Advanced Sampling Theory with Applications

Advanced Sampling Theory with Applications

How Michael 'selected' Amy
Volume I

by

Sarjinder Singh
St. Cloud State University,
Department of Statistics,
St. Cloud, MN, U.S.A.

KLUWER ACADEMIC PUBLISHERS
DORDRECHT / BOSTON / LONDON

A C.I.P. Catalogue record for this book is available from the Library of Congress.

ISBN Vol. 1: 1-4020-1706-5
ISBN Vol. 2: 1-4020-1707-3
ISBN set: 1-4020-1689-1

Published by Kluwer Academic Publishers,
P.O. Box 17, 3300 AA Dordrecht, The Netherlands.

Sold and distributed in North, Central and South America
by Kluwer Academic Publishers,
101 Philip Drive, Norwell, MA 02061, U.S.A.

In all other countries, sold and distributed
by Kluwer Academic Publishers,
P.O. Box 322, 3300 AH Dordrecht, The Netherlands.

Printed on acid-free paper

Printed in the Netherlands.

Dedicated to

GOD

with whose grace...

TABLE OF CONTENTS

2 SIMPLE RANDOM SAMPLING

3 USE OF AUXILIARY INFORMATION: SIMPLE RANDOM SAMPLING

4 USE OF AUXILIARY INFORMATION: PROBABILITY PROPORTIONAL TO SIZE AND WITH REPLACEMENT (PPSWR) SAMPLING

5 USE OF AUXILIARY INFORMATION: PROBABILITY PROPORTIONAL TO SIZE AND WITHOUT REPLACEMENT (PPSWOR) SAMPLING

6 USE OF AUXILIARY INFORMATION: MULTI-PHASE SAMPLING

VOLUME II

7 SYSTEMATIC SAMPLING

8 STRATIFIED AND POST-STRATIFIED SAMPLING

9 NON-OVERLAPPING, OVERLAPPING, POST, AND ADAPTIVE CLUSTER SAMPLING

10 MULTI-STAGE, SUCCESSIVE, AND RE-SAMPLING STRATEGIES

11 RANDOMIZED RESPONSE SAMPLING: TOOLS FOR SOCIAL SURVEYS

12 NON-RESPONSE AND ITS TREATMENTS

13 MISCELLANEOUS TOPICS

APPENDIX

TABLES

POPULATIONS

PREFACE

Advanced Sampling Theory with Applications: How Michael 'Selected' Amy is a comprehensive exposition of basic and advanced sampling techniques along with their applications in the diverse fields of science and technology.

PURPOSE

This book is a multi-purpose document. It can be used as a text by teachers, as a reference manual by researchers, and as a practical guide by statisticians. It covers 1179 references from different research journals through almost 2158 citations across 1248 pages, a large number of complete proofs of theorems, important results such as corollaries, and 335 unsolved exercises from several research papers. It includes 162 solved, data based, real life numerical examples in disciplines such as Agriculture, Demography, Social Science, Applied Economics, Engineering, Medicine, and Survey Sampling. These solved examples are very useful for an understanding of the applications of advanced sampling theory in our daily life and in diverse fields of science. An additional 177 unsolved practical problems are given at the ends of the chapters. University and college professors may find these useful when assigning exercises to students. Each exercise gives exposure to several complete research papers for researchers/students. For example, by referring to Exercise 3.1 at the back of Chapter 3, different types of estimators of a population mean studied by Chakrabarty (1968), Vos (1980), Adhvaryu and Gupta (1983), Walsh (1970), Sahai and Sahai (1985) and Sisodia and Dwivedi (1981) are examined. Thus, this single exercise discusses about six research papers. Similarly, Exercise 5.7 explains the other possibilities in the calibration approach considered by Deville and Särndal (1992) and their followers.

The data based problems show statisticians how to select a sample and obtain estimates of parameters from a given population by using different sampling strategies like SRSWR, SRSWOR, PPSWR, PPSWOR, RHC, systematic sampling, stratified sampling, cluster sampling, and multi-stage sampling. Derivations of calibration weights from the design weights under single phase and two-phase sampling have been provided for simple numerical examples. These examples will be useful to understand the meaning of benchmarks to improve the design weights. These examples also explain the background of well known scientific computer packages like CALMAR, GES, SAS, STATA, and SUDAAN, etc., some of which are very expensive, used to generate calibration weights by most organizations in the public and private sectors. The ideas of hot deck, cold deck, mean method of imputation, ratio method of imputation, compromised imputation, and multiple imputation have been explained with very simple numerical examples. Simple examples are also provided to understand Jackknife variance estimation under single phase, two-phase [or random non-response by following Sitter (1997)] and multi-stage stratified designs.

I have provided a summary of my book from which a statistician can reach a fruitful decision by making a comparison in his/her mind with the existing books in the international market.

Summary Statistic of different components						
General Information	Pages	Solved examples	Unsolved exercises	Practical problems	No. of citations	Figures, tables, maps and graphs.
Title (s)	4	-	-	-	-	-
Dedication	2	-	-	-	-	-
Table of contents	14	-	-	-	-	-
Preface	8	-	-	-	9	1
1	70	13	11	20	2	58
2	66	20	22	19	58	24
3	158	36	68	38	307	61
4	54	9	15	10	84	26
5	180	13	43	15	651	43
6	86	10	29	10	170	21
7	34	8	17	9	72	23
8	116	21	24	19	112	70
9	64	12	11	14	61	57
10	60	3	31	4	162	13
11	86	3	33	5	216	7
12	90	8	24	9	154	28
13	40	6	7	5	100	15
Appendix	26	-	-	-	-	12
Bibliography	62	-	-	-	-	-
Author Index	22	-	-	-	-	-
Subject Index	4	-	-	-	-	-
Related Books	2	-	-	-	-	-
Total	1248	162	335	177	2158	459

This book also covers, in a very simple and compact way, many new topics not yet available in any book on the international market. A few of these interesting topics are: median estimation under single phase and two-phase sampling, difference between low level and higher level calibration approach, calibration weights and design weights, estimation of parametric functions, hidden gangs in finite populations, compromised imputation, variance estimation using distinct units, general class of estimators of population mean and variance, wider class of estimators of population mean and variance, power transformation estimators, estimators based on the mean of non-sampled units of the auxiliary character, ratio and regression type estimators for estimating finite population variance similar to proposed by Isaki (1982), unbiased estimators of mean and variance under Midzuno's scheme of sampling, usual and modified jackknife variance estimator, estimation of regression coefficient, concept of revised selection probabilities, multi-character surveys sampling, overlapping, adaptive, and post cluster sampling, new techniques in systematic sampling, successive sampling, small area estimation, continuous populations, and estimation of measurement errors.

This book has 459 tables, figures, maps, and graphs to explain the exercises and theory in a simple way. The collection of 1179 references (assembled over more than ten years from journals available in India, Australia, Canada, and the USA) is a vital resource for researcher. The most interesting part is the method of notation along with complete proofs of the basic theorems. From my experience and discussion with several research workers in survey sampling, I found that most people dislike the form or method of notation used by different writers in the past. In the book I have tried to keep these notations simple, neat, and understandable. I used data relating to the United States of America and other countries of the world, so that international students should find it interesting and easy to understand. I am confident that the book will find a good place and reputation in the international market, as there is currently no book which is so thorough and simple in its presentation of the subject of survey sampling.

TIMELINESS AND AUDIENCE

The objective, style, and pattern of this book are quite different from other books available in the market. This book will be helpful to:

(a) Graduates and undergraduates majoring in statistics and programs where sampling techniques are frequently used;
(b) Graduates currently involved in M.Sc. or Ph.D. programs in sampling theory or using sampling techniques in their research;
(c) Government organizations such as the US Bureau of Statistics, the Statistics Canada, the Australian Bureau of Statistics, the New Zealand Bureau of Statistics, and the Indian Statistical Institute, in addition to private organizations such as RAND and WESTSTAT, etc.

APPROACH

In this book I have begun each chapter with basic concepts and complete derivations of the theorems or results. I ended each chapter by filling the gap between the origin of each topic and the recent references. In each chapter I provided exercises which summarize the research papers. Thus this book not only gives the basic techniques of sampling theory but also reviews most of the research papers available in the literature related to sampling theory. It will also serve as an umbrella of references under different topics in sampling theory, in addition to clarifying the basic mathematical derivations. In short, it is an advanced book, but provides an exposure to elementary ideas too. It is a much better restatement of the existing knowledge available in journals and books. I have used data, graphs, tables, and pictures to make sampling techniques clear to the learners.

EXERCISES

At the end of each chapter I have provided exercises and their solutions are given through references to the related research papers. Exercises can be used to clarify or relate the classroom work to the other possibilities in the literature.

PRACTICAL PROBLEMS

At the end of each chapter I have provided practical problems which enable students and teachers to do additional exercises with real data.

DATA

I have taken real data related to the United States of America and many other countries around the world. This data is freely available in libraries for public use and it has been provided in the Appendix of this book for the convenience of the readers. This will be interesting to the international students.

NEW TECHNOLOGIES

This provides to students or researchers new formulae available in the literature, which can be used to develop new computer programs for estimating parameters in survey sampling and to learn basic statistical techniques.

SOLUTION MANUAL

I am working on a complete solution manual to the practical problems and selected theoretical exercises given at the end the chapters.

SOME MEMORIES

I was born in the village of Ajnoud, in the district of Ludhiana, in the state of Punjab, India in 1963. My primary education is from the Govt. Primary School, Ajnoud; the Govt. Middle School, Bilga; and Govt. High School, Sahnewal, which are near my birthplace. I did my undergraduate work at Govt. College Karamsar, Rarra Sahib. Still I remember that I used to bicycle my way to college, about 15 km, daily on the bank of canals. It was fun and that life has never come back. M.Sc. and Ph.D. degrees in statistics were completed at the Punjab Agricultural University (PAU), Ludhiana, and most of the time spent in room no. 46 of hostel no. 5.

I attended conferences of the Indian Society of Agricultural Statistics held at Gujarat, Haryana, Orissa, and Kerala, and was a winner of the Gold Medal in 1994 for the Young Scientist Award. I attended conferences of the Australian Statistical Society in Sydney and the Gold Coast. I attended a conference of the International Indian Statistical Association at Hamilton, and the Statistical Society of Canada conferences at Hamilton, Regina, and Halifax in addition to the Concordia University conference. I also attended the Joint Statistical Meetings (JSM-2001, 2002) at Atlanta and New York.

At present I am an Assistant Professor at St. Cloud State University, St. Cloud, MN, USA, and recently introduced the idea of obtaining exact traditional linear regression estimator using calibration approach. From 2001 to 2002 I did post doctoral work at Carleton University, Canada. From 2000 to 2001 I was a Visiting Instructor at the University of Saskatchewan, Canada. From 1999 to 2000 I was a Visiting Instructor at the University of Southern Maine, USA, where I taught several courses to undergraduate and graduate students, and introduced the idea of compromised imputation in survey sampling. From 1998 to 1999 I was Visiting Scientist at the University of Windsor Canada. From 1996 to 1998 I was Research Officer–II in the Methodology Division of the Australian Bureau of Statistics where I developed higher order calibration approach for estimating the variance of the GREG, and introduced the concept of hidden gangs in finite populations. From 1995 to 1996 I was Research Assistant at Monash University, Australia. From 1991 to 1995 I was Research Fellow, Assistant Statistician and then Assistant Professor at PAU, Ludhiana, India and was also awarded a Ph.D. in statistics in 1991. I have published over 80 research papers in reputed journals of statistics and energy science. I am also co-author of a monograph entitled, *Energy in Punjab Agriculture*, published by the Indian Council of Agricultural Research, New Delhi.

Advanced Sampling Theory with Applications is my additional achievement. In this book you can enjoy my new ideas such as:

"How did Michael select Amy?"

"How can you weigh elephants in a circus?"
and

"How many girls like Bob?"

in addition to higher order calibration, bias filtration, hybridising imputation and calibration techniques, hidden gangs, median estimation using two-phase sampling, several new randomised response models, and exact traditional linear regression using calibration technique etc..

ACKNOWLEDGEMENTS

Indeed the words at my command are not adequate to convey the feelings of gratitude toward the late Prof. Ravindra Singh for his constant, untiring and ever encouraging support since 1996 when I started writing this book. Prof. Ravindra Singh passed away Feb. 4, 2003, which is a great loss to his erstwhile students and colleagues, including me. He was my major advisor in my Ph.D. and was closely associated in my research work. Since 1996 Mr. Stephen Horn, supervisor at the Australian Bureau of Statistics, always encouraged to me to complete this book and I appreciate his sincere co-operation, contribution and kindness in joint research papers as well guidance to complete this book. The help of Prof. M.L. King, Monash University is also appreciated. I started writing this book while staying with Dr. Jaswinder Singh, his wife Dr. Rajvinder Kaur, and their daughter Miss

Jasraj Kaur in Australia during 1996. Almost seven years I worked day and night on this book, and during May–July, 2003, I rented a room near an Indian restaurant in Malton, Canada to save cooking time and spent most of the time on this book

Thanks are due to Prof. Ragunath Arnab, University of Durban--Westville, for help in completing the work in Chapter 10 related to his contribution in successive sampling, and completing some joint research papers. The help of Prof. H.P. Singh, Vikram University in joint publications is also duly acknowledged.

The contribution of late Prof. D.S. Tracy, University of Windsor, of reading a few chapters of the very early draft of the manuscript has also been duly acknowledged. The contribution of Ms. Margot Siekman, University of Southern Maine in reading a few chapters has also been duly acknowledged. Thanks are also due to a professional editor Kathlean Prenderqast, University of Saskatchewan, for critically checking the grammar and punctuation of a few chapters. Prof. M. Bickis, University of Saskatchewan, really helped me in my career when I was on the road and looking for a job by going from university to university in Canada. Prof. Silvia Valdês and Ms. Laurie McDermott's help, University of Southern Maine, has been much appreciated. Thanks are also due to Professor Patrick Farrell, Carleton University, for giving me a chance to work with him as a post doctoral fellow. Thanks are also due to Prof. David Robinson at SCSU for providing a very peaceful work environment in the department. The aid of one Stat 321 student, Miss Kok Yuin Ong in cross checking all the solved numerical examples, and a professional English editor Mr. Eric Westphal in reading the entire manuscript at SCSU is much appreciated. Thanks are also due to a professional editor Dr. M. Cole from England for editing the complete manuscript, and to bring it in the present form. Mary Shrode and Mitra Sangrovla, Learning Resources and Technology Service, SCSU, for help in drawing a few illustrations using NOVA art explosion 600,000 images collection is duly acknowledged.

I am also thankful to the galaxy of my friends/colleagues, *viz.*, Dr. Inderjit Grewal (PAU), Dr. B.R. Garg (PAU), Dr. Sukhjinder Sidhu (PAU), Prof. L. N. Upadhyaya (Indian School of Mines), Er. Amarjot Singh (Australia), Mr. Qasim Shah (Australia), Mr. Kuldeep Virdi (Canada), Mr. Kulwinder Channa (Canada), Prof. Balbinder Deo (Canada), Er. Mohan Jhajj (Canada), Mr. Gurbakhash Ubhi (Canada), Mr. Gurmeet Ghatore (USA), Dr. Gurjit Sidhu (USA), Prof. Balwant Singh (USA), Prof. Munir Mahmood (USA), and Mr. Suman Kumar (USA). All cannot be listed, but none is forgotten. I met uncle Mr. Trilochan Singh at Ottawa, who changed my style of living a bit and taught me to get involved with other things, not only sampling theory, and I appreciate his advice. I sincerely appreciate Dr. Joginder Singh's advice at Ottawa, who taught me to do meditation imagining the writing of the name of God with eyes closed and I found it helps when under pressure from work. I am most grateful to my teachers and colleagues for their help and co-operation. Special thanks are due to my father Mr. Sardeep Ubhi, my mother Mrs. Ranjit Ubhi for making this book possible, my brothers Jatinder and Kulwinder, and my late sister Sarjinder.

The permission of Dimitri Chappas, NOAA/ National Climatic Data Center to print a few maps is also duly acknowledged. Free access to data given in the Appendix by Agricultural Statistics and Statistical Abstracts of the United States are also duly acknowledged. I would also like to extend my thanks to the Editor James Finlay, Associate Editor Inge Hardon, and reviewers for bringing the original version of the manuscript into the present form and into the public domain.

Note that I used EXCEL to solve the numerical examples, and while using a hand calculator there may be some discrepancies in the results after one or two decimal places. Further note that the names used in the examples such as Amy, Bob, Mr. Bean, etc., are generic, and are not intended to resemble any real people. I would also like to submit that all opinions and methods of presentation of results in this book are solely the author's and are not necessarily representative of any institute or organization. I tried to collect all recent and old papers, but if you have any published related paper and would like that to be highlighted in the next volume of my book, please feel free to mail a copy to me, and it will be my pleasure to give a suitable place to your paper. To my knowledge this will be the first book, in survey sampling, open to everyone to share contribution irrespective your designation, status, group of scientists, journals names, or any other discriminating character existing in this world, you feel. Your opinions are most welcome and any suggestion for improvement will be much appreciated via e-mail.

Sarjinder Singh (B.Sc., M.Sc., Ph.D., Gold Medalist, and Post Doctorate)
Assistant Professor, Department of Statistics, St. Cloud State University,
St. Cloud, MN, 56301-4498, USA **E-mail:** sarjinder@yahoo.com

1. BASIC CONCEPTS AND MATHEMATICAL NOTATION

1.0 INTRODUCTION

In this chapter we introduce some basic concepts and mathematical notation, which should be known to every survey statistician. The meaning and the use of these terms is supported by using them in the subsequent chapters.

1.1 POPULATION

In statistical language the term population is applied to any finite or infinite collection of individuals or units. It has displaced the older term 'universe'. It is practically synonymous with 'aggregate'. A population is a collection of objects or units about which we want to know something or draw an inference. The population may be finite or infinite. Assume a population consists of electric bulbs produced by a plant. We may want to estimate the average life of the bulbs. The number of bulbs produced by the plant may be finite or infinite.

1.1.1 FINITE POPULATION

If the number of objects or units in the population is countable, it is said to be a finite population. For example, the number of houses in a suburb is a finite population.

1.1.2 INFINITE POPULATION

If the number of objects or units in the population is infinite, it is said to be an infinite population. For example, the number of stars in the sky forms an infinite population. In general, the population is denoted by Ω and its size is denoted by N. In the case of infinite population, $N \to \infty$.

1.1.3 TARGET POPULATION

A finite or infinite population about which we require information is called target population. For example, all 18 year old girls in the United States.

1.1.4 STUDY POPULATION

This is the basic finite set of individuals we intend to study. For example, all 18 year old girls whose permanent address is in New York.

1.2 SAMPLE

A subset of the population, which represents the entire population, is called a sample. The sample is denoted by s and its size by n.

1.3 EXAMPLES OF POPULATIONS AND SAMPLES

We provide here a few examples of populations and samples as follows:

(a) All bulbs manufactured in a plant constitute a population. Now consider we want to estimate the average lifetime of all the bulbs. Instead of taking the whole population for testing purpose into consideration, we take 50 bulbs. Then the collection of 50 bulbs will be called a sample;

(b) Consider we want to find the percentage of ticketless travellers in the TTC buses of Toronto. Then all persons travelling in all the busses of Toronto will constitute the population and the persons checked by a particular checker(s) will form a sample.

1.4 CENSUS

A census is a particular case of a sample. If we take a whole population as the sample then the sampling survey is called a census.

1.5 RELATIVE ASPECTS OF SAMPLING VERSUS CENSUS

The following table provides some of the major differences between a sample and a census.

Aspect	Sample	Census
Cost	Less	More
Effort	Less	More
Time consumed	Less	More
Errors	May be predicted with certain confidence	No such errors
Accuracy of measurements	More	Less

1.6 STUDY VARIABLE

The variable of interest or the variable about which we want to draw some inference is called a study variable. Its value for the ith unit is generally denoted by y_i. For example, the life of the bulbs produced by certain plant can be taken as a study variable.

1.7 AUXILIARY VARIABLE

A variable having a direct or indirect relationship to the study variable is called an auxiliary variable. The value of an auxiliary variable for the ith unit is generally denoted by x_i or z_i, etc.. For example, the time or money spent on producing each bulb by the plant to maintain the quality can be taken as an auxiliary variable.

1.8 DIFFERENCE BETWEEN STUDY VARIABLE AND AUXILIARY VARIABLE

The main differences between the study variable and auxiliary variable are as follows:

Factors	Study Variable	Auxiliary Variable
Cost	More	Less
Effort	More	Less
Sources of availability	Current Surveys or Experiments	Current or Past Survey, Books or Journals etc.
Interest of an investigator	More	Less
Error in measurement	More	Less
Sources of error	More	Fewer
Notation	Y	X, Z

1.9 PARAMETER

An unknown quantity, which may vary over different sets of values forming population is called a parameter. Any function of population values of a variable is called a parameter. It is generally denoted by θ.
Mathematically, suppose a population Ω consists of N units and the value of its ith unit is Y_i. Then any function of Y_i values is a parameter, *i.e.*,

$$\text{Parameter} = f(Y_1, Y_2, ..., Y_N). \tag{1.9.1}$$

For example, if Y_i denotes the total life time of the ith bulb, then the average life time of the bulbs produced by the company is a parameter and is given by

$$\text{Parameter} = \frac{1}{N}(Y_1 + Y_2 + ... + Y_N). \tag{1.9.2}$$

1.10 STATISTIC

A summary value calculated from a sample of observations, usually but not necessarily as an estimator of some population parameter is called a statistic and is generally denoted by $\hat{\theta}$. Mathematically, suppose a sample s consists of n units and the value of the ith unit of the sample is denoted by y_i. Any function of y_i values will be a statistic, *i.e.*,

$$\text{Statistic} = f(y_1, y_2, ..., y_n). \tag{1.10.1}$$

For example, if y_i denotes the total life time of the ith bulb, then the average life time of the bulbs produced by the company is estimated by the statistic, defined as

$$\text{Statistic} = \frac{1}{n}(y_1 + y_2 + ... + y_n). \tag{1.10.2}$$

1.11 STATISTICS

Statistics is a science of collecting, analysing and interpreting numerical data relating to an aggregate of individuals or units.

1.12 SAMPLE SELECTION

A sample can be selected from a population in many ways. In this chapter, we will discuss only two simple methods of sample selection. As the readers get familiar with sample selection, more complicated schemes will be discussed in following chapters.

1.12.1 CHIT METHOD OR LOTTERY METHOD

Suppose we have $N = 10{,}000$ blocks in New York City. We wish to draw a sample of $n = 100$ blocks to draw an inference about a character under study, *e.g.*, average amount of alcohol used or number of bulbs used in each block produced by a certain company. Assign numbers to the 10,000 blocks and write these numbers on chits and fold them in such way that all chits look identical. Put all the chits in a box. Then there are two possibilities:

1.12.1.1 WITH REPLACEMENT SAMPLING

Select one chit out of 10,000 chits in the box and note the number of the block written on it. This is the first unit selected in the sample. Before selecting the second chit, we replace the first chit in the box and mix with the other chits thoroughly. Then select the second chit and note the name of the block written on it. This is called the second unit selected in the sample. Go on repeating the process, until 100 chits have been selected. Note that the chits are selected after replacing the previous chit in the box some chits may be selected more than once. Such a sampling procedure is called Simple Random Sampling With Replacement or simply SRSWR sampling. Let us explain with a few numbers of blocks in a population as follows:

Suppose a population consists of $N = 3$ blocks, say A, B and C. We wish to draw all possible samples of size $n = 2$ using SRSWR sampling. The possible ordered samples are: *AA*, *AB*, *AC*, *BA*, *BB*, *BC*, *CA*, *CB*, *CC*. Thus a total 9 samples of size 2 can be drawn from the population of size 3, which in fact is given by $3^2 = 9$.

In general, the total number of samples of size n drawn from a population of size N in with replacement sampling is N^n and is denoted by $s(\Omega)$.
Thus

$$s(\Omega) = N^n. \tag{1.12.1}$$

Now imagine the situation, 'How many WR samples, each of $n = 100$ blocks, are possible out of $N = 10{,}000$ blocks?'

1.12.1.2 WITHOUT REPLACEMENT SAMPLING

In case of without replacement sampling, we do not replace the chit while selecting the next chit; *i.e.*, the number of chits in the box goes on decreasing as we go on selecting chits. Hence, there is no chance for a chit to be selected more than once. Such a sampling procedure is called Simple Random Sampling and Without Replacement or simply SRSWOR sampling. Let us explain it as follows: Suppose a population consists of $N = 3$ blocks A, B and C. We wish to draw all possible unordered samples of size $n = 2$. Evidently, the possible samples are: AB, AC, BC. Thus a total of 3 samples of size 2 can be drawn from the population of size 3, which in fact is given by ${}^3C_2 = 3$. In general, the total number of samples of size n drawn without replacement from a population of size N is given by ${}^NC_n$ or $\binom{N}{n}$.

Thus

$$s(\Omega) = {}^NC_n = \frac{N!}{n!(N-n)!} \tag{1.12.2}$$

where $n! = n(n-1)(n-2)\ldots\ldots 2.1$, and $0! = 1$.

Now think again, 'How many WOR samples, each of $n = 100$ blocks, are possible out of $N = 10{,}000$ blocks?'

Note that it is a very cumbersome job to make identical chits if the size of the population is very large. In such situations, another method of sample selection is based on the use of a random number table. A random number table is a set of numbers used for drawing random samples. The numbers are usually compiled by a process involving a chance element, and in their simplest form, consist of a series of digits 0 to 9 occurring at random with equal probability.

1.12.2 RANDOM NUMBER TABLE METHOD

As mentioned above, in this table the numbers from 0 to 9 are written both in columns and rows. For the purpose of illustrations, we used Pseudo-Random Numbers (PRN), generated by using the UNIF subroutine following Bratley, Fox, and Schrage (1983), as given in Table 1 of the Appendix. We generally apply the following rules to select a sample:

Rule 1. First we write all random numbers into groups of columns as already done in Table 1 of the Appendix. We take as many columns in each group as the number of digits in the population size.
Rule 2. List all the individuals or units in the population and assign them numbers $1, 2, 3, ..., N$.
Rule 3. Randomly select any starting point in the table of random numbers. Write all the numbers less than or equal to N that follow the starting point until we obtain n numbers. If we are using SRSWOR sampling discard any number that is repeated in the random number table. If we are using SRSWR sampling retain the repeated numbers.
Rule 4. Select those units that are assigned the numbers listed in Rule 3. This will constitute a required random sample.

Let us explain these rules as follows: Suppose we are given a population of $N = 225$ units and we want to select a sample of say $n = 36$ units from it. To pickup a random sample of 36 units out of a population of 225 units, use any three columns from the random number table. For example, use column 1 to 3, 4 to 6, etc., rejecting any number greater than 225 (and also the number 000). As an example, the following table lists the 36 units selected using SRSWR sampling procedure with the use of Pseudo-Random Numbers (PRN) given in Table 1 of the Appendix.

Units selected in the sample											
014	049	053	**039**	196	183	171	225	179	153	142	138
070	083	001	209	222	075	219	092	155	012	099	211
027	**039**	**048**	**048**	080	161	006	059	199	150	025	173

In the case of SRSWOR sampling, the figures **039**, **048** would not get repeated; *i.e.*, we would take every unit only once, so we will continue to select two more distinct random numbers as **078** and **163**.

Although the above method of selecting a sample by using a random number table is very efficient, may make a lot of rejections of the random numbers, therefore we would like to discuss a shortcut method called the remainder method.

1.12.2.1 REMAINDER METHOD

Using the above example, if any three digit selected random number is greater than 225 then divide it by 225. We choose the serial number from 1 through 224 corresponding to the remainder when it is not zero and the serial number 225 when the remainder is zero. However, it is necessary to reject the numbers from 901 to 999 (besides 000) in adopting this procedure as otherwise units with serial number 1 to 99 will have a larger probability (5/999) of selection, while those with serial number 100 to 225 will have probability only equal to 4/999. If we use this procedure and also the same three figure random numbers as given in columns 1 to 3, 4 to 6, etc., we obtain the sample of units which are assigned numbers given below. Again in SRSWR sampling the number that gives rise to the same remainder are not discarded while in SRSWOR sampling procedure such numbers are discarded. Thus an SRSWR sample is as given below:

Units selected in the sample											
138	151	099	025	014	022	197	176	11	209	042	194
015	049	095	040	027	124	116	097	126	142	073	158
108	053	046	001	207	156	201	027	111	209	065	184

Note that in the SRSWR sample, only one unit **209** is repeated, thus for SRSWOR sampling, we continue to apply remainder approach until another distinct unit is selected, which is **089** in this case. Further note that the first random number **992** was discarded due to requirement of this rule.

1.13 PROBABILITY SAMPLING

Probability sampling is any method of selection of a sample based on the theory of probability. At any stage of the selection process, the probability of a given set of units being selected must be known.

1.14 PROBABILITY OF SELECTING A SAMPLE

Every sample selected from the population has some known probability of being selected at any occasion. It is generally denoted by the symbol, p_t or $p(t)$. For example the probability of selecting a sample using

with replacement sampling, $p_t = 1/N^n$, $t = 1, 2, ..., N^n$, (1.14.1)

and

without replacement sampling, $p_t = 1/{}^NC_n$, $t = 1, 2, ..., {}^NC_n$. (1.14.2)

The following table describes the difference between with replacement and without replacement sampling procedures.

With replacement sampling	Without replacement sampling
Cheaper	Costly
Few units may be selected more than once.	A unit can get selected only once.
Less efficient.	More efficient.
Number of possible samples $s(\Omega) = N^n$	Number of possible samples $s(\Omega) = {}^NC_n$
Probability of selecting a particular sample $p_t = 1/N^n$, $t = 1,2,....,N^n$.	Probability of selecting a particular sample $p_t = 1/{}^NC_n$, $t = 1,2,...,{}^NC_n$.
Probability of selecting i^{th} unit in a sample $P_i = 1/N$, $i = 1,2,..., N$.	Probability of selecting i^{th} unit in a sample $P_i = 1/N$, $i = 1,2,....,N$.

1.15 POPULATION MEAN/TOTAL

Let Y_i, $i = 1,2,....,N$, denote the value of the ith unit in a population, then the population mean is defined as

$$\overline{Y} = \frac{1}{N}\left(Y_1 + Y_2 + ... + Y_N\right) = \frac{1}{N}\sum_{i=1}^{N} Y_i \tag{1.15.1}$$

and population total is defined as

$$Y = \left(Y_1 + Y_2 + ... + Y_N\right) = \sum_{i=1}^{N} Y_i = N\,\overline{Y}\,. \tag{1.15.2}$$

The units of measurements of population mean are the same as those for the actual data. For example, if the ith unit, Y_i, $\forall i$, is measured in dollars, then the population mean, $\overline{Y}$, has the same units as dollars.

1.16 POPULATION MOMENTS

The rth order central population moments are defined as

$$\mu_r = \frac{1}{(N-1)}\sum_{i=1}^{N}\left(Y_i - \overline{Y}\right)^r,\ r = 2,\ 3, ... \tag{1.16.1}$$

If $r = 2$ then μ_2 represents the second order population moment, given by

$$\mu_2 = S_y^2 = \frac{1}{N-1}\sum_{i=1}^{N}\left(Y_i - \overline{Y}\right)^2 \tag{1.16.2}$$

and is named the population mean square.

Note that the population variance is defined as

$$\sigma_y^2 = \frac{1}{N}\sum_{i=1}^{N}\left(Y_i - \overline{Y}\right)^2 = \frac{(N-1)}{N}S_y^2\,. \tag{1.16.3}$$

If the data is in dollars, then the units of measurement of σ_y^2 will be dollars^2.

1.17 POPULATION STANDARD DEVIATION

The positive square root of the population variance is called the population standard deviation and it is denoted by σ_y. The units of measurements of σ_y will again be the same as that of actual data. For instance, in the above example, the units of measurements of σ_y will be dollars.

1.18 POPULATION COEFFICIENT OF VARIATION

The ratio of standard deviation to population mean is called the coefficient of variation. It is denoted by C_y that is

$$C_y = \frac{\sigma_y}{\overline{Y}}\,. \tag{1.18.1}$$

Evidently C_y is a unit free number. It is useful to compare the variability in two different populations having different units of measurements, *e.g.*, \$ and kg. It is also called the relative standard error (RSE). Sometimes we also consider $C_y \approx S_y / \overline{Y}$.

1.19 RELATIVE MEAN SQUARE ERROR

The relative mean square error is defined as the square of the coefficient of variation C_y and is generally written by RMSE.

Mathematically

$$\text{RMSE} = C_y^2 = \frac{\sigma_y^2}{\overline{Y}^2}. \tag{1.19.1}$$

Sometimes it is also denote it by ϕ^2.

1.20 SAMPLE MEAN

Let y_i, $i = 1,2,...,n$, denote the value of the ith unit selected in the sample, then the sample mean is defined as

$$\overline{y} = \frac{1}{n}\sum_{i=1}^{n} y_i. \tag{1.20.1}$$

1.21 SAMPLE VARIANCE

The sample variance s_y^2 is defined as

$$s_y^2 = \frac{1}{n-1}\sum_{i=1}^{n}(y_i - \overline{y})^2. \tag{1.21.1}$$

Remark 1.1. The population mean $\overline{Y}$ and population variance σ_y^2 etc., are unknown quantities (parameters) and can be denoted by the symbol θ. The sample mean $\overline{y}$ and sample variance s_y^2 etc., are known after sampling and are called statistic and can be denoted by $\hat{\theta}$. Also note that sample standard deviation (or standard error) and sample coefficient of variation can also be defined as $s_y = \sqrt{s_y^2}$ and $\hat{C}_y = \frac{s_y}{\overline{y}}$, respectively. Note that standard error is a statistic where as standard deviation is a parameter.

1.22 ESTIMATOR

A statistic $\hat{\theta}_t$ obtained from values in the sample s is also called an estimator of the population parameter θ. Note that the notation $\hat{\theta}_t$, or $\hat{\theta}$, or $\hat{\theta}_n$ have same meaning. For example the notation $\bar{y}_t$, or $\bar{y}$, or $\bar{y}_n$ have the same meaning, and s_y^2 or $s_{y(t)}^2$ have the same meaning. We choose according to our requirements for a given topic or exercise.

1.23 ESTIMATE

Any numeric value obtained from the sample information is called the estimate of the population parameter. It is also called a statistic.

1.24 SAMPLE SPACE

A sample space is a set of all possible values of a variable of interest. It is denoted by ψ or S. For example, if we throw a pair of fair coins, each having two faces, then the sample space will consist of all possible 4 outcomes as:

$$\psi = \{HH,\ HT,\ TH,\ TT\}.$$

A pictorial representation of such a sample space is given in Figure 1.24.1.

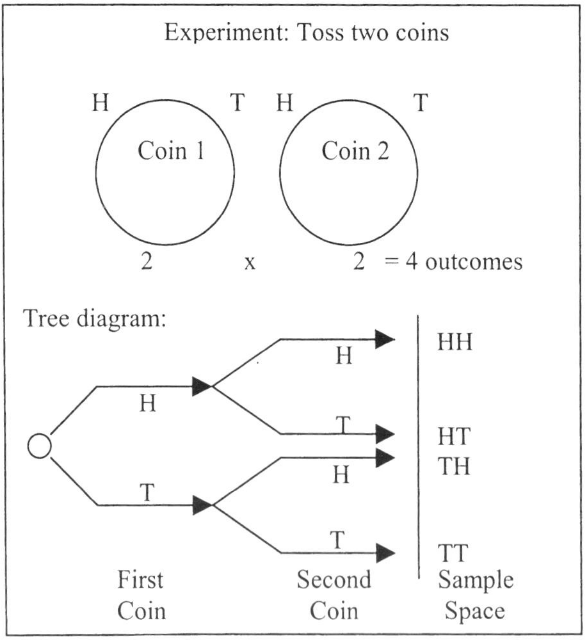

Fig.1.24.1 Sample space while tossing two coins.

1.25 UNIVARIATE RANDOM VARIABLE

A random variable is a real valued function defined on the sample space ψ. It is generally of two types:

(i) Qualitative random variable; (ii) Quantitative random variable.

Let us discuss these random variables in more details as follows:

1.25.1 QUALITATIVE RANDOM VARIABLES

Qualitative random variables assume values that are not necessarily numerical, but can be categorized. For example, Gender has two possible values: Male and Female. These two can be arbitrary coded numerically as Female = 0 and Male = 1. Such coded variables are called Nominal variables. In another example, consider Grades that can take five possible values: A, B, C, D, and F. These five categories can be arbitrarily coded numerically as: $A = 4$, $B = 3$, $C = 2$, $D = 1$, and $F = 0$. Note that here the magnitude of coding tells us quality of Grade that if code is 3 then the Grade is better than the Grade if code is 2. Such a coded variable is called Ordinal variable. Also note that in the case of the Nominal variable, code Male = 1 and Female = 0, does not mean that males are superior to females. Adding, subtracting or averaging such qualitative variables has no meaning. Thus qualitative variables are of two types: (a) Nominal variables; (b) Ordinal variables. Pie charts or Bar charts are generally used to present qualitative variables.

1.25.2 QUANTITATIVE RANDOM VARIABLES

Quantitative random variables can take numerical values for which adding, subtracting or averaging such variables does have meaning. Examples of continuous variables are weight, height, number of students, etc.. In general, two types of quantitative random variables are available: (a) Discrete random variable; (b) Continuous random variable.

1.25.2.1 DISCRETE RANDOM VARIABLE

If a random variable takes a countable number of values, it is called a discrete random variable. In other words, a real valued function defined on a discrete sample space is called a discrete random variable. For example, the number of students can be 0, 1, 2 etc..

1.25.2.2 CONTINUOUS RANDOM VARIABLE

A random variable is said to be continuous if it can take all possible values between certain limits. For example, height a student can be 5.6 feet.

A pictorial representation to differentiate between Qualitative and Quantitative variables is given in Figure 1.25.1.

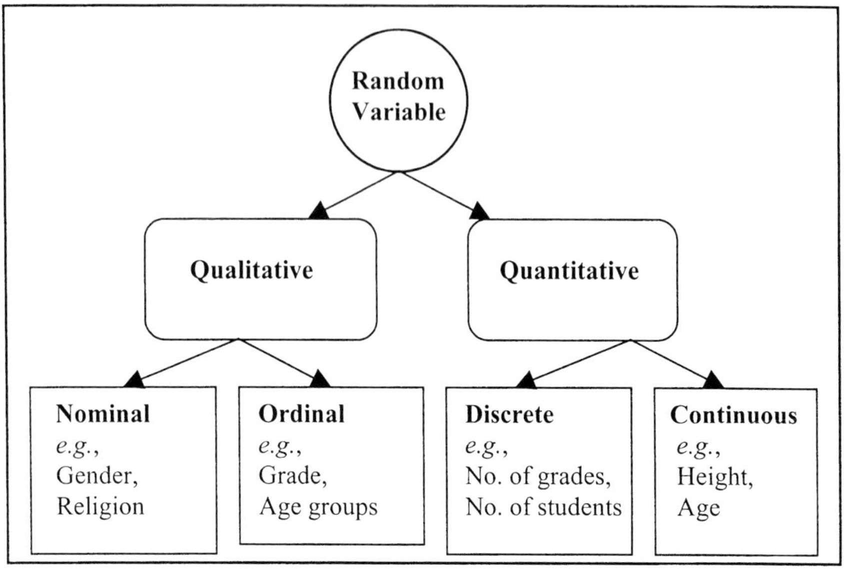

Fig. 1.25.1 Forms of a random variable or data.

Note that Age itself is a quantitative variable whereas Age Groups is a qualitative variable. Pie charts, bar charts, dot plots, line charts, stem and leaf plots, histograms and box plots are generally used to present quantitative variables.

1.26 PROBABILITY MASS FUNCTION (p.m.f.) OF A UNIVARIATE DISCRETE RANDOM VARIABLE

Let X be a discrete random variable taking at most a countable infinite number of values $x_1, x_2, \ldots,$ and with each possible outcome x_i, we associate a number $p_i = P(X = x_i) = p(x_i)$, called the probability of x_i. Then $p(x_i)$ is called the p.m.f. of x_i if it satisfy the following two conditions:

$$\text{(a) } p(x_i) \geq 0 \quad \text{and} \quad \text{(b) } \sum_{i=1}^{\infty} p(x_i) = 1 . \tag{1.26.1}$$

1.27 PROBABILITY DENSITY FUNCTION (p.d.f.) OF A UNIVARIATE CONTINUOUS RANDOM VARIABLE

Let X be a continuous random variable on an interval $x \in [a, b]$, where $-\infty < a \leq x \leq b < +\infty$. Then a function $f(x)$ is said to be a probability density function (p.d.f.) if it satisfies the following two conditions:

$$(\text{a})\ f(x) \geq 0 \quad \text{and} \quad (\text{b})\ \int_a^b f(x)dx = 1. \tag{1.27.1}$$

1.28 EXPECTED VALUE AND VARIANCE OF A UNIVARIATE RANDOM VARIABLE

If a discrete random variable X takes all possible values x_i with probability mass function, $p(x_i)$, in the sample space, ψ, then its expected value is

$$E(x) = \sum_{\psi} x_i p(x_i) \tag{1.28.1}$$

and, the variance of the random variable X is given by

$$V(x) = \sum_{\psi} (x_i - E(x))^2 p(x_i) \tag{1.28.2}$$

or, equivalently

$$V(x) = \sum_{\psi} x_i^2 p(x_i) - \{E(x)\}^2. \tag{1.28.3}$$

Sometimes (1.28.2) is called a formula by definition and that in (1.28.3) is called a computing formula.

If X is a continuous random variable with $x \in [a,b]$ and $f(x)$ be the probability density function, then its expected value is

$$E(x) = \int_a^b x f(x)dx \tag{1.28.4}$$

and the variance of the random variable X is given by

$$V(x) = \int_a^b (x - E(x))^2 f(x)dx \tag{1.28.5}$$

or equivalently

$$V(x) = \int_a^b x^2 f(x)dx - \{E(x)\}^2. \tag{1.28.6}$$

1.29 DISTRIBUTION FUNCTION OF A UNIVARIATE RANDOM VARIABLE

Let X be a random variable, then the function, $F(x) = P(X \leq x)$ is called a distribution function of the random variable and it has the following properties:

(i) If $a \le X \le b$ then $P(a \le X \le b) = F(b) - F(a)$;

(ii) If $a \le b$ then $F(a) \le F(b)$;

(iii) $0 \le F(x) \le 1$;

(iv) The distribution of $F(x)$ is uniform between 0 and 1.

Again it can be of two-types:

1.29.1 DISCRETE DISTRIBUTION FUNCTION

In this case there are a countable number of points $x_1, x_2, ...$ along with associated probabilities $p(x_1), p(x_2), ...$, where $p(x_i) \ge 0$ and $\sum_{i=1}^{\infty} p(x_i) = 1$ such that $F(x) = P(X \le x) = \sum_{\{i: x_i \le x\}} p(x_i)$. For example if x_i takes integral values $x_i = \{1, 2, 3, 4, 5\}$ with probabilities $p(x_i) = 0.2$, then function $F(x)$ is a step function as shown in the Figure 1.29.1.

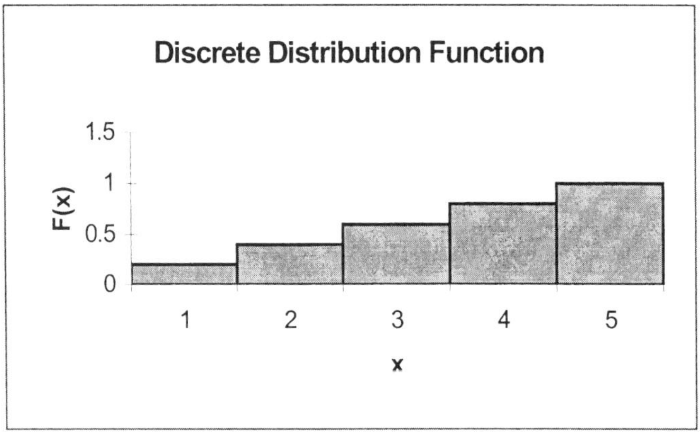

Fig. 1.29.1 Discrete distribution function.

1.29.2 CONTINUOUS DISTRIBUTION FUNCTION

If X is a continuous random variable with the probability density function (p.d.f.) $f(x)$, then the function

$$F(x) = P(X \le x) = \int_{-\infty}^{x} f(t)dt \ , \ -\infty < x < +\infty \qquad (1.29.1)$$

is called the distribution function or sometimes the cumulative distribution function (c.d.f.) of the random variable X. The relationship between $F(x)$ and $f(x)$ is given by $f(x)=\dfrac{dF(x)}{dx}$. The c.d.f. $F(x)$ is a non-decreasing function of x and is continuous on the right. Also note that $F(-\infty)=0$, $F(+\infty)=1$, $0 \le F(x) \le 1$, and $P(a \le x \le b)=\int_a^b f(x)dx = F(b)-F(a)$. For example, if x is a continuous random variable with probability density function (p.d.f.)

$$f(x)=\begin{cases} 1 & 0 < x < 1, \\ 0 & \text{otherwise}, \end{cases} \tag{1.29.2}$$

then its cumulative distribution function (c.d.f.) is given by

$$F(x)=\begin{cases} 0 & \text{if } \; x < 0, \\ x & \text{if } \; 0 \le x \le 1, \\ 1 & \text{if } \; x > 1, \end{cases} \tag{1.29.3}$$

and its graphical representation is given in Figure 1.29.2.

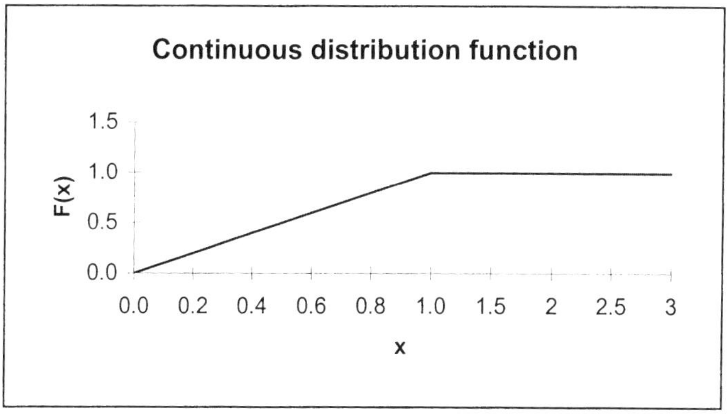

Fig. 1.29.2 Continuous distribution function.

1.30 SELECTION OF A SAMPLE USING KNOWN UNIVARIATE DISTRIBUTION FUNCTION

There are two cases:

1.30.1 DISCRETE RANDOM VARIABLE

If x_i is a discrete random variable with probability mass function $p(x_i)$ and distribution function, $F(x)=P[X \le x]=\sum_{i:x_i \le x} p(x_i)$. Let $0 \le F(x) \le 1$ be any random number drawn from the Pseudo-Random Number (PRN) Table 1 given in the Appendix.

Then we can write $F(x)$ as

$$F(x) = \sum_{i:x_{i-1}<x} p(x_{i-1}) + p(x) \text{ (say)} \tag{1.30.1}$$

Then the integral value of the random variable x selected in the sample is given by

$$x = p^{-1}\left[F(x) - \sum_{i:x_{i-1}<x} p(x_{i-1})\right] \tag{1.30.2}$$

where p^{-1} denotes the inverse function.

Example 1.30.1. A discrete random variable X has the following probability mass function:

x	0	1	2	3	4	5	6	7
$p(x)$	0	0.1	0.2	0.2	0.3	0.01	0.02	0.17

Select a random sample of three units using the method of random numbers.

Solution: The cumulative distribution function of the random variable X is given by

x	0	1	2	3	4	5	6	7
$F(x)$	0	0.1	0.3	0.5	0.8	0.81	0.83	1.00

We used the first six columns of the Pseudo-Random Number (PRN) Table 1 given in the Appendix multiplied by 10^{-6} as the randomly selected values of $F(x)$. Then the integral value of the random variable x selected in the sample is obtained using the inverse relationship $x = p^{-1}\left[F(x) - \sum_{i:x_{i-1}<x} p(x_{i-1})\right]$ as follows:

Value of $F(x)$ using PRN Table 1	Value of x observed in the sample
0.992954	6
0.588183	3
0.601448	3

In case of with replacement sampling, the value of $x = 3$ has been selected twice, as otherwise for WOR sampling we have to continue the process until three distinct values of x are not selected.

Example 1.30.2. If x follows a binomial distribution with parameters N and p, that is, $x \sim B(N, p)$, say $N = 10$ and $p = 0.4$. Select an SRSWR sample of $n = 4$ units by using the random number method.

Solution. Since x follows a binomial distribution, so $P(X=x)=C_x^N p^x(1-p)^{N-x}$, for $x=0,1,2...,N$. Here $N=10$ and $p=0.4$, so the cumulative distribution function $F(x)=P[X\le x]=\sum_{x\le x} C_x^N p^x(1-p)^{N-x}$ is:

x	0	1	2	3	4	5	6	7	8	9	10
$F(x)$	.0060	.0403	.1672	.3822	.6330	.8337	.9452	.9877	.9983	.9999	1.000

We used three columns from 7th to 9th of the Pseudo-Random Number (PRN) Table 1 given in the Appendix multiplied by 10^{-3} as the randomly selected values of $F(x)$. Then the integral value of the random variable x selected in the sample is obtained using the inverse relationship $x=p^{-1}\left[F(x)-\sum_{i:x_{i-1}<x} p(x_{i-1})\right]$ as follows:

Value of $F(x)$ using PRN Table 1	Value of x observed in the sample
0.622	3
0.771	4
0.917	5
0.675	4

1.30.2 CONTINUOUS RANDOM VARIABLE

If x is a continuous random variable with probability density function $f(x)$ and distribution function, $F(x)=P[X\le x]=\int_{-\infty}^{x} f(t)dt$. Let $0\le F(x)\le 1$ be any random number drawn from the Pseudo-Random Number (PRN) Table 1 given in the Appendix.

Then we can write $F(x)$ as

$$F(x)=\int_{-\infty}^{x_0} f(t)dt+f(x) \quad \text{(say)} \tag{1.30.3}$$

where $x_0<x$ but very close to x.

Then the value of the random variable x selected in the sample is given by

$$x=f^{-1}\left(F(x)-\int_{-\infty}^{x_0} f(t)dt\right) \tag{1.30.4}$$

where f^{-1} denotes the inverse function.

Example 1.30.3. A continuous random variable X has the cumulative distribution function as

$$F(x)=\begin{cases}0 & \text{if} \quad x<1,\\ \frac{1}{16}(x-1)^4 & \text{if} \ 1\le x\le 3,\\ 1 & \text{if} \quad x>3.\end{cases} \tag{1.30.5}$$

Select a sample of $n = 10$ units by using SRSWR sampling.

Solution. We are given $F(x)=\frac{1}{16}(x-1)^4$ which implies that $x = 2[F(x)]^{1/4}+1$. By using the first three columns of the Pseudo-Random Numbers (PRN) Table 1 given in the Appendix multiplied by 10^{-3} we obtain the observed values of $F(x)$ and the sampled values of x as:

$F(x)$	x
0.992	2.995988
0.588	2.751356
0.601	2.760956
0.549	2.721563
0.925	2.961397
0.014	1.687958
0.697	2.827419
0.872	2.932676
0.626	2.778990
0.236	2.393985

Example 1.30.4. A continuous random variable x has density function

$$f(x)=\begin{cases}\pi^{-1}\left(1+x^2\right)^{-1} & \text{if} -\infty < x < +\infty,\\ 0 & \text{otherwise.}\end{cases} \tag{1.30.6}$$

Select a sample of $n = 5$ units by using with replacement sampling.

Solution. The cumulative distribution function (c.d.f.) of x is given by

$$F(x)=P(X\le x)=\int_{-\infty}^{x} f(x)dx=\frac{1}{\pi}\int_{-\infty}^{x}\frac{1}{1+x^2}dx=\frac{1}{\pi}\left[\tan^{-1}(x)\right]_{-\infty}^{x}=\frac{1}{\pi}\left[\tan^{-1}(x)+\frac{\pi}{2}\right]$$

which implies that

$$x=\tan\left[\pi F(x)-\frac{\pi}{2}\right]. \tag{1.30.7}$$

Using the three columns multiplied by 10^{-3}, say 7th to 9th, of the Pseudo-Random Numbers (PRN) Table 1 given in the Appendix, the first five observed values of $F(x)$ are given by 0.622, 0.771, 0.917, 0.675 and 0.534. Thus the sampled five values from the above distribution are

$F(x)$	x
0.622	0.403214
0.771	1.141487
0.917	3.747745
0.675	0.612801
0.534	0.107222

Note that we have used the *tan* function in radians and $\pi = 4\tan^{-1}(1)$.

Example 1.30.5. A continuous random variable x has density function

$$f(x) = \begin{cases} \frac{1}{5} & \text{if } 5 < x < 10, \\ 0 & \text{otherwise.} \end{cases} \tag{1.30.8}$$

Select a sample of $n = 5$ units by using with replacement sampling.

Solution. The distribution of x is uniform between 5 and 10, so its probability distribution function is

$$F(x) = P[X \le x] = \int_5^x f(x)dx = \frac{1}{5}(x-5) \tag{1.30.9}$$

which implies that

$$x = 5[F(x)+1]. \tag{1.30.10}$$

Using the three columns multiplied by 10^{-3}, say 7th to 9th, of the Pseudo-Random Number Table 1 given in the appendix, the first five observed values of $F(x)$ are given by 0.622, 0.771, 0.917, 0.675 and 0.534. Thus the sampled five values from the above distribution are given by

$F(x)$	x
0.622	8.110
0.771	8.855
0.917	9.585
0.675	8.375
0.534	7.670

1.31 DISCRETE BIVARIATE RANDOM VARIABLE

If X and Y are discrete random variables, the probability that X will take on the value x and Y will take on the value y as $P(X = x, Y = y) = p(x, y)$, is called the joint probability distribution function of a bivariate random variable.

1.32 JOINT PROBABILITY DISTRIBUTION FUNCTION OF BIVARIATE DISCRETE RANDOM VARIABLES

A bivariate function can serve as the joint probability distribution of a pair of discrete random variables X and Y if and only if its values, $p(x, y)$, satisfy the conditions:

(a) $p(x,y) \geq 0$ for each pair of values (x, y) within its domain.

and

(b) $\sum_{x}\sum_{y} p(x,y) = 1$, where the sum extends over all possible pairs (x,y).

1.33 JOINT CUMULATIVE DISTRIBUTION FUNCTION OF BIVARIATE DISCRETE RANDOM VARIABLES

If X and Y are discrete random variables, the function given by

$$F(x, y) = P(X \leq x, Y \leq y) = \sum_{s \leq x} \sum_{t \leq y} p(s,t) \tag{1.33.1}$$

for $-\infty < x < +\infty$, $-\infty < y < +\infty$, where $p(s,t)$ is the value of the joint probability distribution of X and Y at the point (s, t), is called the Joint distribution function or the Joint cumulative distribution, of X and Y.

1.34 MARGINAL DISTRIBUTIONS OF A BIVARIATE DISCRETE RANDOM VARIABLE

If X and Y are discrete random variables and $p(x,y)$ is the value of the joint probability distribution at (x, y), the function given by

$$p_X(x) = \sum_{y} p(x,y) \tag{1.34.1}$$

for each x within the range of X is called the **marginal distribution** of X, and the function,

$$p_Y(y) = \sum_{x} p(x, y) \tag{1.34.2}$$

for each y within the range of Y is called the **marginal distribution** of Y.

1.35 SELECTION OF A SAMPLE USING KNOWN DISCRETE BIVARIATE DISTRIBUTION FUNCTION

Let $p(x,y)$ denote the joint probability mass function (p.m.f.) of two random variables x and y. Also, let $F(x,y)$ denote the cumulative mass function (c.m.f.) of x and y. It is well known that, the distribution of the marginal distribution function (m.d.f.) $p_y(y)$ for any joint probability density function of x and y is rectangular (or uniform) in the range [0, 1]. Random numbers in the random number table also follow the same distribution. Then to find out the value of y one solves the equation (1.35.1) below.

$$p_y(y) = \sum_x \sum_{y=0}^{y} p(x, y) = R_2. \tag{1.35.1}$$

The known form of the joint density function $p(x, y)$ [one can choose any suitable form for $p(x_1, y)$] can be substituted in (1.35.1). The value y^* of y so obtained is used to find the value x^* of x. For this we use the conditional mass function of x given $y = y^*$ since the distribution of the conditional mass function will also be uniform in [0, 1]. Thus another random number R_1 is drawn and the value x^* of X is determined from the equation

$$\sum_0^x p\left(x,\ y^*\right) = R_1 \tag{1.35.2}$$

where $p\left(x,\ y^*\right)$ is the conditional density function of X given $y = y^*$. The values x^* and y^* of x and y, respectively, so obtained will follow the joint probability mass function $p(x, y)$. The equations (1.35.1) and (1.35.2) may be solved either through usual methods or through iteration procedures.

1.36 CONTINUOUS BIVARIATE RANDOM VARIABLE

A bivariate function with values $f(x, y)$, defined over the two-dimensional plane is called a joint probability density function of the continuous random variables X and Y if and only if

$$P[(X, Y) \in S] = \iint_S f(x, y) dx dy. \tag{1.36.1}$$

1.37 JOINT PROBABILITY DISTRIBUTION FUNCTION OF BIVARIATE CONTINUOUS RANDOM VARIABLE

A bivariate function can serve as the joint probability distribution of a pair of continuous random variables X and Y if and only if its values, $f(x, y)$, satisfy the conditions:

(a) $f(x, y) \geq 0$ for each pair of values (x, y) within its domain; (1.37.1)

(b) $\int_{-\infty}^{+\infty} \int_{-\infty}^{+\infty} f(x, y) dx dy = 1$. (1.37.2)

1.38 JOINT CUMULATIVE DISTRIBUTION FUNCTION OF A BIVARIATE CONTINUOUS RANDOM VARIABLE

If X and Y are continuous random variables, the function given by

$$F(x, y) = P(X \le x, Y \le y) = \int_{-\infty}^{y}\int_{-\infty}^{x} f(s,t)\,ds\,dt \tag{1.38.1}$$

for $-\infty < x < +\infty$, $-\infty < y < +\infty$, where $f(s,t)$ is the value of the joint probability distribution of X and Y at the point (s, t), is called the Joint distribution function or the Joint cumulative distribution, of X and Y.

1.39 MARGINAL CUMULATIVE DISTRIBUTIONS OF BIVARIATE CONTINUOUS RANDOM VARIABLE

If X and Y are continuous random variables and $f(x,y)$ is the value of the joint probability density function, then cumulative marginal probability distribution function of y is given by

$$F_Y(y) = \int_{-\infty}^{y}\int_{-\infty}^{+\infty} f(x,y)\,dx\,dy \tag{1.39.1}$$

for $-\infty < y < +\infty$, and the cumulative marginal probability distribution function of x is given by

$$F_X(x) = \int_{-\infty}^{x}\int_{-\infty}^{+\infty} f(x,y)\,dy\,dx \tag{1.39.2}$$

for $-\infty < x < +\infty$.

1.40 SELECTION OF A SAMPLE USING KNOWN BIVARIATE CONTINUOUS DISTRIBUTION FUNCTION

In general, let $f(x, y)$ denote the joint probability density function (p.d.f.) of two continuous random variables x and y. Also let $F(x, y)$ denote the cumulative density function (c.d.f.) of x and y. It is well known that the distribution of the marginal distribution function (m.d.f.) $F_2(y)$ for any joint probability density function of x and y is rectangular (or uniform) in the range [0, 1]. Random numbers in the random number table also follow the same distribution. To find out the value of y, one solves the equation (1.40.1) below.

$$F_2(y) = \int_0^y\left\{\int_{-\infty}^{+\infty} f(x, y)\,dx\right\}dy = R_2. \tag{1.40.1}$$

The known form of the joint density function $f(x, y)$ [one can choose any suitable form for $f(x_1, y)$] can be substituted in (1.40.1). The value y^* of y so obtained is used to find the value x^* of x. For this we use the conditional density of x given

$y = y^*$ since the distribution of the conditional distribution function will also be uniform in [0,1]. Thus another random number R_1 is drawn and the value x^* of x is determined from the equation:

$$\int_0^x f_1\left(x \mid y = y^*\right) dx = R_1 \tag{1.40.2}$$

where $f_1\left(x \mid y = y^*\right)$ is the conditional density function of x given $y = y^*$. The values x^* and y^* of x and y, respectively, so obtained will follow the joint probability density function $f(x, y)$. The equations (1.40.1) and (1.40.2) may be solved either through usual methods or through iteration procedures.

Example 1.40.1. If the joint probability density function of two continuous random variables x and y is given by,

$$f(x,y) = \begin{cases} \frac{2}{3}(x + 2y) & \text{for } 0 < x < 1,\ 0 < y < 1, \\ 0 & \text{otherwise,} \end{cases}$$

then, select six pairs of observations (x, y) by using the Random Number Table method.

Solution. We have

$$F_y(y) = \int_0^y \left\{ \int_{-\infty}^{+\infty} f(x, y) dx \right\} dy = \int_0^y \left\{ \frac{2}{3} \int_0^1 (x + 2y) dx \right\} dy = \frac{y + 2y^2}{3}.$$

Let $F_y(y) = R_2$ be a Pseudo-Random Number (PRN) selected from Table 1 (say, by using first three columns) is $R_2 = 0.992$ then by solving the quadratic equation $2y^2 + y - 3R_2 = 0$, we obtain one real root $0 < y < 1$ by solving

$$y = \frac{-1 \mp \sqrt{(1)^2 + 24R_2}}{2} \quad \text{as} \quad y = \frac{-1 + \sqrt{1 + 24 \times 0.992}}{2} = 0.995.$$

Now, given $y = y^* = 0.995$ we have

$$f\left(x \mid y = y^*\right) = f(x \mid y = 0.995) = \frac{2}{3}(1.99 + x).$$

Let $0 < R_1 < 1$ be any other random number, say obtained by using 7th to 9th columns, of the Pseudo-Random Numbers given in Table 1 of the Appendix, then the value of x is given by solving the integral

$$\int_0^x f\left(x \mid y = y^*\right) dx = R_1 \quad \text{or} \quad \frac{2}{3} \int_0^x (1.99 + x) dx = 0.622$$

or, equivalently solving a quadratic equation $x^2 + 3.98x - 3R_1 = 0$, which implies

$$x = \frac{-3.98 \pm \sqrt{3.98^2 + 12R_1}}{2},$$

and the real root between the range of x as 0 to 1 is given by

$$x = \frac{-3.98 + \sqrt{3.98^2 + 12R_1}}{2} = \frac{-3.98 + \sqrt{3.98^2 + 12 \times 0.622}}{2} = 0.423\,.$$

On repeating the above process we obtain a sample of $n = 6$ observations as given below:

R_2	y	R_1	x
0.992	0.995	0.622	0.423
0.588	0.722	0.771	0.514
0.601	0.732	0.917	0.600
0.549	0.691	0.675	0.456
0.925	0.954	0.534	0.368
0.014	0.039	0.513	0.355

1.41 PROPERTIES OF A BEST ESTIMATOR

An estimator $\hat{\theta}_t$ of a population parameter, θ, is said to be best if it has the following properties:

(a) Unbiasedness; (b) Consistency; (c) Sufficiency ; and (d) Efficiency.

Let us now briefly explain the meaning of these terms.

1.41.1 UNBIASEDNESS

An estimator $\hat{\theta}_t$ is an unbiased estimator of a population parameter θ if

$$E\left(\hat{\theta}_t\right) = \sum_{t=1}^{s(\Omega)} p_t \hat{\theta}_t = \theta \tag{1.41.1}$$

where, p_t, denote the probability of selecting the tth sample from the population, Ω, and $\sum_{t=1}^{s(\Omega)} p_t = 1$. Note that total number of possible samples, in case of SRSWR sampling are, $s(\Omega) = N^n$, and in case of SRSWOR sampling are, $s(\Omega) = {}^N C_n$.

For example:

(i) Sample mean $\bar{y}_t$ is an unbiased estimator of population mean $\bar{Y}$ under both SRSWR and SRSWOR sampling

$$E(\bar{y}_t) = \sum_{t=1}^{s(\Omega)} p_t \bar{y}_t = \bar{Y}. \tag{1.41.2}$$

(ii) Sample variance s_y^2 is an unbiased estimator of population mean squared error S_y^2 under SRSWOR sampling

$$E\left(s_y^2\right) = \sum_{t=1}^{{}^N C_n} p_t \left(s_y^2\right)_t = S_y^2 \tag{1.41.3}$$

and also s_y^2 is an unbiased estimator of population variance σ_y^2 under SRSWR sampling

$$E\left(s_y^2\right) = \sum_{t=1}^{N^n} p_t \left(s_y^2\right)_t = \sigma_y^2 . \tag{1.41.4}$$

Example 1.41.1. Suppose a population consists of $N = 4$ units. The variable Y_i takes values 1, 2, 3, 4 and are distinguished as A, B, C, and D, respectively. Then the population mean is given by

$$\bar{Y} = \frac{1}{N}\sum_{i=1}^{N} Y_i = \frac{1}{4}(1+2+3+4) = 2.5$$

the population mean squared error is given by

$$S_y^2 = \frac{1}{N-1}\sum_{i=1}^{N}\left(Y_i - \bar{Y}\right)^2 = \frac{1}{4-1}\left[(1-2.5)^2 + (2-2.5)^2 + (3-2.5)^2 + (4-2.5)^2\right] = \frac{5}{3}$$

and population variance is given by

$$\sigma_y^2 = \frac{1}{N}\sum_{i=1}^{N}\left(Y_i - \bar{Y}\right)^2 = \frac{1}{4}\left[(1-2.5)^2 + (2-2.5)^2 + (3-2.5)^2 + (4-2.5)^2\right] = \frac{5}{4} = 1.25 .$$

Show that the sample mean $\bar{y}_t$ is an unbiased estimator of population mean $\bar{Y}$ under both SRSWR and SRSWOR sampling. The sample variance s_y^2 is an unbiased estimator of population mean squared error S_y^2 under SRSWOR sampling, and the population variance σ_y^2 under SRSWR sampling, respectively.

Solution. There are two cases:

Case I. Suppose we are drawing all possible samples of size $n = 2$ by using SRSWOR sampling.

The total number of all possible samples is $s(\Omega) = {}^N C_n = {}^4 C_2 = 6$ and $p_t = 1/6$.

Now we have the following table.

Sample No.	Sampled units	Sample mean $\bar{y}_t$	Sample variance $s^2_{y(t)}$	Probability of selecting the sample p_t
1	(A, B) or $(1, 2)$	$\bar{y}_1 = (1+2)/2 = 1.5$	0.5	1/6
2	(A, C) or $(1, 3)$	$\bar{y}_2 = (1+3)/2 = 2.0$	2.0	1/6
3	(A, D) or $(1, 4)$	$\bar{y}_3 = (1+4)/2 = 2.5$	4.5	1/6
4	(B, C) or $(2, 3)$	$\bar{y}_4 = (2+3)/2 = 2.5$	0.5	1/6
5	(B, D) or $(2, 4)$	$\bar{y}_5 = (2+4)/2 = 3.0$	2.0	1/6
6	(C, D) or $(3, 4)$	$\bar{y}_6 = (3+4)/2 = 3.5$	0.5	1/6

Thus the expected value of the sample mean, $\bar{y}_t$, is given by

$$E(\bar{y}_t) = \frac{1}{{}^N C_n} \sum_{t=1}^{{}^N C_n} \bar{y}_t = \frac{1}{6} \sum_{s=1}^{6} \bar{y}_t = \frac{1}{6}(1.5 + 2 + 2.5 + 2.5 + 3 + 3.5) = 2.5 = \bar{Y},$$

and that of sample variance, s_y^2, is given by

$$E\left[s^2_{y(t)}\right] = \frac{1}{{}^N C_n} \sum_{t=1}^{{}^N C_n} s^2_{y(t)} = \frac{1}{6}(0.5 + 2.0 + 4.5 + 0.5 + 2.0 + 0.5) = \frac{10}{6} = \frac{5}{3} = S_y^2 .$$

Distribution of sample mean and sample variance using WOR sampling:

Distribution of sample means		Distribution of sample variances	
Sample mean	Frequency	Sample variance	Frequency
1.5	1	0.5	3
2.0	1	2.0	2
2.5	2	4.5	1
3.0	1		
3.5	1		

The above table shows that the distribution of sample means is symmetric and that of sample variance is skewed to the right in the case of without replacement sampling.

Case II. Suppose we are drawing all possible samples of size $n = 2$ by using SRSWR sampling.

The total number of all possible samples is $s(\Omega) = N^n = 4^2 = 16$ and $p_t = 1/16$ for all $t = 1, 2, ..., 16$.

Now we have the following table:

Sample No. t	Sampled units	Sample means $\bar{y}_t$	Sample variance $s^2_{y(t)}$	Probability of selecting a sample p_t
1	(A, A) or (1, 1)	$\bar{y}_1 = (1+1)/2 = 1.0$	0.0	1/16
2	(A, B) or (1, 2)	$\bar{y}_2 = (1+2)/2 = 1.5$	0.5	1/16
3	(A, C) or (1, 3)	$\bar{y}_3 = (1+3)/2 = 2.0$	2.0	1/16
4	(A, D) or (1, 4)	$\bar{y}_4 = (1+4)/2 = 2.5$	4.5	1/16
5	(B, A) or (2, 1)	$\bar{y}_5 = (2+1)/2 = 1.5$	0.5	1/16
6	(B, B) or (2, 2)	$\bar{y}_6 = (2+2)/2 = 2.0$	0.0	1/16
7	(B, C) or (2, 3)	$\bar{y}_7 = (2+3)/2 = 2.5$	0.5	1/16
8	(B, D) or (2, 4)	$\bar{y}_8 = (2+4)/2 = 3.0$	2.0	1/16
9	(C, A) or (3, 1)	$\bar{y}_9 = (3+1)/2 = 2.0$	2.0	1/16
10	(C, B) or (3, 2)	$\bar{y}_{10} = (3+2)/2 = 2.5$	0.5	1/16
11	(C, C) or (3, 3)	$\bar{y}_{11} = (3+3)/2 = 3.0$	0.0	1/16
12	(C, D) or (3, 4)	$\bar{y}_{12} = (3+4)/2 = 3.5$	0.5	1/16
13	(D, A) or (4, 1)	$\bar{y}_{13} = (4+1)/2 = 2.5$	4.5	1/16
14	(D, B) or (4, 2)	$\bar{y}_{14} = (4+2)/2 = 3.0$	2.0	1/16
15	(D, C) or (4, 3)	$\bar{y}_{15} = (4+3)/2 = 3.5$	0.5	1/16
16	(D, D) or (4, 4)	$\bar{y}_{16} = (4+4)/2 = 4.0$	0.0	1/16

Distribution of sample mean and sample variance using SRSWR sampling: Again for the SRSWR sampling the sample mean has symmetric distribution and the distribution of sample variance is skewed to the right.

Distribution of sample means		Distribution of sample variances	
Sample mean	Frequency	Sample variance	Frequency
1.0	1	0.0	4
1.5	2	0.5	6
2.0	3	2.0	4
2.5	4	4.5	2
3.0	3		
3.5	2		
4.0	1		

Thus the expected value of the sample mean $\bar{y}_t$ is given by

$$E(\bar{y}_t) = \frac{1}{N^n}\sum_{t=1}^{N^n}\bar{y}_t = \frac{1}{16}\sum_{s=1}^{16}\bar{y}_t = \frac{1}{16}(1+1.5+....+4) = \frac{40}{16} = 2.5 = \bar{Y}$$

and that of the sample variance s_y^2 is given by

$$E\left[s_{y(t)}^2\right] = \frac{1}{N^n}\sum_{t=1}^{N^n} s_{y(t)}^2 = \frac{1}{16}\sum_{s=1}^{16} s_{y(t)}^2 = \frac{1}{16}(0+0.5+...+2+0.5) = \frac{20}{16} = 1.25 = \sigma_y^2 .$$

Then we have the following new term.

1.41.1.1 BIAS

It is the difference between the expected value of a statistic $\hat{\theta}_t$ and the actual value of the parameter θ that is

$$B(\hat{\theta}_t) = E(\hat{\theta}_t) - \theta. \tag{1.41.5}$$

Thus an estimator $\hat{\theta}_t$ is unbiased if $E(\hat{\theta}_t) = \theta$, which is obvious by setting $B(\hat{\theta}_t) = 0$.

1.41.2 CONSISTENCY

There are several definitions for the consistency of any statistic, but we will use the simplest. An estimator $\hat{\theta}_t$ of the population parameter θ is said to be consistent if

$$\operatorname*{Lim}_{n\to\infty}(\hat{\theta}_t) = \theta. \tag{1.41.6}$$

For example:

(i) The sample mean $\bar{y}_t$ (or simply $\bar{y}$) is a consistent estimator of the finite population mean, $\bar{Y}$.

(ii) The sample mean squared error s_y^2 is a consistent estimator of the population mean squared error, S_y^2.

Remark 1.2. An unbiased estimator need not necessarily be consistent, *e.g.*, the sample mean based on a sample of size one is unbiased but not consistent.

1.41.3 SUFFICIENCY

An estimator $\hat{\theta}_t$ is said to be sufficient for a parameter θ if the distribution of a sample $y_1, y_2,..., y_n$ given $\hat{\theta}_t$ does not depend on θ. The distribution of $\hat{\theta}_t$ then contains all the information in the sample relevant to the estimation of θ and knowledge of $\hat{\theta}_t$ and its sampling distribution is 'sufficient' to give that information. In general, a set of estimators or statistics $\hat{\theta}_1, \hat{\theta}_2,........,\hat{\theta}_k$ are 'jointly sufficient' for parameters $\theta_1, \theta_2,.....,\theta_k$ if the distribution of sample values given $\hat{\theta}_1, \hat{\theta}_2,........,\hat{\theta}_k$ does not depend on these $\theta_1, \theta_2,...,\theta_k$.

1.41.4 EFFICIENCY

Before defining the term efficiency, we shall discuss two more terms, *viz.*, variance and mean square error of the estimator.

1.41.4.1 VARIANCE

The variance of an estimator $\hat{\theta}_t$ of a population parameter θ is defined as

$$V(\hat{\theta}_t) = E[\hat{\theta}_t - E(\hat{\theta}_t)]^2 = E\left[(\hat{\theta}_t)^2\right] - [E(\hat{\theta}_t)]^2. \quad (1.41.7)$$

It is generally denoted by the symbol $\sigma_{\hat{\theta}}^2$.

1.41.4.2 MEAN SQUARE ERROR

The mean square error (MSE) of an estimator $\hat{\theta}_t$ of a parameter θ is defined as:

$$\text{MSE}(\hat{\theta}_t) = E[\hat{\theta}_t - \theta]^2 = V(\hat{\theta}_t) + \{B(\hat{\theta}_t)\}^2 \quad (1.41.8)$$

where $B(\hat{\theta}_t)$ denotes the bias in the estimator $\hat{\theta}_t$ of θ.
Evidently if $B(\hat{\theta}_t) = 0$ then

$$\text{MSE}(\hat{\theta}_t) = V(\hat{\theta}_t).$$

Thus if $\hat{\theta}_1$ and $\hat{\theta}_2$ are two different estimators of the parameter θ then the estimator $\hat{\theta}_1$ is said to be more efficient than the estimator $\hat{\theta}_2$ if and only if

$$\text{MSE}(\hat{\theta}_1) < \text{MSE}(\hat{\theta}_2).$$

1.42 RELATIVE EFFICIENCY

In general, the relative efficiency of an estimator $\hat{\theta}_1$ with respect to another estimator $\hat{\theta}_2$ is expressed as a percentage and is defined as:

$$\text{RE} = \text{MSE}(\hat{\theta}_2) \times 100 / \text{MSE}(\hat{\theta}_1). \quad (1.42.1)$$

1.43 RELATIVE BIAS

The ratio of the absolute value of the bias in an estimator to the square root of the mean square error of the estimator is called the relative bias.
It is defined as:

$$\text{RB} = \left| B(\hat{\theta}_t) \right| / \sqrt{\text{MSE}(\hat{\theta}_t)} \quad (1.43.1)$$

where $B(\hat{\theta}_t) = E(\hat{\theta}_t) - \theta$ and the relative bias is independent of the units of measurement of the original data.

1.44 VARIANCE ESTIMATION THROUGH SPLITTING

If $\hat{\theta}_1,\ \hat{\theta}_2,....,\hat{\theta}_n$ are independently distributed random variables with $E(\hat{\theta}_j)=\theta\ \ \forall\ j$, and $\hat{\theta}=\frac{1}{n}\sum_{j=1}^{n}\hat{\theta}_j$, then

$$\hat{v}(\hat{\theta})=\frac{1}{n(n-1)}\sum_{j=1}^{n}(\hat{\theta}_j-\hat{\theta})^2 \tag{1.44.1}$$

is an unbiased estimator of $V(\hat{\theta})$. If $\hat{\theta}_j=\hat{\theta}_{(j)}$ is the jth estimator of θ obtained by dropping the jth unit from the sample of size n, then such a method of variance estimation are also called Jackknife method of variance estimation, and the estimator of variance takes the form

$$\hat{v}_{\text{Jack}}(\hat{\theta})=\frac{(n-1)}{n}\sum_{j=1}^{n}(\hat{\theta}_{(j)}-\hat{\theta})^2 \tag{1.44.2}$$

where $\hat{\theta}=\frac{1}{n}\sum_{j=1}^{n}\hat{\theta}_{(j)}$.

For example, if $\hat{\theta}=\bar{y}=\frac{1}{n}\sum_{i=1}^{n}y_i$ is an estimator of the population mean $\bar{Y}$ under SRSWR sampling, then $\hat{\theta}_{(j)}=\bar{y}_{(j)}=\frac{1}{n-1}\sum_{i\neq j=1}^{n}y_i$, denote the estimator of the population mean $\bar{Y}$ obtained by dropping j^{th} unit from the sample. Clearly, we can write

$$\hat{\theta}_{(j)}=\bar{y}_{(j)}=\frac{1}{n-1}\sum_{i\neq j=1}^{n}y_i=\frac{1}{n-1}\left[\sum_{i=1}^{n}y_i-y_j\right]=\frac{1}{n-1}\left[\sum_{i=1}^{n}y_i-\bar{y}+\bar{y}-y_j\right]$$

$$=\frac{1}{n-1}\left[n\,\bar{y}-\bar{y}+\bar{y}-y_j\right]=\bar{y}-\frac{1}{n-1}(y_j-\bar{y}).$$

Also

$$\hat{\theta}=\frac{1}{n}\sum_{j=1}^{n}\bar{y}_{(j)}=\frac{1}{n}\sum_{j=1}^{n}\left[\frac{1}{n-1}\sum_{i\neq j=1}^{n}y_i\right]=\frac{1}{n(n-1)}\sum_{j=1}^{n}\sum_{i\neq j=1}^{n}y_i$$

$$=\frac{1}{n(n-1)}\sum_{j=1}^{n}\left[\sum_{i=1}^{n}y_i-y_j\right]=\frac{1}{n(n-1)}\sum_{j=1}^{n}\left[n\,\bar{y}-y_j\right]$$

$$=\frac{1}{n(n-1)}\left[\sum_{j=1}^{n}n\bar{y}-\sum_{j=1}^{n}y_j\right]=\frac{1}{n(n-1)}\left[n^2\bar{y}-n\bar{y}\right]=\frac{n(n-1)}{n(n-1)}\bar{y}=\bar{y}.$$

Thus the Jackknife estimator of variance of $\hat{\theta}=\bar{y}$ is given by

$$\hat{v}_{\text{Jack}}(\bar{y})_{\text{srswr}}=\frac{(n-1)}{n}\sum_{j=1}^{n}(\bar{y}_{(j)}-\bar{y})^2=\frac{(n-1)}{n}\sum_{j=1}^{n}\left[\left\{\bar{y}-\frac{1}{n-1}(y_j-\bar{y})\right\}-\bar{y}\right]^2$$

$$=\frac{(n-1)}{n}\sum_{j=1}^{n}\left[\frac{1}{n-1}(y_j-\bar{y})\right]^2=\frac{1}{n(n-1)}\sum_{j=1}^{n}(y_j-\bar{y})^2=\frac{s_y^2}{n},$$

where $s_y^2 = \frac{1}{n-1}\sum_{i=1}^{n}(y_i - \bar{y})^2$.

Thus $\hat{v}_{\text{Jack}}(\bar{y})_{\text{srswr}}$ is an unbiased estimator of $V(\bar{y})$ under SRSWR sampling. The Jackknife technique provides good estimate of variance under SRSWR sampling, but for other sampling schemes we need to adjust it to obtain unbiased estimator of variance. For example, under SRSWOR sampling, an unbiased estimator of $V(\bar{y})$ will be

$$\hat{v}_{\text{Jack}}(\bar{y})_{\text{srswor}} = \frac{(1-f)(n-1)}{n}\sum_{j=1}^{n}\left(\bar{y}_{(j)} - \bar{y}\right)^2 \tag{1.44.3}$$

where $f = n/N$.

Note that this is not always possible to adjust Jackknife estimator of variance to make it unbiased for other sampling schemes available in the literature.

1.45 LOSS FUNCTION

The risk or loss associated with an estimator $\hat{\theta}_t$ of θ, of order r is defined as

$$R(\hat{\theta}_t) = E\left[\hat{\theta}_t - \theta\right]^r \tag{1.45.1}$$

for $r = 2,3,....$ If $r = 2$ then $R(\hat{\theta}_t) = \text{MSE}(\hat{\theta}_t)$ which is generally called quadratic loss function.

1.46 ADMISSIBLE ESTIMATOR

Let Γ be a class of estimators of a population parameter θ. For a given loss function, let $R(\hat{\theta}_t)$ represent the risk or expected loss associated with the estimator $\hat{\theta}_t$ of θ. Out of two estimators $\hat{\theta}_1$ and $\hat{\theta}_2$ of the population parameter θ, the estimator $\hat{\theta}_1$ will be said to be uniformly better than $\hat{\theta}_2$ if, for a given loss function, the following inequality

$$R(\hat{\theta}_1) < R(\hat{\theta}_2) \tag{1.46.1}$$

holds for all possible values of the characteristic under study. Now an estimator $\hat{\theta}_t$ belonging to Γ is said to be admissible in Γ if there exists no other estimator in Γ which is better than $\hat{\theta}_t$.

1.47 SAMPLE SURVEY

A sample survey is a survey which is carried out using sampling methods, *i.e.*, in which only a portion and not the whole population is surveyed.

1.48 SAMPLING DISTRIBUTION

A sampling distribution is a distribution of a statistic in all possible samples which can be chosen according to a specified sampling scheme. The expression almost always relates to a sampling scheme involving random selection, and most usually concerns the distribution of a function of a fixed number n of independent variables.

Example 1.48.1. Select all possible SRSWR samples each of two units from the population consisting of four units 1,3,5 and 7.

(a) Construct the sampling distribution of the sample means.
(b) Construct the sampling distribution of the sample variances.

Solution. The list of 16 samples of size 2 from the population and the mean of each sample is given in the following table.

Sample	1,1	1,3	1,5	1,7	3,1	3,3	3,5	3,7	5,1	5,3	5,5	5,7	7,1	7,3	7,5	7,7
Means	1	2	3	4	2	3	4	5	3	4	5	6	4	5	6	7
Variances	0	2	8	18	2	0	2	8	8	2	0	2	18	8	2	0

(a) The relative frequency distribution of the sample means is

Sample means	Frequency	Relative frequency
1	1	0.0625
2	2	0.1250
3	3	0.1875
4	4	0.2500
5	3	0.1875
6	2	0.1250
7	1	0.0625

A histogram of the above sampling distribution of sample means is given below.

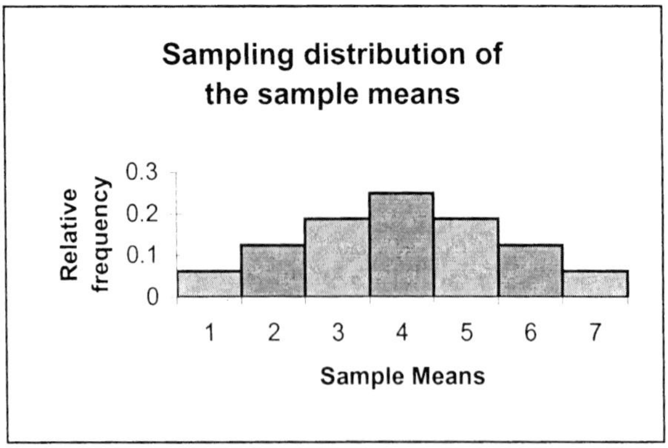

Fig. 1.48.1 Distribution of the sample means.

(b) The relative frequency distribution of the sample variances is given in the following table.

Sample variance	Frequency	Relative frequency
0	4	0.250
2	6	0.375
8	4	0.250
18	2	0.125

A relative histogram of the above sampling distribution of sample variances is given below.

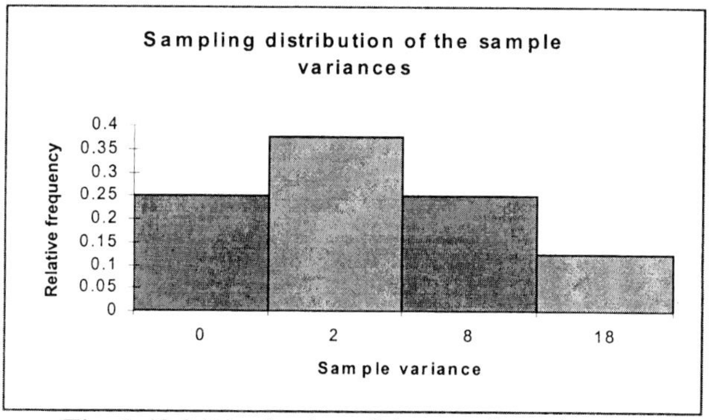

Fig. 1.48.2 Distribution of the sample variances.

1.49 SAMPLING FRAME

A sample space ψ (or S) of identifiable units or elements of population to be surveyed is called a sampling frame. It may be a discrete space such as households, individuals or a continuous space such as area under a particular crop.

1.50 SAMPLE SURVEY DESIGN

Let $\psi = \{ t \}$, $t = 1, 2, ..., s(\Omega)$ be a specified space of samples, B_t be a Borel set in ψ and p_t be the probability measure defined on B_t, then the triplet (ψ, B_t, p_t) is called a sample survey design.

1.51 ERRORS IN THE ESTIMATORS

In general, two types of errors, which arise during the process of sampling, have been observed in actual practice in the estimators:

(a) Sampling errors; (b) Non-sampling errors.

Let us briefly explain these errors.

1.51.1 SAMPLING ERRORS

An error which arises due to sampling is called a sampling error. Let us explain this with the help of the following example. For a population of size $N = 4$, let the units be $A = 1$, $B = 2$, $C = 3$, and $D = 4$. The population mean is given by, $\overline{Y} = 2.5$. There are ${}^{N}C_{n} = {}^{4}C_{2} = 6$ possible samples each of size $n = 2$. The units selected in the six samples are: (A, B), (A, C), (A, D), (B, C), (B, D), and (C, D). Thus six sample means are given by:

Samples	1	2	3	4	5	6
Units Selected	(A, B) or (1, 2)	(A, C) or (1, 3)	(A, D) or (1, 4)	(B, C) or (2, 3)	(B, D) or (2, 4)	(C, D) or (3, 4)
Sample Means ($\overline{y}_t$)	1.5	2.0	2.5	2.5	3.0	3.5

If we take each of the sample means and population mean separately, then, we have the following cases: error of $(A, B) = |1.5 - 2.5| = 1.0$; error of $(A, C) = |2.0 - 2.5| = 0.5$; error of $(A, D) = |2.5 - 2.5| = 0.0$; error of $(B, C) = |2.5 - 2.5| = 0.0$; error of $(B, D) = |3.0 - 2.5| = 0.5$; error of $(C, D) = |3.5 - 2.5| = 1.0$. Note that we are measuring only two units out of four units, *i.e.*, we have only partial information in the sample therefore sampling error arises. One of the measurements for the sampling error is the variance of the estimator. For example, the variance of the sample mean estimator, $\overline{y}_t$, is

$$V(\overline{y}_t) = E[\overline{y}_t - E(\overline{y}_t)]^2 = \sum_{t=1}^{6} p_t (\overline{y}_t - \overline{Y})^2 = \frac{1}{6}\sum_{t=1}^{6}(\overline{y}_t - \overline{Y})^2$$

$$= \frac{1}{6}\left[(1.5 - 2.5)^2 + (2.0 - 2.5)^2 + (2.5 - 2.5)^2 + (2.5 - 2.5)^2 + (3.0 - 2.5)^2 + (3.5 - 2.5)^2\right] = 0.41$$

because $p_t = \frac{1}{6}\ \forall\ t = 1,..,6$, and $E(\overline{y}_t) = \overline{Y}$.

Also note that

$$\left(\frac{N - n}{Nn}\right) S_y^2 = \left(\frac{4 - 2}{4 \times 2}\right) \times \frac{5}{3} = 0.41 = V(\overline{y}_t).$$

For its theoretical derivations refer to Chapter 2.

1.51.2 NON-SAMPLING ERRORS

These errors are of four types:

(i) Non-response errors; (ii) Measurement errors;

(iii) Tabulation errors; and (iv) Computational errors.

Let us discuss each of these errors in brief as follows:

1.51.2.1 NON-RESPONSE ERRORS

The people from whom we get the information are called the respondents and the people in the sample from whom we do not get information are called non-respondents. The error which arises, when we fail to get the information is called non-response error and the phenomenon is called non-response. This error arises because of the fact that we are not able to cover the whole sample. For example, if we want to interview 100 farmers and suppose 5 out of them do not allow us to interview them. Then we are interviewing only 95. So the sample is not complete. Such errors are called non-response errors.

1.51.2.2 MEASUREMENT ERRORS

The errors that we bring in measuring the characters are called measurement errors. For example, suppose we want to measure the age of the respondents. Among the respondents, some may report their age less than their actual age. These types of errors are called measurement errors.

1.51.2.3 TABULATION ERRORS

The errors which arise due to missing some numbers due to non availability of data or recording some numbers wrongly, while making a table is called a tabulation error.

1.51.2.4 COMPUTATIONAL ERRORS

After the table is formed, we start our calculations. The errors committed in calculations are known as computational errors.

1.52 POINT ESTIMATOR

A point estimator endeavours to give the best single estimated value of the parameter. For example, the average height of school children is 5.3 feet.

1.53 INTERVAL ESTIMATOR

An interval estimator of a population parameter which specifies a range of values bounded by an upper and a lower limit, within which the true value is asserted to lie. For example, the average height of school children lies between 4.9 feet and 5.5 feet.

1.54 CONFIDENCE INTERVAL

If it is possible to define two statistics $\hat{\theta}_1$ and $\hat{\theta}_2$ (functions of sample values only) to estimate a population parameter θ, such that

$$P\left(\hat{\theta}_1 < \theta < \hat{\theta}_2\right) = (1-\alpha) \quad (1.54.1)$$

where $1-\alpha$ is some fixed probability, the interval between $\hat{\theta}_1$ and $\hat{\theta}_2$ is called a confidence interval. The assertion that θ lies in this interval will be true, on the average, in a proportion $1-\alpha$ of the cases when the assertion is made. Note that an interval estimate at the same level of confidence with smaller width is considered as better estimate. For example, if someone says with 95% confidence that the average marks of a particular class lies between 65% to 85%, then this estimate is better than the estimate if someone says with same confidence that the average marks lies between 0% to 100%. We saw for SRSWOR sampling, the sample mean $\bar{y}_t$ is unbiased for population mean with

$$V(\bar{y}_t) = \left(\frac{N-n}{Nn}\right)S_y^2, \quad (1.54.2)$$

and an unbiased estimator of variance is given by

$$\hat{v}(\bar{y}_t) = \left(\frac{N-n}{Nn}\right)s_y^2. \quad (1.54.3)$$

Thus there are two cases: If $V(\bar{y}_t)$ is known then for a large sample, a $(1-\alpha)100\%$ confidence interval estimate for the population mean $\bar{Y}$ is given by

$$\bar{y}_t \pm Z_{\alpha/2}\sqrt{V(\bar{y}_t)} \quad (1.54.4)$$

where $Z_{\alpha/2}$ values are given in Table 3 of the Appendix, and if $V(\bar{y}_t)$ is unknown then for a small sample, a $(1-\alpha)100\%$ confidence interval estimate for the population mean $\bar{Y}$ is given by

$$\bar{y}_t \pm t_{\alpha/2}(\text{df} = n-1)\sqrt{\hat{v}(\bar{y}_t)} \quad (1.54.5)$$

where $t_{\alpha/2}(\text{df} = n-1)$ values are given in Table 2 of the Appendix, and df stands for degree of freedom. Note that if $\alpha - 0.05$ it represents $(1-\alpha)100\% = (1-0.05)100\%$ $= 95\%$ confidence interval.

Let us illustrate it with the following example:

Example 1.54.1. Consider a population consisting of $N = 7$ units, *viz.* $A = 1$, $B = 1$, $C = 8$, $D = 8$, $E = 8$, $F = 9$, and $G = 9$. Consider all possible SRSWOR samples of $n = 2$ units, and compute confidence interval estimates of population mean based on each sample under the following two different situations:

(a) Variance or mean square error S_y^2 is known;

(b) Variance or mean square error S_y^2 is unknown.

Find the proportion of confidence interval estimates in which the true population mean is included.

Solution. Note that for this population, we have population mean $\bar{Y} = 6.29$ and population mean square $S_y^2 = 13.24$.

Now the total number of SRSWOR samples will be

$$^{N}C_n = {}^{7}C_2 = \frac{7!}{2!\,(7-2)!} = \frac{7!}{2!\times 5!} = \frac{7\times 6\times 5\times 4\times 3\times 2\times 1}{2\times 1\times 5\times 4\times 3\times 2\times 1} = 21 .$$

(a) When population variance is known: The lower and upper limits are

$$L_1 = \bar{y}_t - z_{\alpha/2}\sqrt{V(\bar{y}_t)} = \bar{y}_t - 1.96\sqrt{\left(\frac{N-n}{Nn}\right)S_y^2} ,$$

and

$$U_1 = \bar{y}_t + z_{\alpha/2}\sqrt{V(\bar{y}_t)} = \bar{y}_t + 1.96\sqrt{\left(\frac{N-n}{Nn}\right)S_y^2} .$$

(b) When population variance is not known: The lower and upper limits are

$$L_2 = \bar{y}_t - t_{\alpha/2}(\text{df} = n-1)\sqrt{\hat{v}(\bar{y}_t)} = \bar{y}_t - 12.71\sqrt{\left(\frac{N-n}{Nn}\right)s_y^2} ,$$

and

$$U_2 = \bar{y}_t + t_{\alpha/2}(\text{df} = n-1)\sqrt{\hat{v}(\bar{y}_t)} = \bar{y}_t + 12.71\sqrt{\left(\frac{N-n}{Nn}\right)s_y^2} .$$

All possible samples, sample means, and variances, lower and upper limits of the 95% confidence interval, and their coverage is given in the following table.

Sample Values				Sample		Variance known			Variance unknown		
				$\bar{y}_t$	s_y^2	L_1	U_1	$\bar{Y} \in CI(1)$	L_2	U_2	$\bar{Y} \in CI(2)$
A	*B*	1	1	1.0	0.00	-3.26	5.26	No	1.0	1.0	No
A	*C*	1	8	4.5	24.50	0.24	8.76	Yes	-33.1	42.1	Yes
A	*D*	1	8	4.5	24.50	0.24	8.76	Yes	-33.1	42.1	Yes
A	*E*	1	8	4.5	24.50	0.24	8.76	Yes	-33.1	42.1	Yes
A	*F*	1	9	5.0	32.00	0.74	9.26	Yes	-38.0	48.0	Yes
A	*G*	1	9	5.0	32.00	0.74	9.26	Yes	-38.0	48.0	Yes
B	*C*	1	8	4.5	24.50	0.24	8.76	Yes	-33.1	42.1	Yes
B	*D*	1	8	4.5	24.50	0.24	8.76	Yes	-33.1	42.1	Yes
B	*E*	1	8	4.5	24.50	0.24	8.76	Yes	-33.1	42.1	Yes
B	*F*	1	9	5.0	32.00	0.74	9.26	Yes	-38.0	48.0	Yes
B	*G*	1	9	5.0	32.00	0.74	9.26	Yes	-38.0	48.0	Yes
C	*D*	8	8	8.0	0.00	3.74	12.26	Yes	8.0	8.0	No
C	*E*	8	8	8.0	0.00	3.74	12.26	Yes	8.0	8.0	No
C	*F*	8	9	8.5	0.50	4.24	12.76	Yes	3.1	13.9	Yes
C	*G*	8	9	8.5	0.50	4.24	12.76	Yes	3.1	13.9	Yes
D	*E*	8	8	8.0	0.00	3.74	12.26	Yes	8.0	8.0	No

Continued……

D F	8 9	8.5	0.50	4.24	12.76	Yes	3.1	13.9	Yes
D G	8 9	8.5	0.50	4.24	12.76	Yes	3.1	13.9	Yes
E F	8 9	8.5	0.50	4.24	12.76	Yes	3.1	13.9	Yes
E G	8 9	8.5	0.50	4.24	12.76	Yes	3.1	13.9	Yes
F G	9 9	9.0	0.00	4.74	13.26	Yes	9.0	9.0	No

Thus we observed that the population mean $\bar{Y} = 6.29$ lies 20 times between L_1 and U_1 out of total 21 times, and hence the observed proportion of confidence intervals containing population mean = 20/21 = 0.9524. In other words, 95.24% cases the population mean lies between the confidence interval estimates when variance is known. Thus the observed percentage is very close to the expected coverage of 95%.

Also we observed, when population variance is not known, then the population mean lies 16 times between L_2 and U_2, and hence the observed proportion of the confidence interval estimates containing population mean = 16/21 = 0.7619, that is, only 76.19% times the population mean lies between the confidence interval estimates when variance is unknown. Here the observed percentage of the coverage is lower than the expected coverage of 95%. This may be due to very small sample and population size. In practice as the sample size becomes large, (How large? Just smile because there is no unique answer), then the observed proportion of coverage in both cases converges to 95%.

1.55 POPULATION PROPORTION

It is defined as the total number of units of a particular interest in a subgroup A (such that $A \cup A^c = \Omega$) of a population divided by the total number of units in the population. In other words, when the variable of interest Y takes only two values 1 and 0 that is $Y_i = 1$ (if i $\in$ A) and $Y_i = 0$ (if i $\in$ A^c), then the population mean $\bar{Y}$ also becomes population proportion P as follows:

$$\bar{Y} = \frac{1}{N}\sum_{i=1}^{N} Y_i = \frac{1}{N}(1+0+1+0+......0+1) = \frac{N_1}{N} = P, \qquad (1.55.1)$$

where N_1 denotes the number of units of the population in the group A, and N denotes the total number of units in the population. Note that the value of population proportion P lies between 0 and 1 that is $0 \le P \le 1$. Further note that here we are dealing with qualitative variables.

1.56 SAMPLE PROPORTION

It is defined as the total number of units of a particular interest in a subgroup A (such that $A \cup A^c = s$) of a sample divided by the total number of units in the sample. In other words, when the variable of interest y takes only two values 1 and

0, that is, $y_i = 1$ (if $i \in A$) and $y_i = 0$ (if $i \in A^c$), then the sample mean $\bar{y}_s$ also becomes sample proportion $\hat{p}$, as follows:

$$\bar{y}_t = \frac{1}{n}\sum_{i=1}^{n} y_i = \frac{1}{n}(1+0+1+0+......0+1) = \frac{n_1}{n} = \hat{p}, \tag{1.56.1}$$

where n_1 denotes the number of units of the sample in the group A, and n denotes the total number of units in the sample. Note that the value of sample proportion also lies between 0 and 1 that is $0 \le \hat{p} \le 1$.

1.57 VARIANCE OF SAMPLE PROPORTION AND CONFIDENCE INTERVAL ESTIMATES

(a) The variance of sample proportion under SRSWOR sampling is give by

$$V_{\text{wor}}(\hat{p}) = \frac{(N-n)}{n(N-1)}P(1-P) \tag{1.57.1}$$

and its estimator is given by

$$\hat{v}_{\text{wor}}(\hat{p}) = \frac{(N-n)}{N(n-1)}\hat{p}(1-\hat{p}). \tag{1.57.2}$$

Thus a $(1-\alpha)100\%$ confidence interval estimate of the population proportion P is given by

$$\hat{p} \pm Z_{\alpha/2}\sqrt{\hat{v}_{\text{wor}}(\hat{p})}. \tag{1.57.3}$$

(b) The variance of sample proportion under SRSWR sampling is give by

$$V_{\text{wr}}(\hat{p}) = \frac{P(1-P)}{n}, \tag{1.57.4}$$

and its estimator is given by

$$\hat{v}_{\text{wr}}(\hat{p}) = \frac{\hat{p}(1-\hat{p})}{(n-1)}. \tag{1.57.5}$$

Thus a $(1-\alpha)100\%$ confidence interval estimate of the population proportion P is given by

$$\hat{p} \pm Z_{\alpha/2}\sqrt{\hat{v}_{\text{wr}}(\hat{p})}. \tag{1.57.6}$$

For detail of derivations of the results related to proportion, refer to Chapter 2.

Example 1.57.1. Consider a class consisting of 6 students. Their names and major are given in the following table:

Name	Major
Amy	Math
Bob	English
Chris	Math
Don	English
Erin	Math
Frank	English

(a) Find the proportion of English students in the class.
(b) How many SRSWOR samples, each of $n = 4$ students, will be there?
(c) What is the sampling distribution of estimate of proportion?

Solution. (a) Count the number of students in the population, $N = 6$.

Count the number of students with major English, $\text{COUNT} = N_1 = 3$.

Compute the proportion of English students:

$$P = \frac{\text{COUNT}}{N} = \frac{N_1}{N} = \frac{3}{6} = 0.5 \text{ (parameter).}$$

Recall that a parameter is an unknown quantity and we try to estimate it by taking a random sample from the population.

(b) How many SRSWOR samples, each of four units, will there be?

The possible combinations of choosing 4 objects out of 6 objects are given by:

$${}^6C_4 = \frac{6!}{4!(6-4)!} = \frac{6!}{4!\times 2!} = \frac{6\times 5\times 4\times 3\times 2\times 1}{4\times 3\times 2\times 1\times 2\times 1} = 15.$$

Note that each combination can be taken as a without replacement sample, so the total number of distinct samples will be 15.

(c) Sampling distribution of estimate of proportion: Let us construct those 15 samples as follows:

Sample No.	Sampled students				Major of the selected students M--Math E--English				Proportion estimate $\hat{p} = \frac{\text{count}}{n}$
1	Amy	Bob	Chris	Don	M	E	M	E	=2/4=0.50
2	Amy	Bob	Chris	Erin	M	E	M	M	=1/4=0.25
3	Amy	Bob	Chris	Frank	M	E	M	E	=2/4=0.50
4	Amy	Bob	Don	Erin	M	E	E	M	=2/4=0.50
5	Amy	Bob	Don	Frank	M	E	E	E	=3/4=0.75
6	Amy	Bob	Erin	Frank	M	E	M	E	=2/4=0.50
7	Amy	Chris	Don	Erin	M	M	E	M	=1/4=0.25
8	Amy	Chris	Don	Frank	M	M	E	E	=2/4=0.50
9	Amy	Chris	Erin	Frank	M	M	M	E	=1/4=0.25
10	Amy	Don	Erin	Frank	M	E	M	E	=2/4=0.50
11	Bob	Chris	Don	Erin	E	M	E	M	=2/4=0.50
12	Bob	Chris	Don	Frank	E	M	E	E	=3/4=0.75
13	Bob	Chris	Erin	Frank	E	M	M	E	=2/4=0.50
14	Bob	Don	Erin	Frank	E	E	M	E	=3/4=0.75
15	Chris	Don	Erin	Frank	M	E	M	E	=2/4=0.50

From the above table we have the following table:

Proportion estimate $\hat{p}$	Tally marking	Frequency f_i	Relative frequency $RF_i = f_i/\sum f_i$
0.25	III	3	3/15
0.50	IIIII IIII	9	9/15
0.75	III	3	3/15
	Sum	15	1

The above table shows that the distribution of estimates of proportion is symmetric, or say normal.

Let $x_i = \hat{p}$ and $p_i = RF_i$. Then the expected value of $x_i = \hat{p}$

$$E(x_i) = \mu = E(\hat{p}) = \sum_{i=1}^{3} p_i x_i = p_1x_1 + p_2x_2 + p_3x_3 = \frac{3}{15}\times 0.25 + \frac{9}{15}\times 0.50 + \frac{3}{15}\times 0.75$$

$$= 0.05 + 0.30 + 0.15 = 0.5 = P \text{ (Population proportion).}$$

Because $E(\hat{p}) = P$, which implies that the sample proportion is an unbiased estimator of the population proportion.

By the computing formula of variance, we have

$$\sigma^2 = V(x_i) = V(\hat{p}) = \left[\sum_{i=1}^{3}\left(p_i x_i^2\right)\right] - (\mu)^2 = \left[p_1x_1^2 + p_2x_2^2 + p_3x_3^2\right] - (\mu)^2$$

$$= \left[\frac{3}{15}\times 0.25^2 + \frac{9}{15}\times 0.50^2 + \frac{3}{15}\times 0.75^2\right] - (0.5)^2$$

$$= [0.0125 + 0.15 + 0.1125] - (0.25) = 0.275 - 0.25 = 0.025.$$

Also note that

$$\left(\frac{N-n}{N-1}\right)\frac{P(1-P)}{n} = \left(\frac{6-4}{6-1}\right)\frac{0.5(1-0.5)}{4} = \frac{2}{5}\times\frac{0.25}{4} = 0.025.$$

Thus the variance of the estimator of population proportion using without replacement sampling is given by

$$\sigma_{\hat{p}}^2 = \left(\frac{N-n}{N-1}\right)\frac{P(1-P)}{n}.$$

Similarly, repeat this example yourself by taking all SRSWR samples, and study the unbiasedness and variance of the estimate of proportion, and also histogram the sampling distribution of the estimate of proportion.

Example 1.57.2. Consider a class of 16 students taking statistics course, and their names, marks, and major subjects are given in the following table:

Sr. No.	Name	Marks	Major
1	Ruth	92	Math
2	Ryan	97	Math
3	Tim	68	English
4	Raul	62	Math
5	Marla	97	English
6	Erin	68	Math
7	Judy	76	English
8	Troy	75	English
9	Tara	51	Math
10	Lisa	94	Math
11	John	70	Math
12	Cher	89	English
13	Lona	62	Math
14	Gina	63	Math
15	Jeff	48	Math
16	Sara	97	Math

1. Compute the following parameters:
   (a) Population mean;
   (b) Population variance;
   (c) Population standard deviation;
   (d) Population mean square error;
   (e) Population coefficient of variation;
2. (a) Select an SRSWOR sample of 4 units using Random Number method;
   (b) Estimate the population mean and population total;
   (c) Compute the variance of the estimator of population mean;
   (d) Estimate s_y^2;
   (e) Estimate the variance of the estimator of population mean;
   (f) Construct 95% confidence interval of the population mean assuming that population mean square is known and sample size is large. Does the population mean falls in it? Interpret it;
   (g) Construct 95% confidence interval assuming that population mean square is unknown and sample size in small. Does the population mean falls in it? Interpret it;
3. (a) Compute the population proportion of major in English students;
   (b) Estimate the proportion of major in English on the basis above sample;
   (c) Compute the variance of the estimator of population proportion;
   (d) Estimate the variance of the estimator of population proportion;
   (e) Construct 95% confidence interval for the proportion.

Solution. We have

Name	Y_i	Y_i^2
Ruth	92	8464
Ryan	97	9409
Tim	68	4624
Raul	62	3844
Marla	97	9409
Erin	68	4624
Judy	76	5776
Troy	75	5625
Tara	51	2601
Lisa	94	8836
John	70	4900
Cher	89	7921
Lona	62	3844
Gina	63	3969
Jeff	48	2304
Sara	97	9409
Sum	1209	95559

From the population information we have

$$N = 16,\ \sum_{i=1}^{N} Y_i = 1209 \text{ and } \sum_{i=1}^{N} Y_i^2 = 95559.$$

1.

(a) Population mean:

$$\bar{Y} = \frac{\sum_{i=1}^{N} Y_i}{N} = \frac{1209}{16} = 75.56 \text{ (parameter)}.$$

(b) Population variance:

$$\sigma_y^2 = \frac{\sum_{i=1}^{N} Y_i^2 - \frac{\left(\sum_{i=1}^{N} Y_i\right)^2}{N}}{N} = \frac{95559 - \frac{(1209)^2}{16}}{16} = 262.75 \text{ (parameter)}.$$

(c) Population standard deviation:

$$\sigma_y = \sqrt{\sigma_y^2} = \sqrt{262.75} = 16.20 \text{ (parameter)}.$$

(d) Population mean square error:

$$S_y^2 = \frac{\sum_{i=1}^{N} Y_i^2 - \frac{\left(\sum_{i=1}^{N} Y_i\right)^2}{N}}{N-1} = \frac{95559 - \frac{(1209)^2}{16}}{16-1} = 280.26 \text{ (parameter)}.$$

(e) Population coefficient of variation: We are using SRSWOR sampling, so

$$C_y = \frac{S_y}{\bar{Y}} = \frac{\sqrt{S_y^2}}{\bar{Y}} = \frac{\sqrt{280.26}}{75.56} = \frac{16.74}{75.56} = 0.2215 \text{ (parameter).}$$

2.

(a) Selection of 4 units using SRSWOR sampling ($n = 4$): Let us start with 1st row and 6th column of the Pseudo-Random Number Table 1 given in the Appendix.

Random Number	Decision: R – Rejection, S -- Selection	Name of the selected student
62	R	
77	R	
92	R	
67	R	
53	R	
51	R	
33	R	
07	S	Judy
62	R	
69	R	
76	R	
48	R	
50	R	
88	R	
37	R	
72	R	
63	R	
21	R	
33	R	
25	R	
76	R	
09	S	Tara
43	R	
80	R	
94	R	
62	R	
68	R	
15	S	Jeff
42	R	
93	R	
29	R	
01	S	Ruth

So our SRSWOR sample consists of four students $=\{\text{Judy, Tara, Jeff, Ruth}\}$.

Now from the sample we have the following information:

Name	y_i	y_i^2
Judy	76	5776
Tara	51	2601
Jeff	48	2304
Ruth	92	8464
Sum	267	19145

Thus

$$n = 4,\quad \sum_{i=1}^{n} y_i = 267 \quad\text{and}\quad \sum_{i=1}^{n} y_i^2 = 19145\,.$$

(b) Sample mean:

$$\bar{y}_t = \frac{\sum_{i=1}^{n} y}{n} = \frac{267}{4} = 66.75 \text{ (statistic)}$$

which is an estimate of population mean.

Note that an estimator of population total $Y = \sum_{i=1}^{N} Y_i$ will be given by

$$\hat{y}_t = N\,\bar{y}_t = 16\times 66.75 = 1068 \text{ (statistic)}.$$

(c) The variance of the sample mean estimator is given by

$$V(\bar{y}_t) = \left(\frac{N-n}{Nn}\right)S_y^2 = \left(\frac{16-4}{16\times 4}\right)\times 280.26 = 52.548 \text{ (parameter)}.$$

(d) An estimator of S_y^2 is given by

$$s_y^2 = \frac{\sum_{i=1}^{n} y_i^2 - \frac{\left(\sum_{i=1}^{n} y_i\right)^2}{n}}{n-1} = \frac{19145 - \frac{(267)^2}{4}}{4-1} = 440.91 \text{ (statistic)}.$$

(e) An estimator of the variance of the estimator of the population mean is

$$\hat{v}(\bar{y}_t) = \left(\frac{N-n}{Nn}\right)s_y^2 = \left(\frac{16-4}{16\times 4}\right)\times 440.91 = 82.67 \text{ (statistic)}.$$

(f) Here 95% confidence interval is given by

$$\bar{y}_t \pm 1.96\sqrt{V(\bar{y}_t)},\ \text{ or }\ 66.75 \pm 1.96\sqrt{52.548},\ \text{ or }\ 66.75\pm 14.20,\ \text{ or }\ [52.55,\ 80.20].$$

Yes, the true population mean $\bar{Y} = 75.26$ lies in the 95% confidence interval estimate. The interpretation of 95% confidence interval is that we are 95% sure that the true mean lies in these two limits of this interval estimate. Note that interval estimate is a statistic.

(g) Here 95% confidence interval estimate is given by

$$\bar{y}_t \pm t_{\alpha/2}(\text{df} = n-1)\sqrt{\hat{v}(\bar{y}_t)}, \quad \text{or} \quad 66.75 \pm t_{0.025}(\text{df} = 3)\sqrt{82.67},$$

or $66.75 \pm 3.182\sqrt{82.67}$, or 66.75 ± 28.93, or $[37.82, 95.68]$

where $t_{\alpha/2}(\text{df} = n-1) = t_{0.025}(\text{df} = 3) = 3.182$ is taken from Table 2 of the Appendix.

Yes, again the true population mean lies in this 95% confidence interval and its interpretation is same as above. Again note that interval estimate is a statistic.

3.

(a) Let us give upper case 'FLAG' of 1 to English majors and 0 to Math major students in the whole population, then we have

Sr. No.	Name	Major	FLAG: E--1, M--0
1	Ruth	Math	0
2	Ryan	Math	0
3	Tim	English	1
4	Raul	Math	0
5	Marla	English	1
6	Erin	Math	0
7	Judy	English	1
8	Troy	English	1
9	Tara	Math	0
10	Lisa	Math	0
11	John	Math	0
12	Cher	English	1
13	Lona	Math	0
14	Gina	Math	0
15	Jeff	Math	0
16	Sara	Math	0
		Sum	5

Population Proportion:

$$P = \frac{\sum_{i=1}^{N} \text{FLAG}_i}{N} = \frac{\text{No. of students with english major}}{\text{Total No. of Students}} = \frac{5}{16} = 0.3125 \text{ (parameter).}$$

(b) Let us now give the same lower case 'flag' to students in the sample.

Name	Major	flag
Judy	English	1
Tara	Math	0
Jeff	Math	0
Ruth	Math	0
	Sum	1

Sample proportion: The sample proportion is given by

$$\hat{p} = \frac{\sum_{i=1}^{n} \text{flag}_i}{n} = \frac{1}{4} = 0.25 .$$

(c) Variance of the estimator of proportion under SRSWOR sampling is given by

$$V_{\text{wor}}(\hat{p}) = \frac{(N-n)}{n(N-1)} P(1-P) = \frac{(16-4)}{4\times(16-1)} \times 0.3125 \times (1-0.3125) = 0.0429 .$$

(d) An estimator of variance of the estimator of proportion under SRSWOR sampling is given by

$$\hat{v}_{\text{wor}}(\hat{p}) = \frac{(N-n)}{N(n-1)} \hat{p}(1-\hat{p}) = \frac{(16-4)}{16(4-1)} \times 0.25 \times (1-0.25) = 0.04687 .$$

(e) A 95% confidence interval estimate for the true population proportion is

$$\hat{p} \pm 1.96\sqrt{\hat{v}_{\text{wor}}(\hat{p})}$$

or $0.25 \pm 1.96\sqrt{0.04687}$, or 0.25 ± 0.424, or $[-0.174, 0.674]$, or $[0.0, \quad 0.674]$.

Note that a proportion can never be negative, so lower limit has been changed to 0.

Caution! It must be noted that we have here a very small sample, but in practice when we deal with the problem of estimation of proportion, the minimum sample size of 30 units is recommended from large populations. Note that instead of using 'FLAG' or 'flag', sometimes we assign codes 0 or 1 directly to the variable Y or X.

Example 1.57.3. For the population considered in the previous example:

(a) John considers a sampling scheme consisting of only 4 samples as follows.

Sample Number	Samples	p_s
1	Cher, John, Marla, Sara	0.25
2	Erin, Judy, Raul, Tara	0.25
3	Gina, Lisa, Ruth, Tim	0.25
4	Jeff, Lona, Ryan, Troy	0.25

(b) Mike considers another sampling plan consisting of 13 samples each of 4 students as given below:

Sample No.	Samples	p_t
1	Cher, Erin, Gina, Jeff	1/13
2	Cher, Erin, Gina, John	1/13
3	Cher, Erin, Gina, Judy	1/13
4	Cher, Erin, Gina, Lisa	1/13
5	Cher, Erin, Gina, Lona	1/13
6	Cher, Erin, Gina, Marla	1/13
7	Cher, Erin, Gina, Raul	1/13
8	Cher, Erin, Gina, Ruth	1/13
9	Cher, Erin, Gina, Ryan	1/13
10	Cher, Erin, Gina, Sara	1/13
11	Cher, Erin, Gina, Tara	1/13
12	Cher, Erin, Gina, Tim	1/13
13	Cher, Erin, Gina, Troy	1/13

Let $\bar{y}_t$ be an estimator of the population mean. Find the following for each one of the above sampling schemes:

$$E(\bar{y}_t);\ V(\bar{y}_t);\ B(\bar{y}_t);\ \text{and MSE}(\bar{y}_t).$$

Comment on the statement, 'Mike's sampling scheme is better than John's sampling scheme'. Justify your logic and discuss the relative efficiency.

Solution. (a) John's sampling plan:

Sample No.	Sampled values				$\bar{y}_t$	p_t	$\{\bar{y}_t - E(\bar{y}_t)\}^2$	$\{\bar{y}_t - \bar{Y}\}^2$
1	89	70	97	97	88.25	0.25	162.56	161.03
2	68	76	62	51	64.25	0.25	126.56	127.91
3	63	94	92	68	79.25	0.25	14.06	13.61
4	48	62	97	75	70.50	0.25	25.00	25.60
				Sum	302.25	1.00	328.18	328.18

where $\bar{Y} = 75.56$ and $E(\bar{y}_t) = \sum_{t=1}^{4} p_t \bar{y}_t = \frac{1}{4} \times 302.25 = 75.5$.

Thus

$$B(\bar{y}_t) = E(\bar{y}_t) - \bar{Y} = 75.50 - 75.56 = -0.06,$$

$$V(\bar{y}_t) = \sum_{t=1}^{4} p_t \{\bar{y}_t - E(\bar{y}_t)\}^2 = \frac{1}{4} \times 328.18 = 82.045,$$

and

$$\text{MSE}(\bar{y}_t) = \sum_{t=1}^{4} p_t \{\bar{y}_t - \bar{Y}\}^2 = \frac{1}{4} \times 328.18 = 82.045 .$$

(b) Mike's sampling plan:

Sample No.	Sampled Values				$\bar{y}_t$	p_t	$\{\bar{y}_t - E(\bar{y}_t)\}^2$	$\{\bar{y}_t - \bar{Y}\}^2$
1	89	68	63	48	67.0	1/13	49.269600	73.316410
2	89	68	63	70	72.5	1/13	2.308062	9.378906
3	89	68	63	76	74.0	1/13	0.000370	2.441406
4	89	68	63	94	78.5	1/13	20.077290	8.628906
5	89	68	63	62	70.5	1/13	12.384990	25.628910
6	89	68	63	97	79.3	1/13	27.360950	13.597660
7	89	68	63	62	70.5	1/13	12.384990	25.628910
8	89	68	63	92	78.0	1/13	15.846520	5.941406
9	89	68	63	97	79.3	1/13	27.360950	13.597660
10	89	68	63	97	79.3	1/13	27.360950	13.597660
11	89	68	63	51	67.8	1/13	39.303250	61.035160
12	89	68	63	68	72.0	1/13	4.077293	12.691410
13	89	68	63	75	73.8	1/13	0.072485	3.285156
				Sum	962.3	1	237.807700	268.769560

where $\bar{Y} = 75.56$.

Thus we have

$$E(\bar{y}_t) = \sum_{t=1}^{13} p_t \bar{y}_t = \frac{1}{13} \times 962.3 = 74.02 ,$$

and

$$B(\bar{y}_t) = E(\bar{y}_t) - \bar{Y} = 74.02 - 75.56 = -1.54 ,$$

$$V(\bar{y}_t) = \sum_{t=1}^{13} p_t \{\bar{y}_t - E(\bar{y}_t)\}^2 = \frac{1}{13} \times 237.8077 = 18.2929,$$

and

$$\text{MSE}(\bar{y}_t) = \sum_{t=1}^{13} p_t \{\bar{y}_t - \bar{Y}\}^2 = \frac{1}{13} \times 268.76956 = 20.675.$$

Although John's sampling scheme is less biased, it has too much mean square error compared to Mike's sampling scheme. Thus we shall prefer Mike's sampling scheme over John's sampling scheme. Also note that the relative efficiency of Mike's sampling scheme over John's sampling scheme is given by

$$\text{RE} = \frac{\text{MSE}(\bar{y}_t)_{\text{John}}}{\text{MSE}(\bar{y}_t)_{\text{Mike}}} \times 100 = \frac{82.045}{20.675} \times 100 = 396.83\% .$$

Thus one can say that Mike's sampling plan is almost four times more efficient than John's sampling scheme.

1.58 RELATIVE STANDARD ERROR

The relative standard error of an estimator $\hat{\theta}$ of population parameter θ is defined as the positive square root of the relative variance of the estimator $\hat{\theta}$. Mathematically

$$\text{RSE}(\hat{\theta}) = \sqrt{\text{RV}(\hat{\theta})} \tag{1.58.1}$$

where $\text{RV}(\hat{\theta}) = V(\hat{\theta}) / [E(\hat{\theta})]^2$ denotes the relative variance of the estimator $\hat{\theta}$. The another famous name for relative standard error is coefficient of variation.

1.59 AUXILIARY INFORMATION

In many sample surveys, it is possible to collect information about some variable(s) in addition to the variable of interest or study variable. The auxiliary information is accurately known from many sources like reference books, journals, administrative records etc. and is cheaper to obtain than the study variable. For example, while estimating the average income of people living in a particular city, the plot area owned by individual may be known from some published sources. Later on we will observe that the known auxiliary information is also helpful in increasing the efficiency of the estimators. Before dealing with two variables, we should be familiar with the following terms. If Y_i and X_i denote the values of ith unit for the study variable Y and auxiliary variable X, then we have:

(a) The covariance between X and Y is

$$\text{Cov}(X,Y) = E[X - E(X)][Y - E(Y)] = E(XY) - E(X)E(Y). \tag{1.59.1}$$

The covariance between X and Y is same as that between Y and X, *i.e.*,

$$\text{Cov}(X,Y) = \text{Cov}(Y,X).$$

For example, for SRSWOR sampling, the covariance between X and Y is given by

$$S_{xy} = \frac{1}{N-1}\sum_{i=1}^{N}(X_i - \bar{X})(Y_i - \bar{Y}),$$

where $\bar{X} = \frac{1}{N}\sum_{i=1}^{N} X_i$ and $\bar{Y} = \frac{1}{N}\sum_{i=1}^{N} Y_i$. Note that an unbiased estimator S_{xy} is given by

$$s_{xy} = \frac{1}{n-1}\sum_{i=1}^{n}(x_i - \bar{x})(y_i - \bar{y})$$

where $\bar{x} = \frac{1}{n}\sum_{i=1}^{n} x_i$ and $\bar{y} = \frac{1}{n}\sum_{i=1}^{n} y_i$.

(b) The population correlation coefficient between X and Y is defined as

$$\rho_{xy} = \frac{\text{Cov}(X,Y)}{\sqrt{V(X)}\sqrt{V(Y)}}. \tag{1.59.2}$$

For simple random sampling it is given by

$$\rho_{xy} = S_{xy} \Big/ \sqrt{S_x^2 S_y^2} \; . \tag{1.59.3}$$

A biased estimator of the correlation coefficient ρ_{xy} is defined as

$$r_{xy} = s_{xy} \Big/ \sqrt{s_x^2 s_y^2} \; . \tag{1.59.4}$$

The value of ρ_{xy} (or r_{xy}) is a unit free number and it lies in the interval $[-1, +1]$. It is also independent of change of origin and scale of the variables X and Y. The linear relationship can also be seen with the help of scatter diagrams as follows:

SCATTER PLOTS

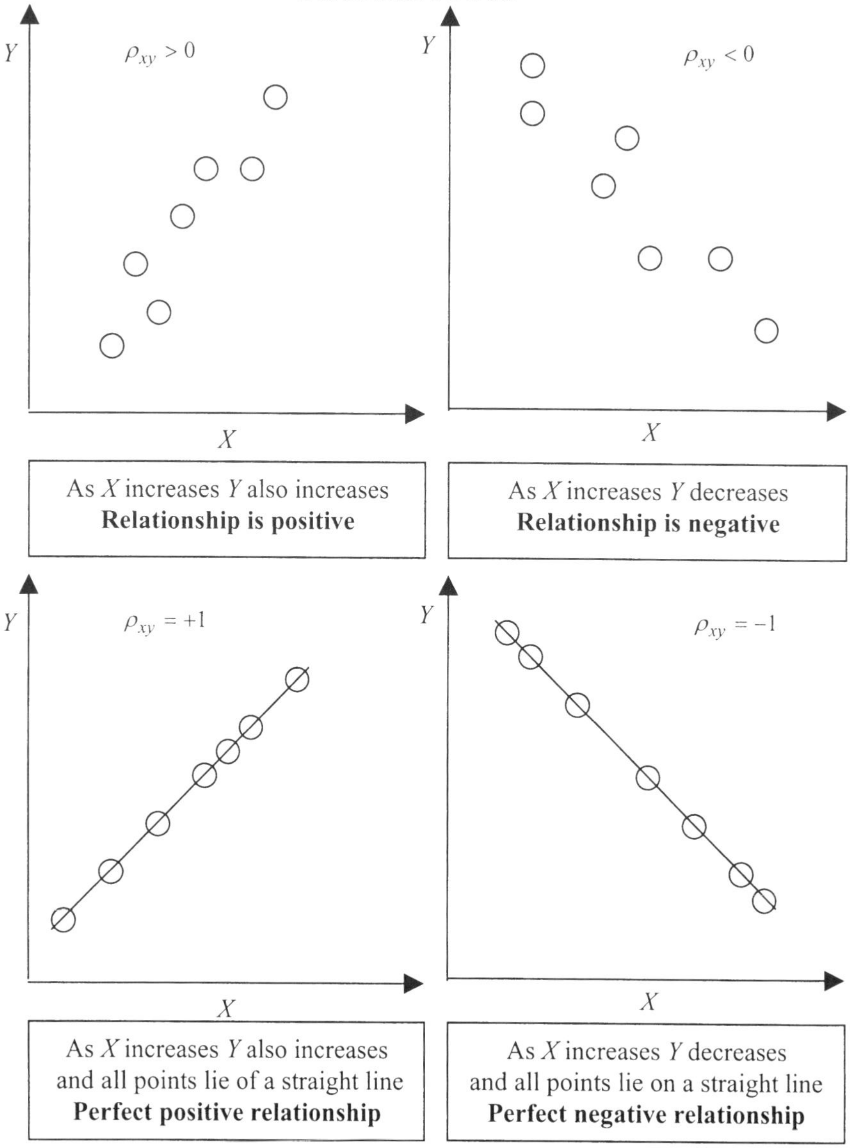

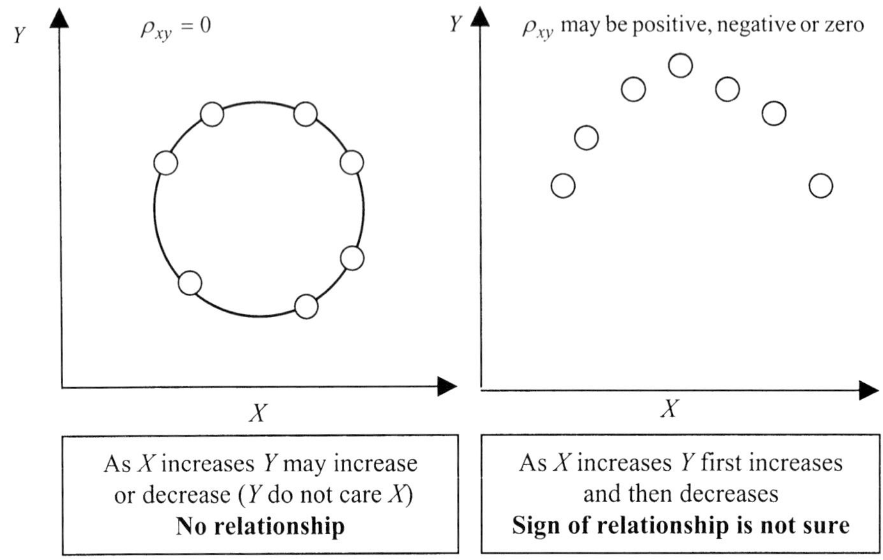

Fig. 1.59.1 Scatter plots.

Note that a similar scatter plot can be made from sample values to find the sign of sample correlation coefficient r_{xy}.

(c) The population regression coefficient of X on Y is defined as

$$\beta = \mathrm{Cov}(X,Y)/V(X). \tag{1.59.5}$$

For simple random sampling, it is given by

$$\beta = S_{xy} \big/ S_x^2 . \tag{1.59.6}$$

A biased estimator of β is given by

$$b = s_{xy} \big/ s_x^2 \tag{1.59.7}$$

which in fact represents a change in the study variable Y with a unit change in the auxiliary variable X. Note that sign of β (or b) is same as that of ρ_{xy} (or $\mathrm{r_{xy}}$).

Example 1.59.1. Consider the following population consisting of five ($N = 5$) units A, B, C, D, and E, where for each one of the unit in the population two variables Y and X are measured.

Units	A	B	C	D	E
Y_i	9	11	13	16	21
X_i	14	18	19	20	24

Find the following parameters:

(a) $\overline{Y}$, $\overline{X}$, S_x^2, S_y^2, S_{xy}, ρ_{xy} and β.

(b) Select all possible SRSWOR samples of $n=3$ units. Show that $\bar{y}$, $\bar{x}$, s_x^2, s_y^2, s_{xy} are unbiased for $\bar{Y}$, $\bar{X}$, S_x^2, S_y^2, S_{xy}, but r_{xy} and b remain biased estimators of ρ_{xy} and β respectively.

(c) Compute $\operatorname{Cov}(\bar{y},\bar{x})$ by using definition.

(d) Compute $\frac{(1-f)}{n}S_{xy}$ and comment on it.

Solution. From the complete population information, we have

Units	Y_i	X_i	$(Y_i-\bar{Y})$	$(X_i-\bar{X})$	$(Y_i-\bar{Y})^2$	$(X_i-\bar{X})^2$	$(Y_i-\bar{Y})(X_i-\bar{X})$
A	9	14	-5	-5	25	25	25
B	11	18	-3	-1	9	1	3
C	13	19	-1	0	1	0	0
D	16	20	2	1	4	1	2
E	21	24	7	5	49	25	35
Sum	70	95	0	0	88	52	65

(a) From the above table we have $\sum_{i=1}^{N} Y_i = 70$, $\sum_{i=1}^{N} X_i = 95$, so that

$$\bar{Y}=\frac{1}{N}\sum_{i=1}^{N} Y_i=\frac{1}{5}\times 70=14, \text{ and } \bar{X}=\frac{1}{N}\sum_{i=1}^{N} X_i=\frac{1}{5}\times 95=19.$$

From the above table, $\sum_{i=1}^{N}(Y_i-\bar{Y})^2=88$, $\sum_{i=1}^{N}(X_i-\bar{X})^2=52$ and $\sum_{i=1}^{N}(Y_i-\bar{Y})(X_i-\bar{X})=65$,

so that

$$S_y^2=\frac{1}{N-1}\sum_{i=1}^{N}\left(Y_i-\bar{Y}\right)^2=\frac{88}{5-1}=22,\quad S_x^2=\frac{1}{N-1}\sum_{i=1}^{N}\left(X_i-\bar{X}\right)^2=\frac{52}{5-1}=13,$$

$$S_{xy}=\frac{1}{N-1}\sum_{i=1}^{N}(Y_i-\bar{Y})(X_i-\bar{X})=\frac{65}{5-1}=16.25,\quad \rho_{xy}=\frac{S_{xy}}{\sqrt{S_x^2S_y^2}}=\frac{16.25}{\sqrt{13\times 22}}=0.960,$$

and $\beta=\frac{S_{xy}}{S_x^2}=\frac{16.25}{13}=1.25$.

(b) Here we have $N=5$ and $n=3$, so the total number of possible SRSWOR samples will be ${}^5C_3=10$.

	y_i	x_i	$(y_i-\bar{y})$	$(x_i-\bar{x})$	$(y_i-\bar{y})^2$	$(x_i-\bar{x})^2$	$(y_i-\bar{y})(x_i-\bar{x})$
Units	**Sample 1**						
A	9	14	-2	-3	4	9	6
B	11	18	0	1	0	1	0
C	13	19	2	2	4	4	4
Sum	33	51	0	0	8	14	10

Continued......

Units	Sample 2						
A	9	14	-3	-3.34	9	11.16	10.02
B	11	18	-1	0.67	1	0.45	-0.67
D	16	20	4	2.67	16	7.13	10.68
Sum	36	52	0	0.00	26	18.73	20.03
Units	Sample 3						
A	9	14	-4.67	-4.67	21.81	21.81	21.81
B	11	18	-2.67	-0.67	7.13	0.45	1.79
E	21	24	7.33	5.34	53.73	28.52	39.14
Sum	41	56	-0.01	0.00	82.67	50.77	62.74
Units	Sample 4						
A	9	14	-3.67	-3.67	13.44	13.44	13.44
C	13	19	0.33	1.33	0.11	1.78	0.44
D	16	20	3.33	2.34	11.11	5.48	7.80
Sum	38	53	0.00	0.00	24.67	20.70	21.69
Units	Sample 5						
A	9	14	-5.33	-5.00	28.44	25.00	26.67
C	13	19	-1.33	0.00	1.78	0.00	0.00
E	21	24	6.67	5.00	44.44	25.00	33.33
Sum	43	57	0.00	0.00	74.67	50.00	60.00
Units	Sample 6						
A	9	14	-6.33	-5.33	40.11	28.44	33.78
D	16	20	0.67	0.67	0.44	0.44	0.44
E	21	24	5.67	4.67	32.11	21.78	26.44
Sum	46	58	0.00	0.00	72.67	50.67	60.67
Units	Sample 7						
B	11	18	-2.33	-1.00	5.44	1.00	2.33
C	13	19	-0.33	0.00	0.11	0.00	0.00
D	16	20	2.67	1.00	7.11	1.00	2.67
Sum	40	57	0.00	0.00	12.67	2.00	5.00
Units	Sample 8						
B	11	18	-4.00	-2.33	16.00	5.44	9.33
C	13	19	-2.00	-1.33	4.00	1.78	2.67
E	21	24	6.00	3.67	36.00	13.44	22.00
Sum	45	61	0.00	0.00	56.00	20.67	34.00
Units	Sample 9						
B	11	18	-5.00	-2.67	25.00	7.11	13.33
D	16	20	0.00	-0.67	0.00	0.44	0.00
E	21	24	5.00	3.33	25.00	11.11	16.67

Continued......

Sum	48	62	0.00	0.00	50.00	18.67	30.00
Units	**Sample 10**						
C	13	19	-3.67	-2.00	13.44	4.00	7.33
D	16	20	-0.67	-1.00	0.44	1.00	0.67
E	21	24	4.33	3.00	18.78	9.00	13.00
Sum	50	63	0.00	0.00	32.67	14.00	21.00

From the above table we have

Sample	$\bar{y}$	$\bar{x}$	s_y^2	s_x^2	s_{xy}	r_{xy}	b
1	11.00	17.00	4.00	7.00	5.00	0.945	0.71
2	12.00	17.33	13.00	9.37	10.02	0.829	1.12
3	13.67	18.67	41.33	25.38	31.37	0.968	1.24
4	12.67	17.67	12.33	10.35	10.84	0.960	1.05
5	14.33	19.00	37.33	25.00	30.00	0.982	1.20
6	15.33	19.33	36.33	25.33	30.33	1.000	1.20
7	13.33	19.00	6.33	1.00	2.500	0.993	2.50
8	15.00	20.33	28.00	10.33	17.00	0.999	1.65
9	16.00	20.67	25.00	9.33	15.00	0.982	1.61
10	16.67	21.00	16.33	7.00	10.50	0.982	1.50
Sum	140.00	190.00	220.00	130.10	162.56	9.641	14.00
Expected Values	14.00	19.00	22.00	13.00	16.25	0.964	1.40

Thus we have the following results:

$E(\bar{y}) = 14 = \bar{Y}$, that is, the sample mean $\bar{y}$ is unbiased for population mean of the study variable;

$E(\bar{x}) = 19 = \bar{X}$, that is, the sample mean $\bar{x}$ is unbiased for population mean of the auxiliary variable;

$E(s_y^2) = 22.00 = S_y^2$, that is, the sample variance s_y^2 is unbiased for population S_y^2 of the study variable;

$E(s_x^2) = 13.00 = S_x^2$, that is, the sample variance s_x^2 is unbiased for population S_x^2 of the auxiliary variable;

$E(s_{xy}) = 16.25 = S_{xy}$, that is, the sample covariance s_{xy} is unbiased for population S_{xy} of both variables;

$E(r_{xy}) = 0.964 \neq \rho_{xy}$, that is, the sample r_{xy} is biased for population ρ_{xy}, and

$B(r_{xy}) = E(r_{xy}) - \rho_{xy} = 0.964 - 0.960 = 0.004$;

and

$E(b) = 1.4 \neq \beta$, that is, the sample b is biased for the population β;

and

$B(b) = E(b) - \beta = 1.40 - 1.25 = 0.15.$

(c) The covariance between $\bar{y}$ and $\bar{x}$ is defined as:

$$\mathrm{Cov}(\bar{y},\bar{x}) = E[\bar{y} - E(\bar{y})][\bar{x} - E(\bar{x})] = E[\bar{y} - \bar{Y}][\bar{x} - \bar{X}] = \sum_{s=1}^{10} p_s[\bar{y}_s - \bar{Y}][\bar{x}_s - \bar{X}]$$

Now we have

$\bar{y}_t$	$\bar{x}_t$	$(\bar{y}_t - \bar{Y})$	$(\bar{x}_t - \bar{X})$	$(\bar{y}_t - \bar{Y})(\bar{x}_t - \bar{X})$
11.00	17.00	-3.00	-2.00	6.0000
12.00	17.33	-2.00	-1.67	3.3400
13.67	18.67	-0.33	-0.33	0.1089
12.67	17.67	-1.33	-1.33	1.7689
14.33	19.00	0.33	0.00	0.0000
15.33	19.33	1.33	0.33	0.4389
13.33	19.00	-0.67	0.00	0.0000
15.00	20.33	1.00	1.33	1.3300
16.00	20.67	2.00	1.67	3.3400
16.67	21.00	2.67	2.00	5.3400
	Sum	0.00	0.00	21.6667

So by definition,

$$\mathrm{Cov}(\bar{y},\ \bar{x}) = \sum_{t=1}^{10} p_t[\bar{y}_t - \bar{Y}][\bar{x}_t - \bar{X}] = \frac{1}{10} \times 21.6667 = 2.1667\,.$$

(d) Now we have

$$\frac{N-n}{Nn} S_{xy} = \frac{(5-3)}{5 \times 3} \times 16.25 = 2.16667.$$

Thus we have

$$\mathrm{Cov}(\bar{y},\ \bar{x}) = \frac{N-n}{Nn} S_{xy} = \frac{(1-f)}{n} S_{xy}, \quad \text{where } f = n/N.$$

For theoretical proof refer to Chapter 2.

1.60 SOME USEFUL MATHEMATICAL FORMULAE

If x and y are two random variable and c and d are two real constants, then

(a) $V(cx) = c^2 V(x)$. (1.60.1)

(b) $\mathrm{Cov}(cx, dy) = cd\,\mathrm{Cov}(x, y)$. (1.60.2)

(c) If $x = \sum_{i=1}^{n} x_i$, where the x_i are also random variables, then we have

$$V(x) = V\left(\sum_{i=1}^{n} x_i\right) = \sum_{i=1}^{n} V(x_i) + \sum_{i \neq j=1}^{n} \mathrm{Cov}(x_i, x_j). \tag{1.60.3}$$

(d) If $x = \sum_{i=1}^{n} c_i x_i$, where c_i are real constants, then we have

$$V(x) = V\left(\sum_{i=1}^{n} c_i x_i\right) = \sum_{i=1}^{n} c_i^2 V(x_i) + \sum_{i \neq j=1}^{n} c_i c_j \mathrm{Cov}(x_i,\ x_j). \tag{1.60.4}$$

Note that if x_i and x_j are independent then $\mathrm{Cov}(x_i, x_j) = 0$ and

$$V(x) = V\left(\sum_{i=1}^{n} c_i x_i\right) = \sum_{i=1}^{n} c_i^2 V(x_i).$$

(e) If $x = \sum_{i=1}^{n} c_i x_i$ and $y = \sum_{i=1}^{n} d_i y_i$, where c_i and d_i are real constants, then we have

$$\mathrm{Cov}(x, y) = \mathrm{Cov}\left(\sum_{i=1}^{n} c_i x_i,\ \sum_{i=1}^{n} d_i y_i\right) = \sum_{i=1}^{n} c_i d_i \mathrm{Cov}(x_i,\ y_i) + \sum_{i \neq j=1}^{n} c_i d_j \mathrm{Cov}(x_i,\ y_j). \tag{1.60.5}$$

Note that if x_i and y_j are independent then $\mathrm{Cov}(x_i, y_j) = 0$ and

$$\mathrm{Cov}(x,\ y) = \mathrm{Cov}\left(\sum_{i=1}^{n} c_i x_i,\ \sum_{i=1}^{n} d_i y_i\right) = \sum_{i=1}^{n} c_i d_i \mathrm{Cov}(x_i,\ y_i).$$

1.61 ORDERED STATISTICS

These are parameters which deal with arranging the data in ascending or descending order, and we introduce a few of them here as follows:

1.61.1 POPULATION MEDIAN

It is a measure which divides the population into exactly two equal parts, and it is denoted by M_y. Its analogous from the sample is called sample median, and is denoted by $\hat{M}_y$. A pictorial representation is given below:

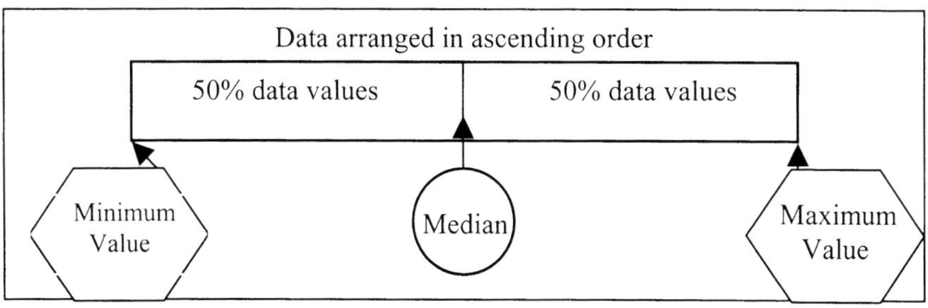

Fig. 1.61.1 Structure of data to find median.

Rules to find sample median: Consider a sample having n observations, and we wish to find the sample median. The first step is to arrange the data in ascending order, and then after that there are two situations:

(i) If the sample size n is odd, then the value at the $\left(\frac{n+1}{2}\right)$th position from the ordered data are called sample median. As an illustration, consider a sample consisting of $n=5$ (odd) observations as 50, 90, 30, 60, and 70. First step is to arrange the data in ascending order as: 30, 50, 60, 70, 90. The second step is to pick up a value at the $\left(\frac{n+1}{2}\right)\text{th}=\left(\frac{5+1}{2}\right)\text{th}=3\text{rd position}=60$, so $\hat{M}_y=60$.

(ii) If the sample size n is even, then the average of the values at the $\left(\frac{n}{2}\right)$th and $\left(\frac{n}{2}+1\right)$th positions from ordered data are called sample median. As an illustration, consider a sample consisting of $n=6$ (even) observations as 50, 90, 30, 60, 70 and 20. First step is to arrange the data in ascending order as: 20, 30, 50, 60, 70, 90. The second step is to pick up two values: one at $\left(\frac{n}{2}\right)\text{th}=\left(\frac{6}{2}\right)\text{th}=3\text{rd position}=50$, and second at $\left(\frac{n}{2}+1\right)\text{th}=\left(\frac{6}{2}+1\right)\text{th}=4\text{th position}=60$. Then the average of these values is called the median, and is given by $\hat{M}_y=(50+60)/2=55$.

1.61.2 POPULATION QUARTILES

These are three measures which divide the population into four equal parts. The ith quartile is represented by $Q_i, i=1,2,3$. A pictorial representation is given below:

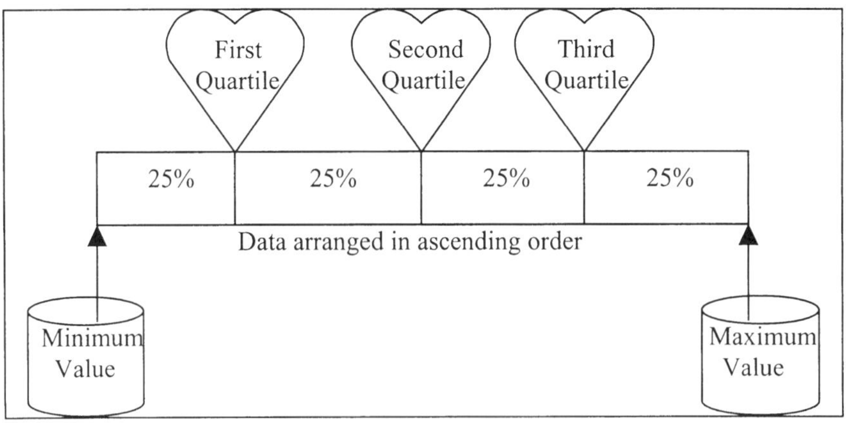

Fig. 1.61.2 Structure of data to find quartiles.

Note that the second quartile Q_2 is a median. The first quartile Q_1 is a median of the data less than or equal to the second quartile Q_2, and third quartile Q_3 is the median of the data more than or equal to the second quartile Q_2. Thus finding three quartiles needs to find median three times from the given ordered data. The population interquartile range is defined as: $\Theta = (Q_3 - Q_1)$. The sample analogous of population quartiles are called sample quartiles and are denoted by $\hat{Q}_i, i = 1,2,3$ and sample interquartile range is defined as: $\hat{\Theta} = (\hat{Q}_3 - \hat{Q}_1)$, which is a measure of variation in the data set.

1.61.3 POPULATION PERCENTILES

These are 99 measures, which divide the population into equal 100 parts. The ith population percentile is represented by $P_i, i = 1,2,....,99$ and its pictorial representation is given below:

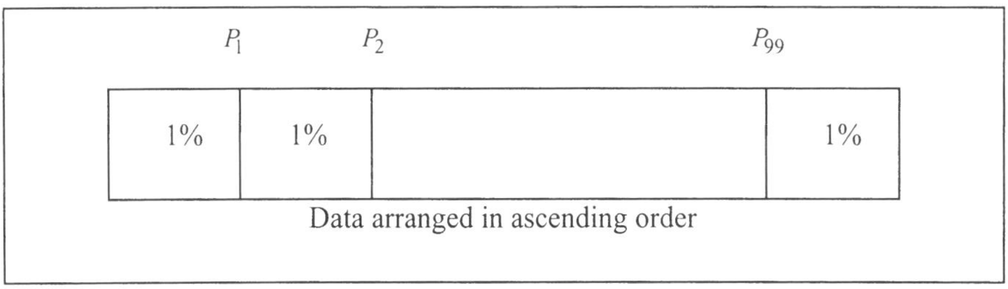

Fig. 1.61.3 Structure of data to find percentile.

Its sample analogous is represented by $\hat{P}_i, i = 1,2,....,99$.

1.61.4 POPULATION MODE

It is value which occurs most frequently in the population and is denoted by M_o, and its sample analogous is called sample mode and is denoted by $\hat{M}_o$. As an illustration, for the data set 60, 70, 30, 60, 30, 30, 80, 30, the mode value is 30, because it occurred most frequently.

1.62 DEFINITION(S) OF STATISTICS

There are several definitions of statistics and we list a few of them are as follows:

(a) It is a science to describe or predict the behaviour of a population based on a random and representative sample drawn from the same population.

(b) The science of statistics is the method of judging collective, natural, or social phenomena from the results obtained from the analysis or enumeration or collection of estimates.
(c) The science which deals with the collection, analysis and interpretation of numerical data.

1.63 LIMITATIONS OF STATISTICS

A few limitations of statistics are:

(a) Statistics does not deal with individual measurements. This is the reason we need police to investigate individuals;
(b) Statistics deals only with quantitative characters or variables, and we have to assign codes to qualitative variables before analysis;
(c) Statistics results are true only on an average;
(d) Statistics can be misused or misinterpreted. For example last year 90% of the pedestrians who died in road accidents were walking on paths, so it is safer to walk in the middle of the road.

1.64 LACK OF CONFIDENCE IN STATISTICS

A few people have the following types of views in their mind about statistics:

(a) Statistics can prove anything;
(b) There are three types of lies --- lies, damned lies, and statistics;
(c) Statistics are like clay of which one can make a God or devil as he/she pleases;
(d) It is only a tool, and cannot prove or disprove anything.

1.65 SCOPE OF STATISTICS

It has scope in almost every kind of category we are divided in this world due to our social setup, for example, Trade, Industry, Commerce, Economics, Biology, Botany, Astronomy, Physics, Chemistry, Education, Medicine, Sociology, Psychology, Religious studies, Meteorology, National defence, and Business: Production, Sale, Purchage, Finance, Accounting, Quality control, etc..

EXERCISES

Exercise 1.1. Define the terms population, parameter, sample, and statistic.

Exercise 1.2. Describe the advantage of a sample survey in comparison with a census survey. Write the circumstances under which census surveys are preferred to sample surveys and vice versa?

Exercise 1.3. Describe the relationship between the variance and mean squared error of an estimator. Hence deduce the term relative efficiency.

Exercise 1.4. You are required to plan a sample survey to study the environment activities of a business in the United States. Suggest a suitable survey plan on the following points: (a) sampling units; (b) sampling frame; (c) method of sampling; and (d) method of collecting information. Prepare a suitable questionnaire which may be used to collect the required information.

Exercise 1.5. Define population, sampling unit and sampling frame for conducting surveys on each of the following subjects. Mention other possible sampling units, if any, in each case and discuss their relative merits.

(a) Housing conditions in the United States.
(b) Study of incidence of lung cancer and heart attacks in the United States.
(c) Measurement of the volume of timber available in the forests of Canberra .
(d) Study of the birth rate in India.
(e) Study of nutrient contents of food consumed by the residents of California.
(f) Labour manpower of large businesses in Canada.
(g) Estimation of population density in India.

Exercise 1.6. What do you understand by the following terms?

(a) Unbiasedness; (b) Consistency; (c) Sufficiency; and (d) Efficiency.

Exercise 1.7. Define the following:

(a) Sampling frame; (b) Sample survey design; and (c) Nonresponse.

Exercise 1.8. Show that the sample variance $s_y^2 = \frac{1}{n-1}\sum_{i=1}^{n}(y_i - \bar{y})^2$ can be put in different ways as

$$s_y^2 = \frac{1}{n-1}\left[\sum_{i=1}^{n} y_i^2 - n\bar{y}^2\right] = \frac{1}{n-1}\left[\sum_{i=1}^{n} y_i^2 - \left(\sum_{i=1}^{n} y_i\right)^2 \Big/ n\right] = \frac{n\left(\sum_{i=1}^{n} y_i^2\right) - \left(\sum_{i=1}^{n} y_i\right)^2}{n(n-1)},$$

and the sample covariance

$$s_{xy} = \frac{1}{n-1}\sum_{i=1}^{n}(x_i - \bar{x})(y_i - \bar{y})$$

can be written as

$$s_{xy} = \frac{1}{n-1}\left[\sum_{i=1}^{n} x_i y_i - n\bar{x}\,\bar{y}\right] = \frac{1}{n-1}\left[\sum_{i=1}^{n} x_i y_i - \frac{\left(\sum_{i=1}^{n} x_i\right)\left(\sum_{i=1}^{n} y_i\right)}{n}\right] = \frac{n\left(\sum_{i=1}^{n} x_i y_i\right) - \left(\sum_{i=1}^{n} x_i\right)\left(\sum_{i=1}^{n} y_i\right)}{n(n-1)}.$$

Exercise 1.9. Show that the population mean square error

$$S_y^2 = \frac{1}{N-1}\sum_{i=1}^{N}(Y_i - \bar{Y})^2$$

can be put in different ways as

$$s_y^2 = \frac{1}{N-1}\left[\sum_{i=1}^{N} Y_i^2 - N\overline{Y}^2\right] = \frac{1}{N-1}\left[\sum_{i=1}^{N} Y_i^2 - \frac{\left(\sum_{i=1}^{N} Y_i\right)^2}{N}\right] = \frac{N\left(\sum_{i=1}^{N} Y_i^2\right) - \left(\sum_{i=1}^{N} Y_i\right)^2}{N(N-1)},$$

and the population covariance

$$S_{xy} = \frac{1}{N-1}\sum_{i=1}^{N}\left(X_i - \overline{X}\right)\left(Y_i - \overline{Y}\right)$$

can be written as

$$S_{xy} = \frac{1}{N-1}\left[\sum_{i=1}^{N} X_iY_i - N\,\overline{X}\,\overline{Y}\right] = \frac{1}{N-1}\left[\sum_{i=1}^{N} X_iY_i - \frac{\left(\sum_{i=1}^{N} X_i\right)\left(\sum_{i=1}^{N} Y_i\right)}{N}\right] = \frac{N\left(\sum_{i=1}^{N} X_iY_i\right) - \left(\sum_{i=1}^{N} X_i\right)\left(\sum_{i=1}^{N} Y_i\right)}{N(N-1)}.$$

Exercise 1.10. Construct a sample space and tree diagram for each one of the following situations:

(a) Toss a fair coin; (b) Toss two fair coins; (c) Toss a fair die; (d) Toss a fair coin and a fair die; (e) Toss two fair dice; and (f) Toss a fair die and a fair coin.

Exercise 1.11. State what type of variable each of the following is. If a variable is quantitative, say whether it is discrete or continuous; and if the variable is qualitative say whether it is nominal or ordinal.

1 Religious preference.
2 Amount of water in a glass.
3 Master card number.
4 Number of students in a class of 32 who turn in assignments on time.
5 Brand of personal computer.
6 Amount of fluid dispensed by a machine used to fill cups with chocolate.
7 Number of graduate applications in statistics each year at the SCSU.
8 Amount of time required to drive a car for 35 miles.
9 Room temperature recorded every half hour.
10 Weight of letters to be mailed.
11 Taste of milk.
12 Occupation list.
13 Coded numbers to different colors, *e.g.*, Red--1, Green--2, and Pink--3.
14 Average daily low temperature per year in the St. Cloud city.
15 Nationality of the students in your University.
16 Phone number.
17 Rent paid by the tenant.
18 Frog Jump in cms.
19 Colors of marbles.

20 Number of mistakes in the examination.
21 Time to finish an examination.
22 Shoe number.
23 Gender of a student.
24 Discipline of a student.
25 Rating of a politician: good, better or best.
26 Sum of two real numbers.
27 Sum of two integers (or whole numbers).
28 Number of passengers in a bus.
29 Age of a patient.
30 Age groups, *e.g.*, 0 to 5 years, 6 to 10 years etc.
31 Area code.
32 Postal code.
33 Product of two positive real numbers.
34 Length of a string in cms.
35 Height of door in feet.
36 Weather conditions: good, better and best.
37 Average of real numbers.
38 Proportion of red balls in a bag.
39 Number of e-mail accounts.
40 Number of questions in an examination.

PRACTICAL PROBLEMS

Practical 1.1. From a population of size 5 how many samples of size 2 can be drawn by using (a) SRSWR and (b) SRSWOR sampling?

Practical 1.2. Mr. Bean selects all possible samples of two units from a population consisting of four units *viz.* 10, 15, 20, 25 by using SRSWOR sampling. He noted that the harmonic mean of this population is given by

$$H_y = N \Big/ \sum_{i=1}^{N} \frac{1}{y_i} = 4 \Big/ \left\{ \frac{1}{10} + \frac{1}{15} + \frac{1}{20} + \frac{1}{25} \right\} = 15.58442.$$

The total number of possible samples $= {}^N C_n = {}^4 C_2 = 6$ and these samples are given by

(10, 15), (10, 20), (10, 25), (15, 20), (15, 25) and (20, 25).

The harmonic means for these samples are, respectively, are

$$\hat{H}_1 = n \Big/ \sum_{i=1}^{n} \frac{1}{y_i} = 2 \Big/ \left\{ \frac{1}{10} + \frac{1}{15} \right\} = 12, \quad \hat{H}_2 = n \Big/ \sum_{i=1}^{n} \frac{1}{y_i} = 2 \Big/ \left\{ \frac{1}{10} + \frac{1}{20} \right\} = 13.33333$$

$$\hat{H}_3 = 14.28571, \quad \hat{H}_4 = 17.14286, \quad \hat{H}_5 = 18.75 \text{ and } \hat{H}_6 = 22.22222.$$

Mr. Bean took the harmonic mean of these six sample harmonic means, as follows:

$$\text{HM} = \frac{6}{\sum_{i=1}^{6}\frac{1}{\hat{H}_i}} = \frac{6}{\left\{\frac{1}{12}+\frac{1}{13.33333}+\frac{1}{14.28571}+\frac{1}{17.14286}+\frac{1}{18.75}+\frac{1}{22.22222}\right\}}$$

$$= 15.58442 = H_y .$$

Then Mr. Bean made the following statements.

(a) Sample harmonic mean is an unbiased estimator of population harmonic mean.

(b) The expected value of sample harmonic mean $\hat{H}_t$ is defined as

$$E\left(\hat{H}_t\right) = \frac{s(\Omega)}{\sum_{t=1}^{s(\Omega)}\frac{1}{\hat{H}_t}} = H_y .$$

Do you agree with him? If not, why?

Hint: Expected value: $E\left(\hat{H}_t\right) = \sum_{t=1}^{s(\Omega)} p_t \hat{H}_t$ with $p_t = \frac{1}{6}$ $\forall\ t = 1,2,...,s(\Omega)$ and $s(\Omega) = {}^N C_n$.

(c) Find the bias, variance, and mean square error in the estimator $\hat{H}_t$.

(d) Does the relation $\text{MSE}\left(\hat{H}_t\right) = V\left(\hat{H}_t\right) + \left\{B\left(\hat{H}_t\right)\right\}^2$ hold?

Practical 1.3. Suppose that a population consists of 5 units given by: 10, 15, 20, 25, and 30.Select all possible samples of 3 units using SRSWR and SRSWOR sampling.

(a) Show that the sample mean is an unbiased estimator of population mean in each case.
(b) The sample variance is unbiased estimator of the population variance under SRSWR sampling, and for population mean squared error under SRSWOR sampling.
(c) Also plot the sampling distribution of sample mean and sample variance in each situation.
(d) Find the variance of sample mean under SRSWOR sampling using the definition of variance? Show all steps.
(e) Also compute $\left(\frac{N-n}{Nn}\right)S_y^2$ and comment on it.

Practical 1.4. Repeat Mr. Bean's exercise with the geometric mean (GM) and comment on the results.

Hint: The GM of n numbers $y_1, y_2, ..., y_n$ is $\text{GM} = \left(\prod_{i=1}^{n} y_i\right)^{1/n}$.

Practical 1.5. If a random variable x follows a Poisson distribution, that is, $x \sim P(\lambda)$ with $\lambda = 0.4$ over $N = 20$ trials. Select a with replacement sample of $n = 5$ units by using the method of cumulative distribution function.

Hint: The p.d.f. of a Poisson random variable x is given by $P[X = x] = \dfrac{e^{-\lambda}\lambda^x}{x!}$.

Practical 1.6. Suppose an urn contains N balls of which Np are black and Nq are white so that $p + q = 1$. The probability that if n balls are drawn (without replacement), exactly x of them will be black, is given by

$$P(X = x) = \left\{ {}^{Np}C_x \, {}^{Nq}C_{n-x} \right\} / \, {}^{N}C_x$$

such that $0 \le x \le Np$; and $0 \le n - x \le Nq$. Using the concept of c.d.f., select a sample of three units by using without replacement sampling.
Hint: Hypergeometric distribution.

Practical 1.7. If a discrete random variable X has a cumulative distribution function:

$$F(x) = \begin{cases} 0 & \text{for} \quad x < 1, \\ 1/3 & \text{for} \quad 1 \le x < 4, \\ 1/2 & \text{for} \quad 4 \le x < 6, \\ 5/6 & \text{for} \quad 6 \le x < 10, \\ 1 & \text{for} \quad x \ge 10. \end{cases}$$

Select a sample of $n = 5$ units by using with replacement sampling.
Hint: Use random number table method.

Practical 1.8. If the distribution function of a population consisting of $N = 5$ units is given by

$$F(x) = \frac{x^2 + 5x}{50} \quad \text{for} \quad x = 1,\ 2, 3, 4, 5.$$

Draw a without replacement sample of $n = 2$ units.
Hint: Use random number table method.

Practical 1.9. If the distribution function of a continuous random variable x is

$$f(x) = \frac{1}{\pi\left[\, 1 + (x - 100)^2 \,\right]}, \quad -\infty < x < +\infty$$

Use the first 6 columns multiplied by 10^{-6} as the values of the cumulative distribution function (c.d.f.) $F(x)$ of the random variable x, and select a random sample of 15 units by using with replacement sampling.
Hint: $F(x) = 100 + \tan\left[\pi\left(F(x) - 0.5\right)\right]$.

Practical 1.10. If the distribution function of a continuous random variable x in a population is given by

$$f(x)=\frac{1}{\sqrt{2\pi}\sigma}\exp\left\{-\left(\frac{x-\mu}{\sigma}\right)^2\right\}, \qquad -\infty < x < +\infty$$

with $\mu = 100$ and $\sigma = 2.5$ .Use the first 6 columns multiplied by 10^{-6} as the values of the cumulative distribution function (c.d.f.) $F(x)$ of the random variable x, and select a random sample of 15 units by using with replacement sampling.
Hint: $x = \mu + z\sigma$ and $z \sim N(0,1)$.

Practical 1.11. Find the value of c so that

$$f(x,y)=\begin{cases} cx(x-y) & \text{for } 0<x<1,\ -x<y<+x, \\ 0 & \text{otherwise,}\end{cases}$$

becomes a joint probability density function. Select a sample of $n = 5$ units by using Random Number Table method.

Hint: $\int\limits_{0-x}^{1+x}\int f(x,y)dydx = 1$ or Freund (2000).

Practical 1.12. In the hope of preventing ecological damage from oil spills, a biochemical company is developing an enzyme to break up oil into less harmful chemicals. The table below shows the time it took for the enzyme to break up oil samples at different temperatures. The researcher plans to use these data in statistical analysis:

Oil Temperature (°C)	3	10	15	21	33	37
Time to Break Up (seconds)	165	98	65	62	26	9

(a) If you are a consultant which variable you will consider dependent and independent? Denote your dependent variable by Y and independent variable with X .
(b) Assuming that these six observations form a population, compute the following parameters:

$$\bar{Y}, \bar{X},\ S_y^2, S_x^2,\ S_y, S_x, C_y=\frac{S_y}{\bar{Y}}, C_x=\frac{S_x}{\bar{X}}, S_{xy}, \rho_{xy}=\frac{S_{xy}}{S_xS_y}, \beta=\frac{S_{xy}}{S_x^2} \text{ and } K=\rho_{xy}\frac{C_y}{C_x}.$$

Practical 1.13. Consider the following population consisting of 5 units $A=10$, $B = 20$, $C = 25$, $D = 50$, and $E = 4$.
(a) Compute the population harmonic mean.
(b) Select all possible SRSWOR samples each consisting of 3 units.
(c) For each sample of 3 units, obtain estimate of the population harmonic mean.
(d) How many sample harmonic means are less than population harmonic mean?
(e) Find the bias in the sample harmonic mean.

(f) Find the variance of sample harmonic mean by the definition.
(g) Find the MSE of the sample harmonic mean by the definition.
(h) Does the relation $\text{MSE}(\hat{H}_s) = V(\hat{H}_s) + \{B(\hat{H}_s)\}^2$ hold?

Practical 1.14. Consider a population consisting of 15 countries as listed in the following table, and also gives the hypothetical suicide rates in these countries per 100,000 persons.

Sr. No.	Country	Suicide rate (%)
1	Australia	22
2	Austria	55
3	Canada	29
4	Denmark	59
5	France	35
6	Ireland	10
7	Israel	12
8	Japan	35
9	Netherlands	20
10	Norway	26
11	Poland	26
12	Sweden	39
13	Switzerland	55
14	United Kingdom	18
15	United States	25

1. Compute the following parameters:
    (a) Population mean;
    (b) Population range;
    (c) Population variance;
    (d) Population standard deviation;
    (e) Population mean square error;
    (f) Population coefficient of variation;
    (g) Proportion of countries having suicide rate more that 25%.

2. (a) Select an SRSWOR sample of 5 units using Random Number Table method (**Rule:** Start from 1st row and 3rd column of the Pseudo-Random Number Table 1 given in the Appendix).
    (b) Estimate the population mean and population total.
    (c) Compute the variance of the estimator of population mean .
    (d) Estimate s_y^2.
    (e) Estimate the variance of the estimator of population mean.
    (f) Construct 95% confidence interval of the population mean assuming that population mean square is known and sample size is large. Does the population mean falls in it? Interpret it.

(g) Construct 95% confidence interval assuming that population mean square is unknown and sample size in small. Does the population mean falls in it? Interpret it.
(h) Find the variance of estimator of proportion of countries having suicide rate more than 25%.

Practical 1.15. Consider a population consisting of the following six units:

Units	*A*	*B*	*C*	*D*	*E*	*F*
Values	718	912	1014	1113	1110	615

Now consider the following sampling plan:

Sample No.	Samples	Probability p_t
1	*A, C,E*	1/9
2	*A, C, F*	1/9
3	*A, D, E*	1/9
4	*A, D, F*	1/9
5	*B, C, E*	1/9
6	*B, C, F*	1/9
7	*B, D, E*	1/9
8	*B, D, F*	1/9
9	*C, D, F*	1/9

Compute the following:
(a) $E(\bar{y}_t)$, $V(\bar{y}_t)$, $B(\bar{y}_t)$, and $\text{MSE}(\bar{y}_t)$.
(b) Does the relation $\text{MSE}(\bar{y}_t) = V(\bar{y}_t) + \{B(\bar{y}_t)\}^2$ holds?

Practical 1.16. For a bivariate data of $n = 10$ pairs of observations we are given

$$\sum_{i=1}^{n} x_i = 57, \quad \sum_{i=1}^{n} y_i = 47, \quad \sum_{i=1}^{n} x_i^2 = 377, \quad \sum_{i=1}^{n} y_i^2 = 263, \text{ and } \sum_{i=1}^{n} x_i y_i = 299.$$

Assume that these 10 observations form a sample compute the following statistic:

$$\bar{y}\,;\bar{x};\ s_y^2;\ s_x^2;\ s_y\,;s_x;\ \hat{C}_y = \frac{s_y}{\bar{y}};\ \hat{C}_x = \frac{s_x}{\bar{x}};\ s_{xy}\,;\ r_{xy} = \frac{s_{xy}}{s_x s_y};\ b = \frac{s_{xy}}{s_x^2} \text{ and}$$

$$\hat{K} = r_{xy}\frac{\hat{C}_y}{\hat{C}_x}.$$

Practical 1.17. The following data shows the daily temperatures in New York over a period of two weeks:

°C	48	50	47	54	49	53	48	48	51	58	53	54	49	52

Find the following: sample size; sample mean; median; mode; first quartile; second quartile; third quartile; minimum value; maximum value; and interquartile range.

Practical 1.18. Construct scatter diagrams and find the linear correlation coefficient in each one of the following five samples each of five units and comment on the different situations will arise:

Data Set-I						
	x	10	20	30	40	50
	y	5	25	30	35	45

Data Set-II						
	x	50	40	30	20	10
	y	5	25	30	35	45

Data Set-III						
	x	10	20	30	40	50
	y	5	10	15	20	25

Data Set-IV						
	x	50	40	30	20	10
	y	5	10	15	20	25

Data set-V						
	x	10	10	30	50	50
	y	5	25	15	5	25

Practical 1.19. The following balloon is filled with five gases with their different atomic number and atomic weights.

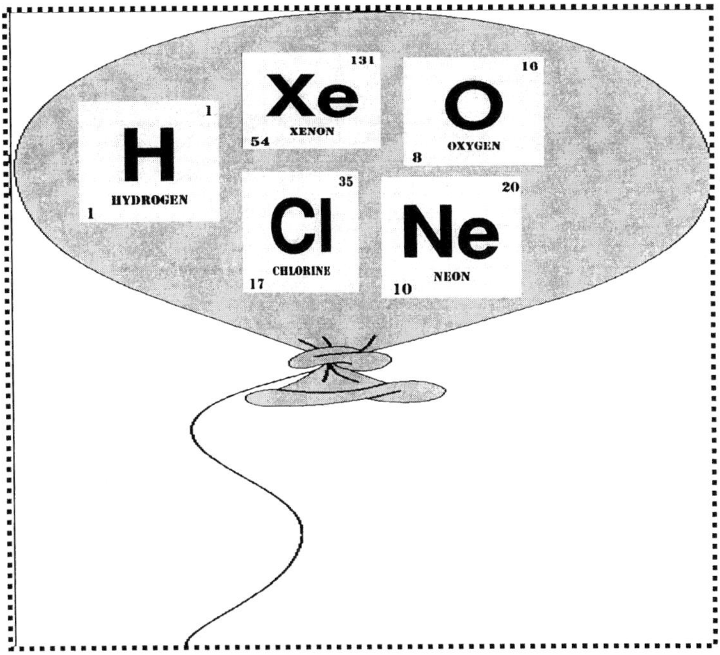

(I) Sampling distribution of atomic weight of gases:

(a) Find the average atomic weight of all the gases in the balloon;
(b) Find the population variance σ^2 of atomic weight of all the gases in the balloon;
(c) Select all possible with replacement samples each consisting of two gases;
(d) Estimate the average atomic weight from each one of the 25 samples;
(e) Construct a frequency distribution table of all possible sample means;
(f) Construct an histogram. Is it symmetric?;
(g) Find the expected value of all sample means of atomic weights from the frequency distribution table you developed;
(h) Find the variance of all the sample means of atomic weights from the frequency distribution table you developed.

(II) Sampling distribution of proportion of inert gases in the balloon:

(a) Find the proportion of inert gases in the balloon, and denote it by P;
(b) Select all possible with replacement samples each consisting of three gases;
(c) Estimate the proportion of inert gases in each sample;
(d) Construct a frequency distribution table of all possible sample proportions of inert gases;
(e) Construct an histogram. Is it symmetric?
(f) Find the expected value of all sample proportions of inert gases from the frequency distribution table you developed;
(g) Find the variance of all the sample proportions of inert gases from the frequency distribution table you developed.

Practical 1.20. Consider a sample $y_1, y_2, ..., y_n$ and let $\bar{y}_k$ and s_k^2 denote the sample mean and variance, respectively, of the first k observations.
(a) Show that

$$s_{k+1}^2 = \frac{(k-1)}{k} s_k^2 + \frac{1}{k+1}(y_{k+1} - \bar{y}_k)^2.$$

(b) Suppose that a sample of 15 observations has sample mean and a sample standard deviation 12.60 and 0.50, respectively. If we consider 16th observation of the data set as 10.2. What will be the values of the sample mean and sample standard deviation for all 16 observations?

2. SIMPLE RANDOM SAMPLING

2.0 INTRODUCTION

Simple Random Sampling (SRS) is the simplest and most common method of selecting a sample, in which the sample is selected unit by unit, with equal probability of selection for each unit at each draw. In other words, simple random sampling is a method of selecting a sample s of n units from a population Ω of size N by giving equal probability of selection to all units. It is a sampling scheme in which all possible combinations of n units may be formed from the population of N units with the same chance of selection.

As discussed in chapter 1:

(a) If a unit is selected, observed, and replaced in the population before the next draw is made and the procedure is repeated n times, it gives rise to a simple random sample of n units. This procedure is known as simple random sampling with replacement and is denoted as SRSWR.

(b) If a unit is selected, observed, and not replaced in the population before making the next draw, and the procedure is repeated until n distinct units are selected, ignoring all repetitions, it is called simple random sampling without replacement and is denoted by SRSWOR. Let us discuss the properties of the estimators of population mean, variance, and proportion in each of these cases.

2.1 SIMPLE RANDOM SAMPLING WITH REPLACEMENT

Suppose we select a sample of $n \geq 2$ units from the population of size N by using SRSWR sampling. Let y_i, $i = 1,2,...,n$, denote the value of the ith unit selected in the sample and Y_i, $i = 1,2,...,N$, be the value of the ith unit in the population. Then we have the following theorems:

Theorem 2.1.1. The sample mean $\bar{y}_n = n^{-1} \sum_{i=1}^{n} y_i$ is an unbiased estimator of the population mean $\bar{Y} = N^{-1} \sum_{i=1}^{N} Y_i$.

Proof. We have to prove that $E(\bar{y}_n) = \bar{Y}$. Now we have

$$E(\bar{y}_n) = E\left[\frac{1}{n}\sum_{i=1}^{n} y_i\right] = \frac{1}{n}\sum_{i=1}^{n} E(y_i). \qquad (2.1.1)$$

Now y_i is a random variable and each unit has been selected by SRSWR sampling, therefore y_i can take values $Y_1, Y_2, ..., Y_N$ with probabilities $1/N$, $1/N$, ..., $1/N$. By the definition of the expected value we have

$$E(y_i) = \frac{1}{N}\sum_{i=1}^{N} Y_i = \bar{Y}.$$

Thus (2.1.1) implies

$$E(\bar{y}_n) = \frac{1}{n}\sum_{i=1}^{n}\left[\frac{1}{N}\sum_{i=1}^{N} Y_i\right] = \frac{1}{n}\sum_{i=1}^{n}\bar{Y} = \bar{Y}. \tag{2.1.2}$$

Hence the theorem.

Corollary 2.1.1. The estimator $\hat{y}_n = N\bar{y}_n$ is an unbiased estimator of population total, Y.

Proof. We have

$$E(\hat{y}_n) = E[N\bar{y}_n] = NE(\bar{y}_n) = N\bar{Y} = Y \tag{2.1.3}$$

which proves corollary.

Theorem 2.1.2. The variance of the estimator $\bar{y}_n$ of the population mean $\bar{Y}$ is

$$V(\bar{y}_n) = n^{-1}\sigma_y^2, \text{ where } \sigma_y^2 = N^{-1}\sum_{i=1}^{N}\left(Y_i - \bar{Y}\right)^2 = N^{-1}\left[\sum_{i=1}^{N} Y_i^2 - N\bar{Y}^2\right] \tag{2.1.4}$$

Proof. Because of independence of draws, we have

$$V(\bar{y}_n) = V\left(\frac{1}{n}\sum_{i=1}^{n} y_i\right) = \frac{1}{n^2}\sum_{i=1}^{n} V(y_i). \tag{2.1.5}$$

By the definition of variance we have

$$V(y_i) = E[y_i - E(y_i)]^2 = E(y_i^2) - \{E(y_i)\}^2 = \frac{1}{N}\sum_{i=1}^{N} Y_i^2 - \bar{Y}^2$$

$$= \frac{1}{N}\left[\sum_{i=1}^{N} Y_i^2 - N\bar{Y}^2\right] = \frac{1}{N}\sum_{i=1}^{N}\left(Y_i - \bar{Y}\right)^2 = \sigma_y^2.$$

Using (2.1.5) we have $V(\bar{y}_n) = \sigma_y^2/n$. Hence the theorem.

Theorem 2.1.3. An unbiased estimator of the variance $V(\bar{y}_n)$ is given by

$$\hat{v}(\bar{y}_n) = \frac{s_y^2}{n}, \tag{2.1.6}$$

where

$$s_y^2 = \frac{1}{n-1}\sum_{i=1}^{n}\left(y_i - \bar{y}_n\right)^2 = \frac{1}{n-1}\left[\sum_{i=1}^{n} y_i^2 - n\bar{y}_n^2\right].$$

Proof. We have to show that $E[\hat{v}(\bar{y}_n)] = V(\bar{y}_n)$. Now we have

$$E[\hat{v}(\bar{y}_n)] = E\left[\frac{s_y^2}{n}\right] = \frac{1}{n}E\left(s_y^2\right). \tag{2.1.7}$$

Note that

$$E\left(\bar{y}_n^2\right) = V(\bar{y}_n) + \bar{Y}^2 = \frac{\sigma_y^2}{n} + \bar{Y}^2$$

and

$$E\left[s_y^2\right] = E\left[\frac{1}{n-1}\left(\sum_{i=1}^{n} y_i^2 - n\bar{y}_n^2\right)\right] = \frac{1}{n-1}E\left[\sum_{i=1}^{n} y_i^2 - n\bar{y}_n^2\right] = \frac{1}{n-1}E\left[\frac{n}{n}\sum_{i=1}^{n} y_i^2 - n\bar{y}_n^2\right]$$

$$= \frac{n}{n-1}\left[\frac{1}{n}\sum_{i=1}^{n} E\left(y_i^2\right) - E\left(\bar{y}_n^2\right)\right] = \frac{n}{n-1}\left[\frac{1}{n}\sum_{i=1}^{n}\left(\frac{1}{N}\sum_{i=1}^{N} Y_i^2\right) - \left(\frac{\sigma_y^2}{n} + \bar{Y}^2\right)\right]$$

$$= \frac{n}{n-1}\left[\frac{1}{N}\sum_{i=1}^{N} Y_i^2 - \bar{Y}^2 - \frac{\sigma_y^2}{n}\right] = \frac{n}{n-1}\left(\sigma_y^2 - \frac{\sigma_y^2}{n}\right) = \sigma_y^2 .$$

From (2.1.7) we have

$$E[\hat{v}(\bar{y}_n)] = \frac{1}{n}E\left(s_y^2\right) = \frac{\sigma_y^2}{n} = V(\bar{y}_n).$$

Hence the theorem.

Corollary 2.1.2. The variance of the estimator $\hat{y}_n = N\bar{y}_n$ of the population total is $V(\hat{y}_n) = N^2 V(\bar{y}_n)$.

Theorem 2.1.4. Under SRSWR sampling, while estimating population mean (or total), the minimum sample size with minimum relative standard error (RSE) equal to ϕ, is given by

$$n \geq \left[\frac{\sigma_y^2}{\bar{Y}^2\phi^2}\right]. \tag{2.1.8}$$

Proof. The relative standard error of the estimator $\bar{y}_n$ is given by

$$\text{RSE}(\bar{y}_n) = \sqrt{V(\bar{y}_n)/\{E(\bar{y}_n)\}^2} = \sqrt{\sigma_y^2/\left(n\bar{Y}^2\right)} . \tag{2.1.9}$$

We need an estimator $\bar{y}_n$ such that $\text{RSE}(\bar{y}_n) \leq \phi$, which implies that

$$\sqrt{\sigma_y^2/\left(n\bar{Y}^2\right)} \leq \phi, \quad \text{or} \quad \frac{\sigma_y^2}{n\bar{Y}^2} \leq \phi^2, \quad \text{or} \quad n \geq \frac{\sigma_y^2}{\phi^2\bar{Y}^2}.$$

Hence the theorem.

Remark 2.1: If $\phi = \left(\dfrac{e}{\bar{Y}Z_{\alpha/2}}\right)$, with $e = Z_{\alpha/2}\dfrac{\sigma_y}{\sqrt{n}}$ then $P\left(\left|\dfrac{(\bar{y}_n - \bar{Y})}{\bar{Y}}\right| \leq e\right) = 1 - \alpha$.

Example 2.1.1. In 1995, a fisherman selected an SRSWR sample of six kinds of fish out of 69 kinds of fish available at Atlantic and Gulf Coasts as given below:

Kind of fish	Saltwater catfish	White perch	Blue runner	Scup	Summer flounder	Scup
No. of fish	13859	3489	2319	3688	16238	3688

(a) Estimate the average number of fish in each species group.
(b) Construct a 95% confidence interval for the average number of fish in each species group
(c) Estimate the total number of fish at Atlantic and Gulf Coasts during 1995.
(d) Construct a 95% confidence interval for the total number of fish at Atlantic and Gulf Coasts during 1995.

Solution. We are given $N = 69$ and $n = 6$. From the sample information we have

Sample Unit	y_i	y_i^2
1	13859	192071881
2	3489	12173121
3	2319	5377761
4	3688	13601344
5	16238	263672644
6	3688	13601344
Sum	43281	500498095

(a) Thus the average number of fish in each species group is given by

$$\bar{y}_n = \frac{1}{n}\sum_{i=1}^{n} y_i = \frac{1}{6}\sum_{i=1}^{6} y_i = \frac{43281}{6} = 7213.5 .$$

(b) A $(1-\alpha)100\%$ confidence interval for the population mean $\bar{Y}$ is given by

$$\bar{y}_n \pm t_{\alpha/2}(\text{df} = n-1)\sqrt{\hat{v}(\bar{y}_n)}$$

where $\hat{v}(\bar{y}_n) = s_y^2/n$. Now we have

$$s_y^2 = \frac{1}{n-1}\left\{\sum_{i=1}^{n} y_i^2 - n^{-1}\left(\sum_{i=1}^{n} y_i\right)^2\right\} = \frac{1}{6-1}\left[500498095 - \frac{(43281)^2}{6}\right] = 37658120.3 .$$

Thus

$$\hat{v}(\bar{y}_n) = \frac{s_y^2}{n} = \frac{37658120.3}{6} = 6276353.38 .$$

Using Table 2 from the Appendix the 95% confidence interval for the average number of fish is given by

$$\bar{y}_n \pm t_{0.05/2}(\text{df} = 6-1)\sqrt{\hat{v}(\bar{y}_n)}, \text{ or } 7213.5 \pm 2.571\sqrt{6276353.38}, \quad \text{or } [772.46,\ 13654.53].$$

(c) An estimate of total number of fish is given by

$$\hat{y} = N\bar{y}_n = 69 \times 7213.5 = 497731.5 .$$

(d) The 95% confidence interval for the total number of fish is given by

$$N \times [772.46,\ 13654.53], \text{ or } 69 \times [772.46,\ 13654.53], \text{ or } [53299.7,\ 942162.5].$$

Example 2.1.2. We wish to estimate the average number of fish in each one of the species groups caught by marine recreational fishermen at the Atlantic and Gulf coasts. There are 69 species groups caught during 1995 as shown in the population 4 in the Appendix. What is the minimum number of species groups to be selected by SRSWR sampling to attain the accuracy of relative standard error 30%?

Given: $S_y^2 = 37199578$ and $Y = 311528$.

Solution. We are given $N = 69$, $S_y^2 = 37199578$ and $Y = 311528$, thus

$$\bar{Y} = \frac{Y}{N} = \frac{311528}{69} = 4514.898$$

and

$$\sigma_y^2 = \frac{(N-1)}{N} S_y^2 = \frac{(69-1)}{69} \times 37199578 = 36660453.68.$$

For $\phi = 0.30$ either we are estimating population total or population mean, the minimum sample size for the required degree of precision is given by

$$n \geq \left(\frac{\sigma_y^2}{\phi^2 \bar{Y}^2} \right) = \frac{36660453.68}{0.3^2 \times (4514.898)^2} = 19.98 \approx 20.$$

Thus a sample of size $n = 20$ units is required to attain 30% relative standard error of the estimator of population mean under SRSWR sampling.

Example 2.1.3. Select an SRSWR sample of twenty units from population 4 given in the Appendix. Collect the information on the number of fish during 1995 in each of the species group selected in the sample. Estimate the average number of fish in each one of the species groups caught by marine recreational fishermen at Atlantic and Gulf coasts during 1995. Construct the 95% confidence interval for the average number of fish in each species group available in the United States.

Solution. The population size is $N = 69$, thus we used the first two columns of the Pseudo-Random Number (PRN) Table 1 given in the Appendix to select 20 random numbers between 1 and 69. The random numbers so selected are 58, 60, 54, 01, 69, 62, 23, 64, 46, 04, 32, 47, 57, 56, 57, 60, 33, 05, 22 and 38.

Random No.	Species group	y_i	$(y_i - \bar{y})$	$(y_i - \bar{y})^2$
01	Sharks, other	2016	-3977.25	15818517.560
04	Eels	152	-5841.25	34120201.560
05	Herrings	30027	24033.75	577621139.100
22	Crevalle jack	3951	-2042.25	4170785.063
23	Blue runner	2319	-3674.25	13500113.060
32	Yellowtail snapper	1334	-4659.25	21708610.560
33	Snappers, others	492	-5501.25	30263751.560

Continued.......

38	Pinfish	16855	10861.75	117977613.100
46	Spot	11567	5573.75	31066689.060
47	Kingfish	4333	-1660.25	2756430.063
54	Tautog	3816	-2177.25	4740417.563
56	Wrasses, other	185	-5808.25	33735768.060
57	Little tunny/ Atl bonito	782	-5211.25	27157126.560
57	Little tunny/ Atl bonito	782	-5211.25	27157126.560
58	Atlantic mackerel	4008	-1985.25	3941217.563
60	Spanish mackerel	2568	-3425.25	11732337.560
60	Spanish mackerel	2568	-3425.25	11732337.560
62	Summer flounder	16238	10244.75	104954902.600
64	Southern flounder	1446	-4547.25	20677482.560
69	Other fish	14426	8432.75	71111272.560
	Sum	119865	0.00	1165943840.000

An estimate of the average number of fish in each species group during 1995 is:

$$\bar{y}_n = \frac{1}{n}\sum_{i=1}^{n} y_i = \frac{119865}{20} = 5993.25.$$

Now

$$s_y^2 = \frac{1}{n-1}\sum_{i=1}^{n}(y_i - \bar{y}_n)^2 = \frac{1165943840}{20-1} = 61365465.26,$$

and the estimate of variance of the estimator $\bar{y}_n$ is given by

$$\hat{v}(\bar{y}_n) = \frac{s_y^2}{n} = \frac{61365465.25}{20} = 3068273.26.$$

A $(1-\alpha)100\%$ confidence interval for the average number of fish in each one of the species groups caught during 1995 by marine recreational fishermen in the United States is

$$\bar{y}_n \mp t_{\alpha/2}(\text{df} = n-1)\sqrt{\hat{v}(\bar{y}_n)}.$$

Using Table 2 from the Appendix the 95% confidence interval is given by

$$\bar{y}_n \mp t_{0.025}(\text{df} = 20-1)\sqrt{\hat{v}(\bar{y}_n)}, \text{ or, } 5993.25 \mp 2.093\sqrt{3068273.26}$$

or $[2327.05, 9659.45]$.

Example 2.1.4. The depth y of the roots of plants in a field is uniformly distributed between 5cm and 8cm with the probability density function

$$f(y) = \frac{1}{3} \quad \forall\, 5 < y < 8.$$

We wish to estimate the average length of roots of the plants with an accuracy of relative standard error of 5%, what is the required minimum with replacement sample size n?

Solution. We know that if y has a uniform distribution function

$$f(y) = \frac{1}{b-a} \;\forall\; a < y < b.$$

Then the population mean

$$\overline{Y} = \frac{a+b}{2} = \frac{5+8}{2} = 6.5,$$

and the population variance

$$\sigma_y^2 = \frac{(b-a)^2}{12} = \frac{(8-5)^2}{12} = \frac{9}{12} = 0.75.$$

We need $\phi = 0.05$, thus the required minimum sample size is given by

$$n \geq \frac{\sigma_y^2}{\phi^2 \overline{Y}^2} = \frac{0.75}{0.05^2 \times 6.5^2} = 7.1 \approx 7.$$

Example 2.1.5. The depth y of the roots of plants in a field is uniformly distributed between 5cm and 8 cm with the probability density function

$$f(y) = \frac{1}{3} \;\forall\; 5 < y < 8.$$

Select a with replacement sample of $n = 7$ units. Construct a 95% confidence interval to estimate the average depth of roots.

Solution. The cumulative distribution function (c.d.f.) is given by

$$F(y) = P[Y \leq y] = \int_5^y f(y)dy = \int_5^y \frac{1}{3}dy = \frac{(y-5)}{3}$$

which implies that $y = 3F(y) + 5$. We used the first three columns multiplied by 10^{-3} of the Pseudo-Random Number (PRN) Table 1 given in the Appendix to select seven values of $F(y)$ and the required sampled values are computed by using the relationship $y = 3F(y) + 5$ as follows:

$F(y)$	y_i	y_i^2
0.992	7.976	63.61658
0.588	6.764	45.75170
0.601	6.803	46.28081
0.549	6.647	44.18261
0.925	7.775	60.45063
0.014	5.042	25.42176
0.697	7.091	50.28228
Sum	48.098	335.98640

Thus an estimate of the average depth of the roots is given by

$$\bar{y}_n = \frac{1}{n}\sum_{i=1}^{n} y_i = \frac{48.098}{7} = 6.8711.$$

To find sample variance s_y^2 we apply here alternative method as

$$s_y^2 = \frac{1}{n-1}\left\{\sum_{i=1}^{n} y_i^2 - n^{-1}\left(\sum_{i=1}^{n} y_i\right)^2\right\} = \frac{1}{7-1}\left[335.9864 - \frac{48.098^2}{7}\right] = 0.9164,$$

and the estimate of variance of the estimator $\bar{y}_n$ is given by

$$\hat{v}(\bar{y}_n) = \frac{s_y^2}{n} = \frac{0.9164}{7} = 0.1309.$$

A $(1-\alpha)100\%$ confidence interval for the average depth of roots in the field is

$$\bar{y}_n \mp t_{\alpha/2}(\text{df} = n-1)\sqrt{\hat{v}(\bar{y}_n)}.$$

Using Table 2 from the Appendix the 95% confidence interval estimate of the average depth of the roots is given by

$$\bar{y}_n \mp t_{0.025}(\text{df} = 7-1)\sqrt{\hat{v}(\bar{y}_n)}, \quad \text{or} \quad 6.8711 \mp 2.447\sqrt{0.1309}, \quad \text{or} \quad [5.9857,\ 7.756].$$

Theorem 2.1.5. The covariance between two sample means $\bar{y}_n$ and $\bar{x}_n$ under SRSWR sampling is:

$$\text{Cov}(\bar{y}_n, \bar{x}_n) = \frac{\sigma_{xy}}{n}, \tag{2.1.10}$$

where

$$\sigma_{xy} = \frac{1}{N}\sum_{i=1}^{N}(Y_i - \bar{Y})(X_i - \bar{X}).$$

Proof. We have

$$\text{Cov}(\bar{y}_n, \bar{x}_n) = \text{Cov}\left(\frac{1}{n}\sum_{i=1}^{n} y_i, \frac{1}{n}\sum_{i=1}^{n} x_i\right) = \frac{1}{n^2}\sum_{i=1}^{n}\text{Cov}(y_i, x_i). \tag{2.1.11}$$

Now

$$\text{Cov}(y_i,\ x_i) = E(y_i x_i) - E(y_i)E(x_i). \tag{2.1.12}$$

The random variables $(y_i x_i)$, y_i and x_i, respectively, can take any one of the value $(Y_i X_i)$, Y_i and X_i for $i = 1,2,...,N$ with probability $1/N$. Thus we have

$$\text{Cov}(y_i,\ x_i) = E(y_i x_i) - E(y_i)E(x_i) = \frac{1}{N}\sum_{i=1}^{N} Y_i X_i - \left(\frac{1}{N}\sum_{i=1}^{N} Y_i\right)\left(\frac{1}{N}\sum_{i=1}^{N} X_i\right)$$

$$= \frac{1}{N}\sum_{i=1}^{N} Y_i X_i - \bar{Y}\,\bar{X} = \frac{1}{N}\sum_{i=1}^{N}(Y_i - \bar{Y})(X_i - \bar{X}) = \sigma_{xy}.$$

On substituting it in (2.1.11) we obtain

$$\text{Cov}(\bar{y}_n,\ \bar{x}_n) = \frac{\sigma_{xy}}{n}.$$

Hence the theorem.

Theorem 2.1.6. An unbiased estimator of $\text{Cov}(\bar{y}_n, \bar{x}_n)$ is given by

$$\hat{\text{cov}}(\bar{y}_n, \bar{x}_n) = \frac{s_{xy}}{n}, \tag{2.1.13}$$

where

$$s_{xy} = \frac{1}{n-1}\sum_{i=1}^{n}(y_i - \bar{y})(x_i - \bar{x}).$$

Proof. We have to prove that $E(s_{xy}) = \sigma_{xy}$. Now we have

$$E(s_{xy}) = E\left[\frac{1}{n-1}\sum_{i=1}^{n}(y_i - \bar{y})(x_i - \bar{x})\right] = E\frac{1}{n-1}\left\{\sum_{i=1}^{n} y_i x_i - n\bar{y}_n\bar{x}_n\right\}$$

$$= \frac{1}{n-1}\left\{\sum_{i=1}^{n} E(y_i x_i) - nE(\bar{y}_n\bar{x}_n)\right\} = \frac{1}{n-1}\left\{\sum_{i=1}^{n}\frac{1}{N}\sum_{i=1}^{N} Y_i X_i - n\left(\text{Cov}(\bar{y}_n, \bar{x}_n) + \bar{Y}\,\bar{X}\right)\right\}$$

$$= \frac{1}{n-1}\left\{\frac{n}{N}\sum_{i=1}^{N} Y_i X_i - n\left(\frac{\sigma_{xy}}{n} + \bar{Y}\,\bar{X}\right)\right\} = \frac{n}{n-1}\left\{\frac{1}{N}\sum_{i=1}^{N} Y_i X_i - \bar{Y}\,\bar{X} - \frac{\sigma_{xy}}{n}\right\}$$

$$= \frac{n}{n-1}\left\{\sigma_{xy} - \frac{\sigma_{xy}}{n}\right\} = \sigma_{xy}.$$

Hence the theorem.

2.2 SIMPLE RANDOM SAMPLING WITHOUT REPLACEMENT

Suppose we selected a sample of $n \geq 2$ units from the population of size N by using SRSWOR sampling. Let y_i, $i = 1,2,...,n$ denote the value of the ith unit selected in the sample and Y_i, $i = 1,2,...,N$ be the value of the ith unit in the population. Then we have the following theorems:

Theorem 2.2.1. The sample mean $\bar{y}_n = n^{-1}\sum_{i=1}^{n} y_i$ is an unbiased estimator of the population mean $\bar{Y} = N^{-1}\sum_{i=1}^{N} Y_i$.

Proof. We have to show that $E(\bar{y}_n) = \bar{Y}$. It is interesting to note that this result can be proved by using three different methods as shown below.

Method I. Note that the estimator $\bar{y}_n = n^{-1}\sum_{i=1}^{n} y_i$ can easily be written as

$$\bar{y}_n = \frac{1}{n}\sum_{i=1}^{N} t_i Y_i \tag{2.2.1}$$

where t_i is a random variable defined as:

$$t_i = \begin{cases} 1 & \text{if the ith unit of the population is in the sample, } i.e., \text{ if } i \in s, \\ 0 & \text{otherwise.} \end{cases}$$

Note that Y_i is a fixed value in the population for the ith unit, therefore, the expected value of (2.2.1) is given by

$$E(\bar{y}_n) = E\left[\frac{1}{n}\sum_{i=1}^{N} t_i Y_i\right] = \frac{1}{n}E\left(\sum_{i=1}^{N} t_i Y_i\right) = \frac{1}{n}\left[\sum_{i=1}^{N} Y_i E(t_i)\right]. \tag{2.2.2}$$

Note that ${}^{(N-1)}C_{(n-1)}$ is the number of samples in which a given population unit can occur out of all ${}^{N}C_n$ SRSWOR samples, and therefore the probability that i^{th} population unit is selected in the sample is $= \frac{{}^{(N-1)}C_{(n-1)}}{{}^{N}C_n} = \frac{n}{N}$. So the random variable t_i takes the value 1 with probability $\frac{n}{N}$ and 0 with probability $\left(1-\frac{n}{N}\right)$.

Thus the expected value of t_i is

$$E(t_i) = \frac{n}{N}\times 1 + \left(1-\frac{n}{N}\right)\times 0 = \frac{n}{N}. \tag{2.2.3}$$

From (2.2.2) we have

$$E(\bar{y}_n) = \frac{1}{n}\left[\sum_{i=1}^{N} E(t_i)Y_i\right] = \frac{1}{n}\sum_{i=1}^{N}\frac{n}{N}Y_i = \frac{1}{N}\sum_{i=1}^{N} Y_i = \bar{Y}.$$

Hence the theorem.

Method II. We can also prove the same result as follows

$$E(\bar{y}_n) = E\left[\frac{1}{n}\sum_{i=1}^{n} y_i\right] = \frac{1}{n}\sum_{i=1}^{n} E(y_i). \tag{2.2.4}$$

In (2.2.4) the sample value y_i is a random variable and can take any population value Y_i, $i = 1,2,...,N$, with probability $1/N$.

Thus we have

$$E(y_i) = \frac{1}{N}\sum_{i=1}^{N} Y_i = \bar{Y}.$$

From (2.2.4) we have

$$E(\bar{y}_n) = \frac{1}{n}\sum_{i=1}^{n} E(y_i) = \frac{1}{n}\sum_{i=1}^{n}\bar{Y} = \bar{Y}.$$

Hence the theorem.

Method III. To prove the above result by another method, let us consider

$(\bar{y}_n)_t$ = sample mean $\bar{y}_t$ based on the tth sample selected from the population.

Note that there are ${}^{N}C_n$ possible samples, the probability of selecting the tth sample is

$$p_t = 1/\left({}^{N}C_n\right).$$

By the definition of expected value, we have

$$E\left[(\bar{y}_n)_t\right] = \sum_{t=1}^{{}^N C_n} p_t (\bar{y}_n)_t = \frac{1}{{}^N C_n} \sum_{t=1}^{{}^N C_n} (\bar{y}_n)_t = \frac{1}{{}^N C_n} \sum_{t=1}^{{}^N C_n} \left(\frac{1}{n} \sum_{i=1}^{n} y_i \right)_t$$

$$= \frac{1}{n\left({}^N C_n\right)} \sum_{t=1}^{{}^N C_n} \left(\sum_{i=1}^{n} y_i \right)_t = \frac{1}{n\left({}^N C_n\right)} \sum_{i=1}^{N} {}^{N-1}C_{n-1} Y_i = \frac{\left({}^{(N-1)}C_{(n-1)}\right)}{n\left({}^N C_n\right)} \sum_{i=1}^{N} Y_i = \bar{Y}.$$

Hence the theorem.

Example 2.2.1. Show that the equality $\sum_{t=1}^{{}^N C_n} \left(\sum_{i=1}^{n} y_i \right)_t = \left\{{}^{(N-1)}C_{(n-1)}\right\} \sum_{i=1}^{N} Y_i$ holds good.

Solution. Let us consider a population consisting of four units A, B, C, and D. The number of possible samples without replacement of size two is ${}^N C_n = {}^4 C_2 = 6$. Let Y_1, Y_2, Y_3 and Y_4 be the values of four units A, B, C, and D respectively, in the population. Let y_i denote the value of the ith unit selected in the sample. Then we have the following situation:

Sample no.	1	2	3	4	5	6
Sampled units	(A, B)	(A, C)	(A, D)	(B, C)	(B, D)	(C, D)
Population units	(Y_1, Y_2)	(Y_1, Y_3)	(Y_1, Y_4)	(Y_2, Y_3)	(Y_2, Y_4)	(Y_3, Y_4)

The values of the units in the sample in all these cases are y_1 and y_2.

Thus we have

$$\sum_{t=1}^{{}^N C_n} \left(\sum_{i=1}^{n} y_i \right)_t = \sum_{t=1}^{6} \left(\sum_{i=1}^{2} y_i \right)_t = \sum_{t=1}^{6} (y_1 + y_2)_t$$

$$= (y_1 + y_2)_1 + (y_1 + y_2)_2 + (y_1 + y_2)_3 + (y_1 + y_2)_4 + (y_1 + y_2)_5 + (y_1 + y_2)_6$$

$$= (Y_1 + Y_2) + (Y_1 + Y_3) + (Y_1 + Y_4) + (Y_2 + Y_3) + (Y_2 + Y_4) + (Y_3 + Y_4)$$

$$= 3Y_1 + 3Y_2 + 3Y_3 + 3Y_4 = {}^3C_1 Y_1 + {}^3C_1 Y_2 + {}^3C_1 Y_3 + {}^3C_1 Y_4$$

$$= {}^{(4-1)}C_{(2-1)} Y_1 + {}^{(4-1)}C_{(2-1)} Y_2 + {}^{(4-1)}C_{(2-1)} Y_3 + {}^{(4-1)}C_{(2-1)} Y_4$$

$$= {}^{(N-1)}C_{(n-1)} Y_1 + {}^{(N-1)}C_{(n-1)} Y_2 + {}^{(N-1)}C_{(n-1)} Y_3 + {}^{(N-1)}C_{(n-1)} Y_4$$

$$= \sum_{i=1}^{N} \left\{ {}^{(N-1)}C_{(n-1)} \right\} (Y_i).$$

Theorem 2.2.2. The probability for any population unit to get selected in the sample at any particular draw is equivalent to inverse of the population size, that is,

$$\text{Probability of selecting the } i^{th} \text{ unit in a sample} = \frac{1}{N}. \tag{2.2.5}$$

Proof. Let us consider that at the rth draw, the ith population unit Y_i is selected. This is possible only if this unit has not been selected in the previous $(r-1)$ draws. Let us now consider the draws one by one.

First draw: The probability for the particular unit 'Y_i' to get selected on the first draw out of N units is $=1/N$. Note that 'the probability' that Y_i is not selected on the first draw, from a population of N units is $=\{1-1/N\}=(N-1)/N$.

Second draw: The probability that a particular unit is selected on the second draw (if it is not already selected on the first draw) is the product of two probabilities, namely

(Probability that Y_i is not selected on the first draw) $\times$ (Probability that Y_i is selected on the second draw)

Therefore the probability that Y_i is selected on the second draw is equal to

$$\frac{(N-1)}{N}\times\frac{1}{(N-1)}=\frac{1}{N}.$$

Note that the probability the Y_i is not selected on the second draw out of the remaining $(N-1)$ population units is equal to

$$1-\frac{1}{(N-1)}=\frac{N-2}{N-1}.$$

Third draw: The probability that a particular population unit is selected on the third draw (if it is not selected on the second draw) is the product of three probabilities.

Probability that Y_i is selected on the third draw (if it is not selected on first or second draw) is a product of three probabilities as

(Probability that Y_i is not selected on first draw) $\times$ (Probability that Y_i is not selected on second draw) $\times$ (Probability that Y_i is selected on the third draw)

$$=\left(\frac{N-1}{N}\right)\times\left(\frac{N-2}{N-1}\right)\times\frac{1}{N-2}=\frac{1}{N}.$$

Note that the probability that Y_i is not selected on the third draw out of $(N-2)$ population units is equal to

$$1-\frac{1}{N-2}=\frac{N-3}{N-2}.$$

This procedure continues up to $(r-1)$ draws.

rth draw: Probability that Y_i is not selected up to $(r-1)$th draw is given by

$$\frac{(N-1)}{N}\times\frac{(N-2)}{(N-1)}\times....\times\frac{N-(r-1)}{N-(r-2)}=\frac{N-r+1}{N}.$$

Probability that Y_i is selected at r^{th} draw [assuming that it is not selected at any of the previous $(r-1)$ draws] is equal to

$$\frac{1}{N-(r-1)}=\frac{1}{N-r+1}.$$

So we obtain the probability of a particular unit Y_i to get selected at the rth draw is

$$\frac{(N-r+1)}{N}\times\frac{1}{(N-r+1)}=\frac{1}{N}.$$

Hence the theorem.

Theorem 2.2.3. The variance of the estimator $\bar{y}_n$ is given by

$$V(\bar{y}_n)=\frac{(1-f)}{n}S_y^2 \tag{2.2.6}$$

where $S_y^2=\frac{1}{N-1}\sum_{i=1}^{N}(Y_i-\bar{Y})^2$ and $f=n/N$ denote the finite population correction factor (f.p.c.).

Proof. We have

$$V(\bar{y}_n)=V\left(\frac{1}{n}\sum_{i=1}^{n}y_i\right)=V\left(\frac{1}{n}\sum_{i=1}^{N}t_iY_i\right), \tag{2.2.7}$$

where t_i is a random variable that takes the value '1' if the ith unit is included in the sample, otherwise it takes the value 0. Note that the Y_i is fixed for the ith unit in the population we have

$$V(\bar{y}_n)=\frac{1}{n^2}V\left(\sum_{i=1}^{N}t_iY_i\right)=\frac{1}{n^2}\left[\sum_{i=1}^{N}V(t_iY_i)+\sum_{i\neq j=1}^{N}\text{Cov}(t_iY_i,\ t_jY_j)\right]$$

$$=\frac{1}{n^2}\left[\sum_{i=1}^{N}Y_i^2V(t_i)+\sum_{i\neq j=1}^{N}Y_iY_j\text{Cov}(t_i,t_j)\right]. \tag{2.2.8}$$

In (2.2.8) we need to determine the $V(t_i)$ and $\text{Cov}(t_i,\ t_j)$. Note that the distributions of t_i and t_i^2 are

$$t_i=\begin{cases}1 & \text{with probability } n/N,\\ 0 & \text{with probability } 1-(n/N),\end{cases} \quad\text{and}\quad t_i^2=\begin{cases}1 & \text{with probability } (n/N),\\ 0 & \text{with probability } \{1-(n/N)\}.\end{cases}$$

We have

$$V(t_i)=E[t_i-E(t_i)]^2=E(t_i^2)-\{E(t_i)\}^2$$

$$=\frac{n}{N}-\left(\frac{n}{N}\right)^2=\frac{n}{N}\left(1-\frac{n}{N}\right)=\frac{n(N-n)}{N^2}. \tag{2.2.9}$$

The probability that both i^{th} and j^{th} units are included in the sample is $\frac{{}^{(N-2)}C_{(n-2)}}{{}^{N}C_n}=\frac{n(n-1)}{N(N-1)}$ and otherwise the probability is $1-\frac{n(n-1)}{N(N-1)}$, therefore

$$t_i t_j = \begin{cases} 1 & \text{with probability } \dfrac{n(n-1)}{N(N-1)}, \\ 0 & \text{with probability } \left\{1 - \dfrac{n(n-1)}{N(N-1)}\right\}. \end{cases}$$

Now we have

$$\text{Cov}(t_i, t_j) = E(t_i t_j) - E(t_i)E(t_j) = \frac{n(n-1)}{N(N-1)} - \left(\frac{n}{N}\right)\left(\frac{n}{N}\right) = \frac{n}{N}\left[\frac{n-1}{N-1} - \frac{n}{N}\right]$$

$$= \frac{n(n-N)}{N^2(N-1)} = -\frac{n}{N}\left(\frac{N-n}{N(N-1)}\right) = -\frac{n(N-n)}{N^2(N-1)}. \tag{2.2.10}$$

Using (2.2.9) and (2.2.10) in (2.2.8) we obtain

$$V(\bar{y}_n) = \frac{1}{n^2}\left[\sum_{i=1}^{N} Y_i^2 \left\{\frac{n(N-n)}{N^2}\right\} + \sum_{i \neq j=1}^{N} Y_i Y_j \left\{\frac{-n(N-n)}{N^2(N-1)}\right\}\right]$$

$$= \frac{(N-n)}{nN^2}\left[\sum_{i=1}^{N} Y_i^2 - \frac{1}{(N-1)}\sum_{i \neq j=1}^{N} Y_i Y_j\right]. \tag{2.2.11}$$

Note that

$$\bar{Y}^2 = \left(\frac{1}{N}\sum_{i=1}^{N} Y_i\right)^2 = \frac{1}{N^2}\left[\sum_{i=1}^{N} Y_i^2 + \sum_{i \neq j=1}^{N} Y_i Y_j\right], \tag{2.2.12}$$

we obtain

$$\sum_{i \neq j=1}^{N} Y_i Y_j = N^2 \bar{Y}^2 - \sum_{i=1}^{N} Y_i^2 .$$

On substituting (2.2.12) in (2.2.11) we obtain

$$V(\bar{y}_n) = \frac{(N-n)}{nN^2}\left[\sum_{i=1}^{N} Y_i^2 - \frac{1}{(N-1)}\left(N^2\bar{Y}^2 - \sum_{i=1}^{N} Y_i^2\right)\right]$$

$$= \frac{(N-n)}{nN^2}\left[\sum_{i=1}^{N} Y_i^2 + \frac{1}{N-1}\sum_{i=1}^{N} Y_i^2 - \frac{N^2}{N-1}\bar{Y}^2\right]$$

$$= \frac{(N-n)}{nN^2}\left[\left(1 + \frac{1}{N-1}\right)\sum_{i=1}^{N} Y_i^2 - \frac{N^2}{N-1}\bar{Y}^2\right]$$

$$= \frac{(N-n)}{nN^2}\left[\frac{N}{(N-1)}\sum_{i=1}^{N} Y_i^2 - \frac{N^2}{N-1}\bar{Y}^2\right] = \frac{(N-n)}{nN^2}\left[\frac{N}{(N-1)}\left\{\sum_{i=1}^{N} Y_i^2 - N\bar{Y}^2\right\}\right]$$

$$= \frac{N-n}{nN} S_y^2 = \frac{(1-f)}{n} S_y^2$$

where

$$S_y^2 = \frac{1}{(N-1)}\left[\sum_{i=1}^{N} Y_i^2 - N\bar{Y}^2\right].$$

Hence the theorem.

Theorem 2.2.4. An unbiased estimator of the variance $V(\bar{y}_n)$ is given by

$$\hat{v}(\bar{y}_n) = \left(\frac{1-f}{n}\right)s_y^2, \tag{2.2.13}$$

where

$$s_y^2 = \frac{1}{n-1}\left(\sum_{i=1}^{n} y_i^2 - n\bar{y}_n^2\right) \text{ and } f = n/N .$$

Proof. We have to show that $E[\hat{v}(\bar{y}_n)] = V(\bar{y}_n)$. Note that

$$E[\hat{v}(\bar{y}_n)] = E\left[\frac{1-f}{n}s_y^2\right] = \left(\frac{1-f}{n}\right)E\left(s_y^2\right). \tag{2.2.14}$$

Now

$$E\left(s_y^2\right) = E\left[\frac{1}{n-1}\left\{\sum_{i=1}^{n} y_i^2 - n\bar{y}_n^2\right\}\right] = \frac{1}{n-1}\left[E\left(\sum_{i=1}^{n} y_i^2\right) - nE\left(\bar{y}_n^2\right)\right]$$

$$= \frac{n}{n-1}\left[\frac{1}{n}\sum_{i=1}^{n} E\left(y_i^2\right) - E\left(\bar{y}_n^2\right)\right]. \tag{2.2.15}$$

Note that $E\left(\bar{y}_n^2\right) = V(\bar{y}_n) + \{E(\bar{y}_n)\}^2 = \frac{N-n}{Nn}S_y^2 + \bar{Y}^2$ and each unit Y_i becomes selected with probability $1/N$, thus (2.2.15) becomes

$$E\left(s_y^2\right) = \frac{n}{n-1}\left[\frac{1}{n}\sum_{i=1}^{n}\left(\sum_{i=1}^{N}\frac{1}{N}Y_i^2\right) - \left\{\frac{(N-n)}{Nn}S_y^2 + \bar{Y}^2\right\}\right]$$

$$= \frac{n}{n-1}\left[\frac{1}{N}\sum_{i=1}^{N} Y_i^2 - \bar{Y}^2 - \frac{(N-n)}{Nn}S_y^2\right] = \frac{n}{n-1}\left[\frac{1}{N}\left\{\sum_{i=1}^{N} Y_i^2 - N\bar{Y}^2\right\} - \frac{(N-n)}{Nn}S_y^2\right]$$

$$= \frac{n}{n-1}\left[\frac{N-1}{N}\left\{\frac{1}{N-1}\left(\sum_{i=1}^{N} Y_i^2 - N\bar{Y}^2\right)\right\} - \frac{(N-n)}{Nn}S_y^2\right]$$

$$= \frac{n}{n-1}\left[\frac{N-1}{N}S_y^2 - \frac{(N-n)}{Nn}S_y^2\right] = S_y^2 .$$

On substituting this value in (2.2.13), we obtain $E[\hat{v}(\bar{y}_n)] = V(\bar{y}_n)$. Hence the theorem.

Theorem 2.2.5. Under SRSWOR sampling, while estimating population mean (or total), the minimum sample size with minimum relative standard error (RSE) equal to ϕ, is

$$n \geq \left[\frac{1}{N} + \frac{\phi^2\bar{Y}^2}{S_y^2}\right]^{-1}. \tag{2.2.16}$$

Proof. The relative standard error of the estimator $\bar{y}_n$ is given by

$$\text{RSE}(\bar{y}_n) = \sqrt{V(\bar{y}_n)/\{E(\bar{y}_n)\}^2} = \sqrt{\frac{(1-f)S_y^2}{n\bar{Y}^2}}. \tag{2.2.17}$$

Note that we need an estimator $\bar{y}_n$ such that $\text{RSE}(\bar{y}_n) \le \phi$, which implies that

$$\sqrt{\left(\frac{1}{n}-\frac{1}{N}\right)\frac{S_y^2}{\bar{Y}^2}} \le \phi, \quad \text{or} \quad \left(\frac{1}{n}-\frac{1}{N}\right)\frac{S_y^2}{\bar{Y}^2} \le \phi^2, \quad \text{or} \quad n \ge \left[\frac{1}{N}+\frac{\phi^2\bar{Y}^2}{S_y^2}\right]^{-1}.$$

Hence the theorem.

Remark 2.2: If $\phi = \frac{e}{(\bar{Y}Z_{\alpha/2})}$, with $e = Z_{\alpha/2}\sqrt{\frac{(1-f)}{n}S_y^2}$ then $P\left(\left|\frac{(\bar{y}_n-\bar{Y})}{\bar{Y}}\right| \le e\right) = 1-\alpha$.

Example 2.2.2. A fishermen recruiting company, XYZ, selected an SRSWOR sample of six kinds of fish out of 69 kinds of fish available at Atlantic and Gulf Coasts as below:

Kind of fish	Pinfish	Bluefish	Searobins	Silver perch	Tautog	Puffers
No. of fish	16855	10940	4793	2146	3816	935

(a) Estimate the average number of fish in each species group.
(b) Construct a 95% confidence interval for the average number of fish in each species group
(c) Estimate the total number of fish at Atlantic and Gulf Coasts during 1995.
(d) Construct a 95% confidence interval for the total number of fish at Atlantic and Gulf Coasts during 1995.

Solution. We are given $N = 69$ and $n = 6$. From the sample information, we have

Sample Unit	y_i	y_i^2
1	16855	284091025
2	10940	119683600
3	4793	22972849
4	2146	4605316
5	3816	14561856
6	935	874225
Sum	39485	446788871

(a) Thus the average number of fish in each species group is given by

$$\bar{y}_n = \frac{1}{n}\sum_{i=1}^{n} y_i = \frac{1}{6}\sum_{i=1}^{6} y_i = \frac{39485}{6} = 6580.83.$$

(b) A $(1-\alpha)100\%$ confidence interval for the population mean $\bar{Y}$ is given by

$$\bar{y}_n \pm t_{\alpha/2}(\mathrm{df} = n-1)\sqrt{\hat{v}(\bar{y}_n)}\ , \text{ where } \hat{v}(\bar{y}_n) = \left(\frac{1-f}{n}\right)s_y^2 .$$

Now we have

$$s_y^2 = \frac{1}{n-1}\left\{\sum_{i=1}^{n} y_i^2 - n^{-1}\left(\sum_{i=1}^{n} y_i\right)^2\right\} = \frac{1}{6-1}\left[446788871 - \frac{(39485)^2}{6}\right] = 37388933.4 .$$

Thus

$$\hat{v}(\bar{y}_n) = \left(\frac{1-f}{n}\right)s_y^2 = \left(\frac{1-0.0869}{6}\right)\times 37388933.4 = 5689972.5 .$$

Using Table 2 from the Appendix the 95% confidence interval for the average number of fish is given by

$$\bar{y}_n \pm t_{0.025}(\mathrm{df} = 6-1)\sqrt{\hat{v}(\bar{y}_n)}, \text{ or } 6580.83 \pm 2.571\sqrt{5689972.5}, \text{ or } [448.05,\ 12713.61].$$

(c) An estimate of total number of fish is given by

$$\hat{y} = N\bar{y}_n = 69\times 6580.83 = 454077.27.$$

(d) The 95% confidence interval for the total number of fish is given by

$$N\times[448.05, 12713.61], \text{ or } 69\times[448.05, 12713.61], \text{ or } [30915.45, 877239.09].$$

Example 2.2.3. We wish to estimate the average number of fish in each one of the species groups caught by marine recreational fishermen at the Atlantic and Gulf coasts. There were 69 species groups caught during 1995 as shown in the population 4 in the Appendix. What is the minimum number of species groups to be selected by SRSWOR sampling to attain the accuracy of relative standard error 30%?

Given: $S_y^2 = 37199578$ and $Y = 311528$.

Solution. We are given $N = 69$, $S_y^2 = 37199578$ and $\bar{Y} = 4514.898$. Thus for $\phi = 0.30$, either we are estimating population total or population mean, the minimum sample size under SRSWOR sampling for the required degree of precision is

$$n \ge \left[\frac{1}{N} + \frac{\phi^2\bar{Y}^2}{S_y^2}\right]^{-1} = \left[\frac{1}{69} + \frac{0.3^2\times 4514.898^2}{37199578}\right]^{-1} = 15.6 \approx 16 .$$

Thus a minimum sample of size $n = 16$ units is required to attain 30% relative standard error of the estimator of population total or mean under SRSWOR sampling.

Example 2.2.4. Select an SRSWOR sample of sixteen units from population 4 given in the Appendix. Collect the information on the number of fish during 1995 in each of the species group selected in the sample. Estimate the average number of fish in each one of the species groups caught by marine recreational fishermen at Atlantic and Gulf coasts during 1995. Construct 95% confidence interval for the average number of fish in each species groups available in the United States.

Solution. The population size is $N = 69$, therefore we used the second and third columns of the Pseudo-Random Number (PRN) Table 1 given in the Appendix to select 16 random numbers between 1 and 69. The random numbers so selected are 01, 49, 25, 14, 26, 36, 42, 44, 65, 20, 40, 66, 17, 08, 33, and 53.

Random No.	Species group	y_i	$(y_i - \bar{y}_n)$	$(y_i - \bar{y}_n)^2$
01	Sharks, other	2016	-937.8130	879492.2852
08	Toadfishes	1632	-1321.8100	1747188.2850
14	Sculpins	71	-2882.8100	8310607.9100
17	Temperate basses, other	23	-2930.8100	8589661.9100
20	Sea basses, other	2068	-885.8130	784663.7852
25	Florida pompano	644	-2309.8100	5335233.7850
26	Jacks, other	1625	-1328.8100	1765742.6600
33	Snappers, other	492	-2461.8100	6060520.7850
36	Grunts, other	3379	425.1875	180784.4102
40	Red porgy	230	-2723.8100	7419154.5350
42	Spotted seatrout	24615	21661.1900	469207043.9000
44	Sand seatrout	4355	1401.1880	1963326.4100
49	Black drum	1595	-1358.8100	1846371.4100
53	Barracuda	908	-2045.8100	4185348.7850
65	Winter flounder	2324	-629.8130	396663.7852
66	Flounders, other	1284	-1669.8100	2788273.7850
	Sum	47261	0.0000	521460078.4000

An estimate of the average number of fish in each species group during 1995 is

$$\bar{y}_n = \frac{1}{n}\sum_{i=1}^{n} y_i = \frac{47261}{16} = 2953.813\,.$$

Now

$$s_y^2 = \frac{1}{n-1}\sum_{i=1}^{n}(y_i - \bar{y}_n)^2 = \frac{521460078.4}{16-1} = 34764005.23\,.$$

and the estimate of variance of the estimator $\bar{y}_n$ is

$$\hat{v}(\bar{y}_n) = \left(\frac{1-f}{n}\right)s_y^2 = \left(\frac{1-16/69}{16}\right)\times 34764005.23 = 1668924.16.$$

A $(1-\alpha)100\%$ confidence interval for the average number of fish in each one of the species groups caught during 1995 by marine recreational fishermen in the United States is

$$\bar{y}_n \mp t_{\alpha/2}(\mathrm{df} = n-1)\sqrt{\hat{v}(\bar{y}_n)}.$$

Using Table 2 from the Appendix the 95% confidence interval is given by

$$\bar{y}_n \mp t_{0.025}(\mathrm{df} = 16-1)\sqrt{\hat{v}(\bar{y}_n)}\,,\text{ or } 2953.813 \mp 2.131\sqrt{1668924.16}\,,\text{ or } [200.842, 5706.784].$$

Example 2.2.5. The distribution of yield (kg/ha) y of a crop in 1000 plots has a Cauchy distribution:

$$f(y) = \frac{1}{\pi\left\{1+(y-10)^2\right\}}, \quad -\infty < y < +\infty .$$

We wish to estimate the average yield with an accuracy of relative standard error of 0.15%. What is the minimum sample size n required while using SRSWOR sampling?

Solution. Since the true mean and variance of a variable having Cauchy distribution are unknown, therefore it is not possible to find the required sample size under such a distribution.

Example 2.2.6. The distribution of yield (kg/ha) y of a crop in 1000 plots has a logistic distribution

$$f(y) = \frac{1}{4\beta}\text{sech}^2\left\{\frac{1}{2}\left(\frac{x-\alpha_\bullet}{\beta_\bullet}\right)\right\}$$

with $\alpha_\bullet = 40$ and $\beta_\bullet = 2.5$.

(a) Find the value of minimum sample size n required to estimate average yield with an accuracy of standard error of 5%

(b) Select a sample of the required size and construct 95% confidence interval for the average yield.

(c) Does the true average yield lies in the 95% confidence interval?

Solution. (a) We know that the mean and variance of a logistic distribution are given by

$$\text{Mean} = \alpha_\bullet = 40$$

and

$$\text{Variance} = \sigma_y^2 = \beta_\bullet^2\frac{\pi^2}{3} = 2.5^2 \times \frac{3.14159^2}{3} = 20.56.$$

Also we are given $N = 1000$ thus

$$S_y^2 = \frac{N}{N-1}\sigma_y^2 = \frac{1000}{1000-1}\times 20.56 = 20.5806 .$$

Thus the minimum sample size required for $\phi = 0.05$ is given by

$$n \ge \left[\frac{1}{N} + \frac{\phi^2\bar{Y}^2}{S_y^2}\right]^{-1} = \left[\frac{1}{1000} + \frac{0.05^2 \times 40^2}{20.5806}\right]^{-1} = 5.11 \approx 5.$$

We know that the cumulative distribution function for the logistic distribution is

$$F(y) = \frac{1}{2}\left[1 + \tanh\left\{\frac{(y-\alpha_\bullet)}{2\beta_\bullet}\right\}\right]$$

which implies that

$$y = \alpha_\bullet + 2\beta_\bullet \tanh^{-1}\{2F(y)-1\} = 40 + 5\tanh^{-1}\{2F(y)-1\} .$$

Using the last three columns of the Pseudo-Random Number (PRN) Table 1 given in the Appendix, multiplied by 10^{-3}, we obtain five values of $F(y)$ and the corresponding values of y as given below:

$F(y)$	y_i	y_i^2
0.072	36.460	1329.344
0.776	42.522	1808.111
0.406	39.071	1526.531
0.565	40.646	1652.128
0.108	36.675	1345.089
Sum	195.375	7661.202

Thus an estimate of the average yield (kg/ha) of the crop is given by

$$\bar{y}_n = \frac{1}{n}\sum_{i=1}^{n} y_i = \frac{195.375}{5} = 39.075 .$$

We use the alternative method to find s_y^2 given by

$$s_y^2 = \frac{1}{n-1}\left[\sum_{i=1}^{n} y_i^2 - n\bar{y}_n^2\right] = \frac{1}{5-1}\left[7661.202 - 5\times 39.075^2\right] = 6.7309 ,$$

and estimate of the variance of the estimator $\bar{y}_n$ given by

$$\hat{v}(\bar{y}_n) = \left(\frac{1-f}{n}\right)s_y^2 = \left(\frac{1-5/1000}{5}\right)\times 6.7309 = 1.339449 .$$

A $(1-\alpha)100\%$ confidence interval for the average yield of the crop is given by

$$\bar{y}_n \mp t_{\alpha/2}(\text{df} = n-1)\sqrt{\hat{v}(\bar{y}_n)} .$$

Using Table 2 from the Appendix the 95% confidence interval is given by

$$\bar{y}_n \mp t_{0.025}(\text{df} = 5-1)\sqrt{\hat{v}(\bar{y}_n)}, \text{ or } 39.075 \mp 2.776\sqrt{1.339449}, \text{ or} \left[35.8622, 42.2877\right].$$

(c) Yes, the resultant 95% confidence interval estimate contains the true average yield $\alpha_\bullet = 40$.

Theorem 2.2.6. The covariance between the two sample means $\bar{y}_n$ and $\bar{x}_n$ under SRSWOR sampling is:

$$\text{Cov}(\bar{y}_n, \bar{x}_n) = \frac{(1-f)}{n} S_{xy} , \tag{2.2.18}$$

where

$$S_{xy} = \frac{1}{N-1}\sum_{i=1}^{N}\left(Y_i - \bar{Y}\right)\left(X_i - \bar{X}\right).$$

Proof. We have

$$\operatorname{Cov}(\bar{y}_n,\ \bar{x}_n) = \operatorname{Cov}\left(\frac{1}{n}\sum_{i=1}^{n} y_i,\ \frac{1}{n}\sum_{i=1}^{n} x_i\right) = \operatorname{Cov}\left(\frac{1}{n}\sum_{i=1}^{N} t_i Y_i,\ \frac{1}{n}\sum_{i=1}^{N} t_i X_i\right) \tag{2.2.19}$$

where t_i is a random variable that takes the value '1' if the ith unit is included in the sample, otherwise it takes the value 0. Note that the pair Y_i and X_i is fixed for the ith unit in the population, we have

$$\operatorname{Cov}(\bar{y}_n, \bar{x}_n) = \frac{1}{n^2}\operatorname{Cov}\left(\sum_{i=1}^{N} t_i Y_i,\ \sum_{i=1}^{N} t_i X_i\right) = \frac{1}{n^2}\left[E\left\{\sum_{i=1}^{N} t_i Y_i\right\}\left\{\sum_{i=1}^{N} t_i X_i\right\} - E\left\{\sum_{i=1}^{N} t_i Y_i\right\}E\left\{\sum_{i=1}^{N} t_i X_i\right\}\right]$$

$$= \frac{1}{n^2}\left[E\left\{\sum_{i=1}^{N} t_i^2 Y_i X_i + \sum_{i\neq j=1}^{N} t_i t_j Y_i X_j\right\} - E\left\{\sum_{i=1}^{N} t_i Y_i\right\}E\left\{\sum_{i=1}^{N} t_i X_i\right\}\right]$$

$$= \frac{1}{n^2}\left[\left\{\sum_{i=1}^{N} E(t_i^2) Y_i X_i + \sum_{i\neq j=1}^{N} E(t_i t_j) Y_i X_j\right\} - \left\{\sum_{i=1}^{N} E(t_i) Y_i\right\}\left\{\sum_{i=1}^{N} E(t_i) X_i\right\}\right]. \tag{2.2.20}$$

In (2.2.20) we need to determine the $E(t_i^2)$ and $E(t_i t_j)$. Note that the distributions of t_i and t_i^2 are:

$$t_i = \begin{cases} 1 \text{ with probability } n/N, \\ 0 \text{ with probability } \{1-(n/N)\}, \end{cases} \quad \text{and} \quad t_i^2 = \begin{cases} 1 \text{ with probability } (n/N), \\ 0 \text{ with probability } \{1-(n/N)\}. \end{cases}$$

We have

$$E(t_i^2) = 1\times\frac{n}{N} + 0\times\left(1-\frac{n}{N}\right) = \frac{n}{N}. \tag{2.2.21}$$

The probability that both i^{th} and j^{th} units are included in the sample is $\dfrac{^{(N-2)}C_{(n-2)}}{^{N}C_n} = \dfrac{n(n-1)}{N(N-1)}$ and otherwise the probability is $1-\dfrac{n(n-1)}{N(N-1)}$, therefore

$$t_i t_j = \begin{cases} 1 \text{ with probability } & \dfrac{n(n-1)}{N(N-1)}, \\ 0 \text{ with probability } & \left\{1-\dfrac{n(n-1)}{N(N-1)}\right\}. \end{cases}$$

Now we have

$$E(t_i t_j) = 1\times\frac{n(n-1)}{N(N-1)} + 0\left\{1-\frac{n(n-1)}{N(N-1)}\right\} = \frac{n(n-1)}{N(N-1)}. \tag{2.2.22}$$

Using (2.2.21) and (2.2.22) in (2.2.20) we obtain

$$\operatorname{Cov}(\bar{y}_n,\ \bar{x}_n) = \frac{1}{n^2}\left[\left\{\sum_{i=1}^{N} E(t_i^2) Y_i X_i + \sum_{i\neq j=1}^{N} E(t_i t_j) Y_i X_j\right\} - \left\{\sum_{i=1}^{N} E(t_i) Y_i\right\}\left\{\sum_{i=1}^{N} E(t_i) X_i\right\}\right]$$

$$= \frac{1}{n^2}\left[\left\{\sum_{i=1}^{N} E(t_i^2) Y_i X_i + \sum_{i\neq j=1}^{N} E(t_i t_j) Y_i X_j\right\} - \left\{\sum_{i=1}^{N} E(t_i) Y_i\right\}\left\{\sum_{i=1}^{N} E(t_i) X_i\right\}\right]$$

$$=\frac{1}{n^2}\left[\left\{\frac{n}{N}\sum_{i=1}^{N}Y_iX_i+\frac{n(n-1)}{N(N-1)}\sum_{i\neq j=1}^{N}Y_iX_j\right\}-\left\{\frac{n}{N}\sum_{i=1}^{N}Y_i\right\}\left\{\frac{n}{N}\sum_{i=1}^{N}X_i\right\}\right]. \quad (2.2.23)$$

Note that

$$\overline{Y}\,\overline{X}=\left(\frac{1}{N}\sum_{i=1}^{N}Y_i\right)\left(\frac{1}{N}\sum_{i=1}^{N}X_i\right)=\frac{1}{N^2}\left(\sum_{i=1}^{N}Y_i\right)\left(\sum_{i=1}^{N}X_i\right)=\frac{1}{N^2}\left[\sum_{i=1}^{N}Y_iX_i+\sum_{i\neq j=1}^{N}Y_iX_j\right]$$

which implies that

$$\sum_{i\neq j=1}^{N}Y_iX_j=N^2\overline{Y}\,\overline{X}-\sum_{i=1}^{N}Y_iX_i\ .$$

Thus from (2.2.23) we have

$$\mathrm{Cov}(\bar{y}_n,\ \bar{x}_n)=\frac{1}{n^2}\left[\left\{\frac{n}{N}\sum_{i=1}^{N}Y_iX_i+\frac{n(n-1)}{N(N-1)}\sum_{i\neq j=1}^{N}Y_iX_j\right\}-\left\{\frac{n}{N}\sum_{i=1}^{N}Y_i\right\}\left\{\frac{n}{N}\sum_{i=1}^{N}X_i\right\}\right]$$

$$=\frac{1}{n^2}\left[\left\{\frac{n}{N}\sum_{i=1}^{N}Y_iX_i+\frac{n(n-1)}{N(N-1)}\left(N^2\overline{Y}\,\overline{X}-\sum_{i=1}^{N}Y_iX_i\right)\right\}-\left\{\frac{n}{N}\sum_{i=1}^{N}Y_i\right\}\left\{\frac{n}{N}\sum_{i=1}^{N}X_i\right\}\right]$$

$$=\frac{1}{n^2}\left[\left\{\frac{n}{N}-\frac{n(n-1)}{N(N-1)}\right\}\sum_{i=1}^{N}Y_iX_i+\overline{Y}\,\overline{X}\left\{\frac{n(n-1)N}{(N-1)}-n^2\right\}\right]$$

$$=\frac{1}{n^2}\left[\frac{n}{N}\left\{\frac{N-n}{N-1}\right\}\sum_{i=1}^{N}Y_iX_i-\overline{Y}\,\overline{X}\left\{\frac{n(N-n)}{(N-1)}\right\}\right]$$

$$=\frac{1}{n^2}\frac{n}{N}\left\{\frac{N-n}{N-1}\right\}\left[\sum_{i=1}^{N}Y_iX_i-N\overline{Y}\,\overline{X}\right]=\frac{N-n}{nN}\frac{1}{N-1}\left[\sum_{i=1}^{N}Y_iX_i-N\overline{Y}\,\overline{X}\right]$$

$$=\left(\frac{1-f}{n}\right)\frac{1}{N-1}\sum_{i=1}^{N}\left(Y_i-\overline{Y}\right)\left(X_i-\overline{X}\right)=\left(\frac{1-f}{n}\right)S_{xy}\ .$$

Hence the theorem.

Theorem 2.2.7. An unbiased estimator of the covariance $\mathrm{Cov}(\bar{y}_n,\ \bar{x}_n)$ is given by

$$\mathrm{c\hat{o}v}(\bar{y}_n,\ \bar{x}_n)=\left(\frac{1-f}{n}\right)s_{xy} \quad (2.2.24)$$

where

$$s_{xy}=\frac{1}{n-1}\left[\sum_{i=1}^{n}y_ix_i-n\bar{y}_n\bar{x}_n\right].$$

Proof. We have to show that

$$E\left[\mathrm{c\hat{o}v}(\bar{y}_n,\ \bar{x}_n)\right]=\mathrm{Cov}(\bar{y}_n,\ \bar{x}_n).$$

We have

$$E\left[\mathrm{c\hat{o}v}(\bar{y}_n,\ \bar{x}_n)\right]=E\left[\frac{1-f}{n}s_{xy}\right]=\frac{1-f}{n}E\left(s_{xy}\right)$$

Now

$$E\left(s_{xy}\right)=E\left[\frac{1}{n-1}\left\{\sum_{i=1}^{n}y_ix_i-n\bar{y}_n\bar{x}_n\right\}\right]=\frac{1}{n-1}\left[E\left(\sum_{i=1}^{n}y_ix_i\right)-nE\left(\bar{y}_n\bar{x}_n\right)\right]$$

$$= \frac{n}{n-1}\left[\frac{1}{n}\sum_{i=1}^{n} E(y_i x_i) - E(\bar{y}_n \bar{x}_n)\right].$$

Now

$$E(\bar{y}_n \bar{x}_n) = \text{Cov}(\bar{y}_n,\ \bar{x}_n) + E(\bar{y}_n)E(\bar{x}_n) = \frac{N-n}{Nn} S_{xy} + \bar{Y}\bar{X},$$

and each pair of units Y_i and X_i gets selected with probability $1/N$, therefore we have

$$E(s_{xy}) = \frac{n}{n-1}\left[\frac{1}{n}\sum_{i=1}^{n}\left(\sum_{i=1}^{N}\frac{1}{N} Y_i X_i\right) - \left\{\frac{(N-n)}{Nn} S_{xy} + \bar{Y}\,\bar{X}\right\}\right]$$

$$= \frac{n}{n-1}\left[\frac{1}{N}\sum_{i=1}^{N} Y_i X_i - \bar{Y}\,\bar{X} - \frac{(N-n)}{Nn} S_{xy}\right]$$

$$= \frac{n}{n-1}\left[\frac{1}{N}\left\{\sum_{i=1}^{N} Y_i X_i - N\bar{Y}\,\bar{X}\right\} - \frac{(N-n)}{Nn} S_{xy}\right]$$

$$= \frac{n}{n-1}\left[\frac{N-1}{N}\left\{\frac{1}{N-1}\left(\sum_{i=1}^{N} Y_i X_i - N\bar{Y}\,\bar{X}\right)\right\} - \frac{(N-n)}{Nn} S_{xy}\right]$$

$$= \frac{n}{n-1}\left[\frac{N-1}{N} S_{xy} - \frac{(N-n)}{Nn} S_{xy}\right] = S_{xy}.$$

Thus we obtain

$$E[\hat{\text{cov}}(\bar{y}_n,\ \bar{x}_n)] = \text{Cov}(\bar{y}_n,\ \bar{x}_n).$$

Hence the theorem.

Example 2.2.7. Consider the joint probability density function of two continuous random variables x and y is

$$f(x,y) = \begin{cases} \frac{2}{3}(x+2y) & 0 < x < 1,\ \ 0 < y < 1, \\ 0 & \text{otherwise.} \end{cases}$$

(a) Select six pairs of observations $(y,\ x)$ by using the Random Number Table method.

(b) Estimate the value of covariance between x and y.

Solution. (a) See Chapter 1.

(b) Estimate of covariance:

R_2	y	R_1	x
0.992	0.995	0.622	0.423
0.588	0.722	0.771	0.514
0.601	0.732	0.917	0.600
0.549	0.691	0.675	0.456
0.925	0.954	0.534	0.368
0.014	0.039	0.513	0.355

So we obtain

	y	$(y-\bar{y})$	x	$(x-\bar{x})$	$(y-\bar{y})(x-\bar{x})$
	0.995	0.306167	0.423	-0.030	-0.009080
	0.722	0.033167	0.514	0.061	0.002034
	0.732	0.043167	0.600	0.147	0.006360
	0.691	0.002167	0.456	0.003	0.000000
	0.954	0.265167	0.368	-0.085	-0.022450
	0.039	-0.649830	0.355	-0.098	0.063467
Sum	4.133	0.000000	2.716	0.000	0.040335

Thus an estimate of the covariance between two sample means is given by

$$\hat{\text{cov}}(\bar{y}_n, \bar{x}_n) = \left(\frac{1-f}{n}\right)s_{xy} = \left(\frac{1-f}{n}\right)\frac{1}{n-1}\sum_{i=1}^{n}(y_i-\bar{y})(x_i-\bar{x})$$

$$= \frac{1-0}{6}\left(\frac{1}{6-1}\right)\times 0.040335 = 0.0013445.$$

Note that $f = 0$ because of infinite population size.

2.3 ESTIMATION OF POPULATION PROPORTION

Let N be the total number of units in the population Ω and N_a be the number of units possessing a certain attribute, A (say). Then population proportion is the ratio of number of units possessing the attribute A to the total number of units in the population, *i.e.*, $P_y = N_a/N$. Thus we have the following theorem:

Theorem 2.3.1. The population proportion P_y is a special case of the population mean $\bar{Y}$.

Proof. We know that population mean is given by

$$\bar{Y} = \frac{1}{N}\sum_{i=1}^{N} Y_i \, . \tag{2.3.1}$$

Let us define

$$Y_i = \begin{cases} 1 & \text{if the } i^{th} \text{ unit possesses the attribute } A, \\ 0 & \text{otherwise.} \end{cases}$$

Then (2.3.1) becomes,

$$\bar{Y} = \frac{1+0+1+.....+1}{N} = \frac{N_A}{N} = P_y \, .$$

Hence the theorem.

We will discuss the problem of estimation of population proportion using SRSWR and SRSWOR sampling.

Case I. When the sample is drawn using simple random sampling with replacement (SRSWR sampling), we have the following theorems.

Theorem 2.3.2. An unbiased estimator of population proportion P_y is given by

$$\hat{p}_y = \frac{r}{n} \tag{2.3.2}$$

where r is the number of units in the sample that possesses the attribute A.

Proof. To prove that $E(\hat{p}_y) = P_y$, let us proceed as follows:

Defining

$$y_i = \begin{cases} 1 & \text{if the } i^{th} \text{ sampled unit possess the attribute } A, \\ 0 & \text{otherwise,} \end{cases}$$

we have

$$\bar{y}_n = \frac{1}{n}\sum_{i=1}^{n} y_i = \frac{r}{n} = \hat{p}_y .$$

Therefore

$$E(\hat{p}_y) = E\left(\frac{r}{n}\right) = E(\bar{y}_n) = \bar{Y} = P_y .$$

Hence the theorem.

Theorem 2.3.3. The variance of the estimator $\hat{p}_y$ is given by

$$V(\hat{p}_y) = \frac{P_y Q_y}{n} \tag{2.3.3}$$

where

$$Q_y = 1 - P_y .$$

Proof. We know that $V(\bar{y}_n) = \frac{\sigma_y^2}{n}$, which implies that $V(\hat{p}_y) = \frac{\sigma_p^2}{n}$.

Now

$$\sigma_y^2 = \frac{1}{N}\left[\sum_{i=1}^{N} Y_i^2 - N\bar{Y}^2\right].$$

Note that

$$Y_i = \begin{cases} 1 & \text{if the } i^{th} \text{ unit possesses the attribute } A, \\ 0 & \text{otherwise,} \end{cases}$$

and

$$Y_i^2 = \begin{cases} 1 & \text{if the } i^{th} \text{ unit possesses the attribute } A, \\ 0 & \text{otherwise.} \end{cases}$$

So that

$$\sigma_p^2 = \frac{1}{N}\left[N_A - NP_y^2\right] = \frac{N_A}{N} - P_y^2 = P_y(1 - P_y) = P_y Q_y .$$

Thus we have

$$V(\hat{p}_y) = \frac{P_y Q_y}{n}.$$

Hence the theorem.

Theorem 2.3.4. An unbiased estimator of $V(\hat{p}_y)$ is given by

$$\hat{v}(\hat{p}_y) = \frac{\hat{p}_y \hat{q}_y}{n-1}$$

where

$$\hat{q}_y = 1 - \hat{p}_y.$$

Proof. We have to show that $E[\hat{v}(\hat{p}_y)] = V(\hat{p}_y)$ or in other words we have to show that

$$E\left(\frac{\hat{p}_y \hat{q}_y}{n-1}\right) = \frac{P_y Q_y}{n}.$$

Now we know that s_y^2/n is an unbiased estimator of σ_y^2/n.

Defining

$$y_i = \begin{cases} 1 & \text{if the } i^{th} \text{ sampled unit } \in A, \\ 0 & \text{otherwise,} \end{cases} \text{ and } y_i^2 = \begin{cases} 1 & \text{if the } i^{th} \text{ sampled unit } \in A, \\ 0 & \text{otherwise.} \end{cases}$$

Hence we will obtain

$$s_y^2 = \frac{1}{n-1}\left[\sum_{i=1}^{n} y_i^2 - n\bar{y}_n^2\right] = \frac{1}{n-1}\left[r - n\hat{p}_y^2\right] = \frac{n}{n-1}\left[\frac{r}{n} - \hat{p}_y^2\right] = \frac{n}{n-1}\hat{p}_y(1-\hat{p}_y) = \frac{n\hat{p}_y\hat{q}_y}{n-1}.$$

So that

$$\frac{s_y^2}{n} = \frac{\hat{p}_y \hat{q}_y}{n-1}.$$

Hence the theorem.

Theorem 2.3.5. Under SRSWR sampling, while estimating population proportion, the minimum sample size with minimum relative standard error (RSE) equal to ϕ, is

$$n \ge Q_y / (\phi^2 P_y). \tag{2.3.4}$$

Proof. The relative standard error of the estimator $\hat{p}_y$ is given by

$$\text{RSE}(\hat{p}_y) = \sqrt{V(\hat{p}_y)/\{E(\hat{p}_y)\}^2} = \sqrt{P_y Q_y / n P_y^2}. \tag{2.3.5}$$

Note that we need an estimator $\hat{p}_y$ such that $\text{RSE}(\hat{p}_y) \le \phi$, which implies that

$$\sqrt{\frac{Q_y}{nP_y}} \le \phi, \text{ or } \frac{Q_y}{nP_y} \le \phi^2, \text{ or } n \ge \frac{Q_y}{\phi^2 P_y}.$$

Hence the theorem.

Note that the relation $n \ge Q_y / (\phi^2 P_y)$ shows that as $P_y \to 0$ then $n \to \infty$.

Example 2.3.1. We wish to estimate the proportion of the number of fish in the group *Herring* caught by marine recreational fishermen at the Atlantic and Gulf coasts. There are 30027 fish out of total 311,528 fish caught during 1995 as shown in the population 4 in the Appendix. What is the minimum number of fish to be selected by SRSWR sampling to attain 5% relative standard error of the estimator of population proportion?

Solution. We have

$$P_y = \frac{30027}{311528} = 0.0964, \quad \text{and} \quad Q_y = 1 - P_y = 1 - 0.0964 = 0.9036.$$

Thus for $\phi = 0.05$, we have

$$n \geq Q_y \Big/ \left(\phi^2 P_y\right) = \frac{0.9036}{0.05^2 \times 0.0964} = 3749.4 \approx 3750.$$

Thus a minimum sample of size $n = 3750$ fish is required to attain 5% relative standard error of the estimator of population proportion under SRSWR sampling.

Example 2.3.2. A fisherman visited the Atlantic and Gulf coast and caught 4000 fish one by one. He noted the species group of each fish caught by him and put back that fish in the sea before making the next catch. He observed that 400 fish belong to the group *Herrings*.

(a) Estimate the proportion of fish in the group *Herrings* living in the Atlantic and Gulf coast.

(b) Construct the 95% confidence interval.

Solution. We are given $n = 4000$ and $r = 400$.

(a) An estimate of the proportion of the fish in the *Herrings* group is given by

$$\hat{p}_y = \frac{r}{n} = \frac{400}{4000} = 0.1.$$

(b) Under SRSWR sampling an estimate of the $V(\hat{p}_y)$ is given by

$$\hat{v}(\hat{p}_y) = \frac{\hat{p}_y \hat{q}_y}{n-1} = \frac{0.1 \times 0.9}{4000 - 1} = 2.2505 \times 10^{-5}.$$

A $(1-\alpha)100\%$ confidence interval for the true proportion P_y is given by

$$\hat{p}_y \mp Z_{\alpha/2} \sqrt{\hat{v}(\hat{p}_y)}.$$

Thus the 95% confidence interval for the proportion of fish belonging to the *Herrings* group is given by

$$\hat{p}_y \mp 1.96 \sqrt{\hat{v}(\hat{p}_y)}, \quad \text{or} \quad 0.1 \mp 1.96 \sqrt{2.2505 \times 10^{-5}}, \quad \text{or} \quad [0.0907,\ 0.1092].$$

Example 2.3.3. The height y of plants in a field is uniformly distributed between 5 cm to 20 cm with the probability density function

$$f(y) = \frac{1}{15} \quad \forall\ 5 < y < 20.$$

We wish to estimate the proportion of plants with height more than 15 cm, what is the minimum required sample size n to have an accuracy of relative standard error of 45%?

(a) Select a sample of the required size, and estimate the proportion of plants with height more than 15 cm.

(b) Construct a 95% confidence interval estimate, assuming that your sample size is large, and interpret your results.

Solution. We know that if y has uniform distribution function

$$f(y) = \frac{1}{b-a} \quad \forall\, a < y < b.$$

Thus the proportion of plants with height more than 15cm is given by

$$P_y = \int_{15}^{20} f(y)dy = \int_{15}^{20} \frac{1}{15}dy = \frac{1}{15}(20-15) = \frac{5}{15} = 0.3333,$$

and the variance

$$\sigma_y^2 = P_y(1-P_y) = 0.3333(1-0.3333) = 0.2222.$$

(a) We need $\phi = 0.45$, thus the required minimum sample size is given by

$$n \geq \frac{\sigma_y^2}{\phi^2 P_y^2} = \frac{0.2222}{0.45^2 \times 0.3333^2} = 9.8 \approx 10.$$

(b) We select a with replacement sample of $n = 10$ units as follows. The cumulative distribution function (c.d.f.) is given by

$$F(y) = P[Y \leq y] = \int_{5}^{y} f(y)dy = \int_{5}^{y} \frac{1}{15}dy = \frac{(y-5)}{15}$$

which implies that $y = 15F(y) + 5$. We used 4th to 6th columns multiplied by 10^{-3} of the Pseudo-Random Number (PRN) Table 1 given in the Appendix to select ten values of $F(y)$ and the required sampled values are computed by using the relationship $y = 15F(y) + 5$ as follows:

$F(y)$	Observed y value	>15 1 -- yes 0 -- no
0.954	19.31	1
0.183	7.75	0
0.448	11.72	0
0.171	7.57	0
0.567	13.51	0
0.737	16.06	1
0.856	17.84	1
0.233	8.50	0
0.895	18.43	1
0.263	8.95	0

Thus an estimate of the proportion P_y is given by,

$$\hat{p}_y = \frac{\text{number of 'yes' answers}}{\text{sample size}} = \frac{4}{10} = 0.4,$$

and an estimate of its variance is given by

$$\hat{v}(\hat{p}_y) = \frac{\hat{p}_y(1-\hat{p}_y)}{n-1} = \frac{0.4(1-0.4)}{10-1} = 0.0267.$$

Thus a 95% confidence interval estimate of the required proportion P_y is given by

$$\hat{p}_y \mp 1.96\sqrt{\hat{v}(\hat{p}_y)}\;, \quad \text{or} \quad 0.4 \mp 1.96\sqrt{0.0267}\,, \quad \text{or} \quad [0.0797,\ 0.7202].$$

Case II. When a sample is drawn using SRSWOR sampling, we have the following theorems.

Theorem 2.3.6. The unbiased estimator of the population proportion P_y is given by

$$\hat{p}_y = \frac{r}{n}$$

where r is the number of units possessing the attribute A.
Proof. Obvious.

Theorem 2.3.7. The variance of the estimator $\hat{p}_y$ is given by

$$V(\hat{p}_y) = \frac{(N-n)}{n(N-1)}P_yQ_y.$$

Proof. We know that

$$V(\bar{y}_n) = \frac{(N-n)}{Nn}S_y^2, \quad \text{where} \quad S_y^2 = \frac{1}{N-1}\left(\sum_{i=1}^{N}Y_i^2 - N\bar{Y}^2\right).$$

Again we define

$$Y_i = \begin{cases} 1 & \text{if the } i^{th} \text{ unit possesses the attribute } A, \\ 0 & \text{otherwise,} \end{cases}$$

and

$$Y_i^2 = \begin{cases} 1 & \text{if the } i^{th} \text{ unit possesses the attribute } A, \\ 0 & \text{otherwise.} \end{cases}$$

So

$$S_y^2 = \frac{1}{N-1}\left(N_A - NP_y^2\right) = \frac{N}{N-1}\left(P_y - P_y^2\right) = \frac{NP_yQ_y}{N-1} = S_p^2.$$

Hence we have

$$V(\hat{p}_y) = \frac{N-n}{Nn}S_p^2 = \frac{N-n}{Nn} \times \frac{N}{N-1}P_yQ_y = \frac{(N-n)}{n(N-1)}P_yQ_y,$$

which proves the theorem.

Theorem 2.3.8. The unbiased estimator of the variance $V(\hat{p}_y)$ is given by

$$\hat{v}(\hat{p}_y) = \frac{(N-n)}{N(n-1)}\hat{p}_y\hat{q}_y.$$

Proof. We will prove that $E[\hat{v}(\hat{p}_y)] = V(\hat{p}_y)$, that is

$$E\left[\frac{N-n}{N(n-1)}\hat{p}_y\hat{q}_y\right] = \frac{N-n}{N(n-1)}P_yQ_y .$$

Now we know that $\frac{(N-n)}{Nn}s_y^2$ is an unbiased estimator of $\frac{N-n}{Nn}S_y^2$.

Changing

$$Y_i = \begin{cases} 1 & \text{if the } i^{th} \text{ population unit} \in \text{A}, \\ 0 & \text{otherwise}, \end{cases} \quad \text{and} \quad Y_i^2 = \begin{cases} 1 & \text{if the } i^{th} \text{ poulation unit} \in \text{A}, \\ 0 & \text{otherwise}, \end{cases}$$

makes

$$\frac{(N-n)}{Nn}S_y^2 = \frac{(N-n)}{n(N-1)}P_yQ_y .$$

Similarly, if we make the changes

$$y_i = \begin{cases} 1 & \text{if the } i^{th} \text{ sampled unit} \in \text{A}, \\ 0 & \text{otherwise}, \end{cases} \quad \text{and} \quad y_i^2 = \begin{cases} 1 & \text{if the } i^{th} \text{ sampled unit} \in \text{A}, \\ 0 & \text{otherwise}, \end{cases}$$

then

$$s_y^2 = \frac{1}{(n-1)}\left[\sum_{i=1}^{n} y_i^2 - n\bar{y}_n^2\right].$$

will reduce to

$$s_p^2 = \frac{1}{n-1}\left[r - n\hat{p}_y^2\right] = \frac{n}{n-1}\left[\frac{r}{n} - \hat{p}_y^2\right] = \frac{n}{n-1}\left(\hat{p}_y - \hat{p}_y^2\right) = \frac{n\hat{p}_y(1-\hat{p}_y)}{n-1} = \frac{n\hat{p}_y\hat{q}_y}{n-1}.$$

Therefore

$$\frac{(N-n)}{Nn}s_p^2 = \left(\frac{N-n}{Nn}\right)\frac{n\hat{p}_y\hat{q}_y}{(n-1)} = \frac{(N-n)}{N(n-1)}\hat{p}_y\hat{q}_y .$$

Hence the theorem.

Theorem 2.3.9. Under SRSWOR sampling, while estimating population proportion, the minimum sample size with minimum relative standard error (RSE) equal to ϕ is

$$n \ge NQ_y \Big/ \left\{\phi^2 (N-1)P_y + Q_y\right\}. \tag{2.3.6}$$

Proof. The relative standard error of the estimator $\hat{p}_y$ is given by

$$\text{RSE}(\hat{p}_y) = \sqrt{V(\hat{p}_y) / \{E(\hat{p}_y)\}^2} = \sqrt{\frac{(N-n)P_yQ_y}{n(N-1)P_y^2}}. \tag{2.3.7}$$

Note that we need an estimator $\hat{p}_y$ such that $\text{RSE}(\hat{p}_y) \le \phi$, which implies that

$$\sqrt{\left(\frac{N}{n}-1\right)\frac{Q_y}{(N-1)P_y}} \le \phi, \quad \text{or} \quad \left(\frac{N}{n}-1\right) \le \frac{(N-1)\phi^2}{Q_y}, \quad \text{or} \quad n \ge \frac{NQ_y}{\phi^2(N-1)P_y + Q_y}.$$

Hence the theorem.

Example 2.3.4. We wish to estimate the proportion of the number of fish in the group *Herrings* caught by marine recreational fishermen at the Atlantic and Gulf coasts. There are 30,027 fish out of total 311,528 of fish caught during 1995 as shown in the population 4 in the Appendix. What is the minimum number of fish to be selected by SRSWOR sampling to attain the accuracy of relative standard error 5%?

Solution. We have

$$P_y = \frac{30027}{311528} = 0.0964, \text{ and } Q_y = 1 - P_y = 1 - 0.0964 = 0.9036.$$

Thus for $\phi = 0.05$, we have

$$n \geq \frac{NQ_y}{\phi^2(N-1)P_y + Q_y} = \frac{311528 \times 0.9036}{0.05^2(311528-1)\times 0.0964 + 0.9036} = 3704.8 \approx 3705 .$$

Thus a minimum sample of size $n = 3705$ fish is required to attain 5% relative standard error of the estimator of population proportion under SRSWOR sampling.

Example 2.3.5. A fisherman visited the Atlantic and Gulf coast and caught 4000 fish. He noted the species group of each fish caught by him. He observed that 400 fish belong to the group *Herrings*.

(a) Estimate the proportion of fish in the group *Herrings* living in the Atlantic and Gulf coast.

(b) Construct the 95% confidence interval.

Given: Total number of fish living in the coast = 311528.

Solution. We are given $N = 311,528$, $n = 4,000$ and $r = 400$.

(a) An estimate of the proportion of the fish in the *Herrings* group is

$$\hat{p}_y = \frac{r}{n} = \frac{400}{4000} = 0.1 .$$

(b) Under SRSWOR sampling, an estimate of the $V(\hat{p}_y)$ is given by

$$\hat{v}(\hat{p}_y) = \frac{(N-n)}{N}\frac{\hat{p}_y\hat{q}_y}{n-1} = \frac{(311528-4000)}{311528} \times \frac{0.1\times 0.9}{4000-1} = 2.2216\times 10^{-5} .$$

A $(1-\alpha)100\%$ confidence interval for the true proportion P_y is given by

$$\hat{p}_y \mp Z_{\alpha/2}\sqrt{\hat{v}(\,\hat{p}_y\,)} .$$

Thus the 95% confidence interval for the proportion of fish belonging to *Herrings* group is given by

$$\hat{p}_y \mp 1.96\sqrt{\hat{v}(\,\hat{p}_y\,)}, \text{ or } 0.1 \mp 1.96\sqrt{2.2216\times 10^{-5}}, \text{ or } [0.0908, 0.1092].$$

Example 2.3.6. In a field there are 1,000 plants and the distribution of their height is given by the probability mass function

y (cm)	50	100	150	200	225	275
$p(y)$	0.1	0.2	0.3	0.1	0.2	0.1

(a) Select a random sample of $n = 10$ units and estimate the proportion of plants with height more than or equal to 225 cm.
(b) Construct a 95% confidence interval, assuming that it is a large sample.

Solution. The cumulative distribution function $F(y)$ is given by

y (cm)	50	100	150	200	225	275
$p(y)$	0.1	0.2	0.3	0.1	0.2	0.1
$F(y)$	0.1	0.3	0.6	0.7	0.9	1.0

Using the first three columns, multiplied by 10^{-3}, of the Pseudo-Random Number (PRN) Table 1 given in the Appendix, we obtain the 10 values of $F(y)$ and y as:

$F(y)$	y	≥ 225 yes--1 no--0
0.992	225	1
0.588	100	0
0.601	150	0
0.549	100	0
0.925	225	1
0.014	Discard this number	
0.697	150	0
0.872	200	0
0.626	150	0
0.236	50	0
0.884	225	1

(a) An estimate of the proportion of plants with height more than 225 cm is

$$\hat{p}_y = \frac{\text{No. of plants with height} \geq 225 \text{ cm}}{\text{No. of plants in the sample}} = \frac{3}{10} = 0.3 .$$

(b) We have $N = 1000$ and $n = 10$, therefore, an estimate the $V(\hat{p}_y)$ is given by

$$\hat{v}(\hat{p}_y) = \frac{(N-n)}{N}\frac{\hat{p}_y\hat{q}_y}{n-1} = \frac{(1000-10)}{1000} \times \frac{0.3 \times 0.7}{10-1} = 0.0231 .$$

Thus a 95% confidence interval for the proportion of plants with height more than or equal to 225 cm is:

$$\hat{p}_y \mp 1.96\sqrt{\hat{v}(\hat{p}_y)}\ , \quad \text{or} \quad 0.3 \mp 1.96\sqrt{0.0231}\ , \quad \text{or} \quad [0.0021, 0.5978] .$$

Note that an estimate of proportion always follow normal distribution, so the use of 1.96 for deriving 95% confidence interval estimate is appropriate irrespective of the sample size.

2.4 SEARLS' ESTIMATOR OF POPULATION MEAN

Searls (1964) considered an estimator of the population mean $\bar{Y}$ defined as

$$\bar{y}_{\text{searl}} = \lambda \bar{y}_n \tag{2.4.1}$$

where λ is a constant such that the MSE of the estimator $\bar{y}_{\text{searl}}$ is minimum. Thus we have the following theorem.

Theorem 2.4.1. The minimum mean squared error of the estimator, $\bar{y}_{\text{searl}}$, is

$$\text{Min.MSE}(\bar{y}_{\text{searl}}) = V(\bar{y}_n)/\left\{1 + V(\bar{y}_n)/\bar{Y}^2\right\}. \tag{2.4.2}$$

Proof. We have

$$\begin{aligned}
\text{MSE}(\bar{y}_{\text{searl}}) &= E\left[\bar{y}_s - \bar{Y}\right]^2 = E\left[\lambda\bar{y}_n - \bar{Y}\right]^2 = E\left[\lambda\bar{y}_n - E(\lambda\bar{y}_n) + E(\lambda\bar{y}_n) - \bar{Y}\right]^2 \\
&= E\left[\lambda\{\bar{y}_n - E(\bar{y}_n)\} + \lambda E(\bar{y}_n) - \bar{Y}\right]^2 = E\left[\lambda\{\bar{y}_n - E(\bar{y}_n)\} + (\lambda - 1)\bar{Y}\right]^2 \\
&= E\left|\lambda^2\{\bar{y}_n - E(\bar{y}_n)\}^2 + (\lambda - 1)^2\bar{Y}^2 + 2(\lambda - 1)\bar{Y}\lambda\{\bar{y}_n - E(\bar{y}_n)\}\right| \\
&= \lambda^2 E\{\bar{y}_n - E(\bar{y}_n)\}^2 + (\lambda - 1)^2\bar{Y}^2 + 0 = \lambda^2 V(\bar{y}_n) + (\lambda - 1)^2\bar{Y}^2.
\end{aligned} \tag{2.4.3}$$

Thus the mean squared error, $\text{MSE}(\bar{y}_{\text{searl}})$, is given by

$$\text{MSE}(\bar{y}_{\text{searl}}) = \lambda^2 V(\bar{y}_n) + (\lambda - 1)^2\bar{Y}^2. \tag{2.4.4}$$

On differentiating (2.4.4) with respect to λ and equating to zero we obtain

$$\lambda V(\bar{y}_n) + (\lambda - 1)\bar{Y}^2 = 0\text{, or}\quad \lambda = 1/\left\{1 + V(\bar{y}_n)/\bar{Y}^2\right\}. \tag{2.4.5}$$

On substituting (2.4.5) in (2.4.4) we obtain

$$\begin{aligned}
\text{Min.MSE}(\bar{y}_{\text{searl}}) &= \lambda^2 V(\bar{y}_n) + (\lambda - 1)^2\bar{Y}^2 = \frac{\bar{Y}^4 V(\bar{y}_n)}{\left\{\bar{Y}^2 + V(\bar{y}_n)\right\}^2} + \left\{\frac{\bar{Y}^2}{\bar{Y}^2 + V(\bar{y}_n)} - 1\right\}^2 \bar{Y}^2 \\
&= \frac{\bar{Y}^4 V(\bar{y}_n)}{\left\{\bar{Y}^2 + V(\bar{y}_n)\right\}^2} + \left\{\frac{\bar{Y}^2 - \bar{Y}^2 - V(\bar{y}_n)}{\bar{Y}^2 + V(\bar{y}_n)}\right\}^2 \bar{Y}^2 \\
&= \frac{\bar{Y}^4 V(\bar{y}_n)}{\left\{\bar{Y}^2 + V(\bar{y}_n)\right\}^2} + \frac{\bar{Y}^2\{V(\bar{y}_n)\}^2}{\left\{\bar{Y}^2 + V(\bar{y}_n)\right\}^2} = \frac{\bar{Y}^2 V(\bar{y}_n)\left\{\bar{Y}^2 + V(\bar{y}_n)\right\}}{\left\{\bar{Y}^2 + V(\bar{y}_n)\right\}^2} \\
&= \frac{\bar{Y}^2 V(\bar{y}_n)}{\bar{Y}^2 + V(\bar{y}_n)} = \frac{V(\bar{y}_n)}{1 + V(\bar{y}_n)/\bar{Y}^2}.
\end{aligned} \tag{2.4.6}$$

Hence the theorem.

Theorem 2.4.2. Under SRSWR sampling, the minimum mean squared error of the Searls' estimator is

$$\text{Min.MSE}(\bar{y}_{\text{searl}}) = n^{-1}\sigma_y^2 / \left\{1 + n^{-1}\sigma_y^2/\bar{Y}^2\right\}. \tag{2.4.7}$$

Proof. Obvious from (2.4.2) because under SRSWR sampling we have

$$V(\bar{y}_n) = n^{-1}\sigma_y^2.$$

Theorem 2.4.3. The relative efficiency of the Searls' estimator $\bar{y}_{\text{searl}}$ with respect to usual estimator $\bar{y}_n$, under SRSWR sampling, is given by

$$\text{RE} = 1 + \sigma_y^2 / \left(n\bar{Y}^2\right). \tag{2.4.8}$$

Thus the relative gain in the Searls' estimator is inversely proportional to the sample size, n. In other words, as $n \to \infty$, the value of $\text{RE} \to 1$.

Proof. It follows from the definition of the relative efficiency. Note that the relative efficiency of Searls' estimator with respect to the usual estimator is given by

$$\text{RE} = \frac{\text{MSE}(\bar{y}_n)}{\text{MSE}(\bar{y}_{\text{searl}})} = \frac{V(\bar{y}_n)}{\text{Min.MSE}(\bar{y}_{\text{searl}})} = n^{-1}\sigma_y^2\left\{\frac{n^{-1}\sigma_y^2}{1 + n^{-1}\bar{Y}^2\sigma_y^2}\right\}^{-1}$$

$$= 1 + n^{-1}\bar{Y}^2\sigma_y^2. \tag{2.4.9}$$

Hence the theorem.

Theorem 2.4.4. Under SRSWOR sampling, the minimum mean squared error of the Searls' estimator is

$$\text{Min.MSE}(\bar{y}_{\text{searl}}) = \left(\frac{1}{n} - \frac{1}{N}\right)S_y^2 \Big/ \left\{1 + \left(\frac{1}{n} - \frac{1}{N}\right)S_y^2 \Big/ \bar{Y}^2\right\}. \tag{2.4.10}$$

Proof. Obvious from (2.4.2) because under SRSWOR sampling we have

$$V(\bar{y}_n) = \left(\frac{1}{n} - \frac{1}{N}\right)S_y^2.$$

Theorem 2.4.5. The relative efficiency of the Searls' estimator $\bar{y}_{\text{searl}}$ with respect to $\bar{y}_n$, under SRSWOR, is given by

$$\text{RE} = 1 + \left(\frac{1}{n} - \frac{1}{N}\right)C_y^2.$$

Thus the relative gain in efficiency of the Searls' estimator is inversely proportional to the sample size n. In other words, as $n \to N$ the value of $\text{RE} \to 1$.

Proof. By the definition of the relative efficiency we have

$$\text{RE} = \frac{V(\bar{y}_n)}{\text{Min.MSE}(\bar{y}_{searl})} = \left(\frac{1}{n} - \frac{1}{N}\right)S_y^2\left\{\frac{\left(\frac{1}{n} - \frac{1}{N}\right)S_y^2}{1 + \left(\frac{1}{n} - \frac{1}{N}\right)\frac{S_y^2}{\bar{Y}^2}}\right\}^{-1}$$

$$= 1 + \left(\frac{1}{n} - \frac{1}{N}\right)\frac{S_y^2}{\bar{Y}^2} = 1 + \left(\frac{1}{n} - \frac{1}{N}\right)C_y^2.$$

Hence the theorem.

Example 2.4.1. We wish to estimate the average number of fish in each one of the species groups caught by marine recreational fishermen at the Atlantic and Gulf coasts. There were 69 species caught during 1995 as shown in the population 4 of the Appendix. We selected a sample of 20 units by SRSWR sampling. What is the gain in efficiency owed to the Searls' estimator over the sample mean?

Given: $S_y^2 = 37199578$ and $Y = 311528$.

Solution. We are given $N = 69$, $S_y^2 = 37199578$ and $Y = 311528$, thus

$$\bar{Y} = \frac{Y}{N} = \frac{311528}{69} = 4514.898$$

and

$$\sigma_y^2 = \frac{(N-1)}{N} S_y^2 = \frac{(69-1)}{69} \times 37199578 = 36660453.68 .$$

The relative efficiency of the Searls' estimator $\bar{y}_{\text{searl}}$ with respect to usual estimator $\bar{y}_n$, under SRSWR, is given by

$$\text{RE} = \left[1 + \frac{\sigma_y^2}{n\bar{Y}^2}\right] \times 100 = \left[1 + \frac{36660453.68}{20 \times 4514.898^2}\right] \times 100 = 108.99\% .$$

Example 2.4.2. On a bank of a river the height of trees is uniformly distributed with the p.d.f. given by

$$f(y) = \frac{1}{300} \quad \forall\ 200 \le y \le 500 \text{ feet.}$$

Find the relative efficiency of Searls' estimator over the usual estimator based on a sample of 5 or 20 units, respectively.

Solution. We are given $a = 200$ and $b = 500$, therefore the population mean

$$\bar{Y} = \frac{a+b}{2} = \frac{200 + 500}{2} = 350 \text{ feet}$$

and population variance

$$\sigma_y^2 = \frac{(b-a)^2}{12} = \frac{(500-200)^2}{12} = 7500 \text{ feet}^2 .$$

Thus the relative efficiency of the Searls' estimator over the usual one is given by

$$\text{If } n = 5 \text{ then RE} = \left(1 + \frac{\sigma_y^2}{n\bar{Y}^2}\right) \times 100 = \left(1 + \frac{7500}{5 \times 350^2}\right) \times 100 = 101.22\% .$$

$$\text{If } n = 20 \text{ then RE} = \left(1 + \frac{\sigma_y^2}{n\bar{Y}^2}\right) \times 100 = \left(1 + \frac{7500}{20 \times 350^2}\right) \times 100 = 100.30\ \%.$$

Searls (1967), Reddy (1978a) and Arnholt and Hebert (1995) studied the properties of this estimator and found that it is useful if C_y is large and sample size is small.

2.5 USE OF DISTINCT UNITS IN THE WR SAMPLE AT THE ESTIMATION STAGE

We shall discuss the problem of estimation of finite population mean and variance by using only distinct units from the SRSWR sample. However, before going further we shall discuss some results, which will be helpful in deriving the result from distinct units for SRSWR sample. Basu (1958) introduced the concept of sufficiency in sampling from finite populations. According to him, for every ordered sample s^{o} there exists an unordered sample s^{uo} which is obtained from s^{o} by ignoring information concerning the order in which the labels occur. The data obtained from the sample s^{uo} can be represented as

$$d^{\text{uo}} = \left(y_i : i \in s^{\text{uo}}\right). \tag{2.5.1}$$

Similarly the data obtained from s^{o} can be represented as:

$$d^{\text{o}} = \left(y_i : i \in s^{\text{o}}\right). \tag{2.5.2}$$

Then the probability of observing the ordered d^{o} given the unordered data d^{uo} is

$$P\left(d^{\text{o}} \mid d^{\text{uo}}\right) = P\left(d^{\text{o}}\right)/P\left(d^{\text{uo}}\right) = p\left(s^{\text{o}}\right)/\sum p\left(s^{\text{o}}\right) \tag{2.5.3}$$

where $\sum$ is the summation over all those ordered samples s^{o} which results in the unordered sample s^{uo}. Since the probability $P\left(d^{\text{o}} \mid d^{\text{uo}}\right)$ is independent of any population parameters and hence the unordered statistics d^{uo} is a sufficient statistic for any population parameter.

Let us now first state the Rao--Blackwell theorem, which is based on Rao (1945) and Blackwell (1947) results.

Theorem 2.5.1. Let $\hat{\theta}_s^{\text{o}} = \hat{\theta}_s\left(d^{\text{o}}\right)$ be an estimator of θ constructed from ordered data d^{o}. Suppose $\hat{\theta}_s = E\left[\hat{\theta}_s^{\text{o}} \mid d^{\text{uo}}\right]$ then:

$$\text{(a)} \quad E\left(\hat{\theta}_s\right) = E\left(\hat{\theta}_s^{\text{o}}\right); \tag{2.5.4}$$

$$\text{(b)} \quad \text{MSE}\left(\hat{\theta}_s^{\text{o}}\right) = \text{MSE}\left(\hat{\theta}_s\right) + E\left(\hat{\theta}_s - \hat{\theta}_s^{\text{o}}\right)^2. \tag{2.5.5}$$

Proof. (a) We have

$$E\left(\hat{\theta}_s\right) = E\left[\hat{\theta}_s^{\text{o}} \mid d^{\text{uo}}\right] = E\left[\sum \hat{\theta}_s^{\text{o}}\left(d^{\text{o}}\right) P\left(d^{\text{o}} \mid d^{\text{uo}}\right)\right] = E\left(\sum \hat{\theta}_s^{\text{o}}\left(d^{\text{o}}\right)\frac{p\left(s^{\text{o}}\right)}{p\left(s^{\text{uo}}\right)}\right)$$

$$= \sum_{s^{\text{uo}}}\left\{\sum \hat{\theta}_s^{\text{o}}\left(d^{\text{o}}\right)\frac{p\left(s^{\text{o}}\right)}{p\left(s^{\text{uo}}\right)}\right\} p\left(s^{\text{uo}}\right) = \sum_{s^{\text{o}}} \hat{\theta}_s^{\text{o}}\left(d^{\text{o}}\right) p\left(s^{\text{o}}\right) = E\left(\hat{\theta}_s^{\text{o}}\right).$$

Hence the part (a) of the theorem.

(b) We have

$$\text{MSE}\left(\hat{\theta}_s^{\text{o}}\right) = E\left[\hat{\theta}_s^{\text{o}} - \theta\right]^2 = E\left[\hat{\theta}_s^{\text{o}} - \hat{\theta}_s + \hat{\theta}_s - \theta\right]^2$$

$$= E\left(\hat{\theta}_s^{\text{o}} - \hat{\theta}_s\right)^2 + E\left(\hat{\theta}_s - \theta\right)^2 + 2E\left(\hat{\theta}_s^{\text{o}} - \hat{\theta}_s\right)\left(\hat{\theta}_s - \theta\right)$$

$$= E\left(\hat{\theta}_s^{\text{o}} - \hat{\theta}_s\right)^2 + \text{MSE}\left(\hat{\theta}_s\right) + 0\,.$$

Hence the theorem.

Now we will discuss the problem of estimation of mean and variance on the basis of distinct units in the sample. Clearly a unit can be repeated only in WR sampling schemes. Hence we are dealing only with SRSWR sampling scheme. Suppose ν denote the number of distinct units in the sample of size n drawn from the population of N units by using SRSWR scheme.
The distribution of distinct units in the sample was first developed by Feller (1957) as follows

$$P(\nu = t) = \frac{1}{N^n}\binom{N}{t}\sum_{r=1}^{t}(-1)^r\binom{t}{r}(t-r)^n, \quad \text{where } t = 1,2,...,\text{Min.}(n, N)\,. \tag{2.5.6}$$

2.5.1 ESTIMATION OF MEAN

For estimating the population mean $\overline{Y}$ by using information only from the distinct units we have the following theorems.

Theorem 2.5.1.1. An unbiased estimator of population mean $\overline{Y}$ is given by

$$\bar{y}_\nu = \frac{1}{\nu}\sum_{i=1}^{\nu} y_i\,. \tag{2.5.1.1}$$

Proof. Following Raj and Khamis (1958), let E_2 and E_1 be the expected values defined for a given sample (fixed number of distinct units) and for all possible samples, respectively, then by taking expected value on both sides of (2.5.1.1), we have

$$E(\bar{y}_\nu) = E_1E_2(\bar{y}_\nu | \nu) = E_1E_2\left(\frac{1}{\nu}\sum_{i=1}^{\nu} y_i\right) = E_1\left(\frac{1}{n}\sum_{i=1}^{n} y_i\right) = \overline{Y}\,. \tag{2.5.1.2}$$

Hence the theorem.

Theorem 2.5.1.2. The variance of the unbiased estimator $\bar{y}_\nu$ based on distinct units is

$$V(\bar{y}_\nu) = \left[E\left(\frac{1}{\nu}\right) - \frac{1}{N}\right]S_y^2\,. \tag{2.5.1.3}$$

Proof. Suppose V_2 and V_1 denote the variance for the given sample (fixed number of distinct units) and over all possible samples, we have

$$V(\bar{y}_v) = E_1V_2(\bar{y}_v \mid v) + V_1E_2(\bar{y}_v \mid v) = E_1V_2\left(\frac{1}{v}\sum_{i=1}^{v} y_i \mid v\right) + V_1E_2\left(\frac{1}{v}\sum_{i=1}^{v} y_i \mid v\right)$$

$$= E_1\left(\frac{1}{v} - \frac{1}{n}\right)s_y^2 + V_1(\bar{y}_n) = \left(E_1\left(\frac{1}{v}\right) - \frac{1}{n}\right)S_y^2 + \left(\frac{1}{n} - \frac{1}{N}\right)S_y^2$$

$$= \left[E\left(\frac{1}{v}\right) - \frac{1}{N}\right]S_y^2.$$

Hence the theorem.

Corollary 2.5.1.1. Pathak (1961) has shown that

$$E\left(\frac{1}{v}\right) = \frac{1}{N^n}\sum_{J=1}^{N} J^{(n-1)}. \tag{2.5.1.4}$$

Thus we have the following theorem

Theorem 2.5.1.3. The variance of the estimator $\bar{y}_v$ is given by

$$V(\bar{y}_v) = \left(\sum_{J=1}^{N-1} J^{(n-1)} \Big/ N^n\right)S_y^2. \tag{2.5.1.5}$$

Proof. We have

$$V(\bar{y}_v) = \left[E\left(\frac{1}{v}\right) - \frac{1}{N}\right]S_y^2 = \left[\sum_{J=1}^{N} J^{(n-1)} \Big/ N^n - 1/N\right]S_y^2$$

$$= \left[\left\{\sum_{J=1}^{N} J^{(n-1)} - N^{(n-1)}\right\} \Big/ N^n\right]S_y^2 = \left[\sum_{J=1}^{N-1} J^{(n-1)} \Big/ N^n\right]S_y^2.$$

Hence the theorem.

It is interesting to note that as the sample size n drawn with SRSWR sampling approaches to the population of size N, the magnitude of the relative efficiency also increases. The reason of increase in the relative efficiency may be that the increase in sample size also increases the probability of repetition of units in SRSWR sampling.

The relative efficiency under the Feller (1957) distribution is given by

$$\text{RE} = \frac{V(\bar{y}_n)}{V(\bar{y}_v)} = \frac{(N-1)N^{(n-1)}}{n\left\{\sum_{J=1}^{N-1} J^{(n-1)}\right\}} \tag{2.5.1.6}$$

which is free from any population parameter but depends upon population size and sample size.

The following table shows the percent relative efficiency of distinct units based estimators with respect to the estimators based on SRSWR sampling for different values of sample sizes n and population sizes $N = 10$.

Benefit of use of distinct units								
	Sample size (n)							
J	2	3	4	5	6	7	8	9
	$J^{(n-1)}$							
1	1	1	1	1	1	1	1	1
2	2	4	8	16	32	64	128	256
3	3	9	27	81	243	729	2187	6561
4	4	16	64	256	1024	4096	16384	65536
5	5	25	125	625	3125	15625	78125	390625
6	6	36	216	1296	7776	46656	279936	1679616
7	7	49	343	2401	16807	117649	823543	5764801
8	8	64	512	4096	32768	262144	2097152	16777216
9	9	81	729	6561	59049	531441	4782969	43046721
Sum	45	285	2025	15333	120825	978405	8080425	67731333
	$\text{RE} = \left[(N-1)N^{(n-1)}\right] \Big/ \left[n\left\{\sum_{J=1}^{N-1} J^{(n-1)}\right\}\right]$							
RE	100.00	105.26	111.11	117.39	124.15	131.41	139.23	147.64

Corollary 2.5.1.2. An approximate expression for $V(\bar{y}_v)$ valid up to order N^{-2} is in Pathak (1962) as

$$V(\bar{y}_v) = \left[\frac{1}{n} - \frac{1}{2N} + \frac{(n-1)}{12N^2}\right] S_y^2 . \tag{2.5.1.7}$$

Theorem 2.5.1.4. (a) Show that an alternative estimator of the population mean based on distinct units is

$$\bar{y}_1 = \frac{v}{E(v)}\bar{y}_v + \bar{X}\left(1 - \frac{v}{E(v)}\right) \tag{2.5.1.8}$$

where $\bar{X}$ is a good guess or a priori estimator of the population mean $\bar{Y}$.

(b) If there is no priori information or good guess about the population mean $\bar{Y}$, then an alternative estimator of population mean is given by

$$\bar{y}_2 = \frac{v}{E(v)}\bar{y}_v \tag{2.5.1.9}$$

where $\bar{y}_v = \frac{1}{v}\sum_{i=1}^{v} y_i$ is a mean of distinct units in the sample.

Proof. Let us consider that an estimator of the population mean is given by

$$\bar{y}_s = f_1(v)\bar{y}_v + f_2(v) \tag{2.5.1.10}$$

where $f_1(v)$ and $f_2(v)$ are suitably chosen constants such that $\bar{y}_s$ is an unbiased estimator of $\bar{Y}$ and its variance is minimum. Now from the property of unbiasedness we have

$$E(\bar{y}_s) = E[f_1(v)\bar{y}_v + f_2(v)] = f_1(v)\bar{Y} + f_2(v) = \bar{Y}. \tag{2.5.1.11}$$

This implies that

$$f_2(v) = [1 - f_1(v)]\bar{Y}. \tag{2.5.1.12}$$

Evidently the value of $f_2(v)$ contains the unknown value $\bar{Y}$, the exact value of $f_2(v)$ is not known unless $f_1(v) = 1$, which implies $f_2(v) = 0$. Thus we chose $f_1(v) = 1$, then $f_2(v) = 0$, which means a better estimator of population mean $\bar{Y}$ would be $\bar{y}_v = v^{-1}\sum_{i=1}^{v} y_i$. In practical situations, sometimes a priori information or knowledge of $\bar{X}$ (say) is available about population mean $\bar{Y}$ from past surveys or pilot surveys. In such situations, the value of $f_2(v)$ is given by

$$f_2(v) = [1 - f_1(v)]\bar{X}. \tag{2.5.1.13}$$

Thus if we will chose $f_2(v)$ as given in (2.5.1.13), then the bias in the estimator $\bar{y}_s$ will be minimum. Unfortunately, $f_2(v)$ depends upon the value of $f_1(v)$ too. The best method to chose $f_1(v)$ is such that the variance of $\bar{y}_s$ is minimum. Now the variance of the estimator $\bar{y}_s$ is given by

$$V(\bar{y}_s) = E_1V_2(\bar{y}_s) + V_1E_2(\bar{y}_s) = E_1V_2[f_1(v)\bar{y}_v + f_2(v)] + V_2E_1[f_1(v)\bar{y}_v + f_2(v)]$$

$$= E_1\left[f_1^2(v)\left(\frac{1}{v} - \frac{1}{N}\right)S_y^2\right] + V_2[f_1(v)\bar{Y} + f_2(v)]$$

$$= E_1\left[f_1^2(v)\left(\frac{1}{v} - \frac{1}{N}\right)S_y^2\right] + V_2[\bar{Y}] = E_1\left[f_1^2(v)\left(\frac{1}{v} - \frac{1}{N}\right)S_y^2\right]. \tag{2.5.1.14}$$

The variance of $\bar{y}_s$ will be minimum if $E_1\left[f_1^2(v)\left(\frac{1}{v} - \frac{1}{N}\right)S_y^2\right]$ is minimum subject to the condition $E_1[f_1(v)] = 1$. Then by Schwartz inequality, we have

$$E_1\left[f_1^2(v)\left(\frac{1}{v} - \frac{1}{N}\right)\right] \geq E_1\left(\frac{1}{v} - \frac{1}{N}\right). \tag{2.5.1.15}$$

In the above inequality, the equality sign holds if and only if

$$f_1(v) = \left(\frac{1}{v} - \frac{1}{N}\right) \Big/ E\left(\frac{1}{v} - \frac{1}{N}\right) = \frac{(Nv)/(N-v)}{E[(Nv)/(N-v)]}. \tag{2.5.1.16}$$

Thus if we have a priori information $\bar{X}$ about $\bar{Y}$, then an optimum estimator of $\bar{Y}$ is given by

$$\bar{y}_1 = \frac{(Nv)/(N-v)}{E[(Nv)/(N-v)]}\bar{y}_v + \bar{X}\left[1 - \frac{(Nv)/(N-v)}{E[(Nv)/(N-v)]}\right]. \tag{2.5.1.17}$$

If no such information about $\overline{Y}$ is available, then we have $\overline{X} = 0$ and the above estimator reduces to

$$\overline{y}_2 = \frac{(Nv)/(N-v)}{E[(Nv)/(N-v)]}\overline{y}_v . \tag{2.5.1.18}$$

Pathak (1961) has shown that

$$E\left[\frac{Nv}{(N-v)}\right] = N^2 \sum_{m=1}^{n} \frac{p_{12...m}}{(N-m)} \tag{2.5.1.19}$$

where

$$p_{12....m} = \begin{cases} 1-\binom{m}{1}\left(1-\frac{1}{N}\right)^n + ... + (-1)^m\binom{m}{m}\left(1-\frac{m}{N}\right)^n & \text{for } m \le n, \\ 0 & \text{otherwise.} \end{cases} \tag{2.5.1.20}$$

The relation (2.5.1.20) shows that the estimators $\overline{y}_1$ and $\overline{y}_2$ given at (2.5.1.17) and (2.5.1.18) are very difficult to compute for large sample sizes. Now if we ignore the sampling fraction n/N and hence v/N, then the above estimators, respectively, reduce to

$$\overline{y}_1 = \frac{v}{E(v)}\overline{y}_v + \overline{X}\left(1 - \frac{v}{E(v)}\right) \tag{2.5.1.21}$$

and

$$\overline{y}_2 = \frac{v}{E(v)}\overline{y}_v . \tag{2.5.1.22}$$

Hence the theorem.

Theorem 2.5.1.5. Show that if the square of the population coefficient of variation $C_y^2 = S_y^2/\overline{Y}^2$ exceeds $(n-1)$, then the estimator $\overline{y}_2 = (v/E(v))\overline{y}_v$ is more efficient than $\overline{y}_v$.

Proof. We know that

$$V(\overline{y}_v) = \left[E_1\left(\frac{1}{v}\right) - \frac{1}{N}\right]S_y^2 . \tag{2.5.1.23}$$

By the definition of variance we have

$$V(\overline{y}_2) = E_1V_2(\overline{y}_2 \mid v) + V_1E_2(\overline{y}_2 \mid v) = E_1V_2\left[\frac{v}{E(v)}\overline{y}_v \mid v\right] + V_1E_2\left[\frac{v}{E(v)}\overline{y}_v \mid v\right]$$

$$= E_1\left[\frac{v^2}{\{E(v)\}^2}\left(\frac{1}{v} - \frac{1}{N}\right)S_y^2\right] + V_1\left[\frac{v}{E(v)}\overline{Y}\right].$$

$$= \frac{S_y^2}{\{E(v)\}^2}\left[E(v) - \frac{E(v^2)}{N}\right] + \frac{\overline{Y}^2}{\{E(v)\}^2}V_1(v) . \tag{2.5.1.24}$$

It is very easy to derive that

$$E(v)=N\left[1-\left(1-\frac{1}{N}\right)^n\right],$$

and

$$E\left(v^2\right)=N\left[1-\left(1-\frac{1}{N}\right)^n\right]+N(N-1)\left[1-2\left(1-\frac{1}{N}\right)^n+\left(1-\frac{2}{N}\right)^n\right],$$

therefore

$$V(v)=E\left(v^2\right)-\{E(v)\}^2=N\left(1-\frac{1}{N}\right)^n-N^2\left(1-\frac{1}{N}\right)^{2n}+N(N-1)\left(1-\frac{2}{N}\right)^n.$$

From (2.5.1.24) we have

$$V(\bar{y}_2)=\frac{S_y^2(N-1)}{N^2\left\{1-\left(1-\frac{1}{N}\right)^n\right\}^2}\left[\left\{1-\left(1-\frac{1}{N}\right)^n\right\}-\left\{1-2\left(1-\frac{1}{n}\right)^n+\left(1-\frac{2}{N}\right)^n\right\}\right]$$

$$+\frac{\bar{Y}^2}{N^2\left[1-\left(1-\frac{1}{N}\right)^n\right]^2}\left[N\left(1-\frac{1}{N}\right)^n-N^2\left(1-\frac{1}{N}\right)^{2n}+N(N-1)\left(1-\frac{2}{N}\right)^n\right]. \tag{2.5.1.25}$$

Now from (2.5.1.23) and (2.5.1.25), we have

$$V(\bar{y}_v)-V(\bar{y}_2)$$

$$=S_y^2\left\{\frac{\sum_{J=1}^{N-1}J^{(n-1)}}{N^n}\;\frac{(N-1)\left\{\left(1-\frac{1}{N}\right)^n-\left(1-\frac{2}{N}\right)^n\right\}}{N^2\left\{1-\left(1-\frac{1}{N}\right)^n\right\}^2}\right\}\;\frac{\bar{Y}^2\left[\left(1-\frac{1}{N}\right)^n-N\left(1-\frac{1}{N}\right)^{2n}+(N-1)\left(1-\frac{2}{N}\right)^n\right]}{N\left[1-\left(1-\frac{1}{N}\right)^n\right]^2}$$

$$=C_1S_y^2-C_2\bar{Y}^2 \quad \text{(say).} \tag{2.5.1.26}$$

Now the estimator $\bar{y}_2$ is better than $\bar{y}_v$ if $V(\bar{y}_v)-V(\bar{y}_2)<0$ or if $\left(S_y^2/\bar{Y}^2\right)>(C_2/C_1)$.

The approximate values of C_1 and C_2 for large populations, correct up to terms of order N^{-2}, are given by

$$C_1=\frac{1}{2nN}+\frac{5(n-1)}{12nN^2} \text{ and } C_2=\frac{(n-1)}{2nN}-\frac{(n-1)(n-2)}{3nN^2}$$

and thus, $(C_2/C_1)\approx(n-1)$. Hence the theorem.

Theorem 2.5.1.6. If squared error be the loss function then show that $\bar{y}_v$ is admissible amongst all functions of $\bar{y}_v$ and v.

Proof. Let $t=\bar{y}_v+f(\bar{y}_v,v)$ be the function of $\bar{y}_v$ and v. Suppose that the estimator t is uniformly better than $\bar{y}_v$. Suppose $R(t)$ be the quadratic loss function for the estimator t. Then the estimator t will be uniformly better than the estimator $\bar{y}_v$ if $R(t) \le V(\bar{y}_v)$, where $V(\bar{y}_v) = E(\bar{y}_v - \bar{Y})^2$.

Also we have

$$R(t) = E[t - \bar{Y}]^2 = E[\bar{y}_v - \bar{Y} + f(\bar{y}_v, v)]^2 = E(\bar{y}_v - \bar{Y})^2 + E\{f(\bar{y}_v, v)\}^2 + 2Ef(\bar{y}_v, v)(\bar{y}_v - \bar{Y}).$$

Thus

$$R(t) \le V(\bar{y}_v) \text{ if } E(\bar{y}_v - \bar{Y})^2 + E\{f(\bar{y}_v, v)\}^2 + 2Ef(\bar{y}_v, v)(\bar{y}_v - \bar{Y}) \le E(\bar{y}_v - \bar{Y})^2 \tag{2.5.1.27}$$

holds for all $Y_1, Y_2, ..., Y_N$. In particular, if $Y_1 = Y_2 = ... = Y_N = C$ (say), where C is an arbitrary chosen constant, then the relation (2.5.1.27) implies that $f(C, v)$ is zero, which proves the theorem.

2.5.2 ESTIMATION OF FINITE POPULATION VARIANCE

Consider the problem of estimation of

$$\sigma_y^2 = N^{-1}\sum_{i=1}^{N}(Y_i - \bar{Y})^2$$

using distinct units in a sample of n units drawn by using SRSWR sampling. The usual estimator of σ_y^2 is given by

$$s_y^2 = (n-1)^{-1}\sum_{i=1}^{n}(y_i - \bar{y})^2 = [2n(n-1)]^{-1}\sum_{i \ne j=1}^{n}(y_i - y_j)^2.$$

If we now construct an estimator based on only distinct units then we have the following theorem.

Theorem 2.5.2.1. A uniformly better estimator of σ_y^2 than s_y^2 is given by

$$s_v^2 = \left[1 - \frac{C_v(n-1)}{C_v(n)}\right]s_d^2 \tag{2.5.2.1}$$

where

$$s_d^2 = \begin{cases}(v-1)^{-1}\sum_{i=1}^{v}(y_i - \bar{y}_v)^2 & \text{if } v > 1,\\ 0 & \text{otherwise,}\end{cases} \tag{2.5.2.2}$$

and

$$C_v(n) = v^n - \binom{v}{1}(v-1)^n + ... + (-1)^{(v-1)}\binom{v}{v-1}1^n. \tag{2.5.2.3}$$

Proof. Suppose we have any convex loss function and T be an ordered sufficient statistic, then by the Rao--Blackwell theorem we have

$$E[s_y^2 \mid T] = E\left[\frac{1}{n-1}\sum_{i=1}^{n}(y_i - \bar{y})^2 \mid T\right] = E\left[\frac{1}{2n(n-1)}\sum_{i=1}^{n}(y_i - y_j)^2 \mid T\right]$$

$$= E\left[\frac{1}{2}(y_1 - y_2)^2 \mid T\right]. \tag{2.5.2.4}$$

To prove that the estimator at (2.5.2.1) is uniformly better than s_y^2 let us consider the following cases:

If $v=1$, *i.e.*, only one unit has been selected in the sample of two units drawn by SRSWR then (2.5.2.4) is obviously zero. Suppose $\Sigma^{/}$ and $\Sigma^{//}$ denote the summations over all integral values of α' such that the following equalities holds:
$\sum_{i=1}^{v}\alpha_{(i)}=n$, $\alpha_{(i)}>0$ for $i=1,2,...,v$ and $\sum_{j=1}^{v}\alpha_{(j)}=(n-2)$, $\alpha_{(j)}\geq 0$, $\alpha_{(j)}^{/}>0$ and

$\alpha_{(k)}^{/}>0$, for $k\neq j\neq j'=1,2,...,v$.

Now if $v>1$ then we have

$$P\left[x_1=x_{(i)},\ x_2=x_{(j)}\mid T\right]=\frac{\frac{1}{N^2}\Sigma^{//}\frac{(n-2)!}{\alpha(1)!\alpha(2)!....\alpha(v)!}\left(\frac{1}{N}\right)^{\alpha(1)}.....\left(\frac{1}{N}\right)^{\alpha(v)}}{\Sigma^{/}\frac{n!}{\alpha(1)!\alpha(2)!....\alpha(v)!}\left(\frac{1}{N}\right)^{\alpha(1)}.....\left(\frac{1}{N}\right)^{\alpha(v)}}. \tag{2.5.2.5}$$

Pathak (1961) has shown some mathematical relations as:

$$\Sigma^{/}\frac{n!}{\alpha(1)!\alpha(2)!.....\alpha(v)!}=C_v(n) \tag{2.5.2.6}$$

and

$$\Sigma^{//}\frac{(n-2)!}{\alpha(1)!\alpha(2)!....\alpha(v)!}=\frac{C_v(n)-C_v(n-1)}{v(v-1)}. \tag{2.5.2.7}$$

Therefore we have

$$P\left[x_1=x_{(i)},x_2=x_{(j)}\mid T\right]=\frac{C_v(n)-C_v(n-1)}{v(v-1)C_v(n)},\ i\neq j=1,2,...,v. \tag{2.5.2.8}$$

Now if $v>1$ then we have

$$E\left[\frac{(y_1-y_2)^2}{2}\mid T\right]=\sum_{i\neq j=1}^{v}\frac{\left(y_{(i)}-y_{(j)}\right)^2}{2}P\left[x_1=x_{(i)},\ x_2=x_{(j)}\mid T\right].$$

On substituting the value of $P\left[x_1=x_{(i)},x_2=x_{(j)}\mid T\right]$ from (2.5.2.8), we obtain

$$E\left[\frac{(y_1-y_2)^2}{2}\mid T\right]=\frac{C_v(n)-C_v(n-1)}{C_v(n)}\frac{1}{2v(v-1)}\sum_{i\neq j=1}^{v}\left(y_{(i)}-y_{(j)}\right)^2=\frac{C_v(n)-C_v(n-1)}{C_v(n)}s_v^2.$$

Hence the theorem.

Example 2.5.2.1. We have selected an SRSWR sample of 20 units from the population 1 by using the 3rd and 4th columns of the Pseudo-Random Numbers (PRN) given in Table 1 in the Appendix. The 20 states corresponding to the serial numbers 29, 14, 47, 22, 42, 23, 48, 06, 07, 42, 21, 31, 31, 36, 16, 27, 10, 18, 26 and 48 were selected in the sample. Later on we observed that the states at serial numbers 42, 31 and 48 have been selected more than once. We reduced our sample size by keeping only 17 states in the sample and collected the information about the real estate farm loans in these states. The data so collected has been given below:

Random Number	State	Real estate farm loans
29	NH	6.044
14	IN	1213.024
47	WA	1100.745
22	MI	323.028
42	TN	553.266
23	MN	1354.768
48	WV	99.277
06	CO	315.809
07	CT	7.130
21	MA	7.590
31	NM	140.582
36	OK	612.108
16	KS	1049.834
27	NE	1337.852
10	GA	939.460
18	LA	282.565
26	MT	292.965

(a) Estimate the average real estate farm loans in the United States using information from distinct units only.
(b) Estimate the finite population variance of the real estate loans in the US using information from distinct units only.
(c) Estimate the average real estate loans and its finite population variance by including repeated units in the sample. Comment on the results.

Solution. Here $n = 20$ and $v = 17$, and on the basis of distinct units information, we have

Random Number	State	Real estate farm loans, y_i	$(y_i - \bar{y}_v)$	$(y_i - \bar{y}_v)^2$
29	NH	6.044	-560.7820	314476.8000
14	IN	1213.024	646.1977	417571.5000
47	WA	1100.745	533.9187	285069.2000
22	MI	323.028	-243.7980	59437.6100
42	TN	553.266	-13.5603	183.8816
23	MN	1354.768	787.9417	620852.1000
48	WV	99.277	-467.5490	218602.3000
06	CO	315.809	-251.0170	63009.6800
07	CT	7.130	-559.6960	313259.9000
21	MA	7.590	-559.2360	312745.2000
31	NM	140.582	-426.2440	181684.2000

Continued......

36	OK	612.108	45.2817	2050.4330
16	KS	1049.834	483.0077	233296.4000
27	NE	1337.852	771.0257	594480.6000
10	GA	939.460	372.6337	138855.9000
18	LA	282.565	-284.2610	80804.4800
26	MT	292.965	-273.8610	75000.0100
	Sum	9636.047	0.0000	3911380.0000

(a) An unbiased estimate of the average real estate farm loans in the United States is given by

$$\bar{y}_v = \frac{1}{v}\sum_{i=1}^{v} y_i = \frac{1}{17}\sum_{i=1}^{17} y_i = \frac{9636.047}{17} = 566.826\,.$$

(b) An estimator of the finite population variance σ_y^2 based on distinct units information is given by

$$s_v^2 = \left[1 - \frac{C_v(n-1)}{C_v(n)}\right] s_d^2\,.$$

Now

$$s_d^2 = \frac{1}{(v-1)}\sum_{i=1}^{v}\left(y_i - \bar{y}_v\right)^2 = \frac{3911380}{17-1} = 244461.25\,,$$

and

$$C_v(n) = v^n - \binom{v}{1}(v-1)^n + ... + (-1)^{(v-1)}\binom{v}{v-1}1^n\,.$$

$$\begin{aligned}C_v(n) = C_{17}(20) = {} & 17^{20} - C_1^{17}(17-1)^{20} + C_2^{17}(17-2)^{20} - C_3^{17}(17-3)^{20} + C_4^{17}(17-4)^{20}\\ & - C_5^{17}(17-5)^{20} + C_6^{17}(17-6)^{20} - C_7^{17}(17-7)^{20} + C_8^{17}(17-8)^{20}\\ & - C_9^{17}(17-9)^{20} + C_{10}^{17}(17-10)^{20} - C_{11}^{17}(17-11)^{20} + C_{12}^{17}(17-12)^{20}\\ & - C_{13}^{17}(17-13)^{20} + C_{14}^{17}(17-14)^{20} - C_{15}^{17}(17-15)^{20} + C_{16}^{17}(17-16)^{20}\\ & = 2.6366\times 10^{20}\,,\end{aligned}$$

and

$$\begin{aligned}C_v(n-1) = C_{17}(19) = {} & 17^{19} - C_1^{17}(17-1)^{19} + C_2^{17}(17-2)^{19} - C_3^{17}(17-3)^{19} + C_4^{17}(17-4)^{19}\\ & - C_5^{17}(17-5)^{19} + C_6^{17}(17-6)^{19} - C_7^{17}(17-7)^{19} + C_8^{17}(17-8)^{19}\\ & - C_9^{17}(17-9)^{19} + C_{10}^{17}(17-10)^{19} - C_{11}^{17}(17-11)^{19} + C_{12}^{17}(17-12)^{19}\\ & - C_{13}^{17}(17-13)^{19} + C_{14}^{17}(17-14)^{19} - C_{15}^{17}(17-15)^{19} + C_{16}^{17}(17-16)^{19}\\ & = 4.4805\times 10^{18}\,.\end{aligned}$$

Hence an estimate of the finite population variance is given by

$$s_v^2 = \left(1 - \frac{4.4805 \times 10^{18}}{2.6366 \times 10^{20}}\right) \times 244461.25 = 240307.004 .$$

(c) From the sample information including repeated units, we have

Random Number	State	Real estate farm loans, y_i	$(y_i - \bar{y}_n)$	$(y_i - \bar{y}_n)^2$
29	NH	6.044	-515.4150	265652.200
14	IN	1213.024	691.5654	478262.700
47	WA	1100.745	579.2864	335572.700
22	MI	323.028	-198.4310	39374.700
42	TN	553.266	31.8074	1011.711
23	MN	1354.768	833.3094	694404.600
48	WV	99.277	-422.1820	178237.300
06	CO	315.809	-205.6500	42291.760
07	CT	7.130	-514.3290	264533.900
21	MA	7.590	-513.8690	264060.900
31	NM	140.582	-380.8770	145067.000
36	OK	612.108	90.6494	8217.314
16	KS	1049.834	528.3754	279180.600
27	NE	1337.852	816.3934	666498.200
10	GA	939.460	418.0014	174725.200
18	LA	282.565	-238.8940	57070.150
26	MT	292.965	-228.4940	52209.330
42	TN	553.266	31.8074	1011.711
31	NM	140.582	-380.8770	145067.000
48	WV	99.277	-422.1820	178237.300
	Sum	10429.172	0.0000	4270686.000

If the repeated units are included in the sample then an estimate of the average real estate farm loans in the United States is given by

$$\bar{y}_n = \frac{1}{n}\sum_{i=1}^{n} y_i = \frac{10429.172}{20} = 521.4586$$

and an estimate of the finite population variance is given by

$$s_y^2 = \frac{1}{n-1}\sum_{i=1}^{n}(y_i - \bar{y}_n)^2 = \frac{4270686}{20-1} = 224772.9474 .$$

Clearly estimates of the average and finite population variance remain under estimate if repeated units are included in the sample. For details on distinct units one can refer to Raj and Khamis (1958), Pathak (1962) and Pathak (1966). Some comparisons of SRSWR and SRSWOR sampling schemes have also been considered by several other researchers *viz.* Deshpande (1980), Ramakrishnan (1969) and Seth and Rao (1964), and Basu (1958).

2.6 ESTIMATION OF TOTAL OR MEAN OF A SUBGROUP (DOMAIN) OF A POPULATION

Sometime we are interested in estimating the total or mean value of a variable of interest within a subgroup or part of the population. Such a part or subgroup of a population is called the domain of interest. For example, in a state wide survey, a district may be considered as a domain. After completing the survey sampling process from the whole population, one may be interested in estimating the mean or total of a particular subgroup of the population. We are interested in estimating population parameters of a subgroup of a population. For example

Population	Subgroup
United States population	Employed
New York	Unemployed
Retailers	Supermarkets
All workers in a firm	Part time workers

Let D be the domain of interest and N_D be the number of units in this domain. Let $Y_D = \sum_{i=1}^{N_D} Y_i$ and $\overline{Y}_D = \frac{1}{N_D}\sum_{i=1}^{N_D} Y_i$ be the total and mean for the domain D respectively. Suppose we selected an SRSWOR sample s of n units from the entire population Ω and $n_D \le n$ units out of the selected units are from the domain D of interest. In certain situations the value of N_D is known and in another situations the value of N_D is unknown. We shall discuss the both situations as follows. Define a variable

$$Y_i^* = \begin{cases} 1 & \text{if } i \in D, \\ 0 & \text{if } i \notin D. \end{cases} \tag{2.6.1}$$

Then we have $\sum_{i=1}^{N} Y_i^* = Y_D = Y^*$ (say).

Case 1. When N_D is unknown. Then we have the following theorems.

Theorem 2.6.1. Under SRS sampling an unbiased estimator of total Y_D of the subgroup (or domain D) is given by

$$\hat{Y}_D = \frac{N}{n}\sum_{i=1}^{n} Y_i^* . \tag{2.6.2}$$

Proof. Taking expected values on both sides of (2.6.2), we have

$$E\left(\hat{Y}_D\right) = E\left(\frac{N}{n}\sum_{i=1}^{n} Y_i^*\right) = \frac{N}{n}\sum_{i=1}^{n} E\left(Y_i^*\right) = \frac{N}{n}\sum_{i=1}^{n}\left[\sum_{j=1}^{N} Y_j P_r\left(Y_i^* = Y_j^*\right)\right]$$

$$= \frac{N}{n}\sum_{i=1}^{n}\frac{1}{N}\sum_{j=1}^{N} Y_j^* = \frac{N}{n}\times\frac{n}{N}\sum_{j=1}^{N} Y_j^* = \sum_{j=1}^{N} Y_j^* = \sum_{j=1}^{N_D} Y_j^* = Y_D .$$

Hence the theorem.

Theorem 2.6.2. The variance of the estimator $\hat{Y}_D$ under SRSWOR sampling is:

$$V(\hat{Y}_D) = \frac{N^2(1-f)}{n} S_D^2, \quad \text{where } S_D^2 = \frac{1}{N-1}\left\{\sum_{i=1}^{N} Y_i^{*2} - N^{-1}\left(\sum_{i=1}^{N} Y_i^*\right)^2\right\}. \tag{2.6.3}$$

Proof. Obvious from the results of SRSWOR sampling.

Theorem 2.6.3. An unbiased estimator of the $V(\hat{Y}_D)$ under SRSWOR sampling is given by

$$\hat{v}(\hat{Y}_D) = \frac{N^2(1-f)}{n} s_D^2, \quad \text{where } s_D^2 = \frac{1}{n-1}\left\{\sum_{i=1}^{n} Y_i^{*2} - n^{-1}\left(\sum_{i=1}^{n} Y_i^*\right)^2\right\}. \tag{2.6.4}$$

Proof. Obvious.

Case II. When N_D is known. Here we need the following lemma.

Lemma 2.6.1. If $i \in S_D$ indicates that the ith unit is in the sample sub group of D then, we have

$$\Pr(i \in S_D \mid n_D > 0) = \frac{n_D}{N_D}. \tag{2.6.5}$$

Proof. We have

$$\Pr(i \in S_D \mid n_D > 0)$$

$$= \frac{\Pr(i \in S_D \cap n_D)}{\Pr(n_D)} = \frac{\text{Number of samples of sizes } n_D \text{ with } i \in S_D}{\text{Number of samples of sizes } n_D}$$

$$= \frac{\text{Number of ways } (n_D - 1) \text{ can be chosen from } (N_D - 1) \text{ and } (n - n_D) \text{ from } (N - N_D)}{\text{Number of ways } n_D \text{ can be chosen from } N_D \text{ and } (n - n_D) \text{ from } (N - N_D)}$$

$$= \frac{C^{(N_D-1)}_{(n_D-1)} C^{(N-N_D)}_{(n-n_D)}}{C^{(N_D)}_{(n_D)} C^{(N-N_D)}_{(n-n_D)}} = \frac{\dfrac{(N_D-1)!}{(n_D-1)!(N-N_D)!}}{\dfrac{N_D!}{n_D!(N-n_D)!}} = \frac{n_D}{N_D}.$$

Hence the lemma.

Then we have the following theorems

Theorem 2.6.4. Under SRS sampling a biased estimator of Y_D is given by

$$\hat{Y}_D = \begin{cases} \dfrac{N_D}{n_D}\sum_{i=1}^{n} Y_i^* & \text{if } n_D > 0, \\ 0 & \text{otherwise.} \end{cases} \tag{2.6.6}$$

Its relative bias is equal to the negative of the probability that n_D is equal to zero.

Proof. Here n_D is not fixed before taking the sample and therefore is a random variable. Thus we have

$$E(\hat{Y}_D) = E_1\left[E_2(\hat{Y}_D \mid n_D)\right]$$

Now when $n_D > 0$ then we have

$$E(\hat{Y}_D) = E_2\left[(\hat{Y}_D \mid n_D > 0)\right] = E_2\left[\frac{N_D}{n_D}\sum_{i=1}^{n} Y_i^* \mid n_D > 0\right]$$

$$= E_2\left[\frac{N_D}{n_D}\sum_{i\in S_D} Y_i \mid n_D > 0\right], \text{ where } S_D \text{ indicates sample subgroup of } D$$

$$= E_2\left[\frac{N_D}{n_D}\sum_{i\in D} I_i Y_i \mid n_D > 0\right], \text{ where } I_i = \begin{cases} 1 & \text{if } i \in S_D, \\ 0 & \text{otherwise,} \end{cases}$$

$$= \frac{N_D}{n_D}\left[\sum_{i\in D} E_2(I_i \mid n_D > 0) Y_i\right] = \frac{N_D}{n_D}\left[\sum_{i\in D} Y_i \Pr(I_i \in S_D \mid n_D > 0)\right]$$

$$= \frac{N_D}{n_D}\left[\sum_{i\in D} Y_i \times \frac{n_D}{N_D}\right] = \sum_{i\in D} Y_i = Y_D .$$

Therefore

$$E(\hat{Y}_D) = \begin{cases} E(\hat{Y}_D \mid n_D > 0) = Y_D, \\ E(\hat{Y}_D \mid n_D = 0) = 0. \end{cases}$$

Thus we have

$$E(\hat{Y}_D) = E_1\left[E_2(\hat{Y}_D \mid n_D)\right] = Y_D \Pr[n_D > 0] + 0\Pr(n_D = 0) = Y_D \Pr[n_D > 0].$$

Thus the relative bias in the estimator $\hat{Y}_D$, when N_D is unknown, is given by

$$\text{RB}(\hat{Y}_D) = \frac{E(\hat{Y}_D) - Y_D}{Y_D} = \frac{Y_D \Pr(n_D > 0) - Y_D}{Y_D} = \Pr(n_D > 0) - 1 = -\Pr(n_D = 0).$$

Hence the theorem.

To find the variance of the estimator $\hat{Y}_D$ we need the following lemma.

Lemma 2.6.2. Show that

$$E\left(\frac{1}{n_D} \mid n_D > 0\right) = \frac{1}{nP_D} + \frac{1-P_D}{n^2 P_D^2}, \quad \text{where } P_D = \frac{N_D}{N}. \tag{2.6.7}$$

Proof. We have

$$\frac{1}{n_D} = \frac{1}{n\left(\dfrac{n_D}{n}\right)} = \frac{1}{n\left(\dfrac{n_D}{n} - P_D\right) + nP_D} = \frac{1}{nP_D + \sqrt{P_D(1-P_D)/n}\left\{\dfrac{\dfrac{n_D}{n} - P_D}{\sqrt{P_D(1-P_D)/n}}\right\} \times n}$$

$$= \frac{1}{nP_D}\left[1 + \frac{\sqrt{P_D(1-P_D)/n}}{nP_D}\left\{\frac{\dfrac{n_D}{n} - P_D}{\sqrt{P_D(1-P_D)/n}}\right\} \times n\right]^{-1}$$

$$\approx \frac{1}{nP_D}\left[1 - n\frac{\sqrt{P_D(1-P_D)/n}}{nP_D}\left\{\frac{\frac{n_D}{n} - P_D}{\sqrt{P_D(1-P_D)/n}}\right\} + n^2\frac{P_D(1-P_D)/n}{(nP_D)^2}\frac{\left(\frac{n_D}{n} - P_D\right)^2}{P_D(1-P_D)/n} + \dots\right].$$

Tanking expected values on both sides, we have

$$E\left(\frac{1}{n_D} \mid n_D > 0\right) \approx \frac{1}{nP_D}\left[1 - n\frac{\sqrt{\frac{P_D(1-P_D)}{n}}}{nP_D}\frac{E\left(\frac{n_D}{n} - P_D\right)}{\sqrt{P_D(1-P_D)/n}} + n^2\frac{\frac{P_D(1-P_D)}{n}}{(nP_D)^2}\frac{E\left(\frac{n_D}{n} - P_D\right)^2}{P_D(1-P_D)/n} + \dots\right]$$

$$\approx \frac{1}{nP_D} 1 - n\frac{\sqrt{P_D(1-P_D)/n}}{n^2P_D^2}\frac{E\left(\frac{n_D}{n} - P_D\right)}{\sqrt{P_D(1-P_D)/n}} + n^2\frac{P_D(1-P_D)/n}{(nP_D)^3}\frac{E\left(\frac{n_D}{n} - P_D\right)^2}{P_D(1-P_D)/n}.$$

Noting that $n_D \sim \text{Binomial}(\text{n}, \text{P}_\text{D})$ and it can be approximated by $\text{Normal}\left(nP_D,\ nP_D(1-P_D)\right)$ for large n. Thus we have

$$E\left(\frac{1}{n_D} \mid n_D > 0\right) \approx \frac{1}{nP_D} + \frac{(1-P_D)}{(nP_D)^2}. \tag{2.6.8}$$

Hence the lemma. For more details about the expected values of an inverse random variable one can refer to Stephen (1945).

Theorem 2.6.5. Show that the variance of the estimator $\hat{Y}_D$, when N_D is known, is

$$V\left(\hat{Y}_D\right) \approx \Pr(n_D > 0)\left[\left\{P_D\frac{N^2(1-f)}{n} + \frac{N^2}{n^2}(1-P_D)\right\}S_D^2 + \Pr(n_D = 0)Y_D^2\right]. \tag{2.6.9}$$

Proof. We have

$$V\left(\hat{Y}_D\right) = E_1\left[V_2\left(\hat{Y}_D \mid n_D\right)\right] + V_1\left[E_2\left(\hat{Y}_D \mid n_D\right)\right]. \tag{2.6.10}$$

Now

$$V_2\left(\hat{Y}_D \mid n_D\right) = \begin{cases}\dfrac{N_D^2}{n_D}\left(1 - \dfrac{n_D}{N_D}\right)S_D^2 & \text{if } n_D > 0, \\ 0 & \text{if } n_D = 0,\end{cases} \quad \text{and } E_2\left(\hat{Y}_D \mid n_D\right) = \begin{cases}Y_D & \text{if } n_D > 0, \\ 0 & \text{if } n_D = 0.\end{cases}$$

From (2.6.10) we have

$$V\left(\hat{Y}_D\right) = E_1\left[\begin{cases}\dfrac{N_D^2}{n_D}\left(1 - \dfrac{n_D}{N_D}\right)S_D^2 & \text{if } n_D > 0 \\ 0 & \text{if } n_D = 0\end{cases}\right] + V_1\left[\begin{cases}Y_D & \text{if } n_D > 0 \\ 0 & \text{if } n_D = 0\end{cases}\right]. \tag{2.6.11}$$

Now

$$V_1\left[\begin{cases}Y_D & \text{if } n_D > 0 \\ 0 & \text{if } n_D = 0\end{cases}\right] = V_1\left[Y_D I(n_D > 0)\right] = Y_D^2 V_1\left[I(n_D > 0)\right] = Y_D^2 \Pr(n_D > 0)(1 - \Pr(n_D > 0))$$

$$= Y_D^2 \Pr(n_D > 0)\Pr(n_D = 0), \tag{2.6.12}$$

and

$$E_1\left[\begin{cases}\dfrac{N_D^2}{n_D}\left(1-\dfrac{n_D}{N_D}\right)S_D^2 & \text{if } n_D>0\\ 0 & \text{if } n_D=0\end{cases}\right]$$

$$=\sum_{j=1}^{\min(N_D,n)}\Pr(n_D=j)\frac{N_D^2}{n_D}\left(1-\frac{n_D}{N_D}\right)S_D^2+\Pr(n_D=0)\times 0$$

$$=\Pr(n_D>0)\sum_{j=1}^{\min(N_D,n)}\frac{\Pr(n_D=j)}{\Pr(n_D>0)}\frac{N_D^2}{n_D}\left(1-\frac{n_D}{N_D}\right)S_D^2$$

$$=\Pr(n_D>0)\sum_{j=1}^{\min(N_D,n)}\Pr(n_D=j|n_D>0)\frac{N_D^2}{n_D}\left(1-\frac{n_D}{N_D}\right)S_D^2$$

$$=\Pr(n_D>0)E\left\{N_D^2\left(\frac{1}{n_D}-\frac{1}{N_D}\right)S_D^2|n_D>0\right\}$$

$$=N_D^2S_D^2\left\{E\left(\frac{1}{n_D}\,|\,n_D>0\right)-\frac{1}{N_D}\right\}\Pr(n_D>0)$$

$$\approx N_D^2S_D^2\left\{\frac{1}{nP_D}+\frac{1-P_D}{n^2P_D^2}-\frac{1}{N_D}\right\}\Pr(n_D>0),\quad \text{where } P_D=\frac{N_D}{N}$$

$$\approx\left\{P_D\frac{N^2}{n}\left(1-\frac{n}{N}\right)S_D^2+N_D^2\frac{1-P_D}{n^2P_D^2}S_D^2\right\}\Pr(n_D>0)\,. \tag{2.6.13}$$

On using (2.6.12) and (2.6.13) in (2.6.11) we have

$$V\left(\hat{Y}_D\right)\approx\Pr(n_D>0)\left[\left\{P_D\frac{N^2(1-f)}{n}+\frac{N^2}{n^2}(1-P_D)\right\}S_D^2+\Pr(n_D=0)Y_D^2\right].$$

Hence the theorem.

Theorem 2.6.6. An estimator for estimating $V\left(\hat{Y}_D\right)$ is given by

$$\hat{v}\left(\hat{Y}_D\right)\approx\left\{\hat{p}_D\frac{N^2(1-f)}{n}+\frac{N^2}{n^2}(1-\hat{p}_D)\right\}s_D^2 \tag{2.6.14}$$

where

$$\hat{p}_D=\frac{n_D}{n}\ \text{and}\ s_D^2=\frac{1}{n_D-1}\left\{\sum_{i=1}^{n}Y_i^{*2}-\left(\sum_{i=1}^{n}Y_i^*\right)^2\Big/n_D\right\}.$$

Proof. Obvious by the method of moments.

2.7 DEALING WITH A RARE ATTRIBUTE USING INVERSE SAMPLING

The problem of estimation of proportion of some rare types of genes or acreage under some special types of plants can not be done with the help of direct binomial distribution. The problem is that if we select a sample of size n by SRSWR or SRSWOR sampling, then the observed number of genes of particular type or interest in the selected sample will be zero due to its rare availability. Thus the traditional SRSWR or SRSWOR sampling schemes cannot be used to estimate the proportion of rare attribute in survey sampling. One of the possible solutions is an Inverse Sampling. Inverse Sampling is a technique in which sampling is continued until a predetermined number of units possessing the attribute occurs in the sample is useful in estimating the proportion of a rare attribute. Now inverse sampling can be done either by using SRSWR or by SRSWOR sampling. If the Inverse Sampling is done by SRSWR sampling then the total number of trials n (say) to get predetermined number of units m (say) possessing attribute A (say) follows the Negative Binomial Distribution. It is also called Binomial waiting-time distribution or Pascal distribution. If the inverse sampling is done by SRSWOR sampling, then the total number of trials n (say) to get predetermined number of units m (say) possessing an attribute A (say) follows Negative Hypergeometric distribution. Figure 2.7.1 has been devoted to differentiate between Negative Binomial and Negative Hypergeometric distributions.

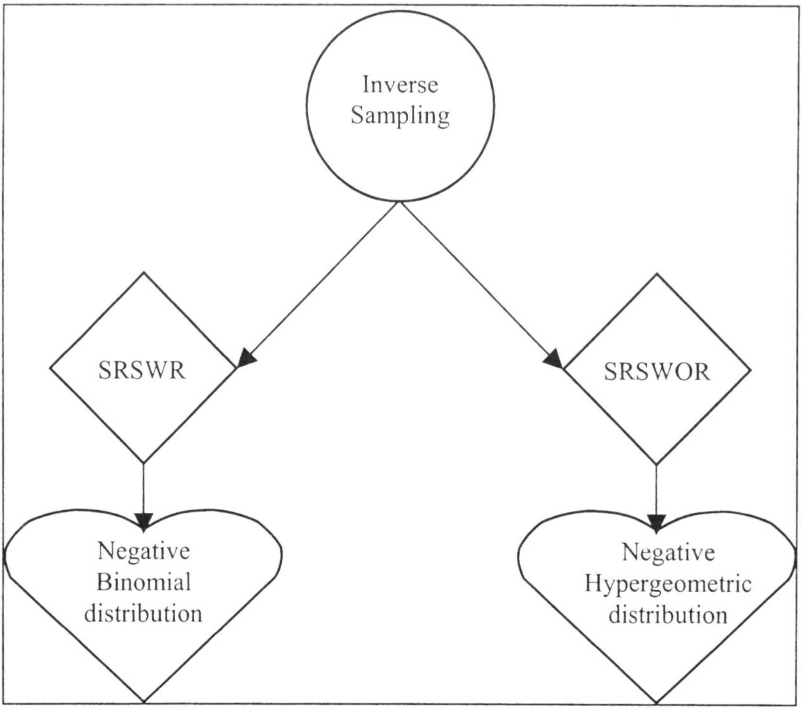

Fig. 2.7.1 Pictorial representation of the Inverse Sampling.

Thus we have the following situations.

Case I. SRSWOR Inverse Sampling: Let N be the population size, P be the proportion of the rare attribute of interest and $Q = 1 - P.$ In case of the Inverse Sampling the sample size n is required to attain m is a random variable and its probability distribution due to SRSWOR sampling is given by

$$P(n = t) = \frac{\binom{NP}{m-1}\binom{NQ}{t-m}}{\binom{N}{t-1}} \times \frac{(NP - m + 1)}{N - t + 1}, \quad t = m, m+1, ..., m + NQ. \tag{2.7.1}$$

Such a distribution is called negative hypergeometric distribution and we have $\sum_{t=m}^{m+NQ} P(n = t) = 1.$ Then we have the following theorem.

Theorem 2.7.1. An unbiased estimator of the proportion of rare attribute P is given by

$$\hat{p} = \frac{m-1}{n-1}. \tag{2.7.2}$$

Proof. We have

$$E(\hat{p}) = E\left(\frac{m-1}{n-1}\right) = \sum_{t=m}^{m+NQ} \frac{m-1}{n-1} P(n = t)$$

$$= \sum_{t=m}^{m+NQ} \left\{\frac{m-1}{n-1}\right\}\left\{\frac{NP - m + 1}{N - t + 1}\right\}\binom{NP}{m-1}\binom{NQ}{t-m}\binom{N}{t-1}^{-1}$$

$$= P \sum_{t=m}^{m+NQ-1} \left\{\frac{NP - m + 1}{N - t + 1}\right\}\binom{NP-1}{m-2}\binom{NQ}{t-m}\binom{N-1}{t-2}^{-1} = P.$$

Hence the theorem.

Theorem 2.7.2. An estimator for estimating the variance of $\hat{p}$ is given by

$$\hat{v}(\hat{p}) = \frac{m-1}{n-1}\left[\frac{m-1}{n-1} - \frac{(N-1)(m-2)}{N(n-2)} - \frac{1}{N}\right]. \tag{2.7.3}$$

Proof. We have

$$E\left\{\frac{(m-1)(m-2)}{(n-1)(n-2)}\right\} = \sum_{t \ge m} \left\{\frac{(m-1)(m-2)}{(t-1)(t-2)}\right\} P(n = t)$$

$$= \left\{\frac{P(NP-1)}{N-1}\right\} \sum_{t \ge m} \left\{\frac{NP - m + 1}{N - t + 1}\right\}\binom{NP-2}{m-3}\binom{NQ}{t-m}\binom{N-2}{t-3}^{-1}$$

$$= \frac{NP^2}{N-1} - \frac{P}{N-1}.$$

By the method of moments an estimator of P^2 is given by

$$\left\{\hat{p^2}\right\}_{\text{estimate}} = \frac{(N-1)(m-1)(m-2)}{N(n-1)(n-2)} + \frac{(m-1)}{N(n-1)}.$$

The variance of the estimator $\hat{p}$ is given by

$$V(\hat{p}) = E(\hat{p}^2) - \{E(\hat{p})\}^2.$$

By the method of moments, an unbiased estimator of $V(\hat{p})$ is given by

$$\hat{v}(\hat{p}) = \{\hat{p}\}^2 - \left\{\hat{p^2}\right\}_{\text{estimate}} = \frac{(m-1)^2}{(n-1)^2} - \frac{(N-1)(m-1)(m-2)}{N(n-1)(n-2)} - \frac{(m-1)}{N(n-1)}$$

$$= \frac{m-1}{n-1}\left[\frac{m-1}{n-1} - \frac{(N-1)(m-2)}{N(n-2)} - \frac{1}{N}\right].$$

Hence the theorem.

Case II. SRSWR Inverse Sampling: In case of large N, the negative hypergeometric probability distribution of total sample size n becomes negative binomial distribution and is given by

$$P(n;\ m,\ P) = \binom{n-1}{m-1} P^m Q^{n-m} \quad \text{for } n = m, m+1, m+2, \ldots$$

Then we have the following theorem.

Theorem 2.7.3. An unbiased estimator of the required proportion P of a rare attribute is

$$\hat{p} = \frac{m-1}{n-1} \tag{2.7.4}$$

and an estimator of the $V(\hat{p})$ is given by

$$\hat{v}(\hat{p}) = \frac{\hat{p}(1-\hat{p})}{n-2}. \tag{2.7.5}$$

Proof. Obvious for large N from the previous theorems.

2.8 CONTROLLED SAMPLING

While using simple random sampling and without replacement (SRSWOR) design, the number of possible samples, ${}^N C_n$, is very large, even for moderate sample and population sizes. For example, if

Number of samples		
	N	
n	30	40
5	142,506	658,008
10	30,045,015	847,660,528
15	155,117,520	40,225,345,056

Some times in the field surveys, all the possible samples are not equally preferable from the operational point of view, because a few of them may be inaccessible, expensive, or inconvenient, etc.. It is therefore advantageous if the sampling design is such that the total number of possible samples is much less than ${}^{N}C_{n}$, retaining the unbiasedness properties of sample mean and sample variance for their respective population parameters. Neyman's (1923) notation for causal effects in randomized experiments and Fisher's (1925) proposal to actually randomize treatments to units. Neyman (1923) appears to have been the first to provide a mathematical analysis for a randomized experiment with explicit notation for the potential outcomes, implicitly making the stability assumption. This notation became standard for work in randomized experiments (e.g., Pitman, 1937; Welch, 1937; McCarthy, 1939; Anscombe, 1948; Kempthorne, 1952; Brillinger, Jones, and Tukey, 1978; Hodges and Lehmann, 1970, and dozens of other places, often assuming constant treatment effects as in Cox, 1958, and sometimes being used quite informally as in Freedman, Pisani, and Purves, 1978). Neyman's formalism was a major advance because it allowed explicit probabilistic inferences to be drawn from data, where the probabilities were explicitly defined by the randomized assignment mechanism. Independently and nearly simultaneously, Fisher (1925) invented a somewhat different method of inference for randomized experiments, also based on the special class of randomized assignment mechanisms. Fisher's test and resulting 'significance levels' (*i.e.*, p values), remain the accepted rigorous standard for the analysis of randomized clinical trials at the end of the twentieth century, so called 'intent to treat analyses. The notions of the central role of randomized experiments seems to have been 'in the air' in the 1920, but Fisher was the first to combine physical randomization with a theoretical analysis tied to it. A review on randomization is also available by Fienberg and Tanur (1987). These ideas were primarily associated with the notion of fairness and objectivity in their earlier work. The role of the International Statistical Institute in the earlier work related to sample surveys, as reviewed by Smith and Sugden (1985). Fienberg and Tanur (1987) explored some of the developments following from their earlier pioneering work with an emphasis on the parallels between the methodologies in the design of experiments and the design of sample surveys. Chakrabarti (1963) initiated the idea that the results on the existence and construction of balanced sampling designs can be easily translated to the language of design theory by using the correspondence between sampling design and block designs. Bellhouse (1984a) also worked on these lines and has shown that a systematic application of the treatments minimises the variance of the treatment constant averaged over the application of the treatment. The lack of cross reference in the review papers by Cox (1984) and Smith (1984) suggested that the specialisation extends even to compartmentalisation within the minds and professional lives of outstanding investigators, for both these authors have been steeped in the tradition of parallels. For example, consider a balanced incomplete block design (BIBD) with standard parameters (b, v, r, k, λ), where v denotes the number of varieties, b the number of blocks, k the block size, r the number of times each treatment occurs and λ the number of times any pair of treatments occur together in a blocks. In practice

$b < \{{}^vC_k\}$. Each treatment represents a population units $v = N$, each block as a sample with $k = n$. Thus each unit occurs in r samples and each pair of units in λ samples. Choose each sample with probability $1/b$ and under such designs, define indicator variables

$$I_{si} = \begin{cases} 1 & \text{if } i \in s \\ 0 & \text{if } i \notin s \end{cases} \quad \text{such that} \quad E(I_{si}) = \frac{r}{b},$$

and

$$I_{sij} = \begin{cases} 1 & \text{if } i, j \in s \\ 0 & \text{if } i, j \notin s \end{cases} \quad \text{such that} \quad E(I_{sij}) = \frac{\lambda}{b}.$$

The sample mean

$$\bar{y} = \frac{1}{n}\sum_{i=1}^{n} y_i$$

can be written as

$$\bar{y} = \frac{1}{n}\sum_{i=1}^{N} Y_i I_{si}$$

such that

$$E(\bar{y}) = E\left[\frac{1}{n}\sum_{i=1}^{N} Y_i I_{si}\right] = \frac{1}{n}\sum_{i=1}^{N} Y_i E(I_{si}) = \frac{1}{n}\sum_{i=1}^{N} Y_i \frac{r}{b} = \frac{r}{nb}\sum_{i=1}^{N} Y_i = \bar{Y}$$

because $vr = bk$.

Similarly using $r(k-1) = \lambda(v-1)$ and $bk = vr$ we have

$$E\left(\sum\sum_{i \neq j}^{n} y_i y_j\right) = E\left(\sum\sum_{i \neq j}^{N} Y_i Y_j I_{sij}\right) = \frac{\lambda}{b}\sum\sum_{i \neq j}^{N} Y_i Y_j = \left(\frac{r(r-1)}{b(v-1)}\right)\sum\sum_{i \neq j}^{N} Y_i Y_j = \frac{k(n-1)}{v(v-1)}\sum\sum_{i \neq j}^{N} Y_i Y_j$$

$$= \frac{n(n-1)}{N(N-1)}\sum\sum_{i \neq j}^{N} Y_i Y_j .$$

Thus, we have the following theorem:

Theorem 2.8.1. Under controlled sampling design, the sample mean and sample variance remain unbiased to their respective parameters.

2.9 DETERMINANT SAMPLING

Subramani and Tracy (1993) used the concept of incomplete block design in sample surveys and introduced a new sampling scheme called determinant sampling. This scheme totally ignores the units close to each other for selection in the sample. In the preceding discussion, the units which are close to each other in some sense are called contiguous units. Chakrabarti (1963) excluded contiguous units when translating the results of sampling designs to experimental designs since these units have a tendency to provide identical information which may be induced by factors like time, category or location. As an example, in socio economic surveys people have a tendency to exhibit similar expenditure patterns on household items during different weeks of the month. Moreover people belonging to the same income category class have a greater tendency to have similar expenditure patterns. With regard to the factor location, residents of a specific area show similar symptoms of a disease caused by environmental pollution as of some infectious disease. Similarly in crop field surveys contiguous farms and fields should be avoided. Because of this limitation, Rao (1975, 1987) has suggested that if contiguous units occur in any observed sample, they may be collapsed into a single unit, with the corresponding response as the average observed response over these units. An estimate of the unknown parameter is then recommended on the basis of such a reduced sample. The situations for getting more information on the population by avoiding pairs of contiguous units in the observed sample are well summarised by Hedayat, Rao, and Stufken (1988). Tracy and Osahan (1994a) further extend their work for other sampling schemes.

EXERCISES

Exercise 2.1. Define simple random sampling. Is the sample mean a consistent or unbiased estimator of the population mean? Derive the variance of the estimator using (a) SRSWR sampling (b) SRSWOR sampling. Also derive an unbiased estimator of variance in each situation.

Exercise 2.2. A population consists of N units, the value of one unit being known to be Y_1. An SRSWOR of $(n-1)$ units is drawn from the remaining $(N-1)$ population units. Show that the estimator

$$\hat{y}_1 = Y_1 + (N-1)\bar{y}_{n-1}$$

where $\bar{y}_{n-1} = (n-1)^{-1}\sum_{i-1}^{n-1} y_i$,is an unbiased estimator of the population total, $Y = \sum_{i=2}^{N} Y_i$, but the variance of the estimator $\hat{y}_1$ is not less than the variance of estimator $\hat{y}_2 = N\bar{y}_n$, where $\bar{y}_n = n^{-1}\sum_{i=1}^{n} y_i$ is an estimator of population mean based on the sample of size n selected from the population of N units. In other words, the estimator $\hat{y}_1$ is no more efficient than $\hat{y}_2$. Give reasons.

Hint: By setting $V(\hat{y}_1) \geq V(\hat{y}_2)$ we obtain $N > n$ which is always true for SRSWOR sampling, where $V(\hat{y}_1) = (N-1)^2\left(\frac{1}{n-1} - \frac{1}{N-1}\right)S_y^2$ and $V(\hat{y}_2) = N^2\left(\frac{1}{n} - \frac{1}{N}\right)S_y^2$.

Exercise 2.3. Suppose in the list on N businesses serially numbered, k businesses are found to be dead and t new businesses came into existence making the total number of business $(N-k+t)$. Give a simple procedure for selecting a businesses with equal probability from $(N-k+t)$ businesses, avoiding renumbering of the original businesses and show that the newly developed procedure achieves equal probability for the new business too.

Hint: Using SRSWOR sampling the probability of selecting each unit will be $(N-k+t)^{-1}$.

Exercise 2.4. Show that the bias in the Searls' estimator defined as, $\bar{y}_{\text{searl}} = \lambda \bar{y}_n$, is $B(\bar{y}_{\text{searl}}) = -\bar{Y}\, V(\bar{y}_n) / \{\bar{Y}^2 + V(\bar{y}_n)\}$. Hence deduce its values under SRSWR and SRSWOR.
Hint: Reddy (1978a).

Exercise 2.5. An analogue to the Searls' estimator for estimating the population proportion is defined as, $\hat{p}_{\text{searl}} = \gamma\, \hat{p}_y$, where γ is a constant. Find the minimum mean square error of the estimator $\hat{p}_{\text{searl}}$ under SRSWR and SRSWOR sampling. Also study the bias of the estimator in each situation.
Hint: Conti (1995).

Exercise 2.6. An estimator of the optimum value of λ in the Searls' estimator of population mean $\bar{Y}$ under SRSWR sampling is given by,

$$\hat{\lambda} = \left[1 + \frac{N}{n(N-1)}\left(s_y^2 / \bar{y}^2\right)\right]^{-1}.$$

Show that $\hat{\lambda}$ is a consistent estimator of the optimum value of λ. Also calculate the bias and mean squared error, to the first order of approximation, in the estimator of population mean defined as, $\bar{y}_0 = \hat{\lambda}\bar{y}_n$. Deduce the results for estimating population proportion with the estimator, $\hat{p}_{\text{searl}} = \hat{\gamma}\, \hat{p}_y$, where $\hat{\gamma}$ is a consistent estimator of γ.
Hint: Mangat, Singh, and Singh (1991).

Exercise 2.7. Show that: (a) under SRSWR sampling s_y^2 is an unbiased estimator of σ_y^2, (b) under SRSWOR sampling s_y^2 is an unbiased estimator of S_y^2.

Exercise 2.8. Define the Searls' estimator of population mean. Show that the relative efficiency of the Searls' estimator is a decreasing function of sample size under (a) SWSWR (b) SRSWOR sampling designs.

Exercise 2.9. Show that the probability of selecting the ith unit in the sth sample remain the same under SRSWR and SRSWOR and is given by $1/N$.

Exercise 2.10. Why is the Searls' estimator not useful in actual practice? Suggest some modifications to make it practicable.
Hint: Use $\hat{\lambda}$ in place of λ.

Exercise 2.11. In case of SWSWR sampling, if there are two characters Y and X, the covariance between Y and X is defined as $\sigma_{xy} = \frac{1}{N}\sum_{i=1}^{N}(Y_i - \bar{Y})(X_i - \bar{X})$. Then the usual estimator of σ_{xy} is given by $s_{xy} = \frac{1}{n-1}\sum_{i=1}^{n}(y_i - \bar{y})(x_i - \bar{x})$. Show that an estimator better than s_{xy} based only on distinct units is:

$$s_{v(xy)} = \begin{cases} \left[1 - \frac{C_v(n-1)}{C_v(n)}\right] s_{d(xy)} & \text{if } v > 1, \\ 0 & \text{otherwise,} \end{cases} \quad \text{where } s_{d(xy)} = (v-1)^{-1}\sum_{i=1}^{v}(y_i - \bar{y}_v)(x_1 - \bar{x}_v).$$

Hint: Pathak (1962).

Exercise 2.12. (a) Show that the usual estimator of the population total (namely $N\bar{y}$) in SRSWOR has average minimum mean squared error, for permutations of values attached to the units, in the general class of linear translation invariant estimators of the population total Y.
(b) Show that for SRSWOR sampling of size n, the estimator which minimises the average mean squared error, for permutations of values attached to the units, in the class of all linear estimators is given by,

$$t_c = \frac{N}{n(1+\delta)}\sum_{i\in s} y_i$$

where $\delta = \left(\frac{N-n}{N-1}\right)\frac{C_y^2}{n}$ and C_y is the known population coefficient of variation.

Hint: Ramakrishnan and Rao (1975).

Exercise 2.13. Let a finite population consist of N units. To every unit there is attached a characteristic y. The characteristics are assumed to be measured on a given scale with distinct points $y_1, y_2, ..., y_t$. Let N_t be the number of units associated with scale point y_t, with $N = \sum_t N_t$. A simple random sample of size n is drawn. Using the likelihood function of ($N_1, N_2, ..., N_t$) and assuming N/n to be an integer, show that a maximum likelihood estimator of the population mean $\bar{Y}$ is

$$\bar{y}_{\text{ml}} = \frac{1}{n}\sum_t n_t y_t$$

where n_t is the number of times the values of y_t are observed in the sample.
Hint: Hartley and Rao (1968).

Exercise 2.14. Suppose we selected a sample of size n such that the ith unit of the population occurs f_i times in the sample. Assume that n_1 of these units $(n_1 < n)$ are selected with frequency one. Evidently $n = n_1 + \sum_{i=1}^{r} f_i$, where r is the number of units occurring f_i times in the sample. Let d $(= n_1 + r)$ be the number of distinct units in the sample. The d units are measured by one set of investigators and the r repeated units by another set, preferably by the supervising staffs. The measurement of the d units be denoted by $x_1, x_2, ..., x_{n_1}$ for the non-repeated ones and $x_{n_1+1}, x_{n_1+2}, ..., x_{n_1+r}$ for the repeated ones. The measurement of the r repeated units be denoted by $z_1, z_2, ..., z_r$. Using the above information and notation, study the asymptotic properties of the following estimators of population mean:

(a) $\bar{x}_d = \frac{1}{d}\left[n_1\bar{x}_{n_1} + r\bar{x}_r\right]$; (b) $\bar{z}_R = \bar{z}_r\left(\frac{\bar{x}_d}{\bar{x}_r}\right)$; (c) $\bar{z}_{lr} = \bar{z}_r + \hat{\beta}(\bar{x}_d - \bar{x}_r)$;

(d) $\bar{y}_1 = (1-w)\bar{x}_d + w\bar{z}_R$; and (e) $\bar{y}_2 = (1-w)\bar{x}_d + w\bar{z}_{lr}$.

Hint: Mohanty (1977).

Exercise 2.15. Discuss the problem of the estimation of domain total in survey sampling. Derive the estimator of domain total and find its variance under different situations.

Exercise 2.16. Under SRSWR sampling, show that the distinct units based unbiased estimators of the finite population variance σ_y^2 are given by

(a) $\hat{v}_1 = \frac{\sum_{J=1}^{(N-1)} J^{(n-1)}}{N^{n-1}(N-1)} s_y^2$; (b) $\hat{v}_2 = \left[\left(\frac{1}{v} - \frac{1}{N}\right) + N^{1-n}\left(1 - \frac{1}{v}\right)\right] s_d^2$;

(c) $\hat{v}_3 = \frac{C_{v-1}(n-1)}{C_v(n)} s_d^2$; (d) $\hat{v}_4 = \frac{\sum_{J=1}^{(N-1)} J^{(n-1)}}{N^{n-1}(N-1)}\left[1 - \frac{C_v(n-1)}{C_v(n)}\right] s_d^2$;

and (e) $\hat{v}_5 = \left[\left(\frac{1}{v} - \frac{1}{N}\right) + \frac{(N-1)}{N^n - N}\right] s_d^2$.

Hint: Pathak (1966).

Exercise 2.17. Discuss the method and theory of the estimation of rare attributes in survey sampling.

Exercise 2.18. Write a program in FORTRAN or SAS to find the values of the coefficients $C_v(n-1)$ and $C_v(n)$. Test the results for $n = 5$ and $v = 3$ with all steps using your desk calculator.

Exercise 2.19. Under an SRSWOR sampling of n units out of a population of N distinct units, consider the following estimator of population mean $\bar{Y}$ as

$$\bar{y}_{new} = \sum_{i=1}^{n} c_i y_i$$

where c_i is a constant depending on the ith draw, y_i is the value of y on the unit selected at the ith draw.

(a) Show that $\bar{y}_{\text{new}}$ is unbiased for population mean $\bar{Y}$ if and only if $\sum_{i=1}^{n} c_i = 1$.

(b) Show also that under this condition $V(\bar{y}_{\text{new}}) = S_y^2\left\{\sum_{i=1}^{n} c_i^2\left(1-\frac{1}{N}\right)\right\}$.

Hint: $V(\bar{y}_{\text{new}}) = V\left(\sum_{i=1}^{n} c_i y_i\right) = \sum c_i^2 V(y_i) + \sum\sum_{i\neq j} c_i c_j \text{Cov}(y_i, y_j) = \sum c_i^2 V(y_i) = V(y_i)\sum c_i^2$.

(c) Show that $V(\bar{y}_{\text{new}})$ is minimised subject to the condition $\sum_{i=1}^{n} c_i = 1$ if and only if $c_i = 1/n$, $i = 1,2,...,n$.

Hint: $\sum c_i^2 \geq (\sum c_i)^2/n = 1/n$ and equality holds if and only if $c_i = 1/n$.

Exercise 2.20. An SRSWR sample of size n is drawn until we have a pre-assigned number m of distinct units in it. Let $\bar{y}_n = \frac{1}{n}\sum_{i=1}^{m} f_i y_i$ and $\bar{y}_m = \frac{1}{m}\sum_{i=1}^{m} y_i$ be the sample means based on the sample including repetitions and without repetitions.

(a) Show that both estimators $\bar{y}_n$ and $\bar{y}_m$ are unbiased for population mean $\bar{Y}$.

(b) Show that $V(\bar{y}_n) = \sigma_y^2 E\left(\frac{1}{n}\right)$.

(c) Show that $V(\bar{y}_m) = \left(\frac{1}{m} - \frac{1}{N}\right)S_y^2$.

(d) Show that the estimator $\bar{y}_m$ is better than $\bar{y}_n$ if $E\left(\frac{1}{n}\right) > \frac{N-m}{m(N-1)}$.

Hint: Raj and Khamis (1958).

Exercise 2.21. Discuss controlled sampling. Show that the sample mean and sample variance remain unbiased to their respective parameters.

Exercise 2.22. Discuss the concept of rare attribute and give a possible solution using inverse sampling.

PRACTICAL PROBLEMS

Practical 2.1. Consider the problem of estimation of the total number of fish caught by marine recreational fishermen at Atlantic and Gulf coasts. We know that there were 69 species caught during 1992 as shown in the population 4 in the Appendix. What is the minimum number of species groups to be selected by SRSWR sampling to attain the accuracy of relative standard error 12%?

Given: $S_y^2 = 31{,}010{,}599$ and $Y = 291{,}882$.

Practical 2.2. Your supervisor has suggested you to think on the problem of estimation of the total number of fish caught by marine recreational fishermen at Atlantic and Gulf coasts. He told you that there were 69 species caught during 1993 as shown in the population 4 in the Appendix. He needs your help in deciding the sample size using SRSWOR design with the relative standard error 25%. How your knowledge in statistics can help him?

Given: $S_y^2 = 39{,}881{,}874$ and $Y = 316{,}784$.

Practical 2.3. The demand for the *Bluefish* has been found to be highest in certain markets. In order to supply these types of fish the estimation of the proportion of bluefish is an important issue. At Atlantic and Gulf coasts, in a large sample of 311,528 fish there were shown to be 10,940 Bluefish caught during 1995. What is the minimum number of fish to be selected by SRSWR sampling to attain the accuracy of relative standard error 12%?

Practical 2.4. John considers the problem of estimation of the total number of fish caught by marine recreational fishermen at Atlantic and Gulf coasts. There were 69 species caught during 1994 as shown in the population 4 in the Appendix. John selected a sample of 20 units by SRSWR sampling. What will be his gain in efficiency if he considers the Searls' estimator instead of usual estimator?

Given: $S_y^2 = 49{,}829{,}270$ and $Y = 341{,}856$.

Practical 2.5. Select an SRSWR sample of twenty units from population 4 given in the Appendix. Collect the information on the number of fish during 1994 in each of the species group selected in the sample. Estimate the average number of fish caught by marine recreational fishermen at the Atlantic and Gulf coasts during 1994. Construct 95% confidence interval for the average number of fish in each species group of the United States.

Practical 2.6. Use population 4 of the Appendix to select an SRSWOR sample of sixteen units. Obtain the information on the number of fish during 1993 in each of the species group selected in the sample. Develop an estimate of the average number of fish caught by marine recreational fishermen at Atlantic and Gulf coasts during 1993. Construct the 95% confidence interval estimate.

Practical 2.7. Select an SRSWR sample of 20 states using Random Number Table method from population 1 of the Appendix. Note the frequency of each state selected in the sample. Construct a new sample by keeping only distinct states and collect the information about the nonreal estate farm loans in these states. From the information collected in the sample:

(a) Estimate the average nonreal estate farm loans in the United States using information from distinct units only.
(b) Estimate the finite population variance of the nonreal estate loans in the United States using distinct units only.

(c) Estimate the average nonreal estate loans and its finite population variance by including repeated units in the sample. Comment on the results.

Practical 2.8. A fisherman visited the Atlantic and Gulf coast and caught 6,000 fish one by one. He noted the species group of each fish caught by him and put back that fish in the sea before making the next caught. He observed that 700 fish belong to the group *Herrings*.
(a) Estimate the proportion of fish in the group *Herrings* living in the Atlantic and Gulf coast.
(b) Construct the 95% confidence interval.

Practical 2.9. During 1995 Michael visited the Atlantic and Gulf coast and caught 7,000 fish. He observed the species group of each one of the fish caught by him using SRSWOR sampling and found that 1,068 fish belong to the group *Red snapper*.
(a) Estimate the proportion of fish in the group *Red snapper* living in the Atlantic and Gulf coast.
(b) Construct the 95% confidence interval.
Given: Total number of fish living in the coast = 311,528.

Practical 2.10. Following the instructions of an *ABC* company, select an SRSWR sample of 25 units from the population 1 by using the 4th and 5th columns of the Pseudo-Random Numbers (PRN) given in Table 1 of the Appendix. Record the states selected more than once in the sample. Reduce the sample size by keeping only each state once in the sample and collect the information about the real estate farm loans in these states. Use this information to:

(a) Estimate the average real estate farm loans in the United States using information from distinct units only.
(b) Estimate the finite population variance of the real estate loans in the US using information from distinct units only.
(c) Estimate the average real estate loans and its finite population variance by including repeated units in the sample. Comment on the results.

Practical 2.11. You think of a practical situation where you have to estimate a total of a variable or characteristic of a subgroup (domain) of a population. Take a sample of reasonable size from the population under study and collect the information from the units selected in the sample. Apply the appropriate formulae to construct the 95% confidence interval estimate.

Practical 2.12. A practical situation arises where you have to estimate a proportion of a rare attribute in a population, *e.g.*, extra marital relations. Collect the information from the units selected in the sample through inverse sampling from the population under study. Apply the appropriate formulae to construct the 95% confidence intervals for the proportion of the rare attribute in the population.

Practical 2.13. A sample of 30 out of 100 managers was taken, and they were asked whether or not they usually take work home. The responses of these managers are given below where 'Yes' indicates they usually take work home and 'No' means they do not.

Yes	Yes	Yes	Yes	No	No	Yes	No	Yes	No
No	Yes	No	Yes	No	Yes	Yes	Yes	Yes	Yes
No	No	Yes	Yes	Yes	Yes	Yes	Yes	No	Yes

Construct 95% confidence intervals for the proportion of all managers who take work home using the following sampling schemes:
(a) Simple Random Sampling and With Replacement;
(b) Simple Random Sampling and Without Replacement.

Practical 2.14. From a list of 80,000 farms in a state, a sample of 2,100 farms was selected by SRSWOR sampling. The data for the number of cattle for the sample were as follows:

$$\sum_{i=1}^{n} y_i = 38{,}000\,,\ \text{and}\ \sum_{i=1}^{n} y_i^2 = 920{,}000\,.$$

Estimate from the sample the total number of cattle in the state, the average number of cattle per farm, along with their standard errors, coefficient of variation and 95% confidence interval.

Practical 2.15. At St. Cloud State University, the length of hairs, y, on the heads of girls is assumed to be uniformly distributed between 5 cm and 25cm with the probability density function

$$f(y) = \frac{1}{20} \quad \forall \quad 5 < y < 25$$

(a) We wish to estimate the average length of hairs with an accuracy of relative standard error of 5%, what is the required minimum number of hairs to be taken from the girls?
(b) Select a sample of the required size, and use it to construct a 95% confidence interval for the average length of hairs?

Practical 2.16. The distribution of weight y shipped to 1000 locations has a logistic distribution

$$f(y) - \frac{1}{4\beta_\bullet}\text{scch}^2\left\{\frac{1}{2}\left(\frac{x-\alpha\bullet}{\beta_\bullet}\right)\right\}$$

with $\alpha_\bullet = 10$ and $\beta_\bullet = 0.5$.
(a) Find the value of the minimum sample size n required to estimate the average weight shipped with an accuracy of standard error of 0.05% .
(b) Select a sample of the required size and construct 95% confidence interval for the average weight shipped.
(c) Does the true weight lies in the 95% confidence interval?

Practical 2.17. Assume that the life of every person is made of an infinite number of good and bad events. Count the total number of good and bad events you remember that have happened to you. Estimate the proportion of good events in your life. Construct a 95% confidence interval estimate. Name the sampling scheme you adopted to estimate proportion of good happenings, and comment.

Practical 2.18. Assuming that every one dreams infinite number times during sleeping hours in the life. Count the number of good and bad dreams in your life you remember. Estimate the proportion of good dreams and construct a 95% confidence interval estimate. Name the sampling scheme you followed to estimate the proportion of good dreams, and comment.

Practical 2.19. Dr. Dreamer believes that if a person takes good dreams during sleeping hours then he/she is mentally more healthy, and pleasant person. You are instructed to report stories of your dreams to the doctor until you are not having 15 good dreams. Find the Dr. Dreamer's 95% confidence interval estimate of the proportion of good dreams in your life. Can you be considered a pleasant person? Comment and list the sampling scheme used.

3. USE OF AUXILIARY INFORMATION: SIMPLE RANDOM SAMPLING

3.0 INTRODUCTION

It is well known that suitable use of auxiliary information in probability sampling results in considerable reduction in the variance of the estimators of population parameters *viz.* population mean (or total), median, variance, regression coefficient, and population correlation coefficient, etc.. In this chapter we will consider the problem of estimation of different population parameters of interest to survey statisticians using known auxiliary information under SRSWOR and SRSWR sampling schemes only. Before proceeding further it is necessary to define some notation and expected values, which will be useful throughout this chapter.

3.1 NOTATION AND EXPECTED VALUES

Assume that a simple random sample (SRS) of size n is drawn from the given population of N units. Let the value of the study variable Y and the auxiliary variable X for the ith unit $(i = 1,2,...,N)$ of the population be denoted by Y_i and X_i and for the ith unit in the sample $(i = 1,2,...,n)$ by y_i and x_i, respectively. From the sample observations we have

$$\bar{y} = \frac{1}{n}\sum_{i=1}^{n} y_i\,,\quad \bar{x} = \frac{1}{n}\sum_{i=1}^{n} x_i\,,\quad s_y^2 = \frac{1}{(n-1)}\sum_{i=1}^{n}(y_i - \bar{y})^2\,,\quad s_x^2 = \frac{1}{(n-1)}\sum_{i=1}^{n}(x_i - \bar{x})^2,$$

and

$$s_{xy} = \frac{1}{(n-1)}\sum_{i=1}^{n}(y_i - \bar{y})(x_i - \bar{x}).$$

For the population observations we have the analogue quantities

$$\bar{Y} = \frac{1}{N}\sum_{i=1}^{N} Y_i\,,\quad \bar{X} = \frac{1}{N}\sum_{i=1}^{N} X_i\,,\quad S_y^2 = \frac{1}{(N-1)}\sum_{i=1}^{N}(Y_i - \bar{Y})^2\,,\quad S_x^2 = \frac{1}{(N-1)}\sum_{i=1}^{N}(X_i - \bar{X})^2,$$

and

$$S_{xy} = \frac{1}{(N-1)}\sum_{i=1}^{N}(Y_i - \bar{Y})(X_i - \bar{X}).$$

In general define the following population parameters

$$\mu_{rs} = \frac{1}{(N-1)}\sum_{i=1}^{N}(Y_i - \bar{Y})^r (X_i - \bar{X})^s\,,\quad \text{and}\quad \lambda_{rs} = \mu_{rs}\Big/\left(\mu_{20}^{r/2}\mu_{02}^{s/2}\right).$$

Note that

$$\mu_{20} = S_y^2\,,\quad \mu_{02} = S_x^2\,,\quad \text{and } \mu_{11} = S_{xy}\,,\quad \text{so that}\quad C_y^2 = S_y^2\big/\bar{Y}^2 = \mu_{20}\big/\bar{Y}^2,$$

$$C_x^2 = S_x^2\big/\bar{X}^2 = \mu_{02}\big/\bar{X}^2\,,\quad \text{and}\quad \rho_{xy} = S_{xy}\big/(S_x S_y) = \mu_{11}\big/\left(\sqrt{\mu_{20}}\sqrt{\mu_{02}}\right).$$

Let us define

$$\varepsilon_0 = \frac{\bar{y}}{\bar{Y}} - 1, \quad \varepsilon_1 = \frac{\bar{x}}{\bar{X}} - 1, \quad \varepsilon_2 = \frac{s_y^2}{S_y^2} - 1, \quad \varepsilon_3 = \frac{s_x^2}{S_x^2} - 1, \text{ and } \varepsilon_4 = \frac{s_{xy}}{S_{xy}} - 1$$

such that

$$E(\varepsilon_0) = E(\varepsilon_1) = E(\varepsilon_2) = E(\varepsilon_3) = E(\varepsilon_4) = 0$$

and

$$E\left(\varepsilon_0^2\right) = \left(\frac{1-f}{n}\right)C_y^2, \quad E\left(\varepsilon_1^2\right) = \left(\frac{1-f}{n}\right)C_x^2, \quad E(\varepsilon_0\varepsilon_1) = \left(\frac{1-f}{n}\right)\rho_{xy}C_yC_x .$$

To the first order of approximation we have

$$E\left(\varepsilon_2^2\right) = \left(\frac{1-f}{n}\right)(\lambda_{40} - 1), \quad E\left(\varepsilon_3^2\right) = \left(\frac{1-f}{n}\right)(\lambda_{04} - 1), \quad E\left(\varepsilon_4^2\right) = \left(\frac{1-f}{n}\right)\left(\frac{\lambda_{22}}{\rho_{xy}^2} - 1\right),$$

$$E(\varepsilon_0\varepsilon_2) = \left(\frac{1-f}{n}\right)C_y\lambda_{30}, \quad E(\varepsilon_0\varepsilon_3) = \left(\frac{1-f}{n}\right)C_y\lambda_{12}, \quad E(\varepsilon_0\varepsilon_4) = \left(\frac{1-f}{n}\right)C_y\frac{\lambda_{21}}{\rho_{xy}},$$

$$E(\varepsilon_1\varepsilon_2) = \left(\frac{1-f}{n}\right)C_x\lambda_{21}, \quad E(\varepsilon_1\varepsilon_3) = \left(\frac{1-f}{n}\right)C_x\lambda_{03}, \quad E(\varepsilon_1\varepsilon_4) = \left(\frac{1-f}{n}\right)C_x\frac{\lambda_{12}}{\rho_{xy}},$$

$$E(\varepsilon_2\varepsilon_3) = \left(\frac{1-f}{n}\right)(\lambda_{22} - 1), \quad E(\varepsilon_2\varepsilon_4) = \left(\frac{1-f}{n}\right)\left(\frac{\lambda_{31}}{\rho_{xy}} - 1\right), \text{ and } E(\varepsilon_3\varepsilon_4) = \left(\frac{1-f}{n}\right)\left(\frac{\lambda_{13}}{\rho_{xy}} - 1\right).$$

where $f = n/N$ denotes the finite population correction (f.p.c.) factor. These expected values can easily be obtained by following Sukhatme (1944), Sukhatme and Sukhatme (1970), Srivastava and Jhajj (1981), and Tracy (1984).
Also define

$$\hat{\mu}_{rs} = \frac{1}{(n-1)}\sum_{i=1}^{n}(y_i - \bar{y})^r(x_i - \bar{x})^s, \quad \hat{\lambda}_{rs} = \frac{\hat{\mu}_{rs}}{\hat{\mu}_{20}^{r/2}\hat{\mu}_{02}^{s/2}}, \text{ and } r_{xy} = \frac{s_{xy}}{s_x s_y} = \frac{\hat{\mu}_{11}}{\sqrt{\hat{\mu}_{20}}\sqrt{\hat{\mu}_{02}}}$$

as unbiased or consistent estimators of μ_{rs}, λ_{rs} and ρ_{xy} respectively.

The next section has been devoted to estimate the population mean in the presence of known auxiliary information.

3.2 ESTIMATION OF POPULATION MEAN

Several estimators of population mean are available in the literature and we will discuss some of them.

3.2.1 RATIO ESTIMATOR

Cochran (1940) was the first to show the contribution of known auxiliary information in improving the efficiency of the estimator of the population mean $\bar{Y}$ in survey sampling. Assuming that the population mean $\bar{X}$ of the auxiliary variable is known, he introduced a ratio estimator of population mean $\bar{Y}$ defined as

$$\bar{y}_R = \bar{y}\left(\frac{\bar{X}}{\bar{x}}\right). \tag{3.2.1.1}$$

Then we have the following theorems:

Theorem 3.2.1.1. The bias in the ratio estimator $\bar{y}_R$ of the population mean $\bar{Y}$, to the first order of approximation, is

$$B(\bar{y}_R) = \left(\frac{1-f}{n}\right)\bar{Y}\left[C_x^2 - \rho_{xy}C_xC_y\right]. \tag{3.2.1.2}$$

Proof. The estimator $\bar{y}_R$ in terms of ε_0 and ε_1, can easily be written as

$$\bar{y}_R = \bar{Y}(1+\varepsilon_0)(1+\varepsilon_1)^{-1}. \tag{3.2.1.3}$$

Assuming $|\varepsilon_1| < 1$ and using the binomial expansion of the term $(1+\varepsilon_1)^{-1}$ we have

$$\bar{y}_R = \bar{Y}(1+\varepsilon_0)\left(1-\varepsilon_1+\varepsilon_1^2+O(\varepsilon_1)\right) = \bar{Y}\left[1+\varepsilon_0-\varepsilon_1+\varepsilon_1^2-\varepsilon_0\varepsilon_1+O(\varepsilon_1)\right]. \tag{3.2.1.4}$$

where $O(\varepsilon_1)$ denotes the higher order terms of ε_1. Note that $|\varepsilon_1| < 1$, $\varepsilon_1^g \to 0$ as $g > 1$ increases. Therefore the terms in (3.2.1.4) with higher powers of ε_1 are negligible and can be ignored. Now taking expected values on both sides of (3.2.1.4) and using the results from section 3.1 we obtain

$$E(\bar{y}_R) = \bar{Y}\left[1+\left(\frac{1-f}{n}\right)\left\{C_x^2 - \rho_{xy}C_xC_y\right\}+O\left(n^{-1}\right)\right]. \tag{3.2.1.5}$$

Thus the bias in the estimator $\bar{y}_R$ to the first order of approximation is given by (3.2.1.2). Hence the theorem.

Theorem 3.2.1.2. The mean squared error of the ratio estimator $\bar{y}_R$ of the population mean $\bar{Y}$, to the first order of approximation, is given by

$$\text{MSE}(\bar{y}_R) = \left(\frac{1-f}{n}\right)\bar{Y}^2\left[C_y^2 + C_x^2 - 2\rho_{xy}C_yC_x\right]. \tag{3.2.1.6}$$

Proof. By the definition of mean squared error (MSE) and using (3.2.1.4) we have

$$\text{MSE}(\bar{y}_R) = E\left[\bar{y}_R - \bar{Y}\right]^2 \approx E\left[\bar{Y}\left(1+\varepsilon_0-\varepsilon_1+\varepsilon_1^2-\varepsilon_0\varepsilon_1+O\left(\varepsilon^2\right)\right)-\bar{Y}\right]^2$$

$$\approx \bar{Y}^2 E\left[\varepsilon_0-\varepsilon_1+\varepsilon_1^2-\varepsilon_0\varepsilon_1\right]^2.$$

Again neglecting higher order terms and using results from section 3.1 the MSE to the first order of approximation is given by

$$\text{MSE}(\bar{y}_R) = \bar{Y}^2 E\left[\varepsilon_0^2+\varepsilon_1^2-2\varepsilon_0\varepsilon_1\right] = \left(\frac{1-f}{n}\right)\bar{Y}^2\left[C_y^2+C_x^2-2\rho_{xy}C_xC_y\right].$$

Hence the theorem.

By substituting the values of C_y, C_x and ρ_{xy} in (3.2.1.6), one can easily see that the mean squared error of the estimator $\bar{y}_R$, to the first order of approximation, can be written as

$$\mathrm{MSE}(\bar{y}_R) = \left(\frac{1-f}{n}\right)\frac{1}{(N-1)}\sum_{i=1}^{N}\left[\left(Y_i - \bar{Y}\right) - R\left(X_i - \bar{X}\right)\right]^2 \tag{3.2.1.7}$$

where $R = \bar{Y}/\bar{X}$ is the ratio of the two population means.

Theorem 3.2.1.3. An estimator of the mean squared error of the ratio estimator $\bar{y}_R$, to the first order of approximation, is

$$\hat{\mathrm{MSE}}(\bar{y}_R) = \left(\frac{1-f}{n}\right)\left[s_y^2 + r^2 s_x^2 - 2r s_{xy}\right] \tag{3.2.1.8}$$

where $r = \bar{y}/\bar{x}$ denotes the estimator of ratio of two sample means.

Proof. Obvious by replacing the population parameters with the corresponding sample statistics in (3.2.1.6). Such a method of obtaining estimators is also called a method of moments.

Theorem 3.2.1.4. Another form of the estimator of the mean squared error of the ratio estimator $\bar{y}_R$, to the first order of approximation, is

$$\hat{\mathrm{MSE}}(\bar{y}_R) = \left(\frac{1-f}{n}\right)\frac{1}{(n-1)}\sum_{i=1}^{n}\left[\left(y_i - \bar{y}\right) - r\left(x_i - \bar{x}\right)\right]^2 \tag{3.2.1.9}$$

where $r = \bar{y}/\bar{x}$ is the ratio of two sample means.

Proof. Obvious by the method of moments.

Theorem 3.2.1.5. The ratio estimator $\bar{y}_R$ is more efficient than sample mean $\bar{y}$ if

$$\rho_{xy}\frac{C_y}{C_x} > \frac{1}{2}. \tag{3.2.1.10}$$

Proof. The proof follows from the fact that the ratio estimator $\bar{y}_R$ is more efficient than the sample mean $\bar{y}$ if

$$\mathrm{MSE}(\bar{y}_R) < V(\bar{y})$$

or if

$$\left(\frac{1-f}{n}\right)\bar{Y}^2\left[C_y^2 + C_x^2 - 2\rho_{xy}C_yC_x\right] < \left(\frac{1-f}{n}\right)\bar{Y}^2 C_y^2$$

or if

$$C_x^2 - 2\rho_{xy}C_yC_x < 0$$

or if

$$\rho_{xy} > \frac{1}{2}\frac{C_x}{C_y}.$$

Hence the theorem.

In the condition (3.2.1.10), if we assume that $C_y \approx C_x$, then it holds for all values of the correlation coefficient ρ_{xy} in the range $(0.5, 1.0]$. A Monte Carlo study of ratio estimator is available from Rao and Beegle (1967). Thus we have the following theorem.

Theorem 3.2.1.6. The ratio estimator $\bar{y}_R$ is more efficient than the sample mean $\bar{y}$ if $\rho_{xy} > 0.5$, *i.e.*, if the correlation between X and Y is positive and high.

Example 3.2.1.1. Mr. Bean was interested in estimating the average amount of real estate farm loans (in $000) during 1997 in the United States. He took an SRSWOR sample of eight states from the population 1 given in the Appendix. From the states selected in the sample he gathered the following information.

State	CA	GA	LA	MS	NM	PA	TX	VT
Nonreal estate farm loans (X) $	3928.732	540.696	405.799	549.551	274.035	298.351	3520.361	19.363
Real estate farm loans (Y) $	1343.461	939.460	282.565	627.013	140.582	756.169	1248.761	57.747

The average amount $878.16 of nonreal estate farm loans (in $000) for the year 1997 is known. Apply the ratio method of estimation for estimating the average amount of the real estate farm loans (in $000) during 1997. Also find an estimator of the mean squared error of the ratio estimator and hence deduce 95% confidence interval.

Solution. From the sample information, we have

Sr. No.	y_i	x_i	$(y_i-\bar{y})^2$	$(x_i-\bar{x})^2$	$(y_i-\bar{y})(x_i-\bar{x})$
1	1343.461	3928.732	447549.2926	7489094.4980	1830775.5040
2	939.460	540.696	70219.8326	424341.5022	-172618.6237
3	282.565	405.799	153589.3331	618286.5613	308159.4078
4	627.013	549.551	2252.1431	412883.3536	30493.8092
5	140.582	274.035	285036.1296	842863.5418	490149.5300
6	756.169	298.351	6674.7674	798806.9376	-73019.5217
7	1248.761	3520.361	329810.4398	5420748.0630	1337093.6030
8	57.747	19.363	380346.9504	1375337.8720	723260.3716
Sum	5395.758	9536.888	1675478.8890	17382362.3300	4474294.0800

Thus we have $n = 8$,

$$s_x^2 = \frac{\sum_{i=1}^{8}(x_i - \bar{x})^2}{8-1} = \frac{17382362.33}{7} = 2483194.6\,,\ \bar{x} = \frac{\sum_{i=1}^{8} x_i}{8} = \frac{9536.888}{8} = 1192.11\,,$$

$$s_y^2 = \frac{\sum_{i=1}^{8}(y_i - \bar{y})^2}{8-1} = \frac{1675478.89}{7} = 239354.1\,,\ \bar{y} = \frac{\sum_{i=1}^{8} y_i}{8} = \frac{5395.758}{8} = 674.469\,,$$

$$s_{xy} = \frac{\sum_{i=1}^{8}(y_i - \bar{y})(x_i - \bar{x})}{8-1} = \frac{4474294.08}{7} = 639184.86\,,\ \text{and}\ r = \frac{\bar{y}}{\bar{x}} = \frac{674.469}{1192.11} = 0.5658\,.$$

We are given $\bar{X} = 878.16$, $N = 50$ and $f = 0.16$.

Thus the ratio estimate of average amount of real estate farm loans during 1997, $\bar{Y}$ (say), is given by

$$\bar{y}_{\mathrm{R}} = \bar{y}\left(\frac{\bar{X}}{\bar{x}}\right) = 674.469\left(\frac{878.162}{1192.11}\right) = 496.86$$

and an estimate of $\mathrm{MSE}(\bar{y}_{\mathrm{R}})$ is given by

$$\mathrm{M\hat{S}E}(\bar{y}_{\mathrm{R}}) = \left(\frac{1-f}{n}\right)\left[s_y^2 + r^2 s_x^2 - 2rs_{xy}\right]$$

$$= \left(\frac{1-0.16}{8}\right)\left[239354.1 + (0.5658)^2 \times 2483194.6 - 2 \times 0.5658 \times 639184.86\right]$$

$$= 32654.65\,.$$

A $(1-\alpha)100\%$ confidence interval for population mean $\bar{Y}$ is given by

$$\bar{y}_{\mathrm{R}} \pm t_{\alpha/2}(\mathrm{df} = n-1)\sqrt{\mathrm{M\hat{S}E}(\bar{y}_{\mathrm{R}})}\,.$$

Using the Table 2 given in the Appendix the 95% confidence interval is given by

$$496.86 \pm 2.365\sqrt{32654.65}\ \text{or}\ [69.490,\quad 924.229].$$

Example 3.2.1.2. After applying the ratio method of estimation, Mr. Bean wants to know if he achieved any gain in efficiency by using the ratio estimator. The amount of real and nonreal estate farm loans (in $000) during 1997 in 50 different states of the United States has been presented in population 1 of the Appendix. Find the relative efficiency of the ratio estimator, for estimating the average amount of real estate farm loans during 1997 by using known information on nonreal estate farm loans during 1997, with respect to the usual estimator of population mean, given the sample size is of eight units.

Solution. From the description of the population, we have $Y_i =$ Amount (in $000) of real estate farm loans in different states during 1997, $X_i =$ Amount (in $000) of nonreal estate farm loans in different states during 1997, $\overline{Y} = 555.43$, $\overline{X} = 878.16$, $S_y^2 = 342021.5$, $C_x^2 = 1.5256$, $C_y^2 = 1.1086$, $\rho_{xy} = 0.8038$, and $N = 50$.

Thus we have

$$\text{MSE}(\bar{y}_\text{R}) = \left(\frac{1-f}{n}\right)\overline{Y}^2\left[C_y^2 + C_x^2 - 2\rho_{xy}C_xC_y\right]$$

$$= \left(\frac{1-0.16}{8}\right)(555.43)^2\left[1.1086 + 1.5256 - 2\times 0.8038\sqrt{1.1086\times 1.5256}\right]$$

$$= 17606.39\,.$$

Also

$$V(\bar{y}) = \left(\frac{1-f}{n}\right)S_y^2 = \left(\frac{1-0.16}{8}\right)\times 342021.5 = 35912.26\,.$$

Thus the percent relative efficiency (RE) of the ratio estimator $\bar{y}_\text{R}$ with respect to the usual estimator $\bar{y}$ is given by

$$\text{RE} = V(\bar{y})\times 100/\text{MSE}(\bar{y}_\text{R}) = \frac{35912.26}{17606.39}\times 100 = 203.97\%$$

which shows that the ratio estimator is more efficient than the usual estimator of population mean. It should be noted that the relative efficiency does not depend upon the sample size.

Theorem 3.2.1.7. The minimum sample size for the relative standard error (RSE) to be less than or equal to a fixed value ϕ is given by

$$n \ge \left[\frac{\phi^2\overline{Y}^2}{S_y^2 + R^2S_x^2 - 2RS_{xy}} + \frac{1}{N}\right]^{-1}. \tag{3.2.1.11}$$

Proof. The relative standard error of the ratio estimator $\bar{y}_\text{R}$ is

$$\text{RSE}(\bar{y}_\text{R}) = \sqrt{V(\bar{y}_\text{R})/\overline{Y}^2} = \sqrt{\left(\frac{1}{n}-\frac{1}{N}\right)\left(C_y^2 + C_x^2 - 2\rho_{xy}C_xC_y\right)}.$$

Now

$$\text{RSE}(\bar{y}_\text{R}) \le \phi \;\text{ if }\; \sqrt{\left(\frac{1}{n}-\frac{1}{N}\right)\left(C_y^2 + C_x^2 - 2\rho_{xy}C_xC_y\right)} \le \phi\,,$$

squaring on both sides we obtain

$$\left(\frac{1}{n}-\frac{1}{N}\right)\left(C_y^2 + C_x^2 - 2\rho_{xy}C_xC_y\right) \le \phi^2$$

$$\text{or}\quad \left(\frac{1}{n}-\frac{1}{N}\right) \le \frac{\phi^2}{C_y^2 + C_x^2 - 2\rho_{xy}C_xC_y} \qquad \text{or}\quad \frac{1}{n} \le \frac{\phi^2}{C_y^2 + C_x^2 - 2\rho_{xy}C_xC_y} + \frac{1}{N}$$

$$\text{or}\quad n > \left[\frac{\phi^2}{C_y^2 + C_x^2 - 2\rho_{xy}C_xC_y} + \frac{1}{N}\right]^{-1} \qquad \text{or}\quad n \ge \left[\frac{\phi^2\overline{Y}^2}{S_y^2 + R^2S_x^2 - 2RS_{xy}} + \frac{1}{N}\right]^{-1}.$$

Hence the theorem.

Example 3.2.1.3. Mr. Bean wishes to estimate the average real estate farm loans in the United States with the help of ratio method of estimation by using known information about the nonreal estate farm loans as shown in population 1 in the Appendix. What is the minimum sample size required for the relative standard error (RSE) to be equal to 12.5%?

Solution. From the description of the population 1 given in the Appendix, we have $N = 50$, $\overline{Y} = 555.43$, $\overline{X} = 878.16$, $S_y^2 = 342021.5$, $S_x^2 = 1176526$, $S_{xy} = 509910.41$, $R = \overline{Y}/\overline{X} = \frac{555.43}{878.16} = 0.63249$, $\phi = 0.125$, thus the minimum sample size is

$$n \ge \left[\frac{\phi^2 \overline{Y}^2}{S_y^2 + R^2 S_x^2 - 2RS_{xy}} + \frac{1}{N}\right]^{-1}$$

$$= \left[\frac{0.125^2 \times (555.43)^2}{342021.5 + 0.63249^2 \times 1176526 - 2 \times 0.63249 \times 509910.41} + \frac{1}{50}\right]^{-1} = 20.51 \approx 21.$$

Example 3.2.1.4. Mr. Bean selected an SRSWOR sample of 21 states listed in population 1. He collected information about the real estate farm loans and nonreal estate farm loans from the selected states. He applied the ratio method of estimation for estimating the average real estate farm loans assuming that the average nonreal estate farm loans in the United States is known and is equal to \$878.16. Discuss his 95% confidence interval.

Solution. Note that the population size is 50. Mr. Bean started with the first two columns of the Pseudo-Random Numbers (PRN) given in Table 1 of the Appendix and selected the following 21 distinct random numbers between 1 and 50 as: 01, 23, 46, 04, 32, 47, 33, 05, 22, 38, 29, 40, 03, 36, 27, 19, 14, 42, 48, 06, and 07.

Random No.	State	x_i	y_i	$(x_i - \overline{x})^2$	$(y_i - \overline{y})^2$	$(y_i - \overline{y})(x_i - \overline{x})$
01	AL	348.334	408.978	303627.6	21302.6	80424.302
03	AZ	431.439	54.633	218948.3	250299.3	234099.570
04	AR	848.317	907.700	2605.2	124445.1	-18005.653
05	CA	3928.732	1343.461	9177106.0	621777.6	2388748.500
06	CO	906.281	315.809	47.9	57179.9	-1655.427
07	CT	4.373	7.130	800998.3	300087.3	490274.840
14	IN	1022.782	1213.024	15233.5	433084.8	81224.255
19	ME	51.539	8.849	718797.2	298206.9	462979.800
22	MI	440.518	323.028	210534.2	53779.6	106406.960
23	MN	2466.892	1354.768	2457163.0	639737.2	1253769.700
27	NE	3585.406	1337.852	7214853.0	612963.4	2102960.000
29	NH	0.471	6.044	807998.0	301278.3	493388.550

Continued......

32	NY	426.274	201.631	223808.6	124821.8	167141.200
33	NC	494.730	639.571	163723.9	7163.7	-34247.221
36	OK	1716.087	612.108	667046.1	3269.1	46697.097
38	PA	298.351	756.169	361209.5	40496.2	-120944.720
40	SC	80.750	87.951	670119.2	218071.5	382274.620
42	TN	388.869	553.266	260599.1	2.8	850.596
46	VA	188.477	321.583	505351.9	54451.9	165883.560
47	WA	1228.607	1100.745	108404.8	297911.6	179708.250
48	WV	29.291	99.277	757016.8	207621.7	396450.630
	Sum	18886.520	11653.577	25645192.0	4667952.0	8858429.300

x – Nonreal estate farm loans, y – Real estate farm loans.

Given $N = 50$ and $\overline{X} = 878.16$. Now from the above table, $n = 21$, $\overline{y} = 554.93223$, $\overline{x} = 899.35809$, $s_x^2 = 1282260$, $s_y^2 = 233397.6$, $s_{xy} = 442921.47$, $r = 0.617$, and $f = 0.42$.

Thus ratio estimate of the average real estate farm loans in the United States is

$$\overline{y}_R = \overline{y}\left(\frac{\overline{X}}{\overline{x}}\right) = 554.93223\left(\frac{878.16}{899.35809}\right) = \$541.85\,.$$

An estimate of $\mathrm{MSE}(\overline{y}_R)$ is given by

$$\hat{\mathrm{MSE}}(\overline{y}_R) = \left(\frac{1-f}{n}\right)\left[s_y^2 + r^2 s_x^2 - 2rs_{xy}\right]$$

$$= \left(\frac{1-0.42}{21}\right)\left[233397.6 + (0.617)^2 \times 1282260 - 2\times 0.617 \times 442921.47\right]$$

$$= 4832.64\,.$$

A $(1-\alpha)100\%$ confidence interval for population mean $\overline{Y}$ is given by

$$\overline{y}_R \pm t_{\alpha/2}(\mathrm{df} = n-1)\sqrt{\hat{\mathrm{MSE}}(\overline{y}_R)}\,.$$

Using Table 2 from the Appendix the 95% confidence interval for the average real estate farm loans is given by

$$541.85 \pm 2.086\sqrt{4832.64}, \text{ or } [396.84, \quad 686.86].$$

3.2.2 PRODUCT ESTIMATOR

Murthy (1964) considered another estimator of population mean $\overline{Y}$ using known population mean $\overline{X}$ of the auxiliary variable as a product estimator

$$\overline{y}_P = \overline{y}\left(\frac{\overline{x}}{\overline{X}}\right). \tag{3.2.2.1}$$

Then we have the following theorems:

Theorem 3.2.2.1. The exact bias in the product estimator $\bar{y}_P$ of the population mean $\bar{Y}$ is given by

$$B(\bar{y}_P)=\left(\frac{1-f}{n}\right)\bar{Y}\rho_{xy}C_yC_x. \tag{3.2.2.2}$$

Proof. The product estimator $\bar{y}_P$ in terms of ε_0 and ε_1 can easily be written as

$$\bar{y}_P=\bar{Y}(1+\varepsilon_0)(1+\varepsilon_1)=\bar{Y}(1+\varepsilon_0+\varepsilon_1+\varepsilon_0\varepsilon_1). \tag{3.2.2.3}$$

Taking expected values on both sides of (3.2.2.3) and using the results from section 3.1 we have

$$E(\bar{y}_P)=\bar{Y}\left[1+\left(\frac{1-f}{n}\right)\rho_{xy}C_xC_y\right]. \tag{3.2.2.4}$$

Thus the bias in the product estimator $\bar{y}_P$ of the population mean is given by

$$B(\bar{y}_P)=E(\bar{y}_P)-\bar{Y}=\left(\frac{1-f}{n}\right)\bar{Y}\rho_{xy}C_xC_y.$$

Hence the theorem.

Theorem 3.2.2.2. The mean squared error of the product estimator $\bar{y}_P$, to the first order of approximation, is given by

$$\text{MSE}(\bar{y}_P)=\left(\frac{1-f}{n}\right)\bar{Y}^2\left[C_y^2+C_x^2+2\rho_{xy}C_yC_x\right]. \tag{3.2.2.5}$$

Proof. By the definition of mean squared error (MSE) using (3.2.2.3) and again neglecting higher order terms and using results from section 3.1 we have

$$\text{MSE}(\bar{y}_P)=E\left[\bar{y}_P-\bar{Y}\right]^2=E\left[\bar{Y}(1+\varepsilon_0+\varepsilon_1+\varepsilon_0\varepsilon_1)-\bar{Y}\right]^2=\bar{Y}^2E\left[\varepsilon_0+\varepsilon_1+\varepsilon_0\varepsilon_1\right]^2.$$

Thus the MSE, to the first order of approximation, is given by

$$\text{MSE}(\bar{y}_P)=\bar{Y}^2E\left[\varepsilon_0^2+\varepsilon_1^2+2\varepsilon_0\varepsilon_1\right]=\left(\frac{1-f}{n}\right)\bar{Y}^2\left[C_y^2+C_x^2+2\rho_{xy}C_xC_y\right].$$

Hence the theorem.

Theorem 3.2.2.3. An estimator of the MSE of the product estimator $\bar{y}_P$, to the first order of approximation, is given by

$$\hat{\text{MSE}}(\bar{y}_P)=\left(\frac{1-f}{n}\right)\left[s_y^2+r^2s_x^2+2rs_{xy}\right]. \tag{3.2.2.6}$$

Proof. It follows by the method of moments.

Theorem 3.2.2.4. The product estimator $\bar{y}_P$ is more efficient than sample mean $\bar{y}$ if

$$\rho_{xy}\frac{C_y}{C_x}<-\frac{1}{2}. \tag{3.2.2.7}$$

Proof. The proof follows from the fact that the product estimator $\bar{y}_P$ is more efficient than the sample mean $\bar{y}$ if

$$\text{MSE}(\bar{y}_P) < V(\bar{y})$$

or if

$$\left(\frac{1-f}{n}\right)\bar{Y}^2\left[C_y^2 + C_x^2 + 2\rho_{xy}C_xC_y\right] < \left(\frac{1-f}{n}\right)\bar{Y}^2 C_y^2$$

or if

$$C_x^2 + 2\rho_{xy}C_xC_y < 0$$

or if

$$\rho_{xy}\frac{C_y}{C_x} < -\frac{1}{2}.$$

Hence the theorem.

In the condition (3.2.2.7), if we assume that $C_y \approx C_x$, then it holds for all values of the correlation coefficient ρ_{xy} in the range $[-1.0, -0.5)$. Thus we have the following theorem.

Theorem 3.2.2.5. The product estimator $\bar{y}_P$ is more efficient than the sample mean $\bar{y}$ if $\rho_{xy} < -0.5$, *i.e.*, if the correlation between X and Y is negative and high.

Remark 3.2.2.1. We observed that the product and ratio estimators are better than sample mean if the value of ρ_{xy} lies in the interval $[-1.0, -0.5)$ and $(+0.5, +1.0]$, respectively. Thus the sample mean estimator remains better than both the ratio and product estimators of the population mean if ρ_{xy} lies in the range $[-0.5, +0.5]$.

Example 3.2.2.1. A psychologist would like to estimate the average duration of sleep (in minutes) during the night for persons 50 years of age and older in a small village in the United States. It is known that there are 30 persons living in the village aged 50 and over. Instead of asking everybody the psychologist selects an SRSWOR sample of six persons of this age group and records the information as given below

Person no.	3	7	10	17	22	29
Age x (years)	55	67	56	78	71	66
Duration of sleep y (in minutes)	408	420	456	345	360	390

Assume that the average age 67.267 years of the subjects is known as shown in the population 2 in the Appendix. Assuming that as the age of a person increases then the sleeping hours decrease, apply the product method of estimation for estimating the average sleep time in the particular village under study. Also find an estimator of the mean squared error of the product estimator and deduce a 95% confidence interval.

Solution. From the sample information, we have

Sr. No.	y_i	x_i	$(y_i-\bar{y})^2$	$(x_i-\bar{x})^2$	$(y_i-\bar{y})(x_i-\bar{x})$
1	408	55	132.25	110.25	-120.75
2	420	67	552.25	2.25	35.25
3	456	56	3540.25	90.25	-565.25
4	345	78	2652.25	156.25	-643.75
5	360	71	1332.25	30.25	-200.75
6	390	66	42.25	0.25	-3.25
Sum	2379	393	8251.50	389.50	-1498.50

Here Y_i = Duration of sleep (in minutes), X_i = Age of subjects $(\geq 50 \text{ years})$, $n=6$, $\bar{y}=396.5$, $\bar{x}=65.5$, $s_x^2=77.9$, $s_y^2=1650.3$, $s_{xy}=-299.7$, and $r=\bar{y}/\bar{x}=6.053$. Also we are given $\bar{X}=67.267$, $N=30$ and $f=0.20$.

Thus product estimate of the average sleep time, $\bar{Y}$ (say), is given by

$$\bar{y}_P=\bar{y}\left(\frac{\bar{x}}{\bar{X}}\right)=396.5\left(\frac{65.5}{67.267}\right)=386.08,$$

and an estimate of $\text{MSE}(\bar{y}_P)$ is given by

$$\hat{\text{MSE}}(\bar{y}_P)=\left(\frac{1-f}{n}\right)\left[s_y^2+r^2s_x^2+2rs_{xy}\right]$$

$$=\left(\frac{1-0.20}{6}\right)\left[1650.3+(6.053)^2\times 77.9-2\times 6.053\times 299.7\right]=116.83\,.$$

A $(1-\alpha)100\%$ confidence interval for population mean $\bar{Y}$ is given by

$$\bar{y}_P \pm t_{\alpha/2}(\text{df}=n-2)\sqrt{\hat{\text{MSE}}(\bar{y}_P)}\,.$$

Using Table 2 from the Appendix the 95% confidence interval for average sleeping time is given by

$$386.08\pm 2.776\sqrt{116.83}\,, \quad \text{or} \quad [356.07,\ 416.08].$$

Example 3.2.2.2. The duration of sleep (in minutes) and age of 30 people aged 50 and over living in a small village of the United States is given in the population 2. Suppose a psychologist selected an SRSWOR sample of six individuals to collect the required information. Find the relative efficiency of the product estimator, for estimating average duration of sleep using age as an auxiliary variable, with respect to the usual estimator of population mean.

Solution. Using the description of the population 2 given in the Appendix we have Y_i = Duration of sleep (in minutes), X_i = Age of subjects $(\geq 50 \text{ years})$, $N=30$, $X=2018$, $Y=11526$, $\bar{X}=67.267$, $\bar{Y}=384.2$, $S_y^2=3582.58$, $S_x^2=85.237$, $C_y^2=0.0243$, $C_x^2=0.0188$, $S_{xy}=-472.607$, and $\rho_{xy}=-0.8552$.

Thus we have

$$\text{MSE}(\bar{y}_P) = \left(\frac{1-f}{n}\right)\bar{Y}^2\left[C_y^2 + C_x^2 + 2\rho_{xy}C_xC_y\right]$$

$$= \left(\frac{1-0.20}{6}\right)(384.2)^2\left[0.0243 + 0.0188 - 2\times 0.8552\sqrt{0.0243\times 0.0188}\right]$$

$$= 128.759.$$

Also

$$V(\bar{y}) = \left(\frac{1-f}{n}\right)S_y^2 = \left(\frac{1-0.20}{6}\right)\times 3582.58 = 477.677\,.$$

Thus the percent relative efficiency (RE) of the product estimator $\bar{y}_P$ with respect to the usual estimator $\bar{y}$ is given by

$$\text{RE} = V(\bar{y})\times 100/\text{MSE}(\bar{y}_P) = \frac{477.677}{128.759}\times 100 = 370.98\%$$

which shows that the product estimator is more efficient than the usual estimator of population mean. The relative efficiency is independent of sample size n.

Corollary 3.2.2.1. The minimum sample size for the relative standard error (RSE) to be less than or equal to a fixed value ϕ is given by

$$n \geq \left[\frac{\phi^2\bar{Y}^2}{S_y^2 + R^2S_x^2 + 2RS_{xy}} + \frac{1}{N}\right]^{-1}. \tag{3.2.2.8}$$

3.2.3 REGRESSION ESTIMATOR

We consider an estimator of the population mean $\bar{Y}$ as

$$\bar{y}_{\text{dif}} = \bar{y} + d(\bar{X} - \bar{x}) \tag{3.2.3.1}$$

where d is a constant to be chosen such that the variance of the estimator $V(\bar{y}_{\text{dif}})$ is minimum. Such an estimator is called difference estimator. The estimator $\bar{y}_{\text{dif}}$ can be written as

$$\bar{y}_{\text{dif}} = \bar{Y}(1+\varepsilon_0) + d\left[\bar{X} - \bar{X}(1+\varepsilon_1)\right] = \bar{Y}(1+\varepsilon_0) - d\bar{X}\varepsilon_1\,. \tag{3.2.3.2}$$

Taking expected value on both sides of (3.2.3.2), we obtain

$$E(\bar{y}_{\text{dif}}) = \bar{Y}\,. \tag{3.2.3.3}$$

Thus the difference estimator $\bar{y}_{\text{dif}}$ is unbiased for the population mean, $\bar{Y}$. The variance of the estimator $\bar{y}_{\text{dif}}$ is given by

$$V(\bar{y}_{\text{dif}}) = E\left[\bar{y}_{\text{dif}} - \bar{Y}\right]^2 = E\left[\bar{Y}(1+\varepsilon_0) - d\bar{X}\varepsilon_1 - \bar{Y}\right]^2 = E\left[\bar{Y}\varepsilon_0 - d\bar{X}\varepsilon_1\right]^2$$

$$= E\left[\ \bar{Y}^2\varepsilon_0^2 + d^2\bar{X}^2\varepsilon_1^2 - 2d\bar{Y}\ \bar{X}\varepsilon_0\varepsilon_1\ \right]$$

$$= \frac{(1-f)}{n}\left[\bar{Y}^2C_y^2 + d^2\bar{X}^2C_x^2 - 2d\bar{Y}\ \bar{X}\rho_{xy}C_xC_y\right]. \tag{3.2.3.4}$$

On differentiating (3.2.3.4) with respect to d and equation to zero we obtain

$$d = \rho_{xy}\frac{C_y}{C_x}\frac{\overline{Y}}{\overline{X}} = \frac{S_{xy}}{S_x^2}. \tag{3.2.3.5}$$

On substituting optimum value of d in (3.2.3.4), the minimum variance of the estimator $\overline{y}_{\text{dif}}$ is given by

$$\text{Min.}V(\overline{y}_{\text{dif}}) = \frac{(1-f)}{n}\left[\overline{X}^2C_y^2 + \left(\rho_{xy}\frac{\overline{Y}}{\overline{X}}\frac{C_y}{C_x}\right)^2\overline{X}^2C_x^2 - 2\rho_{xy}\frac{\overline{Y}}{\overline{X}}\frac{C_y}{C_x}\overline{X}\,\overline{Y}\rho_{xy}C_xC_y\right]$$

$$= \frac{(1-f)}{n}\left(S_y^2 - \frac{S_{xy}^2}{S_x^2}\right) = \frac{(1-f)}{n}S_y^2\left(1 - \frac{S_{xy}^2}{S_y^2S_x^2}\right) = \frac{(1-f)}{n}S_y^2\left(1-\rho_{xy}^2\right). \tag{3.2.3.6}$$

For the optimum value of $d = \rho_{xy}\dfrac{C_y}{C_x}\dfrac{\overline{Y}}{\overline{X}} = \dfrac{S_{xy}}{S_x^2} = \beta$ (regression coefficient, say) the difference estimator becomes

$$\overline{y}_{\text{dif}} = \overline{y} + \left(\frac{S_{xy}}{S_x^2}\right)\left(\overline{X} - \overline{x}\right). \tag{3.2.3.7}$$

Thus the difference estimator becomes non-functional if the value of the regression coefficient $\beta = S_{xy}/S_x^2$ is unknown. In such situations, Hansen, Hurwitz, and Madow (1953) consider the linear regression estimator of the population mean $\overline{Y}$ as

$$\overline{y}_{\text{LR}} = \overline{y} + \hat{\beta}\left(\overline{X} - \overline{x}\right), \tag{3.2.3.8}$$

where $\hat{\beta} = s_{xy}/s_x^2$ denotes the estimator of the regression coefficient $\beta = S_{xy}/S_x^2$. Then we have the following theorems:

Theorem 3.2.3.1. The bias in the linear regression estimator $\overline{y}_{\text{LR}}$ of population mean $\overline{Y}$ is given by

$$B(\overline{y}_{\text{LR}}) = \left(\frac{1-f}{n}\right)\beta\overline{X}C_x\left(\lambda_{03} - \frac{\lambda_{12}}{\rho_{xy}}\right). \tag{3.2.3.9}$$

Proof. The linear regression estimator $\overline{y}_{\text{LR}}$, in terms of ε_0, ε_1, ε_3 and ε_4, can easily be written as

$$\overline{y}_{\text{LR}} = \overline{Y}(1+\varepsilon_0) + \frac{S_{xy}(1+\varepsilon_4)}{S_x^2(1+\varepsilon_3)}\left[\overline{X} - \overline{X}(1+\varepsilon_1)\right]$$

$$= \overline{Y}(1+\varepsilon_0) + \beta(1+\varepsilon_4)(1+\varepsilon_3)^{-1}\left[\overline{X} - \overline{X}(1+\varepsilon_1)\right].$$

Using the binomial expansion $(1+\varepsilon_3)^{-1} = 1 - \varepsilon_3 + \varepsilon_3^2 + O(\varepsilon_3)$ we obtain

$$\overline{y}_{\text{LR}} = \overline{Y}(1+\varepsilon_0) - \beta\overline{X}\left[\varepsilon_1 + \varepsilon_1\varepsilon_4 - \varepsilon_1\varepsilon_3 + O(\varepsilon)\right]. \tag{3.2.3.10}$$

Taking expected value on both sides of (3.2.3.10) and neglecting higher order terms, we obtain

$$E(\bar{y}_{LR}) = \bar{Y} - \beta\bar{X}\left(\frac{1-f}{n}\right)\left(C_x\frac{\lambda_{12}}{\rho_{xy}} - C_x\lambda_{03}\right) = \bar{Y} + \left(\frac{1-f}{n}\right)\beta\bar{X}C_x\left(\lambda_{03} - \frac{\lambda_{12}}{\rho_{xy}}\right).$$

Thus the bias is given by

$$B(\bar{y}_{LR}) = E(\bar{y}_{LR}) - \bar{Y} = \left(\frac{1-f}{n}\right)\beta\bar{X}C_x\left(\lambda_{03} - \frac{\lambda_{12}}{\rho_{xy}}\right).$$

Hence the theorem.

Theorem 3.2.3.2. The mean squared error of the linear regression estimator $\bar{y}_{LR}$, to the first order of approximation, is

$$\text{MSE}(\bar{y}_{LR}) = \left(\frac{1-f}{n}\right)S_y^2\left(1 - \rho_{xy}{}^2\right). \tag{3.2.3.11}$$

Proof. By the definition of mean squared error (MSE) and using (3.2.3.10) and neglecting higher order terms we have

$$\text{MSE}(\bar{y}_{LR}) = E\left[\bar{y}_{LR} - \bar{Y}\right]^2 = E\left[\bar{Y}(1+\varepsilon_0) - \beta\bar{X}\{\varepsilon_1 + \varepsilon_1\varepsilon_4 - \varepsilon_1\varepsilon_3 + O(\varepsilon_i)\} - \bar{Y}\right]^2$$

$$= E\left[\bar{Y}\varepsilon_0^2 + \beta^2\bar{X}^2\varepsilon_1^2 - 2\beta\bar{Y}\bar{X}\varepsilon_0\varepsilon_1\right].$$

Thus the MSE, to the first order of approximation, is given by

$$\text{MSE}(\bar{y}_{LR}) = \left(\frac{1-f}{n}\right)\left[\bar{Y}^2C_y^2 + \beta^2\bar{X}^2C_x^2 - 2\beta\bar{Y}\bar{X}\rho_{xy}C_xC_y\right]$$

$$= \left(\frac{1-f}{n}\right)\left[\bar{Y}^2\frac{S_y^2}{\bar{Y}^2} + \left(\frac{S_{xy}}{S_x^2}\right)^2\bar{X}^2\frac{S_x^2}{\bar{X}^2} - 2\frac{S_{xy}}{S_{xy}^2}\bar{Y}\,\bar{X}\frac{S_{xy}}{S_xS_y}\frac{S_x}{\bar{X}}\frac{S_y}{\bar{Y}}\right]$$

$$= \left(\frac{1-f}{n}\right)\left[S_y^2 + \frac{S_{xy}^2}{S_x^2} - 2\frac{S_{xy}^2}{S_x^2}\right] = \left(\frac{1-f}{n}\right)\left[S_y^2 - \frac{S_{xy}^2}{S_x^2}\right]$$

$$= \left(\frac{1-f}{n}\right)S_y^2\left(1 - \rho_{xy}^2\right).$$

Hence the theorem.

Theorem 3.2.3.3. An estimator of the mean squared error of the linear regression estimator $\bar{y}_{LR}$, to the first order of approximation, is given by

$$\hat{\text{MSE}}(\bar{y}_{LR}) = \left(\frac{1-f}{n}\right)s_y^2\left[1 - r_{xy}^2\right]. \tag{3.2.3.12}$$

Proof. It follows by the method of moments.

Theorem 3.2.3.4. The linear regression estimator $\bar{y}_{LR}$ is always more efficient than the sample mean $\bar{y}$ if $\rho_{xy} \neq 0$.

Remark 3.2.3.1. If $\hat{\beta} = \bar{y}/\bar{x}$ then the linear regression estimator $\bar{y}_{\text{LR}}$ reduces to the usual ratio estimator $\bar{y}_{\text{R}}$ and if $\hat{\beta} = -\bar{y}/\bar{X}$, then the linear regression estimator $\bar{y}_{\text{LR}}$ reduces to the usual product estimator $\bar{y}_{\text{P}}$.

Example 3.2.3.1. A bank manager in the United States of America is interested in estimating the average amount of real estate farm loans (in $000) during 1997. A statistician took an SRSWOR sample of eight states from the population 1 as given in the Appendix and collected the following information.

State	AZ	CO	DE	LA	MT	NC	VT	WA
Nonreal estate farm loans (X) $	431.439	906.281	43.229	405.799	722.034	494.730	19.363	1228.607
Real estate farm loans (Y) $	54.633	315.809	42.808	282.565	292.965	639.571	57.747	1100.745

Apply the regression method of estimation for estimating the average amount of the real estate farm loans (in 000) during 1997. Also find an estimator of the mean squared error of the regression estimator and deduce a 95% confidence interval. Assume that the average amount $878.16 of nonreal estate farm loans (in $000) for the year 1997 is known.

Solution. From the sample information, we have

Sr. No.	y_i	x_i	$(y_i - \bar{y})^2$	$(x_i - \bar{x})^2$	$(y_i - \bar{y})(x_i - \bar{x})$
1	54.633	431.439	86272.833580	9999.250014	29371.136040
2	315.809	906.281	1059.266526	140509.336300	-12199.870350
3	42.808	43.229	93359.198370	238345.342500	149170.138100
4	282.565	405.799	4328.373443	15784.467310	8265.656001
5	292.965	722.034	3068.093643	36327.883500	-10557.336240
6	639.571	494.730	84806.540240	1347.275378	-10689.142320
7	57.747	19.363	84453.227620	262217.989200	148812.484500
8	1100.745	1228.607	566090.147800	486048.449000	524544.791500
Sum	2786.843	4251.482	923437.681200	1190579.993000	826717.857300

Here $n = 8$, $\bar{y} = 348.3554$, $\bar{x} = 531.4353$, $s_x^2 = 170082.85$, $s_y^2 = 131919.67$, $s_{xy} = 118102.55$, $\hat{\beta} = s_{xy}/s_x^2 = 0.6943$ and $r_{xy} = s_{xy}/(s_x s_y) = 0.7884$. Also we are given $\bar{X} = 878.16$, $N = 50$ and $f = 0.16$.

Thus the regression estimate of average amount of real estate farm loans during 1997, $\bar{Y}$ (for example), is given by

$$\bar{y}_{\text{LR}} = \bar{y} + \hat{\beta}(\bar{X} - \bar{x}) = 348.3554 + 0.6943 \times (878.16 - 531.4353) = 589.08$$

and an estimator of $\operatorname{MSE}(\bar{y}_{LR})$ is given by

$$\hat{\operatorname{MSE}}(\bar{y}_{LR}) = \left(\frac{1-f}{n}\right)s_y^2\left[1 - r_{xy}^2\right] = \left(\frac{1-0.16}{8}\right)\times 131919.67\times\left[1-(0.7884)^2\right] = 5241.78 .$$

A $(1-\alpha)100\%$ confidence interval for population mean $\bar{Y}$ is given by

$$\bar{y}_{LR} \pm t_{\alpha/2}(\text{df} = n-2)\sqrt{\hat{\operatorname{MSE}}(\bar{y}_{LR})} .$$

Using Table 2 from the Appendix the 95% confidence interval is given by

$$589.08 \pm 2.447\sqrt{5241.78} \quad \text{or} \quad [411.916,\ 766.24].$$

Example 3.2.3.2. Suppose a bank manager selects an SRSWOR sample of eight states to collect the required information on real estate and nonreal estate farm loans during 1997. Find the relative efficiency of the regression estimator, for estimating the average amount of real estate farm loans during 1997 by using data on nonreal estate farm loans during 1997 with respect to the ratio estimator of population mean. The amounts of real and nonreal estate farm loans (in $000) during 1997 in the 50 states of the United States have been given in population 1 of the Appendix.

Solution. From the description of the population 1 given the Appendix we have $\bar{Y} = 555.43$, $\bar{X} = 878.16$, $S_y^2 = 342021.5$, $\rho_{xy} = 0.8038$, and $N = 50$.

Also from example 3.2.1.2 we have

$$\operatorname{MSE}(\bar{y}_R) = 17606.39 .$$

Now

$$\operatorname{MSE}(\bar{y}_{LR}) = \left(\frac{1-f}{n}\right)S_y^2\left(1-\rho_{xy}^2\right) = \left(\frac{1-0.16}{8}\right)\times 342021.5\times\left[1-(0.8038)^2\right] = 12709.55 .$$

Thus the percent of relative efficiency (RE) of the regression estimator $\bar{y}_{LR}$ with respect to the ratio estimator $\bar{y}_R$ is given by

$$\text{RE} = V(\bar{y}_R)\times 100/\operatorname{MSE}(\bar{y}_{LR})\times 100 = \frac{17606.39}{12709.55}\times 100 = 138.53\%$$

which shows that the regression estimator is more efficient than the ratio estimator of population mean. The relative efficiency is independent of sample size, n.

Corollary 3.2.3.1. The minimum sample size for the relative standard error (RSE) of the linear regression estimator to be less than or equal to a fixed value ϕ is

$$n \ge \left[\frac{\phi^2\bar{Y}^2}{S_y^2\left(1-\rho_{xy}^2\right)} + \frac{1}{N}\right]^{-1} . \tag{3.2.3.13}$$

Example 3.2.3.3. Suppose a bank manager in the United States of America is interested in making future plans about the selection of sample size while estimating average real estate farm loans. The manager would also like to apply the regression method of estimation using known information about nonreal estate farm loans.

What is the minimum sample size required for relative standard error (RSE) to be equal to 12.5%? Use that data as shown in population 1 of the Appendix.

Solution. The population 1 given in the Appendix shows $N = 50$, $\bar{Y} = 555.43$, $S_y^2 = 342021.5$, and $\rho_{xy} = 0.8038$. Here $\phi = 0.125$, thus the minimum sample size is:

$$n \geq \left[\frac{\phi^2 \bar{Y}^2}{S_y^2\left(1-\rho_{xy}^2\right)} + \frac{1}{N}\right]^{-1} = \left[\frac{0.125^2 \times (555.43)^2}{342021.5\left(1-0.8038^2\right)} + \frac{1}{50}\right]^{-1} = 16.7 \approx 18.$$

Example 3.2.3.4. A bank manager selects an SRSWOR sample of eighteen states from population 1 of the Appendix and collects information about real estate farm loans and nonreal estate farm loans. Estimate the average real estate farm loans by using the regression method of estimation, given that the average amount of nonreal estate farm loans in the United States is known to be equal to \$878.16.

Solution. The bank manager used the 19th and 20th columns of the Pseudo-Random Numbers (PRN) given in Table 1 of the Appendix to select the following 18 distinct random numbers between 1 and 50 as:16, 31, 50, 29, 08, 33, 19, 28, 11, 07, 27, 37, 48, 22, 24, 46, 41, and 32.

Random No	State	x_i	y_i	$\left(x_i - \bar{x}\right)^2$	$\left(y_i - \bar{y}\right)^2$	$\left(y_i - \bar{y}\right)\left(x_i - \bar{x}\right)$
07	CT	4.373	7.130	393759.3178	88444.6782	186617.0307
08	DE	43.229	42.808	346504.6366	68496.5732	154059.6645
11	HI	38.067	40.775	352608.4687	69564.8537	156617.8679
16	KS	2580.304	1049.834	3796373.8360	555483.2696	1452178.4160
19	ME	51.539	8.849	336790.3888	87425.1840	171592.4291
22	MI	440.518	323.028	36617.6715	342.3055	-3540.3997
24	MS	549.551	627.013	6777.3141	103997.5427	-26548.5219
27	NE	3585.406	1337.852	8723342.7430	1067761.5890	3051958.4380
28	NV	16.710	5.860	378428.5240	89201.6782	183729.3102
29	NH	0.471	6.044	398671.5725	89091.8028	188463.1771
31	NM	274.035	140.582	128049.7837	26877.7990	58665.9727
32	NY	426.274	201.631	42271.9539	10587.4839	21155.4634
33	NC	494.730	639.571	18808.8729	112254.8170	-45949.8268
37	OR	571.487	114.899	3646.7642	35958.5887	11451.3097
41	SD	1692.817	413.777	1125596.9840	11935.6717	115908.3954
46	VA	188.477	321.583	196602.1805	290.9242	-7562.8256
48	WV	29.291	99.277	363108.01270	42127.3572	123680.1559
50	WY	386.479	100.964	60219.4149	41437.6914	49953.5137
	Sum	11373.76	5481.477	16708178.4400	2501279.8100	5842429.5700

$x-$ Nonreal estate farm loans, $y-$ Real estate farm loans.

Here $N = 50$ and $\overline{X} = 878.16$. The above table shows $n = 18$, $\overline{y} = 304.5265$, $\overline{x} = 631.8754$, $s_x^2 = 982834.03$, $s_y^2 = 147134.11$, and $s_{xy} = 343672.33$. Thus $\hat{\beta} = 0.3496$, $r_{xy} = 0.9037$ and $f = 0.36$.

Thus the regression estimate of the average real estate farm loans in the United States is

$$\overline{y}_{\text{LR}} = \overline{y} + \hat{\beta}\left(\overline{X} - \overline{x}\right) = 304.5265 + 0.3496(878.16 - 631.8754) = 390.627.$$

An estimate of $\text{MSE}(\overline{y}_{\text{LR}})$ is given by

$$\hat{\text{MSE}}(\overline{y}_{\text{LR}}) = \left(\frac{1-f}{n}\right)s_y^2\left(1 - r_{xy}^2\right) = \left(\frac{1-0.36}{18}\right) \times 147134.11 \times \left(1 - 0.9037^2\right) = 959.059.$$

A $(1-\alpha)100\%$ confidence interval for population mean $\overline{Y}$ is given by

$$\overline{y}_{\text{LR}} \pm t_{\alpha/2}(\text{df} = n-2)\sqrt{\hat{\text{MSE}}(\overline{y}_{\text{LR}})}.$$

Using Table 2 from the Appendix the 95% confidence interval for the average real estate farm loans is given by

$$390.627 \pm 2.120\sqrt{959.059} \text{ or } [324.973, \quad 456.280].$$

Example 3.2.3.5. Consider the following population consisting of five ($N = 5$) units A, B, C, D, and E where for each one of the units in the population two variables Y and X are measured.

Units	A	B	C	D	E
Y_i	9	11	13	16	21
X_i	14	18	19	20	24

Do the following:

(a) Select all possible SRSWOR samples each of $n = 3$ units;
(b) Find the variance of the sample mean estimator by definition;
(c) Find the variance of the sample mean estimator using the formula. Comment;
(d) Find the exact mean square error of the ratio estimator by definition;
(e) Find the approximate mean square error of the ratio estimator using first order approximation;
(f) Find the ratio of approximate mean square error to that of exact mean square error of the ratio estimator and comment;
(g) Find the exact mean square error of the regression estimator using definition;
(h) Find the approximate mean square error of the regression estimator using first order approximation;
(i) Find the ratio of approximate mean square error of the regression estimator to that of the exact mean square error and comment;
(j) Find the exact relative efficiency of the ratio estimator with respect to sample mean estimator;

(k) Find the approximate relative efficiency of the ratio estimator with respect to the sample mean estimator and comment;
(l) Find the exact relative efficiency of the regression estimator with respect to the sample mean estimator;
(m) Find the approximate relative efficiency of the regression estimator with respect to the sample mean and comment.

Solution. (a) From Chapter 1 we have following information for this population $\overline{Y}=14$, $\overline{X}=19$, $S_y^2=22$, $S_x^2=13$, $S_{xy}=16.25$, $\rho_{xy}=0.96$, and $\beta=1.25$. Also from the all possible 10 samples of $n=3$ units taken from the population of $N=5$ units.

Sample No.	$\bar{y}_t$	$\bar{x}_t$	b_t	$p_t\{\bar{y}_t-\overline{Y}\}^2$	$\bar{y}_R(t)$	$p_t\{\bar{y}_R(t)-\overline{Y}\}^2$	$\bar{y}_{LR}(t)$	$p_t\{\bar{y}_{LR}(t)-\overline{Y}\}^2$
1	11.00	17.00	0.71	0.900	12.29	0.291	12.42	0.250
2	12.00	17.33	1.12	0.400	13.16	0.071	13.87	0.002
3	13.67	18.67	1.24	0.011	13.91	0.001	14.08	0.001
4	12.67	17.67	1.05	0.177	13.62	0.014	14.07	0.000
5	14.33	19.00	1.20	0.011	14.33	0.011	14.33	0.011
6	15.33	19.33	1.20	0.177	15.07	0.114	14.93	0.087
7	13.33	19.00	2.50	0.045	13.33	0.045	13.33	0.045
8	15.00	20.33	1.65	0.100	14.02	0.000	12.81	0.143
9	16.00	20.67	1.61	0.400	14.71	0.050	13.31	0.047
10	16.67	21.00	1.50	0.713	15.08	0.117	13.67	0.011
Sum	140.00	190.00	14.00	2.933		0.714		0.596

where

$$\bar{y}_t=n^{-1}\sum_{i=1}^{n}y_i,\quad \bar{x}_t=n^{-1}\sum_{i=1}^{n}x_i,\quad b_t=\frac{s_{xy}}{s_x^2},\quad \bar{y}_{LR}(t)=\bar{y}_t+b_t\left(\overline{X}-\bar{x}_t\right),\quad \bar{y}_R(t)=\bar{y}_t\left(\frac{\overline{X}}{\bar{x}_t}\right),$$

$$p_t=1\Big/\binom{N}{n}=1\Big/\binom{5}{3}=1/10,\ \text{and}\ t=1,2,....10.$$

Now with this information, we can answer all of the above questions:
(b) The exact variance of the sample mean $\bar{y}_t$ is given by

$$\text{Exact V}(\bar{y}_t)=\sum_{t=1}^{\binom{N}{n}}p_t\left(\bar{y}_t-\overline{Y}\right)^2=2.933\,.$$

(c) The variance of the sample mean $\bar{y}_t$ with formula is given by

$$\text{V}(\bar{y}_t)=\left(\frac{1-f}{n}\right)S_y^2=\left(\frac{1-3/5}{3}\right)\times 22=2.933\,.$$

We can see from (b) and (c) that the exact variance and variance by the formula are same.

(d) The exact mean square error of the ratio estimator $\bar{y}_R(t) = \bar{y}_t\left(\frac{\bar{X}}{\bar{x}_t}\right)$ is given by

$$\text{Exact MSE}\{\bar{y}_R\} = \sum_{t=1}^{\binom{N}{n}} p_t\left(\bar{y}_R(t) - \bar{Y}\right)^2 = 0.714.$$

(e) The approximate mean square error of the ratio estimator is given by

$$\text{Approx.MSE}\{\bar{y}_R\} = \left(\frac{1-f}{n}\right)\left[S_y^2 + R^2 S_x^2 - 2RS_{xy}\right]$$

$$= \left(\frac{1-3/5}{3}\right)\left[22 + (14/19)^2 \times 13 - 2 \times (14/19) \times 16.25\right] = 0.681.$$

(f) The ratio of approximate mean square error to the exact mean square error is given by

$$\text{Ratio of Mean Square Errors} = \frac{\text{Approx. MSE}(\bar{y}_R)}{\text{Exact. MSE}(\bar{y}_R)} = \frac{0.681}{0.714} = 0.953.$$

Note that this ratio of the mean square errors approaches unity if sample size and population size are such that $f = n/N \to 0$.

(g) The exact mean square error of the linear regression estimator

$\bar{y}_{LR}(t) = \bar{y}_t + b_t(\bar{X} - \bar{x}_t)$

is given by

$$\text{Exact MSE}\{\bar{y}_{LR}\} = \sum_{t=1}^{\binom{N}{n}} p_t\left(\bar{y}_{LR}(t) - \bar{Y}\right)^2 = 0.596.$$

(h) The approximate mean square error of the linear regression estimator is

$$\text{Approx.MSE}\{\bar{y}_{LR}\} = \left(\frac{1-f}{n}\right)S_y^2\left[1 - \rho_{xy}^2\right] = \left(\frac{1-3/5}{3}\right) \times 22 \times \left[1 - 0.96^2\right] = 0.230.$$

(i) The ratio of approximate mean square error to the exact mean square error of the linear regression estimator is given by

$$\text{Ratio of Mean Square Errors} = \frac{\text{Approx. MSE}(\bar{y}_{LR})}{\text{Exact. MSE}(\bar{y}_{LR})} = \frac{0.230}{0.596} = 0.386.$$

Note that, for this particular example, the ratio of approximate mean square to the exact mean square is far away from one, but if $f = n/N \to 0$ then this ratio approaches to unity.

(j) The exact relative efficiency (RE) of the ratio estimator with respect to sample mean estimator is

$$\text{Exact RE of the Ratio Estimator} = \frac{V(\bar{y}_t) \times 100}{\text{Exact. MSE}(\bar{y}_R)} = \frac{2.933 \times 100}{0.714} = 410.78\%.$$

(k) The approximate relative efficiency of the ratio estimator with respect to the sample mean estimator is

$$\text{Approximate RE the Ratio Estimator} = \frac{V(\bar{y}_t)\times 100}{\text{Approx. MSE}(\bar{y}_R)} = \frac{2.933\times 100}{0.681} = 430.69\%.$$

It shows that the approximate relative efficiency expression for the ratio estimator gives a slightly higher efficiency than in reality.

(l) The exact relative efficiency of the regression estimator with respect to the sample mean estimator is

$$\text{Exact RE of the Regression Estimator} = \frac{V(\bar{y}_t)\times 100}{\text{Exact MSE}(\bar{y}_{LR})} = \frac{2.933\times 100}{0.596} = 492.11\%.$$

(m) The approximate relative efficiency of the linear regression estimator with respect to sample mean estimator is

$$\text{Approx. RE of the Regression Estimator} = \frac{V(\bar{y}_t)\times 100}{\text{Approx. MSE}(\bar{y}_{LR})} = \frac{2.933\times 100}{0.230} = 1275.21\%.$$

It also shows that the approximate relative efficiency expression for the regression estimator gives higher efficiency than in reality.

Caution: Be careful while using approximate expression for mean square error of the linear regression estimator or approximate expression for estimating the mean square error of the linear regression estimator. The interval estimate of the population mean may be bigger than you are constructing with the approximate results.

Note the following graphical situations in the Figure 3.2.1 for the use of ratio, product, and regression estimator in actual practice.

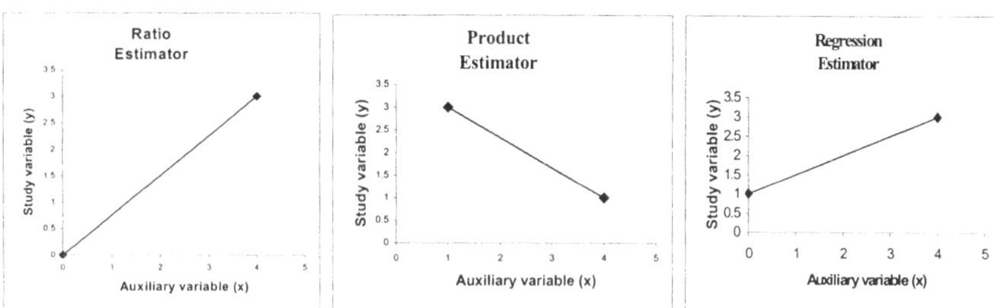

Fig. 3.2.1 Situations for using of ratio, product or regression estimators.

The following table is used to collect some information about these three estimators, which will be useful to the readers:

	Ratio Estimator	Product estimator	Regression estimator
1	The correlation between y and x must be positive and high (within +0.5 and +1.0).	The correlation between y and x must be negative and high (within -1.0 and -0.5).	The correlation between y and x must be non-zero within the range [-1.0, +1.0].
2	The regression lines between y and x should pass through the origin.	The regression line between y and x may or may not pass though the origin. **Note:** If the regression line with two negatively co-related variables will pass through the origin, then one of the variables among y and x will be negative, which may not be practicable.	The regression line may have both parameters *viz.* Intercept and slope.
3	The approximate mean square error expression will be small if the (a) f.p.c. $f = n/N$ is large, (b) sample size n is large, (c) correlation between y and x is very close to plus one, (d) error terms, say $\varepsilon_i = \left(Y_i - \bar{Y}\right) - R\left(X_i - \bar{X}\right)$ are small.	The approximate mean square error expression will be small if the (a) f.p.c. $f = n/N$ is large, (b) sample size n is large, (c) correlation between y and x is very close to minus one, (d) error terms, say $\varepsilon_i = \left(Y_i - \bar{Y}\right) + R\left(X_i - \bar{X}\right)$ are small.	The approximate mean square error expression will be small if the (a) f.p.c. $f = n/N$ is large, (b) sample size n is large, (c) correlation between y and x is very close to plus or minus one, (d) error terms, say $\varepsilon_i = \left(Y_i - \bar{Y}\right) - \beta\left(X_i - \bar{X}\right)$ are small.
4	The usual estimator of the approximate mean square error may be low if sample size is large, which may provide us the smaller confidence interval estimate than the actual one.	The usual estimator of the approximate mean square error may be low if sample size is large, which may provide us the smaller confidence interval estimate than the actual one.	The usual estimator of the approximate mean square error may be low if sample size is large, which may provide us the smaller confidence interval estimate than the actual one.

Continued.....

5	We have to estimate only one model parameter, so the degree of freedom for constructing confidence interval estimates will be $\text{df} = (n-1)$.	If both variables are positive ($x > 0$, and $y > 0$) but the correlation is negative, then we have both intercept and slope, and then we should must use $\text{df} = (n-2)$.	Here we have two unknown parameters, *viz.*: intercept and slope, thus we must use $\text{df} = (n-2)$. Its more justification is given in the Section 3.6.

3.2.4 POWER TRANSFORMATION ESTIMATOR

Srivastava (1967) considered another estimator of population mean, $\overline{Y}$, using the known population mean, $\overline{X}$, of the auxiliary variable, as a power transformation estimator given by

$$\bar{y}_{\text{PW}} = \bar{y}\left(\frac{\bar{x}}{\overline{X}}\right)^{\alpha} \tag{3.2.4.1}$$

where α is a suitably chosen constant, if $\alpha = 1$ then $\bar{y}_{\text{PW}}$ reduces to $\bar{y}_{\text{P}}$ and if $\alpha = -1$ then $\bar{y}_{\text{PW}}$ reduces to $\bar{y}_{\text{R}}$.

Then we have the following theorems.

Theorem 3.2.4.1. The bias in the power transformation estimator $\bar{y}_{\text{PW}}$, to the first order of approximation, is given by

$$B(\bar{y}_{\text{PW}}) = \left(\frac{1-f}{n}\right)\overline{Y}\left[\frac{\alpha(\alpha-1)}{2}C_x^2 + \alpha\rho_{xy}C_xC_y\right]. \tag{3.2.4.2}$$

Proof. The power transformation estimator $\bar{y}_{\text{PW}}$, in terms of ε_0 and ε_1, can easily be written as

$$\bar{y}_{\text{PW}} = \overline{Y}(1+\varepsilon_0)(1+\varepsilon_1)^{\alpha} = \overline{Y}(1+\varepsilon_0)\left(1+\alpha\varepsilon_1+\frac{\alpha(\alpha-1)}{2}\varepsilon_1^2 + O(\varepsilon_1)\right)$$

$$= \overline{Y}\left[1+\varepsilon_0+\alpha\varepsilon_1+\frac{\alpha(\alpha-1)}{2}\varepsilon_1^2+\alpha\varepsilon_0\varepsilon_1+O(\varepsilon_i)\right]. \tag{3.2.4.3}$$

Taking expected values on both sides of (3.2.4.3) and using results from section 3.1, we obtain (3.2.4.2). Hence the theorem.

Theorem 3.2.4.2. The minimum mean squared error of the power transformation estimator $\bar{y}_{\text{PW}}$, to the first order of approximation, is given by

$$\text{Min.MSE}(\bar{y}_{\text{PW}}) = \left(\frac{1-f}{n}\right)S_y^2\left[1-\rho_{xy}^2\right]. \tag{3.2.4.4}$$

Proof. By the definition of mean squared error (MSE), using (3.2.4.3) and again neglecting the higher order terms we have

$$\text{MSE}(\bar{y}_{\text{PW}}) = E\left[\bar{y}_{\text{PW}}-\overline{Y}\right]^2 = E\left[\overline{Y}(1+\varepsilon_0+\alpha\varepsilon_1+O(\varepsilon_i))-\overline{Y}\right]^2$$

$$= \overline{Y}^2E\left[\ \varepsilon_0^2+\alpha^2\varepsilon_1^2+2\alpha\varepsilon_0\varepsilon_1\ \right].$$

Thus the MSE, to the first order of approximation, is given by

$$\mathrm{MSE}(\bar{y}_{\mathrm{PW}})=\left(\frac{1-f}{n}\right)\bar{Y}^2\left[C_y^2+\alpha^2C_x^2+2\alpha\rho_{xy}C_xC_y\right]. \tag{3.2.4.5}$$

On differentiating (3.2.4.5) with respect to α and equating to zero we obtain the optimum value of α as

$$\alpha=-\rho_{xy}\frac{C_y}{C_x}. \tag{3.2.4.6}$$

On substituting the optimum value of α in (3.2.4.5) we obtain

$$\mathrm{Min.MSE}(\bar{y}_{\mathrm{PW}})=\left(\frac{1-f}{n}\right)S_y^2\left[1-\rho_{xy}^2\right]. \tag{3.2.4.7}$$

Hence the theorem.

The power α depends upon the optimum values of unknown parameters. Thus the estimator $\bar{y}_{\mathrm{PW}}$ is not practicable. Thus we have the following corollary.

Corollary 3.2.4.1. A practically useful power transformation type estimator $\bar{y}_{\mathrm{PW(pract)}}$ of population mean $\bar{Y}$ is given by

$$\bar{y}_{\mathrm{PW(pract)}}=\bar{y}\left(\frac{\bar{x}}{\bar{X}}\right)^{\hat{\alpha}} \tag{3.2.4.8}$$

where

$$\hat{\alpha}=-\left(\bar{x}s_{xy}\right)/\left(\bar{y}s_x^2\right)$$

is a consistent estimator of α. Note that while making confidence interval estimate with the power transformation estimator the degree of freedom will be $(n-2)$.

Remark 3.2.4.1. The difference estimator $\bar{y}_{\mathrm{dif}}$ of the population mean, $\bar{Y}$, given as

$$\bar{y}_{\mathrm{dif}}=\bar{y}+d\left(\bar{X}-\bar{x}\right) \tag{3.2.4.9}$$

has the same variance equal to the mean squared error of the linear regression estimator for the optimum value of $d=S_{xy}/S_x^2=\beta$. Again note that the degree of freedom for constructing confidence interval estimates will be $\mathrm{df}=(n-1)$, because the slope is assumed to be known, but we estimate the intercept.

3.2.5 A DUAL OF RATIO ESTIMATOR

Srivenkataramana and Tracy (1980) considered the following estimator of the population mean $\bar{Y}$ based on the use of the mean value of the non-sampled information of the auxiliary variable defined as

$$\bar{y}_{\mathrm{nsu}}=\bar{y}\frac{\left(N\bar{X}-n\bar{x}\right)}{(N-n)\bar{X}} \tag{3.2.5.1}$$

or

$$\bar{y}_{\text{nsu}} = \bar{y}\left(\frac{\bar{x}^*}{\bar{X}}\right) \tag{3.2.5.2}$$

where $\bar{x}^* = \dfrac{(N\bar{X} - n\bar{x})}{(N-n)} = \dfrac{1}{N-n}\sum_{i=1}^{N-n} x_i$ denotes the mean of non-sampled units of the auxiliary variable.

Then we have the following theorems.

Theorem 3.2.5.1. The estimator $\bar{y}_{\text{nsu}}$ is an inconsistent estimator of the population mean, $\bar{Y}$.

Proof. The estimator $\bar{y}_{\text{nsu}}$ in terms of ε_0 and ε_1 can be written as

$$\bar{y}_{\text{nsu}} = \bar{y}\frac{N\bar{X} - n\bar{x}}{(N-n)\bar{X}} = \bar{Y}(1+\varepsilon_0)\frac{N\bar{X} - n\bar{X}(1+\varepsilon_1)}{(N-n)\bar{X}} = \bar{Y}(1+\varepsilon_0)\left(1 - \frac{n}{N-n}\varepsilon_1\right)$$

$$= \bar{Y}\left(1 + \varepsilon_0 - \frac{n}{N-n}\varepsilon_1 - \frac{n}{N-n}\varepsilon_0\varepsilon_1\right).$$

Taking expected values on both sides we have

$$E(\bar{y}_{\text{nsu}}) = \bar{Y}E\left(1 + \varepsilon_0 - \frac{n}{N-n}\varepsilon_1 - \frac{n}{N-n}\varepsilon_0\varepsilon_1\right)$$

$$= \bar{Y}\left[1 + E(\varepsilon_0) - \frac{n}{N-n}E(\varepsilon_1) - \frac{n}{N-n}E(\varepsilon_0\varepsilon_1)\right] = \bar{Y}\left(1 - \frac{n}{N-n}\times\frac{1-f}{n}\rho_{xy}C_xC_y\right)$$

$$= \bar{Y} - \frac{n}{N-n}\times\frac{1-f}{n}\bar{Y}\rho_{xy}C_xC_y = \bar{Y} - \frac{\bar{Y}\rho_{xy}C_xC_y}{N}.$$

Thus the bias in the estimator $\bar{y}_{\text{nsu}}$ is given by

$$B(\bar{y}_{\text{nsu}}) = E(\bar{y}_{\text{nsu}}) - \bar{Y} = -\frac{\bar{Y}\rho_{xy}C_xC_y}{N} = -\frac{S_{xy}}{N\bar{X}}$$

which proves the theorem.

Theorem 3.2.5.2. The mean squared error of the estimator $\bar{y}_{\text{nsu}}$ is given by

$$\text{MSE}(\bar{y}_{\text{nsu}}) = \left(\frac{1-f}{n}\right)\bar{Y}^2\left[C_y^2 + g^2C_x^2 - 2g\rho_{xy}C_xC_y\right] \tag{3.2.5.3}$$

where $g = \dfrac{n}{N-n}$ and $N^2 \to \infty$.

Proof. We have

$$\text{MSE}(\bar{y}_{\text{nsu}}) = E\left[\bar{y}_{\text{nsu}} - \bar{Y}\right]^2 = E\left[\bar{Y}\left(1 + \varepsilon_0 - \frac{n}{N-n}\varepsilon_1 - \left(\frac{n}{N-n}\right)\varepsilon_0\varepsilon_1\right) - \bar{Y}\right]^2$$

$$\approx \bar{Y}^2E\left[\varepsilon_0 - \frac{n}{N-n}\varepsilon_1\right]^2 = \bar{Y}^2E\left[\varepsilon_0^2 + \left(\frac{n}{N-n}\right)^2\varepsilon_1^2 - 2\left(\frac{n}{N-n}\right)\varepsilon_0\varepsilon_1\right]$$

$$= \left(\frac{1-f}{n}\right)\bar{Y}^2\left[C_y^2 + \left(\frac{n}{N-n}\right)^2 C_x^2 - 2\left(\frac{n}{N-n}\right)\rho_{xy}C_xC_y\right]$$

$$= \left(\frac{1-f}{n}\right)\bar{Y}^2\left[C_y^2 + g^2C_x^2 - 2g\rho_{xy}C_xC_y\right].$$

Hence the theorem.

Theorem 3.2.5.3. The estimator $\bar{y}_{\text{nsu}}$ is more efficient than the ratio estimator $\bar{y}_{\text{R}}$ if

$$n < \frac{N}{2}, \text{ and } \rho_{xy} < \frac{N}{2(N-n)}, \tag{3.2.5.4}$$

assuming that the correlation coefficient ρ_{xy} is positive .

Proof. The estimator $\bar{y}_{\text{nsu}}$ will be more efficient than the ratio estimator $\bar{y}_{\text{R}}$ if

$$\text{MSE}(\bar{y}_{\text{nsu}}) < \text{MSE}(\bar{y}_{\text{R}})$$

or $\left(\frac{1-f}{n}\right)\bar{Y}^2\left[C_y^2 + g^2C_x^2 - 2g\rho_{xy}C_xC_y\right] < \left(\frac{1-f}{n}\right)\bar{Y}^2\left[C_y^2 + C_x^2 - 2\rho_{xy}C_xC_y\right]$

or $(g^2-1)C_x^2 - 2(g-1)\rho_{xy}C_xC_y < 0$ or $(g-1)(g+1)C_x^2 - 2(g-1)\rho_{xy}C_xC_y < 0$

or $$(g-1)\left[(g+1)C_x^2 - 2\rho_{xy}C_xC_y\right] < 0. \tag{3.2.5.5}$$

Now there are two cases:

Case 1. The inequality (3.2.5.5) will be satisfied if

$g-1<0$ and $(g+1)C_x^2 - 2\rho_{xy}C_xC_y > 0$

or $\frac{n}{N-n} - 1 < 0$ and $(g+1)C_x > 2\rho_{xy}C_y$

or $\frac{n-N+n}{N-n} < 0$ and $\rho_{xy}\frac{C_y}{C_x} < \frac{(g+1)}{2}$

or $n < \frac{N}{2}$ and $\rho_{xy}\frac{C_y}{C_x} < \frac{n+N-n}{2(N-n)} = \frac{N}{2(N-n)}$

For $C_y \approx C_x$ we have $n < \frac{N}{2}$ and $\rho_{xy} < \frac{N}{2(N-n)}$.

This condition holds in practice. For example, if $N = 100$ and $n = 30$ then ρ_{xy} is supposed to be less that 0.714.

Case 2. The inequality (3.2.5.5) will be satisfied if

$g-1>0$ and $(g+1)C_x^2 - 2\rho_{xy}C_xC_y < 0$

or $\frac{n}{N-n} - 1 > 0$ and $(g+1)C_x < 2\rho_{xy}C_y$

or $\dfrac{n-N+n}{N-n} > 0$ and $\rho_{xy}\dfrac{C_y}{C_x} > \dfrac{(g+1)}{2}$

or $n > \dfrac{N}{2}$ and $\rho_{xy}\dfrac{C_y}{C_x} > \dfrac{n+N-n}{2(N-n)} = \dfrac{N}{2(N-n)}$.

For $C_y \approx C_x$ we have

$$n > \frac{N}{2} \text{ and } \rho_{xy} > \frac{N}{2(N-n)}.$$

This condition will not hold in practice. For example, if $N = 100$ and $n = 70$ then the value of ρ_{xy} needs to be more that 1.667, which is not possible. Hence the theorem.

3.2.6 GENERAL CLASS OF ESTIMATORS

Srivastava (1971) proposed a general class of estimators to estimate the population mean $\bar{Y}$ of the study variable which in the case of single known mean $\bar{X}$ of the auxiliary variable is given by

$$t_g = \bar{y}H(u) \tag{3.2.6.1}$$

where $u = \bar{x}/\bar{X}$ and $H(\bullet)$ is a parametric function, such that, it satisfies the following conditions:

(a) $H(1) = 1$; (3.2.6.2)

(b) The first and second order partial derivatives of H with respect to u exist and are known constants at a given point $u = 1$.

Expanding $H(u)$ about the value 1 in a second order Taylor's series we have

$$H(u) = H[1+(u-1)] = H(1) + (u-1)\frac{\partial H}{\partial u}\Big|_{u=1} + (u-1)^2\frac{1}{2}\frac{\partial^2 H}{\partial u^2}\Big|_{u=1} + \ldots\ldots \tag{3.2.6.3}$$

Note that $|u-1| < 1$ thus the higher order terms can be neglected. Using (3.2.6.2) and (3.2.6.3) in (3.2.6.1) we obtain

$$t_g = \bar{y}[\ 1 + (u-1)H_1 + (u-1)^2 H_2 + \ldots\ldots\ ] \tag{3.2.6.4}$$

where $H_1 = \dfrac{\partial H}{\partial u}\Big|_{u=1}$ and $H_2 = \dfrac{1}{2}\dfrac{\partial^2 H}{\partial u^2}\Big|_{u=1}$ denote the first and second order partial derivatives of H with respect to u and are the known constants. Evidently the class of estimators t_g given at (3.2.6.4) can easily be written in terms of ε_0 and ε_1 as

$$\begin{aligned} t_g &= \bar{Y}(1+\varepsilon_0)\left[1 + \varepsilon_1 H_1 + \varepsilon_1^2 H_2 + \ldots\right] \\ &= \bar{Y}\left[1 + \varepsilon_0 + \varepsilon_1 H_1 + \varepsilon_1^2 H_2 + \varepsilon_0\varepsilon_1 H_1 + O(\varepsilon)\right]. \end{aligned} \tag{3.2.6.5}$$

Thus we have the following theorems:

Theorem 3.2.6.1. The bias in the general class of estimators t_g defined at (3.2.6.1), to the first order of approximation, is

$$B(t_g) = \left(\frac{1-f}{n}\right)\bar{Y}\left[H_2C_x^2 + H_1\rho_{xy}C_yC_x\right]. \tag{3.2.6.6}$$

Proof. Taking the expected value on both sides of (3.2.6.5) we obtain

$$E(t_g) = \bar{Y}\left[1 + \left(\frac{1-f}{n}\right)\left(H_2C_x^2 + H_1\rho_{xy}C_yC_x\right)\right].$$

Thus the bias in the class of estimators t_g is given by

$$B(t_g) = E(t_g) - \bar{Y} = \left(\frac{1-f}{n}\right)\bar{Y}\left[H_2C_x^2 + H_1\rho_{xy}C_yC_x\right].$$

Hence the theorem.

Theorem 3.2.6.2. The minimum mean squared error of the general class of estimators t_g defined at (3.2.6.1), to the first order of approximation, is given by

$$\text{Min.MSE}(t_g) = \left(\frac{1-f}{n}\right)\bar{Y}^2C_y^2\left(1 - \rho_{xy}^2\right). \tag{3.2.6.7}$$

Proof. By the definition of the mean squared error we have

$$\text{MSE}(t_g) = E\left[t_g - \bar{Y}\right]^2 \approx E\left[\bar{Y}\left\{1 + \varepsilon_0 + \varepsilon_1H_1 + \varepsilon_1^2H_2 + ...\right\} - \bar{Y}\right]^2.$$

Again neglecting higher order terms we have

$$\text{MSE}(t_g) = \bar{Y}^2E\left[\varepsilon_0^2 + H_1^2\varepsilon_1^2 + 2H_1\varepsilon_0\varepsilon_1\right]$$

$$= \left(\frac{1-f}{n}\right)\bar{Y}^2\left[C_y^2 + H_1^2C_x^2 + 2H_1\rho_{xy}C_yC_x\right]. \tag{3.2.6.8}$$

On differentiating (3.2.6.8) with respect to H_1 and equating to zero we obtain

$$H_1 = -\rho_{xy}\frac{C_y}{C_x}. \tag{3.2.6.9}$$

On substituting the optimum value of H_1 from (3.2.6.9) in (3.2.6.8) we obtain (3.2.6.7). Hence the theorem.

Remark 3.2.6.1. If we attach any function of $\bar{x}/\bar{X}$ to the sample mean $\bar{y}$ the asymptotic minimum mean squared error of the resultant estimator cannot be reduced further than that given in (3.2.6.7). Thus the usual ratio estimator, product estimator, and power transformation estimator are the special cases of the class of estimators defined in (3.2.6.1).

3.2.7 WIDER CLASS OF ESTIMATORS

One may note here that regression estimator and difference estimator are not special cases of the general class of estimators defined in (3.2.6.1). Srivastava (1980) defined another class of estimators and named a wider class of estimators as

$$t_{\mathrm{w}} = H[\bar{y},\ u] \tag{3.2.7.1}$$

where $H[\bar{y},\ u]$ is a function of $\bar{y}$ and u and satisfies the following regularity conditions:

(a) The point $(\bar{y}, u)$ assumes the value in a closed convex subset R_2 of two-dimensional real space containing the point $(\bar{Y}, 1)$;
(b) The function $H(\bar{y}, u)$ is continuous and bounded in R_2;
(c) $H(\bar{Y},\ 1) = \bar{Y}$ and $H_0(\bar{Y},\ 1) = 1$, where $H_0(\bar{Y},\ 1)$ denotes the first order partial derivative of H with respect to $\bar{y}$;
(d) The first and second order partial derivatives of $H(\bar{y},\ u)$ exist and are continuous and bounded in R_2.

Expanding $H(\bar{y},\ u)$ about the point $(\bar{Y},\ 1)$ in a second order Taylor series we have

$$t_{\mathrm{w}} = H(\bar{y}, u) = H\left[\bar{Y} + (\bar{y} - \bar{Y}), 1 + (u-1)\right]$$

$$= H(\bar{Y}, 1) + (\bar{y} - \bar{Y})\frac{\partial H}{\partial \bar{y}}\Big|_{\bar{y}=\bar{Y}, u=1} + (u-1)\frac{\partial H}{\partial u}\Big|_{\bar{y}=\bar{Y}, u=1} + (u-1)^2\frac{1}{2}\frac{\partial^2 H}{\partial u^2}\Big|_{\bar{y}=\bar{Y}, u=1} +$$

$$+ (\bar{y} - \bar{Y})^2\frac{1}{2}\frac{\partial^2 H}{\partial^2 \bar{y}^2}\Big|_{\bar{y}=\bar{Y}, u=1} + (\bar{y} - \bar{Y})(u-1)\frac{1}{2}\frac{\partial^2 H}{\partial \bar{y}\partial u}\Big|_{\bar{y}=\bar{Y}, u=1} + \ldots\ldots \tag{3.2.7.2}$$

Using the above regularity conditions and $\frac{\partial H}{\partial \bar{y}}\Big|_{\bar{y}=\bar{Y}, u=1} = 1$, we have

$$t_{\mathrm{w}} = \bar{y} + (u-1)H_1 + (u-1)^2 H_2 + (\bar{y} - \bar{Y})(u-1)H_3 + (\bar{y} - \bar{Y})^2 H_4 + .. \tag{3.2.7.3}$$

where

$$H_1 = \frac{\partial H}{\partial u}\Big|_{\bar{y}=\bar{Y}, u=1},\quad H_2 = \frac{1}{2}\frac{\partial^2 H}{\partial u^2}\Big|_{\bar{y}=\bar{Y}, u=1},\quad H_3 = \frac{1}{2}\frac{\partial^2 H}{\partial \bar{y}\partial u}\Big|_{\bar{y}=\bar{Y}, u=1},\ \text{and}$$

$$H_4 = \frac{1}{2}\frac{\partial^2 H}{\partial^2 \bar{y}^2}\Big|_{\bar{y}=\bar{Y}, u=1}.$$

Thus we have the following theorems.

Theorem 3.2.7.1. The asymptotic bias in the wider class of estimators t_{w} of the population mean $\bar{Y}$ is:

$$B(t_{\mathrm{w}}) = \left(\frac{1-f}{n}\right)\left[\bar{Y}\rho_{xy}C_x C_y H_3 + C_x^2 H_2 + \bar{Y}^2 C_y^2 H_4\right]. \tag{3.2.7.4}$$

Proof. The wider class of estimators t_{w}, in terms of ε_0 and ε_1, can easily be written as

$$t_{\mathrm{w}}=\bar{Y}(1+\varepsilon_0)+\varepsilon_1 H_1+\varepsilon_1^2 H_2+\bar{Y}\varepsilon_0\varepsilon_1 H_3+\bar{Y}^2\varepsilon_0^2 H_4+\ldots. \tag{3.2.7.5}$$

Taking expected values on both sides of (3.2.7.5) and using the definition of bias, we obtain (3.2.7.4). Hence the theorem.

Theorem 3.2.7.2. The minimum mean squared error of the wider class of estimators, t_{w}, is given by

$$\text{Min.MSE}(t_{\mathrm{w}})=\left(\frac{1-f}{n}\right)\bar{Y}^2 C_y^2\left(1-\rho_{xy}^2\right). \tag{3.2.7.6}$$

Proof. By the definition of mean squared error, we have

$$\text{MSE}(t_{\mathrm{w}})=E\left[t_{\mathrm{w}}-\bar{Y}\right]^2\approx E\left[\bar{y}+H_1\varepsilon_1+O(\varepsilon)-\bar{Y}\right]^2=E\left[\bar{Y}\left(\frac{\bar{y}}{\bar{Y}}-1\right)+H_1\varepsilon_1\right]^2$$

$$=E\left[\bar{Y}\varepsilon_0+H_1\varepsilon_1\right]^2=E\left[\bar{Y}^2\varepsilon_0^2+H_1^2\varepsilon_1^2+2H_1\bar{Y}\varepsilon_0\varepsilon_1\right]$$

$$=\left(\frac{1-f}{n}\right)\left[\bar{Y}^2C_y^2+H_1^2C_x^2+2H_1\bar{Y}\rho_{xy}C_xC_y\right]. \tag{3.2.7.7}$$

On differentiating (3.2.7.7) with respect to H_1 and equating to zero we obtain

$$H_1=-\rho_{xy}\bar{Y}\frac{C_y}{C_x}. \tag{3.2.7.8}$$

On substituting (3.2.7.8) in (3.2.7.7), we obtain (3.2.7.6). Hence the theorem.

Remark 3.2.7.1. If we have any function of $\bar{y}$, $\bar{x}$ and $\bar{X}$ to estimate the population mean, $\bar{Y}$, the asymptotic minimum mean squared error of the resultant estimator again cannot be reduced further than that given in (3.2.6.7). Thus the usual linear regression estimator and difference estimator are special cases of the wider class of estimators defined at (3.2.7.1).

3.2.8 USE OF KNOWN VARIANCE OF THE AUXILIARY VARIABLE AT ESTIMATION STAGE OF POPULATION MEAN

In this section, we will show that the known variance of the auxiliary variable can also be used as a benchmark, in addition to the known population total or mean of the auxiliary variable, to improve the estimators of the finite population mean of the study variable under certain circumstances.

3.2.8.1 A CLASS OF ESTIMATORS

Srivastava and Jhajj (1981) introduced the use of known value of the variance of the auxiliary variable to improve the efficiency of the estimators of population mean. They considered a general class of ratio type estimators as

$$\bar{y}_{\mathrm{SJ}}=\bar{y}H(u,v) \tag{3.2.8.1}$$

where $u=\bar{x}/\bar{X}$, $v=s_x^2/S_x^2$ and $H(u,v)$ is a function of u and v such that:

where $u = \bar{x}/\bar{X}$, $v = s_x^2/S_x^2$ and $H(u,v)$ is a function of u and v such that:
(a) The point (u, v) assumes the value in a closed convex subset R_2 of two-dimensional real space containing the point $(1, 1)$;
(b) The function $H(u, v)$ is continuous and bounded in R_2;
(c) $H(1,1) = 1$;
(d) The first and second order partial derivatives of $H(u,v)$ exist and are continuous and bounded in R_2.

Thus all ratio and product type estimators of population mean $\bar{Y}$ defined as

$$\bar{y}_1 = \bar{y}\left(\frac{\bar{X}}{\bar{x}}\right)\left(\frac{S_x^2}{s_x^2}\right), \quad \bar{y}_2 = \bar{y}\left(\frac{\bar{X}}{\alpha\bar{x} + (1-\alpha)\bar{X}}\right)\left(\frac{S_x^2}{\gamma s_x^2 + (1-\gamma)S_x^2}\right), \quad \text{and} \quad \bar{y}_3 = \bar{y}\left(\frac{\bar{X}}{\bar{x}}\right)^{\alpha}\left(\frac{S_x^2}{s_x^2}\right)^{\gamma}$$

are the special cases of the class of estimators defined in (3.2.8.1).

Expanding $H(u, v)$ about the point $(1, 1)$ in a second order Taylor's series we obtain

$$\begin{aligned}
\bar{y}_{SJ} &= \bar{y}H(u,v) = \bar{y}H[1 + (u-1), 1 + (v-1)] \\
&\approx \bar{y}\Bigg[H(1,1) + (u-1)\frac{\partial H}{\partial u}\Big|_{(1,1)} + (v-1)\frac{\partial H}{\partial v}\Big|_{(1,1)} + (u-1)^2\frac{1}{2}\frac{\partial^2 H}{\partial u^2}\Big|_{(1,1)} \\
&\quad + (v-1)^2\frac{1}{2}\frac{\partial^2 H}{\partial v^2}\Big|_{(1,1)} + (u-1)(v-1)\frac{1}{2}\frac{\partial^2 H}{\partial u\partial v}\Big|_{(1,1)} + .. \Bigg] \\
&= \bar{Y}(1+\varepsilon_0)\left[1 + \varepsilon_1 H_1 + \varepsilon_3 H_2 + \varepsilon_1^2 H_3 + \varepsilon_3^2 H_4 + \varepsilon_1\varepsilon_3 H_5 + ...\right] \\
&\approx \bar{Y}\left[1 + \varepsilon_0 + \varepsilon_1 H_1 + \varepsilon_3 H_2 + \varepsilon_1^2 H_3 + \varepsilon_3^2 H_4 + \varepsilon_1\varepsilon_3 H_5 \right. \\
&\quad \left. + \varepsilon_0\varepsilon_1 H_1 + \varepsilon_0\varepsilon_3 H_2 + ...\right]
\end{aligned} \tag{3.2.8.2}$$

where

$$H_1 = \frac{\partial H}{\partial u}\Big|_{(1,1)}, \quad H_2 = \frac{\partial H}{\partial v}\Big|_{(1,1)}, \quad H_3 = \frac{1}{2}\frac{\partial^2 H}{\partial u^2}\Big|_{(1,1)}, \quad H_4 = \frac{1}{2}\frac{\partial^2 H}{\partial v^2}\Big|_{(1,1)}, \quad \text{and}$$

$$H_5 = \frac{1}{2}\frac{\partial^2 H}{\partial u\partial v}\Big|_{(1,1)}.$$

Thus we have the following theorems:

Theorem 3.2.8.1. The asymptotic bias in the class of ratio type estimators $\bar{y}_{SJ}$ is

$$B(\bar{y}_{SJ}) = \left(\frac{1-f}{n}\right)\bar{Y}\left[C_x^2 H_3 + (\lambda_{04} - 1)H_4 + C_x\lambda_{03}H_5 + \rho_{xy}C_yC_xH_1 + C_y\lambda_{12}H_2\right]. \tag{3.2.8.3}$$

Proof. It follows by taking expected values on both sides (3.2.8.2) we have

$$\begin{aligned}
E(\bar{y}_{SJ}) &= \bar{Y}E\left[1 + \varepsilon_0 + \varepsilon_1 H_1 + \varepsilon_3 H_2 + \varepsilon_1^2 H_3 + \varepsilon_3^2 H_4 + \varepsilon_1\varepsilon_3 H_5 + \varepsilon_0\varepsilon_1 H_1 + \varepsilon_0\varepsilon_3 H_2 + ...\right] \\
&= \bar{Y}\left[1 + 0 + 0H_1 + 0H_2 + E(\varepsilon_1^2)H_3 + E(\varepsilon_3^2)H_4 + E(\varepsilon_1\varepsilon_3)H_5 + E(\varepsilon_0\varepsilon_1)H_1 + E(\varepsilon_0\varepsilon_3)H_2\right]
\end{aligned}$$

$$= \overline{Y} + \left(\frac{1-f}{n}\right)\overline{Y}\left[C_x^2 H_3 + (\lambda_{04}-1)H_4 + C_x\lambda_{03}H_5 + \rho_{xy}C_yC_xH_1 + C_y\lambda_{12}H_2\right].$$

Thus the bias is given by

$$B(\overline{y}_{SJ}) = E(\overline{y}_{SJ}) - \overline{Y}$$

$$= \left(\frac{1-f}{n}\right)\overline{Y}\left[C_x^2 H_3 + (\lambda_{04}-1)H_4 + C_x\lambda_{03}H_5 + \rho_{xy}C_yC_xH_1 + C_y\lambda_{12}H_2\right].$$

Hence the theorem.

Theorem 3.2.8.2. The minimum MSE of the class of estimators $\overline{y}_{SJ}$ is given by

$$\text{Min.MSE}(\overline{y}_{SJ}) = \left(\frac{1-f}{n}\right)\overline{Y}^2 C_y^2\left[1-\rho_{xy}^2 - \frac{(\lambda_{03}\rho_{xy}-\lambda_{12})^2}{\lambda_{04}-1-\lambda_{03}^2}\right]. \tag{3.2.8.4}$$

Proof. By the definition of the mean squared error we have

$$\text{MSE}(\overline{y}_{SJ}) = E\left[\overline{y}_{SJ}-\overline{Y}\right]^2 = \overline{Y}^2 E\left[\varepsilon_0 + \varepsilon_1 H_1 + \varepsilon_3 H_2 + O(\varepsilon)\right]^2$$

$$= \overline{Y}^2 E\left[\varepsilon_0^2 + \varepsilon_1^2 H_1^2 + \varepsilon_3^2 H_2^2 + 2\varepsilon_0\varepsilon_1 H_1 + 2\varepsilon_0\varepsilon_3 H_2 + 2\varepsilon_1\varepsilon_3 H_1 H_2\right]$$

$$= \left(\frac{1-f}{n}\right)\overline{Y}^2\Big[C_y^2 + C_x^2 H_1^2 + (\lambda_{04}-1)H_2^2 + 2\rho_{xy}C_xC_yH_1 + 2C_y\lambda_{12}H_2$$

$$+2C_x\lambda_{03}H_1H_2\Big]. \tag{3.2.8.5}$$

On differentiating (3.2.8.5) with respect to H_1 and H_2, and equating to zero, respectively we obtain

$$H_1C_x + H_2\lambda_{03} = -\rho_{xy}C_y, \quad \text{and} \quad H_1C_x\lambda_{03} + H_2(\lambda_{04}-1) = -C_y\lambda_{12}. \tag{3.2.8.6}$$

Solving equations in (3.2.8.6) for H_1 and H_2 we obtain

$$H_1 = -\frac{C_y\{\rho_{xy}(\lambda_{04}-1) - \lambda_{12}\lambda_{03}\}}{C_x(\lambda_{04}-1-\lambda_{03}^2)}, \quad \text{and} \quad H_2 = -\frac{C_y\{\lambda_{12}-\rho_{xy}\lambda_{03}\}}{\lambda_{04}-1-\lambda_{03}^2}. \tag{3.2.8.7}$$

On substituting these optimum values of H_1 and H_2 in (3.2.8.5) we have (3.2.8.4). Hence the theorem.

3.2.8.2 A WIDER CLASS OF ESTIMATORS

Srivastava and Jhajj (1981) also considered a wider class of estimators of population mean $\overline{Y}$ as

$$\overline{y}_{SJ(w)} = H(\overline{y},\ u,\ v) \tag{3.2.8.8}$$

where $u = \overline{x}/\overline{X}$, $v = s_x^2/S_x^2$ and $H(\overline{y},u,v)$ is a function of $\overline{y}$, u and v, such that:

(a) The point $(\overline{y}, u, v)$ assumes the value in a closed convex subset R_3 of three-dimensional real space containing the point $(\overline{Y}, 1, 1)$;

(b) The function $H(\bar{y}, u, v)$ is continuous and bounded in R_3 ;

(c) $H(\bar{Y}, 1, 1) = \bar{Y}$;

(d) The first and second order partial derivatives of $H(\bar{y}, u, v)$ exist, and are continuous, and bounded in R_3.

Expanding $H(\bar{y}, u, v)$ about the point $(\bar{Y}, 1, 1)$ in a second order Taylor's series we have

$$\begin{aligned}
\bar{y}_{SJ(w)} &= H(\bar{y},u,v) = H\left[\bar{Y} + (\bar{y} - \bar{Y}),\ 1 + (u-1),\ 1 + (v-1)\right] \\
&= H(\bar{Y},1,1) + (\bar{y} - \bar{Y})\frac{\partial H}{\partial \bar{y}}\Big|_{(\bar{Y},1,1)} + (u-1)\frac{\partial H}{\partial u}\Big|_{(\bar{Y},1,1)} + (v-1)\frac{\partial H}{\partial v}\Big|_{(\bar{Y},1,1)} \\
&\quad + (\bar{y} - \bar{Y})^2 \frac{1}{2}\frac{\partial^2 H}{\partial \bar{y}^2}\Big|_{(\bar{Y},1,1)} + (u-1)^2 \frac{1}{2}\frac{\partial^2 H}{\partial u^2}\Big|_{(\bar{Y},1,1)} + (v-1)^2 \frac{1}{2}\frac{\partial^2 H}{\partial v^2}\Big|_{(\bar{Y},1,1)} \\
&\quad + (\bar{y} - \bar{Y})(u-1)\frac{1}{2}\frac{\partial^2 H}{\partial \bar{y}\partial u}\Big|_{(\bar{Y},1,1)} + (u-1)(v-1)\frac{1}{2}\frac{\partial^2 H}{\partial u \partial v}\Big|_{(\bar{Y},1,1)} + \ldots \\
&\quad + (v-1)(\bar{y} - \bar{Y})\frac{1}{2}\frac{\partial^2 H}{\partial v \partial \bar{y}}\Big|_{(\bar{Y},1,1)} + \ldots\ldots \\
&= \bar{Y}(1 + \varepsilon_0) + \varepsilon_1 H_1 + \varepsilon_3 H_2 + \bar{Y}\varepsilon_0^2 H_3 + \varepsilon_1^2 H_4 + \varepsilon_3^2 H_5 \\
&\quad + \bar{Y}\varepsilon_0\varepsilon_1 H_6 + \varepsilon_1\varepsilon_3 H_7 + \bar{Y}\varepsilon_0\varepsilon_3 H_8 + \ldots
\end{aligned} \tag{3.2.8.9}$$

where

$\frac{\partial H}{\partial \bar{y}}\Big|_{(\bar{Y},1,1)} = 1$, $H_1 = \frac{\partial H}{\partial u}\Big|_{(\bar{Y},1,1)}$, $H_2 = \frac{\partial H}{\partial v}\Big|_{(\bar{Y},1,1)}$, $H_3 = \frac{1}{2}\frac{\partial^2 H}{\partial \bar{y}^2}\Big|_{(\bar{Y},1,1)}$, $H_4 = \frac{1}{2}\frac{\partial^2 H}{\partial u^2}\Big|_{(\bar{Y},1,1)}$,

$H_5 = \frac{1}{2}\frac{\partial^2 H}{\partial v^2}\Big|_{(\bar{Y},1,1)}$, $H_6 = \frac{1}{2}\frac{\partial^2 H}{\partial \bar{y}\partial u}\Big|_{(\bar{Y},1,1)}$, $H_7 = \frac{1}{2}\frac{\partial^2 H}{\partial u \partial v}\Big|_{(\bar{Y},1,1)}$, and $H_8 = \frac{1}{2}\frac{\partial^2 H}{\partial v \partial \bar{y}}\Big|_{(\bar{Y},1,1)}$.

Thus we have the following theorems:

Theorem 3.2.8.3. The asymptotic bias in the wider class estimators $\bar{y}_{SJ(w)}$ is

$$B(\bar{y}_{SJ(w)}) = \left(\frac{1-f}{n}\right)\left[\bar{Y}C_y^2 H_3 + C_x^2 H_4 + (\lambda_{04} - 1)H_5 + \bar{Y}\rho_{xy}C_y C_x H_6 + C_x\lambda_{03}H_7 + \bar{Y}C_y\lambda_{12}H_8\right]. \tag{3.2.8.10}$$

Proof. It follows by taking expected value on both sides of (3.2.8.9).

Theorem 3.2.8.4. The minimum MSE of the class of estimators $\bar{y}_{SJ(w)}$ is given by

$$\text{Min.MSE}(\bar{y}_{SJ(w)}) = \left(\frac{1-f}{n}\right)\bar{Y}^2 C_y^2\left[1 - \rho_{xy}^2 - \frac{(\lambda_{03}\rho_{xy} - \lambda_{12})^2}{\lambda_{04} - 1 - \lambda_{03}^2}\right]. \tag{3.2.8.11}$$

Proof. By the definition of the mean squared error we have

$$\text{MSE}\left(\bar{y}_{\text{SJ(w)}}\right)=E\left[\bar{y}_{\text{SJ(w)}}-\bar{Y}\right]^2=E\left[\bar{Y}\varepsilon_0+\varepsilon_1 H_1+\varepsilon_3 H_2\right]^2$$

$$=\left(\frac{1-f}{n}\right)\left[\bar{Y}^2C_y^2+C_x^2H_1^2+(\lambda_{04}-1)H_2^2+2\bar{Y}\rho_{xy}C_xC_yH_1\right.$$

$$\left.+2\bar{Y}C_y\lambda_{12}H_2+2C_x\lambda_{03}H_1H_2\right]. \tag{3.2.8.12}$$

On differentiating (3.2.8.12) with respect to H_1 and H_2, and equating to zero, respectively we obtain

$$H_1C_x+H_2\lambda_{03}=-\bar{Y}\rho_{xy}C_y, \tag{3.2.8.13}$$

and

$$H_1C_x\lambda_{03}+H_2(\lambda_{04}-1)=-\bar{Y}C_y\lambda_{12}. \tag{3.2.8.14}$$

Solving (3.2.8.13) and (3.2.8.14) for H_1 and H_2 we have

$$H_1=-\frac{\bar{Y}C_y\{\rho_{xy}(\lambda_{04}-1)-\lambda_{12}\lambda_{03}\}}{C_x\left(\lambda_{04}-1-\lambda_{03}^2\right)}, \quad \text{and} \quad H_2=-\frac{\bar{Y}C_y\{\lambda_{12}-\rho_{xy}\lambda_{03}\}}{\lambda_{04}-1-\lambda_{03}^2}. \tag{3.2.8.15}$$

On substituting these optimum values of H_1 and H_2 in (3.2.8.12), we obtain (3.2.8.11). Hence the theorem.

Remark 3.2.8.1.

(a) The difference type of estimator $\bar{y}_1=\bar{y}+\gamma_1\left(\bar{X}-\bar{x}\right)+\gamma_2\left(S_x^2-s_x^2\right)$, where γ_1 and γ_2 are real constants, is a special case of the wider class of estimators. Sahoo, Sahoo, and Espejo (1998) have presented an empirical investigation on the performance of five strategies for estimating the finite population mean using parameters such as mean or variance or both of an auxiliary variable. They considered the problems of comparison of bias, efficiency and approach to normality or asymmetry.

(b) The asymptotic minimum mean squared error of the ratio type and the wider class of estimators remains the same.

(c) Note that λ_{12} and λ_{03} are odd ordered moments. In case X and Y follow the bivariate normal distribution then both λ_{12} and λ_{03} are zero. In such situations the minimum mean squared error of the class of estimators proposed by Srivastava and Jhajj (1981) reduces to the mean squared error of the usual linear regression estimator. Thus there is no advantage in using the known variance of auxiliary variable for the construction of the estimator of the population mean $\bar{Y}$ if the joint distribution of the study variable Y and auxiliary variable X is a bivariate normal distribution.

(d) As the use of number of known parameters of an auxiliary variable in the construction of ratio or regression type estimators increases, no doubt the mean square error of the resultant estimators decreases, but note that at the same time the stability of the estimators decreases.

(e) There are large number of estimators belonging to the same class of estimators with the same minimum asymptotic mean square error, so it is difficult to select an estimator for a particular survey, and there is no theoretical technique available in the literature to select an estimator.

Example 3.2.8.1. Use information given in the population 1 of the Appendix to show the relative efficiency of the general class of estimators over the linear regression estimator while estimating real estate farm loans with the help of known nonreal estate farm loans.

Solution. From the description of the population 1 given in the Appendix we have $\overline{Y} = 555.43$, $\overline{X} = 878.16$, $C_y^2 = 1.1086$, $\lambda_{03} = 1.5936$, $\rho_{xy} = 0.8038$, $N = 50$, $\lambda_{12} = 1.0982$, and $\lambda_{04} = 4.5247$.

Now

$$\text{Min.MSE}(\bar{y}_{\text{LR}}) = \left(\frac{1-f}{n}\right)\overline{Y}^2 C_y^2\left(1-\rho_{xy}^2\right) = \left(\frac{1-0.16}{8}\right)(555.43)^2 \times 1.1086 \times \left(1-0.8038^2\right)$$

$$= 12709.55,$$

and the value of minimum $\text{MSE}(\bar{y}_{\text{SJ}})$ is given by

$$\text{Min.MSE}(\bar{y}_{\text{SJ}}) = \left(\frac{1-f}{n}\right)\overline{Y}^2 C_y^2\left[1-\rho_{xy}^2 - \frac{(\lambda_{03}\rho_{xy}-\lambda_{12})^2}{\lambda_{04}-1-\lambda_{03}^2}\right]$$

$$= \left(\frac{1-0.16}{8}\right)(555.43)^2 \times 1.1086 \times \left[1-(0.8038)^2 - \frac{(1.5936\times 0.8038-1.0982)^2}{4.5247-1-(1.5936)^2}\right]$$

$$= 11491.74.$$

Thus percent relative efficiency (RE) of the general class of estimators, $\bar{y}_{\text{SJ}}$, with respect to the linear regression estimator, $\bar{y}_{\text{LR}}$, is given by

$$\text{RE} = V(\bar{y}_{\text{LR}})\times 100/\text{Min.MSE}(\bar{y}_{\text{SJ}}) = \frac{12709.55}{11491.74}\times 100 = 110.59\%.$$

It should be noted that in this case the relative efficiency is independent of sample size n.

The next section of this chapter has been devoted to constructing the unbiased ratio and product type of estimators of the population mean. We will discuss Quenouille's method, interpenetrating sampling method, exactly unbiased ratio and product type estimators, and bias filtration techniques.

3.2.9 METHODS TO REMOVE BIAS FROM RATIO AND PRODUCT TYPE ESTIMATORS

We have observed that the ratio and product type of estimators are biased. Several researchers have attempted to reduce the bias from these estimators. We should also like to discuss a few methods to construct unbiased ratio and product type estimators of population mean before going on to the problems of estimation of finite population variance, correlation coefficient, and regression coefficient.

3.2.9.1 QUENOUILLE'S METHOD

In this method we draw a sample of size $2n$ units from a population of N units by SRSWOR sampling. We divide the sample of $2n$ units into two equal halves each of size n. The sample based on $2n$ units is called the pooled sample. Then we have three biased ratio estimators of the population mean as:

(a) $\bar{y}_{R_1} = \bar{y}_1\left(\dfrac{\bar{X}}{\bar{x}_1}\right)$, where $\bar{y}_1 = n^{-1}\sum_{i=1}^{n} y_i$ and $\bar{x}_1 = n^{-1}\sum_{i=1}^{n} x_i$ are the first half sample means for Y and X variables, respectively;

(b) $\bar{y}_{R_2} = \bar{y}_2\left(\dfrac{\bar{X}}{\bar{x}_2}\right)$, where $\bar{y}_2 = n^{-1}\sum_{i=1}^{n} y_i$ and $\bar{x}_2 = n^{-1}\sum_{i=1}^{n} x_i$ are the second half sample means for Y and X variables, respectively;

(c) $\bar{y}_{R} = \bar{y}\left(\dfrac{\bar{X}}{\bar{x}}\right)$, where $\bar{y} = (2n)^{-1}\sum_{i=1}^{2n} y_i$ and $\bar{x} = (2n)^{-1}\sum_{i=1}^{2n} x_i$ are the sample means for the Y and X variables, respectively, based on the pooled sample.

By following the ratio method of estimation, we have

$$E\left(\bar{y}_{R_1}\right) = \bar{Y} + \left(\frac{1}{n} - \frac{1}{N}\right)\bar{Y}\left(C_x^2 - \rho_{xy}C_xC_y\right), \tag{3.2.9.1}$$

$$E\left(\bar{y}_{R_2}\right) = \bar{Y} + \left(\frac{1}{n} - \frac{1}{N}\right)\bar{Y}\left(C_x^2 - \rho_{xy}C_xC_y\right), \tag{3.2.9.2}$$

and

$$E\left(\bar{y}_{R}\right) = \bar{Y} + \left(\frac{1}{2n} - \frac{1}{N}\right)\bar{Y}\left(C_x^2 - \rho_{xy}C_xC_y\right). \tag{3.2.9.3}$$

Quenouille (1956) considered an estimator of the population mean $\bar{Y}$ as

$$\bar{y}_Q = a\left(\bar{y}_{R_1} + \bar{y}_{R_2}\right) + (1 - 2a)\bar{y}_R \tag{3.2.9.4}$$

where a is a suitably chosen constant such that the bias in the estimator $\bar{y}_Q$ is zero. Thus we have the following theorem:

Theorem 3.2.9.1. The Quenouille's estimator $\bar{y}_Q$ is an unbiased estimator of population mean $\bar{Y}$ if

$$a = -\frac{(N-2n)}{2N}. \tag{3.2.9.5}$$

Proof. We have

$$E(\bar{y}_Q) = E\left[a(\bar{y}_{R_1} + \bar{y}_{R_2}) + (1-2a)\bar{y}_R\right] = a\left[E(\bar{y}_{R_1}) + E(\bar{y}_{R_2})\right] + (1-2a)E(\bar{y}_R)$$

$$= a\left[\bar{Y} + \left(\frac{1}{n} - \frac{1}{N}\right)\bar{Y}\left(C_x^2 - \rho_{xy}C_xC_y\right) + \bar{Y} + \left(\frac{1}{n} - \frac{1}{N}\right)\bar{Y}\left(C_x^2 - \rho_{xy}C_xC_y\right)\right]$$

$$+ (1-2a)\left[\bar{Y} + \left(\frac{1}{2n} - \frac{1}{N}\right)\bar{Y}\left(C_x^2 - \rho_{xy}C_xC_y\right)\right]$$

$$= \bar{Y} + \bar{Y}\left(C_x^2 - \rho_{xy}C_xC_y\right)\left[2a\left(\frac{1}{n} - \frac{1}{N}\right) + (1-2a)\left(\frac{1}{2n} - \frac{1}{N}\right)\right].$$

Evidently the bias in the estimator $\bar{y}_Q$ will be zero if

$$2a\left(\frac{1}{n} - \frac{1}{N}\right) + (1-2a)\left(\frac{1}{2n} - \frac{1}{N}\right) = 0, \quad \text{or if} \quad a = -\frac{(N-2n)}{2N}.$$

Hence the theorem.

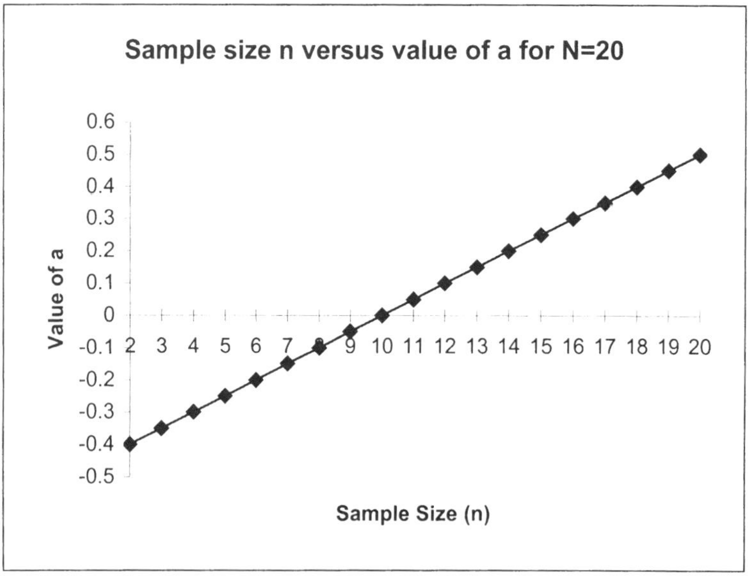

Fig. 3.2.2 Value of Quenouille's constant.

For more details, one can refer to Singh and Singh (1993), Murthy (1962) and Rao (1965a). The reduction in bias to the desired degree by using the method of Quenouille (1956) has also been discussed by Singh (1979).

3.2.9.2 INTERPENETRATING SAMPLING METHOD

Let us first present an idea about the interpenetrating samples. If we want to select n units with SRSWOR sampling, we can select k independent samples each of size $m = n/k$, where we assume that n/k is an integer. We draw m units out of N units, then put back these m units so as to make the population size the same. To make the k samples independent, each individual sample of m units is selected with SRSWOR sampling. Now we have k samples each of size m. From the jth sample, a ratio type estimator to estimate the population mean $\overline{Y}$ is

$$\bar{y}_{\mathrm{R_j}} = \bar{y}_j\left(\frac{\overline{X}}{\bar{x}_j}\right)$$

where $\bar{y}_j = m^{-1}\sum_{i=1}^{m} y_i$ and $\bar{x}_j = m^{-1}\sum_{i=1}^{m} x_i$ denote the jth sample means for the Y and X variables, respectively, for $j = 1, 2, ..., k$. Let us define a new estimator of the population mean $\overline{Y}$ as

$$\bar{y}_{\mathrm{RK}} = \frac{1}{k}\sum_{j=1}^{k} \bar{y}_{\mathrm{R_j}} = \frac{1}{k}\sum_{j=1}^{k} \bar{y}_j\left(\frac{\overline{X}}{\bar{x}_j}\right). \tag{3.2.9.6}$$

Also from the full sample information, we have the usual ratio estimator of population mean $\overline{Y}$ given by

$$\bar{y}_{\mathrm{R}} = \bar{y}\left(\frac{\overline{X}}{\bar{x}}\right)$$

where $\bar{y} = n^{-1}\sum_{i=1}^{n} y_i$ and $\bar{x} = n^{-1}\sum_{i=1}^{n} x_i$ are the sample means based on full sample information. By following the ratio method of estimation, we have

$$E(\bar{y}_{\mathrm{RK}}) = \overline{Y} + \left(\frac{1}{m} - \frac{1}{N}\right)\overline{Y}\left(C_x^2 - \rho_{xy}C_yC_x\right), \tag{3.2.9.7}$$

and

$$E(\bar{y}_{\mathrm{R}}) = \overline{Y} + \left(\frac{1}{n} - \frac{1}{N}\right)\overline{Y}\left(C_x^2 - \rho_{xy}C_yC_x\right). \tag{3.2.9.8}$$

Note that m units are drawn k times from a population of size N which is equivalent to a sample of size $n = km$ is drawn from a population of size kN. Thus we have the following theorem:

Theorem 3.2.9.2. An unbiased estimator of the population mean $\overline{Y}$ is given by

$$\bar{y}_{\mathrm{u}} = \frac{k\bar{y}_{\mathrm{R}} - \bar{y}_{\mathrm{RK}}}{k-1}. \tag{3.2.9.9}$$

Proof. We have

$$E(\bar{y}_{\mathrm{u}}) = E\left[\frac{k\bar{y}_{\mathrm{R}} - \bar{y}_{\mathrm{RK}}}{k-1}\right] = \frac{kE(\bar{y}_{\mathrm{R}}) - E(\bar{y}_{\mathrm{RK}})}{k-1}$$

$$= \frac{k\left[\overline{Y} + \left(\frac{1}{n} - \frac{1}{kN}\right)\overline{Y}\left(C_x^2 - \rho_{xy}C_xC_y\right)\right] - \frac{1}{k}\sum_{j=1}^{k}\left[\overline{Y} + \left(\frac{1}{m} - \frac{1}{N}\right)\overline{Y}\left(C_x^2 - \rho_{xy}C_xC_y\right)\right]}{(k-1)}$$

$$= \frac{(k-1)\overline{Y} + \left(\frac{k}{n} - \frac{1}{N}\right)\overline{Y}\left(C_x^2 - \rho_{xy}C_xC_y\right) - \left(\frac{1}{m} - \frac{1}{N}\right)\overline{Y}\left(C_x^2 - \rho_{xy}C_xC_y\right)}{(k-1)}$$

$$= \frac{(k-1)\overline{Y} + \left(\frac{1}{m} - \frac{1}{N}\right)\overline{Y}\left(C_x^2 - \rho_{xy}C_xC_y\right) - \left(\frac{1}{m} - \frac{1}{N}\right)\overline{Y}\left(C_x^2 - \rho_{xy}C_xC_y\right)}{(k-1)} = \overline{Y}.$$

Hence the theorem.

Theorem 3.2.9.3. The variance of the unbiased estimator $\bar{y}_u$ of the population mean $\overline{Y}$ is

$$V(\bar{y}_u) = \left(\frac{1}{n} - \frac{1}{kN}\right)\left[S_y^2 + R^2S_x^2 - 2RS_{xy}\right]. \tag{3.2.9.10}$$

Proof. We have

$$V(\bar{y}_u) = V\left[\frac{k\bar{y}_R - \bar{y}_{Rk}}{k-1}\right] = \frac{1}{(k-1)^2}\left[k^2V(\bar{y}_R) + V(\bar{y}_{Rk}) - 2k\mathrm{Cov}(\bar{y}_R,\ \bar{y}_{Rk})\right]$$

$$= \frac{1}{(k-1)^2}\left[k^2V\left(\bar{y}\frac{\overline{X}}{\bar{x}}\right) + V\left(\frac{1}{k}\sum_{j=1}^{k}\bar{y}_j\frac{\overline{X}}{\bar{x}_j}\right) - 2k\mathrm{Cov}\left(\frac{1}{k}\sum_{j=1}^{k}\bar{y}\frac{\overline{X}}{\bar{x}},\ \frac{1}{k}\sum_{j=1}^{k}\bar{y}_j\frac{\overline{X}}{\bar{x}_j}\right)\right]$$

$$= \frac{1}{(k-1)^2}\left[k^2V\left(\bar{y}\frac{\overline{X}}{\bar{x}}\right) + \sum_{j=1}^{k}V\left(\bar{y}_j\frac{\overline{X}}{\bar{x}_j}\right)\Big/k^2 - 2\frac{k}{k^2}\sum_{j=1}^{k}\mathrm{Cov}\left(\bar{y}\frac{\overline{X}}{\bar{x}},\ \bar{y}_j\frac{\overline{X}}{\bar{x}_j}\right)\right]$$

$$= \frac{1}{(k-1)^2}\left[k^2\left(\frac{1}{n} - \frac{1}{kN}\right)\left(S_y^2 + R^2S_x^2 - 2RS_{xy}\right) + \frac{k\left(\frac{1}{m} - \frac{1}{N}\right)\left(S_y^2 + R^2S_x^2 - 2RS_{xy}\right)}{k^2}\right.$$
$$\left. - 2\left(\frac{1}{m} - \frac{1}{N}\right)\left(S_y^2 + R^2S_x^2 - 2RS_{xy}\right)\right]$$

$$= \frac{1}{(k-1)^2}\left[\left\{k\left(\frac{1}{m} - \frac{1}{N}\right) + \frac{1}{k}\left(\frac{1}{m} - \frac{1}{N}\right) - 2\left(\frac{1}{m} - \frac{1}{N}\right)\right\}\left(S_y^2 + R^2S_x^2 - 2RS_{xy}\right)\right]$$

$$= \frac{1}{(k-1)^2}\left[\left\{k + \frac{1}{k} - 2\right\}\left(\frac{1}{m} - \frac{1}{N}\right)\left(S_y^2 + R^2S_x^2 - 2RS_{xy}\right)\right]$$

$$= \left(\frac{1}{mk} - \frac{1}{kN}\right)\left(S_y^2 + R^2S_x^2 - 2RS_{xy}\right) = \left(\frac{1}{n} - \frac{1}{kN}\right)\left(S_y^2 + R^2S_x^2 - 2RS_{xy}\right).$$

Hence the theorem.

Note that $k > 1$, thus the unbiased estimator $\bar{y}_{\mathrm{u}}$ is less efficient than the ratio estimator $\bar{y}_{\mathrm{R}}$ in case of finite populations.

If $kN \to \infty$ then $V(\bar{y}_{\mathrm{u}}) \approx \mathrm{MSE}(\bar{y}_{\mathrm{R}})$.

Sengupta (1981a, 1982a) considered the problem of interpenetrating sub-sampling with unequal sizes of the samples and compared with an equicost procedure based on equal sized samples. He observed that unequal sized samples lead to more precise estimates of the finite population mean in almost all cases. He considered the simple random sampling design only and assumed that the cost of the survey is proportional to the number of distinct units in the sample by following Koop (1967), Singh and Bansal (1975, 1978), Singh and Singh (1974) and Srikantan (1963). Schucany, Gray, and Owen (1971) considered the problem of higher order bias reduction in estimating general parameter in survey sampling.

Example 3.2.9.1. Select three different samples each of five units by using SRSWOR sampling from the population 1 given in the Appendix. Collect the information for the real and nonreal estate farm loans from the states selected in each sample. The average nonreal estate farm loan is assumed to be known. Obtain three different ratio estimates of the average real estate farm loans from the information collected in the three samples. Pool the information collected in three samples to obtain a pooled ratio estimate of the average real estate farm loans.
(a) Derive an unbiased estimate of the average real estate farm loans.
(b) Construct 95% confidence interval.
Given: Average nonreal estate farm loans \$878.16.

Solution. Here $N = 50$, $k = 3$, $m = 5$ and $n = mk = 5 \times 3 = 15$. We selected the following three independent samples each of size 5 units. The first sample is selected by using first two columns, the second sample is selected by using 3rd and 4th columns and the third sample is selected by using 5th and 6th columns of the Pseudo-Random Numbers (PRN) given in Table 1 of the Appendix.

Sample I

Random Number $1 \le R_{1i} \le 50$	State	Real estate farm loans, y_i	Nonreal estate farm loans, x_i
01	AL	408.978	348.334
23	MN	1354.768	2466.892
46	VA	321.583	188.477
04	AR	907.700	848.317
32	NY	201.631	426.274
	Sum	3194.660	4278.294

Thus $\bar{y}_1 = 638.932$ and $\bar{x}_1 = 855.6588$.

Sample II

Random Number $1 \le R_{2i} \le 50$	State	Real estate farm loans, y_i	Nonreal estate farm loans, x_i
29	NH	6.044	0.471
14	IN	1213.024	1022.782
47	WA	1100.745	1228.607
22	MI	323.028	440.518
42	TN	553.266	388.869
	Sum	3196.107	3081.247

Thus $\bar{y}_2 = 639.2214$ and $\bar{x}_2 = 616.2494$.

Sample III

Random Number $1 \le R_{3i} \le 50$	State	Real estate farm loans, y_i	Nonreal estate farm loans, x_i
48	WV	99.277	29.291
37	OR	114.899	571.487
33	NC	639.571	494.730
18	LA	282.565	405.799
25	MO	1579.686	1519.994
	Sum	2715.998	3021.301

Thus $\bar{y}_3 = 543.1996$ and $\bar{x}_3 = 604.2602$. It is given that $\bar{X} = 878.16$. Thus three different ratio estimates of the average real estate farm loans in the United States are

$$\bar{y}_{R_1} = \bar{y}_1 \frac{\bar{X}}{\bar{x}_1} = \frac{638.932 \times 878.16}{855.6588} = 655.7339, \quad \bar{y}_{R_2} = \bar{y}_2 \frac{\bar{X}}{\bar{x}_2} = \frac{639.2214 \times 878.16}{616.2494} = 910.8953,$$

and

$$\bar{y}_{R_3} = \bar{y}_3 \frac{\bar{X}}{\bar{x}_3} = \frac{543.1996 \times 878.16}{604.2602} = 789.4218.$$

Thus a pooled estimate from the above three ratio estimates of the average real estate farm loans is given by

$$\bar{y}_{RK} = \frac{1}{3}\sum_{j=1}^{3} \bar{y}_{R_j} = \frac{655.7339 + 910.8953 + 789.4218}{3} = 785.3503.$$

Now we have the pooled sample information as follows:

Pooled Sample

State	y_i	x_i	$(y_i - \bar{y})$	$(x_i - \bar{x})$	$(y_i - \bar{y})^2$	$(x_i - \bar{x})^2$	$(y_i - \bar{y})(x_i - \bar{x})$
AL	408.978	348.334	-198.1400	-343.722	39259.3	118144.9	68104.989
MN	1354.768	2466.892	747.6503	1774.836	558981.0	3150042.0	1326956.627
VA	321.583	188.477	-285.5350	-503.579	81530.1	253591.9	143789.300
AR	907.700	848.317	300.5823	156.261	90349.7	24417.5	46969.256

Continued.....

NY	201.631	426.274	-405.4870	-265.782	164419.4	70640.1	107771.111
NH	6.044	0.471	-601.0740	-691.585	361289.6	478290.0	415693.612
IN	1213.024	1022.782	605.9063	330.726	367122.5	109379.6	200388.897
WA	1100.745	1228.607	493.6273	536.551	243667.9	287886.8	264856.174
MI	323.028	440.518	-284.0900	-251.538	80706.9	63271.4	71459.384
TN	553.266	388.869	-53.8517	-303.187	2900.0	91922.4	16327.132
WV	99.277	29.291	-507.8410	-662.765	257902.1	439257.6	336579.087
OR	114.899	571.487	-492.2190	-120.569	242279.2	14536.9	59346.378
NC	639.571	494.730	32.45333	-197.326	1053.2	38937.6	-6403.891
LA	282.565	405.799	-324.5530	-286.257	105334.4	81943.1	92905.516
MO	1579.686	1519.994	972.5683	827.939	945889.2	685481.1	805226.151
Sum	9106.765	10380.842	0.0000	0.000	3542685.0	5907743.0	3949969.724

Thus $\bar{y} = 607.1176$, $\bar{x} = 692.056$, $s_y^2 = 253048.93$, $s_x^2 = 421981.64$, $s_{xy} = 282140.69$ and $r = \bar{y}/\bar{x} = 0.8773$.

A ratio estimate of the average real estate farm loans from the pooled sample information is given by

$$\bar{y}_R = \bar{y}\frac{\bar{X}}{\bar{x}} = 607.1176 \times \frac{878.16}{692.056} = 770.380 .$$

An unbiased estimate of the average real estate farm loans in the United States is

$$\bar{y}_u = \frac{k\bar{y}_R - \bar{y}_{RK}}{k-1} = \frac{3 \times 770.380 - 785.3503}{3-1} = 762.894 .$$

An estimate for estimating the $V(\bar{y}_u)$ is given by

$$\hat{v}(\bar{y}_u) = \left(\frac{1}{n} - \frac{1}{kN}\right)\left(s_y^2 + r^2 s_x^2 - 2rs_{xy}\right)$$

$$= \left(\frac{1}{15} - \frac{1}{3 \times 50}\right)\left(253048.93 + 0.8773^2 \times 421981.64 - 2 \times 0.8773 \times 282140.69\right)$$

$$= 4967.1166.$$

A $(1-\alpha)100\%$ confidence interval of the population mean $\bar{Y}$ is given by

$$\bar{y}_u \mp t_{\alpha/2}(\text{df} = n-1)\sqrt{\hat{v}(\bar{y}_u)} .$$

Using Table 2 from the Appendix the 95% confidence interval of the average amount of the real estate farm loans in the United States is

$$762.894 \mp 2.145\sqrt{4967.1166}, \quad \text{or} \quad [611.71, \ 914.06].$$

3.2.9.3 EXACTLY UNBIASED RATIO TYPE ESTIMATOR

Hartley and Ross (1954) suggested an ingenious method of proposing an exactly unbiased ratio type estimator of population mean. Suppose we draw a sample of n units from a population of N units. For each unit in the sample, let us define the ratio $r_i = y_i/x_i$, where $i = 1,2,...,n$ and sample mean of the ratios as

$$\bar{r} = \frac{1}{n}\sum_{i=1}^{n} r_i = \frac{1}{n}\sum_{i=1}^{n}\frac{y_i}{x_i}. \tag{3.2.9.11}$$

We will define a ratio estimator of the population mean as

$$\bar{y}_{\mathrm{R}} = \bar{r}\bar{X}. \tag{3.2.9.12}$$

Taking expected values on both sides of (3.2.9.12) we obtain

$$E(\bar{y}_{\mathrm{R}}) = E(\bar{r}\,\bar{X}) = \bar{X}E(\bar{r}) = \bar{X}E\left[\frac{1}{n}\sum_{i=1}^{n}\frac{y_i}{x_i}\right] = \bar{X}\left(\frac{1}{N}\sum_{i=1}^{N}\frac{Y_i}{X_i}\right) \neq \bar{Y}. \tag{3.2.9.13}$$

Note that

$$\bar{R} = \frac{1}{N}\sum_{i=1}^{N}\frac{Y_i}{X_i} \quad \text{and} \quad R_i = \frac{Y_i}{X_i}.$$

Therefore the bias in the estimator $\bar{y}_{\mathrm{R}}$ is given by

$$B(\bar{y}_{\mathrm{R}}) = E(\bar{y}_{\mathrm{R}}) - \bar{Y} = \frac{\bar{X}}{N}\sum_{i=1}^{N}\frac{Y_i}{X_i} - \bar{Y} = \frac{\bar{X}}{N}\sum_{i=1}^{N}\frac{Y_i}{X_i} - \frac{1}{N}\sum_{i=1}^{N}\frac{Y_i}{X_i}X_i = \bar{X}\,\bar{R} - \frac{1}{N}\sum_{i=1}^{N}R_iX_i$$

$$= -\frac{1}{N}\sum_{i=1}^{N}R_iX_i + \bar{R}\,\bar{X} = -\frac{1}{N}\left[\sum_{i=1}^{N}R_iX_i - N\bar{R}\,\bar{X}\right] = -\frac{(N-1)}{N}S_{rx}$$

where

$$S_{rx} = \frac{1}{(N-1)}\left[\sum_{i=1}^{N}R_iX_i - N\bar{R}\,\bar{X}\right] = \frac{1}{N-1}\left[\sum_{i=1}^{N}Y_i - N\bar{R}\,\bar{X}\right] = \frac{N}{(N-1)}\left[\bar{Y} - \bar{R}\,\bar{X}\right].$$

Thus an estimator of S_{rx} is given by

$$s_{rx} = \frac{n}{(n-1)}\left[\bar{y} - \bar{r}\,\bar{x}\right]$$

and hence an estimator of $B(\bar{y}_R)$ is

$$\hat{B}(\bar{y}_R) = -\left(\frac{N-1}{N}\right)\frac{n}{(n-1)}(\bar{y} - \bar{r}\,\bar{x}). \tag{3.2.9.14}$$

Thus we have the following theorem:

Theorem 3.2.9.4. An unbiased ratio type estimator of the population mean $\bar{Y}$ is given by

$$\bar{y}_{\mathrm{HR}} = \bar{r}\,\bar{X} + \frac{(N-1)}{N}\frac{n}{(n-1)}(\bar{y} - \bar{r}\,\bar{x}). \tag{3.2.9.15}$$

Proof. We have

$$E(\bar{y}_{\mathrm{HR}}) = E(\bar{r}\,\bar{X}) + \left(\frac{N-1}{N}\right)E\left[\frac{n}{n-1}(\bar{y} - \bar{r}\,\bar{x})\right] = \bar{X}E\left(\frac{1}{n}\sum_{i=1}^{n}\frac{y_i}{x_i}\right) + \frac{(N-1)}{N}E(s_{rx})$$

$$= \bar{X}\left(\frac{1}{N}\sum_{i=1}^{N}\frac{Y_i}{X_i}\right) + \frac{(N-1)}{N}S_{rx} = \frac{\bar{X}}{N}\sum_{i=1}^{N}\frac{Y_i}{X_i} + \frac{(N-1)}{N} \times \frac{1}{(N-1)}\left[\sum_{i=1}^{N}R_iX_i - N\bar{R}\,\bar{X}\right]$$

$$= \frac{\bar{X}}{N}\sum_{i=1}^{N}\frac{Y_i}{X_i} + \frac{1}{N}\left[\sum_{i=1}^{N}Y_i - N\frac{\bar{X}}{N}\sum_{i=1}^{N}\frac{Y_i}{X_i}\right] = \frac{\bar{X}}{N}\sum_{i=1}^{N}\frac{Y_i}{X_i} + \frac{1}{N}\sum_{i=1}^{N}Y_i - \frac{\bar{X}}{N}\sum_{i=1}^{N}\frac{Y_i}{X_i} = \bar{Y}.$$

Hence the theorem.

Theorem 3.2.9.5. For large samples, the variance of the estimator $\bar{y}_{\text{HR}}$ is given by

$$V(\bar{y}_{\text{HR}}) = \frac{1}{n}\left[\sigma_y^2 + \bar{R}^2\sigma_x^2 - 2\bar{R}\sigma_{xy}\right], \quad \text{where } \bar{R} = \frac{1}{N}\sum_{i=1}^{N}\frac{Y_i}{X_i}. \tag{3.2.9.16}$$

Proof. For large values of n and N, the estimator $\bar{y}_{\text{HR}}$ can be written as

$$\bar{y}_{\text{HR}} = \bar{r}\,\bar{X} + \frac{(N-1)}{N}\frac{n}{(n-1)}(\bar{y} - \bar{r}\,\bar{x}) \approx \bar{r}\,\bar{X} + (\bar{y} - \bar{r}\,\bar{x}) = \bar{y} + \bar{r}\left(\bar{X} - \bar{x}\right).$$

Defining $\tau = \frac{\bar{r}}{\bar{R}} - 1$ such that $E(\tau) \approx 0$. The estimator $\bar{y}_{\text{HR}}$ in terms of ε_0, ε_1 and τ can be written as

$$\bar{y}_{\text{HR}} = \bar{Y}(1+\varepsilon_0) - \bar{R}\bar{X}\varepsilon_1(1+\tau).$$

Now we have

$$V(\bar{y}_{\text{HR}}) = E[\bar{y}_{\text{HR}} - E(\bar{y}_{\text{HR}})]^2 \approx E\left[\bar{Y}(1+\varepsilon_0) - \bar{R}\bar{X}\varepsilon_1 - \bar{Y}\right]^2$$

$$= E\left[\bar{Y}^2\varepsilon_0^2 + \bar{R}^2\bar{X}^2\varepsilon_1^2 - 2\bar{R}\,\bar{X}\varepsilon_0\varepsilon_1\right] \approx \frac{1}{n}\left[\bar{Y}^2C_y^2 + \bar{R}^2\bar{X}^2C_x^2 - 2\bar{R}\,\bar{X}\rho_{xy}C_xC_y\right]$$

$$= \frac{1}{n}\left[S_y^2 + \bar{R}^2S_x^2 - 2\bar{R}S_{xy}\right] \approx \frac{1}{n}\left[\sigma_y^2 + \bar{R}^2\sigma_x^2 - 2\bar{R}\sigma_{xy}\right].$$

Hence the theorem.

Remark 3.2.9.1. Exact variance of the estimator $\bar{y}_{\text{HR}}$ is available in Robson (1957).

Theorem 3.2.9.6. In case of infinite population, the unbiased ratio estimator $\bar{y}_{\text{HR}}$ is more efficient than the usual ratio estimator $\bar{y}_{\text{R}}$ if either

$$\left(\bar{R} - R\right) < 0 \quad \text{and} \quad \beta < \frac{\bar{R}+R}{2} \quad \text{or} \quad \left(\bar{R} - R\right) > 0 \text{ and } \beta > \frac{\bar{R}+R}{2}.$$

Proof. For large values of n and N we have

$$\text{MSE}(\bar{y}_{\text{R}}) = \frac{1}{n}\left[\sigma_y^2 + R^2\sigma_x^2 - 2R\sigma_{xy}\right] \text{ and } V(\bar{y}_{\text{HR}}) = \frac{1}{n}\left[\sigma_y^2 + \bar{R}^2\sigma_x^2 - 2\bar{R}\sigma_{xy}\right].$$

Now the estimator $\bar{y}_{\text{HR}}$ is more efficient than $\bar{y}_{\text{R}}$ if

$$V(\bar{y}_{\text{HR}}) < MSE(\bar{y}_{\text{R}})$$

$$\text{or if } \frac{1}{n}\left[\sigma_y^2 + \bar{R}^2\sigma_x^2 - 2\bar{R}\sigma_{xy}\right] < \frac{1}{n}\left[\sigma_y^2 + R^2\sigma_x^2 - 2R\sigma_{xy}\right]$$

$$\text{or if } \left(\bar{R}^2 - R^2\right)\sigma_x^2 - 2\left(\bar{R} - R\right)\sigma_{xy} < 0$$

or if $\left(\overline{R}+R\right)\left(\overline{R}-R\right)\sigma_x^2-2\left(\overline{R}-R\right)\sigma_{xy}<0$

or if $\left(\overline{R}-R\right)\sigma_x^2\left[\left(\overline{R}+R\right)-2\beta\right]<0$.

The above condition will hold if either

$$\left(\overline{R}-R\right)<0 \text{ and } \beta<\frac{\overline{R}+R}{2} \text{ or } \left(\overline{R}-R\right)>0 \text{ and } \beta>\frac{\overline{R}+R}{2}.$$

Hence the theorem.

Example 3.2.9.2. The estimation of the average amount of real estate farm loans (in \$000) is an important issue for the banks operating in the United States of America. A statistician suggests that they consider the terminology of unbiased estimation proposed by Hartley and Ross (1954). Select an SRSWR sample of eight states from the population 1 given in the Appendix and derive the appropriate estimate of the real estate farm loans, given that the average amount \$878.16 of nonreal estate farm loans (in \$000) for the year 1997 is known.

Solution. An SRSWR sample of eight states is selected by using the 31st and 32th columns of the Pseudo-Random Numbers (PRN) given in Table 1 of the Appendix as: 17, 36, 50, 50, 31, 05, 18, and 50. Note that the state WY has been selected three times in the sample.

Random No.	State	x_i	y_i	$(x_i-\bar{x})^2$	$(y_i-\bar{y})^2$	$(y_i-\bar{y})(x_i-\bar{x})$	$r_i=\frac{y_i}{x_i}$
5	CA	3928.732	1343.461	8546932.65	770219.936	2565739.2530	0.341958
17	KY	557.656	1045.106	200311.97	335549.968	-259257.9300	1.874105
18	LA	405.799	282.565	359303.44	33589.451	109858.1135	0.696318
31	NM	274.035	140.582	534628.95	105792.279	237822.6531	0.513007
36	OK	1716.087	612.108	505334.38	21394.547	103977.8835	0.356688
50	WY	386.479	100.964	382838.26	133133.948	225762.6385	0.261241
50	WY	386.479	100.964	382838.26	133133.948	225762.6385	0.261241
50	WY	386.479	100.964	382838.26	133133.948	225762.6385	0.261241
	Sum	8041.746	3726.714	11295026.15	1665948.025	3435427.8890	4.565798

From the above table:

$n=8$, $\bar{y}=465.8393$, $\bar{x}=1005.218$, $s_x^2=1613575.16$, $s_y^2=237992.57$,

$s_{xy}=490775.41$, and $\bar{r}=\frac{1}{n}\sum_{i=1}^{n}\frac{y_i}{x_i}=\frac{4.565798}{8}=0.57072$.

Also $\overline{X}=878.16$, $N=50$ and $f=0.16$ are given.

Thus the Hartley and Ross (1954) unbiased ratio estimate of average amount of real estate farm loans during 1997, $\overline{Y}$ (say), is given by

$$\bar{y}_{\text{HR}} = \bar{r}\bar{X} + \frac{(N-1)}{N}\frac{n}{(n-1)}(\bar{y} - \bar{r}\ \bar{x})$$

$$= 0.57072 \times 878.16 + \frac{8(50-1)}{50(8-1)}(465.8393 - 0.57072 \times 1005.218)$$

$$= 501.183 - 120.802 = 380.381\ .$$

A rough estimate of $V(\bar{y}_{\text{HR}})$ is given by

$$\hat{v}(\bar{y}_{\text{HR}}) = \frac{1}{n}\left[s_y^2 + \bar{r}^2 s_x^2 - 2\bar{r} s_{xy}\right]$$

$$= \frac{1}{8}\left[237992.57 + (0.57072)^2 \times 1613575.16 - 2 \times 0.57072 \times 490775.41\right]$$

$$= 25422.21\ .$$

A $(1-\alpha)100\%$ confidence interval for population mean $\bar{Y}$ is given by

$$\bar{y}_{\text{HR}} \mp t_{\alpha/2}(\text{df} = n-1)\sqrt{\hat{v}(\bar{y}_{\text{HR}})}\ .$$

Using Table 2 from the Appendix the 95% confidence interval is given by

$$380.381 \pm t_{0.05/2}(\text{df} = 8-1)\sqrt{25422.21}$$

or $380.381 \pm t_{0.025}(\text{df} = 7)\sqrt{25422.21}$

or $380.381 \pm 2.365\sqrt{25422.21}$ or $[3.29,\ 757.46]$.

3.2.9.4 UNBIASED PRODUCT TYPE ESTIMATOR

The usual product estimator of the population mean $\bar{Y}$ is defined as

$$\bar{y}_{\text{p}} = \bar{y}\left(\frac{\bar{x}}{\bar{X}}\right). \tag{3.2.9.17}$$

The bias in the estimator $\bar{y}_{\text{p}}$ is given by

$$B(\bar{y}_{\text{p}}) = \left(\frac{1-f}{n}\right)\bar{Y}\rho_{xy}C_yC_x = \left(\frac{1-f}{n}\right)\bar{Y}\frac{S_{xy}}{S_xS_y}\frac{S_y}{\bar{Y}}\frac{S_x}{\bar{X}} = \left(\frac{1-f}{n}\right)\frac{S_{xy}}{\bar{X}}. \tag{3.2.9.18}$$

Thus an unbiased estimator of bias in the product estimator is given by

$$\hat{B}(\bar{y}_{\text{p}}) = \left(\frac{1-f}{n}\right)\frac{s_{xy}}{\bar{X}}. \tag{3.2.9.19}$$

Thus we have the following theorem:

Theorem 3.2.9.7. A product type unbiased estimator of the population mean is:

$$\bar{y}_{\text{pu}} = \bar{y}\left(\frac{\bar{x}}{\bar{X}}\right) - \left(\frac{1-f}{n}\right)\frac{s_{xy}}{\bar{X}}. \tag{3.2.9.20}$$

Proof. The estimator $\bar{y}_{\text{pu}}$ in terms of ε_0, ε_1 and ε_4 can be written as

$$\bar{y}_{\text{pu}} = \bar{y}\left(\frac{\bar{x}}{\bar{X}}\right) - \left(\frac{1-f}{n}\right)\frac{s_{xy}}{\bar{X}} = \bar{Y}(1+\varepsilon_0)\frac{\bar{X}(1+\varepsilon_1)}{\bar{X}} - \left(\frac{1-f}{n}\right)\frac{S_{xy}(1+\varepsilon_4)}{\bar{X}}$$

$$= \bar{Y}(1+\varepsilon_0)(1+\varepsilon_1) - \left(\frac{1-f}{n}\right)\frac{S_{xy}(1+\varepsilon_4)}{\bar{X}}$$

$$= \bar{Y}(1+\varepsilon_0+\varepsilon_1+\varepsilon_0\varepsilon_1) - \left(\frac{1-f}{n}\right)\frac{S_{xy}(1+\varepsilon_4)}{\bar{X}}.$$

Taking expected values on both sides we obtain

$$E(\bar{y}_{pu}) = \bar{Y}\left[1+0+0+\left(\frac{1-f}{n}\right)\rho_{xy}C_xC_y\right] - \left(\frac{1-f}{n}\right)\frac{S_{xy}(1+0)}{\bar{X}}$$

$$= \bar{Y} + \left(\frac{1-f}{n}\right)\bar{Y}\rho_{xy}C_xC_y - \left(\frac{1-f}{n}\right)\frac{S_{xy}}{\bar{X}}$$

$$= \bar{Y} + \left(\frac{1-f}{n}\right)\bar{Y}\frac{S_{xy}}{S_xS_y}\frac{S_x}{\bar{X}}\frac{S_y}{\bar{Y}} - \left(\frac{1-f}{n}\right)\frac{S_{xy}}{\bar{X}}$$

$$= \bar{Y} + \left(\frac{1-f}{n}\right)\frac{S_{xy}}{\bar{X}} - \left(\frac{1-f}{n}\right)\frac{S_{xy}}{\bar{X}} = \bar{Y}.$$

Hence the theorem.

Theorem 3.2.9.8. The variance $V(\bar{y}_{pu})$ to the first order of approximation is same as

$$V(\bar{y}_{pu}) = \left(\frac{1-f}{n}\right)\left[S_y^2 + R^2S_x^2 + 2RS_{xy}\right]. \tag{3.2.9.21}$$

Proof. The variance of the estimator $\bar{y}_{pu}$ is given by

$$V(\bar{y}_{pu}) = E\left[\bar{y}_{pu} - \bar{Y}\right]^2 = E\left[\bar{Y}(1+\varepsilon_0+\varepsilon_1+\varepsilon_0\varepsilon_1) - \left(\frac{1-f}{n}\right)\frac{S_{xy}(1+\varepsilon_4)}{\bar{X}} - \bar{Y}\right]^2$$

$$\approx \left(\frac{1-f}{n}\right)\left[S_y^2 + R^2S_x^2 + 2RS_{xy}\right] - \text{MSE}(\bar{y}_p).$$

Hence the theorem.

Example 3.2.9.3. It is a well known phenomenon that as people become older their sleeping time reduces. A psychologist wants to study the average duration of sleep (in minutes) during the night for the persons 50 years of age or over in a small village of the United States. Instead of asking everybody, the psychologist selects an SRSWOR sample of six persons and records the information as given below

Person no.	7	12	15	19	24	29
Age x (years)	67	70	53	77	87	66
Duration of sleep y (in minutes)	420	360	510	330	270	390

The average age of 67.267 years of the subjects is known as shown in the population 2 in the Appendix. Apply the unbiased product method of estimation for estimating the average sleep time in the particular village under study. Also find an estimator of the variance of the unbiased product estimator and hence deduce a 95% confidence interval.

Solution. From the sample information, we have

Sr. No.	y_i	x_i	$(y_i-\bar{y})^2$	$(x_i-\bar{x})^2$	$(y_i-\bar{y})(x_i-\bar{x})$
7	420	67	1600	9	-120
12	360	70	400	0	0
15	510	53	16900	289	-2210
19	330	77	2500	49	-350
24	270	87	12100	289	-1870
29	390	66	100	16	-40
Sum	2280	420	33600	652	-4590

We have Y_i = Duration of sleep (in minutes), X_i = Age of subjects $(\geq 50$ years$)$, $n=6$, $\bar{y}=380$, $\bar{x}=70$, $s_x^2=130.4$, $s_y^2=6720$, $s_{xy}=-918$ and $r=\bar{y}/\bar{x}=5.4286$. Also we are given $\bar{X}=67.267$, $N=30$ and $f=0.20$.

Thus unbiased product estimate of average sleep time, $\bar{Y}$ (say), is given by

$$\bar{y}_{\text{pu}}=\bar{y}\left(\frac{\bar{x}}{\bar{X}}\right)-\left(\frac{1-f}{n}\right)\frac{s_{xy}}{\bar{X}}=380\left(\frac{70}{67.267}\right)-\left(\frac{1-0.20}{6}\right)\frac{(-918)}{67.267}=397.26$$

and an estimate of approximate $V(\bar{y}_{\text{pu}})$ is given by

$$\hat{v}(\bar{y}_{\text{pu}})=\left(\frac{1-f}{n}\right)\left[s_y^2+r^2s_x^2+2rs_{xy}\right]$$

$$=\left(\frac{1-0.20}{6}\right)\left[6720+(5.4286)^2\times 130.4-2\times 5.4286\times 918\right]=79.458\,.$$

A $(1-\alpha)100\%$ confidence interval for population mean $\bar{Y}$ is given by

$$\bar{y}_{\text{pu}}\pm t_{\alpha/2}(\text{df}=n-2)\sqrt{\hat{v}(\bar{y}_{\text{pu}})}\ .$$

Using Table 2 from the Appendix the 95% confidence interval is given by

$$397.26\pm 2.776\sqrt{79.458}\quad\text{or}\quad[372.51,\ 422.00].$$

3.2.9.5 CLASS OF ALMOST UNBIASED ESTIMATORS OF POPULATION RATIO AND PRODUCT

Singh and Sahoo (1989) have considered the problem of estimation of the parameter $D=Y\bar{Y}/\bar{X}^{\delta}$, and defined a generalized estimator d_g of the form

$$d_g=\left(\bar{y}/\bar{x}^{\delta}\right) \tag{3.2.9.22}$$

where δ is a constant which takes the value $+1$ and -1 according as we wish to estimate population ratio or product. The estimator d_g in terms of ε_0 and ε_1 can be written as:

$$d_g = \frac{\bar{Y}(1+\varepsilon_0)}{\bar{X}^{\delta}(1+\varepsilon_1)^{\delta}} = D(1+\varepsilon_0)(1+\varepsilon_1)^{-\delta} = D(1+\varepsilon_0)\left(1-\delta\varepsilon_1+\frac{\delta(\delta+1)}{2}\varepsilon_1^2+..\right)$$

$$= D\left[1+\varepsilon_0-\delta\varepsilon_1+\frac{\delta(\delta+1)}{2}\varepsilon_1^2-\delta\varepsilon_0\varepsilon_1+...\right]. \tag{3.2.9.23}$$

Taking expected values on both sides of (3.2.9.23) we obtain

$$E(d_g) = D\left[1+\delta\left\{\frac{(\delta+1)}{2}C_x^2-\rho_{xy}C_xC_y\right\}\right] = D\left[1+\delta\left\{\frac{(\delta+1)}{2}\left(\frac{S_x^2}{\bar{X}^2}\right)-\frac{S_{xy}}{\bar{X}\,\bar{Y}}\right\}\right]$$

or

$$E(d_g) = D\left[1+\delta\left\{\frac{(\delta+1)}{2}C_x^2-C_{xy}\right\}\right], \quad \text{where } C_{xy} = S_{xy}/(\bar{X}\,\bar{Y}). \tag{3.2.9.24}$$

Employing the techniques developed by Beale (1962), Tin (1965) and Sahoo (1983), we may construct three almost unbiased estimators, up to terms of order $O(n^{-1})$, for D as

$$d_B = d[1+\delta c_{yx}]\left[1+\frac{\delta(\delta+1)}{2}c_x^2\right]^{-1}, \tag{3.2.9.25}$$

$$d_T = d\left[1+\delta\left(c_{yx}-\frac{(\delta+1)}{2}c_x^2\right)\right], \tag{3.2.9.26}$$

and

$$d_S = d\left[1+\delta\left(\frac{\delta+1}{2}c_x^2-c_{yx}\right)\right]^{-1}, \tag{3.2.9.27}$$

respectively, where $c_x^2 = s_x^2/\bar{x}^2$ and $c_{xy} = s_{xy}/(\bar{x}\,\bar{y})$ are the estimators of C_x^2 and C_{xy}, respectively. Singh and Sahoo (1989) proposed a class of almost unbiased ratio and product estimators. According to them, whatever the sample chosen let $t = (\delta,\, c_{xy},\, c_x^2)$ assume values in a bounded closed convex subset S of the three dimensional real space containing the point $T = (\delta,\, C_{xy},\, C_x^2)$. Let $f(t)$ be a function of t satisfying the following conditions:

(a) The function $f(t)$ is continuous in S;
(b) first and second order partial derivatives of $f(t)$ exist and are continuous in S;
(c) After the expansion under the given conditions

$$f(t) = 1+\delta\left(c_{xy}-\frac{(\delta+1)}{2}c_x^2\right) + \text{higher order terms}.$$

These conditions taken together are called regularity conditions and any situation where these conditions hold is called a regular estimation case. Then we have the following theorem:

Theorem 3.2.9.9. Let $f(t)$ be any function of δ, c_{xy} and c_x^2 satisfying the above conditions. Then a class of almost unbiased estimators of D may be defined as

$$d_{ss} = df(t). \tag{3.2.9.28}$$

The class of estimators represented by d_{ss} gives us an infinite number of almost unbiased ratio and product estimators by substituting a proper choice of $f(t)$. It is easy to see that the estimators suggested by Beale (1962), Tin (1965), Sahoo (1983), Robson (1957) and the estimators of the type:

$$d_1 = d\left(1+\delta c_{xy}\right)\left[1-\frac{\delta(\delta+1)}{2}c_x^2\right]; \qquad d_2 = d\left(1-\delta c_{xy}\right)^{-1}\left[1-\frac{\delta(\delta+1)}{2}c_x^2\right];$$

$$d_3 = d\left(1-\delta c_{xy}\right)^{-1}\left[1+\frac{\delta(\delta+1)}{2}c_x^2\right]^{-1}; \quad d_4 = d\left[1+\frac{\delta(\delta+1)}{2}c_x^2\right]^{-1}\exp\left(\delta c_{xy}\right);$$

$$d_5 = d\left(1+\delta c_{xy}\right)\exp\left[-\frac{\delta(\delta+1)}{2}c_x^2\right]; \quad d_6 = d\exp\left[\delta c_{yx}-\frac{\delta(\delta+1)}{2}c_x^2\right];$$

$$d_7 = d\left[1+\log\left(1+dc_{xy}-\frac{\delta(\delta+1)}{2}c_x^2\right)\right]; \text{ and } d_8 = d\left[1+\log\left(1-\delta\left(\frac{\delta(\delta+1)}{2}c_x^2-c_{xy}\right)\right)\right]$$

etc. for $\delta = \pm 1$ are the members of the class d_{ss}.

3.2.9.6 FILTRATION OF BIAS

Singh and Singh (1991, 1993a, 1993b, 1993c) have suggested a new method to separate the bias precipitates from the ratio and product type estimators by using a funnel connected with a filter paper. The apparatus consists of a linear variety of estimators and linear constraints. Consider

$\hat{R}_i = \bar{y}\left(\bar{X}/\bar{x}\right)^i$, such that $\hat{R}_i \in G$ for $i = 1,2,3$

where G denotes the set of all possible ratio type estimators for estimating the population mean $\bar{Y}$.

By definition the set G will be a linear variety if

$$\hat{R}_s = \sum_{i=1}^{3} g_i \hat{R}_i \in G \tag{3.2.9.29}$$

for

$$\sum_{i=1}^{3} g_i = 1 \text{ and } g_i \in R \tag{3.2.9.30}$$

where g_i $(i = 1,2,3)$ denotes the amount of chemicals used for separating bias precipitates and R denotes the set of real numbers. Using (3.2.9.30), the relation (3.2.9.29) in terms of ε_0 and ε_1 may be written as

$$\hat{R}_s = \bar{Y} + \bar{Y}\left[\varepsilon_0 - \left(g_1 + 2g_2 + 3g_3\right)\varepsilon_1 + O\left(\varepsilon^2\right)\right]. \tag{3.2.9.31}$$

Let us choose

$$g_1 + 2g_2 + 3g_3 = K \text{ (say, another constant).} \tag{3.2.9.32}$$

Then the mean square error (MSE) to the first order of approximation is given by

$$\mathrm{MSE}\left(\hat{R}_s\right)=\left(\frac{1-f}{n}\right)\bar{Y}^2\left[C_y^2+K^2C_x^2-2K\rho_{xy}C_xC_y\right]. \tag{3.2.9.33}$$

The MSE (up to terms of order $O\left(n^{-1}\right)$) is minimum for

$$K=\rho_{xy}\,C_y/C_x \tag{3.2.9.34}$$

and minimum mean squared error is given by

$$\mathrm{Min.MSE}\left(\hat{R}_s\right)=\left(\frac{1-f}{n}\right)S_y^2\left(1-\rho_{xy}^2\right). \tag{3.2.9.35}$$

From (3.2.9.30), (3.2.9.32) and (3.2.9.34), Singh and Singh (1991, 1993a, 1993b, 1993c) made a funnel consisting of two equations given by

$$\sum_{i=1}^{3} g_i=1 \tag{3.2.9.36}$$

and

$$\sum_{i=1}^{3} i g_i=\rho_{xy}\frac{C_y}{C_x}. \tag{3.2.9.37}$$

In the above two equations, we have three unknowns to be determined. It is therefore not possible to find out unique values for the amount of chemicals g_1, g_2 and g_3 to be used for filtration. They claimed that their funnel is handicapped and cannot be used without a filter paper. In order to get unique values of these chemicals, they suggested using a filter paper with the funnel by imposing a linear constraint as:

$$\sum_{i=1}^{3} g_i B\left(\hat{R}_i\right)=0 \tag{3.2.9.38}$$

where $B\left(\hat{R}_i\right)$ denotes the bias in the $\hat{R}_i$, $i=1,2,3$, of population mean and the above three equations can be written as

$$\begin{bmatrix}1, & 1, & 1\\ 1, & 2, & 3\\ B\left(\hat{R}_1\right), & B\left(\hat{R}_2\right), & B\left(\hat{R}_3\right)\end{bmatrix}\begin{bmatrix}g_1\\ g_2\\ g_3\end{bmatrix}=\begin{bmatrix}1\\ K\\ 0\end{bmatrix} \text{ or } A_{3\times 3}G_{3\times 1}=K_{3\times 1}. \tag{3.2.9.39}$$

The values of g_i, $i=1,2,3$ obtained by solving the system of equations (3.2.8.39) separates the bias precipitates from the linear variety of estimators if $|A|\neq 0$, *i.e.* if

$$B\left(\hat{R}_2\right)\neq\frac{B\left(\hat{R}_1\right)+B\left(\hat{R}_3\right)}{2}. \tag{3.2.9.40}$$

Thus we have the following theorem:

Theorem 3.2.9.10. The values of g_i, $i=1,2,3$ which filter the bias up to terms of first order are given by

$$g_1=3-3K+K^2,\quad g_2=-3+5K-2K^2,\quad \text{and}\quad g_3=1-2K+K^2. \tag{3.2.9.41}$$

Proof. To the first order of approximation we have

$$B(\hat{R}_1) = \left(\frac{1-f}{n}\right)\bar{Y}\left[C_x^2 - \rho_{xy}C_yC_x\right], \tag{3.2.9.42}$$

$$B(\hat{R}_2) = \left(\frac{1-f}{n}\right)\bar{Y}\left[3C_x^2 - 2\rho_{xy}C_yC_x\right], \tag{3.2.9.43}$$

and

$$B(\hat{R}_3) = \left(\frac{1-f}{n}\right)\bar{Y}\left[6C_x^2 - 3\rho_{xy}C_yC_x\right]. \tag{3.2.9.44}$$

On substituting these values of biases in (3.2.9.39), we obtain the values of g_i given in the theorem.

Remark 3.2.9.2. If someone is interested in reducing the bias of second order, then $B(\hat{R}_i), i = 1,2,3$ can be obtained to the second order of approximation by following Gupta and Kothwala (1990).

Example 3.2.9.4. Ms. Stephanie Singh wishes to estimate the average amount of real estate farm loans (in \$000) during 1997 in the United States using known information about the nonreal estate farm loans. She would like to apply the method of filtration of bias from the ratio method of estimation. Suggest her the values of the required real constants $g_i, i = 1,2,3$.

Given: $\rho_{xy} = 0.8038$, $C_x^2 = 1.5256$ and $C_y^2 = 1.1086$.

Solution. Here $\rho_{xy}\dfrac{C_y}{C_x} = 0.8038\sqrt{\dfrac{1.1086}{1.5256}} = 0.6852$, thus values of g_1, g_2 and g_3 are:

$$g_1 = 3 - 3\rho_{xy}\frac{C_y}{C_x} + \rho_{xy}^2\frac{C_y^2}{C_x^2} = 3 - 3\times 0.6852 + 0.6852^2 = 1.414,$$

$$g_2 = -3 + 5\rho_{xy}\frac{C_y}{C_x} - 2\rho_{xy}^2\frac{C_y^2}{C_x^2} = -3 + 5\times 0.6852 - 2\times 0.6852^2 = -0.513,$$

and

$$g_3 = 1 - 2\rho_{xy}\frac{C_y}{C_x} + \rho_{xy}^2\frac{C_y^2}{C_x^2} = 1 - 2\times 0.6852 + 0.6852^2 = 0.099.$$

Example 3.2.9.5. Ms. Stephanie Singh considers the problem of estimating the average amount of real estate farm loans (in \$000) during 1997 in the United States. She takes an SRSWOR sample of eight states from the population 1 of the Appendix and collects the following information:

State	UT	NV	ME	IL	CT	CA	AZ	OH
Nonreal estate farm loans (X)\$	197.244	16.710	51.539	2610.572	4.373	3928.732	431.439	635.774
Real estate farm loans (Y)\$	56.908	5.860	8.849	2131.048	7.130	1343.461	54.633	870.720

The average amount \$878.16 of nonreal estate farm loans (in \$000) for the year 1997 is known. She wishes to use three estimators $\hat{R}_i = \bar{y}\left(\bar{X}/\bar{x}\right)^i$, such that $\hat{R}_i \in G$ for $i = 1,2,3$, where G denotes the set of all possible ratio type estimators for estimating the population mean $\bar{Y}$. Discuss her estimate based on the estimator

$$\hat{R}_s = \sum_{i=1}^{3} g_i \hat{R}_i \in G \text{ for } \sum_{i=1}^{3} g_i = 1 \text{ and } g_i \in R .$$

Also find an estimator of the mean squared error of the estimator $\hat{R}_s$ and hence deduce a 95% confidence interval.

Given: $g_1 = 1.414$, $g_2 = -0.513$ and $g_3 = 0.099$.

Solution. From the sample information we have

States	x_i	y_i	$(x_i - \bar{x})^2$	$(y_i - \bar{y})^2$	$(y_i - \bar{y})(x_i - \bar{x})$
UT	197.244	56.908	619847.3916	252926.6405	395949.3886
NV	16.710	5.860	936710.1523	306878.4676	536149.3972
ME	51.539	8.849	870505.5608	303575.7923	514066.5475
IL	2610.572	2131.048	2643954.4550	2468738.1800	2554844.6740
CT	4.373	7.130	960742.7856	305473.0066	541738.8552
CA	3928.732	1343.461	8668220.1620	614083.6173	2307165.3590
AZ	431.439	54.633	305929.4276	255220.0935	279426.8010
OH	635.774	870.720	121643.2159	96655.0015	-108431.6615
Sum	7876.383	4478.609	15127553.1500	4603550.8000	7020909.3610

Here $n = 8$, $\bar{y} = 559.8261$, $\bar{x} = 984.5479$, $s_x^2 = 2161079.02$, $s_y^2 = 657650.11$, $s_{xy} = 1002987.05$, and $r_{xy} = s_{xy}/\left(s_x s_y\right) = 0.8413$. Also given $\bar{X} = 878.16$, $N = 50$ and $f = 0.16$.

Thus the unbiased ratio type estimate of average amount of real estate farm loans during 1997, $\bar{Y}$ (for example), is given by

$$\hat{R}_s = \sum_{i=1}^{3} g_i \bar{y}\left(\frac{\bar{X}}{\bar{x}}\right)^i$$

$$= 1.414 \times 559.8261\left(\frac{878.16}{984.5479}\right) - 0.513 \times 559.8261\left(\frac{878.16}{984.5479}\right)^2$$

$$+ 0.099 \times 559.8261\left(\frac{878.16}{984.5479}\right)^3$$

$$= 706.05 - 228.47 + 39.33 = 516.91 \; .$$

and an estimator of $\text{MSE}\left(\hat{R}_s\right)$ is given by

$$\hat{\text{MSE}}\left(\hat{R}_s\right) = \left(\frac{1-f}{n}\right)s_y^2\left(1-r_{xy}^2\right) = \left(\frac{1-0.16}{8}\right)\times 657650.11\times\left[1-(0.8413)^2\right] = 20178.35\,.$$

A $(1-\alpha)100\%$ confidence interval for population mean $\bar{Y}$ is given by

$$\hat{R}_s \pm t_{\alpha/2}(\text{df} = n-1)\sqrt{\hat{\text{MSE}}\left(\hat{R}_s\right)}\,.$$

Using Table 2 from the Appendix the 95% confidence interval is given by

$$516.91 \pm 2.365\sqrt{20178.35} \text{ or } [180.96, 852.86]\,.$$

Unbiased ratio and product type estimators have also been discussed by Tukey (1958), Durbin (1959), Murthy and Nanjamma (1959), Nieto de Pascual (1961), Rao (1966), Rao and Webster (1966), Murthy (1967), Rao (1969), Hutchison (1971), Schucany, Gray, and Owen (1971), Sharot (1976) and Rao (1981). A complete review on the reduction of bias using auxiliary information can be had from Sahoo, Sahoo, and Wywial (1997). Williams (1958, 1961) compared the two unbiased regression estimators, one suggested by himself and other by Mickey (1959). Williams (1963) compared the precision of some unbiased regression estimators. Rao (1967) investigated in a variety of natural populations the performance of Mickey's estimator and concluded that it is usually inferior to standard ratio and regression estimators. Sahoo (1994) compared the efficiency of William's (1963) estimator with the standard regression estimator and found through an empirical study that the former is usually inferior to the latter. Rao and Rao (1971) investigated the exact efficiency of Mickey's unbiased ratio type estimator.

The next section has been devoted to estimating the finite population variance in the presence of auxiliary information.

3.3 ESTIMATION OF FINITE POPULATION VARIANCE

We consider the problem of estimation of finite population variance (or population mean square error) of the study variable Y defined by

$$S_y^2 = (N-1)^{-1}\sum_{i=1}^{N}\left(Y_i - \bar{Y}\right)^2 \qquad (3.3.1)$$

in the presence of the known population mean

$$\bar{X} = N^{-1}\sum_{i=1}^{N} X_i$$

and the known population variance

$$S_x^2 = (N-1)^{-1}\sum_{i=1}^{N}\left(X_i - \bar{X}\right)^2$$

of the auxiliary variable, X. Note that the admissibility of the usual estimator of finite population variance has been studied by Strauss (1982), but we will discuss a few estimators of S_y^2.

3.3.1 RATIO TYPE ESTIMATOR

Isaki (1983) proposed a ratio type estimator of finite population variance S_y^2 as

$$s_1^2 = s_y^2\left(S_x^2 / s_x^2\right). \tag{3.3.1.1}$$

The estimator s_1^2 in terms of ε_2 and ε_3 can easily be written as

$$s_1^2 = S_y^2(1+\varepsilon_2)(1+\varepsilon_3)^{-1} = S_y^2(1+\varepsilon_2)(1-\varepsilon_3+\varepsilon_3^2+...)$$

$$= S_y^2\left[1+\varepsilon_2-\varepsilon_3+\varepsilon_3^2-\varepsilon_2\varepsilon_3+....\right]. \tag{3.3.1.2}$$

Thus we have the following theorems:

Theorem 3.3.1.1. The bias, up to terms of order $O(n^{-1})$, in the estimator s_1^2 is

$$B\left(s_1^2\right) = \left(\frac{1-f}{n}\right)S_y^2(\lambda_{04}-\lambda_{22}). \tag{3.3.1.3}$$

Proof. Taking expected values on both sides of (3.3.1.2) we have

$$E\left(s_1^2\right) = S_y^2 E\left[1+\varepsilon_2-\varepsilon_3+\varepsilon_3^2-\varepsilon_2\varepsilon_3\right] = S_y^2\left[1+\left(\frac{1-f}{n}\right)\{(\lambda_{04}-1)-(\lambda_{22}-1)\}\right]$$

and using the result $B\left(s_1^2\right) = E\left(s_1^2\right) - S_y^2$ we have (3.3.1.3).
Hence the theorem.

Theorem 3.3.1.2. The mean squared error of the estimator s_1^2, up to the first order of approximation, is

$$\text{MSE}\left(s_1^2\right) = \left(\frac{1-f}{n}\right)S_y^4[\lambda_{40}+\lambda_{04}-2\lambda_{22}]. \tag{3.3.1.4}$$

Proof. We have

$$\text{MSE}\left(s_1^2\right) = E\left[s_1^2 - S_y^2\right]^2 \approx E\left[S_y^2\left(1+\in_2-\in_3+\in_3^2-\in_2\in_3+....\right)-S_y^2\right]^2$$

$$\approx S_y^4 E[\in_2-\in_3]^2 = S_y^4 E\left[\in_2^2+\in_3^2-2\in_2\in_3\right]$$

$$= \left(\frac{1-f}{n}\right)S_y^4[(\lambda_{40}-1)+(\lambda_{04}-1)-2(\lambda_{22}-1)].$$

Hence the theorem.

Theorem 3.3.1.3. A consistent estimator of the mean squared error of the estimator, s_1^2, up to the first order of approximation, is given by

$$\hat{\text{MSE}}\left(s_1^2\right) = \left(\frac{1-f}{n}\right)s_y^4\left[\hat{\lambda}_{40}+\hat{\lambda}_{04}-2\hat{\lambda}_{22}\right]. \tag{3.3.1.5}$$

Proof. It follows from (3.3.1.4.) by the methods of moments.

Corollary 3.3.1.1. The limiting distribution of the estimator s_{I}^2 is asymptotically normal under certain regularity conditions described by Swain and Mishra (1994).

Example 3.3.1.1. Consider the following population consisting of five ($N = 5$) units A, B, C, D, and E where for each one of the unit in the population two variables Y and X are measured.

Units	A	B	C	D	E
Y_i	9	11	13	16	21
X_i	14	18	19	22	24

(a) Find population mean squares S_y^2 and S_x^2 of the study variable (Y) and auxiliary variable (X), respectively.
(b) Select all possible samples of three units ($n = 3$) with SRSWOR sampling.
(c) Find the sample variance s_y^2 and s_x^2 from each sample.
(d) Find the exact mean square error of the estimator s_y^2 using the definition.
(e) Assuming that the population mean square S_x^2 of the auxiliary variable is known, find the ratio estimates of the S_y^2 as

$$s_{\mathrm{I}}^2 = s_y^2\left(\frac{S_x^2}{s_x^2}\right)$$

from each sample, and its exact mean square error by definition.
(f) Find the relative efficiency of the ratio estimator s_{I}^2 with respect to sample estimator s_y^2.

Solution. (a) The population mean squares of Y and X variables are given by
$S_y^2 = 22$ and $S_x^2 = 14.8$.

(b) and (c) All the 10 possible samples, estimates of population mean square errors from a given sample $s_y^2|_t$, $s_x^2|_t$ and related results are given as:

Sample No.		Sampled Values			$s_y^2\|_t$	$s_x^2\|_t$	$s_I^2\|_t$	p_t	$\{s_I^2\|_t - S_y^2\}^2$	$\{s_y^2\|_t - S_y^2\}^2$
1	y:	9	11	13	4.000	7.000	8.457	0.1	183.409	324.000
	x:	14	18	19						
2	y:	9	11	16	13.000	16.000	12.025	0.1	99.501	81.000
	x:	14	18	22						
3	y:	9	11	21	41.333	25.333	24.147	0.1	4.611	373.778
	x:	14	18	24						
4	y:	9	13	16	12.333	16.333	11.176	0.1	117.170	93.444
	x:	14	19	22						

Continued ......

5	y:	9	13	21	37.333	25.000	22.101	0.1	0.010	235.111
	x:	14	19	24						
6	y:	9	16	21	36.333	28.000	19.205	0.1	7.813	205.444
	x:	14	22	24						
7	y:	11	13	16	6.333	4.333	21.631	0.1	0.136	245.444
	x:	18	19	22						
8	y:	11	13	21	28.000	10.333	40.103	0.1	327.727	36.000
	x:	18	19	24						
9	y:	11	16	21	25.000	9.333	39.643	0.1	311.270	9.000
	x:	18	22	24						
10	y:	13	16	21	16.333	6.333	38.168	0.1	261.418	32.111
	x:	19	22	24						
								Sum	1313.065	1635.333

(d) The mean square error of the estimator s_y^2 is given by

$$\text{MSE}\left(s_y^2\right)=\sum_{t=1}^{10} p_t\left\{s_y^2\big|_t - S_y^2\right\}^2 = 163.5333\,.$$

(e) The ratio estimates of $s_\text{I}^2\big|_t$, $t = 1,2,....,10$ are given in the above table and the mean square error of the ratio estimator s_I^2 is given by

$$\text{MSE}\left(s_\text{I}^2\right)=\sum_{t=1}^{10} p_t\left\{s_I^2\big|_t - S_y^2\right\}^2 = 131.3065\,.$$

(f) The relative efficiency of the ratio estimator s_I^2 with respect to usual estimator s_y^2 is given by

$$\text{RE} = \frac{\text{MSE}\left(s_y^2\right)}{\text{MSE}\left(s_I^2\right)}\times 100 = \frac{163.5333}{131.3065}\times 100 = 124.54\%\,.$$

Note that this small example will demonstrate that the higher level calibration estimators of variance given in Chapter 5 are better than the low level calibration estimators of variance and we will study calibration approaches in Chapter 5.

Example 3.3.1.2. Mr. Jack Allen wishes to estimate the finite population variance of real estate farm loans in the United States using the known benchmark as the variance of the auxiliary variable nonreal estate farm loans. Using the information given in population 1 of the Appendix, discuss the relative efficiency of the ratio type estimator

$$s_\text{I}^2 = s_y^2\left(\frac{S_x^2}{s_x^2}\right)$$

with respect to the usual estimator s_y^2.

Given: $\lambda_{40} = 3.5822$, $\lambda_{04} = 4.5247$ and $\lambda_{22} = 2.8411$.

Solution. The percent relative efficiency of the estimator s_1^2 with respect to s_y^2 is given by

$$\mathrm{RE} = \frac{V(s_y^2)\times 100}{\mathrm{MSE}(s_I^2)} = \frac{(\lambda_{40}-1)\times 100}{\lambda_{40}+\lambda_{04}-2\lambda_{22}} = \frac{(3.5822-1)\times 100}{3.5822+4.5247-2\times 2.8411} = 106.49\% .$$

Theorem 3.3.1.4. The minimum sample size for the relative standard error of the estimator s_1^2 to be less than or equal to ϕ is given by

$$n \ge \left[\frac{\phi^2}{\lambda_{40}+\lambda_{04}-2\lambda_{22}}+\frac{1}{N}\right]^{-1}. \tag{3.3.1.6}$$

Proof. We have

$$\mathrm{RSE} = \sqrt{\mathrm{MSE}(s_1^2)/(S_y^2)^2} \le \phi$$

which implies

$$\left(\frac{1}{n}-\frac{1}{N}\right)(\lambda_{40}+\lambda_{04}-2\lambda_{22}) \le \phi^2$$

or

$$\frac{1}{n} \le \frac{\phi^2}{\lambda_{40}+\lambda_{04}-2\lambda_{22}}+\frac{1}{N}, \text{ or } n \ge \left[\frac{\phi^2}{\lambda_{40}+\lambda_{04}-2\lambda_{22}}+\frac{1}{N}\right]^{-1}.$$

Hence the theorem.

Example 3.3.1.3. Mr. Jack Allen is a bank consultant in the United States of America. He considers the problem of estimation of finite population variance of the real estate farm loans in the United States. Based on the information given in the population 1 of the Appendix, what is the minimum sample size required for the estimator s_1^2 to have minimum relative standard deviation 45%?

Solution. Here $\phi = 0.45$. From the description of the population 1 we have $\lambda_{40} = 3.5822,\ \lambda_{04} = 4.5247$ and $\lambda_{22} = 2.8411$. Thus using (3.3.1.6), we have

$$n \ge \left[\frac{\phi^2}{\lambda_{40}+\lambda_{04}-2\lambda_{22}}+\frac{1}{N}\right]^{-1} = \left[\frac{0.45^2}{3.5822+4.5247-2\times 2.8411}+\frac{1}{50}\right]^{-1} = 9.66 \approx 10 .$$

Thus a minimum sample of 10 states is required to maintain the required level of accuracy of the estimator.

Example 3.3.1.4. Select an SRSWOR sample of 10 states from the population 1 given in the Appendix. Apply the ratio method of estimation for estimating the variance of the real estate farm loans (in \$000) during 1997. Use the known variance of nonreal estate farm loans, $S_x^2 = 1176526$. Find an estimator of the mean squared error of the ratio estimator and hence deduce a 95% confidence interval.

Solution. Using the 23rd and 24th columns of the Pseudo-Random Numbers Table 1 given in the Appendix, we selected the 10 distinct random numbers between 1 and 50 as : 44, 32, 33, 09, 10, 50, 31, 38, 06, and 13. The results obtained are as given below.

State	x_i	y_i	$(x_i-\bar{x})^2$	$(y_i-\bar{y})^2$	$(x_i-\bar{x})^4$	$(y_i-\bar{y})^4$	$(y_i-\bar{y})^2(x_i-\bar{x})^2$
CO	906.281	315.809	60694.826	87013.200	3683861941	7571297044	5281251085
FL	464.516	825.748	38181.863	46207.372	1457854696	2135121198	1764283556
GA	540.696	939.460	14213.838	108024.626	202033179.2	11669319875	1535444494
IL	2610.572	2131.048	3805051.808	2311187.427	1.44784E+13	5.34159E+12	8.79419E+12
NM	274.035	140.582	148905.535	221094.623	22172858454	48882832253	32922213175
NY	426.274	201.631	54589.425	167410.269	2980005352	28026198155	9138830368
NC	494.730	639.571	27287.009	828.403	744580874.8	686252.3986	22604654.64
PA	298.351	756.169	130730.551	21135.344	17090476929	446702782.9	2763035216
UT	197.244	56.908	214067.045	306784.162	45824699843	94116522153	65672379110
WY	386.479	100.964	74768.777	259921.531	5590370066	67559202082	19434015051
Sum	6599.178	6107.890	4568490.679	3529606.958	1.45782E+13	5.602E+12	8.93272E+12

From the above table:

$$n=10,\ \ N=50,\ \ f=0.20,\ \ \hat{\mu}_{02}=s_x^2=\frac{\sum_{i=1}^{10}(x_i-\bar{x})^2}{10-1}=507610.07,$$

$$\hat{\mu}_{04}=\frac{\sum_{i=1}^{10}(x_i-\bar{x})^4}{10-1}=1.6198\times10^{12},\qquad \hat{\mu}_{20}=s_y^2=\frac{\sum_{i=1}^{10}(y_i-\bar{y})^2}{10-1}=392178.55,$$

$$\hat{\mu}_{40}=\frac{\sum_{i=1}^{10}(y_i-\bar{y})^4}{10-1}=6.2244444\times10^{11},\ \ \hat{\mu}_{22}=\frac{1}{10-1}\sum_{i=1}^{10}(y_i-\bar{y})^2(x_i-\bar{x})^2=9.925244\times10^{11},$$

$$\hat{\lambda}_{40}=\frac{\hat{\mu}_{40}}{\hat{\mu}_{20}^2}=\frac{6.2244444\times10^{11}}{392178.55^2}=4.04699,\ \hat{\lambda}_{04}=\frac{\hat{\mu}_{04}}{\hat{\mu}_{02}^2}=\frac{1.6198\times10^{12}}{507610.07^2}=6.28638,\ \text{and}$$

$$\hat{\lambda}_{22}=\frac{\hat{\mu}_{22}}{\hat{\mu}_{20}\hat{\mu}_{02}}=\frac{9.925244\times10^{11}}{392178.55\times507610.07}=4.985711.$$

Thus the ratio estimate of variance, S_y^2, of the real estate farm loans is

$$s_1^2=s_y^2\left(\frac{S_x^2}{s_x^2}\right)=392178.55\left(\frac{1176526}{507610.07}\right)=908981.69$$

and an estimate of $\text{MSE}(s_I^2)$ is given by

$$\hat{\text{MSE}}(s_1^2)=\left(\frac{1-f}{n}\right)s_y^4\left[\hat{\lambda}_{40}+\hat{\lambda}_{04}-2\hat{\lambda}_{22}\right]$$

$$=\left(\frac{1-0.20}{10}\right)(392178.55)^2[6.28638+4.04699-2\times 4.985711]=4453524452\,.$$

A $(1-\alpha)100\%$ confidence interval for population variance S_y^2 is given by

$$s_{\mathrm{I}}^2 \pm t_{\alpha/2}^2(\mathrm{df}=n-1)\sqrt{\hat{\mathrm{MSE}}(s_{\mathrm{I}}^2)}\,.$$

Using Table 2 from the Appendix and the relationship $t^2 = F$, the 95% confidence interval is given by

$$908981.69 \pm (2.262)^2\sqrt{4453524452} \text{ or } [567523.82,\, 1250439.56].$$

3.3.2 DIFFERENCE TYPE ESTIMATOR

Isaki (1983) also considered the difference type estimator of finite population variance S_y^2 given by

$$s_2^2 = s_y^2 + k_0\left(S_x^2 - s_x^2\right) \tag{3.3.2.1}$$

where k_0 is a real constant. The estimator s_2^2 in terms of ε_2 and ε_3 can be written as

$$s_2^2 = S_y^2(1+\varepsilon_2)+k_0\left[S_x^2 - S_x^2(1+\varepsilon_3)\right] = S_y^2(1+\varepsilon_2)-k_0 S_x^2\varepsilon_3\,. \tag{3.3.2.2}$$

Thus we have the following theorems:

Theorem 3.3.2.1. The estimator s_2^2 is an unbiased estimator of S_y^2.

Proof. Taking expected values on both sides of (3.3.2.2), we have

$$E\left(s_2^2\right) = S_y^2\,.$$

Hence the theorem.

Theorem 3.3.2.2. The minimum variance of the estimator s_2^2 is given by

$$\mathrm{Min}.V\left(s_2^2\right)=\left(\frac{1-f}{n}\right)S_y^4\left[\lambda_{40}-1-\frac{(\lambda_{22}-1)^2}{\lambda_{04}-1}\right]. \tag{3.3.2.3}$$

Proof. By the definition of variance, we have

$$V\left(s_2^2\right)=E\left[s_2^2-S_y^2\right]^2=E\left[s_y^2+k_0\left(S_x^2-s_x^2\right)-S_y^2\right]^2=E\left[S_y^2(1+\varepsilon_2)-k_0S_x^2\varepsilon_3-S_y^2\right]^2$$

$$=E\left(S_y^2\varepsilon_2-k_0S_x^2\varepsilon_3\right)^2=E\left[S_y^4\varepsilon_2^2+k_0^2S_x^4\varepsilon_3^2-2k_0S_y^2S_x^2\varepsilon_2\varepsilon_3\right]$$

$$=\left(\frac{1-f}{n}\right)\left[S_y^4(\lambda_{40}-1)+k_0^2S_x^4(\lambda_{04}-1)-2k_0S_y^2S_x^2(\lambda_{22}-1)\right]. \tag{3.3.2.4}$$

On differentiating (3.3.2.4) with respect to k_0 and equating to zero we obtain

$$k_0 = S_y^2(\lambda_{22}-1)/S_x^2(\lambda_{04}-1)\,. \tag{3.3.2.5}$$

Substituting of the optimum value of k_0 in (3.3.2.4) proves the theorem.

Example 3.3.1.5. Using the information given in population 1 of the Appendix, find the relative efficiency of the regression type estimator $s_2^2 = s_y^2 + k_0\left(S_x^2 - s_x^2\right)$ with respect to the ratio type estimator $s_1^2 = s_y^2\left(S_x^2/s_x^2\right)$ while estimating the variance of real estate farm loans using know variance of nonreal estate farm loans.
Given: $\lambda_{40} = 3.5822$, $\lambda_{04} = 4.5247$ and $\lambda_{22} = 2.8411$.

Solution. The percent relative efficiency of the estimator s_2^2 with respect to s_1^2 is

$$\mathrm{RE} = \frac{V\left(s_1^2\right)}{\mathrm{MSE}(\ s_2^2\ )}\times 100 = \frac{(3.5822 + 4.5247 - 2\times 2.8411)}{\left\{3.5822 - 1 - \dfrac{(2.8411-1)^2}{4.5247-1}\right\}}\times 100 = 149.63\%\ .$$

3.3.3 POWER TRANSFORMATION TYPE ESTIMATOR

We consider here an estimator of finite population variance S_y^2 as

$$s_3^2 = s_y^2\left(\frac{\bar{x}}{\bar{X}}\right)^{\alpha_0}. \tag{3.3.3.1}$$

The estimator s_3^2 in terms of ε_1 and ε_2 can easily be written as

$$s_3^2 = S_y^2(1+\varepsilon_2)(1+\varepsilon_1)^{\alpha_0}$$

which, after ignoring the higher order terms, becomes

$$= S_y^2(1+\varepsilon_2)\left[1+\alpha_0\varepsilon_1+\frac{\alpha_0(\alpha_0-1)}{2}\varepsilon_1^2+..\right] = S_y^2\left[1+\varepsilon_2+\alpha_0\varepsilon_1+\frac{\alpha_0(\alpha_0-1)}{2}\varepsilon_1^2+\alpha_0\varepsilon_1\varepsilon_2+..\right].$$

Obviously the estimator s_3^2 is a biased estimator of S_y^2. Thus we have the following theorems:

Theorem 3.3.3.1. The bias in the estimator s_3^2 of S_y^2 to the first order of approximation is given by

$$B\left(s_3^2\right) = \left(\frac{1}{n} - \frac{1}{N}\right)S_y^2\left[\frac{\alpha_0(\alpha_0-1)}{2}C_x^2 + \alpha_0 C_x\lambda_{21}\right]. \tag{3.3.3.2}$$

Proof. We have

$$B\left(s_3^2\right) = E\left(s_3^2\right) - S_y^2 = S_y^2 E\left[1+\varepsilon_2+\alpha_0\varepsilon_1+\frac{\alpha_0(\alpha_0-1)}{2}\varepsilon_1^2+\alpha_0\varepsilon_1\varepsilon_2\right] - S_y^2$$

$$= S_y^2\left[1+0+0+\frac{\alpha_0(\alpha_0-1)}{2}E\left(\varepsilon_1^2\right)+\alpha_0 E(\varepsilon_1\varepsilon_2)\right] - S_y^2$$

$$= \left(\frac{1-f}{n}\right)S_y^2\left[\frac{\alpha_0(\alpha_0-1)}{2}C_x^2 + \alpha_0\rho_{xy}C_x C_y\right].$$

Hence the theorem.

Theorem 3.3.3.2. The minimum mean squared error of the estimator s_3^2 is

$$\text{Min.MSE}\left(s_3^2\right)=\left(\frac{1-f}{n}\right)S_y^4\left[\lambda_{40}-1-\lambda_{21}^2\right]. \tag{3.3.3.3}$$

Proof. We have

$$\text{MSE}\left(s_3^2\right)=E\left[s_3^2-S_y^2\right]\approx E\left[S_y^2(1+\varepsilon_2+\alpha_\circ\varepsilon_1+\alpha_\circ\varepsilon_1\varepsilon_2)-S_y^2\right]^2=S_y^4E\left[\varepsilon_2+\alpha_\circ\varepsilon_1\right]^2$$

$$=S_y^4E\left[\varepsilon_2^2+\alpha_\circ^2\varepsilon_1^2+2\alpha_\circ\varepsilon_1\varepsilon_2\right]$$

$$=\left(\frac{1-f}{n}\right)S_y^4\left[(\lambda_{40}-1)+\alpha_\circ^2C_x^2+2\alpha_\circ C_x\lambda_{21}\right]. \tag{3.3.3.4}$$

On differentiating (3.3.3.4) with respect to $\alpha_\circ$ and equating to zero, we obtain

$$\alpha_\circ=-\frac{\lambda_{21}}{C_x}. \tag{3.3.3.5}$$

On substituting the optimum value of $\alpha_\circ$ from (3.3.3.5) in (3.3.3.4), we have (3.3.3.3). Hence the theorem.

Remark 3.3.3.1. If the population follows the bivariate normal distribution, then $\lambda_{21}=0$, hence the minimum mean square error of the proposed estimator becomes

$$\text{Min.MSE}\left(s_3^2\right)=\left(\frac{1-f}{n}\right)S_y^4(\lambda_{40}-1)=V\left(s_y^2\right). \tag{3.3.3.6}$$

The expression (3.3.3.6) conveys the message to survey statisticians that if the study variable and auxiliary variable follow bivariate normal distribution then use of known population mean or total of auxiliary variable is not helpful in improving the usual estimator of variance.

3.3.4 GENERAL CLASS OF ESTIMATORS

Srivastava and Jhajj (1980) proposed a general class of ratio type estimators for estimating the finite population variance S_y^2 as

$$s_{SJ}^2=s_y^2H(u,v) \tag{3.3.4.1}$$

where $u=\bar{x}/\bar{X}$, $v=s_x^2/S_x^2$ and $H(u,v)$ is a parametric function such that $H(1,1)=1$ and satisfies certain regularity conditions as defined earlier. Thus all ratio or product type estimators of population variance considered by Das and Tripathi (1978) and Kaur and Singh (1982), for example

$$s_1^2=s_y^2\left(\frac{\bar{X}}{\bar{x}}\right)\left(\frac{S_x^2}{s_x^2}\right),\quad s_2^2=s_y^2\left(\frac{\bar{X}}{\bar{x}}\right)^{\alpha}\left(\frac{S_x^2}{s_x^2}\right)^{\gamma},\ \text{and}\ s_3^2=s_y^2\left(\frac{\bar{X}}{\alpha\bar{x}+(1-\alpha)\bar{X}}\right)\left(\frac{S_x^2}{\gamma s_x^2+(1-\gamma)S_x^2}\right)$$

are special cases of the class of estimators defined in (3.3.4.1). Now the class of estimators s_{SJ}^2 defined in (3.3.4.1) can be written as

$$s_{SJ}^2 = S_y^2(1+\varepsilon_2)\left[1+(u-1)\frac{\partial H}{\partial u}|_{(1,1)}+(v-1)\frac{\partial H}{\partial v}|_{(1,1)}+(u-1)^2\frac{\partial^2 H}{\partial u^2}|_{(1,1)}\right.$$

$$\left.+(v-1)^2\frac{\partial^2 H}{\partial v^2}|_{(1,1)}+(u-1)(v-1)\frac{\partial^2 H}{\partial u \partial v}+....\right]$$

$$= S_y^2\left[1+\varepsilon_2+\varepsilon_1 H_1+\varepsilon_3 H_2+\varepsilon_1^2 H_3+\varepsilon_3^2 H_4+\varepsilon_1\varepsilon_3 H_5+\varepsilon_1\varepsilon_2 H_1+\varepsilon_2\varepsilon_3 H_2+...\right]. \quad (3.3.4.2)$$

Thus we have the following theorems:

Theorem 3.3.4.1. The bias in this class of estimators of variance, to the first order of approximation, is given by

$$B\left(s_{SJ}^2\right) = \left(\frac{1-f}{n}\right)S_y^2\left[C_x^2 H_3+(\lambda_{04}-1)H_4+C_x\lambda_{03}H_5+C_x\lambda_{21}H_1+(\lambda_{22}-1)H_2\right]. \quad (3.3.4.3)$$

Proof. Obvious by taking expected values on both sides of (3.3.4.2).

Theorem 3.3.4.2. The minimum mean squared error of the class of estimators s_{SJ}^2, to the first order of approximation, is

$$\text{Min.MSE}\left(s_{SJ}^2\right) = \left(\frac{1-f}{n}\right)S_y^4\left[(\lambda_{40}-1)-\lambda_{21}^2-\frac{\{\lambda_{21}\lambda_{03}-(\lambda_{22}-1)\}^2}{\lambda_{04}-1-\lambda_{03}^2}\right]. \quad (3.3.4.4)$$

Proof. By the definition of mean squared error and by ignoring higher order terms of ε_1, ε_2, and ε_3 we have

$$\text{MSE}\left(s_{SJ}^2\right) = E\left[s_{SJ}^2 - S_y^2\right]^2 \approx E\left[S_y^2(\varepsilon_2+\varepsilon_1 H_1+\varepsilon_3 H_2)\right]^2$$

$$= S_y^4 E\left[\varepsilon_2^2+\varepsilon_1^2 H_1^2+\varepsilon_3^2 H_2^2+2H_1\varepsilon_1\varepsilon_2+2H_2\varepsilon_2\varepsilon_3+2\varepsilon_1\varepsilon_3 H_1 H_2\right]$$

$$= \left(\frac{1-f}{n}\right)S_y^4\left[(\lambda_{40}-1)+C_x^2 H_1^2+(\lambda_{04}-1)H_2^2+2C_x\lambda_{21}H_1+2(\lambda_{22}-1)H_2\right.$$

$$\left.+\ 2C_x\lambda_{03}H_1 H_2\ \right]. \quad (3.3.4.5)$$

On differentiating it with respect to H_1 and H_2, respectively, and equating to zero we obtain

$$C_x H_1 + \lambda_{03}H_2 = -\lambda_{21}, \quad (3.3.4.6)$$

and

$$C_x\lambda_{03}H_1 + (\lambda_{04}-1)H_2 = -(\lambda_{22}-1). \quad (3.3.4.7)$$

On solving equations in (3.3.4.6) and (3.3.4.7) for H_1 and H_2 we have

$$H_1 = -\frac{\lambda_{21}(\lambda_{04}-1)-\lambda_{03}(\lambda_{22}-1)}{C_x\left(\lambda_{04}-1-\lambda_{03}^2\right)}, \text{ and } H_2 = -\frac{(\lambda_{22}-1)-\lambda_{03}\lambda_{21}}{\lambda_{04}-1-\lambda_{03}^2}. \quad (3.3.4.8)$$

On substituting these optimum values of H_1 and H_2 in (3.3.4.5) we will obtain (3.3.4.4). Hence the theorem.

Theorem 3.3.4.3. The minimum MSE, to the first order of approximation, of the wider class of estimators defined as

$$s^2_{SJ(w)} = H\left(s^2_y,\ u,\ v\right) \tag{3.3.4.9}$$

where $H(.,.,.)$ is a parametric function such that $H\left(S^2_y, 1, 1\right) = S^2_y$, $\frac{\partial H}{\partial s^2_y}\Big|_{\left(S^2_y,1,1\right)} = 1$

and satisfies the regularity conditions, is

$$\text{Min.MSE}\left(s^2_{SJ(w)}\right) = \left(\frac{1-f}{n}\right)S^4_y\left[(\lambda_{40}-1) - \lambda^2_{21} - \frac{\{\lambda_{21}\lambda_{03} - (\lambda_{22}-1)\}^2}{\lambda_{04} - 1 - \lambda^2_{03}}\right]. \tag{3.3.4.10}$$

Proof. Expanding $H\left(s^2_y, u, v\right)$ around the point $\left(S^2_y, 1, 1\right)$ in a first order Taylor's series we have

$$\begin{aligned}
s^2_{SJ(w)} &= H\left(s^2_y, u, v\right) = H\left[S^2_y + \left(s^2_y - S^2_y\right),\ 1 + (u-1),\ 1 + (v-1)\right] \\
&= H\left(S^2_y, 1, 1\right) + \left(s^2_y - S^2_y\right)\frac{\partial H}{\partial s^2_y}\Big|_{\left(S^2_y,1,1\right)} + (u-1)\frac{\partial H}{\partial u}\Big|_{\left(S^2_y,1,1\right)} + (v-1)\frac{\partial H}{\partial v}\Big|_{\left(S^2_y,1,1\right)} + \ldots \\
&= S^2_y + \left(s^2_y - S^2_y\right) + (u-1)\frac{\partial H}{\partial u}\Big|_{\left(S^2_y,1,1\right)} + (v-1)\frac{\partial H}{\partial v}\Big|_{\left(S^2_y,1,1\right)} + \ldots \\
&= s^2_y + (u-1)H_1 + (v-1)H_2 + \ldots = S^2_y(1+\varepsilon_2) + \varepsilon_1 H_1 + \varepsilon_3 H_2 + \ldots
\end{aligned} \tag{3.3.4.11}$$

Thus the mean squared error, to the first order of approximation, of the wider class of estimators is

$$\begin{aligned}
\text{MSE}\left[s^2_{SJ(w)}\right] &= E\left[s^2_{SJ(w)} - S^2_y\right]^2 = E\left[S^2_y(1+\varepsilon_2) + \varepsilon_1 H_1 + \varepsilon_3 H_2 - S^2_y\right]^2 \\
&= E\left[S^2_y\varepsilon_2 + \varepsilon_1 H_1 + \varepsilon_3 H_2\right]^2 \\
&= E\left[S^4_y\varepsilon^2_2 + \varepsilon^2_1 H^2_1 + \varepsilon^2_3 H^2_2 + 2S^2_y H_1\varepsilon_1\varepsilon_2 + 2H_2 S^2_y\varepsilon_2\varepsilon_3 + 2H_1 H_2\varepsilon_1\varepsilon_3\right] \\
&= \left(\frac{1-f}{n}\right)\Big[S^4_y(\lambda_{40}-1) + H^2_1 C^2_x + H^2_2(\lambda_{04}-1) + 2S^2_y C_x\lambda_{21}H_1 \\
&\qquad + 2H_2 S^2_y(\lambda_{22}-1) + 2H_1 H_2 C_x\lambda_{03}\Big].
\end{aligned} \tag{3.3.4.12}$$

The mean squared error of the wider class of estimators is minimum for the set of normal equations

$$\begin{bmatrix} C_x, & \lambda_{03} \\ C_x\lambda_{03}, & (\lambda_{04}-1) \end{bmatrix}\begin{bmatrix} H_1 \\ H_2 \end{bmatrix} = \begin{bmatrix} -S^2_y\lambda_{21} \\ -S^2_y(\lambda_{22}-1) \end{bmatrix}. \tag{3.3.4.13}$$

On solving (3.3.4.13) for H_1 and H_2 we have

$$H_1 = -\frac{S^2_y[\lambda_{21}(\lambda_{04}-1) - \lambda_{03}(\lambda_{22}-1)]}{C_x\left(\lambda_{04} - 1 - \lambda^2_{03}\right)}, \text{ and } H_2 = -\frac{S^2_y[(\lambda_{22}-1) - \lambda_{21}\lambda_{03}]}{\lambda_{04} - 1 - \lambda^2_{03}}. \tag{3.3.4.14}$$

On substituting H_1 and H_2 from (3.3.4.14) in (3.3.4.12) we prove the theorem.

Example 3.3.4.1. Suppose we selected an SRSWOR sample of eight states from population 1 given in the Appendix and collected the required information. Find the relative efficiency of the general class estimators which makes use of known variance of the auxiliary variable at the estimation stage, for estimating the finite population variance of the amount of real estate farm loans during 1997 by using information from nonreal estate farm loans during 1997, with respect to the ratio and regression estimators of finite population variance.

Solution. From the description of the population 1 given in the Appendix we have
$Y_i =$ Amount (in \$000) of real estate farm loans in different states during 1997,
$X_i =$ Amount (in \$000) of nonreal estate farm loans in different states during 1997,
$S_y^2 = 342021.5$, $\lambda_{03} = 1.5936$, $N = 50$, $\lambda_{21} = 0.9387$, $\lambda_{40} = 3.5822$, $\lambda_{04} = 4.5247$, $\lambda_{22} = 2.8411$ and $f = 0.16$.
Now we have

$$\mathrm{MSE}\left(s_1^2\right) = \left(\frac{1-f}{n}\right) S_y^4\left[\lambda_{40} + \lambda_{04} - 2\lambda_{22}\right]$$

$$= \left(\frac{1-0.16}{8}\right)(342021.5)^2\left[3.5822 + 4.5247 - 2\times 2.8411\right] = 2.97820183\times 10^{10},$$

$$\mathrm{MSE}\left(s_2^2\right) = \left(\frac{1-f}{n}\right) S_y^4\left[\lambda_{40} - 1 - \frac{(\lambda_{22}-1)^2}{\lambda_{04}-1}\right]$$

$$= \left(\frac{1-0.16}{8}\right)(342021.5)^2\left[3.5822 - 1 - \frac{(2.8411-1)^2}{4.5247-1}\right] = 1.990441025\times 10^{10},$$

and

$$\mathrm{Min.MSE}\left(s_{SJ}^2\right) = \left(\frac{1-f}{n}\right) S_y^4\left[\lambda_{40} - 1 - \lambda_{21}^2 - \frac{\{\lambda_{21}\lambda_{03} - (\lambda_{22}-1)\}}{\lambda_{04} - 1 - \lambda_{03}^2}\right]$$

$$= \left(\frac{1-0.16}{8}\right)(342021.5)^2\left[3.5822 - 1 - (0.9387)^2 - \frac{(0.9387\times 1.5936 - 1.8411)^2}{4.5247 - 1 - (1.5936)^2}\right]$$

$$= 1.940787669\times 10^{10}.$$

Thus the percent relative efficiency (RE) of s_2^2 with respect to s_1^2 is given by

$$\mathrm{RE} = \frac{\mathrm{MSE}\left(s_1^2\right)}{\mathrm{MSE}\left(s_2^2\right)}\times 100 = \frac{2.97820183\times 10^{10}}{1.990441025\times 10^{10}}\times 100 = 149.63\%$$

and the percent relative efficiency (RE) of s_{SJ}^2 with respect to s_2^2 is given by

$$\mathrm{RE} = \frac{\mathrm{MSE}\left(s_2^2\right)}{\mathrm{MSE}\left(s_{SJ}^2\right)}\times 100 = \frac{1.990441025\times 10^{10}}{1.940787669\times 10^{10}}\times 100 = 102.58\%.$$

Thus the general class of estimators remains more efficient than both ratio and regression type estimators. It is noted that the relative efficiency, to the first order of approximation, in each case remains independent of the sample size.

We have seen that the minimum mean square errors of the class of estimators of population mean and variance depend upon the values of the unknown population parameters H_1 and H_2 etc.. Srivastava and Jhajj (1983a) have shown that consistent estimators of the unknown parameters H_1 and H_2 can be used in place of actual values. The minimum mean square error of the resultant class of estimators with estimated optimum values remains the same as that with the actual optimum values to the first order of approximation. We are not discussing this procedure in detail, but the interested reader may refer to Srivastava and Jhajj (1983a) and Singh and Singh (1984a). Singh and Zaidi (2000) estimated square of population mean and variance.

The next section has been devoted to studying the asymptotic properties of the various estimators of regression coefficient

$$\beta = \frac{S_{xy}}{S_x^2}.$$

3.4 ESTIMATION OF REGRESSION COEFFICIENT

The meaning and purpose of the estimation of regression coefficient is well known to survey statisticians. Some like to define it as the slope of the linear regression of X on Y or change in the study variable for per unit change in the auxiliary variable. Here we will discuss a few estimators for estimating the regression coefficient in survey sampling.

3.4.1 USUAL ESTIMATOR

Srivastava, Jhajj, and Sharma (1986) considered the usual estimator of regression coefficient β as

$$b_1 = \frac{s_{xy}}{s_x^2} \tag{3.4.1.1}$$

which in terms of ε_3 and ε_4 can easily be written as

$$b_1 = \frac{S_{xy}(1+\varepsilon_4)}{S_x^2(1+\varepsilon_3)}$$

$$= \beta(1+\varepsilon_4)(1+\varepsilon_3)^{-1} = \beta(1+\varepsilon_4)(1-\varepsilon_3+\varepsilon_3^2+...)$$

$$= \beta\left[1+\varepsilon_4-\varepsilon_3+\varepsilon_3^2-\varepsilon_3\varepsilon_4+....\right]. \tag{3.4.1.2}$$

Thus we have following theorems:

Theorem 3.4.1.1. The bias, to the first order of approximation, in the estimator b_1 of β is

$$B(b_1)=\left(\frac{1-f}{n}\right)\beta\left[(\lambda_{04}-1)-\left(\frac{\lambda_{13}}{\rho_{xy}}-1\right)\right]. \quad (3.4.1.3)$$

Proof. Taking expected values on both sides of (3.4.1.2) we have

$$E(b_1)=\beta E\left[1+\varepsilon_4-\varepsilon_3+\varepsilon_3^2-\varepsilon_3\varepsilon_4+....\right]$$

$$=\beta\left[1+E(\varepsilon_4)-E(\varepsilon_3)+E\left(\varepsilon_3^2\right)-E(\varepsilon_3\varepsilon_4)\right]$$

$$=\beta\left[1+0-0+\left(\frac{1-f}{n}\right)\left\{(\lambda_{04}-1)-\left(\frac{\lambda_{13}}{\rho_{xy}}-1\right)\right\}\right].$$

Thus the bias in the estimator b_1 of β to the first order of approximation is

$$B(b_1)=E(b_1)-\beta=\left(\frac{1-f}{n}\right)\beta\left[(\lambda_{04}-1)-\left(\frac{\lambda_{13}}{\rho_{xy}}-1\right)\right].$$

Hence the theorem.

Theorem 3.4.1.2. The mean squared error of the usual estimator b_1 of the regression coefficient β is given by

$$\text{MSE}(b_1)=\left(\frac{1-f}{n}\right)\beta^2\left[\frac{\lambda_{22}}{\rho_{xy}^2}+\lambda_{04}-2\frac{\lambda_{13}}{\rho_{xy}}\right]. \quad (3.4.1.4)$$

Proof. By using the definition of mean square error (MSE) we have

$$\text{MSE}(b_1)=E(b_1-\beta)^2\approx E[\beta(1+\varepsilon_4-\varepsilon_3+...)-\beta]^2\approx\beta^2E[\varepsilon_4-\varepsilon_3]^2$$

$$=\beta^2E\left[\varepsilon_4^2+\varepsilon_3^2-2\varepsilon_3\varepsilon_4\right]=\left(\frac{1-f}{n}\right)\beta^2\left[\left(\frac{\lambda_{22}}{\rho_{xy}^2}-1\right)+(\lambda_{04}-1)-2\left(\frac{\lambda_{13}}{\rho_{xy}}-1\right)\right].$$

Hence the theorem.

3.4.2 UNBIASED ESTIMATOR

We are also considering an unbiased estimator of the regression coefficient β as

$$b_2=\frac{s_{xy}}{s_x^2}. \quad (3.4.2.1)$$

Thus we have the following theorem:

Theorem 3.4.2.1. The variance of the unbiased estimator of the regression coefficient β is given by

$$V(b_2) = \left(\frac{1-f}{n}\right)\beta^2\left(\frac{\lambda_{22}}{\rho_{xy}^2} - 1\right). \tag{3.4.2.2}$$

Proof. By the definition of variance we have

$$V(b_2) = E[b_2 - \beta]^2 = E[\beta(1+\varepsilon_4) - \beta]^2 = \beta^2 E(\varepsilon_4^2) = \left(\frac{1-f}{n}\right)\beta^2\left(\frac{\lambda_{22}}{\rho_{xy}^2} - 1\right).$$

Hence the theorem.

Example 3.4.2.1. For estimating the regression coefficient of the amount of the real estate farm loans (in \$000) on the nonreal estate farm loans, Mr. Nelson used the 53rd and 54th columns of the Pseudo-Random Numbers given in the Table 1 of the Appendix to select eight distinct random numbers between 1 and 50 as:06, 08, 45, 15, 22, 39, 43 and 34. He collected the following sample information:

State	CO	DE	VT	IA	MI	RI	TX	ND
Nonreal estate farm loans (X)\$	906.281	43.229	19.363	3909.738	440.518	0.233	3520.361	1241.369
Real estate farm loans (Y)\$	315.809	42.808	57.747	2327.025	323.028	1.611	1248.761	449.099

The population variance $S_x^2 = 1176526$ of nonreal estate farm loans (in \$000) for the year 1997 is known from population 1 of the Appendix.

(a) Estimate the regression coefficient β with two different methods.

(b) Also find an estimate of the mean squared errors in each case and hence derive the 95% confidence intervals.

Solution. From the sample information, we have

Sr. No.	$(y_i-\bar{y})(x_i-\bar{x})$	$(x_i-\bar{x})^2$	$(y_i-\bar{y})^2$	$(x_i-\bar{x})^4$	$(y_i-\bar{y})^4$	$(x_i-\bar{x})^2(y_i-\bar{y})^2$	$(x_i-\bar{x})^3(y_i-\bar{y})$
1	99053.71	125213.72	78359.12	15678474394	6140152522	9811637177	12402882820
2	672862.23	1480863.86	305729.37	2.19296E+12	93470449627	4.52744E+11	9.96417E+11
3	667522.49	1539518.88	289432.16	2.37012E+12	83770977628	4.45586E+11	1.02766E+12
4	4587225.90	7020388.11	2997361.60	4.92858E+13	8.98418E+12	2.10426E+13	3.22041E+13
5	223516.52	671774.48	74369.65	4.51281E+11	5530845327	49959635561	1.50153E+11
6	748540.17	1587356.83	352984.51	2.5197E+12	1.24598E+11	5.60312E+11	1.1882E+12
7	1475983.10	5108614.79	426441.65	2.60979E+13	1.81852E+11	2.17853E+12	7.54023E+12
8	2752.01	352.22	21502.41	124058.2636	462353625.9	7573558.476	969310.3289
Sum	8477456.20	17534082.89	546180.49	8.29335E+13	9.48E+12	2.47396E+13	4.31192E+13

Thus from the above table we have

$$n=8,\quad N=50,\quad f=0.16,\quad \bar{y}=\frac{1}{8}\sum_{i=1}^{8}y_i=595.736,\quad \bar{x}=\frac{1}{8}\sum_{i=1}^{8}x_i=1260.137,$$

$$\hat{\mu}_{02}=s_x^2=\frac{1}{8-1}\sum_{i=1}^{8}(x_i-\bar{x})^2=2504868.98,\quad \hat{\mu}_{04}=\frac{1}{8-1}\sum_{i=1}^{8}(x_i-\bar{x})^4=1.184764\times10^{13},$$

$$\hat{\mu}_{20}=s_y^2=\frac{1}{8-1}\sum_{i=1}^{8}(y_i-\bar{y})^2=649454.36,\quad \hat{\mu}_{11}=\frac{1}{8-1}\sum_{i=1}^{8}(y_i-\bar{y})(x_i-\bar{x})=1211065.171,$$

$$\hat{\mu}_{22}=\frac{\sum_{i=1}^{8}(y_i-\bar{y})^2(x_i-\bar{x})^2}{8-1}=3.534228\times10^{12},\quad \hat{\mu}_{13}=\frac{\sum_{i=1}^{8}(y_i-\bar{y})(x_i-\bar{x})^3}{8-1}=6.15988\times10^{12},$$

$$\hat{\lambda}_{04}=\frac{\hat{\mu}_{04}}{\hat{\mu}_{02}^2}=\frac{1.184764\times10^{13}}{2504868.98^2}=1.88826,\quad \hat{\lambda}_{22}=\frac{\hat{\mu}_{22}}{\hat{\mu}_{20}\hat{\mu}_{02}}=\frac{3.534228\times10^{12}}{649454.36\times2504868.98}=2.1725,$$

$$\hat{\lambda}_{13}=\frac{\hat{\mu}_{13}}{\sqrt{\hat{\mu}_{20}}(\hat{\mu}_{02})^{\frac{3}{2}}}=\frac{6.15988\times10^{12}}{\sqrt{649454.36}\times(2504868.98)^{\frac{3}{2}}}=1.9280,$$

and

$$r_{xy}=\frac{\hat{\mu}_{11}}{\sqrt{\hat{\mu}_{20}\hat{\mu}_{02}}}=\frac{1211065.171}{\sqrt{649454.36\times2504868.98}}=0.9495.$$

(a) Usual estimator of regression coefficient

The usual estimator of the regression coefficient β is given by

$$b_1=\frac{s_{xy}}{s_x^2}=\frac{\hat{\mu}_{11}}{\hat{\mu}_{02}}=\frac{1211065.171}{2504868.98}=0.4835$$

and an estimator of $\text{MSE}(b_1)$ is given by

$$\hat{\text{MSE}}(b_1)=\left(\frac{1-f}{n}\right)b_1^2\left[\frac{\hat{\lambda}_{22}}{r_{xy}^2}+\hat{\lambda}_{04}-2\frac{\hat{\lambda}_{13}}{r_{xy}}\right]$$

$$=\left(\frac{1-0.16}{8}\right)(0.4835)^2\left[\frac{2.1725}{0.9495^2}+1.88826-2\times\frac{1.9280}{0.9495}\right]$$

$$=0.005815.$$

The $(1-\alpha)100\%$ confidence interval for regression coefficient β is given by

$$b_1\pm t_{\alpha/2}(\text{df}=n-2)\sqrt{\hat{\text{MSE}}(b_1)}.$$

Using Table 2 of the Appendix the 95% confidence interval of β is given by

$$0.4835\pm t_{0.05/2}(\text{df}=8-2)\sqrt{0.005815}$$

$$0.4835\pm2.447\sqrt{0.0057815}\ \text{ or }\ [0.2974,\ 0.6695].$$

(b) Unbiased estimator of regression coefficient

The unbiased estimator of the regression coefficient β is given by

$$b_2 = \frac{s_{xy}}{s_x^2} = \frac{\hat{\mu}_{11}}{s_x^2} = \frac{1211065.141}{1176526} = 1.029$$

and an estimator of $\text{MSE}(b_2)$ is given by

$$\hat{\text{MSE}}(b_2) = \left(\frac{1-f}{n}\right) b_2^2 \left[\frac{\hat{\lambda}_{22}}{r_{xy}^2} - 1\right] = \left(\frac{1-0.16}{8}\right)(1.029)^2 \left[\frac{2.1725}{0.9495^2} - 1\right] = 0.1568.$$

The $(1-\alpha)100\%$ confidence interval for regression coefficient β using b_2 is

$$b_2 \pm t_{\alpha/2}(\text{df} = n-2)\sqrt{\hat{\text{MSE}}(b_2)}\,.$$

Using Table 2 from the Appendix the 95% confidence interval is given by

$$1.029 \pm t_{0.05/2}(\text{df} = 8-2)\sqrt{0.1568}$$

$$1.029 \pm 2.447\sqrt{0.1568} \quad \text{or} \quad [0.060, 1.997].$$

Note that we used $(n-2)$ degree of freedom while constructing confidence interval estimate for the regression coefficient.

3.4.3 IMPROVED ESTIMATORS OF REGRESSION COEFFICIENT

Singh and Singh (1988) introduced a general class of estimators to estimate the regression coefficient as

$$b_{ss} = \frac{s_{xy}}{s_x^2} H(u) \tag{3.4.3.1}$$

where $H(\bullet)$ is a parametric function such that $H(1) = 1$ satisfying certain regularity conditions as defined earlier. Thus we have the following theorem:

Theorem 3.4.3.1. The minimum mean square error, to the first order of approximation, of the general class of estimators b_{ss} is given by

$$\text{Min.MSE}(b_{ss}) = \text{MSE}(b_1) - \left(\frac{1-f}{n}\right)\beta^2\left(\frac{\lambda_{12}}{\rho_{xy}} - \lambda_{03}\right)^2. \tag{3.4.3.2}$$

Proof. The estimator b_{ss} can be written as

$$b_{ss} = \frac{s_{xy}}{s_x^2} H(u) = \frac{s_{xy}}{s_x^2} H[1 + (u-1)] = \frac{s_{xy}}{s_x^2}\left[H(1) + (u-1)\frac{\partial H}{\partial u}\Big|_{u=1} + (u-1)^2 \frac{\partial^2 H}{\partial u^2}\Big|_{u=1} + ..\right]$$

$$= \frac{s_{xy}}{s_x^2}\left[1 + (u-1)H_1 + (u-1)^2 H_2 + ...\right] = \frac{S_{xy}(1+\varepsilon_4)}{S_x^2(1+\varepsilon_3)}\left[1 + \varepsilon_1 H_1 + \varepsilon_1^2 H_2 + ...\right]$$

$$= \beta(1+\varepsilon_4)(1+\varepsilon_3)^{-1}\left[\ 1 + \varepsilon_1 H_1 + \varepsilon_1^2 H_2 + ...\ \right]$$

$$= \beta(1+\varepsilon_4)\left(1 - \varepsilon_3 + \varepsilon_3^2 + ...\right)\left[1 + \varepsilon_1 H_1 + \varepsilon_1^2 H_2 + ...\right]$$

$$= \beta(1+\varepsilon_4-\varepsilon_3+\varepsilon_3^2-\varepsilon_3\varepsilon_4+...)[1+\varepsilon_1 H_1+\varepsilon_1^2 H_2+....]$$

$$= \beta(1+\varepsilon_4-\varepsilon_3+\varepsilon_3^2-\varepsilon_3\varepsilon_4+\varepsilon_1 H_1+\varepsilon_1^2 H_2+\varepsilon_1\varepsilon_4 H_1-\varepsilon_1\varepsilon_3 H_1+.....).$$

The mean squared error of the estimator b_{ss}, to the first order of approximation, is

$$\text{MSE}(b_{ss}) = E[b_{ss}-\beta]^2$$

$$\approx E\left[\beta(1+\varepsilon_4-\varepsilon_3+\varepsilon_3^2-\varepsilon_3\varepsilon_4+\varepsilon_1 H_1+\varepsilon_1^2 H_2+\varepsilon_1\varepsilon_4 H_1-\varepsilon_1\varepsilon_3 H_1+.....)-\beta\right]^2$$

$$\approx E[\beta(\varepsilon_4-\varepsilon_3+\varepsilon_1 H_1)]^2$$

$$= \beta^2 E\left[\varepsilon_4^2+\varepsilon_3^2-2\varepsilon_3\varepsilon_4+\varepsilon_1^2 H_1^2+2H_1(\varepsilon_1\varepsilon_4-\varepsilon_1\varepsilon_3)\right]$$

$$= \left(\frac{1-f}{n}\right)\beta^2\left[\left(\frac{\lambda_{22}}{\rho_{xy}^2}-1\right)+(\lambda_{04}-1)-2\left(\frac{\lambda_{13}}{\rho_{xy}}-1\right)+H_1^2 C_x^2 \right.$$

$$\left. +2H_1 C_x\left(\frac{\lambda_{12}}{\rho_{xy}}-\lambda_{03}\right)\right]. \tag{3.4.3.3}$$

On differentiating (3.4.3.3) with respect to H_1 and equating to zero, we obtain

$$H_1 = -C_x^{-1}\left(\frac{\lambda_{12}}{\rho_{xy}}-\lambda_{03}\right). \tag{3.4.3.4}$$

On substituting the optimum value of H_1 in (3.4.3.3) we obtain

$$\text{Min.MSE}(b_{ss}) = \left(\frac{1-f}{n}\right)\beta^2\left[\frac{\lambda_{22}}{\rho_{xy}^2}+\lambda_{04}-2\frac{\lambda_{13}}{\rho_{xy}}-\left(\frac{\lambda_{12}}{\rho_{xy}}-\lambda_{03}\right)^2\right]$$

$$= \text{MSE}(b_1)-\left(\frac{1-f}{n}\right)\beta^2\left(\frac{\lambda_{12}}{\rho_{xy}}-\lambda_{03}\right)^2. \tag{3.4.3.5}$$

Hence the theorem.

Now we have the following remark:

Remark 3.4.3.1. Several improved estimators of regression coefficient β can be suggested for example

$$b_{s1} = \frac{s_{xy}}{s_x^2}\left(\frac{\bar{x}}{\bar{X}}\right)^{\alpha},\quad b_{s2} = \frac{s_{xy}}{s_x^2}\left(\frac{\bar{x}}{\bar{X}}\right)^{\alpha}\left(\frac{s_x^2}{S_x^2}\right)^{\gamma},\text{ and } b_{s3} = \frac{s_{xy}}{s_x^2}\left(\frac{\bar{X}}{\alpha\bar{x}+(1-\alpha)\bar{X}}\right)\left(\frac{s_x^2}{S_x^2}\right)^{\gamma} \text{ etc..}$$

It is interesting to note that the use of any one of these estimators of regression coefficient β will result in different types of regression estimators of the population mean $\bar{Y}$ defined as

$$\bar{y}_J = \bar{y} + b_{sj}\left(\bar{X} - \bar{x}\right), \quad j = 1, 2, 3, ...etc. \tag{3.4.3.6}$$

Sampath (1989) has shown that the resultant regression type of estimators of the population mean have the same minimum mean squared error, to the first order of approximation, as that of the usual regression estimator. In other words, the construction of an improved estimator of regression coefficient β is not helpful in improving the estimator of the population mean $\bar{Y}$

Example 3.4.3.1. The real and nonreal estate farm loans (in \$000) during 1997 in the different 50 states of the United States have been given in population 1 of the Appendix. If we selected an SRSWOR sample of eight states, find the relative efficiency of the usual estimator b_1 with respect to the unbiased estimator b_2 of regression coefficient for this population.

Solution. Using the results from the description of the population 1 we have

$$\text{MSE}(b_1) = \left(\frac{1-f}{n}\right)\beta^2\left[\frac{\lambda_{22}}{\rho_{xy}^2} + \lambda_{04} - 2\frac{\lambda_{13}}{\rho_{xy}}\right]$$

$$= \left(\frac{1-0.16}{8}\right)(0.4334)^2\left[\frac{2.8411}{(0.8038)^2} + 4.5247 - 2\times\frac{3.2561}{0.8038}\right] = 0.016178,$$

and

$$\text{MSE}(b_2) = \left(\frac{1-f}{n}\right)\beta^2\left[\frac{\lambda_{22}}{\rho_{xy}^2} - 1\right] = \left(\frac{1-0.16}{8}\right)(0.4334)^2\left[\frac{2.8411}{(0.8038)^2} - 1\right] = 0.067005.$$

Thus the percent relative efficiency of b_1 with respect to b_2 is given by

$$\text{RE} = \frac{\text{MSE}(b_2)}{\text{MSE}(b_1)}\times 100 = \frac{0.067005}{0.016178}\times 100 = 414.17\%.$$

3.5 ESTIMATION OF FINITE POPULATION CORRELATION COEFFICIENT

For a bivariate normal population, an estimator of population correlation coefficient ρ_{xy} due to Pearson (1896) has been defined as

$$r_{xy} = \frac{s_{xy}}{s_x s_y}. \tag{3.5.1}$$

Wakimoto (1971), Gupta, Singh, and Lal (1978, 1979), Rana (1989) and Biradar and Singh (1992a) studied the behaviour of the estimator r_{xy} under SRSWOR sampling. Singh, Mangat, and Gupta (1996) have shown that the class of estimators, to estimate the population correlation coefficient, proposed by Srivastava and Jhajj (1986) as

$$r_{SJ} = r_{xy}H(u, v) \tag{3.5.2}$$

where $u = \bar{x}/\bar{X}$, $v = s_x^2/S_x^2$ and $H(.,.)$ is a parametric function such that $H(1, 1) = 1$ satisfying certain regularity conditions can take an inadmissible value, *i.e.*, outside the range $[-1.0, +1.0]$ from a given sample. In other words, the ratio type or regression type estimators cannot be made to estimate the correlation coefficient. The estimator r_{xy} can easily be written in terms of ε_2, ε_3 and ε_4 as follows:

$$r_{xy} = \frac{s_{xy}}{\left(s_x^2\right)^{1/2}\left(s_y^2\right)^{1/2}}$$

$$= \frac{S_{xy}(1+\varepsilon_4)}{S_x(1+\varepsilon_3)^{1/2}S_y(1+\varepsilon_2)^{1/2}} = \frac{S_{xy}}{S_xS_y}(1+\varepsilon_4)(1+\varepsilon_3)^{-\frac{1}{2}}(1+\varepsilon_2)^{-\frac{1}{2}}$$

$$= \rho_{xy}(1+\varepsilon_4)\left(1-\frac{1}{2}\varepsilon_2+\frac{3}{8}\varepsilon_2^2+....\right)\left(1-\frac{1}{2}\varepsilon_3+\frac{3}{8}\varepsilon_3^2+...\right)$$

$$= \rho_{xy}\left[1+\varepsilon_4-\frac{1}{2}\varepsilon_2-\frac{1}{2}\varepsilon_3+\frac{3}{8}\varepsilon_2^2+\frac{3}{8}\varepsilon_3^2+\frac{1}{4}\varepsilon_2\varepsilon_3-\frac{1}{2}\varepsilon_2\varepsilon_4-\frac{1}{2}\varepsilon_3\varepsilon_4+...\right]. \tag{3.5.3}$$

Thus we have the following theorems:

Theorem 3.5.1. The bias, up to terms of order $O(n^{-1})$, in the estimator r_{xy} of ρ_{xy} is

$$B(r_{xy}) = \left(\frac{1-f}{n}\right)\rho_{xy}\left[\frac{3}{8}(\lambda_{40}+\lambda_{04}-2)-\frac{1}{2}\left(\frac{\lambda_{31}}{\rho_{xy}}+\frac{\lambda_{13}}{\rho_{xy}}-2\right)+\frac{1}{4}(\lambda_{22}-1)\right]. \tag{3.5.4}$$

Proof. It follows by taking expected values on both sides of (3.5.3). Hence the theorem.

Theorem 3.5.2. The mean squared error, up to terms of order $O(n^{-1})$, of the usual estimator r_{xy} is

$$\text{MSE}(r_{xy}) = \left(\frac{1-f}{n}\right)\rho_{xy}^2\left[\left(\frac{\lambda_{22}}{\rho_{xy}^2}-1\right)+\frac{1}{2}(\lambda_{22}-1)+\frac{1}{4}(\lambda_{40}+\lambda_{04}-2)-\left(\frac{\lambda_{31}}{\rho_{xy}}+\frac{\lambda_{13}}{\rho_{xy}}-2\right)\right]. \tag{3.5.5}$$

Proof. By the definition of the mean squared error (MSE) we have

$$\text{MSE}(r_{xy}) = E[r_{xy}-\rho_{xy}]^2 \approx E\left[\rho_{xy}\left(1+\varepsilon_4-\frac{1}{2}\varepsilon_2-\frac{1}{2}\varepsilon_3+...\right)-\rho_{xy}\right]^2$$

$$= \rho_{xy}^2 E\left[\varepsilon_4-\frac{1}{2}\varepsilon_2-\frac{1}{2}\varepsilon_3\right]^2$$

$$= \rho_{xy}^2 E\left[\varepsilon_4^2+\frac{1}{4}\varepsilon_2^2+\frac{1}{4}\varepsilon_3^2-\varepsilon_2\varepsilon_4-\varepsilon_3\varepsilon_4+\frac{1}{2}\varepsilon_2\varepsilon_3\right].$$

Using expected values we obtain (3.5.5). Hence the theorem.

Example 3.5.1. The real and nonreal estate farm loans (in $000) during 1997 in the 50 states of the United States have been presented in population 1 of the Appendix. If we selected an SRSWOR sample of eight states to collect the required information. Study the relative bias of the usual estimator of the correlation coefficient.

Solution. Using the results from the description of the population 1 given in the Appendix, we have

$$B(r_{xy}) = \left(\frac{1-f}{n}\right)\rho_{xy}\left[\frac{3}{8}(\lambda_{40}+\lambda_{04}-2) - \frac{1}{2}\left(\frac{\lambda_{31}}{\rho_{xy}} + \frac{\lambda_{13}}{\rho_{xy}} - 2\right) + \frac{1}{4}(\lambda_{22}-1)\right]$$

$$= \left(\frac{1-0.16}{8}\right)(0.8038)\left[\frac{3}{8}(3.5822+4.5247-2) - \frac{1}{2}\left(\frac{2.9287}{0.8038} + \frac{3.2561}{0.8038} - 2\right)\right.$$

$$\left. + \frac{1}{4}(2.8411-1)\right] = -0.00817.$$

Also

$$\text{MSE}(r_{xy}) = \left(\frac{1-f}{n}\right)\rho_{xy}^2\left[\left(\frac{\lambda_{22}}{\rho_{xy}^2} - 1\right) + \frac{1}{2}(\lambda_{22}-1) + \frac{1}{4}(\lambda_{40}+\lambda_{04}-2) - \left(\frac{\lambda_{31}}{\rho_{xy}} + \frac{\lambda_{13}}{\rho_{xy}} - 2\right)\right]$$

$$= \left(\frac{1-0.16}{8}\right)(0.8038)^2\left[\left(\frac{2.8411}{(0.8038)^2} - 1\right) + \frac{(2.8411-1)}{2} + \frac{(4.5247+3.5822-2)}{4}\right.$$

$$\left. - \left(\frac{2.9287}{0.8038} + \frac{3.25611}{0.8038} - 2\right)\right] = 0.01018.$$

Thus the required relative bias is given by

$$\text{RB} = \frac{|B(r_{xy})|}{\sqrt{\text{MSE}(r_{xy})}} = \frac{|-0.00817|}{\sqrt{0.01018}} = 0.081\,.$$

Example 3.5.2. Consider the problem of estimation of the correlation coefficient between the amount of the real and nonreal estate farm loans in the United States. An SRSWOR sample of eight states has been drawn from the population 1 given in the Appendix and gathered the following information.

State	MA	NC	MO	FL	TX	IA	TN	NH
Nonreal estate farm loans (X)$	56.471	494.730	1519.994	464.516	3520.361	3909.738	388.869	0.471
Real estate farm loans (Y)$	7.590	639.571	1579.686	825.748	1248.761	2327.025	553.266	6.044

(a) Estimate the correlation coefficient ρ_{xy} between the real estate farm loans and the nonreal estate farm loans.

(b) Also find an estimate of the mean squared error and hence deduce the 80% confidence interval for the correlation coefficient.

Solution. (a) Estimation of correlation coefficient: From the sample information we have

State	y_i	x_i	$(y_i-\bar{y})^2$	$(x_i-\bar{x})^2$	$(y_i-\bar{y})(x_i-\bar{x})$
MA	7.590	56.471	793651.8068	1532452.7350	1102829.9420
NC	639.571	494.730	67024.2263	639462.1131	207025.2481
MO	1579.686	1519.994	464066.9897	50895.4728	153684.4457
FL	825.748	464.516	5287.2349	688697.0799	60343.2120
TX	1248.761	3520.361	122709.8273	4954930.1980	779755.4929
IA	2327.025	3909.738	2040794.0310	6840025.5460	3736185.6620
TN	553.266	388.869	119159.8469	819975.0729	312582.9556
NH	6.044	0.471	796408.7712	1674236.0830	1154719.1440
Sum	7187.691	10355.150	4409102.7340	17200674.3000	7507126.1030

From the above table, we have $n = 8$, $N = 50$, $f = 0.16$,

$$\hat{\mu}_{20} = s_y^2 = \frac{1}{8-1}\sum_{i=1}^{8}(y_i-\bar{y})^2 = 629871.82,$$

$$\hat{\mu}_{02} = s_x^2 = \frac{1}{8-1}\sum_{i=1}^{8}(x_i-\bar{x})^2 = 2457239.19,$$

and

$$\hat{\mu}_{11} = \frac{1}{8-1}\sum_{i=1}^{8}(y_i-\bar{y})(x_i \quad \bar{x}) - 1072446.59\,.$$

Thus the usual estimator of the correlation coefficient, ρ_{xy}, is given by

$$r_{xy} = \hat{\rho}_{xy} = \frac{\hat{\mu}_{11}}{\sqrt{\hat{\mu}_{20}\hat{\mu}_{02}}} = \frac{1072446.59}{\sqrt{629871.82 \times 2457239.19}} = 0.8620\,.$$

(b) Estimation of mean squared error and confidence interval: From the sample information we have

Sr. No.	$(y_i-\bar{y})^2(x_i-\bar{x})^2$	$(y_i-\bar{y})^4$	$(x_i-\bar{x})^4$	$(y_i-\bar{y})(x_i-\bar{x})^3$	$(y_i-\bar{y})^3(x_i-\bar{x})$
1	1.21623E+12	6.29883E+11	2.34841E+12	1.69003E+12	8.75263E+11
2	42859453356	4492246907	4.08912E+11	1.32385E+11	13875707073
3	23618908852	2.15358E+11	2590349152	7821842526	71319878084

Continued.......

4	3641303239	27954852.93	4.74304E+11	41558193926	319048736.9
5	6.08019E+11	15057701710	2.45513E+13	3.86363E+12	95683661855
6	1.39591E+13	4.16484E+12	4.67859E+13	2.55556E+13	7.62479E+12
7	97708104162	14199069118	6.72359E+11	2.5631E+11	37247337145
8	1.33338E+12	6.34267E+11	2.80307E+12	1.93327E+12	9.19628E+11
Sum	1.72845E+13	5.67813E+12	7.80469E+13	3.34806E+13	9.63812E+12

From the above table we have

$$\hat{\mu}_{04} = \frac{1}{8-1}\sum_{i=1}^{8}(x_i-\bar{x})^4 = 1.114955\times 10^{13},\quad \hat{\mu}_{22} = \frac{1}{8-1}\sum_{i=1}^{8}(y_i-\bar{y})^2(x_i-\bar{x})^2 = 2.469214\times 10^{12},$$

$$\hat{\mu}_{40} = \frac{1}{8-1}\sum_{i=1}^{8}(y_i-\bar{y})^4 = 8.111614\times 10^{11},\quad \hat{\mu}_{13} = \frac{\sum_{i=1}^{8}(y_i-\bar{y})(x_i-\bar{x})^3}{8-1} = 4.78294\times 10^{12},$$

$$\hat{\mu}_{31} = \frac{\sum_{i=1}^{8}(y_i-\bar{y})^3(x_i-\bar{x})}{8-1} = 1.37687\times 10^{12},\quad \hat{\lambda}_{40} = \frac{\hat{\mu}_{40}}{\hat{\mu}_{20}^2} = 2.04457,$$

$$\hat{\lambda}_{22} = \frac{\hat{\mu}_{22}}{\hat{\mu}_{20}\hat{\mu}_{02}} = \frac{2.469214\times 10^{12}}{629871.82\times 2457239.19} = 1.59536,\quad \hat{\lambda}_{04} = \frac{\hat{\mu}_{04}}{\hat{\mu}_{02}^2} = 1.84655,$$

$$\hat{\lambda}_{13} = \frac{\hat{\mu}_{13}}{\sqrt{\hat{\mu}_{20}}(\hat{\mu}_{02})^{3/2}} = \frac{4.78294\times 10^{12}}{\sqrt{629871.82}\times 2457239.19^{3/2}} = 1.56457, \text{ and}$$

$$\hat{\lambda}_{31} = \frac{\hat{\mu}_{31}}{\hat{\mu}_{20}^{3/2}\sqrt{\hat{\mu}_{02}}} = \frac{1.376874\times 10^{12}}{629871.82^{3/2}\sqrt{2457239.19}} = 1.75708.$$

Thus an estimator of $\mathrm{MSE}(r_{xy})$ is given by

$$\mathrm{M\hat{S}E}(r_{xy}) = \left(\frac{1-f}{n}\right)r_{xy}^2\left[\left(\frac{\hat{\lambda}_{22}}{r_{xy}^2}-1\right)+\frac{1}{2}(\hat{\lambda}_{22}-1)+\frac{1}{4}(\hat{\lambda}_{40}+\hat{\lambda}_{04}-2)-\left(\frac{\hat{\lambda}_{31}}{r_{xy}}+\frac{\hat{\lambda}_{13}}{r_{xy}}-2\right)\right]$$

$$= \left(\frac{1-0.16}{8}\right)(0.8620)^2\left[\left(\frac{1.59536}{0.8620^2}-1\right)+\frac{1}{2}(1.59536-1)\right.$$
$$\left.+\frac{1}{4}(2.04457+1.84655-2)-\left(\frac{1.75708}{0.8620}+\frac{1.56457}{0.8620}-2\right)\right]$$

$$= \left(\frac{1-0.16}{8}\right)(0.8620)^2[1.14706+0.29768+0.47278-1.85342] = 0.005001.$$

The $(1-\alpha)100\%$ confidence interval for the correlation coefficient ρ_{xy} is given by

$$r_{xy} \pm t_{\alpha/2}(\mathrm{df}=n-2)\sqrt{\mathrm{M\hat{S}E}(r_{xy})}.$$

Using Table 2 from the Appendix the 80% confidence interval of the correlation coefficient is given by

$$0.8620 \pm t_{0.20/2}(\text{df} = 8-2)\sqrt{0.005001}\,, \quad \text{or} \quad 0.8620 \pm t_{0.10}(\text{df} = 6)\sqrt{0.005001}$$

$$0.8620 \pm 1.440\sqrt{0.005001}\,, \quad \text{or} \quad [0.7601,\ 0.9638].$$

Again note the use of $(n-2)$ degree of freedom while constructing the confidence interval estimate of the population correlation coefficient.

3.6 SUPERPOPULATION MODEL APPROACH

We shall first introduce the concept of a superpopulation model and its role in survey sampling. Estimation of population mean from a sample is equivalent to predicting the mean of the non-sampled values of the study variable. Thus model-based sampling theory considers the problem of estimating finite population parameters can be expressed as prediction problem. The model based strategies have been proved to overcome the gap between finite population problems and rest of the statistics. We discuss here two types of super population models for regression and ratio estimator of the population mean.

3.6.1 RELATIONSHIP BETWEEN LINEAR MODEL AND REGRESSION ESTIMATOR

In fact under a superpopulation model a relation between a study variable Y_i and auxiliary variable X_i is:

$$Y_i = \alpha + \beta X_i + \varepsilon_i \tag{3.6.1.1}$$

where ε_i is a random variable such that $E(\varepsilon_i) = 0$, $E(\varepsilon_i^2) = \sigma^2$ and $E(\varepsilon_i \varepsilon_j) = 0$ for $i \neq j$. In short, under the superpopulation model, the population itself becomes a large random sample. Royall (1970a, 1970b, 1970c) was the first to show that the linear regression estimator of population mean defined as

$$\bar{y}_{\text{LR}} = \bar{y} + b(\bar{X} - \bar{x}) \tag{3.6.1.2}$$

is the best linear unbiased predictor of population mean $\bar{Y}$. Following Brewer (1963a), Royall (1970a, 1970b, 1970c), and Scott and Smith (1969), by imposing a superpopulation model on the actual finite population, inference about the characteristics of the finite population can be made via the structure of the model. In this section we shall only introduce the idea of superpopulation model under SRSWOR design. Thus we have the following theorems:

Theorem 3.6.1.1. Under the superpopulation model (3.6.1.1) the leading term of the variance of the linear regression estimator $\bar{y}_{\text{LR}}$ is given by

$$V(\bar{y}_{\text{LR}}) = \left(\frac{1-f}{n}\right)\frac{1}{(N-1)}\sum_{i=1}^{N}\varepsilon_i^2 \tag{3.6.1.3}$$

where

$$\varepsilon_i = (Y_i - \bar{Y}) - \beta(X_i - \bar{X}) \tag{3.6.1.4}$$

denotes the residual of Y_i to the fitted values $\bar{Y} + \beta(X_i - \bar{X})$.

Proof. We have

$$Y_i = \alpha + \beta X_i + \varepsilon_i \text{ for } i = 1,2,...,N. \tag{3.6.1.5}$$

Taking the sum on both sides of (3.6.1.5) and dividing by N we obtain

$$\frac{1}{N}\sum_{i=1}^{N} Y_i = \alpha + \beta \frac{1}{N}\sum_{i=1}^{N} X_i$$

which in fact implies that

$$\alpha = \bar{Y} - \beta \bar{X}. \tag{3.6.1.6}$$

On substituting the value of α in (3.6.1.5) and solving for ε_i, we obtain (3.6.1.4). Following the same steps as for MSE in the Theorem 3.2.3.2 we can easily see that variance, up to terms of $O(n^{-1})$, of the usual linear regression estimator is

$$V(\bar{y}_{\text{LR}}) = \left(\frac{1-f}{n}\right)\left[S_y^2 + \beta^2 S_x^2 - 2\beta S_{xy}\right]$$

$$= \left(\frac{1-f}{n}\right)\frac{1}{(N-1)}\left[\sum_{i=1}^{N}(Y_i - \bar{Y})^2 + \beta^2 \sum_{i=1}^{N}(X_i - \bar{X})^2 - 2\beta \sum_{i=1}^{N}(Y_i - \bar{Y})(X_i - \bar{X})\right]$$

$$= \left(\frac{1-f}{n}\right)\frac{1}{(N-1)}\sum_{i=1}^{N}\left[(Y_i - \bar{Y})^2 + \beta^2 (X_i - \bar{X})^2 - 2\beta (Y_i - \bar{Y})(X_i - \bar{X})\right]$$

$$= \left(\frac{1-f}{n}\right)\frac{1}{(N-1)}\sum_{i=1}^{N}\left[(Y_i - \bar{Y}) - \beta(X_i - \bar{X})\right]^2$$

$$= \left(\frac{1-f}{n}\right)\frac{1}{(N-1)}\sum_{i=1}^{N}\varepsilon_i^2.$$

Hence the theorem.

Theorem 3.6.1.2. An unbiased estimator of approximate variance of the linear regression estimator is:

$$\hat{v}(\bar{y}_{LR}) = \left(\frac{1-f}{n}\right)\frac{1}{(n-2)}\sum_{i=1}^{n} e_i^2 \tag{3.6.1.7}$$

where $e_i = (y_i - \bar{y}) - b(x_i - \bar{x})$ and $b = \sum_{i=1}^{n}(y_i - \bar{y})(x_i - \bar{x}) \Big/ \sum_{i=1}^{n}(x_i - \bar{x})^2$.

Proof. It follows from the well known Gauss—Markov Theorem. We have

$$y_i = \alpha + \beta x_i + e_i \tag{3.6.1.8}$$

and

$$\bar{y} = \alpha + \beta \bar{x} + \bar{e} \tag{3.6.1.9}$$

where $E(e_i) = 0$, $E(e_i^2) = \sigma^2$, $E(e_i e_j) = 0$ and e_i and x_i are independent.

By subtracting (3.6.1.9) from (3.6.1.8), then

$$(y_i - \bar{y}) = \beta(x_i - \bar{x}) + (e_i - \bar{e}) \tag{3.6.1.10}$$

or $y_i^* = \beta x_i^* + (e_i - \bar{e}).$ (3.6.1.11)

Remember that $E(e_i) = 0$ but $\bar{e} \neq 0$. In other words, in taking a very large number of samples we expect e_i to have a mean value of zero by assumption, but in any particular sample $\bar{e}$ is not necessarily zero.

Similarly we have

$$\hat{y}_i^* = \hat{\beta}\, x_i^* . \tag{3.6.1.12}$$

One can easily observe that

$$e_i = y_i^* - \hat{y}_i^* = (e_i - \bar{e}) - (\hat{\beta} - \beta)x_i^* .$$

Thus we have

$$\sum_{i=1}^{n} e_i^2 = \sum_{i=1}^{n}\left[(e_i - \bar{e}) - (\hat{\beta} - \beta)x_i^*\right]^2$$

$$= \sum_{i=1}^{n}\left[(e_i - \bar{e})^2 + (\hat{\beta} - \beta)^2 x_i^{*2} - 2(e_i - \bar{e})(\hat{\beta} - \beta)x_i^*\right]$$

$$= \sum_{i=1}^{n}(e_i - \bar{e})^2 + (\hat{\beta} - \beta)^2 \sum_{i=1}^{n} x_i^{*2} - 2\sum_{i=1}^{n}(e_i - \bar{e})(\hat{\beta} - \beta)x_i^* .$$

Taking expected values on both sides, we have

$$E\left(\sum_{i=1}^{n} e_i^2\right) = E\sum_{i=1}^{n}(e_i - \bar{e})^2 + E(\hat{\beta} - \beta)^2 \sum_{i=1}^{n} x_i^{*2} - 2E\sum_{i=1}^{n}(e_i - \bar{e})(\hat{\beta} - \beta)x_i^* . \tag{3.6.1.13}$$

Now we have

$$E\sum_{i=1}^{n}(e_i - \bar{e})^2 = E\left[\sum_{i=1}^{n} e_i^2 - n^{-1}\left(\sum_{i=1}^{n} e_i\right)^2\right] = \sum_{i=1}^{n} E(e_i^2) - \frac{1}{n}E\left(\sum_{i=1}^{n} e_i\right)^2$$

$$= \sum_{i=1}^{n} E(e_i^2) - \frac{1}{n}E\left(\sum_{i=1}^{n} e_i^2 + 2\sum_{i \neq j=1}^{n} e_i e_j\right) = \sum_{i=1}^{n}\sigma^2 - \frac{1}{n}\left(\sum_{i=1}^{n}\sigma^2 + 0\right)$$

$$= n\sigma^2 - \frac{n\sigma^2}{n} = (n-1)\sigma^2 . \tag{3.6.1.14}$$

$$E\left[(\hat{\beta} - \beta)^2 \sum_{i=1}^{n} x_i^{*2}\right] = \sum_{i=1}^{n} x_i^{*2} E(\hat{\beta} - \beta)^2 = \sum_{i=1}^{n} x_i^{*2} \times \frac{\sigma^2}{\sum_{i=1}^{n} x_i^{*2}} = \sigma^2 . \tag{3.6.1.15}$$

Now

$$\hat{\beta} - \beta = \sum_{i=1}^{n} w_i e_i , \quad \text{where } w_i = x_i^* \Big/ \sum_{i=1}^{n} x_i^{*2} \text{ and using } \sum_{i=1}^{n} x_i^* = 0 ,$$

therefore we have

$$E\left[(\hat{\beta} - \beta)\sum_{i=1}^{n} x_i^*(e_i - \bar{e})\right] = E\left[(\hat{\beta} - \beta)\left(\sum_{i=1}^{n} x_i^* e_i - \sum_{i=1}^{n} x_i^* \bar{e}\right)\right] = E\left[\frac{\sum_{i=1}^{n} x_i^* e_i}{\sum_{i=1}^{n} x_i^{*2}}\left(\sum_{i=1}^{n} x_i^* e_i - \sum_{i=1}^{n} x_i^* \bar{e}\right)\right]$$

$$= E\left[\frac{\sum_{i=1}^{n} x_i^* e_i}{\sum_{i=1}^{n} x_i^{*2}} \sum_{i=1}^{n} x_i^* e_i\right] = E\left[\frac{\sum_{i=1}^{n} x_i^{*2} e_i^2 + 2\sum_{i \neq j=1}^{n} x_i^* x_j^* e_i e_j}{\sum_{i=1}^{n} x_i^{*2}}\right]$$

$$= \frac{\sum_{i=1}^{n} x_i^{*2} E\left(e_i^2\right) + 2\sum_{i \neq j=1}^{n} x_i^* x_j^* E\left(e_i e_j\right)}{\sum_{i=1}^{n} x_i^{*2}} = \frac{\sigma^2 \sum_{i=1}^{n} x_i^{*2} + 2\sum_{i \neq j=1}^{n} x_i^* x_j^* \times 0}{\sum_{i=1}^{n} x_i^{*2}}$$

$$= \sigma^2 . \tag{3.6.1.16}$$

Using these expected values in (3.6.1.13) we have

$$E\left(\sum_{i=1}^{n} e_i^2\right) = E\sum_{i=1}^{n}(e_i - \bar{e})^2 + E\left\{\left(\hat{\beta} - \beta\right)^2 \sum_{i=1}^{n} x_i^{*2}\right\} - 2E\sum_{i=1}^{n}(e_i - \bar{e})\left(\hat{\beta} - \beta\right)x_i^*$$

$$= (n-1)\sigma^2 + \sigma^2 - 2\sigma^2 = (n-2)\sigma^2$$

which implies that

$$\sigma^2 = \frac{1}{n-2} E\left(\sum_{i=1}^{n} e_i^2\right). \tag{3.6.1.17}$$

By the method of moments an unbiased estimator of σ^2 is given by

$$\hat{\sigma}^2 = \frac{1}{n-2} \sum_{i=1}^{n} e_i^2 . \tag{3.6.1.18}$$

Under SRSWOR sampling we have

$$\sigma^2 = \frac{1}{n-2} E\left(\sum_{i=1}^{n} e_i^2\right) = \frac{1}{N-1} \sum_{i=1}^{N} \varepsilon_i^2 . \tag{3.6.1.19}$$

Therefore

$$V(\bar{y}_{\text{LR}}) = \left(\frac{1-f}{n}\right) \frac{1}{(N-1)} \sum_{i=1}^{N} \varepsilon_i^2 = \left(\frac{1-f}{n}\right) \sigma^2 . \tag{3.6.1.20}$$

Hence an unbiased estimator of $V(\bar{y}_{LR})$ is given by

$$\hat{v}(\bar{y}_{\text{LR}}) = \left(\frac{1-f}{n}\right) \frac{1}{(n-2)} \sum_{i=1}^{n} e_i^2 . \tag{3.6.1.21}$$

Hence the theorem.

3.6.2 IMPROVED ESTIMATOR OF VARIANCE OF LINEAR REGRESSION ESTIMATOR

Deng and Wu (1987) proposed an improved estimator of $V(\bar{y}_{\text{LR}})$ as follows:

$$\hat{v}(\bar{y}_{\text{LR}})_{\text{DW}} = \frac{(1-f)}{n} \frac{1}{(n-2)} \sum_{i=1}^{n} e_i^2 \left(\frac{\bar{x}}{\bar{X}}\right)^g . \tag{3.6.2.1}$$

Now we have

$$\frac{1}{(n-2)}\sum_{i=1}^{n} e_i^2 = \frac{1}{n\left(1-\frac{2}{n}\right)}\sum_{i=1}^{n} e_i^2 = \frac{1}{n}\sum_{i=1}^{n} e_i^2\left[1-\frac{2}{n}\right]^{-1} \approx \frac{1}{n}\sum_{i=1}^{n} z_i\left[1+\frac{2}{n}+\frac{4}{n^2}+....\right] \quad (3.6.2.2)$$

where $z_i = e_i^2,\ i = 1, 2, ..., n$ and

$$\left(\frac{\bar{x}}{\bar{X}}\right)=\left[1+\left(\frac{\bar{x}}{\bar{X}}-1\right)\right]^{g} \approx \left[1+g\left(\frac{\bar{x}}{\bar{X}}-1\right)+\frac{g(g-1)}{2}\left(\frac{\bar{x}}{\bar{X}}-1\right)^2+....\right]. \quad (3.6.2.3)$$

Using (3.6.2.2) and (3.6.2.3) the estimator (3.6.2.1) can be written as

$$\hat{v}(\bar{y}_{\text{LR}})_{\text{DW}} \approx \left(\frac{1-f}{n}\right)\bar{z}\left[1+g\left(\frac{\bar{x}}{\bar{X}}-1\right)+\frac{g(g-1)}{2}\left(\frac{\bar{x}}{\bar{X}}-1\right)^2+...\right]. \quad (3.6.2.4)$$

Defining

$$\varepsilon_1 = \frac{\bar{x}}{\bar{X}}-1 \text{ and } \delta_1 = \frac{\bar{z}}{\bar{Z}}-1$$

where $\bar{Z} = \frac{1}{N}\sum_{i=1}^{N} Z_i$ for $Z_i = \varepsilon_i^2$, we have

$$E\left(\varepsilon_1^2\right)=\left(\frac{1-f}{n}\right)C_x^2, \quad E\left(\delta_1^2\right)=\left(\frac{1-f}{n}\right)C_z^2, \text{ and } E(\varepsilon_1\delta_1)=\left(\frac{1-f}{n}\right)\rho_{xz}C_xC_z$$

where

$$C_z^2 = S_z^2/\bar{Z}^2, \quad S_z^2=(N-1)^{-1}\sum_{i=1}^{N}\left(Z_i-\bar{Z}\right)^2, \quad S_{xz}=(N-1)^{-1}\sum_{i=1}^{N}\left(Z_i-\bar{Z}\right)\left(X_i-\bar{X}\right), \text{ and}$$

$\rho_{xz} = S_{xz}/(S_xS_z)$.

Thus we have the following theorem.

Theorem 3.6.2.1. The minimum mean squared error of the estimator of approximate variance of the usual regression estimator is given by

$$\text{Min.MSE}\left[\hat{v}(\bar{y}_{\text{LR}})_{\text{DW}}\right]=\left(\frac{1-f}{n}\right)^3 S_z^2\left[1-\rho_{xz}^2\right]. \quad (3.6.2.5)$$

Proof. By the definition of mean squared error we have

$$\text{MSE}\left[\hat{v}(\bar{y}_{\text{LR}})_{\text{DW}}\right]=E\left[\hat{v}(\bar{y}_{\text{LR}})_{\text{DW}}-V(\bar{y}_{\text{LR}})\right]^2$$

$$\approx E\left[\left(\frac{1-f}{n}\right)\bar{Z}(1+\delta_1)\left\{1+g\varepsilon_1+\frac{g(g-1)}{2}\varepsilon_1^2+...\right\}-\left(\frac{1-f}{n}\right)\bar{Z}\right]^2$$

$$\approx \left(\frac{1-f}{n}\right)^2 \bar{Z}^2 E\left[\delta_1^2+g^2\varepsilon_1^2+2g\varepsilon_1\delta_1\right]$$

$$=\left(\frac{1-f}{n}\right)^3 \bar{Z}^2\left[C_z^2+g^2C_x^2+2g\rho_{xz}C_xC_z\right]. \quad (3.6.2.6)$$

On differentiating (3.6.2.6) with respect to g and equating to zero we obtain

$$g=-\rho_{xz}\frac{C_z}{C_x}. \quad (3.6.2.7)$$

On substituting (3.6.2.7) in (3.6.2.6) we obtain (3.6.2.5). Hence the theorem.

Example 3.6.2.1. Apply the regression method of estimation for estimating the average amount of the real estate farm loans (in $000) during 1997. Also find an estimator of the variance of the regression estimator assuming that the relationship between Y and X is given by $Y_i = \beta_0 + \beta_1 X_i + \varepsilon_i$. An SRSWOR sample of eight states selected from the population 1 of in the Appendix is given below.

State	AK	CA	CT	ME	NE	NY	VA	WI
Nonreal estate farm loan (X)$	3.433	3928.732	4.373	51.539	3585.406	426.274	188.477	1372.439
Real estate farm loan (Y)$	2.605	1343.461	7.130	8.849	1337.852	201.631	321.583	1229.572

The average amount $878.16 of nonreal estate farm loan (in $000) for the year 1997 is known.

(a) Derive the 95% confidence intervals using unbiased estimator of variance under the linear model.

(b) Derive the 95% confidence intervals using the estimator proposed by Deng and Wu (1987).

(c) Compare both confidence interval estimates in (a) and (b), and comment.

Given: $Z_i = \varepsilon_i^2$, $C_z^2 = 1.5097$, $C_x^2 = 1.5256$ and $\rho_{xz} = 0.4762$.

Solution. From the sample information we have

Sr. No.	y_i	x_i	$(y_i - \bar{y})^2$	$(x_i - \bar{x})^2$	$(y_i - \bar{y})(x_i - \bar{x})$
1	2.605	3.433	306894.26	1420032.4	660151.3371
2	1343.461	3928.732	619173.25	7472830.7	2151040.8800
3	7.130	4.373	301901.21	1417793.0	654242.6277
4	8.849	51.539	300015.14	1307695.5	626361.2614
5	1337.852	3585.406	610377.54	5713638.7	1867478.7040
6	201.631	426.274	125992.61	591069.0	272892.5174
7	321.583	188.477	55226.12	1013257.9	236555.0651
8	1229.572	1372.439	452911.00	31454.7	119357.4588
Sum	4452.683	9560.673	2772491.10	18967772.0	6588079.8520

From the above table we have

$n = 8$, $\bar{y} = 556.585$, $\bar{x} = 1195.084$, $s_x^2 = 2709682$, $s_y^2 = 396070.2$, $s_{xy} = 941154.3$, $\hat{\beta} = s_{xy}/s_x^2 = 0.3473$, $r_{xy} = s_{xy}/(s_x s_y) = 0.9085$, $e_i = y_i - 141.496 - 0.347 x_i$, and $\sum_{i=1}^{8} e_i^2 = 484252.1$. Also $\bar{X} = 878.16$, $N = 50$ and $f = 0.16$.

Thus the regression estimate of the average amount of the real estate farm loans during 1997, $\bar{Y}$ (say), is given by

$$\bar{y}_{\text{LR}} = \bar{y} + \hat{\beta}\left(\bar{X} - \bar{x}\right) = 556.585 + 0.3473 \times (878.16 - 1195.084) = 446.517\ .$$

(a) Usual estimator of variance: The usual estimate of $V(\bar{y}_{\text{LR}})$ under the superpopulation model is given by

$$\hat{v}(\bar{y}_{\text{LR}}) = \left(\frac{1-f}{n}\right)\frac{1}{n-2}\sum_{i=1}^{n} e_i^2 = \left(\frac{1-0.16}{8}\right) \times \frac{1}{8-2} \times 484252.1 = 8474.41.$$

A $(1-\alpha)100\%$ confidence interval for population mean $\bar{Y}$ is given by

$$\bar{y}_{\text{LR}} \pm t_{\alpha/2}(\text{df} = n-2)\sqrt{\hat{v}(\bar{y}_{\text{LR}})}\ .$$

Using Table 2 from the Appendix the 95% confidence interval is given by

$$446.517 \pm 2.447\sqrt{8474.41} \text{ or } [221.255, 671.779].$$

(b) Deng and Wu's estimator: The Deng and Wu (1987) estimate of $V(\bar{y}_{\text{LR}})$ under the superpopulation model is given by

$$\hat{v}(\bar{y}_{\text{LR}})_{\text{DW}} = \left(\frac{1-f}{n}\right)\frac{1}{n-2}\sum_{i=1}^{n} e_i^2\left(\frac{\bar{x}}{\bar{X}}\right)^g$$

$$= \left(\frac{1-0.16}{8}\right) \times \frac{1}{8-2} \times 484252.1 \times \left(\frac{1195.084}{878.162}\right)^{-0.4737} = 7323.475.$$

Given $C_z^2 = 1.5097$, $C_x^2 = 1.5256$ and $\rho_{xz} = 0.4762$, therefore

$$g = -\rho_{xz}\frac{C_z}{C_x} = -0.4762 \times \sqrt{\frac{1.5097}{1.5256}} = -0.4737.$$

A $(1-\alpha)100\%$ confidence interval for the population mean $\bar{Y}$ is given by

$$\bar{y}_{\text{LR}} \pm t_{\alpha/2}(\text{df} = n-2)\sqrt{\hat{v}(\bar{y}_{\text{LR}})_{\text{DW}}}\ .$$

Using Table 2 from the Appendix the 95% confidence interval is given by

$$446.517 \pm 2.447\sqrt{7323.475} \text{ or } [237.109, 655.924].$$

(c) Interpretation: Note that in this case the length of the confidence interval by the usual estimator is more than that of Deng and Wu (1987) which states that their estimator perform better than usual estimator of variance in this situation.

Deng and Wu (1987) have shown that the estimator at (3.6.2.1) remains better than the estimators proposed by Royall and Eberhardt (1975), Royall and Cumberland (1978, 1981a, 1981b, 1985) and Rao (1968a, 1969). It is to be noted that Deng and Wu (1987) have taken $(n-2)$ in the denominator of the usual estimator of variance of the regression estimator, which comes only from Gauss--Markov Theorem discussed above. Following Deville and Särndal (1992) we have the following theorem.

Theorem 3.6.2.2. An estimator for estimating the approximate variance of the usual linear regression estimator is given by

$$\hat{v}(\bar{y}_{\text{LR}})_{\text{DS}} = \frac{(1-f)}{n}\frac{1}{(n-1)}\sum_{i=1}^{n} e_i^2 \tag{3.6.2.8}$$

where $e_i = y_i - \hat{\beta}x_i$ with $\hat{\beta} = \sum_{i=1}^{n} x_i y_i \Big/ \sum_{i=1}^{n} x_i^2$. This technique is called model assisted, not model based, technique.

Proof. Following Deville and Särndal (1992), the regression estimator can be written as

$$\bar{y}_{\text{LR}} = \bar{y} + \hat{\beta}(\bar{X} - \bar{x}) = \frac{1}{n}\sum_{i=1}^{n}\left[y_i - \hat{\beta}x_i\right] + \hat{\beta}\bar{X} = \frac{1}{n}\sum_{i=1}^{n} e_i + \hat{\beta}\bar{X} = \bar{e} + \hat{\beta}\bar{X}\,. \tag{3.6.2.9}$$

Thus the variance of the linear regression estimator can be approximated as

$$V(\bar{y}_{\text{LR}}) = V(\bar{e} + \hat{\beta}\bar{X}) \approx V(\bar{e}) = \left(\frac{1-f}{n}\right)V(e_i) = \left(\frac{1-f}{n}\right)\frac{1}{N-1}\sum_{i=1}^{N}\varepsilon_i^2 \tag{3.6.2.10}$$

where $Y_i = \beta X_i + \varepsilon_i$ is the true superpopulation model passing through the origin. By the sample analogue an estimator of the variance $V(\bar{y}_{\text{LR}})$ given in (3.6.2.10) is given by (3.6.2.8). Hence the theorem.

Remark 3.6.2.1. In the above theorem if we estimate $\beta = \bar{Y}/\bar{X}$ with $\hat{\beta} = \bar{y}/\bar{x}$, then the results are true for the usual ratio estimator, and if we estimate it with $\hat{\beta} = -\bar{y}/\bar{X}$, then results are true for the product estimator.

3.6.3 RELATIONSHIP BETWEEN LINEAR MODEL AND RATIO ESTIMATOR

Consider a finite population of N units taken from a superpopulation can be described with a linear model as

$$Y_i = R\,X_i + \varepsilon_i\,,\quad i = 1,2,\ldots,N \tag{3.6.3.1}$$

where $E(\varepsilon_i) = 0$, $E(\varepsilon_i^2) = \sigma^2$ and $E(\varepsilon_i\varepsilon_j) = 0$, $i \neq j$. Clearly this model is a special case of (3.6.1.1) with zero intercept and $\beta = R$. A sample analogous of the above model is given by

$$y_i = r\,x_i + e_i\,,\quad i = 1,2,\ldots,n \tag{3.6.3.2}$$

where $E(e_i) = 0$, $E(e_i^2) = \sigma^2$ and $E(e_i e_j) = 0$, $i \neq j$. The value of R from model (3.6.3.1) and its estimate from (3.6.3.2) can be obtained under different assumptions:

Method I. From model (3.6.3.1) on setting $E(\varepsilon_i) = 0$ we obtain

$$\sum_{i=1}^{N}\varepsilon_i = 0\,. \tag{3.6.3.3}$$

From (3.6.3.1) we have

$$\varepsilon_i = Y_i - RX_i \,. \tag{3.6.3.4}$$

On using (3.6.3.4) in (3.6.3.3) we obtain

$\sum_{i=1}^{N}(Y_i - RX_i) = 0$ or $\sum_{i=1}^{N} Y_i - R\sum_{i=1}^{N} X_i = 0$ or $\frac{1}{N}\sum_{i=1}^{N} Y_i - R\frac{1}{N}\sum_{i=1}^{N} X_i = 0$ or $\overline{Y} - R\overline{X} = 0$ or $R = \overline{Y}/\overline{X}$, which is a ratio of two population means.

Using the sample analogous model (3.6.3.2) and setting $E(e_i) = 0$ we obtain

$$r = \frac{\bar{y}}{\bar{x}} \,. \tag{3.6.3.5}$$

If we multiply r in (3.6.3.5) with $\overline{X}$ then we obtain a traditional ratio estimator of the population mean as

$$\bar{y}_{\mathrm{R}} = r\,\overline{X} = \frac{\bar{y}}{\bar{x}}\overline{X} \,. \tag{3.6.3.6}$$

Method II. Again from the model (3.6.3.1) on setting $E(e_i^2) = 0$ we obtain sum of square due to errors (SSE) as

$$\mathrm{SSE} = \sum_{i=1}^{N}(Y_i - RX_i)^2 \,. \tag{3.6.3.7}$$

On setting $\frac{\partial\,\mathrm{SSE}}{\partial\,R} = 0$ we obtain $\sum_{i=1}^{N}(Y_i - RX_i)X_i = 0$ and it gives us

$$R = \frac{\sum_{i=1}^{N} Y_i X_i}{\sum_{i=1}^{N} X_i^2} \,. \tag{3.6.3.8}$$

Note that we are not interested in this ratio, but the sample analogous model of (3.6.3.8) will give us

$$r = \frac{\sum_{i=1}^{n} y_i x_i}{\sum_{i=1}^{n} x_i^2} \,. \tag{3.6.3.9}$$

If we multiply r in (3.6.3.9) with the known population mean $\overline{X}$, we obtain a new ratio type estimator of the population mean

$$\bar{y}_{\mathrm{R}_{\mathrm{e}^2}} = \left(\frac{\sum_{i=1}^{n} y_i x_i}{\sum_{i=1}^{n} x_i^2}\right)\overline{X} \,. \tag{3.6.3.10}$$

Note that $\sum_{i=1}^{n} y_i x_i = n\bar{x}\,\bar{y} + (n-1)s_{xy}$ and $\sum_{i=1}^{n} x_i^2 = n\bar{x}^2 + (n-1)s_x^2$, then (3.6.3.10) becomes

$$\bar{y}_{\mathrm{R}_{e2}} = \frac{\bar{y}}{\bar{x}}\left[\frac{1+\dfrac{(n-1)}{n}\dfrac{s_{xy}}{\bar{x}\,\bar{y}}}{1+\dfrac{(n-1)}{n}\dfrac{s_x^2}{\bar{x}^2}}\right]\bar{X} \tag{3.6.3.11}$$

which is called Beale (1962) ratio estimator of the population mean $\bar{Y}$.

Method III. The weighted sum of square of errors (WSSE) is given by

$$\mathrm{WSSE} = \sum_{i=1}^{N}\frac{1}{X_i}\varepsilon_i^2 = \sum_{i=1}^{N}\frac{1}{X_i}(Y_i - RX_i)^2 . \tag{3.6.3.12}$$

On setting

$$\frac{\partial \mathrm{WSSE}}{\partial R} = 0$$

we obtain the ratio as

$$R = \bar{Y}/\bar{X}$$

which is again a ratio of two population means, and its sample analogous will again lead to the traditional ratio estimator of the population mean. An interesting discussion on imposing linear structures in conventional sampling theory and related work is also available in Pokropp (2001).

3.7 JACKKNIFE VARIANCE ESTIMATOR

Quenouille's (1951) method of bias reduction, popularly known as the Jackknife procedure, has been successfully applied for estimating the variance of estimators. We shall discuss the idea of Jackknife variance estimator for ratio and regression estimator of the population mean, although it can be used to estimate the variance of any linear or non-linear estimator $\hat{\theta}$ of a parameter θ.

3.7.1 RATIO ESTIMATOR

Consider $(y_i, x_i), i = 1,2,...,n$ denotes a simple random sample of size n from a bivariate infinite population with means $(\bar{Y}, \bar{X})$. Let $\bar{y} = n^{-1}\sum_{i=1}^{n} y_i$ and $\bar{x} = n^{-1}\sum_{i=1}^{n} x_i$ be the sample means based of n observations. Divide the sampled data into g groups each having $n_j = n/g$, $j = 1, 2, ..., g$ observations. Let $\bar{y}_{(j)} = \frac{1}{n-n_j}\sum_{i=1}^{n-n_j} y_i$, $\bar{x}_{(j)} = \frac{1}{n-n_j}\sum_{i=1}^{n-n_j} x_i$ be the sample means after dropping j^{th} group from the sample. From the jth subsample obtained by dropping n_j units, a ratio type estimator to estimate the population mean $\bar{Y}$ is given by

$$\bar{y}_{\mathrm{R}(j)} = \bar{y}_{(j)}\left(\frac{\bar{X}}{\bar{x}_{(j)}}\right), \text{ for } j = 1,2,...,g. \tag{3.7.1.1}$$

Let us define a new estimator of the population mean $\overline{Y}$ as

$$\overline{y}_{Rg} = \frac{1}{g}\sum_{j=1}^{g}\overline{y}_{R(j)} = \frac{1}{g}\sum_{j=1}^{g}\overline{y}_{(j)}\left(\frac{\overline{X}}{\overline{x}_{(j)}}\right). \qquad (3.7.1.2)$$

Then a usual Jackknife estimator of variance $\hat{v}_u$ is given by

$$\hat{v}_u = \frac{g-1}{g}\sum_{j=1}^{g}\left[\overline{y}_{R(j)} - \overline{y}_{Rg}\right]^2 . \qquad (3.7.1.3)$$

Also from the full sample information we have the usual ratio estimator of population mean defined as

$$\overline{y}_R = \overline{y}\left(\frac{\overline{X}}{\overline{x}}\right)$$

where $\overline{y} = n^{-1}\sum_{i=1}^{n} y_i$ and $\overline{x} = n^{-1}\sum_{i=1}^{n} x_i$ are the sample means based on full sample information.

A modified Jackknife estimator of variance is given by

$$\hat{v}_m = \frac{g-1}{g}\sum_{j=1}^{g}\left[\overline{y}_{R(j)} - \overline{y}_R\right]^2 . \qquad (3.7.1.4)$$

For $g = n$ it reduces to the situation of dropping one unit at a time while making groups.

Example 3.7.1.1. People Bank took an SRSWOR sample of eight states from the population 1 given in the Appendix and collected the following information:

State	AR	NY	WA	NC	CA	MI	PA	SD
Nonreal estate farm loan (X)\$	848.317	426.274	1228.607	494.730	3928.732	440.518	298.351	1692.817
Real estate farm loan (Y)\$	907.700	201.631	1100.745	639.571	1343.461	323.028	756.169	413.777

(a) Apply the ratio method of estimation for estimating the average amount of the real estate farm loans (in \$000) during 1997.

(b) Estimates the variance of the ratio estimator using the usual Jackknife estimator of variance and derive the 95% confidence interval estimate.

(c) Estimates the variance of the ratio estimator using the modified Jackknife estimator of variance and derive the 95% confidence interval estimate.

(d) Which estimate of variance gives smaller confidence interval estimate?

Given: The known average amount \$878.16 of nonreal estate farm loans.

Solution. From the sample information we have

State	x_i	y_i
AR	848.317	907.700
NY	426.274	201.631
WA	1228.607	1100.745
NC	494.730	639.571
CA	3928.732	1343.461
MI	440.518	323.028
PA	298.351	756.169
SD	1692.817	413.777
Sum	9358.346	5686.082

Thus $n = 8$, $\bar{y} = 710.7603$ and $\bar{x} = 1169.793$. Note that $\bar{X} = 878.16$.

(a) The ratio estimate of the average amount of real estate farm loans during 1997, $\bar{Y}$ (say), is given by

$$\bar{y}_{\mathrm{R}} = \bar{y}\left(\frac{\bar{X}}{\bar{x}}\right) = 710.7603\left(\frac{878.162}{1169.793}\right) = 533.5667 .$$

Taking $g = n$ we have the following table:

Jackknife mechanism						
x_j	y_j	$\bar{x}(j)$	$\bar{y}(j)$	$\bar{y}_{R(j)}$	$\left(\bar{y}_{R(j)} - \bar{y}_{Rn}\right)^2$	$\left(\bar{y}_{R(j)} - \bar{y}_{R}\right)^2$
848.317	907.700	1215.718	682.6260	493.0869	2440.14489	1638.50906
426.274	201.631	1276.010	783.4930	539.2059	10.75108	31.81452
1228.607	1100.745	1161.391	655.0481	495.3000	2226.40154	1464.24376
494.730	639.571	1266.231	720.9301	499.9815	1806.52439	1127.87878
3928.732	1343.461	775.659	620.3744	702.3549	25558.47260	28489.89390
440.518	323.028	1273.975	766.1506	528.1128	206.55132	29.73054
298.351	756.169	1294.285	704.2733	477.8427	4178.59394	3105.02188
1692.817	413.777	1095.076	753.1864	603.9932	3783.29055	4960.07229
			Sum	4339.8780	40210.73030	40847.16470

where

$$\bar{x}(j) = \frac{n\bar{x} - x_j}{n-1}, \quad \bar{y}(j) = \frac{n\bar{y} - y_j}{n-1}, \text{ and } \bar{y}_{R(j)} = \frac{\bar{y}(j)}{\bar{x}(j)}\bar{X} .$$

Thus an average of all Jackknife ratio estimates is given by

$$\bar{y}_{\mathrm{R}n} = \frac{1}{n}\sum_{j=1}^{n} \bar{y}_{\mathrm{R}(j)} = \frac{4339.878}{8} = 542.4847 .$$

(b) Usual Jackknife estimator of variance: The usual Jackknife estimator of variance $\hat{v}_{\mathrm{u}}$ is given by

$$\hat{v}_{\mathrm{u}} = \frac{n-1}{n}\sum_{j=1}^{n}\left[\bar{y}_{\mathrm{R}(j)} - \bar{y}_{\mathrm{R}n}\right]^2 = \frac{8-1}{8}\times 40210.7303 = 35184.38 .$$

A $(1-\alpha)100\%$ confidence interval for population mean $\bar{Y}$ is given by

$$\bar{y}_{\mathrm{R}} \pm t_{\alpha/2}(\mathrm{df} = n-1)\sqrt{\hat{v}_{\mathrm{u}}} .$$

Using Table 2 from the Appendix the 95% confidence interval is given by

$$533.5667 \pm 2.365\sqrt{35184.38}, \text{ or } [89.9518, \quad 977.1816].$$

(c) Modified Jackknife estimator of variance: The modified Jackknife estimator of variance is given by

$$\hat{v}_{\mathrm{m}} = \frac{n-1}{n}\sum_{j=1}^{n}\left[\bar{y}_{\mathrm{R}(j)} - \bar{y}_{\mathrm{R}}\right]^2 = \frac{8-1}{8}\times 40847.1647 = 35741.269 .$$

A $(1-\alpha)100\%$ confidence interval for population mean $\bar{Y}$ is given by

$$\bar{y}_{\mathrm{R}} \pm t_{\alpha/2}(\mathrm{df} = n-1)\sqrt{\hat{v}_{\mathrm{m}}} .$$

Using Table 2 of the Appendix the 95% confidence interval is given by

$$533.5667 \pm 2.365\sqrt{35741.269}, \text{ or } [86.4549, \quad 980.6785].$$

(d) In this particular example the usual Jackknife estimator of the variance of the ratio estimate provides smaller confidence interval estimate at the same level of confidence.

3.7.2 REGRESSION ESTIMATOR

Assume that (y_i, x_i), $i = 1, 2, ..., n$ denotes a simple random sample of size n from a bivariate infinite population with means $(\bar{Y}, \bar{X})$. Let $\bar{y} = n^{-1}\sum_{i=1}^{n} y_i$ and $\bar{x} = n^{-1}\sum_{i=1}^{n} x_i$ be the sample means based on n observations. Let $b = s_{xy}/s_x^2$ be an estimator of the regression coefficient. Then the linear regression estimator of population mean $\bar{Y}$ is

$$\bar{y}_{\mathrm{lr}} = \bar{y} + b(\bar{X} - \bar{x}). \tag{3.7.2.1}$$

Divide the sampled data into n sub-samples each having $(n-1)$ observations. Let

$$\bar{y}(j) = (n-1)^{-1}\sum_{i=1}^{n-1} y_i = \frac{(n\bar{y} - y_j)}{(n-1)}, \quad \text{and} \quad \bar{x}(j) = (n-1)^{-1}\sum_{i=1}^{n-1} x_i = \frac{(n\bar{x} - x_j)}{(n-1)}$$

be the sample means after dropping j^{th} unit from the sample. From the jth sub-sample obtained by dropping one unit a regression estimator to estimate the population mean $\overline{Y}$ is given by

$$\bar{y}_{lr}(j) = \bar{y}(j) + b(j)\left(\overline{X} - \bar{x}(j)\right) \qquad (3.7.2.2)$$

where

$$b(j) = b - \frac{(x_j - \bar{x})d_j}{(n-1)s_x^2(1-k_j)}, \quad k_j = \frac{1}{n} + \frac{(x_j - \bar{x})^2}{(n-1)s_x^2}, \text{ and } d_j = (y_j - \bar{y}) - b(x_j - \bar{x}). \qquad (3.7.2.3)$$

Let us define a new regression estimator of the population mean $\overline{Y}$ as

$$\bar{y}_{lrn} = \frac{1}{n}\sum_{j=1}^{n}\bar{y}_{lr}(j) = \frac{1}{n}\sum_{j=1}^{n}\left[\bar{y}(j) + b(j)\left(\overline{X} - \bar{x}(j)\right)\right]. \qquad (3.7.2.4)$$

Then a usual Jackknife estimator of variance $\hat{v}_u(\bar{y}_{lr})$ is given by

$$\hat{v}_u(\bar{y}_{lr}) = \frac{n-1}{n}\sum_{j=1}^{n}\left[\bar{y}_{lr}(j) - \bar{y}_{lrn}\right]^2. \qquad (3.7.2.5)$$

A modified Jackknife estimator of variance is given by

$$\hat{v}_m(\bar{y}_{lr}) = \frac{n-1}{n}\sum_{j=1}^{n}\left[\bar{y}_{lr}(j) - \bar{y}_{lr}\right]^2. \qquad (3.7.2.6)$$

For details see Miller (1974), Rao (1969, 1974, 1979), Rao and Rao (1971) and Krewski and Chakrabarty (1981).

Example 3.7.2.1. A key bank took an SRSWOR sample of eight states from the population 1 given in the Appendix and collected the following information:

State	CA	FL	MO	IN	NJ	MA	OK	ME
Nonreal estate farm loan (X)$	3928.732	464.516	1519.994	1022.782	27.508	56.471	1716.087	51.539
Real estate farm loan (Y)$	1343.461	825.748	1579.686	1213.024	39.860	7.590	612.108	8.849

A statistician advised them to apply the regression method of estimation for estimating the average amount of the real estate farm loans.

(a) Apply the regression method of estimation for estimating the average amount of the real estate farm loans (in $000) during 1997.
(b) Estimates the variance of the regression estimator using the usual Jackknife estimator of variance and derive the 95% confidence interval estimate.
(c) Estimates the variance of the regression estimator using the modified Jackknife estimator of variance and derive the 95% confidence interval estimate.
(d) Which confidence interval estimate is smaller?

Given: The known average amount $878.16 of nonreal estate farm loans.

Solution. Using the sample information we have

State	x_i	y_i	$(x_i-\bar{x})^2$	$(y_i-\bar{y})^2$	$(y_i-\bar{y})(x_i-\bar{x})$
CA	3928.732	1343.461	8010476.0	409178.0	1810444.900
FL	464.516	825.748	401876.9	14873.6	-77313.289
MO	1519.994	1579.686	177696.3	767192.5	369225.210
IN	1022.782	1213.024	5726.2	259318.5	-38534.508
NJ	27.508	39.860	1146925.0	440804.0	711033.730
MA	56.471	7.590	1085728.0	484695.5	725429.090
OK	1716.087	612.108	381471.0	8405.7	-56626.326
ME	51.539	8.849	1096030.0	482944.0	727544.680
Sum	8787.629	5630.326	12305929.0	2867412.0	4171203.500

From the above table, $n=8$, $\bar{y}=703.7908$, $\bar{x}=1098.454$, $s_x^2=1757989.86$, $s_y^2=409630.28$, $s_{xy}=595886.21$, $b=s_{xy}/s_x^2=0.3389$. Given $\bar{X}=878.16$, $N=50$ and $f=0.16$.

(a) The regression estimate of average amount of the real estate farm loans is

$$\bar{y}_{\text{lr}}=\bar{y}+b(\bar{X}-\bar{x})=703.7908+0.3389\times(878.16-1098.454)=629.133.$$

Jackknife mechanism							
$\bar{x}(j)$	$\bar{y}(j)$	$k(j)$	d_j	$b(j)$	$\bar{y}_{lr}(j)$	$\{\bar{y}_{lr}(j)-\bar{y}_{lrn}\}^2$	$\{\bar{y}_{lr}(j)-\bar{y}_{lr}\}^2$
694.128	612.4093	0.775944	-319.511	0.666878	735.1361	8855.95961	11236.6590
1189.016	686.3683	0.157657	336.798	0.359497	574.6163	4410.78061	2972.0715
1038.234	578.6629	0.139440	733.035	0.309721	529.0847	12531.75270	10009.6650
1109.264	631.0431	0.125465	534.878	0.342661	551.8529	7952.55919	5972.2173
1251.446	798.6380	0.218201	-300.987	0.305395	684.6383	1901.68053	3080.8341
1247.308	803.2480	0.213228	-343.073	0.301978	691.7733	2574.88106	3923.8054
1010.220	716.8883	0.155999	-300.999	0.356799	669.7693	825.94493	1651.3054
1248.013	803.0681	0.214065	-340.142	0.302081	691.3426	2531.35713	3870.0336
				Sum	5128.2130	41584.91570	42716.5910

(b) Usual Jackknife estimator of variance: The usual Jackknife estimator of the variance $\hat{v}_{\text{u}}(\bar{y}_{\text{lr}})$ is given by

$$\hat{v}_{\text{u}}(\bar{y}_{\text{lr}})=\frac{n-1}{n}\sum_{j=1}^{n}\left[\bar{y}_{lr}(j)-\bar{y}_{\text{lrn}}\right]^2=\frac{8-1}{8}\times 41584.9157=36386.80$$

where

$$\bar{y}_{\text{lrn}}=\frac{1}{n}\sum_{j=1}^{n}\bar{y}_{\text{lr}}(j)=\frac{1}{n}\sum_{j=1}^{n}\left[\bar{y}(j)+b(j)(\bar{X}-\bar{x}(j))\right]=\frac{5128.213}{8}=641.03.$$

A $(1-\alpha)100\%$ confidence interval for population mean $\bar{Y}$ is given by

$$\bar{y}_{\text{lr}}\pm t_{\alpha/2}(\text{df}=n-2)\sqrt{\hat{v}_{\text{u}}(\bar{y}_{\text{lr}})}.$$

Using Table 2 from the Appendix the 95% confidence interval is given by

$$629.133 \pm 2.447\sqrt{36386.80}\,, \quad \text{or} \quad [162.36, 1095.90].$$

(c) Modified Jackknife estimator of variance: The modified Jackknife estimator of variance is given by

$$\hat{v}_{\mathrm{m}}(\bar{y}_{\mathrm{lr}}) = \frac{n-1}{n}\sum_{j=1}^{n}\left[\bar{y}_{\mathrm{lr}}(j) - \bar{y}_{\mathrm{lr}}\right]^2 = \frac{8-1}{8}\times 42716.591 = 37377.02\,.$$

A $(1-\alpha)100\%$ confidence interval for population mean $\bar{Y}$ is given by

$$\bar{y}_{\mathrm{lr}} \pm t_{\alpha/2}(\mathrm{df} = n-2)\sqrt{\hat{v}_{\mathrm{m}}(\bar{y}_{\mathrm{lr}})}\,.$$

Using Table 2 of the Appendix the 95% confidence interval for the average real estate farm loans is given by

$$629.133 \pm 2.447\sqrt{37377.02}\,, \quad \text{or} \quad [156.05,\ 1102.21].$$

(d) In this particular example, the usual Jackknife estimator of the variance of the regression estimate provides smaller confidence interval estimate at the same level of confidence.

While using the known single population mean of an auxiliary variable at the estimation stage or for the construction of ratio, product, and regression type estimators, a natural question arises of how to utilise the information if available on more than one auxiliary variable. The next section studies such situations.

3.8 ESTIMATION OF POPULATION MEAN USING MORE THAN ONE AUXILIARY VARIABLE

Let us first consider the case where the information on two auxiliary variables is available. Suppose Y_i, X_{1i} and X_{2i} are respectively values of ith unit of the study variable Y and auxiliary variables X_1 and X_2 from a finite population Ω. Let $\bar{Y}$, $\bar{X}_1$ and $\bar{X}_2$ be the population means of the study variable Y and auxiliary variables X_1 and X_2. Let the auxiliary variables X_1 and X_2 be correlated with Y with correlation coefficients ρ_{yx_1} and ρ_{yx_2}, respectively. A simple random sample of size n is drawn by SRSWOR sampling from the population Ω and let $\bar{y}$, $\bar{x}_1$ and $\bar{x}_2$ denote the corresponding sample means.

Now define

$$\varepsilon_0 = \frac{\bar{y}}{\bar{Y}} - 1\,,\ \eta_1 = \frac{\bar{x}_1}{\bar{X}_1} - 1 \ \text{ and } \ \eta_2 = \frac{\bar{x}_2}{\bar{X}_2} - 1$$

such that

$$E(\varepsilon_0) = E(\eta_1) = E(\eta_2) = 0$$

and

$$E(\varepsilon_0^2)=\left(\frac{1-f}{n}\right)C_y^2,\ \ E(\eta_1^2)=\left(\frac{1-f}{n}\right)C_{x1}^2,\ \ E(\eta_2^2)=\left(\frac{1-f}{n}\right)C_{x2}^2,$$

$$E(\varepsilon_0\eta_1)=\left(\frac{1-f}{n}\right)\rho_{yx_1}C_yC_{x_1},\ \ E(\varepsilon_0\eta_2)=\left(\frac{1-f}{n}\right)\rho_{yx_2}C_yC_{x_2},$$

$$E(\eta_1\eta_2)=\left(\frac{1-f}{n}\right)\rho_{x_1x_2}C_{x_1}C_{x_2}.$$

where

$\rho_{yx_1}=S_{yx_1}/(S_yS_{x_1})$, $\rho_{yx_2}=S_{yx_2}/(S_yS_{x_2})$, $\rho_{x_1x_2}=S_{x_1x_2}/(S_{x_1}S_{x_2})$, $C_y=S_y/\bar{Y}$, $C_{x_1}=S_{x_1}/\bar{X}_1$, $C_{x_2}=S_{x_2}/\bar{X}_2$, $s_y^2=(N-1)^{-1}\sum_{i=1}^{N}(Y_i-\bar{Y})^2$, $s_{x_1}^2=(N-1)^{-1}\sum_{i=1}^{N}(X_{1i}-\bar{X}_1)^2$, $s_{x2}^2=(N-1)^{-1}\sum_{i=1}^{N}(X_{2i}-\bar{X}_2)^2$, $S_{yx_1}=(N-1)^{-1}\sum_{i=1}^{N}(Y_i-\bar{Y})(X_{1i}-\bar{X}_1)$, $S_{yx_2}=(N-1)^{-1}\sum_{i=1}^{N}(Y_i-\bar{Y})(X_{2i}-\bar{X}_2)$, and $S_{x_1x_2}=(N-1)^{-1}\sum_{i=1}^{N}(X_{1i}-\bar{X}_1)(X_{2i}-\bar{X}_2)$ have their usual meanings.

We have redefined these terms only to avoid any kind of confusion in learning the estimation strategies with more than one auxiliary variable. Note that if we have p auxiliary variables, say $X_1, X_2, ..., X_p$ and their population means $\bar{X}_1, \bar{X}_2, ..., \bar{X}_p$ are known, then the above results can easily be extended.

3.8.1 MULTIVARIATE RATIO ESTIMATOR

Olkin (1958) proposed a weighted ratio type estimator of population mean $\bar{Y}$ as

$$\bar{y}_{R_0}=w\bar{y}_{R_1}+(1-w)\bar{y}_{R_2} \tag{3.8.1.1}$$

where $\bar{y}_{R_j}=\bar{y}(\bar{X}_j/\bar{x}_j)$, for $j=1,2$, are the two usual ratio estimators of population mean using different auxiliary variables at the estimation stage.

Then the estimator $\bar{y}_{R_0}$ in terms of ε_0, η_1 and η_2, can easily be expressed as:

$$\bar{y}_{R_0}=\bar{Y}\left(1+\varepsilon_0-\eta_2+\eta_2^2-\varepsilon_0\eta_2+..\right)+w\bar{Y}\left(\eta_2-\eta_1+\eta_1^2-\eta_2^2+\varepsilon_0\eta_2-\varepsilon_0\eta_1+...\right). \tag{3.8.1.2}$$

Thus we have the following theorems:

Theorem 3.8.1.1. The bias, up to terms of order $O(n^{-1})$, in the multivariate ratio estimator, $\bar{y}_{R_0}$, is

$$B(\bar{y}_{R_0})=\left(\frac{1-f}{n}\right)\bar{Y}\left[\left(C_{x_2}^2-\rho_{yx_2}C_yC_{x_2}\right)+w\left(C_{x_1}^2-C_{x_2}^2+\rho_{yx_2}C_yC_{x_2}-\rho_{yx_1}C_yC_{x_1}\right)\right] \tag{3.8.1.3}$$

Proof. It follows by taking expected values on both sides of (3.8.1.2). Hence the theorem.

Theorem 3.8.1.2. The minimum mean squared error of the multivariate ratio estimator $\bar{y}_{R_0}$ is given by

$$\text{Min.MSE}\left(\bar{y}_{R_0}\right)=\left(\frac{1-f}{n}\right)\bar{Y}^2\Bigg[C_y^2+C_{x2}^2-2\rho_{yx_2}C_yC_{x2}-\frac{\left(\rho_{yx_2}C_yC_{x_2}-C_{x2}^2-\rho_{yx_1}C_yC_{x_1}+\rho_{x_1x_2}C_{x_1}C_{x_2}\right)^2}{C_{x2}^2+C_{x1}^2-2\rho_{x_1x_2}C_{x_1}C_{x_2}}\Bigg]. \quad (3.8.1.4)$$

Proof. By the definition of mean squared error we have

$$\text{MSE}\left(\bar{y}_{R_0}\right)=E\left[\bar{y}_{R_0}-\bar{Y}\right]^2\approx\bar{Y}^2E\left[(\varepsilon_0-\eta_2)+w(\eta_2-\eta_1)\right]^2$$

$$=E\left[\left(\varepsilon_0^2+\eta_2^2-2\varepsilon_0\eta_2\right)+w^2\left(\eta_2^2+\eta_1^2-2\eta_1\eta_2\right)+2w\left(\varepsilon_0\eta_2-\eta_2^2-\varepsilon_0\eta_1+\eta_1\eta_2\right)\right]$$

$$=\left(\frac{1-f}{n}\right)\bar{Y}^2\Big[C_y^2+C_{x_2}^2-2\rho_{yx_2}C_yC_{x_2}+w^2\left(C_{x_1}^2+C_{x_2}^2-2\rho_{x_1x_2}C_{x_1}C_{x_2}\right)+2w\left(\rho_{yx_2}C_yC_{x_2}-C_{x_2}^2-\rho_{yx_1}C_yC_{x_1}+\rho_{x_1x_2}C_{x_1}C_{x_2}\right)\Big]. \quad (3.8.1.5)$$

On differentiating (3.8.1.5) with respect to w and equating to zero we obtain

$$w=-\frac{\left(\rho_{yx_2}C_yC_{x_2}-C_{x_2}^2-\rho_{yx_1}C_yC_{x_1}+\rho_{x_1x_2}C_{x_1}C_{x_2}\right)}{\left(C_{x_1}^2+C_{x_2}^2-2\rho_{x_1x_2}C_{x_1}C_{x_2}\right)}. \quad (3.8.1.6)$$

On substituting this value of w in (3.8.1.5) we obtain (3.8.1.4). Hence the theorem.

3.8.2 MULTIVARIATE REGRESSION TYPE ESTIMATORS

Raj (1965a) proposed a multivariate regression type estimator, which in the case of two auxiliary variables can be written as

$$\bar{y}_{\text{Raj}}=\bar{y}+\beta_1\left(\bar{X}_1-\bar{x}_1\right)+\beta_2\left(\bar{X}_2-\bar{x}_2\right). \quad (3.8.2.1)$$

The above estimator $\bar{y}_{\text{Raj}}$ in terms of ε_0, η_1 and η_2 can be easily expressed as:

$$\bar{y}_{\text{Raj}}=\bar{Y}(1+\varepsilon_0)-\beta_1\bar{X}_1\eta_1-\beta_2\bar{X}_2\eta_2. \quad (3.8.2.2)$$

Thus we have the following theorems:

Theorem 3.8.2.1. The estimator $\bar{y}_{\text{Raj}}$ is an unbiased estimator of the population mean $\bar{Y}$.

Proof. On taking expected values on both sides of (3.8.2.2) we have

$$E\left(\bar{y}_{\text{Raj}}\right)=E\left[\bar{Y}(1+\varepsilon_0)-\beta_1\bar{X}_1\eta_1-\beta_2\bar{X}_2\eta_2\right]=\bar{Y}.$$

Hence the theorem.

Theorem 3.8.2.2. The minimum variance of the estimator $\bar{y}_{\text{Raj}}$ is given by

$$\text{Min.}V(\bar{y}_{\text{Raj}}) = \left(\frac{1-f}{n}\right)\bar{Y}^2 C_y^2\left[1 - R^2_{y\bullet x_1 x_2}\right] \tag{3.8.2.3}$$

where $R^2_{y\bullet x_1x_2} = \dfrac{\rho_{yx_1}^2 + \rho_{yx_2}^2 - 2\rho_{yx_1}\rho_{yx_2}\rho_{x_1x_2}}{1-\rho_{x_1x_2}^2}$ denotes the multiple correlation between Y, X_1 and X_2.

Proof. By the definition of variance of an estimator we have

$$V(\bar{y}_{\text{Raj}}) = E\left[\bar{y}_{\text{Raj}} - \bar{Y}\right]^2 = E\left[\bar{Y}(1+\varepsilon_0) - \beta_1\bar{X}_1\eta_1 - \beta_2\bar{X}_2\eta_2 - \bar{Y}\right]^2$$

$$= E\left[\bar{Y}\varepsilon_0 - \beta_1\bar{X}_1\eta_1 - \beta_2\bar{X}_2\eta_2\right]^2$$

$$= E\Big[\ \bar{Y}^2\varepsilon_0^2 + \beta_1^2\bar{X}_1^2\eta_1^2 + \beta_2^2\bar{X}_2^2\eta_2^2 - 2\bar{Y}\ \bar{X}_1\beta_1\varepsilon_0\eta_1$$

$$- 2\bar{Y}\ \bar{X}_2\beta_2\varepsilon_0\eta_2 + 2\beta_1\beta_2\bar{X}_1\bar{X}_2\eta_1\eta_2\ \Big]$$

$$= \left(\frac{1-f}{n}\right)\Big[\bar{Y}^2C_y^2 + \beta_1^2\bar{X}_1^2C_{x_1}^2 + \beta_2^2\bar{X}_2^2C_{x_2}^2 - 2\beta_1\bar{Y}\ \bar{X}_1\rho_{yx_1}C_yC_{x_1}$$

$$- 2\beta_2\bar{Y}\ \bar{X}_2\rho_{yx_2}C_yC_{x_2} + 2\beta_1\beta_2\bar{X}_1\bar{X}_2\rho_{x_1x_2}C_{x_1}C_{x_2}\ \Big]. \tag{3.8.2.4}$$

On differentiating (3.8.2.4) with respect to β_1 and β_2, respectively, and equating to zero we obtain

$$\beta_1\bar{X}_1C_{x_1} + \beta_2\rho_{x_1x_2}\bar{X}_2C_{x_2} = \bar{Y}\rho_{yx_1}C_y, \tag{3.8.2.5}$$

and

$$\beta_1\rho_{x_1x_2}\bar{X}_1C_{x_1} + \beta_2\bar{X}_2C_{x_2} = \bar{Y}\rho_{yx_2}C_y. \tag{3.8.2.6}$$

The above system of equations can be written as

$$\begin{bmatrix} \bar{X}_1C_{x_1}, & \bar{X}_2C_{x_2}\rho_{x_1x_2} \\ \bar{X}_1C_{x_1}\rho_{x_1x_2}, & \bar{X}_2C_{x_2} \end{bmatrix}\begin{bmatrix} \beta_1 \\ \beta_2 \end{bmatrix} = \begin{bmatrix} \bar{Y}\rho_{yx_1}C_y \\ \bar{Y}\rho_{yx_2}C_y \end{bmatrix}.$$

Using Cramer's rule we have

$$\Delta = \det\begin{bmatrix} \bar{X}_1C_{x_1}, & \bar{X}_2C_{x_2}\rho_{x_1x_2} \\ \bar{X}_1C_{x_1}\rho_{x_1x_2}, & \bar{X}_2C_{x_2} \end{bmatrix} = \bar{X}_1\bar{X}_2C_{x_1}C_{x_2}\left(1-\rho_{x_1x_2}^2\right),$$

$$\Delta_1 = \det\begin{bmatrix} -\bar{Y}\rho_{yx_1}C_y, & \bar{X}_2C_{x_2}\rho_{x_1x_2} \\ -\bar{Y}\rho_{yx_2}C_y, & \bar{X}_2C_{x_2} \end{bmatrix} = \bar{Y}\ \bar{X}_2C_yC_{x_2}\left(\rho_{yx_1} - \rho_{yx_2}\rho_{x_1x_2}\right)$$

and

$$\Delta_2 = \det\begin{bmatrix} \bar{X}_1C_{x_1}, & \bar{Y}\rho_{yx_1}C_y \\ \bar{X}_1C_{x_1}\rho_{x_1x_2}, & \bar{Y}\rho_{yx_2}C_y \end{bmatrix} = \bar{Y}\ \bar{X}_1C_yC_{x_1}\left(\rho_{yx_2} - \rho_{yx_1}\rho_{x_1x_2}\right).$$

On solving (3.8.2.5) and (3.8.2.6) for β_1 and β_2 we obtain

$$\beta_1 = \frac{\Delta_1}{\Delta} = \frac{\overline{Y}C_y\left(\rho_{yx_1} - \rho_{x_1x_2}\rho_{yx_2}\right)}{\overline{X}_1 C_{x_1}\left(1-\rho_{x_1x_2}^2\right)}, \quad \text{and} \quad \beta_2 = \frac{\Delta_2}{\Delta} = \frac{\overline{Y}C_y\left(\rho_{yx_2} - \rho_{x_1x_2}\rho_{yx_1}\right)}{\overline{X}_2 C_{x_2}\left(1-\rho_{x_1x_2}^2\right)}.$$

On substituting these values of β_1 and β_2 in (3.8.2.4) we obtain

$$\text{Min.}V\left(\bar{y}_{\text{Raj}}\right)$$

$$= \left(\frac{1-f}{n}\right)\Big[\overline{Y}^2C_y^2 + \beta_1^2\overline{X}_1^2C_{x_1}^2 + \beta_2^2\overline{X}_2^2C_{x_2}^2 - 2\beta_1\overline{YX}_1\rho_{yx_1}C_yC_{x_1} - 2\beta_2\overline{YX}_2\rho_{yx_2}C_yC_{x_2}$$

$$+ 2\beta_1\beta_2\overline{X}_1\overline{X}_2\rho_{x_1x_2}C_{x_1}C_{x_2}\Big]$$

$$= \left(\frac{1-f}{n}\right)S_y^2\left[1 + \frac{\left(\rho_{yx_1} - \rho_{yx_2}\rho_{x_1x_2}\right)^2}{\left(1-\rho_{x_1x_2}^2\right)^2} + \frac{\left(\rho_{yx_2} - \rho_{yx_1}\rho_{x_1x_2}\right)^2}{\left(1-\rho_{x_1x_2}^2\right)^2} - \frac{2\rho_{yx_1}\left(\rho_{yx_1} - \rho_{yx_2}\rho_{x_1x_2}\right)}{\left(1-\rho_{x_1x_2}^2\right)}\right.$$

$$\left. - \frac{2\rho_{yx_2}\left(\rho_{yx_2} - \rho_{yx_1}\rho_{x_1x_2}\right)}{\left(1-\rho_{x_1x_2}^2\right)} + \frac{2\rho_{x_1x_2}\left(\rho_{yx_1} - \rho_{yx_2}\rho_{x_1x_2}\right)\left(\rho_{yx_2} - \rho_{yx_1}\rho_{x_1x_2}\right)}{\left(1-\rho_{x_1x_2}^2\right)^2}\right]$$

$$= \left(\frac{1-f}{n}\right)S_y^2\left[1 + \frac{1}{\left(1-\rho_{x_1x_2}^2\right)^2}\left\{\rho_{yx_1}^2 + \rho_{yx_2}^2\rho_{x_1x_2}^2 - 4\rho_{yx_1}\rho_{yx_2}\rho_{x_1x_2} + \rho_{yx_2}^2 + \rho_{yx_1}^2\rho_{x_1x_2}^2\right.\right.$$

$$\left. + 2\rho_{x_1x_2}\left(\rho_{yx_1}\rho_{yx_2} - \rho_{yx_2}^2\rho_{x_1x_2} - \rho_{yx_1}^2\rho_{x_1x_2} + \rho_{yx_1}\rho_{yx_2}\rho_{x_1x_2}^2\right)\right\}$$

$$\left. - \frac{2}{\left(1-\rho_{x_1x_2}^2\right)}\left\{\rho_{yx_1}^2 - \rho_{yx_1}\rho_{yx_2}\rho_{x_1x_2} + \rho_{yx_2}^2 - \rho_{yx_1}\rho_{yx_2}\rho_{x_1}\right\}\right]$$

$$= \left(\frac{1-f}{n}\right)S_y^2\left[1 + \frac{1}{\left(1-\rho_{x_1x_2}^2\right)^2}\left\{\rho_{yx_1}^2 - \rho_{yx_1}^2\rho_{x_1x_2}^2 + \rho_{yx_2}^2 - \rho_{yx_2}^2\rho_{x_1x_2}^2 - 2\rho_{yx_1}\rho_{yx_2}\rho_{x_1x_2} + 2\rho_{yx_1}\rho_{yx_2}\rho_{x_1x_2}^3\right\}\right.$$

$$\left. - \frac{2}{\left(1-\rho_{x_1x_2}^2\right)}\left\{\rho_{yx_1}^2 + \rho_{yx_2}^2 - 2\rho_{yx_1}\rho_{yx_2}\rho_{x_1x_2}\right\}\right]$$

$$= \left(\frac{1-f}{n}\right)S_y^2\left[1 + \frac{1}{\left(1-\rho_{x_1x_2}^2\right)^2}\left\{\rho_{yx_1}^2\left(1-\rho_{x_1x_2}^2\right) + \rho_{yx_2}^2\left(1-\rho_{x_1x_2}^2\right) - 2\rho_{yx_1}\rho_{yx_2}\rho_{x_1x_2}\left(1-\rho_{x_1x_2}^2\right)\right\}\right.$$

$$\left. - \frac{2}{\left(1-\rho_{x_1x_2}^2\right)}\left\{\rho_{yx_1}^2 + \rho_{yx_2}^2 - 2\rho_{yx_1}\rho_{yx_2}\rho_{x_1x_2}\right\}\right]$$

$$=\left(\frac{1-f}{n}\right)S_y^2\left[1-\frac{\rho_{yx_1}^2+\rho_{yx_2}^2-2\rho_{yx_1}\rho_{yx_2}\rho_{x_1x_2}}{1-\rho_{x_1x_2}^2}\right]=\left(\frac{1-f}{n}\right)S_y^2\left(1-R_{y\bullet x_1x_2}^2\right).$$

Hence the theorem.

Example 3.8.2.1. The season average price (in \$) per pound amount of the commercial apple crop in 36 different states of the United States has been given in population 3. Suppose we selected an SRSWOR sample of nine states to collect the required information from 1996. Find the relative efficiency of the regression type estimator of average price in the United States that makes use of past information from two years with respect to the estimator that makes use of past information only from one year.

Solution. From the description of the population we have

$Y_i =$ Season average price per pound during 1996,

$X_{1i} =$ Season average price per pound during 1995,

$X_{2i} =$ Season average price per pound during 1994,

$N = 36$, $S_y^2 = 0.006488$, $\rho_{yx_1} = 0.8775$, $\rho_{yx_2} = 0.8577$, $\rho_{x_1x_2} = 0.8788$, $n = 9$ and $f = n/N = 9/36 = 0.25$.

Now we have

(a) Use of one auxiliary variable

Regression estimator: $\bar{y}_{l_1} = \bar{y} + \hat{\beta}\left(\bar{X}_1 - \bar{x}_1\right)$.

Mean square error:

$$\mathrm{MSE}\left(\bar{y}_{l_1}\right)=\left(\frac{1-f}{n}\right)S_y^2\left(1-\rho_{yx_1}^2\right)=\left(\frac{1-0.25}{9}\right)\times 0.006488\times\left[1-0.8775^2\right]=0.0001243.$$

(b) Use of two auxiliary variables

Regression estimator: $\bar{y}_{l_2} = \bar{y} + \hat{\beta}_1\left(\bar{X}_1 - \bar{x}_1\right) + \hat{\beta}_2\left(\bar{X}_2 - \bar{x}_2\right)$.

Mean square error:

$$\mathrm{MSE}\left(\bar{y}_{l_2}\right)=\left(\frac{1-f}{n}\right)S_y^2\left[1-\frac{\rho_{yx_1}^2+\rho_{yx_2}^2-2\rho_{yx_1}\rho_{yx_2}\rho_{x_1x_2}}{1-\rho_{x_1x_2}^2}\right]$$

$$=\left(\frac{1-0.25}{9}\right)(0.006488)\left[1-\frac{0.8775^2+0.8577^2-2\times 0.8775\times 0.8577\times 0.8788}{1-0.8788^2}\right]$$

$$=0.0001066.$$

Thus the required relative efficiency is given by

$$\mathrm{RE}=\mathrm{MSE}\left(\bar{y}_{l_1}\right)\times 100/\mathrm{MSE}\left(\bar{y}_{l_2}\right)=\frac{0.00012343}{0.0001066}\times 100=116.60\%\ .$$

Now we have the following corollaries:

Corollary 3.8.2.1. A practically useful linear regression estimator of the population mean $\bar{Y}$ in the presence of two auxiliary variables X_1 and X_2 is given by

$$\bar{y}_{\text{Raj(p)}} = \bar{y} + \hat{\beta}_1\left(\bar{X}_1 - \bar{x}\right) + \hat{\beta}_2\left(\bar{X}_2 - \bar{x}_2\right) \tag{3.8.2.7}$$

where

$$\hat{\beta}_1 = \left\{s_y\left(r_{yx_1} - r_{x_1x_2}r_{yx_2}\right)\right\} \Big/ \left\{s_{x_1}\left(1 - r^2_{x_1x_2}\right)\right\}$$

and

$$\hat{\beta}_2 = \left\{s_y\left(r_{yx_2} - r_{x_1x_2}r_{yx_1}\right)\right\} \Big/ \left\{s_{x_2}\left(1 - r^2_{x_1x_2}\right)\right\}$$

denote the estimators of the partial regression coefficients β_1 and β_2, respectively.

Corollary 3.8.2.2. Under the superpopulation model

$$y_i = \hat{\beta}_0 + \hat{\beta}_1 x_{1i} + \hat{\beta}_2 x_{2i} + e_i \tag{3.8.2.8}$$

where $E(e_i) = 0$, $E\left(e_i^2\right) = \sigma^2$, and $E\left(e_i e_j\right) = 0 \;\; i \neq j = 1,2,....,n.$. Then the values of $\hat{\beta}_1$ and $\hat{\beta}_2$ in the estimator $\bar{y}_{\text{Raj(p)}} = \bar{y} + \hat{\beta}_1\left(\bar{X}_1 - \bar{x}_1\right) + \hat{\beta}_2\left(\bar{X}_2 - \bar{x}_2\right)$ can be obtained by solving the following set of normal equations:

$$\begin{bmatrix} \sum\limits_{i=1}^{n}\left(x_{1i} - \bar{x}_1\right)^2, & \sum\limits_{i=1}^{n}\left(x_{1i} - \bar{x}_1\right)\left(x_{2i} - \bar{x}_2\right) \\ \sum\limits_{i=1}^{n}\left(x_{1i} - \bar{x}_1\right)\left(x_{2i} - \bar{x}_2\right), & \sum\limits_{i=1}^{n}\left(x_{2i} - \bar{x}_2\right)^2 \end{bmatrix} \begin{bmatrix} \hat{\beta}_1 \\ \hat{\beta}_2 \end{bmatrix} = \begin{bmatrix} \sum\limits_{i=1}^{n}\left(y_i - \bar{y}\right)\left(x_{1i} - \bar{x}_1\right) \\ \sum\limits_{i=1}^{n}\left(y_i - \bar{y}\right)\left(x_{2i} - \bar{x}_2\right) \end{bmatrix} \tag{3.8.2.9}$$

or

$$\begin{bmatrix} s^2_{x_1}, & s_{x_1x_2} \\ s_{x_1x_2}, & s^2_{x_2} \end{bmatrix} \begin{bmatrix} \hat{\beta}_1 \\ \hat{\beta}_2 \end{bmatrix} = \begin{bmatrix} s_{yx_1} \\ s_{yx_2} \end{bmatrix}. \tag{3.8.2.10}$$

Corollary 3.8.2.3. Under the superpopulation model

$$y_i = \hat{\beta}_0 + \hat{\beta}_1 x_{1i} + \hat{\beta}_2 x_{2i} + e_i \tag{3.8.2.11}$$

where $E(e_i) = 0$, $E\left(e_i^2\right) = \sigma^2$ and $E\left(e_i e_j\right) = 0$, $i \neq j = 1,2,....,n.$ An unbiased estimator of the $V\left(\bar{y}_{\text{Raj(p)}}\right)$ is given by

$$\hat{v}\left(\bar{y}_{\text{Raj(p)}}\right) = \left(\frac{1-f}{n}\right)\frac{1}{n-3}\sum_{i=1}^{n} e_i^2 . \tag{3.8.2.12}$$

Corollary 3.8.2.4. An estimator of the $V\left(\bar{y}_{\text{Raj(p)}}\right)$ is given by

$$\hat{v}\left(\bar{y}_{\text{Raj(p)}}\right) = \left(\frac{1-f}{n}\right)s_y^2\left(1 - \hat{R}^2_{y\bullet x_1x_2}\right) \tag{3.8.2.13}$$

where

$$\hat{R}^2_{y\bullet x_1x_2} = \frac{r^2_{yx_1} + r^2_{yx_2} - 2r_{yx_1}r_{yx_2}r_{x_1x_2}}{1 - r^2_{x_1x_2}}. \tag{3.8.2.14}$$

Theorem 3.8.2.3. The minimum sample size required to achieve the minimum relative standard error ϕ with the estimator $\bar{y}_{\text{Raj}}$ is given by

$$n \geq \left\{ \frac{\phi^2}{C_y^2\left(1 - R_{y.x_1x_2}^2\right)} + \frac{1}{N} \right\}^{-1}. \tag{3.8.2.15}$$

Proof. We have $\text{RSE} = \sqrt{\frac{\text{MSE}(\bar{y}_{\text{Raj}})}{\bar{Y}^2}} \leq \phi$ which implies that

$$\sqrt{\left(\frac{1}{n} - \frac{1}{N}\right) C_y^2\left(1 - R_{y.x_1x_2}^2\right)} \leq \phi, \text{ or } \left(\frac{1}{n} - \frac{1}{N}\right) C_y^2\left(1 - R_{y.x_1x_2}^2\right) \leq \phi^2,$$

$$\text{or } \frac{1}{n} \leq \frac{\phi^2}{C_y^2\left(1 - R_{y.x_1x_2}^2\right)} + \frac{1}{N}, \text{ or } n \geq \left\{ \frac{\phi^2}{C_y^2\left(1 - R_{y.x_1x_2}^2\right)} + \frac{1}{N} \right\}^{-1}.$$

Hence the theorem.

Example 3.8.2.2. We wish to estimate the season's average price per pound (Y) of the commercial apple crop during 1996 in the United States. The correlation between the price during 1996 (Y) with that during 1995 (X_2) and 1994 (X_1) are assumed to be known. Find the minimum sample size, n, required to estimate the average price with relative standard deviation 5.6%.

Given: $R_{y \bullet x_1x_2}^2 = 0.8029$, $C_y^2 = 0.1563$, and $N = 69$.

Solution. The minimum sample size is given by

$$n \geq \left[\frac{\phi^2}{C_y^2\left(1 - R_{y \bullet x_1x_2}^2\right)} + \frac{1}{N} \right]^{-1} = \left[\frac{0.056^2}{0.1563 \times \left(1 - 0.8029^2\right)} + \frac{1}{69} \right]^{-1} = 14.09 \approx 14.$$

Example 3.8.2.3. Select an SRSWOR sample of 14 states from the population 3 given in the Appendix. Collect the information about the season's average price per pound of the apple crop during 1996, 1995, and 1994. Estimate the average per pound price (\$) of the commercial apple crop during 1996 in the United States. Assume that the average prices per pound of the commercial apple crop during 1995 and 1994 are accurately known.

(a) Apply the regression estimator of population mean with two auxiliary variables.

(b) Derive the 95% confidence intervals using superpopulation model approach

(c) Derive the 95% confidence intervals using design based approach.

Given: $\bar{X}_1 = 0.1856$ and $\bar{X}_2 = 0.1708$.

Solution. We started with the first two columns of the Pseudo-Random Numbers (PRN) given in Table 1 of the Appendix to select 14 distinct random numbers $1 \le R \le 36$ as 01, 23, 04, 32, 33, 05, 22, 29, 03, 36, 27, 19, 14 and 06.

Sr. No.	State and Territory	Year 1994 x_{2i}	Year 1995 x_{1i}	Year 1996 y_i
01	AZ	0.078	0.071	0.122
03	CA	0.133	0.183	0.160
04	CO	0.157	0.145	0.223
05	CT	0.283	0.276	0.292
06	DE	0.168	0.125	0.173
14	ME	0.174	0.179	0.185
19	MO	0.198	0.160	0.228
22	NM	0.219	0.298	0.306
23	NY	0.118	0.121	0.130
27	PA	0.104	0.095	0.133
29	SC	0.130	0.126	0.136
32	VT	0.165	0.181	0.194
33	VA	0.090	0.099	0.101
36	WI	0.230	0.241	0.133
	Sum	2.247	2.300	2.516

From the sample information we have

$\bar{y} = 0.1797$, $\bar{x}_1 = 0.1642$, $\bar{x}_2 = 0.1605$, $r_{yx_1} = 0.7641$, $r_{yx_2} = 0.7722$, $r_{x_1x_2} = 0.8893$, $s_y = 0.063063$, $s_{x_1} = 0.06801$, and $s_{x_2} = 0.057778$.

The estimates of the partial regression coefficients β_1 and β_2 are

$$\hat{\beta}_1 = \frac{s_y\left(r_{yx_1} - r_{x_1x_2}r_{yx_2}\right)}{s_{x_1}\left(1 - r_{x_1x_2}^2\right)} = \frac{0.063063(0.7641 - 0.8893 \times 0.7722)}{0.068009(1 - 0.8893^2)} = 0.3431,$$

and

$$\hat{\beta}_2 = \frac{s_y\left(r_{yx_2} - r_{x_1x_2}r_{yx_1}\right)}{s_{x_2}\left(1 - r_{x_1x_2}^2\right)} = \frac{0.063063(0.7722 - 0.8893 \times 0.7641)}{0.057778(1 - 0.8893^2)} = 0.4837.$$

(a) **Regression estimator:** An estimate of the average price of the commercial apple crop during 1996 is given by

$$\bar{y}_{\text{Raj(p)}} = \bar{y} + \hat{\beta}_1\left(\bar{X}_1 - \bar{x}_1\right) + \hat{\beta}_2\left(\bar{X}_2 - \bar{x}_2\right)$$

$$= 0.1797 + 0.3431(0.1856 - 0.1642) + 0.4837(0.1708 - 0.1605) = 0.19202.$$

(b) **Superpopulation model approach:** The linear relationship between y, x_1, and x_2 is $\hat{y}_i = 0.04571 - 0.343x_{1i} - 0.4837x_{2i}$, thus the residual terms are given by

$e_i = y_i - 0.04571 - 0.343x_{1i} - 0.4837x_{2i}$	
e_i	e_i^2
0.0142013	0.000201677
-0.0128294	0.000164594
0.0515996	0.002662519
0.0147073	0.000216305
0.0031409	9.86525E-06
-0.0062887	3.95477E-05
0.0316214	0.000999913
0.0521159	0.002716067
-0.0143017	0.000204539
0.0043907	1.92782E-05
-0.0158216	0.000250323
0.0063784	4.0684E-05
-0.0222099	0.00049328
-0.1066481	0.011373817

One can see that $\sum_{i=1}^{14} e_i = 0.0$ and $\sum_{i=1}^{14} e_i^2 = 0.019392408$. Thus an estimate of $V(\bar{y}_{\text{Raj(p)}})$ is

$$\hat{v}(\bar{y}_{\text{Raj(p)}}) = \left(\frac{1-f}{n}\right)\frac{1}{n-3}\sum_{i=1}^{n} e_i^2 = \left(\frac{1-0.39}{14}\right)\times\frac{0.019392408}{14-3} = 0.00007694.$$

A $(1-\alpha)100\%$ confidence interval for estimating the population mean $\bar{Y}$ is

$$\bar{y}_{\text{Raj(p)}} \mp t_{\alpha/2}(\text{df} = n \quad 3)\sqrt{\hat{v}(\bar{y}_{\text{Raj(p)}})} \,.$$

Using Table 2 from the Appendix the 95% confidence interval for the average price of the apple crop during 1996 is given by

$$0.19202 \mp t_{0.05/2}(\text{df} = 14-3)\sqrt{0.00007694}$$

or $0.19202 \mp 2.201\sqrt{0.00007694}$, or $[0.1727, 0.2113]$.

(c) Design based approach: An estimator of the multiple correlation coefficient is

$$\hat{R}^2_{y\bullet x_1x_2} = \frac{r_{yx_1}^2 + r_{yx_2}^2 - 2r_{yx_1}r_{yx_2}r_{x_1x_2}}{1-r_{x_1x_2}^2}$$

$$= \frac{0.7641^2 + 0.7722^2 - 2\times 0.7641\times 0.7722\times 0.889}{1-0.8893^2} = 0.6249\,.$$

An estimator of the $V(\bar{y}_{\text{Raj(p)}})$ is given by

$$\hat{v}(\bar{y}_{\text{Raj(p)}}) = \left(\frac{1-f}{n}\right)s_y^2\left(1-\hat{R}^2_{y\bullet x_1x_2}\right) = \left(\frac{1-039}{14}\right)\times 0.003977\times(1-.6249) = 0.00006499.$$

A $(1-\alpha)100\%$ confidence interval for estimating the population mean $\overline{Y}$ is given by

$$\bar{y}_{\text{Raj(p)}} \mp t_{\alpha/2}(\text{df}=n-3)\sqrt{\hat{v}(\bar{y}_{\text{Raj(p)}})} \ .$$

Using Table 2 from the Appendix the 95% confidence interval for the average price of the apple crop during 1996 is

$$0.19202 \mp t_{0.05/2}(\text{df}=14-3)\sqrt{0.00006499}$$

or $0.19202 \mp 2.201\sqrt{0.00006499}$, or $[0.1743, 0.2097]$.

Corollary 3.8.2.5. If we have p auxiliary variables, for example $X_1, X_2,..., X_p$, and their population means $\overline{X}_1, \overline{X}_2,..., \overline{X}_p$ are known, then the multivariate regression type estimator proposed by Raj (1965) is

$$\bar{y}_{\text{m}} = \bar{y} + \sum_{i=1}^{p} \beta_i (\overline{X}_i - \bar{x}_i). \tag{3.8.2.16}$$

Proceeding as in the above theorem the minimum variance of the multivariate regression type estimator $\bar{y}_m$ is given by

$$\text{Min}.V(\bar{y}_m) = \left(\frac{1-f}{n}\right)\overline{Y}^2 C_y^2 \left(1 - R^2_{y \bullet x_1 x_2 .... x_p}\right) \tag{3.8.2.17}$$

where $R^2_{y \bullet x_1 x_2 ... x_p}$ denotes the multiple correlation coefficient between Y and X_1, X_2, ..., X_p.

3.8.3 GENERAL CLASS OF ESTIMATORS

Srivastava (1971) proposed a general class of ratio type estimators for estimating the population mean $\overline{Y}$ as

$$\bar{y}_s = \bar{y}H(u_1, u_2, ..., u_p) = \bar{y}H(\underline{u}) \tag{3.8.3.1}$$

where

$u_j = \bar{x}_j / \overline{X}_j$, $j = 1, 2, ..., p$, and $H(\underline{u})$ is a parametric function such that $H(\underline{\in}) = 1$ for $\underline{\in} = (1, 1, ..., 1)_{1\times p}$, satisfying certain regularity condition such as the first and second order partial derivatives of H with respect to $\underline{u}$ exist and are known. Expanding $H(\underline{u})$ around the point $\underline{\in}$ by using second order Taylor's series we have

$$H(\underline{u}) = H[\underline{\in} + (\underline{u} - \underline{\in})] = H(\underline{\in}) + (\underline{u} - \underline{\in})\frac{\partial H}{\partial \underline{u}}\Big|_{\underline{u}=\underline{\in}} + \frac{1}{2}(\underline{u} - \underline{\in})' \frac{\partial^2 H}{\partial \underline{u}' \underline{u}}\Big|_{\underline{u}=\underline{\in}} (\underline{u} - \underline{\in}) + ... \tag{3.8.3.2}$$

Thus the general class of estimators can be written as

$$\bar{y}_{\text{s}} = \bar{y}\left[1 + (\underline{u} - \underline{\in})\frac{\partial H}{\partial \underline{u}}\Big|_{\underline{u}=\underline{\in}} + \frac{1}{2}(\underline{u} - \underline{\in})' \frac{\partial^2 H}{\partial \underline{u}' \underline{u}}\Big|_{\underline{u}=\underline{\in}} (\underline{u} - \underline{\in}) + ...\right]$$

$$= \bar{Y}\left[1+\varepsilon_0+(\underline{u}-\underline{\in})\frac{\partial H}{\partial \underline{u}}|_{\underline{u}=\underline{\in}}+\frac{1}{2}(\underline{u}-\underline{\in})'\frac{\partial^2 H}{\partial \underline{u}'\underline{u}}|_{\underline{u}=\underline{\in}}(\underline{u}-\underline{\in})+\varepsilon_0(\underline{u}-\underline{\in})\frac{\partial H}{\partial \underline{u}}|_{\underline{u}=\underline{\in}}+....\right]$$

$$= \bar{Y}\left[1+\varepsilon_0+(\underline{u}-\underline{\in})\underline{H_1}+\frac{1}{2}(\underline{u}-\underline{\in})'\underline{H_2}(\underline{u}-\underline{\in})+\varepsilon_0(\underline{u}-\underline{\in})\underline{H_1}+....\right] \tag{3.8.3.3}$$

where $\underline{H_1}=\frac{\partial H}{\partial \underline{u}}|_{\underline{u}=\underline{\in}}$ and $\underline{H_2}==\frac{1}{2}\frac{\partial^2 H}{\partial \underline{u}'\underline{u}}|_{\underline{u}=\underline{\in}}$ are the matrices consisting of first and second order partial derivatives of the function H with respect to $\underline{u}$ and evaluated at $\underline{u}=\underline{\in}$. Thus we have the following theorems:

Theorem 3.8.3.1. The bias, to the first order of approximation, in the general class of estimators of population mean is

$$B(\bar{y}_s)\approx\left(\frac{1-f}{n}\right)\bar{Y}\left[\sum_{j=1}^{p}C_{x_j}^2 H_{2j}+\sum_{j=1}^{p}\rho_{yx_j}C_yC_{x_j}H_{1j}\right] \tag{3.8.3.4}$$

where H_{2j} and H_{1j} denote the jth diagonal element of H_2 and j^{th} component of H, respectively. Also $C_{x_j}^2=S_{x_j}^2/\bar{X}_j^2$ and $\rho_{yx_j}=(S_{yx_j})/(S_yS_{x_j})$ for $j=1,2,...,p$.

Proof. Follows by taking expected values on both sides of (3.8.3.3).

Theorem 3.8.3.2. The minimum mean squared error, up to the terms of $O(n^{-1})$, of the general class of estimators of population mean is given by

$$\text{Min.MSE}(\bar{y}_s)=\left(\frac{1-f}{n}\right)\bar{Y}^2C_y^2\left[1-R^2_{y\bullet x_1x_2....x_p}\right]. \tag{3.8.3.5}$$

Proof. By the definition of mean squared error, we have

$$\text{MSE}(\bar{y}_s)=E\left[\bar{y}_s-\bar{Y}\right]^2$$

$$\approx E\left[\bar{Y}\left(1+\varepsilon_0+(\underline{u}-\underline{\in})\underline{H_1}+(\underline{u}-\underline{\in})'\underline{H_2}(\underline{u}-\underline{\in})+\varepsilon_0(\underline{u}-\underline{\in})\underline{H_1}+....\right)-\bar{Y}\right]^2$$

$$=\bar{Y}^2E\left[\varepsilon_0+(\underline{u}-\underline{\in})\underline{H_1}\right]^2=\bar{Y}^2E\left[\left\{\varepsilon_0+(\underline{u}-\underline{\in})\underline{H_1}\right\}'\left\{\varepsilon_0+(\underline{u}-\underline{\in})\underline{H_1}\right\}\right]$$

$$=\bar{Y}^2E\left[\varepsilon_0^2+\underline{H_1}'(\underline{u}-\underline{\in})'(\underline{u}-\underline{\in})\underline{H_1}+2\varepsilon_0(\underline{u}-\underline{\in})\underline{H_1}\right]$$

$$=\left(\frac{1-f}{n}\right)\bar{Y}^2\left[C_y^2+\sum_{t=1}^{p}\sum_{j=1}^{p}H_{1t}H_{1j}\rho_{x_tx_j}C_{x_t}C_{x_j}+2\sum_{j=1}^{p}\rho_{yx_j}C_yC_{x_j}H_{1j}\right]. \tag{3.8.3.6}$$

On differentiating (3.8.3.6) with respect to $\underline{H_1}=(H_{11},...,H_{1p})'$ and equating to zero, we will obtain a set of p equations, as

$$\begin{bmatrix} C_{x_1}^2, & \rho_{x_1x_2}C_{x_1}C_{x_2},..., \rho_{x_1x_p}C_{x_1}C_{x_p} \\ \rho_{x_2x_1}C_{x_2}C_{x_2}, & C_{x_2}^2, \quad ...,\rho_{x_2x_p}C_{x_2}C_{x_p} \\ \\ \rho_{x_px_1}C_{x_p}C_{x_2}, & \rho_{x_px_2}C_{x_p}C_{x_2},..., \quad C_{x_p}^2 \end{bmatrix} \begin{bmatrix} H_{11} \\ H_{12} \\ \\ H_{1p} \end{bmatrix} = \begin{bmatrix} -\bar{Y}\rho_{yx_1}C_yC_{x_1} \\ -\bar{Y}\rho_{yx_2}C_yC_{x_2} \\ \\ -\bar{Y}\rho_{yx_p}C_yC_{x_p} \end{bmatrix} \quad (3.8.3.7)$$

or $\underline{A}\,\underline{H_1} = C$. (3.8.3.8)

The set of equations given by (3.8.3.8) can easily be solved for unknown parameters as $\underline{H_1} = \underline{A^{-1}}\ C$. As we saw in the Theorem 3.8.2.2, one can easily see that by substituting the optimum values so obtained in (3.8.3.6) we obtain (3.8.3.5). Hence the theorem.

Corollary 3.8.3.1. A wider class of estimators for estimating population mean, $\bar{Y}$, using p auxiliary variables $X_1, X_2, ..., X_p$ can easily be defined as

$$\bar{y}_W = H(\bar{y}, \underline{u}) \quad (3.8.3.9)$$

where $H[.,.]$ is a parametric function such that $H(\bar{Y}, \underline{\in}) = 1$, satisfying certain regularity conditions. It is easy to show that the wider class of estimators $\bar{y}_W$ has the same asymptotic mean squared error as that of the general class of ratio type estimators of population mean defined at (3.8.3.1).

Corollary 3.8.3.2. Let $X = (X_{ij})_{n\times p}$ be the matrix of p variates and an equation of the plane of regression of X_1 on $X_2, X_3, ..., X_p$ be defined as

$$X_1 = b_{12.34...p}X_2 + b_{13.24...p}X_3 + b_{14.235...p}X_4 + .. + b_{1p.23...(p-1)}X_p. \quad (3.8.3.10)$$

Let s_i^2 denote the sample variance of the ith variable and r_{ij} be the simple correlation between the ith and j^{th} variables. Define the correlation matrix as

$$\Delta = \begin{bmatrix} 1, & r_{12}, & r_{13}, & ..., & r_{1p} \\ r_{21}, & 1, & r_{23}, & ..., & r_{2p} \\ \dots & & & & \\ \dots & & & & \\ r_{p1}, & r_{p2}, & r_{p3}, & ... & 1 \end{bmatrix}_{p\times p} . \quad (3.8.3.11)$$

Let Δ_{ij} be the cofactor of the element in the ith row and j^{th} column of the correlation matrix Δ, then the partial regression coefficients b_{1j}, $j = 2,3,..., p$ are given by

$$b_{1j} = -\frac{s_1}{s_j}\frac{\det(\Delta_{1j})}{\det(\Delta_{11})}, \quad j = 2,3,..., p \quad (3.8.3.12)$$

and the estimator of multiple correlation coefficient $R^2_{1\bullet 2,3,4,...,p}$ is given by

$$\hat{R}^2_{1\bullet 2,3,\ldots,p} = 1 - \frac{\det(\Delta)}{\det(\Delta_{11})}. \tag{3.8.3.13}$$

Example 3.8.3.1. An estimate of total number of fish at the Atlantic and Gulf Coats helps in making decision to recruit labour by fishermen contractors. The average number of fish during 1994, 1993, and 1992 are known to be 4954.435, 4591.072, and 4230.174 respectively. Apply the following estimator to estimate the total numbers of fish in all 69 types of groups

$$\bar{y}_{\text{lr}} = \bar{y} + \hat{H}_{11}(u_1 - 1) + \hat{H}_{12}(u_2 - 1) + \hat{H}_{13}(u_3 - 1)$$

where $u_j = \bar{x}_j / \bar{X}_j$ and the estimates $\hat{H}_{1j}$, $j = 1,2,3$ are obtained by solving the of normal equations given by

$$\begin{bmatrix} \hat{C}^2_{x_1}, & r_{x_1x_2}\hat{C}_{x_1}\hat{C}_{x_2}, & r_{x_1x_3}\hat{C}_{x_1}\hat{C}_{x_3} \\ r_{x_2x_1}\hat{C}_{x_1}\hat{C}_{x_2}, & \hat{C}^2_{x_2}, & r_{x_2x_3}\hat{C}_{x_2}\hat{C}_{x_3} \\ r_{x_3x_1}\hat{C}_{x_1}\hat{C}_{x_3}, & r_{x_2x_3}\hat{C}_{x_2}\hat{C}_{x_3}, & \hat{C}^2_{x_3} \end{bmatrix} \begin{bmatrix} \hat{H}_{11} \\ \hat{H}_{12} \\ \hat{H}_{13} \end{bmatrix} = \begin{bmatrix} -\bar{y} r_{yx_1}\hat{C}_y\hat{C}_{x_1} \\ -\bar{y} r_{yx_2}\hat{C}_y\hat{C}_{x_2} \\ -\bar{y} r_{yx_3}\hat{C}_y\hat{C}_{x_3} \end{bmatrix}.$$

To estimate the number of fish during 1995, a consultant takes an SRSWOR sample of 16 types of fish as given in the following table:

Sr. No.	Species group	1992 x_3	1993 x_2	1994 x_1	1995 y
3	Skates/rays	2152	1981	2939	2353
5	Saltwater catfishes	13466	12690	14441	13859
4	Eels	138	222	186	152
16	Striped bass	3840	4799	8521	10758
21	Bluefish	11990	10301	12405	10940
43	Weakfish	1668	2219	4929	5739
55	Cunner	1931	1876	1255	1375
58	Atlantic mackerel	1045	2307	4860	4008
33	Snappers, other	746	861	462	492
69	Other fishes	12249	14953	20488	14426
39	Sheepshead	5933	5593	4383	5118
10	Pollock	168	397	862	832
59	King mackerel	1289	1023	1148	1252
45	Silver perch	1198	1034	1729	2146
62	Summer flounder	11918	22919	17741	16238
66	Flounders, other	1103	999	918	897

Solution. From the sample information we have

Sample	x_3	x_2	x_1	y
Means	4427.1250	5260.8750	6079.1875	5661.5625
Variances	24711755.32	43283631.72	44114004.70	31798379.33
Coefficient of variations	1.1228	1.2505	1.0925	0.9960

and sample correlation matrix is given by

Correlation matrix

Variables	x_3	x_2	x_1	y
x_3	1.0000			
x_2	0.9133	1.0000		
x_1	0.9304	0.9347	1.0000	
y	0.9176	0.9246	0.9723	1.0000

Here

$n = 16$, $p = 3$, $\bar{y} = 5661.56$, $\bar{x}_1 = 6079.18$, $\bar{x}_2 = 5260.87$, $\bar{x}_3 = 4427.12$,

$\hat{C}_y = 0.9960$, $\hat{C}_{x1} = 1.0925$, $\hat{C}_{x2} = 1.2505$, $\hat{C}_{x3} - 1.1229$, $r_{yx_1} = 0.9723$,

$r_{yx2} = 0.9246$, $r_{yx3} = 0.9176$, $r_{x_1x_2} = 0.9347$, $r_{x_1x_3} = 0.9305$ and $r_{x_2x_3} = 0.9133$.

The set of normal equations becomes

$$\begin{bmatrix} 1.1935, 1.2769, 1.1415 \\ 1.2769, 1.5637, 1.2824 \\ 1.1415, 1.2824, 1.2609 \end{bmatrix} \begin{bmatrix} \hat{H}_{11} \\ \hat{H}_{12} \\ \hat{H}_{13} \end{bmatrix} = \begin{bmatrix} -5989.87 \\ -6519.78 \\ -5810.18 \end{bmatrix}.$$

A solution of these normal equations is given by

$$\begin{bmatrix} \hat{H}_{11} \\ \hat{H}_{12} \\ \hat{H}_{13} \end{bmatrix} = \begin{bmatrix} 1.1935, 1.2769, 1.1415 \\ 1.2769, 1.5637, 1.2824 \\ 1.1415, 1.2824, 1.2609 \end{bmatrix}^{-1} \begin{bmatrix} -5989.87 \\ -6519.78 \\ -5810.18 \end{bmatrix}$$

$$= \begin{bmatrix} 9.234, -4.127, -4.163 \\ -4.127, 5.702, -2.059 \\ -4.163, -2.059, 6.657 \end{bmatrix} \begin{bmatrix} -5989.87 \\ -6519.78 \\ -5810.18 \end{bmatrix} = \begin{bmatrix} -4215.54 \\ -492.43 \\ -318.31 \end{bmatrix}.$$

It is given that $\bar{X}_1 = 4954.435$, $\bar{X}_2 = 4591.072$,and $\bar{X}_3 = 4230.174$, therefore, $u_1 = \bar{x}_1 / \bar{X}_1 = 1.2270$, $u_2 = \bar{x}_2 / \bar{X}_2 = 1.1458$ and $u_3 = \bar{x}_3 / \bar{X}_3 = 1.0465$.

Hence a point estimate of the average number of fish during 1995 is given by

$$\bar{y}_{lr} = \bar{y} + \hat{H}_{11}(u_1 - 1) + \hat{H}_{12}(u_2 - 1) + \hat{H}_{13}(u_3 - 1)$$

$$= 5661.56 - 4215.54(1.2270 - 1) - 492.43(1.1458 - 1) - 318.31(1.0465 - 1) = 4618.03.$$

In this case we have

$$\Delta = \det \begin{bmatrix} 1.0000, & 0.9723, & 0.9246, & 0.9176 \\ 0.9723, & 1.0000, & 0.9347, & 0.9305 \\ 0.9246, & 0.9347, & 1.0000, & 0.9133 \\ 0.9176, & 0.9305, & 0.9133, & 1.0000 \end{bmatrix} = 0.000794,$$

and

$$\Delta_{11} = \det \begin{bmatrix} 1.0000, & 0.9347, & 0.9305 \\ 0.9347, & 1.0000, & 0.9133 \\ 0.9305, & 0.9133, & 1.0000 \end{bmatrix} = 0.0150528,$$

therefore the value of estimate of multiple correlation coefficient is given by

$$\hat{R}^2_{y \bullet x_1, x_2, x_3} = 1 - \frac{\Delta}{\Delta_{11}} = 1 - \frac{0.000794}{0.0150528} = 0.9473.$$

An estimate of the variance of the estimator $\bar{y}_{\text{lr}}$ is given by

$$\hat{v}(\bar{y}_{\text{lr}}) = \frac{(1-f)}{n} s_y^2 \left(1 - \hat{R}^2_{y \bullet x_1, x_2, x_3}\right)$$

$$= \frac{1 - 0.32}{16} \times 31798379.33\left(1 - 0.9473^2\right) = 138687.52.$$

A $(1-\alpha)100\%$ confidence interval of the population mean $\bar{Y}$ is given by

$$\bar{y}_{\text{lr}} \pm t_{\alpha/2}(\text{df} = n - p - 1)\sqrt{\hat{v}(\bar{y}_{\text{lr}})}.$$

Using Table 2 of the Appendix the 99% confidence interval of the average number of fish during 1995 is given by

$$\bar{y}_{\text{lr}} \pm t_{0.01/2}(\text{df} = 16 - 3 - 1)\sqrt{\hat{v}(\bar{y}_{\text{lr}})}, \quad \text{or} \quad \bar{y}_{\text{lr}} \pm t_{0.01/2}(\text{df} = 16 - 3 - 1)\sqrt{\hat{v}(\bar{y}_{\text{lr}})}$$

or

$4618.03 \pm t_{0.005}(\text{df}=12)\sqrt{138687.52}$, or $4618.03 \pm 3.055 \times 372.407$, or $[3480.32, 5755.73]$.

A point estimate of the total number of fish during 1995 is given by

$$\hat{y}_{\text{lr}} = N\bar{y}_{\text{lr}} = 69 \times 4618.03 = 318644.07$$

and a 99% interval estimate for the same is given by

$$N\left[\bar{y}_{\text{lr}} \pm t_{0.01/2}(\text{df} = 16-3-1)\sqrt{\hat{v}(\bar{y}_{\text{lr}})}\right]$$

or

$$69[3480.32, 5755.73], \text{ or } [240142.08, 397145.37].$$

The next section has been devoted to study the general class of estimators to estimate any population parameter (*e.g.*, population mean, population variance, population correlation coefficient, population coefficient of variation, population regression coefficient, etc.) by making use of p auxiliary variables at the estimation stage.

3.9 GENERAL CLASS OF ESTIMATORS TO ESTIMATE ANY POPULATION PARAMETER

Singh, Mangat, and Mahajan (1995) considered the problem of estimation of any population parameter F_0 of the study variable Y by using known supplementary information on p auxiliary variables $F_1, F_2, ..., F_p$. Let $f_0, f_1, ..., f_p$ be the unbiased or consistent estimators of $F_0, F_1, ..., F_p$, respectively, each based on a sample of size $n > p$. Let $\underline{u} = (u_1, u_2, ..., u_p)$, where $u_i = f_i / F_i$, $i = 1,2,..., p$, assume values in a bounded closed, convex subset R_p of p-dimensional space containing the point $\underline{\in} = (1, 1, ..., 1)$. Let $\underline{b}^{/} = (b_1, b_2, ..., b_p)$, where $b_i = \{F_0 \text{Cov}(f_0, f_i)\}/\{F_i V(f_0)\}$, $i = 1,2,..., p$, and $\underline{A} = [a_{ij}]_{p\times p}$, where $a_{ij} = \{F_0^2 \text{Cov}(f_i, f_j)\}/\{F_i F_j V(f_0)\}$, $i, j = 1, 2, ..., p$ and the matrix $\underline{A}$ is assumed to be positive definite. Define

$$e_0 = (f_0 / F_0) - 1, \quad \varepsilon^{/} = (e_1, e_2, ..., e_p)$$

where

$$e_i = u_i - 1, \quad C_0^2 = V(f_0)/F_0^2, \quad C_i^2 = V(f_i)/F_i^2, \text{ and } C_{0i} = \text{Cov}(f_0, f_i)/(F_0 F_i).$$

Then we have

$$E(e_i) = B(f_i)/F_i, \quad i = 1,2,\ldots, p,$$

where $B(f_i)$ denotes the bias in the estimator f_i of F_i and

$$E(e_0^2) = C_0^2, \quad E(e_i^2) = C_i^2, \quad E(e_0 e_i) = C_{0i}, \quad E(e_0\varepsilon) = C_0^2 b, \text{ and } E(\varepsilon\varepsilon^{/}) = C_0^2 \underline{A}$$

for $i = 1,2,..., p$.

A wider class of estimators to estimate any population parameter F_0 of the study variable is then defined as

$$t_h = h(f_0, \underline{u}) \tag{3.9.1}$$

where $h(f_0, \underline{u})$ is a function of f_0 and u_i, $i = 1,2,..., p$, such that

$$h(F_0,\ \underline{e}) = F_0 \tag{3.9.2}$$

and h is bounded and continuous with bounded and continuous first and second order partial derivatives in R^{p+1}. Expanding $h(F_0,\ \underline{u})$ about the point $(F_0,\ \underline{e})$ in a second order Taylor's series we obtain

$$t_h = h(F_0,\ \underline{e}) + (f_0 - F_0)h_0^{/}(F_0,\ \underline{e}) + \sum_{i=1}^{p}(u_i - 1)h_i^{/}(F_0,\ \underline{e})$$
$$+ \sum_{i<j}^{p}(u_i - 1)(u_j - 1)h_{ij}^{/}(F_0,\ \underline{e}) + \sum_{i=1}^{p}(u_i - 1)^2 h_{ii}^{//}(F_0,\ \underline{e})$$
$$+ \sum_{i=1}^{p}(u_i - 1)(f_0 - F_0)h_{0i}^{//}(F_0,\ \underline{e}) + (f_0 - F_0)^2 h_{00}^{//}(F_0,\ \underline{e}) + \ldots\ . \tag{3.9.3}$$

Using (3.9.2) and further assuming that $h_0^{/}(F_0,\ \underline{e}) = 1$ we obtain

$$t_h = f_0 + \sum_{i=1}^{p}(u_i - 1)h_i^{/}(F_0, \underline{e}) + \sum_{i<j}^{p}(u_i - 1)(u_j - 1)h_{ij}^{/}(F_0,\ \underline{e}) + \frac{1}{2}\sum_{i=1}^{p}(u_i - 1)^2 h_{ii}^{//}(F_0, \underline{e})$$
$$+ \frac{1}{2}\sum_{i=1}^{p}(u_i - 1)(f_0 - F_0)h_{0i}^{//}(F_0, \underline{e}) + \frac{1}{2}(f_0 - F_0)^2 h_{00}^{//}(F_0, \underline{e}) + .. \tag{3.9.4}$$

where $h_i^{/}(F_0,\ \underline{e})$ and $h_{ij}^{//}(F_0,\ \underline{e})$ denote the first and second order partial derivatives of $h(f_0,\ \underline{u})$ with respect to f_0 and u_i, respectively. Thus we have the following theorem.

Theorem 3.9.1. The bias in the wider class of estimators of any population parameter is of the order $O(n^{-1})$, *i.e.*, prove that

$$E(t_h) = F_0 + O(n^{-1}). \tag{3.9.5}$$

Proof. Taking expected values on both sides of (3.9.4) we obtain

$$E(t_h) = F_0 + \sum_{i<j}^{p} C_{ij} h_{ij}^{/}(F_0, \underline{e}) + \frac{1}{2}\sum_{i=1}^{p} C_{ii} h_{ii}^{//}(F_0,\ \underline{e}) + \frac{1}{2}F_0 C_{00} h_{00}^{//}(F_0,\ \underline{e})$$
$$+ \frac{1}{2}F_0 \sum_{i=1}^{p} C_{0i} h_{0j}^{//}(F_0,\ \underline{e}) + B(f_0) + \sum_{i=1}^{p} B(f_i)h_i^{/}(F_0,\ \underline{e})/F_i. \tag{3.9.6}$$

Note that $f_0,\ f_1,\ f_2, \ldots,\ f_p$ are either unbiased or consistent estimators of $F_0,\ F_1,\ F_2, \ldots,\ F_p$, respectively, by the definition of consistency from Gujarati (1978), C_{ij}, $(i, j = 0, 1, 2, \ldots, p)$ representing the variance--covariance terms and $B(f_i)$ will tend to zero as sample size $n \to \infty$. Thus (3.9.6) can be expressed as (3.9.5).

Hence the theorem.

Theorem 3.9.2. The minimum mean squared error, up to terms of order $O(n^{-1})$, of the class of estimators t_h is given by

$$\text{Min.MSE}(t_h) = V(f_0)\left(1 - R^2_{f_0 \bullet f_1, f_2, \ldots, f_p}\right) \tag{3.9.7}$$

where $R^2_{f_0 \bullet f_1, f_2, ..., f_p}$ denotes the multiple correlation coefficient between f_0 and $f_1, f_2, ..., f_p$.

Proof. By the definition of mean squared error we have

$$\text{MSE}(t_h) = E[t_h - F_0]^2 = E\left[f_0 + \sum_{i=1}^{p}(u_i - 1)h_i(F_0, \underline{\in}) - F_0\right]^2$$

$$= F_0^2 C_0^2 + \{h_m^{/}(F_0, \underline{\in})\}^{/} C_0^2 \underline{A} h_m^{/}(F_0, \underline{\in}) + 2C_0^2 b^{/} h_m^{/}(F_0, \underline{\in}) h_0(F_0, \underline{\in}) F_0 . \quad (3.9.8)$$

On differentiating (3.9.8) and equating to zero we obtain the optimum values of the parameters as

$$h_m(F_0, \underline{\in}) = F_0 \underline{A^{-1}}\ \underline{b} . \quad (3.9.9)$$

On substituting the optimum values from (3.9.9) in (3.9.8), we obtain

$$\text{Min.MSE}(t_h) = V(f_0)\left[1 - \underline{b}^{/}\ \underline{A^{-1}}\ \underline{b}\right] . \quad (3.9.10)$$

where $\underline{b}^{/}\ \underline{A^{-1}}\ \underline{b} = R^2_{f_0 \bullet f_1, f_2, ..., f_p}$. Hence the theorem.

Remark 3.9.1. (a) Note that the multiple correlation coefficient increases with the number of secondary variables, it follows from (3.9.10) that the minimum mean square error of t_h is a monotone decreasing function of the number of secondary variables.

(b) The value of $R^2_{f_0 \bullet f_1, f_2, ..., f_p}$ also increases if there is high correlation between two auxiliary variables and such high correlation between the auxiliary variables may bring artificial reduction in the variance of the estimator of population mean or total. Such a problem can be addressed as a problem of multicollinearity in survey sampling. For example suppose there are three variables Y, X_1, and X_2. Then the minimum variance of the linear regression estimator

$$\bar{y}_{\text{lr}} = \bar{y} + \beta_1\left(\bar{X}_1 - \bar{x}_1\right) + \beta_2\left(\bar{X}_2 - \bar{x}_2\right)$$

is given by

$$V\left(\bar{y}_{\text{lr}}\right) = \frac{1-f}{n} S_y^2 \left(1 - \frac{\rho_{yx_1}^2 + \rho_{yx_2}^2 - 2\rho_{yx_1}\rho_{yx_2}\rho_{x_1x_2}}{1 - \rho_{x_1x_2}^2}\right).$$

Consider the following two situations:

Situation I. Let $\rho_{yx_1} = 0.6$, $\rho_{yx_2} = 0.8$, and $\rho_{x_1x_2} = 0.3$;

Situation II. Let $\rho_{yx_1} = 0.6$, $\rho_{yx_2} = 0.8$, and $\rho_{x_1x_2} = 0.95$.

Then the ratio of $V\left(\bar{y}_{\text{lr}}\right)$ under case 1 to case 2 is given by

$$\text{Ratio} = \frac{V(\bar{y}_{lr})_{\text{situation I}}}{V(\bar{y}_{lr})_{\text{situation II}}} = \frac{\left(1 - \dfrac{\rho_{yx_1}^2 + \rho_{yx_2}^2 - 2\rho_{yx_1}\rho_{yx_2}\rho_{x_1x_2}}{1-\rho_{x_1x_2}^2}\right)_{\text{situation I}}}{\left(1 - \dfrac{\rho_{yx_1}^2 + \rho_{yx_2}^2 - 2\rho_{yx_1}\rho_{yx_2}\rho_{x_1x_2}}{1-\rho_{x_1x_2}^2}\right)_{\text{situation II}}} = 2.233\,.$$

Clearly the reduction in variance $V(\bar{y}_{lr})$ in situation II is due to high correlation between x_1 and x_2.

Biradar and Singh (1992a), Singh and Kataria (1990), Singh (1988), and Singh and Upadhyaya (1986) have suggested some methods to improve the general class of estimators. Srivastava (1992) has shown that those methods are not valid for improving the general class of estimators. Biradar and Singh (1997, 1998) considered a class of estimators based on a general sampling design for a population parameter ϕ_0 utilizing the information on two parameters ϕ_1 and ϕ_2 of an auxiliary variable.

Remark 3.9.2. It is to be noted that although Srivastava's (1971) general class of estimators and its ramifications fetched man followers of course leading to numerous published papers in world wide journals but they are of dubious theoretical merit and practical use according to the views of a few well known scientists.

3.10 ESTIMATION OF RATIO OR PRODUCT OF TWO POPULATION MEANS

Suppose there are two variables Y_1 and Y_2 under study in a finite population Ω of size N. Let Y_{1i} and Y_{2i} denote the values of the ith unit in the population. Suppose we want to estimate the ratio of two population means defined as

$$R_{y_1y_2} = \frac{\bar{Y}_1}{\bar{Y}_2} \tag{3.10.1}$$

where $\bar{Y}_1 = N^{-1}\sum_{i=1}^{N} Y_{1i}$ and $\bar{Y}_2 = N^{-1}\sum_{i=1}^{N} Y_{2i}$ denote, respectively, the population means of the two variables. Suppose a sample of n units is drawn by using SRSWOR and both (y_{1i}, y_{2i}) for $i = 1, 2, ..., n$ paired observations are observed from the sample. Let $\bar{y}_1 = n^{-1}\sum_{i=1}^{n} y_{1i}$ and $\bar{y}_2 = n^{-1}\sum_{i=1}^{n} y_{2i}$ denote the sample means. Obviously an estimator of the population ratio, $R_{y_1y_2}$, is given by

$$\hat{R}_{y_1y_2} = \frac{\bar{y}_1}{\bar{y}_2}\,. \tag{3.10.2}$$

To find the bias and mean squared error of the estimator $\hat{R}_{y_1y_2}$ let us define

$$\varepsilon_0 = \frac{\bar{y}_1}{\bar{Y}_1} - 1, \text{ and } \delta_0 = \frac{\bar{y}_2}{\bar{Y}_2} - 1,$$

such that

$$E(\varepsilon_0) = E(\delta_0) = 0$$

and

$$E(\varepsilon_0^2) = \left(\frac{1-f}{n}\right)C_{y1}^2, \quad E(\delta_0^2) = \left(\frac{1-f}{n}\right)C_{y2}^2, \text{ and } E(\varepsilon_0\delta_0) = \left(\frac{1-f}{n}\right)\rho_{y1y2}C_{y1}C_{y2}.$$

Now the estimator $\hat{R}_{y1y2}$ in terms of ε_0 and δ_0 can easily be written as

$$\hat{R}_{y1y2} = R_{y1y2}\left[\,1 + \varepsilon_0 - \delta_0 + \delta_0^2 - \varepsilon_0\delta_0 + ...\,\right]. \tag{3.10.3}$$

Then we have the following theorems.

Theorem 3.10.1. The bias, to the first order of approximation, in the estimator $\hat{R}_{y1y2}$ is given by

$$B(\hat{R}_{y1y2}) = \left(\frac{1-f}{n}\right)R_{y1y2}\left[C_{y2}^2 - \rho_{y1y2}C_{y1}C_{y2}\right]. \tag{3.10.4}$$

Proof. Taking expected values of (3.10.3) we obtain

$$E(\hat{R}_{y1y2}) = R_{y1y2}E\left[\,1 + \varepsilon_0 - \delta_0 + \delta_0^2 - \varepsilon_0\delta_0 + ...\,\right]$$

$$= R_{y1y2}\left[\,1 + E(\varepsilon_0) - E(\delta_0) + E(\delta_0^2) - E(\varepsilon_0\delta_0)\,\right]$$

$$= R_{y1y2}\left[1 + 0 - 0 + \left(\frac{1-f}{n}\right)\left\{C_{y2}^2 - \rho_{y1y2}C_{y1}C_{y2}\right\}\right].$$

Thus the bias, to the first order of approximation, is given by

$$B(\hat{R}_{y1y2}) = E(\hat{R}_{y1y2}) - R_{y1y2} = \left(\frac{1-f}{n}\right)R_{y1y2}\left\{C_{y2}^2 - \rho_{y1y2}C_{y1}C_{y2}\right\}.$$

Hence the theorem.

Theorem 3.10.2. The mean square error, to the first order of approximation, of the estimator $\hat{R}_{y1y2}$, is given by

$$\text{MSE}(\hat{R}_{y1y2}) = \left(\frac{1-f}{n}\right)R_{y1y2}^2\left[C_{y1}^2 + C_{y2}^2 - 2\rho_{y1y2}C_{y1}C_{y2}\right]. \tag{3.10.5}$$

Proof. By the definition of mean squared error, we have

$$\text{MSE}(\hat{R}_{y1y2}) = E\left[\hat{R}_{y1y2} - R_{y1y2}\right]^2 \approx R_{y1y2}^2E\left[\varepsilon_0 - \delta_0 + \delta_0^2 - \varepsilon_0\delta_0\right]^2$$

$$\approx R_{y1y2}^2E\left[\varepsilon_0^2 + \delta_0^2 - 2\varepsilon_0\delta_0\right] = \left(\frac{1-f}{n}\right)R_{y1y2}^2\left[C_{y1}^2 + C_{y2}^2 - 2\rho_{y1y2}C_{y1}C_{y2}\right].$$

Hence the theorem.

Remark 3.10.1. We can also estimate product of two population means, $P_{y1y2} = \bar{Y}_1\bar{Y}_2$, with the help of estimator $\hat{P}_{y1y2} = \bar{y}_1\bar{y}_2$.

Remark 3.10.2. In the case of the presence of known auxiliary information, general class of estimators of the form, $\hat{R}_h = \left\{\dfrac{\bar{y}_1}{\bar{y}_2}\right\}H(u)$, $\hat{P}_h = \{\bar{y}_1\bar{y}_2\}H(u)$ where $u = \bar{x}/\bar{X}$ and $H(\bullet)$ is a parametric function can also be constructed to estimate the ratio $R_{y_1y_2}$ and product $P_{y_1y_2}$ of two population means. The interested reader may refer to Singh (1982a).

3.11 MEDIAN ESTIMATION IN SURVEY SAMPLING

The median is often regarded as a more appropriate measure of location than the mean when variables with a highly skewed distribution, such as income, are studied. As we have seen in the previous sections of this chapter, there is extensive literature available on the estimation of mean and totals in sample surveys. Relatively few efforts have been made to develop an efficient estimator of the median. Gross (1980), Sedransk and Meyer (1978), and Smith and Sedransk (1983) have considered the problem of estimation of the median using simple random sampling. Kuk and Mak (1989) are the first researchers to attempt the estimation of the median using auxiliary information. Francisco and Fuller (1991) have also considered the problem of estimation of the median as a part of estimation of finite population distribution function. In this chapter we shall restrict ourself to the discussion of the ratio type estimator developed by Kuk and Mak (1989). Let Y_i and X_i, $i = 1, 2, ..., N$, be the values of the population units for the study variable Y and auxiliary variable X respectively. Furthermore, let y_i and x_i, $i = 1,2,...,n$, be the values of the units included in an SRSWOR sample of size n. Assuming the median M_x of the variable X is known we have the following theorem:

Theorem 3.11.1. The ratio type estimator to estimate the median M_y of the study variable is given by

$$\hat{M}_{\mathrm{R}} = \hat{M}_y\left(\frac{M_x}{\hat{M}_x}\right) \tag{3.11.1}$$

where $\hat{M}_y$ and $\hat{M}_x$ are the sample median estimators.

Proof. Follows from the usual ratio method of estimation.

Suppose $y_{(1)}, y_{(2)}, ..., y_{(n)}$ are the Y values of sample units in the ascending order. Furthermore let $p = \dfrac{t}{n}$ be the proportion of Y values in the sample which are less than or equal to the median value M_y which is an unknown parameter and is to be estimated from sample observations and so is the case of p. If $\hat{p}$ is an estimator of p, the sample median $\hat{M}_y$ in terms of quantiles can be written as $\hat{Q}_y(\hat{p})$, where $\hat{p} = 0.5$. Kuk and Mak (1989) defined a matrix of proportions $[\ P_{ij}\ ]$ as:

	$Y \le M_y$	$Y > M_y$	Total
$X \le M_x$	P_{11}	P_{21}	$P_{.1}$
$X > M_x$	P_{12}	P_{22}	$P_{.2}$
Total	$P_{1.}$	$P_{2.}$	1

Thus we have the following theorems.

Theorem 3.11.2. The variances of the estimators $\hat{M}_y$ and $\hat{M}_x$ are, respectively, given by

$$V\left(\hat{M}_y\right) = \left(\frac{1-f}{n}\right)\frac{\{f_y(M_y)\}^{-2}}{4}, \tag{3.11.2}$$

and

$$V\left(\hat{M}_x\right) = \left(\frac{1-f}{n}\right)\frac{\{f_x(M_x)\}^{-2}}{4}, \tag{3.11.3}$$

where f_y and f_x are the density functions of y and x respectively. The covariance between $\hat{M}_y$ and $\hat{M}_x$ is given by

$$\text{Cov}\left(\hat{M}_y,\ \hat{M}_x\right) = \left(\frac{1-f}{n}\right)(P_{11} - 0.25)\{f_x(M_x)f_y(M_y)\}^{-1}. \tag{3.11.4}$$

Proof. Let F_y be the cumulative distribution function of y. Then under some assumptions given by Kuk and Mak (1989) and Francisco and Fuller (1991) the Taylor's series expansion gives

$$F_y\left(\hat{M}_y\right) = F_y\left[M_y + \left(\hat{M}_y - M_y\right)\right] = F_y\left(M_y\right) + \left(\hat{M}_y - M_y\right)f_y\left(M_y\right) + o_p\left(n^{-1/2}\right).$$

So we have

$$\hat{M}_y - M_y = \{f_y(M_y)\}^{-1}\left[F_y\left(\hat{M}_y\right) - F_y\left(M_y\right)\right] + o_p\left(n^{-1/2}\right).$$

Furthermore it may be shown that

$$F_y\left(\hat{M}_y\right) - F_y\left(M_y\right) = \hat{F}_y\left(\hat{M}_y\right) - \hat{F}_y\left(M_y\right) + o_p\left(n^{-1/2}\right),$$

where $\hat{F}_y$ is an estimator of F_y. So we have

$$\hat{M}_y - M_y = \left\{f_y\left(M_y\right)\right\}^{-1}\left[\hat{F}_y\left(\hat{M}_y\right) - \hat{F}_y\left(M_y\right)\right] + o_p\left(n^{-1/2}\right)$$
$$= \{f_y(M_y)\}^{-1}\left[0.5 - p_y\right] + o_p\left(n^{-1/2}\right), \quad \text{with } p_y = \hat{F}_y\left(M_y\right).$$

Similarly

$$\hat{M}_x - M_x = \{f_x(M_x)\}^{-1}\left[0.5 - p_x\right] + o_p\left(n^{-1/2}\right), \quad \text{with } p_x = \hat{F}_x\left(M_x\right).$$

Then we have

$$V(\hat{M}_y) = V(\hat{M}_y - M_y) = \{f_y(M_y)\}^{-2} V(p_y) = (1-f)(4n)^{-1}\{f_y(M_y)\}^{-2},$$

$$V(\hat{M}_x) = V(\hat{M}_x - M_x) = \{f_x(M_x)\}^{-2} V(p_x) = (1-f)(4n)^{-1}\{f_x(M_x)\}^{-2},$$

and

$$\begin{aligned}\text{Cov}(\hat{M}_x, \hat{M}_y) &= \text{Cov}[\ \hat{M}_x - M_x,\ \hat{M}_y - M_y\ ] \\ &= \text{Cov}[\ \{f_x(M_x)\}^{-1}(0.5 - p_x),\ \{f_y(M_y)\}^{-1}(0.5 - p_y)\ ] \\ &= \{f_x(M_x) f_y(M_y)\}^{-1}\text{Cov}(p_x, p_y) \\ &= \{f_x(M_x) f_y(M_y)\}^{-1}[E(p_x p_y) - E(p_x)E(p_y)] \\ &= \{f_x(M_x) f_y(M_y)\}^{-1}[P_{11} - 0.25].\end{aligned}$$

Hence the theorem.

Now we have the following theorems:

Theorem 3.11.3. The variance of the ratio estimator $\hat{M}_R$ of population median M_y is given by

$$V(\hat{M}_R) = \left(\frac{1-f}{n}\right)\left[\frac{\{f_y(M_y)\}^{-2}}{4} + \left(\frac{M_y}{M_x}\right)^2 \frac{\{f_x(M_x)\}^{-2}}{4} - 2\left(\frac{M_y}{M_x}\right)(P_{11} - 0.25)\{f_x(M_x) f_y(M_y)\}^{-1}\right]. \tag{3.11.5}$$

Proof. By the ratio method of estimation we have

$$V(\hat{M}_R) = V(\hat{M}_y) + \left\{\frac{E(\hat{M}_y)}{E(\hat{M}_x)}\right\}^2 V(\hat{M}_x) - 2\left\{\frac{E(\hat{M}_y)}{E(\hat{M}_x)}\right\}\text{Cov}(\hat{M}_y, \hat{M}_x). \tag{3.11.6}$$

On substituting the values of $V(\hat{M}_y)$, $V(\hat{M}_x)$ and $\text{Cov}(\hat{M}_y, \hat{M}_x)$ from theorem 3.11.2 in (3.11.6), we have (3.11.5). Hence the theorem.

Theorem 3.11.4. The ratio estimator of population median $\hat{M}_R$ is more efficient than the sample median estimator $\hat{M}_y$ if

$$\rho_c > \frac{\{f_x(M_x)\}^{-1} M_x^{-1}}{2\{f_y(M_y)\}^{-1} M_y^{-1}}, \tag{3.11.7}$$

where $\rho_c = 4(P_{11} - 0.25)$ goes from -1 to +1 as P_{11} increases from 0.0 to +0.5. This condition is analogous to the condition under which the ratio estimator of population mean remains superior to the sample mean.

Proof. By setting $V(\hat{M}_R) < V(\hat{M}_y)$ we have

$$\left(\frac{M_y}{M_x}\right)^2 \frac{\{f_x(M_x)\}^{-2}}{4} - 2\left(\frac{M_y}{M_x}\right)(P_{11} - 0.25)\{f_x(M_x)f_y(M_y)\}^{-1} < 0$$

or

$$\left(\frac{M_y}{M_x}\right)\{f_x(M_x)\}^{-2} - 2\rho_c\{f_x(M_x)f_y(M_y)\}^{-1} < 0, \quad \text{where } \rho_c = 4(P_{11} - 0.25)$$

or

$$M_x^{-1}\{f_x(M_x)\}^{-1} - 2\rho_c M_y^{-1}\{f_y(M_y)\}^{-1} < 0$$

or

$$\rho_c > \frac{M_x^{-1}\{f_x(M_x)\}^{-1}}{2M_y^{-1}\{f_y(M_y)\}^{-1}}.$$

Hence the theorem.

Example 3.11.1. The amounts of the real and nonreal estate farm loans (in \$000) during 1997 in the US have been presented in population 1. Suppose we selected an SRSWOR sample of eight states to collect the required information. Find the relative efficiency of the ratio estimator of median for estimating median of the amount of real estate farm loans during 1997 by using information on nonreal estate farm loans during 1997 with respect to the usual estimator of population median. Assume that both real and nonreal estate farm loans follow independent normal distributions.

Solution. From the description of the population 1 given in the Appendix we have

Y_i = Amount (in \$000) of the real estate farm loans in different states during 1997,

X_i = Amount (in \$000) of the nonreal estate farm loans in different states during 1997,

$N = 50$, $M_y = 322.305$, $M_x = 452.517$, $\mu_y = 555.434$, $\mu_x = 878.162$, $\sigma_x = 1073.776$, $\sigma_y = 578.948$, and $P_{11} = 0.42$.

Note that we are given

$$f(x) = \frac{1}{\sqrt{2\pi}\sigma_x} e^{-\frac{1}{2}\left(\frac{x-\mu_x}{\sigma_x}\right)^2}, \quad \text{and} \quad f(y) = \frac{1}{\sqrt{2\pi}\sigma_y} e^{-\frac{1}{2}\left(\frac{y-\mu_y}{\sigma_y}\right)^2}$$

which implies that

$$f_x(M_x) = \frac{1}{\sqrt{2\pi}\times 1073.776} e^{-\frac{1}{2}\left(\frac{452.517-878.162}{1073.776}\right)^2} = 3.4345\times 10^{-4},$$

and

$$f_y(M_y) = \frac{1}{\sqrt{2\pi}\times 578.948} e^{-\frac{1}{2}\left(\frac{322.305-555.434}{578.948}\right)^2} = 6.3541\times 10^{-4}.$$

Therefore we have

$$V\left(\hat{M}_y\right)=\left(\frac{1-f}{n}\right)\frac{\left\{f_y\left(M_y\right)\right\}^{-2}}{4}=\left(\frac{1-0.16}{8}\right)\frac{\left\{6.3541\times10^{-4}\right\}^{-2}}{4}=65016.15$$

and

$$V\left(\hat{M}_{\mathrm{R}}\right)=\left(\frac{1-f}{n}\right)\left[\frac{\left\{f_y\left(M_y\right)\right\}^{-2}}{4}+\left(\frac{M_y}{M_x}\right)^2\frac{\left\{f_x\left(M_x\right)\right\}^{-2}}{4}-2\left(\frac{M_y}{M_x}\right)\left(P_{11}-0.25\right)\left\{f_x\left(M_x\right)f_y\left(M_y\right)\right\}^{-1}\right]$$

$$=\left(\frac{1-0.16}{8}\right)\left[\frac{\left\{6.3541\times10^{-4}\right\}^{-2}}{4}+\left(\frac{322.305}{452.517}\right)^2\frac{\left\{3.4345\times10^{-4}\right\}^{-2}}{4}\right.$$

$$\left.-2\left(\frac{322.305}{452.517}\right)\left(0.42-0.25\right)\left\{6.3541\times10^{-4}\times3.4345\times10^{-4}\right\}^{-1}\right]$$

$$=61393.76\,.$$

Thus the percent relative efficiency (RE) of the ratio estimator $\hat{M}_{\mathrm{R}}$ with respect to the usual estimator $\hat{M}_y$ is given by

$$\mathrm{RE}=\frac{V\left(\hat{M}_y\right)}{\mathrm{MSE}\left(\hat{M}_{\mathrm{R}}\right)}\times100=\frac{65016.15}{61393.76}\times100=105.90\%$$

which shows that the ratio estimator is more efficient than the usual estimator of population median.

Remark 3.11.1. In the above example, for simplicity we have considered univariate normal distributions for X and Y separately, but a more interesting example may be considered by assuming bivariate joint normal distribution of X and Y.

Theorem 3.11.5. The minimum variance of the regression type estimator of median defined as

$$\hat{M}_{\mathrm{lr}}=\hat{M}_y+\gamma\left(M_x-\hat{M}_x\right) \tag{3.11.8}$$

is given by

$$\mathrm{Min.}V\left(\hat{M}_{\mathrm{lr}}\right)=\left(\frac{1-f}{n}\right)2P_{11}\left(1-2P_{11}\right)\left\{f_y\left(M_y\right)\right\}^{-2}. \tag{3.11.9}$$

Proof. We have

$$V\left(\hat{M}_{\mathrm{lr}}\right)=V\left(\hat{M}_y\right)+\gamma^2V\left(\hat{M}_x\right)-2\gamma\mathrm{Cov}\left(\hat{M}_y,\hat{M}_x\right)$$

$$=\left(\frac{1-f}{n}\right)\left[\frac{\left\{f_y\left(M_y\right)\right\}^{-2}}{4}+\gamma^2\frac{\left\{f_x\left(M_x\right)\right\}^{-2}}{4}-2\gamma\left(P_{11}-0.25\right)\left\{f_x\left(M_x\right)f_y\left(M_y\right)\right\}^{-1}\right]. \tag{3.11.10}$$

On differentiating (3.11.10) with respect to γ and equating to zero we obtain

$$\gamma=\frac{\left[4\left(P_{11}-0.25\right)\left\{f_y\left(M_y\right)f_x\left(M_x\right)\right\}^{-1}\right]}{\left[\left\{f_x\left(M_x\right)\right\}^{-2}\right]}. \tag{3.11.11}$$

On substituting the optimum value of γ in (3.11.10) we obtain

$$\text{Min.}V\left(\hat{M}_{lr}\right) = \left(\frac{1-f}{n}\right)\left[\frac{\{f_y(M_y)\}^{-2}}{4} - \frac{(P_{11}-0.25)^2}{\{f_x(M_x)\}^{-2}/4}\{f_x(M_x)f_y(M_y)\}^{-2}\right]$$

$$= \left(\frac{1-f}{n}\right)\frac{\{f_y(M_y)\}^{-2}}{4}\left[1-16(P_{11}-0.25)^2\right]$$

$$= \left(\frac{1-f}{n}\right)2P_{11}(1-2P_{11})\{f_y(M_y)\}^{-2}.$$

Hence the theorem.

Example 3.11.2. Consider the following population consisting of five ($N=5$) units A, B, C, D and E where for each one of the unit in the population two variables Y and X are measured.

Units	A	B	C	D	E
Y_i	9	11	13	16	21
X_i	14	18	19	20	24

(a) Find the population medians M_y and M_x of the study variable and auxiliary variable, respectively.
(b) Select all possible samples of three units ($n=3$) with SRSWOR sampling.
(c) Find the estimates of the medians M_y and M_x from each sample.
(d) Find the exact bias in the estimator $\hat{M}_y$ using the definition.
(e) Find the exact mean square error of the estimator $\hat{M}_y$ using the definition.
(f) Assuming that the median M_x of the auxiliary variable is known, find the ratio estimate of the median $\hat{M}_R = \hat{M}_y\left(M_x/\hat{M}_x\right)$ from each sample.
(g) Find the exact bias in the ratio estimator $\hat{M}_R$ using the definition.
(h) Find the exact mean square error of the ratio estimator $\hat{M}_R$ using the definition.
(i) Find the relative efficiency of the ratio estimator $\hat{M}_R$ with respect to $\hat{M}_y$.

Solution. (a) The population medians of Y and X variables are given by

$$M_y = 13 \text{ and } M_x = 19.$$

(b) and (c) All 10 possible samples, estimates of medians from a given sample $\hat{M}_y|_s$, $\hat{M}_x|_s$ and related results are given in the following table:

Sample No.		Sampled Values			$\hat{M}_y \mid_s$	$\hat{M}_x \mid_s$	p_s	$\hat{M}_R \mid_s$	$\{\hat{M}_y \mid_s - M_y\}^2$	$\{\hat{M}_R \mid_s - M_y\}^2$
1	y_i :	9	11	13	11	18	0.1	11.61	4	1.93
	x_i :	14	18	19						
2	y_i :	9	11	16	11	18	0.1	11.61	4	1.93
	x_i :	14	18	20						
3	y_i :	9	11	21	11	18	0.1	11.61	4	1.93
	x_i :	14	18	24						
4	y_i :	9	13	16	13	19	0.1	13.00	0	0.00
	x_i :	14	19	20						
5	y_i :	9	13	21	13	19	0.1	13.00	0	0.00
	x_i :	14	19	24						
6	y_i :	9	16	21	16	20	0.1	15.20	9	4.84
	x_i :	14	20	24						
7	y_i :	11	13	16	13	19	0.1	13.00	0	0.00
	x_i :	18	19	20						
8	y_i :	11	13	21	13	19	0.1	13.00	0	0.00
	x_i :	18	19	24						
9	y_i :	11	16	21	16	20	0.1	15.20	9	4.84
	x_i :	18	20	24						
10	y_i :	13	16	21	16	20	0.1	15.20	9	4.84
	x_i :	19	20	24						
				Sum	133	190			39	20.31

(d) The bias in the estimator $\hat{M}_y$ is given by

$$B(\hat{M}_y) = \sum_{s=1}^{10} p_s \hat{M}_y \mid_s - M_y = 13.3 - 13 = 0.3.$$

(e) The mean square error of the estimator $\hat{M}_y$ is given by

$$\text{MSE}(\hat{M}_y) = \sum_{s=1}^{10} p_s \{\hat{M}_y \mid_s - M_y\}^2 = 3.9.$$

(f) The ratio estimates of $\hat{M}_R \mid_s$, $s = 1,2,....,10$ are given in the above table.

(g) The mean square error of the ratio estimator $\hat{M}_R$ is given by

$$\text{MSE}(\hat{M}_R) = \sum_{s=1}^{10} p_s \{\hat{M}_R \mid_s - M_y\}^2 = 2.031.$$

(h) The relative efficiency of the ratio estimator $\hat{M}_{\mathrm{R}}$ with respect to the usual estimator $\hat{M}_y$ is given by

$$\mathrm{RE} = \frac{\mathrm{MSE}\left(\hat{M}_y\right)}{\mathrm{MSE}\left(\hat{M}_{\mathrm{R}}\right)} \times 100 = \frac{3.90}{2.031} \times 100 = 192.05\% .$$

Remark: 3.11.1. It is not clear if the estimator $\hat{M}_{\mathrm{lr}}$ of median can work as efficiently as the usual linear regression estimator of population mean, $\bar{y}_{\mathrm{LR}}$. Kuk and Mak (1989) studied two more estimators under the names of 'position estimator' and 'stratification estimator', these were found to be as efficient as the $\hat{M}_{\mathrm{lr}}$ from the variance point of view. Graf (2002) also pointed out that estimators of median developed by Kuk and Mak (1989, 1994) and Ren (2000) deserve practical investigations. A bootstrap method for smoothed estimators of median has been discussed by Brown, Hall, and Young (2001). Nelson and Meeden (1998) used prior information about the population quartiles of the auxiliary variable to improve estimator of median.

EXERCISES

Exercise 3.1. Under SRSWOR sampling, find the first order approximations of bias and mean squared error in each of the following estimators:

(a) $\bar{y}_1 = \alpha\bar{y} + (1-\alpha)\bar{y}\left(\frac{\bar{X}}{\bar{x}}\right)$; (b) $\bar{y}_2 = \alpha\bar{y} + (1-\alpha)\bar{y}\left(\frac{\bar{x}}{\bar{X}}\right)$;

(c) $\bar{y}_3 = \bar{y}\left[\frac{\bar{X}}{\alpha\bar{x} + (1-\alpha)\bar{X}}\right]$; (d) $\bar{y}_4 = \bar{y}\left[\frac{(1+\alpha)\bar{X} + (1-\alpha)\bar{x}}{(1-\alpha)\bar{X} + (1+\alpha)\bar{x}}\right]$;

and

(e) $\bar{y}_5 = (1-\alpha)\bar{y} + \alpha\bar{y}\left(\frac{\bar{X}}{\bar{x}}\right)^{\gamma}$;

where α and γ are suitably chosen constants such that $\mathrm{MSE}(\bar{y}_t)$, $t = 1,2,3,4,5$, is minimum. Show that $\mathrm{Min.MSE}(\bar{y}_t)$, $t = 1,2,3$, to the first order of approximation, is the same as that of the usual linear regression estimator.

Hint: (a) and (b) Chakrabarty (1968), Vos (1980), Adhvaryu and Gupta (1983); (c) Walsh (1970) ; (d) Sahai and Sahai (1985); (e) Sisodia and Dwivedi (1981).

Exercise 3.2. Compare the following estimators with the usual ratio estimator under SRSWOR design

(a) $\bar{y}_1 = \bar{y}\left(\dfrac{\bar{X}+C_x}{\bar{x}+C_x}\right)$, where $C_x = S_x/\bar{X}$ denotes the known coefficient of variation of the auxiliary variable X,

(b) $\bar{y}_2 = \bar{y}\left(\dfrac{N\bar{X}-n\bar{x}}{(N-n)\bar{X}}\right)$ or $\bar{y}_2 = \bar{y}\dfrac{\bar{x}^*}{\bar{X}}$, where $\bar{x}^* = \dfrac{1}{N-n}\sum_{i=1}^{N-n} X_i$ denotes the mean of non-sampled units of the x variable,

(c) $\bar{y}_3 = \bar{y}^*\left(\dfrac{\bar{X}}{\bar{x}}\right)$, where $\bar{y}^* = \lambda\bar{y}$ denotes the Searls' estimator.

(d) $\bar{y}_4 = \dfrac{\bar{y}\bar{x}+s_{xy}/n}{\bar{x}^2+s_x^2/n}$, where $s_x^2 = \dfrac{\sum_{i=1}^{n}(x_i-\bar{x})^2}{(n-1)}$ and $s_{xy} = \dfrac{\sum_{i=1}^{n}(x_i-\bar{x})(y_i-\bar{y})}{(n-1)}$.

(e) $\bar{y}_5 = \dfrac{\bar{y}}{\bar{x}}\left[1+\left(\dfrac{1}{n}-\dfrac{1}{N}\right)\left(\dfrac{s_{xy}}{\bar{y}\,\bar{x}}-\dfrac{s_x^2}{\bar{x}^2}\right)\right]$,

and

(f) $\bar{y}_6 = \dfrac{\bar{y}}{\bar{x}}\left[1+\left(\dfrac{1}{n}-\dfrac{1}{N}\right)\left(\dfrac{s_x^2}{\bar{x}^2}-\dfrac{s_{xy}}{\bar{y}\,\bar{x}}\right)\right]^{-1}$.

Hint: (a) Sisodia and Dwivedi (1981); (b) Srivenkataramana (1980), Srivenkataramana and Tracy (1980, 1981); (c) Prasad (1989); (d) Srivastava, Dwivedi, Chaubey, and Bhatnagar (1983); (e) and (f) Swain and Sahoo (1982).

Exercise 3.3. Find the minimum MSE of the estimator of population mean, given by

$$\bar{y}_1 = \bar{y}\left(\frac{S_x^2}{s_x^2}\right)^{\beta}$$

and study the behaviour of the resultant MSE under bivariate normal distribution and discuss your views.

Exercise 3.4. Show that the minimum MSE of the estimator of population mean $\bar{Y}$ defined as

$$\bar{y}_{sl} = \alpha\bar{y} + \beta(\bar{X}-\bar{x})$$

is

$$\text{Min.MSE}(\bar{y}_{sl}) = \frac{\left(\frac{1-f}{n}\right)S_y^2(1-\rho_{xy}^2)}{1+\left(\frac{1-f}{n}\right)C_y^2(1-\rho_{xy}^2)}$$

for the optimum values of α and β.

Exercise 3.5. Suppose a class of ratio type estimators to estimate the population mean $\overline{Y}$ is defined as

$$\bar{y}_c = \bar{y} f(u, v, w)$$

where $u = \bar{x}/\overline{X}$, $v = s_x^2/S_x^2$, and $w = r_{xy}/\rho_{xy}$, and $r_{xy} = s_{xy}/(s_x s_y)$ is an estimator of population correlation coefficient, ρ_{xy}.

(a) Find the minimum mean squared error of $\bar{y}_c$ by defining certain regularity conditions.

(b) Also show that the minimum mean squared error of the ratio type class of estimators, $\bar{y}_c$, is equivalent to that of the wider class of estimators, defined as

$$\bar{y}_w = f(\bar{y}, u, v, w).$$

Hint: Srivastava and Jhajj (1983b, 1995).

Exercise 3.6. Consider $\hat{P}_i = \bar{y}(\bar{x}/\overline{X})^i$, such that $\hat{P}_i \in H, i = 1,2,3$, where H denotes the set of all possible product type estimators of population mean. Construct the following terms:

(a) Linear variety of estimators;

(b) Funnel to filter the bias precipitates;

(c) Filter paper to filter the bias precipitates;

(d) Amount of chemicals to reduce the bias of first order of approximation.

Hint: Singh and Singh (1991, 1993a, 1993b, 1993c).

Exercise 3.7. Study the bias and MSE of the predictive product estimators, given by

(a) $\bar{y}_1 = \bar{y}\left(\dfrac{\tilde{x}}{\widetilde{X}}\right)$, and (b) $\bar{y}_2 = \dfrac{1}{n}\sum\limits_{i=1}^{n}\dfrac{y_i x_i}{\widetilde{X}}$,

where $\bar{y} = n^{-1}\sum\limits_{i=1}^{n} y_i$ is the usual sample mean, $\tilde{x} = n\left(\sum\limits_{i=1}^{n}\dfrac{1}{x_i}\right)^{-1}$, and $\widetilde{X} = N\left(\sum\limits_{i=1}^{N}\dfrac{1}{X_i}\right)^{-1}$ are the sample and population harmonic means respectively, under the super population model defined as , $y_i = \beta/x_i + e_i$, where $E(e_i) = 0$, $E(e_i^2) = \lambda x_i^{-1}$.

Hint: Agarwal and Jain (1989), Agrawal and Sthapit (1996).

Exercise 3.8. Study the asymptotic bias and MSE of the following almost unbiased product type estimators of population mean

(a) $\bar{y}_1 = \bar{y}\left(\dfrac{\bar{x}}{\overline{X}}\right) - \dfrac{(1-f)}{n}\dfrac{s_{xy}}{\overline{X}}$; (b) $\bar{y}_2 = \bar{y}\left(\dfrac{\bar{x}}{\overline{X}}\right)\left[1 + \dfrac{(1-f)}{n}\dfrac{s_{xy}}{\bar{x}\,\bar{y}}\right]^{-1}$;

(c) $\bar{y}_3 = \bar{y}\left(\dfrac{\bar{x}}{\overline{X}}\right) - \dfrac{(1-f)}{n}\dfrac{s_{xy}}{\bar{x}}$; and (d) $\bar{y}_4 = \bar{y}\left[1 - \dfrac{(1-f)s_x^2}{n\bar{x}^2 + \alpha(1-f)s_x^2}\right]$;

where α is the characterising scalar. Also show that $\bar{y}_1$ and $\bar{y}_2$ are the special cases of $\bar{y}_4$ for certain choices of α and the estimator $\bar{y}_4$ remains better in the sense of smaller MSE than the other estimators, $\bar{y}_t, t = 1,2,3$.

Hint: (a) Robson (1957); (b) Singh (1989); (c) Dubey (1993); (d) Srivastava and Bhatnagar (1981), Bhatnagar (1996).

Exercise 3.9. Let X_m and X_M denote the minimum and maximum values of a known positive variate X respectively. Using these values, let us transform the auxiliary variable X to create two new variables Z and U such that

$$Z_i = \frac{X_i + X_m}{X_M + X_m} \text{ and } U_i = \frac{X_i + X_M}{X_M + X_m} \text{ for } i = 1,2,\ldots,N.$$

The same transformations are applied on the x_i values in the sample as $z_i = \dfrac{x_i + X_m}{X_M + X_m}$ and $u_i = \dfrac{x_i + X_M}{X_M + X_m}$, for $i = 1,2,\ldots,n$. Find the conditions under which the estimators

$$(\text{a})\ \bar{y}_1 = \bar{y}\left(\frac{\bar{Z}}{\bar{z}}\right), \text{ and } (\text{b})\ \bar{y}_2 = \bar{y}\left(\frac{\bar{U}}{\bar{u}}\right)$$

where $\bar{z} = n^{-1}\sum_{i=1}^{n} z_i$, $\bar{u} = n^{-1}\sum_{i=1}^{n} u_i$, $\bar{Z} = N^{-1}\sum_{i=1}^{N} Z_i$ and $\bar{U} = N^{-1}\sum_{i=1}^{N} U_i$, of population mean $\bar{Y}$ are better than the usual ratio estimator.

Hint: Mohanty and Sahoo (1995), Sahoo, Sahoo, and Mohanty (1995a), Sahoo (1986).

Exercise 3.10. Consider the transformation on the auxiliary information of sample observations as, $x_i^* = \left(N\bar{X} - nx_i\right)/(N-n)$.

(a) Adjust the bias in the estimator of population mean given by

$$\bar{y}_1 = \bar{r}^*\bar{X}$$

where $\bar{r}^* = n^{-1}\sum_{i=1}^{n} r_i^*$ for $r_i^* = y_i / x_i^*$.

(b) Show that an unbiased estimator of population mean is given by

$$\bar{y}_2 = \bar{r}^*\bar{X} + \frac{(N-1)n}{N(n-1)}\left(\bar{y} - \bar{r}^*\bar{x}^*\right).$$

Hint: Singh and Singh (1993).

Exercise 3.11. Find the asymptotic bias and mean squared error expressions for three estimators of the parameter $K = \rho_{xy}\dfrac{C_y}{C_x} = \dfrac{\bar{X}}{\bar{Y}}\dfrac{S_{xy}}{S_x^2}$, given by

$$(\text{a})\ k_1 = \frac{\bar{x}}{\bar{y}}\frac{s_{xy}}{s_x^2};\quad (\text{b})\ k_2 = \frac{\bar{X}}{\bar{y}}\frac{s_{xy}}{s_x^2};\ \text{and}\quad (\text{c})\ k_3 = \frac{\bar{X}}{\bar{y}}\frac{s_{xy}}{S_x^2}.$$

Also study the properties of the six general classes of estimators defined as:

$$k_{aJ} = k_J H(u) \text{ and } k_{bJ} = k_J H(u,v) \text{ for } J = 1,2,3$$

where $u = \bar{x}/\bar{X}$ and $v = s_x^2/S_x^2$. Construct the wider classes of estimators to estimate the parameter K and study the properties. Comment on the results.

Hint: Singh and Singh (1988), Srivastava, Jhajj, and Sharma (1986), Reddy (1978a).

Exercise 3.12. Study the properties of the almost unbiased product type estimators of population mean, $\bar{Y}$ given by

(a) $\bar{y}_1 = \bar{y}\left(\frac{\bar{x}}{\bar{X}}\right) - \frac{(N-n)}{nN}\left(\frac{s_{xy}}{\bar{X}}\right)$; (b) $\bar{y}_2 = \frac{n(N-1)}{N(n-1)}\bar{y}\left(\frac{\bar{x}}{\bar{X}}\right) - \frac{(N-n)}{N(n-1)}\left(\frac{\bar{p}}{\bar{X}}\right)$;

(c) $\bar{y}_3 = \frac{1}{(n-1)}\left[\bar{y}\left(\frac{\bar{x}}{\bar{X}}\right) - \left(\frac{\bar{p}}{\bar{X}}\right)\right]$; and (d) $\bar{y}_4 = \bar{y}\left(\frac{\bar{X}}{\bar{x}}\right)\left[1 + \left(\frac{1-f}{n}\right)\left(\frac{s_{xy}}{\bar{X}\bar{y}} - \frac{s_x^2}{\bar{x}^2}\right)\right]$;

where $s_{xy} = (n-1)^{-1}\sum_{i=1}^{n}(y_i - \bar{y})(x_i - \bar{x})$ and $\bar{p} = n^{-1}\sum_{i=1}^{n} y_i x_i$.

Hint: Shah and Shah (1979); Murthy (1964); Pandey and Dubey (1989).

Exercise 3.13. Suppose a ratio type estimator of the finite population variance S_y^2 given by

$$s_1^2 = s_y^{*2}\left(S_x^2/s_x^2\right)$$

where $s_y^{*2} = \lambda s_y^2$ denotes the Searls (1964) type estimator. Find the minimum mean squared error of the estimator s_1^2 and find the condition under which it is more efficient than the ratio type estimator of variance proposed by Isaki (1983) for $\lambda = 1$.

Hint: Prasad and Singh (1990).

Exercise 3.14. (a) Show that the power type of estimator of variance S_y^2 given by

$$s_\alpha^2 = s_y^2\left(S_x^2/s_x^2\right)^\alpha$$

is always more efficient than the ratio type estimator $s_l^2 = s_y^2\left(S_x^2/s_x^2\right)$, for the optimum choice of real constant α.

(b) In case of multi-auxiliary information (say, k-variables), study the asymptotic properties of the estimator, defined as

$$s_m^2 = s_y^2\prod_{j=1}^{k}\left(S_{x_j}^2/s_{x_j}^2\right)^{\alpha_j}$$

where α_j, $j = 1, 2, ..., k$ are real constants to be chosen.

Hint: Garcia and Cebrian (1996); Cebrian and Garcia (1997).

Exercise 3.15. Show that an improved and admissible estimator of population correlation coefficient ρ_{xy} is given by

$$r_s = \frac{s_{xy}^*}{s_x s_y}$$

where $s_{xy}^* = \lambda s_{xy}$ denotes the Searls (1964) type estimator of S_{xy}.

Hint: Singh, Mangat, and Gupta (1996).

Exercise 3.16. Consider

$$\hat{\theta}_1 = \frac{1}{\bar{y}}, \ \hat{\theta}_2 = \frac{1}{\bar{y}}\left(\frac{\hat{c}_y}{C_y}\right), \text{ and } \hat{\theta}_3 = \frac{1}{\bar{y}}\left(\frac{C_y}{\hat{c}_y}\right)$$

where $\hat{c}_y = (s_y/\bar{y})$, $C_y = (S_y/\bar{Y})$ and such that $\hat{\theta}_i \in H, i = 1,2,3$, where H denotes the set of all possible estimators for estimating the 'Inverse of population mean $\theta = 1/\bar{Y}$'.

Define the following terms in statistical language:

(a) Linear Variety of estimators;

(b) Funnel to filter the bias precipitates;

(c) Filter paper to filter the bias precipitates;

(d) Amount of chemicals to reduce the bias of first order of approximation.

Hint: Singh and Gangele (1995, 1997), Singh and Singh (1991, 1993a, 1993b, 1993c).

Exercise 3.17. (a) Show that the MSE, to the first order of approximation, of the usual ratio estimator of population mean can be expressed as

$$\text{MSE}(\bar{y}_R) = \left(\frac{1-f}{n}\right)\frac{1}{N-1}\sum_{i=1}^{N}\varepsilon_i^2,$$

where $\varepsilon_i = (Y_i - \bar{Y}) - R(X_i - \bar{X})$ and $R = \bar{Y}/\bar{X}$ have their usual meanings.

(b) Study the asymptotic properties of an estimator of $\text{MSE}(\bar{y}_R)$ defined as:

$$\hat{\text{MSE}}(\bar{y}_R) = \left(\frac{1-f}{n}\right)\frac{1}{(n-1)}\sum_{i=1}^{n}e_i^2\left(\frac{\bar{x}}{\bar{X}}\right)^g,$$

where g is a suitably chosen constant and $e_i = (y_i - \bar{y}) - r(x_i - \bar{x})$ for $r = \bar{y}/\bar{x}$.

Hint: Wu (1982).

Exercise 3.18. Consider $\hat{c}_y = (s_y/\bar{y})$ as an estimator of the population coefficient of variation given by $C_y = (S_y/\bar{Y})$.

(a) Derive the bias and mean squared error of the estimator $\hat{c}_y$ to the first order of approximation.

(b) By suggesting an estimator of the bias in the estimator $\hat{c}_y$, propose an almost unbiased estimator of C_y.

(c) Obtain the mean square error of the almost unbiased estimator in (b) to the first order of approximation and compare it with that of the estimator in (a).
(d) Find the bias and MSE of the general class of estimators defined as:

$$\hat{c}_h = \left(s_y/\bar{y}\right)H(u)$$

where $u = \bar{x}/\bar{X}$ and $H(.)$ is a parametric function such that $H(1) = 1$ and define certain regularity conditions required.

Hint: Expand the ratio $s_y/\bar{y}$ in terms of ε_0 and ε_2 by using binomial expansion and use the results from the section 3.1 to proceed further.

Exercise 3.19. Study the second order asymptotic properties of the estimators of population mean $\bar{Y}$ given by

(a) $t_1^* = \bar{y}\left(\dfrac{\bar{X}}{\bar{x}}\right)^{\hat{a}}$, and (b) $t_2^* = \dfrac{\bar{y}\bar{X}}{\hat{a}\bar{x} + (1-\hat{a})\bar{X}}$,

where $\hat{a} = \left(\bar{y}\, s_{xy}\right)/\left(\bar{x}\, s_x^2\right)$ and compare the results.

Hint: Sahoo and Swain (1987).

Exercise 3.20. (a) Show that the minimum mean squared error of the estimator of population mean as:

$$\bar{y}_{\text{gen}} = \alpha\bar{y} + \beta\left(\bar{X} - \bar{x}\right)$$

where $\alpha + \beta = 1$ is given by

$$\text{MSE}\left(\bar{y}_{\text{gen}}\right) = \left(\frac{1-f}{n}\right)\frac{S_y^2 S_x^2\left(1-\rho_{xy}^2\right)}{S_x^2 + 2\rho_{xy}S_xS_y + S_y^2}.$$

Hint: Jain (1987).

(b) Find the bias and mean square error of the estimator

$$\bar{y}_{\text{ds}} = w_1\bar{y} + w_2\bar{x} + (1 - w_1 - w_2)\bar{X},$$

where w_1 and w_2 are suitably chosen constants, such that the MSE of $\bar{y}_{\text{ds}}$ is minimum.

Hint: Dubey and Singh (2001).

Exercise 3.21. Consider an estimator of population mean $\bar{Y}$ is defined as:

$$\bar{y}_{\text{sk}} = \bar{y}\left[\frac{(A+C)\bar{X} + f\,B\,\bar{x}_\alpha}{(A+f\,B)\bar{X} + C\,\bar{x}_\alpha}\right],$$

where

$\bar{x}_\alpha = \alpha\bar{x} + (1-\alpha)\bar{X}$, $\alpha = n/(N+n)$, $f = n/N$ and A, B and C are the functions of k and $A = (k-1)(k-2)$, $B = (k-1)(k-4)$, $C = (k-2)(k-3)(k-4)$ such that $k \in (0, \infty)$.

(a) Show that several estimators are special cases of the class of estimators defined as $\bar{y}_{\text{sk}}$ for different choices of A, B and C.

(b) Find the mean square error of $\bar{y}_{\text{sk}}$ over the sample mean estimator for different choices of parameters involved, and comment.

Hint: Singh and Shukla (1993).

Exercise 3.22. Suppose Y is the variable under study and X_1, X_2, ...,X_p are the p auxiliary variables correlated with it. A sample of size n is drawn from the finite population of N units with SRSWR. If prior information about the coefficient of variation C_y of Y along with information about p auxiliary variables is available, study the properties of the estimator of population mean $\bar{Y}$ defined as

$$\bar{y}_m = W\left[\bar{y} + \underline{\beta}''\left(\underline{\bar{X}} - \underline{\bar{x}}\right)\right] ,$$

where $\underline{\beta}'' = \left[\beta_1, \beta_2, ..., \beta_p\right]_{p\times 1}$, $\underline{\bar{X}}'' = \left[\bar{X}_1, \bar{X}_2, ..., \bar{X}_p\right]_{p\times 1}$, $\underline{\bar{x}}'' = \left[\bar{x}_1, \bar{x}_2, ..., \bar{x}_p\right]_{p\times 1}$ and W is a suitably chosen positive constant such that the mean square error of the estimator $\bar{y}_m$ is minimum.

Hint: Kothwala and Gupta (1989).

Exercise 3.23. Study the asymptotic properties of the multivariate estimator of population mean, $\bar{Y}$ defined as:

(a) $\bar{y}_1 = \bar{y}\sum_{t=1}^{p} w_t\left(\frac{\bar{x}_t}{\bar{X}_t}\right)$, (b) $\bar{y}_2 = \prod_{t=1}^{p}\left[\bar{y}\left(\frac{\bar{x}_t}{\bar{X}_t}\right)\right]^{w_t}$ and (c) $\bar{y}_3 = \bar{y}\left[\sum_{t=1}^{p} w_t\left(\frac{\bar{X}_t}{\bar{x}_t}\right)\right]^{-1}$

where $\sum_{t=1}^{p} w_t = 1$, and w_t are real constants, and $\bar{x}_t = n^{-1}\sum_{i=1}^{n} x_{ti}$ denote the sample mean unbiased estimator of the known population mean $\bar{X}_t = N^{-1}\sum_{i=1}^{N} X_{ti}$ of the tth auxiliary variable, $t = 1, 2, ..., p$.

Hint: Singh (1967b), Tuteja and Bahl (1991).

Exercise 3.24. Suppose there are two auxiliary variables X_i and Z_i on which information is available and we wish to estimate the population mean, $\bar{Y}$,of the study variable, Y. Then study the behaviour of the estimators

(a) $\bar{y}_0 = \bar{y}\left(\frac{\bar{X}}{\bar{x}}\right)\left(\frac{\bar{z}}{\bar{Z}}\right)$, and (b) $\bar{y}_1 = w\bar{y}\left(\frac{\bar{X}}{\bar{x}}\right)\left(\frac{\bar{z}}{\bar{Z}}\right)$

assuming that $\rho_{yx} > 0$ and $\rho_{yz} < 0$, w is a constant.

(c) Obtain an expression for bias in the estimator of the population mean given by

$$\bar{y}_2 = \frac{\bar{X}}{n}\sum_{i=1}^{n}\left(\frac{y_i}{x_i}\right)\left(\frac{z_i}{\bar{Z}}\right)$$

and show that an estimator after adjusting the bias is given by

$$\bar{y}_3 = \bar{y}_2 + \frac{n(N-1)}{N(n-1)}\left[\left(\bar{y} - \bar{r}\bar{x}\right) - \left(\frac{\bar{X}}{\bar{Z}}\right)\left(\bar{r}_m\bar{Z} - \bar{r}\,\bar{z}\right)\right], \text{ where } \bar{r} = \frac{1}{n}\sum_{i=1}^{n}\frac{y_i}{x_i} .$$

(d) Show that the estimator $\bar{y}_0$ defined in (a) is a special case of another estimator defined as:

$$\bar{y}_4 = w_1\bar{y}\left(\frac{\bar{X}}{\bar{x}}\right) + w_2\bar{y}\left(\frac{\bar{z}}{\bar{Z}}\right) \quad \text{where } w_1 + w_2 = 1.$$

(e) Find the bias and variance of the estimator

$$\bar{y}_5 = \bar{y}\left(\frac{\bar{u}}{\bar{U}}\right)\left(\frac{\bar{z}}{\bar{Z}}\right),$$

where $\bar{u} = n^{-1}\sum_{i=1}^{n} u_i$, $\bar{U} = N^{-1}\sum_{i=1}^{N} u_i$, for $u_i = L - x_i$ and L is a scalar to be chosen so that the mean squared error of $\bar{y}_5$ is minimum.

(f) Find the bias and variance of the product-cum-difference estimator, defined as

$$\bar{y}_{\text{pd}} = \frac{\bar{y}}{\bar{X}}\left[\bar{x} + k\left(\bar{z} - \bar{Z}\right)\right]$$

for the optimum value of k .

(g) Study the generalized regression ratio estimator

$$\bar{y}_{\text{grr}} = \left[\bar{y} + b\left(\bar{X} - \bar{x}\right)\right]\left\{\frac{\bar{z}}{\bar{Z}}\right\}^{\alpha}$$

for the optimum values of b and α .

(h) Study the asymptotic properties of the following estimators of the population mean:

(1) $\bar{y}_1 = \bar{y}\left(\frac{\bar{x}}{\bar{X}}\right)^{\alpha_1}\left(\frac{\bar{z}}{\bar{Z}}\right)^{\alpha_2}$; (2) $\bar{y}_2 = \bar{y}\left[w_1\left(\frac{\bar{x}}{\bar{X}}\right)^{\frac{\alpha_1}{w_1}} + w_2\left(\frac{\bar{z}}{\bar{Z}}\right)^{\frac{\alpha_2}{w_2}}\right]$;

and

(3) $\bar{y}_3 = \bar{y}\left[w_1\left(\frac{\bar{x}}{\bar{X}}\right) + w_2\left(\frac{\bar{z}}{\bar{Z}}\right)\right]^{\alpha}$;

where $w_1 + w_2 = 1$, α_1 and α_2 are real constants.

(i) Discuss the optimum choice of k in the estimator of population mean:

$$\bar{y}_k = \bar{y}\left[\frac{k\bar{X} + \bar{Z}}{k\bar{x} + \bar{z}}\right].$$

Hint: (a) and (b) Singh (1967a, 1967b, 1969), Tracy and Singh (1998); (c) Sahoo and Swan (1980); (d) Biradar and Singh(1992b), Tracy, Singh, and Singh (1996); (e) Singh and Ray (1981), (f) and (g) Khare and Srivastava (1981); (h) Singh and Singh (1984b); (i) Agarwal (1980).

Exercise 3.25. Suppose $\hat{M}_y$ and $\hat{M}_x$ denote the estimators of the population medians M_y and M_x of the study variable and auxiliary variables, respectively. Find the condition under which the product estimator of population median

$$\hat{M}_{\mathrm{p}} = \hat{M}_y\left(\frac{\hat{M}_x}{M_x}\right)$$

remains more efficient than the sample median estimator $\hat{M}_y$.

Hint: Singh and Joarder (2002).

Exercise 3.26. Use the power transformation type estimator to estimate the population median, M_y, defined as follows:

$$\hat{M}_{\mathrm{pw}} = \hat{M}_y\left(\frac{\hat{M}_x}{M_x}\right)^{\alpha}.$$

(a) Show that the ratio and product type estimators are special cases of $\hat{M}_{\mathrm{pw}}$.

(b) Find the optimum value of α such that MSE of the $\hat{M}_{\mathrm{pw}}$ is minimum.

(c) Discuss the difficulties in the use of optimum value of α.

(d) Compare the minimum MSE of $\hat{M}_{\mathrm{pw}}$ with that of usual ratio and product estimator of median.

Exercise 3.27. In case of SRSWR sample of n units, the well known Hartley and Ross (1954) unbiased ratio type estimator of population mean is given by

$$\bar{y}_{\mathrm{HR}} = \bar{r}\bar{X} - \frac{n}{(n-1)}(\bar{y} - \bar{r}\ \bar{x})$$

where $\bar{r} = n^{-1}\sum_{i=1}^{n}(y_i/x_i)$ and $\bar{X} = N^{-1}\sum_{i=1}^{N}X_i$, etc., have their usual meanings. Show that an estimator better than $\bar{y}_{\mathrm{HR}}$ based only on the distinct units v in the sample is:

$$\bar{y}_{v(\mathrm{HR})} = \begin{cases} \bar{r}_v\bar{X} + \left[1 - \dfrac{C_v(n-1)}{C_v(n)}\right]\left(\dfrac{v}{v-1}\right)(\bar{y}_v - \bar{r}_v\bar{x}_v) & \text{if } v > 1, \\ \bar{r}_v\bar{X} & \text{otherwise,} \end{cases}$$

where $\bar{r}_v = \frac{1}{v}\sum_{i=1}^{v}\frac{y_{(i)}}{x_{(i)}}$ and $\bar{x}_v = \frac{1}{v}\sum_{i=1}^{v}x_{(i)}$.

Hint: Pathak (1962).

Exercise 3.28. Find choices of δ, ω, η and ε such that the class of estimators defined as

$$\bar{y}_{\mathrm{c}} = \bar{y}\left(\frac{\bar{x}}{\bar{X}}\right)^{\delta}\left[\omega + (1-\omega)\left(\frac{\bar{x}}{\bar{X}}\right)^{\varepsilon}\right]^{\eta}$$

reduces to the following estimators:

(a) $\bar{y}_{\mathrm{Sr}} = \bar{y}\left(\frac{\bar{x}}{\bar{X}}\right)^{\varepsilon}$ [Srivastava (1967)]

(b) $\bar{y}_{\text{RW}} = \bar{y}\left[\dfrac{\bar{X}}{\omega\bar{x} + (1-\omega)\bar{X}}\right]$ [Reddy (1973, 1974), Walsh (1970)]

(c) $\bar{y}_{\text{Gu}} = \bar{y}\left[\omega\dfrac{\bar{X}}{\bar{x}} + (1-\omega)\left(\dfrac{\bar{X}}{\bar{x}}\right)^2\right]$ [Gupta (1978)]

(d) $\bar{y}_{\text{RS}} = \bar{y}\left[\dfrac{(1-\omega)\bar{x} + \omega\bar{X}}{\bar{x}}\right]$ [Ray and Sahai (1979), Chakrabarty (1979)]

(e) $\bar{y}_{\text{SR}} = \bar{y}\left[2 - \left(\dfrac{\bar{x}}{\bar{X}}\right)^{\varepsilon}\right]$ [Sahai and Ray (1980)]

(f) $\bar{y}_{\text{Tr}} = \bar{y}\dfrac{(1-\omega)\bar{X} + \omega\bar{x}}{\bar{X}}$ [Tripathi (1980)]

(g) $\bar{y}_{\text{AG}} = \omega\bar{y} + (1-\omega)\bar{y}\dfrac{\bar{X}}{\bar{x}}$ [Adhvaryu and Gupta (1983)]

(h) $\bar{y}_{\text{AG}} = \omega\bar{y}\dfrac{\bar{X}}{\bar{x}} + (1-\omega)\bar{y}\dfrac{\bar{x}}{\bar{X}}$ [Adhvaryu and Gupta (1983)]

(i) $\bar{y}_{\text{MS}_1} = \bar{y}\dfrac{\bar{x}}{(1-\omega)\bar{x} + \omega\bar{X}}$ [Mohanty and Sahoo (1987)]

(j) $\bar{y}_{\text{MS}_2} = \bar{y}\dfrac{\bar{X}^2}{(1-\omega)\bar{x}^2 + \omega\bar{x}\bar{X}}$ [Mohanty and Sahoo (1987)]

(k) Replace $\bar{x}$ by $\tilde{x} = (1+d)\bar{X} - d\bar{x}$ for some real constant d in $\bar{y}_{\text{c}}$ and discuss the different members of the resultant class of estimators.

Hint: (a) to (j) Ceccon, Diana, and Salvan (1991); (k) Diana (1992); David and Sukhatme (1974).

Exercise 3.29. Suppose $\bar{x}$ and $\bar{y}$ denote the sample means in an SRSWOR sample of size n and let $\bar{w} = \bar{x}/\bar{X}$, $\bar{z} = \bar{y}/Y_k$, where Y_k be some Y_i $i = 1,2,...,N$, but less than population mean $\bar{Y}$ be known from past surveys. Find the bias and variance of the estimator of population mean $\bar{Y}$ defined as

$$\bar{y}_d = Y_k\bar{z}_d$$

where $\bar{z}_d = 1 + \bar{z} - \bar{w}$.

Hint: Srivenkataramana (1978).

Exercise 3.30. Discuss the efficiency conditions of the following estimators of population mean $\bar{Y}$ as

(a) $\bar{y}_1 = \bar{y}_p - \left(\dfrac{s_{xy}}{n\bar{X}}\right)$, and (b) $\bar{y}_2 = \bar{y}_p - \left(\dfrac{s_{xy}}{n\bar{x}}\right)$

where $\bar{y}_p = \bar{y}\left(\frac{\bar{x}}{\bar{X}}\right)$, under the super population model, $y_i = \alpha + \beta x_i + e_i$ with $\beta < 0$, $E(e_i \mid x_i) = 0$, $E(e_i e_j \mid x_i x_j) = 0 \forall i \neq j$ and $V(e_i \mid x_i) = n\delta$, where δ is a constant of order n^{-1} and the variate x_i/n have a gamma distribution with the parameter $m = nh$.

Hint: Singh and Singh (1997), Dalabehera and Sahoo (1995).

Exercise 3.31. Suppose $r = \bar{y}/\bar{x}$ and $\bar{r} = n^{-1}\sum_{i=1}^{n}(y_i/x_i)$.

(a) Find the values of θ_t, $t = 1, 2, 3$ such that the estimator of $\bar{Y}$ defined as

$$\bar{y}_1 = \theta_1\bar{y} + \theta_2\,\bar{r}\bar{X} + \theta_3\, r\,\bar{X}$$

is unbiased under SRSWOR sampling for $\sum_{t=1}^{3}\theta_t = 1$.

(b) Show that the estimator $\bar{y}_2 = \lambda r\bar{X} + \{1-\lambda\}\bar{r}\bar{X}$ is almost unbiased for population mean for the optimum value of λ given by

$$\lambda_{\text{opt}} = k + (1-ck)\left(\frac{\bar{x}}{\bar{X}}\right)$$

where $c = (n-1)/n$ and k is some constant.

Hint: Rao (1981).

Exercise 3.32. Compare the predictive ratio estimator

$$\bar{y}_1 = \bar{y}\left(\frac{\bar{X}}{\bar{x}}\right) + \left(\frac{1}{n}-\frac{1}{N}\right)\bar{y}\left(\frac{\bar{X}_r}{\bar{x}}\right)\left(\hat{c}_{xy}-\hat{c}_x^2\right)$$

where

$$\bar{X}_r = \left(N\bar{X}-n\bar{x}\right)/(N-n),\ \ \hat{c}_{xy} = \frac{1}{(n-1)}\sum_{i=1}^{n}\left(\frac{x_i}{\bar{x}}-1\right)\left(\frac{y_i}{\bar{y}}-1\right),\ \text{and}\ \ \hat{c}_x^2 = \frac{1}{(n-1)}\sum_{i=1}^{n}\left(\frac{x_i}{\bar{x}}-1\right)^2$$

with the usual ratio estimator and with the estimator, given by

$$\bar{y}_2 = \bar{y}\left(\frac{\bar{X}}{\bar{x}}\right)\left[1+\left(\frac{1}{n}-\frac{1}{N}\right)\left(\hat{c}_{xy}-\hat{c}_x^2\right)\right].$$

Hint: Sahoo, Sahoo, and Mohanty (1995b), Srivastava (1983), and Tin (1965).

Exercise 3.33. Consider the problem of estimation of median of a study variable y as

$$\hat{M}_{\text{new}} = \hat{M}_y\prod_{i=1}^{3}\left(\frac{Q_{ix}}{\hat{Q}_{ix}}\right)^{\alpha_i},$$

where Q_{ix}, $i = 1, 2, 3$ denote the ith known quartiles of the auxiliary variable and $\hat{Q}_{ix}$ denotes its sample analogue. Find the minimum MSE of $\hat{M}_{\text{new}}$ for the optimum values of α_i.

Hint: Singh, Singh, and Puetas (2003c).

Exercise 3.34. (a) Assume that the mode M_o of the auxiliary variable is known, find the bias and variance of the estimator of median M_y defined as

$$\hat{M}_{\text{new}} = \hat{M}_y\left(\frac{M_x + M_o}{\hat{M}_x + M_o}\right).$$

Compare the mean squared error of $\hat{M}_{\text{new}}$ with that of the usual ratio estimator

$$\hat{M}_{\text{R}} = \hat{M}_y\left(\frac{M_x}{\hat{M}_x}\right),$$

and develop a condition of it being efficient estimator.

(b) Study the following estimator for population median M_y defined as

$$\hat{M}_y^{(a)} = \hat{M}_y\left(\frac{A - \hat{M}_x}{A - M_x}\right)$$

where A is a suitably chosen scalar.

Hint: Singh, Singh, and Joarder (2003), Singh, Singh, and Puertas (2003a).

Exercise 3.35. Study the asymptotic properties of an estimator of the median M_y defined as

$$\hat{M}_{\text{new}} = \hat{M}_y \prod_{i=1}^{99}\left(\frac{P_{ix}}{\hat{P}_{ix}}\right)^{\alpha_i}$$

where P_{ix}, $i = 1, 2, ..., 99$ denotes the ith known percentile of the auxiliary variable and $\hat{P}_{ix}$ denotes its sample analogue. Find the minimum MSE of $\hat{M}_{\text{new}}$ for the optimum values of α_i.

Hint: Singh (2002a).

Exercise 3.36. An SRSWOR sample is selected from a finite population of N units from a superpopulation model

$$M : y_i = \alpha + \beta x_i + e_i$$

with $E_M(e_i) = 0$, $V_M(e_i) = \sigma_y^2(1 - \rho_{xy}^2)$, and $\text{Cov}_M(e_i, e_j) = 0$ for $i \neq j$.

Show that under this model, the variance of the linear regression estimator

$$\bar{y}_{\text{LR}} = \bar{y} + \left(s_{xy}/s_x^2\right)\left(\bar{X} - \bar{x}\right)$$

can be written as

$$E_M[V(\bar{y}_{\text{LR}})] = \frac{1}{n}\left[(1-f)\sigma_y^2(1-\rho_{xy}^2)\left\{1 + \frac{1}{n-3}\right\}\right]$$

Hint: Sukhatme, Sukhatme, Sukhatme, and Asok (1984).

Exercise 3.37. Study the asymptotic properties of the multivariate estimator of population mean $\overline{Y}$ given by

$$\overline{y}_1 = \sum_{k=1}^{p} w_k \left[\overline{y} + \alpha_k \left(\overline{X}_k - \overline{x}_k\right)\right].$$

Discuss the choice of w_k and α_k such that the estimator $\overline{y}_1$ reduces to the well known multivariate ratio type of estimator proposed by Olkin (1958) as

$$\overline{y}_0 = \sum_{k=1}^{p} w_k \left(\frac{\overline{X}_k}{\overline{x}_k}\right).$$

Also discuss the nature of the estimator defined as

$$\overline{y}_{\text{HRM}} = \sum_{k=1}^{p} w_k \overline{y}_{\text{HR}(k)},$$

where $\overline{y}_{\text{HR}(k)} = \overline{r}_k \overline{X}_k + \dfrac{n(N-1)}{N(n-1)}\left(\overline{y} - \overline{r}_k \overline{x}_k\right)$, $\sum_{k=1}^{p} w_k = 1$, $\overline{r}_k$, $\overline{x}_k$ and $\overline{X}_k$ have the same meaning as $\overline{r}$, $\overline{x}$ and $\overline{X}$ for the kth auxiliary variable.

Hint: Raj (1965), Agarwal, Sharma, and Kashyap (1997), Ramachandran and Pillai (1976), Goodman and Hartley (1958).

Exercise 3.38. Study the asymptotic properties of the following two estimators of population total Y defined as

(a) $T_1 = \hat{Y}\left(\dfrac{\widetilde{X}}{X}\right)$, and (b) $T_2 = \hat{Y}\left(\dfrac{X}{\widetilde{X}}\right)$,

where $\widetilde{X} = (1+g)\left(X - f\hat{X}\right)$ where g is suitably chosen constant and $f = n/N$.

Hint: Bandyopadhyay (1980).

Exercise 3.39. Show that under the transformation $u_i = a + bx_i$, where a and b are pre-assigned constants, a product type estimator of population mean, $\overline{Y}$, defined as

$$\overline{y}_{(1)} = \overline{y}\left(\frac{\overline{u}}{\overline{U}}\right)$$

can be expressed as

$$\overline{y}_{(1)} = \overline{y}\left(1 + \frac{b\left(\overline{x} - \overline{X}\right)}{a + b\overline{X}}\right).$$

(a) Find the bias and variance of the resultant estimator of population mean.
(b) If $a = C_x$ and $b = \beta_2(x)$, where $\beta_2(x)$ denotes the kurtosis and C_x denotes the coefficient variation of the auxiliary variable, then find the bias and MSE expressions for the ratio type estimator

$$\overline{y}_r = \overline{y}\left(\frac{\overline{U}}{\overline{u}}\right).$$

Hint: (a) Tripathi and Singh (1992), (b) Upadhyaya and Singh (1999).

Exercise 3.40. Show that an unbiased estimator of finite population variance, S_y^2, can be defined as

$$d_a = s_y^2 - k\left(s_x^2 / S_x^2\right) + k .$$

Deduce its minimum variance for the optimum value of real constant k.
Hint: Prasad and Singh (1992).

Exercise 3.41. Let $\hat{Y}$ and $\hat{X}$ be the unbiased estimators of population totals Y and X of the two variables y, x respectively, based on a sample of size n drawn by SRSWOR sampling from a finite population of size N. Consider a transformation

$$\hat{W} = \hat{X} - \lambda ,$$

where λ is a scalar. Obtain the minimum mean squared error of the estimator,

$$\hat{Y}_1 = \hat{Y}\left(\frac{\hat{W}}{W}\right)$$

for the optimum value of λ.
Hint: Srivenkataramana and Tracy (1983, 1986).

Exercise 3.42. Suppose $y_i = \begin{cases} 1 & \text{if } \ i \in A, \\ 0 & \text{otherwise,} \end{cases}$ and $x_i = \begin{cases} 1 & \text{if } i \in B, \\ 0 & \text{otherwise,} \end{cases}$ for $\forall\ i$, where A and B represent two groups in the population.

(a) Show that the usual difference estimator

$$\bar{y}_P = \bar{y} + k\left(\bar{X} - \bar{x}\right)$$

reduces to the estimator of population proportion of the units belonging to A.
(b) Assume that the proportion of units possessing the attribute B are known. Discuss the optimum choice of k.
Hint: Das (1982).

Exercise 3.43. Suppose $\bar{y}^*_{(j)} = \dfrac{n\bar{y}_n - m\bar{y}_j}{n-m}$ and $\bar{x}^*_{(j)} = \dfrac{n\bar{x}_n - m\bar{x}_j}{n-m}$, where $\bar{y}_n = n^{-1}\sum_{i=1}^{n} y_i$ and $\bar{x}_n = n^{-1}\sum_{i=1}^{n} x_i$, are the sample means based on a sample of $(n-m)$ units obtained by omitting the jth group and $\bar{y}_j$, $\bar{x}_j$ are the sample means based on the jth sub-sample of size $m = n/g$, then show that a general class of almost unbiased ratio cum product estimator is given by

$$\bar{y}_{\text{au}} = \{1 - \lambda(1-\theta)\}\bar{y} + \lambda\bar{y}\left(\frac{\bar{X}}{\bar{x}}\right)\left(\frac{\bar{z}}{\bar{Z}}\right) - \frac{\lambda}{g}\sum_{j=1}^{g}\bar{y}^*_{(j)}\left(\frac{\bar{X}}{\bar{x}^*_{(j)}}\right)\left(\frac{\bar{z}^*}{\bar{Z}}\right),$$

with $\theta = \dfrac{(N-n)(n-m)}{n(N-n+m)}$ and λ is a real constant.
Hint: Biradar and Singh (1992).

Exercise 3.44. Study the asymptotic properties of an estimator of, $\bar{Y}$ given by,

$$\bar{y}_p^* = \bar{y}\left(\frac{\bar{x}}{\bar{X}}\right) - \beta\left\{1 - \left(\frac{\bar{x}}{\bar{X}}\right)\right\}$$

where β is a suitably chosen constant, under the super population model

$$Y_i = \alpha + \beta X_i + \varepsilon_i ,$$

with α and β are unknown real constants, ε_i are random errors distributed with $E_c(\varepsilon_i \mid X_i) = 0$, $E_c(\varepsilon_i^2 \mid X_i) = \delta X_i^g$ and $E_c(\varepsilon_i \varepsilon_j \mid X_i X_j) = 0$ for every $i \neq j$. Also assume that $0 < \delta < \infty$, $0 \le g \le 2$ and the X_i are independently identically distributed with a common gamma density

$$G(\theta) = \frac{1}{\Gamma(\theta)} e^{-x} x^{\theta-1} \text{ for } x > 0 \text{ and } 2 \le \theta < \infty .$$

Hint: Singh, Singh, and Espejo (1998).

Exercise 3.45. Let the variables y, x, z take real values (y_i, x_i, z_i) on the ith unit $(i = 1,2,,...,N)$ in a finite population. Assuming that the population means $\bar{X}$ and $\bar{Z}$ of the auxiliary variable are known and we are to estimate population mean $\bar{Y}$ of the study variable y. Assuming that y is positively correlated with x and negatively with z, study the asymptotic properties of the ratio cum product estimators of population mean defined as

$$\bar{y}_1 = \bar{y}\left(\frac{\bar{X}}{\bar{x}}\right)\left(\frac{\bar{z}}{\bar{Z}}\right)$$

where $(\bar{y}, \bar{x}, \bar{z})$ are the unbiased estimators of the population means $(\bar{Y}, \bar{X}, \bar{Z})$ respectively based on a simple random sample of size n drawn without replacement. Define $u_i = A - x_i$ and $v_i = B + z_i$, $i = 1,2,\ldots,N$ where A and B are suitably chosen scalars. Then $\bar{u} = A - \bar{x}$ and $\bar{v} = B + \bar{z}$ are the unbiased estimators of $\bar{U} = A - \bar{X}$ and $\bar{V} = B + \bar{Z}$, respectively. Study the asymptotic bias and variance of the estimator,

$$\bar{y}_2 = \bar{y}\left(\frac{\bar{u}}{\bar{U}}\right)\left(\frac{\bar{v}}{\bar{V}}\right).$$

Compare the estimator $\bar{y}_2$ with the estimator $\bar{y}_1$ and discuss your opinions.

Hint: Singh (1967a, 1967b), Tracy, Singh, and Singh (1998), Chang and Huang (2001a).

Exercise 3.46. Suppose a simple random sample of $n \le N$ units is drawn without replacement from the population under consideration and observations are made on the selected units. Let the totals of y, x variates for the n units in the sample be $Y^{(1)}$ and $X^{(1)}$ and similar totals of the $(N-n)$ units not included in the sample are $Y^{(2)}$ and $X^{(2)}$. Study the asymptotic properties of the following two estimators of population ratio $R = Y/X$, given by

(a) $\hat{R}_1 = Y^{(1)}\left[1 + \frac{X^{(2)}}{fX}\right]$ and (b) $\hat{R}_2 = Y^{(1)}\left[1 + \frac{gX}{X^{(1)}}\right]$

with $f = n/N$ and $g = 1 - f$.

Hint: Srivenkataramana and Tracy (1979).

Exercise 3.47. Consider a ratio estimator

$$\hat{\bar{Y}} = \bar{y}\left(\frac{\bar{X}}{\bar{x}}\right)$$

based on a SRSWOR sampling of n units, where $\bar{y}, \bar{x}$ and $\bar{X}$ have their usual meanings. Under the model

$$E_m(Y_i) = \beta X_i\,,\; V_m(Y_i) = \sigma^2 X_i\,,\text{ and }\; C_m(Y_i, Y_j) = 0\,.$$

Show that:

(a) The ratio estimator can be written as:

$$\tilde{\bar{Y}}_{\text{R}} = \bar{y} + r(\bar{X} - \bar{x})$$

where $r = \bar{y}/\bar{x}$;

(b) $\text{Cov}\left(\bar{x}, \tilde{\bar{Y}}\right) = \frac{1-f}{n} S_x^2[Q - R]$, where $f = n/N$ and $Q = S_{xy}/S_x^2$;

(c) When the model holds true we expect the negligible gain in efficiency from the optimum estimator, $\tilde{\bar{Y}}_{\text{MU}} = \bar{y} + Q(\bar{X} - \bar{x})$;

(d) If the model is wrong the relative gain in efficiency of $\tilde{\bar{Y}}_{\text{MU}} = \bar{y} + Q(\bar{X} - \bar{x})$ over $\tilde{\bar{Y}}_{\text{R}}$ is expected to be substantial.

Hint: Montanari (1998, 1999).

Exercise 3.48. Suppose $(\bar{Y}, \bar{X})$ denote the population means of the two variables y and x , and $(\bar{y}, \bar{x})$ denote the means of a random sample of size n. Let $\bar{y}(j)$ and $\bar{x}(j)$ be the means obtained by deleting n/g observations of the jth group.

(a) Compare the classical estimator $\hat{R} = \bar{y}/\bar{x}$ and the Jackknife estimator

$$\hat{R}^* = g\,\hat{R} - \frac{(g-1)}{g}\sum_{j=1}^{g}\frac{\bar{y}(j)}{\bar{x}(j)}$$

with $g = 2$.

(b) Study the following three estimators of population mean given by

$$\bar{y}_1 = \hat{R}\bar{X} = \bar{y} + \hat{R}(\bar{X} - \bar{x}), \qquad \bar{y}_2 = \hat{R}^*\bar{X}\,, \qquad \text{and} \qquad \bar{y}_3 = \bar{y} + \hat{R}^*(\bar{X} - \bar{x})$$

under the super-population model

$$y_i = \alpha + \beta x_i + \varepsilon_i\,,$$

where ε_i has mean zero and variance δx^l $(0 \le l \le 2)$ and is uncorrelated with x_i .

Hint: Rao (1979).

Exercise 3.49. Compare the usual Jackknife variance estimator

$$\hat{v}_{\text{u}} = \frac{1}{g(g-1)} \sum_{j=1}^{g} \left[\hat{\theta}_j(r) - \hat{\theta}(r)\right]^2$$

where $\hat{\theta}_j(r) = gr - (1-g)r_j$ and $\hat{\theta}(r) = \frac{1}{g}\sum_{j=1}^{g} \hat{\theta}_j(r)$, with the modified Jackknife variance estimator

$$\hat{v}_{\text{m}} = \frac{1}{g(g-1)} \sum_{j=1}^{g} \left[\hat{\theta}_j(r) - \hat{\theta}\right]^2$$

where $\hat{\theta} = r$.
Hint: Krewski and Chakrabarty (1981).

Exercise 3.50. In SRSWOR sampling scheme consider the following class of estimators of the population mean as

$$\bar{y}_{\text{c}} = \sum_{i=1}^{n} w_i y_i$$

where w_i is a constant depending upon the ith draw and y_i is the value of y on the unit selected at the ith draw.

(a) Show that $\bar{y}_c$ is unbiased for $\bar{Y}$ if and only if $\sum_{i=1}^{n} w_i = 1$.

(b) Show that the variance of the estimator $\bar{y}_{\text{c}}$ is given by

$$V(\bar{y}_{\text{c}}) = S_y^2\left(\sum_{i=1}^{n} w_i^2 - \frac{1}{N}\right).$$

(c) Show that $V(\bar{y}_c)$ is minimum if and only if $w_i = 1/n \;\forall i$.

Exercise 3.51. Let $\bar{Y}_s$ and $\bar{Y}_{\bar{s}}$ denote, respectively, the means of observed and unobserved y values, while $S^2_{y(s)}$ and $S^2_{y(\bar{s})}$ denote the corresponding variances.

(a) Show that an estimator of population variance S_y^2 can be written as

$$\hat{S}_y^2 = \frac{1}{N-1}\left[(n-1)s_y^2 + (N-n-1)T + \frac{n(N-n)}{N}\left(\bar{y} - T^*\right)^2\right]$$

where T and T^* are, respectively, the predictors of $S^2_{y(s)}$ and $\bar{Y}_{\bar{s}}$.

(b) Show that if $T = s_y^2\left(S^2_{x(\bar{s})}/s_x^2\right)$ and $T^* = \bar{y} + \left(s_y/s_x\right)\left(\bar{X}_{\bar{s}} - \bar{x}\right)$ then the estimator $\hat{S}_y^2$ reduces to $\hat{S}_I^2 = s_y^2\left(S_x^2/s_x^2\right)$

Hint: Biradar and Singh (1998).

Exercise 3.52. Let $\hat{M}_y$, $\hat{M}_x$ and $\hat{M}_z$ denote, respectively, the median estimators of M_y, M_x and M_z for the y, x, and z values. Study the asymptotic properties of the multivariate ratio type estimator

(a) $\hat{M}_{\text{R}} = \hat{M}_y\left(\frac{M_x}{\hat{M}_x}\right)\left(\frac{M_z}{\hat{M}_z}\right)$, and (b) $\hat{M}_{\text{PW}} = \hat{M}_y\left(\frac{M_x}{\hat{M}_x}\right)^{\alpha}\left(\frac{M_z}{\hat{M}_z}\right)^{\beta}$

were α and β are real constants to be chosen.
Hint: Garcia and Cebrian (2001).

Exercise 3.53. If the relationship between the two variates x and y is given by the relationship $y = a + bx$, $b \neq 0$, then show that in SRSWOR sampling, the sample estimator $\hat{y} = N\bar{y}$ is more efficient than the ratio estimator $\hat{y}_{\text{R}} = \hat{y}(X/\hat{x})$, where $\hat{x} = N\bar{x}$ and X denotes the known total if

$$\left(a^2\overline{X}^2/b^2 s_x^2\right)V\left(\bar{x}^{-1}\right) > \left(\frac{1-f}{n}\right).$$

Hint: Raj (1968)

Exercise 3.54. (a) An SRSWOR sample s of n units is divided at random into g subgroups each of size, rounded to the nearest integer. Let $\hat{y}_{\text{R}}(j) = \hat{y}[j](X/\hat{x}[j])$ and $\hat{y}_{\text{R}} = \hat{y}(X/\hat{x})$ denote, respectively, the ratio estimator of population total based on j^{th} sub-sample and the complete sample, $j = 1, 2, ..., g$. If $\alpha = -(1-f)/(n-g)$, where $f = n/N$, then an unbiased estimator of the population total is given by

$$\hat{y}_{\alpha} = \alpha \sum_{j=1}^{g} \hat{y}_R(j) + (1 - g\alpha)\hat{y}_R$$

(b) Suppose g be the number of independent interpenetrating samples each of n units drawn from the population using SRSWOR sampling. Let y_i and x_i be unbiased estimators of Y and X based on the ith sample. Show that an unbiased estimator of the population mean $\overline{Y}$ is given by

$$\bar{y}_u = \bar{r}\overline{X} + \frac{g(\bar{y} - \bar{r}\,\bar{x})}{g-1}$$

where $\bar{y} = g^{-1}\sum_{i=1}^{g} y_i$, $\bar{x} = g^{-1}\sum_{i=1}^{g} x_i$ and $\bar{r} = g^{-1}\sum_{i=1}^{g}(y_i/x_i)$.

(c) An unbiased regression estimator based on splitting the sample into g groups each of size $m = n/g$ is given by

$$\hat{\bar{y}}_{\text{lru}} = \bar{y} + \bar{b}_g\left(\overline{X} - \bar{x}\right) + \frac{1-f}{g}\left\{\sum_{j=1}^{g} \bar{x}_j b_{-j} - n\bar{x}\bar{b}_g\right\}$$

where $\bar{b}_g = g^{-1}\sum_{j=1}^{g} b_{-j}$, b_{-j} is the sample regression coefficient computed from the sample after omitting the jth group and $\bar{x}_j$ is the sample mean for the jth group.
Hint: Mickey (1959), Rao (1969).

Exercise 3.55. Let $\bar{y}_1 = \bar{y}$, $\bar{y}_2 = \bar{y}(\bar{X}/\bar{x})$ and $\bar{y}_3 = \bar{y}(\bar{x}/\bar{X})$ denote the usual, ratio and product estimator of population mean $\bar{Y}$. Consider a new estimator $\bar{y}_{\text{new}} = \sum_{i=1}^{3}\alpha_t\bar{y}_t$, such that $\sum_{t=1}^{3}\alpha_t = 1$ and $\sum_{t=1}^{3}\alpha_t B(\bar{y}_t) = 0$, where $B(\bar{y}_t)$ denotes the bias in the tth estimator of population mean. Choose α_t such that the $\text{MSE}(\bar{y}_{\text{new}})$ is minimum.
Hint: Singh and Singh (1993a).

Exercise 3.56. An SRSWOR sample of n units is drawn from a finite population of N units, show that the bias in the estimator of the ratio $R = \bar{Y}/\bar{X}$ is given by,

$$B(\hat{R}) = -\left\{\text{Cov}(\hat{R},\bar{x})/\bar{X}\right\}$$

where $\hat{R} = \bar{y}/\bar{x}$, $\bar{y} = n^{-1}\sum_{i=1}^{n} y_i$ and $\bar{x} = n^{-1}\sum_{i=1}^{n} x_i$.

Exercise 3.57. Let y_{ji} be the value of the ith unit in the population $\Omega = [U_1, U_2, ..., U_N]$ for j^{th} variable y_j , $(j = 0, 1, 2)$ of which y_0 is the variable under study and y_1 and y_2 are the auxiliary variables defined over Ω. Let $\bar{y}_j$ be the conventional unbiased estimator of population mean $\bar{Y}_j$ $(j = 0,1,2)$, based on a simple random sample of size n drawn without replacement from the population. When the correlation between the variable under study y_0 and the auxiliary variable y_1 is positive (high), the ratio estimator $\hat{\bar{y}}_1 = \bar{y}_0(\bar{Y}_1/\bar{y}_1)$ has been widely used. If the correlation between study variable y_0 and auxiliary variable y_2 is negative (high), a product estimator $\hat{\bar{y}}_2 = \bar{y}_0(\bar{y}_2/\bar{Y}_2)$ can be used effectively. It is assumed that the population means $\bar{Y}_1$ and $\bar{Y}_2$ of auxiliary variables y_1 and y_2 are known. In large scale sample surveys, we often collect data on more than one auxiliary variable and some of these may be correlated positively with the study variable while others may be negatively correlated. Let the correlation between y_0 and y_1 be positive and between y_0 and y_2 be negative.

(a) Study the following estimators for $\bar{Y}_0$ as

$$\hat{\bar{Y}}_s = u\hat{\bar{y}}_1 + v\hat{\bar{y}}_2 \qquad \text{[Srivastava (1965)]}$$

where u and v are constants such that $u + v = 1$, and

$$\hat{\bar{Y}}_m = \bar{y}_0\left(\bar{Y}_1/\bar{y}_1\right)\left(\bar{y}_2/\bar{Y}_2\right) \qquad \text{[Singh (1967a, 1967b)]}$$

known as ratio cum product estimator of $\bar{Y}_0$.

(b) Consider $\hat{\bar{y}}_0 = \bar{y}_0$, $\hat{\bar{y}}_1 = \bar{y}_0\left(\bar{Y}_1/\bar{y}_1\right)$, $\hat{\bar{y}}_2 = \bar{y}_0\left(\bar{y}_2/\bar{Y}_2\right)$ and $\hat{\bar{y}}_3 = \bar{y}_0\left(\bar{Y}_1/\bar{y}_1\right)\left(\bar{y}_2/\bar{Y}_2\right)$ such that $\hat{\bar{y}}_t \in H$, for $t = 0,1,2,3$, where H denotes the set of all possible estimators of population mean $\bar{Y}_0$. The set H is a linear variety if $\hat{\bar{y}}_h = \sum_{t=0}^{3} h_t \hat{\bar{y}}_t \in H$ for $\sum_{t=0}^{3} h_t = 1$ and $h_t \in R$, where h_t $(t = 0,1,2,3)$ denote the real constants used for reducing the bias and R stands for the set of real numbers. Find the values of h_t $(t = 0,1,2,3)$ such that the variance of the estimator $\hat{\bar{y}}_h$ is minimum and bias is zero.

Hint: Tracy, Singh, and Singh (2001).

Exercise 3.58. Consider any variable $V = \{V : Y \text{ or } X\}$ in a finite population having values : $\Omega = \{V_i : 1 \le i \le N\}$ and let the values of the variable in a sample s be denoted as $s = \{v_i : 1 \le i \le n\}$. Then α^{th} quantile is defined as $V_\alpha = \inf\{v : F_v(v) \ge \alpha\}$, where F_v is the population cumulative distribution function of the variable V. Consider $\hat{F}_Y(y) = n^{-1}\sum_{i=1}^{n}\Delta(y - y_i)$, where $\Delta(a) = 1$ if $a > 0$ and $\Delta(a) = 0$ if $a \le 0$. The inverse of $\hat{F}_Y$ is $\hat{F}_Y^{-1}(\alpha) = \inf\{y : \hat{F}_Y(y) \ge \alpha\} = y_\alpha$. Note that $\alpha = \hat{F}_Y(y_\alpha)$ and that $Y_\alpha = \hat{F}_Y^{-1}\{\hat{F}_Y(Y_\alpha)\}$. Define a new variable $Z_i = \Delta(Y_\alpha - Y_i)$, $i = 1,2,..,N$. If we estimate Y_α as $\hat{z} = \alpha + b(\bar{w} - \bar{W})$, where $W = g(X)$ is correlated to Z variable. Find the minimum variance of the resultant estimator.

Hint: Mak and Kuk (1993)

Exercise 3.59. Let y and x_{ki}, $k = 1, 2, ..., p$, respectively, be the survey variable and the auxiliary variables related to y and the information about the quantiles of the auxiliary variables or distribution functions are known. From the sample of n units from a population of size N we observe (x_{ki}, y_i) where $i \in s$. Consider $Q_{X_k}(\alpha_k)$ for $\alpha_k \in (0.0,0.5) \cup (05,1.0)$ are known and we wish to estimate $Q_Y(\beta)$ with $\beta = 1/2$. Study the asymptotic properties of the following estimators of $Q_Y(\beta)$ as

$$\hat{F}_R = \hat{F}_Y(Q_Y(\beta))\sum_{k=1}^{p} W_k \left\{\frac{F_{X_k}\left(Q_{X_k}(\alpha_k)\right)}{\hat{F}_{X_k}\left(Q_{X_k}(\alpha_k)\right)}\right\}$$

and

$$\hat{F}_D = \hat{F}_Y(Q_Y(\beta)) + \sum_{k=1}^{p} b_k \left\{F_{X_k}\left(Q_{X_k}(\alpha_k)\right) - \hat{F}_{X_k}\left(Q_{X_k}(\alpha_k)\right)\right\}, \quad \text{where } \sum_{k=1}^{p} W_k = 1.$$

Hint: Rueda and Arcos (2002)

Exercise 3.60. Let y and x, respectively, be the survey variable and the auxiliary variable related to each other and the information about the quantiles of the auxiliary variables or distribution functions are known. From the sample of n units from a population of size N we observe (x_i, y_i) where $i \in s$. Consider $F_X(M_x)$ be a known and we wish to estimate $F_Y(M_y)$. Study the asymptotic properties of the following estimators of $F_Y(M_y)$ as

$$\hat{F}_R = \hat{F}_Y(\hat{M}_y)\frac{F_X(M_x)}{\hat{F}_X(\hat{M}_x)}$$

and

$$\hat{F}_D = \hat{F}_Y(\hat{M}_y) + b\{F_X(M_X) - \hat{F}_X(\hat{M}_X)\}.$$

where b is a real constant.

Hint: Rueda , Arcos, and Artes (1998).

Exercise 3.61. Show that under simple random without replacement (SRSWOR) sampling, the following estimator:

$$\hat{y}_{ss} = \frac{n(N-1)}{(n-1)}\left[\bar{y} - \frac{\sum_{i=1}^{n} y_i(x_i - \bar{X})}{\sum_{i=1}^{n}(x_i - \bar{X})^2}(\bar{x} - \bar{X})\right]$$

is unbiased for the population total.

Hint: Singh and Srivastava (1980).

Exercise 3.62. For the following situations, discuss how you may use ratio, product, difference or regression estimators:

(a) Estimate the average number of fish caught per month by marine recreational fishermen at Atlantic and Gulf coasts, assuming that the number of employees per month are known.
(b) Estimate the average amount that graduate students spent on stationary in your class, assuming that weekly sale of stationary at a local shop is known.
(c) Estimate the proportion of time devoted to politics in the television on the national channel of your country, and no further auxiliary information is available.
(d) Estimate the total weight of bones discarded at the time of shipment of usable meat of chickens, assuming that the number of shipments are known.

Exercise 3.63. Consider a population of N identifiable units on which a study variable y is associated with p auxiliary variables $x_1,...,x_p$ whose population variances σ_k^2, $k = 1, 2, ..., p$ are assumed to be known. Let (y_i, x_{ik}), $i = 1, 2, ..., n$ be the observed values in an SRSWR sample on the $(p+1)$ variables. Consider the problem of estimation of finite population variance

$$\sigma_0^2 = \frac{1}{N}\sum_{i=1}^{N}\left(Y_i - \overline{Y}\right)^2, \text{ where } \overline{Y} = \frac{1}{N}\sum_{i=1}^{N} Y_i .$$

Find the bias and variance of the following estimators of σ_0^2 as:

(a) $\hat{\sigma}_1^{*2} = \sum_{ki=1}^{p} w_k \hat{r}_k \sigma_k^2$ where $\hat{r}_k = s_0^2 / s_k^2$, $k = 1, 2, ..., p$, $0 < w_k < 1$ and $\sum_{k=1}^{p} w_k = 1$.

(b) $\hat{\sigma}_2^{*2} = s_0^2\left[\sum_{k=1}^{p} w_k \sigma_k^2 \Big/ \sum_{k=1}^{p} w_k s_k^2\right]$; (c) $\hat{\sigma}_3^{*2} = \prod_{k=1}^{p}\left(w_k \hat{r}_k \sigma_k^2\right)$ (d) $\hat{\sigma}_4^{*2} = \left(\sum_{k=1}^{p} \frac{w_k}{\hat{r}_k \sigma_k^2}\right)^{-1}$

(e) $\hat{\sigma}_5^{*2} = \sum_{k=1}^{p}\left(w_k \hat{r}_k \sigma_k^2\right) + w_{p+1} s_0^2$, where $\sum_{k=1}^{p+1} w_k = 1$.

Hint: Isaki (1983), John (1969), Shukla (1996), Mohanty and Pattanaik (1984), Singh and Singh (2001).

Exercise 3.64. Let Y_i and X_i, $i = 1,2,..,N$ denote the values of the population units for the study variable Y and the auxiliary variable X, respectively. Further, let y_i and x_i, $i = 1,2,..,n$ denote the values of the units included in a sample s_n of size n drawn by simple random sampling without replacement (SRSWOR). The parameter under interest is the population interquartile range of the study variable Y defined by

$$\Theta_y = Q_{3y} - Q_{1y}$$

where Q_{1y} and Q_{3y} denote the first and third population quartiles of Y respectively. The conventional estimator of Θ_y is

$$\hat{\Theta}_y = \hat{Q}_{3y} - \hat{Q}_{1y}$$

where $\hat{Q}_{1y}$ and $\hat{Q}_{3y}$ are sample estimates of Q_{1y} and Q_{3y} respectively. Let $\Theta_x = Q_{3x} - Q_{1x}$ be the known population interquartile range of the auxiliary variable X and its conventional estimator is

$$\hat{\Theta}_x = \hat{Q}_{3x} - \hat{Q}_{1x}$$

where $\hat{Q}_{1x}$ and $\hat{Q}_{3x}$ are sample estimates of first quartile Q_{1x} and third quartile Q_{3x} of X respectively.

(a) Study the asymptotic properties of the following estimators of Θ_y defined as

$$\hat{\Theta}_y^{(1)} = \hat{\Theta}_y\left(\Theta_x / \hat{\Theta}_x\right), \text{ and } \hat{\Theta}_y^{(2)} = \hat{\Theta}_y + \beta_{iqr}\left(\Theta_x - \hat{\Theta}_x\right)$$

where β_{iqr} is a suitably chosen constant such that the variance of $\hat{\Theta}_y^{(2)}$ is minimum.

(b) Study the asymptotic properties of the following estimators of M_y defined as

$$\hat{M}_y^{(1)} = \hat{M}_y\left(\Theta_x / \hat{\Theta}_x\right) \text{ and } \hat{M}_y^{(2)} = \hat{M}_y + \beta_{iqr}\left(\Theta_x - \hat{\Theta}_x\right)$$

where β_{iqr} is a suitably chosen constant such that the variance of $\hat{\Theta}_y^{(2)}$ is minimum.

Hint: Singh and Singh (2002), Singh, Singh, and Puetas (2003b).

Exercise 3.65. Study the asymptotic properties of the following estimators of the population mean, $\bar{Y}$, defined as

(a) $\bar{y}_{\text{SD}} = \bar{y}\left(\dfrac{\bar{X}+C_x}{\bar{x}+C_x}\right)$, where C_x denotes the known population coefficient of variation of the auxiliary variable.

(b) $\bar{y}_{\text{SK}} = \bar{y}\left(\dfrac{\bar{X}+\beta_2(x)}{\bar{x}+\beta_2(x)}\right)$, where $\beta_2(x)$ is the population coefficient of kurtosis of the auxiliary variable.

(c) $\bar{y}_{\text{US1}} = \bar{y}\left(\dfrac{\bar{X}\beta_2(x)+C_x}{\bar{x}\beta_2(x)+C_x}\right)$, and (d) $\bar{y}_{\text{US2}} = \bar{y}\left(\dfrac{\bar{X}C_x+\beta_2(x)}{\bar{x}C_x+\beta_2(x)}\right)$.

Hint: Sisodia and Dwivedi (1981), Singh and Kakran (1993), Upadhyaya and Singh (1999).

Exercise 3.66. Consider a finite population of N units $\Omega:\{u_1, u_2, ..., u_N\}$. Let y and x be the variables taking value y_i and x_i respectively on u_i $(i = 1,2,...,N)$. For estimating $\bar{Y}$, Srivenkataramana (1980) and Bandyopadhyay (1980) proposed a dual to product estimator as

$$\hat{\bar{Y}}_T = \bar{y}\left(\frac{\bar{X}}{\bar{x}^*}\right), \text{ where } \bar{x}^* = \frac{N\bar{X}-n\bar{x}}{N-n}.$$

Using predictive approach advocated by Basu (1971), Srivastava (1983) envisaged another estimator for $\bar{Y}$ as

$$\hat{\bar{Y}}_s = \bar{y}\frac{\left[n\bar{X}+(N-2n)\bar{x}\right]}{\left(N\bar{X}-n\bar{x}\right)} = \bar{y}\left[1-\left(\frac{\bar{X}-\bar{x}}{\bar{x}^*}\right)\right]$$

Let a sample of size n be drawn without replacement from the population and let it be split into g sub-samples each of size $m = n/g$ each, where m is an integer. Let $(\bar{x}_j, \bar{y}_j)$, $j = 1,2,...,g$ be the unbiased estimators of $(\bar{X}, \bar{Y})$ based on j^{th} sub-sample of size m. The Jackknife versions of the above estimators are given by

$$\hat{\bar{Y}}_{s_j} = \bar{y}_j\left(\frac{\bar{X}}{\bar{x}_j^*}\right), \text{ and } \hat{\bar{Y}}_{T_j} = \bar{y}_j\left[1-\frac{\bar{X}-\bar{x}_j}{\bar{x}_j^*}\right], \text{ where } \bar{x}_j^* = \frac{N\bar{X}-n\bar{x}_j}{N-n}, \; j = 1,2,...,g.$$

Let $\hat{\bar{y}}_1 = \dfrac{1}{g}\sum\limits_{j=1}^{g}\hat{\bar{Y}}_{T_j}$, $\hat{\bar{y}}_2 = \dfrac{1}{g}\sum\limits_{j=1}^{g}\hat{\bar{Y}}_{s_j}$, $\hat{\bar{y}}_3 = \hat{\bar{Y}}_T$, $\hat{\bar{y}}_4 = \hat{\bar{Y}}_s$ and $\hat{\bar{y}}_5 = \bar{y}$. Study the asymptotic properties of an unbiased estimator of $\bar{Y}$ defined as

$$\hat{\bar{Y}}_\theta = \sum_{i=1}^{5}\theta_i\hat{\bar{y}}_i, \text{ such that } \sum_{i=1}^{5}\theta_i = 1 \text{ and where } \theta_i,\; i = 1,2,3,4,5 \text{ are real constants.}$$

Hint: Upadhyaya, Singh, and Singh (2003).

Exercise 3.67. Consider an estimator of population mean $\overline{Y}$ as

$$\overline{y}_{\text{ilu}} = \frac{1}{N}\sum_{i\in\Omega}\hat{y}_{i\bullet}$$

where

$$\hat{y}_{i\bullet} = \begin{cases} y_i & \text{if } i \in s, \\ a + bx_i & \text{if } i \in (\Omega - s), \end{cases}$$

and a and b are constants to be chosen such that the mean square error of the estimator is minimum.

(a) Show that the estimator $\overline{y}_{\text{ilu}}$ can be written as

$$\overline{y}_{\text{ilu}} = f\ \overline{y} + (1-f)a + b(\overline{X} - f\ \overline{x})$$

where $f = n/N$.

(b) Show that the minimum mean square error of the estimator $\overline{y}_{\text{ilu}}$ is given by

$$\text{Min.MSE}(\overline{y}_{\text{ilu}}) = \left(\frac{1-f}{n}\right)f^2 S_y^2\left(1-\rho_{xy}^2\right).$$

Hint: Singh (2003a).

Exercise 3.68. Find two complex numbers such that the variance of the difference estimator $\overline{y}_{\text{dif}} = \overline{y} + k(\overline{X} - \overline{x})$ is zero.

Hint: Set $V(\overline{y}_{\text{dif}}) = 0$ and solve the quadratic equation for two complex values of k.

PRACTICAL PROBLEMS

Practical 3.1. A private company ABC was interested in estimating the average amount of real estate farm loans (in \$000) during 1997 in the United States. The company collected information from six states included in an SRSWOR sample as shown below:

State	CT	ME	NE	NY	VA	WI
Nonreal estate farm loans (X)\$	4.373	51.539	3585.406	426.274	188.477	1372.439
Real estate farm loans (Y) \$	7.130	8.849	1337.852	201.631	321.583	1229.572

(a) Given that the average amount of nonreal estate farm loans is \$878.16 (in \$000) for the year 1997, obtain the ratio estimate for the average amount of the real estate farm loans (in \$000) during 1997. Develop an estimator of the mean squared error of the ratio estimator and hence deduce the 95% confidence interval. Verify using information given in population 1 of the Appendix, if the true mean lies in the confidence interval you suggested. Interpret your findings in two lines.

(b) Estimate the average of the real estate farm form loans during 1997 by using Beale's estimator defined as

$$\hat{\bar{y}}_b = \frac{\bar{y}\left\{1 + \frac{(n-1)}{n}\hat{c}_{yx}\right\}}{\bar{x}\left\{1 + \frac{(n-1)}{n}\hat{c}_{xx}\right\}}\bar{X}, \text{ where } \hat{c}_{xy} = \frac{s_{xy}}{\bar{x}\,\bar{y}}, \text{ and } \hat{c}_{xx} = \frac{s_x^2}{\bar{x}^2}.$$

Construct 95% confidence interval for estimating the average real estate farm loans by assuming that the Beale's estimator has the same mean squared error as that of usual ratio estimator.

(c) Also apply Tin's estimator of the population mean given by

$$\hat{\bar{y}}_t = \bar{y}\left(\frac{\bar{X}}{\bar{x}}\right)\left[1 + \frac{(1-f)}{n}\left(\hat{c}_{xy} - \hat{c}_{xx}\right)\right].$$

Construct 95% confidence interval estimate of the average real estate farm loans assuming that the mean squared error of Tin's estimator is same as that of usual ratio estimator.

(d) Estimate the average of the real estimate farm loans using a new estimator of population mean, defined as

$$\hat{\bar{y}}_{\text{new}} = \left(\frac{\sum_{i=1}^{n} x_i y_i}{\sum_{i=1}^{n} x_i^2}\right)\bar{X}.$$

Construct 95% confidence interval estimate for the real estate farm loans assuming that the mean squared error of this new estimator is same as that of usual ratio estimator.

(e) Compare your results obtained in part (a), (b) , (c) and (d), and comment.

Answer:

(a) 484.69, [120.86, 848.52]; (b) $\hat{c}_{xy} = 1.5306$, $\hat{c}_{xx} = 2.1997$, $\hat{\bar{y}}_b = 389.30$;

(c) $\hat{\bar{y}}_t = 437.129$; (d) $\hat{\bar{y}}_{\text{new}} = 389.30$.

Practical 3.2. An instructor suggested to the class to use ratio estimator while estimating the average nonreal estate farm loans by making use of known real estate farm loans. Use the information given in population 1 of the Appendix to support the instructor's statement.

Hint: Discuss the relative efficiency of the ratio estimator with respect to usual estimator.

Practical 3.3. A team of doctors wishes to estimate the average duration of sleep (in minutes) during the night for persons aged 50 years and over in a small village in the United States. It is known that there are 30 persons living in the village aged 50 years and over. Instead of asking everybody, the psychologist selects an SRSWOR sample of five of these people and records the information as given below:

Person no.	18	24	9	12	24
Age X (years)	63	87	84	70	87
Duration of sleep Y (minutes)	405	270	276	360	270

It is well known fact that as the age of a person increases, the sleeping hours decrease. Apply the appropriate method of estimation for estimating the average sleep time in the particular village under study. Find an estimator of the mean squared error of the estimator you used and derive the 95% confidence interval estimate. Assume that the average age 67.267 years of the subjects is known as shown in the population 2 in the Appendix.
Hint: Apply product method of estimation.

Practical 3.4. The age and sleeping time (in minutes) of 30 persons aged 50 and over living in a small village in the United States is given in population 2 of the Appendix. Discuss the relative efficiency of the product estimator under SRSWOR design.

Practical 3.5. The regression method of estimation has been found to be the most efficient method among others. Use it to estimate the average amount of the real estate farm loans (in $000) during 1997 based on an SRSWOR sample of six states selected from the population 1 in the Appendix and is given below

State	AL	FL	MD	OH	TX	VT
Nonreal estate farm loans (X)$	348.334	464.516	57.684	635.774	3520.361	19.363
Real estate farm loans (Y) $	408.978	825.748	139.628	871.720	1248.761	57.747

The average amount of nonreal estate farm loans $878.16 ($000) for the year 1997 is known. Construct a 95% confidence interval estimate and interpret it in non-technical language.

Practical 3.6. Find the relative efficiency of the regression estimator for estimating the average amount of the nonreal estate farm loans during 1997 by using data on the real estate farm loans during 1997 as an auxiliary variable with respect to the ratio estimator of population mean. The real and nonreal estate farm loans (in $000) during 1997 in the 50 states of the United States have been presented in population 1 of the Appendix.

Practical 3.7. The amounts of the real and nonreal estate farm loans (in $000) during 1997 in the 50 states of the United States have been given in population 1 in the Appendix. If we select an SRSWOR sample of six states to collect the required information, find the relative efficiency of the general class estimators which makes use of the known variance of the auxiliary variable at the estimation stage, for estimating average amount of nonreal estate farm loans during 1997 by using information from real estate farm loans during 1997 as an auxiliary variable, with respect to the regression estimator of population mean.

Practical 3.8. The study of the relationship between age and duration of sleep helps a local hospital in developing future policies. If hospital researchers consider 10 patients to collect the information from the population 2 of the Appendix, then what will be the relative bias of the usual estimator of the correlation coefficient under SRSWOR design?

Practical 3.9. A bank manager raised the issue that the variation in the nonreal estate farm loans effects their customers. The bank selected an SRSWOR sample of six states. The manager decides to pick up an estimator from the general class of estimators. Discuss the relative efficiency of the general class estimators which makes use of known variance of the auxiliary variable at the estimation stage, for estimating the finite population variance of the amount of the nonreal estate farm loans during 1997 by using known information about the real estate farm loans during 1997, with respect to the ratio and regression estimators of finite population variance.
Hint: Use information from population 1 of the Appendix.

Practical 3.10. A private company Kitty Management believes that real and nonreal estate farm loans have a cause and effect relationship between them. They want to know the effect of unit change in real estate farm loans on nonreal estate farm loans. A statistician suggests to them that there are two different measuring tools to estimate regression coefficient. Study the relative efficiency of the usual estimator b_1 with respect to the unbiased estimator b_2 of regression coefficient by using complete information available in population 1 of the Appendix.

Practical 3.11. Your instructor provided you an SRSWOR sample of six states from the population 3 of the Appendix as:

State	DE	MD	NC	VT	WA	WI
Year 1994	0.168	0.173	0.088	0.165	0.138	0.230
Year 1996	0.173	0.158	0.117	0.194	0.204	0.241

Find the mistake made by the instructor during the collection of data. Correct the data accordingly. Apply the regression method of estimation for estimating the average price of apple crop during 1996, assuming that average price during 1994 ($0.1708) is known.

Practical 3.12. Jackknife variance estimation technique has become popular due to its simplicity. Suppose we took an SRSWOR sample of six states from the population 1 in the Appendix and gathered the following information:

State	AR	KY	MN	OK	UT	WA
Nonreal estate farm loans (X)$	848.317	557.656	2466.892	1716.087	197.244	1228.607
Real estate farm loans (Y)$	907.700	1045.106	1354.768	612.108	56.908	1100.745

Apply the ratio method of estimation for estimating the average amount of the real estate farm loans (in \$000) during 1997. Also find an estimate of variance of the ratio estimator using Jackknife technique and deduce the 95% confidence intervals. Assume that the average amount \$878.16 of nonreal estate farm loans (in \$000) for the year 1997 is known.

Answers: 95% CIs are $\left[210.830, 1060.399\right]$, and $\left[209.063, 1062.166\right]$.

Practical 3.13. Sometimes the estimation of variance of the regression estimator is difficult, but the Jackknife technique has been found to be the best solution in such situations. Apply the regression method of estimation for estimating the average amount of the real estate farm loans (in \$000) during 1997. Also find estimates of variance of the regression estimator using Jackknife technique and hence deduce the 95% confidence intervals by using the data given below:

State	KY	NC	SD	UT	AL	FL
Nonreal estate farm loans (X)\$	557.656	494.73	1692.817	197.244	348.334	464.516
Real estate farm loans (Y)\$	1045.106	639.571	413.777	56.908	408.978	825.784

The average amount \$878.16 of nonreal estate farm loans (in \$000) for the year 1997 is known.

Practical 3.14. The season average price (in \$) per pound of the commercial apple crop in 36 states of the United States has been given in population 3 of the Appendix. Suppose we selected an SRSWOR sample of six states to collect the required information for 1996. Find the relative efficiency of the regression type estimator of average price in the United States that makes use of past information from two years with respect to the estimator that makes use of past information from only from one year.

Practical 3.15. The real and nonreal estate farm loans (in \$000) during 1997 in 50 states of the United States have been presented in population 1 of the Appendix. Suppose we selected an SRSWOR sample of six states to collect the required information. Find the relative efficiency of the ratio estimator of median, for estimating median of the amount of the nonreal estate farm loans during 1997 by using information from real estate farm loans during 1997, with respect to the usual estimator of population median. Assume that both real and nonreal estate farm loans follow normal distributions.

Practical 3.16. Consider the problem of estimating the average nonreal estate farm loan in the United States. We wish to apply the ratio method of estimation using known information about the real estate farm loans as shown in population 1 in the Appendix. What is the minimum sample size required for relative standard error (RSE) to be equal to 25%?

Practical 3.17. Select an SRSWOR sample of ten states listed in population 1 of the Appendix. Collect information about the real estate farm loans and nonreal estate farm loans from the selected states. Apply the ratio method of estimation for estimating the average nonreal estate farm loans, assuming that the average real estate farm loans in the United States is known.

Practical 3.18. Consider the population under study consists of the students today present in the class. Construct a list of names of the students and assign a number to each of them. Use random number table to select a sample of 20% of the students present in the class. Collect information about their GPA from the students selected in the sample. Assume that the average number of lectures attended by the whole class are known (or can be found from the register). Also collect information about the number of lectures attended by the students selected in the sample. Assuming that the relationship between the average number of lectures attended and GPA of the students is positive, apply the appropriate method(s) to estimate the average GPA of the class, and derive 95% confidence interval estimate(s).

Practical 3.19. John needs an estimate of the average nonreal estate farm loan in the United States. His supervisor has advised him to apply the regression method of estimation using known information about the real estate farm loans as shown in population 1 of the Appendix with minimum relative standard error equal to 10%. What will be John's sample size to meet his supervisor's conditions?

Practical 3.20. Select an SRSWOR sample of the size obtained in Practical 3.19 from population 1 of the Appendix. Collect information about the real estate farm loans and nonreal estate farm loans from the selected states. Apply the regression method of estimation for estimating the average nonreal estate farm loans, assuming that the average real estate farm loans in the United States is known.

Practical 3.21. For estimating the regression coefficient of the amount of the real estate farm loans (in \$000) on the nonreal estate farm loans during 1997 in the United States, we took an SRSWOR sample of six states from the population 1 in the Appendix. From the states selected in the sample, we collected the following information:

State	AK	CA	CT	ME	VA	WI
Nonreal estate farm loans (X)\$	3.433	3928.732	4.373	51.539	188.477	1372.439
Real estate farm loans (Y)\$	2.605	1343.461	7.130	8.849	321.583	1229.572

Assume that the population variance $S_x^2 = 1176526$ of the nonreal estate farm loans (in \$000) for the year 1997 is known. Estimate the regression coefficient β with two different methods. Also find an estimate of the mean squared errors in each case and hence deduce 95% confidence intervals.

Practical 3.22. The amounts of the real and nonreal estate farm loans (in \$000) during 1997 in 50 states of the United States have been given in population 1 in the Appendix. Suppose we selected an SRSWOR sample of six states to collect the required information. Study the relative efficiency of the usual estimator b_1 with respect to the unbiased estimator b_2 of regression coefficient for this population.

Practical 3.23. A team of medical doctors claims that there is a strong negative relationship between the age of a person and the hours of sleep required. Justify their statement based on the information given in population 2 of the Appendix. Also study the relative bias of the usual estimator of the correlation coefficient based on a sample of 10 units.

Practical 3.24. A student in medical college studies the statement made by the team of medical doctors about the negative relationship between age and duration of sleep, and takes an SRSWOR sample of 6 persons from population 2 of the Appendix as given below:

Person No.	17	08	24	30	02	16
Age (x)	78	74	87	72	72	66
Duration of sleep	345	381	270	345	384	420

Estimate the correlation coefficient ρ_{xy} between the age and the duration of sleep and comment. Find an estimate of the mean squared error and derive a 75% confidence interval for the correlation coefficient.

Practical 3.25. Select four different samples each of four units by using SRSWOR sampling from the population 1 of the Appendix. Collect the information for the real and nonreal estate farm loans from the states selected in each sample. The average nonreal estate farm loan is assumed to be known. Obtain four different ratio estimates of the average real estate farm loans from the information collected in the four samples. Pool the information collected in four samples to obtain a pooled ratio estimate of the average real estate farm loans.
(a) Derive an unbiased estimate of the average real estate farm loans.
(b) Construct 95% confidence interval.
Given: Average nonreal estate farm loans \$878.16.
Rules: Use the Pseudo-Random Numbers (PRN) given in Table 1 of the Appendix to select four samples with starting columns as:

Sample Number	Starting Columns
1	3 and 4
2	8 and 9
3	5 and 6
4	4 and 5

Answer: The 95% CI is $[545.36, 904.80]$.

Practical 3.26. Consider the problem of estimating the finite population variance of the duration of sleep in the United States by using the known benchmark as the variance of the auxiliary variable, age. Using information given in population 2 of the Appendix, find the relative efficiency of the product type of estimator $s_p^2 = s_y^2\left(s_x^2 / S_x^2\right)$ with respect to the estimator s_y^2.

Practical 3.27. Estimate the finite population variance of the duration of sleep in the United States based on the information given in the population 2 of the Appendix. What is the minimum sample size required for the estimator $s_p^2 = s_y^2\left(s_x^2 / S_x^2\right)$ to have minimum relative standard deviation 30%?

Practical 3.28. A pilot survey related to population 1 of the Appendix indicates that the values of certain parameters of interest are given as $\lambda_{40} = 3.5822$, $\lambda_{04} = 4.5247$ and $\lambda_{22} = 2.8411$. Use this information to study the relative efficiency of the regression type estimator $s_2^2 = s_y^2 + k\left(S_x^2 - s_x^2\right)$ with respect to the ratio type of estimator $s_I^2 = s_y^2 S_x^2 / s_x^2$ under two situations:

(a) estimate finite population variance of the nonreal estate farm loans using real estate farm loans as known benchmark.

(b) estimate finite population variance of the real estate farm loans using nonreal estate farm loans as known benchmark.

Practical 3.29. A supermarket is worried about the average price of the commercial apple crop during 1996. The correlation between the price during 1996 (Y) with that during 1995 (X_2) and 1994 (X_1) is assumed to be known. Find the minimum sample size, n, required to estimate the average price with relative standard deviation 15% from the population 3 given in the Appendix.

Given: $R^2_{y\bullet x_1,x_2} = 0.8029$.

Practical 3.30. Select an SRSWOR sample of the size developed in Practical 3.29 and select so many states from population 3 given in the Appendix. Collect the information about the season's average price per pound of the apple crop during 1996, 1995, and 1994. Estimate the average price per pound (\$) of the commercial apple crop during 1996 in the United States. Assume that the average price per pound of the commercial apple crop during 1995 and 1994 are accurately known. Apply the regression estimator of population mean with two auxiliary variables. Derive the 95% confidence intervals estimates using (a) superpopulation model approach, and (b) design based approach.

Given: $\overline{X}_1 = 0.1856$ and $\overline{X}_2 = 0.1708$.

Practical 3.31. A private organisation PQR was interested in estimating the average amount of real estate farm loans (in \$000) during 1997 in the United States. The organisation collected information from 15 states, the real estate (y) and nonreal estate (x) farm loans, included in an SRSWOR sample taken from a list of 50 states and observed the following results:

$$\sum_{i=1}^{n} x_i = 18867.089\,,\quad \sum_{i=1}^{n} y_i = 12525.246\,,\quad \sum_{i=1}^{n} x_i^2 = 48780336.98\,,\quad \sum_{i=1}^{n} y_i^2 = 18001501.38\,,\ \text{and}$$

$$\sum_{i=1}^{n} x_i y_i = 26591710.56\,.$$

The average amount of nonreal estate farm loans is \$878.16 (in \$000) for the year 1997 is known.
(a) Obtain a ratio estimate for the average amount of the real estate farm loans (in \$000) during 1997. Develop an estimator of the mean squared error of the ratio estimator and hence deduce the 95% confidence interval.
(b) Obtain a regression estimate for the average amount of the real estate farm loans (in \$000) during 1997. Develop an estimator of the mean squared error of the regression estimator and hence deduce the 95% confidence interval.
(c) Comment on the interval estimates.
Answer: (a) [329.37, 836.58], (b) [460.08, 881.46] .

Practical 3.32. From a population consisting of 100 individuals with their weekly income (y) and age (x), we have the following information:

$$\sum_{i=1}^{N} x_i = 35{,}00\,,\quad \sum_{i=1}^{N} y_i = 50{,}000\,,\quad \sum_{i=1}^{N} x_i^2 = 85{,}0000\,,\quad \sum_{i=1}^{N} y_i^2 = 90{,}000000\,,\ \text{and}$$

$$\sum_{i=1}^{N} x_i y_i = 7500000\,.$$

If we consider a sample of $n = 20$ units, find the following:

(a) Compute the variance of the sample mean estimator;
(b) Compute the mean square error of the ratio estimator;
(c) Find the relative efficiency of the ratio estimator with respect to usual sample mean estimator;
(d) Find the mean square error of the regression estimator;
(e) Find the relative efficiency of the regression estimator with respect to both ratio and sample mean estimator.
Answer: (a) 26262.63; (b) 19872.19; (c) 132.15%; (d) 7900.31; (e) 251.53%, and 332.42%;

Practical 3.33. An estimate of total fertility rate (TFR) in the world is helpful to the policy makers in the world for each country. The fertility rate has been found to have relationship with crude birth rate (CBR), crude death rate (CDR) and infant mortality rate (IMR). The average CBR, CDR and IMR for 96 countries during 1997 have been found to be 26.011, 10.872 and 50.138, respectively. To estimate the total fertility rate (TFR) in the world, a team of consultants takes an SRSWOR sample of 20 countries out of a list of 90 countries as given in the following table:

Sr. No.	Country or area	CBR	CDR	IMR	TFR
3	Algeria	26.5	5.4	42.2	3.16
9	Bangladesh	27.4	10.1	93.0	3.08
14	Bulgaria	12.5	13.6	14.8	1.74
19	Cameroon	41.4	13.9	74.1	5.73
23	China	15.0	6.8	32.6	1.81
28	Czech Republic	13.3	11.0	8.0	1.74
32	Ethiopia	44.3	17.6	117.7	6.75
36	Greece	10.7	9.5	6.6	1.52
37	Guatemala	31.2	6.5	44.6	4.00
42	India	23.5	8.9	63.5	2.90
44	Iran	29.2	5.8	45.1	3.90
48	Kazakhstan	18.8	9.8	59.6	2.29
51	Korea, South	15.8	5.7	7.4	1.80
57	Morocco	25.1	5.1	32.9	3.13
62	Nigeria	41.5	11.7	63.7	5.95
66	Poland	14.2	10.2	11.8	1.93
69	Russia	14.5	14.8	21.6	1.95
76	Spain	12.0	9.1	5.7	1.51
82	Taiwan	14.4	5.7	6.3	1.75
95	Zambia	43.8	25.8	97.6	6.28

Apply the following estimator to estimate the TFR all 96 countries listed in population 8 of the Appendix:

$$\bar{y}_{lr} = \bar{y} + \hat{H}_{11}(u_1 - 1) + \hat{H}_{12}(u_2 - 1) + \hat{H}_{13}(u_3 - 1),$$

where $u_j = \bar{x}_j / \bar{X}_j$ and the estimates $\hat{H}_{1j}$, $j = 1,2,3$ are obtained by solving the of normal equations:

$$\begin{bmatrix} \hat{C}_{x_1}^2, & r_{x_1x_2}\hat{C}_{x_1}\hat{C}_{x_2}, & r_{x_1x_3}\hat{C}_{x_1}\hat{C}_{x_3} \\ r_{x_2x_1}\hat{C}_{x_1}\hat{C}_{x_2}, & \hat{C}_{x_2}^2, & r_{x_2x_3}\hat{C}_{x_2}\hat{C}_{x_3} \\ r_{x_3x_1}\hat{C}_{x_1}\hat{C}_{x_3}, & r_{x_2x_3}\hat{C}_{x_2}\hat{C}_{x_3}, & \hat{C}_{x_3}^2 \end{bmatrix} \begin{bmatrix} \hat{H}_{11} \\ \hat{H}_{12} \\ \hat{H}_{13} \end{bmatrix} = \begin{bmatrix} -\bar{y}r_{yx_1}\hat{C}_y\hat{C}_{x_1} \\ -\bar{y}r_{yx_2}\hat{C}_y\hat{C}_{x_2} \\ -\bar{y}r_{yx_3}\hat{C}_y\hat{C}_{x_3} \end{bmatrix}$$

Practical 3.34. Consider the following population consisting of five ($N = 5$) units P, Q, R, S and T where for each one of the units in the population two variables Y and X are measured.

Units	P	Q	R	S	T
Y_i	9	11	13	16	21
X_i	24	20	19	18	14

(a) Find the population medians M_y and M_x of the study variable and auxiliary variables, respectively.
(b) Select all possible samples of three units ($n = 3$) with SRSWOR sampling.
(c) Find the estimates of the medians $\hat{M}_y$ and M_x from each sample.
(d) Find the exact bias in the estimator $\hat{M}_y$ using the definition.
(e) Find the exact mean square error of the estimator $\hat{M}_y$ using the definition.
(f) Assuming that the median M_x of the auxiliary variable is known, find the product estimates of the median $\hat{M}_{\mathrm{P}} = \hat{M}_y\left(\frac{\hat{M}_x}{M_x}\right)$ from each sample.
(g) Find the exact bias in the product estimator $\hat{M}_P$ using the definition.
(h) Find the exact mean square error of the product estimator $\hat{M}_{\mathrm{P}}$ using the definition.
(i) Find the relative efficiency of the product estimator $\hat{M}_{\mathrm{P}}$ with respect to sample estimator $\hat{M}_y$.

Practical 3.35. Consider the following population consisting of five ($N = 5$) units A, B, C, D, and E where for each one of the units in the population two variables Y and X are measured.

Units	A	B	C	D	E
Y_i	190	210	230	260	290
X_i	142	182	192	202	242

(a) Select all possible samples each of $n = 3$ units.
(b) Estimate the population mean using ratio estimator from each sample. Find the exact bias and exact mean square error of the ratio estimator
(c) Estimate the population mean using the Hartley and Ross (1954) estimator from each sample. Find the exact bias and exact mean square error.
(d) Find the loss in relative efficiency of the Hartley and Ross estimator at the cost of reducing bias.
(e) Give your views on the results.

Practical 3.36. Consider the following population consisting of seven ($N = 7$) units P, Q, R, S, T, U and V, where for each one of the unit in the population two variables Y and X are measured.

Units	P	Q	R	S	T	U	V
Y_i	9	11	13	19	21	26	29
X_i	10	26	28	25	36	37	40

(a) Find the population medians M_y and M_x of the study variable and auxiliary variables, respectively.
(b) Select all possible samples of three units ($n = 3$) with SRSWOR sampling.
(c) Find the estimates of the medians $\hat{M}_y$ and M_x from each sample.
(d) Find the exact bias in the estimator $\hat{M}_y$ using the definition.
(e) Find the exact mean square error of the estimator $\hat{M}_y$ using the definition.
(f) Assuming that the median M_X of the auxiliary variable is known, find the linear regression type estimator of the median $\hat{M}_{\text{LR}} = \hat{M}_y + \gamma\left(M_x - \hat{M}_x\right)$ from each sample, where $\gamma = \dfrac{4(P_{11} - 0.25)\{f_y(M_y)\}}{f_x(M_x)}$ and given that $f(x) = \dfrac{1}{30}$ $\forall\, 10 \le x \le 40$, $f(y) = \dfrac{1}{20}$ $\forall\, 9 \le y \le 29$ and $f(x, y) = \dfrac{1}{961}$ $\forall\, 9 \le x, y \le 40$.
(g) Find the exact bias in the linear regression type estimator $\hat{M}_{LR}$ using the definition.
(h) Find the exact mean square error of the linear regression type estimator $\hat{M}_{LR}$ using the definition.
(i) Find the relative efficiency of the linear regression type estimator $\hat{M}_{LR}$ with respect to sample estimator $\hat{M}_y$.

Hint: $P_{11} = \int_9^{M_y} \int_9^{M_x} f(x, y)dxdy$, Singh and Joarder (2002)

Practical 3.37. From a population consisting of 200 individuals with their weekly income (y) and age (x), we have the following information:

$\sum_{i=1}^{N} x_i = 18867.089$, $\sum_{i=1}^{N} y_i = 12525.246$, $\sum_{i=1}^{N} x_i^2 = 48780336.98$, $\sum_{i=1}^{N} y_i^2 = 18001501.38$, and $\sum_{i=1}^{N} x_i y_i = 26591710.56$.

(I) If we consider 10% of the population as an SRSWOR sample of units, then find the following:

(a) Compute the variance of the sample mean estimator;

(b) Compute the mean square error of the ratio estimator;

(c) Find the relative efficiency of the ratio estimator with respect to usual sample mean estimator;

(d) Compute the mean square error of the regression estimator;

(e) Find the relative efficiency of the regression estimator with respect to the ratio estimator;

(f) Find the relative efficiency of the regression estimator with respect to the sample mean estimator.

(II) If we consider 20% of the population as an SRSWOR sample of units, then again repeat (a) to (f), and report the change in relative efficiencies of the ratio and regression estimator with respect to the usual sample mean estimator. Comment on your findings.

Practical 3.38. The following map shows states in the USA that have beaches. We wish to estimate the average of certain parameters of interest with the beaches across the USA.

Beaches in the USA

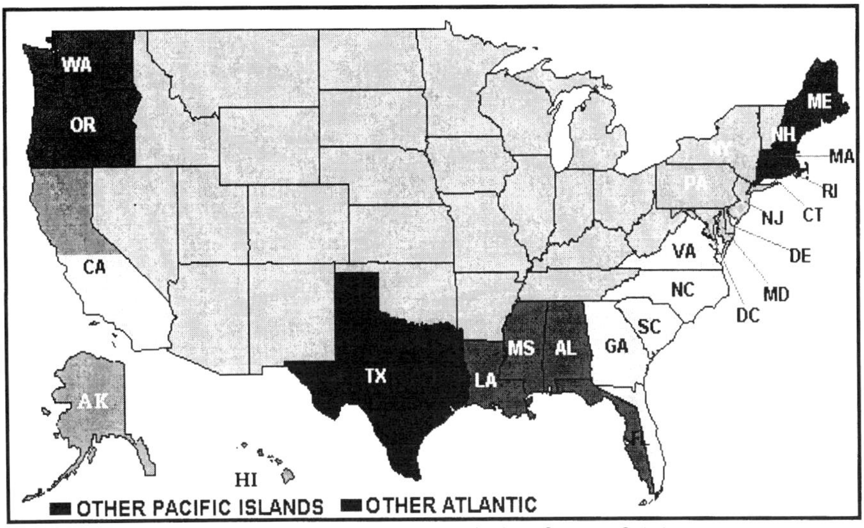

Source: Printed with permission from NOAA

(a) Make a list of all the states that have beaches in the USA and arrange them in alphabetical order. (**Rule:** Use two letter abbreviations for sorting the states)

(b) Select an SRSWOR sample of 5 states from the sorted list of states. (Rule: Start from the first two columns of the Pseudo-Random Number Table given in the Appendix). Collect the information on the number of immigrants admitted during 1996 in these states from the population 9 given in the Appendix.

(c) Estimate the total number of immigrants admitted to these states during 1996 and construct a 95% confidence interval estimate.

(d) Assume that the number of immigrants admitted to all the states shown in the above map are known for the year 1994. Using this information, apply the ratio estimator to estimate the total number of immigrants during 1996. Also construct 95% confidence interval estimate for it.

(e) Apply the regression estimator to estimate the total number of immigrants during 1996, and compare the resultant 95% confidence interval estimate with other two cases.

(f) Use Jackknife to estimate the variance of the ratio estimator, and construct 95% confidence interval estimate for the total number of immigrants during 1996.

(g) Use Jackknife to estimate the variance of the regression estimator, and construct 95% confidence interval estimate for the total number of immigrants during 1996.

Optional:

(h) Collect information from the internet about the current temperature in the states selected in your sample. Estimate the average temperature on the beaches in the USA, and construct a 95% confidence interval estimate.

(i) Collect information about the precipitation (or any other auxiliary variable related to temperature, for example number of visitors) of all the beaches in states of the USA. Use this information to use the ratio (or product) estimator to find the average temperature on the beaches, and construct a 95% confidence interval estimate.

(j) Use the regression estimator to estimate the average temperature and develop 95% confidence interval estimate.

(k) Use the Jackknife to estimate the variance of the ratio (or product) estimator, and construct 95% confidence interval estimate of the average temperature across all beaches in the USA.

(l) Use the Jackknife to estimate the variance of the regression estimator, and construct 95% confidence interval estimate of the average temperature across all beaches in the USA.

(m) Discuss your interval estimates in each case.

4. USE OF AUXILIARY INFORMATION: PROBABILITY PROPORTIONAL TO SIZE AND WITH REPLACEMENT (PPSWR) SAMPLING

4.0 INTRODUCTION

Through the previous chapter we have seen that the proper use of auxiliary information at the estimation stage for estimating any population parameter results in a gain in the efficiency of the resultant estimators. For example, the product and ratio estimators of the population mean remain better than the sample mean when the correlation between the study variable and the auxiliary variable lies in the interval $[-1.0, -0.5)$ and $(+0.5, +1.0]$, respectively. In this chapter we shall show that the auxiliary information can also be used to select a sample which can provide better estimators of population parameters. In other words, the auxiliary information can be used at the sample selection stage as well as at the estimation stage. A sampling scheme with replacement in which each sampling unit has unequal probability of selection, the probability being proportional to the size of the auxiliary variable associated with the particular unit, is called probability proportional to size and with replacement (PPSWR) sampling scheme.

4.1 WHAT IS PPSWR SAMPLING?

Let Y be a study variable and X be an auxiliary variable. For example, consider we want to estimate the population in the villages of a particular district. Then we would choose as our auxiliary variable a variable on which we have information, *e.g.*:

(a) Area of each village of the district (correlation with a study variable = 0.70, say);
(b) Number of households in each village of the district (correlation with a study variable = 0.85, say).

On the basis of the above information, we would choose the auxiliary variable which has maximum correlation with the study variable. Thus the variable at (b) may be a more useful auxiliary variable when selecting a sample using PPSWR sampling.

Let us explain the method of PPSWR sampling with the help of a simple example. Consider a population consists of $N = 4$ units, *viz.*, A, B, C, and D. Consider there are two variables Y and X associated with each unit as follows:

Unit number or identifier	A	B	C	D
Values of a study variable (Y_i)	2	4	5	6
Values of an auxiliary variable (X_i)	4	8	10	12

Obviously we have $X = \sum_{i=1}^{4} X_i = 34$ and $Y = \sum_{i=1}^{4} Y_i = 17$. Consider we wish to draw a sample of $n = 2$ units by using PPSWR sampling. Note that we are using WR sampling, the total number of possible samples will be $N^n = 4^2 = 16$ and are listed in Table 4.1.1. Evidently we have

$$E(\hat{Y}) = \sum_{j=1}^{N^n} p(j)\hat{Y}_j = \frac{16}{1156}\times 17 + \frac{32}{1156}\times 17 + ... + \frac{144}{1156}\times 17 = 17 = Y . \tag{4.1.1}$$

Hence

$$\hat{Y} = \frac{X}{n}\sum_{i=1}^{n}\frac{y_i}{x_i}$$

is an unbiased estimator of population total Y. Now the variance of this estimator is given by

$$V(\hat{Y}) = E\left[\hat{Y} - E(\hat{Y})\right]^2 = \sum_{j=1}^{N^n} p(j)\left(\hat{Y}_j - E(\hat{Y})\right)^2$$

$$= \frac{16}{1156}(17-17)^2 + ... + \frac{144}{1156}(17-17)^2 = 0 . \tag{4.1.2}$$

The possible 16 samples are shown in the following table:

Table 4.1.1. Possible 16 with replacement samples.

Sample number (j)	Sampled units	Sampled Y values	Sampled X Values	Ratio y_i/x_i $i = 1, 2$	Probability of selecting the jth sample, $p(j)$, with PPSWR design	$\hat{Y}_j = \frac{X}{n}\sum_{i=1}^{n}\frac{y_i}{x_i}$
1	(A, A)	$y_1 = 2$ $y_2 = 2$	$x_1 = 4$ $x_2 = 4$	0.5 0.5	$\frac{4}{34}\times\frac{4}{34} = \frac{16}{1156}$	17
2	(A, B)	$y_1 = 2$ $y_2 = 4$	$x_1 = 4$ $x_2 = 8$	0.5 0.5	$\frac{4}{34}\times\frac{8}{34} = \frac{32}{1156}$	17
3	(A, C)	$y_1 = 2$ $y_2 = 5$	$x_1 = 4$ $x_2 = 10$	0.5 0.5	$\frac{4}{34}\times\frac{10}{34} = \frac{40}{1156}$	17
4	(A, D)	$y_1 = 2$ $y_2 = 6$	$x_1 = 4$ $x_2 = 12$	0.5 0.5	$\frac{4}{34}\times\frac{12}{34} = \frac{48}{1156}$	17
5	(B, A)	$y_1 = 4$ $y_2 = 2$	$x_1 = 8$ $x_2 = 4$	0.5 0.5	$\frac{8}{34}\times\frac{4}{34} = \frac{32}{1156}$	17
6	(B, B)	$y_1 = 4$ $y_2 = 4$	$x_1 = 8$ $x_2 = 8$	0.5 0.5	$\frac{8}{34}\times\frac{8}{34} = \frac{64}{1156}$	17
7	(B, C)	$y_1 = 4$ $y_2 = 5$	$x_1 = 8$ $x_2 = 10$	0.5 0.5	$\frac{8}{34}\times\frac{10}{34} = \frac{80}{1156}$	17

Continued......

8	(B, D)	$y_1 = 4$ $y_2 = 6$	$x_1 = 8$ $x_2 = 12$	0.5 0.5	$\frac{8}{34} \times \frac{12}{34} = \frac{96}{1156}$	17
9	(C, A)	$y_1 = 5$ $y_2 = 2$	$x_1 = 10$ $x_2 = 4$	0.5 0.5	$\frac{10}{34} \times \frac{4}{34} = \frac{40}{1156}$	17
10	(C, B)	$y_1 = 5$ $y_2 = 4$	$x_1 = 10$ $x_2 = 8$	0.5 0.5	$\frac{10}{34} \times \frac{8}{34} = \frac{80}{1156}$	17
11	(C, C)	$y_1 = 5$ $y_2 = 5$	$x_1 = 10$ $x_2 = 10$	0.5 0.5	$\frac{10}{34} \times \frac{10}{34} = \frac{100}{1156}$	17
12	(C, D)	$y_1 = 5$ $y_2 = 6$	$x_1 = 10$ $x_2 = 12$	0.5 0.5	$\frac{10}{34} \times \frac{12}{34} = \frac{120}{1156}$	17
13	(D, A)	$y_1 = 6$ $y_2 = 2$	$x_1 = 12$ $x_2 = 4$	0.5 0.5	$\frac{12}{34} \times \frac{4}{34} = \frac{48}{1156}$	17
14	(D, B)	$y_1 = 6$ $y_2 = 4$	$x_1 = 12$ $x_2 = 8$	0.5 0.5	$\frac{12}{34} \times \frac{8}{34} = \frac{96}{1156}$	17
15	(D, C)	$y_1 = 6$ $y_2 = 5$	$x_1 = 12$ $x_2 = 10$	0.5 0.5	$\frac{12}{34} \times \frac{10}{34} = \frac{120}{1156}$	17
16	(D, D)	$y_1 = 6$ $y_2 = 6$	$x_1 = 12$ $x_2 = 12$	0.5 0.5	$\frac{12}{34} \times \frac{12}{34} = \frac{144}{1156}$	17

The following graph shows the relationship between Y and X as $X = 2Y$ and $\theta = \tan^{-1}(1/2)$.

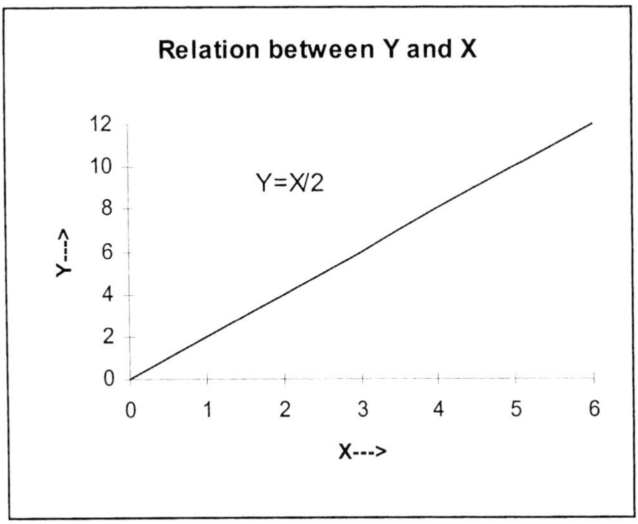

Fig. 4.1.1 Linear relationship with zero intercept.

We observed that the variance of the estimator reduces to zero if an exact linear relationship exist between the study variable and the auxiliary variable. As the direct proportionality between Y and X deviates, the variance of the estimator increases. If Y and X are perfectly correlated and the regression line passes through the origin, then the relationship is of the type as shown in the Fig. 4.1.1. Note that if Y and X are perfectly correlated but the regression line does not pass through the origin then the accuracy of the estimators under a PPSWR design will be decreased. Thus the conditions determining the use of the PPSWR sampling scheme being more efficient than SRSWR sampling are:

(a) there is high positive correlation between Y and X ; and

(b) the line of regression of Y on X passes through the origin.

Let us now consider the following graph:

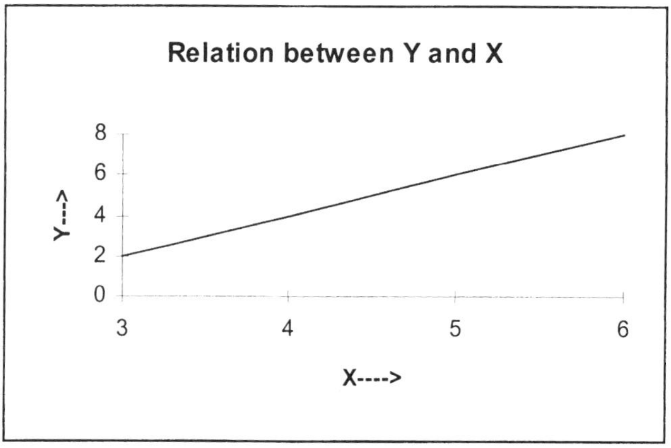

Fig. 4.1.2 Linear relationship with non-zero intercept.

The above regression line shows a perfect positive correlation but it does not pass through the origin and therefore the second condition is not satisfied. In this case, the accuracy of estimators will decrease. To discuss the effect on accuracy, let us again consider a population consisting of $N = 4$ units, *viz.*, A, B, C and D. Assume there are two variables Y and X associated with each other through the relation $Y = 2 + 0.5X$ as follows:

Unit number or identifier	A	B	C	D
Values of a study variable (Y_i)	3	4	5	6
Values of an auxiliary variable (X_i)	2	4	6	8

Obviously we have $X = \sum_{i=1}^{4} X_i = 20$ and $Y = \sum_{i=1}^{4} Y_i = 18$. Consider we wish to draw a sample of $n = 2$ units by using PPSWR sampling. Note that we are using with replacement (WR) sampling, therefore the total number of possible samples will be $N^n = 4^2 = 16$. The possible 16 samples are shown in the Table 4.1.2.

Evidently we have

$$E\left(\hat{Y}\right) = \sum_{j=1}^{N^n} p(j)\hat{Y}_j = \frac{4}{400}\times 30 + \frac{8}{400}\times 25 + \dots\dots + \frac{64}{400}\times 15 = 18 = Y\,. \tag{4.1.3}$$

Hence $\hat{Y} = \frac{X}{n}\sum_{i=1}^{n}(y_i/x_i)$ is again an unbiased estimator of population total Y. However the variance of this estimator is

$$V\left(\hat{Y}\right) = E\left[\hat{Y} - E\left(\hat{Y}\right)\right]^2 = \sum_{j=1}^{N^n} p(j)\left[\hat{Y}_j - E\left(\hat{Y}\right)\right]^2$$

$$= \frac{4}{400}\times(30-18)^2 + \frac{8}{400}\times(25-18)^2 + \dots\dots + \frac{64}{400}(15-18)^2 = 9.667\,. \tag{4.1.4}$$

which is obviously greater than zero.

Thus both conditions (a) and (b) must be satisfied for PPSWR sampling to provide more efficient estimators.

Table 4.1.2. Possible with replacement samples

Sample No. (j)	Sample Units	Sample Y values	Sample X Values	Ratio $= y_i/x_i$ $i = 1,2$	Probability of selecting the jth sample, $p(j)$	$\hat{Y}_j = \frac{X}{n}\sum_{i=1}^{n}\frac{y_i}{x_i}$
1	(A, A)	$y_1 = 3$ $y_2 = 3$	$x_1 = 2$ $x_2 = 2$	1.5 1.5	$\frac{2}{20}\times\frac{2}{20} = \frac{4}{400}$	30.00
2	(A, B)	$y_1 = 3$ $y_2 = 4$	$x_1 = 2$ $x_2 = 4$	1.5 1.0	$\frac{2}{20}\times\frac{4}{20} = \frac{8}{400}$	25.00
3	(A, C)	$y_1 = 3$ $y_2 = 5$	$x_1 = 2$ $x_2 = 6$	1.5 0.833	$\frac{2}{20}\times\frac{6}{20} = \frac{12}{400}$	23.33
4	(A, D)	$y_1 = 3$ $y_2 = 6$	$x_1 = 2$ $x_2 = 8$	1.5 0.75	$\frac{2}{20}\times\frac{8}{20} = \frac{16}{400}$	22.50
5	(B, A)	$y_1 = 4$ $y_2 = 3$	$x_1 = 4$ $x_2 = 2$	1.0 1.5	$\frac{4}{20}\times\frac{2}{20} = \frac{8}{400}$	25.00
6	(B, B)	$y_1 = 4$ $y_2 = 4$	$x_1 = 4$ $x_2 = 4$	1.0 1.0	$\frac{4}{20}\times\frac{4}{20} = \frac{16}{400}$	20.00
7	(B, C)	$y_1 = 4$ $y_2 = 5$	$x_1 = 4$ $x_2 = 6$	1.0 0.833	$\frac{4}{20}\times\frac{6}{20} = \frac{24}{400}$	18.33
8	(B, D)	$y_1 = 4$ $y_2 = 6$	$x_1 = 4$ $x_2 = 8$	1.0 0.75	$\frac{4}{20}\times\frac{8}{20} = \frac{32}{400}$	17.50

Continued.......

9	$(C,\ A)$	$y_1 = 5$ $y_2 = 3$	$x_1 = 6$ $x_2 = 2$	0.833 1.5	$\frac{6}{20}\times\frac{2}{20}=\frac{12}{400}$	23.33
10	$(C,\ B)$	$y_1 = 5$ $y_2 = 4$	$x_1 = 6$ $x_2 = 4$	0.833 1.0	$\frac{6}{20}\times\frac{4}{20}=\frac{24}{400}$	18.33
11	$(C,\ C)$	$y_1 = 5$ $y_2 = 5$	$x_1 = 6$ $x_2 = 6$	0.833 0.833	$\frac{6}{20}\times\frac{6}{20}=\frac{36}{400}$	16.33
12	$(C,\ D)$	$y_1 = 5$ $y_2 = 6$	$x_1 = 6$ $x_2 = 8$	0.833 0.75	$\frac{6}{20}\times\frac{8}{20}=\frac{48}{400}$	15.83
13	$(D,\ A)$	$y_1 = 6$ $y_2 = 3$	$x_1 = 8$ $x_2 = 2$	0.75 1.5	$\frac{8}{20}\times\frac{2}{20}=\frac{16}{400}$	22.50
14	$(D,\ B)$	$y_1 = 6$ $y_2 = 4$	$x_1 = 8$ $x_2 = 4$	0.75 1.0	$\frac{8}{20}\times\frac{4}{20}=\frac{32}{400}$	17.50
15	$(D,\ C)$	$y_1 = 6$ $y_2 = 5$	$x_1 = 8$ $x_2 = 6$	0.75 0.833	$\frac{8}{20}\times\frac{6}{20}=\frac{48}{400}$	15.83
16	$(D,\ D)$	$y_1 = 6$ $y_2 = 6$	$x_1 = 8$ $x_2 = 8$	0.75 0.75	$\frac{8}{20}\times\frac{8}{20}=\frac{64}{400}$	15.00

Remark 4.1.1. (a) The probability of selecting the ith unit in a particular sample by using PPSWR sampling is given by $P_i = X_i/X$ where $X = \sum_{i=1}^{N} X_i$ for $i = 1,2,...,N$.

(b) The probability of selecting any particular sample s of n units from a population Ω of N units by using PPSWR sampling is given by

$$p(s) = P_1 P_2 ....P_n = \prod_{i=1}^{n} P_i\ , \quad \text{where } s = 1,2,3,...,N^n . \tag{4.1.5}$$

There are several methods of selecting a sample by using PPSWR sampling, but we will discuss here only two methods:

(a) Cumulative total method; (b) Lahiri's method.

4.1.1 CUMULATIVE TOTAL METHOD

Consider we have N serial numbers or identifiable labels of the units in a population Ω. Consider we wish to select a sample s of size n with probability proportional to the size of the auxiliary variable X. The first step is to prepare a table of cumulative totals on the auxiliary variable as shown below:

Cumulative Total Table			
Sr. No.	Auxiliary Variable (X_i)	Cumulative Totals	Abbreviation
1	X_1	X_1	T_1
2	X_2	$X_1 + X_2$	T_2
3	X_3	$X_1 + X_2 + X_3$	T_3
.			
N	X_N	$X_1 + X_2 + ... + X_N$	T_N
Total	$X = \sum_{i=1}^{N} X_i$	**Remark:** Note that $X = T_N$	

The second step is to select n random numbers $(R_i,\ i = 1, 2, ..., n)$, say, between 0 and T_N. Consider the first random number selected is R_1 and now if $T_{i-1} < R_1 \le T_i$, then the ith population unit X_i is selected to be included in the sample. Then a second random number R_2 is selected and again tested in the same manner and the process continued till n units are selected. These n selected units will form a sample s of size n with the PPSWR sampling scheme. A pictorial representation of the method is shown below:

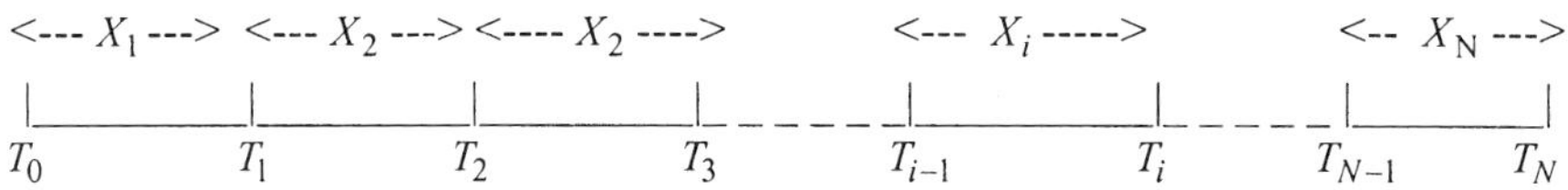

Fig. 4.1.1.1 Cumulative total method.

The probability of any random number falling in any particular interval is proportional to its length, that is,

Probability for the ith unit getting selected in a sample is

$$P_i = \frac{(T_i - T_{i-1})}{T_N} = \left(\frac{X_i}{X}\right) \propto X_i . \quad (4.1.1.1)$$

Note that X is constant, thus the probability of selection of the ith population unit is proportional to the size of the ith unit of the auxiliary variable. Hence this method is called a method of probability proportional to size.

Example 4.1.1. Use the cumulative total method to select a sample of eight units from population 1 given in the Appendix by using nonreal estate farm loans as an auxiliary variable.

Solution. The cumulative totals of the auxiliary variable 'nonreal estate farm loans' are given in column 2 of the Table 4.1.1.1.

Table 4.1.1.1. Cumulative totals (C.T.), random numbers, and states selected.

Sr. No.	C.T.	Random Numbers	Units Selected	Sr. No.	C.T.	Random Number	Units Selected
1	348.334			26	25514.990		
2	351.767			27	29100.390		
3	783.206			28	29117.100		
4	**1631.523**	**01473**	**AR**	29	29117.570		
5	**5560.255**	**04981, 05365**	**CA, CA**	30	29145.080		
6	6466.536			31	29419.120		
7	6470.909			32	29845.390		
8	6514.138			33	30340.120		
9	6978.654			34	31581.490		
10	7519.350			**35**	**32217.260**	**32063**	**OH**
11	7557.417			**36**	**33933.350**	**33313**	**OK**
12	8563.453			37	34504.840		
13	11174.030			38	34803.190		
14	12196.810			39	34803.420		
15	16106.550			40	34884.170		
16	18686.850			41	36576.990		
17	19244.510			42	36965.860		
18	19650.300			**43**	**40486.220**	**38107**	**TX**
19	19701.840			44	40683.460		
20	19759.530			45	40702.830		
21	19816.000			46	40891.300		
22	20256.520			47	42119.910		
23	**22723.410**	**22650**	**MN**	48	42149.200		
24	23272.960			49	43521.640		
25	**24792.950**	**23626**	**MO**	50	43908.120		

We used the first five columns of the Pseudo-Random Numbers (PRN) given in Table 1 of the Appendix to select eight random numbers between 1 and $T_N = 43908$. These selected random numbers came in the sequence as 01473, 23626, 04981, 32063, 33313, 05365, 22650 and 38107. These random numbers have been shown in the column 3 of the Table 4.1.1.1. The last column of this table shows the states selected in the sample from the population 1 given in the Appendix. It is to be noted that the state CA has been selected twice.

Remark 4.1.1.1. The difficulty in this method is in calculating the cumulative totals when the population size N is too large. Thus for large populations we shall discuss another method called Lahiri's method.

4.1.2 LAHIRI'S METHOD

Lahiri (1951) introduced a new method, which does not need cumulative totals, for selecting a PPSWR sample, but in this method we need to know the maximum value of X_i which we denote by X_0. Sometimes it is not possible to know the maximum value of X_i, *e.g.*, $X =$ Number of errors in a book. In such cases, we choose X_0 to be more than the maximum among all values of X_i. In other words, we choose X_0 such that $X_0 \geq \text{Max}(X_1, X_2, ..., X_N)$. Consider we expect (depending upon the quality of printing) that the maximum number of errors in a book may be 100. Then if we choose any value of X_0 greater than or equal to 100, for example $X_0 = 200$, it will not affect the sampling procedure too much. The steps for selecting a sample by using Lahiri's method are as follows:

(a) Select a random number R_i such that $1 \leq R_i \leq N$.

(b) Select another random number R_j such that $1 \leq R_j \leq X_0$.

(c) Compare the magnitude of the random number R_j with that of X_i. If the magnitude of $R_j \leq X_i$, then the unit with serial number R_i is selected to be included in the sample, otherwise it is rejected. The process is continued till we select a random sample of the desired size.

Remarks 4.1.2.1.

(i) Any draw on which a unit is selected is called the effective draw otherwise it is called ineffective draw.
(ii) One should note that the larger the difference between the exact maximum and X_0, the larger the number of rejections.

Now we have the following theorem:

Theorem 4.1.2.1. With the Lahiri's method of PPSWR sampling, the probability for the ith population unit to get selected on the first effective draw is proportional to the size of the ith unit of the auxiliary variable. In other words, the probability of selecting i^{th} unit in the sample by using PPSWR sampling is given by

$$P_i = \frac{X_i}{X}, \quad \text{where } X = \sum_{i=1}^{N} X_i . \tag{4.1.2.1}$$

Proof. We shall not take into account those draw which are ineffective because units are selected only on the effective draws. There are two possibilities, *i.e.*, either the draw is effective or not effective. Thus if the first draw is effective then the corresponding unit is selected and if the first draw is ineffective then we go to the second draw and so on.

The probability for the ith unit to be selected on the first draw if it is effective is

$$\frac{X_i}{(N X_0)},$$

where $1/N$ is the probability for a particular unit to be selected and X_i/X_0 is the probability of an effective draw.

The probability of the first draw not being effective and consequently the probability for the ith unit not being selected on the first draw is

$$1-\left(\frac{X_i}{X_0}\right).$$

Therefore the probability that any unit among the N units not being selected on the first draw is

$$1-\sum_{i=1}^{N}\frac{1}{N}\frac{X_i}{X_0}=1-\frac{\overline{X}}{X_0}, \quad \text{where } \overline{X}=N^{-1}\sum_{i=1}^{N}X_i .$$

Now the probability for the ith unit to be selected on the second draw (in case the first draw was ineffective and second draw is effective) is

$$\left(1-\frac{\overline{X}}{X_0}\right)\left(\frac{X_i}{NX_0}\right). \tag{4.1.2.2}$$

Similarly, if the first and second draws are ineffective and the third is effective, then the probability for the ith unit to be selected on the third draw

$$\left(1-\frac{\overline{X}}{X_0}\right)^2\left(\frac{X_i}{NX_0}\right)$$

and so on.

Hence the probability that the ith unit will be selected on the first effective draw is

$$\frac{X_i}{NX_0}+\left(1-\frac{\overline{X}}{X_0}\right)\frac{X_i}{NX_0}+\left(1-\frac{\overline{X}}{X_0}\right)^2\frac{X_i}{NX_0}+...$$

$$=\frac{X_i}{NX_0}\left\{1-\left(1-\frac{\overline{X}}{X_0}\right)\right\}^{-1}=\frac{X_i}{X}. \tag{4.1.2.3}$$

Hence the theorem.

Example 4.1.2.1. Use Lahiri's method to select a sample of eight units from population 1 given in the Appendix by using nonreal estate farm loans as an auxiliary variable.

Solution. The maximum value of nonreal estate farm loans is \$3928.732. Therefore we decided to choose $X_0 = 4000$. We started with the first two columns of the Pseudo-Random Numbers (PRN) given in Table 1 of the Appendix to choose a random number R_i between 1 and $N = 50$ and reported it in the second column of

Table 4.1.2.1. We started with 7th to 10th columns to choose another random number R_j between 1 and 4000, and the random numbers so obtained have been presented in the third column of the Table 4.1.2.1. The value of the nonreal estate farm loans (X_i) corresponding to the first random number R_i has been coded in the 4th column. The 5th column of the Table 4.1.2.1 has been devoted for making the decision following Lahiri's instructions.

(a) If $R_j > X_i$ then the pair of selected random numbers (R_i, R_j) is rejected and marked 'R' in the 5th column.

(b) If $R_j < X_i$ then the pair of random numbers (R_i, R_j) is selected and marked 'S' in the 5th column.

Table 4.1.2.1. Lahiri's method of sample selection.

Trial Number	$1 \le R_i \le N$	$1 \le R_j \le X_0$	Value of the nonreal estate farm loans X_i ($)	**Decision** S--Selected R--Rejected
1	01	3399	348.334	R
2	23	0757	2466.892	**S**
3	46	3707	188.477	R
4	04	2194	848.317	R
5	32	3331	426.274	R
6	47	2597	1228.607	R
7	33	0926	494.730	R
8	05	1557	3928.732	**S**
9	22	2958	440.518	R
10	38	0122	298.351	**S**
11	29	3235	0.471	R
12	40	0990	80.750	R
13	46	2111	188.477	R
14	03	0275	431.439	**S**
15	36	3363	1716.083	R
16	27	3900	3585.406	R
17	19	3548	51.539	R
18	29	2605	0.471	R
19	29	0395	0.471	R
20	14	1574	1022.782	R
21	47	2207	1228.607	R
22	22	2013	440.518	R
23	42	0607	388.869	R
24	23	2782	2466.892	R
25	48	2579	29.291	R

Continued......

26	06	3849	906.281	R
27	07	3466	4.373	R
28	42	0270	388.869	S
29	21	3490	56.471	R
30	31	1064	274.035	R
31	31	1101	274.035	R
32	36	3770	1716.087	R
33	16	2300	2580.304	S
34	27	1036	3585.406	S
35	10	3688	540.696	R
36	18	3591	405.799	R
37	26	0747	722.034	R
38	48	2486	29.291	R
39	02	0949	3.433	R
40	44	0635	197.244	R
41	12	2000	1006.036	R
42	49	0905	1372.439	S

Thus this method will include the following states in the sample.

Sampled units	1	2	3	4	5	6	7	8
Population unit	23	05	38	03	42	16	27	49
State	MN	CA	PA	AZ	TN	KS	NE	WI

We have seen that every unit within the sample has a different probability of being selected. Also it follows from (4.1.5) that in the PPSWR sampling scheme, unlike SRSWR or SRSWOR sampling, every sample has different probability of selection from a given population. Now we shall discuss the problem of estimation of the population total using PPSWR sampling.

4.2 ESTIMATION OF POPULATION TOTAL

Assume we selected a sample s of size n using PPSWR sampling from a finite population Ω of N units. Let (y_i, x_i), $i = 1,2,...,n$, denote the values of the study variable Y, and auxiliary variable X, for the n units selected in the sample. Let $P_i = X_i/X$, where $X = \sum_{i=1}^{N} X_i$, denotes the known and positive probability of selecting the ith population unit in the sample for $i = 1,2,...,N$. Following Hansen and Hurwitz (1943) we have the following theorems:

Theorem 4.2.1. An unbiased estimator of population total $Y = \sum_{i=1}^{N} Y_i$ is given by

$$\hat{Y}_{\text{HH}} = \frac{1}{n}\sum_{i=1}^{n}(y_i/P_i). \tag{4.2.1}$$

Proof. This can be proved by two different methods as given below

Method I. We have

$$E\left(\hat{Y}_{\text{HH}}\right) = E\left[\frac{1}{n}\sum_{i=1}^{n}\left(y_i/P_i\right)\right] = \frac{1}{n}\sum_{i=1}^{n}E\left(y_i/P_i\right). \tag{4.2.2}$$

Now y_i/P_i is a random variable and can take the values Y_1/P_1, $Y_2/P_2, ..., Y_N/P_N$. Note that the draws are independent, the probability for the unit with value Y_1/P_1 to be selected is P_1, Y_2/P_2 to be selected is P_2, and so on.

Therefore

$$E\left(y_i/P_i\right) = \sum_{i=1}^{N}P_i\left(Y_i/P_i\right) = \sum_{i=1}^{N}Y_i = Y. \tag{4.2.3}$$

Using (4.2.3) in (4.2.2) we obtain

$$E\left(\hat{Y}_{\text{HH}}\right) = \frac{1}{n}\sum_{i=1}^{n}Y = Y. \tag{4.2.4}$$

Hence the theorem.

Method II. We have

$$E\left(\hat{Y}_{\text{HH}}\right) = E\left[\frac{1}{n}\sum_{i=1}^{n}y_i/P_i\right] = \frac{1}{n}E\left[\sum_{i=1}^{N}r_i\left(Y_i/P_i\right)\right] = \frac{1}{n}\sum_{i=1}^{N}E\left(r_i\right)\left(Y_i/P_i\right) \tag{4.2.5}$$

where r_i is such that

$$r_i = \begin{cases} 1 & \text{if the } i^{th} \text{ unit is selected in the sample,} \\ 0 & \text{otherwise.} \end{cases} \tag{4.2.6}$$

Obviously r_i is a binomial variate and it can take any value between 0 and n depending upon how many times i^{th} unit is being selected in the sample under PPSWR sampling scheme with probability of success P_i.

Thus

$$E\left(r_i\right) = nP_i.$$

Therefore (4.2.5) reduces to

$$E\left(\hat{Y}_{\text{HH}}\right) = \frac{1}{n}\sum_{i=1}^{N}E\left(r_i\right)\left(Y_i/P_i\right) = \frac{1}{n}\sum_{i=1}^{N}nP_i\left(Y_i/P_i\right) = \sum_{i=1}^{N}Y_i = Y. \tag{4.2.7}$$

Hence the theorem.

Theorem 4.2.2. The variance of the estimator $\hat{Y}_{\text{HH}}$ is given by

$$V\left(\hat{Y}_{\text{HH}}\right) = \frac{1}{n}\left[\sum_{i=1}^{N}\left(Y_i^2/P_i\right) - Y^2\right]. \tag{4.2.8}$$

Proof. Again there are two methods to prove this theorem as follows:

Method I. Note that the draws are independent in PPSWR sampling, therefore

$$V\left(\hat{Y}_{\text{HH}}\right) = V\left[\frac{1}{n}\sum_{i=1}^{n}\frac{y_i}{P_i}\right] = \frac{1}{n^2}\sum_{i=1}^{n}V\left(\frac{y_i}{P_i}\right). \tag{4.2.9}$$

Now we have

$$V(y_i/P_i) = E(y_i/P_i)^2 - [E(y_i/P_i)]^2 = \sum_{i=1}^{N} P_i\left(Y_i^2/P_i^2\right) - \left[\sum_{i=1}^{N} P_i(Y_i/P_i)\right]^2$$

$$= \sum_{i=1}^{N}\left(Y_i^2/P_i\right) - Y^2. \tag{4.2.10}$$

Further note that y_i^2/P_i^2 takes value Y_i^2/P_i^2 with probability P_i and y_i/P_i takes value Y_i/P_i with probability P_i. On using (4.2.10) in (4.2.9) we have the theorem.

Method II. Again using the binomial variate r_i defined in (4.2.6) we have

$$V\left(\hat{Y}_{\text{HH}}\right) = V\left(\frac{1}{n}\sum_{i=1}^{n} y_i/P_i\right) = V\left(\frac{1}{n}\sum_{i=1}^{N} r_i Y_i/P_i\right) = \frac{1}{n^2}V\left(\sum_{i=1}^{N} r_i Y_i/P_i\right)$$

$$= \frac{1}{n^2}\left[\sum_{i=1}^{N} V\left(r_i Y_i/P_i\right) + \sum_{i\neq j=1}^{N} \text{Cov}\left(r_i Y_i/P_i,\ r_j Y_j/P_j\right)\right]$$

$$= \frac{1}{n^2}\left[\sum_{i=1}^{N} \frac{Y_i^2}{P_i^2} V(r_i) + \sum_{i\neq j=1}^{N} \frac{Y_i Y_j}{P_i P_j}\text{Cov}\left(r_i,\ r_j\right)\right]. \tag{4.2.11}$$

Now we have

$$V(r_i) = nP_i(1 - P_i) \tag{4.2.12}$$

and

$$\text{Cov}\left(r_i,\ r_j\right) = E\left(r_i r_j\right) - E(r_i)E\left(r_j\right) = E\left(r_i r_j\right) - nP_i nP_j. \tag{4.2.13}$$

Also if E_1 and E_2 denote the expected values for all possible values of r_i and for the given values of r_i, respectively, then

$$E\left(r_i r_j\right) = E_1 E_2\left[r_i r_j | r_i\right] = E_1\left[r_i E_2\left(r_j \mid r_i\right)\right] = E_1\left[r_i(n - r_i)\frac{P_j}{1 - P_i}\right] = \frac{P_j}{(1 - P_i)}E_1\left[nr_i - r_i^2\right]$$

$$= \frac{P_j}{(1 - P_i)}\left[nE_1(r_i) - E_1\left(r_i^2\right)\right] = \frac{P_j}{(1 - P_i)}\left[n.nP_i - \left\{V(r_i) + (nP_i)^2\right\}\right] = n(n-1)P_i P_j. \tag{4.2.14}$$

From (4.2.13) and (4.2.14) we have

$$\text{Cov}\left(r_i,\ r_j\right) = n(n-1)P_i P_j - nP_i nP_j = -nP_i P_j. \tag{4.2.15}$$

Note that

$$Y^2 = \left(\sum_{i=1}^{N} Y_i\right)^2 = \sum_{i=1}^{N} Y_i^2 + \sum_{i\neq j=1}^{N} Y_i Y_j,$$

and putting $V(r_i)$ and $\text{Cov}\left(r_i,\ r_j\right)$ in (4.2.11) we have

$$V\left(\hat{Y}_{\text{HH}}\right) = \frac{1}{n^2}\left[\sum_{i=1}^{N} \frac{Y_i^2}{P_i^2} V(r_i) + \sum_{i\neq j=1}^{N} \frac{Y_i Y_j}{P_i P_j}\text{Cov}\left(r_i,\ r_j\right)\right] = \frac{1}{n^2}\left[\sum_{i=1}^{N} \frac{Y_i^2}{P_i^2} nP_i(1 - P_i) - \sum_{i\neq j=1}^{N} \frac{Y_i Y_j}{P_i P_j} nP_i P_j\right]$$

$$= \frac{1}{n}\left[\sum_{i=1}^{N} \frac{Y_i^2}{P_i}(1 - P_i) - \sum_{i\neq j=1}^{N} Y_i Y_j\right] = \frac{1}{n}\left[\sum_{i=1}^{N} \frac{Y_i^2}{P_i} - Y^2\right].$$

Hence the theorem.

Theorem 4.2.3. An unbiased estimator of $V(\hat{Y}_{HH})$ is given by

$$\hat{v}(\hat{Y}_{HH}) = \frac{1}{n(n-1)}\left[\sum_{i=1}^{n}\frac{y_i^2}{P_i^2} - n\hat{Y}_{HH}^2\right]. \qquad (4.2.16)$$

Proof. We have

$$E[\hat{v}(\hat{Y}_{HH})] = E\left[\frac{1}{n(n-1)}\left\{\sum_{i=1}^{n}\frac{y_i^2}{P_i^2} - n\hat{Y}_{HH}^2\right\}\right] = \frac{1}{(n-1)}E\left[\frac{1}{n}\left(\sum_{i=1}^{n}\frac{y_i^2}{P_i^2}\right) - \frac{n}{n}\hat{Y}_{HH}^2\right]$$

$$= \frac{1}{(n-1)}\left[\frac{1}{n}E\left(\sum_{i=1}^{n}\frac{y_i^2}{P_i^2}\right) - E\left(\hat{Y}_{HH}^2\right)\right] = \frac{1}{(n-1)}\left[\left(\sum_{i=1}^{N}\frac{Y_i^2}{P_i}\right) - \left\{V(\hat{Y}_{HH}) + \left(E(\hat{Y}_{HH})\right)^2\right\}\right]$$

$$= \frac{1}{(n-1)}\left[\left(\sum_{i=1}^{N}\frac{Y_i^2}{P_i} - Y^2\right) - \frac{1}{n}\left(\sum_{i=1}^{N}\frac{Y_i^2}{P_i} - Y^2\right)\right] = \frac{1}{n}\left(\sum_{i=1}^{N}\frac{Y_i^2}{P_i} - Y^2\right).$$

Hence the theorem.

Remark 4.2.1. If $P_i = 1/N \quad \forall \quad i = 1,2,..., N$, then the results of PPSWR sampling reduce to those of the SRSWR sampling.

Example 4.2.1. For estimating the total amount of the real estate farm loans (in \$000) during 1997 in the United States we took a PPSWR sample of eight states from the population 1 given in the Appendix by using Lahiri's method. From the states selected in the sample we gathered the following information.

State	AZ	CA	KS	MN	NE	PA	TN	WI
Nonreal estate farm loans, X	431.439	3928.732	2580.304	2466.892	3585.406	298.351	388.869	1372.439
Real estate farm loans, Y	54.633	1343.461	1049.834	1354.768	1337.852	756.169	553.266	1229.572

Assume that the total amount \$43908.12 of nonreal estate farm loans (in \$000) for the year 1997 is known. Apply the Hansen and Hurwitz (1943) estimator for estimating the total amount of the real estate farm loans (in \$000) during 1997 in the United States. Also find an estimator of variance of the used estimator and hence deduce a 95% confidence interval.

Solution. We are given $X = 43908.12$ and $P_i = x_i/X$, therefore we have

Sampled Units	y_i	x_i	$P_i = \frac{x_i}{X}$	$\frac{y_i}{P_i}$	$\frac{y_i^2}{P_i^2}$
AZ	54.633	431.439	0.009825950	5560.072965	30914411.4
CA	1343.461	3928.732	0.089476206	15014.729130	225442090.8
KS	1049.834	2580.304	0.058765987	17864.653640	319145849.8

Continued......

MN	1354.768	2466.892	0.056183048	24113.465820	581459233.7
NE	1337.852	3585.406	0.081657015	16383.797580	268428823.3
PA	756.169	298.351	0.006794894	111284.893300	12384327470.0
TN	553.266	388.869	0.008856426	62470.574720	3902572706.0
WI	1229.572	1372.439	0.031257066	39337.409480	1547431784.0
			Sum	292029.596600	19259722369.0

Thus an estimate of population total Y (say) is given by

$$\hat{Y}_{\text{HH}} = \frac{1}{n}\sum_{i=1}^{n}\frac{y_i}{P_i} = \frac{1}{8}\times 292029.5966 = 36503.7\,.$$

An estimator of $V\left(\hat{Y}_{\text{HH}}\right)$ is given by

$$\hat{v}\left(\hat{Y}_{\text{HH}}\right) = \frac{1}{n(n-1)}\left[\sum_{i=1}^{n}\frac{y_i^2}{P_i^2} - n\hat{Y}_{\text{HH}}^2\right] = \frac{1}{8(8-1)}\left[19259722369 - 8\times(36503.7)^2\right] = 153563601.9\,.$$

Using Table 2 from the Appendix the 95% confidence interval is given by

$$\hat{Y}_{\text{HH}} \mp t_{\alpha/2}(\text{df} = n-1)\sqrt{\hat{v}\left(\hat{Y}_{\text{HH}}\right)}$$

or $36503.7 \mp 2.365\sqrt{153563601.9}$, or $[7196.43,\ 65810.96]$.

Example 4.2.2. The amounts of the real and nonreal estate farm loans (in \$000) during 1997 in different 50 states of the United States have been presented in population 1 of the Appendix. Consider we selected a PPSWR sample of eight states to collect the required information. Find the relative efficiency of the PPSWR sampling estimator for estimating total amount of the real estate farm loans during 1997 by using information on the nonreal estate farm loans during 1997 with respect to the ratio estimator of population total based on SRSWR sample.

Solution. From the description of the population we have $N = 50$,
Y_i = Amount (in \$000) of the real estate farm loans in different states during 1997,
X_i = Amount (in \$000) of the nonreal estate farm loans in different states during 1997,
$\bar{Y} = 555.43$, $\bar{X} = 878.16$, $S_y^2 = 342021.5$, $C_x^2 = 1.5256$, $C_y^2 = 1.1086$, and $\rho_{xy} = 0.8038$.
From Chapter 3 under SRSWR sampling $(f = 0)$ the MSE of the ratio estimator

$$\hat{Y}_{\text{R}} = N\bar{y}\left(\frac{\bar{X}}{\bar{x}}\right)$$

of population total Y is given by

$$\text{MSE}\left(\hat{Y}_{\text{R}}\right) = \frac{N^2}{n}\bar{Y}^2\left[C_y^2 + C_x^2 - 2\rho_{xy}C_xC_y\right]$$

$$= \frac{(2500)(555.43)^2}{8}\left[1.1086 + 1.5256 - 2\times 0.8038\sqrt{1.1086\times 1.5256}\right] = 52399977.81\,.$$

The variance of the estimator $\hat{Y}_{\text{HH}}$ of the population total under PPSWR sampling is given by

$$V\left(\hat{Y}_{\text{HH}}\right)=\frac{1}{n}\left[\sum_{i=1}^{N}\frac{Y_i^2}{P_i}-Y^2\right]=\frac{1}{8}\left[1139152442-(27771.726)^2\right]=45985459.59\,.$$

Thus the percent relative efficiency (RE) of the PPSWR sampling estimator $\hat{Y}_{\text{HH}}$ with respect to the ratio estimator $\hat{Y}_{\text{R}}$ under SRSWR sampling is given by

$$\text{RE}=\frac{\text{MSE}\left(\hat{Y}_{\text{R}}\right)}{V\left(\hat{Y}_{\text{HH}}\right)}\times 100=\frac{52399977.81}{45985459.59}\times 100=113.95\%$$

which shows that the PPSWR estimator is more efficient than the ratio estimator under SRSWR design in this situation.

Theorem 4.2.4. The variance $V\left(\hat{Y}_{\text{HH}}\right)$ is minimum if $P_i \propto Y_i$ or the variance $V\left(\hat{Y}_{\text{HH}}\right)=0$ if $P_i = Y_i/Y$, $i=1,2,....,N$.

Proof. We like to choose P_i to minimise

$$V\left(\hat{Y}_{\text{HH}}\right)=\frac{1}{n}\left(\sum_{i=1}^{N}\frac{Y_i^2}{P_i}-Y^2\right)$$

under the constraint $\sum_{i=1}^{N}P_i=1$.

The Lagrange function is given by

$$L=\sum_{i=1}^{N}\frac{Y_i^2}{P_i}-Y^2+\lambda^2\left(\sum_{i=1}^{N}P_i-1\right).$$

Now

$$\frac{\partial L}{\partial P_i}=0$$

implies that

$$-\frac{Y_i^2}{P_i^2}+\lambda^2=0\,,\quad\text{or}\quad P_i=\left|\frac{Y_i}{\lambda}\right|=\frac{Y_i}{\lambda}\quad\text{if }\ Y_i\geq 0\ \forall\, i\,.$$

Note that $\sum_{i=1}^{N}P_i=1$ which implies that $P_i=Y_i/Y$ and

$$V\left(\hat{Y}_{\text{HH}}\right)=\frac{1}{n}\left[\sum_{i=1}^{N}\frac{Y_i^2}{P_i}-Y^2\right]=\frac{1}{n}\left[\sum_{i=1}^{N}\frac{Y_i^2}{Y_i/Y}-Y^2\right]=\frac{1}{n}\left[Y\sum_{i=1}^{N}Y_i-Y^2\right]=\frac{1}{n}\left[Y^2-Y^2\right]=0\,.$$

Hence the theorem.

Thus the optimum values of the size measures are normed values of Y_i. However, note that the values of Y_i are not known for all the units in the population, thus the values of an auxiliary variable X which is expected to have high positive correlation with the study variable Y can be taken as optimum values of size measures.

As noted it is not necessarily the case that PPSWR sampling is always more efficient than SRSWR the next section has been devoted to study the efficiency conditions.

4.3 RELATIVE EFFICIENCY OF PPSWR SAMPLING WITH RESPECT TO SRSWR SAMPLING

Here we shall discuss two different methods based on superpopulation and cost aspects in survey sampling.

4.3.1 SUPERPOPULATION MODEL APPROACH

We have the following theorem.

Theorem 4.3.1. The PPSWR sampling remains more efficient that SRSWR sampling if the regression line should pass through or near the origin.

Proof. In order to study the relative efficiency of PPSWR sampling with respect to SRSWR sampling we have to consider the superpopulation model approach by following Foreman and Brewer (1971). The superpopulation model is defined by the linear relationship between Y and X as follows

$$Y_i = a + bX_i + e_i\,, \tag{4.3.1.1}$$

where e_i is an error term satisfying the following assumptions:

(a) $E(e_i|X_i) = 0$;

(b) $E(e_i^2|X_i) = a'X_i^g$, $a' > 0$ and g is a constant and in most applications it takes value between 1 and 2. The value of a' is a constant for a given value of g and determines the coefficient of correlation;

(c) $E(e_ie_j|X_iX_j) = 0,\ \forall\ i \neq j$.

To study the relative efficiency let us define

$$V_1 = \text{Variance of the estimator of total under SRSWR} = N^2\frac{\sigma_y^2}{n}\,, \tag{4.3.1.2}$$

and

$$V_2 = \text{Variance of the estimator of total under PPSWR } = \frac{\sigma_z^2}{n}\,, \tag{4.3.1.3}$$

where

$$\sigma_y^2 = \frac{1}{N}\sum_{i=1}^{N}\left(Y_i - \bar{Y}\right)^2 \text{ and } \sigma_z^2 = \sum_{i=1}^{N}\frac{Y_i^2}{P_i} - Y^2\,.$$

Now we have

$$E(V_1) = \frac{N^2}{n}E\left(\sigma^2\right) = \frac{N^2}{n}E\left[\frac{1}{N}\sum_{i=1}^{N}\left(Y_i - \bar{Y}\right)^2\right] = \frac{N}{n}E\sum_{i=1}^{N}\left(Y_i - \bar{Y}\right)^2\,.$$

Also

$$\overline{Y} = \left(\frac{1}{N}\sum_{i=1}^{N} Y_i\right) = \frac{1}{N}\sum_{i=1}^{N}\left(a + bX_i + e_i\right) = a + b\overline{X} + \frac{1}{N}\sum_{i=1}^{N} e_i .$$

Thus we have

$$E(V_1) = \frac{N}{n}\sum_{i=1}^{N} E\left[Y_i - \overline{Y}\right]^2 = \frac{N}{n}\sum_{i=1}^{N} E\left[a + bX_i + e_i - \left(a + b\overline{X} + \frac{1}{N}\sum_{i=1}^{N} e_i\right)\right]^2$$

$$= \frac{N}{n}\sum_{i=1}^{N} E\left[b\left(X_i - \overline{X}\right) + \left(e_i - \overline{e}\right)\right]^2, \quad \text{where } \overline{e} = \frac{1}{N}\sum_{i=1}^{N} e_i$$

$$= \frac{N}{n}\sum_{i=1}^{N}\left[b^2 E\left(X_i - \overline{X}\right)^2 + E\left(e_i - \overline{e}\right)^2 + 2bE\left(X_i - \overline{X}\right)\left(e_i - \overline{e}\right)\right]. \tag{4.3.1.4}$$

Now we have

$$E\left(X_i - \overline{X}\right)^2 = \sigma_x^2, \quad E\left(e_i - \overline{e}\right)^2 = E\left(e_i^2\right) + E\left(\overline{e}^2\right) - 2E\left(\overline{e}e_i\right),$$

$$E\left(e_i^2\right) = E_1E_2\left[e_i^2 | X_i\right] = E_1\left[V\left(e_i\right)| X_i\right] = E_1\left[a' X_i^g\right] = a' X_i^g ,$$

$$E\left(\overline{e}^2\right) = E\left[\frac{1}{N}\sum_{i=1}^{N} e_i\right]^2 = \frac{1}{N^2}\left[\sum_{i=1}^{N} E\left(e_i^2\right) + \sum_{i \ne j=1}^{N} E\left(e_i e_j\right)\right] = \frac{a'}{N^2}\sum_{i=1}^{N} X_i^g = \frac{a'}{N} X_i^g ,$$

$$E\left(e_i\overline{e}\right) = \frac{1}{N} E\left[e_i \sum_{i=1}^{N} e_i\right] = \frac{1}{N} E\left[e_i^2 + \sum_{i \ne j=1}^{N} e_i e_j\right] = \frac{a'}{N} X_i^g .$$

Thus we have

$$E\left(e_i - \overline{e}\right)^2 = a' X_i^g + \frac{1}{N} a' X_i^g - \frac{2}{N} a' X_i^g = a' X_i^g - \frac{1}{N} a' X_i^g$$

and

$$E\left(X_i - \overline{X}\right)\left(e_i - \overline{e}\right) = E\left[e_i\left(X_i - \overline{X}\right) - \overline{e}\left(X_i - \overline{X}\right)\right]$$

$$= E\left[e_i\left(X_i - \overline{X}\right)| X_i\right] = E\left[\left(X_i - \overline{X}\right)E\left(e_i | X_i\right)\right] = 0 .$$

Putting these values in (4.3.1.4) we obtain

$$E(V_1) = \frac{N}{n}\sum_{i=1}^{N}\left[b^2\sigma_x^2 + a' X_i^g - \frac{a'}{N} X_i^g + 2b.0\right]$$

$$= \frac{N^2}{n}\left[b^2\sigma_x^2 + \frac{a'}{N}\sum_{i=1}^{N} X_i^g - \frac{a'}{N^2}\sum_{i=1}^{N} X_i^g\right]. \tag{4.3.1.5}$$

We now consider

$$E(V_2) = \frac{1}{n} E\left[\sum_{i=1}^{N}\frac{Y_i^2}{P_i} - Y^2\right]. \tag{4.3.1.6}$$

Obviously

$$E\left[\sum_{i=1}^{N}\frac{Y_i^2}{P_i}-Y^2\right]=E\left[X\sum_{i=1}^{N}\frac{Y_i^2}{X_i}-\left(\sum_{i=1}^{N}Y_i\right)^2\right]=E\left[X\sum_{i=1}^{N}\frac{(a+bX_i+e_i)^2}{X_i}-\left\{\sum_{i=1}^{N}(a+bX_i+e_i)\right\}^2\right]$$

$$=E\left[\left\{X\sum_{i=1}^{N}\frac{(a+bX_i)^2+e_i^2+2e_i(a+bX_i)}{X_i}\right\}\right.$$

$$\left.-\left\{\sum_{i=1}^{N}(a+bX_i+e_i)^2+\sum_{i\neq j=1}^{N}(a+bX_i+e_i)(a+bX_j+e_j)\right\}\right]$$

$$=X\sum_{i=1}^{N}\frac{(a+bX_i)^2+a'X_i^g}{X_i}-\sum_{i=1}^{N}\left[(a+bX_i)^2+a'X_i^g\right]-\sum_{i\neq j=1}^{N}(a+bX_i)(a+bX_j)$$

$$=X\sum_{i=1}^{N}\frac{(a+bX_i)^2+a'X_i^g}{X_i}-\sum_{i=1}^{N}a'X_i^g-\left\{\sum_{i=1}^{N}(a+bX_i)\right\}^2$$

$$=a^2X\sum_{i=1}^{N}\frac{1}{X_i}+a'X\sum_{i=1}^{N}X_i^{g-1}-a'\sum_{i=1}^{N}X_i^g-N^2a^2. \tag{4.3.1.7}$$

Now we know that the harmonic mean of the values of X_i given by

$$\text{HM}=N\Big/\sum_{i=1}^{N}\frac{1}{X_i}=\tilde{X}(say)$$

which implies that

$$\sum_{i=1}^{N}\frac{1}{X_i}=\frac{N}{\tilde{X}},\quad\text{and}\quad X=N\overline{X}.$$

Therefore (4.3.1.7) becomes

$$E\left[\sum_{i=1}^{N}\frac{Y_i^2}{P_i}-Y^2\right]=N^2a^2\left(\frac{\overline{X}}{\tilde{X}}-1\right)-a'\sum_{i=1}^{N}X_i^g+a'N\overline{X}\sum_{i=1}^{N}X_i^{g-1}. \tag{4.3.1.8}$$

On using (4.3.1.8) in (4.3.1.6) we obtain

$$E(V_2)=\frac{1}{n}\left[N^2a^2\left(\frac{\overline{X}}{\tilde{X}}-1\right)-a'\sum_{i=1}^{N}X_i^g+a'N\overline{X}\sum_{i=1}^{N}X_i^{g-1}\right]. \tag{4.3.1.9}$$

From (4.3.1.9) and (4.3.1.5) the PPSWR sampling will be more efficient than SRSWR if

$$E(V_1)-E(V_2)>0 \tag{4.3.1.10}$$

or if

$$\frac{N^2}{n}\left[b^2\sigma_x^2+\frac{a'}{N}\sum_{i=1}^{N}X_i^g-\frac{a'}{N^2}\sum_{i=1}^{N}X_i^g\right]-\frac{1}{n}\left[N^2a^2\left(\frac{\overline{X}}{\tilde{X}}-1\right)-a'\sum_{i=1}^{N}X_i^g+a'N\overline{X}\sum_{i=1}^{N}X_i^{g-1}\right]>0$$

or if

$$\frac{a'}{N}\left[\sum_{i=1}^{N} X_i^g - \overline{X}\sum_{i=1}^{N} X_i^{g-1}\right] > a^2\left(\frac{\overline{X}}{\tilde{\overline{X}}} - 1\right) - b^2\sigma_x^2 . \tag{4.3.1.11}$$

Now we know that

$$\text{Cov}(X,Y) = \frac{1}{N}\left[\sum_{i=1}^{N} X_i Y_i - N\,\overline{X}\,\overline{Y}\right]. \tag{4.3.1.12}$$

In the present case we have $X = X_i$ and $Y = X_i^{g-1}$, so that

$$\overline{Y} = \frac{1}{N}\sum_{i=1}^{N} X_i^{g-1} .$$

Using this value in (4.3.1.12) we obtain

$$\text{Cov}\left(X_i,\ X_i^{g-1}\right) = \frac{1}{N}\left[\sum_{i=1}^{N} X_i^g - \overline{X}\sum_{i=1}^{N} X_i^{g-1}\right]. \tag{4.3.1.13}$$

From (4.3.1.11) and (4.3.1.13) we find that the condition under which PPSWR sampling is better than SRSWR is

$$a'\text{Cov}\left(X_i,\ X_i^{g-1}\right) > a^2\left(\frac{\overline{X}}{\tilde{\overline{X}}} - 1\right) - b^2\sigma_x^2 . \tag{4.3.1.14}$$

From the relation (4.3.1.14) the following cases are obvious:

(a) If the line passes through the origin, *i.e.*, $a = 0$ and $g > 1$ then the condition (4.3.1.14) is always satisfied;

(b) If the intercept a increases then the value of a' may be found such that the inequality (4.3.1.14) may not be satisfied.

In such situations PPSWR sampling will be less efficient than SRSWR sampling. Hence the theorem.

4.3.2 COST ASPECT

The relative efficiency of PPSWR sampling with respect to SRSWR sampling is

$$\text{RE} = \frac{V\left(\hat{Y}_{\text{SRSWR}}\right)}{V\left(\hat{Y}_{\text{PPSWR}}\right)} = \left\{\frac{N^2}{n}\sigma^2\right\}\Bigg/\left\{\frac{1}{n}\sum_{i=1}^{N} P_i\left(\frac{Y_i}{P_i} - Y\right)^2\right\} = \left\{N^2\sigma^2\right\}\Bigg/\sum_{i=1}^{N} P_i\left(\frac{Y_i}{P_i} - Y\right)^2 \tag{4.3.2.1}$$

Consider the total cost of the survey depends upon the number of draws and total sample size. For simplicity, let us consider the following cost function

$$C = C_0 + nC_1 + C_2\sum_{i=1}^{n} x_i \tag{4.3.2.2}$$

where C_0 is the overhead cost, C_1 is the cost per sampled unit and C_2 is the cost of collecting data per unit size of the sample. In (4.3.1.2) the values of C_0, C_1, C_2 and n are fixed, but $\sum_{i=1}^{n} x_i = n\overline{x}$ changes from sample to sample. Thus

(a) For SRSWR sampling $E\left(\sum_{i=1}^{n} x_i\right) = E(n\bar{x}) = n\bar{X}$, thus the expected cost of the survey is given by

$$E(C) = C^* = C_0 + nC_1 + n\bar{X}C_2. \tag{4.3.2.3}$$

The Lagrange function is then given by

$$L_1 = \frac{N^2}{n}\sigma^2 - \lambda_1\left[C^* - C_0 - n\left(C_1 + \bar{X}C_2\right)\right]. \tag{4.3.2.4}$$

Setting $\partial L_1/\partial n = 0$ implies $n = (N\sigma)\Big/\left(\sqrt{\lambda}\sqrt{C_1 + C_2\bar{X}}\right)$ and substituting it in (4.3.2.3), we have

$$n = \left(C^* - C_0\right)\Big/\left(C_1 + C_2\bar{X}\right). \tag{4.3.2.5}$$

Thus the minimum variance under SRSWR sampling is given by

$$\text{Min.}V\left(\hat{Y}_{\text{SRSWR}}\right) = \frac{N^2}{n}\sigma^2 = \frac{N^2\left(C_1 + C_2\bar{X}\right)}{C^* - C_0}\sigma^2. \tag{4.3.2.6}$$

(b) For PPSWR sampling $E\left(\sum_{i=1}^{n} x_i\right) = n\sum_{i=1}^{N} X_i P_i = nX^{-1}\sum_{i=1}^{N} X_i^2$, thus the expected cost of the survey is given by

$$E(C) = C^* = C_0 + nC_1 + nC_2X^{-1}\sum_{i=1}^{N} X_i^2. \tag{4.3.2.7}$$

The Lagrange function is then given by

$$L_2 = \frac{1}{n}\sum_{i=1}^{N} P_i\left(\frac{Y_i}{P_i} - Y\right)^2 - \lambda_2\left[C^* - C_0 - n\left(C_1 + C_2\sum_{i=1}^{N} X_i^2\Big/X\right)\right]. \tag{4.3.2.8}$$

Setting $\partial L_2/\partial n = 0$ implies

$$n = \sqrt{\sum_{i=1}^{N} P_i\left(\frac{Y_i}{P_i} - Y\right)^2}\Bigg/\left\{\sqrt{\lambda}\sqrt{C_1 + C_2\frac{\sum_{i=1}^{N} X_i^2}{X}}\right\}$$

and substituting it in (4.3.2.7) we obtain

$$n = \frac{\left(C^* - C_0\right)}{\left(C_1 + C_2\sum_{i=1}^{N} X_i^2\Big/X\right)}. \tag{4.3.2.9}$$

Thus the minimum variance under PPSWR sampling is given by

$$\text{Min.}V\left(\hat{Y}_{\text{PPSWR}}\right) = \frac{1}{n}\sum_{i=1}^{N} P_i\left(\frac{Y_i}{P_i} - Y\right)^2 = \left\{\frac{C_1 + C_2\sum_{i=1}^{N} X_i^2\Big/X}{C^* - C_0}\right\}\sum_{i=1}^{N} P_i\left(\frac{Y_i}{P_i} - Y\right)^2. \tag{4.3.2.10}$$

Using (4.3.2.6) and (4.3.2.10) the relative efficiency of PPSWR sampling with respect to SRSWR sampling for the fixed cost of the survey is given by

$$\mathrm{RE}_{\mathrm{Cost}} = \frac{\mathrm{Min.}V\left(\hat{Y}_{\mathrm{SRSWR}}\right)}{\mathrm{Min.}V\left(\hat{Y}_{\mathrm{PPSWR}}\right)} = \left(C_1 + C_2\overline{X}\right)\left(C_1 + C_2\sum_{i=1}^{N} X_i^2 \Big/ X\right)^{-1} \times \mathrm{RE}. \qquad (4.3.2.11)$$

Case I. If $C_2 = 0$ and $C_1 > 0$ then $\mathrm{RE}_{\mathrm{Cost}} = \mathrm{RE}$.

Case II. If $C_1 = 0$ and $C_2 > 0$ then

$$\mathrm{RE}_{\mathrm{Cost}} = \frac{\mathrm{RE}}{\frac{1}{N}\sum_{i=1}^{N} X_i^2 / \overline{X}^2} = \frac{\mathrm{RE}}{\frac{1}{N}\sum_{i=1}^{N}\left(X_i/\overline{X} - 1 + 1\right)^2}$$

$$= \frac{\mathrm{RE}}{\frac{1}{N}\sum_{i=1}^{N}\left[\left(X_i/\overline{X} - 1\right)^2 + 1 + 2\left(X_i/\overline{X} - 1\right)\right]} = \frac{\mathrm{RE}}{1 + C_x^2} \qquad (4.3.2.12)$$

where C_x denotes the coefficient of variation of the auxiliary variable X. Large the variation in the auxiliary variable, less will be the efficiency of PPSWR sampling for the expected cost. If there is no variation in the auxiliary variable, *i.e.*, $C_x = 0$, then $\mathrm{RE}_{\mathrm{Cost}} = \mathrm{RE}$.

4.4 PPSWR SAMPLING: MORE THAN ONE AUXILIARY VARIABLE IS AVAILABLE

We have observed in the previous chapter that for a simple random sample (SRS) of size n drawn with replacement from a population of size N, the usual estimators $\hat{R} = \bar{y}/\bar{x}$ and $\hat{P} = \bar{y}\,\bar{x}$ for estimating the ratio $R = \overline{Y}/\overline{X}$ and product $P = \overline{Y}\,\overline{X}$, respectively, are well known. Prasad (1989) has increased the efficiency of the estimator of R by certain scalar multiplication. The estimators of the product P proposed by Beale (1962), Robson (1957), Tin (1965) and Sahoo (1983) are found to be special cases of the class of estimators suggested by Singh and Sahoo (1989). The efficiency of the estimators of the ratio and product available in the literature is generally low. It is well known that suitable use of auxiliary information in probability sampling results in considerable reduction in variance of the estimators used for estimating a population parameter. Deng and Wu (1987) have increased the efficiency of the estimator of variance of linear regression estimator by making use of prior information about the population mean of the auxiliary variable. Srivastava and Jhajj (1980, 1986) have shown that the efficiency of the estimators for estimating finite population variance and the finite population correlation coefficient, ρ_{xy}, may be increased by making use of some known population parameters of the auxiliary variable at the estimation stage. If there is more than one auxiliary variable, the problem remains as to how the entire information can be utilised in a better way. Multi-variate ratio and regression methods of estimation provide a solution to the problem. Singh (1967a, 1969) has also suggested a method of using two auxiliary variates and the estimators suggested by him would be more efficient than the usual ratio estimator under some conditions. Agarwal and Kumar

(1980) have used two auxiliary variables: one at the stage of selection of the sample and the other at the estimation stage. Then taking the best linear combination of the probability proportional to size (PPS) estimator and the ratio estimator so obtained to estimate the population mean of the study variable. Jhajj and Srivastava (1983) have proposed a class of estimators of the population mean of the study variable when the sampling is done by the method of probability proportional to suitable size variable which is different from the auxiliary variable used at estimation stage. We would like to first discuss the class of estimators proposed by Jhajj and Srivastava (1983) and to show that other estimators available in the literature are the special cases of their class of estimators.

4.4.1 NOTATION AND EXPECTATIONS

Assume that information on two auxiliary variables x_i and z_i (say) for $i = 1,2,...,N$, highly correlated with the study variable y is available. Let a sample of size n be drawn with PPSWR sampling. Let $P_i = z_i \Big/ \sum_{i=1}^{N} z_i$ denote the probability of selection on the basis of known auxiliary variable z_i, $i = 1, 2,..., N$. Let y_i and x_i denote the value of the variable under study y and the auxiliary variable x for the ith unit of the population, and $\overline{Y}$ and $\overline{X}$ denote their population means respectively. Let

$$u_i = \frac{y_i}{NP_i}, \quad v_i = \frac{x_i}{NP_i}, \quad \overline{u}_n = n^{-1}\sum_{i=1}^{n} u_i, \quad \text{and} \quad \overline{v}_n = n^{-1}\sum_{i=1}^{n} v_i .$$

Defining

$$\delta_0 = \frac{\overline{u}_n}{\overline{Y}} - 1, \text{ and } \delta = \frac{\overline{v}_n}{\overline{X}} - 1$$

we have

$$E(\delta_0) = E(\delta) = 0$$

and

$$E(\delta_0^2) = n^{-1}C_u^2, \quad E(\delta^2) = n^{-1}C_v^2, \quad E(\delta_0\delta) = n^{-1}\rho_{uv}C_uC_v,$$

where

$$C_u^2 = \sigma_u^2/\overline{Y}^2 = \text{Relative variance of } \overline{u}_n \text{ with } \sigma_u^2 = \sum_{i=1}^{N} P_i(u_i - \overline{Y})^2,$$

$$C_v^2 = \sigma_v^2/\overline{X}^2 = \text{Relative variance of } \overline{v}_n \text{ with } \sigma_v^2 = \sum_{i=1}^{N} P_i(v_i - \overline{X})^2,$$

and

$$\rho_{uv} = \frac{\sigma_{uv}}{\sigma_u\sigma_v} = \frac{\text{Cov}(\overline{u}_n, \overline{v}_n)}{\sqrt{V(\overline{u}_n)\, V(\overline{v}_n)}}$$ correlation between y_i and x_i for PPSWR sampling,

where $\sigma_{uv} = \sum_{i=1}^{N} P_i(u_i - \overline{Y})(v_i - \overline{X})$.

It may be noted that the results for SRSWR sampling are straightforward if $P_i = 1/N \ \forall \ i$, (*i.e.*, if $z_i = 1 \ \forall \ i$). It is worth noting that the same auxiliary variable cannot be used at both selection stage and estimation stage. In other words if $z_i = x_i \quad \forall \ i = 1, 2, ..., N$, then $E(\delta^2) = 0$ and ρ_{uv} will take indeterminate form. Thus it is compulsory to use one auxiliary variable at selection stage and other at estimation stage.

4.4.2 CLASS OF ESTIMATORS

Jhajj and Srivastava (1983) defined a class of ratio type estimators with PPSWR sampling for estimating population mean $\overline{Y}$ as

$$\bar{y}_{\text{JS}} = \bar{u}_n H(w), \tag{4.4.2.1}$$

where $w = \bar{v}_n / \overline{X}$ and $H(\bullet)$ is a parametric function such that $H(1) = 1$ satisfying certain regularity conditions. Expanding $H(w)$ around the point 1 by using Taylor's second order series we have

$$\bar{y}_{\text{JS}} = \bar{u}_n H(w) = \bar{u}_n H[1 + (w-1)]$$

$$= \bar{u}_n \left[H(1) + (w-1)\frac{\partial H}{\partial w}\Big|_{w=1} + (w-1)^2 \frac{1}{2}\frac{\partial^2 H}{\partial w^2}\Big|_{w=1} + \ldots.. \right]. \tag{4.4.2.2}$$

The expression (4.4.2.2) in terms of δ_0 and δ can easily be written as:

$$\bar{y}_{\text{JS}} = \overline{Y}(1+\delta_0)[\ 1 + \delta H_1 + \delta^2 H_2 + \ldots\ ] = \overline{Y}[\ 1 + \delta_0 + \delta H_1 + \delta^2 H_2 + \delta_0 \delta H_1 + ..\ ] \tag{4.4.2.3}$$

where $H_1 = \frac{\partial H}{\partial w}\Big|_{w=1}$ and $H_2 = \frac{1}{2}\frac{\partial^2 H}{\partial w^2}\Big|_{w=1}$ denote the first and second order partial derivatives of H with respect to w and are assumed to be known and constants at the given point. Thus we have the following theorems.

Theorem 4.4.2.1. The bias, up to terms of order $O(n^{-1})$, in the class of ratio type of estimators in PPSWR sampling is

$$B(\bar{y}_{\text{JS}}) = \frac{\overline{Y}}{n}\left[C_u^2 H_2 + H_1 \rho_{uv} C_u C_v\right]. \tag{4.4.2.4}$$

Proof. On taking expected values on both sides of (4.4.2.3) we have

$$E(\bar{y}_{\text{JS}}) = \overline{Y} E\left[1 + \delta_0 + \delta H_1 + \delta^2 H_2 + \delta_0 \delta H_1 + \ldots.\right]$$

$$= \overline{Y}\left[1 + E(\delta_0) + E(\delta) H_1 + E(\delta^2) H_2 + E(\delta_0 \delta) H_1\right]$$

$$= \overline{Y}\left[1 + \frac{1}{n}\left\{H_2 C_u^2 + H_1 \rho_{uv} C_u C_v\right\}\right].$$

Thus the required bias is given by

$$B(\bar{y}_{\text{JS}}) = E(\bar{y}_{\text{JS}}) - \overline{Y} = \frac{\overline{Y}}{n}\left[C_u^2 H_2 + H_1 \rho_{uv} C_u C_v\right].$$

Hence the theorem.

Theorem 4.4.2.2. The minimum mean squared error of the general class of ratio type estimators $\bar{y}_{\mathrm{JS}}$ for estimating population mean $\bar{Y}$ is given by

$$\text{Min.MSE}(\bar{y}_{\mathrm{JS}}) = \frac{\bar{Y}^2}{n} C_u^2 \left[1 - \rho_{uv}^2\right]. \tag{4.4.2.5}$$

Proof. By the definition of the mean squared error we have

$$\text{MSE}(\bar{y}_{\mathrm{JS}}) = E\left[\bar{y}_{\mathrm{JS}} - \bar{Y}\right]^2 \approx E\left[\bar{Y}\left(1 + \delta_0 + \delta H_1 + ...\right) - \bar{Y}\right]^2 \approx \bar{Y}^2 E\left[\delta_0 + \delta H_1\right]^2$$

$$= \bar{Y}^2 E\left[\delta_0^2 + \delta^2 H_1^2 + 2\delta_0 \delta H_1\right] = \frac{\bar{Y}^2}{n}\left[C_u^2 + H_1^2 C_v^2 + 2\rho_{uv} C_u C_v H_1\right]. \tag{4.4.2.6}$$

On differentiating (4.4.2.6) with respect to H_1 and equating to zero we have

$$H_1 = -\rho_{uv} \frac{C_u}{C_v}. \tag{4.4.2.7}$$

On substituting the value of H_1 from (4.4.2.7) in (4.4.2.6) we obtain the minimum mean squared error given by (4.4.2.5). Hence the theorem.

Remark 4.4.2.1. One can easily observe that the following estimators of the population mean $\bar{Y}$ are special cases of the general class of ratio type estimators under PPSWR sampling:

$$\bar{y}_1 = \alpha \bar{u}_n + (1-\alpha)\bar{u}_n \left(\frac{\bar{X}}{\bar{v}_n}\right); \quad \bar{y}_2 = \bar{u}_n \left(\frac{\bar{v}_n}{\bar{X}}\right)^{\alpha}; \quad \text{and} \quad \bar{y}_3 = \bar{u}_n \left[\frac{\bar{X}}{\alpha \bar{v}_n + (1-\alpha)\bar{X}}\right].$$

The estimator $\bar{y}_1$ was proposed by Agarwal and Kumar (1980), but the estimators $\bar{y}_2$, and $\bar{y}_3$ are the analogues of the estimators proposed by Srivastava (1967) and Walsh (1970) for PPSWR sampling strategies. Estimation of the population mean using supplementary information has been considered by Unam (1995) which shows the comparison of ratio type estimators with PPSWR sampling through simulation.

Thus all ratio type estimators are special cases of the class of estimators $\bar{y}_{\mathrm{JS}}$. In order to cover the regression type estimators, Jhajj and Srivastava (1983) also proposed a wider class of estimators as discussed in the following section.

4.4.3 WIDER CLASS OF ESTIMATORS

The wider class of estimators of population mean $\bar{Y}$ is defined as

$$\bar{y}_{\mathrm{JSW}} = H(\bar{u}_n, w) \tag{4.4.3.1}$$

where $w = \bar{v}_n / \bar{X}$ and $H(\bullet, \bullet)$ is a parametric function such that $H(\bar{Y}, 1) = \bar{Y}$ satisfying certain regularity conditions as:

(a) The first and second order partial derivatives of the function H with respect to $\bar{u}_n$ and w exist and are known constants;

(b) $\frac{\partial H}{\partial \bar{u}_n}|_{(\bar{Y},1)} = 1$.

Expanding $H(\bar{u}_n, w)$ around the point $(\bar{Y}, 1)$ by Taylor's second order series we have

$$\bar{y}_{\text{JSW}} = H\left[\bar{Y} + (\bar{u}_n - \bar{Y}), 1 + (w-1)\right]$$

$$= H(\bar{Y},1) + (\bar{u}_n - \bar{Y})\frac{\partial H}{\partial \bar{u}_n}|_{(\bar{Y},1)} + (w-1)\frac{\partial H}{\partial w}|_{(\bar{Y},1)} + (w-1)^2\frac{\partial^2 H}{2\partial w^2}$$

$$+ (\bar{u}_n - \bar{Y})^2\frac{\partial^2 H}{2\partial \bar{u}_n^2} + (\bar{u}_n - \bar{Y})(w-1)\frac{\partial^2 H}{2\partial \bar{u}_n \partial w} + \ldots$$

Using the regularity conditions we have

$$\bar{y}_{\text{JSW}} = \bar{u}_n + (w-1)H_{01} + (w-1)^2 H_{02} + (\bar{u}_n - \bar{Y})^2 H_{20} + (\bar{u}_n - \bar{Y})(w-1)H_{11} + \ldots \tag{4.4.3.2}$$

where H_{jk} denote the jth and k^{th} derivative of the function H with respect to $\bar{u}_n$ and w, respectively. Then (4.4.3.2) in terms of δ_0 and δ can easily be written as

$$\bar{y}_{\text{JSW}} = \bar{Y}(1+\delta_0) + \delta H_{01} + \delta^2 H_{02} + \bar{Y}^2\delta_0^2 H_{20} + \bar{Y}\delta_0\delta H_{11} + \ldots. \tag{4.4.3.3}$$

Thus we have the following theorems:

Theorem 4.4.3.1. The bias, up to terms of $o(n^{-1})$, in the wider class of estimators $\bar{y}_{\text{JSW}}$ is given by

$$B(\bar{y}_{\text{JSW}}) = \frac{1}{n}\left[C_v^2 H_{02} + \bar{Y}^2 C_u^2 H_{20} + \bar{Y}\rho_{uv}C_u C_v H_{11}\right]. \tag{4.4.3.4}$$

Proof. Taking expected value on both sides of (4.4.3.3) we have

$$E(\bar{y}_{\text{JSW}}) = E\left[\bar{Y}(1+\delta_0) + \delta H_{01} + \delta^2 H_{02} + \bar{Y}^2\delta_0^2 H_{20} + \bar{Y}\delta_0\delta H_{11} + \ldots\right]$$

$$= \bar{Y}\left(1 + E(\delta_0)\right) + E(\delta)H_{01} + E(\delta^2)H_{02} + \bar{Y}^2 E(\delta_0^2)H_{20} + \bar{Y}E(\delta_0\delta)H_{11}$$

$$= \bar{Y} + \frac{1}{n}\left[C_v^2 H_{02} + \bar{Y}^2 C_u^2 H_{20} + \bar{Y}\rho_{uv}C_u C_v H_{11}\right].$$

Thus the required bias is given by

$$B(\bar{y}_{\text{JSW}}) = E(\bar{y}_{\text{JSW}}) - \bar{Y} = \frac{1}{n}\left[C_v^2 H_{02} + \bar{Y}^2 C_u^2 H_{20} + \bar{Y}\rho_{uv}C_u C_v H_{11}\right].$$

Hence the theorem.

Theorem 4.4.3.2. The minimum mean squared error of the wider class of estimators $\bar{y}_{\text{JSW}}$ is given by

$$\text{Min.MSE}(\bar{y}_{\text{JSW}}) = \frac{\bar{Y}^2 C_u^2}{n}\left(1 - \rho_{uv}^2\right). \tag{4.4.3.5}$$

Proof. By the definition of the mean squared error we have

$$\text{MSE}(\bar{y}_{\text{JSW}}) = E\left[\bar{y}_{JSW} - \bar{Y}\right]^2 \approx E\left[\bar{Y}\delta_0 + H_1\delta\right]^2 = E\left[\bar{Y}^2\delta_0^2 + H_1^2\delta^2 + 2H_1\bar{Y}\delta_0\delta\right]$$

$$= \frac{1}{n}\left[\bar{Y}^2 C_u^2 + H_1^2 C_v^2 + 2H_1\rho_{uv}C_u C_v\right]. \tag{4.4.3.6}$$

On differentiating (4.4.3.6) with respect to H_1 and equating to zero we have

$$H_1 = -\overline{Y}\rho_{uv}\frac{C_u}{C_v}. \tag{4.4.3.7}$$

On substituting the value of H_1 in (4.4.3.6) we have

$$\text{Min.MSE}\left(\overline{y}_{\text{JSW}}\right) = \frac{1}{n}\left[\overline{Y}^2C_u^2 + \left(-\overline{Y}\rho_{uv}\frac{C_u}{C_v}\right)^2 C_v^2 + 2\overline{Y}\left(-\overline{Y}\rho_{uv}\frac{C_u}{C_v}\right)\rho_{uv}C_uC_v\right]$$

$$= \frac{1}{n}\left[\overline{Y}^2C_u^2 + \overline{Y}^2\rho_{uv}^2C_u^2 - 2\overline{Y}^2\rho_{uv}^2C_u^2\right] = \frac{\overline{Y}^2C_u^2}{n}\left[1-\rho_{uv}^2\right].$$

Hence the theorem.

Remark 4.4.3.1. It may be noted that estimators of population mean $\overline{Y}$ of the form

$$\overline{y}_{10} = \overline{u}_n + \alpha\left(\overline{X} - \overline{v}_n\right), \quad \text{and} \quad \overline{y}_{11} = \overline{u}_n + \hat{\alpha}\left(\overline{X} - \overline{v}_n\right)$$

where $\hat{\alpha} = \sum_{i=1}^{n} P_i\left(u_i - \overline{u}_n\right)\left(v_i - \overline{v}_n\right) \Big/ \sum_{i=1}^{n} P_i\left(v_i - \overline{v}_n\right)^2$ is an estimator of α, are special cases of the wider class of estimators.

Example 4.4.3.1. Select a PPSWR sample of six units from population-3 given in the Appendix by Lahiri's method using the season average price during 1994 of apple crop as a benchmark or auxiliary variable in the United States. Collect the information on the prices of the apple crop during 1995 and 1996 from population 3 for the units selected in the sample. Use the estimator

$$\overline{y}_{11} = \overline{u}_n + \hat{\alpha}\left(\overline{X} - \overline{v}_n\right), \quad \text{where} \quad \hat{\alpha} = \sum_{i=1}^{n} P_i\left(u_i - \overline{u}_n\right)\left(v_i - \overline{v}_n\right) \Big/ \sum_{i=1}^{n} P_i\left(v_i - \overline{v}_n\right)^2$$

for estimating the average price of the apple crop during 1996 by assuming that the average price $\overline{X} = 0.1856$ during 1995 is known. Also construct the 95% confidence interval for the average price of the apple crop during 1996 in the United States.

Solution. We applied Lahiri's method to select a sample of $n = 6$ units from the population of $N = 36$ units. The maximum value of the price of apples during 1994 is 0.332. Taking $X_0 = 0.333$ We used first two columns of the Pseudo-Random Numbers (PRN) given in Table 1 of the Appendix to select a random number $1 \le R_i \le 36$ and the three columns from 7th to 9th to select a random number $1 \le R_j \le 0.333$.

Then the first seven effective pairs to select a sample of seven units are $(01, .075)$, $(05, .155)$, $(29, .012)$, $(36, .099)$, $(19, .027)$, and $(29, .039)$. Ultimately we have the following sampled information.

Sr. No.	1	2	3	4	5	6	Sum
State	AZ	CT	MO	SC	SC	WI	
Z_i	0.0780000	0.2830000	0.1980000	0.1300000	0.1300000	0.2300000	
P_i	0.0126788	0.0460013	0.0321847	0.0211313	0.0211313	0.0373862	
x_i	0.0710000	0.2760000	0.1600000	0.1260000	0.1260000	0.2410000	
y_i	0.1220000	0.2920000	0.2280000	0.1260000	0.1260000	0.1330000	
u_i	0.2672877	0.1763235	0.1967811	0.1656308	0.1656308	0.0988184	1.0704723
v_i	0.1555527	0.1666620	0.1380920	0.1656308	0.1656308	0.1790618	0.9706301
$P_i u_i^* v_i^*$	-0.0000070	-0.0000005	-0.0000140	-0.0000010	-0.0000010	-0.0000515	-0.0000750
$P_i u_i^{*2}$	0.0001001	0.0000002	0.0000109	0.0000035	0.0000035	0.0002368	0.0003550
$P_i v_i^{*2}$	0.0000005	0.0000011	0.0000180	0.0000003	0.0000003	0.0000112	0.0000314

where

$u_i^* = (u_i - \bar{u}_n)$, $v_i^* = (v_i - \bar{v}_n)$, $P_i = z_i \Big/ \sum_{i=1}^{N} z_i$, with $\sum_{i=1}^{N} z_i = 6.152$ (given), $u_i = y_i/(NP_i)$, and $v_i = x_i/(NP_i)$.

Then we have

$$\bar{u}_n = n^{-1}\sum_{i=1}^{n} u_i = 0.178412\,,\ \bar{v}_n = n^{-1}\sum_{i=1}^{n} v_i = 0.161772\,,\text{ and }\ \hat{\alpha} = \frac{\sum_{i=1}^{n} P_i u_i^* v_i^*}{\sum_{i=1}^{n} P_i v_i^{*2}} = -2.38563\,.$$

Thus an estimator of the average price of the apple crop during 1996 is given by

$$\bar{y}_{11} = \bar{u}_n + \hat{\alpha}\left(\bar{X} - \bar{v}_n\right) = 0.178412 - 2.385636 \times (0.1856 - 0.161772) = 0.121567.$$

Taking

$$\hat{\text{MSE}}(\bar{y}_{11}) = \frac{s_u^2}{n}\left(1 - r_{uv}^2\right)$$

as an estimator of $\text{MSE}(\bar{y}_{11})$ where

$$s_u^2 = \sum_{i=1}^{n} P_i (u_i - \bar{u}_n)^2 = 0.000355$$

and

$$r_{uv} = \frac{\sum_{i=1}^{n} P_i (u_i - \bar{u}_n)(v_i - \bar{v}_n)}{\sqrt{\sum_{i=1}^{n} P_i (u_i - \bar{u}_n)^2 \sum_{i=1}^{n} P_i (v_i - \bar{v}_n)^2}} = \frac{-0.0000750}{\sqrt{0.0003550 \times 0.0000314}} = -0.71036,$$

we have

$$\hat{\text{MSE}}(\bar{y}_{11}) = \frac{0.000355}{6}\left[1 - (-0.71036)^2\right] = 0.00002931.$$

Using Table 2 from the Appendix the 95% confidence interval of the average price of the apple crop during 1996 in the United States is

$$\bar{y}_{11} \mp t_{\alpha/2}(\text{df} = n-2)\sqrt{\hat{\text{MSE}}(\bar{y}_{11})}$$

$$0.12156 \mp 2.776\sqrt{0.00002931}, \quad \text{or} \quad [0.1065, \ 0.1366].$$

4.4.4 PPSWR SAMPLING WITH NEGATIVELY CORRELATED VARIABLES

Consider y and x are highly negatively correlated and $X = \sum_{i=1}^{N} X_i > n.\text{Max}(x_i)$. Srivenkataramana (1980), Sahoo, Sahoo, and Mohanty (1994) considered a transformed auxiliary variable z whose value for X_i is defined as

$$Z_i = \frac{X - nX_i}{N - n} \quad \forall \ X_i \in \Omega \tag{4.4.4.1}$$

and proposed an unbiased estimator of population total Y as

$$\hat{Y}_{\text{SM}} = \frac{1}{n}\sum_{i=1}^{n}\frac{y_i}{P_i^-}, \quad \text{where } P_i^- = Z_i \Big/ \sum_{i=1}^{N} Z_i = Z_i \Big/ \sum_{i=1}^{N} X_i. \tag{4.4.4.2}$$

Thus we have the following theorem:

Theorem 4.4.4.1. The variance of the estimator $\hat{Y}_{\text{SM}}$ is given by

$$V(\hat{Y}_{\text{SM}}) = \frac{1}{n}\left(\sum_{i=1}^{N}\frac{Y_i^2}{P_i^-} - Y^2\right). \tag{4.4.4.3}$$

Proof. Obvious by following Hansen and Hurwitz (1943).

The limitation of this strategy is that in some situations the condition $X > n\text{Max}(x_i)$ may not be satisfied, but this condition is reasonably satisfied in most of the practical situations. A practical example in which this condition holds is given below:

Example 4.4.4.1. Consider that the ranks of the auxiliary variable x_i are used to select the sample of n units with PPSWR sampling from a population of size N. The condition $X > n.\text{Max}(x_i)$ will be satisfied if

$$n < 1 + \left\{\sum_{i=1}^{N-1} X_i\right\} \Big/ \{\text{Max}(x_i)\}. \tag{4.4.4.4}$$

Evidently $\text{Max}(x_i) = N$ and $\sum_{i=1}^{N-1} X_i = \frac{N(N-1)}{2}$, therefore (4.4.4.4) reduces to the condition on sample size

$$n < \frac{(N+1)}{2}. \tag{4.4.4.5}$$

In other words, the condition $X > n.\text{Max}(x_i)$ remains satisfied if the size of the selected sample is less than 50% of the population size. Thus the condition $X > n.\text{Max}(x_i)$ may be satisfied in most practical situations. The literature related to the use of ranks and its benefits in selecting the sample can be had from Wright (1990).

We observe that if the correlation between Y_i and X_i is positive and high and the regression line passes through the origin then PPSWR sampling remains better than the SRSWR sampling. If the correlation between Y_i and X_i is negative then the transformation suggested by Sahoo, Sahoo, and Mohanty (1994) works well.

Example 4.4.4.2. In population 2 the age and duration (minutes) of sleep are negatively correlated. Assuming that the whole population information is known. Discuss the gain in efficiency due to Sahoo, Sahoo, and Mohanty (1994) method of estimation over the Hansen and Hurwitz (1943) estimator. Assume that the sample size consist of five units.

Solution. Defining $P_i = X_i/X$, $Z_i = (X - nX_i)/(N - n)$, and $P_i^- = Z_i/X$. Then from population 2 we have

Person	Age X_i	Duration of sleep, Y_i	$P_i = X_i/X$	Y_i^2/P_i	Z_i	$P_i^- = Z_i/X$	Y_i^2/P_i^-
1	60	492	0.029732408	8141419.200	68.72	0.034053518	7108340.396
2	72	384	0.035678890	4132864.000	66.32	0.032864222	4486824.608
3	55	408	0.027254708	6107715.491	69.72	0.034549058	4818192.083
4	56	465	0.027750248	7791822.321	69.52	0.034449950	6276496.692
5	82	312	0.040634291	2395612.098	64.32	0.031873142	3054107.463
6	78	315	0.038652131	2567128.846	65.12	0.032269574	3074877.918
7	67	420	0.033201189	5313062.687	67.32	0.033359762	5287807.487
8	74	381	0.036669970	3958579.703	65.92	0.032666006	4443793.962
9	84	276	0.041625372	1830037.714	63.92	0.031674926	2404930.663
10	56	465	0.027750248	7791822.321	69.52	0.034449950	6276496.692
11	68	420	0.033696729	5234929.412	67.12	0.033260654	5303563.766
12	70	360	0.034687810	3736182.857	66.72	0.033062438	3919856.115
13	59	435	0.029236868	6472136.441	68.92	0.034152626	5540569.501
14	64	405	0.031714569	5171913.281	67.92	0.033657086	4873416.519
15	53	510	0.026263627	9903430.189	70.12	0.034747275	7485479.179
16	66	420	0.032705649	5393563.636	67.52	0.033458870	5272144.550
17	78	345	0.038652131	3079390.385	65.12	0.032269574	3688458.999
18	63	405	0.031219029	5254007.143	68.12	0.033756194	4859108.191
19	77	330	0.038156591	2854028.571	65.32	0.032368682	3364363.135
20	73	285	0.036174430	2245370.548	66.12	0.032765114	2479008.621
21	55	438	0.027254708	7038930.764	69.72	0.034549058	5552799.656
22	71	360	0.035183350	3683560.563	66.52	0.032963330	3931641.612

Continued……

23	63	390	0.031219029	4872028.571	68.12	0.033756194	4505839.695
24	87	270	0.043111992	1690944.828	63.32	0.031377602	2323313.329
25	61	375	0.030227948	4652151.639	68.52	0.033954410	4141582.750
26	58	375	0.028741328	4892780.172	69.12	0.034251734	4105631.510
27	60	390	0.029732408	5115630.000	68.72	0.034053518	4466498.836
28	70	360	0.034687810	3736182.857	66.72	0.033062438	3919856.115
29	66	390	0.032705649	4650572.727	67.52	0.033458870	4545879.739
30	72	345	0.035678890	3336006.250	66.32	0.0328642220	3621719.692
Sum	2018	11526	1.000000000	143043805.2	2018	1.0000000000	135132599.5

From the above table we have

$X = Z = 2018$, and $Y = 11526$.

Also we have

$$V\left(\hat{Y}_{\text{HH}}\right)=\frac{1}{n}\left(\sum_{i=1}^{N}\frac{Y_i^2}{P_i}-Y^2\right)=\frac{1}{5}\left[143043805.25-11526^2\right]=2039025.84,$$

and

$$V\left(\hat{Y}_{\text{SM}}\right)=\frac{1}{n}\left(\sum_{i=1}^{N}\frac{Y_i^2}{P_i^-}-Y^2\right)=\frac{1}{5}\left[135132599.5-11526^2\right]=456784.69.$$

Thus the percent relative efficiency (RE) of the estimator $\hat{Y}_{\text{SM}}$ with respect to the estimator $\hat{Y}_{\text{HH}}$ is given by

$$\text{RE}=\frac{V\left(\hat{Y}_{\text{HH}}\right)}{V\left(\hat{Y}_{\text{SM}}\right)}\times 100=\frac{2039025.84}{456784.69}\times 100=446.38\%.$$

The next section has been devoted to the situation in which in large scale multipurpose surveys, where the cost involved is quite high, the information collected on several variables of interest is available. Some of these study variables may be positively and highly correlated with the auxiliary information used at the selection stage while others may be poorly correlated. Such a situation is called 'multi-character survey'.

4.5 MULTI-CHARACTER SURVEY

We shall here discuss two cases:

(a) Study variables have poor positive correlation with the selection probabilities.
(b) Study variables have poor positive as well as poor negative correlation with the selection probabilities.

4.5.1 STUDY VARIABLES HAVE POOR POSITIVE CORRELATION WITH THE SELECTION PROBABILITIES

Bansal and Singh (1985) proposed an estimator of population total Y given by

$$\hat{Y}_{\text{BS}} = \frac{1}{n}\sum_{i=1}^{n}\frac{y_i}{P_i^*} \tag{4.5.1.1}$$

where

$$P_i^* = \left(1+\frac{1}{N}\right)^{1-\rho_{xy}}\left(1+P_i\right)^{\rho_{xy}} - 1 . \tag{4.5.1.2}$$

(a) If $\rho_{xy} = 0$ then (4.5.1.2) becomes $P_i^* = 1/N$ and the estimator (4.5.1.1) reduces to the well known estimator owed to Rao (1966a, 1966b), that is $\hat{Y}_{\text{Rao}} = \frac{N}{n}\sum_{i=1}^{n} y_i$.

(b) If $\rho_{xy} = 1$ then (4.5.1.2) becomes $P_i^* = P_i$ and the estimator (4.5.1.1) reduces to the well known estimator of Hansen and Hurwitz (1943), that is $\hat{Y}_{\text{HH}} = \frac{1}{n}\sum_{i=1}^{n}\frac{y_i}{P_i}$.

We now study some further properties of the estimator $\hat{Y}_{\text{BS}}$ in the following theorems:

Theorem 4.5.1.1. The estimator $\hat{Y}_{\text{BS}}$ is not consistent.

Proof. Taking expected values on both sides of (4.5.1.1) we have

$$E\left(\hat{Y}_{\text{BS}}\right) = E\left[\frac{1}{n}\sum_{i=1}^{n}\frac{y_i}{P_i^*}\right] = \sum_{i=1}^{N}\frac{Y_i P_i}{P_i^*} . \tag{4.5.1.3}$$

Thus the bias in the estimator $\hat{Y}_{\text{BS}}$ is given by

$$B\left(\hat{Y}_{\text{BS}}\right) = E\left(\hat{Y}_{\text{BS}}\right) - Y = \sum_{i=1}^{N}\left(\frac{P_i}{P_i^*} - 1\right)Y_i . \tag{4.5.1.4}$$

Evidently $B\left(\hat{Y}_{\text{BS}}\right)$ is not converging to zero as sample size increases. Hence the theorem.

Theorem 4.5.1.2. The variance of the estimator $\hat{Y}_{\text{BS}}$ is given by

$$V\left(\hat{Y}_{\text{BS}}\right) = \frac{1}{n}\left[\sum_{i=1}^{N}\frac{Y_i^2 P_i}{P_i^{*2}} - \left(\sum_{i=1}^{N}\frac{Y_i P_i}{P_i^*}\right)^2\right] . \tag{4.5.1.5}$$

Now we have the following remarks:

Remark 4.5.1.1. On putting $P_i^* = 1/N$ in (4.5.1.5) we have the variance of the estimator of population total proposed by Rao (1966a, 1966b) as follows:

$$V\left(\hat{Y}_{\text{Rao}}\right) = \frac{N^2}{n}\left\{\sum_{i=1}^{N} Y_i^2 P_i - \left(\sum_{i=1}^{N} Y_i P_i\right)^2\right\} . \tag{4.5.1.6}$$

Remark 4.5.1.2. On putting $P_i^* = P_i$ in (4.5.1.5) we have the variance of the estimator of population total proposed by Hansen and Hurwitz (1943) as follows:

$$V\left(\hat{Y}_{\text{HH}}\right)=\frac{1}{n}\left(\sum_{i=1}^{N}\frac{Y_i^2}{P_i}-Y^2\right). \tag{4.5.1.7}$$

Theorem 4.5.1.3. An unbiased estimator for estimating the $V\left(\hat{Y}_{\text{BS}}\right)$ is given by

$$\hat{v}\left(\hat{Y}_{\text{BS}}\right)=\frac{1}{n(n-1)}\left(\sum_{i=1}^{n}\frac{y_i^2}{P_i^{*2}}-n\hat{Y}_{\text{BS}}^2\right). \tag{4.5.1.8}$$

Proof. Taking expected values on both sides of (4.5.1.8) we obtain

$$E\left[\hat{v}\left(\hat{Y}_{\text{BS}}\right)\right]=\frac{E}{n(n-1)}\left(\sum_{i=1}^{n}\frac{y_i^2}{P_i^{*2}}-n\hat{Y}_{\text{BS}}^2\right)=\frac{1}{n(n-1)}\left\{E\left(\sum_{i=1}^{n}\frac{y_i^2}{P_i^{*2}}\right)-nE\left(\hat{Y}_{\text{BS}}^2\right)\right\}$$

$$=\frac{1}{n(n-1)}\left[n\sum_{i=1}^{N}\frac{Y_i^2}{P_i^{*2}}P_i-n\left\{V\left(\hat{Y}_{\text{BS}}\right)+\left(\sum_{i=1}^{N}\frac{Y_iP_i}{P_i^{*2}}\right)^2\right\}\right]$$

$$=\frac{1}{n(n-1)}\left[n\sum_{i=1}^{N}\frac{Y_i^2}{P_i^{*2}}P_i-n\left\{V\left(\hat{Y}_{\text{BS}}\right)+\left(\sum_{i=1}^{N}\frac{Y_iP_i}{P_i^{*2}}\right)^2\right\}\right]$$

$$=\frac{1}{(n-1)}\left[\sum_{i=1}^{N}\frac{Y_i^2}{P_i^{*2}}P_i-\left(\sum_{i=1}^{N}\frac{Y_iP_i}{P_i^{*2}}\right)^2-V\left(\hat{Y}_{\text{BS}}\right)\right]$$

$$=\frac{1}{(n-1)}\left[nV\left(\hat{Y}_{\text{BS}}\right)-V\left(\hat{Y}_{\text{BS}}\right)\right]=V\left(\hat{Y}_{\text{BS}}\right).$$

Hence the theorem.

Theorem 4.5.1.4. An unbiased estimator of bias $B\left(\hat{Y}_{\text{BS}}\right)$ is given by

$$\hat{B}\left(\hat{Y}_{\text{BS}}\right)=\frac{1}{n}\sum_{i=1}^{n}\left(\frac{P_i}{P_i^*}-1\right)\frac{y_i}{P_i}. \tag{4.5.1.9}$$

Proof. Taking expected values on both sides of (4.5.1.9) we have

$$E\left[\hat{B}\left(\hat{Y}_{\text{BS}}\right)\right]=E\left[\frac{1}{n}\sum_{i=1}^{n}\left(\frac{P_i}{P_i^*}-1\right)\frac{y_i}{P_i}\right]=\sum_{i=1}^{N}\left(\frac{P_i}{P_i^*}-1\right)Y_i.$$

Hence the theorem.

Remark 4.5.1.3. An estimator for estimating the $\text{MSE}\left(\hat{Y}_{\text{BS}}\right)$ is given by

$$\text{M}\hat{\text{S}}\text{E}\left(\hat{Y}_{\text{BS}}\right)=\hat{v}\left(\hat{Y}_{\text{BS}}\right)+\left\{\hat{B}\left(\hat{Y}_{\text{BS}}\right)\right\}^2. \tag{4.5.1.10}$$

In order to compare the three estimators let us consider a superpopulation model as follows:

Let Y_i and P_i denote respectively the value of variable Y and the relative measure of size X for the ith ($i = 1,2,...,N$) unit in the population so that the values $P_i = X_i/X$ will serve the purpose of selection probabilities with $\sum_{i=1}^{N} P_i = 1$. A general superpopulation model following Cochran (1963) is given by

$$Y_i = \beta P_i + e_i\ ,\ i = 1,2,....,N\ . \tag{4.5.1.11}$$

where e_i are the error terms such that:

$$E_m[e_i \mid P_i] = 0;$$

$$E_m\left(e_i^2 \mid P_i\right) = aP_i^g,\ \ a > 0,\ \ g \geq 0; \tag{4.5.1.12}$$

and $E_m\left(e_i e_j \mid P_i P_j\right) = 0$, where E_m denotes the expected value under the super-population model. Here $E_m\left(e_i^2 \mid P_i\right)$ is the residual variance of Y for $P = P_i$. The expected value of this residual variance is given by,

$$E\left(aP_i^g\right) = \frac{a}{N}\sum_{i=1}^{N} P_i^g\ ,$$

when the infinite superpopulation is simulated by the finite population of N units having the same characteristics as that of the superpopulation. Also, this expected value of residual variance is known to be given by

$$\sigma_y^2\left(1 - \rho_{xy}^2\right),$$

where ρ_{xy} is the correlation coefficient between Y and P. Thus we have

$$\frac{a}{N}\sum_{i=1}^{N} P_i^g = \sigma_y^2\left(1 - \rho_{xy}^2\right). \tag{4.5.1.13}$$

Now we have the following theorems:

Theorem 4.5.1.5. The expected value of the variance $V\left(\hat{Y}_{\text{Rao}}\right)$ under the superpopulation model (4.5.1.11) is

$$E_m\left[V\left(\hat{Y}_{\text{Rao}}\right)\right] = \frac{N^2}{n}\left[a\sum_{i=1}^{N} P_i^{g+1}\left(1 - P_i\right) + \beta^2\left\{\sum_{i=1}^{N} P_i^3 - \left(\sum_{i=1}^{N} P_i^2\right)^2\right\}\right]. \tag{4.5.1.14}$$

Proof. The $V\left(\hat{Y}_{\text{Rao}}\right)$ in (4.5.1.6) under the model (4.5.1.11) can easily be written as

$$V\left(\hat{Y}_{\text{Rao}}\right) = \frac{N^2}{n}\left[\sum_{i=1}^{N}\left(\beta P_i + e_i\right)^2 P_i - \left\{\sum_{i=1}^{N}\left(\beta P_i + e_i\right)P_i\right\}^2\right]$$

$$= \frac{N^2}{n}\left[\beta^3\sum_{i=1}^{N} P_i^3 + \sum_{i=1}^{N} P_i e_i^2 + 2\beta\sum_{i=1}^{N} P_i^2 e_i - \beta^2\left(\sum_{i=1}^{N} P_i^2\right)^2\right.$$

$$\left.- \sum_{i=1}^{N} P_i^2 e_i^2 - \sum_{i \neq j=1}^{N} P_i P_j e_i e_j - 2\beta\left(\sum_{i=1}^{N} P_i^2\right)\left(\sum P_i e_i\right)\right]. \tag{4.5.1.15}$$

On taking expected values on both sides of (4.5.1.15) and using (4.5.1.12) we have (4.5.1.14). Hence the theorem.

Theorem 4.5.1.6. The expected value of the variance $V(\hat{Y}_{BS})$ under the superpopulation model (4.5.1.11) is

$$E_m\left[V(\hat{Y}_{BS})\right] = \frac{1}{n}\left[a\sum_{i=1}^{N}\frac{P_i^{g+1}(1-P_i)}{P_i^{*2}} + \beta^2\left\{\sum_{i=1}^{N}\frac{P_i^3}{P_i^{*2}} - \left(\sum_{i=1}^{N}\frac{P_i^2}{P_i^*}\right)^2\right\}\right]. \tag{4.5.1.16}$$

Proof. Follows on the lines of the proof of the above theorem.

Theorem 4.5.1.7. Under the superpopulation model (4.5.1.11) the estimator $\hat{Y}_{BS}$ of population total will be more efficient then Rao (1966a, 1966b) estimator if

$$\rho_{xy}^2 > (1-\delta)^{-1}. \tag{4.5.1.17}$$

where

$$\delta = \frac{\left[\sum_{i=1}^{N}P_i^3\left(N^2 - 1/P_i^*\right) - \left(N\sum_{i=1}^{N}P_i^2\right)^2 + \left(\sum_{i=1}^{N}P_i^2/P_i^*\right)^2\right]\left(\sum_{i=1}^{N}P_i^g\right)}{N\sigma_p^2\sum_{i=1}^{N}P_i^{g+1}\left(1-P_i^{g+1}\right)\left(1/P_i^{*2} - N^2\right)} \tag{4.5.1.18}$$

for $\sigma_p^2 = \frac{1}{N}\left\{\sum_{i=1}^{N}P_i^2 - N^{-1}\left(\sum_{i=1}^{N}P_i\right)^2\right\}$.

Proof. It follows by setting $E_m\left[\hat{Y}_{BS}\right] \le E_m\left[\hat{Y}_{Rao}\right]$ and using (4.5.1.14) and (4.5.1.16). Hence the theorem.

Example 4.5.1.1. A team of medical doctors wishes in estimating the total of three variables 'Crude Birth Rate (CBR)', 'Crude Death Rate (CDR)' and 'Infant Mortality Rate (IMR)' in the world. We know that CBR and IMR have positive and high correlation with the 'Total Fertility Rate (TFR)', where as CDR has low positive correlation with TFR. Select a sample of 10 countries from the list given in population 8 of the Appendix by using PPSWR sampling and using TFR as known auxiliary information. The correlation coefficients values of TFR with TBR, CDR and IMR are +0.9855, +0.5492 and +0.8525 respectively. Apply the appropriate transformations on the selection probabilities using know values of correlation coefficient to obtain estimates of total CBR, CDR and IMR. Also construct 95% confidence interval in each situation.

Solution. (a) Selection of PPSWR Sample:

The population 8 in the Appendix consists of $N = 96$ countries. The maximum value of TFR is 7.17. Thus in the Lahiri's method we selected random numbers $1 \le R_i \le 96$ and $1 \le R_j \le 8$. We used first two columns of the Pseudo-Random Number (PRN) Table 1 given in the Appendix to select the random number R_i whereas the random number R_j selected using the 13th column. Then we performed the following trials:

Lahiri's method				
Trial No.	$1 \le R_i \le 96$	$1 \le R_j \le 8$	Value of TFR	Decision S--Selected R--Rejected
1	**58**	**3**	**5.76**	**S**
2	60	7	1.53	R
3	**54**	**3**	**3.10**	**S**
4	92	5	2.53	R
5	01	4	2.07	R
6	**69**	**1**	**1.95**	**S**
7	**87**	**1**	**2.35**	**S**
8	62	8	5.95	R
9	23	8	1.81	R
10	**88**	**4**	**6.24**	**S**
11	94	8	6.86	R
12	64	8	2.66	R
13	46	2	1.55	R
14	04	7	6.05	R
15	**32**	**5**	**6.75**	**S**
16	**94**	**6**	**6.86**	**S**
17	47	4	1.50	R
18	57	8	3.13	R
19	56	6	2.79	R
20	77	7	1.95	R
21	57	7	3.13	R
22	**81**	**4**	**5.19**	**S**
23	60	4	1.53	R
24	33	6	1.73	R
25	05	6	2.64	R
26	**72**	**2**	**5.87**	**S**
27	22	6	1.80	R
28	88	6	1.80	R
29	**38**	**3**	**3.16**	**S**

Thus we have the following sample information collected from 10 countries selected in the PPSWR sample out of 96 countries or area listed in population 8. Assuming that the total of the TFR is 336.15 and is known, the selection probabilities P_i have been listed in the last column of the table given below.

Sr. No.	Country or area	Crude birth rate	Crude death rate	Infant mortality rate	Total fertility rate	Selection Probability P_i
32	Ethiopia	44.3	17.6	117.7	6.75	0.020080321
38	Guinea	40.0	16.9	123.7	5.46	0.016242749
54	Malaysia	24.1	5.2	20.9	3.10	0.009222073
58	Mozambique	42.0	16.8	114.9	5.76	0.017135207
69	Russia	14.5	14.8	21.6	1.95	0.005800982
72	Senegal	43.4	10.4	58.4	6.04	0.017968169
81	Syria	36.1	5.3	35.2	5.19	0.015439536
87	Turkey	20.4	5.2	33.3	2.35	0.006990927
88	Uganda	43.1	21.8	95.6	6.24	0.018563141
94	Yemen	43.3	7.9	57.9	6.86	0.020407556

(b) Estimation of total Crude Birth Rate (CBR):

We have $N = 96$, $\rho_{xy(1)} = 0.9855$ and $P_{1i}^{*} = \left(1+\dfrac{1}{N}\right)^{1-\rho_{xy(1)}}\left(1+P_i\right)^{\rho_{xy(1)}} - 1$ thus we have

Sr. No.	Country or area	CBR y_{1i}	P_i	P_{1i}^{*}	y_{1i}/P_{1i}^{*}	$\left(y_{1i}/P_{1i}^{*}\right)^2$
32	Ethiopia	44.3	0.020080321	0.019939540	2221.716	4936023.195
38	Guinea	40.0	0.016242749	0.016158031	2475.549	6128343.431
54	Malaysia	24.1	0.009222073	0.009239385	2608.399	6803744.595
58	Mozambique	42.0	0.017135207	0.017037470	2465.155	6076988.131
69	Russia	14.5	0.005800982	0.005867759	2471.131	6106488.384
72	Senegal	43.4	0.017968169	0.017858270	2430.247	5906098.555
81	Syria	36.1	0.015439536	0.015366526	2349.262	5519033.079
87	Turkey	20.4	0.006990927	0.007040517	2897.514	8395589.856
88	Uganda	43.1	0.018563141	0.018444549	2336.734	5460324.512
94	Yemen	43.3	0.020407556	0.020261985	2137.007	4566798.293
				Sum	24392.710	59899432.030

An estimate of total CBR is given by

$$\hat{Y}_{\text{CBR}} = \frac{1}{n}\sum_{i=1}^{n}\frac{y_{1i}}{P_{1i}^{*}} = \frac{24392.71}{10} = 2439.271$$

and an estimate of the $V\left(\hat{Y}_{\text{CBR}}\right)$ is given by

$$\hat{v}\left(\hat{Y}_{\text{CBR}}\right) = \frac{1}{n(n-1)}\left[\sum_{i=1}^{n}\frac{y_{1i}^{2}}{P_{1i}^{*2}} - n\hat{Y}_{\text{CBR}}^{2}\right]$$

$$= \frac{1}{10(10-1)}\left[59899432.03 - 10 \times 2439.271^2\right] = 4433.35 .$$

A $(1-\alpha)100\%$ confidence interval of total CBR is given by

$$\hat{Y}_{\text{CBR}} \pm t_{\alpha/2}(\text{df} = n-1)\sqrt{\hat{v}(\hat{Y}_{\text{CBR}})} .$$

Using Table 2 from the Appendix the 95% Confidence Interval estimate of the total CBR is given by

$$2439.271 \pm 2.262\sqrt{4433.35} , \text{ or } [2288.66, 2589.88] .$$

(c) Estimation of total Crude Death Rate (CDR):

Here $N = 96$, $\rho_{xy(2)} = 0.5492$, and $P_{2i}^* = \left(1+\frac{1}{N}\right)^{1-\rho_{xy(2)}}(1+P_i)^{\rho_{xy(2)}} - 1$ thus we have:

Sr. No.	Country or area	CDR y_{2i}	Selection Probability P_i	P_{2i}^*	y_{2i}/P_{2i}^*	$(y_{2i}/P_{2i}^*)^2$
32	Ethiopia	17.6	0.020080321	0.015712557	1120.123	1254676.010
38	Guinea	16.9	0.016242749	0.013612204	1241.533	1541404.074
54	Malaysia	5.2	0.009222073	0.009760421	532.764	283837.384
58	Mozambique	16.8	0.017135207	0.014100977	1191.407	1419450.260
69	Russia	14.8	0.005800982	0.007879117	1878.383	3528323.172
72	Senegal	10.4	0.017968169	0.014556991	714.433	510415.059
81	Syria	5.3	0.015439536	0.013172143	402.364	161897.024
87	Turkey	5.2	0.006990927	0.008533810	609.341	371296.383
88	Uganda	21.8	0.018563141	0.014882611	1464.797	2145629.387
94	Yemen	7.9	0.020407556	0.015891492	497.121	247129.634
				Sum	9652.267	11464058.390

An estimate of total CDR is given by

$$\hat{Y}_{\text{CDR}} = \frac{1}{n}\sum_{i=1}^{n}\frac{y_{2i}}{P_{2i}^*} = \frac{9652.267}{10} = 965.2267$$

and an estimate of the $V(\hat{Y}_{\text{CDR}})$ is given by

$$\hat{v}(\hat{Y}_{\text{CDR}}) = \frac{1}{n(n-1)}\left[\sum_{i=1}^{n}\frac{y_{2i}^2}{P_{2i}^{*2}} - n\hat{Y}_{\text{CDR}}^2\right]$$

$$= \frac{1}{10(10-1)}\left[11464058.39 - 10 \times 965.2267^2\right] = 23860.36 .$$

A $(1-\alpha)100\%$ confidence interval of total CDR is given by

$$\hat{Y}_{\text{CDR}} \pm t_{\alpha/2}(\text{df} = n-1)\sqrt{\hat{v}(\hat{Y}_{\text{CDR}})} .$$

Using Table 2 from the Appendix the 95% Confidence Interval estimate of the total CDR is given by

$$965.2267 \pm 2.262\sqrt{23860.36}, \quad \text{or} \quad [810.75, 1119.69].$$

(d) Estimation of Infant Mortality Rate (IMR):

Here $N = 96$, $\rho_{xy(3)} = 0.8528$, and $P_{3i}^{*} = \left(1+\dfrac{1}{N}\right)^{1-\rho_{xy(3)}}(1+P_i)^{\rho_{xy(3)}} - 1$ thus we have:

Sr. No.	Country or area	IMR y_{3i}	Selection Probability P_i	P_{3i}^{*}	y_{3i}/P_{3i}^{*}	$(y_{3i}/P_{3i}^{*})^2$
32	Ethiopia	117.7	0.020080321	0.018652051	6310.298	39819860.21
38	Guinea	123.7	0.016242749	0.015383046	8041.320	64662833.81
54	Malaysia	20.9	0.009222073	0.009397829	2223.918	4945811.29
58	Mozambique	114.9	0.017135207	0.016143441	7117.442	50657976.27
69	Russia	21.6	0.005800982	0.006479085	3333.804	11114249.47
72	Senegal	58.4	0.017968169	0.016853056	3465.247	12007936.64
81	Syria	35.2	0.015439536	0.014698605	2394.785	5734995.29
87	Turkey	33.3	0.006990927	0.007494466	4443.279	19742724.66
88	Uganda	95.6	0.018563141	0.017359870	5506.954	30326538.49
94	Yemen	57.9	0.020407556	0.018930720	3058.521	9354549.37
				Sum	45895.570	248367475.50

An estimate of total IMR is given by

$$\hat{Y}_{\text{IMR}} = \frac{1}{n}\sum_{i=1}^{n}\frac{y_{3i}}{P_{3i}^{*}} = \frac{45895.57}{10} = 4589.557$$

and an estimate of the $V(\hat{Y}_{\text{IMR}})$ is given by

$$\hat{v}(\hat{Y}_{\text{IMR}}) = \frac{1}{n(n-1)}\left[\sum_{i=1}^{n}\frac{y_{3i}^2}{P_{3i}^{*2}} - n\hat{Y}_{\text{IMR}}^2\right]$$

$$= \frac{1}{10(10-1)}\left[248367475.5 - 10\times 4589.557^2\right] = 419190.45 .$$

A $(1-\alpha)100\%$ confidence interval of total IMR is given by

$$\hat{Y}_{\text{IMR}} \pm t_{\alpha/2}(\text{df} = n-1)\sqrt{\hat{v}(\hat{Y}_{\text{IMR}})} .$$

Using Table 2 from the Appendix the 95% Confidence Interval estimate of the total IMR is given by

$$4589.557 \mp 2.262\sqrt{419190.45}, \text{ or } [3125.02, 6054.08].$$

The problem of estimation of population total using multi-character survey has also been considered by Pathak (1966), Kumar and Herzel (1988), Amahia, Chaubey, and Rao (1989) and Rao (1993a, 1993b). They considered the following types of transformations on the selection probabilities.

$$P_{i0}^{*} = \frac{1}{N} \qquad \text{[Rao (1966a, 1966b)]}$$

$$P_{i1}^{*} = \left(1+\frac{1}{N}\right)^{(1-\rho_{xy})}(1+p_i)^{\rho_{xy}} - 1 \qquad \text{[Bansal and Singh (1985)]}$$

$$P_{i2}^{*} = \frac{(1-\rho_{xy})}{N} + \rho_{xy}\, p_i \qquad \text{[Amahia, Chaubey, and Rao (1989)]}$$

$$P_{i3}^{*} = \left[N(1-\rho_{xy}) + \frac{\rho_{xy}}{p_i}\right]^{-1} \qquad \text{[Amahia, Chaubey, and Rao (1989)]}$$

$$P_{i4}^{*} = \left(\frac{1}{N}\right)^{(1-\rho_{xy})}(p_i)^{\rho_{xy}} \qquad \text{[Mangat and Singh (1992-93)]}$$

4.5.1.1 GENERAL CLASS OF ESTIMATORS

Singh, Grewal, and Joarder (2002) proposed a general class of estimators as

$$\hat{Y}_g = \frac{1}{n}\sum_{i=1}^{n} y_i H(p_i) \qquad (4.5.1.1.1)$$

where $H(p_i)$ is a function of p_i and satisfies certain regularity conditions such as:

(i) $H\left(\frac{1}{N}\right) = N$;

(ii) The first and second order partial derivatives of H with respect to p_i exist and are assumed to be known constants.

Expanding $H(p_i)$ around $1/N$ with Taylor's series and noting $|p_i - 1/N| < 1$ we have

$$\hat{Y}_g = \frac{1}{n}\sum_{i=1}^{n} y_i\left[N + \left(p_i - \frac{1}{N}\right)H' + \left(p_i - \frac{1}{N}\right)^2 H'' + ...\right], \qquad (4.5.1.1.2)$$

where

$$H' = \frac{\partial H}{\partial p_i}\Big|_{p_i=\frac{1}{N}} \quad \text{and} \quad H'' = \frac{\partial^2 H}{\partial^2 p_i}\Big|_{p_i=\frac{1}{N}}$$

denote the first and second order partial derivatives of H with respect to p_i and are known constants for $p_i = 1/N$.

Thus we have the following cases:

Case I. If we choose $H' = -\rho_{xy}N^2$ and $H'' = \{N^3/(N+1)\}\rho_{xy}[\rho_{xy}(1+2N)+1]$ then the class of estimators $\hat{Y}_g$ reduces to estimator of Bansal and Singh (1985).

Case II. If we choose $H^{'} = -\rho_{xy}N^2$ and $H^{''} = 2\rho_{xy}^2 N^3$ then the class of estimators $\hat{Y}_g$ reduces to the first estimator of Amahia, Chaubey, and Rao (1989).

Case III. If we choose $H^{'} = -\rho_{xy}N^2$ and $H^{''} = 2\rho_{xy}N^3$ then the class of estimators $\hat{Y}_g$ reduces to the second estimator of Amahia, Chaubey, and Rao (1989).

Thus there are several choices of $H^{'}$ and $H^{''}$ exist depending upon the form of the function $H(p_i)$, but still no one knows which is the best or optimum.

4.5.2 STUDY VARIABLES HAVE POOR POSITIVE AS WELL AS POOR NEGATIVE CORRELATION WITH THE SELECTION PROBABILITIES

A natural question arises that in multi-character survey, some variables may be poorly negatively correlated and some variables may be poorly positively correlated. Singh and Horn (1998) have made an attempt to suggest a transformation in such situations. They suggested an alternative estimator of population total Y as

$$\hat{Y}_{\text{P}} = \frac{1}{n}\sum_{i=1}^{n}\frac{y_i}{p_i^*}, \tag{4.5.2.1}$$

where

$$p_i^* = \left(p_i^+\right)^{\rho_{xy}(1+\rho_{xy})/2}\left(p_i^-\right)^{-\rho_{xy}(1-\rho_{xy})/2}\left(\frac{1}{N}\right)^{(1-\rho_{xy})(1+\rho_{xy})} \tag{4.5.2.2}$$

with $p_i^- = z_i \Big/ \sum_{i=1}^{N} z_i = z_i \Big/ \sum_{i=1}^{N} x_i$ for $z_i = \dfrac{X - nx_i}{N - n}$, $p_i^+ = x_i \Big/ \sum_{i=1}^{N} x_i$ and ρ_{xy} is the correlation between y and x.

Thus there are the following cases:

Case I. If $\rho_{xy} = 1$ then $p_i^* = p_i^+$ and $\hat{Y}_{\text{P}}$ reduces to $\hat{Y}_{\text{HH}}$ owed to Hansen and Hurwitz (1943).

Case II. If $\rho_{xy} = 0$ then $p_i^* = 1/N$ and $\hat{Y}_{\text{P}}$ reduces to the estimator proposed by Rao (1966a, 1966b).

Case III. If $\rho_{xy} = -1$ then $p_i^* = p_i^-$ and $\hat{Y}_{\text{P}}$ reduces to $\hat{Y}_{\text{SM}}$ owed to Sahoo, Sahoo, and Mohanty (1994).

The transformation at (4.5.2.2) makes use of the known correlation coefficient ρ_{xy} between the study variable y and the auxiliary variable x. The study variable y can be any one among the k variables of interest, say $y_1, y_2, ..., y_k$, some of them have low positive and others having low negative correlation with the auxiliary variable x. In actual practice, the value of ρ_{xy} is not known in most of the surveys. Thus Singh and Horn (1998) have advised to use the estimator of ρ_{xy} in (4.5.2.2). Suppose r_{xy} is an estimator of the correlation coefficient ρ_{xy} defined as

$$r_{xy} = \frac{\sum_{i=1}^{n} p_i (u_i - \bar{u})(v_i - \bar{v})}{\sqrt{\sum_{i=1}^{n} p_i (u_i - \bar{u})^2 \sum_{i=1}^{n} p_i (v_i - \bar{v})^2}} \tag{4.5.2.3}$$

where $u_i = \dfrac{y_i}{Np_i}$, $v_i = \dfrac{x_i}{Np_i}$, $\bar{u} = n^{-1} \sum_{i=1}^{n} u_i$, $\bar{v} = n^{-1} \sum_{i=1}^{n} v_i$, and $p_i = \alpha p_i^{+} + (1-\alpha) p_i^{-}$ denotes the probability of selecting the ith unit in the sample for a suitable choice of α. In practice α can be taken as 0.5. Note that if $\alpha = 1$ then the estimator r_{xy} will not work and in such situation the size measure is supposed to be different than x variable. Then an estimator of the transformation p_i^* at (4.5.2.2) is given by

$$\hat{p}_i^* = \left(p_i^{+}\right)^{r_{xy}(1+r_{xy})/2} \left(p_i^{-}\right)^{-r_{xy}(1-r_{xy})/2} \left(\frac{1}{N}\right)^{(1-r_{xy})(1+r_{xy})}. \tag{4.5.2.4}$$

We have the following theorems:

Theorem 4.5.2.1. The estimator $\hat{p}_i^*$ at (4.5.2.4) is a consistent estimator of p_i^*.

Proof. Note that

$$\left|\frac{1}{N} - 1\right| < 1, \quad \left|p_i^{+} - 1\right| < 1, \quad \text{and} \quad \left|p_i^{-} - 1\right| < 1,$$

therefore using the binomial expansion and neglecting the higher order terms the estimator (4.5.2.4) can easily be rewritten as

$$\hat{p}_i^* = \left(1 - r_{xy}^2\right)\left(\frac{1}{N}\right) + \frac{r_{xy}}{2}\left(p_i^{+} - p_i^{-}\right) + \frac{r_{xy}^2}{2}\left(p_i^{+} + p_i^{-}\right) + O\left(p_i^{+2}, p_i^{-2}, \left(\frac{1}{N}\right)^2\right). \tag{4.5.2.5}$$

Let E_1 denotes the expected value over all possible values of the correlation coefficient, r_{xy}, then

$$E_1\left(\hat{p}_i^*\right) = p_i^* + O\left(n^{-1}\right). \tag{4.5.2.6}$$

Hence the theorem.

Theorem 4.5.2.2. A practically useful estimator of population total in multi-character surveys is obtained by replacing p_i^* by $\hat{p}_i^*$ in (4.5.2.1) as

$$\hat{Y}_S = \frac{1}{n} \sum_{i=1}^{n} \frac{y_i}{\hat{p}_i^*}. \tag{4.5.2.7}$$

Theorem 4.5.2.3. The expected value of the estimator $\hat{Y}_S$ is given by

$$E\left(\hat{Y}_S\right) = \sum_{i=1}^{N} \frac{y_i}{p_i^*} p_i + O\left(n^{-1}\right). \tag{4.5.2.8}$$

Proof. Let E_2 denote the expected value for the given value of r_{xy}, then we have

$$E\left(\hat{Y}_S\right)=E_1E_2\left(\hat{Y}_S \mid r\right)=E_1E_2\left[\frac{1}{n}\sum_{i=1}^{n}\frac{y_i}{\hat{p}_i^*} \mid r\right]=E_1\left[\sum_{i=1}^{n}\frac{y_i}{\hat{p}_i^*}p_i\right]=\sum_{i=1}^{N}\frac{y_i}{p_i^*}p_i+O\left(n^{-1}\right).$$

Hence the theorem.

In order to find the variance of the estimator $\hat{Y}_S$, let us define

$$\mu_{rs}=\sum_{i=1}^{N}p_i\left(u_i-\bar{Y}\right)^r\left(v_i-\bar{X}\right)^s,\ \hat{\mu}_{rs}=\sum_{i=1}^{n}p_i\left(u_i-\bar{u}\right)^r\left(v_i-\bar{v}\right)^s \text{ and } \lambda_{rs}=\frac{\mu_{rs}}{\mu_{20}^{r/2}\mu_{02}^{s/2}}.$$

Defining

$$\delta=\left(\hat{\mu}_{02}/\mu_{02}\right)-1,\quad \eta=\left(\hat{\mu}_{11}/\mu_{11}\right)-1,\quad \text{and}\quad \varepsilon=\left(\hat{\mu}_{20}/\mu_{20}\right)-1$$

we have to the first order of approximation

$$E_1\left(\delta^2\right)=n^{-1}\left(\lambda_{04}-1\right),\quad E_1\left(\eta^2\right)=n^{-1}\left(\frac{\lambda_{22}}{\rho_{xy}^2}-1\right),\quad E_1\left(\varepsilon^2\right)=n^{-1}\left(\lambda_{40}-1\right)$$

$$E_1\left(\delta\eta\right)=n^{-1}\left(\frac{\lambda_{13}}{\rho_{xy}}-1\right),\quad E_1\left(\varepsilon\eta\right)=n^{-1}\left(\frac{\lambda_{31}}{\rho_{xy}}-1\right),\quad \text{and}\quad E_1\left(\delta\varepsilon\right)=n^{-1}\left(\lambda_{22}-1\right)$$

Let V_1 denote the variance of the estimator of the correlation coefficient, then we have the following lemmas:

Lemma 4.5.2.1. The value of $V_1\left(r_{xy}\right)$ is given by

$$V_1\left(r_{xy}\right)=n^{-1}\rho_{xy}^2\left[\lambda_{22}\left(\frac{2+\rho_{xy}^2}{2\rho_{xy}^2}\right)+\frac{1}{4}\left(\lambda_{40}+\lambda_{04}\right)-\left(\frac{\lambda_{13}}{\rho_{xy}}+\frac{\lambda_{31}}{\rho_{xy}}\right)\right]. \tag{4.5.2.9}$$

Lemma 4.5.2.2. The value of $V_1\left(r_{xy}^2\right)$ is given by

$$V_1\left(r_{xy}^2\right)=n^{-1}\rho_{xy}^4\left[2\lambda_{22}\left(\frac{2+\rho_{xy}^2}{\rho_{xy}^2}\right)+\lambda_{04}+\lambda_{40}-4\left(\frac{\lambda_{13}+\lambda_{31}}{\rho_{xy}}\right)\right]. \tag{4.5.2.10}$$

Thus we have the following theorem:

Theorem 4.5.2.4. The variance of the estimator $\hat{Y}_S$ to the first order of approximation is given by

$$V\left(\hat{Y}_S\right)\approx\frac{1}{n}\left[\sum_{i=1}^{N}\frac{y_i^2p_i}{p_i^{*2}}-\left(\sum_{i=1}^{N}\frac{y_ip_i}{p_i^*}\right)^2\right]$$

$$+\frac{1}{n}\sum_{i=1}^{N}\frac{y_i^2 p_i^2}{p_i^{*2}}\left[\left(\frac{1}{N^2}+\frac{\left(p_i^+ + p_i^-\right)^2}{4}\right)\rho_{xy}^4\left\{\frac{2\lambda_{22}\left(2+\rho_{xy}^2\right)}{\rho_{xy}^2}+\lambda_{40}+\lambda_{04}-\frac{4(\lambda_{13}+\lambda_{31})}{\rho_{xy}^2}\right\}\right.$$

$$\left.+\frac{\rho_{xy}^2\left(p_i^+ - p_i^-\right)^2}{4}\left\{\frac{\lambda_{22}\left(2+\rho_{xy}^2\right)}{\rho_{xy}^2}+\frac{1}{4}(\lambda_{40}+\lambda_{04})-\frac{1}{\rho_{xy}}(\lambda_{13}+\lambda_{31})\right\}\right] \quad (4.5.2.11)$$

Proof. Let V_1 and V_2 denote the variance for all possible values of r_{xy} and for the given value of r_{xy}, respectively, then we have

$$V\left(\hat{Y}_S\right)=E_1V_2\left(\hat{Y}_S|r_{xy}\right)+V_1E_2\left(\hat{Y}_S|r_{xy}\right)$$

$$=E_1V_2\left[\frac{1}{n}\sum_{i=1}^{n}\frac{y_i}{\hat{p}_i^*}\,|\,r_{xy}\right]+V_1E_2\left[\frac{1}{n}\sum_{i=1}^{n}\frac{y_i}{\hat{p}_i^*}\,|\,r_{xy}\right]$$

$$=E_1\left[\frac{1}{n}\left\{\sum_{i=1}^{N}\frac{y_i^2 P_i}{\hat{p}_i^{*2}}-\left(\sum_{i=1}^{N}\frac{y_i P_i}{\hat{p}_i^*}\right)^2\right\}\right]+V_1\left[\sum_{i=1}^{N}\frac{y_i P_i}{\hat{p}_i^*}\right]$$

$$\approx\frac{1}{n}\left[\sum_{i=1}^{N}\frac{y_i^2 P_i}{p_i^{*2}}-\left(\sum_{i=1}^{N}\frac{y_i P_i}{p_i^*}\right)^2\right]+\sum_{i=1}^{N}\frac{y_i^2 p_i^2}{p_i^{*2}}V_1\left(\hat{p}_1^*\right),$$

which after using above expected values and on simplification reduces to (4.5.2.11). Hence the theorem.

4.6 CONCEPT OF REVISED SELECTION PROBABILITIES

We observed that SRSWR will be better than PPSWR sampling if the line of regression passes far away from the origin. Thus linearity between Y and X is not a sufficient condition for PPSWR scheme to be better than SRSWR scheme. So in practice it is very difficult to have an idea whether PPSWR sampling scheme can be preferred over SRSWR scheme or not. In order to overcome this problem, Reddy and Rao (1977) suggested that the sample be selected by probability proportional to revised sizes scheme and with replacement. The revised sizes are obtained through a location shift in the auxiliary variable as

$$X_i' = X_i + \left(L^{-1}-1\right)\overline{X} \quad \text{where } 0 < L < 1 .$$

This can also be treated as a compromise selection probability between PPSWR and SRSWR leading to new selection probabilities given by

$$P_i' = X_i' \Big/ \sum_{i=1}^{N} X_i' = LP_i + (1-L)/N . \quad (4.6.1)$$

Thus we have the following theorem:

Theorem 4.6.1. An unbiased estimator of population total Y is given by

$$\hat{Y}_{RR} = \frac{1}{n}\sum_{i=1}^{n}\frac{y_i}{P_i'} \tag{4.6.2}$$

with variance

$$V\left(\hat{Y}_{RR}\right) = \frac{1}{n}\left[\sum_{i=1}^{N}\frac{Y_i^2}{P_i'} - Y^2\right]. \tag{4.6.3}$$

Proof. Obvious by following Hansen and Hurwitz (1943).

For details one can refer to Reddy (1974), Singh, Kumar, and Chandak (1983) and Bedi and Rao (1996).

Example 4.6.1. One can observe for population 1, as the values of L changes from 0.1 to 0.9 with a step of 0.2, the percent relative efficiency of the estimator $\hat{Y}_{RR}$ with respect to the estimator $\hat{Y}_{HH}$ changes as shown in the following table.

L	0.10	0.30	0.50	0.70	0.90
RE	57.45	85.26	110.49	125.20	119.65

Thus a proper choice of L may lead to efficient estimators of population total.

The next section has been devoted to define the estimator of finite population correlation coefficient ρ_{xy} under PPSWR sampling.

4.7 ESTIMATION OF CORRELATION COEFFICIENT USING PPSWR SAMPLING

Assume (y_i, x_i), $i = 1,2,...,n$, denotes a pair of observations on the variables Y and X for the sample of size n drawn with varying probability P_i and with replacement procedure from a population of size N. Following Gupta, Singh, and Kashani (1993), we have the following theorem.

Theorem 4.7.1. An estimator of $\rho_{xy} = S_{xy}\left(S_x^2 S_y^2\right)^{-1/2}$ under PPSWR sampling is given by

$$r_{xy} = \hat{\theta}_1\left(\hat{\theta}_2\hat{\theta}_3\right)^{-1/2}, \tag{4.7.1}$$

where

$$\hat{\theta}_1 = \sum_{i=1}^{n}\frac{x_i y_i}{P_i^2} - \left(\sum_{i=1}^{N}\frac{x_i}{P_i}\right)\left(\sum_{i=1}^{N}\frac{y_i}{P_i}\right) + N(n-1)\sum_{i=1}^{N}\frac{x_i y_i}{P_i}, \tag{4.7.2}$$

$$\hat{\theta}_2 = \sum_{i=1}^{n}\frac{x_i^2}{P_i^2} - \left(\sum_{i=1}^{n}\frac{x_i}{P_i}\right) + N(n-1)\sum_{i=1}^{n}\frac{x_i^2}{P_i}, \tag{4.7.3}$$

and

$$\hat{\theta}_3 = \sum_{i=1}^{n} \frac{y_i^2}{P_i^2} - \left(\sum_{i=1}^{n} \frac{y_i}{P_i}\right) + N(n-1)\sum_{i=1}^{n} \frac{y_i^2}{P_i}. \tag{4.7.4}$$

It is remarkable here that if $P_i = x_i/X$, then the above estimator of correlation coefficient becomes non-functional. Thus one can use the second auxiliary variable, say Z, for selecting a sample while using above estimator of correlation coefficient. In order to see the bias and variance of the estimator r_{xy}, one can refer to Gupta, Singh, and Kashani (1993). The bias and variance expressions are complicated and hence are not reported here. Gupta (2002) suggested a new estimator of correlation coefficient under PPSWR sampling which guarantee admissible estimate under such designs.

EXERCISES

Exercise 4.1. Define an estimator of the regression coefficient $\beta = S_{xy}\left(S_x^2\right)^{-1}$ in PPSWR sampling.

Exercise 4.2. Write a short note on PPSWR sampling? Discuss at least one method of selecting a sample by PPSWR sampling.

Exercise 4.3. Consider a population consists of $N = 5$ units. Consider the population values are given by:

Y_i : 3, 4, 5, 6, 8

X_i : 2, 4, 7, 9, 6

Draw all possible samples of size $n = 3$ units. Construct the sample space. Show that the estimator

$$\hat{Y}_{\text{HH}} = \frac{1}{n}\sum_{i=1}^{n} \frac{y_i}{P_i}, \quad \text{where } P_i = X_i \Big/ \sum_{i=1}^{N} X_i,$$

is an unbiased estimator of population total. Also show that an unbiased estimator of the variance

$$V\left(\hat{Y}_{\text{HH}}\right) = \frac{1}{n}\left(\sum_{i=1}^{N} \frac{Y_i}{P_i} - Y^2\right) \text{ is given by } \hat{v}\left(\hat{Y}_{\text{HH}}\right) = \frac{1}{n(n-1)}\left[\sum_{i=1}^{n} \frac{y_i^2}{P_i^2} - \left(\hat{Y}_{\text{HH}}\right)^2\right].$$

Exercise 4.4. Show that the probability of selection of the ith unit in the sample by using Lahiri's PPSWR sampling scheme is given by

$$P_i = X_i \Big/ \sum_{i=1}^{N} X_i.$$

What is the probability of selecting s^{th} sample of n units using PPSWR sampling?

Exercise 4.5. Show that the estimator $\hat{Y}_{\text{HH}} = \frac{1}{n}\sum_{i=1}^{n} \frac{y_i}{P_i}$ is an unbiased estimator of population total. Find its variance and deduce the unbiased estimator of variance.

Exercise 4.6. Write a short note on the relative efficiency of the PPSWR sampling with respect to SRSWR sampling. Also discuss the concept of revised selection probabilities.

Exercise 4.7. Find the asymptotic bias and variance of the following estimators:

$$\bar{y}_1 = \alpha\bar{u}_n + (1-\alpha)\bar{u}_n\left(\frac{\bar{X}}{\bar{v}_n}\right); \qquad \bar{y}_2 = \bar{u}_n\left(\frac{\bar{v}_n}{\bar{X}}\right)^{\alpha};$$

$$\bar{y}_3 = \bar{u}_n\left(\frac{\bar{X}}{\alpha\bar{v}_n + (1-\alpha)\bar{X}}\right); \text{ and } \qquad \bar{y}_4 = \bar{u}_n + \alpha\left(\bar{X} - \bar{v}_n\right);$$

where $\bar{u}_n = n^{-1}\sum_{i=1}^{n}\frac{y_i}{P_i}$, $\bar{v}_n = n^{-1}\sum_{i=1}^{n}\frac{x_i}{P_i}$ and $P_i = Z_i \Big/ \sum_{i=1}^{N} Z_i$ have their usual meaning under PPSWR sampling.

(a) Discuss your views.

(b) Suggest an estimator of the optimum value of α and study the effect of replacement of α by its estimator in each case.

Hint: Pandey and Singh (1984).

Exercise 4.8. Is it possible to apply PPSWR sampling scheme if the study variable and auxiliary variable are negatively correlated?

Hint: Sahoo, Sahoo, and Mohanty (1994).

Exercise 4.9. Under multi-character surveys, consider an estimator of population total Y is

$$\hat{Y}_{\text{M}} = \frac{1}{n}\sum_{i=1}^{n}\frac{y_i}{P_i^*},$$

where $P_i^* = \rho_{xy}P_i + (1-\rho_{xy})/N$ or $P_i^* = \left[\rho_{xy}/P_i + (1-\rho_{xy})N\right]^{-1}$

(a) Find the values of correlation coefficient, ρ_{xy}, under which the estimator $\hat{Y}_{\text{M}}$ reduces to $\hat{Y}_{\text{Rao}} = \frac{N}{n}\sum_{i=1}^{n} y_i$ and PPSWR estimator, $\hat{Y}_{\text{HH}} = \frac{1}{n}\sum_{i=1}^{n}\frac{y_i}{P_i}$.

(b) Find the variance of the estimator $\hat{Y}_{\text{M}}$ and derive the value of $V\left(\hat{Y}_{\text{Rao}}\right)$ and $V\left(\hat{Y}_{\text{HH}}\right)$.

Hint: Amahia, Chaubey, and Rao (1989), Kumar and Agarwal (1997), Arnab (2001).

(c) Consider a general class of estimators of the form $\hat{Y}_g = \frac{1}{n}\sum_{i=1}^{n} y_i H(p_i)$, where $H(p_i)$ is a function of p_i and satisfies certain regularity conditions defined as:

$$H\left(\frac{1}{N}\right) = N;$$

and the first and second order partial derivatives of H with respect to p_i exist and are assumed to be known constants.

Show that the estimators in (a) and (b) are special cases of (c) for certain choice of parameters in the function $H(p_i)$. Study the bias and variance of the general class of estimators.

Hint: Singh, Grewal, and Joarder (2002), Espejo, Pineda, and Nadarajah (2003).

Exercise 4.10. (a) List a few situations where PPSWR sampling can be used in actual practice.

(b) Pickup one situation of interest to you and collect information on a variable(s) of your interest from a reasonable good number of respondents.

(c) Later you found that a few of the variable(s) have very low correlation with the selection probabilities, what kind of analysis might be helpful to you?

Exercise 4.11. Let $\hat{Y} = N\bar{y}$ be the usual unbiased estimator of Y based on a SRSWR sampling scheme of size n. Let

$$\hat{Y}_{\text{PPS}} = \frac{X}{n}\sum_{i=1}^{n}\frac{y_i}{x_i}$$

be the usual unbiased estimator of Y based on a PPSWR sample of size n and let

$$\hat{Y}^{*}_{\text{PPS}} = \frac{X^{*}}{n}\sum_{i=1}^{n}\frac{y_i}{x_i + d\bar{X}}$$

be an alternative unbiased estimator of Y based on a PPSWR sample of size n, where the size measure is the transformed variate

$$X^{*} = \sum_{i=1}^{N}\left(X_i + d\bar{X}\right) = X(1+d),$$

show that

$$V\left(\hat{Y}^{*}_{\text{PPS}}\right) < \frac{d}{(1+d)}V\left(\hat{Y}\right) + \frac{1}{(1+d)}V\left(\hat{Y}_{\text{PPS}}\right).$$

Hint: Reddy and Rao (1977), Agarwal, Singh, and Goel (1979).

Exercise 4.12. (a) Under multi-character surveys, an estimator of population total Y is

$$\hat{Y}_{\text{P}} = \frac{1}{n}\sum_{i=1}^{n}\frac{y_i}{P_i^{*}},$$

where

$$P_i^{*} = \left(p_i^{+}\right)^{\rho_{xy}\left(1+\rho_{xy}\right)/2}\left(p_i^{-}\right)^{-\rho_{xy}\left(1-\rho_{xy}\right)/2}\left(\frac{1}{N}\right)^{\left(1-\rho_{xy}\right)\left(1+\rho_{xy}\right)}$$

and ρ_{xy} is the correlation between y and x, $p_i^{+} = \dfrac{x_i}{X}$ and $p_i^{-} = \left(X - nx_i\right)/\{(N-n)X\}$ with $X = \sum_{i=1}^{N}x_i$. Discuss the cases for $\rho_{xy} = 1$, $\rho_{xy} = 0$ and $\rho_{xy} = -1$.

(b) If the value of ρ_{xy} is unknown, then use its consistent estimator

$$r_{xy} = \sum_{i=1}^{n} p_i(u_i - \bar{u})(v_i - \bar{v}) \Big/ \sqrt{\sum_{i=1}^{n} p_i(u_i - \bar{u})^2 \sum_{i=1}^{n} p_i(v_i - \bar{v})^2} ,$$

where $u_i = y_i/(Np_i)$, $v_i = x_i/(Np_i)$, $\bar{u} = n^{-1}\sum_{i=1}^{n} u_i$, $\bar{v} = n^{-1}\sum_{i=1}^{n} v_i$ and $p_i = \alpha p_i^+ + (1-\alpha)p_i^-$ for some real constant $0 \le \alpha \le 1$.

Study the asymptotic properties of the practical estimator

$$\hat{Y}_{\text{PC}} = \frac{1}{n}\sum_{i=1}^{n}\frac{y_i}{P_i^{**}}$$

where $P_i^{**} = \left(p_i^+\right)^{r_{xy}(1+r_{xy})/2}\left(p_i^-\right)^{-r_{xy}(1-r_{xy})/2}\left(\frac{1}{N}\right)^{(1-r_{xy})(1+r_{xy})}$.

Hint: Singh and Horn (1998).

Exercise 4.13. If y is perfectly linearly correlated with the auxiliary variable x so that $y = \alpha + \beta x$, where α, $\beta(\neq 0)$ are constants. Show that sample mean estimator $\hat{y} = N\,\bar{y}$, with $\bar{y} = n^{-1}\sum_{i=1}^{n} y_i$ under SRSWR sampling is more efficient than the Hansen and Hurwitz (1943) estimator

$$\hat{y}_{\text{hh}} = \frac{1}{n}\sum_{i=1}^{n} y_i/p_i , \text{ with } p_i = x_i/X \text{ if } \left\{\beta^2/\alpha^2\right\} < \left\{\left(\bar{X} - H\right)/\left(H\sigma_x^2\right)\right\},$$

where $H = N\left(\sum_{i=1}^{N}\frac{1}{x_i}\right)^{-1}$ and $\bar{X} = N^{-1}\sum_{i=1}^{N} x_i$.

Hint: Raj (1968)

Exercise 4.14. Under the superpopulation model $M : Y_i = \beta\, X_i + e_i$ such that $E_M(e_i \mid X_i) = 0$, $E_M\left(e_i^2 \mid X_i\right) = \sigma^2 X_i^g$, and $E\left(e_i e_j \mid X_i, X_j\right) = 0$, $g \ge 0$, show that

$$E_M\left[V\left(\hat{y}_{\text{hh}}\right)\right] = \frac{\sigma^2}{n}\left[\sum_{i=1}^{N} P_i^{g-1} - \sum_{i=1}^{N} P_i^g\right], \text{ where } \hat{y}_{\text{hh}} = \frac{1}{n}\sum_{i=1}^{n}\frac{y_i}{p_i}.$$

Hint: Rao (1966a, 1966b).

Exercise 4.15. Consider a situation of having information on two auxiliary variables x_i and z_i (say), for $i = 1,2,....,N$. Consider a sample s of size n be drawn with PPSWR sampling. Let $P_i = z_i \Big/ \sum_{i=1}^{N} z_i$ denote the probability of selection on the basis of known auxiliary variable z_i for $i = 1,2,....,N$. Let y_i and x_i denote the value of the variable under study y and the second auxiliary variable x to be used at estimation stage. Let $u_i = y_i/(NP_i)$ and $v_i = x_i/(NP_i)$ for $i = 1,2,...,n$. Obviously $\hat{R}_1 = \bar{u}_n/\bar{v}_n$, $\hat{R}_2 = \left(\bar{u}_n/\bar{X}\right)$ and $\hat{R}_3 = \left(\bar{u}_n/\bar{v}_n\right)\left(\bar{X}/\bar{v}_n\right)$ can be considered as three estimators of the ratio $R = \bar{Y}/\bar{X}$. Find the values of the real constants g_t, $t = 1,2,3$ such that the linear variety of the estimators of R defined as

$$\hat{R}_{\text{ms}} = \sum_{t=1}^{3} g_t \hat{R}_t, \quad \text{for} \quad \sum_{t=1}^{3} g_t = 1$$

remains unbiased with minimum variance.
Hint: Mahajan and Singh (1997).

PRACTICAL PROBLEMS

Practical 4.1. Use the cumulative total method to select a sample of ten units from population 1 given in the Appendix by using nonreal estate farm loans as an auxiliary variable.

Practical 4.2. Use Lahiri's method to select a sample of ten units from population 1 given in the Appendix by using nonreal estate farm loans as auxiliary variable.

Practical 4.3. Your supervisor used PPSWR sampling to select a sample of six states from the population 1 in the Appendix and collected the following information.

State	KY	NE	ME	OH	OK	TX
Nonreal estate farm loan (X)$	557.656	3585.406	51.539	635.774	1716.087	3520.361
Real estate farm loan (Y) $	1045.106	1337.852	8.849	870.720	612.108	1248.761

The total amount $43908.12 of nonreal estate farm loans (in $000) for the year 1997 is known. Use the Hansen and Hurwitz (1943) estimator for estimating the total amount of the real estate farm loans (in $000) during 1997 in the United States. Also find an estimator of variance of the estimator and hence derive the 95% confidence interval.

Practical 4.4. Discuss the relative efficiency of the PPSWR sampling estimator for estimating total amount of the real estate farm loans during 1997 by using information on nonreal estate farm loans during 1997 with respect to the usual and ratio estimator of population mean based on SRSWOR sample. Use the information given in population 1 of the Appendix.

Practical 4.5. Consider the problem of estimation of average price of the commercial apple crop. Select a PPSWR sample of seven units from population 3 in the appendix by Lahiri's method using the season average price during 1994 of apple crop as a benchmark or auxiliary variable in the United States. Collect the information on the prices of the apple crop during 1995 and 1996 from population 3 for the units included in the sample.
Use the estimator

$$\bar{y}_{11} = \bar{u}_n + \hat{\alpha}\left(\bar{X} - \bar{v}_n\right),$$

where

$$\hat{\alpha} = \left\{ \sum_{i=1}^{n} P_i (u_i - \bar{u}_n)(v_i - \bar{v}_n) \right\} \Big/ \sum_{i=1}^{n} P_i (v_i - \bar{v}_n)^2 ,$$

for estimating the average price of the apple crop during 1997 by assuming that the average price during 1995 is known. Also construct 95% confidence interval for estimating the average price of the apple crop during 1996 in the United States.

Practical 4.6. It is a well known fact that age and duration of sleep are negatively correlated. Assuming that the whole population information is known as presented in population 2 of the Appendix. Discuss the gain in efficiency due to Sahoo, Sahoo, and Mohanty (1994) method of estimation over the Hansen and Hurwitz (1943) estimator. Assume that the sample size consist of ten units.

Practical 4.7. Your boss Ms. Stephanie Singh used TFR to select a PPSWR sample of 5 units from the population 8 given in the Appendix. Her interest was to estimate the total 'Crude Birth Rate (CBR)', 'Crude Death Rate (CDR)', 'Infant Mortality Rate (IMR)' and 'Expectation of Life at Birth (ELB)' in the world as shown in the following table.

Sr. No.	Country or area	Crude birth rate	Crude death Rate	Infant mortality Rate	Expectation of life at birth	Total fertility rate
42	India	23.5	8.9	63.5	61.5	2.90
06	Australia	13.0	6.9	5.0	80.4	1.80
20	Canada	12.3	7.2	5.5	80.0	1.80
01	United states	14.2	8.8	6.2	76.3	2.07
90	United Kingdom	12.1	10.9	6.0	77.1	1.79

Later she found that CBR and IMR have high positive correlation with the 'Total Fertility Rate (TFR)', where as CDR has low positive correlation with TFR and ELB has high negative correlation. She was worried as she came to know the correlation coefficients values of TFR with TBR, CDR, IMR and ELB are +0.9855, +0.5492, +0.8525 and -0.8152 respectively. Help her by suggesting an appropriate transformation on the selection probabilities using know values of correlation coefficients to obtain good estimates of total CBR, CDR, IMR, and ELB. Also suggest her to construct 95% confidence interval in each situation.
Hint: Singh and Horn (1998).

Practical 4.8. The following table gives the weekly wages and expenditure of all the seven households on a drive way:

Household No.	1	2	3	4	5	6	7
Weekly wages ($)	358	540	225	188	579	150	530
Weekly expenditure ($)	225	329	198	160	357	130	329

(I) John considers the problem of estimation of weekly expenditure using an SRSWR sampling design and selects only two households, the second and seventh household.
(a) Estimate the total weekly expenditure based on John's sampling scheme.
(b) Estimate the variance of the estimator of total weekly expenditure and derive the 95% confidence interval estimate for John's SRSWR design.
(II) Michael considers the same problem of estimation of weekly expenditure using the probability proportional to weekly wages design and again selects the same second and seventh households in the sample.
(d) Estimate the total expenditure based on Michael's sampling scheme.
(e) Estimate the variance of the estimator of total weekly expenditure and derive the 95% confidence interval estimate for Michael's PPSWR design.
(III) Who is more accurate : John or Michael?

Practical 4.9. Consider a class with six girls, their height and response to the question, 'Do you like Bob?' are given in the following table:

Girls	Amy	Cathy	Eileen	Heather	Jennifer	Kelly
Girl likes Bob	Yes	No	Yes	No	Yes	No
Height (cm)	167	165	169	164	171	162

(a) Find the true proportion of girls who likes Bob.

(b) Not all girls want to disclose their response, but agree to give four responses. Select a sample of four girls using SRSWR sampling and estimate the proportion of girls who likes Bob. (**Rule:** Use the first column of the Pseudo-Random Number Table 1 given in the Appendix to select the sample)

(c) Estimate the variance of the estimate of proportion and construct 95% confidence interval estimate.

(d) Does the true proportion lie in your confidence interval estimate?

(e) Select a sample of four girls using probability proportional to height of the girls. (**Rule:** Use Cumulative Total Method and first three columns of the Pseudo-Random Number Table 1 given in the Appendix)

(f) Estimate the proportion of girls who like Bob using PPSWR sampling.

(g) Estimate the variance of the estimator of proportion of girls who likes Bob and construct 95% confidence interval estimate.

(h) Does the true proportion lie in your new confidence interval estimate?

Practical 4.10. Consider a class with six students, their marks in the assignments and final examination are given in the following table:

Student	Amy	Bob	Cathy	Don	Erika	Frank
Assignments	78	58	87	75	56	69
Final examination	72	56	94	67	58	75

(a) Find the true average marks for the final examination.
(b) Select a sample of three students using SRSWR sampling and estimate the average marks for the final examination. (**Rule:** Use third column of the Pseudo-Random Number Table 1 given in the Appendix to select the sample)
(c) Estimate the variance of the estimate of average and construct 95% confidence interval estimate.
(d) Does the true average lie in your confidence interval estimate?
(e) Select a sample of three students using probability proportional to marks on assignments. (**Rule:** Use Lahiri's method and specify the columns used from the Pseudo-Random Number Table 1 given in the Appendix)
(f) Estimate the average marks for the final examination using PPSWR sampling.
(g) Estimate the variance of the estimator of average marks and construct a 95% confidence interval estimate.
(h) Do the true average marks from the final examination lie in your new confidence interval estimate?

5. USE OF AUXILIARY INFORMATION: PROBABILITY PROPORTIONAL TO SIZE AND WITHOUT REPLACEMENT (PPSWOR) SAMPLING

5.0 INTRODUCTION

In probability proportional to size and without replacement (PPSWOR) sampling scheme, we will discuss the Horvitz and Thompson (1952) estimator, two forms of the variance of the Horvitz and Thompson (1952) estimator and their estimators, superpopulation model, construction of inclusion probabilities, calibrated estimators of population total and calibrated estimators of variance of the estimators of total like ratio estimator, linear regression estimator, regression predictor, distribution function, Rao, Hartley, and Cochran (1962) sampling scheme, unbiased estimation strategies under IPPS sampling and unified approach. At the end, we will celebrate Golden Jubilee Year 2003 of the traditional linear regression estimator owed to Hansen, Hurwitz, and Madow (1953). Before going further we would like to define a few important symbols and mathematical relations.

5.0.1 USEFUL SYMBOLS

(a) $P_i = X_i/X$, where $X = \sum_{i\in\Omega} X_i$, denotes the probability of selecting the ith population unit at any particular draw.

(b) π_i : Probability of including the ith population unit in the sample.

(c) π_{ij} : Probability of including both i^{th} and j^{th} (such that $i \neq j$) population units in the sample, and so on.

5.0.2 SOME MATHEMATICAL RELATIONS

To discuss some mathematical relations, we make use of the following theorems:

Theorem 5.0.1. The sum of the probabilities of including the ith population unit in the sample over the whole population is equal to the sample size, *i.e.*,

$$\sum_{i\in\Omega} \pi_i = n \tag{5.0.1}$$

where $i \in \Omega$ implies that i^{th} unit belongs to population Ω of size N.

Proof. Let us define a variable t_i which takes the value 1 if the ith population unit is included in the sample and zero otherwise, *i.e.*,

$$t_i = \begin{cases} 1 \text{ with probability } \pi_i, \\ 0 \text{ with probability } (1-\pi_i). \end{cases} \tag{5.0.2}$$

The expected value of the random variable t_i is

$$E(t_i) = 1\times \pi_i + (1-\pi_i)\times 0 = \pi_i . \tag{5.0.3}$$

Note that we are using without replacement (WOR) sampling, so there is no chance of getting any unit repeated. The n selected units will be distinct and we have

$$\sum_{i\in\Omega} t_i = n . \tag{5.0.4}$$

Taking expected values on both sides of (5.0.4) we have

$$\sum_{i\in\Omega} E(t_i) = n . \tag{5.0.5}$$

On using (5.0.3) in (5.0.5) we have

$$\sum_{i\in\Omega} \pi_i = n . \tag{5.0.6}$$

Hence the theorem.

Theorem 5.0.2. The sum of the probabilities of including different pairs of units from the whole population in the sample is given by product of the probability of including i^{th} population unit and the remaining sample size. In other words

$$\sum_{j(\neq i)\in\Omega} \pi_{ij} = (n-1)\pi_i . \tag{5.0.7}$$

Proof. Let us consider a random variable t_{ij} which takes the value 1 if both i^{th} and j^{th} population units are included in the sample and 0 otherwise, *i.e.*,

$$t_{ij} = \begin{cases} 1 \text{ with probability } \pi_{ij}, \\ 0 \text{ with probability } (1-\pi_{ij}) \end{cases} \tag{5.0.8}$$

Thus the expected value of t_{ij} is given by

$$E(t_{ij}) = \pi_{ij}\times 1 + (1-\pi_{ij})\times 0 = \pi_{ij}. \tag{5.0.9}$$

Now we have

$$E\left[\sum_{j(\neq i)\in\Omega} t_{ij}\right] = \sum_{j(\neq i)\in\Omega} E(t_{ij}) = \sum_{j(\neq i)\in\Omega} \pi_{ij} . \tag{5.0.10}$$

Note that π_{ij} is the probability for both i^{th} and j^{th} $(i \neq j)$ units to be included in the sample s we have

$$\begin{aligned}\sum_{j(\neq i)\in\Omega} \pi_{ij} &= \sum_{j(\neq i)\in\Omega}\left[\text{Probability for both } i^{th} \text{ and } j^{th}(i\neq j) \text{ units are included in the sample}\right] \\ &= \sum_{j(\neq i)\in\Omega} P[j\in s \mid i\in s]P(i\in s) \\ &= \sum_{j(\neq i)\in\Omega} P(j\in s\mid i\in s)\pi_i = \pi_i \sum_{j(\neq i)\in\Omega} P(j\in s\mid i\in s). \end{aligned} \tag{5.0.11}$$

In Theorem 5.0.1 we proved that the sum of the probabilities of including i^{th} population unit in the sample is $\sum_{i\in\Omega}\pi_i = n$, where Ω denotes the set of all N units in the population and n is the number of units in the sample s. Now in $\sum_{j(\neq i)\in\Omega} P(j\in s\mid i\in s)$ the number of population units is $(N-1)$ (since i^{th} unit has been already selected) and in the sample we have to select a further $(n-1)$ units. Thus

$$\sum_{j(\neq i)\in\Omega} P(j\in s\mid i\in s) = (n-1). \tag{5.0.12}$$

Hence the theorem.

5.1 HORVITZ AND THOMPSON ESTIMATOR AND RELATED TOPICS

The Horvitz and Thompson (1952) estimator (or HT estimator) of the population total Y is a linear estimator of the sample observations. The Horvitz and Thompson (a universal estimator), on the basis of n sample observations $y_i, i = 1, 2, ..., n$, can be defined as

$$\hat{Y}_{\text{HT}} = \sum_{i \in s} d_i y_i \tag{5.1.1}$$

where d_i, $i = 1, 2, ..., n$ are predetermined real constants or design weights. Thus we have the following theorems:

Theorem 5.1.1. For the estimator $\hat{Y}_{\text{HT}}$ to be unbiased for the population total, Y, the design weights are given by

$$d_i = 1/\pi_i . \tag{5.1.2}$$

Proof. Taking expected values on both sides of (5.1.1), we have

$$E(\hat{Y}_{\text{HT}}) = E\left[\sum_{i \in s} d_i y_i\right] = E\left[\sum_{i \in \Omega} t_i d_i Y_i\right]$$

where t_i is a random variable and takes a value 1 if the ith population unit is included in the sample and 0 otherwise. Note that here both d_i and Y_i (i^{th} population value) are constants. Therefore we have

$$E(\hat{Y}_{\text{HT}}) = \sum_{i \in \Omega} E(t_i) d_i Y_i = \sum_{i \in \Omega} \pi_i d_i Y_i . \tag{5.1.3}$$

Now $\sum_{i \in \Omega} \pi_i d_i Y_i$ is equal to $\sum_{i \in \Omega} Y_i$ if $d_i = 1/\pi_i$. Therefore the condition that $\hat{Y}_{\text{HT}}$ is an unbiased estimator of population total Y is that the design weights d_i are equal to $1/\pi_i$. Thus we have the following theorem:

Theorem 5.1.2. An unbiased estimator of the population total Y is given by

$$\hat{Y}_{\text{HT}} = \sum_{i \in s} \frac{y_i}{\pi_i} . \tag{5.1.4}$$

Proof. Follows by substituting $d_i = 1/\pi_i$ in (5.1.1).

Theorem 5.1.3. Under SRSWOR sampling, the estimator (5.1.4) reduces to the estimator of population total Y given by

$$\hat{Y}_{\text{HT}} = N\, \bar{y} . \tag{5.1.5}$$

Proof. Under SRSWOR sampling, the probability of including the ith population unit in the sample s is

$$\pi_i = \binom{N-1}{n-1} \Big/ \binom{N}{n} = \frac{n}{N} . \tag{5.1.6}$$

On substituting the value of π_i from (5.1.6) in (5.1.4) we have

$$\hat{Y}_{\text{HT}} = \sum_{i \in s} \frac{y_i}{n/N} = \frac{N}{n} \sum_{i=1}^{n} y_i = N\ \bar{y}\ . \tag{5.1.7}$$

Hence the theorem.

Theorem 5.1.4. The variance of the Horvitz and Thompson (1952) estimator, $\hat{Y}_{\text{HT}} = \sum_{i \in s} y_i/\pi_i$, of the population total, Y, is

$$V\left(\hat{Y}_{\text{HT}}\right) = \sum_{i \in \Omega} \frac{(1-\pi_i)}{\pi_i} Y_i^2 + \sum_{i \in \Omega} \sum_{j(\neq i) \in \Omega} \frac{\left(\pi_{ij} - \pi_i \pi_j\right)}{\pi_i \pi_j} Y_i Y_j\ . \tag{5.1.8}$$

Proof. Let us define a random variable t_i as in (5.0.2), i.e.,

$$t_i = \begin{cases} 1 \text{ with probability } \pi_i, \\ 0 \text{ with probability } (1-\pi_i). \end{cases} \tag{5.1.9}$$

Now we have

$$V\left(\hat{Y}_{\text{HT}}\right) = V\left(\sum_{i \in s} \frac{y_i}{\pi_i}\right) = V\left(\sum_{i \in \Omega} \frac{t_i Y_i}{\pi_i}\right). \tag{5.1.10}$$

Because the sample has been taken by using WOR sampling, the random variable values t_i and t_j for $i \neq j$ are not independent, thus

$$V\left(\hat{Y}_{\text{HT}}\right) = \sum_{i \in \Omega} V\left(\frac{t_i Y_i}{\pi_i}\right) + \sum_{i \in \Omega} \sum_{j(\neq i) \in \Omega} \text{Cov}\left(\frac{t_i Y_i}{\pi_i}, \frac{t_j Y_j}{\pi_j}\right). \tag{5.1.11}$$

Now we have

$$V(t_i) = E\left(t_i^2\right) - \{E(t_i)\}^2\ . \tag{5.1.12}$$

From (5.1.9) we have

$$t_i = \begin{cases} 1 \text{ if the } i^{th} \text{ population unit is included in the sample}, \\ 0 \text{ otherwise.} \end{cases} \tag{5.1.13}$$

Thus the expected value of t_i is given by

$$E(t_i) = \pi_i \times 1 + (1-\pi_i) \times 0 = \pi_i\ . \tag{5.1.14}$$

Also from (5.1.13) we have

$$t_i^2 = \begin{cases} 1 \text{ if the } i^{th} \text{ population unit is included in the sample}, \\ 0 \text{ otherwise.} \end{cases} \tag{5.1.15}$$

Therefore

$$E\left(t_i^2\right) = \pi_i \times 1 + (1-\pi_i) \times 0 = \pi_i \tag{5.1.16}$$

and

$$V(t_i) = E\left(t_i^2\right) - \{E(t_i)\}^2 = \pi_i - \pi_i^2 = \pi_i(1-\pi_i)\ . \tag{5.1.17}$$

Now

$$\text{Cov}(t_i, t_j) = E(t_i t_j) - E(t_i)E(t_j) \tag{5.1.18}$$

and

$$t_i t_j = \begin{cases} 1 \text{ if both } i^{th} \text{ and } j^{th} \text{ population units are included in the sample}, \\ 0 \text{ otherwise.} \end{cases} \tag{5.1.19}$$

Thus

$$E(t_i t_j) = 1 \times \pi_{ij} + 0 \times (1 - \pi_{ij}) = \pi_{ij}.$$

Therefore we get

$$\operatorname{Cov}(t_i, t_j) = E(t_i t_j) - E(t_i)E(t_j) = \pi_{ij} - \pi_i \pi_j. \tag{5.1.20}$$

Now plugging the values of $V(t_i)$ and $\operatorname{Cov}(t_i, t_j)$ in (5.1.11) we have

$$V(\hat{Y}_{\text{HT}}) = \sum_{i\in\Omega} \frac{Y_i^2}{\pi_i^2} V(t_i) + \sum_{i\in\Omega} \sum_{j(\neq i)\in\Omega} \frac{Y_i Y_j}{\pi_i \pi_j} \operatorname{Cov}(t_i, t_j)$$

$$= \sum_{i\in\Omega} \frac{Y_i^2}{\pi_i^2} \pi_i (1 - \pi_i) + \sum_{i\in\Omega} \sum_{j(\neq i)\in\Omega} \frac{Y_i Y_j}{\pi_i \pi_j} (\pi_{ij} - \pi_i \pi_j),$$

which on simplifying reduces to (5.1.8). Hence the theorem.

Theorem 5.1.5. In case of SRSWOR sampling, the variance of the Horvitz and Thompson (1952) estimator reduces to

$$V(\hat{Y}_{\text{HT}})_{\text{SRSWOR}} = \frac{N^2(1-f)}{n} s_y^2 \tag{5.1.21}$$

where $f = n/N$ denotes the finite population correction (f.p.c.) factor.

Proof. We know that the probability of including the ith population unit in the sample by using SRSWOR sampling is

$$\pi_i = \binom{N-1}{n-1} \Big/ \binom{N}{n} = \frac{n}{N} \tag{5.1.22}$$

and the probability of including both i^{th} and j^{th} population units in the sample is

$$\pi_{ij} = \binom{N-2}{n-2} \Big/ \binom{N}{n} = \frac{n(n-1)}{N(N-1)}. \tag{5.1.23}$$

On substituting the values of π_i and π_{ij} in (5.1.8) we have

$$V(\hat{Y}_{\text{HT}}) = \sum_{i\in\Omega} \frac{(1 - n/N)}{n/N} Y_i^2 + \sum_{i\in\Omega} \sum_{j(\neq i)\in\Omega} \frac{\frac{n(n-1)}{N(N-1)} - \frac{n}{N}\frac{n}{N}}{\frac{n}{N}\frac{n}{N}} Y_i Y_j$$

$$= \frac{N-n}{n} \sum_{i=1}^{N} Y_i^2 - \frac{N-n}{n(N-1)} \sum_{i=1}^{N} \sum_{j\neq i=1}^{N} Y_i Y_j. \tag{5.1.24}$$

Note that

$$Y^2 = \left(\sum_{i=1}^{N} Y_i \right)^2 = \sum_{i=1}^{N} Y_i^2 + \sum_{i=1}^{N} \sum_{j\neq i=1}^{N} Y_i Y_j. \tag{5.1.25}$$

Therefore

$$\sum_{i=1}^{N} \sum_{j\neq i=1}^{N} Y_i Y_j = Y^2 - \sum_{i=1}^{N} Y_i^2. \tag{5.1.26}$$

On substituting (5.1.26) in (5.1.24) we have

$$V(\hat{Y}_{\mathrm{HT}}) = \frac{(N-n)}{n}\left[\sum_{i=1}^{N} Y_i^2 - \frac{1}{N-1}\left\{Y^2 - \sum_{i=1}^{N} Y_i^2\right\}\right] = \frac{(N-n)}{n(N-1)}\left[N\sum_{i=1}^{N} Y_i^2 - Y^2\right]$$

$$= \frac{(N-n)}{n(N-1)}\left[N\sum_{i=1}^{N} Y_i^2 - (N\bar{Y})^2\right] = \frac{N(N-n)}{n}S_y^2 = \frac{N^2(1-f)}{n}S_y^2. \tag{5.1.27}$$

Hence the theorem.

Theorem 5.1.6. Another form for the $V(\hat{Y}_{\mathrm{HT}})$, developed by Sen (1953), and Yates and Grundy (1953) independently, is given by

$$V(\hat{Y}_{\mathrm{HT}})_{\mathrm{SYG}} = \frac{1}{2}\sum_{i\in\Omega}\sum_{j(\neq i)\in\Omega}(\pi_i\pi_j - \pi_{ij})\left(\frac{Y_i}{\pi_i} - \frac{Y_j}{\pi_j}\right)^2. \tag{5.1.28}$$

Proof. We have

$$V(\hat{Y}_{\mathrm{HT}})_{\mathrm{SYG}} = \frac{1}{2}\sum_{i\in\Omega}\sum_{j(\neq i)\in\Omega}(\pi_i\pi_j - \pi_{ij})\left(\frac{Y_i}{\pi_i} - \frac{Y_j}{\pi_j}\right)^2$$

$$= \frac{1}{2}\sum_{i\in\Omega}\sum_{j(\neq i)\in\Omega}(\pi_i\pi_j - \pi_{ij})\left[\frac{Y_i^2}{\pi_i^2} + \frac{Y_j^2}{\pi_j^2} - 2\frac{Y_i}{\pi_i}\frac{Y_j}{\pi_j}\right]$$

$$= \frac{1}{2}\sum_{i\in\Omega}\sum_{j(\neq i)\in\Omega}(\pi_i\pi_j - \pi_{ij})\left(\frac{Y_i^2}{\pi_i^2} + \frac{Y_j^2}{\pi_j^2}\right) - \frac{1}{2}\times 2\sum_{i\in\Omega}\sum_{j(\neq i)\in\Omega}(\pi_i\pi_j - \pi_{ij})\frac{Y_i}{\pi_i}\frac{Y_j}{\pi_j}$$

$$= \frac{1}{2}\sum_{i\in\Omega}\sum_{j(\neq i)\in\Omega}(\pi_i\pi_j - \pi_{ij})\left(\frac{Y_i^2}{\pi_i^2} + \frac{Y_j^2}{\pi_j^2}\right) + \sum_{i\in\Omega}\sum_{j(\neq i)\in\Omega}(\pi_{ij} - \pi_i\pi_j)\frac{Y_i}{\pi_i}\frac{Y_j}{\pi_j}. \tag{5.1.29}$$

Note that the probabilities for Y_i^2/π_i^2 and Y_j^2/π_j^2 are the same, therefore $\left(Y_i^2/\pi_i^2 + Y_j^2/\pi_j^2\right) = 2Y_i^2/\pi_i^2$ and (5.1.29) becomes

$$V(\hat{Y}_{\mathrm{HT}})_{\mathrm{SYG}} = \sum_{i\in\Omega}\sum_{j(\neq i)\in\Omega}(\pi_i\pi_j - \pi_{ij})\left(\frac{Y_i^2}{\pi_i^2}\right) + \sum_{i\in\Omega}\sum_{j(\neq i)\in\Omega}(\pi_{ij} - \pi_i\pi_j)\frac{Y_i}{\pi_i}\frac{Y_j}{\pi_j}$$

$$= \sum_{i\in\Omega}\frac{Y_i^2}{\pi_i^2}\pi_i\sum_{j(\neq i)\in\Omega}\pi_j - \sum_{i\in\Omega}\frac{Y_i^2}{\pi_i^2}\sum_{j(\neq i)\in\Omega}\pi_{ij} + \sum_{i\in\Omega}\sum_{j(\neq i)\in\Omega}(\pi_{ij} - \pi_i\pi_j)\frac{Y_i}{\pi_i}\frac{Y_j}{\pi_j}. \tag{5.1.30}$$

Note that $\sum_{j(\neq i)\in\Omega}\pi_j = n - \pi_i$, the ith unit with probability π_i is not there in the sum because the sum of all the inclusion probabilities is n. Also we know that $\sum_{j(\neq i)\in\Omega}\pi_{ij} = (n-1)\pi_i$. Using these results in (5.1.30) we have

$$V(\hat{Y}_{\mathrm{HT}})_{\mathrm{SYG}} = \sum_{i\in\Omega}\frac{Y_i^2}{\pi_i}(n - \pi_i) - \sum_{i\in\Omega}\frac{Y_i^2}{\pi_i^2}(n-1)\pi_i + \sum_{i\in\Omega}\sum_{j(\neq i)\in\Omega}(\pi_{ij} - \pi_i\pi_j)\frac{Y_i}{\pi_i}\frac{Y_j}{\pi_j}$$

$$= \sum_{i\in\Omega}\frac{Y_i^2}{\pi_i} - \sum_{i\in\Omega}Y_i^2 + \sum_{i\in\Omega}\sum_{j(\neq i)\in\Omega}\left(\frac{\pi_{ij}-\pi_i\pi_j}{\pi_i\pi_j}\right)Y_iY_j$$

$$= \sum_{i\in\Omega}\left(\frac{1-\pi_i}{\pi_i}\right)Y_i^2 + \sum_{i\in\Omega}\sum_{j(\neq i)\in\Omega}\left(\pi_{ij}-\pi_i\pi_j\right)\frac{Y_i}{\pi_i}\frac{Y_j}{\pi_j} = V\left(\hat{Y}_{\text{HT}}\right). \quad (5.1.31)$$

Hence the theorem.

Corollary 5.1.1. The variance $V\left(\hat{Y}_{\text{HT}}\right)_{\text{SYG}}$ reduces to zero if $\pi_i \propto Y_i$.

Proof. We have $\pi_i \propto Y_i$. Let α be the constant of proportionality, then $\pi_i = \alpha Y_i$. Taking sum on both sides of it we have $\sum_{i\in\Omega}\pi_i = \alpha\sum_{i\in\Omega}Y_i$, which implies that $\alpha = n/Y$. Thus we have $\pi_i = nY_i/Y$. On substituting $\pi_i = nY_i/Y$ in (5.1.28) we obtain $V\left(\hat{Y}_{\text{HT}}\right)_{\text{SYG}} = 0$. Hence the corollary.

Theorem 5.1.7. An unbiased estimator of the variance $V\left(\hat{Y}_{\text{HT}}\right)$ of the Horvitz and Thompson (1952) estimator of the population total Y is given by

$$\hat{v}\left(\hat{Y}_{\text{HT}}\right) = \sum_{i\in s}\frac{(1-\pi_i)}{\pi_i^2}y_i^2 + \sum_{i\in s}\sum_{j(\neq i)\in s}\left(\frac{\pi_{ij}-\pi_i\pi_j}{\pi_{ij}}\right)\frac{y_i}{\pi_i}\frac{y_j}{\pi_j}. \quad (5.1.32)$$

Proof. We know that the variance of the Horvitz and Thompson (1952) estimator of the population total, Y, is given by

$$V\left(\hat{Y}_{\text{HT}}\right) = \sum_{i\in\Omega}\frac{(1-\pi_i)}{\pi_i}Y_i^2 + \sum_{i\in\Omega}\sum_{j(\neq i)\in\Omega}\left(\frac{\pi_{ij}-\pi_i\pi_j}{\pi_i\pi_j}\right)Y_iY_j. \quad (5.1.33)$$

The estimator for the first term in (5.1.33) will be of the type

$$\sum_{i\in s}\left(\frac{1-\pi_i}{\pi_i}\right)y_i^2\alpha_i. \quad (5.1.34)$$

Now chose α_i in such a manner that

$$E\left[\sum_{i\in s}\left(\frac{1-\pi_i}{\pi_i}\right)y_i^2\alpha_i\right] = \sum_{i\in\Omega}\left(\frac{1-\pi_i}{\pi_i}\right)Y_i^2. \quad (5.1.35)$$

Now

$$E\left[\sum_{i\in s}\left(\frac{1-\pi_i}{\pi_i}\right)y_i^2\alpha_i\right] = E\left[\sum_{i\in\Omega}\left(\frac{1-\pi_i}{\pi_i}\right)Y_i^2\alpha_i t_i\right]$$

$$= \sum_{i\in\Omega}\left(\frac{1-\pi_i}{\pi_i}\right)Y_i^2\alpha_i E(t_i) = \sum_{i\in\Omega}\left(\frac{1-\pi_i}{\pi_i}\right)Y_i^2\alpha_i\pi_i. \quad (5.1.36)$$

Note that (5.1.35) will be true if

$$\sum_{i\in\Omega}\left(\frac{1-\pi_i}{\pi_i}\right)Y_i^2\alpha_i\pi_i = \sum_{i\in\Omega}\left(\frac{1-\pi_i}{\pi_i}\right)Y_i^2 \quad (5.1.37)$$

or if $\alpha_i\pi_i = 1$, *i.e.*, $\alpha_i = 1/\pi_i$. Thus an unbiased estimator of the first term of (5.1.33) is

$$\sum_{i\in s}\left(\frac{1-\pi_i}{\pi_i}\right)\frac{y_i^2}{\pi_i} = \sum_{i\in s}\left(\frac{1-\pi_i}{\pi_i^2}\right)y_i^2 . \tag{5.1.38}$$

Similarly an estimator for the second term of (5.1.33) will be of the type

$$\sum_{i\in s}\sum_{j(\neq i)\in s}\left(\frac{\pi_{ij}-\pi_i\pi_j}{\pi_i\pi_j}\right)y_i y_j \alpha_{ij} , \tag{5.1.39}$$

where α_{ij} is chosen in such a manner that

$$E\left[\sum_{i\in s}\sum_{j(\neq i)\in s}\left(\frac{\pi_{ij}-\pi_i\pi_j}{\pi_i\pi_j}\right)y_i y_j \alpha_{ij}\right] = \sum_{i\in\Omega}\sum_{j(\neq i)\in\Omega}\left(\frac{\pi_{ij}-\pi_i\pi_j}{\pi_i\pi_j}\right)Y_i Y_j . \tag{5.1.40}$$

Again we have

$$E\left[\sum_{i\in\Omega}\sum_{j(\neq i)\in\Omega}\left(\frac{\pi_{ij}-\pi_i\pi_j}{\pi_i\pi_j}\right)y_i y_j \alpha_{ij}\right] = E\left[\sum_{i\in\Omega}\sum_{j(\neq i)\in\Omega}\left(\frac{\pi_{ij}-\pi_i\pi_j}{\pi_i\pi_j}\right)Y_i Y_j \alpha_{ij} t_{ij}\right]$$

$$= \sum_{i\in\Omega}\sum_{j(\neq i)\in\Omega}\left(\frac{\pi_{ij}-\pi_i\pi_j}{\pi_i\pi_j}\right)Y_i Y_j \alpha_{ij} E(t_{ij})$$

$$= \sum_{i\in\Omega}\sum_{j(\neq i)\in\Omega}\left(\frac{\pi_{ij}-\pi_i\pi_j}{\pi_i\pi_j}\right)Y_i Y_j \alpha_{ij}\pi_{ij} . \tag{5.1.41}$$

Thus (5.1.40) will be true if

$$\sum_{j(\neq i)\in\Omega}\left(\frac{\pi_{ij}-\pi_i\pi_j}{\pi_i\pi_j}\right)Y_i Y_j \alpha_{ij}\pi_{ij} = \sum_{j(\neq i)\in\Omega}\left(\frac{\pi_{ij}-\pi_i\pi_j}{\pi_i\pi_j}\right)Y_i Y_j . \tag{5.1.42}$$

or in other words $\alpha_{ij} = 1/\pi_{ij}$. Therefore an estimator of the second term of (5.1.33) is given by

$$\sum_{i\in s}\sum_{j(\neq i)\in s}\left(\frac{\pi_{ij}-\pi_i\pi_j}{\pi_i\pi_j}\right)y_i y_j \frac{1}{\pi_{ij}} = \sum_{i\in s}\sum_{j(\neq i)\in s}\left(\frac{\pi_{ij}-\pi_i\pi_j}{\pi_{ij}}\right)\frac{y_i y_j}{\pi_i\pi_j} . \tag{5.1.43}$$

On combining (5.1.38) and (5.1.43) an unbiased estimator of $V(\hat{Y}_{HT})$ is given by

$$\hat{v}(\hat{Y}_{HT}) = \sum_{i\in s}\frac{(1-\pi_i)}{\pi_i^2}y_i^2 + \sum_{i\in s}\sum_{j(\neq i)\in s}\left(\frac{\pi_{ij}-\pi_i\pi_j}{\pi_{ij}}\right)\frac{y_i}{\pi_i}\frac{y_j}{\pi_j} . \tag{5.1.44}$$

Hence the theorem.

Theorem 5.1.8. An unbiased estimator of the variance of the Horvitz and Thompson (1952) estimator of the population total Y in the Sen--Yates--Grundy (1953) form is given by

$$\hat{v}_{\text{SYG}}(\hat{Y}_{\text{HT}}) = \frac{1}{2}\sum_{i\in s}\sum_{j(\neq i)\in s}\left(\frac{\pi_i\pi_j-\pi_{ij}}{\pi_{ij}}\right)\left(\frac{y_i}{\pi_i}-\frac{y_j}{\pi_j}\right)^2 . \tag{5.1.45}$$

Proof. We know that the Sen--Yates--Grundy (1953) form of the variance of the Horvitz and Thompson (1952) estimator of the population total Y is given by

$$V_{\text{SYG}}\left(\hat{Y}_{\text{HT}}\right)=\frac{1}{2}\sum_{i\in\Omega}\sum_{j(\neq i)\in\Omega}\left(\pi_i\pi_j-\pi_{ij}\right)\left(\frac{Y_i}{\pi_i}-\frac{Y_j}{\pi_j}\right)^2 . \tag{5.1.46}$$

The estimator for (5.1.46) will be of the type

$$\frac{1}{2}\sum_{i\in s}\sum_{j(\neq i)\in s}\left(\pi_i\pi_j-\pi_{ij}\right)\left(\frac{y_i}{\pi_i}-\frac{y_j}{\pi_j}\right)^2\alpha_{ij} . \tag{5.1.47}$$

If we choose α_{ij} in such a manner that

$$E\left[\frac{1}{2}\sum_{i\in s}\sum_{j(\neq i)\in s}\left(\pi_i\pi_j-\pi_{ij}\right)\left(\frac{y_i}{\pi_i}-\frac{y_j}{\pi_j}\right)^2\alpha_{ij}\right]=\frac{1}{2}\sum_{i\in\Omega}\sum_{j(\neq i)\in\Omega}\left(\pi_i\pi_j-\pi_{ij}\right)\left(\frac{Y_i}{\pi_i}-\frac{Y_j}{\pi_j}\right)^2 \tag{5.1.48}$$

.

Now we have

$$E\left[\frac{1}{2}\sum_{i\in s}\sum_{j(\neq i)\in s}\left(\pi_i\pi_j-\pi_{ij}\right)\left(\frac{y_i}{\pi_i}-\frac{y_j}{\pi_j}\right)^2\alpha_{ij}\right]=E\left[\frac{1}{2}\sum_{i\in\Omega}\sum_{j(\neq i)\in\Omega}\left(\pi_i\pi_j-\pi_{ij}\right)\left(\frac{Y_i}{\pi_i}-\frac{Y_j}{\pi_j}\right)^2\alpha_{ij}t_{ij}\right]$$

$$=\frac{1}{2}\sum_{i\in\Omega}\sum_{j(\neq i)\in\Omega}\left(\pi_i\pi_j-\pi_{ij}\right)\left(\frac{Y_i}{\pi_i}-\frac{Y_j}{\pi_j}\right)^2\alpha_{ij}E\left(t_{ij}\right)$$

$$=\frac{1}{2}\sum_{i\in\Omega}\sum_{j(\neq i)\in\Omega}\left(\pi_i\pi_j-\pi_{ij}\right)\left(\frac{Y_i}{\pi_i}-\frac{Y_j}{\pi_j}\right)^2\alpha_{ij}\pi_{ij} , \tag{5.1.49}$$

which implies that

$$\frac{1}{2}\sum_{i\in\Omega}\sum_{j(\neq i)\in\Omega}\left(\pi_i\pi_j-\pi_{ij}\right)\left(\frac{Y_i}{\pi_i}-\frac{Y_j}{\pi_j}\right)^2\alpha_{ij}\pi_{ij}=\frac{1}{2}\sum_{i\in\Omega}\sum_{j(\neq i)\in\Omega}\left(\pi_i\pi_j-\pi_{ij}\right)\left(\frac{Y_i}{\pi_i}-\frac{Y_j}{\pi_j}\right)^2 \tag{5.1.50}$$

if $\alpha_{ij}\pi_{ij}=1$, *i.e.*, $\alpha_{ij}=1/\pi_{ij}$.

Therefore an unbiased estimator of the variance in the Sen--Yates--Grundy (1953) form is given by (5.1.45). Hence the theorem.

The condition for the estimator of variance of the Horvitz and Thompson (1952) estimator of population total Y to be non-negative is given by

$$\pi_i\pi_j \geq \pi_{ij} . \tag{5.1.51}$$

Example 5.1.1. Consider a population of $N=5$ units $\Omega=\{A,B,C,D,E\}$ and if we wish to select a sample of $n=3$ units using without replacement sampling out of total 10 sample $s_t=1,2,3,...10$ given by

$s_1=\{A,B,C\}$, $s_2=\{A,B,D\}$, $s_3=\{A,B,E\}$, $s_4=\{A,C,D\}$, $s_5=\{A,C,E\}$,
$s_6=\{A,D,E\}$, $s_7=\{B,C,D\}$, $s_8=\{B,C,E\}$, $s_9=\{B,D,E\}$, and $s_{10}=\{C,D,E\}$.

(a) Find the first order and second order inclusion probabilities provided that all samples have equal chance of selection that is $P(s_t) = 1/10, \ \forall \ t$.

(b) Compare your results with SRSWOR sampling scheme.

(c) $P(s_1) = 0.50$, $P(s_2) = 0.30$, $P(s_3) = 0.2$, $P(s_t) = 0.00, t = 4, 5, 6, 7, 8, 9, 10$ then find the new first order and second order inclusion probabilities.

Solution. (a) When all samples have equal chance of selection, that is, $P(s_t) = 0.1$ $\forall \ t$ then the first order inclusion probabilities π_i, $i = 1, 2, ..., 5$ are given by

$$\begin{aligned}
\pi_1 &= P[\text{First unit } A \text{ from the population is included in the sample}] \\
&= P(s_1) + P(s_2) + P(s_3) + P(s_4) + P(s_5) + P(s_6) \\
&= 0.1 + 0.1 + 0.1 + 0.1 + 0.1 + 0.1 = 0.6,
\end{aligned}$$

$$\begin{aligned}
\pi_2 &= P[\text{Second unit } B \text{ from the population is included in the sample}] \\
&= P(s_1) + P(s_2) + P(s_3) + P(s_7) + P(s_8) + P(s_9) \\
&= 0.1 + 0.1 + 0.1 + 0.1 + 0.1 + 0.1 = 0.6,
\end{aligned}$$

$$\begin{aligned}
\pi_3 &= P[\text{Third unit } C \text{ from the population is included in the sample}] \\
&= P(s_1) + P(s_4) + P(s_5) + P(s_7) + P(s_8) + P(s_{10}) \\
&= 0.1 + 0.1 + 0.1 + 0.1 + 0.1 + 0.1 = 0.6,
\end{aligned}$$

$$\begin{aligned}
\pi_4 &= P[\text{Fourth unit } D \text{ from the population is included in the sample}] \\
&= P(s_2) + P(s_4) + P(s_6) + P(s_7) + P(s_9) + P(s_{10}) \\
&= 0.1 + 0.1 + 0.1 + 0.1 + 0.1 + 0.1 = 0.6,
\end{aligned}$$

and

$$\begin{aligned}
\pi_5 &= P[\text{Fifth unit } E \text{ from the population is included in the sample}] \\
&= P(s_3) + P(s_5) + P(s_6) + P(s_8) + P(s_9) + P(s_{10}) \\
&= 0.1 + 0.1 + 0.1 + 0.1 + 0.1 + 0.1 = 0.6.
\end{aligned}$$

The second order inclusion probabilities π_{ij}, $i \neq j = 1, 2, 3, 4, 5$ (note that $\pi_{ij} = \pi_{ji}$) are given by

$$\begin{aligned}
\pi_{12} &= P[\text{First and Second units } A \text{ and } B \text{ from the population are included in the sample}] \\
&= P(s_1) + P(s_2) + P(s_3) = 0.1 + 0.1 + 0.1 = 0.3,
\end{aligned}$$

$$\begin{aligned}
\pi_{13} &= P[\text{First and Third units } A \text{ and } C \text{ from the population are included in the sample}] \\
&= P(s_1) + P(s_4) + P(s_5) = 0.1 + 0.1 + 0.1 = 0.3,
\end{aligned}$$

$$\begin{aligned}
\pi_{14} &= P[\text{First and Fourth units } A \text{ and } D \text{ from the population are included in the sample}] \\
&= P(s_2) + P(s_4) + P(s_6) = 0.1 + 0.1 + 0.1 = 0.3,
\end{aligned}$$

$$\begin{aligned}
\pi_{15} &= P[\text{First and Fifth units } A \text{ and } E \text{ from the population are included in the sample}] \\
&= P(s_3) + P(s_5) + P(s_6) = 0.1 + 0.1 + 0.1 = 0.3,
\end{aligned}$$

$$\begin{aligned}
\pi_{23} &= P[\text{Second and Third units } B \text{ and } C \text{ from the population are included in the sample}] \\
&= P(s_1) + P(s_7) + P(s_8) = 0.1 + 0.1 + 0.1 = 0.3,
\end{aligned}$$

$$\begin{aligned}
\pi_{24} &= P[\text{Second and Fourth units } B \text{ and } D \text{ from the population are included in the sample}] \\
&= P(s_2) + P(s_7) + P(s_9) = 0.1 + 0.1 + 0.1 = 0.3,
\end{aligned}$$

$$\begin{aligned}\pi_{25} &= P[\text{Second and Fifth units } B \text{ and } E \text{ from the population are included in the sample}]\\ &= P(s_3)+P(s_8)+P(s_9) = 0.1+0.1+0.1 = 0.3,\end{aligned}$$

$$\begin{aligned}\pi_{34} &= P[\text{Third and Fourth units } C \text{ and } D \text{ from the population are included in the sample}]\\ &= P(s_4)+P(s_7)+P(s_{10}) = 0.1+0.1+0.1 = 0.3,\end{aligned}$$

$$\begin{aligned}\pi_{35} &= P[\text{Third and Fifth units } C \text{ and } E \text{ from the population are included in the sample}]\\ &= P(s_5)+P(s_8)+P(s_{10}) = 0.1+0.1+0.1 = 0.3,\end{aligned}$$

and

$$\begin{aligned}\pi_{45} &= P[\text{Fourth and Fifth units } D \text{ and } E \text{ from the population are included in the sample}]\\ &= P(s_6)+P(s_9)+P(s_{10}) = 0.1+0.1+0.1 = 0.3.\end{aligned}$$

Thus we observed that if all the samples have the same chance of selection then $\pi_i = 0.6, \ \forall i = 1,2,..,5$ and $\pi_{ij} = 0.3, \ \forall i \neq j = 1,2,..,5$.

(b) Also note that under SRSWOR sampling

$$\pi_i = \frac{n}{N} = \frac{3}{5} = 0.6 \text{, and } \pi_{ij} = \frac{n(n-1)}{N(N-1)} = \frac{3(3-1)}{5(5-1)} = \frac{6}{20} = 0.3.$$

Thus if all samples have the same chance of selection then PPSWOR and SRSWOR sampling schemes are equivalent.

(c) In this case we have

$$\begin{aligned}\pi_1 &= P[\text{First unit } A \text{ from the population is included in the sample}]\\ &= P(s_1)+P(s_2)+P(s_3)+P(s_4)+P(s_5)+P(s_6)\\ &= 0.5+0.3+0.2+0.0+0.0+0.0 = 1.0,\end{aligned}$$

$$\begin{aligned}\pi_2 &= P[\text{Second unit } B \text{ from the population is included in the sample}]\\ &= P(s_1)+P(s_2)+P(s_3)+P(s_7)+P(s_8)+P(s_9)\\ &= 0.5+0.3+0.2+0.0+0.0+0.0 = 1.0,\end{aligned}$$

$$\begin{aligned}\pi_3 &= P[\text{Third unit } C \text{ from the population is included in the sample}]\\ &= P(s_1)+P(s_4)+P(s_5)+P(s_7)+P(s_8)+P(s_{10})\\ &= 0.5+0.0+0.0+0.0+0.0+0.0 = 0.5,\end{aligned}$$

$$\begin{aligned}\pi_4 &= P[\text{Fourth unit } D \text{ from the population is included in the sample}]\\ &= P(s_2)+P(s_4)+P(s_6)+P(s_7)+P(s_9)+P(s_{10})\\ &= 0.3+0.0+0.0+0.0+0.0+0.0 = 0.3,\end{aligned}$$

and

$$\begin{aligned}\pi_5 &= P[\text{Fifth unit } E \text{ from the population is included in the sample}]\\ &= P(s_3)+P(s_5)+P(s_6)+P(s_8)+P(s_9)+P(s_{10})\\ &= 0.2+0.0+0.0+0.0+0.0+0.0 = 0.2.\end{aligned}$$

The second order inclusion probabilities π_{ij}, $i \neq j = 1,2,3,4,5$, (note that $\pi_{ij} = \pi_{ji}$) are given by

$$\begin{aligned}\pi_{12} &= P[\text{First and Second units } A \text{ and } B \text{ from the population are included in the sample}]\\ &= P(s_1)+P(s_2)+P(s_3) = 0.5+0.3+0.2 = 1.0,\end{aligned}$$

$$\pi_{13} = P[\text{First and Third units } A \text{ and } C \text{ from the population are included in the sample}]$$
$$= P(s_1) + P(s_4) + P(s_5) = 0.5 + 0.0 + 0.0 = 0.5,$$

$$\pi_{14} = P[\text{First and Fourth units } A \text{ and } D \text{ from the population are included in the sample}]$$
$$= P(s_2) + P(s_4) + P(s_6) = 0.3 + 0.0 + 0.0 = 0.3,$$

$$\pi_{15} = P[\text{First and Fifth units } A \text{ and } E \text{ from the population are included in the sample}]$$
$$= P(s_3) + P(s_5) + P(s_6) = 0.2 + 0.0 + 0.0 = 0.2,$$

$$\pi_{23} = P[\text{Second and Third units } B \text{ and } C \text{ from the population are included in the sample}]$$
$$= P(s_1) + P(s_7) + P(s_8) = 0.5 + 0.0 + 0.0 = 0.5,$$

$$\pi_{24} = P[\text{Second and Fourth units } B \text{ and } D \text{ from the population are included in the sample}]$$
$$= P(s_2) + P(s_7) + P(s_9) = 0.3 + 0.0 + 0.0 = 0.3,$$

$$\pi_{25} = P[\text{Second and Fifth units } B \text{ and } E \text{ from the population are included in the sample}]$$
$$= P(s_3) + P(s_8) + P(s_9) = 0.2 + 0.0 + 0.0 = 0.2,$$

$$\pi_{34} = P[\text{Third and Fourth units } C \text{ and } D \text{ from the population are included in the sample}]$$
$$= P(s_4) + P(s_7) + P(s_{10}) = 0.0 + 0.0 + 0.0 = 0.0,$$

$$\pi_{35} = P[\text{Third and Fifth units } C \text{ and } E \text{ from the population are included in the sample}]$$
$$= P(s_5) + P(s_8) + P(s_{10}) = 0.0 + 0.0 + 0.0 = 0.0,$$

and

$$\pi_{45} = P[\text{Fourth and Fifth units } D \text{ and } E \text{ from the population are included in the sample}]$$
$$= P(s_6) + P(s_9) + P(s_{10}) = 0.0 + 0.0 + 0.0 = 0.0.$$

The next example explains *How Michael selected Amy*?

Example 5.1.2. John and Michael were appointed to select three players ($n = 3$) out of five players ($N = 5$) from the list $\Omega = \{\text{Amy, Bob, Chris, Don, Eric}\}$ with their scores 125, 126, 128, 90 and 127, respectively.

(I) John considers the following sampling scheme consisting of only four possibilities:

$$s_1 = \{\text{Amy, Bob, Chris}\},\ s_2 = \{\text{Bob, Chris, Don}\},$$

$$s_3 = \{\text{Chris, Don, Eric}\}, \text{ and } s_4 = \{\text{Amy, Don, Eric}\},$$

such that

$$P(s_1) = 0.25,\ P(s_2) = 0.25,\ P(s_3) = 0.25 \text{ and}$$
$$P(s_4) = 0.25.$$

(a) Find the first order inclusion probabilities for John's sampling scheme.
(b) Find the estimates of total score from each sample using John's sampling .
(c) Find the bias in John's sampling scheme.
(d) Find the second order inclusion probabilities for John's sampling scheme.

(e) Find the variance of John's sampling scheme using the Sen--Yates--Grundy formula.
(f) Find the variance of John's sampling plan using usual formula.
(g) Find the variance of John's sampling plan using definition of variance.
(h) Are three variances equal for John's sampling scheme?

(II) Michael likes Amy and cleverly suggests the following changes in John's sampling scheme as: $P(s_1)=0.50$, $P(s_2)=0.00$, $P(s_3)=0.00$ and $P(s_4)=0.50$.
(i) Find the first order inclusion probabilities for Michael's sampling scheme.
(j) Find the estimates of total score from each sample using Michael's sampling.
(k) Find the bias in Michael's sampling scheme.
(l) Find the second order inclusion probabilities for Michael's sampling scheme.
(m) Find the variance of Michael's sampling scheme using the Sen--Yates--Grundy formula.
(n) Find the variance of Michael's sampling plan using usual formula.
(o) Find the variance of Michael's sampling plan using definition of variance.
(p) Are three variances equal for Michael's sampling scheme?

(III) Discussion on John and Michael's schemes:

(q) Find the relative efficiency of Michael's sampling scheme over John's sampling.
(r) Would you like to comment on the results?

Solution: (I) John's Sampling Scheme:

(a) The first order inclusion probabilities for John's sampling scheme are

$$\pi_1 = P[\text{Amy is included in the sample}] = P(s_1) + P(s_4) = 0.25 + 0.25 = 0.50,$$

$$\pi_2 = P[\text{Bob is included in the sample}] = P(s_1) + P(s_2) = 0.25 + 0.25 = 0.50,$$

$$\pi_3 = P[\text{Chris is included in the sample}] = P(s_1) + P(s_2) + P(s_3) = 0.25 + 0.25 + 0.25 = 0.75,$$

$$\pi_4 = P[\text{Don is included in the sample}] = P(s_2) + P(s_3) + P(s_4) = 0.25 + 0.25 + 0.25 = 0.75,$$

and

$$\pi_5 = P[\text{Eric is included in the sample}] = (s_3) + P(s_4) = 0.25 + 0.25 = 0.50\,.$$

Note that $\sum_{i=1}^{5} \pi_i = 3$.

(b) Let $\hat{Y}_{\text{HT}(t)}$, $t = 1, 2, 3, 4$ be the estimates of total score based on the first, second, third and fourth sample respectively, then we have

$$\hat{Y}_{\text{HT}(1)\text{John}} = \sum_{i \in s_1} \frac{y_i}{\pi_i} = \frac{\text{Amy}}{\pi_1} + \frac{\text{Bob}}{\pi_2} + \frac{\text{Chris}}{\pi_3} = \frac{125}{0.50} + \frac{126}{0.50} + \frac{128}{0.75} = 672.66\overline{6},$$

$$\hat{Y}_{\text{HT}(2)\text{John}} = \sum_{i \in s_2} \frac{y_i}{\pi_i} = \frac{\text{Bob}}{\pi_2} + \frac{\text{Chris}}{\pi_3} + \frac{\text{Don}}{\pi_4} = \frac{126}{0.50} + \frac{128}{0.75} + \frac{90}{0.75} = 542.66\overline{6},$$

$$\hat{Y}_{\text{HT}(3)\text{John}} = \sum_{i\in s_3}\frac{y_i}{\pi_i} = \frac{\text{Chris}}{\pi_3}+\frac{\text{Don}}{\pi_4}+\frac{\text{Eric}}{\pi_5} = \frac{128}{0.75}+\frac{90}{0.75}+\frac{127}{0.50} = 544.66\overline{6},$$

and

$$\hat{Y}_{\text{HT}(4)\text{John}} = \sum_{i\in s_4}\frac{y_i}{\pi_i} = \frac{\text{Amy}}{\pi_1}+\frac{\text{Don}}{\pi_4}+\frac{\text{Eric}}{\pi_5} = \frac{125}{0.50}+\frac{90}{0.75}+\frac{127}{0.50} = 624.000\,.$$

(c) The bias in John's sampling scheme: Note that true total score of the five players is

$$Y = \sum_{i=1}^{5} Y_i = \text{Amy}+\text{Bob}+\text{Chris}+\text{Don}+\text{Eric} = 125+126+128+90+127 = 596.$$

Now the bias in John's sampling scheme is given by

$$B\left(\hat{Y}_{\text{HT}}\right)_{\text{John}} = E\left(\hat{Y}_{\text{HT}}\right) - Y = \sum_{t=1}^{4}\left\{P(s_t)\hat{Y}_{\text{HT(t)}_{\text{John}}}\right\} - Y$$

$$= \left\{0.25\times 672.66\overline{6} + 0.25\times 542.66\overline{6} + 0.25\times 544.66\overline{6} + 0.25\times 624.000\right\} - 596$$

$$= 596.000 - 596.000 = 0.000\,.$$

Thus John's sampling scheme is unbiased for estimating the total score.

(d) The second order inclusion probabilities for John's sampling scheme are

$$\pi_{12} = P[\text{Amy and Bob are included in the sample}] = P(s_1) = 0.25,$$

$$\pi_{13} = P[\text{Amy and Chris are included in the sample}] = P(s_1) = 0.25,$$

$$\pi_{14} = P[\text{Amy and Don are included in the sample}] = P(s_4) = 0.25,$$

$$\pi_{15} = P[\text{Amy and Eric are included in the sample}] = P(s_4) = 0.25,$$

$$\pi_{23} = P[\text{Bob and Chris are included in the sample}] = P(s_1) + P(s_2) = 0.25 + 0.25 = 0.50,$$

$$\pi_{24} = P[\text{Bob and Don are included in the sample}] = P(s_2) = 0.25,$$

$$\pi_{25} = P[\text{Bob and Eric are included in the sample}] = 0.00,$$

$$\pi_{34} = P[\text{Chris and Don are included in the sample}] = P(s_2) + P(s_3) = 0.25 + 0.25 = 0.50,$$

$$\pi_{35} = P[\text{Chris and Eric are included in the sample}] = P(s_3) = 0.25,$$

and

$$\pi_{45} = P[\text{Don and Eric are included in the sample}] = P(s_3) + P(s_4) = 0.25 + 0.25 = 0.50\,.$$

(e) The variance of John's sampling plan using the Sen --Yates --Grundy formula is

$$V_{\text{SYG}}\left(\hat{Y}_{\text{HT}}\right)_{\text{John}}$$

$$= \frac{1}{2}\sum\sum_{i\neq j\in\Omega}\left(\pi_i\pi_j - \pi_{ij}\right)\left(\frac{Y_i}{\pi_i} - \frac{Y_j}{\pi_j}\right)^2$$

$$= \left(\pi_1\pi_2 - \pi_{12}\right)\left(\frac{Y_1}{\pi_1} - \frac{Y_2}{\pi_2}\right)^2 + \left(\pi_1\pi_3 - \pi_{13}\right)\left(\frac{Y_1}{\pi_1} - \frac{Y_3}{\pi_3}\right)^2$$

$$+\left(\pi_1\pi_4-\pi_{14}\right)\left(\frac{Y_1}{\pi_1}-\frac{Y_4}{\pi_4}\right)^2+\left(\pi_1\pi_5-\pi_{15}\right)\left(\frac{Y_1}{\pi_1}-\frac{Y_5}{\pi_5}\right)^2$$

$$+\left(\pi_2\pi_3-\pi_{23}\right)\left(\frac{Y_2}{\pi_2}-\frac{Y_3}{\pi_3}\right)^2+\left(\pi_2\pi_4-\pi_{24}\right)\left(\frac{Y_2}{\pi_2}-\frac{Y_4}{\pi_4}\right)^2$$

$$+\left(\pi_2\pi_5-\pi_{25}\right)\left(\frac{Y_2}{\pi_2}-\frac{Y_5}{\pi_5}\right)^2+\left(\pi_3\pi_4-\pi_{34}\right)\left(\frac{Y_3}{\pi_3}-\frac{Y_4}{\pi_4}\right)^2$$

$$+\left(\pi_3\pi_5-\pi_{35}\right)\left(\frac{Y_3}{\pi_3}-\frac{Y_5}{\pi_5}\right)^2+\left(\pi_4\pi_5-\pi_{45}\right)\left(\frac{Y_4}{\pi_4}-\frac{Y_5}{\pi_5}\right)^2$$

$$=\left(0.50\times0.50-0.25\right)\left(\frac{125}{0.50}-\frac{126}{0.50}\right)^2+\left(0.50\times0.75-0.25\right)\left(\frac{125}{0.50}-\frac{128}{0.75}\right)^2$$

$$+\left(0.50\times0.75-0.25\right)\left(\frac{125}{0.50}-\frac{90}{0.75}\right)^2+\left(0.50\times0.50-0.25\right)\left(\frac{125}{0.50}-\frac{127}{0.50}\right)^2$$

$$+\left(0.50\times0.75-0.50\right)\left(\frac{126}{0.50}-\frac{128}{0.75}\right)^2+\left(0.50\times0.75-0.25\right)\left(\frac{126}{0.50}-\frac{90}{0.75}\right)^2$$

$$+\left(0.50\times0.50-0.00\right)\left(\frac{126}{0.50}-\frac{127}{0.50}\right)^2+\left(0.75\times0.75-0.50\right)\left(\frac{128}{0.75}-\frac{90}{0.75}\right)^2$$

$$+\left(0.75\times0.50-0.25\right)\left(\frac{128}{0.75}-\frac{127}{0.50}\right)^2+\left(0.75\times0.50-0.50\right)\left(\frac{90}{0.75}-\frac{127}{0.50}\right)^2$$

$$=3035.33\overline{3}\,.$$

(f) The variance of John's sampling plan using usual formula is given by

$$V\left(\hat{Y}_{\text{HT}}\right)_{\text{John}}$$

$$=\sum_{i\in\Omega}\left(\frac{1-\pi_i}{\pi_i}\right)Y_i^2+\sum_{i\in\Omega}\sum_{j(\neq i)\in\Omega}\left(\frac{\pi_{ij}-\pi_i\pi_j}{\pi_i\pi_j}\right)Y_iY_j$$

$$=\left(\frac{1-\pi_1}{\pi_1}\right)Y_1^2+\left(\frac{1-\pi_2}{\pi_2}\right)Y_2^2+\left(\frac{1-\pi_3}{\pi_3}\right)Y_3^2+\left(\frac{1-\pi_4}{\pi_4}\right)Y_4^2+\left(\frac{1-\pi_5}{\pi_5}\right)Y_5^2$$

$$+2\left[\left(\frac{\pi_{12}-\pi_1\pi_2}{\pi_1\pi_2}\right)Y_1Y_2+\left(\frac{\pi_{13}-\pi_1\pi_3}{\pi_1\pi_3}\right)Y_1Y_3+\left(\frac{\pi_{14}-\pi_1\pi_4}{\pi_1\pi_4}\right)Y_1Y_4\right.$$

$$+\left(\frac{\pi_{15}-\pi_1\pi_5}{\pi_1\pi_5}\right)Y_1Y_5+\left(\frac{\pi_{23}-\pi_2\pi_3}{\pi_2\pi_3}\right)Y_2Y_3+\left(\frac{\pi_{24}-\pi_2\pi_4}{\pi_2\pi_4}\right)Y_2Y_4$$

$$\left.+\left(\frac{\pi_{25}-\pi_2\pi_5}{\pi_2\pi_5}\right)Y_2Y_5+\left(\frac{\pi_{34}-\pi_3\pi_4}{\pi_3\pi_4}\right)Y_3Y_4+\left(\frac{\pi_{35}-\pi_3\pi_5}{\pi_3\pi_5}\right)Y_3Y_5+\left(\frac{\pi_{45}-\pi_4\pi_5}{\pi_4\pi_5}\right)Y_4Y_5\right]$$

$$= \left(\frac{1-0.50}{0.50}\right)(125)^2 + \left(\frac{1-0.50}{0.50}\right)(126)^2 + \left(\frac{1-0.75}{0.75}\right)(128)^2$$

$$+ \left(\frac{1-0.75}{0.75}\right)(90)^2 + \left(\frac{1-0.50}{0.50}\right)(127)^2$$

$$+ 2\left[\left(\frac{0.25-0.50\times 0.50}{0.50\times 0.50}\right)(125\times 126) + \left(\frac{0.25-0.50\times 0.75}{0.50\times 0.75}\right)(125\times 128)\right.$$

$$+ \left(\frac{0.25-0.50\times 0.75}{0.50\times 0.75}\right)(125\times 90) + \left(\frac{0.25-0.50\times 0.50}{0.50\times 0.50}\right)(125\times 127)$$

$$+ \left(\frac{0.50-0.50\times 0.75}{0.50\times 0.75}\right)(126\times 128) + \left(\frac{0.25-0.50\times 0.75}{0.50\times 0.75}\right)(126\times 90)$$

$$+ \left(\frac{0.00-0.50\times 0.50}{0.50\times 0.50}\right)(126\times 127) + \left(\frac{0.50-0.75\times 0.75}{0.75\times 0.75}\right)(128\times 90)$$

$$\left. + \left(\frac{0.25-0.75\times 0.50}{0.75\times 0.50}\right)(128\times 127) + \left(\frac{0.50-0.75\times 0.50}{0.75\times 0.50}\right)(90\times 127)\right]$$

$$= 55791.33\overline{3} + 2(-26378) = 3035.33\overline{3}.$$

(g) Variance by definition for John's sampling scheme is given by

$$V_{\text{Def}}\left(\hat{Y}_{\text{HT}}\right)_{\text{John}} = \sum_{t=1}^{4} P(s_t)\left\{\hat{Y}_{\text{HT}(t)\text{John}} - Y\right\}^2$$

$$= 0.25\left(672.66\overline{6} - 596\right)^2 + 0.25\left(542.66\overline{6} - 596\right)^2$$

$$+ 0.25\left(544.66\overline{6} - 596\right)^2 + 0.25(624 - 596)^2$$

$$= 3035.33\overline{3}.$$

(h) Yes, $V_{\text{SYG}}\left(\hat{Y}_{\text{HT}}\right)_{\text{John}} = V\left(\hat{Y}_{\text{HT}}\right)_{\text{John}} = V_{\text{Def}}\left(\hat{Y}_{\text{HT}}\right)_{\text{John}}$.

(II) Michael's sampling scheme:

(i) The first order inclusion probabilities for Michael's sampling scheme are

$$\pi_1 = P[\text{Amy is included in the sample}] = P(s_1) + P(s_4) = 0.50 + 0.50 = 1.00,$$

$$\pi_2 = P[\text{Bob is included in the sample}] = P(s_1) + P(s_2) = 0.50 + 0.00 = 0.50,$$

$$\pi_3 = P[\text{Chris is included in the sample}] = P(s_1) + P(s_2) + P(s_3) = 0.50 + 0.00 + 0.00 = 0.50,$$

$$\pi_4 = P[\text{Don is included in the sample}] = P(s_2) + P(s_3) + P(s_4) = 0.00 + 0.00 + 0.50 = 0.50,$$

and

$$\pi_5 = P[\text{Eric is included in the sample}] = P(s_3) + P(s_4) = 0.00 + 0.50 = 0.50.$$

Note that $\sum_{i=1}^{5} \pi_i = 3$.

(j) Let $\hat{Y}_{HT(t)_{Michael}}$, $t = 1, 2, 3, 4$ be the estimates of total score based on the first, second, third and fourth sample respectively, then we have

$$\hat{Y}_{HT(1)_{Michael}} = \sum_{i \in s_1} \frac{y_i}{\pi_i} = \frac{Amy}{\pi_1} + \frac{Bob}{\pi_2} + \frac{Chris}{\pi_3} = \frac{125}{1.00} + \frac{126}{0.50} + \frac{128}{0.50} = 633.000,$$

$$\hat{Y}_{HT(2)_{Michael}} = \sum_{i \in s_2} \frac{y_i}{\pi_i} = \frac{Bob}{\pi_2} + \frac{Chris}{\pi_3} + \frac{Don}{\pi_4} = \frac{126}{0.50} + \frac{128}{0.50} + \frac{90}{0.50} = 688.000,$$

$$\hat{Y}_{HT(3)_{Michael}} = \sum_{i \in s_3} \frac{y_i}{\pi_i} = \frac{Chris}{\pi_3} + \frac{Don}{\pi_4} + \frac{Eric}{\pi_5} = \frac{128}{0.50} + \frac{90}{0.50} + \frac{127}{0.50} = 690.000,$$

and

$$\hat{Y}_{HT(4)_{Michael}} = \sum_{i \in s_4} \frac{y_i}{\pi_i} = \frac{Amy}{\pi_1} + \frac{Don}{\pi_4} + \frac{Eric}{\pi_5} = \frac{125}{1.00} + \frac{90}{0.50} + \frac{127}{0.50} = 559.000.$$

(k) The bias in Michael's sampling scheme is given by

$$B\left(\hat{Y}_{HT}\right)_{Michael} = E\left(\hat{Y}_{HT}\right) - Y = \sum_{t=1}^{4} \left\{P(s_t)\hat{Y}_{HT(t)_{Michael}}\right\} - Y$$

$$= \{0.50 \times 633.000 + 0.00 \times 688.000 + 0.00 \times 690.000 + 0.50 \times 559.000\} - 596$$

$$= 596.000 - 596 = 0.000 .$$

Thus Michael's sampling scheme is also unbiased for estimating the total score.

(l) The second order inclusion probabilities for Michael's sampling scheme are

$\pi_{12} = P[\text{Amy and Bob are included in the sample}] = P(s_1) = 0.50,$

$\pi_{13} = P[\text{Amy and Chris are included in the sample}] = P(s_1) = 0.50,$

$\pi_{14} = P[\text{Amy and Don are included in the sample}] = P(s_4) = 0.50,$

$\pi_{15} = P[\text{Amy and Eric are included in the sample}] = P(s_4) = 0.50,$

$\pi_{23} = P[\text{Bob and Chris are included in the sample}] = P(s_1) + P(s_2) = 0.50 + 0.00 = 0.50,$

$\pi_{24} = P[\text{Bob and Don are included in the sample}] = P(s_2) = 0.00,$

$\pi_{25} = P[\text{Bob and Eric are included in the sample}] = 0.00,$

$\pi_{34} = P[\text{Chris and Don are included in the sample}] = P(s_2) + P(s_3) = 0.00 + 0.00 = 0.00,$

$\pi_{35} = P[\text{Chris and Eric are included in the sample}] = P(s_3) = 0.00,$

and

$\pi_{45} = P[\text{Don and Eric are included in the sample}] = P(s_3) + P(s_4) = 0.00 + 0.50 = 0.50 .$

(m) The variance of Michael's sampling plan using Sen--Yates--Grundy formula is given by

$$V_{\text{SYG}}\left(\hat{Y}_{\text{HT}}\right)_{\text{Michael}}$$

$$=\frac{1}{2}\sum_{i\neq}\sum_{j\in\Omega}\left(\pi_i\pi_j-\pi_{ij}\right)\left(\frac{Y_i}{\pi_i}-\frac{Y_j}{\pi_j}\right)^2$$

$$=\left(\pi_1\pi_2-\pi_{12}\right)\left(\frac{Y_1}{\pi_1}-\frac{Y_2}{\pi_2}\right)^2+\left(\pi_1\pi_3-\pi_{13}\right)\left(\frac{Y_1}{\pi_1}-\frac{Y_3}{\pi_3}\right)^2$$

$$+\left(\pi_1\pi_4-\pi_{14}\right)\left(\frac{Y_1}{\pi_1}-\frac{Y_4}{\pi_4}\right)^2+\left(\pi_1\pi_5-\pi_{15}\right)\left(\frac{Y_1}{\pi_1}-\frac{Y_5}{\pi_5}\right)^2$$

$$+\left(\pi_2\pi_3-\pi_{23}\right)\left(\frac{Y_2}{\pi_2}-\frac{Y_3}{\pi_3}\right)^2+\left(\pi_2\pi_4-\pi_{24}\right)\left(\frac{Y_2}{\pi_2}-\frac{Y_4}{\pi_4}\right)^2$$

$$+\left(\pi_2\pi_5-\pi_{25}\right)\left(\frac{Y_2}{\pi_2}-\frac{Y_5}{\pi_5}\right)^2+\left(\pi_3\pi_4-\pi_{34}\right)\left(\frac{Y_3}{\pi_3}-\frac{Y_4}{\pi_4}\right)^2$$

$$+\left(\pi_3\pi_5-\pi_{35}\right)\left(\frac{Y_3}{\pi_3}-\frac{Y_5}{\pi_5}\right)^2+\left(\pi_4\pi_5-\pi_{45}\right)\left(\frac{Y_4}{\pi_4}-\frac{Y_5}{\pi_5}\right)^2$$

$$=\left(1.00\times0.50-0.50\right)\left(\frac{125}{1.00}-\frac{126}{0.50}\right)^2+\left(1.00\times0.50-0.50\right)\left(\frac{125}{1.00}-\frac{128}{0.50}\right)^2$$

$$+\left(1.00\times0.50-0.50\right)\left(\frac{125}{1.00}-\frac{90}{0.50}\right)^2+\left(1.00\times0.50-0.50\right)\left(\frac{125}{1.00}-\frac{127}{0.50}\right)^2$$

$$+\left(0.50\times0.50-0.50\right)\left(\frac{126}{0.50}-\frac{128}{0.50}\right)^2+\left(0.50\times0.50-0.00\right)\left(\frac{126}{0.50}-\frac{90}{0.50}\right)^2$$

$$+\left(0.50\times0.50-0.00\right)\left(\frac{126}{0.50}-\frac{127}{0.50}\right)^2+\left(0.50\times0.50-0.00\right)\left(\frac{128}{0.50}-\frac{90}{0.50}\right)^2$$

$$+\left(0.50\times0.50-0.00\right)\left(\frac{128}{0.50}-\frac{127}{0.50}\right)^2+\left(0.50\times0.50-0.50\right)\left(\frac{90}{0.50}-\frac{127}{0.50}\right)^2$$

$$=1369.000.$$

(n) The variance of Michael's sampling plan using usual formula is given by

$$V\left(\hat{Y}_{\text{HT}}\right)_{\text{Michael}}$$

$$=\sum_{i\in\Omega}\left(\frac{1-\pi_i}{\pi_i}\right)Y_i^2+\sum_{i\in\Omega}\sum_{j(\neq i)\in\Omega}\left(\frac{\pi_{ij}-\pi_i\pi_j}{\pi_i\pi_j}\right)Y_iY_j$$

$$=\left(\frac{1-\pi_1}{\pi_1}\right)Y_1^2+\left(\frac{1-\pi_2}{\pi_2}\right)Y_2^2+\left(\frac{1-\pi_3}{\pi_3}\right)Y_3^2+\left(\frac{1-\pi_4}{\pi_4}\right)Y_4^2+\left(\frac{1-\pi_5}{\pi_5}\right)Y_5^2$$

$$+2\left[\left(\frac{\pi_{12}-\pi_1\pi_2}{\pi_1\pi_2}\right)Y_1Y_2+\left(\frac{\pi_{13}-\pi_1\pi_3}{\pi_1\pi_3}\right)Y_1Y_3+\left(\frac{\pi_{14}-\pi_1\pi_4}{\pi_1\pi_4}\right)Y_1Y_4\right.$$

$$+\left(\frac{\pi_{15}-\pi_1\pi_5}{\pi_1\pi_5}\right)Y_1Y_5+\left(\frac{\pi_{23}-\pi_2\pi_3}{\pi_2\pi_3}\right)Y_2Y_3+\left(\frac{\pi_{24}-\pi_2\pi_4}{\pi_2\pi_4}\right)Y_2Y_4$$

$$\left.+\left(\frac{\pi_{25}-\pi_2\pi_5}{\pi_2\pi_5}\right)Y_2Y_5+\left(\frac{\pi_{34}-\pi_3\pi_4}{\pi_3\pi_4}\right)Y_3Y_4+\left(\frac{\pi_{35}-\pi_3\pi_5}{\pi_3\pi_5}\right)Y_3Y_5+\left(\frac{\pi_{45}-\pi_4\pi_5}{\pi_4\pi_5}\right)Y_4Y_5\right]$$

$$=\left(\frac{1-1.00}{1.00}\right)(125)^2+\left(\frac{1-0.50}{0.50}\right)(126)^2+\left(\frac{1-0.50}{0.50}\right)(128)^2$$

$$+\left(\frac{1-0.50}{0.50}\right)(90)^2+\left(\frac{1-0.50}{0.50}\right)(127)^2$$

$$+2\left[\left(\frac{0.50-1.00\times0.50}{1.00\times0.50}\right)(125\times126)+\left(\frac{0.50-1.00\times0.50}{1.00\times0.50}\right)(125\times128)\right.$$

$$+\left(\frac{0.50-1.00\times0.50}{1.00\times0.50}\right)(125\times90)+\left(\frac{0.50-1.00\times0.50}{1.00\times0.50}\right)(125\times127)$$

$$+\left(\frac{0.50-0.50\times0.50}{0.50\times0.50}\right)(126\times128)+\left(\frac{0.00-0.50\times0.50}{0.50\times0.50}\right)(126\times90)$$

$$+\left(\frac{0.00-0.50\times0.50}{0.50\times0.50}\right)(126\times127)+\left(\frac{0.00-0.50\times0.50}{0.50\times0.50}\right)(128\times90)$$

$$\left.+\left(\frac{0.00-0.50\times0.50}{0.50\times0.50}\right)(128\times127)+\left(\frac{0.50-0.50\times0.50}{0.50\times0.50}\right)(90\times127)\right]$$

$$=41989+2(-27560)=1369.000.$$

(o) Variance by definition for Michael's sampling scheme is given by

$$V_{\text{Def}}\left(\hat{Y}_{\text{HT}}\right)_{\text{Michael}}=\sum_{t=1}^{4}P(s_t)\left\{\hat{Y}_{\text{HT}(t)\text{Michael}}-Y\right\}^2$$

$$=0.50(633-596)^2+0.00(688-596)^2$$

$$+0.00(690-596)^2+0.50(559-596)^2$$

$$=1369.000.$$

(p) Yes, $V_{\text{SYG}}\left(\hat{Y}_{\text{HT}}\right)_{\text{Michael}}=V\left(\hat{Y}_{\text{HT}}\right)_{\text{Michael}}=V_{\text{Def}}\left(\hat{Y}_{\text{HT}}\right)_{\text{Michael}}$.

(III) Comparison among schemes:

(q) The percent relative efficiency of Michael's sampling scheme over John's sampling scheme is given by

$$\text{RE}=\frac{V\left(\hat{Y}_{\text{HT}}\right)_{\text{John}}}{V\left(\hat{Y}_{\text{HT}}\right)_{\text{Michael}}}\times100=\frac{3035.33\overline{3}}{1369.000}\times100=221.71\%$$

(r) **Comment:** When we read this example we thought John's scheme would be more efficient and it looks that Michael is doing unnecessary favor to Amy, but from the bias and variance point of view we were surprised that Michael's scheme is more efficient than John's sampling scheme. There may be some different reason behind it, but two reasons apparently comes to our minds:

(i) Why did John not consider all possible samples?

If John did not consider all possible samples we cannot do anything. If he selects all possible samples, then we do not know how much probability he was going to assign to different samples? How was Michael going to react? We may get an answer to this question by doing an unsolved practical at the back of this chapter, so let us think more here!

(ii) What is the correlation between inclusion probabilities and the study variable?

This seems to be a good point. We observed that the value of correlation coefficient between the John's first order inclusion probabilities and the study variable is negative that is $\rho_{\pi y(\text{John})} = -0.569$ and that for Michael's sampling scheme is positive that is $\rho_{\pi y(\text{Michael})} = +0.198$.

Caution: While using PPSWOR sampling, we should make sure that the inclusion probabilities have positive correlation with the study variable. Note that the estimation of variance from each sample for John's sampling scheme is possible, but remains biased. Further note that estimation of variance from each sample using Michael's sampling scheme is not possible, because certain useful second order inclusion probabilities are zero.

Moral: Note that three of the candidates (Bob, Chris and Eric) have higher scores than Amy and John did not consider them together. Michael took the benefit because John was trying to break the merit. Message for the future generations, "Do not attempt to break a merit and be honest if you get a chance to be an administrator as otherwise someone, like Michael, may take benefit of your limitation"

The next example has been taken from Ghosh (1998) which compares the two estimators of variance of the Horvitz and Thompson (1952) estimators.

Example 5.1.3. A computer salesman wishes to estimate the number of left handed students in a town having eight elementary schools. The salesman must make a decision to ensure enough left hand mice are ordered to accommodate his expected sale. He used prior information on the number of registered students to select a sample of $n = 3$ units by using PPSWOR sampling. The information collected by him is given in the following table:

Sample Information					
Sampled Unit	No. of left handed students	First order inclusion probabilities	Second order inclusion probabilities for the sampled units (π_{ij}) j		
i	y_i	π_i	1	2	3
1	10	0.45	-	0.20	0.20
2	4	0.40	0.20	-	0.20
3	2	0.50	0.20	0.20	-

Discuss two 75% confidence intervals based on two estimators of variance of the Horvitz and Thompson estimator.

Solution: An estimate of the total number of left handed students in all schools of the town is given by

$$\hat{Y}_{\text{HT}} = \sum_{i\in s}\frac{y_i}{\pi_i} = \frac{y_1}{\pi_1}+\frac{y_2}{\pi_2}+\frac{y_3}{\pi_3} = \frac{10}{0.45}+\frac{4}{0.40}+\frac{2}{0.50} = 36.2 \approx 36 .$$

Now the variance of the estimator $\hat{Y}_{\text{HT}}$ can be estimated two different ways:

Case 1. The usual estimator of variance of $\hat{Y}_{\text{HT}}$ is given by

$$\hat{v}\left(\hat{Y}_{\text{HT}}\right) = \sum_{i\in s}\left(\frac{1-\pi_i}{\pi_i}\right)\frac{y_i^2}{\pi_i} + \sum\sum_{i\neq j\in s}\left(\frac{\pi_{ij}-\pi_i\pi_j}{\pi_i\pi_j}\right)\frac{y_iy_j}{\pi_{ij}}$$

$$= \left(\frac{1-\pi_1}{\pi_1^2}\right)y_1^2 + \left(\frac{1-\pi_2}{\pi_2^2}\right)y_2^2 + \left(\frac{1-\pi_3}{\pi_3^2}\right)y_3^2$$

$$+2\left(\frac{\pi_{12}-\pi_1\pi_2}{\pi_1\pi_2}\right)\frac{y_1y_2}{\pi_{12}} + 2\left(\frac{\pi_{13}-\pi_1\pi_3}{\pi_1\pi_3}\right)\frac{y_1y_3}{\pi_{13}} + 2\left(\frac{\pi_{23}-\pi_2\pi_3}{\pi_2\pi_3}\right)\frac{y_2y_3}{\pi_{23}}$$

$$= \left(\frac{1-0.45}{0.45^2}\right)\times 10^2 + \left(\frac{1-0.40}{0.40^2}\right)\times 4^2 + \left(\frac{1-0.50}{0.50^2}\right)\times 2^2$$

$$+2\left(\frac{0.20-0.45\times 0.40}{0.45\times 0.40}\right)\frac{10\times 4}{0.20} + 2\left(\frac{0.20-0.45\times 0.50}{0.45\times 0.5}\right)\frac{10\times 2}{0.20}$$

$$+2\left(\frac{0.20-0.40\times 0.50}{0.40\times 0.50}\right)\frac{4\times 2}{0.20}$$

$$= 271.6049 + 60 + 8 + 44.4444 - 22.2222 + 0 = 361.82 .$$

Thus a $(1-\alpha)100\%$ confidence interval for the number of left handed students will be given by

$$\hat{Y}_{\text{HT}} \pm t_{\alpha/2}(\text{df} = n-1)\sqrt{\hat{v}\left(\hat{Y}_{\text{HT}}\right)} .$$

Using Table 2 from the Appendix the 75% confidence interval estimate of the number of left handed students in the town is given by

$$\hat{Y}_{HT} \pm t_{0..25/2}(\text{df} = 3-1)\sqrt{\hat{v}(\hat{Y}_{HT})}$$

or

$$36.22 \pm 1.604\sqrt{361.82}\text{ , or } [5.709, 66.73]\text{, or } [6,\ \ 67].$$

Case II. The Sen--Yates--Grundy estimator of variance of $\hat{Y}_{HT}$ is given by

$$\hat{v}_{SYG}(\hat{Y}_{HT}) = \frac{1}{2}\sum_{i \neq j \in s}\sum\left(\frac{\pi_i\pi_j - \pi_{ij}}{\pi_{ij}}\right)\left(\frac{y_i}{\pi_i} - \frac{y_j}{\pi_j}\right)^2 = \sum_{i<j \in s}\sum\left(\frac{\pi_i\pi_j - \pi_{ij}}{\pi_{ij}}\right)\left(\frac{y_i}{\pi_i} - \frac{y_j}{\pi_j}\right)^2$$

$$= \left(\frac{\pi_1\pi_2 - \pi_{12}}{\pi_{12}}\right)\left(\frac{y_1}{\pi_1} - \frac{y_2}{\pi_2}\right)^2 + \left(\frac{\pi_1\pi_3 - \pi_{13}}{\pi_{13}}\right)\left(\frac{y_1}{\pi_1} - \frac{y_3}{\pi_3}\right)^2$$

$$+ \left(\frac{\pi_2\pi_3 - \pi_{23}}{\pi_{23}}\right)\left(\frac{y_2}{\pi_3} - \frac{y_3}{\pi_3}\right)^2$$

$$= \left(\frac{0.45 \times 0.4 - 0.2}{0.2}\right)\left(\frac{10}{0.45} - \frac{4}{0.4}\right)^2 + \left(\frac{0.45 \times 0.5 - 0.2}{0.2}\right)\left(\frac{10}{0.45} - \frac{2}{0.5}\right)^2$$

$$+ \left(\frac{0.4 \times 0.5 - 0.2}{0.2}\right)\left(\frac{4}{0.4} - \frac{2}{0.5}\right)^2$$

$$= -14.938 + 41.506 + 0 = 26.568\,.$$

A $(1-\alpha)100\%$ confidence interval for the number of left handed students will be

$$\hat{Y}_{HT} \pm t_{\alpha/2}(\text{df} = n-1)\sqrt{\hat{v}_{SYG}(\hat{Y}_{HT})}$$

Using Table 2 from the Appendix the 75% confidence interval estimate of the number of left handed students in the town is given by

$$\hat{Y}_{HT} \pm t_{0.25/2}(\text{df} = 3-1)\sqrt{\hat{v}_{SYG}(\hat{Y}_{HT})},\text{ or } 36.22 \pm 1.604\sqrt{26.568}$$

$$\text{or } [27.95, 44.488]\text{ ,or } [28,\ 44].$$

Following Ghosh (1998), it seems difficult to conclude which estimator of variance is better. In this example the Sen--Yates--Grundy estimator performs better because it provides a smaller interval estimate at the same level of confidence. Ghosh (1998) found that it is difficult to decide to select an estimator of variance in actual practice because Rao and Singh (1973) have shown that the usual estimator of variance is hyper-admissible. Ghosh (1998) ends his paper with a criticism similar to Basu (1971). Basu's famous circus elephant example is available in recent books

written by Thompson (1997) and Brewer (2002). Basu considers the problem of estimation of the weight of 50 elephants in a circus. Instead of weighing all elephants, the owner of the circus weighs only one elephant named *Sambo* by using three year old information about the size of *Sambo* as a middle sized elephant in the circus. If w is the weight of *Sambo* then an estimate of the total weight of 50 elephants in the circus is $\hat{Y} = 50w$. Later a circus statistician suggests that the owner should apply random sampling in place of purpose sampling. Both the owner and statistician decide to use a random sampling device which gives a 99/100 chance of selection to *Sambo* and 1/4900 chance of selection to rest of the 49 elephants in the circus. Thus the first order inclusion probabilities are given by

$$\pi_i = \begin{cases} \dfrac{99}{100} & \text{if Sambo is included in the sample,} \\ \dfrac{1}{4900} & \text{otherwise.} \end{cases}$$

Clearly $\sum_{i=1}^{50} \pi_i = 1$ indicates selection of a sample of only one unit. Owing to the high probability of selection, *Sambo* is selected in the sample and the circus statistician reports an estimate of the total weight of 50 elephants as

$$\hat{Y}_{\text{HT}} = \sum_{i \in s} \frac{y_i}{\pi_i} = \frac{100}{99} w$$

which is approximately the weight of *Sambo*. Then the owner asks the statistician "If a large elephant named *Jumbo* would have been selected what would have been the estimate of total weight?"

Let W be the weight of *Jumbo*. Then the Horvitz and Thompson estimate of the total weight of all elephants is given by

$$\hat{Y}_{\text{HT}} = \sum_{i \in s} \frac{y_i}{\pi_i} = 4900W$$

which clearly is an over estimation of the true weight of 50 elephants. The main mistake made by the circus statistician was to give a small selection probability to *Jumbo* and a large selection probability to *Sambo*. This ignore the fact that PPS sampling works only if the correlation between selection probabilities and the study variable is positive and high. In fact the selection method made by the circus statistician shows that he might be a circus clown instead of a statistician or he may not have understood the meaning of Horvitz and Thompson (1952). Thus aim of Basu (1971) is to show that if some wrong prior (or Bayes) information is used in the estimation process the results may be too biased, and gives us a caution while using Bayes estimates!. Let us make the circus statistician's problem clear with the help of following examples.

Example 5.1.4. Suppose there are 10 elephants in a circus and their weights and diets are given in the following table:

Elephant No.	1	2	3	4	5	6	7	8	9	10
Name	Jumbo	Jumbo	Sambo	Sambo	Niko	Niko	Niko	Niko	Niko	Niko
Weight (kg)	5000	5000	1000	1000	500	500	500	500	500	500
Diet (kg)	300	250	75	75	50	50	50	50	50	50

Based on a sample of one elephant, estimate the total weight of the 10 elephants using the Horvitz and Thompson estimator.

Solution. In this case we are considering a sample of $n = 1$ unit. Then we can consider that the inclusion probabilities are the same as the selection probabilities because $\sum_{i \in \Omega} \pi_i = \sum_{i=1}^{N} P_i = n = 1$. Assuming that the diet of each elephant is known, the selection probabilities are given by

Elephant No.	1	2	3	4	5	6	7	8	9	10
P_i	0.30	0.25	0.075	0.075	0.050	0.050	0.050	0.050	0.050	0.050

Now if the first elephant *Jumbo* is selected in the sample, the Horvitz and Thompson's estimate of the total weight of the 10 elephants is given by

$$\hat{Y}_{\mathrm{HT}(1)} = \frac{5000}{0.30} = 16666.67 \text{ kg.}$$

Similarly, if the second, third or fifth elephant, that is either *Jumbo*, *Sambo*, *Niko*, is included in the sample, then the respective estimates of the weight of all elephants are given by

$$\hat{Y}_{\mathrm{HT}(2)} = \frac{5000}{0.25} = 20000\text{kg}, \; \hat{Y}_{\mathrm{HT}(3)} = \frac{1000}{0.075} = 13333.33\text{kg}, \text{ and } \hat{Y}_{\mathrm{HT}(5)} = \frac{500}{0.050} = 10000\text{kg}.$$

The true weight of the 10 elephants is given by $Y = 15000\text{kg}$. Thus one can easily observe that in Basu's (1971) example, the Horvitz and Thompson (1952) estimator was correct. Rather it was misused by the circus statistician by assigning incorrect inclusion probabilities. The circus statistician problem is illustrated in the next example.

Example 5.1.5. Suppose there are 10 elephants in a circus and their weights and diets are given in the following table:

Elephant No.	1	2	3	4	5	6	7	8	9	10
Elephant name	Jumbo	Jumbo	Sambo	Sambo	Niko	Niko	Niko	Niko	Niko	Niko
Weight (kg)	5000	5000	1000	1000	500	500	500	500	500	500
Diet (kg)	1	1	81	1	1	1	1	1	1	1

Based on a sample of one elephant, estimate the total weight of the 10 elephants using the Horvitz and Thompson estimator.

Solution. Again if we consider a sample of $n = 1$ unit then the inclusion probabilities are the same as the selection probabilities because

$$\sum_{i \in \Omega} \pi_i = \sum_{i=1}^{N} P_i = n = 1.$$

Assuming that the diet of each elephant is known, the selection probabilities are

Elephant No.	1	2	3	4	5	6	7	8	9	10
P_i	1/90	1/90	81/90	1/90	1/90	1/90	1/90	1/90	1/90	1/90

Now if the first elephant *Jumbo* is selected in the sample the Horvitz and Thompson's estimate of the total weight of the 10 elephants is given by

$$\hat{Y}_{\text{HT}}(1) = \frac{5000}{1/90} = 450000 \text{ kg.}$$

The true weight of the 10 elephants is only $Y = 15000\text{kg}$. Similar to Basu (1971), the total weight of the elephants is highly over estimated. The circus statistician never considered that *Jumbo*'s diet was heavier than *Sambo*'s diet. Thompson (1997) also accepted that the circus statistician lost his job and perhaps became a teacher of statistics, but he cannot be a teacher. Brewer (2002) attempted to weigh elephants under SRSWOR sampling by taking a sample of 5 elephants, but Basu's (1971) case is very serious. No doubt a sample of 5 units may represent a heterogeneous population of 50 units and it may be possible to weigh an elephant in a spring balance, but how can a sample of one unit represent a heterogeneous population? There is a very basic assumption in sampling that the sample has to be random and representative of the population. If the circus statistician does not know about these two requirements in sampling theory, then it would have been better if the circus statistician would have become a clown rather than a teacher. Caution! The use of the wrong Bayes information is more dangerous than using no information at all. A layman can understand this in more simple language as follows: Consider a policeman following a thief in going north, and the thief is also going north, but on the way, while following the thief, the policeman received a phone call from the police station that the thief went south, and now the policeman changes his direction towards the south, but due to inaccurate information from the police station, the policeman is now going away from the thief and will never reach the thief. Basu (1971) alerts survey statisticians to be careful while using Bayes estimates. More details about Basu's contribution to the foundation of survey sampling theory can be found by Meeden (1992) and a decent monograph by Ghosh and Meeden (1997).

5.2 GENERAL CLASS OF ESTIMATORS

As we discussed in the previous chapters, ratio and regression type estimators under simple random sampling have been studied by a number of researchers, including Cochran (1963), Srivastava (1967), Reddy (1974), Gupta (1978), Vos (1980), Srivenkataramana and Tracy (1980,1981), and Singh and Singh (1993a). Most of them are special cases of the class of estimators proposed by Srivastava (1971) in which the efficiency of the optimum estimators is the same as that of the linear regression estimator. There are a number of estimators *viz.* Srivenkataramana and Tracy (1980) and Ray and Singh (1981) estimators which do not belong to the Srivastava (1971) class. These are more efficient than the optimum estimators in the Srivastava (1971) class. Das and Tripathi (1980) have used the coefficient of variation of the auxiliary variable to form a class of estimators which are more efficient than linear regression estimator. Srivastava and Jhajj (1981) have proposed a general class of estimators in which, along with the ratio of sample mean to population mean of the auxiliary variable, the ratio of sample variance to population variance of the auxiliary variable has also been used, and the optimum estimator of the proposed class was shown to be better than the linear regression estimator. For the general sampling design, an unbiased estimator of S_x^2 cannot be easily derived but an unbiased estimator of $\sum_{i=1}^{N} X_i^r$ can be easily developed for any $r \ge 1$. Following Cassel, Särndal, and Wretman (1977), under any sampling design, the population total can be estimated unbiasedly if and only if the first order inclusion probabilities π_i are positive for all the units in the population. For any such sampling design, obviously

$$\hat{Y} = \sum_{i \in s} \frac{y_i}{\pi_i}, \quad \hat{X}_1 = \sum_{i \in s} \frac{x_i}{\pi_i} \text{ and } \hat{X}_2 = \sum_{i \in s} \frac{x_i^2}{\pi_i}$$

are the unbiased estimators of

$$Y = \sum_{i=1}^{N} Y_i, \quad X_1 = \sum_{i=1}^{N} X_i \text{ and } X_2 = \sum_{i=1}^{N} X_i^2, \text{ respectively.}$$

Sampath and Chandra (1990) assume that whatever sample is chosen, (u_1, u_2) assumes values only in the closed convex subset D of the two-dimensional real space containing the point $(1, 1)$, where $u_1 = \hat{X}_1 / X_1$ and $u_2 = \hat{X}_2 / X_2$. They considered the general class of estimators of population total Y as

$$\hat{Y}_g = \hat{Y} g(u_1, u_2) \tag{5.2.1}$$

where $g(u_1, u_2)$ is a function satisfying the following conditions:

(a) $g(1,1) = 1$;
(b) $g(u_1, u_2)$ is a continuous and bounded in D; and
(c) The first and second order partial derivatives of the function g exist and are continuous and bounded.

Then we have the following theorem:

Theorem 5.2.1. (a) The lower bound of the asymptotic mean squared error (AMSE) of the general class of estimators $\hat{Y}_g$ defined in (5.2.1) is given by

$$\text{MSE}(\hat{Y}_g) = V_0 - \theta\{V_1 C_{02}^2 + V_2 C_{01}^2 - 2C_{01}C_{02}C_{12}\} \tag{5.2.2}$$

where $\theta = (V_1 V_2 - C_{12}^2)^{-1}$, $V_0 = V(\hat{Y})$, $V_1 = V(\hat{X}_1)$, $V_2 = V(\hat{X}_2)$, $C_{01} = \text{Cov}(\hat{Y}, \hat{X}_1)$, $C_{02} = \text{Cov}(\hat{Y}, \hat{X}_2)$ and $C_{12} = \text{Cov}(\hat{X}_1, \hat{X}_2)$.

(b) Also show that, $\text{MSE}(\hat{Y}_g) \le \text{MSE}(\hat{Y}_{lr})$, where $\text{MSE}(\hat{Y}_{lr}) = V_0\left\{1 - \frac{C_{01}^2}{V_0 V_1}\right\}$ denotes the mean squared error of the general linear regression type estimator studied by Särndal (1980a) and given by $\hat{Y}_{lr} = \hat{Y} + \hat{\beta}(X_1 - \hat{X}_1)$.

Proof. (a) Following the previous chapter, expanding the function $g(u_1, u_2)$ around the point $(1, 1)$ in a Taylor's second order series, the general class of estimators can be approximately written as

$$\hat{Y}_g = \hat{Y}\{1 + (u_1 - 1)g_1(1,1) + (u_2 - 1)g_2(1,1) + ...\},$$

where $g_1(1,1)$ and $g_2(1,1)$ denote the first order partial derivatives of the function g with respect to u_1 and u_2, respectively. Then by the definition of MSE we have

$$\text{MSE}(\hat{Y}_g) = Y^2\left[\frac{V_0}{Y^2} + \frac{V_1}{X_1^2}g_1^2(1,1) + \frac{V_2}{X_2^2}g_2^2(1,1) + 2\frac{C_{01}}{YX_1}g_1(1,1)\right.$$
$$\left. + 2\frac{C_{02}}{YX_2}g_2(1,1) + 2\frac{C_{12}}{X_1X_2}g_1(1,1)g_2(1,1)\right]. \tag{5.2.3}$$

On differentiating (5.2.3) with respect to $g_1(1,1)$ and $g_2(1,1)$ and equating to zero we obtain their optimum values as

$$g_1(1,\ 1) = \frac{X_1}{Y}\theta(V_2C_{01} - C_{12}C_{02}), \text{ and } g_2(1,\ 1) = \frac{X_2}{Y}\theta(V_1C_{01} - C_{12}C_{01}). \tag{5.2.4}$$

On putting (5.2.4) in (5.2.3) we have the minimum mean squared error given by (5.2.2).

(b) Now if

$$\text{MSE}(\hat{Y}_g) - \text{MSE}(\hat{Y}_{\text{lr}}) = V_0 - \theta\left[V_1C_{02}^2 + V_2C_{01}^2 - 2C_{01}C_{02}C_{12}\right] - V_0\left[1 - \frac{C_{01}^2}{V_0V_1}\right] \le 0$$

then we have

$$\frac{C_{01}^2}{V_1} \le \frac{V_1C_{02}^2 + V_2C_{01}^2 - 2C_{01}C_{02}C_{12}}{V_1V_2 - C_{12}^2}, \text{ or } \frac{C_{01}^2}{V_0V_1} \le \frac{V_1C_{02}^2 + V_2C_{01}^2 - 2C_{01}C_{02}C_{12}}{V_0(V_1V_2 - C_{12}^2)}$$

or,

$$\rho_{01}^2 \le \frac{\rho_{02}^2 + \rho_{01}^2 - 2\rho_{01}\rho_{02}\rho_{12}}{1 - \rho_{12}^2}, \text{ or } (\rho_{02} - \rho_{01}\rho_{12})^2 \ge 0$$

which is always true. Hence the theorem.

5.3 MODEL BASED ESTIMATION STRATEGIES

Before defining the model we should define a few notation which will be helpful in understanding the model based estimation strategies. Consider we wish to estimate the population mean, $\bar{Y} = N^{-1}\sum_{i=1}^{N} Y_i$, based on a sample s of n observations drawn with probability $p(s)$ from a population of N units. The function $p(s)$, defined for all samples s, is called the sampling design. We shall consider the problem of estimation of population mean using fixed effective sample size, *i.e.*, all the units in the sample are distinct. Let π_i and π_{ij} be the probabilities of including i^{th} and i^{th}

and j^{th} population units in the sample and are called the first and second order inclusion probabilities. These inclusion probabilities can also be defined as

$$\pi_i = \sum_{s \ni i} p(s) \tag{5.3.1}$$

and

$$\pi_{ij} = \sum_{s \ni i,j} p(s). \tag{5.3.2}$$

Definition 5.3.1. An estimator $\hat{\theta}_s$ is said to be model unbiased for population parameter θ if

$$E_m(\hat{\theta}_s) = \theta \tag{5.3.3}$$

where E_m denotes the expected value over the model M.

Definition 5.3.2. An estimator $\hat{\theta}_s$ is said to be design unbiased for the population parameter θ if

$$E_p(\hat{\theta}_s) = \sum_s p(s)\hat{\theta}_s = \theta \tag{5.3.4}$$

where E_p denotes the expected value with respect to design $p(s)$.

Definition 5.3.3. Under design based approach the mean squared error of the estimator $\hat{\theta}_s$ is defined as

$$\text{MSE}(\hat{\theta}_s) = E_p(\hat{\theta}_s - \theta)^2 . \tag{5.3.5}$$

Under the model M we have two situations:

(a) The finite population mean squared error averaged over the super population model by Cochran (1963) is defined as

$$\text{MSE}(\hat{\theta}_s) = E_m E_p[\hat{\theta}_s - \theta]^2 . \tag{5.3.6}$$

(b) The mean squared error of the estimator and finite population parameter over the super population model M by Royall (1971) is defined as

$$\text{MSE}(\hat{\theta}_s) = E_m[\hat{\theta}_s - \theta]^2 . \tag{5.3.7}$$

The above definitions lead to several possibilities depending upon the interest of the investigators or survey statisticians.

For example:

(a) Rao (1979) minimized (5.3.6) subject to (5.3.4) and found that it leads to the conventional sampling strategies involving randomization;

(b) Rao (1979) also noted that minimization of (5.3.7) subject to (5.3.3) provides purposive selection strategies;

(c) Minimization of (5.3.6) or (5.3.7) is also possible subject to some cost function.

Definition 5.3.4. A cost function is any linear or non-linear function of the costs of selecting a unit in the sample and a number of units in the sample.

Definition 5.3.5. A general superpopulation model can be defined as

$$M \,:\, Y_i = \sum_{k=0}^{m} \beta_k X_i^k + \epsilon_i \,. \tag{5.3.8}$$

The regressors X_i, $i = 1,2,...,N$, are the non-stochastic auxiliary variables and ϵ_i is a random error such that

$$E_m\left(\epsilon_i \mid X_i^k\right) = 0\,,\quad E_m\left(\epsilon_i^2 \mid X_i^k\right) = \sigma^2 f(X_i),\quad \text{and}\quad E_m\left(\epsilon_i \epsilon_j \mid X_i^k X_j^k\right) = 0\,,\ i \neq j\,,$$

where $f(X_i)$ is a function of X_i.

5.3.1 A BRIEF HISTORY OF THE SUPERPOPULATION MODEL

Various strategies have been proposed for estimating a finite population mean or population total under a superpopulation model M that relates the variable of interest to one or more auxiliary variables. Brewer (1963a) and Royall (1970a, 1970b, 1970c, 1971, 1976) have adapted a linear model prediction theory to the finite population situation and have derived the best linear M unbiased (BLU) predictor. Cassel, Särndal, and Wretman (1976, 1977) and Särndal (1980b) have proposed a generalized regression predictor that is asymptotically design unbiased (ADU). Brewer (1979) suggested a predictor that blends aspects of the BLU and generalized regression predictors and retains the ADU property by using a single auxiliary variable. Isaki and Fuller (1982) proposed some ADU predictors involving several auxiliary variables or characters. Wright (1983) examines strategies that are approximately design unbiased and nearly optimal, assuming a large sample survey and a regression superpopulation model and suggested a new class of predictors to link certain features of optimal design-unbiased and model-unbiased predictors.

Thus we have the following theorem:

Theorem 5.3.1.1. Under the superpopulation model

$$Y_i = \beta_0 + \beta_1 X_{1i} + \beta_2 X_{2i} + \ldots\ldots + \beta_k X_{ki} + \epsilon_i \tag{5.3.1.1}$$

with the assumptions

$$E(\epsilon_i) = 0\,,\quad E\left(\epsilon_i^2\right) = \sigma^2 v(x_i) \text{ and } E\left(\epsilon_i \epsilon_j\right) = 0\quad \forall i \neq j\,, \tag{5.3.1.2}$$

a predictor of population mean proposed by Wright (1983) is given by

$$\bar{y}_p = \underline{X}^t \underline{\hat{\beta}} + N^{-1} \sum_{i=1}^{n} \hat{\epsilon}_i \tag{5.3.1.3}$$

where $\underline{\hat{\beta}} = \left(\underline{X}^t \underline{X}\right)^{-1} \underline{X}^t \underline{Y}$

$$\underline{X} = \begin{bmatrix} 1, X_{11}, X_{21}, \ldots\ldots, X_{k1} \\ 1, X_{12}, X_{22}, \ldots\ldots, X_{k2} \\ \\ 1, X_{1n}, X_{2n}, \ldots\ldots, X_{kn} \end{bmatrix}, \quad \underline{Y} = \begin{bmatrix} Y_1 \\ Y_2 \\ Y_3 \\ \\ Y_n \end{bmatrix}, \quad \underline{\beta} = \begin{bmatrix} \beta_1 \\ \beta_2 \\ \\ \\ \beta_k \end{bmatrix}, \text{ and } \hat{\epsilon} = \underline{Y} - \underline{X}'\underline{\beta}.$$

Royall (1971) showed that the choice of design $p(s)$ which minimizes $E_m E_p(\bar{y}_s - \bar{Y})^2$ leads to the purposive design. If $f(x_i)$ is a non-decreasing function of x_i and $f(x_i)/x_i^2$ is non-increasing, then the sampled n units with largest values provide the optimal sample selection. Brewer (1963a) noted that the purposive design which minimizes $E_m E_p(\bar{y}_s - \bar{Y})^2$, the ratio method of estimation is optimal when $f(X_i) = aX_i^g$ for $a = 1$ and $0 \le g \le 1$. Cassel and Särndal (1974) pointed out that the study of Brewer (1963a) holds well even for continuous distributions of the auxiliary characters. Royall's (1970a, 1970b, 1970c) result and his relevant work has been the subject of much controversy for statisticians. The criticisms were noted by Royall (1971), Cox (1971), and Wynn (1977a, 1977b). Neyman (1971) was somewhat stronger in his criticism, saying that it would be dangerous to draw a sample based on an unverified model. The optimal result is dependent on the model assumed. In other words, different models could lead to different kinds of results such as bias, etc.. Related criticism and interesting results can also be seen from Royall and Herson (1973a, 1973b). We will discuss the robust estimation procedure of Scott, Brewer, and Ho (1978) which is in fact the extension of the work of Royall and Herson (1973a, 1973b), ensuring the robustness of the standard ratio estimator against polynomial superpopulation models by choosing balanced samples, to the case of more general regression estimator.

5.3.2 SCOTT, BREWER, AND HO'S ROBUST ESTIMATION STRATEGY

Scott, Brewer, and Ho (1978) have shown that the requirement for robustness is a relationship between the moments of the sample units and those of the remainder of the population and it can be achieved approximately by an unequal probability sampling scheme. Let us first introduce the Royall and Herson (1973a) notations for the purpose of clarity. Royall and Herson (1973a) found that the efficiency and robustness can be combined by choosing an optimal estimator of population total under a superpopulation model and a selection procedure so that the resultant estimator is a BLU estimator under a more general family of polynomial models. They used the notation $\xi[\delta_0, \delta_1, \ldots, \delta_p : V(x)]$ to represent the superpopulation model, given by

$$y_i = \sum_{j=1}^{p} \delta_j \beta_j x_i^j + e_i[V(x_i)]^{\frac{1}{2}}, \quad i = 1,2,\ldots,N, \tag{5.3.2.1}$$

where

$$\delta_j = \begin{cases} 1 & \text{if the term } \beta_j x^j \text{ appearsin the model,} \\ 0 & \text{otherwise,} \end{cases}$$

$$E(e_i) = 0\,,\; E(e_i e_l) = \begin{cases} \sigma^2 V(x) & \text{if } i = j, \\ 0 & \text{otherwise,} \end{cases}$$

with $V(x)$ assumed to be known. Royall and Herson (1979) considered the model $\xi[0,\ 1\ :\ x]$ under which the ratio estimator is the best linear unbiased (BLU). They proved that the ratio estimator is optimal under any p^{th} degree polynomial regression model with variance function, $V(x) = \sum_{j=1}^{p} \delta_j a_j x^j$, provided that the sample is balanced up to degree p, that is, $\bar{x}_s^{(j)} = \bar{X}^{[j]}$ for $j = 1, 2, ..., p$, where $\bar{x}_s^{(j)} = n^{-1} \sum_{i\in s} x_i^j$ and $\bar{X}^{(j)} = N^{-1} \sum_{i\in\Omega} x_i^j$. The superpopulation models of the form $\xi\left[0,\ 1,\ x^{2\gamma}\right]$ have been used by the several researchers including Smith (1938), Jessen (1942), Raj (1958), Rao and Bayless (1969), and Bayless and Rao (1970). It is interesting to note that a more general form of the superpopulation model given by $\xi[0,\ 1\ :\ V(x)]$ has been considered by Scott, Brewer, and Ho (1978). Following them, regardless of the way the sample observations have been obtained, the BLU estimator of population total Y is given by

$$\hat{Y}_0 = \hat{Y}[0, 1 : V(x)] = \sum_{i\in s} y_i + \frac{\sum_{i\in s} V^{-1}(x_i) y_i x_i}{\sum_{i\in s} V^{-1}(x_i) x_i^2}\left(\sum_{i\in\Omega} x_i - \sum_{i\in s} x_i\right). \tag{5.3.2.2}$$

Thus we have the following theorems.

Theorem 5.3.2.1. Let $s^*(p)$ be a particular sample for which

$$\frac{\sum_{i\in(\Omega-s)} x_i^j}{\sum_{i\in(\Omega-s)} x_i} = \frac{\sum_{i\in s} V^{-1}(x_i) x_i^{j+1}}{\sum_{i\in s} V^{-1}(x_i) x_i^2}\,, \text{ for } j = 1, 2 ..., p\,.$$

Then the estimator $\hat{Y}_0$ is ξ unbiased (Royall, 1970a, 1970b, 1070c) under the model $\xi\left(\delta_0,\ \delta_1,\ ...,\ \delta_p\ :\ V^*(x)\right)$.

Proof. We have

$$E_\xi\left(\hat{Y}_0 - Y\right) = \frac{\sum_{i\in(\Omega-s)} x_i}{\sum_{i\in s} \frac{x_i^2}{V(x_i)}} \sum_{i\in s}\left(\frac{\delta_j \beta_j x_i^{j+1}}{V(x_i)}\right) - \sum_{i\in\Omega-s}\left(\sum_{j=0}^{p} \delta_j \beta_j x_i^j\right)$$

$$= \sum_{j=1}^{p} \delta_j \beta_j \left\{ \frac{\sum_{i\in(\Omega-s)} x_i}{\sum_{i\in s} \frac{x_i^2}{V(x_i)}} \sum_{i\in s}\left(\frac{\delta_j \beta_j x_i^{j+1}}{V(x_i)}\right) - \sum_{i\in\Omega-s} x_i^j \right\},$$

since $\sum_{i\in\Omega-s}\left(\sum_{j=0}^{p}\delta_j\beta_j x_i^j\right)=0$.

On simplifying and using the condition of balanced sampling we have

$$E_\xi\left(\hat{Y}_0-Y\right)=0 \quad \text{if} \quad s=s^*(p).$$

Hence the theorem.

Theorem 5.3.2.2. If $s=s^*(p)$, then $\hat{Y}_0$ is the BLU estimator of the population total under the model $\xi\left(\delta_0,...,\delta_p : V^*(x)\right)$ for any variance function of the form

$$V^*(x)=V(x)\sum_{j=0}^{p}\delta_j a_j x^{j-1}. \tag{5.3.2.3}$$

Proof. Under the model $\xi\left(0,...,0,\ \delta_j=1 : V(x)x^{j-1}\right)$, the BLU estimator of population total Y is given by

$$\hat{Y}_{(j)}=\sum_{i\in s}y_i+\left\{\sum_{i\in s}\frac{y_i x_i^j}{V(x_i)x_i^{j-1}}\Big/\sum_{i\in s}\frac{y_i x_i^{2j}}{V(x_i)x_i^{j-1}}\right\}\left(\sum_{i\in\Omega}x_i^j-\sum_{i\in s}x_i^j\right)=\hat{Y}_0 \tag{5.3.2.4}$$

for $j=0,1,...,p$ and $s=s^*(p)$.

Now consider the more general model $\xi_j\left(\delta_0,...,\delta_j=1,...,\delta_p : V(x)x^{j-1}\right)$. For any linear unbiased estimator $\hat{Y}$, $E_{\xi_j}\left(\hat{Y}-Y\right)^2$ depends only on the variance $V(x)x^{j-1}$ and not on the coefficients $\beta_0,...,\beta_p$. Since $\hat{Y}_0$ is unbiased when $s=s^*(p)$ and is the BLU estimator when $\beta_k=0$ for $k\neq j$, this implies that $\hat{Y}_0$ is the BLU estimator under ξ_j.

Thus if $\hat{Y}$ is an unbiased linear estimator under the model $\xi\left(\delta_0,...,\delta_p : V^*(x)\right)$ with

$$V^*(x)=\sum_{j=1}^{p}\delta_j a_j V(x)x^{j-1}$$

then its mean squared error is:

$$E_\xi\left(\hat{Y}-Y\right)^2=\sum_{j=0}^{p}\delta_j a_j E_{\xi_j}\left(\hat{Y}-Y\right)^2. \tag{5.3.2.5}$$

Hence the theorem.

If $V(x)=x$ then the model considered by Scott, Brewer, and Ho (1978) reduces to the Royall and Herson (1973a, 1973b) result leading to balanced sampling and the estimator $\hat{Y}_0$ reduces to the traditional ratio estimator of population total Y defined as

$$\hat{Y}_1=\frac{\sum_{i\in s}y_i}{\sum_{i\in s}x_i}\sum_{i\in\Omega}x_i.$$

In this case the reduction of bias through balanced sampling may be expensive in terms of efficiency under the model $\xi(0,\ 1:x)$. Thus the most efficient sampling strategy is to choose the n units with the largest x_i values. The relative efficiency of the balanced sampling is given by

$$\text{Min.}\left(\frac{\bar{x}_{\Omega-s}}{\bar{x}_s}\right).$$

Royall and Herson (1973a) have given numerical values of relative efficiency for a variety of distributions.

If $V(x)=x^2$ then $\hat{Y}_0$ becomes

$$\hat{Y}_2 = \sum_{i\in s} y_i + \left(\frac{1}{n}\sum_{i\in s}\frac{y_i}{x_i}\right)\left(\sum_{i\in\Omega} x_i - \sum_{i\in s} x_i\right)$$

and the condition of balanced sampling

$$\frac{\sum\limits_{i\in(\Omega-s)} x_i^j}{\sum\limits_{i\in(\Omega-s)} x_i} = \frac{\sum\limits_{i\in s} V^{-1}(x_i)x_i^{j+1}}{\sum\limits_{i\in s} V^{-1}(x_i)x_i^2}$$

reduces to

$$\frac{1}{n}\sum_{i\in s} x_i^{j-1} = \frac{\sum\limits_{i\in\Omega-s} x_i^j}{\sum\limits_{i\in\Omega-s} x_i}, \quad j = 0,1,\ldots,p,$$

which is always true for $j=1$. A sample satisfying this condition is called 'overbalanced'.

The MSE of $\hat{Y}_2$ under the model $\xi(0,1:x^2)$ is given by

$$\text{MSE}\left(\hat{Y}_2\right) = \sigma^2\left\{\sum_{i\in\Omega-s} x_i^2 + \frac{\left(\sum\limits_{i\in\Omega-s} x_i\right)^2}{n}\right\}. \tag{5.3.2.6}$$

Thus (5.3.2.6) shows that if the sampling fraction is small and no single x_i dominates the others, the MSE is affected very little by the choice of sample and little efficiency is lost by choosing an overbalanced sample.

As suggested by Scott, Brewer, and Ho (1978), in many real life situations $V(x)$ increases more quickly than x but less quickly than x^2, so that $\xi(\delta_0,\ldots,\delta_p : V(x))$ with $V(x)=\sigma_1^2 x + \sigma_2^2 x^2$ is often a fairly realistic model.

Thus we have

$$\mathrm{MSE}\left(\hat{Y}_1\right) = \frac{N(N-n)}{n}\left[\sigma_1^2\bar{x} + \sigma_2^2\bar{x}^{(2)}\right], \tag{5.3.2.7}$$

and

$$\mathrm{MSE}\left(\hat{Y}_2\right) = \frac{N(N-n)}{n}\bar{x}_{(\Omega-s)}\left[\sigma_1^2 + \sigma_2^2\bar{x}\right], \tag{5.3.2.8}$$

where $\bar{x}_{(\Omega-s)}$ is the mean of the elements not included in the overbalanced sample, $\bar{x}^{(2)} = N^{-1}\sum_{i=1}^{N} x_i^2$ and $\bar{x} = N^{-1}\sum_{i=1}^{N} x_i$.

Thus it follows from the condition

$$\frac{1}{n}\sum_{i\in s} x_i^{j-1} = \frac{\sum_{i\in\Omega-s} x_i^j}{\sum_{i\in\Omega-s} x_i}$$

with $j = 0$ that $\bar{x}_{(\Omega-s)}$ is smaller than $\bar{x}_s$ and hence less than $\bar{x}$, which implies that $\mathrm{MSE}\left(\hat{Y}_2\right) < \mathrm{MSE}\left(\hat{Y}_1\right)$.

In other words, the ratio estimator with balanced sampling is less efficient than $\hat{Y}_2$ with overbalanced sampling. It may be noted that the loss in efficiency will be less if σ_1^2 dominates σ_2^2, but can be substantial if σ_2^2 is relatively large. Thus to form an efficient estimator, we have to look for overbalanced sampling.

Note that

$$E\left(\frac{1}{n}\sum_{i\in s} x_i^{j-1}\right) = \sum_{i=1}^{N} x_i^j \Big/ \sum_{i=1}^{N} x_i$$

the selection of the sample with probability proportional to x yields an approximately overbalanced sample if the sample size is large and the sampling fraction is small. For example, let the first order inclusion probability π_i be given by

$$\pi_i = \lambda x_i/(1+\lambda x_i) \tag{5.3.2.9}$$

where λ is obtained from the condition $\sum_{i=1}^{N}\pi_i = n$, that is, λ is a solution of the equation

$$\sum_{i=1}^{N}\lambda x_i/(1+\lambda x_i) = n\,. \tag{5.3.2.10}$$

The solution may be obtained iteratively by taking the initial value as

$$\lambda^{(0)} = n\Big/\sum_{i=1}^{N} x_i \quad\text{and}\quad \lambda^{(k+1)} = n\Big/\sum_{i=1}^{N}\left\{x_i\Big/\left(1+\lambda^{(k)}x_i\right)\right\}.$$

Now the probability of not selecting the ith unit in the sample, say π_i^c is:

$$\pi_i^c = 1-\pi_i = 1-\lambda x_i/(1+\lambda x_i) = 1/(1+\lambda x_i)\,. \tag{5.3.2.11}$$

Thus for all j we have

$$E\left(\sum_{i\in\Omega-s} x_i^j\right) = \sum_{i=1}^{N}\pi_i^c x_i^j = \sum_{i=1}^{N} x_i^j \Big/ (1+\lambda x_i). \tag{5.3.2.12}$$

Also we have

$$E\left(\sum_{i\in s} x_i^{j-1}\right) = \sum_{i=1}^{N}\pi_i x_i^{j-1} = \sum_{i=1}^{N}\lambda x_i^j \Big/ (1+\lambda x_i). \tag{5.3.2.13}$$

From (5.3.2.12) and (5.3.2.13) we have

$$E\left(\sum_{i\in\Omega-s} x_i^j\right) = E\left(\sum_{i\in s} x_i^{j-1}\right)\Big/\lambda . \tag{5.3.2.14}$$

Hence it follows that

$$E\left(\sum_{s} x_i^{j-1}\Big/ n\right) = E\left(\sum_{i\in\Omega-s} x_i^j\right)\Big/ E\left(\sum_{i\in\Omega-s} x_i\right). \tag{5.3.2.15}$$

Thus an approximate overbalanced sample can be obtained if the sample size is large enough.

5.3.3 DESIGN VARIANCE AND ANTICIPATED VARIANCE OF LINEAR REGRESSION TYPE ESTIMATOR

Consider a superpopulation model m as

$$m : y_i = \beta x_i + e_i \tag{5.3.3.1}$$

such that $E_m(e_i) = 0$, $E_m(e_i^2) = \sigma^2 x_i^g$ and $E_m(e_i e_j) = 0$ for $i \neq j$ and g is any real number. In the previous sections, we have seen that if the sample size is fixed at n the design variance of the Horvitz and Thompson (1952) type estimator of population total (called generalised linear regression estimator or GREG) is defined as

$$\hat{Y}_g = \sum_{i\in s}\frac{y_i}{\pi_i} + \hat{\beta}\left(X - \sum_{i\in s}\frac{x_i}{\pi_i}\right) \tag{5.3.3.2}$$

where $\hat{\beta}$ is a prediction unbiased and prediction consistent estimator of β. If (5.3.3.1) holds then the above estimator can be written as

$$\hat{Y}_g = \sum_{i\in s}\frac{y_i - \hat{\beta}x_i}{\pi_i} + \hat{\beta}X = \sum_{i\in s}\frac{e_i}{\pi_i} + \hat{\beta}X . \tag{5.3.3.3}$$

If the sample size n is fixed and sampling design is p, then the design variance of the estimator $\hat{Y}_g$ becomes

$$V_p\left(\hat{Y}_g\right) = \frac{1}{2}\sum_{i\neq j\in s}\sum\left(\pi_i\pi_j - \pi_{ij}\right)\left(d_i e_i - d_j e_j\right)^2 . \tag{5.3.3.4}$$

Consider the random variable $\hat{Y}_g - Y$, then the variance of this random variable under the model m and the sampling design p are called Anticipated Variance (AV).

Thus AV of the estimator $\hat{Y}_g$ is given by

$$\mathrm{AV}\left(\hat{Y}_g\right) = E_m\left[V_p\left(\hat{Y}_g\right)\right] = E_m\left[\frac{1}{2}\sum_{i \neq j \in \Omega}\sum\left(\pi_i\pi_j - \pi_{ij}\right)\left(d_ie_i - d_je_j\right)^2\right]$$

$$= E_m\left[\frac{1}{2}\sum_{i \neq j \in \Omega}\sum\left(\pi_i\pi_j - \pi_{ij}\right)\left(d_i^2e_i^2 + d_j^2e_j^2 - 2d_ie_id_je_j\right)\right]$$

$$= \frac{1}{2}\sum_{i\in\Omega}\sum_{j(\neq i)\in\Omega}\left(\pi_i\pi_j - \pi_{ij}\right)\left(d_i^2E_m\left(e_i^2\right) + d_j^2E_m\left(e_j^2\right) - 2d_id_jE_m\left(e_ie_j\right)\right)$$

$$= \frac{1}{2}\sum_{i\in\Omega}\sum_{j(\neq i)\in\Omega}\left(\pi_i\pi_j - \pi_{ij}\right)\left(d_i^2\sigma^2x_i^g + d_j^2\sigma^2x_j^g - 2d_id_j \times 0\right)$$

$$= \frac{\sigma^2}{2}\sum_{i\in\Omega}\sum_{j(\neq i)\in\Omega}\left(\pi_i\pi_j - \pi_{ij}\right)\left(d_i^2x_i^g + d_j^2x_j^g\right) = \sigma^2\sum_{i\in\Omega}\left[\left(n-\pi_i\right)\pi_i^{-1} - (n-1)\pi_i^{-1}\right]x_i^g$$

$$= \sigma^2\sum_{i\in\Omega}\left(\frac{1}{\pi_i} - 1\right)x_i^g$$

which is the same expression as was shown by Godambe (1955) to be the minimum possible anticipated variance for any design unbiased estimator of population total. Similar results can also be had from Brewer (1979) and Särndal and Wright (1984). The Anticipated Variance (AV) can also be derived with an alternative method as follows:

$$\mathrm{AV}\left(\hat{Y}_g\right) = E_mV_p\left(\hat{Y}_g\right) + V_mE_p\left(\hat{Y}_g\right) \approx E_m\left[\sum_{i\in\Omega}\sum_{j(\neq i)\in\Omega}\left(\frac{\pi_{ij} - \pi_i\pi_j}{\pi_i\pi_j}\right)e_ie_j\right]$$

$$= \sum_{i\in\Omega}\sum_{j\in\Omega}\left(\frac{\pi_{ij} - \pi_i\pi_j}{\pi_i\pi_j}\right)E_m\left(e_ie_j\right) = \sigma^2\sum_{i\in\Omega}\left(\frac{\pi_{ij} - \pi_i\pi_i}{\pi_i\pi_i}\right)x_i^g . = \sigma^2\sum_{i\in\Omega}\left(\frac{1}{\pi_i} - 1\right)x_i^g .$$

We have seen in Corollary 5.1.1 that if $\pi_i \propto Y_i$, the variance of the Horvitz and Thompson (1952) estimator of the population total Y reduces to zero. Unfortunately the values of Y_i are not usually known in practice, but the value X of the auxiliary variable correlated with Y may be known. Thus if π_i is chosen proportional to X_i, there may be a substantial reduction in the variance. Now we will consider the construction of the first and second order inclusion probabilities in the following section.

5.4 CONSTRUCTION AND OPTIMAL CHOICE OF INCLUSION PROBABILITIES

Theorem 5.4.1. The inclusion probability of selecting the ith population unit using PPSWOR sampling scheme is given by

$$\pi_i = P\{1 + S - P_i/(1-P_i)\}, \text{ where } S = \sum_{j=1}^{N} P_j/(1-P_j). \tag{5.4.1}$$

Proof. For a sample of size $n = 2$ we have

π_i = Probability of selecting the ith unit in the sample $= P_i + \sum_{j \neq i=1}^{N} P_i P_j/(1-P_j)$

where P_i is the probability of ith unit being selected in the sample at the first draw. If i^{th} unit is not selected at the first draw, then some other unit will be selected. Let the other unit selected be the jth unit. Its probability of getting selected is P_i, so that at second draw the probability of the ith unit getting selected (if i^{th} unit is not selected on the first draw) is the product of two probabilities, namely

(a) the probability that on the first draw the ith unit is not selected but some other unit is selected

and

(b) the probability that the ith unit is selected on the second draw and is not selected on the first draw.

In other words, on the second draw, the probability for selecting the ith unit is

$$P_j\{P_i/(1-P_j)\} \tag{5.4.2}$$

where P_j = probability of first ineffective draw, and $P_i/(1-P_j)$= probability of second effective draw. The summation sign is taken with the second factor for the reason that any of the remaining units (different from the first selected unit) can get selected on the second draw. Therefore

$$\pi_i = P_i\left(1 + \sum_{j \neq i=1}^{N} P_j/(1-P_j)\right) = P_j\left(1 + \sum_{j=1}^{N} P_j/(1-P_j) - P_i/(1-P_i)\right)$$

$$= P_i\left(1 + S - P_i/(1-P_i)\right) \tag{5.4.3}$$

where $P_i/(1-P_i)$ is the probability of selecting the ith unit and $S = \sum_{j=1}^{N} P_j/(1-P_j)$.

Now the factor $P_i/(1-P_i)$ is not constant because it depends upon P_i, but S is a constant since it is a sum over all the units. Therefore π_i is proportional not only to P_i but also to the factor $P_i/(1-P_i)$. To achieve minimum variance, the probability chosen should be such that π_i may become proportional to P_i. Narain (1951), Yates and Grundy (1953), Fellegi (1963), and Hanurav (1967) have suggested different methods to obtain the values of π_i proportional to P_i. Chaudhuri (1981) has shown that the application of the sampling method of Fellegi (1963) to the first occasion units and the unmatched second occasion units provides the required inclusion probabilities. Now we shall discuss in detail the methods for the construction and optimal choice of inclusion probabilities.

Assume π_i^* , $i = 1,2,...,N,$ be the set of desired inclusion probabilities for a sample of size n drawn by using WOR sampling, say the target inclusion probabilities or πps probabilities. Note that $\sum_{i=1}^{N}\pi_i = n$, the πps probabilities should be such that

$$\pi_i^* \cong \pi_i \quad \forall\ i . \tag{5.4.4}$$

Let us now consider the construction of the π_i^* based on a size variable

$$X_i > 0,\ i = 1, 2, ..., N\ ,\ i.e.,\ \pi_i^* = nX_i \,/\, \sum_{i=1}^{N} X_i\ .$$

The first problem with these inclusion probabilities is that some size variables lead to problems, since one or more of the π_i^* exceeds one. This problem can partially be solved by using a revised selection probability discussed in Section 5.4.5. The second problem concerns the exhibition of a good, preferably a best scheme. To solve this problem, Rosen (1997a) has stipulated the following requirements:

(a) The sample selection should be simple to implement.
(b) The scheme should lead to good estimation precision.
(c) The scheme should have good variance estimation properties.

We will now discuss below some target inclusion probability sampling schemes:

5.4.1 PARETO πps SAMPLING ESTIMATION SCHEME

As discussed by Rosen (1997a), the main-steps in this scheme are as follows:
(a) A sampling frame $\Omega = (1,2,....,N)$ with sizes X_i >0, $i = 1,2,...,N$ is at hand, and a sample size n is specified;
(b) Compute the target inclusion probabilities $\pi_i^* = nX_i/X$, $i = 1, 2, ..., N$. It is presumed that $\pi_i^* < 1\ \forall\ i$. If not, then modify the size measures;
(c) Realize independent standard uniform random variables $U_1, U_2,...,U_N$, thereby realizing the ranking variables

$$Q_i = \{U_i(1 - \pi_i^*)\} / \{\pi_i^*(1 - U_i)\},\ i = 1,2,...,N\ . \tag{5.4.1.1}$$

The sample consists of the n units with labels $(J_1, J_2,...,J_n)$ that are determined by $Q_{j1}, Q_{j2},...,Q_{jn}$ being the n smallest values among the realized $Q_1, Q_2,...,Q_N$. The sampled values so obtained result in the non-negative estimator of variance following Rosen (1997b). The general structure of the Pareto πps sampling estimation procedure applies to all order sampling with fixed distribution shape (OSFS) sampling schemes. Sunter (1977) also introduced a πps scheme with control of variance estimation. Aires (2000) has considered the problem of comparison of Pareto π ps sampling with the conditional Poisson sampling by introducing algorithms to calculate the first and second order exact inclusion probabilities.

5.4.2 HANURAV'S METHOD

Keeping in view the stability of the Sen--Yates--Grundy (1953) estimator of variance of the Horvitz and Thompson (1952) estimator of population total, Hanurav (1967) proposed a method such that $\pi_{ij} > \gamma\pi_i\pi_j$ for all pairs (i, j), where γ is a positive constant. Hanurav's (1967) sampling scheme for sample of size $n = 2$ is as follows:

Without loss of generality let us assume that

$$0 < X_1 \leq X_2 \leq .... \leq X_N < \frac{1}{2}\sum_{i=1}^{N} X_i .$$

Conduct a Bernoulli trial with the probability of success

$$\phi = 2(1 - P_N)(P_N - P_{N-1})/(1 - P_N - P_{N-1}).$$

Then there are two possibilities:

(a) If the trial results in success, select one of the units $U_1, U_2, ..., U_{N-1}$ with probabilities proportional to $P_1, P_2, ..., P_{N-1}$. Thus for example, if j^{th} unit is selected, accept U_N and U_j as the sample;

(b) If the trial results in failure, draw two units from the population with replacement and with probabilities $\phi_i^* = P_i/(1 - P_N + P_{N-1})$ for the ith unit $1 \leq i \leq N-1$ and $\phi_N^* = (P_N - \phi/2)/(1-\phi)$.

In case the selections do not coincide then accept the sample; otherwise reject the sample and select two units from the population with replacement and with probabilities proportional to ϕ_i^{*2}. If the sample consists of distinct units, then accept it, otherwise reject it and repeat the process. Hanurav (1967) listed the following obviously desirable properties of a sampling design to base Horvitz and Thompson (1952) estimator:

(a) $\pi_i = nP_i$, $i = 1, 2, ..., N$; (b) No. of distinct units, $\nu = n \;\forall\; s : p(s) > 0$;

(c) $\pi_{ij} = \pi_{ji} > 0 \;\forall\; i \neq j = 1, 2, ..., N$; (d) $\pi_{ij} < \pi_i\pi_j$, $\forall\; i \neq j = 1,2,....,N$;

(e) $\pi_{ij}/(\pi_i\pi_j) > \varepsilon$, where ε is not too close to zero;

(f) π_{ij} should be easily derivable from simple formulae.

Several other researchers have also worked on similar procedures including Fellegi (1963), Vijayan (1968), Fuller (1971), Brewer (1967, 1975), and Asok and Sukhatme (1975, 1976a).

5.4.3 BREWER'S METHOD

Suppose a population consists of N units labelled $i = 1,2,...,N$, with measures of size P_i such that $\sum_{i=1}^{N} P_i = 1$ and $\pi_i = nP_i$ with $0 < \pi_i < 1$. Here we select the $(n-r+1)$th unit (r^{th} last draw), from among those not already selected, with working probabilities proportional to $P_i(1-P_i)/(1-rP_i)$. Following Brewer (1963a), the probabilities of including i^{th} and j^{th} units for $n = 2$ is given by

$$\pi_{ij} = 2P_iP_j\left\{\frac{1}{1-2P_i} + \frac{1}{1-2P_j}\right\} \Big/ \left\{1 + \sum_{k=1}^{N}\frac{P_k}{1-2P_k}\right\}. \tag{5.4.3.1}$$

For $n > 2$, Brewer (1975) provided a recursive formula for the inclusion probabilities of inclusion in the sample as follows. The working selection probability for the ith unit at the $(n-r+1)$th (r^{th} last draw), given that it was not selected in the first $(n-r)$ draws is

$$P_i(r) = rP_i \Big/ \left(1 - \sum_{i=1}^{n-r} P_i\right), \tag{5.4.3.2}$$

where $\sum_{i=1}^{n-r} P_i$ denotes the sum of the selection probabilities corresponding to the selected $(n-r)$ units in the sample. Thus the conditional probabilities of inclusion in the last r draws are proportional to the measures of size of the remaining units. The method of induction can be applied to derive a formula for the working probabilities. Consider a method is available for selecting $(n-1)$ units. Then if the first unit is selected with working probabilities $P_i(n)$, the last $(n-1)$ can be selected from the remaining units with probabilities proportional to P_i values so that the probability of including the ith unit in the last $(n-1)$ units, given that the first unit selected is the jth will be $(n-1)P_i/(1-P_j)$. Then the condition required on $P_i(n)$ will be

$$nP_i = P_i(n) + \sum_{j\neq i=1} P_j(n)\{(n-1)P_i/(1-P_j)\}. \tag{5.4.3.3}$$

Putting $S_i(n) = P_i(n)/(1-P_i)$ and $S(n) = \sum_{j=1}^{N} S_j(n)$ and eliminating $P_i(n)$ from (5.4.3.3) we have

$$S_i(n) = P_i[n-(n-1)S(n)]/(1-nP_i). \tag{5.4.3.4}$$

Putting (5.4.3.4) in $S(n) = \sum_{j=1}^{N} S_j(n)$ we have

$$S(n) = nT(n)/\{1+(n-1)T(n)\}. \tag{5.4.3.5}$$

where $T(n) = \sum_{i=1}^{N} P_i/(1-nP_i)$. Using (5.4.3.5) in (5.4.3.4) we have the required working probabilities as

$$P_i(n) = nP_i(1-P_i)/[(1-nP_i)\{1+(n-1)T(n)\}]. \tag{5.4.3.6}$$

It is obvious that $\sum_{i=1}^{N} P_i(n) = 1$, $0 < P_i(n) < 1$ since $0 < \pi_i = nP_i < 1$. Keeping in view the Sen--Yates--Grundy estimator of variance, Brewer (1975) also suggested a formula for calculating the second order inclusion probabilities given by

$$\pi_{ij}(n) = (n-1)[S_i(n)P_j + S_j(n)P_i] + \sum_{k(\neq i,j)=1}^{N} P_k(n)\pi_{ij}^{(k)}(n-1) \tag{5.4.3.7}$$

where $\pi_{ij}^{(k)}(n-1)$ denote the joint probability of inclusion of the ith and j^{th} units in the remaining $(n-1)$ units given that the kth unit was selected first. Thus the joint inclusion probabilities can also be calculated by using the above recursive formula. Chromy (1974) found that the second order inclusion probabilities π_{ij} given by this procedure asymptotically minimized the expected variance of the Horvitz and Thompson (1952) estimator when $\gamma = 1/2$. Rao (1963a, 1963b) showed that the Horvitz and Thompson (1952) estimator is always more efficient than the corresponding Hansen and Hurwitz (1943) estimator for multinomial sampling, and that its variance estimator was never negative. Thus in this case, the joint probabilities of inclusion, and hence also the variance estimator, are simple functions of size.

5.4.4 SAMPFORD'S METHOD

Sampford (1967) proposed a method for inclusion probabilities proportional to size (IPPS) scheme for selecting a sample of n units, which in fact is an extension of the methods proposed by Brewer (1963a), Rao (1965b), and Durbin (1967). For this scheme of sampling, the first order inclusion probability is given by

$$\pi_i = \text{Prob}\left[i^{th} \text{ unit is selected at the first draw}\right] + \text{Prob}\left[i^{th} \text{ unit is selected at the second draw}\right]$$

$$= P_i + \sum_{j\neq i=1}^{N} P_j P_{i.j} = 2P_i \tag{5.4.4.1}$$

where $P_{i.j}$ denotes the conditional probability $P(i \mid j)$. If P_i denotes the probability of selecting the ith unit at the first draw, then the probability of selecting the second unit from the remaining ($N-1$) units in the population is given by

$$P(j \mid i) = P_{j.i} = P_j\left[\frac{1}{1-2P_i} + \frac{1}{1-2P_j}\right] \Big/ \left[1 + \sum_{i=1}^{N} \frac{P_i}{1-2P_i}\right], \quad j(\neq i) = 1,2,\ldots,N. \tag{5.4.4.2}$$

Furthermore the second order inclusion probability is given by

$$\pi_{ij} = P_i P_{j.i} + P_j P_{i.j} = 2P_iP_j\left[\frac{1}{1-2P_i} + \frac{1}{1-2P_j}\right] \Big/ \left[1 + \sum_{j=1}^{N} \frac{P_j}{(1-P_j)}\right]. \tag{5.4.4.3}$$

It can be easily verified that $\pi_i\pi_j - \pi_{ij} \geq 0$ for Samphord's sampling procedure. Midha (1980), Hartley and Rao (1962), and Asok and Sukhatme (1976b) have also suggested some procedures to approximate these inclusion probabilities. The Durbin (1967) procedure has also been used by Brewer and Hanif (1970) for developing a new multistage estimator of variance.

5.4.5 NARAIN'S METHOD

Narain (1951) proposed another method of sample selection, which is free from any restriction on the set of initial selection probabilities and leads to a more efficient estimator of the population parameters than with replacement sampling. This method consists of making revised selection probabilities $P_i^*(i=1,2,...,N)$ such that the resulting inclusion probabilities π_i are proportional to the original probabilities of selection P_i, $i=1,2,...,N$. For a sample of two units, the revised selection probabilities are given by

$$\pi_i = P_i^*\left(S^* + 1 - \frac{P_i^*}{1-P_i^*}\right) = 2P_i \text{ where } S^* = \sum_{i=1}^{N} P_i^*/\left(1-P_i^*\right). \tag{5.4.5.1}$$

Rao (1989), Yates and Grundy (1953) and Brewer and Undy (1962) have done further work in these directions.

5.4.6 MIDZUNO--SEN METHOD

Midzuno (1952) and Sen (1952) introduced an interesting method for selecting the sample, which is useful in many ways. Using this method, the unit at the first draw is selected with unequal probability, while the rest of the units are selected with equal probability and without replacement. Defining a random variable, t_i, such that

$$t_i = \begin{cases} 1 & \text{if } i^{th} \text{ unit is selected in the sample,} \\ 0 & \text{otherwise.} \end{cases} \tag{5.4.6.1}$$

We have

$$E(t_i) = \pi_i = P_i + \text{P}\begin{bmatrix} i^{th} \text{ unit is not selected at the first draw and is selected at any of the} \\ \text{remaining } (n-1) \text{ draws} \end{bmatrix}$$

$$= P_i + (1-P_i)\left(\frac{n-1}{N-1}\right) = \left(\frac{N-n}{N-1}\right)P_i + \left(\frac{n-1}{N-1}\right) \tag{5.4.6.2}$$

and

$$E(t_i t_j) = \pi_{ij}$$

$$= P\begin{bmatrix} i^{th} \text{ unit is selected at the first draw and } j^{th} \text{ unit is selected} \\ \text{at any of the remaining } (n-1) \text{ draws} \end{bmatrix}$$

$$+ P\begin{bmatrix} j^{th} \text{ unit is selected at the first draw and } i^{th} \text{ unit is selected} \\ \text{at any of the remaining } (n-1) \text{ draws} \end{bmatrix}$$

$$+ P\begin{bmatrix} \text{Neither } i^{th} \text{ nor } j^{th} \text{ units are selected at the first draw but} \\ \text{both of them are selected in the subsequent } (n-1) \text{ draws} \end{bmatrix}$$

$$= P_i\left(\frac{n-1}{N-1}\right) + P_j\left(\frac{n-1}{N-1}\right) + \frac{(1-P_i-P_j)(n-1)(n-2)}{(N-1)(N-2)}$$

$$= \left(\frac{n-1}{N-1}\right)\left[\left(\frac{N-n}{N-2}\right)(P_i + P_j) + \left(\frac{n-2}{N-2}\right)\right]. \tag{5.4.6.3}$$

For this sampling scheme, one can easily see that $\pi_i\pi_j - \pi_{ij} > 0$, which guarantees non-negativity of the estimator of variance proposed by Sen (1953) and Yates and Grundy (1953). At the same time the limitation is that the first order inclusion probabilities are not proportional to the selection probabilities. This can be rectified by finding a new set of revised selection probabilities P_i^* by applying a suitable transformation on the selection probabilities P_i such that the resultant inclusion probabilities π_i, $i = 1, 2, .., N$ are proportional to the value of P_i. Note that P_i^* are the revised selection probabilities we have

$$\pi_i = \left(\frac{N-n}{N-1}\right)P_i^* + \left(\frac{n-1}{N-1}\right). \tag{5.4.6.4}$$

Simplifying (5.4.5.4) for P_i^* and using $\pi_i = nP_i$ we have

$$P_i^* = nP_i\left(\frac{N-1}{N-n}\right) - \left(\frac{n-1}{N-n}\right). \tag{5.4.6.5}$$

Note that the revised selection probabilities P_i^* must always be positive, the initial selection probabilities P_i must satisfy

$$P_i > \left(1-\frac{1}{n}\right)\Big/\left\{N\left(1-\frac{1}{N}\right)\right\} \quad \forall\, i. \tag{5.4.6.6}$$

Thus the use of revised selection probabilities for deriving efficient estimates of population total through Horvitz and Thompson estimator will be possible only in those cases where the original selection probabilities satisfy the above condition. This condition on the initial probabilities usually does not hold, and hence limits the use of the scheme in practice. Some more relevant work related to the discussion of the validity of these inclusion probabilities can be found in Rao (1963a, 1963b), Asok (1974,1980), and Asok and Sukhatme (1978). A generalization of Midzuno--Sen sampling scheme has been given by Prasad and Srivenkataramana (1980), Deshpande and Ajgaonkar (1987), and Kumar and Srivenkataramana (1994). Bedi and Agarwal (1999) suggested a new set of revised probabilities under the Midzuno (1952) sampling scheme. The revised probabilities are functions of the location shift factor L between 0 and 1 and it is remarkable that the optimum value of L is free from any knowledge of unknown population parameter.

5.4.7 KUMAR--GUPTA--NIGAM SCHEME

Following the concept of Gupta, Nigam, and Kumar (1982), Kumar, Gupta, and Nigam (1985) introduced a family of inclusion probabilities proportional to size (IPPS) sampling scheme based on balanced incomplete block design by following Raghavarao (1971) for selecting a sample of n units. The main steps of the proposed scheme are:

(a) Consider two balanced incomplete block designs D_1 ($v = N$, b_1, r_1, $k_1 \ge n$, λ_1) and D_2 ($v = N$, b_2, r_2, $k_2 \ge k_1$, λ_2) where the parameters have their usual meaning as in the case of incomplete block designs;
(b) Perform a random experiment which selects the design D_1 with probability W $(0 \le W \le 1)$ and the design D_2 with probability $(1-W)$;
(c) From the design D_j selected at step (b), select one block $(s,\ D_j)$ with a preassigned probability

$$P(s,D_j) = \sum_{t \in (s,D_j)} P_t\, r_j \quad (s = 1, 2, ..., b_j;\ j = 1 \text{ or } 2), \tag{5.4.7.1}$$

where

$$P_t = [k_1 k_2 (v-1) p_t - \{k_1(k_2-1) - W(k_2-k_1)\}][k_1(v-k_2) + W(k_2-k_1)v]^{-1} \tag{5.4.7.2}$$

$t = 1, 2, .., N$. If $k_j = n$ the elements of the selected block constitute the required sample of size n. Otherwise go to step (d);
(d) Select a sub-sample of size n by equal probability sampling without replacement from the k_j units belonging to the block $(s,\ D_j)$ selected at step (c).
For this scheme, the first and second order inclusion probabilities are

$$\pi_i = nW \sum_{(s,D_1)\supset i} P(s,D_1)/k_1 + n(1-W) \sum_{(s,D_2)\supset i} P(s,D_2), \tag{5.4.7.3}$$

and

$$\pi_{ij} = n(n-1)\left[W \sum_{(s,D_1)\supset i,j} P(s,D_1)/\{k_1(k_1-1)\} + (1-W) \sum_{(s,D_2)\supset i,j} P(s,D_2)/\{k_2(k_2-1)\}\right]. \tag{5.4.7.4}$$

Kumar, Srivastava, and Agarwal (1986) have discussed a general class of unequal probability sampling.

5.4.8 DEY AND SRIVASTAVA SCHEME FOR EVEN SAMPLE SIZE

Dey and Srivastava (1987) considered the following IPPS sampling scheme. Consider a population of N units with y as the study variable and x, an auxiliary variable, as the size. It is assumed that x values are known for all the population units. A sample of size n (>2) is to be selected. To start with, it is assumed that n is even. Divide the population into m $(> n/2)$ groups so that the ith group contains N_i (>2) units $(i = 1, 2, ..., m)$, and for each group

$$X_i / X > (n-2)/\{n(m-1)\} \tag{5.4.8.1}$$

where $X_i = \sum_{u=1}^{N_i} x_{iu}$ is the value of x for the uth unit in the ith group and $X = \sum_{i=1}^{m} X_i$.
Equation (5.4.8.1) is satisfied if the X_i $(\mathrm{i} = 1,2,...,\mathrm{m})$ are made nearly equal. It has been seen in actual populations, considered by Rao and Bayless (1969), that this condition is satisfied for quite a few values of m. Rao and Lanke (1984) suggested a grouping procedure in which N units are divided into R groups so that group totals X_i are nearly equal and group sizes are either $[N/R]$ or $[N/R]+1$, where $[\bullet]$ is

the largest integer. Having formed the m groups, the suggested sampling procedure consists of the following steps:

Step I. Select $n/2$ groups out of m groups using Midzuno (1952) sampling procedure, *i.e.*, select one group with probability

$$P_i' = \{n(m-1)P_i - (n-2)\}/(2m-n), \quad \text{with } P_i = X_i/X \tag{5.4.8.2}$$

and the remaining $(n/2)-1$ groups with equal probabilities without replacement.

Step II. From each of the selected groups, select two units by any IPPS procedure, say by Durbin's (1967) procedure; that is, from the ith selected group $(i = 1, 2, ..., n/2)$ select one unit with probability

$$P_{iu} \mid i = x_{iu}/X_i \tag{5.4.8.3}$$

and the second unit with revised probability

$$P_{iu} \mid i_v = x_{iv}\left[(X_i - 2x_{iv})^{-1} + (X_i - 2x_{iu})^{-1}\right] / D_i\,, \tag{5.4.8.4}$$

where $D_i = \left\{1 + \sum_{i=1}^{N_i} x_{iu}/(X_i - 2x_{iu})\right\}$.

For this sampling scheme the inclusion probability for the i_u^{th} unit is evidently given by

$$\pi_{i_u} = nP_{i_u} \tag{5.4.8.5}$$

where $P_{i_u} = x_{i_u}/X$ and the joint inclusion probabilities for a pair of units are given by

$$\pi_{i_u i_v} = nP_{i_u}P_{i_v}\left(P_i - P_{i_u} - P_{i_v}\right)/\left\{D_i\left(P_i - 2P_{i_u}\right)\left(P_i - 2P_{i_v}\right)\right\}\,. \tag{5.4.8.6}$$

5.4.9 SSS SAMPLING SCHEME

Saxena, Singh, and Srivastava (1986) suggested the following IPPS sampling strategy. Consider a population of size N with Y_i and X_i as the study and the auxiliary variable values, respectively, for the ith unit. We further assume that $X_i > 0$ for all units in the population. The following steps make the IPPS sampling strategy:

Step I. Select a sample, s_1 (say), of size n from the population by simple random sampling without replacement. Let s_1' be its complement, that $s_1' = \Omega - s_1$. Perform independent Bernoulli trials on each unit of s_1 with probability of success p_i for the ith unit in the population. Let the number of successes be r.

Step II. If $r < n$, select $n - r$ units from s_1' by simple random sampling without replacement.

The ultimate sample s will consist of r units selected at Step I and $(n-r)$ units selected at Step II. For this sampling scheme, the first and second order inclusion probabilities are given by

$$\pi_i = n\left(NP_i + \sum_{j=1}^{N} Q_j - 1\right)\Big/\left(N(N-1)\right)$$

and

$$\pi_{ij} = \{n(n-1)/N(N-1)\}-\left[P_iP_j + \left\{(P_i+P_j)\sum_{j=1}^{N} Q_j\right\}\Big/(N-2) + \sum_{j\neq k=1}^{N} Q_jQ_k \Big/\{(N-2)(N-3)\}\right]$$

respectively, where $Q_j = 1-P_j$. Several other research workers such as Chaudhuri (1975a, 1975b), Choudhry and Singh (1979) and Singh (1978) have also suggested some IPPS sampling schemes. As reported by Deville and Tillê (1998), 50 different sampling procedures have been reviewed by Brewer and Hanif (1983). Interested readers may refer to Brewer and Hanif (1983) to study in depth the construction of inclusion probabilities. In addition to that, Chen (1998) has discussed the weighted polynomial models (WPM) for generating the first order inclusion probabilities. Deville and Tillê (2000) have extended the problem of unequal probability sampling of fixed size from a finite population to the generalised case in a partition problem of a population into subsets with unequal inclusion probabilities in each subset.

5.4.10 OPTIMAL CHOICE OF FIRST ORDER INCLUSION PROBABILITIES

Cassel, Särndal, and Wretman (1976) have shown that the average variance of any estimator $\hat{\theta}_s$ satisfying the condition of unbiasedness may be:

$$E_mE_p\left[\hat{\theta}_s-\theta\right]^2 = E_p\left\{B_m\left(\hat{\theta}_s\right)\right\}^2 + E_p\left[V_m\left(\hat{\theta}_s\right)\right] - V_m(\theta), \tag{5.4.10.1}$$

where under the model m

$$V_m\left(\hat{\theta}_s\right) = E_m\left(\hat{\theta}_s^2\right) - \left[E_m\left(\hat{\theta}_s\right)\right]^2, \tag{5.4.10.2}$$

and

$$B_m\left(\hat{\theta}_s\right) = E_m\left(\hat{\theta}_s\right) - E_m[\theta] \tag{5.4.10.3}$$

are respectively the variance and bias in the estimator $\hat{\theta}_s$. Note that $V_m(\theta)$ is constant, it does not enter into the minimization process. Consider an estimator $\hat{\theta}_s^*$ and a set of inclusion probabilities that yield $E_p\left(\hat{\theta}_s^*\right) = \theta$. If $\hat{\theta}_s$ minimizes $E_pV_m\left(\hat{\theta}_s\right)$ subject to the condition $\sum_s p(s)\hat{\theta}_s = \theta$, then the choice of $\hat{\theta}_s^*$ also minimizes (5.4.10.1) if $B_m\left(\hat{\theta}_s^*\right) = 0$. For example, a model unbiased estimator in the case of a linear homogeneous class of estimators that satisfies the above criterion is defined as

$$\hat{y}_s = \sum_{i \in s} d_i y_i \tag{5.4.10.4}$$

which is the well known Horvitz and Thompson estimator. Hàjek (1958) obtained the optimal estimator and design pair for estimating population mean $\overline{Y}$ and introduced a general cost constraint defined as

$$\sum_{i \in \Omega} c_i \pi_i = C \tag{5.4.10.5}$$

to obtain the optimal choice of the first order inclusion probabilities given by $\pi_i \propto \sigma_i / \sqrt{c_i}$ under the super population model $m : y_i = \alpha_i + e_i$, where $E_m(e_i) = 0$, $E(e_i^2) = \sigma_i^2$ and $E(e_i e_j) = 0$. The restriction $B_m(\hat{y}_s) = 0$ implies that

$$\sum_{i=1}^{N} \alpha_i = \frac{1}{n} \sum_{i \in s} \sqrt{\left(c_i / \sigma_i^2\right)} \alpha_i \sum_{i=1}^{N} \sqrt{\left(\sigma_i^2 / c_i\right)}. \tag{5.4.10.6}$$

Several other researchers have also suggested methods to find the optimal first order inclusion probabilities. For example Godambe (1955) also obtained results similar to Hàjek (1958). Rao (1975), Cassel, Särndal, and Wretman (1976) and Rao, and Bellhouse (1978) have also suggested optimal choices of the inclusion probabilities. Maxmin πps sampling designs have been discussed by Hanurav (1967), Rao and Bayless (1969), Sinha (1973), Chao (1982), Gabler (1984), Herzel (1986), Chaudhuri and Vos (1988), and Herzel (1993). Comparisons of PPSWR and Brewer's πps WOR procedures have been done by Sampford (1967), Chaudhuri (1974), Gabler (1981), and Sengupta (1986).

Example 5.4.1. We wish to estimate the total number of fish of all kinds caught by marine recreational fishermen of the Atlantic and Gulf coasts during 1995. Population 4 in the Appendix shows that information on the number of different kinds of fish caught during 1992 is available. Use the known information on the number of fish caught during 1992 to select a sample of eight units by using PPSWOR sampling. Collect the required information from population 4 to estimate the total number of fish caught during 1995. Apply the Sen--Yates--Grundy estimator of variance to construct a 95% confidence interval.

Solution. Although we discussed several methods to select a sample of required size by using PPSWOR sampling, let us restrict ourselves to the case of the Midzuno--Sen sampling scheme. In the Midzuno--Sen method the first unit is selected with probability proportional to the number of fish caught during 1992 and the remaining seven units are selected by SRSWOR sampling. We have used Lahir's method to select the first unit and a random number table method to select the remaining seven units. Using the Pseudo-Random Numbers (PRN) given in Table 1 of the Appendix, we used the first two columns to select a random number $1 \le R_i \le 69$ and columns 7 to 11 to select another random number $1 \le R_j \le 28933$. We performed the following trials to select one unit by using Lahiri's method.

Trial Number	$1 \le R_i \le 69$	$1 \le R_j \le 2893$	No of fish caught in 1992 (X_i)	Decision R--Rejected S--Selected
1	58	07572	1045	R
2	60	21945	5575	R
3	54	25975	4195	R
4	01	09266	1467	R
5	69	15573	12249	R
6	62	01228	11918	S

Thus the species **'Summer flounder'** at serial number 62 has been selected as the first unit in the sample. Out of the remaining 68 species, now we have to select seven units by SRSWOR sampling. We used the first two columns of the Pseudo-Random Numbers (PRN) given in Table 1 of the Appendix to select seven random numbers between 1 and 68. These random numbers came in the sequence as: 58, 60, 54, 01, 62, 23, 64. Thus the ultimate sample is as shown below.

Table 5.4.1. Sample selected using Midzuno--Sen sampling scheme.

Sample units	PRN No.	Population units	Species group	1992 x_i	1995 y_i	P_i	π_i	y_i/π_i
1	01	01	Sharks, other	1467	2016	0.005026	0.107450	18762.21
2	23	23	Blue runner	2371	2319	0.008123	0.110228	21038.21
3	54	54	Tautog	4195	3816	0.014372	0.115834	32943.70
4	58	58	Atlantic mackerel	1045	4008	0.003580	0.106153	37756.82
5	60	60	Spanish mackerel	5575	2568	0.019100	0.120075	21386.63
6	62	62	Summer flounder	11918	16238	0.040832	0.139569	116343.90
7	**62**	63	Gulf flounder	216	163	0.000740	0.103605	1573.28
8	64	65	Winter flounder	1544	2324	0.005290	0.107686	21581.26
								271386.01

From population 4 we have the total number of fish caught during 1992, $X = 291882$. Thus the selection probabilities $P_i = x_i/X$ have been calculated as shown in the Table 5.4.1. The values of the first order inclusion probabilities based on the Midzuno--Sen sampling scheme given by

$$\pi_i = \frac{N-n}{N-1}P_i + \frac{n-1}{N-1}$$

have also been presented in the above table. For example, the value of π_1 is calculated as

$$\pi_1 = \frac{N-n}{N-1}P_1 + \frac{n-1}{N-1} = \frac{69-8}{69-1}\times 0.005026 + \frac{8-1}{69-1} = 0.10745$$

and so on. The values of the second order inclusion probabilities based on Midzuno--Sen sampling scheme given by

$$\pi_{ij} = \frac{n-1}{N-1}\left[\frac{N-n}{N-2}(P_i + P_j) + \frac{n-2}{N-2}\right]$$

have been given in Table 5.4.2. For example the value of π_{12} is calculated as

$$\pi_{12} = \frac{n-1}{N-1}\left[\frac{N-n}{N-2}(P_1 + P_2) + \frac{n-2}{N-2}\right] = \frac{8-1}{69-1}\left[\frac{69-8}{69-2}(0.005026 + 0.008123) + \frac{8-2}{69-2}\right] = 0.010451$$

and so on.

Table 5.4.2. Second order inclusion probabilities for the units selected in the sample.

π_{ij}	j						
$i > j$	1	2	3	4	5	6	7
2	0.010451						
3	0.011037	0.011327					
4	0.010025	0.010315	0.010901				
5	0.011480	0.011770	0.012356	0.011344			
6	0.013517	0.013807	0.014392	0.013381	0.014836		
7	0.009759	0.010049	0.010635	0.009623	0.011078	0.013115	
8	0.010185	0.010476	0.011061	0.010050	0.011505	0.013541	0.009784

The values of Sen--Yates--Grundy weights, $(\pi_i\pi_j - \pi_{ij})/\pi_{ij}$, are given below:

Table 5.4.3. Sen--Yates--Grundy weights.

	$(\pi_i\pi_j - \pi_{ij})/\pi_{ij}$						
$i > j$	1	2	3	4	5	6	7
2	0.133289						
3	0.127726	0.127238					
4	0.137742	0.134317	0.127962				
5	0.123894	0.124520	0.125694	0.123590			
6	0.109511	0.114270	0.123288	0.107219	0.129638		
7	0.140723	0.136417	0.128445	0.142822	0.122973	0.102578	
8	0.136022	0.133105	0.127683	0.137442	0.123948	0.109925	0.140346

Using information from Table 5.4.1, the HT estimate of the total number of fish caught during 1997 is:

$$\hat{Y}_{HT} = \sum_{i \in s} \frac{y_i}{\pi_i} = 271386.01 .$$

Now using information from Table 5.4.1 and Table 5.4.3, the Sen--Yates--Grundy estimator of variance of the HT estimator of total is given by

$$\hat{v}_{\text{SYG}}\left(\hat{Y}_{\text{HT}}\right)$$

$$=\frac{1}{2}\sum_{i\in s}\sum_{j(\neq i)\in s}\left(\frac{\pi_i\pi_j-\pi_{ij}}{\pi_{ij}}\right)\left(\frac{y_i}{\pi_i}-\frac{y_j}{\pi_j}\right)^2=\sum_{i(<j)=1}^{8}\sum_{j=1}^{8}\left(\frac{\pi_i\pi_j-\pi_{ij}}{\pi_{ij}}\right)\left(\frac{y_i}{\pi_i}-\frac{y_j}{\pi_j}\right)^2$$

$$=\left(\frac{\pi_1\pi_2-\pi_{12}}{\pi_{12}}\right)\left(\frac{y_1}{\pi_1}-\frac{y_2}{\pi_2}\right)^2+\left(\frac{\pi_1\pi_3-\pi_{13}}{\pi_{13}}\right)\left(\frac{y_1}{\pi_1}-\frac{y_3}{\pi_3}\right)^2+\left(\frac{\pi_1\pi_4-\pi_{14}}{\pi_{14}}\right)\left(\frac{y_1}{\pi_1}-\frac{y_4}{\pi_4}\right)^2$$

$$+\left(\frac{\pi_1\pi_5-\pi_{15}}{\pi_{15}}\right)\left(\frac{y_1}{\pi_1}-\frac{y_5}{\pi_5}\right)^2+\left(\frac{\pi_1\pi_6-\pi_{16}}{\pi_{16}}\right)\left(\frac{y_1}{\pi_1}-\frac{y_6}{\pi_6}\right)^2+\left(\frac{\pi_1\pi_7-\pi_{17}}{\pi_{17}}\right)\left(\frac{y_1}{\pi_1}-\frac{y_7}{\pi_7}\right)^2$$

$$+\left(\frac{\pi_1\pi_8-\pi_{18}}{\pi_{18}}\right)\left(\frac{y_1}{\pi_1}-\frac{y_8}{\pi_8}\right)^2+\left(\frac{\pi_2\pi_3-\pi_{23}}{\pi_{23}}\right)\left(\frac{y_2}{\pi_2}-\frac{y_3}{\pi_3}\right)^2+\left(\frac{\pi_2\pi_4-\pi_{24}}{\pi_{24}}\right)\left(\frac{y_2}{\pi_2}-\frac{y_4}{\pi_4}\right)^2$$

$$+\left(\frac{\pi_2\pi_5-\pi_{25}}{\pi_{25}}\right)\left(\frac{y_2}{\pi_2}-\frac{y_5}{\pi_5}\right)^2+\left(\frac{\pi_2\pi_6-\pi_{26}}{\pi_{26}}\right)\left(\frac{y_2}{\pi_2}-\frac{y_6}{\pi_6}\right)^2+\left(\frac{\pi_2\pi_7-\pi_{27}}{\pi_{27}}\right)\left(\frac{y_2}{\pi_2}-\frac{y_7}{\pi_7}\right)^2$$

$$+\left(\frac{\pi_2\pi_8-\pi_{28}}{\pi_{28}}\right)\left(\frac{y_2}{\pi_2}-\frac{y_8}{\pi_8}\right)^2+\left(\frac{\pi_3\pi_4-\pi_{34}}{\pi_{34}}\right)\left(\frac{y_3}{\pi_3}-\frac{y_4}{\pi_4}\right)^2+\left(\frac{\pi_3\pi_5-\pi_{35}}{\pi_{35}}\right)\left(\frac{y_3}{\pi_3}-\frac{y_5}{\pi_5}\right)^2$$

$$+\left(\frac{\pi_3\pi_6-\pi_{36}}{\pi_{36}}\right)\left(\frac{y_3}{\pi_3}-\frac{y_6}{\pi_6}\right)^2+\left(\frac{\pi_3\pi_7-\pi_{37}}{\pi_{37}}\right)\left(\frac{y_3}{\pi_3}-\frac{y_7}{\pi_7}\right)^2+\left(\frac{\pi_3\pi_8-\pi_{38}}{\pi_{38}}\right)\left(\frac{y_3}{\pi_3}-\frac{y_8}{\pi_8}\right)^2$$

$$+\left(\frac{\pi_4\pi_5-\pi_{45}}{\pi_{45}}\right)\left(\frac{y_4}{\pi_4}-\frac{y_5}{\pi_5}\right)^2+\left(\frac{\pi_4\pi_6-\pi_{46}}{\pi_{46}}\right)\left(\frac{y_4}{\pi_4}-\frac{y_6}{\pi_6}\right)^2+\left(\frac{\pi_4\pi_7-\pi_{47}}{\pi_{47}}\right)\left(\frac{y_4}{\pi_4}-\frac{y_7}{\pi_7}\right)^2$$

$$+\left(\frac{\pi_4\pi_8-\pi_{48}}{\pi_{48}}\right)\left(\frac{y_4}{\pi_4}-\frac{y_8}{\pi_8}\right)^2+\left(\frac{\pi_5\pi_6-\pi_{56}}{\pi_{56}}\right)\left(\frac{y_5}{\pi_5}-\frac{y_6}{\pi_6}\right)^2+\left(\frac{\pi_5\pi_7-\pi_{57}}{\pi_{57}}\right)\left(\frac{y_5}{\pi_5}-\frac{y_7}{\pi_7}\right)^2$$

$$+\left(\frac{\pi_5\pi_8-\pi_{58}}{\pi_{58}}\right)\left(\frac{y_5}{\pi_5}-\frac{y_8}{\pi_8}\right)^2+\left(\frac{\pi_6\pi_7-\pi_{67}}{\pi_{67}}\right)\left(\frac{y_6}{\pi_6}-\frac{y_7}{\pi_7}\right)^2+\left(\frac{\pi_6\pi_8-\pi_{68}}{\pi_{68}}\right)\left(\frac{y_6}{\pi_6}-\frac{y_8}{\pi_8}\right)^2$$

$$+\left(\frac{\pi_7\pi_8-\pi_{78}}{\pi_{78}}\right)\left(\frac{y_7}{\pi_7}-\frac{y_8}{\pi_8}\right)^2.$$

The above 28 values of $\left(\frac{\pi_i\pi_j-\pi_{ij}}{\pi_{ij}}\right)\left(\frac{y_i}{\pi_i}-\frac{y_j}{\pi_j}\right)^2$ are given in the following table.

Table 5.4.4. The values of $\left(\frac{\pi_i\pi_j - \pi_{ij}}{\pi_{ij}}\right)\left(\frac{y_i}{\pi_i} - \frac{y_j}{\pi_j}\right)^2$ for different values of i and j.

	j						
$i > j$	1	2	3	4	5	6	7
2	690440.4						
3	25687680.6	18035050.2					
4	49697155.8	37543877.8	2964455.1				
5	853303.5	15116.2	16788659.6	33120629.9			
6	1042769336.0	1037919851.0	857526233.2	662159238.9	1168914287.0		
7	41578123.1	51686249.9	126403473.1	186991336.4	48275318.4	1351165614.0	
8	1080852.7	39235.0	16485132.9	35962754.1	4689.3	987107399.0	56182541.0

Thus we have

$$\hat{v}_{\text{SYG}}\left(\hat{Y}_{\text{HT}}\right) = \sum_{i(<j)=1}^{8}\sum_{j=1}^{8}\left(\frac{\pi_i\pi_j - \pi_{ij}}{\pi_{ij}}\right)\left(\frac{y_i}{\pi_i} - \frac{y_j}{\pi_j}\right)^2 = 7857648034\,.$$

Note that the $(1-\alpha)100\%$ confidence interval of population total Y is given by

$$\hat{Y}_{\text{HT}} \mp t_{\alpha/2}(\text{df} = n-1)\sqrt{\hat{v}_{\text{SYG}}\left(\hat{Y}_{\text{HT}}\right)}\,.$$

Using Table 2 from the Appendix the 95% confidence interval of the total number of fish during 1997 is given by

$$271386.01 \mp 2.365\sqrt{7857648034}\,, \quad \text{or} \quad [61744.43,\ 481027.59]\,.$$

The next section has been devoted to discuss calibration approach in sampling theory.

5.5 CALIBRATION APPROACH

Statisticians are often interested in the precision of survey estimates. The most commonly used estimator of population total or population mean is the generalized linear regression (GREG) estimator. Let us consider the simplest case of the GREG where information on only one auxiliary variable is available. Consider a population $\Omega = \{1, 2, .., i, .., N\}$, from which a probability sample s $(s \subset \Omega)$ is drawn with a given sampling design $p(.)$. The inclusion probabilities $\pi_i = P(i \in s)$ and $\pi_{ij} \in P(i \in s,\ j \in s)$ are assumed to be strictly positive and known. Let y_i be the value of the study variable, y, for the ith population unit and let x_i be the value of the ith unit of the associated auxiliary variable. The population total $X = \sum_{i\in\Omega} x_i$ of the auxiliary variable x is assumed to be accurately known. The objective is to estimate the population total $Y = \sum_{i\in\Omega} y_i$. The concept of linear weighting of sample survey data can be found in Bethlehem and Keller (1987). Deville and Särndal (1992) used calibration on the known population total, X, to modify the basic sampling design weights, $d_i = 1/\pi_i$, that appear in the Horvitz and Thompson (1952) estimator

$$\hat{Y}_{\mathrm{HT}} = \sum_{i\in s}\frac{y_i}{\pi_i} = \sum_{i\in s} d_i y_i \,. \tag{5.5.1}$$

A new estimator

$$\hat{Y}_{\mathrm{G}} = \sum_{i\in s} w_i y_i \tag{5.5.2}$$

was proposed by Deville and Särndal (1992), with weights w_i as close as possible in an average sense to the d_i for a given measure and subject to the calibration constraint

$$\sum_{i\in s} w_i x_i = X \,. \tag{5.5.3}$$

Now we have the following theorems:

Theorem 5.5.1. Minimization of chi square (CS) distance between the new weights w_i and selection weights or design weights d_i leads to a general regression type estimator (GREG) of population total, Y, given by

$$\hat{Y}_{\mathrm{G}} = \sum_{i\in s} d_i y_i + \hat{\beta}_{\mathrm{ds}}\left(X - \sum_{i\in s} d_i x_i\right) \tag{5.5.4}$$

where $\hat{\beta}_{\mathrm{ds}} = \sum_{i\in s} d_i q_i x_i y_i \Big/ \sum_{i\in s} d_i q_i x_i^2$.

Proof. Let us define the chi square (CS) type distance function D as

$$D = \sum_{i\in s}(w_i - d_i)^2 (d_i q_i)^{-1} \tag{5.5.5}$$

where q_i are suitably chosen constants such that the estimator depends upon its choice. The Lagrange function L for minimizing D in (5.5.5) subject to the constraint in (5.5.3) is then given by

$$L = \sum_{i\in s}(w_i - d_i)^2 (d_i q_i)^{-1} - 2\lambda\left(\sum_{i\in s} w_i x_i - X\right). \tag{5.5.6}$$

On differentiating (5.5.6) with respect to w_i and equating to zero we have

$$w_i = d_i + \lambda d_i q_i x_i \,. \tag{5.5.7}$$

On substituting (5.5.6) in (5.5.3) and solving for λ we have

$$\lambda = \left(\sum_{i\in s} d_i q_i x_i^2\right)^{-1}\left(X - \sum_{i\in s} d_i x_i\right). \tag{5.5.8}$$

On substituting (5.5.8) in (5.5.7) we have

$$w_i = d_i + \left(d_i q_i x_i \Big/ \sum_{i\in s} d_i q_i x_i^2\right)\left(X - \sum_{i\in s} d_i x_i\right). \tag{5.5.9}$$

Substitution of the value of w_i from (5.5.9) in (5.5.2) leads to the general regression estimator (GREG) of total given by (5.5.4). Hence the theorem.

Corollary 5.5.1. If $q_i = 1/x_i$, then the optimal weight w_i becomes,

$$w_i = d_i X \Big/ \sum_{i \in s} d_i x_i$$

and the resultant estimator reduces to the ratio estimator of population total as

$$\hat{y}_{\text{R}} = \sum_{i \in s} d_i y_i \left(X \Big/ \sum_{i \in s} d_i x_i \right). \tag{5.5.10}$$

Remark 5.5.1. Singh, Horn, and Yu (1998) reported that there is no choice of q_i such that the resultant estimator (5.5.4) reduces to the product estimator of population total discussed by Cochran (1963).

The main difficulty with the calibrated weights given in (5.5.9) is that they do not satisfy the desired constraint of weights being non-negative. Deville and Särndal (1992) have considered several distance functions which guarantee the non-negativity of the weights. Let us discuss a new distance function, which also guarantees the non-negativity of the weights, in the following theorem.

Theorem 5.5.2. The optimal weights obtained by minimizing the distance function

$$D = \sum_{i \in s} w_i \ln(1/d_i) - \sum_{i \in s} w_i \ln(1/w_i). \tag{5.5.11}$$

subject to calibration constraint (5.5.3) lead to non-negative weights.

Proof. In this situation, the Lagrange function L is given by

$$L = \sum_{i \in s} w_i \ln(w_i/d_i) - \lambda\left(\sum_{i \in s} w_i x_i - X \right) = \sum_{i \in s} \{ w_i \ln(w_i) - w_i \ln(d_i) \} - \lambda \left\{ \sum_{i \in s} w_i x_i - X \right\}. \tag{5.5.12}$$

On differentiating (5.5.12) with respect to w_i and equating to zero we have

$$\frac{\partial L}{\partial w_i} = 1 + \ln(w_i) - \ln(d_i) - \lambda x_i = 0\,, \tag{5.5.13}$$

which implies

$$w_i = \exp[\ln(d_i) + \lambda x_i - 1], \tag{5.5.14}$$

where the value of λ can be obtained by solving

$$\sum_{i \in s} x_i \exp[\ln(d_i) + \lambda x_i - 1] = X. \tag{5.5.15}$$

Thus (5.5.14) shows that the calibrated weights are always non-negative if the distance function (5.5.11) is minimized for the calibration constraint (5.5.3). Hence the theorem.

Thus we conclude that the calibration approach can guarantee to yield non-negative estimators of population total depending upon the choice of the distance function to be minimized.

Example 5.5.1. Find the calibration weights for the units selected in the sample by using PPSWOR sampling by making use of known information about the number of fish caught during 1994 as auxiliary variable at estimation stage.

Sample selected using Midzuno--Sen sampling scheme						
Sample units	PRN No.	Population units	Species group	1992 X_3	1994 X_1	1995 Y
1	01	01	Sharks, other	1467	2001	2016
2	23	23	Blue runner	2371	5692	2319
3	54	54	Tautog	4195	2653	3816
4	58	58	Atlantic mackerel	1045	4860	4008
5	60	60	Spanish mackerel	5575	3850	2568
6	62	62	Summer flounder	11918	17741	16238
7	**62**	63	Gulf flounder	216	776	163
8	64	65	Winter flounder	1544	2300	2324

Use the chi square distance function between the design weights and calibrated weights. Discuss two cases where these weights lead to the ratio and GREG estimator for estimating the total number of fish during 1995. Deduce the value of the estimate in each case, provided that the sample has been selected using Midzuno--Sen's scheme of sampling as in the Example 5.4.1 by using information on the number of fish caught during 1992.

Solution. The chi square distance function gives the calibration weights as

$$w_i = d_i + \left(d_i q_i x_{1i} \Big/ \sum_{i\in s} d_i q_i x_{1i}^2\right)\left(X_1 - \sum_{i\in s} d_i x_{1i}\right).$$

From population 4 we have the value of $X_1 = 341856$.

Case I. If $q_i = 1/x_{1i}$ then these weights becomes $w_i = d_i\, X_1 \Big/ \sum_{i\in s} d_i x_{1i}$.

Species group	1994 x_{1i}	1995 y_i	π_i	$d_i = \pi_i^{-1}$	$d_i x_{1i}$	w_i	$w_i y_i$
Sharks, other	2001	2016	0.107450	9.306654258	18622.6152	9.73030	19616.3
Blue runner	5692	2319	0.110228	9.072105091	51638.4222	9.48508	21995.9
Tautog	2653	3816	0.115834	8.633043839	22903.4653	9.02603	34443.3
Atlantic mackerel	4860	4008	0.106153	9.420364945	45782.9736	9.84919	39475.6
Spanish mackerel	3850	2568	0.120075	8.328128253	32063.2938	8.70723	22360.2
Summer flounder	17741	16238	0.139569	7.164914845	127112.7543	7.49107	121640.0
Gulf flounder	776	163	0.103605	9.652043820	7489.9860	10.09141	1644.9
Winter flounder	2300	2324	0.107686	9.286258195	21358.3939	9.70898	22563.7
				Sum	326971.9042		283739.8

Note: To find π_i refer to Example 5.4.1.

Thus we have $\sum_{i=1}^{8} d_i x_{1i} = 340963.6378$ and the ratio estimate of the total number of fish caught during 1997 is given by

$$\hat{Y}_R = \sum_{i \in s} w_i y_i = 283739.8 .$$

Case II. If $q_i = 1$ then these weights become

$$w_i = d_i + \left(d_i x_{1i} \Big/ \sum_{i \in s} d_i x_{1i}^2 \right) \left(X_1 - \sum_{i \in s} d_i x_{1i} \right)$$

and are given in the following table:

x_{1i}	y_i	π_i	$d_i = \pi_i^{-1}$	$d_i x_{1i}$	$d_i x_{1i}^2$	w_i	$w_i y_i$
2001	2016	0.107450	9.306654	18622.6152	37263852.9	9.397594	18945.5
5692	2319	0.110228	9.072105	51638.4222	293925899.0	9.324272	21622.9
2653	3816	0.115834	8.633044	22903.4653	60762893.5	8.744889	33370.5
4860	4008	0.106153	9.420365	45782.9736	222505251.9	9.643938	38652.9
3850	2568	0.120075	8.328128	32063.2938	123443681.0	8.484704	21788.7
17741	16238	0.139569	7.164915	127112.7543	2255107373.0	7.785647	126423.3
776	163	0.103605	9.652044	7489.9860	5812229.1	9.688620	1579.2
2300	2324	0.107686	9.286258	21358.3938	49124305.8	9.390558	21823.6
			Sum	326971.9042	3047945487.0		284206.9

From the above table we have

$$\sum_{i=1}^{8} d_i x_{1i} = 326971.9042 , \quad \text{and} \quad \sum_{i=1}^{8} d_i x_{1i}^2 = 3047945487 .$$

Thus a GREG estimate of the total number of fish caught during 1995 is

$$\hat{Y}_G = \sum_{i \in s} w_i y_i = 284206.90 .$$

Now the question arises of how the calibration can be done if there are two or more auxiliary variables. To answer this question we have the following theorem:

Theorem 5.5.3. Suppose X_1 and X_2 are the known totals of two auxiliary characters X_{1i} and X_{2i}, for $i = 1, 2, ..., N$. The minimization of the CS distance function (5.5.5) subject to the two linear calibration constraints given by

$$\sum_{i \in s} w_i x_{1i} = X_1 \tag{5.5.16}$$

and

$$\sum_{i \in s} w_i x_{2i} = X_2 \tag{5.5.17}$$

leads to the regression type estimator

$$\hat{y}_G = \sum_{i\in s} d_i y_i + \hat{\beta}_1\left(X_1 - \sum_{i\in s} d_i x_{1i}\right) + \hat{\beta}_2\left(X_2 - \sum_{i\in s} d_i x_{2i}\right) \tag{5.5.18}$$

where $\hat{\beta}_1$ and $\hat{\beta}_2$ have their usual meanings.

Proof. In this case the Lagrange function L is given by

$$L = \sum_{i\in s}(w_i - d_i)^2(d_i q_i)^{-1} - 2\lambda_1\left[\sum_{i\in s} w_i x_{1i} - X_1\right] - 2\lambda_2\left[\sum_{i\in s} w_i x_{2i} - X_2\right]. \tag{5.5.19}$$

On differentiating (5.5.19) with respect to w_i and equating to zero we have

$$w_i = d_i + \lambda_1 d_i q_i x_{1i} + \lambda_2 d_i q_i x_{2i}. \tag{5.5.20}$$

On substituting (5.5.20) in (5.5.16) and (5.5.17), respectively we have

$$\lambda_1\left(\sum_{i\in s} d_i q_i x_{1i}^2\right) + \lambda_2\left(\sum_{i\in s} d_i q_i x_{1i} x_{2i}\right) = \left(X_1 - \sum_{i\in s} d_i x_{1i}\right), \tag{5.5.21}$$

and

$$\lambda_1\left(\sum_{i\in s} d_i q_i x_{1i} x_{2i}\right) + \lambda_2\left(\sum_{i\in s} d_i q_i x_{2i}^2\right) = \left(X_2 - \sum_{i\in s} d_i x_{2i}\right). \tag{5.5.22}$$

The system of equations given by (5.5.21) and (5.5.22) can be written as

$$\begin{bmatrix} \sum_{i\in s} d_i q_i x_{1i}^2, & \sum_{i\in s} d_i q_i x_{1i} x_{2i} \\ \sum_{i\in s} d_i q_i x_{1i} x_{2i}, & \sum_{i\in s} d_i q_i x_{2i}^2 \end{bmatrix}\begin{bmatrix} \lambda_1 \\ \lambda_2 \end{bmatrix} = \begin{bmatrix} X_1 - \sum_{i\in s} d_i x_{1i} \\ X_2 - \sum_{i\in s} d_i x_{2i} \end{bmatrix}. \tag{5.5.23}$$

Solving the above system of equations for λ_1 and λ_2 we have

$$\lambda_1 = \frac{\left(X_1 - \sum_{i\in s} d_i x_{1i}\right)\sum_{i\in s} d_i q_i x_{2i}^2 - \left(X_2 - \sum_{i\in s} d_i x_{2i}\right)\sum_{i\in s} d_i q_i x_{1i} x_{2i}}{\left(\sum_{i\in s} d_i q_i x_{1i}^2\right)\left(\sum_{i\in s} d_i q_i x_{2i}^2\right) - \left(\sum_{i\in s} d_i q_i x_{1i} x_{2i}\right)^2}$$

and

$$\lambda_2 = \frac{\left(X_2 - \sum_{i\in s} d_i x_{2i}\right)\sum_{i\in s} d_i q_i x_{1i}^2 - \left(X_1 - \sum_{i\in s} d_i x_{1i}\right)\sum_{i\in s} d_i q_i x_{1i} x_{2i}}{\left(\sum_{i\in s} d_i q_i x_{1i}^2\right)\left(\sum_{i\in s} d_i q_i x_{2i}^2\right) - \left(\sum_{i\in s} d_i q_i x_{1i} x_{2i}\right)^2}.$$

On substituting these values in (5.5.20) we obtain the optimal weights given by

$$w_i = d_i + \frac{d_i q_i x_{1i}\left[\left(X_1 - \sum_{i\in s} d_i x_{1i}\right)\sum_{i\in s} d_i q_i x_{2i}^2 - \left(X_2 - \sum_{i\in s} d_i x_{2i}\right)\sum_{i\in s} d_i q_i x_{1i} x_{2i}\right]}{\left(\sum_{i\in s} d_i q_i x_{1i}^2\right)\left(\sum_{i\in s} d_i q_i x_{2i}^2\right) - \left(\sum_{i\in s} d_i q_i x_{1i} x_{2i}\right)^2}$$

$$+\frac{d_i q_i x_{2i}\left[\left(X_2-\sum_{i\in s} d_i x_{2i}\right)\sum_{i\in s} d_i q_i x_{1i}^2-\left(X_1-\sum_{i\in s} d_i x_{1i}\right)\sum_{i\in s} d_i q_i x_{1i}x_{2i}\right]}{\left(\sum_{i\in s} d_i q_i x_{1i}^2\right)\left(\sum_{i\in s} d_i q_i x_{2i}^2\right)-\left(\sum_{i\in s} d_i q_i x_{1i}x_{2i}\right)^2}. \tag{5.5.24}$$

On inserting this value of w_i in (5.5.2) we obtain

$$\hat{y}_G = \sum_{i\in s} d_i y_i + \hat{\beta}_{1}\left(X_1 - \sum_{i\in s} d_i x_{1i}\right) + \hat{\beta}_2\left(X_2 - \sum_{i\in s} d_i x_{2i}\right) \tag{5.5.25}$$

where

$$\hat{\beta}_1 = \frac{\sum_{i\in s} d_i q_i x_{1i} y_i \sum_{i\in s} d_i q_i x_{2i}^2 - \sum_{i\in s} d_i q_i x_{2i} y_i \sum_{i\in s} d_i q_i x_{1i} x_{2i}}{\left(\sum_{i\in s} d_i q_i x_{1i}^2\right)\left(\sum_{i\in s} d_i q_i x_{2i}^2\right)-\left(\sum_{i\in s} d_i q_i x_{1i}x_{2i}\right)^2}$$

and

$$\hat{\beta}_2 = \frac{\sum_{i\in s} d_i q_i x_{2i} y_i \sum_{i\in s} d_i q_i x_{1i}^2 - \sum_{i\in s} d_i q_i x_{1i} y_i \sum_{i\in s} d_i q_i x_{1i} x_{2i}}{\left(\sum_{i\in s} d_i q_i x_{1i}^2\right)\left(\sum_{i\in s} d_i q_i x_{2i}^2\right)-\left(\sum_{i\in s} d_i q_i x_{1i}x_{2i}\right)^2}.$$

Hence the theorem.

Example 5.5.2. Find the calibration weights for the units selected in the sample by using PPSWOR sampling by making use of known information about the number of fish caught during 1993 and 1994 as auxiliary variables. Use the chi square distance function between the design weights and calibrated weights. Discuss the cases when these weights lead to the regression estimator for estimating the total number of fish during 1995. Deduce the value of the estimate.

Sample selected using Midzuno--Sen sampling scheme using 1992 data							
Sample units	PRN No.	Population units	Species group	1992 X_3	1993 X_2	1994 X_1	1995 Y
1	01	01	Sharks, other	1467	1385	2001	2016
2	23	23	Blue runner	2371	3800	5692	2319
3	54	54	Tautog	4195	4215	2653	3816
4	58	58	Atlantic mackerel	1045	2307	4860	4008
5	60	60	Spanish mackerel	5575	3653	3850	2568
6	62	62	Summer flounder	11918	22919	17741	16238
7	**62**	63	Gulf flounder	216	189	776	163
8	64	65	Winter flounder	1544	3582	2300	2324

Solution. If $q_i = 1 \;\; \forall\, i$ the calibration weights in case of two auxiliary variables become

$$w_i = d_i + d_i x_{1i} \frac{\left[\left(X_1 - \sum_{i=1}^{n} d_i x_{1i}\right)\sum_{i=1}^{n} d_i x_{2i}^2 - \left(X_2 - \sum_{i=1}^{n} d_i x_{2i}\right)\sum_{i=1}^{n} d_i x_{1i} x_{2i}\right]}{\left\{\left(\sum_{i=1}^{n} d_i x_{1i}^2\right)\left(\sum_{i=1}^{n} d_i x_{2i}^2\right) - \left(\sum_{i=1}^{n} d_i x_{1i} x_{2i}\right)^2\right\}}$$

$$+ d_i x_{2i} \frac{\left[\left(X_2 - \sum_{i=1}^{n} d_i x_{2i}\right)\sum_{i=1}^{n} d_i x_{1i}^2 - \left(X_1 - \sum_{i=1}^{n} d_i x_{1i}\right)\sum_{i=1}^{n} d_i x_{1i} x_{2i}\right]}{\left\{\left(\sum_{i=1}^{n} d_i x_{1i}^2\right)\left(\sum_{i=1}^{n} d_i x_{2i}^2\right) - \left(\sum_{i=1}^{n} d_i x_{1i} x_{2i}\right)^2\right\}}.$$

To calculate these weights we proceed as follows:

π_i	d_i	$d_i x_{1i}$	$d_i x_{2i}$	$d_i x_{1i}^2$	$d_i x_{2i}^2$	$d_i x_{1i} x_{2i}$
0.107450	9.306654	18622.62	12889.72	37263853.0	17852256.9	25792322.0
0.110228	9.072105	51638.42	34474.00	293925899.0	131001197.5	196226004.3
0.115834	8.633044	22903.47	36388.28	60762893.5	153376599.3	96538106.3
0.106153	9.420365	45782.97	21732.78	222505252.0	50137527.9	105621320.2
0.120075	8.328128	32063.29	30422.65	123443681.0	111133949.6	117127212.2
0.139569	7.164915	127112.75	164212.70	2255107373.0	3763590489.0	2913297215.0
0.103605	9.652044	7489.99	1824.24	5812229.1	344780.6	1415607.3
0.107686	9.286258	21358.39	33263.38	49124305.9	119149415.9	76505766.8
	Sum	326971.90	335207.70	3047945487.0	4346586217.0	3532523554.0

Note: To find π_i refer to Example 5.4.1.

Using $X_1 = 341856$, $X_2 = 316784$ and the information from the above table we have

$$\frac{\left\{\left(X_1 - \sum_{i=1}^{n} d_i x_{1i}\right)\sum_{i=1}^{n} d_i x_{2i}^2 - \left(X_2 - \sum_{i=1}^{n} d_i x_{2i}\right)\sum_{i=1}^{n} d_i x_{1i} x_{2i}\right\}}{\left\{\left(\sum_{i=1}^{n} d_i x_{1i}^2\right)\left(\sum_{i=1}^{n} d_i x_{2i}^2\right) - \left(\sum_{i=1}^{n} d_i x_{1i} x_{2i}\right)^2\right\}} = \frac{1.29777 \times 10^{14}}{7.69435 \times 10^{17}} = 0.000168666$$

and

$$\frac{\left\{\left(X_2 - \sum_{i=1}^{n} d_i x_{2i}\right)\sum_{i=1}^{n} d_i x_{1i}^2 - \left(X_1 - \sum_{i=1}^{n} d_i x_{1i}\right)\sum_{i=1}^{n} d_i x_{1i} x_{2i}\right\}}{\left\{\left(\sum_{i=1}^{n} d_i x_{1i}^2\right)\left(\sum_{i=1}^{n} d_i x_{2i}^2\right) - \left(\sum_{i=1}^{n} d_i x_{1i} x_{2i}\right)^2\right\}} = \frac{-1.08733 \times 0^{14}}{7.69435 \times 10^{17}} = -0.000141315.$$

Thus the calibrated weights are given by

$$w_i = d_i + 1.68666 \times 10^{-4} \times d_i x_{1i} - 1.41315 \times 10^{-4} \times d_i x_{2i}.$$

Thus we have the following results:

w_i	$w_i y_i$
10.62613538	21422.3
12.91002878	29938.4
7.35385154	28062.3
14.07120429	56397.4
9.43691815	24234.0
5.39870687	87664.2
10.65755444	1737.2
8.18806195	19029.1
Sum	268484.8

Hence an estimate of total number of fish caught during 1995 is given by

$$\hat{Y}_G = \sum_{i\in s} w_i y_i = 268484.80 .$$

Remark 5.5.2: Note that there is no choice of weights q_i such that the estimator $\hat{y}_G$ in (5.5.25) will reduce to a multivariate ratio type estimator with two auxiliary variables.

Now if there are p auxiliary variables X_{ji}, $j = 1, 2, ..., p$, and the population totals $X_j = \sum_{i\in\Omega} X_{ji}$ are known, then we have to minimize the CS type of distance function (5.5.5), subject to the p calibration constraints:

$$\sum_{i\in s} w_i x_{ji} = X_j, \quad \text{for } j = 1, 2, ..., p . \tag{5.5.26}$$

Under such situations, the system of linear equations by minimizing the Lagrange function is given by

$$\begin{bmatrix} \sum_{i\in s} d_i q_i x_{1i}^2, & \sum_{i\in s} d_i q_i x_{1i} x_{2i}, ..., & \sum_{i\in s} d_i q_i x_{1i} x_{pi} \\ \sum_{i\in s} d_i q_i x_{2i} x_{1i}, & \sum_{i\in s} d_i q_i x_{2i}^2, ..., & \sum_{i\in s} d_i q_i x_{2i} x_{pi} \\ \\ \sum_{i\in s} d_i q_i x_{pi} x_{1i}, & \sum_{i\in s} d_i q_i x_{pi} x_{2i}, ..., & \sum_{i\in s} d_i q_i x_{pi}^2 \end{bmatrix} \begin{bmatrix} \lambda_1 \\ \lambda_2 \\ \\ \lambda_p \end{bmatrix} = \begin{bmatrix} X_1 - \sum_{i\in s} d_i x_{1i} \\ X_2 - \sum_{i\in s} d_i x_{2i} \\ X_p - \sum_{i\in s} d_i x_{pi} \end{bmatrix} . \tag{5.5.27}$$

If the system of equations given by (5.5.27) is expressed as

$$A\lambda = C \tag{5.5.28}$$

and its solution is given by

$$\lambda = A^{-1}C . \tag{5.5.29}$$

Then we have the following theorem.

Theorem 5.5.4. The optimal weights in case of p auxiliary variables are given by

$$w_i = d_i + d_i q_i A^{-1} C X \qquad (5.5.30)$$

where $X = (x_{ji})_{p \times n}$ has its usual meaning.

The resultant estimator of the population total in the presence of p auxiliary variables is given by

$$\hat{y}_G = \sum_{i \in s} d_i y_i + \sum_{i \in s} d_i q_i y_i A^{-1} C X . \qquad (5.5.31)$$

Now we will study some properties of the above estimator in the following theorems:

Theorem 5.5.5. The variance of the estimator $\hat{y}_G$ based on p auxiliary characters is

$$V_1(\hat{y}_G) = V(\hat{y}_{HT})\left(1 - R^2_{y \bullet x_1, x_2, ..., x_p}\right), \qquad (5.5.32)$$

where $R^2_{y \bullet x_1, x_2, .., x_p}$ denotes the multiple correlation coefficient of y on $x_1, x_2, ..., x_p$.

Theorem 5.5.6. The Sen--Yates--Grundy form of the variance of the estimator $\hat{y}_G$ can also be written as

$$V_2(\hat{y}_G) = \frac{1}{2} \sum_{i \neq j \in \Omega} \sum (\pi_i \pi_j - \pi_{ij})(d_i e_i - d_j e_j)^2 \qquad (5.5.33)$$

where $\underline{e} = \underline{y} - \underline{\beta} X$ denotes the vector of error terms in the model m .

Theorem 5.5.6. Another form of the variance of the estimator $\hat{y}_G$ is given by

$$V_3(\hat{y}_G) = \frac{1}{2} \sum_{i,j \in s} \sum (\pi_{ij} - \pi_i \pi_j) d_i d_j e_i e_j . \qquad (5.5.34)$$

In the following section we shall discuss the calibration of the estimator of variance of the HT estimator of population total, Y, defined as

$$\hat{Y}_{HT} = \sum_{i \in s} d_i y_i$$

and then the estimation of variance of GREG defined as

$$\hat{Y}_G = \sum_{i \in s} d_i y_i + \left(\sum_{i \in s} d_i q_i x_i y_i \Big/ \sum_{i \in s} d_i q_i x_i^2 \right) \left(X - \sum_{i \in s} d_i x_i \right)$$

in section 5.7.

Remark 5.5.3: The drawback with the above approach is that it may produce negative calibrated weights or large calibrated weights for some units. This can lead to unstable parameter estimates in some domains. We may also produce implausible estimates such as a negative total for a variable which is strictly positive. In order to alleviate these problems we would like to introduce restrictions on the values that the calibrated weights can take. Consider we specify a lower bound l_i and an upper bound u_i for each unit $i \in s$ and bounds may differ from each unit. Thus the above problem can be restated as

$$\text{Minimize : } \sum_{i\in s}(w_i - d_i)^2(d_i q_i)^{-1}$$

subject to

$$\sum_{i\in s} w_i x_i = X$$

and

$$l_i \le w_i \le u_i \text{ for } i \in s .$$

Now it becomes a quadratic programming problem with equality and inequality constraints. The solution to this problem is more complex because of the presence of inequality constraints and there is no closed form expression for the solution. For detail on can refer to Estevao (1994). Singh and Mohl (1996) have also tried to understand calibration approaches and to explain other people in a simple way.

5.6 CALIBRATED ESTIMATOR OF THE VARIANCE OF THE ESTIMATOR OF POPULATION TOTAL

The Sen--Yates--Grundy (1953) form of the variance of the estimator $\hat{Y}_{\text{HT}}$ for a fixed sample size is given by

$$V_{\text{SYG}}\left(\hat{Y}_{\text{HT}}\right) = \frac{1}{2}\sum_{i\neq}\sum_{j\in\Omega}\left(\pi_i\pi_j - \pi_{ij}\right)\left(d_i y_i - d_j y_j\right)^2 \tag{5.6.1}$$

and its unbiased estimator is given by

$$\hat{v}_{\text{SYG}}\left(\hat{Y}_{\text{HT}}\right) = \frac{1}{2}\sum_{i\neq}\sum_{j\in s} D_{ij}\left(d_i y_i - d_j y_j\right)^2 \tag{5.6.2}$$

where $D_{ij} = \left(\pi_i\pi_j - \pi_{ij}\right)/\pi_{ij}$ denotes the design weights. Singh, Horn, Chowdhury and Yu (1999) consider an estimator of variance of the Horvitz and Thompson (1952) estimator

$$\hat{v}_1\left(\hat{Y}_{HT}\right) = \frac{1}{2}\sum_{i\neq}\sum_{j\in s} w_{ij}\left(d_i y_i - d_j y_j\right)^2 \tag{5.6.3}$$

where w_{ij} are the modified weights and are as close as possible in an average sense to the D_{ij} for a given measure and subject to the calibration constraint

$$\frac{1}{2}\sum_{i\neq}\sum_{j\in s} w_{ij}\left(d_i x_i - d_j x_j\right)^2 = V_{\text{SYG}}\left(\hat{X}_{\text{HT}}\right) \tag{5.6.4}$$

where $V_{\text{SYG}}\left(\hat{X}_{\text{HT}}\right)=\frac{1}{2}\sum_{i\neq j\in\Omega}\sum\left(\pi_i\pi_j-\pi_{ij}\right)\left(d_ix_i-d_jx_j\right)^2$ denotes the known variance of the estimator of auxiliary total $X=\sum_{i\in\Omega}X_i$ given by $\hat{X}_{\text{HT}}=\sum_{i\in s}d_ix_i$. The idea of adjusting the denominator of weights D_{ij} has also been discussed by Fuller (1970), but his method may have a limitation of not guarantying the non-negativity of the estimates of variance.

For simplicity they restricted themselves to the two dimensional CS type of distance D between two $n\times n$ grids formed by the weights w_{ij} and D_{ij} for $i,\ j=1,\ 2,\ ...,\ n$, defined as

$$D=\frac{1}{2}\sum_{i\neq j\in s}\sum\left(w_{ij}-D_{ij}\right)^2\left(D_{ij}Q_{ij}\right)^{-1}. \tag{5.6.5}$$

In most situations $Q_{ij}=1$ but other types of weights can also be used. It is shown that the ratio type estimator proposed by Isaki (1983) is a special case for a particular choice of Q_{ij}. Minimization of (5.6.5) subject to (5.6.4) leads to modified optimal weights

$$w_{ij}=D_{ij}+\frac{D_{ij}Q_{ij}\left(d_ix_i-d_jx_j\right)^2}{\left\{\frac{1}{2}\sum_{i\neq j\in s}\sum D_{ij}Q_{ij}\left(d_ix_i-d_jx_j\right)^4\right\}}\left[V_{\text{SYG}}\left(\hat{X}_{\text{HT}}\right)-\frac{1}{2}\sum_{i\neq j\in s}\sum D_{ij}\left(d_ix_i-d_jx_j\right)^2\right]. \tag{5.6.6}$$

Substitution of w_{ij} from (5.6.6) in (5.6.3) leads to the following regression type estimator:

$$\hat{v}_1\left(\hat{Y}_{\text{HT}}\right)=\hat{v}_{\text{SYG}}\left(\hat{Y}_{\text{HT}}\right)+\hat{B}\left[V_{\text{SYG}}\left(\hat{X}_{\text{HT}}\right)-\hat{v}_{\text{SYG}}\left(\hat{X}_{\text{HT}}\right)\right] \tag{5.6.7}$$

where

$$\hat{B}=\sum_{i\neq j\in s}\sum D_{ij}Q_{ij}\left(d_iy_i-d_jy_j\right)^2\left(d_ix_i-d_jx_j\right)^2\Big/\sum_{i\neq j\in s}\sum D_{ij}Q_{ij}\left(d_ix_i-d_jx_j\right)^4=\hat{\mu}_{22}/\hat{\mu}_{04} \tag{5.6.8}$$

and

$$\hat{v}_{\text{SYG}}\left(\hat{X}_{\text{HT}}\right)=\frac{1}{2}\sum_{i\neq j\in s}\sum D_{ij}\left(d_ix_i-d_jx_j\right)^2. \tag{5.6.9}$$

The leading term of the mean squared error of the regression type estimator (5.6.7) is

$$\text{MSE}\left[\hat{v}_1\left(\hat{Y}_{\text{HT}}\right)\right]=V\left[\hat{v}_{\text{SYG}}\left(\hat{Y}_{\text{HT}}\right)\right]+B^2V\left[\hat{v}_{\text{SYG}}\left(\hat{X}_{\text{HT}}\right)\right]-2B\text{Cov}\left[\hat{v}_{\text{SYG}}\left(\hat{Y}_{\text{HT}}\right),\hat{v}_{\text{SYG}}\left(\hat{X}_{\text{HT}}\right)\right] \tag{5.6.10}$$

where

$$B=\sum_{i\neq j\in\Omega}\sum Q_{ij}\left(\pi_i\pi_j-\pi_{ij}\right)\left(d_iy_i-d_jy_j\right)^2\left(d_ix_i-d_jx_j\right)^2\Big/\sum_{i\neq j\in\Omega}\sum Q_{ij}\left(\pi_i\pi_j-\pi_{ij}\right)\left(d_ix_i-d_jx_j\right)^4$$

and

$$\text{Cov}\left[\hat{v}_{\text{SYG}}\left(\hat{Y}_{\text{HT}}\right), \hat{v}_{\text{SYG}}\left(\hat{X}_{\text{HT}}\right)\right] \approx \frac{1}{4}\sum\sum\sum_{i\neq j\neq k\neq l\in\Omega}\sum \left(\pi_{ijkl} - \pi_{ij}\pi_{kl}\right)\left(d_i y_i - d_j y_j\right)^2\left(d_k x_k - d_l x_l\right)^2 .$$

Here π_{ijkl} denotes the positive probability of including four units in the sample, *i.e.*, $\pi_{ijkl} = P(i,\ j,\ k,\ l\ \in s)$.

Expression (5.6.10) shows that the estimator $\hat{v}_1\left(\hat{Y}_{\text{HT}}\right)$ of variance is not always more efficient than the estimator given by Sen-Yates-Grundy (1953), but the estimator (5.6.7) is consistent because the ratio of modified weights to original weights, *i.e.*, w_{ij}/D_{ij} converges in design probability to unity. This condition of consistency is an analogue of the condition given by Särndal, Swensson, and Wretman (1989). Ramakrishnan (1975a) has also modified the Sen-Yates-Grundy (1953) estimator to suit varying sample size designs. It is also possible to calibrate the denominator $d_{ij} = \pi_{ij}^{-1}$ of D_{ij} similar to Fuller (1970) and has the limitation of not guaranteeing the non-negativity of estimates of variance. Sitter and Wu (2002) claim that the two-dimensional chi square distance function can take negative value, and hence calibrates only design weights $d_{ij} = \pi_{ij}^{-1}$. Note that the calibration of $d_{ij} = \pi_{ij}^{-1}$ will not guarantee the non-negative estimate of variance, but the calibration of D_{ij} can guarantee the non-negativity of the variance estimate if calibrated under bounded conditions by using quadratic programming. Further note that distance can be negative, but the magnitude of distance cannot be negative. Stukel, Hidiroglou, and Särndal (1996) have attempted to compare Jackknife and linearization forms of the estimators of variance.

Here we would like to discuss the cases where the estimator $\hat{v}_1\left(\hat{Y}_{\text{HT}}\right)$ reduces to usual estimators in simple designs as follows:

Case I. Under SRSWOR, $\pi_i = \pi_j = n/N$, $\pi_{ij} = n(n-1)/N(N-1)$ and if $q_{ij} = 1$ then (5.6.7) becomes

$$\hat{v}_1\left(\hat{Y}_{\text{srswor}}\right) = \frac{N^2(1-f)}{n}\left[s_y^2 + \hat{b}\left(S_x^2 - s_x^2\right)\right], \tag{5.6.11}$$

where

$\hat{Y}_{\text{srswor}}$ denotes the estimator of population total under SRSWOR design,

$s_y^2 = \dfrac{1}{2n(n-1)}\sum_{i=1}^{n}\sum_{j=1}^{n}\left(y_i - y_j\right)^2$ is an unbiased estimator of S_y^2,

$s_x^2 = \dfrac{1}{2n(n-1)}\sum_{i=1}^{n}\sum_{j=1}^{n}\left(x_i - x_j\right)^2$ is an unbiased estimator of S_x^2,

and

$\hat{b} = \hat{\mu}_{22}/\hat{\mu}_{04}$

where

$$\hat{\mu}_{22} = \frac{N^4(1-f)}{n^4(n-1)} \sum_{i=1}^{n} \sum_{j=1}^{n} (y_i - y_j)^2 (x_i - x_j)^2, \text{ and } \hat{\mu}_{04} = \frac{N^4(1-f)}{n^4(n-1)} \sum_{i=1}^{n} \sum_{j=1}^{n} (x_i - x_j)^4.$$

The ratio

$$\hat{v}_1\left(\hat{Y}_{\text{srswor}}\right) \Big/ \left\{N^2\left(\frac{1-f}{n}\right)\right\}$$

is a regression type estimator of finite population variance S_y^2 as proposed by Isaki (1983).

Case II. For an IPPS sampling scheme for which $\pi_i = np_i$ and π_{ij} is a second order inclusion probability such that $\sum_{i\in\Omega} \pi_i = n$ and $\sum_{j(\neq i)\in\Omega} \pi_{ij} = (n-1)\pi_i$, then the Horvitz and Thompson (1952) estimator becomes

$$\hat{Y}_{\text{HT}} = \frac{1}{n} \sum_{i=1}^{n} \frac{y_i}{p_i}$$

with

$$\hat{v}_1\left(\hat{Y}_{\text{HT}}\right) = \hat{v}_y\left(\hat{Y}_{\text{HT}}\right) + \hat{b}\left[V\left(\hat{X}_{\text{HT}}\right) - \hat{v}_x\left(\hat{X}_{\text{HT}}\right)\right] \tag{5.6.12}$$

where

$$\hat{v}_y\left(\hat{Y}_{\text{HT}}\right) = \frac{1}{2n^2} \sum_{i \neq j \in s} \sum k_{ij} \left(\frac{y_i}{p_i} - \frac{y_j}{p_j}\right)^2, \quad \hat{v}_x\left(\hat{Y}_{\text{HT}}\right) = \frac{1}{2n^2} \sum_{i \neq j \in s} \sum k_{ij} \left(\frac{x_i}{p_i} - \frac{x_j}{p_j}\right)^2$$

where

$$k_{ij} = \frac{n^2 p_i p_j}{\pi_{ij}} - 1, \quad \hat{b} = \hat{\mu}_{22} / \hat{\mu}_{04}$$

with

$$\hat{\mu}_{22} = \sum_{i \neq j \in s} \sum k_{ij} \left(\frac{y_i}{P_i} - \frac{y_j}{P_j}\right)^2 \left(\frac{x_i}{P_i} - \frac{x_j}{P_j}\right)^2, \quad \text{and} \quad \hat{\mu}_{04} = \sum_{i \neq j \in s} \sum k_{ij} \left(\frac{x_i}{P_i} - \frac{x_j}{P_j}\right)^4$$

have their usual meanings under IPPS sampling.

Case III. If $q_{ij} = (d_i x_i - d_j x_j)^{-2}$ then we have ratio type estimator given by

$$\hat{v}_1\left(\hat{Y}_{\text{HT}}\right)_{\text{Ratio}} = \hat{v}_{\text{SYG}}\left(\hat{Y}_{\text{HT}}\right)\left[\frac{V_{\text{SYG}}\left(\hat{X}_{\text{HT}}\right)}{\hat{v}_{\text{SYG}}\left(\hat{X}_{\text{HT}}\right)}\right]. \tag{5.6.13}$$

Under SRSWOR sampling

$$\hat{v}_1\left(\hat{Y}_{\text{HT}}\right)_{\text{Ratio}} \Big/ \left\{N^2\left(\frac{1-f}{n}\right)\right\}$$

leads to the ratio type of estimator

$$s_1^2 = s_y^2\left(S_x^2 / s_x^2\right)$$

proposed by Isaki (1983).

5.7 ESTIMATION OF VARIANCE OF GREG

Following Särndal, Swensson, and Wretman (1989), Deville and Särndal (1992), Särndal (1996), and Rao (1997), the GREG can be written as

$$\hat{Y}_{\text{G}} = \sum_{i\in s}\frac{e_i}{\pi_i} + \hat{\beta}_{\text{ds}}X \tag{5.7.1}$$

and the Sen --Yates --Grundy (1953) form of estimator of variance of the GREG is

$$\hat{v}_{\text{SYG}}\left(\hat{Y}_{\text{G}}\right) = \frac{1}{2}\sum_{i\neq j\in s}\sum D_{ij}\left(w_i e_i - w_j e_j\right)^2 \tag{5.7.2}$$

where $D_{ij} = \left(\pi_i\pi_j - \pi_{ij}\right)/\pi_{ij}$, $i \neq j$, and $e_i = y_i - \hat{\beta}_{\text{ds}}x_i$. This estimator can easily be written as

$$\hat{v}_{\text{SYG}}\left(\hat{Y}_{\text{G}}\right) = \frac{1}{2}\sum_{i\neq j\in s}\sum D_{ij}\left(d_i e_i - d_j e_j\right)^2 + \hat{\psi}_1\left(X - \sum_{i=1}^{n} d_i x_i\right) + \hat{\psi}_2\left(X - \sum_{i=1}^{n} d_i x_i\right)^2 \tag{5.7.3}$$

where

$$\hat{\psi}_1 = \left\{\sum_{i\in s} d_i q_i x_i^2\right\}^{-1} \sum_{i\neq j\in s}\sum D_{ij}\left(d_i e_i - d_j e_j\right)\left(d_i q_i x_i e_i - d_j q_j x_j e_j\right)$$

and

$$\hat{\psi}_2 = 0.5\left\{\sum_{i\in s} d_i q_i x_i^2\right\}^{-2} \sum_{i\neq j\in s}\sum D_{ij}\left(d_i q_i x_i e_i - d_j q_j x_j e_j\right)^2 .$$

The estimator in (5.7.2) covers a variety of estimators of variance. Let us consider SRSWOR design, *i.e.*, $\pi_i = \pi_j = n/N$ and $\pi_{ij} = n(n-1)/N(N-1)$.

Then we have the following cases:

Case I. If $q_i = 1$, then $\hat{Y}_{\text{G}}$ reduces to the usual regression estimator of total. Now if $w_i = d_i$ in (5.7.2), it reduces to

$$\hat{v}\left(\hat{Y}_{\text{G}}\right) = \frac{N^2(1-f)}{n(n-1)}\sum_{i=1}^{n} e_i^2 \tag{5.7.4}$$

which is the usual estimator of variance of the regression estimator.

Case II. If $q_i = 1/x_i$ then the estimator $\hat{Y}_{\text{G}}$ reduces to the ratio estimator $\hat{Y}_{\text{R}}$ of the population total. The estimator of variance (5.7.2) reduces to

$$\hat{v}\left(\hat{Y}_{\text{R}}\right) = \frac{N^2(1-f)}{n(n-1)}\sum_{i=1}^{n} e_i^2\left\{\frac{X}{\hat{X}}\right\}^2 \tag{5.7.5}$$

where $e_i = y_i - \left(\frac{\bar{y}}{\bar{x}}\right)x_i$ and $\hat{X} = \frac{N}{n}\sum_{i=1}^{n} x_i$. The estimator at (5.7.5) is a special case of a class of estimators of variance of the ratio estimator proposed by Wu (1982) as

$$\hat{v}_{\text{W}}\left(\hat{Y}_{\text{R}}\right) = \frac{N^2(1-f)}{n(n-1)}\sum_{i=1}^{n} e_i^2\left\{\frac{X}{\hat{X}}\right\}^g \quad \text{for } g = 2. \tag{5.7.6}$$

Case III. If $q_i = 1$ then the estimator (5.7.2) becomes

$$\hat{v}_{\text{SYG}}\left(\hat{Y}_{\text{G}}\right) = \frac{N^2(1-f)}{n(n-1)}\sum_{i=1}^{n} e_i^2 + \hat{\psi}_1\left(X - \hat{X}\right) + \hat{\psi}_2\left(X - \hat{X}\right)^2 \tag{5.7.7}$$

where,

$$\hat{\psi}_1 = \frac{(N-n)}{n(n-1)} \frac{\sum\limits_{i \neq j=1}^{n}\sum\limits^{n}\left(e_i - e_j\right)\left(x_i e_i - x_j e_j\right)}{\left(\sum\limits_{i=1}^{n} x_i^2\right)}, \text{ and } \hat{\psi}_2 = \frac{(N-n)}{2N(n-1)} \frac{\sum\limits_{i \neq j=1}^{n}\sum\limits^{n}\left(x_i e_i - x_j e_j\right)^2}{\left(\sum\limits_{i=1}^{n} x_i^2\right)^2}.$$

Deng and Wu (1987) have defined a general class of estimators of the variance of the regression estimator:

$$\hat{v}_{\text{DW}}\left(\hat{Y}_{\text{G}}\right) = \frac{N^2(1-f)}{n(n-1)}\sum_{i=1}^{n} e_i^2 \left\{\frac{X}{\hat{X}}\right\}^g. \tag{5.7.8}$$

The linear form of the class of estimators (5.7.6) takes the form

$$\hat{v}_{\text{DW}}\left(\hat{Y}_{\text{G}}\right) = \frac{N^2(1-f)}{n(n-1)}\sum_{i=1}^{n} e_i^2 \left[1 + g\left(\frac{X}{\hat{X}} - 1\right) + \frac{g(g-1)}{2}\left(\frac{X}{\hat{X}} - 1\right)^2 + \ldots\right] \tag{5.7.9}$$

which is again similar to (5.7.3).

Following Singh, Horn, and Yu (1998), the estimators of variance of the estimators of total considered so far belong to the low level calibration approach. The estimators studied by Chaudhuri and Mitra (1992) are also special cases of the low level calibration approach. As noted earlier, there is no choice of q_i which reduces $\hat{Y}_{\text{G}}$ to the product method of estimation considered by Cochran (1963). Thus the estimation of variance of the product estimator has not been discussed here. To discuss the efficiency of such estimators consider an analogue of the general class of estimators of the variance of GREG by following Srivastava (1971) given as

$$\hat{v}_{\text{S}}\left(\hat{Y}_{\text{G}}\right) = \left(\frac{N^2(1-f)}{n(n-1)}\sum_{i=1}^{n} e_i^2\right) H\left(\frac{X}{\hat{X}}\right) \tag{5.7.10}$$

where $H(.)$ is a parametric function such that $H(1) = 1$ satisfying certain regularity conditions defined earlier in Chapter 3. Following Srivastava (1971), it is easy to see that analogues of the general class of estimators (5.7.10) attain the minimum variance of the class of estimators proposed by Deng and Wu (1987) for regression estimators and Wu (1982) for a ratio estimator. If we attach any function of the ratio $X/\hat{X}$ to the usual estimator of variance given by $\frac{N^2(1-f)}{n(n-1)}\sum_{i=1}^{n} e_i^2$ as given in (5.7.10), the asymptotic variance of the resultant estimator remains the same. In other words, the efficiency of the estimators of variance of the regression estimator GREG of total obtained through low level calibration remains less than or equal to the class of estimators proposed by Wu (1982) and Deng and Wu (1987). The weights w_i used to construct the estimator of variance of GREG in (5.7.2) were obtained while estimating the population total and hence named as low level calibration weights for variance estimation.

Example 5.7.1. We wish to estimate the total number of fish of all kinds caught by recreational fisherman on the Atlantic and Gulf coasts during 1995. Population 4 in the Appendix shows that information on the number of different kinds of fish caught during 1992 and 1994 is available. Use the known information on the number of fish caught during 1992 to select a sample of eight units by using PPSWOR sampling. Collect the required information from population 4 to estimate the total number of fish caught during 1995 using a regression estimator by using information during 1994 as an the auxiliary variable. Apply the Sen--Yates--Grundy form of the estimator of variance of the regression estimator to construct a 95% confidence interval. Also use the estimator of variance of the GREG proposed by Särndal, Swenson and Wretman (1989).

Solution. Again referring to example 5.4.1, the ultimate sample is as shown below.

Sample selected using Midzuno-Sen sampling scheme						
Sample units	PRN No.	Population units	Species group	1992 X_3	1994 X_1	1995 Y
1	01	01	Sharks, other	1467	2001	2016
2	23	23	Blue runner	2371	5692	2319
3	54	54	Tautog	4195	2653	3816
4	58	58	Atlantic mackerel	1045	4860	4008
5	60	60	Spanish mackerel	5575	3850	2568
6	62	62	Summer flounder	11918	17741	16238
7	**62**	63	Gulf flounder	216	776	163
8	64	65	Winter flounder	1544	2300	2324

As before, the Sen--Yates--Grundy weights based on first and second order inclusion probabilities are:

Sen-Yates-Grundy weights $\left(\pi_i\pi_j - \pi_{ij}\right)/\pi_{ij}$							
$i > j$	1	2	3	4	5	6	7
2	0.133289						
3	0.127726	0.127238					
4	0.137742	0.134317	0.127962				
5	0.123894	0.124520	0.125694	0.123590			
6	0.109511	0.114270	0.123288	0.107219	0.129638		
7	0.140723	0.136417	0.128445	0.142822	0.122973	0.102578	
8	0.136022	0.133105	0.127683	0.137442	0.123948	0.109925	0.140346

If $q_i = 1$ then these weights become

$$w_i = d_i + \left(d_i x_{1i} \Big/ \sum_{i=1}^{n} d_i x_{1i}^2 \right)\left(X_1 - \sum_{i=1}^{n} d_i x_{1i} \right)$$

and are given in the following table:

x_{1i}	y_i	π_i	$d_i = \pi_i^{-1}$	$d_i x_{1i}$	$d_i x_{1i}^2$	w_i	$w_i y_i$
2001	2016	0.107450	9.306654	18622.61517	37263852.9	9.397594	18945.5
5692	2319	0.110228	9.072105	51638.42218	293925899.0	9.324272	21622.9
2653	3816	0.115834	8.633044	22903.46530	60762893.4	8.744889	33370.5
4860	4008	0.106153	9.420365	45782.97363	222505251.9	9.643938	38652.9
3850	2568	0.120075	8.328128	32063.29377	123443681.0	8.484704	21788.7
17741	16238	0.139569	7.164915	127112.75430	2255107373.0	7.785647	126423.3
776	163	0.103605	9.652044	7489.98600	5812229.1	9.688620	1579.2
2300	2324	0.107686	9.286258	21358.39385	49124305.8	9.390558	21823.6
			Sum	326971.90420	3047945487.0		284206.9

From the above table we have

$$\sum_{i=1}^{8} d_i x_{1i} = 326971.9042 \text{ and } \sum_{i=1}^{8} d_i x_{1i}^2 = 3047945487 .$$

Thus the regression estimate of the total number of fish caught during 1995 is

$$\hat{Y}_G = \sum_{i \in s} w_i y_i = 284206.90 .$$

Now we assume the following superpopulation model passing through the origin as

$$y_i = \beta x_{1i} + e_i .$$

Estimates of error term					
d_i	y_i	x_{1i}	$d_i x_{1i}^2$	$d_i x_{1i} y_i$	e_i
9.306671806	2016	2001	37263923.22	37543262.97	292.3759
9.072095548	2319	5692	293925589.80	119749375.10	-2583.9800
8.633049291	3816	2653	60762931.83	87399678.80	1530.7550
9.420379236	4008	4860	222505589.40	183498436.70	-178.3130
8.328116785	2568	3850	123443511.00	82338425.03	-748.3180
7.164889400	16238	17741	2255099365.00	2064049574.00	956.2333
9.652041724	163	776	5812227.87	1220867.45	-505.4320
9.286219716	2324	2300	49124102.30	49636701.63	342.8229
		Sum	3047937240.00	2625436321.00	-893.8559

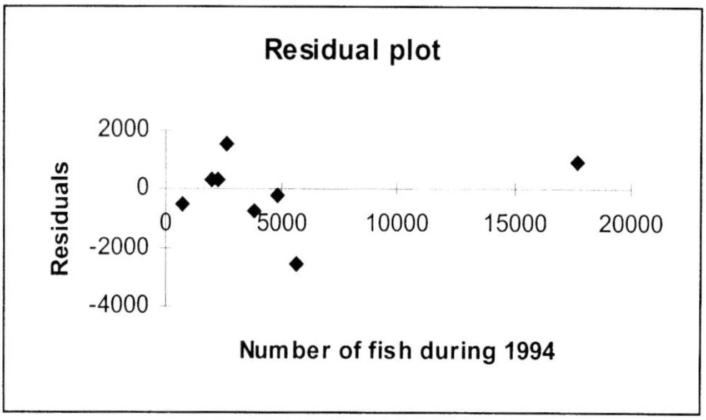

Fig. 5.7.1 A residual plot.

Note that we used

$$e_i = y_i - \hat{\beta}_{ds} x_i \text{ with } \hat{\beta}_{ds} = \sum_{i\in s} d_i x_{1i} y_i \Big/ \sum d_i x_{1i}^2 = 0.861381359,$$

therefore, the sum of the estimates of these residuals may or may not be equal to zero.

The Sen--Yates--Grundy estimator of variance of the GREG estimator of population total is

$$\hat{v}_{\text{SYG}}\left(\hat{Y}_{\text{G}}\right)$$

$$= \frac{1}{2}\sum_{i\in s}\sum_{j(\neq i)\in s}\left(\frac{\pi_i\pi_j - \pi_{ij}}{\pi_{ij}}\right)\left(\frac{e_i}{\pi_i} - \frac{e_j}{\pi_j}\right)^2 = \sum_{i(<j)=1}^{8}\sum_{j=1}^{8}\left(\frac{\pi_i\pi_j - \pi_{ij}}{\pi_{ij}}\right)\left(\frac{e_i}{\pi_i} - \frac{e_j}{\pi_j}\right)^2$$

$$= \left(\frac{\pi_1\pi_2 - \pi_{12}}{\pi_{12}}\right)\left(\frac{e_1}{\pi_1} - \frac{e_2}{\pi_2}\right)^2 + \left(\frac{\pi_1\pi_3 - \pi_{13}}{\pi_{13}}\right)\left(\frac{e_1}{\pi_1} - \frac{e_3}{\pi_3}\right)^2 + \left(\frac{\pi_1\pi_4 - \pi_{14}}{\pi_{14}}\right)\left(\frac{e_1}{\pi_1} - \frac{e_4}{\pi_4}\right)^2$$

$$+ \left(\frac{\pi_1\pi_5 - \pi_{15}}{\pi_{15}}\right)\left(\frac{e_1}{\pi_1} - \frac{e_5}{\pi_5}\right)^2 + \left(\frac{\pi_1\pi_6 - \pi_{16}}{\pi_{16}}\right)\left(\frac{e_1}{\pi_1} - \frac{e_6}{\pi_6}\right)^2 + \left(\frac{\pi_1\pi_7 - \pi_{17}}{\pi_{17}}\right)\left(\frac{e_1}{\pi_1} - \frac{e_7}{\pi_7}\right)^2$$

$$+ \left(\frac{\pi_1\pi_8 - \pi_{18}}{\pi_{18}}\right)\left(\frac{e_1}{\pi_1} - \frac{e_8}{\pi_8}\right)^2 + \left(\frac{\pi_2\pi_3 - \pi_{23}}{\pi_{23}}\right)\left(\frac{e_2}{\pi_2} - \frac{e_3}{\pi_3}\right)^2 + \left(\frac{\pi_2\pi_4 - \pi_{24}}{\pi_{24}}\right)\left(\frac{e_2}{\pi_2} - \frac{e_4}{\pi_4}\right)^2$$

$$+ \left(\frac{\pi_2\pi_5 - \pi_{25}}{\pi_{25}}\right)\left(\frac{e_2}{\pi_2} - \frac{e_5}{\pi_5}\right)^2 + \left(\frac{\pi_2\pi_6 - \pi_{26}}{\pi_{26}}\right)\left(\frac{e_2}{\pi_2} - \frac{e_6}{\pi_6}\right)^2 + \left(\frac{\pi_2\pi_7 - \pi_{27}}{\pi_{27}}\right)\left(\frac{e_2}{\pi_2} - \frac{e_7}{\pi_7}\right)^2$$

$$+ \left(\frac{\pi_2\pi_8 - \pi_{28}}{\pi_{28}}\right)\left(\frac{e_2}{\pi_2} - \frac{e_8}{\pi_8}\right)^2 + \left(\frac{\pi_3\pi_4 - \pi_{34}}{\pi_{34}}\right)\left(\frac{e_3}{\pi_3} - \frac{e_4}{\pi_4}\right)^2 + \left(\frac{\pi_3\pi_5 - \pi_{35}}{\pi_{35}}\right)\left(\frac{e_3}{\pi_3} - \frac{e_5}{\pi_5}\right)^2$$

$$+ \left(\frac{\pi_3\pi_6 - \pi_{36}}{\pi_{36}}\right)\left(\frac{e_3}{\pi_3} - \frac{e_6}{\pi_6}\right)^2 + \left(\frac{\pi_3\pi_7 - \pi_{37}}{\pi_{37}}\right)\left(\frac{e_3}{\pi_3} - \frac{e_7}{\pi_7}\right)^2 + \left(\frac{\pi_3\pi_8 - \pi_{38}}{\pi_{38}}\right)\left(\frac{e_3}{\pi_3} - \frac{e_8}{\pi_8}\right)^2$$

$$+\left(\frac{\pi_4\pi_5-\pi_{45}}{\pi_{45}}\right)\left(\frac{e_4}{\pi_4}-\frac{e_5}{\pi_5}\right)^2+\left(\frac{\pi_4\pi_6-\pi_{46}}{\pi_{46}}\right)\left(\frac{e_4}{\pi_4}-\frac{e_6}{\pi_6}\right)^2+\left(\frac{\pi_4\pi_7-\pi_{47}}{\pi_{47}}\right)\left(\frac{e_4}{\pi_4}-\frac{e_7}{\pi_7}\right)^2$$

$$+\left(\frac{\pi_4\pi_8-\pi_{48}}{\pi_{48}}\right)\left(\frac{e_4}{\pi_4}-\frac{e_8}{\pi_8}\right)^2+\left(\frac{\pi_5\pi_6-\pi_{56}}{\pi_{56}}\right)\left(\frac{e_5}{\pi_5}-\frac{e_6}{\pi_6}\right)^2+\left(\frac{\pi_5\pi_7-\pi_{57}}{\pi_{57}}\right)\left(\frac{e_5}{\pi_5}-\frac{e_7}{\pi_7}\right)^2$$

$$+\left(\frac{\pi_5\pi_8-\pi_{58}}{\pi_{58}}\right)\left(\frac{e_5}{\pi_5}-\frac{e_8}{\pi_8}\right)^2+\left(\frac{\pi_6\pi_7-\pi_{67}}{\pi_{67}}\right)\left(\frac{e_6}{\pi_6}-\frac{e_7}{\pi_7}\right)^2+\left(\frac{\pi_6\pi_8-\pi_{68}}{\pi_{68}}\right)\left(\frac{e_6}{\pi_6}-\frac{e_8}{\pi_8}\right)^2$$

$$+\left(\frac{\pi_7\pi_8-\pi_{78}}{\pi_{78}}\right)\left(\frac{e_7}{\pi_7}-\frac{e_8}{\pi_8}\right)^2.$$

The above 28 values of $\left(\frac{\pi_i\pi_j-\pi_{ij}}{\pi_{ij}}\right)\left(\frac{e_i}{\pi_i}-\frac{e_j}{\pi_j}\right)^2$ for different combinations i and j are given in the following table:

	j						
i>j	1	2	3	4	5	6	7
2	91238120.44						
3	14065768.94	170976306.70					
4	2667685.08	63612617.02	28389310.20				
5	9931124.16	36881063.74	47536545.93	2561215.89			
6	1868160.75	104865300.70	4992906.747	7803347.57	22190860.32		
7	8127085.79	47010740.63	42049622.94	1461282.77	225325.07	14113424.37	
8	29093.82	94361304.94	12849024.92	3250757.81	10988465.09	1478773.16	9121878.27

Thus we have

$$\hat{v}_{\text{SYG}}\left(\hat{Y}_{\text{G}}\right)=\sum_{i(<j)=1}^{8}\sum_{j=1}^{8}\left(\frac{\pi_i\pi_j-\pi_{ij}}{\pi_{ij}}\right)\left(\frac{e_i}{\pi_i}-\frac{e_j}{\pi_j}\right)^2=854647113.8\,.$$

A $(1-\alpha)100\%$ confidence interval of population total Y is given by

$$\hat{Y}_{\text{G}}\mp t_{\alpha/2}(\text{df}=n-2)\sqrt{\hat{v}_{\text{SYG}}\left(\hat{Y}_{\text{G}}\right)}.$$

Using Table 2 from the Appendix the 95% confidence interval of the total number of fish caught during 1997 is given by

$$284206.9\mp 2.447\sqrt{854647113.8}\,,\quad\text{or}\quad[212670.45,\ 355743.35].$$

An estimator of variance of the GREG owed to Särndal, Swenson, and Wretman (1989) is given by

$$\hat{v}_{\text{SYG}}\left(\hat{Y}_{\text{G}}\right)=\frac{1}{2}\sum_{i\in s}\sum_{j(\neq i)\in s}\left(\frac{\pi_i\pi_j-\pi_{ij}}{\pi_{ij}}\right)\left(w_ie_i-w_je_j\right)^2=\sum_{i(<j)=1}^{8}\sum_{j=1}^{8}\left(\frac{\pi_i\pi_j-\pi_{ij}}{\pi_{ij}}\right)\left(w_ie_i-w_je_j\right)^2.$$

The above 28 values of $\left(\dfrac{\pi_i\pi_j - \pi_{ij}}{\pi_{ij}}\right)(w_ie_i - w_je_j)^2$ are given in the following table.

	j						
i>*j*	1	2	3	4	5	6	7
2	96029625.10						
3	14456136.50	178738153.50					
4	2748863.48	67239252.44	29199645.52				
5	10252616.95	39207263.34	48956859.12	2648940.95			
6	2416283.22	113663403.70	4352131.18	9005203.82	24667396.60		
7	8223784.93	50272113.22	42935981.61	1441813.32	259376.84	15624764.01	
8	30258.71	99296317.46	13198368.82	3352643.45	11348317.19	1962788.15	9121878.27

Thus we have

$$\hat{v}_{\text{SYG}}\left(\hat{Y}_{\text{G}}\right) = \sum_{i(<j)=1}^{8}\sum_{j=1}^{8}\left(\frac{\pi_i\pi_j - \pi_{ij}}{\pi_{ij}}\right)(w_ie_i - w_je_j)^2 = 900650181.40\,.$$

A $(1-\alpha)100\%$ confidence interval of population total Y is given by

$$\hat{Y}_{\text{G}} \mp t_{\alpha/2}(\text{df} = n-2)\sqrt{\hat{v}_{\text{SYG}}\left(\hat{Y}_{\text{G}}\right)}\,.$$

Using Table 2 of the Appendix the 95% confidence interval of the total number of fish caught during 1997 is

$$284206.9029 \mp 2.447\sqrt{900650181.40}\,, \quad \text{or} \quad [210770.40,\, 357643.40].$$

The next section is devoted to the higher level calibration approach introduced by Singh, Horn, and Yu (1998), where the variance of the auxiliary variable is assumed to be known. Several new estimators are shown as special cases of the higher level calibration approach.

5.8 IMPROVED ESTIMATOR OF VARIANCE OF THE GREG:THE HIGHER LEVEL CALIBRATION APPROACH

Singh, Horn, and Yu (1998) applied the calibration approach to estimate the variance of the GREG estimator. The weights D_{ij} of the Sen--Yates--Grundy (1953) for an estimator of variance given in (5.7.2) are calibrated so that the estimator of variance for the auxiliary variable has the exact variance and suggested an estimator of the variance of GREG

$$\hat{v}\left(\hat{Y}_{\text{G}}\right) = \frac{1}{2}\sum_{i \neq j \in s}\sum \Omega_{ij}(w_ie_i - w_je_j)^2 \tag{5.8.1}$$

where Ω_{ij} are the modified weights attached to the quadratic expression by the Sen--Yates-- Grundy (1953) type of estimator and are as close as possible in an average sense for a given measure to the D_{ij} with respect to the calibration equation, defined as

$$\frac{1}{2}\sum_{i\neq j\in s}\sum \Omega_{ij}\left(d_i x_i - d_j x_j\right)^2 = V_{\text{SYG}}\left(\hat{X}_{\text{HT}}\right) \tag{5.8.2}$$

where $V_{\text{SYG}}\left(\hat{X}_{\text{HT}}\right)=\frac{1}{2}\sum_{i\neq j\in\Omega}\sum\left(\pi_i\pi_j-\pi_{ij}\right)\left(d_i x_i - d_j x_j\right)^2$ denotes the known variance of the estimator of the auxiliary total $X\left(=\sum_{i\in\Omega} x_i\right)$ given by $\hat{X}_{\text{HT}}=\sum_{i\in s} d_i x_i$. To compute the right hand side of (5.8.2), we need either information on every unit of the auxiliary variable in the population, or only $V_{\text{SYG}}\left(\hat{X}_{\text{HT}}\right)$ obtained from a past survey or pilot survey. The examples of a situation where information on every unit of the auxiliary variable is known are the establishment turnover recorded from census or administrative records or Business Register (BR) or Internal Revenue Service, etc.. The use of a known variance of the auxiliary variable has also been supported by Das and Tripathi (1978), Singh and Srivastava (1980), Srivastava and Jhajj (1980, 1981), Isaki (1983), Singh and Singh (1988), Swain and Mishra (1992), Shah and Patel (1996), and Garcia and Cebrian (1996). Singh, Mangat, and Mahajan (1995) have reviewed classes of estimators of unknown population parameters making use of the known variance of an auxiliary variable. For simplicity, Singh, Horn, and Yu (1998) have been restricted themselves to the two-dimensional CS type distance, D, between two $n\times n$ grids formed by the weights Ω_{ij} and D_{ij} for $i, j = 1,2,...,n$, given by

$$D=\frac{1}{2}\sum_{i\neq j\in s}\sum\left(\Omega_{ij}-D_{ij}\right)^2\left(D_{ij}Q_{ij}\right)^{-1}. \tag{5.8.3}$$

In most of the situations $Q_{ij}=1$ but other types of weights can also be used. They have shown that the ratio type adjustment using a known variance of the auxiliary variable is a special case for a particular choice of Q_{ij}. Minimization of (5.8.3) subject to (5.8.2) leads to modified optimal weights as

$$\Omega_{ij}=D_{ij}+\frac{D_{ij}Q_{ij}\left(d_i x_i - d_j x_j\right)^2}{\frac{1}{2}\sum_{i\neq j\in s}\sum D_{ij}Q_{ij}\left(d_i x_i - d_j x_j\right)^4}\left[V_{\text{SYG}}\left(\hat{X}_{\text{HT}}\right)-\frac{1}{2}\sum_{i\neq j\in s}\sum D_{ij}\left(d_i x_i - d_j x_j\right)^2\right] \tag{5.8.4}$$

for the optimal choice of the Lagrange multiplier λ given by

$$\lambda=\left\{V_{\text{SYG}}\left(\hat{X}_{\text{HT}}\right)-\frac{1}{2}\sum_{i\neq j\in s}\sum D_{ij}\left(d_i x_i - d_j x_j\right)^2\right\}\Big/\frac{1}{2}\sum_{i\neq j\in s}\sum D_{ij}Q_{ij}\left(d_i x_i - d_j x_j\right)^4. \tag{5.8.5}$$

Substituting Ω_{ij} from (5.8.4) in (5.8.1) leads to the following regression type of estimator

$$\hat{v}\left(\hat{Y}_{\text{G}}\right)=\hat{v}_{\text{SYG}}\left(\hat{Y}_{\text{G}}\right)+\hat{B}_1\left[V_{\text{SYG}}\left(\hat{X}_{HT}\right)-\hat{v}_{\text{SYG}}\left(\hat{X}_{\text{HT}}\right)\right] \tag{5.8.6}$$

where

$$\hat{B}_1=\sum_{i\neq j\in s}\sum D_{ij}Q_{ij}\left(d_i x_i - d_j x_j\right)^2\left(w_i e_i - w_j e_j\right)^2\Big/\sum_{i\neq j\in s}\sum D_{ij}Q_{ij}\left(d_i x_i - d_j x_j\right)^4=\hat{\mu}_{22}/\hat{\mu}_{04},$$

$$\hat{v}_{\text{SYG}}\left(\hat{X}_{\text{HT}}\right)=\frac{1}{2}\sum_{i\neq j\in s}\sum D_{ij}\left(d_i x_i - d_j x_j\right)^2, \text{ and } \hat{v}_{\text{SYG}}\left(\hat{Y}_{\text{G}}\right)=\frac{1}{2}\sum_{i\neq j\in s}\sum D_{ij}\left(w_i e_i - w_j e_j\right)^2.$$

The regression coefficient $\hat{B}_1$ makes use of the known total X of the auxiliary variable and hence can be treated as an improved estimator of the regression coefficient by following Singh and Singh (1988). Under the higher level calibration approach, Singh, Horn, and Yu (1998) have discussed the following cases:

Case I. Under SRSWOR sampling design, if $q_i = x_i^{-1}$ and $Q_{ij} = (d_i x_i - d_j x_j)^{-2}$ are, respectively, the weights attached in the low level and higher level calibration approach, then the estimator (5.8.6) reduces to the estimator of the variance of the ratio estimator as

$$\hat{v}(\hat{Y}_{\text{R}}) = \frac{N^2(1-f)}{n(n-1)} \sum_{i=1}^{n} e_i^2 \left(\frac{X}{\hat{X}}\right)^2 \left(\frac{S_x^2}{s_x^2}\right). \tag{5.8.7}$$

Case II. If $q_i = 1$ and $Q_{ij} = 1 \ \forall \ i, \ j$ then we have

$$\hat{v}_{\text{SYG}}(\hat{Y}_{\text{G}}) = \frac{N^2(1-f)}{n(n-1)} \sum_{i=1}^{n} e_i^2 + \hat{\psi}_1(X - \hat{X}) + \hat{\psi}_2(X - \hat{X})^2 + \hat{\psi}_3(S_x^2 - s_x^2) \tag{5.8.8}$$

where $\hat{\psi}_1$ and $\hat{\psi}_2$ are given earlier respectively and

$$\hat{\psi}_3 = \frac{N^2(1-f)}{n \sum\sum_{i \neq j \in s} (x_i - x_j)^4} \left[\sum\sum_{i \neq j \in s} (x_i - x_j)^2 \left\{ (e_i - e_j) + \frac{n(X - \hat{X})(e_i x_i - e_j x_j)}{N \sum_{i \in s} x_i^2} \right\}^2 \right]. \tag{5.8.9}$$

Without loss of generality, the estimators of variance of ratio and GREG given in (5.8.7) and (5.8.8) are neither members of a low level calibration nor of the class of estimators by Deng and Wu (1987). These estimators are members of the analogues of classes of estimators for estimating variance of GREG given by Srivastava and Jhajj (1981) as

$$\hat{v}_{\text{SJ}}(\hat{Y}_{\text{G}}) = \left(\frac{N^2(1-f)}{n(n-1)} \sum_{i=1}^{n} e_i^2 \right) H\left(\frac{X}{\hat{X}}, \frac{S_x^2}{s_x^2} \right) \tag{5.8.10}$$

where $H(\bullet, \ \bullet)$ is a parametric function such that $H(1, \ 1) = 1$ which satisfies certain regularity conditions defined in Chapter 3. Following Srivastava and Jhajj (1981) and Deng and Wu (1987), it is easy to see that the class of estimators in (5.8.10) remains better than the class of estimators defined in (5.7.6) and hence (5.7.10). A difficult issue in using (5.8.1) is how to get non-negative estimates of variance using calibration. The simplest way is to optimise the CS distance function (5.8.3) subject to calibration constraint (5.8.2) along with the conditions $\Omega_{ij} \geq 0$ $\forall \ i, \ j = 1, 2, ..., n$. Constraint calibration weights can be obtained by following Estevao (1994). While it is difficult to develop a solution to this problem theoretically, well known quadratic programming techniques can yield useful numerical results. Straightforward extension, using other distance functions as discussed by Deville and Särndal (1992) for instance, to the two-dimensional problem is not possible owed to the unpredictable nature of the weights D_{ij}.

Padmawar (1994, 1998a) has suggested sampling strategies admitting the non-negative unbiased estimators of the variance following the lines of Rao and Vijayan (1977) and Rao (1979). Chaudhuri (1981) also discussed the methods for constructing non-negative estimators of the variance by following Sharma (1970), Rao (1972, 1977a), Vijayan (1975) and Chaudhuri (1976) and Bandopadhyaya, Chattopadhyaya, and Kundu (1977). Arnab (1992) and Chaudhuri and Roy (1997a) have also considered the problem of estimation of population total under a super population model.

Example 5.8.1. We want to estimate the total real estate farm loans of all the operating banks in the United States. Take an SRSWOR sample of 15 states from population 1 and note the records of the real estate farm loans as well as nonreal estate farm loans. Given that information on the nonreal estate farm loans is available for all states, apply the ratio estimator for estimating the total real estate farm loans in the United States. Construct the 95% confidence intervals using low and higher level calibration estimators.

Given: $N = 50$, $X = 43908.12$ and $S_x^2 = 1176526$.

Solution. We used the first two columns of the Pseudo-Random Numbers (PRN) given in Table 1 of the Appendix to select 15 distinct random numbers $1 \le R_i \le 50$. We observed the random numbers in the sequence as 01, 23, 46, 04, 32, 47, 33, 05, 22, 38, 29, 40, 03, 36 and 27.

Selected sample and analysis

Pop. No	State	Nonreal estate farm loans, x_i	Real estate farm loans y_i	$(x_i - \bar{x})^2$	$(y_i - \bar{y})^2$	e_i	e_i^2
01	AL	348.334	408.978	563337.81	49034.26	209.145	43741.647
23	MN	2466.892	1354.768	1871423.08	524687.56	-60.444	3653.434
46	VA	188.477	321.583	828856.08	95377.08	213.457	45563.946
04	AR	848.317	907.700	62787.99	76887.08	421.036	177271.119
32	NY	426.274	201.631	452415.42	183855.54	-42.915	1841.675
47	WA	1228.607	1100.745	16825.89	221210.49	395.915	156748.865
33	NC	494.730	639.571	365012.12	83.84	355.753	126560.397
05	CA	3928.732	1343.461	8007992.54	508434.88	-910.382	828795.394
22	MI	440.518	323.028	433456.76	94486.64	70.311	4943.599
38	PA	298.351	756.169	640866.43	15814.12	585.010	342237.154
29	NH	0.471	6.044	1206529.42	389838.89	5.774	33.337
40	SC	80.750	87.951	1036613.81	294266.97	41.626	1732.738
03	AZ	431.439	54.633	445493.95	331524.68	-192.876	37201.077
36	OK	1716.087	612.108	380929.26	335.14	-372.380	138667.086
27	NE	3585.406	1337.852	6182750.22	500467.39	-719.031	517005.641
	Sum	16483.385	9456.222	22495290.81	3286304.58	0.000	2425997.112

From the above table we have

$\bar{y} = 630.414$, $\bar{x} = 1098.892$, $s_x^2 = 1606806.48$, $s_y^2 = 234736.042$, $r = \bar{y}/\bar{x} = 0.57368$,

$e_i = (y_i - \bar{y}) - r(x_i - \bar{x})$ and $\sum_{i=1}^{n} e_i^2 = 2425997.11$.

An estimate of the total nonreal estate farm loans in the United States during 1997 is given by

$$\hat{X} = N\bar{x} = 50 \times 1098.892 = 54944.6 .$$

Also we are given $\bar{X} = 878.1624$ and $f = 0.3$.

Thus a ratio estimate of the total real estate farm loans in the United States during 1997 is given by

$$\hat{Y}_{\mathrm{R}} = N\bar{y}\left(\frac{\bar{X}}{\bar{x}}\right) = 50 \times 630.414 \times \left(\frac{878.1624}{1098.892}\right) = 25189.276 .$$

(a) Low level calibration

$$\hat{v}_l\left(\hat{Y}_{\mathrm{R}}\right) = \frac{N^2(1-f)}{n(n-1)}\sum_{i=1}^{n} e_i^2\left(\frac{X}{\hat{X}}\right)^2 = \frac{50^2(1-0.3)}{15(15-1)} \times 2425997.11 \times \left(\frac{43908.12}{54944.6}\right)^2$$

$$= 12910667.19.$$

Making use of Table 2 from the Appendix the 95% confidence interval based on the low level calibration estimator for estimating the total real estate farm loans in the United States is given by

$$\hat{Y}_{\mathrm{R}} \mp t_{\alpha/2}(\mathrm{df} = n-1)\sqrt{\hat{v}_1\left(\hat{Y}_{\mathrm{R}}\right)}, \quad \text{or} \quad 25189.27 \mp t_{0.025}(\mathrm{df} = 14)\sqrt{12910667.19}$$

$$\text{or} \quad 25189.27 \mp 2.145\sqrt{12910667.19}, \quad \text{or} \quad [17481.98,\ 32896.56].$$

(b) Higher level calibration

$$\hat{v}_{\mathrm{h}}\left(\hat{Y}_{\mathrm{R}}\right) = \frac{N^2(1-f)}{n(n-1)}\sum_{i=1}^{n} e_i^2\left(\frac{X}{\hat{X}}\right)^2\left(\frac{S_x^2}{s_x^2}\right) = \hat{V}_1\left(\hat{Y}_{\mathrm{R}}\right)\left(\frac{S_x^2}{s_x^2}\right)$$

$$= 12910667.19 \times \frac{1176526}{1606806.486} = 9453369.624 .$$

Using Table 2 of the Appendix, the 95% confidence interval based on higher level calibration estimator for estimating the total real estate farm loans in the United States is given by

$$\hat{Y}_{\mathrm{R}} \mp t_{\alpha/2}(\mathrm{df} = n-1)\sqrt{\hat{v}_{\mathrm{h}}\left(\hat{Y}_{\mathrm{R}}\right)}, \quad \text{or} \quad 25189.27 \mp t_{0.025}(\mathrm{df} = 14)\sqrt{9453369.624}$$

$$\text{or} \quad 25189.27 \mp 2.145\sqrt{9453369.624}, \quad \text{or} \quad [18594.18,\ 31784.36].$$

This example shows that the width of the 95% confidence interval obtained from a higher level calibrated estimator is smaller than that obtained from a low level calibration estimator.

5.8.1 RECALIBRATED ESTIMATOR OF THE VARIANCE OF GREG

Farrell and Singh (2002a) considered a new estimator of the variance of GREG as

$$\hat{v}_2\left(\hat{Y}_G\right)=\frac{1}{2}\sum_{i\neq j\in s}\sum\Omega_{ij}^{\circ}\left(w_ie_i-w_je_j\right)^2 \tag{5.8.1.1}$$

where Ω_{ij}° are recalibrated weights such that chi square type of distance function

$$D^{\circ}=\frac{1}{2}\sum_{i\neq j\in s}\sum\left(\Omega_{ij}^{\circ}-\Omega_{ij}\right)^2\left(c_{ij}\Omega_{ij}\right)^{-1} \tag{5.8.1.2}$$

is minimized subject to the calibration constraint

$$E_m\left[\frac{1}{2}\sum_{i\neq j\in s}\sum\Omega_{ij}^{\circ}\left(w_i\hat{e}_i-w_j\hat{e}_j\right)^2\right]=E_mV_{\text{SYG}}\left(\hat{Y}_G\right) \tag{5.8.1.3}$$

or equivalently

$$\sum_{i\neq j\in s}\sum\Omega_{ij}^{\circ}\left(w_i^2v(x_i)+w_j^2v(x_j)-2\rho_e\,w_iw_j\sqrt{v(x_i)v_j(x_j)}\right)$$

$$=\sum_{i\neq j\in\Omega}\sum\Theta_{ij}\left(d_i^2v(x_i)+d_j^2v(x_j)-2\rho_ed_id_j\sqrt{v(x_i)v(x_j)}\right)$$

where c_{ij} are real constants, $e_i=y_i-\beta x_i$ such that $E_m(e_i)=0$, $V_m(e_i)=\sigma^2v(x_i)$ and $E_m(e_ie_j)=\rho_e\,\sigma^2\sqrt{v(x_i)v(x_j)}$ for $i\neq j$, $\sigma^2>0$. Here ρ_e is the correlation coefficient between successive error terms that are related according to $e_i=\rho_e\,e_{i-1}+u_i$ where the $u_i\sim i.i.d.\ N(0,1)$.

Optimization of (5.8.1.2) subject to (5.8.1.3) yields the recalibrated weights

$$\Omega_{ij}^{\circ}=\Omega_{ij}+\frac{c_{ij}\Omega_{ij}\left(w_i^2v(x_i)+w_j^2v(x_j)-2\rho_e\,w_iw_j\sqrt{v(x_i)v(x_j)}\right)}{\frac{1}{2}\sum_{i\neq j\in s}\sum c_{ij}\Omega_{ij}\left(w_i^2v(x_i)+w_j^2v(x_j)-2\rho_e\,w_iw_j\sqrt{v(x_i)v(x_j)}\right)^2}\left\{V_{\text{ma}}-\hat{v}_{\text{ma}1}\right\} \tag{5.8.1.4}$$

Note that optimization means the distance may be minimum or maximum. The optimisation of a distance function always leads to a new estimator in calibration approach, but may or may not be efficient one. Further note that total of distances between design and calibration weights can be negative, but the magnitude of distance cannot be negative. If a few weights are negative and others are positive, the resultant estimator of a parameter in calibration approach can still be positive.

Substitution of (5.8.1.4) into (5.8.1.1) provides a new estimator of the variance of GREG as

$$\hat{v}_2\left(\hat{Y}_G\right)=\hat{v}_0\left(\hat{Y}_G\right)+\hat{B}\left[V_{\text{YG}}\left(\hat{X}_{\text{HT}}\right)-\hat{v}_{\text{SYG}}\left(\hat{X}_{\text{HT}}\right)\right]+\hat{D}\left[V_{\text{ma}}-\hat{v}_{\text{ma}1}\right] \tag{5.8.1.5}$$

where

$$\hat{B}=\sum_{i\neq j\in s}\sum D_{ij}q_{ij}\left(d_ix_i-d_jx_j\right)^2\left(w_ie_i-w_je_j\right)^2\Big/\sum_{i\neq j\in s}\sum D_{ij}q_{ij}\left(d_ix_i-d_jx_j\right)^4,$$

$$\hat{D}=\frac{\sum\sum_{i\neq j\in s} c_{ij}D_{ij}\left(w_i^2 v(x_i)+w_j^2 v(x_j)-2\rho_e\, w_i w_j\sqrt{v(x_i)v(x_j)}\right)(w_i e_i-w_j e_j)^2}{\sum\sum_{i\neq j\in s} c_{ij}\Omega_{ij}\left(w_i^2 v(x_i)+w_j^2 v(x_j)-2\rho_e\, w_i w_j\sqrt{v(x_i)v(x_j)}\right)^2},$$

$$\hat{v}_{\text{SYG}}\left(\hat{X}_{\text{HT}}\right)=\frac{1}{2}\sum\sum_{i\neq j\in s} D_{ij}(d_i x_i-d_j x_j)^2,\quad \hat{v}_0\left(\hat{Y}_{\text{G}}\right)=\frac{1}{2}\sum\sum_{i\neq j\in s} D_{ij}(w_i e_i-w_j e_j)^2,$$

$$V_{\text{ma}}=\frac{1}{2}\sum\sum_{i\neq j\in}\Theta_{ij}\left(d_i^2 v(x_i)+d_j^2 v(x_j)-2\rho_e\, d_i d_j\sqrt{v(x_i)v(x_j)}\right),$$

and

$$\hat{v}_{\text{mal}}=\frac{1}{2}\sum\sum_{i\neq j\in s}\Omega_{ij}\left(w_i^2 v(x_i)+w_j^2 v(x_j)-2\rho_e\, w_i w_j\sqrt{v(x_i)v(x_j)}\right).$$

The class (5.8.1.5) is wider than Wu (1982), Deville and Sarndal (1992), Deng and Wu (1987) Singh, Horn, and Yu (1998), and Wu and Sitter (2001) and named it higher order model assisted calibration. Note that the estimator suggested by Shah and Patel (1996) as

$$\hat{Q}_{\text{greg}}=s_y^2+\hat{\gamma}_1\left(S_x^2-s_x^2\right)+\hat{\gamma}_2\left(\frac{1}{N}\sum_{i\in\Omega}v(x_i)-\frac{1}{n}\sum_{i\in s}v(x_i)\right)\tag{5.8.1.6}$$

where $\hat{\gamma}_1$ and $\hat{\gamma}_2$ are sample dependent constants, is also a special case of (5.8.1.5). We illustrate some special cases of (5.8.1.5) here, but a large number of estimators can be derived as special cases.

Case I. Improved estimator of the variance of the ratio predictor under SRSWOR sampling:

If $q_i=1/x_i$, $\pi_i=n/N$, $\pi_{ij}=n(n-1)/N(N-1)$, $v(x_i)=x_i^g$ for $0\leq g\leq 2$ and if

$Q_{ij}=\left(w_i^2 x_i^g+w_j^2 x_j^g-2\rho_e w_i w_j\sqrt{x_i^g x_j^g}\right)^{-1}$, then

$$\hat{v}\left(\hat{Y}_{\text{ratio}}\right)=\hat{v}_0\left\{\frac{\frac{1}{N}\sum_{i\in\Omega}x_i^g-\frac{\rho_e}{N(N-1)}\left(\sum_{i\in\Omega}\sqrt{x_i^g}\right)^2}{\frac{1}{n}\sum_{i\in\Omega}x_i^g-\frac{\rho_e}{n(n-1)}\left(\sum_{i\in\Omega}\sqrt{x_i^g}\right)^2}\right\}\tag{5.8.1.7}$$

where $\hat{v}_0=\frac{N^2(1-f)}{n(n-1)}\sum_{i\in s}e_i^2$ and $e_i=y_i-(\bar{y}/\bar{x})x_i$. Assuming that $\left|(x_i-\bar{X})/\bar{X}\right|<1$, $\left|(x_i-\bar{x})/\bar{x}\right|<1$ and autocorrelation $\rho_e=0$, then (5.8.1.7) can be approximated as

$$\hat{v}\left(\hat{Y}_{\text{ratio}}\right) = \hat{v}_0\left(\frac{\overline{X}}{\overline{x}}\right)^g \left[\frac{1+\dfrac{1}{N}\sum\limits_{i\in\Omega}\sum\limits_{j=0}^{\infty}\left(\dfrac{x_i-\overline{X}}{\overline{X}}\right)^{j+1}\prod\limits_{k=0}^{j}\dfrac{(g-k)}{(j+1)!}}{1+\dfrac{1}{n}\sum\limits_{i\in\Omega}\sum\limits_{j=0}^{\infty}\left(\dfrac{x_i-\overline{x}}{\overline{x}}\right)^{j+1}\prod\limits_{k=0}^{j}\dfrac{(g-k)}{(j+1)!}}\right]. \tag{5.8.1.8}$$

If $j=0$ then (5.8.1.8) becomes

$$\hat{v}\left(\hat{Y}_{\text{ratio}}\right) = \hat{v}_0\left(\frac{\overline{X}}{\overline{x}}\right)^g$$

which is the same as the class of estimators defined by Wu (1982).

If $j=1$ then (5.8.1.8) becomes

$$\hat{v}\left(\hat{Y}_{\text{ratio}}\right) = \hat{v}_0\left(\frac{\overline{X}}{\overline{x}}\right)^g \left\{\frac{1+\dfrac{g(g-1)(N-1)}{2N}\dfrac{S_x^2}{\overline{X}^2}}{1+\dfrac{g(g-1)(n-1)}{2n}\dfrac{s_x^2}{\overline{x}^2}}\right\}$$

which is a member of Singh, Horn, and Yu (1998).

However, if $j=2$ then

$$\hat{v}\left(\hat{Y}_{\text{ratio}}\right) = \hat{v}_0\left(\frac{\overline{X}}{\overline{x}}\right)^g \left\{\frac{1+\dfrac{g(g-1)(N-1)}{2N}\dfrac{S_x^2}{\overline{X}^2}+\dfrac{g(g-1)(g-2)}{6N}\dfrac{\mu_{03}}{\overline{X}^3}}{1+\dfrac{g(g-1)(n-1)}{2n}\dfrac{s_x^2}{\overline{x}^2}+\dfrac{g(g-1)(g-2)}{6N}\dfrac{\hat{\mu}_{03}}{\overline{x}^3}}\right\},$$

where

$\mu_{03} = \dfrac{1}{N-1}\sum\limits_{i=1}^{N}\left(X_i-\overline{X}\right)^3$ and $\hat{\mu}_{03} = \dfrac{1}{n-1}\sum\limits_{i=1}^{n}\left(x_i-\overline{x}\right)^3$ are the third order population and sample moments of the auxiliary variable. This estimator does not belong to the class of estimators proposed by Singh, Horn, and Yu (1998), implying that the recalibration method is more general than that of these authors.

5.8.2. RECALIBRATION USING OPTIMAL DESIGNS FOR THE GREG

Consider the case of autocorrelation ρ_e not necessarily zero and note that if $\pi_i \propto \sqrt{v(x_i)}$ then (5.8.1.3) becomes

$$\sum_{i\neq j\in s}\sum \Omega_{ij}^{\circ}\left(h_i^2+h_j^2-2\rho_e h_i h_j\right) = 2(1-\rho_e)\sum_{i\neq j\in\Omega}\sum \Theta_{ij} \tag{5.8.2.1}$$

where $h_i = 1 + \left(q_i x_i \Big/ \sum_{i \in s} d_i q_i x_i^2 \right)\left(X - \hat{X}_{\mathrm{HT}}\right)$. The recalibrated estimator of variance takes the form

$$\hat{v}_2\left(\hat{Y}_{\mathrm{G}}\right) = \frac{1}{2}\sum_{i \neq j \in s}\sum \Omega_{ij}\left(w_i e_i - w_j e_j\right)^2 + \hat{\beta}_{\mathrm{opt}}\left[\left(1-\rho_e\right)\sum_{i \neq j \in s}\sum \Theta_{ij} - \frac{1}{2}\sum_{i \neq j \in s}\sum \Omega_{ij}\left(h_i^2 + h_j^2 - 2\rho_e h_i h_j\right)^2\right] \tag{5.8.2.2}$$

where

$$\hat{\beta}_{\mathrm{opt}} = \frac{\sum_{i \neq j \in s}\sum \Omega_{ij} c_{ij}\left(h_i^2 + h_j^2 - 2\rho_e h_i h_j\right)\left(w_i e_i - w_j e_j\right)^2}{\sum_{i \neq j \in s}\sum c_{ij}\Omega_{ij}\left(h_i^2 + h_j^2 - 2\rho_e h_i h_j\right)^2}.$$

This result illustrates that the Farrell and Singh (2002a) technique works to recalibrate the Yates and Grundy (1953) form of the estimator of the variance of GREG under the condition of minimum variance for the estimator of the total under the true model. It is often the case that $c_{ij} = 1$. If this is so, then (5.8.2.2) is a new estimator of variance of the GREG estimator. Three other special cases of (5.8.2.2) are also worthy of note.

Case I. If $\rho_e = 0$ and $c_{ij} = \left(h_i^2 + h_j^2\right)^{-1}$, then the estimator (5.8.2.2) reduces to

$$\hat{v}_2\left(\hat{Y}_{\mathrm{G}}\right) = \frac{1}{2}\sum_{i \neq j \in s}\sum w_{ij}\left(w_i e_i - w_j e_j\right)^2\left\{\frac{\sum_{i \neq j \in \Omega}\sum \Theta_{ij}}{\sum_{i \neq j \in s}\sum w_{ij}\left(h_i^2 + h_j^2\right)}\right\}. \tag{5.8.2.3}$$

Case II. If $\rho_e \in [-1, 0) \cup (0, 1)$ and $c_{ij} = \left(h_i^2 + h_j^2 - 2\rho_e h_i h_j\right)^{-1}$ then (5.8.2.2) takes the form

$$\hat{v}_2\left(\hat{Y}_{\mathrm{G}}\right) = \frac{1}{2}\sum_{i \neq j \in s}\sum w_{ij}\left(w_i e_i - w_j e_j\right)^2\left\{\frac{2\left(1-\rho_e\right)\sum_{i \neq j \in \Omega}\sum \Theta_{ij}}{\sum_{i \neq j \in s}\sum w_{ij}\left(h_i^2 + h_j^2 - 2\rho_e h_i h_j\right)}\right\}. \tag{5.8.2.4}$$

Case III. If $\rho_e = +1$, then for $c_{ij} = \left(h_i - h_j\right)^{-2}$ the estimator of variance in (5.8.2.2) becomes

$$\hat{v}_2\left(\hat{Y}_{\mathrm{G}}\right) = 0. \tag{5.8.2.5}$$

Note also that the condition $\pi_i \propto \sqrt{v(x_i)}$ corresponds to the Godambe and Joshi (1965) lower bound of variance, so the variance for fixed sample design under the true model may be equal to zero. This demonstrates the usefulness of the recalibration method of the estimator of the variance of the GREG estimator. Although different choices of the function, $v(x_i)$, leads to different estimators of the variance of GREG and a most common form of the function can be considered as $v(x_i) = x_i^g$, where g is any known model parameter.

5.9 CALIBRATED ESTIMATORS OF VARIANCE OF ESTIMATOR OF TOTAL AND DISTRIBUTION FUNCTION

Rao (1994) has considered the problem of estimation of the general parameters of interest, given by

$$H_y = \sum_{j\in\Omega} h(y_j), \quad \text{and} \quad \overline{H}_y = N^{-1}H_y \tag{5.9.1}$$

for a specified function h. The choice of $h(y)=y$ gives the population total $H_y = Y$ and the population mean $\overline{H} = \overline{Y}$, while the choice $h(y)=\Delta(t-y)$ with $\Delta(a)=1$ when $a \geq 0$ and $\Delta(a)=0$ otherwise gives the distribution function

$$\overline{H}_y = F(t) = N^{-1}\sum_{j\in\Omega}\Delta(t-y_j) \tag{5.9.2}$$

for each t. Rao (1994) has suggested a general class of estimators of H_y given by

$$\hat{H}_y = \sum_{i\in s} d_i(s)h(y_i) \tag{5.9.3}$$

where the basic weights $d_i(s)$ can depend both on s and i $(i \in s)$ and satisfy the design unbiasedness condition. The choice $h(y)=y$ in (5.9.3) gives Godambe's (1955) class of estimators of total. If $d_i(s)=\pi^{-1}$ then (5.9.3) reduces to the Horvitz and Thompson (1952) estimator of population total. If $d_i(s)=w_i$ and $h(y_i)=I(y_i \leq t)$, then (5.9.3) reduces to the estimator $\hat{F}(t)$ suggested by Silva and Skinner (1995). Rao (1979) has suggested an estimator to estimate the variance of the estimator $\hat{H}_y$ as

$$\hat{v}(\hat{H}_y) = \sum_{\substack{i<j\\ i,j\in s}}\sum D_{ij}(s)w_iw_j(z_i - z_j)^2 \tag{5.9.4}$$

where $z_i = h(y_i)/w_i$ and weights $D_{ij}(s)$ can depend both on s and $(i,j)\in s$, and satisfy the unbiasedness condition. For example, the Sen--Yates--Grundy (1953) estimator of the variance of Horvitz and Thompson estimator is a special case of (5.9.4) with $w_i = \pi_i$ and $D_{ij}(s) = (\pi_i\pi_j - \pi_{ij})/(\pi_{ij}\pi_i\pi_j)$ for any design with fixed sample size n. Singh (2001) proposed an estimator of the variance of $\hat{H}_y$ as

$$\hat{v}_s(\hat{H}_y) = \sum_{\substack{i<j\\ i,j\in s}}\sum w_{ij}(s)w_iw_j(z_i - z_j)^2, \tag{5.9.5}$$

where $w_{ij}(s)$ are the modified weights and are as close as possible in an average sense for a given measure to the $d_{ij}(s)$ with respect to the calibration equation:

$$\sum_{\substack{i<j\\ i,j\in s}}\sum w_{ij}(s)w_iw_j(q_i - q_j)^2 = V(\hat{H}_x), \tag{5.9.6}$$

where $q_i = h(x_i)/w_i$ and

$$V(\hat{H}_x) = \sum_{\substack{i<j\\ i,j\in\Omega}}\sum D_{ij}(\Omega)w_iw_j(q_i - q_j)^2, \quad \text{for} \quad D_{ij}(\Omega) = (\pi_i\pi_j - \pi_{ij})/\pi_i\pi_j$$

denotes the known second order moment of the estimator $\hat{H}_x = \sum_{i\in s} d_i(s)h(x_i)$ of the auxiliary parameter H_x.

The minimisation of the two-dimensional CS type distance D between two lower triangular $n \times n$ grids formed by the weights $w_{ij}(s)$ and $D_{ij}(s)$ for $i, j = 1,2,...,n$ is defined as

$$\sum_{\substack{i<j \\ i,j\in s}}\sum \left[\{w_{ij}(s) - D_{ij}(s)\}^2 \left(D_{ij}(s)Q_{ij}\right)^{-1}\right] . \tag{5.9.7}$$

Minimization of (5.9.7) subject to (5.9.6) leads to the modified optimal weights given by

$$w_{ij}(s) = D_{ij}(s) + \frac{w_i w_j D_{ij}(s)Q_{ij}(q_i - q_j)^2}{\sum\sum_{\substack{i<j \\ i,j\in s}} w_i^2 w_j^2 D_{ij}(s)Q_{ij}(q_i - q_j)^4}\left[V(\hat{H}_x) - \sum\sum_{\substack{i<j \\ i,j\in s}} D_{ij}(s)w_i w_j (q_i - q_j)^2\right]. \tag{5.9.8}$$

On substituting the value of $w_{ij}(s)$ from (5.9.8) in (5.9.5) we obtain a regression type estimator for the variance of $\hat{H}_y$ as

$$\hat{v}_s(\hat{H}_y) = \hat{v}(\hat{H}_y) + \hat{B}\left[V(\hat{H}_x) - \hat{v}(\hat{H}_x)\right] \tag{5.9.9}$$

where

$$\hat{B} = \sum\sum_{\substack{i<j \\ i,j\in s}} w_i^2 w_j^2 D_{ij}(s)Q_{ij}(q_i - q_j)^2 (z_i - z_j)^2 \Big/ \sum\sum_{\substack{i<j \\ i,j\in s}} w_i^2 w_j^2 D_{ij}(s)Q_{ij}(q_i - q_j)^4$$

and

$$\hat{v}(\hat{H}_x) = \sum\sum_{\substack{i<j \\ i,j\in s}} D_{ij}(s)w_i w_j (q_i - q_j)^2 .$$

The approximate leading term of the mean squared error of the regression type estimator (5.9.9) is given by

$$\text{MSE}\left[\hat{v}_s(\hat{H}_y)\right] = V\left[\hat{v}(\hat{H}_y)\right] + B^2 V\left[\hat{v}(\hat{H}_x)\right] - 2B\text{Cov}\left[\hat{v}(\hat{H}_y), \hat{v}(\hat{H}_x)\right] \tag{5.9.10}$$

where

$$B = \left\{\sum\sum_{\substack{i<j \\ i,j\in \Omega}} w_i^2 w_j^2 D_{ij}(\Omega)Q_{ij}(q_i - q_j)^2 (z_i - z_j)^2\right\} \Big/ \left\{\sum\sum_{\substack{i<j \\ i,j\in \Omega}} w_i^2 w_j^2 D_{ij}(\Omega)Q_{ij}(q_i - q_j)^4\right\}$$

and

$$\text{Cov}\left[\hat{v}(\hat{H}_y), \hat{v}(\hat{H}_x)\right] \approx \sum\sum_{\substack{i<j<k<l \\ i,\, j,\, k,l\in \Omega}}\sum\sum D_{ij}(\Omega)D_{kl}(\Omega)(\pi_{ijkl} - \pi_{ij}\pi_{kl})(q_i - q_j)^2 (z_i - z_j)^2 .$$

Note that D_{ij}, $D_{ij}(s)$, and $D_{ij}(\Omega)$ have their different meanings. Dorfman and Hall (1993) considers the estimator of the distribution function of a variable over a finite population, when a sample of units is available and the values of a related auxiliary variable are known for the whole population and developed several estimators based on nonparametric regression models. Wu and Sitter (2001) also considered the problem of estimation of variance of the estimator of distribution function in the presence of complete auxiliary information. Rueda and Arcos (2002) make use of quantiles of the auxiliary variable while estimating distribution function and hence median.

5.9.1 UNIFIED SETUP

An estimator for estimating the variance of an estimator of population total, *e.g,*, a ratio estimator, regression predictor, or estimator of distribution function, can be written as

$$\hat{v}_y = \sum\sum D_{ij} f(y_i, y_j, x_i, x_j) \tag{5.9.1.1}$$

where D_{ij} are some design weights and $f(\bullet,\bullet,\bullet,\bullet)$ is a quadratic function of y_i and x_i and $\sum\sum$ denotes the sum over sampled values. A calibrated estimator of variance is then defined as

$$\hat{v}_{\rm p} = \sum\sum w_{ij} f(y_i, y_j, x_i, x_j) \tag{5.9.1.2}$$

where w_{ij} are calibrated weights subject to a function and calibration constraint.

Consider $|w_{ij} - d_{ij}| < 1$. Let $g(w_{ij})$ be any function of the new weights w_{ij} satisfying the regularity conditions

(a) $g(d_{ij}) = 0$,

(b) The first and second order partial derivatives of the function g exist and are known.

The weights w_{ij} are obtained such that the function $g(w_{ij})$ has the minimum value subject to the constraint

$$\sum\sum w_{ij} h(x_i, x_j) = V_x\,. \tag{5.9.1.3}$$

Expanding the function $g(w_{ij})$ around the point D_{ij} by using second order Taylor series, we have

$$g(w_{ij}) = g[D_{ij} + (w_{ij} - D_{ij})]$$

$$= g(D_{ij}) + (w_{ij} - D_{ij})\frac{\partial g}{\partial w_{ij}}|_{w_{ij}=g_{ij}} + \frac{1}{2}(w_{ij} - D_{ij})^2 \frac{\partial^2 g}{\partial w_{ij}^2}|_{w_{ij}=g_{ij}} + ... \tag{5.9.1.4}$$

Using regularity condition (a) we have

$$g(w_{ij}) = (w_{ij} - D_{ij})\psi_{ij} + \frac{1}{2}(w_{ij} - D_{ij})^2 \psi'_{ij} + ..., \tag{5.9.1.5}$$

where ψ_{ij} and ψ'_{ij} denote the first and second order derivatives respectively of the function g with respect to w_{ij} and known constants. For example, if $\psi_{ij} = 0$ and $\psi'_{ij} = 1/(D_{ij}Q_{ij})$, then $\sum\sum g(w_{ij})$ reduces to the CS distance function discussed in the previous sections. Minimization of $\sum\sum g(w_{ij})$ subject to the calibration constraint (5.9.1.3) leads to optimum weights given by

$$w_{ij} = D_{ij}+\frac{1}{\psi_{ij}}\left[h(x_i,x_j)\left\{\sum\sum\frac{h^2(x_i,x_j)}{2\psi_{ij}}\right\}^{-1}\left\{V_x-\sum\sum D_{ij}h(x_i,x_j)+\sum\sum\frac{\psi_{ij}}{2\psi_{ij}}\right\}-\psi_{ij}\right]. \quad (5.9.1.6)$$

On substituting (5.9.1.6) in (5.9.1.2), one can easily get the higher order calibrated estimator of variance.

5.10 CALIBRATION OF ESTIMATOR OF VARIANCE OF REGRESSION PREDICTOR

The Horvitz and Thompson (1952) estimator and the generalized regression (GREG) predictor for the population total, Y, are given respectively

$$\hat{Y}_{\text{HT}} = \sum_i d_i y_i I_{si} \quad (5.10.1)$$

and

$$\hat{Y}_g = \sum_i d_i y_i I_{si} + \hat{\beta}_Q\left(X - \sum_i d_i x_i I_{si}\right) \quad (5.10.2)$$

where $d_i = \dfrac{1}{\pi_i}$, $I_{si} = \begin{cases}1 & \text{if } i \in s,\\ 0 & \text{if } i \notin s,\end{cases}$ $\hat{\beta}_Q = \left\{\sum_i Q_i x_i y_i I_{si}\right\}\Big/\left\{\sum_i Q_i x_i^2 I_{si}\right\}$, and Q_i is a suitably chosen constant. An approximate expression of the variance of $\hat{Y}_g$ was obtained by Särndal (1982) and is given by

$$V_y = V_p(\hat{Y}_g) = \frac{1}{2}\sum_{i\neq j}\sum\Theta_{ij}(d_i \in_i - d_j \in_j)^2$$

where $\in_i = y_i - \beta_Q x_i$, $\beta_Q = \{\sum Q_i x_i y_i \pi_i\}/\sum(Q_i x_i^2 \pi_i)$ and $\Theta_{ij} = (\pi_i\pi_j - \pi_{ij})$. Särndal (1982) introduce two estimators of variance as

$$\hat{v}_1 = \frac{1}{2}\sum_{i\neq j}\sum D_{ij}(d_i e_i - d_j e_j)^2 I_{sij} \quad (5.10.3)$$

and

$$\hat{v}_2 = \frac{1}{2}\sum_{i\neq j}\sum D_{ij}(g_{si}d_i e_i - g_{sj}d_j e_j)^2 I_{sij} \quad (5.10.4)$$

where $D_{ij} = d_{ij}\Theta_{ij}$ with $d_{ij} = \pi_{ij}^{-1}$, $g_{si} = 1 + (X - \sum d_i I_{si})\dfrac{Q_i \pi_i x_i}{\sum Q_i x_i^2 I_{si}}$ and

$$I_{sij} = \begin{cases}1 & \text{if } (i \text{ and } j) \in s,\\ 0 & \text{otherwise.}\end{cases}$$

Kott (1990) proposed the following estimators:

$$\hat{v}_{k_j} = w_j \hat{v}_j, \; j = 1, 2 \quad (5.10.5)$$

where w_j is the calibration weight satisfies

$$E_m(t_g - Y)^2 = w_j E_m(\hat{v}_j). \quad (5.10.6)$$

The estimators are practicable for the model $M(f)$ with $\sigma_i^2 = \sigma^2 f_i$. Chaudhuri and Roy (1997a) have shown that V_y, the variance of the regression predictor, can be written as

$$V_y = \sum_i \alpha_i y_i^2 + \sum_{i \neq j}\sum \alpha_{ij} y_i y_j, \tag{5.10.7}$$

where

$$\alpha_i = (d_i - 1) + \left(Q_i^2 x_i^2 \pi_i^2\right)\frac{V_p\left(\sum d_i x_i I_{s_i}\right)}{\left(\sum Q_i x_i^2 \pi_i\right)^2} + 2Q_i x_i \pi_i \frac{\sum_{k=1}^{N} x_k d_i d_k \Theta_{ik}}{\left(\sum Q_i x_i^2 \pi_i\right)}, \tag{5.10.8}$$

and

$$\alpha_{ij} = -d_i d_j \Theta_{ij} + Q_i Q_j \pi_i \pi_j x_i x_j \frac{V_p\left(\sum d_i x_i I_{si}\right)}{\left(\sum Q_i x_i^2 \pi_i\right)^2} + Q_j x_j \pi_j \frac{\sum_{k=1}^{N} d_i d_k \Theta_{ik} x_k}{\sum Q_i x_i^2 \pi_i} + Q_i x_i \pi_i \frac{\sum_{k=1}^{N} d_j d_k \Theta_{jk} x_k}{\sum Q_i x_i^2 \pi_i} \tag{5.10.9}$$

with $Q_i > 0$ being an assignable constant to form different estimators of regression coefficient, and $V_p\left(\sum_{i \in s} d_i x_i I_{si}\right) = \frac{1}{2}\sum_{i \neq j \in \Omega}\sum \Theta_{ij}\left(d_i x_i - d_j x_j\right)^2$. They considered a class H of non-homogeneous quadratic unbiased estimator as:

$$\hat{v}_y = a_s + \sum_i b_{si} y_i^2 I_{si} + \sum_{i \neq j}\sum b_{sij} y_i y_j I_{sij} \tag{5.10.10}$$

where a_s, b_{si} and b_{sij} are constants free from the y_i values and satisfy the unbiasedness conditions as follows:

$$E_p(a_s) = 0,\ E_p(b_{si} I_{si}) = \alpha_i,\ \text{and}\ E_p(b_{sij} I_{sij}) = \alpha_{ij}.$$

Chaudhuri and Roy (1997a) derived the lower bound of the variance of an estimator belonging to the class H under the following superpopulation model M: $E_m(y_i) = \mu_i$, $V_m(y_i) = \sigma_i^2$ and $C_m(y_i, y_j) = 0$ for $i \neq j$ and showed that the variance estimator

$$\hat{v}_0 = \sum_i d_i \alpha_i \left(y_i^2 - \sigma_i^2 - \mu_i^2\right) I_{si} + \sum_{i \neq j}\sum \alpha_{ij} d_{ij}\left(y_i y_j - \mu_i \mu_j\right) I_{sij} + \sum_i \alpha_i \left(\sigma_i^2 + \mu_i^2\right) + \sum_{i \neq j}\sum \alpha_{ij} \mu_i \mu_j \tag{5.10.11}$$

is optimal within H in the sense that its variance attains the lower bound. In the next section, Arnab and Singh (2002a) have shown that the result concerning the lower bound is incorrect and hence the optimality property of $\hat{v}_0$ cannot be acceptable.

The estimator $\hat{v}_0$ at (5.10.11) can not be used in practice since it involves unknown parameters μ_i and σ_i^2. So Chaudhuri and Roy (1997a) proposed the following alternative estimators when $\mu_i = \beta x_i$ and $\sigma_i^2 = \sigma^2 x_i^g$ by replacing the unknown β and σ^2 with their suitable estimators as:

$$\hat{v}_3 = \sum_i \alpha_i \left(y_i^2 - \hat{\theta} x_i^2 - \hat{\phi} x_i^g\right)\frac{I_{si}}{\pi_i} + \hat{\phi}\sum_i \alpha_i x_i^g$$

$$+\hat{\theta}\sum_i \alpha_i x_i^2 + \sum_{i\neq j}\sum \alpha_{ij}\left(y_i y_j - \hat{\theta} x_i x_j\right)\frac{I_{sij}}{\pi_{ij}} + \hat{\theta}\sum_{i\neq j}\sum \alpha_{ij} x_i x_j \tag{5.10.12}$$

and

$$\hat{v}_4 = \left(\sum_i \alpha_i x_i^2\right)\left[\left(\frac{\sum \alpha_i y_i^2 (I_{si}/\pi_i)}{\sum \alpha_i x_i^2 (I_{si}/\pi_i)}\right) - \hat{\phi}\left(\frac{\sum \alpha_i x_i^g (I_{si}/\pi_i)}{\sum \alpha_i x_i^2 (I_{si}/\pi_i)}\right)\right]$$

$$+\left(\sum_{i\neq j}\sum \alpha_{ij} x_i x_j\right)\left(\frac{\sum_{i\neq j}\sum \alpha_{ij} y_i y_j \left(I_{sij}/\pi_{ij}\right)}{\sum_{i\neq j}\sum \alpha_{ij} x_i x_j \left(I_{sij}/\pi_{ij}\right)}\right) + \hat{\phi}\sum_i \alpha_i x_i^g \tag{5.10.13}$$

where

$$\hat{\theta} = \left(\frac{\sum_{i\in s} x_i^{1-g} y_i}{\sum_{i\in s} x_i^{2-g}}\right)^2 - \frac{1}{(n-1)}\left[\frac{\sum_{i\in s} \frac{y_i^2}{x_i^g}}{\sum_{i\in s} x_i^{2-g}} - \left(\frac{\sum_{i\in s} x_i^{1-g} y_i}{\sum_{i\in s} x_i^{2-g}}\right)^2\right], \hat{\phi} = \frac{1}{(n-1)}\left[\sum_{i\in s} \frac{y_i^2}{x_i^g} - \frac{\left(\sum_{i\in s} x_i^{1-g} y_i\right)^2}{\sum_{i\in s} x_i^{2-g}}\right].$$

5.10.1. CHAUDHURI AND ROY'S RESULTS

The lower bound of the estimator of variance by Chaudhuri and Roy (1997a) has been restated in the following theorem:

Theorem 5.10.1.1.Under model M, and $\hat{v}_y \in H$

$$V_m(\hat{v}_y) \geq \sum \alpha_i^2 (d_i - 1)\eta_i^2 + \sum\sum \alpha_{ij}^2 (\psi_{ij} - 1)\eta_{ij} \tag{5.10.1.1}$$

where $\eta_i^2 = \delta_i - \left(\sigma_i^2 + \mu_i^2\right)^2$ and $\eta_{ij} = \left(\sigma_i^2 + \mu_i^2\right)\left(\sigma_j^2 + \mu_j^2\right) - \mu_i^2 \mu_j^2$.

The equality is attained in the above if the estimator of variance $\hat{v}_y$ takes the form

$$\hat{v}_0 = \sum d_i \alpha_i \left(y_i^2 - \sigma_i^2 - \mu_i^2\right) I_{si} + \sum\sum \alpha_{ij}\psi_{ij}\left(y_i y_j - \mu_i \mu_j\right) I_{sij} + \sum \alpha_i \left(\sigma_i^2 + \mu_i^2\right) + \sum\sum \alpha_{ij}\mu_i\mu_j. \tag{5.10.1.2}$$

Thus we have the following theorem:

Theorem 5.10.1.2. By relaxing the assumptions of Chaudhuri and Roy (1997a), a new expression for the variance of $\hat{v}_0$ is given by

$$M(\hat{v}_0) = E_m E_p (\hat{v}_0 - V)^2$$

$$= \sum_i \alpha_i^2 \eta_i^2 \left(\frac{1}{\pi_i} - 1\right) + 2\sum_{i\neq j}\sum \alpha_{ij}^2 \eta_{ij}\left(\frac{1}{\pi_{ij}} - 1\right) + 4\sum_{i\neq j\neq k}\sum\sum \sigma_i^2 \alpha_{ij}\alpha_{ik}\mu_j\mu_k \left(\frac{\pi_{ijk}}{\pi_{ij}\pi_{ik}} - 1\right)$$

$$+4\sum_{i\neq j}\sum \alpha_i \alpha_{ij}\mu_j \left\{\gamma_i - \mu_i\left(\sigma_i^2 + \mu_i^2\right)\right\}\left(\frac{1}{\pi_i} - 1\right) \tag{5.10.1.3}$$

where $\gamma_i = E_m\left(y_i^3\right) < \infty$.

Proof. Let $E_p(E_m)$, $V_p(V_m)$, and $C_p(C_m)$ be the expectation, variance and covariance with respect to design p (model M).

Then

$$M(\hat{v}_0) = E_m E_p(\hat{v}_0 - V)^2 = E_m V_p(\hat{v}_0), \text{ since } E_p(\hat{v}_0) = V_y .$$

Now

$$\begin{aligned}
V_p(\hat{v}_0) &= V_p\left[\sum d_i\alpha_i\left(y_i^2 - \sigma_i^2 - \mu_i^2\right)I_{si} + \sum\sum \alpha_{ij}d_{ij}\left(y_i y_j - \mu_i\mu_j\right)I_{sij}\right] \\
&= V_p\left[\sum d_i\alpha_i\left(y_i^2 - \sigma_i^2 - \mu_i^2\right)I_{si}\right] + V_p\left[\sum\sum \alpha_{ij}d_{ij}\left(y_i y_j - \mu_i\mu_j\right)I_{sij}\right] \\
&\quad + 2C_p\left[\sum d_i\alpha_i\left(y_i^2 - \sigma_i^2 - \mu_i^2\right)I_{si}, \sum\sum \alpha_{ij}d_{ij}\left(y_i y_j - \mu_i\mu_j\right)I_{sij}\right] \\
&= \sum_i \alpha_i^2\left\{y_i^2 - \left(\sigma_i^2 + \mu_i^2\right)\right\}^2 V_p\left(\frac{I_{si}}{\pi_i}\right) + \sum_{i\neq j}\sum \alpha_i\alpha_j\left\{y_i^2 - \left(\sigma_i^2 + \mu_i^2\right)\right\}\left\{y_j^2 - \left(\sigma_j^2 + \mu_j^2\right)\right\}C_p\left(\frac{I_{si}}{\pi_i}, \frac{I_{sj}}{\pi_j}\right) \\
&\quad + 2\sum_{i\neq j}\sum \alpha_{ij}^2\left\{y_i y_j - \mu_i\mu_j\right\}^2 V_p\left(\frac{I_{sij}}{\pi_{ij}}\right) + \sum_{i\neq j\neq k\neq j}\sum\sum\sum \alpha_{ij}\alpha_{kl}\left\{y_i y_j - \mu_i\mu_j\right\}\left\{y_k y_l - \mu_k\mu_l\right\}C_p\left(\frac{I_{sij}}{\pi_{ij}}, \frac{I_{skl}}{\pi_{kl}}\right) \\
&\quad + 4\sum_{i\neq j\neq k}\sum\sum \alpha_{ij}\alpha_{ik}\left\{y_i y_j - \mu_i\mu_j\right\}\left\{y_i y_k - \mu_i\mu_k\right\}C_p\left(\frac{I_{sij}}{\pi_{ij}}, \frac{I_{sik}}{\pi_{ik}}\right) \\
&\quad + 4\sum_{i\neq j}\sum \alpha_i\alpha_{ij}\left\{y_i^2 - \left(\sigma_i^2 + \mu_i^2\right)\right\}\left\{y_i y_j - \mu_i\mu_j\right\}C_p\left(\frac{I_{si}}{\pi_i}, \frac{I_{sij}}{\pi_{ij}}\right).
\end{aligned} \tag{5.10.1.4}$$

Further noting

(i) $V_p\left(\dfrac{I_{si}}{\pi_i}\right) = \dfrac{\pi_i - \pi_i^2}{\pi_i^2} = \dfrac{1-\pi_i}{\pi_i}$, (ii) $C_p\left(\dfrac{I_{si}}{\pi_i}, \dfrac{I_{sj}}{\pi_j}\right) = \dfrac{\pi_{ij} - \pi_i\pi_j}{\pi_i\pi_j}$ for $i \neq j$,

(iii) $V_p\left(\dfrac{I_{sij}}{\pi_{ij}}\right) = \dfrac{\pi_{ij} - \pi_{ij}^2}{\pi_{ij}^2} = \dfrac{1-\pi_{ij}}{\pi_{ij}}$ for $i \neq j$, (iv) $C_p\left(\dfrac{I_{si}}{\pi_i}, \dfrac{I_{sij}}{\pi_{ij}}\right) = \dfrac{\pi_{ij} - \pi_i\pi_{ij}}{\pi_i\pi_{ij}}$ for $i \neq j$,

(v) $C_p\left(\dfrac{I_{sij}}{\pi_{ij}}, \dfrac{I_{sik}}{\pi_{ik}}\right) = \dfrac{\pi_{ijk} - \pi_{ij}\pi_{ik}}{\pi_{ij}\pi_{ik}}$ for $i \neq j \neq k$,

(vi) $C_p\left(\dfrac{I_{sij}}{\pi_{ij}}, \dfrac{I_{skl}}{\pi_{kl}}\right) = \dfrac{\pi_{ijkl} - \pi_{ij}\pi_{kl}}{\pi_{ij}\pi_{kl}}$ for $i \neq j \neq k \neq l$,

(vii) $E_m\left\{y_i^2 - \left(\sigma_i^2 + \mu_i^2\right)\right\}^2 = E\left(y_i^4\right) - \left(\sigma_i^2 + \mu_i^2\right)^2 = \eta_i^2$,

(viii) $E_m\left\{y_i^2 - \left(\sigma_i^2 + \mu_i^2\right)\right\}\left\{y_j^2 - \left(\sigma_j^2 + \mu_j^2\right)\right\} = 0$ for $i \neq j$,

(ix) $E_m\left\{y_i y_j - \mu_i\mu_j\right\}^2 = \left(\sigma_i^2 + \mu_i^2\right)\left(\sigma_j^2 + \mu_j^2\right) - \mu_i^2\mu_j^2 = \eta_{ij}$ for $i \neq j$,

(x) $E_m\left\{y_i y_j - \mu_i\mu_j\right\}\left\{y_k y_l - \mu_k\mu_l\right\} = 0$ for $i \neq j \neq k \neq l$,

(xi) $E_m\{y_iy_j-\mu_i\mu_j\}\{y_iy_k-\mu_i\mu_k\}=\alpha_i^2\mu_j\mu_k$,

and

(xii) $E_m\{y_i^2-(\sigma_i^2+\mu_i^2)\}\{y_iy_j-\mu_i\mu_j\}=\mu_j\{\gamma_i-\mu_i(\sigma_i^2+\mu_i^2)\}$,

where $\gamma_i=E_m(y_i^3)<\infty$.

Hence we have

$$M(\hat{v}_0)=E_mV_p(\hat{v}_0)=\sum_i\alpha_i^2\eta_i^2(d_i-1)+2\sum\sum_{i\neq j}\alpha_{ij}^2\eta_{ij}(\psi_{ij}-1)+4\sum\sum\sum_{i\neq j\neq k}\sigma_i^2\alpha_{ij}\alpha_{ik}\mu_j\mu_k(\pi_{ijk}\psi_{ij}\psi_{ik}-1)$$
$$+4\sum\sum_{i\neq j}\alpha_i\alpha_{ij}\mu_j\{\gamma_i-\mu_i(\sigma_i^2+\mu_i^2)\}(d_i-1). \tag{5.10.1.5}$$

Note that the results in (viii), (ix), (x) , and (xi) are derived under the assumption $C_m(y_i,\ y_j)=0$, $C_m(y_i^2,\ y_j^2)=0$ and $C_m(y_i^2,\ y_j)=0$ for $i\neq j$, but these assumptions may also be relaxed. The following theorem states a new lower bound of variance as:

Theorem 5.10.1.3. A new lower bound of variance $M(\hat{v}_0)$ is given by

$$\sum\alpha_i^2(d_i-1)\eta_i^2+\sum\sum\alpha_{ij}^2(\psi_{ij}-1)\eta_{ij}. \tag{5.10.1.6}$$

Proof. Following Chaudhuri and Roy (1997a) we can write

$$M(\hat{v}_y)=E_mE_p(\hat{v}_y-V)^2=E_pV_m(\hat{v}_y)+E_p\{E_m(\hat{v}_y-V)\}^2-V_m(V).$$

Now we have

$$V_m(\hat{v}_y)=\sum_i b_{si}^2I_{si}V_m(y_i^2)+2\sum\sum_{i\neq j}b_{sij}^2I_{sij}V_m(y_iy_j)+4\sum\sum\sum_{i\neq j\neq k}b_{sij}b_{sik}I_{sij}I_{sik}C_m(y_iy_j,y_iy_k)$$
$$+4\sum\sum_{i\neq j}b_{si}^2b_{sij}I_{sij}C_m(y_i^2,y_iy_j)$$

and

$$V_m(V)=\sum_i\alpha_i^2V_m(y_i^2)+2\sum\sum_{i\neq j}\alpha_{ij}^2V_m(y_iy_j)+4\sum\sum\sum_{i\neq j\neq k}\alpha_{ij}\alpha_{ik}C_m(y_iy_j,y_iy_k)+4\sum\sum_{i\neq j}\alpha_i^2\alpha_{ij}C_m(y_i^2,y_iy_j)$$

Now noting

$$V_m(y_i^2)=\eta_i^2,\quad V_m(y_iy_j)=\eta_{ij},\quad C_m(y_iy_j,y_iy_k)=\sigma_i^2\mu_j\mu_k,$$

and

$$C_m(y_i^2,y_iy_j)=\mu_j\{\gamma_i-\mu_i(\sigma_i^2+\mu_i^2)\}.$$

Thus we have

$$M(\hat{v}_y)=A_0+B_0 \tag{5.10.1.7}$$

where

$$A_0=E_p(a_s-r_s)^2+\sum_i\{E_p(b_{si}^2I_{si}-\alpha_i^2)\}\eta_i^2+2\sum\sum_{i\neq j}\{E_p(b_{sij}^2I_{sij})-\alpha_{ij}^2\}\eta_{ij} \tag{5.10.1.8}$$

and

$$B_0 = 4\sum_{i\neq j\neq k}\sum\sum\left\{E_p\left(b_{sij}b_{sik}I_{sij}I_{sik}\right)-\alpha_{ij}\alpha_{ik}\right\}\sigma_i^2\mu_j\mu_k$$
$$+4\sum_{i\neq j}\sum\left\{E_p\left(b_{si}^2b_{sij}I_{sij}\right)-\alpha_i^2\alpha_{ij}\right\}\mu_j\left\{\gamma_i-\mu_i\left(\sigma_i^2+\mu_i^2\right)\right\} \quad (5.10.1.9)$$

with

$$r_s = -\sum_i\left(\sigma_i^2+\mu_i^2\right)\left(b_{si}I_{si}-\alpha_i\right)+\sum_{i\neq j}\sum\mu_i\mu_j\left(b_{sij}I_{sij}-\alpha_{ij}\right).$$

Now from Chaudhuri and Roy (1997a) we note that A_0 attains a minimum when $b_{si}=\pi_i^{-1}$, $b_{sij}=\pi_{ij}^{-1}$ and $a_s=r_s$, and the minimum value

$$A_0=\sum_i\alpha_i^2\left(\frac{1}{\pi_i}-1\right)\eta_i^2+\sum_{i\neq j}\sum\alpha_{ij}^2\left(\frac{1}{\pi_{ij}}-1\right)\eta_{ij}$$

which Chaudhuri and Roy (1997a) claimed as the lower bound of $M(\hat{v}_y)$. Obviously the claim is not justifiable under the new assumptions because B_0 does not attains a minimum when $b_{si}=\pi_i^{-1}$, and $b_{sij}=\pi_{ij}^{-1}$.

5.10.2 CALIBRATED ESTIMATORS OF VARIANCE OF REGRESSION PREDICTOR

In this section, we discuss Arnab and Singh (2002a) calibration techniques under the following situations:

(i) Model Assisted Calibration (MAC) when the regression of y on x is passing through the origin.

(ii) The regression is unknown but the population variance V_x of the GREG predictor for the auxiliary variable x is known.

5.10.2.1 MODEL ASSISTED CALIBRATION

The Horvitz and Thompson (1952) type estimator of the variance of the GREG predictor is given by

$$\hat{v}_{\text{ht}}(y)=\sum_i\alpha_id_iy_i^2I_{si}+\sum_{i\neq j}\sum\alpha_{ij}\psi_{ij}y_iy_jI_{sij}. \quad (5.10.2.1.1)$$

Now we consider a calibrated estimator of variance of the regression predictor as

$$\hat{v}_c=\sum_i\alpha_iw_iy_i^2I_{si}+\sum_{i\neq j}\sum\alpha_{ij}w_{ij}y_iy_jI_{sij}, \quad (5.10.2.1.2)$$

where w_i and w_{ij} are calibration weights again obtained by minimising a compromised distance function

$$D_{\text{sp}}=\sum_i\frac{\left(w_i\alpha_i-d_i\alpha_i\right)^2}{d_i\alpha_iq_i}I_{si}+\sum_{i\neq j}\sum\frac{\left(\alpha_{ij}w_{ij}-\psi_{ij}\alpha_{ij}\right)^2}{\psi_{ij}\alpha_{ij}q_{ij}}I_{sij} \quad (5.10.2.1.3)$$

where q_i and q_{ij} are suitably chosen constants to form different kinds of estimators, subject to a model assisted calibration (MAC) constraint given by

$$E_m(\hat{v}_c(y)) = E_m(V_y) \tag{5.10.2.1.4}$$

or equivalently

$$\sum_i \alpha_i w_i E_m(y_i^2) I_{si} + \sum_{i \neq j}\sum \alpha_{ij} w_{ij} E_m(y_i y_j) I_{sij} = \sum_i \alpha_i E_m(y_i^2) + \sum_{i \neq j}\sum \alpha_{ij} E_m(y_i y_j). \tag{5.10.2.1.5}$$

For the superpopulation model

$$M \; : \; y_i = \beta x_i + e_i \tag{5.10.2.1.6}$$

such that $E_m(y_i) = \beta x_i$, $V_m(y_i) = \sigma^2 x_i^g$, and $C_m(y_i, y_j) = 0$, the calibration constraint (5.10.2.1.5) reduces to

$$\sum_i \alpha_i w_i(\sigma^2 x_i^g + \beta^2 x_i^2) I_{si} + \sum_{i \neq j}\sum \alpha_{ij} w_{ij} \beta^2 x_i x_j I_{sij} = \sum_i \alpha_i(\sigma^2 x_i^g + \beta^2 x_i^2) + \sum_{i \neq j}\sum \alpha_{ij} \beta^2 x_i x_j$$

or

$$\sigma^2 \sum_i \alpha_i w_i x_i^g I_{si} + \beta^2 \left\{ \sum_i \alpha_i w_i x_i^2 I_{si} + \sum_{i \neq j}\sum \alpha_{ij} w_{ij} x_i x_j I_{sij} \right\}$$

$$= \sigma^2 \sum_i \alpha_i x_i^g + \beta^2 \left\{ \sum_i \alpha_i x_i^2 + \sum_{i \neq j}\sum \alpha_{ij} x_i x_j \right\}. \tag{5.10.2.1.7}$$

On comparing the coefficients of σ^2 and β^2 on both sides of (5.10.2.1.7), the system of calibration equations becomes

$$\sum_i \alpha_i w_i x_i^2 + \sum_{i \neq j}\sum \alpha_{ij} w_{ij} x_i x_j I_{sij} = \sum_i \alpha_i x_i^2 + \sum_{i \neq j}\sum \alpha_{ij} x_i x_j \tag{5.10.2.1.8}$$

and

$$\sum_i \alpha_i w_i x_i^g I_{si} = \sum_i \alpha_i x_i^g . \tag{5.10.2.1.9}$$

Note that a slightly new set of calibration constraints can also developed by considering an autocorrelated model. Then we have the following theorems:

Theorem 5.10.2.1.1. For $V_x > 0$, that is the first and second order inclusion probabilities are the functions of another auxiliary variable, then the calibrated weights are given by

$$w_i \alpha_i = d_i \alpha_i + \frac{C\Delta_1 - B\Delta_2}{AC - B^2} d_i q_i \alpha_i x_i^2 + \frac{A\Delta_2 - B\Delta_1}{AC - B^2} d_i q_i \alpha_i x_i^g \tag{5.10.2.1.10}$$

and

$$w_{ij}\alpha_{ij} = \psi_{ij}\alpha_{ij} + \frac{C\Delta_1 - B\Delta_2}{AC - B^2} \psi_{ij} q_{ij} \alpha_{ij} x_i x_j , \tag{5.10.2.1.11}$$

where

$A = \sum_i \alpha_i d_i q_i x_i^4 I_{si} + \sum_{i \neq j}\sum \alpha_{ij} \psi_{ij} q_{ij} x_i^2 x_j^2 I_{sij}$, $B = \sum_i \alpha_i d_i q_i x_i^{g+2} I_{si}$, $\Delta_1 = V_x - \hat{v}_{ht}(x)$,

$C = \sum_i \alpha_i d_i q_i x_i^{2g} I_{si}$, $\Delta_2 = \sum_i \alpha_i x_i^g - \sum_i \alpha_i d_i x_i^g I_{si}$, $V_x = \sum_i \alpha_i x_i^2 + \sum_{i \neq j}\sum \alpha_{ij} x_i x_j$,

and $\hat{v}_{ht}(x) = \sum_i \alpha_i d_i x_i^2 I_{si} + \sum_{i \neq j}\sum \alpha_{ij} \psi_{ij} x_i x_j I_{sij}$.

Proof. In order to minimise (5.10.2.1.3) subject to (5.10.2.1.8) and (5.10.2.1.9) consider the Lagrange function as

$$\phi = \sum_i \frac{(w_i\alpha_i - d_i\alpha_i)^2}{d_i\alpha_i q_i} I_{si} + \sum_{i\neq j}\sum \frac{(\alpha_{ij}w_{ij} - \psi_{ij}\alpha_{ij})^2}{\psi_{ij}\alpha_{ij}q_{ij}} I_{sij}$$

$$- 2\lambda\left\{\sum_i \alpha_i w_i x_i^2 I_{si} + \sum_{i\neq j}\sum \alpha_{ij}w_{ij}x_i x_j I_{sij}\right\} - 2\mu\left\{\sum_i \alpha_i w_i x_i^g I_{si}\right\}. \qquad (5.10.2.1.12)$$

Now

$$\frac{\partial\phi}{\partial w_i} = 0 \quad \Rightarrow \quad w_i\alpha_i = d_i\alpha_i + d_i\alpha_i q_i\left(\lambda x_i^2 + \mu x_i^g\right) \qquad (5.10.2.1.13)$$

and

$$\frac{\partial\phi}{\partial w_{ij}} = 0 \quad \Rightarrow \quad w_{ij}\alpha_{ij} = \psi_{ij}\alpha_{ij} + \lambda\psi_{ij}\alpha_{ij}q_{ij}x_i x_j\,. \qquad (5.10.2.1.14)$$

On substituting (5.10.2.1.13) and (5.10.2.1.14) in (5.10.2.1.8) and (5.10.2.1.9) we have

$$\lambda = \frac{C\Delta_1 - B\Delta_2}{AC - B^2}, \quad \text{and} \quad \mu = \frac{A\Delta_2 - B\Delta_1}{AC - B^2}.$$

On substituting these vales of λ and μ in (5.10.2.1.13) and (5.10.2.1.14) we have the theorem.

Theorem 5.10.2.1.2. The new calibrated estimator of the variance is given by

$$\hat{v}_c = \hat{v}_{ht}(y) + \hat{\beta}^2\left(V_x - \hat{v}_{ht}(x)\right) + \hat{\sigma}^2\left(\sum_i \alpha_i x_i^g - \sum_i \alpha_i d_i x_i^g I_{si}\right), \qquad (5.10.2.1.15)$$

where

$$\hat{\beta}^2 = \frac{CP - BQ}{AC - B^2} \quad \text{and} \quad \hat{\sigma}^2 = \frac{AQ - BP}{AC - B^2}$$

with

$$P = \sum_i \alpha_i d_i q_i x_i^2 y_i^2 I_{si} + \sum_{i\neq j}\sum \alpha_{ij}\psi_{ij}q_{ij}x_i x_j y_i y_j I_{sij}\,, \text{ and } Q = \sum_i \alpha_i d_i q_i x_i^g y_i^2 I_{si}\,.$$

Proof. Use (5.10.2.1.10) and (5.10.2.1.11) in (5.10.2.1.2) we have the theorem.

Case I. For an SRSWOR sampling with $Q_i = q_i = 1/x_i\,, q_{ij} = 1/(x_i x_j)$ we have $\pi_i = n/N\,, \pi_{ij} = n(n-1)/N(N-1)$,

$$\alpha_i = \frac{N(1-f)}{n} + \frac{(1-f)}{n}C_x^2 - 2\frac{N(1-f)}{n(N-1)}\left(\frac{x_i}{\overline{X}} - 1\right) = \alpha_{i\bullet} \qquad (5.10.2.1.16)$$

and

$$\alpha_{ij} = -\frac{N(1-f)}{n(N-1)} + \frac{(1-f)}{n}C_x^2 - \frac{N(1-f)}{n(N-1)\overline{X}}\left\{\left(x_i - \overline{X}\right) + \left(x_j - \overline{X}\right)\right\} = \alpha_{ij\bullet}\,, \qquad (5.10.2.1.17)$$

where $f = n/N$ and $\overline{X} = N^{-1}\sum_{i=1}^{N} x_i\,.$

In this situation $\hat{Y}_g$ reduces to the ratio estimator, *viz.*

$$\hat{Y}_g = N\,\bar{y}\left(\frac{\bar{X}}{\bar{x}}\right) = \hat{y}_{\text{ratio}} .$$

The conventional Horvitz and Thompson (1952) and the calibrated variance estimators came as

$$\hat{v}_{\text{ht}}\left(\hat{y}_{\text{ratio}}\right)_{\text{stage=1}} = \frac{N}{n}\sum_{i\in s}\alpha_{i\bullet}y_i^2 + \frac{N(N-1)}{n(n-1)}\sum_{i\neq}\sum_{j\in s}\alpha_{ij\bullet}y_iy_j . \tag{5.10.2.1.18}$$

The second stage calibrated estimator of variance of the ratio estimator becomes

$$\hat{v}_{\text{ht}}\left(\hat{y}_{\text{ratio}}\right)_{\text{stage=2}} = \hat{v}_{\text{ht}}\left(\hat{y}_{\text{ratio}}\right)_{\text{stage=1}} + \hat{\gamma}_{1g\bullet}\left\{V_x - \hat{v}_x\right\}$$

$$+\hat{\gamma}_{2g\bullet}\left\{\sum_{i\in\Omega}\alpha_{i\bullet}x_i^g - \frac{N}{n}\sum_{i\in s}\alpha_{i\bullet}x_i^g\right\} \tag{5.10.2.1.19}$$

where

$$\hat{v}_x = \frac{N}{n}\sum_{i\in s}\alpha_{i\bullet}x_i^2 + \frac{N(N-1)}{n(n-1)}\sum_{i\neq}\sum_{j\in s}\alpha_{ij\bullet}x_ix_j ,$$

$$\hat{\gamma}_{1g\bullet} = \frac{\left(\sum_{i\in s}\alpha_{i\bullet}x_i^{2g-1}\right)\left(\sum_{i\in s}\alpha_{i\bullet}x_iy_i^2 + \frac{N-1}{n-1}\sum_{i\neq}\sum_{j\in s}\alpha_{ij\bullet}y_iy_j\right) - \left(\sum_{i\in s}\alpha_{i\bullet}x_i^{g+1}\right)\left(\sum_{i\in s}\alpha_{i\bullet}x_i^{g-1}y_i^2\right)}{\left(\sum_{i\in s}\alpha_{i\bullet}x_i^3 + \frac{N-1}{n-1}\sum_{i\neq}\sum_{j\in s}\alpha_{ij\bullet}x_ix_j\right)\left(\sum_{i\in s}\alpha_{i\bullet}x_i^{2g-1}\right) - \left(\sum_{i\in s}\alpha_{i\bullet}x_i^{g+1}\right)^2}$$

and

$$\hat{\gamma}_{2g\bullet} = \frac{\left(\sum_{i\in s}^{s}\alpha_{i\bullet}x_i^3 + \frac{N-1}{n-1}\sum_{i\neq}\sum_{j\in s}\alpha_{ij\bullet}x_ix_j\right)\left(\sum_{i\in s}\alpha_{i\bullet}x_i^{g-1}y_i^2\right) - \left(\sum_{i\in s}\alpha_{i\bullet}x_i^{g+1}\right)\left(\sum_{i\in s}\alpha_{i\bullet}x_iy_i^2 + \frac{N-1}{n-1}\sum_{i\neq}\sum_{j\in s}\alpha_{ij\bullet}y_iy_j\right)}{\left(\sum_{i\in s}\alpha_{i\bullet}x_i^3 + \frac{N-1}{n-1}\sum_{i\neq}\sum_{j\in s}\alpha_{ij\bullet}x_ix_j\right)\left(\sum_{i\in s}\alpha_{i\bullet}x_i^{2g-1}\right) - \left(\sum_{i\in s}\alpha_{i\bullet}x_i^{g+1}\right)^2} .$$

Case II. For SRSWOR sampling with $q_i = Q_i = 1$, $q_{ij} = 1$ yield

$$\pi_i = \frac{n}{N}, \quad \pi_{ij} = \frac{n(n-1)}{N(N-1)},$$

$$\alpha_i = N\left(\frac{1-f}{n}\right)\left[1 + \frac{NS_x^2}{\left(\sum_{i\in\Omega}x_i^2\right)^2}x_i^2 - 2\frac{Nx_i\left(x_i - \bar{X}\right)}{(N-1)\sum_{i\in\Omega}x_i^2}\right] = \alpha_{i\circ} , \tag{5.10.2.1.20}$$

and

$$\alpha_{ij} = N\left(\frac{1-f}{n}\right)\left[-\frac{1}{N-1} + \frac{NS_x^2}{\left(\sum_{i\in\Omega}x_i^2\right)^2}x_ix_j \quad \frac{N\left\{x_i\left(x_j - \bar{X}\right) + x_j\left(x_i - \bar{X}\right)\right\}}{(N-1)\sum_{i\in\Omega}x_i^2}\right] = \alpha_{ij\circ} . \tag{5.10.2.1.21}$$

In this situation $\hat{Y}_g$ reduces to the regression estimator

$$\hat{Y}_{\text{reg}} = N\left[\bar{y} + \hat{\beta}_{\text{ds}}\left(\bar{X} - \bar{x}\right)\right], \quad \text{where } \hat{\beta}_{\text{ds}} = \left\{\sum_{i\in s} x_i y_i\right\} \Big/ \left\{\sum_{i\in s} x_i^2\right\}.$$

The Horvitz and Thompson (1952) and calibrated variance estimators are respectively given by

$$\hat{v}_{\text{ht}}\left(\hat{y}_{\text{reg}}\right)_{\text{stage=1}} = \frac{N}{n}\sum_{i\in s}\alpha_{i\circ}y_i^2 + \frac{N(N-1)}{n(n-1)}\sum_{i\neq j\in s}\sum \alpha_{ij\circ}y_i y_j \tag{5.10.2.1.22}$$

and

$$\hat{v}_{\text{ht}}\left(\hat{y}_{\text{reg}}\right)_{\text{stage=2}} = \hat{v}_{\text{ht}}\left(\hat{y}_{\text{reg}}\right)_{\text{stage=1}} + \hat{\gamma}_{1g\circ}\left\{V_x - \hat{v}_x\right\}$$

$$+ \hat{\gamma}_{2g\circ}\left\{\sum_{i\in\Omega}\alpha_{i\circ}x_i^g - \frac{N}{n}\sum_{i\in s}\alpha_{i\circ}x_i^g\right\} \tag{5.10.2.1.23}$$

where

$$\hat{v}_x = \frac{N}{n}\sum_{i\in s}\alpha_{i\circ}x_i^2 + \frac{N(N-1)}{n(n-1)}\sum_{i\neq j\in s}\sum \alpha_{ij\circ}x_i x_j,$$

$$\hat{\gamma}_{1g\circ} = \frac{\left(\sum_{i\in s}\alpha_{i\circ}x_i^{2g}\right)\left(\sum_{i\in s}\alpha_{i\circ}x_i^2 y_i^2 + \frac{N-1}{n-1}\sum_{i\neq j\in s}\sum \alpha_{ij\circ}x_i x_j y_i y_j\right) - \left(\sum_{i\in s}\alpha_{i\circ}x_i^g y_i^2\right)\left(\sum_{i\in s}\alpha_{i\circ}x_i^{g+2}\right)}{\left(\sum_{i\in s}\alpha_{i\circ}x_i^4 + \frac{N-1}{n-1}\sum_{i\neq j\in s}\sum \alpha_{ij\circ}x_i^2 x_j^2\right)\left(\sum_{i\in s}\alpha_{i\circ}x_i^{2g}\right) - \left(\sum_{i\in s}\alpha_{i\circ}x_i^{g+2}\right)^2},$$

and

$$\hat{\gamma}_{2g\circ} = \frac{\left(\sum_{i\in s1}\alpha_{i\circ}x_i^4 + \frac{N-1}{n-1}\sum_{i\neq j\in s}\sum \alpha_{ij\circ}x_i^2 x_j^2\right)\left(\sum_{i\in s}\alpha_{i\circ}x_i^g y_i^2\right) - \left(\sum_{i\in s}\alpha_{i\circ}x_i^{g+2}\right)\left(\sum_{i\in s}\alpha_{i\circ}x_i^2 y_i^2 + \frac{N-1}{n-1}\sum_{i\neq j\in s}\sum \alpha_{ij\circ}x_i x_j y_i y_j\right)}{\left(\sum_{i\in s}\alpha_{i\circ}x_i^4 + \frac{N-1}{n-1}\sum_{i\neq j\in s}\sum \alpha_{ij\circ}x_i^2 x_j^2\right)\left(\sum_{i\in s}\alpha_{i\circ}x_i^{2g}\right) - \left(\sum_{i\in s}\alpha_{i\circ}x_i^{g+2}\right)^2}.$$

5.10.2.2 CALIBRATION ESTIMATORS WHEN VARIANCE OF AUXILIARY VARIABLE IS KNOWN

Arnab and Singh (2002a) proposed alternative calibrated estimator of the variance of the regression predictor as

$$\hat{v}_p = a_s + \sum_i b_{si}^* y_i^2 I_{si} + \sum_{i\neq j}\sum b_{sij}^* y_i y_j I_{sij} \tag{5.10.2.2.1}$$

where b_{si}^* and b_{sij}^* are the modified weights. Now the variance of the regression predictor of the auxiliary variable is obtained by replacing y_i with x_i in (5.10.7) and is given by

$$V_x = \sum_i \alpha_i x_i^2 + \sum_{i\neq j}\sum \alpha_{ij}x_i x_j\,. \tag{5.10.2.2.2}$$

An estimator of the variance of the regression predictor of the auxiliary variable is obtained by substituting y_i for x_i, given by

$$\hat{v}_x = a_s + \sum_i b_{si}x_i^2 I_{si} + \sum\sum_{i\neq j} b_{sij}x_i x_j I_{sij}\ . \tag{5.10.2.2.3}$$

Thus the choice of these weights can be made in several ways:

5.10.2.2.1 EACH COMPONENT OF V_x IS KNOWN

Consider each component of the decomposition of $V_x = \sum_i \alpha_i x_i^2 + \sum\sum_{i\neq j}\alpha_{ij}x_i x_j$ is known. Let $\sum_i \alpha_i x_i^2 = T_1(x)$ and $\sum\sum_{i\neq j}\alpha_{ij}x_i x_j = T_2(x)$. The calibrated weights b_{si}^* and b_{sij}^* are obtained by minimizing the CS distance functions $\sum_i \frac{\left(b_{si}^* - b_{si}\right)^2 I_{si}}{b_{si}q_{si}}$ and $\sum\sum_{i\neq j}\frac{\left(b_{sij}^* - b_{sij}\right)^2 I_{sij}}{b_{sij}q_{sij}}$ subject to $\sum_i b_{si}^* x_i^2 I_{si} = T_1(x)$ and $\sum\sum_{i\ j} b_{sij}^* x_i x_j I_{sij} = T_2(x)$, respectively. Minimisation yields

$$b_{si}^* = b_{si} + \frac{b_{si}q_{si}x_i^2}{\sum_i b_{si}q_{si}x_i^4 I_{si}}\left[\sum_i \alpha_i x_i^2 - \sum_i b_{si}x_i^2 I_{si}\right] \tag{5.10.2.2.4}$$

and

$$b_{sij}^* = b_{sij} + \frac{b_{sij}q_{sij}x_i x_j}{\sum\sum_{i\neq j} b_{sij}q_{sij}x_i^2 x_j^2 I_{sij}}\left[\sum\sum_{i\neq j}\alpha_{ij}x_i x_j - \sum\sum_{i\neq j} b_{sij}x_i x_j I_{sij}\right]. \tag{5.10.2.2.5}$$

On putting the values of b_{si}^* and b_{sij}^* from (5.10.2.2.4) and (5.10.2.2.5) in (5.10.2.2.1), we obtain an another calibrated estimator of the variance of the regression predictor as

$$\hat{v}_{p1} = \hat{v}_y + b_1\left[T_1(x) - \sum_i b_{si}x_i^2 I_{si}\right] + b_2\left[T_2(x) - \sum\sum_{i\neq j} b_{sij}x_i x_j I_{sij}\right], \tag{5.10.2.2.6}$$

where

$b_1 = \sum_i b_{si}q_{si}x_i^2 y_i^2 I_{si} \Big/ \sum_i b_{si}q_{si}x_i^4 I_{si}$ and $b_2 = \sum\sum_{i\neq j} b_{sij}q_{sij}x_i x_j y_i y_j I_{sij} \Big/ \sum\sum_{i\neq j} b_{sij}q_{sij}x_i^2 x_j^2 I_{sij}$.

Case I. Consider an SRSWOR design with $a_s = 0$, $b_{si} = \frac{N}{n}\alpha_{i\bullet}$ and $b_{sij} = \frac{N(N-1)}{n(n-1)}\alpha_{ij\bullet}$, $q_i = Q_i = x_i^{-1}$, $q_{sij} = \left(x_i x_j\right)^{-1}$, where $\alpha_{i\bullet}$ and $\alpha_{ij\bullet}$ are given in (5.10.1.16) and (5.10.1.17). In this case the variance estimator for the ratio estimator $\hat{y}_{\text{ratio}}$ is given by

$$\hat{v}_{p1}\left(\hat{y}_{\text{ratio}}\right) = \frac{N}{n}\sum_{i\in s}\alpha_{i\bullet}y_i^2 + \frac{N(N-1)}{n(n-1)}\sum_{i\neq j\in s}\sum \alpha_{ij\bullet}y_i y_j + \hat{\gamma}_{1\bullet}\left[T_1(x) - \frac{N}{n}\sum_{i\in s}\alpha_{i\bullet}x_i^2\right]$$
$$+ \hat{\gamma}_{2\bullet}\left[T_2(x) - \frac{N(N-1)}{n(n-1)}\sum_{i\neq j\in s}\sum \alpha_{ij\bullet}x_i x_j\right] \tag{5.10.2.2.7}$$

where

$$\hat{\gamma}_{1\bullet} = \sum_{i\in s}\alpha_{i\bullet}x_i y_i^2 \Big/ \sum_{i\in s}\alpha_{i\bullet}x_i^3\,,\quad \hat{\gamma}_{2\bullet} = \sum_{i\neq j\in s}\sum \alpha_{ij\bullet}y_i y_j \Big/ \sum_{i\neq j\in s}\sum \alpha_{ij\bullet}x_i x_j\,,\quad \sum_{i\in\Omega}\alpha_{i\bullet}x_i^2 = T_1(x)$$

and $\sum_{i\neq j\in\Omega}\sum \alpha_{ij\bullet}x_i x_j = T_2(x)$.

Case II. Consider an SRSWOR design with $a_s = 0$, $b_{si} = (N/n)\alpha_{i\circ}$ and $b_{sij} = \{N(N-1)\}/\{n(n-1)\}\alpha_{ij\circ}$, $q_i = Q_i = 1$, $q_{sij} = 1$, where $\alpha_{i\circ}$ and $\alpha_{ij\circ}$ are as given in (5.10.2.1.20) and (5.10.2.1.21).
In this case we have

$$\hat{V}_{p1}(\hat{y}_{greg}) = \frac{N}{n}\sum_{i\in s}\alpha_{i\circ}y_i^2 + \frac{N(N-1)}{n(n-1)}\sum_{i\neq j\in s}\sum \alpha_{ij\circ}y_i y_j + \hat{\gamma}_{1\circ}\left[T_1(x) - \frac{N}{n}\sum_{i\in s}\alpha_{i\circ}x_i^2\right]$$

$$+\hat{\gamma}_{2\circ}\left[T_2(x) - \frac{N(N-1)}{n(n-1)}\sum_{i\neq j\in s}\sum \alpha_{ij\circ}x_i x_j\right] \tag{5.10.2.2.8}$$

where

$$\hat{\gamma}_{1\circ} = \left\{\sum_{i\in s}\alpha_{i\circ}x_i^2 y_i^2\right\}\Big/\left\{\sum_{i\in s}\alpha_{i\circ}x_i^4\right\},\quad \hat{\gamma}_{2\circ} = \left\{\sum_{i\neq j\in s}\sum \alpha_{ij\circ}x_i x_j y_i y_j\right\}\Big/\left\{\sum_{i\neq j\in s}\sum \alpha_{ij\circ}x_i^2 x_j^2\right\},$$

$\sum_{i\in\Omega}\alpha_{i\circ}x_i^2 = T_1(x)$, and $\sum_{i\neq j\in\Omega}\sum \alpha_{ij\circ}x_i x_j = T_2(x)$.

5.10.2.2.2 COMPROMIZED CALIBRATION

Arnab and Singh (2002a) further considered the situation when V_x, the variance of the regression predictor for the auxiliary variable is known but $T_1(x)$ and $T_2(x)$ are unknown, and determine the calibrated weights b_{si}^* and b_{sij}^* by minimizing

$$D = \sum_i \frac{\left(b_{si}^* - b_{si}\right)^2}{b_{si}q_{si}} + \sum_{i\neq j}\sum \frac{\left(b_{sij}^* - b_{sij}\right)^2}{b_{sij}q_{sij}} \tag{5.10.2.2.9}$$

subject to the calibration constraint

$$a_s + \sum_i b_{si}^* x_i^2 I_{si} + \sum_{i\neq j}\sum b_{sij}^* x_i x_j I_{sij} = V_x\,. \tag{5.10.2.2.10}$$

The calibrated weights are given by

$$b_{si}^* = b_{si} + \frac{b_{si}q_{si}x_i^2 I_{si}\{V_x - \hat{v}_x\}}{\sum_i b_{si}q_{si}x_i^4 I_{si} + \sum_{i\neq j}\sum b_{sij}q_{sij}x_i^2 x_j^2 I_{sij}} \tag{5.10.2.2.11}$$

and

$$b_{sij}^* = b_{sij} + \frac{b_{si}q_{sij}x_i x_j I_{si}\{V_x - \hat{v}_x\}}{\sum_i b_{si}q_{si}x_i^4 I_{si} + \sum_{i\neq j}\sum b_{sij}q_{sij}x_i^2 x_j^2 I_{sij}}\,. \tag{5.10.2.2.12}$$

The resultant estimator of variance of the regression predictor is given by

$$\hat{v}_{p2} = \hat{v}_y + \left[\frac{\sum_i b_{si}q_{si}y_i^2x_i^2I_{si} + \sum\sum_{i\neq j} b_{sij}q_{sij}x_ix_jy_iy_jI_{sij}}{\sum_i b_{si}q_{si}x_i^4I_{si} + \sum\sum_{i\neq j} b_{sij}q_{sij}x_i^2x_j^2I_{sij}}\right](V_x - \hat{v}_x). \tag{5.10.2.2.13}$$

If we assume b_{sij} (and hence b_{sij}^*) equal to zero for all values of i and j in the sample, then the proposed strategy is an improved version of the estimator studied by Särndal (1996).

Case I. Consider an SRSWOR design with $a_s = 0$, $b_{si} = (N/n)\alpha_{i\bullet}$ and $b_{sij} = \{N(N-1)/n(n-1)\}\alpha_{ij\bullet}$, $q_i = Q_i = x_i^{-1}$, $q_{sij} = (x_ix_j)^{-1}$, where $\alpha_{i\bullet}$ and $\alpha_{ij\bullet}$ are as given earlier. In this case the estimator of the variance of the ratio estimator $\hat{y}_{\text{ratio}}$ is

$$\hat{v}_{p2}(\hat{y}_{\text{ratio}}) = \frac{N}{n}\sum_{i\in s}\alpha_{i\bullet}y_i^2 + \frac{N(N-1)}{n(n-1)}\sum_{i\neq j\in s}\sum \alpha_{ij\bullet}y_iy_j$$

$$+\hat{\gamma}_{3\bullet}\left[V_x - \left\{\frac{N}{n}\sum_{i\in s}\alpha_{i\bullet}x_i^2 + \frac{N(N-1)}{n(n-1)}\sum_{i\neq j\in s}\sum \alpha_{ij\bullet}x_ix_j\right\}\right] \tag{5.10.2.2.14}$$

where

$$\hat{\gamma}_{3\bullet} = \frac{\sum_{i\in s}\alpha_{i\bullet}x_iy_i^2 + \frac{N-1}{n-1}\sum_{i\neq j\in s}\sum \alpha_{ij\bullet}y_iy_j}{\sum_{i\in s}\alpha_{i\bullet}x_i^3 + \frac{N-1}{n-1}\sum_{i\neq j\in s}\sum \alpha_{ij\bullet}x_ix_j}, \text{ and } V_x = \sum_{i\in\Omega}\alpha_{i\bullet}x_i^2 + \sum_{i\neq j\in\Omega}\sum \alpha_{ij\bullet}x_ix_j\,.$$

Case II. Consider an SRSWOR design with $a_s = 0$, $b_{si} = (N/n)\alpha_{i\circ}$ and $b_{sij} = \{N(N-1)/n(n-1)\}\alpha_{ij\circ}$, $q_i = Q_i = 1$, $q_{sij} = 1$, where $\alpha_{i\circ}$ and $\alpha_{ij\circ}$ are given earlier. In this case the estimator of the variance of the regression estimator $\hat{y}_{\text{reg}}$ is

$$\hat{v}_{p2}(\hat{y}_{greg}) = \frac{N}{n}\sum_{i\in s}\alpha_iy_i^2 + \frac{N(N-1)}{n(n-1)}\sum_{i\neq j\in s}\sum \alpha_{ij\circ}y_iy_j$$

$$+\hat{\gamma}_{3\circ}\left[V_x - \left\{\frac{N}{n}\sum_{i\in s}\alpha_\circ x_i^2 + \frac{N(N-1)}{n(n-1)}\sum_{i\neq j\in s}\sum \alpha_{ij\circ}x_ix_j\right\}\right] \tag{5.10.2.2.15}$$

where

$$\hat{\gamma}_{3\circ} = \frac{\sum_{i\in s}\alpha_{i\circ}y_i^2x_i^2 + \frac{N-1}{n-1}\sum_{i\neq j\in s}\sum \alpha_{ij\circ}x_ix_jy_iy_j}{\sum_{i\in s}\alpha_{i\circ}x_i^4 + \frac{N-1}{n-1}\sum_{i\neq j\in s}\sum \alpha_{ij}x_\circ^2x_j^2}, \text{ and } V_x = \sum_{i\in\Omega}\alpha_{i\circ}x_i^2 + \sum_{i\neq j\in\Omega}\sum \alpha_{ij\circ}x_ix_j\,.$$

Brewer (1999) said, "It is appropriate to estimate the anticipated variance for sample design purposes, but for the analysis of any particular sample the prediction variance is more logical choice".

Thus the next section has been devoted to find the prediction variance of the calibrated estimator of variance.

5.10.2.3 PREDICTION VARIANCE

The prediction variance is given by

$$E_m\left[\hat{v}_c - V_y\right]^2$$

$$= E_m\left[\sum_{i\in s} w_i y_i^2 + \sum_{i\neq j\in s}\sum w_{ij} y_i y_j - \left\{\sum_{i\in\Omega}\alpha_i y_i^2 + \sum_{i\neq j\in\Omega}\sum \alpha_{ij} y_i y_j\right\}\right]^2$$

$$= E_m\left[\sum_{i\in s} w_i y_i^2 + \sum_{i\neq j\in s}\sum w_{ij} y_i y_j - \left\{\sum_{i\in s}\alpha_i y_i^2 + \sum_{i\neq j\in s}\sum \alpha_{ij} y_i y_j + \sum_{i\notin s}\alpha_i y_i^2 + \sum_{i\neq j\notin s}\sum \alpha_{ij} y_i y_j\right\}\right]^2$$

$$= E_m\left[\sum_{i\in s}(w_i - \alpha_i)y_i^2 + \sum_{i\neq j\in s}\sum (w_{ij} - \alpha_{ij})y_i y_j - \left\{\sum_{i\notin s}\alpha_i y_i^2 + \sum_{i\neq j\notin s}\sum \alpha_{ij} y_i y_j\right\}\right]^2$$

$$= E_m\left[\sum_{i\in\Omega} w_i I_{si} y_i^2 + \sum_{i\neq j\in\Omega}\sum w_{ij} I_{sij} y_i y_j - \left\{\sum \alpha_i y_i^2 + \sum_{i\neq j\in\Omega}\sum \alpha_{ij} y_i y_j\right\}\right]^2$$

$$= \sum_{i\in s}\left(w_i^2 - 2w_i\alpha_i\right)m_{4i} + \sum_{i\in\Omega}\alpha_i^2 m_{4i} + \sum_{i\neq j\in s}\sum w_i w_j m_{2i} m_{2j} - 2A\sum_{i\in s} w_i m_{2i} + 2\sum_{i\in s} w_i \alpha_i m_{2j}^2 + \sum_{i\neq j\in\Omega}\sum \alpha_i \alpha_j m_{2i} m_{2j}$$

$$+ 2\left[\sum_{i\neq j\in s}\sum\left(w_{ij}^2 - 2\alpha_{ij} w_{ij}\right)m_{2i} m_{2j} + \sum_{i\neq j\in\Omega}\sum \alpha_{ij}^2 m_{2i} m_{2j}\right]$$

$$+ 4\left[\beta^2 \sum_{i\neq j\neq k\in s}\sum\sum w_{ij} w_{ik} m_{2i} x_j x_k - 2\beta^2 \sum_{i\neq j\in s}\sum w_{ij} m_{2i} x_j A(i,j) + \beta^2 \sum_{i\neq j\neq k\in\Omega}\sum\sum w_{ij} w_{ik} x_j x_k\right]$$

$$+ \beta^4\left[\sum_{i\neq j\neq k\neq l\in s}\sum\sum\sum w_{ij} w_{kl} x_i x_j x_k x_j + \sum_{i\neq j\neq k\neq l\in\Omega}\sum\sum\sum \alpha_{ij}\alpha_{kl} x_i x_j x_k x_l - 2\sum_{i\neq j\in s}\sum w_{ij} x_i x_j B(i,j)\right]$$

$$+ 4\beta\left[\sum_{i\neq j\in s}\sum w_i w_{ij} m_{3i} x_j - \sum_{i\in s} w_i m_{3i}\sum_{j(\neq i)\in\Omega}\alpha_{ij} x_j - \sum_{i\in s}\alpha_i m_{3i}\sum_{j(\neq i)\in s} w_{ij} x_j + \sum_{i\neq j\in\Omega}\sum \alpha_i \alpha_{ij} m_{3i} x_j\right]$$

$$+ 2\beta^2\left[\sum_{i\neq j\neq k\in s}\sum\sum w_i w_{ik} m_{2i} x_j x_k - \sum_{i\in s} w_i m_{2i}\sum_{j\neq k(\neq i)\in\Omega}\sum \alpha_{jk} x_j x_k - \sum_{j\neq k\in s}\sum w_{jk} x_j x_k \sum_{i(\neq j,k)\in\Omega}\alpha_i m_{2i}\right.$$

$$\left.+ \sum_{i\neq j\neq k\in\Omega}\sum\sum \alpha_i \alpha_{ik} m_{2i} x_j x_k\right].$$

5.11 ORDERED AND UNORDERED ESTIMATORS

We will now discuss the estimators which take into account the order of the units selected in the sample as well as those that ignore the order of the units. It is remarkable here that:

(a) For each ordered estimator, we can find an unordered estimator;
(b)The unordered estimator is more efficient than its corresponding ordered estimators.

Let us discuss each of these situations:

5.11.1 ORDERED ESTIMATORS

Suppose that n units are selected in the sample in the order $y_1, y_2, ..., y_n$ with selection probabilities $p_1, p_2, ..., p_n$, respectively. Then an estimator of population total can be defined as

$$\hat{T}_i = \sum_{j=1}^{i-1} y_j + \frac{y_i}{p_i}\left[1 - \sum_{j=1}^{i-1} p_j\right] \quad \text{for } i = 1, 2, ..., n. \tag{5.11.1.1}$$

Thus we have n possible estimators of population total and each estimator is found to be unbiased.

Theorem 5.11.1.1. Each of the estimators $\hat{T}_i$, $i = 1, 2, ..., n$, is an unbiased estimator of the population total Y.

Proof. Here we assume that the results of the first $(i-1)$ draws are known. The theorem is proved if we show that $E(\hat{T}_i) = Y$. Now we have

$$E(\hat{T}_i) = E_1E_2\left[\hat{T}_i \mid \text{Results of first } (i-1) \text{ draws are known}\right] \tag{5.11.1.2}$$

since $E_2\left[\hat{T}_i \mid \text{Results of first } (i-1) \text{ draws are known}\right]$ is a sum of two terms. The first term is the sum of the units selected in the first $(i-1)$ draws whereas the second term is obtained as follows. The number of units in the population are N and we know that $(i-1)$units are drawn in the $(i-1)$ draws. At the time of ith draw, the number of units will be $N-(i-1) = N-i+1$. Now suppose that the jth unit is selected out of $(N-i+1)$ units at i^{th} draw. The probability for the jth unit (P_j) at the ith draw out of $(N-i+1)$units being selected is given by

$$P_j = P_i \Big/ \left\{1 - \sum_{j=1}^{i-1} P_j\right\} \tag{5.11.1.3}$$

where $1 - \sum_{j=1}^{i-1} P_j$ is the sum of the probabilities for $(N-i+1)$units. Therefore

$$E_2\left[\hat{\text{T}}_\text{i} \mid \text{Results for first } (i-1) \text{ draws are known}\right] = \sum_{j=1}^{i-1} y_j + E_2\left[\frac{y_i}{P_i}\left\{1 - \sum_{j=1}^{i-1} P_j\right\}\right]$$

$$= \sum_{j=1}^{i-1} y_j + E_2\left(\frac{y_i}{P_j}\right)$$

$$= \sum_{j=1}^{i-1} y_j + \left[Y - \sum_{j=1}^{i-1} y_j\right] = Y$$

note that $P_j = P_i\left(1 - \sum_{j=1}^{i-1} P_j\right)^{-1}$.

Therefore

$$E\left(\hat{T}_i\right) = E_1(Y) = Y. \tag{5.11.1.4}$$

Hence the theorem.

Corollary 5.11.1.1. The estimator of population total, Y, given by

$$\hat{T} = \frac{1}{n}\sum_{i=1}^{n}\hat{T}_i \tag{5.11.1.5}$$

is an unbiased estimator of thc population total.
Proof. It follows from (5.11.1.4).

Theorem 5.11.1.2. The ordered estimators $\hat{T}_i$ and $\hat{T}_j$ for $i \neq j$ are uncorrelated.
Proof. We have to show that $\operatorname{Cov}\left(\hat{T}_i,\ \hat{T}_j\right) = 0\ \forall\ i \neq j$. For simplicity let us assume that $i < j$ and results up to the ith draw are known. Now we know that

$$\operatorname{Cov}\left(\hat{T}_i,\ \hat{T}_j\right) = E_1C_2\left[\hat{T}_i,\ \hat{T}_j|A_{j-1}\right] + C_1\left[E_2\left(\hat{T}_i \mid A_{j-1}\right),\ E_2\left(\hat{T}_j \mid A_{j-1}\right)\right] \tag{5.11.1.6}$$

where

$E_1,\ E_2$ = Conditional expected values,

$C_1,\ C_2$ = Conditional covariance terms,

A_{j-1} = Results for the first $(j-1)$ draws which are known,

$\hat{T}_j$ = Involves the outcome of jth draw and is a random variable,

and

$\hat{T}_i$ = Involves the outcome of ith draw and is constant for a given A_{j-1}.

The covariance between a constant and a random variable is zero; therefore

$$C_2\left(\hat{T}_i,\ \hat{T}_j \mid A_{j-1}\right) = 0. \tag{5.11.1.7}$$

Now

$$C_1\left[E_2\left(\hat{T}_i \mid A_{j-1}\right), E_2\left(\hat{T}_j \mid A_{j-1}\right)\right] = C_1\left[\hat{T}_i, Y\right] = 0. \tag{5.11.1.8}$$

Using (5.11.1.7) and (5.11.1.8) in (5.11.1.6) we have

$$\operatorname{Cov}\left(\hat{T}_i,\ \hat{T}_j\right) = 0. \tag{5.11.1.9}$$

Hence the theorem.

Theorem 5.11.1.3. The unbiased estimator for the variance $V\left(\hat{T}\right)$ is given by

$$\hat{v}\left(\hat{T}\right) = \frac{1}{n(n-1)}\left[\sum_{i=1}^{n}\hat{T}_i^2 - n\hat{T}^2\right]. \tag{5.11.1.10}$$

Proof. The theorem is proved if we can show

$$E\left[\hat{v}\left(\hat{T}\right)\right] = V\left(\hat{T}\right), \tag{5.11.1.11}$$

where

$$V\left(\hat{T}\right) = V\left[\frac{1}{n}\sum_{i=1}^{n}\hat{T}_i\right] = \frac{1}{n^2}\sum_{i=1}^{n}V\left(\hat{T}_i\right). \tag{5.11.1.12}$$

Then we have

$$E\left[\hat{v}\left(\hat{T}\right)\right]=\frac{1}{n(n-1)}E\left[\sum_{i=1}^{n}\hat{T}_i^2-n\hat{T}^2\right]=\frac{1}{n(n-1)}\left[\sum_{i=1}^{n}E\left(\hat{T}_i^2\right)-nE\left(\hat{T}^2\right)\right]. \tag{5.11.1.13}$$

Now

$$E\left(\hat{T}^2\right)=V\left(\hat{T}\right)+\left\{E\left(\hat{T}\right)\right\}^2=V\left(\hat{T}\right)+Y^2.$$

Using it in (5.11.1.13) we have

$$E\left[\hat{v}\left(\hat{T}\right)\right]=\frac{1}{n(n-1)}\left[\sum_{i=1}^{n}\left\{V\left(\hat{T}_i\right)+Y^2\right\}-n\left\{V\left(\hat{T}\right)+Y^2\right\}\right]=V\left(\hat{T}\right).$$

Hence the theorem.

For simplicity we restrict ourselves to the case of sample size of two units. Let y_i and y_j be the values of the units selected with varying probabilities and without replacement in a sample of size two and let P_i and P_j be the corresponding initial selection probabilities at the first and second draw respectively. If y_i and y_j are the values of the units drawn at the first and second draw respectively, then Raj's ordered estimator of population total, Y, corresponding to ordered sample $s_1^o=(y_i, y_j)$ is given by

$$\hat{Y}_1^o=\frac{1}{2}\left\{(1+P_i)\frac{y_i}{P_i}+(1-P_i)\frac{y_j}{P_j}\right\} \tag{5.11.1.14}$$

and an estimator of the $V\left(\hat{Y}_1^o\right)$ is given by

$$\hat{v}\left(\hat{Y}_1^o\right)=\frac{1}{4}(1-P_i)^2\left(\frac{y_i}{P_i}-\frac{y_j}{P_j}\right)^2. \tag{5.11.1.15}$$

If y_j and y_i are the values of the units drawn at the first and second draw respectively, then Raj's ordered estimator of population total, Y, corresponding to the ordered sample $s_2^o=(y_j, y_i)$ is given by

$$\hat{Y}_2^o=\frac{1}{2}\left\{(1+P_j)\frac{y_j}{P_j}+(1-P_j)\frac{y_i}{P_i}\right\} \tag{5.11.1.16}$$

and an estimator of the $V\left(\hat{Y}_2^o\right)$ is given by

$$\hat{v}\left(\hat{Y}_2^o\right)=\frac{1}{4}(1-P_j)^2\left(\frac{y_i}{P_i}-\frac{y_j}{P_j}\right)^2. \tag{5.11.1.17}$$

The probabilities of the estimators $\hat{Y}_1^o$ and $\hat{Y}_2^o$ are respectively given by the probabilities of the selecting the ith ordered sample as

$$p\left(s_1^o\right)=PiP_j/(1-P_i), \quad \text{and} \quad p\left(s_2^o\right)=PiP_j/(1-P_j).$$

Then Raj's (1956) unordered estimator of population total, Y, is defined as

$$\hat{T}_{\text{raj}} = \frac{\hat{Y}_1^o P\left(s_1^o\right) + \hat{Y}_2^o P\left(s_2^o\right)}{P\left(s_1^o\right) + P\left(s_2^o\right)} = \frac{1}{\left(2 - P_i - P_j\right)}\left[\frac{y_i}{P_i}\left(1 - P_j\right) + \frac{y_j}{P_j}\left(1 - P_i\right)\right]. \quad (5.11.1.18)$$

Theorem 5.11.1.4. An expression for the variance of the estimator $\hat{T}$ for a sample of two units is given by

$$V\left(\hat{T}_{\text{raj}}\right) = \left(1 - \frac{1}{2}\sum_{i=1}^{N} P_i^2\right)\left\{\frac{1}{2}\sum_{i=1}^{N} P_i\left(\frac{Y_i}{P_i} - Y\right)^2\right\} - \frac{1}{4}\sum_{i=1}^{N} P_i^2\left(\frac{Y_i}{P_i} - Y\right)^2. \quad (5.11.1.19)$$

The variance expression (5.11.1.19) becomes too complicated for $n > 2$.

Pathak (1967a, 1967b) provides an upper bound for the variance of Raj's estimator for any sample size n and is given by

$$V\left(\hat{T}_{\text{raj}}\right) \le \frac{1}{n}\sum_{i=1}^{N} P_i\left(\frac{Y_i}{P_i} - Y\right)^2 - \left\{1 + O\left(\frac{n}{N}\right)\right\}\frac{(n-1)}{2n}\left[\sum_{i=1}^{N} P_i^2 \sum_{i=1}^{N} P_i\left(\frac{Y_i}{P_i} - Y\right)^2 + \sum_{i=1}^{N} P_i^2\left(\frac{Y_i}{P_i} - Y\right)^2\right].$$

Under the assumption of large population size, Mukhopadhyay (1977) showed that

$$V\left(\hat{T}_{\text{raj}}\right) = \frac{1}{2n}\sum_{i \ne j = 1}^{N}\left(y_i/p_i - y_j/p_j\right)^2 p_i p_j\left[1 - \frac{n-1}{2}\left(p_i + p_j\right) + \frac{(n-1)(n-2)}{n}\left(p_i^2 + p_j^2 + p_i p_j\right) - \frac{(n-1)(n-2)}{2n}\left(p_i + p_j\right)\sum_{l=1}^{N} p_l^2\right].$$

The admissibility of the estimator $\hat{T}_{\text{raj}}$ for PPSWOR samples of two units within the class of all unbiased estimators of population total was claimed to be proved by Joshi (1970). While indicating that this claim is incorrect, Patel and Dharmadhikari (1978) proved its admissibility when restricted to the class of linear unbiased estimators only. Sengupta (1980, 1982b) generalized the results of Joshi (1966) by showing that an estimator identical to $\hat{T}_{\text{raj}}$ remains admissible within the class of all estimators of population total for any fixed size sampling design of size two. Sengupta (1983) also provided a sufficient condition for the admissibility of unbiased estimators of finite population parameters when sample size is two at most. The result is used to check admissibility of several unbiased estimators of population total. Rosen (1997b) has provided an asymptotic theory for order sampling and introduced a novel general class of varying probabilities sampling schemes, called order sampling schemes. The main result concerns asymptotic distributions of linear statistics. Even if the results are theoretical, they provide the ground work for applications of practical sampling interest. Rosen showed that order sampling yields interesting contributions to the problem of finding simple and good πps schemes. Bhargava (1978) considered some applications of the technique of combined unordering of different estimators which enables us to obtain various new results in addition to those given by Basu (1958) and Pathak (1967a, 1967b). Rosen (1998) has discussed in detail the methods for calculating the inclusion probabilities for ordered sampling. Some order relations between the selection and the inclusion probabilities for PPSWOR sampling scheme have also been discussed by Rao, Sengupta, and Sinha (1991). Das (1951) proposed an estimator based on ordered sample s, however his estimator does not have a non-negative variance estimator. Mukhopadhyay (1977) considered the comparison of ordered estimators under Midzuno--Sen's and probability proportional to with replacement size based on samples of two units. Andreatta and Kaufman (1986) studied these estimators under informative design, that is, their selection probabilities depend upon the values of the study variable.

5.11.2 UNORDERED ESTIMATORS

If we have a sample of n units then there can be $n!$ arrangements of the units, that is, there will be $n!$ ordered samples. For example, if $n = 2$, we have two ordered samples and one unordered sample. In other words, AB, BA are two ordered samples, but if we ignore the order then there is only one sample of two units. For a sample of $n = 3$ units, the number of ordered samples is $3! = 6$, namely ABC, ACB, BAC, BCA, CAB, and CBA. Suppose s_u denotes the uth unordered sample, then the number of unordered samples is ${}^{N}C_n$ and if s_o denotes the oth ordered sample of the s_u unordered samples then the number of ordered sample s_o is $n!$.

Defining

$\hat{T}(s_o, o)$= estimate of population total based on o^{th} ordered sample corresponding to the $s_u{}^{th}$ unordered sample.

$\hat{T}(s_u)$ = Unordered estimator of population total based on $s_u{}^{th}$ unordered sample.

$P(s_o, o)$= Probability of selecting the oth ordered sample of the $s_u{}^{th}$ unordered sample.

$P(s_u) = \sum_{o=1}^{n!} P(s_o, o)$ = Probability for selecting $s_u{}^{th}$ unordered sample.

Then we have the following theorems:

Theorem 5.11.2.1. An unordered estimator of population total is given by

$$\hat{T}(s_u) = \sum_{o=1}^{n!} \left\{ P(s_o, o)\hat{T}(s_o, o) / P(s_u) \right\}. \tag{5.11.2.1}$$

Proof. It follows from the result given in (5.11.2.10) based on next two theorems.

For n=2 the estimator (5.11.2.1) reduces to Murthy's (1957) unordered estimator of population total which we discuss in the following theorem.

Theorem 5.11.2.2. Murthy's (1957) unordered estimator of population total based on two units is given by

$$\hat{T}_{\text{M}} = \frac{1}{(2 - P_i - P_j)}\left[\frac{y_i}{P_i}(1 - P_j) + \frac{y_j}{P_j}(1 - P_i)\right]. \tag{5.11.2.2}$$

Proof. Suppose u_i, u_j and u_j, u_i constitutes the units in the sample and the probabilities of selection attached to these units are P_i, P_j and P_j, P_i, respectively. Now if P_i is the probability of selecting the ith unit at the first draw, then the probability of selecting j^{th} unit in the second draw given that i^{th} unit has already been selected in the first draw $= P_j/(1 - P_i)$.

Therefore the probability for selecting u_i and $u_j = P_i P_j/(1 - P_i) = P(s_o, 1)$.

Similarly the probability for selecting u_i and $u_j = P_i P_j/(1 - P_j) = P(s_o, 2)$.

Thus the sum of these probabilities is given by

$$P(s_u) = P(s_o, 1) + P(s_o, 2) = \frac{P_i P_j}{1 - P_i} + \frac{P_i P_j}{1 - P_j} = \frac{P_i P_j (2 - P_i - P_j)}{(1 - P_i)(1 - P_j)}. \tag{5.11.2.3}$$

Now the first ordered estimator based on the units u_i and u_j is given by

$$\hat{T}(s_o, 1) = y_i + \frac{y_j}{P_j}(1 - P_i) \tag{5.11.2.4}$$

with probability $P(s_o, 1)$, and the second ordered estimator based on the units u_j and u_i is given by

$$\hat{T}(s_o, 2) = y_j + \frac{y_i}{P_i}(1 - P_j) \tag{5.11.2.5}$$

with probability $P(s_o, 2)$. Then we have

$$\hat{T}(s_u) = \sum_{o=1}^{n!} \frac{P(s_o, o)\hat{T}(s_o, o)}{P(s_u)} = \sum_{o=1}^{2!} \frac{P(s_o, o)\hat{T}(s_o, o)}{P(s_u)} = \frac{P(s_o, 1)\hat{T}(s_o, 1) + P(s_o, 2)\hat{T}(s_o, 2)}{P(s_u)}$$

$$= \left[\left\{y_i + \frac{y_j}{P_j}(1 - P_i)\right\}\frac{P_i P_j}{(1 - P_i)} + \left\{y_j + \frac{y_i}{P_i}(1 - P_j)\right\}\frac{P_i P_j}{(1 - P_j)}\right] \Big/ \left[\frac{P_i P_j (2 - P_i - P_j)}{(1 - P_i)(1 - P_j)}\right]$$

$$= (2 - P_i - P_j)^{-1}\left[\frac{y_i}{P_i}(1 - P_j) + \frac{y_j}{P_j}(1 - P_i)\right].$$

Thus we have the following theorem:

Theorem 5.11.2.3. The ordered estimator of population total is always less efficient than the corresponding unordered estimator.

Proof. We have to show that

$$V\left[\hat{T}(s_o, o)\right] \geq V\left[\hat{T}(s_u)\right]. \tag{5.11.2.6}$$

Now

$$V\left[\hat{T}(s_o, o)\right] = E_1 V_2\left[\hat{T}(s_o, o) \mid s_u\right] + V_1 E_2\left[\hat{T}(s_o, o) \mid s_u\right], \tag{5.11.2.7}$$

where $V\left[\hat{T}(s_o, o)\right]$ is based on the oth ordered of the s_u^{th} unordered sample. E_2, V_2 are the conditional expectation and variance for a given unordered sample s_u and E_1, V_1 are the expectation and variance over all the unordered samples s_u.
Therefore we have

$$V\left[\hat{T}(s_o, o)\right] = E_1 V_2\left[\hat{T}(s_o, o) \mid s_u\right] + V_1\left[\hat{Q}\right], \tag{5.11.2.8}$$

where

$$\hat{Q} = E_2\left[\hat{T}(s_o, o) \mid s_u\right]. \tag{5.11.2.9}$$

Now to find the value of $\hat{Q}$ in (5.11.2.9), we proceed as follows: The ordered estimators $\hat{T}(s_o, o)$ for a given unordered sample s_u have values corresponding to each order, that is $\hat{T}(s_o, 1)$, $\hat{T}(s_o, 2)$, ..., $\hat{T}(s_o, n!)$, where

$\hat{T}(s_o, 1)$ = Estimator of the population total based on the first order sample from the uth unordered sample s_u,

$\hat{T}(s_o, 2)$ = Estimator of the population total based on the second sample from the the uth unordered sample s_u,

and

$\hat{T}(s_o, n!)$ = Estimator of population total based on $n!^{\text{th}}$ ordered sample from the uth unordered sample s_u

with probabilities $P(s_o, 1)$, $P(s_o, 2)$, and $P(s_o, n!)$, respectively. Now to find the ordered estimator $\hat{T}(s_o, o)$ we adjust the probabilities such that sum of these probabilities equals one.

Thus we use the adjusted probabilities given by $P(s_o, 1)/P(s_u)$, $P(s_o, 2)/P(s_u)$, .., $P(s_o, n!)/P(s_u)$, respectively, where $P(s_u) = \sum_{o=1}^{n!} P(s_o, o)$.

Thus we have

$$\hat{Q} = E_2\left[\hat{T}(s_o, o) \mid s_u\right] = \sum_{o=1}^{n!} \frac{\hat{T}(s_o, o) P(s_o, o)}{P(s_u)} = \hat{T}(s_u). \tag{5.11.2.10}$$

Using (5.11.2.10) in (5.11.2.8), we have

$$V\left[\hat{T}(s_o, o)\right] = E_1 V_2\left[\hat{T}(s_o, o) \mid s_u\right] + V_1\left[\hat{T}(s_u)\right]. \tag{5.11.2.11}$$

Now the first term, namely $E_1 V_2\left[\hat{T}(s_o, o) \mid s_u\right]$, will be zero if for each unordered sample s_u, the estimator takes the same values as for the ordered samples; otherwise, it is positive.

Therefore we have

$$V\left[\hat{T}(s_o, o)\right] \geq V_1\left[\hat{T}(s_u)\right].$$

Hence the theorem.

5.12 RAO--HARTLEY--COCHRAN (RHC) SAMPLING STRATEGY

Sampling Scheme: Suppose a population consists of N units and we wish to draw a sample of n units. First of all, divide randomly the N units into n groups as shown below:

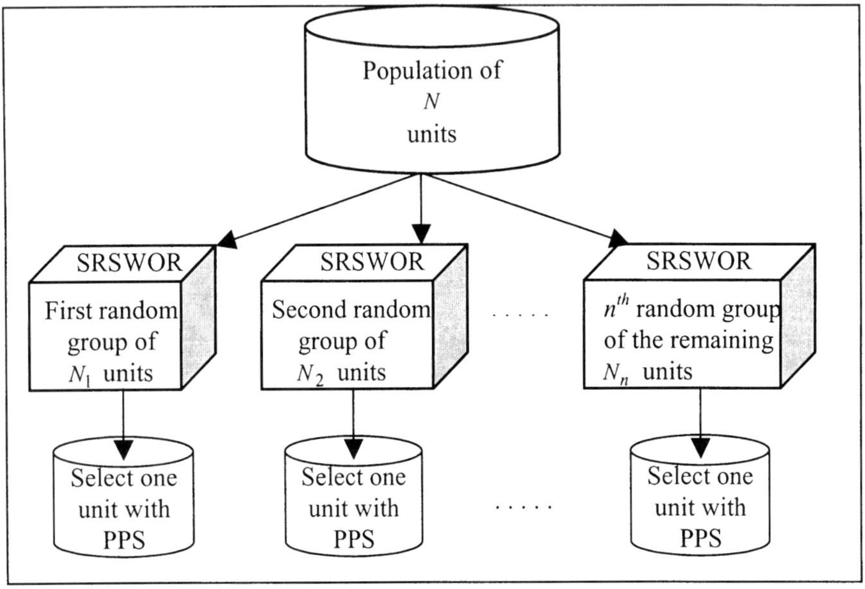

Fig. 5.12.1 Rao --Hartley--Cochran strategy.

First random group: Out of N units, select N_1 units by using SRSWOR sampling.

Second random group: Out of $(N - N_1)$ units, select N_2 units by SRSWOR sampling and so on such that

$$\sum_{i=1}^{n} N_i = N. \tag{5.12.1}$$

The allocation of units to different groups is done randomly and we select one unit from each of the n groups with probability proportional to size (PPS) and thus we obtain a sample of size n.

Suppose $P_1, P_2, ..., P_N$ are the probabilities associated with the N units in the population and $\sum_{i=1}^{N} P_i = 1.$

Further suppose that P_{ij} denotes the probability corresponding to the jth unit in the ith group, G_i, $\forall\, i = 1, 2, ..., n.$.

Thus the Rao, Hartley, and Cochran (1962) mechanism can be better understood from the following table, which gives the structure of population units after making random groups, as follows:

Structure of data in RHC-Sampling Strategy									
1st Group (G_1)		2nd Group (G_2)			i^{th} Group (G_i)			n^{th} Group (G_n)	
Value	Prob.	Value	Prob.		Value	Prob.		Value	Prob.
Y_{11}	P_{11}	Y_{21}	P_{21}		Y_{i1}	P_{i1}		Y_{n1}	P_{n1}
Y_{12}	P_{12}	Y_{22}	P_{22}		Y_{i2}	P_{i2}		Y_{n2}	P_{n2}
.	.	.	.		.	.		.	.
.	.	.	.		.	.		.	.
Y_{1N_1}	P_{1N_1}	Y_{2N_2}	P_{2N_2}		Y_{iN_i}	P_{iN_i}		Y_{nN_n}	P_{nN_n}
	τ_1		τ_2			τ_i			τ_n

where $\tau_i = \sum_{j \in G_i} Pij$, $i = 1, 2, ..., n$,denotes the sum of selection probabilities of the ith random group.

Now we have the following theorems:

Theorem 5.12.1. The unbiased estimator of population total Y is given by

$$\hat{Y}_{\text{RHC}} = \sum_{i=1}^{n} \frac{y_{i1}}{(P_{i1}/\tau_i)}. \tag{5.12.2}$$

Proof. Suppose E_2 denotes the expected value for the given random group G_i and E_1 denotes the expected value over all possible random groups. Then we have

$$E\left(\hat{Y}_{\text{RHC}}\right) = E_1 E_2\left[\hat{Y}_{\text{RHC}} \mid G_i\right] = E_1 E_2\left[\sum_{i=1}^{n} \frac{y_{i1}}{(P_{i1}/\tau_i)} \mid G_i\right] = E_1\left[\sum_{i=1}^{n}\left(\sum_{j=1}^{N_i} Y_{ij}\right)\right] = E_1\left(\sum_{i=1}^{N} Y_i\right) = Y .$$

Hence the theorem.

Theorem 5.12.2. The variance of the estimator $\hat{Y}_{\text{RHC}}$ is given by

$$V\left(\hat{Y}_{\text{RHC}}\right) = \frac{\left(\sum_{i=1}^{n} N_i^2 - N\right)}{N(N-1)}\left[\sum_{j=1}^{N} \frac{Y_j^2}{P_j} - Y^2\right]. \tag{5.12.3}$$

Proof. We have

$$V\left(\hat{Y}_{\text{RHC}}\right) = E_1 V_2\left(\hat{Y}_{\text{RHC}} \mid G_i\right) + V_1 E_2\left(\hat{Y}_{\text{RHC}} \mid G_i\right)$$

$$= E_1 V_2\left[\sum_{i=1}^{n} \frac{y_{i1}}{(P_{i1}/\tau_i)} \mid G_i\right] + V_1 E_2\left[\sum_{i=1}^{n} \frac{y_{i1}}{(P_{i1}/\tau_i)} \mid G_i\right]$$

$$= E_1 V_2\left[\sum_{i=1}^{n} \frac{y_{i1}}{(P_{i1}/\tau_i)} \mid G_i\right] + V_1(Y) = E_1 V_2\left[\sum_{i=1}^{n} \frac{y_{i1}}{(P_{i1}/\tau_i)} \mid G_i\right]. \tag{5.12.4}$$

Note that we have selected independent samples from each group, therefore

$$E_1V_2\left[\sum_{i=1}^{n}\frac{y_{i1}}{(P_{i1}/\tau_i)}\mid G_i\right]=E_1\left[\sum_{i=1}^{n}V_2\left(\frac{y_{i1}}{P_{i1}/\tau_i}\mid G_i\right)\right]. \tag{5.12.5}$$

Thus we have

$$V_2\left(\frac{y_{i1}\tau_i}{P_{i1}}\mid G\right)=E_2\left[\frac{y_{i1}}{(P_{i1}/\tau_i)}\right]^2-\left\{E\left(\frac{y_{i1}}{P_{i1}/\tau_i}\right)\right\}^2. \tag{5.12.6}$$

In a given random group of N_i units, the random variable $\frac{y_{i1}}{(P_{i1}/\tau_i)}$ can take any of the values $\frac{Y_{i1}}{(P_{i1}/\tau_i)}, \frac{Y_{i2}}{(P_{i2}/\tau_i)},......,\frac{Y_{iN_i}}{(P_{iN_i}/\tau_i)}$ with probabilities $\frac{P_{i1}}{\tau_i}, \frac{P_{i2}}{\tau_i},......,\frac{P_{iN_i}}{\tau_i}$, respectively.

Thus we have

$$E_2\left[\frac{y_{i1}}{(P_{1i}/\tau_i)}\right]=\frac{Y_{i1}}{(P_{i1}/\tau_i)}\times(P_{i1}/\tau_i)+\frac{Y_{i2}}{(P_{i2}/\tau_i)}\times(P_{i2}/\tau_i)+........+\frac{Y_{iN_i}}{(P_{iN_i}/\tau_i)}\times(P_{iN_i}/\tau_i)$$

$$=Y_{i1}+Y_{i2}+.......+Y_{iN_i}=\sum_{j=1}^{N_i}Y_{ij}=Y_{i\bullet}. \tag{5.12.7}$$

Also we have

$$E\left[\frac{y_{i1}}{(P_{i1}/\tau_i)}\right]^2=\sum_{j=1}^{N_i}\frac{Y_{ij}^2}{(P_{ij}/\tau_i)} \tag{5.12.8}$$

because $\left[\frac{y_{i1}}{(P_{1i}/\tau_i)}\right]^2$ takes the values $\left[\frac{Y_{i1}}{(P_{i1}/\tau_i)}\right]^2, \left[\frac{Y_{i2}}{(P_{i2}/\tau_i)}\right]^2,......,\left[\frac{Y_{iN_i}}{(P_{iN}/\tau_i)}\right]^2$ with probabilities $\frac{P_{i1}}{\tau_i}, \frac{P_{i2}}{\tau_i},......,\frac{P_{iN_i}}{\tau_i}$, respectively. Using (5.12.7) and (5.12.8) in (5.12.6) we have

$$V_2\left(\frac{y_{i1}\tau_i}{P_{i1}}\mid G\right)=\sum_{j=1}^{N_i}\frac{Y_{ij}^2}{(P_{ij}/\tau_i)}-Y_{i\bullet}^2. \tag{5.12.9}$$

Note that

$$\tau_i=\sum_{j=1}^{N_i}P_{ij}$$

therefore (5.12.5) implies that

$$E_1V_2\left[\sum_{i=1}^{n}\frac{y_{i1}}{(P_{i1}/\tau_i)}\mid G\right]=E_1\left[\sum_{i=1}^{n}V_2\left(\frac{y_{i1}}{P_{i1}/\tau_i}\mid G\right)\right]=E_1\left[\sum_{i=1}^{n}\left\{\sum_{j=1}^{N_i}\frac{Y_{ij}^2}{(P_{ij}/\tau_i)}-Y_{i\bullet}^2\right\}\right]$$

$$=\sum_{i=1}^{n}\left[E_1\left(\sum_{j=1}^{N_i}\frac{Y_{ij}^2}{(P_{ij}/\tau_i)}-Y_{i\bullet}^2\right)\right]=\sum_{i=1}^{n}\left[E_1\left(\sum_{j=1}^{N_i}\frac{Y_{ij}^2}{(P_{ij}/\tau_i)}\right)-E_1\left(Y_{i\bullet}^2\right)\right]$$

$$= \sum_{i=1}^{n}\left[E_1\left(\tau_i \sum_{j=1}^{N_i}\frac{Y_{ij}^2}{P_{ij}}\right) - E_1\left(Y_{i\bullet}^2\right)\right]$$

$$= \sum_{i=1}^{n}\left[E_1\left(\sum_{j=1}^{N_i} P_{ij}\right)\left(\sum_{j=1}^{N_i}\frac{Y_{ij}^2}{P_{ij}}\right) - E_1\left(Y_{i\bullet}^2\right)\right]$$

$$= \sum_{i=1}^{n}\left[E_1\left(\sum_{j=1}^{N_i} Y_{ij}^2 + \sum_{j\neq l=1}^{N_i}\frac{P_{il}Y_{ij}^2}{P_{ij}}\right) - E_1\left(Y_{i\bullet}^2\right)\right]$$

$$= \sum_{i=1}^{n}\left[E_1\left(\frac{N_i}{N_i}\sum_{j=1}^{N_i} Y_{ij}^2 + \sum_{j\neq l=1}^{N_i}\frac{P_{il}Y_{ij}^2}{P_{ij}}\right) - E_1\left(Y_{i\bullet}^2\right)\right]$$

$$= \sum_{i=1}^{n}\left[E_1\left(\frac{N_i}{N_i}\sum_{j=1}^{N_i} Y_{ij}^2\right) + E_1\left(\sum_{j\neq l=1}^{N_i}\frac{P_{il}Y_{ij}^2}{P_{ij}}\right) - E_1\left(Y_{i\bullet}^2\right)\right]$$

$$= \sum_{i=1}^{n}\left[\frac{N_i}{N}\sum_{j=1}^{N} Y_j^2 + \frac{N_i(N_i-1)}{N(N-1)}\sum_{j\neq l=1}^{N}\frac{P_lY_j^2}{P_j} - E_1\left(Y_{i\bullet}^2\right)\right]. \quad (5.12.10)$$

Furthermore

$$E_1\left(Y_{i\bullet}^2\right) = V_1(Y_{i\bullet}) + \{E_1(Y_{i\bullet})\}^2 = V_1\left[N_i\frac{Y_{i\bullet}}{N_i}\right] + \{E_1(Y_{i\bullet})\}^2 = N_i^2V_1\left(\frac{Y_{i\bullet}}{N_i}\right) + \left[N_i\bar{Y}\right]^2$$

$$= N_i^2\left(\frac{N-N_i}{NN_i}\right)S_y^2 + N_i^2\bar{Y}^2. \quad (5.12.11)$$

Now consider

$$\sum_{j=1}^{N}\frac{Y_j^2}{P_j} = \sum_{j=1}^{N}\frac{Y_j^2}{P_j}\times 1 = \sum_{j=1}^{N}\frac{Y_j^2}{P_j}\times\sum_{j=1}^{N}P_j = \sum_{j=1}^{N}Y_j^2 + \sum_{j\neq l=1}^{N}\frac{P_lY_j^2}{P_j} \quad (5.12.12)$$

which implies that

$$\sum_{j\neq l=1}^{N}\frac{P_lY_j^2}{P_j} = \sum_{j=1}^{N}\frac{Y_j^2}{P_j} - \sum_{j=1}^{N}Y_j^2. \quad (5.12.13)$$

Therefore (5.12.10) implies that

$$E_1\left[V_2\left(\sum_{i=1}^{n}\frac{y_{i1}}{(P_{i1}/\tau_i)}\mid G\right)\right]$$

$$= \sum_{i=1}^{n}\left\{\frac{N_i}{N}\sum_{j=1}^{N}Y_j^2 + \frac{N_i(N_i-1)}{N(N-1)}\left(\sum_{j=1}^{N}\frac{Y_j^2}{P_j} - \sum_{j=1}^{N}Y_j^2\right) - N_i^2\left(\frac{N-N_i}{NN_i}\right)S_y^2 - N_i^2\bar{Y}^2\right\}$$

$$= \sum_{i=1}^{n}\left[\frac{N_i}{N}\sum_{j=1}^{N}Y_j^2 + \frac{N_i(N_i-1)}{N(N-1)}\left(\sum_{j=1}^{N}\frac{Y_j^2}{P_j} - \sum_{j=1}^{N}Y_j^2\right) - \frac{N_i(N-N_i)}{N}\frac{1}{(N-1)}\left\{\sum_{j=1}^{N}Y_j^2 - N\bar{Y}^2\right\} - N_i^2\bar{Y}^2\right]$$

$$= \sum_{i=1}^{n}\left[\frac{N_i}{N}\sum_{j=1}^{N} Y_j^2 + \frac{N_i(N_i-1)}{N(N-1)}\sum_{j=1}^{N}\frac{Y_j^2}{P_j} - \frac{N_i(N_i-1)}{N(N-1)}\sum_{j=1}^{N} Y_j^2 - \frac{N_i}{(N-1)}\sum_{j=1}^{N} Y_j^2 + \frac{NN_i}{(N-1)}\overline{Y}^2 \right.$$
$$\left. + \frac{N_i^2}{N(N-1)}\sum_{j=1}^{N} Y_j^2 - \frac{N_i^2}{(N-1)}\overline{Y}^2 - N_i^2\overline{Y}^2 \right]$$
$$= \sum_{i=1}^{n}\left[\frac{N_i(N_i-1)}{N(N-1)}\sum_{j=1}^{N}\frac{Y_j^2}{P_j} + \sum_{j=1}^{N} Y_j^2\left\{\frac{N_i}{N} - \frac{N_i(N_i-1)}{N(N-1)} - \frac{N_i}{N-1} + \frac{N_i^2}{N(N-1)}\right\} + \frac{Y^2}{N^2}\left\{\frac{NN_i}{N-1} - \frac{N_i^2}{N-1} - N_i^2\right\}\right]$$
$$= \sum_{i=1}^{n}\left[\frac{N_i(N_i-1)}{N(N-1)}\sum_{j=1}^{N}\frac{Y_j^2}{P_j} + \sum_{j=1}^{N} Y_j^2\left\{\frac{N_i(N-1) - N_i(N_i-1) - N_iN + N_i^2}{N(N-1)}\right\} + \frac{Y^2}{N^2}\left\{\frac{NN_i - N_i^2 - N_i^2(N-1)}{N-1}\right\}\right]$$
$$= \sum_{i=1}^{n}\left[\frac{N_i(N_i-1)}{N(N-1)}\sum_{j=1}^{N}\frac{Y_j^2}{P_j} - \frac{N_i(N_i-1)}{N(N-1)}Y^2\right] = \sum_{i=1}^{n}\left[\frac{N_i(N_i-1)}{N(N-1)}\left\{\sum_{j=1}^{N}\frac{Y_j^2}{P_j} - Y^2\right\}\right]$$
$$= \frac{\left(\sum_{i=1}^{n} N_i^2 - N\right)}{N(N-1)}\left[\sum_{j=1}^{N}\frac{Y_j^2}{P_j} - Y^2\right]. \tag{5.12.14}$$

Hence the theorem.

Theorem 5.12.3. The RHC scheme is more efficient than PPSWR sampling if $N_i = \frac{N}{n}, \ \forall\ i = 1, 2, ..., n.$

Proof. Without loss of generality we have

$$V\left(\hat{Y}_{\text{PPSWR}}\right) = \frac{1}{n}\left[\sum_{j=1}^{N}\frac{Y_j^2}{P_j} - Y^2\right], \quad \text{and} \quad V\left(\hat{Y}_{\text{RHC}}\right) = \frac{\left(\sum_{i=1}^{n} N_i^2 - N\right)}{N(N-1)}\left[\sum_{j=1}^{N}\frac{Y_j^2}{P_j} - Y^2\right].$$

Combining these results we have

$$V\left(\hat{Y}_{\text{RHC}}\right) = \frac{n\left(\sum_{i=1}^{n} N_i^2 - N\right)}{N(N-1)} V\left(\hat{Y}_{\text{PPSWR}}\right). \tag{5.12.15}$$

To find the minimum value of the variance $V\left(\hat{Y}_{\text{RHC}}\right)$ with respect to N_i we have the Lagrange function

$$L = \sum_{i=1}^{n} N_i^2 - \lambda\left[\sum_{i=1}^{n} N_i - N\right]. \tag{5.12.16}$$

On differentiating (5.12.16) with respect to N_i and equating to zero we have

$$N_i = \lambda/2\,. \tag{5.12.17}$$

On substituting (5.12.17) in the equation $\sum_{i=1}^{n} N_i = N$ we have

$$\lambda = 2N/n\,. \tag{5.12.18}$$

Plugging back this value of λ in (5.12.17) we have the optimum value of N_i that is

$$N_i = N/n\,. \tag{5.12.19}$$

On substituting the optimum value of N_i in (5.12.15) we have

$$V\left(\hat{Y}_{\text{RHC}}\right)=\left[\frac{\sum\limits_{i=1}^{n}\left(\frac{N}{n}\right)^2-N}{N(N-1)}\right]\left[nV\left(\hat{Y}_{\text{PPSWR}}\right)\right]=\frac{(N-n)}{N-1}V\left(\hat{Y}_{\text{PPSWR}}\right). \tag{5.12.20}$$

Note that $(N-n)/(N-1)<1\ \ \forall\ n\geq 2$, $V\left(\hat{Y}_{\text{RHC}}\right)<V\left(\hat{Y}_{\text{PPSWR}}\right)$. Hence the theorem.

Theorem 5.12.4. An unbiased estimator of the variance $V\left(\hat{Y}_{\text{RHC}}\right)$ is given by

$$\hat{v}\left(\hat{Y}_{\text{RHC}}\right)=\frac{\left(\sum\limits_{i=1}^{n}N_i^2-N\right)}{\left(N^2-\sum\limits_{i=1}^{n}N_i^2\right)}\left[\sum_{i=1}^{n}\frac{y_{i1}^2}{\left(P_{i1}^2/\tau_i\right)}-\hat{Y}_{\text{RHC}}^2\right]. \tag{5.12.21}$$

Proof. We know that

$$\hat{v}\left(\hat{Y}_{\text{RHC}}\right)=\frac{\left(\sum\limits_{i=1}^{n}N_i^2-N\right)}{N(N-1)}\times n\times\hat{v}\left(\hat{Y}_{\text{PPSWR}}\right). \tag{5.12.22}$$

Also we know that

$$nV\left(\hat{Y}_{\text{PPSWR}}\right)=\sum_{j=1}^{N}\frac{Y_j^2}{P_j}-Y^2. \tag{5.12.23}$$

Now the estimator of the right hand side of (5.12.23) is given by

$$\sum_{i=1}^{n}\frac{y_{i1}^2}{\left(P_{i1}/\tau_i\right)}-\left[\left(\sum_{i=1}^{n}\frac{y_{i1}}{\left(P_{i1}/\tau_i\right)}\right)^2-\hat{v}\left(\hat{Y}_{\text{RHC}}\right)\right]. \tag{5.12.24}$$

Note that the estimator of $\sum\limits_{j=1}^{N}Y_j$ is $\sum\limits_{i=1}^{n}\frac{y_{i1}}{\left(P_{i1}/\tau_i\right)}$, the estimator of $\sum\limits_{j=1}^{N}\frac{Y_j^2}{P_j}$ is $\sum\limits_{i=1}^{n}\left(\frac{y_{i1}^2}{P_{i1}}\Big/P_{i1}\right)\tau_i=\sum\limits_{i=1}^{n}\frac{y_{i1}^2}{P_{i1}^2}\tau_i$ and, that of Y^2 is $\left\{\left(\sum\limits_{i=1}^{n}\frac{y_{i1}}{\left(P_{i1}/\tau_i\right)}\right)^2-\hat{v}\left(\hat{Y}_{\text{RHC}}\right)\right\}$.

Thus, we have

$$n\,\hat{v}\left(\hat{Y}_{\text{PPSWR}}\right)=\sum_{i=1}^{n}\frac{y_{i1}^2}{P_{i1}^2}\tau_i-\left[\left(\sum_{i=1}^{n}\frac{y_{i1}}{P_{i1}}\tau_i\right)^2-\hat{v}\left(\hat{Y}_{\text{RHC}}\right)\right]. \tag{5.12.25}$$

Therefore

$$\hat{v}\left(\hat{Y}_{\text{RHC}}\right)=\frac{\left(\sum\limits_{i=1}^{n}N_i^2-N\right)}{N(N-1)}\left[\sum_{i=1}^{n}\frac{y_{i1}^2}{\left(P_{i1}^2/\tau_i\right)}-\left\{\left(\sum_{i=1}^{n}\frac{y_{i1}}{\left(P_{i1}/\tau_i\right)}\right)^2-\hat{v}\left(\hat{Y}_{\text{RHC}}\right)\right\}\right]$$

or

$$\hat{v}\left(\hat{Y}_{\text{RHC}}\right)\left[1-\frac{\left(\sum\limits_{i=1}^{n}N_i^2-N\right)}{N(N-1)}\right]=\frac{\left(\sum\limits_{i=1}^{n}N_i^2-N\right)}{N(N-1)}\left[\sum_{i=1}^{n}\frac{y_{i1}^2}{\left(P_{i1}^2/\tau_i\right)}-\hat{Y}_{\text{RHC}}^2\right]$$

which on simplifying reduces to (5.12.21). Hence the theorem.

Some improvements in the RHC strategy have also been suggested by Hartley, Rao, and Kiefer (1969), Gabler and Horst (1995), Mangat (1993), Bansal and Singh (1986) and Singh and Kishore (1975). Padmawar (1996) has considered an interesting extension of the RHC strategy for the case of continuous populations. Its comparison with Midzuno's scheme of sampling has been discussed by Chaudhuri (1977) under a superpopulation model setup.

Example 5.12.1. From the population 1 select a sample of five units by using the RHC scheme. Estimate the total real estate farm loans using RHC estimator and making use of nonreal estate farm loans as an auxiliary variable. Also find 95% confidence interval for the total real estate farm loans in the United States.

Solution. We are to select a sample of size 5 by using RHC scheme. Thus the population must be divided into five random groups. To do this we selected 50 distinct random numbers between 1 and 50 by starting with the first two columns of the Pseudo-Random Numbers (PRN) given in Table 1 of the Appendix.

These random numbers came in the sequence 01, 23, 46, 04, 32, 47, 33, 05, 22, 38, 29, 40, 03, 36, 27, 19, 14, 42, 48, 06, 07, 21, 31, 16, 10, 18, 26, 02, 44, 12, 37, 25, 50, 30, 09, 11, 49, 43, 15, 39, 35, 20, 34, 13, 24, 41, 17, 08, 28 and 45.

The states bearing serial numbers corresponding to the first ten selected random numbers constitute the first random group, whereas the next ten form the second random group and so on.

Let

Y_{ij} = Real estate farm loans ($000) for the jth state in the ith random group,

X_{ij} = Nonreal estate farm loans ($000) for the jth state in the ith random group,

and

P_{ij} = the initial selection probability of the jth unit in the ith random group.

We are given $X = 43908.12$, thus the following are the 5 random groups of units along with initial selection probabilities.

Random Group I					
Sr. No.	Random No.	State and Territory	Y_{1j}	X_{1j}	P_{1j}
1	01	AL	408.978	348.334	0.007933
2	23	MN	1354.768	2466.892	0.056183
3	46	VA	321.583	188.477	0.004293
4	04	AR	907.700	848.317	0.019320
5	32	NY	201.631	426.274	0.009708
6	47	WA	1100.745	1228.607	0.027981
7	33	NC	639.571	494.730	0.011267
8	05	CA	1343.467	3928.732	0.089476
9	22	MI	323.028	440.518	0.010033
10	38	PA	756.169	298.351	0.006795
			Sum	10669.230	0.242990

Random Group II					
Sr. No.	Random no.	State and Territory	Y_{2j}	X_{2j}	P_{2j}
1	29	NH	6.044	0.471	0.0000107
2	40	SC	87.951	80.750	0.0018391
3	03	AZ	54.633	431.439	0.0098260
4	36	OK	612.108	1716.087	0.0390836
5	27	NE	1337.852	3586.406	0.0816798
6	19	ME	8.849	57.684	0.0013137
7	14	IN	1213.024	1022.782	0.0232937
8	42	TN	553.266	388.869	0.0088564
9	48	WV	99.277	29.291	0.0006671
10	06	CO	315.809	906.281	0.0206404
			Sum	8220.060	0.1872100

Random Group III					
Sr. No.	Random no.	State and Territory	Y_{3j}	X_{3j}	P_{3j}
1	07	CT	7.130	4.373	0.000099
2	21	MA	7.590	56.471	0.001286
3	31	NM	140.582	274.035	0.006241
4	16	KS	1049.834	2580.304	0.058766
5	10	GA	939.460	540.696	0.012314
6	18	LA	282.565	405.799	0.009242
7	26	MT	292.965	722.034	0.016444
8	02	AK	2.605	3.433	0.000078
9	44	UT	56.908	197.244	0.004492
10	12	ID	53.753	1006.036	0.022912
			Sum	5790.425	0.131876

Random Group IV					
Sr. No.	Random no.	State and Territory	Y_{4j}	X_{4j}	P_{4j}
1	37	OR	114.899	1519.487	0.034606
2	25	MO	1579.686	1519.994	0.034618
3	50	WY	100.964	386.479	0.008802
4	30	NJ	39.860	27.508	0.000626
5	09	FL	825.748	464.516	0.010579
6	11	HI	40.775	38.067	0.000867
7	49	WI	1229.752	1372.479	0.031258
8	43	TX	1248.761	3520.361	0.080176
9	15	IA	2327.025	3909.738	0.089044
10	39	RI	1.611	0.233	0.000005
			Sum	12758.860	0.290580

Random Group V					
Sr. No.	Random no.	State and Territory	Y_{5j}	X_{5j}	P_{5j}
1	35	OH	870.720	635.774	0.014480
2	20	MD	139.628	57.684	0.001314
3	34	ND	449.099	1241.369	0.028272
4	13	IL	2131.048	2610.572	0.059455
5	24	MS	627.013	549.551	0.012516
6	41	SD	413.777	1692.817	0.038554
7	17	KY	1045.106	557.656	0.012701
8	08	DE	42.808	43.229	0.000985
9	28	NV	5.860	16.710	0.000381
10	45	VT	57.747	19.363	0.000441
			Sum	7424.730	0.169100

Now apply Lahiri's method of selection of sample in each group to select one unit independently. The first random group consists of $N_1 = 10$ units and maximum value of the auxiliary variable X_{1j} is 3928.732. Choosing $X_0 = 4000$ we select random number $1 \le R_i \le 10$ by starting with the first two columns and another random number $1 \le R_j \le 4000$ by starting from the 7th to 10th columns of the Pseudo-Random Numbers. Then the first effective pair of random numbers is (04, 0757). Thus from the first random group the unit with Sr. No. 4, that is the state **AR,** will be included in the sample.

The second random group consists of $N_2 = 10$ units and the maximum value of the auxiliary variable X_{2j} is 3586.406. Choosing $X_0 = 3600$ we select random number $1 \le R_i \le 10$ by starting with the 7th and 8th columns and another random number $1 \le R_j \le 3600$ by starting from the 13th to 16th columns of the Pseudo-Random Numbers. Then the first effective pair of random numbers is (07, 0536). Thus from the second random group, the unit with Sr. No. 7, that is the state **IN,** will be included in the sample.

The third random group consists of $N_3 = 10$ units and the maximum value of the auxiliary variable X_{3j} is 2580.304. Choosing $X_0 = 2600$ we select random number $1 \le R_i \le 10$ by starting with the 13th and 14th columns and another random number $1 \le R_j \le 2600$ by starting from the 19th to 22th columns of the Pseudo-Random Numbers. Then the first effective pair of random numbers is (07, 0705). Thus from the third random group the unit with Sr. No. 7, that is the state **MT,** will be included in the sample.

The fourth random group consists of $N_4 = 10$ units and the maximum value of the auxiliary variable X_{4i} is 3909.738. Choosing $X_0 = 4000$ we select random number $1 \le R_i \le 10$ by starting with the 19th and 20th columns and another random number $1 \le R_j \le 4000$ by starting from the 25th to 28th columns of the Pseudo-Random Numbers. Then the first effective pair of random numbers is (08, 1230). Thus from the fourth random group the unit with Sr. No. 8, that is the state **TX,** will be included in the sample.

The fifth random group consist of $N_5 = 10$ units and the maximum value of the auxiliary variable X_{5j} is 1610.572. Choosing $X_0 = 27$ we select random number $1 \le R_i \le 10$ by starting with the 25th and 26th columns and another random number $1 \le R_j \le 1700$ by starting from the 31st to 34th columns of the Pseudo-Random Numbers. Then the first effective pair of random numbers is (06, 0599). Thus from the fifth random group the unit with Sr. No. 6, that is the state **SD,** will be included in the sample.

After combining all above steps, the ultimate sample consists of the following information.

Sr. No.	State and Territory	Nonreal estate loan x_{i1}	Real estate loan y_{i1}	P_{i1}	τ_i	$\frac{y_{i1}}{P_{i1}/\tau_i}$	$\frac{y_{i1}^2}{P_{i1}^2/\tau_i}$
1	AR	848.317	907.700	0.019320	0.242990	11416.090	536347692.4
2	IN	1022.782	1213.024	0.023294	0.187210	9749.003	507681529.1
3	MT	722.034	292.965	0.016444	0.131876	2349.463	41857340.6
4	TX	3520.361	1248.761	0.080176	0.290580	4525.876	70491971.3
5	SD	1692.817	413.777	0.038554	0.169100	1814.867	19478071.8
					Sum	29855.300	1175856605.0

Note that here $\tau_i = \sum_{j \in G_i} P_{ij}$, $i = 1, 2, 3, 4, 5$.

The estimate of the total real estate farm loans during 1997 in the US is given by

$$\hat{Y}_{\text{RHC}} = \sum_{i=1}^{n} \frac{y_{i1}}{(P_{i1}/\tau_i)} = 29855.30$$

and an estimate of variance of the estimator $\hat{Y}_{RHC}$ is given by

$$\hat{v}\left(\hat{Y}_{\text{RHC}}\right) = \frac{\left(\sum_{i=1}^{n} N_i^2 - N\right)}{\left(N^2 - \sum_{i=1}^{n} N_i^2\right)} \left[\sum_{i=1}^{n} \frac{y_{i1}^2}{\left(P_{i1}^2/\tau_i\right)} - \hat{Y}_{\text{RHC}}^2\right]$$

$$= \frac{\left(\sum_{i=1}^{5} 10^2 - 50\right)}{\left(50^2 - \sum_{i=1}^{5} 10^2\right)} \left[1175856605 - (29855.30)^2\right] = 64016475.05 .$$

Using table 2 from the Appendix the 95% confidence interval of the total farm real estate loan is given by

$$\hat{Y}_{\text{RHC}} \mp t_{\alpha/2}(\text{df} = n-1)\sqrt{\hat{v}\left(\hat{Y}_{\text{RHC}}\right)}\ , \quad \text{or} \quad 29855.30 \mp 2.776\sqrt{64016475.05}$$

or

$$[7644.44,\ 52066.16].$$

5.13 UNBIASED STRATEGIES USING IPPS SAMPLING SCHEMES

We shall discuss a few sampling schemes under which the usual ratio type estimators of population mean and variance become unbiased.

5.13.1 ESTIMATION OF POPULATION MEAN USING A RATIO ESTIMATOR

Consider a population consists of N units and we want to draw a sample of n units. Assume that X values are known for all units and $P_i = X_i/X$, for $X = \sum_{i=1}^{N} X_i$, denotes the probability of selecting the ith unit. Following Midzuno--Sen's sampling scheme:

(a) The first unit is selected with probability proportional to X_i value

(b) The remaining $(n-1)$ units are selected with SRSWOR from the remaining $(N-1)$ units with equal probability.

Let the tth sample s consists of n units such that $(y_i,\ x_i)$, $i \in s$ and $t = 1, 2, ..., \binom{N}{n}$ is the total number of samples using SRSWOR sampling. Our aim is to find the probability $p(t)$ of selecting any given sample. Consider the first unit x_i is selected on the first draw.

Then the probability of selecting the first unit on the first draw is $\left(\frac{x_1}{X}\right)\binom{N-1}{n-1}^{-1}$, where $\binom{N-1}{n-1}^{-1}$ denotes the probability of selecting the remaining $(n-1)$ units out of $(N-1)$ units.

Similarly the probability of selecting the second unit on the first draw is $\left(\frac{x_2}{X}\right)\binom{N-1}{n-1}^{-1}$, and so on.

The probability of selecting the nth unit on the first draw is $\left(\frac{x_n}{X}\right)\binom{N-1}{n-1}^{-1}$.

Hence the probability of selecting the tth sample on the first draw is given by

$$P(t) = \left(\frac{x_1}{X}\right)\binom{N-1}{n-1}^{-1} + \left(\frac{x_2}{X}\right)\binom{N-1}{n-1}^{-1} + \ldots\ldots + \left(\frac{x_n}{X}\right)\binom{N-1}{n-1}^{-1}$$

$$= \binom{N-1}{n-1}^{-1} \frac{\sum_{i=1}^{n} x_i}{X}. \tag{5.13.1.1}$$

Let $\bar{y}_t = n^{-1}\sum_{i=1}^{n} y_i$, $\bar{x}_t = n^{-1}\sum_{i=1}^{n} x_i$ be the sample means obtained from the tth sample for $t = 1,2,\ldots.,\binom{N}{n}$.

Also assuming that the population mean $\bar{X}$ of the auxiliary variable is known, the usual ratio estimator of population mean is given by

$$\bar{y}_R = \bar{y}_t\left(\frac{\bar{X}}{\bar{x}_t}\right). \tag{5.13.1.2}$$

Then we have the following theorem:

Theorem 5.13.1. The ratio estimator $\bar{y}_R$ is unbiased for the population mean.

Proof. We have

$$E(\bar{y}_R) = \sum_{t=1}^{\binom{N}{n}}\left(\bar{y}_t \frac{\bar{X}}{\bar{x}_t}\right)P(t) = \sum_{t=1}^{\binom{N}{n}}\left(\frac{\bar{y}_t}{\bar{x}_t}\bar{X}\right)\left[\frac{\sum_{i=1}^{n} x_i}{\binom{N-1}{n-1}X}\right] = \sum_{t=1}^{\binom{N}{n}}\left(\frac{\bar{y}_t}{\bar{x}_t}\right)\frac{\bar{X}\, n\, \bar{x}_t}{X\binom{N-1}{n-1}}$$

$$= \frac{1}{N\binom{N-1}{n-1}}\sum_{t=1}^{\binom{N}{n}}(n\,\bar{y}_t) = \frac{n}{N\binom{N-1}{n-1}}\sum_{t=1}^{\binom{N}{n}}(\bar{y}_t) = \frac{1}{\binom{N}{n}}\sum_{t=1}^{\binom{N}{n}}(\bar{y}_t) = \bar{Y}.$$

Hence the theorem.

5.13.2 ESTIMATION OF FINITE POPULATION VARIANCE

Nanjamma, Murthy, and Sethi (1959), Singh and Srivastava (1980) and Swain and Mishra (1992) have considered the use of a Midzuno type of sampling scheme as follows:

Step I. Select two units, for example, i^{th} and j^{th}, with the probability of their joint selection being proportional to $(X_i - X_j)^2$;

Step II. Select $(n-2)$ units from the remaining units of the population by simple random sampling and without replacement.

Note that the first step may be performed by considering all possible pairs of units and selecting a pair with the assigned probability.

Evidently the probability of selecting the tth sample is given by

$$P(t) = s_x^2 \Big/ \left\{\binom{N}{n} S_x^2\right\}, \quad \text{for } t = 1,2,\ldots,\binom{N}{n}.$$

Thus we have the following theorem:

Theorem 5.13.2.1. The ratio type of estimator

$$s_1^2 = s_y^2\left(\frac{S_x^2}{s_x^2}\right)$$

is unbiased for S_y^2.

Proof. We have

$$E\left(s_1^2\right) = E\left[s_y^2 \frac{S_x^2}{s_x^2}\right] = \sum_{t=1}^{\binom{N}{n}} S_x^2 \left(\frac{s_y^2}{s_x^2}\right)_t P(t) = S_x^2 \sum_{t=1}^{\binom{N}{n}} \left(\frac{s_y^2}{s_x^2}\right)_t \frac{\left(s_x^2\right)_t}{\binom{N}{n} S_x^2} = \frac{1}{\binom{N}{n}} \sum_{t=1}^{\binom{N}{n}} \left(s_y^2\right)_t = S_y^2 .$$

Hence the theorem.

Singh and Srivastava (1980) proposed two unbiased regression type strategies which depend on only one auxiliary variable. Wywial (1999) has considered an elegant extension of such sampling schemes to the case of multi-auxiliary information. Further, Wywial (2000) consider a study of Horvitz and Thompson (1952) type estimators under such sampling schemes. Chen (1998) has also proposed weighted polynomial models and weighted sampling schemes for finite populations.

5.14 GODAMBE'S STRATEGY: ESTIMATION OF PARAMETERS IN SURVEY SAMPLING

Godambe (1955) has considered the problem of estimation strategies in survey sampling through the concept of a unified approach. His way of estimating the parameters in survey sampling is unique and can be studied from Godambe (1955, 1960, 1976, 1980, 1984, 1989, 1991, 1995), Godambe and Heyde (1987), Godambe and Kale (1991), Godambe and Thompson (1984, 1986, 1989, 1996--97) and Binder and Patak (1994). We will discuss Godambe's strategy only briefly as his detailed work is available in a book by Thompson (1997). Following Godambe's strategy, let us consider the problem of estimation of population parameters which can be implicitly defined by a population equation of the form

$$\sum_{j\in\Omega} \phi_j(y_j, x_j, \theta) = 0 \tag{5.14.1}$$

where y_j and x_j are values of observed variables Y and X respectively. The ϕ_j are known real valued functions and θ is a real valued parameter of interest. Some special cases of (5.14.1) are as follows:

(a) the population mean μ_y defined by

$$\sum_{j\in\Omega} (y_j - \theta) = 0; \tag{5.14.2}$$

(b) the population total t_y defined by

$$\sum_{j\in\Omega} (y_j - t_y/N) = 0; \tag{5.14.3}$$

(c) the population ratio R of Y to X defined by

$$\sum_{j\in\Omega} (y_j - Rx_j) = 0; \tag{5.14.4}$$

(d) the population c.d.f. evaluated at y defined as

$$\sum_{j\in\Omega} [I(y_j \le y) - \theta] = 0, \quad \text{where } I(y_j \le y) = \begin{cases} 1 & \text{if } y_j \le y, \\ 0 & \text{otherwise;} \end{cases} \tag{5.14.5}$$

(e) the population median defined as the least value of θ such that

$$\sum_{j\in\Omega} \left[I(y_j \le \theta) - \frac{1}{2} \right] = 0. \tag{5.14.6}$$

When a population function (or parameter) is defined by (5.14.1), its estimator can be defined as a solution of the sample estimating equation

$$\sum_{j\in s} \phi_j(y_j, x_j, \theta) / \pi_j = 0 \tag{5.14.7}$$

where π_j denotes the probability of including the jth unit in the sample. An estimator of the population mean can easily be obtained by solving $\sum_{j\in s}(y_j-\theta)/\pi_j=0$, that is,

$$\bar{y}_s=\left(\sum_{j\in s}\frac{y_j}{\pi_j}\right)\Bigg/\left(\sum_{j\in s}\frac{1}{\pi_j}\right). \tag{5.14.8}$$

An estimator of the population total can be obtained by solving $\sum_{j\in s}(y_j-\theta/N)/\pi_j=0$, that is,

$$\hat{y}_s=\left(N\sum_{j\in s}\frac{y_j}{\pi_j}\right)\Bigg/\left(\sum_{j\in s}\frac{1}{\pi_j}\right). \tag{5.14.9}$$

Similarly estimators of ratio R of y to x, median of y and c.d.f. of y can be obtained by following Thompson (1997). Godambe (1995) also considered the problem of estimation of two types of parameters defined below:

(a) The parameters of the superpopulation from which the present survey population may be assumed to have been drawn as a random sample.
(b) The parameters of the survey population itself.

The parameters (a) and (b) may or may not be correlated. A survey may be conducted with the intention of estimation of both (a) and (b) types of parameters. On the other hand, if the survey's emphasis is on the estimation of either (a) or (b), the former is a case of analytical surveys and the latter is a case of enumerative surveys. If $\in_i(\theta)$ denotes the robust optimal estimating function for the ith unit of the survey population, then the induced parameter solves the equation

$$\sum_{i\in\Omega}\in_i(\theta)=0. \tag{5.14.10}$$

Thus the underlying concept of the term 'target parameter' can most simply be explicated by 'induced parameter'. To explain it let $m_i(\theta)$ and $v_i(\theta)$ be the mean and variance of the random variable y_i associated with the ith unit of the survey population $(i=1,2,...,N)$. In equation (5.14.10), let

$$\in_i(\theta)=\left\{\frac{y_i-m_i(\theta)}{v_i(\theta)}\right\}\frac{\partial m_i(\theta)}{\partial\theta} \tag{5.14.11}$$

or its trimmed version if one wants to exclude the tails. The score function of the logistic model discussed by Skinner, Holt, and Smith (1989) can easily be seen as a special case of (5.14.11) and hence (5.14.10). The robust optimality is as compelling as the optimality of the sample mean, or its weighted version, for the Gaussian linear models. The joint distribution of the study variable Y_Ω and known auxiliary variable X_Ω is assumed to be such that $(y_i,x_i), i=1,2,..,N$, are independent with density

$$f(y_i,x_i;\theta,n_i)=f_2(x_i;\theta,n_i)f_1(y_i\mid x_i;\theta),\quad i=1,2,...,N \tag{5.14.12}$$

where θ is a parameter of interest and $n_{\Omega} = (n_1, n_2, ..., n_N)$ is a set of nuisance parameters. The data consist of the known values of the auxiliary variate X_{Ω}, the sample s drawn with sampling design $P(\bullet \mid X_{\Omega})$, and the variable $y_i, i \in s$, can be defined as

$$d = \{X_{\Omega}, s, (i, y_i) : i \in s\}. \tag{5.14.13}$$

Then the probability density of the data d is given by

$$\text{Prob}(d;\theta,n_{\Omega}) = \left\{\prod_{i=1}^{N} f_2(x_i;\theta,n_i)\right\} P(s \mid X_{\Omega}) \left\{\prod_{i \in s} f_1(y_i \mid x_i;\theta)\right\}. \tag{5.14.14}$$

The nuisance parameter n_i can vary here. In (5.14.14), for every fixed θ the class of distributions of X_{Ω} obtained for the possible variations of the nuisance parameter is called a complete class of distributions. Following Godambe (1991), an estimator $g = g(d,\theta)$ of the parameter θ such that $E(g) = g^*$ is optimal if

$$\left[\frac{g}{E(\partial g / \partial \theta)}\right]^2 \tag{5.14.15}$$

is its a minimum for $g = g^*$. Then the optimal estimating function for estimating θ, based on data d, is

$$g^*(d,\theta) = \sum_{i \in s} \frac{\partial \log f_1(y_i \mid x_i;\theta)}{\partial \theta} \tag{5.14.16}$$

that is, the optimal estimate for θ is the solution of the equation

$$g^*(d,\theta) = 0 \tag{5.14.17}$$

which, in fact, is a special case of the general result of Godambe (1976). The interesting result is that the estimate of θ is free from the sampling design $P(\bullet \mid X_{\Omega})$, which, in fact supports the result of Royall (1970a, 1970b, 1970c). The estimates of parameters of interest in general can be obtained by solving the maximum likelihood equations given by

$$\frac{\partial \log[\text{Prob}(d;\, \theta, n_{\Omega})]}{\partial \theta} = 0, \tag{5.14.18}$$

and

$$\frac{\partial \log[\text{Prob}(d;\, \theta, n_{\Omega})]}{\partial n_{\Omega}} = 0. \tag{5.14.19}$$

If $(\hat{\theta}, \hat{n}_{\Omega})$ is the solution of the equations (5.14.18) and (5.14.19), then the estimate of μ is given by

$$\hat{\mu} = \mu(\hat{\theta}, \hat{n}_{\Omega}). \tag{5.14.20}$$

It is to be noted that the equations (5.14.18) and (5.14.19) are independent of the sampling design $p(\bullet \mid X)$, and so are the estimates $\hat{\theta}$, $\hat{n}_{\Omega}$ and $\hat{\mu}$.. The estimating equations (5.14.18) and (5.14.19), and hence the implied estimate $\hat{\mu}$ of μ, are dependent on the full model (5.14.12). Godambe raised the following interesting question: Can this dependence of the estimate on the entire model be reduced by some alternative procedure of estimation? This query is particularly meaningful in that, as remarked earlier, the modelling of the design or auxiliary variate X must in practice be very tentative. An alternative estimating procedure that utilizes only the conditional distribution in (5.14.12), namely f_1, would be very desirable. It would be more helpful if this alternative procedure depends only on some semiparametric relationship underlying the conditional distribution f_1. Such an alternative procedure is given below. The main concept here is that of the 'induced parameter' defined by (5.14.10). Since the variates $y_i, i=1,...,N_0$ are *i.i.d.* with the unknown mean μ, the induced parameter is given by the solution μ_0 of the equation

$$\sum_{i=1}^{N_0}(y_i-\mu_0)=0, \tag{5.14.21}$$

that is

$$\mu_0=\frac{1}{N_0}\sum_{i=1}^{N_0}y_i\,. \tag{5.14.22}$$

Assume that in the conditional distribution f_1, θ is a regression parameter and

$$E_m\{(y_i-a-\theta x_i)\,|\,x_i\}=0\,,\quad i=1,...,N_0\,, \tag{5.14.23}$$

where a is a known constant. The theory of estimating functions given by Godambe and Thompson (1986) can be seen to provide optimal estimation of the induced parameter μ_0 in (5.14.22) by using (15.14.23) and the sampling design $p(\bullet\,|\,X_\Omega)$. Note that this optimality is both conditional on holding the design variable X_Ω fixed and unconditional. It should be noted that the unconditional optimality is important here, for it is in the unconditional distribution that

$$\mu_0\to\mu \text{ as } N_0\to\infty\,. \tag{5.14.24}$$

In other words, for survey populations with large size N_0, we have $\mu_0\cong\mu$. This also provides justification for using the estimate which is optimal or approximately optimal for the induced parameter μ_0 for the parent parameter μ. The conditional and unconditional optimal results are given below. Let

$$H=\sum_{i=1}^{N_0}(y_i-a-\theta x_i). \tag{5.14.25}$$

It is remarkable here that for given (a,θ,X_Ω), H is a function of μ_0. Using the data d in (5.14.13) and parameter θ, we can obtain an optimal estimating function $h^*(d,\theta)$ for H. For the given sampling design $p(\bullet\,|\,X_\Omega)$, let the first order inclusion probabilities be

$$\pi_i=\sum_{s\ni i}p(z\,|\,X_\Omega),\quad i=1,...,N_0\,. \tag{5.14.26}$$

For a fixed θ, let $h(d,\theta)$ denote an estimating function based on data d, which is design unbiased for the function H because $E_p(h)=\sum hp(s\,|\,X_\Omega)=H$. Let E_m denote the expectation with respect to the model (15.14.23) both conditionally on X_Ω and unconditionally. In the class of all design unbiased estimating functions h, h^* is said to be optimal if, in the class, $E_m E_p(h-H)^2$ is minimized for $h = h^*$; in other words, if

$$E_m E_p\left(h^* - H\right)^2 \le E_m E_p(h-H)^2 . \tag{5.14.27}$$

Following Godambe and Thompson (1986), one such optimum estimating function is defined as

$$h^*(d,\theta) = \sum_{i \in s} \frac{y_i - a - \theta x_i}{\pi_i} \tag{5.14.28}$$

which is also called a design weighted optimum estimating function. Similarly, for a fixed θ, the optimal estimating function or estimate $\hat{\mu}_0$ for μ_0 is given by the solution of the equation $h^* - H = 0$ or equivalently

$$\hat{\mu}_0 = \frac{1}{N_0} + \left(a + \frac{\theta}{N_0} \sum_{i=1}^{N_0} x_i \right). \tag{5.14.29}$$

It is clear that $\hat{\mu}_0$ in (15.14.29) depends upon the unknown parameter θ. Let $\tilde{\theta}$ be an estimator of θ obtained by setting $g_1^*\left(\tilde{\theta}\right) = 0$. Then an approximation to $\hat{\mu}_0$ is given by

$$\tilde{\mu}_0 = \frac{1}{N_0} h\left(d, \tilde{\theta}\right) + \left(a + \frac{\tilde{\theta}}{N_0} \sum_{i=1}^{N_0} x_i \right). \tag{5.14.30}$$

In the case of unconditional distribution, where X_Ω is allowed to vary, $\mu_0 \to \mu$. This is an analog of the condition given by Godambe (1976) for the conditional score function g^*. The approximation $\tilde{\mu}_0$ of $\hat{\mu}_0$ can easily be justified by solving the equation $h^* - H = 0$ for a given θ. Godambe (1991) has shown that for large samples the difference $h^* - H$ is almost free from θ because $\frac{\partial}{\partial \theta} E\left(h^* - H\right) = 0$. Thus replacement of θ by its estimator $\tilde{\theta}$ can be expected to provide an *approximately optimal* or *near optimal* estimate. The optimality property of h^*, given by

$$E_m E_p\left(h^* - H\right)^2 \le E_m E_p(h-H)^2 ,$$

is restricted to design unbiased estimating functions. Here it is important to point out that design unbiased estimating functions is a wider class of estimates than the design unbiased estimates. For many sampling designs which admit a 'natural' sequence of sample sizes and (survey) population sizes $\to \infty$, design unbiased estimating functions in common use provide 'consistent' estimates under appropriate conditions. Godambe and Thompson (1986) have pointed out that an obvious finite sample explication of consistent estimation is provided by the notion of a design unbiased estimating function. It is notable that the appropriate design weights can lead to optimal estimation for a less restrictive model (15.14.23) than the one we discussed above. For example, the estimating function $g_1^* = \sum_{i\in s} \epsilon_i(\theta)$, and hence the estimate $\tilde{\theta}$ depend upon the conditional variance of y_i, namely σ_i^2, $i = 1,...,N_0$. Hence these conditional variances can affect the estimate $\tilde{\mu}_0$ of the population parameter of interest μ_0. Consider a sampling design of fixed sample size with inclusion probabilities $\pi_i \propto X_i$, $i = 1,...,N_0$. In such situations, the estimating function $h^* - H$ and hence the solution $\hat{\mu}_0$ of the equation $h^* - H = 0$ are independent of θ, rendering estimation of θ irrelevant. If the above mentioned sampling design is implemented with an appropriate stratification, $h^* - H$ and hence $\hat{\mu}_0$ will also be independent of intercept a; in other words, the use of appropriate design weights can robust the optimality. Although, it is assumed in the above discussion that the entire design vector X_Ω is known, it is to be noted that the estimating function h^* is also optimal if only the sample values of x_i, $i \in s$, and the population mean $\overline{X} = N_0^{-1}\sum_{i=1}^{N_0} x_i$ of the auxiliary variable is known in addition to the design inclusion probabilities π_i, $i = 1,...,N_0$.

5.14.1 OPTIMAL ESTIMATING FUNCTION

Consider y_i, $i = 1,...,N$, are drawn from a superpopulation model with a parameter θ, possibly a vector. Assume that the model is such that y_i, $i = 1,...,N$, are independent and that for some specified real functions $\phi_i(y_i, \theta)$ of the indicated variables

$$E_\theta \phi_i(y_i, \theta) = 0, \tag{5.14.1.1}$$

where $i = 1,...,N$, and the choice ϕ_i depends upon the problem of estimation under consideration. Godambe (1995) considered the problem of estimation of the general function, defined as

$$H = \sum_{i\in\Omega} \phi_i(y_i, \theta) \tag{5.14.1.2}$$

with the estimator defined as

$$h^* = \sum_{i\in s} \frac{\phi_i}{\pi_i} \tag{5.14.1.3}$$

where $\pi_i = \sum_{s \ni i} P(s \mid X_\Omega)$ denotes the first order inclusion probability. The function h^* is optimal in the sense of having minimum variance.

Assume the function ϕ_i can be

$$\phi_i(y_i,\theta) = f_i(y_i) - \alpha_i(\theta), \tag{5.14.1.4}$$

where $f_i(y_i)$ and $\alpha_i(\theta)$ are special functions of the indicated variables. Then it follows from the optimality of h^* that for any given θ, the optimal estimate of

$$\sum_{i\in\Omega} f_i(y_i) \tag{5.14.1.5}$$

is given by

$$h = h^* + \Sigma\alpha_i(\theta) = \sum_{i\in s}\frac{f_i(y_i) - \alpha_i(\theta)}{\pi_i} + \sum_{i\in\Omega}\alpha_i(\theta). \tag{5.14.1.6}$$

It is to be noted that the replacement of θ in (5.14.1.6) by its plausible estimate can be expected to affect the optimality of the estimating functions h only a little in large samples. The reason behind this is that its optimality is in particular independent of any assumption concerning the variance functions $E_m(y_i - E_m(y_i))^2$, $i = 1,...,N$, in the superpopulation model. The optimal estimate h given in (5.14.1.6) is itself an estimating function and contains the unknown superpopulation parameter θ. Also, we have $\frac{\partial}{\partial\theta}E\left(h^* - H\right) = 0$.

The estimating function (5.14.1.6) is calibrated in so that for $f_i(y_i) = \alpha_i(\theta)$, $i = 1,...,N$, we have

$$\sum_{i\in s}\frac{f_i(y_i) - \alpha_i(\theta)}{\pi_i} + \sum_{i\in\Omega}\alpha_i(\theta) = \sum_{i\in\Omega} f_i(y_i). \tag{5.14.1.7}$$

This will be true if θ is replaced by its estimate. We list here a few assumptions about the functions f_i and α_i, which have important implications for estimation of robustness. For the superpopulation model satisfying the relations $E_\theta\phi_i(y_i,\theta) = 0$ and $\phi_i(y_i,\theta) = f_i(y_i) - \alpha_i(\theta)$, there exists a survey population based estimate θ_N, with the following properties:

(a) θ_N is consistent for θ;

(b) $\sum_{i\in\Omega} f_i(y_i) = \sum_{i\in\Omega}\alpha_i(\theta_N)$ for all $y_1,...,y_N$; (5.14.1.8)

(c) $\left|\frac{\partial\alpha_i(\theta)}{\partial\theta}\right| < C$, a fixed constant, for all θ and $i = 1,...,N$.

Using first order Taylor Series expansion of $\alpha_i(\theta)$ around the point θ_N, under the assumption 2 and for large N if $N^{-1}\sum_i \alpha_i$ is of $O(1)$, then the optimal estimating function h becomes

$$\frac{1}{N}\sum_{i\in s} d_i\{f_i(y_i)-\alpha_i(\theta)\}\cong 0\,. \tag{5.14.1.9}$$

We know that the replacement of θ by its plausible estimate can be expected to affect the optimality of the estimating function h only moderately in large samples. Let $\hat{\theta}$ be a consistent sample based estimate of the population parameter θ and the approximation given in (5.14.1.9) holds good by replacing θ with $\hat{\theta}$. Then under assumption 2, $\sum_{i\in\Omega}\alpha_i(\hat{\theta})$ is a near optimal estimate for the population parameter $\sum_{i\in\Omega} f_i(y_i)$.

5.14.2 REGRESSION TYPE ESTIMATORS

If $f_i(y_i)=y_i$, $i=1,2,...,N$, then the problem of estimation of $\sum_{i\in\Omega} f_i(y_i)$ reduces to the problem of estimation of population total, $Y=\sum_{i\in\Omega} y_i$. Assume that $\alpha_i(\theta)=\beta x_i$, $i=1,2,...,N$, such that $\sum_{i\in\Omega}\alpha_i(\theta)=\beta\sum_{i\in\Omega} x_i=\beta X$ is known.

Let $[f_i(y_i)=y_i,\ \alpha_i(\theta)=\beta x_i;\ i=1,2,...,n]$ be the n sampled observed values on the study and auxiliary variable.

Then the estimator

$$h=\sum_{i\in s}\frac{f_i(y_i)-\alpha_i(\theta)}{\pi_i}+\sum_{i\in\Omega}\alpha_i(\theta)=\sum_{i\in}\frac{y_i-\beta x_i}{\pi_i}+\beta X=\sum_{i\in s}\frac{y_i}{\pi_i}+\beta\left(X-\sum_{i\in s}\frac{x_i}{\pi_i}\right) \tag{5.14.2.1}$$

which is the usual difference estimator if β is known; otherwise it is the usual regression estimator.

If $\beta=\left(\sum_{i\in s} y_i/\pi_i\right)\Big/\left(\sum_{i\in s} x_i/\pi_i\right)$ then the estimator (5.14.2.1) reduces to the usual ratio estimator. Godambe (1995) has also discussed ratio and regression type estimators under stratified random sampling and over different occasions. For details in stratified random sampling and on sampling different occasions, one can refer to Chapter 8 and Chapter 10. Godambe's paradox and the ancillary principle has also been discussed by Bhave (1987), and its resolution is discussed by Godambe (1987). Godambe and Thompson (1999) have shown that the theory of estimating functions can also be used to construct confidence intervals for population parameters in survey sampling. Godambe (1998) also considered the problem of estimation of parameters in survey sampling.

5.14.3 SINGH'S STRATEGY IN TWO-DIMENSIONAL SPACE

Singh (2000a) consider the problem of estimation of the general parameter of interest in two dimensional space, defined as

$$\frac{1}{2}\sum_{i\in\Omega}\sum_{j\in\Omega}\Psi\left[y_i, y_j, x_i, x_j, \Theta\right] = 0 . \tag{5.14.3.1}$$

The indices i and j may or may not be equal. An unbiased estimator of (5.14.3.1) for a measurable design will be

$$\frac{1}{2}\sum_{i\in s}\sum_{j\in s} d_{ij}\Psi\left[y_i, y_j, x_i, x_j, \Theta\right] = 0 \tag{5.14.3.2}$$

with $d_{ij} = \pi_{ij}^{-1}$. Singh (2000a) developed a new estimator of variance by considering the following function to be estimated:

$$\frac{1}{2}\sum_{i\in\Omega}\sum_{j(\neq i)\in\Omega}\left[\Theta_{ij}\left(d_i y_i - d_j y_j\right)^2 - 2V\{N(N-1)\}^{-1}\right] = 0 , \tag{5.14.3.3}$$

where $d_i = \pi_i^{-1}$, $\Theta_{ij} = \pi_i\pi_j - \pi_{ij}$, $V = \frac{1}{2}\sum_{i\in\Omega}\sum_{j(\neq i)\in\Omega}\Theta_{ij}\left(d_i y_i - d_j y_j\right)^2$ and $2/N(N-1)$ denotes an adjustment factor. Using (5.14.3.2), an unbiased estimator of (5.14.3.3) is given by

$$\frac{1}{2}\sum_{i\in s}\sum_{j(\neq i)\in s} d_{ij}\left\{\Theta_{ij}\left(d_i y_i - d_j y_j\right)^2 - 2V\left(N(N-1)\right)^{-1}\right\} = 0 , \tag{5.14.3.4}$$

which leads to the following new estimator of variance of Horvitz and Thompson (1952) estimator of total in Sen--Yates--Grundy (1953) form given by

$$\hat{v}_2 = \frac{\dfrac{N(N-1)}{2}\displaystyle\sum_{i\in s}\sum_{j(\neq i)\in s}\frac{\Theta_{ij}}{\pi_{ij}}\left(\frac{y_i}{\pi_i} - \frac{y_j}{\pi_j}\right)^2}{\displaystyle\sum_{i\in s}\sum_{j(\neq i)\in s}\frac{1}{\pi_{ij}}} = \hat{V}_1\left(N(N-1)\Big/\sum_{i\in s}\sum_{j(\neq i)\in s}\frac{1}{\pi_{ij}}\right), \tag{5.14.3.5}$$

where

$\hat{v}_1 = \frac{1}{2}\sum_{i\in s}\sum_{j(\neq i)\in s} d_{ij}\Theta_{ij}\left(d_i y_i - d_j y_j\right)^2$ denotes the usual Sen--Yates--Grundy (1953) estimator of variance for the fixed sample and a measurable design.

Under Midzuno--Sen's scheme of sampling with revised selection probabilities

$$\pi_i = nP_i , \quad \text{and} \quad \pi_{ij} = \frac{n(n-1)}{(N-2)}\left\{P_i + P_j - \frac{1}{N-1}\right\}$$

the estimator in (5.14.3.5) reduces to

$$\hat{v}_2 = \frac{\dfrac{1}{2n(N-2)}\displaystyle\sum_{i\in s}\sum_{j(\neq i)\in s}\left\{n(N-2)P_iP_j\left(P_i + P_j - \frac{1}{N-1}\right)^{-1} - (n-1)\right\}\left(\frac{y_i}{P_i} - \frac{y_j}{P_j}\right)^2}{\{N(N-1)\}^{-1}\displaystyle\sum_{i\in s}\sum_{j(\neq i)\in s}\left(P_i + P_j - \frac{1}{N-1}\right)^{-1}} . \tag{5.14.3.6}$$

Under PPSWR sampling, $\pi_i = nP_i$, $\pi_{ij} = n(n-1)P_iP_j$ and therefore (5.14.3.5) reduces to

$$\hat{v}_2 = \frac{1}{2n}\sum_{i\in s}\sum_{j(\neq i)\in s}\left(y_i/P_i - y_j/P_j\right)^2 \bigg/ \left\{\{N(N-1)\}^{-1}\sum_{i\in s}\sum_{j(\neq i)\in s}\left(1/P_iP_j\right)\right\}. \quad (5.14.3.7)$$

Following Singh (2000a) one can easily get an improved estimator of variance of GREG given by

$$\hat{v}_2\left(\hat{Y}_{\text{greg}}\right) = \frac{1}{2}\sum_{i\in s}\sum_{j(\neq i)\in s}\Omega_{ij}^{*}\left(w_ie_i - w_je_j\right)^2 \bigg/ \{N(N-1)\}^{-1}\sum_{i\in s}\sum_{j(\neq i)\in s} d_{ij}\,, \quad (5.14.3.8)$$

where

$$\Omega_{ij}^{*} = d_{ij}\Theta_{ij} + \frac{d_{ij}\Theta_{ij}Q_{ij}\left(d_ix_i - d_jx_j\right)^2}{\frac{1}{2}\sum_{i\in s}\sum_{j(\neq i)\in s} d_{ij}\Theta_{ij}Q_{ij}\left(d_ix_i - d_jx_j\right)^4}\left[V\left(\hat{X}_{HT}\right) - \frac{1}{2}\sum_{i\in s}\sum_{j(\neq i)\in s} d_{ij}\Theta_{ij}\left(d_ix_i - d_jx_j\right)^2\right]$$

and

$$w_i = d_i + \left(d_iq_ix_i\right)\left\{\sum_{i=1}^{n} d_iq_ix_i^2\right\}^{-1}\left(X - \hat{X}_{HT}\right)$$

with q_i and Q_{ij} are suitably chosen constants, which are the higher and lower order calibrated weights as reported by Singh, Horn, and Yu (1998). Singh(2000a) considered the problem of estimation of the ratio of variance of the study variable to that of the auxiliary variable as

$$\frac{1}{2}\sum_{i\in\Omega}\sum_{j(\neq i)\in\Omega}\Theta_{ij}\left[\left(d_iy_i - d_jy_j\right)^2 - V_R\left(d_ix_i - d_jx_j\right)^2\right] = 0\,. \quad (5.14.3.9)$$

Obviously for a measurable design, an estimator of (5.14.3.9) is given by

$$\frac{1}{2}\sum_{i\in s}\sum_{j(\neq i)\in s} d_{ij}\Theta_{ij}\left[\left(d_iy_i - d_jy_j\right)^2 - V_R\left(d_ix_i - d_jx_j\right)^2\right] = 0\,. \quad (5.14.3.10)$$

Thus an estimator of the variance ratio is given by

$$\hat{v}_R = \frac{1}{2}\sum_{i\in s}\sum_{j(\neq i)\in s} d_{ij}\Theta_{ij}\left(d_iy_i - d_jy_j\right)^2 \bigg/ \frac{1}{2}\sum_{i\in s}\sum_{j(\neq i)\in s} d_{ij}\Theta_{ij}\left(d_ix_i - d_jx_j\right)^2. \quad (5.14.3.11)$$

He then considered the problem of estimation of the regression coefficient as

$$\frac{1}{2}\sum_{i\in\Omega}\sum_{j(\neq i)\in\Omega}\Theta_{ij}\left[\left(d_iy_i - d_jy_j\right)\left(d_ix_i - d_jx_j\right) - \beta\left(d_ix_i - d_jx_j\right)^2\right] = 0\,. \quad (5.14.3.12)$$

For a measurable design, an unbiased estimator of (5.14.3.12) is given by

$$\frac{1}{2}\sum_{i\in s}\sum_{j(\neq i)\in s} d_{ij}\Theta_{ij}\left[\left(d_iy_i - d_jy_j\right)\left(d_ix_i - d_jx_j\right) - \beta\left(d_ix_i - d_jx_j\right)^2\right] = 0\,. \quad (5.14.3.13)$$

Thus an estimator of the regression coefficient is given by

$$\hat{\beta} = \sum_{i\in s}\sum_{j(\neq i)\in s} d_{ij}\Theta_{ij}\left(d_ix_i - d_jx_j\right)\left(d_iy_i - d_jy_j\right) \bigg/ \sum_{i\in s}\sum_{j(\neq i)\in s} d_{ij}\Theta_{ij}\left(d_ix_i - d_jx_j\right)^2. \quad (5.14.3.14)$$

In Chapter 3 we considered the problem of estimation of the median M_y of the study variable in the presence of a known median M_x of the auxiliary variable as considered by Kuk and Mak (1989) by defining a matrix of proportions $\left(P_{ij}\right)$as

	$Y \le M_y$	$Y > M_y$	*Total*
$X \le M_x$	P_{11}	P_{21}	$P_{.1}$
$X > M_x$	P_{12}	P_{22}	$P_{.2}$
Total	$P_{1.}$	$P_{2.}$	1

Kuk and Mak (1989) considered three estimators $\hat{M}_R$, $\hat{M}_s$ and $\hat{M}_p$ under the names of ratio, stratified and position estimators of median of y, respectively, in the presence of a known median of variable x. It is interesting to note that the variances of all three estimators are a function of P_{11} for which $X \le M_x$ and $Y \le M_y$ as defined above. The value of P_{11} can be determined by

$$\frac{1}{2}\sum_{i\in\Omega}\sum_{j\in\Omega}\left[I\left(y_i \le M_y \cap x_j \le M_x\right) - P_{11}\right] = 0 \tag{5.14.3.15}$$

where

$$I\left(y_i \le M_y \cap x_j \le M_x\right) = \begin{cases} 1 & \text{if } y_i \le M_y \cap x_j \le M_x, \\ 0 & \text{otherwise.} \end{cases}$$

An unbiased estimator of (5.14.3.15) is given by

$$\frac{1}{2}\sum_{i\in s}\sum_{j\in s} d_{ij}\left[I\left(y_i \le M_y \cap x_j \le M_x\right) - P_{11}\right] = 0 \tag{5.14.3.16}$$

or equivalently

$$\hat{P}_{11} = \frac{1}{2}\sum_{i\in s}\sum_{j\in s} d_{ij}\left[I\left(y_i \le M_y \cap x_j \le M_x\right)\right] \Big/ \frac{1}{2}\sum_{i\in s}\sum_{j\in s} d_{ij} \,. \tag{5.14.3.17}$$

Singh (2000a) suggested that this procedure would be useful for estimating the variances of the estimators of median proposed by Kuk and Mak (1989). Finally he considered the problem of poststratification in the two dimensional plane. Let y and x be the two variables used for post stratifying the given data set. Let us define

$$y_i = \begin{cases} 1 & \text{if } a \le y \le b, \\ 0 & \text{otherwise,} \end{cases} \quad \text{and} \quad x_i = \begin{cases} 1 & \text{if } c \le y \le d, \\ 0 & \text{otherwise,} \end{cases}$$

where a, b, c and d are predefined limits used for defining the stratum boundaries. For example, the strata defined as

$$\frac{1}{2}\sum_{i\in\Omega}\sum_{j\in\Omega}\left[I\left(a \le y_i \le b, c \le x_j \le d\right) - W\right] = 0, \tag{5.14.3.18}$$

where

$$I\left(a \le y_i \le b,\ c \le x_j \le d\right) = \begin{cases} 1 & \text{if } a \le y_i \le b,\ c \le x_j \le d, \\ 0 & \text{otherwise,} \end{cases}$$

contains the number in the sample with $a \le y_i \le b$ and $c \le x_j \le d$. In case of post-stratification, the sample size n is a random variable; therefore an estimator of (5.14.3.18) is given by

$$\frac{1}{2}\sum_{i\in s}\sum_{j\in s} d_{ij}\left[I\left(a \le y_i \le b, c \le x_j \le d\right) - W\right] = 0, \tag{5.14.3.19}$$

or equivalently

$$\hat{W} = \frac{1}{2}\sum_{i\in s}\sum_{j\in s} d_{ij}\left\{I\left(a \le y_i \le b, c \le x_j \le d\right)\right\} \Big/ \frac{1}{2}\sum_{i\in s}\sum_{j\in s} d_{ij}\,. \tag{5.14.3.20}$$

5.14.4 GODAMBE'S STRATEGY FOR LINEAR BAYES AND OPTIMAL ESTIMATION

Godambe (1999) has discussed very thoroughly the optimal estimating function in linear Bayes form. He followed the linear Bayes methodology introduced by Hartigan (1969) to handle the Bayesian semi-parametric models based on fewer moments. As in the case of non-Bayesian statistics, it is common practice to replace a full distributional assumption by a much weaker assumption about its first few moments such as mean and variance. In the Bayesian approach one may similarly consider the replacement of a completely specified prior distribution by an assumption about just a few moments of the distribution. Interestingly, Godambe (1999) proposed an alternative methodology based on the theory of optimum estimating functions and showed that his strategy is more readily applicable and efficient in common problems than the linear Bayes methodology. Godambe (1999) generalized results of Godambe and Thompson (1989) for extending the theory of optimum estimating functions to semi-parametric Bayesian models. Following Godambe (1999), let $X = \{x\}$ be an abstract sample space, $P = \{p\}$ be a class of probability distribution of X, $\theta = (\theta_1, \theta_2, ..., \theta_m)$ is a real valued m dimensional parameter of interest defined on P, and $\Omega = \{\theta(p) : p \in \mathrm{P}\}$. Here we wish to estimate θ based on sample information $X = \{x\}$ through elementary estimating functions. The elementary estimating function h_j is a real valued function defined on $X \times \Omega$ such that, under the distribution $p \in P$, X,

$$E\left(h_j \mid \mathrm{X}_j\right) = 0, \quad j = 1,2,...,k \tag{5.14.4.1}$$

where X_j is some *partition* of the sample space or a σ field generated by a partition. Such an elementary estimating function h_j is called *unbiased* conditionally on X_j. Let us now consider

$$g = (g_1, ..., g_m), \tag{5.14.4.2}$$

to be an estimating function, such that $g_r = \sum_{j=1}^{k} h_j a_{jr}$, where a_{jr} are any functions of (x, θ) which are measurable with respect to X_j, $j = 1,2,...,k$, $r = 1,2,...,m$. Suppose a class G of all such estimating functions g is given by

$$G = \{g\}. \tag{5.14.4.3}$$

Now if we introduce another estimating function

$$g^* = \left(g_1^*, ..., g_m^*\right) \tag{5.14.4.4}$$

where $g_r^* = \sum_{j=1}^{k} h_j a_{jr}^*$ with $a_{jr}^* = E\left\{\frac{\partial h_j}{\partial \theta_r} \mid X_j\right\} \Big/ E\left\{h_j^2 \mid X_j\right\}$ exists for $r = 1,2,...,m$. One may note here that the estimating function g^* is also a member of the class G.

Denoting $h_j a_{jr}^*$ by h_{jr}, the elementary estimating functions $h_j,\ j = 1,2,...,k$ are said to be *mutually orthogonal* if

$$E\left(h_{jr} h_{j'r'} \mid X_j\right) = 0 \tag{5.14.4.5}$$

for $j \neq j';\ j,\ j' = 1, 2, ..., k;\ r, r' = 1, 2, ..., m.$ Corresponding to the elementary estimating function g, define the two matrices as:

$$J = E\begin{bmatrix} g_1^2, g_1g_2, ..., g_1g_m \\ g_2g_1, g_2^2, ..., g_2g_m \\ .... \\ .... \\ g_mg_1, g_mg_2, ..., g_m^2 \end{bmatrix} = \left\| E\left(g_r g_{r'}\right) \right\| \quad \text{and} \quad H = E\begin{bmatrix} \frac{\partial g_1}{\partial \theta_1}, \frac{\partial g_1}{\partial \theta_2}, ..., \frac{\partial g_1}{\partial \theta_m} \\ \frac{\partial g_2}{\partial \theta_1}, \frac{\partial g_2}{\partial \theta_2}, ..., \frac{\partial g_2}{\partial \theta_m} \\ .... \\ \frac{\partial g_m}{\partial \theta_1}, \frac{\partial g_m}{\partial \theta_2}, ..., \frac{\partial g_m}{\partial \theta_m} \end{bmatrix} = \left\| E\left(\frac{\partial g_r}{\partial \theta_{r'}}\right) \right\|.$$

Similarly for the elementary estimating function g^* define the two matrices J^* and H^* as follows

$$J^* = E\begin{bmatrix} g_1^{*2}, g_1^* g_2^*, ..., g_1^* g_m^* \\ g_2^* g_1^*, g_2^{*2}, ..., g_2^* g_m^* \\ .... \\ .... \\ g_m^* g_1^*, g_m^* g_2^*, ..., g_m^{*2} \end{bmatrix} = \left\| E\left(g_r^* g_{r'}^*\right) \right\| \quad \text{and} \quad H^* = E\begin{bmatrix} \frac{\partial g_1^*}{\partial \theta_1}, \frac{\partial g_1^*}{\partial \theta_2}, ..., \frac{\partial g_1^*}{\partial \theta_m} \\ \frac{\partial g_2^*}{\partial \theta_1}, \frac{\partial g_2^*}{\partial \theta_2}, ..., \frac{\partial g_2^*}{\partial \theta_m} \\ .... \\ \frac{\partial g_m^*}{\partial \theta_1}, \frac{\partial g_m^*}{\partial \theta_2}, ..., \frac{\partial g_m^*}{\partial \theta_m} \end{bmatrix} = \left\| E\left(\frac{\partial g_r^*}{\partial \theta_r}\right) \right\|.$$

Thus we have the following theorem.

Theorem 5.14.4.1. If the elementary estimating function $h_j,\ j = 1, 2, ..., k$, are 'mutually orthogonal' in the class G, then the estimating function g^* is optimal in the sense that the matrix

$$D = J - H\left(H^*\right)^{-} J^* \left\{\left(H^*\right)^t\right\}^{-} H^t \tag{5.14.4.6}$$

is positive semi-definite for all $g \in G$, where A^t and A^- denote the transpose and generalized inverse of the corresponding matrix A. An estimate of θ is obtained by solving the equation $g^* = 0$ for the observed value of x. The details about this theorem can be found in Godambe and Thompson (1989).

In order to generalize the above theorem Godambe (1999) first considered X_j to be the partition of $X \times \Omega$ with the parameter θ having some distribution. Let p denote a joint distribution of (x, θ) and $P = \{p\}$. Thus in the above discussion, all the 'expected values' are to be replaced by the 'expected values with respect to the joint distribution of p of (x, θ)'. In other words, $E(\bullet | \theta)$ reduces to the expected value in the above theorem, when the partition of $X \times \Omega$ is given by holding θ fixed. Thus the elementary estimating functions h_j, $j = 1,2,...,k$, would be functions of the parameter θ. Following Godambe (1999), the estimating functions g including g^* in the class G, under the new setup are obtained from the functions h_j with coefficients a_{jr} that are measurable functions with respect to the new partition of $X \times \Omega$ corresponding to h_j, $j = 1,...,k$; $r = 1,...,m$. Assuming the orthogonality of the estimating functions h_j, $j = 1,2,...,k$, defining the class $G = \{g\}$ and the matrices J_1 and H_1 similar to J and H, respectively, we have the following theorem due to Godambe (1999).

Theorem 5.14.4.2. Assume the interpretation of estimating function g^*, the class G, the orthogonality and the matrices J_1 and H_1. Now if the elementary estimating functions h_j, $j = 1,...,k$, are mutually orthogonal, then in the class G, the estimating function g^* is optimal in the sense that the matrix

$$D^* = J_1 - H_1 \left(H_1^*\right)^- J_1^* \left\{\left(H_1^*\right)'\right\}^- H_1' \tag{5.14.4.7}$$

is positive semi-definite for all $g \in G$.

Proof. Because of the orthogonality of the estimating functions h_j, $j = 1,2,...,k$, we have

$$\left\| E\left(\frac{\partial g_r}{\partial \theta_{r'}}\right) \right\| = \left\| E\left(g_r g_{r'}^*\right) \right\|, \quad r, \ r' = 1,...,m. \tag{5.14.4.8}$$

Following Godambe and Thompson (1987), the above equality shows the positive semi-definiteness of the matrix D^*. Hence the theorem.

Again we may note here that the estimate of θ is obtained by solving the equation $g^* = 0$ for given x. In case of the scalar parameter $\theta = \theta_1$ the optimality of the corresponding estimating function g^* in the above theorem is equivalent to the inequality $E\left\{g^* \Big/ E\left(\frac{\partial g^*}{\partial \theta_1}\right)\right\}^2 \le E\left\{g \Big/ E\left(\frac{\partial g}{\partial \theta_1}\right)\right\}^2$. For more details on this topic one can refer to Basawa, Godambe, and Taylor (1997).

5.15 UNIFIED THEORY OF SURVEY SAMPLING

In this section we consider the problem of the estimating a population total or mean through a unified approach of survey sampling. The concepts of uniform admissibility, hyper-admissibility, 'uniformly minimum variance unbiased estimator' and the concept of sufficiency in survey sampling are discussed:

5.15.1 CLASS OF ADMISSIBLE ESTIMATORS

We first discuss possible definitions of best and admissible estimators of population total $Y = \sum_{i \in \Omega} Y_i$ for a given sampling design P. Let s be a sample of units drawn according to a probability scheme P : $\{P(s)\}$ from the totality of possible samples S, where $p(s) \geq 0$ for all $s \in S$ and $\sum_{s \in S} p(s) = 1$. Let $y = (y_1, y_2, ..., y_N)$ be a vector of variate values associated with different units in a population of size N, be an element of the Euclidean space R_N.

5.15.2 ESTIMATOR

An estimator of a population parameter is a function defined on $S \times R_N$, whose value for any $s \in S$, depends upon y, through only those $y_i \in s$. Murthy and Singh (1969) discussed the following definitions on admissibility of estimators in a class $C(P)$ of estimators of population total. Suppose $e_1(s, y)$ and $e_2(s, y)$ be any two estimators of the population total Y based on sample s under design P.

Then we have the following definitions:

5.15.3 ADMISSIBLE ESTIMATOR

In a class $C(P)$ of estimators of population total, Y, an estimator $e_1 \in C(P)$ is admissible if for every other estimator $e \in C(P)$ either $\text{MSE}(e_1) = \text{MSE}(e)$ for all y or $\text{MSE}(e) < \text{MSE}(e_1)$ for at least one y.

5.15.4 STRICTLY ADMISSIBLE ESTIMATOR

An estimator $e_1 \in C(P)$ is strictly admissible if for every other estimator $e \in C(P)$ such that $\text{MSE}(e) < \text{MSE}(e_1)$ for at least one y.

Now we have the following theorem:

Theorem 5.15.1. If an estimator is strictly admissible, it is necessarily admissible, but the reverse is not true.

Proof. Consider a class $C(P)$ containing four estimators e_1, e_2, e_3 and e_4. Let $\text{MSE}(e_1) = \text{MSE}(e_2)$ for all y and $\text{MSE}(e_1) \underset{<}{\overset{>}{-}} \text{MSE}(e_i)$, $i = 3,4$ for all y with inequalities holding for some of the y values. In this case both e_1 and e_2 are admissible, but none of them are strictly admissible. Now if e_1 is excluded then e_2 becomes strictly admissible. This proves the theorem.

Godambe (1955) had proven the non-existence of the uniformly best estimator in a class of homogeneous linear unbiased estimators of population total. Following Godambe (1955) a most general type of linear estimator of population total Y may be defined as $e_s = \sum_{i \in s} \beta_{si} y_i$. It may be worth while to mention that $\sum_{i \in s} y_i$ stands for the summation over all different individuals in the sample s (these may be, in number, $\le n$). The estimator e_s will be unbiased if and only if $\sum_{s \ni i} \beta_{si} p_s = 1$ for $i = 1, 2, ..., N$ and $\sum_{s \ni i}$ denotes the sum over all samples which include the ith unit.

Let us first explain the meaning of this sum with the help of a numerical example.

Example 5.15.1. Consider a population of $N = 5$ units as 10, 20, 30, 40 and 50. Select all possible samples of size $n = 3$ units by using SRSWOR sampling. Find the following:

(i) $\sum_{s \ni 2} y_{s2}$; (ii) find β_{s2} such that $\sum_{s \ni 2} \beta_{s2} p_s = 1$.

Solution. Let $A = 10$, $B = 20$, $C = 30$, $D = 40$ and $E = 50$ be the units in the population. Here $N = 5$ and $n = 3$. Therefore the total number of possible without replacement samples will be given by $\Omega(s) = {}^N C_n = {}^5 C_3 = 10$ and the sample space S is given as below:

Table 5.15.1 Sample space.

Sample Number	Units included in the samples	Values of the units	Totals
1	A, B, C	10, 20, 30	60
2	A, B, D	10, 20, 40	70
3	A, B, E	10, 20, 50	80
4	A, C, D	10, 30, 40	80
5	A, C, E	10, 30, 50	90
6	A, D, E	10, 40, 50	100
7	B, C, D	20, 30, 40	90
8	B, C, E	20, 30, 50	100
9	B, D, E	20, 40, 50	110
10	C, D, E	30, 40, 50	120

(i) In sum $\sum_{s \ni 2} y_{s2}$, the value y_{s_2} corresponds to the sum over all possible samples having the second unit $y_{s_2} = B$ in the samples. Clearly the unit B is included in 6 samples with numbers 1,2,3,7, 8 and 9 in the above table.
Thus we have

$$\sum_{s \ni 2} y_{s2} = 60 + 70 + 80 + 90 + 100 + 110 = 510 .$$

(ii) To find β_{s2} such that $\sum_{s \ni 2} \beta_{s2} p_s = 1$ and under SRSWOR,

$$p_s = \frac{1}{{}^N C_n} = \frac{1}{{}^5 C_3} = \frac{1}{10} \quad \text{for all } s \in S .$$

Thus

$$\sum_{s \ni 2} \beta_{s2} p_s = 1 \text{ implies } \sum_{s \ni 2} \beta_{s2} \frac{1}{10} = 1 \text{ or } \sum_{s \ni 2} \beta_{s2} = 10 .$$

Keep in mind that the second unit belongs to six samples as shown above in Table 5.15.1. In other words we are looking for six constants such that their sum is 10. The following table shows that such a choice of constants is not unique unless we give equal weight to all samples in the sample space.

Table 5.15.2. Choice of weights.

Sample no.	1	2	3	7	8	9	Sum
β_{s2}	5/3	5/3	5/3	5/3	5/3	5/3	10
β_{s2}	2	2	3	2	0.5	0.5	10
β_{s2}	1.176471	1.372549	1.568627	1.764706	1.960784	2.156863	10

Thus all these choices of β_{s2} provide unbiased estimators of population total or mean under SRSWOR sampling. We shall see in the next theorem that the second row of the Table 5.15.2 provides the correct choice for unbiasedness, that is $\beta_{s2} = N/n = 5/3$.

Now let B_p denote the class of all β for which the corresponding estimates are unbiased. Then one would usually define an estimator e_{sb} as the best linear estimate of population total such that the variance is minimum, which is not possible. We have the following theorem:

Theorem 5.15.2. Prove that there does not exist any $\beta_0 \in B_p$ such that variance of the estimator $e(s, \beta_0)$ is always less than or equal to all other estimators $e(s, \beta)$.
Proof. If e_s is unbiased then

$$V(e_s) = \sum_{s \in S} e_s^2 p_s - Y^2 .$$

The Lagrange function is given by

$$L = V(e_s) - \lambda\left(\sum_{s \ni i} \beta_{si} p_s - 1\right).$$

Suppose β_0 exists then for a given y and $i \in s$ we have

$$\left\{\frac{\partial V(e_s)}{\partial \beta_{si}} - \lambda_i \frac{\partial}{\partial \beta_{si}} \sum_{s \ni i} \beta_{si} p_s\right\}\beta_0 = 0,$$

where λ_i is a Lagrange multiplier. Upon differentiating, we obtain $2y_i\{e_s\}\beta_0 = \lambda_i$ for all $s \ni i$ and where $p_s \neq 0$. Moreover, this must hold for all y since β_0 is supposed to give the minimum for all values of y. That is for any s_1 and s_2 containing i^{th} unit, p_{s_1} and p_{s_2} being positive, we have $\{e_{s1}\}\beta_0 = \{e_{s2}\}\beta_0$ for all values of y. For example if we take $y_i = 1$ and $y_j = 0$ for all $i \neq j$ then we have $\beta_{si} = \beta_0$ which implies that, $\beta_0 = \beta_{s0} = \frac{1}{\sum_{s \ni i} p_s} = \frac{1}{\pi_i}$, for all s and i, $\pi_i = \sum_{s \ni i} p_s$ denoting the first order inclusion probability. Hence if β_0 exists it must satisfy the above condition. (Under SRSWOR sampling, $\beta_0 = N/n$.)

Conversely, we have to choose β_0 such that the variance of the estimator e_s is minimum. The choice of $\beta_0 = \beta_{s0} = 1\big/\sum_{s \ni i} p_s = d_i$ holds but does not provide minimum variance. Hence an optimum β_0 does not exist in the class of linear unbiased estimators.

Thus Godambe (1955) has proven the non-existence of the uniformly best estimator in the class of homogeneous linear unbiased estimators of population total. The linearity restriction has been removed by Godambe and Joshi (1965) and has proven the non-existence of the best estimator in the entire unbiased class of estimators. Murthy and Singh (1969) extends the non-existence of such a result as wide classes of estimators of the population total and we have the following theorems:

Theorem 5.15.3. For a given design P if an estimator e_1 is admissible in the class $C(P)$ and if there is another estimator e_2 in the class $C(P)$ such that e_1 is not as good as e_2, then e_1 is not a best estimator in that class. This implies non-existence of a best estimator in the wider class of estimators $C(P)$.

Proof. Note that the estimator e_1 is admissible, for any other estimator $e \in C(P)$ either $\text{MSE}(e_1) = \text{MSE}(e)$ for all y or $\text{MSE}(e_1) < \text{MSE}(e)$ for at least one y holds. Further the estimator e_1 is not at least as good as e_2 therefore $\text{MSE}(e_1) > \text{MSE}(e_2)$ for at least one y. Hence the theorem by contradiction.

Theorem 5.15.4. If there exist two (or more) admissible estimators in a class $C(P)$, with unequal MSEs for at least one y for a given design P, then that class does not contain a best estimator.

Proof. Let e_1 and e_2 be two admissible estimators in $C(P)$. Then the estimator e_1 is not at least as good as e_2 and e_2 is not at least as good as e_1 if either $\text{MSE}(e_1) = \text{MSE}(e_2)$ for all y or $\text{MSE}(e_1) \gtrless \text{MSE}(e_2)$ for all y with inequalities for at least one y. Clearly the relation $\text{MSE}(e_1) = \text{MSE}(e_2)$ does not hold if the relation $\text{MSE}(e_1) \gtrless \text{MSE}(e_2)$ holds. That is, neither e_1 nor e_2 is a best estimator. Hence the theorem.

5.15.5 LINEAR ESTIMATORS OF POPULATION TOTAL

Let $y = (y_1, y_2, ..., y_N)$ be a vector of variate values associated with different units in a population of size N and be an element of the Euclidean space R_N. Then a linear estimator e_b is a function on $S \times R_N$ such that

$$e_b(s, y) = \sum_{i \in \Omega} b(s,i) y_i \quad \text{where } b(s,i) = \begin{cases} 1 & \text{if } i \in s, \\ 0 & \text{if } i \notin s. \end{cases}$$

It is noted here that the estimator is unbiased for population total Y if

$$\sum_{s \in S} e_b(s, y) p(s) = Y$$

for all $y \in R_N$.

Theorem 5.15.5. For any given design P, any constant is strictly admissible in the class L of linear estimators of the population total, except in trivial cases.

Proof. For a design P consider an estimator e_0 as $e_0 = \delta$, for all y, where δ is a constant not equal to zero. Let $e_1 \in L$ be any other estimator such that

$$e_1 = a_s + \sum_{i \in s} b_{si} y_i .$$

Suppose $y_{(\delta)}$ is the set of vectors in R_N for which the population total $Y = \delta$ for all $y \in y_{(\delta)}$. Clearly the mean squared error of the constant estimator e_0 is given by

$$\text{MSE}(e_0) = \begin{cases} [\delta - Y]^2 & \text{if } y \notin y_{(\delta)}, \\ 0 & \text{if } y \in y_{(\delta)}. \end{cases}$$

Keep in mind that $\text{MSE}(e_1) \geq 0$ for all $y \in \text{R}_N$. In other words, the estimator e_0 is strictly admissible unless $\text{MSE}(e_1) = 0$ for all $y \in y_{(\delta)}$, which shows that both estimators e_0 and e_1 are identical for all $y \in y_{(\delta)}$. One can easily see that, in fact, e_0 and e_1 are identical for all $y \in R_N$. Hence the theorem.

Now we have the following theorem to state a very useful result:

Theorem 5.15.6. There does not exist a best estimator, and hence the best and the uniformly best estimator, in the class of linear (L), linear unbiased (L_u), homogeneous linear (L^h), homogeneous linear unbiased (L_u^h), all (both linear and non-linear) unbiased (A_u) and all (A) estimators of the population total for any sampling design P.

Proof. In the previous theorem, we have seen that any constant belongs to the class of linear estimators (L), that it is strictly admissible in L and hence there does not exist any best estimator in this class. Its proof also follows from the fact that if there are at least two strictly admissible estimators in $C(P)$ then a best estimator does not exist in that class of estimators.

Godambe and Joshi (1965) have shown that the Horvitz and Thompson (1952) estimator of the population total defined as

$$e_{ht} = \sum_{i \in s} d_i y_i,$$

where $\pi_i = \sum_{s \ni i} p_s$ is admissible in the class of all unbiased estimators A_u and hence in the class of homogeneous linear and unbiased estimators (L_u^h). Following Murthy (1967) and Rao and Bayless (1969), if the design is not unicluster, then it is possible to find another estimator e_{ht}^* in the class of homogeneous, linear and unbiased estimators (L_u^h) such that e_{ht} is not as good as e_{ht}^* for a given design P. We know that for a given design P, if an estimator e_{ht} is admissible in the class $C(P)$ and if there is another estimator e_{ht}^* in the class $C(P)$ such that e_{ht} is not as good as e_{ht}^*, then e_{ht} is not a best estimator in that class. This implies non-existence of a best estimator in the wider class of estimators $C(P)$, which proves the non-existence of a best estimator in (L_u^h). The results are obvious for linear unbiased L_u and all other unbiased A_u class of estimators. Now consider a class A consisting of all estimators of the population total. Then obviously two admissible estimators with unequal mean square errors for at least some y are the usual and ratio estimators, given by

$$e_u = \frac{N}{n}\sum_{i \in s} y_i, \quad \text{and} \quad e_r = \frac{\sum_{i \in s} y_i}{\sum_{i \in s} x_i} X$$

where $X = \sum_{i \in \Omega} x_i$. We know that if there exist two (or more) admissible estimators in a class $C(P)$, with unequal MSEs for at least one y, for a given design P, then that class does not contain a best estimator. This shows the non-existence of a best estimator in the class of all estimators (A). Hence the theorem.

5.15.6 ADMISSIBLE ESTIMATORS OF VARIANCES OF ESTIMATORS OF TOTAL

Let $\Omega = \{u_1, u_2, ..., u_N\}$ denote a finite population. With each unit u_i, $i = 1,2,...,N$, is an associated variate value y_i. Let $y = (y_1, y_2, ..., y_N)$ denote a point in the N dimensional Euclidean space R_N. Let s denote any non-empty subset of Ω and S denotes the set of all possible samples s. Then a sampling design is defined by attaching a selection probability $p(s)$ to each $s \in S$ so that $p(s) \geq 0$, $\sum_{s\in S} p(s) = 1$. The size $n(s)$ of a sample s denotes the number of units u_i included in s. Let $\pi_i = \sum_{s \ni i} p(s)$ and $\pi_{ij} = \sum_{s \ni i,j} p(s)$ denote the positive first and second order inclusion probabilities.

5.15.6.1 CONDITION FOR THE UNBIASED ESTIMATOR OF VARIANCE

An estimator $e(s, y)$ is said to be an unbiased estimator of a function $V(y)$ defined on R_N, if and only if $\sum_{s\in S} p(s)e(s, y) = V(y)$ for all $y \in R_N$.

5.15.6.2 ADMISSIBLE AND UNBIASED ESTIMATOR OF VARIANCE

An estimator $e(s, y)$ is said to be admissible in the class of all unbiased estimators of a function $V(y)$ if there exists no other unbiased estimator $e_1(s, y)$ of $V(y)$ such that

$$\sum_{s\in S} p(s)[e_1(s, y) - V(y)]^2 \leq \sum_{s\in S} p(s)[e(s, y) - V(y)]^2,$$

for all $y \in R_N$, with the strict inequality holding for at least one $y \in R_N$.

5.15.6.3 FIXED SIZE SAMPLING DESIGN

A sampling design is said to be of fixed size v if $p(s) = 0$ whenever $n(s) \neq v$.

5.15.6.4 HORVITZ AND THOMPSON ESTIMATOR AND ITS VARIANCE IN TWO FORMS

The Horvitz and Thompson (1952) estimator of the population total, Y, defined as

$$e_{ht}(s, y) = \sum_{i\in s} d_i y_i$$

which is unbiased if and only if the first order inclusion probabilities are known and positive. Then the first form of the variance, which is valid under any kind of sampling design, of the Horvitz and Thompson estimator is given by

$$V(y) = \sum_{i\in\Omega} (d_i - 1) y_i^2 + \sum_{i\in\Omega} \sum_{j(\neq i)\in\Omega} (d_i d_j \pi_{ij} - 1) y_i y_j$$

which we can refer to as a class of unbiased estimators of variance.

On the other hand, for a fixed sample size design the Sen--Yates--Grundy form of variance is give by

$$V_{\text{syg}}(y)=\frac{1}{2}\sum_{i\in\Omega}\sum_{j(\neq i)\in\Omega}\left(\pi_i\pi_j-\pi_{ij}\right)\left(d_iy_i-d_jy_j\right)^2 .$$

Assuming that the second order inclusion probabilities are known and positive, an unbiased estimator of $V(y)=V_{\text{syg}}(y)$ based on a fixed sample design is given by

$$\hat{v}_{\text{syg}}\{e_{ht}(s,y)\}=\frac{1}{2}\sum_{i\in s}\sum_{j(\neq i)\in s}d_{ij}\left(\pi_i\pi_j-\pi_{ij}\right)\left(d_iy_i-d_jy_j\right)^2 .$$

Then following Joshi (1970) we have the following theorem.

Theorem 5.15.6.1. For a fixed sampling design of size two the Sen--Yates--Grundy estimator of variance is admissible in the entire unbiased class of estimators of variance.

Proof. Suppose the theorem is not true. In this situation, there exist an unbiased and admissible estimator $\hat{v}_u\{e_{\text{ht}}(s,y)\}$ such that

$$\sum_{s\in S}p(s)\left[\hat{v}_u\{e_{\text{ht}}(s,y)\}-V(y)\right]^2\leq\sum_{s\in S}p(s)\left[\hat{v}_{\text{syg}}\{e_{\text{ht}}(s,y)\}-V(y)\right]^2$$

for all $y\in R_N$ and for strict inequality with at least one $y\in R_N$. Suppose the new unbiased estimator has a relationship with the Sen--Yates--Grundy estimator of variance shown by

$$\hat{v}_u\{e_{\text{ht}}(s,y)\}=\hat{v}_{\text{syg}}\{e_{\text{ht}}(s,y)\}+H(s,y)$$

where $H(s,y)$ is any function of sample values.

Then we have

$$\sum_{s\in S}p(s)\left[\hat{v}_{\text{syg}}\{e_{\text{ht}}(s,\ y)\}-V(y)+H(s,y)\right]^2\leq\sum_{s\in S}p(s)\left[\hat{v}_{\text{syg}}\{e_{\text{ht}}(s,\ y)\}-V(y)\right]^2$$

or

$$\sum_{s\in S}p(s)\left[H(s,\ y)\right]^2\leq-2\sum_{s\in S}p(s)H(s,y)\left[\hat{v}_{\text{syg}}\{e_{\text{ht}}(s,\ y)\}-V(y)\right].$$

The estimators $\hat{v}_u\{e_{\text{ht}}(s,y)\}$ and $\hat{v}_{\text{syg}}\{e_{\text{ht}}(s,y)\}$ are unbiased.

Therefore

$$\sum_{s\in S}p(s)H(s,y)=0$$

and the above inequality becomes,

$$\sum_{s\in S}p(s)\left[H(s,y)\right]^2\leq-2\sum_{s\in S}p(s)\left[H(s,y)\right]\left[\hat{v}_{\text{syg}}\{e_{\text{ht}}(s,y)\}\right].$$

Considering $y_i\propto\pi_i$, then one can easily see that the above inequality does not hold and is a contradiction to our assumption. Hence the theorem.

We already discussed Murthy (1957, 1963) estimators in detail based on a sample of two units. Joshi (1970) used the above mechanism to prove the admissibility of Murthy's estimators. Patel and Dharmadhikari (1978) have shown that Joshi's method is not applicable. They discussed the admissibility (within the class of linear unbiased estimators) of Murthy's estimators with an effective sample of two units.

Now a linear estimator of population total defined as

$$e_l(s) = \sum_{i \in s} b(s,i) y_i$$

is unbiased in linear manner if $\sum_{s \ni i} b(s,i)p(s) = 1$, $i = 1,2,..., N$.

Suppose V_i denotes the variance of an estimator e_l when $Y_i = 1$ and $Y_j = 0$ for $j \neq i$. Assuming the estimator e_l is unbiased we have

$$V_i = \sum_{s \ni i} b^2(s,i)p(s) - 1 .$$

Let $k_1, k_2,..., k_N$ be non-zero constants such that $\sum_{i \in \Omega} k_i = 1$ and let $q_1, q_2,..., q_N$ be positive numbers such that the weighted variance $\sum_{i \in \Omega} q_i V_i$ is minimum subject to conditions of unbiasedness and zero variance at a given point $(k_1, k_2,..., k_N)$ defined as

$$\sum_{i \in s} b(s,i) k_i = 1, \quad s \in S .$$

Consider a Lagrange function defined as

$$L = \sum_{i \in \Omega} q_i V_i - 2 \sum_{i \in \Omega} \lambda_i \sum_{s \ni i} b(s,i)p(s) - 2 \sum_{s \in S} \mu_s \sum_{i \in s} b(s,i) k_i$$

$$= \sum_{i \in \Omega} q_i \left\{ \sum_{s \ni i} b^2(s,i)p(s) - 1 \right\} - 2 \sum_{i \in \Omega} \lambda_i \sum_{s \ni i} b(s,i)p(s) - 2 \sum_{s \in S} \mu_s \sum_{i \in s} b(s,i) k_i$$

where λ_i and μ_s are the Lagrange multipliers. Upon differentiating L with respect to $b(s, i)$ and equating to zero we have

$$\frac{\partial L}{\partial b(s,i)} = 2 q_i b(s,i) p(s) - 2 \lambda_i p(s) - 2 \mu_s k_i = 0 ,$$

which implies that

$$q_i b(s,i) p(s) = \lambda_i p(s) + \mu_s k_i , \quad i = 1,2,..., N,$$

or

$$b(s,i) = \frac{\lambda_i}{q_i} + \frac{\mu_s}{p(s)} \frac{k_i}{q_i} = \xi_i + a_s \eta_i$$

where $\xi_i = \lambda_i / q_i$, $a_s = \mu_s / p(s)$ and $\eta_i = k_i / q_i$..

We know that

$$\sum_{i \in s} b(s,i) k_i = 1 ,$$

which implies that

$$\sum_{i \in s} (\xi_i + a_s \eta_i) k_i = 1$$

or

$$\sum_{i\in s}\xi_i k_i + a_s \sum_{i\in s}\eta_i k_i = 1,$$

which implies that

$$a_s = \frac{1}{\delta(s)}\left(1-\sum_{i\in s}\xi_i k_i\right),$$

where $\delta(s)=\sum_{i\in s}\eta_i k_i$.

Also we have

$$\sum_{s\ni i} b(s,i)p(s) = 1,$$

which implies that

$$\sum_{s\ni i}\left(\xi_i + a_s\eta_i\right)p(s) = 1$$

or

$$\sum_{s\ni i}\xi_i p(s) + \sum_{s\ni i}\eta_i a_s p(s) = 1$$

or

$$\xi_i \sum_{s\ni i} p(s) + \eta_i \sum_{s\ni i}\frac{p(s)}{\delta(s)}\left(1-\sum_{i\in s}\xi_i k_i\right) = 1$$

or

$$\xi_i\pi_i - \eta_i \sum_{s\ni i}\frac{p(s)}{\delta(s)}\sum_{i\in s}\xi_i k_i = 1-\eta_i\sum_{s\ni i}\frac{p(s)}{\delta(s)} \quad \text{for } i=1,2,...,N.$$

Remember that k_i and q_i are known, and hence the η_i and $\delta(s)$ are known. For a given design, $p(s)$ and π_i are also known. Thus the unknown quantities are only values of ξ_i which is, in fact, a function of Lagrange multipliers. The above system of equations can be written as

$$\begin{bmatrix} \left(\pi_1-\eta_1 k_1\sum\limits_{s\ni 1}\frac{p(s)}{\delta(s)}\right), -\eta_2 k_1 \sum\limits_{s\ni(2,1)}\frac{p(s)}{\delta(s)}, \ldots\ldots, -\eta_N k_1\sum\limits_{s\ni(N,1)}\frac{p(s)}{\delta(s)} \\ -\eta_1 k_2\sum\limits_{s\ni(1,2)}\frac{p(s)}{\delta(s)}, \left(\pi_2-\eta_2 k_2\sum\limits_{s\ni 2}\frac{p(s)}{\delta(s)}\right) \ldots, -\eta_N k_2\sum\limits_{s\ni(N,2)}\frac{p(s)}{\delta(s)} \\ ------------------------ \\ ------------------------ \\ -\eta_1 k_N\sum\limits_{s\ni(1,N)}\frac{p(s)}{\delta(s)}, -\eta_{21}k_N\sum\limits_{s\ni(2,N)}\frac{p(s)}{\delta(s)}, \ldots\ldots \left(\pi_N-\eta_N k_N\sum\limits_{s\ni N}\frac{p(s)}{\delta(s)}\right) \end{bmatrix} \begin{bmatrix}\xi_1\\ \xi_2\\ \\ \\ \xi_N\end{bmatrix} = \begin{bmatrix} 1-\eta_1\sum\limits_{s\ni 1}\frac{p(s)}{\delta(s)} \\ 1-\eta_2\sum\limits_{s\ni 1}\frac{p(s)}{\delta(s)} \\ \\ \\ 1-\eta_{N1}\sum\limits_{s\ni 1}\frac{p(s)}{\delta(s)} \end{bmatrix}$$

Clearly a solution $(\xi_1,\xi_2,...,\xi_N)$ to the above system of equations can be obtained to find the weights $b(s,i)$ subject to the condition of unbiasedness and admissibility.

In other words, to show the admissibility of an estimator of the form $e_l(s) = \sum_{i\in s} b(s,i)y_i$, we have to show that the quantity $b(s,i)$ given by the above system of equations is independent of the choice of a particular solution.

Consider a situation where the sample consist of only two unit, that is, $s = \{i, j\}$. Now if we take $q_i = k_i^2/(1-p_i)$, $\eta_i = (1-p_i)/k_i$ and $\eta_i k_i = (1-p_i)$, then $\delta(s) = 2 - p_i - p_j$ and the weights $b(s,i)$ become

$$b(s,i) = \left(1-p_j\right)\Big/\left\{k_i\left(2-p_i-p_j\right)\right\} + \left(k_i p_j - p_i k_j\right)\Big/\left(k_i p(s)\right).$$

Now if we take $k_i = p_i$ then $b(s,i) = (1-p_j)/\{p_i(2-p_i-p_j)\}$, $i = 1,2$ and the estimator of population total defined as

$$e_l(s) = \sum_{i\in s} b(s,i)y_i = \frac{1}{2-p_i-p_j}\left[\frac{(1-p_j)y_i}{p_i} + \frac{(1-p_i)y_j}{p_j}\right]$$

which is same estimator proposed by Murthy (1957, 1963).

Again if we take $k_i = \pi_i - p_i$ such that $\sum_{i\in s} k_i = 1$, we can consider a trivial solution $(\xi_1, \xi_2, ..., \xi_N) = (0,0,...,0)$. Then the weights $b(s,i)$ can be written as

$$b(s,i) = (1-p_i)/\{(2-p_i-p_j)(\pi_i - p_i)\}$$

and the estimator of the population total defined as

$$e_l(s) = \sum_{i\in s} b(s,i)y_i = \frac{1}{2-p_i-p_j}\left[\frac{(1-p_i)y_i}{\pi_i - p_i} + \frac{(1-p_j)y_j}{\pi_j - p_j}\right]$$

which is again the same estimator discussed by Murthy (1957, 1963).

Thus we have the following theorem.

Theorem 5.15.6.2 The estimators of population total proposed by Murthy (1957, 1963) are admissible.

5.15.7 POLYNOMIAL TYPE ESTIMATORS

Hanurav (1966) starts with the basic concepts in sampling theory for finite populations and considers a fundamental problem of optimum estimation procedures to estimate the population total. For unicluster designs Hanurav (1966) has shown that any estimator e_0 in $M^*(P)$, where $M^*(P)$ is a class of all polynomial unbiased estimators of total, is admissible in $M^*(P)$ if and only if

$$e_0 = g_s + \hat{Y}_{HT}(P),$$

where $\hat{Y}_{\text{HT}}(P)$ denotes the Horvitz and Thompson (1952) type of estimator under design P and values of g_s are constants independent of the study variable, the following condition is satisfied $\sum_{s \in S} g_s p_s = 0$. A given parametric function $g(y)$ is said to be estimatable in a given design P if and only if there exist a statistic e_s such that

$$E(e) = \sum_{s \in S} e_s p_s = g(y).$$

Theorem 5.15.7.1. A set of necessary and sufficient conditions for the estimability of the quadratic parametric function

$$Q = t_0 + \sum_{i \in \Omega} t_i y_i + \sum_{i \in \Omega} q_{ii} y_i^2 + \sum_{i \in \Omega} \sum_{j(\neq i) \in \Omega} q_{ij} y_i y_j$$

in a design P is given by

(i) $\pi_i > 0$ if $t_i^2 + q_{ii}^2 > 0$, (ii) $\pi_{ij} > 0$ if $q_{ij} + q_{ji} \neq 0$.

Theorem 5.15.7.2. Given a design P, an estimable parametric function $g(y)$ and s_1, s_2 are effectively equivalent. If e is any unbiased estimator of $g(y)$ with respect to any design P such that for any two samples s_1 and s_2 the condition $p_{s_1} > 0$ $p_{s_2} > 0$ get violated, then the estimator, $e^* = E(e \mid (s, y))$, is also unbiased for $g(y)$ and $V(e^*) \le V(e)$ $\forall y \in R_N$ with strict inequality occurs at least once. The above theorem basically restates the well known Basu (1958) result.

5.15.8 ALTERNATIVE OPTIMALITY CRITERION

While looking for an optimal estimator in survey sampling, we have to keep the following criteria in mind:

(i) Bayesian approach; (ii) Linear invariance;
(iii) Regular estimators; (iv) Hyper-admissibility.

(i) Bayesian approach: This approach makes use of prior information about the known distribution of the study variable. In this case we generally prefer to minimize a loss function in place of variance or mean squared error.

(ii) Linear Invariance: Roy and Chakravorty (1960) introduced the concept that an estimator should remain invariant under linear transformation of the study variable y.

(iii) Regular estimator: An estimator e is said to be a regular estimator if $V(e) = k\sigma^2$, where k is a constant and σ^2 is a finite population variance.

(iv) Hyper-admissibility: An estimator e is said to be hyper-admissible if it is admissible not only in the whole $\mathbf{R_N}$ but also in each of its principal hyperplanes.

For example, for any design P which is not a unicluster design the class of polynomial, $M^*(P)$, unbiased estimators of the population total Y admits just one estimator which is hyper-admissible. This optimum estimator is only the Horvitz and Thompson (1952) estimator. A class of hyper-admissible estimators for unicluster designs is given by $e_0 = g_s + \hat{Y}_{\text{HT}}(P)$ as defined earlier.

5.15.9 SUFFICIENT STATISTIC IN SURVEY SAMPLING

An estimator is said to be sufficient for a parameter, if it contains all information in the sample regarding the parameter. More precisely, if $e = e_s(y_1, y_2, ..., y_n)$ is an estimator of a parameter θ, based on sample information $y_1, y_2, ..., y_n$ of size n from a population with density $f(y, \theta)$ such that the conditional distribution of $y_1, y_2, ..., y_n$ given e, is independent of parameter θ, then e is a sufficient estimator of θ. For example, let $y_1, y_2, ..., y_n$, be a random sample from a Bernoulli population with parameter p such that,

$$y_i = \begin{cases} 1 & \text{with probability } p, \\ 0 & \text{with probability } (1\text{-}p) = q, \end{cases}$$

where $0 < p < 1$.

Then the statistic $e = e_s(y_1, y_2, ..., y_n) = \sum_{i=1}^{n} y_i$ follows a binomial distribution with parameters n and p, and

$$P(e = x) = {}^nC_x p^x q^{(n-x)}, \quad x = 0,1,2,...,n.$$

The conditional distribution of $y_1, y_2, ..., y_n$ given e is

$$P(y_1 \cap y_2 \cap .... \cap y_n \mid e = x) = \frac{P[(y_1 \cap y_2 \cap .... \cap y_n) \cap (e = x)]}{P(e = x)} = \frac{p^x q^{n-x}}{{}^nC_x p^x q^{n-x}} = \frac{1}{{}^nC_x},$$

which is free from p, indicates that $e = e_s(y_1, y_2, ..., y_n) = \sum_{i=1}^{n} y_i$ is a sufficient statistic for p.

There is another way to find a sufficient statistic called Factorization Theorem. The statistic $e = e_s(y_1, y_2, ..., y_n)$ is sufficient for θ if and only if the likelihood function L of the sample values can be expressed in the form, $L = g_\theta(e(y))h(y)$ where $g_\theta(e(y))$ depends on θ and y only through the value of $e(y)$ and $h(y)$ is independent of θ. For example, $y_1, y_2, ..., y_n$ is a random sample from a population with normal density $N(\mu, \sigma^2)$. Then the statistic $e_s = \sum_{i=1}^{n} y_i$ is sufficient for μ and the statistic $e_s^* = \sum_{i=1}^{n} y_i^2$ is sufficient for σ^2 because the likelihood function can be written as

$$L = \prod_{i=1}^{n} f(y_i, \theta) = \left(\frac{1}{\sqrt{2\pi}\sigma}\right)^n \exp\left[-\frac{1}{2\sigma^2}\left(\sum_{i=1}^{n} y_i^2 - 2\mu \sum_{i=1}^{n} y_i + n\mu^2\right)\right] = g_\theta(e(y))h(y),$$

where

$$e(y) = \left(\sum_{i=1}^{n} y_i, \sum_{i=1}^{n} y_i^2\right) \text{ and } h(y) = 1.$$

Thus $\left(\sum_{i=1}^{n} y_i, \sum_{i=1}^{n} y_i^2\right)$ is sufficient for $\theta = (\mu, \sigma^2)$.

It is easy to recognize the concept of sufficiency in general from the above discussion. Yamada and Morimoto (1992) have been able to compare the following four definitions of sufficiency in a very elegant way.

(i) The conditional distribution of a sample when given the estimator e does not depend on the parameter.

(ii) The distribution of a sample can be reconstructed from that of the estimator e through randomisation or mathematically using a stochastic kernal.

(iii) For every decision problem, given a decision function based on the sample, there exists a decision function based on e which is at least as good as the former.

(iv) For any prior distribution of the parameter, the posterior is a function of the sample through e.

The first definition is owed to Fisher (1920, 1922), the second and third are due to Blackwell (1951) and the fourth is due to Kolmogorov (1942). Based on the first definition of sufficiency, Halmos and Perlman (1974), and Bahadur (1954) introduced its new definition as follows.

"Let $E = (X, \mathrm{A}, P)$ be a statistical experiment and B be a sub-field (or a sub-σ-field of A) . Then B is called sufficient if for every $A \in \mathrm{A}$ there exist a B-measurable function $P(A|B)(y)$ which satisfies for all B in $\mathbf{B}$ and p in P such that $p(A \cap B) = \int_B P(A \mid \mathrm{B})(y)dp$ holds."

Further Yamada and Morimoto (1992) have developed relationships between above definitions of sufficiency in more descriptive way. The readers must follow the lecture notes of Ghosh and Pathak (1992) to have better understanding of sufficiency in survey sampling.

5.16 ESTIMATORS BASED ON CONDITIONAL INCLUSION PROBABILITIES

Following Tillê (1998) let $\eta = \eta(x_i,\ i \in s)$ be a statistic based on the auxiliary information. Since the population is finite the statistic η takes a finite number of possible values denoted $\{\eta_1,\ \eta_2,\ ...,\ \eta_I\}$.

Define a indicator variable

$$I_i = \begin{cases} 1 & \text{if } i \in s, \\ 0 & \text{otherwise.} \end{cases} \tag{5.16.1}$$

Then the first order conditional inclusion probabilities are given by

$$\pi_{i|\eta} = E(I_i \mid \eta) \ \text{ for all } i \in \Omega . \tag{5.16.2}$$

and the second order conditional inclusion probabilities are given by

$$\pi_{ij|\eta} = E(I_i I_j \mid \eta) \text{ for all } i, j \in \Omega \text{ and } i \neq j . \tag{5.16.3}$$

Then we have the following theorem.

Theorem 5.16.1. Show that the simple conditionally weighted (SCW) estimator of population mean $\bar{Y}$ defined as

$$\bar{y}_{\text{SCW}} = \frac{1}{N}\sum_{i \in s}\frac{y_i}{\pi_{i|\eta}} \tag{5.16.4}$$

is virtually conditionally unbiased.

Proof. An estimator $\hat{\theta}$ of θ is said to be virtually conditionally unbiased (VCU) with respect to auxiliary statistic η if

$$B(\hat{\theta} \mid \eta) = \sum_{i \in \Omega}\theta_i \alpha_i(\eta) I(\pi_{i|\eta} = 0) \tag{5.16.5}$$

for all $(\theta_1, \theta_2, ..., \theta_N) \in R^N$, where the coefficients $\alpha_i(\eta)$ depends on η and $I(\bullet)$ is an indicator function such that

$$I(\pi_{i|k} = 0) = \begin{cases} 1 & \text{if } \pi_{i|\eta} = 0, \\ 0 & \text{if } \pi_{i|\eta} > 0. \end{cases}$$

Now the bias in the estimator $\bar{y}_{\text{SCW}}$ is given by

$$B[\bar{y}_{\text{SCW}}] = E(\bar{y}_{\text{SCW}} \mid \eta) - \bar{Y} = E\left[\frac{1}{N}\sum_{i \in s}\frac{y_i}{\pi_{i|\eta}}\right] - \bar{Y} = E\left(\frac{1}{N}\sum_{\substack{i \in \Omega \\ \pi_{i|\eta} > 0}}\frac{Y_i I_i}{\pi_{i|\eta}} \mid \eta\right) - \bar{Y}$$

$$= \frac{1}{N}\sum_{\substack{i \in \Omega \\ \pi_{i|\eta} > 0}} E\left(\frac{Y_i I_i}{\pi_{i|\eta}} \mid \eta\right) - \frac{1}{N}\sum_{i \in \Omega} Y_i = -\frac{1}{N}\sum_{i \in \Omega} Y_i I[\pi_{i|\eta} = 0].$$

Hence the theorem.

Theorem 5.16.2. Show that a corrected conditionally weighted (CCW) estimator of population mean $\bar{Y}$ is

$$\bar{y}_{\text{CCW}} = \frac{1}{N}\sum_{i \in s} y_i / \left(h_i \pi_{i|\eta}\right) \tag{5.16.6}$$

where $h_i = E\left[I\left(\pi_{i|\eta} > 0\right)\right] = \Pr\left(\pi_{i|\eta} > 0\right)$ is an unbiased estimator of population mean.

Proof. The bias in the estimator $\bar{y}_{\text{CCW}}$ is given by

$$B\left(\bar{y}_{\text{CCW}} \mid \eta\right) = E\left[\frac{1}{N}\sum_{i \in s}\frac{y_i}{h_i \pi_{i|\eta}}\right] - \bar{Y} = E\left[\frac{1}{N}\sum_{i \in \Omega}\frac{Y_i I_i}{h_i \pi_{i|\eta}}\right] - \frac{1}{N}\sum_{i \in \Omega} Y_i$$

$$= \frac{1}{N}\sum_{i \in \Omega} Y_i\left(\frac{I\left(\pi_{i|\eta} > 0\right)}{h_i} - 1\right) = 0 .$$

Hence the theorem.

Note that Rao (1985) also discussed conditional inference in survey sampling.

Remarks 5.16.1. Note the following points carefully:

(a) The CCW estimator is not VCW but it is unconditionally unbiased;

(b) The estimators CCW and SCW are not invariant.

5.17 CURRENT TOPICS IN SURVEY SAMPLING

Rao (1996) discussed some current topics in survey sampling at the Golden Jubilee conference of the Indian Society of Agricultural Statistics, New Delhi. In particular, inferential issues were studied and the advantage of conditional design based approach were demonstrated. Practically useful estimators for dual frame surveys were presented. The jackknife method was shown to provide a unified, but computer intensive, approach to variance estimation and analysis of survey data. Finally, small area estimation was considered and model based indirect estimators that borrow information from related small areas were introduced. Moors, Smeets, and Boekema (1998) have considered an interesting problem of sampling with probabilities proportional to the variable of interest. Brewer (1994) has discussed the past and present prospects in survey sampling inference. Again Rao (1999a, 1999b) has considered the problem of review of some current trends in sample survey theory and methods. He provides a brief discussion on developments in survey design and data collection and processing, issues related to inference from survey data, re-sampling methods for analysis of survey data, and small area estimation. Following him the principal steps in sample survey have been

(a) Survey design,

(b) Data collection and Processing,

(c) Estimation and analysis of data.

We would like to discuss these issues briefly here. They are covered in more depth by Rao (1999a, 1999b).

5.17.1 SURVEY DESIGN

Researchers have paid much attention to sampling errors, and have developed numerous methods for optimal allocation of resources to minimze the sampling variance associated with the estimators of total or mean. Much less attention has been paid to reducing the total survey error arising from both sampling and non-sampling errors. Many researchers, including Fellegi and Sunter (1974), Linacare and Trewin (1993), and Smith (1995), have emphasized the need for a total survey design approach in which resources are allocated to those sources of error where error reduction is most effective thus resulting in superior survey designs. Linacare and Trewin (1993) applied this approach to the design of the Construction Industry Surveys. Smith (1995) proposed the sum of component MSEs, rather than the MSE of the estimated total, as a measure of the total error which may be written as the sum of the errors from different sources. While estimating the population mean or total, it is customary to study the effect of measurement errors. It has been found that the usual estimators are design unbiased and consistent under the assumption of zero mean measurement errors. Following Mahalanobis (1946), the traditional variance estimators remain valid provided the sample is in the form of interpenetrating sub-samples. It is interesting to note that this useful feature no longer holds in the case of distribution functions, quantiles, and some other complex parameters as shown by Fuller (1995). The usual estimators are biased and inconsistent and thus can lead to erroneous inferences. Fuller (1995) obtained bias adjusted estimators under the assumption of independent and normally distributed errors. Eltinge (1999) extended Fuller's results to the case of non-normal errors using small standard deviation approximations. Singh, Gambino, and Mantel (1994) illustrated the use of compromise allocation to redesign the Canadian Labour Force Survey. As mentioned by Rao (1999a, 1999b), the interpenetrating sub-samples provide a valid estimate of the total variance of an estimated total in the presence of measurement errors, but such designs are not often used, at least in North America, due to cost and operational considerations. Hartley and Rao (1968) and Hartley and Biemer (1978) provided interview and coder assignment conditions that permit the estimation of total variance and its components, such as sampling, interviewer, and coder variances, directly from stratified multistage surveys that satisfy the estimability conditions. Groves (1996) noted that the customary one way random effect interviewer variance model may be unrealistic because it fails to reflect 'fixed' effects of interviewer attributes such as race, age, and gender that can affect the responses.

5.17.2 DATA COLLECTION AND PROCESSING

In comparison with the face to face interviewing technique of data collection, telephone surveys have become popular in developed countries and are growing very rapidly in developing countries too. Computer Assisted Telephone Interview (CATI) has also been found helpful in reducing both measurement and data processing errors. A combination of both face to face interviews and telephone surveys has also been found useful in monthly surveys. There are several more methods of collecting data like random digit dialing (RDD), which provides coverage of both listed and unlisted telephone households. The two stage Mitofsky--Waksberg technique and its refinements are designed to increase the proportion of eligible numbers in the sample and thus reduce data collection cost by following Casady and Lepkowski (1993). Following Groves and Lepkowski (1986) the dual frame approaches are also useful in obtaining more efficient estimates by combining a sample selected from an incomplete directory list frame with another sample selected by random digit dialing. Many large scale surveys, especially surveys on family expenditure and health, use long questionnaires for data collection. Such surveys can lead to high rates of non-response and decreases in the quality of response, but this problem may be reduced by splitting the long questionnaire into two or more parts. For example, Wretman (1995) splits his long questionnaire into five non-overlapping parts. For surveys dealing with sensitive questions, the quality of responses and response rates might depend on the ordering of the questions in the list. It is also important to keep in mind that data on sensitive characters or variables can also be collected through a Randomized Response Technique, which makes use of a device to collect indirect answers to the questions in the surveys. Data collected from survey or census by any one of the above methods needs editing. The purpose of editing is to see which records are unacceptable, outliers, or missing. Then imputation of the missing records and proper treatment of the outliers or unacceptable values is most important, if care is taken to assume that the editing procedure changes as few values as possible. For example, Fellegi and Holt (1976) developed a method for automatic editing of survey data with the help of computers under certain assumptions.

5.17.3 ESTIMATION AND ANALYSIS OF DATA

Suppose we are interested in estimating the population total of a characteristic of interest y. A sample s is selected according to a specified sampling design $p(s)$ from a population of size N and the sampled data $\{(i, y_i): i \in s\}$ are collected, assuming non-sampling errors are absent.

The basic idea of the inference is to obtain an estimator $\hat{Y}$, its standard error $s(\hat{Y})$ or coefficient of variation,

$$c(\hat{Y}) = s(\hat{Y})/\hat{Y},$$

and associated normal theory intervals

$$\hat{Y} \mp Z_{\alpha/2} s(\hat{Y})$$

on Y from the sampled data for large sample size n. As we noted in this chapter, there are three approaches:

(a) Design based approach;

(b) Model based approach;

and

(c) Model assisted approach.

The traditional work in sampling including the Horvitz and Thompson estimator comes under design approach. The work related to Royall (1970a, 1970b, 1970c) and Brewer (1963a) comes under the model based approach. The contribution related to Särndal, Swensson, and Wretman (1991) comes under the concept of model assisted approach. Following Rao (1999), the methods of re-sampling and small area estimation, which we shall discuss in subsequent chapters, are also under the current topics in survey sampling.

5.18 MISCELLANEOUS DISCUSSIONS/TOPICS

In this section we introduce some topics which exist in the literature and may be useful for the research oriented readers.

5.18.1 GENERALIZED IPPS DESIGNS

Generalized πps designs were defined by Rao (1972). The term IPPS stands for Inclusion Probabilities Proportional to Size. Working with a general superpopulation model $\theta(g)$, the strategy consisting of $G\pi PS$ design together with the associated HT estimator of population total has proved to be better than two other well known strategies of Rao (1971, 1972). Following Rao (1972), if a design is such that $\pi_i \propto X_i^{g/2}$ $(i = 1,2,...,N)$ and $\sum\limits_{i \in s} x_i^{1-(g/2)} = k$, a constant for any sample s with $p_s > 0$ then that design is called generalized πps design. For $g = 2$, the $G\pi ps$ design is a πps design with fixed sample size. Ramachandran (1982) has shown that the $g(\theta)$ optimality of the strategy consists of $G\pi ps$ design together with the associated HT estimator in the entire class of design based unbiased strategies of the population total with expected sample size fixed. Ramachandran generalized the results of Rao (1971,1972) by following Godambe and Joshi (1965). Pedgaonkar and Prabhu--Ajgaonkar (1978) have shown that the $G\pi ps$ strategy is better than the RHC strategy with fixed sample size. McLeod and Bellhouse (1983) gave a simple and useful algorithm for drawing a simple random sample without replacement in a single pass through a sampling frame for a finite population whose size is unknown. Richardson (1989) extended their method to probability proportional to size sampling. Korwar (1996) provides a method, which is an adoption of the method of McLeod and Bellhouse for the simple random sample without replacement, for drawing a sample with probability proportional to aggregate size in a single pass through a sampling frame of a finite population whose size is unknown. Some discussion on the question of availability of a unique best estimator in the Horvitz and Thompson (1952) class of estimators has been given by Chaudhuri (1975a, 1975b).

5.18.2 TAM'S OPTIMAL STRATEGIES

Tam (1984) has provided necessary and sufficient conditions for an estimator design pair to be optimal under a regression superpopulation model with correlated residuals. Tam (1986) has provided necessary and sufficient conditions for the optimality of an arbitrary linear predictor of the total of a finite population in survey sampling under a general linear model with a symmetric and positive definite covariance matrix. He extended the work of Pfeffermann (1984) and Tallis (1978) by following Cassel, Särndal, and Wretman (1977), Royall (1970a, 1970b, 1970c, 1976), Royall and Herson (1973a, 1973b), Royall and Pfeffermann (1982), Särndal (1980a, 1980b), Scott, Brewer and Ho (1978), and Zyskind (1967).

5.18.3 USE OF RANKS IN SAMPLE SELECTION

Wright (1990) has shown that there can be a gain in estimation strategies over equal probability sampling methods when one makes use of auxiliary information for probability proportional to size with replacement sampling methods. When a suitable variable X is not available, one may know how to rank units reasonably well relative to the unknown y values before sample selection. When such ranking is possible Wright (1990) has introduced a simple and efficient sampling plan using the ranks as the unknown X measure of size. He showed that the resultant sampling plan is similar to, has the simplicity of, and has no greater sampling variance than with replacement sampling, but is without replacement. Kumar, Srivenkataramana, and Srinath (1996) have also shown the use of ranks in unequal probability sampling for sample selection and stratification including determining the strata boundaries. They also suggested a few sampling schemes. For samples of size two, two sampling schemes and their IPPS versions have been discussed along with their extension to large sample situations. Non-negative unbiased estimators of the variance have also been suggested.

5.18.4 PREDICTION APPROACH

Mukhopadhyay (1994) has considered the optimal prediction of finite population total and variance under a location model with measurement errors. Bayes predictors of population total and variance under a class of priors have been derived and a minimax predictor for population total has been obtained. Under a regression superpopulation model with measurement errors, an optimal predictor of total has been derived. Srivastava and Ouyang (1992) have studied a general estimator in sampling by utilising extraneous information through a sample weight function. They discussed the use of extraneous information (including that collected during the survey) to obtain guesses. Mukhopadhyay and Bhattacharyya (1990, 1991) have provided optimal estimation of a finite population variance under some exchangeable general linear models. They examined the robustness of the optimal strategies under a class of alternative models and sampling designs ensuring near unbiasedness. Samiuddin and Kattan (1991) have suggested a new unequal probability sampling scheme. Samiuddin, Kattan, Hanif, and Asad (1992) have contributed some useful remarks on the use of models and sampling schemes while using unequal probability sampling strategies.

5.18.5 TOTAL OF BOTTOM (OR TOP) PERCENTILES OF A FINITE POPULATION

Reddy and Rao (1990) have considered the problem of estimation of population total of bottom (top) P percentiles of a finite population using the Horvitz and Thompson (1952) type estimator.

5.18.6 GENERAL FORM OF ESTIMATOR OF VARIANCE

Let Q denote a scalar quantity to be estimated. A point estimator of Q will of the form $\hat{Q} = \hat{Q}(\underline{X}, y)$ and variance $V(\hat{Q}) = V(\underline{X}, y)$, where $\underline{X} = (X_{ij})_{n\times p}$ is matrix of auxiliary information and $y = (y_i)_{n\times 1}$ is a vector of study variable. The estimator $\hat{Q}$ can be considered

$$\hat{Q} = g(T_{x1}, T_{x2}, ..., T_{xp}, T_y) \tag{5.18.6.1}$$

where $T_{x_j} = n^{-1}\sum_{i=1}^{n} X_{ij}$, $j = 1,2,...,p$, $T_y = n^{-1}\sum_{i=1}^{n} y_i$, and g is a smooth function. It is to be noted that estimation of mean, proportion and ratio of means is a special case of the general estimator $\hat{Q}$, but the estimation of median, variance, or correlation is not a member of this general estimator. The estimand Q will be the same smooth function g of the expectations of the means $Q = g(E(T_{x1}), E(T_{x2}), ..., E(T_{xp}), E(T_y))$, where the expectations are taken over repeated sampling and estimator $\hat{Q}$ is called an estimator based on the method of moments. Thus the general form of the estimator of the variance $V(\hat{Q})$ is given by

$$\hat{v}(\hat{Q}) = n^{-1}[\partial g(\underline{T})/\partial \underline{T}]^t \underline{S}[\partial g(\underline{T})/\partial \underline{T}] \tag{5.18.6.2}$$

where $\underline{T} = (T_{x1}, T_{x2}, ..., T_{xp}, T_y)^t$ and $S = (n-1)^{-1}\left[\underline{Z}^t \underline{Z} - n\underline{T}\,\underline{T}^t\right]$ with $\underline{Z} = (\underline{X}, y)$. Schafer and Schenker (2000) have developed a model for missing data based on this procedure.

5.18.7 POISSON SAMPLING

Hajek (1964) suggested a sampling procedure in which a Bernoulli trial is conducted with probability of success π_i such that the ith trial results in success then the ith unit is included in the sample. Clearly the sample size m is a random variable for this sampling procedure. The second order inclusion probabilities are given by $\pi_{ij} = \pi_i \pi_j$. Under such a sampling procedure an unbiased estimator of population total is given by

$$\hat{Y}_{\text{P}} = \sum_{i\in s}\frac{y_i}{\pi_i} \tag{5.18.7.1}$$

with variance

$$V\left(\hat{Y}_{\text{P}}\right) = \sum_{i\in\Omega}(1-\pi_i)\frac{Y_i^2}{\pi_i} \tag{5.18.7.2}$$

and an unbiased estimator of the $V\left(\hat{Y}_{\text{P}}\right)$ is given by

$$\hat{v}\left(\hat{Y}_P\right) = \sum_{i\in s}\frac{(1-\pi_i)}{\pi_i^2}y_i^2 \,. \tag{5.18.7.3}$$

For more depth in Poisson sampling, one can refer to Ogus and Clark (1971) and Brewer, Early, and Hanif (1984).

5.18.8 COSMETIC CALIBRATION

Cosmetic calibration was introduced by Särndal and Wright (1984). Brewer (1995) suggested a procedure for constructing a cosmetic estimator. Brewer (1999) has shown that cosmetic estimators are by definition interpretable both as design based and as prediction based estimators. Formulae for them can be obtained directly by equating these two estimators or indirectly by a simple form of calibration. Note that they constitute a subset of GREG, their design variances cannot be estimated without knowing the relevant second order inclusion probabilities, but under the prediction model to which they are calibrated those probabilities do not affect their anticipated variances, so it is more appropriate to estimate these and/or their prediction variances. Brewer (1999) has shown that cosmetic calibration is a simple and effective method for eliminating negative and unacceptably small positive sample weights. Interestingly he suggests here to estimate the anticipated variance of any calibrated estimator of population total under superpopulation model $m\,:\; y_i = \beta x_i + e_i$, such that $E_m(e_i)=0$, $E_m\left(e_i^2\right)=\sigma^2 x_i^g$, and $E_m\left(e_i e_j\right)=0$ for $i\neq j$. We have seen that a calibrated estimator of population total Y can be written as

$$\hat{Y}_c = \sum_{i\in s} w_i y_i, \tag{5.18.8.1}$$

where w_i are the calibration weights. The by the definition of Anticipated Variance

$$\text{AV}\left(\hat{Y}_c\right) = E_m\left[\hat{Y}_c - Y\right]^2 = E_m\left[\sum_{i\in s} w_i y_i - \sum_{i\in\Omega} y_i\right]^2 = E_m\left[\sum_{i\in s} w_i y_i - \left\{\sum_{i\in s} y_i + \sum_{i\notin s} y_i\right\}\right]^2$$

$$= E_m\left[\sum_{i\in s}(w_i - 1)y_i - \sum_{i\notin s} y_i\right]^2$$

$$= \sigma^2\left[\sum_{i\in s} w_i(w_i-1)^2 x_i^g + \left(\sum_{i\in\Omega} x_i^g - \sum_{i\in s}\pi_i^{-1}x_i^g\right) - \left(\sum_{i\in s} w_i x_i^g - \sum_{i\in s}\pi_i^{-1}x_i^g\right)\right]. \tag{5.18.8.2}$$

Brewer (1999) discusses the following advantages of the cosmetically calibrated estimator:

(a) The estimator is clearly interpretable as prediction based as well as design based;

(b) Its anticipated variance and its prediction variance can both be estimated more easily and more efficiently than the design variance of the standard GREG;
(c) Design based estimation has a tendency to be more reliable for large samples, and prediction based estimation for small samples and small domains. Thus the estimators used for large domains are typically design based while those for small domains are often purely prediction based or synthetic. If the large domain estimators are calibrated, the estimates for their component small domains automatically sum them without forcing;
(d) As an unexpected spin off, the elimination of negative and other unacceptably small weights is streamlined by the use of cosmetic calibration.

5.18.9 MIXING OF NON-PARAMETRIC MODELS IN SURVEY SAMPLING

In order to estimate the population total $t_y = \Sigma_{i\in\Omega} y_i$, the Horvitz and Thompson (1952) estimator is given by

$$\hat{t}_y = \sum_{i\in s}\frac{y_i}{\pi_i} = \sum_{i\in\Omega}\frac{y_i I_i}{\pi_i}, \tag{5.18.9.1}$$

where

$$I_i = \begin{cases} 1 & \text{if } i\in s, \\ 0 & \text{if } i\notin s, \end{cases} \tag{5.18.9.2}$$

such that $E_p(I_i) = \pi_i$. The Sen--Yates--Grundy form of the variance of $\hat{t}_y$ is

$$V_{\text{syg}}(\hat{t}_y) = \frac{1}{2}\sum_{i\neq j\in\Omega}\sum \Theta_{ij}\left(\frac{y_i}{\pi_i} - \frac{y_j}{\pi_j}\right)^2, \tag{5.18.9.3}$$

where $\Theta_{ij} = (\pi_i\pi_j - \pi_{ij})$ for $i \neq j$ such that $\lim_{N\to\infty} \sup(n) \max_{i,j\in\Omega} |\Theta_{ij}| < \infty$. An estimator for (5.18.9.3) is

$$\hat{v}_{\text{syg}}(\hat{t}_y) = \frac{1}{2}\sum_{i\neq j\in s}\sum D_{ij}\left(\frac{y_i}{\pi_i} - \frac{y_j}{\pi_j}\right)^2 \tag{5.18.9.4}$$

where $D_{ij} = (\pi_i\pi_j - \pi_{ij})/\pi_{ij}$ are called the Sen--Yates--Grundy (SYG) design weights. In order to derive an alternative estimator of the population total, Breidt and Opsomer (2000) consider modelling y_i and x_i under a superpopulation model ξ through the relationship

$$y_i = m(x_i) + \varepsilon_i \tag{5.18.9.5}$$

where $m(x_i)$ is a smooth function of x, such that $E_\xi(y_i) = m(x_i)$, $V_\xi(y_i) = v(x_i)$, and $v(x_i)$ is smooth and strictly positive. Let K denote a continuous kernel function and let h denote the bandwidth. Breidt and Opsomer (2000) proposed a local polynomial kernel estimator of degree q for $m(x_i)$ that is based on the entire finite population as follows. Let $y_\Omega = [y_i]_{i\in\Omega}$ be the N-dimensional vector of y_i values in the finite population, e_r a vector with a 1 in the rth position and 0 elsewhere, $\in' = (\varepsilon_i, \varepsilon_2, ..., \varepsilon_N)_{1\times N}$,

$$X_{\Omega i} = \begin{bmatrix} 1, (x_1 - x_i), (x_1 - x_i)^2, ..., (x_1 - x_i)^q \\ 1, (x_2 - x_i), (x_2 - x_i)^2, ..., (x_2 - x_i)^q \\ \\ 1, (x_N - x_i), (x_N - x_i)^2, ..., (x_N - x_i)^q \end{bmatrix}_{N\times(q+1)}$$

and

$$W_{\Omega i} = \begin{bmatrix} \frac{1}{h}K\left(\frac{x_1 - x_i}{h}\right), & 0, & ...., & 0 \\ 0, & \frac{1}{h}K\left(\frac{x_2 - x_i}{h}\right), & ...0 \\ .... \\ 0, & 0, & ,\frac{1}{h}K\left(\frac{x_N - x_i}{h}\right) \end{bmatrix}_{N\times N}.$$

Minimization of $e_r \in' W_{\Omega i} \in$ leads to the local polynomial kernel estimator of the regression function at x_i given by

$$m_i = e_i'\left(X_{\Omega i}' W_{\Omega i} X_{\Omega i}\right)^{-1} X_{\Omega i}' W_{\Omega i} y_\Omega = w_{\Omega i}' y_\Omega . \tag{5.18.9.6}$$

This estimator is well defined if $\left(X_{\Omega i}' W_{\Omega i} X_{\Omega i}\right)$ is non-singular. If the m_i are known, then a design unbiased estimator of t_y would be the generalized difference estimator

$$t_y^* = \sum_{i\in s} \frac{y_i - m_i}{\pi_i} + \sum_{i\in U} m_i . \tag{5.18.9.7}$$

The Sen--Yates--Grundy form of the variance of (5.18.9.7) is

$$V\left(t_y^*\right) = \frac{1}{2} \sum_{i\neq j\in\Omega}\sum \Theta_{ij}\left(\frac{y_i - m_i}{\pi_i} - \frac{y_j - m_j}{\pi_j}\right)^2 \tag{5.18.9.8}$$

for $i \neq j$. If the m_i are known, a design-unbiased estimator of $V\left(t_y^*\right)$ is

$$\hat{v}\left(t_y^*\right) = \frac{1}{2} \sum_{i\neq j\in s}\sum D_{ij}\left(\frac{y_i - m_i}{\pi_i} - \frac{y_j - m_j}{\pi_j}\right)^2 . \tag{5.18.9.9}$$

In practice, the values of m_i are unknown and must be estimated using sample information. Let the n dimension vector $y_s = \{y_i\}_{i\in s}$, the n by $(q+1)$ matrix X_{si}, and the n by n matrix W_{si}, be the sample equivalents of y_Ω, X_Ω, and W_Ω. Minimization of $e_r \varepsilon' W_{si} \varepsilon$ leads to design based estimators of m_i given by

$$\hat{m}_i^\circ = e_i'\left(X_{si}' W_{si} X_{si}\right)^{-1} X_{si}' W_{si} y_s = w_{si}^{\circ'} y_s . \tag{5.18.9.10}$$

If l_{ij} denotes the (i, j)th element of the inverse $\left(X_{si}' W_{si} X_{si}\right)^{-1}$, then (5.18.9.10) can be written as

$$\hat{m}_i^{\circ} = e_1' \sum_{t=1}^{n} \frac{\sum_{j=0}^{q} l_{(t-1)j}(x_t - x_i)^j}{\pi_t h} K\left(\frac{x_t - x_i}{h}\right) y_t \tag{5.18.9.11}$$

for $i = 1,2,...n$.

Note that x_i is known for the entire population, the $\hat{m}_i^{\circ}$ can be calculated for all $i \in \Omega$. Using this result Breidt and Opsomer (2000) proposed the estimator

$$\tilde{t}_y^{\circ} = \sum_{i \in s} \frac{y_i - \hat{m}_i^{\circ}}{\pi_i} + \sum_{i \in U} \hat{m}_i^{\circ} \tag{5.18.9.12}$$

for t_y. Following Särndal, Swenson, and Wretman (1992), the variance of $\tilde{t}_y^{\circ}$ can be approximated as

$$V\left(\tilde{t}_y^{\circ}\right) \approx V\left(t_y^{*}\right) = \frac{1}{2} \sum_{i \neq} \sum_{j \in \Omega} \Theta_{ij} \left(\frac{y_i - m_i}{\pi_i} - \frac{y_j - m_j}{\pi_j}\right)^2. \tag{5.18.9.13}$$

An estimator for the variance of (5.18.9.13) in Sen--Yates--Grundy form is given by

$$\hat{v}_{\text{syg}}\left(\hat{t}_y^{\circ}\right) = \frac{1}{2} \sum_{i \neq} \sum_{j \in s} D_{ij} \left(\frac{y_i - \hat{m}_i^{\circ}}{\pi_i} - \frac{y_j - \hat{m}_j^{\circ}}{\pi_j}\right)^2. \tag{5.18.9.14}$$

If $\left(X_{si}' W_{si} X_{si}\right)^{-1}$ is not invertible, one possibility is to replace it with $\left(X_{si}' W_{si} X_{si} + \text{diag}(R)_{(q+1)\times(q+1)}\right)^{-1}$, where R stands for a ridge constant. Fan (1993) made use of this approach, choosing $\text{diag}(\text{R})_{(q+1)\times(q+1)} = \text{diag}\left(\frac{\delta}{N^2}\right)_{(q+1)\times(q+1)}$, with δ being a small positive number.

Let $\hat{m}_i$ be a new set of new estimates for m_i obtained by either one of these two methods. Using these estimates, the method of Breidt and Opsomer (2000) yields

$$\tilde{t}_y = \sum_{i \in s} \frac{y_i - \hat{m}_i}{\pi_i} + \sum_{i \in U} \hat{m}_i \tag{5.18.9.15}$$

as an estimator for the population total, and an estimator for estimating the $V\left(\tilde{t}_y\right)$ in Sen--Yates--Grundy form is given by

$$\hat{v}_{\text{syg}}\left(\tilde{t}_y\right) = \frac{1}{2} \sum_{i \neq} \sum_{j \in s} D_{ij} \left(\frac{y_i - \hat{m}_i}{\pi_i} - \frac{y_j - \hat{m}_j}{\pi_j}\right)^2. \tag{5.18.9.16}$$

Still more remain seems to be done on these lines. Note that one thing is very obvious by following Singh, Horn and Yu (1998) that one can calibrate the estimator of variance $\hat{v}_{\text{syg}}\left(\tilde{t}_y\right)$ by constructing model assisted calibration constraints $\hat{v}_{\text{syg}}\left(\tilde{t}_y\right)$ under the model (5.18.9.5) such that

$$E_{\xi}\left\{\hat{v}_{\text{syg}}\left(\tilde{t}_y\right)\right\} = E_{\xi}\left\{V_{\text{syg}}\left(\tilde{t}_y\right)\right\},$$

and the chi square distance is minimum between the design weights and calibrated weights. Further note that Bredit and Opsomer (2000) have reported second form of the variance of Horvitz and Thompson estimator.

5.19 GOLDEN JUBILEE YEAR 2003 OF THE LINEAR REGRESSION ESTIMATOR

In this section a question has been addressed raised by Deville and Särndal's (1992) calibration approach to the eminent survey statisticians working at Govt. Organizations such as the U.S. Bureau of Statistics, Statistics Canada, Australian Bureau of Statistics, and private organizations such as RAND, WESTSTAT etc., and their consultants from different universities across the world. Note that during the last decade the survey statisticians working for these institutes have tried a lot to construct a traditional linear regression estimator through calibration approach, but there was no success. For example, as reported by Farrell and Singh (2002b) that Wu and Sitter (2001) rediscovered the Deville and Särndal (1992) estimator by setting one of the auxiliary variable at constant level while in the search of a traditional linear regression estimator using calibration approach.

Singh (2003c) discovered that the traditional linear regression estimator can also be shown as a special case of calibration approach, and pointed out that all the papers related to minimizing chi square distance function in survey sampling need modifications. There is a series of such papers by many followers of Deville and Särndal (1992) and it seems that everyone skipped a very important point while using chi square distance function. The technique developed by Singh (2003c) is logically more accurate than whatever is done by survey statisticians during the last decade. The traditional linear regression estimator due to Hansen, Hurwitz, and Madow (1953) is shown to be unique in its class of estimators, and celebrates Golden Jubilee Year 2003 for its outstanding performance. Singh (2003c) considers an estimator of the population total Y as

$$\hat{Y}_s = \sum_{i \in s} w_i^{\oplus} y_i \tag{5.19.1}$$

where $w_i^{\oplus}$ are called the calibrated $\oplus$ (read as plus) weights such that the chi square distance function defined as

$$D^{\oplus} = \frac{1}{2} \sum_{i \in s} \frac{\left(w_i^{\oplus} - d_i\right)^2}{d_i q_i^{\oplus}} \tag{5.19.2}$$

is minimum subject to the two constraints, given by

$$\sum_{i \in s} w_i^{\oplus} = \sum_{i \in s} d_i \, , \tag{5.19.3}$$

and

$$\sum_{i \in s} w_i^{\oplus} x_i = X . \tag{5.19.4}$$

The choice of weights $q_i^{\oplus}$ makes different forms of estimators. Note that the condition (5.19.3) is a requirement of the chi square test given by Sir R.A. Fisher, and is ignored by all the followers of Deville and Särndal (1992). Obviously the Lagrange function is given by

$$L = \frac{1}{2}\sum_{i\in s}\frac{\left(w_i^{\oplus} - d_i\right)^2}{d_i q_i^{\oplus}} - \lambda_1\left\{\sum_{i\in s} w_i^{\oplus} - \sum_{i\in s} d_i\right\} - \lambda_2\left\{\sum_{i\in s} w_i^{\oplus} x_i - X\right\}.$$

On setting $\partial L/\partial w_i^{\oplus} = 0$ we have

$$w_i^{\oplus} = d_i + \lambda_1 d_i q_i^{\oplus} + \lambda_2 d_i q_i^{\oplus} x_i . \tag{5.19.5}$$

On using (5.19.5) in (5.19.3) we have

$$\lambda_1\left(\sum_{i\in s} d_i q_i^{\oplus}\right) + \lambda_2\left(\sum_{i\in s} d_i q_i^{\oplus} x_i\right) = 0 . \tag{5.19.6}$$

On using (5.19.5) in (5.19.4) we have

$$\lambda_1\left(\sum_{i\in s} d_i q_i^{\oplus} x_i\right) + \lambda_2\left(\sum_{i\in s} d_i q_i^{\oplus} x_i^2\right) = \left(X - \hat{X}_{\text{HT}}\right). \tag{5.19.7}$$

where

$$\hat{X}_{\text{HT}} = \sum_{i\in s} d_i x_i .$$

On solving (5.19.6) and (5.19.7) for λ_1 and λ_2 we have

$$\lambda_1 = \frac{-\left(\sum_{i\in s} d_i q_i^{\oplus} x_i\right)\left(X - \hat{X}_{\text{HT}}\right)}{\left(\sum_{i\in s} d_i q_i^{\oplus}\right)\left(\sum_{i\in s} d_i q_i^{\oplus} x_i^2\right) - \left(\sum_{i\in s} d_i q_i^{\oplus} x_i\right)^2} \text{ and } \lambda_2 = \frac{\left(\sum_{i\in s} d_i q_i^{\oplus}\right)\left(X - \hat{X}_{\text{HT}}\right)}{\left(\sum_{i\in s} d_i q_i^{\oplus}\right)\left(\sum_{i\in s} d_i q_i^{\oplus} x_i^2\right) - \left(\sum_{i\in s} d_i q_i^{\oplus} x_i\right)^2}$$

On substituting these values in (5.19.5) the calibrated plus weights are

$$w_i^{\oplus} = d_i + \left\{\frac{d_i q_i^{\oplus} x_i\left(\sum_{i\in s} d_i q_i^{\oplus}\right) - d_i q_i^{\oplus}\left(\sum_{i\in s} d_i q_i^{\oplus} x_i\right)}{\left(\sum_{i\in s} d_i q_i^{\oplus}\right)\left(\sum_{i\in s} d_i q_i^{\oplus} x_i^2\right) - \left(\sum_{i\in s} d_i q_i^{\oplus} x_i\right)^2}\right\}\left(X - \hat{X}_{\text{HT}}\right). \tag{5.19.8}$$

On substituting (5.19.8) in (5.19.1) we have

$$\hat{Y}_{\text{s}} = \sum_{i\in s} d_i y_i + \hat{\beta}_{\text{ols}}\left(X - \sum_{i\in s} d_i x_i\right) \tag{5.19.9}$$

where

$$\hat{\beta}_{\text{ols}} = \frac{\left(\sum_{i\in s} d_i q_i^{\oplus} x_i y_i\right)\left(\sum_{i\in s} d_i q_i^{\oplus}\right) - \left(\sum_{i\in s} d_i q_i^{\oplus} y_i\right)\left(\sum_{i\in s} d_i q_i^{\oplus} x_i\right)}{\left(\sum_{i\in s} d_i q_i^{\oplus}\right)\left(\sum_{i\in s} d_i q_i^{\oplus} x_i^2\right) - \left(\sum_{i\in s} d_i q_i^{\oplus} x_i\right)^2} \tag{5.19.10}$$

which is clearly usual traditional linear regression estimator.

Note that if $q_i^{\oplus} = 1$ and under SRSWOR sampling where $d_i = N/n$, the estimator (5.19.9) reduces to

$$\hat{Y}_s = N\left[\bar{y} + \left(s_{xy}/s_x^2\right)\left(\bar{X} - \bar{x}\right)\right] \tag{5.19.11}$$

where

$\bar{y} = n^{-1}\sum_{i=1}^{n} y_i$, $\bar{x} = n^{-1}\sum_{i=1}^{n} x_i$, $s_x^2 = (n-1)^{-1}\sum_{i=1}^{n}(x_i - \bar{x})^2$ and $s_{xy} = (n-1)^{-1}\sum_{i=1}^{n}(x_i - \bar{x})(y_i - \bar{y})$,

and it is the famous traditional linear regression estimator of Hansen, Hurwitz, and Madow (1953), and celebrates Golden Jubilee Year 2003 for its outstanding performance. Note that there is no choice of weights $q_i^{\oplus}$ which reduces the estimator (5.19.9) into ratio or product methods of estimation. The only choice of weights $q_i^{\oplus} = 1$ reduces it to the traditional linear regression estimator, and leads to the following theorem.

Theorem 5.19.1. The traditional linear regression estimator due to Hansen, Hurwitz, and Madow (1953) is unique in its class of estimators.

A new estimator of the variance of the traditional linear regression estimator $\hat{Y}_s$ has also been suggested by Singh (2003c) as

$$\hat{v}_s\left(\hat{Y}_s\right) = \frac{1}{2}\sum_{i \neq j \in s}\sum D_{ij}\left(w_i^{\oplus}e_i^{\oplus} - w_j^{\oplus}e_j^{\oplus}\right)^2 \tag{5.19.12}$$

where $e_i^{\oplus} = y_i - \hat{\alpha}_{\text{ols}} - \hat{\beta}_{\text{ols}}x_i$. Note that $\hat{\alpha}_{\text{ols}}$ and $\hat{\beta}_{\text{ols}}$ are the least square estimates of α and β in the model $y_i = \alpha + \beta\, x_i + e_i$ obtained by minimizing $\sum_{i \in s} d_i q_i^{\oplus} e_i^{\oplus 2}$.

Singh (2003c) also studied a further calibrated estimator of variance given by

$$\hat{v}_{ss}\left(\hat{Y}_s\right) = \frac{1}{2}\sum_{i \neq j \in s}\sum \Omega_{ij}^{\oplus}\left(w_i^{\oplus}e_i^{\oplus} - w_j^{\oplus}e_j^{\oplus}\right)^2 \tag{5.19.13}$$

where $\Omega_{ij}^{\oplus}$ are two dimensional calibrated plus weights and are chosen such that two dimensional chi square distance

$$D^{\oplus} = \frac{1}{2}\sum_{i \neq j \in s}\sum \frac{\left(\Omega_{ij}^{\oplus} - D_{ij}\right)^2}{D_{ij}Q_{ij}^{\oplus}}, \tag{5.19.14}$$

is minimum subject to two calibration constraints given by

$$\sum_{i \neq j \in s}\sum \Omega_{ij}^{\oplus} = \sum_{i \neq j \in s}\sum D_{ij}, \tag{5.19.15}$$

and

$$\frac{1}{2}\sum_{i \neq j \in s}\sum \Omega_{ij}^{\oplus}\left(d_i x_i - d_j x_j\right)^2 = v\left(\hat{X}_{\text{HT}}\right). \tag{5.19.16}$$

Singh (2003c) suggested that the statistical package GES developed by Statistics Canada could be modified to obtain the traditional linear regression type estimates of population total, and to estimate its variance using the modified calibration approach discussed in this section. Similar changes in other statistical packages such as SUDAAN, CALMAR, SAS, and STATA etc. are also suggested.

Example 5.19.1. Continuing from Example 5.5.1, find the calibrated plus weights which leads to traditional linear regression estimate of the number of fish caught during 1995.

Solution. Continuing from Example 5.5.1 and for $q_i^{\oplus} = 1$ we have

$$w_i^{\oplus} = d_i + \left\{ \frac{d_i x_i \left(\sum_{i \in s} d_i \right) - d_i \left(\sum_{i \in s} d_i x_i \right)}{\left(\sum_{i \in s} d_i \right)\left(\sum_{i \in s} d_i x_i^2 \right) - \left(\sum_{i \in s} d_i x_i \right)^2} \right\} \left(X - \hat{X}_{\text{HT}} \right)$$

Thus we have the following table:

x_{1i}	y_i	π_i	$d_i = \pi_i^{-1}$	$d_i x_{1i}$	$d_i x_{1i}^2$	$w_i^{\oplus}$	$w_i^{\oplus} y_i$
2001	2016	0.107450	9.307	18622.615	37263852.955	9.071	18288.135
5692	2319	0.110228	9.072	51638.422	293925899.046	9.167	21257.489
2653	3816	0.115834	8.633	22903.465	60762893.451	8.469	32318.978
4860	4008	0.106153	9.420	45782.974	222505251.853	9.443	37846.597
3850	2568	0.120075	8.328	32063.294	123443681.033	8.267	21228.615
17741	16238	0.139569	7.165	127112.754	2255107373.414	8.074	131111.675
776	163	0.103605	9.652	7489.986	5812229.140	9.294	1514.894
2300	2324	0.107686	9.286	21358.394	49124305.852	9.078	21098.349
		Sum	70.864	326971.904	3047945486.743		284664.732

Thus a traditional linear regression estimate of the number of fish caught during 1995 is given by

$$\hat{y}_{\text{lr}} = \sum_{i \in s} w_i^{\oplus} y_i = 284664.732.$$

EXERCISES

Exercise 5.1. Find the bias and variance of the estimators of population total, Y, defined as

$$\hat{Y}_{g1} = \hat{Y} u_1^{\alpha} u_2^{\beta}, \quad \hat{Y}_{g2} = \hat{Y}\{1 + \alpha(u_1 - 1) + \beta(u_2 - 1)\}, \quad \text{and} \quad \hat{Y}_{g3} = \hat{Y}[1 + \alpha(u_1 - 1) + \beta(u_2 - 1)]^{-1}$$

where $u_1 = \hat{X}_1 / X_1$ and $u_2 = \hat{X}_2 / X_2$ for $\hat{X}_r = \sum_{i=1}^{n} \frac{x_i^r}{\pi_i}$ and $X_r = \sum_{i=1}^{N} X_i^r$.

Hint: Sampath and Chandra (1990).

Exercise 5.2. Find the bias and mean squared error of the estimators of population total Y defined as

$$\hat{Y}_{kg1} = \alpha \hat{Y} \left(\frac{X_1}{\hat{X}_1} \right) + \beta \hat{Y} \left(\frac{X_1}{\hat{X}_1} \right)^2 \quad \text{and} \quad \hat{Y}_{kg2} = \hat{Y}_{kg1} + \gamma \hat{Y} \left(\frac{X_1}{\hat{X}_1} \right)^3$$

where α, β and γ are suitably chosen constants such that MSEs of the estimators are minimum.
Hint: Kapadia and Gupta (1984).

Exercise 5.3. Show that for any sampling design with positive first order inclusion probabilities for all the units in the population, the covariance between $\hat{Y} = \sum_{i=1}^{n} d_i y_i$ and $\hat{X}_r = \sum_{i=1}^{n} d_i x_i^r$ is given by

$$\text{Cov}\left(\hat{Y}, \hat{X}_r\right) = \sum_{i=1}^{N} Y_i X_i^r (1 - \pi_i)\pi_i^{-1} + \sum_{i \neq j=1}^{N} Y_i X_j^r \left(\pi_{ij} - \pi_i \pi_j\right)\left\{\pi_{ij}\pi_i\pi_j\right\}^{-1}$$

where $d_i = \pi_i^{-1}$, for $r = 1, 2, ..., m$. Derive its value under SRSWOR sampling design.
Hint: Sampath and Chandra (1990).

Exercise 5.4. Study the asymptotic properties of the estimator of population total Y defined as

$$\hat{Y}_m = \hat{Y}\left[\sum_{r=1}^{m} \lambda_r \left(\frac{\hat{X}_r}{X_r}\right)\right],$$

where $\hat{X}_r = \sum_{i=1}^{n} \frac{x_i^r}{\pi_i}$ has its usual meaning.

Exercise 5.5. Show that the minimization of $E_m E_p\left[\bar{y}_s - \bar{Y}\right]$ for any design $p(s)$ under the model $m : Y_i = \beta X_i + \epsilon_i$, where $E_m\left(\epsilon_i^2 | X_i\right) = \sigma^2 f(X_i)$ leads to the estimator of total, Y, given by

$$\hat{y}_1 = \sum_{i \in s} y_i + \left(\sum_{i \in s} \frac{x_i y_i}{f(x_i)}\right) \Bigg/ \left(\sum_{i \in s} \frac{x_i^2}{f(x_i)}\right) \sum_{i \in \bar{s}} x_i .$$

Hint: Royall (1970a, 1970b, 1970c, 1971), Bellhouse (1984).

Exercise 5.6. Consider $\bar{y}_s = \frac{1}{N}\sum_{i \in s} \frac{y_i}{\pi_i}$, $\bar{x}_{sj} = \frac{1}{N}\sum_{i \in s} \frac{x_{ij}}{\pi_i}$ and $\bar{X}_j = \frac{1}{N}\sum_{i=1}^{N} X_{ij}$ for $j = 1, 2, ..., p$ have their usual meanings. Study the bias and variance of the estimator of the population mean $\bar{Y}$ defined as

$$\bar{y}_1 = \bar{y}_s + \sum_{j=1}^{p} \hat{\beta}_{sj}\left(\bar{X}_j - \bar{x}_{sj}\right),$$

where $\hat{\beta}_{sj}$ denotes the jth partial regression coefficient, under the superpopulation model.
Hint: Särndal (1980b).

Exercise 5.7. (a) Minimize each of the following distance functions:

$$D_1 = \sum_{i=1}^{n}\frac{(w_i - d_i)^2}{2d_i}; \quad D_2 = \sum_{i=1}^{n}\left[w_i \ln\left(\frac{w_i}{d_i}\right) - w_i + d_i\right]; \quad D_3 = 2\sum_{i=1}^{n}\left(\sqrt{w_i} - \sqrt{d_i}\right)^2;$$

$$D_4 = \sum_{i=1}^{n}\left[-d_i \ln\left(\frac{w_i}{d_i}\right) + w_i - d_i\right]; \text{ and } D_5 = \sum_{i=1}^{n}\frac{(w_i - d_i)^2}{2w_i}$$

subject to the calibration constraint

$$\sum_{i=1}^{n} w_i x_i = X,$$

where d_i are the known design weights and w_i are the calibrated weights to be found. Discuss the nature of calibrated weights in each situation.

(b) Optimise the generalized distance function

$$D = \sum_{i=1}^{n}\frac{(w_i - d_i)^{\alpha}}{d_i q_i}$$

subject to

$$\sum_{i=1}^{n} w_i x_i = X.$$

Find α such that the variance of the resultant estimator is minimum. Is optimum $\alpha = 2$?

(c) In each one of the above cases (a) and (b), study the following two estimators of the population total, Y, defined as

$$\hat{Y}_{\text{ds}} = \sum_{i=1}^{n} w_i y_i \text{ and } \hat{Y}_{\text{es}} = \sum_{i=1}^{n} w_i \{y_i + e_i\} - \sum_{i=1}^{n} d_i e_i$$

where $e_i = y_i - \hat{\beta}_{\text{ds}} x_i$.

Hint: Deville and Särndal (1992), Estevao and Särndal (2000).

Exercise 5.8. Find the bias and variance of the Rao, Hartley, and Cochran (1962) strategy and the Horvitz and Thompson (1952) strategy for a multi-character survey for estimating population totals given by

$$\hat{Y}_{\text{RHC}} = \sum_{i=1}^{n}\frac{y_i}{P_i^*}\tau_i \text{ and } \hat{Y}_{\text{HT}} = \frac{1}{n}\sum_{i=1}^{n}\frac{y_i}{P_i^*\pi_i}$$

where

$$P_i^* = \left(1 + \frac{1}{N}\right)^{1-\rho_{xy}} (1 + P_i)^{\rho_{xy}} - 1$$

and ρ_{xy} is the known correlation between the selection probabilities P_i and the variable under study, under the superpopulation model, $Y_i = \beta P_i + e_i$, where e_i are the error terms such that $E(e_i \mid P_i) = 0$, $E(e_i^2 \mid P_i) = aP_i^g$ with a>0 and $g \ge 0$ and $E(e_i e_j \mid P_i P_j) = 0 \ \forall i \ne j$. Can you suggest two more transformations P_i^* ?

Hint: Bansal and Singh (1989,1990) , Bedi (1995), Mangat and Singh (1992-93).

Exercise 5.9. Let there be a population of N units and we want to select a sample of n units. For this selection, the population is randomly divided into $(n+k)$ groups of sizes $N_1, N_2, ..., N_{n+k}$ such that $\sum_{i=1}^{n+k} N_i = N$. For the first group we select N_1 units out of N units with SRSWOR sampling. Then for the second group, select N_2 units out of $(N - N_1)$ units with the same sampling scheme and so on. From these $(n+k)$ groups we then select a sample of n random groups using SRSWOR sampling. Now for selecting the ultimate sample of n units, we select one unit with probability proportional to the original probabilities P_i such that $\sum_{i=1}^{N} P_i = 1$ from each of the groups independently. For this scheme, show that an unbiased estimator of population total is given by

$$\hat{Y}_1 = \frac{(n+k)}{n} \sum_{i \in s} \frac{y_i}{(P_i/\tau_i)} \quad \text{where } \tau_i = \sum_{j \in G_i} P_{ij} .$$

Find its variance and compare with the usual RHC scheme.
Hint: Bansal and Singh (1986).

Exercise 5.10. Suppose a selection scheme consists of the following steps:

Step I. Select the first unit with probability for the ith unit proportional to

$$(X_i - \bar{X})^2 + N^{-1} \sum_{i=1}^{N} (X_i - \bar{X})^2 .$$

Step II. Select the second unit from the remaining $(N-1)$ units with conditional probability for the jth unit being proportional to $(X_j - X_i)^2$.

Step III. Select $(n-2)$ units from the remaining units of the population by simple random sampling and without replacement.
Show that under such a sampling scheme, the probability of selecting s^{th} sample is

$$P(s) = s_x^2 \Big/ \left\{ \binom{N}{n} S_x^2 \right\} .$$

Deduce that the regression estimator of population mean defined as,

$$\bar{y}_{\text{lr}} = \bar{y}_s + \left(s_{xy}/s_x^2\right)\left(\bar{X} - \bar{x}_s\right)$$

and the ratio type estimator of finite population variance defined as

$$s_1^2 = s_y^2\left(S_x^2/s_x^2\right)$$

are unbiased for population mean and population variance.
Hint: Singh and Srivastava (1980), Swain and Mishra (1992).

Exercise 5.11. Show that the difference between the variance of the estimator of population total Y under PPSWR sampling and PPSWOR sampling schemes is:

$$\text{Difference} = V\left(\hat{Y}_{\text{HT}}\right) - V\left(\hat{Y}_{\text{HH}}\right) = \sum_{i \neq j = 1}^{N} \left(Y_i - P_i Y\right)\left(Y_j - P_j Y\right)\frac{\pi_{ij}^2}{n^2 P_i P_j}.$$

Show that $\text{Difference} \geq 0$ for Midzuno (1952) and Sen (1952) sampling schemes.
Hint: Prabhu--Ajgaonkar (1975).

Exercise 5.12. If the probability of selecting the sth sample is given by

$$P(s) = m \Big/ \left\{ \binom{N}{n} \bar{x} \right\}.$$

Show that a product type estimator of population mean, $\bar{y}_1 = (\bar{y}\bar{x})/m$, where m is the equiprobable harmonic mean of x values, is unbiased.
Hint: Ruiz and Santos (1990).

Exercise 5.13. In the RHC sampling scheme, let the first random group be made by using Midzuno--Sen's scheme of sampling while the remaining $(n-1)$ groups are constructed as usual. Show that if the random groups are of equal size then the resultant strategy and the usual RHC strategy are equally efficient; otherwise, find the condition under which the resultant strategy fares better than the usual RHC strategy.
Hint: Singh and Lal (1978).

Exercise 5.14. Write the FORTRAN codes to generate the first and second order inclusion probabilities to estimate the variance of the Horvitz and Thompson estimator of population total.
Hint: Bandyopadhyay, Chattopadhyaya, and Kundu (1977).

Exercise 5.15. In the RHC sampling scheme let the sample be selected in such a way that:
(a) The cost is proportional to the expected number of distinct units in the sample; and
(b) The cost is proportional to the total expected size of the sample where $\{P_j\}$ are taken as the relative measures of sizes for the respective units in the population.

Then show that the RHC scheme remains less efficient than the PPSWR sampling scheme for the fixed cost of surveys.
Hint: Singh and Kishore (1975).

Exercise 5.16. Show that under Midzuno--Sen's scheme of sampling, the variance of the Horvitz and Thompson (1952) estimator (HTE) of total may not generally decrease with an increase in sample size or average effective sample size. Suggest an improvement so that the resultant πps property yields a variance of HTE which decreases with increasing sample size.
Hint: Chaudhuri and Arnab (1978).

Exercise 5.17. Discuss the efficiency of the usual ratio estimator under Midzuno-Sen's scheme of sampling.
Hint: Singh (1975a).

Exercise 5.18. Show that the general formula for estimating the variance of the estimator of population total under the RHC scheme is

$$\hat{v}_b = \sum_{i=1}^{n}\sum_{j=1}^{n} a_{ij} d_{ij},$$

where $a_{ij} = b_{ij}/(N_i N_j)$ $i \neq j$, $d_{ij} = \dfrac{Q_i Q_j}{2}\left(\dfrac{y_i}{P_i} - \dfrac{y_j}{P_j}\right)^2$ and $\sum_{i=1}^{n}\sum_{j=1}^{n} b_{ij} = \sum_{i=1}^{n} N_i(N_i - 1)$.

Hint: Chaudhuri and Mitra (1992), Ohlsson (1989).

Exercise 5.19. Discuss the asymptotic properties of the two estimators of the variance of the linear regression estimators:

$$\hat{v}_1 = \sum_{j \neq i=1}^{n}\sum^{n} D_{ij}\left(d_i e_i - d_j e_j\right)^2 \quad \text{and} \quad \hat{v}_2 = \sum_{j \neq i=1}^{n}\sum^{n} D_{ij}\left(d_{i_i} g_i e_i - d_j g_j e_j\right)^2$$

where $g_i = \pi_i w_i$ and the other symbols have their usual meanings.
Hint: Chaudhuri and Maiti (1994), Valliant (2002).

Exercise 5.20. Show that an estimator of variance of the estimator of population total can be written as

$$e(s, y) = \sum_{k \in s} b(s, kk) y_k^2 + \sum_{k \neq l \in s} b(s, kl) y_k y_l$$

where $b(s, kk)$ and $b(s, kl)$ are suitably chosen constants. Find the bias and variance of this estimator.
Hint: Mukhopadhyay (1982), Hanif, Mukhopadhyay, and Bhattacharyya (1993)

Exercise 5.21. Discuss the relative efficiencies of the strategies due to Horvitz and Thompson (1952), Rao, Hartley, and Cochran (1962), Sen (1952) and Midzuno (1952) sampling schemes in estimating a finite population total under the assumptions of a superpopulation model.
Hint: Chaudhuri and Arnab (1979).

Exercise 5.22. Is there any model free evaluation technique to compare the bias and the mean squared error of the regression estimator?
Hint: Konijn (1979).

Exercise 5.23. Assume that the sampling design is such that when the population size N and sample size n are large enough, the joint probability sampling distribution of the vector $\hat{z} = \left(\hat{\bar{Y}}, \hat{\bar{X}}^T\right)^T$ can be assumed normal with mean vector $z = \left(\bar{Y}, \bar{X}^T\right)^T$ and variance matrix

$$V(\hat{z}) = \begin{bmatrix} V\left(\hat{\bar{Y}}\right), C\left(\hat{\bar{X}}, \hat{\bar{Y}}\right) \\ C\left(\hat{\bar{X}}, \hat{\bar{Y}}\right), V\left(\hat{\bar{X}}\right) \end{bmatrix}$$

where $V\left(\hat{\bar{Y}}\right)$ and $V\left(\hat{\bar{X}}\right)$ are the sampling variance matrices of $\hat{\bar{Y}}$ and $\hat{\bar{X}}$, respectively, and $C\left(\hat{\bar{X}}, \hat{\bar{Y}}\right)$ is the sampling covariance vector. Then show that the estimator $\hat{\bar{Y}}$ is normal with mean vector

$$E\left(\hat{\bar{Y}} \mid \hat{\bar{X}}\right) = \bar{Y} + B^T\left(\hat{\bar{X}} - \bar{X}\right)$$

and variance matrix

$$V\left(\hat{\bar{Y}} \mid \hat{\bar{X}}\right) = V\left(\hat{\bar{Y}}\right) - B^T V\left(\hat{\bar{X}}\right) B,$$

where $B = V\left(\hat{\bar{X}}\right)^{-1} C\left(\hat{\bar{X}}, \hat{\bar{Y}}\right)$.

Also show that the conditional mean squared error of the estimator, $\hat{\bar{Y}}$, can be written as

$$\text{MSE}\left(\hat{\bar{Y}} \mid \hat{\bar{X}}\right) = V\left(\hat{\bar{Y}}\right) + B^T\left[\left(\hat{\bar{X}} - \bar{X}\right)\left(\hat{\bar{X}} - \bar{X}\right)^T - V\left(\hat{\bar{X}}\right)\right]B.$$

Hint: Montanari (1999).

Exercise 5.24. If the estimating function g^* is optimal, that is if it satisfies the condition that $E\left\{g^* \Big/ E\left(\frac{\partial g^*}{\partial \theta_1}\right)\right\}^2 \le E\left\{g \Big/ E\left(\frac{\partial g}{\partial \theta_1}\right)\right\}^2$, then prove the following results:

(a) $\text{Corr}^2\left\{g^*, \frac{\partial \log p(\theta|x)}{\partial \theta}\right\} \ge \text{Corr}^2\left\{g, \frac{\partial \log p(\theta|x)}{\partial \theta}\right\}$

(b) $E\left\{g^* - \frac{\partial \log p(\theta|x)}{\partial \theta}\right\}^2 \ge E\left\{g - \frac{\partial \log p(\theta|x)}{\partial \theta}\right\}^2$

for all distributions $p \in \mathrm{P}$ and estimating functions $g \in G$.

Hint: Godambe (1999).

Exercise 5.25. Consider a πps sampling design with $\pi_i = nx_i/X = np_i$, $i = 1,2,...,N$. Let $y_i, i = 1,2,...,n$ be the values of the n units in the sample s drawn by the above πps sampling design. Suppose that the finite population $\{Y_1, Y_2,..., Y_N\}$ is a random sample from the following superpopulation model, $y_i = \beta x x_i + e_i$, with $E(e_i) = 0$, $E\left(e_i^2\right) = \sigma^2 x_i^2$, $E\left(e_i e_j\right) = 0$, $i \ne j$, $E\left(e_i^3\right) = 0$ and

$E(e_i^4) = 3\sigma^4 x_i^4$, where $i = 1,2,...,N$, β and $\sigma^2 > 0$ are unknown parameters. An unbiased estimator of population mean $\bar{Y}$ is

$$\bar{y}_{\text{HT}} = \frac{1}{N}\sum_{i=1}^{n}\frac{y_i}{\pi_i} = \frac{1}{N}\frac{1}{n}\sum_{i=1}^{n}\frac{y_i}{p_i}$$

Assuming PPSWR sampling designs, compare the following estimators of variance $V(\bar{y}_{\text{HT}})$, given by

(a) $\hat{v}_{\text{WR}} = \dfrac{\bar{X}^2}{n(n-1)}\displaystyle\sum_{i=1}^{n}\left(\frac{y_i}{x_i} - \frac{1}{n}\sum_{i=1}^{n}\frac{y_i}{x_i}\right)^2$, (b) $\hat{v}_{\text{M}} = \left(1 - 2\displaystyle\sum_{i=1}^{n} p_i + n\sum_{i=1}^{N} p_i^2\right)\hat{v}_{\text{WR}}$,

(c) $\hat{v}_{\text{M}}^{*} = \left(1 - n\displaystyle\sum_{i=1}^{N} p_i^2\right)\hat{V}_{\text{WR}}$, (d) $\hat{v}_{\text{Z}} = \dfrac{n-1}{n+1}\left(1 - 2\displaystyle\sum_{i=1}^{n} p_i + n\sum_{i=1}^{N} p_i^2\right)\hat{v}_{\text{WR}}$,

and

(e) $$\hat{v}_{\text{Z}}^{*} = \frac{n-1}{n+1}\left[1 - 2\sum_{i=1}^{n} p_i + n\sum_{i=1}^{N} p_i^2 + \frac{2}{n-1}\left\{n\sum_{i=1}^{n} p_i^2 - \left(\sum_{i=1}^{n} p_i\right)^2\right\}\right]\hat{v}_{\text{WR}}.$$

Hint: Kott (1988), Zou (1999).

Exercise 5.26. If a simple random sample is drawn without replacement from a population of size N with random sample size n_s where $n_s > 0$, then if $\eta = n_s$ then show that the first order conditional inclusion probability is given by

$$\pi_{i|\eta} = n_s/N, \ \forall\, i \in \Omega$$

Hint: Tillê (1998).

Exercise 5.27. Suppose σ^2 is known. Then the necessary and sufficient conditions for the estimator $\hat{Y} = w_{0s} + \sum_{k\in s} w_{ks} y_k$ of the linear function $\sum_{k=1}^{N} p_k Y_k$ to be admissible are that there exists λ_s such that $w_{ks} = \lambda_s a_k + p_k \ (k \in s)$, and one of the following two conditions are satisfied:

(i) $0 \le \lambda_s \le \dfrac{c_s}{d_s}$, where $c_s \approx \displaystyle\sum_{k \notin s} p_k a_k$ and $d_s \approx \displaystyle\sum_{k\in s} a_k^2$;

(ii) $\lambda_s = \dfrac{c_s}{d_s}$, and $w_{0s} = -\dfrac{c_s}{d_s}\displaystyle\sum_{k\in s} a_k b_k + \sum_{k\notin s} p_k b_k - \frac{a\sigma^2}{2}\left[\left(\frac{c_s}{d_s}\right)^2 d_s + \sum_{k\notin s} p_k^2\right]$

under the superpopulation model $y_k = a_k\beta + b_k + \varepsilon_k$, $k = 1,2,..., N$.

Hint: Zou (1997).

Exercise 5.28. A necessary and sufficient condition for the estimator

$$\hat{v}_s = \frac{1-f}{n}s_y^2$$

to be admissible in the class of all quadratic estimators is $f \ge 2/(n+1)$.

Hint: Zou and Liang (1997).

Exercise 5.29. Consider a design with $p(\bar{s}) = \binom{N-1}{n}^{-1} \sum_{\bar{s}} p_i$. Show that an unbiased estimator of population total is given by $\hat{\bar{y}} = \bar{y}\left(\frac{\bar{X}}{\bar{x}_{\bar{s}}}\right)$, where $\bar{x}_{\bar{s}} = \frac{1}{N-n} \sum_{\bar{s}} x_i$.

Hint: Deshpande (1978).

Exercise 5.30. For any sampling design p and any real number α, show that

$$E\left(\bar{y}^{\alpha}\right) = \frac{1}{n^{\alpha}} \sum_{i_1=1}^{N} \sum_{i_2=1}^{N} \dots \sum_{i_\alpha=1}^{N} y_{i_1} y_{i_2} \dots y_{i\alpha} \pi(i_1, i_2, \dots, i_\alpha),$$

where $\bar{y} = n^{-1} \sum_{i=1}^{n} y_i$ is the sample mean, and $\pi(i_1, i_2, \dots, i_\alpha)$ is the αth order inclusion probability of including units $(i_1, i_2, \dots, i_\alpha)$. Also for any real number β show that

$$E\left\{\left(\sum_{i \in s} y_i^{\beta}\right)\left(\bar{y}^{\alpha}\right)\right\} = \frac{1}{n^{\alpha}} \sum_{i_0=1}^{N} \sum_{i_1=1}^{N} \sum_{i_2=1}^{N} \dots \sum_{i_\alpha=1}^{N} y_{i_0}^{\beta} y_{i_1} y_{i_2} \dots y_{i\alpha} \pi(i_0, i_1, i_2, \dots, i_\alpha).$$

Deduce the result that for any non-negative real number β if $M_{\beta} = \sum_{i \in s} (y_i - \bar{y})^{\beta}$ then show that

$$E\{M_{\beta}\} = \sum_{t=1}^{\beta-1} (-1)^t \binom{\beta}{t} \frac{1}{n^t} \left\{ \sum_{i_0=1}^{N} \sum_{i_1=1}^{N} \dots \sum_{i_t=1}^{N} y_{i_0}^{\beta-t} y_{i_1} \dots y_{i_t} \pi(i_0, i_1, \dots, i_t) \right\}$$

$$+ (-1)^{\beta} \frac{1}{n^{\beta-1}} \left\{ \sum_{i_1=1}^{N} \dots \sum_{i_\beta=1}^{N} y_{i_1} \dots y_{i_\beta} \pi(i_1, \dots, i_\beta) + \sum_{i=1}^{N} y_i^{\beta} \pi(i) \right\}$$

Hint: Srivastava and Saleh (1985)

Exercise 5.31. Find the bias and variance of the following estimator of the population total Y as

$$\hat{Y}_{\mathrm{H}} = \sum_{i \in s} d_i y_i \left(\frac{X}{\sum_{i \in s} d_i x_i} \right), \quad \text{where } d_i = \pi_i^{-1}.$$

Hint: Hàjek (1959).

Exercise 5.32. Find the bias and variance of the estimator of the population total Y defined as

$$\hat{Y}_{\mathrm{H}} = \sum_{i \in s} d_i y_i + b\left(X - \sum_{i \in s} d_i x_i \right)$$

where $d_i = \pi_i^{-1}$. Discuss the optimum choice of the constant b.

Hint: Cassel, Särndal, and Wretman (1977).

Exercise 5.33. Justify the statement 'For each sampling design there exists a rejective method of drawing a sample and terminates with selection of a sample with probability one.' Also show that for Hãjek's (1964) method the expected number of trials for drawing a sample is $\left[\sum_{s\in\Omega}\left\{\prod_{i\in s} p_i \prod_{j\notin s}\left(1-p_j\right)\right\}\right]^{-1}$.

Hint: Hãjek (1959), Deshpande and Ajgaonkar (1977).

Exercise 5.34. In the non-parametric model $y_i = m(x_i) + e_i$, where $E_m(e_i) = 0$ and $E_m\left(e_i^2\right) = a x_i^g$, $g \geq 0$, discuss different method of estimating $m(x_i)$. Show that if $m(x_i) = \beta x_i$ the non-parametric model reduces to parametric one.

Hint: Breidt and Opsomer (2000).

Exercise 5.35. Let y and x_i, $i = 1,2,...,l$, respectively, be the survey variable and the auxiliary variables related to y and the information about the quantiles of the auxiliary characters or distribution functions are known. From the sample of n units from a population of size N we observe (x_{ik}, y_k) where $k \in s$. Consider $Q_{X_i}(\alpha_i)$ for $\alpha_i \in (0.0,0.5)\cup(05,1.0)$ are known and we wish to estimate $Q_Y(\beta)$ with $\beta = 1/2$. Study the asymptotic properties of the following estimators of $Q_Y(\beta)$ as

$$\hat{F}_{\text{R}} = \hat{F}_{\text{HTy}}(Q_Y(\beta))\sum_{i=1}^{l} W_i\left\{\frac{F_{\text{HTx}_i}\left(Q_{X_i}(\alpha_i)\right)}{\hat{F}_{\text{HTx}_i}\left(Q_{X_i}(\alpha_i)\right)}\right\}, \quad \text{with } \sum_{i=1}^{l} W_i = 1$$

and

$$\hat{F}_{\text{D}} = \hat{F}_{\text{HTy}}(Q_Y(\beta)) + \sum_{i=1}^{l} b_i\left\{F_{\text{HTx}_i}\left(Q_{X_i}(\alpha_i)\right) - \hat{F}_{\text{HTx}_i}\left(Q_{X_i}(\alpha_i)\right)\right\}$$

where

$$\hat{F}_{\text{HTy}} = \sum_{k\in s}\frac{\Delta(Q_Y(\beta) - y_k)}{N\,\pi_k} \text{ and } \hat{F}_{\text{HTx}_i} = \sum_{k\in s}\frac{\Delta\left(Q_{X_i}(\alpha_i) - x_{ik}\right)}{N\,\pi_k}.$$

Hint: Rueda and Arcos (2002).

Exercise 5.36. Consider a penalized chi square distance function between design weights d_i and calibrated weights w_i^* as

$$D = \frac{1}{2}\sum_{i\in s}\frac{\left(w_i^* - d_i\right)^2}{d_i q_i^*} + \frac{1}{2}\Phi^2\sum_{i\in s}\frac{w_i^{*2}}{d_i q_i^*},$$

where q_i^* are weights and Φ is a positive quantity that reflects a penalty to be decided by the investigator based on prior knowledge, or the desire for certain levels of efficiency and bias. Minimize D subject to the following two situations:

(i) No auxiliary information is available;

and

(ii) Calibration constraint of Deville and Särndal (1992).

Discuss the different estimators of $\hat{Y}_{\text{new}} = \sum_{i\in s} w_i^* y_i$ as special cases for different sampling schemes and choices of weights q_i. What choice of penalty and which situation leads to Searls' estimator?
Hint: Farrell and Singh (2002b).

Exercise 5.37. Under the linear model $Y_i = \alpha + \beta X_i + \varepsilon_i$, such that $E(\varepsilon_i) = 0$, $E(\varepsilon_i^2) = 0$ and $E(\varepsilon_i\varepsilon_j) = 0$, $i \neq j$, consider the problem of estimation of population regression coefficient, defined as

$$\beta = \left\{N\sum_{i=1}^{N} X_i Y_i - \sum_{i=1}^{N} X_i \sum_{i=1}^{N} Y_i\right\} \Bigg/ \left\{N\sum_{i=1}^{N} X_i^2 - \left(\sum_{i=1}^{N} X_i\right)^2\right\}.$$

Show that the usual OLS estimator of β defined as

$$\hat{\beta}_{\text{ols}} = \left\{n\sum_{i=1}^{n} x_i y_i - \sum_{i=1}^{n} x_i \sum_{i=1}^{n} y_i\right\} \Bigg/ \left\{n\sum_{i=1}^{n} x_i^2 - \left(\sum_{i=1}^{n} x_i\right)^2\right\}$$

is unbiased, with variance

$$V\left(\hat{\beta}_{\text{ols}}\right) = \sigma^2 \Bigg/ \left\{n\sum_{i=1}^{n} x_i^2 - \left(\sum_{i=1}^{n} x_i\right)^2\right\}.$$

Consider the sample s is taken from a finite population Ω with sampling design p such that π_i and π_{ij} denote the first and second order inclusion probabilities. Show that the variance of the design consistent estimator of β, given by

$$\hat{\beta}_{\text{dc}} = \left\{\sum_{i\in s} d_i \sum_{i\in s} d_i x_i y_i - \sum_{i\in s} d_i x_i \sum_{i\in s} d_i y_i\right\} \Bigg/ \left\{\sum_{i\in s} d_i \sum_{i\in s} d_i x_i^2 - \left(\sum_{i\in s} d_i x_i\right)^2\right\},$$

where $d_i = 1/\pi_i$, has the variance

$$V\left(\hat{\beta}_{\text{dc}}\right) = \left\{\sum_{i}\sum_{j\in\Omega} \Delta_{ij} d_i d_j \left(X_i - \bar{X}\right)\left(X_j - \bar{X}\right)\varepsilon_i\varepsilon_j\right\} \Bigg/ \left\{\sum_{i\in\Omega} d_i \left(X_i - \bar{X}\right)^2\right\}^2,$$

where $\Delta_{ij} = \left(\pi_{ij} - \pi_i\pi_j\right)$ and $\varepsilon_i = \left(Y_i - \bar{Y}\right) - \beta\left(X_i - \bar{X}\right)$.
Hint: Sharma, Singh, Rai ,and Verma (2000).

Exercise 5.38. Consider an auxiliary variable x has a negative correlation with the study variable y. Let $X = \sum_{i=1}^{N} X_i$ and consider a transformed variable $X_i^* = (X - X_i)$, $i = 1,2,...,N$, so that $\sum_{i=1}^{N} X_i^* = (N-1)X$ and the probabilities of selection are

$$p_i^* = \frac{(1-p_i)}{(N-1)}, \quad i = 1,2,...,N.$$

Then study the following estimators of the population total

$$\hat{Y}^{*}_{\text{HH}} = \frac{1}{n}\sum_{i=1}^{n}\frac{y_i}{p_i^{*}} \quad \text{and} \quad \hat{Y}^{*}_{\text{HT}} = \sum_{i=1}^{n}\frac{y_i}{\pi_i^{*}},$$

where π_i^{*} is the first order inclusion probability with probability set p_i^{*}.

Hint: Bedi and Rao (2001).

Exercise 5.39. Find the condition on the real constants c_i, $i = 1, 2, ..., N$ such that the estimator

$$\hat{Y}_{\text{B}} = \sum_{i\in s}\frac{y_i - c_i}{\pi_i} + C,$$

where $C = \sum_{i=1}^{N} c_i$, is unbiased for estimating population total Y.

Hint: Basu (1971).

Exercise 5.40. An estimator of population total, Y, due to Deville and Sarndal (1992) is given by

$$\hat{Y}_{\text{ds}} = \sum_{i\in s} d_i y_i + \hat{\beta}_{\text{ds}}\left(X - \sum_{i\in s} d_i x_i\right)$$

where $\hat{\beta}_{\text{ds}} = \sum_{i\in s} d_i q_i x_i y_i \Big/ \sum_{i\in s} d_i q_i x_i^2$,with approximate variance

$$V\left(\hat{Y}_{\text{ds}}\right) = \sum_{i\in\Omega} d_i(1-\pi_i)e_i^2 - \sum_{i\in\Omega}\sum_{j(\neq i)\in\Omega} d_i d_j \Theta_{ij} e_i e_j$$

and consider an estimator of the variance is given by

$$\hat{v}_1\left(\hat{Y}_{\text{ds}}\right) = \sum_{i\in s} w_i^{*} d_i(1-\pi_i)e_i^2 - \sum_{i\in s}\sum_{j(\neq i)\in s} d_i d_j D_{ij} e_i e_j .$$

(a) Consider a model m such that $E_m(e_i \mid x_i) = 0$, $E_m\left(e_i^2 \mid x_i\right) = \sigma^2 v(x_i)$ and $E_m\left(e_i e_j \mid x_i x_j\right) = 0$. Assuming that $v(x_i)$ is known, show that the calibration equation obtained by equating

$$E_m\left[\hat{v}_1\left(\hat{Y}_{\text{ds}}\right)\right] = E_m\left[V\left(\hat{Y}_{\text{ds}}\right)\right]$$

is given by

$$\sum_{i\in s} w_i^{*} d_i(1-\pi_i)v(x_i) = \sum_{i\in\Omega} d_i(1-\pi_i)v(x_i).$$

(b) Consider an estimator of the population total Y as

$$\hat{Y}_{\text{ss}} = \sum_{i\in s} w_i^{*} y_i$$

where w_i^{*} are recalibrated weights, and are obtained by the minimization of a penalized chi square distance function

$$D_p = \frac{1}{2}\sum_{i\in s}\frac{\left(w_i^{*} - w_i\right)^2}{w_i q_i^{*}} + \frac{1}{2}\Phi^2\sum_{i\in s}\frac{\left(w_i^{*}\right)^2}{w_i q_i^{*}},$$

assuming $w_i > 0$ for all $i = 1, 2, ..., n$, and are already calibrated weights (or design weights), and Φ is a known real constant and called penalty to the function.

(c) If no information on model parameter $v(x_i)$ is available, then show that the minimization of D_p leads to an estimator of total given by

$$\hat{Y}_{ss} = \frac{1}{1+\Phi^2}\sum_{i\in s} w_i y_i .$$

Also deduce the Searls' (1964) estimator as a special case for certain choice of Φ.
(d) Show that minimization of D_p subject to the calibration constraint developed in (a) leads to a new estimator is given by

$$\hat{Y}_{\text{SS}} = \frac{1}{\left(1+\Phi^2\right)}\left[\sum_{i=1}^{n} w_i y_i + \hat{\beta}_{\text{SS}}\left\{\sum_{i=1}^{N} d_i(1-\pi_i)v(x_i) - \sum_{i=1}^{n} w_i d_i(1-\pi_i)v(x_i)\right\}\right].$$

Further deduce the estimator owed to Prasad (1989), $\bar{y}_{p(\text{r})} = N\,\bar{y}_{\text{searl}}\left(\bar{X}/\bar{x}\right)$, as a special case of it for certain choice of Φ.
(e) Suggest some improved estimators of variance through calibration.
Hint: Singh (2002b), Singh and Horn (1999).

Exercise 5.41. (a) Let p_i be the probability mass at, y_i, for $i \in s$ and $\bar{X}$ be the know population means for the (vector valued) auxiliary variable, x. Study the asymptotic properties of the empirical maximum likelihood estimator of the population mean, $\bar{Y}$, defined as

$$\hat{\bar{Y}}_{\text{EL}} = \sum_{i\in s} \hat{p}_i y_i$$

where the values of $\hat{p}_i$ maximize the empirical likelihood function $L(p) = \prod_{i\in s} p_i$

subject to $\sum_{i\in s} p_i = 1$, $p_i > 0$ and $\sum_{i\in s} p_i x_i = \bar{X}$.

(b) Consider the estimation of a quadratic parameter, $V = \sum_{i=1}^{N}\sum_{j=i+1}^{N} \Phi(y_i,\ y_j)$, which includes population variance and co-variance as a special cases.
Study the asymptotic properties of an estimator of V defined as

$$\hat{V}_{\text{wu}} = N(N-1)\sum_{i\in s}\sum_{j>i} \hat{p}_{ij}\Phi(y_i, y_j),$$

where the p_{ij} 's maximizes the pseudo empirical likelihood function defined as

$$\hat{l}(p) = \sum_{i\in s}\sum_{j>i} d_{ij}\log(p_{ij}),$$

subject to the conditions

$$\sum_{i\in s}\sum_{j>i} p_{ij} = 1 \quad (p_{ij} > 0), \text{ and } \sum_{i\in s}\sum_{j>i} p_{ij}\Phi(\hat{y}_i, \hat{y}_j) = \frac{1}{N(N-1)}\sum_{i=1}^{N}\sum_{j=i+1}^{N} \Phi(\hat{y}_i, \hat{y}_j)$$

where $\hat{y}_i$ are the fitted values.
Hint: Wu (2001), Wu and Sitter (2001), Sitter and Wu (2002), Chen, Sitter, and Wu (2002).

Exercise 5.42. Study the asymptotic properties of an unbiased ratio estimator of the population total defined as

$$\hat{T}_{\text{Rao}} = \left(\frac{T_x}{n}\right)\sum_{i\in s}\left(\frac{Y_i}{X_i}\right) + \frac{1}{N}\sum_{i<j\in s}\sum\left(B_iY_i - B_jY_j\right)\left(\frac{1}{B_i} - \frac{1}{B_j}\right) - \left(1 - \frac{1}{N}\sum_{i\in\Omega}\frac{1}{B_i}\right)\left(\sum_{i\in s}\frac{B_iY_i}{\pi_i}\right)$$

Show that if $B_i = T_x/(NX_i)$, and $\sum_{i\in\Omega} B_i^{-1} = N$ then $\hat{T}_{\text{Rao}}$ reduces to Hartley and Ross (1954) unbiased ratio estimator under SRSWOR sampling.
Hint: Rao (2002).

Exercise 5.43. Suppose X_1 and X_2 are the known totals of two auxiliary characters X_{1i} and X_{2i}, for $i = 1, 2, ..., N$. Consider an estimator of the population total Y as

$$\hat{y}_{\text{lr}} = \sum_{i\in s} w_i^{\oplus} y_i$$

Then show that the minimization of the CS distance function

$$D = \sum_{i\in s}\frac{\left(w_i^{\oplus} - d_i\right)^2}{d_i q_i^{\oplus}}$$

subject to the three linear calibration constraints, given by

$$\sum_{i\in s} w_i = \sum_{i\in s} d_i\,, \quad \sum_{i\in s} w_i x_{1i} = X_1\,, \text{ and } \sum_{i\in s} w_i x_{2i} = X_2\,.$$

leads to the traditional linear regression estimator with two auxiliary variables as

$$\hat{y}_{\text{lr}} = \sum_{i\in s} d_i y_i + \hat{\beta}_{1(\text{ols})}\left(X_1 - \sum_{i\in s} d_i x_{1i}\right) + \hat{\beta}_{2(\text{ols})}\left(X_2 - \sum_{i\in s} d_i x_{2i}\right)$$

where $\hat{\beta}_{1(\text{ols})}$ and $\hat{\beta}_{2(\text{ols})}$ are the ordinary least square estimates.
Hint: Singh (2003c).

PRACTICAL PROBLEMS

Practical 5.1. John and Michael were appointed to select three players ($n = 3$) out of five players ($N = 5$) from the list $\Omega = \{\text{Amy, Bob, Chris, Don, Eric}\}$ with their scores 125, 126, 128, 90 and 127, respectively.

(I) John considers the following sampling scheme consisting of all possibilities:

$s_1 = \{\text{Amy, Bob, Chris}\}$, $s_2 = \{\text{Amy, Bob, Don}\}$, $s_3 = \{\text{Amy, Bob, Eric}\}$,
$s_4 = \{\text{Amy, Chris, Don}\}$, $s_5 = \{\text{Amy, Chris, Eric}\}$, $s_6 = \{\text{Amy, Don, Eric}\}$,
$s_7 = \{\text{Bob, Chris, Don}\}$, $s_8 = \{\text{Bob, Chris, Eric}\}$, $s_9 = \{\text{Bob, Don, Eric}\}$,
and $s_{10} = \{\text{Chris, Don, Eric}\}$ such that
$P(s_t) = 0.1, \ \forall\ t = 1, 2, 3, ..., 10.$

(a) Find the first order inclusion probabilities for John's sampling scheme.

(b) Find the estimates of total score from each sample using John's sampling scheme.
(c) Find the bias in John's sampling scheme.
(d) Find the second order inclusion probabilities for John's sampling scheme.
(e) Find the variance of John's sampling scheme using Sen--Yates--Grundy formula.
(f) Find the variance of John's sampling plan using usual formula.
(g) Find the variance of John's sampling plan using definition of variance.
(h) Are three variances equal for John's sampling scheme?

(II) Michael likes Amy and cleverly suggests the following changes in John's sampling scheme as:

$P(s_t) = 1/6$, $\forall$ t = 1, 2, 3, 4, 5, 6 , and $P(s_t) = 0.00$, $\forall$ t = 7, 8, 9, 10.

(i) Find the first order inclusion probabilities for Michael's sampling scheme.
(j) Find the estimates of total score from each sample using Michael's sampling scheme.
(k) Find the bias in Michael's sampling scheme.
(l) Find the second order inclusion probabilities for Michael's sampling scheme.
(m) Find the variance of Michael's sampling scheme using Sen--Yates--Grundy formula.
(n) Find the variance of Michael's sampling plan using usual formula.
(o) Find the variance of Michael's sampling plan using definition of variance.
(p) Are three variances equal for Michael's sampling scheme?

(III) Comparison among techniques:
(q) Find the relative efficiency of Michael's sampling scheme over John's sampling scheme.
(r) Would you like to comment on your results?

Practical 5.2. Use the known information on the number of fish caught during 1992 to select a sample of seven units by using PPSWOR sampling. Collect the required information from population 4 to estimate the total number of fish caught during 1995 through a regression estimator by using information during 1994 as the auxiliary variable. Apply the Sen--Yates--Grundy form of the estimator of variance of the regression estimator to construct a 95% confidence interval. Also use the calibrated estimators of variance of the general linear regression estimator. Population 4 in the Appendix shows that the information on the number of different kinds of fish caught during 1992 and 1994 is available.

Practical 5.3. Develop the calibration weights for the units selected in the sample using PPSWOR sampling by making use of known information about the number of fish caught during 1994 as the auxiliary variable.

Sampled Units	Population units	Species group	First order inclusion probabilities	1994 X	1995 Y
1	10	Pollock	0.003453	862	832
2	15	White perch	0.075421	4648	3489
3	28	Gray snapper	0.110243	4845	4552
4	44	Sand seatrout	0.077703	5665	4355
5	63	Gulf flounder	0.00444	776	163
6	68	Puffers	0.043168	1141	935

For simplicity use the chi square distance function between the design weights and calibrated weights. Discuss the three cases where these weights lead to the ratio, GREG and traditional linear regression estimator for estimating the total number of fish caught during 1995. Derive the value of the estimate in each case.
Given: Total number of fish caught during 1994 is 341856.

Practical 5.4. Take an SRSWOR sample of 10 states from population 1 and note the records of real estate farm loans as well as nonreal estate farm loans. Given that information on the nonreal estate farm loans is available for all states, apply the ratio estimator for estimating the total real estate farm loans in the United States. Construct the 95% confidence intervals using the lower and higher level calibration approach.

Given: $N = 50$, $X = 43908.12$ and $S_x^2 = 1176526$. Repeat the exercise by taking PPSWOR sample.

Practical 5.5. Select a sample of 4 units by using the RHC scheme from population 1 of the Appendix. Estimate the total real estate farm loans using the RHC estimator and making use of nonreal estate farm loans as an auxiliary variable. Find a 95% confidence interval for the total real estate farm loans in the United States.
Hint: Divide the population into four random groups of unequal sizes.

Practical 5.6. The demand of fish for human consumption creates the need to estimate the total number of fish of all kinds caught by recreational fishermen of the Atlantic and Gulf coasts during 1994. Population 4 in the Appendix shows that the information on the number of different kinds of fish caught during 1992 is available. Use the known information on the number of fish caught during 1992 to select a sample of seven units by using PPSWOR sampling (Midzuno--Sen Sampling Scheme). Collect the required information from population 4 to estimate the total number of fish caught during 1994. Apply the Sen--Yates--Grundy estimator of variance to construct a 95% confidence interval.

Practical 5.7. Mr. Mario was interested to estimate the total value of production of the soybeans for beans in all the 30 states of the United States of America growing this particular crop. Mr. Mario selected a sample of $n = 5$ states based on some prior information and listed the first order (π_i) and second order (π_{ij}) inclusion probabilities based on PPSWOR sampling as

Data 1,000 cwt	First order inclusion probabilities		Second order inclusion probabilities π_{ij}				
y_i	π_i	$i\backslash j$	1	2	3	4	5
5430	0.55	1	?	?	?	?	?
7009	0.58	2	0.25	?	?	?	?
7180	0.60	3	0.27	0.29	?	?	?
6062	0.63	4	0.28	0.30	0.32	?	?
4054	0.65	5	0.30	0.32	0.32	0.36	?

(a) Apply the Horvitz and Thompson estimator to estimate the total value of production of the crop.
(b) Construct a 95% confidence interval of the total value of production based on Sen--Yates--Grundy estimator of variance.
(c) Construct a 95% confidence interval of the total value of production based on usual estimator of variance.
(d) Comment on the confidence interval estimates.

Practical 5.8. Calibration weights are found to be useful in survey sampling. Find the calibration weights for the units selected in the sample using PPSWOR sampling by making use of known information about the number of fish caught during 1992 and 1993 as auxiliary variables. Use the chi square distance function between the design weights and calibrated weights. Discuss situations where these weights lead to the GREG and traditional linear regression estimator for estimating the total number of fish caught during 1995. Construct a 95% confidence interval.

Sampled	Population units	Species group	1992	1993	1994	1995
			X_3	X_2	X_1	Y
1	05	Herrings	28933	34060	38007	30027
2	11	Red hake	559	216	369	184
3	17	Temperate basses	5	35	32	23
4	33	Snappers, others	746	861	462	492
5	42	Spotted seatrout	22304	21538	22181	24615
6	52	Mullets	5571	4186	4386	4657
7	61	Tunas/mackerels, oth	1190	794	1018	1029
8	69	Other fish	12249	14953	20488	14426

Practical 5.9. The following table provides a set of first order (π_i) and second order (π_{ij}) inclusion probabilities based on a population of size $N = 5$ units with values X and selection probabilities P_i .

Data	Selection Probabilities	First order inclusion probabilities	Second order inclusion probabilities π_{ij}					
X	P_i	π_i	$i \backslash j$	1	2	3	4	5
50	0.10	0.55	1	d_1	?	?	?	?
75	0.15	0.58	2	0.25	d_2	?	?	?
100	0.20	0.60	3	0.27	0.29	d_3	?	?
125	0.25	0.63	4	0.28	0.30	0.32	d_4	?
150	0.30	0.65	5	0.30	0.32	0.32	0.36	d_5

(a) Find the value of sample size n .
(b) Based on the sample size found in part (a), show that the second order inclusion probabilities are correct.
(c) Complete the missing entries (?) in the above table with the appropriate second order inclusion probabilities.
(d) Are these section probabilities correct? Why?
(e) What will be diagonal entries d_i, $i = 1, 2, 3, 4, 5,$ in the above table?
(f) Compute $V(\hat{Y}_{\text{HT}})$ and $V_{\text{SYG}}(\hat{Y}_{\text{HT}})$ if $Y_1 = Y_2 = 25$, $Y_3 = 40$, $Y_4 = 55$, and $Y_5 = 75$.
(g) Does the equality $V(\hat{Y}_{\text{HT}}) = V_{\text{SYG}}(\hat{Y}_{\text{HT}})$ holds?

Hint: $\sum_{i=1}^{N} \pi_i = n$, $\sum\sum_{j \neq i=1}^{N} \pi_{ij} = (n-1)\pi_i$, $\pi_{ij} = \pi_{ji}$, $\forall\ i \neq j$, $\sum_{i=1}^{N} P_i = 1$, and $\pi_{ii} = \pi_i$.

Practical 5.10. A psychologist wants to know if the birth rate of left handed girls and left handed boys is same in a town having eight elementary schools. The psychologist used the prior information on the number of registered boys and girls to select two samples (one of boys and another of girls) each consisting $n = 3$ units by using PPSWOR sampling. The information collected by him is given the following table:

Sample Information

Sampled Unit	No. of left handed girls	No. of left handed boys	First order inclusion probabilities	Second order inclusion probabilities for sampled units (π_{ij}) j		
i	y_i	y_i	π_i	1	2	3
1	6	4	0.45	-	0.20	0.20
2	1	3	0.40	0.20	-	0.20
3	0	2	0.50	0.20	0.20	-

(a) Derive a 75% confidence interval estimate for total number of left handed boys using the Horvitz and Thompson estimator and usual estimator of its variance.
(b) Derive a 75% confidence interval estimate for total number of left handed girls using the Horvitz and Thompson estimator and usual estimator of its variance.
(c) Assuming that both samples are independent, use (a) and (b) to estimate the difference between total number of left handed boys and girls, and derive 75% confidence interval estimate for the same.
(d) Derive a 75% confidence interval estimate for total number of left handed boys using the Horvitz and Thompson estimator and the Sen--Yates--Grundy estimator of its variance.
(e) Derive a 75% confidence interval estimate for total number of left handed girls using the Horvitz and Thompson estimator and the Sen--Yates--Grundy estimator of its variance.
(f) Assuming that both samples are independent, use (d) and (e) to estimate the difference between total number of left handed boys and girls, and derive 75% confidence interval estimate for the same.
(g) Comment on the confidence interval estimates obtained.
Hint: Ghosh (1998).

Practical 5.11. Consider a population of $N = 6$ units $\Omega = \{A, B, C, D, E, F\}$ and if we wish to select a sample of $n = 4$ units using without replacement sampling out of total 15 sample $s_t = 1,2,3,...15$ given by
$s_1 = \{A, B, C, D\}$, $s_2 = \{A, B, C, E\}$, $s_3 = \{A, B, C, F\}$, $s_4 = \{A, B, D, E\}$, $s_5 = \{A, B, D, F\}$, $s_6 = \{A, B, E, F\}$, $s_7 = \{A, C, D, E\}$, $s_8 = \{A, C, D, F\}$, $s_9 = \{A, C, E, F\}$, $s_{10} = \{A, D, E, F\}$, $s_{11} = \{B, C, D, E\}$, $s_{12} = \{B, C, D, F\}$, $s_{13} = \{B, C, E, F\}$, $s_{14} = \{B, D, E, F\}$, and $s_{15} = \{C, D, E, F\}$.
(a) Find the first order and second order inclusion probabilities provided that all samples have equal chance of selection that is $P(s_t) = 1/15 \quad \forall t$.

(b) If $P(s_t) = 0.050$, $t = 1,2,3,4,5$ and $P(s_t) = 0.075, t = 6,7,8,9,10,11,12,13,14,15$, then find the first order and second order inclusion probabilities.

(c) If $P(s_t) = 0.10$, $t = 1,2,3,4,5,6,7,8,9,10$ and $P(s_t) = 0.00, t = 11,12,13,14,15$, then find the first order and second order inclusion probabilities.

(d) If $P(s_t) = 0.050$, $t = 1,2,3,4,5$, $P(s_t) = 0.10$, $t = 6,7,8,9,10$, and $P(s_t) = 0.05$, $t = 11, 12, 13, 14, 15$, then find the first order and second order inclusion probabilities.

(e) For each one of the above situation find the usual $V(\hat{Y}_{\text{HT}})$ and Sen--Yates--Grundy $V_{\text{SYG}}(\hat{Y}_{\text{HT}})$ given that $A = 10$, $B = 20$, $C = 45$ $D = 40$, $E = 50$, and $F = 75$.

(f) Does the relation $V(\hat{Y}_{\text{HT}}) = V_{\text{SYG}}(\hat{Y}_{\text{HT}})$ holds?

(g) Which sampling scheme is most efficient?

Miscellaneous:

(h) For each one of the sampling plan (a) to (d), from each sample estimate population total using Horvitz and Thompson estimator and test for unbiasedness.
(i) For each one of the sampling plan (a) to (d), from each sample estimate the variance using the usual estimator $\hat{v}(\hat{Y}_{\text{HT}})$, and test for unbiasedness.
(j) For each one of the sampling plan (a) to (d), from each sample estimate the variance using the Sen--Yates--Grundy estimator $\hat{v}_{\text{SYG}}(\hat{Y}_{\text{HT}})$, and test for unbiasedness.

Practical 5.12. Consider a population of $N = 4$ units $\Omega = \{A, B, C, D\}$ with $A = 10$, $B = 15$, $C = 12$ and $D = 25$. Consider a sampling scheme which gives $\pi_i = 0.50$, $i = 1,2,3,4$ and $\pi_{ij} = 0.20, \forall i \neq j = 1,2,3,4$.
(a) Estimate the population total, Y, with the Horvitz and Thompson estimator, $\hat{Y}_{\text{HT}}$, assuming that the units A and D are included in the sample.
(b) Find $\hat{v}(\hat{Y}_{\text{HT}})$ and construct 95% confidence interval estimate.
(c) Find $\hat{v}_{\text{SYG}}(\hat{Y}_{\text{HT}})$ and construct 95% confidence interval estimate.
(d) Comment on the interval estimates.
(e) Are $\hat{v}(\hat{Y}_{\text{HT}})$ and $\hat{v}_{\text{SYG}}(\hat{Y}_{\text{HT}})$ are equal?
(f) Find the population total Y, $V(\hat{Y}_{\text{HT}})$ and $V_{\text{SYG}}(\hat{Y}_{\text{HT}})$.
(g) Does the relation $V(\hat{Y}_{\text{HT}}) = V_{\text{SYG}}(\hat{Y}_{\text{HT}})$ holds?

Practical 5.13. Consider a bivariate population of $N = 6$ units

$$\Omega = \left\{\frac{Y_i:\ A,\ B,\ C,\ D,\ E,\ F}{X_i:\ P,\ Q,\ R,\ S,\ T,\ U}\right\}$$

with $A = 10$, $B = 15$, $C = 12$, $D = 25$, $E = 27$, $F = 30$, $P = 8$, $Q = 12$, $R = 18$, $S = 23$, $T = 30$, and $U = 28$.
Consider a sample of $n = 4$ units

$$s = \left\{\frac{y_i : A,\ B,\ C,\ F}{x_i : P,\ Q,\ R,\ U}\right\}.$$

and a sampling scheme which gives
$\pi_i = 0.75$, $i = 1,2,3,4$, $\pi_5 = 0.2$ and $\pi_6 = 0.8$ and $\pi_{ij} = 0.15, \forall i \neq j = 1,2,3,4,5,6$.
(a) Find the design weights.
(b) Derive the calibrated weights with $q_i = 1$ using chi-square distance function, assuming population total of the auxiliary variable X is known.
(c) Estimate the population total, Y, with the GREG estimator ($\hat{Y}_G$).
(d) Find $\hat{v}_{\text{SYG}}(\hat{Y}_G)$ using usual estimator and construct 95% confidence interval estimates.

(e) Find $\hat{v}_{\text{SYG}}\left(\hat{Y}_G\right)$ using low level calibration approach and construct 95% confidence interval estimates.
(f) Assume that $V_{\text{SYG}}\left(\hat{X}_{\text{HT}}\right)$ is known and construct 95% confidence interval estimate using higher order calibration approach.
(g) Discuss the estimates.

Practical 5.14. Consider a population of $N = 4$ students

$$\Omega = \{\text{Raghunath, Mohan, Kulwinder, Sarjinder}\}$$

with their weights 110, 135, 120, and 145 lbs, respectively.

Two methodologist; Melissa and Stephanie, were asked to suggest a sampling scheme to select one student to estimate the total weight.

(I) Melissa likes every one and considers the following sampling scheme consisting of four possibilities:

$$s_1 = \{\text{Kulwinder}\},\ s_2 = \{\text{Mohan}\},\ s_3 = \{\text{Raghunath}\} \text{ and } s_4 = \{\text{Sarjinder}\}$$

such that

$$P(s_1) = 0.25,\quad P(s_2) = 0.25,\ P(s_3) = 0.25,\ P(s_4) = 0.25$$

(a) Find the first order inclusion probabilities for the Melissa's sampling scheme.
(b) Find the second order inclusion probabilities for the Melissa's sampling scheme
(c) Find the variance of the Melissa's sampling plan using the Sen--Yates--Grundy formula.
(d) Find the variance of the Melissa's sampling plan using the usual formula of the variance.
(e) Does the relation $V\left(\hat{Y}_{\text{HT}}\right) = V_{\text{SYG}}\left(\hat{Y}_{\text{HT}}\right)$ holds for Melissa's sampling plan?
(II) Stephanie likes Sarjinder and cleverly suggests the following changes in the above sampling plan:

$$P(s_1) = 0.20,\quad P(s_2) = 0.25,\ P(s_3) = 0.24,\ P(s_4) = 0.31$$

(f) Find the first order inclusion probabilities for Stephanie's sampling scheme.
(g) Find the second order inclusion probabilities for Stephanie's sampling scheme.
(h) Find the variance of the Stephanie's sampling plan using the usual formula of the variance.
(i) Does the relation $V\left(\hat{Y}_{\text{HT}}\right) = V_{\text{SYG}}\left(\hat{Y}_{\text{HT}}\right)$ holds for Stephanie's sampling plan?
(j) Find the relative efficiency of the Stephanie's cleverness over Melissa's sampling plan?
(k) What is your opinion about the Stephanie's sampling scheme?
(l) Can you estimate the variance using any formula either $\hat{v}\left(\hat{Y}_{\text{HT}}\right)$ or $\hat{v}_{\text{SYG}}\left(\hat{Y}_{\text{HT}}\right)$ for Melissa's (or Stephanie's) sampling scheme? Give you opinion.

Practical 5.15. Stephen and Sarjinder were appointed to select four students ($n = 4$) out of a multicultural list of six students ($N = 6$) with different cultural backgrounds as: $\Omega = \{$Poonam, Quang, Ryan, Stephanie, Tom, Udeesh$\}$ and GPA 3.70, 3.20, 3.80, 3.72, 3.90 and 3.92, respectively.

(I) Stephen considers the following sampling scheme consisting of only four possibilities:

$$s_1 = \{\text{Poonam, Quang, Ryan, Stephanie}\}, \quad s_2 = \{\text{Quang, Ryan, Tom, Udeesh}\},$$

$$s_3 = \{\text{Ryan, Stephanie, Tom, Udeesh}\}, \text{ and } s_4 = \{\text{Poonam, Stephanie, Tom, Udeesh}\},$$

such that

$$P(s_t) = 0.25 \ \forall t = 1, 2, 3, 4.$$

(a) Find the first order inclusion probabilities for Stephen's sampling scheme, and derive the estimates of average GPA from each sample.
(b) Find the bias in Stephen's sampling scheme.
(c) Find the second order inclusion probabilities for Stephen's sampling scheme, and derive the variance using:
(i) Sen--Yates--Grundy formula, (ii) Usual formula, and (iii) By definition.
(d) Are three variances equal for Stephen's sampling scheme?

(II) Sarjinder likes both Poonam and Stephanie and suggests the following changes in Stephen's sampling scheme as:

$$P(s_t) = 0.50 \text{ for } t = 1, 4, \text{ and } P(s_t) = 0.00 \text{ for } t = 2, 3.$$

(h) Find the first order inclusion probabilities for Sarjinder's sampling scheme, and estimates of average GPA from each sample.
(i) Find the bias in Sarjinder's sampling scheme.
(j) Find the second order inclusion probabilities for Sarjinder's sampling scheme, and derive the variance using:
(i) Sen--Yates--Grundy formula, (ii) Usual formula, and (iii) By definition.
(k) Are three variances equal for Sarjinder's sampling scheme?

(III) Comparison among techniques:
(o) Find the relative efficiency of Sarjinder's sampling scheme over Stephen's sampling scheme.
(p) Would you like to comment on your results?

6. USE OF AUXILIARY INFORMATION: MULTI-PHASE SAMPLING

6.0 INTRODUCTION

In the previous chapters we have seen that use of known auxiliary information at the estimation stage as well as at the selection stage leads to improved estimation strategies in survey sampling. When such information is not completely known or lacking and it is relatively cheaper to obtain information on the auxiliary variable(s), one can consider taking a large preliminary sample for estimating population mean(s) of the auxiliary variable(s) to be used at the estimation or selection stage of the ultimate estimation strategies. For example, in the case of single auxiliary variable X, since it is cheaper to obtain information on X, we consider taking a large preliminary sample for estimating population mean $\overline{X}$ or distribution of X as the case may be, and only a small sample (some times a sub-sample) for measuring the study variable Y.

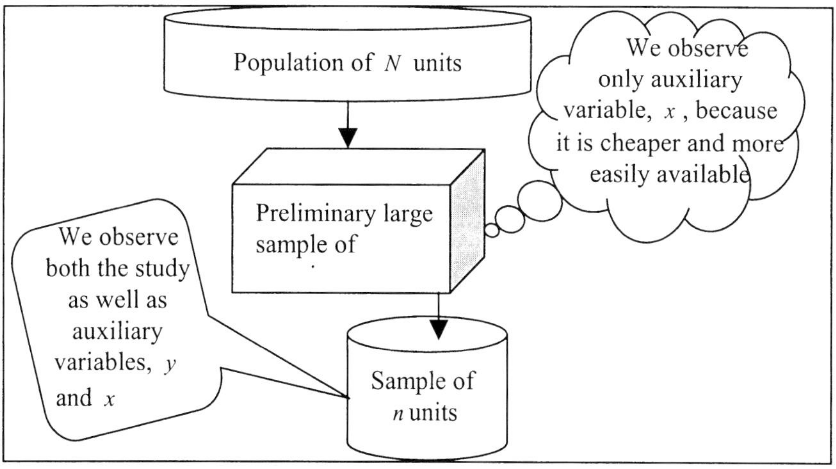

Fig.6.0.1 Two-phase sampling scheme.

This could mean devoting a part of the resources to this large preliminary sample and, therefore, reduction in sample size for measuring the study variable. This sampling technique is called double sampling or two-phase sampling and was invented by Neyman (1938). In cases in which the sample for the main surveys is selected in three or more phases, the sampling procedure is called three-phase or multi-phase sampling. This procedure is advantageous when the gain in precision is substantial as compared to the increase in cost owed to collection of information on the auxiliary variate for large samples.

In this chapter, we will discuss two-phase sampling estimation strategies under different situations such as:

(a) SRSWOR scheme at the first as well as second phases of the sample selection;
(b) SRSWOR scheme at the first phase and PPSWR sampling at the second phase;
(c) PPSWOR scheme at first as well as second phases of the sample selection.

In the following section, we will discuss a situation when simple random without replacement sampling is applied at both phases of the sample selection.

6.1 SRSWOR SCHEME AT THE FIRST AS WELL AS AT THE SECOND PHASES OF THE SAMPLE SELECTION

Under this strategy we would like to discuss the usual ratio and regression type strategies using one and two auxiliary variables. Before proceeding further, it is necessary to define notation and expected values, which will remain useful throughout this chapter.

6.1.0 NOTATION AND EXPECTED VALUES

Let $\left(x_1^*, x_2^*, ..., x_m^*\right)$ be the first phase sample s_1 (say) drawn by simple random sampling from the population of N units and let only the auxiliary variable X be measured. Also, let $\left(y_1, y_2, ..., y_n\right)$ and $\left(x_1, x_2, ..., x_n\right)$ denote, respectively, the second phase sample s_2 (say) drawn by simple random sampling from the first phase sample for the study variable Y and auxiliary variable X.

Let

$$\bar{y} = n^{-1}\sum_{i=1}^{n} y_i\,,\ \bar{x} = n^{-1}\sum_{i=1}^{n} x_i\,,\ \bar{x}^* = m^{-1}\sum_{i=1}^{m} x_i^*\,,\ s_x^2 = (n-1)^{-1}\sum_{i=1}^{n}(x_i - \bar{x})^2\,,$$

$$s_y^2 = (n-1)^{-1}\sum_{i=1}^{n}(y_i - \bar{y})^2 \text{ and } s_x^{*2} = (m-1)^{-1}\sum_{i=1}^{m}\left(x_i^* - \bar{x}^*\right)^2 .$$

Defining

$$\epsilon_0 = \frac{\bar{y}}{\bar{Y}} - 1\,,\ \epsilon_1 = \frac{\bar{x}}{\bar{x}^*} - 1\,,\ \epsilon_2 = \frac{\bar{x}}{\bar{X}} - 1\,,\ \epsilon_3 = \frac{\bar{x}^*}{\bar{X}} - 1\,,\ \delta_0 = \frac{s_y^2}{S_y^2} - 1\,,\text{ and } \delta_1 = \frac{s_x^2}{s_x^{*2}} - 1$$

such that

$$E(\epsilon_0) = E(\epsilon_1) = E(\epsilon_2) = E(\epsilon_3) = E(\delta_0) = E(\delta_1) = 0$$

and

$$E\left(\epsilon_0^2\right) = \left(\frac{1}{n} - \frac{1}{N}\right)C_y^2\,,\ E\left(\epsilon_1^2\right) = \left(\frac{1}{n} - \frac{1}{m}\right)C_x^2\,,\ E\left(\epsilon_2^2\right) = \left(\frac{1}{n} - \frac{1}{N}\right)C_x^2\,,$$

$$E\left(\epsilon_3^2\right) = \left(\frac{1}{m} - \frac{1}{N}\right)C_x^2\,,\ E\left(\delta_0^2\right) = \left(\frac{1}{n} - \frac{1}{N}\right)(\lambda_{40} - 1)\,,\ E\left(\delta_1^2\right) = \left(\frac{1}{n} - \frac{1}{m}\right)(\lambda_{04} - 1)\,,$$

$$E(\epsilon_0\epsilon_1)=\left(\frac{1}{n}-\frac{1}{m}\right)\rho_{xy}C_yC_x\,,\ E(\epsilon_0\epsilon_2)=\left(\frac{1}{n}-\frac{1}{N}\right)\rho_{xy}C_yC_x\,,\ E(\epsilon_0\epsilon_3)=\left(\frac{1}{m}-\frac{1}{N}\right)\rho_{xy}C_yC_x,$$

$$E(\epsilon_2\epsilon_3)=\left(\frac{1}{m}-\frac{1}{N}\right)C_x^2,\quad E(\epsilon_0\ \delta_1)=\left(\frac{1}{n}-\frac{1}{m}\right)C_y\lambda_{12},\quad E(\epsilon_1\ \delta_1)=\left(\frac{1}{n}-\frac{1}{m}\right)C_x\lambda_{03},$$

$$E(\epsilon_0\ \delta_0)=\left(\frac{1}{n}-\frac{1}{N}\right)C_y\lambda_{30}\,,\ E(\delta_0\ \epsilon_1)=\left(\frac{1}{n}-\frac{1}{m}\right)C_x\lambda_{21}\,,\text{ and }E(\delta_0\delta_1)=\left(\frac{1}{n}-\frac{1}{m}\right)(\lambda_{22}-1)$$

where

$$\lambda_{rs}=\frac{\mu_{rs}}{\mu_{20}^{r/2}\mu_{02}^{s/2}}\,,\quad C_y^2=\frac{\mu_{20}}{\overline{Y}^2}\,,\quad C_x^2=\frac{\mu_{02}}{\overline{X}^2}\,,\quad \rho_{xy}=\frac{\mu_{11}}{\sqrt{\mu_{20}\mu_{02}}}$$

for

$$\mu_{rs}=(N-1)^{-1}\sum_{i=1}^{N}\left(Y_i-\overline{Y}\right)^r\left(X_i-\overline{X}\right)^s .$$

Note that some of these expected values are true only up to the first order of approximation.

The above results can easily be proven on the lines of the following two theorems:

Theorem 6.1.0.1. $E\left(\epsilon_0^2\right)=\left(\frac{1}{n}-\frac{1}{N}\right)C_y^2$.

Proof. Note that

$$\bar{y}^*=m^{-1}\sum_{i=1}^{m}y_i^*\quad\text{and}\quad s_y^{*2}=(m-1)^{-1}\sum_{i=1}^{m}\left(y_i^*-\bar{y}^*\right)^2 .$$

Then we have

$$V(\epsilon_0)=E\left(\epsilon_0^2\right)-\{E(\epsilon_0)\}^2=E\left(\epsilon_0^2\right)$$

$$=E_1V_2\left[\epsilon_0\mid\text{first phase sample}\right]+V_1E_2\left[\epsilon_0\mid\text{first phase sample}\right]$$

$$=E_1V_2\left[\left(\frac{\bar{y}}{\overline{Y}}-1\right)\mid\text{first phase sample}\right]+V_1E_2\left[\left(\frac{\bar{y}}{\overline{Y}}-1\right)\mid\text{first phase sample}\right]$$

$$=\frac{E_1\left[V_2(\bar{y})\mid\text{first phase sample}\right]}{\overline{Y}^2}+V_1\left[\left\{\frac{E_2(\bar{y})}{\overline{Y}}-1\right\}\mid\text{first phase sample}\right]$$

$$=E_1\left[\frac{\left(\frac{1}{n}-\frac{1}{m}\right)s_y^{*2}}{\overline{Y}^2}\right]+V_1\left[\frac{\bar{y}^*}{\overline{Y}}-1\right]=\left(\frac{1}{n}-\frac{1}{m}\right)C_y^2+\left(\frac{1}{m}-\frac{1}{N}\right)C_y^2=\left(\frac{1}{n}-\frac{1}{N}\right)C_y^2 .$$

Hence the theorem.

Theorem 6.1.0.2. $E(\epsilon_0\epsilon_1)=\left(\frac{1}{n}-\frac{1}{m}\right)\rho_{xy}C_yC_x$.

Proof. We have

$$\mathrm{Cov}(\epsilon_0,\epsilon_1)=E(\epsilon_0\epsilon_1)-E(\epsilon_0)E(\epsilon_1)=E(\epsilon_0\epsilon_1)$$

$$=E_1[C_2(\epsilon_0,\epsilon_1)\mid \text{first phase sample}]+C_1[\{E_2(\epsilon_0),E_2(\epsilon_1)\}\mid \text{first phase sample}]$$

$$=E_1\left[C_2\left\{\frac{\bar{y}}{\bar{Y}}-1,\frac{\bar{x}}{\bar{x}^*}-1\right\}\mid \text{first phase sample}\right]$$

$$+C_1\left[\left\{E_2\left(\frac{\bar{y}}{\bar{Y}}-1\right),E_2\left(\frac{\bar{x}}{\bar{x}^*}-1\right)\right\}\mid \text{first phase sample}\right]$$

$$=E_1\left[C_2(\bar{y},\bar{x})/\left(\bar{Y}\,\bar{x}^*\right)\mid \text{first phase sample}\right]+C_1\left[\left(\frac{\bar{y}^*}{\bar{Y}}-1\right),0\right]$$

$$=E_1\left[\left(\frac{1}{n}-\frac{1}{m}\right)\frac{s_{xy}^*}{\bar{Y}\,\bar{x}^*}\right]\approx\left(\frac{1}{n}-\frac{1}{m}\right)\frac{S_{xy}}{\bar{Y}\,\bar{X}}=\left(\frac{1}{n}-\frac{1}{m}\right)\rho_{xy}C_yC_x.$$

Hence the theorem.

6.1.1 RATIO ESTIMATOR

The ratio estimator $\bar{y}_{\text{rd}}$ of $\bar{Y}$ in two phase or double sampling takes the form

$$\bar{y}_{\text{rd}}=\bar{y}\left(\frac{\bar{x}^*}{\bar{x}}\right). \tag{6.1.1.1}$$

Assuming that $|\epsilon_i|<1,\ i=0,1$ and using the binomial expansion

$$(1+\epsilon_1)^{-1}=1-\epsilon_1+\epsilon_1^2+...$$

the above estimator can easily be written as

$$\bar{y}_{\text{rd}}=\bar{Y}(1+\epsilon_0)(1+\epsilon_1)^{-1}=\bar{Y}(1+\epsilon_0)\left(1-\epsilon_1+\epsilon_1^2+......\right)$$

$$=\bar{Y}\left[1+\epsilon_0-\epsilon_1+\epsilon_1^2-\epsilon_0\epsilon_1+.....\right]. \tag{6.1.1.2}$$

Thus we have the following theorems:

Theorem 6.1.1.1. The bias in the ratio estimator $\bar{y}_{\text{rd}}$, to the first order of approximation, is given by

$$B(\bar{y}_{\text{rd}})=\left(\frac{1}{n}-\frac{1}{m}\right)\bar{Y}\left[C_x^2-\rho_{xy}C_yC_x\right]. \tag{6.1.1.3}$$

Proof. Taking expected values on both sides of (6.1.1.2) we have

$$E(\bar{y}_{\text{rd}})=\bar{Y}E\left[1+\epsilon_0-\epsilon_1+\epsilon_1^2-\epsilon_0\epsilon_1+.....\right]=\bar{Y}\left[1+E(\epsilon_0)-E(\epsilon_1)+E\left(\epsilon_1^2\right)-E(\epsilon_0\epsilon_1)\right]$$

$$=\bar{Y}\left[1+0-0+\left(\frac{1}{n}-\frac{1}{m}\right)\left(C_x^2-\rho_{xy}C_xC_y\right)\right].$$

Thus the bias in the estimator $\bar{y}_{rd}$, to the first order of approximation, is given by

$$B(\bar{y}_{rd}) = E(\bar{y}_{rd}) - \bar{Y} = \left(\frac{1}{n} - \frac{1}{m}\right)\bar{Y}\left(C_x^2 - \rho_{xy}C_xC_y\right).$$

Hence the theorem.

Theorem 6.1.1.2. The mean squared error of the estimator $\bar{y}_{rd}$, to the first order of approximation, is

$$\text{MSE}(\bar{y}_{rd}) = \left(\frac{1}{m} - \frac{1}{N}\right)S_y^2 + \left(\frac{1}{n} - \frac{1}{m}\right)\left[S_y^2 + R^2S_x^2 - 2RS_{xy}\right] \quad (6.1.1.4)$$

where $R = \bar{Y}/\bar{X}$ has its usual meaning.

Proof. By the definition of mean squared error we have

$$MSE(\bar{y}_{rd}) = E\left[\bar{y}_{rd} - \bar{Y}\right]^2 \approx E\left[\bar{Y}(1 + \epsilon_0 - \epsilon_1 + ...) - \bar{Y}\right]^2 \approx \bar{Y}^2E\left[\epsilon_0^2 + \epsilon_1^2 - 2\epsilon_0\epsilon_1\right]$$

$$= \bar{Y}^2\left[\left(\frac{1}{n} - \frac{1}{N}\right)C_y^2 + \left(\frac{1}{n} - \frac{1}{m}\right)\left\{C_x^2 - 2\rho_{xy}C_xC_y\right\}\right]$$

$$= \left(\frac{1}{m} - \frac{1}{N}\right)S_y^2 + \left(\frac{1}{n} - \frac{1}{m}\right)\left[S_y^2 + R^2S_x^2 - 2RS_{xy}\right].$$

Hence the theorem.

Corollary 6.1.1.1. An estimator of the mean squared error of the estimator $\bar{y}_{rd}$, to the first order of approximation, is

$$\hat{\text{MSE}}(\bar{y}_{rd}) = \left(\frac{1}{m} - \frac{1}{N}\right)s_y^2 + \left(\frac{1}{n} - \frac{1}{m}\right)\left[s_y^2 + r^2s_x^2 - 2rs_{xy}\right] \quad (6.1.1.5)$$

where $r = \bar{y}/\bar{x}$ has its usual meaning.

Example 6.1.1.1. From population 1 in the Appendix select a first phase sample of 10 units by SRSWOR sampling and note only the nonreal estate farm loans from the selected units in the sample. From the selected first phase sample of 10 units, select a sub-sample of 5 units and note the real estate farm loans as well as nonreal estate farm loans. Estimate the average real estate farm loans by using ratio estimator in two-phase sampling. Deduce the 95% confidence interval.

Solution. We used the first two columns of the Pseudo-Random Number (PRN) Table 1 given in the Appendix to select an SRSWOR sample of $m = 10$ units. The following 10 distinct random numbers 01, 23, 46, 04, 32, 47, 33 05, 22, and 38 between 1 and 50 resulted in the following first phase sample.

First phase sample information										
Sr. No.	1	2	3	4	5	6	7	8	9	10
Pop Units	01	23	46	04	32	47	33	05	22	38
States	AL	MN	VA	AR	NY	WA	NC	CA	MI	PA
x_i^*	348.334	2466.892	188.477	848.317	426.274	1228.607	494.73	3928.732	440.518	298.351

where x_i^* = Nonreal estate farm loans, and.

We used the 7th and 8th columns of the Pseudo-Random Numbers to select a second-phase sample of $n = 5$ units from the above list of selected first phase sample units. The following five distinct random numbers between 1 and 10 were observed: 07, 09, 01, 02 and 03. Thus the second phase sample consists of the following information:

Second phase sample information					
Sr. No.	First phase unit number	Population unit number	State and Territory	Nonreal estate farm loans	Real estate farm loans
1	07	33	NC	494.730	639.571
2	09	22	MI	440.518	323.028
3	01	01	AL	348.334	408.978
4	02	23	MN	2466.892	1354.768
5	03	46	VA	188.477	321.583

From the second phase sample information we have

Sr. No.	State and Territory	Nonreal estate farm loans, x	Real estate farm loans y	$(x_i - \bar{x})^2$	$(y_i - \bar{y})^2$	$(y_i - \bar{y})(x_i - \bar{x})$
1	NC	494.730	639.571	85884.281	899.12	-8787.5273
2	MI	440.518	323.028	120597.980	82115.26	99513.4880
3	AL	348.334	408.978	193121.750	40243.42	88158.2540
4	MN	2466.892	1354.768	2819382.900	555296.80	1251237.1000
5	VA	188.477	321.583	359176.310	82945.50	172603.7600
	Sum	3938.951	3047.928	3578163.221	761500.10	1602725.1193

Note that

$$\bar{x} = \frac{1}{n}\sum_{i=1}^{n} x_i = \frac{3938.951}{5} = 787.7902\,, \quad \bar{y} = \frac{1}{n}\sum_{i=1}^{n} y_i = \frac{3047.928}{5} = 609.5856\,,$$

$$r = \frac{\bar{y}}{\bar{x}} = \frac{609.5856}{787.7902} = 0.7738\,, \quad s_x^2 = (n-1)^{-1}\sum_{i=1}^{n}(x_i - \bar{x})^2 = \frac{3578163.221}{4} = 894540.79\,,$$

$$s_y^2 = (n-1)^{-1}\sum_{i=1}^{n}(y_i - \bar{y})^2 = \frac{761500.1193}{4} = 190375.025,$$

and

$$s_{xy} = (n-1)^{-1}\sum_{i=1}^{n}(y_i - \bar{y})(x_i - \bar{x}) = \frac{1602725.1}{4} = 400681.275.$$

Thus ratio estimate of the real estate farm loans is given by

$$\bar{y}_{\text{rd}} = \bar{y}\left(\frac{\bar{x}^*}{\bar{x}}\right) = 606.5856\left(\frac{1066.9232}{787.7902}\right) = 825.576\,.$$

An estimate of the $\text{MSE}(\bar{y}_{\text{rd}})$ is given by

$$\hat{\text{MSE}}(\bar{y}_{\text{rd}}) = \left(\frac{1}{m} - \frac{1}{N}\right)s_y^2 + \left(\frac{1}{n} - \frac{1}{m}\right)\left[s_y^2 + r^2 s_x^2 - 2rs_{xy}\right]$$

$$= \left(\frac{1}{10} - \frac{1}{50}\right) \times 190375.025$$

$$+ \left(\frac{1}{5} - \frac{1}{10}\right)\left[190375.025 + 0.7738^2 \times 894540.79 - 2 \times 0.7738 \times 400681.275\right]$$

$$= 25820.171 .$$

The $(1-\alpha)100\%$ confidence interval estimate for the average real estate farm loans in the US during 1997 is

$$\bar{y}_{\text{rd}} \mp t_{\alpha/2}(\text{df} = n-1)\sqrt{\hat{\text{MSE}}(\bar{y}_{\text{rd}})} .$$

Using Table 2 from the Appendix, the 95% confidence interval for the average real estate farm loans in the US during 1997 is given by

$$825.576 \mp 2.776\sqrt{25820.171} , \quad \text{or } [379.51, \quad 1271.63].$$

6.1.1.1 COST FUNCTION

Suppose C_0 is the overhead cost, C_1 is the cost of selection and processing of a single unit in the first phase and C_2 is the cost of selection and processing of a single unit in the second phase, then the total cost function C for selecting m units in the first phase and n units in the second phase is given by

$$C = C_0 + mC_1 + nC_2 . \tag{6.1.1.6}$$

Theorem 6.1.1.3. The minimum MSE of the ratio estimator $\bar{y}_{\text{rd}}$ for the fixed cost C given by (6.1.1.6) is

$\text{Min.MSE}(\bar{y}_{\text{rd}})$

$$= \left[\frac{\sqrt{C_1}D}{(C-C_0)\sqrt{S_y^2 - V_{\text{R}}}} - \frac{1}{N}\right]S_y^2 + \frac{D}{(C-C_0)}\left[\frac{\sqrt{C_2 V_{\text{R}}\left(S_y^2 - V_R\right)} - \sqrt{C_1}V_{\text{R}}}{\sqrt{S_y^2 - V_{\text{R}}}}\right], \tag{6.1.1.7}$$

where

$$V_{\text{R}} = S_y^2 + R^2 S_x^2 - 2RS_{xy} \tag{6.1.1.8}$$

and

$$D = \sqrt{C_1\left(S_y^2 - V_{\text{R}}\right)} + \sqrt{C_2 V_{\text{R}}} . \tag{6.1.1.9}$$

Proof. The Lagrange function is

$$L = \left(\frac{1}{m} - \frac{1}{N}\right)S_y^2 + \left(\frac{1}{n} - \frac{1}{m}\right)V_{\text{R}} - \lambda\left[C - C_0 - mC_1 - nC_2\right]. \tag{6.1.1.10}$$

On differentiating (6.1.1.10) with respect to m and equating to zero we have

$$\frac{\partial L}{\partial m} = -\frac{S_y^2}{m^2} + \frac{V_R}{m^2} + \lambda C_1 = 0,$$

which implies that

$$m = \sqrt{S_y^2 - V_R} \Big/ \left\{\sqrt{\lambda}\sqrt{C_1}\right\}. \tag{6.1.1.11}$$

On differentiating (6.1.1.10) with respect to n and equating to zero we have

$$\partial L / \partial n = -V_R / n^2 + \lambda C_2 = 0$$

which implies that

$$n = \sqrt{V_R} \Big/ \left\{\sqrt{\lambda}\sqrt{C_2}\right\}. \tag{6.1.1.12}$$

On substituting these values of m and n in equation (6.1.1.6) we have

$$\sqrt{\lambda} = \left\{\sqrt{C_1\left(S_y^2 - V_R\right)} + \sqrt{C_2 V_R}\right\} \Big/ (C - C_0) = D/(C - C_0) \text{ (say)}. \tag{6.1.1.13}$$

On substituting this value of λ in (6.1.1.11) and (6.1.1.12) we obtain the optimum sample sizes

$$m = \left\{(C - C_0)\sqrt{S_y^2 - V_R}\right\} \Big/ \left\{\sqrt{C_1} D\right\} \tag{6.1.1.14}$$

and

$$n = \left\{(C - C_0)\sqrt{V_R}\right\} \Big/ \left\{\sqrt{C_2} D\right\}. \tag{6.1.1.15}$$

On substituting (6.1.1.14) and (6.1.1.15) in (6.1.1.4) we obtain (6.1.1.7). Hence the theorem.

Theorem 6.1.1.4. The minimum cost C for the fixed $\text{MSE}(\bar{y}_{rd}) = V_0$ (say) is

$$\text{Min.}C = C_0 + D^2\left(V_0 + S_y^2/N\right)^{-1}. \tag{6.1.1.16}$$

where $D = \sqrt{C_1\left(S_y^2 - V_R\right)} + \sqrt{C_2 V_R}$ and $V_R = S_y^2 + R^2 S_x^2 - 2RS_{xy}$.

Proof. Consider the Lagrange function

$$L = C - C_0 - mC_1 - nC_2 - \lambda\left[\left(\frac{1}{m} - \frac{1}{N}\right)S_y^2 + \left(\frac{1}{n} - \frac{1}{m}\right)V_R - V_0\right]. \tag{6.1.1.17}$$

On differentiating (6.1.1.17) with respect to m and equating to zero we have

$$\frac{\partial L}{\partial m} = -C_1 + \frac{\lambda S_y^2}{m^2} - \frac{\lambda V_R}{m^2} = 0,$$

which implies that

$$m = \left\{\sqrt{\lambda}\sqrt{S_y^2 - V_R}\right\} \Big/ \sqrt{C_1}. \tag{6.1.1.18}$$

On differentiating (6.1.1.17) with respect to n and equating to zero we have

$$\frac{\partial L}{\partial n} = -C_2 + \frac{\lambda V_R}{n^2} = 0$$

which implies that

$$n = \left\{\sqrt{\lambda}\sqrt{V_R}\right\} \Big/ \left\{\sqrt{C_2}\right\}. \tag{6.1.1.19}$$

On substituting these values of m and n in equation $V(\bar{y}_{rd}) = V_0$ we have

$$\sqrt{\lambda} = \left(V_0 + S_y^2/N\right)^{-1}\left\{\sqrt{C_1\left(S_y^2 - V_R\right)} + \sqrt{C_2 V_R}\right\} = D\left(V_0 + S_y^2/N\right)^{-1}. \tag{6.1.1.20}$$

On substituting this value of λ in (6.1.1.18) and (6.1.1.19) we obtain the optimum sample sizes

$$m = \left\{D\sqrt{S_y^2 - V_R}\right\}\Big/\left\{\sqrt{C_1}\left(V_0 + S_y^2/N\right)\right\} \tag{6.1.1.21}$$

and

$$n = D\sqrt{V_R}\Big/\left\{\sqrt{C_2}\left(V_0 + S_y^2/N\right)\right\}. \tag{6.1.1.22}$$

On substituting (6.1.1.21) and (6.1.1.22) in $C = C_0 + mC_1 + nC_2$ we have (6.1.1.16). Hence the theorem.

Theorem 6.1.1.5. Then minimum variance under single-phase and SRSWOR scheme for the fixed cost of survey is

$$V(\bar{y}) = \left(\frac{C_2}{C - C_0} - \frac{1}{N}\right)S_y^2. \tag{6.1.1.23}$$

Solution. Under SRSWOR sampling we know that

$$V(\bar{y}) = \left(\frac{1}{n} - \frac{1}{N}\right)S_y^2. \tag{6.1.1.24}$$

The cost function in this situation will be

$$C = C_0 + nC_2 \tag{6.1.1.25}$$

which implies that

$$n = (C - C_0)/C_2. \tag{6.1.1.26}$$

On substituting (6.1.1.26) in (6.1.1.24), we have (6.1.1.23). Hence the theorem.

Example 6.1.1.2. The amounts of the real and nonreal estate farm loans (in \$000) during 1997 in the 50 states of the United States have been presented in population 1 of the Appendix. Suppose we selected first phase and second phase samples each of size 10 and 5 respectively.

(a) Find the relative efficiency of the ratio estimator, for estimating the average amount of real estate farm loans during 1997 by using information selected in the first phase sample only on the nonreal estate farm loans during 1997, with respect to the usual estimator of population mean.

(b) Suppose a budget of US\$5000 is available to spend on the survey, \$2000 of which will be the overhead cost. Suppose selection, compilation, and analysis of one unit in the first phase sample cost \$50, whereas for the second phase unit is \$500. Find the optimum values of the first phase and second phase sample sizes. Also find the relative efficiency of the ratio estimator over the sample mean for the fixed cost.

(c) What will be the minimum cost for attaining a 30% relative standard deviation?

Solution. From the description of the population we have

Y_i = Amount (\$000) of the real estate farm loans during 1997.

X_i = Amount ($000) of the Nonreal estate farm loans during 1997.

$\overline{Y} = 555.43$, $\overline{X} = 878.16$, $R = 0.6325$, $S_y^2 = 342021.5$, $S_x^2 = 1176526$, $C_x^2 = 1.5256$, $C_y^2 = 1.1086$, $S_{xy} = 509910.41$, $\rho_{xy} = 0.8038$, and $N = 50$.

(a) We have $m = 10$ and $n = 5$, therefore

$$\mathrm{MSE}(\overline{y}_{\mathrm{rd}}) = \left(\frac{1}{m} - \frac{1}{N}\right)S_y^2 + \left(\frac{1}{n} - \frac{1}{m}\right)\left[S_y^2 + R^2 S_x^2 - 2RS_{xy}\right]$$

$$= \left(\frac{1}{10} - \frac{1}{50}\right) \times 342021.5$$

$$+ \left(\frac{1}{5} - \frac{1}{10}\right)\left[342021.5 + 0.6325^2 \times 1176526 - 2 \times 0.6325 \times 509910.41\right]$$

$$= 44127.86.$$

Also

$$V(\overline{y}) = \left(\frac{1}{n} - \frac{1}{N}\right)S_y^2 = \left(\frac{1}{5} - \frac{1}{50}\right) \times 342021.5 = 61563.87.$$

Thus relative efficiency of the ratio estimator over the sample mean is given by

$$\mathrm{RE} = \frac{V(\overline{y})}{\mathrm{MSE}(\overline{y}_{\mathrm{rd}})} \times 100 = \frac{61563.87}{44127.86} \times 100 = 139.51\%.$$

(b) We have

$$C = 5000,\ C_0 = 2000,\ C_1 = 50,\ C_2 = 500,$$

$$V_{\mathrm{R}} = S_y^2 + R^2 S_x^2 - 2RS_{xy}$$

$$= 342021.5 + 0.6325^2 \times 1176526 - 2 \times 0.6325 \times 509910.41 = 167661.41$$

and

$$D = \sqrt{C_1\left(S_y^2 - V_{\mathrm{R}}\right)} + \sqrt{C_2 V_{\mathrm{R}}} = \sqrt{50(342021.5 - 167661.41)} + \sqrt{500 \times 167661.41} = 12108.53.$$

The optimum first phase and second phase sample sizes are, respectively, given by

$$m = \left\{(C - C_0)\sqrt{S_y^2 - V_{\mathrm{R}}}\right\} \Big/ \left\{\sqrt{C_1}D\right\}$$

$$= \left\{(5000 - 2000)\sqrt{342021.5 - 167661.41}\right\} \Big/ \left\{\sqrt{50} \times 12108.53\right\} = 14.6 \approx 15,$$

and

$$n = \left\{(C - C_0)\sqrt{V_{\mathrm{R}}}\right\} \Big/ \left\{\sqrt{C_2}D\right\} = \frac{(5000 - 2000)\sqrt{167661.41}}{\sqrt{500} \times 12108.53} = 4.5 \approx 5.$$

Thus we have

$$\mathrm{MSE}(\overline{y}_{\mathrm{rd}}) = \left(\frac{1}{m} - \frac{1}{N}\right)S_y^2 + \left(\frac{1}{n} - \frac{1}{m}\right)\left[S_y^2 + R^2 S_x^2 - 2RS_{xy}\right]$$

$$= \left(\frac{1}{15} - \frac{1}{50}\right) \times 342021.5$$

$$+ \left(\frac{1}{5} - \frac{1}{15}\right)\left[342021.5 + 0.6325^2 \times 1176526 - 2 \times 0.6325 \times 509910.41\right]$$

$$= 38315.86 .$$

For single-phase sampling, the optimum sample size for the fixed cost will be

$$n = (C - C_0)/C_2 = (5000 - 2000)/500 = 6,$$

and

$$V(\bar{y}) = \left(\frac{1}{n} - \frac{1}{N}\right) S_y^2 = \left(\frac{1}{6} - \frac{1}{50}\right) \times 342021.5 = 50163.15 .$$

Thus percent relative efficiency of the ratio estimator over the sample mean for the fixed cost is given by

$$\text{RE} = \frac{V(\bar{y})}{\text{MSE}(\bar{y}_{\text{rd}})} \times 100 = \frac{50163.15}{38315.85} \times 100 \approx 130.92\% .$$

(c) By the definition of relative standard deviation we have

$$\sqrt{\text{MSE}(\bar{y}_{\text{rd}})/\bar{Y}^2} = 0.30 , \quad \text{or} \quad \sqrt{\text{MSE}(\bar{y}_{\text{rd}})/555.43^2} = 0.30 \text{, or}$$

$$\text{MSE}(\bar{y}_{\text{rd}}) = 27765.22 = V_0 \text{ (say).}$$

Thus the minimum cost of the survey for the fixed precision will be

$$\text{Min.}C = C_0 + D^2\left(V_0 + S_y^2/N\right)^{-1} = 2000 + 12108.53^2(27765.22 + 342021.5/50)^{-1} = 6236.77 .$$

Thus a minimum of \$6236.77 instead of \$5000 is required for attaining the 30% relative standard deviation of the estimator.

6.1.2 DIFFERENCE ESTIMATOR

The difference estimator of the population mean $\bar{Y}$ in two-phase sampling is

$$\bar{y}_{\text{dd}} = \bar{y} + d\left(\bar{x}^* - \bar{x}\right) \tag{6.1.2.1}$$

where d is a known constant and can be chosen such that the variance of the estimator $\bar{y}_{\text{dd}}$ is minimum. The above estimator can easily be written as

$$\bar{y}_{\text{dd}} = \bar{Y}(1 + \epsilon_0) + d\bar{X}(\epsilon_3 - \epsilon_2) . \tag{6.1.2.2}$$

Thus we have the following theorems:

Theorem 6.1.2.1. The difference estimator $\bar{y}_{\text{dd}}$ is an unbiased estimator of the population mean $\bar{Y}$.

Proof. Taking expected values on both sides of (6.1.2.2) we have

$$E(\bar{y}_{\text{dd}}) = \bar{Y}(1 + E(\epsilon_0)) + d\bar{X}(E(\epsilon_3) - E(\epsilon_2)) = \bar{Y}(1 + 0) + d\bar{X}(0 - 0) = \bar{Y} .$$

Hence the theorem.

Theorem 6.1.2.2. The minimum variance of the difference estimator $\bar{y}_{\text{dd}}$ of population mean $\bar{Y}$ is

$$\text{Min.}V(\bar{y}_{\text{dd}}) = \left(\frac{1}{m} - \frac{1}{N}\right)S_y^2 + \left(\frac{1}{n} - \frac{1}{m}\right)S_y^2\left(1 - \rho_{xy}^2\right). \tag{6.1.2.3}$$

Proof. We have

$$V(\bar{y}_{\text{dd}}) = E\left[\bar{y}_{\text{dd}} - E(\bar{y}_{\text{dd}})\right]^2 = E\left[\bar{Y}(1+\epsilon_0) + d\bar{X}(\epsilon_3 - \epsilon_2) - \bar{Y}\right]^2 = E\left[\bar{Y}\epsilon_0 + d\bar{X}(\epsilon_3 - \epsilon_2)\right]^2$$

$$= E\left[\bar{Y}^2\epsilon_0^2 + d^2\bar{X}^2\left(\epsilon_3^2 + \epsilon_2^2 - 2\epsilon_2\epsilon_3\right) + 2d\bar{Y}\bar{X}\left(\epsilon_0\epsilon_3 - \epsilon_0\epsilon_2\right)\right]$$

$$= \left(\frac{1}{n} - \frac{1}{N}\right)\bar{Y}^2C_y^2 + \left(\frac{1}{n} - \frac{1}{m}\right)\left[d^2\bar{X}^2C_x^2 - 2d\bar{Y}\bar{X}\rho_{xy}C_xC_y\right]. \tag{6.1.2.4}$$

On differentiating (6.1.2.4) with respect to d and equating to zero we have

$$d = \rho_{xy}\frac{C_y}{C_x}\left(\frac{\bar{Y}}{\bar{X}}\right) = \frac{S_{xy}}{S_x^2}. \tag{6.1.2.5}$$

On substituting (6.1.2.5) in (6.1.2.4) we obtain (6.1.2.3). Hence the theorem.

6.1.3 REGRESSION ESTIMATOR

The exact distribution of the regression estimator in two-phase sampling has been derived by Causeur (1999). We consider the regression estimator of population mean in two-phase sampling in the following theorem:

Theorem 6.1.3.1. Find the asymptotic MSE of the linear regression estimator of population mean $\bar{Y}$ in two-phase sampling

$$\bar{y}_{\text{lrd}} = \bar{y} + \hat{\beta}\left(\bar{x}^* - \bar{x}\right) \tag{6.1.3.1}$$

where $\hat{\beta} = s_{xy}/s_x^2$ is an estimator of the regression coefficient $\beta = S_{xy}/S_x^2$.

Proof. Let us define $\eta = \hat{\beta}/\beta - 1$, such that $E(\eta) \approx 0$. Then the estimator $\bar{y}_{\text{lrd}}$ of the population mean $\bar{Y}$ can be written as

$$\bar{y}_{\text{lrd}} = \bar{Y}(1+\epsilon_0) + \beta\bar{X}(1+\eta)(\epsilon_3 - \epsilon_2) = \bar{Y}(1+\epsilon_0) + \beta\bar{X}(\epsilon_3 - \epsilon_2)(1+\eta).$$

Thus the MSE of the estimator $\bar{y}_{\text{lrd}}$ is given by

$$\text{MSE}(\bar{y}_{\text{lrd}}) = E\left[\bar{Y}(1+\epsilon_0) + \beta\bar{X}(\epsilon_3 - \epsilon_2)(1+\eta) - \bar{Y}\right]^2$$

$$\approx E\left[\bar{Y}^2\epsilon_0^2 + \beta^2\bar{X}^2\left(\epsilon_3^2 + \epsilon_2^2 - 2\epsilon_2\epsilon_3\right) - 2\beta\bar{Y}\bar{X}\left(\epsilon_0\epsilon_3 - \epsilon_0\epsilon_2\right)\right]$$

$$= \left(\frac{1}{n} - \frac{1}{N}\right)\bar{Y}^2C_y^2 + \beta^2\bar{X}^2\left\{\left(\frac{1}{m} - \frac{1}{N}\right)C_x^2 + \left(\frac{1}{n} - \frac{1}{N}\right)C_x^2 - 2\left(\frac{1}{m} - \frac{1}{N}\right)C_x^2\right\}$$

$$+ 2\beta\bar{Y}\bar{X}\left\{\left(\frac{1}{m} - \frac{1}{N}\right)\rho_{xy}C_xC_y - \left(\frac{1}{n} - \frac{1}{N}\right)\rho_{xy}C_xC_y\right\}$$

$$= \left(\frac{1}{n} - \frac{1}{N}\right)S_y^2 + \left(\frac{1}{n} - \frac{1}{m}\right)\left\{\beta^2\bar{X}^2C_x^2 - 2\beta\bar{Y}\bar{X}\rho_{xy}C_yC_x\right\}$$

$$=\left(\frac{1}{n}-\frac{1}{m}+\frac{1}{m}-\frac{1}{N}\right)S_y^2+\left(\frac{1}{n}-\frac{1}{m}\right)\left\{\beta^2 S_x^2-2\beta S_{xy}\right\}$$

$$=\left(\frac{1}{m}-\frac{1}{N}\right)S_y^2+\left(\frac{1}{n}-\frac{1}{m}\right)\left\{S_y^2+\beta^2 S_x^2-2\beta S_{xy}\right\}$$

$$=\left(\frac{1}{m}-\frac{1}{N}\right)S_y^2+\left(\frac{1}{n}-\frac{1}{m}\right)\left\{S_y^2+\left(\frac{S_{xy}}{S_x^2}\right)^2 S_x^2-2\left(\frac{S_{xy}}{S_x^2}\right)S_{xy}\right\}$$

$$=\left(\frac{1}{m}-\frac{1}{N}\right)S_y^2+\left(\frac{1}{n}-\frac{1}{m}\right)S_y^2\left(1-\rho_{xy}^2\right).$$

Hence the theorem.

Example 6.1.3.1. The amounts of real and nonreal estate farm loans (in \$000) during 1997 in the United States have been presented in population 1 of the Appendix. Suppose we selected first phase and second phase samples each of size 10 and 5 respectively. Find the relative efficiency of the regression estimator, for estimating the average amount of the real estate farm loans during 1997 using information selected in the first phase sample only on the nonreal estate farm loans during 1997, with respect to the ratio estimator of population mean.

Solution. Continuing from example 6.1.1.2, we have $\text{MSE}(\bar{y}_{\text{rd}}) = 44127.86$. Also the mean squared error of the regression estimator is given by

$$\text{MSE}(\bar{y}_{\text{lrd}})=\left(\frac{1}{m}-\frac{1}{N}\right)S_y^2+\left(\frac{1}{n}-\frac{1}{m}\right)S_y^2\left(1-\rho_{xy}^2\right)$$

$$=\left(\frac{1}{10}-\frac{1}{50}\right)\times 342021.5+\left(\frac{1}{5}-\frac{1}{10}\right)\times 342021.5\left(1-0.8038^2\right)$$

$$= 39466.05\,.$$

Thus percent relative efficiency of the linear regression estimator $\bar{y}_{\text{lrd}}$ over the ratio estimator $\bar{y}_{\text{rd}}$ is

$$\text{RE}=\frac{V(\bar{y}_{\text{rd}})}{\text{MSE}(\bar{y}_{\text{lrd}})}\times 100=\frac{44127.86}{39466.05}\times 100=111.8\%\,.$$

6.1.4 GENERAL CLASS OF ESTIMATORS OF POPULATION MEAN

We discuss the results of the general class of estimators, which is in fact an analogue in two-phase sampling of the general class of estimators proposed by Srivastava and Jhajj (1981) for single-phase sampling. Srivastava (1981a, 1981b) and Srivastava and Jhajj (1987) have also considered some generalized estimators of population mean in two-phase sampling. We have the following theorem:

Theorem 6.1.4.1. Consider a general class of ratio type estimators $\bar{y}_{\text{gd}}$ of the population mean $\bar{Y}$ in two-phase sampling is

$$\bar{y}_{\text{gd}} = \bar{y}H(u,\ v) \tag{6.1.4.1}$$

where $H(\bullet,\bullet)$ is a parametric function such that it satisfies the following regularity conditions:

(a) $H(1,1)=1$;

(b) The first and second order partial derivatives of H with respect to $u = \bar{x}/\bar{x}^*$ and $v = s_x^2/s_x^{*2}$ exist and are continuous and known constants.

Then show that its minimum MSE is given by

$$\text{Min.MSE}\left(\bar{y}_{\text{gd}}\right) = \left(\frac{1}{m} - \frac{1}{N}\right)S_y^2 + \left(\frac{1}{n} - \frac{1}{m}\right)S_y^2\left[1 - \rho_{xy}^2 - \frac{\left(\lambda_{12} - \rho_{xy}\lambda_{03}\right)^2}{\lambda_{04} - 1 - \lambda_{03}^2}\right]. \tag{6.1.4.2}$$

Proof. Expanding the function $H(u,v)$ around the point $(1,1)$ up to the first order Taylor's series we have

$$\bar{y}_{\text{gd}} = \bar{y}\left[H(1,1) + (u-1)\frac{\partial H}{\partial u}\Big|_{(1,1)} + (v-1)\frac{\partial H}{\partial v}\Big|_{(1,1)} + \dots\dots\dots\right]$$

$$= \bar{Y}(1+\epsilon_0)(1+\epsilon_1 H_{10} + \delta_1 H_{01} + \dots.) \tag{6.1.4.3}$$

where $H_{10} = \frac{\partial H}{\partial u}\Big|_{(1,1)}$ and $H_{01} = \frac{\partial H}{\partial v}\Big|_{(1,1)}$ denote the known first order partial derivatives of the function H with respect to u and v respectively. By definition, we have

$$\text{MSE}\left(\bar{y}_{\text{gd}}\right) = E\left[\bar{y}_{\text{gd}} - \bar{Y}\right]^2 = E\left[\bar{Y}\{1+\epsilon_0 + \epsilon_1 H_{10} + \delta_1 H_{01} + \dots\} - \bar{Y}\right]^2$$

$$= \bar{Y}^2 E\left[\epsilon_0^2 + H_{10}^2\epsilon_1^2 + H_{01}^2\delta_1^2 + 2\epsilon_0\epsilon_1 H_{10} + 2\epsilon_0\delta_1 H_{01} + 2\epsilon_1\delta_1 H_{10}H_{01}\right]$$

$$= \bar{Y}^2\left[\left(\frac{1}{n} - \frac{1}{N}\right)C_y^2 + \left(\frac{1}{n} - \frac{1}{m}\right)\left\{H_{10}^2 C_x^2 + H_{01}^2(\lambda_{04}-1) + 2H_{10}\rho_{xy}C_yC_x\right.\right.$$

$$\left.\left. + 2H_{01}C_y\lambda_{12} + 2H_{01}H_{10}C_x\lambda_{03}\ \right\}\right]. \tag{6.1.4.4}$$

On differentiating (6.1.4.4) with respect to H_{10} and H_{01}, respectively, and equating to zero we have

$$H_{10}C_x + H_{01}\lambda_{03} = -\rho_{xy}C_y, \tag{6.1.4.5}$$

and

$$H_{10}C_x\lambda_{03} + H_{01}(\lambda_{04}-1) = -C_y\lambda_{12}. \tag{6.1.4.6}$$

Solving (6.1.4.5) and (6.1.4.6) for H_{10} and H_{01} we have

$$H_{10} = C_y\left[\lambda_{03}\lambda_{12} - \rho_{xy}(\lambda_{04}-1)\right]\Big/\left[C_x\left\{\lambda_{04} - 1 - \lambda_{03}^2\right\}\right], \tag{6.1.4.7}$$

and

$$H_{01} = C_y\left[\rho_{xy}\lambda_{03} - \lambda_{12}\right]\Big/\left\{\lambda_{04} - 1 - \lambda_{03}^2\right\}. \tag{6.1.4.8}$$

On substituting (6.1.4.7) and (6.1.4.8) in (6.1.4.4) we obtain the minimum mean squared error given by (6.1.4.2). Hence the theorem.

Remark 6.1.4.1. The optimum first phase and second phase sample sizes for the fixed cost (or variance) can also be derived for the difference, regression, and general class of estimators of mean. The optimum first phase and second phase sample sizes from the above theorems can easily be obtained by replacing V_R with

$$V_R = S_y^2\left(1-\rho_{xy}^2\right)$$

for the difference and regression estimator and with

$$V_R = S_y^2\left\{1-\rho_{xy}^2 - \frac{\left(\lambda_{12}-\rho_{xy}\lambda_{03}\right)^2}{\lambda_{04}-1-\lambda_{03}^2}\right\}$$

for the general class of estimators.

Example 6.1.4.1. The amounts of the real and nonreal estate farm loans (in $000) during 1997 in the United States have been presented in population 1 of the Appendix. Suppose we selected first phase and second phase samples each of size 10 and 5 respectively. Find the relative efficiency of the general class of estimators, for estimating average amount of the real estate farm loans during 1997 by using information selected in the first phase sample only on the nonreal estate farm loans during 1997, with respect to the regression estimator in two-phase sampling of the population mean.

Proof. We have

Y_i = Amount of the real estate farm loans in different states during 1997.

X_i = Amount of the nonreal estate farm loans in different states during 1997.

$N = 50$, $\bar{Y} = 555.43$, $S_y^2 = 342021.5$, $\lambda_{03} = 1.5936$, $\rho_{xy} = 0.8038$, $\lambda_{12} = 1.0982$, and $\lambda_{04} = 4.5247$.

So we have

$$\mathrm{MSE}\left(\bar{y}_{\mathrm{lrd}}\right) = \left(\frac{1}{m}-\frac{1}{N}\right)S_y^2 + \left(\frac{1}{n}-\frac{1}{m}\right)S_y^2\left(1-\rho_{xy}^2\right)$$

$$= \left(\frac{1}{10}-\frac{1}{50}\right)\times 342021.5 + \left(\frac{1}{5}-\frac{1}{10}\right)\times 342021.5\left(1-0.8038^2\right) = 39466.05\,.$$

Also for the general class of estimators, we have

$$\mathrm{MSE}\left(\bar{y}_{\mathrm{gd}}\right) = \left(\frac{1}{m}-\frac{1}{N}\right)S_y^2 + \left(\frac{1}{n}-\frac{1}{m}\right)S_y^2\left[1-\rho_{xy}^2 - \frac{\left(\lambda_{12}-\rho_{xy}\lambda_{03}\right)^2}{\lambda_{04}-1-\lambda_{03}^2}\right]$$

$$= \left(\frac{1}{10}-\frac{1}{50}\right)\times 342021.5$$

$$+\left(\frac{1}{5}-\frac{1}{10}\right)\times 342021.5\times\left[1-0.8038^2 - \frac{\left(1.0983-0.8038\times 1.5936\right)^2}{4.5247-1-1.5936^2}\right]$$

$$= 38308.00\,.$$

Thus the percent relative efficiency (RE) of the general class of estimators $\bar{y}_{gd}$ with respect to the regression estimator $\bar{y}_{lrd}$ is given by

$$RE = \frac{MSE(\bar{y}_{lrd})}{MSE(\bar{y}_{gd})} \times 100 = \frac{39466.05}{38308.00} \times 100 = 103.02\% .$$

6.1.5 ESTIMATION OF FINITE POPULATION VARIANCE

Singh (1991) defined a general class of estimators for estimating the finite population variance S_y^2 given by

$$s_h^2 = s_y^2 H(u,v) \tag{6.1.5.1}$$

where $u = \bar{x}/\bar{x}^*$, $v = s_x^2/s_x^{*2}$, and $H(\bullet,\bullet)$ is a parametric function and it satisfies some regularity conditions such that:

(a) $H(1,1) = 1$;

(b) The first and second order partial derivatives of H with respect to $u = \bar{x}/\bar{x}^*$ and $v = s_x^2/s_x^{*2}$ exist and are continuous and known constants.

Thus we have the following theorem:

Theorem 6.1.5.1. The minimum mean squared error of the general class of estimators, s_h^2, is given by

$$\text{Min.MSE}(s_h^2)$$
$$= S_y^4\left[\left(\frac{1}{m}-\frac{1}{N}\right)(\lambda_{40}-1)+\left(\frac{1}{n}-\frac{1}{m}\right)\left\{(\lambda_{40}-1)-\lambda_{21}^2-\frac{(\lambda_{22}-1-\lambda_{03}\lambda_{21})^2}{\lambda_{04}-1-\lambda_{03}^2}\right\}\right]. \tag{6.1.5.2}$$

Proof. Expanding $H(u,v)$ around the point $(1,1)$ up to the first order Taylor's series, we have

$$s_h^2 = s_y^2 H(u,v) = s_y^2 H[1+(u-1),1+(v-1)]$$
$$= s_y^2\left[1+(u-1)H_{10}+(v-1)H_{01}+(u-1)^2 H_{20}+(v-1)^2 H_{02}+(u-1)(v-1)H_{11}+....\right]$$

where

$$H_{10} = \frac{\partial H}{\partial u}|_{(1,1)},\ H_{01} = \frac{\partial H}{\partial v}|_{(1,1)},\ H_{20} = \frac{\partial^2 H}{\partial u^2}|_{(1,1)},\ H_{02} = \frac{\partial^2 H}{\partial v^2}|_{(1,1)},\ \text{and } H_{11} = \frac{\partial^2 H}{\partial u \partial v}.$$

Thus the mean squared error of the general class of estimators s_h^2, to the first order of approximation, is given by

$$MSE(s_h^2) = E[s_h^2 - S_y^2]^2 = E[s_y^2\{1+(u-1)H_{10}+(v-1)H_{01}\} - S_y^2]^2$$
$$= E[(s_y^2 - S_y^2) + s_y^2(u-1)H_{10} + s_y^2(v-1)H_{01} + ...]^2$$

$$= E\left[S_y^2(1+\delta_0) - S_y^2 + S_y^2 \in_1 (1+\delta_0)H_{10} + S_y^2\delta_1(1+\delta_0)H_{01}\right]^2$$

$$\approx S_y^4 E[\delta_0 + \in_1 H_{10} + \delta_1 H_{01}]^2$$

$$= S_y^4 E\left[\delta_0^2 + \in_1^2 H_{10}^2 + \delta_1^2 H_{01}^2 + 2\delta_0 \in_1 H_{10} + 2\delta_0\delta_1 H_{01} + 2 \in_1 \delta_1 H_{10} H_{01}\right]$$

$$= S_y^4\left[\left(\frac{1}{n}-\frac{1}{N}\right)(\lambda_{40}-1) + \left(\frac{1}{n}-\frac{1}{m}\right)\left\{H_{10}^2 C_x^2 + H_{01}^2(\lambda_{04}-1) + 2C_x\lambda_{21}H_{10}\right.\right.$$

$$\left.\left. + 2(\lambda_{22}-1)H_{01} + 2C_x\lambda_{03}H_{10}H_{01}\right\}\right]. \tag{6.1.5.3}$$

On differentiating (6.1.5.3) with respect to H_{10} and H_{01}, equating to zero each derivative, we have the following set of minimal equations

$$\begin{bmatrix} C_x, & \lambda_{03} \\ C_x\lambda_{03}, & (\lambda_{04}-1) \end{bmatrix}\begin{bmatrix} H_{10} \\ H_{01} \end{bmatrix} = \begin{bmatrix} -\lambda_{21} \\ -(\lambda_{22}-1) \end{bmatrix}. \tag{6.1.5.4}$$

On solving for H_{10} and H_{01} we obtain

$$H_{10} = -\frac{\{\lambda_{21}(\lambda_{04}-1) - \lambda_{03}(\lambda_{22}-1)\}}{\left\{C_x\left(\lambda_{04}-1-\lambda_{03}^2\right)\right\}}, \text{ and } H_{01} = -\frac{\{(\lambda_{22}-1) - \lambda_{21}\lambda_{03}\}}{\left\{\lambda_{04}-1-\lambda_{03}^2\right\}}.$$

On substituting these values of H_{10} and H_{01} in (6.1.5.3) we have the theorem.

The optimum first phase and second phase sample sizes for the fixed cost (or variance) can also be obtained for estimating the finite population variance.

6.1.6 CALIBRATION APPROACH IN TWO-PHASE SAMPLING

We introduce here the concept of two-phase sampling calibration approach and its generalisation which has been studied by Dupont (1995), Hidiroglou and Särndal (1995,1998) and Estevao and Särndal (2002). At the moment we are using only one auxiliary variable to keep the procedure simple. Suppose a first phase probability sample s_1 is drawn from the population Ω using a sampling design that generates the selection probabilities π_{1i}. From the given first phase sample s_1, the second phase sample s_2 (subset of s_1) is drawn with a sampling design with the selection probabilities $\pi_{2i} = \pi_{i|s_1}$. Evidently from the first phase sample s_1 the Horvitz and Thompson (1952) type estimator of population total X is given by

$$\hat{x}^* = \sum_{i \in s_1} d_{1i} x_i$$

where $d_{1i} = 1/\pi_{1i}$. From the second phase sample unbiased estimators of Y and X are, respectively, given by

$$\hat{y} = \sum_{i \in s_2} d_{2i} y_i, \text{ and } \hat{x} = \sum_{i \in s_2} d_{2i} x_i$$

where $d_{2i} = (1/\pi_{1i}) \times (1/\pi_{2i})$.

Our interest is in the estimator given by

$$\hat{y}_w = \sum_{i\in s_2} w_{2i} y_i \,. \tag{6.1.6.1}$$

Following the simplest case of Deville and Särndal (1992), we choose the weights w_{2i} such that the chi square (CS) distance given by

$$D = \sum_{i\in s_2} \left(w_{2i} - d_{2i}\right)^2 \left(d_{2i} q_{2i}\right)^{-1} \tag{6.1.6.2}$$

is minimum subject to the calibration constraint given by

$$\sum_{i\in s_2} w_{2i} x_i = \hat{x}^* \tag{6.1.6.3}$$

where q_{2i} are suitable weights as discussed in Chapter 5.

Thus we have the following theorem:

Theorem 6.1.6.1. The calibration estimator $\hat{y}_w$ of population total Y in two-phase sampling is given by

$$\hat{y}_w = \hat{y} + \hat{\beta}_{\text{ds}}\left(\hat{x}^* - \hat{x}\right), \quad \text{where } \hat{\beta}_{\text{ds}} = \left(\sum_{i\in s_2} d_{2i} q_{2i} x_i y_i\right) \Big/ \left(\sum_{i\in s_2} d_{2i} q_{2i} x_i^2\right). \tag{6.1.6.4}$$

Proof. From (6.1.6.2) and (6.1.6.3), the Lagrange function L is

$$L = \sum_{i\in s_2} \left(w_{2i} - d_{2i}\right)^2 \left(d_{2i} q_{2i}\right)^{-1} - 2\lambda\left(\sum_{i\in s_2} w_{2i} x_i - \hat{x}^*\right). \tag{6.1.6.5}$$

On differentiating (6.1.6.5) with respect to w_{2i} and equating to zero we have

$$w_{2i} = d_{2i} + \lambda d_{2i} q_{2i} x_i \,. \tag{6.1.6.6}$$

On substituting (6.1.6.6) in (6.1.6.3) we have

$$\lambda = \left[\hat{x}^* - \sum_{i\in s_2} d_{2i} x_i\right] \Big/ \sum_{i\in s_2} d_{2i} q_{2i} x_i^2 \,. \tag{6.1.6.7}$$

On substituting (6.1.6.7) in (6.1.6.6), we obtain the second phase calibrated weights

$$w_{2i} = d_{2i} + d_{2i} q_{2i} x_i \left(\sum_{i=1}^{n} d_{2i} q_{2i} x_i^2\right)^{-1} \left(\hat{x}^* - \sum_{i=1}^{n} d_{2i} x_i\right). \tag{6.1.6.8}$$

On putting these calibrated weights in (6.1.6.1) we have the estimator (6.1.6.4). Hence the theorem.

Corollary 6.1.6.1. If $q_{2i} = 1/x_i$, then the regression estimator $\hat{y}_w$ reduces to the ratio estimator

$$\hat{y}_{\text{rd}} = \hat{y}\left(\frac{\hat{x}^*}{\hat{x}}\right)$$

of the population total in two-phase sampling.

Example 6.1.6.1. Select a preliminary large sample of 10 units by PPSWOR sampling using the Midzuno--Sen sampling scheme and using the number of species groups during 1992 as an auxiliary variable given in the population 4 of the Appendix. Select a second phase sample of 5 units from the given first phase sample by using the Midzuno--Sen sampling scheme. Find the calibration weights for the units selected in the ultimate sample by making use of known information about the number of fish caught during 1994 as an auxiliary variable. Use the chi square distance function between the design weights and calibration weights. Discuss two cases when these weights lead to the ratio and generalized regression estimator (GREG) for estimating the total number of fish during 1995. Deduce the estimate in each case.

Solution. In the Midzuno--Sen sampling scheme the first unit is to be selected with probability proportional to the number of fish during 1992 and the remaining 9 units with SRSWOR sampling. We used the first two columns of the Pseudo-Random Numbers (PRN) given in Table 1 of the Appendix to select a random number $1 \le R_i \le 69$ and another random number $1 \le R_j \le 28933$ by using the 7th to 11th columns. The first effective pair of the random numbers is (62, 01228). Thus the unit at serial number 62 is included as a first unit in the preliminary large sample of 10 units. The remaining 9 units are selected by SRSWOR sampling from the remaining 58 units in the population. We used the 13th and 14th columns of the Pseudo-Random Numbers to draw 9 distinct random numbers between 1 and 58. The random number came in the sequence as 05, 34, 30, 55, 46, 07, 13, 19 and 44.

First phase sample information							
Sr. No.	Random No.	Species group	1992 z_i	P_{1i}	π_{1i}	1994 x_i	$\frac{x_i}{\pi_{1i}}$
1	62	Summer flounder	11918	0.0408	0.1678	17741	105727.1
2	05	Herrings	28933	0.0991	0.2184	38007	174024.7
3	34	Pigfish	2955	0.0101	0.1411	4918	34854.71
4	30	Lane snapper	919	0.0031	0.1351	1088	8053.3
5	55	Cunner	1931	0.0066	0.1381	1255	9087.6
6	46	Spot	14974	0.0513	0.1769	18491	104528.0
7	07	Saltwater catfish	13466	0.0461	0.1724	14441	83764.5
8	13	Searobins	4768	0.0163	0.1465	4707	32129.7
9	19	Groupers	4661	0.0159	0.1462	4583	31347.5
10	44	Sand seatrout	3780	0.0129	0.1436	5665	39449.9
		Sum	88305				622966.9

In the above table the values of π_{1i} are given by

$$\pi_{1i} = (N - m)P_{1i}/(N-1) + (m-1)/(N-1)$$

for $N = 69$, $m = 10$, and $P_{1i} = z_i/Z$ with $Z = 291882$ (given).

Thus an estimate of the total number of species during 1992 is

$$\hat{x}^* = \sum_{i \in s_1} d_{1i} x_i = 622966.9\ .$$

Again applying the Midzuno--Sen sampling scheme to the units selected in the first phase sample, we selected one unit with probability proportional to the number of fish during 1992 and the remaining 4 units with SRSWOR sampling. We used the first two columns of the Pseudo-Random Numbers to select a random number $1 \le R_i \le 10$ and another random number $1 \le R_j \le 28933$ by using the 7^{h} to 11th columns. The first effective pair of the random numbers is (01, 07572). Thus the unit at serial number 01 from the given first phase sample is included as a first unit in the second phase sample of $n = 5$ units. The remaining 4 units are selected by SRSWOR sampling from the remaining 9 units in the given first phase sample. We used the 13th column of the Pseudo-Random Numbers to draw 4 distinct random numbers between 1 and 9. The random number came in the sequence as 3, 7, 5, and 4. Thus the ultimate sample consists of the following information.

Second phase sample information									
Second-phase	First-phase.	Pop. Unit	π_{1i}	1992 z_i	P_{2i}	π_{2i}	d_{2i}	1994 x_i	1994 y_i
1	1	62	0.1678	11918	0.1349	0.5194	11.4737	17741	16238
3	4	30	0.1351	919	0.0104	0.4502	16.4414	1008	859
7	8	13	0.1465	4768	0.0539	0.4744	14.3886	4707	4793
5	6	46	0.1769	14974	0.1696	0.5387	10.4936	18491	11567
4	5	55	0.1381	1931	0.0219	0.4566	15.8588	1255	1375

In the above table we used

$$d_{2i} = 1/(\pi_{1i}\pi_{2i}), \quad \text{and} \quad \pi_{2i} = \frac{m-n}{m-1}p_{2i} + \frac{n-1}{m-1} \quad \text{with } p_{2i} = z_i \Big/ \sum_{i=1}^{m} z_i .$$

Under the chi square distance function the second phase calibration weights are

$$w_{2i} = d_{2i} + d_{2i}q_{2i}x_i\left(\sum_{i\in s_2} d_{2i}q_{2i}x_i^2\right)^{-1}\left(\hat{x}^* - \sum_{i\in s_2} d_{2i}x_i\right)$$

where

$$\hat{x}^* = \sum_{i\in s_1} d_{1i}x_i .$$

Case 1. If $q_{2i} = 1/x_i$, then these weights become

$$w_{2i} = d_{2i}\left(\sum_{i\in s_2} d_{2i}x_i\right)^{-1}\hat{x}^* = \frac{622966.9}{501796.3}d_{2i} .$$

Sr. No.	d_{2i}	x_i	y_i	$x_i d_{2i}$	w_{2i}	$w_{2i}y_i$
1	11.4737	17741	16238	203554.91	14.24429	231298.78
2	16.4414	1008	859	16572.93	20.41156	17533.53
3	14.3885	4707	4793	67726.67	17.86673	85635.23
4	10.4936	18491	11567	194037.16	13.02752	150689.32
5	15.8588	1255	1375	19902.79	19.68828	27071.38
			Sum	501794.46		512228.24

Thus a ratio estimate of the total number of fish during 1995 in the United States is

$$\hat{y}_{rd} = \sum_{i \in s_2} w_{2i} y_i = 512228.24 .$$

Case II. If $q_i = 1$ then these weights become

$$w_{2i} = d_{2i} + d_{2i} x_i \left(\sum_{i \in s_2} d_{2i} x_i^2 \right)^{-1} \left(\hat{x}^* - \sum_{i \in s_2} d_{2i} x_i \right).$$

Sr. No.	d_{2i}	x_i	y_i	$d_{2i} x_i^2$	w_{2i}	$w_{2i} y_i$
1	11.4737	17741	16238	3611267688.00	14.73639	239289.40
2	16.4414	1008	859	16705514.65	16.70704	14351.35
3	14.3885	4707	4793	318789433.30	15.47406	74167.17
4	10.4936	18491	11567	3587941081.00	13.60373	157354.40
5	15.8588	1255	1375	24978006.47	16.17781	22244.49
			Sum	7559681724.00		507406.80

Thus the generalized regression estimate (GREG) of the total number of fish in the United States during 1995 is given by

$$\hat{y}_w = \sum_{i \in s_2} w_{2i} y_i = 507406.8 .$$

6.2 TWO-PHASE SAMPLING USING TWO AUXILIARY VARIABLES

In the case of two auxiliary variables, Rao (1972) and Singh (1984) have considered the following sampling strategies. Consider a finite population of size N. Let X and Z denote the two auxiliary variables and let Y denote the variable under study. Now when two auxiliary variables are used in a survey, there may be three possible sampling schemes, namely:

Scheme I. A first phase sample of size m is selected by SRSWOR for observing the auxiliary variables X and Z. A smaller sub-sample of size n for observing Y, X, and Z is selected from the first phase sample;

Scheme II. The first phase sample is selected as in Scheme I but the smaller sample of size n is selected independently from the whole population;

Scheme III. Often the auxiliary information may be collected by two different agencies and hence two independent preliminary samples of sizes m_1 and m_2 are selected for observing X and Z, and the small sample of size n is also selected independently from the population by SRSWOR. For simplicity we will consider $m_1 = m_2$.

Before proceeding further let us define

$$\epsilon_0 = \frac{\bar{y}}{\bar{Y}} - 1, \quad \epsilon_1 = \frac{\bar{x}}{\bar{X}} - 1, \quad \epsilon_2 = \frac{\bar{x}^*}{\bar{X}} - 1, \quad \epsilon_3 = \frac{\bar{z}}{\bar{Z}} - 1, \quad \text{and} \quad \epsilon_4 = \frac{\bar{z}^*}{\bar{Z}} - 1,$$

such that

$$E(\epsilon_j) = 0, \quad j = 0,1,2,3,4.$$

Under Scheme I.

$$E(\epsilon_0^2) = \left(\frac{1}{n} - \frac{1}{N}\right)C_y^2, \quad E(\epsilon_1^2) = \left(\frac{1}{n} - \frac{1}{N}\right)C_x^2, \quad E(\epsilon_2^2) = \left(\frac{1}{m} - \frac{1}{N}\right)C_x^2, \quad E(\epsilon_3^2) = \left(\frac{1}{n} - \frac{1}{N}\right)C_z^2,$$

$$E(\epsilon_4^2) = \left(\frac{1}{m} - \frac{1}{N}\right)C_z^2, \quad E(\epsilon_0\epsilon_1) = \left(\frac{1}{n} - \frac{1}{N}\right)\rho_{xy}C_xC_y, \quad E(\epsilon_0\epsilon_2) = \left(\frac{1}{m} - \frac{1}{N}\right)\rho_{xy}C_xC_y,$$

$$E(\epsilon_0\epsilon_3) = \left(\frac{1}{n} - \frac{1}{N}\right)\rho_{yz}C_yC_z, \quad E(\epsilon_0\epsilon_4) = \left(\frac{1}{m} - \frac{1}{N}\right)\rho_{yz}C_yC_z, \quad E(\epsilon_1\epsilon_2) = \left(\frac{1}{m} - \frac{1}{N}\right)C_x^2,$$

$$E(\epsilon_1\epsilon_3) = \left(\frac{1}{n} - \frac{1}{N}\right)\rho_{xz}C_xC_z, \quad E(\epsilon_1\epsilon_4) = \left(\frac{1}{m} - \frac{1}{N}\right)\rho_{xz}C_xC_z, \quad E(\epsilon_2\epsilon_3) = \left(\frac{1}{m} - \frac{1}{N}\right)\rho_{xz}C_xC_z,$$

$$E(\epsilon_2\epsilon_4) = \left(\frac{1}{m} - \frac{1}{N}\right)\rho_{xz}C_xC_z, \quad \text{and} \quad E(\epsilon_3\epsilon_4) = \left(\frac{1}{m} - \frac{1}{N}\right)C_z^2.$$

Under Scheme II.

$$E(\epsilon_1\epsilon_2) = E(\epsilon_3\epsilon_4) = E(\epsilon_0\epsilon_2) = E(\epsilon_0\epsilon_4) = E(\epsilon_2\epsilon_3) = E(\epsilon_1\epsilon_4) = 0.$$

Under Scheme III.

$$E(\epsilon_1\epsilon_2) = E(\epsilon_3\epsilon_4) = E(\epsilon_0\epsilon_2) = E(\epsilon_0\epsilon_4) = E(\epsilon_2\epsilon_3) = E(\epsilon_1\epsilon_4) = E(\epsilon_2\epsilon_4) = 0.$$

Now we have the following theorem:

Theorem 6.2.1. The difference type and unbiased estimator of population mean $\bar{Y}$ is given by

$$\bar{y}_1 = \bar{y} + \beta_1(\bar{x}^* - \bar{x}) + \beta_2(\bar{z}^* - \bar{z}). \tag{6.2.1}$$

Under scheme I the minimum variance of $\bar{y}_1$ is given by

$$\text{Min.}V(\bar{y}_1)_\text{I} = \frac{S_y^2}{n}\left[1 - \left(\frac{m-n}{m}\right)\left\{\frac{\rho_{xy}^2 + \rho_{yz}^2 - 2\rho_{xy}\rho_{yz}\rho_{xz}}{1 - \rho_{xz}^2}\right\}\right] - \frac{S_y^2}{N}. \tag{6.2.2}$$

Under scheme II and as $N \to \infty$, the minimum variance of $\bar{y}_1$ is given by

$$\text{Min.}V(\bar{y}_1)_\text{II} = \frac{S_y^2}{n}\left[1 - \left(\frac{m}{n+m}\right)\left\{\frac{\rho_{xy}^2 + \rho_{yz}^2 - 2\rho_{xy}\rho_{yz}\rho_{xz}}{1 - \rho_{xz}^2}\right\}\right]. \tag{6.2.3}$$

Under scheme III and as $N \to \infty$, the minimum variance of $\bar{y}_1$ is given by

$$\text{Min.}V(\bar{y}_1)_{\text{III}} = \frac{S_y^2}{n}\left[1-\left(\frac{m}{n+m}\right)\left\{\frac{\rho_{xy}^2+\rho_{yz}^2-2\rho_{xy}\rho_{yz}\rho_{xz}}{1-\left(\frac{m}{m+n}\right)^2\rho_{xz}^2}\right\}\right]. \tag{6.2.4}$$

Proof. The estimator $\bar{y}_1$ in terms of $\in_j$, $j = 0,1,2,3,4$ can easily be written as

$$\bar{y}_1 = \bar{Y}(1+\in_0)+\beta_1\bar{X}(\in_2-\in_1)+\beta_2\bar{Z}(\in_4-\in_3). \tag{6.2.5}$$

Taking expected values on both sides of (6.2.5) we have

$$E(\bar{y}_1) = \bar{Y}, \tag{6.2.6}$$

which proves the unbiasedness.

By the definition of variance we have

$$\begin{aligned} V(\bar{y}_1) &= E[\bar{y}_1-\bar{Y}]^2 = E[\bar{Y}\in_0+\beta_1\bar{X}(\in_2-\in_1)+\beta_2\bar{Z}(\in_4-\in_3)]^2 \\ &= E\Big[\bar{Y}^2\in_0^2+\beta_1^2\bar{X}^2(\in_2^2+\in_1^2-2\in_1\in_2)+\beta_2^2\bar{Z}^2(\in_4^2+\in_3^2-2\in_3\in_4) \\ &\quad +2\bar{Y}\bar{X}\beta_1(\in_0\in_2-\in_0\in_1)+2\beta_2\bar{Y}\bar{Z}(\in_0\in_4-\in_0\in_3) \\ &\quad +2\beta_1\beta_2\bar{X}\bar{Z}(\in_2\in_4-\in_2\in_3-\in_1\in_4+\in_1\in_3)\Big]. \end{aligned} \tag{6.2.7}$$

Under Scheme I the variance at (6.2.7) reduces to

$$V(\bar{y}_1)_{\text{I}} = \left(\frac{1}{n}-\frac{1}{N}\right)S_y^2+\left(\frac{1}{n}-\frac{1}{m}\right)\left[\beta_1^2S_x^2+\beta_2^2S_z^2-2\beta_1S_{xy}-2\beta_2S_{yz}+2\beta_1\beta_2S_{xz}\right]. \tag{6.2.8}$$

On differentiating (6.2.8) with respect to β_1 and β_2, respectively, and equating to zero we have

$$\beta_1S_x^2+\beta_2S_{xz} = S_{xy}, \tag{6.2.9}$$

and

$$\beta_1S_{xz}+\beta_2S_z^2 = S_{yz}. \tag{6.2.10}$$

On solving (6.2.9) and (6.2.10) for β_1 and β_2 we have

$$\beta_1 = \frac{S_y(\rho_{xy}-\rho_{yz}\rho_{xz})}{S_x(1-\rho_{xz}^2)}, \quad \text{and} \quad \beta_2 = \frac{S_y(\rho_{yz}-\rho_{xz}\rho_{xy})}{S_z(1-\rho_{xz}^2)}. \tag{6.2.11}$$

On substituting β_1 and β_2 from (6.2.11) in (6.2.8), we have (6.2.2). Hence the first part of the theorem.

Under Scheme II and as $N\to\infty$ then (6.2.7) becomes

$$V(\bar{y}_1)_{\text{II}} = \frac{S_y^2}{n}+\left(\frac{1}{m}+\frac{1}{n}\right)(\beta_1^2S_x^2+\beta_2^2S_z^2)-\frac{2\beta_1S_{xy}}{n}-\frac{2\beta_2S_{yz}}{n}+2\beta_1\beta_2S_{xz}\left(\frac{1}{m}+\frac{1}{n}\right). \tag{6.2.12}$$

On differentiating (6.2.12) with respect to β_1 and β_2, respectively and equating to zero and solving for β_1 and β_2, we obtain

$$\beta_1 = \frac{\left(\frac{m}{m+n}\right)\left\{S_y\left\{\rho_{xy} - \rho_{yz}\rho_{xz}\right\}\right\}}{\left\{S_x\left(1-\rho_{xz}^2\right)\right\}}, \text{ and } \beta_2 = \frac{\left(\frac{m}{m+n}\right)\left\{S_y\left\{\rho_{yz} - \rho_{xy}\rho_{xz}\right\}\right\}}{\left\{S_z\left(1-\rho_{xz}^2\right)\right\}}. \quad (6.2.13)$$

On substituting β_1 and β_2 from (6.2.13) in (6.2.12), we have (6.2.3). Hence the second part of the theorem.

Under Scheme III and as $N \to \infty$ then (6.2.7) becomes

$$V(\bar{y}_1)_{\text{III}} = \frac{S_y^2}{n} + \left(\frac{1}{m} + \frac{1}{n}\right)\left(\beta_1^2 S_x^2 + \beta_2^2 S_z^2\right) - \frac{2}{n}\left(\beta_1 S_{xy} + \beta_2 S_{yz} - \beta_1\beta_2 S_{xz}\right) \quad (6.2.14)$$

which in fact reduces to (6.2.4) for the optimum values of β_1 and β_2 obtained from (6.2.14). Hence the theorem.

The estimation procedure based on two-phase sampling schemes, when none of the auxiliary variable population means are known, has been considered by Khan and Tripathi (1967), Tripathi (1970, 1976, 1987), and Adhvaryu (1978). However, in many socio-economic and agricultural surveys, the population means (totals) of some of the auxiliary variables may be known while those of the others may not be readily available. For example, to estimate the total number of agricultural labourers in a rural county, the information about the area and population of the village may be known from the recent county records while the information about the number of cultivators and cultivated areas of the village in the county may not be readily available. The estimation of population mean of a survey variable under the partial knowledge of the auxiliary means has been considered by Singh (1969), Chand (1975), Kiregyera (1980, 1984), Mukerjee, Rao, and Vijayan (1987), and Srivastava, Khare, and Srivastava (1990). These estimators and their modifications are popular in survey sampling under the name 'Chain Ratio Type Estimators' which will be discussed in the next section, but let us first do an example.

Example 6.2.1. The season average per pound prices (in \$) of the commercial Apple crop in 36 different American states have been given in population 3.

Scheme I. Select a first phase sample of 10 units by SRSWOR for observing the auxiliary variables:

X_{1i} = Season average price per pound during 1995;

X_{2i} = Season average price per pound during 1994.

Select a second phase sample from the first phase sample of 5 units and observe:

Y_i = Season average price per pound during 1996;

X_{1i} = Season average price per pound during 1995;

X_{2i} = Season average price per pound during 1994.

Scheme II. Select the first phase sample of 10 units as in Scheme I and collect information on two variables. Select the second phase sample of 5 units independently from the whole population and collect information on three variables.

Scheme III. Suppose the information on 10 units selected by SRSWOR sampling about 'Season price per pound average during 1994' is collected by a company XYZ. The information about 'Season price per pound average during 1995' on an independent sample of 10 units is collected by another company ABC in the United States. A small sample of 5 units is also selected independently from the population by SRSWOR sampling to collect information on the 'Season average price per pound during 1996.'

We wish to estimate the average season price per pound during 1996 by making proper use of information under different schemes. Which sampling scheme would you prefer to recommend for the future?

Solution. From the description of the population we have $Y_i =$ Season average price per pound during 1996, $X_{1i} =$ Season average price per pound during 1995, and $X_{2i} =$ Season average price per pound during 1994. Here $Y = 7.317$, $X_1 = 6.683$, $X_2 = 5.9222$, $N = 36$, $\bar{Y} = 0.2033$, $\bar{X}_1 = 0.1856$, $\bar{X}_2 = 0.1645$, $S_y^2 = 0.15633$, $S_{x1}^2 = 0.16406$, $S_{x2}^2 = 0.17396$, $C_y^2 = 0.1563$, $C_{x1}^2 = 0.1641$, $C_{x2}^2 = 0.1739$, $\rho_{yx_1} = 0.8775$, $\rho_{yx_2} = 0.8759$, and $\rho_{x_1x_2} = 0.74135$. Also we have $m = 10$, $n = 5$.

Under scheme I the minimum variance of $\bar{y}_1$ is given by

$$\text{Min.}V(\bar{y}_1)_{\text{I}} \approx \frac{S_y^2}{n}\left[1-\left(\frac{m-n}{m}\right)\left\{\frac{\rho_{x_1y}^2+\rho_{yx_2}^2-2\rho_{x_1y}\rho_{yx_2}\rho_{x_1x_2}}{1-\rho_{x_1x_2}^2}\right\}\right]$$

$$= \frac{0.15633}{5}\left[1-\left(\frac{10-5}{10}\right)\left\{\frac{0.8775^2+0.8759^2-2\times0.8775\times0.8759\times0.7413}{1-0.7413^2}\right\}\right]$$

$$= 0.017465.$$

Under scheme II the minimum variance of $\bar{y}_1$ is given by

$$\text{Min.}V(\bar{y}_1)_{\text{II}} \approx \frac{S_y^2}{n}\left[1-\left(\frac{m}{n+m}\right)\left\{\frac{\rho_{x_1y}^2+\rho_{yx_2}^2-2\rho_{x_1y}\rho_{yx_2}\rho_{x_1x_2}}{1-\rho_{x_1x_2}^2}\right\}\right]$$

$$= \frac{0.15633}{5}\left[1-\left(\frac{10}{10+5}\right)\left\{\frac{0.8775^2+0.8759^2-2\times0.8775\times0.8759\times0.7413}{1-0.7413^2}\right\}\right]$$

$$= 0.012865.$$

Under scheme III the minimum variance of $\bar{y}_1$ is given by

$$\text{Min.}V(\bar{y}_1)_{\text{III}} \approx \frac{s_y^2}{n}\left[1-\left(\frac{m}{n+m}\right)\left\{\frac{\rho_{x_1y}^2+\rho_{yx_2}^2-2\rho_{x_1y}\rho_{yx_2}\rho_{x_1x_2}}{1-\left(\frac{m}{m+n}\right)^2\rho_{x_1x_2}^2}\right\}\right]$$

$$= \frac{0.15633}{5}\left[1-\left(\frac{10}{10+5}\right)\left\{\frac{0.8775^2+0.8759^2-2\times0.8775\times0.8759\times0.7413}{1-\left(\frac{10}{10+5}\right)^2\times0.7413^2}\right\}\right]$$

$$= 0.021389.$$

In this situation the estimator of population mean under scheme II has minimum variance. Hence from the efficiency point of view scheme II will be preferred.

6.3 CHAIN RATIO TYPE ESTIMATORS

Following the notation of the previous section and assuming that population mean $\bar{Z}$ of the second auxiliary variable is known and $\rho_{xy} > \rho_{yz}$ (variable Z is closely related to X; however, it is not as closely related to Y as X is related to Y), Chand (1975) proposed a chain ratio type estimator of population mean $\bar{Y}$ as

$$\bar{y}_C = \bar{y}\left(\frac{\bar{x}^*}{\bar{x}}\right)\left(\frac{\bar{Z}}{\bar{z}^*}\right). \tag{6.3.1}$$

Thus we have the following theorems:

Theorem 6.3.1. Under scheme I the bias in the estimator $\bar{y}_C$ to the first order of approximation is

$$B(\bar{y}_C) = \left(\frac{1}{n}-\frac{1}{N}\right)\bar{Y}\left(C_x^2-\rho_{xy}C_xC_y\right)+\left(\frac{1}{m}-\frac{1}{N}\right)\bar{Y}\left(C_z^2-C_x^2+\rho_{xy}C_xC_y-\rho_{yz}C_yC_z\right). \tag{6.3.2}$$

Proof. The estimator $\bar{y}_C$ in terms of ϵ_0, ϵ_1, ϵ_2, and ϵ_4 can be written as

$$\bar{y}_C = \bar{Y}\left[1+\epsilon_0-\epsilon_1+\epsilon_2-\epsilon_4+\epsilon_0\epsilon_2-\epsilon_0\epsilon_1-\epsilon_1\epsilon_2+\epsilon_1^2-\epsilon_0\epsilon_4-\epsilon_2\epsilon_4+\epsilon_1\epsilon_4+\epsilon_4^2+\ldots\right]. \tag{6.3.3}$$

Taking expected values on both sides of (6.3.3) and taking its deviation from population mean $\bar{Y}$, we have (6.3.2). Hence the theorem.

Theorem 6.3.2. Under scheme I the mean squared error of the estimator $\bar{y}_C$, to the first order of approximation, is

$$\mathrm{MSE}(\bar{y}_C)=\left(\frac{1}{n}-\frac{1}{N}\right)\bar{Y}^2\left[C_y^2+C_x^2-2\rho_{xy}C_xC_y\right]$$

$$+\left(\frac{1}{m}-\frac{1}{N}\right)\bar{Y}^2\left[C_z^2-C_x^2+2\rho_{xy}C_xC_y-2\rho_{yz}C_yC_z\right]. \qquad (6.3.4)$$

Proof. We have

$$\mathrm{MSE}(\bar{y}_C)=E\left[\bar{y}_C-\bar{Y}\right]^2\approx E\left[\bar{Y}(\epsilon_0-\epsilon_1+\epsilon_2-\epsilon_4)\right]^2$$

$$=\bar{Y}^2E\left[\epsilon_0^2+\epsilon_1^2+\epsilon_2^2+\epsilon_4^2-2\epsilon_0\epsilon_1+2\epsilon_0\epsilon_2-2\epsilon_0\epsilon_4-2\epsilon_1\epsilon_2+2\epsilon_1\epsilon_4-2\epsilon_2\epsilon_4\right].$$

On substituting the expected values and after simplification we obtain (6.3.4). Hence the theorem.

Khare and Srivastava (1998) and Tracy and Singh (1999) have considered the study of chain ratio type estimators in more detail by proposing a few classes of estimators of population mean.

6.4 CALIBRATION USING TWO AUXILIARY VARIABLES

In this case a first phase probability sample s_1 of size m is drawn from the population Ω, using a sampling design that generates the selection probabilities π_{1i}. Given that sample s_1 has been drawn, the second phase sample s_2 (subset of s_1) of size n is selected from s_1 using a sampling design with the selection probabilities $\pi_{2i}=\pi_{i|s_1}$. The first phase sampling weight of ith unit is denoted by $d_{1i}=1/\pi_{1i}$, and the second phase sampling weight by $d_{2i}=(\pi_{1i}\pi_{2i})^{-1}$.

Table 6.4.1. Relationship between set of units and available data at different levels.

Set of units	Data available	HT estimators	Calibrated estimators
Population	$\{z_i : i\in\Omega\}$ or $Z=\sum_{i=1}^{N}Z_i$ is known.		
First phase sample	$\{(x_i,z_i) : i\in s_1\}$	$\hat{x}^*=\sum_{i\in s_1}d_{1i}x_i$, $\hat{z}^*=\sum_{i\in s_1}d_{1i}z_i$	$\hat{x}_c^*=\sum_{i\in s_1}w_{1i}x_i$
Second phase sample	$\{(x_i,y_i,z_i) : i\in s_2\}$	$\hat{y}=\sum_{i\in s_2}d_{2i}y_i$, $\hat{x}=\sum_{i\in s_2}d_{2i}x_i$, $\hat{z}=\sum_{i\in s_2}d_{2i}z_i$	$\hat{y}_c=\sum_{i\in s_2}w_{2i}y_i$

Furthermore we assume that:

(a) Z_i is known for all $i \in \Omega$ (or $Z = \sum_{i=1}^{N} Z_i$ is known) and z_i is observed for all $i \in s_1$;

(b) x_i is observed for all $i \in s_1$;

(c) y_i is observed for all $i \in s_2$.

The Table 6.4.1 summarises our assumptions on the auxiliary information available for estimation.

Now we have the following theorems:

Theorem 6.4.1. Under chi square (CS) type of distance function for the first phase data set defined as

$$D_1 = \sum_{i \in s_1} (w_{1i} - d_{1i})^2 (d_{1i} q_{1i})^{-1} \tag{6.4.1}$$

subject to the calibration equation

$$\sum_{i \in s_1} w_{1i} z_i = Z \tag{6.4.2}$$

the first phase calibrated estimator

$$\hat{x}_c^* = \sum_{i \in s_1} w_{1i} x_i \tag{6.4.3}$$

becomes

$$\hat{x}_c^* = \sum_{i \in s_1} d_{1i} x_i + \left(\sum_{i \in s_1} d_{1i} q_{1i} x_i z_i \Big/ \sum_{i \in s_1} d_{1i} q_{1i} z_i^2 \right) \left(Z - \sum_{i \in s_1} d_{1i} z_i \right). \tag{6.4.4}$$

Proof. Consider the Lagrange function L_1 given by

$$L_1 = \sum_{i \in s_1} (w_{1i} - d_{1i})^2 (d_{1i} q_{1i})^{-1} - 2\lambda_1 \left[\sum_{i \in s_1} w_{1i} z_i - Z \right]. \tag{6.4.5}$$

On differentiating (6.4.5) with respect to w_{1i} and equating to zero we have

$$w_{1i} = d_{1i} + \lambda_1 \, d_{1i} q_{1i} z_i \,. \tag{6.4.6}$$

On substituting (6.4.6) in (6.4.2) we have

$$\lambda_1 = \left(Z - \sum_{i \in s_1} d_{1i} z_i \right) \Big/ \sum_{i \in s_1} d_{1i} q_{1i} z_i^2 \,. \tag{6.4.7}$$

On substituting (6.4.7) in (6.4.6) we obtain the optimal weights given by

$$w_{1i} = d_{1i} + \left(d_{1i} q_{1i} z_i \Big/ \sum_{i \in s_1} d_{1i} q_{1i} z_i^2 \right) \left(Z - \sum_{i \in s_1} d_{1i} z_i \right). \tag{6.4.8}$$

On substituting (6.4.8) in (6.4.3) we have (6.4.4). Hence the theorem.

Theorem 6.4.2. Under the CS type of distance function for the second phase data set defined as

$$D_2 = \sum_{i \in s_2} (w_{2i} - d_{2i})^2 (d_{2i} q_{2i})^{-1} \tag{6.4.9}$$

subject to the calibration equation

$$\sum_{i\in s_2} w_{2i}x_i = \hat{x}_c^* \tag{6.4.10}$$

the second phase calibrated estimator

$$\hat{y}_c = \sum_{i\in s_2} w_{2i}y_i \tag{6.4.11}$$

becomes

$$\hat{y}_c = \sum_{i\in s_2} d_{2i}y_i + \hat{\beta}_{2(ds)}\left[\left(\sum_{i\in s_1} d_{1i}x_i - \sum_{i\in s_2} d_{2i}x_i\right) + \hat{\beta}_{1(ds)}\left(Z - \sum_{i\in s_1} d_{1i}z_i\right)\right] \tag{6.4.12}$$

where $\hat{\beta}_{1(ds)} = \sum_{i\in s_1} d_{1i}q_{1i}x_i z_i \Big/ \sum_{i\in s_1} d_{1i}q_{1i}z_{1i}^2$ and $\hat{\beta}_{2(ds)} = \sum_{i\in s_2} d_{2i}q_{2i}x_i y_i \Big/ \sum_{i\in s_2} d_{1i}q_{1i}x_{1i}^2$.

Proof. The Lagrange function L_2 is

$$L_2 = \sum_{i\in s_2}(w_{2i} - d_{2i})^2(d_{2i}q_{2i})^{-1} - 2\lambda_2\left(\sum_{i\in s_2} w_{2i}x_i - \hat{x}_c^*\right). \tag{6.4.13}$$

On differentiating (6.4.13) with respect to w_{2i} and equating to zero we have

$$w_{2i} = d_{2i} + \lambda_2 d_{2i}q_{2i}x_i\,. \tag{6.4.14}$$

On substituting (6.4.14) in (6.4.10) we have

$$\lambda_2 = \left[\hat{x}_c^* - \sum_{i\in s_2} d_{2i}x_i\right] \Big/ \sum_{i\in s_2} d_{2i}q_{2i}x_i^2 \tag{6.4.15}$$

which on substitution in (6.4.14) leads to the second phase optimal weights

$$w_{2i} = d_{2i} + \left(d_{2i}q_{2i}x_i \Big/ \sum_{i\in s_2} d_{2i}q_{2i}x_i^2\right)\left[\hat{x}_c^* - \sum_{i\in s_2} d_{2i}x_i\right]. \tag{6.4.16}$$

Substituting (6.4.16) in (6.4.11), we have (6.4.12). Hence the theorem.

Example 6.4.1. Select a preliminary large sample of 15 units by using the Midzuno--Sen sampling scheme and using the number of fish caught during 1992 in the United States as a selection variable given in the population 4 in the Appendix. Collect the information on the number of fish caught during 1993 and 1994 from the units selected in the sample. Assume that the total number fish caught during 1993 is known, derive the first phase calibration weights and hence estimate the total number of fish caught during 1994 in the United States. Select a second phase sample of 10 units from the given first phase sample by using the Midzuno--Sen sampling scheme. Collect the information on the number of fish caught during 1994 and 1995 for the selected units in the second phase sample. Derive the second phase calibration weights, and hence deduce the estimate of the number of fish caught during 1995 in the United States.

Solution. In the Midzuno--Sen sampling scheme the first unit is to be selected with probability proportional to the number of fish caught during 1992 and the remaining 14 units with SRSWOR sampling. We used the first two columns of the Pseudo- Random Numbers (PRN) given in Table 1 of the Appendix to select a random number $1 \le R_i \le 69$ and another random number $1 \le R_j \le 28933$ by using the 7[h] to 11th columns. The first effective pair of random numbers is (62, 01228). Thus the unit at serial number 62 is included as a first unit in the preliminary large sample of 15 units. The remaining 14 units are selected by SRSWOR sampling from the remaining 58 units in the population. We used the 13th and 14th columns of the Pseudo-Random Numbers to draw 14 distinct random numbers between 1 and 58. The random numbers came in the sequence as 05, 34, 30, 55, 46, 07, 13, 19, 44, 25, 58, 68, 47, and 67.

First phase sample information										
Sr. No.	PRN No.	Species group	1992 x_i^*	P_{1i}	π_{1i}	d_{1i}	1993 z_i	$d_{1i}z_i$	$d_{1i}z_i^2$
1	62	Summer flounder	11918	0.0408	0.2383	4.1962	22919	96174.090	2204214018.00
2	05	Herrings	28933	0.0991	0.2846	3.5137	34060	119676.800	4076192604.00
3	34	Pigfish	2955	0.0101	0.2139	4.6746	2691	12579.350	33851040.24
4	30	Lane snapper	919	0.0031	0.2083	4.7988	1079	5177.974	5587034.02
5	55	Cunner	1931	0.0066	0.2111	4.7362	1876	8885.269	16668764.30
6	46	Spot	14974	0.0513	0.2466	4.0547	14263	57833.490	824879029.40
7	07	Saltwater catfish	13466	0.0461	0.2425	4.1233	12690	52325.790	664014268.20
8	13	Searobins	4768	0.0163	0.2188	4.5692	7726	35301.980	272743125.60
9	19	Groupers	4661	0.0159	0.2185	4.5753	4236	19381.100	82098340.12
10	44	Sand Seatrout	3780	0.0129	0.2161	4.6260	4068	18818.820	76554980.01
11	25	Florida pompano	498	0.0017	0.2072	4.8253	641	3093.073	1982659.95
12	58	Atlantic mackerel	1045	0.0035	0.2087	4.7909	2307	11052.800	25498800.66
13	68	Other fish	2100	0.0071	0.2115	4.7259	1323	6252.488	8272041.08
14	47	King fish	3778	0.0129	0.2161	4.6261	3304	15284.900	50501301.24
15	67	Puffers	1103	0.0037	0.2088	4.7873	999	4782.576	4777793.03
							Sum	466620.500	8347835800.00

In the above table the values of π_{1i} are given by

$$\pi_{1i} = \frac{N-m}{N-1}P_{1i} + \frac{m-1}{N-1},$$

with $N = 69$, $m = 15$, and $P_{1i} = x_i^*/X^*$ with $X^* = 291882$ (given, the total number of fish caught during 1992).

Thus an estimate of the total number of species caught during 1993 is given by

$$\hat{z} = \sum_{i\in s_1} d_{1i}z_i = 466620.5.$$

For $q_{1i} = 1 \forall i$ and the total number of fish during 1993 are known to be $Z = 316784$, the first phase calibration weights are given by

$$w_{1i} = d_{1i} + \left(d_{1i}z_i \Big/ \sum_{i\in s_1} d_{1i}z_i^2\right)\left(Z - \sum_{i\in s_1} d_{1i}z_i\right)$$

$$= d_{1i} + \frac{d_{1i}z_i}{8347835800}(316784 - 466620.5).$$

First phase calibration weights			
Sr. No.	w_{1i}	x_i	$w_{1i}x_i$
1	2.470018	17741	43820.590
2	1.365609	38007	51902.720
3	4.448813	4918	21879.260
4	4.705924	1088	5120.045
5	4.576801	1255	5743.885
6	3.016730	18491	55782.350
7	3.184185	14441	45982.810
8	3.935604	4707	18524.890
9	4.227456	4583	19374.430
10	4.288281	5665	24293.110
11	4.769869	425	2027.194
12	4.592594	4860	22320.010
13	4.613765	1141	5264.306
14	4.351829	4805	20910.540
15	4.701520	918	4315.995
		Sum	347262.100

Thus an estimate of the total number of fish based on first phase sample during 1994 in the US is given by

$$\hat{x}^* = \sum_{i \in s_1} w_{1i}x_i = 347262.1\,.$$

Again applying the Midzuno--Sen sampling scheme on the units selected in the first phase sample, we selected one unit with probability proportional to the number of fish caught during 1992 and remaining 9 units with SRSWOR sampling. We used the first two columns of the Pseudo-Random Numbers (PRN) given in Table 1 of the Appendix to select a random number $1 \le R_i \le 15$ and another random number $1 \le R_j \le 28933$ by using the 7th to 11th columns. The first effective pair of random numbers is (01, 07572). Thus the unit at serial number 01 in the given first phase sample is included as a first unit in the second phase sample of $n = 10$ units. The remaining 9 units are selected by SRSWOR sampling from the remaining 14 units in the given first phase sample. We used the 13th and 14th column of the Pseudo-Random Numbers to draw 9 distinct random numbers between 1 and 14. The random numbers came in the sequence as 05, 07, 13, 09, 06, 11, 12, 04, and 03.

Thus the second phase sample consists of the following information:

Second phase sample information											
Species group	P_{1i}	π_{1i}	p_{2i}	π_{2i}	d_{2i}	1994 x_i	$d_{2i}x_i$	$d_{2i}x_i^2$	1995 y_i	w_{2i}	$w_{2i}y_i$
Summer flounder	0.041	0.238	0.040	0.657	6.38	17741	113236.0	2008919805.0	16238	3.82	62051.80
Lane snapper	0.003	0.208	0.003	0.644	7.45	1088	8107.6	8821100.0	859	7.26	6243.62
Cunner	0.007	0.211	0.006	0.645	7.34	1255	9212.4	11561586.8	1375	7.13	9806.76
Spot	0.051	0.247	0.051	0.661	6.13	18491	113399.0	2096863507.0	11567	3.56	41266.91
Saltwater catfish	0.046	0.243	0.046	0.659	6.25	14441	90312.1	1304197188.0	13859	4.21	58361.07
Searobins	0.016	0.219	0.016	0.648	7.04	4707	33155.1	156061147.0	4793	6.29	30166.37
Sand Seatrout	0.013	0.216	0.013	0.647	7.14	5665	40474.7	229289141.0	4355	6.22	27128.08
Atlantic mackerel	0.004	0.209	0.003	0.644	7.43	4860	36147.9	175678928.0	4008	6.62	26533.76
Other fish	0.007	0.212	0.007	0.645	7.32	1141	8354.72	9532731.7	935	7.13	6669.63
King fish	0.013	0.216	0.012	0.647	7.14	4805	34331.3	164961659.0	4333	6.36	27594.05
						Sum	486731.0	6165886793.0			295822.10

In the above table we used

$$\pi_{2i} = \frac{m-n}{m-1} p_{2i} + \frac{n-1}{m-1} \quad \text{with} \quad p_{2i} = x_i^* \Big/ \sum_{i=1}^{m} x_i^* .$$

Under the chi square distance function the second phase calibration weights are

$$w_{2i} = d_{2i} + \left(d_{2i} q_{2i} x_i \Big/ \sum_{i \in s_2} d_{2i} q_{2i} x_i^2 \right) \left(\hat{x}^* - \sum_{i \in s_2} d_{2i} x_i \right)$$

where $\hat{x}^* = \sum_{i=1}^{m} d_{1i} x_i$.

If $q_{2i} = 1$ the second phase calibration weights are

$$w_{2i} = d_{2i} + \left(d_{2i} x_i \Big/ \sum_{i \in s_2} d_{2i} x_i^2 \right) \left(\hat{x}^* - \sum_{i=1}^{n} d_{2i} x_i \right) = d_{2i} + \frac{d_{2i} x_i}{6165886793} (347262.1 - 486731) .$$

Thus the regression estimate of the total number of fish caught in the United States during 1995 is given by

$$\hat{y}_w = \sum_{i \in s_2} w_{2i} y_i = 295822.1 .$$

6.5 ESTIMATION OF VARIANCE OF CALIBRATED ESTIMATOR IN TWO-PHASE SAMPLING: LOW AND HIGHER LEVEL CALIBRATION

In general calibrated estimator in two-phase sampling can be written as

$$\hat{y}_c = \sum_{i \in s_2} w_{2i} y_i . \tag{6.5.1}$$

Following Särndal, Swensson, and Wretman (1992) the two required set of residuals are given by

$$e_{1i} = y_i - \hat{\beta}_0 z_i \quad \forall \; i \in s_2 \tag{6.5.2}$$

and

$$e_{2i} = y_i - \hat{\beta}_1 z_i - \hat{\beta}_2 x_i \quad \forall \; i \in s_2 . \tag{6.5.3}$$

Using the concept of two-phase sampling, an unbiased estimator of $V(\hat{y}_c)$ is given by

$$\hat{v}_u(\hat{y}_c) = \sum_{i \in s_2} \sum_{j \in s_2} \frac{(\pi_{2ij} - \pi_{2i}\pi_{2j})e_{2i}e_{2j}}{\pi_{2i}\pi_{2j}\pi_{2ij}} + \frac{1}{2} \sum_{i \in s_2} \sum_{j \in s_2} \frac{(\pi_{1ij} - \pi_{1i}\pi_{1j})e_{1i}e_{1j}}{\pi_{1i}\pi_{1j}\pi_{1ij}\pi_{2ij}} . \tag{6.5.4}$$

Following Hidiroglou and Särndal (1995, 1998), the low level calibrated estimator of variance of $\hat{y}_c$ is

$$\hat{v}_{ll}(\hat{y}_c) = \sum_{i \in s_2} \sum_{j \in s_2} W_{2ij} \tilde{w}_i^* \tilde{w}_j^* e_{2i} e_{2j} + \frac{1}{2} \sum_{i \in s_2} \sum_{j \in s_2} W_{1ij} \tilde{w}_{1i} \tilde{w}_{1j} e_{1i} e_{1j} \tag{6.5.5}$$

where $W_{2ij} = (\pi_{2ij} - \pi_{2i}\pi_{2j})(\pi_{2ij})^{-1}$ and $W_{1ij} = (\pi_{1ij} - \pi_{1i}\pi_{1j})(\pi_{1ij}\pi_{2ij})^{-1}$.

A large number of estimators of variance can be shown as special cases of the estimator considered at (6.5.5). For simplicity, if $q_{1i} = 1/z_i$ and $q_{2i} = 1/x_i$ then (6.5.5) takes the form

$$\hat{y}_{\text{Ratio}}(\hat{y}_c) = \sum_{i \in s_2} \sum_{j \in s_2} W_{2ij} e_{2i} e_{2j} \left[\frac{\sum_{i \in s_1} d_{1i} x_i}{\sum_{i \in s_2} w_i^* x_i} \right]^{g_1} \left[\frac{Z}{\sum_{i \in s_1} d_{1i} z_i} \right]^{g_2} + \sum_{i \in s_2} \sum_{j \in s_2} W_{1ij} e_{1i} e_{1j} \left[\frac{Z}{\sum_{i \in s_1} d_{1i} z_i} \right]^{g_3}$$

which is an analogue of the class of ratio type estimators proposed by Wu (1982) for $g_1 = g_2 = g_3 = 2$.

Similarly from (6.5.5) a regression type estimator for $q_{1i} = q_{2i} = 1$ can be developed for estimating the variance of the chain regression type estimators.

Singh (2000b) suggested a higher order calibration estimator of the variance in two-phase sampling as

$$\hat{v}_{\text{ho}}(\hat{y}_c) = \sum_{i \in s_2} \sum_{j \in s_2} \Omega_{2ij} \tilde{w}_i^* \tilde{w}_j^* e_{2i} e_{2j} + \frac{1}{2} \sum_{i \in s_2} \sum_{j \in s_2} \Omega_{1ij} \tilde{w}_{1i} \tilde{w}_{1j} e_{1i} e_{1j}, \tag{6.5.6}$$

where Ω_{1ij} and Ω_{2ij} are the weights such that the distance between Ω_{1ij} and W_{1ij} and that between Ω_{2ij} and W_{2ij} is minimum. Define two chi square type of distance functions as

$$D_1 = \frac{1}{2} \sum_{i \in s_1} \sum_{j \in s_1} (\Omega_{1ij} - W_{1ij})^2 (Q_{1ij} W_{1ij})^{-1} \tag{6.5.7}$$

and

$$D_2 = \frac{1}{2}\sum_{i\in s_2}\sum_{j\in s_2}\left(\Omega_{2ij} - W_{2ij}\right)^2\left(Q_{2ij}W_{2ij}\right)^{-1}. \tag{6.5.8}$$

Also let us define the first calibration constraint as follows:

$$\sum_{i\in s_1}\sum_{j\in s_1}\Omega_{1ij}d_{1i}d_{ij}z_iz_j = V(\hat{z}), \tag{6.5.9}$$

where $V(\hat{z})$ denotes the known variance of the estimator of the total of the cheaper auxiliary character, Z.

The second calibration constraint is

$$\sum_{i\in s_2}\sum_{j\in s_2}\Omega_{2ij}d_{2i}d_{2j}x_ix_j = \hat{v}(\hat{x}) \tag{6.5.10}$$

where

$$\hat{v}(\hat{x}) = \sum_{i\in s_1}\sum_{j\in s_1}\frac{\left(\pi_{1ij} - \pi_{1i}\pi_{1j}\right)x_ix_j}{\pi_{1ij}\pi_{1i}\pi_{1j}}.$$

Minimisation of (6.5.7) subject to (6.5.9) leads to second order calibrated weights obtained from first phase sample information given by

$$\Omega_{1ij} = W_{1ij} + \frac{Q_{1ij}W_{1ij}d_{1i}d_{1j}z_iz_j}{\sum_{i\in s_1}\sum_{j\in s_1}Q_{1ij}W_{1ij}d_{1i}^2d_{1j}^2z_i^2z_j^2}\left[V(\hat{z}) - \sum_{i\in s_1}\sum_{j\in s_2}W_{1ij}d_{1i}d_{1j}z_iz_j\right]. \tag{6.5.11}$$

Similarly, minimisation of (6.5.8) subject to (6.5.10) leads to second order calibrated weights obtained from second phase sample information, given by

$$\Omega_{2ij} = W_{2ij} + \frac{Q_{2ij}W_{2ij}d_{2i}d_{2j}x_ix_j}{\sum_{i\in s_2}\sum_{j\in s_2}Q_{2ij}W_{2ij}d_{2i}^2d_{2j}^2x_i^2x_j^2}\left[\hat{v}(\hat{x}) - \sum_{i\in s_2}\sum_{j\in s_2}W_{2ij}d_{2i}d_{2j}x_ix_j\right]. \tag{6.5.12}$$

Use of (6.5.11) and (6.5.12) in (6.5.6) forms the higher order calibration estimator of variance. Several estimators can be shown as special cases of the higher order calibration approach.

For example, if $Q_{1ij} = 1/\left(d_{1i}d_{1j}x_{1i}x_{1j}\right)$ and $Q_{2ij} = 1/\left(d_{2i}d_{2j}x_{2i}x_{2j}\right)$, then an estimator of the variance of the chain ratio type estimator becomes

$$\hat{v}_{\text{Ratio}}\left(\hat{y}_c\right)_{\text{ho}} = \sum_{i\in s_2}\sum_{j\in s_2}W_{2ij}d_{2i}d_{2j}e_{2i}e_{2j}\left[\frac{\sum_{i\in s_1}d_{1i}x_i}{\sum_{i\in s_2}w_i^*x_i}\right]^2\left[\frac{Z}{\sum_{i\in s_1}d_{1i}z_i}\right]^2\left[\frac{\hat{v}(\hat{x})}{\sum_{i\in s_2}\sum_{j\in s_2}W_{2ij}d_{2i}d_{2j}x_ix_j}\right]$$

$$+\sum_{i\in s_2}\sum_{j\in s_2}W_{1ij}d_{1i}d_{1j}e_{1i}e_{1j}\left[\frac{Z}{\sum_{i\in s_1}d_{1i}z_i}\right]^2\left[\frac{V(\hat{z})}{\sum_{i\in s_1}\sum_{j\in s_1}d_{1i}d_{1j}z_iz_j}\right]$$

Similarly higher order calibration estimators for estimating the variance of the chain regression type estimators can be developed by choosing

$q_{1i} = q_{2i} = Q_{1i} = Q_{2i} = 1 \;\forall\; i$ and j.

6.6 TWO-PHASE SAMPLING USING MULTI-AUXILIARY VARIABLES

We have described the estimation methods of two-phase sampling through simple examples. Complicated situations, with more than two auxiliary characters, have been considered by Dupont (1995) and Hidiroglou and Särndal (1995, 1998). Dupont (1995) has used two different approaches for deriving the estimators: a regression model assisted approach, which seeks to adapt the idea of the regression estimator, and a calibration approach, which attempts to adapt the idea of calibration. The estimators obtained by the two approaches may be linked together. Hidiroglou and Särndal (1995, 1998) considered multiplicative and additive type weights obtained through the combination of first phase and second phase calibration weights or selection weights. Tripathi and Ahmed (1995) have considered a class of estimators for a finite population mean for the situations where population means of some auxiliary variables are known while those of others are unknown. Hidiroglou (1995) has applied the two-phase calibration approach in the survey of employment, payrolls, and hours in a panel survey. Singh and Singh (1997--98) considered the problem estimating the ratio of two finite population means using two auxiliary variables in double sampling. Dalabehera and Sahoo (2000) proposed a class of unbiased estimators in two-phase sampling.

6.7 UNIFIED APPROACH IN TWO-PHASE SAMPLING

Chaudhuri and Adhikary (1983) showed the admissibility of an estimator in two-phase sampling with varying probabilities following the original work by Rao and Bellhouse (1978). Later Chaudhuri, and Adhikary (1985) modified their results and extended it for two-phase sampling by following Godambe and Joshi (1965). The results for single-phase sampling discussed by Godambe (1969), Sekkappan (1973), Sekkappan and Thompson (1975) and Ramakrishnan (1975b) can also be extended for two-phase sampling. Chaudhuri and Adhikary (1985) have defined additional concepts for deducing the single-phase Bayesian results of Meeden and Ghosh (1981, 1983) and Verdeman and Meeden (1983) for two-phase sampling.

Suppose a first phase sample s_1 of size m is drawn from the population of N units with sampling design p_1 and values x_i $(i \in s_1)$ on a auxiliary variable X are ascertained. A sub-sample s_2 of size n is drawn from the first phase sample s_1 with design p_2 and the values y_i, x_i for $i \in s_2$ of the variable Y and X. The resulting two-phase sample is $s = (s_1, s_2)$, the two-phase sampling design is p, and the sample selection probabilities are given by $p(s) = p_1(s_1)p_2(s_2 \mid s_1)$. Such a class of design p will be denoted by D.

The first phase inclusion probabilities will be denoted as P_j and second phase inclusion probabilities will be denoted as Q_j and are assumed to be positive.
Define

$$P_j(s_1) = \sum_{s_2 \ni j} p_2(s_2 \mid s_1).$$

Let E_p denote the design expectation and V_p denote the design variance.

Let $\underline{X} = (X_1, X_2, ..., X_N)$, $\underline{Y} = (Y_1, Y_2, ..., Y_N)$, $\underline{Z} = (X, Y)$, $U_i = Y_i - X_i$ and $\underline{U} = (U_1, U_2, ..., U_N)$ such that X, Y, and U are the total of X_i, Y_i, and U_i. Also define $x = \{(i, x_i) \mid i \in s_1\}$, $y = \{(i, y_i) \mid i \in s_2\}$ and $u = \{(i, u_i = (y_i - x_i)) \mid i \in s_1\}$.

Thus we can define two sub-classes of the main class D as follows:

(a) A sub-class D_1 can be defined for which the sizes of the first phase sample are fixed at m;
(b) Another sub-class D_2 can be defined for which the sizes of the second phase sample are fixed at $n < m$.

Then we have the following results:

Result 6.7.1. A class C of design unbiased estimators of population total, Y, is given by

$$e_0 = \sum_{j \in s_2} u_j / Q_j + h_1(s, \underline{Y}, \underline{X}) + \sum_{j \in s_1} x_j / P_j + h_2(s_1, \underline{X}) \tag{6.7.1}$$

based on a design in the restrictive class D_3 such that:

(a) h_1 is free of y_i, x_i for $i \notin s_2$ but may involve them and i when $i \in s_2$;
(b) h_2 is free of x_i for $i \notin s_1$ but may involve them and i when $i \in s_1$;
(c) $E_{p2}[h_1 + h_2] = 0$ if $p_1(s_1) > 0$.

Result 6.7.2. Another admissible estimator of population total in class C using two-phase sampling is

$$e_1 = \sum_{j \in s_2} u_j / Q_j + \sum_{j \in s_1} x_j / P_j, \text{ based on any } p \text{ in } D_3. \tag{6.7.2}$$

Result 6.7.3. An admissible estimator within a wider class C_1 of p unbiased estimators is given by

$$e_2 = e(u) + e(x) \tag{6.7.3}$$

where
(a) $e(u)$ is free of y_i, x_i for $i \notin s_2$ but may involve them, in the form u_i and i for $i \in s_2$,
(b) $e(x)$ is free of x_i for $i \in s_1$ but may involve them and i for $i \in s_1$,
and
(c) $E_p\{e(u)\} = U$ and $E_p\{e(x)\} = X$.

Result 6.7.4. Suppose a prior $\xi = \xi(x)$ for Y may be postulated prior to second phase sampling. Let E_ξ denote the prior mean operator and the posterior mean by $E_\xi(\bullet \mid d)$, where $d = (s, y, x)$. Then under a square error loss function, a Bayes estimator of population total in two-phase sampling is given by

$$e_B = \sum_{i\in s_2} y_i + E_\xi\left[\left(\sum_{j\in s_1 - s_2} y_j + \sum_{j\in\Omega - s_1} y_j\right) \| d\right]. \tag{6.7.4}$$

Result 6.7.5. For every design p in class D the estimator

$$e_g = \sum_{i\in s_2} y_i + \sum_{i\in s_1 - s_2} a_i + \sum_{i\in\Omega - s_2} b_i \tag{6.7.5}$$

is admissible for Y with a_i, b_i arbitrary.

Mukerjee and Chaudhuri (1990) considered the problem of sampling a finite population in two-phases with varying probabilities, using the first phase sample with known size measures and the second phase sub-sample there from utilising, in addition, ascertained auxiliary variate values for the initial sample.

6.8 CONCEPT OF THREE-PHASE SAMPLING

Let $\left(x_1^{\#}, x_2^{\#}, ..., x_l^{\#}\right)$ be the first phase sample s_1 (say) drawn by simple random sampling from a population of N units where only auxiliary variable X is measured. Let $\left(x_1^*, x_2^*, ..., x_m^*\right)$ be the second phase sample s_2 (say) drawn by simple random sampling from the first phase sample s_1 units and again only auxiliary variable X is measured. Let $(y_1, y_2, ..., y_n)$ and $(x_1, x_2, ..., x_n)$ denote the third phase sample s_3 (say) drawn by simple random sampling from the second phase sample for the study variable Y and auxiliary variable X.

Let

$$\bar{y} = n^{-1}\sum_{i=1}^{n} y_i\,, \quad \bar{x} = n^{-1}\sum_{i=1}^{n} x_i\,, \quad \bar{x}^* = m^{-1}\sum_{i=1}^{m} x_i^* \ \text{and}\ \bar{x}^{\#} = l^{-1}\sum_{i=1}^{l} x_i^{\#}\,.$$

Define

$$\epsilon_0 = \frac{\bar{y}}{\bar{Y}} - 1, \quad \epsilon_1 = \frac{\bar{x}}{\bar{X}} - 1, \quad \epsilon_2 = \frac{\bar{x}^*}{\bar{X}} - 1 \ \text{and}\ \epsilon_3 = \frac{\bar{x}^{\#}}{\bar{X}} - 1$$

so that

$$E(\epsilon_j) = 0, \ j = 0, 1, 2, 3\,.$$

Now we have the following theorem:

Theorem 6.8.1.

$$E\left(\epsilon_0^2\right) = \left(\frac{1}{n} - \frac{1}{N}\right)C_y^2\,. \tag{6.8.1}$$

Proof. Let E_1, E_2, and E_3 denote the expected values over all possible first, second and third phase samples, respectively, and let V_1, V_2 and V_3 denote the variances over all possible first, second, and third phase samples, respectively. Then we have

$$E(\epsilon_0^2) = V(\epsilon_0) = E_1E_2V_3(\epsilon_0) + E_1V_2E_3(\epsilon_0) + V_1E_2E_3(\epsilon_0)$$

$$= E_1E_2V_3\left(\frac{\bar{y}}{\bar{Y}} - 1\right) + E_1V_2E_3\left(\frac{\bar{y}}{\bar{Y}} - 1\right) + V_1E_2E_3\left(\frac{\bar{y}}{\bar{Y}} - 1\right)$$

$$= E_1E_2\left[\left(\frac{1}{n} - \frac{1}{m}\right)\frac{s_y^2}{\bar{Y}^2}\right] + E_1V_2\left(\frac{\bar{y}^*}{\bar{Y}} - 1\right) + V_1E_2\left(\frac{\bar{y}^*}{\bar{Y}} - 1\right)$$

$$= E_1\left[\left(\frac{1}{n} - \frac{1}{m}\right)\frac{s_y^{*2}}{\bar{Y}^2}\right] + E_1\left[\left(\frac{1}{m} - \frac{1}{l}\right)\frac{s_y^{*2}}{\bar{Y}^2}\right] + V_1\left(\frac{\bar{y}^{\#}}{\bar{Y}} - 1\right)$$

$$= \left(\frac{1}{n} - \frac{1}{m}\right)C_y^2 + \left(\frac{1}{m} - \frac{1}{l}\right)C_y^2 + \left(\frac{1}{l} - \frac{1}{N}\right)C_y^2$$

$$= \left(\frac{1}{n} - \frac{1}{N}\right)C_y^2 .$$

Hence the theorem.

Now we have the following corollary:

Corollary 6.8.1. One can easily prove that

$$E(\epsilon_1^2) = \left(\frac{1}{n} - \frac{1}{N}\right)C_x^2, \quad E(\epsilon_2^2) = \left(\frac{1}{m} - \frac{1}{N}\right)C_x^2, \quad \text{and} \quad E(\epsilon_3^2) = \left(\frac{1}{l} - \frac{1}{N}\right)C_x^2 .$$

Theorem 6.8.2. Prove that under the concept of three phase sampling

$$E(\epsilon_0\epsilon_1) = \left(\frac{1}{n} - \frac{1}{N}\right)\rho_{xy}C_xC_y . \tag{6.8.2}$$

Proof. Let C_1, C_2 and C_3 denote the co-variances over all possible first, second, and third phase samples, respectively. Then

$$E(\epsilon_0\epsilon_1)$$

$$= \text{Cov}(\epsilon_0, \epsilon_1) = E_1E_2\{\, C_3(\epsilon_0,\, \epsilon_1)\,\} + E_1C_2\{E_3(\epsilon_0),\, E_3(\epsilon_1)\} + C_1\{E_2E_3(\epsilon_0),\, E_2E_3(\epsilon_1)\}$$

$$= \left\{E_1E_2C_3\left(\frac{\bar{y}}{\bar{Y}} - 1, \frac{\bar{x}}{\bar{X}} - 1\right)\right\} + E_1C_2\left\{E_3\left(\frac{\bar{y}}{\bar{Y}} - 1\right), E_3\left(\frac{\bar{x}}{\bar{X}} - 1\right)\right\} + C_1\left\{E_2E_3\left(\frac{\bar{y}}{\bar{Y}} - 1\right), E_2E_3\left(\frac{\bar{x}}{\bar{X}} - 1\right)\right\}$$

$$= E_1E_2\left(\frac{1}{n} - \frac{1}{m}\right)\left(\frac{s_{xy}}{\bar{X}\,\bar{Y}}\right) + E_1C_2\left\{\left(\frac{\bar{y}^*}{\bar{Y}} - 1\right), \left(\frac{\bar{x}^*}{\bar{X}} - 1\right)\right\} + C_1\left\{E_2\left(\frac{\bar{y}^*}{\bar{Y}} - 1\right), E_2\left(\frac{\bar{x}^*}{\bar{X}} - 1\right)\right\}$$

$$= E_1\left(\frac{1}{n}-\frac{1}{m}\right)\left(\frac{\overset{*}{s}_{xy}}{\overline{X}\,\overline{Y}}\right)+E_1\left(\frac{1}{m}-\frac{1}{l}\right)\frac{\overset{*}{s}_{xy}}{\overline{X}\,\overline{Y}}+C_1\left\{\left(\frac{\overline{y}^{\#}}{\overline{Y}}-1\right),\left(\frac{\overline{x}^{\#}}{\overline{X}}-1\right)\right\}$$

$$= \left(\frac{1}{n}-\frac{1}{m}\right)\rho_{xy}C_xC_y+\left(\frac{1}{m}-\frac{1}{l}\right)\rho_{xy}C_xC_y+\left(\frac{1}{l}-\frac{1}{N}\right)\rho_{xy}C_xC_y$$

$$= \left(\frac{1}{n}-\frac{1}{N}\right)\rho_{xy}C_xC_y\,.$$

Hence the theorem.

Corollary 6.8.2. Following the above theorem we can easily prove that

$$E(\epsilon_0\epsilon_2)=\left(\frac{1}{m}-\frac{1}{N}\right)\rho_{xy}C_xC_y,\quad E(\epsilon_0\epsilon_3)=\left(\frac{1}{l}-\frac{1}{N}\right)\rho_{xy}C_xC_y,$$

$$E(\epsilon_1\epsilon_2)=\left(\frac{1}{m}-\frac{1}{N}\right)C_x^2,\ E(\epsilon_1\epsilon_3)=\left(\frac{1}{m}-\frac{1}{N}\right)C_x^2,\ \text{and}\ \ E(\epsilon_2\epsilon_3)=\left(\frac{1}{l}-\frac{1}{N}\right)C_x^2.$$

Remark 6.8.1. Under three-phase sampling the following types of estimators can be studied

$$\overline{y}_{t1}=\overline{y}\left(\frac{\overline{x}^*}{\overline{x}}\right)^{\alpha}\left(\frac{\overline{x}}{\overline{x}^{\#}}\right)^{\beta},\ \text{ and }\ \overline{y}_{t2}=\overline{y}+\alpha\left(\overline{x}-\overline{x}^*\right)+\beta\left(\overline{x}^{\#}-\overline{x}\right).$$

The bias and variance expressions for the above estimators can easily be obtained using three-phase expected values.

Remark 6.8.2. The three-phase sampling can be extended to four-phase sampling and so on. Such a sampling plan is called multi-phase sampling.

6.9 ESTIMATION OF VARIANCE OF REGRESSION ESTIMATOR UNDER TWO-PHASE SAMPLING

Suppose that a first phase sample s_1 of fixed size m is taken by SRSWOR from a population of N units and the auxiliary variable X is observed for all $i \in s_1$. A simple random sub-sample s_2 of size n is taken by SRSWOR from sample s_1, and the variable of interest y_i and auxiliary variable x_i are observed for all $i \in s_2$. The simple linear regression estimator for two-phase sampling is given by

$$\overline{y}_{\mathrm{lr}}=\overline{y}+b\left(\overline{x}^*-\overline{x}\right) \tag{6.9.1}$$

where $\overline{y}=n^{-1}\sum_{i=1}^{n}y_i$, $\overline{x}=n^{-1}\sum_{i=1}^{n}x_i$ are the means for the second phase sample s_2, $\overline{x}^*=m^{-1}\sum_{i=1}^{m}x_i$ is the mean for the first phase sample s_1 and $b=s_{xy}/s_x^2$. The variance of the estimator (6.9.1) is given by

$$V(\bar{y}_{\mathrm{lr}})=\left(\frac{1}{m}-\frac{1}{N}\right)S_y^2+\left(\frac{1}{n}-\frac{1}{m}\right)\left[S_y^2+\beta^2S_x^2-2\beta S_{xy}\right]. \tag{6.9.2}$$

The variance expression in (6.9.2) can be easily written as

$$V(\bar{y}_{\mathrm{lr}})=\left(\frac{1}{m}-\frac{1}{N}\right)S_y^2+\left(\frac{1}{n}-\frac{1}{m}\right)S_D^2 \tag{6.9.3}$$

where

$$S_D^2=(N-1)^{-1}\sum_{i=1}^{N}\left[\left(Y_i-\bar{Y}\right)-\beta\left(X_i-\bar{X}\right)\right]^2=(N-1)^{-1}\sum_{i=1}^{N}D_i^2 \text{ with } D_i=\left(Y_i-\bar{Y}\right)-\beta\left(X_i-\bar{X}\right).$$

By the method of moments an obvious estimator of $V(\bar{y}_{lr})$ is given by

$$\hat{v}(\bar{y}_{\mathrm{lr}})=\left(\frac{1}{m}-\frac{1}{N}\right)s_y^2+\left(\frac{1}{n}-\frac{1}{m}\right)s_d^2 \tag{6.9.4}$$

where

$$s_d^2=(n-1)^{-1}\sum_{i=1}^{n}\left[(y_i-\bar{y})-b(x_i-\bar{x})\right]^2=(n-1)^{-1}\sum_{i=1}^{n}d_i^2 \quad \text{with } d_i=(y_i-\bar{y})-b(x_i-\bar{x}).$$

Following Sitter (1997) we have

$$S_y^2=S_D^2+\beta^2S_x^2. \tag{6.9.5}$$

A sample analogue of (6.9.5) leads to

$$s_y^2=s_d^2+b^2s_x^2. \tag{6.9.6}$$

On the basis of the relationship (6.9.6) Sitter (1997) considered the following two estimators of variance of the regression estimator in two-phase sampling as

$$\hat{v}_1(\bar{y}_{\mathrm{lr}})=\left(\frac{1}{n}-\frac{1}{N}\right)s_d^2+\left(\frac{1}{m}-\frac{1}{N}\right)b^2s_x^{*2} \tag{6.9.7}$$

and

$$\hat{v}_0(\bar{y}_{\mathrm{lr}})=\left(\frac{1}{n}-\frac{1}{N}\right)s_d^2+\left(\frac{1}{m}-\frac{1}{N}\right)b^2s_x^2. \tag{6.9.8}$$

Sitter (1997) also considered the problem of estimation of variance of the regression estimator in two-phase sampling as follows.
Defining

$$\bar{x}^*(j)=\frac{m\bar{x}^*-x_j}{m-1} \quad \text{if } j\in s_1, \tag{6.9.9}$$

$$\bar{x}(j)=\begin{cases}\dfrac{n\bar{x}-x_j}{n-1} & \text{if } j\in s_2,\\ \bar{x} & \text{if } j\in s_1-s_2,\end{cases} \tag{6.9.10}$$

$$\bar{y}(j)=\begin{cases}\dfrac{n\bar{y}-y_j}{n-1} & \text{if } j\in s_2,\\ \bar{y} & \text{if } j\in s_1-s_2,\end{cases} \tag{6.9.11}$$

and

$$b(j)=\begin{cases}b-\left\{(x_j-\bar{x})d_j\right\}/\left\{(n-1)s_x^2(1-k_j)\right\} & \text{if } j\in s_2,\\ b & \text{if } j\in s_1-s_2,\end{cases} \tag{6.9.12}$$

where $k_j = 1/n + \left(x_j - \bar{x}\right)^2 / \left\{(n-1)s_x^2\right\}$.

The Jackknife variance estimator of variance in two-phase sampling is given by

$$\hat{v}_{\mathrm{J}}\left(\bar{y}_{\mathrm{lr}}\right) = \frac{(m-1)}{m}\sum_{j=1}^{m}\left[\bar{y}_{lr}(j) - \bar{y}_{lr}\right]^2 \tag{6.9.13}$$

where

$$\bar{y}_{\mathrm{lr}}(j) = \bar{y}(j) + b(j)\left[\bar{x}^*(j) - \bar{x}(j)\right] \tag{6.9.14}$$

denotes the jth regression estimator of population mean in two-phase sampling. Due to computer facilities available now, the problem of estimation of variance of parameters of interest following the Jackknife technique has become quite popular. The usual ratio and product estimators in two-phase sampling have been studied by Sengupta (1981b) and stability of the Jackknife variance estimator in ratio estimation has been considered by Krewski and Chakrabarty (1981). Sitter (1997) also considered the problem of estimation of variance of the general linear regression estimator using two-phase sampling in the presence of multi-auxiliary information. Fuller (1998) considered the problem of estimation of variance of the regression estimator for two-phase sampling. Given the covariance matrix of the first phase control variables, a replication variance estimator which uses only the second phase sample has been suggested.

Example 6.9.1. From population 1 in the Appendix, select a first phase sample of 10 units by SRSWOR sampling and note only the nonreal estate farm loans for the units selected in the sample. From the selected 10 units select a sub-sample of 5 units and note the real estate farm loans as well as nonreal estate farm loans. Estimate the average real estate farm loans by using the regression estimator. Construct the 95% confidence intervals by using three different estimators of the variance.

Solution. We used the first two columns of the Pseudo-Random Number (PRN) Table 1 given in the Appendix to select an SRSWOR sample of $m = 10$ units. The following 10 distinct random numbers 01, 23, 46, 04, 32, 47, 33 05, 22, and 38 between 1 and 50 resulted in the following first phase sample.

First phase sample information										
Sr. No.	1	2	3	4	5	6	7	8	9	10
Pop. Units	01	23	46	04	32	47	33	05	22	38
States	AL	MN	VA	AR	NY	WA	NC	CA	MI	PA
x_i^*	348.334	2466.892	188.477	848.317	426.274	1228.607	494.73	3928.732	440.518	298.351

where x_i^* = Nonreal estate farm loans. From the first phase sample, $\bar{x}^* = 1066.9232$, and $s_x^{*2} = 1470305.988$.

We used the 7th and 8th columns of the Pseudo-Random Numbers to select a second phase sample of $n = 5$ units from the above list of selected first phase sample units. The following five distinct random numbers between 1 and 10 were observed: 07, 09, 01, 02 and 03. Thus the second phase sample consists of the following information.

Second phase sample information					
Sr. No.	First-phase unit number	Population unit number	State and Territory	Nonreal estate farm loan, x_i	Real estate farm loan y_i
1	07	33	NC	494.730	639.571
2	09	22	MI	440.518	327.028
3	01	01	AL	348.334	408.978
4	02	23	MN	2466.892	1354.768
5	03	46	VA	188.477	321.583
			Sum	3938.951	3051.928

From the second phase sample we have

Sr. No.	State and Territory	$(x_i - \bar{x})^2$	$(y_i - \bar{y})^2$	$(y_i - \bar{y})(x_i - \bar{x})$	d_i	d_i^2
1	NC	85884.28	851.7875	-8553.079	160.338	25708.49
2	MI	120597.98	80291.5294	98402.217	-127.943	16369.36
3	AL	193121.75	40565.0213	88509.818	-4.737	22.45
4	MN	2819382.85	554105.1574	1249893.828	-7.0663	49.93
5	VA	359176.31	83406.9417	173083.210	-20.592	424.02
	Sum	3578163.18	759220.4376	1601335.995		42574.26

Now

$$\bar{x} = \frac{1}{n}\sum_{i=1}^{n} x_i = \frac{3938.951}{5} = 787.7902\,,\quad \bar{y} = \frac{1}{n}\sum_{i=1}^{n} y_i = \frac{3051.928}{5} = 610.3856\,,$$

$$r = \frac{\bar{y}}{\bar{x}} = \frac{610.3856}{787.8082} = 0.7748\,,\quad s_x^2 = (n-1)^{-1}\sum_{i=1}^{n}(x_i - \bar{x})^2 = \frac{3578163.18}{4} = 894540.795\,,$$

$$s_y^2 = (n-1)^{-1}\sum_{i=1}^{n}(y_i - \bar{y})^2 = \frac{759220.4}{4} = 189805.1\,,$$

$$s_{xy} = (n-1)^{-1}\sum_{i=1}^{n}(y_i - \bar{y})(x_i - \bar{x}) = \frac{1601335.995}{4} = 400333.9988\,,$$

$$b = s_{xy}/s_x^2 = 0.4475\,,\quad \text{and}\quad s_d^2 = \frac{1}{n-1}\sum_{i=1}^{n} d_i^2 = \frac{42574.26}{4} = 10643.56\,.$$

Thus the regression estimate of the real estate farm loans is:

$$\bar{y}_{\text{lr}} = \bar{y} + b\left(\bar{x}^* - \bar{x}\right) = 610.3856 + 0.4475(1066.9232 - 787.7902) = 735.297\,.$$

Case I. We have

$$\hat{v}_0\left(\bar{y}_{\text{lr}}\right) = \left(\frac{1}{n} - \frac{1}{N}\right)s_d^2 + \left(\frac{1}{m} - \frac{1}{N}\right)b^2 s_x^2$$

$$= \left(\frac{1}{5} - \frac{1}{50}\right)\times 10643.56 + \left(\frac{1}{10} - \frac{1}{50}\right)\times 0.4475^2 \times 894540.795 = 16246.83\,.$$

Using Table 2 from the Appendix, the 95% confidence interval for the average real estate farm loans in the United States during 1997 is

$$\bar{y}_{\text{lr}} \mp t_{\alpha/2}(\text{df} = n-2)\sqrt{\hat{v}_0\left(\bar{y}_{\text{lr}}\right)}\ , \quad \text{or } 735.297 \mp 3.182\sqrt{16246.83}$$

or $\left[329.70, 1140.89\right]$.

Case II. We have

$$\hat{v}_1\left(\bar{y}_{\text{lr}}\right) = \left(\frac{1}{n} - \frac{1}{N}\right)s_d^2 + \left(\frac{1}{m} - \frac{1}{N}\right)b^2 s_x^{*2}$$

$$= \left(\frac{1}{5} - \frac{1}{50}\right)\times 10643.56 + \left(\frac{1}{10} - \frac{1}{50}\right)\times 0.4475^2 \times 1470305.988$$

$$= 25470.878\,.$$

The $(1-\alpha)100\%$ and hence using Table 2 from the Appendix the 95% confidence interval for the average real estate farm loans in the United States during 1997 is given by

$$\bar{y}_{\text{lr}} \mp t_{\alpha/2}(\text{df} = n-2)\sqrt{\hat{v}_1\left(\bar{y}_{\text{lr}}\right)}\ , \text{ or } 735.297 \mp 3.182\sqrt{25470.878} \text{ or}\left[227.46, 1243.13\right].$$

Case III. We have

First phase		Second phase				Required results				
x_j^*	$\bar{x}^*(j)$	x_j	$\bar{x}(j)$	y_j	$\bar{y}(j)$	d_j	$k(j)$	$b(j)$	$\bar{y}_{lr}(j)$	$\{\bar{y}_{lr}(j)-\bar{y}_{lr}\}^2$
494.730	1130.50	494.73	861.077	639.57	602.09	161.25	0.224	0.464	727.364	52.5628
440.518	1136.52	440.52	874.630	323.03	681.25	-131.01	0.234	0.431	794.196	3549.9610
348.334	1146.77	348.33	897.676	408.98	659.74	-3.78	0.254	0.447	771.165	1335.9420
2466.892	911.37	2466.89	368.014	1354.77	423.29	-6.92	0.988	0.717	812.953	6136.9170
188.477	1164.53	188.48	937.641	321.58	681.58	-19.56	0.300	0.443	782.161	2260.6620
848.317	1091.21		787.810		609.58			0.448	745.485	118.1760
426.274	1138.12		787.810		609.58			0.448	766.489	1016.0420
1228.607	1048.96		787.810		609.58			0.448	726.558	64.8937
3928.732	748.950		787.810		609.58			0.448	592.177	20288.4100
298.351	1152.32		787.810		609.58			0.448	772.856	1462.4490
									Sum	36286.0200

Thus Jackknife estimator of the variance of the linear regression estimator $\bar{y}_{\mathrm{lr}}$ is

$$\hat{v}_{\mathrm{J}}(\bar{y}_{\mathrm{lr}}) = \frac{(m-1)}{m}\sum_{j=1}^{m}\left[\bar{y}_{\mathrm{lr}}(j) - \bar{y}_{lr}\right]^2 = \frac{(10-1)}{10}\times 36286.02 = 32657.42\,.$$

A $(1-\alpha)100\%$ confidence interval for the average real estate farm loans in the United States during 1997 is given by

$$\bar{y}_{\mathrm{lr}} \mp t_{\alpha/2}(\mathrm{df} = n-2)\sqrt{\hat{v}_{\mathrm{J}}(\bar{y}_{\mathrm{lr}})}\ .$$

Using Table 2 from the Appendix the 95% confidence interval for the average real estate farm loans in the United States during 1997 is given by

$$735.297 \mp 3.182\sqrt{32657.42} \quad \text{or} \quad [160.266,\ 1310.327].$$

6.10 TWO-PHASE SAMPLING USING PPSWR SAMPLING

Raj (1964) and Singh and Singh (1965) considered the problem of estimation of population mean using the concept of probability proportional to size and with replacement sampling of second phase sample of n units from the given first phase sample of m units selected with SRSWOR sampling. A pictorial representation of such a two-phase sampling is shown in Figure 6.10.1.

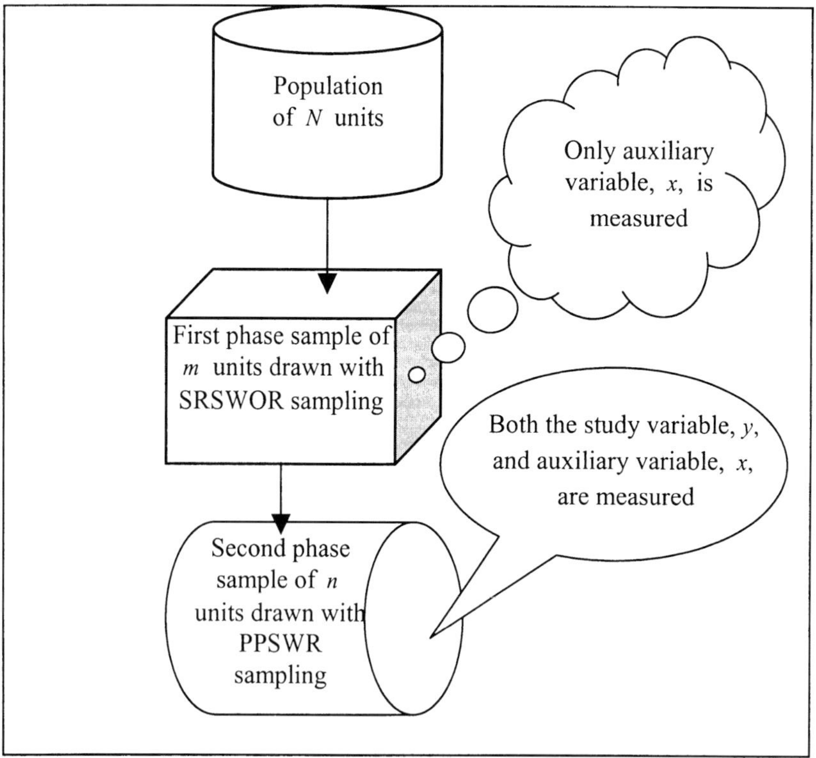

Fig. 6.10.1 Two-phase strategy using PPSWR sampling.

It means that the probability of selecting one unit in the first phase sample of size m is given by $1/N$ and that of selecting a second phase sample from the given first phase sample is given by $p_i = x_i/x^*$, where $x^* = \sum_{i=1}^{m} x_i$.

Then we have the following theorem:

Theorem 6.10.1. An unbiased estimator of the population mean $\bar{Y}$ under PPSWR two-phase sampling is given by

$$\bar{y}_{\text{ppsd}} = \frac{1}{mn}\sum_{i=1}^{n}\frac{y_i}{p_i}. \tag{6.10.1}$$

Proof. Defining E_1 and E_2 as the expected values for the given first phase sample and all possible first phase samples respectively, we have

$$E(\bar{y}) = E_1E_2\left[\bar{y}_{\text{ppsd}} \mid \text{first phase}\right] = E_1E_2\left[\frac{1}{mn}\sum_{i=1}^{n}\frac{y_i}{p_i} \mid \text{first phase}\right] = E_1\left[\frac{1}{m}\sum_{i=1}^{m} y_i\right] = \bar{Y}.$$

Hence the theorem.

Theorem 6.10.2. The variance of the estimator $\bar{y}_{\text{ppsd}}$ is given by

$$V\left(\bar{y}_{\text{ppsd}}\right) = \left(\frac{1}{m}-\frac{1}{N}\right)S_y^2 + \frac{(m-1)}{mnN(N-1)}S_z^2 \tag{6.10.2}$$

where

$$S_y^2 = \frac{1}{N-1}\sum_{i=1}^{N}\left(Y_i-\bar{Y}\right)^2 \quad \text{and} \quad S_z^2 = \frac{1}{N-1}\sum_{i=1}^{N}\left(Z_i-\bar{Z}\right)^2 \quad \text{with } Z_i = Y_i/P_i.$$

Proof. The variance of the estimator $\bar{y}_{\text{ppsd}}$ is given by

$$V\left(\bar{y}_{\text{ppsd}}\right) = E_1V_2\left(\bar{y}_{\text{ppsd}} \mid \text{first phase}\right) + V_1E_2\left(\bar{y}_{\text{ppsd}} \mid \text{first phase}\right)$$

$$= E_1V_2\left(\frac{1}{mn}\sum_{i=1}^{n}\frac{y_i}{p_i} \mid \text{first phase}\right) + V_1E_2\left(\frac{1}{mn}\sum_{i=1}^{n}\frac{y_i}{p_i} \mid \text{first phase}\right)$$

$$= E_1\left(\frac{1}{m^2n}\sum_{i=1}^{m}p_i\left(\frac{y_i}{p_i}-\bar{y}^*\right)^2\right) + V_1\left(\frac{1}{m}\sum_{i=1}^{m}y_i\right)$$

$$= \frac{m(m-1)}{m^2nN(N-1)}\sum_{i=1}^{N}\sum_{j>i}^{m}p_ip_j\left(\frac{y_i}{p_i}-\frac{y_j}{p_j}\right)^2 + \left(\frac{1}{m}-\frac{1}{N}\right)S_y^2$$

$$= \frac{(m-1)}{mnN(N-1)}\sum_{i=1}^{N}p_i\left(\frac{y_i}{p_i}-\bar{Y}\right)^2 + \left(\frac{1}{m}-\frac{1}{N}\right)S_y^2$$

$$= \frac{(m-1)}{mnN(N-1)}\sum_{i=1}^{N}p_i\left(Z_i-\bar{Y}\right)^2 + \left(\frac{1}{m}-\frac{1}{N}\right)S_y^2 = \left(\frac{1}{m}-\frac{1}{N}\right)S_y^2 + \frac{(m-1)}{mnN}S_z^2.$$

Hence the theorem.

Theorem 6.10.3. An unbiased estimator of the variance of the estimator $\bar{y}_{\text{ppsd}}$ is

$$\hat{v}\left(\bar{y}_{\text{ppsd}}\right)=\left(\frac{1}{m}-\frac{1}{N}\right)\frac{m\sum_{i=1}^{n}p_i z_i^2-s_z^2}{nm(m-1)}+\frac{s_z^2}{m^2} \tag{6.10.3}$$

where

$$s_z^2=(n-1)^{-1}\left\{\sum_{i=1}^{n}z_i^2-n^{-1}\left(\sum_{i=1}^{n}z_i\right)^2\right\} \quad \text{with} \quad z_i=y_i/Np_i\ .$$

Proof. Obvious by taking expected values on both sides.

Example 6.10.1. From population 1 given in the Appendix select a first phase sample of 10 units by SRSWOR sampling and note only the nonreal estate farm loans from the units selected in the sample. From the selected 10 units select a sub-sample of 5 units using PPSWR sampling and note the real estate farm loans as well as nonreal estate farm loans. Estimate the average real estate farm loans by using PPSWR two-phase sampling estimator. Construct the 95% confidence intervals.

Solution. We used the first two columns of the Pseudo-Random Number (PRN) Table 1 given in the Appendix to select an SRSWOR sample of $m=10$ units. The following 10 distinct random numbers 01, 23, 46, 04, 32, 47, 33 05, 22, and 38 between 1 and 50 resulted in the following first phase sample.

First phase sample information										
Sr. No.	1	2	3	4	5	6	7	8	9	10
Pop. Units	01	23	46	04	32	47	33	05	22	38
States	AL	MN	VA	AR	NY	WA	NC	CA	MI	PA
x_i^*	348.334	2466.892	188.477	848.317	426.274	1228.607	494.73	3928.732	440.518	298.351

where x_i^* = Nonreal estate farm loans.

Cumulative Total Method							
First phase unit no.	Pop-unit no.	State	x_i	Cumulative x_i	Random No.	Units Selected	$p_i=\frac{x_i}{x^*}$
1	01	AL	348.334	348.334	01473	S	0.032649
2	23	MN	2466.982	2815.316			0.231226
3	46	VA	188.317	3003.633			0.017651
4	04	AR	848.317	3851.950	03965	S	0.079511
5	32	NY	426.274	4278.224	04981, 05365	S, S	0.039954
6	47	WA	1228.607	5506.831			0.115155
7	33	NC	494.730	6001.561	07673	S	0.046370
8	05	CA	3928.732	9930.293			0.368233
9	22	MI	440.518	10370.810			0.041289
10	38	PA	298.351	10669.160			0.027964

We used the cumulative total method to select the second phase sample as follows. The cumulative totals of the auxiliary variable, nonreal estate farm loans, is given in the fifth column of the above table. We used the first five columns of Pseudo-Random Numbers given Table 1 in the Appendix to select eight random numbers between 1 and 10,670. These random numbers came in the sequence as 01473, 04981, 05365, 03965, and 07673. These random numbers have been shown in the sixth column of the above table. The seventh column shows the states selected in the second phase sample with PPSWR sampling. Note that the state **NY** has been selected twice. Now, from the ultimate sample selected in the second phase, we have the following table:

Second phase sample information										
Sr. No.	First phase unit no.	State Territory	x_i	y_i	p_i	z_i	z_i^2	$p_i z_i^2$	$\frac{y_i}{p_i}$
1	1	AL	348.334	408.978	0.03264	250.60	62799.99	2049.79	12529.960
2	4	AR	848.317	907.700	0.07951	228.32	52131.61	4144.98	11416.170
3	5	NY	426.274	201.631	0.03995	100.94	10189.22	407.06	5047.084
4	5	NY	426.274	201.631	0.03995	100.94	10189.22	407.06	5047.084
5	7	NC	494.730	639.571	0.04637	275.86	76096.26	3528.58	13792.78
					Sum	956.66	211406.30	10537.48	47833.080

where $z_i = y_i / Np_i$.

Thus an estimate of average real estate farm loans is given by

$$\bar{y}_{\text{ppsd}} = \frac{1}{mn}\sum_{i=1}^{n}\frac{y_i}{p_i} = \frac{1}{10\times 5}\times 47833.08 = 956.66 .$$

Now we have

$$s_z^2 = \frac{\sum_{i=1}^{n} z_i^2 - n^{-1}\left(\sum_{i=1}^{n} z_i\right)^2}{n-1} = \frac{211406.31 - \frac{956.66^2}{5}}{5-1} = 7091.66 .$$

Thus an estimate of the variance of $\bar{y}_{\text{ppsd}}$ is given by

$$\hat{v}\left(\bar{y}_{\text{ppsd}}\right) = \left(\frac{1}{m} - \frac{1}{N}\right)\frac{m\sum_{i=1}^{n} p_i z_i^2 - s_z^2}{nm(m-1)} + \frac{s_z^2}{m^2}$$

$$= \left(\frac{1}{10} - \frac{1}{50}\right)\frac{10\times 10537.48 - 7091.659}{10\times 5\times (10-1)} + \frac{7091.659}{10^2} = 88.389.$$

A $(1-\alpha)100\%$ confidence interval estimate of the population mean is given by

$$\bar{y}_{\text{ppsd}} \mp t_{\alpha/2}(\text{df} = n-1)\sqrt{\hat{v}\left(\bar{y}_{\text{ppsd}}\right)} .$$

Using Table 2 from the Appendix the 95% confidence interval is given by

or $956.66 \mp 2.776\sqrt{88.389}$ or $[930.56,\ 982.76]$.

6.11 CONCEPT OF DUAL FRAME SURVEYS

The generalized linear regression (GREG) estimator has been found to be most commonly used as an estimator of population total/mean in survey sampling. Let us consider the simplest case of the GREG where information on only one auxiliary variable is available. Consider two populations $\Omega_t = \{1_t, 2_t, ..., i_t, ..., N_t\}$, for $t = 1, 2$ from which two independent probability samples $s_t (s_t \subset \Omega)$ are drawn with a given sampling design, $p_t(.)$. The inclusion probabilities $\pi_{i_t} = \Pr(i_t \in s_t)$ and $\pi_{ij}(t) \in \Pr(i_t \,\&\, j_t \in s_t)$ are assumed to be strictly positive and known. Let y_{i_t} be the value of the variable of interest, y, for the ith population element, with which also is associated an auxiliary variable x_{ti} for the tth population. For the elements $i_1 \in s_1$, we observe $(y_{i_1}, x_{i_1}, z_{i_1})$ in the main or first survey. In the second independent survey, we observe (x_{i_2}, z_{i_2}). The population total of the common auxiliary variable z_{i_2}, $i = 1,2,...,n_2$; $X = \sum_{i_t=1}^{N_t} x_{i_t}$, $t = 1,2$, is assumed to be accurately known in both surveys. The objective is to estimate the population total $Y = \sum_{i_1=1}^{N} y_{i_1}$ using auxiliary information available in both surveys.

6.11.1 COMMON VARIABLES USED FOR FURTHER CALIBRATION OF WEIGHTS

Suppose that two sample surveys have one variable in common. If the population total of this variable is known, then it can be used as a control variable in GREG. Suppose the population total of this common variable is unknown. Let it in the first survey or sample s_1 of n_1 units take values z_{i_1}, $i = 1,2,...,n_1$. In the second independent survey or sample of n_2 units it takes values z_{i_2}, $i = 1,2,...,n_2$, defining estimators of unknown common total as $\hat{Z}_1 = \sum_{i \in s_t} d_{i_1} z_{i_1}$ and $\hat{Z}_2 = \sum_{i \in s_t} d_{i_2} z_{i_2}$, where $d_{i2} = 1/\pi_{i_2}$ such that $E(\hat{Z}_t) = Z$, the unknown total for the common variable z_i. Let us define $\hat{Z} = \sum_{t=1}^{2} \alpha_t \hat{Z}_t$, where $\sum_{t=1}^{2} \alpha_t = 1$, is an unbiased estimator of Z based on the estimators obtained in both surveys. Let us consider a new estimator of population total, Y, as

$$\hat{Y}_{\text{P}} = \sum_{i_1 \in s_1} \gamma_{i_1} y_{i_1} \tag{6.11.1}$$

with weights γ_{i_1} as close as possible for a given metric to the w_{i_1}, while respecting the calibration equation

$$\sum_{i_1 \in s_1} \gamma_{i_1} z_{i_1} = \hat{Z}. \tag{6.11.2}$$

A simple case is the minimization of chi square type of distance function given by

$$\sum_{i_1 \in s_1} \left(\gamma_{i_1} - w_{i_1}\right)^2 \left(w_{i_1} h_{i_1}\right)^{-1}, \tag{6.11.3}$$

where h_{i_1} are suitably chosen weights. By minimizing (6.11.3) subject to calibration equation (6.11.2) we obtain weights

$$\gamma_{i_1} = w_{i_1} + \left(w_{i_1} h_{i_1} z_{i_1} \Big/ \sum_{i_1 \in s_1} w_{i_1} h_{i_1} z_{i_1}^2 \right) \left(\hat{Z} - \sum_{i_1 \in s_1} w_{i_1} z_{i_1} \right). \tag{6.11.4}$$

Substitution of the value of γ_{i_1} from (6.11.4) in (6.11.1) leads to a new estimator of population total from the tth survey as

$$\hat{Y}_{\text{P}} = \sum_{i_1 \in s_1} d_{i_1} y_{i_1} + \hat{\beta}_{1_1}\left(X - \hat{X}_{\text{HT}}\right) + \hat{\beta}_{2_1}\left(\hat{Z} - \hat{Z}_{\text{GREG}}\right) \tag{6.11.5}$$

where

$$\hat{X}_{\text{HT}} = \sum_{i_1 \in s_1} d_{i_1} x_{i_1}, \quad \hat{\beta}_{1_1} = \sum_{i_1 \in s_1} d_{i_1} q_{i_1} x_{i_1} y_{i_1} \Big/ \sum_{i_1 \in s_1} d_{i_1} q_{i_1} x_{i_1}^2,$$

$$\hat{\beta}_{2_1} = \sum_{i_1 \in s_1} w_{i_1} h_{i_1} z_{i_1} y_{i_1} \Big/ \sum_{i_{1t} \in s_1} w_{i_1} h_{i_1} z_{i_1}^2, \quad \text{and} \quad \hat{Z}_{\text{GREG}} = \sum_{i_1 \in s_1} w_{i_1} z_{i_1}.$$

Now from Renssen and Nieuwenbroek (1997) we have the following corollary:

Corollary 6.11.1.1. An adjusted estimator of population total, Y, is given by

$$\hat{Y}_{\text{AD}} = \sum_{i_1 \in s_1} d_{i_1} y_{i_1} + \hat{\beta}_{11}^{*}\left(X - \hat{X}_{\text{HT}}\right) + \hat{\beta}_{21}^{*}\left(\hat{Z} - \hat{Z}_{\text{HT}}\right) \tag{6.11.6}$$

where $\hat{\beta}_{11}^{*}$ and $\hat{\beta}_{21}^{*}$ have their usual meanings. On comparing (6.11.5) with (6.11.6), one can easily see that we are using the regression type estimator for estimating unknown total, Z, given by $\hat{Z}_{\text{GREG}}$, whereas in (6.11.6), Renssen and Nieuwenbroek (1997) used the Horvitz and Thompson (1952) type estimator, given by $\hat{Z}_{\text{HT}}$.

6.11.2 ESTIMATION OF VARIANCE USING DUAL FRAME SURVEYS

Conditioning on the calibrated weights, γ_{i_1} as fixed or given, an expression for the variance of the proposed estimator $\hat{Y}_{\text{P}} = \sum_{i_1 \in s_1} \gamma_{i_1} y_{i_1}$ is given by

$$V\left(\hat{Y}_{\text{P}}\right) = \frac{1}{2} \sum_{i \in \Omega} \sum_{j(\neq i) \in \Omega} \left[D_{ij}(2)\pi_{ij}^2(2)\left(\gamma_{i_1} e_{i_2} - \gamma_{j_1} e_{j_2}\right)^2 + D_{ij}(1)\pi_{ij}(1)\left(\frac{\gamma_{i_1} e_{i_1}}{d_{i_1}} - \frac{\gamma_{j_1} e_{j_1}}{d_{j_1}} \right)^2 \right]. \tag{6.11.7}$$

An estimator for estimating the variance is given by

$$\hat{v}_0\left(\hat{Y}_{\text{P}}(t)\right) = \frac{1}{2} \sum_{i \in s_t} \sum_{j \in s_t} D_{ij}(t)\left[\pi_{ij}(t)\left(\gamma_{i_t} e_{i_t} - \gamma_{j_t} e_{j_t}\right)^2 + \left(\frac{\gamma_{i_t} e_{i_t}}{d_{i_t}} - \frac{\gamma_{j_t} e_{j_t}}{d_{j_t}} \right)^2 \right]. \tag{6.11.8}$$

Several estimators of variance can be shown as special cases of this estimator. Let us consider the following case:

Case I. Suppose both surveys used simple random sampling and without replacement (SRSWOR) scheme for selecting the same number, n, of the units. Under such situations, $\pi_{i_t} = n/N$ and $\pi_{ij}(t) = n(n-1)/N(N-1)$. If we choose the weights $q_{i_t} = 1/x_{i_t}$ and $h_{i_t} = 1/z_{i_t}$, then the calibrated weights are $\gamma_{i_t} = \frac{N}{n}\left(\hat{z}/\hat{Z}_{\text{HT}}\right)$ and the estimator of variance becomes

$$\hat{v}_0\left(\hat{Y}_{\text{P}}(t)\right) = \frac{N^2(1-f)}{n} \sum_{i \in s_t} \sum_{j \in s_t} D_{ij}(t)\left[\pi_{ij}(t)\left(\gamma_{i_t} e_{i_t} - \gamma_{j_t} e_{j_t}\right)^2 + \left(\frac{\gamma_{i_t} e_{i_t}}{d_{i_t}} - \frac{\gamma_{j_t} e_{j_t}}{d_{j_t}}\right)^2\right]. \quad (6.11.9)$$

Hartley (1962, 1974) pointed out that a dual frame design can result in considerable cost savings over a single frame design with comparable design. For example, if the frame Ω_1 is an area frame and frame Ω_2 is a list frame, the frame Ω_1 may be complete but expensive to sample, the frame Ω_2 may be incomplete but has a lower cost per unit sampling. Bankier (1986), Fuller and Burmeister (1972), Kalton and Anderson (1986), Skinner (1991), and Skinner and Rao (1996) have considered the problem of estimation of population total using dual frame surveys. Lohr and Rao (2000) have considered several estimators of population total in dual frame surveys and compared them under a unified setup. They also considered the problem of estimation of variance of the estimator of population total under dual frame survey through the concept of Jackknifing. Deville and Goga (2002) studied Horvitz and Thompson estimator when information is gathered on two samples.

6.12 ESTIMATION OF MEDIAN USING TWO-PHASE SAMPLING

In two-phase sampling, in the first phase we select a preliminary large sample s' of n' units by simple random sampling without replacement (SRSWOR) and only the auxiliary character X is measured. Let $\hat{M}_x'$ be the estimator of median M_x of the auxiliary character X based on the first phase sample. In the second phase a sub-sample s of n units is drawn from the preliminary large sample by SRSWOR and both the study variable Y and the auxiliary variable X are measured. Let $\hat{M}_x$ and $\hat{M}_y$ denote the estimators of M_x and M_y respectively, based on the sample drawn at the second phase.

6.12.1. GENERAL CLASS OF ESTIMATORS

Following Srivastava and Jhajj (1981) and Srivastava (1971), Singh, Joarder, and Tracy (2001) proposed a general class of estimators for estimating the median as

$$\hat{M}_1 = H\left(\hat{M}_y, U\right) \quad (6.12.1)$$

where $U = \hat{M}_x / \hat{M}'_x$. Whatever may be the sample chosen, $H(\hat{M}_y, U)$ assumes the value in a bounded closed convex sub-set R_2 of two-dimensional real space containing the point $(M_y, 1)$ such that $H(M_y, 1) = M_y$. Using a first order Taylor's series expansion around the point $(M_y, 1)$ we have

$$\hat{M}_1 = H(M_y,1) + (\hat{M}_y - M_y)H_1(M_y,1) + (U-1)H_2(M_y,1) + o(n^{-1}) \tag{6.12.2}$$

where $H_1(M_y,1)$ and $H_2(M_y,1)$ denote the first order partial derivatives of $H(\hat{M}_y, U)$ with respect to $\hat{M}_y$ and U, respectively.

Furthermore, under the assumptions that $H_1(M_y,1) = 1$ and that $H_2(M_y,1)$ is an unknown constant, we have the following theorems:

Theorem 6.12.1. The general class of estimators, $\hat{M}_1$, is asymptotically unbiased and normal.

Proof. Following Kuk and Mak (1989), we have

$$\hat{M}_y - M_y = \{f_y(M_y)\}^{-1}\left(\frac{1}{2} - p_y\right) + o_p\left(n^{-0.5}\right),$$

$$\hat{M}_x - M_x = \{f_x(M_x)\}^{-1}\left(\frac{1}{2} - p_x\right) + o_p\left(n^{-0.5}\right),$$

and

$$\hat{M}'_x - M_x = \{f_x(M_x)\}^{-1}\left(\frac{1}{2} - p_x\right) + o_p\left(n'^{-0.5}\right)$$

which proves the theorem.

Theorem 6.12.2. The minimum variance of $\hat{M}_1$, up to terms of order $o(n^{-1})$, is given by

$$\text{Min}.V(\hat{M}_1) = \{f_y(M_y)\}^{-2}\left[\left(\frac{1}{n} - \frac{1}{N}\right)\frac{1}{4} - 4\left(\frac{1}{n} - \frac{1}{n'}\right)(P_{11} - 0.25)^2\right], \tag{6.12.3}$$

where $f_y(M_y)$ denotes the values of the marginal density $f_y(y)$ at the median value of y.

Proof. Follows from Chapter 3.

6.12.2 REGRESSION TYPE ESTIMATOR

We also consider the following regression type estimator

$$\hat{M}_2 = \hat{M}_y + d(\hat{M}_x' - \hat{M}_x) \tag{6.12.4}$$

where d is a suitably chosen constant such that the variance of the proposed estimator $\hat{M}_2$ is minimum.

The minimum variance of the estimator $\hat{M}_2$ up to terms of order $o(n^{-1})$ is given by

$$\text{Min.}V(\hat{M}_2)=\{f_y(M_y)\}^{-2}\left[\left(\frac{1}{n}-\frac{1}{N}\right)\frac{1}{4}-4\left(\frac{1}{n}-\frac{1}{n'}\right)(P_{11}-0.25)^2\right] \tag{6.12.5}$$

for the optimum value of d given by

$$d=f_x(M_x)f_y^{-1}(M_y)(4P_{11}-1). \tag{6.12.6}$$

An estimator of the optimum value of d is given by

$$\hat{d}=\hat{f}_x(\hat{M}_x)\hat{f}_y^{-1}(\hat{M}_y)(4p_{11}-1), \tag{6.12.7}$$

where $\hat{f}_y(\hat{M}_y)$ and $\hat{f}_x(\hat{M}_x)$ may be obtained by following Silverman (1986). The regression type of estimator after replacing d with $\hat{d}$ becomes

$$\hat{M}_{2\text{lr}}=\hat{M}_y+\hat{d}(\hat{M}_x'-\hat{M}_x). \tag{6.12.8}$$

Theorem 6.12.3. The estimators $\hat{M}_2$ and $\hat{M}_{2\text{lr}}$ have the same asymptotic variance because

$$E(\hat{d})=d+O(n^{-1}).$$

Proof. Define

$$\delta_0=\frac{\hat{f}_x(\hat{M}_x)}{f_x(M_x)}-1,\quad \delta_1=\frac{\hat{f}_y(\hat{M}_y)}{f_y(M_y)}-1,\quad \text{and}\quad \delta_3=\frac{(p_{11}-P_{11})}{(P_{11}-1)}$$

such that

$$E(\delta_0)=E(\delta_1)=E(\delta_2)\approx 0$$

which proves the lemma.

The estimators of the form

$$\hat{M}_3=\hat{M}_y\left(\frac{\hat{M}_x'}{\hat{M}_x}\right),\quad \hat{M}_4=\hat{M}_y\left(\frac{\hat{M}_x}{\hat{M}_x'}\right)^d,\quad \text{and } \hat{M}_5=\hat{M}_y\left[\frac{\hat{M}_x'}{d\hat{M}_x+(1-d)\hat{M}_x'}\right],$$

which are in fact analogous of the ratio estimator and the estimators proposed by Srivastava (1967) and Walsh (1968) in double sampling, respectively, are also special cases of the class of estimators $\hat{M}_1$.

Theorem 6.12.4. The general class of estimators, $\hat{M}_1$, is always more efficient than the ratio estimator $\hat{M}_3$.

Proof. If we set $V(\hat{M}_1)<V(\hat{M}_3)$, where $V(\hat{M}_3)$ denotes the variance of the ratio estimator $\hat{M}_3$, then it reduces to

$$\left[M_y\{f_x(M_x)\}^{-1} - 4(P_{11} - 0.25)\{f_y(M_y)\}^{-1}M_x\right]^2 > 0, \tag{6.12.9}$$

which is always true. Hence the theorem.

6.12.3 POSITION ESTIMATOR

Suppose n_x^* is the number of units in the second phase sample S_n with $X \le \hat{M}_x'$. Thus if the values of P_{ij} $(i, j = 1,2)$ are known we can predict p by

$$\hat{p}_0' = n^{-1}\left[n_x^* P_{11}/P_{.1} + \left(n - n_x^*\right)P_{12}/P_{.2}\right] \approx \left(2n^{-1}\right)\left[n_x^* P_{11} + \left(n - n_x^*\right)\left(0.5 - P_{11}\right)\right] \tag{6.12.10}$$

because $P_{.j} = P_{1j} + P_{2j} \approx 0.5$, $j = 1,2$. If we replace P_{ij} by p_{ij} from the sample in (6.12.10) we obtain an estimator of p as

$$\hat{p}_1' = n^{-1}\left[n_x^* p_{11}/p_{.1} + \left(n - n_x^*\right)p_{12}/p_{.2}\right] \approx \left(2n^{-1}\right)\left[n_x^* p_{11} + \left(n - n_x^*\right)\left(0.5 - p_{11}\right)\right]. \tag{6.12.11}$$

Then a position estimator in two-phase sampling is given by

$$\hat{M}_6 = \hat{Q}_y(\hat{p}_1'). \tag{6.12.12}$$

Thus we have the following theorems:

Theorem 6.12.5. The estimator $\hat{M}_6$ is asymptotically unbiased and normal.

Proof. We have

$$\hat{M}_6 - M_y = \left\{f_y\left(M_y\right)\right\}^{-1}\left(\hat{p}'_1 - p_y\right) + o_p\left(n^{-0.5}\right)$$

which proves the theorem.

Theorem 6.12.6. The variance of the estimator $\hat{M}_6$, up to terms of order $o\left(n^{-1}\right)$, is

$$V\left(\hat{M}_6\right) = \left\{f_y\left(M_y\right)\right\}^{-2}\left[\left(\frac{1}{n} - \frac{1}{N}\right)\frac{1}{4} - 4\left(\frac{1}{n} - \frac{1}{n'}\right)\left(P_{11} - 0.25\right)^2\right]. \tag{6.12.13}$$

Proof. We have $\hat{M}_6 = \hat{Q}_y(\hat{p}'_1)$. Following Kuk and Mak (1989), we have

$$F_y\left(\hat{M}_6\right) = F_y\left[M_y + \left(\hat{M}_6 - M_y\right)\right] = F_y\left(M_y\right) + f_y\left(M_y\right)\left[\hat{M}_6 - M_y\right] + o_p\left(n^{-0.5}\right).$$

This implies

$$\hat{M}_6 - M_y = \left\{f_y\left(M_y\right)\right\}^{-1}\left[F_y\left(\hat{M}_6\right) - M_y\right] + o_p\left(n^{0.5}\right).$$

Also we have

$$F_y\left(\hat{M}_6\right) - F_y\left(M_y\right) = F_y\left(\hat{M}_6\right) - F_y\left(M_y\right) + o_p\left(n^{0.5}\right).$$

So we have

$$\hat{M}_6 - M_y = \left\{f_y\left(M_y\right)\right\}^{-1}\left[F_y\left(\hat{M}_6\right) - F_y\left(M_y\right)\right] + o_p\left(n^{0.5}\right)$$
$$= \left\{f_y\left(M_y\right)\right\}^{-1}\left[p_1' - p_y\right] + o_p\left(n^{0.5}\right).$$

Now we have

$$p'_0 - p_y = 2\left[p_x^* P_{11} + \left(1 - p_x^*\right)\left(0.5 - P_{11}\right)\right] - p_y = \left(4P_{11} - 1\right)\left(p_x^* - 0.5\right) - \left(p_y - 0.5\right).$$

Also we have

$$p_1' - p_0' = 2\left[p_x^* p_{11} + \left(1 - p_x^*\right)\left(0.5 - p_{11}\right)\right] - 2\left[p_x^* P_{11} + \left(1 - p_x^*\right)\left(0.5 - P_{11}\right)\right]$$
$$= 2\left(p_{11} - P_{11}\right)\left(2p_x^* - 1\right).$$

Clearly $(p_1' - p_0')\sqrt{n} \to 0$ in probability, since $p_x^* \to 0.5$ in probability and $(p_{11} - P_{11})$ is of order $o_p\left(n^{-0.5}\right)$ and we may replace p_1' with p_0'. The details of these assumptions are available in Kuk and Mak (1989) and we have

$$\hat{M}_6 - M_y = \left\{f_y\left(M_y\right)\right\}^{-1}\left(p_0' - p_y\right) + o_p\left(n^{-0.5}\right).$$

Hence $\hat{M}_6$ is approximately normal with mean M_y. To find the variance of the estimator $\hat{M}_6$ let E_2 and E_1 denote the expected values for the given first phase sample and over all possible first phase samples respectively. Similarly defining the variances V_2 and V_1, we have

$$V\left(\hat{M}_6\right) = E_1 V_2\left(\hat{M}_6\right) + V_1 E_2\left(\hat{M}_6\right). \tag{6.12.14}$$

Now we have

$$V_2\left(\hat{M}_6\right) = V_2\left(\hat{M}_6 - M_y\right) = \left\{f_y\left(M_y\right)\right\}^{-2} V_2\left(p_0' - p_y\right). \tag{6.12.15}$$

Again

$$V_2\left(p_0' - p_y\right) = V_2\left[\left(4P_{11} - 1\right)\left(p_x^* - 0.5\right) - \left(p_y - 0.5\right)\right]$$
$$= \left(4P_{11}^* - 1\right)^2 V\left(p_x^*\right) + V\left(p_y\right) - 2\left(4P_{11}^* - 1\right)\text{Cov}_2\left(p_x^*, p_y\right)$$
$$= \left(\frac{1}{n} - \frac{1}{n'}\right)\left(4P_{11}^* - 1\right)^2 / 4 + \left(\frac{1}{n} - \frac{1}{n'}\right)\Big/ 4 - 2\left(\frac{1}{n} - \frac{1}{n'}\right)\left(4P_{11}^* - 1\right)\left(P_{11}^* - 0.25\right)$$
$$= \left(\frac{1}{n} - \frac{1}{n'}\right)\Big/ 4 + \left(\frac{1}{n} - \frac{1}{n'}\right)\left[\left(4P_{11}^* - 1\right)^2 / 4 - 2\left(4P_{11}^* - 1\right)\left(P_{11}^* - 0.25\right)\right]$$
$$= \left(\frac{1}{n} - \frac{1}{n'}\right)\Big/ 4 - 4\left(\frac{1}{n} - \frac{1}{n'}\right)\left(P_{11}^* - 0.25\right)^2 \tag{6.12.16}$$

where P_{11}^* is defined by a cross classification of the first phase sample using $\hat{M}_x'$ and $\hat{M}_y'$ as the cut-offs. Cov_2 denotes the covariance term for the given first phase sample. Using (6.12.16) in (6.12.15) and then using the result in (6.12.14), we have the theorem.

6.12.4 STRATIFICATION ESTIMATOR

Suppose $\tilde{F}_{y1}(y)$ and $\tilde{F}_{y2}(y)$ denote the proportion of units in the second phase sample for which $X \le \hat{M}_x'$ and $X > \hat{M}_x'$, respectively, for the value Y that have y values less than or equal to y. Then $F_y(y)$ can be estimated by

$$\tilde{F}_y(y)=\{n_x'\tilde{F}_{y1}(y)\}/n'+\{(n-n_x')\tilde{F}_{y2}(y)\}/n'\approx 0.5\left[\tilde{F}_{y1}(y)+\tilde{F}_{y2}(y)\right] \tag{6.12.17}$$

where n_x' is the number of units in the preliminary large sample with $X\leq \hat{M}_x'$. Then we may consider an analogue of the stratification estimator in two-phase sampling which is of the form

$$\hat{M}_7=\text{Inf.}\{y:\tilde{F}_y(y)\geq 0.5\}. \tag{6.12.18}$$

The estimator $\hat{M}_7$ is asymptotically unbiased and normal. Its variance is given in the following theorem:

Theorem 6.12.7. The variance of the estimator $\hat{M}_7$, up to terms of order $o(n^{-1})$, is

$$V(\hat{M}_7)=\{f_y(M_y)\}^{-2}\left[\left(\frac{1}{n}-\frac{1}{N}\right)\frac{1}{4}-4\left(\frac{1}{n}-\frac{1}{n'}\right)(P_{11}-0.25)^2\right]. \tag{6.12.19}$$

Proof. We have

$$F_y(\hat{M}_7)=F_y\left[M_y+(\hat{M}_7-M_y)\right]=F_y(M_y)+f_y(M_y)\left[\hat{M}_7-M_y\right]+o_p(n^{-0.5}).$$

This implies

$$\begin{aligned}\hat{M}_7-M_y&=\{f_y(M_y)\}^{-1}\left[F_y(\hat{M}_7)-F_y(M_y)\right]+o_p(n^{-0.5})\\&=\{f_y(M_y)\}^{-1}\left[0.5-p_y\right]+o_p(n^{-0.5}).\end{aligned}$$

For large N we have $F_y(M_y)\approx F_{y1}(M_y)+F_{y2}(M_y)$, so that

$$V[F_y(M_y)]=E_1V_2[F_y(M_y)]+V_1E_2[F_y(M_y)]\ . \tag{6.12.20}$$

Now we have

$$\begin{aligned}V_2[F_y(M_y)]&=V_2[F_{y1}(M_y)]+V_2[F_{y2}(M_y)]\\&=0.5\left[\left(\frac{1}{n}-\frac{1}{n'}\right)\frac{1}{4}-4\left(\frac{1}{n}-\frac{1}{n'}\right)(P_{11}^*-0.25)^2\right].\\&\quad+0.5\left[\left(\frac{1}{n}-\frac{1}{n'}\right)\frac{1}{4}-4\left(\frac{1}{n}-\frac{1}{n'}\right)(P_{11}^*-0.25)^2\right]\end{aligned}$$

On substituting it in (6.12.20) we have the theorem

The regression type estimator, position estimator, stratification estimator and general class of estimators attain the same minimum variance to the first order of approximation. An estimator to estimate the variance of the estimators, $\hat{M}_i$, $i=1,2,4,5,6,7$, is

$$\hat{V}(\hat{M}_i)=\left\{\hat{f}_y(\hat{M}_y)\right\}^{-2}\left[\left(\frac{1}{n}-\frac{1}{N}\right)\frac{1}{4}-4\left(\frac{1}{n}-\frac{1}{n'}\right)(p_{11}-0.25)^2\right].$$

6.12.5 OPTIMUM FIRST AND SECOND-PHASE SAMPLES FOR MEDIAN ESTIMATION

We obtain the optimum first phase and second phase sample sizes for the fixed cost as well as for the fixed variance cases.

6.12.5.1 COST IS FIXED

Let C_1 and C_2 denote the cost per unit in the second phase and first phase, respectively, then the fixed cost C is given by

$$C = nC_1 + n'C_2 . \tag{6.12.21}$$

The variance of $\hat{M}_i$, $i = 1, 2, 4, 5, 6, 7$ will be minimum for the fixed cost C given by (6.12.21) if the optimum values of n and n' are, respectively, given by

$$n = \frac{C\sqrt{2P_{11}(1-2P_{11})}}{C_1\sqrt{2P_{11}(1-2P_{11})} + 2|P_{11} - 0.25|\sqrt{C_1C_2}} \tag{6.12.22}$$

and

$$n' = \frac{2C|P_{11} - 0.25|}{\sqrt{2C_1C_2P_{11}(1-2P_{11})} + 2C_2|P_{11} - 0.25|} . \tag{6.12.23}$$

6.12.5.2 VARIANCE IS FIXED

Also this variance V_0 can be achieved by $\hat{M}_i$ for $i = 1, 2, 4, 5, 6, 7$ for the minimum cost function (6.12.21) if the optimum values of n and n' are, respectively

$$n = \frac{\{f_y(M_y)\}^{-2}\sqrt{2P_{11}(1-2P_{11})}\left[\sqrt{2P_{11}(1-2P_{11})C_1} + 2\sqrt{C_2}|P_{11} - 0.25|\right]}{\sqrt{C_1}\ \left[\ V_0 + \{f_y(M_y)\}^{-2}/(4N)\ \right]}, \tag{6.12.24}$$

and

$$n' = \frac{2|P_{11} - 0.25|\{f_y(M_y)\}^{-2}\left[\sqrt{2P_{11}(1-2P_{11})C_1} + 2\sqrt{C_2}|P_{11} - 0.25|\right]}{\sqrt{C_2}\left[\ V_0 + \{f_y(M_y)\}^{-2}/(4N)\ \right]} . \tag{6.12.25}$$

6.12.6 KUK AND MAK'S TECHNIQUE IN TWO-PHASE SAMPLING

Kuk and Mak (1994) proposed another ingenious method which in double sampling can also be extended for estimating the finite population distribution function of Y defined as:

$$F(y) = N^{-1}\sum_{i=1}^{N}\Delta(y - y_i),$$

where $\Delta(a) = 1$ when $a \ge 0$ and $\Delta(a) = 0$ otherwise. Note that F simply puts probability N^{-1} at each y_i, $i = 1, 2, ..., N$ assuming them to be distinct. The naive estimator of F is the sample distribution function defined as:

$$\hat{F}(y) = n^{-1} \sum_{i \in s} \Delta(y - y_i) \tag{6.12.26}$$

which puts the probability mass n^{-1} at each sample point. Before we introduce a new estimator in double sampling, let us restrict ourselves reviewing to the definition of the customary ratio estimator of population mean of Y in double sampling given by

$$\hat{y}_r = \left(\frac{\bar{x}'}{\bar{x}}\right)\bar{y} = h(\bar{y}), \tag{6.12.27}$$

where $\bar{y}$ and $\bar{x}$ are the second phase sample means of y and x, respectively, and $\bar{x}'$ is the first phase sample mean of x variable. Also $h(a) = \left(\frac{\bar{x}'}{\bar{x}}\right)a$ is a function satisfying $h(\bar{x}) = \bar{x}'$ for the given first phase sample. Let G denote the population distribution function of the auxiliary variable X. Also let $\hat{G}'$ and $\hat{G}$ be the sample distribution functions of the auxiliary variable X based on the first and second phase sample information respectively. An obvious choice of $h(\bullet)$ is

$$h(\bullet) = \hat{G}' \circ \hat{G}^{-1}(\bullet), \tag{6.12.28}$$

where $\hat{G}^{-1}$ denotes the inverse of the function $\hat{G}$ and $\circ$ denotes the composition of two functions. Based on this information define an estimator of $F(y)$ in double sampling as:

$$\hat{F}_{t1}(y) = \hat{G}' \circ \hat{G}^{-1} \circ F(y) = \hat{G}'\left[\hat{G}^{-1}\{F(y)\}\right]. \tag{6.12.29}$$

Let the variables y_i and x_i be related to each other through a model defined as:

$$\hat{y}_{ij} = \hat{B}x_i + v(x_i)\hat{u}_j, \quad j \in s, \; i \in s' \subset \Omega, \tag{6.12.30}$$

where s' and s denote the first phase and second phase samples from the population Ω. Here aim is to redistribute the point mass at each x_i of the given first phase sample evenly to the n points. Also $\hat{B} = \left\{\sum_{i \in s} \frac{y_i x_i}{v^2(x_i)}\right\}\left\{\sum_{i \in s} \frac{x_i^2}{v^2(x_i)}\right\}^{-1}$ is the weighted least square estimator of the regression coefficient. After the redistribution of mass based on a given first phase sample, we obtain $\hat{Q}'$ and $\hat{Q}$ as

$$\hat{Q}'(y) = (n'n)^{-1} \sum_{i \in s'} \sum_{j \in s} \Delta(y - \hat{y}_{ij}) = n'^{-1} \sum_{i \in s'} \hat{P}_i(y), \tag{6.12.31}$$

$$\hat{Q}(y) = n^{-2} \sum_{i \in s} \sum_{j \in s} \Delta(y - \hat{y}_{ij}) = n^{-1} \sum_{i \in s} \hat{P}_i(y), \tag{6.12.32}$$

where $\hat{P}_i(y) = n^{-1} \sum_{j \in s} \Delta(y - \hat{y}_{ij})$ is an estimator of $\Pr[y' \le y | x' = x]$ under the model $y_i = \beta x_i + v(x_i)u_i$ for a given first phase sample.

Replacing $\hat{G}'$ and $\hat{G}^{-1}$ with $\hat{Q}'$ and $\hat{Q}^{-1}$ we obtain another estimator of $F(y)$ as

$$\hat{F}_{t2}(y) = \hat{Q}' \circ \hat{Q}^{-1} \circ F(y) = \hat{Q}'\left[\hat{Q}^{-1}\{\hat{F}(y)\}\right]. \tag{6.12.33}$$

Now we have the following theorem:

Theorem 6.12.8. The estimator $\hat{F}_{t1}(y)$ is asymptotically normal with mean $F(y)$ and variance given by

$$V\left[\hat{F}_{t1}(y)\right]=\left(\frac{1}{n'}-\frac{1}{N}\right)F(y)[1-F(y)]$$

$$+\left(\frac{1}{n}-\frac{1}{n'}\right)\left[F(y)\{1-F(y)\}+G(x)\{1-G(x)\}-2\{H(x,y)-F(y)G(x)\}\right]. \quad (6.12.34)$$

where $x=G^{-1}(\alpha)$ with $\alpha=F(y)$ and $H(x,y)$ denotes the finite population joint distribution of X and Y.

Proof. Its proof follows from the relation

$$\hat{F}_{t1}(y)-F(y)\approx\left\{\hat{F}(y)-F(y)\right\}-\left\{\hat{G}(x)-G(x)\right\}+\left\{\hat{G}'(x)-G(x)\right\}. \quad (6.12.35)$$

6.12.7 CHEN AND QIN TECHNIQUE IN TWO-PHASE SAMPLING

Chen and Qin (1993) suggested an interesting empirical likelihood method for quantile estimation in the presence of auxiliary information. Following their notation, they considered the problem of estimation of $\theta=E\{T(X,Y)\}$ subject to the constraint $0=E\{w(X)\}$. An estimator of any population parameter θ based on the method of empirical likelihood function is given by

$$\hat{\theta}_n=\frac{1}{n}\sum_{i\in s}T(x_i,y_i)/\left\{1+\hat{\lambda}w(x_i)\right\} \quad (6.12.36)$$

with $\hat{\lambda}$ satisfying $\sum_{i\in s}w(x_i)/\{1+\lambda w(x_i)\}=0$ for some suitable weights $w(x_i)$, $i=1,2,\ldots,n$. A few special cases of the estimator (6.12.36) are as follows:

Case I. If $w(x)=\bar{x}-\bar{x}'$, where $\bar{x}=n^{-1}\sum_{i\in s}x_i$ and $\bar{x}'=n'^{-1}\sum_{i\in s'}x_i$ are the second phase and first phase sample means, respectively, of the auxiliary character X then in such situations the estimator (6.12.36) will be an analogue of the Hartley and Rao (1968) estimator in double sampling.

Case II. If $w(x)=I\left[x\le\hat{M}'_x\right]-0.5$ and θ is the median of X, then the estimator (6.12.36) will lead to the position estimator in double sampling.

Case III. If $w(x_i)=x_i-\bar{x}'$ and x_i is a $0-1$ variable and $\bar{x}$ is the estimator of the population proportion $\bar{X}$ (say) based on the first phase sample information, then the estimator (6.12.36) leads to an analogue of the post stratified estimator of Silva and Skinner (1995) in double sampling.

Case IV. The ratio type estimator proposed by Garcia and Cebrian (1998) is also a special case of it.

Example 6.12.7.1. The amounts of the real and nonreal estate farm loans (in \$000) during 1997 in the United States have been presented in population 1 given in the Appendix. Suppose we selected an SRSWOR first phase sample of ten states to collect the information on nonreal estate farm loans only. From the given first phase sample of ten units, we selected a second phase sample of eight states and both the real and nonreal estate farm loans were observed. Find the relative efficiency of the ratio estimator of median, for estimating median of the amount of the real estate farm loans during 1997 by using information on the nonreal estate farm loans during 1997 collected in the first phase sample, with respect to the usual estimator of population median. Assume that both the real and nonreal estate farm loans follow independent normal distributions.

Solution. From the description of the population, we have Y_i = Amount (in \$000) of the real estate farm loans in different states during 1997, X_i = Amount (in \$000) of the nonreal estate farm loans in different states during 1997, $N = 50$, $M_y = 322.305$, $M_x = 452.517$, $\mu_y = 555.434$, $\mu_x = 878.162$, $\sigma_x = 1073.776$, $\sigma_y = 578.948$, and $P_{11} = 0.42$.

Note that we are given

$$f(x) = \frac{1}{\sqrt{2\pi}\sigma_x} e^{-\frac{1}{2}\left(\frac{x-\mu_x}{\sigma_x}\right)^2} \quad \text{and} \quad f(y) = \frac{1}{\sqrt{2\pi}\sigma_y} e^{-\frac{1}{2}\left(\frac{y-\mu_y}{\sigma_y}\right)^2}$$

which implies that

$$f(M_x) = \frac{e^{-\frac{1}{2}\left(\frac{452.517-878.162}{1073.776}\right)^2}}{\sqrt{2\pi}\times 1073.776} = 3.4345\times 10^{-4},$$

and

$$f(M_y) = \frac{e^{-\frac{1}{2}\left(\frac{322.305-555.434}{578.948}\right)^2}}{\sqrt{2\pi}\times 578.948} = 6.3541\times 10^{-4}.$$

Under single phase sampling we have

$$V(\hat{M}_y) = \left(\frac{1-f}{n}\right)\frac{\{f_y(M_y)\}^{-2}}{4} = \left(\frac{1-0.16}{8}\right)\frac{\{6.35419\times 10^{-4}\}^{-2}}{4} = 65014.30 .$$

Under two-phase sampling the ratio estimator of median is given by

$$\hat{M}_{\text{RD}} = \hat{M}_y\left(\frac{\hat{M}'_x}{\hat{M}_x}\right)$$

with variance

$$v\left(\hat{M}_{\text{RD}}\right)=\left(\frac{1}{n'}-\frac{1}{N}\right)\frac{1}{4}\left\{f_y\left(M_y\right)\right\}^{-2}+\left(\frac{1}{n}-\frac{1}{n'}\right)\left[\frac{1}{4}\left\{f_y\left(M_y\right)\right\}^{-2}+\left(\frac{M_y}{M_x}\right)^2\frac{1}{4}\left\{f_x\left(M_x\right)\right\}^{-2}\right.$$

$$\left.+\ 2\left(\frac{M_y}{M_x}\right)\left(P_{11}-0.25\right)\left\{f_x\left(M_x\right)f_y\left(M_y\right)\right\}^{-1}\ \right]$$

$$=\left(\frac{1}{10}-\frac{1}{50}\right)\frac{\left\{6.3541\times10^{-4}\right\}^{-2}}{4}$$

$$+\ \left(\frac{1}{8}-\frac{1}{10}\right)\left[\frac{\left\{6.3541\times10^{-4}\right\}^{-2}}{4}+\left(\frac{322.305}{452.517}\right)^2\frac{\left\{3.4345\times10^{-4}\right\}^{-2}}{4}\right.$$

$$\left.-\ 2\left(\frac{322.305}{452.517}\right)\left(0.42-0.25\right)\left\{6.3541\times10^{-4}\times3.4345\times10^{-4}\right\}^{-1}\ \right]$$

$$=49536.111+0.025\left[619201.38+1075170.28-1109669.104\right]\ =64153.67\,.$$

Thus the percent relative efficiency (RE) of the ratio estimator $\hat{M}_{\text{RD}}$ with respect to the usual estimator $\hat{M}_y$ is given by

$$\text{RE}=\frac{V\left(\hat{M}_y\right)}{\text{MSE}\left(\hat{M}_{\text{RD}}\right)}\times100=\frac{65014.30}{64153.67}\times100=101.34\%$$

which shows that the ratio estimator is more efficient than the usual estimator of population median.

6.13 DISTRIBUTION FUNCTION WITH TWO-PHASE SAMPLING

Singh and Joarder (2002) consider an estimator of distribution function in two-phase sampling as

$$\hat{F}_{\text{RD}}(t)=\hat{F}_y(t)\left\{\frac{\hat{F}'_x(t)}{\hat{F}_x(t)}\right\} \tag{6.13.1}$$

where $\hat{F}'_x(t)=\frac{1}{n'}\sum_{i=1}^{n'}\Delta(t-x_i)$, $\hat{F}_x(t)=\frac{1}{n}\sum_{i=1}^{n}\Delta(t-x_i)$ and $\hat{F}_y(t)=\frac{1}{n}\sum_{i=1}^{n}\Delta(t-y_i)$. The preliminary large sample consists of n' units using SRSWOR sampling where only auxiliary variable X is measured. In the second phase, a sub-sample of n units is drawn from the preliminary sample of n' units through SRSWOR sampling and both the study variable Y and the auxiliary variable X are measured on the selected units. The following lemmas are needed to find the variance of the above estimator.

Lemma 6.31.1. The variance of $\hat{F}_y(t)$ is given by

$$V\left[\hat{F}_y(t)\right]=\frac{(N-n)}{nN^2}\left[\sum_{i=1}^{N}\left\{\Delta(t-y_i)\right\}^2-\frac{1}{N-1}\sum_{i\neq j=1}^{N}\Delta(t-y_i)\Delta(t-y_j)\right]. \tag{6.13.2}$$

Proof. Let E_1 and E_2 denote the expected values over all possible first phase samples and for a given first phase sample, respectively. Also let the variances V_1 and V_2 be similarly defined. Then we have

$$V\left[\hat{F}_y(t)\right] = E_1 V_2\left\{\hat{F}_y(t)\right\} + V_1 E_2\left\{\hat{F}_y(t)\right\}$$

$$= E_1 V_2\left\{\frac{1}{n}\sum_{i=1}^{n}\Delta(t-y_i)\right\} + V_1 E_2\left\{\frac{1}{n}\sum_{i=1}^{n}\Delta(t-y_i)\right\}$$

$$= \frac{(N-n)}{nN^2}\left[\sum_{i=1}^{N}\{\Delta(t-y_i)\}^2 - \frac{1}{N-1}\sum_{i\neq j=1}^{N}\Delta(t-y_i)\Delta(t-y_j)\right].$$

Hence the lemma.

Lemma 6.13.2. The variance of $\hat{F}_x(t)$ is given by

$$V\left[\hat{F}_x(t)\right] = \frac{(N-n)}{nN^2}\left[\sum_{i=1}^{N}\{\Delta(t-x_i)\}^2 - \frac{1}{N-1}\sum_{i\neq j=1}^{N}\Delta(t-x_i)\Delta(t-x_j)\right]. \tag{6.13.3}$$

Lemma 6.13.3. The variance of $\hat{F}'_x(t)$ is given by

$$V\left[\hat{F}'_x(t)\right] = \frac{(N-n')}{n'N^2}\left[\sum_{i=1}^{N}\{\Delta(t-x_i)\}^2 - \frac{1}{N-1}\sum_{i\neq j=1}^{N}\Delta(t-x_i)\Delta(t-x_j)\right]. \tag{6.13.4}$$

Lemma 6.13.4. The covariance between $\hat{F}_y(t)$ and $\hat{F}_x(t)$ is given by

$$\text{Cov}\left[\hat{F}_y(t),\ \hat{F}_x(t)\right] = \frac{(N-n)}{nN^2}\left[\sum_{i=1}^{N}\Delta(t-x_i)\Delta(t-y_i) + \frac{1}{N-1}\sum_{i\neq j=1}^{N}\Delta(t-x_i)\Delta(t-y_j)\right]. \tag{6.13.5}$$

Proof. We have

$$\text{Cov}\left(\hat{F}_y(t),\ \hat{F}_x(t)\right)$$

$$= E_1\left[C_2\left\{\hat{F}_y(t),\ \hat{F}_x(t)\right\}\right] + C_1\left[E_2\left\{\hat{F}_y(t)\right\},\ E_2\left\{\hat{F}_x(t)\right\}\right]$$

$$= E_1\left[C_2\left\{\frac{1}{n}\sum_{i=1}^{n}\Delta(t-y_i), \frac{1}{n}\sum_{i=1}^{n}\Delta(t-x_i)\right\}\right] + C_1\left[E_2\left\{\frac{1}{n}\sum_{i=1}^{n}\Delta(t-x_i)\right\}, E_2\left\{\frac{1}{n}\sum_{i=1}^{n}\Delta(t-y_i)\right\}\right]$$

$$= \frac{(n'-n)}{n(n'-1)}E_1\left\{\frac{1}{n'}\sum_{i=1}^{n'}\Delta(t-y_i)\sum_{i=1}^{n'}\Delta(t-x_i)\right\} + C_1\left\{\frac{1}{n'}\sum_{i=1}^{n'}\Delta(t-x_i), \frac{1}{n'}\sum_{i=1}^{n'}\Delta(t-y_i)\right\}$$

which proves the lemma.

Lemma 6.13.5. The covariance between $\hat{F}_x(t)$ and $\hat{F}'_x(t)$ is given by

$$\text{Cov}\left[\hat{F}_x(t),\ \hat{F}'_x(t)\right] = V\left[\hat{F}'_x(t)\right]. \tag{6.13.6}$$

Proof. We have

$$\text{Cov}\left[\hat{F}_x(t),\ \hat{F}'_x(t)\right] = E_1\left[C_2\left\{\hat{F}_x(t),\ \hat{F}'_x(t)\right\}\right] + C_1\left[E_2\left(\hat{F}_x(t)\right),\ E_2\left(\hat{F}'_x(t)\right)\right]$$

$$= 0 + C_1\left[\frac{1}{n'}\sum_{i=1}^{n'}\Delta(t-x_i), \frac{1}{n'}\sum_{i=1}^{n'}\Delta(t-y_i)\right] = V\left\{\hat{F}'_x(t)\right\}.$$

Hence the lemma.

Lemma 6.13.6. The covariance between $\hat{F}_y(t)$ and $\hat{F}_x^{'}(t)$ is given by

$$\mathrm{Cov}\left[\hat{F}_y(t),\ \hat{F}_x^{'}(t)\right] = \frac{(N-n')}{n'N^2}\left[\sum_{i=1}^{N}\Delta(t-y_i)\Delta(t-x_i)+\frac{1}{N-1}\sum_{i\neq j}^{N}\Delta(t-y_i)\Delta(t-x_j)\right]. \quad (6.13.7)$$

Proof. We have

$$\mathrm{Cov}\left[\hat{F}_y(t),\ \hat{F}_x^{'}(t)\right] = E_1\left[C_2\left\{\hat{F}_y(t),\ \hat{F}_x^{'}(t)\right\}\right] + C_1\left[E_2\left\{\hat{F}_y(t)\right\},\ E_2\left\{\hat{F}_x^{'}(t)\right\}\right]$$

$$= 0 + C_1\left[\frac{1}{n'}\sum_{i=1}^{n'}\Delta(t-y_i),\ \frac{1}{n'}\sum_{i=1}^{n'}\Delta(t-x_i)\right]$$

$$= \frac{(N-n')}{n'N^2}\left[\sum_{i=1}^{N}\Delta(t-y_i)\Delta(t-x_i)+\frac{1}{N-1}\sum_{i\neq j}^{N}\Delta(t-y_i)\Delta(t-x_j)\right]$$

which proves the lemma.

Thus we have the following theorem.

Theorem 6.13.1. The variance of the estimator $\hat{F}_{\mathrm{RD}}(t)$ at (6.13.1) is approximately

$$V\left[\hat{F}_{\mathrm{RD}}(t)\right] = V\left\{\hat{F}_y(t)\right\} + \left\{\frac{F_y(t)}{F_x(t)}\right\}^2\left[V\left\{\hat{F}_x^{'}(t)\right\} + V\left\{\hat{F}_x(t)\right\} - 2\mathrm{Cov}\left\{\hat{F}_y(t), \hat{F}_x(t)\right\}\right]$$

$$+2\left\{\frac{F_y(t)}{F_x(t)}\right\}\left[\mathrm{Cov}\left\{\hat{F}_y(t), \hat{F}_x^{'}(t)\right\} - \mathrm{Cov}\left\{\hat{F}_y(t), \hat{F}_x(t)\right\}\right].$$

Proof. The proof of this theorem follows directly from elementary concepts. The algebraic expression for $V\left\{\hat{F}_{\mathrm{RD}}(t)\right\}$ may be obtained by using lemmas.

6.14 IMPROVED VERSION OF TWO-PHASE CALIBRATION APPROACH

This section has been especially designed to improve the work of Hidiroglou and Särndal (1995, 1998), and hence that Singh (2000b), and is based on Golden Jubilee Year 2003 celebration by Singh (2003c) of the traditional linear regression estimator owed to Hansen, Hurwitz, and Madow (1953) for its outstanding performance in the literature. It has been shown that chain regression estimator is unique in its class of estimators.

6.14.1 IMPROVED FIRST PHASE CALIBRATION

Singh (2003c) considers an improved calibrated estimator of population total X_2 given by

$$\hat{X}_2^{\oplus} = \sum_{i\in s_1}\tilde{d}_{1i}^{\oplus}x_{2i} \quad (6.14.1.1)$$

where $\tilde{d}_{1i}^{\oplus}$ are the calibrated plus weights obtained from the first phase sample information. Although these weights can be chosen in many ways, but we will discuss here only the simple case. We choose the first phase calibrated weights $\tilde{d}_{1i}^{\oplus}$ such that the chi square distance defined as

$$D_1^{\oplus} = \sum_{i\in s_1} \frac{\left(\tilde{d}_{1i}^{\oplus} - d_{1i}\right)^2}{q_{1i}^{\oplus} d_{1i}} \tag{6.14.1.2}$$

is minimum subject to the two first phase calibration constraint defined as

$$\sum_{i\in s_1} \tilde{d}_{1i}^{\oplus} = \sum_{i\in s_1} d_{1i}\,, \tag{6.14.1.3}$$

and

$$\sum_{i\in s_1} \tilde{d}_{1i}^{\oplus} x_{1i} = X_1\,. \tag{6.14.1.4}$$

Note that the condition (6.14.1.3) is a requirement of the chi square test owed to Sir R.A. Fisher and is ignored by all the followers of Deville and Särndal (1992).

The choice of $q_{1i}^{\oplus}$ decides the form of the estimator. The improved first phase Lagrange function is then defined as

$$L_1 = \sum_{i\in s_1} \frac{\left(\tilde{d}_{1i}^{\oplus} - d_{1i}\right)^2}{q_{1i}^{\oplus} d_{1i}} - 2\lambda_{11}^{\oplus}\left[\sum_{i\in s_1} \tilde{d}_{1i}^{\oplus} x_{1i} - X_1\right] - 2\lambda_{12}^{\oplus}\left[\sum_{i\in s_1} \tilde{d}_{1i}^{\oplus} - \sum_{i\in s_1} d_{1i}\right], \tag{6.14.1.5}$$

where $\lambda_{11}^{\oplus}$ and $\lambda_{12}^{\oplus}$ denote the first phase Lagrange multipliers.

On differentiating (6.14.1.5) with respect to $\tilde{d}_{1i}^{\oplus}$ and equating to zero we have

$$\tilde{d}_{1i}^{\oplus} = d_{1i} + \lambda_{11}^{\oplus} q_{1i}^{\oplus} d_{1i} x_{1i} + \lambda_{12}^{\oplus} d_{1i} q_{1i}^{\oplus} \tag{6.14.1.6}$$

On substituting (6.14.1.6) in (6.14.1.3) we have

$$\lambda_{11}^{\oplus}\left(\sum_{i\in s_1} d_{1i} q_{1i}^{\oplus} x_{1i}\right) + \lambda_{12}^{\oplus}\left(\sum_{i\in s_1} d_{1i} q_{1i}^{\oplus}\right) = 0\,, \tag{6.14.1.7}$$

and on substituting (6.14.1.6) in (6.14.1.4) we have

$$\lambda_{11}^{\oplus}\left(\sum_{i\in s_1} d_{1i} q_{1i}^{\oplus} x_{1i}^2\right) + \lambda_{12}^{\oplus}\left(\sum_{i\in s_1} d_{1i} q_{1i} x_{1i}\right) = \left(X_1 - \sum_{i\in s_1} d_{1i} x_{1i}\right). \tag{6.14.1.8}$$

On solving (6.14.1.7) and (6.14.1.8) for $\lambda_{11}^{\oplus}$ and $\lambda_{12}^{\oplus}$, and substituting back in (6.14.1.6), we have the modified first phase calibrated weights as

$$\tilde{d}_{1i}^{\oplus} = d_i + \frac{\left[\left(x_{1i} d_{1i} q_{1i}^{\oplus}\right)\left(\sum_{i\in s_1} d_{1i} q_{1i}^{\oplus}\right) - \left(d_{1i} q_{1i}\right)\left(\sum_{i\in s_1} d_{1i} q_{1i}^{\oplus}\right)\left(\sum_{i\in s_1} d_{1i} q_{1i}^{\oplus} x_{1i}\right)\right]}{\left(\sum_{i\in s_1} d_{1i} q_{1i}^{\oplus}\right)\left(\sum_{i\in s_1} d_{1i} q_{1i}^{\oplus} x_{1i}^2\right) - \left(\sum_{i\in s_1} d_{1i} q_{1i}^{\oplus} x_{1i}\right)^2}\left[X_1 - \sum_{i\in s_1} d_{1i} x_{1i}\right] \tag{6.14.1.9}$$

Thus a modified first phase calibrated estimator of X_2 is given by

$$\hat{X}_2^{\oplus} = \sum_{i\in s_2} d_i x_{2i} + \hat{\beta}_{1(\text{ols})}\left[X_1 - \sum_{i\in s_1} d_{1i} x_{1i}\right], \tag{6.14.1.10}$$

where

$$\hat{\beta}_{1(\text{ols})} = \frac{\left(\sum_{i\in s_1} d_{1i}q_{1i}^{\oplus}x_{1i}x_{2i}\right)\left(\sum_{i\in s_1} d_{1i}q_{2i}^{\oplus}\right) - \left(\sum_{i\in s_1} d_{1i}q_{1i}^{\oplus}x_{2i}\right)\left(\sum_{i\in s_1} d_{1i}q_{1i}^{\oplus}x_{1i}\right)}{\left(\sum_{i\in s_1} d_{1i}q_{1i}^{\oplus}\right)\left(\sum_{i\in s_1} d_{1i}q_{1i}^{\oplus}\right) - \left(\sum_{i\in s_1} d_{1i}q_{1i}^{\oplus}x_{1i}\right)^2}. \tag{6.14.1.11}$$

Clearly the estimator (6.14.1.10) is a traditional linear regression estimator due to Hansen, Hurwitz and Madow (1953), and hence is improved version of first phase calibrated estimator owed to Hidiroglou and Särndal (1995, 1998) and hence that Singh (2000b).

Note that there is no choice of $q_{1i}^{\oplus}$ which reduces the estimator (6.14.1.10) into ratio or product method of estimation, and leads to the following theorem.

Theorem 6.1.4.1. The traditional first phase calibrated linear regression estimator of the population total X_2 is unique in its class of estimators.

6.14.2 IMPROVED SECOND PHASE CALIBRATION

Singh (2003c) considers an improved second phase calibrated estimator of population total, Y, defined as

$$\hat{Y}_c^{\oplus} = \sum_{i\in s_2} \tilde{d}_i^{\oplus} y_i, \tag{6.14.2.1}$$

where $\tilde{d}_i^{\oplus}$ are called the second phase calibrated plus weights. Let us choose the second phase calibrated weights $\tilde{d}_i^{\oplus}$ such that the chi square distance function defined as

$$D_2^{\oplus} = \sum_{i\in s_2} \frac{\left(\tilde{d}_i^{\oplus} - d_{1i}d_{2i}\right)^2}{d_{1i}d_{2i}q_{2i}^{\oplus}}, \tag{6.14.2.2}$$

is minimum subject to the two calibration constraints defined as

$$\sum_{i\in s_2} \tilde{d}_i^{\oplus} = \sum_{i\in s_2} d_{1i}d_{2i}, \tag{6.14.2.3}$$

and

$$\sum_{i\in s_2} \tilde{d}_i^{\oplus} x_{2i} = \hat{X}_2^{\oplus}, \tag{6.14.2.4}$$

where $\hat{X}_2^{\oplus}$ is given by (6.14.1.1) after fixing first phase calibration. The choice of $q_{2i}^{\oplus}$ makes different forms of estimators in two-phase sampling. The modified second phase Lagrange function is given by

$$L_2^{\oplus} = \sum_{i\in s_2} \frac{\left(\tilde{d}_i^{\oplus} - d_{1i}d_{2i}\right)^2}{d_{1i}d_{2i}q_{2i}^{\oplus}} - 2\lambda_{21}^{\oplus}\left[\sum_{i\in s_2} \tilde{d}_i^{\oplus} x_{2i} - \hat{X}_2^{\oplus}\right] - 2\lambda_{22}^{\oplus}\left[\sum_{i\in s_2} \tilde{d}_i^{\oplus} - \sum_{i\in s_2} d_{1i}d_{2i}\right], \tag{6.14.2.5}$$

where $\lambda_{21}^{\oplus}$ and $\lambda_{22}^{\oplus}$ are Lagrange multipliers. On setting $\dfrac{\partial L_2^{\oplus}}{\partial \tilde{d}_i^{\oplus}} = 0$ we have

$$\tilde{d}_i^{\oplus} = d_{1i}d_{2i} + \lambda_{21}^{\oplus}d_{1i}d_{2i}q_{2i}^{\oplus}x_{2i} + \lambda_{22}^{\oplus}d_{1i}d_{2i}q_{2i}^{\oplus}. \tag{6.14.2.6}$$

From (6.14.2.3) and (6.14.2.6) we have

$$\lambda_{21}^{\oplus}\left(\sum_{i\in s_2} d_{1i}d_{2i}q_{2i}^{\oplus}x_{2i}\right) + \lambda_{22}^{\oplus}\left(\sum_{i\in s_2} d_{1i}d_{2i}q_{2i}^{\oplus}\right) = 0, \tag{6.14.2.7}$$

and from (6.14.2.4) and (6.14.2.6) we have

$$\lambda_{21}^{\oplus}\left(\sum_{i\in s_2} d_{1i}d_{2i}q_{2i}^{\oplus}x_{2i}^2\right) + \lambda_{22}^{\oplus}\left(\sum_{i\in s_2} d_{1i}d_{2i}q_{2i}^{\oplus}x_{2i}\right) = \left(\hat{X}_2^{\oplus} - \sum_{i\in s_2} d_{1i}d_{2i}x_{2i}\right). \tag{6.14.2.8}$$

On solving (6.14.2.7) and (6.14.2.8) for $\lambda_{21}^{\oplus}$ and $\lambda_{22}^{\oplus}$, and substituting them in (6.14.2.) we obtain the improved second phase calibrated weights as

$$\tilde{d}_i^{\oplus} = d_{1i}d_{2i} + \frac{\left(d_{1i}d_{2i}q_{2i}^{\oplus}x_{2i}\right)\left(\sum_{i\in s_2} d_{1i}d_{2i}q_{2i}^{\oplus}\right) - \left(d_{1i}d_{2i}q_{2i}^{\oplus}\right)\left(\sum_{i\in s_2} d_{1i}d_{2i}q_{2i}^{\oplus}x_{2i}\right)}{\left(\sum_{i\in s_2} d_{1i}d_{2i}q_{2i}^{\oplus}x_{2i}^2\right)\left(\sum_{i\in s} d_{1i}d_{2i}q_{2i}^{\oplus}\right) - \left(\sum_{i\in s_2} d_{1i}d_{2i}q_{2i}^{\oplus}x_{2i}\right)^2}\left[\hat{X}_2^{\oplus} - \sum_{i\in s_2} d_{1i}d_{2i}x_{2i}\right]. \tag{6.14.2.9}$$

On substituting (6.14.2.9) in (6.14.2.1), the improved second phase calibrated estimator of the population total Y is given by

$$\hat{Y}_c^{\oplus} = \sum_{i\in s_2} d_{1i}d_{2i}y_i + \hat{\beta}_{2(\text{ols})}\left[\hat{X}_2^{\oplus} - \sum_{i\in s_2} d_{1i}d_{2i}x_{2i}\right], \tag{6.14.2.10}$$

where

$$\hat{\beta}_{2(\text{ols})} = \frac{\left(\sum_{i\in s_2} d_{1i}d_{2i}q_{2i}^{\oplus}x_{2i}y_i\right)\left(\sum_{i\in s_2} d_{1i}d_{2i}q_{2i}^{\oplus}\right) - \left(\sum_{i\in s_2} d_{1i}d_{2i}q_{2i}^{\oplus}y_i\right)\left(\sum_{i\in s_2} d_{1i}d_{2i}q_{2i}^{\oplus}x_{2i}\right)}{\left(\sum_{i\in s_2} d_{1i}d_{2i}q_{2i}^{\oplus}x_{2i}^2\right)\left(\sum_{i\in s_2} d_{1i}d_{2i}q_{2i}^{\oplus}\right) - \left(\sum_{i\in s_2} d_{1i}d_{2i}q_{2i}^{\oplus}x_{2i}\right)^2}. \tag{6.14.2.11}$$

Note that if $q_{1i}^{\oplus} = 1$ and $q_{2i}^{\oplus} = 1$, then the resultant calibrated estimator (6.14.2.10) can be claimed as a traditional chain regression type estimator in survey sampling.

Thus we have the following theorem:

Theorem 6.14.2.1. The traditional chain regression estimator is unique in its class of estimators.

Note that in the same way all the ten cases (See Exercise 6.27) considered by Estevao and Särndal (2002) can be improved and we are leaving it to the readers as an exercise. The problem of estimation of variance with the help of modified two-dimensional methodology as discussed in Chapter 5 can also be extended for two-phase sampling, and will be discussed in the next volume of this book.

EXERCISES

Exercise 6.1. Assume that a sample of size m is selected using SRSWR sampling out of N units to observe the variable X, while a sub-sample of size n is selected out of m units to observe the study variable Y and auxiliary variable X with the same sampling strategy. Suppose $\bar{y}_n$, $\bar{x}_n$ denote the second phase sample means, and $\bar{x}_m$ is the first phase sample mean.

(a) Find the first order bias and mean squared error of the estimators of population mean $\bar{Y}$ defined as

$$\bar{y}_{\text{B}} = \bar{y}_n\left(\frac{\bar{x}_m}{\bar{x}_n}\right), \qquad \text{[Bose (1943)]}$$

$$\bar{y}_{\text{P}} = \bar{r}_n\bar{x}_m, \qquad \text{[Patterson (1950)]}$$

and

$$\bar{y}_{\text{yr}} = \bar{r}_n\bar{x}_n + \frac{n(m-1)}{m(n-1)}\left(\bar{y}_n - \bar{r}_n\bar{x}_n\right), \qquad \text{[Yates (1949), Rao (1975)]}$$

where $\bar{r}_n = \frac{1}{n}\sum_{i=1}^{n}\frac{y_i}{x_i}$.

(b) Now suppose

c_1 = Cost per unit of observation on the auxiliary variable X,

c_2 = Cost per unit of observation on the variable Y,

so that the total cost C is

$$C = mc_1 + nc_2 .$$

Find the minimum mean squared errors of the above estimators defined in part (a) for the fixed cost. Comment on the resultant mean squared errors.

Hint: Sukhatme (1962), Tukey (1956).

Exercise 6.2. Suppose two samples of sizes n_1 and n_2 are drawn independently from a finite population of size N. Let $\bar{x}_1$ denote the sample mean of auxiliary variable X of the first phase sample of size n_1, and let $\bar{y}_2$ and $\bar{x}_2$ denote the sample means of study variable Y and auxiliary variable X respectively, based on the second phase sample of size n_2. Assuming the two samples are drawn independently and the regression coefficient β is known, consider the following two estimators of population mean $\bar{Y}$ as

$$\bar{y}_0 = \bar{y}_2 + \beta(\bar{x}_1 - \bar{x}_2)$$

and

$$\bar{y}_1 = \bar{y}_2 + \beta(\bar{x} - \bar{x}_2), \quad \text{where } \bar{x} = a\bar{x}_1 + b\bar{x}_2 \text{ with } a + b = 1.$$

Show that

$$V(\bar{y}_0) - V(\bar{y}_1) \geq 0 .$$

Hint: Shah and Gupta (1986).

Exercise 6.3. Suppose a sample of size m is selected using SRSWOR sampling out of N units to observe the variates X and Z, while a sub-sample of size n is selected out of m units to observe the variates Y and X with the same sampling strategy. Suppose $\bar{y}_n$ and $\bar{x}_n$ denote the second phase sample means, and $\bar{x}_m$ and $\bar{z}_m$ are the first phase sample means for the associated variables. Assuming that population mean $\bar{Z}$ of the second auxiliary variable Z is known. Find the first order bias and mean squared error of the estimators of population mean $\bar{Y}$ defined as

$$\bar{y}_1 = \bar{y}_n\left[\frac{\bar{x}_m + b\left(\bar{Z} - \bar{z}_m\right)}{\bar{x}_n}\right] \qquad \text{[Kiregyera (1980, 1984)]}$$

$$\bar{y}_2 = \bar{y}_n + \beta\left[\frac{\bar{x}_m}{\bar{z}_m}\bar{Z} - \bar{x}_n\right] \qquad \text{[Kiregyera (1980, 1984)]}$$

$$\bar{y}_3 = \bar{y}_n + b\left[\bar{x}_{\text{frd}} - \bar{x}_n\right] \qquad \text{[Singh and Singh (1991)]}$$

where

$$\bar{x}_{\text{frd}} = \bar{x}_m\left[\frac{(A+C)\bar{Z} + fB\bar{z}_m}{(A+fB)\bar{Z} + C\bar{z}_m}\right].$$

$$\bar{y}_4 = \bar{y}_n\left(\frac{\bar{x}_m}{\bar{x}_n}\right)\left[\frac{\bar{Z} + C_z}{\bar{z}_m + C_z}\right]^{\alpha} \qquad \text{[Singh and Upadhyaya (1995)]}$$

$$\bar{y}_5 = \bar{y}_n\left[\frac{\bar{x}_m + b\left(\bar{Z} - \bar{z}_m\right)}{\bar{x}_m + \lambda\left(\bar{x}_n - \bar{x}_m\right)}\right], \qquad \text{[Upadhyaya, Kushwaha, and Singh (1990)]}$$

and

$$\bar{y}_6 = C_1\bar{y}_n + C_2\bar{x}_n + C_3\bar{x}_m + C_4\bar{z}_n + C_5\bar{z}_m + C_6\bar{Z}$$

where C_j, $j = 1, 2, 3, 4, 5, 6$, are suitable chosen constants such that bias in $\bar{y}_6$ is equal to zero.

Hint: Mishra and Rout (1997).

Exercise 6.4. Using the concept of two-phase sampling study the asymptotic properties of the following estimators of population mean $\bar{Y}$ defined as

$$\bar{y}_1 = \bar{y}_n\left(\frac{\bar{x}_m}{\bar{x}_n}\right)^{\alpha} \quad \text{and} \quad \bar{y}_2 = \bar{y}_n\left[\frac{\bar{x}_m}{\alpha\bar{x}_n + (1-\alpha)\bar{x}_m}\right].$$

Hint: Sahoo and Swain (1989).

Exercise 6.5. In a finite population Ω of size N, let the value of the variable Y_j, $(j = 0, 1, 2)$ on the ith unit be y_{ji}. Let $\bar{Y}_j$ and S_j^2, respectively, denote the population mean and mean square error of the jth variable. Consider we are interested in estimating the ratio and product of population means of the first two variables, defined as $R = \bar{Y}_0/\bar{Y}_1$ and $P = \bar{Y}_0\bar{Y}_1$, respectively. Suppose a preliminary large sample of m units is drawn from the given population by SRSWOR scheme and only the auxiliary variable Y_2 is measured on it. The main sample or second phase sample of size n is drawn either from the preliminary sample of size $m\ (> n)$ or independently from the population using SRSWOR scheme and variables $Y_j,\ (j = 0,1,2)$ are measured.

Suppose

$$\bar{y}_{nj} = n^{-1}\sum_{i=1}^{n} y_{ji} \quad \text{and} \quad s_{nj}^2 = (n-1)^{-1}\sum_{i=1}^{n}\left(y_{ji} - \bar{y}_{nj}\right)^2$$

denote the second phase sample means and sample variance of the jth variable. Also let

$$\bar{y}_{m2} = m^{-1}\sum_{i=1}^{m} y_{2i} \quad \text{and} \quad s_{m2}^2 = (m-1)^{-1}\sum_{i=1}^{m}\left(y_{2i} - \bar{y}_{m2}\right)^2$$

denote the first phase sample mean and sample variance.

(a) Study the asymptotic properties of the general class of estimators of R and P defined as

$$\hat{R}_{\mathrm{H}} = \hat{R}H(u,v) \qquad \text{And} \qquad \hat{P}_{\mathrm{H}} = \hat{P}H(u,v)$$

where $\hat{R} = \bar{y}_{n0}/\bar{y}_{n1}$, $\hat{P} = \bar{y}_{n0}\bar{y}_{n1}$, $u = \bar{y}_{n2}/\bar{y}_{m2}$, $v = s_{n2}^2/s_{m2}^2$ and $H(\bullet, \bullet)$ is a parametric function such that $H(1,1) = 1$, satisfying certain regularity conditions. Also find the optimum sample sizes for the fixed costs.

Hint: Singh (1982b), Khare (1991), Singh, Singh, and Singh (1994), Ahmed (1997).

(b) Assuming that population mean of another auxiliary variable Y_3 is known, study the asymptotic properties of the estimators:

$$\hat{R}^* = w\hat{R}\left(\frac{\bar{y}_{n2}}{\bar{y}_{m2}}\right) + (1-w)\hat{R}\left(\frac{\bar{y}_{n2}}{\bar{y}_{m2}}\right)\left(\frac{\bar{y}_{m3}}{\bar{Y}_3}\right).$$

Hint: Singh, Singh, and Singh (1994), Prasad, Singh, and Singh (1996).

Exercise 6.6. Show that the following are unbiased estimators of the population mean, $\bar{Y}$,

$$t_1 = \bar{r}(\bar{x}_1 - \bar{x}) + M(\bar{y} - \bar{r}\,\bar{x}) + \bar{y}, \quad t_2 = \bar{r}(\bar{x}^* - \bar{x}) + aM(\bar{y} - \bar{r}\,\bar{x}) + \bar{y},$$

and

$$t_3 = \bar{r}(\bar{x}_w - \bar{x}) + M_w(\bar{y} - \bar{r}\,\bar{x}) + \bar{y}$$

where $M = (N-n)/\{N(n-1)\}$, $M_w = (w-n)/\{w(n-1)\}$, $n_1 \le w < N$, $\bar{x}^* = a\bar{x}_1 + b\bar{x}$, $a + b = 1$; $\bar{x}_1$ is mean of x based on n_1 units and w is the number of distinct units in the first and second-phase samples, $\bar{y}$ and $\bar{x}$ are the means based on second phase n units, and $\bar{x}_w$ is the mean based on only distinct units.

(a) Show that t_1 is a special case of t_2 for $a = 1$ and $b = 0$.

(b) Compare these three estimators under a superpopulation model. Use the Quenouille's method to construct unbiased regression type estimators in two-phase sampling.

Hint: Rao (1975), Singh, Katyar, and Gangwar (1996).

Exercise 6.7. Show that an unbiased estimator of variance of the regression estimator in two-phase sampling is given by

$$\hat{v}(\bar{y}_{\text{d}}) = \left(\frac{1}{n} - \frac{1}{m}\right)\frac{1}{(n-3)}\sum_{i=1}^{n}\left[(y_i - \bar{y}) - \hat{\beta}_{\text{ols}}(x_i - \bar{x})\right]^2 + \frac{s_y^2}{m}.$$

Hint: Tikkiwal (1960).

Exercise 6.8. Let a finite population of N units be represented by the set of $(p+1)$ component vectors $(y_i,\, x_{1i},\, x_{2i},\ldots, x_{pi})$, $i = 1, 2, \ldots, N$. Let $\mu_y = N^{-1}\sum_{i=1}^{N} y_i$ and $\mu_{x_j} = N^{-1}\sum_{i=1}^{N} x_{ji}$, $j = 1, 2, \ldots, p$. Suppose a sample of n units is taken by SRSWOR sampling. Let $\bar{y} = n^{-1}\sum_{i=1}^{n} y_i$ and $\bar{x}_{ji} = n^{-1}\sum_{i=1}^{n} x_{ji}$, $j = 1, 2, \ldots, p$.

(a) Show that for some real constants α_j the estimator

$$\bar{y}_{m_1} = \bar{y} - \sum_{j=1}^{p}\alpha_j\left(\bar{x}_j - \mu_{x_j}\right)$$

is an unbiased estimator of the population mean.

(b) Suppose that the sample elements are drawn once at a time, so the sample is ordered set, the order being that of the order of drawn. Let Z_α denote the ordered set of observations on the first α sample elements, $1 < \alpha < n$, and let $\alpha_i(Z_\alpha)$ denote functions of these observations. Let $Y(\alpha)$, $X_j(\alpha)$ denote the sums and $\bar{y}(\alpha)$, $\bar{x}_j(\alpha)$, $j = 1, 2, \ldots, p$ denote means of the indicated observations on the first α sample elements. Also let $E\{\bullet \mid Z_\alpha\}$ denote the conditional expectation given Z_α. Show that the following three estimators are unbiased estimators of population mean $\bar{Y}$, defined as

$$t_1 = \bar{y} - \sum_{j=1}^{p}\alpha_j(Z_\alpha)\left[\bar{x}_j - \mu_{x_j}\right] - \frac{\alpha(N-n)}{(n-\alpha)N}\left\{\bar{y}(\alpha) - \bar{y} - \sum_{j=1}^{p}\alpha_j(Z_\alpha)\left(\bar{x}_j(\alpha) - \bar{x}_j\right)\right\}$$

$$t_2 = \frac{(N-\alpha)n}{(n-\alpha)N}\left\{\bar{y} - \sum_{j=1}^{p}\alpha_j(Z_\alpha)\left[\bar{x}_j - \mu_{x_j}\right]\right\} - \frac{\alpha(N-n)}{(n-\alpha)N}\left\{\bar{y}(\alpha) - \bar{y} - \sum_{j=1}^{p}\alpha_j(Z_\alpha)\left(\bar{x}_j(\alpha) - \bar{x}_j\right)\right\}$$

and

$$t_2 = \sum_{j=1}^{p}\alpha_j(Z_\alpha)\mu_{x_j} + \frac{(N-\alpha)}{(n-\alpha)N}\left\{Y(\alpha) - \sum_{j=1}^{p}\alpha_j(Z_\alpha)X_j(n)\right\} - \frac{N-n}{N(n-\alpha)}\left[Y(\alpha) - \sum_{j=1}^{p}\alpha_j(Z_\alpha)X_j(\alpha)\right]$$

(c) For $p = 1$ show that the minimisation of

$$\sum_{i=1}^{n} w(x_i)\left[y_i - \frac{\mu_y}{\mu_x}x_i\right]^2$$

where $w(x_i)$ is some suitably chosen weight to form different kinds of estimators, leads to the biased ratio estimator

$$\hat{\mu}_y = \mu_x\alpha(Z_\varepsilon)$$

where

$$\alpha(Z_\alpha) = \sum_{i=1}^{n} y_i x_i w(x_i) \Big/ \sum_{i=1}^{n} x_i^2 w(x_i).$$

(d) Show that if $w(x_i) = 1/x_i^2$, $\alpha(Z_\alpha) = \frac{1}{\alpha}\sum_{i=1}^{n}\frac{y_i}{x_i}$ and the estimator t_3 becomes

$$t_3^* = \bar{r}\mu_x + \frac{(N-1)n}{N(n-1)}(\bar{y} - \bar{r}\,\bar{x}), \quad \text{with } \bar{r} = \frac{1}{n}\sum_{i=1}^{n}\frac{y_i}{x_i}.$$

(e) Show that if $w(x_i) = 1/x_i$, then the estimator t_3 becomes,

$$t_3^{**} = \bar{R}_\alpha \mu_x + \frac{(N-\alpha)n}{N(n-\alpha)}(\bar{y} - \bar{R}_\alpha \bar{x})$$

for $\alpha = 1$ and $\alpha = (n-1)$. Also show that if $\alpha = 1$ the estimator t_3^{**} reduces to t_3^* and if $\alpha = (n-1)$ then $\bar{R}_{n-1} = \frac{1}{n}\sum_{i=1}^{n}\frac{n\bar{y} - y_i}{n\bar{x} - x_i}$.

Hint: Mickey (1959).

Exercise 6.9. Suppose an initial sample s_1 of size m is selected with replacement with probabilities p_i proportional to Z_i ($i = 1, 2, ..., N$), where N is the population size. Information on the variable X is gathered on the first-phase sample s_1. The second phase sample s_2 of size n is a sub-sample of s_1, selected with SRSWOR in which information on Y and X is collected. Find the variance of the unbiased difference type estimator of population total Y given by

$$\hat{y} = n^{-1}\sum_{i=1}^{n} y_i/p_i + \beta\left[m^{-1}\sum_{i=1}^{m} x_i/p_i - n^{-1}\sum_{i=1}^{n} x_i/p_i\right].$$

Hint: Srivenkataramana and Tracy (1989), Särndal and Swensson (1987), Raj (1964, 1965b).

Exercise 6.10. Consider a sequence of finite populations such that its tth member U_t consists of N_t units labelled $i = 1, 2, ..., N_t$, $t = 1, 2, ...$, and $N_t \to \infty$ as $t \to \infty$. Let $\xi(t)$ be a quantity that tends to zero as $t \to \infty$. Let $w(\geq 0)$, X, Y denote respectively a size measure, an auxiliary variable, and the variable of interest. Let their respective values for the ith unit $(i = 1, 2, ..., N_t)$ of U_t $(t = 1, 2, ...)$ be w_{it} (known for each i), x_{it} and y_{it}. The corresponding totals (means) over U_t are W_t, X_t, Y_t($\bar{W}_t, \bar{X}_t, \bar{Y}_t$). Let

$$W_t = (w_{1t}, ..., w_{N_t t}),\ X_t = (X_{1t}, ..., X_{N_t t}) \text{ and } Y_t = (Y_{1t}, ..., Y_{N_t t}).$$

From U_t an initial sample s_{1t} (each possible s_{1t} with a fixed number n_{1t} of distinct units) is supposed to be drawn with a probability $p_{1t}(s_{1t})$ that may involve the w_{it} for $i \in U_t$. For the units in s_{1t} the x values are supposed to be ascertained. From s_{1t} a sub-sample s_{2t} (say, each possible s_{2t} with a given number $n_{2t} < n_{1t}$ of distinct units) is to be chosen with a conditional probability $p_{2t}(s_{2t}) = p_{2t}(s_{2t} \mid s_{1t})$ given that

$p_{1t}(s_{1t}) > 0$. These probabilities may involve w_{it}, for $i \in U_t$ and also x_{it} for $i \in s_{it}$. The overall two-phase sample $s_t = (s_{1t}, s_{2t})$ has the selection probability $p_t(s_t) = p_{1t}(s_{1t})p_{2t}(s_{2t} \mid s_{1t})$, where p_t, p_{1t} and p_{2t} denote respectively the design corresponding to double, first phase, and second phase sampling plan. Define E_{p_t}, $E_{p_{1t}}$ and $E_{p_{2t}}$ be the design expectation operators with respect to p_t, p_{1t}, and p_{2t} where $E_{p_{2t}}$ is a conditional expectation for each fixed s_{1t} with $p_{1t} > 0$; $I_{kit}(I_{kit}) = 1$ if $i \in s_{kt}$, and $= 0$ otherwise, for $k = 1, 2$. Let π_{1it} and π_{1ijt} be the first and second order inclusion probabilities according to design p_{1t} and let $r_{1it} = 1/\pi_{1it}$ be the design weight for every s_{1t} with $p_{1t} > 0$. Let π_{2it} and π_{2ijt} be the first and second order conditional inclusion probabilities with respect to design $p_{2t}(\bullet \mid s_{1t})$ and let design weights be $r_{2it} = 1/\pi_{2it}$.

Consider two estimators such as

$$e_{1t}(w) = \sum_{i \in s_{1t}} w_{it} r_{1it}$$

and

$$e_{2t}(w) = \sum_{i \in s_{2t}} w_{it} r_{1it} r_{2it} .$$

Similarly define $e_{1t}(x)$, $e_{2t}(x)$ and $e_{2t}(y)$ based on the information available from survey data.

Find the bias and variance of the regression type estimator of population total Y given by

$$\hat{y}_t = N_t^{-1}\left[e_{2t}(y) - \hat{\beta}_{1t}\{e_{2t}(x) - R_t\} - \hat{\beta}_{2t}\{e_{2t}(w) - W_t\}\right]$$

where $R_t = e_{1t}(x) - \hat{\beta}_{3t}\{e_{1t}(w) - W_t\}$ and $\hat{\beta}_{jt}\,(j = 1,2)$ are quantities in terms of known W and sampled x and y values.

Hint: Mukerjee and Chaudhuri (1990).

Exercise 6.11. Compare the following three estimators in two-phase sampling:

(a) Ratio estimator : $\bar{y}_1 = \bar{y}_n(\bar{x}_m/\bar{x}_n)$,

(b) Regression estimator : $\bar{y}_2 = \bar{y}_n + \beta(\bar{x}_m - \bar{x}_n)$, where $\beta = S_{xy}/S_x^2$, and

(c) PPSWR sampling : $\bar{y}_3 = \dfrac{N\,\bar{x}_m}{m \quad n} \sum_{i=1}^{n} \dfrac{y_i}{x_i}$.

Hint: Prabhu--Ajgaonkar (1975).

Exercise 6.12. Find the asymptotic bias and variance of the chain ratio type estimators of finite population variance S_y^2 as

(a) $s_1^2 = s_y^2\left(\dfrac{\bar{x}^*}{\bar{x}}\right)^{\alpha}\left(\dfrac{\bar{Z}}{\bar{z}}\right)^{\beta}$, (b) $s_2^2 = s_y^2\left(\dfrac{s_x^2}{s_x^{*2}}\right)^{\alpha}\left(\dfrac{s_z^{*2}}{S_z^2}\right)^{\beta}$, and

(c) $s_3^2 = s_y^2\left(\frac{\bar{x}}{\bar{x}^*}\right)^{\alpha}\left(\frac{s_x^2}{s_x^{*2}}\right)^{\beta}\left(\frac{\bar{z}^*}{\bar{Z}}\right)^{\phi}\left(\frac{s_z^{*2}}{S_z^2}\right)^{\delta}$.

Hint: Gupta, Singh, and Mangat (1992-93).

Exercise 6.13. Suppose a simple random sample s_1 of size m is taken without replacement from the population of N elements and X_i alone is observed for all elements $i \in s_1$. A simple random sub-sample s_2 of size n is then drawn without replacement from s_1 and Y_i, X_i are observed for all $i \in s_2$. A ratio estimator of population mean $\bar{Y}$ is

$$\bar{y}_r = \bar{y}\left(\frac{\bar{x}^*}{\bar{x}}\right)$$

where $\bar{y}$ and $\bar{x}$ are the means for s_2 and $\bar{x}^*$ is the mean for s_1. Define

$$\bar{y}_r(j) = \bar{y}(j)\left\{\frac{\bar{x}^*(j)}{\bar{x}(j)}\right\},$$

where

$$\bar{x}(j) = \begin{cases} \dfrac{n\bar{x} - x_j}{n-1} & \text{if } j \in s_2, \\ \bar{x} & \text{if } j \in (s_1 - s_2), \end{cases} \qquad \bar{y}(j) = \begin{cases} \dfrac{n\bar{y} - y_j}{n-1} & \text{if } j \in s_2, \\ \bar{y} & \text{if } j \in (s_1 - s_2), \end{cases}$$

and $\bar{x}^*(j) = \dfrac{m\bar{x} - x_j}{m-1}$ if $j \in s_1$. Show that the linear form of the modified Jackknife estimator

$$\hat{v}_J = \frac{m-1}{m}\sum_{j \in s_1}\{\bar{y}_r(j) - \bar{y}_r\}^2$$

is identical to usual estimator of variance under mild conditions.

Hint: Rao and Sitter (1995), Sitter and Rao (1997), Rao (1996b).

Exercise 6.14. Suppose that two sample surveys have one variable in common, say Z_i that is two agencies are collecting information on it. Suppose the population total of this variable is unknown to both agencies. Let X be the known auxiliary variable and Y be the variable under study. Suppose $\hat{z}_1 = \sum_{i \in s_1} z_i/\pi_{1i}$ and $\hat{z}_2 = \sum_{i \in s_2} z_i/\pi_{2i}$ are the two different estimators of unknown total Z of the common variable obtained by both independent surveys agencies. Study the properties of the regression type estimators of population total Y defined as

$$\hat{y}_0(t) = \hat{y}_t + b_1(X - \hat{x}_t) + b_2(\hat{z}_1 - \hat{z}_2)$$

where $\hat{y}_t = \sum_{i \in s_t} y_{ti}/\pi_{ti}$ and $\hat{x}_t = \sum_{i \in s_t} x_{ti}/\pi_{ti}$ where $t = 1, 2$ denotes the tth agency.

Hint: Renssen and Nieuwenbroek (1997).

Exercise 6.15. Consider the problem of estimating the population mean $\overline{Y}$ of the study variable, Y, from a finite population of size N. Suppose the information on an auxiliary variable, X, highly correlated with Y, is not available, but values of X are assumed known over a large random sample of m units. Suppose that information on another auxiliary variable, Z, is available on all units of the population with population mean $\overline{Z}$. Let (y_i, x_i, z_i) for $i = 1, 2, ..., n$ denote the information collected on the second phase sample. Then study the asymptotic properties of the following four estimators:

(a) $t_1 = \bar{y}_n + b_{yx}\left[(\bar{x}_m - \bar{x}_n) - b_{xz}(\bar{z}_m - \overline{Z})\right]$; (b) $t_2 = \bar{y}_n + b_{yx}(\bar{x}_m - \bar{x}_n) + b_{yz}(\overline{Z} - \bar{z}_n)$;

(c) $t_3 = \bar{y}_n + b_{yx}(\bar{x}_m - \bar{x}_n) + b_{yz}(\overline{Z} - \bar{z}_n)$;

and

(d) $t_4 = \bar{y}_n + b_{yx}(\bar{x}_m - \bar{x}_n) + b_{yx}b_{xz}(\overline{Z} - \bar{z}_m) + b_{yz}(\overline{Z} - \bar{z}_n)$.

Hint: Ahmed (1998), Mukerjee, Rao, and Vijayan (2000), Sahoo and Sahoo (1999a, 1999b), Pradhan (2001).

(e) Develop a test statistic to test a hypothesis that known population mean $\overline{Z}$ can be included or not in the regression type estimators using two-phase sampling.

Hint: Das and Bez (1995).

(f) Study the asymptotic properties of these four estimators (a) to (d) under the superpopulation models:

$$y_k = q_1 x_k + e_{1k}$$

with

$E(e_{1k} \mid x_k) = 0$, $E(e_{1k}^2 \mid x_k) = a x_k^g$, $a > 0$, $g \ge 0$, $E(e_{1k}e_{1j} \mid x_k, x_j) = 0$, $k \ne j = 1,2,...,N$,

and e_{1k} s are independent of x, and

$$x_k = q_2 z_k + e_{2k}$$

with

$E(e_{2k} \mid x_k) = 0$, $E(e_{2k}^2 \mid x_k) = c x_k^h$, $c > 0$, $h \ge 0$, $E(e_{2k}e_{2j} \mid x_k, x_j) = 0$, $k \ne j = 1, 2, ..., N$,

and e_{2k} s are independent of z. Assume e_1 and e_2 are also independent.

Hint: Sahoo and Sahoo (1999a, 1999b).

Exercise 6.16. Let $s = \{i_1, i_2, ..., i_n\}$ where $i_j \in \Omega$ $(j = 1,2,3,...,n)$ denotes a random sample of n units drawn from Ω using the sampling design

$$P = \left\{P_s : P_s \ge 0, \sum_{s \in S} P_s = 1\right\},$$

where S denotes the set of all possible samples. Assume that paired observations $\{(y_i, z_i), i \in s\}$ are available. Suppose $\overline{Z} = \overline{Y}$, that is, both variables have common population mean. Let

$$\bar{y}_{\mathrm{HT}} = \frac{1}{N}\sum_{i \in s}\frac{y_i}{\pi_i} \text{ and } \bar{z}_{\mathrm{HT}} = \frac{1}{N}\sum_{i \in s}\frac{z_i}{\pi_i},$$

where $\pi_i = \sum_{s \in S(i)} P_s$ and $S(i)$ denotes the set of all samples containing the ith unit.

(a) Find the bias and variance of the estimator

$$\hat{\mu}_1 = w\bar{y}_{HT} + (1-w)\bar{z}_{HT}.$$

(b) Also find the optimum value of w so that the variance of the estimator $\hat{\mu}_1$ is minimum.

Hint: Tripathi and Chaubey (1992).

Exercise 6.17. In the first phase we select a preliminary large sample s' of n' units by SRSWOR and only the auxiliary variable X is measured. Let $\hat{M}_x{}'$ be the estimator of median M_x of the auxiliary variable X based on the first-phase sample. In the second phase, a sub-sample s of n units is drawn from the preliminary large sample by SRSWOR and both the study variable Y and the auxiliary variable X are measured. Let $\hat{M}_x$ and $\hat{M}_y$ denote the estimators of M_x and M_y respectively, based on the sample drawn at the second phase.

(a) Study the asymptotic properties of a general class of estimators for estimating the median M_y as

$$\hat{M}_1 = H(\hat{M}_y, U)$$

where $U = \hat{M}_x / \hat{M}'_x$, $H(\hat{M}_y, U)$ assumes the value in a bounded closed convex sub-set R_2 of two dimensional real space containing the point $(M_y, 1)$ such that $H(M_y, 1) = M_y$ and satisfies certain regularity conditions.

(b) Show that the following estimators

$$\hat{M}_2 = \hat{M}_y \frac{\hat{M}_x{}'}{\hat{M}_x}, \quad \hat{M}_3 = \hat{M}_y\left(\hat{M}_x / \hat{M}'_x\right)^d, \text{ and } \hat{M}_4 = \hat{M}_y\left[\hat{M}'_x / \left\{d\hat{M}_x + (1-d)\hat{M}'_x\right\}\right]$$

are the special cases of the general class of estimators.

Hint: Singh, Joarder, and Tracy (2001).

Exercise 6.18. Consider a two-phase sampling design in which second phase SRSWOR sample of r units is selected from an infinite population and the regression estimator of the mean of characteristic y be

$$\hat{\mu}_y = \bar{y}_2 + \hat{\beta}_{\text{ols}}(\bar{x}_1 - \bar{x}_2)$$

where $(\bar{y}_2, \bar{x}_2) = r^{-1}\sum_{i=1}^{r}(y_i, x_i)$, $\bar{x}_1 = n^{-1}\sum_{i=1}^{n}x_i$, $\hat{\beta}_{\text{ols}} = \sum_{i=1}^{r}(x_i - \bar{x}_2)(y_i - \bar{y}_2) \Big/ \sum_{i=1}^{r}(x_i - \bar{x}_2)^2$.

(a) Then show that

$$V(\hat{\mu}_y) = \left[n^{-1}\rho_{xy}^2 + r^{-1}\left(1 - \rho_{xy}^2\right)\right]\sigma_y^2,$$

where ρ_{xy} is the population correlation between y and x and σ_y^2 is the population variance of y.

(b) Also show that an estimator of variance is given by

$$\hat{v}(\hat{\mu}_y) = n^{-1}(n-1)^{-1}\sum_{i=1}^{n}(\hat{y}_i - \overline{y}_1)^2 + r^{-1}(r-2)^{-1}\sum_{i=1}^{r}(y_i - \hat{y}_i)^2,$$

where $\hat{y}_i = \overline{y}_2 + (x_i - \overline{x}_2)\hat{\beta}_{\text{ols}}$ for $i = 1, 2, ..., n$ and $\overline{y}_1 = n^{-1}\sum_{i=1}^{n}\hat{y}_i$.

Hint: Kim (2001).

Exercise 6.19. Let the first phase sample s_1 of n_1 units is taken with SRSWOR sampling, and the second phase sample s_2 of n_2 units is taken as PPSWR sampling.

(a) Differentiate between nested and non-nested two-phase sampling designs.
(b) Consider an estimator of population total Y as:

$$\hat{Y} = \frac{N}{n_1 n_2}\sum_{s_2} y_i / p_{1i} .$$

Show that: If the design is nested then

$$V(\hat{Y}) = \frac{N^2(1-f_1)}{n_1}S_y^2 + \frac{V(\hat{Y}_\text{p})}{n_2},$$

and if the design is non-nested then:

$$V(\hat{Y}) = \frac{N^2(1-f_1)}{n_1}R^2 S_x^2 + \frac{V(\hat{Y}_\text{p})}{n_2}\left[1 + \frac{(1-f_1)}{n_1}\frac{S_x^2}{\overline{X}^2}\right],$$

where

$$V(\hat{Y}_\text{p}) = \frac{1}{n_1}\sum_{i=1}^{N} p_i (y_i / p_i - Y)^2, \quad f_1 = \frac{n_1}{N}, \quad p_{1i} = x_i \Big/ \sum_{i \in s_1} x_i , \text{ and } p_i = x_i \Big/ \sum_{i \in \Omega} x_i .$$

(c) Consider an estimator of Y as

$$\hat{Y}_\text{Rat} = \sum_{i \in s_2} y_i \left(\sum_{i \in s_1} x_i \Big/ \sum_{i \in s_2} x_i \right).$$

Show that: If the design is nested then

$$V(\hat{Y}_\text{Rat}) = \frac{N^2(1-f_1)}{n_1}\left\{2RS_{xy} - R^2 S_x^2\right\} + \frac{N^2(1-f_2)}{n_2}\frac{1}{(N-1)}\sum_{i=1}^{N}(y_i - Rx_i)^2,$$

and if the design is non-nested then

$$V(\hat{Y}_\text{Rat}) = \frac{N^2(1-f_1)}{n_1}R^2 S_x^2 + \frac{N^2(1-f)}{n_2}\frac{1}{(N-1)}\sum_{i=1}^{N}(y_i - Rx_i)^2.$$

Hint: Hidiroglou (2001).

Exercise 6.20. In the first phase sample s_1 consisting of m units measure the values of the p auxiliary variables $(x_{i1}, x_{i2}, ..., x_{ip})$, $i = 1,2,...,m$. In the second phase sample $s_2 \subset s_1$ consisting of n units measure the study variable and auxiliary variables as $(y_i, x_{i1}, x_{i2}, ..., x_{ip})$, $i = 1,2,...,n$. Let

$$\bar{x}_j^* = m^{-1}\sum_{i=1}^{m} x_{ij}\ ,\ j = 1, 2, ..., p$$

denote the means of p auxiliary variables based on the first phase sample. Also let

$$\bar{y} = n^{-1}\sum_{i=1}^{n} y_i \text{ and } \bar{x}_j = n^{-1}\sum_{i=1}^{n} x_{ij}$$

denote the means of study variable y and p auxiliary variates in the second phase sample. Based on the jth auxiliary variate define a difference estimator of the population mean in two-phase sampling as

$$\bar{y}_{d(j)} = \bar{y} + b_j\left(\bar{x}_j^* - \bar{x}_j\right)$$

Now consider a weighted estimator of population mean $\bar{Y}$ as

$$\bar{y}_{\text{wd}} = \sum_{j=1}^{p} w_j \bar{y}_{d(j)},\ \text{ such that } \sum_{j=1}^{p} w_j = 1,\ 0 < w_j < 1.$$

(a) Show that $\bar{y}_{\text{wd}}$ is an unbiased for population mean $\bar{Y}$ with variance $V(\bar{y}_{\text{wd}}) = WBW'$

where $b_{jk} = \left(\frac{1}{m} - \frac{1}{N}\right)S_y^2 + \left(\frac{1}{n} - \frac{1}{m}\right)\left[b_j b_k S_{jk} - b_j S_{yj} - d_k S_{yk}\right]$, $S_y^2 = (N-1)^{-1}\sum_{i=1}^{N}\left(Y_i - \bar{Y}\right)^2$,

$S_{jk} = (N-1)^{-1}\sum_{i=1}^{N}\left(X_{ji} - \bar{X}_j\right)\left(X_{ki} - \bar{X}_k\right)$, $W = \left(w_1, w_2, ..., w_p\right)$, $B = \left(b_{jk}\right)_{p\times p}$, and

$S_{yj} = (N-1)^{-1}\sum_{i=1}^{N}\left(Y_i - \bar{Y}\right)\left(X_{ji} - \bar{X}_j\right)$.

(b) Also show that an unbiased estimator of the $V(\bar{y}_{\text{wd}})$ is given by

$$v(\bar{y}_{\text{wd}}) = \left(\frac{1}{m} - \frac{1}{N}\right)s_y^2 + \left(\frac{1}{n} - \frac{1}{m}\right)\frac{1}{n-1}\sum_{i=1}^{n}\left[(y_i - \bar{y}) - \sum_{j=1}^{p} w_j b_j\left(x_{ij} - \bar{x}_j\right)\right]^2$$

where $s_y^2 = (n-1)^{-1}\sum_{i=1}^{n}(y_i - \bar{y})^2$.

Hint: Raj (1964)

Exercise 6.21. Consider the problem of estimation of population mean using two-phase sampling. Consider an initial first phase sample of m units is selected at random and information on auxiliary variable x is measure. A second phase sample of n units in selected with PPSWR sampling with probability proportional to x. Find the variance of the unbiased estimator of the population total Y given by

$$\hat{y}_{\text{uds}} = \frac{N}{nm}\sum_{i=1}^{n}\frac{y_i}{p_i}\ ,\ \text{where } p_i = x_i \Big/ \sum_{i=1}^{m} x_i\ .$$

Hint: Raj (1964).

Exercise 6.22. Under a superpopulation model, consider $E_m(y_i \mid x_i) = \alpha + \beta x_i$, $V_m(y_i \mid x_i) = \sigma^2 x_i^g$ and $C_m(y_i,\ y_j \mid x_i,\ x_j) = 0$, $i \neq j$, and let the cost functions be

$C = nC_1$ for single phase SRSWOR sampling and $C = nC_1 + mC_2$ for two-phase SRSWOR sampling, where $C_2 < C_1$.

(a) Find the optimum first phase and second phase sample size such that the variance of the usual ratio estimator in two-phase sampling under the superpopulation model is minimum for the fixed cost of surveys.

(b) Compare the results with single phase sampling under SRSWOR sampling for the fixed cost.

Hint: Ajgaonkar (1975).

Exercise 6.23. In the first phase we select a preliminary large sample s' of n' units by SRSWOR and only the auxiliary variables X and Z are measured. Let $\hat{M}_x'$ and $\hat{M}_z'$ be the estimators of medians M_x (unknown) and M_z (known) of the auxiliary variables X and Z based on the first phase sample. In the second phase, a sub-sample s of n units is drawn from the preliminary large sample by SRSWOR and both the study variable Y and the auxiliary variable X are measured. Let $\hat{M}_x$ and $\hat{M}_y$ denote the estimators of M_x and M_y respectively, based on the sample drawn at the second phase.

(a) Study the asymptotic properties of a general class of estimators for estimating the median M_y as

$$\hat{M}_a = H\left(\hat{M}_y, U, V\right)$$

where $U = \hat{M}_x / \hat{M}'_x$, $V = \hat{M}'_z / M_z$ and $H\left(\hat{M}_y, U, V\right)$ assumes the value in a bounded closed convex sub-set R_3 of three dimensional real space containing the point $\left(M_y, 1, 1\right)$ such that $H\left(M_y, 1, 1\right) = M_y$ and satisfies certain regularity conditions.

(b) Show that the following estimators

$$\hat{M}_2 = \hat{M}_y\left(\frac{\hat{M}_x'}{\hat{M}_x}\right)\left(\frac{\hat{M}_z'}{M_z}\right); \quad \hat{M}_3 = \hat{M}_y\left(\frac{\hat{M}_x}{\hat{M}_x'}\right)^{\alpha}\left(\frac{\hat{M}_Z'}{M_z}\right)^{\beta};$$

and

$$\hat{M}_4 = \hat{M}_y \frac{\hat{M}_x'}{\left\{d\hat{M}_x + (1-d)\hat{M}_x'\right\}}\left[\frac{M_z}{\left\{g\hat{M}_z' + (1-g)M_z\right\}}\right]$$

are the special cases of the general class of estimators.

Hint: Allen, Saxena, Singh, Singh, and Smarandache (2002).

Exercise 6.24. (I) Assume that a sample of size m is selected using SRSWR sampling out of N units to observe the variable X, while a sub-sample of size n is selected out of m units to observe the study variable Y and auxiliary variable X with the same sampling strategy. Suppose $\bar{y}_n$, $\bar{x}_n$ denote the second phase sample means, and $\bar{x}_m$ is the first phase sample mean.

(a) Find the first order bias and mean squared error of the estimators of population mean $\bar{Y}$ defined as

$$\bar{y}_R = \bar{y}_n\left(\frac{\bar{x}_m}{\bar{x}_n}\right) \qquad \text{[Bose (1943)]}$$

$$\bar{y}_{HR} = \bar{r}\,\bar{x}_m + \left(\frac{m-1}{m}\right)s_{rx}, \qquad \text{[Sukhatme (1962)]}$$

where

$$r_i = \frac{y_i}{x_i}, \quad \bar{r} = \frac{1}{n}\sum_{i=1}^{n} r_i, \text{ and } s_{rx} = \frac{1}{n-1}\sum_{i=1}^{n} r_i\left(x_i - \bar{x}_n\right).$$

(II) Consider another cheaper auxiliary variable z is available for all the units in the population and hence $\bar{Z} = N^{-1}\sum_{i=1}^{N} Z_i$ is known and $\bar{z}_m$ be its estimator obtained from the first phase sample information.

(a) Find the bias and mean square error of the following estimators:

$$\bar{y}_c = \bar{y}_n\left(\frac{\bar{x}_m}{\bar{x}_n}\right)\left(\frac{\bar{Z}}{\bar{z}_m}\right); \qquad \text{[Chand (1975)]}$$

$$\bar{y}_{ds} = \bar{r}\,\bar{g}\,\bar{Z}, \text{ where } \bar{g} = \frac{1}{m}\sum_{i=1}^{m}\frac{x_i}{z_i};$$

$$\bar{y}_{dsu} = \left(\bar{r}\,\bar{g} + \frac{m-1}{m}s_{rg}\right)\bar{Z} + \frac{N-1}{N}s_{hz},$$

where $s_{hz} = (n-1)^{-1}\sum_{i\in s} h_i\left(z_i - \bar{z}\right)$ and $s_{rg} = (n-1)^{-1}\sum_{i\in s} g_i\left(r_i - \bar{r}\right)$.

Hint: Dalabehera and Sahoo (2000).

$$\bar{y}_t = \bar{y} + b_{yx}\left[\frac{\{\bar{x}_m - t_1(\bar{z}_m - \bar{Z})\}}{\{\bar{z}_m - t_2(\bar{z}_m - \bar{Z})\}^{\alpha}}\left(\bar{Z}\right)^{\alpha} - \bar{x}_n\right],$$

where α is a suitably chosen constant, and t_1 and t_2 are suitably chosen statistics such that their means exist, and

$$b_{yx} = \sum_{i=1}^{n}(y_i - \bar{y}_n)(x_i - \bar{x}_n)\Big/\sum_{i=1}^{n}(x_i - \bar{x}_n)^2.$$

(b) Further show that the following estimators :

$$\bar{y}_t^{(1)} = \bar{y} + b_{yx}\left\{\bar{x}_m\left(\frac{\bar{Z}}{\bar{z}_m}\right) - \bar{x}_n\right\}; \quad \bar{y}_t^{(2)} = \bar{y} + b_{yx}\left\{\bar{x}_m\left(2 - \left(\frac{\bar{Z}}{\bar{z}_m}\right)^{\alpha_1}\right) - \bar{x}_n\right\};$$

$$\bar{y}_t^{(3)} = \bar{y} + b_{yx}\left\{\bar{x}_m\left(\frac{\bar{Z}}{\bar{Z} + \alpha_1\left(\bar{z}_m - \bar{Z}\right)}\right) - \bar{x}_n\right\};$$

and

$$\bar{y}_t^{(4)} = \bar{y} + b_{yx}\left\{\alpha_1\bar{x}_m + (1-\alpha_1)\bar{x}_m\left(\frac{\bar{Z}}{\bar{z}_m}\right)^{\alpha_2} - \bar{x}_n\right\},$$

where α_1 and α_2 are suitably chosen constants, are the special cases of the estimator $\bar{y}_t$.

Hint: Das and Tripathi (1979), Singh, Singh, and Upadhyaya (2001).

Exercise 6.25. Consider a population of N identifiable units on which a study variable y is associated with k auxiliary variables $x_1, x_2, .., x_k$. Let x_{hi}^*, $h = 1, 2, ..., m$ be the information collected on the k auxiliary variables in the first phase sample of m units. Let (y_h, x_{hi}), $h = 1, 2, ..., n$ be the observed values in an SRSWR sample on the $(k+1)$ variables in the second phase sample.
Consider the problem of estimation of finite population variance

$$\sigma_0^2 = \frac{1}{N}\sum_{h=1}^{N}\left(Y_h - \bar{Y}\right)^2, \quad \text{where } \bar{Y} = \frac{1}{N}\sum_{h=1}^{N} Y_h .$$

Let $s_i^{*2} = (m-1)^{-1}\sum_{h=1}^{m}\left(x_{hi}^* - \bar{x}_{hm}^*\right)^2$, with $\bar{x}_{hm}^* = m^{-1}\sum_{h=1}^{m} x_{hi}^*$, be an estimator of σ_i^2 based on the preliminary large sample of m units. Further, let $s_i^2 = (n-1)^{-1}\sum_{h=1}^{n}(x_{hi} - \bar{x}_{hn})^2$, with $\bar{x}_{hn} = n^{-1}\sum_{h=1}^{n} x_{hi}$ and $s_0^2 = (n-1)^{-1}\sum_{h=1}^{n}(y_{hi} - y_{hn})^2$, with $\bar{y}_{hn} = n^{-1}\sum_{h=1}^{n} y_{hi}$, be the estimators of σ_i^2, $i = 0, 1, 2, .., k$ based on the second phase sample. Find the bias and variance of the following estimators of σ_0^2 as

(a) $\hat{\sigma}_1^{*2} = \sum_{i=1}^{k} w_i \hat{r}_i s_i^{*2}$ where $\hat{r}_i = \dfrac{s_0^2}{s_i^2}, = 1, 2, ..., k$, $0 < w_i < 1$ and $\sum_{i=1}^{k} w_i = 1$,

(b) $\hat{\sigma}_2^{*2} = s_0^2\left[\dfrac{\sum_{i=1}^{k} w_i s_i^{*2}}{\sum_{i=1}^{k} w_i s_i^2}\right]$, (c) $\hat{\sigma}_3^{*2} = \prod_{i=1}^{k}\left(\hat{r}_i s_i^{*2}\right)^{w_i}$, (d) $\hat{\sigma}_4^{*2} = \left(\sum_{i=1}^{k}\dfrac{w_i}{\hat{r}_i s_i^{*2}}\right)^{-1}$,

(e) $\hat{\sigma}_5^{*2} = \sum_{i=1}^{k}\left(w_i \hat{r}_i s_i^{*2}\right) + w_{k+1}s_0^2$, where $\sum_{i=1}^{k+1} w_i = 1$.

Hint: Singh and Singh (2001).

Exercise 6.26. Consider a two-phase design in which in each phase the sampling scheme is as follows:

(i) The first phase sample s^* of m ($m < N$) units is drawn from the population Ω to observe two auxiliary variables x and z.

(ii) The second phase sample s of size n $(n < m)$ is drawn from s^* to observe $y, x,$ and z.

Consider the population mean $\bar{Z}$ of one auxiliary variable z is known.
Let

$$\bar{y} = n^{-1}\sum_{i=1}^{n} y_i, \quad \bar{x} = n^{-1}\sum_{i=1}^{n} x_i, \quad \bar{x}^* = m^{-1}\sum_{i=1}^{m} x_i, \quad \text{and} \quad \bar{z}^* = m^{-1}\sum_{i=1}^{m} z_i .$$

After decomposing the whole population Ω into three mutually exclusive domains as s, $r_2 = \bar{s} \cap s^*$ and $r_1 = (\Omega - s^*)$ of n, $(m-n)$ and $(N-m)$ units respectively, where $\bar{s} = \Omega - s$, then the population mean can be written as

$$\bar{Y} = \frac{1}{N}\left[\sum_{i \in s} y_i + \sum_{i \in r_2} y_i + \sum_{i \in r_1} y_i\right] = f\,\bar{y} + (f^* - f)\bar{Y}_2 + (1 - f^*)\bar{Y}_1$$

where $f = n/N$ and $f^* = m/N$.

Study the properties of the following estimator of the population mean, $\bar{Y}$, defined as

$$\hat{\bar{Y}} = f\,\bar{y} + (f^* - f)\,T_2 + (1 - f^*)\,T_1$$

under the following situations:

(a) $T_2 = \bar{y}\left(\frac{\bar{x}^*}{\bar{x}}\right)$ and $T_1 = \bar{y}\left(\frac{\bar{Z}}{\bar{z}^*}\right)$, (b) $T_2 = \bar{y}\left(\frac{\bar{x}^*}{\bar{x}}\right)$ and $T_1 = \bar{y}\left(\frac{\bar{x}^*}{\bar{x}}\right)\left(\frac{\bar{Z}}{\bar{z}^*}\right)$

Hint: Sahoo and Sahoo (2001).

Exercise 6.27. Consider the problem of estimation of population total Y in the presence of two auxiliary variables x and z. Consider the auxiliary information is available at three levels as follows:

(a) At the level of population Ω of size N: the vector total $\sum_{i \in \Omega} x_i$

(b) At the level of first phase sample s_1 of size m : x_i and z_i are known for $i \in s_1$

(c) At the level of second phase sample s_2 of size n : y_i, x_i and z_i are known for every $i \in s_2$;

Study the following calibrated estimator of the population total Y defined as

$$\hat{Y}_{ES} = \sum_{i \in s_2} w_{2i} y_i,$$

where w_{2i}, $i = 1, 2, ..., n$, are the second phase calibrated weights obtained under the following situations:

Situation Number	Sequence of calibration from first phase to second phase sample	Calibration constraints
1	Calibrate on x from s_1 to Ω to obtain w_{1i} then both auxiliary variables x and z from s_2 to s_1 to find w_{2i}.	$\sum_{s_1} w_{1i} x_i = \sum_{i \in \Omega} x_i$, $\sum_{s_2} w_{2i} x_i = \sum_{i \in s_1} w_{1i} x_i$ and $\sum_{s_2} w_{2i} z_i = \sum_{i \in s_1} w_{1i} z_i$

Continued......

2	Calibrate x from s_1 to Ω to obtain w_{1i} and then x from s_2 to s_1 to obtain w_{2i}.	$\sum_{s_1} w_{1i}x_i = \sum_{i\in\Omega} x_i$, $\sum_{s_2} w_{2i}x_i = \sum_{i\in s_1} w_{1i}x_i$ *Note:* it is a special case of the first situation.
3	Calibrate x from s_1 to Ω to obtain w_{1i} and the z from s_2 to s_1 to find w_{2i}.	$\sum_{s_1} w_{1i}x_i = \sum_{i\in\Omega} x_i$, $\sum_{s_2} w_{2i}z_i = \sum_{i\in s_1} w_{1i}z_i$ *Note:* it is a special case of the first situation.
4	Calibrate x from s_1 to Ω to obtain w_{1i}, and the $w_{2i} = d_{2i}w_{1i}$, where d_{2i} are the known design weights at the second phase sample.	$\sum_{s_1} w_{1i}x_i = \sum_{i\in\Omega} x_i$ *Note:* it is a special case of the first situation
5	Calibrate on x from s_2 to Ω and z from s_2 to s_1 to find direct w_{2i} in a single shot of calibration.	$\sum_{s_2} w_{2i}x_i = \sum_{i\in\Omega} x_i$, $\sum_{s_2} w_{2i}z_i = \sum_{i\in s_1} d_{1i}z_i$ where d_{1i} are the known first phase design weights.
6	Calibrate on x from s_2 to Ω to find w_{2i}.	$\sum_{s_2} w_{2i}x_i = \sum_{i\in\Omega} x_i$ *Note:* it is a special case of the fifth situation.
7	Calibrate both x and z from s_2 to s_1 to obtain w_{2i}	$\sum_{s_2} w_{2i}x_i = \sum_{i\in s_1} d_{1i}x_i$, $\sum_{s_2} w_{2i}z_i = \sum_{i\in s_1} d_{1i}z_i$
8	Calibrate on x from s_2 to s_1 to find w_{2i}.	$\sum_{s_2} w_{2i}x_i = \sum_{i\in s_1} d_{1i}x_i$ *Note*: it is a special case of the seventh situation.
9	Calibrate on z from s_2 to s_1 to find w_{2i}.	$\sum_{s_2} w_{2i}z_i = \sum_{i\in s_1} d_{1i}z_i$ *Note*: it is a special case of the seventh situation.
10	Calibrate from s_2 to s_1 to find w_{2i}.	$\sum_{s_2} w_{2i}f(x_i, z_i) = \sum_{i\in s_1} d_{1i}f(x_i, z_i)$ where $f(x_i, z_i)$ is any joint function of both auxiliary variables, $f(x_i, z_i) = x_i^{\alpha} z_i^{\beta}$, where α and β are know. *Note*: it is a special case of the seventh situation.

Hint: Estevao and Särndal (2002).

Exercise 6.28. (I) Consider a population is represented by $\Omega = \{1, 2, ..., i, ..., N\}$. A first phase probability sample s_1 $(s_1 \subset \Omega)$ is drawn from the population Ω using a sampling design that generates the selection probabilities π_{1i}. Given that the first phase sample s_1 has been drawn, the second phase sample s_2 $(s_2 \subset s_1 \subset \Omega)$ is selected from s_1 with a sampling design with the selection probabilities $\pi_{2i} = \pi_{i|s_1}$. Evidently the first phase and second phase sampling weights are defined as $d_{1i} = 1/\pi_{1i}$ and $d_{2i} = 1/\pi_{2i}$, respectively. The overall sampling weights for the selected i^{th} unit in the second phase sample s_2 will be $d_i^* = d_{1i}d_{2i}$. Considered the problem of estimation of general parameters of interest as,

$$H_y = \sum_{i \in \Omega} H(y_i) \quad \text{and} \quad \overline{H}_y = N^{-1} \sum_{i \in \Omega} H(y_i)$$

for a specified function h.

(a) Study the asymptotic properties the estimator of H_y in two phase sampling, defined as

$$\hat{H}_1 = \sum_{i=1}^{n} \tilde{d}_i^* H(y_i),$$

where $\tilde{d}_i^*$ are the ultimate calibrated weights obtained by minimizing chi square distance between second phase design weights.

(b) Show that the chain ratio type estimator of population parameter H_y is

$$\hat{H}_R = \sum_{i=1}^{n} d_{1i}d_{2i}H(y_i)\left(\frac{\sum_{i=1}^{m} d_{1i}H(x_i)}{\sum_{i=1}^{n} d_{1i}d_{2i}H(x_i)}\right)\left(\frac{H_z}{\sum_{i=1}^{m} d_{1i}H(z_i)}\right)$$

and the chain regression type of estimator is

$$\hat{H}_G = \sum_{i=1}^{n} d_{1i}d_{2i}H(y_i) + \hat{\beta}_1\left[\sum_{i=1}^{m} d_{1i}H(x_i) - \sum_{i=1}^{n} d_{1i}d_{2i}H(x_i)\right] + \hat{\beta}_2\left[H_z - \sum_{i=1}^{m} d_{1i}H(z_i)\right]$$

where

$$\hat{\beta}_1 = \sum_{i=1}^{n} d_{1i}d_{2i}H(x_i)H(y_i) \Big/ \sum_{i=1}^{n} d_{1i}d_{2i}\{H(z_i)\}^2 \quad \text{and} \quad \hat{\beta}_2 = \hat{\beta}_1\left[\sum_{i=1}^{m} d_{1i}H(x_i)h(z_i) \Big/ \sum_{i=1}^{m} d_{1i}\{H(z_i)\}^2\right]$$

are the special cases of the calibrated estimator.

Hint: Singh and Puetas (2002).

(II) Consider $\hat{H}_y^c(j)$ be a calibrated estimator of population distribution function obtained by dropping j^{th} units from the sample s_1 of m units by following Singh and Puetas (2002). Then the Jackknifed estimator of population distribution function in two-phase sampling can be written as

$$\hat{H}_y{}^c(j) = \begin{cases} \hat{H}^o_{2y}(j) + \hat{\beta}_2(j)\{\hat{H}^o_{1x}(j) - \hat{H}^o_{2x}(j)\} + \hat{\beta}_1(j)\hat{\beta}_2(j)\{H_z - \hat{H}^o_{1z}(j)\} & \text{if } j \in s_2, \\ \hat{H}^o_{2y} + \hat{\beta}_2\{\hat{H}^o_{1x}(j) - \hat{H}^o_{2x}\} + \hat{\beta}_1(j)\hat{\beta}_2\{H_z - \hat{H}^o_1(j)\} & \text{if } j \in (s_1 - s_2), \end{cases}$$

where s_2 be the second phase sample of n units,

$$\hat{H}^o_{1z}(j) = \hat{H}^o_1 + w^o_{1j}\left(1 - w^o_{1j}\right)^{-1}\left(\hat{H}^o_1 - H(z_j)\right), \hat{H}^o_{1x}(j) = \hat{H}^o_{1x} + w^o_{1j}\left(1 - w^o_{1j}\right)^{-1}\left(\hat{H}^o_{1x} - H(x_j)\right),$$

$$\hat{H}^o_{2x}(j) = \hat{H}^o_{2x} + w^o_{2j}\left(1 - w^o_{2j}\right)^{-1}\left(\hat{H}^o_{2x} - H(x_j)\right), \hat{H}^o_{2y}(j) = \hat{H}^o_{2y} + w^o_{2j}\left(1 - w^o_{2j}\right)^{-1}\left(\hat{H}^o_{2y} - H(y_j)\right),$$

$\hat{\beta}_1$, $\hat{\beta}_1(j)$, $\hat{\beta}_2$, and $\hat{\beta}_2(j)$ have their usual meanings.

(a) Show that

$$\hat{H}^c_y(j) - \hat{H}^c_y = \begin{cases} \in_2(j) + \hat{\beta}_2 \in_1(j) + \hat{\beta}_2(j)d_2(j) + \hat{\beta}_2\delta_2(j) & \text{if } \ j \in s_2, \\ \hat{\beta}_2 \in_1(j) & \text{if } \ j \in (s_1 - s_2), \end{cases}$$

where

$$\in_2(j) = \left(\hat{H}^o_{2y}(j) - \hat{H}^o_{2y}\right) - \hat{\beta}_2(j)\left(\hat{H}^o_{2x}(j) - \hat{H}^o_{2x}\right) - \hat{\beta}_1(j)\hat{\beta}_2(j)\left(\hat{H}^o_{1z}(j) - H_z\right),$$

$$\in_1(j) = \left(\hat{H}^o_{1x}(j) - \hat{H}^o_{1x}\right) - \hat{\beta}_1(j)\left(\hat{H}^o_{1z}(j) - H_z\right), \quad d_2(j) = \left(\hat{H}^o_{1x}(j) - \hat{H}^o_{2x}(j)\right),$$

and

$$\delta_2(j) = \left(\hat{H}^o_{2x}(j) - \hat{H}^o_{1x}(j)\right) - \hat{\beta}_1(j)\left(H_z - \hat{H}^o_{1z}(j)\right) - \hat{\beta}_1\left(H_z - \hat{H}^o_{1z}\right).$$

(b) Show that the modified Jackknife estimator of variance of $\hat{H}^c_y$ given by

$$\hat{v}_J\left(\hat{H}^c_y\right) = \frac{(m-1)}{m} \sum_{j \in s_1} \left\{\hat{H}^c_y(j) - \hat{H}^c_y\right\}^2$$

is approximately equivalent to the variance of the chain ratio and regression estimators.

Hint: Singh (2003b).

Exercise 6.29. Under the concept of two-phase sampling compare the mean squared error of the ratio estimator $\hat{M}_R$ defined as

$$\hat{M}_R = \hat{M}_y\left(\frac{\hat{M}^*_x}{\hat{M}_x}\right),$$

with the following estimator for population median M_y defined as

$$\hat{M}^{(a)}_y = \hat{M}_y\left(\frac{A - \hat{M}^*_x}{A - \hat{M}_x}\right),$$

where A is a suitably chosen real constant.

Hint: Singh, Singh, and Puertas (2003a).

PRACTICAL PROBLEMS

Practical 6.1. Select a first phase sample of 15 units by SRSWOR sampling from population 1 of the Appendix and record only the nonreal estate farm loans from the units selected in the preliminary sample. Select a sub-sample of 5 units from the preliminary large sample and note the real and nonreal estate farm loans. Estimate the average real estate farm loans by using the ratio estimator. Construct the 95% confidence intervals by estimating the variance of the ratio estimator with two different estimators: (a) Jackknifing; (b) method of moments.

Practical 6.2. A key bank in the United States of America is interested in the average of real estate farm loans. The bank manager has information about nonreal estate farm loans in 15 states selected by SRSWOR sampling. From these selected states select a sub-sample of 5 states and note the real estate farm loans as well as nonreal estate farm loans from population 1 given in the Appendix. Use the following estimators to estimate the average real estate farm loans

$$\bar{y}_1 = \bar{y}\left(\bar{x}^*/\bar{x}\right)\left(s_x^{*2}/s_x^2\right) \text{ and } \bar{y}_2 = \bar{y} + \alpha\left(\bar{x}^* - \bar{x}\right) + \beta\left(s_x^{*2} - s_x^2\right).$$

Suggest estimators for optimum values of α and β. Construct the 95% confidence interval in each case. Explain the difference in the estimates based on these two estimators to the bank manager.

Practical 6.3. A private consultant selected first phase and second phase samples of sizes 20 and 10 respectively. Discuss the relative efficiency of the general class of estimators for estimating average amount of the real estate farm loans during 1997 by using information selected in the first phase sample only on the nonreal estate farm loans during 1997, with respect to the regression estimator of population mean.

Practical 6.4. People Bank has information about the real and nonreal estate farm loans (in \$000) during 1997 in the United States has been presented in population 1. If the bank select first phase and second phase samples of size 10 and 5 respectively, then:

(a) Find the relative efficiency of the ratio estimator, for estimating the average amount of the real estate farm loans during 1997 by using information selected in the first phase sample only on the nonreal estate farm loans during 1997, with respect to the usual estimator of population mean;
(b) Suppose a budget of US\$5000 is available to spend on the survey, out of which, \$2000 will be the overhead cost. Suppose selection, compilation, and analysing of one unit in the first phase sample costs \$50, while for the second phase unit the cost is \$500. Find the optimum values of the first phase and second phase sample sizes. Also find the relative efficiency of the ratio estimator over the sample mean for the fixed cost;
(c) What will be the minimum cost for attaining a 20% relative standard deviation?

Practical 6.5. A private company XYZ selected first phase and second phase samples of size 20 and 10, respectively, from population 1 given in the Appendix. Find the relative efficiency of the regression estimator, for estimating average amount of real estate farm loans during 1997 by using information selected in the first phase sample only on the nonreal estate farm loans during 1997, with respect to the ratio estimator of population mean.

Practical 6.6. Mr. Nelson wishes to make a future strategy of selection of estimator while estimating the average real estate farm loans. Suppose he selected first phase and second phase samples each of size 10 and 5 respectively. Suggest a few estimators as a member of the general class of estimators. Find the relative efficiency of these members for estimating average amount of real estate farm loans during 1997 by using information selected in the first phase sample only on the nonreal estate farm loans during 1997 with respect to the regression estimator in two-phase sampling of the population mean. Has Mr. Nelson any hope in finding a new estimator?

Practical 6.7. Ms. Stephanie Singh selects a preliminary large sample of 20 units by PPSWOR sampling using the Midzuno--Sen sampling scheme and using the number of species groups during 1992 as an auxiliary variable and given in population 4 of the Appendix. Select a second-phase sample of 10 units from the given first phase sample by using the Midzuno--Sen sampling scheme. Find the calibration weights for the units selected in the ultimate sample by making use of known information about the number of fish caught during 1994 as an auxiliary variable. Use the chi square distance function between the design weights and calibration weights. Discuss three cases when these weights leads to the GREG, ratio, and traditional linear regression estimator for estimating the total number of fish caught during 1995. Deduce the estimates of the total number of fish in each case.

Practical 6.8. Use the Midzuno--Sen sampling scheme to select a preliminary large sample of 15 units by using the number of fish caught during 1992 in the United States as a selection variable given in the population 4. Collect the information on the number of fish caught during 1993 and 1994 from the units selected in the sample. Assume that the total number of fish caught during 1993 are known, derive the first phase calibration weights and hence estimate the total number of fish caught during 1994 in the United States. Select a second phase sample of 10 units from the given first phase sample by using Midzuno--Sen sampling scheme. Collect the information on the number of fish caught during 1994 and 1995 for the selected units in the second phase sample. Derive the second phase calibration weights, and hence deduce the estimate of the number of fish caught during 1995 in the United States.

Practical 6.9. Take a first phase sample of 15 units by SRSWOR sampling and note only the nonreal estate farm loans from the units selected in the sample given in population 1 of the Appendix. Select a sub-sample of 10 units from the given preliminary large sample and note the real estate farm loans as well as nonreal estate farm loans. Estimate the average real estate farm loans by using regression estimator. Deduce the 95% confidence interval.

Practical 6.10. The real and nonreal estate farm loans (in $000) during 1997 in different 50 states of the United States have been presented in population 1 of the Appendix. Suppose we selected an SRSWOR first phase sample of ten states to collect the information on nonreal estate farm loans only. From the given first phase sample of ten units, we selected a second phase sample of seven units and both the real and nonreal estate farm loans are observed. Find the relative efficiency of the ratio estimator of the median, for estimating median of the amount of the real estate farm loans during 1997 by using information of the nonreal estate farm loans during 1997 collected in the first phase sample with respect to the usual estimator of population median. Assume that both the real and nonreal estate farm loans follows a bivariate normal distribution.

Hint: The joint p.d.f. for the bivariate normal distribution is given by

$$f(x,y)=\frac{\exp\left\{-\dfrac{1}{2\left(1-\rho_{xy}^2\right)}\left[\left(\dfrac{x-\mu_x}{\sigma_x}\right)^2+\left(\dfrac{y-\mu_y}{\sigma_y}\right)^2-2\rho_{xy}\left(\dfrac{x-\mu_x}{\sigma_x}\right)\left(\dfrac{y-\mu_y}{\sigma_y}\right)\right]\right\}}{\sqrt{2\pi}\sigma_x\sigma_y\sqrt{\left(1-\rho_{xy}^2\right)}}.$$

Advanced Sampling Theory with Applications

Advanced Sampling Theory with Applications

How Michael 'selected' Amy
Volume II

by

Sarjinder Singh
St. Cloud State University,
Department of Statistics,
St. Cloud, MN, U.S.A.

KLUWER ACADEMIC PUBLISHERS
DORDRECHT / BOSTON / LONDON

A C.I.P. Catalogue record for this book is available from the Library of Congress.

ISBN Vol. 1: 1-4020-1706-5
ISBN Vol. 2: 1-4020-1707-3
ISBN set: 1-4020-1689-1

Published by Kluwer Academic Publishers,
P.O. Box 17, 3300 AA Dordrecht, The Netherlands.

Sold and distributed in North, Central and South America
by Kluwer Academic Publishers,
101 Philip Drive, Norwell, MA 02061, U.S.A.

In all other countries, sold and distributed
by Kluwer Academic Publishers,
P.O. Box 322, 3300 AH Dordrecht, The Netherlands.

Printed on acid-free paper

Printed in the Netherlands.

Dedicated to

GOD

with whose grace...

TABLE OF CONTENTS

2 SIMPLE RANDOM SAMPLING

3 USE OF AUXILIARY INFORMATION: SIMPLE RANDOM SAMPLING

4 USE OF AUXILIARY INFORMATION: PROBABILITY PROPORTIONAL TO SIZE AND WITH REPLACEMENT (PPSWR) SAMPLING

5 USE OF AUXILIARY INFORMATION: PROBABILITY PROPORTIONAL TO SIZE AND WITHOUT REPLACEMENT (PPSWOR) SAMPLING

6 USE OF AUXILIARY INFORMATION: MULTI-PHASE SAMPLING

VOLUME II

7 SYSTEMATIC SAMPLING

8 STRATIFIED AND POST-STRATIFIED SAMPLING

9 NON-OVERLAPPING, OVERLAPPING, POST, AND ADAPTIVE CLUSTER SAMPLING

10 MULTI-STAGE, SUCCESSIVE, AND RE-SAMPLING STRATEGIES

11 RANDOMIZED RESPONSE SAMPLING: TOOLS FOR SOCIAL SURVEYS

12 NON-RESPONSE AND ITS TREATMENTS

13 MISCELLANEOUS TOPICS

APPENDIX

TABLES

POPULATIONS

7. SYSTEMATIC SAMPLING

7.0 INTRODUCTION

Sampling scheme in which only the first unit is selected at random, the rest being automatically selected according to a predetermined pattern is known as systematic sampling. Systematic sampling provides a very simple sampling design in practice to select a sample of size n from a population of size N. Systematic sampling is both operationally convenient and efficient in sampling some natural populations like forest areas for estimating the volume of timber and hardwood seedlings, etc..

7.1 SYSTEMATIC SAMPLING

There are two possibilities:

(I) When $N = nk$ and k is an integer **(II)** When $N \neq nk$.

Let us discuss each of these cases in detail.

Case I. If $N = nk$. The N units in the population can be arranged in n rows as shown in Table 7.1.1 and can be named as a sequential list of population units. Such a list can be prepared only if we have a finite number of units in the population.

Table 7.1.1. Sequential list of population units.

Row 1	1	2	3		r		k
Row 2	$k+1$	$k+2$	$k+3$		$k+r$		$2k$
Row 3	$2k+1$	$2k+2$	$2k+3$		$2k+r$		$3k$
•	•	•	•		•		•
Row n	$(n-1)k+1$	$(n-1)k+2$	$(n-1)k+3$		$(n-1)k+r$		$nk = N$
Means	$\bar{y}_1$	$\bar{y}_2$	$\bar{y}_3$		$\bar{y}_r$		$\bar{y}_k$

The first step is to select a random number from 1 to k, that is in the range of integers listed in the first row. Let the first selected random number is 2. Then first unit selected in the sample is number 2 in the sequential list. After selecting the second unit from the population, every k^{th} unit is automatically included in the sample. Thus the units in the sample of size n are at the serial numbers

$$2,\ k+2,\ 2k+2,\ \ldots,\ (n-1)k+2\,.$$

The random number selected from 1 to k is called a *random start*. The number k is called sampling interval. Corresponding to each random number from 1 to k, there is only one possible sample of size n. Thus in systematic sampling the total number of samples will be k. If r denotes the random start, then the systematic sample consists of the units at the serial numbers given by the sequence

$$\{r + ik,\ i = 0,1,2.....,(n-1)\}\ .$$

The sequential list of population units can either be made using known magnitude of auxiliary information or by just numbering the list of population units. Assume $\bar{y}_r$ denotes the sample mean corresponding to random start ' r '.

Then

$$\bar{y}_r = \frac{1}{n}\sum_{i=0}^{n-1} y_{r+ik} . \tag{7.1.1}$$

Thus we have following theorem.

Theorem 7.1.1. The sample mean $\bar{y}_r$ is an unbiased estimator of the population mean $\bar{Y}$.

Proof. Note that we have k possible samples in the whole population, thus the probability of selecting one sample is given by $1/k$. By the definition of expected value we have

$$E(\bar{y}_r) = \frac{1}{k}\sum_{r=1}^{k}\bar{y}_r = \frac{1}{k}\sum_{r=1}^{k}\left(\frac{1}{n}\sum_{i=0}^{n-1} y_{r+ik}\right) = \frac{1}{nk}\sum_{r=1}^{k}\sum_{i=1}^{n-1} y_{r+ik} = \frac{1}{N}\sum_{i=1}^{N} Y_i = \bar{Y} . \tag{7.1.2}$$

Hence the theorem.

Theorem 7.1.2. The variance between k sample means is given by

$$V(\bar{y}_r) = \sigma_b^2 = \frac{1}{k}\sum_{r=1}^{k}(\bar{y}_r - \bar{Y})^2 . \tag{7.1.3}$$

Proof. By the definition of variance we have

$$V(\bar{y}_r) = E(\bar{y}_r)^2 - \{E(\bar{y}_r)\}^2 = \frac{1}{k}\sum_{r=1}^{k}\bar{y}_r^2 - \bar{Y}^2 = \frac{1}{k}\sum_{r=1}^{k}(\bar{y}_r - \bar{Y})^2 .$$

Hence the theorem.

From (7.1.3) one can easily calculate the variance between the sample means if all possible sample means and population mean are known. Unfortunately we generally have the knowledge of only one sample mean and only one sample mean cannot be used to estimate the variance between all possible sample means. In (7.1.3) σ_b^2 stands for the variance between all possible samples selected by using systematic sampling. Till now we have discussed the variation between different column means of the Table 7.1.1. A natural question arises that there will be variation within units of each column. Such a variation is called within sample variation. Suppose y_{ri} denotes the ith ($i = 1,2,3,...,n$) observation in the rth sample. Then the rth sample mean can also be defined as

$$\bar{y}_r = \frac{1}{n}\sum_{i=1}^{n} y_{ri} . \tag{7.1.4}$$

The variation within the rth sample can be defined as

$$\sigma_r^2 = \frac{1}{n}\sum_{i=1}^{n}(y_{ri} - \bar{y}_r)^2 . \tag{7.1.5}$$

Thus the overall measure of within sample variation can be defined as the average of these values across all k samples as

$$\sigma_w^2 = \frac{1}{k}\sum_{r=1}^{k}\sigma_r^2 = \frac{1}{k}\sum_{r=1}^{k}\frac{1}{n}\sum_{i=1}^{n}(y_{ri}-\bar{y}_r)^2 = \frac{1}{N}\sum_{i=1}^{n}\sum_{r=1}^{k}(y_{ri}-\bar{y}_r)^2. \tag{7.1.6}$$

Now we have the following theorem.

Theorem 7.1.3. The total variance of the population is the sum of between and within sample variances, that is

$$\sigma^2 = \sigma_b^2 + \sigma_w^2. \tag{7.1.7}$$

Proof. Note that

$$\sum_{r=1}^{k}\sum_{i=1}^{n}(y_{r,i}-\bar{y}_r)(\bar{y}_r-\bar{Y}) = 0.$$

Now we know that

$$N\sigma^2 = \sum_{i=1}^{n}\sum_{r=1}^{k}(y_{ri}-\bar{Y})^2 = \sum_{i=1}^{n}\sum_{r=1}^{k}(y_{ri}-\bar{y}_r+\bar{y}_r-\bar{Y})^2$$

$$= \sum_{i=1}^{n}\sum_{r=1}^{k}(y_{ir}-\bar{y}_r)^2 + 2\sum_{i=1}^{n}\sum_{r=1}^{k}(y_{ir}-\bar{y}_r)(\bar{y}_r-\bar{Y}) + n\sum_{r=1}^{k}(\bar{y}_r-\bar{Y})^2$$

$$= N\sigma_w^2 + N\sigma_b^2.$$

Hence the theorem.

Remark 7.1.1. In systematic sampling we have

within sample variation $\sigma_w^2 = \dfrac{1}{N}\sum_{i=1}^{n}\sum_{r=1}^{k}(y_{r,i}-\bar{y}_r)^2$, (7.1.8)

between sample variation $\sigma_b^2 = \dfrac{1}{N}\sum_{r=1}^{k}(\bar{y}_r-\bar{Y})^2$, (7.1.9)

total variation $\sigma^2 = \dfrac{1}{N}\sum_{i=1}^{n}\sum_{r=1}^{k}(y_{r,i}-\bar{Y})^2$, (7.1.10)

and

$$\sigma^2 = \sigma_w^2 + \sigma_b^2. \tag{7.1.11}$$

We now discuss the situations under which systematic sampling is better than simple random sampling without replacement. The variance of the estimator of population mean under systematic sampling is

$$V(\bar{y}_{sy}) = \sigma_b^2 = \sigma^2 - \sigma_w^2 \tag{7.1.12}$$

and that under SRSWOR is

$$V(\bar{y}_{srs}) = \frac{(N-n)}{n(N-1)}\sigma^2. \tag{7.1.13}$$

The design effect (DE) or relative efficiency of SRSWOR with respect to systematic sampling is defined as

$$\mathrm{DE} = \frac{V(\bar{y}_{\text{sy}})}{V(\bar{y}_{\text{srs}})} = \frac{n(N-1)}{(N-n)}\left(1-\frac{\sigma_w^2}{\sigma^2}\right). \tag{7.1.14}$$

Now if DE= 1 then

$$\sigma_w^2 = \frac{\left(1-\frac{1}{n}\right)}{\left(1-\frac{1}{N}\right)}\sigma^2. \tag{7.1.15}$$

Evidently if DE< 1, *i.e.*,

$$\sigma_w^2 > \left(1-\frac{1}{n}\right)\left(1-\frac{1}{N}\right)^{-1}\sigma^2$$

then systematic sampling for a sample of size n remains better than SRSWOR sample of the same size. The greater the variation σ_w^2 within samples, greater will be the gain due to systematic sampling. It is obvious that if σ_w^2 will increase then σ_b^2 will decrease, since the total variation σ^2 remains fixed. Thus for large gain in efficiency due to systematic sampling, the units within each systematic sample should be as heterogeneous as possible.

Case II. When $N \neq nk$, *i.e.*, it is not possible to find k such that $k = N/n$ is an integer. For example when $N = 5$ and $n = 2$, $k = N/n = 5/2 = 2.5$ is not an integer. In such situations, we can either take $k = 2$ or 3. Suppose the population consists of following units.

Table 7.1.2. Sequential list of population units.

Serial No.	1	2	3	4	5
Value of Y	2	3	2	5	6

Evidently the population mean $\bar{Y}$ is given by

$$\bar{Y} = \frac{1}{N}\sum_{i=1}^{N} Y_i = \frac{18}{5} = 3.6\,.$$

Now there are following possibilities:

(a) If we take $k = 2$ then we have to select a random number between 1 and 2 and every second unit will be included in the sample. Consider the random start is $r = 1$. Then the units selected will be first, third and fifth or in other words, the sample values will be $\{2, 2, 6\}$. Thus the first sample mean

$$\bar{y}_1 = \frac{2+2+6}{3} = 3.33\,.$$

If the selected random start $r = 2$ then the units at serial numbers 2 and 4 will be selected. The resultant sample will be $\{3, 5\}$. The mean of the second sample will be

$$\bar{y}_2 = \frac{3+5}{2} = 4.$$

While discussing the above example we observe the following difficulties:

(i) We had a sample of 3 units, while we were interested in a sample of size 2.
(ii) There are only two possible sample means

$$\bar{y}_1 = \frac{10}{3} \text{ and } \bar{y}_2 = 4.$$

The mean of these sample means is given by

$$\frac{1}{2}\left(\frac{10}{3} + 4\right) = 3.667$$

which is not equal to the true population mean $\bar{Y} = 3.6$.
(b) If we take $k = 3$, then we have to select a random start r between 1 and 3. Then we have the following possibilities:

(*i*) If $r = 1$ then the sample is $\{2, 5\}$ with mean $\bar{y}_1 = \dfrac{2+5}{2} = 3.5,$

(*ii*) If $r = 2$ then the sample is $\{3, 6\}$ with mean $\bar{y}_2 = \dfrac{3+6}{2} = 4.5,$

(*iii*) If $r = 3$ then the sample is $\{2\}$ with mean $\bar{y}_3 = \dfrac{2}{1} = 2.0.$

Again we have the mean of three sample means as

$$\text{Mean} = \frac{3.5 + 4.5 + 2.0}{3} = 3.3333 \neq 3.6 = \bar{Y}.$$

Thus we have the following corollaries for Case II.

Corollary 7.1.1. The sample size is a random variable, that is the sample size is not fixed.
Corollary 7.1.2. The sample mean is not an unbiased estimator of the population mean.

Theorem 7.1.4. Let y_r denote the total corresponding to random start r and let units be taken in the sample skipping k of them. Then the estimator

$$\bar{y}_u = \frac{k}{N} y_r \tag{7.1.16}$$

is unbiased for the population mean, $\bar{Y}$.
Proof. We have

$$E(\bar{y}_u) = E\left(\frac{k}{N} y_r\right) = \frac{1}{k}\sum_{r=1}^{k} \frac{k}{N} y_r = \frac{1}{N}\sum_{r=1}^{k} y_r = \bar{Y}. \tag{7.1.17}$$

Hence the theorem.

The above theorem can also be justified with the help of following example.

Example 7.1.1. Case I. If $k=2$, then the possible samples are $\{2,2,6\}$ and $\{3,5\}$. Thus $y_1 = 2+2+6 = 10$ and $y_2 = 3+5 = 8$. Now the mean of $\bar{y}_u = \frac{k}{N} y_r$, $r = 1,2$, will be

$$E(\bar{y}_u) = \frac{1}{2}\sum_{r=1}^{2}\frac{k}{N}y_r = \frac{1}{2}\left(\frac{2\times 10}{5}+\frac{2\times 8}{5}\right) = \frac{18}{5} = 3.6 = \bar{Y}.$$

Case II. If $k=3$ then possible samples are $\{2,\ 5\}$, $\{3,\ 6\}$ and $\{2\}$. Obviously $y_1 = 2+5 = 7$, $y_2 = 3+6 = 9$, and $y_3 = 2$. Again mean value of $\bar{y}_u = \frac{k}{N} y_r$, $r = 1,2,3$, will be

$$E(\bar{y}_u) = \frac{1}{3}\sum_{r=1}^{3}\frac{k}{N}y_r = \frac{1}{3}\left(\frac{3\times 7}{5}+\frac{3\times 9}{5}+\frac{3\times 2}{5}\right) = \frac{1}{3}\times\frac{54}{5} = \frac{18}{5} = 3.6 = \bar{Y}.$$

The sample means as such are not unbiased for population mean. A sampling scheme providing unbiased estimators is called *Modified Systematic Sampling.*

7.2 MODIFIED SYSTEMATIC SAMPLING

In case of modified systematic sampling, instead of selecting a random start 'r' from 1 to k, we have to select a random number from 1 to N, then every k^{th} unit on the right and left of it is included in the sample. Thus, for example, if $k=2$ then the Table 7.2.1 shows the possible samples for population given in Table 7.1.2 for different random starts.

Table 7.2.1. Possible samples using modified systematic sampling.

Random Start 'r'	Serial number of sampled units	Sampled values	Sample means
1	1, 3, 5	2, 2, 6	10/3
2	2, 4	3, 5	4
3	1, 3, 5	2, 2, 6	10/3
4	4, 2	5, 3	4
5	5, 3, 1	6, 2, 2	10/3

Evidently the expected value of sample means is

$$E(\bar{y}_{ms}) = \frac{1}{5}\left[\frac{10}{3}+4+\frac{10}{3}+4+\frac{10}{3}\right] = 3.6 = \bar{Y}.$$

Hence under modified systematic sampling, the sample mean $\bar{y}_{ms}$ is an unbiased estimator of the population mean $\bar{Y}$, but still the sample size is a random variable as shown in Table 7.2.1.

This difficulty can also be removed by following *Circular Systematic Sampling* scheme as discussed in the next section.

7.3 CIRCULAR SYSTEMATIC SAMPLING

Murthy (1961), Sukhatme and Sukhatme (1970), and Konijn (1973) have suggested using the circular systematic sampling (CSS) design in the situations when N is not a multiple of n. In this sampling scheme, the sequential list of the population units is first prepared on the circle as shown in Figure 7.3.1.

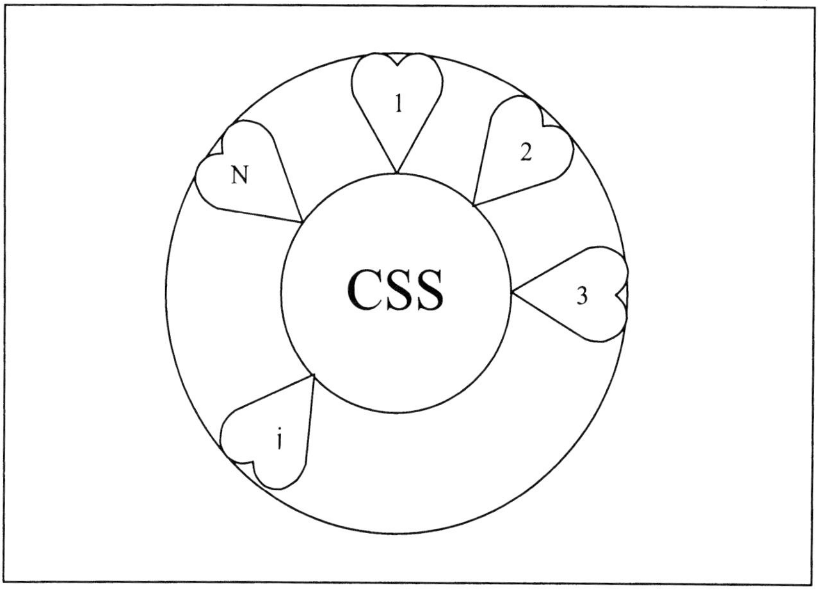

Fig. 7.3.1 Circular Systematic Sampling (CSS).

The main steps involved in selecting a sample using CSS scheme are as follows:

(a) Select a random number from 1 to N and name it as 'random start';

(b) Chose some integer value of $k = N/n$ or rounded to nearest integer and name it as skip;

(c) Select all units in the sample with serial numbers

$$r + jk \quad \text{if} \quad r + jk \le N,$$
$$(r + jk - N) \quad \text{if} \quad r + jk > N\ ; \quad j = 0, 1, 2, ..., (n-1).$$

Example 7.3.1. Suppose a population consists of 10 ministers sitting on a round table. We wish to select a sample of 3 ministers, to study corruption level in their country, using CSS scheme with skip $k = 10/3 \approx 3$. The sequential list of 10 ministers sitting on the round table can be prepared as shown in Figure 7.3.2. Now if the random start is $r = 3$, then the ministers included in the sample would be with serial numbers, $3 + 0 \times 3 = 3$, $3 + 1 \times 3 = 6$, and $3 + 2 \times 3 = 9$. Similarly, if $r = 8$ then the ministers in the sample would be with serial numbers, $8 + 0 \times 3 = 8$, $8 + 1 \times 3 - 10 = 1$, and $8 + 2 \times 3 - 10 = 4$.

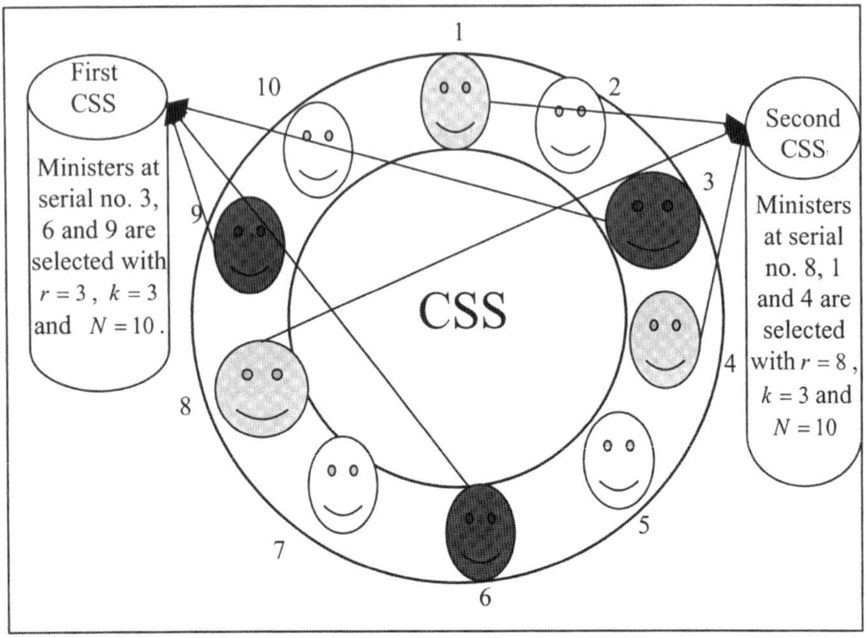

Fig.7.3.2 Pictorial representation of two systematic samples each of size 3 from a population of 10 units with two different random starts, $r=3$ and $r=8$.

Now we have the following theorem.

Theorem 7.3.1. Under the a CSS scheme an unbiased estimator of population mean $\bar{Y}$ is given by

$$\bar{y}_r = \frac{1}{n}\sum_{j=0}^{n-1} y_{r+jk} \quad . \tag{7.3.1}$$

Proof. Taking expected values on both sides of (7.3.1) we have

$$E(\bar{y}_r) = E\left[\frac{1}{n}\sum_{j=0}^{n-1} y_{r+jk}\right] = \frac{1}{Nn}\sum_{r=1}^{N}\left(\sum_{j=0}^{n-1} y_{r+jk}\right) = \frac{1}{Nn}\times n\times\sum_{i=1}^{N} Y_i = \frac{1}{N}\sum_{i=1}^{N} Y_i = \bar{Y}.$$

Hence the theorem.

Difficulty in CSS scheme: Sudakar (1978) pointed out that the use of the skip or span of sampling as the integer nearest to N/n in CSS does not always draw a sample of desired size as shown in the following example.

Example 7.3.2. Let $N=15$, $n=6$, $k=3$ and the starting point be $r=3$. The sample in this case has only five (instead of six) distinct units with serial numbers 3, 6, 9, 12, 15.

Sudakar (1978) has mentioned that if we take the span of sampling as the largest integer smaller than or equal to N/n, then we do not encounter the above difficulty, although this choice depends upon n.

Thus we have the following theorem.

Theorem 7.3.2. A necessary and sufficient condition for all units of the sample of size n selected by circular systematic sampling with random start r to be distinct for all $r \le N$ and $n \le N$, is that N and k are relatively coprime.

Proof. Let N and k be integers with $k < N$, $r \le N$ and $n < N$. Also let the sample s be consist of with serial numbers $s = \{i_1, i_2, ..., i_n\}$ where $i_j = \{r + jk\}\text{mod}(N)$, $j = 0, 1, ..., (n-1)$.

Then the necessary and sufficient (n.s.) conditions are proved as follows:

(a) Sufficiency: Assume N and k are relatively coprime and there exists a random start r and sample size n such that two elements of the sample are same. Without loss of generality, suppose that i_1 and i_{j+1} are same. Now $i_1 = r$ and $i_{j+1} = [r + jk]\text{mod}(N)$. Thus $i_1 = i_{j+1}$ implies that jk is a multiple of N where $j < n < N$. This contradicts that k and N are coprimes. Hence the first part of theorem.

(b) Necessity: Assume for all $r \le N$ and $n < N$, all elements of the sample are distinct and N and k are not coprimes. Let the greatest common divisor (g.c.d.) of $(k, N) = a$, with $k = b.a - N = c.a$, where b and c are both smaller than N. For any random start r, let us take $n \ge c+1$.

Then we have

$$i_{c+1} = [r + ck]\text{mod}(N) = [c.b.a + r]\text{mod}(N) = [b.N + r]\text{mod}(N) = r = i_1,$$

which again contradicts our assumption that all elements of the sample s are distinct. Hence the theorem.

7.4 PPS CIRCULAR SYSTEMATIC SAMPLING

Consider a finite population of N units with positive integer size-measure X_i and cumulative totals $C_i = X_1 + X_2 + ... + X_i$ $(i = 1, 2, ..., N,\ C_0 = 0,\ C_N = X)$. Then we have the following definitions.

Definition 7.4.1. A PPS circular systematic sample (CSS) of size n is drawn by selecting a random number r between 1 and X and choosing the units i $(i = 1, 2, ..., N)$ for which $C_{i-1} < r + jk\text{mod}(X) \le C_i$, $j = 0, 1, 2, ..., (n-1)$.

Then we have the following theorem.

Theorem 7.4.1. A necessary condition for a sample to be selected according to the definition 7.4.1 to contain all distinct units is that

$$n \le \min\{I, [X/M]\} = n_0 \quad \text{(say)}, \tag{7.4.1}$$

where $M = \underset{1 \le i \le N}{\text{Max}}\, X_i$, $[x]$ = the largest integer contained in x and I is the smallest positive integer for which $(I \times k)/X$ is an integer.

Proof. Follows from Chaudhuri and Adhikary (1987).

Sengupta (1988) pointed out that condition (7.4.1) is not sufficient and resultant sample will not contain all distinct units even if (7.4.1) holds. He provided the following example.

Example 7.4.1. Suppose $N = 10$, $X = 300$, $M = 65$ and $[X/M] = 4$. The data is given in the following table:

Table 7.4.1. Difficulty in PPS Circular Systematic Sampling.

i	1	2	3	4	5	6	7	8	9	10
X_i	12	14	24	65	13	16	29	31	46	50
C_i	12	26	50	115	128	144	173	204	250	300

Let $k = 120$, $n = 3$. Here $I = 5$, $n_0 = 4$, so that (7.4.1) holds. But for $r = 112$ the sample contains only two distinct units 4, 9. Furthermore Sengupta (1988) gave the following theorem.

Theorem 7.4.2. Let $M = \underset{1 \le i \le N}{\text{Max}}(X_i)$. Then a necessary and sufficient condition for a PPS Circular Systematic Sample of size n with sampling interval k to always contain all distinct units is that, $n \le n_1$, where n_1 is the smallest positive integer j for which

$$jk \bmod(X) < M \text{ or } > (X - M).$$

Brewer (1963a) suggested a method of selecting a systematic sample with unequal probability sampling. Some new circular systematic sampling (CSS) schemes have also been studied by Uthayakumaran (1998). Hartley (1966) studied PPSWOR systematic sampling. Although we have discussed many difficulties and their solutions in systematic sampling in the preceding sections, the more serious difficulty is in the estimation of variance of the estimator of mean/total using systematic sampling schemes.

7.5 ESTIMATION OF VARIANCE UNDER SYSTEMATIC SAMPLING

We will discuss here certain methods to estimate the variance of the estimators of mean (total) under systematic sampling scheme.

7.5.1 SUB-SAMPLING OR REPLICATED SUB-SAMPLING SCHEME

In this method, rather than choosing one systematic sample of size n, choose m sub-samples each of size n/m by selecting m random starts from 1 to $k = (Nm)/n$ using without replacement sampling scheme. Compute the m sample means $\bar{y}_j$, $j = 1, 2, ..., m$. Also compute the full sample mean defined as

$$\bar{y}_p = \frac{1}{m}\sum_{j=1}^{m}\bar{y}_j . \qquad (7.5.1)$$

An unbiased estimator of $V(\bar{y}_p)$ is

$$\hat{v}(\bar{y}_p) = \hat{\sigma}_b^2 = \frac{1-f}{m(m-1)}\sum_{j=1}^{m}(\bar{y}_j - \bar{y}_p)^2 \qquad (7.5.2)$$

where $f = n/N$. Such a method of estimating the variance under systematic sampling is also called interpenetrating systematic sampling.

Example 7.5.1. Select three sub-samples each consisting of 5 states from population 1 having 50 states as given in the Appendix by using systematic sampling. Collect the information on the real estate farm loans from the states selected in the sample. Obtain a pooled estimate of the average real estate farm loans in the United States. Use an appropriate method for estimating the variance of the resultant pooled estimator.

Solution. We are given $N = 50$, $m = 3$, and $n = 5 \times m = 5 \times 3 = 15$ is the total sample size. Select $m = 3$ distinct random numbers between 1 and $k = \dfrac{N}{n/m} = \dfrac{50}{15/3} = 10$.

Use these three distinct random numbers to form three different systematic sub-samples with a skip of $k = 10$. We used 6th and 7th columns of the Pseudo-Random Numbers (PRN) given in Table 1 of the Appendix to select three distinct random numbers between 1 and 10. We observed the three distinct random numbers as 07, 09, and 01. Then we have the following sub-samples:

First Sub-Sample with random start 07			Second Sub-Sample with random start 09			Third Sub-Sample with random start 01		
PRN No.	State	Real estate farm loans	PRN No.	State	Real estate farm loans	PRN No.	State	Real estate farm loans
07	CT	7.1300	09	FL	825.748	01	AL	408.9780
17	KY	1045.1060	19	ME	8.849	11	HI	40.7750
27	NE	1337.8520	29	NH	6.044	21	MA	7.5900
37	OR	114.8990	39	RI	1.611	31	NH	6.0440
47	WA	1100.7450	49	WI	1229.752	41	SD	413.7770
	$\bar{y}_1$	721.1464		$\bar{y}_2$	414.401		$\bar{y}_3$	175.4328

Thus the pooled sample mean estimate $\bar{y}_p$ is given by

$$\bar{y}_p = \frac{\bar{y}_1 + \bar{y}_2 + \bar{y}_3}{3} = \frac{721.1464 + 414.401 + 175.4328}{3} = 436.993 .$$

An unbiased estimate of $V(\bar{y}_p)$ is

$$\hat{v}(\hat{y}_p) = \hat{\sigma}_b^2 = \frac{1-f}{m(m-1)} \sum_{j=1}^{m} (\bar{y}_j - \bar{y}_p)^2$$

$$= \frac{1-0.3}{3(3-1)}\left[(721.146 - 436.993)^2 + (414.401 - 436.993)^2 + (175.433 - 436.993)^2\right]$$

$$= 17461.184 .$$

7.5.2 SUCCESSIVE DIFFERENCES

If y_i denotes the ith unit selected in the sample, then an estimator of the variance of sample mean can also be obtained as

$$\hat{\sigma}_b^2 = \frac{(1-f)}{2n(n-1)} \sum_{i=1}^{n-1} (y_i - y_{i+1})^2 . \tag{7.5.3}$$

The assumption here is that each successive pair of units, *i.e.*, y_i and y_{i+1} are drawn using SRSWOR sampling from the $2k$ eligible units.

Example 7.5.2. Select a sample of 10 states from population 1 consisting of 50 states by using the systematic sampling scheme. Collect the information on the real estate farm loans from the states selected in the sample. Use an appropriate method for estimating the variance of the estimator of population mean.

Solution. We have $N = 50$ and $n = 10$, therefore $k = N/n = 50/10 = 5$. We used the 8th column of the Pseudo-Random Numbers (PRN) given in Table 1 of the Appendix to select a random number between 1 and 5. We observed random number 2. Thus the systematic sample consists of the following 10 distinct units as 02, 07, 12, 17, 22, 27, 32, 37, 42 and 47.

Random No.	State	Real estate farm loan	$(y_i - y_{i+1})$	$(y_i - y_{i+1})^2$
2	AK	2.605	-4.525	20.475
7	CT	7.130	-46.623	2173.704
12	ID	53.753	-991.353	982780.771
17	KY	1045.106	722.078	521396.638
22	MI	323.028	-1014.82	1029867.750
27	NE	1337.852	1136.221	1290998.160
32	NY	201.631	86.732	7522.439
37	OR	114.899	-438.367	192165.627
42	TN	553.266	-547.479	299733.255
47	WA	1100.745		
	Sum	4740.015		4326658.820

Thus an estimate of variance of the estimator of mean for the systematic sampling is

$$\hat{\sigma}_b^2 = \frac{(1-f)}{2n(n-1)}\sum_{i=1}^{n-1}(y_i - y_{i+1})^2 = \frac{(1-0.2)}{2\times 10(10-1)}\times 4326658.82 = 19229.595\,.$$

7.5.3 VARIANCE OF CIRCULAR SYSTEMATIC SAMPLING

In case of CSS scheme, we have N possible samples instead of k samples and we can define

$$\sigma_w^2 = \frac{1}{nN}\sum_{i=1}^{n}\sum_{r=1}^{N}(y_{r,i} - \bar{y}_r)^2\,, \quad \sigma_b^2 = V(\bar{y}_{\text{css}}) = \frac{1}{N}\sum_{r=1}^{N}(\bar{y}_r - \bar{Y})^2 \quad \text{and} \quad \sigma^2 = \sigma_w^2 + \sigma_b^2\,.$$

Ray and Das (1995, 1997) have proposed circular systematic sampling schemes which provide unbiased estimator of population mean and also estimator of variance of the mean without putting any restriction on the population size.

7.6 SYSTEMATIC SAMPLING IN POPULATION WITH LINEAR TREND

Quite often situation arises when the study variable Y_i is related to the random start i through a linear relationship, usually known as a linear trend. We discuss below some results when such a trend is present.

7.6.1 ESTIMATORS WITH LINEAR TREND

In the following theorem, we show that the usual estimator of population mean under systematic sampling scheme remains more efficient than simple random sampling in the presence of linear trend.

Theorem 7.6.1. If Y_i has linear relation with the random start i, that is

$$Y_i = a + bi \tag{7.6.1.1}$$

under systematic sampling strategy while drawing a sample of n units, and if another sample of n units is drawn by SRSWOR or SRSWR, then for large N we have

$$V(\bar{y})_{\text{SRS}} / V(\bar{y}_i)_{\text{SYS}} = n\,. \tag{7.6.1.2}$$

Proof. Under SRSWR sampling we have

$$V(\bar{y}_{\text{WR}}) = \frac{\sigma_y^2}{n}, \quad \text{where } \sigma_y^2 = \frac{1}{N}\sum_{i=1}^{N}(Y_i - \bar{Y})^2 \text{ and } \bar{Y} = \frac{1}{N}\sum_{i=1}^{N}Y_i\,. \tag{7.6.1.3}$$

Under model (7.6.1.1) we have

$$\bar{Y} = \frac{1}{N}\sum_{i=1}^{N}Y_i = \frac{1}{N}\sum_{i=1}^{N}[a + bi] = \frac{1}{N}\left[Na + b\sum_{i=1}^{N}i\right]$$

$$= a + \frac{b}{N}\sum_{i=1}^{N}i = a + \frac{b}{N}\left\{\frac{N(N+1)}{2}\right\} = a + \frac{b(N+1)}{2} \tag{7.6.1.4}$$

and

$$\sigma_y^2 = \frac{1}{N}\sum_{i=1}^{N}\left(Y_i - \bar{Y}\right)^2 = \frac{1}{N}\sum_{i=1}^{N}\left[a + bi - \left\{a + \frac{b(N+1)}{2}\right\}\right]^2 = \frac{b^2}{N}\sum_{i=1}^{N}\left[i - \frac{(N+1)}{2}\right]^2$$

$$= \frac{b^2}{N}\sum_{i=1}^{N}\left[i^2 + \left(\frac{N+1}{2}\right)^2 - 2\left(\frac{N+1}{2}\right)i\right] = \frac{b^2}{N}\left[\sum_{i=1}^{N} i^2 + N\left(\frac{N+1}{2}\right)^2 - (N+1)\sum_{i=1}^{N} i\right]$$

$$= \frac{b^2}{N}\left[\frac{N(N+1)(2N+1)}{6} + \frac{N(N+1)^2}{4} - \frac{N(N+1)^2}{2}\right] = \frac{b^2\left(N^2-1\right)}{12}. \tag{7.6.1.5}$$

From (7.6.1.3) we have

$$V\left(\bar{y}_{\text{WR}}\right) = \frac{\sigma_y^2}{n} = \frac{b^2\left(N^2-1\right)}{12n}. \tag{7.6.1.6}$$

Under SRSWOR sampling we have

$$V\left(\bar{y}_{\text{WOR}}\right) = \left(\frac{N-n}{Nn}\right)S_y^2 = \left(\frac{N-n}{Nn}\right)\frac{N}{(N-1)}\sigma_y^2 = \frac{(N-n)b^2(N+1)}{12n}. \tag{7.6.1.7}$$

If a systematic sample of size n is selected with random start i and skip k, then the units selected in the sample will be listed at serial numbers $i, k+i, 2k+i, ..., (n-1)k+i$. Using (7.6.1.1) we have

$$\bar{y}_i = \frac{1}{n}\sum_{j=0}^{n-1} y_{i+jk} = \frac{1}{n}\left[y_i + y_{k+i} + y_{2k+i} + ... + y_{(n-1)k+i}\right]$$

$$= \frac{1}{n}\left[(a+bi) + \{a + b(k+i)\} + \{a + b(2k+i)\} + ... + \{a + b((n-1)k+i)\}\right]$$

$$= a + \frac{b}{n}\left[i + (k+i) + (2k+i) + ... + \{(n-1)k+i\}\right] = a + b\left[i + \frac{(n-1)k}{2}\right]. \tag{7.6.1.8}$$

Thus we have

$$E\left(\bar{y}_i\right) = \frac{1}{k}\sum_{i=1}^{k}\bar{y}_i = \frac{1}{k}\left[\sum_{i=1}^{k}\left\{a + b\left(i + \frac{(n-1)k}{2}\right)\right\}\right]$$

$$= \frac{1}{k}\left[ka + b\sum_{i=1}^{k} i + \sum_{i=1}^{k}\frac{(n-1)k}{2}\right] = \frac{1}{k}\left[ka + \frac{bk(k+1)}{2} + \frac{(n-1)bk^2}{2}\right].$$

$$= a + \frac{b}{2}\left[k + 1 + k(n-1)\right] = a + \frac{b}{2}(nk+1) = a + \frac{b}{2}(N+1) = \bar{Y}. \tag{7.6.1.9}$$

Thus the sample mean is an unbiased estimator for the population mean under linear trend. The variance of the estimator under systematic sampling reduces to

$$V\left(\bar{y}_i\right)_{\text{SYS}} = E\left[\bar{y}_i - \bar{Y}\right]^2 = \frac{1}{k}\sum_{i=1}^{k}\left(\bar{y}_i - \bar{Y}\right)^2 = \frac{1}{k}\sum_{i=1}^{k}\left[a + b\left\{i + \frac{(n-1)k}{2}\right\} - \left\{a + b\left(\frac{N+1}{2}\right)\right\}\right]^2$$

$$= \frac{b^2}{k}\sum_{i=1}^{k}\left[i + \frac{(n-1)k}{2} - \frac{(N+1)}{2}\right]^2 = \frac{b^2}{k}\sum_{i=1}^{k}\left[i - \left(\frac{k+1}{2}\right)\right]^2$$

$$= \frac{b^2}{k}\sum_{i=1}^{k}\left[i^2 + \left(\frac{k+1}{2}\right)^2 - 2\left(\frac{k+1}{2}\right)i\right] = \frac{b^2}{k}\left[\sum_{i=1}^{k} i^2 + \frac{k(k+1)^2}{4} - 2\left(\frac{k+1}{2}\right)\sum_{i=1}^{k} i\right]$$

$$= \frac{b^2}{k}\left[\frac{k(k+1)(2k+1)}{6} + \frac{k(k+1)^2}{4} - \frac{k(k+1)^2}{2}\right] = \frac{b^2(k^2-1)}{12}. \qquad (7.6.1.10)$$

From (7.6.1.6) for large N we have

$$V(\bar{y}_i)_{\text{WR}} = \frac{b^2(N^2-1)}{12n} = \frac{b^2N^2\left(1-\frac{1}{N^2}\right)}{12n} = \frac{b^2n^2k^2}{12n} = \frac{nb^2k^2}{12}. \qquad (7.6.1.11)$$

Again from (7.6.1.7) for large N we have

$$V(\bar{y}_i)_{\text{WOR}} = \frac{b^2(N-n)(N+1)}{12n} = \frac{b^2N^2\left(1-\frac{n}{N}\right)\left(1+\frac{1}{N}\right)}{12n}$$

$$= \frac{b^2N^2}{12n} = \frac{b^2n^2k^2}{12n} = \frac{nb^2k^2}{12}. \qquad (7.6.1.12)$$

From (7.6.1.11) and (7.6.1.12) we have

$$V(\bar{y}_i)_{\text{WR}} = V(\bar{y}_i)_{\text{WOR}} = V(\bar{y})_{\text{SRS}} = \frac{nb^2k^2}{12}. \qquad (7.6.1.13)$$

Now from (7.6.1.10) we have

$$V(\bar{y}_i)_{\text{SYS}} = \frac{b^2(k^2-1)}{12} = \frac{b^2}{12}\left(\frac{N^2}{n^2}-1\right)$$

$$= \frac{b^2}{12}\left(\frac{N^2-n^2}{n^2}\right) = \frac{b^2}{12}\left[1-\left(\frac{n}{N}\right)^2\right]\left(\frac{N^2}{n^2}\right) = \frac{b^2k^2}{12}. \qquad (7.6.1.14)$$

Comparing (7.6.1.13) with (7.6.1.14) we have

$$\text{RE} = \frac{V(\bar{y})_{\text{SRS}}}{V(\bar{y}_i)_{\text{SYS}}} = n \qquad (7.6.1.15)$$

which proves the theorem.

In systematic sampling it is interesting to note that it is possible to adjust the weights given to the units of the sample observations in such a way that the variance of the resultant estimator of population mean is zero. Such an adjustment is called the Yates (1948) end correction.

7.6.2 MODIFICATION OF ESTIMATES

Here we adjust the weights given to the sampled observations in the estimator of population mean in such a way so that the variance of the estimator is zero. The estimator of population mean with random start i and skip k in the systematic sampling is given by

$$\bar{y}_i = \frac{1}{n}\sum_{j=0}^{n-1} y_{i+jk} \,. \tag{7.6.2.1}$$

Let us change the weights as

$$\left(\frac{1}{n}-x\right), \frac{1}{n}, \frac{1}{n}, ..., \frac{1}{n}, \frac{1}{n}, \left(\frac{1}{n}+x\right). \tag{7.6.2.2}$$

Clearly we have only changed the weights for the first and the last unit in the sample. The value of x is determined suitably such that the modified estimator $\bar{y}_i^*$ (with new weights) matches the population mean, $\bar{Y}$ for all i. Then the systematic sampling estimator with new weights is given by

$$\bar{y}_i^* = \left(\frac{1}{n}-x\right)y_i + \frac{1}{n}y_{k+i} + \frac{1}{n}y_{2k+i} + ... + \frac{1}{n}y_{(n-2)k+i} + \left(\frac{1}{n}+x\right)y_{(n-1)k+i} \,. \tag{7.6.2.3}$$

Note that the linear trend is given by,

$$y_i = a + bi \,. \tag{7.6.2.4}$$

The estimator (7.6.2.3) becomes

$$\bar{y}_i^* = \left(\frac{1}{n}-x\right)(a+bi) + \frac{1}{n}\left[\sum_{j=1}^{n-2}\{a+b(jk+i)\}\right] + \left(\frac{1}{n}+x\right)\left[a+b\{(n-1)k+i\}\right]$$

$$= x\left[-a-bi+a+b\{(n-1)k+i\}\right] + \frac{1}{n}\sum_{j=0}^{n-1}\{a+b(jk+i)\}$$

$$= b(n-1)kx + \frac{1}{n}\left[na+bni+\frac{bkn(n-1)}{2}\right].$$

Now if $\bar{y}_i^* = \bar{Y}\ \forall\ i$ then $V\left(\bar{y}_i^*\right) = 0$, therefore we have

$$b(n-1)kx + \frac{1}{n}\left[na+bni+\frac{bkn(n-1)}{2}\right] = a + \frac{b(N+1)}{2}, \text{ or } (n-1)kx + i + \frac{k(n-1)}{2} = \frac{(N+1)}{2},$$

which implies that

$$x = \frac{1}{k(n-1)}\left[\frac{(N+1)}{2} - i - \frac{k(n-1)}{2}\right] = \frac{1}{k(n-1)}\left[\frac{k+1}{2} - i\right]. \tag{7.6.2.5}$$

Thus if we use this value of x in the weights of the first and last unit in the sample, then the variance of the usual estimator of population mean in systematic sampling reduces to zero. This type of adjustment or correction of weights is called Yates (1948) end correction. Bellhouse and Rao (1975) extended the Yates end corrections to the case $N \neq nk$ for CSS scheme. The performance of systematic sampling in the presence of a linear trend can also be improved by following Singh, Jindal and Garg (1968), where they suggested to choose pairs of units equidistant from the ends of the population. If i is the random start chosen from 1 to k then the units selected will be $\{i+(j-1)k,\ nk-(j-1)k-i+1\}$, $j = 1, 2, ..., n/2$ if n is even and $\{i+(j-1)k,\ nk-(j-1)k-i+1\}$, $i+(n-1)k/2$, $j = 1, 2, ..., (n-1)/2$ if n is odd. For the linear trend $Y_i = a + bi$, $i = 1, 2, ..., N$, the sample mean will be exactly equal to the population mean when n is even. The accuracy in the estimates of systematic sampling can also be increased by following any of the following two methods.

7.6.3 ESTIMATORS BASED ON CENTRALLY LOCATED SAMPLES

Suppose we have to take a sample of n units from a population of N units such that $N = nk$, where k is a sampling span. Then there are two cases:

(i) k is odd; (ii) k is even.

If k is even then the variance of the systematic sampling mean does not become exactly zero, but if k is odd then it may be zero. We shall discuss each case separately.

Case I. k is odd: We chose $\left(\frac{k+1}{2}\right)$th unit as random start. The value of y which corresponds to the random start $\left(\frac{k+1}{2}\right)$ is $a+b\left(\frac{k+1}{2}\right)$. Thus the first unit in the sample corresponds to serial number $\left(\frac{k+1}{2}\right)$, the second unit corresponds to $k+\left(\frac{k+1}{2}\right)$, the third unit corresponds to $2k+\left(\frac{k+1}{2}\right)$ and so on. Thus the sample mean corresponding to the random start $\frac{k+1}{2}$ is given by

$$\bar{y}_{\left(\frac{k+1}{2}\right)} = \frac{1}{n}\sum_{j=0}^{n-1}\left[a+b\left\{jk+\frac{(k+1)}{2}\right\}\right] = \frac{1}{n}\left[na+\frac{nb(k+1)}{2}+\frac{bkn(n-1)}{2}\right]$$

$$= a+\frac{bk}{2}+\frac{b}{2}+\frac{nbk}{2}-\frac{bk}{2} = a+\frac{b(nk+1)}{2}$$

$$= a+\frac{b(N+1)}{2} = \bar{Y} = \text{Population mean.}$$

Thus with $\left(\frac{k+1}{2}\right)$ as random start we have the sample mean equal to the population mean. Since no randomness is involved, the variance of the estimator $\bar{y}_{\left(\frac{k+1}{2}\right)}$ is zero. A pictorial representation of centrally located systematic sampling when k is odd is given in the Figure 7.6.3.1.

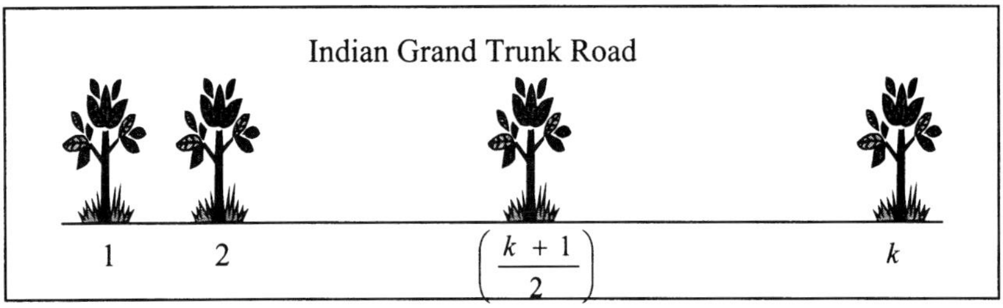

Fig. 7.6.3.1 Centrally located systematic sampling for odd random start.

Case II. k is even: In this situation we chose two random starts, as shown in the figure below.

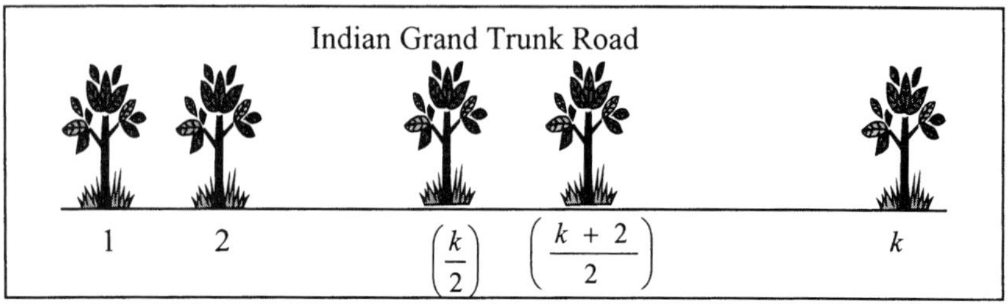

Fig. 7.6.3.2 Centrally located systematic sampling for even random start.

In this case there are two possibilities of choosing random start point, *i.e.*, either $\frac{k}{2}$ or $\left(\frac{k}{2}+1\right)$. Now, in order to decide which random start should be considered, we perform an experiment with an unbiased coin, *e.g.*, head for $\frac{k}{2}$ and tail for $\left(\frac{k}{2}+1\right)$. If we take $\frac{k}{2}$ as random start then from the linear trend, the y value corresponding to the first unit is $a+\frac{bk}{2}$, second unit $a+b\left(k+\frac{k}{2}\right)$, and $(n-1)$th unit $a+b\left[(n-1)k+\frac{k}{2}\right]$. Thus the sample mean with random start $\frac{k}{2}$ will be

$$\bar{y}_{\left(\frac{k}{2}\right)} = \frac{1}{n}\sum_{j=0}^{n-1}\left[a+b\left(kj+\frac{k}{2}\right)\right] = \frac{1}{n}\left[na+\frac{nbk}{2}+\frac{nbk(n-1)}{2}\right]$$

$$= a+\frac{bk}{2}+\frac{nbk}{2}-\frac{bk}{2} = a+\frac{Nb}{2}. \tag{7.6.3.1}$$

Similarly, if the random start is $\left(\frac{k}{2}+1\right)$ then we have

$$\bar{y}_{\left(\frac{k}{2}+1\right)} = \frac{1}{n}\sum_{j=1}^{n-1}\left[a+b\left\{kj+\frac{(k+2)}{2}\right\}\right] = \frac{1}{n}\left[na+\frac{nb(k+2)}{2}+\frac{nbk(n-1)}{2}\right]$$

$$= a+\frac{b(N+2)}{2}. \tag{7.6.3.2}$$

Note that we are choosing each of the two random starts with probability 1/2, so

$$\bar{y} = \frac{1}{2}y_{\left(\frac{k}{2}\right)} + \frac{1}{2}y_{\left(\frac{k}{2}+1\right)} = \frac{1}{2}\left[a+\frac{bN}{2}+a+\frac{b(N+2)}{2}\right] = a+\frac{b(N+1)}{2} = \bar{Y}.$$

Furthermore

$$V(\bar{y}) = E[\bar{y} - E(\bar{y})]^2 = \frac{1}{2}\left[\left\{\left(a + \frac{bN}{2}\right) - \left(a + \frac{b(N+1)}{2}\right)\right\}^2 + \left\{\left(a + \frac{b(N+2)}{2}\right) - \left(a + \frac{b(N+1)}{2}\right)\right\}^2\right]$$

$$= \frac{1}{2}\left[\frac{b^2}{4} + \frac{b^2}{4}\right] = \frac{2b^2}{8} = \frac{b^2}{4} \neq 0. \qquad (7.6.3.3)$$

7.6.4 ESTIMATORS BASED ON BALANCED SYSTEMATIC SAMPLING

We assume that $N = nk$ and n is even. Instead of taking the sampling span from 1 to k, we take the sampling span from 1 to $2k$. In other words, instead of taking n groups of k units each, we take $n/2$ groups of $2k$ units each and the sampling span is $2k$ instead of k. A pictorial representation of Balanced Systematic Sampling (BSS) is presented below:

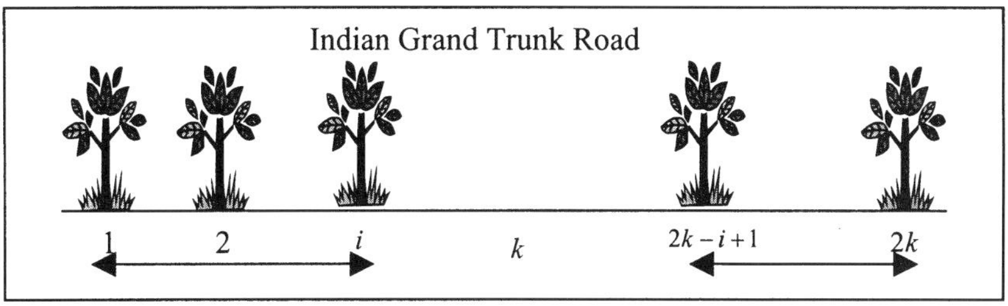

Fig. 7.6.4.1 Balanced Systematic sampling.

Furthermore, as shown in the above diagram, we select two random starting points i and $2k-i+1$ (say), one between 1 and k and another between k and $2k$ such that both points are at equal distance from k. Then we have the following data in the sample. The first unit is selected at serial number i, the second unit is selected at serial number $2k-i+1$, the third unit is selected at serial number $i+2k$, the fourth unit is selected at serial number $4k-i+1$ and so on. Hence using the linear trend we have

$$\bar{y}_{\text{BSS}} = \frac{1}{n}\sum_{j=0}^{\frac{n}{2}-1}[a + b\{2kj + i\}] + \frac{1}{n}\sum_{j=0}^{\frac{n}{2}-1}[a + b(2kj + 2k - i + 1)]$$

$$= \frac{1}{n}\sum_{j=0}^{\frac{n}{2}-1}[a + 2bkj + bi + a + 2bkj + 2bk - bi + b] = \frac{1}{n}\sum_{j=0}^{\frac{n}{2}-1}[2a + 4bkj + 2kb + b]$$

$$= \frac{2}{n}\left[\frac{na}{2} + \frac{2bk}{2}\left(\frac{n}{2} - 1\right)\left(\frac{n}{2}\right) + \frac{nbk}{2} + \frac{nb}{4}\right] = a + bk\left(\frac{n}{2} - 1\right) + bk + \frac{b}{2}$$

$$= a + \frac{nbk}{2} - bk + bk + \frac{b}{2} = a + \frac{Nb}{2} + \frac{b}{2} = a + \frac{b(N+1)}{2} = \bar{Y}. \qquad (7.6.4.1)$$

Thus irrespective of the random start i, the balanced systematic sample (BSS) mean is equal to population mean and equivalently $V(\bar{y}_{\text{BSS}}) = 0$.

Singh and Garg (1979) have proposed a sampling scheme named as *Balanced Random Sampling Scheme*, which has the advantage of both simple random sampling and systematic sampling in the sense that a part of the sampling variance depends upon the arrangements of the units in the population and the other part is independent of the arrangements. The resultant sampling scheme performs the best for the populations showing linear trend or periodicity.

Example 7.6.1. Select a samples of 10 states from population 1 consisting of 50 states by using systematic sampling scheme. Collect the information on the real estate farm loans from the states selected in the sample. Assume that there is a linear trend between the real estate farm loans and the serial numbers. Apply the Yates end correction to obtain unbiased estimate of population mean.

Solution. We have $N = 50$ and $n = 10$, therefore $k = N/n = 50/10 = 5$. We used the 8th column of the Pseudo-Random Numbers (PRN) given in Table 1 of the Appendix to select a random number between 1 and 5. We observed random number 2. Thus the systematic sample consists of the following 10 distinct units such as 02, 07, 12, 17, 22, 27, 32, 37, 42 and 47.

According to the Yates end correction the adjustment factor in the first and last weights is given by

$$x = \frac{1}{k(n-1)}\left[\frac{k+1}{2} - i\right] = \frac{1}{5(10-1)}\left[\frac{5+1}{2} - 2\right] = \frac{1}{45}.$$

Thus we have the following table:

Random No.	State	Real estate farm loans y_i	Old weights d_i	New weights w_i	$w_i y_i$	$d_i y_i$
02	AK	2.605	0.1	0.0778	0.202	0.261
07	CT	7.130	0.1	0.1000	0.713	0.713
12	ID	53.753	0.1	0.1000	5.375	5.375
17	KY	1045.106	0.1	0.1000	104.510	104.511
22	MI	323.028	0.1	0.1000	32.302	32.303
27	NE	1337.852	0.1	0.1000	133.785	133.785
32	NY	201.631	0.1	0.1000	20.163	20.163
37	OR	114.899	0.1	0.1000	11.489	11.489
42	TN	553.266	0.1	0.1000	55.326	55.326
47	WA	1100.745	0.1	0.1222	134.535	110.074
				Sum	498.404	474.002

From the above calculations the biased estimate of the average real estate farm loans is 474.002. The Yates end corrected and unbiased estimate is 498.404.

7.7 SINGH AND SINGH'S SYSTEMATIC SAMPLING SCHEME

Singh and Singh (1977) suggested a new systematic sampling scheme which provides an unbiased estimator of variance of the sample mean. Here we shall discuss their technique briefly. Suppose a population consists of N distinct units and a sample of size n has to be drawn. Let $u(\le n)$ and d be two predetermined integers which are chosen in such a way that: (i) Every sample contains distinct units; (ii) The inclusion probability for each pair of units is non-zero, and starting with a random number $r(\le N)$, select u units continuously, and thereafter the remaining $n-u=v$ (say) units with span d such that $d \le u$ and $u+vd \le N$. With these conditions, a sample of size n can be drawn in two or more phases. The condition for the number of phases $p(\text{say})$, required for selecting a sample of n units from the population of N units is given by

$$p \ge \frac{\log\{\log(N/2)\} - \log\{\log(n/2)\}}{\log(2)}. \tag{7.7.1}$$

The sample space corresponding to the selection scheme contains N possible samples. Thus the probability of selecting each sample is $1/N$ and the probability measure associated with the selection procedure is

$$P = P(s_r)\left\{P(s_r) = \frac{1}{N},\ r = 1, 2, 3, \dots, N\right\}, \tag{7.7.2}$$

where the sample s_r corresponding to random start r is given by $s_r = (s_r', s_r'')$, where s_r' consists of units with indices $r+m$, $(m = 0, 1, 2, \dots, u-1)$ and s_r'' consists of units with indices $(r+u-1)+td$, $(t = 1, 2, \dots, v)$. Assuming that the first and second order inclusion probabilities π_i and π_{ij} are known. Then we have the following results.

Result 7.7.1. The Horvitz and Thompson (1952) type estimator of population mean is given by

$$\bar{y}_{\text{sys}} = \frac{1}{N}\sum_{i=1}^{n}\frac{y_i}{\pi_i}. \tag{7.7.3}$$

Result 7.7.2. The variance of the estimator $\bar{y}_{\text{sys}}$ is given by

$$V(\bar{y}_{\text{sys}}) = \frac{1}{N^2}\sum_{i=1}^{N}\sum_{j>i}^{N}\left(1 - \pi_{ij} - \frac{N^2}{n^2}\right)(y_i - y_j)^2. \tag{7.7.4}$$

Result 7.9.3. An unbiased estimator of the $V(\bar{y}_{\text{sys}})$ is given by

$$\hat{V}(\bar{y}_{\text{sys}}) = \frac{1}{N^2}\sum_{i=1}^{n}\sum_{j>i}^{n}\left(\frac{1}{\pi_{ij}} - \frac{N^2}{n^2}\right)(y_i - y_j)^2. \tag{7.7.5}$$

Example 7.7.1. Select a sample of 10 states from population 1 consisting of 50 states by using Singh and Singh's systematic sampling scheme. Calculate the minimum number of phases to select a sample of 10 states. Collect the information on the real estate farm loans and nonreal estate farm loans from the states selected in the sample. Assume that information on nonreal estate farm loans is known for the whole population. Estimate the total real estate farm loans.

Given: $X = 43908.12$.

Solution. We have $N = 50$ and $n = 10$, therefore, $k = N/n = 50/10 = 5$. Let us decide $u = 7$ and $d = 4$. We used 8th column of the Pseudo-Random Numbers (PRN) given in Table 1 of the Appendix to select a random number between 1 and 5. We observed random number 2. Select first seven units with skip 5 with random start number 2. The remaining 3 units were selected with a skip 4. Thus we have the following situation.

Singh and Singh's scheme										
	Skip = 5							Skip = 4		
Random no.	02	07	12	17	22	27	32	36	40	44
State	AK	CT	ID	KY	MI	NE	NY	OK	SC	UT

The minimum number of phases to select the sample are

$$p \geq \frac{\log\{\log(N/2)\} - \log\{\log(n/2)\}}{\log(2)} = \frac{\log\{\log(25)\} - \log\{\log(5)\}}{\log(2)} = 1.$$

Random No.	State	Real estate farm loans y_i	Nonreal estate farm loans x_i	p_i	$\pi_i = np_i$	$\frac{y_i}{\pi_i}$
02	AK	2.605	3.433	0.000078	0.00078186	3331.79
07	CT	7.130	4.373	0.000099	0.00099594	7159.04
12	ID	53.753	1006.036	0.022912	0.22912300	234.60
17	KY	1045.106	557.656	0.012701	0.12700521	8228.84
22	MI	323.028	440.518	0.010033	0.10032723	3219.74
27	NE	1337.852	3585.406	0.081657	0.81657015	1638.37
32	NY	201.631	426.274	0.009708	0.09708318	2076.89
36	OK	612.108	1716.087	0.039084	0.39083591	1566.15
40	SC	87.951	80.750	0.001839	0.01839068	4782.37
44	UT	56.908	197.244	0.004492	0.04492199	1266.82
					Sum	33504.64

where $p_i = x_i/X$ and $\pi_i = np_i$.

Thus an estimate of the average amount of the real estate farm loans in the United States is

$$\bar{y}_{\text{sys}} = \frac{1}{N}\sum_{i=1}^{n}\frac{y_i}{\pi_i} = \frac{33504.64}{50} = 670.093.$$

The values of $(y_i - y_j)^2$, $i \neq j$ are given below.

i>j	1	2	3	4	5	6	7	8	9
2	20.48								
3	2616.12	2173.70							
4	1086808.34	1077394.18	982780.77						
5	102670.90	99791.55	72509.03	521396.64					
6	1782884.55	1770821.04	1648910.24	85700.22	1029867.75				
7	39611.35	37830.64	21867.90	711450.08	14737.23	1290998.16			
8	371493.91	365998.38	311760.31	187487.27	83567.25	526704.35	168491.37		
9	7283.94	6532.03	1169.50	916145.69	55261.20	1562252.51	12923.14	274740.56	
10	2948.82	2477.85	9.95	976535.29	70819.85	1640817.53	20944.75	308247.04	963.67

From the above table we have

$$\sum_{i=1}^{n}\sum_{j>i}^{n}(y_i - y_j)^2 = 20258417.01.$$

For simplicity, if we take $\pi_{ij} = n(n-1)/\{N(N-1)\}$ then

$$\left(\frac{1}{\pi_{ij}} - \frac{N^2}{n^2}\right) = \frac{N(N-1)}{n(n-1)} - \frac{N^2}{n^2} = \frac{50\times 49}{10\times 9} - \frac{50^2}{10^2} = 2.2222.$$

Using the above information the estimator of the variance of the estimator $\bar{y}_{\text{sys}}$ is

$$\hat{v}(\bar{y}_{\text{sys}}) = \frac{1}{N^2}\sum_{i=1}^{n}\sum_{j>i}^{n}\left(\frac{1}{\pi_{ij}} - \frac{N^2}{n^2}\right)(y_i - y_j)^2 = \frac{2.2222\times 20258417.01}{50^2} = 18007.48.$$

Thus $(1-\alpha)100\%$ and hence, using Table 2 from the Appendix, the 95% confidence interval of the average amount of real estate farm loans in the United States is

$\bar{y}_{\text{sys}} \mp t_{\alpha/2}(\text{df} = n-1)\sqrt{\hat{v}(\bar{y}_{\text{sys}})}$, or $670.093 \mp 2.626\sqrt{18007.48}$, or $[317.704,\ 1022.481]$.

7.8 ZINGER STRATEGY IN SYSTEMATIC SAMPLING

Given the finite population Ω of N distinct and identifiable units with variable of interest, Y. Then the Zinger (1980) sampling design exists if the following conditions hold:

(a) First a sample s of size $m < N$ is selected from Ω using design p_s. The strategy $(p_s, \bar{y}_s)$ is unbiased for the population mean $\bar{Y}$;

(b) Once the sample s is selected and held fixed, another sample r of size $n < (N-m)$ is selected from $(\Omega - s)$ with probability design p_r. The strategy $\{(p_r, t_r) \mid s \text{ fixed}\}$ is unbiased for $\bar{y}_{(\Omega-s)}$, and in turn $(p_s, \bar{y}_{\Omega-s})$ is unbiased for population mean $\bar{Y}$. Espejo (1997) used the above defined Zinger design to propose a Zinger statistic as

$$t_E = \beta\bar{y}_s + (1-\beta)(t_r \mid s \text{ fixed}). \tag{7.8.1}$$

where β is a non-zero real constant. Then we have the following theorems.

Theorem 7.8.1. The estimator t_E is unbiased for population mean $\overline{Y}$.

Proof. We know that

$$E(\overline{y}_s)=\overline{Y}, \text{ and that } E\{E(t_r \mid s)\}=E(\overline{y}_{\Omega-s})=E\left[\frac{N\overline{Y}-m\overline{y}_s}{N-m}\right]=\overline{Y}.$$

Thus we have

$$E(t_E)=\beta E(\overline{y}_s)+(1-\beta)E\{E(t_r \mid s)\}=\beta\overline{Y}+(1-\beta)\overline{Y}=\overline{Y}.$$

Hence the theorem.

Theorem 7.8.2. The variance of the estimator t_E is given by

$$V(t_E)=\left\{\beta-(1-\beta)\frac{m}{N-m}\right\}^2 V(\overline{y}_s)+(1-\beta)^2 E\{V(t_r \mid s)\}. \tag{7.8.2}$$

Proof. It follows from Espejo (1997).

Theorem 7.8.3. An estimator of $V(\overline{y}_s)$, where $\overline{y}_s=m^{-1}\sum_{i=1}^{m} y_i$, is given by

$$\hat{v}(\overline{y}_s)=\left\{\beta-(1-\beta)\frac{m}{N-m}\right\}^{-2}\left\{\hat{V}(t_E)-(1-\beta)^2\hat{V}(t_r \mid s \text{ fixed})\right\}. \tag{7.8.3}$$

Espejo (1997) has shown that the estimators proposed by Gautschi (1957), Heilbron (1978), Wolter (1984), Wu (1984), Ruiz and Santos (1992) and Rana and Singh (1989) are also special cases of the estimator (7.8.1). Systematic sampling has also been discussed by Madow and Madow (1944), Madow (1949, 1953), Finney (1948, 1995), Sukhatme, Panse, and Sastri (1958), and Reddy (1980).

7.9 POPULATIONS WITH CYCLIC OR PERIODIC TRENDS

The precision of systematic sampling from populations having cyclic or periodic trend depends upon the periodicity of the trend.

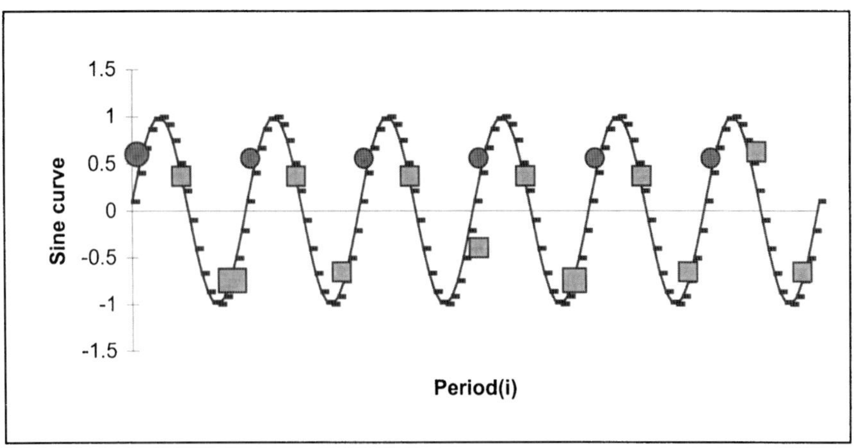

Fig.7.9.1 Population with regular cyclic trend.

The population may consist of a periodic trend given by the sine curve as

$$y_i = \alpha + \sin(\pi i/n + \beta) \tag{7.9.1}$$

where i varies from 0 to an integral multiple of $2n$. In Figure 7.9.1, we have taken $\alpha = 0$, $\beta = 0.1$, $n = 10$, $\pi = 22/7$ and $i = 0,1,2,3,....,120$. Here $2 \times n = 20$, which means successive sampling units will repeat themselves after every 20th value. It can be easily observed that a 5% systematic sample from such a population will form sampling units drawn from the same position of each cycle. An estimate from such a sample will be as good as a single value. On the other hand, a 5% random sample will contain units from different parts of the population and estimates from such samples will be more precise for the effect of a periodic trend. In Figure 7.9.1, the height of the curve is equal to the value of the study variable Y. Now if the skip k is equal to the period of the sine curve or an integral multiple of it, then the units marked with the circles are in the sample. Since all circles are at the same height from the x axis, therefore every observation within the systematic sample is exactly the same. In other words, the sample has the same information as from any single unit of the population. Thus this case can be considered as the least favourable for systematic sampling. In contrast, the most favourable situation happens when the span k is an odd multiple of half-period. In this case, every systematic sample mean is equal to the population mean, which results in zero variance of estimator of population mean. Such a situation of half-periodicity are indicated the Figure 7.9.1 with squares. The choice between these two cases depends upon the relationship between k and wavelength. Increase in periodicity in cyclic populations, while selecting the sample, increases the efficiency of the estimates. Madow (1949), Finney (1948, 1950), and Milne (1959) have observed linear and periodic type trends in the natural populations.

7.10 MULTI-DIMENSIONAL SYSTEMATIC SAMPLING

So for we have discussed only one-dimensional systematic sampling in which the population units are supposed to be in ascending or descending order on a line or boundary line of a circle based on the serial number or magnitude of any known auxiliary variable. A natural question arises, that if there are two auxiliary variables (say X_1 and X_2) then what is the best method to use both auxiliary variable at selection stage in systematic sampling. In such situations it is possible to arrange the population units on a plane or area instead of on a line. Such a systematic sampling scheme is known as two-dimensional systematic sampling scheme. Quenouille (1949), Das (1950), Manwani and Singh (1978) and Singh and Chaudhary (1986) have discussed the problem of two-dimensional systematic sampling. The simplest extension of a linear systematic sampling scheme to two-dimensional scheme is also known as squared grid or aligned sampling scheme. To explain two-dimensional systematic sampling scheme, we consider population consisting of N square grid areas of equal size. Now a sample of n grid areas has to be selected. Let there be r rows and s columns, such that $r.s = N$ as shown in the following Figure 7.10.1.

GRID		X_1 (say) $\rightarrow$					
		1	2	3			s
X_2 $\downarrow$	1						
	2						
	3						
	r						

Fig. 7.10.1 Two-dimensional systematic sample.

The simplest way is to select a pair of random numbers (i, j) such that $i \le r$ and $j \le s$. Thus the random location of a grid is unique. For example, let us suppose the population consists of $N = 100$ units and we wish to select a sample of $n = 9$ units. The simplest way to create 100 grid areas is to form a plane with $r = 10$ rows and $s = 10$ columns as shown in Figure 7.10.2.

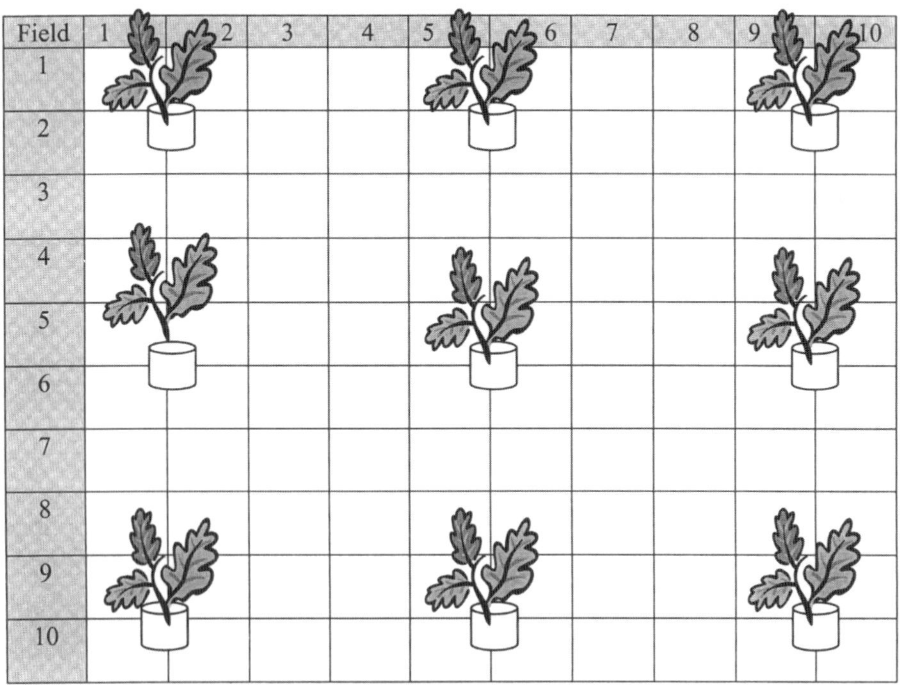

Fig. 7.10.2 Square grid or Aligned sample.

Another method of selecting a sample from a two-dimensional space is called Unaligned Sampling Method. In this method the grids are arranged in the form of $r \times l$ rows and $m \times c$ columns and it is required to select a systematic sample of size $n = rc$ grids. A pictorial representation of such a method is given in Figure 7.10.3.

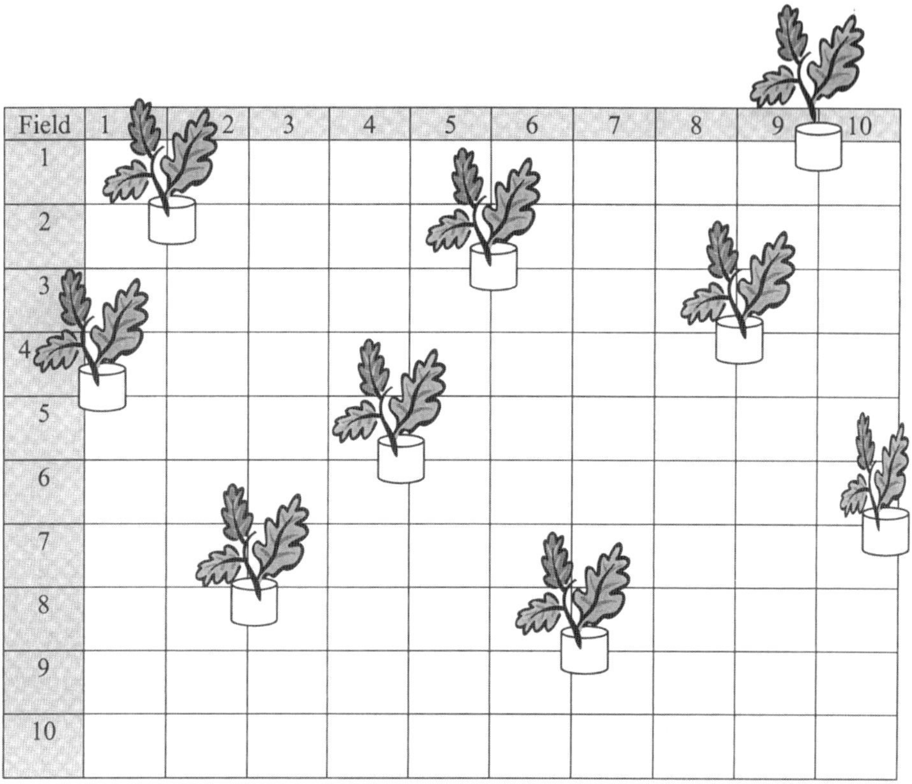

Fig. 7.10.3 Unaligned systematic sampling.

Select r independent random numbers $i_1, i_2, ..., i_r$ each less than or equal to l, and c independent random numbers $j_1, j_2, ..., j_c$ each less than or equal to m. The grids included in the unaligned systematic sample are $\{i_{x+1} + xl,\ j_{x+1} + ym\}$ for $x = 0,1,2,....,(r-1)$ and $y = 0,1,2,....,(c-1)$. Quenouille (1949) and Das (1950) have shown that an unaligned systematic sampling remains better than square grid or aligned two dimensional systematic sampling. Milne (1959) has suggested a central square grid technique for which the first sampling point starts from the centre of the square. If there are more than two auxiliary variables to be used for selecting the sample using systematic sampling, such a sampling scheme is called Multi-dimensional systematic sampling. Yates (1960) has considered three dimension lattice sampling strategy. Decent reviews on systematic sampling can be had from Buckland (1951) and Iachan (1982).

EXERCISES

Exercise 7.1. Give the circumstances under which systematic sampling is preferred over simple random sampling. Discuss the difficulties in estimating the variance of the estimator of population mean under systematic sampling design.

Exercise 7.2. What was the need of circular systematic sampling? Is it always preferable than usual systematic sampling? Give a practical example of using circular systematic sampling.

Exercise 7.3. Discuss a procedure of drawing a sample of n units from a population of N units using (a) systematic sampling, (b) circular systematic sampling, (c) any modified procedure without difficulties in part (a) and (b).

Exercise 7.4. What is Yates end correction in Systematic Sampling? How it effect the variance of the estimator of population mean under systematic sampling? Does there any difficulty in this method while implementing in actual practice?

Exercise 7.5. Discuss the systematic sampling strategies under (a) a linear trend, (b) a cyclic trend.

Exercise 7.6. What is Balanced Systematic Sampling (BSS)? Show that the variance of the estimator of the population mean under BSS reduces to zero.

Exercise 7.7. In a population with quadratic trend $y_i = i^2$, $i = 1,2,...,25$, as shown in the following figure.

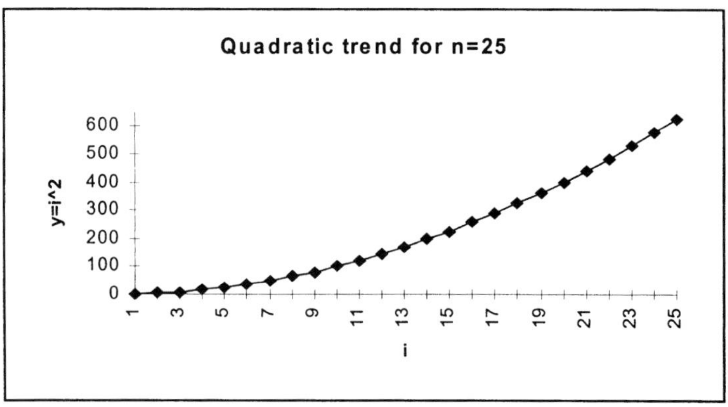

compare the value of $V(\bar{y}_{sys}) = E(\bar{y}_{sys} - \bar{Y})^2$ given by k^{th} systematic sample of size 5 by Yates end correction method and usual method of systematic sampling.

Exercise 7.8. Mr. Bean studied that the alternating current follows sine curve as shown in the Figure 7.9.1. Mr. Bean's interest was to estimate the average amount of current. He used systematic sampling to select a sample of $n = 4$ units with random start $r = 3$ from a population of $N = 40$ as shown below.

Random Start	3	13	23	33
Observed Value	+0.86	-0.86	+0.86	-0.86

From the sample information, he obtained that the average current is zero. Do you agree with him? If not, why?

Exercise 7.9. Write a short note on two-dimensional systematic sampling. Is it possible to use three-dimensional systematic sampling? Can you suggest a practical situation?

Exercise 7.10. If the units in the population can be arranged in increasing (or decreasing) order of magnitude of the study variable. Show that the value of intraclass correlation coefficient is negative (Refer to Chapter 9). Also show that both balanced systematic sampling and modified systematic sampling remain superior to the usual systematic sampling for estimating population total or mean.
Hint: Reddy (1980).

Exercise 7.11. Suggest an estimator of correlation coefficient in systematic sampling. Derive the expression of bias and variance.
Hint: Gupta and Singh (1990).

Exercise 7.12. Under the superpopulation model:

$$E_m(y_i \mid x_i) = \alpha + \beta x_i , \quad V_m(y_i \mid x_i) = \sigma^2 x_i^g , \quad \text{and } \mathrm{Cov}_m(y_i, y_j) = 0 , \ i \neq j ,$$

assuming that the units in the population are arranged in the ascending order of the auxiliary variable x. Let V_{srswr}, V_{srswor}, V_{sys}, V_{bss} and V_{mss} denote the variance of the sample mean estimator under SRSWR sampling, SRSWOR sampling, Systematic Sampling, Balanced Systematic Sampling, and Modified Systematic Sampling, respectively, then show that

$$E_m\left[V_{\text{srswor}} - V_{\text{srswr}}\right] < 0 , \ E_m\left[V_{\text{srswor}} - V_{\text{bss}}\right] > 0, \text{ and } E_m\left[V_{\text{srswor}} - V_{\text{mss}}\right] > 0 .$$

Given that balanced systematic sampling is due to Murthy (1977) and modified systematic sampling is due to Singh, Jindal, and Garg (1968).
Hint: Reddy (1980).

Exercise 7.13. Consider a superpopulation model: $m : y_i = \alpha + \beta\left(i - \dfrac{N+1}{2}\right) + e_i$, $i = 1,2,...,N$, where e_i is a random variable with $E_m(e_i \mid x_i) = 0$, $V_m(e_i \mid x_i) = \sigma^2$, $\mathrm{Cov}_m(e_i, e_j \mid x_i, x_j) = 0$, $i \neq j$. Let us consider the case such that $N = nk$, k is an integer and n is sample size. Let the expected value of the mean square error of the of the estimator $\bar{y}_s$ be defined as $E_m E_p\left[\bar{y}_s - \bar{Y}\right]^2$ for $s = \text{sys, mss, css}$ and bss over the design p. Show that:

(a) For systematic sampling

$$E_m\left[\text{MSE}\left(\bar{y}_{\text{sys}}\right)\right] = \frac{\sigma^2(1-f)}{n} + \frac{b^2\left(k^2-1\right)}{12};$$

(b) For modified systematic sampling (Singh, Jindal, and Garg,1968))

$$E_m\left[\text{MSE}\left(\bar{y}_{\text{mss}}\right)\right] = \begin{cases} \dfrac{\sigma^2(1-f)}{n} + \dfrac{b^2\left(k^2-1\right)}{12n^2} & \text{if } n \text{ is odd,} \\ \dfrac{\sigma^2(1-f)}{n} & \text{if } n \text{ is even;} \end{cases}$$

(c) For centred sampling

$$E_m\left[\text{MSE}\left(\bar{y}_{\text{css}}\right)\right] = \begin{cases} \dfrac{\sigma^2(1-f)}{n} + \dfrac{b^2}{4} & \text{if } k \text{ is even,} \\ \dfrac{\sigma^2(1-f)}{n} & \text{if } k \text{ is odd;} \end{cases}$$

(d) For Balanced Systematic Sampling

$$E_m\left[\text{MSE}\left(\bar{y}_{\text{bss}}\right)\right] = \begin{cases} \dfrac{\sigma^2(1-f)}{n} + \dfrac{b^2\left(k^2-1\right)}{12n^2} & \text{if } n \text{ is odd,} \\ \dfrac{\sigma^2(1-f)}{n} & \text{if } n \text{ is even.} \end{cases}$$

Hint: Fountain and Pathak (1989).

Exercise 7.14. Consider a superpopulation model :

$$m: \ y_i = \alpha + \beta\left(i - \frac{N+1}{2}\right) + e_i, i = 1, 2, ..., N$$

where e_i is a random variable with

$$E_m(e_i \mid x_i) = 0, \ V_m(e_i \mid x_i) = \sigma^2, \ \text{Cov}_m(e_i, e_j \mid x_i, x_j) = 0, \ i \neq j.$$

Let us consider the case such that $N = nk$, k is an integer and n is sample size. Now, consider the linear regression estimator defined as

$$\bar{y}_{\text{lr}} = \bar{y} + \beta\left[\frac{N+1}{2} - \frac{1}{n}\sum_{i=1}^{n} i\right].$$

(a) Show that $\bar{y}_{\text{lr}}$ is the best linear unbiased estimator of the population mean $\bar{Y}$. Also show the following relations:

(b) under SRSWOR sampling $E_m\left[\text{MSE}\left(\bar{y}_{\text{lr}}\right)_{\text{srswor}}\right] = \dfrac{\sigma^2(1-f)}{n} + \dfrac{\sigma^2(k-1)}{nk(n-1)};$

(c) under systematic sampling $E_m\left[\text{MSE}\left(\bar{y}_{\text{lr}}\right)_{\text{sys}}\right] = \dfrac{\sigma^2(1-f)}{n} + \dfrac{\sigma^2(k-1)(k+1)}{nk^2(n-1)(n+1)};$

(d) under centred sampling,

$$E_m\left[\mathrm{MSE}(\bar{y}_{\mathrm{lr}})_{\mathrm{css}}\right]=\begin{cases}\dfrac{\sigma^2(1-f)}{n} & \text{if } k \text{ is odd,}\\[2ex] \dfrac{\sigma^2(1-f)}{n}+\dfrac{3\sigma^2}{nk^2(n-1)(n+1)} & \text{if } k \text{ is even.}\end{cases}$$

Hint: Fountain and Pathak (1989).

Exercise 7.15. (a) Under the model $m\ :\ y_i=\alpha+\beta i+\gamma x_i+e_i$, where α, β and γ are constants, $e_i\sim N(0,\ \sigma^2)$, and x_i is a periodic function of i with a period of $2l$, and $N=2lQ=nk$ Compare the expected variances given by $E_m[V_{\mathrm{srswr}}]$, $E_m[V_{\mathrm{sys}}]$, $E_m[V_{\mathrm{srswor}}]$, $E_m[V_{\mathrm{bss}}]$, and $E_m[V_{\mathrm{mss}}]$.

Hint: Madow and Madow (1944).

(b) After eliminating the linear trend from the model : m: $y_i=\alpha+\beta i+\gamma x_i+e_i$, show that the following relations holds:

(i) $E_m[V_{\mathrm{sys}}]\le E_m[V_{\mathrm{bss}}]$

(ii) $E_m[V_{\mathrm{bss}}]=E_m[V_{\mathrm{sys}}]\le E_m[V_{\mathrm{srswor}}]$ for k/l and n both are even.

(iii) $E_m[V_{\mathrm{sys}}]\approx E_m[V_{\mathrm{srswor}}]$ for k/l being odd.

and

(iv) $E_m[V_{\mathrm{srswor}}]\approx E_m[V_{\mathrm{mss}}]$ for k/l, k and $n/2$ are odd.

Hint: Bellhouse and Rao (1975).

Exercise 7.16. Under the model $m\ : y_i=\rho y_{i-1}+e_i$, $i=1,2,...,N$, where $e_i\sim(0,\ \sigma^2)$ and are independent and identically distributed then show that the centrally located systematic sample (c.s.s) is optimum.

Hint: Blight (1973).

Exercise 7.17. Let N $(=kn,\ \text{where}\ n\le k)$ be the population size. The population units $U_1, U_2,...,U_N$ are arranged in an $n\times k$ matrix M (say) and the jth row of the matrix M is denoted by R_j, $j=1,2,...,n$. Note that the elements of the row $R_j=\{U_{(j-1)k+i},\ i=1,2,...,k\}$. Consider a sampling scheme to select n units from the diagonal of the matrix M and hence belongs to different rows and columns. Let y_{ij} be the value corresponding to i^{th} row and j^{th} column then the sample will consist of the observations, given by $S_i=\{y_{1r},\ y_{2(r+1)},...,\ y_{n(r+n-1)}\}$, for $i=1,2,...,\ k$. Note that if $r+n-1>k$ then it has to be reduced to mod k.

(a) In the hypothetical situation of linear tread $Y_i=a+bX_i$, $i=1,2,....,N$, show that the variances of the simple random sample mean $(\bar{y}_r)$, systematic sample mean $(\bar{y}_{\mathrm{sy}})$ and the diagonal systematic sampling mean $(\bar{y}_{\mathrm{dsy}})$ are given by

$$V(\bar{y}_r) = \frac{(k-1)(N+1)b^2}{12}, \quad V(\bar{y}_{sy}) = \frac{(k-1)(k+1)b^2}{12}, \text{ and } V(\bar{y}_{dsy}) = \frac{(k-n)(n(k-n)+2)b^2}{12n}$$

(b) Also show that

$$V(\bar{y}_{dsy}) \le V(\bar{y}_{sy}) \le V(\bar{y}_r) .$$

(c) If $k = n$, then $V(\bar{y}_{dsy}) = 0$ that is the diagonal systematic sampling becomes a completely trend free sampling.

Hint: Subramani (2000), Mukerjee and Sengupta (1990).

PRACTICAL PROBLEMS

Practical 7.1. Select a sample of 10 states from population 1 consisting of 50 states by using systematic sampling scheme. Collect the information on the nonreal estate farm loans from the states selected in the sample. Assume that there is a linear trend between the real estate farm loans and the serial numbers. Apply the Yates end correction to obtain unbiased estimate of population mean.

Practical 7.2. Select a sample of 10 states from population 1 consisting of 50 states by using systematic sampling scheme. Collect the information on the nonreal estate farm loans from the states selected in the sample. Use an appropriate method for estimating the variance of the estimator of population mean.

Practical 7.3. Use two letter abbreviations to sort the 50 states listed in population 1 of the Appendix, and then write them on a circle in the clock wise direction as shown below:

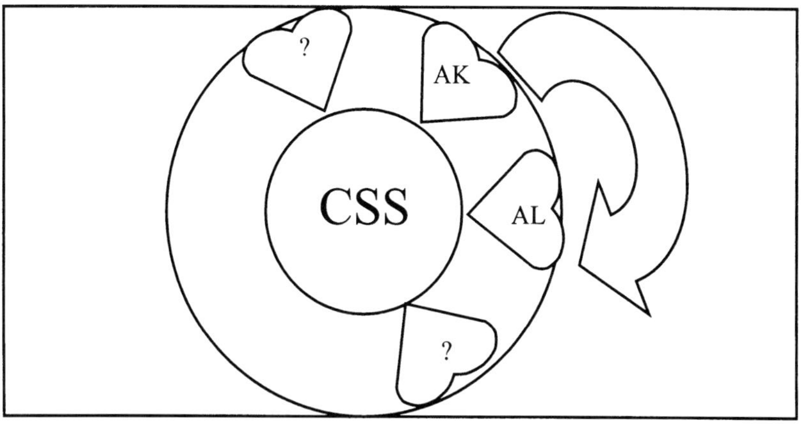

Select a circular systematic sample of 10 states out of 50 states, and collect the information from population 1 given in the Appendix on the real estate farm loans from the states selected in the CS sample.

(a) Estimate the average real estate farm loans in the United States.

(b) Select another independent circular systematic sample of 10 states. Use the information from the both samples to estimate the variance of the estimator of average real estate farm loans based on circular systematic sampling.

Practical 7.4. Select a sample of 15 states from population 1 consisting of 50 states by using Singh and Singh's systematic sampling scheme. Calculate the minimum number of phases to select a sample of 15 states. Collect the information on the real estate farm loans and nonreal estate farm loans from the states selected in the sample. Assume that information on nonreal estate farm loans is known for the whole population. Estimate the total real estate farm loans and find estimator of its variance.

Given: $X = 43908.12$.

Practical 7.5. Select three sub-samples each consisting of 5 states from population 1 with 50 states by using systematic sampling. Collect the information on the nonreal estate farm loans from the states selected in the sample. Obtain a pooled estimate of the average nonreal estate farm loans in the United States. Use an appropriate method for estimating the variance of the resultant pooled estimator.

Practical 7.6. From the population 1 given in the appendix, apply the following methods to select a sample of 10 units and compare the estimate.

(a) Select a sample with random start $1 \le i \le N$ with the units:

$$s(n) = \begin{cases} \{i + jk,\ N - i - jk + 1,\ j = 0,1,2,...,(n/2 - 1)\} & \text{if } n \text{ is even,} \\ \{i + jk,\ N - i - jk + 1,\ i + (n-1)k/2,\ j = 0,1,...,\{(n-1)/2 - 1\}\} & \text{if } n \text{ is odd,} \end{cases}$$

which is called modified systematic sampling.

Hint: Singh, Jindal and Garg (1968).

(b) Select a sample with random start $1 \le i \le N$ with the units:

$$s(n) = \begin{cases} \{i + 2jk,\ 2(j+1)k - i + 1,\ j = 0,1,2,...,(n/2 - 1)\} & \text{if } n \text{ is even,} \\ \{i + 2jk,\ 2(j+1)k - i + 1,\ i + (n-1)k,\ j = 0,1,...,\{(n-3)/2\}\} & \text{if } n \text{ is odd,} \end{cases}$$

which is also called balanced systematic sampling.

Hint: Murthy (1977), Sethi (1965).

(c) Apply the following formulae to estimate the variance in each situation and construct 95% confidence intervals:

(i) $\hat{v}_y = \dfrac{1-f}{6n(n-2)} \sum_{j=3}^{n} \left(y_{i,j} - 2y_{i,j-1} + y_{i,j-2}\right)^2$, called Yates' method;

(ii) $\hat{v}_{sd} = \dfrac{1-f}{2n(n-1)} \sum_{j=2}^{n} \left(y_{i,j} - y_{i,j-1}\right)^2$, , called the method of successive differences.

Hint: Yates (1948).

(d) Divide the sample into two parts, say s_1 and s_2 (each consisting of 5 units). Estimate the variance using the following formulae:

$$\hat{v}_k = \frac{1}{4}\left(\bar{y}_{s_1} - \bar{y}_{s_2}\right)^2 .$$

Hint: Koop (1971).

Practical 7.7. Select all possible five samples each of 10 units with their respective random start $1 \le i \le 5$ from the population 1 given in the Appendix. Calculate the following estimators of the variance, given by

$$\hat{v}_1 = \frac{(1-f)}{n} s_y^2 \,, \quad \text{where } s_y^2 = \frac{1}{n-1}\sum_{j=1}^{n}\left(y_{i,j} - \bar{y}_i\right)^2 \text{ and } \bar{y}_i = \frac{1}{n}\sum_{j=1}^{n} y_{i,j} \,,$$

$$\hat{v}_2 = \frac{1-f}{2n(n-1)}\sum_{j=1}^{n}\left(y_{i,j} - y_{i,j-1}\right)^2 \,, \quad \hat{v}_3 = \frac{1-f}{6n(n-2)}\sum_{j=3}^{n}\left(y_{i,j} - 2y_{i,j-1} + y_{i,j-2}\right)^2 \,,$$

and

$$\hat{v}_4 = \frac{1-f}{35n(n-4)}\sum_{j=5}^{n}\left(\frac{y_{i,j}}{2} - y_{i,j-1} + y_{i,j-2} - y_{i,j-3} + \frac{y_{i,j-4}}{2}\right)^2 .$$

Also find their empirically expected values, biases, and mean square errors. Suggest an efficient estimator based on the empirical results.

Practical 7.8. Select $t = 2$ systematic samples, each of $n = 10$ units, with two different random starts from population 1 consisting of $N = 50$ units. Let $k = 5$, so that the condition $N = nk = 10 \times 5 = 50$ is satisfied. Also note that the sample size $n = 10$ is dividable by $t = 2$. Take an SRS of $t = 2$ units from the first block of $tk = 2 \times 5 = 10$ units and select every $tk = 2 \times 5 = 10^{\text{th}}$ unit thereafter. This method divides the population into tk samples each of size $m = N/(tk) = n/t$ units and selecting $t = 2$ systematic samples (or clusters) by an SRSWOR sampling. From the selected t samples, each consisting of m units, find the sample means as $\bar{y}_j$, $j = 1, 2, ..., t$. Estimate the population mean on the basis of pooled estimator

$$\bar{\bar{y}}_{\text{pooled}} = \frac{1}{m}\sum_{j=1}^{t} \bar{y}_j \,, \text{ and estimate its variance as } \hat{v} = \frac{(k-1)}{kt}\frac{1}{(t-1)}\sum_{j=1}^{t}\left(\bar{y}_j - \bar{\bar{y}}_{\text{pooled}}\right)^2 .$$

Derive 95% confidence interval estimate.
Hint: Iachan (1982), Gautschi (1957), Shiue (1966).

Practical 7.9. The following figure shows nine trees on a portion of Indian Grand Trunk Road with their heights (feet) and timbre weights (kgs) as

Height:	10.5	11.5	12.2	8.4	15.9	12.7	14.3	11.7	10.5
Weight:	202	125	219	107	187	198	210	213	209

(a) Select all the possible three samples each of three trees using systematic sampling, and estimate the average height (and weight) from each sample.
(b) Use these estimates to estimate the variance. (**Rule:** Use first column of the Pseudo-Random Number Table 1 given in the Appendix.)

8. STRATIFIED AND POST-STRATIFIED SAMPLING

8.0 INTRODUCTION

Stratified and post-stratified sampling schemes are useful survey techniques commonly used by government agencies, private consultants, and applied statisticians. Let us differentiate between stratified and post-stratified sampling.

Stratified Sampling: In this sampling scheme, the population of N units is subdivided into L homogeneous sub-groups called strata, such that the hth stratum consists of N_h units, where $h = 1,2...,L$ and $\sum_{h=1}^{L} N_h = N$. From the hth population stratum consisting of N_h units, a sample of size n_h is drawn using any sampling design such that $\sum_{h=1}^{L} n_h = n$, the required sample size. A pictorial representation of stratified sampling scheme is given in Figure 8.0.1.

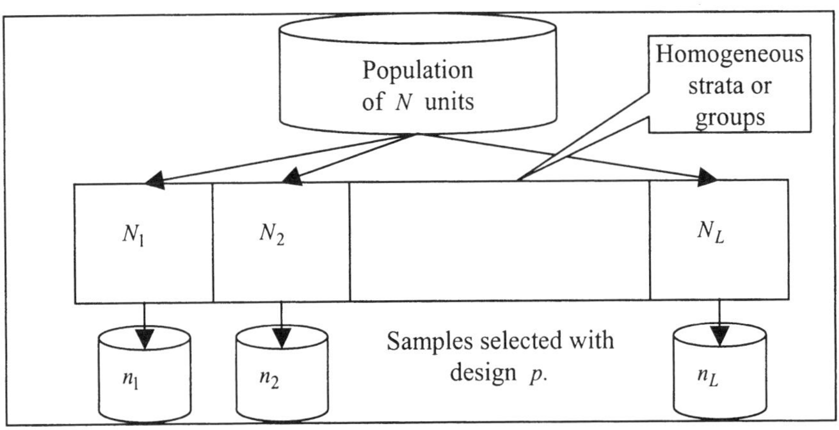

Fig. 8.0.1 Stratified Sampling Scheme.

Post-stratified sampling: In a post-stratified sampling scheme a sample of n units is first selected from the population of N units using any sampling design. The population is stratified into L strata on the basis of some auxiliary information available. In post-stratified sampling, the values of N_h where $h = 1, 2..., L$ and $\sum_{h=1}^{L} N_h = N$ may or may not be known. Each sample unit selected with the chosen design is then post-stratified or placed in the hth stratum based on the auxiliary information associated with each sampled unit such that $\sum_{h=1}^{L} n_h = n$.

A pictorial representation of post-stratified sampling scheme is given in Figure 8.0.2.

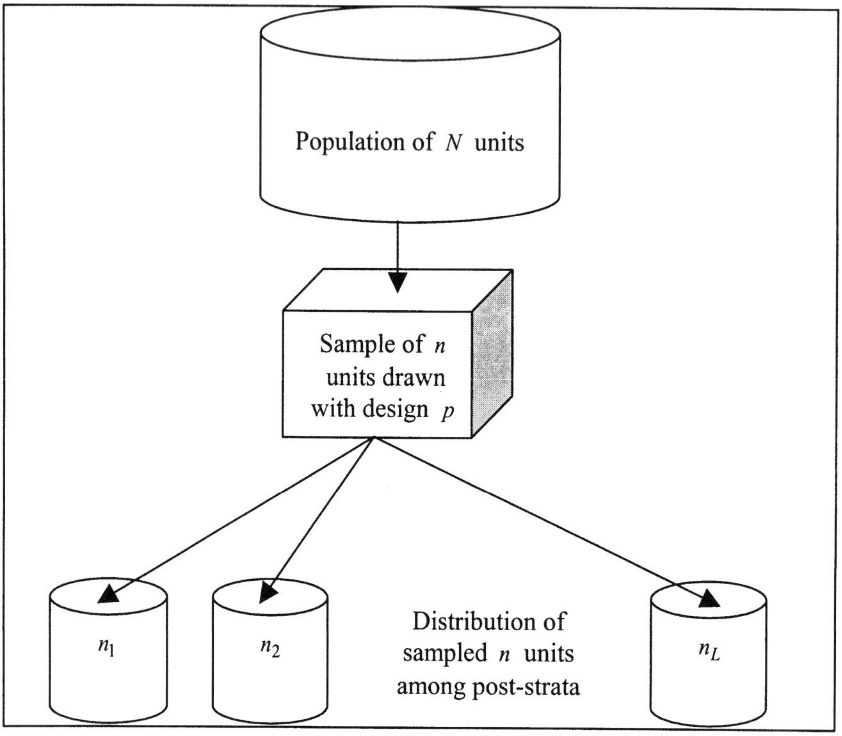

Fig. 8.0.2. Post-stratified Sampling Scheme

Thus the difference between stratified and post-stratified sampling schemes is that in stratified sampling the sub-sample size n_h is a fixed or predecided number, whereas in post-stratified sampling it is a random variable

Now we will discuss each case in detail.

8.1 STRATIFIED SAMPLING

The population of N units is first subdivided into L homogeneous subgroups called strata, such that the hth stratum consists of N_h units, where $h = 1, 2, ..., L$ and $\sum_{h=1}^{L} N_h = N$. Let Y_{hi} be the ith population value of the study variable in the hth stratum, $i = 1, 2, ..., N_h$ such that the hth stratum population mean is given by

$$\bar{Y}_h = N_h^{-1} \sum_{i=1}^{N_h} Y_{hi}, \quad \text{for } h = 1, 2, ..., L.$$

Obviously, using the concept of weighted average the true population mean of the whole population can be written as:

$$\bar{Y} = \frac{N_1\bar{Y}_1 + N_2\bar{Y}_2 + ...+ N_L\bar{Y}_L}{N_1 + N_2 + ...+ N_L} = \frac{N_1\bar{Y}_1 + N_2\bar{Y}_2 + ...+ N_L\bar{Y}_L}{N}$$

$$= \left(\frac{N_1}{N}\right)\bar{Y}_1 + \left(\frac{N_2}{N}\right)\bar{Y}_2 + ...+ \left(\frac{N_L}{N}\right)\bar{Y}_L = W_1\bar{Y}_1 + W_2\bar{Y}_2 + ...+ W_L\bar{Y}_L = \sum_{h=1}^{L} W_h\bar{Y}_h .$$

Consider a sample of size n_h is drawn using SRSWOR sampling from the hth population stratum consisting of N_h units such that $\sum_{h=1}^{L} n_h = n$, the required sample size. Assume the value of the ith unit of the study variable selected from the hth stratum is denoted by y_{hi}, where $i = 1, 2, ..., n_h$ and $W_h = N_h/N$ is the known proportion of population units falling in the hth stratum.

Then we have the following theorems:

Theorem 8.1.1. An unbiased estimator of population mean $\bar{Y}$ is given by

$$\bar{y}_{\text{st}} = \sum_{h=1}^{L} W_h\bar{y}_h \tag{8.1.1}$$

where $\bar{y}_h = n_h^{-1}\sum_{i=1}^{n_h} y_{hi}$ denotes the hth stratum sample mean.

Proof. We have

$$E(\bar{y}_{\text{st}}) = E\left[\sum_{h=1}^{L} W_h\bar{y}_h\right] = \sum_{h=1}^{L} W_h E(\bar{y}_h) = \sum_{h=1}^{L} W_h\bar{Y}_h = \sum_{h=1}^{L} W_h\left[\frac{1}{N_h}\sum_{i=1}^{N_h} Y_{hi}\right]$$

$$= \sum_{h=1}^{L}\frac{N_h}{N}\left[\frac{1}{N_h}\sum_{i=1}^{N_h} Y_{hi}\right] = \frac{1}{N}\sum_{h=1}^{L}\sum_{i=1}^{N_h} Y_{hi} = \frac{Y}{N} = \bar{Y} .$$

Hence the theorem.

Theorem 8.1.2. Under SRSWOR sampling, the variance of the estimator $\bar{y}_{\text{st}}$ is given by

$$V(\bar{y}_{\text{st}}) = \sum_{h=1}^{L} W_h^2\left(\frac{1 - f_h}{n_h}\right)S_{hy}^2 \tag{8.1.2}$$

where $S_{hy}^2 = (N_h - 1)^{-1}\sum_{i=1}^{N_h}\left(Y_{hi} - \bar{Y}_h\right)^2$ denotes the hth stratum population variance, $\bar{Y}_h = N_h^{-1}\sum_{i=1}^{N_h} Y_{hi}$ denotes the hth stratum population mean and $f_h = n_h/N_h$.

Proof. Note that the strata are independent and under SRSWOR sampling we have

$$V(\bar{y}_h) = \left(\frac{1 - f_h}{n_h}\right)S_{hy}^2 .$$

Thus we have

$$V(\bar{y}_{st}) = V\left[\sum_{h=1}^{L} W_h \bar{y}_h\right] = \sum_{h=1}^{L} W_h^2 V(\bar{y}_h) = \sum_{h=1}^{L} W_h^2 \left(\frac{1-f_h}{n_h}\right) S_{hy}^2 .$$

Hence the theorem.

Theorem 8.1.3. An unbiased estimator of $V(\bar{y}_{st})$ under SRSWOR sampling is

$$\hat{v}(\bar{y}_{st}) = \sum_{h=1}^{L} W_h^2 \left(\frac{1-f_h}{n_h}\right) s_{hy}^2, \tag{8.1.3}$$

where $s_{hy}^2 = (n_h - 1)^{-1} \sum_{h=1}^{n_h} (y_{hi} - \bar{y}_h)^2$ and $\bar{y}_h = n_h^{-1} \sum_{i=1}^{n_h} y_{hi}$.

Proof. Obvious on using

$$E(s_{hy}^2) = S_{hy}^2 .$$

Remark 8.1.1. The $(1-\alpha)100\%$ confidence interval estimate is stratified random sampling will be given by

$$\bar{y}_{st} \ \pm \ t_{\alpha/2}(\mathrm{df} = n - L)\sqrt{\hat{v}(\bar{y}_{st})} . \tag{8.1.4}$$

Example 8.1.1. The following data shows daily temperatures in London and New York cities in °F as follows:

City	Temp
NY	48
London	54
NY	52
NY	47
London	57
NY	54
NY	49
London	59
NY	53
NY	50
NY	52
NY	57
London	55
NY	54
London	68
NY	49
NY	51
London	61
NY	55
NY	53
London	50

(a) If we select an SRSWOR sample of 4 units, find the variance of estimator of population mean.
(b) If we stratify the population on the basis of location or cities, and then select two units from each city, find the variance of the estimator of mean in stratified sampling.
(c) Find the relative efficiency of the stratified sampling over the simple random sampling.
(d) Select an SRSWOR sample of four units from the above population and construct 95% confidence interval estimate of the population mean.
(e) Select two units using SRSWOR sampling from London and two units from New York using stratified random sampling, and construct 95% confidence interval estimate of the population mean.

Solution (a) SRSWOR Sampling: From the above data we are given $N = 21$, $n = 4$, $f = n/N = 4/21$, and

City	Y_i	Y_i^2
NY	48	2304
London	54	2916
NY	52	2704
NY	47	2209
London	57	3249
NY	54	2916
NY	49	2401
London	59	3481
NY	53	2809
NY	50	2500
NY	52	2704
NY	57	3249
London	55	3025
NY	54	2916
London	68	4624
NY	49	2401
NY	51	2601
London	61	3721
NY	55	3025
NY	53	2809
London	50	2500
Sum	1128	61064

which implies

$$\sum_{i=1}^{N} Y_i = 1128, \quad \sum_{i=1}^{N} Y_i^2 = 61064$$

and

$$s_y^2 = (N-1)^{-1}\left\{\sum_{i=1}^{N} Y_i^2 - N^{-1}\left(\sum_{i=1}^{N} Y_i\right)^2\right\} = \frac{1}{21-1}\left\{61064 - \frac{(1128)^2}{21}\right\} = 23.71 .$$

Thus the variance of the sample mean estimator is given by

$$V_{\text{srswor}}(\bar{y}) = \frac{(1-f)}{n} S_y^2 = \frac{(1-4/21)}{4} \times 23.71 = 4.715 .$$

(b) Stratified sampling: We stratify the population into two strata based on location as follows:

	Stratum 1 LONDON	
	Y_{1i}	Y_{1i}^2
	54	2916
	57	3249
	59	3481
	55	3025
	68	4624
	61	3721
	50	2500
Sum	404	23516

	Stratum 2 NEW YORK	
	Y_{2i}	Y_{2i}^2
	48	2304
	52	2704
	47	2209
	54	2916
	49	2401
	53	2809
	50	2500
	52	2704
	57	3249
	54	2916
	49	2401
	51	2601
	55	3025
	53	2809
Sum	724	37548

From Stratum 1 we have

$$N_1 = 7 , \quad W_1 = \frac{N_1}{N} = \frac{7}{21} = 0.3333333 , \quad f_1 = \frac{n_1}{N_1} = \frac{2}{7} , \quad \sum_{i=1}^{N_1} Y_{1i} = 404 , \quad \sum_{i=1}^{N_1} Y_{1i}^2 = 23516$$

and

$$S_{1y}^2 = (N_1 - 1)^{-1}\left\{\sum_{i=1}^{N_1} Y_{1i}^2 - N_1^{-1}\left(\sum_{i=1}^{N_1} Y_{1i}\right)^2\right\} = \frac{23516 - (404)^2/7}{7-1} = 33.24 .$$

From Stratum 2 we have

$$N_2 = 14 , \quad W_2 = \frac{N_2}{N} = \frac{14}{21} = 0.6666667 , \quad f_2 = \frac{n_2}{N_2} = \frac{2}{14} , \quad \sum_{i=1}^{N_2} Y_{2i} = 724 , \quad \sum_{i=1}^{N_2} Y_{2i}^2 = 37548$$

and

$$S_{2y}^2 = (N_2-1)^{-1}\left\{\sum_{i=1}^{N_2} Y_{2i}^2 - N_2^{-1}\left(\sum_{i=1}^{N_2} Y_{2i}\right)^2\right\} = \frac{37548-(724)^2/14}{14-1} = 8.22.$$

Thus the variance of the estimator of the population mean in the stratified random sampling is given by

$$V(\bar{y}_{\text{st}}) = \sum_{h=1}^{L} W_h^2\left(\frac{1-f_h}{n_h}\right)S_{hy}^2 = \sum_{h=1}^{2} W_h^2\left(\frac{1-f_h}{n_h}\right)S_{hy}^2 = W_1^2\left(\frac{1-f_1}{n_1}\right)S_{1y}^2 + W_2^2\left(\frac{1-f_2}{n_2}\right)S_{2y}^2$$

$$= \left(\frac{7}{21}\right)^2\left(\frac{1-2/7}{2}\right)\times 33.24 + \left(\frac{14}{21}\right)^2\left(\frac{1-2/14}{2}\right)\times 8.22$$

$$= 1.319246 + 1.565714 = 2.88496.$$

(c) Relative efficiency: The percent relative efficiency of the stratified random sampling over SRSWOR sampling is given by

$$\text{RE} = \frac{V_{\text{srswor}}(\bar{y})}{V(\bar{y}_{\text{st}})}\times 100 = \frac{4.715}{2.88496}\times 100 = 163.45\,.$$

(d) 95% CI estimate using SRSWOR sampling: We started from first two columns of the Pseudo-Random Number Table 1 given in the Appendix to select 4 distinct random numbers between 01 and $N = 21$ (both inclusive) as follows:

Random No.	City Selected	Temp ^{0}F (y_i)	y_i^2
01	NY	48	2304
04	NY	47	2209
05	London	57	3249
03	NY	52	2704
	Sum	204	10466

Thus the sample mean estimate of the population mean based on SRSWOR is

$$\bar{y} = \frac{1}{n}\sum_{i=1}^{n} y_i = \frac{204}{4} = 51\,.$$

The sample variance is given by

$$s_y^2 = (n-1)^{-1}\left\{\sum_{i=1}^{n} y_i^2 - \left(\sum_{i=1}^{n} y_i\right)^2 \Big/ n\right\} = \frac{10466-(204)^2/4}{4-1} = 20.66667\,.$$

An estimate of the variance of the sample mean is given by

$$\hat{v}_{\text{srswor}}(\bar{y}) = \left(\frac{1-f}{n}\right)s_y^2 = \left(\frac{1-4/21}{4}\right)\times 20.66667 = 4.182\,.$$

A $(1-\alpha)100\%$ Confidence Interval estimate of the population mean is given by

$$\bar{y} \;\pm\; t_{\alpha/2}(\text{df} = n-1)\sqrt{\hat{v}_{\text{srswor}}(\bar{y})}\,.$$

Thus a 95% Confidence Interval estimate of the population mean is given by

$$51 \;\pm\; t_{0.025}(\text{df} = 4-1)\sqrt{4.182}\,.$$

Using Table 2 from the Appendix we have

$$51 \pm 3.182\sqrt{4.182}\,, \quad \text{or} \qquad [44.49, 57.51].$$

Note that the true population mean $\bar{Y} = 53.71$ lies in the 95% confidence interval. The interpretation of this confidence interval estimate is that we are 95% sure that the true population mean lies between 44.49 °F to 57.51 °F.

(e) 95% CI estimate using stratified random sampling: We selected two units from Stratum 1 and two units from Stratum 2 using lottery methods:

	Stratum 1 LONDON		Stratum 2 NEW YORK	
	y_{1i}	y_{1i}^2	y_{2i}	y_{2i}^2
	57	3249	52	2704
	55	3025	57	3249
Sum	112	6274	109	5953

From sample stratum 1

$$\sum_{i=1}^{n_1} y_{1i} = 112\,, \quad \bar{y}_1 = \frac{1}{n_1}\sum_{i=1}^{n_1} y_{1i} = \frac{112}{2} = 56\,, \quad \sum_{i=1}^{n_1} y_{1i}^2 = 6274$$

and

$$s_{1y}^2 = (n_1 - 1)^{-1}\left\{\sum_{i=1}^{n_1} y_{1i}^2 - n_1^{-1}\left(\sum_{i=1}^{n_1} y_{1i}\right)^2\right\} = \frac{6274 - (112)^2/2}{2-1} = 2.0\,.$$

From sample stratum 2

$$\sum_{i=1}^{n_2} y_{2i} = 109\,, \quad \bar{y}_2 = \frac{1}{n_2}\sum_{i=1}^{n_2} y_{2i} = \frac{109}{2} = 54.5\,, \quad \sum_{i=1}^{n_2} y_{2i}^2 = 5953$$

and

$$s_{2y}^2 = (n_2 - 1)^{-1}\left\{\sum_{i=1}^{n_2} y_{2i}^2 - n_2^{-1}\left(\sum_{i=1}^{n_2} y_{2i}\right)^2\right\} = \frac{5953 - (109)^2/2}{2-1} = 12.5\,.$$

Thus an estimator of the variance of the estimator of the population mean in stratified random sampling is

$$\hat{v}(\bar{y}_{st}) = \sum_{h=1}^{L} W_h^2\left(\frac{1-f_h}{n_h}\right)s_{hy}^2 = \sum_{h=1}^{2} W_h^2\left(\frac{1-f_h}{n_h}\right)s_{hy}^2 = W_1^2\left(\frac{1-f_1}{n_1}\right)s_{1y}^2 + W_2^2\left(\frac{1-f_2}{n_2}\right)s_{2y}^2$$

$$= \left(\frac{7}{21}\right)^2\left(\frac{1-2/7}{2}\right)\times 2.0 + \left(\frac{14}{21}\right)^2\left(\frac{1-2/14}{2}\right)\times 12.5$$

$$= 0.079365 + 2.380952 = 2.460317\,.$$

Also the stratified estimate of the population mean is given by

$$\bar{y}_{st} = \sum_{h=1}^{L} W_h\bar{y}_h = \sum_{h=1}^{2} W_h\bar{y}_h = W_1\bar{y}_1 + W_2\bar{y}_2 = \left(\frac{7}{21}\right)\times 56 + \left(\frac{14}{21}\right)\times 54.5 = 55\,.$$

Now a $(1-\alpha)100\%$ confidence interval estimate of the population mean $\bar{Y}$ using stratified random sampling is given by

$$\bar{y}_{st} \pm t_{\alpha/2}(\text{df} = n - L)\sqrt{\hat{v}(\bar{y}_{st})}, \quad \text{or} \quad \bar{y}_{st} \pm t_{0.025}(\text{df} = 4 - 2)\sqrt{\hat{v}(\bar{y}_{st})}.$$

Using Table 2 from the Appendix the required 95% CI estimate is given by

$$55 \pm 4.303\sqrt{2.460317}, \quad \text{or} \quad 55 \pm 4.303 \times 1.56854, \quad \text{or} \quad [48.251, \; 61.749].$$

Again note that the true population mean $\bar{Y} = 53.71$ lies in 95% confidence interval. The interpretation of this confidence interval estimate is that we are 95% sure that the true population mean lies between 48.251 °F to 61.749 °F.

Conclusion: The length of the CI estimate obtained using stratified sampling is $L_{st} = 61.749 - 48.251 = 13.498$, is more than that for SRSWOR sampling $L_{srs} = 57.51 - 44.49 = 13.02$, thus we may conclude that stratified sampling does not perform better in this particular situation, perhaps because of small sample size.

Example 8.1.2. Select two units by SRSWOR sampling from each stratum of the population 5 given in the Appendix. Collect the information on the yield/hectare of the tobacco crop from the countries selected in the sample. Assuming that the total number of countries in each continent are known, estimate the average yield/hectare of the tobacco crop in the world. Construct the 95% confidence interval.

Solution. Using Pseudo-Random Number (PRN) Table 1 given in the Appendix we have the following sample information and some results.

Stratum No.	Column no. used from PRN table	Random No.	Countries selected	Yield/ hectare y_{hi}	W_h	$\bar{y}_h$	s^2_{hy}
1	1	5	Nicaragua	2.03	0.0566	2.015	0.00045
		6	Panama	2.00			
2	2	4	Jamaica	1.99	0.0566	1.750	0.11520
		2	Dominican Rep	1.51			
3	3	2	Belgium--Lux	3.69	0.0755	3.240	0.40500
		8	Spain	2.79			
4	4 and 5	07	Macedonia	1.36	0.0943	1.850	0.48020
		08	Poland	2.34			
5	6 and 7	07	Moldova	2.05	0.1132	1.155	1.60210
		02	Armenia	0.26			
6	8	2	Libya	1.61	0.0377	1.785	0.06130
		1	Algeria	1.96			
7	9 and 10	20	Nigeria	2.10	0.2830	2.305	0.08410
		12	Kenya	2.51			
8	11 and 12	15	Thailand	1.33	0.1604	1.090	0.11520
		06	Indonesia	0.85			
9	13 and 14	05	Lebanon	1.33	0.0944	1.240	0.01620
		07	Syria	1.15			
10	15	3	New Zealand	2.58	0.0283	1.765	1.32850
		2	Solomon Islands	0.95			

Then we have

Stratum no.	N_h	f_h	W_h	$\bar{y}_h$	$W_h\bar{y}_h$	s_{hy}^2	$\hat{v}_h$
1	6	0.33	0.0566	2.015	0.1140	0.00045	0.000000
2	6	0.33	0.0566	1.750	0.0990	0.11520	0.000124
3	8	0.25	0.0754	3.240	0.2446	0.40500	0.000863
4	10	0.20	0.0943	1.850	0.1744	0.48020	0.001708
5	12	0.17	0.1132	1.155	0.1307	1.60210	0.008520
6	4	0.50	0.0377	1.785	0.0672	0.06130	0.000022
7	30	0.07	0.2830	2.305	0.6523	0.08410	0.003132
8	17	0.12	0.1603	1.090	0.1748	0.11520	0.001302
9	10	0.20	0.0943	1.240	0.1171	0.01620	0.000057
10	3	0.67	0.0283	1.765	0.0499	1.32850	0.000176
				Sum	1.8243		0.015905

where

$$\hat{v}_h = W_h^2\left(\frac{1-f_h}{n_h}\right)s_{hy}^2 .$$

Thus a stratified estimate of the average yield/hectare of the Tobacco crop in the world is

$$\bar{y}_{st} = \sum_{h=1}^{L} W_h\bar{y}_h = 1.8243$$

and an estimate of the variance of the stratified estimator is given by

$$\hat{v}(\bar{y}_{st}) = \sum_{h=1}^{L} W_h^2\left(\frac{1-f_h}{n_h}\right)s_{hy}^2 = 0.015905 .$$

A $(1-\alpha)100\%$ confidence interval estimate of the average yield/hectare of the Tobacco crop during 1998 in the world is

$$\bar{y}_{st} \mp t_{\alpha/2}(\text{df} = n-L)\sqrt{\hat{v}(\bar{y}_{st})} .$$

Using Table 2 from the Appendix the 95% confidence interval estimate is

$$\bar{y}_{st} \mp t_{0.025}(\text{df} = 20-10)\sqrt{\hat{v}(\bar{y}_{st})}, \text{ or } 1.8243 \mp 2.228\sqrt{0.015905}\text{ , or } [1.5433,\ 2.1052].$$

We know now that the variance of the estimator of population mean under stratified sampling is given by formula in (8.1.2). Furthermore, we known that the choice of sample size n_h for stratified sampling has to be decided once the sample size n is chosen. A natural question arises: What choice of n_h will make the variance of the estimator $\bar{y}_{st}$ a minimum? There are several ways to answer this question, but we will discuss only a few of them here.

8.2 DIFFERENT METHODS OF SAMPLE ALLOCATION

Below we discuss different methods of sample allocation namely:

(a) equal allocation; (b) proportional allocation; and (c) optimum allocation.

Let us discuss each one of the above methods in detail:

8.2.1 EQUAL ALLOCATION

As the name of the method suggests the sub sample sizes are equal, *i.e.*, $n_h = n/L$. Under this choice of sample allocation, the variance of the estimator $\bar{y}_{\text{st}}$ reduces to

$$V(\bar{y}_{\text{st}})_{\text{E}} = \sum_{h=1}^{L} W_h^2\left(\frac{1-f_h}{n_h}\right)S_{hy}^2 = \sum_{h=1}^{L} W_h^2\left(\frac{N_h - n_h}{N_h n_h}\right)S_{hy}^2 = \frac{1}{N^2}\sum_{h=1}^{L}\frac{N_h(N_h - n/L)}{n/L}S_{hy}^2$$

$$= \frac{1}{nN^2}\sum_{h=1}^{L} N_h(LN_h - n)S_{hy}^2 . \quad (8.2.1)$$

An unbiased estimator of $V(\bar{y}_{\text{st}})_{\text{E}}$ is given by

$$\hat{v}(\bar{y}_{\text{st}})_{\text{E}} = \frac{1}{nN^2}\sum_{h=1}^{L} N_h(LN_h - n)s_{hy}^2 . \quad (8.2.2)$$

8.2.2 PROPORTIONAL ALLOCATION

Under this allocation the sub sample size from each stratum is proportional to the size of the subpopulation in the stratum. That is

$$n_h \propto N_h . \quad (8.2.3)$$

To find the constant of proportionality we have

$$n_h = KN_h . \quad (8.2.4)$$

Taking the sum on both sides of (8.2.4) over all possible strata we have

$$\sum_{h=1}^{L} n_h = K\sum_{h=1}^{L} N_h , \quad \text{or} \quad n = KN$$

which implies that

$$K = n/N . \quad (8.2.5)$$

On substituting (8.2.5) in (8.2.4) we have the proportional allocation in the hth stratum as

$$n_h = \left(\frac{n}{N}\right)N_h = n\left(\frac{N_h}{N}\right) = nW_h . \quad (8.2.6)$$

Under this allocation the variance of the estimator $\bar{y}_{\text{st}}$ becomes

$$V(\bar{y}_{\text{st}})_{\text{p}} = \frac{1}{N^2}\sum_{h=1}^{L}\frac{N_h(N_h - n_h)}{n_h}S_{hy}^2 = \frac{1}{N^2}\sum_{h=1}^{L}\frac{N_h(N_h - nW_h)}{nW_h}S_{hy}^2$$

$$= \frac{1}{nN}\sum_{h=1}^{L} N_h\left(1-\frac{n}{N}\right)S_{hy}^2 = \frac{(1-f)}{n}\sum_{h=1}^{L} W_h S_{hy}^2 . \quad (8.2.7)$$

If $f = n/N$ is negligible then we have

$$V(\bar{y}_{st})_p = \frac{1}{n}\sum_{h=1}^{L} W_h S_{hy}^2 . \tag{8.2.8}$$

The unbiased estimator of $V(\bar{y}_{st})_p$ can easily be obtained from (8.2.7) and (8.2.8) by replacing S_{hy}^2 with its sample analogue s_{hy}^2.

Example 8.2.1. Select a sample of 40 countries from population 5 using the method of proportional allocation. Record the yield/hectare of the tobacco crop from the selected countries. Estimate the average yield/hectare of the tobacco crop in the world. Estimate the variance under the method of proportional allocation. Construct a 95% confidence interval.

Solution. By the method of proportional allocation, the number of units to be selected from the hth stratum is given by, $n_h = n N_h / N$. Thus we have

Strata No.	1	2	3	4	5	6	7	8	9	10	Total
N_h	6	6	8	10	12	4	30	17	10	3	106
n_h	3	3	3	3	4	2	11	6	3	2	40

Using Pseudo-Random Number (PRN) Table 1 given in the Appendix we collect the following sample information.

Stratum No.	Column no. used from PRN table	Random no.	Country	Yield/hectare
1	1	5	Nicaragua	2.03
		6	Panama	2.00
		2	El Salvador	1.79
2	2	4	Jamaica	1.99
		2	Dominican Rep.	1.51
		1	Cuba	0.63
3	3	2	Belgium-lux	3.69
		8	Spain	2.79
		1	Austria	1.90
4	4 and 5	07	Macedonia	1.36
		08	Poland	2.34
		02	Albania	0.64
5	6 and 7	07	Moldova	2.05
		02	Armenia	0.26
		10	Turkmenistan	2.36
		09	Tajikistan	2.48
6	8	2	Libya	1.61
		1	Algeria	1.96
7	9 and 10	20	Nigeria	2.10
		12	Kenya	2.51
		30	Zambia	1.29

Continued......

		15	Malawi	1.22
		07	Central African Rep	0.87
		26	Togo	0.50
		22	Zimbabwe	2.06
		01	Angola	0.99
		11	Cote d'lvoire	0.26
		05	Zaire	1.11
		06	Cameroon	1.62
8	11 and 12	15	Thailand	1.33
		06	Indonesia	0.85
		02	Burma	1.22
		10	Korea, South	2.02
		11	Laos	0.75
		05	China	1.75
9	13 and 14	05	Lebanon	1.33
		07	Syria	1.15
		01	Cyprus	1.50
10	15	3	New Zealand	2.58
		2	Solomon Islands	0.95

Thus we have

Stratum no.	$\bar{y}_h$	s_{hy}^2	N_h	n_h	f_h	W_h	$W_h\bar{y}_h$	$W_h s_{hy}^2$
1	1.9400	0.017100	6	3	0.5000	0.0566	0.1098	0.0009
2	1.3766	0.475733	6	3	0.5000	0.0566	0.0779	0.0269
3	2.7933	0.801033	8	3	0.3750	0.0754	0.2108	0.0604
4	1.4466	0.728133	10	3	0.3000	0.0943	0.1364	0.0686
5	1.7875	1.069825	12	4	0.3333	0.1132	0.2023	0.1211
6	1.7850	0.061250	4	2	0.5000	0.0377	0.0673	0.0023
7	1.3209	0.482449	30	11	0.3666	0.2830	0.3738	0.1365
8	1.3200	0.246160	17	6	0.3529	0.1603	0.2116	0.0394
9	1.3266	0.030633	10	3	0.3000	0.0943	0.1251	0.0028
10	1.7650	1.328450	3	2	0.6666	0.0283	0.0499	0.0375
		Sum	106	40		1.0000	1.5654	0.4970

Thus an estimate of yield/hectare of the Tobacco crop during 1998 in the world is

$$\bar{y}_{st} = \sum_{h=1}^{L} W_h\bar{y}_h = 1.5654$$

and

$$\hat{v}(\bar{y}_{st})_p = \frac{(1-f)}{n}\sum_{h=1}^{L} W_h s_{hy}^2 = \frac{1-40/106}{40}\times 0.4970 = 0.007736.$$

Using Table 2 from the Appendix the 95% confidence interval estimate of the yield/hectare in the world during 1998 is

$$\bar{y}_{st} \mp t_{\alpha/2}(\mathrm{df} = n - L)\sqrt{\hat{v}(\bar{y}_{st})_p}\ , \text{ or } \bar{y}_{st} \mp t_{0.025}(\mathrm{df} = 40 - 10)\sqrt{\hat{v}(\bar{y}_{st})_p}$$

or $1.5654 \mp 2.042\sqrt{0.007736}$, or $[1.3857, 1.7450]$.

8.2.3 OPTIMUM ALLOCATION METHOD

This method is based on the cost aspect of the survey. Let C_h be the cost of observing the variable y in the hth stratum and let C_t be the total fixed cost of the survey, then

$$C_t = C_0 + \sum_{h=1}^{L} n_h C_h \tag{8.2.9}$$

where C_0 stands for the known overhead cost. From (8.1.2), the variance of the estimator $\bar{y}_{st}$ is

$$V(\bar{y}_{st}) = \sum_{h=1}^{L} W_h^2 \left(\frac{1-f_h}{n_h}\right) S_{hy}^2 . \tag{8.2.10}$$

We now discuss two cases:

(i) the total cost is fixed; (ii) the variance is fixed.

Case I. Total cost is fixed: Minimization of (8.2.10) subject to (8.2.9) leads to the Lagrange function

$$L_1 = \sum_{h=1}^{L}\left(\frac{1}{n_h} - \frac{1}{N_h}\right) W_h^2 S_{hy}^2 + \lambda\left[C_0 + \sum_{h=1}^{L} n_h C_h - C_t\right]. \tag{8.2.11}$$

On differentiating (8.2.11) with respect to n_h and equating to zero we have

$$n_h = W_h S_{hy} \Big/ \left(\sqrt{\lambda}\sqrt{C_h}\right). \tag{8.2.12}$$

Note that $\sum_{h=1}^{L} n_h = n$, from (8.2.12) we have

$$\frac{1}{\sqrt{\lambda}} = n \Big/ \sum_{h=1}^{L} \frac{W_h S_{hy}}{\sqrt{C_h}} . \tag{8.2.13}$$

On substituting (8.2.13) in (8.2.12) we have

$$n_h = \frac{W_h S_{hy}}{\sqrt{\lambda}\sqrt{C_h}} = n W_h S_{hy} \Bigg/ \left\{\sqrt{C_h} \sum_{h=1}^{L} \frac{W_h S_{hy}}{\sqrt{C_h}}\right\} = n \frac{W_h S_{hy}}{\sqrt{C_h}} \Bigg/ \sum_{h=1}^{L} \frac{W_h S_{hy}}{\sqrt{C_h}} . \tag{8.2.14}$$

In a particular case if $C_1 = C_2 = ... = C_L = C$, that is the cost of sampling in each stratum is the same then (8.2.14) becomes

$$n_h = n\left(\frac{W_h S_{hy}}{\sum_{h=1}^{L} W_h S_{hy}}\right). \tag{8.2.15}$$

In other words the optimum allocation reduces to the famous Neyman (1934) allocation. On substituting (8.2.14) in (8.2.10), the variance of the estimator $\bar{y}_{st}$ under optimum allocation is given by

$$V(\bar{y}_{\text{st}})_{\text{Opt}} = \sum_{h=1}^{L} W_h^2 S_{hy}^2 \left[\sum_{h=1}^{L} \frac{W_h S_{hy}}{\sqrt{C_h}} \bigg/ n\left(\frac{W_h S_{hy}}{\sqrt{C_h}}\right) - \frac{1}{N_h} \right]$$

$$= \frac{1}{n}\left(\sum_{h=1}^{L} W_h S_{hy}\sqrt{C_h}\right)\left(\sum_{h=1}^{L} \frac{W_h S_{hy}}{\sqrt{C_h}}\right) - \sum_{h=1}^{L} \frac{W_h^2 S_{hy}^2}{N_h}. \tag{8.2.16}$$

On substituting (8.2.15) in (8.2.10), the variance of the estimator $\bar{y}_{\text{st}}$ under Neyman allocation is given by

$$V(\bar{y}_{\text{st}})_{\text{N}} = \frac{1}{n}\left(\sum_{h=1}^{L} W_h S_{hy}\right)^2 - \sum_{h=1}^{L} \frac{W_h^2 S_{hy}^2}{N_h}. \tag{8.2.17}$$

It can be easily shown that if f_h is negligible then

$$V(\bar{y}_{\text{st}})_{\text{Opt}} = \frac{1}{n}\left(\sum_{h=1}^{L} W_h S_{hy}\sqrt{C_h}\right)\left(\sum_{h=1}^{L} \frac{W_h S_{hy}}{\sqrt{C_h}}\right) \tag{8.2.18}$$

and

$$V(\bar{y}_{\text{st}})_{\text{N}} = \frac{1}{n}\left(\sum_{h=1}^{L} W_h S_{hy}\right)^2. \tag{8.2.19}$$

Case II. Total variance is fixed: In this case we minimize the cost given by (8.2.9) subject to the fixed variance V_0.

$$V(\bar{y}_{\text{st}}) = \sum_{h=1}^{L}\left(\frac{1}{n_h} - \frac{1}{N_h}\right) W_h^2 S_{hy}^2 = V_0 . \tag{8.2.20}$$

In such situations, the Lagrange function L_2 is given by

$$L_2 = C_0 + \sum_{h=1}^{L} n_h C_h + \lambda\left[\sum_{h=1}^{L}\left(\frac{1}{n_h} - \frac{1}{N_h}\right) W_h^2 S_{hy}^2 - V_0\right]. \tag{8.2.21}$$

On differentiating (8.2.21) with respect to n_h and equating to zero we have

$$n_h = W_h S_{hy}\sqrt{\lambda} \big/ \sqrt{C_h} . \tag{8.2.22}$$

Now there are two possibilities.

(a) Total sample size is fixed: Adding (8.2.22) over all possible strata we have

$$\sum_{h=1}^{L} n_h = \sqrt{\lambda}\sum_{h=1}^{L} W_h S_{hy} \big/ \sqrt{C_h} \quad \text{or} \quad \sqrt{\lambda} = n \bigg/ \sum_{h=1}^{L}\left(W_h S_{hy} \big/ \sqrt{C_h}\right). \tag{8.2.23}$$

Thus the optimum value of n_h is given by substituting (8.2.23) in (8.2.22) as

$$n_h = n\left(W_h S_{hy} \big/ \sqrt{C_h}\right) \bigg/ \sum_{h=1}^{L}\left(W_h S_{hy} \big/ \sqrt{C_h}\right). \tag{8.2.24}$$

On substituting (8.2.24) in (8.2.9) the minimum cost is given by

$$C_{\min} = C_0 + n\left[\sum_{h=1}^{L} W_h S_{hy}\sqrt{C_h} \bigg/ \sum_{h=1}^{L}\left(W_h S_{hy} \big/ \sqrt{C_h}\right)\right]. \tag{8.2.25}$$

(b) Minimum sample size for a fixed value of variance: From (8.2.20) we have

$$\sum_{h=1}^{L} W_h^2 S_{hy}^2 \Big/ n_h = V_0 + \sum_{h=1}^{L} W_h^2 S_{hy}^2 \Big/ N_h \,. \tag{8.2.26}$$

On substituting the value of n_h from (8.2.22) we have

$$\sqrt{\lambda} = \sum_{h=1}^{L} W_h S_{hy} \sqrt{C_h} \Big/ \left\{ V_0 + \sum_{h=1}^{L} W_h^2 S_{hy}^2 \Big/ N_h \right\}. \tag{8.2.27}$$

On substituting this value of λ in (8.2.22) we have

$$n_h = W_h S_{hy} \sum_{h=1}^{L} W_h S_{hy} \sqrt{C_h} \Big/ \sqrt{C_h} \left[V_0 + \sum_{h=1}^{L} W_h^2 S_{hy}^2 \Big/ N_h \right]. \tag{8.2.28}$$

Summing over all possible strata we obtain the minimum sample size n for the fixed variance as

$$n = \sum_{h=1}^{L} n_h = \left(\sum_{h=1}^{L} W_h S_{hy} \Big/ \sqrt{C_h} \right) \left(\sum_{h=1}^{L} W_h S_{hy} \sqrt{C_h} \right) \Big/ \left[V_0 + \sum_{h=1}^{L} W_h^2 S_{hy}^2 \Big/ N_h \right]. \tag{8.2.29}$$

Case III. Total variance and sample size are fixed: Now the Lagrange function is

$$L_3 = C_0 + \sum_{h=1}^{L} n_h C_h + \lambda_1 \left[\sum_{h=1}^{L} W_h^2 \left(\frac{1}{n_h} - \frac{1}{N_h} \right) S_{hy}^2 - V_0 \right] + \lambda_2 \left[\sum_{h=1}^{L} n_h - n \right]. \tag{8.2.30}$$

On differentiating (8.2.30) with respect to n_h and equating to zero we obtain

$$n_h = \sqrt{\lambda_1} W_h S_{hy} \Big/ \sqrt{C_h + \lambda_2} \,. \tag{8.2.31}$$

Note that $\sum_{h=1}^{L} n_h = n$, which implies that

$$\sqrt{\lambda_1} = n \Big/ \sum_{h=1}^{L} \frac{W_h S_{hy}}{\sqrt{C_h + \lambda_2}}, \quad \text{and} \quad n_h = \frac{n W_h S_{hy}}{\sqrt{C_h + \lambda_2}} \Big/ \sum_{h=1}^{L} \frac{W_h S_{hy}}{\sqrt{C_h + \lambda_2}} \tag{8.2.32}$$

the value of λ_2 can be obtained by solving the following equation through iterative procedure

$$\frac{1}{n} \left(\sum_{h=1}^{L} W_h S_{hy} \sqrt{C_h + \lambda_2} \right) \sum_{h=1}^{L} \frac{W_h S_{hy}}{\sqrt{C_h + \lambda_2}} = V_0 + \sum_{h=1}^{L} \frac{W_h^2 S_{hy}^2}{N_h} \,. \tag{8.2.33}$$

On substituting the value of λ_2 from (8.2.33) in (8.2.32) we obtain the allocation of the given sample of n units into different strata for the fixed level of variance.

Example 8.2.2. Select a sample of 40 countries from population 5 using the Neyman method of allocation. Record the yield/hectare of the tobacco crop from the selected countries. Merge the strata if required. Estimate the average yield/hectare of the tobacco crop in the world. Estimate the variance under the Neyman allocation. Construct a 95% confidence interval estimate.

Solution. By the Neyman method of allocation, the number of units to be selected from the hth stratum are given by $n_h = n W_h S_{hy} \Big/ \sum_{h=1}^{L} W_h S_{hy}$. Thus we have

Stratum no.	N_h	s_{hy}^2	S_{hy}	$N_h S_{hy}$	n_h	n_h
1	6	0.02682	0.163788	0.98270	0.5673	1
2	6	0.21809	0.467001	2.80200	1.6177	1
3	8	0.34699	0.589065	4.71250	2.7208	3
4	10	0.23456	0.484314	4.84310	2.7962	3
5	12	0.58214	0.762981	9.15570	5.2861	5
6	4	0.15310	0.391280	1.56510	0.9036	1
7	30	0.34385	0.586387	17.59220	10.1567	10
8	17	0.37855	0.615264	10.45950	6.0389	6
9	10	2.01830	1.420669	14.20670	8.2023	8
10	3	0.97460	0.987218	2.96160	1.7099	2
			Sum	69.28074		40

Using the Pseudo-Random Number (PRN) Table 1 given in the Appendix we collect the following sample information.

Stratum No.	Column no used from PRN table	Random no.	Country	Yield/ Hectare
1	1	5	Nicaragua	2.03
2	2	4	Jamaica	1.99
3	3	2	Belgium--Lux	3.69
		8	Spain	2.79
		1	Austria	1.90
4	4 and 5	07	Macedonia	1.36
		08	Poland	2.34
		02	Albania	0.64
5	6 and 7	07	Moldova	2.05
		02	Armenia	0.26
		10	Turkmenistan	2.36
		09	Tajikistan	2.48
		04	Georgia	1.63
6	8	2	Libya	1.61
7	9 and 10	20	Nigeria	2.10
		12	Kenya	2.51
		30	Zambia	1.29
		15	Malawi	1.22
		07	Central African Rep	0.87
		26	Togo	0.50
		22	Zimbabwe	2.06
		01	Angola	0.99
		11	Cote d'Ivoire	0.26

Continued……..

		05	Zaire	1.11
8	11 and 12	15	Thailand	1.33
		06	Indonesia	0.85
		02	Burma	1.22
		10	Korea, South	2.02
		11	Laos	0.75
		05	China	1.75
9	13 and 14	05	Lebanon	1.33
		07	Syria	1.15
		01	Cyprus	1.50
		09	Turkey	0.91
		06	Oman	1.11
		04	Jordan	1.29
		03	Iraq	1.09
		02	Iran	1.39
10	15	3	New Zealand	2.58
		2	Solomon Islands	0.95

This leads to the pooled strata and related results given below:

Pooled strata	Old strata	N_h	Pooled sample n_h	$\bar{y}_h$	s^2_{hy}	f_h	W_h	$W_h\bar{y}_h$	$\hat{v}_h$
1	1,2,6	16	3	1.8766	0.0537	0.1875	0.15094	0.2832	0.0003313
2	3	8	3	2.7933	0.8010	0.3750	0.07547	0.2108	0.0009505
3	4	10	3	1.4467	0.7281	0.3000	0.09434	0.1364	0.0015120
4	5	12	5	1.7560	0.8073	0.4166	0.11320	0.1987	0.0012070
5	7	30	10	1.2910	0.5251	0.3333	0.28301	0.3654	0.0028040
6	8	17	6	1.3200	0.2462	0.3529	0.16037	0.2116	0.0006829
7	9	10	8	1.2213	0.0363	0.8000	0.09434	0.1152	0.0000081
8	10	3	2	1.7650	1.3285	0.6666	0.02830	0.0499	0.0001773
							Sum	1.5717	0.0076734

where $\hat{v}_h = W_h^2\left(\frac{1-f_h}{n_h}\right)s^2_{hy}$.

Thus an estimate of yield/hectare of the world tobacco crop during 1998 is

$$\bar{y}_{st} = \sum_{h=1}^{L} W_h\bar{y}_h = 1.5717, \quad \text{and} \quad \hat{v}(\bar{y}_{st})_N = \sum_{h=1}^{L} W_h^2\left(\frac{1-f_h}{n_h}\right)s^2_{hy} = 0.0076734.$$

Using Table 2 from the Appendix the 95% confidence interval estimate of the yield/hectare in the world during 1998 is

$$\bar{y}_{st} \mp t_{\alpha/2}(\text{df} = n-L)\sqrt{\hat{v}(\bar{y}_{st})_N}, \quad \text{or} \quad \bar{y}_{st} \mp t_{0.025}(\text{df} = 40-8)\sqrt{\hat{v}(\bar{y}_{st})_N}$$

$$\text{or} \quad 1.5717 \mp 2.037\sqrt{0.0076734}, \quad \text{or} \quad [1.39306,\ 1.75014].$$

Example 8.2.3. Select a sample of 40 countries from population 5 using the method of optimum allocation. Record the yield/hectare of the tobacco crop from the selected countries. Estimate the average yield/hectare of the tobacco crop in the world. Estimate the variance under the method of optimum allocation. Construct a 95% confidence interval.

Given: $C_1 = \$0.5$, $C_2 = \$2.0$, $C_3 = \$3.0$, $C_4 = \$5.0$, $C_5 = \$7.0$, $C_6 = \$1.5$, $C_7 = \$10.0$, $C_8 = \$5.0$, $C_9 = \$5.0$, and $C_{10} = \$3.0$.

Solution. By the method of optimum allocation, the number of units to be selected from the hth stratum is

$$n_h = n\left(W_h S_{hy} / \sqrt{C_h}\right) \Big/ \left(\sum_{h=1}^{L} W_h S_{hy} / \sqrt{C_h}\right)$$

Thus we have

Stratum no.	N_h	C_h	S_{hy}	$N_h S_{hy} / \sqrt{C_h}$	n_h	
1	6	0.5	0.1637	1.3897	1.7760	≈ 2
2	6	2	0.4670	1.9813	2.5320	≈ 3
3	8	3	0.5890	2.7207	3.4770	≈ 3
4	10	5	0.4843	2.1659	2.7679	≈ 3
5	12	7	0.7629	3.4605	4.4224	≈ 4
6	4	1.5	0.3912	1.2779	1.6331	≈ 2
7	30	10	0.5863	5.5629	7.1091	≈ 7
8	17	5	0.6152	4.6776	5.9777	≈ 6
9	10	5	1.4206	6.3534	8.1193	≈ 8
10	3	3	0.9872	1.7099	2.1851	≈ 2
					Sum	40

Using the Pseudo-Random Number (PRN) Table 1 from the Appendix we have:

Stratum No.	Column no used from PRN Table	Random no.	Country	Yield/ hectare
1	1	5	Nicaragua	2.03
		6	Panama	2.00
2	2	4	Jamaica	1.99
		2	Dominican Rep.	1.51
		1	Cuba	0.63
3	3	2	Belgium--Lux	3.69
		8	Spain	2.79
		1	Austria	1.90
4	4 and 5	07	Macedonia	1.36
		08	Poland	2.34
		02	Albania	0.64
5	6 and 7	07	Moldova	2.05

Continued......

		02	Armenia	0.26
		10	Turkmenistan	2.36
		09	Tajikistan	2.48
6	8	2	Libya	1.61
		1	Algeria	1.96
7	9 and 10	20	Nigeria	2.10
		12	Kenya	2.51
		30	Zambia	1.29
		15	Malawi	1.22
		07	Central African Rep	0.87
		26	Togo	0.50
		22	Zimbabwe	2.06
8	11 and 12	15	Thailand	1.33
		06	Indonesia	0.85
		02	Burma	1.22
		10	Korea, South	2.02
		11	Laos	0.75
		05	China	1.75
9	13 and 14	05	Lebanon	1.33
		07	Syria	1.15
		01	Cyprus	1.50
		09	Turkey	0.91
		06	Oman	1.11
		04	Jordan	1.29
		03	Iraq	1.09
		10	Yemen	1.73
10	15	3	New Zealand	2.58
		2	Solomon Islands	0.95

Thus we have the following results:

Stratum no.	$\bar{y}_h$	s_{hy}^2	N_h	n_h	f_h	W_h	$W_h\bar{y}_h$	$\hat{v}_h$
1	2.015	0.00045	6	2	0.33333	0.0566	0.1140	4.80536E-07
2	1.376	0.47573	6	3	0.50000	0.0566	0.0778	0.000254005
3	2.793	0.80103	8	3	0.37500	0.0754	0.2106	0.000948747
4	1.447	0.72813	10	3	0.30000	0.0943	0.1364	0.001510807
5	1.787	1.06982	12	4	0.33333	0.1132	0.2023	0.002284833
6	1.785	0.06125	4	2	0.50000	0.0377	0.0673	2.17635E-05
7	1.507	0.53546	30	7	0.23333	0.2830	0.4265	0.004696889
8	1.320	0.24616	17	6	0.35294	0.1603	0.2116	0.000682147
9	1.263	0.06717	10	8	0.80000	0.0943	0.1191	1.49327E-05
10	1.765	1.32845	3	2	0.66666	0.0283	0.0499	0.000177327
		Sum	106	40		1.0000	1.6155	0.010591932

where $\hat{v}_h = W_h^2\left(\frac{1-f_h}{n_h}\right)s_{hy}^2$. Thus an estimate of the yield/hectare of the tobacco crop during 1998 in the world is

$$\bar{y}_{st} = \sum_{h=1}^{L} W_h \bar{y}_h = 1.6155 \quad \text{and} \quad \hat{v}(\bar{y}_{st})_{Opt} = \sum_{h=1}^{L} W_h^2\left(\frac{1-f_h}{n_h}\right)s_{hy}^2 = 0.010591932.$$

Using Table 2 from the Appendix the 95% confidence interval estimate of the yield/hectare in the world during 1998 is

$$\bar{y}_{st} \mp t_{\alpha/2}(\text{df} = n - L)\sqrt{\hat{v}(\bar{y}_{st})_{\text{Opt}}} \quad \text{or} \quad \bar{y}_{st} \mp t_{0.025}(\text{df} = 40-10)\sqrt{\hat{v}(\bar{y}_{st})_{\text{Opt}}}$$

or $1.6155 \mp 2.042\sqrt{0.010591932}$, or $[1.4053,\ 1.8256]$.

Example 8.2.4. Find the minimum sample size from population 5 to obtain estimates of population mean with different levels of relative variance. Plot relative variance versus sample size.

Given: $C_1 = \$0.5$, $C_2 = \$2.0$, $C_3 = \$3.0$, $C_4 = \$5.0$, $C_5 = \$7.0$, $C_6 = \$1.5$, $C_7 = \$10.0$, $C_8 = \$5.0$, $C_9 = \$5.0$, $C_{10} = \$3.0$ and $\bar{Y} = 1.5507$.

Solution. The minimum sample size n for the fixed variance V_0 is given by

$$n = \frac{\left(\sum_{h=1}^{L} W_h S_{hy}/\sqrt{C_h}\right)\left(\sum_{h=1}^{L} W_h S_{hy}\sqrt{C_h}\right)}{\left[V_0 + \sum_{h=1}^{L} W_h^2 S_{hy}^2/N_h\right]}.$$

Thus we have the following table

Strata No	N_h	C_h	S_{hy}	W_h	$W_h S_{hy}\sqrt{C_h}$	$W_h S_{hy}/\sqrt{C_h}$	$W_h^2 S_{hy}^2/N_h$
1	6	0.5	0.1637	0.0566	0.006552078	0.013104156	0.00001431
2	6	2.0	0.4670	0.0566	0.037383268	0.018691634	0.00011646
3	8	3.0	0.5890	0.0754	0.076994560	0.025664853	0.00024701
4	10	5.0	0.4843	0.0943	0.102162993	0.020432599	0.00020875
5	12	7.0	0.7629	0.1132	0.228503058	0.032643294	0.00062159
6	4	1.5	0.3912	0.0377	0.018080007	0.012053338	0.00005448
7	30	10	0.5863	0.2830	0.524729262	0.052472926	0.00091780
8	17	5.0	0.6152	0.1603	0.220619748	0.044123950	0.00057262
9	10	5.0	1.4206	0.0943	0.299675299	0.059935060	0.00179611
10	3	3.0	0.9872	0.0283	0.048392846	0.016130949	0.00026021
				Sum	1.563093119	0.295252759	0.00480933

The relative variance is defined as

$$\text{RV} = \left\{V(\bar{y}_{st})/\bar{Y}^2\right\} \times 100.$$

Therefore we have the following table.

RV	$V(\bar{y}_{st})$	n	
0.1	0.002405	63.97384098	≈ 64
0.2	0.004809	47.98036859	≈ 48
0.3	0.007214	38.38428905	≈ 39
0.4	0.009619	31.98690430	≈ 32
0.5	0.012023	27.41734456	≈ 27
0.6	0.014428	23.99017519	≈ 24
0.7	0.016833	21.32459927	≈ 21
0.8	0.019237	19.19213870	≈ 19
0.9	0.021642	17.44739833	≈ 17
1.0	0.024047	15.99344810	≈ 16
1.5	0.036070	11.28949194	≈ 11

The following graph in Figure 8.2.1 shows that as the desired level of relative variance decreases the sample size increases and vice versa. A more precise estimate can be obtained with a large sample.

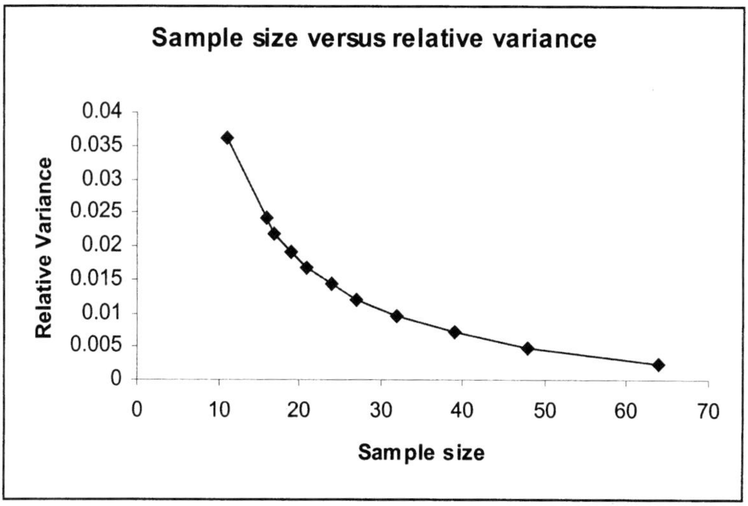

Fig. 8.2.1 Relative variance in stratified sampling.

Now we have the following theorem.

Theorem 8.2.1. Show that

$$V(\bar{y}) \geq V(\bar{y}_{st})_{p} \geq V(\bar{y}_{st})_{N} \tag{8.2.34}$$

where $V(\bar{y}_{st})_{p}$ and $V(\bar{y}_{st})_{N}$ indicate that the variance of the estimator of population mean in stratified sampling under Proportional and Neyman allocation respectively and $V(\bar{y})$ denotes the variance of the usual estimator under SRSWR sampling.

Proof. We know that:

In SRSWR sampling: $V(\bar{y}) = n^{-1}\sigma_y^2$, where $\sigma_y^2 = N^{-1}\sum_{i=1}^{N}(Y_i - \bar{Y})^2$. (8.2.35)

In stratified random sampling:

(a) Under proportional allocation

$$V(\bar{y}_{\text{st}})_{\text{p}} = \frac{1}{n}\sum_{h=1}^{L} W_h S_{hy}^2 , \tag{8.2.36}$$

and

(b) Under Neyman allocation

$$V(\bar{y}_{\text{st}})_{\text{N}} = \frac{1}{n}\left(\sum_{h=1}^{L} W_h S_{hy}\right)^2 . \tag{8.2.37}$$

Now we have to show that

$$V(\bar{y}) - V(\bar{y}_{\text{st}})_{\text{N}} \geq 0 . \tag{8.2.38}$$

Clearly from (8.2.35) and (8.2.36) we have

$$\frac{\sigma_y^2}{n} - \frac{1}{n}\sum_{h=1}^{L} W_h S_{hy}^2 \geq 0 , \quad \text{or} \quad \sigma_y^2 - \sum_{h=1}^{L} W_h S_{hy}^2 \geq 0 . \tag{8.2.39}$$

Now we have

$$\sigma_y^2 = \frac{1}{N}\sum_{i=1}^{N}(Y_i - \bar{Y})^2 = \frac{1}{N}\sum_{h=1}^{L}\sum_{i=1}^{N_h}(Y_{hi} - \bar{Y})^2 = \frac{1}{N}\sum_{h=1}^{L}\sum_{i=1}^{N_h}\left\{(Y_{hi} - \bar{Y}_h) + (\bar{Y}_h - \bar{Y})\right\}^2$$

$$= \frac{1}{N}\sum_{h=1}^{L}\sum_{i=1}^{N_h}\left[(Y_{hi} - \bar{Y}_h)^2 + (\bar{Y}_h - \bar{Y})^2 + 2(Y_{hi} - \bar{Y}_h)(\bar{Y}_h - \bar{Y})\right]$$

$$= \frac{1}{N}\left[\sum_{h=1}^{L}\sum_{i=1}^{N_h}(Y_{hi} - \bar{Y}_h)^2 + \sum_{h=1}^{L}\sum_{i=1}^{N_h}(\bar{Y}_h - \bar{Y})^2 + 0\right]$$

$$= \frac{1}{N}\left[\sum_{h=1}^{L}\frac{(N_h - 1)}{(N_h - 1)}\sum_{i=1}^{N_h}(Y_{hi} - \bar{Y}_h)^2 + \sum_{h=1}^{L} N_h(\bar{Y}_h - \bar{Y})^2\right]$$

$$= \frac{1}{N}\left[\sum_{h=1}^{L}(N_h - 1)S_{hy}^2 + \sum_{h=1}^{L} N_h(\bar{Y}_h - \bar{Y})^2\right],$$

note that $S_{hy}^2 = (N_h - 1)^{-1}\sum_{i=1}^{N_h}(Y_{hi} - \bar{Y}_h)^2$. Assuming that $(N_h - 1)/N \approx W_h$ we have

$$\sigma_{hy}^2 = \sum_{h=1}^{L} W_h S_{hy}^2 + \sum_{h=1}^{L} W_h(\bar{Y}_h - \bar{Y})^2 . \tag{8.2.40}$$

Thus we have

$$\sigma_y^2 - \sum_{h=1}^{L} W_h S_{hy}^2 = \sum_{h=1}^{L} W_h(\bar{Y}_h - \bar{Y})^2 \geq 0 . \tag{8.2.41}$$

This proves that the inequality (8.2.38) holds. If, however, $\bar{Y}_h = \bar{Y}$ for all h then the two variances are equal.

Now we prove the second part of the inequality (8.2.34), that is

$$V(\bar{y}_{\text{st}})_{\text{p}} \geq V(\bar{y}_{\text{st}})_{\text{N}} \tag{8.2.42}$$

$$\text{or} \quad V(\bar{y}_{\text{st}})_{\text{p}} - V(\bar{y}_{\text{st}})_{\text{N}} \ge 0 \quad \text{or} \quad \frac{1}{n}\sum_{h=1}^{L} W_h S_{hy}^2 - \frac{1}{n}\left(\sum_{h=1}^{L} W_h S_{hy}\right)^2 \ge 0$$

$$\text{or} \quad \sum_{h=1}^{L} W_h S_{hy}^2 - \left(\sum_{h=1}^{L} W_h S_{hy}\right)^2 \ge 0 \quad \text{or} \quad \sum_{h=1}^{L} W_h \left[S_{hy} - \sum_{h=1}^{L} W_h S_{hy}\right]^2 \ge 0. \tag{8.2.43}$$

Clearly the inequality (8.2.43) holds because

$$\sum_{h=1}^{L} W_h \left[S_{hy} - \sum_{h=1}^{L} W_h S_{hy}\right]^2 = \sum_{h=1}^{L} W_h \left[S_{hy}^2 + \left(\sum_{h=1}^{L} W_h S_{hy}\right)^2 - 2S_{hy}\left(\sum_{h=1}^{L} W_h S_{hy}\right)\right]$$

$$= \sum_{h=1}^{L} W_h S_{hy}^2 + \sum_{h=1}^{L} W_h \left(\sum_{h=1}^{L} W_h S_{hy}\right)^2 - 2\left(\sum_{h=1}^{L} W_h S_{hy}\right)^2$$

$$= \sum_{h=1}^{L} W_h S_{hy}^2 - \left(\sum_{h=1}^{L} W_h S_{hy}\right)^2. \tag{8.2.44}$$

Thus the inequality (8.2.34) holds. Hence the theorem.

Theorem 8.2.2. Show that the estimator of gain in efficiency $\left(\hat{\text{GE}}\right)$ owed to stratification with respect to SRSWOR sampling is

$$\hat{\text{GE}} = \frac{\left[\frac{N-n}{Nn} s_y^2 - \hat{V}(\bar{y}_{\text{st}})\right]}{\hat{V}(\bar{y}_{\text{st}})} \times 100\%. \tag{8.2.45}$$

Proof. We have

$$\hat{v}(\bar{y}_{\text{st}}) = \sum_{h=1}^{L} W_h^2 \left(\frac{1-f_h}{n_h}\right) s_{hy}^2, \tag{8.2.46}$$

and

$$V(\bar{y}) = \left(\frac{1-f}{n}\right) S_y^2, \quad \text{where } S_y^2 = \frac{1}{N-1}\left(\sum_{i=1}^{N} Y_i^2 - N\bar{Y}^2\right). \tag{8.2.47}$$

To find an estimator of $V(\bar{y})$, we need an estimator of S_y^2, which is given by

$$s_y^2 = \frac{1}{N-1}\left[\sum_{h=1}^{L} \frac{N_h}{n_h} \sum_{i=1}^{n_h} y_{hi}^2 - N\left\{\bar{y}_{\text{st}}^2 - \hat{V}(\bar{y}_{\text{st}})\right\}\right]. \tag{8.2.48}$$

Note that

$$E\left[\sum_{h=1}^{L} \frac{N_h}{n_h} \sum_{i=1}^{n_h} y_{hi}^2\right] = \sum_{h=1}^{L} N_h E\left(\frac{1}{n_h}\sum_{i=1}^{n_h} y_{hi}^2\right) = \sum_{h=1}^{L} N_h \frac{1}{N_h}\sum_{i=1}^{N_h} Y_{hi}^2 = \sum_{h=1}^{L}\sum_{i=1}^{N_h} Y_{hi}^2$$

and

$$E\left[\bar{y}_{\text{st}}^2 - \hat{V}(\bar{y}_{\text{st}})\right] = E\left(\bar{y}_{\text{st}}^2\right) - E\left\{\hat{V}(\bar{y}_{\text{st}})\right\} = V(\bar{y}_{\text{st}}) + \left\{E(\bar{y}_{\text{st}})\right\}^2 - V(\bar{y}_{\text{st}}) = \bar{Y}^2. \tag{8.2.49}$$

The percentage gain in efficiency (GE) owed to stratification can be defined as

$$\text{GE} = \frac{\left\{V(\bar{y}) - V(\bar{y}_{\text{st}})\right\}}{V(\bar{y}_{\text{st}})} \times 100\%. \tag{8.2.50}$$

Therefore the theorem follows by using method of moments.

Example 8.2.5. In example 8.1.2 find an estimate of the percentage gain in efficiency (GE) owed to stratification.

Solution. We have $\hat{v}(\bar{y}_{\text{st}}) = 0.015905$ and $\hat{v}(\bar{y}) = \frac{N-n}{Nn}s_y^2 = 0.024405$. Thus an estimate of the percentage gain in efficiency (GE) is given by

$$\hat{\text{GE}} = \frac{\left[\frac{N-n}{Nn}s_y^2 - \hat{v}(\bar{y}_{\text{st}})\right]}{\hat{v}(\bar{y}_{\text{st}})} \times 100 = \frac{(0.024405 - 0.015905)}{0.015905} \times 100 = 53.44\% . \tag{8.2.51}$$

Reddy (1978b) has considered the case of a finite population of size N divided into L strata of sizes N_h, $h = 1, 2, ..., L$. Let Y_{hi} denote the value of the variable Y for the ith unit in the hth stratum. For estimating the population total $Y = \sum_{h=1}^{L}\sum_{i=1}^{N_h} Y_{hi}$ it is shown in several books (*e.g.*, Cochran 1963) that stratified random sampling with proportional allocation is superior to unstratified random sampling provided the finite population correction factor (f.p.c.) in each stratum is ignored. Reddy (1978b) has shown that the above result is true even without ignoring the f.p.c, under proportional allocation with superpopulation model approach, if the variable Y satisfies the following condition:

$$\max_i(Y_{hi}) \le \min_i(Y_{h+1,i}) \quad \text{for} \quad h = 1, 2, ..., L-1 . \tag{8.2.52}$$

Example 8.2.6. In a circus there are three types of elephants, *viz.*, Light, Medium, and Heavy in weight and some information about them is listed in the following table.

	Light $(h=1)$	Medium $(h=2)$	Heavy $(h=3)$
N_h	250	150	100
n_h	20	12	8
$\bar{y}_h$ (wt., kg)	1800	3200	5200
$\bar{x}_h$ (food, kg/day)	95	155	244.8
$\bar{X}_h$ (food, kg/day)	100	150	250
s_{hy}	490	410	220
s_{hx}	45	25	30
s_{hxy}	17200	8000	5028

(a) How many elephants, N, are available in the circus?

(b) Find W_1, W_2 and W_3.

(c) Select an SRSWOR sample of $n = 40$ elephants using proportional allocation.

(d) Estimate the average weight, $\bar{Y}$, of all elephants in the circus using usual estimator in stratified sampling.

(e) Estimate the variance of $\bar{y}_{st}$.
(f) Construct the 95% CI estimate using usual estimator in stratified sampling.
(g) Find average diet of all the elephants in the circus.
(h) Find pooled sample mean $\bar{y}$ from all the three strata.
(i) Find pooled sample mean $\bar{x}$ from all the three strata.
(j) Find pooled sample variance s_y^2 from all the three strata.
(k) Find pooled sample variance s_x^2 from all the three strata.
(l) Find pooled sample covariance s_{xy} from all the three strata.
(m) Apply ratio estimator to estimate the average weight of the elephants.
(n) Construct the 95% confidence interval estimate using the ratio estimator.
(o) Compare the interval estimates.

Solution: (a) Total number of elephants

$$N = \sum_{h=1}^{L} N_h = 250 + 150 + 100 = 500 .$$

(b) Strata weights are given by

$$W_1 = \frac{N_1}{N} = \frac{250}{500} = 0.50 , \quad W_2 = \frac{N_2}{N} = \frac{150}{500} = 0.30, \text{ and } W_3 = \frac{N_3}{N} = \frac{100}{500} = 0.20 .$$

(c) Proportional allocation of $n = 40$ units

$$n_1 = nW_1 = 40 \times 0.50 = 20 , \quad n_2 = nW_2 = 40 \times 0.30 = 12 \text{ and } n_3 = nW_3 = 40 \times 0.20 = 8 .$$

(d) Estimate of the average weight of all the elephants using stratified sampling is

$$\bar{y}_{st} = \sum_{h=1}^{L} W_h \bar{y}_h = 0.50 \times 1800 + 0.30 \times 3200 + 0.20 \times 5200 = 2900 .$$

(e) Estimate of the variance of $\bar{y}_{st}$ is given by

$$\hat{v}(\bar{y}_{st}) = \sum_{h=1}^{L} W_h^2 \left(\frac{1 - f_h}{n_h} \right) s_{hy}^2$$

$$= 0.5^2 \times \left(\frac{1 - 0.08}{20} \right) \times 490^2 + 0.3^2 \times \left(\frac{1 - 0.08}{12} \right) \times 410^2 + 0.2^2 \times \left(\frac{1 - 0.08}{8} \right) \times 220^2$$

$$= 4143.68 .$$

Note that we used $f_1 = n_1/N_1 = 20/250 = 0.08$, $f_2 = n_2/N_2 = 12/150 = 0.08$ and $f_3 = n_3/N_3 = 8/100 = 0.08$

(f) Using Table 2 from the Appendix the 95% confidence interval estimate using stratified sampling is given by

$$\bar{y}_{st} \mp t_{\alpha/2}(\text{df} = n - L)\sqrt{\hat{v}(\bar{y}_{st})}, \quad \text{or} \quad 2900 \mp t_{\alpha/2}(\text{df} = 40 - 3)\sqrt{4143.68} ,$$

$$\text{or } 2900 \mp 2.026\sqrt{4143.68} , \quad \text{or} \quad [2769.72, 3030.27].$$

(g) The average diet of all the elephants in the circus is

$$\bar{X} = \sum_{h=1}^{L} W_h \bar{X}_h = 0.50 \times 100 + 0.30 \times 150 + 0.20 \times 250 = 145 .$$

(h) The pooled sample mean $\bar{y}$ from all strata is given by

$$\bar{y} = \frac{n_1\bar{y}_1 + n_2\bar{y}_2 + n_3\bar{y}_3}{n} = \frac{20\times1800+12\times3200+8\times5200}{40} = 2900 .$$

(i) The pooled sample mean $\bar{x}$ from all strata is given by

$$\bar{x} = \frac{n_1\bar{x}_1 + n_2\bar{x}_2 + n_3\bar{x}_3}{n} = \frac{20\times95+12\times155+8\times244.8}{40} = 142.96 .$$

(j) The pooled sample variance s_y^2 is given by

$$s_y^2 = \frac{\sum_{h=1}^{L}\left\{(n_h-1)s_{hy}^2\right\} + \sum_{h=1}^{L}\left\{n_h(\bar{y}_h-\bar{y})^2\right\}}{n-1}$$

$$= \frac{(20-1)490^2 + (12-1)410^2 + (8-1)220^2 + 20(1800-2900)^2 + 12(3200-2900)^2 + 8(5200-2900)^2}{40-1}$$

$$= 1906405.128 .$$

(k) The pooled sample variance s_x^2 is given by

$$s_x^2 = \frac{\sum_{h=1}^{L}\left\{(n_h-1)s_{hx}^2\right\} + \sum_{h=1}^{L}\left\{n_h(\bar{x}_h-\bar{x})^2\right\}}{n-1}$$

$$= \frac{(20-1)\times45^2 + (12-1)\times25^2 + (8-1)\times30^2 + 20(95-142.96)^2 + 12(155-142.96)^2 + 8(244.8-142.96)^2}{40-1}$$

$$= 4675.996 .$$

(l) The pooled sample covariance s_{xy} is given by

$$s_{xy} = \frac{\sum_{h=1}^{L}\left\{(n_h-1)s_{hxy}\right\} + \sum_{h=1}^{L}\left\{n_h(\bar{x}_h-\bar{x})(\bar{y}_h-\bar{y})\right\}}{n-1}$$

$$= \frac{1}{(40-1)}[(20-1)\times17200 + (12-1)\times8000 + (8-1)\times5028$$
$$+ 20(1800-2900)(95-142.96) + 12(3200-2900)(155-142.96)$$
$$+ 8(5200-2900)(244.8-142.96)]$$

$$= 87751.692.$$

(m) The ratio estimate of the average weight of elephants is given by

$$\bar{y}_R = \bar{y}\left(\frac{\bar{X}}{\bar{x}}\right) = 2900\left(\frac{145.00}{142.96}\right) = 2941.38 \text{ kg} .$$

(n) Confidence Interval estimate:

Note that $r = \frac{\bar{y}}{\bar{x}} = \frac{2900}{142.96} = 20.285$ and $f = \frac{n}{N} = \frac{40}{500} = 0.08.$

Thus an estimate of the mean square error of the ratio estimator is given by

$$\hat{\text{MSE}}(\bar{y}_R) = \left(\frac{1-f}{n}\right)\left[s_y^2 + r^2 s_x^2 - 2rs_{xy}\right]$$

$$= \left(\frac{1-0.08}{40}\right)\left[1906405.128 + (20.285)^2(4675.996) - 2(20.285)(87751.692)\right]$$

$$= 6219.281.$$

Using Table 2 from the Appendix the 95% confidence interval estimate of the average weight of elephants using ratio estimator is given by

$$\bar{y}_R \pm t_{0.025}(df = 40-1)\sqrt{\hat{MSE}(\bar{y}_R)}, \quad \text{or} \quad 2941.38 \pm 2.023\sqrt{6219.281}$$

or $[\ 2781.841,\ 3100.918\ ]$.

(o) **Comment on the CI estimates:** Although we are losing an extra two degrees of freedom in stratified sampling, still the length of the confidence interval estimate obtained through stratified sampling is smaller than that of a ratio estimator at the same level of confidence. Thus we conclude that stratified sampling performs better than the ratio estimator in this particular situation.

8.3. USE OF AUXILIARY INFORMATION AT ESTIMATION STAGE

The population of N units is first subdivided into L homogeneous sub-groups called strata, such that the hth stratum consists of N_h units, where $h = 1, 2, ..., L$ and $\sum_{h=1}^{L} N_h = N$. Assume from the hth population stratum consisting of N_h units, a sample of size n_h is drawn using SRSWOR sampling, such that $\sum_{h=1}^{L} n_h = n$. Let the ith sample unit of the study variable and auxiliary variable in the hth stratum be denoted by y_{hi} and x_{hi} respectively. Let $\bar{y}_h = n_h^{-1}\sum_{i=1}^{n_h} y_{hi}$ and $\bar{x}_h = n_h^{-1}\sum_{i=1}^{n_h} x_{hi}$ denote the hth stratum sample means for the study and auxiliary variable. Then for the hth stratum let us define

$$\epsilon_{h0} = \frac{\bar{y}_h}{\bar{Y}_h} - 1 \quad \text{and} \quad \epsilon_{h1} = \frac{\bar{x}_h}{\bar{X}_h} - 1$$

such that

$$E(\epsilon_{h0}) = E(\epsilon_{h1}) = 0$$

and

$$E(\epsilon_{h0}^2) = \left(\frac{1-f_h}{n_h}\right)C_{hy}^2, \quad E(\epsilon_{h1}^2) = \left(\frac{1-f_h}{n_h}\right)C_{hx}^2 \quad \text{and} \quad E(\epsilon_{h0}\epsilon_{h1}) = \left(\frac{1-f_h}{n_h}\right)\rho_{hxy}C_{hx}C_{hy}$$

where

$$\bar{Y}_h = N_h^{-1}\sum_{i=1}^{N_h} Y_{hi}, \quad \bar{X}_h = N_h^{-1}\sum_{i=1}^{N_h} X_{hi}, \quad S_{hy}^2 = (N_h - 1)^{-1}\sum_{i=1}^{N_h}(Y_{hi} - \bar{Y}_h)^2,$$

$$S_{hx}^2 = (N_h - 1)^{-1}\sum_{i=1}^{N_h}(X_{hi} - \bar{X}_h)^2, \quad S_{hxy} = (N_h - 1)^{-1}\sum_{i=1}^{N_h}(X_{hi} - \bar{X}_h)(Y_{hi} - \bar{Y}_h),$$

$$C_{hy}^2 = S_{hy}^2/\bar{Y}_h^2, \quad C_{hx}^2 = S_{hx}^2/\bar{X}_h^2 \quad \text{and} \quad \rho_{hxy} = S_{hxy}/(S_{hx}S_{hy}).$$

8.3.1 SEPARATE RATIO ESTIMATOR

The separate ratio estimator $\bar{y}_{\text{sr}}$ of population mean $\bar{Y}$ is given by

$$\bar{y}_{\text{sr}} = \sum_{h=1}^{L} W_h \bar{y}_h \left(\frac{\bar{X}_h}{\bar{x}_h}\right) \tag{8.3.1.1}$$

where $W_h = N_h / N$.

Then we have the following theorems:

Theorem 8.3.1.1. Bias in the estimator $\bar{y}_{\text{sr}}$, to the first order of approximation, is

$$B(\bar{y}_{\text{sr}}) = \sum_{h=1}^{L} W_h \left(\frac{1-f_h}{n_h}\right) \bar{Y}_h \left[C_{hx}^2 - \rho_{hxy} C_{hx} C_{hy}\right]. \tag{8.3.1.2}$$

Proof. The estimator $\bar{y}_{\text{sr}}$, in terms of ϵ_{h0} and ϵ_{h1}, can easily be written as

$$\bar{y}_{\text{sr}} = \sum_{h=1}^{L} W_h \bar{Y}_h (1+\epsilon_{h0})(1+\epsilon_{h1})^{-1} \approx \sum_{h=1}^{L} W_h \bar{Y}_h \left[1+\epsilon_{h0} - \epsilon_{h1} + \epsilon_{h1}^2 - \epsilon_{h0}\epsilon_{h1} + \ldots\right]. \tag{8.3.1.3}$$

Taking expected values on both sides of (8.3.1.3) we have

$$E(\bar{y}_{\text{sr}}) = \sum_{h=1}^{L} W_h \bar{Y}_h \left[1 + \left(\frac{1-f_h}{n_h}\right)\left\{C_{hx}^2 - \rho_{hxy} C_{hx} C_{hy}\right\}\right]. \tag{8.3.1.4}$$

Taking the deviation of (8.3.1.4) from population mean $\bar{Y} = \sum_{h=1}^{L} W_h \bar{Y}_h$ we have (8.3.1.2). Hence the theorem.

Theorem 8.3.1.2. The variance of the separate ratio estimator $\bar{y}_{\text{sr}}$, to the first order of approximation, is

$$V(\bar{y}_{\text{sr}}) = \sum_{h=1}^{L} W_h^2 \left(\frac{1-f_h}{n_h}\right) \bar{Y}_h^2 \left[C_{hy}^2 + C_{hx}^2 - 2\rho_{hxy} C_{hx} C_{hy}\right]. \tag{8.3.1.5}$$

Proof. Note that the strata are independent we have

$$V(\bar{y}_{\text{sr}}) = V\left[\sum_{h=1}^{L} W_h \bar{y}_h \left(\frac{\bar{X}_h}{\bar{x}_h}\right)\right] = \sum_{h=1}^{L} W_h^2 V\left[\bar{y}_h \left(\frac{\bar{X}_h}{\bar{x}_h}\right)\right]. \tag{8.3.1.6}$$

Now we have

$$V\left\{\bar{y}_h \frac{\bar{X}_h}{\bar{x}_h}\right\} = E\left[\bar{y}_h \frac{\bar{X}_h}{\bar{x}_h} - E\left(\bar{y}_h \frac{\bar{X}_h}{\bar{x}_h}\right)\right]^2 \approx E\left[\bar{Y}_h (\epsilon_{h0} - \epsilon_{h1})\right]^2$$

$$= \bar{Y}_h^2 \left(\frac{1-f_h}{n_h}\right)\left(C_{hy}^2 + C_{hx}^2 - 2\rho_{hxy} C_{hx} C_{hy}\right). \tag{8.3.1.7}$$

On substituting (8.3.1.7) in (8.3.1.6) we have the theorem.

Theorem 8.3.1.3. An estimator to estimate the variance of the separate ratio estimator, $\bar{y}_{\text{sr}}$, is given by

$$\hat{v}_1(\bar{y}_{\text{sr}}) = \sum_{h=1}^{L} W_h^2 \left(\frac{1-f_h}{n_h}\right)\left[s_{hy}^2 + r_h^2 s_{hx}^2 - 2 r_h s_{hxy}\right]. \tag{8.3.1.8}$$

where

$\bar{y}_h = n_h^{-1}\sum_{i=1}^{n_h} y_{hi}$, $\bar{x}_h = n_h^{-1}\sum_{i=1}^{n_h} x_{hi}$, $s_{hy}^2 = (n_h - 1)^{-1}\sum_{i=1}^{n_h}(y_{hi} - \bar{y}_h)^2$, $s_{hx}^2 = (n_h - 1)^{-1}\sum_{i=1}^{n_h}(x_{hi} - \bar{x}_h)^2$

$s_{hxy} = (n_h - 1)^{-1}\sum_{i=1}^{n_h}(y_{hi} - \bar{y}_h)(x_{hi} - \bar{x}_h)$, and $r_h = \bar{y}_h/\bar{x}_h$ are the estimators of respective population parameters in the hth stratum.

Another estimator to estimate the variance of the separate ratio estimator $\bar{y}_{sr}$ is

$$\hat{v}_2(\bar{y}_{sr}) = \sum_{h=1}^{L} W_h^2\left(\frac{1 - f_h}{n_h}\right)\frac{1}{(n_h - 1)}\sum_{i=1}^{n_h} e_{hi}^2 \tag{8.3.1.9}$$

where $e_{hi} = (y_{hi} - \bar{y}_h) - r_h(x_{hi} - \bar{x}_h)$ denotes the ith residual term in the hth stratum.

The third improved estimator of the variance of the separate ratio estimator $\bar{y}_{sr}$, owed to Wu (1985), is

$$\hat{v}_3(\bar{y}_{sr}) = \sum_{h=1}^{L} W_h^2\left(\frac{1 - f_h}{n_h}\right)\left(\frac{\bar{X}_h}{\bar{x}_h}\right)^{g_h}\frac{1}{(n_h - 1)}\sum_{i=1}^{n_h} e_{hi}^2 \tag{8.3.1.10}$$

where g_h denotes the suitably chosen constant in the hth stratum such that the variance of the estimator $\hat{v}_3(\bar{y}_{sr})$ is minimum. It is shown by Wu (1985) that the optimum value of g_h leads to an efficient estimator.

Example 8.3.1. Select a sample of 40 countries from population 5 using the method of proportional allocation. Record the yield/hectare and area under the tobacco crop from the countries selected in the sample. Apply ratio cum product estimator to estimate the average yield/hectare of the tobacco crop in the world. Assuming that the total area in each continent under the tobacco crop is known, estimate the variance of the estimator used. Construct a 95% confidence interval.

Solution. By the method of proportional allocation, the number of units to be selected from the hth stratum are given by $n_h = nN_h/N$. Thus we have

Stratum No.	N_h	n_h	f_h	W_h
1	6	3	0.500	0.0566
2	6	3	0.500	0.0566
3	8	3	0.375	0.0755
4	10	3	0.300	0.0943
5	12	4	0.333	0.1132
6	4	2	0.500	0.0377
7	30	11	0.367	0.2830
8	17	6	0.353	0.1604
9	10	3	0.300	0.0943
10	3	2	0.667	0.0283
Sum	106	40		0.9999

Using the Pseudo-Random Number (PRN) Table 1 given in the Appendix we have the following sample information and some results.

Stratum No.	Column no used from PRN Table	Random no.	Country	Yield/ hectare	Area Hectare
1	1	5	Nicaragua	2.03	2240
		6	Panama	2.00	1094
		2	El Salvador	1.79	580
2	2	4	Jamaica	1.99	1175
		2	Dominican Rep.	1.51	27050
		1	Cuba	0.63	59000
3	3	2	Belgium-Lux	3.69	320
		8	Spain	2.79	15150
		1	Austria	1.90	105
4	4 and 5	07	Macedonia	1.36	22000
		08	Poland	2.34	19100
		02	Albania	0.64	24000
5	6 and 7	07	Moldova	2.05	18600
		02	Armenia	0.26	4304
		10	Turkmenistan	2.36	1100
		09	Tajikistan	2.48	3228
6	8	2	Libya	1.61	900
		1	Algeria	1.96	2700
7	9 and 10	20	Nigeria	2.10	10000
		12	Kenya	2.51	8805
		30	Zambia	1.29	4882
		15	Malawi	1.22	116700
		07	Central African Rep	0.87	750
		26	Togo	0.50	4000
		22	Zimbabwe	2.06	103110
		01	Angola	0.99	3950
		11	Cote d'lvoire	0.26	10000
		05	Zaire	1.11	3700
		06	Cameroon	1.62	3400
8	11 and 12	15	Thailand	1.33	51500
		06	Indonesia	0.85	206625
		02	Burma	1.22	36000
		10	Korea, South	2.02	25730
		11	Laos	0.75	4000
		05	China	1.75	1445000
9	13 and 14	05	Lebanon	1.33	3750

Continued….

		07	Syria	1.15	15000
		01	Cyprus	1.50	161
10	15	3	New Zealand	2.58	600
		2	Solomon Islands	0.95	100

Thus we have:

St. no.	$\bar{y}_h$	$\bar{x}_h$	$\bar{X}_h$	r_h	s_{hy}^2	s_{hx}^2	s_{hxy}	$\hat{\bar{y}}_h$	$\hat{v}_h$
1	1.9400	1304.7	3194.5	0.00149	0.0171	722185.3	90.1	0.2688	0.000721
2	1.3767	29075.0	14660.0	4.7E-05	0.4757	839008125.0	-19863.0	0.1545	0.000246
3	2.7933	5191.7	18309.4	0.00054	0.8010	74387858.3	71.3	0.7434	0.026599
4	1.4467	21700.0	14923.5	6.7E-05	0.7281	6070000.0	-2102.0	0.1984	0.000983
5	1.7875	6808.0	5987.8	0.00026	1.0698	63572981.3	391.1	0.1779	0.011204
6	1.7850	1800.0	3450.0	0.00099	0.0613	1620000.0	315.0	0.1291	0.000365
7	1.3209	24481.5	11682.7	5.4E-05	0.4824	1801653230.0	6724.0	0.1784	0.023125
8	1.3200	294809.2	145162.3	4.5E-06	0.2462	3.227E+11	107376.5	0.1042	0.016126
9	1.3267	6303.7	33976.1	0.00021	0.0306	59939890.3	-1304.8	0.0232	0.004414
10	1.7650	350.0	1333.3	0.00504	1.3284	125000.0	407.5	0.1903	0.000053
							Sum	2.1687	0.083673

where

$$\hat{\bar{y}}_h = \begin{cases} W_h \bar{y}_h \dfrac{\bar{X}_h}{\bar{x}_h} & \text{if } s_{hxy} > 0, \\ W_h \bar{y}_h \dfrac{\bar{x}_h}{\bar{X}_h} & \text{if } s_{hxy} < 0, \end{cases}$$

and

$$\hat{v}_h = \begin{cases} W_h^2 \left(\dfrac{1-f_h}{n_h}\right)\left[s_{hy}^2 + r_h^2 s_{hx}^2 - 2 r_h s_{hxy}\right] & \text{if } s_{hxy} > 0, \\ W_h^2 \left(\dfrac{1-f_h}{n_h}\right)\left[s_{hy}^2 + r_h^2 s_{hx}^2 + 2 r_h s_{hxy}\right] & \text{if } s_{hxy} < 0. \end{cases}$$

We have to use ratio and product estimators in different strata because the correlation between yield/hectare and total number of hectare is uncertain. An estimate of the average yield/hectare of the tobacco crop in the world is given by a new separate ratio cum product estimator defined as

$$\bar{y}_{\text{srp}} = \sum_{h=1}^{L} \hat{\bar{y}}_h = 2.1687.$$

An estimate of the variance of the estimator used for estimation is

$$\hat{v}(\bar{y}_{\text{srp}}) = \sum_{h=1}^{L} \hat{v}_h = 0.083673.$$

Using Table 2 from the Appendix the 95% confidence interval is

$$\bar{y}_{\text{srp}} \mp t_{\alpha/2}(\text{df} = n - L - 3)\sqrt{\hat{v}(\bar{y}_{\text{srp}})} \quad \text{or} \quad \bar{y}_{\text{srp}} \mp t_{0.025}(\text{df} = 40 - 13)\sqrt{\hat{v}(\bar{y}_{\text{srp}})}$$

$$\text{or} \quad 2.1687 \mp 2.052\sqrt{0.083673}, \text{ or } [1.5751, 2.7623].$$

Note that in three strata the estimates of correlation were negative, and in the remaining seven strata the estimates of correlation were positive. Therefore for the strata where ratio estimator was used the loss of degree of freedom was one, and where the product estimator was used the loss of degree of freedom was taken as two.

8.3.2 SEPARATE REGRESSION ESTIMATOR

The separate linear regression estimator $\bar{y}_{\text{slr}}$ of population mean $\bar{Y}$ is given by

$$\bar{y}_{\text{slr}} = \sum_{h=1}^{L} W_h\left[\bar{y}_h + \hat{\beta}_{hxy}\left(\bar{X}_h - \bar{x}_h\right)\right] \tag{8.3.2.1}$$

where $\hat{\beta}_{hxy} = s_{hxy}/s_{hx}^2$.

Then we have the following theorems:

Theorem 8.3.2.1. The bias in the separate regression estimator $\bar{y}_{\text{slr}}$, to the first order of approximation, is

$$B(\bar{y}_{\text{slr}}) = \sum_{h=1}^{L} W_h\left(\frac{1-f_h}{n_h}\right)\beta_{hxy}\bar{X}_h C_{hx}\left(\lambda_{h03} - \frac{\lambda_{h12}}{\rho_{hxy}}\right) \tag{8.3.2.2}$$

where

$$\lambda_{hrs} = \frac{\mu_{hrs}}{\mu_{h20}^{r/2}\mu_{h02}^{s/2}} \quad \text{and} \quad \mu_{hrs} = (N_h - 1)^{-1}\sum_{i=1}^{N_h}\left(Y_{hi} - \bar{Y}_h\right)^r\left(X_{hi} - \bar{X}_h\right)^s \text{ for } h = 1, 2, ..., L.$$

Proof. Follows from the bias expression of the regression estimator in Chapter 3.

Theorem 8.3.2.2. The variance of the separate regression estimator $\bar{y}_{\text{slr}}$, to the first order of approximation, is given by

$$V(\bar{y}_{\text{slr}}) = \sum_{h=1}^{L} W_h^2\left(\frac{1-f_h}{n_h}\right)S_{hy}^2\left(1 - \rho_{hxy}^2\right). \tag{8.3.2.3}$$

Proof. Note that the strata are independent, we have

$$V(\bar{y}_{\text{slr}}) = V\left[\sum_{h=1}^{L} W_h\left\{\bar{y}_h + \hat{\beta}_{hxy}\left(\bar{X}_h - \bar{x}_h\right)\right\}\right] = \sum_{h=1}^{L} W_h^2 V\left\{\bar{y}_h + \hat{\beta}_{hxy}\left(\bar{X}_h - \bar{x}_h\right)\right\}. \tag{8.3.2.4}$$

Applying the concept of the usual linear regression estimator in each stratum, we have

$$V(\bar{y}_{\text{slr}}) = \sum_{h=1}^{L} W_h^2\left(\frac{1-f_h}{n_h}\right)\left[S_{hy}^2 + \beta_{hxy}^2 S_{hx}^2 - 2\beta_{hxy}S_{hxy}\right] = \sum_{h=1}^{L} W_h^2\left(\frac{1-f_h}{n_h}\right)S_{hy}^2\left(1 - \rho_{hxy}^2\right).$$

Hence the theorem.

Theorem 8.3.2.3. An estimator to estimate the variance of the separate regression estimator $\bar{y}_{\text{slr}}$ is

$$\hat{v}_1(\bar{y}_{\text{slr}}) = \sum_{h=1}^{L} W_h^2 \left(\frac{1-f_h}{n_h}\right)\left[s_{hy}^2 + \hat{\beta}_{hxy}^2 s_{hx}^2 - 2\hat{\beta}_{hxy} s_{hxy}\right] \tag{8.3.2.5}$$

where $\hat{\beta}_{hxy} = s_{hxy} / s_{hx}^2$ denotes the estimator of the regression coefficient in the hth stratum.

Another estimator of the variance of the separate regression estimator $\bar{y}_{\text{slr}}$ is

$$\hat{v}_2(\bar{y}_{\text{slr}}) = \sum_{h=1}^{L} W_h^2 \left(\frac{1-f_h}{n_h}\right)\frac{1}{(n_h-1)}\sum_{i=1}^{n_h} e_{hi}^2 \tag{8.3.2.6}$$

where $e_{hi} = (y_{hi} - \bar{y}_h) - \hat{\beta}_{hxy}(x_{hi} - \bar{x}_h)$ denotes the ith residual term in the hth stratum. The third improved estimator of the variance of the separate regression estimator $\bar{y}_{\text{slr}}$, proposed by Wu (1985), is given by

$$\hat{v}_3(\bar{y}_{\text{slr}}) = \sum_{h=1}^{L} W_h^2 \left(\frac{1-f_h}{n_h}\right)\left(\frac{\bar{X}_h}{\bar{x}_h}\right)^{g_h} \frac{1}{(n_h-1)}\sum_{i=1}^{n_h} e_{hi}^2 \tag{8.3.2.7}$$

where g_h denotes the suitably chosen constant in the hth stratum such that the variance of the estimator $\hat{v}_3(\bar{y}_{\text{slr}})$ is a minimum. It is shown by Wu (1985) that the optimum value of g_h leads to efficient estimator of variance.

Example 8.3.2. Select a sample of 40 countries from population 5 using the method of proportional allocation. Record the yield/hectare and area under the tobacco crop from the countries selected in the sample. Apply the regression estimator to estimate the average yield/hectare of the tobacco crop in the world. Assuming that the total area in each continent under the tobacco crop is known, estimate the variance of the estimator used for estimation purpose. Construct a 95% confidence interval.

Solution. Using information from the previous example 8.3.1, for the case of a separate regression estimator we derive following table.

St. No.	s_{hy}^2	s_{hx}^2	s_{hxy}	$\hat{\beta}_{hxy}$	r_{hxy}	$\hat{\bar{y}}_h$	$\hat{v}_h$
1	0.0171	722185.3	90.12	0.000125	0.810959	0.123160	0.0000031
2	0.4757	839008125.0	-19863.00	-0.000024	-0.994250	0.097243	0.0000029
3	0.8010	74387858.0	71.32	0.000000	0.009239	0.211764	0.0009500
4	0.7281	6070000.0	-2102.00	-0.000350	-0.999870	0.357864	0.0000003
5	1.0698	63572981.0	391.09	0.000006	0.047423	0.201787	0.0022800
6	0.0613	1620000.0	315.00	0.000194	0.999592	0.079465	0.0000000
7	0.4824	1801653230.0	6724.01	0.000003	0.228080	0.360304	0.0021070
8	0.2462	3.23E+11	107376.40	0.000000	0.424896	0.020372	0.0005840
9	0.0306	59939890.0	-1304.80	0.000022	-0.146320	0.068331	0.0000045
10	1.3284	125000.0	407.50	0.003260	1.000000	0.140679	0.0000000
					Sum	1.660900	0.0059316

where

$$\hat{\bar{y}}_h = W_h\left[\bar{y}_h + \hat{\beta}_{hxy}\left(\bar{X}_h - \bar{x}_h\right)\right] \quad \text{and} \quad \hat{v}_h = W_h^2\left(\frac{1-f_h}{n_h}\right)s_{hy}^2\left[1 - r_{hxy}^2\right].$$

Thus an estimate of the average yield/hectare of the tobacco crop in the world is given by

$$\bar{y}_{\text{slr}} = \sum_{h=1}^{L}\hat{\bar{y}}_h = 1.6609.$$

An estimate of the variance of the estimator used for estimation is

$$\hat{v}\left(\bar{y}_{\text{slr}}\right) = \sum_{h=1}^{L}\hat{v}_h = 0.0059316.$$

Using Table 2 from the Appendix the 95% confidence interval estimate is given by

$$\bar{y}_{\text{slr}} \mp t_{\alpha/2}(\text{df} = n - 2L)\sqrt{\hat{v}\left(\bar{y}_{\text{slr}}\right)}\ , \quad \text{or} \quad 1.6609 \mp 2.086\sqrt{0.0059316}\ , \quad \text{or} \quad [1.5002,\ 1.8215].$$

Note that we are using separate regression estimator in each stratum, so we are loosing two degree of freedom in each stratum.

Sometimes it is not possible to know the population means $\bar{X}_h,\ h = 1,2,...,L$ of the auxiliary variable in each stratum, but the combined population mean, $\bar{X} = \sum_{h=1}^{L} W_h \bar{X}_h$, is known. In such situations it is not possible to use separate ratio and regression type estimators, but we can use a combined ratio or combined regression estimator.

For deriving the expressions of bias and variance of the combined ratio or regression estimator in stratified sampling, let us define,

$$\delta_0 = \frac{\bar{y}_{\text{st}}}{\bar{Y}} - 1 \quad \text{and} \quad \delta_1 = \frac{\bar{x}_{\text{st}}}{\bar{X}} - 1$$

where $\bar{y}_{\text{st}} = \sum_{h=1}^{L} W_h \bar{y}_h$ and $\bar{x}_{\text{st}} = \sum_{h=1}^{L} W_h \bar{x}_h$ are the unbiased estimators of population mean $\bar{Y}$ and $\bar{X}$ respectively.

Obviously we have

$$E(\delta_0) = E(\delta_1) = 0\ .$$

Assuming that the strata are independent, we have

$$E\left(\delta_0^2\right) = \sum_{h=1}^{L} W_h^2\left(\frac{1-f_h}{n_h}\right)S_{hy}^2 \Big/ \bar{Y}^2\ , \qquad E\left(\delta_1^2\right) = \sum_{h=1}^{L} W_h^2\left(\frac{1-f_h}{n_h}\right)S_{hx}^2 \Big/ \bar{X}^2$$

and

$$E(\delta_0\delta_1) = \sum_{h=1}^{L} W_h^2\left(\frac{1-f_h}{n_h}\right)S_{hxy} \Big/ \left(\bar{X}\bar{Y}\right).$$

8.3.3. COMBINED RATIO ESTIMATOR

The usual combined ratio estimator $\bar{y}_{cr}$ of population mean $\bar{Y}$ is defined as

$$\bar{y}_{cr} = \bar{y}_{st}\left(\frac{\bar{X}}{\bar{x}_{st}}\right). \tag{8.3.3.1}$$

Then we have the following theorems.

Theorem 8.3.3.1. The bias in the combined ratio estimator $\bar{y}_{cr}$, to the first order of approximation, is

$$B(\bar{y}_{cr}) = R\sum_{h=1}^{L} W_h^2\left(\frac{1-f_h}{n_h}\right)\left\{\frac{S_{hx}^2}{\bar{X}} - S_{hxy}\right\} \tag{8.3.3.2}$$

where $R = \bar{Y}/\bar{X}$.

Proof. The estimator $\bar{y}_{cr}$ in terms of δ_0 and δ_1 can be approximately written as

$$\bar{y}_{cr} = \bar{Y}\left(1+\delta_0-\delta_1+\delta_1^2-\delta_0\delta_1+....\right). \tag{8.3.3.3}$$

Then we have

$$B(\bar{y}_{cr}) = E(\bar{y}_{cr}) - \bar{Y} = \bar{Y}\left[\frac{1}{\bar{X}^2}\sum_{h=1}^{L} W_h^2\left(\frac{1-f_h}{n_h}\right)S_{hx}^2 - \frac{1}{\bar{X}\,\bar{Y}}\sum_{h=1}^{L} W_h^2\left(\frac{1-f_h}{n_h}\right)S_{hxy}\right]$$

which on simplification reduces to (8.3.3.2). Hence the theorem.

Theorem 8.3.3.2. The variance of the combined ratio estimator $\bar{y}_{cr}$, to the first order of approximation, is

$$V(\bar{y}_{cr}) = \sum_{h=1}^{L} W_h^2\left(\frac{1-f_h}{n_h}\right)\left[S_{hy}^2 + R^2S_{hx}^2 - 2RS_{hxy}\right]. \tag{8.3.3.4}$$

Proof. By the definition of variance we have

$$V(\bar{y}_{cr}) = E\left[\bar{y}_{cr} - E(\bar{y}_{cr})\right]^2 \approx E\left[\bar{Y}\left(1+\delta_0-\delta_1+\delta_1^2-\delta_0\delta_1+....\right)-\bar{Y}\right]^2$$

$$= \bar{Y}^2E\left[\delta_0^2+\delta_1^2-2\delta_0\delta_1\right]$$

$$= \bar{Y}^2\left[\frac{\sum_{h=1}^{L} W_h^2\left(\frac{1-f_h}{n_h}\right)S_{hy}^2}{\bar{Y}^2} + \frac{\sum_{h=1}^{L} W_h^2\left(\frac{1-f_h}{n_h}\right)S_{hx}^2}{\bar{X}^2} - 2\frac{\sum_{h=1}^{L} W_h^2\left(\frac{1-f_h}{n_h}\right)S_{hxy}}{\bar{X}\,\bar{Y}}\right]$$

$$= \sum_{h=1}^{L} W_h^2\left(\frac{1-f_h}{n_h}\right)\left[S_{hy}^2 + R^2S_{hx}^2 - 2RS_{hxy}\right].$$

Hence the theorem.

Theorem 8.3.3.3. An estimator to estimate the variance of the combined ratio estimator $\bar{y}_{cr}$ is given by

$$\hat{v}_1(\bar{y}_{cr}) = \sum_{h=1}^{L} W_h^2 \left(\frac{1-f_h}{n_h}\right)\left[s_{hy}^2 + r^2 s_{hx}^2 - 2r s_{hxy}\right] \tag{8.3.3.5}$$

where $r = \bar{y}_{st}/\bar{x}_{st}$ denotes the estimator of population ratio $R = \bar{Y}/\bar{X}$ across the L strata. Another estimator of the variance of the combined ratio estimator $\bar{y}_{cr}$ is

$$\hat{v}_2(\bar{y}_{cr}) = \sum_{h=1}^{L} W_h^2 \left(\frac{1-f_h}{n_h}\right)\frac{1}{(n_h-1)}\sum_{i=1}^{n_h} e_{hi}^2 \tag{8.3.3.6}$$

where $e_{hi} = (y_{hi} - \bar{y}_h) - r(x_{hi} - \bar{x}_h)$ denotes the ith residual term in the hth stratum, but using r obtained from all strata. The third improved estimator of the variance of the combined ratio estimator $\bar{y}_{cr}$, due to Wu (1985), is given by

$$\hat{v}_3(\bar{y}_{cr}) = \sum_{h=1}^{L} W_h^2 \left(\frac{1-f_h}{n_h}\right)\frac{1}{(n_h-1)}\sum_{i=1}^{n_h} e_{hi}^2 \left(\frac{\bar{X}}{\bar{x}_{st}}\right)^g \tag{8.3.3.7}$$

where g denotes the suitable chosen constant such that the variance of the estimator $\hat{v}_3(\bar{y}_{cr})$ is minimum. It has been shown by Wu (1985) that the optimum value of g leads to efficient estimator of variance.

Example 8.3.3. Select a sample of 40 countries from population 5 using the method of proportional allocation. Record the production and area of the tobacco crop from the countries selected in the sample. Apply the combined ratio estimator to estimate the average production of the tobacco crop in the world. Assuming that the total area in the world under the Tobacco crop is known. Estimate the variance of the estimator used. Construct a 95% confidence interval.

Given: Total area is 3,650,492.66 hectares.

Solution. By the method of proportional allocation, the number of units to be selected from the hth stratum is given by, $n_h = nN_h/N$. Thus we have

Stratum No.	N_h	n_h	f_h	W_h
1	6	3	0.500	0.0566
2	6	3	0.500	0.0566
3	8	3	0.375	0.0755
4	10	3	0.300	0.0943
5	12	4	0.333	0.1132
6	4	2	0.500	0.0377
7	30	11	0.367	0.2830
8	17	6	0.353	0.1604
9	10	3	0.300	0.0943
10	3	2	0.667	0.0283
Sum	106	40		1.0000

Using the Pseudo-Random Number (PRN) Table 1 from the Appendix we have the following sample information and some results.

St. No.	Column no used from PRN Table	Random no.	Country	Production Metric tons	Area Hectares
1	1	5	Nicaragua	4550	2240
		6	Panama	2188	1094
		2	El Salvador	1038	580
2	2	4	Jamaica	2339	1175
		2	Dominican Rep.	40950	27050
		1	Cuba	37000	59000
3	3	2	Belgium--Lux	1800	320
		8	Spain	42300	15150
		1	Austria	199	105
4	4 and 5	07	Macedonia	30000	22000
		08	Poland	44700	19100
		02	Albania	15000	24000
5	6 and 7	07	Moldova	38150	18600
		02	Armenia	1100	4304
		10	Turkmenistan	2600	1100
		09	Tajikistan	8000	3228
6	8	2	Libya	1450	900
		1	Algeria	5300	2700
7	9 and 10	20	Nigeria	21000	10000
		12	Kenya	22120	8805
		30	Zambia	6300	4882
		15	Malawi	142300	116700
		07	Central African Rep	650	750
		26	Togo	2000	4000
		22	Zimbabwe	212050	103110
		01	Angola	3900	3950
		11	Cote d'lvoire	2600	10000
		05	Zaire	4110	3700
		06	Cameroon	5500	3400
8	11 and 12	15	Thailand	68600	51500
		06	Indonesia	175631	206625
		02	Burma	44000	36000
		10	Korea, South	25000	25730
		11	Laos	3000	4000
		05	China	2524500	1445000
9	13 and 14	05	Lebanon	5000	3750
		07	Syria	17200	15000
		01	Cyprus	241	161
10	15	3	New Zealand	1550	600
		2	Solomon Islands	95	100

Thus we have the following two tables:

Stratum No.	$\bar{y}_h$	$\bar{x}_h$	s_{hy}^2	s_{hx}^2	s_{hxy}
1	2592.00	1304.67	3205948.0	722185.3	1521312.0
2	26763.00	29075.00	451299457.0	839008125.0	479521575.0
3	14766.33	5191.67	569217900.3	74387858.3	205728126.7
4	29900.00	21700.00	220530000.0	6070000.0	-36360000.0
5	12462.50	6808.00	302045625.0	63572981.3	134543200.0
6	3375.00	1800.00	7411250.0	1620000.0	3465000.0
7	38411.82	24481.54	5002786596.0	1801653230.0	2864602080.0
8	473455.20	294809.20	1.013E+12	3.23E+11	5.7049E+11
9	7480.33	6303.67	76515960.3	59939890.3	67664108.7
10	822.50	350.00	1058512.5	125000.0	363750.0

Stratum no.	$W_h\bar{y}_h$	$W_h\bar{x}_h$	$\hat{v}_h$
1	146.7170	73.84925	109.08364
2	1514.8870	1645.75500	546967.57000
3	1114.4400	391.82420	125156.94000
4	2820.7550	2047.17000	727333.48000
5	1410.8490	770.71700	76564.22600
6	127.3585	67.92453	182.52455
7	10870.5400	6928.27000	2065671.30000
8	75931.4900	47280.72000	47255953.90000
9	705.6915	594.68580	25195.84800
10	23.2783	9.90566	29.68217
Sum	94666.0100	59810.82000	50823164.50000

From the above table we have

$$\bar{y}_{\text{st}} = \sum_{h=1}^{L} W_h\bar{y}_h = 94666.01\,,\quad \bar{x}_{\text{st}} = \sum_{h=1}^{L} W_h\bar{x}_h = 59810.82 \quad \text{and} \quad r = \bar{y}_{\text{st}}/\bar{x}_{\text{st}} = 1.582757.$$

We are given the average area under the tobacco crop in the world, $\bar{X} = 34438.61$.

Thus an estimate of average production of tobacco crop based on combined ratio estimator is given by

$$\bar{y}_{\text{cr}} = \bar{y}_{\text{st}}\frac{\bar{X}}{\bar{x}_{\text{st}}} = 94666.01\times\frac{34438.61}{59810.82} = 54507.96$$

and an estimator of its variance is given by

$$\hat{v}(\bar{y}_{\rm cr}) = \sum_{h=1}^{L} W_h^2\left(\frac{1-f_h}{n_h}\right)\left[s_{hy}^2 + r^2 s_{hx}^2 - 2rs_{hxy}\right] = \sum_{h=1}^{L}\hat{v}_h = 50823164.5.$$

Using Table 2 from the Appendix the 95% confidence interval of the average production of the tobacco crop in the world is given by

$$\bar{y}_{\rm cr} \mp t_{\alpha/2}(\text{df} = n-1)\sqrt{\hat{v}(\bar{y}_{\rm cr})}\ , \quad \text{or} \quad 54507.96 \mp 2.023\sqrt{50823164.5}$$

or

$$[40085.92,\quad 68930.00]\ .$$

8.3.4 COMBINED REGRESSION ESTIMATOR

The usual combined linear regression estimator $\bar{y}_{\rm clr}$ of population mean, $\bar{Y}$, is defined as

$$\bar{y}_{\rm clr} = \bar{y}_{\rm st} + \hat{\beta}\left(\bar{X} - \bar{x}_{\rm st}\right) \tag{8.3.4.1}$$

where $\hat{\beta} = \sum_{h=1}^{L} W_h^2\left(\frac{1-f_h}{n_h}\right)s_{hxy} \Big/ \sum_{h=1}^{L} W_h^2\left(\frac{1-f_h}{n_h}\right)s_{hx}^2$. Then we have the following theorems:

Theorem 8.3.4.1. The asymptotic variance of the combined regression estimator $\bar{y}_{\rm clr}$ is

$$V(\bar{y}_{\rm clr}) = V(\bar{y}_{\rm st})\left(1 - \rho_{xy}^2\right) \tag{8.3.4.2}$$

where $\rho_{xy} = \sum_{h=1}^{L} W_h^2\left(\frac{1-f_h}{n_h}\right)S_{hxy} \Big/ \left\{\sqrt{\sum_{h=1}^{L} W_h^2\left(\frac{1-f_h}{n_h}\right)S_{hx}^2}\sqrt{\sum_{h=1}^{L} W_h^2\left(\frac{1-f_h}{n_h}\right)S_{hy}^2}\right\}$ denotes the correlation coefficient in stratified sampling across all strata.

Proof. We have

$$V(\bar{y}_{\rm clr}) = V\left[\bar{y}_{\rm st} + \hat{\beta}\left(\bar{X} - \bar{x}_{\rm st}\right)\right] \approx V(\bar{y}_{\rm st}) + \beta^2 V(\bar{x}_{\rm st}) - 2\beta\,\text{Cov}(\bar{y}_{\rm st},\ \bar{x}_{\rm st}) \tag{8.3.4.3}$$

where $\beta = \text{Cov}(\bar{y}_{\rm st},\ \bar{x}_{\rm st})/V(\bar{x}_{\rm st})$. Hence the theorem.

Theorem 8.3.4.2. An estimator to estimate the variance of the combined regression estimator $\bar{y}_{\rm clr}$ is

$$\hat{v}_1(\bar{y}_{\rm clr}) = \sum_{h=1}^{L} W_h^2\left(\frac{1-f_h}{n_h}\right)\left[s_{hy}^2 + \hat{\beta}^2 s_{hx}^2 - 2\hat{\beta}s_{hxy}\right] \tag{8.3.4.4}$$

where $\hat{\beta} = \sum_{h=1}^{L} W_h^2\left(\frac{1-f_h}{n_h}\right)s_{hxy} \Big/ \sum_{h=1}^{L} W_h^2\left(\frac{1-f_h}{n_h}\right)s_{hx}^2$ denotes the estimator of population regression coefficient β.

Another estimator of the variance of the combined regression estimator $\bar{y}_{\rm clr}$ is

$$\hat{v}_2(\bar{y}_{\rm clr}) = \sum_{h=1}^{L} W_h^2\left(\frac{1-f_h}{n_h}\right)\frac{1}{(n_h-1)}\sum_{i=1}^{n_h} e_{hi}^2 \tag{8.3.4.5}$$

where $e_{hi} = (y_{hi} - \bar{y}_h) - \hat{\beta}(x_{hi} - \bar{x}_h)$ denotes the ith residual term in the hth stratum, but using $\hat{\beta}$ obtained from all strata.
The third estimator of the variance of the combined regression estimator $\bar{y}_{\text{clr}}$, due to Wu (1985), is

$$\hat{v}_3(\bar{y}_{\text{clr}}) = \sum_{h=1}^{L} W_h^2\left(\frac{1-f_h}{n_h}\right)\frac{1}{(n_h-1)}\sum_{i=1}^{n_h} e_{hi}^2\left(\frac{\bar{X}}{\bar{x}_{\text{st}}}\right)^g \tag{8.3.4.6}$$

where g denotes the suitable chosen constant such that the variance of the estimator $\hat{v}_3(\bar{y}_{\text{clr}})$ is minimum. It has been shown by Wu (1985) that the optimum value of g leads to an efficient estimator of variance.

Example 8.3.4. Select a sample of 40 countries from population 5 using the method of proportional allocation. Record the production and area of the tobacco crop from the countries selected in the sample. Apply the combined regression estimator to estimate the average production of the tobacco crop in the world. Assuming that the total area in the whole world under the tobacco crop is known, estimate the variance of the estimator used. Construct a 95% confidence interval.
Given: Total area is 3,650,492.66 hectares.
Solution. Continuing information from the example 8.3.3 we have

Stratum no.	D_{1h}	D_{2h}	$W_h\bar{y}_h$	$W_h\bar{x}_h$	$\hat{v}_h$
1	812.4	385.6	146.72	73.85	45.88
2	256063.5	448028.5	1514.88	1645.76	731844.21
3	244129.7	88273.2	1114.44	391.82	88840.02
4	-75507.3	12605.3	2820.75	2047.17	763600.10
5	287383.1	135791.3	1410.85	770.717	53810.33
6	1233.5	576.7	127.36	67.92	81.01
7	13202257.5	8303383.5	10870.54	6928.27	2441804.40
8	1582437849.0	895944582.9	75931.49	47280.72	14947955.11
9	140515.2	124474.7	705.69	594.68	50504.06
10	48.5	16.7	23.28	9.91	21.91
Sum	1596494785.0	905058118.7	94666.01	59810.82	19078507.03

From the above table we have

$$\bar{y}_{\text{st}} = \sum_{h=1}^{L} W_h\bar{y}_h = 94666.01\,,\ \bar{x}_{\text{st}} = \sum_{h=1}^{L} W_h\bar{x}_h = 59810.82$$

and

$$\hat{\beta} = \sum_{h=1}^{L} D_{1h}\Big/\sum_{h=1}^{L} D_{2h} = \sum_{h=1}^{L} W_h^2\left(\frac{1-f_h}{n_h}\right)s_{hxy}\Big/\sum_{h=1}^{L} W_h^2\left(\frac{1-f_h}{n_h}\right)s_{hx}^2 = \frac{1596494785}{905058118.7} = 1.7639.$$

We are given the average area under the tobacco crop in the world, $\bar{X} = 34438.61$.

Thus an estimate of average production of the tobacco crop based on the combined regression estimator is given by

$$\bar{y}_{\text{clr}} = \bar{y}_{\text{st}} + \hat{\beta}\left(\bar{X} - \bar{x}_{\text{st}}\right) = 94666.01 + 1.7639(34438.61 - 59810.82) = 49911.96$$

and an estimate of its variance is given by

$$\hat{v}\left(\bar{y}_{\text{clr}}\right) = \sum_{h=1}^{L} W_h^2\left(\frac{1-f_h}{n_h}\right)\left[s_{hy}^2 + \hat{\beta}^2 s_{hx}^2 - 2\hat{\beta}s_{hxy}\right] = 19078507.03\,.$$

The $(1-\alpha)100\%$ confidence interval of the average production of the tobacco crop in the world is given by

$$\bar{y}_{\text{clr}} \mp t_{\alpha/2}(\text{df} = n-2)\sqrt{\hat{v}\left(\bar{y}_{\text{clr}}\right)}$$

Using Table 2 from the Appendix the required 95% confidence interval estimate is

$$49911.96 \mp 2.024\sqrt{19078507.03}, \quad \text{or} \quad [41071.34,\ 58752.57].$$

Example 8.3.5. A private company ABC selects a sample of 40 countries, to estimate the average production of the tobacco crop in the world, from population 5 using the method of proportional allocation as given tin the following table.

Stratum No.	1	2	3	4	5	6	7	8	9	10	Total
N_h	6	6	8	10	12	4	30	17	10	3	106
n_h	3	3	3	3	4	2	11	6	3	2	40

(a) Using full information from the description of the population given in the Appendix, find the relative efficiency of separate ratio estimator with respect to combined ratio estimator while estimating average production using known area under the crop.

(b) Using full information from the description of the population given in the Appendix, find the relative efficiency of separate regression estimator with respect to combined regression estimator while estimating average production using know area under the crop.

Solution. From the description of the population 5 given in the appendix we have

St. No.	f_h	W_h	$\bar{Y}_h$	$\bar{X}_h$	R_h	S_{hy}^2	S_{hx}^2	S_{hxy}
1	0.50	0.057	6515.83	3194.50	2.040	51856391.4	10899652.8	23619714.4
2	0.50	0.057	13545.66	14660.00	0.924	390206386.6	584984730.0	420480877.0
3	0.38	0.075	43629.38	18309.37	2.383	3172330153.0	635958094.9	1387271318.0
4	0.30	0.094	21928.10	14923.50	1.469	569396673.9	209817189.2	319262930.2
5	0.33	0.113	11788.00	5987.83	1.969	153854582.5	27842810.4	62846173.1
6	0.50	0.038	4653.00	3450.00	1.349	7232769.3	5876666.6	6066866.6
7	0.37	0.283	16862.27	11682.73	1.443	2049296094.0	760238523.4	1190767859.0
8	0.35	0.160	227371.50	145162.30	1.566	3.72E+11	1.24E+11	2.14E+11
9	0.30	0.094	32854.10	33976.10	0.967	6802998890.0	8340765245.0	7529100326.0
10	0.67	0.028	3548.33	13333.33	0.266	22819758.3	2963333.3	8223083.3

(a) Separate ratio and combined ratio estimators: From the above table we have

$$R = \sum_{h=1}^{L} W_h \bar{Y}_h \Big/ \sum_{h=1}^{L} W_h \bar{X}_h = 52444.56/34438.61 = 1.522838 .$$

Let

$$V_h = W_h^2\left(\frac{1-f_h}{n_h}\right)\left[S_{hy}^2 + R_h^2 S_{hx}^2 - 2R_h S_{hxy}\right]$$

and

$$V_h(c) = W_h^2\left(\frac{1-f_h}{n_h}\right)\left[S_{hy}^2 + R^2 S_{hx}^2 - 2R S_{hxy}\right],$$

then we have the following table

Stratum Number	V_h	$V_h(c)$
1	453.24	2774.25
2	60128.59	248921.34
3	204079.04	500736.86
4	174795.28	173606.04
5	30581.64	57702.68
6	554.50	848.41
7	902438.84	856060.71
8	16174803.72	21597997.70
9	85250.55	6674800.88
10	2490.15	620.38
Sum	17635575.55	30114069.26

Thus the variance of the separate ratio estimator is given by

$$V(\bar{y}_{\text{sr}}) = \sum_{h=1}^{L} V_h = 17635575.55$$

and, that of the combined ratio estimator is given by

$$V(\bar{y}_{\text{cr}}) = \sum_{h=1}^{L} V_h(c) = 30114069.26 .$$

Hence the relative efficiency of the separate ratio estimator with respect to combined ratio estimator is:

$$\text{RE} = \frac{V(\bar{y}_{\text{cr}})}{V(\bar{y}_{\text{sr}})}\times 100 = \frac{30114069.26}{17635575.55}\times 100 = 170.77\% .$$

This particular example shows that the separate ratio estimator is better than combined ratio estimator in such situations.

(b) Separate regression and combined regression estimators: Let us first calculate the combined correlation coefficient across all strata to find the variance of the combined regression estimator.

Stratum Number	$W_h^2\left(\frac{1-f_f}{n_h}\right)S_{hy}^2$	$W_h^2\left(\frac{1-f_f}{n_h}\right)S_{hx}^2$	$W_h^2\left(\frac{1-f_f}{n_h}\right)S_{hxy}$
1	27691.20	5820.39	12612.88
2	208369.37	312380.59	224535.89
3	3764483.39	754667.25	1646222.05
4	1182442.36	435718.62	663000.03
5	328632.07	59472.01	134238.89
6	2574.86	2092.08	2159.79
7	9450943.18	3506067.82	5491582.89
8	1031861873.00	343953957.50	593597958.90
9	14127504.52	17320919.88	15635369.14
10	3046.43	395.60	1097.78
Sum	1060957560.00	366351491.80	617408778.30

We know that the correlation coefficient ρ_{xy} across all strata is given by

$$\rho_{xy} = \sum_{h=1}^{L} W_h^2\left(\frac{1-f_h}{n_h}\right)S_{hxy} \Bigg/ \left\{\sqrt{\sum_{h=1}^{L} W_h^2\left(\frac{1-f_h}{n_h}\right)S_{hy}^2}\sqrt{\sum_{h=1}^{L} W_h^2\left(\frac{1-f_h}{n_h}\right)S_{hx}^2}\right\}$$

$$= \frac{617408778.3}{\sqrt{1060957560}\sqrt{366351491.8}} = 0.9903.$$

Thus the variance of the combined regression estimator is given by

$$V(\bar{y}_{\text{clr}}) = \sum_{h=1}^{L} W_h^2\left(\frac{1-f_h}{n_h}\right)S_{hy}^2\left(1-\rho_{xy}^2\right) = 1060957560 \times \left(1-0.9903^2\right) = 20482751.17\,.$$

Let

$$V_h(l) = W_h^2\left(\frac{1-f_h}{n_h}\right)S_{hy}^2\left(1-\rho_{hxy}^2\right), \quad \text{with } \rho_{hxy} = S_{hxy}/\left(S_{hx}S_{hy}\right)$$

then we have following table:

S_{hy}^2	S_{hx}^2	S_{hxy}	ρ_{hxy}	$V_h(l)$
51856391.4	10899652.8	23619714.4	0.99350	358.90
390206386.6	584984730.0	420480877.0	0.88009	46975.34
3172330153.0	635958094.9	1387271318.0	0.97669	173434.47
569396673.9	209817189.2	319262930.2	0.92368	173605.41
153854582.5	27842810.4	62846173.1	0.96021	25631.04
7232769.3	5876666.7	6066866.7	0.93056	345.16
2049296094.0	760238523.4	1190767859.0	0.95400	849431.67
3.72E+11	1.24E+11	2.14E+11	0.99639	7426685.35
6802998890.0	8340765245.0	7529100326.0	0.99952	13660.12
22819758.3	2963333.3	8223083.3	0.99997	0.16
			Sum	8710127.62

Thus the variance of the separate linear regression estimator is given by

$$V(\bar{y}_{\text{slr}}) = \sum_{h=1}^{L} V_h(l) = \sum_{h=1}^{L} W_h^2 \left(\frac{1-f_h}{n_h}\right) S_{hy}^2 \left(1-\rho_{hxy}^2\right) = 8710127.62 .$$

Thus the relative efficiency of separate linear regression estimator with respect to combined linear regression estimator is given by

$$\text{RE} = \frac{V(\bar{y}_{\text{clr}})}{V(\bar{y}_{\text{slr}})} \times 100\% = \frac{20482751.17}{8710127.618} \times 100\% = 235.16\% .$$

Thus separate ratio and regression estimators remain better than combined ratio and regression estimator in this situation. If the population means are known for each stratum, then one should must go for separate ratio or regression estimator. If only overall population mean of the auxiliary variable is known then we do not have choice and we have to go for combined ratio or regression estimators.

8.3.5 ON DEGREE OF FREEDOM IN STRATIFIED RANDOM SAMPLING

When we stratify the given population into L homogeneous strata or groups each with N_h units, then for each stratum we estimate population mean $\bar{Y}_h$, $h = 1,2,...,L$, as shown below:

Stratum No.	1	2	3		h		L
Stratum population Mean	$\bar{Y}_1$	$\bar{Y}_2$	$\bar{Y}_3$		$\bar{Y}_h$		$\bar{Y}_L$

Then we select n_h, $h = 1,2,...,L$ units from the hth stratum using SRSWOR sampling (or any other sampling scheme). There are five cases:

Case I. When no model or auxiliary information is available, then we estimate the population mean in each stratum independently and every time we loose one degree of freedom. Thus when we consider simple pooled estimator of the population mean $\bar{Y}$ as

$$\bar{y}_{\text{st}} = \sum_{h=1}^{L} W_h \bar{y}_h \qquad (8.3.5.1)$$

then the degree of freedom $df = (n-L)$.

Case II. When we consider separate ratio estimator

$$\bar{y}_{\text{sr}} = \sum_{h=1}^{L} W_h \bar{y}_h \left(\frac{\bar{X}_h}{\bar{x}_h}\right) \qquad (8.3.5.2)$$

as an estimator of the population mean, $\bar{Y}$, then basically we are assuming a linear relationship between the study and auxiliary variables in the hth stratum through the model

$$Y_{hi} = R_h X_{hi} + \varepsilon_{hi}, \qquad (8.3.5.3)$$

where $i = 1, 2, ..., N_h$ and $h = 1, 2, ..., L$ such that $E(\varepsilon_{hi}) = 0$, $E(\varepsilon_{hi}^2) = \sigma_h^2$ and $E(\varepsilon_{hi}\varepsilon_{hj}) = 0$ for $i \neq j$.

Then on setting $\sum_{i=1}^{N_h} \varepsilon_{hi} = 0$ we are estimating only one ratio $R_h = \overline{Y}_h / \overline{X}_h$ in the hth stratum and we are loosing one degree of freedom in each stratum. The for the separate ratio estimator, the degree of freedom should be again as $\text{df} = (n - L)$, because here we assume that the regression line in each stratum passes through the origin, as shown in the Figure 8.3.1.

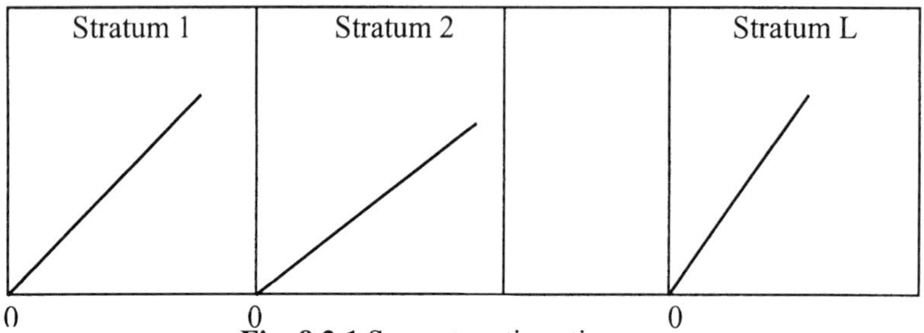

Fig. 8.3.1 Separate ratio estimator.

Case III. When we consider separate regression estimator

$$\bar{y}_{\text{slr}} = \sum_{h=1}^{L} W_h \left[\bar{y}_h + \hat{\beta}_{hxy} \left(\overline{X}_h - \bar{x}_h \right) \right] \tag{8.3.5.4}$$

as an estimator of the population mean $\overline{Y}$ then basically we are assuming a linear relationship between the study and auxiliary variables in the hth stratum through the model

$$Y_{hi} = \alpha_h + \beta_h X_{hi} + \varepsilon_{hi} \tag{8.3.5.5}$$

where $i = 1,2,...,N_h$ and $h = 1,2,...,L$ such that $E(\varepsilon_{hi}) = 0$, $E(\varepsilon_{hi}^2) = \sigma_h^2$ and $E(\varepsilon_{hi}\varepsilon_{hj}) = 0$ for $i \neq j$. Then we are estimating two parameters in each stratum by minimizing $\sum_{i=1}^{N_h} \varepsilon_{hi}^2$ in the hth stratum, and hence we are losing two degrees of freedom in each stratum. Thus for the separate regression estimator the degree of freedom should be taken as $\text{df} = (n - 2L)$, because here we assume that the regression line in each stratum does not pass through the origin and in each stratum we have two parameters *viz.*, intercept and slope, as shown in the Figure 8.3.2.

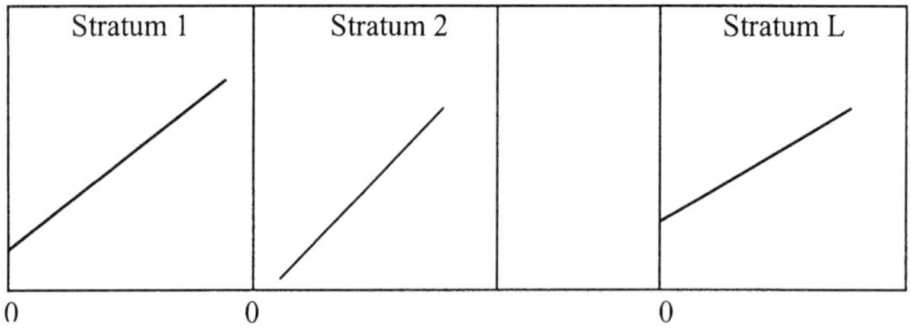

Fig. 8.3.2 Separate regression estimator.

Case IV. When we consider combined ratio estimator

$$\bar{y}_{cr} = \bar{y}_{st}\left(\frac{\bar{X}}{\bar{x}_{st}}\right) \tag{8.3.5.6}$$

as an estimator of the population mean $\bar{Y}$, then basically we are assuming a linear relationship between the study and auxiliary variables across all the L strata through the model

$$Y_{hi} = RX_{hi} + \varepsilon_{hi} \tag{8.3.5.7}$$

where $i = 1,2,...,N_h$ and $h = 1,2,...,L$ such that $E(\varepsilon_{hi}) = 0$, $E(\varepsilon_{hi}^2) = \sigma_h^2$ and $E(\varepsilon_{hi}\varepsilon_{hj}) = 0$ for $i \neq j$. Then on setting $\sum_{h=1}^{L}\sum_{i=1}^{N_h}\varepsilon_{hi} = 0$ we are estimating only one ratio $R = \bar{Y}/\bar{X}$ across all the strata in the population, and we are loosing only one degree of freedom across all strata. The for the combined ratio estimator, the degree of freedom should be taken as df $= (n-1)$. In practice one can think that if we arrange all the strata in ascending order on the basis of auxiliary information, and the regression line passes across all the strata as shown in the Figure 8.3.3, and we have to estimate only one parameter across all strata.

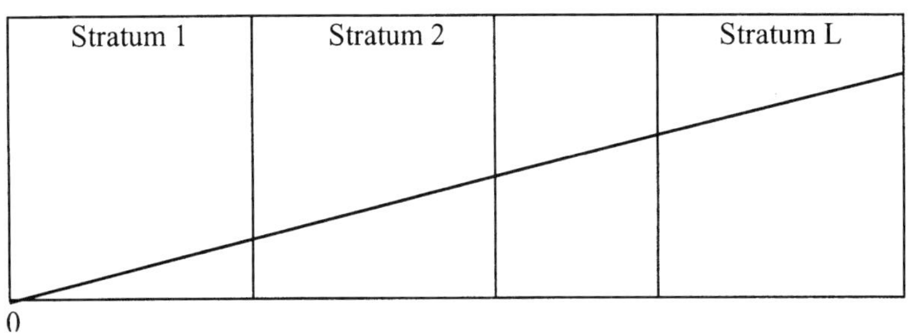

Fig. 8.3.3 Combined ratio estimator.

Case V. When we consider combined regression estimator

$$\bar{y}_{slr} = \bar{y}_{st} + \hat{\beta}(\bar{X} - \bar{x}_{st}) \tag{8.3.5.8}$$

as an estimator of the population mean $\bar{Y}$, then basically we are assuming a linear relationship between the study and auxiliary variables across all the strata through the model

$$Y_{hi} = \alpha + \beta X_{hi} + \varepsilon_{hi} \tag{8.3.5.9}$$

where $i = 1,2,...,N_h$ and $h = 1,2,...,L$ such that $E(\varepsilon_{hi}) = 0$, $E(\varepsilon_{hi}^2) = \sigma_h^2$ and $E(\varepsilon_{hi}\varepsilon_{hj}) = 0$ for $i \neq j$. Then we are estimating two parameters by minimizing $\sum_{h=1}^{L}\sum_{i=1}^{N_h}\varepsilon_{hi}^2$ across all the strata in the population and hence we are loosing two degrees of freedom across all strata. Thus for the combined regression estimator, the degree of freedom should be taken as df $= (n-2)$. In practice one can think that if we arrange the all the strata in ascending order on the basis of auxiliary information, and the regression line passes across all the strata as shown in the Figure 8.3.4 then we have to estimate only two parameters across all strata.

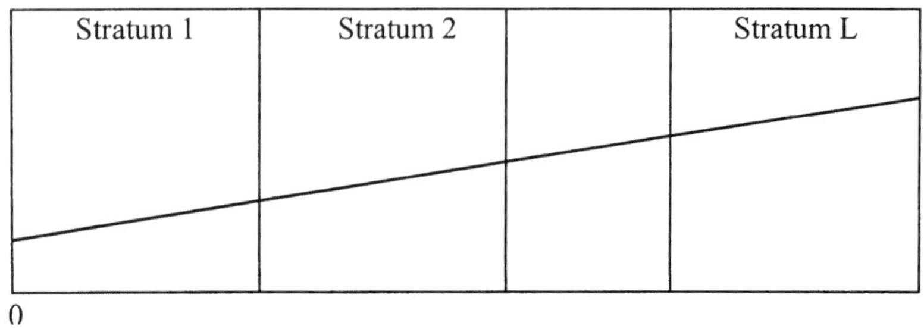

Fig. 8.3.4 Combined regression estimator.

8.4 CALIBRATION APPROACH FOR STRATIFIED SAMPLING DESIGN

Singh, Horn, and Yu (1998) extended the higher level calibration approach for stratified sampling design as follows. Assume the population consists of L strata with N_h units in the hth stratum from which a simple random sample of size n_h is taken without replacement such that total population size is $N = \sum_{h=1}^{L} N_h$ and sample size is $n = \sum_{h=1}^{L} n_h$. Associated with the ith unit of the hth stratum there are two value y_{h_i} and x_{h_i} with $x_{h_i} > 0$ being the auxiliary variable. For the hth stratum, let $W_h = N_h/N$ be the stratum weights, $f_h = n_h/N_h$ the sampling fraction and $\bar{y}_h$, $\bar{x}_h$; $\bar{Y}_h$, $\bar{X}_h$ the sample and population means, respectively. Assume that $\bar{X} = \sum_{h=1}^{L} W_h \bar{X}_h$ is known. The purpose is to estimate population mean $\bar{Y} = \sum_{h=1}^{L} W_h \bar{Y}_h$ by using the auxiliary information X. The usual estimator of population mean $\bar{Y}$ is given by

$$\bar{y}_{\text{st}} = \sum_{h=1}^{L} W_h \bar{y}_h \,. \tag{8.4.1}$$

A new estimator is given by

$$\bar{y}^*_{\text{St}} = \sum_{h=1}^{L} W^*_h \bar{y}_h \tag{8.4.2}$$

with new weights W^*_h which are chosen such that the chi square (CS) type of distance

$$\sum_{h=1}^{L} \left(W^*_h - W_h\right)^2 \left(W_h q_h\right)^{-1} \tag{8.4.3}$$

is minimum, subject to the condition

$$\sum_{h=1}^{L} W^*_h \bar{x}_h = \bar{X} \,. \tag{8.4.4}$$

Note that q_h in (8.4.3) is a suitably chosen weight which determines the form of the estimator. Minimization of (8.4.3), subject to the calibration equation (8.4.4), leads to the combined generalized regression estimator (GREG) given by

$$\bar{y}_{st}^{*} = \sum_{h=1}^{L} W_h \bar{y}_h + \left(\sum_{h=1}^{L} W_h q_h \bar{x}_h \bar{y}_h \Big/ \sum_{h=1}^{L} W_h q_h \bar{x}_h^2 \right) \left[\bar{X} - \sum_{h=1}^{L} W_h \bar{x}_h \right] \tag{8.4.5}$$

for the optimum choice of weights

$$W_h^{*} = W_h + \left(W_h q_h \bar{x}_h \Big/ \sum_{h=1}^{L} W_h q_h \bar{x}_h^2 \right) \left(\bar{X} - \sum_{h=1}^{L} W_h \bar{x}_h \right). \tag{8.4.6}$$

If $q_h = 1/\bar{x}_h$ then the estimator (8.4.5) reduces to the well known combined ratio estimator defined earlier in (8.3.3.1). Note that there is no choice of q_h such that the estimator (8.4.5) reduces to well known combined linear regression estimator in stratified sampling as discussed in Section (8.3.4).

An estimator of variance of combined generalized regression estimator (GREG) is given by

$$\hat{v}\left(\bar{y}_{st}^{*}\right) = \sum_{h=1}^{L} \frac{W_h^2 (1 - f_h)}{n_h} s_{eh}^2 \tag{8.4.7}$$

where $s_{eh}^2 = (n_h - 1)^{-1} \sum_{i=1}^{n_h} e_{hi}^2$ is the hth stratum sample variance and

$e_{hi} = (y_{hi} - \bar{y}_h) - b(x_{hi} - \bar{x}_h)$ with $b = \sum_{h=1}^{L} W_h q_h \bar{y}_h \bar{x}_h \Big/ \sum_{h=1}^{L} W_h q_h \bar{x}_h^2$.

The lower level calibration approach yields an estimator of variance of the combined generalized regression estimator (GREG) as

$$\hat{v}\left(\bar{y}_{st}^{*}\right)_{greg} = \sum_{h=1}^{L} \frac{D_h W_h^{*2}}{W_h^2} s_{eh}^2 \tag{8.4.8}$$

where $D_h = W_h^2 (1 - f_h)/n_h$ and W_h^{*} is given by (8.4.6).

If $q_h = 1/\bar{x}_h$ then (8.4.8) reduces to an estimator given by

$$\hat{v}\left(\hat{y}_{st}^{*}\right)_{ratio} = \left(\frac{\bar{X}}{\bar{x}_{st}}\right)^2 \sum_{h=1}^{L} \frac{W_h^2 (1 - f_h)}{n_h} s_{eh}^2 \tag{8.4.9}$$

which is a special case of a class of estimators for estimating the variance of the combined ratio estimator given by Wu (1985) as

$$\hat{v}\left(\bar{y}_{st}^{*}\right)_{w} = \left(\frac{\bar{X}}{\bar{x}_{st}}\right)^g \sum_{h=1}^{L} \frac{W_h^2 (1 - f_h)}{n_h} s_{eh}^2 \tag{8.4.10}$$

for $g = 2$.

The properties of the estimators of the variance of the combined ratio estimator are also studied by Saxena, Nigam, and Shukla (1995). Using higher level calibration, a new estimator is given by

$$\hat{v}_{ho}\left(\hat{y}_{st}^{*}\right)_{greg} = \sum_{h=1}^{L} \frac{\Omega_h W_h^{*2}}{W_h^2} s_{eh}^2 \tag{8.4.11}$$

where Ω_h are suitably chosen weights such that the chi square distance function

$$\sum_{h=1}^{L}(\Omega_h - D_h)^2 (D_h Q_h)^{-1} \tag{8.4.12}$$

is minimum subject to higher level calibration equation

$$\sum_{h=1}^{L}\Omega_h s_{hx}^2 = V(\bar{x}_{st}) \tag{8.4.13}$$

where

$V(\bar{x}_{st}) = \sum_{h=1}^{L} W_h^2 \frac{(1-f_h)}{n_h} S_{hx}^2$ is assumed to be known, and Q_h, like q_h, is again a suitably chosen weight to form different kind of estimators. This procedure leads to a new estimator for the variance of the combined generalized regression estimator (GREG) as

$$\hat{v}_{ho}\left(\hat{\bar{y}}_{st}^*\right)_{greg} = \hat{v}\left(\hat{\bar{y}}_{st}^*\right)_{greg} + \hat{\beta}_{st}\left[V(\bar{x}_{st}) - \hat{v}(\bar{x}_{st})\right] \tag{8.4.14}$$

where

$\hat{\beta}_{st} = \sum_{h=1}^{L} \frac{W_h^{*2}(1-f_h)}{n_h} Q_h s_{hx}^2 s_{\hat{e}h}^2 \Big/ \sum_{h=1}^{L} \frac{W_h^2(1-f_h)}{n_h} s_{hx}^4$ denotes the combined improved estimator of regression coefficient in stratified sampling and $\hat{v}(\bar{x}_{st}) = \sum_{h=1}^{L} W_h^2 \frac{(1-f_h)}{n_h} s_{hx}^2$ is an unbiased estimator of $V(\bar{x}_{st})$.

If $q_h = 1/\bar{x}_h$ and $Q_h = 1/s_{hx}^2$, then the estimator of (8.4.14) reduces to a new estimator of variance of the combined ratio estimator given by

$$\hat{v}_{st}\left(\hat{\bar{Y}}_{ratio}\right) = \sum_{h=1}^{L} \frac{W_h^2(1-f_h)}{n_h} s_{\hat{e}h}^2 \left(\frac{\bar{X}}{\bar{x}_{st}}\right)^2 \left(\frac{V(\bar{x}_{st})}{\hat{v}(\bar{x}_{st})}\right) \tag{8.4.15}$$

which is again a ratio type estimator proposed by Wu (1985) for estimating the variance of the combined ratio estimator. Note that it makes use of the extra knowledge of the known variance of the auxiliary variable at the estimation stage. Several more new estimators can be constructed for new choices of weights q_h and Q_h. Defining $u = X \Big/ \sum_{i \in s} d_i x_i$ and $v = V\left(\hat{X}_{ht}\right) / \hat{v}\left(\hat{X}_{ht}\right)$. Singh, Horn, and Yu (1998) suggested a wider class of estimators as

$$\hat{v}_{ss}\left(\hat{\bar{y}}_{st}^*\right)_{greg} = \left\{\frac{1}{2}\sum_{i \in s}\sum_{j(\neq i) \in s} D_{ij}\left(d_i e_i - d_j e_j\right)^2\right\} H(u,v) \tag{8.4.16}$$

where $H(u,v)$ is a parametric function of u and v such that $H(1,1) = 1$ satisfying certain regularity conditions. Then all estimators obtained from the functions, $H(u,v) = u^{\alpha} v^{\beta}$, $H(u,v) = \{1+\alpha(u-1)\}/\{1+\beta(v-1)\}$, $H(u,v) = 1+\alpha(u-1)+\beta(v-1)$ and $H(u,v) = \{1+\alpha(u-1)+\beta(v-1)\}^{-1}$ are special cases of the higher level calibration approach, where α and β are unknown parameters involved in the function $H(u, v)$. Replacement of these parameters with their respective consistent estimators in the class of estimators at (8.4.16) yields estimators which possess the same asymptotic variance as shown by Srivastava and Jhajj (1983a), Singh and Singh (1984a), and Mahajan and Singh (1996).

Example 8.4.1. Select a sample of 40 countries from population 5 using the method of proportional allocation. Record the production and area of the tobacco crop from the countries selected in the sample. Assume that the area under the tobacco crop in different continents is known. Obtain the calibration weights which forms combined ratio and combined GREG type estimators. Obtain the values of the estimates.

Solution. Using the proportional allocation, we have the same values of n_h as in the previous example. Further from the sample information, we have

St. no.	$\bar{y}_h$	$\bar{x}_h$	$W_h\bar{x}_h$	W_h^{ratio}	$W_h\bar{x}_h^2$	W_h^{greg}	$\bar{y}_r$	$\bar{y}_{\text{greg}}$
1	2592.0	1304.7	73.84432	0.032587	96342.5	0.056468	84.46	146.36
2	26763.0	29075.0	1645.64500	0.032587	47847128.4	0.053662	872.12	1436.15
3	14766.0	5191.7	391.97108	0.043468	2034984.5	0.074800	641.85	1104.49
4	29900.0	21700.0	2046.31000	0.054292	44404927.0	0.090646	1623.33	2710.33
5	12463.0	6808.0	770.66560	0.065174	5246691.4	0.111824	812.26	1393.66
6	3375.0	1800.0	67.86000	0.021705	122148.0	0.037579	73.26	126.82
7	38411.8	24481.5	6928.27582	0.162934	169614861.6	0.270630	6258.61	10395.40
8	473455.0	294809.0	47287.36360	0.092349	13940740376.0	0.075972	43722.94	35969.48
9	7480.3	6303.7	594.43608	0.054292	3747128.9	0.093239	406.12	697.45
10	822.5	350.0	9.90500	0.016293	3466.7	0.028282	13.40	23.26
		Sum	59816.27651		14213858055.0		54508.36	54003.43

In general the calibration weights in stratified random sampling are given by

$$W_h^* = W_h + \left(W_h q_h \bar{x}_h \Big/ \sum_{h=1}^{L} W_h q_h \bar{x}_h^2\right)\left(\bar{X} - \sum_{h=1}^{L} W_h \bar{x}_h\right).$$

If $q_h = 1/\bar{x}_h$ then the calibration weights which leads to the combined ratio estimator

$$W_h^{\text{ratio}} = W_h\left(\bar{X} \Big/ \sum_{h=1}^{L} W_h \bar{x}_h\right)$$

and an estimate of the average production of tobacco using the combined ratio estimator is given by

$$\bar{y}_{\text{r}} = \sum_{h=1}^{L} W_h^{\text{ratio}} \bar{y}_h = 54508.36.$$

If $q_h = 1$ then the calibration weights leading to the combined GREG estimator are

$$W_h^{\text{greg}} = W_h + \left(W_h \bar{x}_h \Big/ \sum_{h=1}^{L} W_h \bar{x}_h^2\right)\left(\bar{X} - \sum_{h=1}^{L} W_h \bar{x}_h\right)$$

and the combined GREG estimate of the average tobacco production is given by

$$\bar{y}_{\text{greg}} = \sum_{h=1}^{L} W_h^{\text{greg}} \bar{y}_h = 54003.43.$$

The true average production level of the tobacco crop in the world is 52444.56 as shown in the description of population 5. Evidently the GREG estimate is more close than ratio estimate. For this particular sample the sampling error is small in case of GREG estimate than the ratio estimate.

8.4.1 EXACT COMBINED LINEAR REGRESSION USING CALIBRATION

Singh (2003c) recently considers a new estimator of the population mean $\bar{Y}$ in stratified sampling as

$$\bar{y}_{st}^{\otimes} = \sum_{h=1}^{L} W_h^{\otimes} \bar{y}_h \tag{8.4.1.1}$$

where $W_h^{\otimes}$ are the calibrated weights such that the chi square distance function

$$D^{\otimes} = \frac{1}{2}\sum_{h=1}^{L} \frac{\left(W_h^{\otimes} - W_h\right)^2}{W_h Q_h^{\otimes}} \tag{8.4.1.2}$$

is minimum subject to two constraints defined as

$$\sum_{h=1}^{L} W_h^{\otimes} = \sum_{h=1}^{L} W_h \tag{8.4.1.3}$$

and

$$\sum_{h=1}^{L} W_h^{\otimes} \bar{x}_h = \bar{X} \tag{8.4.1.4}$$

where $Q_h^{\otimes}$ are some suitably chosen weights. Minimization of (8.4.1.2) with respect to (8.4.1.3) and (8.4.1.4) leads to new calibrated weights are given by

$$W_h^{\otimes} = W_h + \frac{\left(W_h Q_h \bar{x}_h\right)\left(\sum_{h=1}^{L} W_h Q_h\right) - \left(W_h Q_h\right)\left(\sum_{h=1}^{L} W_h Q_h \bar{x}_h\right)}{\left(\sum_{h=1}^{L} W_h Q_h\right)\left(\sum_{h=1}^{L} W_h Q_h \bar{x}_h^2\right) - \left(\sum_{h=1}^{L} W_h Q_h \bar{x}_h\right)^2}\left(X - \sum_{h=1}^{L} W_h \bar{x}_h\right) \tag{8.4.1.5}$$

and thus a new calibrated estimator of the population mean $\bar{Y}$ becomes

$$\bar{y}_{st}^{\oplus} = \sum_{h=1}^{L} W_h \bar{y}_h + \hat{\beta}_{st(ols)}\left[\bar{X} - \sum_{h=1}^{L} W_h \bar{x}_h\right] \tag{8.4.1.6}$$

where

$$\hat{\beta}_{st(ols)} = \frac{\left(\sum_{h=1}^{L} W_h Q_h^{\otimes} \bar{x}_h \bar{y}_h\right)\left(\sum_{h=1}^{L} W_h Q_h^{\otimes}\right) - \left(\sum_{h=1}^{L} W_h Q_h^{\otimes} \bar{y}_h\right)\left(\sum_{h=1}^{L} W_h Q_h^{\otimes} \bar{x}_h\right)}{\left(\sum_{h=1}^{L} W_h Q_h^{\otimes}\right)\left(\sum_{h=1}^{L} W_h Q_h^{\otimes} \bar{x}_h^2\right) - \left(\sum_{h=1}^{L} W_h Q_h^{\otimes} \bar{x}_h\right)^2} \tag{8.4.1.7}$$

If $Q_h^{\otimes} = 1$ then the estimator (8.4.1.6) reduces to the traditional combined stratified linear regression estimator, and hence better than the estimators developed by Singh, Horn, and Yu (1998), and Tracy, Singh, and Arnab (2003), and leads to the following theorem:

Theorem 8.4.1.1. The combined linear regression estimator in stratified random sampling is unique in its class of estimators.

Note that more work related to this topic will be discussed in the next volume.

8.5 CONSTRUCTION OF STRATA BOUNDARIES

Let us study how to make the strata boundaries which divide the scale of study or auxiliary variable for the strata. The range of the auxiliary variable X is a to b subject to the condition that $(b-a)<\infty$. For example, let $X \in (a,b)$ and let the initial rough four strata boundaries be given by a to B_1; B_1 to B_2; B_2 to B_3; and B_3 to b. It is to be noted here that to find two boundary points B_1 and B_2 will result in three strata, three boundary points B_1, B_2, and B_3 will result in four strata and so on. These boundary points can either be obtained using information on study variable Y or auxiliary variable X. While doing stratification on the basis of the study variable Y we assume that:

(i) The study variable Y has a continuous distribution;
(ii) The first and second order derivatives of the probability density function $f(y)$ exist for all y in the range (a, b);
(iii) The stratum boundary points $y_1, y_2,..., y_{(L-1)}$, which will result into L strata, based on the study variable, take positions as shown in Figure 8.5.1.

|-----|	-----|	-----|	---------|	-----|	---------|	-----|
a	y_1	y_2	y_3	y_{h-1}	y_h	y_{L-1}	y_L

Fig. 8.5.1 Strata Boundary Points.

(iv) The hth stratum weight, mean, and variance are defined as

$$W_h = \int_{y_{h-1}}^{y_h} f(y)dy\,,\quad \mu_{hy} = \frac{1}{W_h}\int_{y_{h-1}}^{y_h} y f(y)dy\,,\quad \text{and}\quad \sigma_{hy}^2 = \frac{1}{W_h}\int_{y_{h-1}}^{y_h} y^2 f(y)dy - \mu_{hy}^2\,;$$

(v) Under assumption (ii) we have the following results:
On differentiating W_h with respect to y_h we get

$$\frac{\partial W_h}{\partial y_h} = f(y_h)\,; \tag{8.5.1}$$

Applying the basic rule of differentiation, for the ith stratum weight, we have

$$\frac{\partial W_i}{\partial y_h} = -f(y_h)\,,\quad i \neq h\,. \tag{8.5.2}$$

On differentiating μ_{hy} with respect to y_h we get

$$\frac{\partial \mu_{hy}}{\partial y_h} = \frac{\partial}{\partial y_h}\left[\frac{1}{W_h}\int_{y_{h-1}}^{y_h} y f(y)dy\right] = \frac{y_h f(y_h)}{W_h} - \frac{\int_{y_{h-1}}^{y_h} y f(y)dy}{W_h^2} \times \frac{\partial W_h}{\partial y_h}$$

$$= \frac{y_h f(y_h)}{W_h} - \frac{\int_{y_{h-1}}^{y_h} y f(y)dy}{W_h^2} \times f(y_h) = \frac{y_h f(y_h)}{W_h} - \frac{\int_{y_{h-1}}^{y_h} y f(y)dy}{W_h} \times \frac{f(y_h)}{W_h}$$

$$= \frac{y_h f(y_h)}{W_h} - \mu_{hy} \times \frac{f(y_h)}{W_h} = \frac{f(y_h)}{W_h}\left[y_h - \mu_{hy}\right].$$

On differentiating σ_{hy}^2 with respect to y_h we have

$$\frac{\partial \sigma_{hy}^2}{\partial y_h} = \frac{\partial}{\partial y_h}\left[\frac{1}{W_h}\int_{y_{h-1}}^{y_h} y^2 f(y)dy - \mu_{hy}^2\right] = \frac{y_h^2 f(y_h)}{W_h} - \frac{\int_{y_{h-1}}^{y_h} y^2 f(y)dy}{W_h^2} \times \frac{\partial W_h}{\partial y_h} - 2\mu_{hy}\frac{\partial \mu_{hy}}{\partial y_h}$$

$$= \frac{y_h^2 f(y_h)}{W_h} - \frac{\int_{y_{h-1}}^{y_h} y^2 f(y)dy}{W_h} \times \frac{f(y_h)}{W_h} - 2\mu_{hy}\frac{f(y_h)}{W_h}\left(y_h - \mu_{hy}\right)$$

$$= \frac{y_h^2 f(y_h)}{W_h} - \left(\sigma_{hy}^2 + \mu_{hy}^2\right) \times \frac{f(y_h)}{W_h} - 2\mu_{hy}\frac{f(y_h)}{W_h}\left(y_h - \mu_{hy}\right)$$

$$= \frac{f(y_h)}{W_h}\left[y_h^2 - \left(\sigma_{hy}^2 + \mu_{hy}^2\right) - 2\mu_{hy}\left(y_h - \mu_{hy}\right)\right]$$

$$= \frac{f(y_h)}{W_h}\left[\left(y_h - \mu_{hy}\right)^2 - \sigma_{hy}^2\right]. \tag{8.5.3}$$

Again from the basic concept of differentiation we have

$$\frac{\partial \sigma_{ih}^2}{\partial y_h} = -\frac{f(y_h)}{W_i}\left[\left(y_h - \mu_{iy}\right)^2 - \sigma_{iy}^2\right]. \tag{8.5.4}$$

8.5.1 STRATA BOUNDARIES FOR PROPORTIONAL ALLOCATION

Under proportional allocation, if finite population correction factor is ignored, then the variance of the estimator $\bar{y}_{st}$ of population mean $\bar{Y}$ is given by

$$V_p\left(\bar{y}_{st}\right) = \frac{1}{n}\sum_{h=1}^{L} W_h \sigma_{hy}^2 . \tag{8.5.1.1}$$

Assuming $i = h+1$ then differentiating (8.5.1.1) with respect to y_h and equating to zero we have

$$n\frac{\partial V_p\left(\bar{y}_{st}\right)}{\partial y_h} = \frac{\partial}{\partial y_h}\left[W_h \sigma_{hy}^2\right] + \frac{\partial}{\partial y_h}\left[W_i \sigma_{iy}^2\right]$$

$$= \sigma_{hy}^2\left(\frac{\partial W_h}{\partial y_h}\right) + W_h\left(\frac{\partial \sigma_{hy}^2}{\partial y_h}\right) + \sigma_{iy}^2\left(\frac{\partial W_i}{\partial y_h}\right) + W_i\left(\frac{\partial \sigma_{iy}^2}{\partial y_h}\right) = 0 . \tag{8.5.1.2}$$

On using first order derivative results from the previous section in (8.5.1.2) we have

$$\sigma_{hy}^2 f(y_h) + \frac{W_h f(y_h)}{W_h}\left[\left(y_h - \mu_{hy}\right)^2 - \sigma_{hy}^2\right] + \sigma_{iy}^2\{-f(y_h)\} + W_i\left\{-\frac{f(y_h)}{W_i}\right\}\left[\left(y_h - \mu_{iy}\right)^2 - \sigma_{iy}^2\right]$$

$$= f(y_h)\left[\sigma_{hy}^2 + \left(y_h - \mu_{hy}\right)^2 - \sigma_{hy}^2 - \sigma_{iy}^2 - \left(y_h - \mu_{iy}\right)^2 + \sigma_{iy}^2\right] = 0 ,$$

or $\left(y_h - \mu_{hy}\right)^2 - \left(y_h - \mu_{iy}\right)^2 = 0$, or $\left(\mu_{hy} - \mu_{iy}\right)\left[\left(\mu_{hy} + \mu_{iy}\right) - 2y_h\right] = 0$. (8.5.1.3)

Note that $\mu_{hy} \neq \mu_{iy}$ thus (8.5.1.3) implies that

$$y_h = \left(\mu_{hy} + \mu_{iy}\right)/2 . \tag{8.5.1.4}$$

Thus we have the following theorem:

Theorem 8.5.1.1. Under proportional allocation, the strata boundary points are given by

$$y_h = \frac{\mu_{hy} + \mu_{iy}}{2}, \quad \text{for } h = 1, 2, ..., L - 1.$$

This set of equations is called the set of minimal equations. It is to be noted that μ_{hy} and μ_{iy} are based on the strata boundaries, therefore it is not possible to obtain y_h directly. In other words, the strata boundaries under proportional allocation technique are the means of the following and preceding strata mean values of the study variable.

8.5.2 STRATA BOUNDARIES FOR NEYMAN ALLOCATION

Under Neyman allocation, if the finite population correction factor is ignored then the variance of the estimator $\bar{y}_{\text{st}}$ of population mean $\bar{Y}$ is given by

$$V\left(\bar{y}_{\text{st}}\right)_{\text{N}} = \frac{1}{n}\left(\sum_{h=1}^{L} W_h \sigma_{hy}\right)^2 . \tag{8.5.2.1}$$

Now minimization of (8.5.2.1) is equivalent to minimization of $\phi = \sum_{h=1}^{L} W_h \sigma_{hy}$. Thus the minimal equations in this case will be

$$\frac{\partial \phi}{\partial y_h} = \frac{\partial}{\partial y_h}\left(W_h \sigma_{hy}\right) + \frac{\partial}{\partial y_h}\left(W_i \sigma_{iy}\right)$$

$$= \sigma_h\left(\frac{\partial W_h}{\partial y_h}\right) + W_h\left(\frac{\partial \sigma_{hy}}{\partial y_h}\right) + \sigma_i\left(\frac{\partial W_i}{\partial y_h}\right) + W_i\left(\frac{\partial \sigma_{iy}}{\partial y_h}\right) = 0 . \tag{8.5.2.2}$$

Now we have $\frac{\partial \sigma_{hy}^2}{\partial y_h} = 2\sigma_{hy}\left(\frac{\partial \sigma_{hy}}{\partial y_h}\right)$, which implies that

$$\frac{\partial \sigma_{hy}}{\partial y_h} = \frac{1}{2\sigma_{hy}}\left(\frac{\partial \sigma_{hy}^2}{\partial y_h}\right) = \frac{f(y_h)}{2\sigma_{hy} W_h}\left[\left(y_h - \mu_{hy}\right)^2 - \sigma_{hy}^2\right]. \tag{8.5.2.3}$$

Similarly we have

$$\frac{\partial \sigma_{iy}}{\partial y_h} = -\frac{f(y_h)}{2\sigma_{iy} W_i}\left[\left(y_h - \mu_{iy}\right)^2 - \sigma_{iy}^2\right]. \tag{8.5.2.4}$$

Thus the set of minimal equations (8.5.2.2) will reduce to

$$\sigma_{hy} f(y_h) + \frac{W_h f(y_h)}{2\sigma_{hy} W_h}\left[\left(y_h - \mu_{hy}\right)^2 - \sigma_{hy}^2\right] - \sigma_{iy} f(y_h) - \frac{W_i f(y_h)}{2\sigma_{iy} W_i}\left[\left(y_h - \mu_{iy}\right)^2 - \sigma_{iy}^2\right]$$

$$= f(y_h)\left[\sigma_{hy} + \frac{1}{2\sigma_{hy}}\left\{(y_h - \mu_{hy})^2 - \sigma_{hy}^2\right\} - \sigma_{iy} - \frac{1}{2\sigma_{iy}}\left\{(y_h - \mu_{iy})^2 - \sigma_{iy}^2\right\}\right] = 0$$

or

$$\sigma_{hy} + \frac{1}{2\sigma_{hy}}\left\{(y_h - \mu_{hy})^2 - \sigma_{hy}^2\right\} = \sigma_{iy} + \frac{1}{2\sigma_{iy}}\left\{(y_h - \mu_{iy})^2 - \sigma_{iy}^2\right\}$$

or

$$\frac{(y_h - \mu_{hy})^2 + \sigma_{hy}^2}{\sigma_{hy}} = \frac{(y_h - \mu_{iy})^2 + \sigma_{iy}^2}{\sigma_{iy}}. \tag{8.5.2.5}$$

Thus we have the following theorem.

Theorem 8.5.2.1. The strata boundary points for the Neyman allocation are given by the solution to the set of minimal equations

$$\frac{(y_h - \mu_{hy})^2 + \sigma_{hy}^2}{\sigma_{hy}} = \frac{(y_h - \mu_{iy})^2 + \sigma_{iy}^2}{\sigma_{iy}} \quad \text{for } i = h+1 \text{ and } h = 1,2,...,L-1. \tag{8.5.2.6}$$

Solving these equations is not easy since μ_h and σ_h are dependent upon y_h.

The methods discussed so for obtaining the strata boundary points are due to Dalenius and Hodges (1957, 1959). In actual practice the following six steps are needed for applying Dalenius and Hodges's method of finding strata boundaries.

Step 1. Sort the population by the auxiliary character X. The value of X is known for each unit in the population and (thought to be) highly correlated with Y.

Step 2. Divide the population into a large number of small and equal width intermediate strata.

Step 3. Calculate the population frequency within each intermediate stratum. Denote the observed frequency in the hth stratum by f_h.

Step 4. Obtain the list of intermediate strata and calculate the cumulative sum of $\sqrt{f_h}$.

Step 5. Divide $T = \sum_{h=1}^{L}\sqrt{f_h}$ by L, the total number of strata; and let $k = T/L$.

Step 6. The stratum boundaries are obtained by finding the value of cumulative sum $\left\{\sqrt{f_h}\right\}$ $h = 1,2,...,L$ that are closest to the multiples of k, and using the right boundary point of the intermediate stratum giving that value of Cum $\left\{\sqrt{f_h}\right\}$.

Some times it may be difficult to use intermediate strata of equal width. For example, when using employment. The Dalenious--Hodges technique proceeds as before, except $\sqrt{W_h f_h}$ has to be used instead of $\sqrt{f_h}$, where W_h is the width of the intermediate stratum h. The above method is also called the Cumulative Square

Root (CSR) method. Unnithan (1978) modified the Newton method used by Shannon (1970), which seems to perform very satisfactorily when adopted for minimizing the variance in the search for the best boundary points of stratification for Neyman allocation. Shiledar--Baxi (1995) discussed a sequel to the modified Kossack and Shiledar--Baxi (1971) procedure proposed to obtain an improved 'unit stratified' design.

Example 8.5.1. Stratify the 50 states listed in population 1 (Appendix) using nonreal estate farm loans as an auxiliary variable. Modify the stratum boundaries into six strata using the cumulative square root method.

Solution. Sort the data in the ascending order of the magnitude of nonreal estate farm loans as follows.

Nonreal estate farm loans	States	Preliminary Strata	Nonreal estate farm loans	States	Preliminary Strata
0.233	RI		540.696	GA	
0.471	NH		549.551	MS	**7**
3.433	AK	**1**	557.656	KY	
4.373	CT		571.487	OR	
16.710	NV		635.774	OH	
19.363	VT		722.034	MT	**8**
27.508	NJ		848.317	AR	
29.291	WV		906.281	CO	
38.067	HI	**2**	1006.036	ID	
43.229	DE		1022.782	IN	**9**
51.539	ME		1228.607	WA	
56.471	MA		1241.369	ND	
57.684	MD		1372.439	WI	
80.750	SC		1519.944	MO	**10**
188.477	VA	**3**	1692.817	SD	
197.244	UT		1716.087	OK	
274.035	NM	**4**	2466.892	MN	
298.351	PA		2580.304	KS	**11**
348.334	AL		2610.572	IL	
386.479	WY	**5**	3520.361	TX	
388.869	TN		3585.406	NE	**12**
405.799	LA		3909.738	IA	
426.274	NY		3928.732	CA	
431.439	AZ	**6**			
440.518	MI				
464.516	FL				
494.730	NC				

From the above table we have

Preliminary Strata	Stratum width W_h	Number of states in each stratum, f_h	$W_h f_h$	$\sqrt{W_h f_h}$	Cum $\sqrt{W_h f_h}$
1	4.140	4	16.560	4.069398	4.069398
2	64.040	10	640.400	25.306130	29.375520
3	8.767	2	17.534	4.187362	33.562890
4	24.316	2	48.632	6.973665	40.536550
5	40.535	3	121.605	11.027470	51.564020
6	88.931	6	533.586	23.099480	74.663500
7	30.791	4	123.164	11.097930	85.761420
8	270.507	4	1082.028	32.894190	118.655600
9	235.333	4	941.332	30.681130	149.336800
10	343.648	4	1374.592	37.075490	186.412200
11	143.680	3	431.040	20.761500	207.173700
12	408.371	4	1633.484	40.416380	247.590100
	Sum	50		247.590100	

Note the use of f_h as frequency. Also note that we need 6 strata, therefore divide the sum $\sum_{h=1}^{12}\sqrt{W_h f_h}$ by 6, and we have the first boundary point at a cumulative frequency of 41.26502. Thus the five boundary points (or six strata) are given in the following table.

Table 8.5.1. Final list of states in different strata obtained by cumulative square root method.

Strata No.	Lower boundary	Upper boundary	Preliminary strata to be merged	States selected in the final strata	Number of states
1	0.00000	41.26502	1, 2, 3, 4	RI, NH, AK, CT, NV,VT, NJ, WV, HI, DE, ME, MA, MD, SC, VA, UT, NM, PA	18
2	41.26502	82.53004	5, 6	AL, WY, TN, LA, NY, AZ, MI, FL, NC	9
3	82.53004	123.79510	7, 8	GA, MS, KY, OR, OH, MT, AR, CO	8
4	123.79510	165.06010	9	ID, IN, WA, ND	4
5	165.06010	206.32510	10	WI, MO, SD, OK	4
6	206.32510	247.59010	11, 12	MN, KS, IL, TX, NE, IA, CA	7
				Sum	50

Example 8.5.2. We wish to estimate the average death rate of the persons living in the United States on the basis of year 2000 projections of the death rate. The projected 191443 persons are to be grouped into five strata on the basis of their age at the death time. The rough distribution of the death rate projected for the year 2000 in the United States has been listed in population 6 in 21 age groups with a gap of four years. We wish to apply the Neyman method of sample allocation for selecting the overall sample. Apply the Cumulative Square Root method to form 5 strata.

Solution. From the information given in the population 6 we have

Sr. No.	Age Group	Strata Boundaries	Year 2000 f_h	$\sqrt{f_h}$	Cum.$\sqrt{f_h}$
1	0—4	0.5--4.5	692	26.305890	26.30589
2	5—9	4.5--9.5	14	3.741657	30.04755
3	10--14	9.5--14.5	15	3.872983	33.92053
4	15--19	14.5--19.5	57	7.549834	41.47037
5	20--24	19.5--24.5	71	8.426150	49.89652
6	25--29	24.5--29.5	68	8.246211	58.14273
7	30--34	29.5--34.5	78	8.831761	66.97449
8	35--39	34.5--39.5	111	10.535650	77.51014
9	40--44	39.5--44.5	182	13.490740	91.00088
10	45--49	44.5--49.5	315	17.748240	108.74910
11	50--54	49.5--54.5	533	23.086790	131.83590
12	55--59	54.5--59.5	853	29.206160	161.04210
13	60--64	59.5--64.5	1357	36.837480	197.87960
14	65--69	64.5--69.5	2010	44.833020	242.71260
15	70--74	69.5--74.5	3022	54.972720	297.68530
16	75--79	74.5--79.5	4423	66.505640	364.19090
17	80--84	79.5--84.5	6921	83.192550	447.38350
18	85--89	84.5--89.5	10609	103.000000	550.38350
19	90--94	89.5--94.5	16915	130.057700	680.44120
20	95--99	94.5--99.5	27188	164.887800	845.32900
21	100--104	99.5--104.5	44053	209.888100	1055.21700
22	105--109	104.5--109.5	71956	268.246200	1323.46300

For making five strata, we need to know four boundary points, say B_1, B_2, B_3, and B_4 by using linear interpolation between the class intervals and the Cum.$\sqrt{f_h}$ values. On dividing 1323.4630 by 5 and taking the cumulative totals, we get the rough five boundary points corresponding to values 264.6926, 529.3853, 794.0779 and 1058.771 as given in the 6th column of the above table. Thus four boundary points obtained through interpolation are given by

$$B_1 = 69.5 + \frac{4(264.6926 - 242.7126)}{54.97272} = 71.09,$$

$$B_2 = 84.5 + \frac{4(529.3853 - 447.3835)}{103} = 87.68,$$

$$B_3 = 94.5 + \frac{4(794.0779 - 680.4412)}{164.8878} = 97.25,$$

and

$$B_4 = 104.5 + \frac{4(1058.771 - 1055.217)}{268.2462} = 104.55.$$

It indicates that the persons with age in the ranges

[0, 71.09], [71.09, 87.68], [87.68, 97.25], [97.25, 104.55], and [104.55, 109]

will form the first, second, third, fourth, and fifth stratum, respectively, according to the cumulative square root method.

8.5.3 STRATIFICATION USING AUXILIARY INFORMATION

Serfling (1968) suggested the use of $\text{Cum}\sqrt{f_h}$ rule for stratification on the auxiliary variable x when the regression of Y on X is linear with uncorrelated homoscedastic errors and nearly perfect correlation. As the $\text{Cum}\sqrt{f_h}$ rule was primarily proposed for stratification on the study variable Y it does not take into account the regression of Y on X and also the form of the conditional variance $V(y|x)$. The $\text{Cum}\sqrt[3]{f_h}$ rule of Singh and Sukhatme (1969) and Singh (1971) though takes into account the regression of Y on X and also the form of the conditional variance function $V(y|x)$ but does not reduce to the rules recommended for optimum stratification on y when $V(y|x) = 0$. While deriving $\text{Cum}\sqrt[3]{f_h}$ rule, Singh (1971) made the assumption that $V(y|x) > 0$ for all x in the range (a, b) of x with $(b - a) < \infty$. Singh (1975b) suggested an improvement in his $\text{Cum}\sqrt[3]{f_h}$ rule such that the form of the conditional variance function $V(y|x)$ also reduces to a rule for optimum stratification on y when $V(y|x) = 0$ for all x in the range (a, b). Schneeberger (1979) commented on the necessary conditions of Dalenius and Hodges (1957, 1959) for optimal stratification points with Neyman's allocation. Bankier (1988) suggested a power allocation method, under which the sample size in the hth stratum takes the optimum value

$$n_h = n\left[\left(\frac{S_h X_h^q}{\overline{Y}_h}\right) \Bigg/ \sum_{h=1}^{L} \frac{S_h X_h^q}{\overline{Y}_h}\right]. \tag{8.5.3.1}$$

The value of q is called the power of this allocation and it can take any value in $[0, 1]$. The choice of q between 0 and 1 can be viewed as a compromise between the Neyman allocation and the almost equal coefficient of variation allocation. Mandowara and Gupta (1999) have obtained optimum points of stratification for two or more stage designs with unequal first stage units and the subsequent units. For numerical illustrations related to Cum.$\sqrt{f}$ and Cum.$\sqrt[3]{f}$ methods, one can also refer to Singh and Mangat (1996). We would like to discuss Singh (1971) method of stratification by using auxiliary information. We have seen that an unbiased estimator of population mean $\bar{Y}$ in stratified random sampling is given by

$$\bar{y}_{st} = \sum_{h=1}^{L} W_h \bar{y}_h \tag{8.5.3.2}$$

with approximate variance of the estimator given by

$$V(\bar{y}_{st}) = \sum_{h=1}^{L} \frac{W_h^2}{n_h} \sigma_{hy}^2 . \tag{8.5.3.3}$$

Obviously for achieving the maximum precision, a stratification design with minimum $V(\bar{y}_{st})$ is desired. From (8.5.3.3) it is clear that the problem of optimum stratification involves the simultaneous determination of (a) Optimum strata boundaries (b) Optimum number of strata, and (c) Sample allocation n_h. Singh (1971) considered the problem of optimum allocation of the sample to different strata when the total expected cost of the survey is fixed. Assume that the cost of observing y on a unit is a function of the value of the variable x for that unit. A function $\theta(x)$ is said to belong to Ω_r if the first r derivatives of $\theta(x)$ exist for all x in the range (a, b) of x. If $C(x)$ $\{C(x) \in \Omega_3,\ C(x) > 0\}$ for all x in (a, b) is a cost function, then the expected cost for observing n_h units in the hth stratum is $n_h \mu_{hc}$, where μ_{hc} is the expected value of $C(x)$ in the hth stratum. The total fixed cost function is then defined as

$$C = C_0 + \sum_{h=1}^{L} n_h \mu_{hc} + \psi(L), \tag{8.5.3.4}$$

where C_0 is the overhead cost and $\psi(L)$ is the cost of constructing L strata with $\psi(1) = 0$. Let the regression of y on x be given by

$$y = \lambda\ (x) + e \tag{8.5.3.5}$$

where $E(e|x) = 0$ and $V(e|x) = \phi(x) > 0\ \ \forall\ x \in (a,b)$ and $(b-a) < \infty$. Assuming that $\lambda(x) \in \Omega_2$ and $\phi(x) \in \Omega_3$. Under the regression model we have

$$\sigma_{hy}^2 = \sigma_{h\lambda}^2 + \mu_{h\phi} . \tag{8.5.3.6}$$

Thus the variance of the estimator under stratified sampling becomes

$$V(\bar{y}_{st}) = \sum_{h=1}^{L} \frac{W_h^2}{n_h} \left(\sigma_{h\lambda}^2 + \mu_{h\phi}\right). \tag{8.5.3.7}$$

For a given number of strata L the Lagrange function is given by

$$L_g = \sum_{h=1}^{L} \frac{W_h^2}{n_h} \left(\sigma_{h\lambda}^2 + \mu_{h\phi}\right) + K\left[C - C_0 - \sum_{h=1}^{L} n_h \mu_{hc} - \psi(L)\right]. \tag{8.5.3.8}$$

The optimum values of n_h are given by

$$n_h = [C - C_0 - \psi(L)] W_h \sqrt{(\sigma_{h\lambda}^2 + \mu_{h\phi}/\mu_{h\phi})} \Big/ \sum_{h=1}^{L} W_h \sqrt{\mu_{hc}(\sigma_{h\lambda}^2 + \mu_{h\phi})}. \quad (8.5.3.9)$$

The minimum variance of the estimator $\bar{y}_{st}$ under stratified sampling is given by

$$\text{Min.}V(\bar{y}_{st}) = \left[\sum_{h=1}^{L} W_h \sqrt{\mu_{hc}(\sigma_{h\lambda}^2 + \mu_{h\phi})}\right]^2 \Big/ \{C - C_0 - \psi(L)\}. \quad (8.5.3.10)$$

If $f(x) \in \Omega_3$ denotes the density function of x then we have

$$W_h = \int_{x_{h-1}}^{x_h} f(x)dx, \quad \mu_{hc} = \frac{1}{W_h}\int_{x_{h-1}}^{x_h} C(x)f(x)dx \quad \text{and} \quad \sigma_{h\lambda}^2 = \frac{1}{W_h}\int_{x_{h-1}}^{x_h} \lambda^2 f(x)dx - (\mu_{h\lambda})^2,$$

where $\mu_{h\lambda}$ is the expected value of $\lambda(x)$ in the hth stratum and (x_{h-1}, x_h) are the stratum boundaries. On differentiating $\sum_{h=1}^{L} W_h \sqrt{\mu_{hc}(\sigma_{h\lambda}^2 + \mu_{h\phi})}$ with respect to x_h and equating to zero we obtain the set of minimal equations given by

$$\frac{\mu_{hc}\left[\phi(x_h) + \{\lambda(x_h) - \mu_{h\lambda}\}^2\right] + C(x_h)(\sigma_{h\lambda}^2 + \mu_{h\phi})}{\sqrt{\mu_{hc}(\sigma_{h\lambda}^2 + \mu_{h\phi})}}$$

$$= \frac{\mu_{ic}\left[\phi(x_h) + \{\lambda(x_h) - \mu_{i\lambda}\}^2\right] + C(x_h)(\sigma_{i\lambda}^2 + \mu_{i\phi})}{\sqrt{\mu_{ic}(\sigma_{i\lambda}^2 + \mu_{i\phi})}}, \quad (8.5.3.11)$$

$i = h+1$, $h = 1, 2, ..., L-1$. Again the exact solution to these equations is not easy. An approximate solution to these equations is

$$K_h^2 \int_{x_{h-1}}^{x_h} g(t)f(t)dt = \text{Constant}, \quad (8.5.3.12)$$

where $K_h = x_h - x_{h-1}$ and $g(t) = \{\phi^2 C'^2 + C^2\phi'^2 + 4C^2\phi\lambda'^2 - 2\phi C\phi' C'\} / (\phi C)^{\frac{3}{2}}$. Following Singh and Sukhatme (1969) an approximate solution to the minimal equations is

$$\int_{x_{h-1}}^{x_h} \sqrt[3]{g(t)f(t)}dt = \text{Constant}. \quad (8.5.3.13)$$

Singh $\text{Cum.}\sqrt[3]{p(x)}$ **rule**: If the function $p(x) = g(x)f(x)$ is bounded and possesses first two derivatives for all x in (a, b), then for a given value of L taking equal intervals of the $\text{Cum.}\sqrt[3]{p(x)}$ yields approximately optimum strata boundaries.

Example 8.5.3. We wish to estimate the average death rate of the persons living in the United States on the basis of year 2000 projections of the death rate. The projected 191443 persons are to be grouped into five strata on the basis of their age at the time of death. The rough distribution of the death rate projected for the year 2000 in the United States has been listed in population 6 in 21 age groups with a gap of four years. We wish to apply the method of proportional allocation for selecting the overall sample required for estimation purpose. Apply the cumulative cube root method to form 5 strata.

Solution. From the information given in the population 6 we have

Sr. No.	Age Group	Strata Boundaries	Year 2000 f_h	$\sqrt[3]{f_h}$	Cum. $\sqrt[3]{f_h}$
1	0--4	0.5--4.5	692	8.845085	8.845085
2	5--9	4.5--9.5	14	2.410142	11.255230
3	10--14	9.5--14.5	15	2.466212	13.721440
4	15--19	14.5--19.5	57	3.848501	17.569940
5	20--24	19.5--24.5	71	4.140818	21.710760
6	25--29	24.5--29.5	68	4.081655	25.792410
7	30--34	29.5--34.5	78	4.272659	30.065070
8	35--39	34.5--39.5	111	4.805896	34.870970
9	40--44	39.5--44.5	182	5.667051	40.538020
10	45--49	44.5--49.5	315	6.804092	47.342110
11	50--54	49.5--54.5	533	8.107913	55.450020
12	55--59	54.5--59.5	853	9.483814	64.933840
13	60--64	59.5--64.5	1357	11.071160	76.005000
14	65--69	64.5--69.5	2010	12.620170	88.625180
15	70--74	69.5--74.5	3022	14.457660	103.082800
16	75--79	74.5--79.5	4423	16.414930	119.497800
17	80--84	79.5--84.5	6921	19.057080	138.554800
18	85--89	84.5--89.5	10609	21.973110	160.528000
19	90--94	89.5--94.5	16915	25.669890	186.197800
20	95--99	94.5--99.5	27188	30.069470	216.267300
21	100--104	99.5--104.5	44053	35.317650	251.585000
22	105--109	104.5--109.5	71956	41.593200	293.178200

For making five strata we need to know four boundary points, say B_1, B_2, B_3, and B_4 by using linear interpolation between the class intervals and the Cum. $\sqrt[3]{f_h}$ values. On dividing 293.1782 by 5 and taking the cumulative totals we have the rough five boundary points corresponding to values 58.6356, 117.2713, 175.9069 and 234.5425 in the 6th column of the above table. Thus four boundary points obtained through interpolation are

$$B_1 = 54.5 + \frac{4(58.6356 - 55.4500)}{9.4838} = 55.84,$$

$$B_2 = 74.5 + \frac{4(117.2713 - 103.0828)}{16.41493} = 77.95,$$

$$B_3 = 89.5 + \frac{4(175.9069 - 160.528)}{25.6699} = 91.89,$$

and

$$B_4 = 99.5 + \frac{4(234.5425 - 216.2673)}{35.31765} = 101.57.$$

It indicates that the persons with age in the ranges

[0, 55.84], [55.84, 77.95], [77.95, 91.89], [91.89, 101.57] and [101.57, 109]

will form the first, second, third, fourth and fifth stratum, respectively, according to cumulative cube root method.

8.6 SUPERPOPULATION MODEL APPROACH

Dayal (1985) considered the superpopulation model, defined as

$$y_j = \alpha + \beta x_j + e_j, \quad j = 1, 2, ..., N \tag{8.6.1}$$

where

$$E(e_j \mid x_j) = 0, \quad V(e_j \mid x_j) = \gamma^2 x_j^g, \quad \text{Cov}(e_j, e_{j'} \mid x_j x_{j'}) = 0, \quad j \neq j'.$$

Rao (1968) has shown that under model (8.6.1), the optimum allocation is given by

$$n_h \propto \sqrt{(T_h^2 - D_h)} \tag{8.6.2}$$

where $T_h = \sum_{i=1}^{N_h} X_{hi}$ and $D_h = N_h \sum_{i=1}^{N_h} (X_{hi}^2 - X_{hi}^g)$. If $g = 2$ then $D_h = 0$ and the optimum allocation reduces to allocation proportional to stratum totals. Then we have the following theorems.

Theorem 8.6.1. Under model (8.6.1) the Neyman optimum allocation reduces to the allocation given by

$$n_h \propto W_h \sqrt{\rho_{xy}^2 S_{hx}^2 + (1 - \rho_{xy}^2) S_x^2 \bar{X}_h^g / \bar{X}^g} \; . \tag{8.6.3}$$

Proof. We have

$$\bar{y}_{\text{st}} = \sum_{h=1}^{L} W_h \bar{y}_h = \sum_{h=1}^{L} W_h \left\{ \frac{1}{n_h} \sum_{i=1}^{n_h} y_{hi} \right\} = \sum_{h=1}^{L} W_h \left\{ \frac{1}{n_h} \sum_{i=1}^{n_h} (\alpha + \beta x_{hi} + e_{hi}) \right\}$$

$$= \sum_{h=1}^{L} W_h (\alpha + \beta \bar{x}_h + \bar{e}_h). \tag{8.6.4}$$

Thus we have

$$V(\bar{y}_{\text{st}}) = E_1 V_2(\bar{y}_{\text{st}} \mid X) + V_1 E_2(\bar{y}_{\text{st}} \mid X) = \frac{\gamma^2}{n_h} \sum_{h=1}^{L} W_h^2 \bar{X}_h^g + \beta^2 V(\bar{x}_{\text{st}}) \tag{8.6.5}$$

where $\bar{X}_h^g = N_h^{-1} \sum_{i=1}^{N_h} X_{hi}^g$ and $V(\bar{x}_{\text{st}}) = \sum_{h=1}^{L} W_h^2 V(\bar{x}_h)$.

Also

$$V(y_j) = E_1 V_2(y_j \mid X) + V_1 E_2(y_j \mid X),$$

which implies

$$S_y^2 = \beta^2 S_x^2 + \gamma^2 \bar{X}^g,$$

where

$$\overline{X}^g = N^{-1}\sum_{h=1}^{L}\sum_{i=1}^{N_h} X_{hi}^g \quad \text{and} \quad \beta^2 = \rho_{xy}^2\left(S_y^2/S_x^2\right).$$

Using these results in (8.6.5) we have

$$V(\bar{y}_{\text{st}}) = \rho_{xy}^2\left(\frac{S_y^2}{S_x^2}\right)V(\bar{x}_{\text{st}}) + \left(1-\rho_{xy}^2\right)\left(\frac{S_y^2}{\overline{X}^g}\right)\sum_{h=1}^{L}\frac{W_h^2\overline{X}_h^g}{n_h}$$

$$= \rho_{xy}^2\left(\frac{S_y^2}{S_x^2}\right)\left\{\sum_{h=1}^{L}\left(\frac{1}{n_h}-\frac{1}{N}\right)W_h^2 S_{hx}^2\right\} + \left(1-\rho_{xy}^2\right)\left(\frac{S_y^2}{\overline{X}^g}\right)\sum_{h=1}^{L}\frac{W_h^2\overline{X}_h^g}{n_h}. \tag{8.6.6}$$

The Lagrange function subject to the condition $n = \sum_{h=1}^{L} n_h$, is then given by

$$L_g = V(\bar{y}_{\text{st}}) + \lambda\left(n - \sum_{h=1}^{L} n_h\right). \tag{8.6.7}$$

On differentiating (8.6.7) with respect to n_h and equating to zero we have (8.6.3). Hence the theorem.

Cassel, Särndal, and Wretman (1977) discussed the correspondence between Horvitz and Thompson (1952) strategy and the stratified random sampling strategy for a special case of the general superpopulation model. Rao (1983c) considered a more general model and discusses the correspondence between these strategies in a more formal way by following Hanurav (1965), Ramachandran and Rao (1974), and Rao (1968, 1977b).

We have seen that in systematic sampling two or more auxiliary variables can be utilised to select a sample in a systematic way. The same question arises here if two or more auxiliary variables are available, whether all the variables can be used at stratification stage or not? In order to answer this question we resort to the next section.

8.7 MULTI-WAY STRATIFICATION

Use of one-way stratification for energy data in agriculture has been discussed by Singh, Singh, Mittal, Pannu, and Bhangu (1994) and Singh, Singh, Pannu, Bhangoo, and Singh (1994). Moses (1978) exhibited the information of multi-way stratification technique in validation of energy data. Chernick and Wright (1983) found that in data validation respondent surveys many variables are candidates for stratification, and since the relationship between these variables and the response is not well understood, two-way or multi-way stratification with proportional allocation along with each variable seems appropriate. Frankel and Stock (1942) discussed the use of multiple stratification techniques in gathering data related to unemployment. They considered the possibility of using sample designs in which the Latin square principle can be used to reduce the number of sample units necessary to represent all strata. For example, suppose two criterion for stratification are used, say A and B, such that p strata can be constructed from the A characteristic, and, within each of these, p from the B characteristic. If one relates the resulting pattern to a single treatment of a $p \times p$ Latin square, it is obvious that in a sample of p sample units each of pA strata will be represented and likewise each of the pB strata. Tepping, Hurwitz and Deming (1943) discussed such designs in some detail and compared their variances with the variances of single stratification sampling. Such a stratification is also called Deep Stratification. Following Bryant, Hartley, and Jessen (1960) the population of size N is stratified along two variables using R rows and C columns. The $R \times C$ sections formed are called cells or strata, the $(i,\ j)$th cell being a typical cell for $i = 1, 2, 3, ..., R$ and $j = 1, 2, 3, ..., C$. Let N_{ij} be number of population units for the $(i, j)th$ cell, Y_{ijk} be the value of kth unit of the study variable for the $(i,\ j)$th cell for $k = 1, 2, ..., N_{ij}$, $P_{ij} = N_{ij}/N$ be the population proportion of units for the (i, j)th cell, $\bar{Y}_{ij} = N_{ij}^{-1}\sum_{k=1}^{N_{ij}} Y_{ijk}$ be the population mean for the (i, j)th cell, $S_{ij}^2 = (N_{ij} - 1)^{-1}\sum_{k=1}^{N_{ij}}(Y_{ijk} - \bar{Y}_{ij})^2$ be the population mean square error for the (i, j)th cell, n_{ij} be the number of units in the sample belonging to the $(i, j)th$ cell, $\bar{y}_{ij} = n_{ij}^{-1}\sum_{k=1}^{n_{ij}} y_{ijk}$ be the sample mean in the (i, j)th cell, $P_{i\bullet} = \sum_{j=1}^{C} P_{ij}$, $P_{\bullet j} = \sum_{i=1}^{R} P_{ij}$, $\bar{Y} = \sum_{i=1}^{R}\sum_{j=1}^{C} P_{ij}\bar{Y}_{ij}$, $n_{i\bullet} = \sum_{j=1}^{C} n_{ij}$ and $n_{\bullet j} = \sum_{i=1}^{R} n_{ij}$ are the marginal totals.

The marginal mean values are given by

$$\bar{Y}_{i\bullet} = \sum_{j=1}^{C} P_{ij}Y_{ij} \Big/ \sum_{j=1}^{C} P_{ij} \quad \text{and} \quad \bar{Y}_{\bullet j} = \sum_{i=1}^{R} P_{ij}Y_{ij} \Big/ \sum_{i=1}^{R} P_{ij}.$$

Table 8.7.1. Two-way stratification based on R rows and C columns.

	Column No.						Marginal
Row No.	1	2	...	j	...	C	Mean
1	P_{11} $\bar{Y}_{11}$	P_{12} $\bar{Y}_{12}$	...	P_{1j} $\bar{Y}_{1j}$	...	P_{1C} $\bar{Y}_{1C}$	$P_{1\bullet}$ $\bar{Y}_{1\bullet}$
2	P_{21} $\bar{Y}_{21}$	P_{22} $\bar{Y}_{22}$	...	P_{2j} $\bar{Y}_{2j}$	...	P_{2C} $\bar{Y}_{2C}$	$P_{1\bullet}$ $\bar{Y}_{1\bullet}$
...	...	...	...	...	...	...	...
i	P_{i1} $\bar{Y}_{i1}$	P_{i2} $\bar{Y}_{i2}$	...	P_{ij} $\bar{Y}_{ij}$	...	P_{iC} $\bar{Y}_{iC}$	$P_{i\bullet}$ $\bar{Y}_{i\bullet}$
...	...	...	...	...	...	...	...
R	P_{R1} $\bar{Y}_{R1}$	P_{R2} $\bar{Y}_{R2}$	...	P_{Rj} $\bar{Y}_{Rj}$	....	P_{RC} $\bar{Y}_{RC}$	$P_{R\bullet}$ $\bar{Y}_{R\bullet}$
Marginal Means	$P_{\bullet 1}$ $\bar{Y}_{\bullet 1}$	$P_{\bullet 2}$ $\bar{Y}_{\bullet 2}$	...	$P_{\bullet j}$ $\bar{Y}_{\bullet j}$	....	$P_{\bullet C}$ $\bar{Y}_{\bullet C}$	1 $\bar{Y}_{\bullet\bullet}$

For simplicity and to achieve the greatest benefit of two-way stratification we assume independence between the variables of stratification, that is, $P_{ij} = P_{i\bullet}P_{\bullet j}$. It is also assumed that $\max(R, C) \le n \le RC$. For proportional allocation, the total sample of n units satisfy the conditions $n_{i\bullet} = nP_{i\bullet}$ and $n_{\bullet j} = nP_{\bullet j}$, where $n_{i\bullet}$ and $n_{\bullet j}$ are rounded to the nearest integer such that sum of $n_{i\bullet}$ as well as $n_{\bullet j}$ is equivalent to n. The values of $n_{i\bullet}$ and $n_{\bullet j}$ are fixed, while n_{ij} is a random variable.

Let us take a hypothetical situation of six villages having four different types of farmers. This table consists of $R = 6$ rows corresponding to their villages and $C = 4$ columns corresponding to different type of farmers as shown in Table 8.7.2. Assume that we require a sample of size $n = 15$ which satisfies the condition $\max(R, C) \le n \le RC$.

Table 8.7.2. Yield in kg/ha of wheat crop in different six regions of the Punjab state in India by different types of farmers and their proportions.

Village	Marginal farmers	Small farmers	Medium farmers	Large farmers	Marginal means	Expected sub-sample sizes $n_{i\bullet}$
1	0.01	0.04	0.05	0.02	0.12	$1.80 \approx 2$
	2831	3095	3762	3741	3458.58	
2	0.03	0.02	0.09	0.05	0.19	$2.85 \approx 3$
	2568	3265	3520	3639	3374.16	
3	0.02	0.03	0.07	0.04	0.16	$2.40 \approx 2$
	2672	3891	3410	3600	3455.44	
4	0.02	0.05	0.08	0.05	0.20	3
	2512	3762	3268	3598	3398.40	
5	0.02	0.03	0.07	0.04	0.16	$2.40 \approx 2$
	2381	3520	3297	3724	3331.06	
6	0.02	0.01	0.06	0.08	0.17	$2.55 \approx 3$
	2381	3210	3542	3842	3527.06	
Marginal	0.12	0.18	0.42	0.28	1.00	
Means	2535.583	3509.056	3448.452	3703.536	3421.24	
$n_{\bullet j}$	$1.8 \approx 2$	$2.7 \approx 3$	$6.3 \approx 6$	$4.2 \approx 4$		$n = 15$

Source: Singh, Pannu, Singh, Singh, and Kaur (1996).

After n, $n_{i\bullet}$, and $n_{\bullet j}$ have been determined, construct a square matrix having n sub-rows $(s = 1, 2, ..., n)$ and n sub-columns $(t = 1, 2, ..., n)$ forming n^2 sub-cells by following Bryant, Hartley, and Jessen (1960). Combining $n_{i\bullet}$ adjacent sub-rows for $i = 1,2,...,R$, form the R rows, and by combining $n_{\bullet j}$ adjacent sub-columns for $j = 1,2,...,C$, form the C columns. Following these notations, the intersection of the

i^{th} row and j^{th} column will contain a single cell consisting of $n_{i\bullet}n_{\bullet j}$ squares or sub-cells. This method of allocation is called random allocation.

Table 8.7.3. A 15×15 grid for allocation of total sample size in a two-way layout.

Row No.	Line No.	Column No.															$n_{i\bullet}$
		$j=1$		$j=2$			$j=3$						$j=4$				
		1	2	3	4	5	6	7	8	9	10	11	12	13	14	15	
$i=1$	1													X			
	2						X										$n_{i\bullet}=2$
$i=2$	3		X														
	4										X						
	5														X		$n_{i\bullet}=3$
$i=3$	6	X															
	7				X												$n_{i\bullet}=2$
$i=4$	8											X					
	9															X	
	10			X													$n_{i\bullet}=3$
$i=5$	11								X								
	12					X											$n_{i\bullet}=2$
$i=6$	13												X				
	14							X									
	15									X							$n_{i\bullet}=3$
$n_{\bullet j}$		$n_{\bullet j}=2$		$n_{\bullet j}=3$			$n_{\bullet j}=6$						$n_{\bullet j}=4$				$n=15$

Steps for two-way random allocation: The following steps are needed for allocating a sample of n units into two-way stratification:

Step 1. Construct $n \times n$ cells as above;
Step 2. In the first sub-row, one sub-column is selected at random and marked;
Step 3. In the second sub-row of the remaining $n-1$ sub-columns, another sub-column is selected at random and marked, and so on;

Step 4. At the end of the marking process, each sub-row and sub-column will contain one mark. It will complete the allocation of the n observations to the RC cells;
Step 5. The number of marks within the boundaries of the (i, j)th cell represents the number of observations to be randomly selected from the cell and n_{ij} denotes the number of observations in it.

Properties of random allocation method: The random allocation method has the following properties:

(a) Every cell has an equal chance of $1/n$ of receiving a mark;
(b) The values of n_{ij} constructed should satisfy $n_{i\bullet} = nP_{i\bullet}$ and $n_{\bullet j} = nP_{\bullet j}$;
(c) $E(n_{ij}) = nP_{i\bullet}P_{\bullet j} = nP_{ij}$.

In such situations we have the following theorem:

Theorem 8.7.1. An unbiased estimator of population mean $\overline{Y}$ under two-way stratification is given by

$$\bar{y}_{\text{twst}} = \frac{1}{n}\sum_{i=1}^{R}\sum_{j=1}^{C} n_{ij}\bar{y}_{ij\bullet} \tag{8.7.1}$$

with variance given by

$$V(\bar{y}_{\text{twst}}) = \frac{1}{n^2}\left[\sum_{i=1}^{R}\sum_{j=1}^{C}\frac{n_{i\bullet}n_{\bullet j}\{n(n-1)N_{ij} - (n-n_{i\bullet})(n-n_{\bullet j}) - (n-1)n_{i\bullet}n_{\bullet j}\}}{n^2(n-1)N_{ij}}S_{ij}^2\right.$$
$$+\sum_{i=1}^{R}\sum_{j=1}^{C}\frac{n_{i\bullet}n_{\bullet j}(n-n_{i\bullet})(n-n_{\bullet j})}{n^2(n-1)}\overline{Y}_{ij}^2 + +\sum_{i=1}^{R}\sum_{j=1}^{C}\sum_{j'=1}^{C}\frac{n_{i\bullet}n_{\bullet j}n_{\bullet j'}(n_{i\bullet}-n)}{n^2(n-1)}\overline{Y}_{ij}\overline{Y}_{ij'}$$
$$\left.+\sum_{i=1}^{R}\sum_{i'=1}^{R}\sum_{j=1}^{C}\frac{n_{i\bullet}n_{i'\bullet}n_{\bullet j}(n_{\bullet j}-n)}{n^2(n-1)}\overline{Y}_{ij}\overline{Y}_{i'j} + \sum_{i=1}^{R}\sum_{i'=1}^{R}\sum_{j=1}^{C}\sum_{j'=1}^{C}\frac{n_{i\bullet}n_{\bullet j}n_{i'\bullet}n_{\bullet j'}}{n^2(n-1)}\overline{Y}_{ij}\overline{Y}_{i'j'}\right]. \tag{8.7.2}$$

Raghunandanan and Bryant (1971) have considered the problem of multi-way stratification for random sample allocation mechanism. Chernick and Wright (1983) have proposed a technique of systematically allocating a sample to the strata formed by double stratification. The method can proportionally allocate the sample along each variable of stratification. If there are R and C strata for the first and second variable of stratification, respectively, the technique requires that the total sample size be at least as large as $\max(R, C)$. They suggested an unbiased estimator of population mean and obtained the variance expression. The technique has been found to be comparable with the random allocation technique proposed by Bryant, Hartley, and Jessen (1960).

8.8 STRATUM BOUNDARIES FOR MULTI-VARIATE POPULATIONS

Sadasivan and Aggarwal (1978) considered the problem of optimum points of stratification with two study variates X and Y (say). The exact equations given by Dalenius (1950) for the univariate case have been extended to the bivariate case. Also for minimizing the generalized variance (variance--covariance matrix) under Neyman allocation, a set of equations giving optimum points of stratification are discussed for two cases:

(a) When the correlation coefficient between the two variables is constant from stratum to stratum;

(b) When the correlation coefficient is varying from stratum to stratum.

While extending the univariate procedure of Dalenius (1950) and Dalenius and Gurney (1951, 1957) to the two-dimensional situation the following assumptions are made:

(a) the variates X and Y have continuous joint p.d.f. $f(x, y)$ in the finite range $x_0 \le x \le x_{L_1}$; $y_0 \le y \le y_{L_2}$;
(b) the population is finite;
(c) divide the population into $L_1 \times L_2$ strata by determining the strata boundaries $x_1, x_2, ..., x_{(L_1-1)}$ for x and $y_1, y_2, ..., y_{(L_2-1)}$ for y such that the generalized variance of the means of the variables for this stratified sample under Neyman's allocation in minimum;
(d) the generalized variance can be set out as

$$\text{GV} = \begin{vmatrix} \sigma_x^2, & \sigma_{xy} \\ \sigma_{xy}, & \sigma_y^2 \end{vmatrix} = \sigma_x^2\sigma_y^2 - \sigma_{xy}^2, \tag{8.8.1}$$

where σ_x^2, σ_y^2 and σ_{xy} denote the variances of x and y and covariance, respectively, under Neyman allocation with fixed cost per unit.

Before proceeding further let us first define some mathematical notation as follows.

Joint p.d.f. of two study variables X and $Y = f(x, y)$.

Area under the (h, k)th cell

$$W_{hk} = \int_{x_{k-1}}^{x_k} \int_{y_{h-1}}^{y_h} f(x, y) dx dy .$$

Marginal density of x_h corresponding to h^{th} cell of y

$$f_{hk}(x_k) = \int_{y_{h-1}}^{y_h} f(x_k, y) dy .$$

Mean value of x in the (h,k)th cell

$$\mu_{hkx} = \frac{1}{W_{hk}} \int_{x_{k-1}}^{x_k} \int_{y_{h-1}}^{y_h} x f(x, y) dx dy .$$

Variance of x in the (h, k)th cell

$$\sigma_{hkx}^2 = \frac{1}{W_{hk}} \int_{x_{k-1}}^{x_k} \int_{y_{h-1}}^{y_h} x^2 f(x, y) dx dy - \mu_{hkx}^2 .$$

Conditional mean of x in (h, k)th cell at a point

$$y_h = \mu_{hkx} \mid y_h = \frac{1}{f_{hk}(y_h)} \int_{x_{k-1}}^{x_k} x f(x, y_h) dx .$$

Covariance between x and y in the (h, k)th cell

$$\sigma_{hkxy} = \frac{1}{W_{hk}} \int_{x_{k-1}}^{x_k} \int_{y_{h-1}}^{y_h} xy f(x, y) dx dy - \mu_{hkx}\mu_{hky} .$$

Second moment of x in (h,k)th cell at a point y_h

$$\mu_{2hkx}^{/} \mid y_h = \frac{1}{f_{hk}(y_h)} \int_{x_{k-1}}^{x_k} x^2 f(x, y_h) dx .$$

Approximate variance of x under the Neyman allocation with cost per unit constant

$$\sigma_x^2 = \frac{1}{n}\left(\sum_{h=1}^{L_2} \sum_{k=1}^{L_1} W_{hk}\sigma_{hkx}\right)^2 .$$

Approximate variance of y under the Neyman allocation with cost per unit constant

$$\sigma_y^2 = \frac{1}{n}\left(\sum_{h=1}^{L_2} \sum_{k=1}^{L_1} W_{hk}\sigma_{hky}\right)^2 .$$

Approximate covariance between x and y under the Neyman allocation with cost per unit constant

$$\sigma_{xy} = \frac{1}{n}\left(\sum_{h=1}^{L_2} \sum_{k=1}^{L_1} W_{hk}\sqrt{\sigma_{hkxy}}\right)^2 .$$

Thus as in the case of univariate strata boundaries, we have the following differentiation results for bi-variate distributions to find the set of minimal equations:

$$\frac{\partial W_{hk}}{\partial x_k} = \int_{y_{h-1}}^{y_h} f(x_k, y) dy = f_{hk}(x_k);$$

$$\frac{\partial \sigma_{hkx}}{\partial x_k} = \frac{f_{hk}(x_k)}{2\sigma_{hxk}W_{hk}}\left[(x_k - \mu_{hkx})^2 - \sigma_{hkx}^2\right];$$

$$\frac{\partial W_{hi}}{\partial x_k} = -f_{hk}(x_k), \; i \neq k ;$$

$$\frac{\partial \sigma_{hix}}{\partial x_k} = -\frac{f_{hk}(x_k)}{2\sigma_{hxk}W_{hk}}\left[(x_k - \mu_{hix})^2 - \sigma_{hix}^2\right].$$

Thus

$$\frac{\partial \sigma_x}{\partial x_k} = \Psi(x,k,k) - \Psi(x,k,l),$$

where

$$\Psi(x,k,l) = \frac{1}{2\sqrt{n}} \sum_h f_{hk}(x_k) \left[\frac{(x_k - \mu_{hlx})^2 + \sigma_{hlx}^2}{\sigma_{hlx}} \right] \quad \text{for } l = i \text{ or } k .$$

Similarly

$$\frac{\partial \sigma_y}{\partial x_k} = \Psi_1(x,k,k) - \Psi_1(x,k,i),$$

where

$$\Psi_1(x,k,l) = \frac{1}{2\sqrt{n}} \sum_h \frac{f_{hk}(x_k)}{\sigma_{hly}} \left[\mu'_{2hky} \mid x_k + \sigma_{hly}^2 + \mu_{hly}^2 - 2\mu_{hly}\mu_{hky} \mid x_k \right],$$

and

$$\frac{\partial \sigma_{xy}}{\partial x_k} = \Psi_2(x,k,k) - \Psi_2(x,k,i)$$

where

$$\Psi_2(x,k,l) = \sqrt{\frac{\sigma_{xy}}{n}} \sum_h \frac{f_{hk}(x_k)}{\sqrt{\sigma_{hlxy}}} \left[x_k \left(\mu_{hky} \mid x_k - \mu_{hly} \right) + \mu_{hly}\mu_{hlx} + \sigma_{hlxy} - \mu_{hlx}\mu_{hky} \mid x_k \right].$$

Now we have the following theorem.

Theorem 8.8.1. The set of minimal equations for minimizing the generalized variance in (8.8.1) is given by

$$\frac{\partial \, \mathrm{GV}}{\partial x_k} = \sigma_x^2 \sigma_y \left(\frac{\partial \sigma_y}{\partial x_k} \right) + \sigma_y^2 \sigma_x \left(\frac{\partial \sigma_x}{\partial x_k} \right) - \sigma_{xy} \left(\frac{\partial \sigma_{xy}}{\partial x_k} \right) = 0 \tag{8.8.2}$$

$$\frac{\partial \mathrm{GV}}{\partial y_k} = \sigma_x^2 \sigma_y \left(\frac{\partial \sigma_y}{\partial y_h} \right) + \sigma_y^2 \sigma_x \left(\frac{\partial \sigma_x}{\partial y_k} \right) - \sigma_{xy} \left(\frac{\partial \sigma_{xy}}{\partial y_h} \right) = 0 . \tag{8.8.3}$$

Using the above results (8.8.2) and (8.8.3) we have the set of minimal equations:

$$\phi(x,k,k) = \phi(x,k,i), \quad \text{for } i = k+1;\ k = 1,2,...,L_1 - 1 \ \text{ and } \ h = 1,2,...,L_2 - 1 \tag{8.8.4}$$

where

$$\phi(x,k,l) = \sigma_x^2 \sigma_y \sum_h \frac{f_{hk}(x_k)}{\sigma_{hly}} \left[\mu'_{2hky} \mid x_k + \sigma_{hly}^2 + \mu_{hly}^2 - 2\mu_{hky}\mu_{hly} \mid x_k \right]$$
$$+ \sigma_y^2 \sigma_x \sum_h \frac{f_{hk}(x_k)}{\sigma_{hlx}} \left[(x_k - \mu_{hlx})^2 + \sigma_{hlx}^2 \right] - 2(\sigma_{xy})^{\frac{3}{2}} \sum_h \frac{f_{hk}(x_k)}{\sqrt{\sigma_{hlxy}}} \left[x_k \left(\mu_{hky} \mid x_k - \mu_{hly} \right) \right.$$
$$\left. + \mu_{hlx}\mu_{hly} + \sigma_{hlxy} - \mu_{hlx}\mu_{hky} \mid x_k \right],$$

and (8.8.4) implies that

$$\phi(y,h,h)=\phi(y,h,j), \quad \text{for } j=h+1, \ h=1,2,...,L_2-1 \text{ and } k=1,2,...,L_1-1, \qquad (8.8.5)$$

where

$$\phi(y,h,l)=\sigma_x^2\sigma_y\sum_k\frac{f_{hk}(y_h)}{\sigma_{lky}}\left[(y_h-\mu_{lky})^2+\sigma_{lky}^2\right]+\sigma_y^2\sigma_x\sum_k\frac{f_{hk}(y_h)}{\sigma_{lkx}}\left[\mu'_{2hkx}\mid y_h+\sigma_{lkx}^2-2\mu_{lkx}\mu_{hkx}\mid y_h\right]$$

$$-2(\sigma_{xy})^{1.5}\sum_k\frac{f_{hk}(y_h)}{\sqrt{\sigma_{lkxy}}}\left[y_h(\mu_{hkx}\mid y_h-\mu_{lkx})+\mu_{lkx}\mu_{lky}+\sigma_{lkxy}-\mu_{lky}\mu_{lkx}\mid y_h\right].$$

Theorem 8.8.2. If the correlation coefficient is assumed to be constant from stratum to stratum, then minimization of generalized variance is equivalent to the minimization of covariance and the set of minimal equations reduces to

$$g(x,k,k)=g(x,k,i), \qquad i=k+1;\ k=1,2,...,L_1-1, \qquad (8.8.6)$$

$$g(y,h,h)=g(y,h,j), \quad j=h+1, \ h=1,2,...,L_2-1, \qquad (8.8.7)$$

where

$$g(x,k,l)=\sum_h\frac{f_{hk}(x_k)}{\sqrt{\sigma_{hlxy}}}\left[x_k\left(\mu'_{2hky}\mid x_k-\mu_{hly}\right)+\mu_{hlx}\mu_{hly}+\sigma_{hlxy}-\mu_{hlx}\mu_{hky}\mid x_k\right]$$

and

$$g(y,h,l)=\sum_k\frac{f_{hk}(y_h)}{\sqrt{\sigma_{lkxy}}}\left[y_h(\mu_{hkx}\mid y_h-\mu_{lkx})+\mu_{lky}\mu_{lkx}+\sigma_{lkxy}-\mu_{lky}\mu_{lkx}\mid y_h\right].$$

The approximate solution to the set of minimal equations given above can also be obtained by following Ekman (1959) method, which reduces to the set of simplified minimal equations given by

$$\sum_h f_{hk}(x_k)\sqrt{(x_k-x_{k-1})(y_h-y_{h-1})} = \text{Constant}$$

and

$$\sum_k f_{hk}(x_k)\sqrt{(x_k-x_{k-1})(y_h-y_{h-1})} = \text{Constant}.$$

Rizvi, Gupta, and Singh (2000) also studied stratification based on two auxiliary variables.

Example 8.8.1. The death rates in the United Sates over the period 1990 to 2065 in nine groups and in twenty two age groups are shown in a 22×9 contingency table as given in population 6 of the Appendix. Construct three homogeneous columns from nine columns and five homogeneous rows from 22 rows.

Solution. From population 6 the frequency distribution is given as shown in the following table.

Age Group	Year 1990	Year 1995	Year 2000	Year 2010	Year 2020	Year 2030	Year 2040	Year 2050	Year 2065	Row Totals f_r	cum $\sqrt{f_r}$
0--4	967	818	692	496	355	255	183	131	80	3977	63.06
5--9	19	16	14	10	7	5	4	3	2	80	72.00
10--14	20	17	15	11	8	6	4	3	2	86	81.28
15--19	67	62	57	48	40	34	28	24	18	378	100.72
20--24	86	78	71	58	48	40	33	27	20	461	122.19
25--29	84	75	68	54	44	35	28	23	16	427	142.85
30--34	97	87	78	62	50	40	32	25	18	489	164.97
35--39	138	124	111	90	72	58	47	38	27	705	191.52
40--44	221	201	182	150	124	102	84	69	52	1185	225.94
45--49	370	341	315	267	227	193	164	139	109	2125	272.04
50--54	613	572	533	464	403	351	305	265	215	3721	333.04
55--59	965	907	853	754	666	589	520	460	382	6096	411.12
60--64	1511	1432	1357	1218	1094	982	882	792	674	9942	510.83
65--69	2233	2119	2010	1810	1629	1466	1320	1188	1015	14790	632.44
70--74	3361	3187	3022	2718	2444	2198	1976	1777	1515	22198	781.43
75--79	4979	4693	4423	3930	3491	3102	2756	2448	2050	31872	959.96
80--84	7748	7323	6921	6182	5523	4933	4407	3936	3323	50296	1184.23
85--89	12267	11687	10609	10108	9177	8331	7564	6868	5942	82553	1471.551
90--94	19099	18341	16915	16246	14987	13827	12758	11774	10439	134386	1838.138
95--99	29744	28864	27188	26390	24869	23442	22102	20844	19095	222538	2309.877
100-104	46334	45554	44053	43329	41933	40600	39325	38104	36364	375596	2922.736
105-109	72195	72100	71956	71906	71845	71831	71861	71930	72097	647721	3727.547
Col. Tot f_c	203118	198598	191443	186301	179036	172420	166383	160868	153455		
cum. $\sqrt{f_c}$	450.686	896.329	1333.87	1765.49	2188.62	2603.85	3011.76	3412.84	3804.57		

From the above table $\sum_r \sqrt{f_r} = 3727.547$ and $\sum_c \sqrt{f_c} = 3804.57$.

We like to make 5×3 contingency table of homogeneous cells.

Divide $\sum_r \sqrt{f_r}$ by number of rows $(R = 5)$ required and $\sum_c \sqrt{f_c}$ by the number of columns $(C = 3)$ required.

Then the five points for grouping 22 rows into five homogeneous groups or rows on the basis of cum $\sqrt{f_r}$ are given by

745.50, 1491.0, 2236.5, 2982.0 and 3727.5.

The three points for grouping nine columns into three homogeneous groups or columns on the basis of cum. $\sqrt{f_c}$ are given by

1268.19, 2536.38 and 3804.57.

The resultant 5×3 contingency table of homogeneous strata is

Age Group	Years 1990 and 1995	Years 2000, 2010 and 2020	Years 2030, 2040, 2050 and 2065
0--69	14240	16615	13607
70--89	55245	68548	63126
90--99	96048	126595	134281
100--104	91888	129315	154393
105--109	144295	215707	287719

8.9 OPTIMUM ALLOCATION IN MULTI-VARIATE STRATIFIED SAMPLING

Ahsan and Khan (1982) considered the problem of allocation to minimize the total budgetary cost of survey subject to the desired precision assigned to the posterior variances of the population means when the sampling is multi-purpose, by assuming that the overhead cost is proportional to the number of individuals contacted in that stratum. Consider a population divided into L strata and let p variables be defined on each unit of population under study. Let $\overline{Y}_{hj}$ $(h = 1, 2,..., L;\ j = 1, 2, ..., p)$ be the unknown population means of the observations on the jth variable within the hth stratum. Let W_h $(h = 1, 2, ..., L)$ be the proportion of the population elements falling in the hth stratum. Let $\bar{y}_{hj}$ be the sample means for the jth variable in the hth stratum. A simple way of representation of the case under consideration is given below.

Stratum Number	Variable or Characteristic Number: 1	2	•	•	p
1	$\overline{Y}_{11}, \bar{y}_{11}, W_1$	$\overline{Y}_{12}, \bar{y}_{12}, W_1$			$\overline{Y}_{1p}, \bar{y}_{1p}, W_1$
2	$\overline{Y}_{21}, \bar{y}_{21}, W_2$	$\overline{Y}_{22}, \bar{y}_{22}, W_2$			$\overline{Y}_{2p}, \bar{y}_{2p}, W_2$
•					
•					
L	$\overline{Y}_{L1}, \bar{y}_{L1}, W_L$	$\overline{Y}_{L2}, \bar{y}_{L2}, W_L$			$\overline{Y}_{Lp}, \bar{y}_{Lp}, W_L$

Assume $\underline{W}$ and $\underline{\overline{Y}}_j$ be the row vectors of the values of W_h and $\overline{Y}_{hj}$, respectively, defined as $\underline{W} = (W_1, W_2, ..., W_L)$ and $\underline{\overline{Y}}_j = (\overline{Y}_{1j}, \overline{Y}_{2j}, ..., \overline{Y}_{Lj})$. The overall population mean $\overline{Y}_j$ for the jth variable is given by

$$\overline{Y}_j = \underline{W}\,\underline{\overline{Y}}_j^t \qquad (8.9.1)$$

where t stands for transpose of the vector. A stratified sample is determined by a set of numbers $(n_1, n_2, ..., n_L)$, where $n_h > 0$ $(h = 1,2,...,L)$ is the number of observations drawn independently from the hth stratum. Let σ_{hj}^2 be the known variance of the jth variable in the hth stratum and for a given population mean $\overline{Y}_j$, the sample mean $\overline{y}_j$ of the jth variable has a conditional L variate normal distribution defined by the mean vector $\overline{Y}_j$ and a diagonal variance--covariance matrix M_j $(j = 1, 2, ..., p)$ given by

$$M_j = \begin{bmatrix} m_{1(1,1)}, \dots\dots\dots\dots \\ \dots\dots, m_{2(2,2)}, \dots\dots\dots \\ \\ \dots\dots\dots\dots\dots m_{m_{p(p,p)}} \end{bmatrix} = \begin{bmatrix} \sigma_{11}^2/n_{11}, \dots\dots\dots\dots \\ \dots\dots, \sigma_{22}^2/n_{22}, \dots\dots\dots \\ \\ \dots\dots\dots\dots\dots \sigma_{pp}^2/n_{pp} \end{bmatrix}. \qquad (8.9.2)$$

A *priori* information about within stratum population means $\overline{Y}_{hj}$ is assumed to be available in terms of the L variate normal distribution of $\overline{Y}_j$ with mean vector $\underline{m}_j = (\overline{Y}_{1j}, \overline{Y}_{2j}, ..., \overline{Y}_{Lj})$ and a non-singular variance--covariance matrix

$$A_j = \begin{bmatrix} a_{11}, 0, \dots\dots, 0 \\ 0,. a_{22}, \dots\dots\dots, 0 \\ \\ 0, \; 0, \dots\dots, \; a_{pp} \end{bmatrix}. \qquad (8.9.3)$$

Raiffa and Schlaifer (1961) have shown that the posterior distribution of population means $\overline{Y}_j$ for a given stratified sample with $n_h > 0$ and observed sample means $\overline{y}_j$ is L variate normal with mean vector

$$\overline{\underline{m}}_j = \left[\underline{\overline{y}}_j W_j + \underline{m}_j \mathrm{V_j}\right]\left[\overline{V}_j\right] \qquad (8.9.4)$$

and variance--covariance matrix

$$\overline{V}_j = \left[W_j + V_j\right]^{-1} \qquad (8.9.5)$$

where $W_j = M_j^{-1}$ and $V_j = A_j^{-1}$.

The overall population mean of the jth variable $\overline{Y}_j$ is a linear combination of the $\overline{Y}_{Lj}$ and has a univariate normal distribution with mean, $b_j = W\underline{m}_j^t$ and variance $a_j = W\underline{A}_j W^t$, and a univariate normal posterior distribution with mean $\overline{b}_j = W\overline{\underline{m}}_j^t$ and variance $\overline{a}_j = W\overline{V}_j W^t$.

Let c_h be the overhead cost of approaching an individual of the hth stratum for measurements and c_{hj} be the cost associated with the measurement of the jth variable of an individual in the hth stratum, then the total cost of the survey is

$$C = \sum_{h=1}^{L} c_h n_h + \sum_{h=1}^{L}\sum_{j=1}^{p} c_{hj} n_{hj}\,, \tag{8.9.6}$$

where $n_h = \max_j (n_{hj}),\ h = 1, 2, ..., L;\ j = 1, 2, ..., p$. Our aim is to draw a sample which attains the desired precision assigned to the posterior variances of $\bar{Y}_j$. For this purpose we require

$$\bar{a}_j \le w_j,\ \ j = 1, 2, ..., p \ \text{ and } \ n_{hj} > 0 \tag{8.9.7}$$

where the values of w_j are the required upper limits on the posterior variances of $\bar{Y}_j,\ j = 1, 2, ..., p$. Note that we have

$$\bar{a}_j = W\bar{V}_j W^t = W\left[W_j + V_j\right]^{-1} W^t = (W_1,..,W_L)\left[\begin{Bmatrix} n_{1j}/\sigma_{1j}^2,...,0 \\ 0,...,n_{Lj}/\sigma_{Lj}^2 \end{Bmatrix} + \begin{Bmatrix} v_{j(1,1)},..,0 \\ 0,...,v_{j(L,L)} \end{Bmatrix}\right]^{-1}\begin{bmatrix} W_1 \\ W_L \end{bmatrix}$$

$$= \sum_{h=1}^{L} W_h^2 \Big/ \left\{v_{j(h,h)} + \left(n_{hj}/\sigma_{hj}^2\right)\right\}, \tag{8.9.8}$$

where $v_{j(h,\,h)}$ are the diagonal elements of V_j. Using (8.9.8) in (8.9.7) we have

$$\sum_{h=1}^{L} W_h^2 \Big/ \left\{v_{j(h,h)} + \left(n_{hj}/\sigma_{hj}^2\right)\right\} \le w_j,\ \ j = 1, 2, ..., \text{p}. \tag{8.9.9}$$

Thus the problem reduces to the following question:

(a) Determine the values of $n_{hj} > 0$ which minimize

$$C = \sum_{h=1}^{L} c_h n_h + \sum_{h=1}^{L}\sum_{j=1}^{p} c_{hj} n_{hj} \tag{8.9.10}$$

subject to

$$\sum_{h=1}^{L} W_h^2 \Big/ \left\{v_{j(h,h)} + \left(n_{hj}/\sigma_{hj}^2\right)\right\} \le w_j\,; \tag{8.9.11}$$

(b) Under the transformations given by

$$n_{hj} = \left[\frac{1}{y_{hj}} - v_{j(hh)}\right]\sigma_{hj}^2 \quad \text{and} \quad n_h = \max_j\left[\left\{\frac{1}{y_{hj}} - v_{j(hh)}\right\}\sigma_{hj}^2\right]$$

the values of c_{hj} which minimizes

$$L = \sum_{h=1}^{L}\sum_{j=1}^{p}\frac{\bar{c}_{hj}}{y_{hj}} + \sum_{h=1}^{L}\sum_{j\in I_j} c_h\left[\frac{1}{y_{hj}} - v_{j(hh)}\right]\sigma_{hj}^2\,, \tag{8.9.12}$$

where I_j is such that the maximum of n_{hj} for each $h \in I_j$ is attained for the jth variable, subject to

$$\sum_{h=1}^{L} W_h^2 y_{hj} \le w_j \quad \text{and} \quad y_{hj} \le \frac{1}{v_{j(hh)}}.$$

Solution to these questions can be obtained by following Kuhn and Tucker (1952) and Kokan and Khan (1967). Sekkappan (1981) also considered a problem of optimum allocation in stratified sampling from a finite population with p variables under study using a superpopulation approach put forth by Ericson (1969). He also studied allocation at second phase using information obtained from the first phase. The results obtained by Khan (1976) and Draper and Guttman (1968a, 1968b) are shown to be special cases.

8.10 STRATIFICATION USING TWO-PHASE SAMPLING

We have seen that in stratified sampling the population has to be divided into L strata, which are homogeneous within them selves, and whose means are widely different. The strata weights W_h can be used to obtain unbiased estimates of population mean or total. Sometime these strata weights are not known then the technique of two-phase sampling can be used to obtain estimates of theses weights. Following the two-phase sampling scheme we have to select a preliminary large sample of m units by SRSWOR sampling to estimate the strata weights. The m units in the first phase sample are then stratified into L strata with m_h units in the hth stratum. Select first phase sample of m_h units from the hth stratum such that $\sum_{h=1}^{L} m_h = m$. A second phase sample of size $n_h < m_h$ is selected from the hth stratum by SRSWOR sampling such that $\sum_{h=1}^{L} n_h = n$. Let $w_h = m_h/m$ be an estimator of original unknown weights $W_h = N_h/N$. In such situations an estimator for estimating the population mean $\bar{Y}$ is given by

$$\bar{y}_{\text{std}} = \sum_{h=1}^{L} w_h \bar{y}_h . \qquad (8.10.1)$$

Then we have the following theorems:

Theorem 8.10.1. The estimator $\bar{y}_{\text{std}}$ is an unbiased estimator of population mean $\bar{Y}$, provided that n_h is free from w_h.

Proof. We have

$$E(\bar{y}_{\text{std}}) = E_1\left[E_2\left\{\sum_{h=1}^{L} w_h \bar{y}_h \mid w_h\right\}\right] = E_1\left[\sum_{h=1}^{L} w_h \bar{Y}_h\right] = \sum_{h=1}^{L} W_h \bar{Y}_h = \bar{Y} .$$

Hence the theorem.

Theorem 8.10.2. The variance of the estimator $\bar{y}_{\text{std}}$ is given by

$$V(\bar{y}_{\text{std}}) = \sum_{h=1}^{L}\left(\frac{1-f_h}{n_h}\right)W_h^2 S_{hy}^2 + \frac{N}{N-1}\left(\frac{1}{m}-\frac{1}{N}\right)\left\{\sum_{h=1}^{L}\left(\frac{1-f_h}{n_h}\right)W_h(1-W_h)S_{hy}^2 + \sum_{h=1}^{L}W_h(\bar{Y}_h-\bar{Y})^2\right\}. \quad (8.10.2)$$

Proof. Note that the random sample size m_h follows multinomial distribution with parameters $(m, W_1, W_2, ..., W_L)$, therefore in particular we have

$$E(w_h) = W_h,\ V(w_h) = \frac{N}{N-1}\left(\frac{1}{m}-\frac{1}{N}\right)W_h(1-W_h),\ \text{Cov}(w_h, w_{h'}) = -\frac{N}{N-1}\left(\frac{1}{m}-\frac{1}{N}\right)W_h W_{h'},$$

thus we have

$$V(\bar{y}_{\text{std}}) = E_1V_2(\bar{y}_{\text{std}} \mid w_h) + V_1E_2(\bar{y}_{\text{std}} \mid w_h) = E_1V_2\left(\sum_{h=1}^{L} w_h\bar{y}_h \mid w_h\right) + V_1E_2\left(\sum_{h=1}^{L} w_h\bar{y}_h \mid w_h\right)$$

$$= E_1\left(\sum_{h=1}^{L} w_h^2 V_2(\bar{y}_h)\right) + V_1\left(\sum_{h=1}^{L} w_h E_2(\bar{y}_h)\right) = \left(\sum_{h=1}^{L} V_2(\bar{y}_h)E_1\left(w_h^2\right)\right) + V_1\left(\sum_{h=1}^{L} w_h\bar{Y}_h\right)$$

$$= \left(\sum_{h=1}^{L} V_2(\bar{y}_h)\left\{V_1(w_h) + (E_1(w_h))^2\right\}\right) + V_1\left(\sum_{h=1}^{L} w_h\bar{Y}_h\right)$$

$$= \sum_{h=1}^{L}\left(\frac{1-f_h}{n_h}\right)S_{hy}^2\left\{\frac{N}{N-1}\left(\frac{1}{m}-\frac{1}{N}\right)W_h(1-W_h) + W_h^2\right\} + \left(\sum_{h=1}^{L}\bar{Y}_h^2 V_1(w_h)\right)$$

$$= \sum_{h=1}^{L}\left(\frac{1-f_h}{n_h}\right)S_{hy}^2\left\{\frac{N}{N-1}\left(\frac{1}{m}-\frac{1}{N}\right)W_h(1-W_h) + W_h^2\right\}.$$

$$+ \sum_{h=1}^{L}\bar{Y}_h^2\frac{N}{N-1}\left(\frac{1}{m}-\frac{1}{N}\right)W_h(1-W_h).$$

which on simplification reduces to (8.10.2). Hence the theorem.

Corollary 8.10.1. For large N and $n_h = nW_h$, the variance $V(\bar{y}_{\text{std}})$ reduces to

$$V(\bar{y}_{\text{std}}) \approx \frac{1}{n}\sum_{h=1}^{L} W_h S_{hy}^2 + \frac{1}{m}\sum_{h=1}^{L} W_h(\bar{Y}_h-\bar{Y})^2. \quad (8.10.3)$$

Corollary 8.10.2. An unbiased estimator of $V(\bar{y}_{\text{std}})$ is given by

$$\hat{v}_u(\bar{y}_{\text{std}}) = \frac{m}{m-1}\sum_{h=1}^{L}\left[\left\{w_h^2 - \left(\frac{N-m}{N-1}\right)\frac{w_h}{m}\right\}\frac{s_{hy}^2}{n_h} + \left(\frac{N-m}{N-1}\right)\frac{w_h(\bar{y}_h-\bar{y}_{\text{std}})^2}{m}\right]. \quad (8.10.4)$$

For more details one can refer to Raj (1968).

Corollary 8.10.3. If n_h are small as compared to the size of preliminary large sample size m, then an unbiased estimator of $V(\bar{y}_{\text{std}})$ is given by

$$\hat{v}_u(\bar{y}_{\text{std}}) = \sum_{h=1}^{L} w_h^2 s_{hy}^2 / n_h \,. \tag{8.10.5}$$

Now we have the following theorem.

Theorem 8.10.3. The minimum variance of the estimator $\bar{y}_{\text{std}}$ under optimum allocation is given by

$$\text{Min.}V(\bar{y}_{\text{std}}) = \left(\sqrt{C_1V_2} + \sqrt{C_2V_1}\right)^2 / (C - C_0) \tag{8.10.6}$$

where $V_1 = \left(\sum_{h=1}^{L} W_h S_{hy}\right)^2$ and $V_2 = \sum_{h=1}^{L} W_h \left(\bar{Y}_h - \bar{Y}\right)^2$.

Proof. The cost function for double sampling, as defined in Chapter 6, is given by

$$C = C_0 + mC_1 + nC_2 \,. \tag{8.10.7}$$

The variance of the estimator $\bar{y}_{\text{std}}$ can be approximated by

$$V(\bar{y}_{\text{std}}) = \frac{V_1}{n} + \frac{V_2}{m} \,. \tag{8.10.8}$$

Thus the Lagrange function is given by

$$L = \frac{V_1}{n} + \frac{V_2}{m} - \lambda(C - C_0 - mC_1 - nC_2) \,.$$

On differentiating L with respect to m and n respectively and equating to zero, the optimum sample sizes, are

$$n = \frac{\sqrt{V_1}(C - C_0)}{\sqrt{C_2}\left(\sqrt{C_1V_2} + \sqrt{C_2V_1}\right)} \quad \text{and} \quad m = \frac{\sqrt{V_2}(C - C_0)}{\sqrt{C_1}\left(\sqrt{C_1V_2} + \sqrt{C_2V_1}\right)} \,. \tag{8.10.9}$$

On substituting (8.10.9) in (8.10.8) it will lead to the minimum variance as

$$\text{Min.}V(\bar{y}_{\text{std}}) = \frac{\sqrt{V_1C_2}\left(\sqrt{C_1V_2} + \sqrt{V_1C_2}\right)}{(C - C_0)} + \frac{\sqrt{C_1V_2}\left(\sqrt{C_1V_2} + \sqrt{V_1C_2}\right)}{(C - C_0)} = \frac{\left(\sqrt{C_1V_2} + \sqrt{V_1C_2}\right)^2}{(C - C_0)} \,.$$

Hence the theorem.

Example 8.10.1. We wish to estimate the yield/hectare of the tobacco crop in the World during 2001. The number of countries on each continent growing this crop during 2001 is unknown. Population 5 in the Appendix shows the yield/hectare of the tobacco crop along with the number of countries on each continent growing this crop. We wish to apply stratified random sampling to estimate the average yield/hectare of the tobacco crop in the world. Using information from 1998 about this crop in different countries, estimate the first phase and second phase sample size for stratification.

Given: $C = \$1500$, $C_0 = \$1000$, $C_1 = \$2$ and $C_2 = \$10$.

Solution. Using information from population 5 we have:

N_h	W_h	$\bar{Y}_h$	S_{hy}^2	S_{hy}	$W_h S_{hy}$	$W_h(\bar{Y}_h - \bar{Y})^2$
6	0.056604	1.913	0.0268266	0.163788	0.009271	0.007429884
6	0.056604	1.388	0.2180900	0.467001	0.026434	0.001498375
8	0.075472	2.556	0.3469980	0.589065	0.044458	0.076273818
10	0.094340	1.549	0.2345600	0.484314	0.045690	2.72642E-07
12	0.113208	1.832	0.5821400	0.762981	0.086375	0.008958078
4	0.037736	1.470	0.1531000	0.391280	0.014765	0.000245754
30	0.283019	1.115	0.3438500	0.586387	0.165959	0.053726742
17	0.160377	1.381	0.3785500	0.615264	0.098674	0.004618562
10	0.094340	1.721	2.0183000	1.420669	0.134025	0.002736046
3	0.028302	2.086	0.9746000	0.987218	0.027940	0.008109795
				Sum	0.653592	0.163597328

where $\bar{Y} = 1.5507$. From the above table we have

$$\sum_{h=1}^{L} W_h S_{hy} = 0.653592 \quad \text{and} \quad \sum_{h=1}^{L} W_h\left(\bar{Y}_h - \bar{Y}\right)^2 = 0.163597328 .$$

Thus we have

$$V_1 = 0.653592^2 = 0.427182 \quad \text{and} \quad V_2 = 0.163597328 .$$

Hence the optimum second and first phase sample sizes, respectively, are given by

$$n = \frac{\sqrt{V_1}(C - C_0)}{\sqrt{C_2}\left(\sqrt{C_1 V_2} + \sqrt{C_2 V_1}\right)} = \frac{\sqrt{0.427182}(1500 - 1000)}{\sqrt{10}\left(\sqrt{2 \times 0.16359} + \sqrt{10 \times 0.42718}\right)} = 39.16 \approx 40$$

and

$$m = \frac{\sqrt{V_2}(C - C_0)}{\sqrt{C_1}\left(\sqrt{C_1 V_2} + \sqrt{C_2 V_1}\right)} = \frac{\sqrt{0.163597}(1500 - 1000)}{\sqrt{2}\left(\sqrt{2 \times 0.16359} + \sqrt{10 \times 0.42718}\right)} = 54.19 \approx 55.$$

8.11 POST-STRATIFIED SAMPLING

Holt and Smith (1979) showed post-stratification is potentially more efficient than stratification. Since in order to maximise the gain in precision the stratification factors can be chosen in different ways for different sets of variables, *e.g.*, age, sex etc., after sampling. This technique is found to be most practicable for surveys where individual responses may be expected to vary with age, sex, occupation, education, state, country, and race etc.. Usually none of these variables are available for stratification at the individual level prior to sampling. Such a situation is called conditional poststratification. In some situations, censuses may provide information on all of these variables at the aggregate level. In some situations, this aggregate level information may not be available. In other such situations this is called unconditional post-stratification. We will discuss both situations.

8.11.1 CONDITIONAL POST-STRATIFICATION

Conditional post-stratification can be defined as a sampling technique in which we select a simple random sample with or without replacement. Once we know the strata sample size $n_h \geq 2$ $(h = 1, 2, ..., L)$ such that $n = \sum_{h=1}^{L} n_h$ we use the traditional stratified estimator. If there is any $n_h < 2$ then we merge the hth stratum with another t^{th} stratum such that the composition is homogeneous. Now we have the following theorems.

Theorem 8.11.1. An unbiased estimator of population mean, $\bar{Y}$, is given by

$$\bar{y}_{pst} = \sum_{h=1}^{L} W_h \bar{y}_h . \tag{8.11.1}$$

Proof. We have

$$E(\bar{y}_{pst}) = E_1 E_2 [\bar{y}_{pst} \mid n_h] = E_1 E_2 \left[\sum_{h=1}^{L} W_h \bar{y}_h \mid n_h \right] = E_1 \left[\sum_{h=1}^{L} W_h E_2 (\bar{y}_h \mid n_h) \right]$$

$$= E_1 \left[\sum_{h=1}^{L} W_h \bar{Y}_h \right] = E_1 (\bar{Y}) = \bar{Y} .$$

Hence the theorem.

Theorem 8.11.2. Conditional variance of the post-stratified estimator $\bar{y}_{pst}$ is

$$V(\bar{y}_{pst}) = \sum_{h=1}^{L} W_h^2 \left\{ \frac{1}{n_h} - \frac{1}{N_h} \right\} S_{hy}^2 . \tag{8.11.2}$$

Proof. We have

$$V(\bar{y}_{pst}) = E_1 V_2 (\bar{y}_{pst} \mid n_h) + V_1 E_2 (\bar{y}_{pst} \mid n_h)$$

$$= E_1 V_2 \left(\sum_{h=1}^{L} W_h \bar{y}_h \mid n_h \right) + V_1 E_2 \left(\sum_{h=1}^{L} W_h \bar{y}_h \mid n_h \right) = E_1 \left[\sum_{h=1}^{L} W_h^2 \frac{(N_h - n_h)}{N_h n_h} S_{hy}^2 \right] + V_1 (\bar{Y})$$

$$= E_1 \left[\sum_{h=1}^{L} W_h^2 \left(\frac{1}{n_h} - \frac{1}{N_h} \right) S_{hy}^2 \right] = \sum_{h=1}^{L} W_h^2 \left\{ \frac{1}{n_h} - \frac{1}{N_h} \right\} S_{hy}^2 . \tag{8.11.3}$$

Hence the theorem.

Espejo and Pineda (1997) proposed an estimator to estimate the variance of $\bar{y}_{pst}$ as

$$\hat{v}(\bar{y}_{pst}) = \frac{(N-1)}{(N-n)} \hat{v}(\bar{y}) - \sum_{h=1}^{L} \frac{W_h}{n_h} \sum_{i=1}^{n_h} y_{hi}^2 + \bar{y}_{pst}^2 \tag{8.11.4}$$

where

$$\hat{v}(\bar{y}) = \sum_{h=1}^{L} \frac{W_h}{n_h} \sum_{i=1}^{n_h} y_{hi}^2 - \bar{y}_{pst}^2 + \sum_{h=1}^{L} W_h^2 \frac{s_{hy}^2}{n_h} .$$

8.11.2 UNCONDITIONAL POST-STRATIFICATION

The unconditional post-stratification is a mechanism for the analysis of the exact variance for the post-stratification estimators when we do not know the stratum sample sizes $n_h \geq 2$, $h = 1, 2, ..., L$, $n = \sum_{h=1}^{L} n_h$ being the sample size from a finite population. We assume that if $n_h < 2$ we collapse the hth stratum with another t^{th} stratum with a homogeneous composition to the hth stratum. Thus n_h are random before selecting the sample and we have the following theorem:

Theorem 8.11.3. The unconditional variance of the post-stratified estimator $\bar{y}_{\text{pst}}$ under SRSWOR is

$$V(\bar{y}_{\text{pst}}) \approx \left(\frac{1}{n} - \frac{1}{N}\right) \sum_{h=1}^{L} W_h S_{hy}^2 + \frac{1}{n^2} \sum_{h=1}^{L} (1 - W_h) S_{hy}^2 . \tag{8.11.5}$$

Proof. We have

$$V(\bar{y}_{\text{pst}}) = E_1 V_2(\bar{y}_{\text{pst}} \mid n_h) + V_1 E_2(\bar{y}_{\text{pst}} \mid n_h) = E_1\left[\sum_{h=1}^{L} W_h^2 \frac{(N_h - n_h)}{N_h n_h} S_{hy}^2\right] + V_1(\bar{Y})$$

$$= E_1\left[\sum_{h=1}^{L} W_h^2 \left(\frac{1}{n_h} - \frac{1}{N_h}\right) S_{hy}^2\right] = \sum_{h=1}^{L} W_h^2 \left\{E_1\left(\frac{1}{n_h}\right) - \frac{1}{N_h}\right\} S_{hy}^2 . \tag{8.11.6}$$

Following Stephan (1945) we have

$$E_1\left(\frac{1}{n_h}\right) = \frac{1}{nW_h} + \frac{(1 - W_h)}{n^2 W_h^2} . \tag{8.11.7}$$

On substituting (8.11.7) in (8.11.6) we have the theorem.

From (8.11.5) we observe that the variance of $\bar{y}_{\text{pst}}$ consists of two components. The first component is the variance of the usual stratified estimator of proportional allocation, and the second component is the additional variance due to post-stratification, since we could not allocate the population owing to lack of information. For large n the second component is smaller compared to the first component. Thus for large samples, the post-stratification technique results in estimates which are nearly as good as with the proportional allocation. It is to be noted here that unconditional formula of variance is useful when planning for a survey (before the sample is selected), or when evaluating one methodology against another. Whereas the conditional formula of variance under post-stratified sampling schemes is useful when constructing the confidence intervals or testing hypotheses using survey results. More work related to poststratification sampling can also be seen from Shukla and Dubey (2001), Agarwal and Panda (1993), Doss, Hartley and Somayajulu (1979), Fuller (1966), Jagers (1986) and Jagers, Oden, and Trulsson (1985). Karlheinz (1990) has discussed the use of two-phase sampling for doing stratification. Chang, Han, and Hawkins (1999) have considered the problem of multiple inverse sampling as a sequential sampling procedure in a stratified population such that random samples are taken population wide and continuously until a specified minimum number of observations are obtained in each stratum. A calibration approach for post-stratified sampling has been considered by Lundstom and Särndal (1999). The multiple inverse sampling procedure can be used in small sample post-stratification to solve the empty post-strata problem. Estimation of population mean for a deeply stratified population under post-stratified random sampling has been discussed by Shukla and Trivedi (2001).

Example 8.11.1. We wish to estimate the average yield/hectare of the tobacco crop in the world. Select an SRSWOR sample of 40 countries from the population 5. Record the yield and area under the crop. Assume each continent as a different strata and number of countries in each continent are known. Post-stratify the countries selected in the SRSWOR sample into different continents. Merge the post-strata if required using total area under crop as an auxiliary variable. Obtain a 95% confidence interval for the average yield of the tobacco crop in the world.

Solution. After applying remainder approach on the first three columns $(N = 106)$ of the Pseudo-Random Number (PRN) Table 1 given in the Appendix we end with the following 40 distinct random numbers as:

038, 058, 071, 019, 077, 014, 061, 024, 096, 036, 094, 008, 041, 049, 002, 092, 053, 044, 030, 046, 075, 078, 015, 085, 034, 063, 066, 091, 009, 039, 086, 021, 074, 006, 101, 035, 090, 079, 054 and 065.

Thus the countries corresponding to these population unit numbers will be included in the sample. Using information about population unit numbers from population 5, we post-stratify the above sampled units as follows:

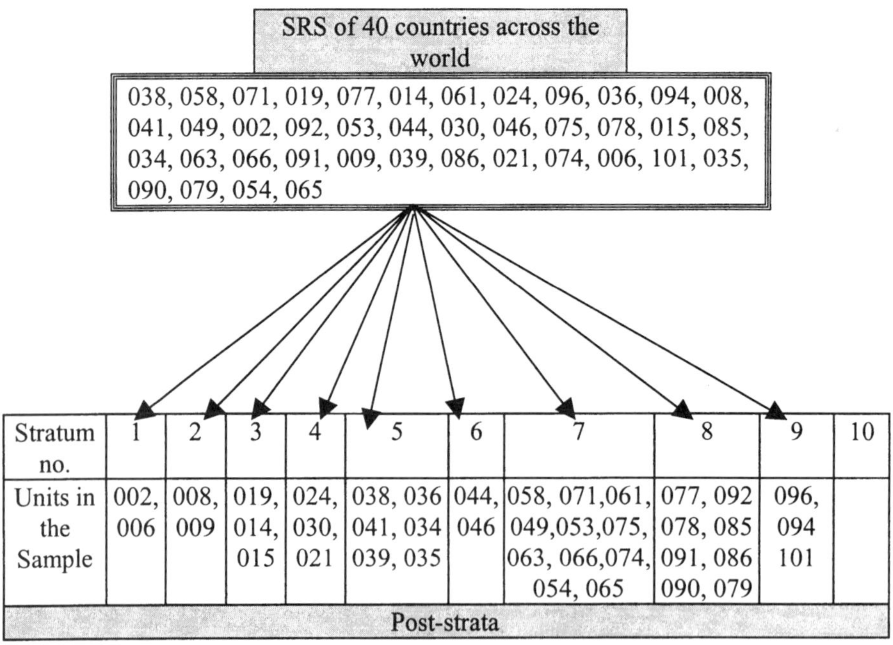

Fig. 8.11.1 Example of post-stratification.

We observed after post-stratification that the 10th stratum remains empty. There are several ways of merging this particular empty stratum with other one. We would like to use the information on the total area under the tobacco crop in ten continents or strata as follows:

Stratum No.	Total Area	N_h	n_h
1	19167.00	6	2
2	87960.00	6	2
3	146474.96	8	3
4	149235.00	10	3
5	65866.13	12	6
6	13800.00	4	2
7	350481.90	30	11
8	2467759.10	17	8
9	339761.00	10	3
10	3999.99	3	0

We also observed each one of the strata 1, 2, and 6 have only two units after post-stratification, which is the minimum requirement for variance estimation. We would prefer to merge stratum 10 with any one of these three. We make the use of known information on the total area under tobacco crop in different continents or strata. We observe that total area in the 10th stratum is 3999.99 hectares which is close to total area of 13800 hectares in the 6th stratum. Thus we shall prefer to merge the 10th empty post-stratum with the 6th post-stratum. After merging these post-strata, we have the following situation.

Post-stratum no.	New N_h values	Sample n_h	New W_h values	f_h	Yield/ hectare y_{hi} .	$\bar{y}_h$	$\sum_{i=1}^{n_h} y_{hi}^2$	s_h^2
1	6	2	0.0566	0.3333	1.79, 2.00	1.895	7.2041	0.0221
2	6	2	0.0566	0.3333	1.51, 1.29	1.400	3.9442	0.0242
3	8	3	0.0755	0.3750	2.14, 3.69, 2.80	2.876	26.0357	0.6050
4	10	3	0.0943	0.3000	1.64, 1.17, 2.09	1.633	8.4266	0.2116
5	12	6	0.1132	0.5000	0.92,1.82, 0.84, 1.63,2.48, 2.50	1.698	19.9217	0.5231
6	7	2	0.0660	0.2857	1.61, 1.18	1.395	3.9845	0.0924
7	30	11	0.2830	0.3666	2.51, 0.00, 1.22, 1.00, 0.87, 1.00, 1.63, 2.10, 0.96, 2.00, 0.93	1.293	23.3988	0.5016
8	17	8	0.1604	0.4706	0.88, 2.50, 1.22, 1.25, 1.33, 2.02, 1.80, 0.56	1.445	19.4782	0.3963
9	10	3	0.0943	0.3000	1.09, 1.50, 5.71	2.767	36.0422	6.5390

Post-Stratum no.	$W_h \bar{y}_h$	$\frac{W_h}{n_h}\sum_{i=1}^{n_h} y_{hi}^2$	$\frac{W_h^2 s_h^2}{n_h}$
1	0.107257	0.203876030	0.000035399
2	0.079240	0.111620860	0.000038760
3	0.217138	0.655231783	0.001149550
4	0.153992	0.264876127	0.000627217
5	0.192214	0.375856073	0.001117188
6	0.092070	0.131488500	0.000201247
7	0.365919	0.601987309	0.003652058
8	0.231778	0.390537910	0.001274509
9	0.260928	1.132926487	0.019382664
Sum	1.700536	3.868401079	0.027478596

Thus an estimate of the average yield/hectare of the tobacco crop in the world is given by

$$\bar{y}_{\text{pst}} = \sum_{h=1}^{L} W_h \bar{y}_h \approx 1.7005 .$$

Now we have

$$\hat{v}(\bar{y}) = \sum_{h=1}^{L} \frac{W_h}{n_h} \sum_{i=1}^{n_h} y_{hi}^2 - \bar{y}_{pst}^2 + \sum_{h=1}^{L} W_h^2 \frac{s_h^2}{n_h}$$

$$= 3.868401079 - 1.7005^2 + 0.027478596 = 1.004179425 .$$

Thus an estimate of variance of $\bar{y}_{\text{pst}}$ is given by

$$\hat{v}(\bar{y}_{\text{pst}}) = \frac{(N-1)}{(N-n)}\hat{v}(\bar{y}) - \sum_{h=1}^{L} \frac{W_h}{n_h} \sum_{i=1}^{n_h} y_{hi}^2 + \bar{y}_{pst}^2$$

$$= \frac{(106-1)}{(106-40)} \times 1.004179425 - 3.868401079 + 1.7005^2 = 0.62089 .$$

Thus a 95% confidence interval of the average yield/hectare of the world tobacco crop is

$$\bar{y}_{\text{pst}} \mp 1.96\sqrt{\hat{v}(\bar{y}_{\text{pst}})}\ , \quad \text{or} \quad 1.7005 \mp 1.96\sqrt{0.6209}\ , \quad \text{or} \quad [0.15608,\ 3.24491].$$

In order to find the percent relative efficiency (PRE) of the post-stratified random sampling with respect to SRSWOR sampling, we have

$$V(\bar{y}_{\text{srswor}}) = \frac{N-n}{Nn} S_y^2 = \frac{106-40}{106 \times 40} \times 0.6323 = 0.0098424 .$$

After post-stratification we have

Post-Strata	New N_h Values	New W_h Values	Sampled n_h Values	New s_{hy}^2 Values	$W_h s_{hy}^2$	$(1-W_h)s_{hy}^2$
1	6	0.0566	2	0.0268266	0.001518	0.0253080
2	6	0.0566	2	0.2180900	0.012344	0.2057460
3	8	0.0755	3	0.3469980	0.026198	0.3207900
4	10	0.0943	3	0.2345600	0.022116	0.2124409
5	12	0.1132	6	0.5821400	0.065898	0.5164210
6	7	0.0660	2	0.5100900	0.033666	0.4764240
7	30	0.2830	11	0.3438500	0.097309	0.2465400
8	17	0.1604	8	0.3785500	0.060719	0.3178300
9	10	0.0943	3	2.0183000	0.190326	1.8279740
Sum	106	1.0000	40		0.510094	4.1494740

Under post-stratified sampling the unconditional variance of the estimator of population mean is given by

$$V(\bar{y}_{\text{pst}}) \approx \left(\frac{1}{n}-\frac{1}{N}\right)\sum_{h=1}^{L} W_h S_{hy}^2 + \frac{1}{n^2}\sum_{h=1}^{L}(1-W_h)S_{hy}^2$$

$$\approx \left(\frac{1}{40}-\frac{1}{106}\right)\times 0.510094 + \frac{4.149474}{40^2} = 0.0105336\,.$$

Thus, the relative efficiency of the post-stratified sampling with respect to SRSWOR sampling or design effect is given by

$$\text{RE} = \frac{V(\bar{y}_{\text{srswor}})}{V(\bar{y}_{\text{pst}})}\times 100 = \frac{0.0098424}{0.0105336}\times 100 = 93.44\%\,.$$

In this particular example the poststratified sampling remains less efficient than simple random sampling.

8.12 ESTIMATION OF PROPORTION USING STRATIFIED RANDOM SAMPLING

Define a variable

$$Y_{hi} = \begin{cases} 1 & \text{if the } i^{th} \text{ unit in the } h^{th} \text{ stratum belongs to group } A, \\ 0 & \text{otherwise,} \end{cases}$$

then

$$\bar{Y}_h = \frac{1}{N_h}\sum_{i=1}^{N_h} Y_{hi} = \frac{N_{ha}}{N_h} = P_h \tag{8.12.1}$$

denote the proportion of units falling in the hth stratum possessing attribute A. Obviously the proportion of the units belonging to group A in the whole population can be written as

$$P=\frac{N_1P_1+N_2P_2+....+N_LP_L}{N}=\sum_{h=1}^{L}W_hP_h\,. \tag{8.12.2}$$

If n_h be the number of units belonging to group A in an SRSWOR sample of n units, then obviously $\hat{p}_h=n_h/n$ be an unbiased estimator of hth stratum population proportion P_h.

Thus we have the following theorems:

Theorem 8.12.1. An unbiased estimator of the overall population proportion P possessing the attribute A is given by

$$\hat{p}_{\text{st}}=\sum_{h=1}^{L}W_h\hat{p}_h\,. \tag{8.12.3}$$

Proof. Taking expected value on both sides of (8.12.3) we have

$$E(\hat{p}_{\text{st}})=\sum_{h=1}^{L}W_hE(\hat{p}_h)=\sum_{h=1}^{L}W_hP_h=P$$

which proves the theorem.

Theorem 8.12.2. The variance of the estimator of population proportion in stratified random sampling is given by

$$V(\hat{p}_{\text{st}})=\sum_{h=1}^{L}W_h^2\left(\frac{1-f_h}{n_h}\right)\left(\frac{N_h}{N_h-1}\right)P_hQ_h \tag{8.12.4}$$

where $Q_h=1-P_h$.

Proof. It follows by using results from Chapter 2 because strata are independent, so

$$V(\hat{p}_{\text{st}})=V\left[\sum_{h=1}^{L}W_h\hat{p}_h\right]=\sum_{h=1}^{L}W_h^2V(\hat{p}_h)=\sum_{h=1}^{L}W_h^2\left(\frac{1-f_h}{n_h}\right)\left(\frac{N_h}{N_h-1}\right)P_h(1-P_h)\,.$$

Hence the theorem.

Theorem 8.12.3. An unbiased estimator of the variance of the estimator of population proportion in stratified random sampling is given by

$$\hat{v}(\hat{p}_{\text{st}})=\sum_{h=1}^{L}W_h^2\left(\frac{1-f_h}{n_h-1}\right)\hat{p}_h\hat{q}_h, \tag{8.12.5}$$

where $\hat{q}_h=1-\hat{p}_h$.

Proof. Again it follows by using results from Chapter 2 because the strata are independent. Hence the theorem.

Note that the methods of equal, proportional, and optimum allocations are still valid while estimating population proportion, and we are considering SRSWOR sampling.

Example 8.12.1. A gardener has 60,000 mango trees in two orchards and over an experience of 4000 years, it has been observed that each tree produces about 100 mangoes. The gardener wishes to estimate the proportion of trees producing more than 100 mangoes out of all 60,000 trees in his two orchards.

(a) If 10% trees are producing more than 100 mangoes, then find the variance of the estimator of population proportion under SRSWOR sampling based on a sample of 50 trees.

(b) Later a statistician found that in the first orchard of 40,000 trees 10% of the trees are producing more than 100 mangoes and in the second orchard of 20,000 trees also 10% of the trees are producing more than 100 mangoes, so he suggested to applying stratified random sampling by selecting 25 units from each orchard.

(c) Do you agree with the statistician?

Solution. (a) Under SRSWOR sampling we have

$$N = 60{,}000\,, \quad n = 50 \ \text{ and } \ P = 0.1$$

so we have

$$V(\hat{p}) = \left(\frac{1-f}{n}\right)\left(\frac{N}{N-1}\right)P(1-P) = \left(\frac{1-50/60{,}000}{50}\right)\left(\frac{60{,}000}{60{,}000-1}\right)\times 0.1\times(1-0.1)$$

$$= 0.01798\,.$$

(b) Under stratified random sampling we have

$$V(\hat{p}_{\text{st}}) = \sum_{h=1}^{L} W_h^2\left(\frac{1-f_h}{n_h}\right)\left(\frac{N_h}{N_h-1}\right)P_h(1-P_h)$$

$$= W_1^2\left(\frac{1-f_1}{n_1}\right)\left(\frac{N_1}{N_1-1}\right)P_1(1-P_1) + W_2^2\left(\frac{1-f_2}{n_2}\right)\left(\frac{N_2}{N_2-1}\right)P_2(1-P_2)$$

$$= \left(\frac{40{,}000}{60{,}000}\right)^2\left(\frac{1-25/40{,}000}{25}\right)\left(\frac{40{,}000}{40{,}000-1}\right)\times 0.1\times(1-0.1)$$

$$+ \left(\frac{20{,}000}{60{,}000}\right)^2\left(\frac{1-25/20{,}000}{25}\right)\left(\frac{20{,}000}{20{,}000-1}\right)\times 0.1\times(1-0.1)$$

$$= 0.003597 + 0.003596 = 0.007193\,.$$

(c) The relative efficiency of the stratified random sampling over SRSWOR sampling will be

$$\text{RE} = \frac{V(\hat{p})}{V(\hat{p}_{\text{st}})}\times 100 = \frac{0.01798}{0.007193}\times 100 = 249.96\%\,.$$

Thus the use of stratified random sampling will be more beneficial than simple random sampling. Yes, we agree with the statistician.

EXERCISES

Exercise 8.1. In some practical situations the correlation of the auxiliary variable x with the study variable y is positive on some units and negative on other units. Evidently, such auxiliary variables cannot be used directly as the auxiliary variate in ratio and product method of estimation, since the population mean and/or the sample mean of the variate may be close to zero and these may occur in the denominator of the estimator. Can you resolve this difficulty by using the technique of (a) Stratified sampling, (b) Post-stratified sampling, (c) Change of scale method? Which method you will prefer and why?
Hint: Srivenkataramana and Tracy (1984).

Exercise 8.2. For stratified random sampling compare the estimator

$$\bar{y}_{\text{st}}^{*} = \sum_{h=1}^{L} \hat{cv}_h \bar{y}_h,$$

where $\hat{cv}_h = s_{hy}/\bar{y}_h$ denotes the estimator of the coefficient of variation in the hth stratum, with the usual estimator

$$\bar{y}_{\text{st}} = \sum_{h=1}^{L} W_h \bar{y}_h$$

under (a) proportional allocation, (b) optimum allocation.
Hint: Bennett (1983).

Exercise 8.3. In case of stratified random sampling, assuming a common fixed sample size n and neglecting the finite population correction factor, the relative precision (RP) of proportional allocation to optimum allocation is given by

$$\text{RP} = \left(\sum_{h=1}^{L} W_h \sqrt{P_h Q_h}\right)^2 \Big/ \sum_{h=1}^{L} W_h P_h Q_h ,$$

where

$$P_h = N_h^{-1} \sum_{i=1}^{N_h} Y_{hi}, \quad Q_h = 1 - P_h \quad \text{and } Y_{hi} = \begin{cases} 1 & \text{if } i \in A \text{ in } h^{th} \text{ stratum,} \\ 0 & \text{otherwise,} \end{cases}$$

and A denotes the attribute of interest.
Hint: Bennett and Islam (1983).

Exercise 8.4. Under the superpopulation model:

$$M : y_j = \alpha + \beta x_j + e_j, \quad j = 1, 2, ..., N$$

where

$$E(e_j \mid x_j) = 0, \quad V(e_j \mid x_j) = \gamma^2 x_j^g, \text{ and } \operatorname{Cov}(e_j, e_{j'} \mid x_j x_{j'}) = 0, \quad j \neq j'$$

the stratum size for optimum allocation in case of combined ratio estimator

$$\bar{y}_{\text{cr}} = \bar{y}_{\text{st}} \left(\frac{\bar{X}}{\bar{x}_{\text{st}}}\right)$$

is given by

$$n_h \propto W_h\left[\left\{\rho_{xy} - \frac{CV(x)}{CV(y)}\right\}^2 S_{hx}^2 + \left(1-\rho_{xy}^2\right)\frac{S_x^2 \overline{X}_h^g}{\overline{X}^g}\right]$$

and that for the combined regression estimator

$$\bar{y}_{\text{clr}} = \bar{y}_{\text{st}} + \hat{\beta}\left(\overline{X} - \bar{x}_{\text{st}}\right)$$

is given by

$$n_h \propto W_h\left(\overline{X}_h^g\right)^{\frac{1}{2}}.$$

Hint: Dayal (1985).

Exercise 8.5. Suppose there are three study variables satisfying the following conditions:

(a) the variates X, Y, and Z have probability density function. $f(x, y, z)$ in the finite range $x_0 \le x \le x_{L_1}$, $y_0 \le y \le y_{L_2}$, and $z_0 \le z \le z_{L_3}$;

(b) population is finite;

(c) divide the population into $L_1 \times L_2 \times L_3$ strata by determining the strata boundaries $x_1, x_2, ..., x_{(L_1-1)}$ for x; $y_1, y_2, ..., y_{(L_2-1)}$ for y; and $z_1, z_2, ..., z_{(L_3-1)}$ such that the generalized variance of the means of the variables for this stratified sample under the Neyman allocation is minimum; and

(d) the generalized variance can be set out as

$$\text{GV} = \begin{vmatrix} \sigma_x^2, & \sigma_{xy}, & \sigma_{xz} \\ \sigma_{xy}, & \sigma_y^2, & \sigma_{yz} \\ \sigma_{xz}, & \sigma_{yz}, & \sigma_z^2 \end{vmatrix},$$

where σ_x^2, σ_y^2, σ_z^2, σ_{xy}, σ_{xz} and σ_{yz} denote the approximate variances and co-variances of x and y under Neyman allocation with cost per unit constant. Find the optimum points of stratification in this tri-variate population.

Exercise 8.6. If m is the size of the first phase sample from which w_h, an estimate of W_h, has been obtained. If n is the size of the second phase sample then show that

$$\bar{y}_{\text{st}} = \sum_{h=1}^{L} w_h \bar{y}_h$$

is an unbiased estimator of population mean $\overline{Y}$ with variance

$$V(\bar{y}_{\text{st}}) = \sum_{h=1}^{L}\left[\left\{W_h^2 + \frac{gW_h(1-W_h)}{m}\right\}\frac{(1-f_h)S_{hy}^2}{n_h} + \frac{gW_h\left(\overline{Y}_h - \overline{Y}\right)^2}{m}\right]$$

where $g = (N-m)/(N-1)$, $f_h = n_h/N_h$ and the value of n_h is independent of w_h.

Hint: Dayal (1979).

Exercise 8.7. For a given sample $s_h \in \Omega_h$ $(h = 1,2,...,L)$ let $(t_{hy},\ t_{hx})$ assume values in a closed convex sub-space R_2 of the two-dimensional real space containing the point $(Y_h,\ X_h)$. Find the bias and variance of the class of estimators defined as

$$\hat{Y}_h = g_h(t_{hy},\ t_{hx})$$

where $g_h(t_{hy},\ t_{hx})$ is a known function of t_{hy} and t_{hx} satisfying certain regularity conditions such that

(a) $g_h(Y_h, X_h) = Y_h$;

(b) the function g_h is continuous in R_2;

and

(c) the first and second order derivatives of g_h exist and are continuous in R_2.

Hint: Dalabehara and Sahoo (1997).

Exercise 8.8. Let $P_{ij} = X_{ij}/X$ be the selection probability for the jth unit of the ith stratum.

(a) Show that an unbiased estimator of population total Y is given by

$$\hat{Y}_{\text{pps}} = \sum_{i=1}^{L} \frac{1}{n_i} \sum_{j=1}^{n_i} \frac{y_{ij}}{p_{ij}}$$

with variance

$$V\left(\hat{Y}_{\text{pps}}\right) = \sum_{i=1}^{L} \frac{1}{n_i} \left(\sum_{j=1}^{N_i} \frac{Y_{ij}^2}{P_{ij}} - Y_i^2 \right).$$

(b) Also derive the expression for optimum sample size n_i^{opt} in the ith stratum.

Hint: Gupta and Rao (1997), Rao (1984).

Exercise 8.9. For the hth stratum define $\bar{x}_h^* = \left(N_h \bar{X}_h - n_h \bar{x}_h\right) / (N_h - n_h)$. Find the bias and variance of the dual to separate ratio estimator of population mean $\bar{Y}$ in stratified sampling, defined as

$$\bar{y}_{\text{dst}} = \sum_{h=1}^{L} W_h \bar{y}_h \left(\frac{\bar{x}_h^*}{\bar{X}_h} \right).$$

Hint: Singh and Singh (1995).

Exercise 8.10. Show that the difference between the variances of the separate ratio and the combined ratio estimator of population total in stratified sampling is

$$V(\hat{y}_{\text{cr}}) - V(\hat{y}_{\text{sr}}) = \sum_{h=1}^{L} \frac{N_h^2 (1 - f_h)}{n_h} \left[\left(R^2 - R_h^2\right) S_{hx}^2 - 2(R - R_h) \rho_{hxy} S_{hx} S_{hy} \right].$$

Discuss the effect of choice of R_h and R from stratum to stratum and across strata.

Exercise 8.11. Consider a population of size N stratified into L strata and the size of the hth stratum being N_h such that $\sum_{h=1}^{L} N_h = N$. A simple random sample of size n is drawn from the population and is classified amongst the L strata such that n_h $(h = 1, 2, ..., L)$ is the number of units in the sample that fall in h^{th} stratum n_h varying from sample to sample.

(a) Compare the usual unbiased estimator

$$\bar{y}_{\text{pst}} = \sum_{h=1}^{L} W_h \bar{y}_h,$$

where $W_h = N_h / N$, with the estimator defined as

$$\bar{y}^*_{\text{pst}} = \sum_{h=1}^{L} W_{h\alpha} \bar{y}_h,$$

where $W_{h\alpha} = \alpha w_h + (1-\alpha) W_h$ for $w_h = n_h / n$.

(b) Discuss the different choices of the real constant α.

Hint: Agrawal and Panda (1993).

Exercise 8.12. Divide the population of size N into L strata, such that the hth stratum has N_h units. From the hth stratum, draw a preliminary large sample of m_h units with SRSWOR sampling and measure the auxiliary variable x_{hi}, $i = 1, 2, ..., m_h$. Out of m_h units selected in the preliminary large sample from the hth stratum select a second phase sample of n_h units using SRSWOR sampling and measure both the study variable y_{hi} and auxiliary variable x_{hi}, $i = 1, 2, ..., n_h$.

(a) Study the bias and variance properties of the ratio and regression type estimators of the population mean $\bar{Y}$ defined as

$$\bar{y}_1 = \sum_{h=1}^{L} W_h \bar{y}_{hn} \left(\frac{\bar{x}_{hm}}{\bar{x}_{hn}} \right), \quad \bar{y}_2 = \sum_{h=1}^{L} W_h \left[\bar{y}_{hn} + k_h (\bar{x}_{hm} - \bar{x}_{hn}) \right],$$

$$\bar{y}_3 = \sum_{h=1}^{L} W_h \bar{y}_{hn} \left[\sum_{h=1}^{L} W_h \bar{x}_{hm} \Big/ \sum_{h=1}^{L} W_h \bar{x}_{hn} \right],$$

and

$$\bar{y}_4 = \sum_{h=1}^{L} W_h \bar{y}_{hn} + k \left[\sum_{h=1}^{L} W_h \bar{x}_{hm} - \sum_{h=1}^{L} W_h \bar{x}_{hn} \right],$$

where $W_h = N_h / N$, $\bar{x}_{hm} = m_h^{-1} \sum_{i=1}^{m_h} x_{hi}$, $\bar{x}_{hn} = n_h^{-1} \sum_{i=1}^{n_h} x_{hi}$ and $\bar{y}_{hn} = n_h^{-1} \sum_{i=1}^{n_h} y_{hi}$ are the first phase and second phase sample means.

(b) Find k_h and k real constants such that the variances of the $\bar{y}_2$ and $\bar{y}_4$ are minimum.

Hint: Lindley and Deely (1993)

Exercise 8.13. Consider we draw a preliminary large sample of m units from the population of N units with SRSWOR sampling. First step is to post-stratify the selected units in the L strata so that the hth stratum contains m_h units and note the associated auxiliary variable x_{hi}, $i = 1, 2, ..., m_h$. From the post-stratified sample with m_h units in the hth stratum, select n_h units with SRSWOR sampling and measure the study variable y_{hi} and auxiliary variable x_{hi}, $i = 1, 2, ..., n_h$.

Study the asymptotic properties of the estimator of population mean $\bar{Y}$ defined as

$$\bar{y}_{\text{pst}} = \sum_{h=1}^{L} \frac{m_h}{m} \bar{y}_{hn} \left(\frac{\bar{x}_{hm}}{\bar{x}_{hn}} \right).$$

Exercise 8.14. Consider P_h be the proportion of units in the hth stratum possessing an attribute of interest, say A. Evidently $P = \sum_{h=1}^{L} W_h P_h$ be the proportion of units in the population possessing the attribute A. Assume n_h units are selected from the hth stratum using SRSWOR sampling. Suggest an unbiased estimator of P. Compare the efficiency of optimal allocation with proportional allocation.

Exercise 8.15. Write a short note on stratification and post-stratification.

Exercise 8.16. Under post-stratification, study the asymptotic properties of the estimator of population mean $\bar{Y}$ given by

$$\bar{y}_{\text{d}} = \frac{\sum_{h=1}^{L} a_h W_h \bar{y}_h / E(a_h)}{\sum_{h=1}^{L} a_h W_h / E(a_h)},$$

where

$$a_h = \begin{cases} 1 - {}^{N-N_h}C_n / {}^{N}C_n & \text{if } N \text{ is finite,} \\ 1 - (1 - W_h)^n & \text{if } N \text{ is infinite.} \end{cases}$$

Hint: Doss, Hartley and Somayajulu (1979).

Exercise 8.17. Assume $\bar{Y} = \sum_{h=1}^{L} W_h \bar{Y}_h$, $W_h = N_h / N$ and $\bar{y}_h = \frac{1}{n_h} \sum_{j=1}^{n_h} y_{hj}$, $h = 1,2,...,L$.

Also let $\bar{y} = \frac{1}{n} \sum_{i=1}^{n} y_i$ and $\bar{y}_p = \frac{1}{n} \sum_{h=1}^{L} W_h \bar{y}_h$ denote the estimators of $\bar{Y}$ obtained from simple random sampling and post-stratification $(n_h \geq 1, h = 1,2,..,L)$, respectively. It should be noted that the sample size n of the first-stage simple random sampling is a fixed number, while the classified sample size n_h thereafter are random variables. In case of small samples a commonly used estimation procedure for $\bar{Y}$ is to collapse the empty post-strata, if there are any, with neighbouring strata. If $\bar{y}_c(L)$ be the estimator of $\bar{Y}$ under such a procedure then define $\bar{y}_c(L)$, $L = 2,3,4$.

Hint: Chang, Liu, and Han (1998), Chang, Han, and Hawkins (1999).

Exercise 8.18. Let the survey population $\Omega = \{i : i = 1,...,N\}$ be stratified into L strata Ω_j, $j = 1,2,...,L$. In the superpopulation model $\phi_i(y_i, \theta) = f_i(y_i) - \alpha_i(\theta)$ let $f_i(y_i) = y_i$ and $\alpha_i(\theta) = \theta_j x_i$ for all $i \in \Omega_j$ and $j = 1,2,...,k$ with x_i, $i = 1,2,...,N$ being a covariate and θ_j some scalar parameter specific to the stratum Ω_j, $j = 1,...,L$.

(a) Show that the estimating function

$$h = \sum_{i \in s}\{f_i(y_i) - \alpha_i(\theta)\}/\pi_i + \sum_{i=1}^{N}\alpha_i(\theta)$$

can be written as

$$\hat{Y}_G = \sum_{j=1}^{L}\sum_{i \in s_j}(y_i - \theta_j x_i)/\pi_i + \sum_{j=1}^{L}\theta_j X_j,$$

where s_j denote the sample from the stratum Ω_j and $X_j = \sum_{\Omega_j} x_i$.

(b) Also show that under SRSWOR sample of fixed size n the estimating function reduces to

$$\hat{Y}_G = \frac{N}{n}\sum_{j=1}^{L} n_j(\bar{y}_j - \theta_j \bar{x}_j) + \sum_{j=1}^{N}\theta_j X,$$

where $\bar{y}_j$ and $\bar{x}_j$ are the means of the y and x over the sample s_j.

(c) If the value of $\theta_j = \bar{y}_j / \bar{x}_j$, $j = 1,2,...,L$, then the estimating function reduces to separate ratio estimator given by

$$\hat{Y}_R = \sum_{j=1}^{L}\frac{\bar{y}_j}{\bar{x}_j}X_j.$$

Hint: Godambe (1995).

Exercise 8.19. Suppose a stratified sampling strategy is defined as follows:
(i) the population is stratified by type, *e.g.*, industry, description or geographic area;
(ii) the type strata are further stratified by size;
(iii) the sample is allocated over all strata in such a fashion as to ensure that the required degree of accuracy will be achieved for each estimate and that, subject to that requirement, the number of sample units is minimized;
(iv) the number of sample units allocated to each particular type size stratum is selected by using SRSWOR.

Under the superpopulation model

$$Y_i = \beta X_i + e_i$$

with

$$E(e_i) = 0, \quad E(e_i e_j) = \begin{cases} \sigma_i^2 & \text{if } i = j, \\ 0 & \text{if } i \neq j. \end{cases}$$

(a) Show that a sensible predictor of the population total Y is given by

$$y^* = \sum_{i\in s} y_i + \hat{\beta}\sum_{i\notin s} x_i,$$

where $\hat{\beta}$ is a sample estimator of β.

(b) If we take $\hat{\beta} = \sum_{i\in s} w_i y_i \Big/ \sum_{i\in s} w_i x_i$, where the weights w_i are bounded but otherwise completely arbitrary, then show that the optimum value of these weights w_i over the repeated sampling design is given by

$$w_i = \alpha\left(n^{-1}\sigma_i^{-1}\sum_{j=1}^{N}\sigma_j - 1\right)$$

with $\alpha = \sum_{j=1}^{N}\pi_j w_j X_j$.

(c) Also show that it leads to an analogue of the Neyman allocation given by

$$\pi_i = n\sigma_i \Big/ \sum_{j=1}^{N}\sigma_j .$$

Hint: Brewer (1979).

Exercise 8.20. If the cost function is of the form $C = C_0 + \sum_{h=1}^{L} C_h n_h^k$ then show that the optimum sample sizes n_h for the minimum variance $V(\bar{y}_{\text{st}})$ is given by

$$n_h = n\left(\frac{W_h^2 S_{hy}^2}{C_h}\right)^{\frac{1}{k+1}} \Bigg/ \sum_{h=1}^{L}\left(\frac{W_h^2 S_{hy}^2}{C_h}\right)^{\frac{1}{k+1}},$$

where k is a constant.

Exercise 8.21. Consider a population Ω of N units divided into independent L strata of sizes N_1, N_2, ..., N_L such that $\sum_{h=1}^{L} N_h = N$. Consider a sampling design that selects n_h units from the hth stratum employing selection probabilities proportional to size sampling within the hth stratum. Define an estimator of population total Y defined as

$$\hat{Y}_{\text{st}} = \sum_{h=1}^{L}\sum_{i\in s_h}\frac{y_{hi}}{\pi_{hi}}$$

where the y_{hi} now denote the y value of the ith unit in the hth stratum, s_h is the sample from the hth stratum, $1 \le h \le L$ and $\pi_{hi} = n_h x_{hi} / X_h$, $1 \le i \le N_h$, and $X_h = \sum_{i=1}^{N_h} x_{hi}$.

Show that the estimator $\hat{Y}_{\text{st}}$ is unbiased for population total, Y, and its variance is given by

$$v\left(\hat{Y}_{st}\right)=-\frac{1}{2}\sum_{h=1}^{L}\sum_{i=1}^{N_h}\sum_{j\neq i=1}^{N_h}\left(\pi_{hij}-\pi_{hi}\pi_{hj}\right)\left(y_{hi}/\pi_{hi}-y_{hj}/\pi_{hj}\right)^2.$$

Hint: Padmawar (1998).

Exercise 8.22. Assuming that reliability ratio δ is known, a necessary and sufficient condition for the estimator

$$\hat{Y}=\left[w_{0s_1}+\sum_{j\in s_1}w_{js_1}y_{1j};\ldots,w_{0s_L}+\sum_{j\in s_L}w_{js_L}y_{Lj}\right]^t$$

to be admissible in the class of linear estimators is that there exist α and a_i satisfying $f_i/\left(1+\delta^{-1}\right)\le a_i\le 1$ and $\frac{1}{n_i}\left(a_i-\frac{f_i}{1+\delta^{-1}}\right)\left(a_j-1\right)=\frac{1}{n_j}\left(a_j-\frac{f_j}{1+\delta^{-1}}\right)\left(a_i-1\right)$ for $1\le i\neq j\le L$ such that $w_{js_i}=a_i/n_i$ for all $j\in s_i$ and $w_{0s_i}=\left(a_i-1\right)\alpha$ for $i=1,2,\ldots,L$.

Hint: Zou and Wan (2000).

Exercise 8.23. (I) Single phase sampling: (a) Consider an estimator in stratified sampling of the form

$$\bar{y}_{st}(\text{new})=\sum_{h=1}^{L}\Omega_h\bar{y}_h,$$

where the weights Ω_h are chosen such that the chi square distance function

$$\sum_{h=1}^{L}\frac{\left(\Omega_h-W_h\right)^2}{W_hQ_h}$$

where Q_h denotes suitable weights, is minimized subject to the following two calibration constraints

$$\sum_{h=1}^{L}\Omega_h\bar{x}_h=\sum_{h=1}^{L}W_h\bar{X}_h \quad\text{and}\quad \sum_{h=1}^{L}\Omega_h s_{hx}^2=\sum_{h=1}^{L}W_hS_{hx}^2,$$

where

$$S_{hx}^2=\left(N_h-1\right)^{-1}\sum_{i=1}^{N_h}\left(X_{hi}-\bar{X}_h\right)^2,\quad s_{hx}^2=\left(n_h-1\right)^{-1}\sum_{i=1}^{n_h}\left(x_{hi}-\bar{x}_h\right)^2,\quad \bar{x}_h=n_h^{-1}\sum_{i=1}^{n_h}x_{hi},\ \text{and}$$

$\bar{X}_h=N_h^{-1}\sum_{i=1}^{N_h}X_{hi}$ have there usual meanings.

Show that the resultant estimator of the population mean is given by

$$\bar{y}_{st}(\text{new})=\sum_{h=1}^{L}W_h\left[\bar{y}_h+\hat{\beta}_1\left(\bar{X}_h-\bar{x}_h\right)+\hat{\beta}_2\left(S_{hx}^2-s_{hx}^2\right)\right]$$

where

$$\hat{\beta}_1=\frac{\sum_{h=1}^{L}W_hQ_h\bar{x}_h\bar{y}_h\left\{\sum_{h=1}^{L}W_h\left(\bar{X}_h-\bar{x}_h\right)\sum_{h=1}^{L}W_hQ_hs_{hx}^4-\sum_{h=1}^{L}W_h\left(S_{hx}^2-s_{hx}^2\right)\sum_{h=1}^{L}W_hQ_h\bar{x}_hs_{hx}^2\right\}}{\sum_{h=1}^{L}W_hQ_h\bar{x}_h^2\sum_{h=1}^{L}W_hQ_hs_{hx}^4-\left(\sum_{h=1}^{L}W_hQ_hs_{hx}^2\right)^2}$$

and

$$\hat{\beta}_2 = \frac{\sum_{h=1}^{L} W_h Q_h s_{hx}^2 \bar{y}_h \left\{ \sum_{h=1}^{L} W_h \left(S_{hx}^2 - s_{hx}^2\right) \sum_{h=1}^{L} W_h Q_h \bar{x}_h^2 - \sum_{h=1}^{L} W_h \left(\bar{X}_h - \bar{x}_h\right) \sum_{h=1}^{L} W_h Q_h \bar{x}_h s_{hx}^2 \right\}}{\sum_{h=1}^{L} W_h Q_h \bar{x}_h^2 \sum_{h=1}^{L} W_h Q_h s_{hx}^4 - \left(\sum_{h=1}^{L} W_h Q_h s_{hx}^2 \right)^2}.$$

(b) The minimum variance of the estimator $\bar{y}_{st}(\text{new})$ is given by

$$V\left(\bar{y}_{st}(\text{new})\right) = \sum_{h=1}^{L} W_h^2 \left(\frac{1-f_h}{n_h}\right) S_{hy}^2 \left\{ 1 - \lambda_{h11}^2 - \frac{\left(\lambda_{h11}\lambda_{ho3} - \lambda_{h12}\right)^2}{\lambda_{h04} - 1 - \lambda_{h03}^2} \right\}$$

where

$$\lambda_{hrs} = \frac{\mu_{hrs}}{\mu_{h20}^{r/2} \mu_{02}^{s/2}} \quad \text{and} \quad \mu_{hrs} = \left(N_h - 1\right)^{-1} \sum_{i=1}^{N_h} \left(Y_{hi} - \bar{Y}_h\right)^r \left(X_{hi} - \bar{X}_h\right)^s.$$

(c) Show that the variance $V\left(\bar{y}_{st}(\text{new})\right)$ can be written as

$$V\left(\bar{y}_{st}(\text{new})\right) = \sum_{h=1}^{L} W_h^2 \left(\frac{1-f_h}{n_h}\right) \frac{1}{N_h - 1} \sum_{i=1}^{N_h} \epsilon_{hi}^2,$$

where

$$\epsilon_{hi} = \left(Y_{hi} - \bar{Y}_h\right) - \beta_1 \left(X_{hi} - \bar{X}_h\right) - \beta_2 \left\{ \left(X_{hi} - \bar{X}_h\right)^2 - \sigma_{hx}^2 \right\} \quad \text{with } \sigma_{hx}^2 = N_h^{-1} \sum_{i=1}^{N_h} \left(X_{hi} - \bar{X}_h\right)^2.$$

(d) Consider an estimator for estimating the variance $V\left(\bar{y}_{st}(\text{new})\right)$ is given by

$$\hat{v}_0\left(\bar{y}_{st}(\text{new})\right) = \sum_{h=1}^{L} W_h^2 \left(\frac{1-f_h}{n_h}\right) \frac{1}{n_h - 3} \sum_{i=1}^{n_h} e_{hi}^2,$$

where

$$e_{hi} = \left(y_{hi} - \bar{y}_h\right) - \hat{\beta}_1 \left(x_{hi} - \bar{x}_h\right) - \hat{\beta}_2 \left\{ \left(x_{hi} - \bar{x}_h\right)^2 - s_{hx}^{*2} \right\}$$

with $s_{hx}^{2*} = n_h^{-1} \sum_{i=1}^{n_h} \left(x_{hi} - \bar{x}_h\right)^2$ being the maximum likelihood estimator of σ_{hx}^2, and compare it with a calibrated estimator of the variance defined as

$$\hat{v}_1\left(\bar{y}_{st}(\text{new})\right) = \sum_{h=1}^{L} \Omega_h^2 \left(\frac{1-f_h}{n_h}\right) \frac{1}{n_h - 3} \sum_{i=1}^{n_h} e_{hi}^2.$$

(II) Double Sampling: Consider the population of N units consists of L strata such that the hth stratum consists of N_h units and $\sum_{h=1}^{L} N_h = N$. From the hth stratum of N_h units, draw a preliminary large sample of m_h units by SRSWOR sampling and measure the auxiliary variable x_{hi} only. Select a sub-sample of n_h units from the given preliminary large sample of m_h units by SRSWOR sampling and measure both the study variable y_{hi} and auxiliary variable x_{hi}.

Let $\bar{x}_h^* = m_h^{-1} \sum_{i=1}^{m_h} x_{hi}$ and $s_{hx}^{*2} = \left(m_h - 1\right)^{-1} \sum_{i=1}^{m_h} \left(x_{hi} - \bar{x}_h^*\right)^2$ denote the first phase sample mean and variance. Also let $\bar{x}_h = n_h^{-1} \sum_{i=1}^{n_h} x_{hi}$, $s_{hx}^2 = \left(n_h - 1\right)^{-1} \sum_{i=1}^{n_h} \left(x_{hi} - \bar{x}_h\right)^2$, and

$\bar{y}_h = n_h^{-1}\sum_{i=1}^{n_h} y_{hi}$, $s_{hy}^2 = (n_h - 1)^{-1}\sum_{i=1}^{n_h}(y_{hi} - \bar{y}_h)^2$ denote the second phase sample mean and variances for the auxiliary and study variables, respectively.

(a) Consider an estimator of the population mean in stratified double sampling as

$$\bar{y}_{\text{st}}(\text{d}) = \sum_{h=1}^{L} W_h^* \bar{y}_h,$$

where the W_h^* are the calibrated weights such that the chi square distance

$$\sum_{h=1}^{L} \frac{\left(W_h^* - W_h\right)^2}{W_h Q_h},$$

where Q_h are prearranged weights, is minimized, subject to the constraints

$$\sum_{h=1}^{L} W_h^* \bar{x}_h = \sum_{h=1}^{L} W_h \bar{x}_h^*, \text{ and } \sum_{h=1}^{L} W_h^* s_{hx}^2 = \sum_{h=1}^{L} W_h s_{hx}^{*2},$$

where the $W_h = N_h/N$ are known stratum weights.

Show that a calibrated estimator of the population mean in stratified double sampling is given by

$$\bar{y}_{\text{st}}(\text{d}) = \sum_{h=1}^{L} W_h \bar{y}_h + \hat{\beta}_1^*\left[\sum_{h=1}^{L} W_h\left(\bar{x}_h - \bar{x}_h^*\right)\right] + \hat{\beta}_2^*\left[\sum_{h=1}^{L} W_h\left(s_{hx}^2 - s_{hx}^{*2}\right)\right],$$

where $\hat{\beta}_1^*$ and $\hat{\beta}_2^*$ have their usual meanings, and work them out

(b) Show that the conditional variance of the stratified double sampling estimator

$$\bar{y}_{\text{st}}(\text{d}) = \sum_{h=1}^{L} W_h^* \bar{y}_h$$

is

$$V\left[\bar{y}_{\text{st}}(\text{d}) \mid W_h^*\right] = \sum_{h=1}^{L} W_h^{*2}\left(\frac{1}{n_h} - \frac{1}{N_h}\right) S_{hy}^2$$

where

$$S_{hy}^2 = (N_h - 1)^{-1}\sum_{i=1}^{N_h}\left(Y_{hi} - \bar{Y}_h\right)^2.$$

(c) Show that a conditionally unbiased estimator of $V\left[\bar{y}_{\text{st}}(\text{d}) \mid W_h^*\right]$ is

$$\hat{v}\left[\bar{y}_{\text{st}}(\text{d}) \mid W_h^*\right] = \sum_{h=1}^{L} W_h^{*2}\left(\frac{1}{n_h} - \frac{1}{N_h}\right) s_{hy}^2$$

where

$$s_{hy}^2 = (n_h - 1)^{-1}\sum_{i=1}^{n_h}(y_{hi} - \bar{y}_h)^2.$$

(d) Show that the minimum variance of the stratified double sampling estimator $\bar{y}_{\text{st}}(\text{d})$, to the first order of approximation, is given by

$$V(\bar{y}_{\text{st}}(\text{d})) = \sum_{h=1}^{L} W_h^2\left[\left(\frac{1}{m_h} - \frac{1}{N_h}\right) S_{hy}^2 + \left(\frac{1}{n_h} - \frac{1}{m_h}\right) S_{hy}^2\left\{1 - \lambda_{h11}^2 - \frac{\left(\lambda_{h11}\lambda_{h03} - \lambda_{h12}\right)^2}{\lambda_{h04} - 1 - \lambda_{h03}^2}\right\}\right].$$

(e) Show that the variance of the stratified double sampling estimator $\bar{y}_{\text{st}}(\text{d})$ is

$$V(\bar{y}_{st}(d)) \approx \sum_{h=1}^{L} W_h^2 \left[\left(\frac{1}{m_h} - \frac{1}{N_h} \right) S_{hy}^2 + \left(\frac{1}{n_h} - \frac{1}{m_h} \right) \frac{1}{N_h} \sum_{i=1}^{N_h} \epsilon_{hi}^2 \right],$$

where

$$\epsilon_{hi} = \left(Y_{hi} - \bar{Y}_h\right) - \beta_1 \left(X_{hi} - \bar{X}_h\right) - \beta_2 \left\{\left(X_{hi} - \bar{X}_h\right)^2 - \sigma_{hx}^2\right\} .$$

(f) Consider an estimator of variance $V(\bar{y}_{st}(d))$ is given by

$$\hat{v}(\bar{y}_{st}(d)) = \sum_{h=1}^{L} W_h^2 \left[\left(\frac{1}{m_h} - \frac{1}{N_h} \right) s_{hy}^2 + \left(\frac{1}{n_h} - \frac{1}{m_h} \right) \frac{1}{n_h} \sum_{i=1}^{n_h} e_{hi}^2 \right],$$

where $e_{hi} = (y_{hi} - \bar{y}_h) - \hat{\beta}_1 (x_{hi} - \bar{x}_h) - \hat{\beta}_2 \left\{(x_{hi} - \bar{x}_h)^2 - s_{hx}^{*2}\right\}$ denotes the estimate of the residual term and $s_{hx}^{*2} = n_h^{-1} \sum_{i=1}^{n_h} (x_{hi} - \bar{x}_h)^2$ denotes the maximum likelihood estimator of σ_{hx}^2. Study an estimator of variance in stratified double sampling as

$$\hat{v}(\bar{y}_{st}(d))_n = \sum_{h=1}^{L} W_h^{*2} \left[\left(\frac{1}{m_h} - \frac{1}{N_h} \right) s_{hy}^2 + \left(\frac{1}{n_h} - \frac{1}{m_h} \right) \frac{1}{n_h - 1} \sum_{i=1}^{n_h} e_{hi}^2 \right].$$

Hint: Tracy, Singh, and Arnab (2003).

Exercise: 8.24. Under stratified random sampling, study the asymptotic properties of the following estimators of the population mean, $\bar{Y}$, defined as

(a) $\bar{y}_{stSD} = \bar{y}_{st} \dfrac{\sum_{h=1}^{L} W_h \left(\bar{X}_h + C_{xh}\right)}{\sum_{h=1}^{L} W_h \left(\bar{x}_h + C_{xh}\right)}$ (b) $\bar{y}_{stSK} = \bar{y}_{st} \dfrac{\sum_{h=1}^{L} W_h \left(\bar{X}_h + \beta_{2h}(x)\right)}{\sum_{h=1}^{L} W_h \left(\bar{x}_h + \beta_{2h}(x)\right)}$

(c) $\bar{y}_{syUS1} = \bar{y}_{st} \dfrac{\sum_{h=1}^{L} W_h \left(\bar{X}_h \beta_{2h}(x) + C_{xh}\right)}{\sum_{h=1}^{L} W_h \left(\bar{x}_h \beta_{2h}(x) + C_{xh}\right)}$ (d) $\bar{y}_{syUS2} = \bar{y}_{st} \dfrac{\sum_{h=1}^{L} W_h \left(\bar{X}_h C_{xh} + \beta_{2h}(x)\right)}{\sum_{h=1}^{L} W_h \left(\bar{x}_h C_{xh} + \beta_{2h}(x)\right)}$

where $\bar{y}_{st} = \sum_{h=1}^{L} W_h \bar{y}_h$ has usual meaning. Discuss the special cases when there is only one stratum ($L = 1$) in the population.

Hint: Kadilar and Cingi (2003).

PRACTICAL PROBLEMS

Practical 8.1. A World Bank manager selects two units by SRSWOR sampling from each stratum of the population 5 given in the Appendix. Later on a statistician collects the information on the production of the tobacco crop from the countries selected in the sample. The total number of countries in each continent are known then discuss the estimate of the average production of the tobacco crop in the world. Derive the 95% confidence interval. Also find an estimate of the percentage gain in efficiency (GE) due to stratification.

Practical 8.2. A statistician suggested to the World Bank to select a sample of 40 countries from population 5 given in the Appendix using the method of proportional allocation. Following his instructions, record the production of tobacco crop from the selected countries. Estimate the average production of the tobacco crop in the world. Estimate the variance under the method of proportional allocation. Construct a 95% confidence interval.

Practical 8.3. The variation in the production of tobacco crops in different continents of the world is known as listed in the descriptive statistics of population 5 given in the Appendix. Use this information to select a sample of 40 countries from population 5 using the Neyman method of allocation. Record the production of tobacco crop from the selected countries. Merge the strata if requited. Estimate the average production of the tobacco crop in the world. Estimate the variance under the Neyman method of allocation. Construct a 95% confidence interval.

Practical 8.4. Owing to budget constraints the management of the World Bank has suggested a condition of spending money in different continents while selecting a sample. Keeping the instructions of the management in your mind, select a sample of 40 countries from population 5 given in the Appendix using the method of optimum allocation. Record the production of Tobacco crop from the selected countries. Estimate the average production of the Tobacco crop in the world. Estimate the variance under the method of Optimum allocation. Construct a 95% confidence interval.
Given: $C_1 = \$0.5$, $C_2 = 2$, $C_3 = 3$, $C_4 = 5$, $C_5 = 7$, $C_6 = 1.5$, $C_7 = 10$, $C_8 = 5$, $C_9 = 5$, and $C_{10} = 3$.

Practical 8.5. Find the minimum sample size from population 5 given in the Appendix to obtain estimates of population mean with different levels of relative standard deviations. Plot relative standard deviation versus sample size.
Given: $C_1 = \$0.5$, $C_2 = 2$, $C_3 = 3$, $C_4 = 5$, $C_5 = 7$, $C_6 = 1.5$, $C_7 = 10$, $C_8 = 5$, $C_9 = 5$, $C_{10} = 3$ and $\overline{Y} = 1.5507$.

Practical 8.6. Stratify the 50 states listed in population 1 given in the Appendix using the real estate farm loans as an auxiliary variable. Modify the stratum boundaries into six strata using cumulative square root method and cumulative cube root method.

Practical 8.7. A team of doctors wish to estimate the average death rate of persons living in the United States on the basis of year 2000 projections. The projected 191,443 persons are to be grouped into five strata on the basis of their age at the time of death. The rough distribution of the death rate projected for the year 2000 in the United States has been listed in population 6 of the Appendix in 21 age groups with a gap of four years. Apply the Neyman method of sample allocation for selecting the overall sample and the cumulative square root method to form 4 strata.

Practical 8.8. Suppose there are eight big cities in a particular country. The following table lists the number of persons and standard deviation of their income in each city with different age groups. We wish to allocate a sample of 1000 persons to collect information for social survey. Suggest the number of persons to be selected from each city with the help of Proportional, Neyman and Optimum allocation.

City	Age group	No. of persons	Standard Deviation	City	Age group	No. of persons	Standard Deviation
1	<24	8500	29.72	5	<24	9500	76.44
	25--34	29500	33.81		25--34	29500	45.89
	35--44	19500	19.33		35--44	32500	18.32
	45--54	4500	11.59		45--54	22500	26.58
	55--64	6500	21.18		55--64	20500	24.88
	65+	15500	21.84		65+	19500	19.14
2	<24	4500	64.39	6	<24	4500	44.63
	25--34	20500	40.89		25--34	27500	13.65
	35--44	27500	20.27		35--44	41500	18.58
	45--54	19500	40.02		45--54	26500	27.99
	55--64	21500	15.90		55--64	16500	12.44
	65+	21500	31.18		65+	23500	15.07
3	<24	6500	15.97	7	<24	5500	4.62
	25--34	12500	33.39		25--34	30500	11.02
	35--44	28500	18.95		35--44	33500	16.58
	45--54	29500	31.38		45--54	21500	23.91
	55--64	29500	22.44		55--64	14500	23.03
	65+	31500	9.85		65+	14500	21.82
4	<24	3500	33.05	8	<24	3500	36.94
	25--34	23500	30.58		25--34	20500	8.74
	35--44	26500	26.56		35--44	13500	12.25
	45--54	16500	66.59		45--54	9500	14.24
	55--64	19500	14.29		55--64	10500	22.09
	65+	20500	11.87		65+	16500	9.97

Which method you will prefer and why?

Given: Cost of processing one unit in the ith the city is $C_1 = \$2$, $C_2 = \$4$, $C_3 = \$2.5$, $C_4 = \$3$, $C_5 = \$4$, $C_6 = \$4.3$, $C_7 = \$2.8$ and $C_8 = \$3.2$.

Practical 8.9. Consider a couple in your class consisting of a husband and wife. Assuming that the life of every couple consists of good and bad events. Ask to both of them to prepare a separate list of good and bad events in their life. Estimate the proportion of good events in for each one of them. Also obtain a pooled estimate of good events among the families assuming that 50% weights to each of them and number of couples are infinitely large.

Practical 8.10. An estimate of the average death rate of persons living in the United States has been found to be useful for making future strategies. The projected 191443 persons are to be grouped into five strata on the basis of their age at the time of death. The rough distribution of the death rate projected for the year 2000 in the United States has been listed in population 6 of the Appendix in 21 age groups with a gap of four years. Apply the method of proportional allocation for selecting the overall sample required for estimation purpose and apply the cumulative cube root method to form 4 strata.

Practical 8.11. The estimation of total production of tobacco crop in the world is an important issue to the health departments. Select an SRSWOR sample of 45 countries from the population 5 given in the Appendix and record the production and area under the crop. Assume each continent as a different strata and number of countries in each continent are known. Post-stratify the countries selected in the SRSWOR sample into different continents. Merge the post-strata if required using total area under crop as an auxiliary variable. Obtain 95% confidence interval for the total production.

Practical 8.12. The following map shows rank of average temperature during December 2001 in different states of the United States of America.

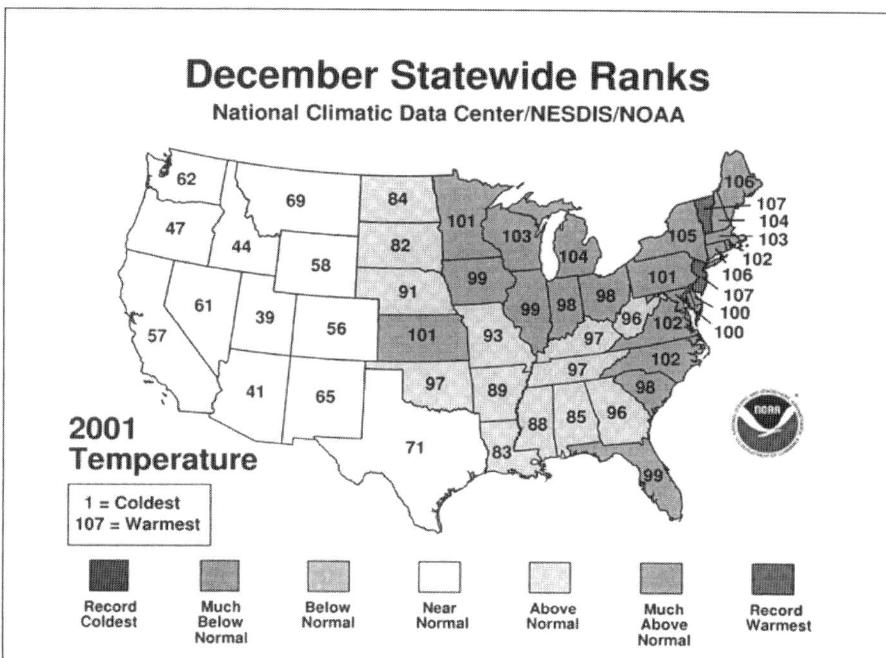

Source: Printed with permission from NOAA/National Climate Data Centre.

(1) SRSWOR sampling

(a) Count the number of states whose rank of temperature is shown in the map.
(b) Compute the average rank temperature in the USA.
(c) List the names of the states in alphabetic order A to Z (**Rule:** Use two letter abbreviations, *e.g.*, use NY for New York).
(d) Select an SRSWOR sample of 15 states (**Rule:** Start with the first two columns of the Pseudo-Random Number Table 1 given in the Appendix).
(e) Construct a 95% confidence interval estimate of the average rank temperature based on the SRSWOR sample information.
(f) Find population mean square S_y^2.
(g) Find the variance of the estimator of the average rank of temperature based on SRSWOR sample of 15 units.

(2) Stratified sampling: Equal allocation

(h) Divide the population into three strata based on information such that the stratum 1 consists of all states having temperature Near Normal, stratum 2 consists of all states having temperature Above Normal, and stratum 3 consists of all states having temperature Much Above Normal or Recorded Warmest. Construct three list of the states falling in different three strata in the alphabetic order A to Z. (Use two letter abbreviations, *e.g.*, NY for New York).
(i) Select sub-samples of five units from each stratum using SRSWOR sampling. (**Rule:** Always start with first row and first column from the Pseudo-Random Number Table 1 given in the Appendix, and make sure the states in each stratum are in alphabet order).
(j) Construct 95% confidence interval estimate of the average rank temperature based on stratified random sample of 15 units.
(k) Find the population mean squares for each one of the three strata S_{hy}^2, $h = 1, 2, 3$.
(l) Find the variance of the estimator of the average rank of temperature based on stratified sampling, while five units are selected from each stratum.
(m) Find the relative efficiency of the stratified sampling over the SRSWOR sampling.

(3) Stratified sampling: Proportional allocation

(n) Select a Stratified Random Sample of 15 units using proportional allocation. (**Rule:** Always start with first row and first column from the Pseudo-Random Number Table 1 given in the Appendix)
(o) Construct a 95% confidence interval estimate of the average rank of temperature.
(p) Find the variance of the estimator of the average rank of temperature using proportional allocation.

(q) Find the relative efficiency of proportional allocation with respect to SRSWOR sampling.
(r) Find the relative efficiency of proportional allocation with respect to equal allocation sampling.

(4) Stratified sampling: Neyman allocation

(s) Assuming that the population mean squares (S_{hy}^2, $h = 1, 2, 3.$) for the three strata are known, select a stratified random sample using Neyman allocation. (**Rule:** Always start with first row and first column from the Pseudo-Random Number Table 1 given in the Appendix, and make sure the states in each stratum are in alphabet order).
(t) Construct the 95% confidence interval estimate of the average rank of temperature.
(u) Find the variance of the estimator of the average rank of temperature using Neyman allocation.
(v) Find the relative efficiency of the Neyman allocation with respect to SRSWOR sampling.
(w) Find the relative efficiency of the Neyman allocation with respect to equal allocation sampling.

(x) Find the relative efficiency of the Neyman allocation with respect to the proportional allocation sampling.

(5) Stratified sampling: Optimum allocation

(y) Let $C_1 = \$4.0$, $C_2 = \$9.0$, and $C_2 = \$25.0$ be the costs of collecting information on a unit in the first, second, and third stratum, respectively. Again assuming that the population mean squares (S_{hy}^2, $h = 1, 2, 3$) for the three strata are known, select a stratified random sample of 15 units using Optimum Allocation. (**Rule:** Always start with first row and first column from the Pseudo-Random Number Table 1 given in the Appendix, and make sure states in each stratum are in alphabet order)
(z) Construct a 95% confidence interval estimate of the average rank of temperature.
(aa) Find the variance of the estimator of the average rank of temperature using Optimum Allocation.
(bb) Find the relative efficiency of the Optimum Allocation with respect to SRSWOR sampling.
(cc) Find the relative efficiency of the Optimum Allocation with respect to equal allocation sampling.
(dd) Find the relative efficiency of the Optimum Allocation with respect to proportional allocation.
(ee) Find the relative efficiency of the Optimum Allocation with respect to Neyman allocation.

(6) Post-stratified sampling

(ff) Collected information for each state included in the SRSWOR sample of 15 units from the whole population as in part (1) and note their status as: Near Normal, Above Normal etc.. (**Rule:** Start with first two columns of the Pseudo-Random Number Table 1 given in the Appendix, and make sure the states are listed in alphabetic order A to Z).
(gg) Post-stratify the sample of 15 units into three strata, *viz.*, Stratum 1 with Near Normal, Stratum 2 with Above Normal, and Stratum-3 with Much Above Normal or Recorded Warmest, temperatures. Is there any empty post-stratum? If so, suggest a possible solution to the issue.
(hh) Estimate the average rank temperature using post-stratification, and deduce 95% CI estimate.
(ii) Find the relative efficiency of the post-stratified sampling with respect to SRSWOR sampling.

(7) Collection of auxiliary information

Experience has shown that the statewide rank of temperature during Dec. 2001 has correlation with the rank of statewide precipitation during Dec. 2000 and a complete information on the precipitation is known. Collect the state wise information on the precipitation using the following map:

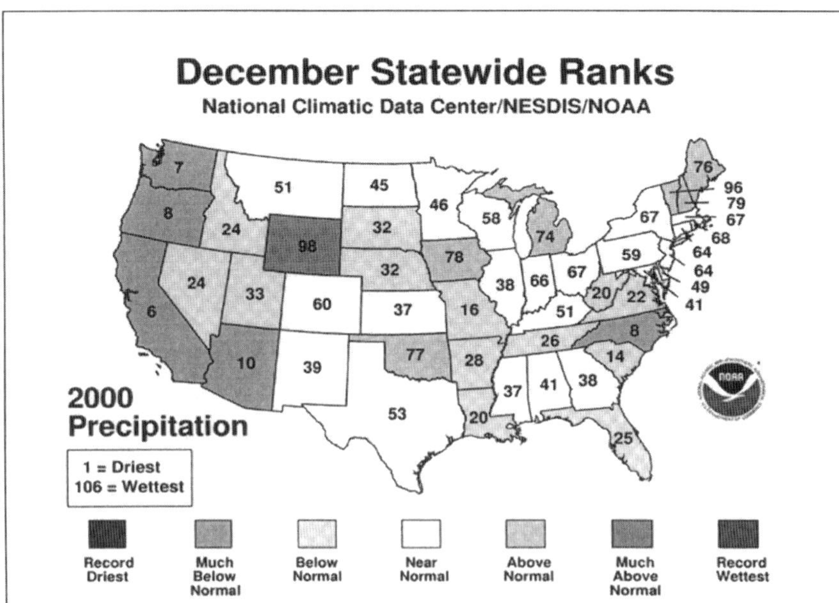

Source: Printed with permission from NOAA/National Climate Data Centre.

(8) Ratio estimator under SRSWOR sampling

(jj) Collected information on precipitation for each state included in the SRSWOR sample of 15 units from the whole population as in part (1). Also calculate the overall average precipitation in the USA based on information from the 48 states listed above.
(kk) Obtain the ratio estimate of average rank temperature.
(ll) Construct a 95% confidence interval estimate of average rank temperature using estimate of mean square error of the ratio estimator based of an SRSWOR sample of 15 units. Interpret it.
(mm) Find mean square error of the ratio estimator.
(nn) Find the relative efficiency of the ratio estimator with respect to the sample mean estimator.

(9) Stratified sampling: Separate ratio estimator

Use the same stratification as used in the previous section and collect information on the precipitation from each stratum. Assume that the information of precipitation in each stratum is known as auxiliary variable.

(oo) Construct a 95% confidence interval estimate of the average rank temperature using separate ratio estimate under equal allocation and proportional allocation. Discuss the comparison among each other.
(pp) Find the relative efficiency of the separate ratio estimator with respect to stratified sampling estimator under equal and proportional allocation.

(10) Stratified sampling: Combined ratio estimator

(qq) Construct a 95% confidence interval estimate of the average rank temperature using combined ratio estimate under equal allocation and proportional allocation. Discuss the interval estimates.
(rr) Find the relative efficiency of the combined ratio estimator with respect to stratified sampling estimator under equal and proportional allocation.

(11) Regression estimator under SRSWOR sampling

(ss) Obtain the regression estimate of average rank temperature from same information in (8).
(tt) Construct a 95% confidence interval estimate of average rank temperature using estimate of mean square error of the regression estimator based on an SRSWOR sample of 15 units. Interpret it.
(uu) Find mean square error of the regression estimator.
(vv) Find the relative efficiency of the regression estimator with respect to the sample mean estimator.
(ww) Find the relative efficiency of the regression estimator with respect to the ratio estimator.

(12) Stratified sampling: Separate regression estimator

(xx) Construct a 95% confidence interval estimate of the average rank temperature using separate regression estimate under equal allocation and proportional allocation. Discuss the interval estimates.
(yy) Find the relative efficiency of the separate regression estimator with respect to stratified sampling estimator under equal and proportional allocation.

(13) Stratified sampling: Combined regression estimator

(zz) Construct a 95% confidence interval estimate of the average rank temperature using combined regression estimate under equal allocation and proportional allocation. Discuss the interval estimates.
(aaa) Find the relative efficiency of the combined regression estimator with respect to stratified sampling estimator under equal and proportional allocation.

(14) Stratified sampling: Miscellaneous

(bbb) Select a sample of 15 units using Neyman allocation for separate ratio estimate and construct a 95% confidence interval estimate, and find its relative efficiency with respect to usual ratio estimator without stratification.
(ccc) Select a sample of 15 units using Neyman allocation for separate regression estimate and construct a 95% confidence interval estimate, and also find its relative efficiency with respect to usual regression estimator without stratification.
(ddd) Select a sample of 15 units using Neyman allocation for combined ratio estimate and construct a 95% confidence interval estimate, and find its relative efficiency with respect to usual ratio estimator without stratification.
(eee) Select a sample of 15 units using Neyman allocation for combined regression estimate and construct a 95% confidence interval estimate, and also find its relative efficiency with respect to usual regression estimator without stratification.
(fff) Select a sample of 15 units using optimum allocation with same costs defined earlier for separate ratio estimate and construct a 95% confidence interval estimate, and find its relative efficiency with respect to usual ratio estimator without stratification.
(ggg) Select a sample of 15 units using optimum allocation with same costs for separate regression estimate and construct a 95% confidence interval estimate, and also find its relative efficiency with respect to usual regression estimator without stratification.
(hhh) Select a sample of 15 units using optimum allocation with same costs for combined ratio estimate and construct a 95% confidence interval estimate, and find its relative efficiency with respect to usual ratio estimator without stratification.
(iii) Select a sample of 15 units using optimum allocation for combined regression estimate and construct 95% confidence interval estimate, and also find its relative efficiency with respect to usual regression estimator without stratification.
(jjj) Which estimator is best among all the above estimators based on stratification? Give your views.

(15) Post-stratified sampling: Miscellaneous

(kkk) Apply the separate ratio estimator for the post-stratification of sample of 15 units in (6) to construct 95% confidence interval estimate, and find its relative efficiency with respect to post-stratified estimator without using auxiliary information.
(lll) Apply the separate regression estimator for the post stratification of sample of 15 units in (6) to construct 95% confidence interval estimate, and find its relative efficiency with respect to post-stratified estimator without using auxiliary information.
(mmm) Apply the combined ratio estimator for the post stratification of sample of 15 units in (6) to construct 95% confidence interval estimate, and find its relative efficiency with respect to post-stratified estimator without using auxiliary information.
(nnn) Apply the combined regression estimator for the post stratification of sample of 15 units in (6) to construct 95% confidence interval estimate, and find its relative efficiency with respect to post stratified estimator without using auxiliary information.
(ooo) Which estimator is best among all the above estimators based on post-stratification? Give your views.

(16) Calibrated weights for stratified sampling: Miscellaneous

(ppp) Derive the calibrated weights using know precipitation rank and deduce the estimates of average rank temperature based on them using combined ratio, combined GREG and combined linear regression estimators.

Practical 8.13. Use SRSWOR sampling to select a sample of 40 countries from population 5 given in the Appendix by using the method of proportional allocation. Record the production and area under the tobacco crop from the countries selected in the sample. Assuming that the total area in each continent under the tobacco crop is known, estimate the variance of the estimator used for estimation purpose. Construct a 95% confidence interval in each case.

(a) Apply separate ratio estimator to estimate the average production of the tobacco crop in the world.
(b) Apply separate regression estimator to estimate the average production of the tobacco crop in the world.
(c) Apply combined ratio estimator to estimate the average production of the tobacco crop in the world.
(d) Apply the combined regression estimator to estimate the average production of the tobacco crop in the world.
(e) Using full information from the description of the population, find the relative efficiency of separate ratio estimator with respect to combined ratio estimator.

(f) Using full information from the description of the population, find the relative efficiency of separate regression estimator with respect to combined regression estimator.

Rules										
St. No.	1	2	3	4	5	6	7	8	9	10
N_h	6	6	8	10	12	4	30	17	10	3
n_h	3	3	3	3	4	2	11	6	3	2
PRN Col.	1	2	3	4 and 5	6 and 7	8	9 and 10	11 and 12	13 and 14	15

Practical 8.14. Consider you are a circus statistician, and in the circus there are three types of elephants, *viz.* Niko, Sambo, and Jumbo, which are of light, medium and heavy in weights. The following table lists the number of elephants in each category along with their average weights (kg) and population standard deviations.

	Niko $(h=1)$	Sambo $(h=2)$	Jumbo $(h=3)$
N_h	250	150	100
$\bar{Y}_h$ (wt.)	2000	3500	5000
S_{hy}	500	400	200

Give the following information to the owner of the circus.

(a) How many elephants, N, are available in the circus?

(b) What is the average weight, $\bar{Y}$, of all elephants in the circus?

(c) Compute the population mean square of the all elephants using the formula

$$S_y^2 = \frac{\sum_{h=1}^{L}\left\{(N_h-1)S_{hy}^2\right\} + \sum_{h=1}^{L}\left\{N_h\left(\bar{Y}_h-\bar{Y}\right)^2\right\}}{N-1}.$$

(d) If we select an SRSWOR sample of $n=40$ elephants from the whole circus, then what will be the variance of the estimator of the average weight of all elephants in the circus?

(e) How many elephants you will select from these strata using proportional allocation to get a sample of $n=40$ units?

(f) What will be variance of the estimator of the population mean under proportional allocation?

(g) What will be the relative efficiency of proportional allocation with respect to SRSWOR sampling?

(h) How many elephants you will select from these strata using the Neyman allocation to get a sample of $n=40$ units?

(i) What will be variance of the estimator of the population mean under the Neyman allocation?

(j) What will be the relative efficiency of the Neyman allocation with respect to SRSWOR sampling?
(k) What will be the relative efficiency of the Neyman allocation with respect to proportional allocation sampling?
(l) Let $C_1 = \$4$, $C_2 = \$9$, and $C_3 = \$25$ be the cost of weighing an elephant in the first, second, and third strata. Find the optimum allocation of a sample of $n = 40$ units over three strata.
(m) What will be variance of the estimator of the population mean under optimum allocation?
(n) What will be relative efficiency of optimum allocation with respect to SRSWOR sampling?
(o) What will be relative efficiency of optimum allocation with respect to proportional allocation?
(p) What will be relation efficiency of optimum allocation with respect to the Neyman allocation?

Practical 8.15. In a circus there are three types of elephants, *viz.*, Light, Medium, and Heavy in weight. The following table lists the number of elephants in each category along with their average weights (kg) and average weight of food (kg/day) along with other information.

	Light $(h=1)$	Medium $(h=2)$	Heavy $(h=3)$
N_h	250	150	100
$\bar{Y}_h$ (wt)	2000	3500	5000
$\bar{X}_h$ (food)	100	150	250
S_{hy}	500	400	200
S_{hx}	50	30	20
S_{hxy}	17500	8400	2800

(a) How many elephants, N, are available in the circus?
(b) What is the average weight, $\bar{Y}$, of all elephants in the circus?
(c) What is the average diet, $\bar{X}$, of all elephants in the circus?
(d) Compute the population parameters of the all elephants using the formulae

$$S_y^2 = \frac{\sum_{h=1}^{L}\left\{(N_h-1)S_{hy}^2\right\} + \sum_{h=1}^{L}\left\{N_h\left(\bar{Y}_h-\bar{Y}\right)^2\right\}}{N-1},$$

$$S_x^2 = \frac{\sum_{h=1}^{L}\left\{(N_h-1)S_{hx}^2\right\} + \sum_{h=1}^{L}\left\{N_h\left(\bar{X}_h-\bar{X}\right)^2\right\}}{N-1},$$

and

$$S_{xy} = \frac{\sum_{h=1}^{L}\left\{(N_h - 1)S_{hxy}\right\} + \sum_{h=1}^{L}\left\{N_h\left(\overline{X}_h - \overline{X}\right)\left(\overline{Y}_h - \overline{Y}\right)\right\}}{N-1}.$$

(e) If we select an SRSWOR sample of $n = 40$ elephants from the whole circus, then what will be the mean square error of the ratio estimator of the average weight of all elephants in the circus?
(f) How many elephants will you select from these strata using proportional allocation to get a sample of $n = 40$ units?
(g) What will be the mean square error of the separate ratio estimator of the population mean under proportional allocation?
(h) What will be the relative efficiency of the separate ratio estimator under proportional allocation with respect to (w.r.t.) usual ratio estimator under SRSWOR sampling?
(i) How many elephants will you select from these strata using Neyman allocation to get a sample of $n = 40$ units?
(j) What will be the mean square error of the separate ratio estimator of the average weight of elephants under Neyman allocation?
(k) What will be the relative efficiency of the separate ratio estimator under the Neyman allocation with respect to the usual ratio estimator under SRSWOR sampling?
(l) What will be the relative efficiency of the separate ratio estimator under the Neyman allocation with respect to the separate ratio estimator under proportional allocation ?
(m) Let $C_1 = \$4$, $C_2 = \$9,$ and $C_3 = \$25$ be the cost of weighing an elephant in the first, second, and third strata. Find the optimum allocation of a sample of $n = 40$ units over three strata.
(n) What will be variance of the separate ratio estimator of the average weight of the elephants under optimum allocation?
(o) What will be relative efficiency of the separate ratio estimator under optimum allocation with respect to the usual ratio estimator under SRSWOR sampling?
(p) What will be relative efficiency of the separate ratio estimator under optimum allocation with respect to separate ratio estimator under proportional allocation?
(q) What will be relative efficiency of the separate ratio estimator under optimum allocation with respect to the separate ratio estimator under the Neyman allocation?
(r) What will be mean square error of the combined ratio estimator of the population mean under proportional allocation?
(s) What will be the relative efficiency of the combined ratio estimator under proportional allocation with respect to the usual ratio estimator under SRSWOR sampling?
(t) What will be the mean square error of the combined ratio estimator of the average weight of elephants under the Neyman allocation?
(u) What is the relative efficiency of the combined ratio estimator under the Neyman allocation with respect to the usual ratio estimator under SRSWOR sampling?

(v) What is the relative efficiency of the combined ratio estimator under the Neyman allocation with respect to the combined ratio estimator under proportional allocation?
(w) What will be variance of the combined ratio estimator of the average weight of the elephants under optimum allocation?
(x) What will be the relative efficiency of the combined ratio estimator under optimum allocation with respect to the usual ratio estimator under SRSWOR sampling?
(y) What will be the relative efficiency of the combined ratio estimator under optimum allocation with respect to combined ratio estimator under proportional allocation?
(z) What will be the relative efficiency of the combined estimator under optimum allocation with respect to the combined ratio estimator under the Neyman allocation?
(aa) If we select an SRSWOR sample of $n = 40$ elephants from the whole circus, then what will be the mean square error of the regression estimator of the average weight of all elephants in the circus?
(bb) What will be the mean square error of the separate regression estimator of the population mean under proportional allocation?
(cc) What will be the relative efficiency of the separate regression estimator under proportional allocation with respect to the usual regression estimator under SRSWOR sampling?
(dd) What will be the mean square error of the separate regression estimator of the average weight of elephants under Neyman allocation?
(ee) What will be the relative efficiency of the separate regression estimator under the Neyman allocation with respect to the usual regression estimator under SRSWOR sampling?
(ff) What is the relative efficiency of the separate regression estimator under the Neyman allocation with respect to the separate regression estimator for proportional allocation?
(gg) Let $C_1 = \$4$, $C_2 = \$9$, and $C_3 = \$25$ be the cost of weighing an elephant in the first, second, and third strata. Find the optimum allocation of a sample of $n = 40$ units over three strata.
(hh) What will be the variance of the separate regression estimator of the average weight of the elephants under optimum allocation?
(ii) What will be the relative efficiency of the separate regression estimator under optimum allocation with respect to the usual regression estimator under SRSWOR sampling?
(jj) What will be the relative efficiency of the separate regression estimator under optimum allocation with respect to the separate regression estimator under proportional allocation?
(kk) What will be relative efficiency of the separate regression estimator under optimum allocation with respect to the separate regression estimator for Neyman allocation?
(ll) What will be mean square error of the combined regression estimator of the average weight of elephants under proportional allocation?

(mm) What is the relative efficiency of the combined regression estimator under proportional allocation with respect to the usual regression estimator under SRSWOR sampling?
(nn) What will be mean square error of the combined regression estimator of the average weight of elephants under the Neyman allocation?
(oo) What will be the relative efficiency of the combined regression estimator under the Neyman allocation with respect to the regression estimator under SRSWOR sampling?
(pp) What will be the relative efficiency of the combined regression estimator under the Neyman allocation with respect to the combined regression estimator under proportional allocation?
(qq) What will be variance of the combined regression estimator of the average weight of the elephants under optimum allocation?
(rr) What will be relative efficiency of the combined regression estimator under optimum allocation with respect to the usual regression estimator under SRSWOR sampling?
(ss) What will be relative efficiency of the combined regression estimator under optimum allocation with respect to the combined regression estimator under proportional allocation?
(tt) What will be the relative efficiency of the combined regression estimator under optimum allocation with respect to the combined regression estimator under Neyman allocation?

Practical 8.16. In a circus there are three types of elephants, *viz.*, Light, Medium, and Heavy in weight and some information about them is listed below:

	Light $(h=1)$	Medium $(h=2)$	Heavy $(h=3)$
N_h	250	150	100
n_h	20	12	8
$\bar{y}_h$ (wt., kg)	1800	3200	5200
$\bar{x}_h$ (food, kg/day)	95	155	244.8
$\bar{X}_h$ (food, kg/day)	100	150	250
s_{hy}	490	410	220
s_{hx}	45	25	30
s_{hxy}	17200	8000	5500

(a) How many elephants, N, are available in the circus?

(b) Select an SRSWOR sample of 40 elephants using proportional allocation.

(c) Estimate the average weight, $\bar{Y}$, of all elephants in the circus using the usual estimator in stratified sampling

$$\bar{y}_{st} = \sum_{h=1}^{L} W_h \bar{y}_h$$

and derive a 95% confidence interval estimate.

(d) Assuming that the average food $\bar{X}_h$, $h = 1, 2, 3$ are known, estimate the average weight, $\bar{Y}$, of all elephants in the circus using separate ratio estimator in stratified sampling and derive a 95% confidence interval estimate.

(e) Assuming that the average food $\bar{X}$ is known, estimate the average weight , $\bar{Y}$, of all elephants in the circus using combined ratio estimator in stratified sampling and derive a 95% confidence interval estimate.

(f) Assuming that the average food $\bar{X}_h$, $h = 1, 2, 3$ are known, estimate the average weight, $\bar{Y}$, of all elephants in the circus using separate regression estimator in stratified sampling and derive a 95% confidence interval estimate.

(g) Assuming that the average food $\bar{X}$ is known, estimate the average weight, $\bar{Y}$, of all elephants in the circus using combined regression estimator in stratified sampling and derive a 95% confidence interval estimate.

(h) Assume that the sample of $n = 40$ elephants is an SRSWOR sample across the whole circus, and then find the following

$$\bar{y} = \frac{n_1\bar{y}_1 + n_2\bar{y}_2 + n_3\bar{y}_3}{n}, \quad \bar{x} = \frac{n_1\bar{x}_1 + n_2\bar{x}_2 + n_3\bar{x}_3}{n},$$

$$s_y^2 = \frac{\sum_{h=1}^{L}\left\{(n_h - 1)s_{hy}^2\right\} + \sum_{h=1}^{L}\left\{n_h\left(\bar{y}_h - \bar{y}\right)^2\right\}}{n-1}, \quad s_x^2 = \frac{\sum_{h=1}^{L}\left\{(n_h - 1)s_{hx}^2\right\} + \sum_{h=1}^{L}\left\{n_h\left(\bar{x}_h - \bar{x}\right)^2\right\}}{n-1},$$

and

$$s_{xy} = \frac{\sum_{h=1}^{L}\left\{(n_h - 1)s_{hxy}\right\} + \sum_{h=1}^{L}\left\{n_h\left(\bar{x}_h - \bar{x}\right)\left(\bar{y}_h - \bar{y}\right)\right\}}{n-1}.$$

(i) Apply the usual ratio estimator

$$\bar{y}_R = \bar{y}\left(\frac{\bar{X}}{\bar{x}}\right)$$

to estimate the average weight of the elephants and derive 95% CI estimate.

(j) Apply the usual regression estimator

$$\bar{y}_{LR} = \bar{y} + \hat{\beta}\left(\bar{X} - \bar{x}\right)$$

to estimate the average weight of the elephants and derive 95% CI estimate.

(k) Compare all your confidence interval estimates and find which one is better and why?.

Practical 8.17. Michael believes that when a soul is born it automatically falls into one of the religious categories, say, Sikh, Hindu, Muslim, Christian etc., existing in the world. Is it an example of stratification or post-stratification? Comment.

Practical 8.18. Ms. Stephanie Singh selects a sample of 50 countries from population 5 given in the Appendix using the method of proportional allocation and records the production and area of the tobacco crop from the countries selected in the sample. Assume that the area under the tobacco crop in different continents is known. Obtain the calibration weights which result in (a) combined ratio, (b) combined GREG, (c) combined linear regression estimators. Deduce the values of the estimates.

Practical 8.19. Divide your class into two groups based on gender. Decide to select a sample of reasonable size with the suggestion of the class instructor. Select a sample of the required size using proportional allocation, and collect information on the GPA of the students selected in the sample from the both strata.
(a) Estimate the average GPA of the class using the usual formula in stratified sampling. Construct the 95% confidence interval estimate.
(b) Collect information about the number of classes attended by the students from the register of the instructor if he/she permits. Use this information to improve your estimates in (a) and comment.

9. NON-OVERLAPPING, OVERLAPPING, POST, AND ADAPTIVE CLUSTER SAMPLING

9.0 INTRODUCTION

In survey sampling the basic assumption is that the population consists of a finite number of distinct and identifiable units. A group of such units is called a *cluster*. If, instead of randomly selecting a unit for sample, a group of units is selected as a single unit in the sample, it is called cluster sampling. If the entire area containing the population under study is divided into smaller segments, and if each unit of the population belongs to only one segment, the procedure is called area sampling or non-overlapping cluster sampling. If one or a few units appears in more than one segment or cluster, then such a procedure is called overlapping cluster sampling. The main purpose of cluster sampling is to divide the population into small groups with each group serving as a sample unit. Clusters are generally made up of neighbouring elements; therefore the elements within a cluster tend to be homogeneous. However at some stage in the research we become interested in heterogeneous clusters rather than homogeneous. More broadly, the concept of forming strata in the previous chapter was to form homogeneous groups, whereas in this chapter the concept of forming clusters will be to form groups of a heterogeneous nature. After dividing the population into clusters the sample of clusters can be selected with either equal or unequal probability. The concept of unequal probability may be based on the size of the cluster; that is, the larger the cluster, the larger the probability of its being selected in the sample. All the units in the selected cluster will be enumerated. As a simple rule the number of units in a cluster should be small and the number of clusters should be large. The main advantage of cluster sampling is that it is cheaper, since the collection of data for neighbouring units is easier and faster. It is also useful when the frame for selecting the sample is not available at the unit level. For example, a list of persons may not be available, whereas a list at a dwelling level may be available. For a given sample size cluster sampling is less efficient than simple random sampling. However, in most situations the loss in efficiency can be balanced by the reduction in cost. Any sampling procedure, for example simple random sampling, stratified sampling, or systematic sampling, may be applied to cluster sampling by using the clusters themselves as sampling units. Smith (1938) and Hansen and Hurwitz (1942) have discussed the efficiency of cluster sampling.

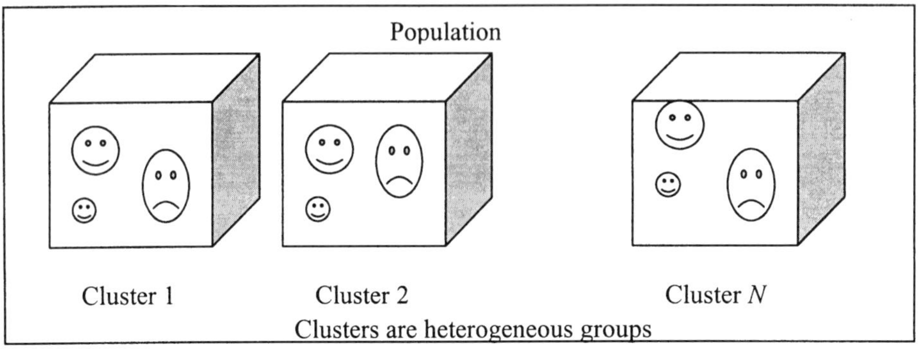

Fig. 9.0.1. Pictorial representation of cluster sampling.

The well known examples of the clusters and their units are given below:

Cluster	School	Dwelling	Employer	Bus	Day
Elements/units in the cluster	Students	Persons	Employee	Passengers	Hours

There are two possibilities: (a) clusters of equal size or (b) clusters of different sizes. It is obvious that if every cluster has the same number of units, then the chance of selection of each cluster in the sample will be the same. On the other hand, if the number of units in the clusters is different and known, then different probabilities proportional to the number of units in the clusters can be assigned before taking the sample. We will discuss both these situations in detail below.

9.1 NON-OVERLAPPING CLUSTERS OF EQUAL SIZE

Let N denote the number of clusters in the population and M the number of units in each cluster. Evidently, the total number of units in the population is NM. We select a sample of n clusters and hence we select nM units from the population. Let us define

Y_{ij} = The value of the study variable y for the jth population element $(j = 1, 2, ..., M)$ for the ith cluster $(i = 1, 2, ..., N)$,

$Y_{i\bullet} = \sum_{j=1}^{M} Y_{ij}$, the ith cluster total for the population y values,

$Y_{\bullet\bullet} = \sum_{i=1}^{N}\sum_{j=1}^{M} Y_{ij}$, the grand total or population total,

$\bar{\bar{Y}} = \frac{1}{NM}\sum_{i=1}^{N}\sum_{j=1}^{M} Y_{ij} = \frac{Y_{\bullet\bullet}}{NM}$, the population mean which we want to estimate from the sample information,

$\bar{Y} = Y_{\bullet\bullet}/N$, population mean per cluster,

$\bar{Y}_{i\bullet} = \dfrac{Y_{i\bullet}}{M} = \dfrac{1}{M}\sum_{j=1}^{M} Y_{ij}$, mean of M population elements in the ith cluster,

$y_{\bullet\bullet} = \sum_{i=1}^{n}\sum_{j=1}^{M} y_{ij}$, the sample total of the study variable y,

$\bar{y} = y_{\bullet\bullet}/n$, sample mean per cluster,

$\bar{\bar{y}} = \dfrac{1}{nM}\sum_{i=1}^{n}\sum_{j=1}^{M} y_{ij} = \dfrac{\bar{y}}{M}$, sample mean per unit,

$S_b^2 = \dfrac{1}{N-1}\sum_{i=1}^{N}\left(\bar{Y}_{i\bullet} - \bar{\bar{Y}}\right)^2$, the variance between the means of different clusters,

$S_i^2 = \dfrac{1}{M-1}\sum_{j=1}^{M}\left(Y_{ij} - \bar{Y}_{i\bullet}\right)^2$, the variance within the ith cluster,

$\bar{s}_w^2 = \dfrac{1}{N}\sum_{i=1}^{N} S_i^2 = \dfrac{1}{N(M-1)}\sum_{i=1}^{N}\sum_{j=1}^{M}\left(Y_{ij} - \bar{Y}_{i\bullet}\right)^2$,mean value of within cluster variances,

$S^2 = \dfrac{1}{NM-1}\sum_{i=1}^{N}\sum_{j=1}^{M}\left(Y_{ij} - \bar{\bar{Y}}\right)^2$, the total population variance,

and

$\rho = \dfrac{\sum_{i=1}^{N}\sum_{j\neq j'=1}^{M}\left(Y_{ij} - \bar{\bar{Y}}\right)\left(Y_{ij'} - \bar{\bar{Y}}\right)}{NM(M-1)S^2}$, the intra-cluster correlation coefficient.

Then we have the following theorems:

Theorem 9.1.1. The sample mean per unit

$$\bar{\bar{y}} = \frac{1}{nM}\sum_{i=1}^{n}\sum_{j=1}^{M} y_{ij} \tag{9.1.1}$$

is unbiased for population mean $\bar{\bar{Y}}$.

Proof. We have

$$E\left(\bar{\bar{y}}\right) = E\left(\frac{1}{nM}\sum_{i=1}^{n}\sum_{j=1}^{M} y_{ij}\right) = \frac{1}{M}E\left(\frac{1}{n}\sum_{i=1}^{n} y_{i\bullet}\right) = \frac{1}{M}\left(\frac{1}{N}\sum_{i=1}^{N} Y_{i\bullet}\right) = \frac{\bar{Y}}{M} = \bar{\bar{Y}}.$$

Hence the theorem.

Theorem 9.1.2. The variance of the estimator $\bar{\bar{y}}$ is given by

$$V\left(\bar{\bar{y}}\right) = \left(\frac{1}{n} - \frac{1}{N}\right)S_b^2. \tag{9.1.2}$$

Proof. We have

$$V\left(\bar{\bar{y}}\right) = V\left[\frac{1}{n}\sum_{i=1}^{n}\bar{y}_{i\bullet}\right] = \left(\frac{1}{n} - \frac{1}{N}\right)\frac{1}{(N-1)}\sum_{i=1}^{N}\left(\bar{Y}_{i\bullet} - \bar{\bar{Y}}\right)^2 = \left(\frac{1}{n} - \frac{1}{N}\right)S_b^2 .$$

Hence the theorem.

Theorem 9.1.3. An unbiased estimator of $V\left(\bar{\bar{y}}\right)$ is given by

$$\hat{v}\left(\bar{\bar{y}}\right)=\left(\frac{1}{n}-\frac{1}{N}\right)s_b^2 \tag{9.1.3}$$

where $s_b^2=(n-1)^{-1}\sum_{i=1}^{n}\left(\bar{y}_{i\bullet}-\bar{\bar{y}}\right)^2$.

Proof. Obvious by using $E\left(s_b^2\right)=S_b^2$.

Remark 9.1.1. A $(1-\alpha)100\%$ confidence interval estimate of the population mean $\bar{\bar{Y}}$ will be given by

$$\bar{\bar{y}} \pm t_{\alpha/2}(\text{df}=n(M-1))\sqrt{\hat{v}\left(\bar{\bar{y}}\right)}\ .$$

Example 9.1.1. Using the MAP of the United States of America construct 10 clusters, from the 50 states, each cluster consisting of 5 states.

Solution. There are several possible ways to construct 10 clusters based on their locations. The number of possibilities may be decreased if we have some additional information, for example, distance to be travelled between states to collect data. In such a situation, we may decide to minimise the distance to be travelled by the data collectors.

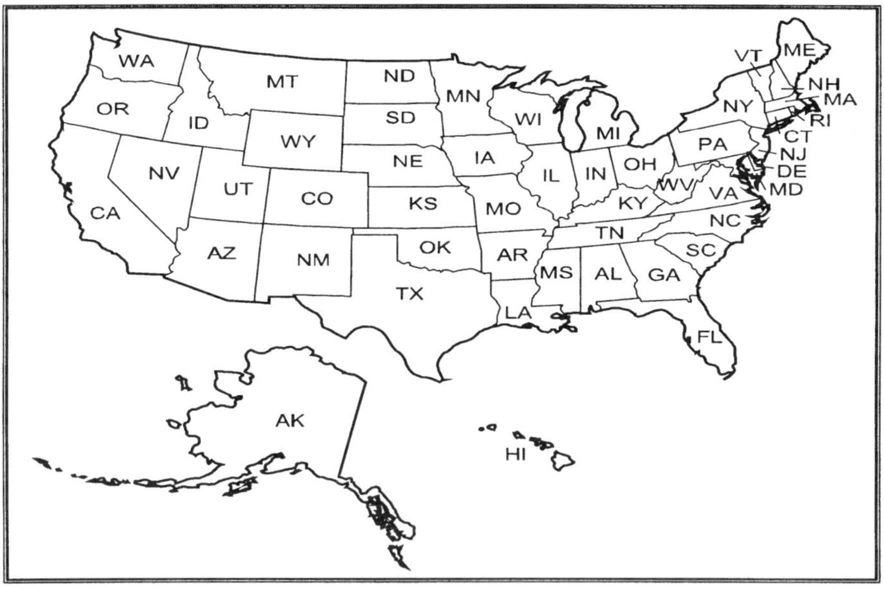

Looking at the MAP of the USA we arrive at the following situation.

Table 9.1.1. Clusters of equal size.

Cluster Number	Names of the states
1	WA, OR, MT, NV, ID
2	CA, AZ, NM, AK, HI
3	UT, WY, CO, SD, NE
4	TX, OK, KS, LA, AR
5	MO, IA, MN, ND, WI
6	IL, IN, MI, OH, PA
7	MS, AL, GA, FL, SC
8	TN, KY, WV, VA, NC
9	NY, VT, NH, ME, MA
10	RI, CT, NJ, DE, MD

Example 9.1.2. Select 6 clusters from Table 9.1.1 using SRSWOR sampling. Record the values of the real estate farm loans for the selected states from population 1 given in the Appendix. Estimate the average real estate farm loans in the United States using the cluster sampling. Estimate the variance of the estimator used for estimating the average real estate farm loans. Construct a 95% confidence interval.

Solution. Starting with the first two columns of the Pseudo-Random Number (PRN) Table 1 given in the Appendix, we select the required 6 distinct random numbers as 01, 04, 05, 03, 06 and 07. Thus the following clusters are included in an SRSWOR sample of 6 units.

PRN	Names of the states	Values of the real estate farm loans y_{ij}	$y_{i\bullet}$
01	WA, OR, MT, NV, ID	1100.745, 114.899, 292.965, 5.860, 53.753	1568.222
03	UT, WY, CO, SD, NE	56.908, 100.964, 315.809, 413.777, 1337.852	2225.310
04	TX, OK, KS, LA, AR	1248.761, 612.108, 1049.834, 282.565, 907.700	4100.968
05	MO, IA, MN, ND, WI	1579.686, 2327.025, 1354.768, 449.099, 1229.752	6940.330
06	IL, IN, MI, OH, PA	2131.048, 1213.024, 323.028, 870.720, 756.169	5293.989
07	MS, AL, GA, FL, SC	627.013, 408.978, 939.460, 825.748, 87.951	2889.150
		Sum	23017.970

Note that $M = 5$ and $n = 6$. Thus an estimate of the average real estate farm loans in the United States is given by

$$\bar{\bar{y}} = \frac{1}{nM}\sum_{i=1}^{n}\sum_{j=1}^{M} y_{ij} = \frac{1}{6\times 5}\times 23017.97 = 767.2656.$$

Now we have

Cluster Number	$\bar{y}_{i\bullet}$	$\left(\bar{y}_{i\bullet} - \bar{\bar{y}}\right)^2$
1	313.644	205772.56
2	445.062	103815.16
3	820.193	2801.31
4	1388.066	385393.14
5	1058.798	84991.14
6	577.830	35885.85
	Sum	818659.16

Thus we have

$$s_b^2 = \frac{\sum_{i=1}^{n}\left(\bar{y}_{i\bullet} - \bar{\bar{y}}\right)^2}{(n-1)} = \frac{818659.16}{6-1} = 163731.832\,.$$

Thus an estimator of $V\left(\bar{\bar{y}}\right)$ is given by

$$\hat{v}\left(\bar{\bar{y}}\right) = \left(\frac{1}{n} - \frac{1}{N}\right)s_b^2 = \left(\frac{1}{6} - \frac{1}{10}\right)\times 163731.832 = 10915.455\,.$$

A $(1-\alpha)100\%$ confidence interval of the population mean $\bar{\bar{Y}}$ is given by

$$\bar{\bar{y}} \mp t_{\alpha/2}(\text{df} = n(M-1))\sqrt{\hat{v}\left(\bar{\bar{y}}\right)}\,.$$

Thus 95% confidence interval for the average real estate farm loans in the United States is given by

$$\bar{\bar{y}} \mp t_{0.025}(\text{df} = 6\times 4)\sqrt{\hat{v}\left(\bar{\bar{y}}\right)} \quad \text{or} \quad \bar{\bar{y}} \mp t_{0.025}(\text{df} = 24)\sqrt{\hat{v}\left(\bar{\bar{y}}\right)}$$

$$\text{or } 767.2656 \mp 2.064\sqrt{10915.45}\,, \quad \text{or} \quad [551.62,\ 982.906]\,.$$

Theorem 9.1.4. The relative efficiency of cluster sampling with respect to SRS is

$$\text{RE} = S^2/\left(MS_b^2\right). \tag{9.1.4}$$

Proof. We know that the variance of estimator $\bar{\bar{y}}$ under cluster sampling is

$$V\left(\bar{\bar{y}}\right) = \left(\frac{1}{n} - \frac{1}{N}\right)S_b^2 \tag{9.1.5}$$

and that of $\bar{y}$ for the SRSWOR estimator is

$$V(\bar{y}) = \left(\frac{1}{nM} - \frac{1}{MN}\right)S^2 = \frac{1}{M}\left(\frac{1}{n} - \frac{1}{N}\right)S^2\,. \tag{9.1.6}$$

The relative efficiency of the estimator in cluster sampling with respect to the one in SRSWOR sampling is

$$\text{RE} = V(\bar{y})/V\left(\bar{\bar{y}}\right) = \left(\frac{1}{nM} - \frac{1}{NM}\right)S^2 \Big/ \left(\frac{1}{n} - \frac{1}{N}\right)S_b^2 = S^2/\left(MS_b^2\right). \tag{9.1.7}$$

Hence the theorem.

Example 9.1.3. Suppose the United States has been divided into 10 neighbouring clusters each consisting of 5 states as shown below.

Cluster Number	Names of the states
1	WA, OR, MT, NV, ID
2	CA, AZ, NM, AK, HI
3	UT, WY, CO, SD, NE
4	TX, OK, KS, LA, AR
5	MO, IA, MN, ND, WI
6	IL, IN, MI, OH, PA
7	MS, AL, GA, FL, SC
8	TN, KY,WV, VA, NC
9	NY, VT, NH, ME, MA
10	RI, CT, NJ, DE, MD

We are interested in estimating the average real estate farm loans in the United States. Is there any gain in efficiency due to clustering as opposed to simple random sampling?

Solution. We have

Cluster No	Values of the real estate farm loans Y_{ij}					$\overline{Y}_{i\bullet}$	$\left(\overline{Y}_{i\bullet}-\overline{\overline{Y}}\right)^2$
1	1100.745	114.899	292.965	5.860	53.753	313.644	58462.46
2	1343.461	54.633	140.582	2.605	40.775	316.411	57132.15
3	56.908	100.964	315.809	413.777	1337.852	445.062	12182.09
4	1248.761	612.108	1049.834	282.565	907.700	820.193	70097.37
5	1579.686	2327.025	1354.768	449.099	1229.752	1388.066	693275.20
6	2131.048	1213.024	323.028	870.720	756.169	1058.798	253374.60
7	627.013	408.978	939.460	825.748	87.951	577.830	501.56
8	553.266	1045.106	99.277	321.583	639.571	531.760	560.45
9	201.631	57.747	6.044	8.849	7.590	56.372	249063.20
10	1.611	7.130	39.860	42.808	139.628	46.207	259312.30
					Sum	5554.345	1653961.00

From the above table we have

$$\overline{\overline{Y}} = \frac{5554.345}{10} = 555.4345 .$$

Thus we have

$$s_b^2 = (N-1)^{-1}\sum_{i=1}^{N}\left(\overline{Y}_{i\bullet}-\overline{\overline{Y}}\right)^2 = \frac{1653961}{10-1} = 183773.444 .$$

Also we have

$$S^2 = 342021.5 .$$

The relative efficiency of the estimator in cluster sampling with respect to the one in SRSWOR sampling is given by

$$\text{RE} = S^2/\left(MS_b^2\right) = \frac{342021.5}{5\times 183773.444} = 0.3722 .$$

In this case, cluster sampling is less efficient than simple random sampling. Thus there is no gain in efficiency through cluster sampling in this example. Let us try to find the reason.

Theorem 9.1.5. The relative efficiency at (9.1.7) in terms of intraclass correlation coefficient ρ is given by

$$\text{RE} = [1+(M-1)\rho]^{-1} . \tag{9.1.8}$$

Thus cluster sampling is more efficient than SRS if $\rho < 0$.

Proof. To find relative efficiency in terms of intraclass correlation coefficient ρ, let us first express

$$\sum_{i=1}^{N}\sum_{j\neq j'=1}^{M}\left(Y_{ij}-\overline{\overline{Y}}\right)\left(Y_{ij'}-\overline{\overline{Y}}\right)$$

in terms of S^2 and S_b^2. To do so we have

$$\sum_{i=1}^{N}\left[\sum_{j=1}^{M}\left(Y_{ij}-\overline{\overline{Y}}\right)\right]^2 = \sum_{i=1}^{N}\left[\sum_{j=1}^{M}Y_{ij}-\sum_{j=1}^{M}\overline{\overline{Y}}\right]^2 = \sum_{i=1}^{N}\left(Y_{i\bullet}-M\overline{\overline{Y}}\right)^2 = \sum_{i=1}^{N}\left(M\overline{Y}_{i\bullet}-M\overline{\overline{Y}}\right)^2$$

$$= M^2\sum_{i=1}^{N}\left(\overline{Y}_{i\bullet}-\overline{\overline{Y}}\right)^2 = M^2(N-1)S_b^2 . \tag{9.1.9}$$

Also we have

$$\sum_{i=1}^{N}\left[\sum_{j=1}^{M}\left(Y_{ij}-\overline{\overline{Y}}\right)\right]^2 = \sum_{i=1}^{N}\left[\sum_{j=1}^{M}\left(Y_{ij}-\overline{\overline{Y}}\right)^2+\sum_{j\neq j'=1}^{M}\left(Y_{ij}-\overline{\overline{Y}}\right)\left(Y_{ij'}-\overline{\overline{Y}}\right)\right]$$

$$= \sum_{i=1}^{N}\sum_{j=1}^{M}\left(Y_{ij}-\overline{\overline{Y}}\right)^2+\sum_{i=1}^{N}\sum_{j\neq j'=1}^{M}\left(Y_{ij}-\overline{\overline{Y}}\right)\left(Y_{ij'}-\overline{\overline{Y}}\right). \tag{9.1.10}$$

Now

$$S^2 = \frac{1}{NM}\sum_{i=1}^{N}\sum_{j=1}^{M}\left(Y_{ij}-\overline{\overline{Y}}\right)^2 \quad \text{and} \quad \rho = \sum_{i=1}^{N}\sum_{j\neq j'=1}^{M}\left(Y_{ij}-\overline{\overline{Y}}\right)\left(Y_{ij'}-\overline{\overline{Y}}\right)\Big/\left\{NM(M-1)S^2\right\}$$

which implies that

$$\sum_{i=1}^{N}\left[\sum_{j=1}^{M}\left(Y_{ij}-\overline{\overline{Y}}\right)\right]^2 = (NM-1)S^2+NM(M-1)\rho S^2. \tag{9.1.11}$$

On comparing (9.1.9) and (9.1.11) we have

$$(NM-1)S^2+NM(M-1)\rho S^2 = M^2(N-1)S_b^2$$

$$\text{or } S^2/\left\{MS_b^2\right\} = M(N-1)/\{(NM-1)+NM(M-1)\rho\} .$$

Note that M and 1 both may be neglected in comparison to NM the term relative efficiency becomes

$$\text{RE} = \frac{S^2}{MS_b^2} = \frac{M(N-1)}{(NM-1)+NM(M-1)\rho}$$

$$= \frac{1}{1+(M-1)\rho} = [1+(M-1)\rho]^{-1} \tag{9.1.12}$$

which proves the first part of the theorem.

To prove the second part we have

$$\text{RE} > 1$$

which implies that

$$[1+(M-1)\rho]^{-1} > 1, \text{ or } 1+(M-1)\rho < 1 \tag{9.1.13}$$

which is possible only if

$$\rho < 0.$$

Hence the theorem.

Remark 9.1.1. Cluster sampling is more efficient than SRS if intraclass correlation coefficient $\rho < 0$.

Remark 9.1.2. If $\rho = 0$, then the cluster sampling and SRS are equally efficient.

Remark 9.1.3. In practice units which are near one another are more similar than units which are apart, therefore ρ is positive and hence in general the efficiency of cluster sampling is less than that of SRS. We will observe later on that cluster sampling is more efficient than SRS for the fixed cost of a given survey.

Example 9.1.4. In Example 9.1.3, find the value of the intraclass correlation coefficient. Use it to find the relative efficiency of cluster sampling over simple random sampling.

Solution. Using information from Example 9.1.3 we have

Cluster Number	$\left(Y_{ij} - \overline{\overline{Y}}\right)$, $j = 1,2,...,5$				
1	545.3105	-440.5360	-262.4700	-549.5750	-501.6820
2	788.0265	-500.8020	-414.8530	-552.8300	-514.6600
3	-498.5270	-454.4710	-239.6260	-141.6580	782.4175
4	693.3265	56.6735	494.3995	-272.8700	352.2655
5	1024.2510	1771.5900	799.3335	-106.3360	674.3175
6	1575.6130	657.5895	-232.4070	315.2855	200.7345
7	71.5785	-146.4570	384.0255	270.3135	-467.4840
8	-2.1685	489.6715	-456.1580	-233.8520	84.1365
9	-353.8040	-497.6880	-549.3910	-546.5860	-547.8450
10	-553.8240	-548.3050	-515.5750	-512.6270	-415.8070

For computing the intra-class correlation coefficient we have

Cluster No. i	$\left(Y_{ij}-\overline{\overline{Y}}\right)\left(Y_{ij'}-\overline{\overline{Y}}\right)$, $j<j'$			
	$j=2$	$j=3$	$j=4$	$j=5$
1	-240229.00	-143127.00	-299689.00	-273572.00
		115627.10	242107.10	221008.50
			144246.60	131676.10
				275711.40
2	-394645.00	-326915.00	-435644.00	-405565.00
		207758.80	276857.90	257742.30
			229342.70	213507.80
				-46239.30
3	226565.60	119459.70	70620.03	-390056.00
		108902.70	64379.17	-355586.00
			33944.76	-187487.00
				-110835.00
4	39293.22	342780.30	-189188.00	244235.00
		28019.34	-15464.50	19964.11
			-134907.00	174159.90
				-96122.50
5	1814554.00	818718.50	-108914.00	690670.70
		1416092.00	-188383.00	1194614.00
			-84997.50	539004.50
				-71703.90
6	1036107.00	-366183.00	496768.10	316280.00
		-152828.00	207328.40	132000.90
			-73274.40	-46652.00
				63288.67
7	-10483.10	27487.96	19348.63	-33461.80
		-56243.00	-39589.20	68466.01
			103807.30	-179526.00
				-126367.00
8	-1061.86	989.19	507.11	-182.45
		-223367.00	-114510.00	41199.23
			106673.10	-38379.50
				-19675.40
9	176083.60	194376.30	193383.90	193829.30
		273424.80	272028.80	272655.40
			300288.90	300980.60
				299443.90
10	303663.90	285537.30	283904.60	230283.40
		282691.80	281075.40	227988.60
			264297.20	214379.20
				213153.40

From the above table we have

Cluster Number	$\sum_{j<j'=1}^{5}\left(Y_{ij}-\overline{\overline{Y}}\right)\left(Y_{ij'}-\overline{\overline{Y}}\right)$
1	173759.80
2	-423798.80
3	-420092.04
4	412769.87
5	6019655.30
6	1612835.67
7	-226560.20
8	-247807.58
9	2476495.50
10	2586974.80
Sum	11964232.30

Thus we have

$$\sum_{i=1}^{N}\sum_{j\neq j'=1}^{M}\left(Y_{ij}-\overline{\overline{Y}}\right)\left(Y_{ij'}-\overline{\overline{Y}}\right)=2\times 11964232.30=23928464.6\,.$$

Thus the value of the intraclass correlation coefficient is given by

$$\rho=\frac{23928464.6}{10\times 5\times(5-1)\times 342021.5}=0.349809\,.$$

Note that because the value of the intraclass correlation coefficient is positive, and hence the cluster sampling is less efficient in this particular situation.

Effect of cluster size (M) on the relative efficiency: The relative efficiency of cluster sampling with respect to SRS is defined as

$$\text{RE}=S^2/\left(MS_b^2\right). \tag{9.1.14}$$

Following Smith (1938) we have

$$S_b^2=S^2/M^g, \qquad 0<g<1. \tag{9.1.15}$$

Thus (9.1.14) implies that

$$\text{RE}=\left(S^2M^g\right)/\left(MS^2\right)=M^{g-1}. \tag{9.1.16}$$

Thus we have the following graph which shows that the increase in cluster size M decreases the efficiency of cluster sampling provided that $(1-g)>0$.

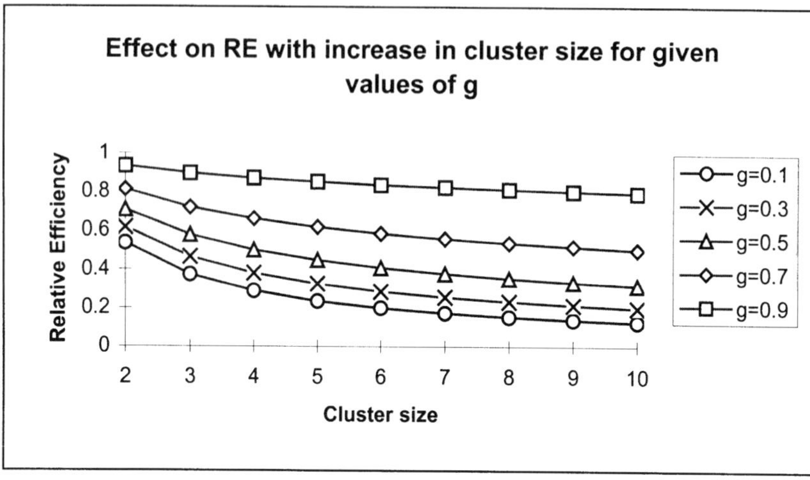

Theorem 9.1.6. An estimator of relative efficiency from the sample information for a large number of clusters is given by

$$\text{Est(RE)} = \left\{s_b^2 + \left(1 - M^{-1}\right)\bar{s}_w^2\right\} / \left\{M s_b^2\right\}. \tag{9.1.17}$$

Proof. In cluster analysis, there are three sources of variations, namely

(a) Total variation,

(b) Between cluster variation,

and

(c) Within cluster variation.

Thus cluster analysis may be expressed in an Analysis of Variance (ANOVA) table form as below:

How population looks after making clusters of equal size?								
Cluster Number	1	2		j		M	Cluster Total	Cluster Variance
1	Y_{11}	Y_{12}		Y_{1j}		Y_{1M}	$Y_{1\bullet}$	S_1^2
2	Y_{21}	Y_{22}		Y_{2j}		Y_{2M}	$Y_{2\bullet}$	$S_{2\bullet}^2$
i	Y_{i1}	Y_{i2}		Y_{ij}		Y_{iM}	$Y_{i\bullet}$	S_i^2
N	Y_{N1}	Y_{N2}		Y_{Nj}		Y_{NM}	$Y_{N\bullet}$	S_N^2
Population grand total and population mean of within cluster variances							$Y_{\bullet\bullet}$	$\bar{S}_w^2$

Thus from the population information we have:

$$\text{Population Grand Total} = Y_{\bullet\bullet} = \sum_{i=1}^{N}\sum_{j=1}^{M} Y_{ij} .$$

Thus population Correction Factor is:

$$\text{CF} = \frac{(\text{Population Grand Total})^2}{\text{Total No. of Observations}} = \frac{(Y_{\bullet\bullet})^2}{NM}.$$

The corrected population total sum of squares (SST) is given by

$$\text{SST} = \sum_{i=1}^{N}\sum_{j=1}^{M} Y_{ij}^2 - CF = \sum_{i=1}^{N}\sum_{j=1}^{M} Y_{ij}^2 - \frac{(Y_{\bullet\bullet})^2}{NM} = \sum_{i=1}^{N}\sum_{j=1}^{M} Y_{ij}^2 - NM\left(\bar{\bar{Y}}\right)^2 = \sum_{i=1}^{N}\sum_{j=1}^{M}\left(Y_{ij} - \bar{\bar{Y}}\right)^2$$

$$= (NM-1)S^2.$$

The corrected between population sum of squares (SSB) owed to cluster totals is

$$\text{SSB} = \frac{\sum_{i=1}^{N} Y_{i\bullet}^2}{M} - \text{CF} = \frac{\sum_{i=1}^{N} Y_{i\bullet}^2}{M} - \frac{(Y_{\bullet\bullet})^2}{NM} = M\sum_{i=1}^{N}\left(\frac{Y_{i\bullet}}{M}\right)^2 - NM\left(\bar{\bar{Y}}\right)^2 = M\sum_{i=1}^{N}\left(\bar{Y}_{i\bullet} - \bar{\bar{Y}}\right)^2$$

$$= M(N-1)S_b^2.$$

The corrected within population sum of squares (SSW) owed to cluster totals is

$$\text{SSW} = \text{SST} - \text{SSB} = \left[\sum_{i=1}^{N}\sum_{j=1}^{M} Y_{ij}^2 - \text{CF}\right] - \left\{\left\{\sum_{i=1}^{N} Y_{i\bullet}^2 \Big/ M\right\} - \text{CF}\right\}$$

$$= \sum_{i=1}^{N}\sum_{j=1}^{M} Y_{ij}^2 - \frac{\sum_{i=1}^{N} Y_{i\bullet}^2}{M} = \sum_{i=1}^{N}\sum_{j=1}^{M} Y_{ij}^2 - M\sum_{i=1}^{N}\bar{Y}_{i\bullet}^2 = \sum_{i=1}^{N}\left[\sum_{j=1}^{M} Y_{ij}^2 - M\bar{Y}_{i\bullet}^2\right] = \sum_{i=1}^{N}\sum_{j=1}^{M}\left(Y_{ij} - \bar{Y}_{i\bullet}\right)^2$$

$$= \sum_{i=1}^{N}\frac{(M-1)}{(M-1)}\sum_{j=1}^{M}\left(Y_{ij} - \bar{Y}_{i\bullet}\right)^2 = (M-1)\sum_{i=1}^{N}\frac{1}{(M-1)}\sum_{j=1}^{M}\left(Y_{ij} - \bar{Y}_{i\bullet}\right)^2$$

$$= (M-1)\sum_{i=1}^{N} S_{i\bullet}^2 = (M-1)\frac{N}{N}\sum_{i=1}^{N} S_{i\bullet}^2 = N(M-1)\bar{S}_w^2.$$

ANOVA for population for equal sized clusters				
Source	Population degree of Freedom (D.F.)	Sum of Squares (SS)	Mean Sum of Squares (MSS)	Population F ratio
Between Clusters	$N-1$	SSB$=M\sum_{i=1}^{N}\left(\bar{Y}_{i\bullet} - \bar{\bar{Y}}\right)^2$	MSB $= \frac{M}{N-1}\sum_{i=1}^{N}\left(\bar{Y}_{i\bullet} - \bar{\bar{Y}}\right)^2$ $= MS_b^2$	$F = \frac{\text{MSB}}{\text{MSW}}$
Within Clusters	$N(M-1)$	SSW$=\sum_{i=1}^{N}\sum_{j=1}^{M}\left(Y_{ij} - \bar{Y}_{i\bullet}\right)^2$	MSW $= \frac{1}{N(M-1)}\sum_{i=1}^{N}\sum_{j=1}^{M}\left(Y_{ij} - \bar{Y}_{i\bullet}\right)^2$ $= \bar{S}_w^2$	
Total	$NM-1$	SST$=\sum_{i=1}^{N}\sum_{j=1}^{M}\left(Y_{ij} - \bar{\bar{Y}}\right)^2$	MST $= \frac{1}{NM-1}\sum_{i=1}^{N}\sum_{j=1}^{M}\left(Y_{ij} - \bar{\bar{Y}}\right)^2$ $= S^2$	

As mentioned earlier, cluster sampling is better if clusters are homogeneous clusters and units are heterogeneous within the given cluster. In other words, the variation of units or elements within the cluster (SSW) should be large, but the variation from cluster to cluster (SSB) should be small for cluster sampling to be more efficient. Thus if the value of population F ratio is large, then the cluster sampling will not be efficient. Similarly, as we saw earlier, if the mean total sum of square $\text{MST} = S^2$, that is the variation among all elements in the population is smaller than the mean between sum of square $\text{MSB} = MS_b^2$, then cluster sampling will be less efficient. Thus for cluster sampling to be effective than SRSWOR sampling, we need $S_b^2 > S^2/M$. Although we already discussed one method to find the values of Intraclass (or Intra-cluster) correlation coefficient ρ, but the ANOVA method can also be used as

$$\rho = 1 - \left(\frac{M}{M-1}\right)\frac{\text{SSW}}{\text{SST}}.$$

Note that $0 \le \frac{\text{SSW}}{\text{SST}} \le 1$, therefore the value of intraclass correlation coefficient lies in $-\frac{1}{(M-1)} \le \rho \le 1$. Again note the conditions of efficiency in terms of intraclass correlation coefficient. Further note that the intra-class correlation coefficient can be easily defined only for clusters of equal size. Similarly the sample analogue of the population looks as given below:

How sample looks after selecting clusters of equal size?								
Sampled Clusters Number	1	2		j		M	Sampled Cluster Total	Sampled Cluster Variance
1	y_{11}	y_{12}		y_{1j}		y_{1M}	$y_{1\bullet}$	s_1^2
2	y_{21}	y_{22}		y_{2j}		y_{2M}	$y_{2\bullet}$	$s_{2\bullet}^2$
i	y_{i1}	y_{i2}		y_{ij}		y_{iM}	$y_{i\bullet}$	s_i^2
n	y_{n1}	y_{n2}		y_{nj}		y_{nM}	$y_{n\bullet}$	s_n^2
Sample grand total and sample mean of within cluster variances							$y_{\bullet\bullet}$	$\bar{s}_w^2$

Thus from the sample information we have

$$\text{Sample Grand Total} = y_{\bullet\bullet} = \sum_{i=1}^{n}\sum_{j=1}^{M} y_{ij}.$$

Thus sample correction factor is

$$\text{cf} = \frac{(\text{Sample Grand Total})^2}{\text{Total No. of Observations}} = \frac{(y_{\bullet\bullet})^2}{nM}.$$

The corrected sample total sum of squares (sst) is given by

$$\text{sst} = \sum_{i=1}^{n}\sum_{j=1}^{M} y_{ij}^2 - \text{cf} = \sum_{i=1}^{n}\sum_{j=1}^{M} y_{ij}^2 - \frac{(y_{\bullet\bullet})^2}{nM} = \sum_{i=1}^{n}\sum_{j=1}^{M} y_{ij}^2 - nM\left(\bar{\bar{y}}\right)^2 = \sum_{i=1}^{n}\sum_{j=1}^{M}\left(y_{ij} - \bar{\bar{y}}\right)^2 = (nM-1)s^2.$$

The corrected between sample sum of squares (ssb) owed to cluster totals is

$$\text{ssb} = \frac{\sum_{i=1}^{n} y_{i\bullet}^2}{M} - \text{cf} = \frac{\sum_{i=1}^{n} y_{i\bullet}^2}{M} - \frac{(y_{\bullet\bullet})^2}{nM} = M\sum_{i=1}^{n}\left(\frac{y_{i\bullet}}{M}\right)^2 - nM\left(\bar{\bar{y}}\right)^2 = M\sum_{i=1}^{n}\left(\bar{y}_{i\bullet} - \bar{\bar{y}}\right)^2 = M(n-1)s_b^2.$$

The corrected within sample sum of squares (ssw) owed to cluster totals is given by

$$\text{ssw} = \text{sst} - \text{ssb} = \left[\sum_{i=1}^{n}\sum_{j=1}^{M} y_{ij}^2 - \text{cf}\right] - \left[\sum_{i=1}^{n} y_{i\bullet}^2 \Big/ M - \text{cf}\right]$$

$$= \sum_{i=1}^{n}\sum_{j=1}^{M} y_{ij}^2 - \frac{\sum_{i=1}^{n} y_{i\bullet}^2}{M} = \sum_{i=1}^{n}\sum_{j=1}^{M} y_{ij}^2 - M\sum_{i=1}^{n}\bar{y}_{i\bullet}^2 = \sum_{i=1}^{n}\left[\sum_{j=1}^{M} y_{ij}^2 - M\bar{y}_{i\bullet}^2\right] = \sum_{i=1}^{n}\sum_{j=1}^{M}\left(y_{ij} - \bar{y}_{i\bullet}\right)^2$$

$$= \sum_{i=1}^{n}\frac{(M-1)}{(M-1)}\sum_{j=1}^{M}\left(y_{ij} - \bar{y}_{i\bullet}\right)^2 = (M-1)\sum_{i=1}^{n}\frac{1}{(M-1)}\sum_{j=1}^{M}\left(y_{ij} - \bar{y}_{i\bullet}\right)^2$$

$$= (M-1)\sum_{i=1}^{n} s_{i\bullet}^2 = (M-1)\frac{n}{n}\sum_{i=1}^{n} s_{i\bullet}^2 = n(M-1)\bar{s}_w^2.$$

ANOVA for sample for equal sized clusters				
Source	Sample degree of freedom (df)	Sample Sum of Squares (ss)	Sample Mean Sum of Squares (mss)	Sample F ratio
Between Clusters	$n-1$	ssb $= M\sum_{i=1}^{n}\left(\bar{y}_{i\bullet} - \bar{\bar{y}}\right)^2$	msb $= \frac{M}{n-1}\sum_{i=1}^{n}\left(\bar{y}_{i\bullet} - \bar{\bar{y}}\right)^2$ $= Ms_b^2$	$F = \frac{\text{msb}}{\text{msw}}$
Within Clusters	$n(M-1)$	ssw $= \sum_{i=1}^{n}\sum_{j=1}^{M}\left(y_{ij} - \bar{y}_{i\bullet}\right)^2$	msw $= \frac{1}{n(M-1)}\sum_{i=1}^{n}\sum_{j=1}^{M}\left(y_{ij} - \bar{y}_{i\bullet}\right)^2$ $= \bar{s}_w^2$	
Total	$nM-1$	sst $= \sum_{i=1}^{n}\sum_{j=1}^{M}\left(y_{ij} - \bar{\bar{y}}\right)^2$	mst $= \frac{1}{nM-1}\sum_{i=1}^{n}\sum_{j=1}^{M}\left(y_{ij} - \bar{\bar{y}}\right)^2$ $= s^2$	

Thus if the value of sample F ratio is large, then we can guess that the cluster sampling may not be efficient. Other guess for the cluster sampling to be effective than SRSWOR sampling, we need $s_b^2 > s^2/M$. Obviously, an estimate of the Intra-class (or Intra-cluster) correlation coefficient ρ can be had from the sample ANOVA method as:

$$\hat{\rho} = 1 - \left(\frac{M}{M-1}\right)\frac{\text{ssw}}{\text{sst}}.$$

Note that $0 \le \frac{\text{ssw}}{\text{sst}} \le 1$, therefore the estimate of intraclass correlation coefficient lies in $-\frac{1}{(M-1)} \le \hat{\rho} \le 1$.

We know that s_b^2 and $\bar{s}_w^2$ are unbiased estimators of S_b^2 and $\bar{S}_w^2$, respectively. But $\frac{1}{nM-1}\sum_{i=1}^{n}\sum_{j=1}^{M}\left(y_{ij}-\bar{\bar{y}}\right)^2$ is not an unbiased estimator of $\frac{1}{NM-1}\sum_{i=1}^{N}\sum_{j=1}^{M}\left(Y_{ij}-\bar{\bar{Y}}\right)^2$. Let us try to put $\frac{1}{NM-1}\sum_{i=1}^{N}\sum_{j=1}^{M}\left(Y_{ij}-\bar{\bar{Y}}\right)^2$ in terms of S^2 and $\bar{S}_w^2$. From the ANOVA for population we have

$$\sum_{i=1}^{N}\sum_{j=1}^{M}\left(Y_{ij}-\bar{\bar{Y}}\right)^2 = \sum_{i=1}^{N}\sum_{j=1}^{M}\left(Y_{ij}-\bar{Y}_{i\bullet}+\bar{Y}_{i\bullet}-\bar{\bar{Y}}\right)^2 = \sum_{i=1}^{N}\sum_{j=1}^{M}\left(Y_{ij}-\bar{Y}_{i\bullet}\right)^2 + M\sum_{i=1}^{N}\left(\bar{Y}_{i\bullet}-\bar{\bar{Y}}\right)^2,$$

which implies that

$$(NM-1)S^2 = N(M-1)\bar{S}_w^2 + M(N-1)S_b^2$$

or

$$S^2 = \frac{N(M-1)\bar{S}_w^2 + M(N-1)S_b^2}{(NM-1)}. \tag{9.1.18}$$

In (9.1.18) the denominator $(NM-1)$ is a constant quantity, so an unbiased estimator of S^2 is given by

$$\text{Est}\left(S^2\right) = \frac{N(M-1)\bar{s}_w^2 + M(N-1)s_b^2}{(NM-1)}. \tag{9.1.19}$$

The formation of ANOVA helps us to find the multipliers of s_b^2 and $\bar{s}_w^2$, which, in fact, are $M(N-1)$ and $N(M-1)$, respectively. Evidently an estimator of the relative efficiency RE is given by

$$\text{Est}(\text{RE}) = \frac{\text{Est}\left(S^2\right)}{Ms_b^2} = \frac{M(N-1)s_b^2 + N(M-1)\bar{s}_w^2}{M(NM-1)s_b^2} = \frac{MN\left(1-\frac{1}{N}\right)s_b^2 + MN\left(1-\frac{1}{M}\right)\bar{s}_w^2}{M^2N\left(1-\frac{1}{NM}\right)s_b^2}$$

$$= \frac{s_b^2 + \left(1 - M^{-1}\right)\bar{s}_w^2}{M s_b^2}. \tag{9.1.20}$$

Hence the theorem.

Example 9.1.5. Mr. Stephen Horn and Mr. Ken Brewer were asked to construct three clusters of 9 regions of the USA using the following two maps:

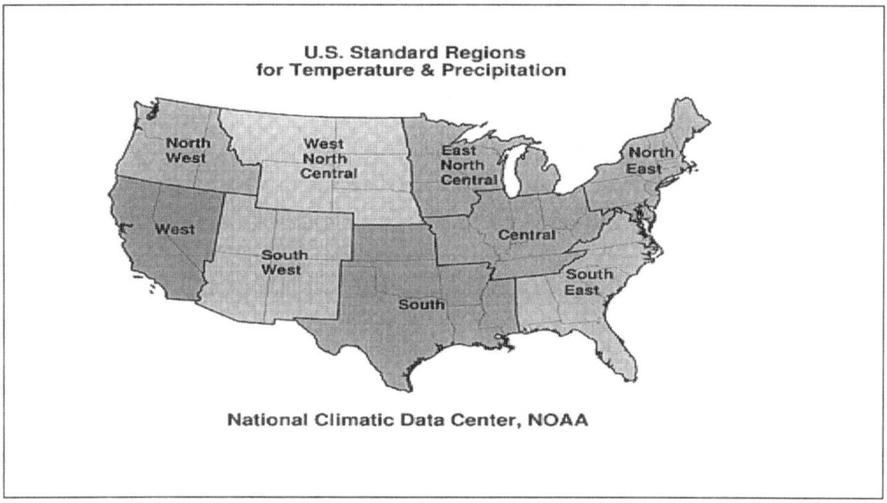

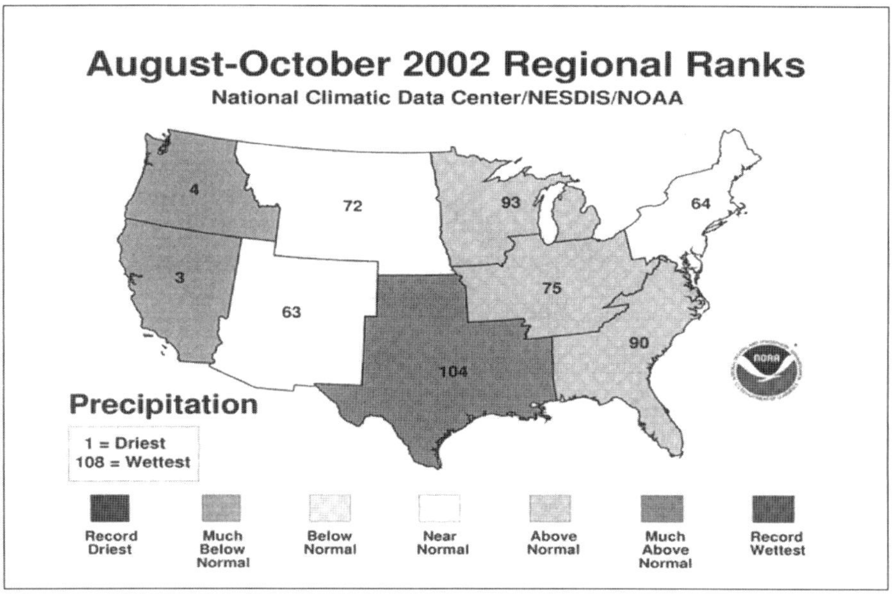

Source: Printed with permission from NOAA.

Mr. Stephen Horn suggested the following three clusters as:

Mr. Stephen Horn's Clusters			
Cluster 1	North West	West North Central	East North Central
Cluster 2	West	South West	South
Cluster 3	North East	Central	South East

Mr. Ken Brewer suggested the following three clusters as:

Mr. Ken Brewer's Clusters			
Cluster 1	North West	West	South West
Cluster 2	West North Central	East North Central	North East
Cluster 3	Central	South East	South

(A) Whose clustering plan is more efficient and why? Apply the ANOVA method.

(B) Find the relative efficiency of cluster sampling with respect to SRSWOR in each case.

(C) Select two clusters out of Mr. Stephen Horn's clusters, and construct 95% confidence interval estimate for the population mean. Apply the sample ANOVA approach to estimate intraclass correlation. (**Rule:** Use first row and first column of the Pseudo-Random Number (PRN) Table 1 given in the Appendix)

(D) Select two clusters out of Mr. Ken Brewer's clusters, and construct 95% confidence interval estimate for the population mean. Apply sample ANOVA approach to estimate intraclass correlation. (**Rule:** Use first row and first column of the Pseudo-Random Number (PRN) Table 1 given in the Appendix)

(E) Does both confidence interval estimates include true average precipitation of the USA?

Solution (A) : **(I) Mr. Stephen Horn's Clustering:** Using information from the second map, we have

Clusters	Stephen's Clustering			Cluster Totals
Cluster I	4	72	93	169
Cluster II	3	63	104	170
Cluster III	64	75	90	229
	Population Grand Total (GT)			568

(a) Population correction factor (CF):

$$\text{CF} = \frac{(\text{GT})^2}{NM} = \frac{568^2}{3\times 3} = 35847.11.$$

(b) Population total corrected sum of squares:

$$\text{SST} = \sum_{i=1}^{N}\sum_{j=1}^{M} Y_{ij}^2 - \text{CF}$$
$$= 4^2 + 72^2 + 93^2 + 3^2 + 63^2 + 104^2 + 64^2 + 75^2 + 90^2 - 35847.11$$
$$= 10616.89.$$

(c) Population between cluster totals corrected sum of squares:

$$\text{SSB} = \sum_{i=1}^{N}\frac{Y_{i\bullet}^2}{M} - \text{CF} = \frac{169^2 + 170^2 + 229^2}{3} - 35847.11 = 786.89.$$

(d) Population within cluster totals corrected sum of squares:

$$\text{SSW} = \text{SST - SSB} = 10616.89 - 786.89 = 9830.$$

Thus we have the following ANOVA table

ANOVA for Mr. Stephen Horn's Clusters				
Sources	D.F.	S.S.	M.S.	F
Between Clusters	2	786.89	393.44	0.2401
Within Clusters	6	9830.00	1638.33	
Total	8	10616.89	1327.11	

The value of intraclass correlation coefficient for Mr. Stephen Horn's clustering is given by

$$\rho_{\text{stephen}} = 1 - \left(\frac{M}{M-1}\right)\frac{\text{SSW}}{\text{SST}} = 1 - \left(\frac{3}{3-1}\right)\frac{9830}{10616.89} = -0.388 \text{ (which is negative).}$$

(II) Mr. Ken Brewer's Clustering: Using information from the second map we have

Clusters	Ken's Clustering			Cluster Totals
Cluster I	4	3	63	70
Cluster II	72	93	64	229
Cluster III	75	90	104	269
	Population Grand Total (GT)			568

(a) Population correction factor (CF):

$$\text{CF} = \frac{(\text{GT})^2}{NM} = \frac{568^2}{3\times 3} = 35847.11.$$

(b) Population total corrected sum of squares:

$$\text{SST} = \sum_{i=1}^{N}\sum_{j=1}^{M} Y_{ij}^2 - \text{CF}$$
$$= 4^2 + 72^2 + 93^2 + 3^2 + 63^2 + 104^2 + 64^2 + 75^2 + 90^2 - (35847.11)$$
$$= 10616.89.$$

(c) Population between cluster totals corrected sum of squares:

$$\text{SSB} = \sum_{i=1}^{N}\frac{Y_{i\bullet}^2}{M} - \text{CF} = \frac{70^2 + 229^2 + 269^2}{3} - (35847.11) = 7386.89.$$

(d) Population within cluster totals corrected sum of squares:

$$\text{SSW} = \text{SST} - \text{SSB} = 10616.89 - 7386.89 = 3230.$$

Thus we have the following ANOVA table

ANOVA for Mr. Ken Brewer's Clusters				
Sources	DF	SS	MS	F
Between Clusters	2	7386.89	3693.45	6.8609
Within Clusters	6	3230.00	538.33	
Total	8	10616.89	1327.11	

The value of intraclass correlation coefficient for Mr. Ken Brewer's clustering is given by

$$\rho_{\text{ken}} = 1 - \left(\frac{M}{M-1}\right)\frac{\text{SSW}}{\text{SST}} = 1 - \left(\frac{3}{3-1}\right)\frac{3230}{10616.89} = 0.5436 \text{ (which is positive).}$$

The value of intraclass correlation coefficient for Mr. Stephen Horn's clustering plan is negative, whereas that for the Mr. Ken Brewer's plan is positive, which indicates that Mr. Horn's clustering plan will perform better than Mr. Brewer's clustering. The reason is that Mr. Horn's clusters have more variation within each cluster and less variation between the clusters.

(B) The relative efficiency of Mr. Stephen Horn's cluster sampling over SRSWOR sampling will be

$$\text{RE}(\text{Mr. Horn}) = \frac{S^2}{MS_b^2}\times 100 = \frac{\text{MST}}{\text{MSB}}\times 100 = \frac{1327.11}{393.44}\times 100 = 337.31\%$$

and the relative efficiency of Mr. Ken Brewer's cluster sampling over SRSWOR sampling will be

$$\text{RE(Mr. Brewer)} = \frac{S^2}{MS_b^2} \times 100 = \frac{\text{MST}}{\text{MSB}} \times 100 = \frac{1327.11}{3693.45} \times 100 = 35.93\% .$$

(C) **Sample from Mr. Stephen Horn's clustering:** Now using the information from second map we have

Clusters	Stephen's Clustering			Cluster Totals
Cluster II	3	63	104	170
Cluster III	64	75	90	229
		Sample grand total (gt)		399

(a) Sample correction factor (cf):

$$\text{cf} = \frac{(\text{gt})^2}{nM} = \frac{399^2}{2 \times 3} = 26533.5.$$

(b) Sample total corrected sum of squares:

$$\text{sst} = \sum_{i=1}^{n} \sum_{j=1}^{M} y_{ij}^2 - \text{cf} = 3^2 + 63^2 + 104^2 + 64^2 + 75^2 + 90^2 - 26533.51 = 6081.49.$$

(c) Sample between cluster totals corrected sum of squares:

$$\text{ssb} = \sum_{i=1}^{n} \frac{y_{i\bullet}^2}{M} - \text{cf} = \frac{170^2 + 229^2}{3} - 26533.51 = 580.16.$$

(d) Sample within cluster totals corrected sum of squares:

$$\text{ssw} = \text{sst - ssb} = 6081.49 - 580.16 = 5501.33.$$

Thus we have the following ANOVA table

Sample ANOVA for Mr. Stephen Horn				
Sources	df	ss	ms	F
Between Clusters	1	580.16	580.16	0.4218
Within Clusters	4	5501.33	1375.33	
Total	5	6081.49	1216.29	

Thus an estimate of the value of intraclass correlation coefficient for Mr. Stephen Horn's cluster sampling is given by

$$\hat{\rho}_{\text{stephen}} = 1 - \left(\frac{M}{M-1}\right)\frac{\text{ssw}}{\text{sst}} = 1 - \left(\frac{3}{3-1}\right)\frac{5501.33}{6081.49} = -0.356 \text{ (which is again negative).}$$

The estimate of average precipitation in the USA is given by

$$\bar{\bar{y}} = \frac{\text{gt}}{n \times M} = \frac{399}{2 \times 3} = 66.5\,.$$

From the above sample ANOVA we have

$$M\, s_b^2 = 580.16$$

which implies

$$s_b^2 = \frac{580.16}{3} = 193.38\,.$$

Thus we have

$$\hat{v}\left(\bar{\bar{y}}\right) = \left(\frac{1}{n} - \frac{1}{N}\right)s_b^2 = \left(\frac{1}{2} - \frac{1}{3}\right) \times 193.38 = 32.23\,.$$

A $(1-\alpha)100\%$ confidence interval estimate of the population mean $\bar{\bar{Y}}$ is given by

$$\bar{\bar{y}} \pm t_{\alpha/2}\left(\text{df} = n(M-1)\right)\sqrt{\hat{v}\left(\bar{\bar{y}}\right)}\,.$$

Thus 95% confidence interval estimate for the average precipitation in the USA is given by

$$66.5 \pm t_{0.025}(\text{df} = 4)\sqrt{32.33}\,.$$

Using Table 2 from the Appendix we have

$$66.5 \pm 2.776 \times 5.69\text{, or } [50.72, \quad 82.28]\,.$$

(D) Sample from Mr. Ken Brewer's Clustering: Now using the information from second map we have

Clusters	Ken's Clustering			Cluster Totals
Cluster II	72	93	64	229
Cluster III	75	90	104	269
		Sample grand total (gt)		498

(a) Sample Correction factor (cf):

$$\text{cf} = \frac{(\text{gt})^2}{nM} = \frac{498^2}{2 \times 3} = 41334\,.$$

(b) Sample total corrected sum of squares:

$$\text{sst} = \sum_{i=1}^{n}\sum_{j=1}^{M} y_{ij}^2 - \text{cf} = 72^2 + 93^2 + 64^2 + 75^2 + 90^2 + 104^2 - 41334 = 1136.$$

(c) Sample between cluster totals corrected sum of squares:

$$\text{ssb} = \sum_{i=1}^{n}\frac{y_{i\bullet}^2}{M} - \text{cf} = \frac{229^2 + 269^2}{3} - 41334 = 266.67.$$

(d) Sample within cluster totals corrected sum of squares:

$$\text{ssw} = \text{sst} - \text{ssb} = 1136 - 266.67 = 869.33.$$

Thus we have the following ANOVA table

Sample ANOVA for Mr. Ken Brewer				
Sources	df	ss	ms	F
Between Clusters	1	266.67	266.67	1.227
Within Clusters	4	869.33	217.33	
Total	5	1136.00	227.20	

Thus an estimate of the value of intraclass correlation coefficient is given by

$$\hat{\rho}_{\text{ken}} = 1 - \left(\frac{M}{M-1}\right)\frac{\text{ssw}}{\text{sst}} = 1 - \left(\frac{3}{3-1}\right)\frac{869.33}{1136} = -0.147 \text{ (which is slightly negative).}$$

The estimate of average precipitation in the USA is give by

$$\bar{\bar{y}} = \frac{\text{gt}}{n \times M} = \frac{498}{2 \times 3} = 83\,.$$

From the above sample ANOVA we have

$$M\, s_b^2 = 266.67$$

which implies

$$s_b^2 = \frac{266.67}{3} = 88.89\,.$$

Thus we have

$$\hat{v}\left(\bar{\bar{y}}\right) = \left(\frac{1}{n} - \frac{1}{N}\right)s_b^2 = \left(\frac{1}{2} - \frac{1}{3}\right) \times 88.89 = 14.82\,.$$

A $(1-\alpha)\,100\%$ confidence interval estimate of the population mean $\bar{\bar{Y}}$ is given by

$$\bar{\bar{y}} \pm t_{\alpha/2}\left(\text{df} = n(M-1)\right)\sqrt{\hat{v}\left(\bar{\bar{y}}\right)}\ .$$

Thus 95% confidence interval estimate for the average precipitation in the USA is given by

$$83 \pm t_{0.025}(\text{df} = 4)\sqrt{14.82}$$

or, using Table 2 from the Appendix

$$83 \pm 2.776 \times 3.85, \quad \text{or} \quad [72.31, \quad 93.68].$$

(e) The true average precipitation in the USA is given by

$$\bar{\bar{Y}} = \frac{\text{GT}}{NM} = \frac{568}{3 \times 3} = 63.11\,.$$

Clearly the true average precipitation in the USA lies in the 95% confidence interval estimate due to Mr. Stephen Horn's clustering, given by $[50.72,\ \ 82.28]$, but not in the 95% confidence interval estimate due to Mr. Ken Brewer's clustering, given by $[72.31,\ \ 93.68]$. Although the estimate of intraclass correlation coefficient for Mr. Ken Brewer's sampled clusters is slightly negative, but it is not performing very well because the true intraclass correlation coefficient in the population is positive.

Theorem 9.1.7. Under the superpopulation model approach, the relative efficiency is free from sample information and is inversely proportional to the size of clusters.

Proof. We know that under the superpopulation model proposed by Smith (1938) the relative efficiency can be written as

$$\mathrm{RE} = S^2/\left(MS_b^2\right) \tag{9.1.21}$$

where $S_b^2 = (N-1)^{-1}\sum_{i=1}^{N}\left(\overline{Y}_{i\bullet} - \overline{\overline{Y}}\right)^2$ and $S^2 = (NM-1)^{-1}\sum_{i=1}^{N}\sum_{j=1}^{M}\left(Y_{ij} - \overline{\overline{Y}}\right)^2$.

Evidently if the cluster size changes then S_b^2 changes but S^2 does not, because Y_{ij} and hence $\overline{\overline{Y}}$ remains the same. Note that NM remains constant, M increases if N decreases, and vice versa. Thus with change in cluster size, only S_b^2 gets altered, because $\overline{Y}_{i\bullet}$ changes with cluster size. Smith (1938) suggested an empirical relation between S_b^2 and S^2 through a superpopulation model parameter g given by

$$S_b^2 = S^2/M^g\ . \tag{9.1.22}$$

Thus the expression for RE reduces to

$$\mathrm{RE} = \frac{S^2}{MS_b^2} = \frac{S^2M^g}{MS^2} = M^{g-1}\ . \tag{9.1.23}$$

Now to find the value of g we make use of data that we selected for one particular value of cluster size M. Thus (9.1.23) implies that RE is constant K (for example), and its estimator is

$$\mathrm{Est.(RE)} = \left\{M(N-1)s_b^2 + N(M-1)\bar{s}_w^2\right\}/M(NM-1)s_b^2\ . \tag{9.1.24}$$

Now we will equate the value of Est.(RE) from the sampled data for a fixed cluster size M with the value of K. For the fixed value of M we can find the value of g given by

$$\mathrm{Est.(RE)} = M^{g-1} \tag{9.1.25}$$

Taking logs on both sides we have

$$g = 1 + \log[\mathrm{Est.(RE)}]/\log(M)\ . \tag{9.1.26}$$

Example 9.1.6. Select 6 clusters from Table 9.1.1 using SRSWOR sampling. Record the values of the real estate farm loans for the selected states from population 1 given in Appendix. Estimate the relative efficiency using the ANOVA approach. From this calculate the value of parameter g.

Solution. Starting with the first two columns of the Pseudo-Random Number (PRN) Table 1 given in the Appendix, we select the required 6 distinct random numbers as 01, 04, 05, 03, 06 and 07. The following clusters are included in an SRSWOR sample of 6 units.

PRN	Names of the states	Values of the real estate farm loans y_{ij}	$y_{i\bullet}$	$\bar{y}_{i\bullet}$	$\left(\bar{y}_{i\bullet} - \bar{\bar{y}}\right)^2$
01	WA, OR, MT, NV, ID	1100.745, 114.899, 292.965, 5.860, 53.753	1568.222	313.644	205772.22
03	UT, WY, CO, SD, NE	56.908, 100.964, 315.809, 413.777, 1337.852	2225.310	445.062	103815.18
04	TX, OK, KS, LA, AR	1248.761, 612.108, 1049.834, 282.565, 907.700	4100.968	820.194	2801.37
05	MO, IA, MN, ND, WI	579.686, 2327.025, 1354.768, 449.099, 1229.752	6940.330	1388.066	385393.10
06	IL, IN, MI, OH, PA	2131.048, 1213.024, 323.028, 870.720, 756.169	5293.989	1058.798	84991.00
07	MS, AL, GA, FL, SC	627.013, 408.978, 939.460, 825.748, 87.951	2889.150	577.830	35885.86
		Sum	23017.970	4603.594	818658.73

From the above table we have

$$\bar{\bar{y}} = \frac{1}{n}\sum_{i=1}^{n}\bar{y}_{i\bullet} = 767.2656$$

and

$$\sum_{i=1}^{n}\left(\bar{y}_{i\bullet} - \bar{\bar{y}}\right)^2 = 818658.73.$$

Now we have

$\left(y_{ij} - \bar{\bar{y}}\right)^2$				
111208.490	425582.220	224961.090	579738.540	509100.278
504607.970	443957.870	203813.090	124954.210	325568.802
231837.790	24073.891	79844.882	234934.700	19721.811
660026.850	2432849.300	345159.030	101230.010	213893.639
1859902.300	198700.520	197347.070	10702.806	123.135
19670.801	128370.030	29650.900	3420.187	461468.371

From the above table we have

$$\sum_{i=1}^{n}\sum_{j=1}^{M}\left(y_{ij} - \bar{\bar{y}}\right)^2 = 10{,}706{,}420.62$$

Also we have

$(y_{ij}-\bar{y}_{i\bullet})^2$				
429920.200	109007.610	23133.497	192898.400	153122.733
582604.910	517291.220	254403.820	165174.450	267970.219
19405.883	602110.820	114400.890	1222132.500	230751.494
271324.520	1608400.200	87598.359	371732.630	29225.338
2412486.200	403471.420	64924.059	85784.552	31804.798
2418.967	28510.998	130776.260	61463.335	239981.435

From the above table, we have

$$\sum_{i=1}^{n}\sum_{j=1}^{M}(y_{ij}-\bar{y}_{i\bullet})^2 = 10{,}714{,}231.64\,.$$

ANOVA for sample			
Source	Degree of freedom df	Sum of Squares SS	Mean Sum of Squares MSS
Between Clusters	$6-1=5$	$M\sum_{i=1}^{n}\left(\bar{y}_{i\bullet}-\bar{\bar{y}}\right)^2 = 4093293.70$	$Ms_b^2 = 818658.74$
Within Clusters	$6(5-1)=24$	$\sum_{i=1}^{n}\sum_{j=1}^{M}(y_{ij}-\bar{y}_{i\bullet})^2 = 10714231.64$	$\bar{s}_w^2 = 446426.32$
Total	$5\times 6-1=29$	$\sum_{i=1}^{n}\sum_{j=1}^{M}(y_{ij}-\bar{\bar{y}})^2 = 10706420.62$	$s^2 = 369186.92$

Thus an estimate of RE is given by

$$\text{Est.}(\text{RE}) = \frac{M(N-1)s_b^2 + N(M-1)\bar{s}_w^2}{M(NM-1)s_b^2} = \frac{(10-1)\times 818658.74 + 10(5-1)\times 446426.32}{(10\times 5-1)\times 818658.74}$$

$$= 0.6288.$$

Thus the value of g is given by

$$g = 1 + \frac{\log[\text{Est.}(\text{RE})]}{\log(M)} = 1 + \frac{\log(0.6288)}{\log(5)} = 0.7117.$$

9.2 OPTIMUM VALUE OF NON-OVERLAPPING CLUSTER SIZE

We have seen that for a given sample size, the sampling variance increases with increase in cluster size and decreases with the number of clusters. It may be noted here that the cost of the survey decreases with the increase in cluster size and increases with the increase in number of clusters. Thus a need arises to compromise between the cluster size and the number of clusters in the sample so that the sampling variance is minimum for the fixed cost of the survey or the cost is minimum for the fixed variance. Mahalanobis (1940, 1942) has considered the problem of determination of optimum cluster size from the point of view of both variance and cost. Singh (1956) studied cluster sizes in cluster sampling and sub-sampling procedures. The best source of discussion of the construction of cost functions in cluster sampling is that of Hansen, Hurwitz, and Madow (1953). In their analysis, they postulate that the cost of the survey using cluster sampling consists of two components in addition to the overhead cost:

(a) Cost of enumerating the elements in the sample and travelling within the cluster, which is proportional to the number of units in the sample.
(b) Cost of travelling between clusters, which is proportional to the distance to be travelled between clusters. Empirical studies show that the expected value of minimum distance between n points located at random is proportional to $\sqrt{n}$.

Thus in general the cost of the survey can be written as

$$C = C_1 nM + C_2\sqrt{n}, \tag{9.2.1}$$

where C_1 is the cost of enumerating a unit within the cluster and C_2 is the cost of travelling per unit distance between clusters.

The variance of the estimator of population mean in cluster sampling is

$$V\left(\bar{\bar{y}}\right) = \left(\frac{1}{n} - \frac{1}{N}\right)S_b^2 = \frac{(1-f)}{n}S_b^2 . \tag{9.2.2}$$

Several researchers, including Jessen (1942), Mahalanobis (1944, 1946) and Hendricks (1944), have shown that there is a log linear relationship between S_w^2 and cluster size M given by

$$S_w^2 = aM^b, \tag{9.2.3}$$

where a and b are positive constants and are to be determined from the survey data. From the ANOVA for population we have

$$S_b^2 = \left\{(MN-1)S^2 - N(M-1)S_w^2\right\} / M(N-1) = S^2 - (M-1)aM^{b-1} . \tag{9.2.4}$$

Ignoring f in (9.2.2), the variance of the estimator of population mean in cluster sampling is given by

$$V\left(\bar{\bar{y}}\right) \approx \frac{1}{n}\left[S^2 - (M-1)aM^{b-1}\right]. \tag{9.2.5}$$

Case I.. Cost is fixed: In this case the Lagrange function L_1 is given by

$$L_1 = \frac{1}{n}\left[S^2 - (M-1)aM^{b-1}\right] + \lambda_1\left[C - C_1 nM - C_2\sqrt{n}\right] \tag{9.2.6}$$

where λ_1 is a Lagrange multiplier. Differentiating (9.2.6) with respect to n, M and λ_1 and equating to zero in each case to find three equations, one has to solve the resultant equations for the optimum values of n and M .

Case II. Variance is fixed: In this case the Lagrange function L_2 is given by

$$L_2 = C_1 nM + C_2\sqrt{n} + \lambda_2\left[V_0 - \frac{1}{n}\left\{S^2 - (M-1)aM^{b-1}\right\}\right], \tag{9.2.7}$$

where λ_2 is a Lagrange multiplier. Differentiating (9.2.7) with respect to n, M and λ_2 and equating to zero in each case to find three equations, one solves the resultant equations for the optimum values of n and M.

9.3 ESTIMATION OF PROPORTION USING NON-OVERLAPPING CLUSTER SAMPLING

Consider an investigator is interested in estimating the proportion of units belonging to a specified class A (for example) when the population consists of N clusters each of size M. An SRSWOR sample of n clusters is selected. Suppose elements in the cluster can be specified into two classes as follows:

$$Y_{ij} = \begin{cases} 1 & \text{if } j^{th} \text{ element of } i^{th} \text{ cluster belongs to class } A, \\ 0 & \text{otherwise.} \end{cases} \tag{9.3.1}$$

Let $a_i = \sum_{j=1}^{M} Y_{ij}$ is the number of units in the ith cluster. Then the investigator's interest will be to estimate the population proportion P defined as

$$P = \frac{1}{MN}\sum_{i=1}^{N} a_i\,. \tag{9.3.2}$$

Then we have the following theorems:

Theorem 9.3.1. An unbiased estimator of population proportion P is given by

$$\hat{p}_{\text{cl}} = \frac{1}{n}\sum_{i=1}^{n} P_i \tag{9.3.3}$$

where $P_i = a_i/M$ is the proportion of units belonging to the ith cluster in the sample.

Theorem 9.3.2. The variance of the estimator $\hat{p}_{\text{cl}}$ is given by

$$V(\hat{p}_{\text{cl}}) = \frac{(1-f)NP(1-P)}{(N-1)nM}\left[1 + (M-1)\rho\right]. \tag{9.3.4}$$

Proof. We have

$$V(\hat{p}_{\text{cl}}) = \frac{(1-f)}{n}S_b^2 \tag{9.3.5}$$

where S_b^2 is the variance between cluster proportions and is given by

$$MS_b^2 = \frac{1}{N}\sum_{i=1}^{N}(P_i - P)^2 = P(1-P) - \frac{1}{N}\sum_{i=1}^{N}P_i(1-P_i). \tag{9.3.6}$$

For large N we have

$$S^2 \approx S_b^2 + S_w^2 = NMP(1-P)/(NM-1)$$

and

$$S_w^2 = \frac{M}{N(M-1)}\sum_{i=1}^{N} P_i(1-P_i).$$

Note that we have

$$(N-1)MS_b^2 = (NM-1)S^2 - N(M-1)S_w^2$$

or

$$S_b^2 = \frac{(NM-1)}{M(N-1)}S^2 - \frac{N(M-1)}{M(N-1)}S_w^2 = \frac{(NM-1)}{M(N-1)} \times \frac{NMP(1-P)}{(NM-1)} - \frac{N(M-1)M}{M(N-1)N(M-1)}\sum_{i=1}^{N} P_i(1-P_i)$$

$$= \frac{NP(1-P)}{(N-1)} - \frac{1}{(N-1)}\sum_{i=1}^{N} P_i(1-P_i) = P(1-P)\left[\frac{N}{(N-1)} - \frac{1}{(N-1)}\sum_{i=1}^{N}\frac{P_i(1-P_i)}{P(1-P)}\right]$$

$$= P(1-P)\left[\frac{N}{(N-1)} - 1 + 1 - \frac{1}{(N-1)}\sum_{i=1}^{N}\frac{P_i(1-P_i)}{P(1-P)}\right]$$

$$= P(1-P)\left[\frac{N-(N-1)}{(N-1)} + 1 - \frac{1}{(N-1)}\sum_{i=1}^{N}\frac{P_i(1-P_i)}{P(1-P)}\right]$$

$$= P(1-P)\left[\frac{1}{(N-1)} + \frac{M-1}{M-1}\left\{1 - \frac{1}{(N-1)}\sum_{i=1}^{N}\frac{P_i(1-P_i)}{P(1-P)}\right\}\right]$$

$$= \frac{P(1-P)}{N-1}\left[1 + (M-1)\frac{N-1}{M-1}\left\{1 - \frac{1}{(N-1)}\sum_{i=1}^{N}\frac{P_i(1-P_i)}{P(1-P)}\right\}\right].$$

Thus the value of the intra-cluster correlation coefficient ρ can easily be written as

$$\rho = \frac{N-1}{M-1}\left\{1 - \frac{1}{(N-1)}\sum_{i=1}^{N}\frac{P_i(1-P_i)}{P(1-P)}\right\}. \tag{9.3.7}$$

Hence the theorem.

Theorem 9.3.3. An unbiased estimator of $V(\hat{p}_{\text{cl}})$ is given by

$$\hat{v}(\hat{p}_{\text{cl}}) = \left(\frac{1-f}{n}\right)s_b^2 = \frac{(1-f)}{n(n-1)}\sum_{i=1}^{n}(P_i - \overline{P})^2 \tag{9.3.8}$$

where $\overline{P} = \frac{1}{n}\sum_{i=1}^{n} P_i$ has its usual meaning.

Theorem 9.3.4. The relative efficiency of cluster sampling as compared to SRSWOR sampling is given by

$$\text{RE} = \frac{V(\hat{p}_{\text{srs}})}{V(\hat{p}_{\text{cl}})} = \frac{\{(N-1)NP(1-P)\}}{\left[(NM-1)\left\{NP(1-P) - \sum_{i=1}^{N} P_i(1-P_i)\right\}\right]}. \tag{9.3.9}$$

Example 9.3.1. Select 6 clusters from table 9.1.1 using SRSWOR sampling. Record the values of the real estate farm loans for the selected states from population 1 given in the Appendix. Estimate the proportion of states having real estate farm loans of more than \$555.4345 in the United States using cluster sampling. Estimate the variance of the estimator used for estimating the required proportion. Construct a 95% confidence interval.

Solution. Starting with the first two columns of the Pseudo-Random Number (PRN) Table 1 given in the Appendix, we select the required 6 distinct random numbers as 01, 04, 05, 03, 06 and 07. Thus the following clusters are included in an SRSWOR sample of 6 units

PRN	Names of the states	Values of the real estate farm loans y_{ij}	$y_{i\bullet}$
01	WA, OR, MT, NV, ID	1100.745, 114.899, 292.965, 5.860, 53.753	1568.222
03	UT, WY, CO, SD, NE	56.908, 100.964, 315.809, 413.777, 1337.852	2225.310
04	TX, OK, KS, LA, AR	1248.761, 612.108, 1049.834, 282.565, 907.700	4100.968
05	MO, IA, MN, ND, WI	1579.686, 2327.025, 1354.768, 449.099, 1229.752	6940.330
06	IL, IN, MI, OH, PA	2131.048, 1213.024, 323.028, 870.720, 756.169	5293.989
07	MS, AL, GA, FL, SC	627.013, 408.978, 939.460, 825.748, 87.951	2889.150
		Sum	23017.970

Here $M = 5$ and $n = 6$. Distinguish the states having real estate farm loans of more than \$555. 4345 in each one of the selected clusters.

Selected Clusters	$Y_{ij} = \begin{cases} 1, & \text{if } Y_{ij} > 555.4345 \\ 0, & \text{otherwise} \end{cases}$					a_i	P_i	$\left(P_i - \overline{P}\right)^2$
1	1	0	0	0	0	1	0.2	0.1225
2	0	0	0	0	1	1	0.2	0.1225
3	1	1	1	0	1	4	0.8	0.0625
4	1	1	1	0	1	4	0.8	0.0625
5	1	1	0	1	1	4	0.8	0.0625
6	1	0	1	1	0	3	0.5	0.0025
						Sum	3.3	0.4350

Using cluster sampling an estimate of population proportion P is given by

$$\hat{p}_{\text{cl}} = \frac{1}{n}\sum_{i=1}^{n} P_i = \frac{3.3}{6} = 0.55$$

where $P_i = \dfrac{a_i}{M}$.

Note that

$$\overline{P} = \frac{1}{n}\sum_{i=1}^{n} P_i = 0.55$$

and an estimate of $V(\hat{p}_{\text{cl}})$ is given by

$$\hat{v}(\hat{p}_{\text{cl}}) = \frac{(1-f)}{n(n-1)}\sum_{i=1}^{n}\left(P_i - \overline{P}\right)^2 = \frac{1-6/50}{6(6-1)} \times 0.435 = 0.01276 .$$

A $(1-\alpha)100\%$ confidence interval of the population proportion, P, is given by

$$\hat{p}_{\text{cl}} \mp t_{\alpha/2}(\text{df} = n(M-1))\sqrt{\hat{v}(\hat{p}_{\text{cl}})} .$$

Using Table 2 from the Appendix the 95% confidence interval for the proportion of states having real estate farm loans of more than \$555.4345 in the United States is given by

$$\hat{p}_{\text{cl}} \mp t_{0.025}(\text{df} = 24)\sqrt{\hat{v}(\hat{p}_{\text{cl}})}\ , \quad 0.55 \mp 2.064\sqrt{0.01276}\ , \text{ or } \quad [0.3168,\ 0.7831].$$

Example 9.3.2. Suppose the United States has been divided into 10 neighbouring clusters consisting of 5 states as shown earlier. We are interested in the proportion of states having real estate farm loans of more than \$555.4345. Is there any expectation for gain in efficiency due to clustering as opposed to simple random sampling?

Solution. Let us test the gain due to cluster sampling through the concept of intraclass correlation coefficient.

Cl. No.	Values of the real estate farm loans Y_{ij}					$Y_{ij} = \begin{cases} 1 \text{ if } Y_{ij} > L \\ 0 \text{ otherwise} \end{cases}$					a_i	P_i	$P_i(1-P_i)$
1	1100.745	114.899	292.965	5.860	53.753	1	0	0	0	0	1	0.20	0.16
2	1343.461	54.633	140.582	2.605	40.775	1	0	0	0	0	1	0.20	0.16
3	56.908	100.964	315.809	413.777	1337.852	0	0	0	0	1	1	0.20	0.16
4	1248.761	612.108	1049.834	282.565	907.700	1	1	1	0	1	4	0.80	0.16
5	1579.686	2327.025	1354.768	449.099	1229.752	1	1	1	0	1	4	0.80	0.16
6	2131.048	1213.024	323.028	870.720	756.169	1	1	0	1	1	4	0.80	0.16
7	627.013	408.978	939.460	825.748	87.951	1	0	1	1	0	3	0.60	0.24
8	553.266	1045.106	99.277	321.583	639.571	1	1	0	0	1	3	0.60	0.24
9	201.631	57.747	6.044	8.849	7.590	0	0	0	0	0	0	0.00	0.00
10	1.611	7.130	39.860	42.808	139.628	0	0	0	0	0	0	0.00	0.00
										Sum	21	4.00	1.44

where limit $L = \$555.4345$.

From the above table we have

$$P = \sum_{i=1}^{N} a_i \Big/ NM = \frac{21}{50} = 0.42$$

and

$$P(1-P) = 0.42 \times (1 - 0.42) = 0.2436 .$$

The value of the intraclass correlation coefficient is given by

$$\rho = \frac{N-1}{M-1}\left\{1 - \frac{1}{(N-1)}\sum_{i=1}^{N}\frac{P_i(1-P_i)}{P(1-P)}\right\} = \frac{10-1}{5-1}\left\{1 - \frac{1}{(10-1)} \times \frac{1.44}{0.2436}\right\} = 0.7722 .$$

which again shows that the cluster sampling is less efficient than simple random sampling. The relative efficiency of cluster sampling over simple random sampling is given by

$$\text{RE} = (N-1)NP(1-P) \Big/ \left\{(NM-1)\left\{NP(1-P) - \sum_{i=1}^{N} P_i(1-P_i)\right\}\right\}$$

$$= \frac{(10-1)\times 10\times 0.4(1-0.4)}{(10\times 5-1)\{10\times 0.4(1-0.4)-1.44\}} = \frac{21.6}{47.04} = 0.4592 .$$

We have assumed that all clusters are of equal size, which restricts the applicability of cluster sampling in actual practice. For example, villages or suburbs which are groups of households, or households which are groups of persons, or schools which are groups of students and teachers, are usually taken as clusters for operational convenience. We see that unequal cluster sampling is a more practical situation. We will now discuss the cluster sampling scheme with clusters of different sizes.

9.4 NON-OVERLAPPING CLUSTERS OF DIFFERENT SIZES

Consider the ith cluster consists of M_i, $i = 1, 2, ..., N$, units and $M_0 = \sum_{i=1}^{N} M_i$ is the total number of units in the population. The population mean $\overline{\overline{Y}}$ can be defined as

$$\overline{\overline{Y}} = \frac{1}{M_0}\sum_{i=1}^{N}\sum_{j=1}^{M_i} Y_{ij} = \frac{1}{M_0}\sum_{i=1}^{N} M_i \overline{Y}_{i\bullet} \tag{9.4.1}$$

where $\overline{Y}_{i\bullet} = \frac{1}{M_i}\sum_{j=1}^{M_i} Y_{ij}$ denotes the ith cluster mean. Suppose n clusters of unequal size are selected with SRSWOR sampling and y_{ij} denotes the value of the jth unit of the variable under study in the ith cluster. The following three estimators of population mean can be suggested.

$$\overline{y}_n = \frac{1}{n}\sum_{i=1}^{n} \overline{y}_{i\bullet}, \tag{9.4.2}$$

where $\overline{y}_{i\bullet} = \frac{1}{M_i}\sum_{j=1}^{M_i} y_{ij}$,

$$\overline{y}_n^{*} = \frac{1}{m_0}\sum_{i=1}^{n} M_i \overline{y}_{i\bullet}, \tag{9.4.3}$$

where $m_0 = \sum_{i=1}^{n} M_i$, and

$$\overline{y}_n^{**} = \frac{1}{n\overline{M}}\sum_{i=1}^{n} M_i \overline{y}_{i\bullet}, \tag{9.4.4}$$

where $\overline{M} = \frac{1}{N}\sum_{i=1}^{N} M_i = \frac{M_0}{N}$.

Theorem 9.4.1. The simple arithmetic mean estimator

$$\overline{y}_n = \frac{1}{n}\sum_{i=1}^{n} \overline{y}_{i\bullet}, \tag{9.4.5}$$

where $\overline{y}_{i\bullet} = \frac{1}{M_i}\sum_{j=1}^{M_i} y_{ij}$, is a biased estimator of population mean.

Proof. Taking the expected value on both sides of (9.4.5) we have

$$E(\bar{y}_n) = E\left[\frac{1}{n}\sum_{i=1}^{n}\bar{y}_{i\bullet}\right] = \frac{1}{n}\sum_{i=1}^{n}E(\bar{y}_{i\bullet}) = \frac{1}{N}\sum_{i=1}^{N}\bar{Y}_{i\bullet} = \bar{Y}_N \neq \bar{\bar{Y}}.$$

Thus the estimator $\bar{y}_n$ is a biased estimator and the bias can be written as

$$B(\bar{y}_n) = E(\bar{y}_n) - \bar{\bar{Y}} = \frac{1}{N}\sum_{i=1}^{N}\bar{Y}_{i\bullet} - \frac{1}{N\bar{M}}\sum_{i=1}^{N}M_i\bar{Y}_{i\bullet} = -\frac{1}{N\bar{M}}\sum_{i=1}^{N}(M_i - \bar{M})\bar{Y}_{i\bullet}$$

$$= -\frac{1}{N\bar{M}}\sum_{i=1}^{N}(M_i - \bar{M})\left(\bar{Y}_{i\bullet} - \bar{\bar{Y}}\right) = -\frac{1}{\bar{M}}\text{Cov}(\bar{Y}_{i\bullet}, M_i). \tag{9.4.6}$$

The expression (9.4.6) indicates that the estimator $\bar{y}_n$ is a biased estimator of population mean $\bar{\bar{Y}}$. It is to be noted that if $\bar{Y}_{i\bullet}$ and M_i are uncorrelated, the bias in the estimator (9.4.5) will be zero.

Theorem 9.4.2. The variance of the estimator $\bar{y}_n$ is given by

$$V(\bar{y}_n) = \frac{(1-f)}{n}S_b^2, \tag{9.4.7}$$

where $S_b^2 = \frac{1}{N-1}\sum_{i=1}^{N}\left(\bar{Y}_{i\bullet} - \bar{\bar{Y}}\right)^2$.

Proof. We have

$$V(\bar{y}_n) = E[\bar{y}_n - E(\bar{y}_n)]^2 = E\left[\bar{y}_n - \bar{Y}_N\right]^2 = \frac{(1-f)}{n}\frac{1}{(N-1)}\sum_{i=1}^{N}\left(\bar{Y}_{i\bullet} - \bar{\bar{Y}}\right)^2.$$

Hence the theorem.

Corollary 9.4.1. An unbiased estimator of $V(\bar{y}_n)$ is given by

$$\hat{v}(\bar{y}_n) = \frac{(1-f)}{n}s_b^2, \tag{9.4.8}$$

where $s_b^2 = (n-1)^{-1}\sum_{i=1}^{n}\left(\bar{y}_{i\bullet} - \bar{\bar{y}}_n\right)^2$.

Theorem 9.4.3. The bias and variance of the second estimator of population mean

$$\bar{y}_n^* = \frac{1}{m_0}\sum_{i=1}^{n}M_i\bar{y}_{i\bullet}, \tag{9.4.9}$$

where $m_0 = \sum_{i=1}^{n}M_i$, are respectively given by

$$B\left(\bar{y}_n^*\right) \approx \frac{1}{nM_0^2}\left[V(m_0) - \text{Cov}\left\{m_0, \sum_{i=1}^{n}M_i\bar{y}_{i\bullet}\right\}\right] \tag{9.4.10}$$

and

$$V\left(\bar{y}_n^*\right) = \frac{(1-f)}{n}S_b^{*2}, \tag{9.4.11}$$

where $S_b^{*2} = \frac{1}{\bar{M}^2(N-1)}\sum_{i=1}^{N}M_i^2\left(\bar{Y}_{i\bullet} - \bar{\bar{Y}}\right)^2$.

Proof. In the estimator $\bar{y}_n^*$, both the numerator $\sum_{i=1}^{n} M_i \bar{y}_{i\bullet}$ and denominator m_0 are random variables and hence form a ratio of two random variables. From the standard ratio method of estimation, the asymptotic bias in the $\bar{y}_n^*$ is given by

$$B(\bar{y}_n^*) \approx \frac{1}{nM_0^2}\left[V(m_0) - \text{Cov}\left\{m_0, \sum_{i=1}^{n} M_i \bar{y}_{i\bullet}\right\}\right]. \tag{9.4.12}$$

Clearly $B(\bar{y}_n^*) \to 0$ as $n \to \infty$, $\bar{y}_n^*$ is an asymptotically unbiased estimator of population mean. One can easily see that the variance of the estimator $\bar{y}_n^*$ is

$$V(\bar{y}_n^*) = E[\bar{y}_n^* - \bar{Y}]^2 = \frac{(1-f)}{n} S_b^{*2}, \tag{9.4.13}$$

where $S_b^{*2} = \dfrac{1}{\bar{M}^2(N-1)} \sum_{i=1}^{N} M_i^2 \left(\bar{Y}_{i\bullet} - \bar{\bar{Y}}\right)^2$.

Corollary 9.4.2. An estimator of the variance of $\bar{y}_n^*$ is given by

$$\hat{v}(\bar{y}_n^*) = \frac{(1-f)}{n} s_b^{*2}, \tag{9.4.14}$$

where

$$s_b^{*2} = \frac{1}{\bar{m}_0^{\,2}(n-1)} \sum_{i=1}^{n} M_i^2 (\bar{y}_{i\bullet} - \bar{y}_n^*)^2 \quad \text{with} \quad \bar{m}_0 = n^{-1}\sum_{i=1}^{n} M_i.$$

Theorem 9.4.4. The third estimator $\bar{y}_n^{**}$ of population mean is unbiased and its variance is given by

$$V(\bar{y}_n^{**}) = \frac{(1-f)}{n} S_b^{**2}, \tag{9.4.15}$$

where $S_b^{**2} = (N-1)^{-1} \sum_{i=1}^{N} \left(\dfrac{M_i}{\bar{M}} \bar{Y}_{i\bullet} - \bar{\bar{Y}}\right)^2$.

Proof. We have

$$E(\bar{y}_n^{**}) = E\left[\frac{1}{n\bar{M}} \sum_{i=1}^{n} M_i \bar{Y}_{i\bullet}\right] = \frac{1}{n\bar{M}} \sum_{i=1}^{n} E(M_i \bar{Y}_{i\bullet}) = \bar{\bar{Y}}.$$

Hence $\bar{y}_n^{**}$ is an unbiased estimator of population mean. By the definition of variance we have

$$V(\bar{y}_n^{**}) = E[\bar{y}_n^{**} - E(\bar{y}_n^{**})]^2 = E\left[\frac{1}{n\bar{M}} \sum_{i=1}^{n} M_i \bar{Y}_{i\bullet} - \bar{\bar{Y}}\right]^2 = \frac{(1-f)}{n} S_b^{**2}.$$

Hence the theorem.

Corollary 9.4.3. An unbiased estimator of the $V(\bar{y}_n^{**})$ is given by

$$\hat{v}(\bar{y}_n^{**}) = \frac{(1-f)}{n} s_b^{**2}, \tag{9.4.16}$$

where $s_b^{**2} = (n-1)^{-1} \sum_{i=1}^{n} \left(\dfrac{M_i}{\bar{M}} \bar{y}_{i\bullet} - \bar{y}_n^{**}\right)^2$.

Theorem 9.4.5. The efficiency of the estimator $\bar{y}_n^{**}$ with respect to SRSWOR sampling is

$$\text{RE} = S^2 / \left(\overline{M} S_b^{**2}\right). \tag{9.4.17}$$

Proof. In unequal cluster sampling, the total number of units in the sample will be $m_0 = \sum_{i=1}^{n} M_i$. The expected value of m_0 is

$$E(m_0) = nE\left(\frac{1}{n}\sum_{i=1}^{n} M_i\right) = \frac{n}{N}\sum_{i=1}^{N} M_i = n\overline{M}.$$

If an equivalent sample of $n\overline{M}$ is directly selected from the population of $N\,\overline{M}$ elements, then the variance of the sample mean estimator under SRSWOR will be

$$V_{\text{srs}}(\bar{y}) = \left(\frac{1}{n\overline{M}} - \frac{1}{N\overline{M}}\right) S^2$$

and that of $\bar{y}_n^{**}$ is

$$V\left(\bar{y}_n^{**}\right) = \left(\frac{1}{n} - \frac{1}{N}\right) S_b^{**2}.$$

By the definition of relative efficiency we have

$$\text{RE} = V_{\text{srs}}(\bar{y}) / V\left(\bar{y}_n^{**}\right) = S^2 / \left(\overline{M} S_b^{**2}\right). \tag{9.4.18}$$

Hence the theorem.

Example 9.4.1. We wish to estimate the average yield/hectare of the world tobacco crop. Select four continents from population 5 by SRSWOR sampling. Collect information about the yield/hectare from all the countries in the selected continents. Estimate the average yield/hectare in the world using three different estimators. Also construct the 95% confidence interval in each situation.

Solution. We used the first two columns of the Pseudo-Random Numbers (PRN) Table 1 given in the Appendix to select the four distinct random numbers as 01, 04, 05 and 03. The continents 'Central America', 'Western & Eastern Europe', 'FSU-12' and 'European Union' will be included in the sample. Thus we have the following sample information.

Random No.	Cluster/ Continent	Cluster Unit No.	Population Unit No.	Country	Yield Metric tons (y_{ij})	$\bar{y}_{i\bullet}$
01	Central America	1	1	Costa Rica	2.03	
		2	2	El Salvador	1.79	
		3	3	Guatemala	2.21	
		4	4	Honduras	1.78	
		5	5	Nicaragua	2.03	
		6	6	Panama	2.00	
					Average	**1.973**

Continued....

03	European Union	1	13	Austria	1.90	
		2	14	Belgium--Lux	3.69	
		3	15	France	2.80	
		4	16	Germany	2.40	
		5	17	Greece	1.96	
		6	18	Italy	2.77	
		7	19	Portugal	2.14	
		8	20	Spain	2.79	
					Average	**2.556**
04	Western and Eastern Europe	1	21	Switzerland	2.09	
		2	22	Albania	0.63	
		3	23	Bulgaria	1.61	
		4	24	Czech Republic	1.64	
		5	25	Croatia	1.67	
		6	26	Hungary	1.75	
		7	27	Macedonia	1.36	
		8	28	Poland	2.34	
		9	29	Romania	1.23	
		10	30	Serbia/Montenegro	1.17	
					Average	**1.549**
05	FSU-12	1	31	Azerbaijan	2.33	
		2	32	Armenia	0.26	
		3	33	Byelarus	2.42	
		4	34	Georgia	1.63	
		5	35	Kyrgyzstan	2.50	
		6	36	Kazakhstan	1.82	
		7	37	Moldova	2.05	
		8	38	Russia	0.92	
		9	39	Tajikistan	2.48	
		10	40	Turkmenistan	2.36	
		11	41	Ukranine	0.84	
		12	42	Uzbekistan	2.37	
					Average	**1.832**

Estimator 1. Here we have

M_i	$\bar{y}_{i\bullet}$	$\left(\bar{y}_{i\bullet}-\bar{y}_n\right)^2$
6	1.973	0.000020
8	2.556	0.334662
10	1.549	0.183612
12	1.832	0.021170
Sum	7.910	0.539465

Thus the first estimate of average yield/hectare of the world tobacco crop is

$$\bar{y}_n = \frac{1}{n}\sum_{i=1}^{n}\bar{y}_{i\bullet} = \frac{7.91}{4} = 1.9775\,.$$

Note that

$$s_b^2 = (n-1)^{-1}\sum_{i=1}^{n}(\bar{y}_{i\bullet} - \bar{y}_n)^2 = 0.179822\,.$$

Thus the estimator of the variance of the estimator $\bar{y}_n$ is given by

$$\hat{v}(\bar{y}_n) = \frac{(1-f)}{n}s_b^2 = \frac{1-4/10}{4}\times 0.179822 = 0.0269733\,.$$

A $(1-\alpha)100\%$ CI of the average yield/hectare of the world tobacco crop is given by

$$\bar{y}_n \mp t_{\alpha/2}\left(\text{df} = \sum_{i=1}^{n}M_i - n\right)\sqrt{\hat{v}(\bar{y}_n)}$$

Using Table 2 from the Appendix the 95% confidence interval is given by

$$1.9775 \mp 2.037\sqrt{0.0269733}\,, \quad \text{or} \quad [1.6429,\ 2.3117].$$

Estimator 2. We have

Cluster no.	M_i	$\bar{y}_{i\bullet}$	$M_i\bar{y}_{i\bullet}$	$M_i^2\left(\bar{y}_{i\bullet} - \bar{y}_n^*\right)^2$
1	6	1.973	11.838	0.0446
2	8	2.556	20.448	24.4590
3	10	1.549	15.490	15.1160
4	12	1.832	21.984	1.6120
Sum	36	7.910	69.760	41.2320

Thus

$$\overline{m}_0 = \frac{1}{n}\sum_{i=1}^{n}M_i = \frac{36}{4} = 9$$

and an estimate of the average yield/hectare of the world tobacco crop is given by

$$\bar{y}_n^* = \frac{1}{m_0}\sum_{i=1}^{n}M_i\bar{y}_{i\bullet} = \frac{69.76}{36} = 1.9378.$$

Note that

$$s_b^{*2} = \frac{1}{\overline{m}^2(n-1)}\sum_{i=1}^{n}M_i^2\left(\bar{y}_{i\bullet} - \bar{y}_n^*\right)^2 = \frac{41.232}{9^2(4-1)} = 0.1697\,.$$

Thus an estimate of the variance of the estimator of population mean is

$$\hat{v}\left(\bar{y}_n^*\right) = \frac{(1-f)}{n}s_b^{*2} = \frac{1-0.4}{4}\times 0.1697 = 0.02545.$$

A $(1-\alpha)100\%$ confidence interval of the average yield/hectare of the world tobacco crop is given by

$$\bar{y}_n^* \mp t_{\alpha/2}\left(\text{df} = \sum_{i=1}^{n}M_i - n\right)\sqrt{\hat{v}\left(\bar{y}_n^*\right)}\,.$$

Using Table 2 from the Appendix the 95% confidence interval is given by

$$1.9378 \mp 2.037\sqrt{0.02545}\,, \quad \text{or} \quad [1.6128,\ 2.2627].$$

Estimator 3. Note that $\overline{M} = 10.6$, and an estimate of the average yield/hectare of the world tobacco crop is given by

$$\bar{y}_n^{**} = \frac{1}{n\overline{M}} \sum_{i=1}^{n} M_i \bar{y}_{i\bullet} = \frac{69.76}{4 \times 10.6} = 1.6453.$$

Now we have

M_i	$\bar{y}_{i\bullet}$	$M_i\,\bar{y}_{i\bullet}$	$\left(\frac{M_i}{\overline{M}}\bar{y}_{i\bullet} - \bar{y}_n^{**}\right)^2$
6	1.973	11.838	0.2793202
8	2.556	20.448	0.0805178
10	1.549	15.490	0.0338484
12	1.832	21.984	0.1522993
Sum	7.910	69.760	0.5459857

Thus we have

$$s_b^{**2} = (n-1)^{-1} \sum_{i=1}^{n} \left(\frac{M_i}{\overline{M}} \bar{y}_{i\bullet} - \bar{y}_n^{**} \right)^2 = \frac{0.5459857}{3} = 0.1819952$$

and an estimate of variance is

$$\hat{v}\left(\bar{y}_n^{**}\right) = \frac{(1-f)}{n} s_b^{**2} = \frac{1-0.4}{4} \times 0.1819952 = 0.027299.$$

A $(1-\alpha)100\%$ confidence interval of the average yield/hectare of the world tobacco crop is given by

$$\bar{y}_n^{**} \mp t_{\alpha/2}\left(\mathrm{df} = \sum_{i=1}^{n} M_i - n \right) \sqrt{\hat{v}\left(\bar{y}_n^{**}\right)}.$$

Using Table 2 from the Appendix the 95% confidence interval is given by

$$1.6453 \mp 2.037\sqrt{0.027299} \quad \text{or} \quad [1.3087,\ 1.9818].$$

Now we would like to discuss how ANOVA works for unequal cluster sampling.

How population looks with clusters of unequal sizes?							
Cluster No.	Units in the ith cluster						Cluster Totals
1	Y_{11}	Y_{12}		Y_{1j}		Y_{1M_1}	$Y_{1\bullet}$
2	Y_{21}	Y_{22}		Y_{2j}		Y_{2M_2}	$Y_{2\bullet}$
i	Y_{i1}	Y_{i2}		Y_{ij}		Y_{iM_i}	$Y_{i\bullet}$
N	Y_{N1}	Y_{N2}		Y_{Nj}		Y_{NM_N}	$Y_{N\bullet}$
				Population Grand Total (GT)			$Y_{\bullet\bullet}$

(a) The population correction factor (CF) is given by

$$\text{CF} = \frac{Y_{\bullet\bullet}^2}{M_\bullet}.$$

(b) The population corrected total sum of squares (SST) is given by

$$\text{SST} = \sum_{i=1}^{N}\sum_{j=1}^{M_i} Y_{ij}^2 - \text{CF} = \sum_{i=1}^{N}\sum_{j=1}^{M_i} Y_{ij}^2 - \frac{Y_{\bullet\bullet}^2}{M_\bullet} = \sum_{i=1}^{N}\sum_{j=1}^{M_i}\left(Y_{ij} - \overline{\overline{Y}}\right)^2 = (M_\bullet - 1)S^2.$$

(c) The population corrected between sum of squares (SSB) is given by

$$\text{SSB} = \sum_{i=1}^{N}\frac{Y_{i\bullet}^2}{M_i} - \text{CF} = \sum_{i=1}^{N}\frac{Y_{i\bullet}^2}{M_i} - \frac{Y_{\bullet\bullet}^2}{M_\bullet} = \sum_{i=1}^{N} M_i\left(\overline{Y}_{i\bullet}^2\right) - M_\bullet \overline{\overline{Y}}^2 = M_\bullet\left\{\sum_{i=1}^{N} P_i \overline{Y}_{i\bullet}^2 - \overline{\overline{Y}}^2\right\}$$

$$= \sum_{i=1}^{N} M_i\left(\overline{Y}_{i\bullet} - \overline{\overline{Y}}\right)^2 = (N-1)S_b'^2, \text{ where } P_i = M_i / M_\bullet$$

(d) The population corrected within sum of squares (SSW) is given by

$$\text{SSW} = \text{SST} - \text{SSW} = \sum_{i=1}^{N}\sum_{j=1}^{M_i} Y_{ij}^2 - \sum_{i=1}^{N}\frac{Y_{i\bullet}^2}{M_i} = \sum_{i=1}^{N}\left[\sum_{j=1}^{M_i}\left(Y_{ij} - \overline{Y}_{i\bullet}\right)^2\right] = (M_\bullet - 1)S^2.$$

Thus the ANOVA for unequal sized clusters will be

ANOVA for population with unequal sized clusters				
Source	Population degree of Freedom DF	Sum of Squares SS	Mean Sum of Squares MSS	Population *F* ratio
Between Clusters	$N-1$	SSB $=\sum_{i=1}^{N} M_i\left(\overline{Y}_{i\bullet} - \overline{\overline{Y}}\right)^2$	MSB $=\frac{1}{N-1}\sum_{i=1}^{N} M_i\left(\overline{Y}_{i\bullet} - \overline{\overline{Y}}\right)^2$ $= S_b'^2$	$F = \frac{\text{MSB}}{\text{MSW}}$
Within Clusters	$N(\overline{M}-1)$	SSW $=\sum_{i=1}^{N}\sum_{j=1}^{M_i}\left(Y_{ij} - \overline{Y}_{i\bullet}\right)^2$	MSW $=\frac{1}{N(\overline{M}-1)}\sum_{i=1}^{N}\sum_{j=1}^{M_i}\left(Y_{ij} - \overline{Y}_{i\bullet}\right)^2$ $= \overline{S}_w^2$	
Total	$N\overline{M}-1$	SST $=\sum_{i=1}^{N}\sum_{j=1}^{M_i}\left(Y_{ij} - \overline{\overline{Y}}\right)^2$	MST $=\frac{1}{N\overline{M}-1}\sum_{i=1}^{N}\sum_{j=1}^{M_i}\left(Y_{ij} - \overline{\overline{Y}}\right)^2$ $= S^2$	

Remark 9.4.1. If clusters are not of equal size, then an alternative measure is adjusted values of coefficient of determination given by

$$R_a^2 = 1 - \frac{\text{MSW}}{\text{MST}}.$$

Further note that the value of adjusted R_a^2 becomes negative if the cluster sampling is efficient. However Zasepa (1962) has shown that the value of intraclass correlation coefficient for unequal clusters can be found as

$$\rho = \frac{\sum_{i=1}^{N}\sum_{j\neq j'=1}^{M_i}\left(Y_{ij}-\overline{\overline{Y}}\right)\left(Y_{ij'}-\overline{\overline{Y}}\right)}{M_{\bullet}\left(\overline{M}-1\right)S^2}.$$

Example 9.4.2. Mr. Ken Brewer learned from Mr. Stephen Horn that the clusters are supposed to be heterogeneous and suggested the following clustering plan:

Cluster No.				
I	North West	West North Central	East North Central	North East
II	West	South West	South	
III	Central	South East		

Verify if Mr. Brewer's new clustering plan may result in efficient estimates? Apply ANOVA approach.

Solution. From the map of Dec. 2002 precipitation (Refer to Example 9.1.5) we have the following information:

Cluster					Cluster Totals
1	4	72	93	64	233
2	3	63	104		170
3	75	90			165
Population Grand Total (GT)					568

(a) The population correction factor is given by

$$\text{CF} = \frac{Y_{\bullet\bullet}^2}{M_{\bullet}} = \frac{568^2}{9} = 35847.11.$$

(b) The population corrected total sum of squares (SST) is given by

$$\text{SST} = \sum_{i=1}^{N}\sum_{j=1}^{M_i} Y_{ij}^2 - \text{CF}$$

$$= 4^2 + 72^2 + 93^2 + 64^2 + 3^2 + 63^2 + 104^2 + 75^2 + 90^2 - 35847.11$$

$$= 10616.89.$$

(c) The population corrected between sum of squares is given by

$$\text{SSB} = \sum_{i=1}^{N}\frac{Y_{i\bullet}^2}{M_i} - \text{CF} = \frac{233^2}{4} + \frac{170^2}{3} + \frac{165^2}{2} - 35847.11 = 970.97.$$

(d) The population corrected within sum of squares is given by

$$\text{SSW} = \text{SST} - \text{SSB} = 10616.89 - 970.97 = 9645.92.$$

Thus the ANOVA for Mr. Ken Brewer's new clustering plan will be:

ANOVA for Mr. Ken Brewer's new clustering plan				
Source	Population degree of Freedom DF	Sum of Squares SS	Mean Sum of Squares MSS	Population F ratio
Between Clusters	2	970.97	485.48	$F = 0.301$
Within Clusters	6	9645.92	1607.65	
Total	8	10616.89	1327.11	

Note that if the clusters are not of equal size then an alternative measure is adjusted values of coefficient of determination given by

$$R_a^2 = 1 - \frac{\text{MSW}}{\text{MST}} = 1 - \frac{1607.65}{1327.11} = -0.214,$$

which is negative, and it shows that Mr. Ken Brewer's new clustering plan will perform better than SRSWOR sampling. Ken! Well done!.

9.5 SELECTION OF NON-OVERLAPPING CLUSTERS WITH UNEQUAL PROBABILITY SAMPLING

In most of the practical situations the study variable is found to be positively and highly correlated with the cluster size. Under such circumstances, it is recommended to select the sample with probability proportional to the size of clusters. Let $P_i = Z_i / Z$ be the probability of selecting the ith cluster in the sample. Following Hansen and Hurwitz (1943), let us define a transformed variable, $u_{ij} = M_i y_{ij} / M_0 P_i$, $j = 1, 2, ..., M_i$; $i = 1, 2, ..., N$. Assuming that n clusters are selected by PPSWR sampling scheme, define

$$\bar{u}_{i\bullet} = \frac{M_i \bar{y}_{i\bullet}}{M_0 P_i} \quad \text{for} \quad i = 1, 2, ..., n\,.$$

Then we have the following theorem:

Theorem 9.5.1. An unbiased estimator of population mean is given by

$$\bar{y}_{\text{pps}} = \frac{1}{n}\sum_{i=1}^{n} \bar{u}_{i\bullet} = \frac{1}{n}\sum_{i=1}^{n} \frac{M_i \bar{y}_{i\bullet}}{M_0 P_i} \tag{9.5.1}$$

and its variance is given by

$$V\left(\bar{y}_{\text{pps}}\right) = \frac{1}{n}\sum_{i=1}^{N} P_i \left(\bar{u}_{i\bullet} - \bar{\bar{Y}}\right)^2 . \tag{9.5.2}$$

Proof. Taking expected values on both sides of (9.5.1) we have

$$E\left(\bar{y}_{\text{pps}}\right)=E\left(\frac{1}{n}\sum_{i=1}^{n}\bar{u}_{i\bullet}\right)=\frac{1}{n}\sum_{i=1}^{n}E\left(\bar{u}_{i\bullet}\right)=\frac{1}{n}\sum_{i=1}^{n}\sum_{i=1}^{N}P_i\frac{M_i\bar{Y}_{i\bullet}}{M_0P_i}=\bar{\bar{Y}}.$$

Following Hansen and Hurwitz (1943) the variance of $\bar{y}_{\text{pps}}$ is given by

$$V\left(\bar{y}_{\text{pps}}\right)=E\left(\bar{y}_{\text{pps}}^2\right)-\bar{\bar{Y}}^2=\frac{1}{n}\sum_{i=1}^{N}P_i\left(\bar{u}_{i\bullet}-\bar{\bar{Y}}\right)^2.$$

Hence the theorem.

Theorem 9.5.2. If $P_i=M_i/M_0$, where $M_0=\sum_{i=1}^{N}M_i$, then an unbiased estimator of population mean $\bar{\bar{Y}}$ is

$$\bar{y}_{\text{pps(cl)}}=\frac{1}{n}\sum_{i=1}^{n}\bar{y}_{i\bullet}. \tag{9.5.3}$$

Proof. We have

$$E\left(\bar{y}_{\text{pps(cl)}}\right)=E\left(\frac{1}{n}\sum_{i=1}^{n}\bar{y}_{i\bullet}\right)=\frac{1}{n}\sum_{i=1}^{n}\left[\sum_{i=1}^{N}P_i\bar{Y}_{i\bullet}\right]=\sum_{i=1}^{N}\frac{M_i}{M_0}\frac{1}{M_i}\sum_{j=1}^{M_i}Y_{ij}=\frac{1}{M_0}\sum_{i=1}^{N}\sum_{j=1}^{M_i}Y_{ij}=\bar{\bar{Y}}.$$

Hence the theorem.

Theorem 9.5.3. The variance of the estimator

$$\bar{y}_{\text{pps(cl)}}=\frac{1}{n}\sum_{i=1}^{n}\bar{y}_{i\bullet}$$

is given by

$$V\left(\bar{y}_{\text{pps(cl)}}\right)=\frac{1}{n}\sum_{i=1}^{N}\frac{M_i}{M_0}\left(\bar{Y}_{i\bullet}-\bar{\bar{Y}}\right)^2. \tag{9.5.4}$$

Proof. We have

$$\bar{y}_{\text{pps(cl)}}=\frac{1}{n}\sum_{i=1}^{n}\bar{y}_{i\bullet}=\frac{1}{n}\sum_{i=1}^{n}\frac{P_i\bar{y}_{i\bullet}}{P_i}$$

Thus

$$V\left(\bar{y}_{\text{pps(cl)}}\right)=V\left[\frac{1}{n}\sum_{i=1}^{n}\bar{y}_{i\bullet}\right]=V\left[\frac{1}{n}\sum_{i=1}^{n}\frac{P_i\bar{y}_{i\bullet}}{P_i}\right]=\frac{1}{n}\sum_{i=1}^{N}P_i\left(\frac{P_i\bar{Y}_{i\bullet}}{P_i}-\bar{\bar{Y}}\right)^2=\frac{1}{n}\sum_{i=1}^{N}P_i\left(\bar{Y}_{i\bullet}-\bar{\bar{Y}}\right)^2.$$

Hence the theorem.

Theorem 9.5.4. The relative efficiency of the estimator $\bar{y}_{\text{pps(cl)}}$ with respect to SRSWR is given by

$$\text{RE}=\left\{\bar{M}\left(1-\frac{\sigma_w^2}{\sigma^2}\right)\right\}^{-1},\quad \text{where } \sigma^2=\frac{1}{\bar{M}M_0}\sum_{i=1}^{N}\sum_{j=1}^{M_i}\left(Y_{ij}-\bar{\bar{Y}}\right)^2 \tag{9.5.5}$$

Proof. Under SRSWR sampling we have

$$V\left(\bar{y}_{\text{srs}}\right)=\frac{1}{n\bar{M}}\left[\frac{1}{M_0}\sum_{i=1}^{N}\sum_{j=1}^{M_i}\left(Y_{ij}-\bar{\bar{Y}}\right)^2\right]=\frac{1}{n\bar{M}}\frac{1}{M_0}\sum_{i=1}^{N}\sum_{j=1}^{M_i}\left(Y_{ij}-\bar{Y}_{i\bullet}+\bar{Y}_{i\bullet}-\bar{\bar{Y}}\right)^2$$

$$= \frac{1}{n\overline{M}} \frac{1}{M_0} \sum_{i=1}^{N} \left[\sum_{j=1}^{M_i} \left\{ \left(Y_{ij} - \overline{Y}_{i\bullet}\right)^2 + \left(\overline{Y}_{i\bullet} - \overline{\overline{Y}}\right)^2 \right\} \right]$$

$$= \frac{1}{n\overline{M}} \frac{1}{M_0} \sum_{i=1}^{N} \left[\frac{M_i}{M_i} \sum_{j=1}^{M_i} \left(Y_{ij} - \overline{Y}_{i\bullet}\right)^2 + M_i \left(\overline{Y}_{i\bullet} - \overline{\overline{Y}}\right)^2 \right]$$

$$= \frac{1}{n\overline{M}} \sum_{i=1}^{N} \left[\frac{M_i \sigma_i^2}{M_0} + \frac{M_i}{M_0} \left(\overline{Y}_{i\bullet} - \overline{\overline{Y}}\right)^2 \right] = \frac{1}{\overline{M}} \left[\frac{\sigma_w^2}{n} + V\left(\overline{y}_{\text{pps(cl)}}\right) \right]$$

where

$$\sigma_w^2 = \sum_{i=1}^{N} \frac{M_i \sigma_i^2}{M_0}.$$

This implies

$$V\left(\overline{y}_{\text{pps(cl)}}\right) = \overline{M} V\left(\overline{y}_{\text{srs}}\right) - \frac{\sigma_w^2}{n} = \frac{\sigma^2}{n\overline{M}} \times \overline{M} - \frac{\sigma_w^2}{n} = \frac{1}{n}\left(\sigma^2 - \sigma_w^2\right).$$

Thus the relative efficiency of cluster sampling is given by

$$\text{RE} = \frac{V\left(\overline{y}_{\text{srs}}\right)}{V\left(\overline{y}_{\text{pps(cl)}}\right)} = \frac{\dfrac{\sigma^2}{n\overline{M}}}{\dfrac{\sigma^2 - \sigma_w^2}{n}} = \left[\overline{M} \left(1 - \frac{\sigma_w^2}{\sigma^2}\right) \right]^{-1}.$$

Hence the theorem.

Example 9.5.1. We wish to estimate the production of the world tobacco crop. Assume we form 10 clusters of the countries in the world based on the continents listed in population 5. We wish to apply PPSWR sampling for selecting the cluster. Should we expect any gain in efficiency over simple random sampling?

Solution. We have

M_i	σ_i^2	$M_i\sigma_i^2$
6	0.022356	0.134133
6	0.181742	1.090450
8	0.303623	2.428986
10	0.211104	2.111040
12	0.533628	6.403540
4	0.114825	0.459300
30	0.332388	9.971650
17	0.356282	6.056800
10	1.816470	18.164700
3	0.649733	1.949200
	Sum	48.769800

Thus we have

$$\sigma_w^2 = \sum_{i=1}^{N} \frac{M_i \sigma_i^2}{M_0} = \frac{48.7698}{106} = 0.46009 .$$

Also we have

$$\sigma^2 = \frac{1}{\overline{M} M_0} \sum_{i=1}^{N} \sum_{j=1}^{M_i} \left(Y_{ij} - \overline{\overline{Y}} \right)^2 = \left(\frac{N-1}{N} \right) S_y^2 = \left(\frac{106-1}{106} \right) \times 0.6323 = 0.626335 .$$

Thus the relative efficiency of PPS cluster sampling over simple random sampling is given by

$$\mathrm{RE} = \left[\overline{M} \left(1 - \frac{\sigma_w^2}{\sigma^2} \right) \right]^{-1} = \left[10.6 \times \left(1 - \frac{0.46009}{0.626335} \right) \right]^{-1} = 0.3554 .$$

Thus for this case the PPS cluster sampling will be less efficient than simple random sampling.

Raj (1954, 1958), Zarcovic (1960), and Foreman and Brewer (1971) have considered the concept of a superpopulation model for comparison purposes. It is to be remarked here that other sampling schemes like PPSWOR sampling, systematic sampling, two-phase sampling and stratified sampling can also be used to construct the estimation strategies under cluster sampling. Madow (1949), Sukhatme (1954), and Sampford (1962) have also suggested estimation strategies under cluster sampling. Singh and Singh (1999) suggested an unbiased class of estimators in cluster sampling.

9.6 OPTIMAL AND ROBUST STRATEGIES FOR NON-OVERLAPPING CLUSTER SAMPLING

Following Royall (1992), Tam (1995) has assumed that the finite population of interest $\underline{y}^t = (y_1, y_2, ..., y_N)$ is a realization of a random vector $\underline{Y}^t$ that is related to the design matrix $\underline{X} = (X_1, X_2, ..., X_N)^t$ via superpopulation model, defined as

$$M(\underline{X}, \underline{V}) : \underline{Y} = \underline{X}\, \underline{\beta} + \underline{\in} , \tag{9.6.1}$$

where $\underline{X}$ is assumed to be of full rank p, $\underline{\in}$ is a null mean vector and covariance matrix is $\sigma^2 \underline{V}$, where $\underline{V}$ is a symmetric positive definite matrix such that the variances of the derived statistics are positive, and σ and $\underline{\beta}$ are unknown quantities. We are interested in estimating the population total defined as

$$T = \underline{1}^t\, \underline{Y} \tag{9.6.2}$$

where $\underline{1}_{N\times 1}$ is a vector of ones. Let s corresponds to the sampled n units and r for the remainder of the population units $(N-n)$. Without loss of generality, the population information can be represented as

$$\underline{1}=\begin{bmatrix}\underline{1}_s\\ \underline{1}_r\end{bmatrix},\quad \underline{Y}=\begin{bmatrix}\underline{Y}_s\\ \underline{Y}_r\end{bmatrix},\quad \underline{X}=\begin{bmatrix}\underline{X}_s\\ \underline{X}_r\end{bmatrix},\quad V=\begin{bmatrix}\underline{V}_s, & \underline{V}_{sr}\\ \underline{V}_{rs}, & \underline{V}_r\end{bmatrix}$$

Then we have the following theorem:

Theorem 9.6.1. The best linear unbiased predictor (BLUP) of population total T is given by

$$\hat{T}(\underline{X},\underline{\mathrm{V}})=\underline{1}_s^t\underline{Y}_s+\underline{1}_r\underline{X}_r\underline{\hat{\beta}}(\underline{X},\underline{\mathrm{V}})+\underline{1}_r^t\underline{V}_{rs}\underline{V}_s^{-1}\left(\underline{Y}_s-\underline{X}_s\underline{\hat{\beta}}(\underline{X},\underline{\mathrm{V}})\right) \tag{9.6.3}$$

where

$$\underline{\hat{\beta}}=\left(\underline{X}_s\underline{V}_s^{-1}\underline{X}_s\right)^{-1}\underline{X}_s^t\underline{V}_s^{-1}\underline{Y}_s$$

is the best linear predictor of β with variance given by

$$V\left[\hat{T}(\underline{X},\underline{\mathrm{V}})\right]=\sigma^2\left(\underline{1}^t\underline{X}-\underline{1}^t\underline{K}^t\underline{V}_s^{-1}\underline{X}_s\right)\underline{A}_s^{-1}\left(\underline{1}^t\underline{X}-\underline{1}^t\underline{K}^t\underline{V}_s^{-1}\underline{X}_s\right)^t$$
$$+\sigma^2\left(\underline{1}^t\underline{V}\underline{1}-\underline{1}\underline{K}^t\underline{V}_s^{-1}\underline{K}\,\underline{1}\right) \tag{9.6.4}$$

where

$$\underline{A}_s=\underline{X}_s\underline{V}_s^{-1}\underline{X}_s \text{ and } K=\left(\underline{V}_s,\ \underline{V}_{\mathrm{sr}}\right).$$

Corollary 9.6.1. If $\underline{V}\,\underline{1}=\underline{X}\underline{\lambda}_1$ for some vector $\underline{\lambda}_1$ then the best linear unbiased predictor (BLUP) of population total Y is

$$\hat{T}_{\mathrm{R}}(\underline{X},\underline{\mathrm{V}})=\underline{1}^t\underline{X}\underline{\hat{\beta}}(\underline{X},\underline{\mathrm{V}}) \tag{9.6.5}$$

with variance given by

$$V\left[\hat{T}_{\mathrm{R}}(\underline{X},\underline{V})\right]=\sigma^2\left(\underline{1}^t\underline{X}\underline{A}_s^{-1}\underline{X}^t-\underline{1}^t\underline{V}\underline{1}\right). \tag{9.6.6}$$

For studying the optimal sampling strategies, Tam (1995) has considered the variance--covariance matrix of the form

$$\underline{V}=\begin{bmatrix}\underline{V}_A, & \underline{O}\\ \underline{O}, & \underline{\mathrm{V}}_{\mathrm{Z}}\end{bmatrix} \tag{9.6.7}$$

where $\underline{V}_A$ is a $(N-z)\times(N-z)$ non-diagonal matrix that has the same correlation coefficient in the off-diagonal elements, $\underline{V}_z$ is a $z\times z$ diagonal matrix, and $0\le z\le N$. In other words, the diagonal matrix V looks like

$$\underline{V}=\begin{bmatrix} v_1^2, & \rho\delta_1\delta_2, & \rho\delta_1\delta_3, & \bullet\ \bullet, & \rho\delta_1\delta_{N-z}, & 0, & 0, & \bullet\bullet, & 0 \\ \rho\delta_1\delta_2, & v_2^2, & \rho\delta_2\delta_3, & \bullet\ \bullet, & \rho\delta_2\delta_{N-z}, & 0, & 0, & \bullet\bullet, & 0 \\ \bullet \\ \bullet \\ \rho\delta_{N-z}\delta_1, & \rho\delta_{N-z}\delta_2, & \rho\delta_{N-z}\delta_3, & \bullet, \bullet, & v_{N-z}^2, & 0, & 0, & \bullet,\bullet, & 0 \\ 0, & 0, & 0, & \bullet\ \bullet, & v_{N-z+1}^2, & 0, & 0, & \bullet,\bullet, & 0 \\ 0, & 0, & 0, & \bullet, \bullet, & 0, & v_{N-z+2}^2, & \bullet,\bullet, & 0 \\ \bullet \\ \bullet \\ 0, & 0, & 0, & \bullet, \bullet, & 0, & 0, & 0, \bullet,\bullet, & 0, v_N^2 \end{bmatrix} \quad (9.6.8)$$

where

$$\delta_k=\begin{cases}1 & \text{if there is correlation } \rho,\\ 0 & \text{otherwise.}\end{cases} \quad (9.6.9)$$

The matrix $\underline{V}$ can also be expressed as

$$\underline{V}=\text{Diag}(\underline{V}_2)+\rho\underline{V}_3\underline{V}_3^t \quad (9.6.10)$$

where

$$\underline{V}_2=\begin{bmatrix}(1-\delta_1\rho)v_1^2, & 0, & 0, \ldots, & 0\\ 0, & (1-\delta_2\rho)v_2^2, & 0, \ldots\ldots & 0\\ \\ 0, & 0, & 0,\ldots\ldots, & (1-\delta_N\rho)v_N^2\end{bmatrix} \text{ and } \underline{V}_3=\begin{bmatrix}\delta_1 v_1^2, & 0, & 0,\ldots, & 0\\ 0, & \delta_2 v_2^2, & 0,\ldots & 0\\ \\ 0, & 0, & 0,\ldots, & \delta_N v_N^2\end{bmatrix}.$$

Letting n_s be the number of zeros in v_{3s}, where $v_3=\left(v_{3s}^{'},\, v_{3r}^{'}\right)$ and defining $n_s=z_s+\dfrac{(n-z_s)(1-\rho)}{1+(n-z_s-1)\rho}$, Tam (1995) extended Royall (1992) results to cluster sampling as given in the following theorems:

Theorem 9.6.2. If $\underline{V}\,\underline{1}=\underline{X}\,\underline{\lambda}_1$ and $\underline{v}_1=\underline{X}\,\underline{\lambda}_2$ for some $\underline{\lambda}_1$ and $\underline{\lambda}_2$ under $M(\underline{X},\ \underline{V})$ defined by (9.6.1) and (9.6.10) we have

$$\operatorname{var}\left\{\hat{T}(\underline{X};\ \underline{V})-T\right\}\geq\left\{\left(\underline{1}^t\,\underline{v}_1\right)^2/n_{\text{opt}}-\underline{1}^t\,\underline{V}\,\underline{1}\right\}\sigma^2 \tag{9.6.11}$$

for all samples s, where

$$n_{\text{opt}}=\begin{cases} n & \text{if} \quad z\geq n,\\ z+\dfrac{(n-z)(1-\rho)}{1+(n-z-1)\rho} & \text{if} \quad z<n. \end{cases} \tag{9.6.12}$$

The lower bound is attained by the estimator

$$\hat{T}(\underline{X};\underline{V})=\sum_{i=1}^{N}\frac{\sqrt{(1-\delta_i\rho)}v_i}{n_{opt}}\sum_{i=1}^{n_s}\left[\frac{1}{\sqrt{(1-\delta_i\rho)}}-\frac{\sqrt{(1-\rho)}\left(n-z_{opt}\right)\rho\delta_i}{(1-\delta_i\rho)\left\{1+\left(n-z_{opt}-1\right)\rho\right\}}\right]\frac{y_i}{v_i} \tag{9.6.13}$$

where

$$z_{\text{opt}}=\begin{cases} n & \text{if} \quad z\geq n,\\ z & \text{if} \quad z<n, \end{cases}$$

if and only if the sample satisfies the condition,

$$n_{\text{opt}}^{-1}\left(\underline{1}^t\,\underline{v}_1\right)\underline{v}_{1s}^t\,\underline{V}_s\,\underline{X}_s=\underline{1}^t\,\underline{X}$$

variance covariance matrix is given by

$$\underline{v}_1=\begin{bmatrix} \sqrt{(1-\delta_1\rho)}v_1, & 0, & 0\\ 0, & \sqrt{(1-\delta_2\rho)}v_2, & 0\\ 0, & 0, & \sqrt{(1-\delta_N\rho)}v_N \end{bmatrix}$$

and $\underline{v}_1$ is given by $\underline{v}_1=\left(\underline{v}_{1s}^t,\underline{v}_{1r}^t\right)$.

Theorem 9.6.3. If $\underline{V}\,\underline{1}=\underline{X}\,\underline{\lambda}_1$, $\underline{v}_1=\underline{X}\,\underline{\lambda}_2$ for some $\underline{\lambda}_1$ and $\underline{\lambda}_2$, and $\delta_i=1$ for $i=1,\ 2,\ldots,N$ under $M(\underline{X},\ \underline{V})$, then

$$\mathrm{V}\left[\hat{T}(\underline{X};\ \underline{V})-T\right]\geq\sigma^2(1-\rho)\left[\frac{1}{n}\left(\sum_{i=1}^{N}v_i\right)^2-\sum_{i=1}^{N}v_i^2\right] \tag{9.6.14}$$

for all samples s.

The estimator given by

$$\hat{T}(\underline{X};\underline{V})=\frac{1}{n}\left(\sum_{i=1}^{N}v_i\right)\left(\sum_{s}\frac{y_i}{v_i}\right) \tag{9.6.15}$$

gives the lower bound of the variance if and only if

$$n^{-1}\left(\sum_i v_i\right)\left(v_i^{-1},\ldots\ldots\ldots,v_n^{-1}\right)\underline{X}_s=\underline{1}^t\,\underline{X}. \tag{9.6.16}$$

9.7 OVERLAPPING CLUSTER SAMPLING

Owing to operational convenience and reduction in cost, sometimes cluster sampling has been found to be useful even if all elements of the population are identifiable. Clusters are formed either before selecting the sample (CBS) or after selecting the sample (CAS). In both situations the clusters may be non-overlapping or overlapping. For non-overlapping cluster sampling the theory is straight forward, as we have discussed in the previous sections of this chapter. However, there may be many practical sampling situations where one gets overlapping clusters. For example, overlapping clusters may exist in some regional epidemiological survey for a contagious disease like mycobacterium tuberculosis (T.B.), which is becoming very prevalent with the spread of AIDS (Gifford--Jones, 1993). For overlapping cluster sampling, one can refer to Goel and Singh (1977), Agarwal and Singh (1982), Amdekar (1985), Singh (1988), and Tracy and Osahan (1994b). Under the overlapping cluster sampling strategies, the population size may or may not be known.

9.7.1 POPULATION SIZE IS KNOWN

Let the population under consideration consist of N distinct and identifiable units. Assume that these N units are expressible in the form of K overlapping clusters with N_i, $(i = 1, 2, ..., K)$ units in the ith cluster and $\sum_{i=1}^{K} N_i = M \geq N$. The equality sign will hold only for non-overlapping clusters. For an overlapping cluster situation, a population unit may be included in more than one cluster and let F_j be the frequency of the jth $(j = 1, 2, ..., N)$ population unit occurring in K clusters. Let Y be the variable of interest and we are interested in estimating the population mean $\bar{Y} = N^{-1} \sum_{j=1}^{N} Y_j$.

Define

$$Z_{ij} = (M/NF_j)Y_{ij}, \quad i = 1, 2, ..., K \quad \text{and} \quad j = 1, 2, ..., N_i,$$

where Y_{ij} denotes the value of Y for the jth unit in the ith cluster. Then we have the following schemes:

Scheme 1. (a) Select k clusters out of K clusters by SRSWR sampling.
(b) From the ith selected cluster of size N_i ($i = 1, 2, ..., k$), select n_i units by SRSWOR sampling.

Under such a sampling scheme we have the following theorems:

Theorem 9.7.1.1. A biased estimator of population mean $\bar{Y}$ is given by

$$\bar{z}_{SRS} = \frac{1}{k}\sum_{i=1}^{k}\frac{1}{n_i}\sum_{j=1}^{n_i} z_{ij} . \tag{9.7.1.1}$$

Proof. Let E_2 denote the conditional expectation for a given sample of clusters and E_1 denote the expectation over all such samples, then we have

$$E(\bar{z}_{SRS}) = E\left(\frac{1}{k}\sum_{i=1}^{k}\frac{1}{n_i}\sum_{j=1}^{n_i} z_{ij}\right) = E_1E_2\left(\frac{1}{k}\sum_{i=1}^{k}\frac{1}{n_i}\sum_{j=1}^{n_i} z_{ij}\right) = E_1\left(\frac{1}{k}\sum_{i=1}^{k} E_2(\bar{z}_{i\bullet})\right)$$

$$= E_1\left(\frac{1}{k}\sum_{i=1}^{k}\bar{Z}_{i\bullet}\right) = \frac{1}{K}\sum_{i=1}^{K}\bar{Z}_{i\bullet} = \bar{Z}_K \neq \bar{Y}.$$

Hence the theorem.

Theorem 9.7.1.2. The variance of the estimator $\bar{z}_{SRS}$ is given by

$$V(\bar{z}_{SRS}) = \frac{1}{k}\left[\sigma_{bz}^2 + K^{-1}\sum_{i=1}^{K}\left(\frac{1}{n_i} - \frac{1}{N_i}\right)S_{iz}^2\right] \tag{9.7.1.2}$$

where $\sigma_{bz}^2 = K^{-1}\sum_{i=1}^{K}\left(\bar{Z}_{i\bullet} - \bar{Z}_K\right)^2$ and $S_{iz}^2 = (N_i - 1)^{-1}\sum_{j=1}^{N_i}\left(Z_{ij} - \bar{Z}_{i\bullet}\right)^2$.

Proof. Let V_2 denote the conditional variance for a given sample of clusters and V_1 denote the expectation over all such samples, then we have

$$V(\bar{z}_{SRS}) = V_1E_2(\bar{z}_{SRS}) + E_1V_2(\bar{z}_{SRS}) = V_1\left[\frac{1}{k}\sum_{i=1}^{k}E_2(\bar{z}_{i\bullet})\right] + E_1\left[\frac{1}{k^2}\sum_{i=1}^{k}V_2(\bar{z}_{i\bullet})\right]$$

$$= V_1\left[\frac{1}{k}\sum_{i=1}^{k}\bar{Z}_{i\bullet}\right] + E_1\left[\frac{1}{k^2}\sum_{i=1}^{k}\left(\frac{1}{n_i} - \frac{1}{N_i}\right)S_{iz}^2\right]$$

$$= \frac{1}{kK}\sum_{i=1}^{K}\left(\bar{Z}_{i\bullet} - \bar{Z}_K\right)^2 + \frac{1}{kK}\sum_{i=1}^{K}\left(\frac{1}{n_i} - \frac{1}{N_i}\right)S_{iz}^2.$$

Hence the theorem.

Scheme-2. (a) Select k clusters out of K clusters by PPSWR sampling, with $P_i = N_i/M$.

(b) From the ith selected cluster of size N_i, $(i = 1, 2, ..., k)$, select n_i units by SRSWOR sampling.

Then we have the following theorems:

Theorem 9.7.1.3. An unbiased estimator of population mean $\bar{Y}$ is given by

$$\bar{z}_{pps} = \frac{1}{k}\sum_{i=1}^{k}\frac{1}{n_i}\sum_{j=1}^{n_i} z_{ij}. \tag{9.7.1.3}$$

Proof. We have

$$E(\bar{z}_{pps}) = E_1E_2\left(\frac{1}{k}\sum_{i=1}^{k}\frac{1}{n_i}\sum_{j=1}^{n_i} z_{ij}\right) = E_1\left[\frac{1}{k}\sum_{i=1}^{k}E_2(\bar{z}_{i\bullet})\right] = E_1\left(\frac{1}{k}\sum_{i=1}^{k}\bar{Z}_{i\bullet}\right) = \sum_{i=1}^{K}\frac{N_i}{M}\bar{Z}_{i\bullet}$$

$$= \frac{1}{N}\sum_{i=1}^{K}\sum_{j=1}^{N_i}\frac{Y_{ij}}{F_j} = \bar{Y}.$$

Hence the theorem.

Theorem 9.7.1.4. The variance of the estimator $\bar{z}_{\text{pps}}$ is given by

$$V(\bar{z}_{\text{pps}}) = \frac{\sigma_{bz}^2}{k} + \frac{1}{k}\sum_{i=1}^{K}P_i\left(\frac{1}{n_i} - \frac{1}{N_i}\right)S_{iz}^2 \qquad (9.7.1.4)$$

where $\sigma_{bz}^2 = \sum_{i=1}^{K}P_i(\bar{Z}_{i\bullet} - \bar{Y})^2$.

Proof. We have

$$V(\bar{z}_{\text{pps}}) = V_1E_2(\bar{z}_{\text{pps}}) + E_1V_2(\bar{z}_{\text{pps}}) = V_1\left[\frac{1}{k}\sum_{i=1}^{k}E_2(\bar{z}_{i\bullet})\right] + E_1\left[\frac{1}{k^2}\sum_{i=1}^{k}V_2(\bar{z}_{i\bullet})\right]$$

$$= V_1\left[\frac{1}{k}\sum_{i=1}^{k}\bar{Z}_{i\bullet}\right] + E_1\left[\frac{1}{k^2}\sum_{i=1}^{k}\left(\frac{1}{n_i} - \frac{1}{N_i}\right)S_{iz}^2\right]$$

$$= \frac{1}{k}\sum_{i=1}^{K}P_i(\bar{Z}_{i\bullet} - \bar{Z})^2 + \frac{1}{k}\sum_{i=1}^{K}P_i\left(\frac{1}{n_i} - \frac{1}{N_i}\right)S_{iz}^2.$$

Hence the theorem.

9.7.2 POPULATION SIZE IS UNKNOWN

It is possible that the population of N units under consideration is expressible in the form of K overlapping clusters with N_i units in the ith cluster such that $\sum_{i=1}^{K}N_i = M \geq N$ is satisfied, but the population size N is unknown. When cluster wise data on units are available on the computer, the values of these frequencies for overlapping clusters may be easily available. Under such situations, define $Z_{ij} = Y_{ij}/F_{ij}$ and $W_{ij} = 1/F_{ij}$ for $i = 1, 2, ..., K$; $j = 1, 2, ..., N_i$, where Y_{ij} is the value of y for the jth unit in the ith cluster and let F_{ij} be its frequency of occurring in K clusters.

Then again we have two schemes as discussed below:

Scheme 1. (a) Select k clusters out of K clusters by SRSWR sample.
(b) From the ith selected cluster of size N_i, $(i = 1, 2, ..., \text{k})$, select n_i units by SRSWOR sampling.

Under such a sampling scheme, we have the following theorem:

Theorem 9.7.2.1. The ratio estimator under scheme 1 given by

$$\bar{z}_{\text{RS}} = \left(\frac{K}{k}\sum_{i=1}^{k}N_i\bar{z}_i\right)\Big/\left(\frac{K}{k}\sum_{i=1}^{k}N_i\bar{w}_i\right) \qquad (9.7.2.1)$$

has relative bias, to the first order of approximation, given by

$$\mathrm{RB}(\bar{z}_{\mathrm{RS}}) \approx \frac{K}{k}\left[\left(\frac{\sigma_{bw}^2}{N^2} - \frac{\sigma_{bzw}}{NY}\right)K + \sum_{i=1}^{K} N_i^2\left(\frac{1}{n_i} - \frac{1}{N_i}\right)\left(\frac{S_{iw}^2}{N^2} - \frac{S_{izw}}{NY}\right)\right] \tag{9.7.2.2}$$

and mean squared error to the first order of approximation is given by

$$\mathrm{MSE}(\bar{z}_{\mathrm{RS}}) \approx \frac{K}{kN^2}\sum_{i=1}^{K} N_i^2\left[\left(\bar{Z}_i - \bar{Y}\ \bar{W}_i\right)^2 + \left(\frac{1}{n_i} - \frac{1}{N}\right)\left(S_{iz}^2 + \bar{Y}^2 S_{iw}^2 - 2\bar{Y}S_{izw}\right)\right]$$

where

$$\sigma_{bzw} = K^{-1}\sum_{i=1}^{K}\left(N_i\bar{Z}_i - K^{-1}Y\right)\left(N_i\bar{W}_i - K^{-1}N\right);\ S_{izw} = (N_i - 1)^{-1}\sum_{j=1}^{N_i}\left(Z_{ij} - \bar{Z}_i\right)\left(W_{ij} - \bar{W}_i\right);$$

$$\bar{Z}_i = N_i^{-1}\sum_{j=1}^{N_i} Z_{ij},\ \bar{z}_i = n_i^{-1}\sum_{j=1}^{n_i} z_{ij},\ \bar{W}_i = N_i^{-1}\sum_{j=1}^{N_i} W_{ij},\ \bar{w}_i = n_i^{-1}\sum_{j=1}^{n_i} w_{ij}\text{ ,etc.,}$$

have their usual meanings.

Proof. Follows by the standard method of moments.

Scheme 2. (a) Select k clusters out of K clusters by PPSWR sampling, with $P_i = N_i/M$.
(b) From the ith selected cluster of size N_i, $(i = 1, 2, ..., k)$, select n_i units by SRSWOR sampling.

Then we have the following theorem:

Theorem 9.7.2.2. A ratio type estimator under scheme 2 is given by

$$\bar{z}_{\mathrm{RP}} = \left(\frac{M}{k}\sum_{i=1}^{k}\bar{z}_i\right)\Big/\left(\frac{M}{k}\sum_{i=1}^{k}\bar{w}_i\right) \tag{9.7.2.3}$$

has relative bias, to the first order of approximation, given by

$$\mathrm{RB}(\bar{z}_{\mathrm{RP}}) \approx \frac{M^2}{k}\left[\left(\frac{\sigma_{bw'}^2}{N^2} - \frac{\sigma_{bzw'}}{YN}\right) + \sum_{i=1}^{K} P_i\left(\frac{1}{n_i} - \frac{1}{N_i}\right)\left(\frac{S_{iw}^2}{N^2} - \frac{S_{izw}}{YN}\right)\right] \tag{9.7.2.4}$$

and has asymptotic mean squared error

$$\mathrm{MSE}(\bar{z}_{\mathrm{RP}}) = \frac{M}{kN^2}\sum_{i=1}^{K} N_i\left[\left(\bar{Z}_i - \bar{Y}\ \bar{W}_i\right)^2 + \left(\frac{1}{n_i} - \frac{1}{N_i}\right)\left(S_{iz}^2 + \bar{Y}^2 S_{iw}^2 - 2\bar{Y}S_{izw}\right)\right], \tag{9.7.2.5}$$

where $\sigma_{bzw}' = \sum_{i=1}^{K} P_i\left(\bar{Z}_i - M^{-1}Y\right)\left(\bar{W}_i - M^{-1}N\right)$, etc., have usual meanings due to Hansen and Hurvitz (1943).

Example 9.7.1. Suppose there are three plots as shown in the figure below and nine partners who are owners of these plots. A few partners have shares in only one plot, a few have shares in two plots and others have shares in all three plots.

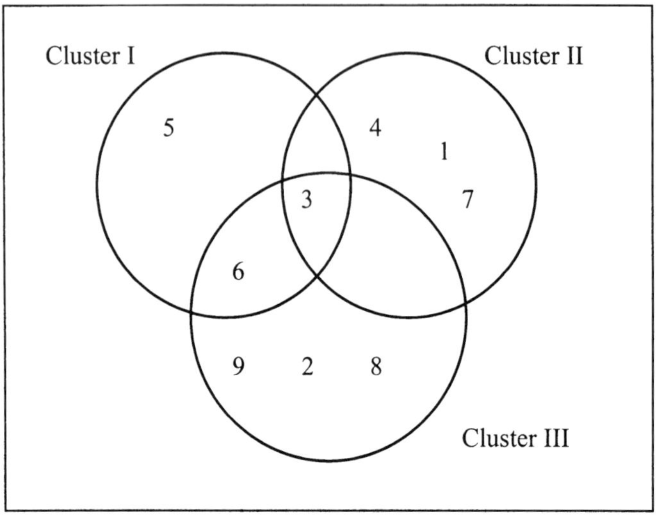

Fig.9.7.2.1 Three overlapping clusters.

The income of the nine partners is given below.

Partner No.	1	2	3	4	5	6	7	8	9
Income $	1000	2000	6000	3000	3000	4000	2000	3000	1000

We wish to estimate the average income of all partners using plots as overlapping clusters. Which sampling scheme would you prefer and why?

Solution. We observed that partner 3 is the owner of all the three plots, partner 6 is owner of two plots and the other partners are owners of only single plots. This means these plots or clusters of the partners have overlapped with each other. Thus we shall use the concept of overlapping cluster sampling. Let us compare here two methods suggested by Tracy and Osahan (1994b).

When clusters are selected with SRSWOR sampling then we have

$$\text{MSE}(\bar{z}_{\text{RS}}) \approx \frac{K}{kN^2}\sum_{i=1}^{K} N_i^2\left[\left(\overline{Z}_i - \overline{Y}\ \overline{W}_i\right)^2 + \left(\frac{1}{n_i} - \frac{1}{N}\right)D_i^2\right]$$

$$= \frac{3}{2\times 9^2}\left[3880211.11 + 15388643.24 + 14404805.5\right] = 623586.29\ .$$

When clusters are selected with PPSWR sampling then we have

$$\text{MSE}(\bar{z}_{\text{RP}}) = \frac{M}{kN^2}\sum_{i=1}^{K} N_i\left[\left(\overline{Z}_i - \overline{Y}\ \overline{W}_i\right)^2 + \left(\frac{1}{n_i} - \frac{1}{N_i}\right)D_i^2\right]$$

$$= \frac{3}{2\times 9^2}\left[1293403.70 + 3847160.81 + 2880961.10\right] = 148546.77\ .$$

We used the following information in the above formulae:

N_i	3			4				5				
n_i	1			2				2				
Partner no.	3	5	7	1	3	4	7	2	3	6	8	9
Y_{ij}	6000	3000	2000	1000	6000	3000	2000	2000	6000	4000	3000	1000
F_{ij}	3	1	2	1	3	1	1	1	3	2	1	1
$Z_{ij} = \frac{Y_{ij}}{F_{ij}}$	2000	3000	1000	1000	2000	3000	2000	2000	2000	2000	3000	1000
$W_{ij} = \frac{1}{F_{ij}}$	0.3333	1	0.5	1	0.3333	1	1	1	0.3333	0.5	1	1
s_{iz}^2	1000,000			666666.67				500,000				
$\overline{W}_i$	0.6111			0.8333333				0.766666				
$\overline{Z}_i$	2000			2000				2000				
s_{iw}^2	0.1203704			0.111111				0.105555				
S_{izw}	250			0.0000				0.0000				
D_i^2	646412.35			1840277.78				1114924.69				
$(\overline{Z}_i - \overline{Y}\overline{W}_i)^2$	193.0015			501720.764				241714.81				

where $D_i^2 = S_{iz}^2 + \overline{Y}^2 S_{iw}^2 - 2\overline{Y}S_{izw}$. Assuming $k = 2$ and we are given $K = 3$.

Thus the relative efficiency of PPSWR sampling strategy with respect to SRSWOR sampling strategy is

$$\text{RE} = \frac{\text{MSE}(\bar{z}_{\text{RS}})}{\text{MSE}(\bar{z}_{\text{RP}})} \times 100 = \frac{623586.29}{148546.77} \times 100 = 419.79\% .$$

Thus we shall prefer to use PPSWR sampling for selecting overlapping clusters in the sample.

9.8 POST-CLUSTER SAMPLING

In cluster sampling the lack of information relating to the composition of clusters poses a serious problem. In such situations, the technique of 'Post-Cluster' sampling introduced by Dalenius (1957) can be adopted. Ghosh (1963) has developed a stochastic model for analysis of post-cluster sampling. The technique of post-cluster sampling differs from two-phase or multi-phase sampling in the sense that it uses a hierarchy of sampling units and differs from ordinary sub-sampling in that sampling units at the second stage are larger than the sampling units at the first stage. Ghosh (1963) proposed a ratio estimate based on post-cluster sampling considering the ratio of the expected value of the variable under study to the expected value of the number of observations. Sadasivan and Srinath (1975) have considered a ratio of the expected value of the variable under study to the expected value of the auxiliary variable. Consider a finite population consisting of N distinct units. Let Y be the variable under study and let it take the values $Y_1, Y_2, ..., Y_N$ in the population. Let X be the auxiliary variable. The N elements in the population may, in principle, be grouped into M clusters, the ith cluster consisting of N_i elements. The population mean, $\overline{X} = N^{-1}\sum_{i=1}^{M}\sum_{j=1}^{N_i} X_{ij}$, is assumed to be known. The ratio estimator to estimate the population mean $\overline{Y}$ under post-cluster sampling is given by

$$\bar{y}_R = \frac{\left(\frac{M}{nm}\sum_{i=1}^{M}\eta_i\sum_{j=1}^{N_i}\varepsilon_{ij}Y_{ij}\right)}{\left(\frac{M}{nm}\sum_{i=1}^{M}\eta_i\sum_{j=1}^{N_i}\varepsilon_{ij}X_{ij}\right)}\overline{X} = \frac{\bar{y}}{\bar{x}}\overline{X} \tag{9.8.1}$$

where n and m denote the size of initial random sample and number of post-clusters selected, η_i and ε_{ij} are the random variables used for selection of post-clusters and the elements from the population. Then we have the following corollary:

Corollary 9.8.1. The covariance between $\bar{y}$ and $\bar{x}$ is given by

$$\operatorname{Cov}(\bar{y},\bar{x}) = \frac{M(N-n)}{mn(N-1)}V_{xy} + \frac{M^2(M-m)(n-1)}{nmN(N-1)(M-1)}V_{bxy} + \frac{(M-m)(N-n)}{mn(N-1)}(\overline{X}\ \overline{Y}) \tag{9.8.2}$$

where $V_{xy} = N^{-1}\sum_{i=1}^{M}\sum_{j=1}^{N_i} X_{ij}Y_{ij} - (\overline{X}\ \overline{Y})$ is the covariance between Y and X in the population, $V_{bxy} = M^{-1}\sum_{i=1}^{M}(X_i - X/M)(Y_i - Y/M)$ is the covariance between cluster totals for X and Y in the population. From (9.8.2), one can easily derive $V(\bar{y})$ and $V(\bar{x})$. Following standard ratio method of estimation, we have the following theorems:

Theorem 9.8.1. The relative bias expression for the ratio estimator $\bar{y}_R$, to the first order of approximation, is given by

$$\operatorname{RB}(\bar{y}_R) = \frac{M(N-n)}{nm(N-1)}\left(C_x^2 - C_{xy}\right) + \frac{(M-m)(n-1)N}{mn(N-1)(M-1)}\left[C_{x_i}^2 - C_{x_iy_i}\right] \tag{9.8.3}$$

where

$C_x^2 = V_x/\overline{X}^2$ = Square of coefficient of variation of X,

$C_{xy} = V_{xy}/(\overline{Y}\ \overline{X})$ = Coefficient of co-variation between X and Y,

$C_{x_i}^2 = \dfrac{V_{bx}}{(X/M)^2}$ = Square of coefficient of variation of values of X_i,

and

$C_{x_iy_i}^2 = \dfrac{V_{bxy}}{(X/M)(Y/M)}$ = Coefficient of co-variation between X_i and Y_i.

Theorem 9.8.2. The variance of the ratio estimator $\bar{y}_{\text{R}}$ is given by

$$V(\bar{y}_{\text{R}})=\bar{Y}^2\left[\left\{\frac{M}{mn}\left(C_y^2+C_x^2-2C_{xy}\right)+\frac{(M-m)(n-1)}{mn(M-1)}\left(C_{y_i}^2+C_{x_i}^2-2C_{x_iy_i}\right)\right\}\right]. \quad (9.8.4)$$

Proof. Follows by the ratio method of estimation.

9.9 ADAPTIVE CLUSTER SMPLING

Thompson (1990) reported that the use of adaptive cluster sampling for patchy populations is an efficient design. Brown (1996) pointed out that adaptive cluster sampling is efficient only for very patchy populations, but can be highly inefficient for other less aggregate populations. Christman (1997) also supported Brown's views about adaptive cluster sampling in case of non-patchy populations. In general, the adaptive cluster sampling can be done in two steps. In the first step a preliminary sample of n units is selected; for example, a random sample of n quadrants from a study area are divided into N evenly sized quadrants. In the second step, for any quadrant in the initial sample for which the variable of interest, y, for example the number of plants in the quadrant, has a value as large as a predefined critical value, the neighbourhood is sampled. The neighbourhood can be defined in several ways, for example, four surrounding quadrants, that is, on the east, west, north, and south sides. If any of the quadrants in the neighbourhood has a value at least as the critical value, the neighbourhood of that quadrant is sampled, and so on. The sampling of the units continues until all the neighbourhoods of sampled quadrants are sampled, similar to inverse sampling. The difference is that here initial sample size remains fixed, but the final sample size is variable and depends on the number of networks selected in the sample. The group of adjacent quadrants whose values are all at least as great as the critical value is known as a network. If we divide a population into K networks then the Horvitz and Thompson (1952) estimator of population total in adaptive cluster sampling becomes

$$\hat{Y}_{\text{adcl}}=\frac{1}{N}\sum_{i=1}^{N}\frac{y_iI_i}{\pi_i} \quad (9.9.1)$$

where

y_i = total of the y values in the ith network.

$$I_i=\begin{cases}1 & \text{if any quadrant in the } i^{th} \text{ net work is in the initial sample,}\\ 0 & \text{otherwise,}\end{cases}$$

and $\pi_i=1-\binom{N-x_i}{n}\Big/\binom{N}{n}$ is the probability that a quadrant in the ith containing x_i quadrants is selected in the initial sample.

Brown (1999) extended the concept of adaptive cluster sampling to stratified sampling design. He considered its application to the ecological data and showed a slight gain in efficiency using stratified design over the adaptive cluster sampling design. He showed that the two-phase adaptive sampling scheme introduced by Francis (1984) may lead to biased estimates. Francis (1984) considered the application of two-phase adaptive sampling design for the estimation of fish biomass from trawl surveys. Jolly and Hampton (1990) also suggested a two-phase adaptive cluster sampling design with the additional advantage that its bias can be estimated very easily. Thompson and Seber (1996) compare the different estimation strategies available in the literature under adaptive cluster sampling.

EXERCISES

Exercise 9.1. Suppose a population consists of N clusters, M_i being the size of the ith cluster $(i = 1, 2, ..., N)$. Clusters are selected with PPSWR sampling with $P_i = M_i/M$. The cluster selected at the $(r+1)$th draw is rejected if the number of distinct clusters selected in the first r draws is a pre-assigned number n. Let the ith cluster occur r_i times in a sample of r draws, $r_i = 0, 1, 2.....;\ i = 1, 2, ..., N$. If $\overline{Y}_{i\bullet}$ is the mean of the ith cluster then show that

$$\bar{y}_s = r^{-1}\sum_{i=1}^{n} r_i \overline{Y}_{i\bullet}$$

is an unbiased estimator of population mean.
Hint: Sampford (1962).

Exercise 9.2. Let Y_{ijk} be the value of Y for the kth element of the ith cluster in the jth stratum. Define

$$\overline{Y}_{i\bullet\bullet} = M_{i\bullet}^{-1}\sum_{j=1}^{K}\sum_{k=1}^{M_{ij}} Y_{ijk} \quad = \text{the per element cluster mean;}$$

$$\hat{\overline{Y}}_{\bullet j\bullet} = \hat{M}_{\bullet j}^{-1}\sum_{i=1}^{n}\sum_{k=1}^{M_{ij}} Y_{ijk} = \text{the per element (large) sample mean in the } j^{th} \text{ stratum,}$$

where $\hat{M}_{\bullet j} = \sum_{i=1}^{n} M_{ij}$ denotes the number of elements in the jth stratum in the sample;

$$\hat{\bar{y}}_{\bullet j\bullet} = \hat{m}_{\bullet j}^{-1}\sum_{i=1}^{n}\sum_{k=1}^{m_{ij}} Y_{ijk} = \text{the per element sample mean in the } j^{th} \text{ stratum,}$$

where $\hat{m}_{\bullet j} = \sum_{i=1}^{n} m_{ij}$ denotes the number of sample elements in the ith cluster of the jth stratum. Study the estimator $\hat{R}_{AS}$ of the population ratio $R = \sum_{i=1}^{N} M_i\overline{Y}_i \Big/ \sum_{i=1}^{N} M_i\overline{X}_i$, defined as

$$\hat{R}_{AS} = \sum_{j=1}^{K}\hat{M}_{\bullet j}\hat{\bar{y}}_{\bullet j} \Big/ \sum_{j=1}^{K}\hat{M}_{\bullet j}\hat{\bar{x}}_{\bullet j}\ .$$

Hint: Akar and Sedransk (1979).

Exercise 9.3. Draw an initial random sample of n elements from N elements in the population with SRSWOR sampling. On the basis of this sample, form M' sub-clusters. Then take a sample of m' sub-clusters from M' with SRSWOR sampling. Again from M super-clusters on the basis of these m' sub-clusters and finally draw a sample of m super-clusters from M with SRSWOR sampling. Let c_{ijk} for $k = 1,2,...,N_{ij},\ j = 1,2,...,M'_i,\ i = 1,2,...,M$ be a random variable such that it takes the value 1 if Y_{ijk} is selected and 0 otherwise. Let b_{ij} be another random variable, taking the value 1 if the jth sub-cluster of the ith super-cluster is selected and 0 otherwise. Let a_i be another random variable taking the value 1 if the ith super-cluster gets selected and 0 otherwise. Then show that an unbiased estimator of population mean is given by

$$\bar{y}_{\mathrm{U}} = \frac{MM'}{mm'n}\sum_{i=1}^{M} a_i \sum_{j=1}^{M'_i} b_{ij} \sum_{k=1}^{N_{ij}} c_{ijk} Y_{ijk} .$$

Hint: Sadasivan and Srinath (1975).

Exercise 9.4. A finite population of M_0 elements is divided into N clusters with the ith cluster having M_i elements. An SRSWOR sample of m elements is selected from the total population elements and the sample elements grouped according to the clusters to which they belong. Show that $\bar{y}_{\mathrm{G}} = (N/mn)y$, where y is the sample total based on the values of the sample elements in the n selected clusters, is an unbiased estimator of population mean.

Hint: Ghosh (1963).

Exercise 9.5. Suppose the overlapping clusters are selected such that if U_i is being selected then $(M_i - 1)$ units are associated with it to form a cluster of M_i units. Then show that an unbiased estimator of population mean $\bar{Y}$ is given by

$$\bar{y}_a = \frac{1}{n}\sum_{i=1}^{n}\sum_{j \in i}\frac{y_j}{M_j} .$$

Hint: Amdekar (1985).

Exercise 9.6. From a population consisting of N clusters each containing M elements, a simple random sample of n clusters is selected to estimate the population mean per element. Consider $S_w^2 = aM^g$, with $g < 0$, be the variance within clusters under the superpopulation model. Find the optimum cluster size such that the variance of the estimator of population mean is minimum for the fixed cost given by $C = nMc_1 + c_2\sqrt{n}$.

Exercise 9.7. If NM elements in a population are grouped at random into N clusters each of size M, then show that an estimator based on a SRSWOR sample of n clusters has the same efficiency as an estimator based on nM elements selected by usual SRSWOR random sample.

Exercise 9.8. Suppose a sample of n clusters, from N clusters each containing M elements, is taken systematically for estimating the population mean, such that N is a multiple of sample size n. Suggest an estimator to estimate the population mean and find its variance in terms of intraclass correlation coefficient ρ_c between pairs of elements in the clusters and that between pairs of clusters ρ_c^* at the samples.
Hint: Madow (1949).

Exercise 9.9. In order to examine the efficiency of sampling households, instead of persons, for estimating the population of males in a given area, it is assumed that:
(a) each household consists of four persons, *viz.*, husband, wife and two children;
(b) the number of boys (or girls) in a family are binomially distributed.
Show that:
(a) The value of intraclass correlation coefficient ρ_c is 1/6;
(b) The efficiency of sampling households compared to that of sampling persons is 200%, that is, the efficiency of cluster sampling is twice the random sampling.
Hint: Sukhatme (1954).

Exercise 9.10. Consider a large library with N distinct titles, each title being present in one or more volumes. A sample of n books is to be drawn to estimate a proportion π (say), for instance, the proportion of Canadian books. Because the sampling the books directly proves physically difficult, the librarian develops a scheme for choosing an SRS from the card catalogue. Since the catalogue contains a card for each volume, not for each title, this is a sampling of titles with probabilities proportional to size, the 'size' of the ith title being the number of volumes M_i. It is assumed that M_i are known at least for the sample units. Define

$$z_i = M_i/M, \quad i = 1,2,...,N, \quad \text{where} \quad M = \sum_{i=1}^{N} M_i, \; y_i = \begin{cases} 1 & \text{if the } i^{th} \text{ title is Spanish,} \\ 0 & \text{otherwise,} \end{cases} \quad \text{then}$$

our interest is to estimate the population proportion, $\pi = \frac{1}{N}\sum_{i=1}^{N} y_i$. Let $u_i = y_i/(Nz_i)$, $v_i = 1/(Nz_i)$, $i = 1, 2, ..., n$, $\bar{u} = n^{-1}\sum_{i=1}^{n} u_i$ and $\bar{v} = n^{-1}\sum_{i=1}^{n} v_i$. Find the bias and variance of the following three estimators of population proportion, defined as

$$\hat{\pi}_1 = \bar{u} \; ; \; \hat{\pi}_2 = \bar{u} + (1 - \bar{v}) \text{ and } \hat{\pi}_3 = \bar{u}/\bar{v}.$$

Discuss their relative efficiencies with respect to one another.
Hint: Alalouf (1996), Mayor (2002).

Exercise 9.11. Discuss balanced cluster sampling.
Hint: Tallis (1991).

PRACTICAL PROBLEMS

Practical 9.1. Using the MAP of the USA, construct 10 clusters of the 50 states, each cluster consisting of 5 states, listed in population 1 based on their locations.

Practical 9.2. Select 6 clusters from the list of clusters you made in practical 9.1 by using SRSWOR sampling. Record the values of the nonreal estate farm loans for the selected states from population 1 given in Appendix. Estimate the average nonreal estate farm loans in the United States using cluster sampling. Estimate the variance of the estimator used for estimating the average nonreal estate farm loans. Construct a 95% confidence interval.

Practical 9.3. Suppose the United States has been divided into neighbouring 10 clusters each consisting of 5 states as shown below.

Cluster Number	Names of the states
1	ME, NH, VT, MA, RI
2	NY, CT, NJ, DE, PA
3	MD,WV,VA, KY, NC
4	WI, MI, IL, IN, OH
5	TN, SC, GA, AL, FL
6	ND, MN, SD, NE, IA
7	MO, OK, AR, MS,LA
8	MT, WY, CO, NM,KS
9	CA, AZ, TX, AK, HI
10	WA, OR, ID, NV, UT

We are interested in estimating the average nonreal estate farm loans in the United States. Is there any gain in efficiency owed to clustering rather than simple random sampling?

Hint: $S_b^2 = 609{,}964.83$ and $S^2 = 1{,}176{,}526.14$.

Practical 9.4. In practical 9.3 find the value of the intraclass correlation coefficient. Use it to find the relative efficiency of the cluster sampling over the simple random sampling.

Answer: The value of intraclass correlation coefficient $\rho = 0.3451$.

Practical 9.5. Select 6 clusters from the table given in Practical 9.3 by using SRSWOR sampling. Record the values of the nonreal estate farm loans for the selected states from population 1 given in the Appendix. Estimate the relative efficiency using the ANOVA approach. Hence, deduce the value of parameter g.

Practical 9.6. Consider the problem of estimation of proportion using cluster sampling. Select 3 clusters from the table given in Practical 9.3 by using SRSWOR sampling. Record the values of the nonreal estate farm loans for the selected states from population 1 given in the Appendix. Estimate the proportion of states having nonreal estate farm loans of more than $878.16 n the United States using cluster sampling. Estimate the variance of the estimator used for estimating the required proportion. Construct a 95% confidence interval.

Practical 9.7. Divide the United States neighbouring 10 clusters each consisting of 5 states as shown in practical 9.3. Consider the problem of estimating the proportion of states having nonreal estate farm loans of more than \$878.16 in the United States. Is there any expectation of gain in efficiency due to clustering rather than simple random sampling?

Practical 9.8. A world level team of doctors believes that the estimation of total tobacco use in the world is an important factor in the persistence of health problems. Select four continents from population 5 by SRSWOR sampling. Collect information about the yield/hectare from all the countries in the selected continents. Estimate the yield/hectare in the world using three different estimators and report your 95% confidence interval to the doctors.

Practical 9.9. A tobacco farmer wishes to find a strategy which provides a better estimate of the production of tobacco crops in the world. He divides the world level tobacco growing countries into 10 clusters as listed in population 5 of the Appendix. He wishes to apply PPSWR sampling for selecting the cluster. Can he expect any gain in efficiency over simple random sampling?

Practical 9.10. For the purpose of comparison in overlapping cluster sampling, suppose the clusters are formed and selected by following two different sampling schemes:

Scheme 1. (a) Select k clusters out of K clusters by SRSWR sampling.
(b) From the ith selected cluster of size N_i, $(i = 1,2,...,k)$, select n_i units by using SRSWOR sampling.

Scheme 2. (a) Select k clusters out of K clusters by PPSWR sampling, with $P_i = N_i/M$, (b) From the ith selected cluster of size N_i, $(i = 1,2,...,k)$, select n_i units by SRSWOR sampling.

Assume that the population size is unknown and three possible clusters are formed as shown in the figure below.

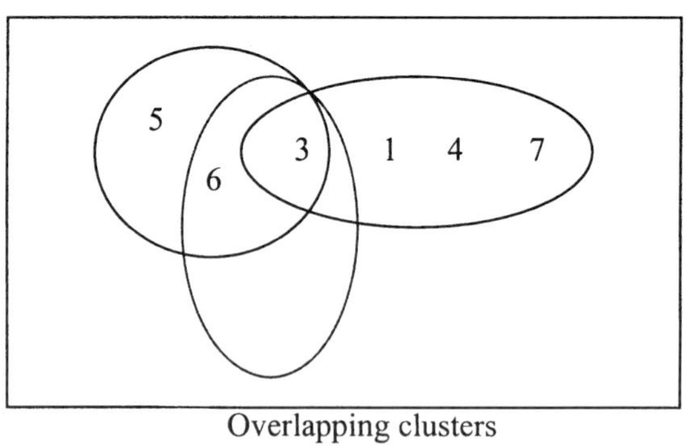

Overlapping clusters

Show that the relative efficiency of the estimator of the population mean based on scheme 2 with respect to that based on scheme 1 is 114.21%.
Hint: Tracy and Osahan (1994b).

Practical 9.11. Ms. Stephanie Singh and Ms. Renee Horn were asked to construct three clusters of 9 regions of the USA using the following two maps:

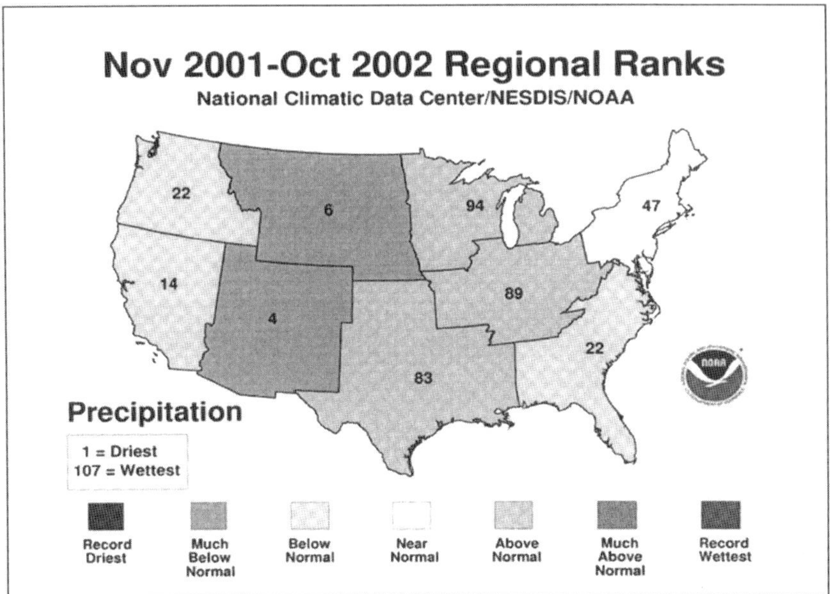

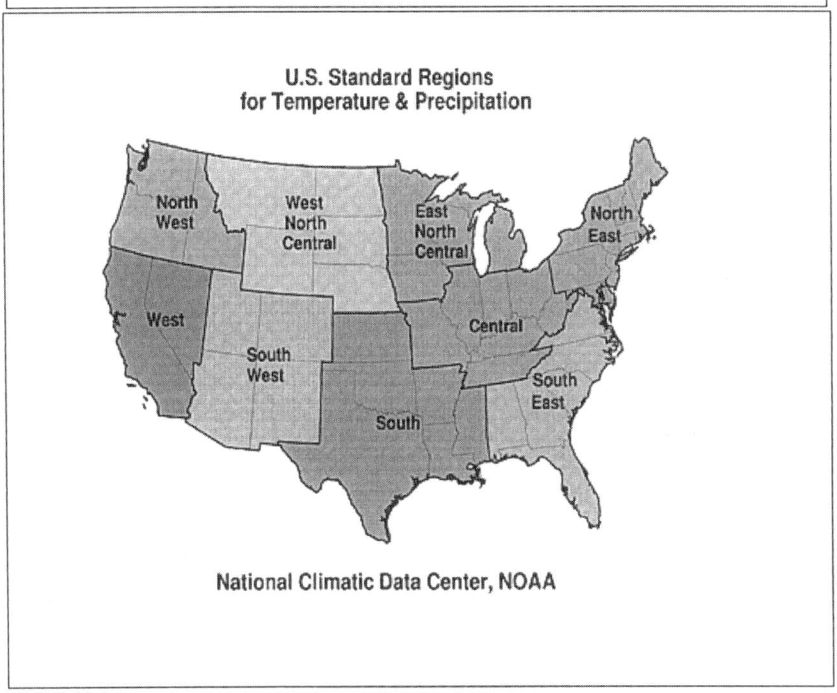

Source: Printed with permission from National Climate Data Center/NESDIS/NOAA.

Ms. Stephanie Singh suggested the following three clusters as:

Ms. Stephanie Singh's Clusters			
Cluster 1	North West	West North Central	East North Central
Cluster 2	West	South West	South
Cluster 3	North East	Central	South East

Ms. Renee Horn suggested the following three clusters as:

Ms. Renee Horn's Clusters			
Cluster 1	North West	West	South West
Cluster 2	West North Central	East North Central	North East
Cluster 3	Central	South East	South

(a) Whose clustering plan is more efficient and why?

(b) Select two clusters using SRSWOR sampling from Ms. Stephanie Singh's clusters and construct 95% confidence interval for the average precipitation. (**Rule:** Always start from first row and first column of the Pseudo-Random Number Table 1 given in the Appendix).

(c) Select two clusters using SRSWOR sampling from Ms. Renee Horn's clusters and construct 95% confidence interval for the average precipitation. (**Rule:** Always start from first row and first column of the Pseudo-Random Number Table 1 given in the Appendix)
Hint: Apply ANOVA approach.

Practical 9.12. Beta Corporation is considering 15 suppliers to secure the large amount of steel pipes that it uses from different five states, *viz.*, NY, VT, NH, ME, and MA of the USA. Among the 15 suppliers, few belong to more than one state, as shown in the following diagram on the next page:

Depending upon the quality and their standard the selling prices of one pipe of these 15 suppliers are given below:

Supplier no.	1	2	3	4	5	6	7	8	9	10	11	12	13	14	15
US$	500	525	400	515	510	490	450	525	410	475	500	405	480	430	420

We wish to estimate the average selling price of the pipe by different suppliers. Instead of asking all the suppliers, we wish to select all the suppliers working in few states. At first-stage select two states and contact all the suppliers within the selected states to collect information about their selling prices. Use the concept of overlapping cluster sampling for estimating average selling price of a pipe.

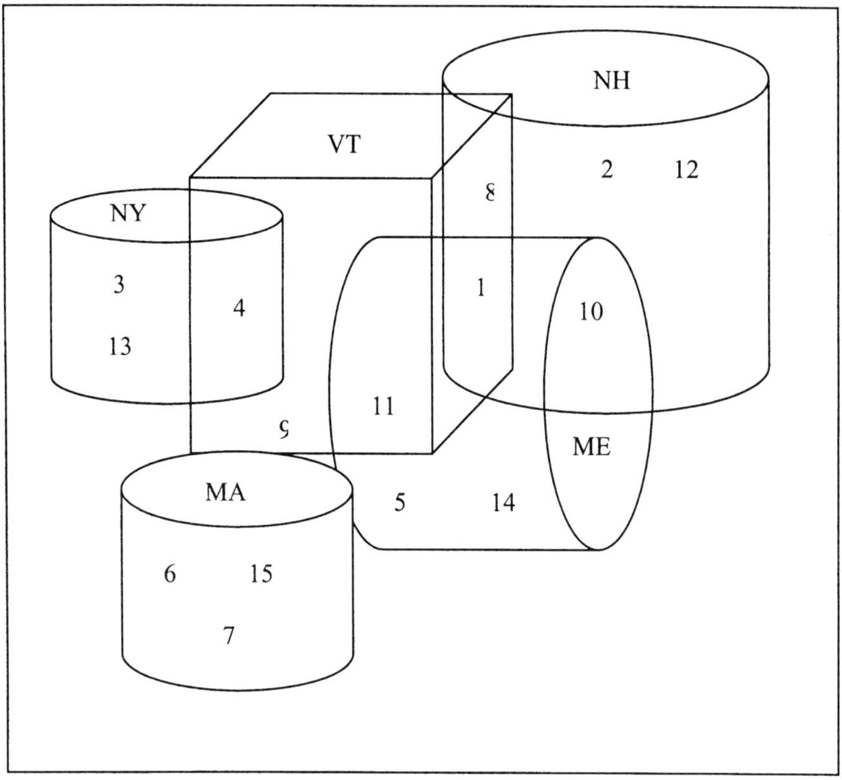

Distribution of 15 suppliers in five different states.

Practical 9.13. Consider a hypothetical situation of a survey on AIDS patients in the NY city, where the residential dwellings were considered as clusters and apartments within dwellings were considered as elements. Nine dwellings were selected by simple random without replacement sampling out of total 1000 dwellings available on a list, and noted the number of AIDS patients in each apartment as listed below:

Cluster	Number of AIDS patients					
1	2	0				
2	4	5	1			
3	2	4	2	0		
4	1	2	4	0	3	
5	2	4	0	0	0	4
6	0	0	0	3	2	
7	0	0	0	0		
8	2	2	1			
9	1	2				

(a) Estimate the total number of AIDS patients in the NY city.

(b) Estimate the relative efficiency of the cluster sampling with respect to simple random without replacement sampling.
(c) Confirm your result in (b) based of the value of coefficient of determination.
Hint: Apply ANOVA approach with unequal number of units in each cluster.

Practical 9.14. The population of interest is the students in your class today. Each of the rows of students in the class will form a cluster.

(I) We wish to estimate the proportion of students in the class 'Who visited theatre last week?'

(a) Using the Pseudo-Random Number Table 1 given in the Appendix select two rows and estimate the proportion of students who visited a theatre last week. Also construct 95% confidence interval estimate.
(b) Now ask the same question to everyone in the class and find the true proportion of such students who visited a theatre last week. Does the true proportion lie in your confidence interval estimate? If yes, it is fine, but if not, then suggest a suitable reason for it.

(II) Consider the problem of estimation of average GPA of the class.

(a) Using the Pseudo-Random Number Table 1 given in the Appendix, select one row and estimate the average GPA of the class. Also construct 95% confidence interval estimate for the same.
(b) Now ask the GPA question to everyone in the class and find the true average GPA. Does the true average value lie in your confidence interval estimate?
Give your opinion.

10. MULTI-STAGE, SUCCESSIVE, AND RE-SAMPLING STRATEGIES

10.0 INTRODUCTION

The meaning of multi-stage sampling is clear from its name. Here we have several stages for the sample selection. In fact it is an extension of the concept of cluster sampling. Similar to cluster sampling first we divide the population into M clusters or heterogeneous groups. We select m clusters and use the estimates of cluster means or totals to form population estimate. For example, in two-stage sampling we again divide our population into M groups and select a sample of groups, which form the first stage sample. The units so selected are called first stage units (FSU).

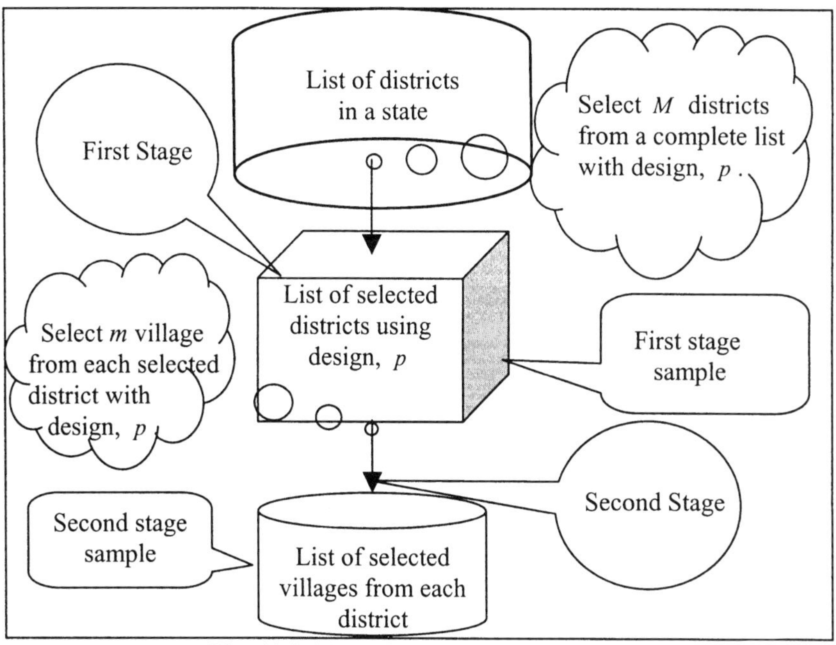

Fig. 10.0.1 Two-Stage Sampling Scheme.

First stage units are sometimes also called preliminary stage units (PSU). From each selected group at first stage, we select a sample of the population units forming the group. The sub-sample of the first stage sample is called the second stage sample and the units so selected are called the second stage units (SSU). Similarly one can think three-stage or four-stage sampling and hence multi-stage sampling. A pictorial representation of two-stage sampling is given in Figure 10.0.1. Thus, a scheme of the type where at each stage we have a selection and go on selecting smaller and smaller units is called multi-stage sampling scheme.

10.1 NOTATION

Let us define a few symbols, which will remain useful for understanding the concept of multi-stage sampling, as follows:

N = Total number of first stage units (FSUs) in the population;

n = Number of FSUs selected in the sample;

M_i = The number of second stage units (SSUs) in the ith selected first stage units (FSUs), ($i = 1, 2, ..., N$);

m_i = The number of second stage units (SSUs) selected from the ith FSU in the sample of n FSUs, ($i = 1, 2, ..., n$);

T_{ij} = Total number of third stage units (TSUs) in the jth SSU of the ith FSU;

t_{ij} = No. of TSUs selected in the sample from the jth SSU of the ith FSU in the sample, $i = 1, 2, ..., n$ and $j = 1,2,...,m_i$;

Y_{ik} = value of the kth TSU of the jth SSU of the ith FSU of the study variable Y in the population, $i = 1, 2, ..., N$; $j = 1,2,...,M_i$, $k = 1,2,...,T_{ij}$;

y_{ik} = value of the kth TSU of the jth SSU of the ith FSU of Y in the sample for $i = 1, 2, ..., n$; $j = 1,2,...,m_i$, $k = 1,2,...,t_{ij}$;

$Y_{ij\bullet} = \sum_{k=1}^{T_{ij}} Y_{ijk}$, total corresponding to all the TSUs in the jth SSU of the ith FSU in the population;

$y_{ij\bullet} = \sum_{k=1}^{t_{ij}} y_{ijk}$, total corresponding to all TSUs in the jth SSU of the ith FSU in the sample;

$S_{ij}^2 = (T_{ij}-1)^{-1}\sum_{k=1}^{T_{ij}}(Y_{ijk} - Y_{ij\bullet}/T_{ij})^2$, population mean square between T_{ij} units of the jth SSU of the ith FSU;

$s_{ij}^2 = (t_{ij}-1)^{-1}\sum_{k=1}^{t_{ij}}(y_{ijk} - y_{ij\bullet}/t_{ij})^2$, the sample mean square between t_{ij} units of the jth SSU of the ith FSU;

$S_i^2 = (M_i-1)^{-1}\left[\sum_{j=1}^{M_i} Y_{ij\bullet}^2 - M_i^{-1}\left(\sum_{j=1}^{M_i} Y_{ij\bullet}\right)^2\right]$, the population mean square error between M_i units of the ith FSU;

$s_i^2 = (m_i-1)^{-1}\left[\sum_{j=1}^{m_i} y_{ij\bullet}^2 - m_i^{-1}\left(\sum_{j=1}^{m_i} y_{ij\bullet}\right)^2\right]$, the sample mean square error between m_i units of the ith FSU;

$S_b^2 = (N-1)^{-1}\left[\sum_{i=1}^{N} Y_{i\bullet\bullet}^2 - N^{-1}\left(\sum_{i=1}^{N} Y_{i\bullet\bullet}\right)^2\right]$, the population mean square error between the totals of FSUs;

and

$s_b^2 = (n-1)^{-1}\left[\sum_{i=1}^{n} y_{i\bullet\bullet}^2 - n^{-1}\left(\sum_{i=1}^{n} y_{i\bullet\bullet}\right)^2\right]$, the sample mean square error between the totals of FSUs.

10.2 PROCEDURE FOR CONSTRUCTION OF ESTIMATORS OF THE TOTAL

Suppose we like to estimate the population total Y. Let $\hat{Y}$ denote its unbiased estimator. At the first stage of sampling, we have N units in the population. Therefore an estimate of per first stage unit (FSU) is given by $\hat{Y}/N$. Note that M_i $i = 1,2,...,N$, units are selected at the second stage from the selected i^{th} first stage unit (FSU), the total number of units at the second stage of sampling will be given by $M = \sum_{i=1}^{N} M_i$ and hence the required unbiased estimator of per second stage units (SSU) is $\hat{Y}/M$. Now let the total number of units at the third stage units (TSU) be $T = \sum_{i=1}^{N}\sum_{j=1}^{M_i} T_{ij}$. Then an unbiased estimator of per third stage unit (TSU) is $\hat{Y}/T$. Our objective is to find an estimator of the population total. For this purpose we proceed as follows:

First Stage	1	2	•	n
Second Stage	$1, 2, ..., m_1$	$1, 2, ..., m_2$	•	$1, 2, ..., m_n$
Third Stage	$t_{1j}, t_{2j}, ..., t_{m_1 j}$ $j = 1,2,...,m_1$	$t_{1j}, t_{2j}, ..., t_{m_2 j}$ $j = 1,2,...,m_2$	•	$t_{1j}, t_{2j}, ..., t_{m_n j}$ $j = 1,2,...,m_n$

Then we have the following corollaries.

Corollary 10.2.1. Suppose SRS has been used at each stage of selection of the sample. Then if the y_i values are known for FSUs an estimator of the population total Y is given by

$$\hat{y}_{\text{ms}} = \frac{N}{n}\sum_{i=1}^{n} y_i . \tag{10.2.1}$$

In the case in which the y_i values are not known an estimator of the population total can be obtained by replacing them by their estimators

$$\hat{y}_i = \frac{M_i}{m_i}\sum_{j=1}^{m_i} y_{ij}\ .$$

Thus

$$\hat{y}_{\text{ms}} = \frac{N}{n}\sum_{i=1}^{n}\frac{M_i}{m_i}\sum_{j=1}^{m_i} y_{ij}\ . \tag{10.2.2}$$

In the case in which the values of y_{ij} are not known they can be estimated as

$$\hat{y}_{ij} = \left(T_{ij}/t_{ij}\right)\sum_{k=1}^{t_{ij}} y_{ijk}$$

and the estimator of population total Y becomes

$$\hat{y}_{\text{ms}} = \frac{N}{n}\sum_{i=1}^{n}\frac{M_i}{m_i}\sum_{j=1}^{m_i}\frac{T_{ij}}{t_{ij}}\sum_{k=1}^{t_{ij}} y_{ijk}\ . \tag{10.2.3}$$

Corollary 10.2.2. Consider FSUs are selected using PPSWR sampling with p_i being the probability of selecting the ith unit and y values are known for FSUs, then an estimator of population total Y is given by

$$\hat{y}_{\text{ms}} = \frac{1}{n}\sum_{i=1}^{n}\frac{y_i}{p_i}\ . \tag{10.2.4}$$

In the case in which the y_i values are not known and SSUs are selected with SRS and they can be replaced by their estimators

$$\hat{y}_i = \frac{M_i}{m_i}\sum_{j=1}^{m_i} y_{ij}\ .$$

Then an estimator of the population total Y is given by

$$\hat{y}_{\text{ms}} = \frac{1}{n}\sum_{i=1}^{n}\frac{1}{p_i}\frac{M_i}{m_i}\sum_{j=1}^{m_i} y_{ij}\ . \tag{10.2.5}$$

In the case in which the values of y_{ij} are not known and TSUs are selected using SRS and then y_{ij} can be estimated as

$$\hat{y}_{ij} = \left(T_{ij}/t_{ij}\right)\sum_{k=1}^{t_{ij}} y_{ijk}$$

and the estimator of population total Y becomes

$$\hat{y}_{\text{ms}} = \frac{1}{n}\sum_{i=1}^{n}\frac{1}{p_i}\frac{M_i}{m_i}\sum_{j=1}^{m_i}\frac{T_{ij}}{t_{ij}}\sum_{k=1}^{t_{ij}} y_{ijk}\ . \tag{10.2.6}$$

Corollary 10.2.3. If SSUs are selected with PPSWR sampling, whereas first and third stage units are selected with SRS, then an estimator of population total is given by

$$\hat{y}_{\text{ms}} = \frac{N}{n}\sum_{i=1}^{n}\frac{1}{m_i}\sum_{j=1}^{m_i}\frac{1}{p_{ij}}\frac{T_{ij}}{t_{ij}}\sum_{k=1}^{t_{ij}} y_{ijk}\ . \tag{10.2.7}$$

In the same fashion the estimator of population total can be obtained under any finite number of stages. It may be remarkable that the lesser the number of stages, the more accurate will be the estimators based on a given sample size, but may not be for fixed cost.

10.3 METHOD FOR CALCULATING THE VARIANCE OF THE ESTIMATORS

Let us first discuss simple cases to find the variance of the estimators of population total under multi-stage sampling. Let us first consider the situation of three-stage sampling, assuming that SRSWOR sampling has been used at each stage of selection of the sample. Evidently an unbiased estimator of population total is

$$\hat{y}_{3s} = \frac{N}{n}\sum_{i=1}^{n}\frac{M_i}{m_i}\sum_{j=1}^{m_i}\frac{T_{ij}}{t_{ij}}\sum_{k=1}^{t_{ij}} y_{ijk}\,. \tag{10.3.1}$$

To find its variance let E_3 and V_3 denote the conditional expectation and variance for the fixed first and second stage samples; E_2 and V_2 denote the conditional expectation and variance for the fixed first stage samples; E_1 and V_1 denote the expected value and variance for all possible first stage samples and we have

$$V(\hat{y}_{3s}) = E_1E_2V_3(\hat{y}_{3s}) + E_1V_2E_3(\hat{y}_{3s}) + V_1E_2E_3(\hat{y}_{3s}) = I_1 + I_2 + I_3 \text{ (say)}. \tag{10.3.2}$$

Note that the sampling has been done independently in each FSU and SSU we have

$$I_1 = E_1E_2V_3(\hat{y}_{3s}) = E_1E_2V_3\left[\frac{N}{n}\sum_{i=1}^{n}\frac{M_i}{m_i}\sum_{j=1}^{m_i}\frac{T_{ij}}{t_{ij}}\sum_{k=1}^{t_{ij}} y_{ijk}\right] = E_1E_2V_3\left[\frac{N}{n}\sum_{i=1}^{n}\frac{M_i}{m_i}\sum_{j=1}^{m_i} T_{ij}\bar{y}_{ij\bullet}\right]$$

$$= E_1E_2\left[\frac{N^2}{n^2}\sum_{i=1}^{n}\frac{M_i^2}{m_i^2}\sum_{j=1}^{m_i} V_3\left(T_{ij}\,\bar{y}_{ij\bullet}\right)\right] = E_1E_2\left[\frac{N^2}{n^2}\sum_{i=1}^{n}\frac{M_i^2}{m_i^2}\sum_{j=1}^{m_i} T_{ij}^2\left(\frac{1}{t_{ij}}-\frac{1}{T_{ij}}\right)S_{ij}^2\right]$$

$$= E_1\left[\frac{N^2}{n^2}\sum_{i=1}^{n}\frac{M_i}{m_i}\sum_{j=1}^{M_i}\frac{T_{ij}\left(T_{ij}-t_{ij}\right)}{t_{ij}}S_{ij}^2\right] = \frac{N}{n}\sum_{i=1}^{N}\frac{M_i}{m_i}\sum_{j=1}^{M_i}\frac{T_{ij}\left(T_{ij}-t_{ij}\right)}{t_{ij}}S_{ij}^2\,, \tag{10.3.3}$$

$$I_2 = E_1V_2E_3(\hat{y}_{3s}) = E_1V_2E_3\left[\frac{N}{n}\sum_{i=1}^{n}\frac{M_i}{m_i}\sum_{j=1}^{m_i}\frac{T_{ij}}{t_{ij}}\sum_{k=1}^{t_{ij}} y_{ijk}\right] = E_1V_2E_3\left[\frac{N}{n}\sum_{i=1}^{n}\frac{M_i}{m_i}\sum_{j=1}^{m_i} T_{ij}\bar{y}_{ij\bullet}\right]$$

$$= E_1V_2\left[\frac{N}{n}\sum_{i=1}^{n}\frac{M_i}{m_i}\sum_{j=1}^{m_i} Y_{ij\bullet}\right] = E_1V_2\left[\frac{N}{n}\sum_{i=1}^{n} M_i\bar{Y}_{ij\bullet}\right] = E_1\left[\frac{N^2}{n^2}\sum_{i=1}^{n} M_i^2 V\left(\bar{Y}_{ij\bullet}\right)\right]$$

$$= E_1\left[\frac{N^2}{n^2}\sum_{i=1}^{n}\frac{M_i\left(M_i-m_i\right)}{m_i}S_i^2\right] = \frac{N}{n}\sum_{i=1}^{N}\frac{M_i\left(M_i-m_i\right)}{m_i}S_i^2 \tag{10.3.4}$$

and

$$I_3 = V_1E_2E_3(\hat{y}_{3s}) = V_1E_2E_3\left[\frac{N}{n}\sum_{i=1}^{n}\frac{M_i}{m_i}\sum_{j=1}^{m_i}\frac{T_{ij}}{t_{ij}}\sum_{k=1}^{t_{ij}} y_{ijk}\right] = V_1E_2E_3\left[\frac{N}{n}\sum_{i=1}^{n}\frac{M_i}{m_i}\sum_{j=1}^{m_i} T_{ij}\bar{y}_{ij\bullet}\right]$$

$$= V_1 E_2\left[\frac{N}{n}\sum_{i=1}^{n}\frac{M_i}{m_i}\sum_{j=1}^{m_i} Y_{ij\bullet}\right] = V_1 E_2\left[\frac{N}{n}\sum_{i=1}^{n} M_i \overline{Y}_{ij\bullet}\right]$$

$$= V_1\left[\frac{N}{n}\sum_{i=1}^{n} Y_{i\bullet\bullet}\right] = \frac{N(N-n)}{n} S_b^2 . \tag{10.3.5}$$

Thus on using (10.3.3), (10.3.4), and (10.3.5) in (10.3.2) we have the resultant variance of the estimator in three stage sampling given in the following theorem:

Theorem 10.3.1. If three-stage sampling is performed using SRSWOR sampling at each stage of sampling, then the variance of the resultant unbiased estimator of population total is given by

$$V(\hat{y}_{3s}) = \frac{N}{n}\sum_{i=1}^{N}\frac{M_i}{m_i}\sum_{j=1}^{M_i}\frac{T_{ij}(T_{ij}-t_{ij})}{t_{ij}} S_{ij}^2 + \frac{N}{n}\sum_{i=1}^{N}\frac{M_i(M_i-m_i)}{m_i} S_i^2 + \frac{N(N-n)}{n} S_b^2 . \tag{10.3.6}$$

The expression (10.3.6) shows that the variance in three-stage sampling consists of three components, which corresponds to the variation at each stage.

10.3.1 SELECTION OF FIRST AND SECOND STAGE UNITS USING SRSWOR SAMPLING

An estimator of population total in two-stage sampling, assuming SRSWOR sampling at each stage, is given by

$$\hat{y}_{2s} = \frac{N}{n}\sum_{i=1}^{n}\frac{M_i}{m_i}\sum_{j=1}^{m_i} y_{ij} \tag{10.3.7}$$

with variance

$$V(\hat{y}_{2s}) = \frac{N(N-n)}{n} S_b^2 + \frac{N}{n}\sum_{i=1}^{N}\frac{M_i(M_i-m_i)}{m_i} S_i^2 \tag{10.3.8}$$

and an unbiased estimator of $V(\hat{y}_{2s})$ is given by

$$\hat{V}(\hat{y}_{2s}) = \frac{N(N-n)}{n} s_b^2 + \frac{N}{n}\sum_{i=1}^{n}\frac{M_i(M_i-m_i)}{m_i} s_i^2 , \tag{10.3.9}$$

where

$$s_b^2 = (n-1)^{-1}\left[\sum_{i=1}^{n}\left(\frac{M_i}{m_i}\sum_{j=1}^{m_i} y_{ij}\right)^2 - n(\hat{y}_{2s}/N)^2\right] \quad \text{and} \quad s_i^2 = (m_i-1)^{-1}\left[\sum_{j=1}^{m_i} y_{ij}^2 - m_i \bar{y}_{i\bullet}{}^2\right].$$

Example 10.3.1. Out of 10 clusters or continents listed in population 5, select 4 continents with SRSWOR sampling as first stage units (FSU). From each one of the selected continents select two countries by SRSWOR sampling and record the yield/hectare of the Tobacco crop. Estimate the total yield/hectare in the world. Also estimate the standard error of the estimator used for estimating the total yield/hectare in the world.

Solution. For selecting first stage units, we used the 7th and 8th columns of the Pseudo-Random Numbers (PRN) Table 1 from the Appendix to select four distinct random numbers between 1 and 10 as 07, 09, 01 and 02. Obviously the four continents selected in an SRSWOR first-stage sample of four units are *Other Africa, Middle East, Central America, and Caribbean.*

Note that there are 30 countries in the continent *Other Africa.* To select two countries out of these 30 countries, we selected two distinct random numbers between 1 and 30 by using first two columns of the Pseudo-Random Numbers (PRN) Table 1 from the Appendix as 01 and 23. Thus at the second-stage the countries *Angola* and *South Africa* will be included in the sample.

Further note that there are 10 countries in the continent of the *Middle East.* To select two countries out of these 10 countries, we selected two distinct random numbers between 1 and 10 by using third and fourth columns of the Pseudo-Random Numbers (PRN) Table 1 from the Appendix as 06 and 07. Thus at second stage, the countries *Oman* and *Syria* will be included in the sample.

Similarly the second stage countries selected from *Central America* are *Honduras* and *Nicaragua*, whereas those from the *Caribbean* are *Haiti* and *Jamaica*, respectively.

The structure of sample information collected takes the following form. Here we are given $N = 10$ and $n = 4$.

First stage unit no. i	First stage sample	M_i	m_i	Second stage unit no.	Second stage sample	y_{ij}	$\hat{y}_i = \frac{M_i}{m_i}\sum_{j=1}^{m_i} y_{ij}$	$\frac{M_i(M_i - m_i)}{m_i} s_i^2$
1	Other Africa	30	2	1	Angola	0.99	44.85	214.2210
				2	South Africa	2.00		
2	Middle East	10	2	1	Oman	1.11	11.30	0.0320
				2	Syria	1.15		
3	Central America	6	2	1	Honduras	1.78	11.43	0.3750
				2	Nicaragua	2.03		
4	Caribbean	6	2	1	Haiti	1.29	9.84	2.9400
				2	Jamaica	1.99		
						Sum	77.42	217.5680

An estimate of total yield/hectare in the world is given by

$$\hat{y}_{2s} = \frac{N}{n}\sum_{i=1}^{n}\frac{M_i}{m_i}\sum_{j=1}^{m_i} y_{ij} = \frac{10}{4}\times 77.42 = 193.55\ .$$

Now we have

$$s_b^2 = \frac{1}{n-1}\left[\sum_{i=1}^{n}\left(\frac{M_i}{m_i}\sum_{j=1}^{m_i} y_{ij}\right)^2 - n\left(\hat{y}_{2s}/N\right)^2\right] = \frac{1}{4-1}\left[77.42^2 - 4\left(\frac{193.55}{10}\right)^2\right]$$

$$= \frac{4495.3923}{3} = 1498.4641 .$$

An unbiased estimate of $V(\hat{y}_{2s})$ is given by

$$\hat{v}(\hat{y}_{2s}) = \frac{N(N-n)}{n} s_b^2 + \frac{N}{n} \sum_{i=1}^{n} \frac{M_i(M_i - m_i)}{m_i} s_i^2 = \frac{10(10-4)}{4} \times 1498.4641 + \frac{10}{4} \times 217.568$$

$$= 23020.8815.$$

A $(1-\alpha)100\%$ confidence interval for estimating population total Y is given by

$$\hat{y}_{2s} \pm t_{\alpha/2}(\text{df} = n(m-1))\sqrt{\hat{v}(\hat{y}_{2s})} .$$

Using Table 2 from the Appendix the 95% confidence interval for total yield/ hectare of tobacco crop in the world is given by

$$193.55 \pm t_{0.025}(\text{df} = 4(2-1))\sqrt{23020.8815}, \text{ or } 193.55 \pm 2.776\sqrt{23020.8815}, \text{ or}$$

$$[-227.64,\ 614.74], \text{ or } [0.00,\ 614.74].$$

10.3.2 OPTIMUM ALLOCATION IN TWO-STAGE SAMPLING

Suppose first stage n units are selected by SRSWOR sampling out of N units in the population. The second stage m units are selected from the given first-stage units each consisting of M units. If y_{ij} denotes the value of the sample unit corresponding to i^{th} first stage unit and j^{th} second stage unit, then an unbiased estimator of population mean is given by

$$\bar{\bar{y}}_{\text{eq}} = \frac{1}{n} \sum_{i=1}^{n} \frac{M}{m} \sum_{j=1}^{m} y_{ij} \tag{10.3.10}$$

with variance given by

$$V(\bar{\bar{y}}_{\text{eq}}) = \frac{N-n}{Nn} S_b^2 + \frac{M-m}{nMm} S_w^2, \tag{10.3.11}$$

where $S_b^2 = \frac{1}{N-1}\sum_{i=1}^{N}\left(\bar{Y}_{i\bullet} - \bar{Y}\right)^2$ and $S_w^2 = \frac{1}{N(M-1)}\sum_{i=1}^{N}\sum_{j=1}^{M}\left(Y_{ij} - \bar{Y}_{i\bullet}\right)^2$.

Consider C_1 be the cost of selecting first stage unit in a sample, C_2 be the cost of selecting a second stage unit, and C_0 be the overhead cost, then the simplest cost function in two-stage sampling can be written as

$$C = C_0 + nC_1 + mnC_2. \tag{10.3.12}$$

One can easily observe that the optimum first stage and second stage sample sizes for the fixed cost of survey are given by

$$n = \left\{(C - C_0)\sqrt{C_2\left(MS_b^2 - S_w^2\right)}\right\} \Big/ \left\{\sqrt{MC_1}\left\{\sqrt{(C_1/M)}\sqrt{\left(MS_b^2 - S_w^2\right)} + S_w\sqrt{C_2}\right\}\right\} \tag{10.3.13}$$

and

$$m = S_w\sqrt{C_1 M} \Big/ \sqrt{C_2\left(MS_b^2 - S_w^2\right)} . \tag{10.3.14}$$

10.4 OPTIMUM ALLOCATION OF SAMPLE IN THREE-STAGE SAMPLING

For simplicity let us assume that m is the number of SSUs selected from each of the selected n FSUs and p is the number of TSUs selected from each of the selected m SSUs. As we saw earlier in three stage sampling, the variance of the estimator consists of three components. Let us assume that it is given by

$$V(\hat{y}_{3s}) = \frac{\sigma_b^2}{n} + \frac{\sigma_w^2}{nm} + \frac{\bar{\sigma}_w^2}{mnp}, \tag{10.4.1}$$

where

$$\sigma_b^2 = \frac{1}{N}\sum_{i=1}^{N}\left(\bar{Y}_{i\bullet\bullet} - \bar{Y}\right)^2, \quad \sigma_w^2 = \frac{1}{NM}\sum_{i=1}^{N}\sum_{j=1}^{M}\left(\bar{Y}_{ij\bullet} - \bar{Y}_{i\bullet\bullet}\right)^2 \text{ and } \bar{\sigma}_w^2 = \frac{1}{NMP}\sum_{i=1}^{N}\sum_{j=1}^{M}\sum_{k=1}^{P}\left(Y_{ijk} - \bar{Y}_{ij\bullet}\right)^2$$

have their usual meanings.

Suppose C_0 is the overhead fixed cost for the survey and C_1, C_2, C_3 are the costs of enumerating a unit at first, second, and third stage of sampling. Obviously, the simplest cost function is

$$C = C_0 + nC_1 + nmC_2 + mnpC_3\,. \tag{10.4.2}$$

The Lagrange function is then given by

$$L = \frac{\sigma_b^2}{n} + \frac{\sigma_w^2}{nm} + \frac{\bar{\sigma}_w^2}{mnp} + \lambda\left[C_0 + nC_1 + mnC_2 + mnpC_3 - C\right]. \tag{10.4.3}$$

On differentiation (10.4.3) with respect to m, (mn) and (mnp) and equating to zero in each case we have

$$\frac{\partial L}{\partial n} = -\frac{\sigma_b^2}{n^2} + \lambda C_1 = 0 \qquad \Rightarrow \quad n = \frac{\sigma_b}{\sqrt{\lambda C_1}}, \tag{10.4.4}$$

$$\frac{\partial L}{\partial (nm)} = -\frac{\sigma_w^2}{(mn)^2} + \lambda C_2 = 0 \;\Rightarrow\; (mn) = \frac{\sigma_w}{\sqrt{\lambda C_2}}, \tag{10.4.5}$$

and

$$\frac{\partial L}{\partial (mnp)} = -\frac{\bar{\sigma}_w^2}{(mnp)} + \lambda C_3 = 0 \;\Rightarrow\; (mnp) = \frac{\bar{\sigma}_w}{\sqrt{\lambda C_3}}. \tag{10.4.6}$$

From (10.4.2), (10.4.4), (10.4.5), and (10.4.6) we have

$$C - C_0 = \frac{\left[\sigma_b\sqrt{C_1} + \sigma_w\sqrt{C_2} + \bar{\sigma}_w\sqrt{C_3}\right]}{\sqrt{\lambda}} \;\Rightarrow\; \sqrt{\lambda} = \frac{C - C_0}{\sigma_b\sqrt{C_1} + \sigma_w\sqrt{C_2} + \bar{\sigma}_w\sqrt{C_3}}. \tag{10.4.7}$$

On substituting (10.4.7) in (10.4.4), (10.4.5), and (10.4.6), and solving for m, n and p, we obtain optimum sample sizes as

$$n_{\text{opt}} = \frac{\sigma_b (C - C_0)}{\sqrt{C_1}\left[\sigma_b C_1 + \sigma_w \sqrt{C_2} + \bar{\sigma}_w \sqrt{C_3}\right]}, \quad m_{\text{opt}} = \frac{\sigma_w}{\sigma_b}\sqrt{\frac{C_1}{C_2}}, \text{ and } p_{\text{opt}} = \frac{\bar{\sigma}_w}{\sigma_w}\sqrt{\frac{C_2}{C_3}}. \quad (10.4.8)$$

The minimum variance can easily be obtained by substituting these optimum values in (10.4.1). The three-stage sampling considered so for is called traditional three stage sampling. If FSUs consists of district, SSUs consist of villages, and TSUs consist of bank accounts of customers, then from each selected village the bank account holding customers are required to be sampled separately and independently across the villages. Raj (1968) and Rao (1975) have discussed the estimation strategies in greater detail.

Chaudhuri, Adhikary, and Dihidar (2000) considered a homogeneous linear unbiased function of the sampled first stage unit (FSU) values taken as an estimator of a survey population total, the sampling variance is expressed as a homogeneous quadratic function of the FSU values. When the FSU values are not ascertainable but unbiased estimators for them are separately available through sampling in later stages and substituted into the estimator, Raj (1968) gave a simple variance estimator formula for this multi-stage estimator of the population total. He requires that the variances of the estimated FSU values in sampling at later stages and their unbiased estimators are available in certain simple forms. Chaudhuri, Adhikary, and Dihidar (2000) illustrate a particular three-stage sampling strategy and present a simulation based numerical exercise showing the relative efficiencies of two alternative variance estimators.

10.5 MODIFIED THREE-STAGE SAMPLING

To draw a sample of n FSUs, Chaudhuri (1997) has used the well known Rao, Hartley, and Cochran (1962) scheme of sampling using $p_i \in \Omega$. First we divide the population Ω at random into n groups, taking N_i FSUs in the ith group such that $\sum_{i=1}^{n} N_i = N$, the population size. From each group, one unit is selected with probability proportional to p_i, and selection is done independently across the groups. Let Q_i be the sum of the probabilities p_i falling in the ith group. The m_i SSUs are also selected using RHC scheme from the ith FSU of size M_i. We split the M_i SSUs into m_i groups at random, taking N_{ij} SSUs in the jth group such that $\sum_{j=1}^{m_i} M_{ij} = M_i$ is satisfied. From every group, say the jth, one unit is chosen with probability proportional to p_{ij} and is repeated across the groups. Let Q_{ij} denote the sum of the p_{ij} values falling in the jth group. Let s_i denote the set of SSUs chosen from the ith selected FSU, $A(s_i)$ be the set of bank accounts of customers with dwelling addresses in s_i and $L_i(s_i)$ be the cardinality of $A(s_i)$. Also let s_i^* be an

SRSWOR sample of $t_i(s_i)$ TSUs chosen from $A(s_i)$ and it is to be repeated independently across the selected FSUs. Thus we have the following theorem:

Theorem 10.5.1. An unbiased estimator of population total Y is given by

$$\hat{y}_{\text{m3s}} = \sum_{i=1}^{n}\frac{Q_i}{p_i}\sum_{j=1}^{m_i}\frac{Q_{ij}}{p_{ij}}\left(\frac{L_i(s_i)}{t_i(s_i)}\sum_{k=1}^{t_i} y_{ijk}\right). \tag{10.5.1}$$

Proof. The estimator (10.5.1) can be written as

$$\hat{y}_{\text{m3s}} = \sum_{i=1}^{n}\frac{Q_i}{p_i}\sum_{j=1}^{m_i}\frac{Q_{ij}}{p_{ij}} w_{ij},$$

where

$$w_{ij} = \frac{L_i(s_i)}{t_i(s_i)}\sum_{k\in s_i^*} y_{ijk} = L_i(s_i)\bar{y}_{ij}.$$

Taking expected values on both sides we have

$$E(\hat{y}_{\text{m3s}}) = E_1E_2E_3\left[\sum_{i=1}^{n}\frac{Q_i}{p_i}\sum_{j=1}^{m_i}\frac{Q_{ij}}{p_{ij}}w_{ij}\right] = E_1\left[\sum_{i=1}^{n}\frac{Q_i}{p_i}E_2\left\{\sum_{j=1}^{m_i}\frac{Q_{ij}}{p_{ij}}w_{ij}\right\}\right]$$

$$= E_1\left[\sum_{i=1}^{n}\frac{Q_i}{p_i}\sum_{j=1}^{M_i}Y_{ij}\right] = E_1\left[\sum_{i=1}^{n}\frac{Q_i}{p_i}y_{i\bullet}\right] = Y.$$

Hence the theorem.

10.6 GENERAL CLASS OF ESTIMATORS IN TWO-STAGE SAMPLING

Let Y_i and X_i be the totals of the ith FSU of the study variable Y and auxiliary variable X, respectively. Assume that the population total $X = \sum_{i=1}^{N} X_i$ of the auxiliary variable is known and we are interested in the estimation of population total $Y = \sum_{i=1}^{N} Y_i$ of the study variable. A sample s of n FSUs is selected according to any design with π_i and π_{ij} as the known first and second order inclusion probabilities, which are in fact function of first stage sample size n. The selected FSUs are sub-sampled independently with suitable selection probabilities at the second stage. When the ith unit becomes selected then it is assumed that from sampling in the second stage, estimators t_{iy} and t_{ix}, respectively, for Y_i and X_i are available such that

$$V_2(t_{iy}) = \sigma_{iy}^2 = M_i^{-1}\sum_{j=1}^{M_i}(Y_{ij}-\bar{Y}_i)^2,\quad V_2(t_{ix}) = \sigma_{ix}^2 = M_i^{-1}\sum_{j=1}^{M_i}(X_{ij}-\bar{X}_i)^2,$$

$$C_2(t_{iy},\, t_{ix}) = \sigma_{ixy} = M_i^{-1}\sum_{j=1}^{M_i}(X_{ij}-\bar{X}_i)(Y_{ij}-\bar{Y}_i),\quad E_2(t_{iy}) = Y_i,\ \text{and}\ E_2(t_{ix}) = X_i.$$

Thus we have the following theorem:

Theorem 10.6.1. The unbiased estimators of Y and X are, respectively, given by

$$t_y = \sum_{i=1}^{n} \frac{t_{iy}}{\pi_i} \quad \text{and} \quad t_x = \sum_{i=1}^{n} \frac{t_{ix}}{\pi_i} \tag{10.6.1}$$

with

$$\text{Cov}(t_y, t_x) = \sigma_{xy} + \sum_{i=1}^{N} \frac{\sigma_{ixy}}{\pi_i}, \tag{10.6.2}$$

where

$$\sigma_{xy} = \frac{1}{2} \sum_{i \neq j=1}^{N} (\pi_i \pi_j - \pi_{ij})(X_i/\pi_i - X_j/\pi_j)(Y_i/\pi_i - Y_j/\pi_j).$$

Proof. By the definition of covariance we have

$$\text{Cov}(t_y, t_x) = E_1 C_2(t_y, t_x) + C_1[E_2(t_y), E_2(t_x)]$$

$$= E_1 C_2\left[\sum_{i=1}^{n} \frac{t_{iy}}{\pi_i}, \sum_{i=1}^{n} \frac{t_{ix}}{\pi_i}\right] + C_1\left[E_2\left(\sum_{i=1}^{n} \frac{t_{iy}}{\pi_i}\right), E_2\left(\sum_{i=1}^{n} \frac{t_{ix}}{\pi_i}\right)\right]$$

$$= E_1\left[\sum_{i=1}^{n} \frac{\sigma_{ixy}}{\pi_i^2}\right] + C_1\left[\sum_{i=1}^{n} \frac{Y_i}{\pi_i}, \sum_{i=1}^{n} \frac{X_i}{\pi_i}\right] = \sigma_{xy} + \sum_{i=1}^{N} \frac{\sigma_{ixy}}{\pi_i}.$$

Hence the theorem.

Corollary 10.6.1. From the above theorem the following results are obvious:

$$V(t_y) = \sigma_y^2 + \sum_{i=1}^{N} \frac{\sigma_{iy}^2}{\pi_i}, \tag{10.6.3}$$

where

$$\sigma_y^2 = \frac{1}{2} \sum_{i \neq j=1}^{N} (\pi_i \pi_j - \pi_{ij})(Y_i/\pi_i - Y_j/\pi_j)^2,$$

and

$$V(t_x) = \sigma_x^2 + \sum_{i=1}^{N} \frac{\sigma_{ix}^2}{\pi_i}, \tag{10.6.4}$$

where

$$\sigma_x^2 = \frac{1}{2} \sum_{i \neq j=1}^{N} (\pi_i \pi_j - \pi_{ij})(X_i/\pi_i - X_j/\pi_j)^2.$$

Sahoo and Panda (1997) first defined a sub-class to estimate the ith unit population total Y_i (say) as

$$\hat{Y}_i = f_{2i}(t_{iy}, t_{ix}) \tag{10.6.5}$$

where $f_{2i}(t_{iy}, t_{ix})$ is a function of t_{iy} and t_{ix}, such that $f_{2i}(Y_i, X_i) = Y_i$ satisfying certain regularity conditions analogue to those defined by Srivastava (1980). Evidently an estimator to estimate Y can be defined as

$$t_y^* = \sum_{i=1}^{n} \frac{\hat{Y}_i}{\pi_i} = \sum_{i=1}^{n} f_{2i}\left(t_{iy}, t_{ix}\right)/\pi_i \tag{10.6.6}$$

and that to estimate X can be defined as

$$t_x^* = \sum_{i=1}^{n} X_i/\pi_i \ . \tag{10.6.7}$$

Using (10.6.6) and (10.6.7), Sahoo and Panda (1997) have defined a main class of estimators of population total in two-stage sampling as

$$\hat{Y}_C = f_1\left[t_y^*,\ t_x^*\right], \tag{10.6.8}$$

where $f_1\left[t_y^*,\ t_x^*\right]$ denotes the function of t_y^* and t_x^* satisfying certain regularity conditions. Then we have the following theorem:

Theorem 10.6.2. The minimum variance of the main class of estimators $\hat{Y}_C$ is

$$\text{Min.}V\left(\hat{Y}_C\right) = \sigma_y^2\left(1-\rho_{xy}^2\right) + \sum_{i=1}^{N} \frac{\sigma_{iy}^2\left(1-\rho_{ixy}^2\right)}{\pi_i}. \tag{10.6.9}$$

Proof. Note that

$$f_1(Y, X) = Y \quad \text{and} \quad \frac{\partial f_1}{\partial t_y^*}\Big|_{(Y,X)} = 1$$

Using the second order Taylor's series, the class of estimators $\hat{Y}_C$ can easily be expanded around the point, (Y, X) as

$$\hat{Y}_C = f_1\left[Y + \left(t_y^* - Y\right),\ X + \left(t_x^* - X\right)\right]$$

$$= f_1(Y,X) + \left(t_y^* - Y\right)\frac{\partial f_1}{\partial t_y^*}\Big|_{(Y,X)} + \left(t_x^* - X\right)\frac{\partial f_1}{\partial t_x^*}\Big|_{(Y,X)} + \ldots..$$

$$= t_y^* + \left(t_x^* - X\right)f_{12}(Y,X) + \ldots.$$

$$= \sum_{i=1}^{n}\frac{\hat{Y}_i}{\pi_i} + \left(t_x^* - X\right)f_{12}(Y,X) + \ldots.$$

$$= \sum_{i=1}^{n}\frac{1}{\pi_i} f_{2i}\left(t_{iy}, t_{ix}\right) + \left(t_x^* - X\right)f_{12}(Y,X) + \ldots.$$

$$= \sum_{i=1}^{n} \frac{f_{2i}(Y_i, X_i) + \left(t_{iy} - Y_i\right)\frac{\partial f_{2i}}{\partial t_{iy}}\Big|_{(Y_i,X_i)} + \left(t_{ix} - X_i\right)\frac{\partial f_{2i}}{\partial t_{ix}}\Big|_{(Y_i,X_i)}}{\pi_i} + \left(t_x^* - X\right)f_{12}(Y,X) + \ldots.$$

$$= \sum_{i=1}^{n} \frac{t_{iy} + \left(t_{ix} - X_i\right)f_{22i}(Y_i, X_i)}{\pi_i} + \left(t_x^* - X\right)f_{12}(Y,X) + \ldots.., \tag{10.6.10}$$

where $f_{12}(Y, X)$ and $f_{22i}(Y_i, X_i)$ denotes the known first order partial derivatives of the respective function used for the construction of the estimator. Then using the definition of variance we have

$$V\left(\hat{\bar{Y}}_C\right)=\sigma_y^2+2f_{12}(Y,X)\sigma_{xy}+f_{12}^2(Y,X)\sigma_x^2+\sum_{i=1}^{N}\frac{\sigma_{iy}^2+2f_{22i}(Y_i,X_i)\sigma_{ixy}+f_{22i}^2(Y_i,X_i)\sigma_{ix}^2}{\pi_i} \quad (10.6.11)$$

The variance $V\left(\hat{\bar{Y}}_C\right)$ is found to be minimum for

$$f_{12}(Y,X)=-\beta_{xy} \text{ and } f_{22i}(Y_i,X_i)=-\beta_{ixy}\,.$$

On substituting these values in (10.6.11) we have (10.6.9). Hence the theorem.

Sahoo and Panda (1999a, 1999b) extended the results to the situation when two auxiliary variables X and Z are available to estimate the population mean of the study variable Y under two-stage sampling.

10.7 PREDICTION ESTIMATOR UNDER TWO-STAGE SAMPLING

Bellhouse and Rao (1986) have pointed out that the prediction estimator may be only marginally better than the classical estimator in PPS sampling under two stage sampling design from the efficiency point of view. Following them, consider a population of N units consisting of L FSUs with M_i units in the ith FSU ($i=1,2,...,L$) such that $N=\sum_{i=1}^{L}M_i$. Let y_{ij} be the value of the jth unit in the ith FSU and we are interested to estimate the population mean

$$\bar{Y}=N^{-1}\sum_{i=1}^{L}\sum_{j=1}^{M_i}y_{ij}$$

by selecting a sample, s, of the FSUs and then choosing, for each $i\in s$, a sample s_i of units from the ith FSU. The probability $p(\tilde{s})$ with which the composite sample $\tilde{s}=\cup_{i\in s}s_i$ is chosen is called a sampling design. The design is said to be of fixed size if the number of distinct FSUs in s is v and the number of distinct units in s_i is m_i, where v and m_i are pre-specified. The first order inclusion probability for i^{th} FSU is $\pi_i=\sum_{s\ni i}p(s)$ and that for j^{th} unit in the ith FSU is $\pi_{ij}=\sum_{\tilde{s}\ni(i,j)}p(\tilde{s})$. Then we have the following theorem:

Theorem 10.7.1. A linear homogeneous estimator of population mean $\bar{Y}$ is

$$\bar{y}_{bs}=\sum_{i\in s}\sum_{j\in s_i}b_{\tilde{s}ij}y_{ij}=N^{-1}\left[\sum_{i\in s}\sum_{j\in s_i}y_{ij}+\sum_{i\in s}\sum_{j\in s_i}d_{\tilde{s}ij}y_{ij}\right] \quad (10.7.1)$$

where $d_{\tilde{s}ij}=N\,b_{\tilde{s}ij}-1$ and $b_{\tilde{s}ij}$ are defined for all $\tilde{s}$ and all $(i,j)\in\tilde{s}$.

Scott and Smith (1969) defined a model by assuming that y_{ij} are uncorrelated, for each i, with mean μ_i and fixed variance σ_i^2 and that μ_i are uncorrelated with mean μ and variance σ^2. They suggested a model as

$$y_{ij}=\mu+e_{ij} \quad (10.7.2)$$

with
$E_m(e_{ij})=0$, $E(e_{ij}^2)=\delta^2+\sigma_i^2$, $E_m(e_{ij}e_{ij'})=\delta^2$, $j\neq j'$ and $E_m(e_{hj}e_{h'j'})=0$, $h\neq h'$.

Then we have the following theorem:

Theorem 10.7.2. If $m_i=m$ and $\sigma_i^2=\sigma^2$ $\forall$ i in the above model, then a prediction estimator in two-stage sampling for a fixed (v, m) design is given by

$$\bar{y}_{\text{pr1}}=\Psi\bar{y}_a+(1-\Psi)\bar{y}_b, \tag{10.7.3}$$

where $\bar{y}_a=\sum_{i\in s}M_i\sum_{j\in s_i}\frac{y_{ij}}{m}\Big/\sum_{i\in s}M_i$, $\bar{y}_b=\frac{1}{mv}\sum_{i\in s}\sum_{j\in s_i}y_{ij}$ and ψ is a constant such that the variance of the estimator $\bar{y}_{\text{pr1}}$ is minimum. It can be easily shown that $V(\bar{y}_{\text{pr1}})$ is minimum if the optimum value of Ψ is given by

$$\Psi_{\text{opt}}=\frac{\lambda}{N}\sum_{i\in s}M_i\,,\quad\text{with}\quad \lambda=\delta^2\left(\delta^2+\frac{\sigma^2}{m}\right)^{-1}. \tag{10.7.4}$$

Theorem 10.7.3. In the case of stratified random sampling with $v=L$ the prediction estimator becomes

$$\bar{y}_{\text{pr2}}=\sum_{i=1}^{L}\frac{M_i}{N}\left[\lambda_i\left(\frac{1}{m}\sum_{j\in s_i}y_{ij}\right)+(1-\lambda_i)\left(\sum_{i=1}^{L}\sum_{j\in s_i}y_{ij}\Big/m\sum_{i=1}^{L}\lambda_i\right)\right], \tag{10.7.5}$$

where $\lambda_i=\delta^2\left(\delta^2+\frac{\sigma_i^2}{m_i}\right)^{-1}$.

Bellhouse and Rao (1986) have proposed a two-stage permutation model:

$$y_{ij}=\bar{\bar{Y}}+e_{ij} \tag{10.7.6}$$

where $E_m(e_{ij})=0$, $E_m(e_{ij}^2)=\sigma_b^2+\sigma_w^2$, $E_m(e_{ij}e_{ij'})=\sigma_b^2-(M_i-1)^{-1}\sigma_w^2$, $j\neq j'$, and $E_m(e_{ij}e_{i'j'})=-\sigma_b^2/(L-1)$, $h\neq h'$.

Then we have the following theorem:

Theorem 10.7.4. The MSE of the estimator $\bar{y}_{\text{bs}}$ under model (10.7.6) is

$$E_m[\text{MSE}(\bar{y}_{\text{bs}})]=\bar{\bar{Y}}^2\left(\sum_{i\in s}\sum_{j\in s_i}b_{\tilde{s}ij}-1\right)^2$$
$$+\frac{\sigma_b^2}{L-1}\left[L\sum_{i\in s}\left(\sum_{j\in s_i}b_{\tilde{s}ij}\right)^2-\left(\sum_{i\in s}\sum_{j\in s_i}b_{\tilde{s}ij}\right)^2-2\sum_{i\in s}\left(\frac{LM_i}{N}-1\right)\sum_{j\in s_i}b_{\tilde{s}ij}\right]$$

$$+\sigma_w^2\left[\sum_{i\in s}\frac{M_i}{M_i-1}\sum_{j\in s_i}b_{\bar{s}ij}^2-\sum_{i\in s}\frac{1}{M_i-1}\left(\sum_{j\in s_i}b_{\bar{s}ij}\right)^2\right]+E_m\left(\bar{E}^2\right) \quad (10.7.7)$$

where $\bar{E}$ is the population mean of e_{ij}.

Theorem 10.7.5. For any fixed size design, the optimal model unbiased predictor of population mean $\bar{Y}$ is

$$\bar{y}_{\text{opt}}=\left\{1-\left(\frac{\gamma L}{L-1}\right)N^{-1}\sum_{i\in s}a_iM_i\right\}\frac{\sum_i a_i\sum_j y_{ij}}{m\sum_i a_i}+\left\{\left(\frac{\gamma L}{L-1}\right)N^{-1}\sum_{i\in s}a_iM_i\right\}\frac{\sum_i a_iM_i\sum_j y_{ij}}{m\sum_i a_iM_i} \quad (10.7.8)$$

where $\gamma=\sigma_b^2/\sigma_w^2$ and $a_i=m_i\left(\frac{Lm_i\gamma}{L-1}+\frac{M_i-m_i}{M_i-1}\right)^{-1}$.

10.8 PREDICTION APPROACH TO ROBUST VARIANCE ESTIMATION IN TWO-STAGE CLUSTER SAMPLING

Royall (1986) has considered the problem of robust variance estimators by following Royall and Cumberland (1978, 1981a, 1981b) for some classes of estimators which are useful for estimating population total and ratios of totals from cluster sampling. In other words, he has considered the problem of estimation of population totals

$$Y=\sum_{i=1}^{N}\sum_{j=1}^{M_i}Y_{ij}\ ,\ Z=\sum_{i=1}^{N}\sum_{j=1}^{M_i}Z_{ij}$$

and the ratio

$$R=\bar{Z}/\bar{Y}$$

where $(Y_{ij},\ Z_{ij})$ are the numbers associated with j^{th} unit in the ith cluster.

The estimators are generally of the form

$$\hat{Y}=\frac{N}{n}\sum_{i\in s}u_i\hat{Y}_i\ ,\quad \hat{Z}=\frac{N}{n}\sum_{i\in s}u_i\hat{Z}_i\ ,\quad \text{and}\quad \hat{R}=\hat{Z}/\hat{Y}$$

where

$$\hat{Y}_i=\frac{M_i}{m_i}\sum_{j=1}^{m_i}y_{ij}\quad \text{and}\quad \hat{Z}_i=\frac{M_i}{m_i}\sum_{j=1}^{m_i}z_{ij}$$

are the simple expansion estimators of the ith cluster totals for Y and Z respectively. The coefficient u_i is free from y or z values. Royall (1986) does not guaranty that the model considered by him is correct, but he adopts it as a tool to use in planning and inference. Now we have the following theorem:

Theorem 10.8.1. If variables in different clusters are uncorrelated, then the error variance of the estimator under two-stage sampling design is given by

$$V\left(\hat{Y}\right)=\frac{N}{f}B_1-2NB_2+NB_3, \quad (10.8.1)$$

where

$B_1 = \frac{1}{n}\sum_{i=1}^{N} u_i^2 M_i^2 V(\bar{y}_{si})$, $\bar{y}_{si} = m_i^{-1}\sum_{j=1}^{m_i} y_{ij}$, $B_2 = n^{-1}\sum_{i=1}^{N} u_i M_i^2 \operatorname{cov}(\bar{y}_{si}, \bar{y}_i)$, $\bar{y}_i = Y_i / M_i$ and

$B_3 = N^{-1}\sum_{i=1}^{N} M_i^2 V(\bar{y}_i)$.

Clearly B_2 and B_3 depend on the variance--covariance structure of units that are not observed, but B_1 depends only on the variances of sample means. Obviously it is possible to produce an unbiased or consistent estimator $\hat{B}_1$ of B_1 under models much more general than the one required for estimating B_2 and B_3. Then the first term of the variance $\frac{N}{f} B_1$ is dominant when $f \to 0$ and it is possible to produce an unbiased or consistent estimator of it under fairly general models.

Now we have the following corollary:

Corollary 10.8.1. The variance of a linear function of independent random variables, given by $\sum l^2 V(X)$ can be estimated by

$$\sum l^2 \left\{ \left(X - \hat{X}\right)^2 + k \right\},$$

where k stands for appropriate bias correction term.

Royall (1986) has considered both the situations when the cluster sizes M_i are known and unknown, to estimate the variance. We will discuss only the situation when the cluster size is known. Royall (1986) considered the following model which depends upon three unknown parameters μ, ρ, and σ^2, given by

$$E\left(Y_{ij}\right) = \mu \text{, and } \operatorname{Cov}\left(Y_{ij}, Y_{kl}\right) = \begin{cases} \sigma^2 & \text{if } i = k, \ j = l, \\ \rho\sigma^2 & \text{if } i = k, \ j \neq l \ = k, \\ 0 & \text{if } i \neq k. \end{cases} \tag{10.8.2}$$

The above model has been also used by Scott and Smith (1969) and Royall (1976). Burdick and Sielken (1979) have studied the same variance estimation problem under the same model but focused on obtaining unbiased variance estimators having chi squared distribution, whereas Royall (1986) has suggested robustness in the sense of consistency under broad conditions. In fact, this model describes populations in which the cluster total Y_i is roughly proportional to cluster size M_i, that is, $E(Y_i) = \mu M_i$, with Y_{ij} correlated within clusters.

Thus we have the following theorem.

Corollary 10.8.2. Under model (10.8.2), the following results hold:

$$V\left(\hat{Y}_i\right)=\frac{M_i^2}{m_i}\sigma^2\left[1-\rho+m_i\rho\right], \tag{10.8.3}$$

and

$$\mathrm{Cov}\left(\hat{Y}_i,\ Y_i\right)=M_i\sigma^2\left[1-\rho+M_i\rho\right]. \tag{10.8.4}$$

Then we have the following corollary:

Corollary 10.8.3. An estimator of variance $V\left(\hat{Y}\right)$ can be obtained as a linear function of two parameters $\theta_1=\sigma^2(1-\rho)$ and $\theta_2=\rho\sigma^2$, which can be estimated by replacing these expressions by their unbiased estimates. Unfortunately such estimates of variance are not robust.

10.8.1 ROYALL'S TECHNIQUE OF VARIANCE ESTIMATION

Royall (1986) suggested the following steps to estimate the three different components of the variance,

$$V\left(\hat{Y}\right)=\frac{N}{f}B_1-2NB_2+NB_3\,.$$

(a) Estimate the third component under the model defined as:

$$M_1:E\left(Y_{ij}\right)=\mu\text{, and } M_1:Cov\left(Y_{ij},Y_{kl}\right)=\begin{cases}\sigma^2 & \text{if } \ i=k,j=l,\\ \rho\sigma^2 & \text{if } \ i=k,j\neq l,\\ 0 & \text{if } \ i\neq k.\end{cases} \tag{10.8.5}$$

(b) Estimate the second component under the model defined as

$$M_2\text{: } E\left(Y_{ij}\right)=\mu,\ i=1,2,...,N;\ j=1,2,...,M_i\text{ ,and } M_2\text{: }\mathrm{Cov}\left(Y_{ij},\ Y_{kl}\right)=0,\ i\neq k\,. \tag{10.8.6}$$

(c) Estimate the first component of variance under the model defined as

$$M_3:E\left(Y_{ij}\right)=\mu,\quad\text{and } M_3:\mathrm{Cov}\left(Y_{ij},Y_{kl}\right)=\begin{cases}\sigma_i^2 & i=k,j=l,\\ \rho_i\sigma_i^2 & i=k,j\neq l,\\ 0 & \text{otherwise,}\end{cases} \tag{10.8.7}$$

for $i=1,2,...,N;\ j=1,2,...,M_i$.

Then we have the following theorem:

Theorem 10.8.2. A robust estimator of variance of estimator of population total in two-stage cluster sampling is given by

$$\hat{v}_{\text{Robust}}=\frac{1}{f^2}\sum_{i=1}^{n}u_i\left(u_i-f\right)\hat{y}_i+\frac{1}{f}\sum_{i=1}^{n}u_iM_i\left(1-f_i\right)\left(\frac{s_i^2}{f_i}\right)+\left\{\sum_{i=1}^{N}M_i^2-\frac{1}{f}\sum_{i=1}^{n}u_iM_i^2\right\}\overset{\bullet}{\theta}_2, \tag{10.8.8}$$

where

$$\hat{v}_i = \frac{r_i^2}{(1-g_i/n)} - \frac{\left\{(M_i/\overline{M})^2/(1-g_i/n)\right\}\sum u_j^2 r_j^2/n^2(1-g_j/n)}{\left\{1+\sum_s (M_j/\overline{M})^2 (u_j/n)^2/(1-g_j/n)\right\}}$$

with $g_i = 2u_i M_i/\overline{M} < n \ \forall \ i \in s$, $\dot{\theta}_2 = \sum_{i=1}^{n}\left(\hat{v}_i - M_i S_i^2/f_i\right)\Big/\sum_{i=1}^{n} M_i^2$, $f_i = n_i/N_i$ and $f = n/N$, etc., have their usual meanings. A general method for estimating the variance in multi-stage sampling has also been discussed by Srinath and Hidiroglou (1980).

10.9 TWO-STAGE SAMPLING WITH SUCCESSIVE OCCASIONS

The theory of two-stage sampling when selected FSUs on an occasion alone are partially replaced on subsequent occasions after the first and by observing the same set of SSUs in the given FSU from one occasion to the other was first developed by Jessen (1942), Yates (1960), Patterson (1950), Eckler (1955), and Tikkiwal (1953, 1958, 1965). Singh (1968) considered the theory of two-stage successive sampling and built up linear unbiased estimators of the population mean on the second and third occasions separately under certain restrictive assumptions when the partial replacement of units is made among the FSUs only. Kathuria and Singh (1971a, 1971b) have shown that under certain circumstances, partial retention of SSUs may be better than the partial retention of FSUs. Agarwal and Tikkiwal (1980) have mentioned the following possibilities:

(a) Replacement among FSUs only,
(b) Replacement among SSUs only,
(c) Replacement among both FSUs and SSUs.

Surveys often have to be repeated on many occasions (over years or seasons or months) for estimating the same characteristic at different points of time. The information collected on previous occasions can be used to study the change or total value for the most recent occasion. An investigator or owner of the industry of cold drinks may be interested in the following type of problems:

(a) The average or total sale of cold drinks for the current season;
(b) The change in average sale of cold drinks for two different seasons;

Note that sampling on successive occasions is done according to a specific rule, with partial replacement of units, is called Rotation Sampling. Hansen, Hurwitz, and Madow (1953) and Rao and Graham (1964) have considered the problems of rotation designs for successive sampling.

Let us discuss here few successive sampling strategies suggested by various researchers.

10.9.1 ARNAB's SUCCESSIVE SAMPLING SCHEME*

To discuss Arnab (1979a) successive sampling scheme let us first define few notation. Let us consider

Y_{it} = Value of the study variable for the ith unit of a finite population Ω of size N at the tth occasion, for $i = 1, 2, \ldots N$ and $t = 1, 2, \ldots, T$.

The following table will be helpful in understanding the estimators proposed by Arnab (1979a).

Table 10.9.1. Description of the sampling procedure on different occasions.

Occasion	Sample (Size) ⊂ Selected From (method of selection)			Total Sample size
1	$s_{11}\ (m_{11}) \subset \Omega$ (PPSWR)		.	m_{11}
2	$s_{21}\ (m_{21}) \subset s_{1,1}$ (SRSWOR)	$s_{22}\ (m_{22}) \subset \Omega$ (PPSWR)	.	$m_{21} + m_{22} = m_{11}$
3	$s_{31}\ (m_{31}) \subset s_{2,1}$ (SRSWOR)	$s_{32}\ (m_{32}) \subset s_{2,2}$ (SRSWOR)	.	$m_{31} + m_{32} + m_{33} = m_{11}$
.....	....	.....		....
t	$s_{t1}\ (m_{t1}) \subset s_{(t-1),1}$ (SRSWOR)	$s_{t2}\ (m_{t2}) \subset s_{(t-1),2}$ (SRSWOR)		$m_{t1} + m_{t2} + \ldots + m_{(t-1),t} = m_{11}$
...	....	.....	.	.....
T	$s_{T1}\ (m_{T1}) \subset s_{(T-1),1}$ (SRSWOR)	$s_{T2}\ (m_{T2}) \subset s_{(T-1),2}$ (SRSWOR)	.	$m_{T1} + m_{T2} + \ldots + m_{(T-1),T} = m_{11}$
Estimators	$\hat{Y}_T(1)$	$\hat{Y}_T(2)$	$\hat{Y}_T(t)$	$\hat{Y}_T(T)$

$Y_T = \sum_{i=1}^{N} Y_{iT}$, population total of the study variable Y at time T, P_i = The probability of selecting the ith unit on any occasion, $\sum_{i=1}^{N} P_i (Y_{ti}/P_i - Y_t)(Y_{t'i}/P_i - Y_{t'}) = \delta^{|t-t'|} V_0$ for every $t, t' = 1,2,\ldots,T$, such that δ and $V_0 = \sum_{i=1}^{N} P_i (Y_{ti}/P_i - Y)^2$ are known quantities.

---

* The proofs of the Theorems in this section are contributed by Prof. R. Arnab, Department of Statistics, University of Durban--Westville, Private Bag X54001, Durban 4000, South Africa.

Arnab's Strategy I. The main steps of Arnab's first strategy are:

(1) On the first occasion, select a sample s_{11} of size m_{11} with PPSWR sampling.

(2) On the second occasion :

(a) Select a sub-sample s_{21} of size m_{21} units from the first sample s_{11} of m_{11} units by SRSWOR sampling;
(b) Select an independent sample s_{22} of m_{22} units, which in fact is equal to the $(m_{11}-m_{21})$ units, by PPSWR sampling from Ω;

(3) The value of the total sample size m_{11} on each occasion is supposed to be fixed and determined from cost consideration.

Thus in general on the tth occasion a sample $s_{tt'}$ of size $m_{tt'}$ is selected by SRSWOR sampling from $s_{(t-1)t'}$ independently for each $t'=1,2,...,(t-1)$. Finally another sample $m_{tt}=\left(m_{11}-\sum\limits_{t'=1}^{t-1} m_{tt'}\right)$ is selected by PPSWR method from the whole population for $t=2,3,...,T$.

Thus we have the following theorems:

Theorem 10.9.1. An unbiased estimator of population total Y_T on occasion T is given by

$$\hat{Y}_T=\sum_{t=1}^{T} C_t \hat{Y}_T(t), \tag{10.9.1}$$

where

$$\hat{Y}_T(t)=\begin{cases}\dfrac{1}{m_{Tt}}\sum\limits_{s_{Tt}}\dfrac{y_{Ti}}{P_i}-\beta_T(t)\left\{\dfrac{1}{m_{Tt}}\sum\limits_{s_{Tt}}\dfrac{y_{(T-1),i}}{p_i}-\hat{Y}_{(T-1)}(t)\right\}, & t=1,2,...,(T-1),\\[2ex] \dfrac{1}{m_{TT}}\sum\limits_{s_{TT}}\dfrac{y_{Ti}}{P_i}, & t=T,\end{cases} \tag{10.9.2}$$

with $\beta_T(t)$ and C_t are being constants such that the estimator $\hat{Y}_T$ is unbiased and has the minimum variance. It can be verified by the method of induction on different occasions that minimum variance of the estimator $\hat{Y}_T$ exists for $\beta_T(t)=\delta$ for all t.

Proof. For $t=1$

$$\hat{Y}_T(1)=\frac{1}{m_{T1}}\sum_{s_{T1}}\frac{y_{Ti}}{p_i}-\beta_T(1)\left\{\frac{1}{m_{T1}}\sum_{s_{T1}}\frac{y_{(T-1)i}}{p_i}-\hat{Y}_{T-1}(1)\right\}.$$

For $T=1$

$$\hat{Y}_1(1) = \frac{1}{m_{T1}} \sum_{s_{T1}} \frac{y_{Ti}}{p_i}$$

and

$$E(\hat{Y}_1) = E\left[\frac{1}{m_{11}} \sum_{s_{11}} \frac{y_{1i}}{p_i}\right] = \sum_{\Omega} Y_{1i} = Y_1 .$$

Now for $T = 2$

$$\hat{Y}_2(1) = \frac{1}{m_{21}} \sum_{s_{21}} \frac{y_{2i}}{p_i} - \beta_2(1)\left\{\frac{1}{m_{21}} \sum_{s_{21}} \frac{y_{1i}}{p_i} - \hat{Y}_1(1)\right\} .$$

Note that s_{21} is selected from s_{11} by SRSWOR sampling, therefore we have

$$E\{\hat{Y}_2(1)\} = E_1E_2\{\hat{Y}_2(1) | s_{11}\} = E_1E_2\left[\frac{1}{m_{21}} \sum_{s_{21}} \frac{y_{2i}}{p_i} - \beta_2(1)\left\{\frac{1}{m_{21}} \sum_{s_{21}} \frac{y_{1i}}{p_i} - \hat{Y}_1(1)\right\}\right],$$

where E_1 and E_2 have their usual meanings.

Thus we have

$$E\{\hat{Y}_2(1)\} = E_1\left[\frac{1}{m_{11}} \sum_{s_{11}} \frac{y_{2i}}{p_i} - \beta_{21}\left\{\frac{1}{m_{11}} \sum_{s_{11}} \frac{y_{1i}}{p_i} - \hat{Y}_1(1)\right\}\right] = Y_2 - \beta_{21}\{Y_1 - Y_1\} = Y_2 .$$

Thus we can assume

$$E(\hat{Y}_{T-1}(t)) = Y_{T-1} .$$

So we have

$$E(\hat{Y}_T(t)) = E_1E_2\{\hat{Y}_T(t) | s_{T-1,1}\}$$

$$= E_1\left[\frac{1}{m_{T-1,1}} \sum_{s_{T-1,1}} \frac{y_{Ti}}{p_i} - \beta_T(t)\left\{\frac{1}{m_{T-1}} \sum_{s_{T-1,1}} \frac{y_{T-1,i}}{p_i} - \hat{Y}_{T-1}(t)\right\}\right]$$

$$= Y_T - \beta_T(t)\{Y_{T-1} - Y_{T-1}\} = Y_T .$$

Note that

$$E\left(\frac{1}{m_{T-1,1}} \sum_{s_{T-1,1}} \frac{y_{Ti}}{p_i}\right) = E\left(\frac{1}{m_{1,1}} \sum_{s_{1,1}} \frac{y_{Ti}}{p_i}\right) = Y_T .$$

Thus we have

$$E(\hat{Y}_T) = \sum_{t=1}^{T} C_t Y_t = Y_T$$

whenever $\sum_{t=1}^{T} C_t = 1$.

Hence the theorem.

Theorem 10.9.2. The variance $V\left[\hat{Y}_T(t)\right]$ with $\beta_T(t)=\delta$ for every $t=1,2,...,(T-1)$ is given by

$$V\left[\hat{Y}_T(t)\right]=\begin{cases}\left[\dfrac{1}{m_{Tt}}\sum\limits_{t'=1}^{T-t}\left(\dfrac{1}{m_{T+1-t',t}}-\dfrac{1}{m_{T-t',t}}\right)\left(\delta^2\right)^{t'}\right]V_0, & \text{for } t=1,2,...,(T-1),\\ \dfrac{V_0}{m_{TT}}, & \text{for } t=T.\end{cases} \tag{10.9.3}$$

Proof. Consider

$$\hat{Y}_T(1)=\frac{1}{m_{T_1}}\sum_{s_{T_1}}\frac{y_{Ti}}{p_i}-\beta_T(1)\left\{\frac{1}{m_{T_1}}\sum_{s_{T_1}}\frac{y_{(T-1),i}}{p_i}-\hat{Y}_{T-1}(1)\right\}.$$

For $T=1$ we have

$$V\left(\hat{Y}_1(1)\right)=V\left(\frac{1}{m_{11}}\sum_{s_{11}}\frac{y_{1i}}{p_i}\right)=\frac{1}{m_{11}}\left(\sum_{i=1}^{N}\frac{y_{1i}^2}{p_i}-Y_1^2\right)=\frac{V_0}{m_{11}}$$

which proves the theorem for $t=1$. Now let us consider

$$\hat{Y}_2(1)=\frac{1}{m_{21}}\sum_{s_{21}}\frac{y_{2i}}{p_i}-\beta_2(1)\left\{\frac{1}{m_{21}}\sum_{s_{21}}\frac{y_{1i}}{p_i}-\hat{Y}_0(1)\right\}.$$

Then

$$V\left(\hat{Y}_2(1)\right)=E_1V_2\left[\hat{Y}_2(1)\mid s_{11}\right]+V_1E_2\left[\hat{Y}_2(1)\mid s_{11}\right]$$

$$=E_1V_2\left[\frac{1}{m_{21}}\sum_{s_{21}}\frac{y_{2i}-\beta_2(i)y_{1i}}{p_i}\right]+V_1\left[\frac{1}{m_{11}}\sum_{s_{11}}\frac{y_{2i}}{p_i}-\beta_2(1)\left\{\frac{1}{m_{11}}\sum_{s_{11}}\frac{y_{1i}}{p_i}-\hat{Y}_0(1)\right\}\right].$$

Note that

$$\hat{Y}_0(1)=\frac{1}{m_{11}}\sum_{s_{11}}\frac{y_{1i}}{p_i},$$

therefore we have

$$V\left(\hat{Y}_2(1)\right)=E_1\left(\frac{1}{m_{21}}-\frac{1}{m_{11}}\right)\frac{1}{m_{11}-1}\sum_{s_{11}}\left(\frac{y_{2i}-\beta_2(i)y_{1i}}{p_i}-\frac{\sum\limits_{s_{11}}\frac{y_{2i}-\beta_2(i)y_{1i}}{p_i}}{m_{11}}\right)^2+V_1\left[\frac{1}{m_{11}}\sum_{s_{11}}\frac{y_{2i}}{p_i}\right]$$

$$=\left(\frac{1}{m_{21}}-\frac{1}{m_{11}}\right)\frac{1}{m_{11}-1}\left[\sum_{\Omega}\frac{\left(y_{2i}-\beta_2(i)y_{1i}\right)^2}{p_i}-\left(Y_2-\beta_2Y_1\right)^2\right]+\frac{V_0}{m_{11}}$$

$$=\left(\frac{1}{m_{21}}-\frac{1}{m_{11}}\right)\left[V_0+\beta_2^2(1)V_0-2\beta_2(1)\delta V_0\right]+\frac{V_0}{m_{11}}.$$

Now setting

$$\frac{\partial V\left(\hat{Y}_2(1)\right)}{\partial\beta_2(1)}=0$$

we have

$$\beta_2(1) = \delta$$

and which leads to the following relation

$$V\left(\hat{Y}_2(1)\right)|_{\beta_2(1)=\delta} = \left(\frac{1}{m_{21}} - \frac{1}{m_{11}}\right)\left(V_0 + \delta^2 V_0 - 2\delta^2 V_0\right) + \frac{V_0}{m_{11}}$$

$$= \left(\frac{1}{m_{21}} - \frac{1}{m_{11}}\right)V_0\left(1 - \delta^2\right) + \frac{V_0}{m_{11}}$$

$$= \left[\left(\frac{1}{m_{21}} - \frac{1}{m_{11}}\right) - \delta^2\left(\frac{1}{m_{21}} - \frac{1}{m_{11}}\right) + \frac{1}{m_{11}}\right]V_0 .$$

Let us suppose that

$$\beta_j(1) = \delta \text{ for } j = 1, 2, ..., (T-1)$$

and

$$V\left[\hat{Y}_j(1)\right] = \left[\frac{1}{m_{j1}} - \sum_{t'=1}^{j-1}\left(\frac{1}{m_{j+1-t',1}} - \frac{1}{m_{j-t'}}\right)\left(\delta^2\right)^{t'}\right]V_0 \text{ for } j = 1, 2, ..., (T-1).$$

Now

$$V\left(\hat{Y}_T(1)\right) = V\left[\left(\frac{1}{m_{T1}}\sum_{s_{T1}}\frac{y_{ti}}{p_i}\right) - \beta_T(1)\left\{\frac{1}{m_{T1}}\sum_{s_{T1}}\frac{y_{t-1,i}}{p_i} - \hat{Y}_{T-1}(1)\right\}\right]$$

$$= V\left(\frac{1}{m_{T1}}\sum_{s_{T1}}\frac{y_{ti}}{p_i}\right) + \beta_T^2(1)V\left\{\frac{1}{m_{T1}}\sum_{s_{T1}}\frac{y_{t-1,i}}{p_i} - \hat{Y}_{T-1}(1)\right\}$$

$$- 2\beta_T(1)\text{Cov}\left\{\frac{1}{m_{T1}}\sum_{s_{T1}}\frac{y_{ti}}{p_i}, \frac{1}{m_{T1}}\sum_{s_{T1}}\frac{y_{ti}}{p_i} - \hat{Y}_{T-1}(1)\right\}. \qquad (10.9.4)$$

To complete the above theorem we need the following results:

Result 10.9.1. $V\left(\frac{1}{m_{T1}}\sum_{s_{T1}}\frac{y_{ti}}{p_i}\right) = \frac{V_0}{m_{T1}}$.

Proof. We have

$$V\left(\frac{1}{m_{T1}}\sum_{s_{T1}}\frac{y_{ti}}{p_i}\right) = E_1\left\{V_2\left(\frac{1}{m_{T1}}\sum_{s_{T1}}\frac{y_{ti}}{p_i} \mid s_{11}\right)\right\} + V_1\left\{E_2\left(\frac{1}{m_{T1}}\sum_{s_{T1}}\frac{y_{ti}}{p_i} \mid s_{11}\right)\right\}.$$

Now noting m_{T1} may be treated as an SRSWOR sample from s_{11} so we have

$$V\left(\frac{1}{m_{T1}}\sum_{s_{T1}}\frac{y_{ti}}{p_i}\right) = E_1\left\{\left(\frac{1}{m_{T1}} - \frac{1}{m_{11}}\right)\frac{1}{m_{11}-1}\sum_{s_{11}}\left(\frac{y_{ti}}{p_i} - \frac{\sum_{s_{11}}\frac{y_{ti}}{p_i}}{m_{11}}\right)^2\right\} + V_1\left\{\frac{1}{m_{11}}\sum_{s_{11}}\frac{y_{ti}}{p_i}\right\}$$

$$=\left(\frac{1}{m_{T1}}-\frac{1}{m_{11}}\right)V_0+\frac{V_0}{m_{11}}=\frac{V_0}{m_{T1}}.$$

Result 10.9.2. $\text{Cov}\left(\frac{1}{m_{T1}}\sum_{s_{T1}}\frac{y_{T-1,i}}{p_i},\ \hat{Y}_{T-1}(1)\right)=V\left(\hat{Y}_{T-1}(1)\right)$ for $T=2,...$

Proof. For $T=2$

$$\text{Cov}\left(\frac{1}{m_{T1}}\sum_{s_{T1}}\frac{y_{T-1,i}}{p_i},\ \hat{Y}_{T-1}(1)\right)=\text{Cov}\left\{\frac{1}{m_{21}}\sum_{s_{21}}\frac{y_{1i}}{p_i},\ \frac{1}{m_{11}}\sum_{s_{11}}\frac{y_{1i}}{p_i}\right\}$$

$$=E_1\left[C_2\left\{\frac{1}{m_{21}}\sum_{s_{21}}\frac{y_{1i}}{p_i},\ \frac{1}{m_{11}}\sum_{s_{11}}\frac{y_{1i}}{p_i}\right\}\mid s_{11}\right]$$

$$+C_1\left[\left\{E_2\left(\frac{1}{m_{21}}\sum_{s_{21}}\frac{y_{1i}}{p_i}\mid s_{11}\right),\ E_2\left(\frac{1}{m_{11}}\sum_{s_{11}}\frac{y_{1i}}{p_i}\mid s_{11}\right)\right\}\right]$$

$$=0+C_1\left[\frac{1}{m_{11}}\sum_{s_{11}}\frac{y_{1i}}{p_i},\ \frac{1}{m_{11}}\sum_{s_{11}}\frac{y_{1i}}{p_i}\mid s_{11}\right]=V\left(\hat{Y}_{T-2}(1)\right).$$

Suppose

$$\text{Cov}\left\{\frac{1}{m_{T-1,1}}\sum_{s_{T-1,1}}\frac{y_{T-2,i}}{p_i},\ \hat{Y}_{T-2}(1)\right\}+V\left(\hat{Y}_{T-2}(1)\right).$$

Then we have

$$\text{Cov}\left\{\frac{1}{m_{T1}}\sum_{s_{T1}}\frac{y_{T-1,i}}{p_i},\ \hat{Y}_{T-1}(1)\right\}$$

$$=C_1\left[E_2\left(\frac{1}{m_{T1}}\sum_{s_{T1}}\frac{y_{T-1,i}}{p_i}\mid s_{T-1,1}\right),\ E_2\left(\hat{Y}_{T-1}(1)\mid s_{T-1,1}\right)\right]$$

$$+E_1\left[C_2\left(\frac{1}{m_{T1}}\sum_{s_{T1}}\frac{y_{T-1,i}}{p_i},\ \hat{Y}_{T-1}(1)\mid s_{T-1}\right)\right]$$

$$=C_1\left[\frac{1}{m_{T-1,1}}\sum_{s_{T-1,1}}\frac{y_{T-1,i}}{p_i},\ \frac{1}{m_{T-1,1}}\sum_{s_{T-1,1}}\frac{y_{T-1,i}}{p_i}-\frac{\delta}{m_{T-1,1}}\sum_{s_{T-1,1}}\frac{y_{T-2,i}}{p_i}+\delta\overline{Y}_{T-2}(1)\right]+0$$

$$=V\left(\frac{1}{m_{T-1,1}}\sum_{s_{T-1,1}}\frac{y_{T-1,i}}{p_i}\right)-\delta^2V\left(\frac{1}{m_{T-1,1}}\sum_{s_{T-1,1}}\frac{y_{T-1,i}}{p_i}\right).$$

$$+\delta\text{Cov}\left\{\frac{1}{m_{T-1,1}}\sum_{s_{T-1,1}}\frac{y_{T-1,i}}{p_i},\hat{Y}_{T-2}(1)\right\}.$$

Noting that

$$\operatorname{Cov}\left(\sum_{s_{T,1}} \frac{y_{T,i}}{p_i}, \sum_{s_{T1}} \frac{y_{T-k,i}}{p_i}\right) = \delta^k V\left(\sum_{s_{T,1}} \frac{y_{T,i}}{p_i}\right).$$

Result 10.9.3. $V\left\{\frac{1}{m_{T1}} \sum_{s_{T1}} \frac{y_{T-1,i}}{p_i} - \hat{Y}_{T-1}(1)\right\} = \frac{V_0}{m_{T1}} - V\left\{\hat{Y}_{T-1}(1)\right\}.$

Proof. Note that

$$\operatorname{Cov}\left(\frac{1}{m_{T1}} \sum_{s_{T1}} \frac{y_{T-1,i}}{p_i}, \hat{Y}_{T-1}(1)\right) = V\left(\hat{Y}_{T-1}(1)\right).$$

We have

$$\begin{aligned} V\left\{\frac{1}{m_{T1}} \sum_{s_{T1}} \frac{y_{T-1,i}}{p_i} - \hat{Y}_{T-1}(1)\right\} &= V\left(\frac{1}{m_{T1}} \sum_{s_{T1}} \frac{y_{T-1,i}}{p_i}\right) + V\left(\hat{Y}_{T-1}(1)\right) - 2\operatorname{Cov}\left(\frac{1}{m_{T1}} \sum_{s_{T1}} \frac{y_{T-1,i}}{p_i}, \hat{Y}_{T-1}(1)\right) \\ &= V\left(\frac{1}{m_{T1}} \sum_{s_{T1}} \frac{y_{T-1,i}}{p_i}\right) + V\left(\hat{Y}_{T-1}(1)\right) - 2V\left(\hat{Y}_{T-1}(1)\right) \\ &= \frac{V_0}{m_{T1}} - V\left\{\hat{Y}_{T-1}(1)\right\}. \end{aligned}$$

Result 10.9.4. $\operatorname{Cov}\left\{\frac{1}{m_{T1}} \sum_{s_{T1}} \frac{y_{Ti}}{p_i}, \frac{1}{m_{T1}} \sum_{s_{T1}} \frac{y_{T-1,i}}{p_i} - \hat{Y}_{T-1}(1)\right\} = \delta\left[\frac{V_0}{m_{T1}} - V\left\{\hat{Y}_{T-1}(1)\right\}\right].$

Proof. We have

$$\begin{aligned} &\operatorname{Cov}\left\{\frac{1}{m_{T1}} \sum_{s_{T1}} \frac{y_{Ti}}{p_i}, \frac{1}{m_{T1}} \sum_{s_{T1}} \frac{y_{T-1,i}}{p_i} - \hat{Y}_{T-1}(1)\right\} \\ &= \operatorname{Cov}\left(\frac{1}{m_{T1}} \sum_{s_{T1}} \frac{y_{Ti}}{p_i}, \frac{1}{m_{T1}} \sum_{s_{T1}} \frac{y_{T-1,i}}{p_i}\right) - \operatorname{Cov}\left(\frac{1}{m_{T1}} \sum_{s_{T1}} \frac{y_{Ti}}{p_i}, \hat{Y}_{T-1}(1)\right) \\ &= \delta V\left(\frac{1}{m_{T1}} \sum_{s_{T1}} \frac{y_{Ti}}{p_i}\right) - \delta \operatorname{Cov}\left(\frac{1}{m_{T1}} \sum_{s_{T1}} \frac{y_{Ti}}{p_i}, \hat{Y}_{T-1}(1)\right). \end{aligned}$$

Noting that

$$\operatorname{Cov}\left\{\frac{1}{m_{T1}} \sum_{s_{T1}} \frac{y_{Ti}}{p_i}, \frac{1}{m_{T1}} \sum_{s_{T1}} \frac{y_{T-k,i}}{p_i}\right\} = \delta^k V\left(\frac{1}{m_{T1}} \sum_{s_{T1}} \frac{y_{Ti}}{p_i}\right)$$

and using Result 10.9.2 we have Result 10.9.4.

Now using these results we note that

$$V\left(\hat{Y}_T(1)\right) = \frac{V_0}{m_{T1}} + \beta_T^2(1)\left\{\frac{V_0}{m_{T1}} - V\left(\hat{Y}_{T-1}(1)\right)\right\} - 2\beta_T(1)\delta\left\{\frac{V_0}{m_{T1}} - V\left(\hat{Y}_{T-1}(1)\right)\right\}.$$

On differentiating it with respect to $\beta_T(1)$ and equating with zero we have

$$\frac{\partial V\left(\hat{Y}_T(1)\right)}{\partial \beta_T(1)} = 0$$

which implies, $\beta_T(1) = \delta$ and

$$V\left(\hat{Y}_T(1)\,|_{\beta_T(1)=\delta}\right) = \frac{V_0}{m_{T1}} - \delta^2\left\{\frac{V_0}{m_{T1}} - V\left(\hat{Y}_{T-1}(1)\right)\right\} = \left[\frac{1}{m_{T1}} - \delta^2\left\{\frac{1}{m_{T1}} - V\left(\hat{Y}_{T-1}(1)\right)\right\}\right]$$

$$= \left[\frac{1}{m_{T1}} - \sum_{t'=1}^{T-1}\left(\frac{1}{m_{T+1-t',1}} - \frac{1}{m_{T-t',1}}\right)\left(\delta^2\right)^{t'} V_0\right].$$

Now we have

$$\hat{Y}_T = \sum_{t=1}^{T} C_T \hat{Y}_T(t).$$

We note that $\hat{Y}_T(t)$ are independent due to the sampling scheme shown in the Table 10.9.1 and $E\{\hat{Y}_T(t)\} = Y$ for every $t = 1, 2, ..., T$, thus the optimum value of $C_T \propto \{V(Y_T(1))\}^{-1}$ and

$$V_{\text{opt}}\left(\hat{Y}_T\right) = \left[\sum_{t=1}^{T} 1/V\left(\hat{Y}_T(t)\right)\right]^{-1} = \left[\sum_{t=1}^{T} 1\Big/\left\{\frac{1}{m_{Tt}} - \sum_{t'=1}^{T-t}\left(\frac{1}{m_{T+1-t',t}} - \frac{1}{m_{T-t',t}}\right)\left(\delta^2\right)^{t'}\right\}V_0\right]^{-1}$$

Now writing $m_{Tt} = \lambda_{Tt} m_{11}$ we have

$$V_{\text{opt}}\left(\hat{Y}_T\right) = \frac{V_0}{m_{11}}\left[\sum_{t=1}^{T} 1\Big/\left\{\frac{1}{\lambda_{Tt}} - \sum_{t'=1}^{T-t}\left(\frac{1}{\lambda_{T+1-t',t}} - \frac{1}{\lambda_{T-t',t}}\right)\left(\delta^2\right)^{t'}\right\}\right]^{-1} = \frac{V_0}{m_{11}}\left[\sum_{t=1}^{T}\frac{1}{\phi(t)}\right]^{-1}$$

where

$$\phi(t) = \frac{1}{\lambda_{Tt}} - \sum_{t'=1}^{T-t}\left(\frac{1}{\lambda_{T+1-t',t}} - \frac{1}{\lambda_{T-t',t}}\right)\left(\delta^2\right)^{t'} = \left(1-\delta^2\right)\frac{1}{\lambda_{Tt}} + \delta^2\phi_{T-1}(t) \text{ for } t = 1, 2, ..., (T-1)$$

and

$$\phi(T) = \frac{1}{\lambda_{TT}} \text{ for } t = T.$$

Thus we have

$$V_{\text{opt}}\left[\hat{Y}_T\right] = \frac{V_0}{m_{11}}\left[\sum_{t=1}^{T-1}\left\{1/\lambda_{Tt}\left(1-\delta^2\right) + \delta^2\phi_{T-1}(t)\right\} + \lambda_{TT}\right].$$

Now equating

$$\frac{\partial V_{\text{opt}}\left(\hat{Y}_T\right)}{\partial \lambda_{Tt}} = 0$$

and writing

$$\lambda_{TT} = m_{11} - \sum_{t=1}^{T}\lambda_{Tt}$$

we get

$$\lambda_{Tt} = \left\{\sqrt{1-\delta^2}\left\{1-\sqrt{1-\delta^2}\right\}\right\} \Big/ \left\{1+\sqrt{1-\delta^2}\right\}\phi_{T-1}(t).$$

Hence the theorem.

Following the proofs of the above theorems, the results given in the following theorems can easily be proved, and avoided to save the space.

Theorem 10.9.3. If $\beta_T(t) = \delta, \ \forall \ t = 1, 2, ..., (T-1)$, then the optimum value of $V(\hat{Y}_T)$ with variation of values of C_t, $t = 1, 2, ..., H$ is

$$V_{\text{opt}}(\hat{Y}_T) = \left[\sum_{t'=1}^{T} \frac{1}{V\{\hat{Y}_T(t')\}}\right]^{-1}$$

$$= \begin{cases} \dfrac{V_0}{m_{11}}\left[\sum\limits_{t'=1}^{T}\left\{\dfrac{1}{\lambda_{Tt'}} - \sum\limits_{t=1}^{T-t'}\left(\dfrac{1}{\lambda_{T+1-t,t'}} - \dfrac{1}{\lambda_{T-t,t'}}\right)(\delta^2)^t\right\}^{-1}\right]^{-1} & t = 1,2,..,(T-1), \\ \dfrac{V_0}{m_{11}}\left[\sum\limits_{t'=1}^{T}\left\{\left(\dfrac{1-\delta^2}{\lambda_{Tt'}} + \delta^2\phi_{T-1}(t')\right)\right\}^{-1}\right]^{-1} & t = T, \end{cases} \tag{10.9.5}$$

where

$$\lambda_{Tt} = \frac{m_{Tt}}{m_{11}} \quad \text{and} \quad \phi_T(t') = \begin{cases} \dfrac{1}{\lambda_{Tt'}} - \sum\limits_{t=1}^{T-t'}\left(\dfrac{1}{\lambda_{T+1-t,t'}} - \dfrac{1}{\lambda_{T-t,t'}}\right)(\delta^2)^t & t' = 1, 2, ..., (T-1), \\ \dfrac{1}{\lambda_{TT}} & t' = T. \end{cases}$$

Theorem 10.9.4. The optimum values of λ_{Tt} (and hence m_{Tt}) are given by

$$\lambda_{Tt(\text{opt})} = \begin{cases} \dfrac{\sqrt{1-\delta^2}}{1+\sqrt{1-\delta^2}}\left(\dfrac{1}{\phi_{(T-1)\text{opt}}(t)}\right) & t = 1, 2, ..., (T-1), \\ 1 - \dfrac{\sqrt{1-\delta^2}}{1+\sqrt{1-\delta^2}}\left(\dfrac{1}{\phi_{(T-1)\text{opt}}}\right) & t = T, \end{cases} \tag{10.9.6}$$

where $\phi_{(1)\text{opt}}(1) = 1$, $g_{(h)\text{opt}} = \dfrac{\left(1+\sqrt{1-\delta^2}\right)g_{(T-1)\text{opt}}}{\left(1-\sqrt{1-\delta^2}\right)+\left(1+\sqrt{1-\delta^2}\right)g_{(T-1)\text{opt}}}$ and $g_{(1)\text{opt}} = 1$.

Theorem 10.9.5. The minimum variance of the estimator of population total on the Tth occasion is given by

$$V_{\text{Min}}(\hat{Y}_T) = \frac{V_0}{m_{11}} g_{(T)\text{opt}} . \tag{10.9.7}$$

The optimum values of C_t are given by

$$C_{t(\text{opt})} = \begin{cases} \left\{ \dfrac{1-\delta^2}{\lambda_{Ti(\text{opt})}} + \delta^2 \phi_{(h-1)}(i)_{\text{opt}} \right\} & t = 1,\ 2, \ldots, (T-1), \\ \lambda_{TT(\text{opt})} g_{(T)\text{opt}} & t = T. \end{cases} \quad (10.9.8)$$

The optimum proportion of matched units on the Tth occasion is given by

$$\lambda_{T(\text{opt})} = \frac{\sqrt{1-\delta^2}}{1+\sqrt{1-\delta^2}} \left(\frac{1}{g_{(T-1)\text{opt}}} \right). \quad (10.9.9)$$

It can be easily shown that the strategies proposed by Tripathi and Srivastava (1979), Ghangurde and Rao (1969), Chotai (1974) and Chaudhuri and Arnab (1977) are the topics related to Arnab's first strategy in successive sampling.

Arnab's Strategy II. The main steps of Arnab's second strategy are:

(1) On the first occasion select a sample s_{11} of size m_{11} with the RHC scheme with size measures p_i.

(2) On the second occasion:

(a) Select a sub-sample s_{21} of size m_{21} units from the first sample s_{11} of m_{11} units by RHC scheme using size measures ${}_1Q_i(2)$, where ${}_1Q_i(2)$ is the sum of the p_i values of the group containing the ith unit formed in selecting s_{11}.
(b) Select an independent sample s_{22} of m_{22} units, which in fact is equal to $(m_{11} - m_{21})$ units, by the RHC scheme from the whole population Ω with p_i size measures.

(3) In general, on the tth occasion a sample $s_{tt'}$ of size $m_{tt'}$ is selected from $s_{(t-1)t'}$ independently for $t' = 1, 2, \ldots, (t-1)$ following RHC scheme using normed size measure ${}_{t'}Q_i(t)$ which, in fact, is the sum of ${}_{t'}Q_i(t-1)$ values containing the ith unit occurring in $s_{(t-1)t'}$ for $t' = 1, 2, \ldots, (t-1)$ and another independent sample s_{tt} is selected from the entire population using p_i as normed size measures.

Then we have the following theorems:

Theorem 10.9.6. An estimator of population total Y_T on the Tth occasion is

$$\hat{Y}_T^* = \sum_{t=1}^{T} c_t^* \hat{Y}_T^*(t) \quad (10.9.10)$$

with

$$\hat{Y}_T^*(t) = \begin{cases} \sum\limits_{i \in s_{Tt}} \frac{y_{Ti}}{p_i}\left({}_tQ_i(T+1)\right) - \hat{\beta}_T(t)\left[\sum\limits_{i \in s_{Tt}} \frac{y_{T-1,t}}{p_i}\left\{{}_tQ_i(T+1)\right\} - \hat{Y}_T^*(t)\right] & t = 1, 2, ..., (T-1), \\ \sum\limits_{i \in s_{TT}} \frac{y_{hi}}{p_i}\left\{{}_TQ_i(T+1)\right\} & t = T, \end{cases}$$

where c_t^* and $\hat{\beta}_T(t)$ are constants chosen such that estimator $\hat{Y}_T^*$ is unbiased and have the minimum variance.

Theorem 10.9.7. The variance of the estimator $\hat{Y}_T^*(t)$ is minimum if $\hat{\beta}_T(t) = \delta$ and minimum variance of the estimator $\hat{Y}_T^*$ is given by

$$V_{\min}\left(\hat{Y}_T^*\right) = \frac{NV_0}{N-1}\left(\frac{1}{m_{12}\lambda_{TT(\text{opt})}} - \frac{1}{N}\right)\left[1 - \frac{1}{\sqrt{1-\delta^2}}\left(1 - \frac{1}{\lambda_{TT(\text{opt})}}\right)\right]^{-1} \tag{10.9.11}$$

for the optimum values of c_t given by

$$c_{t(\text{opt})}^* = \begin{cases} \frac{1}{\phi_T(t)_{opt}}\left(\frac{1}{m_{11}\lambda_{TT(\text{opt})}} - \frac{1}{N}\right)\left[1 - \frac{1}{\sqrt{1-\delta^2}}\left(1 - \frac{1}{\lambda_{TT(\text{opt})}}\right)\right]^{-1} & t = 1, 2, ..., (T-1), \\ \left[1 - \frac{1}{\sqrt{1-\delta^2}}\left(1 - \frac{1}{\lambda_{TT(\text{opt})}}\right)\right]^{-1}, & t = T, \end{cases}$$

and

$$\lambda_{TT(\text{opt})} = \frac{1}{m_{11}\left\{1 - \sqrt{1-\delta^2}\,B(T-1)\right\}}\left[m_{11} - N\sqrt{1-\delta^2}\left\{1 - \sqrt{1-\delta^2}\right\}B(T-1)\right]$$

with

$$B(T-1) = \sum_{t=1}^{T-1}\left\{\delta^2 N\phi_{T-1}(t) - \left(1-\delta^2\right)\right\}^{-1}.$$

Arnab's Strategy III. The main steps of Arnab's third strategy are:

(1) On the first occasion, select a sample s_{11} of size m_{11} with the RHC scheme with size measures p_i.

(2) On the second occasion:

(a) Select a sub-sample s_{21} of size m_{21} units from the first sample s_{11} of m_{11} units by SRSWOR sampling;

(b) Select independent sample s_{22} of m_{22} units, which in fact equal to $(m_{11} - m_{21})$ units, by RHC scheme from the whole population Ω with p_i size measures.

(3) In general, on the tth occasion sub-samples $s_{tt'}$ of size $m_{tt'}$ are selected by WOR sampling from $s_{(t-1),t'}$, $(t'=1,2,...,(t-1))$ and an independent sample of size $m_{tt}\left(=m_{11}-\sum_{t'=1}^{t-1} m_{tt'}\right)$ from the entire population by RHC method using p_i.

Under such a sampling scheme we have the following theorem:

Theorem 10.9.8. An estimator to estimate the population total Y_T, on occasion T, is given by

$$\hat{Y}_T^{**}=\sum_{t=1}^{T} c_t^{**}\hat{Y}_T^{**}(t) \tag{10.9.12}$$

with

$$\hat{Y}_T^{**}(t)=\begin{cases}\dfrac{m_{tt}}{m_{Tt}}\sum_{i\in s_{Tt}}\dfrac{y_{Ti}}{p_i}\{{}_tQ_i(t+1)\}-\left\{\dfrac{m_{tt}}{m_{Tt}}\sum_{i\in s_{Tt}}\dfrac{y_{T-1,i}}{p_i}({}_tQ_i(t+1))-\hat{Y}_{T-1}^{**}(t)\right\} \\ \qquad\qquad\qquad\qquad\qquad\qquad \text{for } t=1,2,...,(T-1), \\ \dfrac{m_{11}}{m_{TT}}\sum_{i\in s_{TT}}\dfrac{y_{Ti}}{p_i}\{{}_TQ_i(T+1)\} \qquad \text{for } t=T.\end{cases}$$

Arnab (1979a) has shown that Strategy II remains always superior than Strategy I from the minimum variance point of view. Comparing Strategies II and III, the efficiency comparison depends upon the values of the parameters involved in the variance expressions. Sekkappan and Thompson (1994) have given thought on multi-phase and successive sampling for a stratified population with unknown stratum sizes. Arnab (1998) has suggested two new sampling strategies for estimating the most recent occasion total on the basis of two samples selected with varying probability at two different occasions. Empirical results shows that Arnab (1998) strategy remains more efficient than one described by Prasad and Graham (1994).

Example 10.9.1. The data in population 9 of the Appendix relates to the number of immigrants coming to 51 states in the United States during 1994--1996. Select a sample (s_1) of 10 states by PPSWR method using number of immigrants in 1994 (z) as measure of size. From the selected sample s_1, select a sub-sample s_m of size $m=4$ from s_1 by SRSWOR method assuming all elements of s_1 are distinct. Finally select an independent sample s_u of size $u=(10-m)=6$ with PPSWR using x as size measure. Estimate the total number of immigrants in 1996 using the composite estimator.

Solution. We want to estimate total number of immigrants in 1996 (Y) using the composite estimator

$$\hat{Y}=\hat{\phi}\hat{Y}_m+\left(1-\hat{\phi}\right)\hat{Y}_u$$

with

$$\hat{\phi} = \left\{ \left(\frac{1}{m} - \frac{1}{n} \right) \left(1 - \hat{\delta}^2 \right) + \frac{1}{n} \right\}^{-1} \Bigg/ \left[\left\{ \left(\frac{1}{m} - \frac{1}{n} \right) \left(1 - \hat{\delta}^2 \right) + \frac{1}{n} \right\}^{-1} + u \right],$$

where an estimate of Y based on the matched sample is given by

$$\hat{Y}_m = \frac{1}{m} \sum_{i \in s_m} \frac{y_i}{p_i} - \frac{\hat{\delta}}{m} \sum_{i \in s_m} \frac{x_i}{p_i} + \frac{\hat{\delta}}{n} \sum_{i \in s_1} \frac{x_i}{p_i}$$

with

$$\hat{\delta} = \sum_{s_m} \left(\frac{y_i}{p_i} - \frac{1}{m} \sum_{s_m} \frac{y_i}{p_i} \right) \left(\frac{x_i}{p_i} - \frac{1}{m} \sum_{s_m} \frac{x_i}{p_i} \right) \Bigg/ \left\{ \sqrt{\sum_{s_m} \left(\frac{y_i}{p_i} - \frac{1}{m} \sum_{s_m} \frac{y_i}{p_i} \right)^2} \sqrt{\sum_{s_m} \left(\frac{x_i}{p_i} - \frac{1}{m} \sum_{s_m} \frac{x_i}{p_i} \right)^2} \right\}$$

and an estimate of Y based on unmatched units is given by

$$\hat{Y}_u = \frac{1}{u} \sum_{i \in s_u} \frac{y_i}{p_i}.$$

(a) Selection of sample s_1 and unmatched sample s_u: We use cumulative total methods to select a PPSWR sample of $n = 10$ units. We used the first six columns of the Pseudo-Random Number (PRN) Table 1 given in the Appendix to select 10 random numbers R_1 (say) between 1 and 789767. The second independent random sample (called unmatched sample s_u) of six states from the population was selected by using columns 7th to 12th of the Pseudo-Random Numbers R_2 (say). The data on the number of immigrants during 1994, cumulative totals CT(z) and the states selected are shown in the following table.

Sr. No.	States	1994 (z)	CT(z)	R_1	States Selected in the sample s_1	R_2	States selected in the sample s_u
1	AL	1837	1937				
2	AK	1129	3066				
3	AZ	9141	12207				
4	AR	1031	13238	014737, 049819	AR, AR		
5	CA	208498	221736				
6	CO	6825	228561				
7	CT	9537	238098	236263	CT		
8	DE	984	239082				
9	DC	3204	242286				
10	FL	58093	300379				
11	GA	10032	310411				
12	HI	7746	318157				
13	ID	1559	319716			339922	ID
14	IL	42400	362116				

Continued......

15	IN	3725	365841				
16	IA	2163	368004				
17	KS	2902	370906				
18	KY	2036	372942				
19	LA	3366	376308				
20	ME	829	377137				
21	MD	15937	393074				
22	MA	22882	415956				
23	MI	12728	428684				
24	MN	7098	435782				
25	MS	815	436597				
26	MO	4362	440959				
27	MT	447	441406				
28	NE	1595	443001				
29	NV	4051	447052				
30	NH	1144	448196	465225	NH		
31	NJ	44083	492279				
32	NM	2936	495215	588183, 601448, 549171, 626895	NM, NM, NM, NM	622048, 534272, 513080	NM, NM, NM
33	NY	144354	639569	644818	NY		
34	NC	6204	645773				
35	ND	635	646408				
36	OH	9184	655592				
37	OK	2728	658320				
38	OR	6784	665104			675048	OR
39	PA	15971	681075				
40	RI	2907	683982				
41	SC	2110	686092				
42	SD	570	686662				
43	TN	3608	690270	697856	TN		
44	TX	56158	746428				
45	UT	2951	749379				
46	VT	658	750037				
47	VA	15342	765379			771280	VA
48	WA	18180	783559				
49	WV	663	784222				
50	WI	5328	789550				
51	WY	217	789767				

Thus the sample s_1 of $n=10$ units is given by

Sr. No.	States	z	X	$p_i = \frac{z_i}{Z}$	$\frac{x_i}{p_i}$
1	AR	1031	934	0.001305	715463.0242
2	AR	1031	934	0.001305	715463.0242
3	CT	9537	9240	0.012076	765172.1799
4	NH	1144	1186	0.001449	818761.9423
5	NM	2936	2758	0.003718	741886.0307
6	NM	2936	2758	0.003718	741886.0307
7	NM	2936	2758	0.003718	741886.0307
8	NM	2936	2758	0.003718	741886.0307
9	NY	144354	128406	0.182780	702514.7997
10	TN	3608	3392	0.004568	742486.0488
				Sum	7427405.1420

Given: Z=789767.

and the unmatched sample s_u of $u=6$ units is given by

Sr. No.	States	z	x	y
1	ID	1559	1612	1825
2	NM	2936	2758	5780
3	NM	2936	2758	5780
4	NM	2936	2758	5780
5	OR	6784	4923	7554
6	VA	15342	16319	21375

(b) Selection of a matched sample s_m

We used the first two columns of the Pseudo-Random Numbers to select a matched sample s_m of $m=4$ units from the given sample s_1 of $n=10$ units by using SRSWOR sampling. The sample s_m selected is given by

Sr. No.	Random no.	States	z	x	y
1	01	AR	1031	934	1494
2	03	CT	9537	9240	10874
3	04	NH	1144	1186	1512
4	05	NM	2936	2758	5780

(c) Calculation of the estimates:

From the matched sample information s_m we have

States	$\frac{x_i}{p_i}$	$\frac{y_i}{p_i}$	$\left(\frac{x_i}{p_i}-\frac{1}{m}\sum_{s_m}\frac{x_i}{p_i}\right)^2$	$\left(\frac{y_i}{p_i}-\frac{1}{m}\sum_{s_m}\frac{y_i}{p_i}\right)^2$	$\left(\frac{x_i}{p_i}-\frac{1}{m}\sum_{s_m}\frac{x_i}{p_i}\right)\left(\frac{y_i}{p_i}-\frac{1}{m}\sum_{s_m}\frac{y_i}{p_i}\right)$
AR	715463.0	1144434.0	2012219533.00	270489496.7	737756225.7
CT	765172.2	900485.1	23535942.68	67806024636.0	-1263280930.0
NH	818761.9	1043818.0	3415367782.00	13703762934.0	-6841300346.0
NM	741886.0	1554787.0	339840510.10	1.55162E+11	-7261555471.0
Sum	3041283.0	4643524.0	5790963768.00	2.36942E+11	-14628380521.0

Thus we have

$$\hat{\delta}=\frac{\sum_{s_m}\left(\frac{y_i}{p_i}-\frac{1}{m}\sum_{s_m}\frac{y_i}{p_i}\right)\left(\frac{x_i}{p_i}-\frac{1}{m}\sum_{s_m}\frac{x_i}{p_i}\right)}{\sqrt{\sum_{s_m}\left(\frac{y_i}{p_i}-\frac{1}{m}\sum_{s_m}\frac{y_i}{p_i}\right)^2}\sqrt{\sum_{s_m}\left(\frac{x_i}{p_i}-\frac{1}{m}\sum_{s_m}\frac{x_i}{p_i}\right)^2}}$$

$$=\frac{-14628380521}{\sqrt{5790963768}\sqrt{2.36942\times10^{11}}}=-0.3949$$

and an estimate of the required total is given by

$$\hat{Y}_m=\frac{1}{m}\sum_{i\in s_m}\frac{y_i}{p_i}-\frac{\hat{\delta}}{m}\sum_{i\in s_m}\frac{x_i}{p_i}+\frac{\hat{\delta}}{n}\sum_{i\in s_1}\frac{x_i}{p_i}$$

$$=\frac{4643524}{4}+\frac{0.3949}{4}\times3041283-\frac{0.3949}{10}\times7427405.14=1167823.44\,.$$

From the unmatched sample information, we have

Sr. No.	States	z	y	p_i	$\frac{y_i}{p_i}$
1	ID	1559	1825	0.001974	924518.7781
2	NM	2936	5780	0.003718	1554786.5330
3	NM	2936	5780	0.003718	1554786.5330
4	NM	2936	5780	0.003718	1554786.5330
5	OR	6784	7554	0.008590	879407.4172
6	VA	15342	21375	0.019426	1100330.4410
				Sum	7568616.2340

An estimate of Y based on unmatched units is given by

$$\hat{Y}_u = \frac{1}{u}\sum_{i\in s_u}\frac{y_i}{p_i} = \frac{7568616.234}{6} = 1261436.04\ .$$

Also we have

$$\hat{\phi} = \frac{\dfrac{1}{\left(\frac{1}{m}-\frac{1}{n}\right)\left(1-\hat{\delta}^2\right)+\frac{1}{n}}}{\dfrac{1}{\left(\frac{1}{m}-\frac{1}{n}\right)\left(1-\hat{\delta}^2\right)+\frac{1}{n}}+u} = \frac{\dfrac{1}{\left(\frac{1}{4}-\frac{1}{10}\right)\left(1-0.3949^2\right)+\frac{1}{10}}}{\dfrac{1}{\left(\frac{1}{4}-\frac{1}{10}\right)\left(1-0.3949^2\right)+\frac{1}{10}}+6} = \frac{4.412905}{10.412905} = 0.4238\ .$$

Hence the composite estimate of total immigration Y during 1996 in the United States is given by

$$\hat{Y} = \hat{\phi}\hat{Y}_m + \left(1-\hat{\phi}\right)\hat{Y}_u = 0.4238\times 1167823.44 + (1-0.4238)\times 1261436.04$$
$$= 1221763.02\ .$$

Note that the estimation of parameters in successive sampling or repeated surveys, the inference can be made in three different ways:

(a) Extracting a new sample on each occasion (Repeated sampling);
(b) Using same sample on every occasion (Panel sampling);
(c) Performing a partial replacement of the units from the one occasion to another (Sampling on successive occasions or Rotation sampling).

The last possibility has been discussed extensively by several research workers, for estimating population mean of the second occasion using information from the first occasion, including Jessen (1942), Tikkiwal (1950, 1955), Rao and Mudholkar (1967), Das (1982), and Artes and Garcia (2000a, 2000b). Ratio estimator in successive sampling was first studied by Avdhani (1968) and his followers Sen, Seller, and Smith (1975), and Artes and Garcia (2001a, 2001b). Gupta (1970) and Artes, Rueda, and Arcos (1998) studied product estimator in successive sampling. The problem of estimation of ratio of two population means in successive sampling over two occasions has been studied by Rao and Mudholkar (1967), Okafor and Arnab (1987), Okafor (1992) and Artes and Garcia (2001c, 2001d). Artes and Garcia (2001e) developed an unbiased ratio cum product estimator of the population mean in successive sampling. Singh and Yadav (1992) have discussed a generalized estimation procedure for successive sampling. Sud, Srivastava, and Sharma (2001a, 2001b) have paid attention for estimating the population variance over repeated surveys. Okafor (2001) discusses some successive sampling estimation strategies in the presence of random non-response.

10.10 ESTIMATION STRATEGIES IN SUPPLEMENTED PANELS

The basic idea of using least squares to incorporate information from a previous occasion into the estimate of the current occasion is that of Jensen (1942). Patterson (1950) considered the use of information from the rotating samples. His genius idea has been spawned in the vast literature owed to Eckler (1955), Rao and Graham (1964), Raj (1965b), Smith (1978), Wolter (1979), Jones (1980), Huang and Ernst (1981), Kumar and Lee (1983), Breau and Ernst (1983), Singh (1996), and Yansaneh and Fuller (1998). Duncan and Kalton (1987), Schreuder, Gregoire, and Wood (1993), Fuller (1990), Lent, Miller, and Cantwell (1996) discussed different kinds of repeated surveys and the objectives. Kasprzyk, Duncan, Kalton, and Singh (1989) have discussed various aspects of panel surveys in an excellent way. A rotation survey is one in which a unit is observed for a partial set of time points and is not observed for the remaining set of time points in the study. The Canadian Labour Force Surveys and the U.S. Current Population Survey are the well known examples of the rotation surveys. The National Resources Inventory (NRI) is nearly a pure panel survey of certain land area with a five year observation interval. A pure panel survey can be defined as a survey in which the same units are observed at each time point of a survey conducted at more than one time point. The longitudinal survey can be defined as a survey conducted at more than two points in time with multiple observations on some units planned as part of the survey design. Fuller and Breidt (1999) introduced generalized least squares estimators for such surveys. Following them, consider a simple three period survey in which one fourth of the units are observed in all three periods and each of the remaining three sets of one fourth of the units is observed in exactly one of the three periods. In other words, if n is the total sample size, then $0.5n$ of the units are observed at each point. Let (Y_1, Y_2, Y_3) denote the value of a characteristic observed at times one, two, and three respectively. Assume that the correlation between observations at time i and time j on the same element is $\rho_{|i-j|}$. For simplicity assume SRS for the selection of all samples. Let $(\bar{y}_{11}, \bar{y}_{21}, \bar{y}_{31})^t$ denote the estimated mean at time one, two and three, of the sample elements that are observed all three periods. Let $(\bar{y}_{12}, \bar{y}_{23}, \bar{y}_{34})^t$ denote the sample means for the three periods for the sample elements that are observed once. These six estimators can be named as elementary estimators. We wish to estimate the population means $\mu = (\mu_1, \mu_2, \mu_3)$ for the three periods.

The above six elementary estimates can be modelled as

$$\bar{y} = X\mu + e, \tag{10.10.1}$$

where $\bar{y} = (\bar{y}_{11}, \bar{y}_{21}, \bar{y}_{31}, \bar{y}_{12}, \bar{y}_{23}, \bar{y}_{34})$ and $X' = (I_3, I_3)$.

The co-variance matrix of e is

$$\Omega = 4n^{-1}\sigma^2 \begin{pmatrix} 1 & \rho_1 & \rho_2 & 0 & 0 & 0 \\ \rho_1 & 1 & \rho_1 & 0 & 0 & 0 \\ \rho_2 & \rho_1 & 1 & 0 & 0 & 0 \\ 0 & 0 & 0 & 1 & 0 & 0 \\ 0 & 0 & 0 & 0 & 1 & 0 \\ 0 & 0 & 0 & 0 & 0 & 1 \end{pmatrix} \tag{10.10.2}$$

It is to be noted that the term $4n^{-1}\sigma^2$ is the variance of the mean of $n/4$ observations. By applying the weighted least squares method the best linear unbiased estimator of $\mu = (\mu_1, \mu_2, \mu_3)$ is given by

$$\hat{\mu} = (\hat{\mu}_1, \hat{\mu}_2, \hat{\mu}_3) = \left(X'\Omega^{-1}X\right)^{-1} X'\Omega^{-1}\bar{y}, \tag{10.10.3}$$

and

$$V(\hat{\mu}) = \left(X'\Omega^{-1}X\right)^{-1}. \tag{10.10.4}$$

Fuller and Breidt (1999) considered the comparison of the variance--covariance matrix in (10.10.4) with the variance--covariance matrix of a pure panel survey in which the same $n/2$ units are observed on all three periods. The best linear unbiased estimator of μ under the pure panel design is

$$\hat{\mu}_{\text{panel}} = (\bar{y}_1, \bar{y}_2, \bar{y}_3)'. \tag{10.10.5}$$

The variance-covariance matrix for the pure panel design is given by

$$V(\hat{\mu}_{\text{panel}}) = 2n^{-1}\sigma^2 \begin{pmatrix} 1 & \rho_1 & \rho_2 \\ \rho_1 & 1 & \rho_1 \\ \rho_2 & \rho_1 & 1 \end{pmatrix}.$$

Fuller and Breidt (1999) found that in most of the practical situations, their generalized least square estimation remains better than the pure panel surveys.

10.11 RE-SAMPLING METHODS

The major benefit of re-sampling methods is that a single standard error formula can be used for all statistics, unlike the linearization method which requires the derivation of separate formula for each statistic. Sometimes the linearization method becomes too cumbersome in handling the situations specially for post-stratification and non-response adjustments. Such situations can be handled very easy through re-sampling methods. Re-sampling methods are found to be more applicable to stratified random sampling or stratified multi-stage sampling. Establishment surveys are the examples of stratified random sampling, whereas large socio economic surveys are the examples of stratified multi-stage sampling. In the case of stratified multi-stage sampling, re-sampling methods are valid if the sample clusters are selected with replacement of the first stage sampling fraction is negligible. Here we consider a stratified multistage design with large number of strata, L, and relatively few primary sampling units or clusters, $m_h \geq 2$, sampled within each stratum, $h = 1, 2, ..., L$, Assuming that sub-sampling within sampled clusters $i = 1, 2, ..., m_h$ is performed to ensure unbiased estimation of cluster totals.

The basic design weights w_{hik} attached to the sample elements hik are adjusted for post-stratification and unit non-response to get adjusted weights w^*_{hik}. The weights w^*_{hik} may also be calibrated weights. As we saw from Godambe's work, many parameters of interest such as mean, total, median, and variance etc., can be derived as a solution to the census equation

$$S(\theta) = \sum_{(hik)\in\Omega} u(y_{hik}, \theta) = 0 . \tag{10.11.1}$$

Obviously the GREG estimator of $S(\theta)$ is

$$\hat{S}(\theta) = \sum_{(hik)\in s} w^*_{hik} u(y_{hik}, \theta). \tag{10.11.2}$$

In this section we would like to discuss different methods of variance of the estimator $\hat{S}(\theta)$ *viz.* Jackknife, Balanced Repeated Replication (BRR), and Balanced Half Sample (BHS) methods. Ahmad (1997) also suggested a re-sampling technique for complex survey data. Willson, Kirnos, Gallagher, and Wanger (2002) considered variance estimation from calibrated samples which improve techniques adopted by statistical packages such as SUDAAN, SAS, and STATA etc..

10.11.1 JACKKNIFE VARIANCE ESTIMATOR

At the first step, keep all the sampled units and solve the equation

$$\hat{S}(\theta) = \sum_{(hik)\in s} w^*_{hik} u(y_{hik}, \theta) = 0 \tag{10.11.1.1}$$

we have an estimator $\hat{\theta}_r$ of θ. Delete one sampled cluster, say $(l\ j)$, and adjust the weights w^*_{hik} accordingly to the following two steps.

Step I. Change the basic weights w_{hik} to the Jackknife weights $w_{hik(lj)} = w_{hik} b_{lj}$ where

$$b_{lj} = \begin{cases} 0 & \text{if } \ (hi) = (lj), \\ \dfrac{m_l}{m_l - 1} & \text{if } \ h = l \ \text{ and } \ i \neq j, \\ 1 & \text{if } \ h \neq l. \end{cases}$$

Step II. Replace w_{hik} by $w_{hik(lj)}$ in the post-stratification process to get adjusted Jackknife weights $w^*_{hik(lj)}$.

Now solve the equation

$$\hat{S}_{(lj)}(\theta) = \sum_{\substack{(hik)\in s \\ i\neq l, k\neq j}} w^*_{hik(lj)} u(y_{hik}, \theta) = 0$$

to obtain the estimator $\hat{\theta}_{(lj)}$ when the sampled cluster (lj) is deleted.

Then a Jackknife variance estimator of the estimator $\hat{\theta}$ is given by

$$\hat{v}_J(\hat{\theta}) = \sum_{l=1}^{L} \frac{m_l - 1}{m_l} \sum_{j=1}^{n_l} \left[\hat{\theta}_{(lj)} - \hat{\theta}\right]\left[\hat{\theta}_{(lj)} - \hat{\theta}\right]^t . \quad (10.11.1.2)$$

The Jackknife variance estimator given in (10.11.1.2) is quite general and is applicable in many practical situations.

Example 10.11.1. Divide the United States of America into four independent strata as Northeast, Midwest, South, and West. Suppose the first stratum Northeast consists of two clusters New England and Mid Atlantic, second stratum Midwest consists of two clusters East North Central and West North Central, third stratum South consists of three clusters South Atlantic, East South Central, and West South Central, and the fourth stratum West consists of two clusters Mountains and Pacific as given in population 7 in the Appendix. From each stratum (or region) select two clusters (or divisions) by SRSWOR sampling and within each selected cluster (or division) select two units (or states) and collect the information on the projected population counts during 1995.
(a) Suggest an estimator for estimating population total in the US using stratified multi-stage design.
(b) Estimate the variance of the estimator of population total by using the technique of Jackknife.
(c) Find the 95% confidence interval for total projected population during 1995 in the USA.

Solution. Let y_{hik} be the value of the kth units (projected counts in a state during 1995) of the study variable in the ith cluster (division) of the hth stratum (region).

Stratum or Region (h)	Cluster or Division (i)	M_{hi}= no. of units in the ith cluster or division	N_h = no. of divisions in the hth region.	n_h= no. of divisions selected from the hth region	m_{hi}= no. of states selected from each cluster or division.
1	1 2	$M_{11} = 6$ $M_{12} = 3$	2	2	2 2
2	1 2	$M_{21} = 5$ $M_{22} = 7$	2	2	2 2
3	1 2	$M_{31} = 9$ $M_{32} = 4$ $M_{33} = 4$	3	2	2 2 ---
4	1 2	$M_{41} = 8$ $M_{42} = 5$	2	2	2 2

Let L be the total number of strata, N_h be the total number of clusters in the hth stratum, n_h be the number of clusters selected from the hth stratum, M_{hi} be the total number of units in the ith cluster of the hth stratum and m_{hi} be the number of units selected from the ith cluster of the hth stratum.

The following table provides the information collected in the sample.

Region (h)	**Division** (i)	**State** (k)	**1995** y_{hik}
Northeast	New England	New Hampshire	1276
		Massachusetts	6032
	Mid Atlantic	New York	17909
		New Jersey	8100
Midwest	East North Central	Ohio	10958
		Michigan	9364
	West North Central	Minnesota	4501
		North Dakota	631
South	South Atlantic	Maryland	5180
		North Carolina	7197
	East South Central	Kentucky	3740
		Alabama	4282
		Idaho	1018
		Colorado	3407
		California	31749
		Hawaii	1253

(a) Estimation of population total: Using these notation an unbiased estimator of population total Y in the United States is

$$\hat{Y} = \sum_{h=1}^{L} \frac{N_h}{n_h} \sum_{i=1}^{n_h} \frac{M_{h_i}}{m_{h_i}} \sum_{k=1}^{m_{h_i}} y_{hik} \,.$$

This estimator can be written as

$$\hat{Y} = \frac{N_1}{n_1} \sum_{i=1}^{n_1} \frac{M_{1i}}{m_{1i}} \sum_{k=1}^{m_{1i}} y_{1ik} + \frac{N_2}{n_2} \sum_{i=1}^{n_2} \frac{M_{2i}}{m_{2i}} \sum_{k=1}^{m_{2i}} y_{2ik} + \frac{N_3}{n_3} \sum_{i=1}^{n_3} \frac{M_{3i}}{m_{3i}} \sum_{k=1}^{m_{3i}} y_{3ik} + \frac{N_4}{n_4} \sum_{i=1}^{n_4} \frac{M_{4i}}{m_{4i}} \sum_{k=1}^{m_{4i}} y_{4ik}$$

$$= \frac{2}{2} \sum_{i=1}^{2} \frac{M_{1i}}{m_{1i}} \sum_{k=1}^{m_{1i}} y_{1ik} + \frac{2}{2} \sum_{i=1}^{2} \frac{M_{2i}}{m_{2i}} \sum_{k=1}^{m_{2i}} y_{2ik} + \frac{3}{2} \sum_{i=1}^{2} \frac{M_{3i}}{m_{3i}} \sum_{k=1}^{m_{3i}} y_{3ik} + \frac{2}{2} \sum_{i=1}^{2} \frac{M_{4i}}{m_{4i}} \sum_{k=1}^{m_{4i}} y_{4ik}$$

$$=\frac{2}{2}\left[\frac{M_{11}}{m_{11}}\sum_{k=1}^{m_{11}} y_{1ik}+\frac{M_{12}}{m_{12}}\sum_{k=1}^{m_{12}} y_{1ik}\right]+\frac{2}{2}\left[\frac{M_{21}}{m_{21}}\sum_{k=1}^{m_{21}} y_{2ik}+\frac{M_{22}}{m_{22}}\sum_{k=1}^{m_{22}} y_{2ik}\right]$$

$$+\frac{3}{2}\left[\frac{M_{31}}{m_{31}}\sum_{k=1}^{m_{31}} y_{3ik}+\frac{M_{32}}{m_{32}}\sum_{k=1}^{m_{32}} y_{3ik}\right]+\frac{2}{2}\left[\frac{M_{41}}{m_{41}}\sum_{k=1}^{m_{41}} y_{4ik}+\frac{M_{42}}{m_{42}}\sum_{k=1}^{m_{42}} y_{4ik}\right]$$

$$=\frac{2}{2}\left[\frac{6}{2}(1276+6032)+\frac{3}{2}(17909+8100)\right]+\frac{2}{2}\left[\frac{5}{2}(10958+9364)+\frac{7}{2}(4501+631)\right]$$

$$+\frac{3}{2}\left[\frac{9}{2}(5180+7197)+\frac{4}{2}(3740+4282)\right]+\frac{2}{2}\left[\frac{8}{2}(1018+3407)+\frac{5}{2}(31749+1253)\right]$$

$$=\frac{2}{2}[21924+39013.5]+\frac{2}{2}[50805+17962]+\frac{3}{2}[55696.5+16044]+\frac{2}{2}[17700+82505]$$

$$=337520.25\,.$$

which is an estimate of the projected population of the United States during 1995 as shown in the appendix.

(b) Estimation of variance using Jackknife estimator: The estimates of population total $\hat{Y}(h,\ i)$ (say) by dropping i^{th} cluster $(i=1,2)$ from the hth stratum $(h=1,2,3,4)$ are given by

$$\hat{Y}(1,1)=\frac{2}{1}[39013.5]+\frac{2}{2}[50805+17962]+\frac{3}{2}[55696.5+16044]+\frac{2}{2}[17700+82505]=354609.75\,,$$

$$\hat{Y}(1,2)=\frac{2}{1}[21924]+\frac{2}{2}[50805+17962]+\frac{3}{2}[55696.5+16044]+\frac{2}{2}[17700+82505]=320430.75\,,$$

$$\hat{Y}(2,1)=\frac{2}{2}[21924+39013.5]+\frac{2}{1}[17962]+\frac{3}{2}[55696.5+16044]+\frac{2}{2}[17700+82505]=304677.25\,,$$

$$\hat{Y}(2,2)=\frac{2}{2}[21924+39013.5]+\frac{2}{1}[50805]+\frac{3}{2}[55696.5+16044]+\frac{2}{2}[17700+82505]=370363.27\,,$$

$$\hat{Y}(3,1)=\frac{2}{2}[21924+39013.5]+\frac{2}{2}[50805+17962]+\frac{3}{1}[16044]+\frac{2}{2}[17700+82505]=278041.5\,,$$

$$\hat{Y}(3,2)=\frac{2}{2}[21924+39013.5]+\frac{2}{2}[50805+17962]+\frac{3}{1}[55696.5]+\frac{2}{2}[17700+82505]=396999\,,$$

$$\hat{Y}(4,1)=\frac{2}{2}[21924+39013.5]+\frac{2}{2}[50805+17962]+\frac{3}{2}[55696.5+16044]+\frac{2}{1}[82505]=402325.25\,,$$

and

$$\hat{Y}(4,2)=\frac{2}{2}[21924+39013.5]+\frac{2}{2}[50805+17962]+\frac{3}{2}[55696.5+16044]+\frac{2}{1}[17700]=272715.25\,.$$

Thus using the concept of Jackknife, an estimate of the variance $V(\hat{Y})$ is given by

$$\hat{v}_J(\hat{Y}) = \sum_{h=1}^{L} \frac{n_h - 1}{n_h} \sum_{i=1}^{n_h} \left[\hat{Y}(h,i) - \hat{Y}\right]^2 = \sum_{h=1}^{4} \frac{n_h - 1}{n_h} \sum_{i=1}^{n_h} \left[\hat{Y}(h,i) - \hat{Y}\right]^2$$

$$= \frac{n_1 - 1}{n_1} \sum_{i=1}^{n_1} \left[\hat{Y}(h,i) - \hat{Y}\right]^2 + \frac{n_2 - 1}{n_2} \sum_{i=1}^{n_2} \left[\hat{Y}(h,i) - \hat{Y}\right]^2$$

$$+ \frac{n_3 - 1}{n_3} \sum_{i=1}^{n_3} \left[\hat{Y}(h,i) - \hat{Y}\right]^2 + \frac{n_4 - 1}{n_4} \sum_{i=1}^{n_4} \left[\hat{Y}(h,i) - \hat{Y}\right]^2$$

$$= \frac{2-1}{2}\left[\left\{\hat{Y}(1,1) - \hat{Y}\right\}^2 + \left\{\hat{Y}(1,2) - \hat{Y}\right\}^2\right] + \frac{2-1}{2}\left[\left\{\hat{Y}(2,1) - \hat{Y}\right\}^2 + \left\{\hat{Y}(2,2) - \hat{Y}\right\}^2\right]$$

$$+ \frac{2-1}{2}\left[\left\{\hat{Y}(3,1) - \hat{Y}\right\}^2 + \left\{\hat{Y}(3,2) - \hat{Y}\right\}^2\right] + \frac{2-1}{2}\left[\left\{\hat{Y}(4,1) - \hat{Y}\right\}^2 + \left\{\hat{Y}(4,2) - \hat{Y}\right\}^2\right]$$

$$= \frac{2-1}{2}\left[\{354609.75 - 337520.25\}^2 + \{320430.75 - 337520.25\}^2\right]$$

$$+ \frac{2-1}{2}\left[\{304677.25 - 337520.25\}^2 + \{370363.27 - 337520.25\}^2\right]$$

$$+ \frac{2-1}{2}\left[\{278041.5 - 337520.25\}^2 + \{396999 - 337520.25\}^2\right]$$

$$+ \frac{2-1}{2}\left[\{402325.25 - 337520.25\}^2 + \{272715.25 - 337520.25\}^2\right]$$

$$= 9108124043.$$

(c) Using Table 2 from the Appendix the $(1-\alpha)100\%$ and hence 95% confidence interval of the population total is given by

$$\hat{Y} \pm t_{\alpha/2}\left(\text{df} = n - \sum_{h=1}^{L} n_h\right)\sqrt{\hat{v}_J(\hat{Y})}, \quad \text{or} \quad \hat{Y} \pm t_{0.025}(\text{df} = 16-8)\sqrt{\hat{v}_J(\hat{Y})}$$

or $337520.25 \pm 2.306\sqrt{9108124043}$, or $[117443.69, 557596.80]$.

The true count for the projected population during 1995 is $Y = 259{,}593$ which clearly lies in the above 95% confidence interval.

10.11.2 BALANCED HALF SAMPLE (BHS) METHOD

A half sample can be formed by selecting one of the two first stage sample clusters from each stratum. A set of R half samples can be defined by an $R \times L$ matrix H with the (r,h)th element defined as

$$\epsilon_{rh} = \begin{cases} +1 \text{ if the cluster of the } h^{th} \text{ stratum in the } r^{th} \text{ half is } 1^{st} \text{ from first stage sample cluster,} \\ -1 \text{ if the cluster of the } h^{th} \text{ stratum in the } r^{th} \text{ half is } 2^{nd} \text{ from first stage sample cluster.} \end{cases}$$

Then a set of R half samples is said to be balanced if $\sum_{r=1}^{R} \epsilon_{rh}\, \epsilon_{rh'} = 0$ for all $h \neq h'$. A balanced H can be made from an $R \times R$ Hadamard matrix by choosing any L columns excluding the column of all values of $+1$, where $L+1 \le R \le L+4$. Assume

$\hat{\theta}_{(r)}$ be the survey estimator of θ obtained by treating the rth half sample as the original data set after adjusting the weights w_{hik} as

$$w_{hik}^{(r)} = \begin{cases} 2w_{hik} & \text{if } \in_{rh} = +1, \\ 0 & \text{if } \in_{rh} = -1. \end{cases} \tag{10.11.2.1}$$

Using these weights in place of the Jackknife weights, a standard Balanced Half Sample (BHS) variance estimator of $\hat{\theta}$ is given by

$$\hat{v}_{\text{BHS}}(\hat{\theta}) = \frac{1}{R}\sum_{r=1}^{R}\left[\hat{\theta}_{(r)} - \hat{\theta}\right]\left[\hat{\theta}_{(r)} - \hat{\theta}\right]^t . \tag{10.11.2.2}$$

A limitation of BHS method is that some of the replicate estimators may become extreme because only a half sample is used. In other words, the weights w_{hik} are sharply perturbed. One well known method to solve this problem is to change the weights w_{hik} as

$$w_{hik}^{(r)} = \begin{cases} 1.5w_{hik} & \text{if } \in_{rh} = +1, \\ 0.5w_{hik} & \text{if } \in_{rh} = -1. \end{cases}$$

In such situations, the full sample information is utilized to obtain estimators $\hat{\theta}_{(r)}\left(\frac{1}{2}\right)$ for $r = 1, 2, ..., R$ of θ.

The usual BHS method of variance estimation becomes

$$\hat{v}_{\text{BHS}\left(\frac{1}{2}\right)}(\hat{\theta}) = \frac{1}{R}\sum_{r=1}^{R}\left[\hat{\theta}_{(r)}\left(\frac{1}{2}\right) - \hat{\theta}\right]\left[\hat{\theta}_{(r)}\left(\frac{1}{2}\right) - \hat{\theta}\right]^t . \tag{10.11.2.3}$$

For practical use, we need a BRR with number of replications (R) as small as possible. It can be made a balance between a desire for small R and the need to have a reasonably stable variance estimator by following Wolter (1985). It may not be trivial to find a BRR, but some methods have been suggested by Gupta and Nigam (1987), Gurney and Jewett (1975), Sitter (1993), and Wu (1981). One may note that the standard BHS variance estimator can also use the weights as

$$w_{hik}^{(r)} = \begin{cases} \left(1 + \sqrt{\dfrac{n_h - m_h}{m_h}}\right) w_{hik} & \text{if cluster } (h,i) \text{ is in replicator,} \\[2ex] \left(1 - \sqrt{\dfrac{n_h - m_h}{m_h}}\right) w_{hik} & \text{if cluster } (h,i) \text{ is not in replicator.} \end{cases}$$

The original BHS was proposed by McCarthy (1969) for $m_h = 2$ and was further studied by Krewski and Rao (1979). It is difficult to construct balanced replicates for arbitrary value of m_h and its limitations can be had from Valliant (1996). Fortunately there is an another solution to deal with arbitrary value of m_h through the method of Bootstrap.

10.11.3 BOOTSTRAP VARIANCE ESTIMATOR

A bootstrap sample can be obtained by drawing $(m_h - 1)$ clusters with replacement from the m_h sample clusters in each stratum independently. Select a large number, R, of bootstrap samples independently which can be represented in terms of the bootstrap frequencies $m_{hi}(r) =$ number of times (h, i)th sample cluster get selected in the rth bootstrap sample $r = 1, 2, ..., R$.
Then the bootstrap design weights are given by

$$w_{hik}(r) = w_{hik}\left[\frac{m_h}{m_h - 1}\right]m_{hi}(r). \tag{10.11.3.1}$$

Using these weights obtain the bootstrap estimators $\hat{\theta}_{(r)}$, $r = 1, 2, ..., R$ of θ.

Then a bootstrap variance estimator of $\hat{\theta}$ is given by

$$\hat{v}_{\text{boot}}\left(\hat{\theta}\right) = \frac{1}{R}\sum_{r=1}^{R}\left[\hat{\theta}_{(r)} - \hat{\theta}\right]\left[\hat{\theta}_{(r)} - \hat{\theta}\right]^t . \tag{10.11.3.2}$$

Several research workers including Kott and Stukel (1998), Lohr and Rao (1998), Rao and Shao (1992), Rao (1996b), Rao and Shao (1999), Shao, Chen, and Chen (1998), and Shao and Chen (1999) have considered the applications of the above standard techniques, with intelligent modifications, to estimate the variance in the presence of non-response or imputed data through different techniques like hot deck, cold deck, ratio imputation, and regression imputation, which we shall discuss in details in Chapter 12.

EXERCISES

Exercise 10.1. A sample of size n is to be taken from a population of size N to estimate the total Y of values y_i $(i = 1, 2, ..., N)$ of a variable y when 'normed' size measures p_i are available. Suppose the population is divided into n random groups of sizes N_g, $(g = 1, 2, ..., N)$ such that $\sum_{g=1}^{n} N_g = N$. If the selected i^{th} unit consists of M_i second stage units (SSU), one may estimate y_i by taking an SRSWOR sample of m_i units instead of ascertaing it. Suggest an estimator of population total on the basis of above two stage sampling schemes and hence deduce an estimator of its variance.
Hint: Chaudhuri, Adhikary, and Seal (1997).

Exercise 10.2. Let Y_i and X_i be the totals of the ith FSU of the study variable Y and auxiliary variable X, respectively. Assume that the population total $X = \sum_{i=1}^{N} X_i$ of the auxiliary variable is known and we are interested in the estimation of population total of the study variable, $Y = \sum_{i=1}^{N} Y_i$. A sample s of n FSUs is selected according to any design with π_i and π_{ij} as the known first and second order inclusion probabilities, which are in fact function of first stage sample size n. The selected FSUs are sub-sampled independently with suitable selection probabilities in the second stage. When the ith unit is being selected then it is assumed that from sampling in the second stage estimators t_{iy} and t_{ix}, respectively, for Y_i and X_i are available such that $E_2(t_{iy}) = Y_i$ and $E_2(t_{ix}) = X_i$. Find the asymptotic properties of the following estimators:

(a) $t_1 = \sum_{i=1}^{n} \frac{1}{\pi_i}\left[t_{iy} - \beta_{ixy}(t_{ix} - X_i)\right] - \beta_{xy}\left(\sum_{i=1}^{n} \frac{t_{ix}}{\pi_i} - X\right)$ [Sahoo (1987)]

when the first and second stage units are selected with SRSWOR sampling.

(b) $t_2 = \sum_{i=1}^{n} \frac{1}{\pi_i}\left[t_{iy} - \beta_{ixy}(t_{ix} - X_i)\right] - \beta_{xy}\left(\sum_{i=1}^{n} \frac{t_{ix}}{\pi_i} - X\right)$ [Sahoo and Panda (1997)]

when the first stage units are selected with PPSWOR sampling and second stage units are selected with SRSWOR sampling.

(c) $t_3 = \sum_{i=1}^{n} M_i \bar{y}_i \left(X \Big/ \sum_{i=1}^{n} X_i\right)$ [Smith (1969)]

(d) $t_4 = \sum_{i=1}^{n} \frac{\bar{y}_i}{\bar{x}_i} X_i \left(X \Big/ \sum_{i=1}^{n} X_i\right)$ [Murthy (1967)]

(e) $t_5 = \left[\frac{N}{n} \sum_{i=1}^{n} \frac{(M_i \bar{y}_i)(M_i \bar{x}_i)}{X_i}\right]\left[\frac{N}{n} \sum_{i=1}^{n} X_i\right] \Big/ X$ [Sahoo and Swain (1986)]

(f) $t_6 = n^{-1} \sum_{i=1}^{n} \frac{M_i}{M}\left[\bar{y}_i - \alpha_i(\bar{x}_i - \bar{X}_i) - \gamma_i(s_{ix}^2 - S_{ix}^2)\right] - d(\bar{x} - \bar{X})$ [Mahajan and Singh (1996)]

(g) $t_7 = \frac{N}{n} \sum_{i=1}^{n} \left[M_i \bar{y}_i + d_i(M_i \bar{x}_i - X_i)\right]\left(\bar{X}/\bar{x}\right)^g$ [Sahoo and Panda (1997)]

Hint: Panda and Sahoo (1999).

Exercise 10.3. Show that Hartley and Ross (1954) unbiased ratio type estimator of population mean in two-stage sampling is

$$t_{\text{u}} = \bar{r}\bar{X} + \frac{(N-1)}{N}\left(s_{brx} - \theta s_{brz}\right) + \frac{1}{n}\sum_{i=1}^{n} \alpha_i^2 \left(\frac{M_i - 1}{M_i} - \frac{N-1}{N}\left(\frac{M_i - m_i}{M_i m_i}\right)\right)\left(s_{irx} - \theta s_{irz}\right),$$

where

$$\alpha_i = NM_i \Big/ \sum_{i=1}^{N} M_i\,,\ \theta = \bar{X}/\bar{Z}\,,\ \bar{r}_i = \sum_{j=1}^{m_i} r_{ij} \Big/ m_i\,,\ \bar{r} = \sum_{i=1}^{n} \alpha_i \bar{r}_i \Big/ n\,,$$

$$s_{brx} = (n-1)^{-1} \sum_{i=1}^{n} (\alpha_i \bar{r}_i - \bar{r})(\alpha_i \bar{x}_i - \bar{x}),\ \ s_{irx} = (m_i - 1)^{-1} \sum_{j=1}^{m_i} (r_{ij} - \bar{r}_i)(x_{ij} - \bar{x}_i),$$

$$s_{brz} = (n-1)^{-1}\sum_{i=1}^{n}(\alpha_i\bar{r}_i - \bar{r})(\alpha_i\bar{z}_i - \bar{z}), \text{ and } s_{irz} = (m_i - 1)^{-1}\sum_{j=1}^{m_i}(r_{ij} - \bar{r}_i)(z_{ij} - \bar{z}_i)$$

Hint: Sahoo (1991).

Exercise 10.4. In two-stage sampling n FSUs are selected with PPSWR sampling. If the ith FSU occurs r_i times in the sample one of the following procedures may be adopted for the second stage:

(a) $r_i\, m_i$ SSUs are selected with SRSWOR sampling;

(b) r_i independent samples of m_i SSUs selected with SRSWOR sampling;

(c) m_i units are selected WOR and observations are weighted by r_i.

Propose unbiased estimators of population total Y and derive the expressions for the variance in each situation. If V_a, V_b, and V_c denote the variance of the estimators under (a), (b), and (c), respectively, then show that the inequality $V_a \le V_b \le V_c$ holds.

Hint: Rao (1961).

Exercise 10.5. A sample of n FSUs is selected with SRSWOR sampling and from each selected FSU a constant fraction f_2 of SSU is selected. If r_i out of the m_i SSUs in the ith FSU possess an attribute, show that the estimator ratio to size $\hat{p} = \sum_{i=1}^{n} r_i \Big/ \sum_{i=1}^{n} m_i$ estimates the population proportion of the attribute. Find the variance of the estimator $\hat{p}$ and suggest an estimator of variance.

Hint: Cochran (1977).

Exercise 10.6. (a) Suppose n FSUs are selected with PPSWR sampling in two stage sampling. From each sampled FSU, m SSUs are selected with SRSWOR sampling. Find the bias and variance of the estimator of population total defined as, $\hat{y}_{\text{R}} = \sum_{i=1}^{n}\alpha_i\hat{Y}_i$, where α_i are the real numbers and $\hat{Y}_i$ is an unbiased estimator of the ith FSU. Deduce an unbiased estimator of variance.

Hint: Raj (1966).

(b) In two stage sampling, n FSUs are selected from N FSUs in the population by Midzuno--Sen's sampling scheme and the m_i SSUs are selected from the ith selected FSU by SRSWOR sampling. Derive an unbiased estimator of population total, and expression for its variance and estimator of variance.

Hint: Raj (1954a, 1954b).

Exercise 10.7. Consider n FSUs are selected with PPSWR sampling. From the ith selected FSU of size M_i, suppose m_i SSUs are selected with SRSWR sampling. For estimating the population total Y the sub-sampling number m_i are to be fixed in two ways:

(a) Expected value of m_i is m;

(b) Total number of SSUs is m_0.

Find the optimum value of m_i in each situation such that the variance of the estimator of population total is minimum. Hence deduce the efficiency comparison.
Hint: Rangarajan (1957).

Exercise 10.8. Show that in the following cases the sample mean is an unbiased estimator of population mean.

(a) Divide the population into N clusters each of size M units. n FSUs are selected with SRSWR sampling. From each selected FSU, m SSUs are selected with SRSWR sampling.
(b) Divide the population into clusters each of m' units. Select a sample of n' clusters with SRSWR sampling.
(c) Deduce the condition such that the efficiency of both schemes will be the same.
Hint: Singh (1956).

Exercise 10.9. Discuss three-stage sampling by proposing an unbiased estimator of total and discuss its variance and estimator of variance. Discuss a method of optimum allocation in three-stage sampling.

Exercise 10.10. Suppose n units are selected with SRSWR sampling on the first occasion and a sub-sample of $n\pi$ units is selected by a similar method and is retained for the second occasion. A supplementary sample of size $n(1-\pi)$ units has been selected independently by the same method. Find the constants α_i, β_i, $(i = 1, 2)$ and π such that the variance of the estimator of population mean on the second occasion defined as

$$\bar{y}_2 = \alpha_1\bar{y}_1 + \alpha_2\bar{y}_2 + \beta_1\bar{y}_1^* + \beta_2\bar{y}_2^*$$

is a minimum.
Hint: Hansen, Hurwitz, and Madow (1953).

Exercise 10.11. From a population of N units, a large sample of m units is drawn by SRSWR sampling scheme to draw information on the auxiliary variable x (say) and named first phase sample. In the second phase, a sub-sample of m_1 units is drawn from the first phase sample and a sample of m_2 units is taken independently from the whole population by SRSWR sampling and the study variable is noted.

(a) Compare the following two estimators of population mean:

(i) Unbiased estimator based on direct sample of m_2 units;
(ii) Regression estimator based on two stage sample of m_1 and m_2 units.

(b) Derive the optimum values of m_1 and m_2 for the fixed cost.
(c) Obtain the best linear estimator from (i) and (ii) and derive its variance expression.
(d) Find the optimum value of a parameter forming the linear variety of estimators from (i) and (ii), such that its resultant variance is minimum.
Hint: Patterson (1950), Cochran (1963).

Exercise 10.12. Consider a population Ω consisting of N units. Let the ith FSU consist of M_i SSUs. On the first occasion a sample of n FSUs is drawn from the population Ω with PPSWR sampling. Within each selected FSU a sample of m SSUs is drawn with SRSWR sampling and order of drawing FSU and SSU is specified. Assume λ to be the proportion of FSUs matched. Reject the first $n\mu = n(1-\lambda)$ units and retain the remaining $n\lambda$ units in the sample as FSUs. Supplement these retained FSUs with a set of new $n\mu$ FSUs selected from the population with PPSWR sampling. Suggest an estimator of population mean. Extend the procedure such that one can estimate the population mean on the hth occasion.
Hint: Kathuria (1975).

Exercise 10.13. In multi-occasion sampling let the first occasion consist of n_i FSUs selected by SRSWOR sampling from a population of N units. For the ith $(i \geq 2)$ occasion a sample of size n_i FSUs consists of the following parts:

(i) The n_i^* FSUs which are also observed from the same variate at least on the previous occasion;
(ii) The n_i^{**} FSUs from the population FSUs not selected so far in the sample such that the condition $n_i = n_i^* + n_i^{**}$ is satisfied;
(iii) The FSU in the sample is studied on different occasions by selecting m SSUs out of M SSUs by SRSWOR sampling again once for all.

Suggest an estimator of population mean based on the above sampling mechanism and derive its variance.
Hint: Agarwal and Tikkiwal (1980).

Exercise 10.14. Let p_j, $j = 1, 2, ..., N$, be the probability of selecting a unit U_j based on a variable z, N being the number of units in the population. On the first occasion select a sample s_1 of n units with probability p_j such that $\sum_{j=1}^{N} p_j = 1$ using PPSWR sampling and observe the study variable y. On the hth occasion

$(h>1)$ the sample s_h of n units consists of a sample s_{hm} of $m_h = n\lambda_h$ units (matched part) selected with SRSWOR sampling from the n units obtained on $(h-1)$th occasion and a sample s_{hu} of $u_h = n - m_h$ units (unmatched part) selected directly from the population with probabilities p_i using PPSWR sampling.
Hint: Tripathi and Srivastava (1979).

Exercise 10.15. Consider a sample s_1 of size n is selected on the first occasion by some suitable sampling design P_1. Let y_{1i} (y_{2i}) be the values of the variate y under study for the ith unit on the first (second) occasion. On the second occasion, a matched sample, say sub-sample s_m of size m $(= n\gamma)$ with $0<\gamma\le 1$, is selected from the sample s_1 selected at the first occasion by some suitable sampling design P_m. Assume the sub-sample s_m is supplemented by another unmatched sample s_u of size u either from the entire population or from the non-sampled units at the first occasion by any suitable sampling design P_u. Assuming that the initial sample s_1 is selected by using the Rao--Hartley--Cochran (RHC) scheme, compare the following two estimator

$$\hat{Y}_2 = \phi\hat{Y}_{2m} + (1-\phi)\hat{Y}_{2u}, \quad \text{and} \quad \hat{Y}_{2m}^{*} = \sum_{i\in s_m}\left(r_i^{*}/p_i^{*}\right)\tilde{P}_i + \beta Z,$$

where

$$\hat{Y}_{2m} = \sum_{i\in s_m}\left(y_{2i}^{*}/p_i^{*}\right)\tilde{P}_i;\ \hat{Y}_{2u} = \sum_{i\in s_u}\left(y_{2i}/p_i\right)P_i^{'}\ ;\ y_{2i}^{*} = y_{2i}\Delta_i/p_i\ ;\ \tilde{P}_i = \sum_{\Omega_m} p_i^{*}\ ;\ P_i^{'} = \sum_{\Omega_u} p_i\ , \text{ with}$$

Ω_m and Ω_u being the groups of those values associated with those units that belong to the random group from which the ith unit was selected in s_m and s_u, respectively. In addition

$$p_i = z_i/Z \quad \text{and} \quad p_i^{*} = \left(y_{1i}\Delta_i p_i\right)\Big/\left(\sum_{i\in s_1} y_{1i}\Delta_i/p_i\right)$$

for the ith unit in the first occasion sample s_1 and Δ_1 be the sum of the p_i values for the group containing the ith unit.
Hint: Arnab (1998).

Exercise 10.16. Study the bias and variance properties of the following estimators of population total in two stage sampling:

$$t_1 = \sum_{i=1}^{n}\frac{1}{\pi_i}\left[t_{iy} - \beta_{ixy}\left(t_{ix} - X_i\right)\right] - \beta_{xy}\left(\sum_{i=1}^{n}\frac{t_{ix}}{\pi_i} - X\right), \quad t_2 = \sum_{i=1}^{n} M_i\bar{y}_i\left(X\Big/\sum_{i=1}^{n}X_i\right),$$

$$t_3 = \sum_{i=1}^{n}\frac{\bar{y}_i}{\bar{x}_i}X_i\left(X\Big/\sum_{i=1}^{n}X_i\right), \quad t_4 = \left[\frac{N}{n}\sum_{i=1}^{n}\frac{\left(M_i\bar{y}_i\right)\left(M_i\bar{x}_i\right)}{X_i}\right]\left[\frac{N}{n}\sum_{i=1}^{n}X_i\right]\Big/X,$$

and

$$t_5 = \frac{N}{n}\sum_{i=1}^{n}\left[M_i\bar{y}_i + d_i\left(M_i\bar{x}_i - X_i\right)\right]\left(nX\Big/N\sum_{i=1}^{n}X_i\right)^{g}.$$

Hint: Sahoo and Panda (1997).

Exercise 10.17. Consider the problem of estimation of change in total on two successive occasions. For a fixed survey population $\Omega = (i : i = 1, 2, ..., N)$, let the population vector on the present occasion be $y = (y_1, y_2, .., y_N)$ and on some previous occasion be $y^* = (y_1^*, y_2^*, .., y_N^*)$. Let the sample data $\{s, (y_i, y_i^*) : i \in s\}$ are available. Show that optimal estimate for the change in total, $\sum_{i\in\Omega} y_i - \sum_{i\in\Omega} y_i^*$, is given by $\sum_{i\in s}(y_i - y_i^*)/\pi_i$.

Hint: Godambe (1995).

Exercise 10.18. Let a general linear unbiased estimator of population total Y under a multi-stage design be defined as

$$\hat{y}_{\text{ms}} = \sum_{i\in s} t_{is} y_i = \sum_{i=1}^{N} t_{is}^* y_i$$

where

$$t_{is}^* = \begin{cases} t_{is} & \text{if } i \in s, \\ 0 & \text{otherwise,} \end{cases} \quad \text{and} \quad E(t_{is}) = \sum_{i\in s} t_{is} p(s) = 1.$$

Then deduce the following results:

(a) An unbiased estimator of variance of $\hat{y}_{\text{ms}}$ is given by

$$\hat{v}_1(\hat{y}_{\text{ms}}) = \sum_{i\in s} a_{is} y_i^2 + \sum_{i,j\in s}\sum b_{ijs} y_i y_j + \sum_{i\in s} t_{is} \hat{v}_2(y_i),$$

where a_{is}, b_{ijs} and t_{is} are predetermined real numbers for every s, and $\hat{v}_2(y_i)$ is an estimator of conditional variance of y_i given s, say $V_{2s}(y_i)$.

Hint: Rao (1966c).

(b) If $V_{2s}(y_i)$ is not independent of s then an estimator of variance of $\hat{y}_{ms}$ can be written as:

$$\hat{v}_2(\hat{y}_{\text{ms}}) = \sum_{i\in s} a_{is} y_i^2 + \sum_{\substack{i,j\in s \\ i<j}}\sum b_{ijs} y_i y_j + \sum_{i\in s}(t_{is}^2 - a_{is})\hat{v}_2(y_i),$$

Hint: Rao (1975).

(c) The following estimators of variance are the special cases of the estimator $\hat{v}_2(\hat{y}_{\text{ms}})$:

(i) Considering $\hat{y}_{\text{ms}}$ as the Horvitz and Thompson estimator $\hat{Y}_{\text{HT}}$ of total in multi-stage sampling, the Sen--Yates--Grundy form of the estimator of variance is given by

$$\hat{v}_{\text{syg}}(\hat{y}_{\text{HT}}) = \sum_{i<j\in s}\sum D_{ij}\left(\frac{y_i}{\pi_i} - \frac{y_j}{\pi_j}\right)^2 + \left\{\frac{1}{n-1} - D_{ij}\right\}\left[\frac{\hat{v}_{2s}(y_i)}{\pi_i^2} + \frac{\hat{v}_{2s}(y_j)}{\pi_j^2}\right],$$

where $D_{ij} = (\pi_i \pi_j - \pi_{ij})/\pi_{ij}$.

(*ii*) In the case of ordered sampling the well known Raj estimator for $n = 2$ becomes

$$\hat{v}_{dr_1} = \frac{1}{4}\left[(1-P_1^2)\left(\frac{y_1}{P_1}-\frac{y_2}{P_2}\right)^2 - (1-P_1)^2\left\{\frac{\hat{v}_{2s}(y_1)}{P_1^2}+\frac{\hat{v}_{2s}(y_2)}{P_2^2}\right\}\right.$$

$$\left. + \left\{(1+P_1)^2\frac{\hat{v}_{2s}(y_1)}{P_1^2}+(1-P_1)^2\frac{\hat{v}_{2s}(y_2)}{P_2^2}\right\}\right],$$

and in the case of unordered sampling, it reduces to

$$\hat{v}_{dr_2} = \frac{1}{4}\left[(1-P_1)(1-P_2)\left\{\left(\frac{y_1}{P_1}-\frac{y_2}{P_2}\right)^2 - \left(\frac{\hat{v}_{2s}(y_1)}{P_1^2}+\frac{\hat{v}_{2s}(y_2)}{P_2^2}\right)\right\}\right.$$

$$\left. + (1+P_1)^2\frac{\hat{v}_{2s}(y_1)}{P_1^2}+(1-P_1)^2\frac{\hat{v}_{2s}(y_2)}{P_2^2}\right].$$

(*iii*) Under the Rao--Hartley--Cochran (RHC) scheme

$$\hat{v}_{\text{RHC}} = \frac{N^2-\sum_{i=1}^{n}N_i^2}{\sum_{i=1}^{n}N_i^2 - N}\sum_{i=1}^{n}\pi_i\left[\left(\frac{y_{it}}{P_{it}}-\hat{y}_{\text{RHC}}\right)^2\right.$$

$$\left. - \left\{(1-\pi_i)^2\frac{\hat{v}_{2s}(y_{it})}{P_{it}^2}+\sum_{\substack{j=1\\ j\neq i}}^{n}\frac{\pi_j^2\hat{v}_{2s}(y_{jt})}{P_{jt}^2}\right\}\right]+\sum_{i=1}^{n}\pi_i^2\frac{\hat{v}_{2s}(y_{it})}{P_{it}^2}.$$

(*iv*) Murthy's estimator of variance for $n = 2$ becomes

$$\hat{v}_m = \frac{(1-P_1)(1-P_2)(1-P_1-P_2)}{(2-P_1-P_2)^2}\left[\left(\frac{y_1}{P_1}-\frac{y_2}{P_2}\right)^2 - \left\{\frac{\hat{v}_{2s}(y_1)}{P_1^2}+\frac{\hat{v}_{2s}(y_2)}{P_2^2}\right\}\right]$$

$$+\frac{(1-P_1)^2(1-P_2)^2}{(2-P_1-P_2)^2}\left\{\frac{\hat{v}_{2s}(y_1)}{(1-P_1)^2P_1^2}+\frac{\hat{v}_{2s}(y_2)}{(1-P_2)^2P_2^2}\right\}.$$

Hints: Brewer and Hanif (1970), Arnab (1988).

Exercise 10.19. Consider a population Ω consists of N first stage units such that the ith first stage unit Ω_i (say) containing M_i second stage units and $M = \sum_{i=1}^{N}M_i$. Let Y_i, X_i and Z_i be the totals of Ω_i for the study variable y and two auxiliary variables x and z, respectively, with corresponding totals $Y = \sum_{i=1}^{N}Y_i$, $X = \sum_{i=1}^{N}X_i$, and $Z = \sum_{i=1}^{N} Z_i$. At the first stage a sample $s \subset \Omega$ of n first stage units is drawn from Ω according to any design with π_i and π_{ij} as the known and positive first and second order inclusion probabilities. For every $i \in s$, a sample s_i of m_i second stage units is drawn from Ω_i with suitable selection probabilities at the second stage. Let $\widetilde{Y}_i$, $\widetilde{X}_i$ and $\widetilde{Z}_i$ be the unbiased estimates of Y_i, X_i and Z_i, respectively, such that

$V_2(\widetilde{Y}_i) = \sigma_{iy}^2$, $V_2(\widetilde{X}_i) = \sigma_{ix}^2$, $V_2(\widetilde{Z}_i) = \sigma_{iz}^2$, $C_2(\widetilde{Y}_i, \widetilde{Z}_i) = \sigma_{iyz}$, $C_2(\widetilde{Y}_i, \widetilde{X}_i) = \sigma_{iyx}$, and $C_2(\widetilde{X}_i, \widetilde{Z}_i) = \sigma_{ixz}$.

From the first phase sample information define

$$\hat{Y} = \sum_{i \in s} \frac{\widetilde{Y}_i}{\pi_i}, \quad \hat{X} = \sum_{i \in s} \frac{\widetilde{X}_i}{\pi_i}, \text{ and } \hat{Z} = \sum_{i \in s} \frac{\widetilde{Z}_i}{\pi_i}$$

such that

$$E(\hat{Y}) = Y, \quad E(\hat{X}) = X, \text{ and } \quad E(\hat{Z}) = Z .$$

Also

$$V(\hat{Y}) = \sigma_y^2 + \sum_{i=1}^{N} \frac{\sigma_{iy}^2}{\pi_i}, \quad V(\hat{X}) = \sigma_x^2 + \sum_{i=1}^{N} \frac{\sigma_{ix}^2}{\pi_i}, \quad V(\hat{Z}) = \sigma_z^2 + \sum_{i=1}^{N} \frac{\sigma_{iz}^2}{\pi_i},$$

$$C(\hat{Y}, \hat{X}) = \sigma_{xy} + \sum_{i=1}^{N} \frac{\sigma_{ixy}}{\pi_i}, \quad C(\hat{Y}, \hat{Z}) = \sigma_{yz} + \sum_{i=1}^{N} \frac{\sigma_{iyz}}{\pi_i}, \text{ and } \quad C(\hat{X}, \hat{Z}) = \sigma_{xz} + \sum_{i=1}^{N} \frac{\sigma_{ixz}}{\pi_i},$$

where

$$\sigma_{xz} = \frac{1}{2} \sum_{i=1}^{N} \sum_{j(\neq i)1}^{N} (\pi_i \pi_j - \pi_{ij}) \left(\frac{X_i}{\pi_i} - \frac{X_j}{\pi_j} \right) \left(\frac{Z_i}{\pi_i} - \frac{Z_j}{\pi_j} \right).$$

(a) Study the asymptotic properties of the following class of estimators of population total in two-stage sampling.

$$\hat{Y}_g = G[\hat{Y}_a, \widetilde{Z}] ,$$

where $\hat{Y}_a = G_i[\widetilde{Y}_i, \widetilde{X}_i]$ for $i \in s$.

(b) Show that the estimator

$$\hat{Y}_{2s} = \sum_{i \in s} (\widetilde{Y}_i - \alpha_i (\widetilde{X}_i - X_i)) / \pi_i - \alpha(\widetilde{Z} - Z)$$

is a special case of the class of estimators in (a).

(c) Defining

$$\hat{Z} = \sum_{i \in s} \frac{\widetilde{Z}_i}{\pi_i}$$

where $\hat{Z}_i = h_i(\widetilde{Z}_i, \widetilde{X}_i)$, study the class of estimators

$$\hat{Y}_{s1} = H[\hat{Y}_a, \hat{Z}_a] ,$$

where $G(.,.)$, $H(.,.)$, $h_i(.,.)$ and $g_i(.,.)$ are the parametric functions as defined by Srivastava (1980).
Hint: Sahoo and Panda (1999a, 1999b).

Exercise 10.20. Let K_h be the number of primary clusters in the hth stratum of the finite population. At the first stage of sampling, $k_h = c_h(K_h)$ clusters are sampled from stratum h as a SRSWOR, where $c_h(1)=1$ and $c_h(t)\geq 2$ for $t \geq 2$. At the second stage of sampling $n_{hi} = g_h(N_{hi})$ observations from a sampled cluster, where N_{hi} is the number of population units in this cluster. Let $\bar{y}_{hi}$ be the mean of these sampled observations.

(a) Show that an unbiased estimator of population mean μ is given by

$$\bar{y} = \sum_{h=1}^{L}\frac{K_h}{k_h}\sum_{i=1}^{k_h}N_{hi}\bar{y}_{hi}\Big/\sum_{h=1}^{L}\frac{K_h}{k_h}\sum_{i=1}^{k_h}N_{hi}\,.$$

(b) Also show that if f.p.c. is ignored, then an estimator of the variance of the estimator $\bar{y}$ under the concept of repeated sampling is given by

$$\hat{v}(\bar{y}) = \sum_{h=1}^{L}\frac{K_h^2}{k_h(k_h-1)}\sum_{i=1}^{k_h}\left[N_{hi}(\bar{y}_{hi}-\bar{y}) - \frac{1}{k_h}\sum_{j=1}^{k_h}N_{hj}(\bar{y}_{hj}-\bar{y})\right]^2\Big/\left(\sum_{h=1}^{L}\frac{K_h}{k_h}\sum_{i=1}^{k_h}N_{hi}\right)^2.$$

Hint: Korn and Graubard (1998).

Exercise 10.21. In exercise 10.20 select the first stage k_h clusters with PPSWOR sampling using 'size' cluster level variable (Z, say) for constructing selection probabilities. At the second stage of sampling, $n_{hi} = g_h(N_{hi})\geq 2$ observations are sampled as a SRSWOR from the ith sampled cluster from the hth stratum. Show that the estimator of the variance of the estimator of population mean $\bar{Y}$ defined as

$$\bar{y}_1 = \sum_{h=1}^{L}\sum_{i=1}^{k_h}\frac{N_{hi}\bar{y}_{hi}}{\pi_{hi}}\Big/\sum_{h=1}^{L}\sum_{i=1}^{k_h}\frac{N_{hi}}{\pi_{hi}}$$

is given by

$$\hat{v}(\bar{y}_1) = \sum_{h=1}^{L}\frac{k_k}{(k_h-1)}\sum_{i=1}^{k_h}\left[\left(\frac{N_{hi}}{\pi_{hi}}\bar{y}_{hi} - \bar{y}\frac{N_{hi}}{\pi_{hi}}\right) - \frac{1}{k_h}\sum_{j=1}^{k_h}\left(\frac{N_{hj}\bar{y}_{hj}}{\pi_{hj}} - \bar{y}\frac{N_{hj}}{\pi_{hj}}\right)\right]^2\Big/\left(\sum_{h=1}^{L}\sum_{i=1}^{k_h}\frac{N_{hi}}{\pi_{hi}}\right)^2.$$

Hint: Korn and Graubard (1998).

Exercise 10.22. Consider a population consists of N FSUs and the ith FSU contains M_i SSUs. Select a sample of n FSUs by PPSWR (and by PPSWOR) sampling. From the ith selected FSU select m_i SSUs by SRSWOR sampling. Derive an expression in each case for optimum allocation of second stage units with the constraint that the total number of sampled SSUs remains fixed.
Hint: Rangarajan (1957), Chaudhuri and Arnab (1982).

Exercise 10.23. Consider s_1 is an SRSWOR sample of n units, $s_2 = s_{2m} \cup s_{2u}$; where s_{2m} is an SRSWOR sample of m matched units from the sample s_1 and s_{2u} is an independent SRSWOR sample of u units from the population Ω. Then show that the following estimators are unbiased estimators of the population mean

$$\bar{y}_1 = \phi \bar{y}_2^{**} + (1-\phi)\left[\bar{y}_2^{*} + b\left(\bar{y}_1 - \bar{y}_1^{*}\right)\right], \quad \text{where } 0 < \phi < 1,$$

and

$$\bar{y}_2 = \phi \frac{m_2 \bar{y}_2^{*} + u_2 \bar{y}_{22}^{**}}{m_2 + u_2} + (1-\phi)\left[\bar{y}_2^{*} + b\left(\bar{y}_1 - \bar{y}_1^{*}\right)\right],$$

where $u_2 = u - m_2$ is the number of units in a sample s_{2u} drawn from unmatched units in the population.
Hint: Pathak and Rao (1967).

Exercise 10.24. Consider a sampling on three occasions in which a constant number $n = m\lambda$ of sampled units is retained from each occasion to the next and a fresh sample of u units is selected from the units not used up to that occasion. Assume that sampling is by SRSWOR and total sample size at each occasion is n. Show that an unbiased estimator of population mean $\bar{Y}$ at the third occasion is given by

$$\bar{y}_s = a_1\left(\bar{y}_1^{*} - \bar{y}_1^{**}\right) + a_2\left(\bar{y}_2^{*} - \bar{y}_2^{**}\right) + a_3\left(\bar{y}_3^{*} - \bar{y}_3^{**}\right).$$

Find the optimum values of a_i, $i = 1, 2, 3$ by assuming that $S_i^2 = S^2$ and $\rho_{ij} = \rho$.
Hint: Singh(1968).

Exercise 10.25. Find the value of θ such that the following estimator based on two occasions is an unbiased estimator of population mean

$$\bar{y}_s = \theta\left[(N-m)\bar{y}_{2u} + m\bar{y}_2^{*}\right] + (1-\theta)\left[\bar{y}_2^{*} + b\left(\bar{y}_1 - \bar{y}^{*}\right)\right].$$

Hint: Singh (1972).

Exercise 10.26. Derive an estimator of the parameter $\tau = \sum_{i=1}^{2} \alpha_i \bar{Y}_i$ where α_i, $i = 1, 2$ are constants. Assume that s_1 is an SRSWOR sample of n units from population Ω, s_{2m} is an SRSWOR sample of m matched units and s_{2u} is an SRSWOR sample of $n-m$ unmatched units from the rest of the population $\Omega - s_1^c$. Obtain the optimum values of the sample sizes n and m such that the variance of the estimator you suggest is minimum for the fixed cost of the survey given by

$$C = C_0 + nC_1 + mC_2 + C_3(n-m).$$

Hint: Kulldoff (1963).

Exercise 10.27. Find the bias and variance of the following estimator of population mean $\bar{Y}$ defined as

$$\hat{\bar{Y}} = \left\{\alpha \frac{\bar{y}_{m-r}}{\bar{z}_{m-r}} + (1-\alpha)\left[\frac{\bar{y}_r}{\bar{z}_r} + \beta_{21}\left(\frac{\bar{x}_n}{\bar{z}_n} - \frac{\bar{x}_r}{\bar{z}_r}\right)\right]\right\}\bar{Z},$$

where $0 \le \alpha \le 1$, $\bar{y}_r(\bar{x}_r)$ denote the means of the matched sub-sample on the second (first) occasion, $\bar{z}_r$ is the mean of the auxiliary variable of the matched sub-sample, and $\beta_{21} = \sum_{i=1}^{N}(Y_i - RZ_i)(X_i - RZ_i) \Big/ \sum_{i=1}^{N}(X_i - RZ_i)^2$ with $R = \bar{Y}/\bar{Z}$.

Hint: Feng and Zou (1997).

Exercise 10.28. Let $(x_i,\ y_i\ :\ i = 1, 2, ..., n)$ be the observed values of the variables on the first and second occasion respectively, and the corresponding true values are $(\eta_i,\ \omega_i\ :\ i = 1, 2, ..., N)$. An SRS sample of n units is obtained on the first occasion. A random sub-sample of $m = n\lambda$, $0 < \lambda < 1$ of units is retained for use on the second occasion. An independent unmatched SRS sample of $u = n - m = n\mu$ units is selected. Consider the measurement errors are defined as $\varepsilon_i = x_i - \eta_i$ and $e_i = y_i - \omega_i$ such that $E_m(\varepsilon_i \mid i) = 0$, $E_m(e_i \mid i) = 0$, $E_m(\varepsilon_i^2 \mid i) = \delta_{i1}^2$ and $E_m(e_i^2 \mid i) = \delta_{i2}^2$, where E_m denote the expectation over the model. Let the single prime indicates units common to two occasions and a double prime indicates the units selected independently, so that we have the following situation:

2nd Occasion	$\bar{y}'$	$\bar{y}''$
$\bar{x}''$	$\bar{x}'$	1st Occasion

Consider the following estimator of the population mean:

$$\bar{y}_{ss} = a\,\bar{y}'' + b\,\bar{y}' + c\bar{x}' + d\bar{x}''.$$

(a) Find the conditions such that $E(\bar{y}_{ss}) = \bar{Y}$ and show that under such conditions, the minimum variance linear unbiased estimator of the population mean is given by

$$\bar{y}_{\text{mvlue}} = d(\bar{x}'' - \bar{x}') + (1-b)\bar{y}'' + b\bar{y}'.$$

(b) Find the minimum variance of $\bar{y}_{\text{mvlue}}$ for the optimum values of the parameters involved in it.

Hint: Sud and Srivastava (2000).

(c) Let σ_x^2 and σ_y^2 be the population variances on the first and second occasions respectively. Find the bias and variance of the following estimator of σ_y^2 given by

$$\hat{\sigma}_y^{*2} = \frac{a}{u}\sum_{k=1}^{u} x_k^2 + \frac{b}{u(u-1)}\sum_{k}^{u}\sum_{j\ne k}^{u} x_k x_j + \frac{c}{m}\sum_{k=1}^{m} x_k^2 + \frac{d}{m(m-1)}\sum_{k}^{m}\sum_{j\ne(k)}^{m} x_k x_j$$

$$+ \frac{e}{m}\sum_{k=1}^{m} y_k^2 + \frac{f}{m(m-1)}\sum_{k}^{m}\sum_{j\ne k}^{m} y_k y_j + \frac{g}{u}\sum_{k=1}^{u} y_k^2 + \frac{h}{u(u-1)}\sum_{k}^{u}\sum_{j\ne(k)}^{u} y_k y_j$$

where a, b, c, d, e, f, g and h are suitably chosen constants.

Hint: Sud, Srivastava, and Sharma (2001a, 2001b).

Exercise 10.29. Consider an SRS s of n units selected on the first occasion from a universe Ω of N units and measurements are taken on two variables y and x in each of the two occasions in bivariate normal population. While selecting the second sample, we assume that $m = pn$, $(0 < p <)$ of the units of the selected sample on the first occasion are retained for the second occasion (matched sample) and the remaining $u = n - m = nq$, $(q = 1 - p)$ units are replaced by a new selection from the universe of $(N - m)$ units left after omitting m units. Let x_i (y_i) be the x(y) variables on the ith occasion, $i = 1, 2$. Consider the problem of estimation of the ratio of two population means. Let $R_i = \overline{Y}_i / \overline{X}_i$, $i = 1, 2$ be the population ratio on the ith occasion. Let $\hat{R}_i = \overline{y}_i / \overline{x}_i$, $i = 1, 2$ be the estimator of the ratio on the ith occasion. Further let $\hat{R}_{im}$ and $\hat{R}_{iu}$ be the estimates of ratio on the ith $(i = 1,2)$ occasion based on matched and unmatched units, respectively. Consider the estimator $\hat{R}_2^*$ of population ratio on the second occasion R_2, given by

$$\hat{R}_2^* = a\,\hat{R}_{1u} + b\hat{R}_{1m} + c\hat{R}_{2m} + d\hat{R}_{2u}\,.$$

(a) Show that the conditions on a, b, c and d for the estimator $\hat{R}_2^*$ to be unbiased are $a + b = 0$ and $c + d = 1$, and the above estimator reduces to

$$\hat{R}_2^* = a\left(\hat{R}_{1u} - \hat{R}_{1m}\right) + c\hat{R}_{2m} + (1 - c)\hat{R}_{2u}\,.$$

(b) The variance of $\hat{R}_2^*$ is given by

$$V\left(\hat{R}_2^*\right) = a^2\left(\frac{1}{q} + \frac{1}{p}\right)\frac{A}{n\overline{X}_1^2} + \frac{c^2 B}{pn\overline{X}_2^2} + \frac{(1-c)^2 B}{qn\overline{X}_2^2} - 2ac\text{Cov}\left(\hat{R}_{1m}, \hat{R}_{2m}\right).$$

where

$$A = S_{y1}^2 + R_1^2 S_{x1}^2 - 2R_1\text{Cov}(y_1, x_1),\ B = S_{y2}^2 + R_2^2 S_{x2}^2 - 2R_2\text{Cov}(y_2, x_2)$$

and

$$\text{Cov}\left(\hat{R}_{1m}, \hat{R}_{2m}\right) = \frac{1}{np\overline{X}_1\overline{X}_2}\left[\text{Cov}(y_1, y_2) - R_1\text{Cov}(y_2, x_1) - R_2\text{Cov}(y_1, x_2) + R_1R_2\text{Cov}(x_1, x_2)\right].$$

(c) Find the optimum values of a and c, which minimize the variance $V\left(\hat{R}_2^*\right)$ and find the minimum variance.

Hint: Artes and Garcia (2001e)

Exercise 10.30. Consider the size N of a population remains same over two occasions, but the values of the units changes over occasions. Let an SRSWOR sample of n_1 units is selected on the first occasion. Out of this sample let n_2^* units are retained on the second occasion while a fresh sample of size n_2^{**} is drawn on the second occasion from the remaining $(N - n_1)$ units of the population so that the total sample size at the second occasion becomes $n_2 = n_2^* + n_2^{**}$. Assume that the information on an auxiliary variable X, which is positively correlated to Y is available at the second occasion.

Let

$\bar{Y}_i$: the population mean of the study variable Y on the ith occasion ($i = 1,2$);

S_{iY}^2 : the population mean square of Y on the ith occasion;

$\bar{y}_1$: the sample mean based on n_1 units drawn on the first occasion;

$\bar{y}_2^*$: the sample mean based on n_2^* units observed on the second occasion and common with the first occasion;

$\bar{y}_2^{**}$: the sample mean based on n_2^{**} units drawn afresh on the second occasion;

$\bar{y}_1^*$: the sample mean based on n_2^* units common to both occasions and observed on the first occasion;

$\bar{X}_2$: the population mean of the auxiliary variable X on the second occasion;

S_{2X}^2 : the population mean square of X on the second occasion;

$\bar{x}_2^*$: the sample mean of X based on n_2^* units common to both occasions and observed on the second occasion; and

$\bar{x}_2^{**}$: the sample mean of X based on n_2^{**} units drawn afresh on the second occasion.

Study the asymptotic properties of the estimator of the population mean on the second occasion $\bar{Y}_2$ as

$$\hat{T}_2 = \phi\ \hat{\bar{Y}}_2^{**} + (1-\phi)\,\hat{\bar{Y}}_2^{*},$$

where ϕ is a constant to be determined

$$\hat{\bar{Y}}_2^{**} = \bar{y}_2^{**} + \beta_{2xy}\left(\bar{X}_2 - \bar{x}_2^{**}\right),$$

$$\hat{\bar{Y}}_2^{*} = \bar{y}_2^{*} + \beta_{1xy}\left(\bar{y}_1 - \bar{y}_1^{*}\right) + \beta_{2xy}\left(\bar{X}_2 - \bar{x}_2^{*}\right),$$

β_{2xy} is a regression coefficient of y on x at the second occasion,

and

β_{21} stands for the regression coefficient of the variable y of the second occasion on the same variate of the first occasion, and both β_{21} and β_{2xy} are known.

Hint: Singh and Singh (2001).

Exercise 10.31. A sample s_1 of size n is selected from a population Ω by RHC scheme using $p_i = \frac{x_i}{X}$, the second sample $s_2 = s_{2m} \cup s_{2u}$, s_{2m} being an SRSWOR sample of m units taken from s_1, and s_{2u} is an independent sample drawn from the population Ω by using RHC scheme using the same p_i measures.

(I) Study the bias and variance of the following estimator:

$$\hat{Y}_1 = \phi\ \hat{Y}_{2m} + (1-\phi)\hat{Y}_{2u},$$

where

$$\hat{Y}_{2m} = \sum_{i\in s_{2m}} \frac{(y_{2i} - y_{1i})\tau_{1i}}{(m/n)p_i} + \sum_{i\in s_1} \frac{y_{1i}\tau_{1i}}{p_i}$$

and

$$\hat{Y}_{2u} = \sum_{i\in s_2} \frac{y_{2i}\tau_{2i}}{p_i}$$

with τ_{1i} and τ_{2i} are the totals of p_i values included in the ith group while selecting s_1 and s_{2u} using RHC scheme.

Hint: Ghangurde and Rao (1969)

(II) Select s_1 and s_{2u} is the same way as above, but select s_{2m} from s_1 following RHC method using τ_i in place of p_i. Let Γ_i denote the total of τ_i over RHC groups while selecting s_{2m} sample. Under this scheme find the bias and variance of the following estimators:

$$\hat{Y}_{c1} = \phi_c\, \hat{Y}_{2m}^c + (1-\phi_c)\hat{Y}_{2u}^c$$

where

$$\hat{Y}_{2m}^c = \sum_{i\in s_{2m}} \frac{(y_{2i} - y_{1i})\Gamma_{1i}}{(m/n)\tau_i} + \sum_{i\in s_1} \frac{y_{1i}\tau_{1i}}{p_i}$$

and

$$\hat{Y}_{2u}^c = \sum_{i\in s_2} \frac{y_{2i}\tau_{2i}}{p_i}.$$

Hint: Chotai (1974).

PRACTICAL PROBLEMS

Practical 10.1. Out of 10 clusters or continents listed in population 5 of the Appendix select 4 continents with SRSWOR sampling as first stage units (FSU). From each one of the selected continents select two countries by SRSWOR sampling and record the production of the tobacco crop. Estimate the total production in the world. Also estimate the standard error of the estimator used for estimating the total production in the world.

Practical 10.2. Divide the United States of America into four independent strata as Northeast, Midwest, South and West. Suppose the first stratum Northeast consists of two clusters New England and Mid Atlantic, the second stratum Midwest consists of two clusters East North Central and West North Central, the third stratum South consists of three clusters South Atlantic, East South Central, and West South Central, and the fourth stratum West consists of two clusters Mountains and Pacific as given in population 7 in the Appendix. From each stratum (or region) select two clusters (or divisions) by SRSWOR sampling and within each selected cluster (or division) select two units (or states) and collect the information on the projected population counts during 2000.

(a) Suggest an estimator for estimating population total in the US using stratified multi-stage design.
(b) Estimate the variance of the estimator of population total by using the technique of the Jackknife.
(c) Find the 95% confidence interval for total projected population during 2000 in the USA.

Practical 10.3. The data in population 9 of the Appendix relates to the number of immigrants coming to 51 states in the United States during 1994--1996. Select a sample (s_1) of 20 cities by PPSWR method using number of immigrants in 1994 (z) as measure of size. From the selected sample s_1, select a sub-sample s_m of size $m = 12$ from s_1 by SRSWOR method assuming all elements of s_1 are distinct. Finally select an independent sample s_u of size $u = (20-12) = 8$ with PPSWR using x as size measure. Estimate the total number of immigrants in 1996 using the composite estimator. Compare your estimate with the estimate obtained in example 10.9.1 and comment on it.

Practical 10.4. The data in population 9 of the Appendix relates to the number of immigrants coming to 51 states in the United States during 1994--1996. Consider a sampling on three occasions in which a constant number m such that $n = m\lambda = 2\times 5$ of sampled units is retained from each occasion to the next and a fresh sample of $u = 5$ units is selected from the units not used up to that occasion. Suppose that sampling is by SRSWOR and total sample size at each occasion is n. Estimate the population mean $\bar{Y}$ at the third occasion by using the estimator

$$\bar{y}_s = a_1\left(\bar{y}_1^{*} - \bar{y}_1^{**}\right) + a_2\left(\bar{y}_2^{*} - \bar{y}_2^{**}\right) + a_3\left(\bar{y}_3^{*} - \bar{y}_3^{**}\right).$$

where the sample means $\bar{y}_j^{*}$, $\bar{y}_j^{**}$, and real constants a_j for $j = 1,2,3$, have their usual meanings.
Hint: Singh(1968).

11. RANDOMIZED RESPONSE SAMPLING: TOOLS FOR SOCIAL SURVEYS

11.0 INTRODUCTION

The randomized response technique (RRT) is useful for reducing response error problems when potentially sensitive questions such as the illegal use of drugs, sexual practice, illegal earning, or incidence of acts of domestic violence are included in surveys of human populations. Direct questioning of respondents about sensitive issues often results in either refusal or falsification of the answers. Social stigma and fear of reprisals sometimes result in untruthful, exaggerated, or misleading responses by respondents when approached with conventional survey methods. Warner (1965) was the first to suggest an ingenious method of counteracting fears in response to sensitive questions.

11.1 PIONEER MODEL

It is widely understood that direct questions on sensitive issues often result in deliberately incorrect answers and/or refusal to respond. The real essence of Warner's (1965) randomized response technique is to reduce the response error of potentially 'sensitive questions'. Warner, while estimating π, the proportion of population possessing a stigmatized character (say) A, used two randomized questions inquiring about the respondents status in relation to the sensitive character only.

(a) Are you a member of group A?
(b) Are you not a member of group A?

Warner (1965) used a spinner (say) whose background consists of two different shapes, *viz.*, circular and heart representing proportion P and $(1-P)$ of the total area as shown in the Figure 11.1.1.

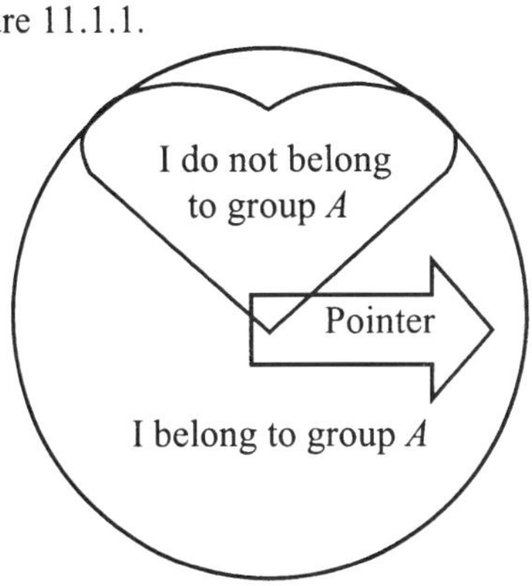

Fig 11.1.1 Spinner as a device.

Every interviewee selected in the sample is requested to rotate the spinner unobserved by the interviewer. If the pointer stops in the circular region and the interviewee possesses the character A then he/she is requested to report 'Yes', and if the pointer stops in the heart region he/she is requested to report 'No'. Similarly, if the spinner stops in the heart region and the interviewee does not possess the character A, again he/she is requested to report, 'Yes' otherwise 'No'. Note that the response 'Yes' comes from two possibilities and also the response 'No' comes from two possibilities, so the respondent's privacy is maintained. One can see from the following table:

<table>
<tr><td colspan="2" rowspan="2">RESPONSE FROM WARNER'S MODEL</td><td colspan="2">Statement pointed by the spinner</td></tr>
<tr><td>Are you member of group A?</td><td>Are you not a member of group A?</td></tr>
<tr><td rowspan="2">Status of the person selected in the sample</td><td>Belongs to group A</td><td>Yes</td><td>No</td></tr>
<tr><td>Do not belong to group A</td><td>No</td><td>Yes</td></tr>
</table>

Evidently the probability of 'Yes' answer is given by

$$\theta_w = P\pi + (1-P)(1-\pi). \tag{11.1.1}$$

Let n_1 be the number of 'Yes' responses out of an SRSWR sample of n respondents. Let y_i be a discrete random variable such that

$$y_i = \begin{cases} 1 & \text{if the } i^{th} \text{ selected person reported 'Yes',} \\ 0 & \text{otherwise.} \end{cases}$$

Obviously the random variable y_i follows Binomial distribution with parameters n and θ_w, that is, $y_i \sim \text{Binomial}(n,\ \theta_w)$. Note that n_1 responses are 'Yes' and $(n-n_1)$ responses are 'No', thus the maximum likelihood function is given by

$$L(\theta_w) = {}^nC_{n_1}(\theta_w)^{n_1}(1-\theta_w)^{(n-n_1)}. \tag{11.1.2}$$

Taking logarithms on both sides of (11.1.2) we have

$$\text{Log}\{L(\theta_w)\} = \text{Log}\{{}^nC_{n_1}\} + n_1\text{Log}(\theta_w) + (n-n_1)\text{Log}\{(1-\theta_w)\}.$$

On setting

$$\frac{\partial\ \text{Log}\{L(\theta_w)\}}{\partial\theta_w} = 0.$$

We get a maximum likelihood estimate of θ_w as

$$\hat{\theta}_w = \frac{n_1}{n}. \tag{11.1.3}$$

Thus we have the following theorem:

Theorem 11.1.1. An unbiased estimator of population proportion π is given by

$$\hat{\pi}_W = \frac{\hat{\theta}_W - (1-P)}{2P-1}, \qquad P \neq 0.5 \tag{11.1.4}$$

where $\hat{\theta}_w$ is the observed proportion of 'Yes' answers in the sample of n units drawn by SRSWR sampling.

Proof. Note that $\hat{\theta}_w \sim B(n,\ \theta_w)$, taking expected values on both sides of (11.1.4) we have

$$E(\hat{\pi}_W) = \frac{E(\hat{\theta}_W) - (1-P)}{2P-1} = \frac{\theta_W - (1-P)}{2P-1} = \pi .$$

Hence the theorem.

Theorem 11.1.2. The variance of the estimator $\hat{\pi}_w$ is given by

$$V(\hat{\pi}_W) = \frac{\pi(1-\pi)}{n} + \frac{P(1-P)}{n(2P-1)^2} . \tag{11.1.5}$$

Proof. Note that $\hat{\theta}_w \cap B(n, \theta_w)$ we have

$$V(\hat{\pi}_W) = V\left[\frac{\hat{\theta}_W - (1-P)}{2P-1}\right] = \frac{V(\hat{\theta}_W)}{(2P-1)^2} = \frac{\theta_w(1-\theta_W)}{n(2P-1)^2}$$

which on simplification reduces to (11.1.5).Hence the theorem.

Theorem 11.1.3. An unbiased estimator of $V(\hat{\pi}_W)$ is given by

$$\hat{v}(\hat{\pi}_W) = \frac{\hat{\theta}_W(1-\hat{\theta}_W)}{n-1} . \tag{11.1.6}$$

Proof. Obvious by taking expected value on both sides of (11.1.6).

Example 11.1.1. Michael believes that 'churches' and 'academic institutes' were made for honest people, but after watching the movie 'Wonder Boys—2000', he felt that due to the increase of unscrupulous people in the world that there may be academic research cheaters (or academic thieves) in the institutes. He selected an SRSWR sample of 20,000 researchers across the world from different institutes by using a randomisation device, like the spinner, producing 70% of the statements, 'Have you ever stolen your colleagues research papers or books?' along with 30% of the statements, 'Have you never stolen your colleagues research papers or books?' Out of 20,000 selected researchers 6060 reported 'Yes' through the above randomization device. Estimate the proportion of 'academic thieves' in the world, and also construct a 95% confidence interval estimate.

Solution. Here $P = 0.7$, $n = 20{,}000$ and $n_1 = 6060$.

Thus the observed proportion of 'academic thieves' is given by

$$\hat{\theta}_W = \frac{6060}{20000} = 0.303 .$$

Using Warner's (1965) estimator, an estimate of true proportion of academic thieves is given by

$$\hat{\pi}_w = \frac{\hat{\theta}_w - (1-P)}{2P-1} = \frac{0.303-(1-0.7)}{1.4-1} = \frac{0.003}{0.4} = 0.0075 .$$

An unbiased estimate of $V(\hat{\pi}_w)$ is given by

$$\hat{v}(\hat{\pi}_w) = \frac{\hat{\theta}_w(1-\hat{\theta}_w)}{n-1} = \frac{0.303(1-0.303)}{20{,}000-1} = \frac{0.211191}{19999} = 0.0000156 .$$

Thus a 95% confidence interval estimate of the true proportion of 'academic thieves' is given by

$$\hat{\pi}_w \pm Z_{\alpha/2}\sqrt{\hat{v}(\hat{\pi}_w)}\ , \quad \text{or} \quad 0.0075 \pm 1.96\sqrt{0.0000156}\ , \quad \text{or} \quad [0.00113,\ 0.01386]\ .$$

Thus Michael claims with 95% confidence that in these days there are 0.113% to 1.386% of the researchers in this world are 'academic thieves', but he is pleased that this proportion is very negligible.

11.2 FRANKLIN'S MODEL

Let π represent the proportion of individuals in a population who belong to group A (*e.g.*, the proportion of women who have had an abortion). Franklin (1989a, 1989b) considers that a with replacement simple random sample of $n > 1$ individuals is chosen. This is usually the case and eliminates concerns about finite population correction factors. A total of $k > 1$ trials per respondent are conducted. For respondent i on trial j, random values are drawn from the densities g_{ij} and h_{ij} (where we assume independence of all densities). The respondent but not the interviewer seeks both values and is asked to report the value from g_{ij} if he/she belongs to group A and the value from h_{ij} otherwise.

Response from Franklin's model		
Status of the respondent		
Belongs to group A		Do not belongs to group A
Report value from g_{ij}		Report value from h_{ij}

The interviewer knows the exact form of g_{ij} and h_{ij} but sees only the value reported by the respondent. This value we denote by Z_{ij}. The interviewer does not know for certain from which of the two densities it comes. From the total of kn observations of Z_{ij} $(i = 1, 2, ..., n;\ j = 1, 2, ..., k)$ inference can be made about π. The conditional density of a k-tuple $(Z_{i1}, Z_{i2}, ..., Z_{ik})$ representing a random observation of the ith respondent given π is

$$\pi \prod g_{ij}(Z_{ij}) + (1-\pi)\prod h_{ij}(Z_{ij}) .$$

Thus $(Z_{i1}, Z_{i2}, ..., Z_{ik})$ can be represented as the mixture of random variables

$$P(X_{i1},\ X_{i2}, ..., X_{ik}) + (1-P)(Y_{i1}, Y_{i2}, ..., Y_{ik}) \tag{11.2.1}$$

where P is Bernoulli (p) and $X_{ij} \sim g_{ij}$ and $Y_{ij} \sim h_{ij}$. The model can be specialised by having $g_{ij} = g_i$ and $h_{ij} = h_i$ for all i. Thus all respondents are observing the same distributions and allows us to perceive the k-tuple response $(Z_{i1}, Z_{i2}, ..., Z_{ik})$ as independent, identically distributed (i.i.d.) random variables with form

$$P(X_1,\ X_2, ..., X_k) + (1-P)(Y_1, Y_2, ..., Y_k) \tag{11.2.2}$$

from the density

$$\pi \prod_{j=1}^{K} g_j(Z_j) + (1-\pi) \prod_{j=1}^{K} h_j(Z_j). \tag{11.2.3}$$

Although any density may be used for g_i or h_i, the Franklin (1989a, 1989b) model consider only both as normal densities with known means μ_{1j} and μ_{2j}, and known standard deviations σ_{1j} and σ_{2j} respectively. The choice of g_i and h_i as Bernoulli (p) with $h_i(Z) = 1 - g_i(Z)$ and $k = 1$ reduces this model to Warner's (1965) original model and the densities h_i and g_i are dependent. Also, if g_i denotes the distribution generated by the first deck of cards with known proportion θ_1 of red cards and h_i denotes the distribution generated by the second deck of cards with known proportion θ_2 $(\neq \theta_1)$ of red cards, then Franklin's model reduces to the model suggested by Kuk (1990). It is a well known result that while the estimators derived from the method of moments (MM) are usually not preferred over maximum likelihood (ML) estimates, but under certain circumstances they might be. Franklin has shown that there is one such possible instance where MM estimators and their associated variances can be found analytically, while those of the ML estimators cannot be obtained. This allows formation of confidence intervals and test of hypothesis to be conducted for these MM estimators. Two obvious MM estimators can be derived, one by concentrating on the row averages of Z_{ij} and other on its column averages.

Let

$$\bar{Z}_{i\bullet} = \frac{1}{k}[Z_{i1} + Z_{i2} + ... + Z_{ik}] \text{ and } \bar{Z}_{\bullet j} = \frac{1}{n}[Z_{1j} + Z_{2j} + ... + Z_{nj}] \tag{11.2.4}$$

for $i = 1, 2, ..., n$ and $j = 1, 2, ..., k$, represent the row and column averages respectively. Concentrating on the row averages gives

$$E(\bar{Z}_{i\bullet}) = \frac{1}{k} \sum_{j=1}^{k} E(Z_{ij}) = \frac{1}{k} \sum_{j=1}^{k} E(Z_j) \tag{11.2.5}$$

since the k-tuple responses are independently and identically distributed (i.i.d.). Now

$$E(Z_j) = \pi\mu_{1j} + (1-\pi)\ \mu_{2j} \quad \text{for } j = 1, 2, ..., k. \tag{11.2.6}$$

Thus

$$E(\overline{Z}_{i\bullet}) = \frac{1}{k}\left[\pi \sum_{j=1}^{k} \mu_{1j} + (1-\pi)\sum_{j=1}^{k} \mu_{2j}\right] = \frac{1}{k}[\pi m_1 + (1-\pi)m_2], \tag{11.2.7}$$

where $m_1 = \sum_{j=1}^{k} \mu_{1j}$ and $m_2 = \sum_{j=1}^{k} \mu_{2j}$.

From these the standard MM approach gives

$$\hat{\pi}_1 = \frac{k\overline{Z} - m_2}{m_1 - m_2} = \frac{\sum_{j=1}^{k} \overline{Z}_{\bullet j} - m_2}{m_1 - m_2}. \tag{11.2.8}$$

Note that the k-tuple responses are i.i.d, it can be easily shown by (11.2.6) that

$$E(\hat{\pi}_1) = \pi$$

with the variance of $\hat{\pi}_1$ given by

$$V(\hat{\pi}_1) = k^2 V(\overline{Z}) / (m_1 - m_2)^2 = V\left(\sum_{j=1}^{k} Z_j\right) \Big/ \left\{n(m_1 - m_2)^2\right\}. \tag{11.2.9}$$

Now the observations $Z_1, Z_2, ..., Z_k$ from a certain single respondent are not independent and need not even be identically distributed. In fact, for example, the joint density of $(Z_i,\ Z_j)$ of two trials from a single respondent is given by

$$f_{ij}(Z_i,\ Z_j) = \pi g_i(Z_i) g_j(Z_j) + (1-\pi) h_i(Z_i) h_j(Z_j). \tag{11.2.10}$$

It can be easily seen that

$$V(Z_i) = \pi(1-\pi)(\mu_{1i} - \mu_{2i})^2 + \pi\sigma_{1i}^2 + (1-\pi)\sigma_{2i}^2, \tag{11.2.11}$$

and

$$\text{Cov}(Z_i, Z_j) = \pi(1-\pi)(\mu_{1i} - \mu_{2i})(\mu_{1j} - \mu_{2j}). \tag{11.2.12}$$

The correlation between Z_i and Z_j is given by

$$\frac{\pi(1-\pi)(\mu_{1i} - \mu_{2i})(\mu_{1j} - \mu_{2j})}{\sqrt{\pi(1-\pi)(\mu_{1i} - \mu_{2i})^2 + \pi\sigma_{1i}^2 + (1-\pi)\sigma_{2i}^2}\sqrt{\pi(1-\pi)(\mu_{1j} - \mu_{2j})^2 + \pi\sigma_{1j}^2 + (1-\pi)\sigma_{2j}^2}}. \tag{11.2.13}$$

Thus as σ_{1i}, σ_{1j}, σ_{2i}, σ_{2j} all $\to 0$, then $\text{Cov}(Z_i, Z_j) \to 1$. Also as $(\mu_{1i} - \mu_{2i})$ and $(\mu_{1j} - \mu_{2j}) \to \infty$, then $\text{Cov}(Z_i,\ Z_j) \to 1$. This is apparent since knowledge of Z_i for a particular respondent under these circumstances gives perfect predictability for Z_i given by the same respondent.

Equations (11.2.9), (11.2.11), and (11.2.12) give the variance of $\hat{\pi}_1$ as

$$V(\hat{\pi}_1) = \frac{1}{n(m_1 - m_2)^2}\left[\sum_{j=1}^{k} V(Z_j) + 2\sum_{i<j} \text{Cov}(Z_i, Z_j)\right]$$

$$= \frac{\pi(1-\pi)}{n} + \frac{\pi \sum_{j=1}^{k} \sigma_{1j}^2 + (1-\pi)\sum_{j=1}^{k} \sigma_{2j}^2}{n(m_1 - m_2)^2}. \tag{11.2.14}$$

Thus the variance of $\hat{\pi}_1$ can be thought of as the variance from the ordinary sampling plus an additional term due to the randomizing effect.

Now concentrating upon the column average gives

$$E(\overline{Z}_{\bullet j}) = \frac{1}{n}\sum_{i=1}^{n} E(Z_{ij}) = \frac{1}{n}\sum_{j=1}^{n} E(Z_j) = E(Z_j) = \pi\mu_{1j} + (1-\pi)\mu_{2j} \tag{11.2.15}$$

for $j = 1, 2, \ldots k$.

Hence a second MM estimator can be formed from the average of the jth column of observations by setting $\overline{Z}_{\bullet j} = E(\overline{Z}_{\bullet j})$ and solving for π to have

$$\hat{\pi}_{\bullet j} = \frac{\overline{Z}_{\bullet j} - \mu_{2j}}{\mu_{1j} - \mu_{2j}}. \tag{11.2.16}$$

Thus a total of k such estimators (not independent of one another) for $j = 1, 2, \ldots, k$ can be formed. Each of these estimators is an unbiased estimator of π . The variance of $\hat{\pi}_{\bullet j}$ is given, since observations are i.i.d. by

$$V(\hat{\pi}_{\bullet j}) = \frac{V(\overline{Z}_{\bullet j})}{(\mu_{1j} - \mu_{2j})^2} = \frac{V(Z_j)}{n(\mu_{1j} - \mu_{2j})^2}. \tag{11.2.17}$$

Using (11.2.11) and (11.2.17) we have

$$V(\hat{\pi}_{\bullet j}) = \frac{\pi(1-\pi)}{n} + \frac{\pi\sigma_{1j}^2 + (1-\pi)\sigma_{2j}^2}{n(\mu_{1j} - \mu_{2j})^2}. \tag{11.2.18}$$

Furthermore

$$\text{Cov}(\hat{\pi}_{\bullet i}, \hat{\pi}_{\bullet j}) = \frac{\text{Cov}(\overline{Z}_{\bullet i}, \overline{Z}_{\bullet j})}{(\mu_{1i} - \mu_{2i})(\mu_{1j} - \mu_{2j})} = \frac{\sum_{m=1}^{n} \text{Cov}(Z_{m_i}, Z_{m_j})}{n^2(\mu_{1i} - \mu_{2i})(\mu_{1j} - \mu_{2j})}. \tag{11.2.19}$$

Note that Z_{pi} and Z_{mi} are independent if $p \neq m$. But since the responses are i.i.d. we have from (11.2.12)

$$\text{Cov}(\hat{\pi}_{\bullet i}, \hat{\pi}_{\bullet j}) = \frac{n\text{Cov}(Z_i, Z_j)}{n(\mu_{1i} - \mu_{2i})(\mu_{1j} - \mu_{2j})} = \frac{\pi(1-\pi)}{n}. \tag{11.2.20}$$

Any weighted average of k estimators $\hat{\pi}_1, \hat{\pi}_2, \ldots, \hat{\pi}_k$ will also be an unbiased estimator for π. An estimator of this type, which (11.2.16) suggests, is one where the weights are proportional to the difference in the means of g_i and h_i. Specifically, let

$$w_j = \frac{\left|\mu_{1j} - \mu_{2j}\right|}{\sum_{j=1}^{k}\left|\mu_{1j} - \mu_{2j}\right|} = \frac{D_j}{D} \tag{11.2.21}$$

where $D_j = \left|\mu_{1j} - \mu_{2j}\right|$ and $D = \sum_{j=1}^{k} D_j$ for $j = 1, 2, ..., k$.

Also let us take second method of moment estimator $\hat{\pi}_2$ given by

$$\hat{\pi}_2 = \sum_{j=1}^{k} w_j \hat{\pi}_{\bullet j} . \tag{11.2.22}$$

Thus by (11.2.18), (11.2.19) and (11.2.21), we have

$$V(\hat{\pi}_2) = \sum_{j=1}^{k} w_j^2 V(\hat{\pi}_{\bullet j}) + 2\sum_{i<j} w_i w_j \text{Cov}(\hat{\pi}_{\bullet i}, \hat{\pi}_{\bullet j})$$

$$= \frac{\pi(1-\pi)}{n} + \frac{\pi\sum_{j=1}^{k}\sigma_{1j}^2 + (1-\pi)\sum_{j=1}^{k}\sigma_{2j}^2}{nD^2} . \tag{11.2.23}$$

Thus this second MM estimator has a variance similar to the first MM estimator. A comparison of the variance of the two MM estimators $\hat{\pi}_1$ and $\hat{\pi}_2$ from (11.2.14) and (11.2.23) reveals that they will be identical if and only if all the terms $(\mu_{1j} - \mu_{2j})$ have the same sign for $j = 1, 2, ..., k$ then

$$\frac{1}{D^2} = \left[\sum_{j=1}^{k}\left|\mu_{1j} - \mu_{2j}\right|\right]^{-2} = \left[\sum_{j=1}^{k}(\mu_{1j} - \mu_{2j})\right]^{-12} = (m_1 - m_2)^{-2} . \tag{11.2.24}$$

In any other case the denominator of the right hand side of (11.2.14) will contain both positive and negative terms and will result in $V(\hat{\pi}_1) \le V(\hat{\pi}_2)$. In fact, when all $(\mu_{1j} - \mu_{2j})$ for $j = 1, 2, ..., k$ have the same sign then we have $\hat{\pi}_1 = \hat{\pi}_2$. Thus to conclude the two MM of estimators will be identical if all terms $(\mu_{1j} - \mu_{2j})$ for all j have the same sign. In any other case, $\hat{\pi}_2$ will have a smaller variance. This aspect of keeping the sign of all $(\mu_{1j} - \mu_{2j})$ the same is the particularly important design consideration in using the $\hat{\pi}_1$ estimator. It is seen that $(m_1 - m_2)$ could not be equal to or near zero. By violating this condition the variance of $\hat{\pi}_1$ could be extremely large. On the other hand, for the estimator $\hat{\pi}_2$ it is seen that the variance can be decreased by making at least one absolute difference large. Singh and Singh (1992a, 1993d) have suggested some improvements over Franklin's model. Arnab (1996) developed a unified setup for Singh and Singh's models. More details about randomized response sampling can be had from Chaudhuri and Mukherjee (1988), Sheers (1992), and Tracy and Mangat (1996a).

11.3 UNRELATED QUESTION MODEL AND RELATED ISSUES

In Warner's model both questions refer to the same sensitive character A or its complement A^c. Greenberg, Abul-Ela, Simmons, and Horvitz (1969) felt that to protect the privacy of the respondents, it is desirable that the two questions be unrelated and suggested an unrelated question model. In Greenberg *et.al.*'s pioneer unrelated question model the data gathering randomization device consists of two questions or statements:

(i) Are you member of group *A?*
(ii) Are you member of group *Y?*

where character Y or complement of it are innocuous and unrelated to A. For example, in estimating the proportion of persons having extra marital relations in a certain community the two questions may be:

(a) Are you having extra marital relations?
(b) Were you born in the month of March?

Clearly the second question has nothing to do with extra marital relations. An SRSWR sample of n is drawn from the population. Greenberg, Abul-Ela, Simmons, and Horvitz (1969), in their theoretical development, dealt with two situations of π_y (proportion of unrelated character, Y, say), being known and unknown.

11.3.1 WHEN PROPORTION OF UNRELATED CHARACTER IS KNOWN

Each interviewee selected in the sample chooses questions (i) and (ii) with probabilities P and $(1-P)$ respectively. Assuming that the proportion π_y of the unrelated question model is known, the probability θ of 'Yes' answer is given by

$$\theta = P\pi + (1-P)\pi_y . \tag{11.3.1.1}$$

Let n_1 be the number of observed 'Yes' answers in a sample of n units, so that

$$\hat{\theta} = \frac{n_1}{n}.$$

Then we have the following theorems:

Theorem 11.3.1. An unbiased estimator of π is given by

$$\hat{\pi}_g = \frac{\hat{\theta} - (1-P)\pi_y}{P} . \tag{11.3.1.2}$$

Proof. It follows because $\hat{\theta} \sim B(n, \theta)$.
Hence the theorem.

Theorem 11.3.2. The variance of the estimator $\hat{\pi}_g$ is given by

$$V(\hat{\pi}_g) = \frac{\theta(1-\theta)}{nP^2}. \tag{11.3.1.3}$$

Proof. Note that $\hat{\theta} \sim B(n, \theta)$, therefore

$$V(\hat{\theta}) = \frac{\theta(1-\theta)}{n}.$$

Hence the theorem.

11.3.2 WHEN PROPORTION OF UNRELATED CHARACTER IS UNKNOWN

Here π_y the proportion of neutral character (say) Y in the population is unknown. In such a situation, Greenberg, Abul-Ela, Simmons, and Horvitz (1969) suggested to take two independent SRSWR samples of sizes n_1 and n_2 from the population. In the ith sample, P_i and $(1-P_i)$ denote the probabilities of selecting the statements regarding the possessing of the sensitive characteristic A and non-sensitive characteristic Y in the randomized response device R_i used for the respondents in the ith sample, $i=1,2$, so that the probability of 'Yes' answer in the ith sample is given by

$$\theta_i = P_i\pi + (1-P_i)\pi_y, \quad i=1,2. \tag{11.3.2.1}$$

Then we have the following theorems:

Theorem 11.3.2.1. An unbiased estimator of π is given by

$$\hat{\pi}_G = \frac{(1-P_2)\hat{\theta}_1 - (1-P_1)\hat{\theta}_2}{P_1 - P_2}, \tag{11.3.2.2}$$

where $\hat{\theta}_1$ and $\hat{\theta}_2$ are the observed proportion of 'Yes' answers in the first and second sample, respectively.

Proof. Solve the two equations given by (11.3.2.1) for π and use the method of moments. The unbiasedness is clear from the fact that $\hat{\theta}_i \sim B(n_i, \theta_i)$.

Theorem 11.3.2.2. The variance of the estimator $\hat{\pi}_G$ is given by

$$V(\hat{\pi}_G) = \frac{1}{(P_1 - P_2)^2}\left[\frac{(1-P_2)^2\theta_1(1-\theta_1)}{n_1} + \frac{(1-P_1)^2\theta_2(1-\theta_2)}{n_2}\right], \tag{11.3.2.3}$$

where $\theta_i = P_i\pi + (1-P_i)\pi_y$, $i=1, 2$, denotes the probability of 'Yes' answer in the first and second sample, respectively.

Proof. It follows from $\hat{\theta}_i \sim B(n_i, \theta_i)$ so that

$$V(\hat{\theta}_i) = \frac{\theta_i(1-\theta_i)}{n_i}, \quad i=1, 2.$$

Hence the theorem.

For the optimum choice of n_1 and n_2

(obtained by setting $\frac{\partial V(\hat{\pi}_G)}{\partial n_i} = 0$)

the minimum variance of the estimator $\hat{\pi}_G$ is

$$\text{Min.}V(\hat{\pi}_G) = \frac{\left[(1-P_2)\sqrt{\theta_1(1-\theta_1)} + (1-P_1)\sqrt{\theta_2(1-\theta_2)}\right]^2}{n(P_1-P_2)^2} \tag{11.3.2.4}$$

where $n = n_1 + n_2$ and n_1 and n_2 are Greenberg, Abul-Ela, Simmons, and Horvitz (1969) optimum sample sizes. They suggested that one of the optimal choices of P_i, $i = 1, 2$, should be close to one and the other close to zero. The value π_y should be chosen close to zero or one according as $\pi < 0.5$ or $\pi > 0.5$ respectively. Moors (1971) claimed that if one of the value of P_i (say, P_2) is chosen to be zero then the model of Greenberg, Abul-Ela, Simmons, and Horvitz (1969) becomes optimal so far as the choice of P_i is concerned. The optimality choice of $P_2 = 0$ (*i.e.*, Moors' model) discloses the privacy of sensitive character whenever any respondent appears in both the independent samples and reports 'Yes' in the first sample and 'No' in the second sample. The probability of repetition of the respondents in both independent samples is quite high when large samples are required for gaining efficiency in the randomized response techniques. This difficulty has been pointed out by Mangat, Singh, and Singh (1997). The probability of a respondent being selected in both samples depends upon the size of stratum or the level of the estimates required. Some organizations require estimates at country, state, district, village, or school level. As the stratum size changes from country to school level, the probability of the respondents being selected in both the samples increases. Another practical situation where the respondent gets selected in both samples is the well-known overlapping cluster sampling. If there is no population list, then in such situations, although an individual/ respondent might not be detected by the interviewer, the interviewee might think that in the first sample, he replied 'Yes' out of choice of two questions, now another interviewer is asking him a direct non-sensitive question. If he understands RRT then he will never respond, as he becomes suspicious. In this case the problem arises from the interviewee's side. Mahmood, Singh, and Horn (1998) observed that the method suggested by Mangat, Singh, and Singh (1997) needs splitting of the population of N units into two random groups. Sometimes it is difficult to do so owing to the population being large or non-availability of a list of individual units at population level. Moreover, the variance expression of the estimator proposed by them is too complicated to implement in actual practice. Mahmood, Singh, and Horn (1998) have provided three simple and alternative survey techniques which parallel the optimality conditions suggested by Moors (1971) and are free from the difficulties in the Moors (1971) as well as in the Mangat, Singh, and Singh (1997) model.

Technique I. Mahmood, Singh, and Horn (1998) suggest a device consisting of three types of statements:

(i) 'I belong to group A'; (ii) 'I belong to group Y^c'

and

(iii) 'I belong to group Y'

with probabilities P_1, P_3 and P_4 respectively, such that $P_1 + P_3 + P_4 = 1$.

Then the probability of 'Yes' answer in the first sample of n_1 respondents is

$$\theta_3 = P_1\pi + P_3(1-\pi_y) + P_4\pi_y . \tag{11.3.2.5}$$

In the second independent sample of n_2 respondents, the question is asked only on the unrelated question 'Y' to estimate the proportion π_y of the unrelated character as discussed in Moors' model. Obviously this technique guarantees the respondents' privacy which is lacking in Moors' model.

By the method of moments, an unbiased estimator of π is given by

$$\hat{\pi}_1 = \frac{\hat{\theta}_3 - P_3(1-\hat{\pi}_y) - P_4\hat{\pi}_y}{P_1} \tag{11.3.2.6}$$

where $\hat{\theta}_3$ and $\hat{\pi}_y$ are the observed proportion of 'Yes' answers in the first and second independent samples.

Thus we have the following theorem:

Theorem 11.3.2.3. The minimum variance of the estimator $\hat{\pi}_1$ is given by

$$\text{Min.}V(\hat{\pi}_1) = \frac{\left[\sqrt{\theta_3(1-\theta_3)} + (P_3 - P_4)\sqrt{\pi_y(1-\pi_y)}\right]^2}{nP_1^2} . \tag{11.3.2.7}$$

Proof. Note that $\hat{\theta}_3 \sim B(n_1, \theta_3)$, $\hat{\pi}_y \sim B(n_2, \pi_y)$ and the samples are independent,

$$V(\hat{\pi}_1) = \frac{V(\hat{\theta}_3) + (P_3 - P_4)^2 V(\hat{\pi}_y)}{P_1^2}$$

$$= \frac{\dfrac{\theta_3(1-\theta_3)}{n_1} + (P_3 - P_4)^2\dfrac{\pi_y(1-\pi_y)}{n_2}}{P_1^2} . \tag{11.3.2.8}$$

On differentiating (11.3.2.8) with respect to n_1, such that $n_1 + n_2 = n$, and equating to zero we have

$$\frac{n_1}{n_2} = \sqrt{\frac{\theta_3(1-\theta_3)}{(P_3 - P_4)\pi_y(1-\pi_y)}} . \tag{11.3.2.9}$$

Hence the minimum variance of the estimator $\hat{\pi}_1$ is given by (11.3.2.7).

Technique II. The randomized response device here differs from technique I in the sense that the second statement 'I belong to Y^c' is simply replaced by the statement 'Try once again.' If this statement appears on the second trial, then the respondent is requested to report 'Yes' irrespective of his actual status. Then the probability of a 'Yes' answer in the first sample of n_1 respondents is given by

$$\theta_4 = (1+P_3)(P_1\pi + P_4\pi_y) + P_3^2 . \quad (11.3.2.10)$$

The second independent sample of n_2 respondents is asked the direct question on Y by following Moors (1971).

By the method of moments an unbiased estimator of π is given by

$$\hat{\pi}_2 = \frac{\hat{\theta}_4 - P_3^2 - P_4(1+P_3)\hat{\pi}_y}{P_1(1+P_3)} . \quad (11.3.2.11)$$

Thus we have the following theorem

Theorem 11.3.2.4. The minimum variance of the estimator $\hat{\pi}_2$ is given by

$$\text{Min.}V(\hat{\pi}_2) = \left[\sqrt{\theta_4(1-\theta_4)} + P_4(1+P_3)\sqrt{\pi_y(1-\pi_y)}\right]^2 \Big/ \left\{n\left(P_1(1+P_3)\right)^2\right\}. \quad (11.3.2.12)$$

Proof. Similar to the proof of Theorem 11.3.2.3.

Technique III. The randomized response device differs from technique I in the sense that the second statement 'I belong to Y^c' is simply replaced by the statement 'I belong to A^c'. Then the probability of 'Yes' answer in the first sample of n_1 respondents is given by

$$\theta_5 = P_1\pi + P_3(1-\pi) + P_4\pi_y . \quad (11.3.2.13)$$

The second independent sample of n_2 respondents is treated by following Moors (1971). By the method of moments, an unbiased estimator of π is given by

$$\hat{\pi}_3 = \frac{\hat{\theta}_5 - P_3 - P_4\hat{\pi}_y}{P_1 - P_3} . \quad (11.3.2.14)$$

Thus we have the following theorem:

Theorem 11.3.2.5. The minimum variance of the estimator $\hat{\pi}_3$ is given by

$$\text{Min.}V(\hat{\pi}_3) = \left[\sqrt{\theta_5(1-\theta_5)} + P_4\sqrt{\pi_y(1-\pi_y)}\right]^2 \Big/ \left\{n(P_1 - P_3)^2\right\}. \quad (11.3.2.15)$$

Proof. It is similar to the proof of Theorem 11.3.2.3.

Mahmood, Singh, and Horn (1998) have shown that the estimator proposed at (11.3.2.6) remains better than the other estimators. A few more strategies in this direction have been suggested by Tracy and Mangat (1995, 1996b), Singh, Singh, and Mangat (2000), and Singh (1999). Note that unfortunately the variance expressions in Singh (1999) and Singh, Singh, and Mangat (2000) are missing small terms, but after making those corrections the suggested strategies still remain efficient than the competitors.

Example 11.3.1. Assume the true proportion of the drug users in a city is 0.1. Consider we used a randomization device to collect information from a large sample of the persons bearing two types of statements:

(i) 'Are you a drug user?' with probability 0.70;
(ii) 'Were you born in spring?' with probability 0.30.

Assuming that the proportion of persons born in spring is 0.2. (In actual practice it is unknown). Find the relative efficiencies of the above three techniques with respect to the Greenberg, Abul-Ela, Simmons, and Horvitz (1969) technique.
Solution. We are given $\pi = 0.1$, $\pi_y = 0.2$, $P_1 = 0.7$ and $P_2 = 0.3$. Thus we have

$$\theta_1 = P_1\pi + (1 - P_1)\pi_y = 0.7 \times 0.1 + (1 - 0.7) \times 0.2 = 0.13,$$

and

$$\theta_2 = P_2\pi + (1 - P_2)\, \pi_y = 0.3 \times 0.1 + (1 - 0.3) \times 0.2 = 0.17\,.$$

Technique I. According to technique I let us split $P_2 = 0.3$ into two parts as $P_3 = 0.15$ and $P_4 = 0.15$. Thus we have

$$\theta_3 = P_1\pi + P_3(1 - \pi_y) + P_4\pi_y = 0.7 \times 0.1 + 0.15(1 - 0.2) + 0.15 \times 0.2 = 0.14\,.$$

Thus the relative efficiency of the estimator $\hat{\pi}_1$ with respect to $\hat{\pi}_G$ is given by

$$\mathrm{RE} = \frac{\left[(1 - P_2)\sqrt{\theta_1(1 - \theta_1)} + (1 - P_1)\sqrt{\theta_2(1 - \theta_2)}\right]^2}{\left[\sqrt{\theta_3(1 - \theta_3)} + (P_3 - P_4)\sqrt{\pi_y(1 - \pi_y)}\right]^2} \frac{P_1^2}{(P_1 - P_2)^2}$$

$$= \frac{\left[(1 - 0.3)\sqrt{0.13(1 - 0.13)} + (1 - 0.7)\sqrt{0.17(1 - 0.17)}\right]^2}{\left[\sqrt{0.14(1 - 0.14)} + (0.15 - 0.15)\sqrt{0.2(1 - 0.2)}\right]^2} \frac{0.7^2}{(0.7 - 0.3)^2} = 3.0822.$$

Technique II. Here we have

$$\theta_4 = (1 + P_3)(P_1\pi + P_4\pi_y) + P_3^2 = (1 + 0.15)(0.15 \times 0.1 + 0.15 \times 0.2) + 0.15^2 = 0.07425\,.$$

Thus the relative efficiency of the estimator $\hat{\pi}_2$ with respect to $\hat{\pi}_G$ is given by

$$\mathrm{RE} = \frac{\left[(1 - P_2)\sqrt{\theta_1(1 - \theta_1)} + (1 - P_1)\sqrt{\theta_2(1 - \theta_2)}\right]^2}{\left[\sqrt{\theta_4(1 - \theta_4)} + P_4(1 + P_3)\sqrt{\pi_y(1 - \pi_y)}\right]^2} \frac{(P_1(1 + P_3))^2}{(P_1 - P_2)^2}$$

$$= \frac{\left[(1 - 0.3)\sqrt{0.13(1 - 0.13)} + (1 - 0.7)\sqrt{0.17(1 - 0.17)}\right]^2}{\left[\sqrt{0.07425(1 - 0.07425)} + 0.15(1 + 0.15)\sqrt{0.2(1 - 0.2)}\right]^2} \frac{(0.7(1 + 0.15))^2}{(0.7 - 0.3)^2} = 4.4747.$$

Technique III. Here we have

$$\theta_5 = P_1\pi + P_3(1-\pi) + P_4\pi_y = 0.7\times 0.1 + 0.15\times(1-0.1) + 0.15\times 0.2 = 0.235\,.$$

The relative efficiency of the estimator $\hat{\pi}_3$ with respect to $\hat{\pi}_G$ is given by

$$\text{RE} = \frac{\left[(1-P_2)\sqrt{\theta_1(1-\theta_1)} + (1-P_1)\sqrt{\theta_2(1-\theta_2)}\right]^2}{\left[\sqrt{\theta_5(1-\theta_5)} + P_4\sqrt{\pi_y(1-\pi_y)}\right]^2}\frac{(P_1-P_3)^2}{(P_1-P_2)^2}$$

$$= \frac{\left[(1-0.3)\sqrt{0.13(1-0.13)} + (1-0.7)\sqrt{0.17(1-0.17)}\right]^2}{\left[\sqrt{0.235(1-0.235)} + 0.15\sqrt{0.2(1-0.2)}\right]^2}\times\frac{(0.7-0.15)^2}{(0.7-0.3)^2} = 0.977.$$

11.4. REGRESSION ANALYSIS

Singh, Joarder, and King (1996) considered the problem of estimation of regression coefficients in the traditional regression model. They assumed that the variable of interest Y_i is related to k non-stochastic regressors via the classical linear regression model

$$Y = X\beta + e, \tag{11.4.1}$$

where Y is an $n\times 1$ vector of Y_i values, X is an $n\times k$ matrix of regressors, β is $k\times 1$ and e is an $n\times 1$ disturbance term such that $e \sim N(0, \sigma^2 I_n)$. Here Y_i is a sensitive variable whose observations have to be obtained by survey methods. Because some respondents are unlikely to respond truthfully to questions about their behaviour which is immoral, unpopular, or unlawful, Eichhorn and Hayre's (1983) scrambling response approach is applied as follows: Each respondent is requested to report the product Y_iS_i where S_i is the value of the scrambling variable drawn by the ith respondent. The privacy of the respondent is protected by the fact that S_i is not known to the interviewer but its distribution, and in particular its mean $E(S_i) = \theta$ and variance $V(S_i) = \gamma^2$ are known. The scrambling device may be a deck of cards, a spinner, etc., following some suitable distribution, *e.g.*, Normal, Weibull, or any discrete distribution. The Y_iS_i value obtained from the ith respondent can be standardized as $Z_i = Y_iS_i/\theta$ after collection. Singh, Joarder, and King (1996) considered the estimation and testing of β under the model

$$Z = X\beta + \eta, \tag{11.4.2}$$

where Z is the $n\times 1$ vector of Z_i values and η is an $n\times 1$ vector of errors whose distribution is unknown.

Theorem 11.4.1. $E(\eta) = 0$ and the unbiased OLS estimator of β is given by

$$\hat{\beta}_* = (X^tX)^{-1}X^tZ\,. \tag{11.4.3}$$

Proof. This theorem follows immediately from the property

$$E(Z) = E_{\mathrm{M}}E_{\mathrm{R}}(Z) = E_{\mathrm{M}}(Y) = X\beta,$$

where E_{M} and E_{R} denote the expected values with respect to model (11.4.1) and the distribution of the randomization device providing the S_i values, respectively.

To find the variance of the OLS estimator $\hat{\beta}$ we need the following lemma.

Lemma 11.4.1. If V_{R} denotes the variance--covariance matrix of Z over the distribution of S_i then we have

$$V_{\mathrm{R}}(Z) = C_\gamma^2 \mathrm{diag}\left(Y_1^2, Y_2^2, ..., Y_n^2\right), \tag{11.4.4}$$

where $C_\gamma = \gamma/\theta$ is the coefficient of variation of the scrambling device.

Theorem 11.4.2. The variance--covariance matrix of the estimator $\hat{\beta}$ is given by

$$V(\hat{\beta}_*) = \sigma^2\left[\left(X'X\right)^{-1} + C_\gamma^2\left(X'X\right)^{-1}\right]$$

$$+ C_\gamma^2\left(X'X\right)^{-1}X'\mathrm{diag}\left[\left(\sum_{j=1}^k X_{1j}\beta_j\right)^2, \left(\sum_{j=1}^k X_{2j}\beta_j\right)^2, ..., \left(\sum_{j=1}^k X_{nj}\beta_j\right)^2\right]X\left(X'X\right)^{-1}. \tag{11.4.5}$$

Proof. Let V_{M} denote the variance-covariance matrix over the model (11.4.1). Then we have

$$V(\hat{\beta}_*) = E_{\mathrm{M}}V_{\mathrm{R}}(\hat{\beta}_*) + V_{\mathrm{M}}E_{\mathrm{R}}(\hat{\beta}_*) = E_{\mathrm{M}}V_{\mathrm{R}}\left[\left(X'X\right)^{-1}X'Z\right] + V_{\mathrm{M}}E_{\mathrm{R}}\left[\left(X'X\right)^{-1}X'Z\right]$$

$$= E_{\mathrm{M}}\left[\left(X'X\right)^{-1}X'C_\gamma^2\mathrm{diag}\left(Y_1^2, Y_2^2, ..., Y_n^2\right)X\left(X'X\right)^{-1}\right] + V_{\mathrm{M}}\left[\left(X'X\right)^{-1}X'Y\right]$$

$$= \sigma^2\left[\left(X'X\right)^{-1} + C_\gamma^2\left(X'X\right)^{-1}\right]$$

$$+ C_\gamma^2\left(X'X\right)^{-1}X'\mathrm{diag}\left[\left(\sum_{j=1}^k X_{1j}\beta_j\right)^2, \left(\sum_{j=1}^k X_{2j}\beta_j\right)^2, ..., \left(\sum_{j=1}^k X_{nj}\beta_j\right)^2\right]X\left(X'X\right)^{-1}.$$

Hence the theorem.

Theorem 11.4.3. An estimator of σ^2 is given by

$$\hat{\sigma}_*^2 = \left\{\hat{\eta}'\hat{\eta} - C_\gamma^2\sum_{i=1}^n\left(\sum_{j=1}^k \hat{\beta}_{*j}X_{i,j}\right)^2 M_{i,i}\right\} \Big/ \left(n - k + C_\gamma^2\sum_{i=1}^n M_{i,i}\right), \tag{11.4.6}$$

where $\hat{\eta} = Z - X\hat{\beta}_*$ is the OLS residual vector from (11.4.3) and $M_{i,i}$ denotes the ith diagonal element of the matrix

$$M = I - X\left(X'X\right)^{-1}X'.$$

Proof. We have

$$E(\hat{\eta}^t\hat{\eta}) = E_M E_R\left[(Z - X\hat{\beta}_*)^t (Z - X\hat{\beta}_*)\right] = E_M E_R\left[Z^t\left(I - X(X^t X)^{-1} X^t\right).Z\right]$$

$$= \left(n - k + C_\gamma^2 \sum_{i=1}^{n} M_{i,i}\right)\sigma^2 + C_\gamma^2 \sum_{i=1}^{n}\left(\sum_{j=1}^{k} \beta_j X_{i,j}\right)^2 M_{i,i}. \qquad (11.4.7)$$

By the method of moments (11.4.7) with unknown β_j values replaced by estimates gives (11.4.6). Hence the theorem.

Theorem 11.4.4. The Wald test statistic to test the null hypothesis $H_0 : \beta = \beta_0$ against the alternative hypothesis $H_a : \beta \neq \beta_0$ for the scrambled response model is given by

$$(\hat{\beta}_* - \beta_0)^t \left(\hat{v}(\hat{\beta}_*)\right)^{-1} (\hat{\beta}_* - \beta_0) \overset{\text{asy}}{\sim} \chi^2(p)$$

under H_0 assuming that $\hat{\beta}_* \overset{a}{\sim} N\left(\beta, V(\hat{\beta}_*)\right)$ and an estimator of $V(\hat{\beta}_*)$ is given by

$$\hat{v}(\hat{\beta}_*) = \hat{\sigma}_*^2\left[(X^t X)^{-1} + C_\gamma^2 (X^t X)^{-1}\right]$$

$$+ C_\gamma^2 (X^t X)^{-1} X^t \text{diag}\left[\left(\sum_{j=1}^{k} X_{1j}\hat{\beta}_{*j}\right)^2, \left(\sum_{j=1}^{k} X_{2j}\hat{\beta}_{*j}\right)^2, \ldots, \left(\sum_{j=1}^{k} X_{n,j}\hat{\beta}_{*j}\right)^2\right] X (X^t X)^{-1}.$$

A detailed empirical study has been carried out to study the nature of estimates of regression coefficients under different levels of respondents lying in direct surveying. Strachan, King, and Singh (1998) have considered likelihood based estimation of the regression model with scrambled responses, where they compared Bayesian estimator achieved through a Markov Chain Monte Carlo (MCMC) sampling scheme with a classical maximum likelihood estimator and the estimator proposed by Singh, Joarder, and King (1996), and later coefficient of determination is studied by Singh and King (1999).

A more serious problem arises when a character under study is sensitive in nature. For example, researchers may be interested in estimating the average income using expenditure and assets of households as independent variables. Obviously high correlation is expected between expenditure and assets of household and will result in matrix $(X'X)$ to become ill conditioned. Singh and Tracy (1999) consider the ridge regression estimator of the regression coefficient under scrambled responses and found more efficient than the ordinary least square estimator.

11.4.1 RIDGE REGRESSION ESTIMATOR

Singh and Tracy (1999) considered a ridge regression estimator under scrambled response as

$$\hat{\beta}_{R_c}^{(s)} = (X'X + R_c I)^{-1} X'z, \qquad (11.4.1.1)$$

where R_c denotes the ridge constant.

Lemma 11.4.1.1. The variance $V_{\mathrm{R}}(z)$ is given by

$$V_{\mathrm{R}}(z) = C_{\gamma}^{2}\mathrm{Diag}\left(Y_1^2,\ Y_2^2,\ \ldots,\ Y_n^2\right). \tag{11.4.1.2}$$

Theorem 11.4.1.1. The bias in the estimator $\hat{\beta}_{\mathrm{R_c}}^{(s)}$ is given by $(A-I)\beta$.

Proof. Let E_{M} and E_{R} denote, respectively, the expected value with respect to model and the distribution of the randomization device providing the S_i values. Then

$$E\left\{\hat{\beta}_{\mathrm{R_c}}^{(s)}\right\} = E_{\mathrm{M}}E_{\mathrm{R}}\left\{\hat{\beta}_{\mathrm{R_c}}^{(s)}\right\} = \left(X'X + R_c I\right)^{-1}X'X\beta .$$

Using

$$B\left(\hat{\beta}_{\mathrm{R_c}}^{(s)}\right) = E\left(\hat{\beta}_{\mathrm{R_c}}^{(s)}\right) - \beta$$

we get the theorem.

Theorem 11.4.1.2. The variance of the proposed estimator $\hat{\beta}_{\mathrm{R_c}}^{(s)}$ is given by

$$V\left\{\hat{\beta}_{\mathrm{R_c}}^{(s)}\right\} = \sigma^2 A\left(X'X\right)^{-1}A' + C_{\gamma}^{2}\left(X'X + R_c I\right)^{-1}X'\left(D + \sigma^2 I\right)X\left(X'X + R_c I\right)^{-1} \tag{11.4.1.3}$$

which is equivalent to trace of the right hand side of (11.4.1.3).

Proof. Let V_{M} and V_{R} denote, respectively, the variance-covariance matrix over the model and the distribution of the randomization device. Thus we have

$$\begin{aligned} V\left(\hat{\beta}_{\mathrm{R_c}}^{(s)}\right) &= E_{\mathrm{M}}V_{\mathrm{R}}\left(\hat{\beta}_{\mathrm{R_c}}^{(s)}\right) + V_{\mathrm{M}}E_{\mathrm{R}}\left(\hat{\beta}_{\mathrm{R_c}}^{(s)}\right) \\ &= E_{\mathrm{M}}V_{\mathrm{R}}\left[\left(X'X + R_c I\right)^{-1}X'z\right] + V_{\mathrm{M}}E_{\mathrm{R}}\left[\left(X'X + R_c I\right)^{-1}X'z\right] \\ &= E_{\mathrm{M}}\left[\left(X'X + R_c I\right)^{-1}X'C_{\gamma}^{2}\mathrm{Diag}\left(Y_1^2, Y_2^2, \ldots, Y_n^2\right)X\left(X'X + R_c I\right)^{-1}\right] \\ &\quad + V_{\mathrm{M}}\left[\left(X'X + R_c I\right)^{-1}X'y\right] \\ &= \sigma^2 A\left(X'X\right)^{-1}A' + C_{\gamma}^{2}\left(X'X + R_c I\right)^{-1}X'\left(D + \sigma^2 I\right)X\left(X'X + R_c I\right)^{-1}. \end{aligned}$$

Hence the theorem.

Theorem 11.4.1.3. The mean square error of the estimator $\hat{\beta}_{\mathrm{R_c}}^{(s)}$ is given by

$$\mathrm{MSE}\left[\hat{\beta}_{\mathrm{R_c}}^{(s)}\right] = V\left[\hat{\beta}_{\mathrm{R_c}}^{(s)}\right] + (A-I)\beta\beta'(A-I)'. \tag{11.4.1.4}$$

Proof. Trivial with the definition of trace of a matrix.

The ridge regression estimator $\hat{\beta}_{\mathrm{R_c}}^{(s)}$ under scrambled responses is more efficient than the estimator $\hat{\beta}^{(s)}$ sue to Singh, Joarder, and King (1996) if

$$V\left[\hat{\beta}^{(s)}\right] - \mathrm{MSE}\left[\hat{\beta}_{\mathrm{R_c}}^{(s)}\right] > 0 \tag{11.4.1.5}$$

or if

$$0 < R_c < \frac{2\left(1 + C_\gamma^2\right)\sigma^2}{\beta'\beta} \tag{11.4.1.6}$$

which can be derived by taking trace of (11.4.1.5). It is interesting to note that the range of ridge constant R_c is wider in the case of scrambled responses than in the case of direct question survey sampling because $C_\gamma > 0$. Also (11.4.1.6) shows that the ridge constant R_c is directly proportional to C_γ.

11. 5 HIDDEN GANGS IN FINITE POPULATIONS

Singh, Horn, and Chowdhury (1998) have introduced a new and interesting model in survey sampling for social sciences. Suppose someone is interested in estimating the proportionate strength (or size) and average of a stigmatizing quantitative character of a particular hidden gang G (defined by that character or otherwise) in the finite population Ω. People are unwilling to admit membership of gang G due to fear of administrative retribution or social embarrassment. Singh, Horn, and Chowdhury (1998) attempted a possible solution to the following types of problems is survey sampling.

(1) Estimation of the proportion of persons in population Ω having income greater than or equal to \$60,000 (say, gang G_1) along with their average income.

(2) Estimation of the proportion of persons having extra marital relations (say, gang G_2) in the whole population and their average income.

(3) Estimation of the proportion of politically active persons in the country (say, gang G_3) and their average income [or the average number of murders committed by them].

(4) More generally, estimation of a proportion of persons involved in a particular crime (say, gang G) along with average value, μ_x, of any stigmatized quantitative character (say, X) of the same gang.

11.5.1 TWO SAMPLE METHOD

In the first method Singh, Horn, and Chowdhury (1998) suggested drawing two independent samples of sizes n_1 and n_2 from the population using SRSWR. The persons appearing in the first sample are provided with a randomization device R_1 (say). The device may be a spinner, deck of cards, or computer grid which generates a random variable taking a value greater than or equal to Ψ_0 (say) from a given probability distribution. Each respondent selected in the sample is requested to draw one random number (which is obviously greater than or equal to Ψ_0) from the randomization device R_1, without showing it to the interviewer. Now he is requested to report the actual value of the stigmatized quantitative variable, say X, if and only if he belongs to gang G, otherwise he is requested to report the random number as drawn. The value of Ψ_0 depends upon the problem under consideration.

For example, in problem 1 the value of Ψ_0 is \$60,000, in problems 2 and 3 the value of Ψ_0 can be taken as zero or any other suitable value. The choice of Ψ_0 can be made such that the respondent's privacy will not be jeopardized if they respond honestly. It is assumed that the distribution of the randomization device, R_1, is known to the interviewer, but not the number drawn by the respondent. Let θ_1 and σ_1^2 denote the known mean and variance of the randomization device R_1. Let π denote the proportion of persons belonging to the gang G in the population Ω. If the respondents are reporting 100% truthfully then the distribution of the ith response, say Z_{1i}, is given by

$$Z_{1i} = \begin{cases} X_i & \text{with probability } \pi, \\ R_{1i} & \text{with probability } (1-\pi), \end{cases} \tag{11.5.1.1}$$

where X_i and R_{1i} denote, respectively, the actual value of the stigmatized quantitative character and the random variable drawn by the ith respondent in the sample. The expected value of the random variable Z_{1i} is given by

$$E(Z_{1i}) = \pi E(X_i) + (1-\pi)E(R_{1i}) = \pi\mu_x + (1-\pi)\theta_1 \tag{11.5.1.2}$$

where μ_x denotes the average value of the stigmatized quantitative variable of the hidden gang G in the finite population Ω. Now in equation (11.5.1.2) we have two unknown parameters of interest, π and μ_x, satisfying the single equation. This difficulty can be removed by taking another independent sample of size n_2 from Ω and using a similar randomization device R_2 with different known distribution. Consider $\theta_2 (\neq \theta_1)$ and σ_2^2 denote the known mean and variance of the second randomization device R_2. Then distribution of the ith response, say Z_{2i}, in the second sample is given by

$$Z_{2i} = \begin{cases} X_i & \text{with probability } \pi, \\ R_{2i} & \text{with probability } (1-\pi), \end{cases} \tag{11.5.1.3}$$

where X_i and R_{2i} denote, respectively, the actual value of the stigmatized quantitative character and the random value drawn by the ith respondent in the second sub-sample. Again the expected value of the random variable Z_{2i} is

$$E(Z_{2i}) = \pi\mu_x + (1-\pi)\theta_2 . \tag{11.5.1.4}$$

We now have the following theorems:

Theorem 11.5.1.1. An unbiased estimator of π is given by

$$\hat{\pi}_s = 1 - \frac{\bar{Z}_1 - \bar{Z}_2}{\theta_1 - \theta_2}, \tag{11.5.1.5}$$

where $\bar{Z}_1 = n_1^{-1}\sum_{i=1}^{n_1} Z_{1i}$ and $\bar{Z}_2 = n_2^{-1}\sum_{i=1}^{n_2} Z_{2i}$ denote the response means in the first and second sub-sample respectively.

Proof. Follows on solving (11.5.1.2) and (11.5.1.4) for π.

Theorem 11.5.1.2. The variance of the estimator $\hat{\pi}_s$ is given by

$$V(\hat{\pi}_s) = \frac{1}{(\theta_1 - \theta_2)^2}\left[\frac{\sigma_{Z_1}^2}{n_1} + \frac{\sigma_{Z_2}^2}{n_2}\right], \tag{11.5.1.6}$$

where

$$\sigma_{Z_1}^2 = \pi\sigma_x^2 + (1-\pi)\sigma_1^2 + \pi(1-\pi)(\mu_x - \theta_1)^2 \tag{11.5.1.7}$$

and

$$\sigma_{Z_2}^2 = \pi\sigma_x^2 + (1-\pi)\sigma_2^2 + \pi(1-\pi)(\mu_x - \theta_2)^2. \tag{11.5.1.8}$$

Proof. Follows from the independence of samples.

Theorem 11.5.1.3. An estimator of μ_x is given by

$$\hat{\mu}_x = \frac{\bar{Z}_2\theta_1 - \bar{Z}_1\theta_2}{(\bar{Z}_2 - \theta_2) - (\bar{Z}_1 - \theta_1)}. \tag{11.5.1.9}$$

Proof. Substitute the value of π obtained from (11.5.1.2) in (11.5.1.4), then (11.5.1.9) follows by the method of moments.

Theorem 11.5.1.4. An expression for the variance of the estimator $\hat{\mu}_x$ is given by

$$V(\hat{\mu}_x) = \frac{1}{\pi^2(\theta_2 - \theta_1)^2}\left[\frac{\sigma_{Z_1}^2}{n_1}(\theta_2 - \mu_x)^2 + \frac{\sigma_{Z_2}^2}{n_2}(\theta_1 - \mu_x)^2\right]. \tag{11.5.1.10}$$

Proof. The estimator (11.5.1.9) can be written as

$$\hat{\mu}_x = \hat{d}_1 / \hat{d}_2,$$

where

$$\hat{d}_1 = \bar{Z}_2\theta_1 - \theta_2\bar{Z}_1 \quad \text{and} \quad \hat{d}_2 = (\bar{Z}_2 - \theta_2) - (\bar{Z}_1 - \theta_1).$$

By the ratio method of estimation, the variance of the estimator $\hat{\mu}_x$ is given by

$$V(\hat{\mu}_x) = \frac{1}{\pi^2(\theta_2 - \theta_1)^2}\left[V(\hat{d}_1) + \left\{\frac{E(\hat{d}_1)}{E(\hat{d}_2)}\right\}^2 V(\hat{d}_2) - 2\left\{\frac{E(\hat{d}_1)}{E(\hat{d}_2)}\right\}\text{Cov}(\hat{d}_1, \hat{d}_2)\right]. \tag{11.5.1.11}$$

One can easily show that

$$V(\hat{d}_1) = \frac{\theta_1^2\sigma_{Z_2}^2}{n_2} + \frac{\theta_2^2\sigma_{Z_1}^2}{n_1}, \quad V(\hat{d}_2) = \frac{\sigma_{Z_1}^2}{n_1} + \frac{\sigma_{Z_2}^2}{n_2},$$

$$\text{Cov}(\hat{d}_1, \hat{d}_2) = \frac{\theta_1\sigma_{Z_2}^2}{n_2} + \frac{\theta_2\sigma_{Z_1}^2}{n_1} \quad \text{and} \quad E(\hat{d}_1)/E(\hat{d}_2) = \mu_x. \tag{11.5.1.12}$$

On using (11.5.1.12) in (11.5.1.11) we have (11.5.1.10). Hence the theorem.

The next section discusses strategies for using the method 1 in actual survey sampling.

Decision Making Strategies: The decision maker may have any of the following objectives:

Case I. When the investigator is interested in obtaining a more precise estimate of population proportion π than mean μ_x, he will seek to minimize the variance of the estimator $\hat{\pi}_s$ at the expense of the variance of $\hat{\mu}_x$. The variance of the estimator $\hat{\pi}_s$ will be minimal if the values of sample sizes n_1 and n_2 are given by

$$n_1 = n\sigma_{Z_1}/(\sigma_{Z_1}+\sigma_{Z_2}) \quad \text{and} \quad n_2 = n\sigma_{Z_2}/(\sigma_{Z_1}+\sigma_{Z_2}) .$$

The minimum variance of the proposed estimator $\hat{\pi}_s$ is then given by

$$\text{Min.}V(\hat{\pi}_s) = (\sigma_{Z_1}+\sigma_{Z_2})^2 / \{n(\theta_1-\theta_2)^2\} .$$

Case II. When the investigator is interested in estimating μ_x more precisely than the size (or proportion) of the gang, then he will be interested in the minimum variance of the estimator $\hat{\mu}_x$ rather than the minimum variance of the estimator $\hat{\pi}_s$. The variance of the estimator $\hat{\mu}_x$ will be minimal if the values of the sample sizes n_1 and n_2 are given by

$$n_1 = \frac{n\sigma_{Z_1}|\theta_2-\mu_x|}{\sigma_{Z_1}|\theta_2-\mu_x|+\sigma_{Z_2}|\theta_1-\mu_x|}, \quad \text{and} \quad n_2 = \frac{n\sigma_{Z_2}|\theta_1-\mu_x|}{\sigma_{Z_1}|\theta_2-\mu_x|+\sigma_{Z_2}|\theta_1-\mu_x|}.$$

The minimum variance of the proposed estimator $\hat{\mu}_x$ is given by

$$\text{Min.}V(\hat{\mu}_x) = \frac{[\sigma_{Z_1}|\theta_2-\mu_x|+\sigma_{Z_2}|\theta_1-\mu_x|]^2}{n\pi^2(\theta_2-\theta_1)^2}.$$

Case III. Let $\sigma^2_{Z_1} = \sigma^2_{Z_2}$ as there is no prior reason to treat the two samples asymmetrically. Then for a given $\delta = |\theta_1-\theta_2|$, we can minimize $V(\hat{\pi}_s)$ and $V(\hat{\mu}_x)$ simultaneously by letting $\theta_1 = \mu_x - \frac{1}{2}\delta$, $\theta_2 = \mu_x + \frac{1}{2}\delta$ and $n_1 = n_2 = \frac{n}{2}$.

Its proof is obvious from that $n_1 = n_2 = \frac{n}{2}$ minimizes $V(\hat{\pi}_s)$ when $\sigma^2_{Z_1} = \sigma^2_{Z_2}$, then the variance becomes

$$\text{Min.}V(\hat{\mu}_x) = \frac{\sigma^2_{Z_1}\{|\theta_2-\mu_x|+|\theta_1-\mu_x|\}^2}{n\pi^2(\theta_2-\theta_1)^2} \geq \frac{\sigma^2_{Z_1}}{n\pi^2}$$

and the lower bound is achieved whenever $\theta_1 \leq \mu_x \leq \theta_2$, and in particular, if $\theta_1 = \mu_x - 0.5\delta$ and $\theta_2 = \mu_x + 0.5\delta$. For this choice of θ_1 and θ_2 it follows that $n_1 = n_2 = n/2$ minimizes $V(\hat{\mu}_x)$.

With $\sigma^2_{Z_1} = \sigma^2_{Z_2}$ the choice $\theta_1 = \mu_x - 0.5\delta$, $\theta_2 = \mu_x + 0.5\delta$ and $n_1 = n_2 = n/2$ minimizes $V(\hat{\pi}_s)$ and $V(\hat{\mu}_x)$ simultaneously for a given $\delta = |\theta_1-\theta_2|$. We should make $|\theta_1-\theta_2|$ as large as possible without threatening the privacy of the respondents.

11.5.2 ONE SAMPLE METHOD

Singh, Horn, and Chowdhury (1998) proposed another method, according to which a sample of n respondents instead of n_1 respondents of the previous method is selected by SRSWR. Each respondent selected in the sample is asked two questions. First question about the value of X_i is asked by using the randomization device R_1 as discussed earlier. The second question is asked about the membership of the gang using the Warner (1965) model. For this situation we define a random variable t_i such that

$$t_i = \begin{cases} 1 & \text{with probability } P\pi + (1-P)(1-\pi), \\ 0 & \text{with probability } P(1-\pi) + \pi(1-P). \end{cases} \tag{11.5.2.1}$$

Then we have the following theorem from Warner (1965).

Theorem 11.5.2.1. An unbiased estimator of π is given by

$$\hat{\pi}_w = \frac{\bar{t} - (1-P)}{2P-1}, \quad P \neq 0.5, \tag{11.5.2.2}$$

where $\bar{t} = n^{-1}\sum_{i=1}^{n} t_i$ and its variance is given by

$$V(\hat{\pi}_w) = \frac{\pi(1-\pi)}{n} + \frac{P(1-P)}{n(2P-1)^2}. \tag{11.5.2.3}$$

Using this estimator of π in (11.5.1.2) we obtain an estimator of μ_x.

This estimator is presented in the following theorem.

Theorem 11.5.2.2. An estimator of μ_x is given by

$$\hat{\mu}_x^* = \frac{\bar{Z} - (1-\hat{\pi}_w)\theta_1}{\hat{\pi}_w}, \tag{11.5.2.4}$$

where $\bar{Z} = n^{-1}\sum_{i=1}^{n} Z_{1i}$ denotes the observed mean of n responses. In order to find the variance of the estimator $\hat{\mu}_x^*$ we need the following lemmas.

Lemma 11.5.2.1. The variance of $\bar{Z}$ is given by

$$V(\bar{Z}) = n^{-1}\sigma_{z_1}^2 = n^{-1}\left[\pi\sigma_x^2 + (1-\pi)\sigma_1^2 + \pi(1-\pi)(\mu_x - \theta_1)^2\right]. \tag{11.5.2.5}$$

Lemma 11.5.2.2. The estimators $\bar{Z}$ and $\hat{\pi}_w$ are correlated.

Proof. By the definition we have

$$\text{Cov}(\bar{Z}, \hat{\pi}_w) = \frac{1}{(2P-1)}\text{Cov}(\bar{Z}, \bar{t}) = \frac{1}{n(2P-1)}[E(Zt) - E(Z)E(t)]$$

where

$$E(Zt) = E\{E(Zt \mid t)\} = E\{tE(Z \mid t)\} = P(t=1)E(Z \mid t=1) = E(t)E(Z \mid t=1).$$

It follows that

$$\text{Cov}(\bar{Z}, \bar{t}) = \frac{E(t)}{n}[E(Z \mid t=1) - E(Z)] \neq 0$$

as

$$E(Z \mid t=1) = P(G \mid t=1)\mu_x + P(\bar{G} \mid t=1)\theta_1 \neq E(Z)$$

since

$$P(G \mid t=1) = \frac{P\pi}{P\pi + (1-P)(1-\pi)} \neq \pi .$$

Hence the lemma.

We now have the following theorem, the proof of which is easily obtained by proceeding on the lines of obtaining the variance for the ratio estimator.

Theorem 11.5.2.3. The variance of the estimator $\hat{\mu}_x^*$, to the first order of approximation, is given by

$$V(\hat{\mu}_x^*) \approx \frac{1}{\pi^2}\left[V(\bar{Z}) + (\mu_x - \theta_1)^2 V(\hat{\pi}_w) - 2(\mu_x - \theta_1)\text{Cov}(\bar{Z}, \hat{\pi}_w)\right] . \tag{11.5.2.6}$$

The result of Theorem 11.5.2.3 can also be expressed as follows:

Theorem 11.5.2.4. From (11.5.2.6), it follows that

$$V(\hat{\mu}_x^*) = \frac{\sigma_{Z1}^2}{n\pi^2} \text{ , if } \theta_1 = \mu_x$$

and

$$V(\hat{\mu}_x^*) = \frac{\sigma_{Z1}^2}{n\pi^2}(1-\rho^2) \text{ , if } \mu_x - \theta_1 = \frac{\text{Cov}(\bar{Z}, \hat{\pi}_w)}{V(\hat{\pi}_w)}$$

where

$$\rho = \frac{\text{Cov}(\bar{Z}, \hat{\pi}_w)}{\sqrt{V(\hat{\pi}_w)}\sqrt{V(\bar{Z})}} .$$

Comparing (11.5.2.6) with (11.5.1.10) we can see that $\hat{\mu}_x^*$ is capable of achieving a variance which is smaller than or at most the same as that of $\hat{\mu}_x$. In this method it is also possible to use a well known unrelated question model in place of Warner's (1965) model.

11.5.3 ESTIMATION OF CORRELATION COEFFICIENT BETWEEN TWO CHARACTERS OF A HIDDEN GANG

We wish to estimate the proportion of persons involved in a particular crime (say, gang G) along with average values, μ_x, μ_y and correlation coefficient ρ_{xy} of any two stigmatized quantitative characters (say, X and Y) of the same gang. The estimator of correlation coefficient r_{xy} may be useful to study the relationship between two characters.

Let

$$\pi = \frac{N_G}{N}$$

be the proportion of persons in the finite population Ω belonging to the hidden gang G. Let Y and X be the two quantitative sensitive characters of a hidden gang G. We wish to estimate the population means

$$\mu_x = \frac{1}{N_G}\sum_{i=1}^{N_G} X_i\ , \quad \text{and} \quad \mu_y = \frac{1}{N_G}\sum_{i=1}^{N_G} Y_i$$

spread of a hidden gang π and $\rho_{xy} = S_{xy}/(S_x S_y)$, where

$$S_x^2 = \{N_G - 1\}^{-1}\sum_{i=1}^{N_G}(X_i - \mu_x)^2\ , S_y^2 = \{N_G - 1\}^{-1}\sum_{i=1}^{N_G}(Y_i - \mu_y)^2,$$

and

$$S_{xy} = \{N_G - 1\}^{-1}\sum_{i=1}^{N_G}(X_i - \mu_x)(Y_i - \mu_y).$$

Here we have selected a sample of n units by SRSWR scheme. Each respondent is provided with three randomization devices, R_x, R_y and R_w (say). The randomization device R_x is provided to each respondent such that if he belongs to hidden gang G then he is requested to report actual value of quantitative character X, otherwise he is assumed to report a random number T_x generated through the device R_x. It is assumed that random variable T_x takes values in the range of actual X values to maintain the privacy and its mean $\theta_x = E(T_x)$ and variance $\gamma_x^2 = V(T_x)$ are assumed to be known. Here π is the probability that he will report value X and $(1-\pi)$ is the probability that he will report value T_x.
Thus the distribution of the first response of the ith respondent is:

$$Z_{1i} = \begin{cases} X & \text{with probability} \quad \pi, \\ T_x & \text{with probability} \quad (1-\pi). \end{cases} \tag{11.5.3.1}$$

Then the expected value of ith response is given by

$$E(Z_{1i}) = \pi\mu_x + (1-\pi)\theta_x\ . \tag{11.5.3.2}$$

The randomization device R_y is provided to each respondent such that if he belongs to hidden gang G then he is requested to report the actual value of the quantitative character Y, otherwise he is supposed to report a random number T_y generated through the device R_y. It is assumed that the random variable T_y takes values in the range of actual Y values to maintain the privacy and its mean $\theta_y = E(T_y)$ and variance $\gamma_y^2 = V(T_y)$ are assumed to be known. Note π is the probability that he will report value Y and $(1-\pi)$ is the probability that he will report value T_y, the distribution of the second response of the ith respondent is given by

$$Z_{2i} = \begin{cases} Y & \text{with probability } \pi, \\ T_y & \text{with probability } (1-\pi). \end{cases} \tag{11.5.3.3}$$

Thus the expected value of the ith response is given by

$$E(Z_{2i}) = \pi\mu_y + (1-\pi)\theta_y . \tag{11.5.3.4}$$

The third device R_w is the same the as invented by Warner (1965), that is, it consists of two outcomes. The statement 'Are you a member of gang G?' occurs with probability P and its complement 'Are you not a member of gang G?' occurs with probability $(1-P)$. Each respondent is also requested to use the device R_w and report 'Yes' or 'No' according to his status and the statement drawn by him from the device, which, in fact, comes from the third response of the ith respondent. Thus the probability of 'Yes' answer is given by

$$\theta = P\pi + (1-P)(1-\pi) . \tag{11.5.3.5}$$

Thus we have the following theorems:

Theorem 11.5.3.1. An unbiased estimator of π is given by

$$\hat{\pi}_w = \frac{\hat{\theta} - (1-P)}{2P-1}, \quad P \neq 0.5, \tag{11.5.3.6}$$

where $\hat{\theta} = n_1/n$ is the observed proportion of 'Yes' answers in the sample.
Proof. Obvious from (11.5.3.5).

Theorem 11.5.3.2. An estimator to estimate the mean μ_x of the quantitative character X of the hidden gang G is given by

$$\hat{\mu}_x = \frac{\overline{Z}_1 - \theta_x}{\hat{\pi}_w} + \theta_x . \tag{11.5.3.7}$$

Proof. Obtained by the method of moments after solving (11.5.3.2) and (11.5.3.5) for μ_x.

Theorem 11.5.3.3. An estimator to estimate the mean μ_y of the quantitative character Y of the hidden gang G is given by

$$\hat{\mu}_y = \frac{\overline{Z}_2 - \theta_y}{\hat{\pi}_w} + \theta_y . \tag{11.5.3.8}$$

Proof. Obtained by the method of moments.

Theorem 11.5.3.4. An estimator of correlation coefficient ρ_{xy} between the two quantitative characters X and Y of a hidden gang G is given by

$$r_{xy} = \frac{s_{z_1 z_2}}{\hat{\pi}_w\sqrt{s_{z_1}^2 - (1-\hat{\pi}_w)\left\{\hat{\pi}_w(\theta_x - \hat{\mu}_x)^2 + \gamma_x^2\right\}}\sqrt{s_{z_2}^2 - (1-\hat{\pi}_w)\left\{\hat{\pi}_w(\theta_y - \hat{\mu}_y)^2 + \gamma_y^2\right\}}} . \tag{11.5.3.9}$$

Proof. The distribution of Z_{1i}^2 is given by

$$Z_{1i}^2 = \begin{cases} X^2 & \text{with probability } \pi, \\ T_x^2 & \text{with probability } (1-\pi). \end{cases} \tag{11.5.3.10}$$

Thus the expected value of Z_{1i}^2 is given by

$$E(Z_{1i}^2) = \pi E(X^2) + (1-\pi)E(T_x^2), \tag{11.5.3.11}$$

which implies that

$$E(X^2) = \frac{E(Z_{1i}^2) - (1-\pi)(\gamma_x^2 + \theta_x^2)}{\pi}. \tag{11.5.3.12}$$

Thus we have

$$V(X) = E(X^2) - \{E(X)\}^2 = \frac{\sigma_{z1}^2 - (1-\pi)\{\pi(\theta_x - \mu_x)^2 + \gamma_x^2\}}{\pi}. \tag{11.5.3.13}$$

Similarly

$$V(Y) = \frac{\sigma_{z2}^2 - (1-\pi)\{\pi(\theta_y - \mu_y)^2 + \gamma_y^2\}}{\pi}. \tag{11.5.3.14}$$

Again we have

$$E(Z_{1i}Z_{2i}) = \pi\pi E(XY) + \pi(1-\pi)E(X)E(T_x) + \pi(1-\pi)E(Y)E(T_y) + (1-\pi)(1-\pi)E(S_x)E(S_y). \tag{11.5.3.15}$$

which implies that

$$E(XY) = \frac{E(Z_{1i}Z_{2i}) - (1-\pi)\{\pi\mu_x\theta_y + \pi\mu_y\theta_x + (1-\pi)\theta_x\theta_y\}}{\pi^2}. \tag{11.5.3.16}$$

Thus we have

$$\mathrm{Cov}(X,Y) = E(XY) - E(X)E(Y) = \frac{\sigma_{z_1z_2}}{\pi^2}. \tag{11.5.3.17}$$

From (11.5.3.13), (11.5.3.14), and (11.5.3.17) the correlation between two characters X and Y of the hidden gang G is given by

$$\rho_{xy} = \frac{\sigma_{z_1z_2}}{\pi\sqrt{\sigma_{z_1}^2 - (1-\pi)\{\pi(\theta_x - \mu_x)^2 + \gamma_x^2\}}\sqrt{\sigma_{z_1}^2 - (1-\pi)\{\pi(\theta_y - \mu_y)^2 + \gamma_y^2\}}}. \tag{11.5.3.18}$$

By the method of moments it proves the theorem.

The difference between the estimator of correlation coefficient developed at (11.5.3.18) and that developed by Bellhouse (1995) and Singh (1991b) is that this formula estimates the correlation between two sensitive characters of a particular hidden gang, whereas Bellhouse's technique estimates the correlation between two sensitive characters of the whole population.

11.6 UNIFIED APPROACH FOR HIDDEN GANGS

Let P be a finite population of size N (known) and N_G (unknown) be the total number of persons belonging to some sensitive group G. Let X_i be the value of a stigmatized quantitative variable, X say, for a person i in the sensitive group (Gang) G. Independent samples $s_k, k = 1,2$ of sizes n_k are selected from population P following the sampling designs p_k. Let $\pi_i(k)$ and $\pi_{ij}(k)$ be the first and second order inclusion probabilities for the sampling design p_k. If the respondent labelled i is selected in the sample s_k, then he/she has to disclose the true response X_i if he belongs to the group G; otherwise he has to produce a randomize response R_{ki} following a certain randomized design. In other words, the persons appearing in the kth sample are provided with a randomization device R_k (say). The device may be a spinner, deck of cards or computer grid which generates a random variable taking the value greater than or equal to Ψ_0 (say) from a given probability distribution (Normal, Weibull or Gamma etc.). Each respondent i selected in the sample is requested to draw one random number (which is obviously greater than or equal to Ψ_0) from the randomization device R_k, without showing it to the interviewer. It is to be noted that for the chosen randomization device R_k, being almost similar in range with X_i is better for collecting sample data, especially in a sensitive survey. Now he/she is requested to report the actual value of the stigmatized quantitative variable, say X_i, iff he/she belongs to gang G; otherwise he/she is requested to report the random number R_{k_i}, say, as drawn. The value of Ψ_0 depends upon the problem under consideration. For example, under problem 1, the value of Ψ_0 is \$80,000; under problems 2 and 3, the value of Ψ_0 can be taken as zero or any other suitable value. The choice of Ψ_0 can be made such that the respondents' privacy will not be jeopardised if they respond honestly. It is assumed that the distribution of the randomization device, R_k, is known to the interviewer but not the number drawn by the respondent. Let θ_k and σ_k^2 denote the known mean and variance of the randomization device R_k. Let π denote the proportion of persons belonging to the gang G in the population. Thus for the ith unit (if it is included in the sample s_k), we obtain a response

$$Z_{ki} = \begin{cases} X_i & \text{if } i \in G \\ R_{ki} & \text{if } i \notin G \end{cases} \quad (11.6.1)$$

where X_i and R_{ki} denote, respectively, the actual value of the stigmatized quantitative character and the random variable drawn by the ith respondent in the kth sample, $k = 1, 2$. Arnab and Singh (2002b) consider the problem of estimation of the population proportion, given by

$$\pi = N_G/N \quad (11.6.2)$$

and the population mean of the desired sub-group or hidden gang given by

$$\mu_x = \frac{1}{N_G}\sum_{i\in G} X_i \,, \tag{11.6.3}$$

where X_i is the value of the quantitative characteristic associated with the respondent. Let G be the population of the gang, $\overline{G} = P - G$, and $I_i = \begin{cases}1 & \text{if } i \in G \\ 0 & \text{if } i \notin G\end{cases}$.

The randomized response from the ith unit for the sample s_k as given in (11.6.1) can therefore be written as

$$Z_{ki} = X_i I_i + (1 - I_i)R_{ki} = X_i I_i + R_{ki} I_i' \,, \tag{11.6.4}$$

where $I_i' = 1 - I_i$. Denoting the expectation and variance as E_R, V_R, respectively with respect to the randomization device, we have

$$E_R(Z_{ki}) = X_i I_i + E_R(R_{ki}) I_i' = X_i I_i + \theta_k I_i' = \gamma_i(k) \text{ (say)} \tag{11.6.5}$$

and

$$V_R(Z_{ki}) = I_i' V_R(R_{ki}) = I_i' \sigma_k^2 \,. \tag{11.6.6}$$

Now consider the following linear homogenous unbiased estimator for $E_R(\overline{Z}_k)$, where $\overline{Z}_k = \frac{1}{N}\sum_{i=1}^{N} Z_{ki}$ based on the sampling design p_k as

$$T_k = \frac{1}{N}\sum_{i\in s_k} b_{s_k i} Z_{ki} \tag{11.6.7}$$

with $b_{s_k i}$ being the known constants satisfying design unbiasedness (p_k unbiasedness) condition given by

$$\sum_{s_k \ni i} b_{s_k i} p_k(s_k) = 1 \,. \tag{11.6.8}$$

Denoting the expectation and variance as by E_p, V_p, respectively, with respect to the sampling design p_k, we have the following theorems:

Theorem 11.6.1. The estimator T_k is an unbiased estimator for

$$E_R(\overline{Z}_k) = \pi \mu_x + (1 - \pi)\theta_k \,. \tag{11.6.9}$$

Proof. We have

$$E(T_k) = E_p E_R(T_k) = \frac{1}{N} E_p \sum_{i\in s_k} b_{s_k i}\gamma_i = \frac{1}{N}\sum_i \gamma_i(k) = \frac{1}{N}\sum_i \left\{X_i I_i + I_i' \theta_k\right\}$$

$$= \frac{1}{N}\sum_{i\in G} X_i + \frac{\theta_k}{N}\sum_i I_i' = \frac{N_G}{N}\frac{1}{N_G}\sum_{i\in G} X_i + \theta_k \frac{(N - N_G)}{N} = \pi\mu_x + (1 - \pi)\theta_k \,.$$

Hence the theorem.

Theorem 11.6.2. The variance of the estimator T_k is given by

$$V_k = V(T_k) = \frac{1}{N^2}\left[\sigma_k^2 \sum_i \alpha_i(k) I_i' + \left\{\sum_i \gamma_i^2(k)(\alpha_i(k)-1) + \sum_{i\neq j}\sum \gamma_i(k)\gamma_j(k)(\alpha_{ij}(k)-1)\right\}\right] \quad (11.6.10)$$

where $\alpha_i(k) = \sum_{s_k \ni i} b_{s_k i}^2 p_k(s_k)$ and $\alpha_{ij}(k) = \sum_{s_k \ni i,j} b_{s_k i} b_{s_k j} p_k(s_k)$.

Proof. We have

$$N^2 V(T_k) = N^2\left[E_p\left[V_R(T_k)\right] + V_p\left[E_R(T_k)\right]\right] = E_p\left[\sum_{i\in s_k} b_{s_k i}^2 \sigma_k^2 I_i'\right] + V_p\left[\sum_{i\in s_k} b_{s_k i}\gamma_i(k)\right]$$

$$= \sigma_k^2 \sum_i I_i' \sum_{s_k \ni i} b_{s_k i}^2 p_k(s_k) + \sum_i \gamma_i^2(k) \sum_{s_k\ni i} b_{s_k i}^2 p_k(s_k)$$

$$+ \sum_{i\neq j}\sum \gamma_i(k)\gamma_j(k) \sum_{s_k \ni i,j} b_{s_k i} b_{s_k j} p_k(s_k) - \left(\sum_{i=1}^{N}\gamma_i(k)\right)^2$$

$$= \sigma_k^2 \sum_i \alpha_i(k) I_i' + \left\{\sum_i \gamma_i^2(\alpha_i(k)-1) + \sum_{i\neq j}\sum \gamma_i(k)\gamma_j(k)(\alpha_{ij}(k)-1)\right\}.$$

Hence the theorem.

Theorem 11.6.3. An unbiased estimator of the variance $V(T_k)$ is given by

$$\hat{\hat{V}}_k = \hat{V}_k + \frac{\sigma_k^2}{N}\left(1-\hat{\pi}(k)\right),$$

where

$$\hat{V}_k = \frac{1}{N^2}\left[\sum_{i\in s_k}\frac{Z_{ki}^2}{\pi_i(k)}(\alpha_i(k)-1) + \sum_{i\neq j \in s_k}\sum \frac{Z_{ki}Z_{kj}}{\pi_{ij}(k)}(\alpha_{ij}(k)-1)\right] \text{ and } \hat{\pi}(k) = \frac{1}{N}\sum_{i\in p_k}\frac{I_i}{\pi_i(k)}.$$

Proof. Consider

$$E(\hat{V}_k) = \frac{1}{N^2}E_p\left[\sum_{i\in s_k}\frac{\gamma_i^2(k)+\sigma_k^2 I_i'}{\pi_i(k)}(\alpha_i(k)-1) + \sum_{i\neq j\in s_k}\sum\frac{\gamma_i(k)\gamma_j(k)}{\pi_{ij}(k)}(\alpha_{ij}(k)-1)\right]$$

$$= \frac{1}{N^2}\left[\sum_i \gamma_i^2(k)(\alpha(k)_i - 1) + \sum_{i\neq j\in s_k}\sum \gamma(k)_i\gamma_j(k)(\alpha_{ij}(k)-1) + \sigma_k^2\sum_i(\alpha_i(k)-1)I_i'\right]$$

$$= V(T_k) - \frac{\sigma_k^2}{N^2}\sum_i I_i' = V(T_k) - \frac{\sigma_k^2(1-\pi(k))}{N}.$$

Hence the theorem.

Corollary 11.6.1. An alternative unbiased estimator for estimating $V(T_k)$ is

$$\hat{\hat{V}} = \hat{V}_k^* + \frac{\sigma_k^2}{N}\left(1-\hat{\pi}(k)\right), \quad (11.6.11)$$

where

$$\hat{V}_k^* = \frac{1}{N^2}\sum_{i<j\in s_k}\sum\left(\frac{\pi_i(k)\pi_j(k)-\pi_{ij}(k)}{\pi_{ij}(k)}\right)\left(\frac{Z_{ki}}{\pi_i(k)} - \frac{Z_{kj}}{\pi_j(k)}\right)^2 + \frac{\sigma_k^2(1-\hat{\pi}(k))}{N}$$

assuming each s_k $(k = 1, 2)$ with positive $p(s_k)$ to have a 'common' number of distinct units.

Theorem 11.6.4. An unbiased estimator of π is given by

$$\hat{\pi} = 1 - \frac{T_1 - T_2}{\theta_1 - \theta_2} \tag{11.6.12}$$

with variance

$$V(\hat{\pi}) = \frac{V(T_1) + V(T_2)}{(\theta_1 - \theta_2)^2} \tag{11.6.13}$$

and an unbiased estimator of $V(\hat{\pi})$ is given by

$$\hat{V}(\hat{\pi}) = \frac{\hat{\hat{V}}_1 + \hat{\hat{V}}_2}{(\theta_1 - \theta_2)^2}. \tag{11.6.14}$$

Proof. Straightforward from previous section.

The estimator of proportion in (11.6.12) becomes non-functional if $\theta_1 - \theta_2 = 0$. It is important to keep in mind while making a pair of randomization device such that $\theta_1 - \theta_2 \neq 0$. The expression (11.6.12) indicates that the variance of the proportion is inversely proportional to the difference $\theta_1 - \theta_2$. Thus one should choose θ_1 and θ_2 such that the difference $\theta_1 - \theta_2$ is maximum without threatening the privacy of the respondents.

We have

$$E(T_k) = \pi \mu_x + (1-\pi)\theta_k, \; k = 1, 2. \tag{11.6.15}$$

This implies that

$$\hat{\mu}_x = \frac{T_1 - (1-\hat{\pi})\theta_1}{\hat{\pi}} = \frac{\theta_1 - \theta_2}{(\theta_1 - \theta_2) - (T_1 - T_2)}\left\{T_1 - \frac{(T_1 - T_2)\theta_1}{\theta_1 - \theta_2}\right\}$$

$$= \frac{T_1\theta_1 - T_1\theta_2 - T_1\theta_1 + T_2\theta_1}{(T_2 - \theta_2) - (T_1 - \theta_1)} = \frac{T_2\theta_1 - T_1\theta_2}{(T_2 - \theta_2) - (T_1 - \theta_1)}. \tag{11.6.16}$$

Thus an estimator of μ_x is given by

$$\hat{\mu}_x = \frac{d_1}{d_2},$$

where $d_1 = T_2\theta_1 - T_1\theta_2$ and $d_2 = (T_2 - \theta_2) - (T_1 - \theta_1)$. Arnab and Singh (2002c) studied its special cases under different sampling designs such as Horvitz and Thompson (1952), and Rao, Hartley, and Cochran (1962) strategies, etc., and applied to real data.

11.7 RANDOMIZED RESPONSE TECHNIQUE FOR A QUANTITATIVE VARIABLE

Eichhorn and Hayre (1983) proposed an ingenious method to collect information on quantitative characters rather than qualitative characters. According to this model each respondent selected in the sample is requested to give the scrambled response $y_i = x_i s$, where x_i are the real values of the quantitative variable and s is the scrambling variable. It is reasonably assumed that the first two moments of s are known, *i.e.*, $E(s) = \theta$ and $V(s) = \gamma^2$ are known. They suggested an unbiased estimator of population mean

$$\bar{y}_0 = \frac{1}{n\theta}\sum_{i=1}^{n} y_i \tag{11.7.1}$$

with variance

$$V(\bar{y}_0) = \frac{1}{n}\left[\sigma_x^2 + \left(\frac{\gamma}{\theta}\right)^2 \left(\sigma_x^2 + \mu_x^2\right)\right], \tag{11.7.2}$$

where $y_i, y_2, ..., y_n$ are the scrambled responses obtained from n sampled respondents and μ_x and σ_x^2 denote the mean and the variance of the sensitive quantitative variable x under study. Mangat (1991), Mangat and Singh (1994) have proposed an optional randomized response technique for qualitative characters and have shown empirically that the proposed optional randomized response technique is superior to Warner's (1965) technique in certain situations. Singh and Joarder (1997a) proposed an optional randomized response technique for estimating the population mean of the quantitative sensitive variable. It is interesting to note that the proposed optional randomized response technique remains always more efficient than the well known technique of Eichhorn and Hayre (1983). In the optional RRT, each respondent selected in the SRSWR sample has either of the following two options:

(i) He can report his actual value of x_i; or

(ii) He can report the scrambled response $t_i = (x_i s)/\theta$

without revealing to the interviewer which mode has been followed for giving the response. Here we should stress the fact that t_i should have the same support as x_i, otherwise one immediately knows the answer is either from option (i) or (ii). Let W be the probability that the respondent selects the option (i) and $(1-W)$ will be the probability that he select the option (ii).

Then response z_i has the distribution:

$$z_i = \begin{cases} x_i & \text{with probability } W, \\ t_i & \text{with probability } (1-W). \end{cases} \tag{11.7.3}$$

Thus we have the following theorem:

Theorem 11.7.1. An unbiased estimator of μ_x is given by

$$\bar{y}_1 = \frac{1}{n}\sum_{i=1}^{n} z_i . \tag{11.7.4}$$

Proof. By the definition of the expected value we have

$$E(z_i) = WE(x_i) + (1-W)E(t_i) = \mu_x . \tag{11.7.5}$$

Hence the theorem.

In order to find the variance of the estimator in (11.7.4) we have the following lemma.

Lemma 11.7.1. The distribution of z_i^2 is given as follows:

$$z_i^2 = \begin{cases} x_i^2 & \text{with probability } W, \\ t_i^2 & \text{with probability } (1-W), \end{cases} \tag{11.7.6}$$

and the expected value of z_i^2 is given by

$$E\left(z_i^2\right) = WE\left(x_i^2\right) + (1-W)E\left(t_i^2\right) = W\left(\sigma_x^2 + \mu_x^2\right) + (1-W)\left(1 + \frac{\gamma^2}{\theta^2}\right)\left(\sigma_x^2 + \mu_x^2\right). \tag{11.7.7}$$

Thus we have the following theorem:

Theorem 11.7.2. The variance of the estimator $\bar{y}_1$ is given by

$$V(\bar{y}_1) = \frac{1}{n}\left[\sigma_x^2 + \left(\frac{\gamma}{\theta}\right)^2 \left(\sigma_x^2 + \mu_x^2\right)(1-W)\right]. \tag{11.7.8}$$

Proof. We have

$$V(\bar{y}_1) = V\left(\frac{1}{n}\sum_{i=1}^{n} z_i\right) = \frac{1}{n^2}\sum_{i=1}^{n} V(z_i) = \frac{\sigma_z^2}{n}. \tag{11.7.9}$$

Note that $\sigma_z^2 = E\left(z_i^2\right) - \{E(z_i)\}^2$, on using (11.7.5) and (11.7.7) we have

$$\sigma_z^2 = \sigma_x^2 + \left(\frac{\gamma}{\theta}\right)^2 \left(\sigma_x^2 + \mu_x^2\right)(1-W). \tag{11.7.10}$$

On putting the value of σ_z^2 from (11.7.10) in (11.7.9) we have (11.7.8). Hence the theorem.

Theorem 11.7.3. An unbiased estimator of the variance of $\bar{y}_1$ is given by

$$s_x^2 = \frac{1}{(n-1)}\sum_{i=1}^{n}(z_i - \bar{y}_1)^2 . \tag{11.7.11}$$

Proof. The estimator s_x^2 can easily be written as

$$s_x^2 = \frac{1}{n-1}\left[\sum_{i=1}^{n} z_i^2 + \bar{y}_1^2 - 2\bar{y}_1\sum_{i=1}^{n} z_i\right] = \frac{1}{n-1}\left[\sum_{i=1}^{n} z_i^2 - n\bar{y}_1^2\right]. \tag{11.7.12}$$

Taking expected values on both sides of (11.7.12) we have

$$E\left(s_x^2\right) = V\left(\bar{y}_1\right).$$

Hence the theorem.

The percent relative efficiency (PRE) of the estimator $\bar{y}_1$ with respect to $\bar{y}_0$ is

$$\text{PRE} = \left\{V\left(\bar{y}_0\right)/V\left(\bar{y}_1\right)\right\} \times 100\% . \tag{11.7.13}$$

On using (11.7.2) and (11.7.8) we have

$$\text{PRE} = \frac{C_x^2 + C_\gamma^2\left(1 + C_x^2\right)}{C_x^2 + C_\gamma^2\left(1 + C_x^2\right)(1 - W)} \times 100\% , \tag{11.7.14}$$

where $C_x = \dfrac{\sigma_x}{\mu_x}$ and $C_\gamma = \dfrac{\gamma}{\theta}$ have their usual meanings. Note that $0 \le W \le 1$, therefore (11.7.14) shows that PRE is always more than 100%.

Thus we have the following theorem:

Theorem 11.7.4. The estimator $\bar{y}_1$ is always more efficient than the estimator $\bar{y}_0$.

Remark 11.7.1. The exact value of W can never be known. In practice, the investigator/ interviewer does not need the actual value of W. The estimator $\bar{y}_1 = \frac{1}{n}\sum_{i=1}^{n} z_i$ at (11.7.4) and estimator of its variance $s_x^2 = (n-1)^{-1}\sum_{i=1}^{n}\left(z_i - \bar{y}_1\right)^2$ at (11.7.11) are free from the value of W.

Remark 11.7.2. The percent relative efficiency (PRE) defined at (11.7.14) is an increasing function of W. In other words, if W approaches one then PRE increases. Thus if the interviewer is interested to know how much he gained by using the optional model, a rough guess can be made by using an unbiased estimator of W. Each respondent selected in the sample can be provided another randomization device R_2 (say) with the following questions:

(a) Have you reported the actual value? with probability T;
(b) Have you reported scrambled response? with probability $(1 - T)$.

If $\hat{\theta}$ is the observed proportion of 'Yes' answers, then following Warner (1965) an unbiased estimator of W is given by

$$\hat{W} = \frac{\hat{\theta} - (1 - T)}{2T - 1}. \tag{11.7.15}$$

Remark 11.7.3. A rough guess can be made about W from a past survey or pilot survey. For example, if out of 100 persons 20 would like to report actual value truthfully then W can be taken as 0.2.

Remark 11.7.4. It is a fact that we have assumed W to be the same for all the respondents in the sample, which is somewhat restrictive. It is worth mentioning here that the case of unequal W for each respondent can be handled by a hierarchical (or empirical) Bayes' approach.

Example 11.7.1. Show graphically that the percent relative efficiency (PRE) of the optional randomized response technique is an increasing function of the proportion of respondents revealing direct answers.

Given: $C_x = 0.1$, and 0.7 .

Solution. The graphical representation of the PRE of the optional randomized response technique is as follows:

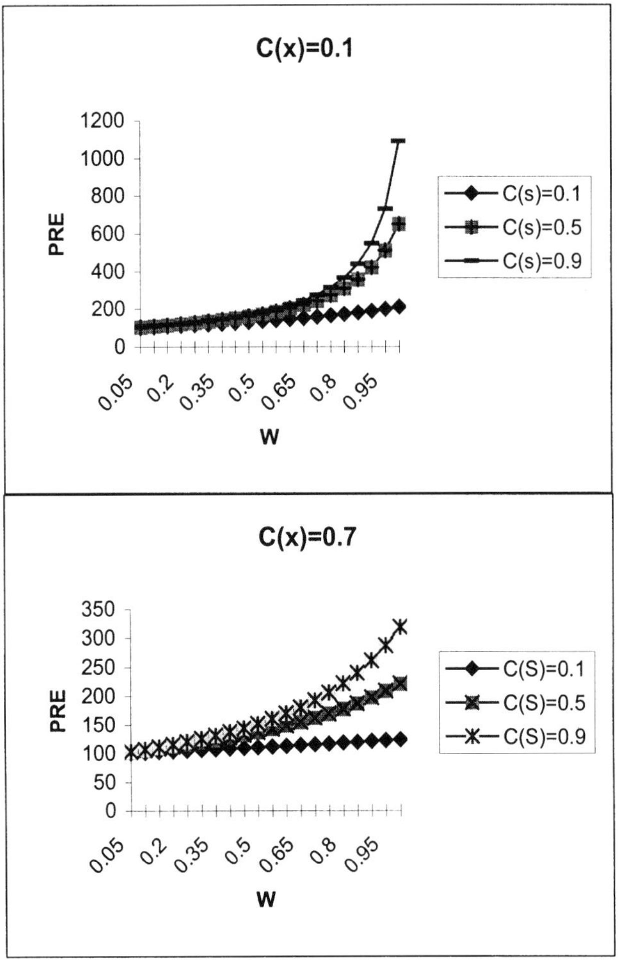

Fig. 11.7.1 Percent relative efficiency.

11.8 GREG USING SCRAMBLED RESPONSES

Following Chaudhuri and Mukherjee (1988), Chaudhuri and Roy (1997b) assumed that randomized response devices are available to produce a response r_i from the ith respondent in the sample such that

$$E_R(r_i) = y_i \quad \text{and} \quad V_i = V(r_i) = \alpha Y_i^2 + \tau Y_i + \theta \quad (\alpha > 0), \tag{11.8.1}$$

where α, τ and θ in (11.8.1) are constants and have their usual meanings. For example, consider a practicable randomization device proposed by Chaudhuri and Adhikary (1990). According to this device the ith respondent in the sample is required to choose independently at random two tickets numbered a_i and b_t out of boxes proposed by the investigator containing the tickets numbered (i) A_1, A_2, ..., A_m with known mean $\overline{A}$ and known variance σ_A^2, and (ii) B_1, B_2, ..., B_t with known mean $\overline{B}$ and variance σ_B^2.

The respondent is required to report the response as

$$Z_i = a_j Y_i + b_t\,.$$

Thus

$$E_R(Z_i) = \overline{A}\,Y_i + \overline{B}\,, \quad R_i = (Z_i - \overline{B})/\overline{A} \quad \text{and} \quad V_R(r_i) = V_i^2 = \left(\sigma_A^2 Y_i^2 + \sigma_B^2\right)/\overline{A}^2$$

where E_R and V_R denote the expected value and variance corresponding to the randomization device. Thus on comparing (11.8.1) with the above randomization device, we get $\alpha = \sigma_A^2/\overline{A}^2$, $\tau = 0$ and $\theta = \sigma_B^2/\overline{A}^2$. Chaudhuri and Mukerjee (1988) have also given an estimator for V_i as

$$\hat{v}_i = \frac{1}{1+\alpha_i}\left(\alpha_i r_i^2 + \beta_i r_i + \theta_1\right) \tag{11.8.2}$$

satisfying $E_R(r_i) = V_i$, $i \in s$ provided $1+\alpha_i \neq 0$. Consider an auxiliary variable x with known positive values x_i, with a total $X = \sum_{i=1}^{N} X_i$, is available. Assume the superpopulation model is given by

$$y_i = \beta x_i + e_i\,, \quad i = 1, 2, ..., N, \tag{11.8.3}$$

where $E_m(e_i) = 0$, $V_m(e_i^2) = \sigma_i^2$ and $C_m(e_i, e_j) = \sigma_{ij}$, $i \neq j$.

Then we have the following theorem:

Theorem 11.8.1. The regression predictor for estimating the population total Y, under randomized response sampling, is

$$t_r = X\hat{\beta}_Q(r) + \sum_{i\in s} R_i\left[r_i - x_i\hat{\beta}_Q(r)\right], \tag{11.8.4}$$

where Q_i and R_i are chosen subject to the condition

$$(1 - R_i\pi_i)/Q_i\pi_i x_i = \text{a constant} \;\; \forall \;\; i \in \Omega$$

$$\hat{\beta}_Q(r) = \sum_{i \in s} r_i x_i Q_i \Big/ \sum_{i \in s} x_i^2 Q_i \text{ and } e_i(r) = r_i - x_i \hat{\beta}_Q(r).$$

Following Chaudhuri and Roy (1997b) we have the following theorem:

Theorem 11.8.2. Two estimators for estimating the variance $V(t_r)$ are given by

$$\hat{V}_k(t_r) = \frac{1}{2} \sum_{i \neq j \in s} \sum d_{ij} \Theta_{ij} \left[\left(d_i a_{ki} e_i(r) - d_j a_{kj} e_j(r) \right)^2 - \left\{ \left(d_i a_{ki} \right)^2 \hat{F}_{ii} + \left(d_j a_{kj} \right)^2 \hat{F}_{jj} - 2 a_{ki} a_{kj} d_i d_j \hat{F}_{ij} \right\} \right]$$
$$+ \sum_{i \in s} \left(d_i a_{ki} \right)^2 \hat{V}_i$$

where $k = 1, 2$, $\hat{F}_{ij}$ are the sample analogue of the parameters obtained as $F_{ij} = E_R\left[e_i(r) - e_i\right]\left[e_j(r) - e_j\right]$, $d_i = 1/\pi_i$, $d_{ij} = 1/\pi_{ij}$, $\Theta_{ij} = \left(\pi_i \pi_j - \pi_{ij}\right)$, $a_{1i} = 1$, and

$$a_{2i} = 1 + \left(X - \sum_s \frac{x_i}{\pi_i} \right) \frac{x_i Q_i \pi_i}{\sum_s x_i^2 Q_i}.$$

The problem of estimation of variance of the linear regression estimator under the randomized response sampling has also been considered by Chaudhuri (1993), and Chaudhuri and Maiti (1994). Tracy and Singh (2000) considered the problem of estimation of variance of the general linear regression estimator under scrambled responses using low level and higher level calibration approach.

11.8.1 CALIBRATION OF SCRAMBLED RESPONSES

Tracy and Singh (2000) consider an estimator of population total Y in randomized response sampling as

$$\hat{Y}_S = \sum_{i=1}^{n} w_i r_i. \tag{11.8.1.1}$$

Then we have the following theorems:

Theorem 11.8.1.1. The conditional bias in the estimator $\hat{Y}_S$ is given by

$$B\left(\hat{Y}_S \mid s\right) = \sum_{i=1}^{N} \left(w_i \pi_i - 1\right) y_i. \tag{11.8.1.2}$$

Proof. Let E_D denote the expected value over the design D. Then

$$E\left(\hat{Y}_S\right) = E_D E_R\left(\hat{Y}_S\right) = E_D E_R\left(\sum_{i=1}^{n} w_i r_i\right) = E_D\left(\sum_{i=1}^{n} w_i y_i\right) = \sum_{i=1}^{N} w_i y_i \pi_i.$$

Taking its difference from the true value we have the theorem.

Theorem 11.8.1.2. The conditional variance of the estimator $\hat{Y}_S$ is given by

$$V\left(\hat{Y}_S \mid s\right) = \frac{1}{2} \sum_{i=1}^{N} \sum_{j(\neq i)=1}^{N} \left(\pi_i \pi_j - \pi_{ij}\right)\left(w_i y_i - w_j y_j\right)^2 + \sum_{i=1}^{N} w_i^2 V_i \pi_i. \tag{11.8.1.3}$$

Proof. Let V_D denote the variance over the design D then we have

$$V\left(\hat{Y}_S \mid s\right) = V_D E_R\left(\hat{Y}_S \mid s\right) + E_D V_R\left(\hat{Y}_S \mid s\right).$$

Hence the theorem.

Theorem 11.8.1.3. A conditionally unbiased estimator of $V(\hat{Y}_S)$ is given by

$$\hat{v}(\hat{Y}_S \mid s) = \frac{1}{2}\sum_{i=1}^{n}\sum_{j(\neq i)=1}^{n} d_{ij}\Theta_{ij}\frac{(w_i r_i - w_j r_j)^2}{\sqrt{(1+\alpha_i)(1+\alpha_j)}} + \sum_{i=1}^{n} w_i^2\,\hat{v}_i, \tag{11.8.1.4}$$

where $\sqrt{(1+\alpha_i)(1+\alpha_j)}$ denotes the bias adjustment factor due to randomized experiments in estimating the variance component. Several estimators are shown to be special cases of the estimator in (11.8.1.4). For simplicity, minimize the chi square distance function

$$\sum_{i=1}^{n}(w_i - d_i)^2 (d_i q_i)^{-1} \tag{11.8.1.5}$$

subject to the condition $\sum_{i=1}^{n} w_i X_i = T_X$ yields $w_i = d_i(1 + q_i X_i'\lambda)$, where λ denotes the Lagrange multiplier and the values of q_i are suitably chosen weights which results in different forms of the estimators. The resulting estimator $\hat{Y}_S$ at (11.8.1.1) becomes

$$\hat{Y}_G = \sum_{i=1}^{n} d_i r_i + \hat{\beta}_S\left(T_X - \sum_{i=1}^{n} d_i X_i\right) \tag{11.8.1.6}$$

where $\hat{\beta}_S = \left(\sum_{i=1}^{n} d_i q_i X_i X_i'\right)^{-1} \sum_{i=1}^{n} d_i q_i X_i r_i$ denotes the weighted estimator of the multiple regression coefficient in the presence of scrambled responses. This type of estimator has also been studied by Chaudhuri and Roy (1997b). The properties of the estimator of regression coefficient $\hat{\beta}_S$ under the general linear regression model have been studied by Singh, Joarder, and King (1996). Thus the general linear regression estimator (GREG) under scrambled responses is a special case of the calibration technique. Following the model assisted survey sampling approach of Chaudhuri and Roy (1997b) for scrambled responses, an estimator for the variance of the regression estimator under scrambled responses can be expressed in the form

$$\hat{v}(\hat{Y}_G \mid s) = \frac{1}{2}\sum_{i \neq j=1}^{n}\sum^{n} d_{ij}\Theta_{ij}\frac{(w_i \eta_i - w_j \eta_j)^2}{\sqrt{(1+\alpha_i)(1+\alpha_j)}} + \sum_{i=1}^{n} w_i^2\,\hat{v}_i, \tag{11.8.1.7}$$

where $\eta_i = r_i - \hat{\beta}_s x_i$ denotes the residual term under scrambled responses studied by Singh, Joarder, and King (1996). We have the following cases:

Case I. If $w_i = 1/\pi_i$ then we have

$$\hat{v}(\hat{Y}_S) = \frac{1}{2}\sum_{i \neq j=1}^{n}\sum^{n} \frac{d_{ij}\Theta_{ij}}{\sqrt{(1+\alpha_i)(1+\alpha_j)}}\left(\frac{r_i}{\pi_i} - \frac{r_j}{\pi_j}\right)^2 + \sum_{i=1}^{n}\frac{\hat{v}_i}{\pi_i} = \hat{V}(\hat{Y}_S)_{HT} \tag{11.8.1.8}$$

leading to the estimator of the variance of the usual Horvitz and Thompson (1952) estimator under RR sampling.

Case II. Under simple random sampling without replacement (SRSWOR) scheme, $\pi_i = \pi_j = n/N$ and $\pi_{ij} = n(n-1)/N(N-1)$ the estimator of variance reduces to

$$\hat{v}(\hat{Y}_{\mathrm{G}} \mid s) = \frac{N^2(1-f)}{n(n-1)} \sum_{i=1}^{n} \frac{\eta_i^2}{(1+\alpha_i)} + \frac{N}{n} \sum_{i=1}^{n} \hat{v}_i^2, \tag{11.8.1.9}$$

where $f = n/N$. Thus (11.8.1.9) denotes the usual estimator of variance of the regression estimator under scrambled responses.

Case III. If we choose $q_i = 1/x_i$ the strategy reduces to the usual ratio estimator of total under scrambled responses, say $\hat{Y}_{\mathrm{R}}$. Under SRSWOR sampling,

$$\hat{Y}_{\mathrm{R}} = N\bar{r}\left(\frac{\bar{X}}{\bar{x}}\right),$$

where $\bar{r} = \frac{1}{n}\sum_{i=1}^{n} r_i$, $\bar{x} = \frac{1}{n}\sum_{i=1}^{n} x_i$, $\bar{X} = \frac{1}{N}\sum_{i=1}^{N} x_i$ and the estimator of variance takes the form given by

$$\hat{v}(\hat{Y}_{\mathrm{R}} \mid s) = \frac{N^2(1-f)}{n(n-1)} \sum_{i=1}^{n} \frac{\eta_i^2}{(1+\alpha_i)}\left(\frac{X}{\hat{X}}\right)^2 + \frac{N}{n} \sum_{i=1}^{n} \hat{v}_i^2, \tag{11.8.1.10}$$

where $\eta_i = r_i - (\bar{r}/\bar{x})x_i$ and $\hat{X} = \frac{N}{n}\sum_{i=1}^{n} x_i$.

Thus we have the following theorems:

Theorem 11.8.1.4. A class of estimators for estimating the variance of the ratio estimator of population total Y, is given by

$$\hat{v}_g(\hat{Y}_{\mathrm{R}} \mid s) = \frac{N^2(1-f)}{n(n-1)} \sum_{i=1}^{n} \frac{\eta_i^2}{(1+\alpha_i)}\left(\frac{X}{\hat{X}}\right)^g + \frac{N}{n} \sum_{i=1}^{n} \hat{v}_i^2, \tag{11.8.1.11}$$

where g is a suitably chosen constant such that the variance of the estimator of variance is minimum.

Theorem 11.8.1.5. A general class of estimators for estimating the variance of the general regression estimator under scrambled responses is given by

$$\hat{v}_g(\hat{Y}_{\mathrm{G}} \mid s) = \frac{N^2(1-f)}{n(n-1)} \sum_{i=1}^{n} \frac{\eta_i^2}{(1+\alpha_i)}\left(\frac{X}{\hat{X}}\right)^g + \frac{N}{n} \sum_{i=1}^{n} \hat{v}_i^2, \tag{11.8.1.12}$$

where g is a suitably chosen constant.

In order to study the efficiency of the above estimators of variance, Tracy and Singh (2000) introduce two classes of estimators for estimating the variance

$$\hat{v}_1(\hat{Y}_{\mathrm{G}} \mid s) = \frac{N^2(1-f)}{n(n-1)} \sum_{i=1}^{n} \frac{\eta_i^2}{(1+\alpha_i)} H\left(\frac{X}{\hat{X}}\right) + \frac{N}{n} \sum_{i=1}^{n} \hat{v}_i^2, \tag{11.8.1.13}$$

and

$$\hat{v}_2(\hat{Y}_{\mathrm{G}} \mid s) = \left[\frac{N^2(1-f)}{n(n-1)} \sum_{i=1}^{n} \frac{\eta_i^2}{(1+\alpha_i)} + \frac{N}{n} \sum_{i=1}^{n} \hat{v}_i^2\right] H\left(\frac{X}{\hat{X}}\right), \tag{11.8.1.14}$$

where $H(\bullet)$ is a parametric function such that $H(1)=1$ satisfying certain regularity conditions.

11.8.2 HIGHER ORDER CALIBRATION OF THE ESTIMATORS OF VARIANCE UNDER SCRAMBLED RESPONSES

Following Singh, Horn, and Yu (1998) (refer to Chapter 5 for details) consider here another estimator of variance

$$\hat{v}_{\mathrm{N}}\left(\hat{Y}_{\mathrm{G}} \mid s\right)=\frac{1}{2}\sum_{i=1}^{n}\sum_{j(\neq i)=1}^{n}\Omega_{ij}\frac{\left(w_i\eta_i-w_j\eta_j\right)^2}{\sqrt{\left(1+\alpha_i\right)\left(1+\alpha_j\right)}}+\sum_{i=1}^{n}\phi_i\hat{v}_i\,, \tag{11.8.2.1}$$

where Ω_{ij} and ϕ_i are the new weights such that the distance between Ω_{ij} and $D_{ij}=d_{ij}\Theta_{ij}$, as well as that between ϕ_i and w_i^2, respectively, are minimum. Tracy and Singh (2000) considered two chi square type of distance functions

$$D_1=\frac{1}{2}\sum_{i=1}^{n}\sum_{j(\neq i)=1}^{n}\left(\Omega_{ij}-D_{ij}\right)^2\left(D_{ij}\Delta_{ij}\right)^{-1} \tag{11.8.2.2}$$

and

$$D_2=\frac{1}{2}\sum_{i=1}^{n}\left(\phi_i-w_i^2\right)^2\left(\delta_i w_i^2\right)^{-1}. \tag{11.8.2.3}$$

They assumes that in many situations the variance of the estimator $\hat{X}_{\mathrm{HT}}$ of population total X given by

$$V_{\mathrm{SYG}}\left(\hat{X}_{\mathrm{HT}}\right)=\frac{1}{2}\sum_{i=1}^{N}\sum_{j(\neq i)=1}^{N}\Theta_{ij}\left(d_i x_i-d_j x_j\right)^2$$

is known either from past surveys or can be calculated. The weights Ω_{ij} are chosen such that the chi square distance (11.8.2.2) is minimum subject to the second order calibration constraint

$$\frac{1}{2}\sum_{i=1}^{n}\sum_{j(\neq i)=1}^{n}\Omega_{ij}\left(d_i x_i-d_j x_j\right)^2=V_{\mathrm{SYG}}\left(\hat{X}_{\mathrm{HT}}\right). \tag{11.8.2.4}$$

Now minimization of (11.8.2.2), subject to (11.8.2.4), leads to the new optimal weights

$$\Omega_{ij}=D_{ij}+\frac{D_{ij}\Delta_{ij}\left(d_i x_i-d_j x_j\right)^2}{\frac{1}{2}\sum_{i=1}^{n}\sum_{j(\neq i)=1}^{n}D_{ij}\Delta_{ij}\left(d_i x_i-d_j x_j\right)^4}\left[V_{\mathrm{SYG}}\left(\hat{X}_{\mathrm{HT}}\right)-\hat{v}_{\mathrm{SYG}}\left(\hat{X}_{\mathrm{HT}}\right)\right] \tag{11.8.2.5}$$

where

$$\hat{v}_{\mathrm{SYG}}\left(\hat{X}_{\mathrm{HT}}\right)=\frac{1}{2}\sum_{i=1}^{n}\sum_{j(\neq i)=1}^{n}D_{ij}\left(d_i x_i-d_j x_j\right)^2.$$

Tracy and Singh (2000) introduced a new calibration constraint

$$\sum_{i=1}^{n}\phi_i\left[\upsilon_1 x_i^2+\upsilon_2 x_i+\upsilon_3\right]=\sum_{i=1}^{N}\left[U_1 X_i^2+U_2 X_i+U_3\right]=Q_X, \tag{11.8.2.6}$$

where $Q_X = \sum_{i=1}^{N}\left[U_1 X_i^2 + U_2 X_i + U_3\right]$ denotes the known quadratic expression or variance of the auxiliary variable and υ_i, $i = 1, 2, 3$ are the random variables such that $E\left[\upsilon_1 x_i^2 + \upsilon_2 x_i + \upsilon_3\right] = Q_X$. Minimization of (11.8.2.3), subject to (11.8.2.6), leads to the new calibration weights given by

$$\phi_i = w_i^2 + \frac{\delta_i w_i^2\left(\upsilon_1 x_i^2 + \upsilon_2 x_i + \upsilon_3\right)}{\sum_{i=1}^{n}\delta_i w_i^2\left(\upsilon_1 x_i^2 + \upsilon_2 x_i + \upsilon_3\right)^2}\left[Q_X - \sum_{i=1}^{n} w_i^2\left(\upsilon_1 x_i^2 + \upsilon_2 x_i + \upsilon_3\right)\right]. \qquad (11.8.2.7)$$

Substitution of (11.8.2.5) and (11.8.2.7) in (11.8.2.1) leads to the calibrated estimator of variance

$$\hat{v}_N\left(\hat{Y}_G \mid s\right) = \frac{1}{2}\sum_{i=1}^{n}\sum_{j(\neq i)=1}^{n} D_{ij}\frac{\left(w_i\eta_i - w_j\eta_j\right)^2}{\sqrt{(1+\alpha_i)(1+\alpha_j)}} + \hat{B}_1\left[V_{YG}\left(\hat{X}_{HT}\right) - \frac{1}{2}\sum_{i=1}^{n}\sum_{j(\neq i)=1}^{n} D_{ij}\left(\frac{x_i}{\pi_i} - \frac{x_j}{\pi_j}\right)^2\right]$$

$$+ \sum_{i=1}^{n} w_i^2\hat{v}_i^2 + \hat{B}_2\left[Q_X - \sum_{i=1}^{n} w_i^2\left(\upsilon_1 x_i^2 + \upsilon_2 x_i + \upsilon_3\right)\right] \qquad (11.8.2.8)$$

where

$$\hat{B}_1 = \frac{\frac{1}{2}\sum_{i=1}^{n}\sum_{j(\neq i)=1}^{n} D_{ij}\Delta_{ij}\left(d_i x_i - d_j x_j\right)^2 \frac{\left(w_i\eta_i - w_j\eta_j\right)^2}{\sqrt{(1+\alpha_i)(1+\alpha_j)}}}{\frac{1}{2}\sum_{i=1}^{n}\sum_{j(\neq i)=1}^{n} D_{ij}\Delta_{ij}\left(d_i x_i - d_j x_j\right)^4}$$

and

$$\hat{B}_2 = \frac{\sum_{i=1}^{n}\frac{\delta_i w_i^2}{1+\alpha_i}\left(\alpha_i r_i^2 + \gamma_i r_i + \theta_i\right)\left(\upsilon_1 x_i^2 + \upsilon_2 x_i + \upsilon_3\right)}{\sum_{i=1}^{n}\delta_i w_i^2\left(\upsilon_1 x_i^2 + \upsilon_2 x_i + \upsilon_3\right)^2}.$$

A large number of estimators can be shown to be special cases of the estimator (11.8.2.8).

Case I. Under the SRSWOR sampling, let $q_i = 1/x_i$, $\Delta_{ij} = \left(d_i x_i - d_j x_j\right)^{-2}$, and $\delta_i = \left(v_1 x_i^2 + v_2 x_i + v_3\right)^{-1}$. Also, for simplicity if we take $\upsilon_1 = 1$ and $\upsilon_2 = \upsilon_3 = 0$ then an estimator for estimating the variance of the ratio estimator under scrambled responses is given by

$$\hat{v}_N\left(\hat{Y}_R \mid s\right) = \frac{N^2(1-f)}{n(n-1)}\sum_{i=1}^{n}\frac{\eta_i^2}{(1+\alpha_i)}\left(\frac{X}{\hat{X}}\right)^2\left(\frac{S_x^2}{s_x^2}\right) + \frac{N}{n}\sum_{i=1}^{n}\hat{v}_i^2\left(\frac{X}{\hat{X}}\right)^2, \qquad (11.8.2.9)$$

where $s_x^2 = (n-1)^{-1}\sum_{i=1}^{n}(x_i - \bar{x})^2$ is an unbiased estimator of $S_x^2 = (N-1)^{-1}\sum_{i=1}^{N}\left(X_i - \bar{X}\right)^2$.

Case II. Under the SRSWOR sampling, a class of estimators for estimating the variance of the GREG under scrambled responses can be defined as follows:

$$\hat{v}_c\left(\hat{Y}_G \mid s\right)=\frac{N^2(1-f)}{n(n-1)}\sum_{i=1}^{n}\frac{\eta_i^2}{(1+\alpha_i)}F\left(\frac{X}{\hat{X}},\frac{S_x^2}{s_x^2}\right)+\frac{N}{n}\sum_{i=1}^{n}\hat{v}_i^2 G\left(\frac{X}{\hat{X}}\right), \tag{11.8.2.10}$$

where $F(\bullet,\bullet)$ and $G(\bullet)$ are parametric functions such that $F(1,1)=1$ and $G(1)=1$, satisfying certain regularity conditions. Under SRSWOR sampling a more general class of estimators has been suggested as

$$\hat{v}_c\left(\hat{Y}_G \mid s\right)=\left[\frac{N^2(1-f)}{n(n-1)}\sum_{i=1}^{n}\frac{\eta_i^2}{(1+\alpha_i)}+\frac{N}{n}\sum_{i=1}^{n}\hat{v}_i^2\right]F\left(\frac{X}{\hat{X}},\frac{S_x^2}{s_x^2}\right). \tag{11.8.2.11}$$

11.8.3 GENERAL CLASS OF ESTIMATORS

Let $u = X/\hat{X}_{\text{HT}}$ and $v = V\left(\hat{X}_{\text{HT}}\right)/\hat{v}\left(\hat{X}_{\text{HT}}\right)$, a wider class of estimators

$$\hat{v}_N\left(\hat{Y}_G \mid s\right)=\left[\frac{1}{2}\sum_{i=1}^{n}\sum_{j=1}^{n}D_{ij}\frac{\left(w_i\eta_i - w_j\eta_j\right)^2}{\sqrt{(1+\alpha_i)(1+\alpha_j)}}+\sum_{i=1}^{n}w_i^2\hat{v}_i^2\right]H(u,v), \tag{11.8.3.1}$$

where $H(u,v)$ is a parametric function of u and v such that $H(1,1)=1$, satisfying certain regularity conditions. All estimators obtained from the functions, $H(u,v)=u^{\alpha}v^{\beta}$, $H(u,v)=\dfrac{1+\alpha(u-1)}{1+\beta(v-1)}$, $H(u,v)=1+\alpha(u-1)+\beta(v-1)$, and $H(u,v)=\{1+\alpha(u-1)+\beta(v-1)\}^{-1}$ are special cases of the higher level calibration approach, with α and β unknown parameters in the function $H(u,v)$.

11.9 ON RESPONDENT'S PROTECTION: QUALITATIVE CHARACTERS

Comparison between different randomized response (RR) models have been performed by several research workers, for example Greenberg, Abul-Ela, Simmons, and Horvitz (1969), Moors (1971), and Dowling and Shachtman (1975). All these comparisons have concentrated solely on variances. The emphasis on variances, however, amounts to considering matters from the statistician's point of view only. In the general public's opinion a much more central question would be: To what extent do the different methods protect the privacy of the interviewees? Since the degree of privacy protection is an essential part of the RR procedures, and since greater privacy, in general, results in more cost in terms of the variance of the estimate, one obvious basis for comparing RR models is to compare variances only if the degree of privacy protection is held constant. First we would like to discuss such two measures for qualitative characters proposed by Leysieffer and Warner (1976), and Lanke (1975, 1976).

11.9.1 LEYSIEFFER AND WARNER'S MEASURE

A population is divided into two complementary sensitive groups, A and A^c with unknown proportions π and $(1-\pi)$, respectively. Let us consider a dichotomous response model where a typical response R is 'Yes' (say, Y) or 'No' (say, N). The conditional probabilities that a response R comes from an individual of groups A and A^c are $P(R \mid A)$ and $P(R \mid A^c)$, respectively. These quantities are at the investigator's disposal and are called design probabilities. The posterior probabilities that a respondent belongs to group A or A^c, respectively, when he/she reports R, are, say, $P(A \mid R)$ and $P(A^c \mid R)$ and are called revealing probabilities by following Lanke (1975) and Anderson (1976, 1977). Using the Bayes' theorem

$$P(A \mid R) = \frac{\pi P(R \mid A)}{\pi P(R \mid A) + (1-\pi)P(R \mid A^c)} \tag{11.9.1.1}$$

and

$$P(A^c \mid R) = 1 - P(A \mid R). \tag{11.9.1.2}$$

According to Leysieffer and Warner (1976), the response R is regarded as jeopardizing with respect to A or A^c if

$$P(A \mid R) > \pi \text{ or } P(A^c \mid R) > 1-\pi \tag{11.9.1.3}$$

respectively. Thus we have

$$\frac{P(A \mid R)}{P(A^c \mid R)}\left(\frac{1-\pi}{\pi}\right) = \frac{P(R \mid A)}{P(R \mid A^c)}. \tag{11.9.1.4}$$

It follows that if the right hand side of (11.9.1.4) is greater (less) than unity, then R is Jeopardizing with respect to A (A^c) in the sense that, with this response, a respondent genuinely of group A (A^c) rather than in group A^c (A) thus tilts the scale against himself/herself if A (A^c) is stigmatizing. From (11.9.1.3) and (11.9.1.4), Leysieffer and Warner (1976) proposed the natural measures of jeopardy carried by R about A and A^c, respectively, which are as follows

$$g(R \mid A) = P(R \mid A)/P(R \mid A^c) \quad \text{and} \quad g(R \mid A^c) = 1/g(R \mid A). \tag{11.9.1.5}$$

These functions are called Jeopardy functions. The response R is non-jeopardizing if and only if

$$g(R \mid A) = 1. \tag{11.9.1.6}$$

Clearly the probability of 'Yes' response is given by

$$\lambda = P(Y \mid A)\pi + (1-\pi)P(Y \mid A^c) = \left[P(Y \mid A) - P(Y \mid A^c)\right]\pi + P(Y \mid A^c). \tag{11.9.1.7}$$

If an SRSWR sample of size n is taken and $\hat{\lambda}$ is the sample proportion of 'Yes' answers, then following Warner (1965) an unbiased estimator of π is given by

$$\hat{\pi} = \frac{\hat{\lambda} - P(Y \mid A^c)}{P(Y \mid A) - P(Y \mid A^c)} \tag{11.9.1.8}$$

which is defined if and only if

$$P(Y \mid A) - P(Y \mid A^c) \neq 0 \tag{11.9.1.9}$$

that is, if and only if the condition $g(R \mid A) = 1$ is violated. In fact, we have proved that the existence of an unbiased estimator for π necessarily makes a response jeopardising with respect to either A or A^c.

The variance of $\hat{\pi}$ becomes

$$V(\hat{\pi})=\frac{\pi(1-\pi)}{n}+\frac{\pi g(Y\mid A)+(1-\pi)g\left(N\mid Y^{c}\right)}{n\{g(Y\mid A)-1\}\left\{g\left(N\mid A^{c}\right)-1\right\}}. \tag{11.9.1.10}$$

Assuming without loss of generality that $P(Y\mid A)>P\left(Y\mid A^{c}\right)$ so that $g(Y\mid A)>1$, $g\left(N\mid A^{c}\right)>1$ and hence 'Y' and 'N' are jeopardizing for A and A^{c}, respectively. On differentiating (11.9.10) partially with respect to $g(Y\mid A)$ and $g\left(N\mid A^{c}\right)$ we have

$$\frac{\partial V(\hat{\pi})}{\partial g(Y\mid A)}<0 \quad \text{and} \quad \frac{\partial V(\hat{\pi})}{\partial g\left(N\mid A^{c}\right)}<0, \text{ respectively.}$$

It indicates that for the sake of efficiency one needs as large magnitudes as possible for $g(Y\mid A)$ and $g\left(N\mid A^{c}\right)$ both above unity. From a practical point of view, regarding protection of privacy, one can fix some maximal allowable levels of $g(Y\mid A)$ and $g\left(N\mid A^{c}\right)$ (say, k_1 and k_2), respectively. After fixing $g(Y\mid A)$ and $g\left(N\mid A^{c}\right)$ at k_1 and k_2 the optimal choice of the design parameters for the particular RR model can be worked out. In this way one can work out the variance expressions for each RR model by substituting the values of design parameters and then can be compared at the same level of protection of privacy.

11.9.2 LANKE'S MEASURE

Lanke (1976) assumed that the member of A may hesitate to reveal which group he/she belongs to. On the other hand, a person who belongs to A^{c} is supposed to be quite willing to acknowledge the fact. It means the membership in A may be embarrassing while membership in A^{c} cannot be considered so. 'Embarrassment' must mean 'suspicion of belonging to A'. A reasonable conclusion is, then, that the larger the conditional probability of belonging to A given a certain answer, the greater the embarrassment caused by giving that response.

The conditional probability

$$\Pr[\text{Interviewee belongs to group } A \mid \text{Interviewee has given the response 'Yes'}]$$

denoted by $P(A\mid Y)$ can be defined as

$$P(A\mid Y)=\frac{\pi P(Y\mid A)}{\pi P(Y\mid A)+(1-\pi)P\left(Y\mid A^{c}\right)}. \tag{11.9.2.1}$$

Similarly when the response is 'No' then

$$P(A\mid N)=\frac{\pi P(N\mid A)}{\pi P(N\mid A)+(1-\pi)P\left(N\mid A^{c}\right)}. \tag{11.9.2.2}$$

Thus for randomized interviewers one method may be considered to be more protective than the other if

$$\psi=\text{Max.}[P(A\mid Y),\ P(A\mid N)] \tag{11.9.2.3}$$

is smaller for the former method than for the latter. The procedure of comparing two strategies (S_1 and S_2, say) at the same level of protection of the privacy can be summarised as follows. First derive the conditional probabilities $P(A|Y)$ and $P(A|N)$ using the design probabilities and also check $P(A|Y) > P(A|N)$ or vice-versa for both strategies. The two strategies, say P_1 and P_2, will then be considered equivalent from the protection of privacy point of view if $P_1(A|Y) = P_2(A|Y)$. On equating these two conditional probabilities a relationship is obtained between the design parameter (say, P_1) of one strategy and the design parameter (say, P_2) of the other strategy. Substituting this value of P_i into the variance expression of the first strategy and on comparing it with the variance of the second strategy, we can be in a position to assess the efficiency of the first strategy with respect to other strategy at the same level of protection of privacy. Using the above two measures Bhargava (1996) and Bhargava and Singh (2002) considered the comparison of Mangat and Singh (1990) and Mangat (1994) strategies with the pioneer model of Warner (1965).

11.9.3 MANGAT AND SINGH'S TWO-STAGE MODEL

In Mangat and Singh (1990) model, each interviewee in an SRSWR sample of n respondents is provided with two randomization devices R_1 and R_2.

The randomization device R_1 consists of two statements, namely,

(i) I belong to sensitive group A,
(ii) Go to randomization device R_2, represented with probabilities T and $(1-T)$, respectively.

The randomization device R_2 which uses two statements, namely,

(i) I belong to sensitive group A,
(ii) I belong to group A^c,

with known probabilities P_2 and $(1-P_2)$ respectively, is exactly the same as used by Warner (1965). The respondent is instructed to experience first the randomization device R_1. He/she is to use R_2 only if directed by the outcome of R_1. The respondent is required to answer 'Yes' if the outcome points to the attribute he/she possesses and answers 'No' if the complement of his/her status is pointed out by the outcome. The whole procedure is completed by the respondent, unobserved by the interviewer.

Then λ_2, the probability of a 'Yes' answer, is given by

$$\lambda_2 = T\pi + (1-T)[P_2\pi + (1-P_2)(1-\pi)] . \tag{11.9.3.1}$$

A flow chart of Mangat and Singh's two-stage model is given below:

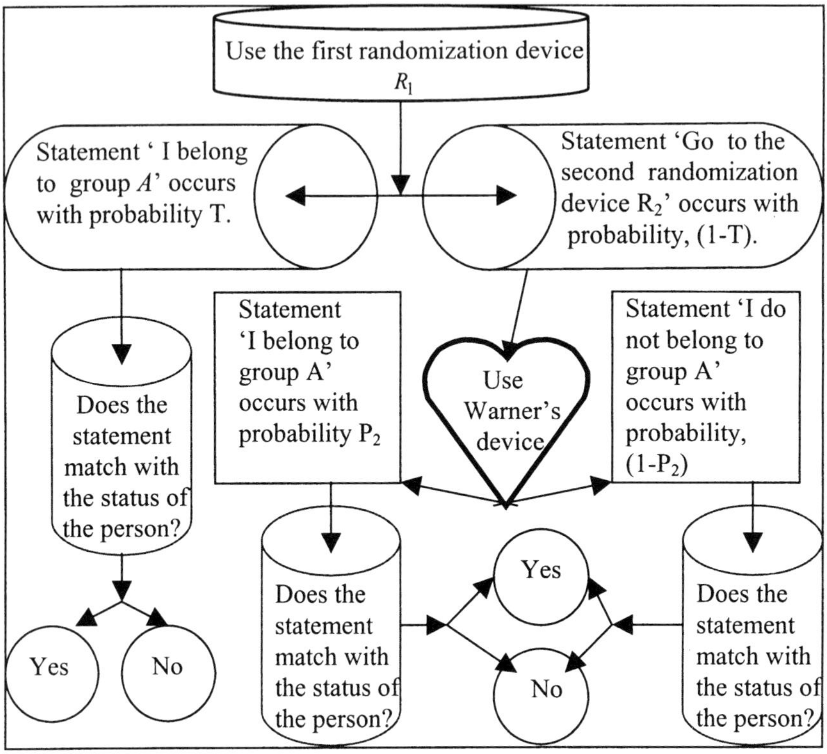

Fig. 11.9.3.1 Mangat and Singh's two-stage model.

Thus we have the following theorems:

Theorem 11.9.3.1. An unbiased estimator of π is given by

$$\hat{\pi}_1 = \frac{\hat{\lambda}_2 - (1-T)(1-P_2)}{2P_2 - 1 + 2T(1-P_2)}, \quad T \neq \frac{1-2P_2}{2(1-P_2)}. \tag{11.9.3.2}$$

Proof. Note that $\hat{\lambda}_2$ is the observed proportion of 'Yes' answers in the sample and it follows a binomial distribution $B(n, \lambda_2)$, which proves the theorem.

Theorem 11.9.3.2. The variance of the estimator $\hat{\pi}_1$ is given by

$$V(\hat{\pi}_1) = \frac{\pi(1-\pi)}{n} + \frac{(1-T)(1-P_2)\{1-(1-T)(1-P_2)\}}{n\{2P_2 - 1 + 2T(1-P_2)\}^2}. \tag{11.9.3.3}$$

Proof. Note that $\hat{\lambda}_2$ is the proportion of 'Yes' answers in the sample and it follows a binomial distribution $B(n, \lambda_2)$, therefore

$$V(\hat{\pi}_1) = V\left[\frac{\hat{\lambda}_2 - (1-T)(1-P_2)}{2P_2 - 1 + 2T(1-P_2)}\right] = \frac{\lambda_2(1-\lambda_2)}{n\{2P_2 - 1 + 2T(1-P_2)\}^2}.$$

On substituting the value of λ_2 and after simplification we obtain (11.9.3.3). Hence the theorem.

Corollary 11.9.3.1. For $P_2 = P_1$ (say) and $T = 0$ the Mangat and Singh (1990) model reduces to Warner (1965) model, or W model.

Corollary 11.9.3.2. The $V(\hat{\pi}_1) < V(\hat{\pi}_w)$ if $T > (1-2P_1)/(1-P_1)$ which shows the estimator $\hat{\pi}_1$ can always be made more efficient than the usual Warner estimator, $\hat{\pi}_w$ (say) by suitably choosing the value of T for any practicable value of P_1.

11.9.4 MANGAT AND SINGH'S TWO-STAGE AND WARNER'S MODEL AT EQUAL LEVEL OF PROTECTION

Here we consider two different measures as follows.

(a) Leysieffer and Warner's measure: For the Warner (1965) model the design probabilities are given by

$$P(Y \mid A) = P(N \mid A^c) = P_1 \quad \text{and} \quad P(N \mid A) = P(Y \mid A^c) = 1 - P_1 . \tag{11.9.4.1}$$

It can be easily checked that the condition

$$P(Y \mid A) > P(Y \mid A^c) \tag{11.9.4.2}$$

holds if $P_1 > 0.5$.

The jeopardy functions are given by

$$g_w(Y \mid A) = \frac{P(Y \mid A)}{P(Y \mid A^c)} = \frac{P_1}{(1-P_1)} \quad \text{and} \quad g_w(N \mid A^c) = \frac{P(N \mid A^c)}{P(N \mid A)} = \frac{P_1}{(1-P_1)}, \tag{11.9.4.3}$$

where the suffix w stands for Warner's model. Let k_1 and k_2 be the maximal allowable values of $g_w(Y \mid A)$ and $g_w(N \mid A^c)$. If $k_1 = k_2 = k$ (say) then maximization of $g_w(Y \mid A)$ and $g_w(N \mid A^c)$ leads to a design with $P_1/(1-P_1) = k$, that is, $P_1 = k/(1+k)$. If, however, $k_1 \neq k_2$, as $g_w(Y \mid A) = g_w(N \mid A^c)$, different upper bounds for them cannot be attained simultaneously. In that case if, without loss of generality, $k_1 < k_2$ one should chose a design such that

$$P_1 = k_1/(k_1 + 1). \tag{11.9.4.4}$$

For the Mangat and Singh (1990) model we have the design probabilities as

$$P(Y \mid A) = P(N \mid A^c) = T + (1-T)P_2 \quad \text{and} \quad P(N \mid A) = P(Y \mid A^c) = (1-T)(1-P_2). \tag{11.9.4.5}$$

Then we have the following theorem.

Theorem 11.9.4.1. The inequality $P(Y \mid A) > P(Y \mid A^c)$ holds if $P_2 > (1-2T)/2(1-T)$.

Proof. On substituting the values of $P(Y \mid A)$ and $P(Y \mid A^c)$ in $P(Y \mid A) > P(Y \mid A^c)$ we have

$$T+(1-T)P_2 > (1-T)(1-P_2) \text{ or } 1-2P_2 < \frac{T}{1-T} \text{ or } \frac{1-2T}{1-T} < 2P_2 \text{ or } P_2 > \frac{1-2T}{2(1-T)}.$$

Hence the theorem.

Now the jeopardy functions for the Mangat and Singh model are given by

$$g_{ms}(Y \mid A) = \frac{T+(1-T)P_2}{(1-T)(1-P_2)} \quad \text{and} \quad g_{ms}(N \mid A^c) = \frac{T+(1-T)P_2}{(1-T)(1-P_2)}. \tag{11.9.4.6}$$

Let k_1 and k_2 be the maximum allowable values of $g_{ms}(Y \mid A)$ and $g_{ms}(N \mid A^c)$. If $k_1 = k_2 = k$ (say) maximization of these jeopardy functions leads to a design with

$$\frac{T+(1-T)P_2}{(1-T)(1-P_2)} = k \text{ or } T+(1-T)P_2 = k(1-T)(1-P_2) \text{ or } P_2 = \frac{k(1-T)-T}{(1-T)(1+k_1)}. \tag{11.9.4.7}$$

Thus $P_1 = k_1/(k_1+1)$ and $P_2 = \{k_1(1-T)-T\}/\{(1-T)(1+k_1)\}$ are the optimal choices for design parameters of the W model and MS model respectively.

Now we have the following theorems.

Theorem 11.9.4.2. With the optimal choice $P_1 = k_1/(k_1+1)$ of the design parameter, the variance of the unbiased estimator $\hat{\pi}_w$ is given by

$$V(\hat{\pi}_w) = \frac{\pi(1-\pi)}{n} + \frac{k_1(k_1-1)^{-2}}{n}. \tag{11.9.4.8}$$

Proof. On substituting the value of P_1 in the expression on variance of W model, we have

$$V(\hat{\pi}_w) = \frac{\pi(1-\pi)}{n} + \frac{P_1(1-P_1)}{n(2P_1-1)^2} = \frac{\pi(1-\pi)}{n} + \frac{\frac{k_1}{k_1+1}\left(1-\frac{k_1}{k_1+1}\right)}{n\left(2\frac{k_1}{k_1+1}-1\right)^2}$$

$$= \frac{\pi(1-\pi)}{n} + \frac{k_1(k_1-1)^{-2}}{n}.$$

Hence the theorem.

Theorem 11.9.4.3. With the optimal choice of design parameter

$$P_2 = \{k_1(1-T)-T\} / \{(1-T)(1+k_1)\}$$

the variance of the unbiased estimator $\hat{\pi}_1$ under the MS model is given by

$$V(\hat{\pi}_1) = \frac{\pi(1-\pi)}{n} + \frac{k_1(k_1-1)^{-2}}{n}. \tag{11.9.4.9}$$

Proof. It follows on substituting the value of P_2 in

$$V(\hat{\pi}_1) = \frac{\pi(1-\pi)}{n} + \frac{(1-T)(1-P_2)\{1-(1-T)(1-P_2)\}}{n\{2P_2-1+2T(1-P_2)\}^2}.$$

Hence the theorem.

Thus for the optimum choices of P_1 and P_2 we have $V(\hat{\pi}_1)=V(\hat{\pi}_w)$, hence we conclude that both MS model and W model are equally efficient at the same level of protection of privacy.

(b) Lanke's measure: According to this measure and considering Warner (1965) model, the revealing probabilities are given by

$$P_w(A \mid Y) = \frac{\pi P_1}{\pi P_1 + (1-\pi)(1-P_1)} \tag{11.9.4.10}$$

and

$$P_w(A \mid N) = \frac{\pi(1-P_1)}{\pi(1-P_1)+(1-\pi)P_1}. \tag{11.9.4.11}$$

To find the $\text{Max}[P_w(A \mid Y), P_w(A \mid N)]$, we have the following theorem.

Theorem 11.9.4.4. Revealing probabilities satisfy the inequalities

(i) $P_w(A \mid Y) > P_w(A \mid N)$ if $P_1 > 0.5$; (ii) $P_w(A \mid N) > P_w(A \mid Y)$ if $P_1 < 0.5$.

Proof. Let us prove first inequality and then second is obvious. On substituting the values of $P_w(A \mid Y)$ and $P_w(A \mid N)$ in the first inequality, we have

$$\frac{\pi P_1}{\pi P_1 + (1-\pi)\,(1-P_1)} > \frac{\pi(1-P_1)}{\pi(1-P_1)+(1-\pi)\,P_1} \quad \text{or} \quad (1-\pi)P_1^2 > (1-\pi)(1-P_1)^2$$

or $P_1 > (1-P_1)$ or $P_1 > 0.5$.

Hence the theorem.

Thus a measure of protection of privacy in W model is given by

$$\psi_w = \begin{cases} P_w(A \mid Y) & \text{if } P_1 > 0.5, \\ P_w(A \mid N) & \text{if } P_1 < 0.5. \end{cases} \tag{11.9.4.12}$$

For the MS model the revealing probabilities are given by

$$P_{\text{ms}}(A \mid Y) = \frac{\pi[T+(1-T)P_2]}{\pi\{T+(1-T)P_2\} + (1-\pi)(1-T)(1-P_2)} \tag{11.9.4.13}$$

and,

$$P_{\text{ms}}(A \mid N) = \frac{\pi(1-T)(1-P_2)}{\pi(1-T)(1-P_2) + (1-\pi)\{T+(1-T)P_2\}}. \tag{11.9.4.14}$$

Theorem 11.9.4.5. The $\text{Max}[P_{\text{ms}}(A \mid Y), P_{\text{ms}}(A \mid N)]$ occurs if the conditional probabilities are satisfying the following inequalities:

(i) $P_{\text{ms}}(A \mid Y) > P_{\text{ms}}(A \mid N)$ if $P_2 > (1-2T)/2(1-T)$; (11.9.4.15)

(ii) $P_{\text{ms}}(A \mid N) > P_{\text{ms}}(A \mid Y)$ if $P_2 < (1-2T)/2(1-T)$. (11.9.4.16)

Proof. On using $P_{\text{ms}}(A \mid Y)$ and $P_{\text{ms}}(A \mid N)$ in the inequality $P_{\text{ms}}(A \mid Y) > P_{\text{ms}}(A \mid N)$, we have

$$\frac{\pi[T+(1-T)P_2]}{\pi\{T+(1-T)P_2\}+(1-\pi)(1-T)(1-P_2)} > \frac{\pi(1-T)(1-P_2)}{\pi(1-T)(1-P_2)+(1-\pi)\{T+(1-T)P_2\}},$$

or

$$\pi(1-\pi)[T+(1-T)P_2]^2 > \pi(1-\pi)(1-T)^2(1-P_2)^2,$$

or

$$[T+(1-T)P_2] > (1-T)(1-P_2),$$

or

$$P_2 > \frac{1-2T}{2(1-T)}.$$

Which proves the first part of the theorem. The second part of the theorem can be similarly proved. Hence the theorem Therefore the measure of protection of privacy in the MS model is given by

$$\psi_{ms} = \begin{cases} P_{ms}(A\mid Y) & \text{if } P_2 > (1-2T)/2(1-T), \\ P_{ms}(A\mid N) & \text{if } P_2 < (1-2T)/2(1-T). \end{cases} \tag{11.9.4.17}$$

Thus there are two measures of protection of privacy in each strategy. We shall consider the following four cases for comparing these two strategies.

Case 1. On setting $\psi_w = \psi_{ms}$ we have

$$P_1(1-\pi)(1-T)(1-P_2) = (1-\pi)(1-P_1)[T+(1-T)P_2],$$

and in turn we have

$$P_1 = T+(1-T)P_2.$$

For such a choice of P_1 the variance of W model reduces to MS model.

Case II. On setting $\psi_w = \psi_{ms}$ we have

$$P_1(1-\pi)[T+(1-T)P_2] = (1-\pi)(1-T)(1-P_1)(1-P_2),$$

which in turn reduces to $P_1 = (1-T)(1-P_2)$.

For such a choice of P_1 the variance of W model reduces to MS model.

Case III. Again on equating $\psi_w = \psi_{ms}$ we have

$$(1-\pi)(1-T)(1-P_1)(1-P_2) = P_1(1-\pi)[T+(1-T)P_2],$$

which again results in

$$P_1 = (1-T)(1-P_2).$$

Case IV. On equating

$$(1-P_1)(1-\pi)[T+(1-T)P_2] = (1-\pi)P_1(1-T)(1-P_2)$$

which on solving reduces to

$$P_1 = T+(1-T)P_2.$$

Thus the variance expression of the MS model and the W model under Leysieffer and Warner's measure remains the same. Hence both the MS model (or two-stage model) and the W model are equally efficient at the same level of protection of the respondents. Nayak (1994) has also reported a similar conclusion.

Remark 11.9.1. Although the MS model and the W model are theoretically performing the same, but are psychologically different to the respondents in a sample, and thus from psychological point of view one may expect more co-operation in the MS model than with the W model.

11.9.5 MANGAT'S MODEL

Mangat (1994) proposed a method in which each of the n respondents selected through SRSWR sampling is instructed to say 'Yes' if he/she belongs to sensitive group, A. If he/she does not belong to group A the respondent is required to use the Warner (1965) randomization device which consists of two statements.

(i) I belong to sensitive group A; (ii) I belong to group A^c

represented with probabilities P_3 and $(1-P_3)$ respectively. The respondent is required to report 'Yes' or 'No' according to the outcome of this device and the actual status that he/she has with respect to the sensitive group A. The whole procedure is completed by the respondent, unobserved by the interviewer. Since the 'Yes' answer may come from respondents in group A and group A^c, the probability of 'Yes' answer for this M model (say) is given by

$$\lambda_3 = \pi + (1-P_3)(1-\pi). \tag{11.9.5.1}$$

Then we have the following theorems:

Theorem 11.9.5.1 An unbiased estimator of π is given by

$$\hat{\pi}_2 = \frac{\hat{\lambda}_3 - (1-P_3)}{P_3}. \tag{11.9.5.2}$$

Proof. Note that $\hat{\lambda}_3$ is the observed proportion of "Yes" answers obtained from the selected sample, $E(\hat{\lambda}_3) = \lambda_3$, which proves the theorem.

Theorem 11.9.5.2. The variance of the estimator $\hat{\pi}_2$ is given by

$$V(\hat{\pi}_2) = \frac{\pi(1-\pi)}{n} + \frac{(1-\pi)(1-P_3)}{nP_3}. \tag{11.9.5.3}$$

Proof. Note that $\hat{\lambda}_3 \sim B(n, \lambda_3)$, thus we have

$$V(\hat{\pi}_2) = V\left[\frac{\hat{\lambda}_3 - (1-P_3)}{P_3}\right] = \frac{V(\hat{\lambda}_3)}{P_3^2} = \frac{\lambda_3(1-\lambda_3)}{nP_3^2},$$

which on substituting the value of λ_3 proves the theorem.

11.9.6 MANGAT'S AND WARNER'S MODEL AT EQUAL LEVEL OF PROTECTION

Again we shall consider here two measures for comparing M model and W model.

(a) Leysieffer and Warner's measure: For the M model we have the design probabilities as follows:

$$P(Y \mid A) = 1, \quad P(N \mid A) = 0, \quad P(Y \mid A^c) = 1 - P_3, \text{ and } P(N \mid A^c) = P_3 .$$

Clearly the condition $P(Y \mid A) > P(Y \mid A^c)$ is true for $P_3 > 0$. Now the jeopardy functions are given by

$$g_2(Y \mid A) = 1/(1 - P_3) \text{ and } g_2(N \mid A^c) = \infty . \tag{11.9.6.1}$$

Note that the second jeopardy function is infinite, we can take maximal allowable limit for $g_2(Y \mid A)$ as k_1, that is,

$$\frac{1}{1-P_3} = k_1, \text{ or } P_3 = [k_1 - 1]/k_1, \tag{11.9.6.2}$$

which is the optimal choice of design parameters for the M model. Then we have the following theorem.

Theorem 11.9.6.1. With the optimal choice of design parameter $P_3 = (k_1 - 1)/k_1$ the variance of the unbiased estimator $\hat{\pi}_2$ is given by

$$V(\hat{\pi}_2) = \frac{\pi(1-\pi)}{n} + \frac{(1-\pi)(k_1 - 1)^{-1}}{n} . \tag{11.9.6.3}$$

Proof. On substituting $P_3 = (k_1 - 1)/k_1$ in $V(\hat{\pi}_2)$ we have

$$V(\hat{\pi}_2) = \frac{\pi(1-\pi)}{n} + \frac{(1-\pi)(1-P_3)}{nP_3} = \frac{\pi(1-\pi)}{n} + (1-\pi)\left(1 - \frac{k_1 - 1}{k_1}\right) \Big/ n\left(\frac{k_1 - 1}{k_1}\right)$$

$$= \frac{\pi(1-\pi)}{n} + \frac{(1-\pi)(k_1 - 1)^{-1}}{n} .$$

Hence the theorem.

The variance expression given in the above theorem is different from the one for W model under the optimum choice of parameters. Thus we have the following theorem:

Theorem 11.9.6.2. The M model is always more efficient that the W model at equal level of protection of the respondents.

Proof. For the optimum choice of parameters we have

$$V(\hat{\pi}_w) = \frac{\pi(1-\pi)}{n} + \frac{k_1(k_1 - 1)^{-2}}{n} ,$$

and

$$V(\hat{\pi}_2) = \frac{\pi(1-\pi)}{n} + \frac{(1-\pi)(k_1 - 1)^{-1}}{n} .$$

Now $V(\hat{\pi}_2) < V(\hat{\pi}_w)$ if

$$(1-\pi)(k_1-1)^{-1} < k_1(k_1-1)^{-2},$$

which reduces to

$$(1-\pi) < k_1/(k_1-1). \tag{11.9.6.4}$$

Note that k_1 is always more than one, therefore the above inequality will always hold. This completes the proof.

(b) Lanke's measure: The revealing probabilities for the M model are given by

$$P_m(A \mid Y) = \frac{\pi}{\pi + (1-\pi)(1-P_3)} \quad \text{and} \quad P_2(A \mid N) = 0\,. \tag{11.9.6.5}$$

Clearly $P(A \mid Y) > P(A \mid N)$, hence the measure of protection of privacy to be used for this strategy is given by

$$\psi_m = P_m(A \mid Y). \tag{11.9.6.6}$$

On equating $\psi_w = P_w(A \mid Y)$ with ψ_m we have

$$P_1(1-\pi)(1-P_3) = (1-P_1)(1-\pi)$$

or

$$P_1 = \frac{1}{2-P_3}. \tag{11.9.6.7}$$

Thus we have the following theorem.

Theorem 11.9.6.3. The variance of the estimator $\hat{\pi}_w$ for $P_1 = 1/(2-P_3)$ is given by

$$V(\hat{\pi}_w) = \frac{\pi(1-\pi)}{n} + \frac{1-P_3}{nP_3^2}\,. \tag{11.9.6.8}$$

Proof. Obvious from the variance expression of the W model.

Theorem 11.9.6.4. The M model is always more efficient than the W model at an equal level of protection of the respondents.

Proof. We have

$$V(\hat{\pi}_w) = \frac{\pi(1-\pi)}{n} + \frac{1-P_3}{nP_3^2} \quad \text{and} \quad V(\hat{\pi}_2) = \frac{\pi(1-\pi)}{n} + \frac{(1-\pi)(1-P_3)}{nP_3}.$$

Now $V(\hat{\pi}_2) < V(\hat{\pi}_w)$ if

$$\frac{\pi(1-\pi)}{n} + \frac{(1-\pi)(1-P_3)}{nP_3} < \frac{\pi(1-\pi)}{n} + \frac{(1-P_3)}{nP_3^2}, \quad \text{or} \quad (1-\pi) < 1/P_3$$

since the right hand side of the above inequality is always greater than one. Hence the theorem.

11.10 ON RESPONDENT'S PROTECTION: QUANTITATIVE CHARACTERS

Let us first introduce a few RR models for quantitative data.

11.10.1 UNRELATED QUESTION MODEL FOR QUANTITATIVE DATA

Greenberg, Kuebler, Abernathy, and Horvitz (1971) suggested an extension of the Unrelated model (U model) to quantitative characters. They considered a sensitive variable X which is supposed to be continuous with true density $g(\bullet)$. The treatment remains similar for a discrete X, if one replaces everywhere a density function by the corresponding probability mass function. The unrelated character is Y with density $h(\bullet)$. The problem is to estimate μ_x, the unknown population mean of X. They considered both the cases when μ_y, the population mean of Y, is known and when it is not known. Here we will consider the case when μ_y is known. For this purpose an SRSWR sample of size n is drawn. Each respondent reports his/her X value with probability P and Y value with probability $Q = 1 - P$. Then a randomized response (RR), say Z, has the density

$$f(Z) = Pg(Z) + Qh(Z). \tag{11.10.1.1}$$

Therefore

$$E(Z) = PE(X) + QE(Y) \quad \text{and} \quad E\left(Z^2\right) = PE\left(X^2\right) + QE\left(Y^2\right) \tag{11.10.1.2}$$

and consequently the population mean μ_z and the population variance σ_z^2 of Z are

$$\mu_z = P\mu_x + Q\mu_y, \tag{11.10.1.3}$$

and

$$\sigma_z^2 = E\left(Z^2\right) - \{E(Z)\}^2 = P\sigma_x^2 + Q\sigma_y^2 + PQ\left(\mu_x - \mu_y\right)^2, \tag{11.10.1.4}$$

where σ_x^2 and σ_y^2 are the population variances of X and Y respectively. If the available responses are $Z_1, Z_2, ..., Z_n$ then we have sample mean and variance as

$$\bar{Z} = \frac{1}{n}\sum_{i=1}^{n} Z_i \quad \text{and} \quad s_z^2 = (n-1)^{-1}\sum_{i=1}^{n}\left(Z_i - \bar{Z}\right)^2$$

respectively, such that

$$E\left(\bar{Z}\right) = \mu_z \quad \text{and} \quad E\left(s_z^2\right) = \sigma_z^2.$$

Thus we have the following theorem:

Theorem11.10.1. An unbiased estimator of μ_x is

$$\hat{\mu}_x = \left(\bar{Z} - Q\mu_y\right)/P$$

with

$$V(\hat{\mu}_x) = \frac{V\left(\bar{Z}\right)}{P^2} = \frac{\sigma_Z^2}{nP^2} = \frac{1}{nP^2}\left[P\sigma_x^2 + Q\sigma_y^2 + PQ\left(\mu_x - \mu_y\right)^2\right]$$

and an unbiased estimator of $V(\hat{\mu}_x)$ is

$$\hat{v}(\hat{\mu}_x) = s_z^2 / \{nP^2\}.$$

Proof. Obvious using above results.

11.10.2 THE ADDITIVE MODEL

Pollock and Bek (1976) considered the additive model. In this model each respondent in an SRSWR sample is asked to sum his/her sensitive attribute (X) to a random value (Y) taken from a known distribution. The observed response, denoted by Z, is

$$Z = X + Y. \tag{11.10.2.1}$$

Here the random variable Y is distributed independently of sensitive attribute X. Then the mean and variance of Z are

$$\mu_z = \mu_x + \mu_y \quad \text{and} \quad \sigma_z^2 = \sigma_x^2 + \sigma_y^2.$$

If $Z_1, Z_2, ..., Z_n$ be the observed responses in a SRSWR sample of size n, then an unbiased estimator of μ_x is obtained as

$$\hat{\mu}_x = \bar{Z} - \mu_y \tag{11.10.2.2}$$

where $\bar{Z}$ denotes the sample mean of the observed responses. The variance of the estimator $\hat{\mu}_x$ is given by

$$V(\hat{\mu}_x) = \frac{\sigma_z^2}{n} = \frac{\sigma_x^2 + \sigma_y^2}{n}. \tag{11.10.2.3}$$

It is easy to obtain an unbiased estimator of $V(\hat{\mu}_x)$ by replacing σ_z^2 with s_z^2.

11.10.3 THE MULTIPLICATIVE MODEL

Pollock and Bek (1976) also considered the multiplicative model. This model was further considered by Eichhorn and Hayre (1983). According to them, each respondent in an SRSWR sample of n units is asked to multiply his/her X value by a random number Y taken from a known distribution and thus giving a scrambled response to the interviewer. They also referred to this model as scrambled RR method. The observed response (Z) is

$$Z = XY. \tag{11.10.3.1}$$

Here also the random variable Y is independent of X. The mean and variances of Z are

$$\mu_z = \mu_x \mu_y \quad \text{and} \quad \sigma_z^2 = \mu_y^2 \sigma_x^2 + \mu_x^2 \sigma_y^2 + \sigma_x^2 \sigma_y^2.$$

Then we have the following theorem.

Theorem 11.10.2. An unbiased estimator of μ_x is given by

$$\hat{\mu}_x = \bar{Z} / \mu_y, \tag{11.10.3.2}$$

where $\bar{Z}$ is the mean of observed sampled scrambled responses. The variance of $\hat{\mu}_x$ is given by

$$V(\hat{\mu}_x)=\frac{\sigma_z^2}{n\mu_y^2}=\frac{1}{n}\left[\sigma_x^2+\frac{\sigma_y^2\left(\mu_x^2+\sigma_x^2\right)}{\mu_y^2}\right]. \tag{11.10.3.3}$$

An unbiased estimator of $V(\hat{\mu}_x)$ can easily be obtained by replacing σ_z^2 with s_z^2.
Proof. Obvious from the above results.

11.10.4 MEASURE OF PRIVACY PROTECTION

Anderson (1977) developed a measure for privacy protection in case of quantitative characters. Following him, consider a population of individuals with some sensitive attribute $X \in \Omega_x$, distributed according to the unknown distribution $F_x(x)$. Our aim is to estimate the mean of X, or some proportion or distribution function of X. For this purpose a random sample of respondents to be interviewed is taken. However, for privacy reasons the sampled individuals may not be interviewed directly. Thus we have to ask questions in such a way that, from the answers, the population parameters can be estimated but the values of X for the respondents can not be inferred. This can be attained by introducing a randomization device known to the interviewer and producing an answer Z with a probability distribution dependent on the X value of the respondent. More precisely, we have the following general RR model. An individual with $X=x$ answers according to the known probability density $h_z(z \mid x)$, $x \in \Omega_x$. The densities $h_z(z \mid x)$, $x \in \Omega_x$, are called response densities and they hold with respect to some measure $d\lambda(z)$ thus including continuous as well as discrete distributions. In general, the family of response densities $\left\{h_z(z|x),\, x \in \Omega_x\right\}$ is chosen by the interviewer and nearly family can be used by choosing an appropriate randomization device. The interviewer can only observe the answer Z from a randomly taken individual. It has the density

$$f_z(z)=\int h_z(z \mid x)dF_x(x)=E[h_z(z \mid X)] \tag{11.10.4.1}$$

which is a mixture of the different response densities with $F_x(x)$ as mixing distribution. The population parameters of X may be inferred from observations of Z, whereas the single X value may not be determined, if the densities $h_z(z \mid x)$, $x \in \Omega_x$, overlap. The family of conditional densities of X given $Z=z$, $z \in \Omega_z$, is

$$h_x(x \mid z)=h_z(z \mid x)f_x(x)/f_z(z), \quad z \in \Omega_z, \tag{11.10.4.2}$$

where Ω_z being the set of possible RR answers. In the following models this family will be called the family of revealing densities. It depends both on the response distributions and on the unknown population distribution of X. Further, the joint density of random variable (X,Z) is given by

$$f_{x,z}(x,z)=h_z(z \mid x)f_x(x). \tag{11.10.4.3}$$

Using (11.10.4.2) and (11.10.4.3) the revealing densities can be defined as

$$h_x(x \mid z) = f_{x,z}(x, z) / f_z(z). \tag{11.10.4.4}$$

The importance of the revealing densities for the degree of protection of the interviewee's privacy is obvious. Therefore we come to the privacy aspect. The choice of the family of response distributions determines the degree of protection of the interviewee's privacy as well as the possibility of obtaining good estimates of the population parameters. Before the interview the density $f_x(x)$ provides knowledge about the variable X of a randomly sampled individual. Assume that he answers $Z = z$. Then the revealing density $h_x(x \mid z)$ yields information about X and the discrepancy between $f_x(x)$ and $h_x(x \mid z)$ reflects the invasion of privacy caused by the answer $Z = z$. If $h_x(x \mid z)$ is very concentrated around some value x, the protection provided by the answer $Z = z$ is very small; conversely, if $h_x(x|z)$ is widely spread, privacy is little affected. Therefore we take $V(X \mid Z = z)$ as a general measure of the privacy protection associated with the answer $Z = z$, and the average $E[V(X \mid Z)]$ as an overall measure of the privacy protection provided by an RR model. In the following section, we shall compare the additive and multiplicative models. Chang and Huang (2001b) also estimated the proportion and sensitivity level of a quantitative character.

11.10.5 COMPARISON BETWEEN ADDITIVE AND MULTIPLICATIVE MODELS

Bharagava (1996) has considered the comparison of additive and multiplicative models when both the sensitive and scrambling random variables are beta distributed.

(a) Multiplicative model: Let us first consider the multiplicative model as $Z = XY$. Let us assume that the sensitive variable X is beta distributed with parameters (α_1, β_1). The interviewees are asked to multiply a random number Y, from a known beta distribution with parameters (α_2, β_2) to their X values. The interviewer thus receives the randomized (scrambled) response Z. Let us also assume that X and Y are independently distributed. The corresponding density functions are

$$f_1(x) = [B(\alpha_1, \beta_1)]^{-1} x^{(\alpha_1 - 1)}(1 - x)^{(\beta_1 - 1)}, \quad 0 < x < 1 \tag{11.10.5.1}$$

$$f_2(y) = [B(\alpha_2, \beta_2)]^{-1} y^{(\alpha_2 - 1)}(1 - y)^{(\beta_2 - 1)}, \quad 0 < y < 1 \tag{11.10.5.2}$$

Note that x and y are independent their joint distribution is given by

$$f(x, y) = [B(\alpha_1, \beta_1)]^{-1}[B(\alpha_2, \beta_2)]^{-1} x^{(\alpha_1 - 1)}(1 - x)^{(\beta_1 - 1)} y^{(\alpha_2 - 1)}(1 - y)^{(\beta_2 - 1)}, \tag{11.10.5.3}$$

where $0 < x < 1$ and $0 < y < 1$. Now making the transformations

$$u = xy \text{ and } v = x. \tag{11.10.5.4}$$

Solving for x and y we have

$$x = v \text{ and } y = u/v. \tag{11.10.5.5}$$

The Jacobian of the transformation is

$$J=\begin{vmatrix}\dfrac{\partial x}{\partial u}, & \dfrac{\partial x}{\partial v}\\ \dfrac{\partial y}{\partial u}, & \dfrac{\partial y}{\partial u}\end{vmatrix}=\begin{vmatrix}0, & 1\\ \dfrac{1}{v}, & -\dfrac{u}{v^2}\end{vmatrix}=-\frac{1}{v},$$

which implies that

$$|J|=\frac{1}{v}.$$

The joint distribution of u and v is given by

$$g(u,v)=f(x,y)|J|$$

$$=[B(\alpha_1,\beta_1)]^{-1}[B(\alpha_2,\beta_2)]^{-1}v^{(\alpha_1-1)}(1-v)^{(\beta_1-1)}\left(\frac{u}{v}\right)^{(\alpha_2-1)}\left(1-\frac{u}{v}\right)^{(\beta_2-1)}$$

$$=[B(\alpha_1,\beta_1)]^{-1}[B(\alpha_2,\beta_2)]^{-1}v^{\alpha_1-(\alpha_2+\beta_2)}(1-v)^{(\beta_1-1)}u^{(\alpha_2-1)}(v-u)^{(\beta_2-1)}. \quad (11.10.5.6)$$

If we assume

$$\alpha_1=\alpha_2+\beta_2$$

we obtain

$$g(u,v)=[B(\alpha_1,\beta_1)]^{-1}[B(\alpha_2,\beta_2)]^{-1}(1-v)^{(\alpha_1-1)}u^{(\alpha_2-1)}(v-u)^{(\beta_2-1)}. \quad (11.10.5.7)$$

The region $(0<x<1,\ 0<y<1)$ in the $(x,\ y)$ plane transforms to the region $(u<v<1,\ 0<u<1)$ in the $(u,\ v)$ plane. Following Rao (1973) the value of integral of $g(u,\ v)$ over v from u to 1, the density of u is also beta with parameters $(\alpha_2,\ \beta_1+\beta_2)$ provided that the condition $\alpha_1=\alpha_2+\beta_2$ is satisfied. Thus the marginal density of u is given by

$$h(u)=[B(\alpha_2,\alpha_1+\alpha_2)]^{-1}u^{(\alpha_2-1)}(1-u)^{(\beta_1+\beta_2-1)},\quad 0<u<1 \quad (11.10.5.8)$$

which reduces to

$$h(u)=B_1u^{(\alpha_2-1)}(1-u)^{(\beta_1+\beta_2-1)},\quad 0<u<1 \quad (11.10.5.9)$$

where $B_1=\dfrac{\Gamma(\alpha_1+\beta_1)}{\Gamma_{\beta_2}\Gamma(\beta_1+\beta_2)}$.

Thus the revealing density function becomes

$$h_x(x\mid z)=h^*(v\mid u)=\frac{g(u,v)}{h(u)}$$

$$=\frac{[B(\alpha_1,\beta_1)]^{-1}[B(\alpha_2,\beta_2)]^{-1}(1-v)^{(\beta_1-1)}u^{\alpha_2-1}(v-u)^{(\beta_2-1)}}{B_1u^{(\alpha_2-1)}(1-u)^{(\beta_1+\beta_2-1)}}$$

$$=\frac{\dfrac{\Gamma(\alpha_1+\beta_1)\Gamma(\alpha_2+\beta_2)}{\Gamma_{\alpha_1}\Gamma_{\beta_1}\Gamma_{\alpha_2}\Gamma_{\beta_2}}(1-v)^{(\beta_1-1)}(v-u)^{(\beta_2-1)}}{\dfrac{\Gamma(\alpha_1+\beta_1)}{\Gamma_{\alpha_2}\Gamma(\beta_1+\beta_2)}(1-u)^{(\beta_1+\beta_2-1)}}$$

$$= \left(\frac{\Gamma(\beta_1+\beta_2)}{\Gamma_{\beta_1}\Gamma_{\beta_2}}\right)\frac{(1-v)^{(\beta_1-1)}(v-u)^{(\beta_2-1)}}{(1-u)^{(\beta_1+\beta_2-1)}}$$

$$= [B(\beta_1,\beta_2)]^{-1}\frac{(1-v)^{(\beta_1-1)}(v-u)^{(\beta_2-1)}}{(1-u)^{(\beta_1+\beta_2-1)}}$$

$$= B_2\frac{(1-v)^{(\beta_1-1)}(v-u)^{(\beta_2-1)}}{(1-u)^{(\beta_1+\beta_2-1)}}, \tag{11.10.5.10}$$

where $B_2 = [B(\beta_1,\beta_2)]^{-1}$, $u < v < 1$ and $0 < u < 1$.

The general rule for working out the measure of privacy protection is given by

$$E(X \mid Z) = E(V \mid U) = \int v h^*(v \mid u)dv \tag{11.10.5.11}$$

$$\text{Var}(X \mid Z) = \text{Var}(V \mid U) = E\left[(V - E(V \mid U))^2 \mid U\right]$$
$$= \int (V - E(V \mid U))^2 h^*(v \mid u)dv \tag{11.10.5.12}$$

and the measure of protection is given by

$$E[\text{Var}(X \mid Z)] = E[\text{Var}(V \mid U)] = \int \text{Var}(V \mid U)h(u)du\ .$$

Therefore we have

$$E(V \mid U) = B_2\int_u^1 \frac{v(1-v)^{(\beta_1-1)}(v-u)^{(\beta_2-1)}}{(1-u)^{(\beta_1+\beta_2-1)}}dv \tag{11.10.5.13}$$

and

$$\text{Var}(V \mid U) = B_2\int_u^1 (v - E(V \mid U))^2\frac{(1-v)^{(\beta_1-1)}(v-u)^{(\beta_2-1)}}{(1-u)^{(\beta_1+\beta_2-1)}}dv, \tag{11.10.5.14}$$

and the measure of privacy protection is

$$E[\text{Var}(V \mid U)] = B_1\int_0^1 \text{Var}(V \mid U)u^{(\alpha_2-1)}(1-u)^{\beta_1+\beta_2-1}du\ . \tag{11.10.5.15}$$

Note that the integral expressions (11.10.5.13) to (11.10.5.15) cannot be evaluated exactly, therefore it is not possible to derive a simple expression for the efficiency of the multiplicative model. Bhargava (1996) has resolved this issue through numerical illustrations with different sets of parameters.

(b) Additive model: The additive model is given by

$$Z = X + Y\ . \tag{11.10.5.16}$$

Assume that the sensitive variable X is beta distributed with parameters (α_1,β_1) and each interviewee in the sample is asked to add a random number Y to his/her X value. The random number Y is from a known beta distribution with parameters (α_2,β_2) and is distributed independently of the sensitive variable X. The interviewer receives the randomized response Z. The corresponding density functions of X and Y and their joint distribution are given by

$$f_1(x) = [B(\alpha_1,\beta_1)]^{-1}x^{(\alpha_1-1)}(1-x)^{(\beta_1-1)},\quad 0 < x < 1, \tag{11.10.5.17}$$

$$f_2(y)=\left[B(\alpha_2,\beta_2)\right]^{-1}y^{(\alpha_2-1)}(1-y)^{(\beta_2-1)}, \quad 0<y<1, \tag{11.10.5.18}$$

and

$$f(x,y)=\left[B(\alpha_1,\beta_1)\right]^{-1}\left[B(\alpha_2,\beta_2)\right]^{-1}x^{(\alpha_1-1)}(1-x)^{(\beta_1-1)}y^{(\alpha_2-1)}(1-y)^{(\beta_2-1)}, \tag{11.10.5.19}$$

where $0<x<1$ and $0<y<1$, respectively. Again let us make the transformations

$$u=x+y \quad \text{and} \quad v=x. \tag{11.10.5.20}$$

Solving for x and y we have

$$x=v \quad \text{and} \quad y=u-v. \tag{11.10.5.21}$$

The Jacobian of the transformation is

$$J=\begin{vmatrix}\dfrac{\partial x}{\partial u}, & \dfrac{\partial x}{\partial v}\\ \dfrac{\partial y}{\partial u}, & \dfrac{\partial y}{\partial v}\end{vmatrix}=\begin{vmatrix}0, & 1\\ 1, & -1\end{vmatrix}=-1$$

therefore

$$|J|=1.$$

Using (11.10.5.21) in (11.10.5.19) and using $|J|=1$ we have the following joint distribution of u and v

$$g(u,v)=\left[B(\alpha_1,\beta_1)\right]^{-1}\left[B(\alpha_2,\beta_2)\right]^{-1}v^{(\alpha_1-1)}(1-v)^{(\beta_1-1)}(u-v)^{(\beta_2-1)}(1-u+v)^{(\beta_2-1)} \tag{11.10.5.22}$$

The region $(0<x<1,\ 0<y<1)$ in the $(x,\ y)$ plane transforms to the region $(0<v<u,\ 0<u<1$ and $u-1<v<1,\ 1<u<2)$ in the $(u,\ v)$ plane. Hence by integrating $g(u,\ v)$ over v, first from 0 to u and then from $(u-1)$ to 1, then the density of u as

$$h_1(u)=\left[B(\alpha_1,\beta_1)\right]^{-1}\left[B(\alpha_2,\beta_2)\right]^{-1}\int_0^u v^{(\alpha_1-1)}(1-v)^{(\beta_1-1)}(u-v)^{(\alpha_2-1)}(1-u+v)^{(\beta_2-1)}dv$$

for $0<u<1$,

$$h_2(u)=\left[B(\alpha_1,\beta_1)\right]^{-1}\left[B(\alpha_2,\beta_2)\right]^{-1}\int_{u-1}^1 v^{(\alpha_1-1)}(1-v)^{(\beta_1-1)}(u-v)^{(\alpha_2-1)}(1-u+v)^{(\beta_2-1)}dv$$

for $1<u<2$.

The revealing density becomes

$$h_1^*(v\mid u)=\frac{v^{(\alpha_1-1)}(1-v)^{(\beta_1-1)}(u-v)^{(\alpha_2-1)}(1-u+v)^{(\beta_2-1)}}{\int_0^u v^{(\alpha_1-1)}(1-v)^{(\beta_1-1)}(u-v)^{(\alpha_2-1)}(1-u+v)^{(\beta_2-1)}dv}, \quad 0<u<1$$

and

$$h_2^*(v\mid u)=\frac{v^{(\alpha_1-1)}(1-v)^{(\beta_1-1)}(u-v)^{(\alpha_2-1)}(1-u+v)^{(\beta_2-1)}}{\int_{u-1}^1 v^{(\alpha_1-1)}(1-v)^{(\beta_1-1)}(u-v)^{(\alpha_2-1)}(1-u+v)^{(\beta_2-1)}dv}, \quad 1<u<2.$$

Thus we have the following results:

$$E_1(V \mid U) = \int_0^u v h_1^*(v \mid u) dv, \qquad 0 < u < 1, \quad |J| = 1,$$

$$E_2(V \mid U) = \int_{u-1}^1 v h_1^*(v \mid u) dv, \quad 1 < u < 2,$$

and

$$\text{Var}_1(V \mid U) = \int_0^u (v - E_1(V \mid U))^2 h_1^*(v \mid u) dv, \ 0 < u < 1,$$

$$\text{Var}_2(V \mid U) = \int_{u-1}^1 (v - E_2(V \mid U))^2 h_2^*(v \mid u) dv, \ 1 < u < 2.$$

The measure of privacy protection is then given by

$$E[\text{Var}(V \mid U)] = \int_0^1 \text{Var}_1(V \mid U) h_1(u) du + \int_1^2 \text{Var}_2(V \mid U) h_2(u) du.$$

Again these expressions cannot be evaluated exactly. Thus Bhargava (1996) obtained the numerical measures of the level of respondents' protection. The following table provides the measure of privacy protection for beta distribution in case of multiplicative and additive models.

Parameters				Multiplicative	Additive
β_1	β_2	α_1	α_2	Model	Model
2	2	4	2	0.0238	0.0133
2	2	5	3	0.0178	0.0155
2	2	3	1	0.0333	0.0217
1	1	3	2	0.0250	0.0222
1	1	4	3	0.0083	0.0155
1	1	2	1	0.0417	0.0321
0.5	0.5	2.5	2	0.0215	0.0191
0.5	0.5	3.5	3	0.0139	0.0117
0.5	0.5	1.5	1	0.0412	0.0352

The above table shows that if the sensitive variable follows a beta distribution, the multiplicative model remains more protective than the additive model in most cases.

11.11 TEST FOR DETECTING UNTRUTHFUL ANSWERING

Assume π be the proportion of people belonging to sensitive group A and $(1-\pi)$ be the proportion of persons belonging to non-sensitive group A^c such that $A \cup A^c = \Omega$. Owing to the sensitive nature of the group A, people do not like to disclose their status to the interviewers. An SRSWR sample of n respondents will be taken and each respondent will be given two random devices R_1 and R_2. Each of the random devices will have two statements:

(i) I belong to group A;
(ii) I do not belong to group A.

Using different probability mechanisms, under each device the respondent chooses statement (i) or (ii) with probability P or $(1-P)$ and simply answers 'Yes' or 'No' depending upon his/her actual status. The responses under the two devices are assumed to be independent. Let l be the probability that a person gives an untruthful answer whether he/she belongs to A or A^c.
Thus we have

$$l = \Pr(\text{Untruthful answer} \mid A) = \Pr(\text{Untruthful answer} \mid A^c). \tag{11.11.1}$$

Let

$$X_i = \begin{cases} 1 & \text{if the } i^{th} \text{ respondent answers 'Yes' with device } R_1, \\ 0 & \text{otherwise,} \end{cases}$$

and

$$Y_i = \begin{cases} 1 & \text{if the } i^{th} \text{ respondent answers 'Yes' with device } R_2, \\ 0 & \text{otherwise.} \end{cases}$$

Then

$$\Pr(X_i = 1, Y_i = 0) = \Pr(X_i = 0, Y_i = 1) = [Pl + (1-P)(1-l)][(1-P)l + P(1-l)]$$
$$= P(1-P) + (1-l)l(1-2P) = \theta_l .$$

Let n_{11}, n_{10}, n_{01} and n_{00} be the number of respondents who answered (Yes, Yes), (Yes, No), (No, Yes), and (No, No) respectively from devices R_1 and R_2 as shown below.

Device	R_2	
R_1	Yes	No
Yes	n_{11}	n_{10}
No	n_{01}	n_{00}

Instead of assuming

$$l = \Pr(\text{Untruthful answer} \mid A) = \Pr(\text{Untruthful answer} \mid A^c)$$

one can assume

$$\Pr(\text{Untruthful answer} \mid A) = l, \text{ and } \Pr(\text{Untruthful answer} \mid A^c) = 0 .$$

Lakshmi and Raghavarao (1992) considered a p-variates $N_p(\mu, \Sigma)$ normal distribution with mean vector μ and positive definite dispersion matrix Σ and by I_n, a column vector of n ones. Lakshmi and Raghavarao (1992) considered the problem of testing the null hypothesis

$$H_0 : l = 0$$

against the alternative hypothesis

$$H_a : l > 0 .$$

Following Rao (1973), the asymptotic distribution of the bivariate random vector

$$\sqrt{n}\left[\left(\frac{n_{10}}{n}, \frac{n_{01}}{n}\right) - \theta_l 1'_2\right]$$

is $N_2(0_2, \Sigma_l)$, where 0_2 is a vector of dimension 2 and

$$\Sigma_l = \begin{bmatrix} \theta_l(1-\theta_l), & -\theta_l^2 \\ -\theta_l^2, & \theta_l(1-\theta_l) \end{bmatrix}.$$

Under the Null hypothesis $H_0 : l = 0$, the statistic

$$T = n\left(\frac{n_{10}}{n} - \theta_0, \frac{n_{01}}{n} - \theta_0\right)\Sigma_0^{-1}\left(\frac{n_{10}}{n} - \theta_0, \frac{n_{01}}{n} - \theta_0\right)^t \tag{11.11.2}$$

is distributed as a central chi square distribution with two degrees of freedom, where $\theta_0 = P(1-P)$ and

$$\Sigma_0 = \begin{bmatrix} \theta_0(1-\theta_0), & -\theta_0^2 \\ -\theta_0^2, & \theta_0(1-\theta_0) \end{bmatrix}.$$

Lakshmi and Raghavarao (1992) developed the following result: A critical region for an α level test

$$H_0 : l = 0$$

against the alternative hypothesis

$$H_a : l > 0$$

is

$$T > \chi^2_{(1-\alpha)}(2). \tag{11.11.3}$$

11.12 STOCHASTIC RANDOMIZED RESPONSE TECHNIQUE

Singh (2002c) suggested another procedure which may result in a greater sense of response confidentiality among the sampled individuals. The procedure can be used in surveys where the respondents selected in the sample assemble at a common place for the conduct of the survey. This could be a situation of collecting data from a small town, community or organization. The procedure invokes K decks of cards (which is named as stochastic randomization device) with different proportions of cards carrying the statement, 'I belong to group A'. After explaining to the respondents how the randomization device provides confidentiality to their responses, the investigator asks one of the assembled respondents to randomly select one deck of cards from the box containing K decks of cards. The deck is then used to collect information on the sensitive attribute from the respondents. Every sampled respondent draws one card from the selected deck of cards and reads the statement on it. In this procedure every respondent is provided with two identical slips of paper with 'Yes' or 'No' printed on them. According to his status in relation to the statement printed on the card drawn, each respondent is requested to put one of the two slips of paper into an empty box. After the survey is completed the number of 'Yes' answers is counted from the box and the proportion p^* for the deck used in the survey is noted. Random selection of one randomization device from several such devices may help in increasing the sense of confidentiality among the respondents. The choice of values of p for preparing K decks of cards for the survey is important in this procedure. These K values of p could either be purposively selected by the investigator or they could be taken as a random sample from a known discrete or continuous density function. Let this density function be denoted by $f(p)$. The value of p corresponding to the deck used in the survey will be selected from this random sample of p values with equal probabilities. Thus the value of p^* used in the survey is a random variable with $f(p)$ as its probability density function. When $f(p)$ is a one point distribution then this procedure reduces to Warner (1965). Assume n_1 persons in the sample answered 'Yes' and $(n-n_1)$ answered 'No'. Note that the probability of 'Yes' answer for a particular choice of p^* is given by

$$\theta = p^*\pi + (1-\pi)(1-p^*). \tag{11.12.1}$$

Consider the following estimator of π as

$$\hat{\pi}_R = \frac{\hat{\theta} - (1-p^*)}{2p^* - 1}, \tag{11.12.2}$$

where $\hat{\theta} = n_1/n$ is the proportion of 'Yes' answers in the sample.

Thus we have the following theorems:

Theorem 11.12.1. The estimator $\hat{\pi}_R$ is unbiased for the population proportion π.

Proof. We have

$$E(\hat{\pi}_R) = E_1E_2(\hat{\pi}_R) = E_1E_2\left\{\frac{\hat{\theta} - (1-p^*)}{2p^* - 1}\right\} = E_1(\pi) = \pi$$

where E_2 denotes the expected value for the fixed value p^* of p and E_1 denotes the expected value for all values of p generated by its distribution. This completes the proof of the theorem.

Theorem 11.12.2. The variance of the estimator $\hat{\pi}_R$ is given by

$$V(\hat{\pi}_R) = \frac{\pi(1-\pi)}{n} + \int_a^b \frac{p(1-p)}{n(2p-1)^2} f(p)dp, \quad a \le p \le b, \quad p \ne 0.5, \tag{11.12.3}$$

where $f(p)$ denotes the probability density function (p.d.f.) of p.

Proof. We have

$$V(\hat{\pi}_R) = E_1V_2(\hat{\pi}_R) + V_1E_2(\hat{\pi}_R)$$

where E_1 and E_2 have been defined earlier. Similarly, V_2 and V_1, respectively, denote the variance for a fixed value p^* of p and for all values of p generated by its distribution. Thus we have

$$V(\hat{\pi}_R) = E_1V_2(\hat{\pi}_R) + V_1E_2(\hat{\pi}_R) = E_1\left[\frac{\pi(1-\pi)}{n} + \frac{p^*(1-p^*)}{n(2p^*-1)^2}\right] + V_1(\pi)$$

$$= \frac{\pi(1-\pi)}{n} + E_1\left\{\frac{p^*(1-p^*)}{n(2p^*-1)^2}\right\} = \frac{\pi(1-\pi)}{n} + \frac{1}{n}\int_a^b \frac{p(1-p)}{(2p-1)^2} f(p)dp .$$

Hence the theorem.

Note that in (11.12.3) $f(p)$ denotes the probability density function of p. We shall now try to choose the probability density function (p.d.f.) $f(p)$ in such a way that the estimator $\hat{\pi}_R$ becomes more efficient than the estimator $\hat{\pi}_w$. Although there could be many choices for $f(p)$, consider the following two such probability density functions:

$$(\ i\ )\ f(p) = \frac{1}{b-a}, \quad a < p < b, \quad p \neq 0.5 ; \tag{11.12.4}$$

$$(\ ii\ )\ f(p) = C\,(2p-1)^2 p^{\alpha-1}(1-p)^{\beta-1}, \quad 0 < p < 1,\ \alpha > 0,\ \ \beta > 0 ; \tag{11.12.5}$$

where

$$C = \frac{\Gamma_{(\alpha+\beta+2)}}{\Gamma_{(\alpha)}\Gamma_{(\beta+2)} + \Gamma_{\beta}\Gamma_{(\alpha+2)} - 2\Gamma_{(\alpha+1)}\Gamma_{(\beta+1)}} .$$

Now we show that the stochastic randomized response procedure in addition to providing greater sense of response confidentiality and is also more efficient than the Warner's (1965) procedure.

Theorem 11.12.3. The variance $V(\hat{\pi}_R)$ of the estimator $\hat{\pi}_R$ with $f(p)$ defined in (11.12.4) becomes

$$V(\hat{\pi}_R) = \frac{\pi(1-\pi)}{n} - \frac{1}{4n} + \frac{1}{4n(2b-1)(2a-1)} . \tag{11.12.6}$$

Proof. On substituting the values of $f(p)$ from (11.12.4) in (11.12.3) we have

$$V(\hat{\pi}_R) = \frac{\pi(1-\pi)}{n} - \frac{1}{4n} + \frac{1}{4n}\int_a^b \frac{1}{(2p-1)^2(b-a)}dp ,$$

which proves the theorem.

Corollary 11.12.1. For $a = p_0$ and $b = p_0 + g(1-p_0)$, with $0 < g < 1$, the estimator $\hat{\pi}_R$ remains always more efficient than Warner's estimator. Here it is possible to find more acceptable choices for a and b.

Theorem 11.12.4. The variance of the estimator $\hat{\pi}_R$ with $f(p)$ defined in (11.12.5) is given by

$$V(\hat{\pi}_R) = \frac{\pi(1-\pi)}{n} + \frac{\alpha\beta}{n\{(\alpha-\beta)^2 + \alpha + \beta\}} . \tag{11.12.7}$$

Proof. After replacing the value of $f(p)$ from (11.12.5) in (11.12.3) we have

$$V(\hat{\pi}_R) = \frac{\pi(1-\pi)}{n} + \frac{1}{n}\int_0^1 p^{(\alpha+1)-1}(1-p)^{(\beta+1)-1}dp,$$

which proves the theorem.

Corollary 11.12.2. If we choose $\alpha = p_0$ and $\beta = 1 - p_0$, where p_0 denotes the value of p used in Warner's model, the variance $V(\hat{\pi}_R)$ under this distribution of p becomes

$$V(\hat{\pi}_R) = \frac{\pi(1-\pi)}{n} + \frac{p_0(1-p_0)}{n\{(2p_0-1)^2+1\}},$$

which is smaller than the variance $V(\hat{\pi}_w)$, making the stochastic procedure more efficient than Warner's (1965) method. Hence one could also possibly find other efficient and more acceptable values of α and β.

Corollary 11.12.3. If we choose $\alpha = (1-p_0)(1-T_0)$ and $\beta = 1-\alpha$, where p_0 and T_0 denote the value of p and T used in Mangat and Singh (1990), the variance $V(\hat{\pi}_R)$ under this distribution becomes

$$V(\hat{\pi}_R) = \frac{\pi(1-\pi)}{n} + \frac{(1-p_0)(1-T_0)\{1-(1-p_0)(1-T_0)\}}{n\{(2p_0-1)^2+2T_0(1-p_0)+1\}},$$

which is smaller than the variance $V(\hat{\pi}_{ms})$ given by

$$V(\hat{\pi}_{ms}) = \frac{\pi(1-\pi)}{n} + \frac{(1-p_0)(1-T_0)\{1-(1-p_0)(1-T_0)\}}{n\{(2p_0-1)^2+2T_0(1-p_0)\}},$$

making the proposed procedure more efficient than the Mangat and Singh (1990) method. Hence in this case as well one could possibly find other efficient and more acceptable values of α and β.

EXERCISES

Exercise 11.1. (I) Consider a social survey has been conducted by an investigator using randomized responses from a device consisting of two devices R_1 and R_2 (say). The device R_1 consists of the following two statements:

(1) 'Do you possess the sensitive attribute A?', with probability T.
(2) 'Go to second randomization device R_2' with probability $(1-T)$.

The second randomization device R_2 is the same as defined by Warner (1965) as follows:

(1) 'Do you possess the sensitive attribute A?' with probability P.
(2) 'Do you not possess the sensitive attribute A?' with probability $(1-P)$.

(a) Show that an unbiased estimator of population proportion π is given by

$$\hat{\pi}_1 = \frac{\hat{\theta}_1 - (1-P)(1-T)}{2P - 1 + 2T(1-P)}.$$

(b) Show that the estimator $\hat{\pi}_1$ reduces to the estimator proposed by Warner (1965) for $T = 0$.

(c) Show that the estimator $\hat{\pi}_1$ is more efficient than $\hat{\pi}_w$ if

$$T > (1-2p)/(1-P).$$

(d) Study the properties of the estimator $\hat{\pi}_1$ under the SRSWR and the SRSWOR sampling designs.

Hint: Mangat and Singh (1990, 1991).

(II) Modify the second randomization device R_2 of Mangat and Singh (1990) model with statements:

(1) 'Do you possess the sensitive attribute A?' with probability P;

(2) 'Say forcibly Yes' with probability $(1-P)W$;

(3) 'Say forcibly No' with probability $(1-P)(1-W)$;

where $W \in [0, 1]$ is a suitably choosen constant. Obviously the probability of 'Yes' answer is given by

$$\theta = T\pi + (1-T)[P\pi + (1-P)W].$$

Develop an estimator of π and discuss its properties.

Hint: Singh, Singh, Mangat, and Tracy (1995), Tracy and Osahan (1999).

(III) Modify the second randomization device R_2 of Mangat and Singh (1990) model with the statements:

(1) 'Do you possess the sensitive attribute A?' with probability P;

(2) 'Do you possess the unrelated attribute Y?' with probability $(1-P)$.

Obviously the probability of 'Yes' answer is given by

$$\theta = T\pi + (1-T)[P\pi + (1-P)\pi_y].$$

Develop an estimator of π and discuss its properties: (a) when π_y is known; (b) when π_y is unknown.

Hint: Mangat (1992), Mangat, Singh, Singh, and Singh (1993).

(IV) Compare Mangat and Singh (1990) model with Warner (1965) model at equal level of protection of the respondents. Discuss your views.

Hint: Nayak (1994), Bhargava (1996), Moors (1997).

Exercise 11.2. Consider we have two mutually exclusive groups G_1 and G_2. In the population of interest the proportion of persons in G_1 which we would like to estimate is π. We select n persons at random from this population. It is convenient to create a random variable T_i, so that $T_i = 1$ if the ith individual is in G_1 and $T_2 = 2$

if it is in G_2. We confront each person with two urns. There are red balls and black balls in each urn. We ask each person to select k balls from each urn (WR sampling), mentally noting the number of red balls obtained from each. We do not observe the drawing of the balls, and the subjects understand that we will not know the results of the separate draws. Then each person is told to reveal the number of red balls obtained from the urn corresponding to his/her group. Persons from G_1 will tell the number of red balls obtained from urn 1, and the persons from G_2 will tell us the number of red balls obtained from urn 2. By such a mechanism, confidentiality of individuals is preserved. Let θ_1 and θ_2 be the proportion of red balls in the two urns. We control the values of θ_1 and θ_2, and we would never consider the case $\theta_1 = \theta_2$. Let r be the observed proportion of 'Yes' answers in a sample of n respondents, then assuming that $k = 1$, show that an unbiased estimator of population proportion π is given by

$$\hat{\pi}_1 = \frac{r - \theta_2}{\theta_1 - \theta_2} .$$

Find its variance and discuss the results and your views.
Hint: Kuk (1990), Chatterjee and Simon (1993).

Exercise 11.3. Let the value Y_i of a sensitive variable y, defined on a finite survey population of N identifiable and labelled persons, be supposed to be unavailable through a direct response survey when one intends to estimate population total Y on choosing a sample s from the population with a probability $p(s)$ according to design p. Instead, let a randomized response R_i be available in independent manner from the respective persons i, on request if sampled, in such a way that their expectations, variance, and covariance (E_R, V_R, C_R) respectively satisfy $E_R(r_i) = Y_i$, $V_R(r_i) = \alpha_i Y_i^2 + \beta_i Y_i + \theta_i = V_i^2$ (say), $C_R(r_i, r_j) = 0$, for $i \neq j$ such that $\alpha_i > 0$, β_i and θ_i are known for every unit in the population.

(a) Show that for PPSWR sampling an unbiased estimator of population total Y is

$$\hat{y}_{wr} = \frac{1}{n}\sum_{i=1}^{n}\frac{r_i}{p_i} .$$

Find its variance and estimator of its variance.
Hint: Chaudhuri and Adhikari (1990), Arnab (1990).

(b) Under multi-character survey and PPSWR sampling design, six estimators of population total Y are given by

$$\hat{y}_{mc} = \frac{1}{n}\sum_{i=1}^{n}\frac{r_i}{P_{ik}^*}, \quad \text{for } k = 0, 1, 2, 3, 4, 5,$$

where

$$P_{i0}^* = \frac{1}{N}, \quad P_{i1}^* = \left(1 + \frac{1}{N}\right)^{(1-\rho_{xy})}(1 + P_i)^{\rho_{xy}} - 1, \quad P_{i2}^* = \frac{(1 - \rho_{xy})}{N} + \rho_{xy} P_i$$

$$P_{i3}^{*}=\left(\frac{1}{N}\right)^{(1-\rho_{xy})}P_i^{\rho_{xy}},\quad P_{i4}^{*}=\left[N(1-\rho_{xy})+\frac{\rho_{xy}}{P_i}\right]^{-1},\ \text{and}\ P_{i5}^{*}=\left(1-\sqrt[3]{\rho_{xy}}\right)/N+\sqrt[3]{\rho_{xy}}P_i$$

with ρ_{xy} being the known correlation coefficient. Find the bias and variance for each one of the estimators.

Hint: Bansal, Singh, and Singh (1994), Grewal, Bansal, and Singh (1999).

(c) Under PPSWOR sampling , an unbiased estimator of population total is

$$\hat{y}_{\text{WOR}}=\sum_{i=1}^{n}\frac{r_i}{\pi_i}.$$

Find its variance and suggest an estimator of variance.

Hint: Godambe (1980b), Arnab (1994).

(d) If population total X of an auxiliary character is known, then find the variance of the generalized regression estimator (GREG) of the population total, defined as

$$\hat{y}_{\text{GREG}}=\sum_{i=1}^{n}\frac{r_i}{\pi_i}+\hat{\beta}_{ds}\left[X-\sum_{i=1}^{n}\frac{x_i}{\pi_i}\right].$$

Suggest at least two estimators of its variance.

Hint: Chaudhuri, Maiti, and Roy (1996) , Tracy and Singh (2000).

(e) Assuming that y_i is a qualitative variable, show that a linear homogeneous unbiased estimator of the population proportion is given by

$$\hat{y}_{\text{p}}=\sum_{i\in s}b_{si}r_i$$

where the b_{si} are constants and are free from y_i values, and satisfy the condition

$$\sum_{s\supset i}b_{si}p(s)=N^{-1}.$$

Hint: Arnab (1996)

Exercise 11.4. Consider a finite population of N first stage units (FSUs) and let the ith , $i=1,2,...,N$, FSU consist of M_i second stage units (SSUs). Let a sample s of n FSUs be selected with probability $p_1(s)$ following some sampling design p_1 and if the ith FSU is selected in the sample, we take a sub-sample s_i of m_i SSUs from M_i SSUs of the ith FSU with probability $p_2(s_i)$ following a sampling design p_2. The sub-sample s_i , $i\in s$, are selected independently. The overall sampling design for selection of sample $(s_i, i\in s)$ is denoted by p. Let $E_p(V_p)$, $E_1(V_1)$, and $E_2(V_2)$ denote expectation (variance) operators over the sampling design p, p_1 and p_2, respectively. Let π_i, π_{ij}, etc., denote the first and second order inclusion probabilities of FSUs. Let Y_{ij} be the value of the character under study for the jth SSU of the ith FSU $(j=1,2,...,M_i;\ i=1,2,...,N)$. Let r_{ij} denote the standardized randomized response, such that

$$E_{\text{R}}(r_{ij})=Y_{ij},\quad V_{\text{R}}(r_{ij})=\sigma_{ij}^2=\theta_{ij}Y_{ij}^2+\beta_{ij}Y_{ij}+\delta_{ij}\quad\text{and}\quad C_{\text{R}}(r_{ij},r_{kl})=0\ \text{for}\ (i,j)\neq(k,l).$$

Find the variance of the estimator of the population total Y defined as

$$\hat{Y}(r) = \sum_{i \in s} b_i(s) \hat{Y}_i(r),$$

where $\hat{Y}_i(r) = \sum_{j \in s_i} b_j(s_i) r_{ij}$, $b_i(s)$ and $b_j(s_i)$ are constants free from r_{ij} values such that $\sum_{s_i \ni j} b_j(s_i) p_2(s_i) = 1$ and $\sum_{s \ni i} b_i(s) p_i(s) = 1$. Suggest an estimator of its variance.
Hint: Arnab (1992b).

Exercise 11.5. (a) Consider a randomization device as defined by Warner (1965) as follows:

(i) 'Do you possess the sensitive attribute A?' with probability P;

(ii) 'Do you not possess the sensitive attribute A?' with probability $(1-P)$;

has been used to conduct a social survey. Show that an unbiased estimator of population proportion π is

$$\hat{\pi}_1 = \frac{\hat{\theta}_1 - (1-P)}{2P-1}$$

under SRSWR and SRSWOR sampling design. Compare the relative efficiency of the above estimator under the two different designs.
Hint: Singh and Kathuria (1995).

(b) Assume the randomized response device consists of a deck having three types of cards as follows:

(i) 'Do you possess the sensitive attribute A?' with probability P_1;

(ii) 'Do you not possess the sensitive attribute A?' with probability P_2 $(P_2 \neq P_1)$;

(iii) 'Forcibly No' with probability P_3;

such that $P_1 + P_2 + P_3 = 1$. Obviously the probability of 'Yes' answer is given by $\theta = P_1\pi + P_2(1-\pi)$. Develop an estimator of π and find its variance expression.
Hint: Mangat, Singh, and Singh (1995).

(c) Replace 'Forcibly No' statement in (b) with 'Forcibly Yes' statement, and construct a new estimator of π. Compare its variance expression with (b) at equal and unequal levels of protection.
Hint: Bhargava and Singh (2000).

Exercise 11.6. Consider a randomization device as defined by Warner (1965) is as follows:

(i) 'Do you possess the sensitive attribute A?' with probability P;

(ii) 'Do you not possess the sensitive attribute A?' with probability $(1-P)$;

has been used to conduct a social survey.

(a) Assume that the interviewer goes on selecting or collecting the response from the respondents from the population until m respondents (pre-specified number of

'Yes' answers) are not reporting 'Yes'. Then show that an unbiased estimator of population proportion π is given by

$$\hat{\pi}_1 = \frac{\hat{\theta}_1 - (1-P)}{2P-1}$$

where $\hat{\theta}_1 = m/n$ follows an Inverse Binomial distribution. Is it possible to list situations where Inverse Binomial Randomized Response (IBRR) can be implemented in actual practice?
Hint: Mangat and Singh (1991b, 1995).

(b) Let $\hat{\theta}_1$ be the proportion of observed 'Yes' answers in a sample of n units drawn with an SRSWR sampling, and it follows Binomial distribution with same parameters as in Warner's pioneer model. Suppose an estimator for estimating the population proportion π is defined as

$$\hat{\pi}_0 = a\hat{\theta}_1 + b .$$

Find the values of a and b such that the estimator $\hat{\pi}_0$ reduces to the estimators defined as

$$\hat{\pi}_1 = \frac{\hat{\theta}_1 - (1-P)}{2P-1} \quad \text{and} \quad \hat{\pi}_2 = \frac{\lambda\hat{\theta}_1 - (1-P)}{2P-1}.$$

Also find the optimum values of a and b such that the mean squared error of the estimator $\hat{\pi}_0$ is minimum.
Hint: Sampath, Uthayakumaran, and Tracy (1995), Mangat, Singh, and Singh (1991), Singh and Singh (1992b).

Exercise 11.7. Consider a population Ω consists of $n(\Omega)$ individuals and a group of individuals in Ω possessing sensitive quantitative variable (*e.g.*, income $\geq$ \$10,000 per month or number of abortions ≥ 2 or number of murders committed $\geq$ 1 etc.) is denoted by G. Our problem is to estimate the proportion $n(G)/n(\Omega)$, where $n(G)$ denotes the number of individuals in the group G. Simultaneously, we are also interested in estimating the average value of the sensitive quantitative variable for $n(G)$ individuals of group G. Because of the fear of the law or of embarrassment in society, the respondents from group G are not likely to reveal their membership of the group and also the value of the sensitive variable for them through a direct survey. Let π denote the proportion of respondents possessing the sensitive variable in group, G (*i.e.*, $\pi = n(G)/n(\Omega)$) and μ_x be the average value of the sensitive quantitative variable (X) for $n(G)$ respondents. Thus we would like to estimate π and μ_x simultaneously. In this randomized response model, two independent sub-samples of sizes n_1 and n_2 are drawn from Ω by using SRSWR such that $n_1 + n_2 = n$, the total sample size required. Each respondent selected in the ith sub-sample, $i = 1,2$ is requested to generate a random number S_i using randomization device R_i and report only the scrambled response $Z_i = XS_i$ if he belongs to group G; otherwise he is required to report a random number $Z_3 = S_3$

generated by another randomization device R_3. The whole process is performed by the interviewee unobserved by the interviewer. Let $E(S_i)=\theta_i$ and $V(S_i)=E(S_i-\theta_i)^2 =\sigma_i^2$, $i=1,2,3$, denote the known mean and variance for the scrambling variable S_i. Also let σ_x^2 denote the variance of the sensitive quantitative variable (X) in group G. Without loss of generality we have

$$E(\bar{Z}_1)=\pi\theta_1\mu_x+(1-\pi)\theta_3 \quad \text{and} \quad E(\bar{Z}_2)=\pi\theta_2\mu_x+(1-\pi)\theta_3,$$

where $\bar{Z}_1$ and $\bar{Z}_2$ are means of responses in the two sub-samples. Then show that an estimator of π is

$$\hat{\pi}_s=\frac{\theta_2\bar{Z}_1-\theta_1\bar{Z}_2-\theta_3(\theta_2-\theta_1)}{\theta_3(\theta_1-\theta_2)} \quad \text{and that for } \mu_x \text{ is } \quad \hat{\mu}_x=\frac{(\bar{Z}_1-\bar{Z}_2)\theta_3}{\theta_2\bar{Z}_1-\theta_1\bar{Z}_2-\theta_3(\theta_2-\theta_1)}.$$

Find the bias and variance of both estimators. Find the values of n_1 and n_2 such that the value of the linear combination of variances of both estimators, that is,

$\text{LC}=\alpha V(\hat{\pi}_s)+(1-\alpha)V(\hat{\mu}_x)$ is minimum.

Hint: Singh, Mangat, and Singh (1997).

Exercise 11.8. Consider a respondent belonging to group A is requested to repeat the trial in the Warner (1965) randomization device if in the first trial he does not receive the statement according to his status. The repetition of the trial is known to the interviewee but remains unknown to the interviewer. Assuming completely truthful responding by the respondents the probability of a 'Yes' answer is given by $\theta=\pi\{P+(1-P)P\}+(1-\pi)(1-P)$. Develop an unbiased estimator of π and find its variance.

Hint: Singh and Joarder (1997b).

Exercise 11.9. In the first phase select a preliminary large sample of m units from the population of N units by using SRSWOR and only auxiliary information X is measured on these m units as $X_1, X_2, \ldots, X_m$. In the second phase a sub-sample of n units is drawn from the preliminary large sample of m units using PPSWR and then the scrambled responses r_i are measured through a randomization device. Study the asymptotic properties of an estimator of the population mean

$$\bar{y}_p=\frac{1}{mn}\sum_{i=1}^{n}\frac{r_i}{p_i^*} \quad \text{where} \quad p_i^*=X_i\Big/\sum_{i=1}^{m}X_i$$

denotes the probability of selecting the ith unit from the given first phase sample.

Hint: Grewal, Bansal, and Singh (2002).

Exercise 11.10. In the direct question survey methods, if there are u distinct units in an SRSWR sample of size n and k_i is the frequency with which the ith distinct unit occurs in the sample, then we have $E(\bar{y}_u)=\bar{Y}$ and $V(\bar{y}_u)\le V(\bar{y}_n)$ where $\bar{y}_u=u^{-1}\sum_{i=1}^{u}y_i$ and $\bar{y}_n=n^{-1}\sum_{i=1}^{n}k_iy_i$. Show that this inequality oscillates in the case of

randomized response with replacement sampling. Deduce the results for qualitative characters also.
Hint: Arnab (1995), Mangat, Singh, Singh, Bellhouse, and Kashani (1995), Singh, Mahmood, and Tracy (2001).

Exercise 11.11. Consider an investigator is interested in estimating the proportion of people who individually belong to two different sensitive groups, called group A and group B. The investigator proposed a randomization device with which the respondent reply 'Yes' or 'No' to whichever of the two statements occur by chance: (i) I am a member of group A. (ii) I am not a member of group B. In order to estimate the proportion of those sensitive groups, two independent non-overlapping samples of sizes n_1 and n_2 are selected. Let P_i denote the probability of the first sensitive question being selected by respondents in the ith sample ($i = 1, 2$). Let n_i be the total number of respondents who replies 'Yes' in i^{th} sample. Let π_a and π_b be the true proportion of persons with attribute A and B, respectively. Obviously, the probability of 'Yes' answer in the ith sample is $\theta_i = P_i\pi_a + (1 - P_i)(1 - \pi_b)$, $i = 1,2$. Deduce the estimators of π_a and π_b and derive their minimum variances subject to total sample size remains fixed $n = n_1 + n_2$.
Hint: Chang and Liang (1996).

Exercise 11.12. Consider π be the proportion of respondents in the population that belong to the sub-group (say) A, each member of which has the value of sensitive variable X that satisfies the inequality $X \geq W$, where W is a known fixed constant. Further let Y denote the random variable that is generated by a randomization device (say R_1). The distribution of the variable Y is known and it takes values equal to or greater than W. Each respondent selected in an SRSWR sample of n respondents is instructed to report actual value for X if he belongs to sub-group A; otherwise, he is instructed to report the random number Y generated through the randomization device R_1 provided to him. The mode followed by the respondents for giving their responses is not revealed to the investigator. Thus the response Z_i from the ith respondent will take value X_i with probability π and a random value Y_i with probability $(1-\pi)$. Thus we have

$$E(Z_i) = \pi\, \mathrm{E}(X_i) + (1-\pi)E(Y_i) = \pi\, \mu_x + (1-\pi)\mu_y$$

where μ_x and μ_y respectively denote the means for the sensitive variable X in the sub-group and for the variable Y in the randomization device R_1. In addition to the above, each respondent in the sample is also provided with the usual Warner's (1965) randomization device R_2 to estimate π. This device may consist of a deck of cards having two types of statements:

(i) 'Do you belong to the sub-group A?' with probability P;

and

(ii) 'Do you belong to the sub-group not A?' with probability $(1-P)$.

Show that an unbiased estimator of π is

$$\hat{\pi} = \{n_1/n - (1-p)\}/(2p-1)$$

where n_1 denotes the number of observed 'Yes' answers from the sample when the randomization device R_2 is used. Also deduce an estimator $\hat{\mu}_x$ of μ_x and show that, to the first order of approximation, its MSE is

$$\text{MSE}(\hat{\mu}_x) = \frac{1}{n\pi^2}\left[\pi\sigma_x^2 + (1-\pi)\sigma_y^2 + \frac{p(1-p)}{(2p-1)^2}(\mu_x - \mu_y)^2\right].$$

Hint: Singh, Singh, and Mangat (1996).

Exercise 11.13. Suppose X denotes the sensitive variable. Let S be the scrambling variable, independent of X and having finite mean and variance. For simplicity, assume that $X \geq 0$ and $S > 0$. The respondent generates S by using some specified randomization device. Each respondent scrambles the response on X by multiplying it with the value taken by the scrambling variable S in his/her case. Only the scrambled result $Y = XS$ is revealed to the interviewer. Note that the particular value taken by S is not known to the interviewer, thus the respondent's privacy is not disclosed. Consider a sample of size n is drawn using simple random with replacement sampling from a population of size N. Let y_i denote the value of the scrambled variable Y for the ith respondent of the sample, $i = 1,2,...,n$. Show that square of coefficient of variation of the sensitive character X is given by

$$C_x^2 = \left(C_y^2 - C_\tau^2\right)/\left(1 + C_\tau^2\right),$$

where $C_x^2 = \mu_2/\overline{X}^2$ and $C_\tau^2 = \tau_2/\theta^2$ are the squares of coefficient of variation of the sensitive character X and scrambling variable S, respectively. Suggest an estimator of C_x^2 and study its asymptotic properties.

Hint: Singh, Singh, and Mangat (1998).

Exercise 11.14. In Franklin's (1989a, 1989b) model $k \geq 1$ responses are obtained from each respondent of a simple random with replacement sample of size n. The response Z_{ij}, $i = 1,2,...,n$; $j = 1,2,...,k$ is a random number drawn from the density g_{ij} if the respondent belongs to the sensitive group A; otherwise, it is drawn from the density h_{ij}. The interviewer does not know the density used by the respondent for drawing the random number. The model can be specialized by having $g_{ij} = g_i$ and $h_{ij} = h_i$ for all $i = 1,2,...,n$. The densities g_i and h_j, respectively, have known means μ_{1i} and μ_{2j} and known variances σ_{1j}^2 and σ_{2j}^2. Suppose an investigator modified the procedure suggested by Franklin (1989b) by using the known proportion of unrelated character π_y in the population and by suitably choosing known parameters of the proposed randomization device as follows:

If a respondent,

(i) belongs to A but not to Y he is instructed to use the density g_{1ij},

(ii) does not belong to both A and Y he is instructed to use the density g_{2ij},

(iii) belongs to both A and Y he is instructed to use the density g_{3ij},

(iv) belongs to Y but not to A he is instructed to use the density g_{4ij}.

On the basis of the above information suggest estimator of finite population proportion of interest. Derive its variance expression.
Hint: Singh (1994).

Exercise 11.15. A sample of n respondents is drawn from a finite population of N units using SRSWR but the information from the d distinct units in the sample, $1 \le d \le n$, is used in the construction of estimator of population proportion based on Warner's model. Let d' denote the respondents reporting "Yes" answer in the interview conducted with the Warner's RR device, then study the properties of the following estimators of π defined as

$$\hat{\pi}_1 = \frac{d'/d - 1 + p}{2p - 1} \quad \text{and} \quad \hat{\pi}_2 = \frac{d'/E(d) - 1 + p}{2p - 1}$$

Also suggest an estimator of π in case of unrelated question model based on the information collected from distinct unit.
Hint: Tracy and Mangat (1998), Arnab (1999).

Exercise 11.16. Develop an estimator of proportion π of the persons possessing the attribute A in the population with the help of the following randomization devices:

(a) Every respondent selected in the sample of size n drawn with SRSWR sampling is requested to report "Yes" if he belongs to sensitive group A, otherwise he is requested to use Warner's (1965) device. Obviously the probability of a 'Yes' answer is $\theta_1 = \pi + (1 - P)(1 - \pi)$.

Hint: Mangat (1994).

(b) Each respondent in a sample of n respondents is instructed to say 'Yes' if he belongs to sensitive group A and to report his membership in Y in terms of 'Yes' or 'No' if he does not belong to group A. Obviously the probability of 'Yes' answer is given by $\alpha = \pi + (1 - \pi)\pi_y$. Can this randomization device be used in mail surveys?
Hint: Mangat, Singh, and Singh (1993), Singh, Mangat, and Singh (1993).

Exercise 11.17. Consider the population under study consists of N units. We assume that the population could be thought of as consisting of two strata, the first stratum having N_1 respondents who will return the completed questionnaire without waiting for any further communication from the investigator while the N_2 members of the second stratum (defined as non-response stratum) will not do so. Thus $N_1 + N_2 = N$.

Let μ_1 and μ_2 denote the population means for the sensitive character X in the first and second strata. Let the sensitive character X in the first and second strata of the population be denoted by X_1 and X_2 respectively, so that

$$E(X_1)=\mu_1 \text{ and } E(X_2)=\mu_2 .$$

Then the actual population mean of the sensitive variable X is given by

$$\mu=\sum_{i=1}^{2} W_i\mu_i ,$$

where $W_i = N_i/N$, $i=1, 2$. Assume we select a sample of size n using SRSWOR method. Questionnaires will then be mailed to each of the selected respondents in the sample with the request that they should be returned after completion. The respondents will also be required to scramble their response on sensitive character X. For scrambling the sensitive character X, each respondent is instructed to select K random natural numbers S_j ($j=1,2,...,K$) out of the random sequence of first N_0 natural numbers sent with the questionnaire by SRSWOR. Then each respondent is instructed to calculate the mean, $S=K^{-1}\sum_{j=1}^{K} S_j$ of the K selected natural numbers and record only the scrambled response $Y=SX$. The value of K and N_0 are same for each respondent. Here S_j is a random variable with $1\le S_j \le N_o$. Using $\sum_{j=1}^{N_o} S_j = N_o(N_o+1)/2$ and $\sum_{j=1}^{N_o} S_j^2 = N_o(N_o+1)(N_o+2)/6$, it can be easily seen that $E(S)=\dfrac{N_o+1}{2}$ and $V(S)=\left\{\dfrac{N_o}{K}-1\right\}\dfrac{N_o+1}{12}$. Let n_1 denote the number of respondents who return the completed questionnaires so that $n_2=(n-n_1)$ is the number of respondents who do not return the questionnaire. We select a sub-sample of h_2 respondents with SRSWOR from n_2 respondents of the non-response stratum such that $n_2 = h_2 g$, ($g \ge 1$). These h_2 respondents will then be interviewed personally. Let $Y_{1i}=X_{1i}S_1$ ($i=1,2,...,n_1$) and $Y_{2i}=X_{2i}S_2$ ($i=1,2,...,h_2$) denote the scrambled responses given by the respondents at the first and second efforts, respectively. Note that X_{1i} and X_{2i} are independent of S_1 and S_2 respectively, therefore $E(X_{1i})=E(Y_{1i})/E(S_1)=\mu_1$ and $E(X_{2i})=E(Y_{2i})/E(S_2)=\mu_2$. On using sample analog of scrambled responses, unbiased estimators of μ_1 and μ_2 are $\hat{\mu}_1 = 2\bar{y}_1/(N_o+1)$ and $\hat{\mu}_2 = 2\bar{y}_2/(N_o+1)$, where $\bar{y}_1 = n_1^{-1}\sum_{i=1}^{n_1} y_{1i}$ and $\bar{y}_2 = h_2^{-1}\sum_{i=1}^{h_2} y_{2i}$. Consider an estimator for the population mean μ as:

$$\hat{\mu}_w = \frac{(n_1\hat{\mu}_1 + n_2\hat{\mu}_2)}{n} .$$

Show that $\hat{\mu}_w$ is unbiased for μ and find its variance for the fixed cost.

Hint: Singh, Singh, and Mangat (1995).

Exercise 11.18. Study the properties of the ridge estimator

$$\hat{\beta}_{R_c}^{(s)} = (X'X + R_c I)^{-1} X'z$$

under scrambled responses. Show that the variance of $\hat{\beta}_{R_c}^{(s)}$ is

$$V\left(\hat{\beta}_{R_c}^{(s)}\right) = \sigma^2 A(X'X)^{-1} A' + C_\gamma^2 (X'X + R_c I)^{-1} X'\left(D + \sigma^2 I\right)X(X'X + R_c I)^{-1},$$

where

$$C_\gamma = \gamma/\theta \text{ and } D = \text{Diag}\left[\left(\sum_{j=1}^{k} X_{1j}\beta_j\right)^2, \dots, \left(\sum_{j=1}^{k} X_{nj}\beta_j\right)^2\right].$$

Hint: Singh and Tracy (1999).

Exercise 11.19. Assume a simple random sample of n people is drawn with replacement from the population. Each interviewee in the sample is then provided with three random devices, say R_0, R_1 and R_2 each consisting of two mutually exclusive outcomes. The random device R_0 is the decision device and the interviewee uses R_0 at the first instance. The random device R_0 has two types of statements namely, (i) Use a random device R_1 (ii) Use a random device R_2. The probabilities with which these two statements occur in the device are T and $(1-T)$, respectively. If the interviewee selects the card from the device R_0 having the statement (i) Use the random device R_1, then the interviewee is instructed to use the density $g_{ij}^*(L_{ij})$, if he/she belongs to the sensitive group A otherwise he/she uses the density $h_{ij}^*(L_{ij})$. On the other hand, if the interviewee selects the card from the device R_0 having the statement (ii) Use the random device R_2, then the interviewee will use the density $g_{ij}(L_{ij})$ if he/she belongs to the sensitive group A, otherwise he/she uses the density $h_{ij}(L_{ij})$. Form an unbiased estimator of the proportion of persons, say π in the sensitive group A based on the information collected through the above randomization device.
Hint: Singh and Singh (1992a, 1993d).

Exercise 11.20. Divide the population into $(k+1)$ groups --- of which the first group consists of responsive group and the remaining k groups belong to non-sensitive group of varying degrees --- each of these non-responsive groups supply $a_1, a_2, \dots, a_k$ levels of responses. Let $\pi_1, \pi_2, \dots, \pi_{k+1}$ be the true probabilities of responses in these $(k+1)$ groups and $p_1, p_2, \dots, p_{k+1}$ be the probabilities that the spinner points to the responsive group. The value of the response variable is 1, $a_1, a_2, \dots, a_k$, that is, the response group supply the full information while the remaining k groups reveals only $a_1, a_2, \dots, a_k$ level of information.
Thus

$$P(x_i = 1) = \pi_1 p_1 + (1-p_1)(1-\pi_1),$$

$$P(x_i = j) = \pi_j p_j + (1 - p_j)(1 - \pi_j), \quad j = 2, 3, ..., (k+1).$$

The likelihood function of the sample is

$$L = \prod_{j=1}^{k+1} \left[p_j \pi_j + (1 - p_j)(1 - \pi_j) \right]^{n_j},$$

where

$$n = \sum_{j=1}^{k+1} n_j .$$

Show that the maximum likelihood estimate of π_j, $j = 1,2,...,(k+1)$ is the solution to the set of singular simultaneous equations

$$\begin{bmatrix} n_1(2p_1 - 1), n_2(1 - p_2), ..., n_{k+1}(1 - p_{k+1}) \\ n_1(1 - p_1), n_2(2p_2 - 1), ..., n_{k+1}(1 - p_{k+1}) \\ \\ n_1(1 - p_1), n_2(1 - p_2), ..., n_{k+1}(2p_{k+1} - 1) \end{bmatrix} \begin{bmatrix} \{\pi_1(2p_1 - 1) + (1 - p_1)\}^{-1} \\ \{\pi_2(2p_2 - 1) + (1 - p_2)\}^{-1} \\ \\ \{\pi_{k+1}(2p_{k+1} - 1) + (1 - p_{k+})\}^{-1} \end{bmatrix} = \begin{bmatrix} 0 \\ 0 \\ \\ 0 \end{bmatrix}$$

Hint: Mishra and Sinha (1999), Bourke (1981), Eriksson (1973).

Exercise 11.21. (a) Assume two independent samples of sizes n_i, $i = 1, 2$ are drawn from the whole population by using an SRSWR method such that $n_1 + n_2 = n$, the total sample size required. Provide a randomization device S_i to the respondents in the ith sample, $i = 1,2$ with two statements: (i) 'I belong to group A' and (ii) 'I belong to group Y', represented with probabilities p_i and $(1 - p_i)$, $i = 1, 2$, respectively. Then for this model $\theta_i = p_i \pi + (1 - p_i)\pi_y$ is the probability of obtaining 'Yes' answer from a person in the ith sample. If $\hat{\theta}_i$ denotes the proportion of 'Yes' answers obtained from the respondents in the ith sample, show that an unbiased estimator of π is

$$\hat{\pi}_G = \frac{(1 - p_2)\hat{\theta}_1 - (1 - p_1)\hat{\theta}_2}{p_1 - p_2}$$

with variance

$$V(\hat{\pi}_G) = \left[(1 - p_2)^2 \theta_1(1 - \theta_1)/n_1 + (1 - p_1)^2 \theta_2(1 - \theta_2)/n_2 \right] / (p_1 - p_2)^2 .$$

Show that the optimal choice of one of the p_i, $i = 1, 2$ is close to one and other close to zero. Show that the choice of the value π_y (unknown) close to zero or 1 according as $\pi < 0.5$ or $\pi > 0.5$ and if $\pi = 0.5$, the minimum variance occurs at the tails of π_y.

(b) In the above part (a) select the respondents in the independent sub-samples by using an SRSWOR sampling. Show that the variance of the estimator $\hat{\pi}_G$ is given by

$$V_k(\hat{\pi}_G) = V(\hat{\pi}_G) - \left[(1 - p_2)^2 (n_1 - 1)\{p_1^2 \pi(1 - \pi) + (1 - p_1)^2 \pi_y(1 - \pi_y)\}\right] / \left[n_1(N - 1)(p_1 - p_2)^2 \right]$$
$$- \left[(1 - p_1)^2 (n_2 - 1)\{p_2^2 \pi(1 - \pi) + (1 - p_2)^2 \pi_y(1 - \pi_y)\}\right] / \left[n_2(N - 1)(p_1 - p_2)^2 \right] .$$

Discus the optimal choice of p_i and π_y under an SRSWOR sampling.

(c) The choice of $p_2 = 0$ under both SRSWR and SRSWOR sampling schemes discloses the privacy of certain kind of respondents.

Hint: Kim (1978), Singh, Singh, and Mangat (2000).

Exercise 11.22. Let π be the true proportion of persons belonging to group A. Let ρ_j, $j = 1,2,3$ be known parameters of a randomization device used for eliciting information in randomized response surveys. Let y_i be the binary response of the ith respondent in the sample consisting of n respondents, then the probabilities for the randomized response sensitive question are given by $\Pr(y_i = 1) = \rho_1 + (1-\rho_1-\rho_2)\pi_{rr}$ and $\Pr(y_i = 0) = \rho_2 + (1-\rho_1-\rho_2)(1-\pi_{rr})$, where π_{rr} denote the proportion of observed 'Yes' answers through the randomization device. Let X_i be a vector of explanatory variable and β is a column vector of unknown parameters, so that the probability for the ith respondent is given by $\pi = e^{\beta' X_i}\left(1+e^{\beta' X_i}\right)^{-1}$. Show that an estimate of β can be obtained by maximizing the likelihood function

$$L = \prod_{i:y_i=1}\left[\rho_1 + (1-\rho_1-\rho_2)\frac{e^{\beta' X_i}}{1+e^{\beta' X_i}}\right]\prod_{i:y_i=0}\left[\rho_2 + (1-\rho_1-\rho_2)\frac{1}{1+e^{\beta' X_i}}\right].$$

Hint: Kerkvliet (1994)

Exercise 11.23. Consider each interviewee in a given study is subjected to the randomization procedure, say R_0, which is similar to the one used by Kuk (1990). The proportion for the red cards in the two decks is represented by θ_1 and θ_2. The values of the probability θ_1 and θ_2 in a particular investigation are determined randomly and depend on the joint distribution of θ_1 and θ_2. Assume r denotes the proportion of red cards reported by a random sample of n respondents. Then for the given values of θ_1 and θ_2 we have

$$\hat{\pi}_s = \frac{r-\theta_2^*}{\theta_1^*-\theta_2^*}, \quad \theta_1^* \neq \theta_2^*,$$

where θ_1^* and θ_2^* are fixed for the given study and their values depend upon the joint distribution of θ_1 and θ_2.

(a) Show that the estimator $\hat{\pi}_s$ is unbiased for the population proportion, π.

(b) The variance $V(\hat{\pi}_s)$ of the estimator $\hat{\pi}_s$ is given by

$$V(\hat{\pi}_s) = \frac{\pi(1-\pi)}{n} + \int_c^d\int_a^b \frac{(1-\pi)\theta_2(1-\theta_2)+\pi\theta_1(1-\theta_1)}{n(\theta_1-\theta_2)^2} f(\theta_1,\theta_2)d\theta_1 d\theta_2$$

where $f(\theta_1, \theta_2)$ denotes the joint p.d.f. of θ_1 and θ_2 and $0 \le a < \theta_1 < b \le 1$, $0 \le c < \theta_2 < d \le 1$.

Hint: Singh (2002c).

Exercise 11.24. Consider an unrelated question model with two non-sensitive characters Y_1 and Y_2. Two independent SRSWR samples of samples of sizes n_1 and n_2 are drawn. In each sample, the respondent answer a direct question about a non-sensitive character, they also answer one of the two questions about A as well as about a non-sensitive character, selected with probability P and $(1-P)$ respectively through a randomization device. In other words, in the first sample, each selected respondent answer through a randomized device question regarding A and Y_1 while question Y_2 without the randomization device. In the second sample each respondent response for A and Y_2 through the randomization device, while they respond to the first natural question Y_1 directly. Let Θ_{1r} and Θ_{2r} be the probabilities of obtaining 'Yes' answer through the randomization device in the first and second sample. Also let $\hat{\Theta}_{1d}$ and $\hat{\Theta}_{2d}$ be the proportion of direct questions obtained in the second and first sample on the neutral characters Y_1 and Y_2, respectively.

(a) Show that the following two estimators of π are possible

$$\hat{\pi}_1 = \frac{\hat{\Theta}_{1r} - (1-P)\hat{\Theta}_{2d}}{P} \quad \text{and} \quad \hat{\pi}_2 = \frac{\hat{\Theta}_{2r} - (1-P)\hat{\Theta}_{1d}}{P}.$$

(b) Consider a linear combination of the above two estimators as $\hat{\pi}_a = \sum_{i=1}^{2} w_i \hat{\pi}_i$ such that $\sum_{i=1}^{2} w_i = 1$. Show that $\hat{\pi}_a$ is unbiased and find the optimum weights such that the variance of $\hat{\pi}_a$ is minimum.

Hint: Folsom, Greenberg, Horvitz, and Abernathy (1973).

Exercise 11.25. Consider that in Warner (1965) model the persons belonging to sensitive group are not reporting truthfully, but only a proportion Δ of them are honest. Assuming that those who are not members of the group A are honest and report truthfully. Show that the probability of a 'Yes' answer becomes $\theta = \pi\,\Delta(2P-1)+(1-P)$. Find the variance of the unbiased estimator of π defined as

$$\hat{\pi}_u = \frac{\hat{\theta} - (1-P)}{\Delta(2P-1)} \quad \text{where } P \neq 0.5.$$

Hint: Bourke and Dalenius (1976).

Exercise 11.26. Show that an optimal estimator of the population total, Y, under scrambled responses is given by

$$e(s,r) = a_s + \sum_{i \in s} b_{s_i} r_i,$$

where r_i denotes the scrambled responses, a_s and b_{s_i} have their usual meanings.

Hint: Arnab (2002).

Exercise 11.27. Consider an unrelated question model (or U model) for the situation when π_y is known. Each sampled respondent is provided with a random device consisting of two statements: (i) I belong to sensitive group A; and (ii) I belong to non-sensitive group Y; represented with probabilities p_1 and $(1-p_1)$, respectively. The respondent selects randomly one of these two statements, unobserved by the interviewer and reports 'Yes' or 'No' with respect to his/her actual status. The probability of 'Yes' answer is $\theta_1 = p_1 \pi + (1-p_1)\pi_y$. Thus, an unbiased estimator of the population proportion π is given by

$$\hat{\pi}_G = \frac{\hat{\theta}_1 - (1-p_1)\pi_y}{p_1}.$$

Consider a two-stage randomized response unrelated question model in which each interviewee is provided with two randomization devices R_1 and R_2. The randomization device R_1 consists of two statements, namely: (i) I belong to sensitive group A; and (ii) Go to randomization device R_2; represented with probabilities T and $(1-T)$, respectively. The randomization device R_2 is the same as used in the U model represented with probabilities p_2 and $(1-p_2)$, respectively. The probability of 'Yes' answer is given by $\theta_2 = T\pi + (1-T)[p_2\pi + (1-p_2)\pi_y]$ and an unbiased estimator of π is given by

$$\hat{\pi}_m = \frac{\hat{\theta}_2 - (1-p_2)(1-T)\pi_y}{T + p_2(1-T)}.$$

(a) Find the variance of $\hat{\pi}_G$ and $\hat{\pi}_m$.

(b) Show that at equal level of protection of the respondents $V(\hat{\pi}_G) = V(\hat{\pi}_m)$.

Hint: Mangat (1992), Mangat, Singh, and Singh (1992), Bhargava and Singh (2001).

Exercise 11.28. Let X denote the response to the first sensitive question (*e.g.*, income) and Y denote the response to the second sensitive question (*e.g.*, expenditure). Further assume S_1 and S_2 be the two scrambling random variables, each independent of X and Y and having finite means and variances. For simplicity also assume that $X \ge 0$, $Y \ge 0$, $S_1 > 0$ and $S_2 > 0$.

Consider the following two cases:

(a) The respondent generates S_1 using some specified method, while S_2 is generated by using the linear relation, $S_2 = \alpha S_1 + \beta$, where α and β are the known constants and therefore S_1 and S_2 are dependent.

(b) S_1 and S_2 are random variables following known distributions. The particular values of S_1 and S_2 to be used by any respondent are obtained from two separate random devices. This way S_1 and S_2 becomes independent.

The interviewee multiplies his response X to the first sensitive question by S_1 and the response Y to the second sensitive question by S_2. The interviewer thus receives two scrambled answers $Z_1 = XS_1$ and $Z_2 = XS_2$. The particular values of S_1 and S_2 are not known to the interviewer, but their joint distribution is known. In this way the respondent's privacy is not violated. Let $E(S_1)=\theta_1$, $E(S_2)=\theta_2$, $V(S_1)=\gamma_{20}$, $V(S_2)=\gamma_{02}$, $\text{Cov}(S_1,S_2)=\gamma_{11}$, $E(X)=\mu_1$, $E(Y)=\mu_2$, $V(X)=\sigma_x^2=m_{20}$, $V(Y)=\sigma_y^2=m_{02}$, $\gamma_{rs}=E[S_1-\theta_1]^r[S_2-\theta_2]^s$ and $m_{rs}=E[X-\mu_1]^r[Y-\mu_2]^s$, where θ_1, θ_2, γ_{20}, γ_{02}, γ_{11}, and γ_{rs} are known to the interviewer but μ_1, μ_2, σ_x^2, σ_y^2 and m_{rs} are known. Also let σ_{z1}^2, σ_{z2}^2 and $\sigma_{Z_1Z_2}$ denote the variance and co-variance of Z_1 and Z_2, respectively.

(a) Show that the correlation coefficient between the two sensitive variables X and Y is then given by

$$\rho_{xy} = \frac{\left(\sigma_{Z_1Z_2} - \gamma_{11}\mu_1\mu_2\right)\sqrt{\gamma_{20}+\theta_1^2}\sqrt{\gamma_{02}+\theta_2^2}}{\left(\gamma_{11}+\theta_1\theta_2\right)\sqrt{\sigma_{Z_1}^2-\gamma_{20}\mu_1^2}\sqrt{\sigma_{Z_2}^2-\gamma_{02}\mu_2^2}}.$$

(b) Develop an estimator of the correlation coefficient using scrambled responses.
Hint: Singh (1991b), Bellhouse (1995).

Exercise 11.29. A randomization device (say, deck of cards) consists of three types of cards bearing statements: (i) I belong to group A; (ii) I belong to group Y; and (iii) Draw one more card. The statements are represented with proportions p, p_1, and p_2, respectively. Note that the characters A and Y are uncorrelated. The respondent is required to draw one card randomly from the above deck and give answer in terms of 'Yes' or 'No' according to his/her actual status if the statements (i) or (ii) are drawn. However if statement (iii) is drawn the respondent is required to repeat the above process without replacing that card. If the statement (iii) is drawn in the second phase, the respondent is directed to report 'No'. If m be the total number of cards in the deck, then the probability of 'Yes' answer is

$$\theta = \left[\pi\, p + p_1\pi_y\right]\left[1 + p_2 m/(m-1)\right].$$

Construct an unbiased estimator of π and study its properties for different values of the parameters involved in it.
Hint: Singh, Singh, Mangat, and Tracy (1994)

Exercise 11.30. Consider a procedure in which the randomised response (RR) device and method for sampling remain same as in Warner (1965) model. However, it differs in the sense that the respondent is free to give an answer in terms of 'Yes' and 'No' either by using RR device or without using it, without revealing to the interviewer which mode has been followed for giving answer. If T be the probability that a respondent gives answer without using the RR device, then the probability of 'Yes' answer is given by

$$\theta_1 = \pi T + (1-T)\theta$$

where $\theta = P\pi + (1-P)(1-\pi)$. If n_1 be the number of 'Yes' answers in an SRSWR sample of n respondents, then study the bias and mean square error of the estimator of π defined as

$$\hat{\pi}_{\text{ms}} = \frac{n_1}{n}.$$

Hint: Mangat and Singh (1994).

Exercise 11.31. Consider a sensitive variable X which is supposed to be continuous with true density $g(\bullet)$ and Y be an unrelated variable with density $h(\bullet)$. Consider the problem of estimation of population mean μ_x of the sensitive variable X using the known mean μ_y of the unrelated character. Each respondent selected in the sample is provided with two randomization devices: The first randomization device consists of a deck of cards having two types of statements: (a) Report the value of sensitive variable X; (b) Go to the second randomization device; with probabilities W and $(1-W)$, respectively. In the second randomization device, each interviewee reports his/her X value with probability P and Y value with probability $Q = (1-P)$. Obviously, the observed randomised response, say Z, has the density function

$$f(z) = Wg(z) + (1-W)[Pg(z) + (1-P)h(z)]$$

and the population mean of the randomised response Z is given by

$$\mu_z = \mu_x[P + W(1-P)] + Q(1-W)\mu_y.$$

Construct an estimator of μ_x and study its properties.

Hint: Singh, Mangat, and Singh (1994).

Exercise 11.32. Consider a randomization device consisting of three types of cards bearing statements: (i) I belong to sensitive category A; (ii) I belong to category Y; and (iii) Blank cards; with probabilities p_1, p_2 and p_3 such that $\sum_{i=1}^{3} p_i = 1$. In case the blank card is drawn by the respondent, he/she will reports 'No'. The rest of the procedure remains as usual. The probability of 'Yes' answer will now be

$$\theta_1 = p_1\pi + p_2\pi_y.$$

Study the properties of the estimator of the proportion of the sensitive attribute π given by

$$\hat{\pi}_1 = \left\{\hat{\theta}_1 - p_2\pi_y\right\}/p_1$$

where $\hat{\theta}_1$ is the proportion of 'Yes' answers obtained from the n sampled respondents.

Hint: Singh, Horn, Singh, and Mangat (2003).

Exercise 11.33. Consider the problem of estimation of population proportion π of individuals possessing a certain sensitive characteristic. We selected an SRSWR sample of n respondents. Each sampled person is provided with a random device which produces integers $1,2,...,L$ with frequencies $p_1, p_2,..., p_L$ respectively. For example, the device might be a deck of M cards with exactly Mp_j of those cards showing the integer j, $j=1,2,...,L$ or perhaps the device is a die with L faces, each one showing one of the integers $1, 2,...,L$ with probabilities $p_1, p_2,..., p_L$ respectively. Using the random device in the absence of the interviewer, the individual produces one of these numbers and he/she reports how far away this number is from $L+1$ if he/she has the characteristic or from 0 if he/she does not have it. Let x_i take on the value $L+1$ if the ith individual possess the characteristic and the value 0 if not. Obviously, $P(x_i = L+1)=\pi$ and $P(x_i = 0)=1-\pi$. Let y_i be the integer produced by the ith individual using the random device. Obviously, the reported random number is $d_i = |x_i - y_i|$ such that $P(d_i = k)=(1-\pi)p_k + \pi\, p_{L+1-k}$, $k = 1,2,...,L$. Study the bias and variance of the estimator of π given by

$$\hat{\pi}_{\text{tcc}} = \frac{\bar{d} - E(y)}{L+1-2E(y)}$$

where $E(y)=\sum_{k=1}^{L} kp_k$ and $\bar{d} = n^{-1}\sum_{i=1}^{n} d_i$.

Hint: Christofides (2003), Chaudhuri (2001).

PRACTICAL PROBLEMS

Practical 11.1. Michael visited libraries of certain developing counties such as India, China, etc., and found that about 20 years ago most of the top class scientific journals were available in the shelves of libraries of the universities of those countries, but he felt that the availability of top class scientific journals is decreasing from the shelves of libraries of developing countries, and there may be some hidden reason behind it. One obvious reason may be that the editors of the journals may not be honestly reviewing papers and scientists from developing countries may have lost interest in the top class journals. Michael decided to do a survey to estimate the proportion of editors and associate editors of the scientific journals to know their review process. Michael selected 5000 editors and associate editors and asked the following questions using a randomization device producing 80% statements, 'Are you reviewing every article honestly?' along with 20% statements, 'Are you not reviewing every article honestly?' Out of 5000 selected persons 3000 reported 'Yes' through the above randomization device. Find Michael's estimate of the proportion of editors reviewing articles honestly. Also construct a 95% confidence interval estimate. Comment!

Practical 11.2. Assume the true proportion of the extramarital relations in the world is 0.3. Suppose we used a randomization device to collect information from a large sample of persons using a randomization device with two types of statements:

(i) 'Are you having extramarital relation?' with probability 0.80;
(ii) 'Were you born in spring?' with probability 0.20.

Assuming that the proportion of persons born spring is 0.4. Find the relative efficiencies of three different techniques of Mahmood, Singh, and Horn (1998) discussed in this chapter with respect to Greenberg, Abul-Ela, Simmons, and Horvitz (1969) technique.

Practical 11.3. Ms. Poonam Singh wishes to estimate the proportion of persons having extra marital relations in the world. Suppose she selected an SRSWR sample of 10000 persons across the world, and took responses through a randomization device producing 80% statements, 'Are you having extra marital relation?' along with 20% statements, 'Are you having no extra marital relation?' Out of 10,000 selected persons 3,000 reported 'Yes' through the above randomization device. Find her estimate of the proportion of extramarital relations in the world. Also construct a 95% confidence interval estimate.

Practical 11.4. A social society of a particular community wishes to study the virginity status of couples before their marriage. In certain communities, the status of virginity before the marriage is an important issue. A researcher selected an SRSWR sample of n couples (each made of one male and one female).

(I) Each male in the sample was asked to respond to one of the two outcomes using randomization device R_1 as follows:

Q1: 'Were you virgin before you got married?' with probability P_1 ;
Q2: 'Were you born during the first 6 months of the year?' with probability $(1-P_1)$.

If π_m denotes the true proportion of virgin males before marriage, and ϕ_m in the probability of males born during the first 6 months of the year, then the probability of a 'Yes' answer for a male in the couple is given by

$$\theta_m = \pi_m P_1 + (1-P_1)\phi_m .$$

(II) Each female in the sample was asked a response to one of the two outcomes using randomization device R_2 as follows:

Q1: 'Were you virgin before you got married?' with probability P_2 ;
Q2: 'Were you born during the first 6 months of the year?' with probability $(1-P_2)$

If π_f denotes the true proportion of virgin females before marriage, and ϕ_f in the probability of female born during the first 6 months of the year, then the probability of 'Yes' answer for a female in the couple is given by

$$\theta_f = \pi_f P_2 + (1-P_2)\phi_f .$$

(III) Assume $\phi_m = \phi_f = \phi$ (say), that is, the proportions of males and females born during the first six months of a year is same, then every couple was asked the following question directly.

'Were either one (not both) of you born during the first 6 months of the year?'

Obviously the probability of a 'Yes' answer from a couple is given by

$$\theta = \pi_f(1-\pi_m)\phi + \pi_m(1-\pi_f)\phi .$$

(IV) Using information from (I) to (III) estimate the proportion of the males and females who were virgin before they were married. Also obtain a pooled estimate.

Given: $n = 5000$, $n_m = 2000$, $n_f = 4000$, $n_c = 2500$ and $P_1 = P_2 = 0.80$, where n_m, n_f, and n_c denote the number of observed 'Yes' responses from males, females, and couples, respectively.

Practical 11.5. Michael visited hospitals and found AIDS is a very serious problem in these days. He selected an SRSWOR sample of 70,000,000 persons across the world and each respondent asked to use the following two randomization devices in the sequence.

The randomization device R_1 consists of two statements *viz.*:
Statement (i): 'Are you suffering from AIDS?' with probability 0.8;
Statement (ii): 'Use second randomization device, R_2 ' with probability 0.2.
The second randomization device R_2 consists of the following two statements:
Statement (i): 'Are you suffering from AIDS?' with probability 0.7;
Statement (ii): 'Are you not suffering from AIDS?' with probability 0.3.
Out of sampled 70,000,000 persons he received 10,360,000 number of 'Yes' answers.

(a) Make a flow chart of the randomization device.
(b) Estimate the proportion of AIDS patients, and derive 95% confidence interval estimate.

12. NON-RESPONSE AND ITS TREATMENTS

12.0 INTRODUCTION

Incompleteness or non-response in the form of absence, censoring, or grouping is a troubling issue of many data sets. Statisticians have recognized for some time that failure to account for the stochastic nature of incompleteness or non-response can spoil the nature of data. There are several factors which effect the non-response rate in any particular inquiry. Some of these factors are the type of information being collected, the official status of the surveying agency, the extent of publicity, the legal obligations of the respondents, the time of visit by the enumerator and length of the schedule, etc.. Hansen and Hurwitz (1946) were the first to deal with the problem of incomplete samples in mail surveys. It is a well known fact that mail surveys or telephone surveys are most commonly used by most of the bureaucratic or business organisations because of their low cost. Rubin (1976) has defined three key concepts: Missing at random (MAR), Observed at random (OAR), and parameter distinctness (PD).

Missing at random (MAR): The data are MAR if the probability of the observed absence pattern given the observed and unobserved data, does not depend on the values of the unobserved data. We shall put all such cases where the data are missing only owing to chance factors. It will therefore include cases where the enumerator is not able to contact the respondents only by chance and had he been able to contact, the data would have been collected. For example, when the information is kept on punched cards the non-response owed to the accidental loss of one or more cards is of this category. This type of non-response is called random non-response.

Observed at random (OAR): The data are OAR if for every possible value of the missing data the probability of the observed absence pattern, given the observed and unobserved data, does not depend on the values of the observed data.

Parameter distinctness (PD): PD holds if there are no a *priori* ties between the parameters of the absence model and those of the data model. In other words, we can possibly classify the non-response with respect to its nature into two broad categories.

Deliberate non-response (DNR): If the respondents are not willing to reveal their response, such cases of non-response will be classified as deliberate non-response. For example, the non-response in surveys in which information is being collected on personal income or on certain personal habits such as drinking, gambling, etc., will come under this category.

Unit and total non-response: Kalton and Kasprzyk (1986) says that it is a common practice to distinguish between total (and unit) non-response, when none of the survey responses are available for a sampled element, and item non-response, when some but not all of the responses are available. Total non-response arises because of refusals, inability to participate, not at homes, and untraced elements. Item non-response arises because of item refusals, do not know, omissions, and answers deleted in editing. For more details about the forms of non-responses the reader is referred to Rubin (1978). We will discuss here a few basic models followed by their recent developments.

12.1 HANSEN AND HURWITZ PIONEER MODEL

Hansen and Hurwitz (1946) proposed a model for mail survey designs to provide unbiased estimators of population mean or total. Their pioneer model consists of the following steps:

(a) Select a sample of respondents and mail a questionnaire to all of them;
(b) After the deadline is over, identify the non-respondents and select a sub-sample of the non-respondents;
(c) Collect data from the non-respondents in the sub-sample by interview;
(d) Combine the data from the two parts of the survey to estimate the population parameters of interest.

Consider that a population consists of N units can be divided into two classes:

(a) those who will respond at the first attempt forming the response class;
(b) those who will not respond at the first attempt forming the non-response class.

Assume that N_1 and N_2 are the number of units in the population that belong to the response and non-response class, respectively. We may regard the sample of n_1 respondents as a simple random sample from the response class and the sample of n_2 as a simple random sample from the non-response class. Let h_2 denote the size of the sub-sample from the n_2 non-respondents to be interviewed so that $n_2 = h_2 g$, where $g > 1$.

Then we have the following lemma:

Lemma 12.1.1. The unbiased estimators of N_1 and N_2 are, respectively,

$$\hat{N}_1 = \frac{n_1}{n} N \quad \text{and} \quad \hat{N}_2 = \frac{n_2}{n} N .$$

We have the following theorems:

Theorem 12.1.1. An unbiased estimator of the population mean $\bar{Y}$ is given by

$$\bar{y}_{\text{hh}} = \frac{n_1\bar{y}_1 + n_2\bar{y}_{h_2}}{n}, \tag{12.1.1}$$

where $\bar{y}_{h_2}$ denotes the mean of h_2 observations from the sub-sample of non-responding units.

Proof. We have

$$E(\bar{y}_{\text{hh}}) = E_1[E_2(\bar{y}_{\text{hh}} \mid n_1, n_2)] = E_1\left[\frac{n_1\bar{y}_1 + n_2\bar{y}_{h_2}}{n}\right] = E_1(\bar{y}_n) = \bar{Y}.$$

Hence the theorem.

Theorem 12.1.2. The variance of the estimator $\bar{y}_{\text{hh}}$ is given by

$$V(\bar{y}_{\text{hh}}) = \left(\frac{1}{n} - \frac{1}{N}\right)S_y^2 + \frac{(g-1)}{n}\left(\frac{N}{N_2}\right)S_{2y}^2 \tag{12.1.2}$$

where S_{2y}^2 is the population mean square error for the non-response class.

Proof. By definition we have

$$V(\bar{y}_{\text{hh}}) = V_1[E_2(\bar{y}_{hh} \mid n_1, n_2)] + E_1[V_2(\bar{y}_{hh} \mid n_1, n_2)]$$

$$= V_1\left[E_2\left(\frac{n_1\bar{y}_1 + n_2\bar{y}_{h_2}}{n} \mid n_1, n_2\right)\right] + E_1\left[V_2\left(\frac{n_1\bar{y}_1 + n_2\bar{y}_{h_2}}{n} \mid n_1, n_2\right)\right]$$

$$= V_1(\bar{y}_n) + E_1\left[\frac{n_2^2}{n^2}V_2(\bar{y}_{h_2})\right] = V_1(\bar{y}_n) + E_1\left[\frac{n_2^2}{n^2}\left(\frac{1}{h_2} - \frac{1}{n_2}\right)s_{2y}^2\right]$$

$$= V_1(\bar{y}_n) + \left[\frac{1}{n}E_1\left(\frac{n_2}{h_2} - 1\right)\frac{n_2}{n}s_{2y}^2\right] = V_1(\bar{y}_n) + \left[\frac{1}{n}E_1(g-1)\frac{n_2}{n}s_{2y}^2\right]$$

$$= V_1(\bar{y}_n) + \left[\frac{1}{n}(g-1)E_1\left(\frac{n_2}{n}\right)E_1(s_{2y}^2)\right] \quad \text{because } g \text{ is constant}$$

$$= \left(\frac{1}{n} - \frac{1}{N}\right)S_y^2 + \frac{(g-1)}{n}\left(\frac{N_2}{N}\right)S_{2y}^2.$$

Hence the theorem.

Now we have the following corollary:

Corollary 12.1.1. The second term in the expression of $V(\bar{y}_{\text{hh}})$ will vanish if $g = 1$. In fact, it is true if it is possible to interview every non-respondent to collect information.

The cost function for this model has been found to be made of three components:

(a) Cost of including a sample unit in the initial survey $= C_0$ (say);

(b) Cost of collecting, editing, and processing per unit in the response class $= C_1$ (say);
(c) Cost of interviewing and processing information per unit in the non-response class $= C_2$ (say).

Thus it is reasonable to consider the cost function given by

$$C^* = nC_0 + n_1C_1 + h_2C_2 . \tag{12.1.3}$$

Note that C^* varies from sample to sample thus it is recommended to use the expected cost function. Taking expected values on both sides of (12.1.3), we have

$$E\left(C^*\right) = E\left[nC_0 + n_1C_1 + h_2C_2\right] = nE\left[C_0 + \frac{n_1}{n}C_1 + \frac{n_2}{ng}C_2\right]$$

$$= n\left[C_0 + E\left(\frac{n_1}{n}\right)C_1 + \frac{1}{g}E\left(\frac{n_2}{n}\right)C_2\right] = n\left[C_0 + \frac{N_1}{N}C_1 + \frac{N_2}{gN}C_2\right]$$

or

$$C = \frac{n}{N}\left[NC_0 + N_1C_1 + \left(\frac{N_2}{g}\right)C_2\right]. \tag{12.1.4}$$

Then we have the following theorem:

Theorem 12.1.3. The optimum values of g and n for the minimum expected cost are, respectively, given by

$$g = \sqrt{C_2\left(S_y^2 - \frac{N_2S_{2y}^2}{N}\right)\Big/ S_{2y}^2\left(C_0 + \frac{N_1C_1}{N}\right)} \tag{12.1.5}$$

and

$$n = \left\{S_y^2 + \frac{N_2(g-1)}{N}S_{2y}^2\right\}\Big/\left\{V_0 + S_y^2/N\right\}. \tag{12.1.6}$$

Proof. Let the variance $V(\bar{y}_{hh})$ has been fixed as V_0, that is, $V(\bar{y}_{hh}) = V_0$. Then the Lagrange function is given by

$$L = \frac{n}{N}\left[NC_0 + N_1C_1 + \left(\frac{N_2}{g}\right)C_2\right] + \lambda\left[V(\bar{y}_{hh}) - V_0\right], \tag{12.1.7}$$

where λ is a Lagrange multiplier. Differentiating (12.1.7) with respect to n and equating to zero we have

$$\frac{\partial L}{\partial n} = \frac{NC_0 + N_1C_1 + \frac{N_2}{g}C_2}{N} - \lambda\frac{S_y^2}{n^2} - \lambda\frac{(g-1)}{n^2}\frac{N_2}{N}S_{y2}^2 = 0$$

which implies

$$n^2 = \lambda\left\{S_y^2 + (g-1)\frac{N_2}{N}S_{y2}^2\right\}N\Big/\left\{NC_0 + N_1C_1 + \frac{N_2}{g}C_2\right\} \tag{12.1.8}$$

or

$$n = \sqrt{\lambda}\sqrt{\left\{S_y^2 + (g-1)\frac{N_2}{N}S_{y2}^2\right\}N}\Big/\sqrt{NC_0 + N_1C_1 + \frac{N_2}{g}C_2} . \tag{12.1.9}$$

Note that

$$V(\bar{y}_{hh}) = \left(\frac{1}{n} - \frac{1}{N}\right)S_y^2 + \frac{(g-1)}{n}\left(\frac{N}{N_2}\right)S_{2y}^2 = V_0$$

On using (12.1.9) we have

$$\sqrt{\lambda} = \left(NC_0 + N_1C_1 + \frac{N_2}{g}C_2\right)\sqrt{\left\{S_y^2 + (g-1)\frac{N_2}{N}S_{y2}^2\right\}} \Bigg/ \sqrt{N}\left(V_0 + \frac{S_y^2}{N}\right). \quad (12.1.10)$$

On substituting (12.1.10) in (12.1.9) we have

$$n = \left\{S_y^2 + (g-1)\frac{N_2}{N}S_{y2}^2\right\} \Big/ \left\{V_0 + S_y^2/N\right\} \quad (12.1.11)$$

which is the required optimum value of sample size n. Now differentiating (12.1.7) with respect to g and equating to zero we have

$$\frac{\partial L}{\partial g} = -\frac{nN_2}{Ng^2}C_2 + \lambda\frac{N_2}{nN}S_{2y}^2 = 0, \quad (12.1.12)$$

which implies

$$g^2 = \frac{n^2C_2}{\lambda S_{2y}^2}. \quad (12.1.13)$$

Using (12.1.8) in (12.1.13) we have

$$g^2 = \frac{\lambda\left\{S_y^2 + (g-1)\frac{N_2}{N}S_{2y}^2\right\}NC_2}{\left\{NC_0 + N_1C_1 + \frac{N_2}{g}C_2\right\}\lambda S_{2y}^2} = \frac{C_2\left(S_y^2 - \frac{N_2}{N}S_{2y}^2\right) + g\frac{N_2C_2}{N}S_{2y}^2}{\left\{C_0 + \frac{N_1}{N}C_1\right\}S_{2y}^2 + \frac{N_2}{gN}C_2S_{2y}^2},$$

or

$$g^2\left(C_0 + \frac{N_1}{N}C_1\right)S_{2y}^2 = C_0\left(S_y^2 - \frac{N_2}{N}S_{2y}^2\right),$$

or

$$g = \sqrt{C_0\left(S_y^2 - \frac{N_2}{N}S_{2y}^2\right)\Big/\left(C_0 + \frac{N_1}{N}\right)S_{2y}^2} \quad (12.1.14)$$

which is the required optimum value of g. Hence the theorem.

Example 12.1.1. Consider a city consists of 1000 persons. We wish to estimate the average income of the persons living in the city. We selected an SRSWOR sample of 100 persons and mailed a questionnaire to each of them regarding their annual income at the beginning of a particular month. Out of 100 questionnaires mailed we received a reply from 70 people. The average annual income from the 70 responses was $\$35,000$. At the end of the month we selected 10 people out of 30 who did not respond through the mail survey and contacted them through personal interviews for collecting information regarding their income. The average income obtained through personal interview survey was $\$38,000$.

(a) If the questionnaire had been mailed to all the 1000 persons in the city, then find the estimate of the number of persons which are expected to respond to it as well as will not respond to it.
(b) Apply the Hansen and Hurwitz technique to estimate the average income of the persons living in the particular city.

Solution. (a) Here $N = 1000$ be the total number of persons living in a particular city. Let N_1 and N_2 be the unknown number of persons belonging to the responding and non-responding class respectively. Then an estimator of N_1 is $\hat{N}_1 = \frac{n_1}{n}N = \frac{70}{100}\times 1000 = 700$, and that of N_2 is $\hat{N}_2 = \frac{n_2}{n}N = \frac{30}{100}\times 1000 = 300$.

(b) According to Hansen and Hurwitz model, an estimate of the average annual income of the persons in the city is given by

$$\bar{y}_{hh} = \frac{n_1\bar{y}_1 + n_2\bar{y}_{h2}}{n} = \frac{70\times 35000 + 30\times 38000}{100} = \$35900\,.$$

Several other research workers including Dalenius (1955), Hendricks (1949), Kish and Hess (1959), Bartholomew (1961), Rao (1968b), and Srinath (1971) discussed different strategies to improve the efficiency of the pioneer model. Deming (1953) has shown that successive recalls are helpful to increase the efficiency in the mail-survey designs. Elliott, Little, and Lewitzky (2000) have shown that sub-sampling call-backs seems to be more cost-effective however their analysis suggests that subsampling should begin at the telephone rather than the face to face interviews. Okafor and Lee (2000) applied Hansen and Hurwitz (1946) technique for ratio and regression estimator. Politz and Simmons (1950) developed a model to reduce the bias due to incomplete sample, without successive call backs. Their technique is popular whenever call backs are not feasible owing to shortage of time or money however the efficiency depends upon the circumstances.

12.2 POLITZ AND SIMMONS MODEL

Following Politz and Simmons (1950) model, the interviewer makes only one call during specific time (such as morning) on six weekdays. The time of calls has been considered as random within interviewing hours. If the respondent is at home the desired information is collected and he is asked how many times in the preceding five days he was at home at the time of the visit. The information so obtained is used to estimate the probability of the respondent's availability. Assume n households have been selected by SRSWR and ψ_i denotes the probability that i^{th} household is available at the time of the first ring. Then we have the following theorem:

Theorem 12.2.1. An estimator of population mean $\bar{Y}$ is given by

$$\bar{y}_{sp} = \frac{1}{n}\sum_{i=1}^{n}\frac{y_i}{\psi_i}\,. \qquad (12.2.1)$$

Assuming $\psi_i = (d+1)/D$, where d denotes the number of times the respondent was at home during the last D days (or hours, or months etc.), $d = 0, 1, 2, ..., D-1$. Show that the bias in the estimator $\bar{y}_{\text{sp}}$ is given by

$$B(\bar{y}_{\text{sp}}) = -\frac{1}{N}\sum_{i=1}^{N} Y_i (1-\psi_i)^D \tag{12.2.2}$$

and variance

$$V(\bar{y}_{\text{sp}}) = \frac{1}{n}\left[\frac{D}{N}\sum_{i=1}^{N}\frac{\alpha_i Y_i^2}{\psi_i} - \left\{\frac{1}{N}\sum_{i=1}^{N} Y_i\left(1-(1-\psi_i)^D\right)\right\}^2\right]. \tag{12.2.3}$$

Proof. Let ψ_{id} $i = 1,2,....,n$ and $d = 0,1,2,....,D-1$ denote the probability that the ith respondent will be at home d times out of $D-1$ attempts or calls fixed by the investigator for collecting information. Assuming ψ_i remains same for every d^{th} day, the probability that the ith person will be available on d^{th} days out of $D-1$ calls is given by the binomial distribution

$$P(i = d) = \binom{D-1}{d}\psi_i^d (1-\psi_i)^{D-1-d}. \tag{12.2.4}$$

Under this distribution we have

$$E\left(\frac{y_i}{\psi_i} \mid i\right) = \sum_{d=0}^{D-1}\frac{y_i}{\psi_i}P(i=d) = \sum_{d=0}^{D-1}\frac{y_i}{\psi_i}\binom{D-1}{d}\psi_i^d(1-\psi_i)^{D-1-d}$$

$$= \sum_{d=0}^{D-1}\frac{y_i}{\frac{d+1}{D}}\binom{D-1}{d}\psi_i^d(1-\psi_i)^{D-1-d} = y_i\sum_{d=0}^{D-1}\frac{D}{d+1}\binom{D-1}{d}\psi_i^d(1-\psi_i)^{D-1-d}$$

$$= \frac{y_i}{\psi_i}\sum_{d=0}^{D-1}\binom{D}{d+1}\psi_i^{d+1}(1-\psi_i)^{D-(1+d)}.$$

Note that

$$\sum_{d=0}^{D-1}\binom{D}{d+1}\psi_i^{d+1}(1-\psi_i)^{D-(1+d)} + \binom{D}{0}(1-\psi_i)^D = 1,$$

which implies that

$$E\left(\frac{y_i}{\psi_i} \mid i\right) = \frac{y_i}{\psi_i}\left[1-(1-\psi_i)^D\right].$$

Note that the probability of the ith person being selected and found at home is given by ψ_i/N, thus we have

$$E(\bar{y}_{\text{sp}}) = E_1E_2(\bar{y}_{sp}) = \frac{1}{n}\sum_{i=1}^{n}E_1E_2\left(\frac{y_i}{\psi_i} \mid i\right) = \frac{1}{n}\sum_{i=1}^{n}\sum_{i=1}^{N}\frac{\psi_i}{N}\frac{Y_i}{\psi_i}\left[1-(1-\psi_i)^D\right]$$

$$= \bar{Y} - \frac{1}{N}\sum_{i=1}^{N}Y_i(1-\psi_i)^D.$$

Hence the bias in $\bar{y}_{\text{sp}}$ is given by

$$B(\bar{y}_{\text{sp}}) = -\frac{1}{N}\sum_{i=1}^{N}Y_i(1-\psi_i)^D$$

which proves the first part of the theorem.

To find the variance of the estimator we have

$$E\left[\left(\frac{y_i}{\psi_i}\right)^2 | i\right] = \sum_{d=0}^{D-1}\left(\frac{y_i}{\psi_i}\right)^2 P(i=d) = y_i^2\sum_{d=0}^{D-1}\left(\frac{D}{d+1}\right)^2\binom{D-1}{d}\psi_i^d(1-\psi_i)^{D-1-d}$$

$$= y_i^2\sum_{d=0}^{D-1}\frac{1}{\psi_i}\left(\frac{D}{d+1}\right)^2\binom{D-1}{d}\psi_i^{d+1}(1-\psi_i)^{D-1-d}$$

$$= \frac{y_i^2 D}{\psi_i}\sum_{d=0}^{D-1}\frac{1}{d+1}\left(\frac{D}{d+1}\right)\binom{D-1}{d}\psi_i^{d+1}(1-\psi_i)^{D-(1+d)} = \frac{y_i^2 D\alpha_i}{\psi_i},$$

where

$$\alpha_i = \sum_{d=0}^{D-1}\frac{1}{d+1}\binom{D}{d+1}\psi_i^{d+1}(1-\psi_i)^{D-(1+d)},$$

which implies

$$E\left(\frac{y_i^2}{\psi_i^2}\right) = \frac{D}{N}\sum_{i=1}^{N}\frac{y_i^2\alpha_i}{\psi_i^2}\times\psi_i = \frac{D}{N}\sum_{i=1}^{N}\frac{\alpha_i Y_i^2}{\psi_i}.$$

Thus the variance of the estimator $\bar{y}_{sp}$ is given by

$$V(\bar{y}_{sp}) = V\left[\frac{1}{n}\sum_{i=1}^{n}\frac{y_i}{\psi_i}\right] = \frac{1}{n^2}\sum_{i=1}^{n}V\left(\frac{y_i}{\psi_i}\right) = \frac{1}{n}\left[E\left(\frac{y_i}{\psi_i}\right)^2 - \left\{E\left(\frac{y_i}{\psi_i}\right)\right\}^2\right]$$

$$= \frac{1}{n}\left[\frac{D}{N}\sum_{i=1}^{N}\frac{\alpha_i Y_i^2}{\psi_i} - \left\{\frac{1}{N}\sum_{i=1}^{N}Y_i\left(1-(1-\psi_i)^D\right)\right\}^2\right].$$

Hence the theorem.

Assuming ψ_i as the probability of selecting i^{th} units under the probability proportional to size and with replacement sampling scheme, an estimator of the variance of the estimator $\bar{y}_{sp}$ can be developed as

$$\hat{v}(\bar{y}_{sp}) = \frac{1}{n(n-1)}\sum_{i=1}^{n}\left(\frac{y_i}{\psi_i} - \bar{y}_{sp}\right)^2. \tag{12.2.5}$$

The Politz and Simmon (1950) estimator adjusts the non-response bias owed to the non-availability of the selected respondents at home during the period of survey by classifying the available respondents into six groups according to their availability at home during the previous week $(D = 6)$ and employing appropriate weighting procedures. Thus this estimator is based on the premise that the selected respondents available at home necessarily co-operate with the enumerator, which however may not be true. Sharma and Sil (1996) considered to study the non-response bias in the Politz--Simmon estimator taking into account the possible non-co-operation from the selected respondents who are, though, available at home yet may be busy otherwise.

Example 12.2.1. Select an SRSWOR sample of sixteen units from population 4 given in the Appendix. Assume that we made 7 attempts to collect information about the different species groups selected in the sample. Information about the number of fish caught of different species is not available every time you contact the fishermen. We collected the information on the number of fish during 1995 in seven visits. It was also noted that out of seven visits, how many times (D) the information about these species was available? Estimate the average number of fish caught by marine recreational fishermen at Atlantic and Gulf coasts during 1995. Construct 95% confidence interval for the average number of fish in the United States.

Solution. The population size $N = 69$. Thus we used the first two columns of the Pseudo-Random Number (PRN) Table 1 given in the Appendix to select 16 random numbers between 1 and 69. The random numbers so selected are 58, 60, 54, 01, 69, 62, 23, 64, 46, 04, 32, 47, 57, 56, 33, and 05. The following information was collected.

PRN	Species group	y_i	Availability during last 6 attempts d	$\psi_i = \frac{d+1}{D}$	$\frac{y_i}{\psi_i}$	$\left(\frac{y_i}{\psi_i} - \bar{y}_{sp}\right)^2$
01	Sharks, other	2016	6	1.000000	2016.00	34479613.30
04	Eels	152	2	0.428571	354.67	56750122.37
05	Herrings	30027	6	1.000000	30027.00	490138226.70
23	Blue runner	2319	6	1.000000	2319.00	31013030.07
32	Yellowtail snapper	1334	6	1.000000	1334.00	42954055.79
33	Snappers, others	492	6	1.000000	492.00	54699845.28
46	Spot	11567	3	0.571429	20242.25	152629114.60
47	King fish	4333	5	0.857143	5055.17	8024572.89
54	Tautog	3816	5	0.857143	4452.00	11805645.03
56	Wrasses, other	185	6	1.000000	185.00	59335197.99
57	Little tunney/Atl bonito	782	6	1.000000	782.00	50494303.34
58	Atlantic mackerel	4008	6	1.000000	4008.00	15053890.75
60	Spanish mackerel	2568	4	0.714286	3595.20	18427568.41
62	Summer flounder	16238	6	1.000000	16238.00	69723595.94
64	Southern flounder	1446	6	1.000000	1446.00	41498518.49
69	Other fish	14426	2	0.428571	33660.67	664233729.80
				Sum	126207.00	1801261031.00

Given: D=7.

Thus an estimate of the average number of fish during 1995 is

$$\bar{y}_{\text{sp}} = \frac{1}{n}\sum_{i=1}^{n}\frac{y_i}{\psi_i} = \frac{126207}{16} = 7887.934 .$$

An estimate of variance of the estimator $\bar{y}_{sp}$ is

$$\hat{v}(\bar{y}_{sp}) = \frac{1}{n(n-1)} \sum_{i=1}^{n} \left(\frac{y_i}{\psi_i} - \bar{y}_{sp} \right)^2 = \frac{1801261031}{16 \times 15} = 7505254.295 .$$

A $(1-\alpha)100\%$ confidence interval for the average number of fish caught during 1995 by marine recreational fishermen in the United States is given by

$$\bar{y}_{sp} \mp t_{\alpha/2}(\text{df} = n-1)\sqrt{\hat{v}(\bar{y}_{sp})} .$$

Using Table 2 given in the Appendix the 95% confidence interval of population mean $\bar{Y}$ is given by

$$\bar{y}_{sp} \mp t_{0.025}(\text{df} = 16-1)\sqrt{\hat{v}(\bar{y}_{sp})}$$

or

$$7887.934 \mp 2.131\sqrt{7505254.295} \text{ or } [2049.907,\ 13725.962].$$

12.3 HORVITZ AND THOMPSON ESTIMATOR UNDER NON-RESPONSE

Let there be a population of N units from which a sample of size n is to be drawn. Let the study variable be denoted by y and the auxiliary variable be denoted by x. The selection probabilities p_i of the population units are taken to be proportional to the corresponding x-values. Let r, $r = 1,2,...,(n-1)$ be the number of units (including repetitions in the case of PPSWR sampling) on which the information on y could not be collected. Singh and Singh (1979) assumed that the value of r is supposed to be less than or equal to $(n-2)$ while estimation of variance of the estimator of population total is concerned. Then we have the following theorem:

Theorem 12.3.1. An unbiased estimator of population total Y is given by

$$\hat{Y}_{HTR} = \frac{n}{n-r} \sum_{i=1}^{n-r} \frac{y_i}{\pi_i}, \tag{12.3.1}$$

where $\pi_i = nP_i$ denotes the first order inclusion probability.

Proof. Let E_3 be the expected value for a given sample of n units from which $(n-r)$ units (on which response has been received) can be treated as an SRSWOR sample, E_2 be the expected value over all such samples for a given r, and E_1 be the expectation over all possible values of r.

Then we have

$$E(\hat{Y}_{HTR}) = E_1E_2E_3(\hat{Y}_{HTR}) = E_1E_2\left[\sum_{i=1}^{n} \frac{y_i}{\pi_i}\right] = E_1(Y) = Y .$$

Hence the theorem.

Theorem 12.3.2. The variance of the estimator $\hat{Y}_{\text{HTR}}$ is given by

$$V\left(\hat{Y}_{\text{HTR}}\right) = V\left(\hat{Y}_{\text{HT}}\right) + \left[\sum_{i=1}^{N}\frac{Y_i^2}{\pi_i} - \frac{1}{n-1}\sum_{i=1}^{N}\sum_{j\neq i=1}^{N}\frac{\pi_{ij}}{\pi_i\pi_j}Y_iY_j\right]E_1\left(\frac{r}{n-r}\right), \tag{12.3.2}$$

where

$$V\left(\hat{Y}_{\text{HT}}\right) = \sum_{i=1}^{N}\frac{1-\pi_i}{\pi_i}Y_i^2 + \sum_{i=1}^{N}\sum_{j\neq i=1}^{N}\frac{\pi_{ij}-\pi_i\pi_j}{\pi_i\pi_j}Y_iY_j \quad \text{for } \hat{Y}_{\text{HT}} = \sum_{i=1}^{n}\frac{y_i}{\pi_i}.$$

Proof. Define the variances V_1, V_2 and V_3 in the same manner as the expectations E_1, E_2, and E_3 we have

$$V\left(\hat{Y}_{\text{HTR}}\right) = E_1E_2V_3\left(\hat{Y}_{\text{HTR}}\right) + E_1V_2E_3\left(\hat{Y}_{\text{HTR}}\right) + V_1E_2E_3\left(\hat{Y}_{\text{HTR}}\right)$$

$$= E_1E_2\left[\left(\frac{nr}{n-r}\right)\frac{1}{n-1}\left\{\sum_{i=1}^{n}\frac{y_i^2}{\pi_i^2} - n^{-1}\left(\sum_{i=1}^{n}\frac{y_i}{\pi_i}\right)^2\right\}\right] + E_1V_2\left[\sum_{i=1}^{n}\frac{y_i}{\pi_i}\right] + V_1E_2\left(\sum_{i=1}^{n}\frac{y_i}{\pi_i}\right)$$

$$= E_1E_2\left[\left(\frac{nr}{n-r}\right)\frac{1}{n-1}\left\{\sum_{i=1}^{n}\frac{y_i^2}{\pi_i^2} - \frac{1}{n}\sum_{i=1}^{n}\frac{y_i^2}{\pi_i^2} - \frac{1}{n}\sum_{i\neq j=1}^{n}\frac{y_iy_j}{\pi_i\pi_j}\right\}\right] + E_1\left[V\left(\hat{Y}_{\text{HT}}\right)\right] + V_1(Y)$$

$$= E_1E_2\left[\left(\frac{nr}{n-r}\right)\frac{1}{n-1}\left\{\frac{n-1}{n}\sum_{i=1}^{n}\frac{y_i^2}{\pi_i^2} - \frac{1}{n}\sum_{i\neq j=1}^{n}\frac{y_iy_j}{\pi_i\pi_j}\right\}\right] + V\left(\hat{Y}_{\text{HT}}\right)$$

$$= E_1\left[\left(\frac{nr}{n-r}\right)\frac{1}{n-1}\left\{\frac{n-1}{n}E_2\sum_{i=1}^{n}\frac{y_i^2}{\pi_i^2} - \frac{1}{n}E_2\sum_{i\neq j=1}^{n}\frac{y_iy_j}{\pi_i\pi_j}\right\}\right] + V\left(\hat{Y}_{\text{HT}}\right)$$

$$= E_1\left[\left(\frac{r}{n-r}\right)\left\{E_2\sum_{i=1}^{n}\frac{y_i^2}{\pi_i^2} - \frac{1}{n-1}E_2\sum_{i\neq j=1}^{n}\frac{y_iy_j}{\pi_i\pi_j}\right\}\right] + V\left(\hat{Y}_{\text{HT}}\right)$$

$$= E_1\left[E_2\sum_{i=1}^{n}\frac{y_i^2}{\pi_i^2} - \frac{1}{n-1}E_2\sum_{i\neq j=1}^{n}\frac{y_iy_j}{\pi_i\pi_j}\right]\left(\frac{r}{n-r}\right) + V\left(\hat{Y}_{\text{HT}}\right)$$

$$= E_1\left[\sum_{i=1}^{N}\frac{Y_i^2}{\pi_i} - \frac{1}{n-1}\sum_{i\neq j=1}^{N}\frac{\pi_{ij}y_iy_j}{\pi_i\pi_j}\right]\left(\frac{r}{n-r}\right) + V\left(\hat{Y}_{\text{HT}}\right)$$

$$= \left(\sum_{i=1}^{N}\frac{Y_i^2}{\pi_i} - \frac{1}{n-1}\sum_{i\neq j=1}^{N}\frac{\pi_{ij}y_iy_j}{\pi_i\pi_j}\right)E_1\left(\frac{r}{n-r}\right) + V\left(\hat{Y}_{\text{HT}}\right)$$

which proves the theorem.

Theorem 12.3.3. An unbiased estimator of the variance $V\left(\hat{Y}_{\text{HTR}}\right)$ is given by

$$\hat{v}\left(\hat{Y}_{\text{HTR}}\right)=\frac{n}{(n-r)^2}\left[\sum_{i=1}^{n-r}\frac{n-(n-r)\pi_i}{\pi_i^2}y_i^2\right.$$
$$\left.+\frac{1}{(n-r-1)}\sum_{i=1}^{n-r}\sum_{j\neq i=1}^{n-r}\frac{\{n(n-r-1)\pi_{ij}-(n-r)(n-1)\pi_i\pi_j\}y_iy_j}{\pi_{ij}\pi_i\pi_j}\right]. \tag{12.3.3}$$

Proof. We have

$$E\left[\hat{v}\left(\hat{Y}_{\text{HTR}}\right)\right]$$
$$=E_1E_2E_3\left[\hat{v}\left(\hat{Y}_{\text{HTR}}\right)\right]$$
$$=E_1E_2E_3\frac{n}{(n-r)^2}\left[\sum_{i=1}^{n-r}\frac{n-(n-r)\pi_i}{\pi_i^2}y_i^2+\frac{1}{(n-r-1)}\sum_{i=1}^{n-r}\sum_{j\neq i=1}^{n-r}\frac{\{n(n-r-1)\pi_{ij}-(n-r)(n-1)\pi_i\pi_j\}y_iy_j}{\pi_{ij}\pi_i\pi_j}\right]$$
$$=E_1E_2\left[\frac{1}{(n-r)}\sum_{i=1}^{n}\frac{n-(n-r)\pi_i}{\pi_i^2}y_i^2+\frac{1}{(n-r)(n-1)}\sum_{i=1}^{n}\sum_{j\neq i=1}^{n}\frac{\{n(n-r-1)\pi_{ij}-(n-r)(n-1)\pi_i\pi_j\}y_iy_j}{\pi_{ij}\pi_i\pi_j}\right]$$
$$=E_1\left[\frac{1}{(n-r)}\sum_{i=1}^{N}\frac{n-(n-r)\pi_i}{\pi_i}Y_i^2+\frac{1}{(n-r)(n-1)}\sum_{i=1}^{N}\sum_{j\neq i=1}^{N}\frac{\{n(n-r-1)\pi_{ij}-(n-r)(n-1)\pi_i\pi_j\}Y_iY_j}{\pi_i\pi_j}\right]$$
$$=E_1\left[\sum_{i=1}^{N}\left\{\frac{1-\pi_i}{\pi_i}+\frac{r}{(n-r)\pi_i}\right\}Y_i^2+\frac{1}{(n-r)(n-1)}\sum_{i=1}^{N}\sum_{j\neq i=1}^{N}\left\{\frac{\pi_{ij}-\pi_i\pi_j}{\pi_i\pi_j}-\frac{r\pi_{ij}}{(n-r)(n-1)\pi_i\pi_j}\right\}Y_iY_j\right]$$
$$=V\left(\hat{Y}_{\text{HT}}\right)+\left[\sum_{i=1}^{N}\frac{Y_i^2}{\pi_i}-\frac{1}{n-1}\sum_{i=1}^{N}\sum_{j\neq i=1}^{N}\frac{\pi_{ij}}{\pi_i\pi_j}Y_iY_j\right]E_1\left(\frac{r}{n-r}\right).$$

Hence the theorem.

Remark 12.3.1. The value of

$$E\left(\frac{r}{n}\right)\approx\frac{E(r)}{n}+\frac{E\left(r^2\right)}{n^2}+\ldots,$$

where $E(r)$ and $E\left(r^2\right)$ can easily be obtained if the distribution of r is known. In practice r may follow Binomial or Negative Binomial distribution.

12.4 RATIO AND REGRESSION TYPE ESTIMATORS

Tracy and Osahan (1994c) studied the effect of random non-response on the usual ratio estimator of the population mean in two situations: (i) non-response in study as well as the auxiliary variable and (ii) non-response in the study variable only. Singh, Joarder, and Tracy (2000) suggested three regression type estimators, which are further studied by Singh and Tracy (2001), in the presence of random non-response in different situations under an assumption that the number of sampling units on which information can not be obtained owing to random non-response follow some distribution.

12.4.1 DISTRIBUTION AND SOME EXPECTED VALUES

Let $\Omega:(v_1, v_2, ..., v_N)$ denote the population of N units from which an SRSWOR sample of size n is drawn. If r $(r = 0, 1, ..., (n-2))$ denotes the number of sampling units on which information could not be obtained owing to random non-response, then the remaining $(n-r)$ units in the sample can be treated as an SRSWOR sample from Ω. Note that we are considering the regression type estimators, therefore we are assuming that r should be less than $(n-2)$. We assume that if p denotes the probability of non-response among the $(n-2)$ possible values of responses, then r has the discrete distribution

$$P(r) = \frac{(n-r)}{nq+2p}\,{}^{n-2}C_r p^r q^{n-2-r}, \tag{12.4.1.1}$$

where $q = 1-p$ and $r = 0, 1, 2, ..., (n-2)$.

Let us define

$$\varepsilon = \frac{\bar{y}_{n-r}}{\bar{Y}} - 1, \quad \delta = \frac{\bar{x}_{n-r}}{\bar{X}} - 1, \quad \text{and} \quad \eta = \frac{\bar{x}}{\bar{X}} - 1,$$

where $\bar{y}_{n-r} = (n-r)^{-1}\sum_{i=1}^{n-r} y_i$, $\bar{x}_{n-r} = (n-r)^{-1}\sum_{i=1}^{n-r} x_i$ and $\bar{x} = n^{-1}\sum_{i=1}^{n} x_i$ have their usual meanings (Tracy and Osahan, 1994c). The probability model defined at (12.4.1.1) is free from actual data values, hence it can be considered as a model suitable for MAR situation. Then under the probability model given by (12.4.1.1), we have the following results:

$$E(\varepsilon) = E(\delta) = E(\eta) = 0,$$

$$E(\varepsilon^2) = \left[\frac{1}{(nq+2p)} - \frac{1}{N}\right]C_y^2, \quad E(\delta^2) = \left[\frac{1}{(nq+2p)} - \frac{1}{N}\right]C_x^2, \quad E(\eta^2) = \left(\frac{1}{n} - \frac{1}{N}\right)C_x^2,$$

$$E(\varepsilon\delta) = \left[\frac{1}{(nq+2p)} - \frac{1}{N}\right]\rho_{xy}C_yC_x, \quad E(\varepsilon\eta) = \left(\frac{1}{n} - \frac{1}{N}\right)\rho_{xy}C_yC_x, \text{ and } E(\delta\eta) = \left(\frac{1}{n} - \frac{1}{N}\right)C_x^2,$$

where $C_y = S_y/\bar{Y}$, $C_x = S_x/\bar{X}$ and $\rho_{xy} = S_{xy}/(S_xS_y)$. It is interesting to note that under the model (12.4.1.1) the above expected values are exact and hence makes valid comparison with the estimators in the absence of non-response. The logic and the practical importance of the distribution defined at (12.4.1.1) has been discussed by Singh and Joarder (1998). In place of (12.4.1.1) we can also use another distribution (*e.g.*, truncated binomial), but these distributions add an extra approximation in the expected values.

12.4.2 ESTIMATION OF POPULATION MEAN

Strategy I. Consider the situation when random non-response exists on both the study variable y and the auxiliary variable x and population mean $\bar{X}$ of the auxiliary variable is known. Thus we consider a regression type estimator as

$$\bar{y}_1 = \bar{y}_{n-r} + \alpha_1\left(\bar{X} - \bar{x}_{n-r}\right). \tag{12.4.2.1}$$

The minimum variance of the estimator at (12.4.2.1) is given by

$$\text{Min.}V(\bar{y}_1) = \left[\frac{1}{(nq+2p)} - \frac{1}{N}\right] S_y^2\left(1-\rho_{xy}^2\right) \tag{12.4.2.2}$$

for the optimum value of α_1 given by

$$\alpha_1 = S_{xy}/S_x^2. \tag{12.4.2.3}$$

If $p=0$ then the variance in (12.4.2.2) reduces to the variance of the usual linear regression estimator. In this situation, we can estimate α_1 by $\hat{\alpha}_1 = s_{xy}^*/s_x^{*2}$, where $s_{xy}^* = (n-r-1)^{-1}\sum_{i=1}^{n-r}(y_i - \bar{y}_{n-r})(x_i - \bar{x}_{n-r})$ and $s_x^{*2} = (n-r-1)\sum_{i=1}^{n-r}(x_i - \bar{x}_{n-r})^2$ are the estimators of S_{xy} and S_x^2 respectively. Thus we have the following theorems:

Theorem 12.4.2.1. The regression estimator, if random non-response exists on both the study and auxiliary variables, is given by

$$\bar{y}_{1\text{lr}} = \bar{y}_{n-r} + \left(s_{xy}^*/s_x^{*2}\right)\left(\bar{X} - \bar{x}_{n-r}\right). \tag{12.4.2.4}$$

Theorem 12.4.2.2. An estimator for estimating the variance of the estimator $\bar{y}_1$ is

$$\hat{v}(\bar{y}_1) = \left[\frac{1}{(n\hat{q}+2\hat{p})} - \frac{1}{N}\right] s_y^{*2}\left(1-r_{xy}^{*2}\right), \tag{12.4.2.5}$$

where

$$s_y^{*2} = (n-r-1)^{-1}\sum_{i=1}^{n-r}(y_i - \bar{y}_{n-r})^2 \quad \text{and} \quad \hat{p} = \frac{(n-1+r) - \sqrt{(n-1+r)^2 - 4nr(n-3)/(n-2)}}{2(n-3)}$$

is a maximum likelihood estimator of p obtained from the distribution given by (12.4.1.1), $\hat{q} = 1-\hat{p}$ and $r_{xy}^* = s_{xy}^*/\left(s_x^* s_y^*\right)$. If $r=0$ then $\hat{p}=0$, and if $r=(n-2)$ then $\hat{p}=1$, *i.e.*, $\hat{p}$ is an admissible estimator of response probability p. The behaviour of the estimator $\hat{p}$ for different values of r and n is given in the following table.

Table 12.4.2.1. The values of $\hat{p}$ for different choices of n and r.

n	r								
	2	4	6	8	10	12	14	16	18
20	0.11687	0.23355	0.34993	0.46586	0.58107	0.69496	0.80624	0.91135	1.00000
40	0.05397	0.10794	0.16190	0.21584	0.26978	0.32369	0.37758	0.43144	0.48527
60	0.03506	0.07013	0.10519	0.14025	0.17531	0.21037	0.24542	0.28047	0.31552
100	0.02061	0.04122	0.06184	0.08245	0.10306	0.12368	0.14429	0.16490	0.18552
1000	0.00200	0.00401	0.00601	0.00802	0.01003	0.01203	0.01404	0.01604	0.01805
2000	0.00100	0.00200	0.00300	0.00400	0.00500	0.00600	0.00701	0.00801	0.00901
5000	0.00040	0.00080	0.00120	0.00160	0.00200	0.00240	0.00280	0.00320	0.00360

Strategy II. Consider the situation in which information on variable y cannot be obtained for r units while information on variable x is available and population mean $\overline{X}$ of the auxiliary variable is known. Under such situations, consider the following estimator

$$\overline{y}_2 = \overline{y}_{n-r} + \alpha_2\left(\overline{X} - \overline{x}\right). \tag{12.4.2.6}$$

The minimum variance of the estimator $\overline{y}_2$ is given by

$$\text{Min.}V\left(\overline{y}_2\right) = \left[\frac{1}{(nq+2p)} - \frac{1}{n}\right]S_y^2 + \left(\frac{1}{n} - \frac{1}{N}\right)S_y^2\left(1 - \rho_{xy}^2\right) \tag{12.4.2.7}$$

for the optimum value of α_2 given by

$$\alpha_2 = S_{xy}/S_x^2\,. \tag{12.4.2.8}$$

If $p = 0$ the variance in (12.4.2.6) reduces to the variance of the usual linear regression estimator. Here α_2 can be estimated by $\hat{\alpha}_2 = s_{xy}^{*}/s_x^2$, where $s_x^2 = (n-1)^{-1}\sum_{i=1}^{n}(x_i - \overline{x})^2$, which leads to the following theorems:

Theorem 12.4.2.3. The regression estimator, if non-response exists only on the study variable, is given by

$$\overline{y}_{2\text{lr}} = \overline{y}_{n-r} + \left(s_{xy}^{*}/s_x^2\right)\left(\overline{X} - \overline{x}\right). \tag{12.4.2.9}$$

Theorem 12.4.2.4. An estimator for estimating the minimum variance of $\overline{y}_2$ is given by

$$\hat{v}\left(\overline{y}_2\right) = \left[\frac{1}{\left(n\hat{q} + 2\hat{p}\right)} - \frac{1}{n}\right]s_y^{*2} + \left(\frac{1}{n} - \frac{1}{N}\right)s_y^{*2}\left(1 - r_{xy}^{*2}\right). \tag{12.4.2.10}$$

Strategy III. Consider the situation when information on variable y cannot be obtained for r units while information on the variable x is obtained for all the sample units. The difference from the previous case is that the population mean $\overline{X}$ of the auxiliary variable is not known. In this case consider another regression type of estimator

$$\overline{y}_3 = \overline{y}_{n-r} + \alpha_3\left(\overline{x}_{n-r} - \overline{x}\right). \tag{12.4.2.11}$$

The minimum variance of $\overline{y}_3$ is given by

$$\text{Min.}V\left(\overline{y}_3\right) = \left[\frac{1}{(nq+2p)} - \frac{1}{n}\right]S_y^2\left(1 - \rho_{xy}^2\right) + \left(\frac{1}{n} - \frac{1}{N}\right)S_y^2 \tag{12.4.2.12}$$

for the optimum value of $\alpha_3 = S_{xy}/S_x^2$. Again α_3 can be estimated by $\hat{\alpha}_2$. Thus we have the following theorems:

Theorem 12.4.2.5. The regression estimator in the absence of population mean $\overline{X}$, but in the presence of non-response, is given by

$$\overline{y}_{3\text{lr}} = \overline{y}_{n-r} + \left(s_{xy}^{*}/s_x^2\right)\left(\overline{x}_{n-r} - \overline{x}\right). \tag{12.4.2.13}$$

Theorem 12.4.2.6. An estimator of the minimum variance of the estimator $\overline{y}_3$ is given by

$$\hat{v}(\overline{y}_3) = \left[\frac{1}{(n\hat{q} + 2\hat{p})} - \frac{1}{n}\right] s_y^{*2}\left(1 - r_{xy}^{*2}\right) + \left(\frac{1}{n} - \frac{1}{N}\right) s_y^{*2}. \tag{12.4.2.14}$$

Example 12.4.2.1. We selected an SRSWOR sample of twenty states from the population 1 given in the Appendix. We collected information about the real estate farm loans and nonreal estate farm loans from the selected states, but unfortunately the information on the real estate farm loans was not available on six states as marked in the table below.

Random No.	State	x_i	y_i	Random No.	State	x_i	y_i
01	AL	348.334	408.978	27	NE	3585.406	1337.852
03	AZ	431.439	54.633	29	NH	0.471	6.044
04	AR	848.317	907.700	32	NY	426.274	201.631
05	CA	3928.732	1343.461	33	NC	494.730	639.571
06	CO	906.281	315.809	36	OK	1716.087	Missing
07	CT	4.373	7.130	38	PA	298.351	Missing
14	IN	1022.782	1213.024	40	SC	80.750	Missing
19	ME	51.539	8.849	42	TN	388.869	Missing
22	MI	440.518	323.028	46	VA	188.477	Missing
23	MN	2466.892	1354.768	47	WA	1228.607	Missing

x -- Nonreal estate farm loans; y -- Real estate farm loans.

Apply the following three estimators

$$\overline{y}_{1\text{lr}} = \overline{y}_{n-r} + \left(s_{xy}^{*}/s_x^{*2}\right)\left(\overline{X} - \overline{x}_{n-r}\right), \quad \overline{y}_{2\text{lr}} = \overline{y}_{n-r} + \left(s_{xy}^{*}/s_x^2\right)\left(\overline{X} - \overline{x}\right)$$

and $\overline{y}_{3\text{lr}} = \overline{y}_{n-r} + \left(s_{xy}^{*}/s_x^2\right)\left(\overline{x}_{n-r} - \overline{x}\right)$ of population mean for estimating the average real estate farm loans assuming that the average nonreal estate farm loans in the United States is known and is equal to \$878.16. Construct 95% confidence interval estimates.

Solution. We are given $N = 50$ and $\overline{X} = 878.16$. From the above table, we have $n = 20$, $\overline{x} = 942.8615$ and $s_x^2 = 1307911.82$.

From the responding states we have

State	x_i	y_i	$(x_i - \bar{x}_{n-r})^2$	$(y_i - \bar{y}_{n-r})^2$	$(y_i - \bar{y}_{n-r})(x_i - \bar{x}_{n-r})$
AL	348.334	408.978	518339.52	29309.09	123256.0896
AZ	431.439	54.633	405581.74	276196.49	334694.2730
AR	848.317	907.700	48389.00	107271.31	-72046.8719
CA	3928.732	1343.461	8182116.94	582602.46	2183328.0850
CO	906.281	315.809	26247.56	69890.43	42830.5240
CT	4.373	7.130	1131923.63	328382.86	609675.5912
IN	1022.782	1213.024	2071.16	400495.32	-28800.8669
ME	51.539	8.849	1033786.66	326415.68	580899.4580
MI	440.518	323.028	394100.19	66125.60	161431.4563
MN	2466.892	1354.768	1956081.96	599991.21	1083342.9730
NE	3585.406	1337.852	6335862.89	574071.40	1907154.3500
NH	0.471	6.044	1140241.68	329628.70	613071.2742
NY	426.274	201.631	412187.11	143297.07	243033.3458
NC	494.730	639.571	328973.36	3527.64	-34066.1414
Sum	14956.090	8122.478	21915903.50	3837205.34	7747803.5390

Thus we have

$\bar{x}_{n-r} = 1068.292$, $\bar{y}_{n-r} = 580.177$, $s_x^{*2} = 1685838.731$, $s_y^{*2} = 295169.6416$,

$s_{xy}^{*} = 595984.887$ and $r_{xy}^{*} = \dfrac{595984.887}{\sqrt{1685838.731 \times 295169.6416}} = 0.84487$.

From Table 12.4.2.1, for $n = 20$ and $r = 6$, an estimate of the probability of non-response is $\hat{p} = 0.34993 \approx 0.35$ and $\hat{q} = 1 - \hat{p} = 0.65$.

Estimator 1. We have

$$\bar{y}_{1\text{lr}} = \bar{y}_{n-r} + \left(s_{xy}^{*} / s_x^{*2}\right)\left(\bar{X} - \bar{x}_{n-r}\right) = 580.177 + \frac{595984.887}{1685838.731}(878.16 - 1068.292) = 512.96$$

and

$$\hat{v}(\bar{y}_{1\text{lr}}) = \left[\frac{1}{(n\hat{q} + 2\hat{p})} - \frac{1}{N}\right] s_y^{*2}\left(1 - r_{xy}^{*\,2}\right)$$

$$= \left[\frac{1}{(20 \times 0.65 + 2 \times 0.35)} - \frac{1}{50}\right] \times 295169.6416 \times \left(1 - 0.84487^2\right) = 4476.61 .$$

A $(1-\alpha)100\%$ confidence interval for the true population mean $\bar{Y}$ is given by

$$\bar{y}_{1\text{lr}} \pm t_{\alpha/2}(\text{df} = n - r - 2)\sqrt{\hat{v}(\bar{y}_{1\text{lr}})} .$$

Using Table 2 from the Appendix the 95% confidence interval for the average real estate farm loans is given by

$$512.96 \pm t_{0.05/2}(\text{df} = 20-6-2)\sqrt{4476.61}$$

or $512.96 \pm 2.179\sqrt{4476.61}$, or $[367.17,\ 658.75]$.

Estimator 2. We have

$$\bar{y}_{2\text{lr}} = \bar{y}_{n-r} + \left(s_{xy}^{*}/s_x^2\right)\left(\bar{X} - \bar{x}\right) = 580.177 + \frac{595984.887}{1307911.82}(878.16 - 942.861) = 550.694$$

and

$$\hat{v}(\bar{y}_{2\text{lr}})$$
$$= \left[\frac{1}{(n\hat{q}+2\hat{p})} - \frac{1}{n}\right]s_y^{*2} + \left(\frac{1}{n} - \frac{1}{N}\right)s_y^{*2}\left(1 - r_{xy}^{*2}\right)$$
$$= \left[\frac{1}{(20\times 0.65 + 2\times 0.35)} - \frac{1}{20}\right]\times 295169.6416 + \left(\frac{1}{20} - \frac{1}{50}\right)\times 295169.6416 \times \left(1 - 0.84487^2\right)$$
$$= 6786.74 + 2534.27 = 9321.02 .$$

A $(1-\alpha)100\%$ confidence interval for the true population mean $\bar{Y}$ is given by

$$\bar{y}_{2\text{lr}} \pm t_{\alpha/2}(\text{df} = n-r-2)\sqrt{\hat{v}(\bar{y}_{2\text{lr}})} .$$

Using Table 2 from the Appendix the 95% confidence interval for the average real estate farm loans is

$$550.694 \pm t_{0.05/2}(\text{df} = 20-6-2)\sqrt{9321.02}$$

or $550.694 \pm 2.179\sqrt{9321.02}$, or $[340.32,\ 761.066]$.

Estimator 3. We have

$$\bar{y}_{3\text{lr}} = \bar{y}_{n-r} + \left(s_{xy}^{*}/s_x^2\right)\left(\bar{x}_{n-r} - \bar{x}\right) = 580.177 + \frac{595984.887}{1307911.82}(1068.292 - 942.8615) = 637.33$$

and

$$\hat{v}(\bar{y}_{3\text{lr}})$$
$$= \left[\frac{1}{(n\hat{q}+2\hat{p})} - \frac{1}{n}\right]s_y^{*2}\left(1 - r_{xy}^{*2}\right) + \left(\frac{1}{n} - \frac{1}{N}\right)s_y^{*2}$$
$$= \left[\frac{1}{(20\times 0.65 + 2\times 0.35)} - \frac{1}{20}\right]\times 295169.6416 \times \left(1 - 0.84487^2\right) + \left(\frac{1}{20} - \frac{1}{50}\right)\times 295169.6416$$
$$= 1942.33 + 8855.09 = 10797.42 .$$

A $(1-\alpha)100\%$ confidence interval for the true population mean $\bar{Y}$ is given by

$$\bar{y}_{3\text{lr}} \pm t_{\alpha/2}(\text{df} = n-r-2)\sqrt{\hat{v}(\bar{y}_{3\text{lr}})} .$$

Using Table 2 from the Appendix the 95% confidence interval for the average farm real estate loans is

$$637.33 \pm t_{0.05/2}(\text{df} = 20-6-2)\sqrt{10797.42}$$

or $637.33 \pm 2.179\sqrt{10797.42}$ or $[410.90,\ 863.75]$.

12.4.3 ESTIMATION OF FINITE POPULATION VARIANCE

Singh and Joarder (1998) consider the problem of estimation of finite variance in the presence of non-response in survey sampling.
Let us define

$$\varepsilon = \left(s_y^{*2}/S_y^2\right)-1, \quad \delta = \left(s_x^{*2}/S_x^2\right)-1, \text{ and } \eta = \left(s_x^2/S_x^2\right)-1,$$

where

$s_y^{*2} = (n-r-1)^{-1}\sum_{i=1}^{n-r}\left(y_i-\bar{y}^*\right)^2$ and $s_x^{*2} = (n-r-1)^{-1}\sum_{i=1}^{n-r}\left(x_i-\bar{x}^*\right)^2$ are conditionally unbiased estimators of S_y^2 and S_x^2, respectively, and $\bar{y}^* = (n-r)^{-1}\sum_{i=1}^{n-r} y_i$ and $\bar{x}^* = (n-r)^{-1}\sum_{i=1}^{n-r} x_i$. Thus under the probability model given by (12.4.1.1) we have the following results.

$$E(\varepsilon) = E(\delta) = E(\eta) = 0,$$

$$E\left(\varepsilon^2\right) = \left[\frac{1}{(nq+2p)}-\frac{1}{N}\right](\lambda_{40}-1), \quad E\left(\delta^2\right) = \left[\frac{1}{(nq+2p)}-\frac{1}{N}\right](\lambda_{04}-1),$$

$$E\left(\eta^2\right) = \left[\frac{1}{n}-\frac{1}{N}\right](\lambda_{04}-1), \quad E(\varepsilon\delta) = \left[\frac{1}{(nq+2p)}-\frac{1}{N}\right](\lambda_{22}-1),$$

$$E(\varepsilon\eta) = \left[\frac{1}{n}-\frac{1}{N}\right](\lambda_{22}-1), \quad \text{and} \quad E(\delta\eta) = \left[\frac{1}{n}-\frac{1}{N}\right](\lambda_{04}-1).$$

It may be noted that if $p=0$, *i.e.*, if there is no non-response, the above expected values coincide with the usual results discussed in Chapter 3.

Strategy I. Consider the situation when random non-response exists on both the study variable y and the auxiliary variable x and population variance S_x^2 of the auxiliary character is known. An estimator of a finite population variance is

$$\hat{v} = s_y^{*2}\left(S_x^2/s_x^{*2}\right). \tag{12.4.3.1}$$

Then we have the following theorems:

Theorem 12.4.3.1. The bias in the estimator $\hat{v}$, up to terms of order n^{-1}, is

$$B(\hat{v}) = \left[\frac{1}{(nq+2p)}-\frac{1}{N}\right]S_y^2(\lambda_{04}-\lambda_{22}). \tag{12.4.3.2}$$

Proof. The estimator $\hat{v}$ in terms of ε and δ can be written as

$$\hat{v} = S_y^2\left(1+\varepsilon-\delta+\delta^2-\varepsilon\delta+....\right). \tag{12.4.3.3}$$

Taking expected value on both sides of (12.4.3.3) and using the results on the expectations from previous section, we get (12.4.3.2). Hence the theorem.

Theorem 12.4.3.2. The mean square error, up to terms of order n^{-1}, of the estimator $\hat{v}$ is given by

$$\text{MSE}(\hat{v}) = \left[\frac{1}{(nq+2p)} - \frac{1}{N}\right] S_y^4(\lambda_{40} + \lambda_{04} - 2\lambda_{22}). \tag{12.4.3.4}$$

Proof. It is easy to check that

$$\text{MSE}(\hat{v}) = E\left(\hat{v} - S_y^2\right)^2 = S_y^4 E\left(\varepsilon - \delta + \delta^2 - \varepsilon\delta\right)^2 = S_y^4 E\left(\varepsilon^2 + \delta^2 - 2\varepsilon\delta\right)$$

$$= \left[\frac{1}{(nq+2p)} - \frac{1}{N}\right] S_y^4(\lambda_{40} + \lambda_{04} - 2\lambda_{22}).$$

Theorem 12.4.3.3. An estimator of the $\text{MSE}(\hat{v})$ is given by

$$\hat{\text{MSE}}(\hat{v}) = \left[\frac{1}{(n\hat{q}+2\hat{p})} - \frac{1}{N}\right] s_y^{*4}\left(\hat{\lambda}_{40}^* + \hat{\lambda}_{04}^* - 2\hat{\lambda}_{22}^*\right). \tag{12.4.3.5}$$

where $\hat{\lambda}_{ls}^* = \hat{\mu}_{ls}^* \Big/ \left\{\left(\hat{\mu}_{20}^*\right)^{l/2}\left(\hat{\mu}_{02}^*\right)^{s/2}\right\}$ and $\hat{\mu}_{ls}^* = (n-r-1)^{-1}\sum_{i=1}^{n-r}\left(y_i - \bar{y}^*\right)^l\left(x_i - \bar{x}^*\right)^s$.

Strategy II. Consider the situation when information on variable y could not be obtained for r units while information on variable x is available and population variance S_x^2 of the auxiliary variable is known, then we have the estimator

$$\hat{v}_1 = s_y^{*2}\left(S_x^2 / s_x^2\right). \tag{12.4.3.6}$$

Thus we have the following theorems and their proofs are obvious.

Theorem 12.4.3.4. The bias in the estimator $\hat{v}_1$, up to terms of order n^{-1}, is

$$B(\hat{v}_1) = \left(\frac{1}{n} - \frac{1}{N}\right) S_y^2(\lambda_{04} - \lambda_{22}). \tag{12.4.3.7}$$

Theorem 12.4.3.5. The mean square error, up to terms of order n^{-1}, is given by

$$\text{MSE}(\hat{v}_1) = \text{MSE}\left(s_I^2\right) + \left[\frac{1}{(nq+2p)} - \frac{1}{N}\right] S_y^4 \tag{12.4.3.8}$$

where $\text{MSE}\left(s_I^2\right)$ denotes the MSE of ratio type estimator of variance proposed by Isaki (1983) as discussed in Chapter 3.

Theorem 12.4.3.6. An estimator of mean square error of $\hat{v}_1$ is given by

$$\hat{\text{MSE}}(\hat{v}_1) = \left(\frac{1}{n} - \frac{1}{N}\right) s_y^{*4}\left(\hat{\lambda}_{40}^* + \hat{\lambda}_{04} - 2\hat{\lambda}_{22}^*\right) + \left[\frac{1}{(n\hat{q}+2\hat{p})} - \frac{1}{n}\right] s_y^{*4} \tag{12.4.3.9}$$

where $\hat{\lambda}_{04} = \hat{\mu}_{04} / \hat{\mu}_{02}^2$ and $\hat{\mu}_{0s} = (n-1)^{-1}\sum_{i=1}^{n}(x_i - \bar{x})^s$.

If information on x is available for all the n units then we can obtain both s_x^2 and s_x^{*2}. Using this information consider another estimator as

$$\hat{v}_2 = s_y^{*2}\left(S_x^2/s_x^2\right)+\alpha\left(s_x^{*2}/S_x^2-1\right) \tag{12.4.3.10}$$

where α is a suitably chosen constant such that the mean square error of $\hat{v}_2$ is minimum. Thus we have the following two theorems:

Theorem 12.4.3.7. The bias in the estimator $\hat{v}_2$ is the same as in the estimator $\hat{v}_1$.

Theorem 12.4.3.8. The minimum mean square error of the estimator $\hat{v}_2$ is given by

$$\text{Min.MSE}(\hat{v}_2)=\text{MSE}(\hat{v}_1)-\frac{S_y^4\left\{\left(\frac{1}{nq+2p}-\frac{1}{N}\right)(\lambda_{22}-1)-\left(\frac{1}{n}-\frac{1}{N}\right)(\lambda_{04}-1)\right\}^2}{\left\{\frac{1}{(nq+2p)}-\frac{1}{N}\right\}(\lambda_{04}-1)}. \tag{12.4.3.11}$$

Proof. We have

$$\text{MSE}(\hat{v}_2)=E\left(\hat{v}_2-S_y^2\right)^2=E\left[S_y^2\left(1+\varepsilon-\eta+\eta^2-\varepsilon\eta\right)+\alpha\delta-S_y^2\right]^2$$

$$=\text{MSE}(\hat{v}_1)+\alpha^2\left\{\frac{1}{(nq+2p)}-\frac{1}{N}\right\}(\lambda_{04}-1)$$

$$+2\alpha S_y^2\left\{\left(\frac{1}{nq+2p}-\frac{1}{N}\right)(\lambda_{22}-1)-\left(\frac{1}{n}-\frac{1}{N}\right)(\lambda_{04}-1)\right\} \tag{12.4.3.12}$$

On differentiating (12.4.3.12) with respect to α and equating to zero we have

$$\alpha=-\frac{\left\{\left(\frac{1}{nq+2p}-\frac{1}{N}\right)(\lambda_{22}-1)-\left(\frac{1}{n}-\frac{1}{N}\right)(\lambda_{04}-1)\right\}S_y^2}{\left\{\frac{1}{(nq+2p)}-\frac{1}{N}\right\}(\lambda_{04}-1)} \tag{12.4.3.13}$$

and then putting the optimum value of α in (12.4.3.12) we have (12.4.3.11). Hence the theorem.

Theorem 12.4.3.9. An estimator of the $\text{Min.MSE}(\hat{v}_2)$ is given by

$$\text{Min.}\hat{\text{MSE}}(\hat{v}_2)=\hat{\text{MSE}}(\hat{v}_1)-\frac{s_y^{*4}\left\{\left(\frac{1}{n\hat{q}+2\hat{p}}-\frac{1}{N}\right)(\hat{\lambda}_{22}^*-1)-\left(\frac{1}{n}-\frac{1}{N}\right)(\hat{\lambda}_{04}-1)\right\}^2}{\left\{\frac{1}{(n\hat{q}+2\hat{p})}-\frac{1}{N}\right\}(\hat{\lambda}_{04}-1)}. \tag{12.4.3.14}$$

If optimum value of α is not known, then it is advisable to replace it with its estimator $\hat{\alpha}$ and then we have the following estimator given by

$$\hat{v}_3 = s_y^{*2}\left(S_x^2/s_x^2\right)+\hat{\alpha}\left(s_x^{*2}/S_x^2-1\right), \tag{12.4.3.15}$$

where

$$\hat{\alpha} = s_y^{*2}\left\{\left(\frac{1}{n\hat{q}+2\hat{p}} - \frac{1}{N}\right)\left(\hat{\lambda}_{22}^{*} - 1\right) - \left(\frac{1}{n} - \frac{1}{N}\right)\left(\hat{\lambda}_{04} - 1\right)\right\} \Big/ \left\{\frac{1}{(n\hat{q}+2\hat{p})} - \frac{1}{N}\right\}\left(\hat{\lambda}_{04} - 1\right)$$

denotes a consistent estimator of α. To find the mean square error of the estimator $\hat{v}_3$ let us define $\kappa = (\hat{\alpha}/\alpha) - 1$, where $E(\kappa) = O(n^{-1})$, then the MSE of $\hat{v}_3$ is given by

$$\mathrm{MSE}(\hat{v}_3) = E\left(\hat{v}_3 - S_y^2\right)^2 = E\left[S_y^2\left(1 + \varepsilon - \eta + \eta^2 - \varepsilon\eta\right) + \alpha(1+\kappa)\delta - S_y^2\right]^2 . \quad (12.4.3.16)$$

This is approximately the same as $\mathrm{MSE}(\hat{v}_2)$. It may be noted here that estimators $\hat{v}_2$ and $\hat{v}_3$ may take inadmissible value, *i.e.*, a negative value. Thus an equally efficient alternative estimator is given below

$$\hat{v}_4 = s_y^{*2}\frac{S_x^2}{s_x^2}\left(\frac{s_x^{*2}}{S_x^2}\right)^{\alpha} \quad (12.4.3.17)$$

for $\alpha \neq 1$. If $\alpha = 1$ then it leads to the following strategy.

Strategy III. Consider the situation when information on variable y could not be obtained for r units while information on the variable x is obtained for all the sample units, but the difference is that the population variance S_x^2 of the auxiliary variable is not known. In this case consider another ratio estimator

$$\hat{v}_5 = s_y^{*2}\left(s_x^2 / s_x^{*2}\right). \quad (12.4.3.18)$$

Thus we have the following theorems:

Theorem 12.4.3.10. The approximate bias in the estimator $\hat{v}_5$, up to terms of order n^{-1}, is given by

$$B(\hat{v}_5) = \left[\frac{1}{n} - \frac{1}{nq+2p}\right]S_y^2(\lambda_{22} - \lambda_{04}). \quad (12.4.3.19)$$

Theorem 12.4.3.11. The asymptotic mean square error of the estimator $\hat{v}_5$, up to terms of order n^{-1}, is

$$\mathrm{MSE}(\hat{v}_5) = \left[\left\{\frac{1}{nq+2p} - \frac{1}{N}\right\}(\lambda_{40} - 1) + \left\{\frac{1}{nq+2p} - \frac{1}{n}\right\}(\lambda_{04} + 1 - 2\lambda_{22})\right]S_y^4 . \quad (12.4.3.20)$$

Theorem 12.4.3.12. An estimator of mean square error of $\hat{v}_5$ is given by

$$\hat{\mathrm{MSE}}(\hat{v}_5) = \left[\left\{\frac{1}{n\hat{q}+2\hat{p}} - \frac{1}{N}\right\}\left(\hat{\lambda}_{40}^{*} - 1\right) + \left\{\frac{1}{n\hat{q}+2\hat{p}} - \frac{1}{n}\right\}\left(\hat{\lambda}_{04} + 1 - 2\hat{\lambda}_{22}^{*}\right)\right]s_y^{*4} . \quad (12.4.3.21)$$

Example 12.4.3.1. A Government organisation selected an SRSWOR sample of twenty states from the population 1 given in the Appendix and collected information about the real estate farm loans and nonreal estate farm loans, but unfortunately the information on the real estate farm loans was not available on four states as marked in the table below.

Sampled Unit	State	x_i	y_i	Sampled Unit	State	x_i	y_i
05	CA	3928.732	1343.461	34	ND	1241.369	Missing
07	CT	4.373	7.130	36	OK	1716.087	612.108
09	FL	464.516	825.748	40	SC	80.750	87.951
13	IL	2610.572	2131.048	42	TN	388.869	553.266
19	ME	51.539	Missing	43	TX	3520.361	Missing
24	MS	549.551	627.013	44	UT	197.244	56.908
25	MO	1519.994	1579.686	46	VA	188.477	Missing
27	NE	3585.406	1337.852	47	WA	1228.607	1100.745
30	NJ	27.508	39.860	48	WV	29.291	99.277
31	NM	274.035	140.582	49	WI	1372.439	1229.752

x -- Nonreal estate farm loans, y -- Real estate farm loans.

Apply the ratio type estimator $\hat{v}_5 = s_y^{*2}\left(s_x^2 / s_x^{*2}\right)$ for estimating the finite population variance of the real estate farm loans in the United States. Construct 75% confidence interval.

Solution. From the Table 12.4.2.1 for $n = 20$ and $r = 4$, we have $\hat{p} = 0.23355$ and $\hat{q} = 1 - \hat{p} = 1 - 0.23355 = 0.76645$. From the responding units in the sample we have

Sr. No.	x_i	y_i	$(x_i - \bar{x}_{n-r})^2$	$(y_i - \bar{y}_{n-r})^2$	$(x_i - \bar{x}_{n-r})^2 (y_i - \bar{y}_{n-r})^2$	$(y_i - \bar{y}_{n-r})^4$
05	3928.732	1343.461	7868634.0	369283.3	2.90575E+12	1.3637E+11
07	4.373	7.130	1252721.0	530922.4	6.65098E+11	2.81879E+11
09	464.516	825.748	434422.5	8095.3	3516775046	65533670.5
13	2610.572	2131.048	2211016.0	1946789.0	4.30438E+12	3.78999E+12
24	549.551	627.013	329559.1	11829.0	3898353147	139925144.2
25	1519.994	1579.686	157109.7	712187.1	1.11891E+11	5.07211E+11
27	3585.406	1337.852	6060374.0	362497.7	2.19687E+12	1.31405E+11
30	27.508	39.860	1201469.0	484296.6	5.81867E+11	2.34543E+11
31	274.035	140.582	721800.4	354253.7	2.557E+11	1.25496E+11
36	1716.087	612.108	351013.1	15293.3	5368158462	233885818.0
40	80.750	87.951	1087585.0	419674.9	4.56432E+11	1.76127E+11
42	388.869	553.266	539864.0	33309.2	17982458456	1109505370.0
44	197.244	56.908	858178.7	460859.3	3.955E+11	2.12391E+11
47	1228.607	1100.745	11021.6	133203.7	1468112708	17743224089.0
48	29.291	99.277	1197563.0	405128.7	4.85167E+11	1.64129E+11
49	1372.439	1229.752	61909.2	244014.1	15106720154	59542870868.0
Sum	17977.970	11772.390	24344241.0	6491637.0	1.2406E+13	5.83837E+12

$$\bar{x}_{n-r} = \frac{\sum_{i=1}^{n-r} x_i}{n-r} = \frac{17977.97}{20-4} = 1123.62\,, \quad \bar{y}_{n-r} = \frac{\sum_{i=1}^{n-r} y_i}{n-r} = \frac{11772.39}{20-4} = 735.77\,,$$

$$\hat{\mu}_{20}^{*} = s_y^{*2} = \frac{\sum_{i=1}^{n-r}\left(y_i - \bar{y}_{n-r}\right)^2}{n-r-1} = \frac{6491637}{20-4-1} = 432778.2\,,$$

$$\hat{\mu}_{02}^{*} = s_x^{*2} = \frac{\sum_{i=1}^{n-r}\left(x_i - \bar{x}_{n-r}\right)^2}{n-r-1} = \frac{24344241}{20-4-1} = 1622949.4\,,$$

$$\hat{\mu}_{22}^{*} = \frac{\sum_{i=1}^{n-r}\left(y_i - \bar{y}_{n-r}\right)^2\left(x_i - \bar{x}_{n-r}\right)^2}{n-r-1} = \frac{1.2406\times 10^{13}}{20-4-1} = 8.270667\times 10^{11}\,,$$

and

$$\hat{\mu}_{40}^{*} = \frac{\sum_{i=1}^{n-r}\left(y_i - \bar{y}_{n-r}\right)^4}{n-r-1} = \frac{5.83837\times 10^{12}}{20-4-1} = 3.8922\times 10^{11}\,.$$

From the complete information on the auxiliary variable we have

Sampled Unit	State	x_i	$(x_i - \bar{x})^2$	$(x_i - \bar{x})^4$
05	CA	3928.732	7726987.825	5.97063E+13
07	CT	4.373	1310138.920	1.71646E+12
09	FL	464.516	468499.180	2.19491E+11
13	IL	2610.572	2136233.635	4.56349E+12
19	ME	51.539	1204389.918	1.45056E+12
24	MS	549.551	359322.319	1.29113E+11
25	MO	1519.994	137646.936	18946679008
27	NE	3585.406	5936142.416	3.52378E+13
30	NJ	27.508	1257712.904	1.58184E+12
31	NM	274.035	765539.252	5.86050E+11
34	ND	1241.369	8534.618	72839716.17
36	OK	1716.087	321603.544	1.03429E+11
40	SC	80.750	1141128.152	1.30217E+12
42	TN	388.869	577777.854	3.33827E+11
43	TX	3520.361	5623419.391	3.16228E+13
44	UT	197.244	905812.835	8.20497E+11
46	VA	188.477	922577.539	8.51149E+11
47	WA	1228.607	6339.504	40189306.41
48	WV	29.291	1253716.893	1.57181E+12
49	WI	1372.439	49931.243	2493129048
	Sum	22979.720	32113454.880	1.41818E+14

$$\bar{x} = \frac{\sum_{i=1}^{n} x_i}{n} = \frac{22979.72}{20} = 1148.986\,,\quad \hat{\mu}_{02} = s_x^2 = \frac{\sum_{i=1}^{n}(x_i - \bar{x})^2}{n-1} = \frac{32113454.88}{20-1} = 1690181.84$$

$$\hat{\mu}_{04} = \frac{\sum_{i=1}^{n}(x_i - \bar{x})^4}{n-1} = \frac{1.41818\times 10^{14}}{20-1} = 7.46411\times 10^{12}\,,\quad \hat{\lambda}_{40}^{*} = \frac{\hat{\mu}_{40}^{*}}{\hat{\mu}_{20}^{*2}} = \frac{3.8922\times 10^{11}}{432778.2^2} = 2.0781$$

$$\hat{\lambda}_{04} = \frac{\hat{\mu}_{04}}{\hat{\mu}_{02}^2} = \frac{7.46411\times 10^{11}}{1690181.84^2} = 2.6128\,,\ \&\ \hat{\lambda}_{22}^{*} = \frac{\hat{\mu}_{22}^{*}}{\hat{\mu}_{20}^{*}\hat{\mu}_{02}^{*}} = \frac{8.240667\times 10^{11}}{432778.2\times 1622949.4} = 1.1775\,.$$

Thus an estimate of the finite population variance of the real estate farm loans is given by

$$\hat{v}_5 = s_y^{*2}\left(s_x^2 / s_x^{*2}\right) = 432778.2\times\frac{1690181.84}{1622949.4} = 450706.50\,.$$

An estimate of the $\text{MSE}(\hat{v}_5)$ is given by

$$\hat{\text{MSE}}(\hat{v}_5) = \left[\left\{\frac{1}{n\hat{q}+2\hat{p}} - \frac{1}{N}\right\}\left(\hat{\lambda}_{40}^{*} - 1\right) + \left\{\frac{1}{n\hat{q}+2\hat{p}} - \frac{1}{n}\right\}\left(\hat{\lambda}_{04} - 2\hat{\lambda}_{22}^{*} + 1\right)\right]s_y^{*4}$$

$$= \left[\left\{\frac{1}{(20\times 0.76645 + 2\times 0.23355)} - \frac{1}{50}\right\}(2.0781-1)\right.$$

$$\left. + \left\{\frac{1}{(20\times 0.23355 + 2\times 0.76645)} - \frac{1}{20}\right\}(2.61281 + 1 - 2\times 1.1775)\right]\times(432778.2)^2$$

$$= 3.4938\times 10^{10}\,.$$

The $(1-\alpha)100\%$ confidence interval for the finite population variance is given by

$$\hat{v}_5 \pm t_{\alpha/2}(\text{df} = n-r-2)\sqrt{\hat{\text{MSE}}(\hat{v}_5)}\,.$$

Using Table 2 from the Appendix the 75% confidence interval of the finite population variance of the real estate farm loans is given by

$$\hat{v}_5 \pm t_{0.25/2}(\text{df} = 20-4-2)\sqrt{\hat{\text{MSE}}(\hat{v}_5)}\,,$$

$$\text{or}\ \ 450706.50 \pm 1.200\sqrt{3.4938\times 10^{10}}\ \ \text{or}\ \ [226404.16,\ 675008.83].$$

The next section has been devoted to discuss a few imputation techniques in brief for handling the non-response in survey sampling.

Important note: Please keep in mind that in the proceeding sections the notation r and $(n-r)$ have different meanings than in the preceding sections.

12.5. CALIBRATED ESTIMATORS OF TOTAL AND VARIANCE IN THE PRESENCE OF NON-RESPONSE

In almost all large scale surveys non-responses are unavoidable. Several methods are available for handling non-response problems in survey sampling. Details are given by Rubin (1987). We consider here the problem of variance estimation in the presence of non-response in a unified setup. Several researchers have suggested several methods of estimating the variance of the estimators of total or mean in the presence of non-response in survey sampling. In some cases the probability of response of the ith individual may be known or may be estimated through logistic regression model. Consider a finite population $\Omega = \{1,2,...,i,...,N\}$ from which a sample s of size n is selected with probability $p(s)$ according to some sampling design p. Let us denote the inclusion probabilities for the ith, and i^{th} and j^{th} $(i \neq j)$ units by π_i and π_{ij}, respectively. Let s_r be the respondent units in s. Here we assume that the set of the response sample $s_r(\subset s)$ selected from s by nature with probability $q(s_r)$. Let us denote $S_{(r)}$ is the collection of all possible samples of size r, those can be selected from s, that is, $S_{(r)}$ will consist of $\binom{n}{r}$ different samples.

Let $\tilde{q} = q(s_r)\left\{\sum_{s_r \in S_{(r)}} q(s_r)\right\}^{-1}$, then $\tilde{q}$ forms a sampling design. Let $\psi_i(s,r)$ and $\psi_{ij}(s,r)$ be the inclusion probabilities for the ith, (i^{th}, j^{th}) $i \neq j$ units. Obviously $\psi_i(s,r)$ and $\psi_{ij}(s,r)$ will depend upon s and r, the number of respondents in the sample for the sampling design q defined on $S_{(r)}$. Arnab and Singh (2001) consider the problem of estimation of population total Y and its variance under two different situations, *viz.*, (i) absence of auxiliary information, and (ii) presence of auxiliary information.

12.5.1 ESTIMATION OF POPULATION TOTAL AND VARIANCE

Let y_i be the value of the variable of interest, y, for the ith population unit. The well known Horvitz and Thompson (1952) estimator of the population total, Y, in the presence of non-response is given by

$$\hat{Y}_1 = \sum_{i \in s_r} \frac{y_i}{\pi_i \psi_{irs}} = \sum_{i \in s_r} \frac{z_i}{\psi_{irs}}, \tag{12.5.1.1}$$

where $\psi_{irs} = \psi_i(r,s)$ and $z_i = y_i/\pi_i$. Let E_s (V_s), $E_{r|s}$ $(V_{r|s})$, and $E_{s_r|r,s}$ $(V_{s_r|r,s})$ denote respectively the unconditional expectation (variance) over the initial sample s conditional expectation (variance) over r given s and conditional expectation (variance) over s_r given r and s. Similarly E_r (V_r) denotes unconditional expectation (variance) of r, the size of the respondent units. Then we have the following theorems.

Theorem 12.5.1.1. The estimator $\hat{Y}_1$ is unbiased for the population total Y.

Proof. We have

$$E(\hat{Y}_1) = E_s E_{r|s} E_{s_r|r,s}[\hat{Y}_1] = E_s E_{r|s} E_{s_r|r,s}\left[\sum_{i\in s_r} \frac{z_i}{\Psi_{irs}}\right] = E_s E_{r|s}\left[\sum_{i\in s} Z_i\right] = E_s\left[\sum_{i\in s} \frac{y_i}{\pi_i}\right] = Y.$$

Hence the theorem.

Theorem. 12.5.1.2. The variance of $\hat{Y}_1$ is given by

$$V(\hat{Y}_1) = E_r\left[\sum_{i\in\Omega} \frac{y_i^2}{\pi_i^2} a_{ir} + \frac{1}{2}\sum_{i\neq}\sum_{j\in\Omega} \delta_{ijr}\left(\frac{y_i}{\pi_i} - \frac{y_j}{\pi_j}\right)^2 + \frac{1}{2}\sum_{i\neq}\sum_{j\in\Omega} \Theta_{ij}\left(\frac{y_i}{\pi_i} - \frac{y_j}{\pi_j}\right)^2\right],$$

where $a_{ir} = \delta_{ir} - \sum_{j(\neq i)} \delta_{ijr}$, $\delta_{ir} = \sum_{s\ni i}\left(\frac{1}{\Psi_{irs}} - 1\right)p(s)$, $\delta_{ijr} = \sum_{s\ni i,j}\left(1 - \frac{\Psi_{ijrs}}{\Psi_{irs}\Psi_{jrs}}\right)p(s)$, and $\Theta_{ij} = (\pi_i\pi_j - \pi_{ij})$.

Proof. Writing $E_{r,s_r|s}$ ($V_{r,s_r|s}$) for the overall expectation (variance) for variation over r and s_r when s is fixed we have

$$V(\hat{Y}_1) = E_s\left[V_{r,s_r|s}(\hat{Y}_1 \mid s)\right] + V_s\left[E_{r,s_r|s}(\hat{Y}_1 \mid s)\right]. \tag{12.5.1.2}$$

Now

$$V_s\left[E_{r,s_r|s}(\hat{Y}_1 \mid s)\right] = V_s\left(\sum_{i\in s} \frac{y_i}{\pi_i}\right) = \frac{1}{2}\sum_{i\neq}\sum_{j\in\Omega} \Theta_{ij}\left(\frac{y_i}{\pi_i} - \frac{y_j}{\pi_j}\right)^2 \tag{12.5.1.3}$$

and

$$V_{r,s_r|s}(\hat{Y}_1 \mid s) = E_{r|s} V_{s_r|r,s}(\hat{Y}_1 \mid s) + V_{r|s} E_{s_r|r,s}(\hat{Y}_1 \mid s) = E_{r|s} V_{s_r|r,s}\left(\sum_{i\in s_r} \frac{Z_i}{\Psi_{irs}}\right) + V_r\left(\sum_{i\in s} \frac{y_i}{\pi_i}\right)$$

$$= E_{r|s}\left[\sum_{i\in s} Z_i^2\left(\frac{1}{\Psi_{irs}} - 1\right) + \sum_{i\neq}\sum_{j\in s} Z_i Z_j\left(\frac{\Psi_{ijrs}}{\Psi_{irs}\Psi_{jrs}} - 1\right)\right] + 0. \tag{12.5.1.4}$$

Noting that $V_r\left(\sum_{i\in s} \frac{y_i}{\pi_i}\right) = 0$. From (12.5.1.4) we have

$$E_s\left[V_{r,s_r|s}(\hat{Y}_1 \mid s)\right] = E_s E_{r|s}\left[\sum_{i\in s} Z_i^2\left(\frac{1}{\Psi_{irs}} - 1\right) + \sum_{i\neq}\sum_{j\in s} Z_i Z_j\left(\frac{\Psi_{ijrs}}{\Psi_{irs}\Psi_{jrs}} - 1\right)\right]$$

$$= E_r E_{s|r}\left[\sum_{i\in s} Z_i^2\left(\frac{1}{\Psi_{irs}} - 1\right) + \sum_{i\neq}\sum_{j\in s} Z_i Z_j\left(\frac{\Psi_{ijrs}}{\Psi_{irs}\Psi_{jrs}} - 1\right)\right].$$

Now noting (i) $E_s E_{r|s} = E_r E_{s|r}$ is valid because y_i values are not involved in the selection of samples and (ii) $E_{s|r} = E_s$, we obtain

$$E_s\left[V_{r,s_r|s}\left(\hat{Y}_1\right)\right] = E_r\left[\sum_{i\in\Omega} Z_i^2\delta_{ir} - \sum_{i\neq}\sum_{j\in\Omega} Z_iZ_j\delta_{ijr}\right]$$

$$= E_r\left[\sum_{i\in\Omega}\frac{y_i^2}{\pi_i^2}a_{ir} + \frac{1}{2}\sum_{i\neq}\sum_{j\in\Omega}\delta_{ijr}\left(\frac{y_i}{\pi_i} - \frac{y_j}{\pi_j}\right)^2\right] \qquad (12.5.1.5)$$

since

$$\sum_{i\neq}\sum_{j\in\Omega} Z_iZ_j\delta_{ijr} = -\frac{1}{2}\sum_{i\neq}\sum_{j}\left\{\left(Z_i - Z_j\right)^2 - \left(Z_i^2 + Z_j^2\right)\right\}\delta_{ijr} = -\frac{1}{2}\sum_{i\neq}\sum_{j}\left(Z_i - Z_j\right)^2 + \sum_i Z_i^2\sum_{j(\neq i)}\delta_{ijr}.$$

From (12.5.1.3) and (12.5.1.5), we have the theorem.

Theorem. 12.5.1.3. An unbiased estimator of $V\left(\hat{Y}_1\right)$ is given by

$$\hat{v}\left(\hat{Y}_1\right) = \sum_{i\in s_r} b_{iis}(r)\frac{y_i^2}{\pi_i^2} + \frac{1}{2}\sum_{i\neq}\sum_{j\in s_r} b_{ijs}(r)\left(\frac{y_i}{\pi_i} - \frac{y_j}{\pi_j}\right)^2 \qquad (12.5.1.6)$$

where $b_{iis}(r) = a_{ir}/\left(\pi_i\psi_i\right)$ and $b_{ijs}(r) = \left(\delta_{ijr} + \Theta_{ij}\right)/\left(\pi_{ij}\psi_{ijrs}\right)$. .

12.5.2. CALIBRATION ESTIMATOR FOR THE TOTAL

The estimator (12.5.1.1) can be written as

$$\hat{Y}_2 = \sum_{i\in s_r} d_iy_i, \quad \text{where } d_i = 1/\left(\pi_i\psi_i\right) \text{ (writing } \psi_{irs} = \psi_i \text{)}. \qquad (12.5.2.1)$$

Note that here d_i is slightly different than that defined in Chapter 5. Let us assume that an auxiliary variable x_i is available and it is positive for every i, then we can estimate ψ_i through x_i. So, in the presence of auxiliary information, we can choose weights in different stages as follows:

Mechanism of adjusting weights							
Study variable	Auxiliary variable	Selection weights	Non-response adjusted weights	Adjusted weights	Weighted study variable	Weighted auxiliary variable for respondents	Weighted auxiliary variable for all.
y_1	x_1	$1/\pi_1$	$1/\psi_1$	d_1	d_1y_1	d_1x_1	x_1/π_1
y_2	x_2	$1/\pi_2$	$1/\psi_2$	d_2	d_2y_2	d_2x_2	x_2/π_2
.	.						
y_r	x_r	$1/\pi_r$	$1/\psi_r$	d_r	d_ry_r	d_rx_r	x_r/π_r
Missing	x_{r+1}	$1/\pi_{r+1}$	$1/\psi_{r+1}$	d_{r+1}	-	-	x_{r+1}/π_{r+1}
.	.	.			-	-	
Missing	x_n	$1/\pi_n$	$1/\psi_n$	d_n	-	-	x_n/π_n

where $d_i = 1/\left(\pi_i\psi_i\right)$.

When the auxiliary variable is available we can propose a calibrated estimator for (12.5.2.1) as

$$\hat{Y}_g = \sum_{i \in s_r} w_i y_i . \tag{12.5.2.2}$$

The calibrated weights w_i are such that the chi square distance function owed to Deville and Särndal (1992) defined as

$$\sum_{i \in s_r} \frac{(w_i - d_i)^2}{d_i q_i} \tag{12.5.2.3}$$

is minimum subject to the constraint

$$\sum_{i \in s_r} w_i x_i = \sum_{i \in s} \frac{x_i}{\pi_i} = \hat{x} \text{ (say)}, \tag{12.5.2.4}$$

where the q_i are suitably chosen weights. The calibration equation (12.5.2.4) is similar as used by Dupont (1995) and Hidiroglou and Särndal (1995, 1998) for two-phase sampling. The choice of q_i leads to different forms of estimators of the population total. Minimization of (12.5.2.3) subject to (12.5.2.4) leads to the calibrated weights

$$w_i = d_i + \frac{d_i q_i x_i}{\sum_{i \in s_r} d_i q_i x_i^2}\left(\hat{x} - \sum_{i \in s_r} d_i x_i\right). \tag{12.5.2.5}$$

On substituting (12.5.2.5) in (12.5.2.2) we obtain a general linear regression estimator (GREG) in the presence on non-response given by

$$\hat{Y}_g = \sum_{i \in s_r} w_i y_i = \sum_{i \in s_r} d_i e_i + b\hat{x}, \tag{12.5.2.6}$$

where $b = \sum_{i \in s_r} d_i q_i x_i y_i \Big/ \sum_{i \in s_r} d_i q_i x_i^2$ and $e_i = y_i - bx_i$. Thus an estimator of $V(\hat{Y}_g)$ is given by

$$\hat{v}(\hat{Y}_g)_{\text{stg}=0} = \sum_{i \in s_r} b_{iis}(r)\frac{e_i^2}{\pi_i^2} + \frac{1}{2}\sum_{i \neq} \sum_{j \in s_r} b_{ijs}(r)\left(\frac{e_i}{\pi_i} - \frac{e_j}{\pi_j}\right)^2 . \tag{12.5.2.7}$$

12.5.3 CALIBRATION OF THE ESTIMATORS OF VARIANCE

If $1/\pi_i = w_i \psi_i$ then we consider first stage calibrated estimator of variance in the presence of non-response as

$$\hat{v}(\hat{Y}_g)_{\text{stg}=1} = \sum_{i \in s_r} b_{iis}(r) w_i^2 \psi_i^2 e_i^2 + \frac{1}{2}\sum_{i \neq} \sum_{j \in s_r} b_{ijs}(r)(w_i \psi_i e_i - w_j \psi_j e_j)^2 . \tag{12.5.3.1}$$

We consider a second stage calibrated estimator of variance of GREG as

$$\hat{v}(\hat{Y}_g)_{\text{stg}=2} = \sum_{i \in s_r} w_{iis}(r) w_i^2 \psi_i^2 e_i^2 + \frac{1}{2}\sum_{i \neq} \sum_{j \in s_r} w_{ijs}(r)(w_i \psi_i e_i - w_j \psi_j e_j)^2 \tag{12.5.3.2}$$

The second stage weights $w_{ijs}(r)$ are obtained such that the distance function

$$\sum_{i\in s_r}\frac{\left(w_{iis}(r)-b_{iis}(r)\right)^2}{b_{iis}(r)Q_{iisr}}+\frac{1}{2}\sum_{i\neq j\in s_r}\sum\frac{\left(w_{ijs}(r)-b_{ijs}(r)\right)^2}{b_{ijs}(r)Q_{ijsr}} \tag{12.5.3.3}$$

is minimum subject to

$$\sum_{i\in s_r} w_{iis}(r)\frac{x_i^2}{\pi_i^2}+\frac{1}{2}\sum_{i\neq j\in s_r}\sum w_{ijs}(r)\left(\frac{x_i}{\pi_i}-\frac{x_j}{\pi_j}\right)^2=\hat{v}_2(x), \tag{12.5.3.4}$$

where

$$\hat{v}_2(x)=\sum_{i\in s}\frac{a_{ir}}{\pi_i^3}x_i^2+\frac{1}{2}\sum_{i\neq j\in s}\sum\frac{\delta_{ijr}+\Theta_{ij}}{\pi_{ij}}\left(\frac{x_i}{\pi_i}-\frac{x_j}{\pi_j}\right)^2. \tag{12.5.3.5}$$

Now we have the following theorem.

Theorem 12.5.3.1. The second stage calibrated weights obtained by minimizing (12.5.3.3) subject to (12.5.3.4) are

$$w_{iis}(r)=b_{iis}(r)+\frac{\hat{v}_2(x)-\hat{v}_1(x)}{R(x)}b_{iis}(r)x_i^2Q_{iisr}, \tag{12.5.3.6}$$

and

$$w_{ijs}(r)=b_{ijs}(r)+\frac{\hat{v}_2(x)-\hat{v}_1(x)}{R(x)}b_{ijs}(r)Q_{ijsr}\left(\frac{x_i}{\pi_i}-\frac{x_j}{\pi_j}\right)^2 \tag{12.5.3.7}$$

with

$$\hat{v}_1(x)=\sum_{i\in s_r} b_{iis}(r)\frac{x_i^2}{\pi_i^2}+\frac{1}{2}\sum_{i\neq j\in s_r}\sum b_{ijs}(r)\left(\frac{x_i}{\pi_i}-\frac{x_j}{\pi_j}\right)^2$$

and

$$R(x)=\sum_{i\in s_r} b_{ijs}(r)Q_{ijs}(r)\frac{x_i^4}{\pi_i^4}+\frac{1}{2}\sum_{i\neq j\in s_r}\sum b_{ijs}(r)Q_{ijs}(r)\left(\frac{x_i}{\pi_i}-\frac{x_j}{\pi_j}\right)^4.$$

Proof. Let

$$\Phi=\sum_{i\in s_r}\frac{\left(w_{iis}(r)-b_{iis}(r)\right)^2}{b_{iis}(r)Q_{iisr}}+\frac{1}{2}\sum_{i\neq j\in s_r}\sum\frac{\left(w_{ijs}(r)-b_{ijs}(r)\right)^2}{b_{ijs}(r)Q_{ijsr}}$$
$$-2\lambda\left[\sum_{i\in s_r} w_{iis}(r)\frac{x_i^2}{\pi_i^2}+\frac{1}{2}\sum_{i\neq j\in s_r}\sum w_{ijs}(r)\left(\frac{x_i}{\pi_i}-\frac{x_j}{\pi_j}\right)^2-\hat{v}_2(x)\right]. \tag{12.5.3.8}$$

Now $\dfrac{\partial\Phi}{\partial w_{iis}(r)}=0$ implies that

$$w_{iis}(r)=b_{iis}(r)+\lambda b_{iis}(r)\frac{x_i^2}{\pi_i^2}Q_{iis}(r),$$

which implies that

$$\sum_{i\in s_r} w_{iis}(r)\frac{x_i^2}{\pi_i^2} = \lambda \sum_{i\in s_r} b_{iis}(r)\frac{x_i^4}{\pi_i^4}Q_{iisr} + \sum_{i\in s_r} b_{iis}(r)\frac{x_i^2}{\pi_i^2}. \tag{12.5.3.9}$$

Again $\dfrac{\partial\Phi}{\partial w_{ijs}(r)} = 0$ implies that

$$w_{ijs}(r) = b_{ijs}(r) + \lambda b_{ijs}(r)Q_{ijs}(r)\left(\frac{x_i}{\pi_i} - \frac{x_j}{\pi_j}\right)^2, \tag{12.5.3.10}$$

which implies that

$$\frac{1}{2}\sum_{i\neq j\in s_r}\sum w_{ijs}(r)\left(\frac{x_i}{\pi_i} - \frac{x_j}{\pi_j}\right)^2 = \frac{\lambda}{2}\sum_{i\neq j\in s_r}\sum b_{ijs}(r)Q_{ijs}(r)\left(\frac{x_i}{\pi_i} - \frac{x_j}{\pi_j}\right)^4 + \frac{1}{2}\sum_{i\neq j\in s_r}\sum b_{ijs}(r)\left(\frac{x_i}{\pi_i} - \frac{x_j}{\pi_j}\right)^2.$$

Adding (12.5.3.9) and (12.5.3.10) we have $\lambda = \dfrac{\hat{v}_2(x) - \hat{v}_1(x)}{R(x)}$, which proves the theorem.

Now we discuss the special cases of the above strategy as follows:

12.5.3.1 PPSWOR SAMPLING

If the sub-sample s_r is selected by SRSWOR sampling, that is, $\psi_{irs} = r/n$ and $\psi_{ijrs} = \{r(r-1)\}/\{n(n-1)\}$, then $\delta_{ijr} = \{(n-r)/r(n-1)\}\pi_{ij}$, $\delta_{ir} = \{(n-r)/r\}\pi_i$, $a_{ir} = 0$, $b_{iis}(r) = 0$, $b_{ijs}(r) = \dfrac{n(n-1)}{r(r-1)}\left\{\dfrac{n-r}{r(n-1)} + \dfrac{\Theta_{ij}}{\pi_{ij}}\right\} = \Delta_{ij}$ (say), $\hat{v}_1(x) = \dfrac{1}{2}\sum_{i\neq j\in s_r}\sum \Delta_{ij}\left(\dfrac{x_i}{\pi_i} - \dfrac{x_j}{\pi_j}\right)^2$,

$\hat{v}_2(x) = \dfrac{r(r-1)}{2n(n-1)}\sum_{i\neq j\in s}\sum \Delta_{ij}\left(\dfrac{x_i}{\pi_i} - \dfrac{x_j}{\pi_j}\right)^2$, and $R(x) = \dfrac{1}{2}\sum_{i\neq j\in s_r}\sum \Delta_{ij}Q_{ijs}(r)\left(\dfrac{x_i}{\pi_i} - \dfrac{x_j}{\pi_j}\right)^4$.

Case I. If $w_i = d_i$, then $e_i = y_i$ and

$$\hat{Y}_g = \hat{Y}_{\text{HT}} = \frac{n}{r}\sum_{i\in s_r}\frac{y_i}{\pi_i}.$$

Then three estimators of variance are

$$\hat{v}\left(\hat{Y}_{\text{HT}}\right)_{\text{stg}=0} = \frac{1}{2}\sum_{i\neq j\in s_r}\sum \Delta_{ij}\left(\frac{y_i}{\pi_i} - \frac{y_j}{\pi_j}\right)^2, \quad \hat{v}\left(\hat{Y}_{\text{HT}}\right)_{\text{stg}=1} = \hat{v}\left(\hat{Y}_{\text{HT}}\right)_{\text{stg}=0}, \text{ and}$$

$$\hat{v}\left(\hat{Y}_{\text{HT}}\right)_{\text{stg}=2} = \frac{1}{2}\sum_{i\neq j\in s_r}\sum w_{ijs}(r)\left(\frac{y_i}{\pi_i} - \frac{y_j}{\pi_j}\right)^2$$

where $w_{ijs}(r) = \Delta_{ij}\left\{1 + \dfrac{\hat{v}_2(x) - \hat{v}_1(x)}{R(x)}Q_{ijs}(r)\left(\dfrac{x_i}{\pi_i} - \dfrac{x_j}{\pi_j}\right)^2\right\}$.

Case II. If $q_i = x_i^{-1}$ and $Q_{ij} = \left(\frac{x_i}{\pi_i} - \frac{x_j}{\pi_j}\right)^{-2}$ we have

$$w_i = \frac{1}{\pi_i}\left\{\sum_{i\in s_r}\frac{x_i}{\pi_i}\right\}^{-1}\left\{\sum_{i\in s}\frac{x_i}{\pi_i}\right\}, \quad e_i = y_i - \sum_{i\in s_r}\frac{y_i}{\pi_i}\left\{\sum_{i\in s}\frac{x_i}{\pi_i}\right\}^{-1} x_i, \text{ and}$$

$$w_{ijs}(r) = \Delta_{ij}\frac{r(r-1)}{n(n-1)}\sum_{i\neq}\sum_{j\in s}\Delta_{ij}\left(\frac{x_i}{\pi_i} - \frac{x_j}{\pi_j}\right)^2\left\{\sum_{i\neq}\sum_{j\in s_r}\Delta_{ij}\left(\frac{x_i}{\pi_i} - \frac{x_j}{\pi_j}\right)^2\right\}^{-1}.$$

In this case $\hat{Y}_g$ becomes an extension of the Horvitz and Thompson (1952) estimator

$$\hat{Y}_{\text{HTR}} = \left(\frac{n}{r}\sum_{i\in s_r}\frac{y_i}{\pi_i}\right)\left(\frac{n}{r}\sum_{i\in s_r}\frac{x_i}{\pi_i}\right)^{-1}\left(\sum_{i\in s}\frac{x_i}{\pi_i}\right).$$

The three estimators of variance of the ratio estimator $\hat{Y}_{\text{HTR}}$ are given by

$$\hat{v}\left(\hat{Y}_{\text{HTR}}\right)_{\text{stg}=0} = \frac{1}{2}\sum_{i\neq}\sum_{j\in s_r}\Delta_{ij}\left(\frac{e_i}{\pi_i} - \frac{e_j}{\pi_j}\right)^2,$$

$$\hat{v}\left(\hat{Y}_{\text{HTR}}\right)_{\text{stg}=1} = \frac{1}{2}\sum_{i\neq}\sum_{j\in s_r}\Delta_{ij}\left(\frac{e_i}{\pi_i} - \frac{e_j}{\pi_j}\right)^2\left\{\left(\sum_{i\in s}\frac{x_i}{\pi_i}\right)\Big/\left(\frac{n}{r}\sum_{i\in s_r}\frac{x_i}{\pi_i}\right)\right\}^2,$$

and $$\hat{v}\left(\hat{Y}_{\text{HTR}}\right)_{\text{stg}=2} = \frac{1}{2}\sum_{i\neq}\sum_{j\in s_r} w_{ijs}(r)\left(\frac{e_i}{\pi_i} - \frac{e_j}{\pi_j}\right)^2\left\{\left(\sum_{i\in s}\frac{x_i}{\pi_i}\right)\Big/\left(\frac{n}{r}\sum_{i\in s_r}\frac{x_i}{\pi_i}\right)\right\}^2.$$

Case III. If $q_i = Q_{ij} = 1$ then $w_i = \frac{n}{r}\frac{1}{\pi_i} + \frac{x_i}{\pi_i}\left\{\sum_{i\in s_r}\frac{x_i^2}{\pi_i}\right\}^{-1}\left(\sum_{i\in s}\frac{x_i}{\pi_i} - \frac{n}{r}\sum_{i\in s_r}\frac{x_i}{\pi_i}\right)$,

$b = \sum_{i\in s_r}\frac{x_i y_i}{\pi_i}\left\{\sum_{i\in s_r}\frac{x_i^2}{\pi_i^2}\right\}^{-1}$, and $e_i = y_i - bx_i$. In this case $\hat{Y}_g$ becomes the regression estimator of the following form

$$\hat{Y}_{\text{HT(lr)}} = \frac{n}{r}\sum_{i\in s_r}\frac{y_i}{\pi_i} + b\left(\sum_{i\in s}\frac{x_i}{\pi_i} - \frac{n}{r}\sum_{i\in s_r}\frac{x_i}{\pi_i}\right).$$

The three estimators of variance of the regression estimator are given by

$$\hat{v}\left(\hat{Y}_{\text{HT(lr)}}\right)_{\text{stg}=0} = \frac{1}{2}\sum_{i\neq}\sum_{j\in s_r}\Delta_{ij}\left(\frac{e_i}{\pi_i} - \frac{e_j}{\pi_j}\right)^2,$$

$$\hat{v}\left(\hat{Y}_{\text{HTR(lr)}}\right)_{\text{stg}=1} = \frac{1}{2}\left(\frac{r}{n}\right)^2\sum_{i\neq}\sum_{j\in s_r}\Delta_{ij}\left(w_i e_i - w_j e_j\right)^2$$

and

$$\hat{v}\left(\hat{Y}_{\text{HTR(lr)}}\right)_{\text{stg}=2} = \frac{1}{2}\left(\frac{r}{n}\right)^2 \sum_{i \neq j \in s_r}\sum w_{ijs}(r)\left(w_i e_i - w_j e_j\right)^2$$

where $w_{ijs}(r) = \left\{1 + \left\{\sum_{i \neq j \in s_r}\sum \Delta_{ij}\left(\frac{x_i}{\pi_i} - \frac{x_j}{\pi_j}\right)^4\right\}^{-1}\left(\frac{x_i}{\pi_i} - \frac{x_j}{\pi_j}\right)^2 \left\{\hat{v}_2(x) - \hat{v}_1(x)\right\}\right\}\Delta_{ij}$.

12.5.3.2 SRSWOR SAMPLING

When both the samples s and s_r are selected by the SRSWOR method, that is, $\pi_i = \frac{n}{N}$, $\pi_{ij} = \frac{n(n-1)}{N(N-1)}$, $\psi_{irs} = \frac{r}{n}$ and $\psi_{ijrs} = \frac{r(r-1)}{n(n-1)}$. Then $b_{iis}(r) = 0$, $b_{ijs}(r) = \Delta = \frac{n^2}{r(r-1)}\left(\frac{1}{r} - \frac{1}{N}\right)$, $\hat{v}_1(x) = N^2\left(\frac{1}{r} - \frac{1}{N}\right)s_{xr}^2$, $\hat{v}_2(x) = N^2\left(\frac{1}{r} - \frac{1}{N}\right)s_{xn}^2$, where $(r-1)s_{xr}^2 = \sum_{i \in s_r}(x_i - \bar{x}_r)^2$, $(n-1)s_{xn}^2 = \sum_{i \in s}(x_i - \bar{x}_n)^2$, $R(x) = \frac{N^4}{2n^4}\Delta\sum_{i \neq j \in s_r}\sum Q_{ijs}(r)\left(x_i - x_j\right)^4$, $\bar{x}_r = \frac{1}{r}\sum_{i \in s_r} x_i$, $\bar{x}_n = \frac{1}{n}\sum_{i \in s} x_i$ and $w_{ijs}(r) = \Delta\left\{1 + N^2\left(\frac{1}{r} - \frac{1}{N}\right)\frac{\left(s_{xn}^2 - s_{xr}^2\right)}{R(x)} Q_{ijs}(r)\left(x_i - x_j\right)^2 \frac{N^2}{n^2}\right\}$.

Case I. If $w_i = d_i$ then

$$\hat{Y}_g = \frac{N}{r}\sum_{i \in s_r} y_i = \hat{Y}_u$$

and

$$\hat{v}\left(\hat{Y}_u\right)_{\text{stg}=0} = N^2\left(\frac{1}{r} - \frac{1}{N}\right)s_{yr}^2 = \hat{v}\left(\hat{Y}_u\right)_{\text{stg}=1}.$$

Now

$$\hat{v}\left(\hat{Y}_u\right)_{\text{stg}=2} = \frac{N^2}{2n^2}\sum_{i \neq j \in s_r}\sum w_{ijs}(r)\left(y_i - y_j\right)^2.$$

If $Q_{ijs}(r) = \left(\frac{x_i}{\pi_i} - \frac{x_j}{\pi_j}\right)^{-2}$ then we have

$$w_{ijs}(r) = \frac{n^2}{r(r-1)}\left(\frac{1}{r} - \frac{1}{N}\right)\frac{s_{xn}^2}{s_{xr}^2}$$

and hence

$$\hat{v}\left(\hat{Y}_u\right)_{\text{stg}=2} = N^2\left(\frac{1}{r} - \frac{1}{N}\right)s_{yr}^2 \frac{s_{xn}^2}{s_{xr}^2},$$

which is a similar ratio type of estimator studied by Isaki (1983), Garcia and Cebrian (1996), and Singh and Joarder (1998).

In the case $Q_{ij} = 1$ we ten have

$$\hat{v}\left(\hat{Y}_u\right)_{\text{stg}=2} = N^2\left(\frac{1}{r} - \frac{1}{N}\right)\left[s_{yr}^2 + \hat{B}\left(s_{xn}^2 - s_{xr}^2\right)\right],$$

where $\hat{B} = \sum_{i \neq j \in s_r}\sum (x_i - x_j)^2 (y_i - y_j)^2 \Big/ \sum_{i \neq j \in s_r}\sum (x_i - x_j)^4$, which is again similar regression type estimator studied by Isaki (1983), Garcia and Cebrian (1996), and Singh and Joarder (1998).

Case II. If we take $q_i = x_i^{-1}$ then $w_i = \dfrac{N}{r}\dfrac{\bar{x}_n}{\bar{x}_r}$ and

$$\hat{Y}_g = \hat{Y}_{\text{R}} = N\,\bar{y}_r \frac{\bar{x}_n}{\bar{x}_r},$$

which is a ratio estimator of population total. In this case $e_i = y_i - (\bar{y}_r/\bar{x}_r)x_i$ and we get three estimators of variance as

$$\hat{v}\left(\hat{Y}_{\text{R}}\right)_{\text{stg}=0} = N^2\left(\frac{1}{r} - \frac{1}{N}\right)\frac{1}{r-1}\sum_{i \in s_r} e_i^2, \qquad \hat{v}\left(\hat{Y}_{\text{R}}\right)_{\text{stg}=1} = N^2\left(\frac{1}{r} - \frac{1}{N}\right)\frac{1}{r-1}\sum_{i \in s_r} e_i^2\left(\frac{\bar{x}_n}{\bar{x}_r}\right)^2$$ and

$$\hat{v}\left(\hat{Y}_{\text{R}}\right)_{\text{stg}=2} = \left(\frac{\bar{x}_n}{\bar{x}_r}\right)^2\left(\frac{N}{n}\right)^2 \sum_{i \neq j \in s_r}\sum w_{ijs}(r)(e_i - e_j)^2 .$$

Further, if we assume $Q_{ijs}(r) = \dfrac{n^2}{N^2}(x_i - x_j)^{-2}$ then we have

$$\hat{v}\left(\hat{Y}_{\text{R}}\right)_{\text{stg}=2} = N^2\left(\frac{1}{r} - \frac{1}{N}\right)\frac{1}{r-1}\sum_{i \in s_r} e_i^2\left(\frac{\bar{x}_n}{\bar{x}_r}\right)^2 \frac{s_{xn}^2}{s_{xr}^2}.$$

Case III. If we take $q_i = 1$ then $w_i = \dfrac{N}{r} + \left(Nx_i \Big/ \sum_{i \in s_r} x_i^2\right)(\bar{x}_n - \bar{x}_r)$, $b = \sum_{i \in s_r} x_i y_i \Big/ \sum_{i \in s_r} x_i^2$,

$e_i = y_i - bx_i$ and

$$\hat{Y}_g = \hat{Y}_{\text{lr}} = N\left[\bar{y}_r + b(\bar{x}_n - \bar{x}_r)\right]$$

is a regression type estimator for the finite population total studied by Singh, Joarder, and Tracy (2000). In this situation a few estimators of variance are

$$\hat{v}\left(\hat{Y}_{\text{lr}}\right)_{\text{stg}=0} = N^2\left(\frac{1}{r} - \frac{1}{N}\right)\sum_{i \in s_r}(e_i - \bar{e})^2,$$

where $\bar{e} = \dfrac{1}{r}\sum_{i \in s_r} e_i$.

$$\hat{v}\left(\hat{Y}_{\text{lr}}\right)_{\text{stg}=1} = N^2\left(\frac{1}{r} - \frac{1}{N}\right)\left[\frac{1}{(r-1)}\sum_{i \in s_r}(e_i - \bar{e})^2 + \hat{\gamma}_1(\bar{x}_n - \bar{x}_r)^2 + \hat{\gamma}_2(\bar{x}_n - \bar{x}_r)\right],$$

where

$$\hat{\gamma}_1 = \frac{r}{2(r-1)} \frac{\sum\sum_{i \neq j \in s_r} (x_i e_i - x_j e_j)^2}{\left(\sum_{i \in s} x_i^2\right)^2}, \text{ and } \hat{\gamma}_2 = \frac{\sum\sum_{i \neq j \in s_r} (e_i - e_j)(x_i e_i - x_j e_j)}{(r-1) \sum_{i \in s_r} x_i^2}.$$

If $Q_{ijs}(r) = \left(\frac{x_i}{\pi_i} - \frac{x_j}{\pi_j}\right)^{-2} = \frac{n^2}{N^2}(x_i - x_j)^2$, then

$$w_{ijs}(r) = \Delta \frac{s_{xn}^2}{s_{xr}^2}, \text{ and } \hat{v}(\hat{\bar{Y}}_{\text{lr}})_{\text{stg}=2} = \hat{v}(\hat{\bar{Y}}_{\text{lr}})_{\text{stg}=1} \frac{s_{xn}^2}{s_{xr}^2}.$$

If $Q_{ijs}(r) = 1$ then

$$w_{ijs}(r) = \Delta\left\{1 + \frac{r(r-1)(s_{xn}^2 - s_{xr}^2)(x_i - x_j)^2}{\frac{1}{2}\sum\sum_{i \neq j \in s_r} (x_i - x_j)^4}\right\}, \quad k = \frac{(\bar{x}_n - \bar{x}_r)}{\sum_{i \in s_r} x_i^2}, \text{ and}$$

then

$$\hat{v}(\hat{\bar{Y}}_{\text{lr}})_{\text{stg}=2} = \hat{v}(\hat{\bar{Y}}_{\text{lr}})_{\text{stg}=1} + \frac{r^2}{n^2} \frac{r(r-1)\Delta(s_{xn}^2 - s_{xr}^2)\sum\sum_{i \neq j \in s_r} (x_i - x_j)^2 \left\{\frac{1}{r}(e_i - e_j) + k(e_i x_i - e_j x_j)\right\}^2}{\sum\sum_{i \neq j \in s_r} (x_i - x_j)^4}.$$

12.6 DIFFERENT TREATMENTS OF NON-RESPONSE

Consider $\bar{Y} = N^{-1} \sum_{i=1}^{N} y_i$ is the mean of the finite population $\Omega = \{1, 2, .., i, .., N\}$. An SRSWOR sample s of size n is drawn from Ω to estimate $\bar{Y}$. Let r be the number of responding units out of the sampled n units. Let the set of responding units be denoted by A and that of non-responding units by $\bar{A}$ such that $s = A \cup \bar{A}$. For every unit $i \in A$ the value y_i is observed. However, for the units $i \in \bar{A}$ the y_i values are missing and imputed values are derived. If $\hat{y}_i$ denotes the imputed value for the ith non-responding unit, then the general method of imputation can be expressed as:

$$y_{\bullet i} = \begin{cases} y_i & \text{if } i \in A, \\ \hat{y}_i & \text{if } i \in \bar{A}. \end{cases} \tag{12.6.1}$$

The general point estimator of population mean takes the form

$$\bar{y}_s = \frac{1}{n}\left[\sum_{i=1}^{r} y_i + \sum_{i=1}^{n-r} \hat{y}_i\right], \tag{12.6.2}$$

where the value of $\hat{y}_i$ is different for each imputation method.

12.6.1 RATIO METHOD OF IMPUTATION

In ratio method of imputation, we assume that imputation is carried out with the aid of an auxiliary variable x, such that x_i the value of x for unit i, is known and positive for every $i \in s$. In other words, the data $x_s = \{x_i : i \in s\}$ are known. Following the notation of Lee, Rancourt, and Särndal (1994, 1995), in the case of single value imputation, if the ith unit requires imputation, the value $\hat{b}\, x_i$ is imputed, where $\hat{b} = \sum_{i=1}^{r} y_i \Big/ \sum_{i=1}^{r} x_i$. The data after imputation becomes

$$y_{\bullet i} = \begin{cases} y_i & \text{if } i \in A, \\ \hat{b} x_i & \text{if } i \in \bar{A}. \end{cases} \tag{12.6.3}$$

This method of imputation is called the ratio method of imputation. Thus we have the following theorem:

Theorem 12.6.1. Under ratio method of imputation the point estimator (12.6.2) of the population mean becomes

$$\bar{y}_{\text{RAT}} = \bar{y}_r \left(\frac{\bar{x}_n}{\bar{x}_r} \right), \tag{12.6.4}$$

where $\bar{x}_n = n^{-1} \sum_{i=1}^{n} x_i$, $\bar{x}_r = r^{-1} \sum_{i=1}^{r} x_i$ and $\bar{y}_r = r^{-1} \sum_{i=1}^{r} y_i$.

12.6.2 MEAN METHOD OF IMPUTATION

Under mean method of imputation the data after imputation becomes

$$y_{\bullet i} = \begin{cases} y_i & \text{if } i \in A, \\ \bar{y}_r & \text{if } i \in \bar{A}. \end{cases} \tag{12.6.5}$$

Thus we have the following theorem:

Theorem 12.6.2. Under mean method of imputation the point estimator (12.6.2) of the population mean is

$$\bar{y}_m = \frac{1}{r} \sum_{i=1}^{r} y_i = \bar{y}_r . \tag{12.6.6}$$

12.6.3 HOT DECK (HD) METHOD OF IMPUTATION

Under the HD method the data after imputation takes the form

$$y_{\bullet i} = \begin{cases} y_i & \text{if } i \in A, \\ y_{g(i)} & \text{if } i \in \bar{A}, \end{cases} \tag{12.6.7}$$

where $y_{g(i)}$ is the y value given by the donor unit $g(i) \in R$, drawn at random (with replacement) from the r responding units. Thus we have the following theorem:

Theorem 12.6.3. Under the HD method of imputation the point estimator (12.6.2) of the population mean becomes

$$\bar{y}_{HD} = \frac{1}{n}\left[\sum_{i=1}^{r} y_i + \sum_{i=1}^{n-r} y_{g(i)}\right]. \tag{12.6.8}$$

12.6.4 NEAREST NEIGHBOUR (NN) METHOD OF IMPUTATION

Under the NN method the data after imputation becomes

$$y_{\bullet i} = \begin{cases} y_i & \text{if } i \in A, \\ y_{g(i)} & \text{if } i \in \bar{A}, \end{cases} \tag{12.6.9}$$

where $y_{g(i)}$ is the y value given by the donor unit $g(i)$ such that $\underset{g \in R_p}{\text{Min.}} |x_g - x_i|$ occurs for $g = g(i)$. If it results in more than one unit, a donor is randomly selected from them. More detail can be had from Chen and Shao (2001).

Thus we have the following theorem:

Theorem 12.6.4. Under the NN method of imputation the point estimator (12.6.2) of the population mean becomes

$$\bar{y}_{NN} = \frac{1}{n}\left[\sum_{i=1}^{r} y_i + \sum_{i=1}^{n-r} y_{g(i)}\right]. \tag{12.6.10}$$

Example 12.6.1. A bank committee selected an SRSWOR sample of twenty states from the population 1 given in the Appendix and collected information about the real and nonreal estate farm loans. Unfortunately the information on the real estate farm loans was not available on six states as marked in the table below.

Sr. No.	State	x_i	y_i	Sr. No.	State	x_i	y_i
02	AK	3.433	2.605	27	NE	3585.406	1337.852
07	CT	4.373	Missing	28	NV	16.710	Missing
08	DE	43.229	42.808	30	NJ	27.508	Missing
11	HI	38.067	40.775	31	NM	274.035	140.582
18	LA	405.799	282.565	33	NC	494.730	639.571
19	ME	51.539	Missing	37	OR	571.487	114.899
20	MD	57.684	139.628	42	TN	388.869	553.266
22	MI	440.518	323.028	47	WA	1228.607	Missing
24	MS	549.551	627.013	48	WV	29.291	99.277
25	MO	1519.994	Missing	49	WI	1372.439	1229.752

x – Nonreal estate farm loans, y – Real estate farm loans.

There is 30% non-response about the real estate farm loans. Impute the missing values with different methods of imputation.

Solution. (a) Ratio method of imputation: We have

$$\hat{b} = \frac{\sum_{i=1}^{r} y_i}{\sum_{i=1}^{r} x_i} = \frac{5573.621}{8254.538} = 0.6752$$

thus the imputed values are given by

State	x_i	$\hat{y}_i = \hat{b}x_i$
CT	4.373	2.95265
ME	51.539	34.79913
MO	1519.994	1026.30000
NV	16.710	11.28259
NJ	27.508	18.57340
WA	1228.607	829.55540

(b) Mean method of imputation: Here we assume that no auxiliary information is available. The missing values are replaced with the mean of the responding units on the study variable only.

State	Imputed values
CT	398.11
ME	398.11
MO	398.11
NV	398.11
NJ	398.11
WA	398.11

(c) Hot deck method of imputation: Here again no auxiliary information is used for imputation. We have to select 6 values from the responding units by using SRSWOR sampling. There are total 14 responses available on the study variable and 6 response are missing. We used first two columns of the Pseudo-Random Number (PRN) Table 1 given in the Appendix to select 6 distinct random numbers between 1 and 14. The distinct random numbers are 01, 04, 05, 03, 14 and 06. Thus the imputed values are given as

State	Imputed values
CT	2.605
ME	282.565
MO	139.628
NV	40.775
NJ	1229.752
WA	323.028

(d) Nearest Neighbourhood method: Here we use the minimum absolute difference between the values of the nonreal estate farm loans of two different states to impute the missing real estate farm loan of the other state.

State	x_i	Imputed values
CT	4.373	2.605
ME	51.539	139.628
MO	1519.994	1229.752
NV	16.710	39.860
NJ	27.508	99.277
WA	1228.607	1229.752

(e) Regression method of imputation: Here we use the model $y_i = a + bx_i + e_i$. The values of a and b are estimated by the method of least squares. In this case the fitted line on the responding units is given by

$$\hat{y}_i = 76.89265 + 0.071471 x_i$$

and the imputed values are given in the following table

State	x_i	Imputed values
CT	4.373	77.2051
ME	51.539	80.5761
MO	1519.994	185.5260
NV	16.710	78.0869
NJ	27.508	78.8586
WA	1228.607	164.7010

12.7 SUPERPOPULATION MODEL APPROACH

The imputation for missing data appeals to one or more model assumptions. The imputed values are assumed, on the average, to be good substitutes for the missing values. The difference between the true value, the unobserved value y_i, and its imputed value $\hat{y}_i$ is assumed to be zero. These assumptions are useful to find the estimator of the variance under different imputation mechanisms. Thus we have three different model components here:

(a) the response mechanism;

(b) the regression relationship;

and

(c) the variance structure for the regression.

Assuming that the response mechanism is uniform, we have considered the following model

$$M : y_i = \beta x_i + \epsilon_i , \tag{12.7.1}$$

where $E_M(\epsilon_i) = 0$, $E(\epsilon_i^2) = \sigma^2 x_i$, and $E(\epsilon_i \epsilon_j) = 0,\ i \neq j$.

12.7.1 DIFFERENT COMPONENTS OF VARIANCE

The variance of the estimator of population total based on the data obtained after imputation consists of three components. An estimator of population total Y based on the imputed data is given by

$$\hat{Y}_{\bullet s} = \frac{N}{n}\sum_{i=1}^{n} y_{\bullet i} = \frac{N}{n}\left[\sum_{i=1}^{r} y_i + \sum_{i=1}^{n-r}\hat{y}_i\right] \tag{12.7.2}$$

and that based on full information is given by

$$\hat{Y}_s = \frac{N}{n}\sum_{i=1}^{n} y_i . \tag{12.7.3}$$

Now we have

$$\hat{Y}_{\bullet s} - Y = \hat{Y}_s - Y + \hat{Y}_{\bullet s} - \hat{Y}_s . \tag{12.7.4}$$

Squaring both sides of (12.7.4) and taking expected values we have

$$E\left[\hat{Y}_{\bullet s} - Y\right]^2 = E_p E_q\left[\hat{Y}_s - Y\right]^2 + E_p E_q\left[\hat{Y}_{\bullet s} - \hat{Y}_s\right]^2 + 2E_p E_q\left[\hat{Y}_s - Y\right]\left[\hat{Y}_{\bullet s} - \hat{Y}_s\right], \tag{12.7.5}$$

where E_p and E_q are the expectation operators with respect to the sampling design p and the response mechanism q respectively. Thus we can say

$$V_{\text{TOT}}\left(\hat{Y}_{\bullet s}\right) = V_{\text{SAM}} + V_{\text{IMP}} + V_{\text{MIX}} = V_{\text{ORD}} + V_{\text{DIF}} + V_{\text{IMP}} + V_{\text{MIX}} \tag{12.7.6}$$

where

$$V_{\text{SAM}} = E_p\left\{\hat{Y}_s - Y\right\}^2,\quad V_{\text{IMP}} = E_p E_q\left\{\hat{Y}_{\bullet s} - \hat{Y}_s\right\}^2, \text{ and } V_{\text{MIX}} = E_p E_q\left\{\hat{Y}_{\bullet s} - \hat{Y}_s\right\}\left\{\hat{Y}_s - Y\right\}.$$

Thus an estimator of $V\left(\hat{Y}_{\bullet s}\right)$ can easily be obtained as

$$\hat{V}\left(\hat{Y}_{\bullet s}\right) = \hat{V}_{\text{ORD}} + \hat{V}_{\text{DIF}} + \hat{V}_{\text{IMP}} + \hat{V}_{\text{MIX}}, \tag{12.7.7}$$

where $\hat{V}_{\text{ORD}} = N^2\left(\frac{1-f}{n}\right)s_{y\bullet s}^2$ with $s_{y\bullet s}^2 = (n-1)^{-1}\sum_{i=1}^{n}\left\{y_{\bullet i} - n^{-1}\sum_{i=1}^{n} y_{\bullet i}\right\}^2$.

Note that an estimator of the sampling variance is also given by

$$\hat{V}_{\text{SAM}} = \hat{V}_{\text{ORD}} + \hat{V}_{\text{DIF}}, \tag{12.7.8}$$

the correction term $\hat{V}_{\text{DIF}}$ should be constructed such that

$$E_m\left\{\hat{V}_{\text{DIF}}\right\} = \frac{N^2(1-f)}{n}E_m\left[s_{ys}^2 - s_{y\bullet s}^2\right] . \tag{12.7.9}$$

We shall now present the three components $\hat{V}_{\text{DIF}}$, $\hat{V}_{\text{IMP}}$ and $\hat{V}_{\text{MIX}}$ for each of the four types of imputation methods. Let us define $s_d = \Omega \cap s$, $r_d = \Omega \cap R$ and $l_d = \Omega \cap R^c$. Then we have the following lemmas:

Lemma 12.7.1. For the ratio method of imputation, under model M the different components of the variance estimators are given by

$$\hat{V}_{\text{DIF}} = \frac{N^2(1-f)}{n^2}\left\{\sum_{R^c} x_i(d)\right\}\hat{\sigma}^2, \quad \hat{V}_{\text{IMP}} = \frac{N^2}{n^2}\left\{\sum_{R^c} x_i(d)\right\}\left\{\frac{\sum_{R^c} x_i(d)}{\sum_{R} x_i(d)} + 1\right\}\hat{\sigma}^2,$$

and

$$\hat{V}_{\text{MIX}} = \frac{N^2(1-f)}{n^2}\left\{\sum_{R^c} x_i(d)\right\}\left\{\frac{\sum_{R} x_i(d)}{\sum_{R} x_i} - 1\right\}\hat{\sigma}^2, \quad \text{with } \hat{\sigma}^2 = \sum_R e_i^2 \Big/ \sum_R x_i$$

and $e_i = y_i - \hat{b}_r x_i$.

Lemma 12.7.2. For mean method of imputation, $x_i = 1 \ \forall i$, then under model M, the different components of the variance estimators are given by

$$\hat{V}_{\text{DIF}} = \frac{N^2(1-f)}{n^2} l_d \hat{\sigma}^2,$$

$$\hat{V}_{\text{IMP}} = \frac{N^2}{n^2} l_d \left\{\frac{l_d}{m} + 1\right\}\hat{\sigma}^2,$$

and

$$\hat{V}_{\text{MIX}} = \frac{N^2(1-f)}{n^2} l_d \left\{\frac{l_d}{m} - 1\right\}\hat{\sigma}^2,$$

with $\hat{\sigma}^2 = \frac{m-1}{m} s_{yr}^2$ and $s_{yr}^2 = (m-1)^{-1} \sum_{R_p} (y_i - \bar{y}_r)^2$ which, in fact, is a special case of ratio method of imputation for $x_i = 1$.

Lemma 12.7.3. Under the NN method of imputation and the model M we have

$$\hat{V}_{\text{DIF}} = 0$$

$$\hat{V}_{\text{IMP}} = \frac{N^2}{n^2} l_d \left\{\left(\frac{l_d}{m} + 1\right)\frac{\sum_{R_p} x_i}{m} + \frac{\sum_{R_p^c} x_i(d)}{l_d}\right\}\hat{\sigma}^2$$

and

$$\hat{V}_{\text{MIX}} = \frac{N^2(1-f)}{n^2} l_d \left\{ \frac{\sum\limits_{R_p} x_i(d)}{m} - \frac{\sum\limits_{R_p^c} x_i(d)}{l_d} \right\} \hat{\sigma}^2,$$

with $\hat{\sigma}^2 = \sum\limits_{R_p} e_i^2 \Big/ \sum\limits_{R_p} x_i$ and $e_i = y_i - \hat{b}_r x_i$.

Lemma 12.7.4. Under Hot Deck method of imputation and the model M we have

$$\hat{V}_{\text{DIF}} = 0,$$

$$\hat{V}_{\text{IMP}} = \frac{N^2}{n^2} l_d \left(\frac{l_d}{m} + 2 \right) \hat{\sigma}^2,$$

and

$$\hat{V}_{\text{MIX}} = \frac{N^2(1-f)}{n^2} l_d \left\{ \frac{m_d}{m} - 1 \right\} \hat{\sigma}^2,$$

with $\hat{\sigma}^2 = \frac{m-1}{m} s_{yr}^2$ and $s_{yr}^2 = (m-1)^{-1} \sum\limits_{R} (y_i - \bar{y}_r)^2$ which in fact is a special case of NN method of imputation for $x_i = 1$.

12.8 JACKKNIFE TECHNIQUE

We know that the principle of the Jackknife technique is to re-calculate the estimator after deleting a unit from the sample. The variance between the re-computed estimators is used to obtain an estimator of the variance of the estimator calculated using the complete sample. After the deletion of the jth unit, the estimator for the whole population is given by

$$\hat{Y}_{\bullet s}^{(j)} = \frac{N}{n-1} \sum_{i \neq j \in s} y_{\bullet i}, \tag{12.8.1}$$

where the superscript (j) denotes that the jth unit was deleted. This is performed for all units $j \in s$. The modified Jackknife estimator of variance is

$$\hat{V} = \frac{n-1}{n} \sum_{j \in s} \left\{ \hat{Y}_{\bullet s}^{(j)} - \hat{Y}_{\bullet s} \right\}^2. \tag{12.8.2}$$

For data sets containing imputed values this estimator does not take the imputation into account. Rao and Shao (1992) proposed a Jackknife variance estimator that corrects the estimator by adjusting the imputed values when the deleted unit is in the response set. For some imputation methods the adjusted values are the re-imputed values based on the reduced response set after deletion of the jth unit. If the jth unit deleted is a non-respondent the imputed values are unchanged. The data set after adjusting the imputed values is

$$y_{\bullet i}^{(aj)} = \begin{cases} y_i & \text{if } i \in R^c, \\ \hat{y}_i + a_i^{(j)} & \text{if } i \in R, \ j \in R, \\ \hat{y}_i & \text{if } i \in R, \ j \in R^c, \end{cases} \tag{12.8.3}$$

where $y_{\bullet i}^{(aj)}$ is the adjusted imputed value and $a_i^{(j)}$ is called the adjustment. The Jackknife variance estimator is then

$$\hat{V}_J = \frac{n-1}{n} \sum_{j \in s} \left\{ \hat{Y}_{\bullet s}^{(aj)} - \hat{Y}_{\bullet s}^{(a)} \right\}^2, \tag{12.8.4}$$

where $\hat{Y}_{\bullet s}^{(aj)} = \frac{N}{n-1} \sum_{i \neq j \in s} y_{\bullet i}^{(aj)}$ and $\hat{Y}_{\bullet s}^{(a)} = \frac{N}{n} \sum_{i \in s} y_{\bullet i}^{(a)}$ are the estimators of the total obtained after dropping the jth unit and from all the sample information, respectively. In (12.8.3), the value of the adjustment factor $a_i^{(j)}$ changes from one method of imputation to the other. For example, for the Ratio and NN method of imputation $a_i^{(j)} = \left(\frac{\bar{y}_r^{(j)}}{\bar{x}_r^{(j)}} - \frac{\bar{y}_r}{\bar{x}_r} \right) x_i$, for the Mean and the HD method of imputation $a_i^{(j)} = \bar{y}_r^{(j)} - \bar{y}_r$. Estimation of variance with Jackknife method for data with imputed values has also been discussed by Lee, Rancourt, and Särndal (1995a, 1995b). Some advanced techniques to estimate the variance of an estimator of total in multi-stage designs is also available from Rao and Shao (1999), Shao, Chen, and Chen (1998), Rao (1996b), Chen, Rao, and Sitter (2000), and Lee and Kim (2002).

12.9 HOT DECK IMPUTATION FOR MULTI-STAGE DESIGNS

Rao and Shao (1992) considered the problem of missing data in a multi-stage survey design. They consider the situation in which the first stage units or clusters are selected with replacement, or so treated and in which independent sub-samples are taken within those clusters selected more than once and followed Krewski and Rao (1981) asymptotic set-up in studying the consistency of the adjusted Jackknife variance estimator. Assume n_h clusters are selected with probabilities p_{hi} and with replacement independently from the hth stratum. Let

$$\hat{Y}_h = \frac{1}{n_h} \sum_{i=1}^{n_h} \frac{\hat{Y}_{hi}}{p_{hi}} \tag{12.9.1}$$

be an unbiased estimator of stratum total Y_h with $\hat{Y}_{hi}$ is a linear unbiased estimator of the total Y_{hi} for a selected cluster based on sampling at second and subsequent stages. If s_n denotes the total sample of n units then if there is no non-response an unbiased estimator of population total Y is given by

$$\hat{Y} = \sum_{(hik) \in s_n} w_{hik} y_{hik} \tag{12.9.2}$$

where y_{hik} and w_{hik} $(k = 1, 2, .., n_{hi};\ i = 1, 2, ..., n_h;\ h = 1, 2, ..., L)$ denote the value of the variable under study and design weights, respectively. Let $\{y^*_{hik}\}$ be the imputed values for the non-respondents, s_m, using a hot deck single imputation class mechanism. Then imputed estimator of population total Y is given by

$$\hat{Y}_{\mathrm{I}} = \sum_{(hik)\in s_r} w_{hik} y_{hik} + \sum_{(hik)\in s_m} w_{hik} y^*_{hik}, \tag{12.9.3}$$

where s_r denotes the sample of respondents. Under SRS sampling if we select the imputed values y^*_{hik} as $y_{gjl}\, w_{gjl}/w_{hik}$ where $(gjl)\in s_r$ denotes the selected donor, then following Platek and Grey (1983) the estimator $\hat{Y}_{\mathrm{I}}$ becomes unbiased, otherwise it remains biased. It is more appropriate for quantitative data, but may not provide good estimates for qualitative data. Rao and Shao (1992) suggested a simple method to select the donors $(gjl)\in s_r$ with replacement with probabilities $w_{gil}\Big/\sum_{s_r} w_{gil}$ and use $y^*_{hik} = y_{gjl}$.

Then we have the following theorem:

Theorem 12.9.1. If E_* denotes the expected value with respect to HD imputation then

$$E_*\left(\hat{Y}_{\mathrm{I}}\right) = \frac{\hat{S}}{\hat{T}}\hat{U} \tag{12.9.4}$$

where $\hat{S} = \sum_{(hik)\in s_r} w_{hik} y_{hik}$, $\hat{T} = \sum_{(hik)\in s_r} w_{hik}$ and $\hat{U} = \sum_{(hik)\in s_n} w_{hik}$.

Proof. Taking expected values on both sides of

$$\hat{Y}_{\mathrm{I}} = \sum_{(hik)\in s_r} w_{hik} y_{hik} + \sum_{(hik)\in s_m} w_{hik} y^*_{hik}$$

we have

$$E_*\left(\hat{Y}_{\mathrm{I}}\right) = \sum_{(hik)\in s_r} w_{hik} y_{hik} + \sum_{(hik)\in s_m} w_{hik} E_*\left(y^*_{hik}\right)$$

$$= \sum_{(hik)\in s_r} w_{hik} y_{hik} + \left(\sum_{(hik)\in s_m} w_{hik}\right)\frac{\sum_{(hik)\in s_r} w_{hik} y_{hik}}{\sum_{(hik)\in s_r} w_{hik}}$$

$$= \sum_{(hik)\in s_r} w_{hik} y_{hik} + \left(\sum_{(hik)\in s_n} w_{hik} - \sum_{(hik)\in s_r} w_{hik}\right)\frac{\sum_{(hik)\in s_r} w_{hik} y_{hik}}{\sum_{(hik)\in s_r} w_{hik}}$$

$$= \left(\sum_{(hik)\in s_n} w_{hik}\right)\frac{\sum_{(hik)\in s_r} w_{hik} y_{hik}}{\sum_{(hik)\in s_r} w_{hik}}.$$

Hence the theorem.

Theorem 12.9.2. Under uniform response mechanism prove that

$$E\left(\hat{S}\right)=pY\ ,\quad E\left(\hat{T}\right)=pN,\ \text{ and }\ E\left(\hat{U}\right)=N\ .$$

Proof. Define a random variable

$$t_{hik}=\begin{cases}1 & \text{if } (hik)\in s_r,\\ 0 & \text{if } (hik)\in s_m,\end{cases}\tag{12.9.5}$$

such that $E_2(t_{hik})=p$ under the uniform response mechanism and E_2 denotes the expected value within a given sample.

Now

$$\hat{S}=\sum_{(hik)\in s_r}w_{hik}y_{hik}$$

can be written as

$$\hat{S}=\sum_{(hik)\in s_r}w_{hik}y_{hik}=\sum_{(hik)\in s_n}t_{hik}w_{hik}y_{hik}\ .$$

Let E_1 denote the expected value over all possible samples, then we have

$$E\left(\hat{S}\right)=E_1E_2\left[\sum_{(hik)\in s_n}t_{hik}w_{hik}y_{hik}\right]=E_1\left[\sum_{(hik)\in s_n}E_2(t_{hik})w_{hik}y_{hik}\right]=E_1\left[\sum_{(hik)\in s_n}p\,w_{hik}y_{hik}\right]$$

$$=pE_1\left[\sum_{(hik)\in s_n}w_{hik}y_{hik}\right]=pE_1\left(\hat{Y}\right)=pY\ .$$

Now the estimator

$$\hat{T}=\sum_{(hik)\in s_r}w_{hik}$$

can be written as

$$\hat{T}=\sum_{(hik)\in s_r}w_{hik}=\sum_{(hik)\in s_n}t_{hik}w_{hik}\ .$$

Taking expected values on both sides, we have

$$E\left(\hat{T}\right)=E_1E_2\left(\hat{T}\right)=E_1E_2\left(\sum_{(hik)\in s_n}t_{hik}w_{hik}\right)=E_1\left(\sum_{(hik)\in s_n}E_2(t_{hik})w_{hik}\right)$$

$$=pE_1\left(\sum_{(hik)\in s_n}w_{hik}\right)=pN\ .$$

Now taking expected values on both sides of

$$\hat{U}=\sum_{(hik)\in s_n}w_{hik}$$

we have

$$E\left(\hat{U}\right)=E\left(\sum_{(hik)\in s_n}w_{hik}\right)=N\ .$$

Hence the theorem.

Defining

$$\in_0 = \frac{\hat{S}}{pY} - 1, \quad \in_1 = \frac{\hat{T}}{pN} - 1, \quad \text{and} \quad \in_2 = \frac{\hat{U}}{N} - 1$$

such that

$$E(\in_i) = 0 \ \text{ for } \ i = 0, 1, 2.$$

Then we have the following theorem.

Theorem 12.9.3. An approximately unbiased estimator of population total Y is given by

$$\hat{Y}_{\text{I}} = \sum_{(hik)\in s_r} w_{hik} y_{hik} + \sum_{(hik)\in s_m} w_{hik} y_{hik}^{*} .$$

Proof. Taking expected value we have

$$E\left(\hat{Y}_{\text{I}}\right) = E\left[E_*\left(\hat{Y}_{\text{I}}\right)\right] = E\left[\frac{\hat{S}}{\hat{T}}\hat{U}\right] = E\left[\frac{pY(1+\in_0)}{pN(1+\in_1)} N(1+\in_2)\right] = YE\left[(1+\in_0)(1+\in_2)(1+\in_1)^{-1}\right]$$

$$\approx YE\left[(1+\in_0)(1+\in_2)\left(1-\in_1+\in_1^2+...\right)\right] = YE\left[\left(1+\in_0+\in_2+\in_0\in_2\right)\left(1-\in_1+\in_1^2+...\right)\right]$$

$$\approx YE\left[\left(1+\in_0+\in_2-\in_1+....\right)\right] = Y.$$

Hence the theorem.

Let n_g be the number of units in the gth stratum. Then estimators corresponding to $\hat{S}$, $\hat{T}$ and $\hat{U}$ by deleting j^{th} cluster from the gth stratum can be defined as

$$\hat{S}_{-gj} = \sum_{(hik)\in s_r, h\neq g} w_{hik} y_{hik} + \frac{n_g}{n_g - 1} \sum_{(gik)\in s_r, i\neq j} w_{gik} y_{gik}, \tag{12.9.6}$$

$$\hat{T}_{-gj} = \sum_{(hik)\in s_r, h\neq g} w_{hik} + \frac{n_g}{n_g - 1} \sum_{(gik)\in s_r, i\neq j} w_{gik}, \tag{12.9.7}$$

and

$$\hat{U}_{-gj} = \sum_{(hik)\in s_n, h\neq g} w_{hik} + \frac{n_g}{n_g - 1} \sum_{(gik)\in s_n, i\neq j} w_{gik}, \tag{12.9.8}$$

such that $E\left(\hat{S}_{-gj}\right) = pY$, $E\left(\hat{T}_{-gj}\right) = pN$ and $E\left(\hat{U}_{-gj}\right) = N$. Also for $(hik)\in s_m$ and $(hi)\neq(gj)$ adjust the imputed values as

$$z_{hik}^{*(gj)} = y_{hik}^{*} + \hat{S}_{-gj}\big/\hat{T}_{-gj} - \hat{S}\big/\hat{T}, \tag{12.9.9}$$

which are called 'synthetic' imputed values such that

$$E_*\left(z_{hik}^{*(gj)}\right) = \hat{S}_{-gj}\big/\hat{T}_{-gj}$$

because $E\left(y_{hik}^{*}\right) = \hat{S}/\hat{T}$. Rao and Shao (1992) used synthetic imputation values to construct an approximately unbiased estimator of population total Y as

$$\hat{Y}_I^a(-gj) = \hat{S}_{-gj} + \sum_{(hik)\in s_m, h\neq g} w_{hik} z_{hik}^{*(gj)} + \frac{n_g}{n_g - 1} \sum_{(gik)\in s_m} w_{gik} z_{gik}^{*(gj)} . \tag{12.9.10}$$

Theorem 12.9.4. The estimator $\hat{Y}_{\rm I}^{a}(-gj)$ is approximately unbiased for the population total Y.

Proof. Taking expected value of $\hat{Y}_{\rm I}^{a}(-gj)$ we have

$$E\left[\hat{Y}_{\rm I}^{a}(-gj)\right] = EE_{*}\left[\hat{Y}_{\rm I}^{a}(-gj)\right] = E\left[\left(\frac{\hat{S}_{-gj}}{\hat{T}_{-gj}}\right)\hat{U}_{-gj}\right] \approx \frac{E\left(\hat{S}_{-gj}\right)}{E\left(\hat{T}_{-gj}\right)}E\left(\hat{U}_{-gj}\right) = \frac{pY}{pN}N = Y .$$

Hence the theorem.

Then a Jackknife estimator of the variance of $\hat{Y}_{\rm I}$ is given by

$$v_J = \sum_{g=1}^{L}\frac{n_g - 1}{n_g}\sum_{j=1}^{n_g}\left\{\hat{Y}_{\rm I}^{a}(-gj) - \hat{Y}_{\rm I}\right\}^2 . \tag{12.9.11}$$

Rao and Shao (1992) have also considered the situation of multiple imputation classes. They showed $\left(\hat{Y}_{\rm I} - Y\right)/\sqrt{v_J} \sim N(0,1)$. Yung and Rao (2000) have considered the post-stratified design for estimating the variance using Jackknife estimator of variance.

12.10 MULTIPLE IMPUTATION

Rubin (1978) introduced mutiple imputation (MI) to account for the inflation in the variance owed to imputation. It requires the construction of $M(\geq 2)$ complete data sets by replacing each missing value by M imputed values using the same imputation procedure. If $\bar{y}_{\rm I1}, \bar{y}_{\rm I2}, ..., \bar{y}_{\rm IM}$ denote the M imputed estimators of the population mean $\bar{Y}$ under SRSWR sampling. The 'final' imputed estimator of population mean $\bar{Y}$ is then given by

$$\bar{y}_{\rm I\bullet} = \frac{1}{M}\sum_{J=1}^{M}\bar{y}_{IJ} \tag{12.10.1}$$

with estimated variance

$$\hat{v}(\bar{y}_{\rm I\bullet}) = \frac{1}{M}\left(\frac{1}{n} - \frac{1}{N}\right)\sum_{J=1}^{M}s_{\rm IJ}^2 + \frac{(M+1)}{M}\left\{\frac{1}{M-1}\sum_{J=1}^{M}\left(\bar{y}_{\rm IJ} - \bar{y}_{\rm I\bullet}\right)^2\right\}, \tag{12.10.2}$$

where s_{IJ}^2 denotes the sample variance for the Jth completed data set and n is the sample size. Rubin and Schenker (1986) pointed out that the variance estimator (12.10.2) leads to valid inference at least when the number of imputations, M is large and imputations are 'proper' in the sense that the imputed values are drawn from the posterior distribution of non-observed values from the given respondent values. The traditional simple random hot deck imputation is not proper in this sense and it may lead to be an underestimate of true variance of $\bar{y}_I$, thus Rubin and Schenker (1986) suggested a new method called Approximate Bayesian Bootstap (ABB) method of variance estimation. Rao (1996b) considered stratified random sampling design for estimating the variance with multiple imputed data. Rubin's method has been found to be applicable in the situations in which the fraction of missing data is large and the user and imputer are the same individual who chooses multiple imputation because of its convenience; for example, see Little and Yao (1996), Paik (1997), Taylor, Muzoz, Bass, Sah, Chmiel, Kingsley et al. (1990), Tu, Meng, and Pagano (1993), and Clayton, Dunn, Pickles, and Spiegelhalter (1998). On the other hand, Fay (1992, 1994, 1996), Meng (1994) and Rubin (1996) have shown that the estimator of variance proposed by Rubin results in upward bias and inconsistent in certain cases. Robins and Wang (2000) derived a general formula for the large sample bias in Rubin's estimator of variance, which not only confirms the findings of the Fay (1992, 1994, 1996), Meng (1994) and Rubin (1996), but also indicates there are other scenarios under which Rubin's estimator of variance is downwardly biased. Robins and Wang (2000) provided an interesting formula which overcomes the deficiencies of Rubin's estimator of variance and they provided a consistent estimator of variance, unlike Rubins estimator, when the imputation and analysis models are mis-specified and incompatible with one another. Schafer and Schenker (2000) used imputed conditional means for drawing inference from it. Let us explain the concept of multiple imputation with the help of a simple example.

Example 12.10.1. Select an SRSWOR sample of twelve units from the population 1 given in the Appendix. Record the values of the real estate farm loans for the states selected in the sample. Observe the random non-response in the selected sample. Impute the missing values three times with the help of hot deck method of imputation. Apply the concept of multiple imputation for estimating population mean and construct the 95% confidence interval.

Solution. We apply the remainder approach on the first two columns of the Pseudo-Random Numbers (PRN) given in Table 1 of the Appendix to select a sample of 12 states from the population 1. The first 12 distinct random numbers between 1 and 50 were selected as: 49, 08, 10, 04, 42, 01, 19, 37, 12, 23, 38 and 44. The information so collected from the selected units in the sample is give below:

Sr. No.	State and Territory	Real estate farm loans
01	AL	408.978
04	AR	907.700
08	DE	42.808
10	GA	939.460
12	ID	Missing
19	ME	8.849
23	MN	1354.768
37	OR	114.899
38	PA	756.169
42	TN	Missing
44	UT	56.908
49	WI	Missing

We observe that data is missing for the three states ID, TN and WI. In the following table we impute these missing values three times with the help of hot deck method of imputation. On the first occasion we used the 3rd column of the PRN to select three random numbers between 1 and 9 to impute the missing values as shown in the fourth column of the Table 12.10.1. First time three distinct random numbers came in the sequence as 2, 8 and 1.

Table 12.10.1. Example of multiple imputation.

Sr. No.	State and Territory	Real estate farm loans	First time imputed data $J=1$	Second time imputed data $J=2$	Third time imputed data $J=3$
1	AL	408.978	408.978	408.978	408.978
2	AR	907.700	907.700	907.700	907.700
3	DE	42.808	42.808	42.808	42.808
4	GA	939.460	939.460	939.460	939.460
5	ME	8.849	8.849	8.849	8.849
6	MN	1354.768	1354.768	1354.768	1354.768
7	OR	114.899	114.899	114.899	114.899
8	PA	756.169	756.169	756.169	756.169
9	UT	56.908	56.908	56.908	56.908
10	ID	Missing	**907.700**	**1354.768**	**939.460**
11	TN	Missing	**756.169**	**114.899**	**756.169**
12	WI	Missing	**408.978**	**56.908**	**408.978**
		After imputation $\bar{y}_{IJ}$	555.282	509.759	557.929
		After imputation s_{IJ}^2	196565.530	275716.920	198684.640

Note: BOLD entries are imputed values.

We use the 7th column of PRN to select three random numbers between 1 and 9 to impute the missing data second time. We find the three distinct random numbers as: 6, 7 and 9. The corresponding imputed data is shown in the fifth column of the Table 12.10.1. We use the 11th column of PRN to select three random numbers between 1 and 9 to impute the missing data third time. We find the three distinct random numbers as: 4, 8 and 1. The corresponding imputed data is shown in the sixth column of the Table 12.10.1. Here $M=3$, $n=12$ and $N=50$. An imputed estimate of the average real estate farm loans during 1997 in the United States is given by

$$\bar{y}_{I\bullet} = \frac{1}{M}\sum_{J=1}^{M}\bar{y}_{IJ} = \frac{555.282+509.759+557.929}{3} = 540.99\,.$$

An estimate of variance is given by

$$\hat{v}(\bar{y}_{I\bullet}) = \frac{1}{M}\left(\frac{1}{n}-\frac{1}{N}\right)\sum_{J=1}^{M}s_{IJ}^2 + \frac{(M+1)}{M}\left\{\frac{1}{M-1}\sum_{J=1}^{M}(\bar{y}_{IJ}-\bar{y}_{I\bullet})^2\right\}$$

$$=\frac{1}{3}\left(\frac{1}{12}-\frac{1}{50}\right)(196565.530+275716.920+198684.640)$$

$$+\frac{(3+1)}{3}\left[\frac{1}{3-1}\left\{(555.282-540.99)^2+(509.759-540.99)^2+(557.929-540.99)^2\right\}\right]$$

$$=14164.86+977.71\ =15142.57\,.$$

A $(1-\alpha)100\%$ confidence interval, assuming large sample size, is

$$\bar{y}_{I\bullet}\pm t_{\alpha/2}(\mathrm{df}=n-1)\sqrt{\hat{v}(\bar{y}_{I\bullet})}\,.$$

Using Table 2 from the Appendix the 95% confidence interval of the average real estate farm loans is

$$540.99\pm 2.201\sqrt{15142.57}\quad\text{or}\quad [270.145,\,811.834].$$

12.10.1 DEGREE OF FREEDOM WITH MULTIPLE IMPUTATION FOR SMALL SAMPLES

Multiple imputation is becoming very popular these days and standard techniques for properly dealing with the missing data are appearing to have easy access to the computers during the forthcoming centuries. Let Y be the complete data set. Let Y_{obs} and Y_{mis} denote the observed and missing components of the complete data, that is $Y\equiv(Y_{\text{obs}},\,Y_{\text{mis}})$. Assuming that with complete data valid inference about a k component quantity β, possibly a model parameter or a finite population characteristic follows the standard large sample statement

$$\left(\hat{\beta}-\beta\right)\sim N(0,\,V) \tag{12.10.3}$$

where $\hat{\beta}\equiv\hat{\beta}(Y)$ is an estimator of β and $V\equiv V(Y)$ is its associated variance. The basic idea of multiple imputation is to fill the missing data multiple times with values drawn from some distribution that predicts the missing values given the observed data and other available information. Each draw of Y_{mis} is an imputation. Thus in case of multiple imputation we have m imputations $Y_{\text{mis}}^{(1)},Y_{\text{mis}}^{(2)},...,Y_{\text{mis}}^{(m)}$ as repeated independent draws of Y_{mis} from Bayesian prediction model. In general we do multiple imputation in three steps:

Step I. Compute the complete data statistics $\hat{\beta}_{*l}\equiv\hat{\beta}\left(Y^{(l)}\right)$ and $V_{*l}\equiv V\left(Y^{(l)}\right)$ for each of the m completed data sets $Y^{(l)}\equiv\left(Y_{\text{obs}},\,Y_{\text{mis}}^{(l)}\right)$ for $l=1,2,...,m.$

Step II. Compute an estimate of β as $\bar{\beta}_m=\frac{1}{m}\sum_{l=1}^{m}\hat{\beta}_{*l}$ and an estimate of variance as

$$V_m=\bar{V}_m+\left(1+\frac{1}{m}\right)B_m$$

where $\bar{V}_m=\frac{1}{m}\sum_{l=1}^{m}V_{*l}$ and $B_m=\frac{1}{m-1}\sum_{l=1}^{m}\left(\hat{\beta}_{*l}-\bar{\beta}_m\right)\left(\hat{\beta}_{*l}-\bar{\beta}_m\right)'$.

Step III. The $(1-\alpha)100\%$ confidence intervals for β are formed on the basis of k component Student's t distribution given by

$$\left(\beta-\overline{\beta}_m\right)\Big/\sqrt{V_m}\sim t_{v_m} \tag{12.10.4}$$

where

$$v_m=(m-1)\hat{\gamma}_m^{-2} \quad \text{with } \hat{\gamma}_m=\left(1+m^{-1}\right)\mathrm{tr}\left(B_m T_m^{-1}\right)\Big/k . \tag{12.10.5}$$

Barnard and Rubin (1999) pointed out that in small data sets, however, it can be unsatisfactory to set the degree of freedom to infinity, especially when there is little missing information, because v_m can then be many times the degree of freedom available if there were no missing data. Thus Barnard and Rubin (1999) thought that there is a need for the new expression for multiple imputation degree of freedom that does not rely on a large complete data sample. They provided a new expression as an adjusted degree of freedom

$$\hat{v}_m=v_{\text{com}}\left[\left\{\lambda\left(v_{\text{com}}\right)\left(1-\hat{\gamma}_m\right)\right\}^{-1}+\frac{v_{\text{com}}}{v_m}\right]^{-1} \tag{12.10.6}$$

where $\lambda(v)=(v+1)/(v+3)$ and $\hat{v}_{\text{obs}}=\lambda\left(v_{\text{com}}\right)v_{\text{com}}\left(1-\hat{\gamma}_m\right)$ denotes the observed degree of freedom with v_{com} being complete degree of freedom.

12.11 COMPROMISED IMPUTATION

Singh and Horn (2000) suggested a compromised imputation, in which the data after imputation becomes

$$y_{\bullet i}=\begin{cases}\alpha n y_i/r+(1-\alpha)\hat{b}x_i & \text{if } i\in A,\\ (1-\alpha)\hat{b}x_i & \text{if } i\in\overline{A},\end{cases} \tag{12.11.1}$$

where α is a suitably chosen constant, such that the variance of the resultant estimator is minimum. Note that Meeden (2000) has also suggested the idea of adjusting responding values in addition non-responding values while doing imputation. They used information from imputed values for the responding units in addition to non-responding units. Thus we have the following theorem:

Theorem 12.11.1. The point estimator of the population mean $\overline{Y}$ under the compromised method of imputation becomes

$$\overline{y}_{\text{comp}}=\alpha\overline{y}_r+(1-\alpha)\overline{y}_r\left(\frac{\overline{x}_n}{\overline{x}_r}\right). \tag{12.11.2}$$

Proof. We have

$$\overline{y}_{\text{comp}}=\frac{1}{n}\sum_{i\in s}y_{\bullet i}=\frac{1}{n}\left[\sum_{i\in A}y_{\bullet i}+\sum_{i\in\overline{A}}y_{\bullet i}\right], \tag{12.11.3}$$

and using (12.11.1) we have (12.11.2). Hence the theorem.

The estimator at (12.11.2) is an analogue of the well known estimator of population mean considered by Chakrabarty (1968), Vos (1980) and Adhvaryu and Gupta (1983) as discussed in Chapter 3, and given by

$$\bar{y}_{\text{cvag}} = \alpha\bar{y}_n + (1-\alpha)\bar{y}_n\left(\bar{X}/\bar{x}_n\right). \tag{12.11.4}$$

Defining

$$\varepsilon = \left(\bar{y}_r/\bar{Y}\right)-1, \quad \delta = \left(\bar{x}_r/\bar{X}\right)-1, \text{ and } \eta = \left(\bar{x}_n/\bar{X}\right)-1.$$

Using the concept of two-phase sampling following Rao and Sitter (1995) and the mechanism of missing completely at random (MCAR), for given r and n, we have

$$E(\varepsilon) = E(\delta) = E(\eta) = 0$$

and

$$E\left(\varepsilon^2\right) = \left(\frac{1}{r}-\frac{1}{N}\right)C_y^2, \quad E\left(\delta^2\right) = \left(\frac{1}{r}-\frac{1}{N}\right)C_x^2, \quad E(\varepsilon\delta) = \left(\frac{1}{r}-\frac{1}{N}\right)\rho_{xy}C_yC_x$$

$$E\left(\eta^2\right) = \left(\frac{1}{n}-\frac{1}{N}\right)C_x^2, \quad E(\delta\eta) = \left(\frac{1}{n}-\frac{1}{N}\right)C_x^2, \quad E(\varepsilon\eta) = \left(\frac{1}{n}-\frac{1}{N}\right)\rho_{xy}C_yC_x$$

where $C_y^2 = S_y^2/\bar{Y}^2$, $C_x^2 = S_x^2/\bar{X}^2$, $\rho_{xy} = S_{xy}/\left(S_xS_y\right)$, S_y^2, S_x^2, and S_{xy} have their usual meanings. Now we have the following theorems.

Theorem 12.11.2. The conditional bias in the estimator $\bar{y}_{\text{comp}}$ at (12.11.2) is

$$B\left(\bar{y}_{\text{comp}}\right) \approx (1-\alpha)\left(\frac{1}{r}-\frac{1}{n}\right)\bar{Y}\left(C_x^2 - \rho_{xy}C_yC_x\right). \tag{12.11.5}$$

Proof. The estimator $\bar{y}_{\text{comp}}$ in terms of ε, δ and η can be written as

$$\bar{y}_{\text{comp}} \approx \alpha\bar{Y}(1+\varepsilon) + (1-\alpha)\bar{Y}\left(1+\varepsilon+\eta-\delta+\varepsilon\eta-\varepsilon\delta-\delta\eta+\delta^2+O\left(\varepsilon^2\right)\right). \tag{12.11.6}$$

Taking expected value on both sides of (12.11.6) and its deviation from actual mean we obtain (12.11.5). Hence the theorem.

Theorem 12.11.3. The minimum MSE the estimator $\bar{y}_{\text{comp}}$ is given by

$$\text{Min.MSE}\left(\bar{y}_{\text{comp}}\right) \approx \text{MSE}\left(\bar{y}_{\text{ratio}}\right) - \left(\frac{1}{r}-\frac{1}{n}\right)\alpha^2\bar{Y}^2C_x^2 \tag{12.11.7}$$

for the optimum value of α given by

$$\alpha = 1-\rho_{xy}C_y/C_x. \tag{12.11.8}$$

Proof. We have

$$\bar{y}_{\text{comp}} \approx \bar{Y} + \bar{Y}\varepsilon + (1-\alpha)\bar{Y}(\eta-\delta) + O\left(\varepsilon^2\right)$$

where $O\left(\varepsilon^2\right)$ indicates terms of higher orders of $\varepsilon, \delta, \eta$ etc.. Thus

$$\text{MSE}\left(\bar{y}_{\text{comp}}\right) = E\left[\bar{y}_{\text{comp}} - \bar{Y}\right]^2 \approx E\left[\bar{Y}\varepsilon + (1-\alpha)\bar{Y}(\eta-\delta)\right]^2$$

$$= \left(\frac{1}{r}-\frac{1}{N}\right)\bar{Y}^2C_y^2 + \left(\frac{1}{r}-\frac{1}{n}\right)\bar{Y}^2\left[(1-\alpha)^2C_x^2 - 2(1-\alpha)\rho_{xy}C_yC_x\right]. \tag{12.11.9}$$

On differentiating (12.11.9) with respect to $(1-\alpha)$ and equating to zero, we obtain $\alpha = 1-\rho_{xy}C_y/C_x$ and on substituting it in (12.11.9), we have the theorem.

12.11.1 PRACTICABILITY OF COMPROMISED IMPUTATION

The main difficulty in using the compromised imputation procedure is the choice of α. It is important to note that the optimum value of α depends only upon the well known parameter $K = \rho_{xy} C_y / C_x$. The value of K is quite stable in the repeated surveys as shown by Reddy (1978a). Thus if the value of K is known then the compromised imputation method can be easily implemented in actual surveys. Some time the value of K is not known. In these situations, Singh and Horn (2000) have suggested two estimators of α. The first estimator is given by

$$\hat{\alpha}_1 = 1 - \left(\bar{x}_r s_{xy}^*\right) / \left(\bar{y} s_x^{*2}\right), \tag{12.11.10}$$

where $s_{xy}^* = (r-1)^{-1} \sum_{i=1}^{r} (y_i - \bar{y}_r)(x_i - \bar{x}_r)$ and $s_x^{*2} = (r-1)^{-1} \sum_{i=1}^{r} (x_i - \bar{x}_r)^2$. The second estimator is given by

$$\hat{\alpha}_2 = 1 - \left(\bar{x}_n s_{xy}^*\right) / \left(\bar{y}_r s_x^2\right), \tag{12.11.11}$$

where $s_x^2 = (n-1)^{-1} \sum_{i=1}^{n} (x_i - \bar{x}_n)^2$. The choice between $\hat{\alpha}_1$ and $\hat{\alpha}_2$ is not very important for the infinite populations, because the asymptotic mean squared error of the resultant estimators of mean remains same by following Sampath (1989). Singh and Horn (2000) have shown that the compromised imputation technique remains better than ratio or mean methods of imputation.

12.11.2. RECOMMENDATIONS OF COMPROMISED IMPUTATION

It is interesting to note that if a strong imputation variable, x_i, is available then $y_i = b x_i$ very nearly holds for all i. Then to very close approximation, $\hat{b} = b$ and $\hat{b} x_i = y_i$, that is, the imputed value is near perfect. Then the imputation rule (12.11.1) reduces to

$$y_{\bullet i} = \begin{cases} y_i[1 + \alpha(n/r - 1)] & \text{if } i \in A, \\ (1-\alpha) y_i & \text{if } i \in \bar{A}. \end{cases} \tag{12.11.12}$$

Under such situations the values of $\rho_{xy} \to 1$ and $C_y \approx C_x$, then the optimum value of α, and hence its estimators $\hat{\alpha}_i$, $i = 1, 2$, will tend to zero. In other words, then the imputed values, using the compromised technique, remain close to the true values in $\bar{A}$. Also the actual values y_i do not have any impact of imputation in. It is remarkable that a bad guess of α may lead to bad results in the compromised imputation. Since the compromised imputation provides better estimator of population mean, therefore, it is recommended to use in future.

12.11.3. WARM DECK IMPUTATION

Singh and Horn (2000) pointed out that this type of compromisation can also be done between other type of imputation methods. For example a compromisation between Hot deck and Cold deck methods of imputation may lead to the 'Warm Deck' method of imputation, defined as

$$\bar{y}_{\text{WD}} = \alpha \bar{y}_{\text{CD}} + (1-\alpha)\bar{y}_{\text{HD}}. \tag{12.11.13}$$

The correlation between estimates obtained via the cold deck and hot deck methods is expected to be high, and hence the resultant estimator (12.11.13) named the "Warm Deck" method of imputation is expected to be efficient for the optimum value of α, given by

$$\alpha_o = \{V(\bar{y}_{\text{HD}}) - \text{Cov}(\bar{y}_{\text{HD}}, \bar{y}_{\text{CD}})\} / \{V(\bar{y}_{\text{HD}}) + V(\bar{y}_{\text{CD}}) - 2\,\text{Cov}(\bar{y}_{\text{HD}}, \bar{y}_{\text{CD}})\}. \tag{12.11.14}$$

A consistent estimator of α_o at (12.11.14) can be obtained by replacing variance - covariance terms by sample analogues.

12.11.4 MEAN CUM NN IMPUTATION

A compromised method of imputation obtained by pooling Mean method and Nearest Neighbourhood (NN) method of imputation, given by

$$\bar{y}_{\text{MN}} = \alpha \bar{y}_r + (1-\alpha)\bar{y}_{\text{NN}} \tag{12.11.15}$$

can be named the 'Mean cum NN' method of imputation. In the same fashion a linear variety of any of two or more imputation procedures can be used to make a compromised imputation procedure. Chen and Shao (2001) have discussed variance estimation techniques for nearest neighbourhood imputation.

Example 12.11.1. Select an SRSWOR sample of twenty units from the population 1 given in the Appendix. Record the values of the real and nonreal estate farm loans for the states selected in the sample. Observe the random non-response in the selected sample. Impute the missing values with the following two methods:
(a) Ratio method of imputation; and (b) Compromised method of imputation.
Estimate the average real estate farm loans with each method and comment on your results.

Solution. We apply the remainder approach on the 3rd and 4th columns of the Pseudo-Random Numbers (PRN) given in Table 1 of the Appendix to select a sample of 20 states from the population 1. The first 20 distinct random numbers between 1 and 50 were selected as: 29, 31, 14, 41, 05, 47, 28, 22,18, 12, 42, 23, 48, 02, 06, 07, 11, 21, 25, and 39. The information so collected from the selected units in the sample is given below.

Sr. No.	State and Territory	Nonreal estate farm loans, x	Real estate farm loans, y	Sr. No.	State and Territory	Nonreal estate farm loans, x	Real estate farm loans, y
02	AK	3.433	2.605	23	MN	2466.892	1354.768
05	CA	3928.732	1343.461	25	MO	1519.994	1579.686
06	CO	906.281	Missing	28	NV	16.710	5.860
07	CT	4.373	7.130	29	NH	0.471	6.044
11	HI	38.067	40.775	31	NM	274.035	140.582
12	ID	1006.036	53.753	39	RI	0.233	Missing
14	IN	1022.782	1213.024	41	SD	1692.817	413.777
18	LA	405.799	282.565	42	TN	388.869	553.266
21	MA	56.471	Missing	47	WA	1228.607	Missing
22	MI	440.518	323.028	48	WV	29.291	Missing

From the above table $\bar{x}_r = 880.6352$ and $\bar{y}_r = 488.0216$, which imply

$$\hat{b} = \bar{y}_r / \bar{x}_r = 0.55417\,.$$

Also $s_x^{*2} = 1264908.658$ and $s_{xy}^{*} = 505875.559$, which imply

$$\hat{\alpha}_1 = 1 - \left(\bar{x}_r s_{xy}^{*}\right) / \left(\bar{y}_r s_x^{*2}\right) = 1 - \frac{880.6352 \times 505875.559}{488.0216 \times 1264908.658} = 0.27832522.$$

The imputed values for the ratio method are then given by

$$\hat{y}_i(r) = \hat{b}x_i = 0.55417 \times x_i$$

if the ith state selected in the sample is not responding to the value of real estate farm loans.

On the other hand, the imputed value for the compromised imputation method is

$$\hat{y}_i(c) = \begin{cases} \dfrac{n\hat{\alpha}}{r} y_i + (1-\hat{\alpha})\hat{b}x_i = 0.3711 y_i + 0.39993 x_i & \text{if the } i^{th} \text{ state is responding,} \\ (1-\hat{\alpha})\hat{b}x_i = 0.39993 x_i & \text{if the } i^{th} \text{ state is not responding.} \end{cases}$$

Thus by the ratio and compromised methods of imputation the data takes the form as shown in the 4th and 5th columns, respectively, of the following table.

Sr. No.	State and Territory	Real estate farm loans, y	Ratio Imputation	Compromised imputation
2	AK	2.605	2.60500	2.339678
5	CA	1343.461	1343.46100	2069.779000
6	CO	Missing	502.23370	362.449400
7	CT	7.130	7.13000	4.394841
11	HI	40.775	40.77500	30.355770
12	ID	53.753	53.75300	422.292200
14	IN	1213.024	1213.02400	859.195300
18	LA	282.565	282.56500	267.151400
21	MA	Missing	31.29453	22.584480
22	MI	323.028	323.02800	296.052400
23	MN	1354.768	1354.76800	1489.340000
25	MO	1579.686	1579.68600	1194.114000
28	NV	5.860	5.86000	8.859487
29	NH	6.044	6.04400	2.431297
31	NM	140.582	140.58200	161.765000
39	RI	Missing	0.12912	0.093184
41	SD	413.777	413.77700	830.561900
42	TN	553.266	553.26600	360.837800
47	WA	Missing	680.85710	491.357400
48	WV	Missing	16.23219	11.714360
		Sum	8551.07100	8887.666700

Thus an estimate of the average of the real estate farm loans in the United States during 1997 by the ratio method of imputation is given by

$$\bar{y}_{\text{ratio}} = \frac{1}{n}\sum_{i=1}^{n}\hat{y}_i(r) = \frac{8551.071}{20} = 427.55$$

and that by the compromised imputation method is given by

$$\bar{y}_{\text{comp}} = \frac{1}{n}\sum_{i=1}^{n}\hat{y}_i(c) = \frac{8887.6667}{20} = 444.38\,.$$

From the description of the population 1 given in the Appendix, the true average real estate farm loans is give by $\bar{Y} = 555.43$. One can see here that the estimate based on the compromised method of imputation is close to the true value of the population mean than the estimate based on the ratio method of imputation.

12.12 ESTIMATION OF RESPONSE PROBABILITIES

Consider a sample s of size n is selected by a given design $p(s)$ and it is assumed that a set r of n_r units respond and a set r^c of $(n-n_r)$ units does not. The information collected in the sample can be represented as $(k,\ I_k,\ Y_k,\ x_k)$, $k \in s,\ N, n$, where I_k is an indicator variable such that $E(I_k) = p_k$ and p_k denotes the individual response probability. Cassel, Särndal, and Wretman (1979) has shown that it is always possible to eliminate non-response bias by estimators of population total or mean by the response probability, p_k, provided that p_k is always positive for all k. Since the response probability p_k is unknown, but it can be estimated. It is generally assumed that the response probability takes the form

$$p_k = F(\theta,\ x_k) \tag{12.12.1}$$

where θ denotes an unknown parameter (or vector of parameters) and $q(\bullet,\ \bullet)$ is a functional form to be explicitly specified, for example $p_k = \text{Exp}(-\theta x_k)$ or $p_k = [1 + \text{Exp}(-\theta x_k)]^{-1}$ etc.. Estimated $\hat{p}_k$ are then obtained by replacing in (12.12.1) estimated values $\hat{\theta}$ of parameter θ. Specification of the functional form as well as estimation of its parameters may be cumbersome. Thus if auxiliary information consists of a grouping or stratification variable(s) and it is reasonable to assume response propensity is homogeneous within strata, then a way to avoid this difficulty is to assume response probabilities constant within strata, thus if $h = 1,2,...,L$ denote the stratum levels, then (12.12.1) becomes

$$p_k = \theta_h\ . \tag{12.12.2}$$

It is natural to use the response rate in each stratum as maximum likelihood estimate of p_k as

$$\hat{p}_k = \hat{\theta}_h = n_{rh}/n_h \tag{12.12.3}$$

where n_{rh} and n_h are the size and respondents in the hth stratum, respectively. Sometimes however no clearly identifiable strata or other groups exists, this may often occur when auxiliary information is represented by the x_k values of a continuous variable, then this method, although still easily applicable, may present same drawbacks. In such situations it seems natural to turn the idea of homogeneous response behaviour within strata into the idea that similar or near x_k are connected with similar p_k. However, the latter does not fit well the strata structure introduced above, where groups are non-overlapping. In fact, x_k values, some distance apart in the same group, give the same estimates of response probabilities, whereas x_k values, near to each other but in different groups, may give very different estimates. Giommi (1984) has given the idea of centring at each unit, for estimating the response probabilities, and using a technique ideally related to the concept of moving average. Thus the response probability of the kth unit can be estimated by the response rate in the group centred at the kth unit or equivalently in the convenient interval of x values centred at $x_k, (k \in s)$. Assuming $2l_k$ be the length of the interval centred at x_k, then p_k can be estimated as

$$\hat{p}_k = \frac{\sum_{j \in s} i_k D(x_k - x_j)}{\sum_{j \in s} D(x_k - x_j)}, \quad \text{where } D(z) = \begin{cases} 1 & \text{if } |z| \le l_k, \\ 0 & \text{if } |z| > l_k, \end{cases} \tag{12.12.4}$$

and i_k is the realised value of I_k, that is $i_k = 1$ if $k \in r$ and 0 otherwise. Using estimates of response probabilities, Cassel, Särndal, and Wretman (1979) proposed the following estimator of population total:

$$\hat{Y}_c = \sum_{k \in s} \frac{y_k}{\pi_k \hat{p}_k} + \hat{\beta}_{ds}\left[X - \sum_{k \in s} \frac{x_k}{\pi_k \hat{p}_k}\right] \tag{12.12.5}$$

where

$$\hat{\beta}_{ds} = \left[\sum_{k \in s} \frac{y_k x_k}{v_k \pi_k \hat{p}_k}\right]\left[\sum_{k \in s} \frac{x_k^2}{v_k \pi_k \hat{p}_k}\right]^{-1}$$

and π_k denotes the probability of including the kth unit in the sample.

Example 12.12.1. We wish to estimate the average duration sleep time (in minutes) of the aged persons living in a small village of the United States as shown in population 2 of the Appendix. Select an SRSWOR of eight persons from the population 2. Collect the information about the duration of sleep and age of the persons selected in the sample. Suppose the response probability of the ith person is inversely proportional to the age of the persons selected. Assuming that an old person will respond less quickly than a matured young person. Estimate the average duration sleep time of the old persons in the particular village.

Solution. The population 2 consists of 30 old persons living in a small village. We used first two columns of the Pseudo-Random Number (PRN) Table 1 given in the Appendix to select eight distinct random numbers between 1 and 30 as : 01, 23, 04, 05, 22, 29, 03, 27.

Person	Age (x)	Duration of sleep (y)	ψ_i	y_i/ψ_i
01	60	492	0.036708	13403.07
03	55	408	0.040045	10188.53
04	56	465	0.039330	11823.04
05	82	312	0.026859	11616.22
22	71	360	0.031021	11605.04
23	63	390	0.034960	11155.61
27	60	390	0.036708	10624.38
29	66	390	0.033371	11686.79
			Sum	92102.68

We assume the response probability

$$\psi_i = x_i^{-1} \Big/ \sum_{i=1}^{N}(1/X_i), \quad \text{where} \quad \sum_{i=1}^{30} 1/X_i = 0.454 \quad \text{is known.}$$

Then an estimate of the average duration sleep time is

$$\bar{y}_{\text{est}} = \frac{1}{Nn}\sum_{i=1}^{n}\frac{y_i}{\psi_i} = \frac{92102.68}{30 \times 8} = 383.76 \text{ minutes.}$$

12.13 ESTIMATORS BASED ON ESTIMATED RESPONSE PROBABILITIES

In almost all large scale surveys non-responses are inevitable. Several methods for handling non-response problems in sample surveys are available in the literature. Good details are given by Rubin (1987). Särndal (1992) developed a method of estimation of the population total and its variance when a single imputation is used for estimating unobserved values under the superpopulation model described below. Let $\Omega = \{1,2,...,N\}$ be a finite population of N identifiable units and y_i be the value of the variate under study for the ith unit of the population Ω. It is assumed that the vector $\underline{y} = (y_1,..,y_i,..,y_N)$ is a random sample from a superpopulation ξ having the following distribution: $E_\xi(y_i) = \beta x_i$, $V_\xi(y_i) = \sigma^2 x_i^g$ and $C_\xi(y_i, y_j) = 0$ for $i \neq j$, $i, j = 1,..,N$, where E_ξ, V_ξ and C_ξ denote respectively expectation, variance and covariance operator with respect to the model ξ; β, $\sigma^2(>0)$, and $g(>0)$ are unknown model parameters, and x_i (>0) is the value of the auxiliary variable for the ith unit. The objective is to estimate the finite population total $Y = \sum_{i\in\Omega} y_i$ on the basis of a sample s, of size n, selected with probability $p(s)$ according to a sampling design p. Let π_i and π_{ij} be the inclusion probabilities for the ith, and i^{th} and j^{th} ($i \neq j$) unit of Ω. Let $s_r (\subset s)$ be the set of respondent units of size m from which responses y_i values are observed and the complement $s - s_r$ (of size $n - m$) be the set of non-response units. Let $\hat{y}_j$ for $j \in s - s_r$ be the imputed value of the jth unit computed according to a certain rule depending on the superpopulation model ξ (for details, see Särndal (1992)).

Let

$$t = \sum_{i\in s} w_i y_i$$

be an unbiased estimator used for estimating Y in case of 100% response $(i.e., m = n)$ where w_i are suitably chosen weights independent of y_i values. In the presence of non-response ($m < n$), Särndal (1992) modified the estimator t as

$$\hat{t} = \sum_{i\in s} w_i y_{i0},$$

where

$$y_{i0} = \begin{cases} y_i & \text{for } i \in s_r, \\ \hat{y}_i & \text{for } i \in s - s_r. \end{cases}$$

Denoting E_p (V_p) and E_r (V_r) as expectation (variance) with respect to the sampling design p and response mechanism respectively, Särndal (1992) derived the overall variance of $\hat{t}$ as

$$V_{\text{tot}} = E_\xi E_p E_r (\hat{t} - Y)^2 = E_\xi V_p(\hat{t}) + E_p V_r E_\xi(\hat{t}) + 2E_\xi E_p [(t - Y)\{E_\xi(\hat{t} - t) \mid s\}],$$

where $V_p(\hat{t}) = E_p(\hat{t} - Y)^2 =$ variance of $\hat{t}$ with respect to the sampling design p. $V_\xi = E_\xi\{(\hat{t} - t)^2 \mid s, r\}$. Now writing $E_p E_r V_\xi(\hat{t}) = V_{\text{imp}} =$ imputation variance and assuming $E_\xi E_p[(t - Y)\{E_\xi(\hat{t} - t) \mid s\}]$ is close to zero. Särndal (1992) derived an approximate expression for V_{tot} as

$$V_{\text{tot}} \cong V_{\text{sam}} + V_{\text{imp}}. \tag{12.13.1}$$

Consider the situation where an SRSWOR sample s of size n is selected and m the size of the response sample s_r, then under the following superpopulation model,

$$E_\xi(y_i) = \beta, \quad V_\xi(y_i) = \sigma^2, \text{ and } C_\xi(y_i, y_j) = 0 \text{ for } i \neq j \tag{12.13.2}$$

the unobserved values of y_i may be imputed by using $\hat{y}_i = \hat{\beta} = \bar{y}_r = \sum_{i \in s_r} y_i / m$ for $i \in s - s_r$, *i.e.* $y_{i0} = y_i$ for $i \in s$ and $y_{i0} = \bar{y}_r$ for $i \in s\text{-}s_r$.

Särndal (1992) proposed an estimator for the population total Y as

$$\hat{t}_{01} = \frac{N}{n} \sum_{i \in s} y_{i0} = N\, \bar{y}_r. \tag{12.13.3}$$

He proposed the estimator for variance of $\hat{t}_{01}$ as

$$\hat{V}_{\text{tot}} = N^2 \left(\frac{1}{m} - \frac{1}{N}\right) S_{yr}^2 = \hat{V}_{t01} \text{ (say), where } (m-1)S_{yr}^2 = \sum_{i \in s_r} (y_i - \bar{y}_r)^2 \tag{12.13.4}$$

For the superpopulation model

$$E_\xi(y_i) = \beta x_i, \quad V_\xi(y_i) = \sigma^2 x_i, \text{ and } C_\xi(y_i, y_j) = 0, \tag{12.13.5}$$

Särndal (1992) used

$$y_{i0} = \begin{cases} y_i & \text{for } i \in s_r, \\ \hat{B} x_i & \text{for } i \in s - s_r, \end{cases} \quad \text{with } \hat{B} = \sum_{i \in s_r} y_i \Big/ \sum_{i \in s_r} x_i$$

and proposed an estimator for Y under SRSWOR as

$$t_{02} = N \bar{x}_s \frac{\bar{y}_r}{\bar{x}_r}, \tag{12.13.6}$$

where $\bar{x}_s = \sum_{i \in s} x_i$, $\bar{x}_r = \sum_{i \in s_r} x_i$, and $\bar{y}_r = \sum_{i \in s_r} y_i$.

Särndal (1992) estimated the variance of t_{02} by

$$\hat{V}_{02} = \hat{V}_{\text{sam}}(2) + \hat{V}_{\text{imp}}(2), \tag{12.13.7}$$

where

$$\hat{V}_{\text{sam}}(2)=N^2\left(\frac{1}{n}-\frac{1}{N}\right)\left\{s_{yos}^2+C_0\hat{\sigma}^2\right\},\quad \hat{V}_{\text{imp}}(2)=N^2\left(\frac{1}{n}-\frac{1}{N}\right)C_1\hat{\sigma}^2$$

where

$$S_{\text{yos}}^2=(n-1)\sum_{i\in s}(y_i-\bar{y}_{\text{so}})^2,\quad \bar{y}_{\text{so}}=\sum_{i\in s}y_{i0}/n,\quad \hat{\sigma}^2=\sum_{i\in s_r}\left(y_i-\hat{B}x_i\right)^2\Big/\left[(m-1)\left\{\bar{x}_r\left(1-cv_{xr}^2/m\right)\right\}\right],$$

$$C_1=\bar{x}_s\bar{x}_{s-r}/\bar{x}_r,\quad cv_{xr}=S_{xr}/\bar{x}_r,\quad (m-1)S_{xr}^2=\sum_{i\in s_r}(x_i-\bar{x}_r)^2$$

and

$$(n-1)C_0=\sum_{i\in s-r}x_i-\frac{\sum_{i\in s-r}x_i^2}{\sum_{i\in s_r}x_i}+\frac{1}{n}\frac{\sum_{i\in s-r}x_i\sum_{i\in s}x_i}{\sum_{i\in s_r}x_i}.$$

Finally, in order to compare the relative efficiency of the estimator $\hat{t}_{02}$, Särndal (1992) conducted a Monte Carlo study with 100,000 repeated response sets s_r, $N=100$, $n=30$. Three different response mechanisms were used:

Mechanism 1: Response probability $\theta_i=1-\exp(a_1y_i)$ decreases with y_i;
Mechanism 2: Response probability $\theta_i=\exp(-a_2y_i)$ increases with y_i;
and
Mechanism 3: Response probability $\theta_i=0.7$ is a constant.

The constants a_1 and a_2 are positive and so chosen to make the average of $\theta_i=\sum_i\theta_i\Big/N=0.7$. Note that Särndal's (1992) estimators cannot be used gainfully when the response probabilities θ_k are known or can be estimated from the available data. So Arnab and Singh (2002d) proposed some alternative estimation procedures assuming that the response probabilities θ_k for the kth unit are either known or estimated through the log linear models:

(i) $\log(\theta_i)=1-c_1p_i$; and (ii) $\log(\theta_i)=c_2p_i$;

respectively, where $p_i=z_i/Z$ and c_1, c_2 are unknown positive constants which are appropriate when response probability increases or decreases with x.

12.13.1 ESTIMATORS BASED ON RESPONSE PROBABILITIES

Arnab and Singh (2002d) assumed that the response probabilities θ_i are independent. Under this assumption, they consider that the response sample s_r (formed by the set of respondent units) is a sub-sample from s selected by the nature according to the Poisson sampling scheme with inclusion probabilities $\pi_{i|s}=\theta_i$ and $\pi_{ij|s}=\theta_{ij}=\theta_i\theta_j$ for $i\neq j$. The Horvitz and Thompson (1952) type estimator for the total Y is

$$\hat{Y}_{\text{ht}} = \sum_{i \in s_r} \frac{y_i}{\pi_i \theta_i}. \tag{12.13.1.1}$$

It can be easily checked that $\hat{Y}_{\text{ht}}$ is unbiased for the total Y. The expression for the variance of $\hat{Y}_{\text{ht}}$ is given in the following theorems:

Theorem 12.13.1. The variance of $\hat{Y}_{\text{ht}}$ is $V(\hat{Y}_{\text{ht}}) = V_1 + V_2$, where

$$V_1 = \frac{1}{2}\sum\sum_{i \neq j}\left(\pi_i \pi_j - \pi_{ij}\right)\left(\frac{y_i}{\pi_i} - \frac{y_j}{\pi_j}\right)^2, \tag{12.13.1.2}$$

and

$$V_2 = \sum_i \frac{y_i^2}{\pi_i}\left(\frac{1}{\theta_i} - 1\right). \tag{12.13.1.3}$$

Proof. We have

$$V(\hat{Y}_{\text{ht}}) = E\{V(\hat{Y}_{\text{ht}} \mid s)\} + V\{E(\hat{Y}_{\text{ht}} \mid s)\} = E\left\{V\left(\sum_{i \in s_r} \frac{y_i}{\pi_i \theta_i} \mid s\right)\right\} + V\left\{E\left(\sum_{i \in s_r} \frac{y_i}{\pi_i \theta_i} \mid s\right)\right\}.$$

Now

$$E\left\{V\left(\sum_{i \in s_r} \frac{y_i}{\pi_i \theta_i} \mid s\right)\right\} = E\left\{\sum_{i \in s} z_i^2\left(\frac{1}{\pi_{i|s}} - 1\right)\right\} + \frac{1}{2}\sum\sum_{i \neq j \in s}\left(\pi_{ij|s} - \pi_{i|s}\pi_{j|s}\right)\frac{z_i}{\pi_{i|s}}\frac{z_j}{\pi_{j|s}}$$

$$= \sum_i \frac{y_i^2}{\pi_i}\left(\frac{1}{\theta_i} - 1\right) = V_2$$

(since $\pi_{ij|s} = \theta_i \theta_j$ and $\pi_{i|s} = \theta_i$) and $V\{E(\sum_{i \in s_r} \frac{y_i}{\pi_i \theta_i} \mid s)\} = V(\sum_{i \in s} \frac{y_i}{\pi_i}) = V_1$

(where $z_i = y_i / \pi_i$).

Theorem 12.13.2. If the sample size is large enough to ensure $\text{Prob}\{m \geq 2\} \cong 1$, then the following two estimators

$$\hat{v}_{\text{ht}}(1) = \hat{V}_{11} + \hat{V}_2 \quad \text{and} \quad \hat{v}_{\text{ht}}(2) = \hat{V}_{12} + \hat{V}_2$$

are unbiased for $V(\hat{Y}_{\text{ht}})$, where

$$\hat{V}_{11} = \frac{1}{2}\sum\sum_{i \neq j \in s_r} \frac{\left(\pi_i \pi_j - \pi_{ij}\right)}{\pi_{ij}\theta_i \theta_j}\left(\frac{y_i}{\pi_i} - \frac{y_j}{\pi_j}\right)^2, \quad \hat{V}_2 = \sum_{i \in s_r} \frac{y_i^2}{\pi_i^2 \theta_i}\left(\frac{1}{\theta_i} - 1\right)$$

and

$$\hat{V}_{12} = \sum_{i \in s_r} \frac{y_i^2}{\pi_i \theta_i}\left(\frac{1}{\pi_i} - 1\right) - \sum\sum_{i \neq j \in s_r} \frac{\left(\pi_i \pi_j - \pi_{ij}\right)}{\pi_{ij}\pi_i \pi_j}\frac{y_i}{\theta_i}\frac{y_j}{\theta_j}.$$

Proof. Noting

$$V_1 = \frac{1}{2}\sum_{i\neq j}\sum(\pi_i\pi_j - \pi_{ij})\left(\frac{y_i}{\pi_i} - \frac{y_i}{\pi_i}\right)^2 = \sum_i y_i^2\left(\frac{1}{\pi_i} - 1\right) + \sum_{i\neq j}\sum(\pi_{ij} - \pi_i\pi_j)\frac{y_i}{\pi_i}\frac{y_j}{\pi_j},$$

we can verify that both the estimators $\hat{V}_{11}$ and $\hat{V}_{12}$ are unbiased for V_1. It can be easily checked that $\hat{V}_2$ is unbiased for V_2.

Case I. Consider an SRSWOR design where $\pi_i = \frac{n}{N} = \pi_{i0}$ and $\pi_{ij} = \frac{n(n-1)}{N(N-1)} = \pi_{ij0}$.

In this case we have $\hat{Y}_{\text{ht}} = \frac{N}{n}\sum_{i\in s_r}\frac{y_i}{\theta_i}$. The variance of $\hat{Y}_{\text{ht}}$ can be estimated by putting $\pi_i = \pi_{i0}$ and $\pi_{ij} = \pi_{ij0}$ in the expressions of the variance estimators given in the Theorem 12.13.2.

Case II. Consider an SRSWOR design where $\pi_i = \pi_{i0}$, $\pi_{ij} = \pi_{ij0}$ and response probabilities θ_i are equal to θ for every i. Further, if we estimate θ by $\hat{\theta} = m/n$, then

$$\hat{Y}_{\text{ht}} = \frac{N}{m}\sum_{i\in s_r} y_i = N\,\bar{y}_r = \hat{t}_{01}. \tag{12.13.1.4}$$

The estimator $\hat{t}_{01}$ was proposed by Särndal (1992) when the method of single imputation is used as described in (12.13.3). Putting $\theta_i = \hat{\theta} = m/n$, $\pi_i = \pi_{i0}$, $\pi_{ij} = \pi_{ij0}$ in the Theorem 12.13.2 we have two approximate variance estimators for $\hat{t}_{01}$ as

$$\hat{v}_{01} = \frac{N(N-n)(m-1)}{m(n-1)}s_{yr}^2 + \frac{N^2(n-m)}{nm^2}\sum_{i\in s_r} y_i^2, \quad \text{and} \quad \hat{v}_{02} = \hat{v}_{01} + \frac{N(N-n)(n-m)}{n(n-1)m^2}\sum_{i\in s_r} y_i^2.$$

Finally, noting that $\hat{v}_o = N^2\left(\frac{1}{m} - \frac{1}{N}\right)s_{yr}^2$, the expression for the estimate of the variance of $\hat{t}_{01}$ given by Särndal in (12.13.1.4), we have the following theorem relating to the magnitude of the three variance estimators of $\hat{t}_{01}$.

Theorem 12.13.3. (i) $\hat{v}_{02} \ge \hat{v}_{01}$ for all the y_i values, and (ii) $\hat{v}_{02} \ge \hat{v}_0$ whenever all the y_i values are positive.

Proof. Straightforward and hence omitted.

12.13.2 CALIBRATION OF RESPONSE PROBABILITIES

We shall discuss estimators of total and variance under the calibrated response probabilities mechanism in the following sections:

12.13.2.1 CALIBRATED ESTIMATOR AND ITS VARIANCE

Consider the x_i, $i \in s$ are known. Then following Deville and Särndal (1992), a calibrated estimator of the population total is given by

$$\hat{Y}_C = \sum_{i \in s_r} w_i \frac{y_i}{\pi_i}, \tag{12.13.2.1}$$

where the w_i are the calibrated weights obtained by minimizing the chi square type distance function

$$D = \sum_{i \in s_r} (w_i - 1/\theta_i)^2 q_i^{-1}$$

subject to the calibration constraint

$$\sum_{i \in s_r} w_i \frac{x_i}{\pi_i} = \sum_{i \in s} \frac{x_i}{\pi_i}.$$

Here the q_i are suitably chosen weights. Minimization of D leads to the calibrated weights

$$w_i = \frac{1}{\theta_i} + \left(\sum_{i \in s_r} \frac{x_i^2 q_i}{\pi_i^2} \right)^{-1} \left(\sum_{i \in s} \frac{x_i}{\pi_i} - \sum_{i \in s_r} \frac{x_i}{\pi_i \theta_i} \right) \frac{x_i q_i}{\pi_i}. \tag{12.13.2.2}$$

On putting (12.13.2.2) in (12.13.2.1), the calibrated estimator $\hat{Y}_C$ becomes

$$\hat{Y}_C = \hat{Y}_{\text{ht}} + \hat{B}_r \left(\sum_{i \in s} \frac{x_i}{\pi_i} - \hat{X}_{\text{ht}} \right) \tag{12.13.2.3}$$

where

$$\hat{X}_{\text{ht}} = \sum_{i \in s_r} \frac{x_i}{\pi_i \theta_i} \text{ and } \hat{B}_r = \sum_{i \in s_r} \frac{y_i x_i q_i}{\pi_i^2} \Bigg/ \sum_{i \in s_r} \frac{x_i^2 q_i}{\pi_i^2}. \tag{12.13.2.4}$$

Now writing

$$B_r = \left(E \sum_{i \in s_r} \frac{y_i x_i q_i}{\pi_i^2} \right) \Bigg/ \left(E \sum_{i \in s_r} \frac{x_i^2 q_i}{\pi_i^2} \right) = \left(\sum_{i \in U} \frac{y_i x_i q_i \theta_i}{\pi_i} \right) \Bigg/ \left(\sum_{i \in U} \frac{x_i^2 q_i \theta_i}{\pi_i} \right)$$

and $E_i = y_i - Bx_i$, an approximate expression for the variance of $\hat{Y}_C$ is obtained as

$$V_C = V(\hat{Y}_C) = E\left\{ V\left(\sum_{i \in s_r} \frac{E_i}{\theta_i \pi_i} \mid s \right) \right\} + V\left\{ E\left(\sum_{i \in s_r} \frac{y_i - Bx_i}{\theta_i \pi_i} + B \sum_{i \in s} \frac{x_i}{\pi_i} \mid s \right) \right\}$$

$$= E\left\{ \sum_{i \in s} \frac{E_i^2}{\pi_i^2} \left(\frac{1}{\theta_i} - 1 \right) \right\} + V\left(\sum_{i \in s} \frac{y_i}{\pi_i} \right) = \sum_i \frac{E_i^2}{\pi_i} \left(\frac{1}{\theta_i} - 1 \right) + \frac{1}{2} \sum_{i \neq j} \sum (\pi_i \pi_j - \pi_{ij}) \left(\frac{y_i}{\pi_i} - \frac{y_i}{\pi_i} \right)^2.$$

12.13.2.2 ESTIMATION OF VARIANCE OF THE CALIBRATED ESTIMATOR

It can be easily checked that

$$Q(r) = \sum_{i \in s_r} \left(e_i^2 / \pi_i^2 \theta_i\right)\left(\frac{1}{\theta_i} - 1\right) \tag{12.13.2.5}$$

is an approximate unbiased estimator of $\sum_i \left(E_i^2 / \pi_i\right)\left(\frac{1}{\theta_i} - 1\right)$ where $e_i = y_i - \hat{B}_r x_i$ and B_r is as given in (12.13.2.4). Hence using the Theorem12.13.2 we obtain the following two design consistent estimators for V_{C} as

$$\hat{V}_{\mathrm{C}}(1) = Q(r) + \hat{V}_{11}, \tag{12.13.2.6}$$

and

$$\hat{V}_{\mathrm{C}}(2) = Q(r) + \hat{V}_{12}, \tag{12.13.2.7}$$

where $Q(r)$ is given in (12.13.2.5), and $\hat{V}_{11}$, $\hat{V}_{12}$ are as defined in the Theorem 12.13.2. Following Deville and Särndal (1992), we obtain two alternative model based as well as design based consistent estimators for V_C when we replace $1/\theta_i$ by the calibrated weight w_i in the expressions of $\hat{V}_{\mathrm{C}}(1)$ and $\hat{V}_{\mathrm{C}}(2)$. The estimators are as follows

$$\hat{V}_C(3) = \tilde{Q}(r) + \hat{V}_{13}, \tag{12.13.2.8}$$

and

$$\hat{V}_C(4) = \tilde{Q}(r) + \hat{V}_{14}, \tag{12.13.2.9}$$

where

$$\tilde{Q}(r) = \sum_{i \in s_r} \frac{e_i^2 w_i}{\pi_i^2}(w_i - 1), \quad \hat{V}_{13} = \frac{1}{2}\sum_{i \neq j \in s_r}\sum w_i w_j \frac{\left(\pi_i \pi_j - \pi_{ij}\right)}{\pi_{ij}}\left(\frac{y_i}{\pi_i} - \frac{y_j}{\pi_j}\right)^2,$$

and $\hat{V}_{14} = \sum_{i \in s_r} w_i \frac{y_i^2}{\pi_i}\left(\frac{1}{\pi_i} - 1\right) - \sum_{i \neq j \in s_r}\sum w_i w_j \frac{\left(\pi_i \pi_j - \pi_{ij}\right)}{\pi_{ij}\pi_i \pi_j} y_i y_j$.

Case I. Consider an SRSWOR design where $\pi_i = n/N$ and $\pi_{ij} = n(n-1)/N(N-1)$ and $q_i = 1/x_i$,then

$$\hat{Y}_{\mathrm{C}} = \frac{N}{n}\sum_{i \in s_r} \frac{y_i}{\theta_i} + b_r\left(\frac{N}{n}\sum_{i \in s_r} x_i - \frac{N}{n}\sum_{i \in s_r} \frac{x_i}{\theta_i}\right), \tag{12.13.2.10}$$

where $b_r = \bar{y}_r / \bar{x}_r$. Denoting $\bar{x}_s = \sum_{i \in s} x_i / n$ and $\bar{x}_r = \sum_{i \in s_r} x_i / m$, and in particular, if we assume $\theta_i = \theta$ for every $i = 1,2,..,N$, then we obtain $w_i = \frac{n}{m}\frac{\bar{x}_s}{\bar{x}_r} = w$ (say) and

$$\hat{Y}_{\mathrm{C}} = N\frac{\bar{x}_s}{\bar{x}_r}\bar{y}_r = t_{02}. \tag{12.13.2.11}$$

Remark 12.13.2.1. It is important to note that the estimator (12.13.2.11) is independent of the response probability θ. Hence one can use t_{02} without estimating the response probability θ. The estimator (12.13.2.11) was obtained by Särndal (1992) with the single imputation method. Now replacing θ by its estimate $\hat{\theta} = m/n$ and putting $\pi_i = n/N$, $\pi_{ij} = n(n-1)/N(N-1)$ and $w_i = w$ in the expression of $E_\xi(\varepsilon_i) = 0$ for $j = 1,2,3,4$, we have the following variance estimators for t_{02}

$$v_{\text{OC}}(1) = \frac{N^2}{n}\frac{n-m}{m^2}(m-1)s_{er}^2 + \frac{N(N-n)(m-1)}{m(n-1)}s_{yr}^2,$$

$$v_{\text{OC}}(2) = \frac{N^2}{n}\frac{n-m}{m^2}(m-1)s_{er}^2 + \frac{N(N-n)}{m(n-1)}\left\{(m-1)s_{yr}^2 + \left(\frac{1}{m}-\frac{1}{n}\right)\sum_{i\in s_r} y_i^2\right\},$$

$$v_{\text{OC}}(3) = \frac{N^2}{n}\frac{(m-1)}{m^2}\left(\frac{\bar{x}_s}{\bar{x}_r}\right)\frac{(n\bar{x}_s - m\bar{x}_r)}{\bar{x}_r}s_{er}^2 + \frac{N(N-n)(m-1)}{m(n-1)}\left(\frac{\bar{x}_s}{\bar{x}_r}\right)^2 s_{yr}^2$$

and

$$v_{\text{OC}}(4) = \frac{N^2}{n}\frac{(m-1)}{m^2}\left(\frac{\bar{x}_s}{\bar{x}_r}\right)\frac{(n\bar{x}_s - m\bar{x}_r)}{\bar{x}_r}s_{er}^2 + \frac{N(N-n)}{nm}\frac{\bar{x}_s}{\bar{x}_r}\left(1-\frac{\bar{x}_s}{\bar{x}_r}\right)\left(\sum_{i\in s_r} y_i^2\right)$$

$$+\left(\frac{\bar{x}_s}{\bar{x}_r}\right)^2\left[\frac{N(N-n)}{m(n-1)}\left\{(m-1)s_{yr}^2 + \left(\frac{1}{m}-\frac{1}{n}\right)\sum_{i\in s_r} y_i^2\right\}\right]$$

where $(m-1)s_{er}^2 = \sum_{i\in s_r}(y_i - b_r x_i)^2$.

Case II. Consider an SRSWOR sampling design, $q_i = 1$ and $\theta_i = \theta$ for every $i = 1,2,..,N$, and if we replace θ by its estimate $\hat{\theta} = m/n$ in (12.13.2.1) and obtain

$$w_i = w_{i0} = \frac{n}{m} + n\left(\sum_{i\in s_r} x_i^2\right)^{-1}(\bar{x}_s - \bar{x}_r)x_i, \text{ and } w_i = w_{i0} \text{ yields}$$

$$\hat{Y}_C = N[\bar{y}_r + b(\bar{x}_s - \bar{x}_r)] = \hat{t}_{03} \text{ (say)}, \tag{12.13.2.12}$$

where $b = \sum_{i\in s_r} y_i x_i \Big/ \sum_{i\in s_r} x_i^2$.

Now writing $(m-1)\tilde{s}_{er}^2 = \sum_{i\in s_r}(y_i - b_r x_i)^2$ and $\tilde{e}_i = y_i - bx_i$, we have the following estimators for the variance of (12.13.2.12) as follows

$$\tilde{v}_{\text{OC}}(1) = \frac{N^2}{n}\left(\frac{n-m}{m^2}\right)(m-1)\tilde{s}_{er}^2 + \frac{N(N-n)(m-1)}{m(n-1)}s_{yr}^2,$$

$$\tilde{v}_{\text{OC}}(2) = \frac{N^2}{n}\left(\frac{n-m}{m^2}\right)(m-1)\tilde{s}_{er}^2 + \frac{N(N-n)}{m(n-1)}\left[(m-1)s_{yr}^2 + \left(\frac{1}{m}-\frac{1}{n}\right)\sum_{i\in s_r} y_i^2\right],$$

$$\tilde{v}_{\rm oC}(3)=\frac{N^2}{n^2}\sum_{i\in s_r} w_{i0}\left(w_{i0}-1\right)\tilde{e}_i^2+\frac{N(N-n)}{n^2(n-1)}\frac{1}{2}\sum_{i\neq j\in s_r}\sum w_{i0}w_{j0}\left(y_i-y_j\right)^2,$$

$$\tilde{v}_{\rm oC}(4)=\frac{N^2}{n^2}\sum_{i\in s_r} w_{i0}\left(w_{i0}-1\right)\tilde{e}_i^2+\frac{N(N-n)}{n^2}\sum_{i\in s_r} w_{i0}y_i^2-\frac{N(N-n)}{n^2(n-1)}\sum_{i\neq j\in s_r}\sum w_{i0}w_{0j}y_iy_j.$$

One can also refer to Lundström (1997) and Lundström and Särndal (1999) and they suggested that calibration can be taken as a standard method for treatment of non-response in survey sampling.

EXERCISES

Exercise 12.1. Distinguish between 'Missing at random' and 'Missing completely at random'. Give examples of random non-response and deliberate non-response in interview surveys.

Exercise 12.2. Let $t=t(s)=\sum_{i\in s} b_{si}y_i$ be an estimator of population total $Y=\sum_{i=1}^{N} Y_i$ in the case of complete response. Let $t^*\left(s^*\right)=\sum_{i\ni s^*}\frac{b_{si}}{\theta_i}y_i$ be an estimator of population total in the presence of non-response and θ_i be the probability of response. Compare the variance of the estimator $t^*\left(s^*\right)$ with that of $t(s)$.
Hint: Arnab (1979b).

Exercise 12.3. Let there be a population of N units from which a sample of size n is to be drawn. Let the study variable be denoted by y and the auxiliary variable be denoted by x. The selection probabilities p_i of the population units are taken to be proportional to the corresponding x values. Let r $(r=0,1,2,...,(n-1))$ be the number of units (including repetitions in the case of PPSWR sampling) on which the information on y could be collected. The value of r is supposed to be less than or equal to $(n-2)$ while estimation of variance of the estimator of population total is concerned. Then show that an unbiased estimator of population total Y is given by

$$\hat{Y}_{\rm HTR}=\frac{n}{n-r}\sum_{i=1}^{n-r} d_iy_i$$

and the variance of the estimator $\hat{Y}_{\rm HTR}$ is

$$V\left(\hat{Y}_{\rm HTR}\right)=\sum_{i=1}^{N}\sum_{j>i=1}^{N}\left(\pi_i\pi_j-\pi_{ij}\right)\left(d_iY_i-d_jY_j\right)^2+\frac{1}{n-1}\sum_{i=1}^{N}\sum_{j>i=1}^{N}\pi_{ij}\left(d_iY_i-d_jY_j\right)^2E_1\left(\frac{r}{n-r}\right),$$

where $d_i=\pi_i^{-1}$.
Hint: Singh and Singh (1979), Bouza (1994).

Exercise 12.4. Study the properties of an estimator of finite population variance σ_y^2 in the presence of random non-response given by

$$\hat{v}_1 = s_y^{*2} + \hat{\beta}^{*2}\left(S_x^2 - s_x^{*2}\right), \quad \hat{v}_2 = s_y^{*2} + k_1\left(S_x^2 - s_x^{*2}\right) + k_2\left(S_z^2 - s_z^{*2}\right)$$

where $\hat{\beta}^* = s_{xy}^* / s_x^{*2}$, k_1 and k_2 are real constants.

Hint: Singh and Joarder (1998), Singh, Chandra, and Singh (2003).

Exercise 12.5. Consider (Y_i, X_i) be the value of the ith unit of the study variable y and the auxiliary variable x for a population of size N with population means $\overline{Y}$ and $\overline{X}$, respectively. Let n be the size of the SRSWOR sample drawn from it. Consider only n_1 selected persons respond and n_2 do not, such that $n_1 + n_2 = n$. From the n_2 non-response, let $r = n_2/k$, $k > 1$ units are selected by making extra efforts. An estimator of population mean $\overline{Y}$ is given by

$$\bar{y}^* = \frac{n_1}{n}\bar{y}_1 + \frac{n_2}{n}\bar{y}_2^*,$$

where $\bar{y}_1$ and $\bar{y}_2^*$ are the sample means based on n_1 and r units, respectively, for the study character. Assuming that population mean $\overline{X}$ of the auxiliary variable is known, study the asymptotic properties of the following estimators:

(a) $m_1 = \bar{y}^*\left(\dfrac{\overline{X}}{\bar{x}^*}\right)$, where $\bar{x}^* = \dfrac{n_1}{n}\bar{x}_1 + \dfrac{n_2}{n}\bar{x}_2^*$, $\bar{x}_1$ and $\bar{x}_2^*$ being the sample means based on n_1 and r units, respectively, for the auxiliary variable;

(b) $m_2 = \bar{y}^*\left(\dfrac{\overline{X}}{\bar{x}}\right)$, where $\bar{x} = n^{-1}\sum_{i=1}^{n} x_i$ denote the sample means based on n units for the auxiliary variable;

(c) $m_3 = \bar{y}^*\left(\dfrac{\overline{X} + C}{\bar{x}^* + C}\right)$ and $m_4 = \bar{y}^*\left(\dfrac{\overline{X} + C_1}{\bar{x} + C_1}\right)$,

with C and C_1 suitably chosen constants such that the variances of m_3 and m_4 are minimum. Compare m_3 and m_4 with the estimator of population mean in the presence of full response defined as

$$\bar{y}_{sd} = \bar{y}\left(\frac{\overline{X} + C_x}{\bar{x} + C_x}\right).$$

Hint: Khare and Srivastava (1997).

Exercise 12.6. Let Y_{ij}, $j = 1,2,...,N_i$, be the value of the study variable for the jth unit in the ith stratum $i = 1,2,...,L$. The population of each stratum is divided into two classes, those who will response at the first attempt and those who will not respond, thus creating the problem of incomplete sample in the mail survey. In order to preserve the advantage of mail survey, an investigator decided to apply the following strategy:

(a) Select a random sample from each stratum;
(b) Send a mail questionnaire to all of the selected units in each stratum;
(c) After the deadline is over, identify the non-respondents in each stratum;
(d) Collect data from the selected non-respondents in the sub-sample by interview and combine data from the two parts of the survey in each stratum to provide the unbiased estimate of population mean.

Let $\bar{y}_i$ and $\bar{y}_{r_i}$ be the sample means of the respondent groups in the ith stratum, obtained at the first attempt and second attempt (mail survey, say), respectively.

(I) Show that an unbiased estimator of the population mean is given by

$$\bar{y}_{\text{st}} = \sum_{i=1}^{L} W_i \left[\frac{n_{i1}\bar{y}_i + n_{i2}\bar{y}_{ri}}{r_i} \right]$$

with $W_i = N_i/N$. Find the variance and suggest an estimator of its variance.

Compare the strategies under:

(a) Optimal allocation; (b) Proportional allocation; and (c) Equal allocation.

(II) If the population mean $\bar{X}$ of the auxiliary variable is known then repeat the above exercise for the estimators, defined as

$$\bar{y}_1 = \sum_{i=1}^{L} W_i \left(\frac{n_{i1}\bar{y}_i + n_{i2}\bar{y}_{ri}}{r_i} \right) \left[\bar{X} \Big/ \sum_{i=1}^{L} W_i \left(\frac{n_{i1}\bar{x}_i + n_{i2}\bar{x}_{ri}}{r_i} \right) \right], \quad \bar{y}_2 = \sum_{i=1}^{L} W_i \left[\frac{n_{i1}\bar{y}_i + n_{i2}\bar{y}_{ri}}{r_i} \right] \left[\frac{\bar{X}}{\bar{x}} \right],$$

$$\bar{y}_3 = \sum_{i=1}^{L} W_i \left(\frac{n_{i1}\bar{y}_i + n_{i2}\bar{y}_{ri}}{r_i} \right) \left[\bar{X}_i \Big/ \left(\frac{n_{i1}\bar{x}_i + n_{i2}\bar{x}_{ri}}{r_i} \right) \right], \quad \text{and} \quad \bar{y}_4 = \sum_{i=1}^{L} W_i \left(\frac{n_{i1}\bar{y}_i + n_{i2}\bar{y}_{ri}}{r_i} \right) \left[\bar{X}_i / \bar{x}_i \right].$$

Hint: Khare (1987), Lindley and Deely (1993).

Exercise 12.7. Discuss the different methods of imputation and suggest estimators of variance.
Hint: Lee, Rancourt, and Särndal (1994, 1995a, 1995b).

Exercise 12.8. Discuss Rao--Shao adjustments for estimating the variance of the estimator of population total using Jackknifing under different methods of imputation.

Exercise 12.9. Consider a finite population of size N. Let y_i be the value of the variable under study, y, for the ith unit. Each unit of the population has a definite probability of providing the necessary information with respect to the variable of interest under the particular given field method. Let p_i denote this probability of obtaining required information from the ith unit and let $q_i = 1 - p_i$. Consider we selected a sample of n units by SRSWOR sampling and only m units provided the required information. In the repeated samples, the value of m will be a random variable taking the values $0,1,2,...,n$. Show that an unbiased estimator of the population total is given by

$$\hat{y}_1 = \begin{cases} 0 & \text{if } m = 0, \\ \dfrac{N}{n}\sum_{i=1}^{N}\dfrac{y_i v_i \varepsilon_i}{P_i} & \text{if } m > 0, \end{cases}$$

where $v_i = 1$ or 0 according as the ith unit is in or is not in the sample and $\varepsilon_i = 1$ or 0 according as the ith unit does or does not provide the required information. Find its variance.

Hint: Singh and Narain (1989).

Exercise 12.10. Let a finite population consist of N units. To every unit there is attached a characteristic y. The characteristics are assumed to be measured on a given scale with distinct points $y_1, y_2,..., y_T$. Let N_t be the number of units associated with scale point y_t, with $N = \sum_t N_t$. A simple random sample of size n is drawn. Using the likelihood function of $(N_1, N_2,..., N_T)$ and assuming N/n to be an integer, a maximum likelihood estimate of population mean $\bar{Y}$ is given by, $\bar{y}_{\text{ml}} = \frac{1}{n}\sum_t n_t y_t$, where n_t is the number of y_t values observed in the sample. When non-response is observed assume that response is obtained from $n(r)$ units in the sample and non-response from $n(\bar{r})$ units, such that $n(r) + n(\bar{r}) = n$. Show that under the likelihood function, defined as

$$L\left[N_{r1}, N_{r2},...., N_{rT(r)}; N(\bar{r})\right] = \left[\binom{N_{r1}}{n_{r1}}\binom{N_{r2}}{n_{r2}}.......\binom{N_{rT(r)}}{n_{rT(r)}}\binom{N(\bar{r})}{n(\bar{r})}\right] \Big/ \binom{N}{n},$$

a maximum likelihood estimator of the population mean is given by

$$\bar{y}_r = \frac{1}{n(r)}\sum_t n_{rt} y_{rt}.$$

Hint: Laake (1986).

Exercise 12.11. (a) Adjust the inclusion probabilities in the Horvitz and Thompson (1952) estimator of population mean (or total) defined as

$$t_{\text{HT}} = N^{-1}\sum_{k \in s}(y_k/\pi_k)$$

where only r out of n units $(r < n)$ are responding. Let $\pi_k^{(j)}$ denotes the jth adjusted inclusion probability, then study the asymptotic properties of the following five estimators

$$t_{\text{cc(j)}} = N^{-1}\sum_{k \in r} y_k / \pi_k^{(j)}, \quad j = 1,2,3,4,5,$$

where

$\pi_k^{(1)} = (\pi_k)^{r/n}$, $\pi_k^{(2)} = \frac{r}{n}\pi_k$, $\pi_k^{(3)} = \pi_k \frac{\Pr(k \in U_R \mid I_k = 1)}{\Pr(k \in U_R)}$, with U_R being a particular set of units in the population that would respond to the survey question, and I_k is an indicator function defined as $I_k = \begin{cases} 1 & \text{if } k \in s, \\ 0 & \text{if } k \notin s, \end{cases}$ such that $\pi_k = \Pr[I_k = 1 \mid k \in U_R]$; $\pi_k^{(4)} = \frac{r}{n}\left[(1 - r_{xy})\frac{n}{N} + r_{xy}\,\pi_k\right]$, where r_{xy} is an estimator of correlation coefficient ρ_{xy} between x and y; and $\pi_k^{(5)} = (1 - r_{xy})(\pi_k)^{r/n} + r_{xy}\frac{r}{n}(\pi_k)$. Study the bias and variance properties of these five estimators.

Hint: Chaubey and Crisalli (1995).

(b) Consider here an estimator of the population total Y as

$$\hat{Y}^{*}_{w\Omega} = \sum_{i \in r} d_i v^{*}_{\Omega i} y_i,$$

where $v^{*}_{\Omega i}$ are the new calibrated response weights obtained by minimizing a new penalized chi square distance function, defined as

$$D_1 = \frac{1}{2}\sum_{i \in r}\frac{\left(v^{*}_{\Omega i} - 1\right)^2}{q_i} + \frac{1}{2}\sum_{i \in r}\frac{\Phi_i v^{*2}_{\Omega i}}{q_i},$$

where q_i are the suitably chosen weights to form different types of estimators, and Φ_i is a relaxed penalty, and it can take any value in the range $(-1, \infty)$.

Under such assumptions study the following situations:

(i) Show that, when no auxiliary information is available, the minimization of D_1 leads to calibrated weights given by

$$v^{*}_{\Omega i} = 1/(1 + \Phi_i).$$

(ii) Show that the resultant estimator of the population total, Y, becomes

$$\hat{Y}^{*}_{w\Omega}(0) = \sum_{i \in r}\frac{d_i}{1 + \Phi_i} y_i .$$

Note that the estimator, $\hat{Y}^{*}_{w\Omega}(0)$, is free from q_i and thus comment on its choice.

Further verify the following cases for different choices of Φ_i:

Case I. If $\Phi_i = V\left(\hat{Y}^{*}_{w\Omega}\right)/Y^2$ and is constant for each respondent in the sample, then $\hat{Y}^{*}_{w\Omega}(0)$ reduces to the Searls (1964) estimator in the presence of non-response.

Case II. If $\Phi_i = (\pi_i)^{((r/n)-1)} - 1$ then the estimator $\hat{Y}^*_{w\Omega}(0)$ becomes

$$\hat{Y}^*_{w\Omega}(0)_{cc(1)} = \sum_{i\in r}\frac{y_i}{(\pi_i)^{(r/n)}}$$

which is the first estimator considered by Chaubey and Crisalli (1995).

Case III. If $\Phi_i = \left(\frac{r}{n} - 1\right)$ then the estimator $\hat{Y}^*_{w\Omega}(0)$ becomes

$$\hat{Y}^*_{w\Omega}(0)_{cc(2)} = \frac{n}{r}\sum_{i\in r} d_i y_i$$

which is the second estimator considered by Chaubey and Crisalli (1995).

Case IV. If $\Phi_i = \dfrac{\Pr(i \in \Omega_R \mid I_i = 1) - \Pr(i \in \Omega_R)}{\Pr(i \in \Omega_R)}$, where Ω_R represent set of units in the population that would respond to the survey question and I_i is an indicator function such that $I_i = \begin{cases} 1 & \text{if } i \in s \\ 0 & \text{if } i \notin s \end{cases}$ and $\pi_i = \Pr[I_i = 1 \mid i \in \Omega_R]$, then the $\hat{Y}^*_{w\Omega}(0)$ becomes

$$\hat{Y}^*_{w\Omega}(0)_{cc(3)} = \sum_{i\in r}\frac{\Pr(i \in \Omega_R)}{\Pr(i \in \Omega_R \mid I_i = 1)} y_i$$

which is the third estimator considered by Chaubey and Crisalli (1995).

Case V. If $\Phi_i = \left[(1 - r_{xy})\pi_i^{(r/n-1)} + r_{xy}(r/n)\right] - 1$ then the estimator $\hat{Y}^*_{w\Omega}(0)$ becomes

$$\hat{Y}^*_{w\Omega}(0)_{cc(4)} = \sum_{i\in r}\frac{y_i}{\left\{(1 - r_{xy})\pi_i^{(r/n)} + r_{xy}\pi_i(r/n)\right\}}$$

which is the fourth estimator considered by Chaubey and Crisalli (1995).

Case VI. If $\Phi_i = \left[(1 - r_{xy})(r/N)(1/\pi_i) + r_{xy}(r/n)\right] - 1$ then the estimator $\hat{Y}^*_{w\Omega}(0)$ becomes

$$\hat{Y}^*_{w\Omega}(0)_{cc(5)} = \frac{n}{r}\sum_{i\in r}\frac{y_i}{\left\{(1 - r_{xy})(n/N) + r_{xy}\pi_i\right\}}$$

which is the fifth estimator considered by Chaubey and Crisalli (1995).

(i) Minimize D_1 subject to the calibration constraint

$$\sum_{i\in r} d_i v_{si} x_i = \sum_{i\in s} x_i$$

and study the resultant estimators.

(ii) Suggest a few new calibration constraints and study the resultant estimators.

Hint: Singh and Arnab (2003).

Exercise 12.12. (I) Assume the data after imputation takes the form

$$y_{\bullet i} = \begin{cases} \dfrac{(n-1)}{(r-1)} y_i & \text{if } i \in A, \\ r^{-1}(r-1)^{-1}\{(n-1)x_i - n\bar{x}_n\}\sum_{i=1}^{r}\dfrac{y_i}{x_i} & \text{if } i \in \bar{A}, \end{cases}$$

where A and $\bar{A}$ denote the responding and non-responding sets of the sample s such that $s = A \cup \bar{A}$.

(a) Show that point estimator

$$\hat{e}_2 = \frac{1}{n}\sum_{i=1}^{n} y_{\bullet i}$$

of the population mean $\bar{Y}$ becomes unbiased estimator of the population mean

$$\hat{e}_2 = \bar{z}_r\bar{x}_n + \frac{r(n-1)}{n(r-1)}\left(\bar{y}_r - \bar{z}_r\bar{x}_r\right),$$

where $\bar{y}_r = r^{-1}\sum_{i \in A} y_i$, $\bar{x}_r = r^{-1}\sum_{i \in A} x_i$, and $\bar{x}_n = n^{-1}\sum_{i \in s} x_i$, $\bar{z}_r = r^{-1}\sum_{i=1}^{r}\frac{y_i}{x_i}$ have the usual meanings.

(b) Find the variance of the estimator $\hat{e}_2$ and estimate it by Jackknife technique.

(II) Consider the data after imputation take the form

$$y_{\bullet i} = \begin{cases} \dfrac{r(n-r+1)}{n} y_i - \dfrac{(n-r)(r-1)}{r}\bar{y}_r(i)\dfrac{\bar{x}_n}{\bar{x}_r(i)} & \text{if } i \in A, \\ \dfrac{r(n-r+1)}{n}\left(\dfrac{\sum_{i=1}^{r} y_i}{\sum_{i=1}^{r} x_i}\right)x_i & \text{if } i \in \bar{A}, \end{cases}$$

where $\bar{y}_r(i) = \dfrac{r\bar{y}_r - y_i}{r-1}$ and $\bar{x}_r(i) = \dfrac{r\bar{x}_r - x_i}{r-1}$.

(a) Show that the point estimator $\hat{e}_2 = \frac{1}{n}\sum_{i=1}^{n} y_{\bullet i}$ of the population mean $\bar{Y}$ under this method of imputation becomes

$$\hat{e}_3 = \frac{r(n-r+1)}{n}\bar{y}_r\frac{\bar{x}_n}{\bar{x}_r} - \frac{(n-r)(r-1)}{rn}\sum_{i=1}^{r}\bar{y}_r(i)\frac{\bar{x}_n}{\bar{x}_r(i)}.$$

(b) Show that the estimator $\hat{e}_3$ is an unbiased estimator of the population mean.

(c) Show that the variance of the estimator $\hat{e}_3$, to the first order of approximation, is equivalent to that of the ratio estimator in two-phase sampling.

Hint: Singh, Horn, and Tracy (2001).

Exercise 12.13. Consider a population consists of N units. We draw an SRSWOR sample of n units and measured the study variable y and auxiliary variable x. We are interested in estimating the population ratio $R = \bar{Y}/\bar{X}$ of the population means $\bar{Y}$ and $\bar{X}$. Assume that after taking the sample only a set of $(n-p-q)$ complete observations (y_i, x_i), $i = 1, 2, ..., (n-p-q)$ is available. Assume that p observations $x_1^*, x_2^*, ..., x_p^*$ in the sample are available for the x variable but the corresponding observations of the y variable are missing. Similarly assume that we have q observations $y_1^{**}, y_2^{**}, ..., y_q^{**}$ on y variable in the sample, but the associated values of the auxiliary variable x are missing. Defining

$$\bar{x} = (n-p-q)^{-1}\sum_{i=1}^{n-p-q} x_i\ ,\ \bar{y} = (n-p-q)^{-1}\sum_{i=1}^{n-p-q} y_i\ ,\ \ \bar{x}^* = p^{-1}\sum_{i=1}^{p} x_i^*,\ \text{ and }\ \bar{y}^{**} = q^{-1}\sum_{i=1}^{p} y_i^{**}\ .$$

Find the bias and variance expressions of the following estimators of R defined as

$$r_1 = \frac{\bar{y}}{\bar{x}},\qquad r_2 = \frac{(n-q)\bar{y}}{(n-p-q)\bar{x} + p\bar{x}^*},\qquad r_3 = \frac{(n-p-q)\bar{y} + q\bar{y}^{**}}{(n-p)\ \bar{x}},$$

$$r_4 = \left(\frac{n-q}{n-p}\right)\frac{(n-p-q)\ \bar{y} + q\bar{y}^{**}}{(n-p-q)\ \bar{x} + p\bar{x}^{**}},\quad \text{and}\quad r_5 = \alpha_0\left\{\left(\alpha_1\bar{y} + \alpha_2\bar{y}^{**}\right)\Big/\left(\alpha_3\bar{x} + \alpha_4\bar{x}^{**}\right)\right\}.$$

Show that the estimators r_i, $i = 1, 2, 3, 4,$ are the special cases of the estimator r_5 for different choices of α_j, $j = 0,1,2,3,$ and 4.

Hint: Toutenburg and Srivastava (1998).

Exercise 12.14. Consider the population of interest has been stratified into L strata with N_h clusters in the hth stratum. At the first stage of sampling, let $n_h \geq 2$ clusters are selected from the hth stratum. Let P_{hi}, $i = 1,2,...,N_h$, $h = 1,2,...,L$ denote the probability of selecting the ith cluster from the hth stratum Assuming that these clusters are selected independently across strata and without replacement and the overall sampling fraction $\sum_h n_h \Big/ \sum_h N_h$ is negligible. Assume y_{hik} be the ultimate population units value, where the index (h, i, k), $k = 1,...,N_{hi}$; $i = 1,...,N_h$; $h = 1,...,L$, have their usual meanings. We wish to estimate the population total

$$Y = \sum_{h=1}^{L}\sum_{i=1}^{N_h}\sum_{k=1}^{N_{h_i}} y_{hik}.$$

If there is no non-response and w_{hik} are the design weights, then show that the unbiased estimator of population total Y is given by

$$\hat{Y} = \sum_{(hik)\in s} w_{hik}\, y_{hik}.$$

In a stratified multi-stage sampling design consider for the ratio method of imputation that the adjusted values are given by:

$$\tilde{y}_{hik}^{(r)}(\varepsilon)=\begin{cases}\hat{\beta}_v^{(r)}(\varepsilon)x_{hik} & \text{if } y_{hik} \text{ is missing,}\\ y_{hik} & \text{if } y_{hik} \text{ is observed,}\end{cases}$$

with

$$\hat{\beta}_v^{(r)}(\varepsilon)=\sum_{(hik)\in s}\delta_{hikv}w_{hik}^{(r)}(\varepsilon)y_{hik}\Big/\sum_{(hik)\in s}\delta_{hikv}w_{hik}^{(r)}(\varepsilon)x_{hik}\,.$$

Then the balanced repeated replication $\text{BRR}(\varepsilon)$, for a fixed ε, method of variance estimation is

$$\hat{v}_{\text{BRR}}(\varepsilon)=\frac{1}{\varepsilon^2 R}\sum_{r=1}^{R}\left\{\hat{\theta}^{(r)}(\varepsilon)-\hat{\theta}\right\}^2$$

where $\hat{\theta}$ is an estimator of any parameter of interest, say θ, and $\hat{\theta}^{(r)}(\varepsilon)$ is computed using the same formula for $\hat{\theta}$ but with the original weights w_{hik} replaced with the new weights, given by

$$w_{hik}^{(r)}(\varepsilon)=\begin{cases}\left\{1+\varepsilon\left(\dfrac{n_h-m_h}{m_h}\right)^{\frac{1}{2}}\right\}w_{hik} & \text{if the } (\text{h},\text{i})\text{th cluster is selected in the } r^{th} \text{ replicate,}\\[2ex] \left\{1+\varepsilon\left(\dfrac{n_h-m_h}{m_h}\right)^{\frac{1}{2}}\right\}w_{hik} & \text{otherwise.}\end{cases}$$

Discuss your views by taking $\hat{\theta}=\tilde{y}_{hik}$ and $\varepsilon=0.5$. Also show that $\hat{v}_{\text{BRR}}(\varepsilon)/\text{var}(\hat{\theta})\to 1$.

Hint: Rao and Shao (1999).

Exercise 12.15. Show that under the multi-stage post-stratified sampling design an estimator of population total Y in the presence of unit non-response is

$$\hat{Y}_{\text{N,PS}}=\sum_c\sum_p\sum_{(hik)\in s}d_{\text{p}\,c}b\,w_{hik}a_{hik\,c}\delta_{hik}\eta_{hik}^p y_{hik}\,,$$

where ${}_cM$ = post-stratum count, y_{hik} denotes the value of the kth unit of the study variable in the ith cluster and h^{th} stratum,

$$d_{\text{p}}=\frac{\sum_{(hik)\in s}w_{hik}\eta_{hik}^p}{\sum_{(hik)\in s}w_{hik}a_{hik}\eta_{hik}^p},\quad a_{hik}=\begin{cases}1 & \text{if } (hik) \text{ responds,}\\ 0 & \text{otherwise,}\end{cases}$$

$${}_cb=\frac{{}_cM}{\sum_p\sum_{(hik)\in s}d_p w_{hik}a_{hik\,c}\delta_{hik}\eta_{hik}^p},\quad \eta_{hik}^p=\begin{cases}1 & \text{if } (hik)\in p^{th} \text{ weighing class,}\\ 0 & \text{otherwise,}\end{cases}$$

and $w_{hik}(>0)$ denotes the design weights. Suggest a method to estimate the variance of the estimator $\hat{Y}_{N,PS}$ using the concept of Jackknife.

Hint: Yung and Rao (2000).

Exercise 12.16. Consider we took a sample s_n of n units using SRSWOR sampling from a population Ω of N units. Let y_i, $i = 1,2,...,n$ denote the value of the ith unit selected in the sample. Unfortunately there is a random non-response, that is, only $r < n$ units selected in the sample are responding. Let y_i^*, $i = r+1,\ r+2,...,n$ denote the imputed values.

The situation is as shown in the following table.

Sampled units	Responding units	Imputed values	Adjusted data
1	y_1	y_1^*	a_1
2	y_2	y_2^*	a_2
3	y_3	y_3^*	a_3
.	.	.	.
r	y_r	y_r^*	a_r
$r+1$	Missing	y_{r+1}^*	a_{r+1}
$r+2$	Missing	y_{r+2}^*	a_{r+2}
.	.	.	.
n	Missing	y_n^*	a_n

Let $\bar{y}^*$ and $\text{var}(\bar{y}^*)$ denote he sample mean and sample variance of the sample which comprises the values making up y^*. Let m and $u > 0$ be the fixed real numbers. Let $a = (a_1, a_2, ..., a_n)$ be the vector of real numbers. Show that the minimization of $\sum_{i=1}^{n}(a_i - y_i^*)^2$ subject to the two constraints $\bar{a} = \frac{1}{n}\sum_{i=1}^{n} a_i = m$ and $\text{var}(a) = \frac{1}{n-1}\sum_{i=1}^{n}(a_i - m)^2 = u$ leads to the adjusted data set given by $a_i^0 = m + \left\{\frac{u}{\text{var}(y^*)}\right\}^{0.5}(y_i^* - \bar{y}^*)$, $i = 1,2,...,n$. Show that the mean method of imputation is a special case of it. Extend the results for ratio method of imputation. Let v_i be another variable corresponding to the auxiliary variable. Show that the minimization of $\sum_{i=1}^{n}(a_i - y_i^*)^2$ subject to the two constraints $\bar{a} = \frac{1}{n}\sum_{i=1}^{n} a_i = m$ and $\text{var}(\bar{a}) = \frac{1}{n-1}\sum_{i=1}^{n}(a_i - v_i)^2 = u$ leads to the adjusted data similar to the ratio method of imputation.

Hint: Meeden (2000).

Exercise 12.17. Consider a census of the population has been undertaken and all units could be classified by the call back index number on which they respond by forming L strata. Each call back stratum h would be of size N_h and the total population size $N = \sum_h N_h$. Let Y_{hi} be the (fixed) value of a variable of interest for the ith unit in the hth call back stratum. Defining $\bar{Y} = N^{-1}\sum_{h=1}^{L}\sum_{i=1}^{N_h} Y_{hi}$, $\sigma^2 = N^{-1}\sum_{h=1}^{L}\sum_{i=1}^{N_h}\left(Y_{hi} - \bar{Y}\right)^2$, $\bar{Y}_h = N_h^{-1}\sum_{i=1}^{N_h} Y_{hi}$ and $\sigma_h^2 = N_h^{-1}\sum_{i=1}^{N_h}\left(Y_{hi} - \bar{Y}_h\right)^2$. Show that under the sub-sampling strategy α an unbiased estimator of $\bar{Y}$ is given by $\bar{y}^{(\alpha)} = \sum_h w_h^{\alpha}\bar{y}_h$ where $\bar{y}_h = \frac{1}{n_h^{\alpha}}\sum_{i\in s_h} Y_{hi}$ for the set of interviewed units s_h in the stratum h , n_h^{α} is the number of interviews obtained in h^{th} call back stratum under the α sub-sampling strategy and $w_h^{\alpha} = \dfrac{n_h^{\alpha}/\alpha_h}{\sum_h\left(n_h^{\alpha}/\alpha_h\right)}$ for $\alpha_h = \begin{cases}1 & \text{if } k < m,\\ \alpha & \text{if } k > m.\end{cases}$ Show that the conditional variance for the fixed number of attempts to be made is given by

$$V\left(\bar{y}^{(\alpha)}\right) \approx n^{\alpha}\left(\sum_h \frac{n^{\alpha}}{\alpha_h}\right)^{-2}\sum_h \frac{p_h}{\alpha_h}\left[\left(\sigma_h^2 + \left(\bar{Y}_h - \bar{Y}\right)^2\right)\right].$$

Hint: Elliott, Little, and Lewitzky (2000).

Exercise 12.18. Consider for estimating the mean of a finite population, a random sample of n units is distributed equally among m enumerators chosen randomly from an infinite population of enumerators. Assume that y_{ij} the value of y on unit i as enumerated by interviewer j, is given by the model $y_{ij} = y_i + a_j + e_{ijk}$, where $E\left(e_{ijk} \mid i, j\right) = 0$, $V\left(e_{ijk} \mid ij\right) = S^2$, and $\text{Cov}\left(e_{ijk}, e_{ijk'}\right) = 0 \;\; \forall\, k \neq k'$.
Find the expected value and variance of the estimator defined as

$$\bar{y}_s = \frac{1}{mn}\sum_{j=1}^{m}\sum_{i=1}^{n} y_{ij} .$$

Hint: Sukhatme (1953).

Exercise 12.19. Assume the data after imputation take the form

$$y_{\bullet i} = \begin{cases} y_i & \text{if } \; i \in A,\\ \bar{y}_r\left[\dfrac{(n-r)\bar{x}_n + \alpha r\left(\bar{x}_n - \bar{x}_r\right)}{\alpha\bar{x}_r + (1-\alpha)\bar{x}_n}\right]\dfrac{x_i}{\sum_{i\in R_p^C} x_i} & \text{if } \; i \in \bar{A}, \end{cases}$$

where α is a suitably chosen constant, such that the variance of the resultant estimator is minimum. The sets A ($\bar{A}$) denote the sets of responding (non-responding) units in the sample s such that $s = A \cup \bar{A}$.

(a) Show that the point estimator

$$\bar{y}_{\text{p}} = \frac{1}{n}\sum_{i=1}^{n} y_{\bullet i}$$

of population mean $\bar{Y}$ under this method of imputation becomes

$$\bar{y}_{\text{p}} = \frac{\bar{y}_r \bar{x}_n}{\{\alpha\bar{x}_r + (1-\alpha)\bar{x}_n\}},$$

where $\bar{y}_r = r^{-1}\sum_{i\in A} y_i$, $\bar{x}_r = r^{-1}\sum_{i\in A} x_i$, and $\bar{x}_n = n^{-1}\sum_{i\in s} x_i$.

Study the asymptotic bias and variance properties of the estimator $\bar{y}_{\text{p}}$.

Hint: Singh (2000c).

Exercise 12.20. (I) Consider the data after imputation becomes

$$y_{\bullet i} = \begin{cases} y_i & \text{if } i \in s, \\ \hat{b}x_i & \text{if } i \in s' - s, \end{cases}$$

where $\hat{b} = \sum_{i\in s} y_i \big/ \sum_{i\in s} x_i$.

(a) Show that the point estimator of population mean given by

$$\bar{y}_s = n^{-1}\sum_{i\in s'} y_{\bullet i}$$

becomes

$$\bar{y}_{\text{R}} = \bar{y}_n \frac{\bar{x}_{n'}}{\bar{x}_n}$$

with variance

$$V(\bar{y}_{\text{R}}) = \left(\frac{1}{n'} - \frac{1}{N}\right)S_y^2 + \left(\frac{1}{n} - \frac{1}{n'}\right)S_d^2$$

where $\bar{x}_{n'} = n'^{-1}\sum_{i\in s'} x_i$, $\bar{x}_n = n^{-1}\sum_{i\in s} x_i$, $\bar{y}_n = n^{-1}\sum_{i\in s} y_i$, $S_y^2 = \frac{1}{N-1}\sum_{i\in\Omega}(Y_i - \bar{Y})^2$,

$S_d^2 = \frac{1}{N-1}\sum_{i\in\Omega} d_i^2$, $\bar{Y} = N^{-1}\sum_{i\in\Omega} Y_i$, $d_i = y_i - Rx_i$ and $R = \sum_{i\in\Omega} Y_i \Big/ \sum_{i\in\Omega} X_i$.

(b) Show that the estimator $\bar{y}_{\text{R}}$ is design consistent for population mean $\bar{Y}$.

(c) A design consistent linearization estimator of the variance of $\bar{y}_{\text{R}}$ is given by the standard formula

$$\hat{v}_0 = \left(\frac{1}{n} - \frac{1}{n'}\right)s_d^2 + \left(\frac{1}{n'} - \frac{1}{N}\right)s_y^2$$

with $s_d^2 = (n-1)^{-1}\sum_{i\in s} d_i^2$, $s_y^2 = (n-1)^{-1}\sum_{i\in s}(y_i - \bar{y})^2$, where $d_i = (y_i - \bar{y}) - \hat{R}(x_i - \bar{x})$.

(d) Show that s_y^2 can be written as

$$s_y^2 = s_d^2 + 2\hat{R}s_{dx} + \hat{R}^2 s_x^2$$

which is an estimator of S_y^2.

(e) Develop a new estimator of S_y^2 by replacing s_x^2 with $s_x^{'2}$ as

$$s_y^2 = s_d^2 + 2\hat{R}s_{dx} + \hat{R}^2 s_x^{'2}.$$

Use it in estimating the variance of the ratio estimator.

(f) Study the following two new estimators for the variance of $\bar{y}_R$ as

$$\hat{v}_1 = \left(\frac{1}{n} - \frac{1}{N}\right)s_d^2 + 2\left(\frac{1}{n'} - \frac{1}{N}\right)\hat{R}s_{dx} + \left(\frac{1}{n'} - \frac{1}{N}\right)\hat{R}^2 s_x^{'2}$$

and

$$\hat{v}_0^* = \left(\frac{1}{n} - \frac{1}{N}\right)s_d^2 + 2\left(\frac{1}{n'} - \frac{1}{N}\right)\hat{R}s_{dx} + \left(\frac{1}{n'} - \frac{1}{N}\right)\hat{R}^2 s_x^2.$$

(g) Consider a Jackknife variance estimator for $\bar{y}_R$ by recalculating $\bar{y}_R$ with the jth element removed for each $j \in s'$ and then using the variance of these n' Jackknife values, $\bar{y}_R(j)$. Define

$$\bar{y}_R(j) = \bar{y}(j)\frac{\bar{x}'(j)}{\bar{x}(j)} \quad \text{for all } j \in s',$$

where

$$\bar{x}(j) = \begin{cases} \dfrac{n\bar{x}_n - x_j}{n-1} & \text{if } j \in s, \\ \bar{x}_n & \text{if } i \in s'-s, \end{cases} \qquad \bar{y}(j) = \begin{cases} \dfrac{n\bar{y}_n - y_j}{n-1} & \text{if } j \in s, \\ \bar{y}_n & \text{if } i \in s'-s, \end{cases}$$

and , $\bar{x}'(j) = \dfrac{n'\bar{x}_{n'} - x_j}{n'-1}$ if $j \in s'$.

Show that the linearization of the Jackknife estimator of variance of $\bar{y}_R$ as

$$\hat{v}_1^J = \frac{n'-1}{n'}\sum_{j \in s'}\{\bar{y}_R(j) - \bar{y}_R\}^2$$

is approximately equal to usual estimators of variance.

Hint: Rao and Sitter (1995).

$$\bar{y}(j)\frac{\bar{x}'(j)}{\bar{x}(j)} - \bar{y}_n\frac{\bar{x}_{n'}}{\bar{x}_n} = \left(\frac{n\bar{y}_n - y_j}{n-1}\right)\frac{\bar{x}'(j)}{\bar{x}(j)} - \bar{y}_n\frac{\bar{x}_{n'}}{\bar{x}_n}$$

$$= -\frac{\bar{x}'(j)}{\bar{x}(j)}\left\{\frac{y_j}{n-1} - \frac{\bar{y}_n x_j}{\bar{x}(n-1)}\right\} - \frac{\bar{x}'(j)\bar{y}_n x_j}{\bar{x}(j)\bar{x}_n(n-1)} - \bar{y}_n\frac{\bar{x}_{n'}}{\bar{x}_n} + \frac{n\bar{y}_n}{n-1}\frac{\bar{x}'(j)}{\bar{x}(j)}$$

$$= -\frac{\bar{x}'(j)}{\bar{x}(j)}\left\{\frac{y_j - \hat{R}x_j}{n-1}\right\} + \frac{\bar{x}'(j)\bar{y}_n}{\bar{x}(j)(n-1)}\left\{n - \frac{x_j}{\bar{x}_n}\right\} - \bar{y}_n\frac{\bar{x}_{n'}}{\bar{x}_n}$$

$$= -\frac{\bar{x}'(j)}{\bar{x}(j)}\left\{\frac{y_j - \hat{R}x_j}{n-1}\right\} + \frac{\bar{x}'(j)\bar{y}_n}{\bar{x}(j)(n-1)}\left\{\frac{n\bar{x}_n - x_j}{\bar{x}}\right\} - \bar{y}_n\frac{\bar{x}_{n'}}{\bar{x}_n}$$

$$= -\frac{\bar{x}'(j)}{\bar{x}(j)}\left\{\frac{y_j - \hat{R}x_j}{n-1}\right\} - \frac{\bar{x}'(j)\bar{y}_n}{\bar{x}(j)\bar{x}_n}\left\{\frac{n\bar{x}_n - x_j}{n-1}\right\} - \bar{y}_n\frac{\bar{x}_{n'}}{\bar{x}_n}$$

$$= -\frac{\bar{x}'(j)}{\bar{x}(j)}\left\{\frac{y_j - \hat{R}x_j}{n-1}\right\} + \frac{\bar{x}'(j)\bar{y}_n}{\bar{x}(j)\bar{x}_n}\left\{\frac{n\bar{x}_n - x_j}{n-1}\right\} - \bar{y}_n\frac{\bar{x}_{n'}}{\bar{x}_n}$$

$$= -\frac{\bar{x}'(j)}{\bar{x}(j)}\left\{\frac{y_j - \hat{R}x_j}{n-1}\right\} + \frac{\bar{x}^*(j)\bar{y}_n}{\bar{x}_n} - \bar{y}_n\frac{\bar{x}_{n'}}{\bar{x}_n}$$

$$= -\frac{\bar{x}'(j)}{\bar{x}(j)}\left\{\frac{y_j - \hat{R}x_j}{n-1}\right\} + \frac{\bar{y}_n}{\bar{x}_n}\left\{\frac{n'\bar{x}_{n'} - x_j}{n'-1} - \bar{x}_{n'}\right\}$$

$$= -\frac{\bar{x}'(j)}{\bar{x}(j)}\left\{\frac{y_j - \hat{R}x_j}{n-1}\right\} + \frac{\bar{y}_n}{\bar{x}_n}\left\{\bar{x}'(j) - \bar{x}_{n'}\right\}$$

$$= -\frac{\bar{x}'(j)}{\bar{x}(j)}\left\{\frac{y_j - \hat{R}x_j}{n-1}\right\} + \hat{R}\left\{\frac{\bar{x}_{n'} - x_j}{n'-1}\right\}.$$

(II) Consider a modified Jackknife estimator of the variance as

$$\hat{v}_1^m = \frac{n'-1}{n'}\sum_{j=1}^{n'}\left[\bar{y}_R^{**}(j) - \bar{y}_R\right]^2,$$

where

$$\bar{y}_R^{**}(j) = \bar{y}_R^*(j)\frac{\bar{x}'(j)}{\bar{x}^*(j)}, \qquad x'(j) = \frac{n'x_{n'} - x_j}{n'-1} \text{ if } j \in s'$$

$$\bar{y}_R^*(j) = \begin{cases} \dfrac{n\bar{y}_n\dfrac{\bar{x}_{n'}}{\bar{x}_n} - y_j}{n-1} & \text{if } j \in s, \\[2ex] \dfrac{n\bar{y}_n\dfrac{\bar{x}_{n'}}{\bar{x}_n} - \dfrac{\bar{y}_n}{\bar{x}_n}x_j}{n-1} & \text{if } j \in s'-s, \end{cases} \quad \text{and } \bar{x}^*(j) = \begin{cases} \dfrac{n\bar{x}_n - x_j}{n-1} & \text{if } j \in s, \\[2ex] \dfrac{n\bar{x}_{n'} - x_j}{n-1} & \text{if } j \in s'-s. \end{cases}$$

(a) Show that the modified Jackknife method of variance estimation becomes

$$\bar{y}_R^{**}(j) - \bar{y}_R = \begin{cases} -\dfrac{\bar{x}'(j)}{\bar{x}(j)}\dfrac{(y_j - \hat{R}x_j)}{n-1} + \hat{R}(\bar{x}'(j) - \bar{x}_{n'}) + \left(\dfrac{n}{n-1}\right)\hat{R}\dfrac{\bar{x}'(j)}{\bar{x}(j)}(\bar{x}_{n'} - \bar{x}_n) & \text{if } j \in s, \\[2ex] \hat{R}(\bar{x}'(j) - \bar{x}_{n'}) & \text{if } j \in s'-s. \end{cases}$$

(b) Show that the modified Jackknife estimator of variance reduces to

$$\hat{v}_1^m \cong \left(\frac{\bar{x}_{n'}}{\bar{x}_n}\right)^2\frac{s_d^2}{n} + 2\left(\frac{\bar{x}_{n'}}{\bar{x}_n}\right)\frac{\hat{R}s_{dx}}{n'} + \frac{\hat{R}^2 s_x^{'2}}{n'} + \frac{n^2(n'-1)}{n'(n-1)^2}\hat{R}^2\left(\frac{\bar{x}_{n'}}{\bar{x}_n}\right)^2(\bar{x}_{n'} - \bar{x}_n)^2.$$

Hint: Arnab and Singh (2002e)

Exercise 12.21. For estimating the mean of a finite population, a random sample of n units is distributed equally among m enumerators chosen randomly from an infinite population of enumerators. Assume that y_{ij}, the value of y on the ith unit as enumerated by the jth interviewer, is

$y_{ij} = y_i + \alpha_j + e_{ijk}$, where $E(e_{ijk} \mid i, j) = 0$, $V(e_{ijk} \mid i, j) = \sigma_e^2$ and $\text{Cov}(e_{ijk}, e_{ijk'} \mid i, j) = 0$.
(a) Find the expected value and variance of the sample mean.
(b) Assuming that the cost function is of the form

$$C = nC_1 + mC_2 + \sqrt{nm}C_3 .$$

Find the optimum number of enumerators for which the variance of the sample mean will be minimum for the fixed cost.
Hint: Sukhatme (1953)

Exercise 12.22. (I) Consider the data after imputation take the form

$$y_{\bullet i} = \begin{cases} y_i & \text{if } i \in A \\ \bar{y}_r\left[n\left(\dfrac{\bar{x}_n}{\bar{x}_r}\right)^{\alpha} - r\right]\dfrac{x_i}{\sum\limits_{i \in R_p^C} x_i} & \text{if } i \in \bar{A} \end{cases}$$

where α is a suitably chosen constant, such that the variance of the resultant estimator is minimum. The sets $A(\bar{A})$ denote the sets of responding (non-responding) units in the sample s such that $s = A \bigcup \bar{A}$.

(a) Show that the point estimator $\bar{y}_s = \dfrac{1}{n} \sum\limits_{i \in s} y_{\bullet i}$ of population mean $\bar{Y}$ becomes:

$$\bar{y}_s = \bar{y}_r\left(\frac{\bar{x}_n}{\bar{x}_r}\right)^{\alpha},$$

where $\bar{y}_r = r^{-1}\sum\limits_{i \in A} y_i$, $\bar{x}_r = r^{-1}\sum\limits_{i \in A} x_i$, and $\bar{x}_n = n^{-1}\sum\limits_{i \in s} x_i$
(b) Study the asymptotic bias and variance properties of the resultant estimator $\bar{y}_s$.
(c) Show that $\alpha = 1$ and $\alpha = -1$ lead to ratio and product methods of imputation.

(II) Let $X = (x_{ij})_{n \times p}, (i = 1,2,...,n;\ j = 1,2,....,p)$ be the $n \times p$ matrix of the p auxiliary vectors associated with the study variable y. It is assumed that full information is available on the auxiliary variables, but responses are missing only for the study variable. Consider a method of imputation given by

$$y_{\bullet i} = \begin{cases} y_i & \text{if } i \in \text{A} \\ \bar{y}_r\left[n\prod\limits_{j=1}^{p}\left(\dfrac{\bar{x}_{nj}}{\bar{x}_{rj}}\right)^{\alpha_j} - r\right]\dfrac{\sum\limits_{j=1}^{p} x_{ij}}{\sum\limits_{i \in \bar{A}}\sum\limits_{j=1}^{p} x_{ij}} & \text{if } i \in \bar{\text{A}} \end{cases}$$

where $\prod\limits_{i=1}^{p} x_i = x_1 x_2 .... x_p$ denote the product of p terms.

(a) Show that the point estimator $\bar{y}_s = n^{-1}\sum\limits_{i \in s} y_{\bullet i}$ becomes

$$\bar{y}_{\text{mult}} = \bar{y}_r \prod_{j=1}^{p} \left(\frac{\bar{x}_{nj}}{\bar{x}_{rj}} \right)^{\alpha_j} .$$

(b) Show that the minimum variance of the estimator $\bar{y}_{\text{mult}}$ is given by

$$V(\bar{y}_{\text{mult}}) \approx \left(\frac{1}{n} - \frac{1}{N} \right) S_y^2 + \left(\frac{1}{r} - \frac{1}{n} \right) S_y^2 \left(1 - R^2_{y \bullet x_1 x_2 \ldots x_p} \right)$$

where $R^2_{y \bullet x_1 x_2 \ldots x_p}$ denotes the multiple correlation coefficient.

Hint: Singh and Deo (2002).

Exercise 12.23. Consider the data after imputation take the form

$$y_{\bullet i} = \begin{cases} y_i & \text{if } i \in A, \\ a + w x_i & \text{if } i \in \bar{A}, \end{cases}$$

where a and w are suitably chosen constants, such that the variance of the resultant estimator is minimum, and the method of imputation becomes optimum. The sets A ($\bar{A}$) denote the sets of responding (non-responding) units in the sample s such that $s = A \cup \bar{A}$.

(a) Show that the point estimator $\bar{y}_s = \frac{1}{n} \sum_{i=1}^{n} y_{\bullet i}$ of population mean $\bar{Y}$ under the above method of imputation becomes

$$\bar{y}_s = p \bar{y}_r + (1 - p) a + w (\bar{x}_n - p \bar{x}_r)$$

where $p = r/n$, $\bar{y}_r = r^{-1} \sum_{i \in A} y_i$, $\bar{x}_r = r^{-1} \sum_{i \in A} x_i$, and $\bar{x}_n = n^{-1} \sum_{i \in s} x_i$.

(b) Show that the estimator $\bar{y}_s$ is unbiased if either $a = (\bar{Y} - w \bar{X})$ for any real value of w , or $w = \left(\frac{\bar{Y}}{\bar{X}} - \frac{a}{\bar{X}} \right)$ for any real value of a .

(c) Discuss choices of a and w such that the above optimum method of imputation reduces to ratio, mean and regression methods of imputation.

(d) The minimum mean square error of the estimator $\bar{y}_s$, for the optimum values of a and w , is given by

$$\text{Min.MSE}(\bar{y}_s) = p^2 S_y^2 \left[\left(\frac{1}{r} - \frac{1}{N} \right) - \frac{ \left\{ \left(\frac{1}{n} - \frac{1}{N} \right) - p \left(\frac{1}{r} - \frac{1}{N} \right) \right\}^2 \rho_{xy}^2 }{ \left(\frac{1}{n} - \frac{1}{N} \right) + p^2 \left(\frac{1}{r} - \frac{1}{N} \right) - 2p \left(\frac{1}{n} - \frac{1}{N} \right) } \right] .$$

(e) Discuss difficulties in using the above optimum method of imputation and suggest some possible solutions.

Hint: Singh and Valdés (2003).

Exercise 12.24. Michael works in a private sector and his boss Harold Mantel considers an imputation technique based on the mechanism of the ratio method of imputation while estimating ratio $R = \bar{Y}/\bar{X}$ of two population means. Harold Mantel shows data to Michael on a spread sheet as shown in the following table in the first two columns, 'data before imputation', and suggests to him to use the following two ratios given by

$$b_{yx} = \sum_{i=1}^{n-p-q} y_i \Big/ \sum_{i=1}^{n-p-q} x_i \text{, and } b_{xy} = \sum_{i=1}^{n-p-q} x_i \Big/ \sum_{i=1}^{n-p-q} y_i$$

for imputing the missing Y variable and the missing X variable, respectively, as shown in the last two columns of the following table.

Data structure before and after imputation with study variable Y and auxiliary variable X.			
Data before imputation		Data after imputation	
y variate	x variate	y variate	x variate
y_1	x_1	y_1	x_1
y_2	x_2	y_2	x_2
.	.	.	.
$y_{(n-p-q)}$	$x_{(n-p-q)}$	$y_{(n-p-q)}$	$x_{(n-p-q)}$
Missing	x_1^*	$b_{yx}x_1^*$	x_1^*
Missing	x_2^*	$b_{yx}x_2^*$	x_2^*
.Missing	.	.	.
Missing	.	.	.
Missing	x_p^*	$b_{yx}x_p^*$	x_p^*
y_1^*	Missing	y_1^*	$b_{xy}y_1^*$
y_2^*	Missing	y_2^*	$b_{xy}y_2^*$
.	Missing	.	.
.	Missing	.	.
y_q^*	Missing	y_q^*	$b_{xy}y_q^*$
	Sum	y_s (say)	x_s (say)

(a) After imputation Michael found the sums y_s and x_s, and took the ratio of these two as

$$\hat{R} = \frac{y_s}{x_s}$$

and reported back to Harold Mantel that his imputation method is not going to work. Justify Michael's claim by showing that

$$\hat{R} = \frac{\bar{y}_{n-p-q}}{\bar{x}_{n-p-q}} = \text{Ratio of observed responses}.$$

Hint: Use $y_s = (n-q)\bar{y}_{n-p-q}\left(\frac{\bar{x}_{n-q}}{\bar{x}_{n-p-q}}\right) + q\bar{y}_q$ and $x_s = (n-p)\bar{x}_{n-p-q}\left(\frac{\bar{y}_{n-p}}{\bar{y}_{n-p-q}}\right) + p\bar{x}_p$.

(b) Michael suggests to his boss Harold Mantel the following efficient class of estimators of population ratio as follows:

$$\hat{R}_w = \left[\nabla_1 \bar{y}_{n-p-q} H\left(\frac{\bar{x}_{n-q}}{\bar{x}_{n-p-q}}\right) + \nabla_2 \bar{y}_q\right] \Big/ \left[\nabla_3 \bar{x}_{n-p-q} G\left(\frac{\bar{y}_{n-p}}{\bar{y}_{n-p-q}}\right) + \nabla_4 \bar{x}_p\right]$$

where ∇_k, $k = 1, 2, 3, 4$ are real constants, and $H(\bullet)$ and $G(\bullet)$ are the parametric functions such that they satisfy the following assumptions:

(a) $H(1) = 1$ and $G(1) = 1$;

(b) first (G_1 and H_1 say) and second order (H_{11} and G_{11}) derivatives of H and G exist and are known constants.

Justify the Michael's claim by comparing the mean square errors of $\hat{R}$ and $\hat{R}_w$.

Hint: Singh, Singh, Tailor, and Allen (2002).

PRACTICAL PROBLEMS

Practical 12.1. John selected an SRSWOR sample of twenty states from the population 1. He collected information about the real estate farm loans and nonreal estate farm loans from the selected states, but unfortunately the information on the real estate farm loans was not available on ten states as marked in the table below.

Random No.	State	x_i	y_i	Random No.	State	x_i	y_i
01	AL	348.334	408.978	27	NE	3585.406	Missing
03	AZ	431.439	54.633	29	NH	0.471	Missing
04	AR	848.317	907.700	32	NY	426.274	Missing
05	CA	3928.732	1343.461	33	NC	494.730	Missing
06	CO	906.281	315.809	36	OK	1716.087	Missing
07	CT	4.373	7.130	38	PA	298.351	Missing
14	IN	1022.782	1213.024	40	SC	80.750	Missing
19	ME	51.539	8.849	42	TN	388.869	Missing
22	MI	440.518	323.028	46	VA	188.477	Missing
23	MN	2466.892	1354.768	47	WA	1228.607	Missing

x – Nonreal estate farm loans, y – Real estate farm loans.

Apply the following three estimators:

$$\bar{y}_{1\text{lr}} = \bar{y}_{n-r} + \left(s_{xy}^{*}/s_x^{*2}\right)\left(\bar{X} - \bar{x}_{n-r}\right),$$

$$\bar{y}_{2\text{lr}} = \bar{y}_{n-r} + \left(s_{xy}^{*}/s_x^{2}\right)\left(\bar{X} - \bar{x}\right),$$

and

$$\bar{y}_{3\text{lr}} = \bar{y}_{n-r} + \left(s_{xy}^{*}/s_x^{2}\right)\left(\bar{x}_{n-r} - \bar{x}\right)$$

of the population mean for estimating the average real estate farm loans assuming that the average nonreal estate farm loans in the United States is known and equal to \$878.16. Construct 95% confidence intervals.

Practical 12.2. Select an SRSWOR sample of sixteen units from population 4 given in the Appendix. Consider you made 7 attempts to collect information about the different species groups selected in the sample. Information about the number of fish caught in different species is not available all the times you contact the fisherman. You collected the information on the number of fish during 1994 in seven visits. It was also noted that out of seven visits, how many times (D) the information about these species was available? Estimate the average number of fish caught by marine recreational fishermen at Atlantic and Gulf coasts during 1994. Construct 95% confidence interval for the average number of fish in the United States.

Practical 12.3. Michael selected an SRSWOR sample of twenty states from the population 1, and tried to collect information about the real estate farm loans and nonreal estate farm loans from the selected states, but unfortunately the information on the real estate farm loans was not available on nine states as marked

Random No.	State	x_i	y_i	Random No.	State	x_i	y_i
01	AL	348.334	408.978	27	NE	3585.406	1337.852
03	AZ	431.439	54.633	29	NH	0.471	6.044
04	AR	848.317	907.700	32	NY	426.274	Missing
05	CA	3928.732	1343.461	33	NC	494.730	Missing
06	CO	906.281	Missing	36	OK	1716.087	Missing
07	CT	4.373	7.130	38	PA	298.351	Missing
14	IN	1022.782	1213.024	40	SC	80.750	Missing
19	ME	51.539	8.849	42	TN	388.869	Missing
22	MI	440.518	323.028	46	VA	188.477	Missing
23	MN	2466.892	1354.768	47	WA	1228.607	Missing

x – Nonreal estate farm loans; y – Real estate farm loans.

Apply the ratio type estimator $\hat{v}_5 = s_y^{*2}\left(s_x^2/s_x^{*2}\right)$ for estimating the finite population variance of the real estate farm loans and construct the 95% confidence interval.

Practical 12.4. Santa Singh and Banta Singh were appointed to select two candidates from a list of four candidates $\Omega = \{\text{Anokha, Banto, Channa, Didar}\}$ with their respective scores 25, 35, 40, and 45, respectively. In the first phase the administration suggested to Santa Singh and Banta Singh to select three candidates for telephone interview, and in the second phase they decided to select two candidates for face to face interview. Santa Singh likes every one whereas Banta Singh likes *Banto* so both suggested the following first phase sampling plan:

First phase sample	Santa Singh	Banta Singh
$s_1' = \{\text{Anokha, Banto, Channa}\}$	$p(s_1') = 1/4$	$p(s_1') = 1/3$
$s_2' = \{\text{Anokha, Banto, Didar}\}$	$p(s_2') = 1/4$	$p(s_2') = 1/3$
$s_3' = \{\text{Anokha, Channa, Didar}\}$	$p(s_3') = 1/4$	$p(s_3') = 0.0$
$s_4' = \{\text{Banto, Channa, Didar}\}$	$p(s_4') = 1/4$	$p(s_4') = 1/3$

(a) Construct the first order and second order inclusion probabilities for the first phase telephone interview.

In the second phase the administration decided to select two candidates for face to face interview out of the selected three candidates during telephone interview. Again Santa Singh and Banta Singh suggested the following possibilities

Selected first phase sample	Second phase sample	Santa Singh $p(s_i \mid s_j'), \forall\ i, j$	Banta Singh $p(s_i \mid s_j'), \forall\ i, j$
s_1'	$s_1 \mid s_1' = \{\text{Anokha, Banto}\}$	1/3	0.5
	$s_2 \mid s_1' = \{\text{Anokha, Channa}\}$	1/3	0.0
	$s_3 \mid s_1' = \{\text{Banto, Channa}\}$	1/3	0.5
s_2'	$s_1 \mid s_2' = \{\text{Anokha, Banto}\}$	1/3	0.5
	$s_2 \mid s_2' = \{\text{Anokha, Didar}\}$	1/3	0.0
	$s_3 \mid s_2' = \{\text{Banto, Didar}\}$	1/3	0.5
s_3'	$s_1 \mid s_3' = \{\text{Anokha, Channa}\}$	1/3	0.0
	$s_2 \mid s_3' = \{\text{Anokha, Didar}\}$	1/3	0.0
	$s_3 \mid s_3' = \{\text{Channa, Didar}\}$	1/3	0.0
s_4'	$s_1 \mid s_4' = \{\text{Banto, Channa}\}$	1/3	0.5
	$s_2 \mid s_4' = \{\text{Banto, Didar}\}$	1/3	0.5
	$s_3 \mid s_4' = \{\text{Channa, Didar}\}$	1/3	0.0

(b) Construct the first order and second order inclusion probabilities for the second phase face to face interview. Find difficulties in Banta Singh's sampling scheme.

(c) Estimate the total score from each one of the second phase sample for the given first phase sample. (Except s'_3 for Banta Singh's scheme).

(d) Find the bias and variance of Santa Singh and Banta Singh's selection schemes by using the definitions of bias and variance.

(e) Discuss the relative efficiency of Banta Singh's sampling scheme over the Santa Singh's sampling scheme and comment.

Practical 12.5. Consider the problem of estimation of number of cattle in a farm using known information about the size of the farms. A preliminary large SRSWOR sample of 70,000 farms out of a total 14,000,000 farms gave the estimated average area per farm as 5.00 hectares. A second phase SRSWOR sample of 3000 farms gave the following information:

$\sum_{i=1}^{n} x_i = 18867.089$, $\sum_{i=1}^{n} y_i = 12525.246$, $\sum_{i=1}^{n} x_i^2 = 48780336.98$, $\sum_{i=1}^{n} y_i^2 = 18001501.38$, and

$\sum_{i=1}^{n} x_i y_i = 26591710.56$.

(a) Estimate the average number of cattle per farm and derive 95% confidence interval estimate using two-phase ratio estimator.

(b) Estimate the average number of cattle per farm and derive 95% confidence interval estimate using two-phase regression estimator.

(c) Comment on the confidence interval estimates obtained.

Practical 12.6. John and Michael were appointed to select two candidates from a list of five candidates $\Omega = \{\text{Amy, Bob, Chris, Don, Eric}\}$ with their scores 125, 126, 128, 90 and 127, respectively. In the first phase, the administration suggested to John and Michael to select three candidates for telephone interview, and in the second phase they decided to select two candidates for face to face interview. Both John and Michael suggested the following first-phase sampling plans for telephone interview:

First phase sample	John and Michael
$s_1' = \{\text{Amy, Bob, Chris}\}$	$p(s_1') = 1/4$
$s_2' = \{\text{Amy, Chris, Don}\}$	$p(s_2') = 1/4$
$s_3' = \{\text{Amy, Don, Eric}\}$	$p(s_3') = 1/4$
$s_4' = \{\text{Amy, Chris, Eric}\}$	$p(s_4') = 1/4$

(a) Construct the first order and second order inclusion probabilities for the first phase telephone interview.

In the second phase the administration decided to select two candidates for face to face interview out of the selected three candidates during telephone interview. John likes every one whereas Michael likes Amy, so they suggested the following possibilities:

Selected first phase sample	Second phase sample	John $p(s_i \mid s_j^{'}), \forall\ i,j$	Michael $p(s_i \mid s_j^{'}), \forall\ i,j$
$s_1^{'}$	$s_1 \mid s_1^{'} = \{\text{Amy, Bob}\}$	1/3	0.5
	$s_2 \mid s_1^{'} = \{\text{Amy, Chris}\}$	1/3	0.5
	$s_3 \mid s_1^{'} = \{\text{Bob, Chris}\}$	1/3	0.0
$s_2^{'}$	$s_1 \mid s_2^{'} = \{\text{Amy, Chris}\}$	1/3	0.5
	$s_2 \mid s_2^{'} = \{\text{Amy, Don}\}$	1/3	0.5
	$s_3 \mid s_2^{'} = \{\text{Chris, Don}\}$	1/3	0.0
$s_3^{'}$	$s_1 \mid s_3^{'} = \{\text{Amy, Eric}\}$	1/3	0.5
	$s_2 \mid s_3^{'} = \{\text{Amy, Don}\}$	1/3	0.5
	$s_3 \mid s_3^{'} = \{\text{Don, Eric}\}$	1/3	0.0
$s_4^{'}$	$s_1 \mid s_4^{'} = \{\text{Amy, Chris}\}$	1/3	0.5
	$s_2 \mid s_4^{'} = \{\text{Amy, Eric}\}$	1/3	0.5
	$s_3 \mid s_4^{'} = \{\text{Chris, Eric}\}$	1/3	0.0

(b) Construct the first order and second order inclusion probabilities for the second phase face to face interview for both sampling schemes.
(c) Estimate the total score from each one of the second phase sample for the given first phase sample for both sampling schemes.
(d) Find the bias and variance of John and Michael's selection schemes by using the definitions of bias and variance.
(e) Discuss the relative efficiency of Michael's sampling scheme over the John's sampling scheme and comment.

Practical 12.7. Select an SRSWOR sample of twenty units from the population 1 given in the Appendix. Record the values of the real estate farm loans for the states selected in the sample. Assume 5% random non-response in the selected sample. Impute the missing values four times with the help of hot deck method of imputation. Apply the concept of multiple imputation for estimating population mean and construct the 95% confidence interval.

Practical 12.8. Select an SRSWOR sample of twenty five units from the population 1 given in the Appendix. Record the values of the real and nonreal estate farm loans for the states selected in the sample. Assume 5% random non-response in the selected sample. Impute the missing values with the following two methods :

(a) ratio method of imputation, and (b) compromised method of imputation.

Estimate the average real estate farm loans with each method and comment on your results.

Practical 12.9. Professor Forgetful (*e.g.*, refer to the film 'The Nutty Professor' directed by Jerry Lewis) believes that the percentage of marks of students in an examination depends upon the number of classes attended by them, and the number of marks in the assignments. Professor Forgetful misplaced mid-term exams of 4 students, but has information about the number of classes attended and marks in the assignments.

Name		Mid-term Marks (%)	Classes Attended	Assignment Marks (%)
Alabi	Ajoke	59.29	30	66
Anderson	Benjamin	72.51	36	79
Anderson	Rebecca	77.66	35	93
Barland	Deborah	Missing	9	85
Beager	Eric	75.26	34	89
Bistodeau	Jill	78.81	36	83
Brendel	Stephanie	86.67	40	84
Cherne,	Tara	73.87	34	85
Danzl	Jennifer	87.20	39	93
Houck	Jennifer	92.55	40	94
Izu	Yohei	55.43	25	75
Kimura	Shigeru	51.17	20	70
Kjeseth	Katie	74.51	32	76
Kladek	William	47.37	20	73
Laliberte	Joseph	Missing	26	54
Montana	Larissa	70.67	33	79
Olson	Cher	64.92	32	89
Rief	Abby	Missing	15	74
Ring	Paula	Missing	16	78
Taylor	Abby	65.56	33	94
Todd	Andrew	82.29	38	81
Tuladhar	Tsering	85.66	39	95

(a) Find the average marks in the class (excluding missing exams).

(b) Impute the missing marks with the mean method of imputation, and again find the average marks in the class.

(c) Assuming the information about the number of classes attended by the students is known, impute the missing marks with the ratio method of imputation and again find the average marks in the class.

(d) Impute the missing marks with the following method of imputation

$$y_{\bullet i}=\begin{cases}\dfrac{(n-1)}{(r-1)}y_i & \text{if } i\in A,\\[2ex] r^{-1}(r-1)^{-1}\{(n-1)x_i-n\bar{x}_n\}\sum\limits_{i=1}^{r}\dfrac{y_i}{x_i} & \text{if } i\in \overline{A},\end{cases}$$

where y_i and x_i denote, respectively, the marks and number of classes attended by the ith student, A and $\overline{A}$ denote the responding and non-responding sets of the sample s such that $s = A\cup\overline{A}$. Again find the average marks in the class after imputation.

(e) Repeat (c) and (d) using known marks in the assignments.

(f) Suggest a new method of imputation to use both of the variables *viz.*, the number of classes attended and marks in the assignments, to impute the missing marks in the exam.

(g) Give your views on the methods of imputation used by Professor Forgetful, and your suggestion in (f).

13. MISCELLANEOUS TOPICS

13.0. INTRODUCTION

The main purpose of this chapter is to keep this book open to the new topics coming in the recent years or which have not been touched upon by the author in the present version of the book. In this chapter we shall introduce a few miscellaneous topics namely:

(a) Estimation of Measurement Errors; (b) Raking Ratio Estimators;
(c) Continuous Populations; and (d) Small Area Estimation.

13.1 ESTIMATION OF MEASUREMENT ERRORS

In many econometric, social studies, agricultural sciences and medical sciences applications the reported values of variables are often not exact, but are only obtained with some uncertainty, herein called measurement error. With knowledge of the measurement error, it is possible to adjust estimates determined from measurements that contain measurement error (see Fuller (1987)).

In other situations more engineering applications, where measuring systems (tools) are used to measure the characteristics of population units and the tools are not perfect, it is the determination of the measurement error itself that is important. Grubbs (1948) estimated the measurement error of three clocks measuring the fuse burning time of shells of guns. Jaech (1981) estimated the measurement error of a device that determined the percentage of an isotope of plutonium in plutonium oxide powder. Morrison, Mangat, Desjardins, and Bhatia (2000) used this methodology to estimate the measurement error of a pipeline inspection tool used to measure corrosion pit depths, and a field tool that obtained data after some portions of the pipeline were excavated for rehabilitation purposes.

The estimation of measurement error is also important where calibration of a tool is required as part of its development procedure. As no instrument can claim to be fully accurate, the instrument used for calibration might also contain some measurement error. The estimation of measurement precision of both tools, the one that is being calibrated and the other that is being used as the standard, is necessary to ensure that calibration is helpful in improving the tools under development.

The statistical methods used for estimation of measurement errors for the tools individually are studied below.

13.1.1 ESTIMATION OF MEASUREMENT ERROR USING A SINGLE MEASUREMENT PER ELEMENT

Grubbs (1948) was the first to suggest a sampling theory methodology for estimating the variance of measurement error of any number of tools separately.

13.1.1.1 MODEL AND NOTATION

In a two tool case for a population of N units the model is

$$Y_i = \Psi_i + U_i, \quad \text{and} \quad X_i = \Psi_i + V_i, \quad (i = 1,2,...,N) \tag{13.1.1.1}$$

where for the ith unit, Y_i and X_i are the measurements reported by Tool 1 and Tool 2, Ψ_i is the true value and U_i and V_i are the random measurement errors in measuring Ψ_i. It is assumed that the errors U_i and V_i are random, normally distributed and have mean zero. They are independent of each other, among themselves and of the true values Ψ_i. The population variances of Ψ_i, U_i and V_i, $(i = 1,2,...,N)$, are denoted by σ_Ψ^2, σ_U^2 and σ_V^2.

13.1.1.2 GRUBBS' ESTIMATORS

For the sample units all variables including the variables used in the model, will be denoted by lower case letters. Assuming the n units are drawn randomly using equal probability with replacement sampling from a population of N units, the usual unbiased estimators of the variances and covariance of the measurements of Tool 1 and Tool 2 are

$$s_y^2 = (n-1)^{-1}\left(\sum_{i=1}^{n} y_i^2 - n\bar{y}^2\right), \; s_x^2 = (n-1)^{-1}\left(\sum_{i=1}^{n} x_i^2 - n\bar{x}^2\right), \text{ and } s_{yx} = (n-1)^{-1}\left(\sum_{i=1}^{n} y_i x_i - n\,\bar{y}\,\bar{x}\right)$$

where $\bar{y} = n^{-1}\sum_{i=1}^{n} y_i$ and $\bar{x} = n^{-1}\sum_{i=1}^{n} x_i$ are the respective sample means.

Theorem 13.1.1.1. The variance of the measurement error of Tool 1 is estimated by

$$\hat{\sigma}_U^2 = s_y^2 - s_{yx}.$$

Proof. If E is the expected value over the units then

$$E\left(s_y^2\right) = E\left[\frac{1}{n-1}\left\{\sum_{i=1}^{n} y_i^2 - n\left(\frac{1}{n}\sum_{i=1}^{n} y_i\right)^2\right\}\right] = E\left[\frac{1}{n}\sum_{i=1}^{n} y_i^2 - \frac{1}{n(n-1)}\sum_{i\neq j=1}^{n} y_i y_j\right]$$

Under model (13.1.1.1) the expected value of s_y^2 becomes

$$E\left(s_y^2\right) = E\left[\frac{1}{n}\sum_{i=1}^{n}\left(\psi_i^2 + u_i^2 + 2\psi_i u_i\right) - \frac{1}{n(n-1)}\sum_{i\neq j=1}^{n}\left(\psi_i + u_i\right)\left(\psi_j + u_j\right)\right]$$

$$= \frac{1}{N}\sum_{i=1}^{N}\Psi_i^2 + \frac{1}{N}\sum_{i=1}^{N} U_i^2 - \frac{n(n-1)}{n(n-1)}\bar{\Psi}\,\bar{\Psi} = \sigma_\Psi^2 + \sigma_U^2 \quad \left(\text{because } \bar{U} = 0\right).$$

Similarly

$$E(s_{yx}) = E\left[\frac{1}{n-1}\left\{\sum_{i=1}^{n} y_i x_i - n\left(\frac{1}{n}\sum_{i=1}^{n} y_i\right)\left(\frac{1}{n}\sum_{i=1}^{n} x_i\right)\right\}\right] = E\left[\frac{1}{n}\sum_{i=1}^{n} y_i x_i - \frac{1}{n(n-1)}\sum_{i\neq j=1}^{n} y_i x_j\right]$$

Under model (13.1.1.1) the expected value of s_{yx} becomes

$$E(s_{yx}) = E\left[\frac{1}{n}\sum_{i=1}^{n}(\psi_i + u_i)(\psi_i + v_i) - \frac{1}{n(n-1)}\sum_{i\neq j=1}^{n}(\psi_i + u_i)(\psi_j + v_j)\right]$$

$$= \frac{1}{N}\sum_{i=1}^{N}\Psi_i^2 - \frac{n(n-1)}{n(n-1)}\overline{\Psi}\;\overline{\Psi} = \sigma_\Psi^2.$$

Using the values of $E(s_y^2)$ and $E(s_{yx})$ we have

$$E\left(\hat{\sigma}_U^2\right) = \sigma_\Psi^2 + \sigma_U^2 - \sigma_\Psi^2 = \sigma_U^2.$$

Hence the proof.

Theorem 13.1.1.2. The estimator of variance of measurement error of Tool 2 is

$$\hat{\sigma}_V^2 = s_x^2 - s_{yx}.$$

Proof. The proof is similar to the one given above for Theorem 13.1.1.1.

Since the variances of measurement errors of tools are estimated from the measurements of sample units, $\hat{\sigma}_U^2$ and $\hat{\sigma}_V^2$ may differ from sample to sample. Under certain conditions of normality

$$V\left(\hat{\sigma}_U^2\right) = \frac{2\sigma_U^4}{n-1} + \frac{\sigma_\Psi^2\sigma_U^2 + \sigma_\Psi^2\sigma_V^2 + \sigma_U^2\sigma_V^2}{n-1}.$$

To determine $V\left(\hat{\sigma}_V^2\right)$ the σ_U^4 in the first term of the right hand side is replaced by σ_V^4. For practical purposes the unknown parameters in the above expression are replaced by their respective sample estimates.

The relationship between total scatter, represented by the root mean square differential/error, and the estimates of variance of measurement error for individual tool, including bias if present, is illustrated in Morrison, Mangat, Carroll, and Riznic (2003).

Example 13.1.1.1. The hypothetical set of 30 true values of a characteristic of a unit given in Table 13.1.1.1 were selected at random from a normal distribution of mean of 40 and standard deviation 15. To these true values input measurement errors from normal distributions equal to mean zero and standard deviations of 10 and 6 were added to generate measurements for Tool 1 and Tool 2 respectively. Using Tool 1 and Tool 2 data estimate the standard deviation (SD) of measurement errors (ME) of Tool 1 and Tool 2 separately.

Solution. Denoting the measurements reported by Tool 1 and Tool 2 by y and x respectively, one has $n = 30$, $s_y^2 = 291.4$, $s_x^2 = 212.1$ and $s_{yx} = 188.8$.

Table 13.1.1.1. Hypothetical data set for two tools.

Sr. No.	True Depth	Tool 1	Tool 2	Sr. No.	True Depth	Tool 1	Tool 2
1	40.80	38.00	28.90	16	9.00	17.60	8.50
2	40.20	49.20	37.40	17	54.60	54.50	57.00
3	20.70	5.00	30.70	18	41.50	37.10	45.50
4	57.50	60.30	56.70	19	32.30	28.50	29.00
5	39.40	50.70	35.90	20	46.30	52.90	32.40
6	29.20	40.60	20.10	21	22.00	14.00	22.10
7	32.10	39.20	37.60	22	45.10	62.90	43.40
8	58.40	65.10	53.10	23	62.40	73.70	72.00
9	21.10	17.40	18.30	24	40.50	43.50	30.50
10	67.80	80.20	60.30	25	36.50	38.00	36.50
11	20.00	27.40	18.90	26	42.70	43.00	59.70
12	24.40	40.60	32.70	27	38.60	41.90	33.50
13	47.80	52.20	52.50	28	34.60	41.00	26.50
14	35.80	48.70	28.90	29	42.80	34.20	34.60
15	38.20	57.70	42.30	30	30.30	36.10	36.80

These values yield estimates of the standard deviations of measurement errors as:

$$\text{SD of ME of Tool 1} = \sqrt{\hat{\sigma}_U^2} = \sqrt{s_y^2 - s_{yx}} = \sqrt{291.4 - 188.8} = 10.1,$$

$$\text{SD of ME of Tool 2} = \sqrt{\hat{\sigma}_V^2} = \sqrt{s_x^2 - s_{yx}} = \sqrt{212.1 - 188.8} = 4.8\,.$$

13.1.2 BHATIA, MANGAT, AND MORRISON'S (BMM) REPEATED MEASUREMENT ESTIMATORS

Grubbs model is based on single measurement made per unit by each instrument. Consider the case where it is possible to measure the characteristic of a unit more than once by each tool. Bhatia, Mangat, and Morrison (1998) presented estimators for this situation using varying probability and equal probability with and without replacement sampling. Here the results are presented for equal probability with replacement sampling only.

13.1.2.1 MODEL AND NOTATION

Let each unit be measured repeatedly using Tool 1 and Tool 2. Using t as a subscript for the tth measurement and i for the ith unit, the model is written as

$$Y_{it} = \Psi_i + U_{it} \text{ and } X_{it} = \Psi_i + V_{it}, \ (i = 1,2,...,N) \tag{13.1.2.1}$$

where

Y_{it} = observation, subject to measurement error, recorded by Tool 1,

X_{it} = observation, subject to measurement error, recorded by Tool 2,

Ψ_i = true value for the ith unit,

U_{it} = error in observing Ψ_i by Tool 1,

and

V_{it} = error in observing Ψ_i by Tool 2.

(i) The errors U_{it} and V_{it} are independent of each other, among themselves, and are independent of the true value of the unit.

(ii) Denoting the sample values by lower case letters $y_i, x_i, \psi_i, u_{it}, v_{it}$, the conditional expectations E_2 over the measurements for the ith sample unit are

$$E_2(u_{it}) = u_i, \ E_2(v_{it}) = 0, \ E(u_{it}^2) = \overline{\overline{u_i^2}}, \ E_2(y_{it}) = \psi_i + u_i, \ E_2(x_{it}) = \psi_i$$

where u_i represents the measurement bias of the Tool 1 over measurements for measuring the ith sample unit selected, but $\text{Cov}(\Psi_i, U_i) = 0$.

The variances σ_Ψ^2 and σ_U^2 are defined as

$$\sigma_\Psi^2 = \frac{1}{N}\sum_{i=1}^{N}\Psi_i^2 - \overline{\Psi}^2 \text{ and } \sigma_U^2 = \frac{1}{N}\sum_{i=1}^{N}\overline{\overline{U_i^2}} - \overline{U}^2$$

where $\overline{\Psi}$ is the mean of the true values for the population units, $\overline{U}$ is the mean over measurements and units in the population, $\overline{\overline{U_i^2}}$ represents the mean of squares of the measurement errors over the infinite large repeated measurements for the ith population unit and σ_U^2 is the variance of the measurement errors.

The actual number of measurements r_1 and r_2 made respectively by Tool 1 and Tool 2 on each unit are considered as independent selections from their respective infinitely large populations of repeated measurements.

Let us define notation for sample units as:

$$\overline{y}_i = \frac{1}{r_1}\sum_{t=1}^{r_1} y_{it}, \ \overline{x}_i = \frac{1}{r_2}\sum_{t=1}^{r_2} x_{it}, \ u_{it} = y_{it} - \overline{x}_i, \ \overline{u}_i = \frac{1}{r_1}\sum_{t=1}^{r_1} u_{it} = \overline{y}_i - \overline{x}_i, \ \overline{y_i^2} = \frac{1}{r_1}\sum_{t=1}^{r_1} y_{it}^2,$$

$$\overline{u_i^2} = \frac{1}{r_1}\sum_{t=1}^{r_1} u_{it}^2, \ s_{ya}^2 = \frac{1}{n}\sum_{i=1}^{n}\overline{y_i^2} - \left(\frac{1}{n}\sum_{i=1}^{n}\overline{y}_i\right)^2, \ s_{\overline{y}}^2 = \frac{1}{n-1}\left[\sum_{i=1}^{n}\overline{y}_i^2 - \frac{1}{n}\left(\sum_{i=1}^{n}\overline{y}_i\right)^2\right],$$

$$s_y^2 = \frac{1}{n-1}\left[\sum_{i=1}^{n} y_i^2 - \frac{1}{n}\left(\sum_{i=1}^{n} y_i\right)^2\right], \ s_{\overline{y}\,\overline{x}} = \frac{1}{n-1}\left[\sum_{i=1}^{n}\overline{y}_i\overline{x}_i - \frac{1}{n}\left(\sum_{i=1}^{n}\overline{y}_i\right)\left(\sum_{i=1}^{n}\overline{x}_i\right)\right],$$

$$s_{y\bar{x}} = \frac{1}{n-1}\left[\sum_{i=1}^{n} y_i\bar{x}_i - \frac{1}{n}\left(\sum_{i=1}^{n} y_i\right)\left(\sum_{i=1}^{n}\bar{x}_i\right)\right], \text{ and } s_{yx} = \frac{1}{n-1}\left[\sum_{i=1}^{n} y_i x_i - \frac{1}{n}\left(\sum_{i=1}^{n} y_i\right)\left(\sum_{i=1}^{n} x_i\right)\right].$$

Expressions for s_{ua}^2, $s_{\bar{u}}^2$, and s_u^2 are obtained by substituting u for y in the expressions for s_{ya}^2, $s_{\bar{y}}^2$, and s_y^2.

Five different situations as given below are considered:

Case I. $r_1 \geq 2$ and r_2 is sufficiently large so that the average of r_2 could be taken as the true value for the ith unit.

Case II. $r_1 \geq 2$ and $r_2 \geq 1$ but not so large that the average of r_2 could be taken as the true value for the ith unit.

Case III. $r_1 = 1$ and r_2 is sufficiently large so that the average of r_2 could be taken as the true value for the ith unit.

Case IV. $r_1 = 1$ and $r_2 \geq 2$ but not so large that the average of r_2 could be treated as the true value for the ith unit.

Case V. $r_1 = 1$ and $r_2 = 1$.

Note that Case V is equal to the Grubbs estimators for two inspection tools.

For equal probability with replacement sampling the estimators of the variance of measurement error σ_U^2 of Tool 1 are given by

$$\hat{\sigma}_U^2 = s_{ua}^2 + \frac{s_{\bar{u}}^2}{n} \qquad \text{(Case I)}$$

$$\hat{\sigma}_U^2 = s_{ya}^2 + \frac{s_{\bar{y}}^2}{n} - s_{\bar{y}\bar{x}} \qquad \text{(Case II)}$$

$$\hat{\sigma}_U^2 = s_u^2 \qquad \text{(Case III)}$$

$$\hat{\sigma}_U^2 = s_y^2 - s_{y\bar{x}} \qquad \text{(Case IV)}$$

$$\hat{\sigma}_U^2 = s_y^2 - s_{yx}. \qquad \text{(Case V)}$$

Theorem 13.1.2.1. For equal probability with replacement sampling the Case II estimator of variance of the measurement error of Tool 1 is given by

$$\hat{\sigma}_U^2 = s_{ya}^2 + \frac{s_{\bar{y}}^2}{n} - s_{\bar{y}\bar{x}}.$$

Proof. Let E_2 be the conditional expectation over measurements for a given i^{th} unit and E_1 the expectation over units. Then

$$E\left[s_{ya}^2 + \frac{s_{\bar{y}}^2}{n}\right] = E_1 E_2\left[s_{ya}^2 + \frac{s_{\bar{y}}^2}{n}\right]$$

$$= E_1E_2\left[\frac{1}{n}\sum_{i=1}^{n}\overline{y_i^2} - \left(\frac{1}{n}\sum_{i=1}^{n}\bar{y}_i\right)^2 + \frac{1}{n(n-1)}\left\{\sum_{i=1}^{n}\bar{x}_i^2 - \frac{1}{n}\left(\sum_{i=1}^{n}\bar{x}_i\right)^2\right\}\right]$$

$$= E_1E_2\left[\frac{1}{n}\sum_{i=1}^{n}\overline{y_i^2} - \frac{1}{n(n-1)}\sum_{i\neq j=1}^{n}\bar{y}_i\bar{y}_j\right].$$

Now

$$E_2\left[\overline{y_i^2}\right] = E_2\left(\frac{1}{r_1}\sum_{t=1}^{r_1}y_{it}^2\right) = E_2\left[\frac{1}{r_1}\sum_{t=1}^{r_1}(\psi_i + u_{it})^2\right] = \psi_i^2 + \overline{\overline{u_i^2}} + 2\psi_i u_i$$

$$E_2\left[\bar{y}_i\,\bar{y}_j\right] = E_2(\bar{y}_i)E_2(\bar{y}_j) = (\psi_i + u_i)(\psi_j + u_j)$$

Substituting the conditional expected values obtained above one gets

$$E_1\left[s_{ya}^2 + \frac{s_{\bar{y}}^2}{n}\right] = E_1\left[\frac{1}{n}\sum_{i=1}^{n}\left(\psi_i^2 + \overline{\overline{u_i^2}} + 2\psi_i u_i\right) - \frac{1}{n(n-1)}\sum_{i\neq j=1}^{n}(\psi_i + u_i)(\psi_j + u_j)\right]$$

$$= \frac{1}{N}\sum_{i=1}^{N}\Psi_i^2 + \frac{1}{N}\sum_{i=1}^{N}\overline{\overline{U_i^2}} + \frac{2}{N}\sum_{i=1}^{N}\Psi_i U_i - \frac{n(n-1)}{n(n-1)}\left(\overline{\Psi} + \overline{U}\right)\left(\overline{\Psi} + \overline{U}\right)$$

$$= \sigma_\Psi^2 + \sigma_U^2\ ,\ \text{because}\ \operatorname{Cov}(\Psi_i,\ U_i) = 0\,.$$

Now

$$E\left(s_{\bar{y}\bar{x}}\right) = E_1E_2\left[\frac{1}{n-1}\left\{\sum_{i=1}^{n}\bar{y}_i\,\bar{x}_i - \frac{1}{n}\left(\sum_{i=1}^{n}y_i\right)\left(\sum_{i=1}^{n}x_i\right)\right\}\right]$$

$$= E_1\left[\frac{1}{n-1}\left\{\sum_{i=1}^{n}(\psi_i + u_i)\ \psi_i - \frac{1}{n}\sum_{i=1}^{n}(\psi_i + u_i)\left(\sum_{i=1}^{n}\psi_i\right)\right\}\right]$$

$$= E_1\left[\frac{1}{n-1}\left\{\sum_{i=1}^{n}\psi_i^2 - \frac{1}{n}\left(\sum_{i=1}^{n}\psi_i\right)^2\right\} + \frac{1}{n-1}\left\{\sum_{i=1}^{n}\psi_i u_i - \frac{1}{n}\sum_{i=1}^{n}\psi_i\sum_{i=1}^{n}u_i\right\}\right]$$

$$= \sigma_\psi^2\quad (\text{since}\ E_1\{\operatorname{Cov}(\psi_i,\ u_i)\} = 0\ ).$$

Thus

$$E\left[s_{ya}^2 + \frac{s_{\bar{y}}^2}{n} - s_{\bar{y}\bar{x}}\right] = \sigma_\Psi^2 + \sigma_U^2 - \sigma_\Psi^2 = \sigma_U^2$$

which proves the theorem.

Corollary 13.1.2.1. Assuming that the average of repeated measurements of Tool 1 does not yield the true value of the units, all other assumptions remaining the same, the BMM Case II estimator for Tool 2 can be written in the same way

$$\hat{\sigma}_V^2 = s_{xa}^2 + \frac{s_{\bar{x}}^2}{n} - s_{\bar{x}\bar{y}}$$

where s_{xa}^2 and $s_{\bar{x}}^2$ can be determined by replacing y by x in the formulae for s_{ya}^2 and $s_{\bar{y}}^2$ and $s_{\bar{x}\bar{y}} = s_{\bar{y}\bar{x}}$.

Example 13.1.2.1. Thirty true values were selected at random from a normal distribution of mean 40 and standard deviation 15. To these values simulated measurement errors of standard deviations 10 and 2 were added three times for Tool 1 and three times for Tool 2 respectively to generate three measurements for each tool. The data is given in Table 13.1.2.1. Using this data estimate the standard deviations of measurement error of Tool 1 and Tool 2 separately assuming the average of neither of the two tools is the true value of the unit.

Table 13.1.2.1. Hypothetical replicated data for two tools.

Sr.No.	True Depth	Tool 1 r1	Tool 1 r2	Tool 1 r3	Tool 2 r1	Tool 2 r2	Tool 2 r3
1	31.29	15.21	25.35	38.14	33.90	29.93	30.63
2	9.89	13.02	13.10	6.84	12.06	8.04	9.20
3	62.40	71.27	49.71	45.83	60.82	61.13	62.28
4	43.68	33.64	33.58	51.04	40.38	44.61	41.66
5	20.37	18.68	27.45	11.98	18.49	19.56	17.95
6	30.24	24.89	11.79	34.26	29.95	31.62	31.10
7	13.81	5.00	32.57	18.61	12.44	12.73	15.38
8	53.25	54.89	49.13	72.18	50.55	52.87	51.65
9	49.65	45.76	36.93	68.38	48.75	52.36	50.16
10	41.17	39.28	38.04	60.11	37.46	44.93	42.64
11	35.04	21.33	22.42	47.47	30.62	33.53	35.23
12	67.61	62.87	55.50	68.14	68.38	67.76	66.67
13	40.70	34.21	49.61	50.09	41.18	39.07	40.30
14	74.77	84.53	71.98	81.88	74.95	76.61	75.50
15	36.56	32.58	38.00	26.39	37.45	36.96	35.55
16	38.63	45.36	42.32	34.87	37.76	37.65	35.30
17	30.17	12.51	31.57	19.99	30.64	28.44	30.29
18	37.68	33.47	29.94	40.02	40.93	36.81	37.56
19	56.79	53.98	56.95	55.76	56.36	55.72	53.23
20	17.36	22.50	7.46	35.98	19.04	21.43	17.98
21	21.92	20.72	5.33	18.63	21.98	24.05	20.22
22	61.95	45.64	73.45	61.77	60.30	60.69	61.24
23	17.19	12.15	40.27	27.27	14.91	18.90	16.60
24	25.38	30.32	7.24	42.56	24.28	26.38	26.06
25	28.82	33.61	28.57	15.90	32.48	27.40	25.31
26	39.40	24.64	38.58	46.49	38.06	38.97	38.97
27	57.37	74.93	44.83	42.40	54.23	60.07	59.31
28	34.29	32.43	44.84	38.70	33.06	37.08	35.23
29	48.17	56.44	52.25	44.10	53.36	48.23	46.27
30	41.26	44.78	46.68	36.75	41.48	39.43	40.96

28	34.29	32.43	44.84	38.70	33.06	37.08	35.23
29	48.17	56.44	52.25	44.10	53.36	48.23	46.27
30	41.26	44.78	46.68	36.75	41.48	39.43	40.96

Solution. The BMM Case II estimator is used: Here $n = 30$, $r_1 = 3$, $r_2 = 3$, and one can easily obtain the following

$$\frac{1}{n}\sum_{i=1}^{n}\overline{y_i^2} = 1811.8\,,\quad \frac{1}{n}\sum_{i=1}^{n}\bar{y}_i = 38.3\,,\quad \frac{1}{n}\sum_{i=1}^{n}\overline{x_i^2} = 1757.5\,,\quad \frac{1}{n}\sum_{i=1}^{n}\bar{x}_i = 38.7\,,$$

$$s_{\bar{y}}^2 = 273.4\,,\quad s_{\bar{x}}^2 = 269.2\,,\quad \text{and}\quad s_{\bar{y}\bar{x}} = 260.9\,.$$

This yield

$$s_{ya}^2 = 1811.8 - 38.3^2 = 344.9,\quad \text{and}\quad s_{xa}^2 = 1757.5 - 38.7^2 = 259.8.$$

Using the Case II formula one gets:

Standard deviation of measurement error of Tool 1:

$$\sqrt{344.9 + \frac{273.4}{30} - 260.9} = \pm 9.6\,.$$

Standard deviation of measurement error of Tool 2:

$$\sqrt{259.8 + \frac{269.2}{30} - 260.9} = \pm 2.8.$$

Remark 13.1.2.1. (a) Estimation of measurement errors for multiple tools: The Grubbs method and the BMM estimators can be extended to more than two tool situations. For this readers are referred to Grubbs (1948) and Bhatia, Mangat, and Morrison (1998).

(b) Remedy for negative measurement error estimates: Since the estimates of measurement error are based on samples of values obtained from the measurements by two tools, and the objective is to split the total scatter into components to be associated with each tool, it is possible for the estimate of measurement error for one of the tools to be negative. This happens when the variance of the measurements obtained by one tool is less than the covariance between the two tools measurements. This can be rectified either using Jaech's (1981, 1985) method or the Bayesian approach suggested by Worthingham, Morrison, Mangat, and Desjardins (2002) and Morrison, Mangat, Carrol, and Riznic (2003).

13.2 RAKING RATIO USING CONTINGENCY TABLES

The use of raking ratio estimator in survey sampling is quite old, which was first introduced by Deming and Stephan (1940). This procedure uses an iterative method of adjusting two-way contingency table so that the row and column sums add to certain preassigned value. The concept is basically similar to the calibration approach. Let Ω be a population of units cross classified into an $R \times C$ table, and let Ω_{rc} be the set of the N_{rc} units in the (r, c)th cell. We draw a sample s, with $s_{rc} = s \cap \Omega_{rc}$, and let d_i be the survey weight attached to the ith unit. Assuming that the variable of interest Y, taking the value y_i for the ith unit and $\sum_{i \in s} w_i y_i$ is a consistent or unbiased estimator of $Y = \sum_{i\in\Omega} Y_i$ .In the presence of non-response, let $\Omega^* \subseteq \Omega$ be a set of units who would respond if they were to be sampled, let $\Omega^*_{rc} = \Omega_{rc} \cap \Omega^*$ be of size N^*_{rc}, let $s^*_{rc} = s_{rc} \cap \Omega^*$ be the size of n^*_{rc}, and let $s^* = s \cap \Omega^*$. Let X be an auxiliary variable associated with the study variable Y. Define $Y_{rc} = \sum_{i\in\Omega_{rc}} y_i$, $X_{rc} = \sum_{i\in\Omega_{rc}} x_i$, $Y^*_{rc} = \sum_{i\in\Omega^*_{rc}} y_i$ and $X^*_{rc} = \sum_{i\in\Omega^*_{rc}} x_i$, $r = 1,2,...,R$ and $c = 1,2,...,C$. The definition of the raking ratio method is to adjust the weights of each of the observation so that the resulting estimators of the auxiliary totals $X_{r\bullet}$ and $X_{\bullet c}$ $(r = 1,2,...,R;\ c = 1,2,...,C)$ correspond to their population values. In order to have a better understanding of it, let us consider the following $R \times C$ contingency table.

Rows	Columns				Totals
	1	2		C	
1	$Y_{11}, Y^*_{11}, X_{11}, X^*_{11}$	$Y_{12}, Y^*_{12}, X_{12}, X^*_{12}$		$Y_{1C}, Y^*_{1C}, X_{1C}, X^*_{1C}$	$Y_{1\bullet}, X_{1\bullet}$
2	$Y_{21}, Y^*_{21}, X_{21}, X^*_{21}$	$Y_{22}, Y^*_{22}, X_{22}, X^*_{22}$		$Y_{2C}, Y^*_{2C}, X_{2C}, X^*_{2C}$	$Y_{2\bullet}, X_{2\bullet}$
R	$Y_{R1}, Y^*_{R1}, X_{R1}, X^*_{R1}$	$Y_{R2}, Y^*_{R2}, X_{R2}, X^*_{R2}$		$Y_{RC}, Y^*_{RC}, X_{RC}, X^*_{RC}$	$Y_{R\bullet}, X_{R\bullet}$
Sum	$Y_{\bullet 1}, X_{\bullet 1}$	$Y_{\bullet 2}, X_{\bullet 2}$		$Y_{\bullet C}, X_{\bullet C}$	$Y_{\bullet\bullet}, X_{\bullet\bullet}$

Define $\hat{Y}^{(t)}_{rc}$ be the raking ratio estimator of Y_{rc} after t^{th} iteration, we have

$$\hat{Y}^{(t)}_{rc} = \begin{cases} \sum_{i\in s^*_{rc}} d_i y_i & \text{if } t = 0, \\ \hat{Y}^{(t-1)}_{rc}\left(\dfrac{X_{r\bullet}}{\hat{X}^{(t-1)}_{r\bullet}}\right) & \text{if } t \text{ is odd}, \\ \hat{Y}^{(t-1)}_{rc}\left(\dfrac{X_{\bullet c}}{\hat{X}^{(t-1)}_{\bullet c}}\right) & \text{if } t > 0 \text{ is even}. \end{cases} \tag{13.2.1}$$

Formulas for the asymptotic variance of the raking ratio estimator are given by Brackstone and Rao (1979) for up to four iterations. Following them we have following relations.

If $t = 0$, $E\left(\hat{Y}^{(0)}_{rc}\right) = E\left[\sum_{i\in s^*_{rc}} d_i y_i\right] = Y$,

$$\text{if } t=2,\quad E\left(\hat{Y}_{rc}^{(2)}\right)=E\left[\hat{Y}_{rc}^{(1)}\left(\frac{X_{\bullet c}}{\hat{X}_{\bullet c}^{(1)}}\right)\right]\approx Y\,.$$

If general if t is odd then

$$E\left(\hat{Y}_{rc}^{(t)}\right)=E\left[\hat{Y}_{rc}^{(t-1)}\left(\frac{X_{r\bullet}}{\hat{X}_{r\bullet}^{(t-1)}}\right)\right]=\sum_{r=1}^{R}\hat{Y}_{r\bullet}^{(t-1)}\left(\frac{X_{r\bullet}}{\hat{X}_{r\bullet}^{(t-1)}}\right),\tag{13.2.2}$$

and if t is even then

$$E\left(\hat{Y}_{rc}^{(t)}\right)=E\left[\hat{Y}_{rc}^{(t-1)}\left(\frac{X_{r\bullet}}{\hat{X}_{r\bullet}^{(t-1)}}\right)\right]=\sum_{c=1}^{C}\hat{Y}_{\bullet c}^{(t-1)}\left(\frac{X_{\bullet c}}{\hat{X}_{\bullet c}^{(t-1)}}\right).\tag{13.2.3}$$

Now using the result that the bias in the usual ratio estimator

$$\bar{y}_{\mathrm{R}}=\bar{y}\left(\frac{\bar{X}}{\bar{x}}\right)$$

is given by

$$B(\bar{y}_{\mathrm{R}})=\frac{1}{\bar{X}}\left[\bar{Y}V(\bar{x})-\mathrm{Cov}(\bar{y},\bar{x})\right].\tag{13.2.4}$$

Thus we have the following theorem:

Theorem 13.2.1. The bias in the raking ratio estimator, to the first order of approximation, is given by

$$B=\begin{cases}\displaystyle\sum_{r=1}^{R}\frac{1}{X_{r\bullet}^{(t-1)}}\left[Y_{r\bullet}^{(t-1)}V\left(\hat{X}_{r\bullet}^{(t-1)}\right)-Cov\left(\hat{X}_{r\bullet}^{(t-1)},\hat{Y}_{r\bullet}^{(t-1)}\right)\right] & \text{if } t \text{ is odd,}\\ \displaystyle\sum_{c=1}^{C}\frac{1}{X_{\bullet c}^{(t-1)}}\left[Y_{\bullet c}^{(t-1)}V\left(\hat{X}_{\bullet c}^{(t-1)}\right)-Cov\left(\hat{X}_{\bullet c}^{(t-1)},\hat{Y}_{\bullet c}^{(t-1)}\right)\right] & \text{if } t \text{ is even.}\end{cases}\tag{13.2.5}$$

Similarly using the result that the variance of the usual ratio estimator $\bar{y}_{\mathrm{R}}$ is

$$V(\bar{y}_{\mathrm{R}})=V(\bar{y})+\left(\frac{\bar{Y}}{\bar{X}}\right)^2V(\bar{x})-2\left(\frac{\bar{Y}}{\bar{X}}\right)\mathrm{Cov}(\bar{y},\bar{x}).\tag{13.2.6}$$

We have the following theorem:

Theorem 13.2.2. The variance of the raking ratio estimator is given by

$$V=\begin{cases}\displaystyle\sum_{r=1}^{R}\left[V\left(\hat{Y}_{r\bullet}^{(t-1)}\right)+\left(\frac{Y_{r\bullet}^{(t-1)}}{X_{r\bullet}^{(t-1)}}\right)V\left(\hat{X}_{r\bullet}^{(t-1)}\right)-\left(\frac{Y_{r\bullet}^{(t-1)}}{X_{r\bullet}^{(t-1)}}\right)\mathrm{Cov}\left(\hat{X}_{r\bullet}^{(t-1)},\hat{Y}_{r\bullet}^{(t-1)}\right)\right] & \text{if } t \text{ is odd,}\\ \displaystyle\sum_{c=1}^{C}\left[V\left(\hat{Y}_{\bullet c}^{(t-1)}\right)+\left(\frac{Y_{\bullet c}^{(t-1)}}{X_{\bullet c}^{(t-1)}}\right)V\left(\hat{X}_{\bullet c}^{(t-1)}\right)-2\left(\frac{Y_{\bullet c}^{(t-1)}}{X_{\bullet c}^{(t-1)}}\right)\mathrm{Cov}\left(\hat{X}_{\bullet c}^{(t-1)},\hat{Y}_{\bullet c}^{(t-1)}\right)\right] & \text{if } t \text{ is even.}\end{cases}\tag{13.2.7}$$

Deming and Stephan (1940) were the first who used raking ratio method of estimation for estimating the cell probabilities π_{rc} in the $r\times c$ contingency table for which the marginal probabilities $\pi_{r\bullet}$ and $\pi_{\bullet c}$ are known. Later on various iterative procedures have been developed by several researchers including Smith (1947), Friedlander (1961), Ireland and Kullback (1968), Fienberg (1970) and Causey

(1972) to find the solutions to this problem. The method originally proposed by Deming and Stephan (1940) is called the Iterative Proportional Fitting Procedure (IPFP) and it minimises the modified ξ^2 distance function as

$$\xi^2 = \sum_{r=1}^{R}\sum_{c=1}^{C}\frac{(n_{rc}-n\pi_{rc})^2}{n_{rc}}, \tag{13.2.8}$$

where $n_{rc} > 0$ denote the sample size in the (r, c)th cell and

$$n = \sum_{r=1}^{R}\sum_{c=1}^{C} n_{rc}.$$

Ireland and Kullback (1968) proved that the IPFP minimises a discrimination function defined as:

$$I(\pi, p) = \sum_{r=1}^{R}\sum_{c=1}^{C}\pi_{rc}\ln\left(\frac{\pi_{rc}}{p_{rc}}\right), \tag{13.2.9}$$

where $p_{rc} = n_{rc}/n$.

Konijn (1981) has derived biases, variances and co-variances for the estimators of the cell and marginal totals and of the corresponding marginal averages in the $R\times C$ contingency table. We shall now like to explain the raking ratio method with the help of a numerical example given by Binder and Theberge (1988) as given below.

Example 13.2.1. Consider $R = C = 2$ and that we have drawn a simple random sample of 80 units with 70 respondents. We obtained the data given in the following table:

$n_{11}^* = 10$	$n_{12}^* = 15$	$N_{1\bullet} = 300$
$n_{21}^* = 20$	$n_{22}^* = 25$	$N_{2\bullet} = 400$
$N_{\bullet 1} = 350$	$N_{\bullet 2} = 350$	$N_{\bullet\bullet} = 700$

Use $w_i = \frac{700}{80} = 8.75$ and taking $y_i = x_i = 1$.

For $t = 0$ we have

$$\hat{Y}_{rc}^{(0)} = \sum_{i\in s_{rc}^*} w_i y_i = \sum_{i\in s_{rc}^*} 8.75 = 8.75\times n_{rc}^* = \hat{N}_{rc}^{(0)} \text{ (say)}. \tag{13.2.10}$$

Thus at the 0th iteration we have the following table.

8.75 × 10 = 87.50	8.75 × 15 = 131.25	218.75
8.75 × 20 = 175.00	8.75 × 25 = 218.75	393.75
262.50	350.00	612.50

For $t = 1$ (odd) the raking ratio estimator takes the form

$$\hat{Y}_{rc}^{(1)} = \hat{Y}_{rc}^{(0)}\frac{X_{r\bullet}}{\hat{X}_{r\bullet}^{(0)}}.$$

Evidently

$$\frac{X_{1\bullet}}{\hat{X}_{1\bullet}^{(0)}} = \frac{300}{218.75} = 1.371429$$

and

$$\frac{X_{2\bullet}}{\hat{X}_{2\bullet}^{(0)}} = \frac{400}{393.75} = 1.015873 .$$

Thus we have following table

87.50 × 1.371429 = 120.00	131.25 × 1.371429 = 180.0	300.00
175.00 × 1.015873 = 177.78	218.75 × 1.015873 = 222.2	399.98
297.78	402.2	699.98

For $t = 2$ (even) the raking ratio estimator takes the form

$$\hat{Y}_{rc}^{(2)} = \hat{Y}_{rc}^{(1)} \frac{X_{\bullet c}}{\hat{X}_{\bullet c}^{(1)}} .$$

Evidently

$$\frac{X_{\bullet 1}}{\hat{X}_{\bullet 1}^{(1)}} = \frac{350}{297.78} = 1.17568, \text{ and } \frac{X_{\bullet 2}}{\hat{X}_{\bullet 2}^{(1)}} = \frac{350}{402.2} = 0.870214 .$$

Thus we have following table

120 × 1.17568 = 141.08	180 × 0.870214 = 156.63	297.67
177.78 × 1.17568 = 208.96	222.2 × 0.870214 = 193.37	402.33
350.04	350.00	700.00

and so on, until convergence is achieved. Such a method of improving the estimates is called racking ratio method of estimation.

13.3 CONTINUOUS POPULATIONS

The literature on the survey sampling of finite populations mainly deals with the populations that consist of sets of discrete units. Here we shall discuss the populations that may be considered as one continuous entity or a continuum of points. Examples of natural continuous populations are air, soil temperature over space and time, pollutant levels in a volume of material such as a river, percent chemical contents along a strip of soil and inches of rainfall over a region and commodity prices over time etc.. Thus a continuous population exist within a support medium such as time or space. Let the function y_α assign the value of a characteristic of interest y to point α of the support region. In geo-statistics this function is called a regionalized variable. In survey sampling, the entire set of labelled pairs $y_p = \{(\alpha, y_\alpha)\}$ that is the subject of inference after sampling is called the population parameter and any real function y_p is called a parametric function. If P denote the support region, then we are interested in estimating the total value of the y characteristic defined as

$$Y = \int_P y_\alpha d\alpha \; . \qquad (13.3.1)$$

In the fixed population sampling strategies, the parameter $\{(\alpha, y_\alpha)\}$ is regarded as fixed but unknown. In case of superpopulation model approach, the parameter $\{(\alpha, y_\alpha)\}$ is treated as a realisation of a random vector or function $\{(\alpha, Y_\alpha)\}$ whose stochastic distribution ξ is specified or partially specified. Let E_ξ denote the expected value with respect to distribution ξ . In the fixed population approach the inference is based on a sample s of n units drawn with probability $p(s)$, called sampling design, from the population. Let E denote the expected value with respect to sampling design.

Design unbiasedness: An estimator $\hat{\theta}$ is said to be p unbiased if

$$E_p(\hat{\theta}) = \sum_s \hat{\theta} p(s) = Q(y_p)$$

if $\hat{\theta}$ denote an estimator of the parametric function $Q(y_p)$.

Model and design unbiasedness: If

$$E_\xi(\hat{\theta}) = Q(Y_p)$$

then $\hat{\theta}$ is called ξ unbiased.
If

$$E_p E_\xi(\hat{\theta}) = Q(y_p)$$

then it is called $p\xi$ unbiased.

Predictive mean squared error: The predictive MSE of $\hat{\theta}$ is defined as

$$E_\xi\left[\hat{\theta} - Q(Y_p)\right]^2 .$$

The superpopulation value Y_α can be partitioned into two components, *viz.*,
(a) The first is a trend component, the values of which usually depend on the location α of the population unit in the support region, defined as

$$f(\alpha) = E_\xi(Y_\alpha) = \sum_{k=0}^{p} c_k f_k(\alpha), \quad \alpha \in P;$$

and $c_0, c_1, ..., c_p$ are $p+1$ independent known functions;

(b) The second component represent the local random fluctuations, denoted as

$$Z(\alpha) = Y_\alpha - f(\alpha) \; ,$$

which is, in fact the vector of residual terms.

The general superpopulation model can be defined as

$$Y_\alpha = f(\alpha) + Z(\alpha) \qquad (13.3.2)$$

with $E_\xi[Z(\alpha)] = 0, \; \alpha \in P$.

We shall now consider a one-dimensional support region, the labelling vector α will now be replaced by a scalar t (say).

Then we have the following assumptions. The error structure $Z(t)$, $t \in P$, has

(a) mean zero,
(b) finite variance, $\sigma_z^2 = V_\xi\{Z(t)\}$,
(c) the covariance, $V_Z(t_i, t_j) = \text{Cov}_\xi\{Z(t_i), Z(t_j)\}$,
(d) second order stationarity in one dimension can be written as a function of a single variable $h = t_j - t_i$, so that, $V_Z(h) = \text{Cov}_\xi\{Z(t), Z(t+h)\}$,

and
(e) linear operations of integration and expectation can be interchanged.

Then, following Bartlett (1986), our objective is to estimate the population total Y as

$$t_c = \int_0^T y(t)dt \tag{13.3.3}$$

over the support region $P = \{t : 0 \le t \le T\}$. Thus t_c can be regarded as a realisation of

$$T_c = \int_0^T Y(t)dt \tag{13.3.4}$$

under the super-population model (13.3.2). Since we have only a finite number of observations over the support region P, therefore, our sample will consist of the set $s = \{t_i : i = 1, 2, ..., n\}$.

Thus if $y_{t_1}, y_{t_2}, ..., y_{t_n}$ are the observed values then we will choose our estimator to be a linear combination

$$\hat{t}_c = \sum_{i=1}^n w_{t_i} y_{t_i}, \tag{13.3.5}$$

which is a realisation of

$$\hat{T}_c = \sum_{i=1}^n w_{t_i} Y_{t_i}, \tag{13.3.6}$$

where w_{t_i} are the weights depending upon the population point t_i selected. Barltett (1986) has discussed the following four criterion to find these weights:

(a) Determine both weights and the sample locations such that the estimator (13.3.5) is model unbiased and has minimum predictive mean square error;
(b) Determine the weights which for any sample s actually obtained the estimator (13.3.5) is model unbiased and has minimum predictive mean square error;
(c) Determine for present weights the sample s for which the estimator is model unbiased and has minimum predictive mean square error;

(d) Determine either the weights or the sample or both such that the estimator has minimum model bias for a given class of trend functions.

Padmawar (1996) considered the problem of estimation of population total of the continuous populations by using following sampling designs.

(i) SRS : Simple random sampling where $p(t) \equiv 1$.

(ii) PPX : design for which $p(t) \propto \prod_{i=1}^{n} t_i^{\alpha}$

(iii) P_M : the analogous of Midzuno--Sen sampling design with

$$p(t) = \frac{1}{n\mu}\sum_{i=1}^{n} t_i,$$

where $\mu = E(t) = \int_0^{\infty} t f(t) dt$.

(iv) P_g : a sampling design with

$$p(t) = A\prod_{i=1}^{n} t_i^{g-1} \sum_{i=1}^{n} x_i^{2-g}$$

and

(v) RHC: the continuous analogous of the RHC strategy.

Padmawar (1996) compared the following four estimators of the population total of a continuous distribution, defined as:

(i) $\bar{y}_R$: the ratio estimator

$$\sum_{i=1}^{n} y(t_i)\left(T \Big/ \sum_{i=1}^{n} t_i\right);$$

(ii) $\bar{y}_{HT}$: the HT estimator

$$\sum_{i=1}^{n} w(t_i) f(t_i) / \pi(t_i),$$

where

$\pi(t_i) = \sum_{j=1}^{n} q_j(t_i)$, $\pi(t_i)$ is assumed to be positive for $t > 0$, and $q_i(t_i) = \int q(t) \prod_{j \neq i = 1}^{n} dt_j$, $1 \le i \le n$;

$$\text{(iii) } \bar{y}_g = \left(T \Big/ \sum_{i=1}^{n} t_i^{2-g}\right) \sum_{i=1}^{n} t_i^{1-g}, \quad g \in [0, 2];$$

and

$$\text{(iv) } t_{RHC} = \sum_{h=1}^{n} W_h \left[\frac{y(t_h)}{t_h / T_h}\right], \text{ where } W_h = \int_{G_h} f(t) dt.$$

Padmawar (1996) has shown that the RHC strategy also remains design unbiased under continuous population.

13.4 SMALL AREA ESTIMATION

The problem of small area estimation is an important technique in survey sampling owed to a heavy demand for reliable small area statistics from public as well as private sectors. Direct survey estimates for small area are likely to yield unacceptably large standard errors owed to the smallness of sample sizes in the area. Thus it make sense to use information from related areas to find accurate estimates for a given area or, simultaneously, for several areas. There are several methods such as Sample size dependent, Synthetic, Empirical best linear unbiased prediction, Empirical Bayes, and Hierarchical Bayes estimation. It is clear from the name 'small area' or 'local area' that we are interested in the estimates of population total and its standard errors in a particular small geographical area, such as a village, district, county, or a census division. Some times we may be interested in 'small domains' rather than 'small area or local area'. For example, we may be interested in the small domain formed by Sex, Age, and Race of group of persons living in a large geographical area. Such a mechanism is also called interchangeable. Brackstone (1987) considered the small area estimation issues based either on a census or using complete enumeration from administrative records. No doubt that the sample survey data can be used to obtain reliable estimators of mean and total for large areas or domains, but it may yield unacceptable large standard errors due to small sample size in the area of interest. Sample sizes for small areas are typically small because the overall sample size in a survey is usually determined to provide specific accuracy at a much higher level of the aggregation than that of small areas. Rao (2003), Ghosh and Rao (1994), and Chaudhuri (1992) have given a complete review on small area estimation strategies in survey sampling. Here we would like to discuss few estimation strategies.

13.4.1 SYMPTOMATIC ACCOUNTING TECHNIQUES

These techniques make use of data from administrative registers in addition to related data from the latest census. This method covers many methods such as: (a) Vital Rates Method (VRM); (b) Census Component Method (CCM); and (c) Housing Unit Method (HUM). We first discuss these methods in brief, but for more details one can refer to Purcell and Kish (1980).

13.4.2 VITAL RATES METHOD (VRM)

Following Bogue (1950), the VRM method makes use of only birth and death data. These data are used as symptomatic variables rather than as components of population change. In the first step, the number of births B_t and deaths D_t in a given t^{th} year are determined for a local area. If $\hat{B}_{10}$ denote the crude birth rate for the local area in the latest census year $(t = 0)$, B_{1t} denote the crude birth rate in the current year for a larger area containing the local area, and B_{10} denote the crude birth rate in the census year for a larger area containing the local area. Then an estimator of the crude birth rate in the current year is given by

$$\hat{B}_{1t} = \hat{B}_{10}(B_{1t}/B_{10}). \tag{13.4.2.1}$$

Similarly, if $\hat{D}_{10}$ denotes the crude death rate for the local area in the latest census year $(t=0)$, D_{1t} denotes the crude death rate in the current year for a larger area containing the local area, and D_{10} denotes the crude death rate in the census year for a larger area containing the local area. Then an estimator of the crude death rate in the current year is given by

$$\hat{D}_{1t} = \hat{D}_{10}(D_{1t}/D_{10}). \tag{13.4.2.2}$$

Then an estimator of total population P_t for the local area in year t is given by

$$\hat{P}_t = \frac{1}{2}\left[B_t/\hat{B}_{1t} + D_t/\hat{D}_{1t}\right]. \tag{13.4.2.3}$$

It is to be noted here that the VRM method works well if $\left(\hat{B}_{1t}/\hat{B}_{10}\right) \approx (B_{1t}/B_{10})$ and $\left(\hat{D}_{1t}/\hat{D}_{10}\right) \approx (D_{1t}/D_{10})$, but sometimes this assumption may be violated and the VRM method becomes non-functional.

Example 13.4.2.1. In a college of a University the total number of the recruitment of new students during 2000 was 450, and the number of students who left the college was 250. The recruitment and departure rate according to 1999 records was 2.5% and 1.5%, respectively, for both the college and the University. The recruitment and departure rate of the students during 2000 for the university is 2.6% and 1.7% respectively. Estimate the total number of students in the college during 2000 by using the VRM method. Assuming that there are a large number of colleges in a University. A particular college is small in comparison to the whole University.

Solution. An estimate of the current recruitment rate is

$$\hat{B}_{1t} = \hat{B}_{10}(B_{1t}/B_{10}) = 0.025 \times (26/25) = 0.026,$$

and an estimate of current departure rate is given by

$$\hat{D}_{1t} = \hat{D}_{10}(D_{1t}/D_{10}) = 0.015 \times (17/15) = 0.017 .$$

Thus an estimate of the total number of students in the college during 2000 is

$$\hat{P}_t = \frac{1}{2}\left[B_t/\hat{B}_{1t} + D_t/\hat{D}_{1t}\right] = \frac{1}{2}\left[450/0.026 + 250/0.017\right] = 16006.78 \approx 16007 .$$

13.4.3 CENSUS COMPONENT METHOD (CCM)

This method takes into account an important factor, migration, while estimating total population in a local area. If M_t denotes the net migration in the local area since last census, then an estimator of total population P_t is given by

$$\hat{P}_t = P_0 + B_t - D_t + M_t, \tag{13.4.3.1}$$

where P_0 denotes the population count of the local area in the census year $t=0$. Migration can be estimated in several ways. For example, the net migration can be subdivided into civilian and military migration. Evidently the military migration can be taken from the administrative records and civilian migration can be taken from school enrolments. If you estimate the net migration from the records for the individuals as opposed to collect units like schools, then such a method is called administrative record method, and can be used for producing local area estimates.

Example 13.4.3.1. In a university let the total number students be 15000 during 1999. During 2000 there is a recruitment of 1500 students according to the schedule and later on 50 students left the university and stopped their study. There was migration +10 students (say, 20 students migrated to other universities and 30 students came from other universities during the academic year 2000). Apply the Census Component Method (CCM) to estimate the total number of students in the university during 2000.

Solution. We have

$$\hat{P}_t = P_0 + B_t - D_t + M_t = 15000 + 1500 - 50 + 10 = 16460.$$

13.4.4 HOUSING UNIT METHOD (HUM)

If $\hat{H}_t$ denotes the estimator of the number of occupied housing units at time t, $P\hat{P}H_t$ is the estimator of the average number of persons per housing unit at time t, and $G\hat{Q}_t$ is the estimator of the number of persons in group quarters at time t, then an estimator of total population P_t is

$$\hat{P}_t = \hat{H}_t \times P\hat{P}H_t + G\hat{Q}_t . \tag{13.4.4.1}$$

Smith and Lewis (1980) have discussed several methods for obtaining the estimates $\hat{H}_t, P\hat{P}H_t$ and $G\hat{Q}_t$ of the respective parameters. Marker (1983) came with an interesting idea that most of the above methods can be derived as a special case of the multiple linear regression model. We will not discuss this method here as it is straightforward by following Marker (1983) or the basic technique of regression analysis, but we will discuss some different methods which makes use of sampling techniques discussed in the previous chapters of this book.

13.4.5 SYNTHETIC ESTIMATOR

Before discussing synthetic estimator, let us try to understand the structure of population under small area estimation process. Let Y_{ij} and X_{ij} be the totals of the study variable and auxiliary variable for the jth unit in the ith area of interest for $i = 1,2,...,K$ and $j = 1,2,...,N_i$. Assume that the population totals X_{ij} bases on N_{ij} units for the auxiliary variable X are accurately known. Now our interest is to draw inference about the study variable Y in a particular area of interest by making use of maximum information. The structure of the population and mechanism of synthetic estimation technique can be easily seen from Table 13.4.5.1. Following Gonzalez (1973), "An unbiased estimate is obtained from a sample survey for a large area; when this estimate is used to derive estimates for sub-areas under the assumption that the small areas have the same characteristics as the large area, we identify these estimates as synthetic estimates". Based on this definition a general synthetics estimator of small area total, Y_i, is defined as

$$\hat{Y}_i^S = \sum_{j=1}^{K} \left(X_{ij}/X_{\bullet j}\right)\hat{Y}_j^*, \tag{13.4.5.1}$$

where $\hat{Y}_j^*$ may be a ratio estimator or difference type estimator or any other direct estimator. As shown in Table 13.4.5.1, the direct estimator $\hat{Y}_j^*$ as a ratio estimator has been discussed by Ghosh and Rao (1994) and a difference estimator has been discussed by Singh, Stukel, and Pfeffermann (1998). It is further investigated by Gershunskaya, Eltinge, and Huff (2002).

Table 13.4.5.1. An easy way to understand the synthetic estimation mechanism.

$i \backslash j$	1	2		N_1	N_2	N_K	Ratio Estimator based on n_i units.	Difference Estimator for given $\hat{\beta}$, based on n_i units selected from a given area.
1	Y_{11} X_{11} N_{11}	Y_{12} X_{12} N_{12}		Y_{1N_1} X_{1N_1} N_{1N_1}			$\hat{Y}_1^* = \hat{Y}_{\bullet 1}\dfrac{X_{\bullet 1}}{\hat{X}_{\bullet 1}}$	$\hat{Y}_1^* = \tilde{Y}_{\bullet 1} + \hat{\beta}\left(X_{\bullet 1} - \hat{X}_{\bullet 1}\right)$
2	Y_{21} X_{21} N_{21}	Y_{22} X_{22} N_{22}			Y_{2N_2} X_{2N_2} N_{2N_k}		$\hat{Y}_2^* = \hat{Y}_{\bullet 2}\dfrac{X_{\bullet 2}}{\hat{X}_{\bullet 2}}$	$\hat{Y}_2^* = \hat{Y}_{\bullet 2} + \hat{\beta}\left(X_{\bullet 2} - \hat{X}_{\bullet 2}\right)$
K	Y_{K1} X_{K1} N_{K1}	Y_{K2} X_{K2} N_{K2}				Y_{KN_K} X_{KN_K} N_{KN_K}	$\hat{Y}_K^* = \hat{Y}_{\bullet K}\dfrac{X_{\bullet K}}{\hat{X}_{\bullet K}}$	$\hat{Y}_K^* = \hat{Y}_{\bullet K} + \hat{\beta}\left(X_{\bullet K} - \hat{X}_{\bullet K}\right)$
Parameter	$Y_{\bullet 1}$ $X_{\bullet 1}$	$Y_{\bullet 2}$ $X_{\bullet 2}$		$Y_{\bullet N_1}$ $X_{\bullet N_1}$	$Y_{\bullet N_2}$ $X_{\bullet N_2}$	$Y_{\bullet N_K}$ $X_{\bullet N_K}$		
Estimator based on n_j units across all areas.	$\hat{Y}_{\bullet 1}$ $\hat{X}_{\bullet 1}$	$\hat{Y}_{\bullet 2}$ $\hat{X}_{\bullet 2}$		$\hat{Y}_{\bullet N_1}$ $\hat{X}_{\bullet N_1}$	$\hat{Y}_{\bullet N_2}$ $\hat{X}_{\bullet N_2}$	$\hat{Y}_{\bullet N_K}$ $\hat{X}_{\bullet N_K}$		

Now we have the following theorems:

Theorem 13.4.5.1. Assuming the direct method of estimation as a ratio estimator, show that the synthetic estimator for the ith area of interest reduces to

$$\hat{Y}_i^S = \sum_{j=1}^{K} \hat{Y}_{\bullet j}\frac{X_{ij}}{\hat{X}_{\bullet j}}. \tag{13.4.5.2}$$

Proof. Substitute $\hat{Y}_{\bullet j}^* = \hat{Y}_{\bullet j}\left(X_{\bullet j}/\hat{X}_{\bullet j}\right)$, $j = 1, 2, ..., K$, into $\hat{Y}_i^S$ in (13.4.5.1), we have (13.4.5.2). Hence the theorem.

Theorem 13.4.5.2. Assuming the direct method of estimation as a difference type estimator, show that the synthetic estimator for the ith area of interest reduces to

$$\hat{Y}_i^S = \sum_{j=1}^{K}\left(X_{ij}/X_{\bullet j}\right)\hat{Y}_{\bullet j} + \hat{\beta}\sum_{j=1}^{K}\left(X_{\bullet j} - \hat{X}_{\bullet j}\right)\left(X_{ij}/\hat{X}_{\bullet j}\right), \tag{13.4.5.3}$$

where $\hat{\beta}$ is an estimator of fixed regression coefficients in the linear regression of y on x.

Proof. It follows by substituting direct estimator $\hat{Y}_j^* = \hat{Y}_{\bullet j} + \hat{\beta}\left(X_{\bullet j} - \hat{X}_{\bullet j}\right)$ in the general synthetic estimator $\hat{Y}_i^S$ in (13.4.5.1). Hence the theorem.

If the direct estimator $\hat{Y}_{\bullet j}^*$ (may be a ratio type or a difference type) is approximately design unbiased, then we have the following theorem.

Theorem 13.4.5.3. The bias in the general synthetic estimator, $\hat{Y}_i^S$, is given by

$$B\left(\hat{Y}_i^S\right) \approx \sum_{j=1}^{K} X_{ij}\left(Y_{\bullet j}/X_{\bullet j} - Y_{ij}/X_{ij}\right) \tag{13.4.5.4}$$

Proof. We know that the true total for the ith small area is given by $Y_i = \sum_{j=1}^{K} Y_{ij}$. Also taking design based expected value on both sides of the general synthetic estimator

$$\hat{Y}_i^S = \sum_{j=1}^{K}\left(X_{ij}/X_{\bullet j}\right)\hat{Y}_j^*$$

we have

$$E\left(\hat{Y}_i^S\right) = E\left[\sum_{j=1}^{K}\left(X_{ij}/X_{\bullet j}\right)\hat{Y}_j^*\right] = \sum_{j=1}^{K}\left(X_{ij}/X_{\bullet j}\right)E\left(\hat{Y}_j^*\right) \approx \sum_{j=1}^{K}\left(X_{ij}/X_{\bullet j}\right)\left(Y_{\bullet j}\right).$$

Thus by the definition of bias we have

$$B\left(\hat{Y}_i^S\right) \approx \sum_{j=1}^{K}\left(X_{ij}/X_{\bullet j}\right)Y_{\bullet j} - \sum_{j=1}^{K} Y_{ij}$$

$$= \sum_{j=1}^{K}\left(X_{ij}/X_{\bullet j}\right)Y_{\bullet j} - \sum_{j=1}^{K}\left(Y_{ij}X_{ij}/X_{ij}\right) = \sum_{j=1}^{K} X_{ij}\left(Y_{\bullet j}/X_{\bullet j} - Y_{ij}/X_{ij}\right).$$

Hence the theorem.

Theorem 13.4.5.4. Under the assumption $\operatorname{Cov}\left(\hat{Y}_i^S, \hat{Y}_i^*\right) \approx 0$ an approximately unbiased estimator of $\operatorname{MSE}\left(\hat{Y}_i^S\right)$ is given by

$$\hat{\operatorname{MSE}}\left(\hat{Y}_i^S\right) = \left(\hat{Y}_i^S - \hat{Y}_i^*\right)^2 + \hat{V}\left(\hat{Y}_i^*\right) \tag{13.4.5.5}$$

where $\hat{V}\left(\hat{Y}_i^*\right)$ is a design unbiased estimator of variance of $\hat{Y}_i^*$.

Proof. The mean squared error of the general synthetic estimator $\hat{Y}_i^S$ is given by

$$\operatorname{MSE}\left(\hat{Y}_i^S\right) = E\left[\hat{Y}_i^S - Y_i\right]^2 = E\left[\hat{Y}_i^S - \hat{Y}_i^* + \hat{Y}_i^* - Y_i\right]^2$$

$$= E\left(\hat{Y}_i^S - \hat{Y}_i^*\right)^2 + E\left(\hat{Y}_i^* - Y_i\right)^2 + 2E\left(\hat{Y}_i^S - \hat{Y}_i^*\right)\left(\hat{Y}_i^* - Y_i\right).$$

Under the assumption

$$E\left(\hat{Y}_i^S - \hat{Y}_i^*\right)\left(\hat{Y}_i^* - Y_i\right) \approx 0$$

we have

$$\text{MSE}\left(\hat{Y}_i^S\right) \approx E\left(\hat{Y}_i^S - \hat{Y}_i^*\right)^2 + E\left(\hat{Y}_i^* - Y_i\right)^2 \approx E\left(\hat{Y}_i^S - \hat{Y}_i^*\right)^2 + V\left(\hat{Y}_i^*\right).$$

Hence the theorem by the method of moments.

It is to be noted here that the condition $\text{Cov}\left(\hat{Y}_i^S,\ \hat{Y}_i^*\right) \approx 0$ may be realistic in practice since the synthetic estimator is much more stable than the direct estimator in small area estimation process.

13.4.6 COMPOSITE ESTIMATOR

A class of estimators can be defined by combining the direct estimator and the synthetic estimator as

$$\hat{Y}_i^C = \gamma_i \hat{Y}_i^* + (1-\gamma_i)\hat{Y}_i^S, \tag{13.4.6.1}$$

where γ_i is called a shrinkage factor between 0 and 1. The estimator $\hat{Y}_i^C$ is expected to have a small prediction mean squared error if γ_i provides a suitable trade off between the instability of direct estimator and the bias of the synthetic estimator. Singh, Stukel, and Pfefferman (1998) reported that most of the work on small area estimation is on the determination of the shrinkage coefficient γ_i.

Theorem 13.4.6.1. The optimum value of γ_i which minimises the $\text{MSE}\left(\hat{Y}_i^C\right)$ is

$$\gamma_i = \text{MSE}\left(\hat{Y}_i^S\right) \Big/ \left\{\text{MSE}\left(\hat{Y}_i^S\right) + V\left(\hat{Y}_i^*\right)\right\}. \tag{13.4.6.2}$$

Proof. We have

$$\text{MSE}\left(\hat{Y}_i^C\right) = E\left[\hat{Y}_i^C - Y_i\right]^2 = E\left[\gamma_i \hat{Y}_i^* + (1-\gamma_i)\hat{Y}_i^S - Y_i\right]^2$$

$$= E\left[\gamma_i \hat{Y}_i^* + (1-\gamma_i)\hat{Y}_i^S - \{\gamma_i Y_i + (1-\gamma_i)Y_i\}\right]^2 = E\left[\gamma_i\left(\hat{Y}_i^* - Y_i\right) + (1-\gamma_i)\left(\hat{Y}_i^S - Y_i\right)\right]^2$$

$$= E\left[\gamma_i^2\left(\hat{Y}_i^* - Y_i\right)^2 + (1-\gamma_i)^2\left(\hat{Y}_i^S - Y_i\right)^2 + 2\gamma_i(1-\gamma_i)\left(\hat{Y}_i^S - Y_i\right)\left(\hat{Y}_i^* - Y_i\right)\right]$$

$$= \gamma_i^2 E\left(\hat{Y}_i^* - Y_i\right)^2 + (1-\gamma_i)^2 E\left(\hat{Y}_i^S - Y_i\right)^2 + 2\gamma_i(1-\gamma_i)E\left(\hat{Y}_i^S - Y_i\right)\left(\hat{Y}_i^* - Y_i\right)$$

$$= \gamma_i^2 V\left(\hat{Y}_i^*\right) + (1-\gamma_i)^2 \text{MSE}\left(\hat{Y}_i^S\right).$$

Now setting

$\partial\text{MSE}\left(\hat{Y}_i^C\right)/\partial\gamma_i = 0$, we get $\gamma_i V\left(\hat{Y}_i^*\right) - (1-\gamma_i)\text{MSE}\left(\hat{Y}_i^S\right) = 0$

which proves the theorem.

The main difficulty in using γ_i is that it depends upon the unknown population parameters. This difficulty was over come by Purcell and Kish (1979) by suggesting as estimator of γ_i as

$$\hat{\gamma}_i = 1 - \left\{1 - \sum_{i=1}^{K}\hat{V}\left(\hat{Y}_i^*\right) \Big/ \sum_{i=1}^{K}\left(\hat{Y}_i^S - \hat{Y}_i^*\right)^2\right\}^{-1}. \tag{13.4.6.3}$$

Drew, Singh, and Choudhry (1982) suggested an interesting method to obtain the shrinkage coefficient γ_i as

$$\hat{\gamma}_i = \begin{cases} 1 & \text{if } \hat{N}_i \geq \delta N_i, \\ \dfrac{\hat{N}_i}{\delta N_i} & \text{otherwise,} \end{cases} \tag{13.4.6.4}$$

where $\hat{N}_i$ is the direct, unbiased estimator of the known domain population size N_i and δ is subjectively chosen to control the contribution of the synthetic estimator. Särndal and Hidiroglou (1989) proposed an alternative method as

$$\hat{\gamma}_i = \begin{cases} 1 & \text{if } \hat{N}_i \geq N_i, \\ \left(\dfrac{\hat{N}_i}{N_i}\right)^{h-1} & \text{otherwise,} \end{cases} \tag{13.4.6.5}$$

where h is subjectively chosen constant. It is to be noted that for $\delta = 1, h = 2$ the Särndal and Hidiroglou (1989) weights reduces to Drew, Singh, and Choudhry (1982) weights.

Example 13.4.6.1. The average area in hectares, number of countries in each continent, and the number of countries to be selected from each one of the 10 continents are as given in the following table.

Continent (i)	No. of countries in the ith continent N_i	Average area in hectares $\overline{X}_i$	No. of units selected from the ith continent (n_i)
1	6	3194.50	2
2	6	14660.00	2
3	8	18309.37	3
4	10	14923.50	4
5	12	5987.83	4
6	4	3450.00	2
7	30	11682.73	11
8	17	145162.30	6
9	10	33976.10	4
10	3	1333.33	2
		Total sample size	40

Select a sample of 40 countries from different continents listed in the population 5 of the Appendix. Estimate the average yield in hectares in each of the continent using the synthetic ratio (SR) estimator and the composite estimator (CE).

Solution. We selected the sub-samples of the required size from each continent by SRSWOR sample and collected the information on the area and yield of tobacco crop as shown below.

Continent	n_i	x_{ij}	$\bar{x}_i$	y_{ij}	$\bar{y}_i$
1	2	9024	7090.50	2.21	1.995
		5157		1.78	
2	2	27050	14112.50	1.51	1.750
		1175		1.99	
3	3	9260	24003.30	2.80	2.780
		47600		2.77	
		15150		2.79	
4	4	24000	14025.00	0.63	1.450
		7000		1.67	
		19100		2.34	
		6000		1.17	
5	4	4500	8226.00	2.33	1.615
		4304		0.26	
		5500		1.82	
		18600		2.05	
6	2	2700	4700.00	1.96	1.570
		6700		1.18	
7	11	705		1.00	
		4000	21939.09	0.45	1.010
		3400		1.62	
		750		0.87	
		10000		0.26	
		10		1.00	
		116700		1.22	
		655		1.63	
		103110		2.06	
		0		0.00	
		2000		1.00	
8	6	36000	325593.16	1.22	1.450
		12165		0.74	
		1445000		1.75	
		445000		1.43	
		11000		1.05	
		4394		2.50	
9	4	18000	78800.00	1.39	1.175
		2100		1.29	
		1800		1.11	
		293300		0.91	
10	2	3300	1700.00	2.73	1.840
		100		0.95	

The overall sample means are given by

$\bar{x} = 68157.725$ and $\bar{y} = 1.486$.

Thus we have the following two cases:

Synthetic Ratio (SR) estimators: The synthetic ratio estimates of average yield in different continents is

Continent i	Synthetic estimator: $\hat{\bar{Y}}_i = \frac{\bar{y}}{\bar{x}}\bar{X}_i$
1	$\frac{1.486}{68157.725}\times 3194.50 = 0.06947$
2	$\frac{1.486}{68157.725}\times 14660.00 = 0.31962$
3	$\frac{1.486}{68157.725}\times 18309.37 = 0.39918$
4	$\frac{1.486}{68157.725}\times 14923.50 = 0.32536$
5	$\frac{1.486}{68157.725}\times 5987.83 = 0.13054$
6	$\frac{1.486}{68157.725}\times 3450.00 = 0.07522$
7	$\frac{1.486}{68157.725}\times 11682.73 = 0.25471$
8	$\frac{1.486}{68157.725}\times 145162.30 = 3.16488$
9	$\frac{1.486}{68157.725}\times 33976.10 = 0.74076$
10	$\frac{1.486}{68157.725}\times 1333.33 = 0.02906$

Composite estimator: The composite estimates of average yield are

				Composite Estimator
Area i	w_i	W_i	Condition for the choice of estimators	$\hat{\bar{Y}}_i = \begin{cases} \bar{y}_i + \frac{\bar{y}}{\bar{x}}(\bar{X}_i - \bar{x}_i) & \text{if } w_i \geq W_i \\ \frac{w_i}{W_i}\left[\bar{y}_i + \frac{\bar{y}}{\bar{x}}(\bar{X}_i - \bar{x}_i)\right] + \left(1 - \frac{w_i}{W_i}\right)\frac{\bar{y}}{\bar{x}}\bar{X}_i & \text{if } w_i < W_i \end{cases}$
1	0.050	0.0566	$w_1 < W_1$	1.695
2	0.050	0.0566	$w_2 < W_2$	1.594
3	0.075	0.0755	$w_3 < W_3$	2.393
4	0.100	0.0933	$w_4 > W_4$	1.469
5	0.100	0.1130	$w_5 < W_5$	1.401
6	0.050	0.0370	$w_6 > W_6$	1.543
7	0.275	0.2830	$w_7 < W_7$	0.771
8	0.150	0.1600	$w_8 < W_8$	-2.131
9	0.100	0.0940	$w_9 > W_9$	0.197
10	0.050	0.0280	$w_{10} > W_{10}$	1.832

The estimate of the average yield in the eighth continent is negative, which is not possible and hence can be taken as zero. It looks this estimator needs improvements, may be due to no correlation between yield and area under the crop.

13.4.7 MODEL BASED TECHNIQUES

The small area estimation techniques are model dependent, and to understand these techniques, one should must know the Henderson (1975) model. We first provide here a complete solution to it.

13.4.7.1 HENDERSON'S MODEL

We give here a complete solution to the general Henderson (1975) model defined as

$$Y = X\beta + Zu + e, \tag{13.4.7.1}$$

where $Y = \underset{1\le i\le t}{\text{col}}\ \underset{1\le j\le n_i}{\text{col}}\left(y_{ij}\right)$ is an $n\times 1$ vector, $X = \left(x_{ijl}\right)_{n\times k}$ matrix, $u = \underset{1\le i\le t}{\text{col}}\left(u_i\right)$, $e = \underset{1\le i\le t}{\text{col}}\ \underset{1\le j\le n_i}{\text{col}}\left(e_{ij}\right)$, and $Z = \left(z_{ij}\right)_{n\times k}$ matrix. Let us partition a matrix as

$$\left[\begin{array}{c|c} A & B \\ \hline C & D \end{array}\right]\left[\begin{array}{c|c} P & Q \\ \hline R & S \end{array}\right] = \left[\begin{array}{c|c} I & O \\ \hline O & I \end{array}\right]. \tag{13.4.7.2}$$

Let D be a non-singular matrix, then by block multiplication we have

$$AP + BR = I, \tag{13.4.7.3}$$

$$AQ + BS = O, \tag{13.4.7.4}$$

$$CP + DR = O, \tag{13.4.7.5}$$

and

$$CQ + DS = I. \tag{13.4.7.6}$$

From (13.4.7.5) we have

$$R = -D^{-1}CP. \tag{13.4.7.7}$$

From (13.4.7.6)

$$S = D^{-1}(I - CQ). \tag{13.4.7.8}$$

From (13.4.7.3) and (13.4.7.7)

$$AP - BD^{-1}CP = I \text{ or } \left(A - BD^{-1}C\right)P = I,$$

which implies

$$P = \left(A - BD^{-1}C\right)^{-1}. \tag{13.4.7.9}$$

From (13.4.7.7) and (13.4.7.9)

$$R = -D^{-1}C\left(A - BD^{-1}C\right)^{-1}. \tag{13.4.7.10}$$

From (13.4.7.4) and (13.4.7.8)

$$AQ + BD^{-1}(I - CQ) = O \text{ or } AQ + BD^{-1} - BD^{-1}CQ = O \text{ or } \left(A - BD^{-1}C\right)Q = -BD^{-1},$$

which implies

$$Q = -\left(A - BD^{-1}C\right)^{-1}BD^{-1}. \tag{13.4.7.11}$$

Put it in (13.4.7.8) we have

$$S = D^{-1} + D^{-1}C\left(A - BD^{-1}C\right)^{-1} BD^{-1}$$

or

$$S = D^{-1}\left[I + C\left(A - BD^{-1}C\right)^{-1} BD^{-1}\right]. \tag{13.4.7.12}$$

From (13.4.7.9), (13.4.7.10), (13.4.7.11) and (13.4.7.12), the inverse of the matrix $\begin{bmatrix} A \mid B \\ C \mid D \end{bmatrix}$ is given by

$$\begin{bmatrix} P & | & Q \\ ----- \\ R & | & S \end{bmatrix} = \begin{bmatrix} \left(A - BD^{-1}C\right)^{-1} & | & -\left(A - BD^{-1}C\right)^{-1} BD^{-1} \\ ------------------------------ \\ -D^{-1}C\left(A - BD^{-1}C\right)^{-1} & | & D^{-1} + D^{-1}C\left(A - BD^{-1}C\right)^{-1} BD^{-1} \end{bmatrix}$$

Now in Henderson (1975) model, the partition of the matrix is given by

$$A = \left(X'R^{-1}X\right), \quad B = \left(X'R^{-1}Z\right), \quad C = \left(Z'R^{-1}X\right), \text{ and } D = Z'R^{-1}Z + G^{-1}. \tag{13.4.7.13}$$

Thus the least square estimates of $\hat{\beta}$ and $\hat{\mu}$ in the Henderson's model

$$y = X\beta + Z\mu + e \tag{13.4.7.14}$$

where

$$V\begin{bmatrix} u \\ e \end{bmatrix} = \sigma^2 \begin{bmatrix} G, & O \\ O, & R \end{bmatrix} \tag{13.4.7.15}$$

are given by

$$\hat{\beta} = \left(A - BD^{-1}C\right)^{-1} X'R^{-1}y - \left(A - BD^{-1}C\right)^{-1} BD^{-1}Z'R^{-1}y$$

or $$\hat{\beta} = \left(A - BD^{-1}C\right)^{-1}\left[X' - BD^{-1}Z'\right]R^{-1}y, \tag{13.4.7.16}$$

and

$$\hat{\mu} = -D^{-1}C\left(A - BD^{-1}C\right)^{-1} X'R^{-1}y + D^{-1}\left[I + C\left(A - BD^{-1}C\right)^{-1} BD^{-1}\right]Z'R^{-1}y$$

$$= D^{-1}\left[Z' + C\left(A - BD^{-1}C\right)^{-1}\left(BD^{-1}Z' - X'\right)\right]R^{-1}y. \tag{13.4.7.17}$$

To simplify these estimates of $\hat{\beta}$ and $\hat{\mu}$ we have

$$\left(A - BD^{-1}C\right)^{-1} = \left[\left(X'R^{-1}X\right) - \left(X'R^{-1}Z\right)\left(Z'R^{-1}Z + G^{-1}\right)^{-1} Z'R^{-1}X\right]^{-1}$$

$$= \left[X'\left(R^{-1} - R^{-1}Z\left(Z'R^{-1}Z + G^{-1}\right)^{-1} Z'R^{-1}\right)X\right]^{-1}$$

$$= \left(X'V^{-1}X\right)^{-1}, \tag{13.4.7.18}$$

where $\left(R^{-1} - R^{-1}Z\left(Z'R^{-1}Z + G^{-1}\right)^{-1} Z'R^{-1}\right) = V^{-1}$ as shown by Henderson (1973).

Now

$$\left(X'-BD^{-1}Z'\right)R^{-1}=\left[X'-\left(X'R^{-1}Z\right)\left(Z'R^{-1}Z+G^{-1}\right)^{-1}Z'\right]R^{-1}$$

$$=X'\left[R^{-1}-R^{-1}Z\left(Z'R^{-1}Z+G^{-1}\right)^{-1}Z'R^{-1}\right]=X'V^{-1}. \qquad (13.4.7.19)$$

On using (13.4.7.18) and (13.4.7.19) in (13.4.7.16) we obtain

$$\hat{\beta}=\left(X'V^{-1}X\right)^{-1}X'V^{-1}y. \qquad (13.4.7.20)$$

Now $\hat{\mu}$ can be written as

$$\hat{\mu}=D^{-1}\left[Z'+C\left(A-BD^{-1}C\right)^{-1}\left(BD^{-1}Z'-X'\right)\right]R^{-1}y$$

$$=\left(Z'R^{-1}Z+G^{-1}\right)^{-1}Z'R^{-1}y$$

$$+\left(Z'R^{-1}Z+G^{-1}\right)^{-1}Z'R^{-1}X\left[X'R^{-1}X-X'R^{-1}Z\left(Z'R^{-1}Z+G^{-1}\right)^{-1}Z'R^{-1}X\right]^{-1}$$

$$\times\left[\left(X'R^{-1}Z\right)\left(Z'R^{-1}Z+G^{-1}\right)^{-1}Z'R^{-1}y-X'R^{-1}y\right]$$

$$=GZ'V^{-1}y+GZ'V^{-1}X\left[\left(X'R^{-1}X\right)-X'R^{-1}ZGZ'V^{-1}X\right]^{-1}\left[\left(X'R^{-1}Z\right)GZ'V^{-1}y-X'R^{-1}y\right]$$

Using the results that $\left(Z'R^{-1}Z+G^{-1}\right)^{-1}Z'R^{-1}=GZ'V^{-1}$ and now using $y=X\hat{\beta}$ in the above expression

$$\hat{\mu}=GZ'V^{-1}y+GZ'V^{-1}X\left[X'\left(R^{-1}-R^{-1}ZGZ'V^{-1}\right)X\right]^{-1}\left[X'\left(R^{-1}-R^{-1}ZGZ'V^{-1}\right)X\hat{\beta}\right]$$

$$=GZ'V^{-1}y-GZ'V^{-1}X\hat{\beta}=GZ'V^{-1}\left(y-X\hat{\beta}\right). \qquad (13.4.7.21)$$

Thus the result follows theorem:

Theorem 13.4.7.1. The best linear unbiased predictor (BLUP) of $\xi'\beta+\zeta'\mu$ is

$$\xi'\hat{\beta}+\zeta'\hat{u}=\xi'\hat{\beta}+\zeta'GZ'V^{-1}\left(y-X\hat{\beta}\right) \qquad (13.4.7.22)$$

Proof. Replace β by $\hat{\beta}$ and μ by $\hat{\mu}$ in $\xi'\beta+\zeta'\mu$ we have the theorem.

Theorem 13.4.7.2. The variance of the best linear unbiased predictor (BLUP),

$$\hat{\mu}_p=\xi'\hat{\beta}+\zeta'\hat{\mu}$$

is

$$V\left(\hat{\mu}_p\right)=\xi'\left(X'V^{-1}X\right)^{-1}\xi-\zeta'\left(Z'R^{-1}Z+G^{-1}\right)^{-1}Z'R^{-1}X\left(X'V^{-1}X\right)^{-1}\xi$$

$$-\xi\left(X'V^{-1}X\right)^{-1}X'R^{-1}Z\left(Z'R^{-1}Z+G^{-1}\right)^{-1}\zeta$$

$$+\zeta'\left(Z'R^{-1}Z+G^{-1}\right)^{-1}\left[I+Z'R^{-1}X\left(X'V^{-1}X\right)^{-1}X'R^{-1}Z\left(Z'R^{-1}Z+G^{-1}\right)^{-1}\right]\zeta$$

$$(13.4.7.23)$$

Proof. Following Henderson (1975) let

$$\begin{bmatrix} C_{11} & | C_{12} \\ --- & | --- \\ C_{12}' & | \ C_{22} \end{bmatrix}$$

be a generalized inverse of the matrix

$$\begin{bmatrix} A & | \ \mathrm{B} \\ --- & | --- \\ C & | \ \mathrm{D} \end{bmatrix}$$

then

$$V\left(\hat{\mu}_p\right) = V\left(\xi'\hat{\beta} + \zeta'\hat{\mu}\right) = [\xi' | \zeta'] \begin{bmatrix} C_{11} & | C_{12} \\ --- & | --- \\ C_{12}' & | \ C_{22} \end{bmatrix} \begin{bmatrix} \xi \\ \zeta \end{bmatrix} \sigma^2$$

$$= \sigma^2 \left[\xi' C_{11}\xi + \zeta' C_{12}'\xi + \xi' C_{12}\zeta + \zeta' C_{22}\zeta\right]$$

where $C_{11} = \left(X'V^{-1}X\right)^{-1}$, $C_{12} = -\left(X'V^{-1}X\right)^{-1} X'R^{-1}Z\left(Z'R^{-1}Z + G^{-1}\right)^{-1}$, and $C_{22} = \left(Z'R^{-1}Z + G^{-1}\right)^{-1}\left[I + Z'R^{-1}X\left(X'V^{-1}X\right)^{-1} X'R^{-1}Z\left(Z'R^{-1}Z + G^{-1}\right)^{-1}\right]$.

Remark 13.4.7.1. Note that BLUP can also be written as:

$$\xi'\hat{\beta} + \zeta'\hat{u} = \xi'\hat{\beta} + \zeta' GZ'V^{-1}\left(y - X\hat{\beta}\right)$$

$$= \xi'\left(X'V^{-1}X\right)^{-1} X'V^{-1}y + \zeta' GZ'V^{-1}\left[y - X\left(X'V^{-1}X\right)^{-1} X'V^{-1}y\right]$$

$$= \left[\xi'\left(X'V^{-1}X\right)^{-1} X'V^{-1} + \zeta' GZ'V^{-1}\left(1 - X\left(X'V^{-1}X\right)^{-1} X'V^{-1}\right)\right]y$$

$$= Wy,$$

where

$$W = \left[\xi'\left(X'V^{-1}X\right)^{-1} X'V^{-1} + \zeta' GZ'V^{-1}\left(1 - X\left(X'V^{-1}X\right)^{-1} X'V^{-1}\right)\right].$$

We first discuss three well known models useful for small area estimation as follows:

(a) Nested error regression model; (b) Random regression coefficient model; and (c) Fay and Herriot model.

13.4.7.2 NESTED ERROR REGRESSION MODEL

Battese, Harter and Fuller (1988) suggested a nested error regression model in the context of estimating mean acreage under a crop for counties (small areas) in Iowa using landsat satellite data in conjunction with survey data with the model

$$y = \underline{x}_{ij}'\beta + v_i + e, \quad i = 1,2,...,t \text{ and } j = 1,2,...,n_i$$

where $y = \underset{1\le i\le t}{\text{col}}\ \underset{1\le j\le n_i}{\text{col}}\left(y_{ij}\right)$ is a vector of the jth sampled unit in the ith sampled area, $\underline{x}_{ij} = \left(x_{ij1}, x_{ij2},...,x_{ijk}\right)'$ is an $n\times k$ matrix, $\beta = \left(\beta_1, \beta_2,..., \beta_k\right)'$, $n = \sum_{i=1}^{t} n_i$, and $e = \underset{1\le i\le t}{\text{col}}\ \underset{1\le j\le n_i}{\text{col}}\left(e_{ij}\right)$. So the nested error regression model can be written as

$$\begin{bmatrix} y_{11} \\ y_{12} \\ . \\ .y_{n_1} \\ .. \\ y_{t1} \\ y_{t2} \\ .. \\ y_{tn_t} \end{bmatrix} = \begin{bmatrix} x_{111}, & x_{112}, & ..., & x_{11k} \\ x_{121}, & x_{122}, & ..., & x_{12k} \\ & & & \\ x_{1n_11}, & x_{1n_12}, & ..., & x_{1n_1k} \\ ..... & ..... & ..... & ..... \\ x_{t11}, & x_{t12}, & ...., & x_{t1k} \\ x_{t21}, & x_{t22}, & ....., & x_{t2k} \\ & & & \\ x_{tn_t1}, & x_{tn_t2}, & ..., & x_{tn_tk} \end{bmatrix} \begin{bmatrix} \beta_1 \\ \beta_2 \\ \\ \beta_k \end{bmatrix} + \begin{bmatrix} v_1 \\ v_1 \\ \\ v_1 \\ \\ v_t \\ v_t \\ \\ v_t \end{bmatrix} + \begin{bmatrix} e_{11} \\ e_{12} \\ \\ e_{1n_1} \\ \\ e_{t1} \\ e_{t2} \\ \\ e_{tn_t} \end{bmatrix}. \tag{13.4.7.24}$$

To find variance under the nested error regression model let us choose

$$Z = \underset{1\le i\le t}{\text{diag}}\ \underset{1\le j\le n_i}{\text{Col}}(1) = I_t \otimes 1_{n_i} = \begin{bmatrix} 1, & 0, & ..., & 0 \\ 0, & 1, & ..., & 0 \\ . & & & \\ 0, & 0, & ..., & 1 \end{bmatrix}_{t\times t} \otimes \begin{bmatrix} 1 \\ 1 \\ . \\ 1 \end{bmatrix}_{n_i\times 1},$$

where $\otimes$ denotes the Krnoker product, $R = \sigma_e^2 I_n$, $G = \sigma_v^2 I_t$,

$$ZGZ' = \sigma_v^2 \begin{bmatrix} J_{n_1}, & 0, & 0 \\ 0, & J_{n_2}, & 0 \\ . & & \\ 0, & 0, & J_{n_t} \end{bmatrix}_{n\times n} \text{ with } J_{n_i} = \begin{bmatrix} 1, & 1, & 1 \\ 1, & 1, & 1 \\ & & \\ 1, & 1, & 1 \end{bmatrix}_{n_i\times n_i}$$ which implies that

$$V = R + ZGZ' = \underset{1\le i\le t}{\text{diag}}\left(V_i\right) = \underset{1\le i\le t}{\text{diag}}\left(\sigma_e^2\right)\left[I_{n_i} + \frac{\sigma_v^2}{\sigma_e^2} J_{n_i}\right] \text{ and } V_i^{-1} = \left(\sigma_e^2\right)^{-1}\left[I_{n_i} - \frac{\gamma_i}{n_i} 1_{n_i} 1'_{n_i}\right]$$ with $1_{n_i} = \underset{1\le j\le n_i}{\text{col}}(1)$. The V_i^{-1} can be obtained by using a standard result $\left(I_\alpha + \alpha J_a\right)^{-1} = I_\alpha - \frac{\alpha}{1+a\alpha} J_a$. One can easily observe that with $\zeta' = (0, 0, ..., 1, ..., 0)$ we have $\zeta' GZ'V^{-1} = \left(\frac{\gamma_i}{n_i}, \frac{\gamma_i}{n_i}, ..., \frac{\gamma_i}{n_i}\right)$, where $\gamma_i = \frac{\sigma_v^2}{\sigma_v^2 + n_i^{-1}\sigma_e^2}$, which further implies that

$$\zeta' GZ'V^{-1}(y - X\beta) = \left(\frac{\gamma_i}{n_i}, \frac{\gamma_i}{n_i}, ..., \frac{\gamma_i}{n_i}\right) \begin{bmatrix} y_{i1} - \hat{\beta} x_{i1} \\ y_{i2} - \hat{\beta} x_{i2} \\ \\ y_{in_i} - \hat{\beta} x_{in_i} \end{bmatrix} = \gamma_i\left(\bar{y}_i - \bar{x}_{i\bullet}\hat{\beta}\right).$$

Again taking $\xi = \left(0, 0, ..., \bar{X}_{i\bullet}, ..., 0\right)$ the best linear unbiased predictor (BLUP) under the nested error regression model becomes

$$\hat{\mu}(\text{nested}) = \xi'\hat{\beta} + \zeta' GZ'V^{-1}\left(y - X\hat{\beta}\right) = \overline{X}_{i\bullet}\beta + \gamma_i\left(\overline{y}_i - \overline{x}_{i\bullet}\hat{\beta}\right). \tag{13.4.7.25}$$

Then we have the following theorem:

Theorem 13.4.7.2.1. The mean squared error under the nested error regression model is given by

$$\text{MSE}[\hat{\mu}(\text{nested})] = \left(\overline{X}_{i\bullet} - \gamma_i\overline{x}_{i\bullet}\right)\left(X'V^{-1}X\right)^{-1}\left(\overline{X}_{i\bullet} - \gamma_i\overline{x}_{i\bullet}\right) + \sigma_v^2(1-\gamma_i). \tag{13.4.7.26}$$

Proof. The true mean under the nested error regression model in the ith area is given by $\mu_i = \overline{X}_{i\bullet}\beta + v_i$, therefore, the mean squared error (MSE) is given by

$$\text{MSE}[\hat{\mu}(\text{nested})] = E[\hat{\mu}(\text{nested}) - \mu_i]^2 = E\left[\left(\overline{X}_{i\bullet} - \gamma_i\overline{x}_{i\bullet}\right)\hat{\beta} + \frac{\gamma_i}{n_i}\sum_{j=1}^{n_i} y_{ij} - \mu_i\right]^2$$

$$= E\left[\left(\overline{X}_{i\bullet} - \gamma_i\overline{x}_{i\bullet}\right)\hat{\beta} + \frac{\gamma_i}{n_i}\sum_{j=1}^{n_i}\left(\beta x_{ij} + v_i + e_{ij}\right) - \overline{X}_{i\bullet}\beta - v_i\right]^2$$

$$= E\left[\left(\overline{X}_{i\bullet} - \gamma_i\overline{x}_{i\bullet}\right)\left(\hat{\beta} - \beta\right) - (1-\gamma_i)v_i + \frac{\gamma_i}{n_i}\sum_{j=1}^{n_i} e_{ij}\right]^2$$

$$= \left(\overline{X}_{i\bullet} - \gamma_i\overline{x}_{i\bullet}\right)V\left(\hat{\beta}\right)\left(\overline{X}_{i\bullet} - \gamma_i\overline{x}_{i\bullet}\right) + (1-\gamma_i)^2\sigma_v^2 + \frac{\gamma_i^2}{n_i}\sigma_e^2 .$$

Note that $\gamma_i = \dfrac{\sigma_v^2}{\sigma_v^2 + n_i^{-1}\sigma_e^2}$, which implies $\dfrac{\gamma_i}{n_i} = (1-\gamma_i)\dfrac{\sigma_v^2}{\sigma_e^2}$ and thus

$$\text{MSE}(\hat{\mu}(\text{nested})) = \left(\overline{X}_{i\bullet} - \gamma_i\overline{x}_{i\bullet}\right)V\left(\hat{\beta}\right)\left(\overline{X}_{i\bullet} - \gamma_i\overline{x}_{i\bullet}\right) + (1-\gamma_i)^2\sigma_v^2 + \sigma_e^2\gamma_i(1-\gamma_i)\frac{\sigma_v^2}{\sigma_e 2}$$

$$= \left(\overline{X}_{i\bullet} - \gamma_i\overline{x}_{i\bullet}\right)\left(X'V^{-1}X\right)^{-1}\left(\overline{X}_{i\bullet} - \gamma_i\overline{x}_{i\bullet}\right) + \sigma_v^2(1-\gamma_i).$$

Hence the theorem.

13.4.7.3 RANDOM REGRESSION COEFFICIENT MODEL

Dempster, Rubin, and Tsutakawa (1981) proposed a more general model with random regression coefficient β, which in case of single concomitant variable x and regression through origin takes the form

$$y_{ij} = \beta_i x_{ij} + e_{ij} = \beta x_{ij} + v_i x_{ij} + e_{ij}, \quad i = 1,2,...,t; \quad j = 1,2,...,n_i, \tag{13.4.7.27}$$

where $\beta_i = \beta + v_i$ and v_i and e_{ij} are independent. The mean of the ith area is given by

$$\mu_i = \overline{X}_i\beta + \overline{X}_i v_i \tag{13.4.7.28}$$

a linear combination of fixed effect β and realized value of random effect v_i. Under random regression model with $k=1, Z_i = X_i$, $\tilde{\gamma}_i = \sigma_e^2\left\{\sigma_v^2 + \sigma_e^2 \Big/ \sum_{j=1}^{n_i} x_{ij}^2\right\}^{-1}$, which implies, $(1-\tilde{\gamma}_i) = \frac{\sigma_e^2}{\sigma_v^2}\left(\tilde{\gamma}_i \Big/ \sum_{j=1}^{n_i} x_{ij}^2\right)$, and we have

$$GZ'V^{-1}\left(y - X\hat{\beta}\right) = \left(\sigma_v^2\right)I_{n_i}x'_i\left(\sigma_e^2\right)^{-1}\left[I_{n_i} - \tilde{\gamma}_i x_i x'_i \Big/ \sum_{j=1}^{n_i} x_{ij}^2\right]\left(y_i - x_i\hat{\beta}\right)$$

$$= \left(\frac{\sigma_v^2}{\sigma_e^2}\right)\left[x'_i e_i - \tilde{\gamma}_i x'_i x_i x'_i e_i \Big/ \sum_{j=1}^{n_i} x_{ij}^2\right] = \left(\frac{\sigma_v^2}{\sigma_e^2}\right)\left[\sum_{j=1}^{n_i} x_{ij}e_{ij} - \tilde{\gamma}_i \sum_{j=1}^{n_i} x_{ij}e_{ij}\right]$$

$$= \left(\frac{\sigma_v^2}{\sigma_e^2}\right)\left[(1-\tilde{\gamma}_i)\sum_{j=1}^{n_i} x_{ij}e_{ij}\right] = \left(\frac{\sigma_v^2}{\sigma_e^2}\right)\left(\frac{\sigma_e^2}{\sigma_v^2}\right)\left(\tilde{\gamma}_i \Big/ \sum_{j=1}^{n_i} x_{ij}^2\right)\left[\sum_{j=1}^{n_i} x_{ij}\left(y_{ij} - \hat{\beta}x_{ij}\right)\right]$$

$$= \tilde{\gamma}_i\left[\sum_{j=1}^{n_i} x_{ij}y_{ij} \Big/ \sum_{j=1}^{n_i} x_{ij}^2 - \hat{\beta}\right] = \tilde{\gamma}_i\left[\hat{\beta}_i - \hat{\beta}\right]. \tag{13.4.7.29}$$

Further if $\xi = \left(0,0,...,\overline{X}_i,...,0,0\right)$ then the best linear unbiased predictor (BLUP) of $\xi'\beta + \zeta'u$ is given by

$$\hat{\mu}(\text{random}) = \overline{X}_i\hat{\beta} + \overline{X}_i\tilde{\gamma}_i\left(\hat{\beta}_i - \hat{\beta}\right). \tag{13.4.7.30}$$

Then we have the following theorem:

Theorem 13.4.7.3.1. The mean squared error of the BLUP under random regression coefficient model with one auxiliary variable is given by

$$\text{MSE}\left[\hat{\mu}(\text{random})\right] = \overline{X}_i^2\left[(1-\tilde{\gamma}_i)^2\left(\sigma_v^2 \Big/ \sum_{i=1}^{t}\tilde{\gamma}_i\right) + \sigma_v^2(1-\tilde{\gamma}_i)\right]. \tag{13.4.7.31}$$

Proof. Note that with one auxiliary variable ($k=1$) we have

$$V\left(\hat{\beta}\right) = \left(X'V^{-1}X\right)^{-1} = \sigma_v^2 \Big/ \sum_{i=1}^{t}\tilde{\gamma}_i$$

$$\hat{\beta}_i = \sum_{j=1}^{n_i} x_{ij}y_{ij} \Big/ \sum_{j=1}^{n_i} x_{ij}^2 = \sum_{j=1}^{n_i} x_{ij}\left(\beta x_{ij} + v_i x_{ij} + e_{ij}\right) \Big/ \sum_{j=1}^{n_i} x_{ij}^2 = \beta + v_i + \sum_{j=1}^{n_i} x_{ij}e_{ij} \Big/ \sum_{j=1}^{n_i} x_{ij}^2$$

which implies

$$E\left[\hat{\beta}_i - \beta\right]^2 = E\left[v_i + \sum_{j=1}^{n_i} x_{ij}e_{ij} \Big/ \sum_{j=1}^{n_i} x_{ij}^2\right]^2 = \sigma_v^2 + \sigma_e^2 \Big/ \sum_{j=1}^{n_i} x_{ij}^2 = \frac{\sigma_v^2}{\tilde{\gamma}_i}.$$

Also

$$E\left[\left(\hat{\beta}_i - \beta\right)v_i\right] = E\left[v_i^2 + v_i\sum_{j=1}^{n_i} x_{ij}e_{ij} \Big/ \sum_{j=1}^{n_i} x_{ij}^2\right] = \sigma_v^2.$$

Thus we have

$$\text{MSE}\left[\hat{\mu}(\text{random})\right] = E\left[\hat{\mu}(\text{random}) - \mu_i\right]^2 = E\left[\overline{X}_i\hat{\beta} + \tilde{\gamma}_i\overline{X}_i\left(\hat{\beta}_i - \hat{\beta}\right) - \overline{X}_i\beta - \overline{X}_i v_i\right]^2$$

$$= \bar{X}_i^2 E\left[(\hat{\beta}-\beta)(1-\tilde{\gamma}_i)+\tilde{\gamma}_i(\hat{\beta}_i-\beta)-v_i\right]^2$$

$$= \bar{X}_i^2 E\left[(\hat{\beta}-\beta)^2(1-\tilde{\gamma}_i)^2+\tilde{\gamma}_i^2(\hat{\beta}_i-\beta)^2+v_i^2-2\tilde{\gamma}_i(\hat{\beta}_i-\beta)v_i\right]$$

$$= \bar{X}_i^2\left[(1-\tilde{\gamma}_i)^2\left(\sigma_v^2\Big/\sum_{i=1}^{t}\tilde{\gamma}_i\right)+\tilde{\gamma}_i^2\sigma_v^2+\sigma_v^2-2\tilde{\gamma}_i\sigma_v^2\right]$$

$$= \bar{X}_i^2\left[(1-\tilde{\gamma}_i)^2\left(\sigma_v^2\Big/\sum_{i=1}^{t}\tilde{\gamma}_i\right)+\sigma_v^2(1-\tilde{\gamma}_i)\right].$$

Hence the theorem.

13.4.7.4 FAY AND HERRIOT MODEL

For estimating per capita income for small areas (population less than 1000), Fay and Herriot (1979) assumed that the k vector of benchmark variables $x_i = (x_{i1}, x_{i2},..., x_{ik})$, related to μ_i, is available for each area i, and that the μ_i are independent $N(x_i'\beta,\ A)$, where β is a k vector of unknown parameters. The sample mean vector $\bar{y} = (\bar{y}_1, \bar{y}_2,..., \bar{y}_t) = \underset{1\le i\le t}{\text{col}}(\bar{y}_i)$, given $\mu = \underset{1\le i\le t}{\text{col}}(\mu_i)$ is $N(\mu, D)$, where $D = \underset{1\le i\le t}{\text{diag}}(D_i)$ with known diagonal element. The model can be stated as a linear model

$$\bar{y}_i = \mu_i + e_i \quad \text{and} \quad \mu_i = x_i'\beta + v_i\ , \quad i = 1,2,...,t, \tag{13.4.7.32}$$

where $e = (e_1, e_2,..., e_t)$ and $v = (v_1, v_2,..., v_t)$ are distributed independently as $N(0, D)$ and $N(0, AI)$, respectively. For the Fay and Herriot model, the best linear unbiased predictor (BLUP) of μ_i is obtained as

$$\hat{\mu}(\text{FH}) = x_i'\hat{\beta} + \left(\frac{A}{A+D_i}\right)(\bar{y}_i - x_i'\hat{\beta}), \tag{13.4.7.33}$$

where $\hat{\beta} = (X'V^{-1}X)^{-1}X'V^{-1}\bar{y}$ with $V = \text{diag}(A+D_1, A+D_2,..., A+D_t)$ and $X = \underset{1\le i\le t}{\text{col}}(x_i')$. Under normality the estimator (13.4.7.33) is also a Bayes estimator, as shown by Fay and Herriot (1979). Note that $\hat{\mu}(\text{FH}) \to \bar{y}_i$ if $D_i/(A_i + D_i) \to 0$, and $\hat{\mu}(\text{FH}) \to x_i'\hat{\beta}$ if $A_i/(A_i + D_i) \to 0$.

13.4.8 FURTHER GENERALIZATIOS

Consider Y_{ijk} denotes the value of the kth unit in the cell (i, j), the μ_j are fixed effects and the error term e_{ijk} are uncorrelated with zero means and variance σ_j^2. Holt, Smith, and Tomberlin (1979) obtain a best linear unbiased prediction (BLUP) estimator of Y_i under the linear model for the finite population

$$y_{ijk} = \mu_j + e_{ijk}, \quad k = 1,...,N_{ij};\ j = 1,2,...,N_k;\ i = 1,2,...,m. \tag{13.4.8.1}$$

Further N_{ij} denotes the number of population elements in the large domain j that belong to the small area i. Let n_{ij} elements in a sample of size n fall in the cell (i, j), and $\bar{y}_{ij}$ and $\bar{y}_{\bullet j}$ denote the sample means for (i, j) and j, respectively. The best linear unbiased estimator of μ_j under the model (13.4.8.1) is $\hat{\mu}_j = \bar{y}_{\bullet j}$ which in turn leads to the BLUP estimator of Y_i given by

$$\hat{Y}_i^B = \sum_{j=1}^{N_k} \hat{Y}_{ij}^C, \tag{13.4.8.2}$$

where $\hat{Y}_{ij}^C$ is a composite estimator of the total Y_{ij} with shrinkage coefficient $\gamma_i = n_{ij}/N_{ij}$ attached to the direct estimator and its complement $(1-\gamma_i) = (N_{ij} - n_{ij})/N_{ij}$ to the synthetic estimator. Thus in this situation if $n_{ij}/N_{ij} \to 0$ for all i, irrespective of the size of between area variation relative to within area variation, the BLUP estimator of Y_i changes to the synthetic estimator. More general models in the presence of auxiliary information may be considered as

$$y_i = X_i^T \beta + v_i z_i, \quad i = 1,2,...,m, \tag{13.4.8.3}$$

where $X_i = (x_{i1}, x_{i2},..., x_{ip})^T$ denotes the set of known auxiliary information, β is the vector of regression parameters and the v_i are independent and identically distributed random variable with $E(v_i) = 0,\ V(v_i) = \sigma_v^2$. It is also possible to consider nested error regression model, for example

$$y_{ij} = X_{ij}^T \beta + v_i + e_{ij}, \quad j = 1,2,..., N_i, \quad i = 1,2,...,m \tag{13.4.8.4}$$

where $e_{ij} = \tilde{e}_{ij} k_{ij}$ and the $\tilde{e}_{ij}$ are independent and identically distributed random variables, independent of the v_i with $E(\tilde{e}_{ij}) = 0$, $V(\tilde{e}_{ij}) = \sigma^2$, the k_{ij} being known constants and N_i the number of elements in the ith area. More details about the small area estimation issues can be had from Singh, Stukel, and Pfefferman (1998) and Rao (1999c), where they have discussed Bayesian versus frequent measures of error in small area estimation in addition to some recent advances in model-based small area estimation. Datta and Ghosh (1991), Datta and Lahiri (1995), Lahiri and Rao (1995), Datta, Day, and Maiti (1998), Datta, Day, and Basawa (1999), Datta, Lahiri, Maiti, and Lu (1999) have suggested robust hierarchical Bayes estimation of small area characteristics in the presence of covariates and outliers etc.. Datta and Lahiri (2000) suggested a unified measure of uncertanity of estimated best linear unbiased predictors. Moura and Holt (1999) have considered the problem of small area estimation using multilevel models. Agarwal and Roy (1999) have also suggested some efficient estimators for small domains. Sisodia and Singh (2001) have shown the application of small area statistic in agriculture is becoming more and more important in the context of Agro Climate Regional Planning process initiated in India. You and Rao (2000a, 2000b), Stukel and Rao (1999), Prasad and Rao (1999), Rao (1999c), Lahiri and Rao (1995), Rao and Yu (1994) are also interesting contributions to small area estimation.

13.4.9 ESTIMATION OF PROPORTION OF A CHARACTERISTIC IN SMALL AREAS OF A POPULATION

The use of estimates of proportion and their accuracy is well known in different disciplines such as economics, criminology, etc., and a practitioner always look for an improved methodology. Fay and Herriot (1979) and their followers, as reviewed by Ghosh and Rao (1994), consider the problem of estimation of mean in small areas of a continuous study variable such as income, tax return and yield of a crop based on the standard normal theory distribution. Unfortunately such a theory is not appropriate for estimating proportion of an attribute in small areas. MacGibbon and Tomberlin (1989) consider, the classical statistical logistic regression theory, that the logit transform of the proportion, not the proportion itself, that has to be modelled in a linear way while estimating proportion in small areas. Dempster and Tomberlin (1980) are the first to consider the problem of inference from a relatively thinly spread complex, multi-stage surveys to small area or domains not necessarily included in the survey by using model based approach. They estimated proportion for small areas and associated uncertainty by making use of random effects, multiple logistic regression model and empirical Bayes techniques. This explicitly model based methods differs substantially from the implicitly model based approach of the synthetic estimation techniques of Gonzalez and Hoza (1976).

Farrell, MacGibbon, and Tomberlin (1994) consider the problem of estimating small area rates and binomial parameters using empirical and hierarchical Bayes approaches by following pioneer empirical Bayes model of Dempster and Tomberlin (1980). Stroud (1994) studied hierarchical Bayes models for univariate natural exponential families with quadratic variance functions. Malec, Sedransk, Moriarity, and LeClere (1997) used a fully Bayes approach to estimate proportions using data from National Health Interview Survey.

Farrell (1997, 2000), Farrell, MacGibbon, and Tomberlin (1994, 1997a, 1997b, 1997c) pointed out that the importance of small area estimation as a facet of survey sampling cannot be over emphasised. Of late there has been an increasing demand for small area statistics in both the public and private sectors. They considered the problem of gain in power of a Bayesian approach by borrowing strength from an ensemble, while simultaneously obtaining desirable frequents operating characteristic and adopted empirical Bayes methodology for the said purpose. Farrell, MacGibbon, and Tomberlin studied the various adjustments in the empirical Bayes interval estimates of small area proportions by following different kinds of bootstrap methodology developed by Laird and Louis (1987) and modifications suggested by Carlin and Gelfand (1990, 1991).

The probabilities associated with individuals in the population as a function of categorical variables, continuous covariates and sample characteristics can be, in particular, modelled as

$$\text{logit}(\pi_{ij}) = \theta + X_{ij}\beta + \phi_i, \tag{13.4.9.1}$$

where π_{ij} denotes the probability of a 'response' for the jth unit in the ith area, the subscript i refers to a set of categorical variable covariates, and the subscript j refers to a set of nested sampling characteristics, indicating primary stage units (PSU), second stage units (SSU), SSU within PSU, and so on. The parameter θ represents a sum of fixed classification effects, the parameter ϕ_i represents a sum of random effects associated with sampling characteristics, the vector X_{ij} represents a vector of quantitative covariates, and the parameter β is a vector of fixed logistic linear regression parameters. The random effects parameters are assumed to have some parametric distribution. The probabilities π_{ij} are obtained by inverting

$$\pi_{ij} = \left[1+\exp\left\{-\left(\theta + X_{ij}\beta + \phi_i\right)\right\}\right]^{-1}. \tag{13.4.9.2}$$

The Bayes estimate for the ith small area is given by

$$\hat{p}_i = \frac{\sum_j \hat{\pi}_{ij}}{N_i}, \tag{13.4.9.3}$$

where

$$\hat{\pi}_{ij} = \left[1+\exp\left\{-\left(\hat{\theta} + X_{ij}\hat{\beta} + \hat{\phi}_i\right)\right\}\right]^{-1} \tag{13.4.9.4}$$

such that

$$\sum_{i,j} y_{ij}X_{ij} = \sum_{i,j} \hat{\pi}_{ij}X_{ij}, \tag{13.4.9.5}$$

$$\sum_{i,j} y_{ij} = \sum_{i,j} \hat{\pi}_{ij}, \tag{13.4.9.6}$$

and

$$\sum_j \left(y_{ij} - \hat{\pi}_{ij}\right) - \hat{\phi}_i / \sigma^2 = 0. \tag{13.4.9.7}$$

The equations (13.4.9.5), (13.4.9.6) and (13.4.9.7) can be solved by the Newton--Raphson algorithm. If Z_{ij} represents a vector of predictor variables, both quantitative and qualitative, associated with the (i, j)th individual and Γ represent a vector of the parameters of the model

$$Z_{ij}^T\Gamma = \theta + X_{ij}\beta + \phi_j, \tag{13.4.9.8}$$

then

$$V(\hat{p}_i) \approx \sum_j Z_{ij}^T \hat{\pi}_{ij}\left(1-\hat{\pi}_{ij}\right)\frac{\hat{\Sigma}_\Gamma}{N_i^2}\sum_j Z_{ij}\hat{\pi}_{ij}\left(1-\hat{\pi}_{ij}\right) \tag{13.4.9.9}$$

where

$$\hat{\Sigma}_\Gamma = \begin{bmatrix} \sum_{ij}\hat{\pi}_{ij}\left(1-\hat{\pi}_{ij}\right)X_{ij}^2, & \sum_{ij}\hat{\pi}_{ij}\left(1-\hat{\pi}_{ij}\right)X_{ij}, & \sum_{j}\hat{\pi}_{ij}\left(1-\hat{\pi}_{ij}\right)X_{ij} \\ \sum_{ij}\hat{\pi}_{ij}\left(1-\hat{\pi}_{ij}\right)X_{ij}, & \sum_{ij}\hat{\pi}_{ij}\left(1-\hat{\pi}_{ij}\right), & \sum_{ij}\hat{\pi}_{ij}\left(1-\hat{\pi}_{ij}\right) \\ \sum_{j}\hat{\pi}_{ij}\left(1-\hat{\pi}_{ij}\right)X_{ij}, & \sum_{ij}\hat{\pi}_{ij}\left(1-\hat{\pi}_{ij}\right), & \sum_{ij}\hat{\pi}_{ij}\left(1-\hat{\pi}_{ij}\right) - 1/\sigma^2 \end{bmatrix}^{-1} \tag{13.4.9.10}$$

and σ^2 is known. If σ^2 is unknown, then it can be estimated by the empirical Bayes EM algorithm of Dempster, Laird, and Rubin (1977).

EXERCISES

Exercise 13.1. What is the raking ratio estimation technique? Explain with the help of a 3×3 contingency table.

Exercise 13.2. Write a short note on the estimation of continuous populations in survey sampling? Give three example where continuous populations exists in actual surveys.

Exercise 13.3. Show that RHC strategy remains design unbiased for estimating population total of a continuous population.
Hint: Padmawar (1996).

Exercise 13.4. Discuss different models for estimating the estimation of measurement errors using ANOVA method. Discuss the estimators of bias and measurements errors under different sampling schemes.
Hint: Bhatia, Mangat, and Morrison (1998).

Exercise 13.5. Discuss the Vital Rate Method and Housing Unit Method techniques for small area estimation. Discuss two methods of choosing optimal weights, γ_i, in case of the composite estimator, $\hat{Y}_i^C$, of small area estimation. Show that if $\gamma_i \to 0$, the composite estimator reduces to synthetic estimator, and if $\gamma_i \to 1$, then the composite estimator reduces to the direct estimator $\hat{Y}_i^*$.
Hint: Ghosh and Rao (1994).

Exercise 13.6. Estimate the variance of the estimator of correlation coefficient using (a) Jackknifing, (b) Bootstrapping, and (c) the Balanced Half sample method. Discuss the effect of non-response on the estimator of correlation coefficient between two variables.

Exercise 13.7. Find the bias and variance of the ratio estimator

$$\bar{y}_R = \bar{y}\left(\frac{\bar{X}}{\bar{x}}\right)$$

and a weighted estimator

$$\bar{y}_\theta = \theta\,\bar{y} + (1-\theta)\bar{y}_R$$

when both variables y and x are subject to error.
Hint: Manisha and Singh (2001).

PRACTICAL PROBLEMS

Practical 13.1. Consider $R = C = 2$, we have drawn a simple random sample of 80 units with 74 respondents, and obtained the data given in the following table:

$n_{11}^{*} = 11$	$n_{12}^{*} = 16$	$N_{1\bullet} = 300$
$n_{21}^{*} = 21$	$n_{22}^{*} = 26$	$N_{2\bullet} = 400$
$N_{\bullet 1} = 350$	$N_{\bullet 2} = 350$	$N_{\bullet\bullet} = 700$

Explain the raking ratio method of estimation with minimum four iterations.

Practical 13.2. The average area in hectares, number of countries in each continent, and the number of countries to be selected from each one of the 10 continent are as given in the following table.

Continent (i)	No. of countries in the ith continent N_i	Average area in hectares $\overline{X}_i$	No. of units selected from the ith continent (n_i)
1	6	3194.50	3
2	6	14660.00	3
3	8	18309.37	4
4	10	14923.50	5
5	12	5987.83	5
6	4	3450.00	3
7	30	11682.73	12
8	17	145162.30	8
9	10	33976.10	5
10	3	1333.33	2

Select a sample of 50 countries from different continents listed in the population 5 of the Appendix. Estimate the average production in each of the continent using the synthetic ratio (SR) estimator and the composite estimator (CE).

Practical 13.3. In a college of a university, the total recruitment of new students during 2000 was 500 and the number of students who left the college was 150. The recruitment and departure rate according to 1999 records was 2.6% and 1.4%, respectively, for both the college and the university. The recruitment and departure rate of the students during 2000 for the university is 2.7% and 1.5% respectively. Estimate the total number of students in the college during 2000 by using VRM method.

Practical 13.4. In a university let the total number students be 20,000 during 1999. During 2000 there is a recruitment of 2,500 students according to the schedule and later on 150 students left the university and stopped their studies. There was migration -10 students (say, 30 students migrated to other universities and 20 students came from other universities during the academic year 2000). Apply the Census Component Method (CCM) to estimate the total number of students in the university during 2000.

Practical 13.5. The following figure shows a plant growing area near the main road of a city. The upward arrow shows a good plant, whereas the two-headed arrow shows a defective plant. The crossing block arrow shows the possibility of footpaths across the plant growing area.

(a) Select a sample of 20 plants across the whole plant growing area.

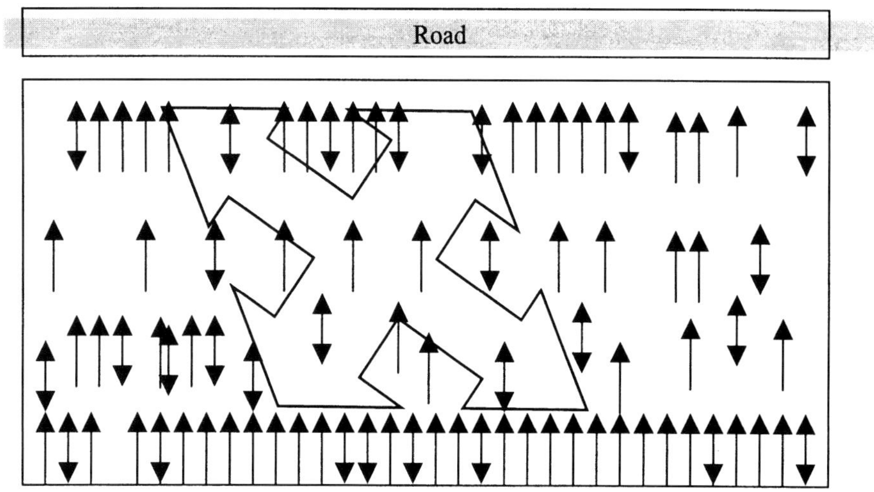

Now divide the plant growing area into four small areas using the four directions of the block arrow. Estimate the proportion of defective plants in each small area.
(b) Consider that the information about the distance of each plant row from the road is known. Suggest if this can be used to improve the estimator of proportion developed above. Note that the growth is less near the footpaths.
(c) Can you think of any information which can improve the estimates?

APPENDIX

Table 1. Pseudo-Random Numbers (PRN).

992954	622048	053688	165744	123095	171617	163786	973077	023072
588183	771280	343591	663332	873618	367912	762966	513710	787776
601448	917515	745302	624833	911823	506723	175641	935065	159406
549171	675048	304073	568609	016631	504253	004851	644863	147565
925567	534272	555153	624063	527207	981904	537885	222215	138108
014737	513080	467607	629610	769730	876103	241252	779302	185096
697856	339922	075453	855050	869127	664305	730911	734215	915553
872233	075720	139834	841453	427838	991093	436525	957477	714145
626895	627161	190747	313897	674490	313503	678962	323374	640289
236263	696957	899853	676831	158751	050072	314718	963637	342154
884270	767387	845366	506386	135152	567334	465516	862593	594974
942383	485006	074816	835762	500252	186142	938443	604162	732781
644818	498059	442486	661170	736683	504168	426933	163153	853715
465225	881559	869629	297738	442775	556942	471196	118548	401922
049819	370781	854896	088106	729287	527940	403024	444583	113867
320636	728960	250949	803613	139839	092083	637599	118688	765193
940767	638199	700635	335070	286080	644274	137531	794310	607639
477870	219452	582000	528396	161651	317053	486002	971898	595671
574258	333150	688670	751336	873796	713099	239312	691403	450906
566179	259758	478480	706027	892073	059970	128034	418305	385466
772153	769327	827373	198803	080205	929413	870198	464626	537843
576891	092665	670905	288755	011477	648964	429105	972735	527691
817574	433974	903417	118562	905366	151992	978704	763841	904615
973142	806267	731170	680235	502065	537687	092620	884410	876879
608932	942025	790592	721489	210155	914388	671936	282358	711834
333138	622638	491253	070561	076166	121776	234911	601944	794614
053651	688769	490430	927370	127476	703745	163014	878087	080662
721677	155734	669360	312921	489308	846283	776593	018163	692605
226505	427675	933935	279406	816254	486369	208397	265881	615237
882750	936294	647940	978020	793783	418755	537679	672050	300690
381070	295889	943669	589360	119593	024805	787483	159494	712930
651853	012286	247554	378621	009225	897628	236542	620496	219186
702670	495896	644947	487506	061240	437818	563445	681374	869224
786451	528864	640408	529911	262859	408046	834394	862338	051535
884886	630173	347278	228907	876472	040491	665615	329032	161707

Continued......

290260	323502	707834	244659	874655	548426	995549	040993	816920
409083	099078	571932	984411	330590	412815	196353	978115	984771
465730	211163	476164	995712	226350	188281	106244	185970	055161
539896	027534	900914	944872	268155	442362	646597	613553	104275
039365	768948	676363	467448	894496	793214	362738	995656	554078
616705	724810	642912	412883	803841	564324	554305	991424	397772
975581	896491	426965	326618	159848	602917	211953	876408	368676
604472	336311	014143	482110	571612	234134	309728	794506	347737
367344	390037	705539	831722	094714	106243	601509	274297	420157
960001	354823	001221	767362	860098	626945	561950	113412	591427
737209	517446	531783	064831	678795	071352	704580	128848	117604
275811	727153	680330	628827	518951	229341	877061	556992	583115
565430	260577	313905	698626	010685	541472	776828	370699	429592
196602	529352	801239	821149	582828	531195	149357	344186	155977
291222	833343	431066	054285	592207	794992	259361	736578	509032
584975	998455	937964	376363	227579	432919	091869	680558	314357
685465	039510	363667	535141	926647	075276	051822	154758	404150

Source: Generated from Uniform distribution.

Table 2. Critical values based on t distribution.

	Level of significance								
	$\frac{\alpha}{2} \rightarrow$	0.25	0.15	0.125	0.10	0.05	0.025	0.01	0.005
df	$\alpha \rightarrow$	0.50	0.30	0.25	0.20	0.10	0.05	0.02	0.01
01		1.000	1.963	2.414	3.078	6.314	12.71	31.82	63.65
02		0.816	1.386	1.604	1.886	2.920	4.303	6.965	9.925
03		0.765	1.250	1.423	1.638	2.353	3.182	4.541	5.841
04		0.741	1.190	1.344	1.533	2.132	2.776	3.747	4.604
05		0.727	1.156	1.301	1.476	2.015	2.571	3.365	4.032
06		0.718	1.134	1.273	1.440	1.943	2.447	3.143	3.707
07		0.711	1.119	1.254	1.415	1.895	2.365	2.998	3.499
08		0.706	1.108	1.240	1.397	1.860	2.306	2.896	3.355
09		0.703	1.100	1.230	1.383	1.833	2.262	2.821	3.250
10		0.700	1.093	1.221	1.372	1.812	2.228	2.764	3.169
11		0.697	1.088	1.214	1.363	1.796	2.201	2.718	3.106
12		0.695	1.083	1.209	1.356	1.782	2.179	2.681	3.055
13		0.694	1.079	1.204	1.350	1.771	2.160	2.650	3.012
14		0.692	1.076	1.200	1.345	1.761	2.145	2.624	2.977
15		0.691	1.074	1.197	1.341	1.753	2.131	2.602	2.947
16		0.690	1.071	1.194	1.337	1.746	2.120	2.583	2.921
17		0.689	1.069	1.191	1.333	1.740	2.110	2.567	2.898
18		0.688	1.067	1.189	1.330	1.734	2.101	2.552	2.878
19		0.688	1.066	1.187	1.328	1.729	2.093	2.539	2.861
20		0.687	1.064	1.185	1.325	1.725	2.086	2.528	2.845
21		0.686	1.063	1.183	1.323	1.721	2.080	2.518	2.831
22		0.686	1.061	1.182	1.321	1.717	2.074	2.508	2.819
23		0.685	1.060	1.180	1.319	1.714	2.069	2.500	2.807
24		0.685	1.059	1.179	1.318	1.711	2.064	2.492	2.797
25		0.684	1.058	1.178	1.316	1.708	2.060	2.485	2.787
26		0.684	1.058	1.177	1.315	1.706	2.056	2.479	2.779
27		0.684	1.057	1.176	1.314	1.703	2.052	2.473	2.771
28		0.683	1.056	1.175	1.313	1.701	2.048	2.467	2.763
29		0.683	1.055	1.174	1.311	1.699	2.045	2.462	2.756
30		0.683	1.055	1.173	1.310	1.697	2.042	2.457	2.750
31		0.682	1.054	1.172	1.309	1.696	2.040	2.453	2.744
32		0.682	1.054	1.172	1.309	1.694	2.037	2.449	2.738

Continued.......

33		0.682	1.053	1.171	1.308	1.692	2.035	2.445	2.733
34		0.682	1.052	1.170	1.307	1.691	2.032	2.441	2.728
35		0.682	1.052	1.170	1.306	1.690	2.030	2.438	2.724
36		0.681	1.052	1.169	1.306	1.688	2.028	2.434	2.719
37		0.681	1.051	1.169	1.305	1.687	2.026	2.431	2.715
38		0.681	1.051	1.168	1.304	1.686	2.024	2.429	2.712
39		0.681	1.050	1.168	1.304	1.685	2.023	2.426	2.708
40		0.681	1.050	1.167	1.303	1.684	2.021	2.423	2.704
Infinity		0.674	1.036	1.150	1.282	1.645	1.960	2.326	2.576

Source: Generated with EXCEL using the $\text{TINV}(\alpha, \text{df})$ function.

Table 3. Area under the standard normal curve for $(0 \le z < \infty)$.

Z	0.00	0.01	0.02	0.03	0.04	0.05	0.06	0.07	0.08	0.09
0.0	0.0000	0.0040	0.0080	0.0120	0.0160	0.0199	0.0239	0.0279	0.0319	0.0359
0.1	0.0398	0.0438	0.0478	0.0517	0.0557	0.0596	0.0636	0.0675	0.0714	0.0753
0.2	0.0793	0.0832	0.0871	0.0910	0.0948	0.0987	0.1026	0.1064	0.1103	0.1141
0.3	0.1179	0.1217	0.1255	0.1293	0.1331	0.1368	0.1406	0.1443	0.1480	0.1517
0.4	0.1554	0.1591	0.1628	0.1664	0.1700	0.1736	0.1772	0.1808	0.1844	0.1879
0.5	0.1915	0.1950	0.1985	0.2019	0.2054	0.2088	0.2123	0.2157	0.2190	0.2224
0.6	0.2257	0.2291	0.2324	0.2357	0.2389	0.2422	0.2454	0.2486	0.2517	0.2549
0.7	0.2580	0.2611	0.2642	0.2673	0.2704	0.2734	0.2764	0.2794	0.2823	0.2852
0.8	0.2881	0.2910	0.2939	0.2967	0.2995	0.3023	0.3051	0.3078	0.3106	0.3133
0.9	0.3159	0.3186	0.3212	0.3238	0.3264	0.3289	0.3315	0.3340	0.3365	0.3389
1.0	0.3413	0.3438	0.3461	0.3485	0.3508	0.3531	0.3554	0.3577	0.3599	0.3621
1.1	0.3643	0.3665	0.3686	0.3708	0.3729	0.3749	0.3770	0.3790	0.3810	0.3830
1.2	0.3849	0.3869	0.3888	0.3907	0.3925	0.3944	0.3962	0.3980	0.3997	0.4015
1.3	0.4032	0.4049	0.4066	0.4082	0.4099	0.4115	0.4131	0.4147	0.4162	0.4177
1.4	0.4192	0.4207	0.4222	0.4236	0.4251	0.4265	0.4279	0.4292	0.4306	0.4319
1.5	0.4332	0.4345	0.4357	0.4370	0.4382	0.4394	0.4406	0.4418	0.4429	0.4441
1.6	0.4452	0.4463	0.4474	0.4484	0.4495	0.4505	0.4515	0.4525	0.4535	0.4545
1.7	0.4554	0.4564	0.4573	0.4582	0.4591	0.4599	0.4608	0.4616	0.4625	0.4633
1.8	0.4641	0.4649	0.4656	0.4664	0.4671	0.4678	0.4686	0.4693	0.4699	0.4706
1.9	0.4713	0.4719	0.4726	0.4732	0.4738	0.4744	0.4750	0.4756	0.4761	0.4767
2.0	0.4772	0.4778	0.4783	0.4788	0.4793	0.4798	0.4803	0.4808	0.4812	0.4817
2.1	0.4821	0.4826	0.4830	0.4834	0.4838	0.4842	0.4846	0.4850	0.4854	0.4857
2.2	0.4861	0.4864	0.4868	0.4871	0.4875	0.4878	0.4881	0.4884	0.4887	0.4890
2.3	0.4893	0.4896	0.4898	0.4901	0.4904	0.4906	0.4909	0.4911	0.4913	0.4916
2.4	0.4918	0.4920	0.4922	0.4925	0.4927	0.4929	0.4931	0.4932	0.4934	0.4936
2.5	0.4938	0.4940	0.4941	0.4943	0.4945	0.4946	0.4948	0.4949	0.4951	0.4952
2.6	0.4953	0.4955	0.4956	0.4957	0.4959	0.4960	0.4961	0.4962	0.4963	0.4964
2.7	0.4965	0.4966	0.4967	0.4968	0.4969	0.4970	0.4971	0.4972	0.4973	0.4974
2.8	0.4974	0.4975	0.4976	0.4977	0.4977	0.4978	0.4979	0.4979	0.4980	0.4981
2.9	0.4981	0.4982	0.4982	0.4983	0.4984	0.4984	0.4985	0.4985	0.4986	0.4986
3.0	0.4987	0.4987	0.4987	0.4988	0.4988	0.4989	0.4989	0.4989	0.4990	0.4990
3.1	0.4990	0.4991	0.4991	0.4991	0.4992	0.4992	0.4992	0.4992	0.4993	0.4993
3.2	0.4993	0.4993	0.4994	0.4994	0.4994	0.4994	0.4994	0.4995	0.4995	0.4995
3.3	0.4995	0.4995	0.4995	0.4996	0.4996	0.4996	0.4996	0.4996	0.4996	0.4997
3.4	0.4997	0.4997	0.4997	0.4997	0.4997	0.4997	0.4997	0.4997	0.4997	0.4998
3.5	0.4998	0.4998	0.4998	0.4998	0.4998	0.4998	0.4998	0.4998	0.4998	0.4998
3.6	0.4998	0.4998	0.4999	0.4999	0.4999	0.4999	0.4999	0.4999	0.4999	0.4999
∞	0.5000									

Source: Generated in Excel using the NORMSDIST(x)

Population 1. All operating banks: Amount (in $000) of agricultural loans outstanding in different states in 1997.

Sr. No.	State and Territory	Nonreal estate farm loans	Real estate farm loans
1	AL	348.334	408.978
2	AK	3.433	2.605
3	AZ	431.439	54.633
4	AR	848.317	907.700
5	CA	3928.732	1343.461
6	CO	906.281	315.809
7	CT	4.373	7.130
8	DE	43.229	42.808
9	FL	464.516	825.748
10	GA	540.696	939.460
11	HI	38.067	40.775
12	ID	1006.036	53.753
13	IL	2610.572	2131.048
14	IN	1022.782	1213.024
15	IA	3909.738	2327.025
16	KS	2580.304	1049.834
17	KY	557.656	1045.106
18	LA	405.799	282.565
19	ME	51.539	8.849
20	MD	57.684	139.628
21	MA	56.471	7.590
22	MI	440.518	323.028
23	MN	2466.892	1354.768
24	MS	549.551	627.013
25	MO	1519.994	1579.686
26	MT	722.034	292.965
27	NE	3585.406	1337.852
28	NV	16.710	5.860
29	NH	0.471	6.044
30	NJ	27.508	39.860
31	NM	274.035	140.582
32	NY	426.274	201.631
33	NC	494.730	639.571
34	ND	1241.369	449.099
35	OH	635.774	870.720
36	OK	1716.087	612.108
37	OR	571.487	114.899
38	PA	298.351	756.169

Continued……

39	RI	0.233	1.611
40	SC	80.750	87.951
41	SD	1692.817	413.777
42	TN	388.869	553.266
43	TX	3520.361	1248.761
44	UT	197.244	56.908
45	VT	19.363	57.747
46	VA	188.477	321.583
47	WA	1228.607	1100.745
48	WV	29.291	99.277
49	WI	1372.439	1229.752
50	WY	386.479	100.964

Source: Agricultural Statistics (1999) Washington, US.

Descriptive parameters of the population 1:

Y_i = Amount (in \$000) of real estate farm loans in different states during 1997.
X_i = Amount (in \$000) of nonreal estate farm loans in different states during 1997.
$Y = 27771.73$, $X = 43908.12$, $\overline{Y} = 555.43$, $\overline{X} = 878.16$, $S_x^2 = 1176526$, $S_y^2 = 342021.5$, $C_x^2 = 1.5256$, $C_y^2 = 1.1086$, $\lambda_{03} = 1.5936$, $\lambda_{30} = 1.1011$, $S_{xy} = 509910.41$, $\rho_{xy} = 0.8038$, $N = 50$, $\beta = 0.4334$, $\lambda_{21} = 0.9387$, $\lambda_{12} = 1.0982$, $\lambda_{40} = 3.5822$, $\lambda_{04} = 4.5247$, $\lambda_{22} = 2.8411$, $\lambda_{31} = 2.9287$, $\lambda_{13} = 3.2561$, $M_y = 322.305$, $M_x = 452.517$, and $P_{11} = 0.42$.

Population 2. Hypothetical situation of a small village having only 30 old persons (age more that 50 years): Approximate duration of sleep (in minutes) and Age (in years) of the persons.

Person	Age (x)	Duration of sleep (y)
1	60	492
2	72	384
3	55	408
4	56	465
5	82	312
6	78	315
7	67	420
8	74	381
9	84	276
10	56	465
11	68	420
12	70	360
13	59	435
14	64	405
15	53	510
16	66	420
17	78	345
18	63	405
19	77	330
20	73	285
21	55	438
22	71	360
23	63	390
24	87	270
25	61	375
26	58	375
27	60	390
28	70	360
29	66	390
30	72	345

Descriptive parameters of the population 2:

Y_i = Duration of sleep (in minutes), X_i = Age of old persons $(\geq 50 \text{ years})$

$N = 30$, $X = 2018$, $Y = 11526$, $\overline{X} = 67.267$, $\overline{Y} = 384.2$, $S_y^2 = 3582.58$, $S_x^2 = 85.237$, $C_y^2 = 0.0243$, $C_x^2 = 0.0188$, $S_{xy} = -472.607$, and $\rho_{xy} = -0.8552$.

Population 3. Apples, commercial crop: Season average price (in $) per pound, by States, 1994--1996.

Sr. No.	State and Territory	Price $		
		Year 1994	Year 1995	Year 1996
1	AZ	0.078	0.071	0.122
2	AR	0.164	0.143	0.180
3	CA	0.133	0.183	0.160
4	CO	0.157	0.145	0.223
5	CT	0.283	0.276	0.292
6	DE	0.168	0.125	0.173
7	GA	0.139	0.164	0.175
8	ID	0.101	0.174	0.157
9	IL	0.209	0.210	0.324
10	IN	0.200	0.197	0.246
11	IA	0.244	0.303	0.312
12	KS	0.206	0.305	0.279
13	KY	0.216	0.255	0.316
14	ME	0.174	0.179	0.185
15	MD	0.173	0.131	0.158
16	MA	0.226	0.208	0.221
17	MI	0.086	0.099	0.125
18	MN	0.332	0.403	0.452
19	MO	0.198	0.160	0.228
20	NH	0.217	0.203	0.198
21	NJ	0.157	0.159	0.145
22	NM	0.219	0.298	0.306
23	NY	0.118	0.121	0.130
24	NC	0.088	0.084	0.117
25	OH	0.181	0.200	0.264
26	OR	0.107	0.116	0.126
27	PA	0.104	0.095	0.133
28	RI	0.310	0.301	0.314
29	SC	0.130	0.126	0.136
30	TN	0.195	0.215	0.235
31	UT	0.121	0.188	0.142
32	VT	0.165	0.181	0.194
33	VA	0.090	0.099	0.101
34	WA	0.138	0.215	0.204
35	WV	0.095	0.110	0.111
36	WI	0.230	0.241	0.133

Source: Agricultural Statistics (1998) Washington, US.

Descriptive parameters of the population 3:

$Y_i =$ Season average price per pound during 1996.
$X_{1i} =$ Season average price per pound during 1995.
$X_{2i} =$ Season average price per pound during 1994.
$Y = 7.317$, $X_1 = 6.683$, $X_2 = 6.152$, $N = 36$, $\overline{Y} = 0.2033$, $\overline{X}_1 = 0.1856$, $\overline{X}_2 = 0.1708$, $S_y^2 = 0.006458$, $S_{x1}^2 = 0.005654$, $S_{x2}^2 = 0.004017$, $C_y^2 = 0.1563$, $C_{x1}^2 = 0.1641$, $C_{x2}^2 = 0.13757$, $\rho_{yx_1} = 0.8775$, $\rho_{yx_2} = 0.8577$, and $\rho_{x_1x_2} = 0.8788$.

Population 4. Fish caught: Estimated number of fish caught by marine recreational fishermen by species group and year, Atlantic and Gulf coasts, 1992--95.

Sr. No.	Species group	Year 1992 X_3	Year 1993 X_2	Year 1994 X_1	Year 1995 Y
1	Sharks, other	1467	1385	2001	2016
2	Sharks, dogfish	1039	1031	993	833
3	Skates/Rays	2152	1981	2939	2353
4	Eels	138	222	186	152
5	Herrings	28933	34060	38007	30027
6	Freshwater catfish	1100	1091	1377	666
7	Saltwater catfish	13466	12690	14441	13859
8	Toadfish	1784	2676	1781	1632
9	Atlantic cod	850	2693	1861	1942
10	Pollock	168	397	862	832
11	Red hake	559	216	369	184
12	Codfish/hakes, other	73	124	130	266
13	Searobins	4768	7726	4707	4793
14	Sculpins	54	698	136	71
15	White perch	3669	5281	4648	3489
16	Striped bass	3840	4799	8521	10758
17	Temperate bass, other	5	35	32	23
18	Black sea bass	11759	12758	11892	17723
19	Groupers	4661	4236	4583	4923
20	Sea bass, other	2797	2690	2138	2068
21	Bluefish	11990	10301	12405	10940
22	Crevalle jack	3542	2569	2978	3951
23	Blue runner	2371	3800	5692	2319
24	Greater amber jack	692	1141	332	164
25	Florida pompano	498	641	425	644
26	Jacks, other	4463	3802	1878	1625
27	Dolphins	1484	1926	2449	2613
28	Gray snapper	5363	5154	4845	4552
29	Red snapper	2024	2546	2011	1608
30	Lane snapper	919	1079	1088	859
31	Vermilion snapper	950	1228	826	1200
32	Yellowtail snapper	1649	2061	1247	1334
33	Snappers, others	746	861	462	492
34	Pigfish	2955	2691	4918	4199
35	White grunt	5593	5356	5784	5678
36	Grunts, other	3039	3521	3186	3379

Continued.......

37	Scup	10078	7077	5662	3688
38	Pinfish	13055	13043	16063	16855
39	Sheepshead	5933	5593	4383	5118
40	Red porgy	207	166	166	230
41	Porgies, other	545	445	434	484
42	Spotted seatrout	22304	21538	22181	24615
43	Weakfish	1668	2219	4929	5739
44	Sand seatrout	3780	4068	5665	4355
45	Silver perch	1198	1034	1729	2146
46	Spot	14974	14263	18491	11567
47	Kingfish	3778	3304	4805	4333
48	Atlantic croaker	16953	21016	26671	17753
49	Black drum	1405	1534	1125	1595
50	Red drum	8682	7649	7609	9236
51	Drums, other	1365	1165	1622	1354
52	Mullet	5571	4186	4386	4657
53	Barracudas	800	788	906	908
54	Tautog	4195	4215	2653	3816
55	Cunner	1931	1876	1255	1375
56	Wrasses, other	275	240	227	185
57	Little tunny/ Atl bonito	996	925	982	782
58	Atlantic mackerel	1045	2307	4860	4008
59	King mackerel	1289	1023	1148	1252
60	Spanish mackerel	5575	3653	3850	2568
61	Tunas/mackerels, other	1190	794	1018	1029
62	Summer flounder	11918	22919	17741	16238
63	Gulf flounder	216	189	776	163
64	Southern flounder	1148	1083	1369	1446
65	Winter flounder	1544	3582	2300	2324
66	Flounders, other	1252	2149	2173	1284
67	Triggerfish / file fish	1103	999	918	897
68	Puffers	2100	1323	1141	935
69	Other fish	12249	14953	20488	14426

Source: Agricultural Statistics (1999), Washington, D.C.

Descriptive parameters of the population 4:

X_{3i} = Estimated number of fish caught during 1992.
X_{2i} = Estimated number of fish caught during 1993.
X_{1i} = Estimated number of fish caught during 1994.
Y_i = Estimated number of fish caught during 1995.

$X_3 = 291882$, $X_2 = 316784$, $X_1 = 341856$, $Y = 311528$, $N = 69$, $\overline{X}_3 = 4230.174$, $\overline{X}_2 = 4591.072$, $\overline{X}_1 = 4954.435$, $\overline{Y} = 4514.899$, $S_{x3}^2 = 31010599$, $S_{x2}^2 = 39881874$, $S_{x1}^2 = 49829270$, $S_y^2 = 37199578$, $C_{x3}^2 = 1.7329$, $C_{x2}^2 = 1.8921$, $C_{x1}^2 = 2.0300$, $C_y^2 = 1.8249$, $\rho_{x2x3} = 0.9632$, $\rho_{x1x3} = 0.9576$, $\rho_{x3y} = 0.9538$, $\rho_{x2x1} = 0.9729$, $\rho_{x2y} = 0.9564$, and $\rho_{x1y} = 0.9601$.

Population 5. Tobacco: Area (hectares), yield and production (metric tons) in specified countries during 1998.

Stratum No.	Continent	Unit No.	Country	Area (Hectare)	Yield (Metric tons)	Production (Metric tons)
1	Central America	1	Costa Rica	1072	2.03	2180
		2	El Salvador	580	1.79	1038
		3	Guatemala	9024	2.21	19962
		4	Honduras	5157	1.78	9177
		5	Nicaragua	2240	2.03	4550
		6	Panama	1094	2.00	2188
2	Caribbean	1	Cuba	59000	0.63	37000
		2	Dominican Rep	27050	1.51	40950
		3	Haiti	565	1.29	730
		4	Jamaica	1175	1.99	2339
		5	Trinidad and	100	1.70	170
		6	St. Vincent	70	1.21	85
3	European Union	1	Austria	105	1.90	199
		2	Belgium--Lux	320	3.69	1180
		3	France	9260	2.80	25930
		4	Germany	3831	2.40	9200
		5	Greece	67300	1.96	132000
		6	Italy	47600	2.77	132000
		7	Portugal	2909	2.14	6226
		8	Spain	15150	2.79	42300
4	West and Eastern Europe	1	Switzerland	635	2.09	1325
		2	Albania	24000	0.63	15000
		3	Bulgaria	49000	1.61	79000
		4	Czech Republic	7500	1.64	12306
		5	Croatia	7000	1.67	11700
		6	Hungary	2000	1.75	3500
		7	Macedonia	22000	1.36	30000
		8	Poland	19100	2.34	44700
		9	Romania	12000	1.23	14750
		10	Sebia/Montenegr	6000	1.17	7000
5	FSU 12	1	Azerbaijan	4500	2.33	10500
		2	Armenia	4304	0.26	1100
		3	Byelarus	1076	2.42	2606
		4	Georgia	5400	1.63	8800
		5	Kyrgyzstan	12000	2.50	30000
		6	Kazakhstan	5500	1.82	10000
		7	Moldova	18600	2.05	38150
		8	Russia	650	0.92	600

Continued......

		9	Tajikistan	3228	2.48	8000
		10	Turkmenistan	1100	2.36	2600
		11	Ukranine	5000	0.84	4200
		12	Uzbekistan	10496	2.37	24900
6	North Africa	1	Algeria	2700	1.96	5300
		2	Libya	900	1.61	1450
		3	Morocco	3500	1.13	3962
		4	Tunisia	6700	1.18	7900
7	Other Africa	1	Angola	3950	0.99	3900
		2	Burundi	705	1.00	705
		3	Chad	200	1.00	200
		4	Congo	4000	0.45	1800
		5	Zaire	3700	1.11	4110
		6	Cameroon	3400	1.62	5500
		7	Central African	750	0.87	650
		8	Benin	200	2.00	400
		9	Ethiopia	3000	1.17	3500
		10	Ghana	3950	0.38	1500
		11	Cote d'lvoire	10000	0.26	2600
		12	Kenya	8805	2.51	22120
		13	Liberia	10	1.00	10
		14	Madagascar	5900	0.93	5500
		15	Malawi	116700	1.22	142300
		16	Mali	1000	0.55	550
		17	Mauritius	655	1.63	1065
		18	Mozambique	2700	1.07	2900
		19	Niger	1000	0.93	930
		20	Nigeria	10000	2.10	21000
		21	Reunion	200	1.00	200
		22	Zimbabwe	103110	2.06	212050
		23	South Africa	15500	2.00	31000
		24	Sierra Leone	540	1.11	600
		25	Somalia	0	0.00	0
		26	Togo	4000	0.50	2000
		27	Tanzania	33900	0.74	25080
		28	Uganda	7525	0.96	7198
		29	Swaziland	200	1.00	200
		30	Zambia	4882	1.29	6300
8	Other Asia	1	Bangladesh	50263	0.88	44000
		2	Burma	36000	1.22	44000
		3	Cambodia	9000	0.56	5000
		4	Sri Lanka	12165	0.74	9000
		5	China	144500	1.75	2524500

Continued......

		6	Indonesia	206625	0.85	175631
		7	India	445000	1.43	635000
		8	Japan	26214	2.56	67100
		9	Korea, North	20000	1.25	25000
		10	Korea, South	25730	2.02	52040
		11	Laos	4000	0.75	3000
		12	Malaysia	11000	1.05	11505
		13	Pakistan	47300	1.91	90450
		14	Philippines	37568	1.80	67500
		15	Thailand	51500	1.33	68600
		16	Taiwan	4394	2.50	10990
		17	Vietnam	36000	0.89	32000
9	Middle East	1	Cyprus	161	1.50	241
		2	Iran	18000	1.39	25000
		3	Iraq	2000	1.09	2180
		4	Jordan	2100	1.29	2700
		5	Lebanon	3750	1.33	5000
		6	Oman	1800	1.11	2000
		7	Syria	15000	1.15	17200
		8	United Arab.	350	5.71	2000
		9	Turkey	293300	0.91	266500
		10	Yemen	3300	1.73	5720
10	Oceania	1	Australia	3300	2.73	9000
		2	Solomon Islands	100	0.95	95
		3	New Zealand	600	2.58	1550

Source: Agricultural Statistics (1999), Washington, D.C.

Stratum wise and overall descriptive parameters of the population 5:
X = Area, Y = Yield, Z = Production

Stratum 1

$N_1 = 6$, $W_1 = 0.0566$, $\overline{X}_1 = 3194.5$, $\overline{Y}_1 = 1.913$, $\overline{Z}_1 = 6515.83$, $S_{1x}^2 = 10899652.8$, $S_{1y}^2 = 0.0268266$, $S_{1z}^2 = 51856391.4$, $S_{1xy} = 249.836$, $S_{1xz} = 23619714.4$, and $S_{1yz} = 639.9066$

Stratum 2

$N_2 = 6$, $W_2 = 0.0566$, $\overline{X}_2 = 14660$, $\overline{Y}_2 = 1.388$, $\overline{Z}_2 = 13545.667$, $S_{2x}^2 = 584984730$, $S_{2y}^2 = 0.21809$, $S_{2z}^2 = 390206386.6$, $S_{2xy} = -8156.1$, $S_{2xz} = 420480877$ and $S_{2yz} = -4340.548$

Stratum 3

$N_3 = 8$, $W_3 = 0.0755$, $\overline{X}_3 = 18309.37$, $\overline{Y}_3 = 2.556$, $\overline{Z}_3 = 43629.38$, $S_{3x}^2 = 635958094.9$, $S_{3y}^2 = 0.346998$, $S_{3z}^2 = 3172330153$, $S_{3xy} = -3667.18$, $S_{3xz} = 1387271318$, and $S_{3yz} = -5300.54$

Stratum 4

$N_4 = 10$, $W_4 = 0.0943$, $\overline{X}_4 = 14923.5$, $\overline{Y}_4 = 1.549$, $\overline{Z}_4 = 21928.1$, $S_{4x}^2 = 209817189.2$, $S_{4y}^2 = 0.23456$, $S_{4z}^2 = 569396673.9$, $S_{4xy} = -1327.096$, $S_{4xz} = 319262930.2$ and

$$S_{4yz} = 1924.369$$

Stratum 5

$N_5 = 12$, $W_5 = 0.1132$, $\overline{X}_5 = 5987.83$, $\overline{Y}_5 = 1.832$, $\overline{Z}_5 = 11788$, $S_{5x}^2 = 27842810.45$, $S_{5y}^2 = 0.58214$, $S_{5z}^2 = 153854582.5$, $S_{5xy} = 892.03$, $S_{5xz} = 62846173.1$ and

$$S_{5yz} = 4252.525$$

Stratum 6

$N_6 = 4$, $W_6 = 0.0377$, $\overline{X}_6 = 3450$, $\overline{Y}_6 = 1.47$, $\overline{Z}_6 = 4653$, $S_{6x}^2 = 5876666.67$, $S_{6y}^2 = 0.1531$, $S_{6z}^2 = 7232769.3$, $S_{6xy} = -561.33$, $S_{6xz} = 6066866.67$ and

$$S_{6yz} = -279.36$$

Stratum 7

$N_7 = 30$, $W_7 = 0.2830$, $\overline{X}_7 = 11682.73$, $\overline{Y}_7 = 1.115$, $\overline{Z}_7 = 16862.27$, $S_{7x}^2 = 760238523.4$, $S_{7y}^2 = 0.34385$, $S_{7z}^2 = 2049296094$, $S_{7xy} = 3983.74$, $S_{7xz} = 1190767859$ and

$$S_{7yz} = 9674.33$$

Stratum 8

$N_8 = 17$, $W_8 = 0.1604$, $\overline{X}_8 = 145162.3$, $\overline{Y}_8 = 1.3817$, $\overline{Z}_8 = 227371.5$, $S_{8x}^2 = 1.24\times10^{11}$, $S_{8y}^2 = 0.37855$, $S_{8z}^2 = 3.72\times10^{11}$, $S_{8xy} = 28823.77$, $S_{8xz} = 2.14\times10^{11}$ and

$$S_{8yz} = 62502.73$$

Stratum 9

$N_9 = 10$, $W_9 = 0.0944$, $\overline{X}_9 = 33976.1$, $\overline{Y}_9 = 1.721$, $\overline{Z}_9 = 32854.1$, $S_{9x}^2 = 8340765245$, $S_{9y}^2 = 2.0183$, $S_{9z}^2 = 6802998890$, $S_{9xy} = -28414.68$, $S_{9xz} = 7529100326$ and

$$S_{9yz} = -25774.2$$

Stratum 10

$N_{10} = 3$, $W_{10} = 0.0283$, $\overline{X}_{10} = 1333.33$, $\overline{Y}_{10} = 2.086$, $\overline{Z}_{10} = 3548.33$, $S_{10x}^2 = 2963333.33$, $S_{10y}^2 = 0.9746$, $S_{10z}^2 = 22819758.33$, $S_{10xy} = 1152.67$, $S_{10xz} = 8223083.33$ and

$$S_{10yz} = 3223.34$$

Overall population

$N = 106$, $\overline{X} = 34438.61$, $\overline{Y} = 1.5507$, $\overline{Z} = 52444.56$, , $S_x^2 = 22340229388$, $S_y^2 = 0.6323$, $S_z^2 = 64196519001$, $S_{xy} = -910.459$, $S_{xz} = 37536712756$ and $S_{yz} = 6122.433$

Population 6. Age specific death rates from 1990 to 2065 (Number per 100, 000 births).

Age Group	Year 1990	Year 1995	Year 2000	Year 2010	Year 2020	Year 2030	Year 2040	Year 2050	Year 2065
0--4	967	818	692	496	355	255	183	131	80
5--9	19	16	14	10	7	5	4	3	2
10--14	20	17	15	11	8	6	4	3	2
15--19	67	62	57	48	40	34	28	24	18
20--24	86	78	71	58	48	40	33	27	20
25--29	84	75	68	54	44	35	28	23	16
30--34	97	87	78	62	50	40	32	25	18
35--39	138	124	111	90	72	58	47	38	27
40--44	221	201	182	150	124	102	84	69	52
45--49	370	341	315	267	227	193	164	139	109
50--54	613	572	533	464	403	351	305	265	215
55--59	965	907	853	754	666	589	520	460	382
60--64	1511	1432	1357	1218	1094	982	882	792	674
65--69	2233	2119	2010	1810	1629	1466	1320	1188	1015
70--74	3361	3187	3022	2718	2444	2198	1976	1777	1515
75--79	4979	4693	4423	3930	3491	3102	2756	2448	2050
80--84	7748	7323	6921	6182	5523	4933	4407	3936	3323
85--89	12267	11687	10609	10108	9177	8331	7564	6868	5942
90--94	19099	18341	16915	16246	14987	13827	12758	11774	10439
95--99	29744	28864	27188	26390	24869	23442	22102	20844	19095
100--104	46334	45554	44053	43329	41933	40600	39325	38104	36364
105--109	72195	72100	71956	71906	71845	71831	71861	71930	72097

Source: Statistical Abstracts of the United States (1993).

Population 7. State population projections, 1995 and 2000.
(Numbers in thousands).

Region	Division	State	Year 1995	Year 2000
Northeast	New England	Maine	1295	1344
		New Hampshire	1276	1410
		Vermont	597	619
		Massachusetts	6032	6159
		Rhode Island	1021	1048
		Connecticut	3354	3422
	Mid Atlantic	New York	17909	17966
		New Jersey	8100	8382
		Pennsylvania	12080	12069
Midwest	East North Central	Ohio	10958	10930
		Indiana	5688	5696
		Illinois	11759	11722
		Michigan	9364	9365
		Wisconsin	4908	4844
	West North Central	Minnesota	4501	4566
		Iowa	2703	2549
		Missouri	5353	5473
		North Dakota	631	596
		South Dakota	719	715
		Nebraska	1582	1539
		Kansas	2546	2534
South	South Atlantic	Delaware	739	802
		Maryland	5180	5608
		District of	592	595
		Virginia	6758	7275
		West Virginia	1749	1651
		North Carolina	7197	7717
		South Carolina	3772	3962
		Georgia	7288	8005
		Florida	14583	16315
	East South Central	Kentucky	3740	3689
		Tennessee	5239	5424
		Alabama	4282	4358
		Mississippi	2717	2772
	West South Central	Arkansas	2473	2509
		Louisiana	4247	4141
		Oklahoma	3072	2924
		Texas	17572	17828
West	Mountain	Montana	774	744
		Idaho	1018	1008
		Wyoming	439	409

Continued......

		Colorado	3407	3424
		New Mexico	1639	1735
		Arizona	4149	4633
		Utah	1807	1845
		Nevada	1283	1409
	Pacific	Washington	5052	5191
		Oregon	2887	2903
		California	31749	33963
		Alaska	560	599
		Hawaii	1253	1362

Source: Statistical Abstracts of the United States (1993).

Population 8. Projected vital statistics by country or area during 2000.

Sr. No.	Country or area	Crude birth rate	Crude death rate	Infant mortality rate	Expectation of life at birth	Total fertility rate
1	United states	14.2	8.8	6.2	76.3	2.07
2	Afghanistan	41.6	16.6	137.5	47.8	5.87
3	Algeria	26.5	5.4	42.2	69.6	3.16
4	Angola	42.6	15.9	125.9	48.9	6.05
5	Argentina	19.9	7.6	17.8	75.0	2.64
6	Australia	13.0	6.9	5.0	80.4	1.80
7	Austria	10.3	10.3	5.8	77.3	1.52
8	Azerbaijan	21.1	8.6	72.3	65.4	2.61
9	Bangladesh	27.4	10.1	93.0	57.5	3.08
10	Belarus	14.0	12.9	12.0	69.4	1.92
11	Belgium	11.2	10.4	6.1	77.7	1.69
12	Bolivia	30.0	9.3	60.2	62.0	3.81
13	Brazil	19.2	10.1	47.7	60.9	2.14
14	Bulgaria	12.5	13.6	14.8	71.6	1.74
15	Burkina Faso	44.9	21.4	112.8	39.8	6.48
16	Burma	28.1	10.7	71.8	58.1	3.57
17	Burundi	41.0	15.3	95.2	48.1	6.25
18	Cambodia	41.0	14.3	100.5	51.5	5.81
19	Cameroon	41.4	13.9	74.1	51.3	5.73
20	Canada	12.3	7.2	5.5	80.0	1.80
21	Chad	42.8	16.3	113.6	48.9	5.64
22	Chile	15.9	5.7	11.9	75.5	2.00
23	China	15.0	6.8	32.6	71.1	1.81
24	Colombia	19.0	4.6	21.3	74.2	2.17
25	Congo (Kinshasa)	46.5	15.6	98.9	48.1	6.39
26	Cote d'lvoire	41.3	17.6	94.9	43.7	5.80
27	Cuba	12.7	7.5	8.6	75.6	1.60
28	Czech Republic	13.3	11.0	8.0	74.2	1.74
29	Dominican Republic	21.1	5.4	40.8	70.4	2.42
30	Ecuador	23.0	5.1	29.3	72.5	2.59
31	Egypt	26.2	8.1	65.7	62.7	3.24
32	Ethiopia	44.3	17.6	117.7	46.0	6.75
33	France	12.1	9.0	5.6	79.1	1.73
34	Germany	10.5	10.8	5.7	76.7	1.56
35	Ghana	30.8	10.2	74.8	57.5	3.95
36	Greece	10.7	9.5	6.6	79.0	1.52
37	Guatemala	31.2	6.5	44.6	66.9	4.00
38	Guinea	40.0	16.9	123.7	47.0	5.46
39	Haiti	32.3	14.7	98.4	50.2	4.50

Continued……

40	Hong Kong	10.2	5.6	4.5	82.8	1.42
41	Hungary	13.1	14.8	11.7	69.6	1.78
42	India	23.5	8.9	63.5	61.5	2.90
43	Indonesia	22.4	8.1	55.4	63.4	2.53
44	Iran	29.2	5.8	45.1	69.1	3.90
45	Iraq	40.8	5.6	50.1	68.7	5.81
46	Italy	11.4	10.1	6.4	78.6	1.55
47	Japan	10.7	8.4	4.3	80.0	1.50
48	Kazakhstan	18.8	9.8	59.6	64.7	2.29
49	Kenya	29.5	12.6	54.9	51.1	3.69
50	Korea, North	19.9	5.4	22.2	71.5	2.21
51	Korea, South	15.8	5.7	7.4	74.7	1.80
52	Madagascar	41.2	13.3	87.6	53.6	5.64
53	Malawi	38.7	27.1	135.8	33.0	5.33
54	Malaysia	24.1	5.2	20.9	71.0	3.10
55	Mali	49.3	17.5	95.6	48.8	6.91
56	Mexico	24.3	4.4	20.7	75.0	2.79
57	Morocco	25.1	5.1	32.9	71.8	3.13
58	Mozambique	42.0	16.8	114.9	46.4	5.76
59	Nepal	35.6	11.2	70.1	55.9	4.68
60	Netherlands	11.5	8.7	4.7	78.2	1.53
61	Niger	51.6	22.2	111.2	42.4	7.17
62	Nigeria	41.5	11.7	63.7	55.6	5.95
63	Pakistan	32.6	10.2	90.3	59.7	4.56
64	Peru	21.8	5.8	44.4	70.8	2.66
65	Philippines	27.3	6.4	33.2	66.8	3.38
66	Poland	14.2	10.2	11.8	72.6	1.93
67	Portugal	11.9	10.3	7.1	76.1	1.53
68	Romania	14.3	12.7	21.7	70.0	1.83
69	Russia	14.5	14.8	21.6	65.4	1.95
70	Rwanda	38.5	23.5	118.8	36.5	5.73
71	Saudi Arabia	37.2	4.7	36.3	71.1	6.30
72	Senegal	43.4	10.4	58.4	58.3	6.04
73	Serbia	14.8	10.5	21.3	72.5	2.10
74	Somalia	41.7	12.0	111.6	57.0	6.53
75	South Africa	25.3	14.4	55.4	51.9	3.02
76	Spain	12.0	9.1	5.7	79.1	1.51
77	Sri Lanka	16.9	5.9	18.7	73.2	1.95
78	Sudan	38.8	10.3	69.2	56.8	5.47
79	Sweden	10.5	11.1	4.4	78.5	1.60
80	Switzerland	11.2	9.5	5.2	78.1	1.59
81	Syria	36.1	5.3	35.2	68.4	5.19
82	Taiwan	14.4	5.7	6.3	77.3	1.75
83	Tajikistan	34.5	8.3	104.1	65.1	4.47

Continued……

84	Tanzania	39.8	20.9	101.4	40.0	5.31
85	Thailand	16.2	7.2	28.3	69.4	1.80
86	Tunisia	22.5	5.1	30.1	73.6	2.68
87	Turkey	20.4	5.2	33.3	73.8	2.35
88	Uganda	43.1	21.8	95.6	38.1	6.24
89	Ukraine	13.2	14.9	21.2	67.8	1.87
90	United Kingdom	12.1	10.9	6.0	77.1	1.79
91	Uzbekistan	28.9	7.7	77.5	65.2	3.56
92	Venezuela	21.5	4.9	25.5	73.3	2.53
93	Vietnam	20.0	6.5	33.7	68.5	2.31
94	Yemen	43.3	7.9	57.9	62.6	6.86
95	Zambia	43.8	25.8	97.6	33.7	6.28
96	Zimbabwe	29.3	21.6	72.2	38.2	3.50

Source: Statistical Abstracts of the United States (1997).

Descriptive parameters of the population 8:

X = Total fertility rate;
Y_1 = Crude birth rate;
Y_2 = Crude death rate;
Y_3 = Infant mortality rate;
Y_4 = Expectation of life at birth in years

Correlation Matrix

	Y_1	Y_2	Y_3	Y_4	X
Y_1	1				
Y_2	0.525913	1			
Y_3	0.882218	0.660721	1		
Y_4	-0.842360	-0.828100	-0.910310	1	
X	0.985520	0.549261	0.852536	-0.815260	1

Population 9. Number of immigrants admitted in the USA.

Sr. No.	States	1994 z	1995 x	1996 y
1	AL	1837	1900	1782
2	AK	1129	1049	1280
3	AZ	9141	7700	8900
4	AR	1031	934	1494
5	CA	208498	166482	201529
6	CO	6825	7713	8895
7	CT	9537	9240	10874
8	DE	984	1051	1377
9	DC	3204	3047	3784
10	FL	58093	62023	79461
11	GA	10032	12381	12608
12	HI	7746	7537	8436
13	ID	1559	1612	1825
14	IL	42400	33898	42517
15	IN	3725	3590	4692
16	IA	2163	2260	3037
17	KS	2902	2434	4303
18	KY	2036	1857	2019
19	LA	3366	3000	4092
20	ME	829	814	1028
21	MD	15937	15055	20732
22	MA	22882	20523	23085
23	MI	12728	14135	17253
24	MN	7098	8111	8977
25	MS	815	757	1073
26	MO	4362	3990	5690
27	MT	447	409	449
28	NE	1595	1831	2150
29	NV	4051	4306	5874
30	NH	1144	1186	1512
31	NJ	44083	39729	63303
32	NM	2936	2758	5780
33	NY	144354	128406	154095
34	NC	6204	5617	7011
35	ND	635	483	606
36	OH	9184	8585	10237
37	OK	2728	2792	3511
38	OR	6784	4923	7554
39	PA	15971	15065	16938
40	RI	2907	2609	3098

Continued……

41	SC	2110	2165	2151
42	SD	570	495	519
43	TN	3608	3392	4343
44	TX	56158	49963	83385
45	UT	2951	2831	4250
46	VT	658	535	654
47	VA	15342	16319	21375
48	WA	18180	15862	18833
49	WV	663	540	583
50	WI	5328	4919	3607
51	WY	217	252	280

Source: Statistical Abstracts of the United States (1999).

BIBLIOGRAPHY

[1] Adhvaryu, D. (1978). Successive sampling using multi-auxiliary information. *Sankhy $\bar{a}$*, C, 40, 167--173.

[2] Adhvaryu, D. and Gupta, P.C. (1983). On some alternative sampling strategies using auxiliary information. *Metrika*, 30, 217--226

[3] Agarwal, C.L. and Tikkiwal, B.D. (1980). Two stage sampling on successive occasions. *Sankhy $\bar{a}$*, 42, C, 31--44.

[4] Agarwal, D.K. and Singh, P. (1982). On cluster sampling strategies using ancillary information. *Sankhy $\bar{a}$*, B, 44, 184--192.

[5] Agarwal, M.C. and Jain, N. (1989). A new predictive product estimator. *Biometrika,* 76, 822--823.

[6] Agarwal, M.C. and Panda, K.B. (1993). An efficient estimator in poststratification. *Metron,* 179--188.

[7] Agarwal, M.C. and Roy, D.C. (1999). Efficient estimators for small domains. *J. Indian Soc. Agric. Statist..*, 52(3), 327--337.

[8] Aagrwal, M.C. and Sthapit, A.B. (1996). Model assisted selection of a product strategy. *J. Indian Soc. Agric. Statist.*, 48(2), 207--215.

[9] Agarwal, S.K. (1980). Two auxiliary variates in ratio method of estimation. *Biom. J.*, 22(7), 569--573.

[10] Agarwal, S.K. and Kumar, P. (1980). Combination of ratio and PPS estimator. *J. Indian Soc. Agric. Statist.*, 32, 81--86.

[11] Agarwal, S.K., Sharma, U.K. and Kashyap, S. (1997). A new approach to use multivariate auxiliary information in sample surveys. *J. Statist. Planning Infer.*, 60, 261--267.

[12] Agarwal, S.K., Singh, M. and Goel, B.B.P.S. (1979). Use of p-auxiliary variates in PPS sampling. *Biom. J.*, 21(8), 781--785.

[13] Ahmed, M.S. (1997). The general class of chain estimators for the ratio of two means using double sampling. *Commun. Statist.--Theory Meth.*, 26(9), 2247--2254.

[14] Ahmed, M.S. (1998). A note on regression type estimators using multiple auxiliary information. *Austral. & New Zealand J. Statist.*, 40(3), 373--376.

[15] Ahmad, T. (1997). A resampling technique for complex survey data. *J. Indian Soc. Agric. Statist.*, 50(3), 364--379.

[16] Ahsan, M.J. and Khan, S.U. (1982). Optimum allocation in multivariate stratified random sampling with overhead cost. *Metrika*, 29, 71--78.

[17] Aires, N. (2000). Comparisons between conditional Poisson sampling and Pareto πps sampling design. *J. Statist. Planning Infer.*, 88, 133--147.

[18] Ajgaonkar, S.G.P. (1975). The efficient use of supplementary information in double sampling procedure. *Sankhy $\bar{a}$*,C,37, 181--189.

[19] Akar, I. and Sedransk, J. (1979). Post-stratified cluster sampling. *Sankhy $\overline{a}$* ,C, 41, 76--83.

[20] Alalouf, I.S. (1996). The estimation of a proportion in cluster sampling. *Commun. Statist.--Theory Meth.*, 25(2), 325--343.

[21] Allen, J., Saxena, S., Singh, H.P., Singh, S. and Smarandache, F. (2002). *Randomness and optimal estimation in data sampling.* American Research Press. 26--43.

[22] Amahia, G.N., Chaubey, Y.P. and Rao, T.J. (1989). Efficiency of a new estimator in PPS sampling for multiple characteristics. *J. Statist. Planning Infer.*, 21,75--84.

[23] Amdekar, S.J. (1985). An unbiased estimator in overlapping clusters. *Calcutta Statist. Assoc. Bull.*, 15, 231--232.

[24] Anderson, H. (1976). Estimation of a proportion through randomized response. *Int. Statist. Rev.* , 44, 213--217.

[25] Anderson, H. (1977). Efficiency versus protection in the general randomized response model. *Scand. J. Statist.*, 4, 11--19.

[26] Andreatta, G. and Kaufman, G.M. (1986). Estimates of finite population when sampling is without replacement and proportional to magnitude. *J. Amer. Statist. Assoc.*, 81, 657--666.

[27] Anscombe, F.J. (1948). The validity of comparative experiments. *J.R. Statist. Soc., A*, 61, 181--211.

[28] Arnab, R. (1979a). On strategies of sampling finite populations on succesive occasions with varying probabilities. *Sankhy $\overline{a}$* , C, 41, 141--155.

[29] Arnab, R. (1979b). An addendum to Singh and Singh's paper on random nonresponse in unequal probability sampling. *Sankhy $\overline{a}$* , C, 41, 138--140.

[30] Arnab, R. (1988). Variance estimation in multi-satge sampling. *Aust. J. Statist.*,30, 107--110.

[31] Arnab, R. (1990). On comutativity of design and model expectations in randomized response surveys. *Commun. Statist.--Theory Meth.*, 19(10), 3751--3757.

[32] Arnab, R. (1992a). Estimation of a finite population mean under superpopulation models. *Commun. Statist. -- Theory Meth.* 21(6), 1717--1724.

[33] Arnab, R. (1992b). Unbiased estimation of variance in Multi-Stage sampling under randomized response surveys. *Statistics,* 23, 357--364.

[34] Arnab, R. (1994). Nonnegative variance estimation in randomized response surveys. *Commun. Statist.--Theory Meth.*, 23(6), 1743--1752.

[35] Arnab, R. (1995). On admissibility and optimality of sampling strategies in randomized response surveys. *Sankhy $\overline{a}$* , B, 57, 385--390.

[36] Arnab, R. (1996). Randomized response trials : A unified approach for qualitative data. *Commun. Statist.--Theory Meth.*, 25(6), 1173--1183.

[37] Arnab, R. (1998). Sampling on two occasions: Estimation of population total. *Survey Methodology*, 24, 185--192.

[38] Arnab, R. (1999). On use of distinct respondents in randomized response surveys. *Biom.J.*,41(4), 507--513.

[39] Arnab, R. (2001). Estimation of a finite population total in varying probability sampling for multi-character surveys. *Metrika*, 54(2), 159--177.

[40] Arnab, R. (2002). Optimum sampling strategies under randomized response surveys. *Biom. J.*, 44(4), 490--495.

[41] Arnab, R. and Singh, S. (2001). On the estimation of population total and variance in the presence of non-response. *Presented on JSM--2001 conference Atlanta, USA.*

[42] Arnab, R. and Singh, S. (2002a). Calibration for variance estimator of generalized regression predictor. *Submitted for possible presentation at JSM--2003, California, USA.*

[43] Arnab, R. and Singh, S. (2002b). Estimation of the size and mean value of a stigmatized characteristic of a hidden gang in a finite population : a unified approach. *Ann. Inst. Math. Stat.*, 54(3), 659--666.

[44] Arnab, R and Singh, S. (2002c). On the estimation of size and mean value of a stigmatized characterstic of a hiden gang in finite populations. *Recent Advances in Statistical Methods--Proceedings of Statistics 2001 Concordia University Conference*, 1--11.

[45] Arnab, R. and Singh, S. (2002d). Estimation of variance form missing data. *Presented at Statistical Society of Canada Conference at Hamilton, Canada.*

[46] Arnab, R. and Singh, S. (2002e). Jackknifing the imputed data in addition to observed data while estimating variance of the ratio estimator. *Working paper.*

[47] Arnholt, A.T. and Hebert, J.L. (1995). Estimating the mean with known coefficient of variation. *American Statistician*, 49(4), 367--369.

[48] Artes, E. and Garcia, A. (2000a). A note on successive sampling using auxiliary information. *Proceedings of the 15th International Workshop on Statistical Modelling*, 376--379.

[49] Artes, E. and Garcia, A. (2000b). Sobre muestreo en ocasiones sucesivas. *Actas del IX congreso sobre ensenanza y aprendizaje de las Matematicas*, 153--155.

[50] Artes, E. and Garcia, A. (2001a). Metodo diferencia multivariate en muestreo en dos ocasiones. *VIII Conferencia Espanola de Biometria*, 199--200.

[51] Artes, E. and Garcia, A. (2001b). Successive sampling for the ratio of population parameters. *Journal of the Portuguese Nacional Statistical Institute*, En prensa.

[52] Artes, E. and Garcia, A. (2001c). Estimating the current mean in successive sampling using a product estimate.*Conference on Agricultural and Environmental Statist. Application in Rome*, XLIII—1--XLIII--2.

[53] Artes, E. and Garcia, A. (2001d). An almost unbiased ratio-cum-product estimator on two occasions. *X International Symposium on Applied Stochastic Models and Data Analysis*, 130--135.

[54] Artes, E. and Garcia, A. (2001e). Estimation of current population ratio in successive sampling. *J. Indian Soc. Agric. Statist.*, 54(3), 342--354.

[55] Artes, E., Rueda, M. and Arcos, A. (1998). Successive sampling using a product estimate. *Applied sciences and the environment, computational mechanics publications*, 85--90.

[56] Asok, C. (1974). *Contribution to the theory of unequal probability sampling without replacement.* Unpublished Ph.D. Thesis, Iowa State University, Ames, Iowa.

[57] Asok, C. (1980). A note on the comparison between simple mean and mean based on distinct units in sampling with replacement. *American Statistician*, 34, 158.

[58] Asok, C. and Sukhatme, B.V. (1975). Unequal probability sampling with random stratification. *Proc. Amer. Statist. Assoc.*, 283--288.

[59] Asok, C. and Sukhatme, B.V. (1976a). On the efficiency comparison of two πps sampling strategies. *Proc. Amer. Statist. Assoc.*, 161--166.

[60] Asok, C. and Sukhatme, B.V. (1976b). On Sampford's procedure of unequal probability sampling without replacement. *J. Amer. Statist. Assoc.*, 71, 912--918.

[61] Asok, C. and Sukhatme, B.V. (1978). A note on Midzuno Scheme of sampling. *Paper presented at the 32nd Annual Conference of the Indian Soc. Agricul. Statist., New Delhi,India.*

[62] Avdhani, M.S. (1968). *Contribution to the theory of sampling from finite population and its application.* Ph.D. thesis, Delhi University.

[63] Bahadur, R.R. (1954). Sufficiency and statistical decision functions. *Ann. Math. Statist.*, 25,423--462.

[64] Bandyopadhyay, S. (1980). Improved ratio and product estimators. *Sankhy* $\bar{a}$, C, 42, 45--49.

[65] Bandyopadhyay, S., Chattopadhyaya, A.K. and Kundu, S. (1977). On estimation of population total. *Sankhy* $\bar{a}$,C, 39, 28--42.

[66] Bankier, M. D. (1986). Estimators based on several stratified samples with application to multiple-frame surveys. *J. Amer. Statist. Assoc.*, 81, 1074--1079.

[67] Bankier, M.D.(1988). Power allocations: Determining sample sizes for sub-national areas. *American Statist.*,42(3),174--178.

[68] Bansal, M.L and Singh, R. (1985). An alternative estimator for multiple characteristics in PPS sampling. *J. Statist. Planning Infer.*, 11, 313--320.

[69] Bansal, M.L. and Singh, R. (1986). On the generalization of Rao, Hartley and Cochran's scheme. *Metrika,* 33, 307--314.

[70] Bansal, M.L. and Singh, R. (1989). An alternative estimator for multiple characteristics corresponding to Horvitz and Thompson estimator in probability proportional to size and without replacement sampling. *Statistica,* anno. XLIX, 3, 447--452.

[71] Bansal, M.L. and Singh, R. (1990). An alternative estimator for multiple characteristics in RHC sampling scheme. *Commun. Statist. --Theory Meth.* 19(5), 1777--1784.

[72] Bansal, M.L., Singh, S. and Singh, R.(1994) Multi-character survey using randomized response technique. *Commun.Statist.--Theory Meth.* 23(6), 1705--1715.

[73] Barnard, J. and Rubin, D.B. (1999). Small sample degree of freedom with multiple imputation. *Biometrika*, 86(4), 948--955.

[74] Bartholomew, D.J. (1961). A method of allowing for not at home. bias in sample surveys. *Applied Statist.*, 10, 52--59.

[75] Bartlett, R.F. (1986). Estimating the total of a continuous populations. *J. Statist. Planning Infer.*, 13, 51--66.

[76] Basawa, I.V., Godambe, V.P. and Taylor, R.L. (1997). *Selected proceedings of the symposium on estimating functions.* Lecture Notes - Monograph Series, Institute of Mathematical Statistics, Hayward, California.

[77] Basu, D. (1958). On sampling with and without replacement. *Sankhy* $\bar{a}$, 20, 287--294.

[78] Basu, D. (1971). *An essay on the logical foundations of survey sampling.* Part one. In: V.P. Godambe and D.A. Sportt (eds.) Foundations of statistical inferences. Holt, Rinehart and Winston, Toronto, 203--242.

[79] Battese, G.E., Harter, R.M. and Fuller, W.A. (1988). An error components model for prediction of county crop areas using surveys and satellite data. *J. Amer. Statist Assoc.*, 83, 28--36.

[80] Bayless, D.L. and Rao, J.N.K. (1970). An empirical study of stabilities of estimators and variance estimators in unequal probability sampling (n=3 or 4). *J. Amer. Statist. Assoc.*, 65, 1645--1667.

[81] Beale, E.M.L. (1962). Some use of computers in operational research. *Industrielle Organ.*, 31, 27--28.

[82] Bedi, P.K. (1995). An alternative estimator in Midzununo scheme for multiple characteristics. *Commun. Statist.--Simula.*, 17--30.

[83] Bedi, P.K. and Agarwal, S.K. (1999). Modified Midzuno scheme of sampling. *J. Statist. Planning Infer.*, 76, 203--214.

[84] Bedi, P.K. and Rao, T.J. (1996). Probability proportional to revised sizes with replacement scheme. *Metron,* 67--82.

[85] Bedi, P.K. and Rao, T.J. (2001). PPS method of estimation under a transformation. *J. Indian Soc. Argic. Statist.*, 54(2), 184--195.

[86] Bellhouse, D.R. (1984a). Optimal randomization for experiments in which autocorrelation is present. *Biometrika*, 71,155--160.

[87] Bellhouse, D.R. (1984b). A review of optimal designs in survey sampling. *Canadian J. Statist.*, 12, 53--65.

[88] Bellhouse, D. R. (1995). Estimation of correlation in randomized response. *Survey Methodology,* 21, 13--19.

[89] Bellhouse, D.R. and Rao, J.N.K. (1975). Systematic sampling in the presence of a trend. *Biometrika*, 62, 694--697.

[90] Bellhouse, D.R. and Rao, J.N.K. (1986). On the efficiency of prediction estimators in two-stage sampling. *J. Statist. Planning Infer.*, 13, 269--281.

[91] Bennett, B.M. (1983). Alternate estimates in stratified sampling. *Metron*, 77--82.

[92] Bennett, B.M. and Islam, M.A. (1983). On relative precision in stratified sampling for proportions. *Metron,* 19--22.

[93] Bethlehem, J.G and Keller, W.J. (1987). Linear weighting of sample survey data. *J.Official Statist.*, 141--153.

[94] Bhargava, M. (1996). *An investigation into the efficiencies of certain randomized response strategies*. Unpublished Ph.D. thesis submitted to Punjab Agricultural University, Ludhiana, India.

[95] Bhargava, M. and Singh, R. (2000). A modified randomization device for Warner's model. *Statistica*, 60, 315--321.

[96] Bhargava, M. and Singh, R. (2001). Efficiency comparison of certain randomized response schemes with U--model. *J. Indian Soc. Agric. Statist.*, 54(1), 19--28.

[97] Bhargava, M. and Singh, R. (2002). On the efficiency comparison of certain randomized response strategies. *Metrika*, 55(3), 191--197.

[98] Bhargava, N.K. (1978). On some applications of the technique of combined unordering. *Sankhy $\overline{a}$* , C, 40, 74--83.

[99] Bhatia, A., Mangat, N.S., and Morrison, T. (1998). Estimation of measurement errors. *Proceedings of the International Pipeline Conference 1998, Calgary, Canada, American Society of Mechanical Engineers, 1, 315--325.*

[100] Bhatnagar, S. (1996). Improved product estimators. *Sankhy $\overline{a}$* , B, 58, 84--89.

[101] Bhave, S.V. (1987). Godambe's paradox and the ancillarity principle. *Statist. Prob. Lett.*, 5, 243--246.

[102] Binder, D.A. and Theberge, A. (1988). Estimating the variance of raking ratio estimators. *Canadian J. Statist.*, 16, 47--55.

[103] Binder, D.A. and Patak, Z. (1994). Use of estimating functions for estimation from complex surveys. *J. Amer. Statist. Assoc.*, 89, 1035--1043.

[104] Biradar, R.S. and Singh, H.P. (1992a). A class of estimators for finite population correlation coefficient using auxiliary information. *J. Indian Soc. Agril. Statist.*, 44, 271--285.

[105] Biradar, R.S. and Singh, H.P. (1992b). A note on an almost unbiased ratio cum product estimator. *Metron,* 249--255.

[106] Biradar, R.S. and Singh, H.P. (1997-98). A class of estimators for population parameter using supplementary information. *Aligarh J. Statist.*, 17/18, 54--71.

[107] Biradar, R.S. and Singh, H.P. (1998). Predictive estimation of finite population variance. *Calcutta Statist. Assoc. Bull.*, 48, 229--235.

[108] Blackwell, D. (1947). Conditional expectation and unbiased sequential estimation. *Ann. Math. Statist.*, 18, 105--110.

[109] Blackwell, D. (1951). Comparison of experiments. *Proc. 2nd Berkeley symp. Math. Stat. Prob.*, 93--102.

[110] Blight, B.J.N. (1973). Sampling from an autocorrelated finite population. *Biometrika*, 60, 375--385.

[111] Bose, C. (1943). Note on the sampling error in the method of double sampling. *Sankhy $\overline{a}$* , 6, 330.

[112] Bogue, D.J. (1950). A technique for making extensive postcensus estimates. *J. Amer. Statist. Assoc.*, 45, 149--163.

[113] Bourke, P.D. (1981). On the analysis of some multivariate randomized response designs for categorical data. *J. Statist. Planing Infer.*, 5, 165--170.

[114] Bourke, P.D. and Dalenious, T. (1976). Some new ideas in the realm of randomized enquiries. *Int. Statist. Rev.*, 44, 219--221.

[115] Bouza, C. (1994). The use of auxiliary information for solving non-response problems. *Test*, 3, 113--122.

[116] Brackstone, G.J. (1987). *Small area data: policy issues and technical challenges. In Small Area Statistics* (R. Platek, J.N.K. Rao, C.E. Särndal and M.P. Singh eds.) 3--20, Wiley New York.

[117] Brackstone, G.J. and Rao, J.N.K. (1979). An investigation of raking ratio estimators. *Sankhyā*, C, 41, 97--114.

[118] Bratley, P., Fox, B. L. and Schrage, L.E. (1983). *A Guide to Simulation*. NY: Springer--Verlag.

[119] Breau, P. and Ernst, L.R. (1983). Alternative estimators to the current composite estimator. *Proc of the section on Survey Research Methods*, Amer. Statist. Assoc., 397--402.

[120] Breidt, F.J. and Opsomer, J.D. (2000). Local polynomial regression estimators in survey sampling. *Ann. Statist.*, 28(4), 1026--1053.

[121] Brewer, K.R.W. (1963a). Ratio estimation and finite populations: Some results deducible from the assumption of an underlying stochastic process. *Austral. J. Statist.*, 5, 93--105.

[122] Brewer, K.R.W. (1963b). A model of systematic sampling with unequal probabilities. *Austral. J. Statist.*, 5, 5--13.

[123] Brewer, K.R.W. (1967). A note on Fellegi's method of sampling without replacement with probabilities proportional to size. *J. Amer. Statist. Assoc.*, 62, 79--85.

[124] Brewer, K.R.W. (1975). A simple procedure for sampling πps wor. *Austral. J. Statist.*, 17, 166--172.

[125] Brewer, K.R.W. (1979). A class of robust sampling designs for large scale surveys. *J. Amer. Statist. Assoc.,* 74, 911--915.

[126] Brewer, K.R.W. (1994). Survey sampling inference : Some past perspectives and present prospects. *Pak. J. Statist.*, A, 10, 213--233.

[127] Brewer, K.R.W. (1995). *Combining design based model based inference.* Chapter 30 in Business Survey Methods (Eds. B.G. Cox, D.A. Binder, B.N. Chinnapa, A. Christianson, MJ. Colledge and P.S. Kott). New York:Wiley, 589--606.

[128] Brewer, K.R.W. (1999a). Cosmetic calibration with unequal probability sampling. *Survey Methodology*, 25(2), 205--212.

[129] Brewer, K.R.W. (1999b). Design based or prediction based inference? Stratified random vs. Stratified balanced sampling. *Int. Statist. Rev.*, 67, 35--47.

[130] Brewer, K.R.W. (2002). *Combined survey sampling inference*. Arnold.

[131] Brewer, K.R.W. and Hanif, M. (1970). Durbin's new multistage variance estimator. *J. R. Statist. Soc.*, B, 32, 302--311.

[132] Brewer, K.R.W. and Hanif, M. (1983). *Sampling with unequal probabilities.* New York : Springer --Verlag.

[133] Brewer, K.R.W. and Undy, G.C. (1962). Samples of two units drawn with unequal probabilities without replacement. *Austral. J. Statist.*, 4, 89--100.

[134] Brewer, K.R.W., Early, L.J. and Hanif, M. (1984). Poisson, modified poisson and collocated sampling. *J. Statist. Planning Infer.*, 10, 15--30.

[135] Brillinger, D.R., Jones, L.V. and Tukey, J.W. (1978). Report of the statistical task force for the weather modification advisory board. *The Management of Western Resources, Vol. II: The Role of Statistics on Weather Resources Management.* Government Printing Office, Washington, DC.

[136] Brown, B.M., Hall, P. and Young, G.A. (2001). The smoothed median and the bootstrap. *Biometrika*, 88(2), 519--534.

[137] Brown, J.A. (1996). *The relative efficiency of adaptive cluster sampling for ecological surveys.* Mathematical and Information Sciences Reports, series B, 96/08, Massey University.

[138] Brown, J.A. (1999). A comparison of two adaptive sampling designs. *Austral. & New Zealand J. Statist.*, 41(4), 395--403.

[139] Bryant, E.C., Hartley, H.O. and Jessen, R.J. (1960). Design and estimation in two-way stratification. *J. Amer. Statist. Assoc.*, 55, 105--124.

[140] Buckland, W.R. (1951). A review of the literature of the systematic sampling. *J. R. Statist. Soc.*, B, 13, 208--215.

[141] Burdick, R.K. and Sielken, R.L. (1979). Variance estimation based on a superpopulation model in two stage sampling. *J. Amer. Statist. Assoc.*, 74, 438--440.

[142] Carlin, B.P. and Gelfand, A.E. (1990). Approaches for empirical Bayes confidence intervals. *J. Amer. Statist. Assoc.*, 85, 105--114.

[143] Carlin, B.P. and Gelfand, A.E. (1991). A sample re-use method for accurate parametric empirical Bayes confidence intervals. *J.R. Statist. Soc.*, B, 53, 189--200.

[144] Casady, R.J. and Lepkowski, J.M. (1993). Stratified telephone survey designs. *Survey Methodology*, 19, 103--113.

[145] Cassel, C.M. and Särndal, C.E. (1974). Evaluation of some sampling strategies using a continuous variable framework. *Commun. Statist.--Theory Meth.*, 3, 373--390.

[146] Cassel, C.M., Särndal, C.E. and Wretman, J.H. (1976). Some results on generalized difference estimation and generalized regression estimation for finite populations. *Biometrika*, 63, 615--620.

[147] Cassel, C.M., Särndal, C.E. and Wretman, J.H. (1977). *Foundations of Inference in Survey Sampling.* John Wiley and Sons, New York.

[148] Cassel, C.M., Särndal, C.E. and Wretman, J.H. (1979). Some uses of statistical models in connection with the non-response problem. *Symposium on Incomplete Data, Preliminary Proc., Washington, D.C.*

[149] Causey, B.D. (1972). Sensitivity of raked contingency table totals to changes in problem conditions. *Ann. Math. Statist.*, 43, 656--658.

[150] Causeur, D. (1999). Exact distribution of the regression estimator in double sampling. *Statistics,* 32, 297--315.

[151] Cebrian, A.A. and Garcia, M.R. (1997). Variance estimation using auxiliary information : An almost unbiased multivariate ratio estimator. *Metrika*, 45, 171--178.

[152] Ceccon, C., Diana, G. and Salvan, A. (1991). Approccio classico al campionamento da popolazioni finite: *Alcuni risultati recenti, CLEUP, Padova.*

[153] Chakrabarty, M.C. (1963). On the use of incidence matrices in sampling from finite populations. *J. Indian Statist. Assoc*, 1, 78--85.

[154] Chakrabarty, R.P. (1968). *Contribution to the theory of ratio type estimators*. Ph.D. Thesis, Texas A and M University.

[155] Chakrabarty, R.P. (1979). Some ratio type estimators. *J. Indian Soc. Agril. Statist,* 31, 49--62.

[156] Chand, L. (1975). *Some ratio type estimators based on two or more auxiliary variables.* Ph.D. thesis submitted to Iowa State University, Ames, Iowa.

[157] Chang, H.J. and Liang, D.H. (1996). A randomized response procedure for two unrelated sensitive questions. *J. Information & Optimization Sci.*, 17(1), 185--198.

[158] Chang, H.J. and Huang, K.C. (2001a). On construction of almost unbiased estimators of finite population mean using transformed auxiliary variable. *Statistical Papers*, 42(4), 505--515.

[159] Chang, H.J. and Huang, K.C. (2001b). Estimation of proportion and sensitivity of a qualitative character. *Metrika*, 53(2), 269--280.

[160] Chang, K.C., Han, C.P. and Hawkins, D.L. (1999). Truncated multiple inverse sampling in post-stratification. *J. Statist. Planning Infer.*, 76, 215--234.

[161] Chang, K.C., Liu, J.F. and Han, C.P. (1998). Multiple inverse sampling in post-stratification. *J. Statist. Planning Infer.*, 69, 209--227.

[162] Chao, M.T. (1982). A general purpose unequal probability sampling plan. *Biometrika*, 69, 653--656.

[163] Chatterjee, S. and Simon, G. (1993). Confidentiality guaranteed : A non-invasive procedure for collecting sensitive information. *Comm. Statist-Theory Meth.*, 22(6), 1629--1651.

[164] Chaubey, Y.P. and Crisalli, A.N. (1995). Adjustment of the inclusion probabilities in case of non-response. *Statistical Society of Canada, Proceedings of the Survey Methods Section*, 75--79.

[165] Chaudhuri, A. (1974). On some properties of sampling scheme due to Midzuno. *Calcutt Statist. Assoc. Bull.*, 23, 1--9.

[166] Chaudhuri, A. (1975a). A simple method of sampling without replacement with inclusion probabilities exactly proportional to size. *Metrika,* 22, 147--152.

[167] Chaudhuri, A. (1975b). Some results concerning Horvitz and Thompson's T_1 -class of estimators. *Metrika*, 217--223.

[168] Chaudhuri, A. (1976). A non-negativity criterion for a certain variance estimator. *Metrika,* 23, 201--205.

[169] Chaudhuri, A. (1977). On some properties of the Horvitz and Thompson estimator based on Midzuno's π ps sampling scheme. *J. Indian Soc. Ag. Statist.*, 47--52.

[170] Chaudhuri, A. (1981). On non-negative variance estimation. *Metrika*, 28, 1-12.

[171] Chaudhuri,A. (1992). *Small domain statistic: a review.* Technical Report ASC /92/2, Indian Statistical Institute, Calcutta.

[172] Chaudhuri, A. (1993). Mean square error estimation in randomized response surveys. *Pak. J. Statist.*, A, 9, 101--104.

[173] Chaudhuri, A. (1997). On a pragmatic modification of survey sampling in three stages. *Commun. Statist.--Theory Meth.*, 26(7), 1805--1810.

[174] Chaudhuri, A. (2001). Using randomized response from a complex survey to eliminate a sensitive proportion in a dichotomous finite population. *J. Statist. Planning Infer.*, 94, 37--42.

[175] Chaudhuri, A. and Adhikary, A.K. (1983). On optimality of double sampling strategies with varying probabilities. *J. Statist. Planning Infer.*, 8, 257--265.

[176] Chaudhuri, A. and Adhikary, A.K. (1985). Some results on admissibility and uniform admissibility in double sampling. *J. Statist. Planning Infer.*, 12, 199--202.

[177] Chaudhuri, A. and Adhikary, A.K. (1987). Circular systematic sampling with varying probabilities. *Calcutta Statist. Assoc. Bull.*, 36, 193--195.

[178] Chaudhuri, A. and Adhikary, A.K. (1990). Variance estimation with randomized response. *Commun. Statist.-- Theory Meth.*, 19(3), 1119--1125.

[179] Chaudhuri, A., Adhikary, A.K., Dihidar, S. (2000). Mean square error estimation in multi-stage sampling. *Metrika*, 52, 2, 115--131.

[180] Chaudhuri, A., Adhikary, A.K. and Seal, A.K. (1997). Small domain estimation by empirical Bayes and Kalman filtering procedures--A case study. *Commun. Statist -- Theory Meth.*, 26(7), 1613--1621.

[181] Chaudhuri, A. and Arnab, R. (1977). On the relative efficiencies of a few strategies of sampling with varying probabilities on two occasions. *Calcutta Statist. Assoc. Bull.*, 26, 25--38.

[182] Chaudhuri, A. and Arnab, R. (1978). On the role of sample size in determining efficiency of Horvitz and Thompson estimators. *Sankhy $\overline{a}$* , C, 40, 104--109.

[183] Chaudhuri, A. and Arnab, R. (1979). On the relative efficiencies of sampling strategies under a superpopulation model. *Sankhy $\overline{a}$* , C, 41, 40--43.

[184] Chaudhuri, A. and Arnab, R. (1982). On unbiased variance estimators with various multi-stage sampling strategies. *Sankhy $\overline{a}$* , B, 44, 92--101.

[185] Chaudhuri, A. and Maiti, T. (1994). Variance estimation in model assisted survey sampling. *Commun. Statist.-- Theory Meth.*, 23(4), 1203--1214.

[186] Chaudhuri, A., Maiti, T. and Roy, D. (1996). A note on competing variance estimators in randomized response surveys. *Austral. J. Statist.*, 38(1), 35--42.

[187] Chaudhuri, A. and Mitra, J. (1992). A note on two variance estimators for Rao--Hartley--Cochran estimator. *Commun. Statist. -- Theory Meth.*, 21(12), 3535--3543.

[188] Chaudhuri, A. and Mukerjee, R. (1988). *Randomized response: Theory and techniques*. Marcel Dekker, New York,.

[189] Chaudhuri, A. and Roy, D. (1997a). Optimal variance estimation for generalized regression predictor. *J. Statist. Planning Infer.*, 60, 139--151.

[190] Chaudhuri, A. and Roy, D. (1997b). Model assisted survey sampling strategies with randomized response. *J. Statist. Planning Infer.*, 60, 61--68.

[191] Chaudhuri, A. and Vos, J.W.E. (1988). *Unified theory and strategies of survey sampling*. North Holand.

[192] Chen, J. and Shao, J. (2001). Jackknife variance estimation for nearest neighbour imputation. *J. Amer. Statist. Assoc.*, 96, 260--269.

[193] Chen, J. and Qin, J. (1993). Empirical likelihood estimation for finite populations and the effective usage of auxiliary information. *Biometrika*, 80, 107--116.

[194] Chen, J., Rao, J.N.K. and Sitter, R.R. (2000). Efficient random imputation for missing data in complex surveys. *Statistica Sinica*, 10(4), 1153--1169.

[195] Chen, J., Sitter, R.R., and Wu, C. (2002). Using empirical likelihood methods to obtain range restricted weights in regression estimators for surveys. *Biometrika*, 89, 230--237.

[196] Chen, S.X. (1998). Weighted polynomial models and weighted sampling schemes for finite population. *Annals of Statistics*, 26, 5, 1894--1515.

[197] Chernick, M.R. and Wright, T. (1983). Estimation of population mean with two-way stratification using a systematic allocation scheme. *J. Statist. Planning Infer.*, 7, 219--231.

[198] Chotai, J. (1974). A note on Rao--Hartley--Cochran method for PPS sampling over two occasions. *Sankhy* $\overline{a}$, C, 36, 173--180.

[199] Choudhry, G.H. and Singh, M.P. (1979). Sampling with unequal probabilities and without replacement--A rejective method. *Survey Methodology*, 5(2), 162--177.

[200] Christman, M. (1997). Efficiency of some sampling designs for spatially clustered populations. *Environmetrics*, 8, 145--166.

[201] Christofides, T.C. (2003). A generalized randomized response technique. *Metrika*, 57, 195--200.

[202] Chromy, J.R. (1974). Pairwise probabilities in probability non-replacement sampling. *Presented at ASA meeting at St. Louis, Missouri, USA*.

[203] Clayton, D., Dunn, G., Pickles, A. and Spiegelhalter, D. (1998). Analysis of longitudinal binary data from multi-phase sampling (with discussion). *J. R. Statist. Soc.*, B, 60, 71--80.

[204] Cochran, W.G. (1940). Some properties of estimators based on sampling scheme with varying probabilities. *Austral. J. Statist.*, 17, 22--28.

[205] Cochran, W.G. (1963). *Sampling Techniques*. John Wiley and Sons : New York.

[206] Cochran, W.G. (1968). Errors of measurement in statistics. *Technometrics*, 10(4), 637--666.

[207] Cochran, W.G. (1977). *Sampling Techniques*. 3rd Ed. John Wiley & Sons, New York.

[208] Conti, P.L. (1995). A note on the estimation of a proportion in sampling finite populations. *Metron*, 35--41.

[209] Cox, D.R. (1958). *The Planning of Experiments*. Wiley, New York.

[210] Cox, D.R. (1971). Discussion of Royall (1971). *Foundations of Statistical Inference* (V.P. Godambe and D.A. Sprott, eds). Holt, Rinehart & Winston, Toronto, 275.

[211] Cox, D.R.(1984). Present position and potential developments: some personal views, design of experiments and regression. *J. Roy. Statist. Soc.*, A, 147, 306--315.

[212] Dalabehera, M. and Sahoo, L.N. (1995). Efficiency of six almost unbiased ratio estimators under a particular model. *Statistical Hefte*, 36, 61--67.

[213] Dalabehara, M. and Sahoo, L.N. (1997). A class of estimators in stratified sampling with two auxiliary variables. *J. Indian Soc. Agril. Statist.,* 50(2), 144--149.

[214] Dalabehera, M. and Sahoo, L.N (2000). An unbiased estimator in two-phase sampling using two auxiliary variables. *J. Indian Soc. Agric. Statist.*, 53(2), 134--140.

[215] Dalenius, T. (1950). The problem of optimum stratification. *Skand Akt.*, 33, 203--213.

[216] Dalenius, T. (1955). The problem of not-at-homes. *Statistisk Tidskrift*, 4, 208--211.

[217] Dalenius, T. (1957). *Sampling in Sweden*. Almquist and Wiksell, Stockholm.

[218] Dalenius, T. and Gurney, M. (1951). The problem of optimum stratification II. *Skand. Akt.,* 34, 133--148.

[219] Dalenius, T. and Gurney, M. (1957). The choice of stratification points. *Skand. and Akt.*,40, 198--203.

[220] Dalenius, T. and Hodges, J.L. (1957). *The choice of stratification points*. Skandinavisk Aktuarietidskrift.

[221] Dalenius, T. and Hodges, J.L. (1959). Minimum variance stratification. *J. Amer. Statist. Assoc.*, 54.

[222] Das, A.C. (1950). Two dimensional systematic sampling. *Sankhy* $\overline{a}$, 10, 95--108.

[223] Das, A.C. (1951). On two-phase sampling and sampling with varying probabilities. *Bull. Int. Statist. Inst.*, 33(2), 105--112.

[224] Das, A.K. (1982). On the use of auxiliary information in estimating proportions. *J. Indian Statist. Assoc.*, 20, 99--108.

[225] Das, A.K. and Tripathi, T.P. (1978). Use of auxiliary information in estimating finite population variance. *Sankhy* $\overline{a}$,C, 40, 139--148.

[226] Das, A.K. and Tripathi, T.P. (1979). A class estimators for population mean when mean of an auxiliary character is known. *Math. Tech. Report No. 22/79*, ISI, Calcutta.

[227] Das, A.K. and Tripathi, T.P. (1980). Sampling strategies for population mean when the coefficient of variation of an auxiliary character is known. *Sankhy* $\overline{a}$ C,42,76--86.

[228] Das, G. and Bez, K. (1995). Preliminary test estimators in double sampling with two auxiliary variables. *Commun. Statist. – Theory Meth.*, 24(5), 1211--1226.

[229] Das, K. (1982). Estimation of population ratio on two occasions. *J. Indian Soc. Agric. Statist.*, 34 (2), 1--9.

[230] Datta, G.S., Day, B. and Basawa, I.V. (1999). Empirical best linear unbiased and empirical Bayes prediction in multivariate small area estimation. *J. Statist. Planning Infer.*, 75, 269--279.

[231] Datta, G.S., Day, B. and Maiti, T. (1998). A nested error regression model for multivariate hierarchical Bayes estimation of small area means. *Sankhy $\overline{a}$*, A, 60, 344--362.

[232] Datta, G.S. and Ghosh, M. (1991). Bayesian prediction in linear models: applications to small area estimation. *Annals of Statistics*, 19, 1748--1770.

[233] Datta, G.S. and Lahiri, P. (1995). Robust hierarchical Bayes estimation of small area characteristics in presence of covariates and outliers. *J. Multtivariate Analysis*, 54, 310--328.

[234] Datta, G.S. and Lahiri, P. (2000). A unified measure of uncertainty of estimated best linear unbiased predictors in small area estimation problems. *Statistica Sinica*, 10, 613--627.

[235] Datta, G.S., Lahiri, P., Maiti, T. and Lu, K.L. (1999). Hierarchical Bayes estimation of unemployment rates for the states of the U.S. *J. Amer. Statist. Assoc.*, 94, 1074--1082.

[236] David, I.P. and Sukhatme, B.V. (1974). On the bias and mean square error of the ratio estimator. *J. Amer. Statist. Assoc.*, 69, 464--466.

[237] Dayal, S. (1979). Use of estimates of proportions of stratum sizes and standard deviations in allocation of sample to different strata under stratified random sampling. *Sankhy $\overline{a}$*, C, 41, 159--175.

[238] Dayal, S. (1985). Allocation of sample using values of auxiliary characteristic. *J. Statist. Planning Infer.*, 11, 321--328.

[239] Deming, W.E. (1953). On a probability mechanism to attain an economic balance between the resulting error of response and bias of non-response. *J. Amer. Statist. Assoc.*, 48, 743--772.

[240] Deming, W.E. and Stephan, F.F. (1940). On a least square adjustment of a sampled frequency table when the expected marginal totals are known. *Ann. Math. Statist.*, 11, 427--444.

[241] Dempster, A.P., Laird, N.M. and Rubin, D.B. (1977). Maximum likelihood from incomplete data via the EM algorithm (with discussion). *J. R. Statist. Soc.*, B, 39, 1--38.

[242] Dempster, A.P., Rubin, D.B., and Tsutakawa, R.K. (1981). Estimation in covariance component models. *J. Amer. Statist. Assoc.*, 76, 341--353.

[243] Dempster, A.P. and Tomberlin, T.J. (1980). The analysis of census undercount from a post-enumeration survey. *Proceedings of the Conference on Census Undercount*, 88-94.

[244] Deng, L.Y. and Wu, C.F.J. (1987). Estimation of variance of the regression estimator. *J. Amer. Statist. Assoc.*, 82, 568--575.

[245] Deshpande, M.N. (1978). A new sampling procedure with varying probabilities. *J. Indian Soc. Agric. Statist.*, 30, 110--114.

[246] Deshpande, M.N. (1980). A note on the comparison between simple random sampling with and without replacement. *Metrika*, 27, 277--279.

[247] Deshpande, M.N. and Ajgoankar, S.G.P. (1977). On multitrial sampling methods. *Biometrika*, 64, 422--424.

[248] Deshpande, M.N. and Ajgoankar, S.G.P. (1987). A generalization of Midzuno sampling procedure. *Aust. J. Stat*, 29, 188--192.

[249] Deville, J.C. and Goga, C. (2002). The Horvitz--Thompson theory for two samples. *International Conference on Improving Surveys, Copenhagen.*

[250] Deville, J.C. and Särndal, C.E. (1992). Calibration estimators in survey sampling. *J. Amer. Statist. Assoc.*, 87, 376--382.

[251] Deville, J.C. and Tillé, Y. (1998). Unequal probability sampling without replacement through a splitting method. *Biometrika*, 85(1), 89--101.

[252] Deville, J.C. and Tillê, Y. (2000). Selection of several unequal probability samples from the same population. *J. Statist. Planning Infer.*, 86, 215--227.

[253] Dey, A. and Srivastava, A.K. (1987). A sampling procedure with inclusion probabilities proportional to size. *Survey Methodology*, 13(1), 85--92.

[254] Diana, G. (1992). A study of kth order approximation of some ratio type strategies. *Metron,* 19--32.

[255] Dorfman, A.H. and Hall, P. (1993). Estimators of the finite population distribution function using non-parametric regression. *Annals of Statistics*, 21(3), 1452--1475.

[256] Doss, D.C., Hartley, H.O. and Somayajulu, G.R. (1979). An exact small sample theory for post-stratification. *J. Statist. Planning Infer.*, 3,235--248.

[257] Dowling, T.A. and Shachtman, R.H. (1975). On the relative efficiency of randomized response models. *J. Amer. Statist. Assoc.*, 70, 84--87.

[258] Draper, N.R. and Guttman, I. (1968a). Some Bayesian stratified two-phase sampling results. *Biometrika*, 55, 131--139.

[259] Draper, N.R. and Guttman, I. (1968b). Bayesian stratified two-phase sampling results: *k* characteristics. *Biometrika,* 55, 587--589.

[260] Drew, D., Singh, M.P. and Choudhry, G.H. (1982). Evaluation of small area estimation techniques for the Canadian labour force surveys. *Survey Methodology*, 8, 17--47.

[261] Dubey, V. (1993). An almost unbiased product estimator. *J. Indian Soc. Agril. Statist.,* 45, 226--229.

[262] Dubey, V. and Singh, S.K. (2001). An improved regression estimator for estimating population mean. *J. Indian Soc. Agric. Statist.*, 54(2), 179--183.

[263] Duncan, G.J. and Kalton, G. (1987). Issue of design and analysis of survey across time. *Int. Statist. Rev.*, 55, 97--117.

[264] Dupont, F. (1995). Alternative adjustments where there are several levels of auxiliary information. *Survey Methodology*, 21, 125--135.

[265] Durbin, J. (1959). A note on the application of Quenouille's method of bias reduction to the estimation of ratios. *Biometrika,* 46, 477--480.

[266] Durbin, J. (1967). Design of multi-stage survey for the estimation of sampling error. *Applied Statist.,* 16, 152--164.

[267] Eckler, A.R. (1955). Rotation sampling. *Ann. Math. Stat.*, 26, 664--685.

[268] Eichhorn, B.H. and Hayre, L.S. (1983). Scrambled randomized response methods for obtaining sensitive quantitative data. *J. Statist. Planning Infer.*,7, 307--316.

[269] Ekman, G. (1959). An approximation useful in univariate stratification. *Ann. Math. Stat.*, 30, 219--229.

[270] Elliott, M.R., Little, R.J.A. and Lewitzky, S. (2000). Subsampling callbacks to improve survey efficiency. *J. Amer. Statist. Assoc.*, 95, 730--738.

[271] Eltinge, J.L. (1999). Accounting for non-Gaussian measurement error in complex survey estimators of distribution functions and quantiles. *Statist. Sinica* ,9 , 425--450.

[272] Ericson, W.A. (1969). Subjective Bayesian models in sampling finite populations. *J. R. Statist. Soc.*, 55, 587--589.

[273] Eriksson, S.A. (1973). A new model for randomized response. *Int. Stat. Rev.*, 41, 101--103.

[274] Espejo, M.R. (1997). Uniqueness of the Zinger strategy with estimable variance : Rana-Singh estimator. *Sankhy $\overline{a}$* , B, 59, 76--83.

[275] Espejo, M.R. and Pineda, M.D.(1997). On variance estimation for poststratification: a review. *Metron*, 209--220.

[276] Espejo, M.R., Pineda, M.D. and Nadarajah, S. (2003). Estimation of finite population parameters with several realizations. *Statistical Papers*, 44 (2), 267--278.

[277] Estevao, V.M. (1994). Calibration of g weights under calibration and bound constraints. *Report, Statistics Canada.*

[278] Estevao, V.M. and Särndal, C.E. (2000). A functional form approach to calibration. *J. Official Statist.*, 16(4), 379--399.

[279] Estevao, V.M. and Särndal, C.E. (2002). The ten cases of auxiliary information for calibration in two-phase sampling. *J. Official Statist.*, 18(2), 233--255.

[280] Farrell, P.J. (1997). Empirical Bayes estimation of small area proportions based on ordinal outcomes variables. *Survey Methodology*, 23, 119--126.

[281] Farrell, P.J. (2000). Bayesian inference for small area proportions. *Sankhy $\overline{a}$* , B, 62, 402--416.

[282] Farrall, P.J., MacGibbon, B. and Tomberlin, T.J. (1994). Protection against outliers in empirical Bayes estimation. *Canad. J. Statist.*, 22,365-376.

[283] Farrell, P.J., MacGibbon, B. and Tomberlin, T.J. (1997a). Bootstrap adjustments for empirical Bayes interval estimates of small-area proportions. *Canad. J. Statist.*, 25(1), 75--89.

[284] Farrell, P.J., MacGibbon, B. and Tomberlin, T.J. (1997b). Empirical Bayes estimators of small area proportions in multistage designs. *Statistica Sinica*, 7, 1065--1083.

[285] Farrell, P.J., MacGibbon, B. and Tomberlin, T.J. (1997c). Empirical Bayes small area estimation using logistic regression models and summary statistics. *J. Business and Econo. Statist.*, 15, 101--108.

[286] Farrell, P.J. and Singh, S. (2002a). Recalibration of higher order calibration weights. *Presented at the Conference of the Statistical Society of Canada, Hamilton,Canada.*

[287] Farrell, P.J. and Singh, S. (2002b). Penalized chi square distance function in survey sampling. *Joint Statistical Meetings, NY-Section on survey research method*, pp. 963--968.

[288] Fan, J. (1993). Local linear regression smoothers and their minimax efficiencies. *Ann. Statist.*, 21, 196--216.

[289] Fay, R. (1992). When are inferences from multiple imputation valid? *Proc. Survey Res. Meth. Sect., Amer Statist. Assoc.*, 227--232.

[290] Fay, R. (1994). Discussion of paper by X.L. Meng. *Statist. Sci.*, 9, 558--560.

[291] Fay, R. (1996). Alternative paradigms for the analysis of imputed survey data. *J. Amer. Statist. Assoc.*, 91, 490--498.

[292] Fay, R.E. and Herriot, R.A. (1979). Estimates of income for small places: An application of James--Stein procedures to census data. *J. Amer. Statist. Assoc.*, 74, 269--277.

[293] Fellegi, I.P. (1963). Sampling with varying probabilities without replacement : rotating and non-rotating samples. *J. Amer. Statist. Assoc.*, 58, 183--201.

[294] Fellegi, I.P. and Holt, D. (1976). A systematic approach to automatic editing and imputation. *J. Amer. Statist. Assoc.*, 71, 17--35.

[295] Fellegi, I.P. and Sunter, A.B. (1974). Balance between different sources of survey errors--some Canadian experiences. *Sankhy $\overline{a}$* , C, 36, 119--142.

[296] Feller, W. (1957). *An introduction to probability theory and its applications.* Vol. 1, John Wiley and Sons, New York.

[297] Feng, S. and Zou, G. (1997). Sample rotation method with auxiliary variable. *Commun. Statist.--Theory Meth.*, 26(6), 1497--1509.

[298] Fienberg, S.E. (1970). An iterative procedure for estimation in contingency tables. *Ann. Math. Statist.*, 41, 907--917.

[299] Fienberg, S.E. and Tanur, J.M. (1987). Experimental and sampling structures : parallel diverging and meeting. *Ins. Statist. Rev.*, 55, 75--96.

[300] Finney, D.J. (1948). Random and systematic sampling in timber surveys. *Forestry*, 22, 1--36.

[301] Finney, D.J. (1950). An example of periodic variation in forest sampling. *Forestry*, 23, 96--111.

[302] Fisher, R.A. (1920). A mathematical examination of the methods of determining the accuracy of an observation by the mean error, and by the mean square error. *Monthly Notices R. Astr. Soc.*, 80, 759--770.

[303] Fisher, R.A. (1922). On the mathematical foundations of theoretical statistics. *Phil. Trans. R. Soc. Lond.*, A, 222, 309--368.

[304] Fisher, R.A. (1925). *Statistical Methods for Research Workers, 1st Edition.* Oliver and Boyd, Edinburgh.

[305] Folsom, R.E., Greenberg, B.G., Horvitz, D.G. and Abernathy, J.R. (1973). The two alternate questions randomized response model for human surveys. *J. Amer. Statist. Assoc.*, 68, 525--530.

[306] Foreman, E.K. and Brewer, K.W.R. (1971). The efficient use of supplementary information in standard sampling procedures. *J. R. Statist. Soc.*, B, 33, 391--400.

[307] Fountain, R.L. and Pathak, P.K. (1989). Systematic and non-random sampling in the presence of linear trends. *Commun. Statist.-- Theory Meth.*, 18, 2511--2526.

[308] Francis, R.I.C. (1984). An adaptive strategy for stratified random trawl surveys. *New Zealand J. Mar. Freshw. Res.*, 18, 59--71.

[309] Francisco, C.A. and Fuller, W.A. (1991). Quantile estimation with a complex survey design. *Ann. Statist.*, 19, 454--469.

[310] Frankel, L.R. and Stock, J.S. (1942). On the sample survey of unemployment. *J. Amer. Statist. Assoc.*, 37, 77--80.

[311] Franklin, L.A. (1989a). A comparison of estimators for randomized response sampling with continuous distributions from a dichotomous population. *Commun. Statist. -- Theory Meth.*, 18, 489--505.

[312] Franklin, L.A. (1989b). Randomized response sampling from dichotomous populations with continuous randomization. *Survey Methodology*, 15, 225--235.

[313] Freedman, D., Pisani, R. and Purves, R. (1978). *Statistics.* Norton, New York.

[314] Freund, J.E. (2000). *Mathematical Statistics.* Fifth Edition. Prentice Hall of India, New Delhi.

[315] Friedlander, D. (1961). A technique for estimating a contingency table, given the marginal totals and some supplementary data. *J. R.. Statist.* A, 124, 412--420.

[316] Fuller, W.A. (1966). Estimation employing post strata. *J. Amer. Statist. Assoc.*, 61,1172--1183.

[317] Fuller, W.A. (1970). Sampling with random stratum boundaries. *J. R. Statist. Soc.*, 32, 209 -- 226.

[318] Fuller, W.A. (1971). *A procedure for selecting non-replacement unequal probability samples.* Unpublished Manuscript, Department of Statistics, Iowa State, University, Ames, Iowa.

[319] Fuller, W.A. (1987). *Measurement error models.* John Wiley and Sons, Inc., New York.

[320] Fuller, W.A. (1995). Estimation in the presence of measurement error. *Int. Statist. Rev.*, 63, 121--147.

[321] Fuller, W.A. (1998). Replication variance estimation for two-phase samples. *Statistica Sinica*, 8, 1153--1164.

[322] Fuller, W.A. and Breidt, F.J. (1999). Estimation for supplemented panels. Sankhy $\overline{a}$,51, 58--70.

[323] Fuller, W.A. and Burmeister, L.F. (1972). Estimators for samples selected from two overlapping frames. *Proceedings of the Social Statistics Section, American Statistical Association*, 245--249.

[324] Gabler, S. (1981). A comparison of Sampford's sampling procedure versus unequal probability sampling with replacement. *Biometrika*, 68, 725--727.

[325] Gabler, S. (1984). On unequal probability sampling: sufficient conditions for the superiority of sampling without replacement. *Biometrika,* 71, 171--175.

[326] Gabler, S. and Horst, S. (1995). Improving the RHC--strategy. *Statistical Hefte*, 36, 327--336.

[327] Garcia, M.R. and Cebrian, A.A. (1996). Repeated substitution method: The ratio estimator for the population variance. *Metrika*, 43, 101--105.

[328] Garcia, M.R. and Cebrian, A.A. (1998). Quantile interval estimation in finite population using a multivariate ratio estimator. *Metrika*, 47, 203--213.

[329] Garcia, M.R. and Cebrian, A.A. (2001). On estimating the median from survey data using multi-auxiliary information. *Metrika*, 54(1), 59--76.

[330] Gautschi, W. (1957). Some remarks on systematic sampling. *Ann. Math. Statist.*, 28, 385--394.

[331] Gershunskaya, J., Eltinge, J.L. and Huff, L. (2002). Use of auxiliary information to evaluate a synthetic estimator in the U.S. current employment statistic program. *Joint Statistical Meetings, NY-Section on survey research methods*, 1149--1154.

[332] Ghangurde, P.D. and Rao, J.N.K. (1969). Some results on sampling over two occasions. *Sankhy $\bar{a}$* , A, 31, 463--472.

[333] Ghosh, M. and Meeden, G. (1997). *Bayesian methods for finite population sampling*. Chapman and Hall.

[334] Ghosh, M. and Pathak, P.K. (1992). *Current Issues in Statistical Inference: Essays in Honor of D. Basu.* Lecture Notes -- Monograph Series, Institute of Mathematical Statistics, Hayward, California.

[335] Ghosh, M. and Rao, J.N.K. (1994). Small area estimation : An appraisal. *Statistical Science*, 9(1), 55--93.

[336] Ghosh, S. (1998). The Horvitz--Thompson vs. Sen--Yates--Grundy variance Estimators: Issues in finite population sampling. *J. Indian Soc. Agric. Statist.*, 50, 2&3, 343--348.

[337] Ghosh, S.P. (1963). Post-cluster sampling. *Ann. Math. Statist.* 34, 587--597.

[338] Giffard--Jones, W. (1993). The doctor game. *The Windsor Star*, April 15, 1993.

[339] Giommi, A. (1984). A simple method for estimating individual response probabilities in sampling from finite populations. *Metrika*, 185--200.

[340] Godambe, V.P. (1995a). Estimation of parameters in survey sampling : Optimality. *Canad. J. Statist.*, 23(3), 227--243.

[341] Godambe, V.P. (1955b). A unified theory of sampling from finite populations. *J. R. Statist. Soc.*, B, 17, 269--278.

[342] Godambe, V.P. (1960). An optimum property of regular maximum likelihood estimation. *Ann. Math. Statist.*, 31,1208--1211.

[343] Godambe, V.P. (1969). Admissibility and Bayes estimation in sampling finite populations--V. *Ann. Math. Statist.*, 40, 672--676.

[344] Godambe, V.P. (1976). Conditional likelihood and unconditional optimum estimating equations. *Biometrika*, 63, 277--284.

[345] Godambe, V.P. (1980a). On the sufficiency and ancillarity in the presence of nuisance parameters. *Biometrika,* 67, 269--276.

[346] Godambe, V.P. (1980b). Estimation in randomized response trials. *Int. Statist. Rev.*, 48, 29--32.

[347] Godambe,V.P. (1984). On ancillarity and Fisher information in presence of a nuisance parameters. *Biometrika*, 71, 626--629.

[348] Godambe, V.P. (1987). Resolution of Godambe's paradox. *Statist. Probab. Lett.*, 5, 239--239.

[349] Godambe, V.P. (1989). *Estimation of cumulative distribution of survey population*. Technical Report STAT : 89--117, University of Waterloo.

[350] Godambe, V.P. (1991). Orthogonality of estimating functions and nuisance parameters. *Biometrika*, 78, 143--151.

[351] Godambe, V.P. (1995). Estimation of parameters in survey sampling : Optimality. *Canad. J. Statist.*, 23(3), 227--243.

[352] Godambe, V.P. (1998). Estimation of parameters in survey sampling. *J. Indian Soc. Agric. Statist.*, 51 (2-3), 315--330.

[353] Godambe, V.P. (1999). Linear Bayes and optimal estimation. *Ann. Inst. Statist. Math.*,51(2), 201--215.

[354] Godambe, V.P. and Heyde, C.C. (1987). Quasi likelihood and optimal estimation. *Int. Statist. Rev.*, 55, 231--244.

[355] Godambe, V.P. and Joshi, V.M. (1965). Admissibility and Bayes estimation in sampling finite populations -- I. *Ann. Math. Statist.*, 36, 1707--1722.

[356] Godambe, V.P. and Kale, B.K. (1991). Estimating functions: an overview. *In estimating Functions* (V.P. Godambe ed.), Clarendon Press, Oxford, 3--20.

[357] Godambe, V.P. and Thompson, M.E. (1984). Robust estimation through estimating equations. *Biometrika*, 71, 115--125.

[358] Godambe, V.P. and Thompson, M.E. (1986). Parameters of superpopulation and survey population, their relationship and estimation. *Int.. Statist. Rev.*, 54, 127--138.

[359] Godambe, V.P. and Thompson, M.E. (1989). An extension of quasi-likelihood estimation (with discussion). *J. Statist. Planning Infer.*, 22, 137--172.

[360] Godambe, V.P. and Thompson, M.E. (1996-97). Optimal estimation in a casual framework. *J. Indian Soc. Agril. Statist.*, 49, 21--46.

[361] Godambe, V.P. and Thompson, M.E. (1999). A new look at confidence intervals in survey sampling. *Survey Methodology*, 25, 161--173.

[362] Goel, B.B.P.S. and Singh, D. (1977). On the formation of clusters. *J. Indian Soc. Agril. Statist.*, 29, 53--68.

[363] Gonzalez, M.E. (1973). Use and evaluation of synthetic estimators. *Proceedings of the Amer. Statist. Assoc., Social Statistics Section,*, 33--36.

[364] Gonzalez, M.E. and Hoza, C. (1976). Small area estimation of unemployment. *Proceedings of the Section on Social Statistics, American Statistical Association*, 437--443.

[365] Gonzalez, M.E. and Hoza, C. (1978). Small area estimation with application to unemployment and housing estimates. *J. Amer. Statist. Assoc.*, 73,7--15.

[366] Goodman, L.A. and Hartley, H.O. (1958). The precision of unbiased ratio type estimators. *J. Amer. Statist. Assoc.*, 53, 491--508.

[367] Graf, M. (2002). Assessing the accuracy of the median in a stratified double stage cluster sampling by means of a nonparametric confidence interval: Application to the swiss earnings structure survey. *Proc. Joint Statistical Meetings, NY--Section on Government Statatistics*, 1223--1228.

[368] Greenberg, B.G., Abul-Ela, A.L.A., Simmons, W.R. and Horvitz, D.G. (1969). The unrelated question randomized response model -- theoretical framework. *J. Amer. Statist. Assoc.*, 64, 520--539.

[369] Greenberg, B.G., Kuebler, R.R., Abernathy, J.R. and Horvitz, D.G. (1971). Application of the randomized response technique in obtaining quantitative data. *J. Amer. Statist. Assoc.*, 66, 243--250.

[370] Grewal, I.S., Bansal, M.L. and Singh, S. (1999). An alternative estimator for multiple characteristics using randomized response technique in PPS sampling. *Aligarh J. Statist.*, 51--65.

[371] Grewal, I.S., Bansal, M.L. and Singh, S. (2002). Estimation of population mean of a stigmatized quantitative variable using double sampling. *Statistica* (Accepted).

[372] Gross, S.T. (1980). Median estimation in sample surveys. *Proc. Surv. Res. Meth. Sect. Amer. Statist. Assoc.,* 181--184.

[373] Groves, R.M. (1996). Non-sampling error in surveys: the journey toward relevance in practice. *Proc. Statist. Can. Symp.*, 96, 7--14.

[374] Groves, R.M. and Lepkowski, J.M. (1986). An experimental implementation of a dual frame telephone sample design. *Proc. Sec. Survey Res. Meth., American Statistical Association*, 340--345.

[375] Grubbs, F.E. (1948). On estimating precision of measuring instruments and product variability. *J. Amer. Statist. Assoc.*, 43, 243--264.

[376] Gujarati, D. (1978). *Basic econometrics (International Student Edition),* Mcgraw – Hill International Book Company, Tokyo.

[377] Gupta, B.K. and Rao, T.J. (1997). Stratified PPS sampling and allocation of sample size. *J. Indian Soc. Agril. Statist.*, 50(2), 199--208.

[378] Gupta, J.P. (2002). Estimation of the correlation coefficient in probability proportional to size with replacement sampling. *Statistical Papers*, 43(4), 525--536.

[379] Gupta, J.P. and Singh, R. (1990). A note on usual correlation coefficient in systematic sampling. *Statistica,* 50, 255--259.

[380] Gupta, J.P., Singh, R. and Kashani, H.B. (1993). An estimator of the correlation coefficient in probability proportional to size with replacement sampling. *Metron,* 165--177.

[381] Gupta, J.P., Singh, R. and Lal, B. (1978). On the estimation of the finite population correlation coefficient--I. *Sankhyā* , C, 41, 38--59.

[382] Gupta, J.P., Singh, R. and Lal, B. (1979). On the estimation of the finite population correlation coefficient--II. *Sankhyā* , C, 42, 1--39.

[383] Gupta, P.C. (1970). *Some estimation problems in sampling using auxiliary information.* Unpublished Ph.D. thesis submitted to IARS, New Delhi.

[384] Gupta, P.C. (1978). On some quadratic and higher degree ratio and product estimator. *J. Indian Soc. Agril. Statist.,* 30, 71--80.

[385] Gupta, P.C. and Kothwala, N.H. (1990). A study of second order approximation for some product type estimators. *J. Indian Soc. Agril. Statist.,* 42, 171--185.

[386] Gupta, R.K., Singh, S. and Mangat, N.S. (1992-93). Some chain ratio type estimators for estimating finite population variance. *Aligarh J. Statist.*, 12&13, 65--69.

[387] Gupta, V.K. and Nigam, A.K. (1987). Mixed orthogonal arrays for variance estimation with unequal number of primary selections per stratum. *Biometrika,* 74, 735--742.

[388] Gupta, V.K., Nigan, A.K. and Kumar, P. (1982). On a family of sampling schemes with inclusion probability proportional to size. *Biometrika,* 69, 191--196.

[389] Gurney, M. and Jewett, R.S. (1975). Constructing orthogonal replications for standard errors. *J. Amer. Statist. Assoc.*, 70, 819--821.

[390] Hājek, J. (1958). Some contribution to the theory of probability sampling. *Bull. Int. Statist. Inst.*, 36, 127--134.

[391] Hājek, J. (1959). Optimum strategies and other problems in probability sampling. *Casopis Pest. Mat.*, 84, 387--423.

[392] Hājek, J. (1964). Asymptotic theory of rejective sampling with varying probability from a finite population. *Ann. Math. Stat.*, 35, 1491--1525.

[393] Halmos, P.R. and Perlman, M.D. (1974). On the existence of a minimal sufficient sub-field. *Ann. Statist..*, 2, 1049--1055.

[394] Hanif, M., Mukhopadhyay, P. and Bhattacharyya, S. (1993). On estimating the variance of Horvitz and Thompson estimator. *Pak. J. Statist.*, A, 9, 123--136.

[395] Hansen, M.H. and Hurwitz, W.N. (1942). Relative efficiencies of various sampling units in population enquiries. *J. Amer. Statist. Assoc.*, 37, 89--94.

[396] Hansen, M.H. and Hurwitz, W.N. (1943). On the theory of sampling from finite populations. *Ann. Math. Stat.*,14, 333--362.

[397] Hansen, M.H. and Hurwitz, W.N. (1946). The problem of non-response in sample surveys. *J. Amer. Statist. Assoc.*,41,517--529.

[398] Hansen, M.H., Hurwitz, W.N. and Madow, W.G. (1953). *Sample survey methods and theory.* New York, John Wiley and Sons, 456--464.

[399] Hanurav, T.V. (1965). *Optimum sampling strategies and some related problems.* Ph.D. Thesis, Indian Statistical Institute.

[400] Hanurav, T.V. (1966). Some aspects of unified sampling theory. *Sankhyā* , A, 28, 175--204.

[401] Hanurav, T.V. (1967). Optimum utilization of auxiliary information: πps sampling of two units from a stratum. *J. R. Statist. Soc.*,B, 29, 374--391.

[402] Hartigan, J.A. (1969). Linear Bayesian methods. *J. R. Statist. Soc.*, B, 31, 440--454.

[403] Hartley, H.O. (1962). Multiple frame surveys. *Proc. of the Social Statistics Section, American Statistical Association*, 203--206.

[404] Hartley, H.O. (1966). Systematic sampling with unequal probability and without replacement. *J. Amer. Statist. Assoc.*, 61, 739--748.

[405] Hartley, H.O. (1974). Multiple frame methodology and selected applications. *Sankhyā*, C, 36, 99--118.

[406] Hartley, H.O. and Biemer, P.P. (1978). The estimation of non-sampling variances in current surveys. *Proc. Sec. Survey Res. Meth.*, American Statistical Association, 257--262.

[407] Hartley, H.O. and Rao, J.N.K. (1962). Sampling with unequal probabilities and without replacement. *Ann. Math. Statist.*, 33, 350--374.

[408] Hartley, H.O. and Rao, J.N.K. (1968). A new estimation theory for sample surveys. *Biometrika*, 55, 547--557.

[409] Hartley, H.O., Rao, J.N.K. and Kiefer, G. (1969). Variance estimation with one unit per stratum. *J. Amer. Statist. Assoc.*, 64, 841--851.

[410] Hartley, H.O. and Ross, A. (1954). Unbiased ratio estimators. *Nature*, 174, 270--271.

[411] Hedayat, A.S., Rao, C.R. and Stufken, J. (1988). Sampling plans excluding contiguous units. *J. Statist. Planning Infer.*, 19, 159--170.

[412] Heilbron, D.C. (1978). Comparison of estimators of the variance of systematic sampling. *Biometrika*, 65, 429--433.

[413] Henderson, C.R. (1975). Best linear unbiased estimation and prediction under a selection model. *Biometrics*, 31, 423--447.

[414] Hendricks, W.A. (1944). The relative efficiencies of groups of farms as sampling units. *J. Amer. Statist. Assoc.*, 39, 366--376.

[415] Hendricks, W.A. (1949). Adjustment for bias caused by non-response in mailed surveys. *Agric. Econo. Res.*, 1, 52--56.

[416] Herzel, A. (1986). Sampling without replacement with unequal probabilities: sample designs with pre-assigned joint inclusion probabilities of any order. *Metron*, 49--68.

[417] Herzel, A. (1993). Maximin πps designs. *Metron*, 5--25.

[418] Hidiroglou, M.A. (1995). Sampling and estimation for stage one of the Canadian survey of employment, payrolls and hours survey redesign. *Statistical Society of Canada, Proc. of the Survey Methods Section*, 123--128.

[419] Hidiroglou, M.A. (2001). Double Sampling. *Survey Methodology*, 27, 143--154.

[420] Hidiroglou, M. A. and Särndal, C.E. (1995). Use of auxiliary information for two-phase sampling. *Proc. Sec. Survey Res. Meth., Amer. Statist. Assoc.*, Vol.II, 873--878.

[421] Hidiroglou, M. A. and Särndal, C.E. (1998). Use of auxiliary information for two-phase sampling. *Survey Methodology*, 24 (1), 11--20.

[422] Hodges, J.L. and Lehmann, E. (1970). *Basic Concepts of Probability and Statistics, 2nd ed.* Holden--Day, San Francisco.

[423] Holt, D. and Smith, T.M.F. (1979). Post-stratification. *J. R. Statist. Soc.*, A, 142 33--46.

[424] Holt, D., Smith, T.M.F. and Tomberlin, T.J. (1979). A model based approach to estimation for small subgroups of a population. *J. Amer. Statist. Assoc.*, 74, 405--410.

[425] Horvitz, D.G. and Thompson, D.J. (1952). A generalisation of sampling without replacement from a finite universe. *J. Amer. Statist. Assoc.*, **47**, 663--685.

[426] Huang, L.R. and Ernst, L.R. (1981). Comparison of an alternative estimator to the current composite estimator in the Current Population Surveys. *Proc. of the Amer. Statist. Assoc., Section on Survey Research Methods*, 303--308.

[427] Hutchison, M.C. (1971). A Monte Carlo comparison of some ratio estimators. *Biometrika*, 58, 313--321.

[428] Iachan, R. (1982). Systematic sampling : A critical review. *Int. Stat. Rev.*, 50, 293--303.

[429] Ireland, C.T. and Kullback, S. (1968). Contingency table with given marginals. B*iometrika*, 55, 179--188.

[430] Isaki, C.T. (1983).Variance estimation using auxiliary information. *J. Amer. Statist. Assoc.* ,78, 117--123.

[431] Isaki, C.T. and Fuller, W.A. (1982). Survey design under a regression superpopulation model. *J. Amer. Statist. Assoc.*, 77, 89--96.

[432] Jaech, J.L. (1981). Constraind expected likelihood estimates of precisions using Grubbs' technique for two dimensional methods. *Nuclear Materials Management Journal*, X(2), 34--39.

[433] Jaech, J.L. (1985). *Statistical analysis of measurement errors.* John Wiley and Sons, New York.

[434] Jagers, P. (1986). Post-stratification against bias in sampling. *Int. Statist. Rev.*, 54, 159--167.

[435] Jagers, P., Oden, A. and Trulsson, L. (1985). Post-stratification and ratio estimation: usages of auxiliary information in survey sampling and opinion polls. *Internat. Statist. Rev.,* 53, 221--238.

[436] Jain, R.K. (1987). Properties of estimators in simple random sampling using auxiliary variable. *Metron,* 265--271.

[437] Jessen, R.J. (1942). Statistical investigation of a sample survey for obtaining farm facts. *Iowa Agricultural Experiment Station Research Bulletin*, 104.

[438] Jhajj, H.S. and Srivastava, S.K. (1983). A class of PPS estimators of population mean using auxiliary information. *J. Indian Soc. Agril. Statist.*, **35**, 57--61.

[439] John, S. (1969). On multivariate ratio and product estimators. *Biometrika*, 56, 533--536.

[440] Jolly, G.M. and Hampton, I. (1990). A stratified random transect design for acoustic surveys of fish stocks. *Canad. J. Fish. Aquat. Sci.*, 47, 1282--1291.

[441] Jones, R.G. (1980). Best linear unbiased estimators for repeated surveys. *J. R. Statist. Soc.*, B, 42, 221--226.

[442] Joshi,V.M. (1966). Admissibility and bayes estimation in sampling finite populations IV. *Ann. Math. Statist.*, 37, 1658--1678.

[443] Joshi, V.M. (1970). Note on the admissibility of the Sen—Yates--Grundy estimator and Murthy's estimator and its variance estimator for samples of size two. *Sankhy* $\overline{a}$, A,32, 431--438.

[444] Kadilar, C. and Cingi, H. (2003). Ratio estimators in stratified random sampling. *Biom. J.*, 45(2), 218--225.

[445] Kalton, G and Anderson, D.W. (1986). Sampling rare populations. *J. R. Statist. Soc.*, A, 65--82.

[446] Kalton, G. and Kasprzyk, J.R. (1986). The treatment of missing data. *Survey Methodology*, 105--110.

[447] Kapadia, S.B. and Gupta, P.C. (1984). A quadratic and higher degree ratio, product estimators in sampling with varying probabilities. *J. Statist. Res.*, 18, 1--18.

[448] Karlheinz, F. (1990). Stratified sampling using double sampling. *Statist. Hefte*, 31, 55--63.

[449] Kasprzyk, D., Duncan, G.J., Kalton, G. and Singh, M.P. *Panel Surveys*. Wiley, New York.

[450] Kathuria, O.P. (1975). Some estimators in two-stage sampling on succesive occasions with partial matching at both stages. *Sankhy* $\overline{a}$, C, 37, 147--162.

[451] Kathuria, O.P. and Singh, D. (1971a). Comparison of estimates in two-stage sampling on successive occasions. *J. Indian Soc. Agril. Statist.*, 23, 31--51.

[452] Kathuria, O.P and Singh, D. (1971b). Relative efficiencies of some alternative replacement procedures in two-stage sampling on successive occasions. *J. Indian Soc. Agril. Statist.*, 23, 101--114.

[453] Kaur, P. and Singh, G. (1982). A note on estimating variance in a finite population. *J. Statist. Res.*, 16(1&2), 51--54.

[454] Kempthorne, O. (1952). *The Design and Analysis of Experiments*. Wiley, New York.

[455] Kerkvliet, J. (1994). Estimating a logit model with randomized data: The case of cocaine use. *Austral. J. Statist.*,36, 9--20.

[456] Khan, S.U. and Tripathi, T.P. (1967). The use of multi-auxiliary information in double sampling. *J. Indian Statist. Assoc.*, 5, 42--48.

[457] Khan, Z. (1976). Optimum allocation in Bayesian stratified two-phase sampling. *J. Indian Soc. Agril. Statist.*, 14, 65--74.

[458] Khare, B.B. (1987). Allocation in stratified sampling in presence of non-response. *Metron*, 213--221.

[459] Khare, B.B. (1991). Determination of sample sizes for a class of two-phase sampling estimators for ratio and product of two population means using auxiliary character. *Metron*, 185--197.

[460] Khare, B.B. and Srivastava, S. (1981). A generalized regression ratio estimator for the population mean using two auxiliary variables. *Aligarh J. Statist.*, 1(1), 43--51.

[461] Khare, B.B. and Srivastava, S. (1997). Transformed ratio type estimators for the population mean in the presence of nonresponse. *Commun. Statist. -- Theory Meth.*, 26(7), 1779--1791.

[462] Khare, B.B and Srivastava, S.R. (1998). Combined generalised chain estimators for ratio and product of two population means using auxiliary characters. *Metron*, 56, 109--116.

[463] Kim, J. (1978). *Randomized response techniques for surveying human populations*. Unpublished Ph.D. dissertation, Temple University, Philadelphia, USA.

[464] Kim, J.K. (2001). Variance estimation after imputation. *Survey Methodology*, 27, 75--83.

[465] Kiregyera, B. (1980). A chain ratio type estimator in finite population: double sampling using two auxiliary variables. *Metrika*, 27, 217--223.

[466] Kiregyera,B.(1984). Regression type estimators using two auxiliary variables and the model of double sampling from finite populations. *Metrika*, 31, 215--226.

[467] Kish, L. and Hess, I. (1959). A replacement procedure for reducing the bias of non-response. *American Statistician*, 13, 17--19.

[468] Kokan, A.R. and Khan, S.U. (1967). Optimum allocation in multivariate surveys: An analytical solution. *J. R. Statist. Soc.*, B, 2, 115--125.

[469] Kolmogorov, A.N. (1942). Sur l'estimation statistique des parameters de la loi de Gauss. *Izv. Akad. Nauk SSSR Ser. Mat.* 6, 3--32.

[470] Konijn, H.S. (1973). *Statistical Theory of Sample Survey Design and Analysis*. North Holand Publishing Company.

[471] Konijn, H.S. (1979). Model free evaluation of the bias and the mean square error of the regression estimator. *Sankhy* $\overline{a}$, C, 41, 69--75.

[472] Konijn, H.S. (1981). Biases, variances and co-variances of raking ratio estimators for marginal and cell totals and averages of observed characteristics. *Metrika*, 28, 109--121.

[473] Koop, J.C. (1967). Replicated (or interpenetrating) samples of unequal sizes. *Ann. Math. Statist.*, 38, 1142--1147.

[474] Koop, J.C. (1971). On splitting a systematic sample for variance estimation. *Ann. Math. Statist.*, 42, 3, 1084--1087.

[475] Korn, E.L. and Graubard, B.I. (1998). Variance estimation for superpopulation parameters. *Statistica Sinica*, 8, 1131--1151.

[476] Kossack, C.F. and Shiledar--Bax, H.R. (1971). On designing of a unit--stratified survey design for discrete set of observations. *Int. Statist. Rev.*, 39, 46--56.

[477] Kothwala, N.H. and Gupta, P.C. (1989). Estimation of population mean with knowledge of coefficient of variation with p--auxiliary variables. *Metron*, 107--119.

[478] Kott, P.S. (1988). Model based finite population correction for the Horvitz and Thompson estimator. *Biometrika*, 75, 797--799.

[479] Kott, P.S. and Stukel, D.M. (1997). Can the Jackknife be used with a two-phase sample? *Survey Methodology*, 23, 81--89.

[480] Kowar, R.M. (1996). One pass selection of a sample with probability proportional to aggregate size. *Sankhy* $\overline{a}$, B, 58, 80--83.

[481] Krewski, D. and Chakrabarty, R.P. (1981). On the stability of the Jackkinfe variance estimator in ratio estimation. *J. Statist. Planning Infer.*, 5, 71--78.

[482] Krewski, D. and Rao, J.N.K. (1981). Inference from stratified samples: properties of the linearization, Jackknife and balanced repeated replication methods. *Ann. Statist.*,9, 1010--1019.

[483] Kuhn, H.W. and Tucker, A.W. (1952). *Non-linear programming.* Proceeding of the second Berkeley Symposium on Mathematical Statistics and Probability.

[484] Kuk, A.Y.C. (1990). Asking sensitive questions indirectly. *Biomerika*, 77(2), 436--438.

[485] Kuk, A.Y.C. and Mak, T.K. (1989). Median estimation in the presence of auxiliary information. *J. R. Statist. Soc.*, B, 51, 261--269.

[486] Kuk, A.Y.C. and Mak, T.K. (1994). A functional approach to estimating finite population distribution functions. *Commun. Statist.-- Theory Meth.*, 23(3), 883--896.

[487] Kulldorff, G. (1963). Some problems of optimum allocation for sampling on two occasions. *Rev. Inter. Statist. Inst.*, 31, 24--57.

[488] Kumar, E.V. and Srivenkataramana, T. (1994). A generalization of Midzuno--Sen sampling scheme for finite populations. *Commun. Statist.--Theory Meth.*, 23(9), 2541--2559.

[489] Kumar, E.V., Srivenkataramana, T. and Srinath, K.P. (1996). Use of ranks in probability proportional to size sampling. *Commun. Statist. -- Theory Meth.*, 25(12), 3195--3215.

[490] Kumar, P. and Agarwal, S.K. (1997). Alternative estimators for the population totals in multiple characteristic survey. *Commun. Statist. -- Theory Meth.*, 26(10), 2527--2537.

[491] Kumar, P., Gupta, V.K. and Nigam, A.K. (1985). On inclusion probability proportional to size sampling scheme. *J. Statist. Planning Infer.*, 12, 127--131.

[492] Kumar, P. and Herzel, A. (1988). Estimating population totals in surveys involving multi-characters. *Metron*, 33--47.

[493] Kumar, P., Srivastava, A.K. and Agarwal, S.K.(1986). A general class of unequal probability sampling schemes. *Statistica*, 46, 67--74.

[494] Kumar, S and Lee, H. (1983). Evaluation of composite estimation for the Canadian Labour Force Survey. *Survey Methodology*, 9, 1--24.

[495] Laake, P. (1986). Optimal estimates and optimal predictors of finite population characteristics in the presence of non-response. *Metrika,* 33, 69--77.

[496] Lahiri, D.B. (1951). A method for sample selection providing unbiased ratio estimates. *Bull. Ins. Statist. Inst.*,33(2), 133--140.

[497] Lahiri, P. and Rao, J.N.K. (1995). Robust estimation of mean squared error of small area estimators. *J. Amer. Statist. Assoc.*, 90, 758--766.

[498] Laird, N.M. and Louis, T.A. (1987). Empirical Bayes confidence intervals based on bootstrap samples. *J. Amer. Statist. Assoc.*, 82, 739--750.

[499] Lakshmi, D.V. and Raghavarao, D. (1992). A test for detecting untruthful answering in randomized response procedure. *J. Statist. Planning Infer.*, 31, 387--390.

[500] Lanke, J. (1975). On the choice of the unrelated question in Simons version of randomised response. *J. Amer. Statist. Assoc.,* 70, 80--83.

[501] Lanke, J. (1976). On the degree of protection in randomized interviews. *Int. Statist. Rev.*, 44, 197--203.

[502] Lee, H. and Kim, J.K. (2002). Jackknife variance estimation for two-phase samples with high sampling fractions. *Joint Statistical Meetings, NY--Section on survey research methods*, 2024--2028

[503] Lee, H., Rancourt, E. and Särndall, C.E. (1994). Experiments with variance estimation from survey data with imputed values. *J. Official Statist.*, 10(3), 231--243.

[504] Lee, H., Rancourt, E. and Särndall, C.E. (1995a). Variance estimation in the presence of imputed data for the generalized estimation system. *Proc. of the American Statist. Assoc. (Social Survey Research Methods Section)*, 384--389.

[505] Lee, H., Rancourt, E. and Särndall, C.E. (1995b). Jackknife variance estimation for data with imputed values. *Statistical Society of Canada, Proceedings of the Survey Methods Section*, 111--115.

[506] Lent, J., Miller, S.M. and Cantwell, P.J. (1996). Effect of composite weight on some estimates from the current population surveys. *Proc. of the section on Survey Research Methods, Amer. Statist. Assoc.*, 130--139.

[507] Leysieffer, F.W. and Warner, S. L. (1976). Respondent jeopardy and optimal designs in randomized response models. *J. Amer. Statist. Assoc.*, 71, 649--656.

[508] Linacre, S.J. and Trewin, D.J. (1993). Total survey design application to a collection of the construction industry. *J. Official Statist.*, 9, 611--621.

[509] Lindley, D.V. and Deely, J.J. (1993). Optimum allocation in stratified sampling with partial information. *Test*, 2(1), 147--160.

[510] Little, R.J.A. and Yao, L. (1996). Intent to treat analysis for longitudinal studies with drop outs. *Biometrics*, 52, 1324--1333.

[511] Lohr, S.L. and Rao, J.N.K. (1998). Jackknife variance estimation in dual frame surveys. *Tech. Rep., Laboratory for research in Statistics and Probability*, Carleton University.

[512] Lohr, S.L. and Rao, J.N.K. (2000). Inference from dual frame surveys. *J. Amer. Statist. Assoc.*, 95, 271-- 280.

[513] Lundström, S. (1997). *Calibration as a standard method for treatment of non-response.* Ph.D. Thesis.

[514] Lundström, S. and Särndal, C.E. (1999). Calibration as a standard method for treatment of non-response. *J. Official Statist.*, 15(2), 305--327.

[515] MacGibbon, B. and Tomberlin, T.J. (1989). Small area estimates of proportion via empirical Bayes techniques. *Survey Methodology*, 15(2), 237--252.

[516] Madow, W.G. (1949). On the theory of systematic sampling -- II. *Ann. Math. Statist.*, 20, 333--354.

[517] Madow, W.G. (1953). On theory of systematic sampling -- III. *Ann. Math. Statist.*, 24, 101--106.

[518] Madow, W.G. and Madow, L.H. (1944). On the theory of systematic sampling -- I. *Ann. Math. Statist.*, 15, 1--24.

[519] Mahajan, P.K. and Singh, S. (1996). On estimation of total in two stage sampling. *J. Statist. Res.*, 30, 127--131.

[520] Mahajan, P.K. and Singh, S. (1997). Almost unbiased ratio and product type estimators: A new approach. *Biom. J.*, 39(3), 509--516.

[521] Mahalanobis, P.C. (1940). A sample survey of acreage under jute in Bengal. *Sankhyā*, 4, 511--530.

[522] Mahalanobis, P.C. (1942). *General report on the sample census of area under jute in Bangal.* Indian Central Jute Committee.

[523] Mahalanobis, P.C. (1944). On large scale sample surveys. *Phil. Transac. Roy. Soc., London*, B, 231, 324--351.

[524] Mahalanobis, P.C. (1946). Recent developments in statistical sampling in the Indian Statistical Institute. *J. R. Statist. Soc.*, 109, 326--378.

[525] Mahmood, M., Singh, S. and Horn, S. (1998). On the confidentiality guaranteed under randomized response sampling : A comparison with several new techniques. *Biom. J.* 40 (2), 237--242.

[526] Mak, T.K. and Kuk, A.Y.C. (1993). A new method for estimating finite population quantiles using auxiliary information. *Canad. J. Statist.*, 21(1), 29--38.

[527] Malec, D., Sedransk, J., Moriarity, C.L. and LeClere, F.B. (1997). Small area inference for binary variables in the National Health Interview Survey. *J. Amer. Statist. Assoc.*, 92, 815--826.

[528] Mandowara, V.L. and Gupta, P.C. (1999). Contribution to optimum points of stratification for multi-stage designs. *Metron*, 57, 51--66.

[529] Mangat, N.S. (1991). An optional randomized response sampling technique using non-stigmatized attribute. *Statistica*, LI, 595--602.

[530] Mangat, N.S. (1992). Two stage randomized response sampling procedure using unrelated question. *J. Indian Soc. Agril. Statist.*, 44 (1), 82—87.

[531] Mangat, N.S. (1993). Estimation of population total using an alternative estimator for RHC scheme. *Statistica*, 53, 251--259.

[532] Mangat, N.S. (1994). An improved randomized response strategy. *J. R. Statist. Soc.*, B, 56, 93--95.

[533] Mangat, N.S. and Singh, R. (1990). An alternative randomized response procedure. *Biometrika*, 77, 439--442.

[534] Mangat, N.S. and Singh, R. (1991a). An alternative randomized response procedure for sampling without replacement. *J. Indian Statist. Assoc.*, 29(2), 127--131.

[535] Mangat, N.S. and Singh, R. (1991b). An alternative approach to randomized response survey. *Statistica*, 51(3), 327--332.

[536] Mangat, N.S. and Singh, R. (1992-93). Sampling with varying probabilities without replacement: A review. *Aligarh J. Statist.*, 12&13, 75--106.

[537] Mangat, N.S. and Singh, R. (1995). A note on the inverse binomial randomized response rocedure. *J. Indian Soc. Agri.l. Statist.*, 47(1), 21--25.

[538] Mangat, N.S., Singh, R. and Singh, S. (1991). Alternative estimators in randomised response technique. *Aligarh J. Statist.*, 11, 75--80.

[539] Mangat, N.S., Singh, R. and Singh, S. (1992). An improved unrelated question randomized response strategy. *Calcutta Statist. Assoc. Bull.*, 42, 277--281.

[540] Mangat, N.S., Singh, R. and Singh, S. (1995). Unrelated question randomised response model without randomization device. *Estadistica*, 47, 59--68.

[541] Mangat, N.S., Singh, R. and Singh, S. (1997). Violation of respondent's privacy in Moor's model -- its rectification through a random group strategy. *Commun Statist. -- Theory Meth.*, 26 (3), 743--754.

[542] Mangat, N.S., Singh, R., Singh, S. Bellhouse, D.R. and Kashani, H.B. (1995). On efficiency of estimator using distinct respondents in randomised response survey. *Survey Methodology*, 21(1), 21--23.

[543] Mangat, N.S., Singh, R., Singh, S. and Singh, B. (1993). On Moors' randomised response model. *Biom. J.*, 35(6), 727--732.

[544] Mangat, N.S. and Singh, S. (1994). An optional randomized response sampling technique. *J. Indian Statist. Assoc.*, 32, 71--75.

[545] Mangat, N.S., Singh, S. and Singh, R. (1993). On the use of a modified randomization device in randomised response inquiries. *Metron*, 51 (1), 211--216.

[546] Mangat, N.S., Singh, S. and Singh, R. (1995). On use of a modified randomization device in Warner's model. *J. Indian Soc. Statist. Opers. Res.*, 16, 65--69.

[547] Manisha and Singh, R.K. (2001). An estimation of population mean in the presence of measurement errors. *J. Indian Soc. Agric. Statist.*, 54(1), 13--18.

[548] Manwani, A.H. and Singh, K.B. (1978). Studies in systematic sampling for two-dimensional finite population with special reference to survey for estimation of guavas. *J. Indian Soc. Agric. Statist.*, 30, 1, 82--93.

[549] Marker, D.A. (1983). Organization of small area estimators. *Proc. Survey Research Method Section, Amer. Statist. Assoc.*, Washington, D.C., 409-414.

[550] Mayor, J.A. (2002). Optimum cluster selection probabilities to estimate the finite population distribution function under PPS cluster sampling. *Test*, 11(1), 73--88.

[551] McCarthy, M.D. (1939). On the application of the z-test to randomized blocks. *Ann. Math. Statist.*, 10, 337.

[552] McCarthy, P.J. (1969). Pseudo replication: Half samples. *Rev. Int. Statist. Inst.*, 37, 239--264.

[553] McLeod, A.I. and Bellhouse, D.R. (1983). A convenient algorithm for drawing a simple random sample. *App. Statist.*, 32, 182--184.

[554] Meeden, G. (1992). Basu's contribution to the foundations of sample survey. *Current issues in statistical inference: Essays in Honor of D. Basu by Ghosh and Pathak. Lecture Notes -- Monograph Series, Institute of Mathematical Statistics, Hayward, California,*17,178--186.

[555] Meeden, G. (2000). A decision theoretic approach to imputation in finite population sampling. *J. Amer. Statist. Assoc.*, 95, 586--595.

[556] Meeden, G. and Gosh, M. (1981). Admissibility in finite problems. *Ann. Statist.* 9, 846--852.

[557] Meeden, G. and Ghosh, M. (1983). Choosing between experiments : applications to finite population sampling. *Ann. Statist.*, 11, 296--305.

[558] Meng, X.L. (1994). Multiple imputation inferences with uncongenial sources of input (with discussion). *Statist. Sci.*, 9, 538--573.

[559] Mickey, M.R. (1959). Some finite population unbiased ratio and regression estimators. *J .Amer. Statist. Assoc.*, 54, 594--612.

[560] Midha, C.K. (1980). *Contribution to survey sampling and design of experiments.* Unpublished Ph.D. thesis, Iowa State University Press, Ames, Iowa.

[561] Midzuno, H. (1952). On the sampling system with probability proportional to sum of sizes. *Ann. Inst. Statist. Math.*, 3, 99--107.

[562] Miller, R.G. (1974). An unbalanced jackknife. *Ann. Statist.*, 2, 880--891.

[563] Milne, A. (1959). The centric systematic area sample treated as a random sample. *Biometrics*, 15, 270--297.

[564] Mishra, G. and Rout, K. (1997). A regression estimator in two-phase sampling in presence of two auxiliary variables. *Metron*, 55, 177--186.

[565] Mishra, R.N. and Sinha, J.N. (1999). Randomized response procedure with multiple statement. *Aligarh J. Statist.*, 19, 1--9.

[566] Mohanty, S. (1977). Sampling with repeated units. *Sankhy $\bar{a}$*, C, 39, 43--46.

[567] Mohanty, S. and Sahoo, L.N. (1987). A class of estimators based on mean per unit ratio estimators. *Statistica,* 47, 473--477.

[568] Mohanty, S. and Sahoo, J. (1995). A note on improving the ratio method of estimation through linear transformation using certain known population parameters. *Sankhy $\bar{a}$*, B, 57, 93--102.

[569] Mohanty, S. and Pattanaik, L.M. (1984). Alternative multivariate ratio estimators using geometric and harmonic means. *J. Indian Soc. Argic. Statist.*, 36, 100--118.

[570] Montanari, G.E. (1998). On regression estimation of finite population means. *Survey Methodology*, 24(1), 69--77.

[571] Montanari, G.E. (1999). A study on the conditional properties of finite population mean estimators. *Metron*, *57*, 21--35.

[572] Moors, J.J.A. (1971). Optimization of the unrelated question randomized response model. *J. Amer. Statist. Assoc.*, 66, 627--629.

[573] Moors, J.J.A. (1997). *A critical evaluation of Mangat's two-step procedure in randomized response*. Discussion paper at Center for Economic Research, Tilburg University, The Netherlands.

[574] Moors, J.J.A., Smeets, R. and Boekema, F.W.M. (1998). Sampling with probabilities proportional to the variable of interest. *Statistica Neerlandica*, 52, 129--140.

[575] Morrison, T., Mangat, N.S., Desjardins, G. and Bhatia, A. (2000). Validation of an in-line inspection metal loss tool. *Proceedings of the International Pipeline Conference 2000, Calgary, Alberta, Canada. The American Society of Mechanical Engineers, New York, Vol 2, 839--844.*

[576] Morrison, T., Mangat, N.S, Carroll, L.B. and Riznic, J. (2002). Statistical estimation of flaw size measurement errors for steam generator inspection tools. *Proceedings of the 4th international Steam Generator Conference, Canadian Nuclear Society, Toronto, Ontario, May 5--8.*

[577] Morrison, T. Mangat, N.S., Carroll, L.B. and Riznic, J. (2003). Statistical estimation of flaw size measurement errors for steam generator tube inspection tools. *Submitted to Nuclear Engineering and Design.*

[578] Moses, L.E. (1978). Energy information validation -- A status report. *Proc. of the 1978 DOE Statistical Symposium*, 33--49.

[579] Moura, F.A.S. and Holt, D. (1999). Small area estimation using multilevel models. *Survey Methodology*, 25(1), 73--80.

[580] Mukerjee, R. and Chaudhuri, A. (1990). Asymptotic optimality of double sampling plans employing generalized regression estimators. *J. Statist. Planning Infer.*, 26, 173--183.

[581] Mukerjee, R. and Sengupta, S. (1990). Optimal estimation of a finite population mean in the presence of linear trend. *Biometrika*, 77., 625--630.

[582] Mukerjee, R., Rao, T.J. and Vijayan, K. (1987). Regression type estimator using multiple auxiliary information. *Austral. J. Statist.*, 29(3), 244--254.

[583] Mukerjee, R., Rao, T.J. and Vijayan, K. (2000). Rejoinder to Ahmed, M.S. (1998): A note on regression type estimators using multiple auxiliary information. *Austral. & New Zealand J. Statist.*, 42(2), 245.

[584] Mukhopadhyay, P. (1977). *Further studies in sampling theory.* Unpublished Ph.D. thesis submitted to the University of Calcutta.

[585] Mukhopadhyay, P. (1982). Optimum strategies for estimating the variance of a finite population under a superpopulation model. *Metrika*, 29, 143--158.

[586] Mukhopadhyay, P. (1994). Prediction in finite population under error in variables superpopulation models. *J. Statist. Planning Infer.*, 41, 151--161.

[587] Mukhopadhyay, P. and Bhattacharyya, S. (1990-91). Estimating a finite population variance under some general linear models with exchangeable errors. *Calcutta Statist. Assoc. Bull.*, 40, 219--228.

[588] Murthy, M.N. (1957). Ordered and unordered estimators in sampling without replacement. *Sankhyā*, 18, 379--390.

[589] Murthy, M.N. (1961). *Introduction to sampling theory: Lecture Notes.* Indian Statistical Institutes.

[590] Murthy, M.N. (1962). Almost unbiased estimators based on interpenetrating sub-samples. *Sankhyā*, 303--314.

[591] Murthy, M.N. (1963). Generalized unbiased estimation for finite populations. *Sankhyā*, B, 25, 245--261.

[592] Murthy, M.N. (1964). Product method of estimation. *Sankhyā*, A, 26, 69--74.

[593] Murthy, M.N. (1967). *Sampling theory and methods.* Statistical Publishing Society, Calcutta.

[594] Murthy, M.N. (1977). *Sampling theory and methods.* Second edition, Statistical Publication Society, Calcutta.

[595] Murthy, M.N. and Nanjamma, N.S. (1959). Almost unbiased ratio estimates based on interpenetrating sub-sample estimates. *Sankhyā*, 21, 381--392.

[596] Murthy, M.N. and Singh, M.P. (1969). On the concepts of best and admissible estimators in sampling theory. *Sankhya*, A, 31, 343--354.

[597] Nanjamma, N.S., Murthy, M.N. and Sethi, V.K. (1959). Some sampling systems providing unbiased ratio estimators. *Sankhyā*, 21, 299--314.

[598] Narain, R.D. (1951). On sampling without replacement with varying probabilities. *J. Indian Soc. Agril. Statist.*, 3, 169--174.

[599] Nayak, T.K. (1994). On randomized response surveys for estimating a proportion. *Commun. Statist.--Theory Meth.*, 23(1), 3303--3321.

[600] Nelson, D. and Meeden, G. (1998). Using prior information about population quantiles in finite population sampling. *Sankhy $\bar{a}$* , A, 60, 426--445.

[601] Neyman, J. (1923). On the application of probability theory to agricultural experiments: essay on principles. *Statistical Science*, 5, 465--480.

[602] Neyman, J. (1934). On two different aspects of the representative methods, the method of stratified sampling and the method of purposive selection. *J. R. Statist. Soc.* 97, 558--606.

[603] Neyman, J. (1938). Contribution to the theory of sampling human populations. *J. Amer. Statist. Assoc.*, 33, 101--116.

[604] Neyman, J. (1971). *Discussion of Royall (1971): Foundations of Statistical Inference* (V.P. Godambe and D.A. Sprott, eds). Holt, Rinehart & Winston, Toronto, 276--278.

[605] Nieto de Pascual, J. (1961). Unbiased ratio estimates in stratified sampling. *J. Amer. Statist. Assoc.*, 56, 70--87.

[606] Ogus, J.K. and Clark, D.F. (1971). *The annual survey of manufacturers*: A report on methodology. Technical Report No. 24, U.S. Bureau of Census, Washington, D.C.

[607] Ohlsson, E. (1989). Variance estimation in Rao--Hartley--Cochran procedure. *Sankhy $\bar{a}$* ,B, 51, 348--361.

[608] Okafor, F.C. (1992). The theory and application of sampling over two occasions for the estimation of current population ratio. *Statistica*, 1, 137--147.

[609] Okafor, F. C. (2001). Treatment of non-response in successive sampling. *Statistica*, 1, 195--204.

[610] Okafor, F.C. and Arnab, R. (1987). Some strategies of two-stage sampling for estimating population ratios over two occasions. *Austrial. J. Statist.*, 29(2), 128--142.

[611] Okafor, F.C. and Lee, H. (2000). Double sampling for ratio and regression estimation with sub-sampling the non-respondents. *Survey Methodology*, 26, 183--188.

[612] Olkin, I. (1958). Multivariate ratio estimation for finite population. *Biometrika*, 43, 154--163.

[613] Padmawar, V.R. (1994). Strategies admitting non-negative unbiased variance estimators. *J. Statist. Planning Infer.*, 40, 81--95.

[614] Padmawar, V.R. (1996). Rao--Hartley--Cochran strategy in survey sampling of continuous populations. *Sankhy $\bar{a}$* , B, 58, 90--104.

[615] Padmawar, V.R. (1998a). On estimating non-negative definite quadratic forms. *Metrika*, 49, 231--244.

[616] Padmawar, V.R. (1998b). On π PS designs and stratification. *J. Indian Statist. Assoc.*, 36, 99--104.

[617] Paik, M.C. (1997). The generalized estimating equation approach when data are not missing completely at random. *J. Amer. Statist. Assoc.*, 92, 1320--1329.

[618] Panda, P. and Sahoo, L.N. (1999). Predictive estimation of finite population mean using a product estimator for two-stage sampling. *Biom. J.*, 41(1), 93--97.

[619] Pandey, B.N. and Dubey, V. (1989). On almost unbiased estimators. *Metron,* 333--338.

[620] Pandey, S.K. and Singh, R.K. (1984). On combination of ratio and PPS estimators. *Biom. J.*, 26 (3), 333--336.

[621] Patel, H.C. and Dharmadhikari, S.W. (1978). Admissibility of Murthy and Midzuno's estimators within the class of linear unbiased estimators of finite population totals. *Sankhy ā* , C, 40, 21--28.

[622] Pathak, P.K. (1961). On the evaluation of moments of distinct units in a sample. *Sankhy ā* , A, 23, 415--420.

[623] Pathak, P.K. (1962). On simple random sampling with replacement. *Sankhy ā* , A, 24, 287--302.

[624] Pathak, P.K. (1966). An estimator in PPS sampling for multiple characteristics. *Sankhy ā* , A, 28, 35--40.

[625] Pathak, P.K. (1967a). Asymptotic efficiency of Des Raj's strategy--I. *Sankhy ā* , A, 29, 283--298.

[626] Pathak, P.K. (1967b). Asymptotic efficiency of Des Raj's strategy--II. *Sankhy ā* , A, 29, 299--304.

[627] Pathak, P.K. and Rao, T.J. (1967). Inadmissibility of customary estimators in sampling over two occasions. *Sankhy ā* , A, 29, 49--54.

[628] Patterson, H.D. (1950). Sampling on successive occassions with partial replacement units. *J. R. Statist. Soc.*, 241--255.

[629] Pearson, K. (1896). Mathematical contribution to the theory of evolution—III--Regression: heredity and panmixia. *Phil Trans (A), Royall Soc. London*, 187, 253--318.

[630] Pedgaonkar, A.M. and Prabhu--Ajaonkar, S.G. (1978). Comparison of sampling strategies. *Metrika,* 25, 149--153.

[631] Pfeffermann, D. (1984). A note on large sample properties of balanced samples. *J. R. Statist. Soc.*, B, 46, 38--41.

[632] Pitman, E.J.G. (1937). Significance tests which can be applied to samples from any population--III: The analysis of variance test. *Biometrika*, 29, 322--335.

[633] Platek, R. and Grey, G.B. (1983). *Imputation methodology: total survey error.* In incomplete data in sample surveys, 2, Ed. W.G. Madow, I. Olkin, and D.B. Rubin, 249-333. New York: Academic Press.

[634] Pokropp, F. (2001). Imposed linear structures in conventional sampling theory. *Allgemeines Statist. Archiv*, 86, 333--352.

[635] Politz, A. and Simmons, W. (1950). Note on an attempt to get not at home into the sample without callbacks. *J. Amer. Statist. Assoc.*, 45, 136--137.

[636] Pollock, K.H. and Bek, Y. (1976). A comparison of three randomized response models for quantitative data. *J. Amer. Statist. Assoc.*, 71, 884--886.

[637] Prabhu--Ajgaonkar, S.G. (1975). The efficient use of supplementary information in double sampling procedures. *Sankhy ā* , C, 37, 181--189.

[638] Pradhan, B.K. (2001). Modified chain regression estimators using multi auxiliary information. *Statistica*, no.1, 249--258.

[639] Prasad, B. (1989). Some improved ratio type estimators of population mean and ratio in finite population sample surveys. *Commun. Statist.--Theory Meth.*, 18(1), 379--392.

[640] Prasad, B. and Singh, H.P. (1990). Some improved ratio type estimators of finite population variance in sample surveys. *Commun. Statist.-- Theory Meth.,* 19, 1127--1139.

[641] Prasad, B. and Singh, H.P. (1992). Unbiased estimators of finite population variance using auxiliary information in sample surveys. *Commun. Statist.-- Theory Meth.*, 21(5), 1367--1376.

[642] Prasad, B., Singh, R.S. and Singh, H.P. (1996). Some chain ratio type estimators for ratio of two population means using two auxiliary characters in two phase sampling. *Metron,* 95--113.

[643] Prasad, N.G.N and Graham, J.B. (1994). PPS sampling over two occasions. *Survey Methodology*, 20, 59--64.

[644] Prasad, N.G.N. and Rao, J.N.K. (1999). On robust small area estimation using a simple random effects model. *Survey Methodology*, 25, 67--72.

[645] Prasad, N.G.N. and Srivenkataramana, T.A. (1980). A modification to Horvitz--Thompson estimator under Midzuno sampling scheme. *Biometrika*, 67, 709--711.

[646] Purcell, N.J. and Kish, L. (1979). Estimation for small domain. *Biometrics*, 35, 365--384.

[647] Purcell, N.J. and Kish, L. (1980). Postcensal estimates for local areas (or domains). *Int. Statist. Rev.*, 48, 3--18.

[648] Quenouille, M.H. (1949). Problems in plane sampling. *Ann. Math. Statist.*, 20, 355--375.

[649] Quenouille, M.H. (1956). Notes on bias in estimation. *Biometrika*, 43, 353--360.

[650] Raghavarao, D. (1971). *Constructions and combinatorial problems in design of experiments.* Wiley, New York.

[651] Raghunandanan, K. and Bryant, E.C. (1971). Variance in multi-way stratification *Sankhyā* , A, 33, 221--226.

[652] Raiffa, H. and Schlaifer, R. (1961). *Applied statistical decision theory.* Boston.

[653] Raj, D. (1954a). On sampling probabilities proportional to size. *Ganita*, 52, 175--182.

[654] Raj, D. (1954b). Ratio estimator in sampling with equal and unequal probabilities. *J. Indian Soc. Agril. Statist.*, 6, 127--138.

[655] Raj, D. (1956). Some estimators in sampling with varying probabilities without replacement. *J. Amer. Statist. Assoc.*, 51, 269--284.

[656] Raj, D. (1958). On the relative accuracy of some sampling techniques. *J. Amer. Statist.Assoc.*,53, 98--101.

[657] Raj, D. (1964). On double sampling for pps estimation. *Ann. Math. Statist.*, 35, 900--902.

[658] Raj, D. (1965a). On a method of using multi-auxiliary in sample surveys. *J. Amer. Statist.Assoc.*, 60, 270--277.

[659] Raj, D. (1965b). On sampling over two occasions with probability proportionate to size. *Ann. Math. Statist.*, 36, 327--330.

[660] Raj, D. (1966).Some remarks on a simple procedure of sampling without replacement. *J. Amer. Statist. Assoc.*, 61,391--397.

[661] Raj, D. (1968). *Sampling Theory*. McGraw-Hill, New York.

[662] Raj, D. and Khamis, S.H. (1958). Some remarks on sampling with replacement. *Ann. Math. Statist.*, 29, 550--557.

[663] Ramachandran, G. (1982). Horvitz and Thompson estimator and generalized πPS designs. *J. Statist. Planning Infer.*, 7, 151--153.

[664] Ramachandran, G. and Rao, T.J. (1974). Allocation to strata and relative efficiencies of stratified and unstratified *πps* sampling schemes. *J. R. Statist. Soc.*, 97, 558--606.

[665] Ramachandran, V. and Pillai, S.S. (1976). Multivariate unbiased ratio type estimation for finite sampling. *J. Indian Soc. Agril. Statist.*, 28, 71--80.

[666] Ramakrishnan, M.K. (1969). Some remarks on the comparison of sampling with and without replacement. *Sankhyā*, A, 31, 333--342.

[667] Ramakrishnan, M.K. (1975a). A generalisation of the Yates--Grundy variance estimator. *Sankhyā*, C, 37, 204--206.

[668] Ramakrishnan, M.K. (1975b). Choice of an optimum sampling strategy -- I. *Ann. Statist.*, 3, 669--679.

[669] Ramakrishnan, M.K. and Rao, V.V. B. (1975). On the sample mean in simple random sampling without replacement. *Sankhyā*, C, 37, 207--210.

[670] Rana, R.S. (1989). Concise estimator of bias and variance of the finite population correlation coefficient. *J. Indian Soc. Agril. Statist.*, 41, 69--76.

[671] Rana, R.S. and Singh, R. (1989). Note on systematic sampling with supplementary observations. *Sankhyā*, B, 51, 205--211.

[672] Rangarajan, R. (1957). A note on two stage sampling. *Sankhyā*, 17, 373--376.

[673] Rao, C.R. (1945). Information and accuracy obtainable in an estimation of a statistical parameter. *Bull. Calcutta Math. Soc.*, 37, 81.

[674] Rao, C.R. (1973). *Linear Statistical Inference and its Applications.* Wiley, New York.

[675] Rao, C.R. (1975). Some problems of sample surveys. *Proceedings of the conference on directions for mathematical statistics, University of Alberta, Edmonton, Canada.*

[676] Rao, C.R. (1987). Strategies of data analyst. *Proceedings of 46th Session of the International Statistical Institute, Tokyo.*

[677] Rao, J.N.K. (1961). On sampling with varying probabilities in sub-sampling designs. *J. Indian Soc. Agric. Statist.*, 13, 211--217.

[678] Rao, J.N.K. (1963a). On two systems of unequal probability sampling without replacement. *Ann. Inst. Statist. Math.*,15, 67--72.

[679] Rao, J.N.K. (1963b). On three procedures of unequal probability sampling without replacement. *J. Amer. Statist. Assoc.*, 58, 202--215.

[680] Rao, J.N.K. (1965a). A note on estimation of ratios by Quenouille's method. *Biometrika*, 52, 647--649.

[681] Rao, J.N.K. (1965b). On two simple schemes of unequal probability sampling without replacement. *J. Indian Soc. Agric. Statist.*, 3, 169--174.

[682] Rao, J.N.K. (1966a). Alternative estimators in PPSWR sampling for multiple characteristics. *Sankhyā*, A, 28, 47--60.

[683] Rao, J.N.K. (1966b). On the relative efficiency of some estimators in PPS sampling for multiple characteristics. *Sankhyā*, A, 28, 61--70.

[684] Rao, J.N.K. (1966c). On the comparison of sampling with and without replacement. *Rev. Int. Statist. Inst.*, 34, 125--138.

[685] Rao, J.N.K. (1967). The precision of Mickey's unbiased ratio estimator. *Biometrika,* 54, 93--108.

[686] Rao, J.N.K. (1968a). Some small sample results in ratio and regression estimation. *J. Indian Statist. Assoc.*, 6, 160--168.

[687] Rao, J.N.K. (1968b). Some non-response sampling theory when the frame contains an unknown amount of duplication. *J. Amer. Statist. Assoc.*, 63, 87--90.

[688] Rao, J.N.K. (1969). *Ratio and regression estimators in new developments in survey sampling*. eds. N. L. Johnson and H. Smith, New York, John Wiley.

[689] Rao, J.N.K. (1975). Unbiased variance estimation for multi-stage designs. *Sankhyā*, C, 37, 133--139.

[690] Rao, J.N.K. (1979). On deriving the mean square errors and their non-negative unbiased estimators in finite population sampling. *J. Indian Soc. Agric. Statist.*, 17, 125--136.

[691] Rao, J.N.K. (1985). Conditional inference in survey sampling. *Survey Methodology,* 11, 15--31.

[692] Rao, J.N.K. (1989). A note on Narain's necessary condition in sampling. *J. Indian Soc. Agric. Statist.* , 41, 316--317.

[693] Rao, J.N.K. (1994). Estimating totals and distribution functions using auxiliary information at the estimation stage. *J. Official Statist.*, 10(2), 153--165.

[694] Rao, J.N.K. (1996a). Some current topics in sample survey theory. *J. Indian Soc. Agril.Statist.*, 50, 244--263.

[695] Rao, J.N.K. (1996b). On variance estimation with imputed survey data. *J. Amer. Statist. Assoc.*, 91, 499--506.

[696] Rao, J.N.K. (1997). Developments in sample survey theory: an appraisal. *Canad. J. Statist.*, 25, 1--21.

[697] Rao, J.N.K. (1999a). Some current trends in sample survey theory and methods. *Sankhyā*, B, 61, 1--57.

[698] Rao, J.N.K. (1999b). Reply to comments on 'some current trends in sample survey theory and methods'. *Sankhyā*, B, 61, 53--57.

[699] Rao, J.N.K. (1999c). Some recent advances in model based small area estimation. *Survey Methodology*, 25, 175--186.

[700] Rao, J.N.K. (2000). Conditional inference for large and small areas. *J. Indian Statist. Assoc.*, 38 (2), 383--398.

[701] Rao, J.N.K. (2002). Discussion of 'Exact linear unbiased estimation in survey sampling'. *J.Statist. Planning Infer.*, 102, 39--40.

[702] Rao, J.N.K. (2003). *Small area estimation.* John Wiley and Sons, NY.

[703] Rao, J.N.K. and Bayless, D.L. (1969). An empirical study of the stabilities of estimators and variance estimators in unequal probability sampling of two units per stratum. *J. Amer. Statist. Assoc.*, 64, 540--559.

[704] Rao, J.N.K. and Beegle, L.D. (1967). A Monte Carlo study of some ratio estimators. *Sankhyā*, B, 29, 47--56.

[705] Rao, J.N.K. and Bellhouse, D.R. (1978). Optimal estimation of a finite population mean under generalized random permutation models. *J. Statist. Planning Infer.*, 2, 125--141.

[706] Rao, J.N.K. and Graham, J.B. (1964). Rotation designs for sampling on repeated occasions. *J. Amer. Statist. Assoc.*, 59, 492--509.

[707] Rao, J.N.K., Hartley,H.O. and Cochran, W.G. (1962). A simple procedure of unequal probability sampling without replacement. *J. R. Statist. Soc.*, B, 24, 482--491.

[708] Rao, J.N.K. and Lanke, J. (1984). Simplified unbiased variance estimation for multistage designs. *Biometrika,* 71, 387--395.

[709] Rao, J.N.K and Rao, P.S.R.S. (1971). Small sample results for ratio estimators. *Biometrika*, 58, 625--630.

[710] Rao, J.N.K. and Shao, J. (1992). Jackknife variance estimation with survey data under hot deck imputation. *Biometrika*, 79, 811--822.

[711] Rao, J.N.K. and Shao, J. (1999). Modified balanced repeated replication for complex survey data. *Biometrika*, 86, 403--415.

[712] Rao, J.N.K. and Singh, M.P. (1973). On the choice of estimator in survey sampling. *Austral. J. Statist.*, 15(2), 95--104.

[713] Rao, J.N.K. and Sitter, R.R. (1995). Variance estimation under two-phase sampling with application to imputation for missing data. *Biometrika*, 82, 453--460.

[714] Rao, J.N.K. and Vijayan, K.(1977). On estimating the variance in sampling with probability proportional to aggregate size. *J. Amer. Statist. Assoc*, 72, 579--584.

[715] Rao, J.N.K.and Webster, J.T.(1966). On two methods of bias reduction in the estimation of ratios. *Biometrika*, 53,571--577.

[716] Rao, J.N.K. and Wu, C.F.J. (1985). Inference from stratified samples: second order analysis of three methods for non-linear statistics. *J. Amer. Statist. Assoc.*, 80, 620--630.

[717] Rao, J.N.K. and Yu, M. (1994). Small area estimation by combining time series and cross sectional data. *Canad. J. Statist.*, 22, 511--528.

[718] Rao, P.S.R.S. (1969). Comparison of four ratio type estimates under a model. *J. Amer. Statist. Assoc.*, 64, 574--580.

[719] Rao, P.S.R.S. (1972). On two phase regression estimator. *Sankhy $\bar{a}$* , A, 34, 373--476.

[720] Rao, P.S.R.S. (1974). Jackknifing the ratio estimator. *Sankhy $\bar{a}$* , 36, 84--97.

[721] Rao, P.S.R.S. (1975). Hartley--Ross type estimators with two-phase sampling. *Sankhy $\bar{a}$* , 37, 140--146.

[722] Rao, P.S.R.S. (1979). On applying the jackknife procedure to the ratio estimator. *Sankhy $\bar{a}$* , 41, 115--126.

[723] Rao, P.S.R.S. and Mudholkar, G.S. (1967). Generalized multivariate estimators for the mean of finite population parameters. *J. Indian Soc. Agric. Statist.*, 62, 1008--1012.

[724] Rao, T.J. (1966). On certain unbiased ratio estimators. *Ann. Inst. Statist. Math.*,18, 117--121.

[725] Rao, T.J. (1967). *Contributions to the theory of sampling strategies*. Ph.D. Thesis, I.S.I. Calcutta.

[726] Rao, T.J. (1968). On the allocation of sample size in stratified sampling. *Ann. Inst. Statist. Math.* , 20, 159--166.

[727] Rao, T.J. (1971). πps sampling designs and Horvitz and Thompson estimator. *J. Amer. Statist. Assoc.*, 66, 872--875.

[728] Rao, T.J. (1972). Horvitz and Thompson and Des Raj estimator revisited. *Austral. J. Statist.*,14, 227--230.

[729] Rao, T.J. (1977a). Estimating the variance of the ratio estimator for the Midzuno--Sen sampling scheme. *Metrika*, 24,203--208.

[730] Rao, T.J. (1977b). Optimum allocation of sample size and prior distributions: a review. *Int. Statist. Rev.*, 45, 173--179.

[731] Rao, T.J. (1981). On a class of almost unbiased ratio estimators. *Ann. Inst. Statist. Math.*, A, 33, 225--231.

[732] Rao, T.J. (1993a). On certain problems of sampling design and estimation for multiple characteristics. *Sankhy $\bar{a}$* , B, 55, 372--381.

[733] Rao, T.J. (1993b). On certain alternative estimators for multiple characteristics in varying probability sampling. *J. Indian Soc. Agril. Statist.* 45(3), 307--318.

[734] Rao, T.J. (1983c). Horvitz--Thompson strategy vs. stratified random sampling strategy. *J. Statist. Planning Infer.*, 8, 43--50.

[735] Rao, T.J. (1984). Allocation of sample size to strata and related problems. *Biom. J.*, 26, 517--526.

[736] Rao, T.J., Sengupta, S. and Sinha, B.K. (1991). Some order relations between selection and inclusion probabilities for PPSWOR sampling scheme. *Metrika*, 38, 335--343.

[737] Ray, S. and Das, M.N. (1995). On systematic sampling allowing estimation of variance of mean. *J. Indian Soc. Agril. Statist.,* 47(2), 192--196.

[738] Ray, S. and Das, M.N. (1997). Circular systematic sampling with drawback. *J. Indian Soc. Agric. Statist.*, 50(1), 70--74.

[739] Ray, S.K. and Sahai, A. (1979). A note on ratio and product type estimators. *Ann. Inst. Statist. Math,* 31, 141--144.

[740] Ray, S.K. and Singh, K. (1981). Difference cum ratio type estimators. *J. Indian Statist. Assoc.,* 19, 147--151.

[741] Reddy, V.N. (1973). On ratio and product method of estimation. *Sankhy $\bar{a}$* ,B, 35, 307--316.

[742] Reddy, V.N. (1974). On a transformed ratio method of estimation. *Sankhy $\bar{a}$* , C, 36, 59--70.

[743] Reddy, V.N. (1978a). A study on the use of prior knowledge on certain population parameters in estimation. *Sankhy $\bar{a}$* , C, 40, 29--37.

[744] Reddy, V.N. (1978b). A comparison between stratified and unstratified random sampling. *Sankhy $\bar{a}$* , C, 40, 99--103.

[745] Reddy, V.N. (1980). Systematic sampling in monotone populations. *Sankhy $\bar{a}$* ,C, 42, 97--108.

[746] Reddy, V.N. and Rao, T.J. (1977). Modified PPS method of estimation. *Sankhy $\bar{a}$* , C, 39, 185--197.

[747] Reddy, V.N. and Rao, T.J. (1990). On estimation of the population total of bottom (top) P percentiles of a finite population. *Metron,* 309--320.

[748] Ren, R. (2000). Estimation de la fonction de repartition et des fractiles d'une population finite. *VIIemes Journees de Methodologies Statistique.* Paris.

[749] Renssen, R.H. and Nieuwenbroek, N.J. (1997). Aligning estimates for common variables in two or more sample surveys. *J. Amer. Statist. Assoc.*, 92, 368--374.

[750] Richardson, S.C. (1989). One pass selection of a sample with probability proportional to size. *Appl. Statist.*, 38, 517--520.

[751] Rizvi, S.E.H., Gupta, J.P. and Singh, R. (2000). Approximately optimum stratification for two study variables using auxiliary information. *J. Indian Soc. Agric. Statist.*, 53(3), 287--298.

[752] Robins, J.M. and Wang, N. (2000). Inference for imputation estimators. *Biometrika*, 87, 113--124.

[753] Robson, D.S. (1957). Applications of multivariate polykays of the theory of unbiased ratio type estimation. *J. Amer. Statist. Assoc.*, 52, 511--522.

[754] Rosen, B. (1997a). On sampling with probability proportional to size. *J. Statist. Planning Infer.,* 62, 159--191.

[755] Rosen, B. (1997b). Asymptotic theory for order sampling. *J. Statist. Planning Infer.,* 62, 135--158.

[756] Rosen, B. (1998). *On inclusion probabilities for order sampling.* R and D Report, Research Methods, Development , 2, 1--23.

[757] Roy, J. and Chakravorty, I.M. (1960). Estimating the mean of a finite population. *Ann. Math. Statist.*, 31, 392--398.

[758] Royall, R.M. (1970a). On finite population sampling theory under certain linear regression models. *Biometrika*,57, 377--387.

[759] Royall, R.M. (1970b). Finite population sampling: on labels in estimation. *Ann. Math. Statist.*, 41, 1774--1779.

[760] Royall, R M. (1970c). On finite population sampling theory under certain linear regression models. *Biometrika*, 57, 377--387.

[761] Royall, R.M. (1971). Linear regression models in finite population sampling theory. *Foundations of Statist. Infer.* (V.P. Godambe and D.A. Sprott, eds). Holt, Rinehart & Winston, Toronto, 259--274.

[762] Royall, R.M. (1976). The linear least squares prediction approach to two-stage sampling. *J. Amer. Statist. Assoc.*,71, 657--664.

[763] Royall, R.M. (1986). The prediction approach to robust variance estimation in two-stage cluster sampling. *J. Amer. Statist. Assoc.*, 81, 119--123.

[764] Royall, R.M. (1992). Robustness and optimal design under prediction models in finite population sampling. *Survey Methodology*, 2, 179--195.

[765] Royall, R.M. and Cumberland, W.G. (1978). Variance estimation in finite population sampling. *J. Amer. Statist. Assoc.*, 73, 351--358.

[766] Royall, R.M. and Cumberland, W.G. (1981a). An empirical study of the ratio estimator and estimators of its variance. *J. Amer. Statist. Assoc.*, 73, 351--358.

[767] Royall, R.M. and Cumberland, W.G. (1981b). The finite population linear regression estimator and estimators of its variance -- An empirical study. *J. Amer. Statist. Assoc.*, 76, 924--930.

[768] Royall, R.M. and Cumberland, W.G. (1985). Conditional coverage properties of finite population confidence intervals. *J. Amer. Statist. Assoc.*, 80, 355--359.

[769] Royall, R.M. and Eberhardt, K.R. (1975). Variance estimates for the ratio estimator. *Sankhy* $\overline{a}$, C, 37, 43--52.

[770] Royall, R.M. and Herson, J. (1973a). Robust estimation in finite populations -- I. *J. Amer. Statist. Assoc.*, 68, 880--889.

[771] Royall, R.M. and Herson, J. (1973b). Robust estimation in finite populations -- II: Stratification on a size variable. *J. Amer. Statist. Assoc.*, 68, 890--893.

[772] Royall, R.M. and Pfeffermann, D. (1982). Balanced samples and robust Bayesian inference in finite population sampling. *Biometrika*, 69, 401-409.

[773] Rubin, D.B. (1976). Inference and missing data. *Biometrika*, 63, 581--592.

[774] Rubin, D.B. (1978). Multiple imputation in sample surveys -- a phenomenological Bayesian approach to nonresponse. *In Proc. Sect. Survey res. Meth.*, pp. 20-34. Washington D.C.:American Statistical Association.

[775] Rubin, R.B.(1987). *Multiple imputation for non-response in surveys.* John Wiley, NewYork.

[776] Rubin, R.B. (1996). Multiple imputation after 18+ years. *J. Amer. Statist. Assoc.*, 91, 473--490.

[777] Rubin, D.B. and Schenker, N. (1986). Multiple imputation for interval estimation from simple random samples with ignorable non-response. *J. Amer. Statist. Assoc.*, *81*, 366--374.

[778] Rueda, M. and Arcos, A. (2002). The use of quantiles of auxiliary variable to estimate medians. *Biom. J.*, 44(5), 619--632.

[779] Rueda, M., Arcos, A. and Artes, E. (1998). Quantile interval estimation in finite population using a multivariate ratio estimator. *Metrika*, 47, 203--213.

[780] Ruiz, M. and Santos, J. (1990). Sampling design providing unbiased new product estimator. *Statistica*, 50, 285--288.

[781] Ruiz, M. and Santos, J. (1992). Variance estimation with systematic sampling. *Rev. Acad. Cienc Zaragoza,* 47, 121--124.

[782] Sadasivan, G. and Aggarwal, R. (1978). Optimum points of stratification in bivariate populations. *Sankhyā*, C, 40, 84--97.

[783] Sadasivan, G. and Srinath, M. (1975). Some contributions to post-cluster sampling. *Sankhyā*, C, 37, 171--180.

[784] Sahai, A. and Ray, S.K. (1980). An efficient estimator using auxiliary information. *Metrika,* 27, 271--275.

[785] Sahai, A. and Sahai, A. (1985). On efficient use of auxiliary information. *J. Statist. Planning Infer.*, 12, 203--212.

[786] Sahoo, J. and Sahoo, L.N. (1999a). A comparative study of some regression type estimators in double sampling procedures. *Aligarh J. Statist.*, 19, 67--76.

[787] Sahoo, J. and Sahoo, L.N. (1999b). An alternative class of estimators in double sampling procedures. *Calcutta Statist. Assoc. Bull.*, 49, 79--83.

[788] Sahoo, J. Sahoo, L.N. and Mohanty, S. (1994). Unequal probability sampling using a transformed auxiliary variable. *Metron,* 71--83.

[789] Sahoo, J., Sahoo, L.N. and Wywial, J. (1997). Some thoughts of reduction of estimation bias using auxiliary information in sample. *Statistics in Transition*, 3(2), 383--401.

[790] Sahoo, L.N (1983). On a method of bias reduction in ratio estimation. *J. Statist. Res.* 17,1--6.

[791] Sahoo, L.N. (1986). On a ratio method of estimation using a transformed auxiliary variable. *Statistica*, 46, 409--413.

[792] Sahoo, L.N. (1987). A regression type estimator in two-stage sampling. *Calcutta Statist. Assoc. Bull.*, 36, 97--100.

[793] Sahoo, L.N. (1991). An unbiased ratio cum product estimator in two-stage sampling. *Metron,* 213--217.

[794] Sahoo, L.N. (1994). *Some estimation problems in finite population sampling using auxiliary information.* Ph. D. Thesis, Utkal University, Bhubaneswar, India.

[795] Sahoo, L.N. and Panda, P. (1997). A class of estimators in two-stage sampling with varying probabilities. *South African Statist.* J., 31, 151--160.

[796] Sahoo, L.N. and Panda, P. (1999a). A class of estimators using auxiliary information in two-stage sampling. *Austral. & New Zealand J. Statist.*, 41(4), 405--410.

[797] Sahoo, L.N. and Panda, P. (1999b). A predictive regression type estimator in two-stage sampling. *J. Indian Soc. Agric. Statist.*, 52(3), 303--308.

[798] Sahoo, L.N., Sahoo, J. and Espejo, M.R. (1998). On some strategies using auxiliary information for estimating finite population mean. *Questiio*, 22, 243--252.

[899] Sahoo, L.N., Sahoo, J. and Mohanty, S. (1995a). Empirical comparison of some regression and regression type strategies. *Statistical Hefte*, 36, 337--347.

[800] Sahoo, L.N., Sahoo, J. and Mohanty, S. (1995b). A new predictive ratio estimator. *J. Indian Soc. Agric. Statist.* 47(3), 240--242.

[801] Sahoo, L.N. and Sahoo, R.K. (2001). Predictive estimation of finite population mean in two-phase sampling using two auxiliary variables. *J. Indian Soc. Agric. Statist.*, 54(2), 258--264.

[802] Sahoo, L.N. and Swain, A.K.P.C. (1980). Unbiased ratio cum product estimator. *Sankhy* $\overline{a}$, C, 42, 56--62.

[803] Sahoo, L.N. and Swain, A.K.P.C. (1986). Chain product estimators. *Aligarh J. Statist.*, 6, 53--58.

[804] Sahoo, L.N. and Swain, A.K.P.C. (1987). Some modified ratio estimators. *Metron*, 285--293.

[805] Sahoo, L.N. and Swain, A.K.P.C. (1989). On two modified ratio estimators in two-phase sampling. *Metron*, 261--266.

[806] Samiuddin, M. and Kattan, A.K.A. (1991). A procedure of unequal probability sampling. *Pak. J. Statist.*, A, 7, 1--7.

[807] Samiuddin, M., Kattan, A.K.A., Hanif, M. and Asad, H. (199 2). Some remarks on models, sampling schemes and estimators in unequal probability sampling. *Pak. J. Statist.*, A, 8, 1--18.

[808] Sampath, S. (1989). On the optimal choice of unknowns in ratio type estimators. *J. Indian Soc. Agric. Statist,* 41, 166--172.

[809] Sampath, S. and Chandra, S.K. (1990). General class of estimators for the population total under unequal probability sampling schemes. *Metron,* 409--419.

[810] Sampath, S., Uthayakumaran, N. and Tracy, D.S. (1995). On an alternative estimator for randomized response technique. *J. Indian Soc. Agric. Statist.* 47(3), 243--248.

[811] Sampford, M.R. (1962). Methods of cluster sampling with and without replacement for clusters of unequal sizes. *Biometrika*, 49, 27--40.

[812] Sampford, M.R. (1967). On sampling without replacement with unequal probabilities of selection. *Biometrika,* 54, 499--513.

[813] Särndal, C.E. (1980a). On π --inverse weighting versus best linear unbiased weighting in probability sampling. *Biometrika,* 67, 639--650.

[814] Särndall, C.E. (1980b). Two model based inference arguments in survey sampling. *Austral. J. Statist.,* 22, 341--348.

[815] Särndal, C.E. (1982). Implications of survey designs for generalized regression estimators of linear functions. *J. Statist. Planning Infer.*, 7, 155--170.

[816] Särndal, C. E.(1992): Methods for estimating the precision of survey estimates when imputation is used. *Survey Methodology*, 18, 241--252.

[817] Särndal, C.E. (1996). Efficient estimators with simple variance in unequal probability sampling. *J. Amer. Statist .Assoc.*,91, 1289--1300.

[818] Särndal, C.E. and Hidiroglou, M.A. (1989). Small domain estimation : a conditional analysis. *J. Amer. Statist. Assoc.*, 84, 266--275.

[819] Särndal, C.E. and Wright, R.L. (1984). Cosmetic form of estimators in survey sampling. *Scand. J. Statist.*, 11, 146--156.

[820] Särndal, C.E. and Swensson, B. (1987). A general view of estimation for two phases of selection with application to two-phase sampling and non-response. *Int. Statist. Rev.*, 55, 279--294.

[821] Särndal, C.E., Swensson, B. and Wretman, J.H. (1989). The weighted residual technique for estimating the variance of the general regression estimator of the finite population total. *Biometrika*, 76 (3), 527--537.

[822] Särndal, C.E., Swensson, B. and Wretman, J.H. (1992). *Model assisted survey sampling.* NewYork: Springer--Verlag.

[823] Saxena, R.R., Singh, P. and Srivastava, A.K. (1986). An unequal probability sampling scheme. *Biometrika*, 73(3), 761--763.

[824] Saxena, S.K., Nigam, A.K. and Shukla, N.D. (1995). Variance estimation for combined ratio estimator. *Sankhyā*, B, 57, 85--92.

[825] Schafer, J.L. and Schenker, N. (2000). Inference with imputed conditional means. *J. Amer. Statist. Assoc.*, 95, 141--154.

[826] Schneeberger, H. (1979). Saddle points of the variance of the sample mean in stratified sampling. *Sankhyā*, C, 41, 92--96

[827] Schreuder, H.T., Gregoire, T.G. and Wood, G.B. (1993). *Sampling methods for multi-resource forest inventory.* Wiley, New York.

[828] Schucany, W.R., Gray, H.L. and Owen, D.B. (1971). Bias reduction in estimation. *J. Amer. Statist. Assoc.*, 66, 524--533.

[829] Scott, A. and Smith, T.M.F. (1969). Estimation in multi-stage surveys. *J. Amer. Statist. Assoc.*, 64, 830--840.

[830] Scott, A.J., Brewer, K.W. and Ho, E.W. (1978). Finite population sampling and robust estimation. *J. Amer. Statist. Assoc.*, 73, 359--361.

[831] Searls, D.T. (1964). The utilization of a known coefficient of variation in the estimation procedure. *J. Amer. Statist. Assoc.*, 59, 1225--1226.

[832] Searls, D.T. (1967). A note on the use of an approximately known co-efficient of variation. *American Statistician*, 21(2), 20--21.

[833] Sedransk, J. and Meyer, J. (1978). Confidence intervals for quantiles of a finite population: simple random and stratified simple random sampling. *J. R. Statist. Soc.*, B, 40, 239--252.

[834] Sekkappan, R.M. (1973). *Bayes estimation and uniform admissibility for sampling from finite population.* Ph. D. Thesis, University of Waterloo.

[835] Sekkappan, R.M. (1981). Subjective Bayesian multivariate stratified sampling from finite populations. *Metrika,* 28, 123--132.

[836] Sekkappan, R.M. and Thompson, M.E. (1975). On a class of uniformly admissible estimators for finite populations. *Ann. Statist.*, 3, 492--499.

[837] Sekkappan, R.M. and Thompson, M.E. (1994). Multi-phase and successive sampling for a stratified population with unknown stratum sizes. *Pak. J. Statist.*, A, 10, 131--142.

[838] Sen, A.R. (1952). Present status of probability sampling and its use in estimation of farm characteristics. *Economet.* ,27, 130.

[839] Sen, A.R. (1953). On the estimate of the variance in sampling with varying probabilities. *J. Indian Soc. Agril. Statist.*, 5, 119--127.

[840] Sen, A.R., Seller, S. and Smith, D.N. (1973). The use of ratio estimate in successive sampling. *Biometrics*, 31, 673--683.

[841] Sengupta, S. (1980). On the admissibility of the symmetrized Des Raj estimator for PPSWOR sample of size two. *Calcutta Statist. Assoc. Bull.*, 29, 35--44.

[842] Sengupta, S. (1981a). On interpenetrating samples of equal and unequal sizes. *Calcutta Statist. Accoc. Bull,* 30, 187--197.

[843] Sengupta, S. (1981b). Jackknifing the ratio and the product estimators in double sampling. *Metrika*, 28, 245--256.

[844] Sengupta, S. (1982a). On interpenetrating samples of unequal sizes. *Metrika*, 29, 175--188.

[845] Sengupta, S. (1982b). Admissibility of the symmetrized Des Raj estimator for fixed size sampling designs of size two. *Calcutta Statist. Assoc. Bull.,* 31, 201--205.

[846] Sengupta, S. (1983). Admissibility of unbiased estimators in finite population sampling for samples of size at most two. *Calcutta Statist. Assoc. Bull.*, 32, 91--102.

[847] Sengupta, S. (1986). A comparison between PPSWR and Brewer's πps WOR procedures. *Calcutta Statist. Assoc. Bull.*, 35, 207--210.

[848] Sengupta, S. (1988). A note on PPS circulation systematic sampling. *Calcutta Statist. Assoc. Bull.*, 37, 111--112.

[849] Serfling, R.J. (1968). Approximate optimum stratification. *J. Amer. Statist. Assoc.*, 63, 1298--1309.

[850] Seth, G.R. and Rao, J.N.K. (1964). On the comparison between simple random sampling with and without replacement. *Sankhy* $\bar{a}$, A, 26, 85--86.

[851] Sethi, V.K. (1965). On optimum pairing of units. *Sankhy* $\bar{a}$, B, 27, 315--320.

[852] Shah, D.N. and Shah, S.M. (1979). Unbiased product type estimators. *Gujarat Statist. Rev.,* 6(2), 34--43.

[853] Shah, D.N. and Gupta, M.R. (1986). Comparison of double sampling estimators. *Metron*, 417--419.

[854] Shah, D.N. and Patel, P.A. (1996). Asymptotic properties of a generalized regression type predictor of a finite population variance in probability sampling. *Canad. J. Statist.*, 24, 373--384.

[855] Shannon, D.F. (1970). Parameter selection for modified Newton methods for function minimisation. *Siam. J. Numerical Analysis*, 7, 102--109.

[856] Shao, J. and Chen, Y. (1999). Approximate balanced half sample and related replication methods for imputed survey data. *Sankhy $\overline{a}$*, 61, 187--201.

[857] Shao, J., Chen, Y and Chen, Y. (1998). Balanced repeated replication for stratified multistage survey data under imputation. *J. Amer. Statist. Assoc.*, 819--831.

[858] Sharma, S.D. and Sil, A. (1996). A study of Politz--Simmon estimator under non-cooperation. *J. Indian Soc. Statist.*, 48(2), 171--184.

[859] Sharma, S.S. (1970). On an estimation in T_3 -- class of linear estimators in sampling with varying probabilities from a finite population. *Ann. Inst. Statist. Math.*, 22, 495--500.

[860] Sharma, Y.K., Singh, R., Rai, A. and Verma, S.S. (2000). Regression estimators from survey data for small sample sizes. *J. Indian Soc. Agric. Statist.*, 53(2), 115--124.

[861] Sharot, T. (1976). Sharpening the Jackknife. *Biometrika*, 63, 315--321.

[862] Sheers, N. (1992). A review of randomized response technique. *Measurement and Evaluation in Counselling and Development.*, 25, 27--41.

[863] Shiledar--Baxi, H.R. (1995). Approximately optimum stratified design for a finite population--II. *Sankhy $\overline{a}$*, 57, 391--404.

[864] Shiue, G.J. (1966). Systematic sampling with multiple random starts. *Forestry Science*, 6, 142--150.

[865] Shukla, G.K. (1996). An alternative multivariate ratio estimate for finite population. *Bull. Calcutta Statist. Assoc.*, 15, 127--134.

[866] Shukla, D. and Dubey, J. (2001). Estimation in mail surveys under PSNR sampling scheme. *J. Indian Soc. Agric. Statist.*, 54(3), 288--302.

[867] Shukla, D. and Trivedi, M. (2001). Mean estimation in deeply stratified population under post-stratification. *J. Indian Soc. Agric. Statist.*, 54(2), 221--235.

[868] Silva, P.L.D.N. and Skinner, C.J. (1995). Estimating distribution functions with auxiliary information using poststratification. *J. Official Statist.*, 11, 277--294.

[869] Silverman, B.W. (1986). *Density estimation for statistics and data analysis.* London:Chapman and Hall.

[870] Singh, A.C. (1996). Combining information in survey sampling by modified regression. *Proc. of the Section on Survey Research Methods*, American Statistical Association, 120--129.

[871] Singh, A.C. ad Mohl, C.A. (1996). Understanding calibration estimators in survey sampling. *Survey Methodology*, 22, 107--115.

[872] Singh, A.C., Stukel, D.M. and Pfeffermann, D. (1998). Bayesian versus frequentist measures of error in small area estimation. *J. R. Statist. Soc.*, B, 60, 377--396.

[873] Singh, A.K. and Singh, H.P. (1997). A note on the efficiencies of three product type estimators under a linear model. *J. Indian Soc. Agric. Statist.*, 50(2), 130--134,

[874] Singh, A.K., Singh, H.P. and Upadhyaya, L.N. (2001). A generalized chain estimator for finite population mean in two-phase sampling. *J. Indian Soc. Agric. Statist.*, 370--375.

[875] Singh, D. (1956). On efficiency of cluster sampling. *J. Indian Soc. Agric. Statist.*, 8, 45--55.

[876] Singh, D. (1968). Estimation in successive sampling using a multi-stage design. *J. Amer. Statist. Assoc.*, 63, 99--112.

[877] Singh, D. and Chaudhary, F.S. (1986). *Theory and analysis of sample survey designs*. Wiley Eastern Limited.

[878] Singh, D., Jindal, K.K. and Garg, J.N. (1968). On modified systematic sampling. *Biometrika*, 55, 541--546.

[879] Singh, D. and Singh, B.D. (1965). Some contribution to two-phase sampling. *Austral. J. Statist.*, 2, 45--67.

[880] Singh, D. and Singh, P. (1977). New systematic sampling. *J. Statist. Planning Infer.*, 1, 163--177.

[881] Singh, G.N. and Upadhyaya, L.N. (1995). A class of modified chain type estimators using two auxiliary variables in two-phase sampling. *Metron*, 117--125.

[882] Singh, G.N. and Singh, V.K. (2001). On the use of auxiliary information in successive sampling. *J. Indian Soc. Agric. Statist.*, 54(1), 1--12.

[883] Singh, H.P. (1988). An improved class of estimators of population mean using auxiliary information. *J. Indian Soc. Agric. Statist.*, 96--104.

[884] Singh, H.P. (1989). A class of unbiased estimators of product of population means. *J. Indian Soc. Agric. Statist.*, 40, 113--118.

[885] Singh, H.P. and Biradar, R.S. (1992). Almost unbiased ratio cum product estimators for the finite population mean. *Test*, 1, 19--29.

[886] Singh, H.P., Chandra, P. and Singh, S. (2003). Variance estimation using multi-auxiliary information for random non-response in survey sampling. *Statistica* (Accepted).

[887] Singh, H.P. and Gangele, R.K. (1995). Almost separation of bias precipitates in the estimator of 'Inverse of population mean' with known coefficient of variation. *J. Indian Soc. Agric. Statist.*, 47, 212--218.

[888] Singh, H.P. and Gangele, R.K. (1997). An approach for almost separation of bias precipitates. *J. Indian Soc. Agric. Statist*, 50, 11--17.

[889] Singh, H.P. and Kakran, M.S. (1993). A modified ratio estimator using known coefficient of kurtosis of an auxiliary character. (Unpublished manuscript).

[890] Singh, H.P., Katyar, N.P. and Gangwar, D.K. (1996). A class of almost unbiased regression type estimators in two phase sampling applying Quenouille's method. *J. Indian Soc. Agric. Statist.*, 48(1), 98--104.

[891] Singh, H.P. and Sahoo, L.N. (1989). A class of almost unbiased estimators for population ratio and product. *Calcutta Statist. Assoc. Bull*, 38, 241--243.

[892] Singh, H.P. and Singh, R. (2001). Improved ratio type estimator for variance using auxiliary information. *J. Indian Soc. Agric. Statist.*, 54(3), 276--287.

[893] Singh, H.P. and Singh, S. (2002). Estimation of median of the study variable using known interquartile range of the auxiliary variable. *Working Paper.*

[894] Singh, H.P., Singh, S. and Joarder, A.H. (2003). Estimation of median using know mode of the auxiliary variable. *J. Statist. Research (To appear).*

[895] Singh, H.P., Singh, S. and Puetas, S.M. (2003a). Ratio type estimators for the median of finite populations. *Allgemeines Statistiches Archiv. (In press)*

[896] Singh, H.P., Singh, S. and Puetas, S.M. (2003b). Estimation of Interquartile range of the study variable using known interquartile range of the auxiliary variable. *Working Paper.*

[897] Singh, H.P., Singh, S. and Puetas, S.M. (2003c). Estimation of median using three known quartiles of the auxiliary variable. *Working Paper.*

[898] Singh, H.P. and Singh, V.P. (1993). A general class of unbiased estimators of a parameter. *Calcutta Statist. Assoc. Bull.*, 43, 169--170.

[899] Singh, H.P. and Singh, V.P. (1995). A class of unbiased dual to ratio estimator in stratified sampling. *J. Indian Soc. Agric. Statist.*, 47(2), 168--175.

[900] Singh, H.P. and Tracy, D.S. (2001). Estimation of population mean in presence of random non-response in sample surveys. *Statistica,* LXI, no.2, 231--248.

[901] Singh, H.P. and Upadhyaya, L.N. (1986) On a class of estimators of the population mean in sampling using auxiliary information. *J. Indian Soc. Agric. Statist.*, 38, 100--104.

[902] Singh, M. (1979). On the reduction of bias of ratio estimator to a desired degree. *Biom. J.*, 21(7), 645--647.

[903] Singh, M., Kumar, P. and Chandak, R. (1983). Use of multi-auxiliary variables as a condensed auxiliary variable in selecting a sample. *Commun. Statist.--Theory Meth.*, 12, 1685--1697.

[904] Singh, M.P. (1967a). Ratio cum product method of estimation. *Metrika,* 12, 34--42.

[905] Singh, M.P. (1967b). Multivariate product method of estimation for finite populations. *J. Indian Soc. Agric. Statist.,* 19, 1--10.

[906] Singh, M.P. (1969). Comparison of some ratio-cum-product estimators. *Sankhyā*, B, 31, 375--378.

[907] Singh, M.P., Gambino, J. and Mantel, H.J. (1994). Issues and strategies for small area data. *Survey Methodology*, 20, 3--22.

[908] Singh, P. (1978). A sampling scheme with inclusion probability proportional to size. *Sankhyā*, C, 40, 122--128.

[909] Singh, P. and Garg, J.N. (1979). On balanced random sampling. *Sankhyā*, C, 41, 60--68.

[910] Singh, P. and Srivastava, A.K. (1980). Sampling scheme providing unbiased regression estimators. *Biometrika,* 67, 205--209.

[911] Singh, P. and Yadav, R.J. (1992). Generalized estimation under successive sampling. *J. Indian Soc. Agric. Statist.*, 44, 27--36.

[912] Singh, R. (1971). Approximately optimum stratification on the auxiliary variable. *J. Amer. Statist. Assoc.*, 66, 829--833.

[913] Singh, R. (1972). A note on successive sampling over two occasions. *Aust. J. Statist.*, 14, 2, 120--122.

[914] Singh, R. (1975a). A note on the efficiency of ratio estimate with Midzuno's scheme of sampling. *Sankhy $\bar{a}$* , C, 37, 211--214.

[915] Singh, R. (1975b). An alternative method of stratification on the auxiliary variable. *Sankhy $\bar{a}$* , C, 37, 100--108.

[916] Singh, R. (1984). Double sampling for two auxiliary characters. *Calcutta Statist. Assoc. Bull.*, 33, 193--197.

[917] Singh, R. and Bansal, M.L. (1975). On the efficiency of interpenetrating sub-samples in simple random sampling. *Sankhy $\bar{a}$* , C, 37, 190--198.

[918] Singh, R. and Bansal, M.L. (1978). A note on the efficiency of interpenetrating sub-samples in simple random sampling. *Sankhy $\bar{a}$* , C, 40, 174--176.

[919] Singh, R. and Kathuria, O.P. (1995). Sampling without replacement in qualitative randomized response model. *J. Indian Soc. Agric. Statist.*, 47(2), 134--141.

[920] Singh, R. and Kishore, L. (1975). On Rao, Hartley and Cochran's method of sampling. *Sankhy $\bar{a}$* , 37, 88--94.

[921] Singh, R. and Lal, M. (1978). On the construction of random groups in the RHC scheme. *Sankhy $\bar{a}$* , C, 40, 129--135.

[922] Singh, R. and Mangat, N.S. (1996). *Elements of survey sampling*. Kluwer Academic Publishers, The Netherlands.

[923] Singh, R., Mangat, N.S. and Singh, S.(1993). A mail survey design for sensitive character without using randomization device. *Commun. Statist. -- Theory Meth.*, 22(9), 2661--2668.

[924] Singh, R. and Narain, P. (1989). Method of estimation from samples with random sample sizes. *J. Statist. Planning Infer.*, 23, 217--225.

[925] Singh, R. and Singh, B. (1974). On replicated samples drawn with Rao, Hartley and Cochran's Scheme. *Sankhy $\bar{a}$* , C, 36, 147--150.

[926] Singh, R. and Singh, H.P. (1993). A Hartley--Ross type estimator for finite population mean when the variables are negatively correlated. *Metron,* 205--216.

[927] Singh, R. and Singh, H.P. (1999). A class of unbiased estimators in cluster sampling. *J. Indian Soc. Agric. Statist.*, 52(3), 299--302.

[928] Singh, R., Singh, H.P. and Espejo, M.R. (1998). The efficiency of an alternative to ratio estimator under a super population model. *J. Statist. Planning Infer.*, 71, 287--301.

[929] Singh, R., Singh, S. and Mangat, N.S.(1995). Mail survey design for sensitive quantitative variable. *Metron* , 53, 43--54.

[930] Singh, R., Singh, S., Mangat, N.S. and Tracy, D.S. (1995). An improved two stage randomized response strategy. *Statistical Papers*, 36, 265--271.

[931] Singh, R. and Sukhatme, B.V. (1969). Optimum stratification. *Ann. Inst. Statist. Math.*, 21, 515-528.

[932] Singh, R.K. (1982a). On estimating ratio and product of population parameters. *Calcutta Statist. Assoc. Bull.*, 31, 69--76.

[933] Singh, R.K. (1982b). Generalized double sampling estimators for the ratio and product of population parameters. *J. Indian Statist. Assoc.*, 20, 39--49.

[934] Singh, R.K. and Ray, S.K. (1981). Product cum difference method of estimation using two auxiliary variables. *Biom. J.*, 23(6), 563--571.

[935] Singh, R.K. and Singh, G. (1984a). A class of estimators with estimated optimum values in sample surveys. *Statist. Prob. Lett.*, 2, 319--321.

[936] Singh, R.K. and Singh, G. (1984b). Improved generalized ratio cum product estimation. *Biom. J.*, 26 (1), 57--61.

[937] Singh, R.K. and Zaidi, S.M.H. (2000). On estimating square of population mean and population variance. *J. Indian Soc. Agric. Statist.*, 53(3), 243--256.

[938] Singh, S. (1988). Estimation in overlapping clusters. *Commun. Statist. - Theory Meth.*, 17(2), 613--621.

[939] Singh, S. (1991a). Estimation of finite population variance using double sample. *Aligarh J.Statist.*, 11, 53--56.

[940] Singh, S. (1991b). *On improved strategies in survey sampling*. Unpublished Ph.D. thesis submitted to Punjab Agricultural University, Ludhiana, India.

[941] Singh, S. (1994). Unrelated question randomized response sampling using continuous distributions. *J. Indian Soc. Agric. Statist.*, 46(3), 349--361. (Gold Medalist presentation).

[942] Singh, S. (1999). An addendum to the confidentiality guaranteed under randomised response sampling by Mahmood, Singh and Horn. *Biom. J.*, 41(8), 955--966.

[943] Singh, S. (2000a). Estimation of parametric functions in two-dimensional space in survey sampling. *South. African Statist. J.*, 34, 51--71.

[944] Singh, S. (2000b). Estimation of variance of regression estimator in two phase sampling. *Calcutta Statist. Assoc. Bull.*, 50, 49--63.

[945] Singh, S. (2000c). A new method of imputation in survey sampling. *Working paper.*

[946] Singh, S. (2001). Generalized calibration approach for estimating variance in survey sampling. *Ann. Inst. Statist. Math.*, 53(2), 404--417.

[947] Singh, S. (2002a). Estimation of median of the study variable using 99 known percentiles of the auxiliary variable. *Working Paper at St. Cloud State University, MN, USA.*

[948] Singh, S. (2002b). On Farrell and Singh's penalized chi square distance functions in survey. *Presented at the SSC--2003 conderence at Halifax, Canada.*

[949] Singh, S. (2002c). A new stochastic randomized response technique. *Metrika*, 56, 131--142.

[950] Singh, S. (2003a). Short note on the linear regression estimator in survey sampling. *Working paper at St. Cloud State University, St. Cloud, MN, USA.*.

[951] Singh, S. (2003b). On Jackknifing the two-phase calibration weights for estimating the variance of estimator of distibution function using two auxiliary variables. *Working paper at St. Cloud State University, St. Cloud, MN, USA.*

[952] Singh, S. (2003c). Golden Jubilee Year--2003 of the linear regression estimator. *Working paper at St. Cloud State University, St. Cloud, MN, USA.*

[953] Singh, S. and Arnab, R. (2003). Penalized chi square distance function for non-response adjustments. *Working paper at St. Cloud State Univerity, and University of Durbin—Westvile(Submitted for presentation at JSM-2003, California, USA).*

[954] Singh, S. and Deo, B. (2002). Imputing with power transformation. *Statistical Papers* (To appear)

[955] Singh, S., Grewal, I.S. and Joarder, A.H. (2002). General class of estimators in multi-character surveys. *Statistical Papers* (To appear).

[956] Singh, S. and Horn, S. (1998). An alternative estimator for multi-character surveys. *Metrika,* 48, 99--107.

[957] Singh, S. and Horn, S. (1999). An improved estimator of the variance of the regression estimator. *Biom. J.*, 41(3), 359--369.

[958] Singh, S. and Horn, S. (2000). Compromised imputation in survey sampling. *Metrika*, 51, 267--276.

[959] Singh, S., Horn, S. and Chowdhury, S. (1998). Estimation of stigmatized characteristics of a hidden gang in finite population. *Austral. & New Zealand J. Statist.*,40(3), 291--297.

[960] Singh, S., Horn, S. Chowdhury, S. and Yu, F. (1999). Calibration of the estimators of variance. *Austral .& New Zealand J. Statist.* , 40(2), 199--212.

[961] Singh, S., Horn, S. and Tracy, D.S. (2001). Hybrid of calibration and imputation : estimation of mean in survey sampling. *Statistica*, LXI (1), 27--41

[962] Singh, S., Horn, S. and Yu, F. (1998). Estimation of variance of general regression estimator : Higher level Calibration Approach. *Survey Methodology,* 24 (1), 41--50.

[963] Singh, S., Horn, S., Singh, R. and Mangat, N.S. (2003). On the use of modified randomization device for estimating the prevalence of a sensitive attribute. *Statistics in Transition* (To appear)

[964] Singh, S. and Joarder, A.H. (1997a). Optional randomized response technique for quantitative sensitive character. *Metron*, LV, 151--157.

[965] Singh, S. and Joarder, A.H. (1997b). Unknown repeated trials in randomized response sampling. *J. Indian Soc. Agric. Statist.*, 50(1), 103--105.

[966] Singh, S. and Joarder, A. (1998). Estimation of finite population variance using random nonresponse in survey sampling. *Metrika,* 241--249.

[967] Singh, S. and Joarder, A.H. (2002). Estimation of distribution function and median in two-phase sampling. *Pak J. Statist.*, 18(2), 301--319.

[968] Singh, S., Joarder, A. H. and King, M. L. (1996). Regression analysis using scrambled responses. *Austral. J. Statist.* 38 (2), 201--211.

[969] Singh, S., Joarder, A.H. and Tracy, D.S. (2000). Regression type estimators for random non-response in survey sampling. *Statistica,* LX, 39--44.

[970] Singh, S., Joarder, A.H. and Tracy, D.S. (2001). Median estimation using double sampling. *Austral. & New Zealand J. Statist.*, 43 (1), 33--46.

[971] Singh, S. and Kataria, P. (1990). An estimator of finite population variance. *J. Indian Soc. Agril. Statist.* 42, 186--188.

[972] Singh, S. and King, M.L. (1999). Estimation of coefficient of determination using scrambled responses. *J. Indian Soc. Agric. Statist.* 52(3), 338--343.

[973] Singh, S., Mahmood, M. and Tracy, D.S. (2001). On the estimation of mean and variance of a sensitive character using distinct units. *Statistical Papers*, 42, 403--411

[974] Singh, S., Mangat, N.S. and Gupta, J.P. (1996). Improved estimator of finite population correlation coefficient. *J. Indian Soc. Agric. Statist.*, 48, 141--149.

[975] Singh, S., Mangat, N.S. and Mahajan, P.K. (1995). General class of estimators. *J. Indian Soc. Agril. Statist.* ,47(2), 129--133.

[976] Singh, S., Mangat, N.S. and Singh, R. (1994). On estimation of mean/total of stigmatised quantitative variable. *Statistica*, 54(3), 383--386.

[977] Singh, S., Mangat, N.S. and Singh, R. (1997). Estimation of size and mean of a sensitive quantitative variable for a sub-group of a population. *Commun. Statist. -- Theory Meth.*, 26(7), 1793--1804.

[978] Singh, S., Pannu, C.J.S., Singh, S., Singh, I.P. and Kaur, S. (1996). *Energy is Punjab Agriculture.* Department of Farm Power and Machinery, Punjab Agricultural University, Ludhiana, India.

[979] Singh, S. and Puetas, S.M. (2002). On the estimation of total, mean and distribution function using two-phase sampling: calibration approach. *J. Indian Soc. Agric. Statist.* (Revised submitted).

[980] Singh, S., Singh, H.P., Tailor, R. and Allen, J. (2002). General class of estimators for estimating ratio of two population means in the presence of random non-response. *Working paper.*

[981] Singh, S. and Singh, R. (1979). On random non-response in unequal probability sampling. *Sankhy* $\bar{a}$, C, 41, 127--137.

[982] Singh, S. and Singh, R. (1991). Almost bias precipitate filtration: A new technique. *Aligarh J. Statist.*, 11, 5--8.

[983] Singh, S and Singh, R. (1992a). Improved Franklin's model for randomized response sampling. *J. Indian Statist. Assoc.* 30, 109--122.

[984] Singh, S. and Singh, R. (1992b). An alternative estimator for randomised response technique. *J. Indian Soc. Agic. Statist.*, 44(2), 149--154.

[985] Singh, S. and Singh, R. (1993a). Almost filtration of bias precipitates:a new approach. *J. Indian Soc. Agril. Statist.* 45, 214--218.

[986] Singh, S. and Singh, R. (1993b). A new method: almost separation of bias precipitates in sample surveys. *J. Indian Statist. Assoc.*, 31, 99--105.

[987] Singh, S. and Singh, R. (1993c). A class of almost unbiased ratio and product type estimators. *J. Indian Soc. Statist. Opers. Res.*, 14, 35--39.

[988] Singh, S. and Singh R (1993d). Generalised Franklin's model for randomised response sampling. *Commun. Statist. --Theory Meth* 22 (3), 741--755.

[989] Singh, S., Singh, R. and Mangat, N.S. (1996). Estimation of mean of a stigmatized quantitative variable for a sub-group of the population. *Metron*, 54 (3-4), 83--91.

[990] Singh, S., Singh, R. and Mangat, N.S. (1998). Estimation of coefficient of variation of a sensitive character. *Metron, 55*, 59--67.

[991] Singh, S., Singh, R and Mangat, N.S. (2000). Some alternative strategies to Moors' model in randomized response sampling. *J. Statist. Planning Infer.*, 83, 243--255.

[992] Singh, S., Singh, R., Mangat, N.S. and Tracy, D.S. (1994). An alternative device for randomised responses. *Statistica*, 54(2), 233--243.

[993] Singh, S. and Singh, S. (1988). Improved estimators of K and B in finite populations. *J. Indian Soc. Agric. Statist.*, 40(2), 121--126.

[994] Singh, S., Singh, S., Mittal, J.P., Pannu, C.J.S. and Bhangoo, B.S. (1994). Energy inputs and crop yield relationships for rice in Punjab. *Energy--International Journal*,19(10), 1061--1065.

[995] Singh, S., Singh, S., Pannu, C.J.S., Bhangoo, B.S. and Singh, M.P. (1994). Energy inputs and crop yield relationships for wheat in Punjab. *Energy Convers. Mgmt*, 35(6), 493--499.

[996] Singh, S. and Tracy, D.S. (1999). Ridge regression using scrambled responses. *Metron*, 57, 147--157.

[997] Singh, S. and Valdés, S.R. (2003). Optimum method of imputation in survey sampling. *Working paper at St. Cloud State University, St. Cloud,MN, USA.*

[998] Singh, V.K. and Shukla, D. (1993). An efficient one parameter family of factor type estimators in sample surveys. *Metron*, 139--159.

[999] Singh, V.K. and Singh, G.N. (1991). Chain type regression estimators with two auxiliary variables under double sampling scheme. *Metron*, 279--289.

[1000] Singh, V.K., Singh, H.P. and Singh, H.P. (1994). Estimation of ratio and product of two finite population means in two-phase sampling. *J. Statist. Planning Infer.* , 41,163--171.

[1001] Singh, V.P. and Singh, H.P. (1997--98). Chain estimators for population ratio in double sampling. *Aligarh J. Statist.*, 17/18, 85--100.

[1002] Sinha, B.K. (1973). On sampling schemes to realize pre-assigned sets of inclusion probabilities of first two orders. *Calcutta Statist. Assoc. Bull.*, 22, 89--100.

[1003] Sisodia, B.V.S. and Dwivedi, V.K. (1981). A modified ratio estimator using coefficient of variation of auxiliary variable. *J. Indian Soc. Agric. Statist.*, 33, 13--18.

[1004] Sisodia, B.V.S. and Singh, A. (2001). On small area estimation--An empirical study. *J. Indian Soc. Agric. Statist.*, 54(3), 303--306.

[1005] Sitter, R.R. (1993). Balanced repeated replications based on orthogonal multi-arrays. *Biometrika*, 80, 211--221.

[1006] Sitter, R.R. (1997). Variance estimation for the regression estimator in two-phase sampling. *J. Amer. Statist. Assoc.*, 92, 780--787.

[1007] Sitter, R.R. and Rao, J.N.K. (1997). Imputation for missing values and corresponding variance estimation. *Canad. J. Statist.*, 25(1), 61--73.

[1008] Sitter, R.R. and Wu, C. (2002). Efficient estimation of quadratic finite population functions. *J. Amer. Statist. Assoc.*, 97, 535--543

[1009] Skinner, C.J. (1991). On the efficiency of raking ratio estimation for multiple frame surveys. *J. Amer. Statist. Assoc.*, 86, 779--784.

[1010] Skinner, C.J., Holt, D. and Smith, T.M.F. (1989). *Analysis of complex surveys*. Wiley New York.

[1011] Skinner, C.J. and Rao, J.N.K. (1996). Estimation in dual frame surveys with complex designs. *J. Amer. Statist. Assoc.*, 91, 349--356.

[1012] Smith, H.F. (1938). An experimental law describing heterogeneity in the yields of agricultural crops. *J. Amer. Statist. Assoc.*, 28, 1--23.

[1013] Smith, J.H. (1947). Estimation of linear functions of cell proportions. *Ann. Math. Statist.*, 18, 231--254.

[1014] Smith, P. and Sedransk, J. (1983). Lower bounds for confidence coefficients for confidence intervals for finite population quantiles. *Commun. Statist.--Theory Meth.*, 12, 1329--1344.

[1015] Smith, S.K. and Lewis, B.B. (1980). Some new techniques for applying the housing unit method of local population estimation. *Demography*, 17, 323--340.

[1016] Smith, T.M.F. (1969). A note on ratio estimates in multi-stage sampling. *J. R. Statist. Soc.*, A, 132, 426--430.

[1017] Smith, T.M.F. (1978). *Principles and problems in the analysis of repeated surveys*. In N. Krishnan Namboodiri, ed. Survey Sampling and Measurement, Acad. Press, NY.

[1018] Smith, T.M.F. (1984). Present position and potential developments: some personal views, sample surveys. *J. R. Statist. Soc.*, A, 147, 208--221.

[1019] Smith, T.M.F. and Sugden, R.A. (1985). Inference and the ignorability of selection for experiments and surveys. *Bull. Int. Statist. Inst.*, 44th Session, Book II, 10.2-1 to 10.2--12.

[1020] Smith, T.M.F. (1995). Problems of resource allocation. *Proc. Statist. Can. Symp.*, 95, *Statistics Canada*, 107--114.

[1021] Srikantan, K.S. (1963). A note on interpenetrating sub-samples of unequal sizes. *Sankhyā*, B, 25, 345--350.

[1022] Srinath, K.P. (1971). Multiphase sampling in non-response problems. *J. Amer. Statist. Assoc.*, 66, 583--586.

[1023] Srinath, K.P. and Hidiroglou, M.A. (1980). Estimation of variance in multi-stage sampling. *Metrika*, 27, 121--125.

[1024] Srivastava, J. and Ouyang, Z. (1992). Studies on a general estimator in sampling, utilizing extraneous information through a sample weight function. *J. Statist. Planning Infer.*, 31, 199--218.

[1025] Srivastava, J.N. and Saleh, F. (1985). Need of t design in sampling theory. *Utilitas Mathematica*, 25, 5--17.

[1026] Srivastava, S.K. (1965). An estimator of the mean of a finite population using several auxiliary variables. *J. Indian Statist. Assoc.*, 3, 189--194.

[1027] Srivastava, S.K. (1967). An estimator using auxiliary information in sample surveys. *Calcutta Statist. Assoc. Bull.*,16,121--132.

[1028] Srivastava, S.K. (1971). A generalized estimator for the mean of a finite population using multi-auxiliary information. *J. Amer. Statist. Assoc.*, 66, 404--407.

[1029] Srivastava, S.K. (1980). A class of estimators using auxiliary information in sample surveys. *Canad. J. Statist.*, 8(2), 253--254.

[1030] Srivastava, S.K. (1981a). A generalized two-phase sampling estimator. *J. Indian Soc. Agric. Statist.*, 33, 38--46.

[1031] Srivastava S.K. (1981b). A note on generalized RPD estimator in double sampling. *J. Indian Soc. Agric. Statist.*, 33, 89--93.

[1032] Srivastava, S.K. (1983). Predictive estimation of finite population mean using product estimator. *Metrika*, 30, 93--99.

[1033] Srivastava, S.K. (1992). A note on improving classes of estimators in sample surveys. *J. Indian Soc. Agric. Statist.*, 44, 267--270.

[1034] Srivastava, S.K. and Jhajj, S.K. (1980). A class of estimators using auxiliary information for estimating finite population variance. *Sankhy* $\overline{a}$, C, 42,87--96.

[1035] Srivastava, S.K. and Jhajj, H.S. (1981). A class of estimators of the population mean in survey sampling using auxiliary information. *Biometrika*, 68, 341--343.

[1036] Srivastava, S.K. and Jhajj, H.S. (1983a). A class of estimators of the population mean using multi-auxiliary information. *Calcutta Statist. Assoc. Bull.*, 32, 47--56.

[1037] Srivastava, S.K. and Jhajj, H.S. (1983b). Class of estimators of mean and variance using auxiliary information when correlation coefficient is also known. *Biom. J.*, 25(4),401--409.

[1038] Srivastava, S.K. and Jhajj, H.S. (1986). On the estimation of finite population correlation coefficient. *J. Indian Soc. Agric. Statist.*, 38, 82--91.

[1039] Srivastava, S.K. and Jhajj, H.S. (1987). Improved estimation in two-phase and successive sampling. *J. Indian Statist. Assoc.*, 25, 71--75.

[1040] Srivastava, S.K. and Jhajj, H.S. (1995). Classes of estimators of finite population mean and variance using auxiliary information. *J. Indian Soc. Agril. Statist.*, 47,119--128.

[1041] Srivastava, S.K., Jhajj, H.S. and Sharma, M.K. (1986). Comparison of some estimators of K and B in finite populations. *J. Indian Soc. Agric. Statist.*, 38(2), 230--236.

[1042] Srivastava, S.R., Khare, B.B. and Srivastava, S.R. (1990). A generalized chain ratio estimator for mean of finite population. *J. Indian Soc. Agric. Statist.*, 42,108--117.

[1043] Srivastava, V.K. and Bhatnagar, S. (1981). Ratio and product methods of estimation when $\overline{X}$ is not known. *J. Statist. Res.*, 15, 29--39.

[1044] Srivastava, V.K., Dwivedi, T.D., Chaubey, Y.P. and Bhatnagar, S. (1983). Finite sample properties of Beale's ratio estimator. *Commun. Statist. -- Theory Meth.*, 12(15), 1795--1805.

[1045] Srivenkataramana, T. (1978). Change of origin and scale in ratio and difference methods of estimation in sampling. *Canad. J. Statist.*, 6, 79--86.

[1046] Srivenkataramana, T. (1980). A dual to ratio estimator in sample surveys. *Biometrika*, 67, 199--204.

[1047] Srivenkataramana, T. and Tracy, D.S. (1979). On ratio and product methods of estimation in sampling. *Statistica Neerlandica*, 33, 37--49.

[1048] Srivenkataramana, T. and Tracy, D.S. (1980). An alternative to ratio method in sample surveys. *Ann. Inst. Statist. Math.*, A, 32, 111--120.

[1049] Srivenkataramana, T. and Tracy, D.S. (1981). Extending product method of estimation to ositive correlation case in surveys. *Austral. J. Statist.*, 23, 95--100.

[1050] Srivenkataramana, T. and Tracy, D.S. (1983). Interchangebility of the ratio and product methods in sample surveys. *Commun. Statist. -- Theory Meth.*, 12(18), 2143--2150.

[1051] Srivenkataramana, T. and Tracy, D.S. (1984). Positive and negative valued auxiliary variates in surveys. *Metron,* 207--319.

[1052] Srivenkataramana, T. and Tracy, D.S. (1986). Transformations after sampling. *Statistics*, 17, 597--608.

[1053] Srivenkataramana, T. and Tracy, D.S. (1989). Two-phase sampling for selection with probability proportional to size in sample surveys. *Biometrika*, 76(4), 818--821.

[1054] Strachan, R., King, M.L. and Singh, S. (1998). Likelihood-based estimation of the regression model with scrambled responses. *Austral. & New Zealand J. Statist.*,40(3), 279--290.

[1055] Strauss, I. (1982). On the admissibility of estimators for the finite population variance. *Metrika,* 29, 195--202.

[1056] Stephan, F.F. (1945). The expected value and variance of the reciprocal and other negative powers of a positive Bernoulli variate. *Ann. Math. Statist.*, 16, 50--61.

[1057] Stroud, T.W.F. (1994). Bayesian analysis of binary survey data. *Canad. J. Statist.*, 22, 33--45.

[1058] Stukel, D.M., Hidiroglou, M.A. and Särndal, C.E. (1996). Variance estimation for calibration estimators: A comparison of Jackknifing versus Taylor linearization. *Survey Methodology*, 22, 107--115.

[1059] Stukel, D.M. and Rao, J.N.K. (1997). Estimation of regression model with nested error structure and unequal error variance under two and three stage cluster sampling. *Statist. Prob. Lett.*, 35, 401--407.

[1060] Stukel, D.M. and Rao, J.N.K. (1999). On small-area estimation under two-fold nested error regression models. *J. Statist. Planning Infer.*, 78, 131--147.

[1061] Subramani, J. (2000). Diagonal systematic sampling scheme for finite populations. *J. Indian Soc. Agric. Statist.*, 53(2), 187--195.

[1062] Subramani, J. and Tracy, D.S. (1993). Determinant sampling scheme for finite populations. *Working paper at University of Windsor, Canada.*

[1063] Sud, U.C. and Srivastava, A.K. (2000). Estimation of population mean in repeated surveys in the presence of measurement errors. *J. Indian Soc. Agric. Statist.*, 53(2), 125--133.

[1064] Sud, U.C., Srivastava, A.K. and Sharma, D.P. (2001a). On a biased estimator in repeated surveys. *J. Indian Soc. Agric. Statist.*, 54(1), 29--42.

[1065] Sud, U.C., Srivastava, A.K. and Sharma, D.P. (2001b). On the estimation of population variance in repeated surveys. *J. Indian Soc. Agric. Statist.*, 54(2), 355--369.

[1066] Sudakar, K. (1978). A note on 'Circular Systematic Sampling Design'. *Sankhyā*, C, 40, 72--73.

[1067] Sukhatme, B.V. (1962). Some ratio type estimators in two-phase sampling. *J. Amer. Statist. Assoc.*, 57, 628--632.

[1068] Sukhatme, P.V. (1944). Moments and product moments of moment statistics for samples of the finite and infinite populations. *Sankhyā*, 6, 363--382.

[1069] Sukhatme, P.V. (1953). *Sampling theory of surveys with applications.* Iowa State College Press, Ames, Iowa.

[1070] Sukhatme, P.V. (1954). *Sampling theory of surveys with applications.* Indian Society of Agricultural Statistics, New Delhi.

[1071] Sukhatme, P.V., Panse, V.G. and Sastri, K.V.R. (1958). Sampling techniques for estimating the catch of sea fish in India. *Biometrics*, 14, 78--96.

[1072] Sukhatme, P.V. and Sukhatme, B.V. (1970). *Sampling theory of surveys with applications.* Second Edition, Asia Publishing House, Bombay, India.

[1073] Sukhatme, P.V., Sukhatme, B.V., Sukhatme, S. and Asok, C. (1984). *Sampling theory of surveys with applications.* Iowa State University Press and Indian Society of Agricultural Statistics, New Delhi.

[1074] Sunter, A.B. (1977). List sequential sampling with equal or unequal probabilities without replacement. *Applied Statist.*, 26, 261--268.

[1075] Swain, A.K.P.C. and Mishra, G. (1992). Unbiased estimators of finite population variance using auxiliary information. *Metron,* 201--215.

[1076] Swain, A.K.P.C. and Mishra, G. (1994). Limiting distribution of ratio estimator of finite population variance. *Sankhyā*, B, 56, 11--17.

[1077] Swain, A.K.P.C. and Sahoo, L.N. (1982). Comparison of three almost unbiased ratio estimators in a survey for quantitative characteristics. *Statistica,* 42(3), 397--401.

[1078] Tallis, G.M. (1978). Note on robust estimation in finite populations. *Sankhyā*, C, 40, 136--138.

[1079] Tallis, G.M. (1991). A note on balanced cluster sampling. *Statist. Prob. Lett.*, 11, 169--172.

[1080] Tam, S.M. (1984). Optimal estimation in survey sampling under a regression superpopulation model. *Biometrika,*71, 645--647.

[1081] Tam, S.M. (1986). Characterization of best model based predictors in survey sampling. *Biometrika,* 73, 232--235.

[1082] Tam, S.M. (1995). Optimal and robust strategies for cluster sampling. *J. Amer. Statist. Assoc.*, 90, 379--382.

[1083] Taylor, J.M., Muzoz, A., Bass, S.M., Sah, A.J., Chmiel, J., Kingsley, L. *et al.* (1990). Estimating the distribution of times from HIV seroconversion to AIDS using multiple imputation. *Statist. Med.*, 9, 505--514.

[1084] Tepping, B.J., Hurwitz, W.N. and Deming, W.E. (1943). On the efficiency of deep stratification in block sampling. *J. Amer. Statist. Assoc.*, 38, 93--100.

[1085] Thompson, M.E. (1997). *Theory of sample surveys.* Chapman & Hall, London, U.K.

[1086] Thompson, S.K. (1990). Adaptive cluster sampling. *J. Amer. Statist. Assoc.*, 85, 1050--1059.

[1087] Thompson, S.K. and Seber, G.A.F. (1996). *Adaptive sampling.* New York: Wiley and Sons.

[1088] Tikkiwal, B.D. (1953). Optimum allocation in succesive sampling. *J. Indian Soc. Agric. Statist.*, 5, 100--102.

[1089] Tikkiwal, B.D. (1955). *Multiphase sampling on successive occasions.* Unpublished Ph.D. thesis submitted to North Carolina State University.

[1090] Tikkiwal, B.D. (1958). Theory of successive two-stage sampling. (Abstract). *Ann. Math. Statist.*, 29, 1291.

[1091] Tikkiwal, B.D. (1960). On the theory of classical regression and double sampling estimation. *J. R. Statist. Soc.*,B, 22, 131--138.

[1092] Tikkiwal, B.D. (1965). The theory of two-stage sampling on succesive occasions. *J. Indian Soc. Agril. Statist.*, 125--136.

[1093] Tikkiwal, B.D. (1979). Succesive sampling – a review. *Bull. Int. Statist. Inst.*, 48, 367--383.

[1094] Tillê, Y. (1998). Estimation is surveys using conditional inclusion probabilities : Simple random sampling. *Int. Statist. Rev.*, 66, 303--322.

[1095] Tin, M. (1965). Comparison of some ratio estimators. *J. Amer. Statist. Assoc.*,60, 294--307.

[1096] Toutenburg, H. and Srivastava, V.K. (1998). Estimation of ratio of population means in survey sampling when some observations are missing. *Metrika*, 48, 177--187.

[1097] Tracy, D.S. (1984). Moments of sample moments. *Commun. Statist.--Theory Meth.*, 3(5), 553--562.

[1098] Tracy, D.S. and Mangat, N.S. (1995). Respondent's privacy hazards in Moors' randomized response model-- A remedial strategy. *Int. J. Math. & Statist. Sci.*, 4(1), 121--130.

[1099] Tracy, D.S. and Mangat, N.S. (1996a). Some developments in randomized response sampling during the last decade – A follow up of review by Chaudhuri and Mukerjee. *J. Applied Statist. Sci.*, 4(2/3) ,147--158.

[1100] Trcay, D.S. and Mangat, N.S. (1996b). On respondent's jeopardy in two alternate question randomized response model. *J. Statist. Planning Infer.*, 55(1), 107--114.

[1101] Tracy, D.S. and Mangat, N.S. (1998). Comparisons of distinct units based estimators in unrelated question randomized response model. Internal. *J. Math. & Statist. Sci.*, 7, 229--240.

[1102] Tracy, D.S. and Osahan, S.S. (1994a). Determinant sampling versus some conventional sampling schemes. *Pak. J. Statist.*, 10(1), 99--121.

[1103] Tracy, D.S. and Osahan, S.S. (1994b). Estimation in overlapping clusters with unknown population size. *Survey Methodology*, 20(1), 53--57.

[1104] Tracy, D.S. and Osahan, S.S. (1994c). Random nonresponse on study variable versus on study as well as auxiliary variables. *Statistica,* 54, 163--168.

[1105] Tracy, D.S. and Osahan, S.S. (1999). A partial randomized response strategy. *Test*, 4(2), 315--322.

[1106] Tracy, D.S. and Singh, H.P. (1998). A modified ratio cum product estimator. *Int.. J. Math. & Statist. Sci.*, 7, 201--212.

[1107] Tracy, D.S. and Singh, H.P. (1999). A general class of chain regression estimators in two-phase sampling. *J. Appl. Statist. Sci.*, 8, 205--216.

[1108] Tracy, D.S., Singh, H.P. and Singh, R. (1996). An alternative to ratio cum product estimator in sample surveys. *J. Statist. Planning Infer.*, 53, 375--387.

[1109] Tracy, D.S., Singh, H.P. and Singh, R. (1998). A class of unbiased estimators alternative to ratio cum product estimator in sample surveys. *Prisankhyan Samikkha*, 5, 43--50.

[1110] Tracy, D.S., Singh, H.P. and Singh, S. (2001). An investigation on the bias reduction in linear variety of ratio cum product estimator. *Allgemeines Statist. Archive*, 85, 323--332.

[1111] Tracy, D.S. and Singh, S. (2000). Calibration estimators in randomized response surveys. *Metron,* 57, 47--68.

[1112] Tracy, D.S., Singh, S. and Arnab, R. (2003). Note on calibration in stratified and double sampling. *Survey Methodology*, June issue (To appear)

[1113] Tripathi, T.P. (1970). *Contributions to the sampling theory using multivariate information.* Ph.D. thesis submitted to Punjabi University, Patiala, India.

[1114] Tripathi, T.P. (1976). On double sampling for multivariate ratio and difference methods of estimation. *J. Indian Soc. Agric. Statist.*, 33, 33--54.

[1115] Tripathi, T.P. (1980). A general class of estimators for population ratio. *Sankhy* $\bar{a}$, C, 42, 63--75.

[1116] Tripathi, T.P. (1987). A class of estimators for population mean using multi-variate auxiliary information under general sampling designs. *Aligarh J. Statist.,* 7,49--62.

[1117] Tripathi, T.P. and Ahmed, M.S. (1995). A class of estimators for a finite population mean based on multivariate information and general two-phase sampling. *Calcutta. Statist. Assoc. Bull.*,45, 203--218.

[1118] Tripathi, T.P. and Chaubey, Y.P. (1992). Improved estimation of a finite population mean based on paired observations. *Commun. Statist.--Theory Meth.*, 21, 3327--3333.

[1119] Tripathi, T.P. and Singh, H.P. (1992). A class of unbiased product type estimators for the mean suitable for positive and negative correlation situations. *Commun. Statist.--Theory Meth.*, 21(2), 507--518.

[1120] Tripathi, T.P. and Srivastava, O.P. (1979). Estimation on successive occasions using PPSWR sampling. *Sankhy* $\bar{a}$, C, 41, 84--91.

[1121] Tu, X.M., Meng, X.L. and Pagano, M. (1993). The AIDS epidemic: Estimating survival after AIDS diagnosis from surveillance data. *J. Amer. Statist. Assoc.*, 88, 26--36.

[1122] Tukey, J.W. (1956). Keeping moments like sampling computations simple. *Ann. Math. Statist.*, 27, 37--54.

[1123] Tukey, J.W. (1958). Bias and confidence in not quite large samples. *Ann. Math. Statist.* (Abstract), 29, 614.

[1124] Tuteja, R.K. and Bahl, S. (1991). Multivariate product estimators. *Calcutta Statist. Assoc. Bull.*, 42, 109--115.

[1125] Unam, I. (1995). Estimating the population mean using supplementary dephased information *Commun. Statist. -- Simula.*, 24, 733--743.

[1126] Unnithan, V.K.G. (1978). The minimum variance boundary points of stratification *Sankhy $\bar{a}$*, C, 40, 60--72.

[1127] Upadhyaya, L.N., Kushwaha, K.S. and Singh, H.P. (1990). A modified chain ratio type estimator in two-phase sampling using multi-auxiliary information. *Metron*, 381--393.

[1128] Upadhyaya, L.N. and Singh, H.P. (1999). Use of transformed auxiliary variable in estimating the finite population mean. *Biom. J.*, 41(5), 627--636.

[1129] Upadhyaya, L.N., Singh, H.P. and Singh, S. (2003). A family of almost unbiased estimators for negatively correlated variables using jackknife technique. *Statistica* (Accepted)

[1130] Uthayakumaran, N. (1998). Additional cicular systematic sampling methods. *Biom. J.*, 40(4), 467--474.

[1131] Valliant, R. (1996). Limitations of balanced half-sampling. *J. Official Statist.*, 12(3), 225--240.

[1132] Valliant, R. (2002). Variance estimation for the general regression estimator. *Survey Methodology*, 28 (1), 103--114.

[1133] Verdeman, S. and Meeden, G.(1983). Admissible estimators in finite population sampling employing various types of prior information. *J. Statist. Planning Infer.*, 7, 329--341.

[1134] Vijayan, K. (1968). An exact πps scheme---generalization of a method of Hanurav. *J. R. Statist. Soc.*, B, 30, 556--566.

[1135] Vijayan, K. (1975). On estimating variance in unequal probability sampling. *J. Amer. Statist. Assoc.*, 70, 713--716.

[1136] Vos, J.W.E. (1980). Mixing of direct, ratio and product method estimators. *Statistica Nearlandica,* 34, 209--213.

[1137] Wakimoto, K. (1971). Stratified random sampling (III): Estimation of the correlation coefficient. *Ann. Inst. Statist. Math*, 23, 339--355.

[1138] Walsh, J.E. (1970). Generalization of ratio estimator for population total. *Sankhy $\bar{a}$*, A, 32, 99--103.

[1139] Warner, S.L. (1965). Randomized response : A survey technique for eliminating evasive answer bias. *J. Amer. Statist. Assoc.*, 60, 63--69.

[1140] Welch, B.L. (1937). On the z test in randomized blocks and Latin squares. *Biometrika*, 29, 21--52.

[1141] Williams, W.H. (1958). *Unbiased regression estimator.* Unpublished Ph.D. Dissertation, Iowa State University, Ames, Iowa.

[1142] Williams, W.H. (1961). Generating unbiased ratio and regression estimators. *Biometrics*, 17, 267--274.

[1143] Williams, W.H. (1963). The precision of some unbiased regression estimators. *Biometrics*, 19, 352--361.

[1144] Willson, D., Kirnos, P., Gallagher, J. and Wagner, A. (2002). Variance estimation from calibrated samples. *Joint Statistiacl Meetings, NY-Section on survey research methods*, 3727--3731.

[1145] Wolter, K.M. (1979). Composite estimation in finite populations. *J. Amer. Statist. Assoc.* 74, 604--613.

[1146] Wolter, K.M. (1984). An investigation of some estimators of variance for systematic sampling. *J. Amer. Statist. Assoc.*,79, 781--790.

[1147] Wolter, K.M. (1985). *Introduction to variance estimation.* New York, Springer--Verlag.

[1148] Worthingham, R., Morrison, T., Mangat, N.S., and Desjardins, G. (2002). Bayesian estimates of measurement error for in--line inspection and field tools. *Paper IPC2002-27263 International Pipeline Conference 2002, Calgary, Alberta, Canada. The American Society of Mechanical Engineers, New York, Proceedings sre on a CD-ROM.*

[1149] Wretman, J.H. (1995). Split questionnaires. *Presented at the Conference on Methodological issues in Official Statistics*, Stockholm.

[1150] Wright, R.L. (1983). Finite population sampling with multivariate auxiliary information. *J. Amer. Statist. Assoc.*, 78, 879--883.

[1151] Wright, T. (1990). Probability proportional to size (πps) sampling using ranks. *Commun. Statist. -- Theory Meth.*, 19(1), 347--362.

[1152] Wu, C. (2001). Empirical Likelihood method for finite populations. *Proceedings of Statistics 2001 Canada, The 4th Conference in Applied Statistics*, 339--350.

[1153] Wu, C. and Sitter, R.R. (2001). A model calibration approach to using complete auxiliary information from survey data. *J. Amer. Statist. Assoc.*, 96, 185--193.

[1154] Wu, C. and Sitter, R.R. (2001). Variance estimation for the finite population distribution function with complete auxiliary information. *Canad. J. Statist.*, 29(2), 289--307.

[1155] Wu, C.F.J. (1981). Balanced repeated replications based on mixed orthogonal arrays. *Biometrika*, 78, 181--188,

[1156] Wu, C.F.J. (1982). Estimation of variance of the ratio estimator. *Biometrika*, 69, 183--189.

[1157] Wu, C.F.J. (1984). Estimation in systematic sampling with supplementary observations. *Sankhy* $\overline{a}$, B, 46, 306--315.

[1158] Wu, C. F. J. (1985). Variance estimation for combined ratio and combined regression estimators. *J. R. Statist. Soc.*, B, 47, 147--154.

[1159] Wynn, H.P. (1977a). Minimax purposive survey sampling design. *J. Amer. Statist. Assoc.*, 72, 655--657.

[1160] Wynn, H.P. (1977b). *Optimum designs for finite populations sampling. Statistica Decision Theory and Related Topic* (S.S. Gupta and D.S. Moore, eds) Academic Press, New York, 471--492.

[1161] Wywial, J. (1999). Generalization of Singh and Srivastava's sampling scheme providing unbiased regression estimators. *Statistics in Transition*, 4(2), 259--281.

[1162] Wywial, J. (2000). On precision of Horvitz--Thompson strategies. *Statistics in Transition*, 4(5), 779--798.

[1163] Yamada, S. and Morimoto, H. (1992). Sufficiency. *Current issues in statistical inference: Essays in Honor of D. Basu by Ghosh and Pathak. Lecture Notes--Monograph Series, Institute of Mathematical Statistics, Hayward, California,* 17, 86--98.

[1164] Yansaneh, I.S. and Fuller, W.A. (1998). Optimal recursive estimation for repeated surveys. *Survey Methodology*, 24,31--40.

[1165] Yates, F. (1948). Systematic Sampling. *Phil. Trans. R. Soc.*, A, 241, 345--377.

[1166] Yates, F. (1949). *Sampling methods for censuses and surveys.* London: Charles Griffin and Co.

[1167] Yates, F. (1960). *Sampling methods for censuses and surveys.* Charles Griffin & Co., London.

[1168] Yates, F. and Grundy, P.M. (1953). Selection without replacement from within strata with probability proportional to size. *J. R. Statist. Soc*, 15(B), 253--261.

[1169] You, Y. and Rao, J.N.K. (2000a). Small area estimation with unmatched sampling and linking models. *Proceedings of Survey Methods Section*, 191--196.

[1170] You, Y. and Rao, J.N.K. (2000b). Hierarchical Bayes estimation of small area means using multi-level models. *Survey Methodology*, 26 (2), 173--181.

[1171] Yung, W. and Rao, J.N.K. (2000). Jackknife variance estimation under imputation for estimators using poststratification information. *J. Amer. Statist. Assoc.*,95, 903--915.

[1172] Zarcovic, S.S. (1960). On the efficiency of sampling with various probabilities and the selection of units with replacement. *Metrika*, 3, 53--60.

[1173] Zasepa, R. (1962). *Badania Statystyezna Metoda Reprezentacying.* Warszawa.

[1174] Zinger, A. (1980). Variance estimation in partially systematic sampling. *J. Amer. Statist. Assoc.*, 75, 89--97.

[1175] Zou, G. (1997). Admissible estimation for finite population under the Linex loss function. *J. Statist. Planning Infer.* 61, 373--384.

[1176] Zou, G. (1999). Variance estimation for unequal probability sampling. *Metrika*, 50, 71--82.

[1177] Zou, G. and Liang, H. (1997). Admissibility of the usual estimators under error in variables superpopulation model. *Statist. Prob. Lett.*, 32, 301--309.

[1178] Zou, G. and Wan, A.T.K. (2000). Simultaneous estimation of several stratum means under error-in-variables superpopulation models. *Ann. Inst. Statist. Meth.*, 52(2), 380--396.

[1179] Zyskind, G. (1967). On canonical forms, non-negative covariance matrices and best and simple least squares linear estimators in linear models. *Ann. Math. Statist.*, 38, 1092--1109.

AUTHOR INDEX

A

B

C

D

E

F

G

H

I

J

K

L

M

N

O

P

S

T

U

V

W

Y

Z

HANDY SUBJECT INDEX

F

G

H

I

J

L

M

N

T

U

V

W

Z

ADDITIONAL INFORMATION

A list of books on survey sampling for further study

1 Barnett, V. (2002). *Sample Survey Principles and Methods.* 3rd Ed., Arnold, London.
2 Biemer, P., Groves, R., Lyberg, L., Mathiowetz, N. and Sudman, S. (1992). *Measurement Errors in Surveys.* Wiley.
3 Brewer, K. (2002). *Combined survey sampling inference.* Arnold.
4 Cassel, C., Särndal, C.E. and Wretman, J.H. (1992). *Foundations of inference in survey sampling.* Krieger Publishing Company.
5 Chaudhuri, A. and Vos, J.W.E..(1988). *Unified theory and strategies of survey sampling.* N. Holand.
6 Chaudhuri, A. and Stenger, H. (1992). *Survey sampling : Theory and Methods.* Marcel Dekker, NY.
7 Cingi, H. (1994). *Sampling Theory.* Hacettepe University Press.
8 Cochran, W.G. (1977). *Sampling Techniques.* 3rd ed., Wiley.
9 Coleman, P.B. (1993). *Practical sampling techniques for infrared analysis.* Marcel Dekker, NY.
10 Cox, B.G., College, M., Binder, D., Kott, P.S. and Christianson, A. (1995). *Business Survey Methods.* Wiley.
11 Deming, W.E. (1950). *Some Theory of Sampling.* Wiley.
12 Groves, R.M. (1988). *Telephone Survey Methodology.* Wiley.
13 Groves, R.M. (1989). *Survey Errors and Survey Costs.* Wiley.
14 Foreman, E.K. (1991). *Survey Sampling Principles.* Marcel Dekker.
15 Hajek, J. (1981). *Sampling from a finite population.* Marcel Dekker.
16 Hansen, M.H., Hurwitz, W.N. and Madow, G. (1953). *Sample Survey Methods and Theory.* Wiley.
17 Hansen, M.H., Hurwitz, W.N. and Madow, G. (1993). *Sample Survey Methods and Theory.* Wiley.
18 Hedayat, A.S. and Sinha, B.K. (1991). *Design and inference in finite population sampling.* Wiley.
19 Jessen, R.J. (1978). *Statistical Survey Techniques.* Wiley.
20 Kish, L. (1995). *Survey Sampling.* Wiley.
21 Lohr, S.L. (1999). *Sampling: Design and Analysis.* Duxbury Press
22 Mandenhall, W. (1979). *Elementary survey sampling.* Duxbury.
23 Moser, C.A. and Kalton, G. (1971). *Survey Methods in Social Investigation.* London:Heineman.
24 Mukhopadhyay, P. (1998). *Small area estimation in survey sampling.* Narosa Publishing House.
25 Mukhopadhyay, P. (1998). *Theory and methods of survey sampling.* Prentice--Hall of India.
26 Mukhopadhyay, P. (2002). *Topics in survey sampling.* Springer--Verlag, NY.
27 Murthy, M.N. (1967). *Sampling theory and methods.* Statistical Publishing Society, Calcutta.
28 Platek, R., Rao, J.N.K. , Särndal, C.E. and Singh, M.P. (1987). *Small area statistics.* Wiley.
29 Raj, D. (1968). *Survey Sampling.* McGraw Hill.
30 Rao, J.N.K. (2003). *Small area estimation : Methods and applications.* Wiley.
31 Rossi, P.H., Wright, J.D. and Anderson, A.B. (1983). *Handbook of Survey Research.* AP.
32 Särndal, C.E., Swenson,B. and Wretman, J.H. (1992). *Model Assisted Survey Sampling.* Springer.
33 Schaefer, R.L., Mendenhall, W. and Ott, R.L. (1996). *Elementary Survey Sampling.* Duxbury.
34 Seber, G.A.F. (1981). *Estimation of Animal Abundance and Related Parameters.* Griffin, London.
35 Singh, R. and Mangat, N.S. (1996). *Elements of survey sampling.* Kluwer Academic Publisher.
36 Som, R.K. (1995). *Practical sampling techniques.* Marcel Dekker, NY.
37 Stuart, A. (1984). *The Ideas of Sampling.* Griffin, London.
38 Sudman, S. (1976). *Applied Sampling.* AP.
39 Sukhatme, P.V., Sukhatme, B.V., Sukhatme, S. and Asok, C. (1984). *Sampling Theory of Surveys With Applications.* 3rd Ed., Iowa State University Press, Ames, Iowa.
40 Thompson, M.E. (1997). *Theory of Sample Surveys.* Chapman & Hall.
41 Thompson, S.K. (1992). *Sampling.* Wiley.
42 Tryfos, P. (1996). *Sampling Methods for Applied Research.* Wiley.
43 Valliant, R., Dorfman, A.H. and Royall, R. (2000). *Finite Population Sampling and Inference: a Prediction Approach.* Wiley.
44 Williams, B. (1978). *A Sampler on Sampling.* Wiley.
45 Wolter, K.M. (1985). *Introduction to Variance Estimation.* Springer.
46 Yates, F. (1960). *Sampling Methods for Censuses and Surveys.* Griffen Publishing Co.

Advanced Sampling Theory with Applications

Multi-purpose	Text book for teachers/students.
	Reference manual for researchers.
	Practical guide for statisticians.
Highlights	Basic concepts to advanced technology, SRSWR, SRSWOR, General class of estimators, Ratio and regression type estimators, Bias filtration, Median estimation, PPSWR sampling, Multi-character survey, PPSWOR sampling, Rao--Hartley--Cochran strategy, Calibration of estimators of total, variance, and distribution function etc., Multi-phase sampling, Systematic sampling, Stratified and Post-stratified sampling, Cluster sampling, Multi-stage sampling, Randomized response sampling, Hidden gangs, Imputation, Measurements errors, Small area estimation, many more topics, and fun.
Unique way to	1179 research papers / references.
	162 solved numerical examples.
	335 unsolved theoretical exercises.
	177 unsolved data based practical problems.
Attractions	Simple notation, easy to understand, and smooth flow to reading.
	Complete proofs to theorems.
	Complete solutions to numerical examples.
	Up to date, and new ideas for future research.

Author: Dr. Sarjinder Singh is an Assistant Professor at St. Cloud State University, St. Cloud, MN, U.S.A.. He has published over 80 research papers. He introduced ideas of higher order calibration, hybridizing imputation and calibration, bias filtration, hidden gangs, several new randomized response models, median estimation using two-phase sampling, and exact traditional linear regression estimator using calibration approach in survey sampling. In this book you can enjoy his new ideas such as: